DICTIONARY
OF
SCIENTIFIC BIOGRAPHY

PUBLISHED UNDER THE AUSPICES OF
THE AMERICAN COUNCIL OF LEARNED SOCIETIES

The American Council of Learned Societies, organized in 1919 for the purpose of advancing the study of the humanities and of the humanistic aspects of the social sciences, is a nonprofit federation comprising forty-five national scholarly groups. The Council represents the humanities in the United States in the International Union of Academies, provides fellowships and grants-in-aid, supports research-and-planning conferences and symposia, and sponsors special projects and scholarly publications.

MEMBER ORGANIZATIONS

AMERICAN PHILOSOPHICAL SOCIETY, 1743
AMERICAN ACADEMY OF ARTS AND SCIENCES, 1780
AMERICAN ANTIQUARIAN SOCIETY, 1812
AMERICAN ORIENTAL SOCIETY, 1842
AMERICAN NUMISMATIC SOCIETY, 1858
AMERICAN PHILOLOGICAL ASSOCIATION, 1869
ARCHAEOLOGICAL INSTITUTE OF AMERICA, 1879
SOCIETY OF BIBLICAL LITERATURE, 1880
MODERN LANGUAGE ASSOCIATION OF AMERICA, 1883
AMERICAN HISTORICAL ASSOCIATION, 1884
AMERICAN ECONOMIC ASSOCIATION, 1885
AMERICAN FOLKLORE SOCIETY, 1888
AMERICAN DIALECT SOCIETY, 1889
AMERICAN PSYCHOLOGICAL ASSOCIATION, 1892
ASSOCIATION OF AMERICAN LAW SCHOOLS, 1900
AMERICAN PHILOSOPHICAL ASSOCIATION, 1901
AMERICAN ANTHROPOLOGICAL ASSOCIATION, 1902
AMERICAN POLITICAL SCIENCE ASSOCIATION, 1903
BIBLIOGRAPHICAL SOCIETY OF AMERICA, 1904
ASSOCIATION OF AMERICAN GEOGRAPHERS, 1904
HISPANIC SOCIETY OF AMERICA, 1904
AMERICAN SOCIOLOGICAL ASSOCIATION, 1905
AMERICAN SOCIETY OF INTERNATIONAL LAW, 1906
ORGANIZATION OF AMERICAN HISTORIANS, 1907
AMERICAN ACADEMY OF RELIGION, 1909
COLLEGE ART ASSOCIATION OF AMERICA, 1912
HISTORY OF SCIENCE SOCIETY, 1924
LINGUISTIC SOCIETY OF AMERICA, 1924
MEDIAEVAL ACADEMY OF AMERICA, 1925
AMERICAN MUSICOLOGICAL SOCIETY, 1934
SOCIETY OF ARCHITECTURAL HISTORIANS, 1940
ECONOMIC HISTORY ASSOCIATION, 1940
ASSOCIATION FOR ASIAN STUDIES, 1941
AMERICAN SOCIETY FOR AESTHETICS, 1942
AMERICAN ASSOCIATION FOR THE ADVANCEMENT OF SLAVIC STUDIES, 1948
METAPHYSICAL SOCIETY OF AMERICA, 1950
AMERICAN STUDIES ASSOCIATION, 1950
RENAISSANCE SOCIETY OF AMERICA, 1954
SOCIETY FOR ETHNOMUSICOLOGY, 1955
AMERICAN SOCIETY FOR LEGAL HISTORY, 1956
AMERICAN SOCIETY FOR THEATRE RESEARCH, 1956
SOCIETY FOR THE HISTORY OF TECHNOLOGY, 1958
AMERICAN COMPARATIVE LITERATURE ASSOCIATION, 1960
AMERICAN SOCIETY FOR EIGHTEENTH-CENTURY STUDIES, 1969
ASSOCIATION FOR JEWISH STUDIES, 1969

DICTIONARY

OF

SCIENTIFIC BIOGRAPHY

CHARLES COULSTON GILLISPIE

Princeton University

EDITOR IN CHIEF

Volume 9

A. T. MACROBIUS – K. F. NAUMANN

CHARLES SCRIBNER'S SONS · NEW YORK

Copyright © 1970, 1971, 1972, 1973, 1974, 1975, 1976, 1978, 1980
American Council of Learned Societies.
First publication in an eight-volume edition 1981.

Library of Congress Cataloging in Publication Data

Main entry under title:

Dictionary of scientific biography.

"Published under the auspices of the American Council
of Learned Societies."
Includes bibliographies and index.
1. Scientists—Biography.   I. Gillispie, Charles
Coulston.   II. American Council of Learned Societies
Devoted to Humanistic Studies.
Q141.D5   1981   509′.2′2 [B]   80-27830
ISBN 0-684-16962-2 (set)

ISBN 0-684-16963-0 Vols. 1 & 2   ISBN 0-684-16967-3 Vols. 9 & 10
ISBN 0-684-16964-9 Vols. 3 & 4   ISBN 0-684-16968-1 Vols. 11 & 12
ISBN 0-684-16965-7 Vols. 5 & 6   ISBN 0-684-16969-X Vols. 13 & 14
ISBN 0-684-16966-5 Vols. 7 & 8   ISBN 0-684-16970-3 Vols. 15 & 16

5 7 9 11 13 15 17 19 B/C 20 18 16 14 12 10 8 6

Printed in the United States of America

Editorial Board

Panel of Consultants

Contributors to Volume 9

The following are the contributors to Volume 9. Each author's name is followed by the institutional affiliation at the time of publication and the names of articles written for this volume. The symbol † indicates that an author is deceased.

S. MAQBUL AHMAD
Aligarh Muslim University
IBN MĀJID; AL-MAQDISĪ; AL-MASʿŪDĪ

GARLAND E. ALLEN
Washington University
T. H. MORGAN

FEDERICO ALLODI †
MASCAGNI

EDOARDO AMALDI
University of Rome
MAJORANA

JEAN ANTHONY
Muséum National d'Histoire Naturelle
MILNE-EDWARDS

WILLIAM H. AUSTIN
University of Houston
MORE

MARGARET E. BARON
NAPIER

ISABELLA G. BASHMAKOVA
Academy of Sciences of the U.S.S.R.
MOLIN

IRINA V. BATYUSHKOVA
Academy of Sciences of the U.S.S.R.
MUSHKETOV

ROBERT P. BECKINSALE
University of Oxford
MARTONNE

JOHN J. BEER
University of Delaware
MITTASCH

WHITFIELD J. BELL, JR.
American Philosophical Society Library
S. G. MORTON

LUIGI BELLONI
University of Milan
MALPIGHI; MARCHI; MENGHINI;
MORGAGNI; MORICHINI

ENRIQUE BELTRÁN
Instituto Mexicano de Recursos Naturales
MAST

OTTO THEODOR BENFEY
Guilford College
J. L. MEYER

ALEX BERMAN
University of Cincinnati
MILLON; MOISSAN

PHILIP W. BISHOP
Smithsonian Institution
MUSHET

BRUNO A. BOLEY
Northwestern University
MENABREA

GERT H. BRIEGER
Duke University
METCHNIKOFF

T. A. A. BROADBENT †
MATHEWS

W. H. BROCK
University of Leicester
MARCHAND; MARIGNAC

THEODORE M. BROWN
City College, City University of New York
MAYOW

JED ZACHARY BUCHWALD
Harvard University
MELLONI; MOSSOTTI

K. E. BULLEN
University of Sidney
J. MILNE; MOHOROVIČIĆ

VERN L. BULLOUGH
California State University, Northridge
MONDINO DE' LUZZI

IVOR BULMER-THOMAS
MENAECHMUS; MENELAUS OF
ALEXANDRIA

W. BURAU
University of Hamburg
W. F. MEYER

JOHN G. BURKE
University of California, Los Angeles
MICHEL-LÉVY; MOHS; K. F. NAUMANN

HAROLD L. BURSTYN
William Paterson College of New Jersey
M. F. MAURY; J. MURRAY

G. V. BYKOV
Academy of Sciences of the U.S.S.R.
MARKOVNIKOV

WILLIAM F. BYNUM
University College London
S. W. MITCHELL

RONALD S. CALINGER
Rensselaer Polytechnic Institute
E. H. MOORE

W. A. CAMPBELL
University of Newcastle upon Tyne
MOND

LUIGI CAMPEDELLI
University of Florence
MAGALOTTI; MAGINI

JULES CARLES
Institut Catholique de Toulouse
MOLIARD

ELOF AXEL CARLSON
State University of New York, Stony Brook
H. J. MULLER

ALBERT CAROZZI
University of Illinois
MAILLET; MUNIER-CHALMAS

CARLO CASTELLANI
MANTEGAZZA; MARCHIAFAVA; MOSSO

W. B. CASTLE
Harvard University
G. R. MINOT

JOHN CHALLINOR
University College of Wales
MANTELL

ROBERT A. CHIPMAN
University of Toledo
MARCONI

EDWIN CLARKE
University College London
C. L. MORGAN

M. J. CLARKSON
Liverpool School of Tropical Medicine
MANSON

I. B. COHEN
Harvard University
A. M. MAYER

WILLIAM COLEMAN
Johns Hopkins University
MENURET DE CHAMBAUD; NAUDIN

GEORGE W. CORNER
American Philosophical Society
MALL; C. S. MINOT

DAVID W. CORSON
University of Arizona
MASSON

ALBERT B. COSTA
Duquesne University
V. MEYER; MICHAEL

PIERRE COSTABEL
École Pratique des Hautes Études
MALEBRANCHE; MILHAUD; MORIN;
MYLON

JULIANA HILL COTTON
MANARDO

MAURICE CRANSTON
London School of Economics
MILL

CONTRIBUTORS TO VOLUME 9

A. C. CROMBIE
University of Oxford
MERSENNE

MICHAEL J. CROWE
University of Notre Dame
A. F. MÖBIUS

GLYN DANIEL
University of Cambridge
MONTELIUS; MORTILLET

ROBERT DARNTON
Princeton University
MESMER

AUDREY B. DAVIS
Smithsonian Institution
MOSS; MURPHY

SUZANNE DELORME
Centre International de Synthèse
METZGER; MONTMOR

SALLY H. DIEKE
Johns Hopkins University
G. MOLL

J. DIEUDONNÉ
MINKOWSKI

YVONNE DOLD-SAMPLONIUS
AL-MĀHĀNĪ

J. DORFMAN
Academy of Sciences of the U.S.S.R.
MANDELSHTAM

SIGALIA C. DOSTROVSKY
Barnard College
MAIRAN

RAYNOR L. DUNCOMBE
*Nautical Almanac Office, U.S. Naval
Observatory*
H. R. MORGAN

CAROLYN EISELE
*Hunter College, City University of New
York*
G. A. MILLER

C. W. F. EVERITT
Stanford University
MAXWELL

JOSEPH EWAN
Tulane University
MICHAUX

W. V. FARRAR
University of Manchester
K. H. MEYER

I. A. FEDOSEYEV
Academy of Sciences of the U.S.S.R.
MAKAROV

MARTIN FICHMAN
York University
MAGNENUS; MALOUIN

C. P. FINLAYSON
University of Edinburgh Library
MONRO (PRIMUS); MONRO (SECUNDUS)

WALTHER FISCHER
MALLET

PIETRO FRANCESCHINI
MENEGHETTI

ERIC G. FORBES
University of Edinburgh
MASKELYNE; J. T. MAYER

H.-CHRIST. FREIESLEBEN
C. MAYER; MOLLWEIDE

O. R. FRISCH
University of Cambridge
MEITNER

JOSEPH S. FRUTON
Yale University
MEYERHOF

A. M. GEIST-HOFMAN
MOLESCHOTT

PATSY A. GERSTNER
*Howard Dittrick Museum of Historical
Medicine*
J. MORTON

OWEN GINGERICH
Smithsonian Astrophysical Observatory
A. C. MAURY; MÉCHAIN; MESSIER

BENTLEY GLASS
*State University of New York, Stony
Brook*
MAUPERTUIS

THOMAS F. GLICK
Boston University
MARKHAM; MARTÍNEZ

NORMAN T. GRIDGEMAN
National Research Council of Canada
VON MISES

A. T. GRIGORIAN
Academy of Sciences of the U.S.S.R.
MESHCHERSKY

M. D. GRMEK
*Archives Internationales d'Histoire des
Sciences*
MAGENDIE

MICHAEL GROSS
Hampshire College
MAREY; MORAT

I. GRATTAN-GUINNESS
Middlesex Polytechnic of Enfield
MATHIEU

JACOB W. GRUBER
Temple University
MIVART

FRANCISCO GUERRA
Laboratorios Abelló
MOLINA; MONARDES

IAN HACKING
University of Cambridge
DE MOIVRE; MONTMOR

SAMI HAMARNEH
Smithsonian Institution
AL-MAJŪSĪ

J. L. HEILBRON
University of California, Berkeley
MOSELY

ROGER HEIM
Muséum National d'Histoire Naturelle
MAIRE

DIETER B. HERRMANN
Archenhold Observatory, Berlin
MADLER; G. MÜLLER

MAURICE HOCQUETTE
MAIGE; MANGIN

DORRIT HOFFLEIT
Maria Mitchell Observatory
M. MITCHELL

WILLIAM T. HOLSER
University of Oregon
MALLARD; MIERS

WŁODZIMIERZ HUBICKI
Marie Curie-Skłodowska University
MAIER; MARCHLEWSKI

THOMAS PARKE HUGHES
University of Pennsylvania
MIDGLEY

ALBERT Z. ISKANDAR
University of California, Los Angeles
IBN AL-NAFĪS

C. DE JAGER
Astronomical Institute at Utrecht
MINNAERT

JULIAN JAYNES
Princeton University
G. E. MÜLLER

DANIEL P. JONES
Oregon State University
MATTHIESSEN

PAUL JOVET
Centre National de Floristique
MAGNOL

HANS KANGRO
University of Hamburg
J. H. J. MÜLLER

ROBERT H. KARGON
Johns Hopkins University
W. MOLYNEAUX

GEORGE B. KAUFFMAN
California State University, Fresno
H. G. MAGNUS

B. M. KEDROV
Academy of Sciences of the U.S.S.R.
MENDELEEV

DANIEL J. KEVLES
California Institute of Technology
D. C. MILLER; MILLIKAN

GEORGE KISH
University of Michigan
MAURO; G. MERCATOR; MÜNSTER

MARC KLEIN
Faculté de Médecine, Strasbourg
MOHL

B. KNASTER
Mazurkiewicz

ZDENĚK KOPAL
University of Manchester
MICHELL

ELAINE KOPPELMAN
Goucher College
MANNHEIM; MOUTARD

SHELDON J. KOPPERL
Grand Valley State College
MOSANDER

V. KRUTA
Purkyne University, Brno
J. G. MENDEL

P. G. KULIKOVSKY
Academy of Sciences of the U.S.S.R.
MAKSUTOV; MOISEEV

NAOITI KUMAGAI
MATUYAMA

GISELA KUTZBACH
University of Wisconsin
MARGULES

YVES LAISSUS
*Bibliothèque Centrale du Muséum
National d'Histoire Naturelle*
J. MARCHANT; N. MARCHANT

WILLIAM LeFANU
Royal College of Surgeons, London
MAYO

HENRY M. LEICESTER
University of the Pacific
MUIR; P. MÜLLER

DONALD J. LE ROY
National Research Council of Canada
W. L. MILLER

JACQUES R. LÉVY
Paris Observatory
MINEUR

J. M. LÓPEZ DE AZCONA
Comisión Nacional de Geología, Madrid
MEDINA

EDWARD LURIE
University of Delaware
MARCOU

DUNCAN McKIE
University of Cambridge
W. H. MILLER

H. LEWIS McKINNEY
University of Kansas
F. MÜLLER

ROGERS McVAUGH
University of Michigan
MOCIÑO

MICHAEL S. MAHONEY
Princeton University
MARIOTTE

J. C. MALLET
Centre National de Floristique
MAGNOL

FREDERICK G. MANN
*University Chemical Laboratory,
Cambridge*
MILLS

F. MARKGRAF
University of Zurich Botanical Garden
MARKGRAF

BRIAN G. MARSDEN
Smithsonian Astrophysical Observatory
J. H. MOORE

A. HUGHLETT MASON
University of Virginia
MASON

ARNALDO MASOTTI
Polytechnic of Milan
MAUROLICO

KIRTLEY F. MATHER
Harvard University
MERRILL

KURT MAUEL
Verein Deutscher Ingenieure
F. R. H. C. MOLL

OTTO MAYR
Smithsonian Institution
C. O. MOHR; MOLLIER; MÜLLER-
BRESLAU

JAGDISH MEHRA
University of Texas
MIE

S. R. MIKULINSKY
Academy of Sciences of the U.S.S.R.
MOROZOV

LORENZO MINIO-PALUELLO
University of Oxford
MICHAEL SCOT; MOERBEKE

CHARLOTTE E. MOORE
National Bureau of Standards
MEGGERS

ELLEN J. MOORE
Natural History Museum, San Diego
E. MITCHELL

GIUSEPPE MORUZZI
University of Pisa
MATTEUCCI

ALIDA M. MUNTENDAM
MARUM

A. NATUCCI
University of Genoa
MALFATTI; MENGOLI

G. NAUMOV
Academy of Sciences of the U.S.S.R.
MIDDENDORF

J. D. NORTH
University of Oxford
MELVILL; W. A. MILLER; NASMYTH

A. NOUGARÈDE
University of Paris VI
MIRBEL

LUBOŠ NOVÝ
Czechoslovak Academy of Sciences
MARCI DE KRONLAND

ROBERT OLBY
University of Leeds
MIESCHER; NAEGELI

C. D. O'MALLEY †
MASSA

V. OREL
Mendelianum, Moravian Museum
J. G. MENDEL

JOHN PARASCANDOLA
University of Wisconsin
MELTZER

LEONARD M. PAYNE
Royal College of Physicians, London
MERRETT

KURT MØLLER PEDERSEN
University of Aarhus
MALUS

OLAF PEDERSEN
University of Aarhus
MOHN

J. PELSENEER
MANSION

G. PETIT
University of Paris
MARION

STUART PIERSON
Memorial University of Newfoundland
MAGELLAN; J. F. MEYER

SHLOMO PINES
The Hebrew University
MAIMONIDES

DAVID PINGREE
Brown University
MAHĀDEVA; MAHĀVĪRA; MAHENDRA
SŪRI; MAKARANDA; MANILIUS;
MĀSHĀ'ALLĀH; MATHURĀNĀTHA
ŚARMAN; MUNĪŚVARA VIŚVARŪPA;
MUÑJĀLA; NĀGĒŚA; NĀRĀYANA

A. F. PLAKHOTNIK
Academy of Sciences of the U.S.S.R.
MESYATSEV

A. F. PLATÉ
Academy of Sciences of the U.S.S.R.
NAMETKIN

HOWARD PLOTKIN
University of Western Ontario
S. MOLYNEAUX

S. PLOTKIN
Academy of Sciences of the U.S.S.R.
MAGNITSKY

DENISE MADELEINE PLOUX,
S.N.J.M.
College of the Holy Names
MATRUCHOT

ARTHUR W. POLLISTER
Columbia University
T. H. MONTGOMERY

CONTRIBUTORS TO VOLUME 9

LORIS PREMUDA
University of Padua
MAGATI; MAGGI; MERCATI

CARROLL PURSELL
University of California, Santa Barbara
J. C. MERRIAM

HANS QUERNER
University of Heidelberg
MEYEN; K. A. MÖBIUS

RHODA RAPPAPORT
Vassar College
MALESHERBES; MONNET

GERHARD REGNÉLL
University of Lund
NATHORST

LADISLAO RETI †
MARTINI

MARIA LUISA RIGHINI BONELLI
Istituto e Museo di Storia della Scienza, Florence
MIELI

GUENTER B. RISSE
University of Wisconsin, Madison
MECKEL

ABRAHAM ROBINSON
Yale University
MÉRAY; MITTAG-LEFFLER

FRANCESCO RODOLICO
University of Florence
MARSILI; MICHELI

JACQUES ROGER
University of Paris
MONTESQUIEU

PAUL LAWRENCE ROSE
New York University
MAGIOTTI; MAGNI; MARLIANI; MONTE

EDWARD ROSEN
City College, City University of New York
MÄSTLIN; MAYR

CHARLES E. ROSENBERG
University of Pennsylvania
H. N. MARTIN

B. A. ROSENFELD
Academy of Sciences of the U.S.S.R.
MALTSEV

BARBARA ROSS
University of Massachusetts
MORAY

K. D. ROTHSCHUH
Universität Münster/Westphalia
MEISSNER

M. J. S. RUDWICK
University of Cambridge
H. MILLER; MURCHISON

A. S. SAIDAN
University of Jordan
AL-NASAWĪ

JULIO SAMSÓ
Universidad Autonoma de Barcelona
MANṢŪR IBN 'ALĪ IBN 'IRĀQ

A. P. M. SANDERS
Biohistorisch Institut der Rijksuniversität, Utrecht
MARTIUS

WILLIAM L. SCHAAF
Florida Atlantic University
MASERES

H. SCHADEWALDT
University of Düsseldorf
R. MARTIN

C. SCHALÉN
University of Lund
MÖLLER

F. SCHMEIDLER
University of Munich
MAURER

CHARLES B. SCHMITT
Warburg Institute
MAIGNAN

A. SEIDENBERG
University of California, Berkeley
MASCHERONI; G. MOHR

E. M. SENCHENKOVA
Academy of Sciences of the U.S.S.R.
MAKSIMOV

ELIZABETH NOBLE SHOR
Scripps Institution of Oceanography
MARSH; C. H. MERRIAM; E. S. MORSE

DIANA M. SIMPKINS
Polytechnic of North London
MALTHUS; MOFFETT; MOTTRAM; G. R. M. MURRAY

W. A. SMEATON
University College London
MONTGOLFIER BROTHERS

CYRIL STANLEY SMITH
Massachusetts Institute of Technology
MERICA

H. A. M. SNELDERS
State University of Utrecht
MULDER; MUNCKE

E. SNORRASON
Rigshospitalet, Copenhagen
O. F. MÜLLER

Y. I. SOLOVIEV
Academy of Sciences of the U.S.S.R.
MENSHUTKIN

LARRY T. SPENCER
Plymouth State College
E. D. MONTGOMERY

PIERRE SPEZIALI
University of Geneva
MOUTON; MYDORGE

ERNEST G. SPITTLER
John Carroll University
MORLEY

FRANS A. STAFLEU
University of Utrecht
MIQUEL

WILLIAM H. STAHL †
MACROBIUS; MARTIANUS CAPELLA

MARTIN S. STAUM
University of Calgary
MARGGRAF

WILLIAM T. STEARN
British Museum (Natural History)
MEDICUS; METTENIUS; P. MILLER; MOENCH

JOHANNES STEUDEL †
J. P. MÜLLER

V. T. STRINGFIELD
U. S. Geological Survey
MEINZER

D. J. STRUIK
Massachusetts Institute of Technology
METIUS FAMILY; MEUSNIER DE LA PLACE; MUSSCHENBROEK

ROGER H. STUEWER
University of Minnesota
MASCART

JUDITH P. SWAZEY
Boston University Medical School
R. MAGNUS; MARIE

LOYD S. SWENSON, JR.
University of Houston
MICHELSON

FERENC SZABADVÁRY
Technical University, Budapest
MARTINOVICS; MITSCHERLICH; C. F. MOHR; F. MÜLLER; A. NAUMANN

GIORGIO TABARRONI
University of Bologna–University of Modena
MANFREDI; MICHELINI; MONTANARI

JULIETTE TATON
MOUCHEZ

RENÉ TATON
École Pratique des Hautes Études
MARALDI FAMILY; MONGE

S. TEKELI
Ankara University
MUḤYI 'L-DĪN AL-MAGHRIBĪ

ANDRZEJ A. TESKE†
NATANSON

ANDRÉE TÉTRY
Faculté des Sciences
MESNIL

JEAN THÉODORIDÈS
Centre National de la Recherche Scientifique
MAUPAS

ROSE THOMASIAN
University of New Hampshire
MORO

CONTRIBUTORS TO VOLUME 9

HEINZ TOBIEN
University of Mainz
MARGERIE; MAYER-EYMAR; C. E. H. VON MEYER

G. J. TOOMER
Brown University
METON

HENRY S. TROPP
Humboldt State University
MOULTON

A. J. TURNER
MILLINGTON

G. L'E. TURNER
University of Oxford
B. MARTIN; MORLAND

R. STEVEN TURNER
University of New Brunswick
J. R. MAYER

JUAN VERNET
University of Barcelona
AL-MAJRĪṬĪ; MARTÍ FRANQUÉS; MELLO

GRAZIELLA FEDERICI VESCOVINI
University of Turin
MARSILIUS OF INGHEN

H. B. VICKERY
Connecticut Agricultural Experiment Station
L. B. MENDEL

KURT VOGEL
University of Munich
MONTUCLA

WILLIAM A. WALLACE, O. P.
Catholic University of America
MAIOR

R. WALZER
University of Oxford
MARINUS

DEBORAH JEAN WARNER
Smithsonian Institution
MAUNDER

J. B. WATERHOUSE
University of Toronto
MAWSON

CHARLES WEBSTER
University of Oxford
MORISON

RODERICK WEBSTER
Adler Planetarium
NAIRNE

JOHN W. WELLS
Cornell University
MATHER

FRANZ WEVER
Max-Planck-Institut für Eisenforschung, Düsseldorf
MARTENS

GEORGE W. WHITE
University of Illinois
J. MORSE

D. T. WHITESIDE
Whipple Science Museum
N. MERCATOR

G. J. WHITROW
Imperial College of Science and Technology
E. A. MILNE

WESLEY C. WILLIAMS
Case Western Reserve University
MÉRY

FRANK H. WINTER
Smithsonian Institution
MONTGÉRY; W. MOORE

JÖRN HENNING WOLF
University of Munich
MOLDENHAWER; MURALT

H. WUSSING
Karl Marx University
C. G. A. MAYER

JEAN WYART
University of Paris
MAUGUIN

ERI YAGI
Toyo University
NAGAOKA

ELLIS L. YOCHELSON
MEEK

ALEXANDER A. YOUSCHKEVITCH
Academy of Sciences of the U.S.S.R.
MARKOV

A. YOUSCHKEVITCH
Academy of Sciences of the U.S.S.R.
MINDING

BRUNO ZANOBIO
University of Pavia
MATTIOLI

DICTIONARY
OF
SCIENTIFIC BIOGRAPHY

DICTIONARY OF
SCIENTIFIC BIOGRAPHY

MACROBIUS — NAUMANN

MACROBIUS, AMBROSIUS THEODOSIUS (*b.* North Africa [?], *fl.* early fifth century A.D.), *Neoplatonic commentary.*

Macrobius bore the title *vir clarissimus et illustris*, indicating that he held high government positions. It has been customary to identify him with one of three officials by that name mentioned in the Codex Theodosianus as serving in 399/400, 410, and 422. A serious objection to these identifications is that Macrobius would thus have been known by the name Theodosius, and not as Macrobius. Moreover internal evidence in the *Saturnalia*, his larger extant work, suggests a date of composition in the 430's, rather than at the close of the fourth century. The only official named Theodosius recorded as holding office during this period was a prefect of Italy in 430, and this identification is therefore accordingly proposed.

The title of Macrobius' other extant work, his commentary on Cicero's *Somnium Scipionis*, thinly disguises its actual contents. Macrobius uses passages of Cicero's work as mere suggestions to construct a treatise on Neoplatonic philosophy—the most satisfactory and widely read Latin compendium on Neoplatonism that existed during the Middle Ages. Like *Somnium Scipionis*, Macrobius' *Commentarii* is in the tradition of Plato's *Timaeus*. Macrobius' main source appears to have been Porphyry's lost commentary on the *Timaeus*; and Cicero himself was probably inspired by Posidonius' lost commentary on the *Timaeus*.

Macrobius' lengthy excursuses on Pythagorean number lore, cosmography, world geography, and the harmony of the spheres established him as one of the leading popularizers of science in the Latin West. His chapters on numbers consist largely of conventional statements about the virtues of the numbers within the sacred Pythagorean decade, but include a fine explanation of the Pythagorean doctrine that numbers underlie all physical objects (*Commentarii*, 1.5.5–13).

Macrobius' excursus on the heavens (1.14.21–1.22.13) presents the stock features of popular handbooks on astronomy. A spherical earth at the center of a spherical universe is encircled by seven planetary spheres and a celestial sphere which rotates diurnally from east to west. The planets have proper motions from west to east in addition to their more apparent motions from east to west, the result of their being "dragged along" by the rotation of the celestial sphere, The celestial circles are defined, with particular attention to the Milky Way, the dramatic setting of Scipio's dream. When Macrobius discusses the order of the planets (1.19.1–10), he is purposely ambiguous because his two infallible authorities, Plato and Cicero, differ about the position of the sun. Macrobius' vague statement about the upper and lower courses of Venus and Mercury has been misinterpreted since the Middle Ages as an exposition of Heraclides' geoheliocentric theory.

Macrobius and Martianus Capella were largely responsible for preserving Crates of Mallos' theory of an equatorial and meridional ocean dividing the earth into four quarters, each of which was assumed to be inhabited, and for the wide adoption of Eratosthenes' figure of 252,000 stades for the circumference of the earth. These concepts dominated scientific thinking on world geography in the Middle Ages.

BIBLIOGRAPHY

I. ORIGINAL WORKS. See the new critical ed. of Macrobius by J. Willis, 2 vols. (Leipzig, 1963; 2nd ed., 1970). W. Stahl's *Macrobius' Commentary on the Dream of Scipio* (New York, 1952; 2nd printing with supp. bibliography, New York–London, 1966) is an English trans. See also P. W. Davies' English trans. of the *Saturnalia* (New York, 1968) and N. Marinone's Italian trans. (Turin, 1967).

II. SECONDARY LITERATURE. In addition to the bibliography in the 2nd printing of Stahl's trans., the following more recent items are pertinent: A. Cameron, "The Date and Identity of Macrobius," in *Journal of Roman Studies*, **56** (1966), 25–38; M. A. Elferink, *La descente de l'âme d'après Macrobe* (Leiden, 1968); J. Flamant, "La technique du banquet dans les Saturnales de Macrobe," in *Revue des études latines*, **46** (1968) [1969], 303–319; H. Görgemanns,

"Die Bedeutung der Traumeinkleidung in *Somnium Scipionis*," in *Wiener Studien*, **81**, n.s. **2** (1968), 46–69; and E. Jeauneau, *Lectio philosophorum* (Amsterdam, 1973); M. H. de Ley, *Macrobius and Numenius* (Brussels, 1972); and "Le traité sur l'emplacement des enfers chez Macrobe," in *L'antiquité classique*, **36** (1967), 190–208. Also see N. Marinone, "Replica Macrobiana," in *Rivista di filologia e di istruzione classica*, **99**, n.s. **59** (1971), 1–4, 367–371; A. R. Sodano, "Porfirio commentatore di Platone," in *Entretiens sur l'antiquité classique*, **12** (1966), 193–228, on Macrobius, 198–211; E. Tuerk, "A propos de la bibliothèque de Macrobe," in *Latomus*, **27** (1968), 433–435; "Macrobe et les *Nuits Attiques*," ibid., **24** (1965), 381–406; J. Willis, "Macrobius," in *Altertum*, **12** (1966), 155–161; and C. Zintzen, "Römisches und neuplatonisches bei Macrobius," in *Palingenesia*, **4** (1969), 357–376.

WILLIAM H. STAHL

MÄDLER, JOHANN HEINRICH (*b.* Berlin, Germany, 29 May 1794; *d.* Hannover, Germany, 14 March 1874), *astronomy.*

After graduating from the Gymnasium, Mädler became, at the age of twenty-three, a seminary teacher in Berlin. His interest in astronomy had been awakened by the appearance of the comet of 1811, but he did not have an opportunity to make extensive astronomical observations until he met the rich Berlin banker Wilhelm Beer, half-brother of the composer Giacomo Meyerbeer. Beer maintained a private observatory in Berlin; and he and Mädler worked there together, mainly on lunar topography. In making their observations they used a Fraunhofer telescope with an aperture of 95 millimeters. Their joint publications gave such a favorable impression of Mädler's abilities that, beginning in 1836, he was an observer at the Berlin observatory, then directed by Encke. Here too he worked on the topography of the moon and the planets, chiefly Mars. The most important achievement from his collaboration with Beer was undoubtedly a map of the moon and accompanying two-volume text: *Der Mond nach seinen kosmischen und individuellen Verhältnissen* (1837). The lunar map (with a diameter of 97.5 centimeters) is in many respects the equal of Lohrmann's representation.

In 1840 Mädler accepted an offer from the observatory at Dorpat; he also obtained a professorship and began to publish his considerable body of work. The observatory possessed an excellent observational instrument, the celebrated Fraunhofer refractor with which Struve had determined the parallax of α Lyrae (1838). Mädler was thus led to undertake a new program of work in Dorpat and thereby to follow the tradition begun by Struve: the observation of

double stars. In his *Die Centralsonne*, Mädler sought to provide evidence that the Milky Way possesses a central constellation. He thought the latter, represented by a center of gravity, was formed by Alcyone in the Pleiades. Mädler vigorously defended this idea, but without success, for further research disproved his views.

Mädler was also a pioneer popularizer of astronomy. After giving popular scientific lectures, in 1841 he published *Populäre Astronomie*, which went through six editions during his lifetime. The book was distinguished by its author's thorough command of the material and pedagogically effective presentation of it. In contrast with most popularizers of science, who believed that "to instruct the public one needs only a superficial knowledge of the subject in question" (preface to the first edition), Mädler incorporated in his book the whole wealth of his knowledge, including that of the most recent literature. He also contributed to the dissemination of astronomical knowledge through articles in journals and newspapers. In 1888–1889—with the active cooperation of Wilhelm Foerster—the popular astronomy movement established its own institution in Berlin (the Urania Observatory); and today there is a large network of popular astronomy journals published throughout the world.

In *Populäre Astronomie*, Mädler had briefly sketched the history of astronomy. Following his departure from Dorpat (1865) and return to Germany, he devoted himself to this subject. The result of his labor, the two-volume *Geschichte der Himmelskunde* (1873), contains an extraordinary treasure of valuable historical data that Mädler had been gathering for decades. It left much to be desired, however, with regard to order and conception—probably a consequence of the author's advanced age, as was pointed out several years later by R. Wolf (1877).

Mädler was an argumentative scientist and thus acquired many enemies. Although he had a knowledge of history, he was often skeptical of new theories. For example, he rejected the progressive developmental ideas introduced into astronomy by Kant, Laplace, and Herschel and, twenty years before the discovery of the first spectroscopic double star, he disputed the validity of the Doppler principle.

BIBLIOGRAPHY

Mädler's writings include *Lehrbuch der Schönschreibekunst* (Berlin, 1825); *Physikalische Beobachtungen des Mars* (Berlin, 1830); *Mappa selenographica*, 4 vols. (Berlin, 1834–1836), prepared with Wilhelm Beer; *Der Mond nach seinen kosmischen und individuellen Verhältnissen oder allgemeine vergleichende Selenographie* (Berlin, 1837), written with Wilhelm Beer; *Populäre Astronomie* (Berlin, 1841 and later); *Astronomische Briefe* (Mitau, Latvia, 1846); *Die Centralsonne* (Dorpat, 1846); *Untersuchungen über die Fixstern-Systeme* (Mitau–Leipzig, 1847–1848). A bibliography of Mädler's scientific papers, which number more than 150 in the period 1829–1851, is in H. Kobold, ed., *Generalregister der Bände 1–40 der Astronomischen Nachrichten Nr. 1–960* (Kiel, 1936), cols. 72–74.

There are 17 letters to J. F. Encke and 8 letters to F. W. Bessel, in the Zentralarchiv der Akademie der Wissenschaften der DDR, Berlin, Bessel Nachlass; and 46 letters to H. C. Schumacher, in the Deutsche Staatsbibliothek, Berlin, Handschriftenabteilung, Schumacher Nachlass, and other MS material in the Staats- und Universitätsbibliothek Göttingen.

Biographical material is in S. Günther, *Allgemeine deutsche Biographie*, XX (1884), 37–39.

Dieter B. Herrmann

MAESTLIN, MICHAEL. See **Mästlin, Michael.**

MAGALHÃES. See **Magellan.**

MAGALOTTI, LORENZO (*b.* Rome, Italy, 13 December 1637; *d.* Florence, Italy, 4 March 1712), *dissemination of science.*

Born to an old and distinguished Florentine family, Magalotti lived during a period of marked contrasts arising from political upheavals, religious wars, and the influence of colonialism. In Italy there flourished, on the one hand, post-Galilean scientific progress, most significantly expressed in the Florentine Accademia del Cimento; on the other, however, was the decline of the Renaissance, which had been characterized by freedom of thought.

Magalotti was one of the first ten members of the Accademia del Cimento, which was founded in Florence in 1657 by Ferdinando II de' Medici and his brother Prince Leopoldo and which lasted only until 1667. Magalotti was the secretary of the Academy and reported its activity in the *Saggi di naturali esperienze fatte nell' Accademia del Cimento* (Florence, 1667), essays on natural experiments mainly carried out by Borelli, Redi, and Vincenzio Viviani. The volume immediately attracted considerable interest and was translated into English and Latin.

Magalotti acquired his scientific skill from studying with Viviani, one of the last pupils of Galileo, and from attending at Pisa, then the major Italian university, the lectures of other scientists, notably Marcello Malpighi, Carlo Renaldini, and Giovanni Alfonso Borelli (1656). In 1667, however, he abandoned scientific studies and embarked on a series of travels as a diplomat in the service of the Medicis, which enabled him to become familiar with much of Europe and inspired many of his writings.

In his time Magalotti acquired considerable fame, but today his work appears not quite worthy of his many gifts and complex personality. He has the distinction, though, of having written the best scientific prose in Italian after that of Galileo; his descriptions of experiments in physics are written in colorful, almost dramatic, language. His contacts with different cultures enriched and gave freedom to his expression, which led his contemporaries to believe that his style showed too much foreign influence.

Magalotti published very little during his lifetime; his writings, which were much in demand, were circulated mostly in manuscript form and were not published until after his death. The most notable were the then celebrated *Lettere contro l'ateismo*, his short stories and poetic works, his many letters on scientific and other scholarly topics, and the singular essays on odors.

Magalotti became so interested in linguistics that he became involved in the Italian dictionary that was being prepared by the Accademia della Crusca. His literary style is characterized by lively prose and brilliant witticisms, which express his character as a man of the world but also may serve to disguise the spiritual disquiet which led him to enter a monastery for a few months in 1691. Magalotti seems in some ways a very modern figure, with acute critical abilities and a questioning mind, characterized also by a certain world-weariness.

BIBLIOGRAPHY

I. Original Works. Among Magalotti's writings are *Saggi di naturali esperienze fatte nell' Accademia del Cimento* (Florence, 1667), translated into English by R. Walter (London, 1684) and into Latin by Petrus van Musschenbroek (Leiden, 1731), photocopy ed. issued by Museum of the History of Science, Florence, and Domus Galileiana, Pisa (Florence–Pisa, 1957); *Notizie varie dell'imperio della China e di qualche altro paese adiacente* (Florence, 1697); the *Lettere familiari del Conte Lorenzo Magalotti gentiluomo fiorentino e accademico della Crusca,* known as *Lettere contro l'ateismo,* 2 pts. (Venice, 1719); *Lettere scientifiche ed erudite del Conte Lorenzo Magalotti gentiluomo . . .* (Florence, 1721); *Delle lettere familiari del*

Conte Lorenzo Magalotti e di altri insigni uomini a lui scritte (Florence, 1769); and *Varie operette del Conte Lorenzo Magalotti, con otto lettere su le terre odorose d'Europa e d'America dette volgarmente buccheri, ora pubblicate per la prima volta* (Milan, 1825).

II. SECONDARY LITERATURE. See Angelo Fabroni, *Vitae italorum doctrina excellentium, qui saec. XVII et XVIII floruerunt* (Rome, 1769), an Italian trans. of which by Cianfrogni is found in the *Lettere familiari* of 1769 (see above); Stefano Fermi, *Lorenzo Magalotti, scienziato e letterato, studio biografico-bibliografico critico* (Florence, 1903); and *Bibliografia magalottiana* (Piacenza, 1904); Cesare Guasti, "Lorenzo Magalotti, diplomatico," in *Giornale storico degli archivi Toscani* (1860–1861); Lorenzo Montano, *Le più belle pagine di Lorenzo Magalotti* (Milan, 1924); and Pompilio Pozzetti, *Lorenzo Magalotti. Elogium, habitum nonis Ianuarii 1787* (Florence, 1787).

LUIGI CAMPEDELLI

MAGATI, CESARE (*b.* Scandiano, Modena, Italy, 1579; *d.* Bologna, Italy, 9 September 1647), *surgery*.

Magati was the son of Giorgio Magati and of the former Claudia Mattacoda; his parents were of modest condition. One of his brothers, Giovanni Battista, was a doctor; his sister Laura was the grandmother of Antonio Vallisnieri. Magati obtained a doctorate in philosophy and medicine at Bologna in 1597. He was a pupil of the military surgeon Flaminio Rota and of Giulio Cesare Claudini, teacher of logic, philosophy, and practical medicine. Magati practiced the treatment of head wounds under Giovanni Battista Cortese, an expert in this field. He later moved to Rome, where he learned the methods of treating wounds used by surgeons there.

When he returned to Scandiano, Magati built up a successful practice and gained the patronage of the Marquis Enzio Bentivoglio, who recognized the earnestness and ability of the young surgeon and took him to Ferrara. There the established physicians received him with hostility, but he overcame their jealousy and in 1613 became a lecturer in surgery. This post became the focal point for the diffusion of his method of treating wounds and at the same time gave Magati useful opportunities to experiment.

In 1616, at the age of thirty-seven, Magati published, in Venice, the book for which he is particularly remembered, the collection of lectures *De rara medicatione vulnerum, seu de vulneribus raro tractandis, libri duo*. A few years later he became seriously ill and decided to join the Capuchin order. The investiture took place on 11 April 1618; he took his final vows a year later at Ravenna as Friar Liberato of Scandiano. But there was little peace for Magati in the monastery,

for he was assailed from all sides with requests for his help and his cures. His superiors granted him permission to practice, and he treated well-known patients throughout the territory of the house of Este.

For years Magati suffered from gallstones, and in 1647 he went to Bologna for an operation. The surgeon very neatly extracted three stones the size of an egg, but failed in the more difficult attempt to extract a fourth one covered with sharp projections; the bladder wall tore and Magati died three days later in great pain.

Magati was one of the forerunners of modern surgery. He was among the first to prescribe a rational treatment of wounds, quite different from contemporary methods, which advocated the frequent replacement of dressings and repeated local applications, on the same day, of various ointments.

Magati's major work, *De rara medicatione vulnerum*, was published in three editions. The book is divided into two parts, general and specialized. The style is prolix, with frequent mentions of Hippocrates and Galen. But the essence of Magati's new ideas is clearly expressed, and may be summarized thus; It is nature, not the doctor and his medicine, that heals wounds, because it is nature that eliminates pus, regenerates the flesh, repairs broken bones with callus, coagulates blood, and expels secretions. Therefore the best method of healing wounds is to give nature the means to do her work in the best way, by eliminating or avoiding obstacles. The frequent exposure of wounds to air is damaging, as is the introduction of probes and bandages, which encourage putrefaction. Magati denied the need to clean and anoint wounds; they should merely be bound with a linen cloth folded several times and left in place for five or six days. The bandage must not be heavy or unevenly distributed, and neither too tight nor too loose.

The validity of Magati's care of wounds was confirmed by Ludovico Settala, a doctor in the hospital at Milan. In the early part of the eighteenth century Dionisio Andrea Sancassani, also from Scandiano, tried to revive Magati's therapeutic methods and wrote three short works on the subject: *Chirone in campo, Lume all'occhio*, and *Magati redivivo*. But the time was not ripe, and minds accustomed to centuries-old methods could not quickly be persuaded to accept such innovations.

Two other works are attributed to Magati: *Tractatus quo raro vulnerum curatio defenditur contra Sennertum*, which was printed at the end of *De rara medicatione*, in the second (1676) edition, of which his brother Giovanni Battista appears as the author; and the *De Re Medica*, which appears to have been printed at the expense of the Este family.

BIBLIOGRAPHY

On Magati or his work, see W. von Brunn, *Kurze Geschichte der Chirurgie* (Berlin, 1928), pp. 219, 225, 265, 276; P. Capparoni, "Cesare Magati (Padre Liberato da Scandiano dei Minori Cappuccini)," in *Profili bio-bibliografici di medici e naturalisti celebri italiani dal secolo XV al secolo XVIII* (Rome, 1932), pp. 70–75; D. Giordano, "Medicazioni strane e medicazioni semplici," in *Scritti e discorsi pertinenti alla storia della medicina e ad argomenti diversi* (Milan, 1930), pp. 25–45; W. von Haberling, F. Hübotter, and H. Vierordt, eds., *Lexicon der hervorragenden Ärzte aller Zeiten und Völker*, 2nd ed., IV (Berlin–Vienna, 1932), 27–28; V. Putti, "Cesare Magati (1579–1647)," in *Biografie di chirurghi dal XVI a XIX secolo* (Bologna, 1941), pp. 9–16; and S. de Renzi, *Storia della medicina in Italia*, IV (Naples, 1845), 484–495.

LORIS PREMUDA

MAGELLAN, JEAN-HYACINTHE (Magalhães, João Jacinto de) (*b.* Aveiro, Portugal, 4 November 1722; *d.* Islington, England, 7 February 1790), *chemistry, physics, scientific instrumentation.*

Little is known about Magellan's youth and early manhood. His family, who made an unproven claim to be descended from Ferdinand Magellan, sent him to an Augustinian monastery in Coimbra when he was eleven years old, and there he lived and studied for about twenty years, first as a novice and then as a monk. There was a scientific tradition among the Coimbra Augustinians (it is reported that they studied the works of Newton), and as a consequence Magellan became well enough versed in astronomy to serve as a guide for, and to gain the friendship of, Gabriel de Bory during the latter's visit to Portugal in 1751 to observe a solar eclipse. A few years later Magellan sought and received permission from Pope Benedict XIV to leave the order. From 1755 to about 1764, Magellan traveled through Europe, finally settling in England, where he resided for the rest of his life. At some point he was converted to Protestantism. He never married.

Magellan produced no scientific work of serious consequence. He did, however, find ways to meet or to write to everyone whose activities interested him, and, as a result, is known chiefly for his wide circle of acquaintances and for acting as an intermediary in disseminating new information. He introduced English scientific instruments into France, edited Cronstedt's *Mineralogy*, and informed the French chemists of Priestley's work. He was a fellow of the Royal Society and a member of several European academies of science. Industrial spy, indefatigable learner of languages, shameless borrower from others' writings, Magellan nevertheless showed little of that malice usually associated with the gossip or the hanger-on. A curious mixture of unoriginality and independence, he had no great ambitions for himself.

Magellan wrote more about scientific instruments than about any other subject. His first work (1775), a description of English octants and sextants of the reflecting or Hadleyan type, was clearly written, detailed, and useful. He also wrote on barometers and other meteorological instruments (although not always with full understanding),[1] and on Atwood's machine. These works were all in French, in keeping with Magellan's role as correspondent of the Academy of Sciences, as agent of Trudaine de Montigny (*intendant* of finances), and as bearer of good news to the Continent from the land of the artisan-scientist coalition.[2]

Through his reading, correspondence, and acquaintances, Magellan kept up with the latest developments in English, Scottish, and Swedish chemistry and experimental physics. His work on "elementary fire" helped to disseminate the new theories of heat being worked out by Black, Irvine, and Crawford, and introduced the term "specific heat" *(chaleur spécifique)* into the language. It also gave the first published table of specific heats, although these were derived from determinations by Richard Kirwan.[3] Magellan early saw how important were the investigations of Priestley, whose good friend he became, and his characteristic response to Priestley's fundamental research was two-fold: he told the French about it, and he produced a pamphlet describing some small improvements in the apparatus for making carbonated water and some refinements in the construction of nitric oxide eudiometers.[4]

Gustav von Engestrom, who had studied mineralogy with Cronstedt, was Magellan's link with Swedish chemistry. Engestrom translated into English, at Magellan's behest, the *Mineralogy* of Cronstedt (1770). Magellan undertook to publish a second edition, which was to have notes by Giovanni Fabbroni and Kirwan; but by the time he was ready, Kirwan's own *Mineralogy* had appeared (1784). Magellan went ahead, and, to Kirwan's great annoyance,[5] borrowed from him where appropriate, and also incorporated recently published findings of Bergman, Scheele, A. Mongez, and M. T. Brünnich. Although he "rearranged" the text to include new developments, Magellan was convinced that much in Cronstedt's system was still valuable. He especially endorsed the latter's combination of chemical analysis with the observation of the external characteristics of minerals.[6]

In his notes to Cronstedt, Magellan gave a good picture of conventional contemporary thinking about the foundations of chemistry. Thus various bodies, he said, although suspected of being compound, "may and even ought to be considered as primitive substances with respect to our knowledge of them, till they shall be experimentally decomposed."[7] Acid and alkaline substances act on other bodies in virtue of an "attraction," about which Magellan says:

> We may complain indeed of the deficiency of our knowledge in regard to the essential cause of this phenomenon which we mean to explain by the word *attraction*; but it being the ultimate effect our knowledge can reach to, after our observation has been driven from cause to cause of all that we can discern in nature, we must rest contented with the simple deductions from such an evident and general principle, whatever may be its original cause.[8]

Magellan also shared the widespread belief that the smallest bit of a chemically reacting substance is some sort of basic unit, and asserted at one and the same time that these smallest parts probably possess polarity, and that we know nothing whatever about them.[9] As Magellan's editing proceeded, the notes on Lavoisier's new theory of combustion and calcination increased,[10] until finally Magellan conceded to the arguments in the *Nomenclature* (1787). He retained the old language of the phlogiston theory for the remainder of the work, however, remarking, "*ut pes et caput uni reddantur formae*, according to the old adage of Horace."

NOTES

1. See W. E. K. Middleton, *A History of the Barometer* (Baltimore, 1964), 102–104, 114, 122–123, 259–260, 377; and *Invention of the Meteorological Instruments* (Baltimore, 1969), 79, 146.
2. See Birembaut, cited below, and M. Daumas, *Les instruments scientifiques aux xviie et xviiie siècles* (Paris, 1953), 138 ff.
3. Robert Fox, *The Caloric Theory of Gases* (Oxford, 1971), 26–29.
4. The story of Magellan's role in the introduction to France of Priestley's discoveries has been masterfully reconstructed in ch. 2 of Guerlac, cited below.
5. See the summary of Kirwan's letter (1788) to Banks in W. R. Dawson, ed., *The Banks Letters* (London, 1958), 493.
6. Cronstedt's *Mineralogy*, I, v–x.
7. *Ibid.*, p. 263 n; the work is continuously paginated.
8. *Ibid.*, p. 328 n.
9. *Ibid.*, pp. 428–431 n.
10. *Ibid.*, notes on pp. 285, 435 ff., 431–432, 444–445, 447, and 491–493, for example.

BIBLIOGRAPHY

I. ORIGINAL WORKS. Among Magellan's works are *Description des octants et sextants anglois ...* (Paris–London, 1775); *Description of a Glass Apparatus, for*

Making Mineral Waters Like Those of Pyrmont, Spa, Seltzer ... Together With the Description of Some New Eudiometers ... in a Letter to the Rev. Dr. Priestley (London, 1777; 2nd ed., rev., 1779; 3rd ed., enl., 1783), German trans. by G. T. Wenzel (Dresden, 1780); *Description des nouveaux instruments à reflection pour observer avec plus de précision les distances angulaires sur mer ...* (London, 1779); *Collection de différents traités sur des instruments d'astronomie, physique ...* (Paris–London, 1775–1780); *Description et usages des nouveaux baromètres, pour mésurer la hauteur des montagnes et la profondeur des mines ...* (London, 1779); *Description et usages des instrumens d'astronomie et de physique faits à Londres, par ordre de la cour de Portugal en 1778 et 1779 ...* (London, 1779); *Notice des instrumens d'astronomie, de géodésie, de physique, etc., faits dernièrement à Londres par ordre de la cour d'Espagne ...* (London, 1780); *Description d'une machine nouvelle de dynamique inventée par M. G. Atwood, au moyen de laquelle on rend très aisement sensible les loix du mouvement des corps en ligne droite, et en rotation ...* (London, 1780); and *Essai sur la nouvelle théorie du feu élémentaire, et de la chaleur des corps ...* (London, 1780).

Magellan contributed to A. G. Lebègue de Presle, *Rélation ou notice des derniers jours de J. J. Rousseau ... avec une addition relative au même sujet, par J. H. Magellan* (London–Paris, 1778). He edited A. F. Cronstedt, *An Essay Towards a System of Mineralogy*, 2 vols. (London, 1781; 2nd ed., 1788); and *Voyages et mémoires de Maurice-Auguste, Comte de Benyowski sur la Pologne*, 2 vols. (Paris, 1791). He also published a number of articles in the *Journal de physique* between 1778 and 1783.

II. SECONDARY LITERATURE. For works about Magellan, see Arthur Birembaut, "Sur les lettres du physicien Magellan conservées aux Archives Nationales," in *Revue d'histoire des sciences et de leurs applications*, **9** (1956), 150–161; J.-P. Brissot, "Mémoires (1754–1793)," in C. Perroud, ed., *Mémoires et documents relatifs aux XVIIIe et XIXe siècles*, I (Paris, 1911), 362–363; Joaquim de Carvalho, "Correspondência científica dirigida a João Jacinto de Magalhães," in *Revista da Faculdade de ciências, Universidade de Coimbra*, **20** (1951), 93–283 and also published separately (Coimbra, 1952); Henry Guerlac, *Lavoisier: The Crucial Year* (Ithaca, N. Y., 1961), esp. ch. 2; John Nichols, *Literary Anecdotes of the Eighteenth Century*, VIII (London, 1814), 48–51n, and Alexandre Alberto de Sousa Pinto, *A vida e a obra de João Jacinto de Magalhães* (Pôrto, 1931).

STUART PIERSON

MAGENDIE, FRANÇOIS (*b*. Bordeaux, France, 6 October 1783; *d*. Sannois, Seine-et-Oise, France, 7 October 1855), *physiology, medicine.*

Magendie was a son of Antoine Magendie, a surgeon, and Marie Nicole de Perey. He had a younger brother, Jean-Jacques, whose name testifies to the admiration their father, an ardent republican, had for the ideas of Rousseau. The two boys were brought up in accord with Rousseau's pedagogical precepts: the emphasis was on their personal independence and not the instruction they received. As Pierre Flourens wrote in his *éloge* of François Magendie: "The new Émile, absolutely given over to himself, went about as he pleased in a liberty that very closely resembled abandonment." In 1791, swept along by the Revolution, the Magendie family moved to Paris, where the father devoted himself more to politics than to medicine. The death of his mother in 1792 and his father's activities on Revolutionary committees threw Magendie still further upon his own intellectual resources. Having reached the age of ten without having attended school or having learned to read and write, at his own wish he entered elementary school, where he made very rapid progress. At the age of fourteen he won the grand prize in a national contest for an essay on knowledge of the rights of man.

At sixteen, too young to be admitted to the École de Santé, Magendie became an apprentice at a Paris hospital, where the surgeon Alexis Boyer, a friend of his father's, accepted him as a pupil and entrusted him with the anatomical dissections. In 1803 Magendie passed the examination required for an *interne des hôpitaux* and entered the Hôpital Saint-Louis as a medical student. In 1807 he became an assistant in anatomy at the École de Médecine and gave courses in anatomy and physiology. He received his medical degree in Paris on 24 March 1808 after defending a dissertation entitled *Essai sur les usages du voile de palais avec quelques propositions sur la fracture du cartilage des côtes*. Magendie's studies reflect the chaotic situation of teaching in France during the period. A liberal education in his childhood, a practically oriented apprenticeship in medicine, the astonishing experience of the successive collapses of academic and doctrinal systems—all these combined to strengthen in Magendie a love of facts and a contempt for words, theories, and social conventions, as well as a rude frankness and a truly exceptional independence of judgment.

After his thesis Magendie's first publication was an article of a theoretical, not to say doctrinal, character. It appeared in the *Bulletin des sciences médicales*, which, published by the Société Médicale d'Émulation, did its utmost to glorify the memory of Bichat; yet Magendie's memoir was a harsh attack on the fundamental ideas of this intellectual master of French physicians. Magendie asserted: "[The] majority of physiological facts must be verified by new experiments and this is the only means of bringing the physics of living bodies out of the state of imperfection in which it lies at present" (*Bulletin des sciences médicales*, **4**

[1809], 147). According to Magendie, the biological sciences had remained behind the physical sciences because they utilized complicated ideas and preconceptions to explain facts which very often were not themselves established with certainty. Magendie still accepted the concept of a vital force (considering it a supposition that served merely to bring together in a single term all the characteristics proper to life), but he proposed "to abolish the two vital properties known under the names of animal sensibility and animal contractility and to consider them as functions" (*op. cit.* p. 166).

Further, he condemned Bichat's attempt to increase the number of vital properties by distinguishing them according to the organic tissues. In Magendie's view, physiology should explain the two phenomena essential to life—nutrition and movement—through reducing them to the organization of living beings and of their parts. Magendie's profession of faith is, in fact, the fundamental dogma of modern biology: "Two living bodies having the same organization will display the same vital phenomena; two living bodies having different organizations will display vital phenomena the diversity of which will always be in direct proportion to the difference in organization" (*op. cit.*, p. 159).

The gnoseological optimism of this youthful piece was rapidly replaced by a certain skepticism and a growing distrust of all theoretical generalization. Although he later honored Bichat by preparing his two major works for publication, Magendie furnished these editions with ample commentaries praising Bichat the experimenter but treating with irony all attempts at a systematic explanation of vital phenomena. The influence of the philosophy of the Idéologues had prevailed over the vitalist doctrines of the Montpellier school; and Magendie, having taken an aversion to all theories, made an extraordinary effort during most of his life to discover and collect the "facts" and refused, to the extent this was possible, to interpret them. "I compare myself," he said to Claude Bernard, "to a ragpicker: with my spiked stick in my hand and my basket on my back, I traverse the field of science and I gather what I find" (C. Bernard, *Magendie*, p. 13). This was, of course, an illusion: Magendie made his discoveries on the basis of certain theoretical considerations and within the framework of a rather well defined philosophy of biology. But this illusion was particularly important at a moment in the history of physiology when it was necessary to replace excessive speculation with recourse to the "facts"—that is, with recourse to the experimental method. In this sense Bernard was perfectly right when he stressed that "M. Magendie is not one of those men concerning whom one can give a sufficient idea simply by enumerating their works or by pointing out the discoveries with which they have enriched science" and that, for the historian of science, Magendie's principal merit consisted in the influence he exerted in orienting physiology toward experimental investigations.

In 1809 Magendie presented to the Académie des Sciences and to the Société Philomatique the results of his first experimental work, which he carried out in collaboration with the botanist and physician A. Raffeneau-Delile. In a series of ingenious experiments on various animals, the two investigators studied the toxic action of several drugs of vegetable origin, particularly of upas, nux vomica, and St.-Ignatius's bean. As Olmsted observes, these experiments mark the beginning of modern pharmacology. For the first time an experimental comparison was made of the similar effects produced by drugs of different botanical origin. Magendie held that the toxic or medicinal action of natural drugs depends on the chemical substances they contain, and it should be possible to obtain these substances in the pure state. As early as 1809 he suspected the existence of strychnine, later isolated, in accord with his predictions, by P. J. Pelletier (1819). Moreover, in 1817, in collaboration with Pelletier, Magendie discovered emetine, the active principle of ipecac. Immediately after the isolation of strychnine he demonstrated that it produces exactly the same type of poisoning as do certain vegetable drugs.

The experiments of 1809 enabled Magendie and Raffeneau-Delile to affirm that upas and nux vomica, which produce generalized convulsions and tetanus, must act on the spinal marrow and, in fact, must stimulate it very strongly. Sectioning the medulla—separating it from the brain—does not suppress the symptoms of the poisoning, whereas destruction of the medulla eliminates them completely. The character of the symptoms was found to be independent of the way in which the poison entered the organism, but the latter circumstance did influence the rapidity with which the first spasms began. Magendie thus formulated the principle of local action: A toxic or medicinal substance acts solely in terms of its direct contact with an effector organ. This principle obliged physiologists to accord great importance to the study of the absorption and transport of poisons and medicines in the organism.

At the beginning of the nineteenth century the generally accepted view was that absorption takes place exclusively through lymphatic vessels. This theory, elaborated by John Hunter and reinforced by Bichat's teaching, had replaced Haller's opinion, according to which food and all other substances are absorbed through the veins. Magendie, however, demonstrated the existence of two absorption paths. He conducted a classic experiment in which a dog was poisoned following the introduction of the toxic substance into a limb that was connected to the body only by a blood vessel, or even only by a quill. Magendie concluded that the absorption of liquids and semiliquids is not a vital and physically inexplicable function of the lymphatics but a simple physicochemical phenomenon of the imbibition of tissues and of passage through vascular walls.

In 1811 Magendie was appointed anatomy demonstrator at the Faculté de Médecine of Paris, and for three years he taught anatomy and surgery. He displayed unusual skill during his operations at the École Pratique. Meanwhile, his rude behavior precipitated a conflict with the professor of anatomy, François Chaussier. The professor of surgery, Guillaume Dupuytren, saw Magendie as a dangerous rival and created difficulties for him at the Faculté de Médecine. In 1813 Magendie resigned from his post as demonstrator, opened an office as a practicing physician, and organized a private course in physiology. In his éloge Flourens speaks of a veritable "volte-face": in his opinion, Magendie suddenly buried all his ambitions as an anatomist and surgeon in order to devote himself to experimental physiology. Whether it was a long-considered project or simply an impulsive act cannot be known; in any case his private courses, featuring experiments on living animals, aroused the curiosity of the medical public and soon enjoyed a large success. It is from this moment that Claude Bernard dated the beginning of the "new physiology." "Magendie," he wrote, "joined example to precept. He undertook private courses of experimental physiology based on vivisections. He attracted numerous students, among whom were a great many foreigners. It was from this center that the young physiologists carried the seeds of the new experimental physiology into the neighboring schools, where it then developed with such prodigious rapidity" (Rapport sur le progrès et la marche de la physiologie en France, p. 7).

Magendie's teaching was not only oral. The interest evoked by his courses led him to write Précis élémentaire de physiologie, in which, just as in his lectures, experimental demonstration replaced theoretical discussion as much as possible. He thus created a new type of physiology textbook: philosophical deductions founded on anatomy and on doctrinal suppositions were greatly reduced in favor of simple and precise descriptions of experimental facts. The first volume of the Précis was published in 1816, the second in 1817.

This work, which went through four French editions and was translated into several other languages, including English and German, exerted a very profound influence on physicians and biologists during the first half of the nineteenth century.

In the introduction to his *Précis*, Magendie explained and justified his methods of investigating vital phenomena. Without abandoning his vitalist position (that is, he still accepted that "corps vivants" differ from "corps bruts" both in their form and composition and in certain supplementary laws that govern them), Magendie criticized the ontological interpretations of soul and of vital principle. He rejected as a dangerous illusion the methodological analogy between the vital force and Newtonian gravitation. For him, vital force would remain an empty term as long as it was impossible to link it, on the example of universal attraction, to a precise law. According to Magendie, the laws of life, even if they possess their own character, cannot be in contradiction with the physicochemical laws. The first task of physiology was to push the physical analysis of vital phenomena as far as possible. In theory and in practice, therefore, Magendie preached an empirical reductionism.

Between 1813 and 1821 Magendie made a great many discoveries in almost all the fields of research that then constituted physiology. Among these were proof of the passive role of the stomach in vomiting; explanation of the mechanism of deglutition; experiments on alimentation with nonnitrogenous substances (demonstration of the mammals' need for a protein supply and the first experimental production of an avitaminosis); experiments on digestive properties of pancreatic juice; proof of the liver's decisive role in detoxification processes; demonstration of the hemodynamic importance of the elasticity of the arteries; discovery of emetine and experiments on the toxic action of hydrocyanic acid; comparative anatomical investigations clarifying the mechanism of absorption; and new observations following vivisections of cranial nerves. Magendie was also the first to make comparative nutrition experiments with chemically pure substances.

Although Magendie was interested primarily in experimental physiology, he did not neglect medical practice. For many years he suffered from not being on a hospital staff, which would have facilitated the clinical study of new medicines. In 1818, following a competitive examination, he was named to the Bureau Central des Hôpitaux Parisiens; but until 1826 he had no official hospital assignment and had to rely on the understanding of his friend Henri Husson in order to observe treatments and to give a clinical

course at the Hôtel-Dieu. In 1819 he was requested to give a course at the Athénée Royal. In 1821 he published the first edition of his *Formulaire pour la préparation et l'emploi de plusieurs nouveaux médicaments*, a therapeutical manual much used by physicians. Magendie introduced into medical practice a series of recently discovered alkaloids: strychnine, morphine, brucine, codeine, quinine, and veratrine. He also generalized the therapeutic use of iodine and bromine salts. Contrary to the dominant opinion among the older physicians, Magendie favored the use of chemical substances over that of natural drugs and, in addition, had great confidence in pharmacological experiments on animals. He did not hesitate to test on himself all the substances that were shown to be harmless in the animal experiments.

In 1821 Magendie was elected to the Académie des Sciences and the Académie Royale de Médecine. In the same year he founded the *Journal de physiologie expérimentale*, the first periodical devoted exclusively to physiology. Starting with the second volume he added the words *et pathologie* to the title. Convinced that pathology is essentially "the physiology of the sick organism," Magendie already envisaged a complete reform of medicine by establishing it upon the experimental study of the vital functions; the idea of this project was later brilliantly defended and developed by his disciple Claude Bernard.

It was in his *Journal* that Magendie published the results of his investigations on the physiology of the nervous system and on the cerebrospinal fluid. The discovery of the Bell-Magendie law (1822) was the source of a distressing dispute over the parts played by Charles Bell and by Magendie in distinguishing the motor and sensory roots of the medulla. The historical documents relating to this subject do not appear to contradict Magendie's final claims:

> In sum, Charles Bell had had, before me, but unknown to me, the idea of separately cutting the spinal roots; he likewise discovered that the anterior influences muscular contractility more than the posterior does. This is a question of priority in which I have, from the beginning, honored him. Now, as for having established that these roots have distinct properties, distinct functions, that the anterior ones control movement, and the posterior ones sensation, this discovery belongs to me [*Comptes rendus . . . des séances de l'Académie des sciences*, **24** (1847), 320].

Although it is possible that, from the start of his research on the spinal nerves, Magendie knew of Bell's general idea through the latter's assistant, John Shaw, it is nonetheless certain that the clear statement and the experimental verification of the law in question belongs to him. This discovery, of fundamental importance for neurophysiology, was completed by Magendie and Claude Bernard with the experimental explanation of an apparent exception known as *sensibilité récurrente* (1847).

In 1823 Magendie produced experimentally and described the rigidity that follows decerebration; he provided the first proof of the cerebellum's role in maintaining the equilibrium of the organism; and he cut the fifth pair of cranial nerves within the cranium itself, demonstrating the direct responsibility of these nerve structures for the sense of touch and their indirect trophic role in the maintenance of the function of the other senses. In 1824 he observed the circular movement ("mouvement de manège") that occurs in the rabbit following the section of the cerebellar peduncle. This experiment was the point of departure for Bernard's discovery of the "piqûre sucrée" and, later, of Jacques Loeb's experiments on the rotary movements of animals. During the period 1824–1828 Magendie made many discoveries concerning the origin, composition, and circulation of the cerebrospinal fluid. He showed that the brain cavities communicate freely with the spinal subarachnoid space and described the medial foramen in the roof of the fourth ventricle *(foramen Magendie)*. Through severing the various branches of the facial nerve, Magendie succeeded in definitively banishing the ancient hypothesis of the "nervous fluid."

During a trip to England in 1824, when he was a guest of William Hyde Wollaston's, Magendie gave several public demonstrations of his method of the experimental section of cranial nerves of living dogs. The cruel side of his experiments provoked an antivivisectionist campaign. Although powerful in Great Britain, this struggle for the protection of animals found no echo in France. Some colleagues, however, reproached Magendie for having experimented on sick people—that is, for having performed operations the goal of which was essentially scientific and not therapeutic. Such proceedings are described in Magendie's publications, but they were never really dangerous or mutilating. Particularly noteworthy are his experiments on the human retina, which could have led him, if he had had a taste for theoretical generalization, to the discovery of the law of the specific energy of the senses.

In 1830 Magendie finally obtained the directorship of a hospital department, the women's ward at the Hôtel-Dieu. Despite everything that one could say concerning his gruff manners toward his colleagues and his cruelty to animals, contemporary testimony agrees on the gentleness, patience, and understanding with which he treated his hospital patients. On 4 April 1831 he replaced J. C. A. Récamier in the chair of

medicine at the Collège de France. It was not without some difficulty that the medical instruction there was changed in style and in substance. Instead of expounding doctrines, Magendie gave public demonstrations of the experimental method; instead of teaching clinical medicine as it was practiced at the patient's bedside, he concentrated on the presentation of physiological and pathological knowledge derived from studies made on animals. Nevertheless, his initial lectures at the Collège de France were devoted to a medical problem of current concern: cholera. Magendie had just made a trip to England, to Sunderland, where he had been able to follow closely an epidemic of this disease. After his return to Paris, cholera broke out there. Magendie fought it courageously and devised a good symptomatic treatment, but he was seriously mistaken in asserting that it was not contagious. He also denied the contagiousness of yellow fever and opposed quarantine.

This error had dire consequences, in particular after 1848, when Magendie was appointed head of the Advisory Committee on Public Hygiene. Even though he belonged to the anticontagionist camp, Magendie had made a positive contribution to the study of infection: he had demonstrated experimentally that the saliva of rabid dogs contains a contagious principle. He also observed the effects of intravenous injections of putrid blood and led B. Gaspard to study the phases of sepsis by the experimental method (1822–1823). Another serious error of Magendie's was his impassioned activity against surgical anesthesia induced by ether (1847).

From 1832 to 1838 Magendie delivered his famous lectures on the physical phenomena of life at the Collège de France. These lectures were dominated by two main ideas: to extend as far as possible the purely physical explanation of vital phenomena and to base medical practice on the certain knowledge of normal and pathological physiology. Among the discoveries belonging to this period, the most interesting is that concerning the phenomenon later called anaphylaxis: Magendie ascertained that a second injection of egg white results in the death of rabbits that had tolerated perfectly well the first injection of the substance.

Beginning in 1838 Magendie's lectures dealt successively with the physiology of the nervous system, the dynamics of the circulation of the blood, the cerebrospinal fluid, and nutrition. In collaboration with Poiseuille, he carried out fundamental studies on arterial pressure and demonstrated the hemodynamic role of the elasticity of the major arteries. He also showed the very poor nutritive value of gelatin, until then utilized in the hospitals as an inexpensive food. In 1846 Magendie demonstrated that the presence of sugar in the blood is not necessarily a pathological phenomenon. These experiments on glycemia served Claude Bernard as the starting point of the research that culminated in his discovery of the glycogenic function of the liver.

Through his marriage to Henriette Bastienne de Puisaye in 1830, Magendie had acquired an estate in Sannois, Seine-et-Oise. There he led a very happy family life. Yielding to the fatigue brought on by approaching old age, he withdrew more and more to his country house, left the Hôtel-Dieu (1845), and had Bernard substitute for him at the Collège de France (1847). At Sannois he undertook experiments in plant physiology with a view to improving agricultural yield. He died probably of a heart ailment.

Balzac masterfully characterized Magendie, under the barely disguised name "docteur Maugredie," in *La peau de chagrin* (1831): "a distinguished intellect, but skeptical and contemptuous, who believed only in the scalpel" and who "claimed that the best medical system was to have none at all and to stick to the facts."

BIBLIOGRAPHY

I. ORIGINAL WORKS. Magendie's principal publications are "Quelques idées générales sur les phénomènes particuliers aux corps vivants," in *Bulletin des sciences médicales*, **4** (1809), 145–170; "Examen de l'action de quelques végétaux sur la moelle épinière," in *Nouveau bulletin scientifique de la Société philomatique*, **1** (1809), 368–405; *Mémoire sur le vomissement* (Paris, 1813); *Précis élémentaire de physiologie*, 2 vols. (Paris, 1816–1817; 2nd ed., rev., 1825; 3rd ed., rev., 1834; 4th ed., 1836); "Mémoire sur les propriétés nutritives des substances qui ne contiennent pas d'azote," in *Bulletin de la Société philomatique*, **4** (1816), 137; "Mémoire sur l'émétine et sur les trois espèces d'ipécacuanha," in *Journal général de médecine, de chirurgie et de pharmacie*, **59** (1817), 223–231, written with P. J. Pelletier; *Formulaire pour la préparation et l'emploi de plusieurs nouveaux médicaments, tels que la noix vomique, la morphine, etc.* (Paris, 1821); "Expériences sur les fonctions des racines des nerfs rachidiens," in *Journal de physiologie expérimentale et de pathologie*, **2** (1822), 276–279; "Expériences sur les fonctions des racines des nerfs que naissent de la moelle épinière," *ibid.*, pp. 366–371; "Mémoire sur les fonctions de quelques parties du système nerveux," *ibid.*, **4** (1824), 399–407; "Mémoire sur un liquide qui se trouve dans le crâne et le canal vertébral de l'homme et des animaux mammifères," *ibid.*, **5** (1825), 27–37, and **7** (1827), 1–29, 66–82; *Lectures on the Blood* (Philadelphia, 1839); *Leçons sur les fonctions et les maladies du système nerveux*, 2 vols. (Paris, 1839–1841); *Phénomènes physiques de la vie*, 4 vols. (Paris, 1842); and *Recherches physiologiques et cliniques sur le liquide céphalo-rachidien ou cérébrospinal* (Paris, 1842).

II. SECONDARY LITERATURE. Among the obituary notices containing information on Magendie's life and work, of special interest are F. Dubois, "Éloge de M. Magendie," in *Mémoires de l'Académie impériale de médecine de Paris*, **22** (1858), 1–36; and P. Flourens, *Éloge historique de François Magendie* (Paris, 1858), see also E. Littré, "Magendie," in *Journal des débats* (30 May and 28 June 1856); and A. E. Serres, *Funérailles de Magendie* (Paris, 1855). In a lecture at the Collège de France, Claude Bernard analyzed the historical influence, philosophical position, and character of his teacher: *François Magendie* (Paris, 1856).

For biographies of Magendie, see M. Genty, "François Magendie," in *Les biographies médicales*, IV (1935), 113–144; P. Menetrier, "Documents inédits concernant Magendie," in *Bulletin de la Société française d'histoire de la médecine*, **20** (1926), 251–258; the best modern biographical study is undoubtedly J. M. D. Olmsted, *François Magendie: Pioneer in Experimental Physiology and Scientific Medicine in XIX Century France* (New York, 1944).

Concerning Magendie's epistemological views and general ideas one should consult T. S. Hall, *Ideas of Life and Matter* (Chicago, 1969), II, 245–251; and O. Temkin, "The Philosophical Background of Magendie's Physiology," in *Bulletin of the History of Medicine*, **20** (1946), 10–35. On the pharmacological experiments see M. P. Earles, "Early Theories of Mode of Action of Drugs and Poisons," in *Annals of Science*, **17** (1961), 97–110. The controversy between Bell and Magendie over the discovery of the properties of the spinal nerves was well analyzed by C. Bernard, *Rapport sur le progrès et la marche de la physiologie en France* (Paris, 1867), pp. 10–14, 154–158; and by A. Flint, Jr., "Considérations historiques sur les propriétés des racines rachidiennes," in *Journal de l'anatomie et de la physiologie . . .*, **5** (1868), 520–538, 577–592. A recent restatement of the issue can be found in E. Clarke and C. D. O'Malley, *The Human Brain and Spinal Cord* (Berkeley–Los Angeles, 1968), pp. 296–303.

M. D. GRMEK

MAGGI, BARTOLOMEO (*b.* Bologna, Italy, 1477; *d.* 1552), *surgery.*

Maggi was professor of surgery at the University of Bologna and was the private physician of Pope Julius III. He did not become internationally renowned until 1550, when, because of his skill in treating wounds, he was summoned to Modena to tend the nephew of Pope Paul III, who had suffered a gunshot wound. Before this, Henry II of France had rewarded him with honors and gifts for his curative treatment of wounded French soldiers. He also had already created a school whose pupils supported and defended him in his controversy with Francesco Rota on the treatment of wounds.

Maggi's great ability is illustrated by his book on the treatment of wounds, which, while reflecting his valuable personal experiences and observations, also recalls a method of treatment already adopted in Italy. This method was also discussed by Paré, who acknowledged his debt to Maggi in the introduction to his own treatise on the subject. Maggi's book, published posthumously at Bologna in 1552 by his brother Giovanni Battista, is entitled *De vulnerum bombardorum et sclopetorum, globulis illatorum, et de eorum symptomatum curatione, tractatus*. This work, which in some ways was avant-garde, was of considerable benefit in the treatment of the war-wounded. Its main thesis can be summarized as follows: The wounds inflicted by firearms neither burn nor poison but are first-degree contusions. The shells propelled by firearms do not burn or scald on touch; do not set clothing on fire; do not produce blisters in the areas hit; do not burn gunpowder, hay, sulfur, straw, or tow; and do not give the wounded a burning sensation. A wax ball produces the same effect as a lead one and, like the lead ball, bounces. Shells, moreover, are not poisonous; the components of gunpowder—charcoal, sulfur, and niter—neither have the characteristics of poisons individually nor become poisonous in combination, since such a mixture can be tasted without ill effects. Wounds are contusions, and the gravity of the contusion determines the symptoms of the victim, who may reach a state of general shock. Maggi's theory became accepted, although it took several years. Shortly after its presentation it was defended by Leonardo Botallo in *De curandis vulneribus sclopetorum* (Lyons, 1560) and argued against by Francesco Rota in *De bellicorum tormentorum vulneribus eorumque curatione liber* (Venice, 1555).

A century before Magati's expounding of strange hemostatic practices (the application of boiled ass's or horse's dung, for example), Maggi, although he knew of other valid hemostatic cures, was treating amputation stumps with clay mixed with vinegar. He recommended the same remedy for those bitten by vipers, because clay is cold and the bite is warm *(contraria contrariis)*, and because earth, from which animals derive their poison, is a healthy medicament. Maggi is further remembered for his method of layered amputation.

Maggi was among the first to teach a rational method of treating gunshot wounds, and therefore his name has a deserved place in the history of surgery.

BIBLIOGRAPHY

On Maggi and his work, see C. Burci, *Storia compendiata della chirurgia italiana dal suo principio fino al secolo XIX* (Florence, 1876), pp. 43–44; D. Giordano, "Medicazioni strane e medicazioni semplici," in *Scritti e discorsi pertinenti alla storia della medicina e ad argomenti diversi* (Milan, 1930), pp. 25–45; A. von Haller, *Bibliotheca chirurgica*, I (Bern–Basel, 1774), 206–207; and S. de Renzi, *Storia della medicina in Italia*, III (Naples, 1845), 660–666.

LORIS PREMUDA

MAGINI, GIOVANNI ANTONIO (*b.* Padua, Italy, 13 June 1555; *d.* Bologna, Italy, 11 February 1617), *mathematics, astronomy, geography, cartography.*

Magini graduated with a degree in philosophy from the University of Bologna in 1579; in 1588 he was appointed to one of the two chairs of mathematics there, having been preferred for that post to his younger contemporary Galileo. (The other chair was held by Pietro Cataldi, a mathematician of great prestige.) Magini alternated lectures on Euclid with classes in astronomy, which, stimulated by his passion for astrology, was actually his chief scholarly interest. Astrology itself had been taught at Bologna since 1125. Its study produced results occasionally useful to astronomers, as, for example, the more accurate calculation of celestial movements. Magini wrote several astrological works that were admired in their time, and also served the Gonzaga prince of Mantua as judicial astrologer (with varying results). For this reason he spent long periods of time in that city.

Like his astrological works, Magini's writings on astronomy remain of only historical interest, due in large part to his adherence to Ptolemaic principles. He rejected the Copernican theory, which was then being vindicated by Galileo; the conservatism of his thought indeed made him Galileo's enemy, and Magini more or less openly lent his support to libels against the younger man. Within the boundaries of his Ptolemaicism, Magini drew up complex theories, among them the multiplication of Ptolemaic spheres and orbits, and also performed some useful calculations. He was, in fact, much more skilled in calculation than in theory, and his ephemerides remained valid for a long time.

Magini's mathematical work was essentially practical. In 1592 he published his *Tabula tetragonica*, a table of the squares of natural numbers which was designed to permit the determination of the products of two factors as the difference between two squares. In 1609 he brought out extremely accurate trigonometric tables, in which he introduced new terms for what are now called cosines, cotangents, and cosecants. Magini's nomenclature enjoyed some currency, and was later adopted by Cavalieri, who succeeded him at Bologna. Magini made further contributions to practical geometry, including works on the geometry of the sphere and the applications of trigonometry, for which he invented certain calculating devices that may be reconstructed from his texts. Of his lectures on Euclid, some notes relating to the third book are extant in the Ambrosian Library in Milan.

Although Magini's fame in his own century rested upon these and other accomplishments (including his studies on mirrors and especially the concave spherical mirrors that he fabricated, one of which he presented to the emperor Rudolf II), he is today remembered chiefly as a geographer and cartographer. One of his earliest works was a commentary on Ptolemaic geography, in which he took up the problem of the topographical representation of the earth. He then embarked upon the ambitious project that, with interruptions, occupied him the rest of his life—an atlas of Italy, providing maps of each region (showing the borders of each state) with exact nomenclature and historical notes. The most complete edition of this atlas was published by his son, Fabio, in 1620, three years after Magini's death. Unfortunately, even this edition represents only a small part of Magini's actual work, since his notes for a greater volume, together with much of his library (particularly astrological works), were confiscated by the Roman Inquisition and apparently lost or destroyed.

BIBLIOGRAPHY

I. ORIGINAL WORKS. Magini wrote in Latin and most of his works were then translated into Italian. The major works are *Ephemerides coelestium motuum* (Venice, 1582); *Novae coelestium orbium Theoricae congruentes cum observationibus N. Copernici* (Venice, 1589); *De planis triangulis liber unicus et de dimitiendi ratione per quadrentem et geometricum quadratum libri quinque* (Venice, 1592); *Tabula tetragonica, seu quadratorum numerorum cum suis radicibus* (Venice, 1592); *Geographiae universae* (Venice, 1596); *Tabulae primi mobilis, quas directionum vulgo dicunt* (Venice, 1604). His later works include *Continuatio Ephemeridum coelestium motuum* (Venice, 1607); *Ephemeridum coelestium motuum, ab anno Domini 1608 usque ad annum 1630* (Frankfurt, 1608); *Tabulae generales ad Primum Mobile spectantes, et primo quidem sequitur magnus canon mathematicus* (Bologna, 1609); *Breve instruttione sopra l'apparenze et mirabili effetti dello specchio concavo sferico* (Bologna, 1611); *Geographiae universae* (Venice, 1616); *Tabulae novae iuxta Tychonis rationes elaboratae* (Bologna, 1619); and his atlas, *Italia* (Bologna, 1620).

II. SECONDARY LITERATURE. The best biography of Magini is A. Favaro, *Carteggio inedito di Ticone Brahe, Giovanni Keplero e di altri celebri astronomi e matematici dei secoli XVI e XVII con Giovanni Antonio Magini* (Bologna, 1886). Other works are R. Almagia, *L'Italia di G. A. Magini e la cartografia dell'Italia nei secoli XVI e XVII* (Naples, 1922); and G. Loria, *Storia delle matematiche* (Milan, 1950), pp. 380, 400, 422–425.

LUIGI CAMPEDELLI

MAGIOTTI, RAFFAELLO (*b.* Montevarchi, Italy, 1597; *d.* Rome, Italy, 1656), *physics, hydrostatics, hydrodynamics.*

Raffaello Magiotti studied in Florence and was one of the three favored pupils, along with Castelli and Torricelli, whom Galileo referred to as his Roman "triumvirate." After becoming a priest in the order of Santa Lucia della Chiavica, he was invited to accompany Cardinal Sacchetti to Rome around 1630 as his houseguest. At Rome Magiotti became well known as a scholar in mathematics, law, medicine, theology, and letters. His wide culture secured for him an appointment as *scrittore* in May 1636 on the scholarly staff of the Vatican Library with a salary of 200 scudi a year. Although Galileo, with whom he maintained a lively correspondence, and Castelli wished to nominate him (1638–1640) for the chair of mathematics at Pisa, Magiotti refused to leave the congenial intellectual life of Rome, where he died in 1656 of the plague.

It was probably Magiotti who, stimulated by Galileo's treatment of siphons in the *Two New Sciences* (1638), encouraged acquaintances at Rome to experiment further on siphons and the vacuum. At any rate, Magiotti was present at an experiment that was devised and staged at Rome by Berti, probably in 1640, but definitely at some time between December 1638 and 2 January 1644. From the description thereof left by Magiotti and other eyewitnesses, this experiment resembled the later "barometric" experiment performed by Torricelli in 1643–1644, although it is not clear whether Berti was trying to demonstrate air pressure. In a letter addressed to Marin Mersenne some years later (12 March 1648) Magiotti says that he had earlier forwarded news of the Berti experiment to Torricelli and had suggested that the use of a liquid such as seawater, which would be heavier than the plain water used by Berti, would make a significant difference in the result. "They [Torricelli and Viviani] then carried out experiments and eventually arrived at [the use of] mercury."

In the field of hydrodynamics, Torricelli openly acknowledged the aid of Magiotti. According to Torricelli's theory of flow (1643), the mean velocities of a liquid flowing out of the bottom outlet of a vessel are proportional to the square root of the head pressure, that is, the column of liquid above the outlet. This hypothesis was borne out experimentally by Magiotti who then determined the rate of flow through various sizes of openings. Magiotti thus anticipated by nearly thirty years the similar experiments of Edme Mariotte (1673). (Not surprisingly Torricelli greatly admired Magiotti and sought the priest's approval of his work on solid cycloids.)

Only one work by Magiotti was printed during his lifetime, the *Renitenza dell'Acqua alla Compressione* (1648). This work, in the form of a letter to Lorenzo de' Medici, embodies the first published announcement of the near incompressibility of water at a constant temperature—although Magiotti errs by insisting that water is absolutely incompressible—and the expansion and contraction of water and air according to changes in temperature. Several thermometers and other devices are mentioned, the most interesting being a "Cartesian devil" or "diver." In Magiotti's description, a stoppered cylinder containing an empty inverted jug is filled with water. As the stopper at the top is pushed into or withdrawn from the cylinder, the varying compression makes the jug fall or rise. The effect is caused by the incompressibility of the water. When the stopper is pushed into the cylinder, touching the water directly with no air space in between, the water forces itself into the jug, compressing the air therein and forcing the jug to descend. When the pressure is relaxed, the decompressed air forces the water out of the jug, which, again being lighter than water, rises to the surface. Although a related effect had been described earlier by Beeckman, Magiotti's is the first thorough—and the first printed—description of the Cartesian devil.

After Magiotti died, Leopold de' Medici commissioned Borelli in 1658 to seek out the late priest's manuscripts in Rome. Borelli reported that the writings had been destroyed by looters two years before. Despite this loss, however, Magiotti holds an important place in the history of science because of his probable connection with the first experiment to produce the vacuum, his experiments on hydrodynamics, and his announcement of the incompressibility of water.

BIBLIOGRAPHY

I. ORIGINAL WORKS. Magiotti's *Renitenza certissima dell'acqua alla compressione* (Rome, 1648) was reprinted in G. Targioni-Tozzetti, *Atti e memorie inedite dell'Accademia del Cimento e Notizie aneddote dei progressi delle scienze in Toscana*, II (Florence, 1780), 182–191; and in

L. Belloni, "Schemi e modelli della macchina vivente nel Seicento, con ristampa della lettera di R. Magiotti ...," in *Physis*, **5** (1963), 259–298. A MS version of the work is among the Galilean MSS at the Biblioteca Nazionale Centrale in Florence. The autograph of Magiotti's letter to Mersenne on the Berti vacuum experiment is in the Nationalbibliothek, Vienna, MS 7049, no. 127; it was printed in C. De Waard, *L'expérience barométrique; ses antécédents et ses explications* (Thouars, 1936), 178–181. The correspondence with Galileo is in A. Favaro, ed., *Opere di Galileo Galilei*, XVIII (Florence, 1890–1909), 525. Correspondence with Torricelli is in G. Loria and G. Loria, eds., *Opere di Evangelista Torricelli*, III (Faenza, 1919), 75 (cf. 37, 43, 102, 109, 150, 165, 204). A letter to Candido del Buono on the comet of December 1652 appears in A. Fabroni, ed., *Lettere inedite di uomini illustri*, II (Florence, 1775), 259–263 (cf. I, 151–152).

II. SECONDARY LITERATURE. A life of Magiotti is in G. Targioni-Tozzetti, *op. cit.*, I, 171–172. Also see *Opere di Galileo*, XX, 472–473. For Magiotti's role in the vacuum experiment and his other work see De Waard, *op. cit.*, 101–117, 132–137; W. E. K. Middleton, *The History of the Barometer* (Baltimore, 1964), 10–18; and *Invention of the Meteorological Instruments* (Baltimore, 1969), 3–18. For Torricelli's opinion of Magiotti, see *Opere di Galileo*, XVIII, 327, 331–332; and *Opere di Evangelista Torricelli*, I, 174, which also cites Magiotti's experimental verification of the theory of flow, II, 190.

Descriptions of the Berti experiment by Zucchi, Kircher, Maignan and K. Schott are reprinted by De Waard, *op. cit.*, 145 ff.

For the Cartesian devil, see G. Govi, "In che tempo e da chi siano inventati i ludioni, detti ordinariamente 'Diavoletti Cartesiani,' " in *Rendiconti dell'Accademia delle scienze fisiche e matematiche*, **18** (1879), 291–296.

PAUL LAWRENCE ROSE

MAGNENUS, JOHANN CHRYSOSTOM (*b.* Luxeuil-les-bains, France, *ca.* 1590; *d.* 1679[?]), *natural philosophy, medicine.*

Little is known of Magnenus' family and early life other than that he received the M.D. from the University of Dôle. He traveled for a period in Italy, becoming well-known as a doctor, and was subsequently appointed professor of medicine at the University of Pavia, where several years later he also secured the chair of philosophy. In 1660 Magnenus was chosen personal physician to the count of Fuensaldagne, ambassador to the French court, whom he accompanied to Paris.

Magnenus' importance in the history of science derives from his attempt to reinstate the Democritean theory of atomism as a respectable part of seventeenth-century natural philosophy. His *Democritus reviviscens* (1646), though marking less of a break with tradition than Gassendi's contemporaneous revival of Epicureanism, was typically regarded (for example, by Boyle) as instrumental in establishing a comprehensive alternative to Aristotelianism. Magnenus adopted Democritus' view that matter is composed of physically indivisible atoms which differ in size and shape for each element. He rejected the concept of *materia prima*, asserting that the elements are not interconvertible but preserve their atomic identity and properties when combined chemically.

Eight fundamental propositions for the existence of atoms were advanced by Magnenus, based on mathematical as well as chemical and other experimental considerations. Much of the experimental evidence was drawn from Daniel Sennert's *Hypomnemata physica* (Frankfurt, 1636) and Jacques Gaffarel's *Curiositez inouyes* (Paris, 1629), although he also cited the work of more prominent scientists of the period, including Galileo. Magnenus countered mathematical objections to the atomic theory by arguing that the continuum could not be built up from mathematical points, whether their number be finite, indeterminate, or infinite. Matter, he averred, consisted of atoms having definite dimensions (unlike mathematical points) which represent the physical limit to material division. There were three elementary atoms: fire, water, and earth. Each possessed specific properties and gave rise, by their various combinations, to all other natural substances. Air, because it had no characteristic properties but could assume, at different times, all primary properties, was not an element. It functioned as the neutral medium for propagating the specific properties of the elements during interaction and served to prevent a vacuum by filling the pores of compound bodies.

Magnenus' restoration of Democritean atomism was limited, to be sure. Unlike Democritus, he denied the existence of a void; his retention of Aristotelian substantial forms (now inherent in individual atoms) and his explanation of combination by an innate tendency to union further separates his system from classical atomism. Moreover, while the widespread reading and citation of his work facilitated the acceptance of atomistic ideas in general, his theory must be distinguished from those corpuscular philosophies, like Gassendi's and Boyle's, which sought to explicate natural phenomena solely on the basis of the size, shape, and movement of imperceptible particles. Magnenus accepted certain of the tenets of seventeenth-century mechanical philosophy but amalgamated them into a broader system incorporating traditional modes of qualitative chemical explanation. Thus he is representative of that atomist school which,

during the seventeenth and eighteenth centuries, posed an alternative to strict mechanism in science.

BIBLIOGRAPHY

I. ORIGINAL WORKS. Magnenus' major scientific publication is *Democritus reviviscens sive de atomis. Addita est Democriti vita* (Pavia, 1646; Leiden, 1648; London–The Hague, 1658). Other writings include: *De tabaco exercitationes quatuordecim* (Pavia, 1648; Pavia–The Hague, 1658), which treats of the medical usage and effects of tobacco; and *De manna liber singularis* (Pavia, 1648; 2nd ed., Pavia–The Hague, 1658).

II. SECONDARY LITERATURE. For Magnenus' life and work, see J. Güsgens, *Die Naturphilosophie des Joannes Chrysostomus Magnenus* (Bonn, 1910). F. Ueberweg, *Grundriss der Geschichte der Philosophie*, rev. ed., III (Berlin, 1924), 171–174, places Magnenus among the French natural philosophers of the first half of the seventeenth century. Other assessments of his atomic theory include G. B. Stones, "The Atomic View of Matter in the XVth, XVIth, and XVIIth Centuries," in *Isis*, **10** (1928), 458–459; and J. R. Partington, *A History of Chemistry*, II (London, 1961), 455–458.

MARTIN FICHMAN

MAGNI, VALERIANO (*b.* Milan, Italy, 15 October 1586; *d.* Salzburg, Austria, 29 July 1661), *physics.*

At the age of two, Magni was taken by his parents, Constantino and Ottavia Magni, from Italy to Prague; he was to spend much of the rest of his life in central Europe. Magni entered the Capuchin order on 25 March 1602, adopting the name Valeriano in place of his original Christian name, Maximilian. After he had gained a reputation as a preacher and instructor at Prague, Linz, and Vienna, he was appointed in 1613 to a chair of philosophy in the Austrian capital. Three years later Magni helped to establish the Franciscan order in Poland at the request of Sigismund III; the king later tried to obtain a cardinal's hat for Magni. During the 1620's Magni was active in various roles: as Hapsburg envoy to Paris (1622–1623); novice-master at Linz; professor of philosophy at Prague; Franciscan provincial of Bohemia (1624); and Hapsburg emissary to Italy (1625). Following the death of his patron, Sigismund III, Magni played a decisive part in the selection of a successor (1632), and later worked in Poland to consolidate the position of the Catholic church. In 1642–1643, and again in 1645, Magni was in Italy, then in Poland (1646–1648), and subsequently in Vienna and Cologne. In 1655 the combative Magni's long-standing feud with the Jesuits (he had incited Urban VIII, a close friend, against them in 1631) led to his being accused of heresy; while trying to reconcile Protestants to the Catholic church, he had admitted the supremacy of the pope to be founded on tradition. Pleading that he was too ill to obey a summons to Rome, Magni was arrested in Vienna at the end of 1655. The emperor's intervention, however, secured his release the following February, whereupon he was sent to Salzburg. There he remained for the rest of his life.

In philosophy Magni was a vehement anti-Aristotelian and an admirer of Galileo and Descartes. In his fight with the Aristotelians, Magni made great use of an experiment designed to demonstrate the existence of the vacuum. Although this was practically identical with the barometric experiment described by Torricelli in 1643–1644, Magni claimed that this idea was conceived independently after reading of Galileo's work on siphons. There is no firm reason for doubting Magni's word on this, although at the time his claim aroused great controversy.

In mid-1647 Magni demonstrated his experiment at Warsaw in the presence of Wenceslas VII, and in July of that year he published an account of it *(Demonstratio ocularis)* which made much of the fact that light could traverse a vacuum, so proving, against Aristotle, that motion was possible in a void. News of the experiment was communicated by a French eyewitness, Des Noyers, to Mersenne at Paris in a letter dated 24 July 1647. Unfortunately for Magni, the French were then pursuing research on the vacuum, and Roberval, who replied on Mersenne's behalf to Des Noyers's letter, implicitly accused Magni of plagiarism. In his reply of 20 September 1647 (apparently printed in Paris in the same year) Roberval stated that the Torricelli experiment had been performed years before and that Magni, who was in Italy in 1645, must have heard of it there; in any case, the experiment had been repeated by Petit and Pascal at Rouen in 1646.

Magni quickly wrote a defense of his work, the "Narratio apologetica," dated 5 November 1647. This he had printed in a collection entitled *Admirando de vacuo*, which also contained reprints of the *Demonstratio ocularis* and of Roberval's letter to Des Noyers. Magni conceded that Torricelli now had the priority, but he strongly denied having heard anything of the Torricelli experiment before the arrival of Roberval's letter. Torricelli's work, he said, was not known to any of his (Magni's) friends at Rome. Rather, it was Galileo's writings that had stimulated him to devise the experiment.

Magni's forthright publication of the damaging letter of Roberval suggests strongly that he was telling the truth. In any event, the *Demonstratio ocularis* of

Magni is certainly important as the first printed account of the barometric experiment, Torricelli having left his description in manuscript. Although it was printed at Warsaw, Magni's treatise became quite well known following its reprinting at Paris in 1647 as part of an edition that also included Petit's account of the Rouen vacuum experiments. Interestingly, it was Magni who acquainted Guericke with the Torricellian barometric experiment when the two men met at Regensburg in 1654. The several editions of Magni's writings, and the controversy that surrounded them, undoubtedly helped to disseminate widely the news of the barometric experiment.

BIBLIOGRAPHY

I. ORIGINAL WORKS. Magni's main work on the vacuum is *Demonstratio ocularis; loci sine locato; corporis successive moti in vacuo; luminis nulli corpori inhaerentis* (Warsaw, 1647); repr. by M. Dominicy as *Observation touchant le vuide faite pour la première fois en France, contenue en une lettre écrite . . . par Monsieur Petit . . . le 10 Novembre 1646. Avec le discours qui en a esté imprimé en Pologne sur le mesme sujet, en Ieuillet 1647* (Paris, 1647). This work is also in *Admirando de vacuo* (Warsaw, 1647), which contains the critical letter of Roberval to Des Noyers as well as the "Narratio apologetica." The treatise also was published in Bologna (1648) and Venice (1649). Magni published a further treatise entitled *Vacuum pleno supletum* (Venice, 1650). His *Principia et specimen philosophiae* (Cologne, 1642) contains the *opuscula* on the vacuum.

Magni's main philosophical work is the *Opus philosophicum: I. Synopsis philosophiae Aristotelis. II. Philosophia Valeriani* (Lithomifflii, 1660). His main attack on the Jesuits is the *Apologia . . . contra imposturas Jesuitarum* (n.d. [1655?], n.p.).

Some of the vacuum materials are reprinted in Pascal, *Oeuvres*, Leon Brunschvicg, ed., II (Paris, 1908–1914), including Des Noyers's letter to Mersenne (15–18); Roberval's critique (21–35); and Magni's defense (503–506).

II. SECONDARY LITERATURE. German Abgottspon, *P. Valerianus Magni, Kapuziner* (Olten, 1939), deals mainly with Magni's political and religious life. Documents and an account of Magni as a scientist are given in Cornelis De Waard, *L'expérience barométrique, ses antécédents et ses explications* (Thouars, 1936); also see W. E. K. Middleton, *The History of the Barometer* (Baltimore, 1964), ch. 3. For the bibliography of Magni's writings on the vacuum see G. Hellmann, "Beiträge zur Erfindungsgeschichte meteorologischer Instrumente," in *Abhandlungen der Preussischen Akademie der Wissenschaften*, Phys.-Math. Kl., no. 1 (1920), 33–34.

An Aristotelian reply to Magni is Jacobus Pierius, *Ad experientiam nuper circa vacuum R. P. Valeriani Magni demonstrationem ocularem . . . responsio ex peripateticae philosophiae principiis desumpta* (Paris, 1648).

Contemporary references to Magni and the experiment are in Otto von Guericke, *Experimenta nova . . . Magde-*

burgica de vacuo spatio (Amsterdam, 1672), 117–118; Honoré Fabri, *Dialogi physici* . . . (Lyons, 1665), 182–183; and Jakub Dobrzenski, *Nova et amaenior . . . fontium . . . philosophia* (Ferrara, 1659), 27–28. Other references are cited in Lynn Thorndike, *A History of Magic and Experimental Science*, VII (New York, 1923–1958), 654, 659.

PAUL LAWRENCE ROSE

MAGNITSKY, LEONTY FILIPPOVICH (*b.* Ostashkov, Russia, 19 June 1669; *d.* Moscow, Russia, 30 October 1739), *mathematics.*

No precise information exists on Magnitsky's origins and early years. It is possible that he studied in Moscow at the Slavonic, Greek, and Latin Academy founded in 1687. It is also possible that he acquired his broad knowledge, which included many foreign languages, independently. In 1701 Peter the Great founded the Navigation School in Moscow, and it soon became the breeding ground for the technical intelligentsia. Peter brought Magnitsky there to teach in 1702. Magnitsky worked there for the rest of his life, and was named director in 1715.

Magnitsky's *Arithmetic* (1703) was the first guide to mathematics published in Russia. Its first edition of 2,400 copies was extraordinarily large for that time and it served as the basic textbook of mathematics in Russia for half a century. The founder of Russian science, Lomonosov, called it, along with one grammar book, "our gateways to learning." Magnitsky's textbook successfully combined the tradition of Russian mathematical literature of the seventeenth century with that of the western European mathematical schools. In the first section a detailed exposition of mathematical problems is given. The second section, almost an encyclopedia of the natural sciences of the time, contains information on algebra and its geometrical applications, the computation of trigonometric tables of sines, tangents, and secants, and information on navigational astronomy, geodesy, and navigation. There are also tables of magnetic declination, tables of latitude of the points of rising and setting of the sun and moon, and coordinates of the most important ports with their times of high and low tide.

Magnitsky also participated in the preparation of a Russian edition (1703) of the logarithmic tables of Vlacq (1628).

BIBLIOGRAPHY

Magnitsky's one published work was *Arifmetika, sirech nauka chislitelnaya. Tablitsy sinusov, tangensov i sekansov i logarifma sinusov i tangensov* ("Arithmetic, Called the Computational Science. Tables of Sines, Tangents, and Secants and Logarithms of Sines and Tangents"; Moscow, 1703).

Works about Magnitsky are: A. P. Denisov, *Leonty Filippovich Magnitsky* (Moscow, 1967); D. D. Galanin, *Leonty Filippovich Magnitsky i yego "Arifmetika"* ("Leonty Filippovich Magnitsky and his 'Arithmetic' "), 3 vols. (Moscow, 1914); and A. P. Youschkevitch, *Istoria matematika v Rossii do 1917 goda* ("History of Mathematics in Russia Until 1917"; Moscow, 1968).

S. PLOTKIN

MAGNOL, PIERRE (*b.* Montpellier, France, 8 June 1638; *d.* Montpellier, 21 May 1715), *botany.*

Magnol's father was an apothecary, and his mother came from a family of physicians. He was interested in botany from his youth; and in 1659, after receiving his medical degree, he decided that (as his son reports) "it would be very advantageous to him to make a serious study of plants" before practicing medicine. He then began to botanize in the area around Montpellier, in Provence, and in the neighboring islands. He was aided by Laugier, a professor of medicine who was the friend of Gaston, duke of Orleans, and who possessed a great knowledge of plants. Magnol's reputation grew rapidly, and people soon competed to join the excursions he led. He established contacts with many French and foreign botanists: John Ray, William Sherard, and James Petiver in London; Herman and Hotton in Leiden; Commelin in Amsterdam; the Rivinuses (Bachmanns) in Leipzig; Breyn in Danzig; Johann Heinrich Lavater in Zurich; Lelio and Giovanni Battista Triumpheti in Rome; Giovanni Ciassi in Venice; Boccone in Palermo; Nappus in Strasbourg; J. Salvador in Barcelona; Jacob Spon in Lyons; and Gui C. Fagon in Paris. In 1663 Magnol obtained, through J. P. de Tournefort, a *brevet de médecin ordinaire du roi*. In 1667 the king opposed his nomination as professor of medicine at the University of Montpellier (he was not appointed until 1694, but in the meantime he renounced Protestantism). Magnol was not disturbed by this rejection, which permitted him to devote his time to botany.

Magnol's *Botanicum Monspeliense*, containing, it is said, the description of 1,354 species, appeared in 1676; it was intended for his students, among whom were Antoine and Bernard de Jussieu. Out of love for botany Magnol agreed in 1687 to substitute for François Chicoyneau, whose sight was starting to fail, as demonstrator of plants at the botanical garden of Montpellier. In 1697 Magnol was named director of the botanical garden. He was called to Paris in 1709 to replace Tournefort at the Académie Royale des Sciences and was particularly warmly received there by Fontenelle. But he soon wished to return to his native city, where, in "his" garden, he cultivated rare plants. In 1697 he had published a catalog of this garden *(Hortus regius Monspeliensis)* in which several new species were described, including *Lonicera pyrenaica* and *Xanthium spinosum*.

An innovator in classification, Magnol was one of the first, in his *Prodromus historiae generalis plantarum* (1689), to classify plants in tables that made possible rapid identifications. In his *Novus caracter plantarum*, posthumously published by his son Antoine in 1720, "he proposed a new classification based on the calyx (a name that he gives even to the unique floral envelope of certain plants)." He demonstrated that the fig contains many flowers but was unable to interpret the fructification of the ferns. He recognized that the coral is a "living body" but thought it was a plant. Magnol also established that desiccation causes the tuber of the arum to lose its "burning acridity." He observed the underground components of the *Bryonia*, the cyclamen, the Jerusalem artichoke, turnips, and other plants and concluded that by drying them, kneading them with wheat or rye flour, and baking them a quite nourishing food could be obtained; it is the root of the creeping wheat-grass that gives bread the most agreeable taste. These observations were published in the *Histoire de la Société des sciences de Montpellier*.

Magnol helped to promote interest in botany, which he thought was excessively neglected by educated people, and attracted attention to the possibility of employing natural classifications. Moreover, he was undoubtedly the first to use the term "family" in the sense of a natural group. The family Magnoliaceae is represented by the genus *Magnolia*, which was dedicated to him by Plumier.

BIBLIOGRAPHY

I. ORIGINAL WORKS. Magnol's works are *Botanicum Monspeliense* (Lyons, 1676); *Prodromus historiae generalis plantarum in quo familiae plantarum per tabulas disponuntur* (Montpellier, 1689); *Hortus regius Monspeliensis* (Montpellier, 1697); and the posthumously published *Novus caracter plantarum, in duos tractatus divisus* (Montpellier, 1720). He also contributed various memoirs to the *Histoire de la Société royale des sciences de Montpellier*.

II. SECONDARY LITERATURE. The principal source is the biography by Magnol's son Antoine, in J. E. Planchon, ed., *La botanique à Montpellier. Notes et documents . . .* (Montpellier, 1884). See also L. Dulieu, "Les Magnol," in *Revue d'histoire des sciences et de leurs applications*, **12** (1959), 209–224; and Robert Zander, "Pierre Magnol," in *Das Gartenamt* (Nov. 1959), 245–246, with portrait.

PAUL JOVET
J. C. MALLET

MAGNUS, HEINRICH GUSTAV (*b*. Berlin, Germany, 2 May 1802; *d*. Berlin, 4 April 1870), *physics, chemistry*.

Magnus' father, Johann Matthias Magnus, the prosperous founder of a large trading firm, was able to provide his son with private instruction in mathematics and natural science. Magnus entered the University of Berlin in 1822; and in 1825 he published his first paper, an investigation of pyrophoric iron, cobalt, and nickel carried out under the direction of Eilhard Mitscherlich, discoverer of the law of isomorphism. After receiving his doctorate in September 1827 with a dissertation on tellurium, Magnus took the advice of Mitscherlich, Heinrich and Gustav Rose, and Friedrich Wöhler, all former students of Berzelius, and in October 1827 went to Stockholm to study with the great Swedish chemist, who became his lifelong friend and adviser.

It was in Berzelius' laboratory that Magnus not only discovered the first platinum-ammine compound (Magnus' green salt $[Pt(NH_3)_4][PtCl_4]$) and its related potassium salt ($K_2[PtCl_4]$) but also worked on the addition compound of ethylene and platinous chloride later described by the Danish chemist W. C. Zeise (Zeise's salt, $K[Pt(C_2H_4)Cl_3]$). In the summer of 1828 Magnus returned to Berlin, where, with the exception of a visit to Paris during 1828 and 1829, he remained until his death. His *Habilitationsschrift* on mineral analysis (1831) permitted him to begin lecturing on technology at the university and on chemistry at the municipal trade school but led to a break with his teacher Mitscherlich, who regarded the young *Privatdozent* as a dangerous competitor. In 1833 Magnus was appointed associate professor and in 1845 professor of technology and physics at the University of Berlin, where he also served as rector during 1861 and 1862. He married Bertha Humblot in 1840. Magnus became a member of the Berlin Academy of Sciences in 1840 and was one of the founding members of the German Chemical Society (1868). A number of his students became famous physicists.

As was true of most chemists of the time, Magnus' research interests were varied. From an initial interest in mineral analysis, he turned to inorganic chemistry, discovering periodic acid and its salts in 1833; organic chemistry, discovering ethionic and isethionic acids in 1833–1839 and the polymerization of hydrocarbons on heating in 1853; physiological chemistry, studying the oxygen and carbon dioxide content of blood in 1837–1845; and agricultural chemistry in 1849. Magnus gradually turned to physicochemical and eventually purely physical investigations, which constitute his most important scientific achievements.

Among these are his contributions to the theory of heat, thermal expansion of gases, boiling of liquids, vapor formation, electrolysis (Magnus' rule), induced and thermoelectric currents, optics, hydrodynamics, magnetism, and mechanics. Although his most important work was in physics, he never ceased investigating chemical problems in his private laboratory. These later chemical works, however, never led to results of general significance but served merely for his own instruction.

Neither a theoretician nor an original thinker, Magnus was, however, an acute, conscientious, and diligent experimenter who uncovered much valuable physical and chemical information—notably the Magnus effect.

BIBLIOGRAPHY

I. ORIGINAL WORKS. Magnus' papers are listed in the Royal Society *Catalogue of Scientific Papers*, IV, 182–184; VIII, 306. See esp. "Ueber die Eigenschaft metallischer Pulver, sich bei der gewöhnlichen Temperatur von selbst in der atmosphärischen Luft zu entzünden," in *Annalen der Physik und Chemie*, **3** (1825), 81–88; "Ueber einige neue Verbindungen des Platinchlorürs," *ibid.*, **14** (1828), 239–242, with English trans. in G. B. Kauffman, ed., *Classics in Coordination Chemistry, Part II. Selected Papers (1798–1935)* (New York, in press); and "Ueber die Weinschwefelsäure, ihren Einfluss auf die Aetherbildung, und über zwei neue Säuren ähnlicher Zusammensetzung," *ibid.*, **27** (1833), 367–387.

II. SECONDARY LITERATURE. See the notices by A. W. Hofmann, in *Berichte der Deutschen Chemischen Gesellschaft*, **3** (1870), 993; and A. W. Williamson, in *Journal of the Chemical Society*, **24** (1871), 610–615.

See also J. J. Berzelius, *Aus Jac. Berzelius und Gustav Magnus' Briefwechsel in den Jahren 1828–1847*, E. Hjelt, ed. (Brunswick, 1900); and W. Prandtl, *Deutsche Chemiker in der ersten Hälfte des neunzehnten Jahrhunderts* (Weinheim, 1956), 303–314.

GEORGE B. KAUFFMAN

MAGNUS, OLAUS. See **Olaus Magnus.**

MAGNUS, RUDOLF (*b.* Brunswick, Germany, 2 September 1873; *d.* Pontresina, Switzerland, 25 July 1927), *neurophysiology, pharmacology.*

Magnus was raised in a rich intellectual environment that embraced medicine, the law, and the humanities. His father practiced law in Brunswick, one grandfather was director of the Hamburg library, and the other was a physician. Magnus' initial career interests lay in literature and philosophy. Partly through the advice of a family friend, the chemist Richard Meyer, however, he decided to study medicine at Heidelberg. There his career was molded by such influential figures as Willy Kühne, the noted physiologist, and the chemist David Meyer; and he began lasting friendships with certain scientific figures, notably Jakob von Uexküll and Otto Cohnheim. In 1895, while still a medical student, Magnus showed his aptitude for original research in a paper presented at the Third International Congress of Physiology at Bern, dealing with a method for measuring blood pressure in an exposed artery. The further development and application of this technique was the subject of his doctoral thesis in 1898.

Magnus began his career in pharmacology, working first at Heidelberg as an assistant in 1898 and as a *Privatdozent* in 1900. At Heidelberg he continued investigations that he had begun as an undergraduate on the cardiovascular, renal, and intestinal systems, earning rapid and wide recognition for studies on the action of arsenic and of various pharmacologic agents in the gut, and on water balance in tissues.

In 1904 Magnus devised a now-standard technique in pharmacology for studying the responses of isolated muscle, suspending a loop of small intestine in warmed, oxygenated Locke-Ringer solution. Using the method to make a series of important observations on responses to alkaloid agents, on local reflexes, and on automatic rhythmicity, he discovered that the degree of stretching of the intestinal muscle determines the direction of stimulus conduction.

During his years at Heidelberg, Magnus made a series of visits to British research laboratories, beginning with a trip to work with E. A. Sharpey-Schafer at Edinburgh in 1900. Together they discovered the diuretic action of pituitary gland extracts. In 1905 Magnus went to Cambridge to learn surgical techniques for studying the autonomic nervous system from J. N. Langley, in order to continue his analysis of the relations between drug effects and nerve supply on the motility of intestinal muscle. The critical event that shaped the direction of his future neurophysiological investigations was his 1908 visit to Charles Sherrington at Liverpool.

Magnus accepted the professorship of pharmacology at the University of Utrecht in 1908, since there was then no vacant chair of pharmacology in Germany. During the next two decades his Utrecht group issued over 300 papers. The major corpus of Magnus' work, and that for which he is best known, deals with the reflex control of posture. He also continued an active program of teaching and research in pure pharmacology.

Magnus was noted as a teacher and speaker as well as a gifted investigator. The lectures he delivered in his pharmacology courses, published as *Pharmakologisches Praktikum*, reported on the latest research projects of his institute, particularly the pharmacology of the pulmonary circulation, and the isolation and identification of choline as the hormonal regulatory agent for intestinal muscle. The wide range of his interests, embracing history, philosophy, and botany, was demonstrated by a lecture series on Goethe as a natural scientist (1906).

Although present concepts of the body's equilibratory system are principally an outgrowth of Magnus' work, data on the role of the cerebellum in maintaining body attitudes had been accumulating slowly since experiments by François Magendie in the 1820's. Toward the end of the nineteenth century, David Ferrier's experiments on various animals led him to the idea of a cerebellovestibular connection, a germinal idea that helped launch the modern period of study of equilibratory functions.

The postural reflex studies by Magnus and his colleagues, particularly A. de Kleijn, were a model of the integrative neurophysiology being pioneered by Sherrington at Liverpool and Oxford. While Magnus was working with him in 1908, one of Sherrington's research interests was muscle tonus in mammals. Analyzing the reflex pathways involved in the production and maintenance of tonus, Sherrington concluded that tone in mammals is due to postural reflexes, as P. Q. Brondgeest in 1860 had shown it to be for frogs and rabbits. In 1910, drawing in part upon his observations of the "reflexe figures" assumed by decerebrate animals, Sherrington published a detailed analysis of the reflex control of stepping and standing.

Under Sherrington's guidance, Magnus in 1908 had begun experiments on mammalian muscle tonus, drawing in part upon Uexküll's work on the changing responses of the muscle bands in a marine worm to varying tensions. Encouraged by Sherrington to continue the investigation of equilibratory phenomena, Magnus and Kleijn published the first of their classic papers on the influence of head position on the tonus of extremities in 1912. The depth of Magnus' subsequent studies is suggested by the fact that one observation he and Sherrington had made independently—that rotation of the head in the decerebrate animal changes the muscle tonus of the limbs—generated a series of eighty-two publications by the Utrecht group.

The investigations of Magnus and his colleagues showed the many automatic reflex actions through which an animal assumes and maintains body postures,

by sequences of coordinated reflexes and by the types of static "figures" demonstrated so clearly in the decerebrate preparation. Their fundamental analysis of equilibratory functions involved, first, a detailed study of tonic neck muscle and labyrinth reflexes and of labyrinth righting reflexes, by which body postures change in response to various stimuli as an animal constantly adjusts to its needs and environment. They then restudied the various reflexes they had cataloged after ablating the cerebellum, cerebrum, brain stem, or cervical spinal cord, in order to localize the brain and spinal cord areas controlling posture.

Their study of the labyrinth organs, which was inspired in part by the work of J. R. Ewald, led Magnus and Kleijn to a fundamental analysis of the responses of the ear's vestibular organs to natural stimuli. In one classic experiment they differentiated between the functions of the otolith organs and the semicircular canals by centrifuging anesthetized guinea pigs at high speed. The otolith membranes were detached by the procedure; but the canals, ampullae, and cristae remained intact. Magnus and Kleijn then found that reflexes resulting from static posture were abolished in the absence of the otolith mechanism but that the animals retained all the labyrinth reactions produced by rectilinear acceleration.

Through such studies Magnus' research group founded our knowledge of the complex integrative reflex system by which the brain stem and cervical spinal cord control musculature. Step by step they documented the functions of the vestibular apparatus, tonic neck and labyrinth reflexes, and other postural and righting reflexes and their neural pathways and control mechanisms. Magnus summarized the work of his Utrecht laboratory, and surveyed the work of others in his field, in *Die Körperstellung* (1924), a monograph justly cited as a classic work in reflex physiology. At the time of his sudden death in 1927, at age fifty-three, he and Kleijn were under consideration for the Nobel Prize in physiology or medicine for their fundamental contributions to neurophysiology.

BIBLIOGRAPHY

I. ORIGINAL WORKS. Magnus' writings include *Goethe als Naturforscher* (Leipzig, 1906); "Welche Teile des Zentralnervensystems müssen für das Zustandekommen der tonischen Hals- und Labyrinthereflexe auf die Körpermuskulatur vorhanden sein," in *Pflügers Archiv für die gesamte Physiologie des Menschen und der Tiere*, **159** (1914), 224–249; *Körperstellung experimentell-physiologische Untersuchungen über die einzelnen beider Körperstellung in Tätigkeit tretenden Reflexe, über ihr Zusammenwirken und ihre Störungen* (Berlin, 1924); and "Animal Posture," in *Pro-*

ceedings of the Royal Society, **98B** (1925), 339–353, the Croonian lecture.

II. SECONDARY LITERATURE. See H. H. Dale, "In Memoriam Rudolf Magnus (1873–1927)," in *Stanford University Publications, Medical Sciences*, **2** (1930), 241–247; J. F. Fulton, "Rudolf Magnus 1873–1927," in *Boston Medical and Surgical Journal*, **197** (1927–1928), 323–324; and I. N. W. Olninck, "Rudolf Magnus," in E. W. Haymaker, ed., *The Founders of Neurology* (Springfield, Ill., 1953), pp. 149–152.

JUDITH P. SWAZEY

MAGNUS, VALERIANUS. See **Magni, Valeriano.**

MAHĀDEVA (*fl.* western India, 1316), *astronomy.*

The scion of a Brahman family of astronomers and astrologers belonging to the Gautamagotra, a family that began with Bhogadeva and extended through successive generations represented by Mādhava, Padmanābha, and his father, Paraśurāma, Mahādeva resided on the banks of the Godāvarī River—probably near its source in Mahārāshtra. He wrote a lengthy set of astronomical tables, the *Mahādevī* (see essay in Supplement), employing the "true linear" arrangement (see D. Pingree, "On the Classification of Indian Planetary Tables," in *Journal for the History of Astronomy*, **1** [1970], 95–108, esp. 103–104) and the parameters of the *Brāhmapakṣa* (see essay in Supplement); their epoch is 28 March 1316. The extreme popularity of these tables in western India is indicated by the fact that over 100 manuscripts of them originating in that area have been identified. They have also been commented on by Nṛsiṃha of Nandipura in Gujarat (1528) and by Dhanarāja of Padmāvatī in Mārwār (Jodhpur) (1635) and have often been imitated by the astronomers of Gujarat and Rajasthan.

BIBLIOGRAPHY

The tables are discussed in detail by O. Neugebauer and D. Pingree, "The Astronomical Tables of Mahādeva," in *Proceedings of the American Philosophical Society*, **111** (1967), 69–92. See also D. Pingree, "Sanskrit Astronomical Tables in the United States," in *Transactions of the American Philosophical Society*, n.s. **58**, no. 3 (1968), 37a–39a; and "Sanskrit Astronomical Tables in England," in *Journal of Oriental Research* (Madras).

DAVID PINGREE

AL-MĀHĀNĪ, ABŪ ʿABD ALLĀH MUḤAMMAD IBN ʿĪSĀ (*b.* Mahān, Kerman, Persia; *fl.* Baghdad, *ca.* 860; *d. ca.* 880), *mathematics, astronomy.*

Our main source of information on al-Māhānī's life consists of quotations from an unspecified work by al-Māhānī in Ibn Yūnus' *Ḥākimite Tables.* Here Ibn Yūnus cites observations of conjunctions and lunar and solar eclipses made by al-Māhānī between 853 and 866. Al-Māhānī remarked, in connection with the lunar eclipses, that he calculated their beginnings with an astrolabe and that the beginnings of three consecutive eclipses were about half an hour later than calculated.

Al-Māhānī's main accomplishments lie in mathematics; in the *Fihrist* he is mentioned only as geometer and arithmetician. Al-Khayyāmī states that al-Māhānī was the first to attempt an algebraic solution of the Archimedean problem of dividing a sphere by a plane into segments the volumes of which are in a given ratio (*On the Sphere and the Cylinder* II, 4). Al-Māhānī expressed this problem in a cubic equation of the form $x^3 + a = cx^2$, but he could not proceed further. According to al-Khayyāmī, the problem was thought unsolvable until al-Khāzin succeeded by using conic sections. In Leiden there exists a manuscript copy of a commentary to al-Māhānī's treatise, probably by al-Qūhī.

Al-Māhānī wrote commentaries to books I, V, X, and XIII of Euclid's *Elements.* Of these, the treatise on the twenty-six propositions of book I that can be proved without a *reductio ad absurdum* has been lost. Part of a commentary on book X, on irrational ratios; an explanation of obscure passages in book XIII; and three (different?) treatises on ratio (book V) are extant. Since book V, on the theory of proportion, was presented in a synthetic form which did not reveal how the doctrine of proportions had come into being, Arabic mathematicians were dissatisfied with definition 5, the fundamental one. They did not deny its correctness, however, and accepted it as a principle. Gradually they replaced the Euclidean "equimultiple" definition by the pre-Eudoxian "anthyphairetic" definition, which compared magnitudes by comparing their expansion in continued fractions. The "anthyphairetic" conception appears in explicit form in al-Māhānī's treatise, in which he referred to Thābit ibn Qurra. Al-Māhānī regarded ratio as "the mutual behavior of two magnitudes when compared with one another by means of the Euclidean process of finding the greatest common measure." Two pairs of magnitudes were for him proportional when "the two series of quotients appearing in that process are identical." Essentially the same theory was worked out later by al-Nayrīzī. Neither established a connection with

Euclid's definition, which was first done by Ibn al-Haytham.

At the request of some geometers al-Māhānī wrote an improved edition of the *Sphaerica* of Menelaus—of book I and part of book II—which has been lost. His improvements consisted of inserting explanatory remarks, modernizing the language (with special consideration given to technical terms), and remodeling or replacing obscure proofs. This edition was revised and finished by Aḥmad ibn Abī Saʿīd al-Harawī in the tenth century. Al-Ṭūsī, who wrote the most widely known Arabic edition, considered al-Māhānī's and al-Harawī's improvements valueless and used the edition by Abū Naṣr Manṣūr ibn ʿIrāq.

BIBLIOGRAPHY

I. ORIGINAL WORKS. C. Brockelmann, *Geschichte der arabischen Literatur*, supp. I (Leiden, 1937), 383, lists the available MSS of al-Māhānī. Information on al-Māhānī is also given in H. Suter's translation of the *Fihrist* in *Das Mathematiker-Verzeichniss im Fihrist des Ibn abī Jaʿkūb al-Nadim*, in *Abhandlungen zur Geschichte der Mathematik*, VI (Leipzig, 1892), 25, 58. Partial translations and discussions of al-Māhānī's work are in M. Krause, *Die Sphärik von Menelaos aus Alexandrien* (Berlin, 1936), 1, 13, 23–26; G. P. Matvievskaya, *Uchenie o chisle na srednevekovom Blizhnem i Srednem Vostoke* ("Studies on Number in the Medieval Near and Middle East"; Tashkent, 1967), ch. 6, which deals with commentaries on Euclid X; and E. B. Plooij, *Euclid's Conception of Ratio* (Rotterdam, 1950), 4, 50, 61.

II. SECONDARY LITERATURE. On al-Māhānī's observations, see "Ibn Yūnus, *Le livre de la grande Table Hakémite*, trans. by J. J. A. Caussin de Perceval in *Notices et extraits de la Bibliothèque nationale*, **7** (1804), 58, 80, 102–112, 164. Information on al-Māhānī as a mathematician, especially his treatment of the Archimedean problem, is in F. Woepcke, *L'algèbre d'Omar Alkhayyāmī* (Paris, 1851), 2, 40–44, 96. On the anthyphairetic theory, see O. Becker, "*Eudoxos Studien I*," in *Quellen und Studien zur Geschichte der Mathematik, Astronomie und Physik*, Abt. B, **2** (1933), 311–333.

YVONNE DOLD-SAMPLONIUS

MAHĀVĪRA (*fl.* Mysore, India, ninth century), *mathematics*.

Mahāvīra, a Jain, wrote during the reign of Amoghavarṣa, the Rāṣṭrakūṭa monarch of Karṇāṭaka and Mahārāṣṭra between 814/815 and about 880. Nothing else of his life is known. His sole work was a major treatise on mathematics, the *Gaṇitasārasaṅgraha* (see essay in Supplement), in nine chapters:

1. Terminology.
2. Arithmetical operations.
3. Operations involving fractions.
4. Miscellaneous operations.
5. Operations involving the rule of three.
6. Mixed operations.
7. Operations relating to the calculations of areas.
8. Operations relating to excavations.
9. Operations relating to shadows.

There is one commentary on this work by a certain Varadarāja, and another in Kannaḍa, entitled *Daivajñavallabha*.

BIBLIOGRAPHY

The *Gaṇitasārasaṅgraha* was edited, with an English trans. and notes, by M. Raṅgācārya (Madras, 1912); and with a Hindī *anuvāda* by Lakṣmīcandra Jaina as *Jīvarāma Jaina Granthamālā* 12 (Solāpura, 1963). There are discussions of various aspects of this work (listed chronologically) by D. E. Smith, "The Ganita-Sara-Sangraha of Mahāvīrācārya," in *Bibliotheca mathematica*, **3**, no. 9 (1908–1909), 106–110; B. Datta, "On Mahāvīra's Solution of Rational Triangles and Quadrilaterals," in *Bulletin of the Calcutta Mathematical Society*, **20** (1932), 267–294; B. Datta, "On the Relation of Mahāvīra to Śrīdhara," in *Isis*, **17** (1932), 25–33; B. Datta and A. N. Singh, *History of Hindu Mathematics*, 2 vols. (Lahore, 1935–1938; repr. in 1 vol., Bombay, 1962), *passim;* E. T. Bell, "Mahavira's Diophantine System," in *Bulletin of the Calcutta Mathematical Society*, **38** (1946), 121–122; and A. Volodarsky, "O traktate Magaviry 'Kratky kurs matematiki,' " in *Fiziko-matematicheskie nauki v stranakh vostoka*, II (Moscow, 1969), 98–130.

DAVID PINGREE

MAHENDRA SŪRI (*fl.* western India, 1370), *astronomy*.

A Jain and a pupil of Madana Sūri of Bhṛgupura (Broach, Gujarat), Mahendra Sūri wrote the first Sanskrit treatise on the astrolabe, the *Yantrarāja* (1370). He evidently used an Islamic source (see essay in Supplement); in it, for instance, $R = 3600' = 60$ parts; $\varepsilon = 1415' = 23;35°$. Furthermore, the commentary by his pupil, Malayendu Sūri, lists the latitudes of Ādane (Aden), Makkā (Mecca), Badaṣasāna (Badakhshan), Balaṣa (Balkh), Nayasāpura (Nīshāpūr), Samarakanda (Samarkand), Kāsagāra (Kashgar), and other Islamic cities, as well as "Hiṃsārapirojāvāda which is inhabited by the king Pīroja" (the king is Fīrūz Shāh Tughlaq [1351–1388], and the place the Hiṣar palace begun by Fīrūz at Firozabad, near Delhi, in 1354), and both the Persian and the Indian (Sanskrit) names of thirty-two stars.

There is another commentary on the *Yantrarāja* by Gopīrāja (1540) and a set of examples for the year 1512.

BIBLIOGRAPHY

The *Yantrarāja* was edited, with Malayendu's commentary, by Sudhākara Dvivedin (Benares, 1882) and by K. K. Raikva (Bombay, 1936). There are notices on Mahendra in S. Dvivedin, *Gaṇakataraṅgiṇī* (Benares, 1933), pp. 48–49, repr. from *The Pandit*, n.s. **14** (1892); and in Ś. B. Dīkṣita, *Bhāratīya Jyotiḥśāstra* (Poona, 1896; repr. Poona, 1931), p. 351 in repr. See also S. L. Katre, "Sultān Fīrūz Shāh Tughluk: Royal Patron of a Contemporary Sanskrit Work," in *Journal of Indian History*, **45** (1967), 357–367.

DAVID PINGREE

MAḤMŪD IBN MASʿŪD AL-SHĪRĀZĪ. See **Quṭb al-Dīn al-Shīrāzī.**

MAIER, MICHAEL (*b.* Rensburg, Holstein, Germany, *ca.* 1568; *d.* Magdeburg, Germany, 1662), *alchemy.*

Maier was probably the son of Johann Maier, an official of the duchy of Holstein. He studied first in either Rensburg or Kiel, and in 1587 he was studying at the University of Rostock. He owed his career to a relation of his mother's, Severin Goebel, a well-known physician of Gdańsk and Königsberg, who financed his studies. In 1589 Maier was in Nuremberg, and he was in Padua with the son of Goebel from 1589 to 1591. He began practicing surgery in 1590 without an academic degree. In 1592 he was at the University of Frankfurt an der Oder, where he had the title of *poeta laureatus caesareus.* He wrote elegant Latin verse, under the anagram "Hermes Malavici."

Next, Maier practiced at Königsberg under the supervision of Severin Goebel. On 24 May 1596 he was enrolled in the University of Bologna as *magister,* and in the same year enrolled himself at the University of Basel and received the doctorate in medicine after presenting his "Theses de epilepsia." It is not known where Maier took his doctorate in philosophy. It seems that before 1600 he was a courtier of Rudolf II and a writer in the German chancellery.

In 1601 Maier was in Königsberg, and on 11 September entered his name on the university rolls as "Michael Meierus Philosophiae et Medicinae Doctor Honoris Gratia," apparently in an attempt to obtain the status of professor or *extraneus* at this university. Obviously this did not occur, for in December 1601 he went to Gdańsk, where in the White Horse Inn he started medical practice, advertising his own remedies, such as frogs dried and then soaked in vinegar.

Before 1612 Maier had returned to Prague as a doctor. He became physician-in-ordinary to Rudolf II, although probably only in an honorary capacity, since his name does not figure in the court accounts. His family coat of arms was augmented, by the grace of the emperor, to include on one half a tree trunk with three branches, and on the other a toad bound by a chain to a flying eagle. The latter symbolized volatile and nonvolatile substances and in all likelihood was taken from an alchemical treatise ascribed to Ibn Sīnā, as can be gathered from Maier's *Symbola aureae mensae.* He was also named *comes palatinus* by the emperor. (A count palatine was an imperial official who exercised a sort of supervision over the universities and had the right to grant doctorates and the title of poet laureate.)

In 1611 Maier was in various cities of Saxony—Torgau, Leipzig, and Mühlhausen—where he met the landgraves Maurice of Hesse and Christian of Anhalt, both of whom shared his passionate devotion to music. During the period 1612–1614 Maier was in England, where he met Robert Fludd, William Paddy, Thomas Smith, and Francis Anthony and translated into Latin a treatise by Thomas Norton under the title of *Crede mihi seu ordinale.* Maier was not favorably impressed by England, as he stated in *Symbola aureae mensae.*

After his return to Germany, Maier helped to organize the publication of the works of Fludd in Frankfurt am Main. He became court physician to Landgrave Maurice, without, however, giving up his private practice. In 1618 he traveled to Stockhausen, where he attended a wealthy nobleman named von Eriedesel. Maier had a house in Frankfurt am Main, where his wife lived, and he bought alchemical works for the landgrave's library at the Frankfurt book fairs. In 1618 Maier moved to Magdeburg to become the physician of Duke Christian Wilhelm. He died there four years later.

Maier is an extremely puzzling figure, both in his works and in his very unsettled life. Without question, as a count palatine he was a political agent of the emperor. Maier was an ardent alchemist, a follower of Paracelsus and neo-Hermetic ideas. He was an implacable enemy of the Roman Catholic church, a defender of the Rosicrucian movement, and probably had a hand in the publication of the *Fama fraternitatis* (1616). Many of his works are written in a very Rosicrucian spirit.

All of Maier's treatises are written with great erudition and display substantial knowledge of mythology and ancient history. They are classic examples of the neo-Hermetic manner, having no clear chemical sense. Yet there appear in his writings sentences and considerations that are sometimes astonishing, as in *Viatorum . . . de montibus planetarium*, in which he deliberates why lead and copper weigh more after being roasted (as Lazarus Ercker had observed). In *Examen fucorum pseudochymiorum* Maier gives examples of the possibility of alchemical fraud and states that it is possible to estimate transmutation truly only by means of docimasy, that is, chemical analysis.

The writings of Maier were highly valued and popular among alchemists. In the history of chemistry they represent a certain regression, however, for Maier was a fervent believer in the transmutation of metals, which was for him a synonym of the word "chymia."

BIBLIOGRAPHY

I. ORIGINAL WORKS. Incomplete bibliographies of Maier's writings are in D. J. Duveen, *Bibliotheca alchemica et chemica* (London, 1965), p. 380; J. Ferguson, *Bibliotheca chimica* (Glasgow, 1906), II, 66; and N. Lenglet du Fresnoy, *Histoire de la philosophie hermétique* (The Hague, 1742), III, 225–230.

Among his works are *Arcana arcanissima* (n.p., n.d. [London, 1614?]); *De circulo physico quadrato* (Oppenheim, 1616); *Apologeticus quo causae clamorum seu revelatiorum Fratrum Rosae Crucis* (Frankfurt, 1617); *Atalanta fugiens* (Oppenheim, 1617); *Examen fucorum pseudochymiorum* (Frankfurt, 1617); *Jocus severus* (Frankfurt, 1617); *Lusus serius* (Frankfurt, 1617); *Silentium post clamores* (Frankfurt, 1617); *Symbola aureae mensae* (Frankfurt, 1617); *Themis aurea* (Frankfurt, 1618); *Tripus aureus* (Frankfurt, 1618), repr. in *Musaeum Hermeticum* (Frankfurt, 1749); *Viatorum hoc est de montibus planetarium* (Frankfurt, 1618); *Verum invectum* (Frankfurt, 1619); *De volucri arboreum* (Frankfurt, 1619); *Septimena philosophia* (Frankfurt, 1620); *Civitas corporis humani* (Frankfurt, 1621); *Cantilenae intellectuales de Phoenice redidivo* (Rostock, 1622); *Tractatus posthumus sive Ulysses* (Frankfurt, 1624); and *Viridarium chymicum* (Frankfurt, 1688).

II. SECONDARY LITERATURE. The literature on Maier's life is scanty. Most of the data in this article are the result of the author's research in various European libraries and archives. The best biography is considered to be J. B. Craven, *Count Michael Maier* (Kirkwall, Scotland, 1910; London, 1968), which really describes the contents of Maier's writings. H. M. E. De Jong, in the new ed. of *Atalanta fugiens* (London, 1969), elucidates the sources of the emblems and allegories in Maier's books. See also L. Thorndike, *A History of Magic and Experimental Science* (New York, 1958), VII, 167, 171–173, 213, and VIII, 113, 194.

A list of older literature on Maier's life and work is in Ferguson (see above), *loc. cit.*

WŁODZIMIERZ HUBICKI

MAIGE, ALBERT (*b.* Auxonne, France, 26 November 1872; *d.* Lille, France, 29 November 1943), *botany.*

Professor at Algiers (1900–1910), Poitiers (1911–1919), and Lille (1920–1943), Maige was a corresponding member of the Institut de France. Having determined the general adaptive characteristics of creeping plants, he determined the tendencies of a tapering evolution toward the morphology and anatomy of either rhizomes or climbing plants for which direct light discourages creeping and diffused light encourages creeping.

Maige's *Flore forestière de l'Algérie* begins with general botanical concepts applied to phytogeography, to silviculture, and to the natural history of the woody plants of Algeria. It also includes four keys designed to assist in the identification of specimens: the reproductive organs and the characteristics of leafy branches, of the bare branches of trees with caducous leaves, and of the principal native woods.

Besides the description of new galls and various anomalies, Maige's works in pathology and teratology include the study of the potato blight (*brunissure*), a physiological disorder caused by a progressive dehydration of the tissues, and, especially, the study of the disease of the cork oak known as "yellow spot," which gives wine the taste of cork.

In physiology, Maige determined that the respiratory rate of the plant decreases regularly from the earliest stages to the time of blooming and that it falls steeply as the flower fades. The respiratory physiology of the flower thus resembles that of the leaf. He found that the influence of variations in turgescence on the respiration of the cell is shown by a notable simultaneous elevation of turgescence and respiration. The diminution of turgescence produces the same effects up to an optimum concentration of the cellular juice, beyond which there occurs a diminution of the respiratory coefficients. Sugar solutions of various concentrations affect the respiration, the turgescence, and the growth of the cell.

Maige conducted research in cytology, the study of pollinic karyokinesis among the Nymphaeaceae. In cytophysiology he used the cytophysiological method of nuclear variations, which consists of depriving cells of nourishment, thus producing a decrease in nuclear volume, and observing whether or not the nucleus grows under the influence of various substances. Combined with the analogous method of plastid variations, it can contribute valuable data concerning the nutritive values of the substances being examined. In particular, these methods enabled Maige to show that the formation of starch causes the different sugars to pass through the same stages as does the breakdown of starch but in the opposite direction—notably through stages of the dextrins and the erythrodextrins. The several enzymes involved in these processes can be arrested at certain stages.

The formation and the digestion of starch in the cells are two distinct phenomena produced by different enzymes, or at least by enzymes localized in different cellular regions. That which governs starch formation is localized, Maige proved, in the leucoplast; that which provokes amylolysis, in the cytoplasm. In addition, the former inhibits the latter during starch formation.

BIBLIOGRAPHY

Maige's works include "Recherches biologiques sur les plantes rampantes," in *Annales des sciences naturelles (Botanique)* (1900), 249–364; *Flore forestière de l'Algérie* (Paris, 1914); and various works on cytophysiology in *Comptes rendus hebdomadaires de l'Académie des sciences* and *Comptes rendus de la Société de biologie.*

MAURICE HOCQUETTE

MAIGNAN, EMANUEL (*b.* Toulouse, France, 17 July 1601[?]; *d.* Toulouse, 29 October 1676), *physics.*

Born of a prominent Armagnac family, Maignan spent his boyhood in Toulouse and entered the order of Minims at an early age, taking his vows in 1619. He first studied philosophy with an Aristotelian named Ruffat but soon rebelled against the Peripatetic system, showing a strong interest and aptitude for mathematical studies. He taught philosophy and theology at the Minim convent of Monte Pincio in Rome from 1636 to 1650, during which time he became interested in the experimental approach to knowledge, coming into contact with Gasparo Berti, Raffaello Magiotti, and Athanasius Kircher. In this group he participated in the important experiments which helped to establish the possibility of artificially creating a void space in nature and which influenced the work of Torricelli and others. His *Cursus philosophicus* (1653), of which more than four-fifths is devoted to natural philosophy, provides one of the fullest accounts of these researches. In 1650 he

returned to Toulouse, where he spent most of the remainder of his life. There he continued his experimental work but devoted much of his energy to the administrative and religious work of his order.

Once described by Pierre Bayle as "one of the greatest philosophers of the seventeenth century," Maignan has largely been forgotten, although he was an original and individualistic thinker of no small merit. His work in optics, instrument making and design, and various branches of physics is in need of reevaluation. His *Perspectiva horaria* (1648) is an extremely detailed and almost exhaustive discussion of sundials, both from a practical and from a theoretical point of view. In this work many optical topics such as sciagraphy are also treated.

Maignan is responsible for introducing a strongly experimental emphasis into the scholastic textbook, turning aside from the bookish Aristotelian tradition, while at the same time remaining critical of Descartes and other contemporary authors. His work, perhaps as well as any of the seventeenth century, shows the marked influence of experimentalism on scholastic thought. After Maignan's death a systematic textbook of his teachings meant for use in the schools was prepared by his follower and biographer, Jean Saguens (*Philosophia Magnani scholastica . . .,* 4 vols. [Toulouse, 1703]).

BIBLIOGRAPHY

I. ORIGINAL WORKS. Maignan's most important works are *Perspectiva horaria sive de horographia gnomonica tum theoretica tum pratica libri quatuor* (Rome, 1648); *Cursus philosophicus* (Toulouse, 1653; 2nd ed., enl., Lyons, 1673); *Philosophia sacra,* 2 pts. (Toulouse, 1661–Lyons, 1672). For more complete lists see Whitmore, listed below; and the catalog of the Bibliothèque Nationale, Paris, CIII, 786–787.

II. SECONDARY LITERATURE. The basic work on Maignan's life is apparently J. Saguens, *De vita . . . Emanuel Maignani* (Toulouse, 1703), not seen. See also (listed chronologically) Pierre Bayle, *Dictionnaire historique et critique,* X (Paris, 1820), 125–133; F. Sander, *Die Auffassung des Raumes bei Emanuel Maignan und Johannes Baptiste Morin* (Paderborn, 1934); C. de Waard, *L'expérience barométrique* (Thouars, 1936), *passim;* R. Ceñal, "Emmanuel Maignan: su vida, su obra, su influencia," in *Revista de estudios politicos* (Madrid), no. 66 (1952), 111–149; and "La filosofía de Emmanuel Maignan," in *Revista de filosofía* (Madrid), **13** (1954), 15–68; J. S. Spink, *French Free-Thought from Gassendi to Voltaire* (London, 1960), 75–84; W. E. K. Middleton, *The History of the Barometer* (Baltimore, 1964), *passim;* and P. J. S. Whitmore, *The Order of Minims in Seventeenth-Century France* (The Hague, 1967), 163–186, with additional bibliography, including MSS sources.

CHARLES B. SCHMITT

MAILLET, BENOÎT DE (*b.* St. Mihiel, France, 12 April 1656; *d.* Marseilles, France, 30 January 1738), *geology, oceanography, cosmogony.*

Maillet belonged to a noble family of Lorraine and received an excellent classical education. Through the favors of his protector, the chancellor Pontchartrain, he was appointed general consul of the king of France at Cairo in 1692, a position he held until 1708. During that time he was chosen by the king as his personal envoy to Ethiopia. Although political circumstances prevented him from accomplishing his mission, Maillet wrote a compilation entitled "Mémoires d'Éthiopie" that was included in Jeronymo Lobo's account of that country (1728).

Between 1708 and 1714 Maillet was consul in Leghorn. He ended his diplomatic career in 1720 as inspector of French establishments in the Levant and the Barbary States. After having spent two years in Paris while the plague was raging at Marseilles, he retired to that city on a handsome pension.

While in Egypt, Maillet completed an important historical and sociological volume entitled *Description de l'Égypte* (1735) and wrote most of his system on the diminution of the sea, *Telliamed ou entretiens d'un philosophe indien avec un missionaire françois sur la diminution de la mer, la formation de la terre, l'origine de l'homme, etc. . . .* This fundamental work, in essence an ultraneptunian theory of the earth, was based largely on his geological field observations made during extensive travels throughout Egypt and other Mediterranean countries. Maillet must have taken full advantage of his fluency in Arabic to gain access to the manuscripts of many ancient Arabic authors, such as al-Khayyāmī, from whom he may have borrowed the original idea of the diminution of the sea.

For the publication of his system Maillet relied on the Abbé J. B. Le Mascrier, who had previously edited the *Description de l'Égypte*. Only ten years after Maillet's death did Le Mascrier reluctantly agree to the publication of the first edition (Amsterdam, 1738), which was followed by a second and third. Two English translations were also published. This unusual delay in publication resulted from the failure of Le Mascrier's editorial work to reduce the dangerous nature of the system; actually, he was willing to be acknowledged only as editor of the third edition. Indeed, even when presented under the name of a fictitious Indian philosopher (his own name spelled backward), Maillet's concepts were unorthodox and highly materialistic.

The proposed system did not admit God as an omnificent ruler, postulating instead an eternal universe undergoing natural changes at random. This universe was based on the Cartesian theory of vortices combined with Fontenelle's concept of the plurality of inhabited worlds. Heavenly bodies were believed to be eternally renovated through a mechanism of alternating luminous phases, during which they were similar to suns, and dark phases, when they were comparable with planets. At present the earth was in a dark phase, and the level of the universal ocean was being lowered by evaporation into outer space, at the rate of three inches per century, until total depletion would occur. This general regression had taken place for at least the past two billion years, an estimate indicating Maillet's acute perception of geological time (as stated in the manuscripts).

The first part of Maillet's system consisted of a study of shoreline features combined with the submarine exploration of the continental shelf by means of a diving machine, the ancestor of modern bathyscaphes. In this survey the sedimentary processes characteristic of rocky coasts, beaches, and deltas were discussed in great detail as a function of the pattern of waves, tides, and bottom currents. This analysis of present-day marine mechanisms was then applied to the geological past, following essentially uniformitarian principles. Maillet used all the aspects of his oceanographical knowledge to explain, by the action of the sea, all the physiographic, lithologic, paleontologic, and structural features of the geological record.

When the sea was in the stage of a universal ocean, its bottom currents' distribution of the sands, silts, and calcinated stones which were the remains of the previous luminous phase built the so-called primitive or primordial mountains, analogous to gigantic submarine bars. They represented the highest mountains, which consisted of a "uniform substance," occasionally displaying horizontal bedding. This description certainly referred to metamorphic rocks, since Maillet stressed the almost complete absence of fossils in them, claiming that at the time of their deposition the universal ocean was too deep to allow the existence of any form of life.

Following a certain amount of evaporation, the summits of the primitive mountains emerged; and in the surrounding shallow waters life appeared. Maillet explained this appearance of life on earth by assuming that the entire universe was filled with the seeds of all living beings. Invisible and imperishable, these seeds were always available for fecundation and the creation of new species whenever the waters of the globes, at certain times and under certain circumstances, became proper for such processes to occur.

The diminution of the sea and the related expansion of the continents compelled all forms of life to undergo a generalized transformism in order to adapt themselves to the new environmental conditions. Marine plants gave rise to land types; sea animals acquired flight and walking capacities and changed into birds and land animals; and mermen and mermaids emerged from the sea, developing into terrestrial human beings. Obviously transformism, but not evolution, was postulated in Maillet's system; and the very early appearance (one billion years ago, as stated in the manuscripts) of mankind, with its "petrified ships," earthenware, weapons, and tools allegedly found with human skeletons in rocks, did not present any particular difficulty of chronology.

Further lowering of the sea level led to the generation of the secondary mountains, or "daughters of the primitive ones." They consisted of fossiliferous and horizontally bedded sands and muds (sedimentary rocks) derived from the erosion of the primitive mountains by waves acting along coastlines. As a consequence of this process, the secondary mountains were younger and less high, and within any geological section of such mountains the beds were superposed from bottom to top in agreement with the law of superposition. This situation was also true with respect to the enclosed fossils, which were assumed to become more abundant at higher levels since the shallowing sea provided an increasingly favorable environment. Consequently, fossils showed important changes through time. This was an early expression of the modern concept of faunal successions.

In Maillet's system all the features of the primitive and secondary mountains dated back to the time of deposition of the soft materials. These were gradually cemented by marine salt and finally exposed through the diminution of the sea as a finished or "congealed" scenery which would remain essentially unaffected by weathering agents. Indeed, in Maillet's system, stream valleys represented ancient channels of marine currents used after emergence by streams only to funnel rainwater to the ocean. In his negation of stream erosion Maillet considered that these channels could be modified by tidal currents only during the gradual retreat of the sea, as shown today by estuaries.

The proposed system was also devoid of any mountain-building mechanism capable of uplifting, tilting, or folding strata. Since Maillet's geological knowledge was limited to essentially undisturbed sedimentary rock areas, he considered tilted and folded beds to be accidental features of mountains. He explained them as well through the action of marine bottom currents and waves he had observed along present coasts. Small-scale folds were interpreted as ripple marks, whereas large-scale tilting and folding of beds were explained by the action of currents which deposit regular layers of sediments over supposed irregularities of an "indurated substratum." Following the same line of reasoning, complicated folds were considered the results of violent storms during which currents flowed in complex patterns around rocky obstacles, eroding and depositing sediments in the shape of highly contorted beds.

Maillet's ideas unquestionably influenced many leading naturalists for almost a century, notably Buffon and Cuvier. The latter, although not in favor of Maillet's system, considered it to be the first systematic presentation of a theory of general transformism. Although Maillet was certainly a forerunner of Lamarck, this position should not obscure his major contribution, which is geological. His ultraneptunian theory of the earth, in which everything was explained by the action of a retreating sea, makes him a marine geologist of the eighteenth century.

BIBLIOGRAPHY

Maillet's "Mémoires d'Éthiopie" is in R. P. Jeronymo Lobo, *Relation historique d'Abissinie, traduite du portugais, continuée et augmentée de plusieurs dissertations, lettres et mémoires par M. Le Grand* (Paris, 1728). His work on Egypt is *Description de l'Égypte contenant plusieurs remarques curieuses sur la géographie ancienne et moderne de ce païs* (Paris, 1735). His major work is *Telliamed . . .* (Amsterdam, 1748; Basel, 1749; new ed., rev., corr., and enl., The Hague–Paris, 1755). English versions of the latter are *Telliamed: Or Discourses Between an Indian Philosopher and a French Missionary on the Diminution of the Sea . . .* (London, 1750); *Telliamed: Or the World Explained Containing Discourses Between an Indian Philosopher and a Missionary on the Diminution of the Sea . . .* (Baltimore, 1797); and *Telliamed: Or Conversations Between an Indian Philosopher and a French Missionary on the Diminution of the Sea,* edited and translated by A. V. Carozzi (Urbana, Ill., 1968), trans. of a 1728 MS and comparison with the final ed. of 1755.

Secondary literature is Fritz Neubert, *Einleitung in eine kritische Ausgabe von B. de Maillets Telliamed. Ein Beitrag zur Geschichte der französischen Aufklärungsliteratur,* Romanische Studien, no. 19 (Berlin, 1920).

                                        ALBERT CAROZZI

MAIMONIDES, RABBI MOSES BEN MAIMON, also known by the acronym **RaMBaM** (*b.* Córdoba, Spain, 1135 or 1138; *d.* Cairo [or Fuṣṭāṭ], Egypt, 1204), *medicine, codification of the Jewish law, philosophy.*

Maimonides was the foremost representative of the school of thought that is designated as Jewish Aristotelianism. In consequence of the invasion of Muslim Spain by the Almohads, his family left Córdoba while he was a child and after an interval settled in 1159/ 1160 in Fez, Morocco, a country which, like Andalusia, was ruled by the Almohads. He lived there until 1165. Maimonides received his philosophical, scientific, and legal training in Spain and the Maghreb and prided himself on belonging to the Andalusian (rather than the Oriental) school of philosophy. It is also probable that the dogmas of the Almohad creed had some influence on his formulation of the thirteen fundamental Jewish religious principles. In 1166 Maimonides settled in Egypt, at first in Alexandria and then in Fuṣtāṭ, near Cairo. In Egypt he was court physician and (either official or unofficial) head of the Jewish community. His works, with very few exceptions, were written in that country, where he spent the rest of his life.

Maimonides' writings may be classed according to their genres:

1. The legal works, the most important of which are his commentary on the Mishnah, written when he was still young, and his codification of the Talmudic law, known as *Mishnah Torah* or *Yad Hazaqa* ("A Strong Hand"). Certain portions of both these works treat of philosophical doctrine.

2. Popular or semipopular theological works destined for the general Jewish reader, such as the "Treatise on Resurrection."

3. A systematic philosophical text, the *Maqāla fī ṣināʿat al-manṭiq*, the only one written by Maimonides; it is a treatise on logic, and perhaps his earliest work.

4. *The Guide of the Perplexed*, completed a short time before Maimonides' death, which is in a class by itself. It deals in an unsystematic way with physics and metaphysics but also is concerned with the presuppositions and the imperatives of politics, religious belief and the religious commandments, and the final end of man. It is intended for the perplexed, that is, for those versed in Jewish lore who also have a smattering of and a capacity for philosophical knowledge and are thus in danger of abandoning the observance of the religious law.

5. A number of medical treatises, written in the last period of his life.

6. A very extensive correspondence, consisting of letters and rabbinical *responsa* addressed to notables of Jewish communities in various countries of the Islamic world and outside it—for instance, in the south of France.

All the main works of Maimonides, except *Mishnah Torah*, which is in Hebrew, were written in Arabic.

Maimonides affirmed that he did not intend to expound novel philosophical views; he attempted to show, *inter alia*, (1) that the teaching of philosophy need not, if the necessary precautions are taken, result in the disruption of society and the destruction of the Jewish religion and (2) that philosophy enables man to attain his final end, which is the perfection of his intellect.

We have firsthand evidence of the esteem in which Maimonides held various philosophers. In a letter to Samuel ibn Tibbon, the translator of the *Guide* into Hebrew, he had very high praise for Aristotle—who should, however, according to him, be read together with his commentators. It may be noted in this connection that Alexander of Aphrodisias appears to have had a significant influence on Maimonides.

The most trustworthy Muslim philosophers were, in Maimonides' opinion, al-Fārābī and Ibn Bājja. Ibn Sīnā was regarded as less reliable, although Maimonides used him freely; Ibn Sīnā sometimes provided him with the theological or semitheological terminology necessary for his purposes. According to the letter to Samuel ibn Tibbon, the study of Plato is much less useful than that of Aristotle; but, like that of al-Fārābī, Maimonides' political philosophy derived to a considerable extent from Plato's writings. The precautions taken by Maimonides in the *Guide* to avoid troubling the religious readers who lacked the capacity for philosophical thought entailed, as he explicitly states, unsystematic exposition and deliberate recourse to self-contradiction. To cite an important example, he set forth three conceptions of God which appear to be mutually incompatible.

The first of these conceptions is the God of Maimonides' brand of negative theology. This theology is different in an important respect from that of most Neoplatonists because, contrary to most of them, Maimonides did not admit mystic union, that is, an ecstatic experience of God which transcends the intellect but which man is able to achieve. Maimonides' negative theology stressed the impossibility of making a correct positive statement about the essence of God. Apparently, positive assertions can be regarded as true only if they are given a negative meaning. For instance, the statement "God is wise" signifies that He is not unwise. Maimonides denied—and was, because of this, taken to task in the fourteenth century by the Jewish philosopher Levi ben Gershon (Gersonides)—that this assertion may, when applied to God, have a positive content. Maimonides' conception of the unknowability of God, of there being nothing in the created world that is similar to or has a trace of Him, and the doctrine of negative attributes that fits in with these other points result in the recognition that it is impossible to transform God into an object of science. Metaphysics is thus deprived of its main object (or, according to another opinion, of one of its main objects).

A philosopher can, nevertheless, acquire the only knowledge of God of which man is capable: a knowledge of His activity. This is tantamount to a knowledge of the natural order of events, the expressions "divine actions" and "natural actions" being interchangeable. It appears to follow that it is in studying natural science and metaphysics that man achieves the only knowledge of God granted to him. It is admittedly a very limited knowledge, for an examination of the "divine actions" or "natural actions" does not legitimate any inference with regard to the divine essence. This impossibility is veiled by a generally accepted linguistic convention, in virtue of which "natural" or "divine" events are regarded as proceeding from certain dispositions in God; for instance, the care of parents for their offspring is said to be due to God's beneficence or mercifulness, and earthquakes and floods to His vengefulness.

Maimonides—who in this matter followed, at least as far as terminology is concerned, a well-established tradition—designated mercifulness, vengefulness, and other terms of this kind, when applied to God, as attributes of action. Such attributes represent an evaluation of the impact of natural (and perhaps also of historical) events on man or human society; they should not be taken as referring to God's essence.

A second conception of God expounded by Maimonides is the Aristotelian one. God is an intellect, that is, the subject, the object, and the act of intellection. Like other Aristotelian philosophers, Maimonides considered that these three form a unity. He follows such predecessors in considering, in disagreement with the tenor of Aristotle's text, that God's knowledge is not confined to Himself only. He may be held to know the specific forms and the natural order—or, in other words, the system of sciences. Since Maimonides adopted the Aristotelian view that the knower and the object of his knowledge are identical, this means that in his view (as in that of other medieval Aristotelian philosophers) God may be equated with a self-cognizant system of sciences, a conception which has a striking similarity to Hegel's interpretation of Aristotle's God (in the concluding portion of his *Encyclopedia*). It may be noted that according to Maimonides, God does not know individuals as such, that is, in their separate existence, but only in virtue of being their cause. It seems probable that this formula, like other theological traits in Maimonides' writings, may be derived from Ibn Sīnā (see Ibn Sīnā's *Kitāb al-Shifāʾ*, in *al-Ilāhīyāt*, II, M. Y. Moussa, S. Dunya, and S. Zayed, eds. [Cairo, 1960], p. 359).

In contradistinction to his Aristotelian predecessors, Maimonides appears to have set store by a comparison which indicates a similarity between God conceived as an intellect and the human intellect. This comparison contrasts—perhaps intentionally—with the extreme negative theology of what has been designated as his first conception of God; the extremism of this theology goes much beyond the analogous views expressed by the Muslim Aristotelian philosophers.

A third conception affirms the existence of a divine will, a notion that had been elaborated by al-Ghazālī and some earlier Mutakallimūn. Hence Maimonides may have been influenced to some extent by these thinkers. As we shall see, however, his views were markedly different from theirs.

A God not endowed with will is, according to Maimonides, an altogether powerless God, who is not able to lengthen the wings of a fly. This idea is wholly unacceptable for religion. In this context the question of whether the world has or has not been created in time becomes crucial. Temporal creation would mean an intervention of God in the course of events or, in other words, a miracle, the greatest of miracles; if this were admitted as possible, there would be no difficulty in accepting lesser ones.

No problem would arise if Maimonides were prepared to follow the example of the Mutakallimūn in denying the existence of a natural order and causality; this would mean complete rejection of Aristotelian physics. This he refused to do, and instead found another solution. He argued that the natural sciences are absolutely correct within certain limits but ought not to go beyond these limits. For there are spheres of knowledge the investigation of which transcends the powers of man; as far as science is concerned, certain questions are insoluble, the question of whether the world has been created in time being one of them. Given this fact, Maimonides chose the hypothesis of temporal creation, for the reason that the religious tradition, including the belief that Israel was chosen by God, can be explained and justified only in the light of this hypothesis. Thus the latter may be considered as a practical postulate required for the preservation of religion, and not as a theoretical truth.

Maimonides' emphasis on the limitations of human science is perhaps his most significant contribution to general—as distinct from Jewish—philosophical thought. Like Kant, he pointed out these limitations in order to make room for belief. He accepted Aristotelian physics insofar as it is concerned with the sublunar world; in his view it provides an example of a perfect scientific theory. It may be noted that in this connection he apparently preferred a mechanistic explanation. He certainly played down the role of final, as compared with efficient, causes in natural science.

Human science cannot, however, provide a satisfactory theory for the world of the heavenly spheres. Following his Muslim predecessors (see, for instance,

Ibn Sīnā, *Risāla fi'l-Ajrām al-'ulwiyya*, in *Tis' Rasā'il* [Cairo, 1908], p. 49), Maimonides posed some questions concerning this celestial world which, according to him, may be insoluble. For instance, he asked whether, given the difference between the stars and the spheres, one should not admit, in opposition to Aristotle's views, the existence of more than one kind of matter in the celestial world. A much more intractable problem was constituted by the flagrant contradiction between the Ptolemaic system, with its recourse to epicycles and eccentrics, and Aristotelian physics.

Unlike his contemporary Ibn Rushd, Maimonides did not believe that a correct system of astronomy was known in Aristotle's time and had since been forgotten, for he considered that at that time knowledge of mathematics was still very imperfect. Nor did he accept any of the attempts made by Muslim philosophers and astronomers to work out an astronomical system compatible with Aristotle's physics. The contradiction between astronomy and physics served his purpose. It proves, according to him, the limitation of human knowledge: man is unable to give a satisfactory scientific account of the world of the spheres.

This line of argument (insofar as it shows that the claim of science to propound an all-embracing, coherent, and true system of nature is untrue) concerns the problem of temporal creation only indirectly. The following reasoning, on the other hand, impinges directly upon this question. Maimonides argued that one should not extrapolate beyond certain limits from the knowledge of the natural order obtaining now, for there may have been a beginning prior to which another order may have existed. In this context Maimonides cited the example of a person who, not knowing the facts of birth, denies the possibility of human beings having first existed as embryos. According to him, the Aristotelian affirmation of the eternity of the world is based on a similar extrapolation.

Maimonides held no brief for Ibn Sīnā's opinion that the individual human soul survives the death of the body and is immortal. Like Alexander of Aphrodisias and other Aristotelians, he considered that in man only the actual intellect—which lacks all individual particularity—is capable of survival. In adopting this view, Maimonides clearly showed that, at least on this point, he preferred the philosophical truth as he saw it, however opposed it may seem to be to the current religious conception, to the sort of halfway house between theology and philosophy, which—in the severe judgment of certain Spanish Aristotelians—Ibn Sīnā, who was the dominant philosophical influence in the Muslim East, sought to establish.

Maimonides did, however, adopt certain conceptions of Ibn Sīnā. Thus, his view that existence is an accident derives from Ibn Sīnā's fundamental tenet that essences per se are neutral with respect to existence, which supervenes on them as an accident.

According to Maimonides, all prophets are philosophers, that is, men whose intellect is actualized. But in contradistinction to other philosophers, prophets have a highly developed imaginative faculty. Prophecy is a natural phenomenon.

This description of prophets does not, according to Maimonides' statement, apply to Moses, whose status is higher. In a popular treatise Maimonides refers to Moses' achieving union with the active intellect; such a union (or, to be more precise, a near union) is, according to Ibn Sīnā, a result of the prophetic faculty (see Ibn Sīnā's *De anima*, F. Rahman, ed. [Oxford, 1959], pp. 248–250), whereas, according to Ibn Bājja, it is attained by the great philosophers without the stimulation of such a faculty.

Religious revelation does not procure any knowledge of the highest truth that cannot be achieved by the human intellect; it does, however, have an educative role—as well as a political one. In Maimonides' words, "The Law as a whole aims at two things: the welfare of the soul and the welfare of the body" (*Guide of the Perplexed*, pt. III, ch. 27).

Because of the great diversity of human character, a common framework for the individuals belonging to one society can be provided only by a special category of men endowed with a capacity for government and for legislation. Those who have only a strong imagination, unaccompanied by proportionate intellectual powers, are not interested in the intellectual education of the members of the state which they found or govern. Moses, on the other hand, is the ideal lawgiver.

The law instituted by Moses had to take into account the historical circumstances—such as the influence of ancient Oriental paganism—and had to avoid too great a break with universal religious usage. To cite one example, sacrifices could not be abolished, because this would have been an excessively violent shock. In spite of these difficulties, Moses succeeded in establishing a polity to which Maimonides in the "Epistle to Yemen" (a popular work) applied the expression *al-madīna al-fāḍila* ("the virtuous city"), used by the Muslim philosophers to designate the ideal state of Plato's *Republic*.

Not only does the Mosaic polity regulate men's actions in the best possible way, but the Scriptures by which this polity is ruled also contain hints toward philosophical truth that may guide such men as are capable of understanding them. Some of these truths are to be discovered in the beliefs taught to all who profess Judaism; these dogmas are, for evident reasons, formulated in language adapted to the understanding of ordinary, unphilosophical people. There are, however, other religious beliefs that, although they are not true, are necessary for the majority of the people, in order to safeguard a tolerable public order and to further morality. Such are the belief that God is angry with those who act in an unjust manner and the belief that He responds instantaneously to the prayers of someone wronged or deceived (*Guide of the Perplexed*, pt. III, ch. 28). The morality suited to men of the common run aims at their exercising a proper restraint over the passions or the appetites; it is an Aristotelian middle-of-the-road morality, not an ascetic one. The ascetic overtones which are occasionally encountered in the *Guide* concern the philosopher rather than the ordinary man.

There is a separate morality for the elite, which rules or should rule (see the *Guide*, pt. I, ch. 54; pt. III, chs. 51, 54). This ethical doctrine is connected with his interpretation of what ought to be man's superior goal, which is to love God and, as far as possible, to resemble Him.

From the point of view of negative theology, love of God can be achieved only through knowledge of divine activity in the world. This appears to signify that the highest perfection can be attained only by a man who leads the theoretical life. Maimonides was at pains, however, to show that the theoretical life can be combined with a life of action, as proved by the examples of the patriarchs and Moses. Moreover, a life of action can constitute an imitation of God. For the prophetic legislators and statesmen endeavor to imitate the operation of nature and God (the two being equivalent). Maimonides emphasized two characteristics that belong to both the actions of God-nature and the actions of the superior statesman. First, however beneficent or destructive—or, in ordinary human parlance, however merciful or vengeful—the actions in question appear to be, neither God nor the prophetic statesman is activated by passions. Second, the activity of nature (or God) tends to preserve the cosmic order, which includes the perpetuity of the species of living beings; but it has no consideration for the individual. In the same way, the prophetic lawgivers and statesmen, who in founding or governing a polity imitate this activity, must have in mind first and foremost the commonweal, the welfare of the majority, and must not be deterred from following a political course of action by the fact that it hurts individuals.

On the whole, Maimonides' medical treatises have been less thoroughly studied than his speculative and legal work. Like other medieval physicians he recognizes Galen as his master. Nevertheless, in a medical treatise entitled *Moses' Chapters on Medicine* he charged Galen with forty contradictions and also taxed him with ignorance in philosophical and theological matters. According to Maimonides, his criticism of Galen was independent of that of al-Rāzī, who wrote a work polemizing against Galen.

Two Hebrew versions of the *Guide* (by Samuel ibn Tibbon and al-Ḥarizi) were prepared a short time after the work was written. It had many Hebrew commentators of various and sometimes conflicting views; and because of its impact, it is certainly the most important work of Jewish medieval philosophy. In the period from 1200 to 1500 it provided most Jewish philosophers with a scheme of reference in relation to which they could formulate their own positions. In the thirteenth and fourteenth centuries it was vehemently denounced as antireligious—and was as vehemently defended.

Spinoza knew Maimonides well, polemizing against him but influenced by him particularly in the *Tractatus theologico-politicus*. Solomon Maimon wrote a commentary on the *Guide*. The *Guide* was translated from Hebrew into Latin in the thirteenth century and exerted, especially with regard to the problem of the eternity of the world but also on many other points, a considerable influence on Scholastic philosophers. This influence is very much in evidence in the works of Thomas Aquinas. In the postmedieval period, Maimonides influenced Jean Bodin and impressed Leibniz.

BIBLIOGRAPHY

I. ORIGINAL WORKS. Bibliographies of Maimonides' works are in J. I. Gorfinkle, "A Bibliography of Maimonides," in *Moses Maimonides 1135–1204*, I. Epstein, ed. (London, 1935), 231–248; L. G. Levy, *Maimonides* (Paris, 1911), supp. in *Cahiers juifs*, **2** (1935), 142–151; and G. Vajda, *Jüdische Philosophie* (Bern, 1950), pp. 20–24. See also M. Steinschneider, *Die arabische Literatur der Juden* (Frankfurt, 1902), pp. 199–221.

Works by Maimonides are *Guide of the Perplexed: Le guide des égarés*, Salomon Munk, ed., 3 vols. (Paris, 1856–1866), Arabic text and French trans. with many detailed notes, French trans. also re-ed. (Paris, 1960), also available in English as *The Guide of the Perplexed*, trans. with intro. and notes by Shlomo Pines (Chicago, 1963), intro. essay by Leo Strauss; "Maqāla fī ṣinā'at al-manṭiq" (Maimonides' treatise on logic), an incomplete Arabic text and the Hebrew versions edited, with an English trans., by I. Efros, in *Proceedings of the American Academy for Jewish Research*, **8** (1937–1938)—the complete text of the Arabic original of his treatise was found and edited, with a Turkish trans., by Mubahat Türker, in *Ankara Üniversitesi Dil ve tarih-coğrafya Fakültesi Dergisi*, **18** (1960), 14–64; "Treatise on Resurrection," original Arabic and Samuel ibn Tibbon's Hebrew trans. edited by J. Finkel, in

Proceedings of the American Academy for Jewish Research, **9** (1939), 1–42, 60–105; "Thamāniyat Fuṣūl," an exposition of ethics, Arabic text edited, with a German trans. by M. Wolff (Leiden, 1903); *Responsen und Briefe des Maimonides*, A. Lichtenberg, ed. (Leipzig, 1859); and *Teshubhot Ha-Rambam* ("Responsa of Maimonides"), 3 vols., J. Blau, ed. (Jerusalem, 1957–1961).

II. SECONDARY LITERATURE. Works on Maimonides are Alexander Altmann, "Das Verhältnis Maimunis zur jüdischen Mystik," in *Monatsschrift für Geschichte und Wissenschaft des Judentums*, **80** (1936), 305–330; Salo Baron, ed., *Essays on Maimonides: An Octocentennial Volume* (New York, 1941); H. Davidson, "Maimonides Shemonah Perakim and Alfarabi's Fuṣūl Al-Madanī," in *Proceedings of the American Academy for Jewish Research*, **31** (1963), 33–50; Z. Diesendruck, "Maimonides' Lehre von der Prophetie," in G. A. Kohut, ed., *Jewish Studies in Memory of Israel Abrahams* (New York, 1927), pp. 74–134; and "Die Teleologie bei Maimonides," in *Hebrew Union College Annual*, **5** (1928), 415–534; I. Epstein, ed., *Moses Maimonides: 1135–1204* (London, 1935); Jakob Guttmann, *Der Einfluss der Maimonideschen Philosophie auf das christliche Abendland* (Leipzig, 1908); S. Pines, "Spinoza's Tractatus Theologica Politicus. Maimonides and Kant," in *Scripta universitatis atque bibliothecae hierosolymitanarum*, **10** (1968), 3–5; A. Rohner, *Das Schöfungsproblem bei Moses Maimonides, Albertus Magnus, und Thomas von Aquin* (Münster, 1913); Leon Roth, *The Guide for the Perplexed, Moses Maimonides* (London, 1948); and Leo Strauss, *Philosophie und Gesetz* (Berlin, 1935); "Quelques remarques sur la science politique de Maimonide et de Farabi," in *Revue des études juives*, **100** (1936), 1–37; and *Persecution and the Art of Writing* (Chicago, 1952), which includes "The Literary Character of *The Guide for the Perplexed*" (pp. 37–94), also in Baron's *Essays on Maimonides* (see above), 37–91.

See also the following works by H. A. Wolfson: "Maimonides and Halevi," in *Jewish Quarterly Review*, **2** (1911–1912), 297–337; "Maimonides on the Internal Senses," *ibid.*, **25** (1934–1935), 441–467; "Hallevi and Maimonides on Design, Chance, and Necessity," in *Proceedings of the American Academy for Jewish Research*, **11** (1941), 105–163; "Hallevi and Maimonides on Prophecy," in *Jewish Quarterly Review*, n.s. **32** (1941–1942), 345–370, and n.s. **33** (1942–1943), 49–82; "The Platonic, Aristotelian, and Stoic Theories of Creation in Hallevi and Maimonides," in I. Epstein, E. Levine, and C. Roth, eds., *Essays in Honor of the Very Rev. Dr. J. H. Hertz* (London, 1942), pp. 427–442; and "Maimonides on Negative Attributes," in A. Marx *et al.*, eds., *Louis Ginzberg Jubilee Volume* (New York, 1945), pp. 419–446.

SHLOMO PINES

MAIOR (or MAIORIS), JOHN (frequently cited as **JEAN MAIR)** (*b.* Gleghornie, near Haddington, Scotland, 1469; *d.* St. Andrews, Scotland, 1550), *logic, mathematics, natural philosophy, history.*

Maior received his early education in Haddington, whence he passed to God's House (later Christ's College), Cambridge, and then to the University of Paris, where he enrolled at the Collège Ste. Barbe about 1492; he completed his education at the Collège de Montaigu. He received the licentiate in arts in 1495 and the licentiate and doctorate in theology in 1506. In 1518 Maior returned to Scotland, where he occupied the first chair of philosophy and theology at Glasgow; in 1522 he was invited to the University of St. Andrews to teach logic and theology. Attracted back to Paris in 1525, he taught there until 1531, when he returned again to St. Andrews. He became provost of St. Salvator's College in 1533 and, as dean of the theological faculty, was invited to the provincial council of 1549, although he could not attend because of advanced age.

Maior spent most of his productive life in Paris, where he formed a school of philosophers and theologians whose influence was unparalleled in its time. Himself taught by nominalists such as Thomas Bricot and Geronymo Pardo and by the Scotist Peter Tartaret, Maior showed a special predilection for nominalism while remaining open to realism, especially that of his *conterraneus* (countryman) John Duns Scotus. To this eclecticism Maior brought a great concern for positive sources, researching and editing with his students many terminist and Scholastic treatises and even contributing to history with his impressive *Historiae Majoris Britanniae, tam Angliae quam Scotiae* (Paris, 1521). His students included the Spaniards Luis Coronel and his brother Antonio and Gaspar Lax; the Scots Robert Caubraith, David Cranston, and George Lokert; and Peter Crokart of Brussels and John Dullaert of Ghent. They and their students quickly diffused Maior's ideals of scholarship through the universities of Spain, Britain, and France, and ultimately throughout Europe. In theology Maior was unsympathetic to the Reformers (he taught the young John Knox while at Glasgow) and remained faithful to the Church of Rome until his death.

Maior's importance for physical science derives from his interest in logic and mathematics and their application to the problems of natural philosophy. He became an important avenue through which the writings of the fourteenth-century Mertonians, especially Bradwardine, Heytesbury, and Swineshead, exerted an influence in the schools of the sixteenth century, including those at Padua and Pisa, where the young Galileo received his education. Among Maior's logical writings the treatise *Propositum de infinito* (1506) is important for its anticipation of modern mathematical treatments of infinity; in it he argues in favor of the existence of actual infinities (*infinita actu)* and discusses the possibilities of motion of an infinite body.

Maior also composed series of questions on all of Aristotle's physical works (Paris, 1526), based on "an exemplar sent to me from Britain" and thus probably written between 1518 and 1525; it is a balanced, if somewhat eclectic, exposition of the main positions that were then being argued by the nominalists and realists. Maior's commentaries on the *Sentences* are significant for their treatment of scientific questions in a theological context; they were used and cited, generally favorably, until the end of the sixteenth century.

BIBLIOGRAPHY

I. ORIGINAL WORKS. Hubert Élie, ed., *Le traité "De l'infini" de Jean Mair* (Paris, 1938), is a Latin ed. of the *Propositum de infinito* with French trans., intro., and notes. Some of Maior's works are listed in the *Dictionary of National Biography*, XII (1921–1922), 830–832. That list has been emended by R. G. Villoslada, S.J., "La universidad de Paris durante los estudios de Francisco de Vitoria, O.P. (1507–1522)," in *Analecta Gregoriana*, **14** (1938), 127–164; and by Élie, *op. cit.*, pp. v–xix. Villoslada also analyzes Maior's philosophical and theological writings and provides a guide to bibliography.

II. SECONDARY LITERATURE. See Hubert Élie, "Quelques maîtres de l'université de Paris vers l'an 1500," in *Archives d'histoire doctrinale et littéraire du moyen âge*, **18** (1950–1951), 193–243, esp. 205–212; and William A. Wallace, O.P., "The Concept of Motion in the Sixteenth Century," in *Proceedings of the American Catholic Philosophical Association*, **41** (1967), 184–195; also A. B. Emden, *A Biographical Register of The University of Cambridge to 1500* (Cambridge, 1963), 384–385.

WILLIAM A. WALLACE, O.P.

MAIR, SIMON. See **Mayr, Simon.**

MAIRAN, JEAN JACQUES D'ORTOUS DE (*b.* Béziers, France, 26 November 1678; *d.* Paris, France, 20 February 1771), *physics.*

Mairan was concerned with a wide variety of subjects, including heat, light, sound, motion, the shape of the earth, and the aurora. He wanted to find physical mechanisms (in the Cartesian sense) to explain phenomena. His theories were generally ingenious descriptions, which were sometimes mathematical and sometimes based on experiment. Despite his enthusiasm and diligence, Mairan often failed to perceive what was trivial and what was crucial about a theory or phenomenon. Nevertheless, Mairan was an important and sometimes controversial figure in the scientific community of his day. Working in the decades during which Newtonian ideas were becoming known in France, Mairan incorporated some of them in his theories; but he remained basically Cartesian.

Mairan's family came from the minor nobility. His parents were François d'Ortous de Mairan and Magdaleine d'Ortous. After studying classics at Toulouse, Mairan studied physics and mathematics in Paris, where Malebranche was one of his teachers. On returning to Béziers in 1704, Mairan continued to be interested in science, and in the years 1715 to 1717 he published his first major works, which received prizes from the Bordeaux Academy. In 1718 Mairan went to Paris and became a member of the Academy of Sciences. (He was made an associate member right away, an unusual procedure.) He later received official lodging in the Louvre. Mairan was secretary of the Academy from 1741 to 1743, succeeding Fontenelle, and he was made *pensionnaire géomètre* in 1746. He also belonged to the Royal Societies of London, Edinburgh, and Uppsala, the Petersburg Academy, and the Institute of Bologna. Mairan was an amateur pianist, he had a serious interest in Chinese culture, and he attended the Paris salons.

Like most of his other work, Mairan's work on heat continued over a long period and was ambitiously conceived. Mairan tried to explain temperature variations and changes of state. His theory of heat was essentially kinetic, but he felt that a subtle matter was necessary to account for the motions of the ultimate particles of ordinary matter and for the changes in these motions. On the basis of observations, experiments, and ingenious estimates, Mairan concluded that the earth has a "central fire" which is an important source of its heat.

Mairan tried to construct a theory of light which would be a Cartesian modification of Newton's theory. In analogy with light, he attempted to understand sound in terms of particles rather than waves. This theory was inspired by the fact that one can distinguish different pitches even when they are all produced together. Mairan postulated different species of particles to carry the different pitches in the propagation of sound.

Mairan's most intense controversies were associated with his ideas in mechanics. He became involved in the notorious *vis viva* controversy, arguing that the "force of a body" depends on its velocity rather than the square of its velocity. Contrary to some, Mairan was aware that velocity should be treated as a vector quantity, but his arguments in general did not provide any special clarification of the problem.

In connection with the shape of the earth, another controversial topic, Mairan tried to reconcile pendulum measurements (indicating that the force of gravity is weaker at the equator) with the Cassinis' (erroneous) measurements of the length of a degree along the meridian (indicating that the earth is elongated at the poles). Mairan proposed that attraction at a point on the earth varies, not according to Newton's law, but inversely as the product of the two principal radii of curvature!

BIBLIOGRAPHY

I. ORIGINAL WORKS. Mairan's works include *Dissertation sur les variations du baromètre* (Bordeaux, 1715); *Dissertation sur la glace* (Bordeaux, 1716; Paris, 1749); *Dissertation sur la cause de la lumière des phosphores et des noctiluques* (Bordeaux, 1717); "Recherches géométriques sur la diminution des degrés terrestres en allant de l'equateur vers les poles," in *Mémoires de l'Académie royale des sciences* (1720); "Dissertation sur l'estimation et la mesure des forces motrices des corps," *ibid.* (1728); "Discours sur la propagation du son dans les différens tons qui le modifient," *ibid.* (1737).

More of Mairan's works are listed in J. C. Poggendorff, *Biographisch-Literarisches Handwörterbuch*, II (Leipzig, 1863), 18, and in the study by Kleinbaum mentioned below.

II. SECONDARY WORKS. Abby R. Kleinbaum, *Jean Jacques Dortous de Mairan (1678–1771): A Study of an Enlightenment Scientist* (Columbia Univ. Ph.D. diss., 1970), is an excellent study of Mairan's scientific work. The *éloge* of Mairan, by Grandjean de Fouchy, is in the *Histoire de l'Académie royale des sciences*, 1771 (Paris, 1774), 89–104. There is a discussion of Mairan's work with respect to Newtonianism and Cartesianism in Pierre Brunet, *L'introduction des théories de Newton en France au XVIII siècle* (Paris, 1928); Mairan's work on the shape of the earth is discussed in I. Todhunter, *A History of the Mathematical Theories of Attraction and the Figure of the Earth* (New York, 1962), I, 59–61; Mairan's role in the *vis viva* controversy is discussed in René Dugas, *A History of Mechanics* (Neuchâtel, 1955); Mairan's work on the aurora is discussed in J. Morton Briggs, "Aurora and Enlightenment," in *Isis*, **58** (1967), 491–503.

SIGALIA C. DOSTROVSKY

MAIRE, RENÉ-CHARLES-JOSEPH-ERNEST (*b.* Lons-le-Saunier, France, 29 May 1878; *d.* Algiers, Algeria, 24 November 1949), *botany.*

The son of a forest ranger, Maire displayed a precocious interest in botany and at the age of fifteen published his first observations on the vegetation of the Jura. A student at the Faculties of Science and Medicine at Nancy, he was encouraged by the botanists Georges Le Monnier and Paul Vuillemin. By the age of twenty he had published about twenty papers. His interests led him to fieldwork as well as to laboratory observations. His favorite objects of study apart from the phanerogams were the fungi.

His doctoral dissertation on the cytology of the Basidiomycetes, which he defended in Paris at the age of twenty-four, is still a basic work. In it he explained why previous authors believed, wrongly, in the existence of acaryotic stages and he specified the nature of the metachromatic corpuscles. In addition he outlined the nuclear evolution of the Ustilaginales (smuts) and the Uredinales and defined the synkaryon. The latter, which is found among the fleshy Basidiomycetes, is a caryologic unit formed from two morphologically distinct but intimately related nuclei. Finally, he demonstrated that among the Ustilaginales the budding basidiospores or sporidia have a structure identical to that of the true blastosporous fungi, such as the *Saccharomyces.*

Maire supported the observations of Pierre-Auguste Dangeard on the cytological characteristics that originate in the spores and the mycelium of these Ascomycetes, and he agreed with Dangeard that among the fungi, fertilization, which is proper to the higher plants, is replaced by the fusion of two nuclei in the mother cell of the basidium and of the ascus.

At age thirty-three, after serving as a *maître de conférences* at the Faculty of Sciences of Caen, Maire was named to the chair of botany at Algiers. He held this post at the French University of North Africa for nearly forty years.

On several voyages in the Mediterranean basin, Maire studied the phanerogams and fungi of Corsica (1902–1904), the Balearic Islands (1905), the Olympus Mountains, and the Taurus Mountains. He demonstrated the phytogeographical heterogeneity of Thessaly and Epirus and identified six stages of vegetation in that region. Maire first went to Africa in 1902. From Tangier he traversed the area south of Oran and the mountains of Tlemcen, visiting Tunisia in 1909. Permanently settled in Algiers, he explored the Djuradjura and Babor mountains, South Oran, Mount Daya and the Tlemcen Mountains. Stationed in Thessaloniki during World War I, he spent his leaves on the island of Skíros and in Pilos. With Braun-Blanquet, who influenced his work, he published a phytogeographical sketch of Morocco (1925), after having climbed the High and Middle Atlas in 1921.

From 1931 to 1936 Maire made twenty-seven trips to Morocco. He described in detail the Mediterranean character of the Sous and examined the flora of the Moroccan coast, the Rif, Mount Zaian, Mount Tichchoukt, the summits of the High Atlas, the fir and oak forests of Tauzin, Ceuta and the Anti-Atlas, the high Dra River, the Tafilalet, and the plateau of the Lakes District. Maire held that the origins of the Moroccan flora and its autonomous evolution since the Pliocene, in conjunction with the penetration of the arcto-Tertiary floral element, explain the Iberian character of this vegetation. Maire was also active in these years in Algeria, notably in the Aurès Mountains, the phytogeographical map of which he helped to establish. From 1932 to 1935 he explored the Western Sahara as far as Tindouf, as well as the Tefedest, the Hoggar, and the Tassili N' Ajjer Mountains; he described three stages of tropical and Mediterranean vegetation in this region. The results of these gigantic botanical labors were set forth in *Contributions à l'étude de la flore de l'Afrique du Nord*; three volumes were prepared before Maire's death, the remaining were completed by his successors.

A first-rate mycologist endowed with an exceptional memory, Maire studied various fungi of Europe and the Maghreb: Laboulbeniales, rusts, Pezizales, Gasteromycetes, and especially the fleshy agarics. In 1908, while traveling in Sweden, Maire encountered the work of Elias Fries, which left a lasting impression on his own work. In his study of the *Russula*, Maire introduced the Ariadne's thread that permitted the discovery of the exact value of the characteristics of this difficult genus. His contributions to the mycology of the cedars of the Atlas Mountains and of Catalonia as well as to toxicology and to phytopathology are also important, and his account of the biology of the Uredinales is a model of clarity.

Maire became correspondent of the Paris Academy of Sciences in 1923 and nonresident member in 1946. He was also honorary president of the Société Mycologique de France, an organization in which he retained a lively interest.

A scientist whose devotion to work consumed all his energy, Maire was egocentric and severe about keeping to a regular schedule. There was no room in his life for anything besides his research. His personality bore the mark of his native Lorraine; he was even-tempered, rigorous, objective, easy to approach, indulgent, and accommodating—traits that made him universally popular.

BIBLIOGRAPHY

I. ORIGINAL WORKS. Maire's writings include his diss., "Recherches cytologiques et taxonomiques sur les basidiomycètes," in *Bulletin de la Société mycologique de France*, **18** (1902), 1–209, with 8 plates; "Les bases de la classification dans le genre *Russula*," *ibid.*, **26** (1910), 49–125, with figures; "La biologie des urédinales," in *Progressus rei botanicae*, **4** (1911), 109–162; "Études sur la végétation et la flore du Grand Atlas et du Moyen Atlas Marocains," which is *Mémoires de la Société des sciences naturelles et physique du Maroc*, no. 7 (1924), with 16 plates; and "Études sur la végétation et la flore marocaines," *ibid.*, no. 8 (1925), with map, plates, and figures.

II. SECONDARY LITERATURE. Articles devoted to Maire include L. Emberger, G. Malençon, and C. Sauvage, "Hommage à René Maire. I. L'Homme. II. Le Mycologue. III. Le Phanérogamiste," in *Bulletin de la Société des sciences naturelles du Maroc*, **1** (1950), 9; J. Feldmann, "René Maire," in *Revue générale de botanique*, **58** (1951), 65; "René Maire. Sa vie et son oeuvre," written with P. Guinier, in *Bulletin de la Société d'histoire naturelle de l'Afrique du Nord*, **41** (1952), contains a complete bibliography; B. P. G. Hochreutiner, "Un grand systématicien et mycologue français, René Maire," in *Mémoires de la Société botanique de France* (1950–1951), 132–136; F. Jelenc, "René Maire (1878–1949)," in *Revue bryologique et lichénologique*, n.s. **19** (1950), 5; and R. Kuhner, "René Maire (1878–1949)," in *Bulletin de la Société mycologique de France*, **49** (1953), 1–49.

ROGER HEIM

IBN MĀJID, SHIHĀB AL-DĪN AḤMAD IBN MĀJID (*fl.* Najd, Saudi Arabia, fifteenth century A.D.), *navigation*.

Ibn Mājid inherited his profession; both his father and grandfather were *mu'allim*, "masters of navigation," and both were known as experts in the navigation of the Red Sea, dreaded by sailors. Of the Arab navigators of the Middle Ages, none surpassed Ibn Mājid himself in the intimate knowledge and experience of both the Red Sea and the Indian Ocean. He knew almost all the sea routes from the Red Sea to East Africa, and from East Africa to China; proud of his achievements, he styled himself "The Successor of the Lions," or "The Lion of the Sea in Fury" (in his *Ḥāwiyat al-ikhtiṣār fī uṣūl 'ilm al-biḥār*, dated A.H. 866, or A.D. 1462, fol. 88b). He became a legend among pious mariners, who called him "Shaykh Mājid" and recited the *Fātiḥa*, the first chapter of the Koran, in his memory before embarking on certain seas.[1]

Ibn Mājid was well versed in the works of a number of both Muslim and Greek geographers, astronomers, and navigators. He considered the study of these sources to be essential to Arab navigators, and is known to have read books by Ptolemy, Abu'l-Ḥasan al-Marrākushī, al-Ṣūfī, al-Ṭūsī, Yāqūt al-Ḥamawī,

Ibn Saʿīd, al-Battānī, Ibn Ḥawqal, and Ulūgh Bēg—as well as the the works of three ʿAbbāsid sailors, Muḥammad ibn Shādān, Sahl ibn Abān, and Layth ibn Kahlān, whom he dismissed as mere compilers.[2]

Ibn Mājid himself wrote at least thirty-eight works, in both prose and poetry, of which twenty-five are extant. In these he took up a wide variety of astronomical and nautical subjects, including the lunar mansions; the stars that correspond to the thirty-eight rhumbs *(khanns)* of the mariner's compass; sea routes of the Indian Ocean and the latitude of harbors; birds as landmarks; coastlines; the "ten large islands" of the Indian Ocean (Arabia, Madagascar, Sumatra, Java, Taiwan, Ceylon, Zanzibar, Bahrein, Ibn Gāwān, and Socotra); a systematic survey of the coastal regions of Asia and Africa (in which he revealed a more detailed knowledge of the coasts of the Indian Ocean than those of the Mediterranean or Caspian Sea); the Red Sea; Arabian, Coptic, Byzantine, and Persian years; *bāshī*, the computation of the elevation of the polestar from its minimum height above the horizon; the proper direction of the Kaaba; landfalls, in particular landfalls on capes during monsoons; certain northern stars; months of the Byzantine calendar; general instructions for navigators; reefs and deeps; signs indicating land; observations of both constellations (Aquarius) and individual stars (Canopus, Arcturus); *majrās* (a course of a journey by sea); and European, especially Portuguese, navigators of the Indian Ocean. Ibn Mājid is, in addition, known to have revised and enlarged a book, called *al-Ḥijāziyya* and written in verse of the *rajaz* form, originally composed by his father.[3]

Of all his works, however, Ibn Mājid's *Kitāb al-Fawāʾid*, dated A.H. 895 (A.D. 1490), was the one most valuable to navigators. Indeed, the Turkish navigator Sīdī ʿAlī Reʾīs (who died in 1562) had, during a stay in Basra, acquired a copy of this book, together with Ibn Mājid's *Ḥāwiya* and some more nearly contemporary works of Sulaymān al-Mahrī, because, according to him, it was extremely difficult to navigate the Indian Ocean without them.[4] A modern scholar, Gabriel Ferrand, correctly described the *Kitāb al-Fawāʾid* as a "compendium of the known knowledge of theoretical and practical navigation" and as "a kind of synthesis of nautical science of the latter years of the Middle Ages." Ferrand described Ibn Mājid himself as the first writer on nautical science in its modern sense, adding that, apart from the inevitable errors in latitudes, his description of the Red Sea for navigational purposes had never been equaled.[5]

The *Kitāb al-Fawāʾid* also makes it clear that Ibn Mājid did not actually invent the mariner's compass,

although others have claimed that invention for him. Indeed, in folio 46b, he specified only that he fixed the needle (*al-maghnāṭīs*, or "magnet") to the case of the instrument. He did boast, however, that the compass used by the Arab navigators of the Indian Ocean was much superior to that employed by their Egyptian or Maghribi (North African) counterparts, since the Arab compass was divided into thirty-two, rather than only sixteen, sections. He further claimed that the Egyptians and Maghribis were unable to sail Arab ships, while Arabs could handle Egyptian and Maghribi vessels with great ease.[6]

Ibn Mājid's *Al-sufāliyya* is also of particular interest, since in it he records (folio 94a) the expeditions of the "Franks," or Portuguese (although the term was also used for Europeans in general). He was aware of the Portuguese circumnavigation of the Cape of Good Hope, an event that took place near the end of his life, and further aware of Portuguese navigation in the Indian Ocean. He wrote that the Franks, having passed through *al-madkhal* ("the place of entry") that lay between the Maghrib and Sofala (Mozambique), reached the latter coast in A.H. 900 (A.D. 1495), and proceeded to India. He gave a further account of their return to Portugal, by way of Zanzibar and through the same "passage of the Franks," and of their second voyage, in A.H. 906 (A.D. 1501) to India, where they purchased houses and settled down, having been befriended by the Sāmrī rulers (the zamorin kings of Kerala).

Al-madkhal was an object of much concern to Arab mariners, who believed it to be a sea channel that lay south of the Mountains of the Moon (the source of the Nile) and connected the Indian Ocean with the Atlantic. According to Ptolemaic tradition, the whole of the southern hemisphere was terra incognita, an extension of the southern coast of Africa; Arab maps of the period show the Indian Ocean as a lake, connected by a sea passage to the Pacific, and lacking any communication with the Atlantic. Al-Bīrūnī, however, had posited a channel between the Indian Ocean and the "Sea of Darkness"—the Atlantic—and had placed it somewhere south of the source of the Nile, between Sofala and the cape al-Raʾsūn (probably in the region of the Agulhas currents on modern maps). Abu ʾl-Fidāʾ quoted him in his own *Taqwīm al-Buldān*, a work known to Ibn Mājid, who considered his predecessor's theory to be proved by the accomplishments of "the experienced ones," the Portuguese.

Camoëns mentioned Ibn Mājid in *The Lusiads*, while a later Arab historian, Quṭb al-Dīn al-Nahrwālī, accused him of having drunkenly confided to the chief of the Franks *(al-amilandī)*, Vasco da Gama, the navigational information that allowed him to sail from East Africa to India.

NOTES

1. Ferrand, *Instructions nautiques*, III, 227–228.
2. *Kitāb al-Fawā'id*, fols. 3b–4a; *cf.* Ferrand, *op. cit.*, 229–233.
3. *Ibid.*, fol. 78a–b.
4. See Ferrand, in *Encyclopaedia of Islam*, 1st ed., IV, 363.
5. *Ibid.*, 365.
6. *Ibid.* and ff.

BIBLIOGRAPHY

I. ORIGINAL WORKS. Twenty-two of Ibn Mājid's works have been published in facsimile by G. Ferrand, *Instructions nautiques et routiers arabes et portugais des XVᵉ et XVIᵉ siècles*, 3 vols. (Paris, 1921–1928). These include both prose and poetical works (*urjūza*): *Kitāb al-Fawā'id fī uṣūl 'ilm al-baḥr wa 'l-qawā'id* (A.H. 895, A.D. 1490); *Ḥāwiyat al-ikhtiṣār fī uṣūl 'ilm al-biḥār* (A.H. 866, A.D. 1462); *Al-Mu'arraba* (A.H. 890, A.D. 1485); *Kiblat al-Islām fī jamī' al-dunyā* (A.H. 893, A.D. 1488); *Urjūza Barr al-'Arab fī Khalīj Fārs*; *Urjūza fī qismat al-jamma 'alā Banāt Na'sh* (A.H. 900, A.D. 1494–1495); *Kanz al-Ma'ālima wa dhakhīratihim fī 'ilm al-najhūlāt fi'l-baḥr wa 'l-nujūm wa 'l-burūj* (not dated, but probably written before A.H. 894, A.D. 1489); *Urjūza fī 'l-natakhāt li-Barr al-Hind wa Barr al-'Arab*; *Mīmiyyāt al-abdāl*; *Urjūza Mukhammasa*; *Urjūza* on the Byzantine months, rhyming in *nūn* (not dated, but probably written before 1475 or 1489); *Ḍaribat al-ḍarā'ib*; *Urjūza* dedicated to the caliph 'Alī ibn Abī Ṭālib (not dated, but written before 1475 or 1489); *Al-Qaṣīda al-Makkiyya*; *Nādirat al-abdāl*; *Al-Qaṣīda al-Bā'iyya*, called *Al-Dhahabiyya* (dated 16 Dhu 1-Ḥijja 882, or 21 March 1478); *Al-Fā'iqa* (not dated, but written before 1475); *Al-Balīgha*; nine short prose sections (*faṣl*); *Urjūza* called *Al-Sab'iyya*; untitled *Qaṣīda* (not dated, but written before 1475, 1478, or 1489); and *Qaṣīda* called *Al-Hādiya* (not dated, written before 1475, 1478, or 1489).

Three further *urjūzas*, *Al-Sufāliyya*, *Al-Ma'laqiyya*, and *Al-Tā'iyya*, have been published, with Russian translations and notes, by T. A. Shumovsky, *Thalāth rāhmānajāt al-majhūla li Aḥmad ibn Mājid: Tri nyeizvyestnioye lotsii Akhmada ibn Madzida arabskogo lotsmana Vasko da Gami* (Moscow–Leningrad, 1957).

Thirteen other works, specifically mentioned by Ibn Mājid in *Kitāb al-Fawā'id*, are no longer known.

II. SECONDARY LITERATURE. See S. Maqbul Ahmad, "The Arabs and the Rounding of the Cape of Good Hope," in *Dr. Zakir Husain Presentation Volume* (New Delhi, 1968), 90–100; M. Reinaud, *Géographie d'Aboulféda*, vol. I of *Introduction générale à la géographie des orientaux* (Paris, 1848); and *Encyclopaedia of Islam*, 2nd ed., III (Leiden, 1968), 856–859.

S. MAQBUL AHMAD

MAJORANA, ETTORE (*b.* Catania, Sicily, 5 August 1906; *d.* at sea, near Naples, 25/26 March 1938), *physics.*

Majorana was the fourth of the five children of Fabio Massimo Majorana, an engineer and inspector general of the Italian ministry of communications, and Dorina Corso. At the age of four he revealed the first signs of a gift for arithmetic. After schooling at home he entered the Jesuit Istituto Massimo in Rome and completed his secondary school education at the Liceo Torquato Tasso, passing his *maturità classica* in the summer of 1923. That fall he entered the School of Engineering of the University of Rome, where his fellow students included his older brother Luciano, Emilio Segrè, and Enrico Volterra, later professor of civil engineering at the University of Houston. Majorana was persuaded by Segrè to take up physics at the beginning of 1928. His lively mind, insight, and the range of his interests immediately impressed the new circle of physicists that had formed around Fermi. He was nicknamed "the Grand Inquisitor" for his exceptionally penetrating and inexorable capacity for scientific criticism, even of his own person and work. He received the doctorate in physics on 6 July 1929 with a thesis on the mechanics of radioactive nuclei sponsored by Fermi.

Fermi convinced Majorana to go abroad financed by a grant from the Consiglio Nazionale delle Ricerche; and Majorana began his journey at the end of January 1933, traveling first to Leipzig and then to Copenhagen. In Leipzig, Heisenberg persuaded Majorana to publish his paper on nuclear forces. He returned to Rome in the autumn of 1933 in poor health aggravated by gastritis, which he had developed in Germany and which was attributed by some to nervous exhaustion. He attended the Istituto di Fisica at intervals but stopped after a few months, despite his friends' attempts to lead him back to a normal life.

Appointed professor of theoretical physics at Naples in November 1937, Majorana soon discovered that his course was too advanced for the majority of students. On 25 March 1938 he wrote from Palermo to his colleague and friend Antonio Carrelli that he found life in general, and his own in particular, useless and had decided to commit suicide. A few hours later he sent a telegram to Carrelli asking him to disregard the letter and boarded a steamer for Naples that evening. Although he was seen at daybreak as the ship entered the Bay of Naples, no trace was ever found of him, despite an inquiry continued for several months and repeated appeals of his family published in the Italian press.

Majorana's total scientific production consists of nine papers, which can be divided into two parts:

six papers on problems of atomic and molecular physics, and three on nuclear physics or the properties of elementary particles. The first group of papers deals with the splitting of Roentgen terms of heavy elements induced by electron spin, the interpretation of recently observed spectral lines in terms of atomic states with two excited electrons, the formation of the molecular ion of helium, the binding of molecular hydrogen through a mechanism different from that of Walter Heitler and Heinz London, and the probability of reversing the magnetic moment of the atoms in a beam of polarized vapor moving through a rapidly varying magnetic field. The last paper remains a classic on nonadiabatic moment-inversion processes. Often quoted, it provides the basis for interpreting the experimental method of flipping neutron spin with a radio-frequency field. The other papers of this period (1928–1932) reveal a thorough knowledge of experimental data and an ease—particularly unusual at the time—in using the symmetry properties of the states to simplify problems or to choose the most suitable approximation for solving each problem quantitatively. The latter ability was at least partly due to Majorana's exceptional gift for calculation.

Majorana's major scientific contribution, however, is found in the last three papers. "Sulla teoria dei nuclei" (1932) concerns the theory of light nuclei under the assumption that they consist solely of protons and neutrons that interact through exchange forces acting only on the space coordinates (and not on the spin), so that the alpha particle—rather than the deuteron—is shown to be, as it is, the system with greatest binding energy per nucleon. The essential work on this paper was completed in the spring of 1932, only two months after the appearance of J. Chadwick's letter to the editor of *Nature* announcing the discovery of the neutron. Fermi and his friends tried in vain to persuade Majorana to publish, but he did not consider his work good enough and even forbade Fermi to mention his results at an international conference that was to take place in July 1932 in Paris. The July 1932 issue of *Zeitschrift für Physik* contains the first of Heisenberg's three famous papers on the same subject. They are based on Heisenberg's exchange forces, which differ from Majorana's forces in that not only the space coordinates but also the spin of the two particles are exchanged.

"Teoria relativistica di particelle con momento intrinseco arbitrario" (1932), the first paper of Majorana's second phase, concerns the relativistic theory of particles with arbitrary intrinsic angular momentum. Although in some ways outside the mainstream of the development of elementary-particle physics, it represents the first attempt to construct a relativistically invariant theory of arbitrary half-integer or integer-spin particles. Majorana's mathematically correct theory contains the first recognition, and the simplest development and application, of the infinite dimensional unitary representations of the Lorentz group. This theory lies outside the mainstream of successive development primarily because, from the outset, Majorana set himself the task of constructing a relativistically invariant linear theory of which the eigenvalues of the mass were all positive. This viewpoint was justified at the time the paper was written (summer 1932), since news of C. D. Anderson's discovery of the positron had not yet reached Rome.

Majorana's last paper was written in 1937 on Fermi's urging, after four years of not publishing because of poor health. It contains a symmetrical theory of the electron and the positron based on the Dirac equation but in which the states of negative energy are avoided and a neutral particle is identical to its antiparticle. The most characteristic point is the discovery of a representation of the Dirac matrices γ_k ($k = 1, 2, 3, 4$), in which the first three components are real, the fourth imaginary, like the vector $x \equiv \vec{r}$, ict (Majorana representation).

At present no neutral particle of the type suggested by Majorana is known, since it has been experimentally established that the neutron, lambda particle, and neutrino differ from their corresponding antiparticles. Nevertheless, Majorana's neutrino, ν_M, characterized by the equality $\nu_M = \bar{\nu}_M$ (the bar indicates the antiparticle), has played an important part in the physics of weak interactions, especially since the discovery by T. D. Lee and C. N. Yang of the nonconservation of parity and the development of the two-component theory of the neutrino. This theory is related to that of Majorana, to which, in certain aspects, it is equivalent. Contrary to the two-component theory, Majorana's does not require the neutrino to have a mass exactly equal to zero, and a small neutrino mass cannot at present be excluded on the basis of available experimental data.

Majorana had an extraordinary gift for mathematics, an exceptionally keen analytic mind, and an acute critical sense. It was perhaps the latter, together with a certain lack of balance on the human side, that interfered with his capacity for creative synthesis and prevented him from reaching a level of scientific productivity comparable to that attained at the same age by major contemporary physicists. Yet his choice of problems and his way—especially his mathematical methods—of attacking them showed that he was naturally in advance of his times and, in some cases, almost prophetic.

BIBLIOGRAPHY

I. ORIGINAL WORKS. Majorana's papers on atomic and molecular physics are "Sullo sdoppiamento dei termini Roentgen e ottici a causa dell'elettrone rotante . . .," in *Atti dell'Accademia nazionale dei Lincei. Rendiconti*, 6th ser., **8** (1928), 229–233, written with G. Gentile; "Sulla formazione dello ione molecolare di Elio," in *Nuovo cimento*, 8th ser., **8** (1931), 22–28; "I presunti termini anomali dell'Elio," *ibid.*, 78–83; "Reazione pseudopolare fra atomi di idrogeno," in *Atti dell'Accademia nazionale dei Lincei. Rendiconti*, **13** (1931), 58–61; "Teoria dei tripletti P′ incompleti," in *Nuovo cimento*, **8** (1931), 107–113; and "Atomi orientati in campo magnetico variabile," *ibid.*, **9** (1932), 43–50.

His papers on elementary particles are "Teoria relativistica di particelle con momento intrinseco arbitrario," *ibid.*, 335–344; "Sulla teoria dei nuclei," in *Ricerca scientifica*, **4** (1933), 559–565; and "Teoria simmetrica dell' elettrone e del positrone," in *Nuovo cimento*, **14** (1937), 171–184. See also the posthumously published "Il valore delle leggi statistiche nella fisica e nelle scienze sociali," in *Scientia* (Bologna), **71** (1942), 58–66.

II. SECONDARY LITERATURE. On Majorana's life and work, see E. Amaldi, *La vita e l'opera di Ettore Majorana* (Rome, 1966); an English trans. of the biographical note in this work is in A. Zichichi, ed., *Strong and Weak Interactions—Present Problems* (New York, 1966), 10–77, which also contains a list of Majorana's MSS at the Domus Galileiana, Pisa.

EDOARDO AMALDI

AL-MAJRĪṬĪ, ABU 'L-QĀSIM MASLAMA IBN AḤMAD AL-FARAḌĪ (*b.* Madrid, Spain, second half of the tenth century; *d.* Córdoba, Spain, *ca.* 1007), *astronomy.*

Little is known of al-Majrīṭī's life. He must have been quite an important personality, for Ibn Ḥazm (*d.* 1064) mentions him in his *Ṭawq al-ḥamāma* ("The Ring of the Dove"). It would appear that he early settled in Córdoba where, as a very young man, he studied with a geometrician named ʿAbd al-Ghāfir ibn Muḥammad; it may also be assumed that he was connected with the group of hellenizing scholars patronized by the Umayyad caliph ʿAbd al-Raḥmān III (A.D. 912–961). It is known that he was engaged in making astronomical observations in about A.D. 979; in this period he must have revised the astronomical tables of al-Khwārizmī. At some later date he also was responsible for making the *Rasāʾil* of the Ikhwān al-Ṣafāʾ known to Andalusian astronomers. He may in addition have served as court astrologer.

Al-Majrīṭī had several important disciples, whose later dispersion into all the provinces of Spain made his work known throughout the peninsula. One of

these, al-Kirmānī (*d.* 1066), continued al-Majrīṭī's work in carrying Ikhwān al-Ṣafāʾ's *Rasāʾil* into Zaragoza and to the northern frontier. Another, Abu 'l-Qāsim Aṣbagh, better known as Ibn al-Samḥ (*d.* 1035), published a two-part treatise of 130 chapters on the construction and use of the astrolabe, as well as some astronomical tables constructed by the Indian methods, and a book, *Libro de las láminas de los siete planetas*, that was translated into Spanish and incorporated into the *Libros del saber de astronomía*. Others of al-Majrīṭī's followers were Abū 'l-Qāsim Aḥmad, nicknamed Ibn al-Ṣaffār (*d.* 1034), whose work on the astrolabe is, in its Latin version, attributed to al-Majrīṭī; the astrologer Ibn al-Khayyāṭ (*d.* 1055), much praised in the *Memoirs* of the zirī king ʿAbd Allāh; al-Zahrāwī; and Abū Muslim ibn Khaldūn of Seville. Through these men al-Majrīṭī exercised a considerable influence on the work of later scientists.

Of al-Majrīṭī's own works, the actual number is in some dispute. In general, it may be assumed that the magical and alchemical works attributed to him are spurious, especially since Ibn Ṣāʿid does not refer to them in his *Ṭabaqāt*. The works that may be considered genuine are the *Commercial Arithmetic (Muʿāmalāt)*, which, according to Ibn Khaldūn, dealt with sales, cadaster, and taxes, using arithmetical, geometrical, and algebraic operations, all of which were apparently used without much distinction; the very brief *Treatise on the Astrolabe* (not to be confused with the longer work by Ibn al-Ṣaffār), which treated both the construction and use of that instrument; his adaptation of al-Khwārizmī's astronomical tables to the longitude of Córdoba and to the Hijra calendar; his revision of some tables by al-Battānī; some notes on the theorem of Menelaus; and the lost *Tasṭīḥ basīṭ al-kura*, an Arabic translation of Ptolemy's *Planisphaerium*, which survives in a Latin version drawn from the Arabic by Hermann of Dalmatia (1143) and in a Hebrew recension (al-Majrīṭī's annotations to the original are also still extant).

Of the works often—but probably wrongly— attributed to al-Majrīṭī, the *Rutbat al-ḥakīm* ("The Rank of the Sage") was composed after 1009; it is alchemical in nature, and gives formulas and instructions for the purification of precious metals and describes the preparation of mercuric oxide on a quantitative basis. *Ghāyat al-ḥakīm* ("The Aim of the Wise") was translated into Spanish in 1256 by order of Alfonso el Sabio; it was widely distributed throughout Europe under the title *Picatrix* (a corruption of Buqrāṭis = Hippocrates), and is a compendium of magic, cosmology, astrological practice, and esoteric wisdom in general. As such, it provides the most complete picture of superstitions current in eleventh-century Islam. Also attributed to al-Majrīṭī are various opuscules which are in fact extracts, including passages on zoology and alchemy, from the *Rasā'il* of the Ikhwān al-Ṣafā', or have a certain relationship with these *Rasā'il* (like the *Risālat al jāmī'a*).

BIBLIOGRAPHY

I. ORIGINAL WORKS. Al-Majrīṭī's writings and those spurious works attributed to him are catalogued in Brockelmann, *Geschichte der arabischen Litteratur*, I (Weimar, 1898), 243, and supp. I (Leiden, 1937), 431.

Of the genuine works, the *Treatise on the Astrolabe* is edited and translated, with commentary, in J. Vernet and M. A. Catalá, "Las obras matemáticas de Maslama de Madrid," in *Al-Andalus*, **30** (1965), 15–45; see *ibid.*, pp. 46–47, an analysis of the position of the fixed stars by M. A. Catalá. Recent publications of the spurious works include Hellmut Ritter, ed., *Ghāyat al-ḥakīm* (Leipzig, 1933), and German trans. with Martin Plessner as "*Picatrix*." *Das Ziehl des Weisen von Pseudo-Maǧriti* (London, 1962); and Jamil Saliba, ed., *Risāla al-jami'a* (Damascus, 1948), which provides a good illustration of eleventh-century Ismā'īlī propaganda.

II. SECONDARY LITERATURE. On al-Majrīṭī's revision of al-Kwārizmī's tables, see G. J. Toomer, in *Dictionary of Scientific Biography*, VII, 360–361; see also Axel Björnbo and H. Suter, *Thabits Werke über den Transversalensatz (liber de figura sectore)* (Erlangen, 1924), 23, 79, and 83. On the works probably falsely attributed to him, see E. J. Holmyard, "Maslama al-Majrīṭī and the Rutbat al-ḥakīm," in *Isis*, **6** (1924), 239–305; also on the *Picatrix* and bibliography related to it, see the index by Willy Hartner, *Oriens, Occidens* (Hildesheim, 1968).

Supplementary material may be found in J. A. Sánchez Pérez, *Biografías de matemáticos árabes que florecieron en España* (Madrid, 1921), no.84; George Sarton, *Introduction to the History of Science*, I (Baltimore, 1927), 668–669; and H. Suter, *Die Mathematiker und Astronomen der Araber und ihre Werke* (Leipzig, 1900), 176.

JUAN VERNET

AL-MAJŪSĪ, ABU'L-ḤASAN 'ALĪ IBN 'ABBĀS (latinized as **Haly Abbas**) (*b*. al-Ahwāz-Khūzistān, near Shiraz, Persia, first quarter of the tenth century; *d*. Shiraz, A.D. 994), *medicine, pharmacology, natural science*.

Nothing is known of al-Majūsī's ancestry except that the nickname Majūsī suggests that he, or most probably his father, was originally a Zoroastrian and that he does not seem to have traveled much outside his native country. Al-Majūsī received his medical training under the physician Abū Māhir Mūsā ibn Sayyār, author of a commentary on phlebotomy. Al-Majūsī served King 'Aḍud al-Dawla (*d*. 983), to whom he dedicated his only medical compendium, *Kāmil al-Ṣinā'ah al-Ṭibbiyyah*, called *al-Malikī (Liber regius)* in honor of his patron, who bore the title *Shāhanshāh* ("king of kings").

The *Kāmil* consists of twenty treatises on the theory and practice of medicine (ten on each). In it the author referred to how he has studied and used indigenous medicinal plants, as well as animal and mineral products, as therapeutics. Although several important physicians and natural scientists appeared in tenth-century Iraq and Persia, only a few seem to have been known to or acknowledged by al-Majūsī. For example, he referred to the two books of al-Rāzī (865–925), the most prolific and original medical author in tenth-century Persia and the leading clinician, social scientist, and alchemist of his time. Yet al-Majūsī did not mention his countryman and contemporary al-Ḥusayn ibn Nūḥ al-Qumrī, author of the famous book *Ghanā wa-Manā* ("On Life and Death"), or Aḥmad ibn Abī al-Ash'ath of Mosul, author of a praiseworthy text on the powers and utility of the materia medica entitled *Quwa 'l-Adwiya 'l-Mufrada* and one of the best medical educators of his time. From the introductory remarks in the *Kāmil*, al-Majūsī seems to have been critical of his predecessors, even those whom he quoted and whose writings influenced him, such as Hippocrates, Galen, Oribasius (fourth century), Ahrun the Priest (sixth century), and Yūḥannā ibn Sirābiyūn (ninth century). He did, however, praise Ḥunayn ibn Isḥāq (*d*. 873) as a reliable translator and fine scholar.

Al-Majūsī gave the following interesting, surprisingly accurate, and almost modern description of pleurisy: "Pleurisy is an inflammation of the pleura, with exudation which pours materials over the pleura from the head or chest .... Following are the four symptoms that always accompany pleurisy: fever, coughing, pricking in the side, and difficult breathing (dyspnea)." In defining theoretical medicine, he recognized three areas:

1. Knowledge of natural (instinctive) matters, such as the elements, temperaments, humors, actions, faculties (or powers), and parts.

2. Knowledge of things not part of human (instinctive) nature. This he apparently copied from Ḥunayn ibn Isḥāq's *Ars medica (al-Masā'il fi 'l-Ṭibb)*, which defined them as the six essential principles: the air we breathe and how to be free from pollution, work and rest, diet, wakefulness and slumber, use of vomit-inducing drugs and laxatives, and psychological impulses.

3. Knowledge of things outside the realm of natural

conditions of the human body and which are concerned with diseases, their causes, and their symptoms.

In describing the arteries and veins, al-Majūsī spoke of their divisions into numerous thin tubules spreading like hairs and of the connection between arteries and veins through tiny pores. He also described the function of the three valves in each of the pulmonary arteries, the aorta, and the two in what he called the veinal artery (most probably referring to the atrioventricular valves).

Al-Majūsī also propagated health measures to preserve normal conditions of body and mind, such as diet, rest and work, bathing, and physical exercises. For example, he cited three advantages of exercise:

1. It awakens and increases innate heat to enable the attraction and digestion of foods for assimilation by body organs (metabolism).

2. It helps relieve the body of its superfluities and cleans and expands its pores.

3. It solidifies and strengthens the body's organs by inducing contacts among them so that the body functions harmoniously and is able to resist disease.

Furthermore, he said of sleep that it helps to relax and refresh the brain and the senses, as well as assisting in digestion and normalizing humors.

Long before Ibn Sīnā, al-Majūsī emphasized the importance of psychotherapy and the relationship between psychology and medicine. Emotional reactions (manifestations, *aʿrāḍ nafsāniyya*), he explained, may cause sickness or promote good health, depending on how they are controlled. He also spoke of passionate love and how it can cause illness if it has no fulfillment.

In addition, al-Majūsī discussed meteorology, hygiene, human behavior, and surgery, recommending frequent use of phlebotomy. In the section on embryology he clearly explained the presently accepted fact that the fetus is pushed out in parturition. His discussion of poisons, their symptoms, and their antidotes is an important chapter in the history of medieval toxicology. Furthermore, al-Majūsī elaborated on the effects of the use of opiates in a manner which is of interest to the history of drug addiction and abuse. His general discussions of materia medica and the therapeutics of crude and compound drugs are based on Dioscorides and Galen, with additions of indigenous, familiar drugs. Like his predecessor al-Rāzī, he used and promoted chemotherapy.

Regarding medical deontology, al-Majūsī emphasized the highest ethical standards and asked his colleagues, as well as all practitioners and medical students, to observe them as ordered and upheld in the Hippocratic writings. He also opposed the use of contraception, or of drugs that cause abortion, except

in cases involving the physical or mental health of the mother, attitudes still heard and commended today.

Al-Majūsī boasted that in his *Kāmil* he covered the three most important points of a medical text: dealing with the most needed and highly honored art of healing; presentation of a much-improved medical compendium; and comprehensive coverage of the topic. In several areas, however, he seems to have fallen short of his objectives. Nonetheless, his diligent studies, personal observations, and detailed coverage of medical matters won al-Majūsī's book the high prestige it deserved in Islam. It was translated more than once into Latin and incunabula copies exist in many libraries, a proof of its wide acceptance and circulation in East and West for almost five centuries.

BIBLIOGRAPHY

I. ORIGINAL WORKS. Al-Majūsī's *Kāmil al-Ṣināʿah ʾl-Ṭibbiyya* in 20 treatises is believed to be his only medical contribution. Numerous Arabic MSS (complete or fragmentary) exist in many libraries. It was published in 2 vols., one on medical theory and one on medical practice (Cairo, 1877). The ninth treatise was also published on its own (Lucknow, 1906). The *Kāmil* was rendered in part into Latin in the *Pantegni* of Constantine the African (*d. ca.* 1085). In 1127 Stephen of Antioch translated the entire work into Latin, with annotations by Michael de Capella. This trans. was edited by Antonius Vitalis Pyrranensis and was first published under the title *Liber regalis dispositio nominatus ex arabico venetiis* (Venice, 1492), repr. under the title *Liber totius medicinae necessariae continens, quem Haly filius Abbas edidit regique inscripsit* (Lyons–Leiden, 1523).

II. SECONDARY LITERATURE. In Arabic the earliest and best biographies of al-Majūsī and accounts of his work are Jamāl al-Dīn ʿAlī al-Qifṭī, *Tārīkh al-Ḥukamāʾ*, Julius Lippert, ed. (Leipzig, 1903), p. 232; and Aḥmad ibn Abī Uṣaybiʿa, *ʿUyūn al-Anbāʾ*, I (Cairo, 1882), 236–237. In the West during the nineteenth century many historians of medicine wrote on al-Majūsī. See (in chronological order) K. P. J. Sprengel, *Versuch einer pragmatische Geschichte der Arzneykunde*, II (Halle, 1823), 412–418; Ferdinand Wüstenfeld, *Geschichte der arabischen Aerzte und Naturforscher* (Göttingen, 1840), p. 59; E. H. F. Meyer, *Geschichte der Botanik*, III (Königsberg, 1856), 176–178; Lucien Leclerc, *Histoire de la médecine arabe*, I (Paris, 1876), 381–388; George J. Fisher, "Biography of Haly Abbas," in *Annals of Anatomy and Surgery*, 7 (1883), 208, 255; and Ernst J. Gurlt, *Geschichte der Chirurgie*, I (Berlin, 1898), 615–618. Twentieth-century works include P. de Koning, *Traité sur le calcul dans les reins* (Leiden, 1898), pp. 124–185; and *Trois traités d'anatomie arabes* (Leiden, 1903), pp. 90–431; Max Neuburger, *Geschichte der Medizin*, II, pt. 1 (Stuttgart, 1911), 210; and Paul Richter,

"Über die spezielle Dermatologie des Ali b. Abbas aus de 10. Jahrhunderts," in *Archiv für Dermatologie und Syphilis*, **113** (1912), 849–864; after the earlier twentieth-century studies, the investigations of Edward G. Browne, *Arabian Medicine* (Cambridge, 1921), pp. 51–57, 123–124, added significant weight to the importance of al-Majūsī's work.

Special studies, besides those of Koning and Richter, include an important comparison and evaluation, by J. Wiberg, "The Anatomy of the Brain in the Works of Galen and 'Alī 'Abbās," in *Janus*, **19** (1914), 17–32, 84–104. For further discussions on al-Majūsī and his compendium see also (in chronological order) George Sarton, *Introduction to the History of Science*, I (Baltimore, 1927), 677–678; Carl Brockelmann, *Geschichte der arabischen Literatur*, I (Leiden, 1943), 273, and supp., I, 423; A. A. Khairallah, *Outline of Arabic Contributions to Medicine* (Beirut, 1946), pp. 116–117; Cyril Elgood, *A Medical History of Persia* (Cambridge, 1951), pp. 99–100, 153–157, 199, 279; A. Z. Iskandar, *Arabic MSS. on Medicine and Sciences* (London, 1967), pp. 119–124; and Sami Hamarneh, *Fihris Makhṭūṭāt al-Ẓāhiriya* (Damascus, 1969), pp. 248–254.

SAMI HAMARNEH

MAKARANDA (*fl.* Benares, India, 1478), *astronomy*.

Makaranda wrote at Kāśī (Benares) an extremely influential set of astronomical tables, entitled *Makaranda* (see essay in Supplement), based on the *Saurapakṣa*. These tables are calendaric (for *tithis*, *nakṣatras*, and *yogas*), planetary, and for eclipses; their epoch is 1478. Their extreme popularity is indicated by the facts that there are almost 100 extant manuscripts (mostly from northern India) and that some twenty commentaries are known; the dated commentaries are by Harikarṇa of Hisāranagara (1610), Viśvanātha of Benares (1612–1630), Divākara (1627), Puruṣottama (1631), Kṛpārāma Miśra of Ahmadabad (1815), Jīvanātha of Patna (1823), and Nīlāmbara Jhā of Koilakh, Mithilā (nineteenth century). The continued popularity of the *Makaranda* at the end of the last century is proven by the several editions.

BIBLIOGRAPHY

The *Makaranda* was published at Benares in 1869 with the commentaries of Gokulanātha, Divākara, and Viśvanātha; the text with the first two commentaries was repub. in *Aruṇodaya*, I, pt. 15 (Calcutta, 1891). The *Makaranda* appeared alone also at Benares in 1880 and again in 1884. The tables in a number of MSS have been analyzed by D. Pingree, "Sanskrit Astronomical Tables in the United States," in *Transactions of the American Philosophical Society*, n.s. **58**, no. 3 (1968), 39b–46b; and in his "Sanskrit Astronomical Tables in England," in *Journal of Oriental Research* (Madras).

DAVID PINGREE

MAKAROV, STEPAN OSIPOVICH (*b.* Nikolayev, Russia, 8 March 1849; *d.* aboard the battleship *Petropavlovsk,* Port Arthur, Russia, 13 April 1904), *oceanography*.

His father, Osip Fyodorovich Makarov, retired from the navy in 1873 with the rank of junior captain. His mother came from a simple family and had no education; she died in 1857, leaving two daughters and three sons, of whom Stepan was the youngest. His officer's rank enabled Makarov's father to enroll his son at the age of ten at the naval school at Nikolayevsk-na-Amure, to which his father had been transferred. The five-month voyage from St. Petersburg instilled in him a love for the ocean and for sea voyages: Makarov often said, "At sea means at home."

Makarov graduated from the naval school in 1865. He served as a cadet and in May 1869 he was commissioned a warrant officer. Assigned to the Black Sea fleet in 1876 as commander of a steamer, he conducted successful military actions during the Russo-Turkish War.

In 1881–1882 as commander of the ambassadorial station ship *Taman* in Constantinople, Makarov conducted hydrological research in the Bosporus. In 1886–1889 he commanded the corvette *Vityaz* on its round-the-world voyage, and in 1896 he became vice-admiral. From the beginning of 1897 Makarov was actively involved in research on icebreakers in the North Atlantic. While planning the icebreaker *Ermak* Makarov studied previous voyages through ice, particularly on the Great Lakes. He supervised the building of the *Ermak* in England and in 1899–1901 completed the first voyages on it in polar latitudes. In December 1899 Makarov was named commander in chief of the Kronshtadt port and military governor of Kronshtadt. Makarov died during the Russo-Japanese War, when his battleship, the *Petropavlovsk*, was sunk by a Japanese mine.

Makarov began his oceanographic studies in 1881–1882 on currents in the Bosporus. In his first experiment Makarov proved the existence of a deep current running counter to the surface current. In the middle of the channel, he let down a barrel which was borne by the surface current toward the Sea of Marmara. At a certain depth the line began to pull in the opposite direction. The force of the deep current was so great that the barrel dragged the boat against the surface current. Makarov organized systematic observations of the water density and temperature at various depths, and of the velocity of the current throughout the strait. The velocity of the current was measured by a rotator, which Makarov invented and which he called a fluctometer. The velocity of the surface current varied from 6 to 3.22 feet per second, and of the lower from 3.22 to 1.84 feet per second. The density of the upper water was 1.015; the lower, 1.028. This difference in density between the less saline Black Sea and the more saline Sea of Marmara appears to be the reason for the existence of contrary currents in the Bosporus. Makarov estimated that the ratio of the volume of inflow to outflow in the Black Sea is 1 : 1.85; the difference is accounted for by fresh water flowing into the Black Sea. The results of Makarov's Bosporus research were published in *Ob obmene vod Chernogo i Sredi-zemnogo morey* ("On the Exchange of Water of the Black and Mediterranean Seas," 1885), which was a major contribution to oceanography.

In his main oceanographic work, *Vityaz i Tikhy okean* ("The *Vityaz* and the Pacific Ocean," 1894), he explained the hydrological observations carried out under his direction aboard the corvette *Vityaz* on its thirty-three-month round-the-world voyage. Although it was undertaken mainly for purposes of military instruction, Makarov began oceanographic observation at the outset. Makarov made more than 250 individual measurements of water density and temperature at depths from twenty-five to 800 meters. After careful analysis of the results of these observations and also descriptions in logs of other voyages, Makarov compiled the first water temperature tables for the North Pacific Ocean. He also considered the origin of the deep waters of the North Pacific, the reason for the homogeneous temperature and density of the water at every depth of the English Channel, the reason for the rising of deep waters near the mouths of large rivers, and the general pattern of ocean currents, with an indication of the primary significance of the action of the Coriolis force on sea currents. This main work of Makarov's, published simultaneously in Russian and French and awarded prizes by the St. Petersburg Academy of Sciences and the Russian Geographical Society, brought Makarov international recognition as a scientific oceanographer.

Makarov conceived the idea of opening up navigation along the northern borders of Siberia with the aid of icebreakers. "Straight Through to the North Pole!" was the expressive title of his report in 1897 to the Russian Geographical Society. On two voyages on the icebreaker *Ermak* to Spitsbergen and to Novaya

Zemlya (1899–1901), Makarov gathered data on the Arctic ice and on the temperatures and salinity of the Arctic basin.

Although the *Ermak*, constructed on Makarov's initiative, did not achieve all the results for which its creator hoped, research in the Arctic Ocean with icebreakers has subsequently been widely realized in the Soviet Union.

Makarov must be credited with a great number of different inventions pertaining to oceanographic research and naval construction. He published several works on naval tactics, the chief of which was *Rassuzhdenia po voprosam morskoy taktiki* ("Considerations on Questions of Naval Tactics"). Several geographical areas including an island in the Nordenskjöld Archipelago and in the Kara Sea were named for him.

BIBLIOGRAPHY

I. ORIGINAL WORKS. Makarov's oceanographic works were published as *Okeanograficheskie raboty* (Moscow, 1950). See also *Ob issledovanii Severnogo Ledovitogo Okeana* ("Research in the Northern Arctic Ocean"; St. Petersburg, 1897), written with F. Vrangel; *Ermak vo ldakh* ("Ermak in the Ice"), 2 pts. (St. Petersburg, 1901); and *Rassuzhdenia po voprosam morskoy taktiki* ("Considerations on Questions of Naval Tactics"; Moscow, 1943).

II. SECONDARY LITERATURE. On Makarov and his work, see D. N. Anuchin, *O lyudyakh russkoy nauki i kultury* ("People of Russian Science and Culture"; Moscow, 1952), 318–328; A. D. Dobrovolsky, *Admiral S. O. Makarov, puteshestvennik i okeanograf* ("Admiral S. O. Makarov, Traveler and Oceanographer"; Moscow, 1948); A. N. Krylov, *Vitse-Admiral Makarov* ("Vice-Admiral Makarov"; Moscow–Leningrad, 1944); B. G. Ostrovsky, *Admiral Makarov* (Moscow, 1954); F. F. Vrangel, *Vitse-Admiral Stepan Osipovich Makarov*, 2 pts. (St. Petersburg, 1911–1913), a biographical sketch.

I. A. FEDOSEYEV

MAKSIMOV, NIKOLAY ALEKSANDROVICH (*b.* St. Petersburg [now Leningrad], Russia, 21 March 1880; *d.* Moscow, U.S.S.R., 9 May 1952), *plant physiology.*

After graduating from the Gymnasium in 1897, Maksimov entered the natural sciences section of the department of physics and mathematics of St. Petersburg University. He graduated in 1902, then remained to prepare for a professorship. In 1905 he became an assistant in the department of botany of the St. Petersburg Forestry Institute. In 1910 Maksimov traveled to Java, where he worked in the Buitenzorg (now Bogor) Botanical Garden. In 1913 he defended his master's thesis, "O vymerzanii i kholodostoykosti rasteny" ("On the Frost Kill and Cold Resistance of Plants"), at St. Petersburg University. The following year he transferred to the Tiflis Botanical Garden, where he organized a laboratory of plant physiology.

Maksimov moved to Leningrad in 1921 and began to work in the main botanical garden of the Academy of Sciences of the U.S.S.R., where he organized a laboratory of experimental plant ecology, which he directed until 1927. From 1925 to 1933 he was also director of the laboratory of plant physiology which he organized in the All-Union Research Institute of Plant Growing. At the same time (1922–1931) he carried out major work in teaching and management of the department of botany at the A. I. Herzen Pedagogical Institute in Leningrad. From 1933 to 1939 Maksimov was in Saratov, at the All-Union Institute of Grain Economy, where he headed the section of plant physiology, and at Saratov University, in the department of plant physiology (1935–1939). In 1936 Maksimov began his work at the Institute of Plant Physiology of the Soviet Academy of Sciences, first as manager of the laboratory for growth and development of plants and, from 1939, as director of the institute. His teaching continued as part of his duties as head of the department of plant physiology of the Timiryazev Agricultural Academy in Moscow (1943–1951).

Maksimov's basic scientific research was connected with the study of frost resistance and drought resistance of plants. He was one of the most important pioneers in the study of the ecological physiology of plants. Maksimov's scientific work began with the study of respiration in fungi and the influence of injury on the respiration coefficient. His results were stated in the article "K voprosu o dykhanii" ("On the Question of Respiration"; 1904), one of the first investigations in which the fermentative nature of respiration was established. In a later study of the respiration of woody plants during the winter, Maksimov became interested in why coniferous needles and winter buds do not die at low temperatures which other plants cannot endure. Studying cold resistance, he spoke against the current idea that this property does not depend on the external environment but is defined only by the plant's inner qualities. Maksimov considered that the damage to or killing of the plant at low temperatures was caused by the formation and accumulation of ice crystals between the cells, which dehydrate and mechanically damage the protoplasm, leading to the coagulation of plasma colloids. He showed that the resistance of the cell to low tempera-

tures can be increased by the use of sugar and mineral salts and by an increase in the cell juice of the amount of other osmotically active substances that decrease the quantity of ice crystals formed. Thus he formulated the first theory of the "chemical defense of the plant against death by frost," which he presented in his master's thesis.

In 1914 Maksimov began a new and more fruitful stage of research—the study of the water system and drought resistance of plants. His first experiments showed that xerophyte plants with enough water supply transpire no less moisture than mesophytes and demonstrated the inadequacy of Schimper's then widely recognized theory. This theory explained the drought resistance of xerophytes as an ability to use water economically because of certain peculiarities in their anatomical-morphological structure that supposedly result in a level of transpiration much curtailed in comparison with that of mesophytes. Maksimov suggested that the basis of drought resistance of xerophytes lies not in their structure but in the biochemical capacity of their protoplasm to bear a prolonged water shortage without harmful consequences. He saw the plant's capacity to sustain prolonged dehydration as a complex of the traits characterizing xerophytes: the peculiarities of their protoplasm, its specific structure at a comparatively high osmotic pressure, and the anatomical-morphological peculiarities of the plant's structure. Maksimov also recognized the variety of adaptations to the conditions of existence in various ecological groups of xerophytes, explaining their origin in nature from an evolutionary position. His works on the water system and drought resistance of plants laid the foundation for a new area in botany —the ecological physiology of plants.

Maksimov also conducted research on photosynthesis, growth, development, photoperiodism, and the natural and artificial stimulators of plant growth. All these investigations were carried out under laboratory conditions and also in a natural situation for wild plants and in the field for cultivated plants. Taken as a whole, this work had great significance for the theory as well as the practice of agriculture: Maksimov developed a series of recommendations for obtaining higher yields in arid regions and in hothouses by the creation of a new regimen of artificial light, for directing the growth and development of plants by means of photoperiodic effects, and for rooting cuttings of cultivated and wild plants under the influence of growth activators.

Besides his research, Maksimov paid much attention to scientific organization and set up laboratories of plant physiology in several institutions. He also devoted a substantial part of his time to teaching. His *Kratky kurs fiziologii rasteny* ("Short Course in Plant Physiology"), which went through nine editions between 1927 and 1958, greatly influenced the development of plant physiology in the Soviet Union. The seventh edition of this text was awarded the K. A. Timiryazev Prize in 1944. Maksimov also wrote *Vvedenie v botaniku* ("Introduction to Botany"; 1915), which went through two editions; a popular book, *Ot chego byvayut zasukhi i mozhno li s nimi borotsya* ("What Causes Droughts and How We Can Fight Them"; 1951); and *Kak zhivet rastenie* ("How a Plant Lives"; 1951). He also published about 250 scientific articles and notes.

For his teaching and research work in scientific organizations Maksimov was elected a corresponding member (1939) and an academician (1946) of the Academy of Sciences of the U.S.S.R. In 1945 he received the Order of the Red Banner of Labor and was elected a vice-president of the All-Union Botanical Society, of which he became an honorary member in 1947. Maksimov was also a corresponding member of the Czechoslovakian Agricultural Academy (1934) and a corresponding member of the Royal Netherlands Botanical Society (1936).

BIBLIOGRAPHY

I. Original Works. Maksimov's writings include "O vymerzanii i kholodostoykosti rasteny. Eksperimentalnye i kriticheskie issledovania" ("On the Frost Resistance and Cold Resistance of Plants. Experimental and Critical Research"), in *Izvestiya Lesnogo instituta*, no. 25 (1913), 1–330, his master's diss.; "Zasukhoustoychivost rasteny s fiziologicheskoy tochki zrenia" ("Drought Resistance of Plants From the Physiological Point of View"), in *Zhurnal opytnoi agronomii*, **22** (1921–1923), 173–186; "Znachenie v zhizni rastenia sootnoshenia mezhdu prodolzhitelnostyu dnya i nochi (fotoperiodizm)" ("The Significance of the Relation Between the Length of Day and of Night in the Life of the Plant [Photoperiodism]"), in *Trudy po prikladnoi botanike, genetike i selektsii*, **14**, no. 5 (1924–1925), 65–90; *Fiziologicheskie osnovy zasukhoustoychivosti rasteny* ("Physiological Bases of Drought Resistance in Plants"; Leningrad, 1936); "Rostovye veshchestva, priroda ikh deystvia i prakticheskoe primenenie" ("Growth Substances, the Nature of Their Effects and Practical Application"), in *Uspekhi sovremennoi biologii*, **22**, no. 2 (1946), 161–180; and *Izbrannye raboty po zasukhoustoychivosti i zimostoykosti rasteny* ("Selected Works in Drought Resistance and Winter Resistance of Plants"), 2 vols. (Moscow, 1952).

II. Secondary Literature. See P. A. Genkel, "Nauchnaya deyatelnost Nikolaya Aleksandrovicha Maksimova i ego rol v sozdanii ekologicheskoy fiziologii rasteny" ("The Scientific Career of N. A. Maksimov and His Role in the Creation of the Ecological Physiology of Plants"), in *Problemy fiziologii rasteny. Istoricheskie ocherki* ("Problems of Plant Physiology. Historical Sketches"; Moscow, 1969), pp. 306–331, literature about Maksimov on p. 326; *Nikolay Aleksandrovich Maksimov, materialy k biobibliografii uchenykh SSSR*, Ser. biol. nauk, fiziol. rast. ("Materials for a Biobibliography of Soviet Scientists, Biological Science Series, plant physiology"), no. 2 (Moscow–Leningrad, 1949), with intro. article by P. A. Genkel and bibliography of Maksimov's works and literature on him compiled by O. V. Isakova; and I. I. Tumanov, "Osnovnye cherty nauchnoy deyatelnosti N. A. Maksimova" ("Basic Outlines of the Scientific Work of N. A. Maksimov"), in *Pamyati akademika N. A. Maksimova* ("Recollections of Academician N. A. Maksimov"; Moscow, 1957), pp. 3–9.

E. M. Senchenkova

MAKSUTOV, DMITRY DMITRIEVICH (*b.* Odessa, Russia, 23 April 1896; *d.* Pulkovo [near Leningrad], U.S.S.R., 12 August 1964), *optics, astronomy.*

Maksutov's father, a seaman, aroused his son's early interest in astronomy; and Maksutov made his first observations with his father's two-inch spyglass, to which he fitted a set of eyepieces. When he was about twelve or thirteen he constructed his first reflector, 180 mm. in diameter. Acquainted with the articles of the well-known Russian optician A. A. Chikin, Maksutov made a Newtonian reflecting telescope of 210 mm. with which he began serious observations. Recognizing his enthusiasm and skill, the Russian Astronomical Society elected the fifteen-year-old optician a member. Before he graduated from the Odessa cadet corps, Maksutov was directing the astronomical observatory from 1909 to 1913 and conducting studies in cosmography with the students of the advanced classes. In 1914 he graduated from the military engineering school and, in 1915, completed courses in radiotelephony at the electrotechnical school. He served briefly in World War I in the Caucasus, and in 1916 transferred to the military aviation school in Tiflis. Having sustained a concussion in an accident he was demobilized and returned to Odessa. In 1917, he decided to go to the United States, hoping to meet the eminent optician G. W. Ritchey, who was then working at the Mount Wilson observatory. He got only as far as Harbin, China, where he lived for a while doing odd jobs. In 1919 he was sent to the radiotelegraph base at Tomsk. In 1920 he transferred together with the base personnel to the side of the Red Army. In Tomsk, Maksutov entered the Polytechnic Institute, where he simultaneously organized an optical workshop and repaired microscopes and telescopes. In 1920 Maksutov was invited to the recently formed Petrograd Optical

Institute, but the following year he returned to Odessa and began serious work in the theory of astronomical optics.

In 1923 Maksutov developed a general theory of aplanatic optical systems. In 1928 he obtained the first of his eighteen patents, for the invention of a photogastrograph, an ingenious instrument for examining the stomach. In 1930 Maksutov was again at the Leningrad Optical Institute, where he organized a laboratory of astronomical optics. Until 1952, when Maksutov transferred all his activities to the Pulkovo observatory, he worked tirelessly on developing a theory of astronomical optics and on manufacturing the optical systems of a series of astronomical instruments. At Pulkovo, where from 1944 he headed the section of astronomical instrument construction, Maksutov devised new methods of construction, improved the meniscus systems that he had invented in 1941 and that are now universally known, and originated new methods of calculating optical systems.

In 1941 Maksutov received the degree of doctor of technical sciences and the State prize; in 1944 he received the title of professor; and in 1946 he was elected corresponding member of the U.S.S.R. Academy of Sciences and received a second State prize. Maksutov was twice awarded the Order of Lenin and the order "Badge of Honor."

Maksutov was surrounded by both Soviet and foreign students, many of whom became eminent opticians. His monographs *Astronomicheskaya optika* ("Astronomical Optics") and *Izgotovlenie i issledovanie astronomicheskoy optiki* ("Preparation and Testing of Astronomical Optics") are basic references for all astronomical instrument makers and have been translated into several languages.

Maksutov's scientific career began in 1923 with his invention of aplanatic optical systems, in which the independently developed systems of H. Chrétien, K. Schwarzschild, and A. Couder proved to be specific cases of his more general solution. In 1932 he published a substantial summary monograph, *"Anaberratsionnye otrazhatelnye poverkhrosti i sistemy i novye metody ikh ispytania"* ("Anaberrational Reflecting Surfaces and Systems and New Methods of Testing Them"). Maksutov significantly improved the shadow method for qualitative verification of optical surfaces and extended it to quantitative applications, as described in the monograph *Tenevye metody issledovania opticheskikh sistem* ("Shadow Methods of Verification of Optical Systems"; 1934). He substantially improved the compensation method of testing the mirrors that he had proposed as far back as 1924 and described it in 1957 in "Novaya metodika issle-

dovania formy zerkal krupnykh teleskopov" ("A New Method for Examining the Forms of Mirrors of Large Telescopes").

As first-class master, Maksutov himself prepared at the Optical Institute the optics of such large instruments as the 381-mm. Schmidt telescope for the Engelhardt observatory, near Kazan, the Pulkovo solar telescope, the 820-mm. Pulkovo refractor, which the firm of Grubb-Parsons declined to make, and certain others. Being interested in replacing glass for the reflector with metal, Maksutov made a number of silvered metal mirrors, of which the largest was a 720-mm. parabolic mirror of high optical efficiency. In 1941 during the war, in dramatic circumstances of which he wrote brilliantly in his monograph *Astronomicheskaya optika* ("Astronomical optics"), Maksutov developed his meniscus system, now used in many photographic as well as optical instruments. Among the numerous instruments produced by Maksutov, the unique dual-meniscus 700-mm. astrometric astrograph of high optical efficiency, constructed for the expedition of the Pulkovo observatory to Chile, occupies a special place. His unfinished monograph "Meniskovye sistemy" ("Meniscus Systems") is the result of his work in this area.

BIBLIOGRAPHY

I. ORIGINAL WORKS. Maksutov's early works include: "Anaberratsionnye otrazhatelnye poverkhnosti i sistemy i novye sposoby ikh ispytania" ("Anaberrational Reflecting Surfaces and Systems and New Methods of Testing Them"), in *Trudy Gosudarstvennogo opticheskogo instituta*, no. 86 (1932), 3–120; "Issledovanie neskolkikh obektivov i zerkal po metodu fokogramm" ("Research on Some Objectives and Mirrors With the Focogram Method"), in *Optiko-Mekhanicheskaya Promyshlennost*, no. 2 (1932), 8–10; *Tenevye metody issledovania opticheskikh sistem* ("Shadow Methods of Research on Optical Systems"; Leningrad–Moscow, 1934); "On the Temperature Coefficient of the Focal Distance of an Object Glass," in *Tsirkulyar Glavnoi astronomicheskoi observatorii v Pulkove*, no. 20 (1936), 37–41; and "Sotovye zerkala iz splavov alyuminia" ("Honeycomb Mirrors From Alloys of Aluminum"), in *Optiko-Mekhanicheskaya Promyshlennost*, no. 3 (1937), 1–3.

Subsequent works are "Novye katadioptricheskie meniskovye sistemy" ("New Catadioptric Meniscus Systems"), in *Doklady Akademii nauk SSSR*, **37**, no. 4 (1942), 147–152, and in *Zhurnal Tekhnicheskoi Fiziki*, **13**, no. 3 (1943), 87–108, translated in *Journal of the Optical Society of America*, **34**, no. 5 (1944), 270–281; "Aplanaticheskie meniskovye teleobektivy" ("Aplanatic Meniscus Teleobjectives"), in *Doklady Akademii nauk SSSR*, Novaja Ser. tekhn. fiz., no. 7 (1945), 504–507; *Astronomicheskaya optika* ("Astronomical Optics"; Moscow, 1946); *Izgotovle-*

nie i issledovanie astronomicheskoy optiki ("Preparation and Testing of Astronomical Optics"; Moscow, 1948); and "Novaya metodika issledovania formy zerkal krupnykh teleskopov" ("A New Method for Examining the Forms of Mirrors of Large Telescopes"), in *Izvestiya glavnoi astronomicheskoi observatorii v Pulkove*, no. 160 (1957), 5–29.

II. SECONDARY LITERATURE. On Maksutov and his work, see the biographies by O. A. Melnikov, in *Astronomichesky kalendar na 1966 god* ("Astronomical Calendar for 1966"; Moscow, 1965), 231–236; N. N. Mikhelson, in *Izvestiya glavnoi astronomicheskoi observatorii v Pulkove*, **24**, no. 178 (1965), 2–7, with portrait and bibliography; and S. A. Shorygin, in *Astronomichesky kalendar na 1944 god* (Moscow, 1943), 125–129.

See also *Sky and Telescope*, **25**, no. 4 (1963), 228, for a list of twenty-two articles on Maksutov telescope constructions published in *Sky and Telescope* after 1956, including Maksutov's "New Catadioptric Meniscus Systems" (see above); and R. Riecker, *Fernrohre und ihre Meister* (Berlin, 1957), pp. 504–507.

P. G. KULIKOVSKY

MALEBRANCHE, NICOLAS (*b.* Paris, France, 5 August 1638; *d.* Paris, 13 October 1715), *philosophy, science.*

Malebranche's life spanned the same years as Louis XIV's, and a famous contemporary, Antoine Arnauld, termed his philosophy "*grand et magnifique,*" adjectives historians often apply to that monarch's reign. The grandeur of his philosophy consists in the way he assimilated the whole of the Cartesian heritage and attempted to elaborate, on theological foundations, an original, rationalist-oriented speculative system. The passage of time and the recently concluded publication of his works have restored to Malebranche the stature of a remarkable intellect, for whom the polemics in which he ceaselessly engaged were merely occasions to buttress his "search for truth." Yet, while his personality can be understood in terms of the profound—and religious—unity of his thought and life, the influence of his work is not free from paradox: Voltaire honored him as one of the greatest speculative thinkers, and d'Alembert placed his portrait above his writing table. A discussion of Malebranche would be incomplete without an attempt to comprehend why Enlightenment philosophers accorded him this praise, suspect as it was in the eyes of theologians.

The youngest son of a large family, Malebranche was born with a delicate constitution. Through his father, a royal counsellor, he was linked to the rural bourgeoisie. His mother, Catherine de Lauson, be-

longed to the minor nobility; her brother, Jean de Lauson, was governor of Canada. His family's modest wealth allowed him to pursue a special program of studies adapted to his physical disability. It was not until age sixteen that he entered the Collège de la Marche of the University of Paris. He received the master of arts degree there in 1656 after having attended the lectures of the renowned Peripatetic M. Rouillard. His piety inclined him toward the priesthood, and for three years he studied theology at the Sorbonne. It seems, however, that he was no more satisfied with this instruction than he had been with commentaries on Aristotle. He entered the Congregation of the Oratory on 20 January 1660, no doubt attracted by its reputation for liberty and culture in the service of the inner life. The impression he made on his new teachers was not altogether favorable. Although he was judged to be suited for the religious life and endowed with the virtues required in communal life, his was considered an "undistinguished intellect."

The explanation of this judgment may well be that, during his four years of Oratorian training, Malebranche, who was ordained priest on 20 September 1664, does not seem to have been sympathetic to the newest elements of the curriculum: an interest in history and erudition, and a passion for positive theology founded on critical study of the Scriptures. Malebranche was taught by the leaders of this tendency, Richard Simon and Charles Lecointe, but did not adopt their views. However that may be, he did become acquainted at the Oratory with the ideas of St. Augustine and Plato.

The stimulus for Malebranche's independent intellectual development came from Descartes, during the first year of his priesthood. It was said that this change resulted from his reading of the newly published *Traité de l'homme*, whose editors had sought to emphasize the broad area of agreement between Descartes and Augustine that was revealed by this posthumous work. Whatever the event that decided Malebranche in favor of this disputed book, it is certain that within three or four years he had completely redone his studies and had made the Cartesian legacy an integral part of his thought. Evidence for this assertion is to be found in *De la recherche de la vérité*, begun as early as 1668. The title itself reveals the inspiration he drew from the manuscripts generously made available to him by the circle around Claude Clerselier. Indeed, the content of the first volume exhibits this inspiration so clearly that Malebranche became involved in difficulties with the censors and had to postpone publication until 1674.

The following year, 1675, saw the publication of a

revised edition of the first volume, the second volume, and Jean Prestet's *Élémens des mathématiques.* The simultaneous appearance of these three books is significant. Prestet, a young man with no resources, owed everything to Malebranche and was evidently his pupil even before the Congregation decided officially in 1674 to recognize Malebranche as professor of mathematics at the seminary. The extremely gifted Prestet rapidly accomplished what Malebranche himself was unable to achieve while he was embroiled in difficulties over his philosophical writings. It was Malebranche, however, who was responsible for the simultaneous publication of 1675, for he wished to place before the public an original philosophical and mathematical synthesis attesting the vitality of Cartesianism.

The general impression given by this synthesis—an impression that accounts for its success—was not deceptive. It was indeed from Descartes that Malebranche attempted to discover a science and a method of reasoning founded on clear and distinct ideas. Later he himself declared that what Augustine lacked was the opportunity to learn from Descartes that bodies are not seen in themselves. From the beginning of his philosophical career Malebranche let it be known that he considered this a fundamental lesson. Rejecting sensible qualities, he held, like Descartes, that things are to be judged solely by the ideas that represent them to us according to their intelligible essence.

All the same the *Recherche de la vérité* touches on various subjects that are not at all Cartesian: primacy of religious goals, refutation of the doctrine of innate ideas, negation of composite substance, union of the problems of error and sin, explanation of the creation by God's love for himself, and affirmation that God acts in the most simple ways, that he is the sole efficient cause, and that natural causes are only "occasional" causes. The list of new branches that Malebranche grafted onto the Cartesian trunk and that corresponded to his hope of establishing a truly Christian philosophy could be expanded; but at this stage of his career it was a matter of possible materials for a new doctrine rather than such a doctrine itself.

Progress toward this goal is represented by *Conversations chrétiennes* (1677) and the third volume of the *Recherche* (1678), "containing several elucidations concerning the principal difficulties of the preceding volumes." But it was with the *Traité de la nature et de la grâce* (Amsterdam, 1680) that Malebranche emerged as the creator of a new system of the world. Inspired by a discussion with Arnauld in 1679, the book's immediate goal was to refute Jansenist ideas concerning grace and predestination. But in order to untangle this essentially religious problem, he transferred the debate to the philosophical plane, thus demonstrating

to what extent he disagreed with Descartes on the value of extending rational reflection to questions of theology.

In examining this book one grasps the essential difference between the two thinkers. A believer and a philosopher, Malebranche did not experience the hyperbolic doubt expressed in the first Cartesian *Méditation;* he did not confront the "Cogito" as the initial indubitable existence; he did not have to seek to escape from a structure of thought closed in upon itself by discovering a God who could guarantee the universality and immutability of truth. For Malebranche, as for Descartes, God was undoubtedly the keystone and foundation of all truth, but for the former he was not the God reached by philosophical speculation whose essence is demonstrated by his existence. Rather, he is Augustine's God *intimior intimo meo,* whose presence in man is the source of the believer's daily meditation and from whom all light descends. He is also the God of wisdom, creator of a universe ordered according to laws that are both perfectly simple and perfectly intelligible—the God who, acting uniquely for his own glory, created man that he might live in union with him and participate in his reason, in his word itself.

Thus, whereas Descartes refused as a vain undertaking any speculation on divine motivations, Malebranche found in this realm something on which he could base the exercise of human reason. In his doctrine the union of man and God is not only the goal of the religious life, it is also the means of attaining a vision, in God, in which there occurs the fullest possible communication of wisdom and intelligibility. Of course, Malebranche does not claim that this communication, the supreme guarantee against error, is a blessing easily obtained or permanently assured. But he does assert that in making the effort to discern the coherence of rational discourse, sinful man, whether Christian or atheist, always obtains some reflection of the universal reason, even if he is unaware of or actually denies its divine nature. Indeed Malebranche contends that attention is a *natural prayer* that God has established as the occasional cause of our knowledge.

The term and the notion of occasional cause are not due to Malebranche, but his use of them and the importance he gave to them were incontestably original. Assigning the source of all effective action to God, he took causality in the strict sense out of the created world. This world is indeed regulated by divine wisdom, but as a function of relationships that carry in themselves no necessity whatever. Moreover, the means that man has received to make it intelligible could only be indirect, that is, occasional. Malebranche thus arrived at a philosophical system that goes far beyond the theological problem that was, so to say, the occasion for its own complete formulation.

When the *Traité* appeared, it was already several years since Malebranche had been assigned any specific duties. Starting in 1680 he devoted all his time to writing and to his role as mediator between theology and Cartesian natural philosophy. He was assailed by polemics that obliged him to review, correct, and improve his system. It is impossible to recount this highly complicated story in a few lines or to discuss in detail the modifications he made in response to a flood of objections and difficulties. However interesting the debates in which Malebranche found himself involved (for example, over the coordination of the two different perfections represented by the divine laws and the divine work) and whatever accusations he was forced to counter (destroying Providence, excluding miracles, minimizing grace to the advantage of liberty), he did not need to modify for the scientific public the basic positions of his philosophy as outlined above.

It should be merely noted in passing that Malebranche, who was more skillful in the art of revising his texts than in that of controversy, rapidly alienated a number of people, even in the Oratory. In Arnauld's opinion he was incapable of maintaining a suitable degree of detachment, and Bossuet judged him severely. Most important, he failed to escape papal censure: the *Traité* was placed on the Index in 1690 while he was in the midst of preparing the third edition. He was sincerely troubled by the decision of the hierarchy, but it did not stop him. The seventh and last edition appeared in 1712, along with the sixth edition of the *Recherche*.

These figures are revealing. Malebranche was read in his own time as much by admirers as by opponents. So much is evident. What most clearly appears in this record of publication, however, is a tireless capacity for modifying his positions and a mind always receptive to suggestions, two rare qualities that testify to his character and intelligence. Malebranche owed his position in the scientific movement of his time to this harmony of his personal qualities with his doctrine of occasionalism.

It is not difficult to understand why occasionalism was a conception particularly conducive to the advance of experimental science. To the degree that nature appeared, to Malebranche, as simply a sphere of relations, the dialogue between reason and experience became for him, inevitably, the fundamental stimulus in the pursuit of knowledge. For when reason was supported by metaphysics, as it was in Descartes, it had much too great a tendency to declare what should be, a priori, and to call upon experience solely for

confirmation. In his view, however, the only means of discovery available to the human mind are occasional causes, that is, causes which could have been totally different and which are the reflection not of some ontological necessity but only of the Creator's will. Consequently, experience is indispensable. Of course, it must be intimately conjoined with the exercise of reason in order to attain knowledge of the relations that God has established in his Creation in fact, and not involuntarily, as it were, to comply with some metaphysical imperative. While Malebranche's philosophy provided, above all, a rationale for the study of physics, what is striking is the way in which he was led to grasp this fact himself and to work simultaneously in very different disciplines.

As noted above, the simultaneous publication of Prestet's *Élémens* and the first edition of the *Recherche* suggests that Malebranche was sufficiently well-versed in Cartesian mathematics to have been capable of inspiring a highly talented disciple and to have worked with him on an up-to-date textbook. John Wallis in his *Treatise of Algebra* (1684) did not hesitate to attribute to Malebranche the authorship of the *Élémens* and to reproach the work for being merely a compilation, one that failed to cite its sources other than Descartes and Viète. In replying to this accusation, Prestet clearly implied that he was not annoyed at the attribution of his book to "a person more skillful than he," but he ironically asserted his astonishment that anyone could have supposed he had read so many specialized works. Dating his own initiation in mathematics to 1671, he artlessly stated that Descartes was virtually his only source and that, moreover, he was completely dissatisfied with the few other books that had come to his attention. These remarks would be as true of Malebranche as of Prestet himself.

It is likely that Malebranche's duties as a professor of mathematics lasted only a short while. In any case they have left no further trace. Moreover, when Leibniz met Prestet at Malebranche's residence during his stay in Paris, he was well aware of their respective roles, as is evident from his later correspondence with Malebranche. The disciple, who clearly surpassed his master in the technical realm, was entrusted with the actual mathematical portion of the work; but the master directed the research, and his orientation of it consisted in giving the greatest possible development to Cartesian mathematics.

The *Élémens* consisted of two parts. The first was devoted to arithmetic and algebra, the second to analysis, that is, the application of the two former disciplines to the resolution of all problems concerned with magnitude *(grandeur)*. By magnitude, the author specified that he meant not only what is susceptible of extension in various dimensions but, more generally, everything "susceptible of more and less" *(de plus et de moins)*—in other words, everything that could enter, according to Archimedean logic, into the formal rules of relations. The plan of the work corresponded to one of the aspects of the intelligibility that Malebranche promised to the exercise of the human mind. Prestet added: "We do not attempt to understand or even to reason about the infinite," a point of view which was in accord with Descartes's thinking and to which he always remained faithful. The authors cited in the section on analysis were Diophantus, Viète, and Descartes. In his view, however, Descartes's method was "the most general, the most fruitful, and the most simple of all." In utilizing this method he completed Descartes's effort, notably with regard to equations of the fourth and fifth degree, an area in which he fancied that he had made a theoretical advance.

As to that, he deluded himself a bit, but he did at least provoke Leibniz' curiosity and interest in the subject. Leibniz was disappointed to learn from Malebranche in 1679 that Prestet, who had entered the Oratory and was busy preparing for the priesthood, had not pursued his investigations. This circumstance explains why the theory of equations and the analytic expression of roots constituted the grounds on which Leibniz chose to attack Malebranche. In telling Malebranche that this was the area that most clearly demonstrated the insufficiency and limitations of the Cartesian method, Leibniz was on the right track. Between 1680 and 1690 Malebranche progressively detached himself from Prestet, whose teaching at the University of Angers during these years was marked by painful conflicts with the Jesuits.

True, a new person in Malebranche's immediate entourage, the Abbé Catelan, lent Prestet a hand in assimilating English mathematics and in attempting to attach Barrow's method and Wallis' arithmetic of the infinitesimals to Cartesian mathematics. But Malebranche also became acquainted with a young gentleman, the Marquis de L'Hospital, whom he considered more receptive to the changes that he suspected might be necessary. Prestet died in 1691 after having published two volumes of *Nouveaux élémens* (1689), leaving in manuscript a third volume on geometry that was never published because of Malebranche's unfavorable opinion. For a few months Catelan sought to continue Prestet's work, but the cause was already lost. From 1690 to 1691 Malebranche devoted all his attention to the compromise that L'Hospital had worked out and then ardently followed what the latter was learning from Johann I Bernoulli in 1692. The arrival in Paris of this messenger of Leibniz' new calculus was the "occasion" that completely rearranged the mathematical landscape. Malebranche left to his Oratorian collaborators the task of completing the fair copy of the manuscript recording the mathematical reform elaborated by L'Hospital the preceding year, and the two of them became converts to the movement emanating from Hannover.

This rapid sequence of events within the space of only two or three years undoubtedly reproduced, in a certain way, the situation of 1671–1675. Malebranche assimilated the innovations, pen in hand, and convinced himself of the necessity of encouraging research in the new direction. L'Hospital was the real mathematician, the one who mastered the material and proceeded faster. He soon asserted his own independence from Malebranche; in 1696 he published his *Analyse des infiniment petits* virtually without consulting him. This independence, moreover, was the sign of a new reality. The rapidity with which mathematics was developing reflected the fruitfulness of analysis, which combined consideration of the infinite with the operational procedures of the differential and integral calculus. And the rapid pace accentuated the distinction between those who truly deserved to be considered mathematicians, and the partisans who could only follow, more or less closely, with greater or less difficulty. Malebranche henceforth belonged in the second category.

All the same, he possessed the valuable assets of freshness and enthusiasm. In this regard Leibniz said he had to laugh to see how Malebranche was so enamored of algebra, so enchanted with its operational effectiveness. The enchantment that Malebranche found in the mathematics of the infinitesimal analysis attests to the same naïveté. He failed to distinguish clearly between the respective roles of logic and calculation. Believing that the new mathematics was within striking distance of perfection, he could not understand what restrained the great masters from placing their discoveries before the public. What diminishes Malebranche's standing as a mathematician in the eyes of the specialists, his naïveté, was the same quality that made his advocacy more effective.

In the fifth edition of the *Recherche de la vérité* (1700), Malebranche replaced all the mathematical references he had previously given with L'Hospital's *Analyse* and a work on integral calculus that his former secretary, Louis Carré, had just compiled from material in the archives of the Oratory. Fully aware of its deficiencies, Malebranche expressed the hope that a better work would shortly appear and hinted that the required effort was under way. He had good reason for doing so, because in 1698 he had, in effect, assigned this task to the Oratorian Charles-René Reyneau, Prestet's successor at Angers. And to the extent that this outstanding teacher encountered great difficulties in absorbing the infinitesimal methods, there were grounds for thinking that the result of his labor would correspond to the conditions required for the dissemination of the new ideas in the schools and would, in short, constitute a good textbook.

The enterprise was marked by many vicissitudes and was not completed until 1708, after the happy conclusion at the Académie des Sciences of the polemic provoked by Michel Rolle against the infinitesimals. (Malebranche played the most active role in bringing about this happy ending.) Although Reyneau's *Analyse démontrée* appeared in 1708, later than expected, it answered all the more fully to the hopes placed in it. The first textbook of the new mathematics, it fulfilled the important social function indispensable to all reform. It was from one of this work's posthumous editions that d'Alembert learned the subject.

It is evident that Malebranche holds no place in the history of mathematics by virtue of any specific discovery, nor any claim to be considered a true mathematician. Nevertheless, the history of mathematics at the end of the seventeenth century—at least in France—cannot be described without referring to his activity. The mainspring of the spread and development of Cartesian mathematics, Malebranche successively insisted on the need for reform and fostered the introduction of Leibnizian mathematics. Throughout these changes, moreover, he was concerned with their implications for teaching.

While the importance of intelligibility in his philosophy accounts for his special interest in mathematics, it was, rather, toward physics and the natural sciences that Malebranche turned his attention. The first edition of the *Recherche* clearly demonstrates that this vast subject attracted Malebranche's interest from the start and that he had already read extensively in it. In the realm of physics, Rohault's recent publication seemed to Malebranche both adequate and faithful to the Cartesian method. The only topic in which Malebranche felt obliged to make a personal contribution was that of the laws of collision. It is also the question to which he returned in 1692 in publishing a small volume entitled *Des loix de la communication des mouvements.*

The date 1692 in itself is significant, but to understand fully Malebranche's statement that this short treatise was written in order to meet Leibniz' criticisms, it is not sufficient to consider only the mathematical developments outlined above. It must also be recalled that in 1686–1687 Leibniz had launched an attack in the *Acta eruditorum* against the Cartesian identification of force with the quantity of motion and had thereby provoked a bitter controversy with Cate-

lan, who was then friendly with Malebranche. Moreover, in 1692 Malebranche was the recipient of a manuscript copy of Leibniz' *Essay de dynamique.* The brief work that he brought out almost simultaneously shows that Malebranche was able to assimilate criticism without capitulating to it.

Although Malebranche agreed to revise the whole of his presentation of the subject, he did not consent to abandon any more of the Cartesian legacy than he had already done in dropping the principle that a force inheres in the state of rest. Further, he assumed that he had answered Leibniz' objections by distinguishing three types of laws, corresponding to the "different suppositions that may be held relating to colliding bodies and to the surrounding medium." On this occasion, moreover, he gave greater importance to the notion of elasticity. Nevertheless, his conclusion, presented with highly interesting remarks on the respective roles of theoretical speculation and experiment, makes clear that he was not satisfied with his work and was ready for a more radical revision. He undertook such a revision in several steps in the years 1698 to 1700, characterizing his own publication of 1692 as a "wretched little treatise."

In the course of this tumultuous development of his ideas Malebranche made his most original contribution to the scientific movement—and did so in his capacity as speculative philosopher. In his exposition of the third law of impact, he invested collision theory with a clarity that was lacking in Mariotte's *Traité de la percussion ou chocq des corps* (1673). After concisely expressing the principles of research, he judiciously chose numerical examples and then stated a position that he firmly maintained in the following years: the scientist's duty is to begin with the diversity of observations and then to establish laws. These laws, when submitted to mathematical operations, should reflect natural effects step by step. It was in this connection that Malebranche was dissatisfied with Mariotte's propositions. The latter had, it is true, clearly distinguished between two operations. First, he disregarded elasticity and treated the bodies as if they were soft. Second, he superimposed the effect of elasticity, which consisted in assigning the respective velocities in inverse ratio to the masses. But in Malebranche's view the first operation was unintelligible, since bodies without elasticity were, he supposed, necessarily hard. And the second operation ran into serious logical difficulties, for taking the force to be the absolute quantity of motion led to paradoxical results. Malebranche satisfied himself with regard to the first point in 1698–1699 by means of a modification of the concept of matter, the subject of his "Mémoire sur la lumière, les couleurs etc." He attempted to overcome the second problem by considering the property of reciprocity, which Mariotte's laws assumed, to be a "revelation" of the experiment, the sort of principle of intelligibility to which all rational effort must be subject. While correcting the proofs for the fifth edition of the *Recherche*, he was rewarded by the discovery that the whole question became clarified if the absolute quantity of motion were replaced by the algebraic quantity, that is, if the sign were taken into account.

This discovery led to the final corrections, which now furnished an original way of demonstrating, without paralogism or *petitio principii*, the laws of elastic collision. Moreover, this method of improving Mariotte's presentation avoided adopting Leibniz' point of view and preserved as much as possible of Descartes's conception.

Convinced that he had found a solution, Malebranche turned his attention all the more resolutely toward other problems. The memoir alluded to above won him membership in the Académie des Sciences at the time of its reorganization in 1699. Henceforth, Malebranche actively participated in scientific life, while gathering the material he was to incorporate in the sixth edition of the *Recherche* (1712), in which he made the necessary revisions, corrections, and additions in those sections devoted to all the topics in which he thought science bore on his philosophy.

It is most important to note that certain authors have erred in ridiculing the patching up of the Cartesian vortices that Malebranche is supposed to have begun. To be sure, he speaks of subtle matter and vortices, but his system arises from a syncretism that borrowed much from recent advances in physics and especially from the work of Huygens. Malebranche's subtle matter is a unique primary substance that, forced to move at high speed in a closed universe, is obliged to whirl in vortices the dimensions of which can decrease without limit, a property predicated on the supposition that no vacuum can exist. The formula for centrifugal force then requires that these small vortices, which are actually the universal material of all physical entities, be not only perfectly elastic but capable, as well, of releasing a "fearful" force upon breaking up. A theoretical model of this sort is not a trivial invention.

Nor is there anything trivial about the manner in which Malebranche utilized this model to study luminous phenomena and to provide an account of universal gravitation, of planetary motion, and of gravity. This model, considered in itself as the seat of action in the universe, inspired his idea that light consists of vibration in a medium under pressure. And considered in all its ramifications, it led him to conceive of the gross matter accessible to our senses as the result of a condensation in the neighborhood of a vortical center. This picture was imposed by the inapplicability to the case of large vortices of a homogeneous mechanical model centered on a point with invariant properties for distances near to or far from the center.

Although all this theoretical effort must be granted a certain originality, none of it was adopted by eighteenth-century science. It was not until much later that scientists again took up the idea that frequency is characteristic of colors or the idea that orthography can help establish the laws of central systems of small diameter—and when they did they were unaware that Malebranche had advocated such views.

Nor did anything come either of the hours that Malebranche spent at the microscope or of his botanical observations. Despite the importance he accorded to the experimental method after 1700, he never considered himself more than an amateur, concerned simply to grasp what it was that the specialization of others was accomplishing. The only experiment that we can confidently attribute to his own efforts—before his reading of Newton's *Opticks*—concerned the virtual equivalence of air and of the vacuum produced in the air pump as mediums for the propagation of light. It was a perfect example of the ambiguity of so-called crucial experiments. Malebranche's improvements in methods for observing generation in eggs in the hatchery were trivial and presupposed confirmation of the ovist theory. Even though the science of life seemed to him a realm apart, incomprehensible without the idea of finality, he applied to it what is now known as the notion of structure, deriving from his mathematical critique of being and extension. That is why he advocated the doctrine of the *emboîtement des germes*.

It has to be admitted that Malebranche came to a scientific career, in the broad sense, too late in life. It was unusual enough that at age sixty he was able to carry out experimental research which showed a greater command of the subject than he could have won from books alone; and more should not be asked of him. Faithful to his speculative temperament, he was ardently concerned to preserve from his Cartesian past those values he thought enduring and to bequeath a system reconciling this past with the science of his day. This arduous enterprise condemned him to be a follower, not a leader, and it is not surprising that his work failed to exhibit intimate knowledge of the most advanced developments of contemporary science. The reformulation of results that have become common knowledge always requires the discovery of new results, if it is to incite interest. Malebranche failed to go beyond the reexpression of either the sine law of refraction of light or the inverse-square law of gravitation, and he left his vibratory theory of colors in only a rudimentary state.

Still, the high level of reflection he demanded from his readers exerted an influence on the most diverse thinkers both in France and abroad. As in the case of mathematics, Malebranche has a claim to be remembered in the history of physics, a science the autonomy of which was scarcely recognized in the last years of the seventeenth century and which had to formulate a charter for itself. In this respect Malebranche indisputably answered to the needs of his time, and his efforts were not in vain.

Thus, to the extent that Malebranche enriched theoretical speculation and worked to fashion a suitable basis for the union of the rational and the experimental, he made a genuine contribution to the autonomy of science. His activity was always inspired by his religious philosophy and, reciprocally, his results appeared to him to provide support for it. Others could complete the separation, retaining the autonomy and discarding the philosophy.

In preparing this account the author has sought to adhere to the facts available to him. This same fidelity, however, obliges him to restore to Malebranche something beyong the authorship of a body of thought that advanced an enlightened rationalism. The restitution concerns the virtues that Malebranche constantly displayed during his life: a capacity to correct himself, a sensitivity to the difficulties of the ordinary reader and to the needs of his time, and a perseverance in educating himself in many fields. In Malebranche, the man is inseparable from the thinker, and the man was wholly imbued with Christian faith. One may, of course, not share this faith, and then the separation of science from belief is easy to effect. But whoever accepts the lesson to be learned in contemplating the total, integrated image of a life will no less easily perceive the violence of such an act. This is why, after several centuries, the message offered by Malebranche endures.

BIBLIOGRAPHY

I. Original Works. The complete works of Malebranche, published under the direction of André Robinet as *Oeuvres complètes*, 20 vols. (Paris, 1958–1968), include correspondence and MSS. For his scientific work, see esp. vol. III, *Éclaircissements de la recherche de la vérité* (1678–1712); and vol. XVII, pt. 1, *Lois du mouvement* (1675–1712); and pt. 2, *Mathematica,* containing unpublished mathematical and other works, with critical annotations by P. Costabel.

Six eds. of *De la recherche de la vérité* were published, at Paris, during Malebranche's lifetime: the first three, in 2 vols. (1674–1675, 1675, 1677–1678); the 4th and 5th, in 3 vols. (1678–1679, 1700); and the 6th, in 4 vols. (1712). The *Traité de la nature et de la grâce* (Amsterdam, 1680) was followed in 1681 by *Éclaircissement, ou la suite du Traité* ....

II. Secondary Literature. Works on Malebranche and his work include V. Delbos, *Étude de la philosophie de Malebranche* (Paris, 1924); G. Dreyfus, *La volonté selon Malebranche* (Paris, 1958); H. Gouhier, *La vocation de Malebranche* (Paris, 1926); and *La philosophie de Malebranche et son expérience religieuse* (Paris, 1926); M. Gueroult, *Malebranche*, 3 vols. (Paris, 1955–1959); A. Robinet, *Malebranche, de l'Académie des sciences* (Paris, 1970); and G. Rodis-Lewis, *Nicolas Malebranche* (Paris, 1963). See also *Malebranche—l'homme et l'oeuvre* (Paris, 1966), published by the Centre International de Synthèse.

Pierre Costabel

MALESHERBES, CHRÉTIEN-GUILLAUME DE LAMOIGNON DE

MALESHERBES, CHRÉTIEN-GUILLAUME DE LAMOIGNON DE (*b.* Paris, France, 6 December 1721; *d.* Paris, 22 April 1794), *agronomy, botany.*

Member of a distinguished family of the *noblesse de robe*, Malesherbes was the son of Guillaume de Lamoignon, chancellor of France (1750–1768), and was related by marriage to the families of Chateaubriand, La Luzerne, Rosanbo, and Tocqueville. A magistrate by profession, he was one of the most enlightened officials of the *ancien régime*, holding at various times the posts of director general of the Librairie, first president of the Cour des Aides, and minister under Turgot (1774–1776) and under Loménie de Brienne (1787–1788). He was an influential spokesman for freedom of the press, religious toleration, and tax reform. A member of the Académie Française and the Société Royale d'Agriculture, he was also honorary member of the Académie des Inscriptions et Belles-Lettres and the Académie Royale des Sciences. Late in 1792 he volunteered to serve as defense counsel at the forthcoming trial of Louis XVI. Subsequently accused of having defended the king and of other "acts of treason," he was tried by the Revolutionary Tribunal and guillotined.

Malesherbes studied botany with Bernard de Jussieu (1746–1749) and chemistry with G.-F. Rouelle, and maintained a lifelong interest in natural history. He wrote little for publication, and accounts of his scientific activities, ideas, and influence must be sought principally in his correspondence and in memoirs often intended for circulation among his friends. These sources show him to have had some competence in botany and especially in agronomy, and reveal his role in supplying scientists with information and patronage.

Among Malesherbes's earliest works, although

published posthumously (1798), was a critique of the first volumes of Buffon's *Histoire naturelle* (1749). Here he not only disagreed with specific details but also replied effectively to Buffon's attack on naturalists who emphasized the accumulation of data and on botanists who believed it possible to discover a natural system of classification.

Malesherbes was concerned with the improvement of breeds of livestock, the cultivation of wastes, and the naturalization in France of such crops as wild rice. From about 1760 his estate at Malesherbes (Loiret) was essentially an experimental farm devoted largely to the cultivation and acclimatization of "exotic" trees. Rather than purely ornamental trees or botanical rarities, useful trees, and especially conifers, were of particular interest to Malesherbes. Varieties of pine, for example, were important for shipbuilding; and he tried to discover the soil and climate most suitable for naturalizing in France the pines of Corsica and the Baltic. While some agronomists advocated a national program of marsh drainage, Malesherbes pointed out that such soil was often sterile for staple crops; instead, he proposed broadening existing afforestation attempts by the introduction of the swamp cypresses of Virginia. These and other trees, he argued, would provide rot-resistant naval timber, alleviate the national fuel shortage, and turn marshes to efficient use. He was able to test some of his ideas during the exceptionally harsh winter of 1788–1789, when he made detailed observations of the survival capabilities of his own forest trees.

Malesherbes's interest in trees came increasingly to center on those of North America when, during the American Revolution, he shared the hope that France could establish close commercial and cultural ties with an independent United States. Although some North American plants had long been available to French naturalists, collection was not systematic and tended to be done through English intermediaries. Malesherbes's own contacts in America were strengthened after 1779 when his nephew, the Marquis de La Luzerne, was sent there as minister plenipotentiary; his American correspondents also came to include Thomas Jefferson and Benjamin Franklin and French diplomats F. Barbé de Marbois, L.-G. Otto, and H. St. Jean de Crèvecoeur. He arranged to have shipments of seeds and seedlings sent to the Paris firm of Vilmorin-Andrieux so that American plants could be widely distributed in France.

Malesherbes's interest in the dissemination of new ideas is apparent in the two agricultural pamphlets published during his lifetime. In the first he emphasized the need for organized communication among agronomists so that experimental results could be verified and made widely known. In the second he attempted to direct the attention of the National Assembly to problems of landholding that were intimately connected with legal, social, and agricultural change.

Malesherbes often contributed to the work of other scientists and served as their patron. He transmitted the results of his own experiments to agronomists H.-L. Duhamel du Monceau and P. Varenne de Fenille for use in their publications, and he did the same for chemist P.-J. Macquer, stipulating that Macquer refrain from publicly acknowledging his aid. During his travels he gathered information useful for the geological maps of J.-E. Guettard, and he donated his collection of minerals to the École des Mines soon after its founding in 1783. It was his patronage that enabled mineralogist A.-G. Monnet to obtain a post with the Bureau du Commerce.

Recognizing the limits of his own training and ability, Malesherbes tried to recruit professional translators and scientists to work on such projects as the translation of English agricultural writings and of Pehr Kalm's *Travels in North America* (first published in Swedish, 1753–1761, and soon afterward in German and English). His role, as he saw it, was that of the scientific amateur, possessed of enough training to understand the work of the professionals and enough ability, wealth, and influence to aid them. The agronomists who knew him best disagreed with part of this evaluation and looked upon Malesherbes as their colleague.

BIBLIOGRAPHY

I. Original Works. Malesherbes's publications are *Mémoire sur les moyens d'accélérer les progrès de l'économie rurale en France. Lu à la Société royale d'agriculture* (Paris, 1790), also published in *Mémoires de la Société nationale d'agriculture de France* (spring trimester 1790); *Idées d'un agriculteur patriote sur le défrichement des terres incultes* (Paris, 1791), repr. in *Annales de l'agriculture françoise*, **10** (*an* X), 9–26; and *Observations de Lamoignon-Malesherbes sur l'histoire naturelle générale et particulière de Buffon et Daubenton*, with intro. and notes by L.-P. Abeille, 2 vols. (Paris, 1798). There are also several works on nonscientific subjects.

Large collections of MSS are extant, some in private hands; see the first work by Grosclaude (below). Also Bibliothèque de l'Institut de France, Paris, MS 997; and Bibliothèque Centrale du Muséum National d'Histoire Naturelle, Paris, MSS 238, 239, 949, 1765. Relevant documents in the possession of the American Philosophical Society, Philadelphia, are described by Gilbert Chinard, "Recently Acquired Botanical Documents," in *Proceedings of the American Philosophical Society*, **101** (1957), 508–522.

II. Secondary Literature. Biographies are numerous and began to appear soon after Malesherbes's death, but these works have in most respects been superseded. See Pierre Grosclaude, *Malesherbes, témoin et interprète de son temps* (Paris, 1961), and *Malesherbes et son temps: Nouveaux documents inédits* (Paris, 1964); J. M. S. Allison, *Lamoignon de Malesherbes, Defender and Reformer of the French Monarchy, 1721–1794* (New Haven, 1938); J. Sabrazès, "Malesherbes, l'homme de bien, le réformateur, le savant," in *Gazette hebdomadaire des sciences médicales de Bordeaux*, no. 39 (25 Sept. 1932); André J. Bourde, *Agronomie et agronomes en France au XVIIIᵉ siècle*, 3 vols. (Paris, 1967); and Joseph Laissus, "Monsieur de Malesherbes et 'la montagne qui cogne' (1782–1783)," in *Comptes rendus du quatre-vingt-douzième Congrès national des sociétés savantes, Strasbourg et Colmar, 1967: Section des sciences*, I (Paris, 1969), 233–254.

RHODA RAPPAPORT

MALFATTI, GIAN FRANCESCO (*b.* Ala, Trento, Italy, 1731; *d.* Ferrara, Italy, 9 October 1807), *mathematics.*

After completing his studies in Bologna under the guidance of Francesco Maria Zanotti, Gabriele Manfredi, and Vincenzo Riccati, Malfatti went to Ferrara in 1754, where he founded a school of mathematics and physics. In 1771, when the University of Ferrara was reestablished, he was appointed professor of mathematics. He held this post for about thirty years, teaching all phases of mathematics from Euclidean geometry to calculus.

Malfatti became famous for his paper "De aequationibus quadrato-cubicis disquisitio analytica" (1770), in which, given an equation of the fifth degree, he constructed a resolvent of the equation of the sixth degree, that is, the well-known Malfatti resolvent. If the root is known, the complete resolution of the given equation may be deduced. The latter, however, cannot be obtained by means of rational root expressions; rather, as Brioschi later demonstrated, it is obtained by means of elliptical transcendents.

Malfatti also demonstrated that a memoir on the theory of probability, published by Lagrange in 1774 and proclaimed by Poisson as "one of Lagrange's most beautiful works," nevertheless required explanation at one point.

In a brief treatise entitled *Della curva cassiniana* (1781), Malfatti demonstrated that a special case of Cassini's curve, the lemniscate, has the property that a mass point moving on it under gravity goes along any arc of the curve in the same time as it traverses the subtending chord.

In 1802 Malfatti gave the first, brilliant solution of the problem that bears his name: "Describe in a triangle three circumferences that are mutually

tangent, each of which touches two sides of the triangle." Many illustrious mathematicians had dealt with this problem. Jacques Bernoulli (1654–1705) had earlier dealt with the special case in which the triangle is isosceles. An elegant geometric solution was supplied by Steiner (*Crelle's Journal*, vol. 1, 1826), while Clebsch, dealing with the same problem in 1857, made an excellent application of the elliptical functions (*Crelle's Journal*, vol. 53, 1857).

In a letter to A. M. Lorgna (27 April 1783), Malfatti gave the polar equation concerning the squaring of the circle.

BIBLIOGRAPHY

I. ORIGINAL WORKS. Among Malfatti's works are "De aequationibus quadrato-cubicis disquisitio analytica," in *Atti dell' Accademia dei Fisiocritici di Siena* (1770); *Memorie della Società italiana delle scienze detta dei XL*, **3**; *Della curva cassiniana* (Pavia, 1781); *Memorie della Società italiana delle scienze detta dei XL*, **10** (1802); and his letter to Lorgna, in *Bullettino di bibliografia e di storia delle scienze matematiche e fisiche*, **9** (1876), 438.

II. SECONDARY LITERATURE. For further information on Malfatti and his work, see G. B. Biadego, "Intorno alla vita e agli scritti di Gianfrancesco Malfatti, matematico del sec. XVIIIᵒ," in *Bullettino di bibliografia e storia delle matematiche del Boncompagni*, **9** (1876); E. Bortolotti, "Sulla risolvente di Malfatti," in *Atti dell'Accademia di Modena*, 3rd ser., **7** (1906); "Commemorazione di G. F. Malfatti," in *Atti della XIX riunione della Società italiana per il progresso delle scienze* (1930). Also see article on Malfatti in *Enciclopedia italiana* (Milan, 1934), XXII, 16; F. Brioschi, "Sulla risolvente di Malfatti," in *Memorie dell'Istituto lombardo di scienze e lettere*, **9** (1863); Gino Loria, *Curve piane speciali: Teoria e storia* (Milan, 1930), I, 265, and II, 23; and *Storia delle matematiche*, 2nd ed. (Milan, 1950), *passim*; A. Procissi, "Questioni connesse al problema di Malfatti e bibliografia," in *Periodico di matematiche*, 4th ser., **12** (1932); and A. Wittstein, *Geschichte des Malfatti'schen Problems* (Munich, 1871).

A. NATUCCI

IBN MALKĀ. See **Abu'l Barakāt.**

MALL, FRANKLIN PAINE (*b.* Belle Plaine, Iowa, 28 September 1862; *d.* Baltimore, Maryland, 17 November 1917), *anatomy, embryology, physiology.*

Mall was the only son of Franz Mall, a farmer born at Solingen, Germany; his mother, the former Louise Christine Miller, was also a native of Germany and died when he was ten years old. He attended a local academy, where an able teacher awakened his interest in science. In 1880, at the age of eighteen, Mall began medical studies at the University of Michigan, where he had three stimulating teachers: Corydon L. Ford, professor of anatomy, a superb lecturer but not an original investigator; Victor C. Vaughan, biochemist and bacteriologist; and Henry Sewall, physiologist. William J. Mayo, later a famous surgeon, was a classmate.

Influenced, no doubt, by Vaughan and Sewall, who had gone to Germany for scientific training, Mall, after taking the M.D. degree in 1883, went to Heidelberg with the intention of becoming a specialist in ophthalmology. Finding himself more interested in anatomical research than in the practice of medicine, he went in 1884 to Leipzig, where he became a student of Wilhelm His, the greatest embryologist of the time, who admitted him to close association in the laboratory and became a lifelong friend and adviser. During the same year Mall met another American postgraduate student, William H. Welch, who later, as dean of the Johns Hopkins University School of Medicine, invited Mall to be its first professor of anatomy. After completing a study of the development of the thymus gland (which, incidentally, contradicted earlier work by His), Mall moved, on the advice of His, to the laboratory of Carl Ludwig at Leipzig. Ludwig set for the young Mall a study of the structure of the small intestine, which he accomplished with such skill and breadth of view as to win the admiration of Ludwig, who personally saw Mall's monograph through the press after the latter left for America in 1886.

Welch, who had returned to America some years earlier, had organized a department of pathology at the Johns Hopkins Hospital preliminary to the creation of a school of medicine. Mall applied to him for a post and was given the first fellowship in the new laboratory. He remained there for three years, studying the structure of the stomach and intestines from both the anatomical and the physiological standpoints, partly in collaboration with the surgeon William S. Halsted. He also studied the microscopic structure of connective tissue by highly original methods. On the basis of his description Halsted developed the method of suturing the intestine in surgical operations that is known by his name.

Because Mall's position with Welch was not permanent, he was much concerned about his future, opportunities for full-time medical research and teaching being very limited at that time; but in 1889 he was offered an adjunct professorship of anatomy at Clark University, Worcester, Massachusetts, then being organized by the psychologist G. Stanley Hall.

The appointment seemed to offer Mall an opportunity to develop anatomical research and to put the teaching of that subject on as high a scientific basis as in Germany; the prevalent approach in American schools was that of didactic instruction, serving only as preparation for surgery. He remained at Clark University for three years, until the faculty's growing dissatisfaction with Hall's administration led to the departure of several young professors, Mall going to the University of Chicago as professor of anatomy at its school of medicine. While at Clark, however, his research led to an important discovery, that of the vasomotor nerves of the portal vein. At this time also Mall constructed a model of an early human embryo by the Born wax-plate method, the first to be made in America, and thus began the program of embryological research on which his reputation is chiefly based.

Mall remained in Chicago only one year, for in 1893, when Johns Hopkins University opened its long-planned school of medicine with Welch as dean, Mall was called to head its department of anatomy. Free to plan instruction without constraint of tradition, Mall at once began to reform the teaching of anatomy in the United States by giving few lectures while providing his students full opportunity to learn for themselves by dissection, with the aid of textbooks and atlases and the advice of instructors engaged in research. Mall designed and maintained quiet and scrupulously clean small laboratory rooms in place of the large dissecting halls of the older schools; he also insisted on accurate work, familiarity with the literature of the field, and scientific rather than purely practical aims. The same principles were applied to the teaching of microscopic anatomy and neurology, conducted largely by his staff. The members of his staff were given full freedom to direct their own researches under his lightly imposed leadership, that of an older, more experienced fellow student rather than a taskmaster. The success of Mall's methods is demonstrated by the fact that they were carried to a score of other universities by those of his pupils and assistants who became professors of anatomy, and many physicians and medical teachers found in his laboratory an intellectual awakening and a stimulus to become independent scientific investigators. Outstanding researchers who worked with him were Ross G. Harrison, Florence R. Sabin, George L. Streeter, Warren H. Lewis, and Herbert M. Evans.

In 1894 Mall married Mabel Glover of Washington, D.C., one of the three women students in his first class at Johns Hopkins. They had two daughters.

Mall's own investigations during the early years in Baltimore dealt chiefly with the structure of the spleen and the liver. His study of the very peculiar arrangement of blood vessels in the spleen underlies the current conception that this organ serves as a storage place for the blood. His work on the liver gave rise to two important generalizations with which he extended ideas of his teacher Wilhelm His. One of these states that the extremely thin tubular tissue (endothelium) of which the capillary blood vessels consist, and which forms the lining of the veins and arteries, is the primary structure of the vascular system and in the larger vessels is reinforced by muscle and connective-tissue cells. Studying the liver, Mall showed that organ to be made up of small structural units each of which contains all the essential tissue elements—hepatic cells, blood vessels, and bile ducts, systematically arranged within the unit; and he demonstrated that the blood vessels supplying the units are so arranged as to distribute the blood equally to each of them. This concept of structural units was extended, by him and by some of his students, to other organs.

While a postgraduate student in Leipzig with Wilhelm His, Mall had begun to collect human embryos; and in Baltimore he continued to build up his collection. With this material he made valuable studies of the development of the intestinal canal, the body cavities, the diaphragm, and the abdominal walls. In 1910–1912 Mall collaborated with Franz Keibel of Freiburg in editing *Handbook of Human Embryology*, written by American and German experts. The two-volume work has not yet been superseded.

The growth of Mall's collection of human embryos and their preparation for research use made so large a demand on the resources of his university department that in 1913 he appealed to the Carnegie Institution of Washington to create a department of embryology at the Johns Hopkins Medical School, to which he gave his collection, by that time the largest in the world. He led this laboratory until his death in 1917, holding its directorship as well as the Johns Hopkins chair of anatomy. The first six volumes of the Carnegie *Contributions to Embryology* were edited by Mall with the highest standards of textual perfection and illustration. A comprehensive program of research laid down by him was continued and largely completed under his successor, George L. Streeter.

Mall's service to anatomy and embryology beyond his own laboratory began in 1900, when, with Charles S. Minot of Harvard and George S. Huntington of Columbia University, he founded the *American Journal of Anatomy* and for eight years, with sound financial judgment, published it from his laboratory with one of his staff, Henry McElderry Knower, as managing editor. Mall and Minot also joined in 1900 in a successful effort to rejuvenate the American

Association of Anatomists, which since its foundation in 1888 had represented the older, relatively unoriginal phase of anatomical study in the United States. Never putting himself forward in its organization, Mall was strongly influential in placing and keeping it on a high scientific level. As its president in 1906–1908 he overcame his reluctance, or inability, to speak in public and made the only formal address of his career, under the characteristically modest title "Some Points of Interest to Anatomists."

As in the councils of his profession, so also in the Johns Hopkins University, Mall exerted a quiet but profound influence. His close friendship since his student days in Germany with William H. Welch, dean of the school of medicine and eloquent, magnetic leader of American medical education, gave Mall a channel for his ideas and a spokesman for the ideals they both cherished. Mall was the chief proponent of the so-called concentration system, under which the medical student's time was no longer divided between several concurrently taught subjects but was allotted to not more than two subjects at a time, which were taken up for long periods, up to two trimesters. Thus the student received a thorough, uninterrupted conspectus of each subject. This system, begun at Johns Hopkins, was generally adopted by the American medical schools.

Mall was also a leader, in his inconspicuous way, in the highly controversial movement for full-time teachers of the clinical subjects, as opposed to use of medical practitioners devoting only part of their time to teaching. Full-time teaching of the preclinical sciences—anatomy, physiology, biochemistry and pathology—was well under way in the American schools by the turn of the century, as Mall's appointment to the Johns Hopkins chair of anatomy shows; but the radical move for which he worked was to put the teaching of internal medicine, surgery, and the major specialties into the hands of able men freed from the necessity of earning their living by private practice and, as salaried professors, devoting their full time to teaching, research, and the care of patients in university hospitals. Mall had gotten the germ of this idea from Carl Ludwig at Leipzig and had discussed it during the early years of the Johns Hopkins Medical School. A former member of his staff, Lewellys F. Barker, first presented the plan publicly in an address at Chicago in 1903. Through Welch, who was at first lukewarm, a program for full-time clinical teaching at Johns Hopkins was drawn up in 1913 and implemented by a grant from the General Education Board of New York.

As the foregoing record has indicated, Mall possessed an unusual combination of far-reaching originality with shy reticence, of scientific detachment with shrewd business sense. He never overran his budget and always had a reserve on hand. He was wise in his selection of young men for his staff and highly successful in placing them, when they matured, in important positions. Admirably fair-minded, he was, however, intolerant of stupidity.

Mall's disapproval of excessive lecturing in anatomical teaching was based not only on his pedagogical ideas but also on lack of talent for formal speaking before classes and scientific societies. When he did, rarely, speak to such groups, he was not outstandingly clear. On the other hand, he always had time for private discussions with students and staff members, in or out of classroom and laboratory. His remarks, however, were not always directly related to the topic of the conversation; they were often very whimsical and seemingly farfetched, but they always included a serious idea which the hearer might miss until after days or months he suddenly perceived what Mall had intended subtly to convey. Without the mediation of his ideas by Minot in the American Association of Anatomists, and in the larger affairs of medical education by Welch and by Abraham Flexner of the General Education Board and the Rockefeller Foundation, Mall would never have seen his reform projects carried out; but indirectly he influenced many pupils and associates to whom he addressed himself privately.

BIBLIOGRAPHY

I. ORIGINAL WORKS. The author of many scientific articles, Mall also was joint editor, with Franz Keibel, of *Handbook of Human Embryology*, 2 vols. (Philadelphia, 1910–1912), to which he contributed 3 chapters.

II. SECONDARY LITERATURE. See "Memorial Services in Honor of Franklin Paine Mall, Professor of Anatomy, Johns Hopkins University, 1893 to 1917," in *Johns Hopkins Hospital Bulletin*, **29** (1918), 109–123, with portrait and complete bibliography; and the following works by Florence R. Sabin: "Franklin Paine Mall, a Review of His Scientific Achievement," in *Science*, n.s. **47** (1918), 254–261; *Franklin Paine Mall, the Story of a Mind* (Baltimore, 1934), with portraits and selected bibliography; and "Franklin Paine Mall, 1862–1917," in *Biographical Memoirs. National Academy of Sciences*, **16** (1936), 65–122, with portrait and complete bibliography.

GEORGE W. CORNER

MALLARD, (FRANÇOIS) ERNEST (*b.* Châteauneuf-sur-Cher, France, 4 February 1833; *d.* Paris, France, 6 July 1894), *mineralogy, mining engineering.*

The son of a lawyer, Mallard was brought up in St.-Amand-Montrond. He studied at the Collège de Bourges, the École Polytechnique, and the École des Mines, from which he graduated in 1853 as *ingénieur des mines*. He began his career as a geologist for the Corps des Mines but in 1859 was transferred to fill the chair of mineralogy at its École des Mineurs at Saint-Étienne. He continued his work for the Corps, first in geological mapping and after 1868 on problems of mining engineering. His work attracted favorable notice from G. A. Daubrée, professor of mineralogy at the École des Mines; and when Daubrée became director of the school in 1872, he chose Mallard to fill the vacated chair of mineralogy. The new post was a decisive change for Mallard, involving a new concentration on more directly mineralogical problems, the area in which he found greatest satisfaction and made his most important contributions. He continued as a member of investigative commissions in the Corps des Mines, for which he was promoted in 1867 to engineer first class and in 1886 to inspector general. In 1869 he received the *croix de chevalier* and in 1888 became an officer of the Legion of Honor. He was proposed, unsuccessfully, for several vacancies in the mineralogy section of the Académie des Sciences and finally was elected in 1890. In 1879 he was the second president of the Société minéralogique de France.

Mallard had great personal modesty but did not hesitate to put into print his thoughtful objections to publications of other scientists. He never wished to marry; he followed his vocations single-mindedly. Termier poignantly describes Mallard's camaraderie with students during field excursions.

Mallard's contributions in crystallography began in 1876, not long after he had assumed the chair of mineralogy at Paris. He took as his starting point the *Études cristallographiques*[1] of Auguste Bravais, who had been professor of physics there from 1845 to 1855. While Sohncke[2] and later A. Schönflies[3] were developing Bravais's concept of a lattice of translationally equivalent points into a complete mathematical description of symmetry of crystals (the 230 space groups), Mallard independently developed other aspects of Bravais's theories. This work had its notable beginning in the memoir *Explication des phénomènes optiques anomaux*, on optically "anomalous" crystals (that is, those crystals the morphology of which seems to be of greater symmetry than their optics), in which the powerful new polarizing microscope showed the importance of twinning in these "crystalline edifices" and of pseudosymmetry as an explanatory concept.

A summary of Mallard's ideas was offered in a two-volume work with atlas entitled *Traité de cristal-*

lographie géométrique et physique, published in 1879 (geometrical crystallography, lattice theory, and morphology) and 1884 (crystal physics). Here, for the first time, the convoluted mathematical apparatus of Bravais was stripped away to reveal in a didactic but complete fashion the essential contributions of the lattice theory. Mallard recognized the way in which this theory corresponded to a special case of Haüy's *molécules intégrantes* while admitting that the unsupportable remainder of Haüy's theories of crystal structure had led to its being completely discredited by the German school of geometrical crystallographers. Wherever possible, he applied Bravais's theory to an understanding of the wide range of physical properties of crystals that had been investigated in the thirty intervening years; and he found in a strict definition of homogeneity a common basis for both the lattice theory and the newer macroscopic description of physical properties of anisotropic crystals by characteristic ellipsoids. The detailed and generally favorable reviews of these volumes in Germany,[4] where a divergent direction of theory had been followed since the time of C. S. Weiss, attests to their persuasive completeness.

A third volume of the *Traité*, in which Mallard planned to discuss isomorphism, polymorphism, twinning, pseudosymmetry, and crystal growth, was never completed. Some of his original work on these subjects was published separately.[5] In this respect his most important contribution was his theory of twins based on the continuation or pseudo-continuation of a lattice between the twinned crystals. Mallard's development and extension of Bravais's lattice theory, especially to the explanation of the importance of crystal faces ("law of Bravais") and of twins, was further developed by observations and refined by Georges Friedel,[6] who was Mallard's student at the École des Mines; Friedel's exposition on these subjects remained the definitive statement until very recent direct structural theories. On the other hand, Mallard's theory of circular polarization in crystals as equivalent to a stack of thin, linearly polarized sheets has not stood the test of time.[7]

Mallard also made notable scientific contributions in his capacity as a mining engineer. In 1878, soon after it was formed, he was appointed to an official commission investigating methods of preventing methane explosions in mines. A series of laboratory and field investigations in collaboration with H. Le Chatelier, then professor of general chemistry at the École des Mines, resulted in a series of joint papers on the design of safety lamps, on combustion temperature, on velocity of flame propagation, and on the importance of mixtures of coal dust. They instituted the

use of ammonium nitrate as an explosive; its low temperature of detonation made it safer (less likely to propagate in mixtures of air with gas or coal), and it has remained the preferred explosive to this day. Mallard continued his collaboration with Le Chatelier in experiments on the thermal properties of crystals.

A key figure in the French school of crystallography and a bridge from Bravais to Friedel, Mallard was described as having ... "definitively displaced the center of gravity of crystallography that, thereafter, could not be cultivated as a descriptive science, but was elevated to the rank of a rational science."[8]

NOTES

1. *Études cristallographiques* (Paris, 1866), reprinting papers published during 1850 and 1851.
2. *Entwickelung einer Theorie der Krystallstruktur* (Leipzig, 1879) and "Erweiterung der Theorie der Krystallstruktur," in *Zeitschrift für Krystallographie und Mineralogie*, **14** (1888), 426–446.
3. The importance of the Mallard-Bravais viewpoint on symmetry is acknowledged by Arthur Schönflies in "Bemerkungen über Theorien der Krystallstruktur," in *Zeitschrift für physikalische Chemie*, **9** (1892), 158–170.
4. C. Klein, in *Neues Jahrbuch für Mineralogie, Geologie und Paläontologie* (1880), **2**, 1–5; *ibid.* (1886), **1**, 1–5.
5. E.g., in Edmond Fremy, *Encyclopédie chimique*, I (Paris, 1882), 610–774; and *Revue scientifique*, **24** (1887), 129–138, 165–171.
6. Summarized in *Leçons de cristallographie* (Paris, 1926).
7. Mallard, *Traité*, II, chs. 8–9; refuted in J. R. Partington, *Advanced Treatise on Physical Chemistry*, IV (London, 1953), p. 355.
8. Translated from Wrybouff's memorial, p. 249.

BIBLIOGRAPHY

I. ORIGINAL WORKS. A complete bibliography is given in Lapparent's memorial (see below). In addition to *Explication des phénomènes optiques anomaux* (Paris, 1877), repr. from *Annales des mines et des carburants*, 7th ser., **10** (1876), 60–196; and *Traité de cristallographie géométrique et physique*, 2 vols. and atlas (Paris, 1879–1884), the most important later crystallographic papers may be "Les groupements cristallins," in *Revue scientifique*, **24** (1887), 129–138, 165–171; "Sur l'isomorphisme des chlorates et des azotates et sur la vraisemblance de la quasi-identité de l'arrangement moléculaire dans toutes les substances cristallisées," in *Bulletin de la Société française de minéralogie*, **7** (1884), 349–401; and "Sur la théorie des macles," *ibid.*, **8** (1885), 452–469. The work with Le Chatelier on mine safety is given in a series of papers in *Annales des mines* (many of which are summarized in short notes in the *Comptes rendus de l'Académie des sciences* [1879–1889]), notably three memoirs in *Annales des mines*, 8th ser., **4** (1883), 276–568; and in a series of commission reports published by the government and listed in the catalog of the *Bibliothèque nationale*, CIV (Paris, 1930), 593–594. One notable paper not listed in Lapparent's bibliography is "De la définition de la température dans la théorie mécanique de la chaleur et de l'interprétation physique du second principle fondamental de cette théorie," in *Comptes rendus de l'Académie des sciences*, **75** (1872), 1479–1484.

II. SECONDARY LITERATURE. Mallard's life and works were well covered from diverse viewpoints in a dozen contemporary memorials, most of which are listed in the Royal Society *Catalogue of Scientific Papers*, XVI, 1028. The most informative are those by A. de Lapparent, in *Annales des mines*, 9th ser., **7** (1895), 267–303, with bibliography; G. Wrybouff, in *Bulletin de la Société française de minéralogie*, **7** (1894), 241–266, with bibliography and portrait; and P. Termier, in *Bulletin de la Société géologique de France*, **23** (1895), 179–191. Mallard is barely mentioned by his contemporary Paul Groth in the latter's *Entwicklungsgeschichte der mineralogischen Wissenschaften* (Berlin, 1926; repr. 1970), and his contributions do not seem to have been reviewed elsewhere.

WILLIAM T. HOLSER

MALLET, ROBERT (*b.* Dublin, Ireland, 3 June 1810; *d.* London, England, 5 November 1881), *technology, seismology.*

Mallet's father, John Mallet, of Devonshire, owned a plumbing business and copper and brass foundry in Dublin, from which he made a large fortune. He married a cousin, also named Mallet. Robert, their only son, was educated at Bective House in Dublin; he entered Trinity College, Dublin, in December 1826, graduating with a B.A. in 1830. After joining his father's company, he built it up into one of the most important engineering works in Ireland; he himself became an inventive designer-engineer and versatile researcher.

In November 1831 Mallet married Cordelia Watson, who died in 1854. He remarried in 1861. He left two daughters and three sons. Mallet's eyesight deteriorated seriously after the winter of 1871–1872.

In 1861 Mallet settled in London as a consulting engineer. He edited four volumes of the *Practical Mechanic's Journal* (1865–1869), for which he prepared the *Journal Record of the Great Exhibition* (London, 1862), and, in collaboration with R. F. Fairlie, *The Safes' Challenge Contest at the International Exhibition of Paris in 1867* (1868).

Mallet was a member of many organizations, including the Royal Irish Academy (1832), the British Association (1835), the Institution of Civil Engineers of Ireland (1839, associate; 1842, member; 1866, president), and the Royal Geological Society of Ireland (1847, president). He became a fellow of the Royal

Society of London in 1854 and later served as a member of its council. Mallet also received an honorary doctor of laws degree from Trinity College and was awarded numerous scientific medals.

Among Mallet's notable technical accomplishments were the erection of the roof of St. George's Church in Dublin with hoists of his own design; his "manumotive" for baggage transport, which, with eight men driving it, covered the five miles from Dublin to Kingstown in twenty minutes (1834); his procedure for the electromagnetic separation of brass and iron filings; and his method of bleaching turf for the manufacture of paper (1835). His multifarious engineering projects also involved work on steam-driven printing plants, bridges, hydraulic presses, ventilators and heaters, brewery machinery, railroads, dock gates, viaducts, lighthouses, and coal mines. Mallet worked out a plan to supply Dublin with water from six reservoirs on the Dodder, for which he carried out the surveying at his own expense in 1841. His name is associated with the "buckled plates," which he patented in 1852, used for flooring and later for railroad ties.

From 1850 to 1856 Mallet worked on the design of heavy guns. With his comprehensive account entitled *The Physical Conditions Involved in the Construction of Artillery, With an Investigation of the Relative and Absolute Values of the Materials Principally Employed, and of Some Hitherto Unexplained Causes of the Destruction of Cannon in Service* (Dublin, 1856), he created a basis for later books on ordnance and on casting and founding.

In view of these many technical achievements and the demands of running such a large business, it is remarkable that Mallet still found time for scientific research in the most varied fields. Among his scientific papers are investigations of the "seed-dispersing apparatus" of *Erodium moschatum* (1836); the blackening of photographic paper due to the radiation of glowing cinders (1837); the photochemical bleaching of caustic potash; the effect of boiling on organic and inorganic substances (1838); the improvement of the manufacture of optical glasses; and the application of the "electrotype process in conducting organic analysis" (1843).

The author of fundamental works on the effect of air and water—especially pure and polluted seawater—on wrought and cast iron and also on steel (1836–1873), Mallet is one of the founders of research on corrosion. In this way and through his studies dealing with the physical properties and electrochemical relationships of copper and tin alloys with tin and zinc, he made important early contributions to the science of materials. He also dealt with the state of aggregation of alloys (1840–1844), with the coefficients of elasticity and with failure in wrought iron (1859), and with the expansion of cast iron (1875).

Completely aside from his profession as an engineer, Mallet was interested in the structure of County Galway trap and the columnar structure of basalts, and in glacier movement, the plasticity of glacial ice, and the lamination of Irish slates.

Stimulated by Charles Lyell's description of the earthquake in Calabria in 1783, Mallet explained its "vorticose movement" by the position of the center of gravity, adhesion, or—as the case may be—friction, and the inertia of the squared stone affected by the tremor (1846). In the same year he presented to the Royal Irish Academy his *On the Dynamics of Earthquakes*. In this work he differentiated the so-called earth wave, great sea wave, forced sea wave, and aerial or sound wave. Mallet considered local elevations of portions of the earth's solid crust to be the cause of earthquakes. He produced an experimental seismograph in 1846. Important elements of his model, which was never actually used, were incorporated in the seismograph that Luigi Palmieri made in 1855. Between 1850 and 1861 Mallet set off explosions in different locations to determine the rate of travel of seismic waves in sand (825 feet per second), solid granite (1,665 feet per second), and quartzite (1,162 feet per second). According to A. Sieberg (1924), Mallet should be considered the founder of the physics of earthquakes. The term "seismology" we owe to Mallet, as well as "seismic focus," "angle of emergence," "isoseismal line," and "meizoseismal area."

Mallet presented his most important seismic results in four *Report[s] to the British Association* (1850, 1851, 1852–1854, 1858) and in four editions of the *Admiralty Manual of Scientific Enquiry* (1849, 1851, 1859, 1871). Between them, they contain an extensive catalog—which he prepared and debated with his son, John W. Mallet—of 6,831 earthquakes reported between 1606 B.C. and A.D. 1858 and his seismic map of the world. In February 1858 he visited the region of the Neapolitan earthquake of 16 December 1857. In 1869, with Oldham, he developed a spherical projection seismograph.

Through his work in seismology, Mallet encountered possible connections with vulcanicity. He eventually studied the problem of the emergence of volcanic foci in particular. In 1862 he published "Proposed Measurement of the Temperatures of Active Volcanic Foci to the Greatest Attainable Depth . . ." (*Report of the British Association for the Advancement of Science* for 1862). In 1872 he submitted his principal paper on vulcanicity, "Volcanic Energy; an Attempt to Develop Its True Origin and Cosmical Relations" (*Philosophical Transactions of the Royal Society*, **163**, no. 1 [1873], 147–227).

In 1870 Mallet made exhaustive analyses of sixteen different types of rock. Through computation he concluded that (1) "the crushing of the earth's solid crust affords a supply of energy sufficient to account for terrestrial vulcanicity"; and (2) that "the necessary amount of crushing falls within the limits that may be admitted as due to terrestrial contraction by secular refrigeration" (*ibid.*, p. 214). Although this theory received little attention in the following years, it nevertheless contained some ingenious conceptions that were taken over by later researchers. In any case, it deserves recognition as an early example of the mathematical utilization of experimental findings in the service of geological speculation.

BIBLIOGRAPHY

I. ORIGINAL WORKS. A list of eighty-five (incorrectly numbered ninety-one) papers by Mallet is in Royal Society *Catalogue of Scientific Papers*, IV, 205–208; VIII, 314; X, 703–704. See also Poggendorff, III, pt. 2, 861–862.

In addition to works cited in the text, Mallet wrote *Great Neapolitan Earthquake of 1857: The First Principles of Observational Seismology*, 2 vols. (London, 1862). Works which he translated and to which he contributed include H. Law, *Civil Engineering* (London, 1869), with Mallet's notes and illustrations; G. Field, *The Rudiments of Colours and of Colouring* (London, 1870), revised and partly rewritten by Mallet; L. L. de Koninck, *A Practical Manual of Chemical Analysis and Assaying* (London, 1872), which he edited and to which he contributed notes; and L. Palmieri, *The Eruption of Vesuvius in 1872* (London, 1873), with notes and intro. by Mallet.

II. SECONDARY LITERATURE. For obituaries of Mallet, see *Engineer*, **52** (1881), 352–353, 371–372, 389–390; *Minutes of Proceedings of the Institution of Civil Engineers*, **68** (1882), 297–304; *Proceedings of the Royal Society*, **33** (1882), 19–20; and *Quarterly Journal of the Geological Society of London*, **38** (1882), 54–56.

See also (listed chronologically) *Dictionary of National Biography*, XXXV (1893), 429–430, which gives Mallet's place of death as Enmore, Surrey; R. Ehlert, "Zusammenstellung, Erläuterung und kritische Beurtheilung der wichtigsten Seismometer," in *Beiträge zur Geophysik*, **3** (1898), 350–475; K. A. von Zittel, *Geschichte der Geologie und Paläontologie bis Ende des 19. Jahrhunderts* (Munich, 1899); Charles Davison, *The Founders of Seismology* (Cambridge, 1927); and L. Mintrop, "100 Jahre physikalische Erdbebenforschung und Sprengtechnik," in *Naturwissenschaften*, **34** (1947), 257–262, 289–295.

WALTHER FISCHER

MALOUIN

MALOUIN, PAUL-JACQUES (*b*. Caen, France, 29 June 1701; *d*. Versailles, France, 3 January 1778), *medicine, chemistry.*

Malouin was born into a venerable Caen family. His parents, N. Malouin and N. Poupart, wanted him to pursue a legal career. He was sent to Paris to study law, but turned instead to scientific pursuits and, after a brief return to his native city from 1730 to 1733, settled in the French capital to teach and practice medicine. A relative of Fontenelle, permanent secretary of the Académie Royale des Sciences, Malouin quickly attracted a prominent clientele, which included members of the royal family. He emphasized the importance of hygiene and the comprehensive application of chemical remedies and theory to medicine, presenting his findings formally as professor of medicine at the Collège Royal from 1767 to 1775. In his will he provided for the establishment of an annual public meeting at the Faculty of Medicine in Paris to apprise the nation of the most recent medical discoveries and advances.

Malouin complemented his medical career with an active interest in the developing science of chemistry. Elected to the Academy as *adjoint chimiste* in 1742, he became *pensionnaire chimiste* in 1766 and director of the Academy for 1772; he was made a fellow of the Royal Society of London in 1753. Malouin's chemical studies are relatively unimportant, although several memoirs read in 1742 and 1743 on zinc and tin were then useful. A contributor to the early volumes of Diderot's *Encyclopédie*, Malouin wrote a number of competent articles on various chemical topics: "Alchimie," "Antimoine," "Acide," and "Alkali." He worked frequently with Bourdelin, professor of chemistry at the Jardin du Roi, and often lectured in his stead. He also contributed important articles on milling and baking in the Academy's series *Description des Arts et Métiers,* applying chemical theory and method to those two trades, vital in the economic and social life of the *ancien régime*. Malouin's methods for grinding wheat and mixing flour yielded bread of higher quality.

BIBLIOGRAPHY

I. ORIGINAL WORKS. Malouin's major works are *Traité de chimie, contenant la manière de préparer les remèdes qui sont les plus en usage dans la pratique de la médecine* (Paris, 1734); and *Chimie médicinale, contenant la manière de préparer les remèdes les plus usités, et la méthode pour la guérison des maladies,* 2 vols. (Paris, 1750; 2nd ed., 1755). His work on milling and baking, entitled *Description et détails des arts du meunier, du vermicelier et du boulanger, avec une histoire abrégée de la boulangerie et un dictionnaire*

de ces arts (Paris, 1767), appeared as a volume in the series *Description des Arts et Métiers, faites ou approuvées par Messieurs de l'Académie Royale des Sciences.*

II. SECONDARY LITERATURE. The best biographical source for Malouin is M. Condorcet, "Éloge de M. Malouin," in A. Condorcet O'Connor and M. F. Arago, eds., *Oeuvres de Condorcet,* II (Paris, 1847), 320–332. Consult also Jean-Charles Des Essartz, *Éloge de Malouin* (Paris, 1778); and F. Hoefer, ed., *Nouvelle biographie générale,* XXXIII (Paris, 1860), 97–98. A brief assessment of Malouin's chemical work is in J. R. Partington, *A History of Chemistry,* III (London, 1961), 72.

MARTIN FICHMAN

MALPIGHI

MALPIGHI, MARCELLO (*b*. Crevalcore, Bologna, Italy, baptized 10 March 1628; *d*. Rome, Italy, 29 November 1694), *medicine, microscopic and comparative anatomy, embryology.*

Malpighi was the son of Marcantonio Malpighi and Maria Cremonini. In 1646 he entered the University of Bologna, where his tutor was the Peripatetic philosopher Francesco Natali. On Natali's advice Malpighi in 1649 began to study medicine. He first attended the school conducted by Bartolomeo Massari, then that of Andrea Mariani; with Carlo Fracassati he was among the nine students allowed to attend the dissections and vivisections that Massari conducted in his own house.

Malpighi graduated as doctor of medicine and philosophy in 1653; three years later, still in Bologna, he began teaching as a lecturer in logic, but toward the end of the year he was called to the chair of theoretical medicine at the University of Pisa. The three years that he spent in Pisa were fundamental to the formation of Malpighi's science. Influenced by Giovanni Alfonso Borelli, who was then professor of mathematics in the same university, Malpighi turned from Peripateticism to a "free and Democritean philosophy." He also participated in animal dissections in Borelli's home laboratory and, through Borelli, entered the scientific orbit of the school of Galileo, which was at that time best represented in Tuscany itself, in the Accademia del Cimento (1657–1667).

By 1659, however, Malpighi was no longer able to tolerate the Pisan climate. He therefore returned to Bologna to become extraordinary lecturer in theoretical medicine. Toward the end of 1660 he assumed the ordinary lectureship at the university in practical medicine. In 1662 he went to the University of Messina, where he held the principal chair of medicine; four years later he returned to Bologna to lecture in practical medicine again. A letter of 28 December 1667 asked him to undertake scientific correspondence with the Royal Society of London; Malpighi agreed, and the society subsequently supervised the printing of all his later works. In 1691 Malpighi was called to Rome as chief physician to Pope Innocent XII. He died there, in his apartments in the Quirinal Palace.

Malpighi's first—and fundamental—work is the *De pulmonibus*, two short letters which he sent to Borelli in Pisa and which were published in Bologna in 1661. After his return to Bologna in 1659 Malpighi, together with Carlo Fracassati, continued to conduct dissections and vivisections. In the course of these he used the microscope to make fundamental discoveries about the lungs, which he quickly announced in the letters to Borelli.

According to the traditional quaternary system, the lungs were fleshy viscera, endowed with a sanguine nature and hot-humid temperament. Having subjected them to microscopical examination, Malpighi found them to be an aggregate of membranous alveoli opening into the ultimate tracheobronchial ramifications and surrounded by a capillary network. He had thus discovered the connections, until then sought in vain, between the arteries and the veins. His observations were of basic significance for two reasons—the pulmonary parenchyma (and subsequently the other parenchymas) for the first time could be seen to have a structure, and the observation of the capillaries confirmed the theory of the circulation of the blood and assured its general acceptance.

Malpighi's mastery of microscopic technique was apparent even in *De pulmonibus*. He used instruments of different magnifying powers and made observations with both reflected and transmitted light. He prepared specimens in a number of ways, including drying, boiling, insufflation (of the tracheobronchial tree or of systems of blood vessels), vascular perfusion, deaeration (by crushing), corrosion, or a combination of these methods. In choosing to examine the frog, Malpighi was able to avail himself of the so-called "microscope of nature." He was able to visualize, with a relatively small magnification, so minute a feature as the capillary (the capillary network itself is so fine in mammals that Malpighi was never able to observe it with the microscopes available to him). Malpighi acutely remarked that nature is accustomed "to undertake its great works only after a series of attempts at lower levels, and to outline in imperfect animals the plan of perfect animals."

Malpighi saw the structure of the lung as air cells surrounded by a network of blood vessels; he interpreted this structure as a well-devised mechanism to insure the mixing of particles of chyle with particles of blood—in other words, for the conversion of chyle

to blood (then called hematosis), a function that the Galenists attributed to the liver. Jean Pecquet had shown in 1647 that the chyle, instead of being conveyed to the liver, was introduced into the blood in the superior vena cava, at a point shortly before that vessel reached the heart, and was then distributed to the lungs through the pulmonary artery.

In the four years, 1662–1666, that he spent at the University of Messina, Malpighi enthusiastically continued his researches on fundamental structures, making use also of marine animals from the Strait of Messina. He published the results of these researches in a series of treatises in 1665–1666. These were devoted mainly to three major topics—neurology, adenology, and hematology.

The short works *De lingua* (Bologna, 1665) and *De externo tactus organo* (Naples, 1665) are closely linked to each other. In *De lingua*, Malpighi reported peeling two layers from the surface of the tongue—the horny layer and the reticular (or mucous) layer that is now named for him—and thus exposing the papillary body, in which he distinguished three orders of papillae. He speculated that these papillae could be reached through pores in the epithelium and thereby stimulated by "sapid" particles dissolved in the saliva, the organismal liquid the significance of which had been recognized just a few years previously by Nicholas Steno. It is easy here to recognize the influence of Galileo, who in *Il saggiatore* (1623) had suggested that the very small taste particles, when "placed on the upper surface of the tongue and, mixed with its moisture, penetrate it and carry the tastes, pleasant or otherwise, according to the differences in the touching of the different shapes of these tiny corpuscles, and according to whether they are few or many, faster or slower."

Malpighi's discovery of the sensory receptors—the papillae of the tongue were followed by the cutaneous (or tactile) papillae—formed part of a wider neuroanatomical research. In the treatise *De cerebro*, which was published in 1665 with *De lingua*, he dealt mainly with the white substance of the central nervous system, which he found to be composed of the same fibers that form the nerves. Malpighi conceived of these fibers as long, fine channels filled with a liquid—the nerve fluid—which was secreted by the cortical gray matter, or, more precisely, by the cortical glands. In his later treatise *De cerebri cortice* (1666), Malpighi claimed to have demonstrated these glands, but his results were in fact due to an artifact.

On the basis of his observations, whether true or false, Malpighi in any event succeeded in constructing a mechanism to encompass the entire neural course from the cortex of the brain to the peripheral endings of the nerves: the neuron, in which the transmission of the nervous impulse could be equated with the transmission of a mechanical impulse through a liquid mass in accordance with Pascal's principle.

During his years in Messina, Malpighi made further investigations into the structure of another mechanism fundamental to his iatromechanical atomism: the gland, or secretion machine. The function of this mechanism was to select specific particles of blood brought by an afferent artery, to separate them from others flowing back through an efferent vein, and to introduce them, as an independent liquid, into an excretory duct. The sieve may thus be used as a convenient model ("cribrum" and "secretio" are even etymologically the same); it offers an a priori explanation of the operation of the secreting mechanism by postulating a proportionality of form and dimension between the pores and the particles to be separated. Malpighi certainly recognized that he could not investigate this "minima simplexque meatuum structura" directly, but he did not abandon his search for the mechanism that might contain the pores. This he localized, a priori, at the point at which the smallest ramifications of the artery, vein, and duct are joined together.

Malpighi continued to search for ever finer and more minute structures within the glandular parenchyma. These investigations were stimulated by the discovery of the pancreatic duct (by Wirsüng in 1642), the testicular duct (by Highmore in 1651), the submandibular duct (by Wharton in 1656), and the parotid duct (by Steno in 1660).

The secreting mechanism devised by Malpighi was based on a follicle that, on one hand, is continuous into the secretory tubule, and, on the other, is surrounded by the ultimate ramifications of the arteries, veins, and nerves. In passing from the artery to the vein, the blood channel and the contiguous glandular follicle are permeable to the particles that must be eliminated and impermeable to the particles of venous blood. By analogy with the sieve, and without invoking vitalistic arguments, secretion can thus be explained in purely mechanical terms. In *De renibus* Malpighi set down a series of convincing observations in support of his system. He skillfully made use of staining techniques by affusion to show the renal tubules, both straight and twisted, while by injecting coloring into the arteries he was able to demonstrate the tufts of vessels attached to the branches of the interlobular arteries. He believed, however, that the ampullar extremities of the renal tubules (the Malpighi corpuscles) were enclosed within the vascular tufts.

Malpighi reiterated and developed his theory of glandular structure in the epistolary dissertation *De structura glandularum conglobatarum consimiliumque*

partium, dated June 1688 and published in London the following year. Although the "conglobate" glands of Sylvius—that is, the lymph nodes—are emphasized in the title, less than half the treatise is devoted to them. For the rest, Malpighi reported additional observations on glands that were already known and considerably expanded his earlier work on the secretory mechanism. He also included remarks on the glandular membranes (later classified by Bichat as serous and mucous).

Having established the capillary circulation and devised a mechanism to explain hematosis; having defined and systematized a nervous mechanism endowed with highly acute sensory receptors; and having postulated a secreting mechanism, Malpighi turned to an analysis of the blood—the universal fluid necessary to all these machines. His chief hematological treatise, *De polypo cordis*, appeared in 1666 (or 1668?) as an appendix to *De viscerum structura*.

"Heart polyps" had been identified for some time and with a certain frequency, especially in patients who had died from severe cardiorespiratory insufficiency. Previous researchers had explained such polyps in various ways, even invoking traditional humoral theory. Malpighi, however, considered these lesions to be the result of an intravitam process of coagulation, which had as its model the coagulation of blood extracted from the organism. The study of coagulum was thus fundamental, and culminated in Malpighi's demonstration that the "phlogistic crust" was, despite its whitish color, derived from the whole blood that "confuses our poor eyes with its purple [color]." To this end he broke the blood down into its component parts (a method that he had successfully employed in his studies of viscera and organs) by continuing artificially in the coagulum the separation (into coagulum and serum) that occurs naturally when blood is extracted from an organism. Malpighi found that the coagulum, after repeated washings, "from being intensely red and black becomes white, while the water is reddened by the extracted particles of color." The phlogistic crust thus corresponds in large part to the bleached-out coagulum; the difference between them is only quantitative (that is, it lies in the amount of coloring material that each contains) and not qualitative, as supposed by the humoral theory.

Microscopic examination of a clot of coagulum also enabled Malpighi to observe, as separate components, the interlacing white fibers that arise from the conglutination of much smaller but similarly shaped filaments (a process similar to that which occurs in the crystallization of salts) and the red fluid that fills the interstices of these meshes of fibers. With the microscope Malpighi could perceive that the red fluid was composed of a host of red "atoms"; it is thus clear that the discovery of the red corpuscles—although variously attributed by a number of authors who would seem to be unaware of their unmistakable description in the *De polypo cordis*—is surely Malpighi's.

Malpighi was able to utilize even a morbid deviation such as the heart polyp toward an investigation of a normal phenomenon. He studied aberrations to cast light upon normal organisms. In the same way, he studied simple animals to understand more complex ones, writing that the

> study of insects, fish, and the first unelaborated outlines of animals has been used in this century . . . to discover much more than was achieved by previous ages, which limited their investigations to the bodies of perfect animals only.

Having stated this methodological formulation, Malpighi applied it in his work on the silkworm, *De bombyce* (London, 1669), and in the later embryological and botanical works that were edited by the Royal Society for publication in London in the 1670's.

Malpighi was led to do embryological research through an analogy with the artisan who "in building machines must first manufacture the individual parts, so that the pieces are first seen separately, which must then be fitted together," as he asserted in *De formatione pulli in ovo* (1673). In *De bombyce*, he had carefully observed the artisan nature construct each of the three stages—larva, chrysalis, and moth—through which the silkworm is formed. He further remarked on the specific apparatuses with which the silkworm is provided, among them the air ducts (tracheae) and the blood duct with a number of pulsating centers (corcula).

With the *De formatione* and the subsequent appendix to it (1675), Malpighi brought a fine structural content to embryology, which became a valuable aid to illustrating the morphology of the adult. So, too, the study of lower forms of life clarifies the morphology of more highly developed ones. Malpighi noted that the study "of the first unelaborated outlines of animals in the course of development" is particularly fruitful because the artisan nature forms them separately before combining them with one another. In the embryo, for example, the miliary glands, which will merge to form the liver, are still distinguishable as the cecal sacs (which in crustaceans remain distinct). From this point on, the paths of embryogenesis and phylogenesis were destined to cross.

The chick fetus develops in a manner similar to that of the plant embryo contained within a plant seed: from being enveloped at the start, it simultaneously "evolves" and grows in size as a result of the influx of food (yolk and albumen) liquefied by the warmth of the nest or by the fermentation process set in motion by fecundation. This notion, that embryogenesis consists of the development of constituents that in some sense existed prior to incubation, but which are nevertheless secondary to fecundation (see Adelmann), since they are induced by the "colliquamentum" of the pellucid area by the aura—or spiritous emanation—of the male seed, gave fuel to the doctrine of preformation, which then became a strong alternative to the traditional doctrine of epigenesis.

Malpighi's chief embryological discoveries were the vascular area embraced by the terminal sinus, the cardiac tube and its segmentation, the aortic arches, the somites, the neural folds and the neural tube, the cerebral vesicles, the optic vesicles, the protoliver, the glands of the prestomach, and the feather follicles.

Malpighi clearly stated his comparative method in the introduction to *Anatomes plantarum idea* (1675):

> The nature of things, enveloped in shadows, is revealed only by the analogical method ["cum solo analogismo pateat"]. Hence the necessity to follow it entirely, so as to be able to analyze the most complex mechanisms by means of simpler ones that are more easily accessible to the experience of the senses. It is the most important and most perfect things, however, that are the most immediately attractive to human genius, since they are the most necessary to human utility and therefore most worthy of consideration.

This had been true in the early work of even Malpighi himself. With youthful ardor, he had flung himself into the investigation of higher animals,

> . . . but these, enveloped in their own shadows, remain in obscurity; hence it is necessary to study them through the analogues provided by simple animals ["simplicium analogismo egent"]. I was therefore attracted to the investigation of insects; but this too has its difficulties. So, in the end, I turned to the investigation of plants, so that by an extensive study of this kingdom I might find a way to return to early studies, beginning with vegetant nature. But perhaps not even this will be enough, since the yet simpler kingdom of minerals and elements should take precedence. At this point the undertaking becomes immense, and absolutely out of all proportion to my strength.

If Malpighi retreated before the demands of making a systematic study of minerals, he nonetheless undertook the study of plants with extraordinary success. *Anatome plantarum*, which appeared in London in two parts (1675 and 1679), earned him acclaim (along with Nehemiah Grew) as the founder of the microscopic study of plant anatomy. In his investigation Malpighi found that plants also have a mechanical structure: he described their ducts (some of which he compared to the tracheae of insects) and their basic "cellular" structure (an aggregate of "utricles"), which Hooke had already described (as "cellulae") in the *Micrographia*.

In his later studies Malpighi used the "microscope of nature" as it was manifest in natural anomalies, and in particular in monstrosities and pathological aberrations. For example, he investigated warts and found the dermic papillae to be strikingly enlarged. Anomalous structures may be not only enlarged but also so arranged as to clarify individual components in the normal state. Thus, in onychogryphosis the lamellar structure of the normal nail is apparent; while in the jugular horn of a calf the reinforced projections of the papillae stand out, whereas in the normal horn they are concealed.

The correspondence between normal and anomalous horn is not only structural, but also morphogenetic, since the metamorphosis of the cutaneous strata into the horn is caused by mechanical stimulation. Under normal conditions this stimulation is exerted by the bony excrescence of the frontal bone: in the jugular horn it is the result of the irritation of the yoke and of the resulting saccate accumulation of fluid in the subcutaneous tissue. Malpighi adduced a similar mechanical morphogenesis in the polycystic kidney: the glandular follicles (Malpighi's corpuscles) appear enlarged and distinct in this condition because they have been dilated by urostasis secondary to a blockage of the outflow channels. Similarly, in the nodules of the cirrhotic liver, the hepatic follicles are enlarged by the "microscope of nature," as are lymphatic follicles that have been altered by disease (usually tuberculosis).

In *De polypo cordis* and subsequent treatises it is possible to identify explicit references to pathological material obtained during autopsy. Malpighi recognized the importance of local lesions, and his pathological investigations were considerably enhanced by the microscopic anatomy of the 1660's, of which he himself was the most important investigator. The discovery of minute functional mechanisms, which in the aggregate give rise to the vital event, gave abnormal structures an added significance. The anatomical investigation of the breakdown of any of these mechanisms—even if only in such macroscopic equivalents as the lesions visible in the dissecting room—demonstrated the effect of such disturbances on the economy of the organism as a whole. Such clinical manifestations are proportional to the place and nature of the lesion; subtle anatomy thus gave rise to the anatomical investigation of the causes and localizations of disease (to paraphrase the title of the later work of Morgagni).

In his medical anatomy (or practical anatomy, as it was then called), in his emphasis on those aspects of anatomy proper to medical practice, and above all, in his use of anatomoclinical parallelism, Malpighi shaped the work of at least two generations. His pupils included Albertini and Valsalva; the *De sedibus et causis morborum per anatomen indagatis* of their pupil Morgagni represents a most important continuation of Malpighi's work.

Malpighi also made considerable contributions to vegetable pathology. In particular he made a study of plant galls, which he found to be a morbid alteration of the structural plan of the infested plant. Finally, Malpighi wrote an important methodological work, *De recentiorum medicorum studio*, in which he supported rational medicine against the empiricists. Rational medicine was also the basis for his many *Consultationes*, which attest to the medical practice that he carried out concurrently with his biological researches.

BIBLIOGRAPHY

I. ORIGINAL WORKS. Malpighi's most important works are *De pulmonibus observationes anatomicae* (Bologna, 1661); *De pulmonibus epistola altera* (Bologna, 1661); *Epistolae anatomicae de cerebro, ac lingua ... Quibus Anonymi accessit exercitatio de omento, pinguedine, et adiposis ductibus* (Bologna, 1665); *De externo tactus organo anatomica observatio* (Naples, 1665); *De viscerum structura exercitatio anatomica ... Accedit dissertatio eiusdem de polypo cordis* (Bologna, 1666); *Dissertatio epistolica de bombyce* (London, 1669); *Dissertatio epistolica de formatione pulli in ovo* (London, 1673); *Anatomes plantarum pars prima. Cui subjungitur appendix iteratas et auctas de ovo incubato observationes continens* (London, 1675), which is prefaced by *Anatomes plantarum idea*, dated November 1671; *Anatomes plantarum pars altera* (London, 1679); "Dissertatio epistolica varii argumenti" [addressed to Jacob Spon], in *Philosophical Transactions of the Royal Society of London*, **14** (1684), 601–608, 630–646; *Opera omnia* (London, 1686; repr. Leiden, 1687); *De structura glandularum conglobatarum consimiliumque partium epistola* (London, 1689); *Opera posthuma* (London, 1697; repr. Amsterdam, 1698); *Consultationum medicinalium centuria prima* (Padua, 1713); and *Consultationum medicarum nonnullarumque dissertationum collectio* (Venice, 1747), written with J. M. Lancisi.

A recent selection is Luigi Belloni, ed., *Opere scelte* (Turin, 1967), with an introduction containing a useful synopsis of Malpighi's work.

II. SECONDARY LITERATURE. A definitive biography is Howard B. Adelmann, *Marcello Malpighi and the Evolution of Embryology* (Ithaca, N. Y., 1966).

LUIGI BELLONI

MALTHUS, THOMAS ROBERT (*b.* near Guildford, Surrey, England, 13 February 1766; *d.* near Bath, England, 23 December 1834), *political economy.*

Malthus is known in the history of science almost exclusively for his influence on Charles Darwin, exerted almost accidentally. His life, work, and friends were mainly centered on social conditions and political economy, and his work on population was part of these. He did have early training in mathematics, however, and based his arguments on the careful analysis of observed data.

Robert Malthus (he appears never to have been called Thomas) was the sixth child of seven born to Daniel Malthus and his wife, the former Henrietta Catherine Graham. Daniel Malthus, a scholar and a friend and admirer of Rousseau, provided a stimulating home life and education for the boy, and later sent him to study with Richard Graves at Claverton and at the Dissenting Academy of Warrington under Gilbert Wakefield.

In 1784 Malthus went up to Jesus College, Cambridge, where his tutor was William Frend. He read for the mathematical tripos and graduated in 1788, being ninth wrangler; but he also read widely in French and English history and literature and in Newtonian physics. He had already shown his interest in the practical rather than the abstract. He played games and lived a full social life, apparently unaffected by his cleft palate and harelip. The friends he made at Cambridge influenced the rest of his life; the most important was William Otter (1768–1840), later bishop of Chichester. Malthus and Otter traveled extensively in Europe and maintained the relationship after their marriages. Malthus' son, Henry, married Otter's daughter Sophia. Otter probably wrote the memorial to Malthus in Bath Abbey, and he certainly wrote the "Memoir" published with the second edition of the *Principles of Political Economy.*

Malthus followed graduation with ordination, but more in the tradition of the younger sons of English gentry entering the Church than as a step consistent with his intellectual development. For some years he held a curacy at Okewood Chapel in Surrey, near the home of his parents at Albury, and was active in his pastoral functions from 1792 to 1794. He showed a genuine interest in and concern for the local people and an understanding of their problems, a sympathy which makes surprising his later references to the laboring class almost as though they were a community apart. From 1803 until his death he held a sinecure as rector of Walesbury in Lincolnshire.

In 1799 Malthus and Otter, together with friends from Jesus College, E. D. Clarke and J. M. Cripps, traveled through northern Germany and Norway.

Afterward Malthus and Otter went on to Sweden, Finland, and Russia. Malthus' detailed diaries of these journeys provided some of the evidence he needed to develop his theory of population growth. Clarke also published a record of his travels. In 1800 Malthus' parents died; by 1802 he was traveling again, this time in France and Switzerland, in a party that included his cousin Harriet Eckersall.

Jesus College elected Malthus to a fellowship in 1793, and he was resident intermittently until he had to resign upon his marriage to Harriet Eckersall in 1804. They had one son, Henry, who followed his father into the ministry, and two daughters: Emily, who married, and Lucy, who died before her father. About the time of Malthus' marriage, the East India Company founded a new college, first at Hertford and then at Haileybury, to give a general education to staff members before they went on service overseas. The first known professorship of history and political economy was established there, and Malthus was invited to fill the post. He took it up in 1805, and it gave him the security of a home and an income that enabled him to spend the rest of his life writing and lecturing.

In order to teach political economy, Malthus needed to extend his knowledge. He wrote two pamphlets on the Corn Laws (1814, 1815); a short, unexceptionable tract on rent (1815); statements on Haileybury (1813, 1817); and a major work, *Principles of Political Economy* (1820). This included an analogy of his population theory with the quantity of funds designed for the maintenance of labor and the prudential habits of the laboring classes.

In 1819 the Royal Society elected Malthus to a fellowship. He was also a member of the French Institute and the Berlin Academy, and a founding member of the Statistical Society (1834). In 1827 he was called upon to give evidence on emigration before a committee of the House of Commons.

Although their life was quiet, Robert and Henrietta Malthus traveled and entertained their many friends, including David Ricardo, Harriet Martineau, Otter, and William Empson, who was also at Haileybury. Malthus managed, in spite of the controversy flowing around him, to keep a reputation as a warm, charming, and lively companion.

Principle of Population. Malthus' first writing was an unpublished pamphlet, *The Crisis* (extracts are quoted by Otter). Stimulated by Pitt's Poor Law Bill of 1796, he supported the proposal for children's allowances, but was already expressing unease at the current idea that an increase in population was desirable.

He was not the first to propound the theory that population tends to increase proportionately faster than the supply of food—and he freely acknowledged that he was not—nor was the first edition of his *Essay on the Principle of Population*, published anonymously in 1798, a fully worked-out thesis. He wrote: "I had for some time been aware that population and food increased in different ratios; and a vague opinion had been floating in my mind that they could only be kept equal by some species of misery or vice."

Stimulated by doubts about Pitt's policy and by publications of William Godwin, Condorcet, and others, Malthus hammered out the *Essay* in discussions with his father, who accepted Godwin's belief in the potential immortality and perfectibility of man. Countering apparently rosy visions, Malthus swung to pessimism about the inevitability of poverty and the irresponsibility of the poor, an attitude which his opponents called inhuman. These observations were based at least partially on experience, for he had, as a curate, seen how in the country many births were registered but few deaths, yet, as he said, "sons of labourers are very apt to be stunted in their growth, and are a long while arriving at maturity."

The central argument of the *Essay* lies in two postulates:

"That food is necessary to the existence of man";

"That the passion between the sexes is necessary, and will remain nearly in its present state" [p. 11]; and four conclusions:

> . . . that the power of population is indefinitely greater than the power in the earth to produce subsistence for man.
>
> Population, when unchecked, increases in a geometrical ratio. Subsistence increases only in arithmetical ratio. A slight acquaintance with numbers will shew the immensity of the first power in comparison with the second.
>
> By that law of our nature which makes food necessary to the life of man, the effects of those two unequal powers must be kept equal.
>
> This implies a strong and constantly operating check on population from the difficulty of subsistence [p. 13].

The postulates are taken as self-evident; the deduced consequences are examined in more detail, including the various checks on population, such as postponed marriage, infant mortality, epidemics, and famine. He presented no numerical data to support either the tendency to geometrical rate of growth of the population or the arithmetical rate of growth of food supply; these suppositions are reasonable but largely intuitive. Malthus seems also to have failed to realize that although the existence of checks is a firm deduction, there is no reason to suppose that they operate constantly.

The style of the essay—short paragraphs, pungent sentences, and an elegant but matter-of-fact air—undoubtedly contributed to the impact of the work on a community already deeply concerned with the social problems of the Industrial Revolution. It was also brief—only some 50,000 words—and the edition seems to have been small, since the work is now rare.

Malthus realized that he needed more evidence to support his views and that he had not taken sufficient account of the effects of rising standards of living. He therefore listened to criticisms and used information gathered on his travels in Europe, information which tended to be observational rather than numerical. For example, he correlated the poverty of fishermen in Drontheim with their earlier marriages and larger families—in contrast with the people of the interior parts of the country—without considering other possible variables.

Malthus' next publication, *The Present High Price of Provisions* (1800), again published anonymously, returned to the problems of poor relief. In it he made the case that linking poor relief to the cost of grain resulted in driving the price even higher. He also pointed out that whereas previously grain had been exported, there was no longer enough to go round; and therefore, assuming that agricultural production had increased or at least not declined, the population must have increased. The first census in Great Britain (1801) tended to confirm this assertion.

The second and greatly expanded edition of the *Principle of Population* was published in 1803 and carried the author's name. It provided the theoretical framework to the conclusions of the first *Essay*, with several additional chapters, including information from China and Japan as well as from countries he had visited. The argument was rewritten in terms more academic if less immediate. He explained that "everything depends on the relative proportions between population and food, and not on the absolute number of people," and that when the absolute quantity of provisions is increased, the number of consumers more than keeps up with it. If, therefore, he argued, it is not possible to maintain the production of food to satisfy the population, then the population must be kept down to the level of food; failure will result in deprivation and misery. He then went on to reexamine positive and preventive checks, introducing the new idea of voluntary "natural restraint" by late marriage and sexual abstinence before marriage. He does not seem to have considered abstinence after marriage and was strongly opposed to both abortion and contraception.

Later editions of the *Essay* were rewritten and included new appendixes of evidence, until the sixth edition (1826) required three volumes and contained some 250,000 words. Malthus' last statement on population was his *Summary View of the Principle of Population* (1830), rewritten from an article he had done for the 1824 supplement to the *Encyclopaedia Britannica*. It was condensed again to some 20,000 words, but by now it contained a greater element of social comment. There is not only the observation of tendencies but also reference to the bad structure of society and the unfavorable distribution of wealth. There have been numerous reprints and translations. Malthus has been widely read, but he has also been widely misquoted or quoted out of context. His observations have been interpreted by both his supposed followers and his enemies with overtones which suggest that his work is prescriptive rather than descriptive.

Influence on the Theory of Evolution. Malthus' *Essay* was a crucial contribution to Darwin's thinking about natural selection when he returned in 1836 from the *Beagle* voyage. In July 1837 Darwin began his "Notebook on Transmutation of Species," in which he wrote:

> In October 1838, that is, fifteen months after I had begun my systematic enquiry, I happened to read for amusement "Malthus on Population," and being well prepared to appreciate the struggle for existence . . . it at once struck me that under these circumstances favourable variations would tend to be preserved and unfavourable ones to be destroyed. The result would be the formation of a new species [*Life and Letters*, I, 83].

Later, in the *Origin of Species*, he wrote that the struggle for existence "is the doctrine of Malthus applied with manifold force to the whole animal and vegetable kingdoms; for in this case there can be no artificial increase of food, and no prudential restraint from marriage" [p. 63].

Alfred Russel Wallace, who arrived at a worked-out formulation of the theory of evolution at almost precisely the same time as Darwin, acknowledged that "perhaps the most important book I read was Malthus's *Principles of Population*" (*My Life*, p. 232).

Although there were four decennial censuses before Malthus' death, he did not himself analyze the data, although he did influence Lambert Quetelet and Pierre Verhulst, who made precise statistical studies on growth of populations in developed countries and showed how the early exponential growth changed to an S curve.

Influence on Social Theory. Notwithstanding the anonymity of the first *Essay*, the authorship soon became known. Godwin wrote to Malthus immediately, and the book loosed a storm of controversy that is still rumbling. It has influenced all demographers since, as well as many students of economic theory and genetic inheritance. The early controversy is described concisely by Leslie Stephen and more fully by Bonar and McCleary. Besides Godwin, Ricardo corresponded lengthily and critically but accepted much of his theory, as did Francis Place. Ricardo and Malthus did not meet until 1811 but formed a valuable friendship. Hazlitt, Cobbett, and Coleridge attacked him for real or supposed views.

The current attitude around the end of the eighteenth century, when need for industrial workers was increasing, was that population growth was desirable in itself and that welfare provisions should encourage large families. Malthus' principle, that population tends to increase up to the limits of the means of subsistence, could be extended to suggest that if the level of subsistence were lowered by reducing state welfare provisions, then the population would naturally settle at a lower level and the working classes could avoid checks due to both misery and vice by planning and observing "prudential restraint." Malthus himself believed that the effects of the Poor Laws were harmful, but he never recommended the withdrawal of benefits and believed it to be "the duty of every individual, to use his utmost efforts to remove evil from himself, and from as large a circle as he can influence."

In 1807 Samuel Whitbread, M.P., introduced a bill to reform the Poor Laws, attempting to reduce misery and vice by a series of proposals which included a national system of education, encouragement of saving, and equalization of county taxes from which the welfare benefits were paid. Malthus wrote him an open letter, published as a pamphlet, in which he supported the plan for general education (he made it clear that the poor should be able to understand both the reason for their condition and the means of alleviating it), but he opposed vigorously the building of tenement cottages on the ground that the rents would increase the number of dependent poor except where there was a high demand for labor. If it were possible, the Poor Laws should be restricted to maintaining only the average number of children that might have been expected from each marriage, and he hoped that "the poor would be deterred from early and improvident marriages more by the fear of dependent poverty than by the contemplation of positive distress."

Malthus appears to have ignored the point that any average must have many examples above the average. Visualizing a progressive increase in the proportion of the dependent population under the laws then in effect, he admitted to being "really unable to suggest any provision which would effectually secure us against an approach to the evils here contemplated, and not be open to the objection of violating our promises to the poor." Probably this pamphlet was widely read and was a main source of the image of Malthus as a pessimist and supporter of laissez-faire political economy. He was an analyst, not a creator of imaginative legislation; and the problems he dealt with are still with us in one form or another. He was at least more clear than some politicians about "our promises to the poor." Nothing came of Whitbread's bill in its original form, and Malthus had produced no constructive amendments, so the law remained unchanged until the Poor Law Amendment Act (1834), which abolished relief outside the workhouses that were to be set up under boards of guardians for those qualified and willing to live there.

More cheerfully and positively, in his *Principles of Political Economy* (1820) Malthus was proposing investment in public works and private luxury as a means of increasing effective demand, and hence as a palliative to economic distress. The nation, he thought, must balance the power to produce and the will to consume.

After all the accretions on Malthusian principles, it was perhaps natural that Marx and Engels should have seen Malthus as an advocate of repressive treatment of the working class, rather than appreciating his anticipation of their own belief that the demand for labor regulates population.

However bitter and distorted the controversy has been, Malthus' achievement was to show that population studies, although overlaid with emotional and often irrational influences, can be examined and analyzed empirically, discussed on a rational basis, and ultimately can form the subject of positive policy making.

BIBLIOGRAPHY

I. ORIGINAL WORKS. Malthus' first major work, published anonymously, was *An Essay on the Principle of Population, as It Affects the Future Improvement of Society, With Remarks on the Speculations of Mr. Godwin, M. Condorcet, and Other Writers* (London, 1798). The 2nd ed., *An Essay on the Principle of Population, or a View of Its Past and Present Effects on Human Happiness, With an Enquiry Into Our Prospects Respecting the Future Removal or Mitigation of the Evils Which It Occasions. . . .* (London, 1803), was signed T. R. Malthus. There was a 3rd ed., with appendixes (1806); a 4th ed. (1807), reprinted with additions, 2 vols. (1817); a 5th ed., with appendixes, 3 vols. (1817); a 6th ed., with appendixes (1826); and the 7th and posthumous ed. (1872). There have been numerous other eds. and trans. It is worth mentioning the facs. repr. of the 1st ed., with notes comparing it with the 2nd, by J. Bonar (London, 1926), and a modern repr. of the

7th ed. (New York, 1969). Extracts from the 1798 and 1803 eds. were reprinted in *Parallel Chapters From the First and Second Editions of "An Essay on the Principle of Population,"* D. Ricardo, ed. (New York, 1895). The last statement was *A Summary View of the Principle of Population* (London, 1830). There was also a repr. of the first *Essay* and *Summary View*, edited, with an intro., by A. Flew (London, 1970).

Malthus' other major work is *Principles of Political Economy Considered With a View to Their Practical Applications* (London, 1820); 2nd ed. (London, 1836), with considerable alterations from the author's own MS, also contains an original memoir by Otter that includes extracts from *The Crisis*; there is also a modern repr. (New York, 1964).

Malthus' journal of his travels is P. James, ed., *The Travel Diaries of Thomas Robert Malthus* (Cambridge, 1966).

Malthus' library of 2,300 volumes is in Jesus College, Cambridge. There have clearly been many letters and other MSS available to students of Malthus, but few can be located now. The travel diaries are in Cambridge University Library. In her introduction James refers briefly to other manuscripts, including the unpublished *Recollections* of Malthus' niece, Louisa Bray.

II. SECONDARY LITERATURE. The most comprehensive bibliography is Library of Congress, *List of References on Malthus and Malthusianism* (Washington, D.C., 1920). Later eds. of Malthus' works, translations, and works on him may be traced through the national bibliographies, and particularly through the *General Catalogue of Printed Books in the British Museum*, CLI (1962), cols. 313–314, and supps. There is also an extensive bibliography for 1793–1880 in Glass (see below).

The best source for Malthus' personal life is P. James, "Biographical Sketch," in *The Travel Diaries* (see above), in which she gives full details of all her sources. Otter's "Memoir," added to the 2nd ed. of *Principles of Political Economy*, and W. Empson's review of this ed. in *Edinburgh Review*, **64** (1837), 469–506, contain much personal information. The standard biography is J. Bonar, *Malthus and His Work* (London, 1885; 2nd ed., 1924). There are also C. R. Drysdale, *The Life and Writings of Thomas R. Malthus*, 2nd ed. (London, 1892); and a short biographical sketch by G. T. Bettany in his ed. of *Principle of Population* (London, 1890). L. Stephen's article in the *Dictionary of National Biography*, XXXVI (1893), 886–890, summarizes the early controversy; and there is an evaluation by J. M. Keynes in his *Essays in Biography*, new ed. (London, 1951), 81–124.

The three works which provoked Malthus' *Essay* are W. Godwin, *An Enquiry Concerning Political Justice, and Its Influence on General Virtue and Happiness* (London, 1793); and *The Enquirer: Reflections on Education, Manners and Literature* (Dublin–London, 1797); and Condorcet, ed., *Outlines of an Historical View of the Progress of the Human Mind* (London, 1795), translated from the French.

Works on Malthus' theories, their influence, and their place in theories of population are numerous. D. V. Glass, ed., *Introduction to Malthus* (London, 1953), contains three essays, reprs. of the *Summary View*, and the letter to Whitbread. A general and appreciative account is G. F. McCleary, *The Malthusian Population Theory* (London, 1953). Ricardo's reactions are published in *The Works and Correspondence of David Ricardo*, P. Staffa, ed., II (London, 1951). One of the most vigorous attacks was W. Hazlitt, *The Spirit of the Age* (London, 1825), 251–276. A detailed study of Malthus' influence on social history is D. Eversley, *Social Theories of Fertility and the Malthusian Debate* (Oxford, 1959). There is an account of the relationship of Malthus' and Darwin's theories in M. T. Ghiselin, *The Triumph of the Darwinian Method* (Berkeley–Los Angeles, 1969), 46–77, which gives further references.

DIANA M. SIMPKINS

MALTSEV (or **Malcev**), **ANATOLY IVANOVICH** (*b*. Misheronsky, near Moscow, Russia, 27 November 1909; *d*. Novosibirsk, U.S.S.R., 7 July 1967), *mathematics.*

The son of a glassblower, Maltsev graduated in 1931 from Moscow University and completed his graduate work there under A. N. Kolmogorov. He received his M.S. in 1937 and the D.S. in 1941 and became professor of mathematics in 1944. He was a corresponding member of the Academy of Sciences of the U.S.S.R. from 1953 and was elected a member in 1958.

From 1932 to 1960 Maltsev taught mathematics at the Ivanovo Pedagogical Institute in Moscow, rising from assistant to head of the department of algebra. He worked at the Mathematical Institute of the Academy in Moscow from 1941 to 1960, when he became head of the department of algebra at the Mathematical Institute of the Siberian branch of the Academy in Novosibirsk as well as head of the chair of algebra and mathematical logic at the University of Novosibirsk. He received the State Prize in 1946 for his work in algebra and, in 1964, the Lenin Prize for his work in the application of mathematical logic to algebra and in the theory of algebraic systems. In 1956 he was named Honored Scientist of the Russian Federation and in 1963 was elected president of the Siberian Mathematical Society.

Maltsev's most important work was in algebra and mathematical logic. In his first publication (1936), which dealt with a general method for obtaining local theorems in mathematical logic, he provided such a theorem for the limited calculus of predicates of arbitrary signature. By means of this theorem an arbitrary set of formulas of this calculus is noncontradictory when—and only when—any finite subset of this set is noncontradictory. In this work the theorem of the extension of infinite models was also proved. Both theorems are important in mathematical logic and in the theory of models, the creation of which Maltsev himself was largely responsible for. His local method enabled him to prove (1941) a series of important theorems of the theory of groups and other algebraic systems. In 1956 he generalized his local theorems to cover many classes of models. Ideas similar to those presented in the last of these works led A. Robinson to formulate his nonstandard analysis, in which actual infinitesimally small and great magnitudes obtained an original substantiation.

Maltsev's most important works in algebra dealt with the theory of Lie groups. He proved (1940, 1943) that for a Lie group to have an exact linear representation, linear representability of the radical of this Lie group and the corresponding factor group constitutes a necessary and sufficient condition. In 1941 he proved that Cartan's theorem of the inclusion of an arbitrary local Lie group into a full Lie group cannot be generalized for local general topological groups. In 1944 he described all semisimple subgroups of simple Lie groups of infinite classes and exceptional classes *G* and *F*, and proved the conjugateness of semisimple factors in Levi's decomposition of Lie groups and algebras.

The following year Maltsev defined the rational submodulus of Lie algebra, characterizing the Lie group by the finite-leaved covering; and he discovered the criteria for a subgroup of a Lie group, corresponding to a given subalgebra of Lie algebra, to be closed. He also proved that maximal compact subgroups of a connected Lie group are conjugate (Cartan's problem) and that a Lie group is homeomorphic to a direct product of such a subgroup by Euclidean space. In 1948 he obtained important results in the theory of nilpotent manifolds, i.e. homogeneous manifolds the fundamental groups of which are nilpotent Lie groups. In 1951 he proved the so-called Maltsev-Kolchin theorem of solvable linear groups and studied properties of solvable groups of integer matrices and new classes of solvable groups. In 1955 he constructed an alternative analogue of Lie groups and a corresponding analogue of Lie algebras that are now called Maltsev algebras. In 1957 he constructed the general theory of free topological algebras as being a generalization of topological groups.

In the last ten years of his life Maltsev obtained important results in the theory of algebraic systems and models and in the synthesis of algebra and mathematical logic, which he described in a series of papers and in the posthumous *Algebraicheskie sistemy* ("Algebraic Systems," 1970). His results in the theory of algorithms are presented in the monograph *Algoritmy i rekursivnye funktsii* ("Algorithms and Recursive Functions," 1965). Maltsev was the author of an important textbook of algebra, *Osnovy lineynoy algebry* ("Foundations of Linear Algebra," 1948), founded the journal *Algebra i logika. Seminar*, and was editor-in-chief of *Sibirskii matematicheskii zhurnal*.

BIBLIOGRAPHY

A bibliography of 96 works follows the obituary of Maltsev by P. S. Aleksandrov *et al.*, in *Uspekhi matematicheskikh nauk*, **23**, no. 3 (1968), 159–170. Works referred to above are "Untersuchungen aus dem Gebiete der mathematischen Logik," in *Matematicheskii sbornik*, **1** (1936), 323–326; "Ob izomorfnom predstavlenii beskonechnykh grupp matritsami" ("On the Isomorphic Representation of Infinite Groups by Means of Matrices"), *ibid.*, **8** (1940), 405–422; "Ob odnom obshchem metode polucheniya lokalnykh teorem teorii grupp" ("On a General Method for Obtaining Local Theorems of the Theory of Groups"), in *Uchenye zapiski Ivanovskogo pedagogicheskogo instituta*, Fiz.-mat. fak., **1**, no. 1 (1941), 3–9; "O lokalnykh i polnykh topologicheskikh gruppakh" ("On Local and Full Topological Groups"), in *Doklady Akademii nauk SSSR*, **32**, no. 9 (1941), 606–608; "O lineyno svyaznykh lokalno zamknutykh gruppakh" ("On Linearly Connected Locally Closed Groups"), *ibid.*, **41**, no. 8 (1943), 108–110; "O poluprostykh podgruppakh grupp Li" ("On Semisimple Subgroups of Lie Groups"), in *Izvestiya Akademii nauk SSSR*, Ser. mat., **8** (1944), 143–174; "On the Theory of the Lie Groups in the Large," in *Matematicheskii sbornik*, **16** (1945), 163–190; **19** (1946), 523–524; *Osnovy lineynoy algebry* ("Foundations of Linear Algebra"; Moscow-Leningrad, 1948; 2nd ed., Moscow, 1956; 3rd ed., Moscow, 1970); "Ob odnom klasse odnorodnykh prostranstv" ("On One Class of Homogenous Spaces"), in *Izvestiya Akademii nauk SSSR*, Ser. mat., **13** (1949), 9–32; "O nekotorykh klassakh beskonechnykh razreshimykh grupp" ("On Certain Classes of Infinite Solvable Groups"), in *Matematicheskii sbornik*, **28** (1951), 567–588; "Analiticheskie lupy" ("Analytical Loops"), *ibid.*, **36** (1955), 569–576; "O predstavleniyakh modeley" ("On Representations of Models"), in *Doklady Akademii nauk SSSR*, **108**, no. 1 (1956), 27–29; "Svobodnye topologicheskie algebry" ("Free Topological Algebras"), in *Izvestiya Akademii nauk SSSR*, Ser. mat., **21** (1957), 171–198; "Modelnye sootvetstviya" ("Model Correspondences"), *ibid.*, **23** (1959), 313–336; "Regulyarnye proizvedenia modeley" ("Regular Products of Models"), *ibid.*, 489–502; "Konstruktivnye algebry" ("Constructive Algebras"), in *Uspekhi matematicheskikh nauk*, **16**, no. 3 (1961), 3–60; *Algoritmy i rekursivnye funktsii* ("Algorithms and Recursive Functions"; Moscow, 1965); and *Algebraicheskie sistemy* ("Algebraic Systems"; Moscow, 1970).

B. A. Rosenfeld

MALUS, ÉTIENNE LOUIS (*b.* Paris, France, 23 July 1775; *d.* Paris, 24 February 1812), *optics*.

The son of Anne-Louis Malus du Mitry and Louise-Charlotte Desboves, Malus was privately educated, mainly in Greek, Latin, and mathematics. He revealed his mathematical skill in 1793 at the entrance examination to the military school in Mézières. His father's position as treasurer of France compromised the family during the Revolution; so Malus served as a simple soldier until 1794, when he was sent to the École Polytechnique. He became sublieutenant of engineers on 20 February 1796 and captain of engineers on 19 June 1796, and he took part in Napoleon's expedition to Egypt and Syria (1798–1801). Malus survived an infection and landed in Marseilles on 14 October 1801. In 1802–1803 he was at Lille; he was subdirector for the fortifications of Anvers (1804–1806) and Strasbourg (1806–1808). In 1808 Malus was called to Paris, where he became major of engineers on 5 December 1810.

Malus was among the first students to enter the École Polytechnique, where he received his basic scientific education (1794–1796). Interested primarily in optics, he composed a memoir in Cairo stating that the constituent principle of light was a particular combination of caloric and oxygen. In September 1802 the Société des Sciences, de l'Agriculture et des Arts de Lille began regular meetings; Malus became vice-president on 11 February 1803 and president the following January. From 1805 he was examiner in geometry and analysis at the École Polytechnique and, from 1806, in physics as well. This post gave Malus the opportunity of long stays in Paris and contacts with other physicists. On 20 April 1807 he presented his first memoir, "Traité d'optique," to the first class of the Institute. He received the mathematical prize of the Institute on 2 January 1810. The greatest event in Malus's career was undoubtedly his election to membership of the first class of the Institute (18 August 1810). Malus was also a member of the Institut d'Égypte (22 August 1798), the Société d'Arcueil (1809), and the Société Philomatique (April 1810). On 22 March 1811 Thomas Young informed Malus that the Royal Society of London had awarded him the Rumford Medal. His last memoir was read to the Institute on 19 August 1811.

At Giessen, just before he was ordered to Egypt, he planned to marry Wilhelmine-Louise Koch, the eldest daughter of the university chancellor, but they were not married until after his return. She died on 18 August 1813. Malus's influential friends included Monge, whom he first met as director of the École Polytechnique; Berthollet, who also was with Napoleon on the Egyptian expedition; and Laplace, who at the beginning of the nineteenth century was particularly interested in optics.

In "Traité d'optique" Malus considered mathematically the properties of a system of contiguous rays of light in three dimensions. He found the equation of the caustic surfaces, and the Malus theorem: Light rays emanating from a point source, after being reflected or refracted from a surface, are all normal to a common surface, but after a second reflection or refraction they will no longer have this property. If the perpendicular surface is identified with a wavefront, it is obvious that this result is false, which Malus did not realize because he adhered to the Newtonian emission theory of light, and the Malus theorem was not proved in its full generality until W. R. Hamilton (1824) and Quetelet and Gergonne (1825). The line of thought and the results of the "Traité d'optique" were continued and generalized by Hamilton in his "Theory of Systems of Rays" (1827).

Double refraction had first been observed in Iceland spar by Erasmus Bartholin. The laws governing it were found by Huygens from the assumption that the wavelets of the extraordinary rays were ellipsoids of revolution with major axes parallel to the axis of the crystal. If one crystal of Iceland spar is placed over another in such a way that the principal sections of the crystals are parallel, then the ordinary rays produced in the upper crystal undergo ordinary refraction only in the lower crystal, while the extraordinary rays undergo only an extraordinary refraction. If the principal sections are perpendicular to each other, the ordinary rays undergo an extraordinary refraction and vice versa. Huygens could not account for these observations, and Newton used them to refute Huygens' wave theory. Newton considered light as particles, and the above-mentioned polarization phenomenon indicated to him that these particles had sides. In Query 25 of the *Optics* he announced his own (false) rule for double refraction, which was adopted for the next century. In 1788 Haüy found experimentally that Huygens' law was true only in certain special cases, but in 1802 Wollaston found experimental evidence for the Huygenian construction. In "Mémoire sur la mesure du pouvoir réfringent" Malus showed that Wollaston's experiments were incomplete, and so the French corpuscularian physicists did not trust Wollaston's results. They thought, moreover, that Wollaston was associated with Thomas Young and therefore with the wave hypothesis.

In this situation the Institute on 4 January 1808 proposed a prize which required an experimental and theoretical explanation of double refraction. The French "Newtonian" scientists hoped that Malus would find a precise and general law for double

refraction within the framework of an emission theory of light. Malus was a skilled mathematician and during 1807 he had carried out experiments on double refraction. By December 1808 Malus had finished his experimental investigations, which verified the Huygenian law. What remained was a theoretical deduction of the law. In January 1809, Laplace published a memoir in which he deduced Huygens' law within the framework of Newtonian mechanics, using the principle of least action, and Malus considered this an insolence which deprived him of the priority. In 1810 Malus won the prize for his "Théorie de la double réfraction," published in 1811. Here he deduced the law following the same method as Laplace, by means of the principle of least action. Malus won the prize therefore mainly because of his original experimental researches and his discussion of the short-range forces that produce double refraction. Also of great importance was his law for the relative intensities of the ordinary and extraordinary rays.

While working on double refraction Malus discovered that a ray of sunlight reflected at a certain angle from a transparent medium behaves in exactly the same manner as if it had been ordinarily refracted by a double refracting medium. He found that each medium had a characteristic angle of reflection for which this happened, 52°45' for water and 35°25' for glass. Malus did not postpone publishing his discovery until the end of the competition, but announced it to the Institute on 12 December 1808. He also showed that if the two rays emanating from a crystal are reflected from a water surface at an angle less than 52°45' and if the principal section of the crystal is parallel to the plane of reflection then the ordinary ray is totally refracted; and if the principal section is perpendicular to the plane of reflection the extraordinary ray is totally refracted. He concluded that these phenomena could be accounted for only by supposing that light consisted of particles which were lined up by reflection and refraction and remained mutually parallel afterward. In "Mémoire sur de nouveaux phénomènes d'optique" Malus said that light particles have sides or poles and used for the first time the word "polarization" to characterize the phenomenon.

All transparent and opaque bodies polarize light more or less by reflection, and for each medium a characteristic angle of reflection will totally polarize the reflected ray in the plane of reflection. The refracted rays will contain light that is polarized perpendicular to the reflected light and light that is not polarized. Malus carried out numerous experiments to determine characteristic angles and relative intensities. If the intensity of the incident, polarized ray is unity, then the intensity I of ray reflected at the characteristic angle will be $I = \cos^2 \alpha$, where α is the angle between the planes of polarization of the incident and the reflected rays. He also found the relative intensities of reflected and ordinarily and extraordinarily refracted light. For instance Malus found that if two double refracting crystals are placed one above the other with parallel refracting surfaces, the relative intensities I_{oo}, I_{oe}, I_{eo}, and I_{ee} of the rays subject to ordinary-ordinary, ordinary-extraordinary, extraordinary-ordinary, and extraordinary-extraordinary refraction will be

$$I_{oo} = I_{ee} = \cos^2 i$$
$$I_{oe} = I_{eo} = \sin^2 i,$$

where i is the angle between the two principal sections. He also found that the ordinary and extraordinary rays are polarized perpendicularly to each other.

All material bodies will, to a certain extent, polarize rays of light by reflection. At first Malus thought that this was not true of metallic surfaces, but he later found that rays reflected from such a surface contain two kinds of mutual perpendicularly polarized light together with light not polarized. By reflection at a certain angle, later called the principal angle of incidence, all the reflected light was circularly polarized (Malus did not use this term). The theory of metallic reflection was developed by Brewster, MacCullagh, and Cauchy.

Malus did not indicate whether his results were found experimentally or theoretically. After his death his researches on polarization were followed up by Arago and Biot in France and Brewster in England. In the wave theory of light polarization was explained from the assumption of the transversality of light waves. This was proposed both by Fresnel and Young (1816), but it was not until 1821 that Fresnel succeeded in laying a mechanical foundation for the theory of transverse waves in an elastic medium.

BIBLIOGRAPHY

I. ORIGINAL WORKS. A bibliography of sixteen memoirs by Malus is in Royal Society *Catalogue of Scientific Papers*, IV, 210–211. Those cited in the text are "Mémoire sur la mesure du pouvoir réfringent des corps opaques," in *Nouveau bulletin des sciences de la Société philomatique de Paris*, **1** (1807), 77–81; "Mémoire sur de nouveaux phénomènes d'optique," *ibid.*, **2** (1811), 291–295; "Traité d'optique," in *Mémoires présentés à l'Institut des sciences par divers savants*, **2** (1811), 214–302; and "Théorie de la double réfraction de la lumière dans les substances cristallines," *ibid.*, 303–508. Malus recorded his activities on scientific expeditions to various parts of Egypt in his diary, published posthumously as *L'agenda de Malus. Souvenirs de l'expédition d'Égypte 1798–1801* (Paris, 1892). Collec-

tions of MSS are at the Bibliothèque de l'Institut and the Archives de l'Académie des Sciences in Paris.

II. SECONDARY LITERATURE. On his life, see the biographies by Jean-Baptiste Biot, in Michaud, ed., *Biographie universelle*, XXVI (Paris, 1820), 410 ff.; and by François Arago in his *Oeuvres*, III (Paris, 1855), 113–155. Much valuable information can be found in M. Crosland, *The Society of Arcueil* (London, 1967). See also Anatole de Norguet, "Malus, fondateur de la Société des sciences de Lille," in *Mémoires de la Société des sciences, de l'agriculture et des arts de Lille*, 3rd ser., **10** (1872), 225–232. The history of the Malus theorem is presented in *The Mathematical Papers of Sir William Rowan Hamilton*, I (Cambridge, 1931), 463 ff. Laplace's paper on double refraction, "Mémoire sur les mouvements de la lumière dans les milieux diaphanes," is in *Oeuvres complètes de Laplace*, XII (Paris, 1898), 267–298.

KURT MØLLER PEDERSEN

MANARDO, GIOVANNI (*b.* Ferrara, Italy, 24 July 1462; *d.* Ferrara, 7 March 1536), *medicine, botany.*

Manardo belonged to a distinguished Ferrarese family; his father, Francesco Manardo, was a notary —as were many of his other relatives—while his great-uncle, Antonio Manardo, was an apothecary. Manardo studied at the University of Ferrara, where his teachers included Battista Guarini, Niccolò Leoniceno, and Francesco Benzi, son of the physician Ugo Benzi. He received his doctorate in arts and medicine on 17 October 1482; that same year he was appointed lecturer at the university. Although Manardo remained at Ferrara for the next ten years, his academic promotion, or a career at court, may have been obviated by his unwillingness to accept the prevalent theoretical and astrological basis assigned to medicine.

From 1493 to 1502 Manardo and his wife, Samaritana da Monte, lived in Mirandola, where he served as tutor and physician to Giovanni Francesco Pico and also assisted him in editing the works of his famous uncle, Pico della Mirandola. In spite of the French invasion of Italy, and under the influence of Pico della Mirandola's theories, Manardo began to concentrate more heavily on separating medicine from astrology, while recognizing astronomy as a discrete science.

It was probably during these years also that Manardo's studies led to his scientific travels in Italy and the brief lectureships attributed to him at Perugia, Padua, and Pavia. In 1507 and 1509 he returned to Ferrara, and again in 1512 when his son Timoteo was included among his pupils. In 1513 Manardo went to Hungary, where, through the influence of Celio Calcagnini and on the recommendation of Cardinal Ippolito d'Este, he was appointed royal physician to Ladislaus Jagellon and to his successor, Louis II. From Hungary Manardo was able to journey to Croatia, Austria, and Poland. Subsequently his son (who had on 31 January 1514 qualified in medicine at Ferrara, under Leoniceno's sponsorship) joined Manardo. The younger man wrote an account of his travels, *Odoiporicon Germanicum et Pannonicum*, which was praised by Calcagnini but is now lost.

In 1518 both Manardo and his son returned to Ferrara; there is no further record of Timoteo Manardo's career, although he may have become a monk. Manardo himself succeeded Leoniceno as professor of medicine at the university in 1524 and also became personal physician to Alfonso I d'Este, duke of Ferrara. He remained in Ferrara for the rest of his life; when he was seventy-three, he married Giulia dei Sassoli da Bergamo, a widow with two children, who survived him, with their daughter, Marietta.

Manardo brought to his science new methods of interpretation, analysis, and classification. He had learned a Galenic, anti-Arabic medicine from Leoniceno; to this he added an empirical, intuitive methodology, firmly based on clinical observation, and a broad knowledge of Greek, Latin, Arabic, and biblical sources. He was thus able to resolve some of the linguistic confusion that surrounded his disciplines, and to devise a consistent nomenclature for his work in both pathology and botany.

In medicine, Manardo divided diseases into groups, according to their natures and cures. In dermatology he distinguished among psoriasis, filariasis, scabies, and syphilis, which he established as a specifically venereal entity. (Leoniceno and others had attributed the spread of syphilis to climatic conditions—humidity, rain, and flood—operating under the influence of the planets.) As an alternative to the widely used and dangerous mercuric treatment of the disease, Manardo proposed the West Indian remedy, *Guaiacum sanctum*, dissolved in wine. In ophthalmology he made a distinction between cataract and glaucoma; he further recognized the relationship between systemic—or internal—health and vision.

Manardo's major medical work, *Epistolae medicinales*, is divided into twenty books which consist of 103 letters based on case histories, professional discussions, and personal observations. Among the epistles is one on external diseases, addressed to the Ferrarese surgeon Santanna, which was followed by a long letter on internal diseases, written in the last years of Manardo's life at the request of A. M. Canano (Manardo lived to complete the discussion of phthisis). In sum, the letters represent a development of Ugo Benzi's *Consilia;* in substance they anticipate the scientific dissertations of the seventeenth century.

As a botanist, Manardo drew upon observations made in the course of his travels to distinguish among the properties of the variants that occur within a single species growing in differing locations. These variations are of practical importance in both pharmacy and dietetics; Manardo made further mention of them in his commentary on the *Simplicia et composita* (sometimes called the *Grabadin*) attributed to Johannes Mesue the Younger and in his criticism of V. M. Adriani's translation of Dioscorides.

Manardo enjoyed considerable contemporary fame. He attended the last illnesses of Cardinal Ippolito d'Este, his early patron, and of Ariosto, who had praised him by name in *Orlando Furioso* (canto XLVI, stanza 14). His pupil L. G. Giraldi wrote a moving poem on the occasion of his departure for Hungary, and G. P. Valeriano dedicated to him book twenty-four of his fifty-eight-book *Geroglifici*. Erasmus owned Manardo's works, S. Champier corresponded with him, and Rabelais (himself a physician) wrote a preface to the 1532 edition of the *Epistolae*. Manardo's writings were plagiarized by Leonhard Fuchs and cited by Vesalius.

BIBLIOGRAPHY

I. Original Works. Manardo's *Epistolae medicinales* were published in a number of editions: books I–VI (Ferrara, 1521; Paris, 1528; Strasbourg, 1529); books VII–XII, with a preface by Rabelais (Lyons, 1532); books I–XVIII (Basel, 1535); and the complete 20-book work (eight eds., Basel, 1540–Hannover, 1611). Manardo also published a partial translation and commentary on Galen's *Ars medicinalis* (Rome, 1525; Basel, 1529, 1536, 1540, 1541; Padua, 1553, 1564), and a commentary on the *Simplicia et composita*, attributed to Johannes Mesue the Younger (Venice, 1558, 1561, 1581, 1589, 1623).

II. Secondary Literature. There is no comprehensive biobibliographical work on Manardo, but three recent publications provide essential references to earlier works about him. See Árpád Herczeg, "Johannes Manardus Hofarzt in Ungarn und Ferrara im Zeitalter der Renaissance," *Janus*, **33** (1929), 52–78, 85–130, with portraits, separately published in Hungarian as *Manardus János, 1462–1536, magyar udvari föorvos élete és müvei* (Budapest, 1929); *Atti del Convegno internazionale per le celebrazione della nascità di Giovanni Manardo, 1462–1536* (Ferrara, 1963); and L. Münster, "Ferrara e Bologna sotto i rapporti delle loro scuole medico—naturalistiche nell' epoca umanistica—rinascimentale," in *Rivista di storia della medicina* (1966), 11–12, 17–18, assesses the value of Manardo's contribution.

Juliana Hill Cotton

MANASSEH. See Māshā'llāh.

MANDELSHTAM, LEONID ISAAKOVICH (*b.* Mogilev, Russia, 5 May 1879; *d.* Moscow, U.S.S.R., 27 November 1944), *physics.*

Mandelshtam's father, Isaak Grigorievich Mandelshtam, was a physician widely known in southern Russia. His mother, Minna Lvovna Kahn, was her husband's second cousin, knew several foreign languages, and was an outstanding pianist. Mandelshtam's uncles, the biologist A. G. Gurvich and the distinguished specialist in petroleum chemistry L. G. Gurvich, greatly influenced his upbringing. Soon after his birth the family moved to Odessa, where Mandelshtam passed his childhood and youth. After graduating from the Gymnasium with honors in 1897, he entered the mathematical section of the Faculty of Physics and Mathematics of Novorossysk University in Odessa; two years later he was expelled for having participated in antigovernment student riots.

He continued his education at the Faculty of Physics and Mathematics of Strasbourg University, where Ferdinand Braun soon attracted him to scientific research in his own laboratory—primarily questions of electromagnetic vibration and their application to radiotelegraphy. In 1902 Mandelshtam defended his dissertation for the doctorate of natural philosophy at Strasbourg University, with highest distinction. He then became Braun's extra-staff personal assistant and, in 1903, his second staff assistant at the Strasbourg Physical Institute. Mandelshtam's friendship and collaboration with the distinguished Russian radiophysicist Nikolay Dmitrievich Papaleksi, which continued until his death, began at this time.

In 1907 Mandelshtam married Lidya Solomonovna Isakovich, the first Russian woman architect. In 1914, just before the beginning of World War I, Mandelshtam returned to Russia with his family and Papaleksi. After working as privatdocent in physics at Novorossysk University, in 1915 he became scientific consultant at the Petrograd radiotelegraph factory and, in 1917, professor of physics at the Polytechnical Institute in Tiflis. From 1918 to 1921 he was scientific consultant of the Central Radio Laboratory in Moscow and later in Petrograd. In 1925 he was appointed professor of theoretical physics at Moscow State University. He settled in Moscow, where he began his long collaboration with the prominent Soviet physicist G. S. Landsberg. In 1928 Mandelshtam was elected a corresponding member of the Academy of Sciences of the U.S.S.R. and in 1929 an active member. From the fall of 1934 Mandelshtam took an active part in the work of the P. N. Lebedev Physical Institute of the Academy of Sciences in Moscow, in addition to his work at the university.

Because of serious heart disease Mandelshtam was evacuated during World War II to Borovoye in Kazakhstan, where he continued his theoretical work. After the war he returned to Moscow and spent the rest of his life at Moscow State University and the Lebedev Physical Institute.

Mandelshtam's scientific research, which embraced extremely varied areas of physics and its practical applications, centered fundamentally on optics and radiophysics. His accomplishments in optics include the discovery of the phenomenon of combination scattering, the study of the effect of the fluctuation scattering of light in a uniform medium, and the theory of the microscope. In an early work (1907) Mandelshtam was the first to show that the scattering of light observed in a uniform medium was caused not by the presence of movement among the molecules, as Rayleigh had asserted, but by the occurrence of irregularities connected, according to Smoluchowski's idea, with the fluctuations of density caused by random heat motion.

In 1918 Mandelshtam proposed the idea that the Rayleigh lines must reveal a fine structure, caused by the scattering of light on adiabatic fluctuations. His work on this question did not appear until 1926, after the publication of an analogous idea of Brillouin. The Mandelshtam-Brillouin effect was first experimentally demonstrated by Mandelshtam and Landsberg in 1930 in crystals and by E. F. Gross in liquids. During this research Mandelshtam and Landsberg discovered in 1928 an essentially new effect in crystals, combination scattering, which consists of a regular variation in the frequency of light scattering. An analogous effect was discovered at the same time in liquids by the Indian physicists C. V. Raman and K. S. Krishnan. A preliminary communication of the discovery of the Indian scientists appeared in print a few months before the communication of Mandelshtam and Landsberg. The effect, known in the Soviet Union as combination scattering, is elsewhere called the Raman effect. The study of these phenomena led Mandelshtam to the discovery and investigation of light scattering in fluctuations originating on the surface of a liquid.

In radiophysics and its applications Mandelshtam's research with Papaleksi on nonlinear vibrations and the creation of radiogeodesy is of especially great significance. Begun in 1918, their research on nonlinear vibrations led to results obtained in the 1930's in the formulation of so-called conditions of discontinuity, which are the basis of "explosive" vibrations and led to the development of the theory of multivibrators. The subsequent discovery of the conditions of appearance of n-type resonance made it possible to stimulate vibrations in a circuit, the frequencies of which are precisely *n* times lower than the frequencies of external electromotive forces. Associated with this work is the research on vibrations in linear systems with parameters changing through time. In 1931 Mandelshtam and Papaleksi constructed the first alternating-current parametrical generator with periodically changing inductivity. One of their most distinguished achievements was the radio-interference method of precise measurement invented in 1938 (radiogeodesy), which was also the most precise method of measuring the velocity of propagation of radio waves. In conjunction with studies in optics and radiophysics, Mandelshtam also conducted theoretical research on the basic problems of quantum mechanics. An outstanding lecturer who loved teaching, Mandelshtam taught a large school of physicists, including a number of distinguished scientists (I. E. Tamm, M. A. Leontovich, A. A. Andronov, S. E. Khaikin, among others).

BIBLIOGRAPHY

Mandelshtam's complete collected works were published as *Polnoe sobranie trudov*, S. M. Rytov and M. A. Leontovich, eds., 5 vols. (Leningrad, 1947–1955). A biographical sketch by N. D. Papaleksi is in I, 7–66. See also N. D. Papaleksi, "Kratky ocherk zhizni i nauchnoy deyatelnosti Leonida Isaakovicha Mandelshtama," in *Uspekhi fizicheskikh nauk*, **27**, no. 2 (1945), 143–158, a short sketch of Mandelshtam's life and scientific work.

J. DORFMAN

MANFREDI, EUSTACHIO (*b.* Bologna, Italy, 20 September 1674; *d.* Bologna, 15 February 1739), *astronomy, hydraulics.*

A well-known poet as well as scientist, Manfredi was the eldest son of Alfonso Manfredi, a notary originally from Lugo (near Ravenna), and Anna Maria Fiorini. Encouraged by his father to study philosophy while attending Jesuit schools, he took a degree in law in 1692 but never practiced it. Having shown an early preference for science, he studied mathematics and hydraulics with Domenico Guglielmini and began to study astronomy by himself. By 1690 he had founded his own scientific academy, the Inquieti, a private institution that in 1714 became the Academy of Sciences of the Institute of Bologna. In 1699 Manfredi became lecturer in mathematics at the University of Bologna; but obliged by family financial difficulties to accept two positions, in 1704 he became head of a pontifical college in Bologna and then

superintendent of waters for the region, a post he retained until his death. He was relieved from the first post in 1711 by his appointment as astronomer of the recently founded Institute of Sciences.

In 1715 Manfredi completed his two-volume *Ephemerides motuum coelestium* for 1715–1725, based on the still unpublished tables of Cassini in Paris, his predecessor in the chair of astronomy at Bologna. Intended, unlike most of its predecessors, not for astrological use but for practical astronomy, the ephemerides were of unusual extent and practicality. They included tables of the meridian crossing of the planets, tables of the eclipses of the satellites of Jupiter and of the conjunction of the moon and the principal stars, as well as maps of the regions of the earth affected by solar eclipses. The ephemerides were preceded by a volume of instructions including tables that were reprinted by Eustachio Zanotti in 1750. In 1725 Manfredi published a similar, highly successful work for the period 1726–1750 that in some ways anticipated the *Nautical Almanac* (1766).

Soon after his appointment as astronomer, Manfredi calculated the latitude and longitude of the new observatory at Bologna by following the polar star with two mobile quadrants and an eight-foot wall semicircle; the three series of observations confirmed his results obtained in other parts of the city. With a team of assistants that included Francesco Algarotti he carefully measured the annual motion of several fixed stars chosen at various ecliptical latitudes, in order to confirm and identify precisely their apparently elliptical orbits. Although he recognized that the phenomenon could not be a parallactic effect—a conclusion he had apparently reached in 1719—he did not publish on it until 1729, the year in which Bradley gave the exact explanation: the first astronomical evidence of the earth's revolution and a confirmation that the value of the velocity of light, although extremely great, is finite. Manfredi regarded these explanations as insufficiently tested hypotheses and remained, like Cassini and certain other contemporary astronomers, a lifelong adherent of the geocentric and geostatic conception of the world. The phenomenon is still known as the annual aberration of fixed stars, the name Manfredi gave it in the title of *De annuis inerrantium stellarum aberrationibus* (1729).

In 1736 Manfredi published *De gnomone meridiano Bononiensi ad Divi Petronii* [*templum*], for which he had been collecting material since his youth. The work also included a history and description of Cassini's meridian and observations made on the solar "species" since the instrument had been introduced in 1655. These observations are of meteorological as well as astronomical interest. The following year Manfredi published *Astronomicae ac geographicae observationes selectae* by Francesco Bianchini of Verona, after patiently organizing and completing his notes.

Most of Manfredi's many publications appeared in Latin in the proceedings of the Academy of Sciences of Bologna and in French in the *Mémoires* of the Académie des Sciences, in which he published his observations and descriptions of solar and lunar eclipses, comets, transits of Mercury, and an aurora borealis. Other works appeared posthumously in Italian and considerably updated older treatises in Latin, which was always less used.

Manfredi had ordered for his observatory the latest astronomical instruments from England, but they did not arrive until two years after his death from kidney and bladder stones. Manfredi was a foreign member of the Académie des Sciences (1726) and a member of the Royal Society of London (1729). Recognition of his mastery of the Italian language was expressed by his membership in the Accademia della Crusca of Florence (1706), an honor then reserved almost entirely for Tuscans.

BIBLIOGRAPHY

I. ORIGINAL WORKS. Manfredi's scientific works include *Ephemerides motuum coelestium*, 2 vols. (Bologna, 1715); *Novissimae ephemerides motuum coelestium*, 2 vols. (Bologna, 1725); *De annuis inerrantium stellarum aberrationibus* (Bologna, 1729); *De gnomone meridiano Bononiensi ad Divi Petronii* (Bologna, 1736); and Francesco Bianchini, *Astronomicae ac geographicae observationes selectae* (Verona, 1737), which Manfredi edited.

Posthumously published works are Domenico Guglielmini, *Della natura de' fiumi* (Bologna, 1739), with Manfredi's annotations; *Elementi della cronologia* (Bologna, 1744); *Istituzioni astronomiche* (Bologna, 1749); and *Elementi della geometria* (Bologna, 1755).

II. SECONDARY LITERATURE. On Manfredi and his work, see Henri Bédaride, "Eustachio Manfredi," in *Études italiennes 1928–1929* (Paris, 1930), 75–124; Paolo Dore, "Origine e funzione dell'Istituto e della Accademia delle scienze di Bologna," in *L'archiginnasio*, **35** (1940), 201, 206; Angelo Fabroni, *Vitae Italorum*, V (Pisa, 1779), 140–225; Fontenelle, "Éloge de M. Manfredi," in *Histoires de l'Académie royale des sciences* for 1739 (Amsterdam, 1743), 80–99; Guido Horn D'Arturo, "Chi fu il primo a parlare di aberrazione?" in *Coelum*, **3** (1933), 279; and *Piccola enciclopedia astronomica*, II (Bologna, 1960), 294–295; D. Provenzal, *I riformatori della bella letteratura italiana* (Rocca S. Casciano, 1900), p. 251–311; P. Riccardi, *Biblioteca matematica italiana*, II (Modena, 1873–1876), cols. 79–88; F. M. Zanotti, *Elogio del dottor Eustachio Manfredi* (Verona, 1739); G. P. Zanotti, *Vita di Eustachio Manfredi* (Bologna, 1745), with portrait.

GIORGIO TABARRONI

MANGIN, LOUIS ALEXANDRE (*b.* Paris, France, 8 September 1852; *d.* Grignon, France, 27 January 1937), *botany*.

Mangin was professor of cryptogamy at the Muséum National d'Histoire Naturelle in Paris from 1906 to 1932; he served as director of that institution concurrently from 1920. He was a member of the Académie des Sciences, its vice-president in 1928, and its president in 1929. His researches were not limited to the cryptogams, although he did write on micromycetes, species of *Penicillium*, and the phylogeny of *Atichiales*, as well as studying the composition and seasonal and geographical variations of the phytoplankton collected on the Antarctic expedition of the *Pourquoi-Pas?* (directed by J.-B. Charcot) and on the North Sea expedition of the *Scotia*. He also did significant work in plant anatomy and physiology and phytopathology, and made important contributions to plant histology.

Mangin's work in plant anatomy included a study of the vascular system of the monocotyledons. He also established that adventitious roots arise from a special meristem (which he called "souche dictogène") which is seated in the pericycle; he thought, however, that only the central core and the bark of the adventitious root are formed by this apical meristem, the root cap being constituted from the internal layers of the bark.

In plant physiology Mangin observed the waxy cuticle and determined thereby the importance and function in respiration of the stomata. In collaboration with Gaston Bonnier he published a series of papers on their joint researches into plant respiration within various experimental environments. Mangin and Bonnier also devised an apparatus—consisting of a gas bubble imprisoned between two columns of mercury and subjected successively to the actions of caustic potash and pyrogallic acid—for the purpose of rapidly analyzing the atmosphere surrounding plants. They were particularly concerned with the ratio that existed between the oxygen absorbed and the carbon dioxide discharged by each species (which they found to be constant).

Mangin may be considered one of the founders of phytopathology; he furthered its study as the guiding spirit of the Société de Pathologie Végétale et d'Entomologie Agricole de France. He published studies on mycorrhiza of fruit trees (1889); on wheat foot-rot (which he showed to be a consequence of the association of the grain with several species of fungus, including *Ophiobolus graminis* and *Leptosphaeria herpotrichoides*); and, with P. Viala, on vine diseases, especially "phtyriose," which they found to be due to a cochineal insect associated with a polypore, *Bornetia corium.* He also did research on root rot in chestnut trees and needle-shedding disease in firs.

As a histologist, Mangin pioneered in the use of color reactives for microscopic investigation. In 1890 to 1910 he employed a whole series of azoic dyes in the work on the composition of plant membranes whereby he established the characteristics of cellulose and pectin materials and showed that the young membranes of vascular plants always contain these compounds. In 1890 he reported his discovery of callose to the Academy, and went on to describe its microchemical properties and to define its diverse forms according to the condensation of the molecule. He showed callose to exist in all membranes and in such special calcified formations as cystoliths; demonstrated gums and mucilages to be the end products of the jellification of cellular membranes; and recorded important observations on the constitution of the membranes of pollen grains. In his final investigations on the subject Mangin ascertained the essentially variable constitution of the cellular membranes in mushrooms—cellulose and callose in Peronosporaceae, cellulose and pectin compounds in Mucoraceae, and various combinations of callose in other groups.

BIBLIOGRAPHY

I. ORIGINAL WORKS. Mangin's report on callose is "Sur la callose, nouvelle substance fondamentale existant dans la membrane," in *Comptes rendus hebdomadaires de l'Académie des sciences*, **110** (1890), 644–647. His collaborative work with Bonnier is summarized in "La fonction respiratoire chez les végétaux," in *Annales des sciences naturelles (botanique)*, 7th series, **2** (1885), 365–380, which draws upon the papers that appeared in *Annales*, 6th series, **17** (1884); **18** (1884), 293–381; and **19** (1885), 217–255.

Mangin was the author of several textbooks, including *Botanique élémentaire* (1883); *Éléments de botanique* (1884); *Cours élémentaire de botanique* (1885); and *Anatomie et physiologie végétale* (1895).

II. SECONDARY LITERATURE. Mangin's work with Bonnier is discussed in M. H. Jumelle, "L'oeuvre scientifique de Gaston Bonnier," in *Revue générale de botanique*, **36** (1924), 289–307.

MAURICE HOCQUETTE

MANILIUS, MARCUS (*fl.* Rome, beginning of the first century A.D.), *astrology.*

Manilius' life is a mystery to us. Even his name is variously presented in the somewhat restricted manuscript tradition of his one surviving work as Manlius or Mallius, to which is often added Boenius or Boevius; this probably reflects some confusion with the philosopher Anicius Manlius Boethius. The fact is that Manilius is known to us only through an incomplete Latin poem on astrology, the *Astronomicôn libri V*.[1] The composition of this poem began while Augustus was still reigning, and book I was written later than A.D. 9; but it has been much debated whether the work as we have it was completed before Augustus' death in A.D. 14[2] or only under his successor, Tiberius.[3] In either case, Manilius intended to write more than the five books preserved in our manuscripts. Not only does he promise to expound the nature of the planets in book II and fail to accomplish this before the end of book V, but the poem as it stands is not adequate for its purpose—the instruction of students in the science of astrology. In fact, its astrological content, while important because of its antiquity (Manilius' is our oldest connected treatise on astrology), is quite rudimentary.

Roughly, the scheme of the *Astronomica* is as follows. Book I treats the sphere, zodiacal and other constellations, great circles, and comets; book II, the zodiacal signs, their classifications, interrelations, and subdivisions, and the *dodecatopus;* book III, the twelve astrological places (here called *athla*), the Lot of Fortune, the rising times of the signs at Alexandria,[4] the lord of the year, and the length of life; book IV, the decans, the *monomoria*, and an astrological geography; and book V, the fixed stars that rise simultaneously with points on the ecliptic. The possessor of only this poem could not hope to cast or to read a horoscope; he would have several thousand Latin hexameters, some of which are very fine, and a curious congeries of strange doctrines, many of which are found in no other extant text in either Greek or Latin.

The sources of Manilius' doctrines are not often evident. Housman has cited those that are in his edition, and also a large number of parallel passages. An attempt at a survey of the sources in book I was made by R. Blum.[5] The evidence which points to his use of Hermetic astrological writings is strong,[6] and the relation of book V to Germanicus' version of Aratus' *Phaenomena* has been studied by H. Wempe.[7] But the fragmentary state of our knowledge of the early stages of the development of astrology in Hellenistic Egypt makes it impossible to pursue the search for Manilius' sources much further. It is even more difficult (though not because of a lack of texts) to discern any influence exercised by Manilius over later astrologers. If he was read at all in antiquity, it was not by the profession.

NOTES

1. But see P. Thielscher, "Ist 'M. Manilii Astronomicon libri V' richtig?" in *Hermes*, **84** (1956), 353–372.

2. See E. Flores, "Augusto nella visione astrologica di Manilio ed il problema della cronologia degli Astronomicon libri," in *Annali della Facoltà di Lettere e Filosofia della Università di Napoli*, **9** (1960–1961), 5–66.
3. See E. Gebhardt, "Zur Datierungsfrage des Manilius," in *Rheinisches Museum für Philologie*, **104** (1961), 278–286.
4. See O. Neugebauer, "On Some Astronomical Papyri and Related Problems of Ancient Geography," in *Transactions of the American Philosophical Society*, n. s. **32** (1942), 251–263.
5. *Manilius' Quelle im ersten Buche der Astronomica* (Berlin, 1934).
6. See G. Villauri, "Gli *Astronomica* di Manilio e le fonti ermetiche," in *Rivista di Filologia e di Istruzione Classica*, n.s. **32** (1954), 133–167; and M. Valvo, "Considerazioni su Manilio e l'ermetismo," in *Siculorum Gymnasium*, **9** (1956), 108–117.
7. "Die literarischen Beziehungen und das chronologischen Verhältnis zwischen Germanicus und Manilius," in *Rheinisches Museum für Philologie*, **84** (1935), 89–96.

BIBLIOGRAPHY

The standard ed. of Manilius is that by A. E. Housman, 2nd ed., 5 vols. (Cambridge, 1937). Numerous articles on the text tradition and certain difficult passages have appeared since 1937; among the most impressive of these are by G. P. Goold, "De fonte codicum Manilianorum," in *Rheinisches Museum für Philologie*, **97** (1954), 359–372, and "Adversaria Maniliana," in *Phoenix*, **13** (Toronto, 1959), 93–112, from whom a new critical ed. is expected. See also E. Flores, *Contributi di filologia maniliana* (Naples, 1966).

DAVID PINGREE

MANNHEIM, VICTOR MAYER AMÉDÉE (*b.* Paris, France, 17 July 1831; *d.* Paris, 11 December 1906), *geometry.*

A follower of the geometric tradition of Poncelet and Chasles, Amédée Mannheim, like his predecessors, spent most of his professional career associated with the École Polytechnique, which he entered in 1848. In 1850 he went to the École d'Application at Metz. While still a student he invented a type of slide rule, a modified version of which is still in use. After graduation as a lieutenant, he spent several years at various provincial garrisons. In 1859 he was appointed *répétiteur* at the École Polytechnique; in 1863, examiner; and in 1864, professor of descriptive geometry. He attained the rank of colonel in the engineering corps, retiring from the army in 1890 and from his teaching post in 1901. He was a dedicated and popular teacher, strongly devoted to the École Polytechnique, and was one of the founders of the Société Amicale des Anciens Élèves de l'École.

Mannheim worked in many branches of geometry. His primary interest was in projective geometry, and he was influenced by Chasles's work on the polar reciprocal transformation, which he further investigated with respect to metric properties. He applied these studies in his work in kinematic geometry, which he defined as the study of motion, independent of force, time, and any elements outside the moving figure. He also made significant contributions to the theory of surfaces, primarily in regard to Fresnel's wave surfaces. Most of his results can be found in his texts, *Cours de géométrie descriptive de l'École Polytechnique* (1880) and *Principes et développements de la géométrie cinématique* (1894), which, although he was an enthusiast for the synthetic method in geometry, contained much differential geometry, as well as a good summary of that subject. In recognition of his contributions to the field of geometry Mannheim was awarded the Poncelet Prize in 1872.

BIBLIOGRAPHY

I. ORIGINAL WORKS. Mannheim's early works on the polar reciprocal transformation include his *Théorie des polaires réciproques* (Metz, 1851); and *Transformation de propriétés métriques des figures à l'aide de la théorie des polaires réciproques* (Paris, 1857). His work in kinematic geometry is found primarily in *Cours de géométrie descriptive de l'École Polytechnique comprenant les éléments de la géométrie cinématique* (Paris, 1880; 2nd ed. 1886); and *Principes et développements de la géométrie cinématique; ouvrage contenant de nombreuses applications à la théorie des surfaces* (Paris, 1894). A complete list of his works is in Poggendorff, III, 865–866; IV, 952; and V, 801; and in the article by Loria cited below. For a list of his important papers in the theory of surfaces, see G. Loria, *Il passato ed il presente delle principali teorie geometriche*, 2nd ed. (Turin, 1896), 115.

II. SECONDARY LITERATURE. For an account of Mannheim's life, see C. A. Laisant, "La vie et les travaux d'Amédée Mannheim," in *L'enseignement mathématique*, **9** (1907), 169–179. A much fuller account of his work is G. Loria, "L'opera geometrica di A. Mannheim," in *Rendiconti de Circolo matematico di Palermo*, **26** (1908), 1–63, and "A. Mannheim—Soldier and Mathematician," in *Scripta Mathematica*, **2** (1934), 337–342. Mannheim's works on the wave surface is considered in C. Niven, "On M. Mannheim's Researches on the Wave Surface," in *Quarterly Journal of Pure and Applied Mathematics*, **15** (1878), 242–257.

ELAINE KOPPELMAN

MANSION, PAUL (*b.* Marchin, near Huy, Belgium, 3 June 1844; *d.* Ghent, Belgium, 16 April 1919), *mathematics, history and philosophy of science.*

Mansion was a professor at the University of Ghent, member of the Royal Academy of Belgium, and director of the Journal *Mathesis*. He entered the École Normale des Sciences at Ghent in 1862; and by the age of twenty-three he was teaching advanced courses. He held an eminent position in the scientific world of Belgium despite his extreme narrow-mindedness. In 1874 he founded, with Eugène-Charles Catalan and J. Neuberg, the *Nouvelle correspondance mathématique*; this title was chosen in memory of the *Correspondance mathématique et physique*, edited by Garnier and Adolphe Quetelet. Through the efforts of Mansion and Neuberg, who were encouraged by Catalan himself, the *Nouvelle correspondance* was succeeded in 1881 by *Mathesis*. Mansion retired in 1910.

Alphonse Demoulin's notice on Mansion (1929) includes a bibliography of 349 items, some of which were published in important foreign compendia. Several others appeared in German translation. Mansion's own French translations of works by Riemann, Julius Plücker, Clebsch, Dante, and even Cardinal Manning attest to the extent of his interests. Among other subjects, he taught the history of mathematics and of the physical sciences, in which field he wrote in particular on Greek astronomy, Copernicus, Galileo, and Kepler. His desire to justify the positions of Catholic orthodoxy is evident.

BIBLIOGRAPHY

A bibliography of Mansion's works is in the notice by A. Demoulin, in *Annuaire de l'Académie royale de Belgique*, **95** (1929), 77–147. On Mansion's life and work see L. Godeaux, in *Biographie nationale publiée par l'Académie royale de Belgique*, XXX (Brussels, 1959), 540–542; and in *Florilège des sciences en Belgique pendant le 19ᵉ siècle et le début du 20ᵉ siècle* (Brussels, 1968), 129–132.

J. Pelseneer

MANSON, PATRICK (*b*. Old Meldrum, Aberdeen, Scotland, 3 October 1844; *d*. London, England, 9 April 1922), *tropical medicine*.

Manson was born at Cromlet Hill, Old Meldrum, in Aberdeenshire, the second son of a family of five sons and four daughters. His father, John Manson, was a bank manager and a local laird. His mother, who exercised a profound influence upon him up to the time of her death at the age of eighty-eight, was Elizabeth Livingstone, a distant cousin of David Livingstone the explorer, and a member of a well-known local family named Blaikie.

In his youth Manson was considered rather dull by his teacher, who complained that he spent too much of his time shooting partridges and rabbits and too little on classical education. At the age of eleven he shot a savage cat and "extracted from its innards a long tapeworm," his first practical exercise in parasitology.

In 1857 the family moved to Aberdeen, where Patrick attended the Gymnasium and the West End Academy. His mother's family owned a large engineering works in the city, to which, at the age of fourteen, Manson became apprenticed. He undertook the heavy work so enthusiastically that he developed curvature of the spine and a partial paresis of the right arm, which for the next six months forced him to spend most of each day lying on his back. Nevertheless, for two hours daily he contrived to study natural history at Marischal College, which so stimulated his interest in science that, upon learning that his work would count as part of the medical curriculum, he decided not to return to engineering but to devote his life to medicine.

He became a student at the Aberdeen Medical School in 1860, graduated M.B., C.M., in 1865, M.D. in 1866, and as his first appointment became assistant medical officer at the Durham County Mental Asylum. He remained in Durham for only a year. Persuaded by his elder brother, then working in Shanghai, to travel overseas, he obtained the post of medical officer for Formosa in the Chinese Imperial Maritime Customs, where his official duties were to inspect ships calling at the port of Takao (now Kaohsiung) and to treat their crews. Of this work he kept a careful diary now preserved at Manson House in London, in which he made detailed descriptions of cases of elephantiasis, leprosy, and "heart disease"— which he later recognized as beriberi. At the end of 1870 he unwittingly became involved in the political struggle between China and Japan, to escape from which, on the advice of the British consul, he left Formosa early in 1871 and settled at Amoy on the mainland of China.

His private practice there and his post at the Baptist Missionary Hospital provided him with an immense number of cases that added greatly to his experience. His special interest in elephantiasis led him to devise surgical procedures for removing the masses of tissue associated with the disease, of which, it is recorded, he removed over a ton in a period of three years.

At the end of 1874 Manson returned to Great Britain on a year's leave, during which he was married to Henrietta Isabella Thurburn on 21 December 1875. He spent much of his leave searching the libraries for literature on elephantiasis, in the course of which, on 25 March 1875, while working in the British Museum, he came across the work of Timothy Lewis, a surgeon in the Indian Army Medical Service, on *Filaria sanguinis hominis* (F.S.H.). Lewis was convinced that F.S.H. was the immature form of a much larger adult worm, which he eventually discovered in 1877, some nine months after its discovery by Thomas Lane Bancroft in Australia. He also believed that the microfilariae or the adult worms were the causative agents of disease, although nothing was known of the method of transmission. Manson pondered long on these discoveries of Lewis, and from them he formulated his theory of mosquito transmission. So great was his interest that upon his return to Amoy he devoted all his spare time to investigating the correctness of his theory.

To this end he enlisted the assistance of two medical students to examine blood for the presence of F.S.H. One of the students could work only at night, and Manson noticed that a significantly higher proportion of positives was obtained by this "night observer" than by the day worker. From this observation he stumbled upon the hitherto unsuspected phenomenon of microfilarial periodicity. By training two men whose blood contained microfilariae to examine each other every three hours for six weeks, he was able conclusively to demonstrate that microfilariae were present in the blood in larger numbers at night than during the day. (The resulting graphs of microfilarial numbers are now preserved in the London School of Hygiene and Tropical Medicine.) He also demonstrated that the microfilariae were surrounded by a sheath, from which they could escape when the blood was cooled in ice. This observation led him to postulate that F.S.H. was an embryo worm that could continue its development outside the human body in the common brown mosquito of Amoy (now identified as *Culex fatigans*). He then persuaded his gardener, who was infected with microfilariae, to allow large numbers of mosquitoes to feed upon him; and by dissecting the fed mosquitoes he was able to trace the development of the worm through the intestine and into the thoracic musculature. In his publications of 1877 and 1878 Manson referred to the mosquitoes as "nurses."

His work, however, was greatly hampered by the absence of literature on the life cycle of mosquitoes. He mistakenly believed that mosquitoes took only a single meal of blood, with the result that most of his mosquitoes died within five days. He also erroneously believed that man became infected by ingesting the larvae in water, into which they were released when the mosquitoes laid their eggs.

Nevertheless, these early experiments provided the first proof of the obligatory involvement of an arthropod vector in the life cycle of a parasite. The almost unending list of parasites now known to require an arthropod as a necessary alternate host is testimony of the fundamental nature of Manson's concept. Yet, when his work on periodicity and on development in the mosquito was presented to the Linnean Society of London by the president, Spencer Cobbold, on 7 March 1878, the only recognition that it received was incredulity and ridicule.

Manson's part in the elucidation of the transmission of malaria by mosquitoes took place some sixteen years later. In 1890, under financial pressure caused by depreciation of the Chinese dollar, he set up practice in London and was appointed physician to the Seamen's Hospital, where he had access to many cases of tropical disease. He carried out prolonged observations on the "exflagellation" of malaria and, in a paper published in 1894, postulated that the process was a normal part of the life cycle of the parasite in the stomach of the mosquito. In the same year he met Ronald Ross, with whom, after showing him the malaria parasite, he spent long hours discussing the mosquito-malaria theory. Largely as a result of pressure on the India Office brought to bear by Manson, Ross was dispatched to India the following year to investigate the theory. Manson's advice was to "follow the flagellum," and Ross soon succeeded in observing exflagellation in the stomach of the mosquito. But the problem of following the parasite into the tissues of the mosquito, of which only one species is suitable for development, proved to be a Herculean task. Throughout the months of investigation that followed, Manson maintained a continuous correspondence with Ross, much of which has now been published. In August 1897 Ross dissected a new type of mosquito (*Anopheles*) that had fed on a malaria patient, and in it he found pigmented round bodies on the stomach wall. The pigmented bodies were sent to Manson, who confirmed their significance. Soon afterward Ross was removed to an area where human malaria was absent, and there he applied himself to the study of *Proteosoma*, a malaria parasite of sparrows. From this study he was able in 1898 to describe its complete life cycle in the mosquito. The discovery was announced by Manson at a meeting of the British Medical Association in Edinburgh. Ross fully acknowledged the part played by Manson; but Manson, with characteristic modesty, disclaimed any credit save that of having "discovered" Ross. Meanwhile, in Italy, Grassi in 1898 transmitted human malaria by the bite of a mosquito and in 1901 described the complete life cycle of the parasite.

Manson was responsible not only for the concept of the mosquito-transmission theory but also for bringing the findings of Ross and Grassi to the attention of the public and thus for spreading the knowledge that eventually led to the practical control of malaria. In 1900, from a consignment of infected mosquitoes sent by Grassi to London, Manson succeeded in infecting his son and a laboratory technician by mosquito bite. In the same year Manson sent two of his pupils, Low and Sambon, to live for three months in a highly malarious area of Italy. By the simple expedient of spending each night in a mosquito-proof hut, they remained healthy and uninfected, while their neighbors lay sick and dying of the disease. Manson was involved in the whole field of tropical medicine, as well as in the discovery of many other pathogenic parasites and in the elucidation of their life cycles, including *Paragonimus westermanni*, *Sparganum mansoni*, *Schistosoma mansoni* and *S. japonicum*, *Oxyspirura mansoni*, *Loa loa*, and *Filaria perstans*.

Manson was a man of deep penetrative mind, original in thought, creative in imagination, careful and patient in experimentation. He was possessed of an overwhelming desire to communicate his knowledge to others, and so stimulating was his enthusiasm that wherever he went he invariably gathered about him a group of eager students. He played a significant part in the foundation of the College of Medicine at Hong Kong (later to form the basis of the University of Hong Kong) and became the first dean. In a letter to *Lancet* published in 1897 he stressed the need for special training of doctors destined for work in the tropics. Despite the disapproval of many of his colleagues, his recommendations led to the foundation of the London School of Tropical Medicine in 1899, some six months after the opening of the world's first tropical school in Liverpool. On many subsequent occasions Manson pleaded for funds for "our tropical schools," which became the models for similar institutions throughout the world. *Tropical Diseases. A Manual of the Diseases of Warm Climates* (1898), his principal work, is now in its seventeenth edition and has become a classic textbook of tropical medicine. He taught continually at the London School of Tropical Medicine and its nearby hospital until his retirement in 1914. He was elected a fellow of the Royal Society in 1900, received a knighthood in 1903, and was medical adviser to the Colonial Office for nearly twenty years. In 1907 he was one of the founders of the Royal Society of Tropical Medicine and was elected its first president. After his retirement he spent most of his time fishing in Ireland, interspersed with visits to Ceylon, Rhodesia, and South Africa. The last of his frequent visits to the London School took place only two weeks before his death at the age of seventy-seven. He is buried in Aberdeen.

BIBLIOGRAPHY

A series of articles in commemoration of Manson's life and work, including a short autobiography, appeared in *Journal of Tropical Medicine and Hygiene*, **25** (1922), 155–206; see also the bibliography of his writings compiled by S. Honeyman, *ibid.*, 206–208. See also the obituaries in *Lancet* (1922), 767–769; *Proceedings of the Royal Society*, **94** (1922), xliii–xlviii; and *Transactions of the Royal Society of Tropical Medicine and Hygiene*, **16** (1922), 1–15; and the article by J. W. W. Stephens in *Dictionary of National Biography 1922–1930* (London, 1937), 560–562.

Full-length studies of Manson's life and work are Philip Manson-Bahr, *Patrick Manson, the Father of Tropical Medicine* (London–New York, 1962); P. Manson-Bahr and A. Alcock, *The Life and Work of Sir Patrick Manson* (London, 1927); and Ronald Ross, *Memories of Sir Patrick Manson* (London, 1930).

M. J. CLARKSON

MANṢŪR IBN ʿALĪ IBN ʿIRĀQ, ABŪ NAṢR (*fl.* Khwarizm [now Kara-Kalpakskaya, A.S.S.R.]; *d.* Ghazna [?] [now Ghazni, Afghanistan], *ca.* 1036), *mathematics, astronomy.*

Abū Naṣr was probably a native of Gīlān (Persia); it is likely that he belonged to the family of Banū ʿIrāq who ruled Khwarizm until it fell to the Maʾmūnī dynasty in A.D. 995. He was a disciple of Abuʾl Wafāʾ al-Būzjānī and the teacher of al-Bīrūnī. Abū Naṣr passed most of his life in the court of the monarchs ʿAlī ibn Maʾmūn and Abuʾl-ʿAbbās Maʾmūn, who extended their patronage to a number of scientists, including al-Bīrūnī and Ibn Sīnā. About 1016, the year in which Abuʾl-ʿAbbās Maʾmūn died, both Abū Naṣr and al-Bīrūnī left Khwarizm and went to the court of Sultan Maḥmūd al-Ghaznawī in Ghazna, where Abū Naṣr spent the rest of his life.

Abū Naṣr's fame is due in large part to his collaboration with al-Bīrūnī. Although this collaboration is generally considered to have begun in about 1008, the year in which al-Bīrūnī returned to Khwarizm from the court of Jurjān (now Kunya-Urgench, Turkmen S.S.R.), there is ample evidence for an earlier date. For example, in his *Al-Āthār al-bāqiya* ("Chronology"), finished in the year 1000, al-Bīrūnī refers to Abū Naṣr as *Ustādhī*—"my master," while Abū Naṣr dedicated his book on the azimuth, written sometime before 998, to his pupil.

This collaboration also presents grave difficulties in assigning the authorship of specific works. A case in point is some twelve works that al-Bīrūnī lists as being written "in my name" *(bismī)*, a phrase that has led scholars to consider them to be of his own composition. Nallino has, however, pointed out that *bismī* might also mean "addressed to me" or "dedicated to me"—by Abū Naṣr—and there is considerable evidence in support of this interpretation. For instance, the phrase is used in this sense in both medieval texts (the *Mafātīḥ al-ʿulūm* of Muḥammad ibn Aḥmad al-Khwārizmī of 977) and modern ones of which there is no doubt of the authorship. The incipits and explicits of the works in question make it clear, moreover, that they were written by Abū Naṣr in response to al-Bīrūnī's request for solutions to specific problems that had arisen in the course of his more general researches. Indeed, in some of al-Bīrūnī's own books he mentioned Abū Naṣr by name and stated that his book incorporates the results of some investigations that the older man carried out at his request. Al-Bīrūnī gave Abū Naṣr full credit for his discoveries—as, indeed, he gave full credit to each of his several collaborators, including Abū Sahl al-Masīḥī, a certain Abū ʿAlī al-Ḥasan ibn al-Jīlī (otherwise unidentified) and Ibn Sīnā, who wrote answers to philosophical questions submitted to him by al-Bīrūnī.

The extent of the collaboration between Abū Naṣr and al-Bīrūnī may be demonstrated by the latter's work on the determination of the obliquity of the ecliptic. Al-Bīrūnī carried out observations in Khwarizm in 997, and in Ghazna in 1016, 1019, and 1020. Employing the classical method of measuring the meridian height of the sun at the time of the solstices, he computed the angle of inclination as 23°35′. On the other hand, however, al-Bīrūnī became acquainted with a work by Muḥammad ibn al-Ṣabbāḥ, in which the latter described a method for determining the position, ortive amplitude, and maximum declination of the sun. Since al-Bīrūnī's copy was full of apparent errors, he gave it to Abū Naṣr and asked him to correct it and to prepare a critical report of Ibn al-Ṣabbāḥ's techniques.

Abū Naṣr thus came to write his *Risāla fi 'l-barāhīn ʿalā ʿamal Muḥammad ibn al-Ṣabbāḥ* ("A Treatise on the Demonstration of the Construction Devised by Muḥammad Ibn al-Ṣabbāḥ"), in which he took up Ibn al-Ṣabbāḥ's method in detail and demonstrated that it must be in error to the extent that it depended on the hypothesis of the uniform movement of the sun on the ecliptic. According to Ibn al-Ṣabbāḥ, the ortive amplitude of the sun at solstice (a_t) may be obtained by making three observations of the solar ortive amplitude (a_1, a_2, a_3) at thirty-day intervals within a single season of the year. He thus reached the formula:

$$2 \sin a_t = \frac{2 \sin a_2 \sqrt{(2 \sin a_2)^2 - 2 \sin a_1\, 2 \sin a_3}}{\sqrt{(2 \sin a_2)^2 - (\sin a + \sin a_3)^2}} .$$

The same result may also be obtained from only two observations (a_1, a_2) if the distance (d) covered by the sun on the ecliptic over the period between the two observations is known:

$$2 \sin a_t = \frac{R \sqrt{\dfrac{R^2(\sin a_1 + \sin a_2)^2}{\cos^2 \dfrac{d}{2}} - 4 \sin a_1 \sin a_2}}{\sin \dfrac{d}{2}} .$$

The value of a_t is thus extractable in two ways, and the value of the maximum declination can then be discovered by applying the formula of al-Battānī and Ḥabash:

$$\sin \text{ort. ampl.} = \frac{\sin \partial x R}{\cos \varphi} .$$

Al-Bīrūnī then took up Abū Naṣr's clarification of Ibn al-Ṣabbāḥ's work, citing it in his own *Al-Qānūn al-Masʿūdī* and *Taḥdīd*. He remained, however, primarily interested in obtaining the angle of inclination, and simplified Ibn al-Ṣabbāḥ's methods to that end. He thus, within the two formulas, substituted three and two, respectively, observations of the declination of the sun for the three and two observations of solar ortive amplitude. By this method he obtained values for the angle of inclination of 23°25′19″ and 23°24′16″, respectively. These values are clearly at odds with that then commonly held (23°35′) and confirmed by al-Bīrūnī's own observations. Al-Bīrūnī then returned to Abū Naṣr's work, and explained the discrepancy as being due to Ibn al-Ṣabbāḥ's supposition of the uniform motion of the sun on the ecliptic, as well as to the continuous use of sines and square roots.

Abū Naṣr's contributions to trigonometry are more direct. He is one of the three authors (the others being Abu'l Wafāʾ and Abū Maḥmūd al-Khujandī) to whom al-Ṭūsī attributed the discovery of the sine law whereby in a spherical triangle the sines of the sides are in relationship to the sines of the opposite angles as

$$\frac{\sin a}{\sin A} = \frac{\sin b}{\sin B} = \frac{\sin c}{\sin C} ,$$

or, in a plane triangle, the sides are in relationship to the sines of the opposite angles as

$$\frac{a}{\sin A} = \frac{b}{\sin B} = \frac{c}{\sin C} .$$

The question of which of these three mathematicians was actually the first to discover this law remains unresolved, however. Luckey has convincingly argued against al-Khujandī, pointing out that he was essentially a practical astronomer, unconcerned with theoretical problems. Both Abū Naṣr and Abu'l Wafā', on the other hand, claimed discovery of the law, and while it is impossible to determine who has the better right, two considerations would seem to corroborate Abū Naṣr's contention. First, he employed the law a number of times throughout his astronomical and geometrical writings; whether or not it was his own finding, he nevertheless dealt with it as a significant novelty. Second, Abū Naṣr treated the demonstration of this law in two of his most important works, the *Al-Majisṭī al-Shāhī* ("Almagest of the Shah") and the *Kitāb fi 'l-sumūt* ("Book of the Azimuth"), as well as in two lesser ones, *Risāla fī maʿrifat al-qisiyy al-falakiyya* ("Treatise on the Determination of Spherical Arcs") and *Risāla fi 'l-jawāb ʿan masāʾil handasiyya suʾila ʿanhā* ("Treatise in Which Some Geometrical Questions Addressed to Him are Answered").

The *Al-Majisṭī al-Shāhī* and the *Kitāb fi 'l-sumūt* have both been lost. It is known that the latter was written at the request of al-Bīrūnī, as well as dedicated to him, and that it was concerned with various procedures for calculating the direction of the *qibla*. Abū Naṣr's other significant work, the most complete Arabic version of the *Spherics* of Menelaus, is, however, still extant (although the original Greek text is lost). Of the twenty-two works that are known to have been written by Abū Naṣr, a total of seventeen remain, of which sixteen have been published.

In addition to the books cited above, the remainder of Abū Naṣr's work consisted of short monographs on specific problems of geometry or astronomy. These lesser writings include *Risāla fī ḥall shubha ʿaraḍat fi 'l-thālitha ʿashar min Kitāb al-Uṣūl* ("Treatise in Which a Difficulty in the Thirteenth Book of the *Elements* is Solved"); *Maqāla fī iṣlāḥ shakl min kitāb Mānālāwus fi 'l-kuriyyāt ʿadala fīhi muṣalliḥū hādha 'l-kitāb* ("On the Correction of a Proposition in the *Spherics* of Menelaus, in Which the Emendators of This Book Have Erred"); *Risāla fī ṣanʿat al-asṭurlāb bi 'l-ṭarīq al-ṣināʿī* ("Treatise on the Construction of the Astrolabe in the Artisan's Manner"); *Risāla fi 'l-asṭurlāb al-sarṭānī al-muŷannaḥ fī ḥaqīqatihi bi 'l-ṭarīq al-sināʿi* ("Treatise on the True Winged Crab Astrolabe, According to the Artisan's Method"); and *Faṣl min kitāb fī kuriyyat al-samāʾ* ("A Chapter From a Book on the Sphericity of the Heavens").

BIBLIOGRAPHY

I. ORIGINAL WORKS. Abū Naṣr's version of the *Spherics* of Menelaus exists in an excellent critical edition, with German trans., by Max Krause, "Die Sphärik von Menelaos aus Alexandrien in der Verbesserung von Abū Naṣr Manṣur ibn ʿAlī ibn ʿIrāq. Mit Untersuchungen zur Geschichte des Textes bei den islamischen Mathematikern," in *Abhandlungen der K. Gesellschaft der Wissenschaften zu Göttingen*, Phil.-hist. Kl., no. 17 (Berlin, 1936). Most of the rest of his extant work has been badly edited as *Rasāʾil Abī Naṣr Manṣūr ilā 'l-Bīrūnī. Dāʾirat al-Maʿārif al-ʿUthmāniyya* (Hyderabad, 1948); six of the same treatises are trans. into Spanish in Julio Samsó, *Estudios sobre Abū Naṣr Manṣūr* (Barcelona, 1969).

II. SECONDARY LITERATURE. On Abū Naṣr and his work, see D. J. Boilot, "L'oeuvre d'al-Beruni: essai bibliographique," in *Mélanges de l'Institut dominicain d'études orientales*, **2** (1955), 161–256; "Bibliographie d'al-Beruni. Corrigenda et addenda," *ibid.*, **3** (1956), 391–396; E. S. Kennedy and H. Sharkas, "Two Medieval Methods for Determining the Obliquity of the Ecliptic," in *Mathematical Teacher*, **55** (1962), 286–290; Julio Samsó, *Estudios sobre Abū Naṣr Manṣūr b. ʿAlī b. ʿIrāq* (Barcelona, 1969); "Contribución a un análisis de la terminología matemático-astronómica de Abū Naṣr Manṣūr b. ʿAlī b. ʿIrāq," in *Pensamiento*, **25** (1969), 235–248; Paul Luckey, "Zur Entstehung der Kugeldreiecksrechnung," in *Deutsche Mathematik*, **5** (1940–1941), 405–446; Muḥammad Shafī, "Abū Naṣr ibn ʿIrāq aur us kā sanah wafāt" ("Abū Naṣr ibn ʿIrāq and the Date of his Death"), in Urdu with English summary, in *60 doğum münasebetyle Zeki Velidi Togan'a armağan. Symbolae in honorem Z. V. Togan* (Istanbul, 1954–1955), 484–492; Heinrich Suter, "Zur Trigonometrie der Araber," in *Bibliotheca Mathematica*, 3rd ser., X (1910), 156–160; and K. Vogel and Max Krause, "Die Sphärik von Menelaus aus Alexandrien in der Verbesserung von Abū Naṣr b. ʿAlī ibn ʿIrāq," in *Gnomon*, **15** (1939), 343–395.

JULIO SAMSÓ

MANTEGAZZA, PAOLO (*b.* Monza, Italy, 31 December 1831; *d.* San Terenzo di Lerici, Italy, 17 August 1910), *medicine, anthropology.*

Mantegazza was born into a rich family who gave him a liberal and sophisticated education. His mother, to whom he dedicated one of his books, was Laura Solari, herself notably well-educated, and famous for her ardent patriotism. Under her inspiration Mantegazza took part in the "Cinque Giornate" of 1848, in which the Milanese were able, after furious street fighting, to repel the Austrian occupying forces. Since he was sixteen, Mantegazza was allowed to serve only as a courier in the insurrection, but it nevertheless marked his baptism of fire.

Mantegazza attended the universities of Pisa and Pavia, graduating from the latter in 1853 with honors in medicine. He began scientific experimentation while he was still a student, and when he was nineteen presented to the Istituto Lombardo Accademia di Scienze e Lettere a memoir on spontaneous generation—a still somewhat controversial topic—that aroused considerable interest.

After his graduation Mantegazza traveled extensively in Europe (he knew seven languages), then moved to South America. In 1856 he established a medical practice in Salto, Argentina, where he was engaged in founding an agricultural colony and where he married an Argentinian. He shortly thereafter abandoned the colonization project and returned with his wife to Italy; in 1858 he became an assistant at the Ospedale Maggiore in Milan. The following year, in spite of his numerous professional commitments, he requested that he be allowed to take part in a competition for the unsalaried post of honorary assistant in the same institution. In 1860 Mantegazza was appointed to the chair of general pathology at the University of Pavia, where he subsequently established the first laboratory of experimental pathology in Europe. Ten years later, in 1870, he went to Florence to fill the first Italian chair of anthropology. Here he built up an important museum of anthropology and ethnology and founded the journal *Archivio per l'antropologia e la etnografia*. Following the death of his first wife, Mantegazza married Maria Fantoni, the daughter of a Florentine aristocrat.

Mantegazza published a great number of books, both popular and scientific. Among the latter, those that record his researches on the physiology of reproduction and on what are today called opotherapy and endocrinology are particularly important. Taking up Spallanzani's work of the preceding century, Mantegazza conducted a series of experiments in which he subjected frog sperm to low temperatures to determine its viability. From the data that he compiled he was able to conclude that it should be possible to preserve sperm by this method; and he went on to speculate on the feasibility of artificial insemination, writing that it might be a practice applicable to man. He also made experiments designed to demonstrate that tuberculosis is contagious, and was the first to show that bacteria reproduce by means of spores. He conducted researches on transplanting amphibian testicles, and did work on animal organ transplants in general.

The abundance and variety of Mantegazza's works written for the layman quite overshadowed his scientific works, however. He was an active popularizer, at a time when science was considered to be the preserve of the initiated few. His works on hygiene are particularly significant; he courageously dealt with a number of then-proscribed topics, including sex education. Indeed, there was almost no medical or social problem to which he did not devote a book, pamphlet, or lecture; his books were highly successful, and a few have had modern editions. (He also wrote several novels in the lachrymose and romantic style popular at the time.)

Mantegazza lived to be nearly eighty. He was much honored for his scientific achievements. He was a member of a number of scientific academies and institutes, and was awarded decorations by his own and foreign governments.

BIBLIOGRAPHY

I. Original Works. Among Mantegazza's original works are *Della vitalità dei zoospermi della rana e del trapiantamento dei testicoli da un animale all' altro* (Milan, 1860); *Della temperatura delle orine in diverse ore del giorno e in diversi climi* (Milan, 1862); *Sugli innesti animali e sulla organizzazione artificiale della fibrina* (Milan, 1864); *Sulla congestione: ricerche di patologia sperimentale* (Milan, 1864); *Degli innesti animali e della produzione artificiale delle cellule*: ricerche sperimentali (Milan, 1865); *Delle alterazioni istologiche prodotte dal taglio dei nervi* (Milan, 1867). Two works on other scientists are *Maurizio Bufalini: biografia* (Turin, 1863); and *Carlo Darwin e il suo ultimo libro* (Milan, 1868). His later scientific books include *Fisiologia dell' amore* (Milan, 1873); *Fisiologia del dolore* (Milan, 1880); and *Fisiologia della donna* (Milan, 1893).

II. Secondary Literature. Works about Mantegazza include F. Bazzi, "Paolo Mantegazza nel cinquantenario della morte," in *Castalia*, **3** (1960), 126; and E. V. Ferrario, "Una lettera inedita di Paolo Mantegazza sulla fecondazione artificiale," *ibid.* (1962), 134.

Carlo Castellani

MANTELL, GIDEON ALGERNON (*b.* Lewes, Sussex, England, 3 February 1790; *d.* London, England, 10 November 1852), *geology*.

The son of a shoemaker in Lewes, Mantell studied medicine in London and in 1811 returned to Lewes, where he became a busy and successful surgeon. Geology was, however, an overmastering passion, and while at Lewes he made great discoveries and amassed an important collection. In 1833 he moved to Brighton, where his practice became largely eclipsed by his interest in geology. His house with his collection of fossils was turned into a public museum, and his distracted wife and children were forced to seek shelter elsewhere. In 1838 he sold the "Mantellian collection" for £5,000 to the British Museum and bought a practice at Clapham, moving to London in 1844. A prolific writer of books and memoirs (as well as letters and verse), he had enormous energy and enthusiasm but in later life suffered from a painful spinal disease. He was a conspicuous member of the Geological Society of London, of which he became a vice-president in 1848.

In 1835 he was the second recipient of its high honor, the Wollaston Medal (the first was William Smith). Mantell was elected a fellow of the Royal Society in 1825 and he received a Royal Medal in 1849.

Mantell's first and most important book was *The Fossils of the South Downs, or Illustrations of the Geology of Sussex* (1822), a large quarto volume with forty-two lithographic plates. It is now known chiefly for the large number of fossils (nearly all invertebrates) from the Cretaceous and Tertiary strata, but particularly from the Chalk, that Mantell described and illustrated. Many were new species, named by him and now familiar. The most notable fossil here fully described for the first time and named by him is the sponge *Ventriculites*. He wrote various papers on other Chalk fossils, particularly on belemnites and the microscopic organisms found in flint nodules. Mantell included a colored geological map that is on a larger scale and is more detailed and accurate than existing maps of the district (the parts of the general maps of England and Wales by William Smith, 1815, and Greenough, 1819), although the succession of the Cretaceous strata below the Chalk was not satisfactorily settled until 1824, by W. H. Fitton and T. Webster.

An important advance in the knowledge of the geology of northwestern Europe was the recognition of the freshwater origin of the Wealden series of the Cretaceous together with the uppermost series of the Jurassic (Purbeckian). This suggestion was first made by Conybeare in *Outlines of the Geology of England and Wales* (1822), in which he looks forward to support by the forthcoming "work of Mr. Mantell on the fossils of Sussex." Mantell, however, following a warning by George Sowerby on some of his fossil shells from the Wealden, deprecates rather than confirms the inference of a freshwater origin. In a letter to Webster of November 1822 Fitton gave his opinion that the whole of the Purbeck-Wealden series was freshwater, and he published this opinion in *Annals of Philosophy* (1824). Thus, it cannot be said that Mantell was the first to establish the freshwater origin of the Wealden beds, as has been stated, although the evidence he had already obtained (1822), and particularly the evidence he later obtained, did in fact support it, as he came to realize.

Mantell is best known for his discovery of the first dinosaur ever to be described properly—a momentous event. During the second and third decades of the nineteenth century remains of aquatic saurians had been found and described in Britain by several leading geologists, particularly by Conybeare and Buckland, and in France by Cuvier, the founder of vertebrate paleontology. But the existence of the great land saurians (named Dinosauria by Richard Owen in 1842) had not even been suspected. Their enormous diversity is now known in great detail, and the extent of their dominance of life during the entire Mesozoic is fully realized. Fossils that were clearly teeth but unlike any known fossil teeth were found in 1822 by Mantell (it was Mrs. Mantell who first noticed them in a pile of stones along the roadside) together with some loosely scattered bones. In 1825 he was shown teeth of the modern lizard iguana, and he saw that his fossil teeth were similar but much larger. Mantell described the fossil teeth in a paper to the Royal Society in that year and called the large herbivorous reptile to which they must have belonged *Iguanodon*. Although bones that could definitely be shown to have belonged to the same animal had not been found, such associations came to light in 1835 in various parts of the Wealden formation of southern England. The fossils were studied by Richard Owen, and *Iguanodon* was reconstructed in a life-size model (together with models of other dinosaurs) in the grounds of the Crystal Palace in south London in 1854. By a curious mistake the reptile was reconstructed with a horned nose, but the bone thus placed was later found to be a large spike at the end of this biped's "thumb."

In 1832 Mantell discovered the first strongly armored group of dinosaurs. He described this fossil, which he named *Hylaeosaurus*, in *The Geology of the South-east of England* (1833). Like the *Iguanodon*, *Hylaeosaurus* was discovered in the Tilgate Forest region of northern Sussex. Meanwhile, Buckland in 1824 had described the remains of the large carnivorous dinosaur *Megalosaurus* from the Jurassic near Oxford. Thus the first three dinosaurs to be known, the *Iguanodon*, the *Megalosaurus*, and the *Hylaeosaurus*, each belonged to a quite distinct group, later called Ornithopoda, Theropoda, and Ankylosauria, respectively. Although Mantell may be said to have been essentially an amateur collector and expounder—although a very expert and extraordinarily industrious one—he was professionally qualified to examine and report on matters of vertebrate paleontology by reason of his anatomical knowledge as a surgeon.

BIBLIOGRAPHY

I. Original Works. Mantell's papers are listed in Royal Society *Catalogue of Scientific Papers*, IV, 219–220. His chief works are *The Fossils of the South Downs, or Illustrations of the Geology of Sussex* (London, 1822); "On the Teeth of the *Iguanodon*, a Newly-discovered Fossil Herbivorous Reptile," in *Philosophical Transactions of the Royal Society*, **115** (1825), 179–186; *Illustrations of the Geology of Sussex* (London, 1827); *The Geology of the South-east of England* (London, 1833); *The Wonders of*

Geology (London, 1838); *The Medals of Creation* (London, 1844); and *Geological Excursions Round the Isle of Wight and the Adjoining Coast of Dorsetshire* (London, 1847).

II. SECONDARY LITERATURE. Obituary notices include Lord Rosse, in *Proceedings of the Royal Society*, **6** (1852), 252–256; *Gentleman's Magazine*, n.s. **38** (1852), 644–647, unsigned; W. Hopkins, in *Proceedings of the Geological Society*, **9** (1853), xxii–xxv; B. Silliman, *American Journal of Science*, **15** (1853), 147–149; *A Reminiscence of G. A. Mantell. By a Member of the Council of the Clapham Museum. To Which is Appended an Obituary by Professor Silliman* (London, 1853); and T. R. Jones, notice prefaced to his edition (7th) of Mantell's *Wonders of Geology* (London, 1857).

See also M. A. Lower, *The Worthies of Sussex* (Lewes, 1865); W. Topley, *The Geology of the Weald*, Memoirs of the Geological Survey of England and Wales (1875), *passim*; T. G. Bonney, in *Dictionary of National Biography*, XXXVI (1893), 99–100; H. B. Woodward, *The History of the Geological Society of London* (London, 1908), 122; S. Spokes, *Gideon Algernon Mantell* (London, 1927); E. C. Curwen, ed., *The Journal of Gideon Mantell, Surgeon and Geologist* (Oxford, 1940); E. H. Colbert, *Dinosaurs: Their Discovery and Their World* (London, 1962), 33–35; W. A. S. Sarjeant, "The Xanthidia," in *Mercian Geologist*, **2** (1967), 249; E. H. Colbert, *Men and Dinosaurs* (London, 1970), *passim;* W. E. Swinton, *The Dinosaurs* (London, 1970), 28–34, 201–208; A. D. Morris, "Gideon Algernon Mantell (1790–1852)," in *Proceedings of the Royal Society of Medicine*, **65** (1971), 215–221; and L. G. Wilson, *Charles Lyell: the Years to 1841* (New Haven–London, 1972), *passim*.

JOHN CHALLINOR

AL-MAQDISĪ (or Muqaddasī), SHAMS AL-DĪN ABŪ ʿABDALLĀH MUḤAMMAD IBN AḤMAD IBN ABĪ BAKR AL-BANNĀʾ AL-SHĀMĪ AL-MAQDISĪ AL-BASHSHĀRĪ (*b.* Bayt al-Maqdis [Jerusalem], *ca.* A.D. 946; *d. ca.* the end of the tenth century), *geography, cartography.*

Al-Maqdisī spent most of his youth in Jerusalem, then traveled throughout the *Mamlakat al-Islām* (the "Kingdom of Islam"), excepting only al-Andalus (southern Spain), Sind, and Sijistān (southern Afghanistan). He also visited Sicily. His great geographical compendium, the *Kitāb aḥsan al-taqāsīm fī maʿrifat al-aqālīm*, which he completed in Shīrāz in A.D. 985, would indicate, among other things, that he was knowledgeable in Islamic jurisprudence and a follower of the Ḥanafī school of Islamic law.

The geographical writers of the Middle Ages—both western and Arab—had mainly dealt in narrow segments of the subject, producing individual works on mathematical, physical, or descriptive geography, or writing of trade routes and kingdoms or toponymy.

Al-Maqdisī was not satisfied with such works; he criticized those of al-Jayhānī, Abū Zayd Aḥmad ibn Sahl al-Balkhī, Ibn al-Faqīh al-Hamadhānī, and Ibn Khurradādhbih, for example, for being directed to the specific needs of specific rulers, or for simply being too brief to be of practical use. He himself therefore planned a work of wider scope, designed to meet the needs and requirements of a wider audience—merchants, travelers, and people of culture. Of geography, he wrote that "It is a science in which kings and nobles take a keen interest, [while] the judges and the jurists seek it and the common people and the leaders love it." Al-Maqdisī's view of the subject embraced a variety of topics, including various sects and schisms, trade and commerce, weights and measures, customs and traditions, coinage and monetary systems, and languages and dialects; to all of these subjects he brought a critical mind and narrative and investigative skills.

Although al-Maqdisī brought a new aim to geography, his method was that of the Balkhī school, of whom the chief adherents were, besides al-Balkhī himself, al-Iṣṭakhrī (who lived in the first half of the tenth century) and Ibn Ḥawqal (who completed his geographical work in A.D. 977). The Balkhī geographers limited their descriptive writings to the *Mamlakat al-Islām* and attempted to align their geographical concepts with those of the Koran and the *Ḥadīth* ("Traditions of the Prophet"). An example of al-Maqdisī's use of the holy books may be found in his discourse on the seas, in which he argued that the Koranic verse describing the "confluence of the two seas between which was situated *al-barzakh* [the interstice]" actually referred to the meeting of the Mediterranean Sea and the Indian Ocean (which most Arab geographers thought to be a lake) at the Isthmus of Suez, since *al-barzakh* was the land between the al-Faramā and al-Qulzum of the Koran. Like the Balkhī geographers, al-Maqdisī also confined himself to the *Mamlakat al-Islām* and began his account with the description of the "Island of Arabia," which must take precedence since it contained both the holy cities of Mecca and Medina. As he noted in the *Kitāb aḥsan al-taqāsīm*, he neither visited the countries of the infidels nor saw any point in describing them.

Al-Maqdisī nonetheless held independent views and differed from the Balkhī geographers on a number of points. He tried to judge every geographical problem independently and in a scholarly manner; he observed, regarding the authenticity of his own work:

> Know that many scholars and ministers have written on the subject [of geography] but most [of their writing], nay, all of it, is based upon hearsay, while we have

entered every region and have acquired knowledge through experience. Moreover, we did not cease investigation, enquiry, and [attempts to gain] insight into the unknown [al-ghayb]. Thus, our book is arranged in three parts: first, what we have observed; second, what we have heard from trustworthy sources; and third, what we have found in books written on this subject and others.

Al-Maqdisī began his *Kitāb aḥsan al-taqāsīm* with general remarks on a number of subjects, among them seas and rivers; place names and their variants (including names common to more than one place); the special characteristics of various regions; the sects of Islam and the non-Muslim inhabitants of the Islamic world; personal travel narratives; and sections entitled "Places About Which There are Differences of Opinion," "Epitome for the Jurists," and "World *Aqālīm* [regions or administrative districts] and the Position of the *Qibla*." These introductory passages embody some of al-Maqdisī's innovations; he was, for example, the first Arab geographer to determine and standardize the meanings and connotations of Arabic geographical terms, and the first to provide a list of towns and other features for quick reference.

According to al-Maqdisī, the Islamic world was not symmetrical, but rather irregular in shape. He divided this world into fourteen regions *(aqālīm)*, of which he designated six—the "Island of Arabia," 'Irāq (southern Mesopotamia), Aqūr (al-Jazīra, or northern Mesopotamia), al-Shām (Syria), Miṣr (Egypt), and al-Maghrib—as *'Arab*. The remaining eight—al-Mashriq (the kingdom of the Samanids), al-Daylam (Gilan and the mountainous regions east of the Caspian Sea), al-Riḥāb (Azerbaydzhan, Arran, and Armenia), al-Jibāl (ancient Media), Khūzistān (the area south of Media and east of Mesopotamia), Fārs (ancient Persia), Kirmān (the region to the south of Fārs), and al-Sind—he called *'Ajam*, Persian. Each of these districts, it may be noted, is demarcated by well-defined physical boundaries, which al-Maqdisī undoubtedly took into account. In commenting upon them, he further divided his remarks on each region into two sections, of which one was dedicated to physical features, toponymy, and political subdivisions, while the other contained a discussion of general features.

Al-Maqdisī drew a map of each *iqlīm*, indicating regional boundaries and trade routes in red, sandy areas in yellow, salt seas in green, rivers in blue, and mountains in ochre. Although most of the maps have been lost, it is possible to reconstruct them to some degree by considering those made by other geographers of the Balkhī school, the conventions of which al-Maqdisī again followed (although his book suggests

that he specifically disagreed with some of the maps drawn by al-Balkhī). The world maps of this school are round, showing the land mass encircled by an ocean from which the Mediterranean Sea and the Indian Ocean flow, almost meeting at the Isthmus of Suez. The boundaries of the various *aqālīm* are then shown within the land mass. Because of this high degree of stylization, these maps are less accurate than the more detailed maps of specific regions, which conform more closely to the geographers' descriptions; since the maps that al-Maqdisī drew for the *Kitāb aḥsan al-taqāsīm* were of the latter type, some fair amount of accuracy may be assumed.

Al-Maqdisī's book is also notable for its literary style. He wrote in an ornamental and varied manner, occasionally framing his comments in rhymed prose *(saj')*. He used the local dialect of each region in describing it, or, when he did not do so, he explained, he used the Syrian dialect that was native to him. Through this imitation, the language of his section on al-Mashriq is the most rhetorical, since the people of this area were perfect in Arabic; but because the language of the people of Egypt and al-Maghrib was weak and unadorned, and that of the inhabitants of al-Baṭā'ih (the swamps of Iraq) ugly, so, too, is the language in which al-Maqdisī wrote of them.

BIBLIOGRAPHY

Al-Maqdisī's geographical work, *Kitāb aḥsan al-taqāsīm fī ma'rifat al-aqālīm* ("The Best Division for Knowledge of Regions"), is in M. J. de Goeje, ed., *Bibliotheca geographorum arabicorum*, 2nd ed., III (Leiden, 1906). An English trans. up to the *iqlīm* of Egypt is G. Ranking and R. Azoo, *Aḥsanu-t-taqāsīm fī ma'rifat-i-l-aqālīm*, in *Bibliotheca indica*, n.s. (Calcutta, 1897–1910); a French trans. is A. Miquel, *Al-Muqaddasī, Aḥsan al-Taqāsīm fī Ma'rifat al-Aqālīm, La meilleure répartition pour la connaissance des provinces* (Damascus, 1963); and an Urdu trans. is Khurshīd Aḥmad Fāriq, *Islāmī dunyā daswin ṣadī 'iswī mēn* (Delhi, 1962).

See I. I. Krakovsky, *Istoria arabskoy geograficheskoy literatury* (Moscow–Leningrad, 1957); translated into Arabic by Salāḥ al-Dīn 'Uthmān Hāshim as *Ta'rīkh al-adab al-jughrāfī al-'Arabī* (Cairo, 1963).

S. MAQBUL AHMAD

MARALDI, GIACOMO FILIPPO (MARALDI I) (*b.* Perinaldo, Imperia, Italy, 21 August 1665; *d.* Paris, France, 1 December 1729), *astronomy, geodesy*.
MARALDI, GIOVANNI DOMENICO (MARALDI II) (*b.* Perinaldo, Imperia, Italy, 17 April 1709; *d.* Perinaldo, 14 November 1788), *astronomy, geodesy*.

Maraldi I was the son of Francesco Maraldi and Angela Cassini, sister of Cassini I, who had helped found the Paris observatory. After finishing his studies of the classics and of mathematics, he was called to Paris in 1687 by his uncle Cassini I. He soon became his devoted collaborator, and eventually assisted his son, Cassini II, as well. He participated in the observatory's work for thirty years.

Upon arriving in France, Maraldi I started producing a new catalog of the fixed stars, a project he continued throughout his career. This important work, which he almost succeeded in completing, unfortunately was never published, with the exception of certain stellar positions utilized by Deslisle, Manfredi, and Brouckner. An active participant in the daily observations made at the observatory, Maraldi I left behind several unpublished journals. He published many notes in the annual volumes of the *Histoire de l'Académie royale des sciences* concerning the planets, their satellites, eclipses and variable stars, as well as some more theoretical memoirs. In one of the latter, "Considérations sur la seconde inégalité du mouvement des satellites de Jupiter et l'hypothèse du mouvement successif de la lumière" (*Histoire de l'Académie pour l'année 1707* [Paris, 1708], 25–32), he defended the point of view of Cassini I, opposing the hypothesis of the finite velocity of light, conceived by Ole Römer to account for certain irregularities in the movement of Jupiter's satellites.

In 1700 and 1701 Maraldi I participated with Cassini II, J. M. de Chazelles and Pierre Couplet in the operations directed by Cassini I to extend the meridian of Paris to France's southern frontier. He then spent two years in Rome, making various astronomical observations—including one on the zodiacal light—and sharing in the determination and construction of the meridian of the Church of the Carthusians. He returned to Paris in 1703 and resumed his observations, interrupting them for several months in 1718, to take part, with Cassini II and G. de La Hire, in the extension as far as Dunkirk of the Paris–Amiens meridian measured by Jean Picard in 1670.

Maraldi I's personal work was overly influenced by the conservative ideas of Cassini I and was only of the second rank. But as a steady and scrupulously careful observer he contributed significantly to the smooth operation of the observatory and to the realization of important programs of astronomical and geodesic research. He was successively student (1694), associate (1699), and pensioner (1702) of the Academy of Sciences of Paris.

In 1726 Maraldi brought his nephew Giovanni Domenico (Maraldi II), the son of his brother Gian Domenico and Angela Francesca Mavena, to Paris. Maraldi II, who had studied in San Remo and Pisa, worked first under Cassini II and then under Cassini

de Thury (Cassini III), until 1771, when he returned to Italy. He carried out regular astronomical and meteorological observations until 1787 and published many notes drawn from them in the *Histoire de l'Académie*. He likewise participated in various geodesic operations directed by Cassini II and Cassini III: the triangulation of the west perpendicular to the meridian of Paris (1733); the partial survey of the Atlantic coast (1735); the verification of the Paris meridian (1739–1740); and also an experiment designed to measure the speed of sound in the air (1738). Lastly, he shared in the fundamental work carried out to establish the map of France. The greater part of Maraldi II's activity, however, was concerned with positional astronomy. Although he published no books of his own, he edited the *Connaissance des temps* from 1735 to 1759 and also the posthumous work of his friend Lacaille: *Coelum australe stelliferum* (Paris, 1763). Among the numerous memoirs which he published, the most important were devoted to the observation and theory of the movements of the satellites of Jupiter, in which he made many improvements. We may note that in 1740 he finally accepted Römer's theory of a finite value of the speed of light.

Named *adjoint* of the Academy of Sciences in 1731, he was promoted to associate in 1733, to pensioner in 1758, and finally to veteran pensioner in 1772. "An industrious and worthy astronomer" and "an assiduous observer of all the phenomena," Maraldi II "was not content to calculate them; he sought to make them serve the development of his theories." This judgment of Delambre, an author who generally had little good to say for the Cassini family and its allies, is sufficient testimony to the quality of his work, especially regarding the improvement of the tables of Jupiter's satellites.

BIBLIOGRAPHY

I. ORIGINAL WORKS. The works of the Maraldis were almost all published in *Histoire de l'Académie royale des sciences;* the list of their works is given in successive vols. of *Table générale des matières contenues dans l'Histoire et dans les Mémoires de l'Académie royale des sciences*, I–IV (1729–1734) for Maraldi I and IV–X (1734–1809) for Maraldi II. The most important of these works are also in Poggendorff, II, cols. 37–38 for Maraldi I and col. 38 for Maraldi II; P. Riccardi, *Biblioteca matematica italiana*, I (Bologna, 1870; repr. Milan, 1952), cols. 98–102 for Maraldi I, cols. 102–105 for Maraldi II; and J. Houzeau and A. Lancaster, *Bibliographie générale de l'astronomie*, 3 vols. (Brussels, 1882–1889; repr. London, 1964), see index.

II. SECONDARY LITERATURE. For studies dealing with both Maraldis see the following (in chronological order): A. Fabroni, *Vitae italorum doctrina excellentium*, VIII (Pisa, 1781), 293–320; J. J. Bailly, *Histoire de l'astronomie moderne*, 3 vols. (Paris, 1779–1782), see III index; J. D. Cassini IV, *Mémoires pour servir à l'histoire des sciences et à celle de l'Observatoire de Paris* (Paris, 1810), 348–357; J. J. Weiss, *Biographie universelle*, new ed., XXVI (Paris, 1861), 410–411; J. B. Delambre, *Histoire de l'astronomie au XVIIIᵉ siècle* (Paris, 1827), 239–250; F. Hoefer, ed., *Nouvelle biographie générale*, XXXIII (Paris, 1860), cols. 348–350; C. Wolf, *Histoire de l'Observatoire de Paris . . .* (Paris, 1902), see index; F. Boquet, *Histoire de l'astronomie* (Paris, 1925), 403–404 and 425–426; and N. Nielsen, *Géomètres français du XVIIIᵉ siècle* (Copenhagen–Paris, 1935), 297–300.

On Maraldi I, see B. Fontenelle, "Éloge de Jacques-Philippe Maraldi," in *Histoire de l'Académie royale des sciences pour l'année 1729* (Paris, 1731), 116–120; and C. G. Jöcher in *Allgemeines Gelehrtenlexikon*, III (Leipzig, 1751), col. 130.

On Maraldi II see J. D. Cassini IV, "Éloge de J. D. Maraldi," in *Magasin encyclopédique*, **1** (1810), 268–282; and J. de Lalande, in *Bibliographie astronomique* (Paris, 1803), see index.

RENÉ TATON

MARCGRAF, GEORG. See **Markgraf, Georg.**

MARCHAND, RICHARD FELIX (*b.* Berlin, Germany, 25 August 1813; *d.* Halle, Germany, 2 August 1850), *chemistry.*

The son of a Berlin lawyer, Marchand had already published a substantial amount of original research on organic chemistry while a medical student at the University of Halle, before he graduated in 1837. This work brought him the friendship of Otto Erdmann, professor of technical chemistry at the neighboring University of Leipzig. In 1838 he became lecturer in chemistry at the Royal Prussian Artillery and Engineering School in Berlin, but not until 1840 was he recognized as privatdocent by the University of Berlin. In 1843 he became extraordinary professor of chemistry at the University of Halle (where he had been licensed to teach since 1838), and in 1846 he succeeded to the permanent chair.

Wöhler, writing in 1839, describes Marchand as a tall, elegant youth with "negroid Mephistopheles features, very free and easy in manner almost to a level of impertinence, full of sudden ideas, wisecracks and satirical remarks; all, however, combined with a sure amiability and genius, so that one cannot help liking him."[1] Nicknamed "Dr. Méchant" by Berlin society, during his short life (ended by cholera) he accomplished an impressive amount of experimental research and popular lecturing. In 1844 he married Marianne Baerensprung. One of their three children, Jacob Felix, became a distinguished physiologist.

Marchand's publications, which are primarily descriptive and analytical, encompass biochemistry, in which he was greatly inspired by Liebig's *Die Thier-Chemie*, organic chemistry, and determinations of atomic weights. In 1837 Marchand published an important analytical paper on the controversial subject of the constitution of ethyl sulfuric acid; in this he supported the view of Serullas that it was a bisulfate of ordinary ether.[2] Although this challenged the interpretation held by Liebig at this time, Liebig soon saw that Marchand's work supported the ethyl-radical concept which had been introduced in 1834. Following Dumas, in 1838, Marchand and Erdmann developed a technique for the estimation of nitrogen in organic compounds by using copper oxide in an inert atmosphere of carbon dioxide. In 1842 they extended Hess's technique for organic analyses, using as oxidizing agents copper oxide and streams of air and oxygen controlled from gasometers. From 1841, following the dramatic reduction of the atomic weight of carbon from 76.43 to 75.08 (O = 100), Marchand and Erdmann devoted their attentions to the accurate redetermination of atomic weights. Their drastic modifications of Berzelian values and their enthusiastic support for Prout's hypothesis that atomic weights were whole numbers on the hydrogen scale, infuriated Berzelius. He dismissed them unkindly—and erroneously—as careless "apes" and bunglers who always echoed Dumas.[3]

NOTES

1. Wöhler to Berzelius, 12 Oct. 1839. See Wallach, p. 138.
2. R. F. Marchand, "Ueber die ätherschwefelsauren Salze," in *Annalen der Physik und Chemie*, **41** (1837), 596–634.
3. Berzelius to Wöhler, 28 Feb. 1843 and 25 Mar. 1845. See Wallach, pp. 393, 530.

BIBLIOGRAPHY

I. ORIGINAL WORKS. A list of Marchand's 132 papers (17 written with Erdmann) is in the Royal Society *Catalogue of Scientific Papers,* IV, 229–233. Marchand also published *Acidium sulphuricum quam vim in alcoholem exerceat quaeque et hinc prodeuntium et similium compositionum natura sit et constitutio* (Leipzig, 1838), his diss.; *Lehrbuch der organischen Chemie* (Leipzig, 1838); *Grundriss der organischen Chemie* (Leipzig, 1839; Berlin, 1838), also in Dutch trans. (Amsterdam, 1840); *Lehrbuch der physiologischen Chemie* (Berlin, 1844); *Chemische Tafeln zur Berechnung der Analysen* (Leipzig, 1847); *Ueber die Alchemie. Ein Vortrag im wissenschaftlichen Vereine zu Berlin am 20 Februar 1847* (Halle, 1847); *Ueber die Luftschifffahrt. Ein Vortrag im wissenschaftlichen Vereine zu Berlin am 12 Januar 1850* (Leipzig, 1850); and *Das Geld* (Leipzig, 1852).

With Erdmann he edited the *Journal für praktische Chemie*, **16–50** (1839–1850).

II. SECONDARY LITERATURE. There are no formal obituaries. Existing biographical information is based entirely on a vague eulogy in B. F. Voigt, ed., *Neuer Nekrolog der Deutschen*, no. 28 (1850), (Weimar, 1852), 452. But note J. Jordan and O. Kern, *Die Universitäten Wittenberg und Halle vor und bei ihrer Vereinigung* (Halle, 1917); and J. Asen, *Gesamtverzeichnis des Lehrkörpers der Universität Berlin* (Leipzig, 1955), 124.

For contemporary references to Marchand, see J. J. Berzelius, *Jahres-Bericht über die Fortschritte der physischen Wissenschaften*, XIV–XXIV (Tübingen, 1835–1845), and *Vollständiges Sach-und-Namen Register zum Jahres-Bericht* (Tübingen, 1847); O. Wallach, *Briefwechsel zwischen J. Berzelius und F. Wöhler*, 2 vols. (Leipzig, 1901), esp. vol. II; and H. G. Söderbaum, ed., *Jac. Berzelius Bref* (Uppsala, 1912–1935), II, 227–231 (V, to Mulder); III, 212 (VII, to Marignac).

For the joint papers with Erdmann on atomic weights, see *Dictionary of Scientific Biography*, IV, 395, notes 7–9.

Marchand's Berlin career may be traced in *Acta der Königlichen Friedrich-Wilhelms-Universität zu Berlin betreffend die Habilitationen der Privatdocenten* (1838–1843), 65–67, 77–82 (Archives of Humboldt-Universität zu Berlin, Philos. Fak. Littr. H, Nro. 1, vol. VI [1203]). See also the lecture registers and faculty records in the archives of Martin Luther-Universität, Halle–Wittenberg. A few letters to Berzelius are preserved at the Royal Swedish Academy of Sciences, Stockholm.

For the context of Marchand's contributions to biochemistry see F. Holmes, ed., *Liebig's Animal Chemistry*, repr. ed. (New York–London, 1964), vii–cxvi; for his atomic weight research see the secondary literature cited for his collaborator O. L. Erdmann in *Dictionary of Scientific Biography*.

W. H. BROCK

MARCHANT, JEAN (*b.* 1650; *d.* Paris, France, 11 November 1738); **MARCHANT, NICOLAS** (*d.* Paris, June 1678), *botany.*

Jean Marchant was the son and successor of Nicolas Marchant. The botanical concerns of father and son were so similar that the works of one have often been attributed to the other. In particular, both men devoted much of their effort—Nicolas, the last ten years of his life, and Jean, almost all his life's work—to the preparation of the *Histoire des plantes*, undertaken in

1667 by the Académie Royale des Sciences at the urging of Claude Perrault. Each prepared a large number of botanical descriptions for this project which was, however, never published, being abandoned by the Academy in 1694.

Nicolas Marchant held a degree in medicine from the University of Padua, and he appears to have become interested in botany at quite an early date. Following his university training he became apothecary to Gaston, duc d'Orléans, the brother of Louis XIII. In this capacity he was often resident at the château of Blois, where Gaston had established a botanical garden. The garden was under the management of Abel Brunyer, who was also first physician to the duke, and of Robert Morison and Jean Laugier. Nicolas Marchant certainly collaborated with all three of these men, and perhaps accompanied Morison on a botanical excursion to the area around La Rochelle in 1657. The duke died on 2 February 1660 and Nicolas Marchant entered the service of the king late in that year, although it is not known what his title or function in the royal household was. (Of the others, the Protestants Brunyer and Laugier received no official reemployment, while Morison rejected an offer made by Nicolas Fouquet, the minister of finance, and returned to his native England.)

The elder Marchant was one of the founding members of the Académie Royale des Sciences, and remained the only botanist in the organization until the election of Denis Dodart in 1673. In addition to his work toward the *Histoire des plantes*, he collaborated in editing the *Mémoires pour servir à l'histoire des plantes*, which Dodart published in 1676. On 9 November 1674 Colbert named Nicolas Marchant "concierge et directeur de la culture des plantes du Jardin Royal," and in this post Marchant had a garden at his disposal for the experimental cultivation of exotic species. He was also one of the first botanists to take up the study of the lower plants; following his death his son named the common liverwort *Marchantia* in his honor.

Jean Marchant was elected to his father's place in the Academy on 18 June 1678 and also succeeded him immediately in his post at the Jardin du Roi. He continued his father's work on the *Histoire* and increased the specimens in the experimental garden until, in the year 1680 alone, he received more than 500 species of seeds and plants sent to him from abroad. The Academy's decision to give up the *Histoire* cost Jean Marchant the royal pension that the government had granted him for this work, and in the same year, 1694, his position at the royal garden was abolished.

Deprived of royal subsidies, of the official support of the Academy, and of his experimental garden, Jean

Marchant nevertheless continued to prepare botanical descriptions, which he now intended to publish as a work of his own. Although the greater part of this work also remained unpublished, some fifteen of his notices did appear in the Academy's *Mémoires*. Among these, his "Observations sur la nature des plantes" is particularly interesting, since in it Jean Marchant dealt with the notion of partial transformism among plants, thus foreshadowing one of the tenets of evolution.

BIBLIOGRAPHY

I. ORIGINAL WORKS. The greater number of the known manuscripts of both Jean and Nicolas Marchant are in the Bibliothèque Centrale du Muséum National d'Histoire Naturelle, Paris, *cotes* MSS 89, 447–451, 1155, 1356, and 2253. Compare *Catalogue général des manuscrits des bibliothèques publiques de France, Paris*, II, *Muséum d'histoire naturelle* by Amédée Boinet (Paris, 1914), 16, 90–91, 192, 226, and LV, *Muséum d'histoire naturelle. Supplément*, by Yves Laissus (Paris, 1965), 53.

The published works of Jean Marchant comprise fifteen memoirs that appeared in *Mémoires de l'Académie royale des sciences depuis 1666 jusqu'à 1699*, **10** (1730), and in *Histoire de l'Académie royale des sciences . . . avec les mémoires de mathématiques et physiques . . .* (1701, 1706, 1707, 1709, 1711, 1713, 1718, 1719, 1723, 1727, 1733, 1735); the complete list is in Abbé Rozier, *Nouvelle table des articles contenus dans les volumes de l'Académie royale de Paris, depuis 1666 jusqu'à 1770 . . .*, IV (Paris, 1776), 245. Two of these fifteen memoirs are erroneously attributed to Nicolas Marchant; see B. de Fontenelle, *Histoire de l'Académie royale des sciences*, 2 vols. (Paris, 1733).

Jean Marchant, "Observations sur la nature des plantes," is in the *Mémoires* for 1719, 59–66; see also *Histoire*, 57–58. "Liste des plantes citées dans les mémoires de l'Académie, dont les descriptions, données par M. Marchant, ont été réservées pour un ouvrage particulier," is in Godin, *Table alphabétique des matières contenues dans l'Histoire et les Mémoires de l'Académie royale des sciences . . .*, 4 vols. (Paris, 1729–1734), II, *années* 1699–1710, 391–392; III, *années* 1711–1720, 216–217; and IV, *années* 1721–1730, 210–211.

II. SECONDARY LITERATURE. On Nicolas Marchant, see Edmond Bonnet, "Gaston de France, duc d'Orléans, considéré comme botaniste," in *Comptes rendus de l'Association française pour l'avancement des sciences, 19ème session, Limoges 1890* (1891), 416–421; and Nicolas-François-Joseph Eloy, *Dictionnaire historique de la médecine ancienne et moderne . . .*, III (Mons, 1778), 159–160.

On both Jean and Nicolas Marchant, see Yves Laissus and Anne-Marie Monseigny, "Les plantes du roi. Note sur un grand ouvrage de botanique préparé au XVIIᵉ siècle par l'Académie royale des sciences," in *Revue d'histoire des sciences*, **22** (1969), 193–236.

YVES LAISSUS

MARCHI, VITTORIO (*b*. Novellara, Reggio nell' Emilia, Italy, 30 May 1851; *d*. Iesi, Ancona, Italy, 12 May 1908), *pathology*.

Marchi studied at the University of Modena, from which he received a degree in chemistry and pharmacy in 1873. He took a further degree in medicine and surgery in 1882; by then he had already conducted significant research, particularly in demonstrating Golgi tendon organs, including those of the motor muscles of the eye. Soon after his second graduation Marchi became a lecturer in anatomy at the university; he simultaneously served as an anatomist and pathologist at the Reggio nell'Emilia lunatic asylum.

In 1883 a government grant allowed Marchi to continue his studies in Golgi's own general pathology laboratory at the University of Pavia. Here he investigated the fine structure of the corpus striatum and optic thalamus by means of Golgi's "black reaction." By this method, nerve fragments are subjected to three processes—being treated with potassium bichromate, osmium chloride and potassium bichromate, and finally silver nitrate—whereby a black precipitate that demonstrates the nerve elements is formed. Marchi refined Golgi's method, omitting a step. By subjecting nerve fragments to only the potassium bichromate and osmium chloride and potassium bichromate, he was able to demonstrate recently degenerated nerve fibers. The destruction of a cell or the interruption of a nerve fiber is followed by the degeneration of the part of the fiber distal to the lesion; one of the concomitant results of such degeneration is the conversion of myelin to droplets of fat, and it is these fat globules ("Marchi's globules") that are stained black by osmium bichlorate in Marchi's method. (Normal fiber remains unstained.)

Marchi described this staining technique in a series of reports that appeared in 1885. These included notes on lesions of the annular protuberance, on the double crossing of the pyramidal fasciculi, and particularly the preliminary note on the descending degeneration secondary to cortical lesions. In the last of these Marchi fully expounded the significance of his method.

More important than these papers, however, was Marchi's experimental work on the descending degeneration that results from entire or partial extirpation of the cerebellum. From 1885 Marchi was in Florence, where he conducted researches at the physiology laboratory of the Istituto di Studi Superiori Pratici e di Perfezionamento, then under the direction of Luciani. His work served to clarify the structure of the cerebellar pedunculi; this in turn led to the recognition of the efferent fibers that run from the cerebellum to the spinal cord. The tractus tectospinalis is known as "Marchi's tract."

Marchi's scientific contributions were not rewarded with the university chair that he sought. He gave up research and in 1888 began to practice medicine at San Benedetto del Tronto. From 1890 until his death he was head physician of the hospital of Iesi.

BIBLIOGRAPHY

I. ORIGINAL WORKS. Among Marchi's works published in journals are "Sulla terminazione della fibra muscolare nella fibra tendinea," in *Lo Spallanzani*, **9** (1880), 194–197; "Sulle terminazioni periferiche dei nervi," in *Rivista sperimentale di freniatria e di medicina legale*, **8** (1882), 477–489, esp. 485–486; "Sugli organi terminali nervosi nei tendini dei muscoli motori dell'occhio," in *Atti della Reale Accademia delle scienze di Torino*, **16** (1880–1881), 206–207; "Sugli organi terminali nervosi (corpi di Golgi) nei tendini dei muscoli motori del bulbo oculare," in *Archivio per le scienze mediche*, **5** (1882), 273–282; and "Ueber di Terminalorgane der Nerven (Golgi's Nervenkörperchen) in den Sehnen der Augenmuskeln," in *Albrecht v. Graefes Archiv für Ophthalmologie*, **28**, pt. 1 (1882), 203–213.

See also "Un caso di sarcoma cerebrale in un alienato," in *Rivista sperimentale di freniatria*, **9** (1883), 114–117; "Sulla fina anatomia dei corpi striati," *ibid.*, 331–334; "Sull'istologia patologica della paralisi progressiva," *ibid.*, 220–221; "Sulla struttura dei talami ottici," *ibid.*, **10** (1884), 329–332; "Sopra un caso di doppio incrociamento dei fasci piramidali," in *Archivio italiano per le malattie nervose*, **22** (1885), 255–266; "Contributo allo studio delle lesioni della protuberanza anulare," in *Rivista sperimentale di freniatria*, **11** (1885), 254–278, written with Giovanni Algeri; "Sulle degenerazioni discendenti consecutive a lesioni della corteccia cerebrale," *ibid.*, 492–494, written with Algeri; "Sulle degenerazioni discendenti consecutive a lesioni sperimentali in diverse zone della corteccia cerebrale," *ibid.*, **12** (1886), 208–252, written with Algeri; "Sulle degenerazioni consecutive all'estirpazione totale e parziale del cervelletto," *ibid.*, 50–56; "Sulla fine struttura dei corpi striati e dei talami ottici," *ibid.*, 285–306; "Sulle degenerazioni consecutive a estirpazione totale e parziale del cervelletto," *ibid.*, 224, and **13** (1887), 446–452; "Sul decorso dei cordoni posteriori nel midollo spinale," *ibid.*, **13** (1887), 206–207; "Ricerche anatomo-patologiche e bacteriologiche sul tifo pellagroso," *ibid.*, **14** (1888), 341–348; and "Sull'origine e decorso dei peduncoli cerebellari e sui loro rapporti cogli altri centri nervosi," *ibid.*, **17** (1891), 357–368.

A separate publication is *Sull'origine e decorso dei peduncoli cerebellari e sui loro rapporti cogli altri centri nervosi* (Florence, 1891).

II. SECONDARY LITERATURE. Works about Marchi include [Arturo Donaggio], *Onoranze nella R. Università di Modena a Vittorio Marchi nel 25° anniversario della sua morte* (Reggio nell'Emilia, 1933); Battista Grassi, *I progressi della biologia e delle sue applicazioni pratiche conseguiti in Italia nell'ultimo cinquantennio* (Rome, 1911), 170; Luigi Luciani, "Vittorio Marchi," in *Archives italiennes de biologie*, **49** (1908), 149–152; Manfredo Manfredi, "Vittorio Marchi e il suo 'metodo,'" in Luigi Barchi, ed., *Medici e naturalisti Reggiani* (Reggio nell'Emilia, 1935), 159–169; P. Petrazzani, "Prof. Vittorio Marchi," in *Rivista sperimentale di freniatria*, **34** (1908), 319–320; and Antonio A. Rizzoli, "An Unusual Case of Meningioma With the Involvement of Russell's Hook Bundle as Described by Vittorio Marchi (1851–1908)," in *Medical History*, **17** (1973), 95–97.

LUIGI BELLONI

MARCHIAFAVA, ETTORE (*b.* Rome, Italy, 3 January 1847; *d.* Rome, 23 October 1935), *pathology, anatomy.*

Marchiafava was the son of Anna Vercelli and Francesco Marchiafava. He began his career at a time of great social and political upheaval. In the scientific world polemics raged—such as those between Bufalini and Tommasini on the ultimate cause of disease—over the direction that science should take in light of the many major discoveries then being made.

The great prevalence of communicable diseases, especially malaria and tuberculosis, exerted a strong influence in determining Marchiafava's line of research. After obtaining a degree at the University of Rome in 1869, he went for a short period to Berlin, where Koch was making progress in the study of tuberculosis. The young scientist returned to Italy with a strong interest in bacteriology and parasitology.

In 1872 Marchiafava was nominated assistant to the professor of pathological anatomy at the University of Rome; he became associate professor in 1881 and full professor in 1883. After his official retirement in 1922 he continued his research and writing in the department he had helped organize.

Marchiafava was not only a great pathological anatomist but also an outstanding clinician and a faithful follower of Morgagni, so that from pathological anatomy and from the data he obtained in studying corpses, he was able to make his clinical interpretations.

Marchiafava's first research was essentially in parasitology. He spent many years studying the morphology and the biological cycle of the malarial parasite. He showed the modifications that the presence of amoeboid bodies causes in the erythrocytes, and demonstrated that these changes were closely related to the growth and multiplication of the parasite. This demonstration derived from the parallel study of microscopic blood data and the clinical pattern of fever peaks. The most important result of the research was Marchiafava's discovery that malarial infection is transmitted through the blood. He spent the entire period from 1880 to 1891 in this intensive study, which enabled him to distinguish between the agent of the estivo-autumnal fever and that of the tertian and quartan fevers. As a senator, elected in 1913, and later as hygiene assessor of Rome (1918), he urged the adoption of antimalarial measures.

In 1884 Marchiafava, in collaboration with A. Celli, identified meningococcus as the etiological agent of epidemic cerebral and spinal meningitis. Another of his findings, which bears his name, was Marchiafava's postpneumonic triad, characterized by the simultaneous presence of a meningitis infection and an endocardial ulcer, which he related to septicemia in the lungs.

His name is also remembered in the Marchiafava-Bignami syndrome, which is a special primitive alteration of nerve fibers caused by chronic alcoholism, affecting in particular the corpus callosum and the frontal commissure.

In 1911 Marchiafava described a form of chronic acquired hemolitic jaundice characterized by a hemoglobinemia with hemoglobinuria and progressive anemia, the Marchiafava-Micheli syndrome. He later conducted more detailed studies on this form of the disease, which he named hemolitic anemia with perpetual hemosiderinuria.

A pioneer in the field of cardiac pathology, Marchiafava showed the importance of coronary sclerosis in the pathogenesis of cardiac infarction and suggested the use of theobromine as a treatment for this disease. Early in his career he made other important studies that showed the bacterial nature of endocardial ulcers. He also did research in angiotic obliteration in interstitial inflammations and particularly in tuberculosis. Marchiafava was especially interested in tuberculosis and examined in detail the structural modifications occurring where the bronchi join the lungs, as well as the clinical epidemiology of the disease. On kidney pathology he studied and described glomerulonephritis related to infections such as scarlet fever.

BIBLIOGRAPHY

I. ORIGINAL WORKS. The main works of Marchiafava are *Sul parasita delle febbri gravi estivo-autunnali* (Rome, 1889); *Sulle febbri malariche estivo-autunnali* (Rome, 1892); *La infezione malarica* (Milan, 1903); *La perniciosità della malaria* (Rome, 1928); and *La eredità in patologia* (Turin, 1930).

II. SECONDARY LITERATURE. Works about Marchiafava include G. Bompiani, "Ettore Marchiafava," in *Pathologica*, **28** (1936), 93–99; L. W. Hackett, "Prof. Ettore Marchiafava," in *Transactions of the Royal Society of Tropical Medicine and Hygiene*, **29** (1936); L. Stroppiana, "Ettore Marchiafava a cento anni dalla nascità," in *Castalia*, **1** (1948), 17–18; and P. Verga, "Ettore Marchiafava," in *Riforma medica*, **51** (1935), 1736–1737.

CARLO CASTELLANI

MARCHLEWSKI, LEON PAWEŁ TEODOR (*b.* Włocławek, Poland, 15 December 1869; *d.* Cracow, Poland, 16 January 1946), *chemistry.*

Marchlewski was the son of Józef Marchlewski, a grain merchant, and Augusta Riksreerend. Having finished his secondary education in Warsaw, he worked there for a year in a chemistry laboratory of the Museum of Agriculture and Industry; then, in 1888, enrolled in the Polytechnical School in Zurich. From 1890 until 1892, the year in which he received the doctorate from the University of Zurich, Marchlewski served as an assistant to Georg Lunge. While working with Lunge he published his first scientific papers; these were largely analytical and technological in nature, and were concerned with inorganic chemistry, the determination of iodine and sulfur in compounds, and the gas-volumetric determination of carbon dioxide. He also drew up tables of the density of hydrochloric and nitric acids that continue to be consulted.

From 1892 until 1898, Marchlewski worked in England in the private laboratory of Edward Schunck at Kersal, near Manchester. Schunck had been Liebig's pupil, and his own area of research was plant pigments. With him, Marchlewski published papers on natural glucosides, including arbutin, phlorizin, and datiscin. Marchlewski offered a new interpretation of the structure of these compounds, which brought him into conflict with E. Fischer; Marchlewski was eventually proved to be right. He also explained the structure of rubiadin and of indican, a compound which had been discovered by Schunck and which Marchlewski showed to be a glucoside of indoxyl. Most important, he began to make his own investigations of plant pigments, including isatin and chlorophyll and its derivatives, a subject upon which he first reported in a paper written with Schunk in 1897.

In 1898 Marchlewski became director of the research laboratory attached to the Claus and Ree factory at Clayton. He was at the same time a lecturer in technology at the Institute of Science and Technology in Manchester. During his stay in England he married Fanny Hargreaves; they had three sons. In 1900 Marchlewski returned with his family to Poland, where he had been appointed to the Food Examination Research Institute in Cracow. He became lecturer in chemical technology at Jagiellonian University there in the same year.

Marchlewski again took up the studies of chlorophyll that he had begun in England. He had already experimentally obtained phylloporphyrin and compared its absorption spectra with those of hematoporphyrin, concluding that the two compounds are closely related, as are chlorophyll and hemoglobin. In Cracow, he obtained phyllocyanin from chlorophyll; he then demonstrated that hemopyrrole could be derived from this compound as well as from hemin. He further obtained phyllophylin, a substance similar to hemin itself, from phylloporphyrin and ferrous salts. Marchlewski published the results of some of these researches with Nencki. His own monograph *Die Chemie der Chlorophylle*, published in 1903, established his authority in this field.

In 1904 Marchlewski declined the offer of a chair of chemistry at the University of Lvov; two years later, he was appointed professor of medical chemistry at Jagiellonian University. He had by then published eighty-four papers—almost half of his life's work—in chemical journals. He received another appointment almost immediately, and left Cracow to become director of the Research Institute in Puławy, where he remained until 1923. He then returned to occupy the chair of medical chemistry at Jagiellonian University, where he remained (except for an interruption during World War II) for the rest of his life. He twice served the university as dean of the medical faculty, and was rector of it in 1932.

Marchlewski's work on chlorophyll involved him in a series of controversies with the German chemist Willstätter. Although Marchlewski was not always correct (it is difficult, for example, to understand how he, an outstanding analyst, could have overlooked the presence of magnesium in chlorophyll), the debates themselves contributed to the growth of chlorophyll research. Marchlewski was also highly critical of the work of the botanist Tsvet, who had devised a technique for separating chlorophyll into its component parts by dissolving it in alcohol, then passing it through a column filled with calcium carbonate, sugar, and inulin. Marchlewski considered Tsvet to be ignorant of chemistry, and Tsvet, as a direct result of Marchlewski's criticism, suspended his researches. Marchlewski fully understood Tsvet's method a few years later, and greatly regretted his interference.

Marchlewski's last important achievement was the discovery of phylloerythrin, a compound that results from the breakdown of chlorophyll during digestion by herbivores. In his last years he devoted himself exclusively to spectral analysis, which he saw as the chief means toward explaining the structure of organic compounds.

BIBLIOGRAPHY

I. Original Works. Marchlewski wrote 201 papers, of which a complete list is given by his student H. Malarski, "Leon Marchlewski. 1869–1946," in *Pamiętniki państwowego naukowego instytutu gospodarstwa wiejskiego*, **18E** (1948), 1–27. His most important books are *Die Chemie der Chlorophylle* (Brunswick, 1903); *Teorye i metody badania współczesnej chemii organicznej* ("Theories and Methods of Contemporary Organic Chemistry"; Lvov, 1905); *Chemia organiczna* ("Organic Chemistry"; Cracow, 1910; repr. 1924); and *Podręcznik do badań fizjologicznochemicznych* ("Handbook of Physiological and Chemical Research"; Cracow, 1916). Marchlewski's personal acta are preserved in the archives of Jagiellonian University, Cracow, S. II, 619; some of his letters may be found in the archives of the Polish Academy of Sciences, Cracow, the Jagiellonian Library, and the Ossolineum, Wrocław.

II. Secondary Literature. See A. Gałecki, "Udział Polaków w uprawianiu i rozwoju chemii" ("Poles Who Participated in the Development of Chemistry"), in *Polska w kulturze powszechnej*, II (Cracow, 1918), 336–337; W. Lampe, "Śp. Leon Marchlewski," in *Rocznik towarzystwa naukowego Warszawskiego*, **39** (1946), 131–134; and B. Skarzyński, "Leon Marchlewski," in *Roczniki Chemii*, **22** (1948), 1–18, repr. in *Polscy badacze przyrody* (Warsaw, 1959), 289–312. On Marchlewski's controversy with Tsvet, see T. Robinson, "Michael Tswett," in *Chymia*, **6** (1960), 146–161.

WŁODZIMIERZ HUBICKI

MARCI OF KRONLAND, JOHANNES MARCUS

(*b.* Lanškroun, Bohemia [now Czechoslovakia], 13 June 1595; *d.* Prague, Bohemia [now Czechoslovakia], 10 April 1667), *physics, mathematics, medicine.*

Marci, whose father was clerk to an aristocrat, received his early education at the Jesuit college in Jindřichův Hradec, then studied philosophy and theology in Olomouc and, from 1618 on, medicine in Prague. He took the M.D. in 1625, then began to lecture at the Prague Faculty of Medicine. He achieved considerable renown as a physician, becoming physician to the Kingdom of Bohemia and personal attendant to two emperors, Ferdinand III and Leopold I.

Although it is recorded that Marci wished to become a priest and a Jesuit, and although he took a staunchly Catholic position during the forced civil re-Catholicization of Bohemia and Moravia (1625–1626), he nevertheless represented the anti-Jesuit party in the affairs of Prague University. To gain support at the Vatican for his party's purpose, which was to prevent the Jesuits from gaining control of the medical and legal faculties (since they already held the faculties of philosophy and theology), Marci undertook a diplomatic trip to Italy, which had important results in his scientific life. During this trip, which he made in 1639, Marci met Paul Guldin and Athanasius Kircher, with whom he corresponded for a long time, and also read Galileo's *Discorsi*, although he did not meet Galileo.

Marci's political activities did not injure his career. He was professor of medicine at Prague University from about 1620 to 1660. In 1648 he took active part in defending the city against the Swedes; he was knighted for merit in 1654. He retained his academic position even after the Prague Charles University merged with the Jesuit institution to become Charles-Ferdinand University, a unification that greatly favored Jesuit pretensions. Marci became rector of the university in 1662; according to Jesuit sources he was admitted to the Society shortly before his death.

As a scientist, Marci worked in considerable isolation. The Catholic Counter-Reformation, exploited by the Hapsburg rulers, had gradually strangled scientific life in Bohemia, and access to the works of foreign scientists was severely limited. Marci's knowledge of the researches of his contemporaries was therefore at best random, and his own work shows evidences of the ideological pressures of his own Prague environment. Marci studied many scientific subjects, including astronomy and mathematics; in the latter he was probably stimulated by the work of the Jesuit Grégoire de Saint-Vincent, who taught at Prague.

Marci's most important work was, however, accomplished in medicine and physics. His 1639 book, *De proportione motus*, contained his theory of the collision of bodies (particularly elastic bodies) and gave an account of the experiments whereby he reached it. Although these experiments are described precisely, Marci was unable to formulate general quantitative laws from them, since his results were not drawn from exact measurements of either of the sizes and weights of the spheres that he employed or of the direction and velocity of their motion. Rather, he was content with simple comparisons of the properties that he investigated, characterizing them as being "smaller," "bigger," or "the same" as each other; his allegations of their proportionalities are thus unproven. Some of his concepts, too (for example that of impulse), lack exact definition, but despite these shortcomings, his observations and conclusions are generally right. He was able to distinguish different qualities of spheres and to state the concepts of solid bodies and of quantity of motion.

The section on the collision of bodies in *De proportione motus* is only one of those in which Marci dealt in problems of mechanics. He also stated the correct relationship between the duration of the oscillation of a pendulum and its length and proposed using a pendulum for measuring short periods of time (for example, for taking the pulse of a patient). He further described the properties of free fall. Here the question of the influence on Marci of Galileo's *Discorsi* must arise. The *Discorsi* was published a year before *De proportione motus*, and Marci certainly read it before publishing his own book, but the exact extent to which he drew upon it remains unknown. Certainly Marci had less skill than Galileo in reducing mechanics to mathematical forms; but if, in later years, he chose to emphasize the divergence of his opinions from Galileo's, he may well have been influenced by the attitude of the church toward the latter's writings.

Marci also carried out research in optics, setting down most of his results in *Thaumantias liber de arcu coelesti* (1648). In his optical experiments, designed to explain the phenomenon of the rainbow, Marci placed himself in the line of such Bohemian and Moravian investigators as Kepler, Christophe Scheiner, Baltasar Konrád, and Melchior Haněl. In his experiments on the decomposition of white light, for which he employed prisms, Marci described the spectral colors and recorded that each color corresponded to a specific refraction angle. He also stated that the color of a ray is constant when it is again refracted through another prism (*Thaumantias liber de arcu coelesti*, pp. 99–100). He did not mention the reconstitution of the spectrum into white light (a result that is first to be found in the work of Newton), although he did study the "mixture" of colored rays. He also made inconclusive experiments on light phenomena on thin films. In general, Marci's optical works are not successful in speculation, since his attempts to deduce the properties of light and to explain the causes of observed phenomena on the basis of his optical knowledge become entangled in the philosophical notions of his time.

Marci's medical works also become involved in philosophical as well as theoretical problems. It is interesting to note that he devoted particular attention to questions of what would now be termed neurology, psychology, and psychophysiology, in treatises that have not yet been fully evaluated. His work on epilepsy is, however, worthy of special note, since in it Marci tried to adopt a purely medical approach to the disease and to analyze critically both previous descriptions of epileptic fits and existing theories of their origin. From these data he drew, in obscure and symbolic terms, the conclusion that epilepsy is, in fact, a nervous disease; this result is in keeping with his theories of perception, memory, and imagination, in which his method was observational and his guiding principle that later formulated by Locke as "nihil est in intellectu quod non prius fuerit in sensu."

It is thus apparent that philosophical considerations figured importantly in Marci's scientific work; it is perhaps less obvious that his philosophy was in turn colored by developments in the natural sciences. Marci's philosophy represented a sometimes incoherent fusion of Aristotelian and Platonic ideas with Catholic mysticism. From these elements he derived a speculative pantheism, based on a "world soul"—uniting the macrocosmos and the microcosmos—and a "virtus plastica sive seminalis," or an "active idea." He attempted to confirm his mystical beliefs by means of then newly established and often subjectively interpreted tenets of natural science; he further called upon these new discoveries to answer such philosophical questions as the relationship between mind and body and to elaborate a general view of the world and nature. (In so doing he drew close to the later systems of *Naturphilosophie*.) Marci's philosophical ideas probably had some influence on such Prague philosophers as Hirnheim (and perhaps even on the young Spinoza), while some of his ideas were taken up by the Cambridge Platonists, among them Ralph Cudworth and, in particular, Francis Glisson.

BIBLIOGRAPHY

Marci's principal works are *De proportione motus figurarum rectilinearum et circuli quadratura ex motu* (Prague, 1639), repr. in *Acta historiae rerum naturalium necnon technicarum*, special issue 3 (1967), 131–258; *Thaumantias liber de arcu coelesti deque colorum apparentium natura, ortu et causis, in quo pellucidi opticae fontes a sua scaturigine, ab his vero colorigeni rivi derivantur* (Prague, 1648; repr. 1968); *Lithurgia mentis seu disceptatio medico-philosophica et optica de natura epilepsiae . . .* (Regensburg, 1678); and *Ortho-Sophia seu philosophia impulsus universalis* (Prague, 1683).

Bibliographies of writings by and about Marci are Dagmar Ledrerová, "Bibliographie de Johannes Marcus Marci," in *Acta historiae rerum naturalium necnon technicarum*, special issue 3 (1967), 39–50; and "Bibliografie Jana Marka Marci," in *Zprávy Čs. společnosti pro dějiny věd a techniky*, nos. 9–10 (1968), 107–119.

LUBOŠ NOVÝ

MARCONI, GUGLIELMO (*b.* Bologna, Italy, 25 April 1874; *d.* Rome, Italy, 20 July 1937), *engineering, physics.*

Marconi was the second son of Giuseppe Marconi, a wealthy landowner, and his second wife, Annie Jameson, the daughter of an Irish whiskey distiller. His limited formal education, of early private tutoring followed by several years at the Leghorn lyceum, included special instruction in physics. His first wife, Beatrice O'Brien, was of an aristocratic Irish family; his second, Maria Bezzi-Scali, belonged to the papal nobility. Marconi was always a devoted citizen of Italy, and frequently acted in an official capacity for his government. Chief among the many honors awarded him was the Nobel Prize for physics, which he shared with K. F. Braun in 1909.

Marconi seems to have first learned in 1894 of Hertz's laboratory experiments with electromagnetic waves. He was immediately curious as to how far the waves might travel, and began to experiment, with the assistance of Prof. A. Righi of Bologna. His initial apparatus resembled Hertz's in its use of a Ruhmkorff-coil spark gap oscillator and dipole antennas with parabolic reflectors, but it replaced Hertz's sparkring detector with the coherer that had been employed earlier by Branly and Lodge. Marconi quickly discovered that increased transmission distance could be obtained with larger antennas, and his first important invention was the use of sizable elevated antenna structures and ground connections at both transmitter and receiver, in place of Hertz's dipoles. With this change he achieved in 1895 a transmission distance of 1.5 miles (the length of the family estate), and at about the same time conceived of "wireless telegraph" communication through keying the transmitter in telegraph code.

Marconi was unable to interest the Italian government in the practical potentialities of his work, however. In February 1896 he moved to London, where one of his Irish cousins, Henry Jameson Davis, helped him prepare a patent application. Davis also arranged demonstrations of the wireless telegraph for government officials and in 1897 helped to form and finance the Wireless Telegraph and Signal Co., Ltd., which in 1900 became Marconi's Wireless Telegraph Co., Ltd. By the latter year Marconi had experimentally increased his signaling distance to 150 miles, and had decided to attempt transatlantic transmission. A powerful transmitter was built at Poldhu, Cornwall, England, and a large receiving antenna placed on Cape Cod, Massachusetts. When the latter blew down in 1901, Marconi, who was anxious to forestall any competitors, sailed for Newfoundland where, using a kiteborne antenna and Solari's carbon-on-steel detector with a telephone receiver, on 12 December he received the first transatlantic wireless communication, the three code dots signifying the letter "S." Already well known, Marconi, at twenty-seven, became world famous overnight.

From 1902 Marconi devoted more of his time to managing his companies, which by 1914 held a commanding position in British and American maritime radio service. (The Radio Corporation of America was formed in 1919, partly to acquire his United States interests.) Throughout his career Marconi was exceptionally fortunate in his ability to attract highly qualified employees and consultants; among them J. A. Fleming, inventor of the thermionic diode; H. J. Round, who developed the triode as a radio-frequency oscillator and amplifier independently of De Forest; R. M. Vyvyan, who installed many of the early spark stations; and C. S. Franklin, designer of directional antennas. It was Franklin who—drawing upon Marconi's earlier notion of exploring the communication potentialities of shortwaves by employing dipole antennas with highly directional reflectors—in 1920 developed such dipole antennas into a beamed radio-telephone circuit between London and Birmingham, operating at 20 MHz. Following a series of discoveries (made by radio amateurs, among others) that indicated the feasibility of establishing a 10,000-mile shortwave communication network, operable by both day and night, Marconi's company completed a globe-girdling system of shortwave beam stations in 1927.

From 1921 on Marconi had used his steam yacht *Elettra* as home, laboratory, and mobile receiving station in propagation experiments. In 1932 he discovered that still higher frequency waves (microwaves) could be received at a point much farther below the optical horizon than had been predicted by any theory. This phenomenon was exploited in later "scatter propagation" circuits, which added new reliability to communications in arctic regions.

BIBLIOGRAPHY

I. ORIGINAL WORKS. Papers by Marconi are "Wireless Telegraphy," in *Proceedings of the Institution of Electrical Engineers*, **28** (1899), 273; "Wireless Telegraphy," in *Proceedings of the Royal Institution of Great Britain*, **16** (1899–1901), 247–256; "Syntonic Wireless Telegraphy," in *Royal Society of Arts. Journal*, **49** (1901), 505; "The Progress of Electric Space Telegraphy," in *Proceedings of the Royal Institution of Great Britain*, **17** (1902–1904), 195–210; "A Note on the Effect of Daylight Upon the Propagation of Electromagnetic Impulses over Long Distances," in *Proceedings of the Royal Society*, **70** (1902), 344; and "Address on Wireless Telegraphy to Annual Dinner," in *Transactions of the American Institute of Electrical Engineers*, **19** (1902), 93–121.

See also "Recent Advances in Wireless Telegraphy," in *Proceedings of the Royal Institution of Great Britain*, **18** (1905–1907), 31–45; "Transatlantic Wireless Telegraphy," *ibid.*, **19** (1908–1910), 107–130; "Radiotelegraphy," *ibid.*, **20** (1911–1913), 193–209; "Radio Telegraphy," in *Proceedings of the Institute of Radio Engineers*, **10** (1922), 215–238; "Results Obtained Over Very Long Distance by Short Wave Directional Wireless Telegraphy, More Generally Referred to as the Beam System," in *Royal Society of Arts. Journal*, **72** (1924), 607; "Radio Communication," in *Proceedings of the Institute of Radio Engineers*, **16** (1928), 40–69; and "Radio Communication by Means of Very Short Electric Waves," in *Proceedings of the Royal Institution of Great Britain*, **27** (1931–1933), 509–544.

II. SECONDARY LITERATURE. For information on Marconi's life and work, see B. L. Jacot de Boinod and D. M. B. Collier, *Marconi—Master of Space* (London, 1935); Douglas Coe, *Marconi, Pioneer of Radio* (New York, 1943); O. E. Dunlap, Jr., *Marconi, The Man and His Wireless* (New York, 1937); Degna Marconi, *My Father, Marconi* (New York, 1962); and W. P. Jolly, *Marconi* (New York, 1972).

ROBERT A. CHIPMAN

MARCOU, JULES (*b.* Salins, France, 20 April 1824; *d.* Cambridge, Massachusetts, 17 April 1898), *geology, paleontology, topography.*

Marcou was born in the Jura, and the natural history of the area had much to do with determining the course of his work in science. Educated in his native Salins and the lycée at Besançon, Marcou went to Paris to study at the Collège Saint-Louis. Ill health caused an interruption in his education, and after returning to Salins he began to explore the geology and paleontology of his native Jura. Marcou's first published work was in mathematics, but his growing knowledge of natural history had become so extensive that by 1845 he was able to publish a highly original analysis of Jurassic fossils ("Recherches géologiques sur le Jura Salinois") in the *Mémoires de la Société d'histoire naturelle de Neuchâtel*. Louis Agassiz, editor of the journal, was impressed by the young man's grasp of a complex subject. The Swiss paleontologist encouraged the young man to do further work in the field, an ambition buttressed by Marcou's appointment as professor of mineralogy at the Sorbonne in 1846 and curator of fossil conchology at the Jardin des Plantes in 1847. With the support of Agassiz, in 1848 Marcou was awarded a traveling fellowship under the auspices of the Jardin des Plantes. He chose to spend this time under the guidance of Agassiz, who had gone to the United States. Thus in a short time, Marcou had established himself as a rising figure in geology and paleontology.

Marcou's outstanding contributions were in stratigraphical geology and geological mapping, most notably a "Geological Map of the World" published in two European editions in 1862 and in 1875, a work one biographer classified as "the point of departure for all subsequent maps of this class." The majority of his more than 180 publications—books, collections of maps, and articles—were in French.

In 1850 Marcou married Jane Belknap, daughter of the historian Jeremy Belknap, and this association with New England lineage and wealth made him independent of material concerns, and he was able to carry on explorations and publish his findings. Marcou did not consider America or its scientists in the light of any permanent physical or intellectual association. After exploring Lake Superior with Agassiz during 1848–1849, he returned to Europe in 1850. He came to the United States again in 1853 as an explorer of the trans-Mississippi West but left the following year. He remained in Europe, chiefly as a professor at the École Polytechnique of Zurich, until 1859, when he returned to the United States to aid Agassiz in the organization and teaching activities of the newly established Museum of Comparative Zoology at Harvard College. After 1864 he returned to France on several occasions but considered Cambridge, Massachusetts, his primary residence. Marcou was physically prepossessing and intellectually dogmatic, with a high opinion of his own abilities—which were significant in areas other than science. His *Derivation of the Name America* (Washington, 1890) remains a highly original piece of scholarship, and his two-volume study of the life of Agassiz was a particularly modern appraisal for its time, especially for its dispassionate presentation of Agassiz's scientific work. In 1867 Marcou was awarded the grand cross of the Legion of Honor, and in 1875 he undertook his last scientific exploration, in the employ of the United States Geographical Surveys West of the One Hundredth Meridian, wherein he did original work in the topography and stratigraphy of southern California.

Marcou did not support the theory of organic evolution. Three years before his death he wrote that Charles Darwin had "failed to give a doctrine well based and acceptable," insisting that natural history progressed through reliance on new facts rather than hypotheses and theories.

This position was in some contrast to Marcou's career in American science. As a field geologist, his experience was limited. Nevertheless, beginning in 1853 he became a party to a series of controversies that demonstrated his disdain for the work of Americans, and at the same time were a witness to the rise of modern American geology. Marcou's role in these disputes took the form of insistence on the correctness of his identification of American stratigraphic topology. This certainty was grounded on little direct knowledge or experience, a condition that infuriated men of the caliber of William P. Blake, James Dwight Dana, James Hall, and William Barton Rogers, who were establishing the character of American professional geology. Early in his American career, Marcou stoutly defended the veracity and general utility of Ebenezer Emmons' so-called Taconic system of New York stratigraphy, pitting himself against Dana and Hall in a matter not fully resolved for nearly fifty years. The *Geology of North America* (Zurich, 1858) contained an entire chapter castigating the scientific work and methods of Blake, Hall, and Dana, as well as "criticisms of the *American Journal of Science and Arts*." Marcou's controversies with American geologists of established reputation were epitomized in his attack on John Wesley Powell and the research orientation of the United States Geological Survey in his *The Geological Map of the United States and the United States Geological Survey* (Cambridge, Mass., 1892).

In 1853 Agassiz's influence had gained Marcou a position as geologist with the United States Army Topographical Corps surveying a Pacific railroad route along the thirty-fifth parallel. Marcou's field experience, extending from Arkansas into California, was sufficient to embolden him to publish a geological map of the United States. This map (published in enlarged editions in 1855 and 1858) was an epitome of the argumentations surrounding the professionalization of American geology. The 1858 edition was especially offensive to American naturalists. Each edition of the map was criticized first by Hall, then by Blake, and finally by Dana, and each condemnation was met by Marcou with greater insistence on his veracity. The points at issue were matters of stratigraphic identification. Marcou was criticized for identifying large portions of the United States as belonging to the Jurassic and Triassic periods, rather than the conventional Cretaceous classification. The matter was made even less agreeable by Marcou's retention of important fossil evidence which belonged to the government; and upon its ultimate return, it was plain that these materials were of European rather than American Jurassic origin. Marcou had identified large portions of the continent as belonging to a period younger than the Jurassic, and, in the view of men such as Blake, all such analyses had been done without benefit of field experience and were of little service to American geology.

Subsequent investigations demonstrated that Marcou was at least partially correct in his nonempirical support of the Taconic system and his definition of the American Jurassic. It is significant that this European geologist, working in the tradition of Georges Cuvier, Jules Thurmann, and Agassiz, was consistent with an earlier period of universalist ambitions to define natural history. It is also noteworthy that his critics demonstrated, by the nature of their disputation, that it was impossible, on the basis of limited knowledge, to construct an American geological map of sufficient detail. By the late 1880's, the work of the early railroad surveys and post-Civil War government geographical and geological efforts had made such contributions possible. Marcou's independence was not unusual in the annals of other aspects of American natural history.

Marcou, whose career began as a promising fieldworker, was at his best as a teacher and descriptive stratigraphic geologist whose observations stimulated Americans to be more critical of their physical history. In this respect, his work served an important purpose in that it helped persuade both Americans and Europeans of the need for careful, comparative research methods and publications.

BIBLIOGRAPHY

I. Original Works. A bibliography of Marcou's publications in invertebrate paleontology is in John Belknap Marcou, "Bibliography of Publications Relating to the Collection of Fossil Invertebrates in the United States National Museum," *Bulletin. United States National Museum*, no. 30 (Washington, 1885–1886), 241–244. Marcou's works published in the United States are listed in Max Meisel, *A Bibliography of American Natural History*, 3 vols. (New York, 1924–1929), II–III.

Among his significant works are *A Geological Map of the United States and the British Provinces of North America ...* (Boston, 1853); *Carte géologique des États Unis ...* (Paris, 1855); *Geology of North America ...* (Zurich, 1858); *Letter to M. Joachim Barrande on the Taconic Rocks of Vermont and Canada* (Cambridge, Mass., 1862); *Carte géologique de la terre* (Zurich, 1875); *American Geological Classification and Nomenclature* (Cambridge, Mass., 1888); and *Life, Letters and Works of Louis Agassiz*, 2 vols. (New York–London, 1895).

II. Secondary Literature. Alpheus Hyatt, "Jules Marcou," in *Proceedings of the American Academy of Arts and Sciences*, **34** (1899), 651–656; Hubert Lyman Clark, "Marcou, Jules," in *Dictionary of American Biography*; William Goetzmann, *Army Exploration in the American West: 1803–1863* (New Haven, 1959), *passim*; Edward Lurie, *Louis Agassiz: A Life in Science* (Chicago, 1960), *passim*; George P. Merrill, *The First One Hundred Years of American Geology* (New Haven, 1924), *passim*.

EDWARD LURIE

MARCUS, JOHANNES. See **Marci of Kronland.**

MAREY, ÉTIENNE-JULES (*b.* Beaune, France, 5 March 1830; *d.* Paris, France, 15 May 1904), *physiology.*

Marey's central significance for the development of physiology in France lies in his adoption and advocacy of two fundamental techniques in experimental physiology: graphical recording and cinematography.

Marey studied medicine at the Faculty of Medicine of Paris and then was an intern at the Hôpital Cochin. His doctoral dissertation (1857) on the circulation of the blood utilized recording instruments that were modified versions of those developed by German physiologists, particularly Karl Ludwig. By installing these instruments in his lodgings on the Rue Cuvier, he established the first private laboratory in Paris for the study of experimental physiology. In 1868 he succeeded Pierre Flourens in the chair of "natural history of organized bodies" at the Collège de France.

During the first decade of his research career (1857–1867) Marey applied the technique of graphical recording to the study of the mechanics and hydraulics of the circulatory system, the heartbeat, respiration, and muscle contraction in general. He analyzed the circulatory and muscular systems in terms of the physical variables, elasticity, resistance, and tonicity. With the graphical trace he established the relationship of heart rate and blood pressure, thus supplementing previous studies of the value of blood pressure in a vessel with traces of its waveform.

After having identified the actions of parts of the organ or system under investigation by means of particular motions of the recording stylus, Marey constructed an artificial model of the organ or system. By manipulating these constructions to obtain wave forms identical to those produced in the living subject, he demonstrated the accuracy of his analyses of the characteristics of his graphical traces.

In this early work Marey sought to apply his methods and results to pathology and to clinical diagnosis. His concern with the greatest possible accuracy in graphical records was matched by his concern for simplifying instruments so that they could be easily used by the clinical diagnostician.

During this first decade Marey's accomplishments depended more upon technical achievement than upon innovative choice of problems. His research topics were in fact fairly straightforward extensions of investigations begun by Bernard, Helmholtz, and Vierordt. Emil du Bois-Reymond, Fick, and Weber were also important influences. After 1868, however, he turned to what was then a more novel area for the application of recording devices—the study of human and animal locomotion. Using traces of the motions of bird and insect wings, Marey showed that changes in the form of the wing modify its air-resistance properties; rather than contracting the wing flexor and extensor muscles, this surface change accounts for much of the upward and forward motion of the flying animal. By this means Marey determined the mechanical requirements for the physiological apparatus of flight. As with his deductions drawn from his circulatory studies, here too he sought to verify his deductions by constructing models that would display the same properties as those of the living specimen. He examined the structure of the muscle and skeletal systems in the light of these mechanical requirements to learn how the size and insertion of muscles, bone length, and joint angles combined to fulfill those requirements.

Marey also studied the length and frequency of steps taken by human beings and quadrupeds under various environmental conditions. Again, he sought the clinical application of his results—in this case to elucidate different locomotor pathologies. This work depended upon the invention of "Marey's tambour," a device for the transmission and recording of subtle motions without seriously limiting the subject's freedom of movement. The tambour is an air-filled metal capsule covered by a rubber membrane. When compression distorts the membrane, air is forced through an opening from the capsule into a fine, flexible tube; at the opposite end of the tube a similar capsule receives these variations in air pressure and its membrane activates the movable lever on the graphical recorder. Marey's tambour was still being used in 1955.

When Marey saw that the pattern of leg motions and hoofbeats of a trotting horse could be depicted clearly by photographs taken in rapid succession, he turned to the perfection of a photographic device that could be used to improve his studies of animal locomotion. Beginning in 1881, his modifications of a camera that had been used by Janssen to record the transit of Venus in 1874 made an important contribution to the development of cinematographic techniques. Also in 1881 he persuaded the municipal council of Paris to annex to his professorial chair land at the Parc-des-Princes, where he constructed a physiological station for the photographic study of animal motion outdoors under the most natural conditions possible. For almost the whole of the following two decades he devoted himself to the application of cinematography to physiology, extending its use to such subjects as photographing water currents produced by the motions of fish and microscopic organisms.

In Marey's view, physiology "is itself but the study of organic movements," and the graph best represents all the variations that such phenomena undergo. Marey believed, however, that these motions ultimately were to be explained by laws of physics and chemistry. Furthermore, while he accepted the application of physiological research to medical problems, he subordinated this utilitarian purpose to a more abstract goal: "analyzing the conditions which modify the functions of life and . . . better determining the laws which regulate these functions." Toward this end medicine served only as one further means of analysis.

Marey's strong desire to see the graph become the language of physiological description led him to fear that confusion and repetition would increase without some standardization of the equipment and parameters used in recording. He therefore proposed to the fourth International Physiological Congress in 1898 that a committee be formed to suggest uniform standards and to perfect the technology of recording devices. When his suggestion was accepted, he solicited and obtained donations from the French government, the municipality of Paris, the Royal Society of London, and other scientific academies for construction at the Parc-des-Princes of an institute where the committee members could work. This institute has since been called the Institut Marey.

In 1895 he became president of the Académie des Sciences, to which he had been elected in 1878.

BIBLIOGRAPHY

I. Original Works. Marey published more than 150 papers, which are indexed in the Royal Society *Catalogue of Scientific Papers*, IV, 237; VIII, 327–328; X, 719–720; XII, 484; XVII, 16–17. Among his major papers are "Recherches hydrauliques sur la circulation du sang," in *Annales des sciences naturelles. Zoologie . . .*, 4th ser., **8** (1857), 329–364; "Études physiologiques sur les caractères du battement du coeur et les conditions qui le modifient," in *Journal de l'anatomie et de la physiologie*, **2** (1865), 276–301, 416–425; "Étude graphique des mouvements respiratoires et des influences qui les modifient," *ibid.*, 425–453; "Études graphiques sur la nature de la contraction musculaire," *ibid.*, **3** (1866), 225–242, 403–416; "Mémoire sur le vol des insectes et des oiseaux," in *Annales des sciences naturelles. Zoologie . . .*, 5th ser., **12** (1869), 49–150, and **15** (1872), art. 13; "De la locomotion terrestre chez les bipèdes et les quadrupèdes," in Robin's *Journal anatomique*, **9** (1873), 42–80; "Emploi de la photographie instantanée pour l'analyse des mouvements chez les animaux," in *Comptes rendus . . . de l'Académie des sciences*, **94** (1882), 1013–1020; "La photochronographie et ses applications à l'analyse des phénomènes physiologiques," in *Archives de physiologie normale et pathologique*, **1** (1889), 508–517; and

"Mesures à prendre pour l'uniformisation des méthodes et le contrôle des instruments employés en physiologie," in *Comptes rendus ... de l'Académie des sciences*, **127** (1899), 375–381.

Among Marey's books are *Physiologie médicale de la circulation du sang* (Paris, 1863); *Du mouvement dans les fonctions de la vie* (Paris, 1868); *La machine animale, locomotion terrestre et aérienne* (Paris, 1873); *La méthode graphique dans les sciences expérimentales* (Paris, 1878); and *Le mouvement* (Paris, 1894).

II. SECONDARY LITERATURE. See Association Internationale de l'Institut Marey, *Travaux de l'Institut Marey* (Paris, 1905–1910), which contain summaries of laboratory work at the institute; A. Chauveau, H. Poincaré, and C. Richet, "Inauguration du monument élevé à la mémoire de Étienne-Jules Marey," in *Mémoires de l'Académie des sciences de l'Institut de France*, 2nd ser., **52** (1915), separately paginated; A. R. Michaelis, *Research Films* (New York, 1955), 4–6, 118–119, and *passim*; "E. J. Marey—Physiologist and First Cinematographer," in *Medical History*, **10** (1966), 2; and "Marey, Étienne-Jules," in Trevor I. Williams, ed., *A Biographical Dictionary of Scientists* (London, 1969), 352–353; "Obituary, É.-J. Marey," in *Lancet* (1904), **1**, 1530–1533; Henri de Parville *et al.*, *Hommage à M. Marey* (Paris, 1902); and C. J. Wiggers, "Some Significant Advances in Cardiac Physiology," in *Bulletin of the History of Medicine*, **34** (1960), 1–15, esp. 9–10.

Background to the development of graphical recording techniques in physiology is in H. E. Hoff and L. A. Geddes, "The Technological Background of Physiological Discovery: Ballistics and the Graphic Method," in *Journal of the History of Medicine and Allied Sciences*, **15** (1960), 345–363.

MICHAEL GROSS

MARGERIE, EMMANUEL MARIE PIERRE MARTIN JACQUIN DE (*b.* Paris, France, 11 November 1862; *d.* Paris, 21 December 1953), *geology, physical geography.*

Margerie came from a cultured Paris family that included several diplomats. He and his brother and sister received an excellent private education, and his childhood vacation travels awakened his interest in geology and geography. At fifteen he attended the lectures of Lapparent at the Institut Catholique in Paris and became a member of the French Geological Society. He took part, in 1878, in the first International Geological Congress in Paris. He did not complete any university training or take any examinations. While very young he began his study of foreign languages, especially German and English, and eventually he was able to read most other European languages as well. This ability was an important factor in his later scientific achievements.

Until the end of World War I, Margerie lived in Paris on an independent income. From 1918 to 1933 he was director of the Service Géologique de la carte d'Alsace et de Lorraine in Strasbourg. After his retirement, he returned to Paris, where he remained until his death. In 1903 Margerie married Renée Ferrer, who survived him. He was sympathetic, with a probing, analytic mind, and he unreservedly made available to his colleagues, especially the younger ones, his extensive knowledge of the international literature on geology and geography. He was a member—and often president—of more than fifty academies and learned societies throughout the world, and received numerous medals, distinctions, and prizes.

Margerie published 265 scientific works, primarily in regional geology, tectonics, and physical geography, as well as geographic and geologic cartography. Most of his publications were designed to make known the work of foreign researchers, to comment upon it, and to synthesize the results; and this was his forte. By virtue of the critical analyses and broad range of subjects, many of these publications were and continue to be of outstanding importance for the study of geology.

Among these works belongs *Les dislocations de l'écorce terrestre* (1888), which Margerie edited with Albert Heim. In this trilingual work (English, French, and German) the editors collected the technical terms, expressions, and concepts employed in geological tectonics, compared them with one another, and listed the corresponding words in the other languages. The lasting importance of this publication for tectonics was reflected in the decision of the 1948 International Geological Congress that the appropriate committee complete the work and bring it up to date.

Also in 1888 Margerie published a work on geomorphology, *Les formes du terrain*, written with General de La Noë, director of the Service Géographique de l'Armée. This book sets forth, for the first time, the causal relationship between the morphology of the earth's surface, and its geological structure and historical development.

In 1896 Margerie published a reference work that is still of value, the *Catalogue des bibliographies géologiques*. But his great reputation among scientists stemmed from his six-volume French translation of Suess's *Das Antlitz der Erde* (1897–1918). Executed with great empathy, and enlarged through many additional illustrations and bibliographical references, this edition enabled scientists who were not proficient in German to study Suess's epochal work.

Margerie was also the author of many specialized geological and geomorphological studies. At the start of his career he published reports (1892, 1893), in collaboration with F. Schrader, on the geomorphological structure of the Pyrenees, which were a combination of his own fieldwork and a critical evaluation of older works. Soon afterward he studied the Swiss and French Jura and the result was the voluminous publication *Le Jura* (1922–1936). This work, which he himself held in especially high regard, is still indispensable for all studies of the Jura.

Margerie devoted a large number of individual studies to the geology and morphology of North America, made from observations gathered in the course of several trips. His last major work, *Études américaines. Géologie et géographie* (1952), is a critical survey of the most important geological works, the topographic and geologic maps, and the history of the United States Geological Survey, as well as a compendium of knowledge about the geological structure of the continent.

Margerie never visited Asia, but through a series of bibliographical analyses he brought the most important publications on Asia to the attention of an international audience. Notable among these was the analysis of the works of the Swedish explorer Hedin on the orography of Tibet (1928).

The 1922 International Geological Congress appointed a committee to prepare a geological map of Africa. Margerie was appointed director of the project because of his wide experience in cartography. The first sheet appeared in 1937, and the publication was concluded in 1952. Margerie was associated with the publication of another cartographic undertaking of international importance—the *Carte générale bathymétrique des océans* (completed in 1931).

Among Margerie's last major publications were the four volumes of *Critique et géologie* (1943–1948). These, together with *Études américaines*, provide a retrospect of the author's career, scientific goals, and accomplishments. Many colleagues and contemporaries, and their correspondence with him, are described in these volumes, which offer a deep insight into the history of geology and geography, and especially into the emergence and development of the leading ideas in these subjects in the late nineteenth and early twentieth centuries.

BIBLIOGRAPHY

I. Original Works. Margerie's works include *Les dislocations de l'écorce terrestre* (Zurich, 1888), written with A. Heim; and *Les formes du terrain* (Paris, 1888), written with G. de La Noë. With F. Schrader he wrote "Aperçu de la structure géologique des Pyrénées," in *Annuaire du Club alpin français*, **18** (1892), 557–619; and "Aperçu de la forme et du relief des Pyrénées," *ibid.*, **19**

(1893), 432–453. He also wrote *Catalogue des bibliographies géologiques* (Paris, 1896); and, with several collaborators, *La face de la terre*, 6 vols. (Paris, 1897–1918), the trans. of Suess's *Das Antlitz der Erde*. Other works include "Le Jura," in *Mémoires pour servir à l'explication de la carte géologique de la France*, 2 vols. (Paris, 1922–1936); "L'oeuvre de Sven Hedin et l'orographie du Tibet," in *Bulletin de la Section de géographie des travaux historiques et scientifiques*, **43** (1928), 1–139; "Les dernières feuilles de la carte générale bathymétrique des océans (panneau du pôle nord)," in *Comptes rendus hebdomadaires des séances de l'Académie des Sciences*, **192** (1931), 1689–1694, and in *Bulletin de l'Institut océanographique*, **580** (1931), 1–6; *Critique et géologie. Contribution à l'histoire des sciences de la terre*, 4 vols. (Paris, 1943–1948); "Carte géologique internationale de l'Afrique," in *Comptes rendus . . . de l'Académie des Sciences*, **235** (1952), 591–592; and *Études américaines. Géologie et géographie*, I (Paris, 1952).

II. Secondary Literature. On Margerie and his work, see the anonymous article in *Comptes rendus du Comité national français de géodésie et géophysique* (1955), 32–34; H. Badoux in *Actes de la société helvétique des sciences naturelles*, **134** (1954), 347–348; P. Fourmarier, in *Bulletin de la Société géologique de France*, 6th ser., **4** (1954), 281–302, with a complete bibliography and portrait; S. Gillet, in *Bulletin du Service de la carte géologique d'Alsace et de Lorraine*, **7** (1954), 5–7, with portrait; C. Jacob, in *Comptes rendus . . . de l'Académie des Sciences*, **238** (1954), 20–23; E. Paréjas, in *Archives des Sciences*, **8** (1955), 69–70; and C. E. Wegmann, in *Geologische Rundschau*, **42** (1954), 314–316.

Heinz Tobien

MARGGRAF, ANDREAS SIGISMUND (*b.* Berlin, Prussia, 3 March 1709; *d.* Berlin, 7 August 1782), *chemistry.*

The few recorded accounts of Marggraf's personal life portray a modest, even-tempered man of precarious health but of single-minded devotion to study and laboratory experimentation. The influence of his mother, Anne Kellner, remains obscure; but it is known that his father, Henning Christian Marggraf, apothecary to the royal court at Berlin and assessor (assistant) at the Collegium Medico-Chirurgicum, introduced him to a circle of pharmacists and chemists. Marggraf's professional apprenticeship comprised several stages: from 1725 to 1730 he was a pupil of Caspar Neumann, his father's colleague at the court pharmacy and medical school and a disciple of Stahl; from 1730 to 1733, he assisted the apothecary Rossler in Frankfurt-am-Main and studied with the chemist Spielmann the elder at the University of Strasbourg; in 1733 at Halle he heard the lectures of Friedrich Hoffmann in medicine and of Johann Juncker in chemistry; in 1734 he traveled to Freiberg, Saxony, to study metallurgy with Henckel. After two years with his father in the Berlin court pharmacy, Marggraf visited Wolfenbüttel, Brunswick, in 1737 but refused an offer of the post of ducal apothecary. He chose to return to Berlin, where he was admitted the following year to the Königlich Preussischen Societät der Wissenschaften (reorganized in 1744–1746 as the Académie Royale des Sciences et Belles-Lettres). Despite the recriminations of a senior academician, J.-H. Pott, Frederick II selected Marggraf as director of the Academy's chemical laboratory in 1753 and as director of its Class of Experimental Philosophy in 1760. Marggraf was also a member of the Kurakademie der Nützlichen Wissenschaften of Mainz and a foreign associate of the Paris Academy of Sciences (1777). Although unable to write after suffering a stroke in 1774, Marggraf confounded attempts to replace him and prepared studies for publication until 1781.

Contemporaries recognized Marggraf as a masterful experimental chemist because of the extraordinary range of his interests and the painstaking nature of his procedures. As an adherent of the Stahl-Juncker phlogiston theory of combustion and calcination, he remained a figure of the "Chemical Ancien Regime." But just as eighteenth-century statecraft sometimes prefigured Revolutionary politics, so Marggraf's interest in chemistry for its own sake, his refinement of analytical tools, and his use of the balance anticipated some facets of the Chemical Revolution.

Marggraf's innovations in analytical methods included an emphasis on "wet methods," or solvent extraction, with careful attention to washing and recrystallization of the end product. His work with certain organic substances, for example the acid extracted from ants (1749) and the "essential oil" of cedar shavings (1753), combined traditional destructive distillation with the sophisticated use of solvents later practiced by G. F. Rouelle. Marggraf's most significant contribution to applied chemistry was his extraction and crystallization of sugar from plants commonly grown in Europe. In 1747 he used boiling rectified alcohol to extract the juice from the dried roots of *Beta alba* (white mangel-wurzel), *Sium sisarum* (skirret), and *Beta radicae rapae* (red mangel-wurzel). When crystals appeared several weeks later, he confirmed their identity with those of cane sugar by microscopic observation—perhaps the first such use of the microscope in the chemical laboratory.

Marggraf also developed a less costly process involving the maceration of roots to obtain the juice and the use of limewater to aid sugar crystallization. Although he envisaged a kind of household industry to assure the poor farmer a new source of sugar, half a century elapsed before any technological application of the

laboratory procedures. Achard, Marggraf's successor as director of the Class of Experimental Philosophy, began experiments on sugar refining in 1786; his first factory became operational under royal patronage in 1802 at Kunern, Silesia. Napoleon's Continental System aroused even greater interest in France and Prussia in a substitute for overseas sugarcane.

On several other occasions Marggraf applied tests significant for modern analytical chemistry. In 1759, to distinguish "cubic niter" (sodium nitrate) from "prismatic niter" (potassium nitrate) crystals, he used, besides the microscope, the flame test—forerunner of modern emission spectroscopy—which differentiated the violet flash of ignition of saltpeter from the yellowish flash of the sodium nitrate. The blowpipe, a tube designed to intensify the flame by directing air upon it, refined this test to reveal characteristic colors and products upon the fusion of a metal. In 1745 Marggraf pioneered the use of the reagent Prussian blue, "fixed alkali ignited with dried cattle blood," as an indicator for the iron content of limestone.

The most notable of Marggraf's isolations of mineral substances were his production of the "acid of phosphorus" and his improved preparation of phosphorus itself. In 1740 he obtained white "flowers" (oxide of phosphorus) from the combustion of phosphorus and recorded, without explanation, the phenomenon so crucial to Lavoisier in 1772, that the calx showed an increase in weight. More remarkable to Marggraf was the hydration of the product in air to form the previously unknown oily phosphoric acid. When heated with coal this acid yielded, in Marggraf's terms, phlogiston and phosphorus.

The preparation of phosphorus had remained a highly prized monopoly of a few German and English chemists (notably Boyle's assistant Hanckwitz) until the French government purchased rights to Kunckel's process in 1737 and permitted Hellot to publish his experiments the same year in the *Mémoires* of the Paris Academy. In 1725 Marggraf had observed Neumann's preparation of phosphorus "with extreme difficulty" from urine and sand. Applying suggestions recorded in Henckel's *Pyritologia* (1725), Marggraf in 1743 evaporated stale urine to obtain a crystallizable "microcosmic salt" that, when heated to redness, yielded a clear glass (sodium metaphosphate), ammonia, and water. A lead calx–sal ammoniac mixture then reduced the "glass" to phosphorus. This method superseded Hellot's preparation, but in 1774 and 1777 Scheele developed a more economical method of obtaining phosphorus from bone ash.

Marggraf attempted to confirm the contention of Stahl and Hellot that phosphorus is a mixture of "acid of sea salt" (hydrochloric acid) and phlogiston. When his efforts to produce phosphorus from hydrochloric acid without urine failed, he cautiously concluded that the acid of phosphorus is distinct and related to the "peculiar salt" in urine necessary for the synthesis of phosphorus. In 1746 he distinguished this salt from Haupt's "sal mirabile perlato" (dodecahydrate of sodium phosphate) by the reducible product it yielded upon heating. Marggraf also substantiated Pott's observation that phosphorus is contained in vegetable matter, and reasoned that the higher yields of phosphorus from urine in the summer are proportional to increased consumption of vegetable foods.

In 1750 Marggraf noted that the earth contained in "Bologna stone" (barium sulfide), another phosphorescent substance, is heavier and more soluble than lime. In the same memoir he anticipated Lavoisier's conclusions by identifying the constituents of gypsum as water, lime, and vitriolic acid.

Until the invention of the Leblanc process, many chemists unsuccessfully sought an inexpensive means of converting common salt into soda for soap manufacture. With that motive Marggraf investigated (1758) the reasoning of H.-L. Duhamel du Monceau that the "alkali" of potash differs from that of rock salt. Besides using microscopic and flame tests, Marggraf recorded the difference in solubility or tendency to deliquescence of the sulfates, chlorides, and carbonates of sodium and potassium. He designated sodium salts as "mineral fixed alkali" and potassium compounds as "vegetable fixed alkali." In 1764 he treated plant parts with acids to establish that the vegetable alkali is an essential plant constituent and not merely a product of distillation.

On at least two occasions Marggraf followed Hoffmann's suggestions concerning the distinctiveness of particular "earths." He showed in 1754 that alumina is a peculiar alkaline earth soluble in acids. Moreover, he refuted the notion of Stahl, Neumann, and Pott that lime is a constituent of alum and, like Lavoisier in 1777, insisted that potash or ammonia is indispensable for alum crystallization. Despite his ignorance of Black's experiments, Marggraf recognized that magnesia, the "bitter earth" related to Epsom salt, is a "genuine and true alkaline earth."

Even Marggraf's less enduring achievements were sometimes remarkable challenges to existing assumptions. His assertion in 1747 that even "pure" commercially available tin contains up to 1/8 arsenic by weight raised doubts about the use of tin in food containers or kitchen utensils until the Bayen commission of the Paris School of Pharmacy concluded in 1781 that the arsenic impurities in various tin samples averaged 1/480 by weight (one grain per ounce). In 1768 Marggraf contradicted the assumption that earths are never volatile by alleging that a distillation of fluorspar with sulfuric acid partially "volatilized" the stone. Only in 1786 did Scheele identify the volatile substance as a mixture of a new acid (prepared by the action of concentrated sulfuric acid on solid fluorite) and glass (now recognized as silicon fluorite).

Marggraf sometimes retained an alchemical outlook—specifically in his memoirs of 1751 and 1756, in which he supported the conviction that water can be transformed into earth. In a 1743 discussion of the crystallization of "microcosmic salt" he had noted his expectation that silver in nitric acid, with phlogiston and a "fine vitrifiable earth," would be subject to "partial transformation"; but he found no trace of a "nobler metal."

Several of Marggraf's pupils were also distinguished analytical chemists. Valentin Rose the elder (1736–1771), a Berlin apothecary, invented a fusible alloy of bismuth, tin, and lead; his son Valentin Rose the younger (1762–1807) was assessor (assistant) at the Berlin Ober-Collegium-Medicum; and their associate Martin Klaproth discovered uranium oxide and became first professor of chemistry at the University of Berlin.

Even without the technical achievements of Achard, Marggraf's work would remain a valuable illustration of the eighteenth-century search for precision in laboratory techniques and for purity in chemical reagents, as well as of the refusal to construct grand theory.

BIBLIOGRAPHY

I. ORIGINAL WORKS. A full list of Marggraf's memoirs was compiled by O. Köhnke in Adolf von Harnack, *Geschichte der Königlich Preussischen Akademie der Wissenschaften zu Berlin*, III (Berlin, 1900), 179–181. The originals appear in the publications of the Berlin Royal Society of Sciences and Royal Academy of Sciences (1740–1781): *Miscellanea Berolinensia ad incrementum scientiarum . . .*, 7 vols. (Berlin, 1710–1743); *Histoires de l'Académie royale des sciences et des belles-lettres de Berlin, . . . avec les mémoires . . .*, 25 vols. (Berlin, 1746–1771); and *Nouveaux mémoires de l'Académie royale des sciences et belles-lettres. . .*, 17 vols. (1772–1788). With the editorial assistance of J.-G. Lehmann, Marggraf collected his memoirs and added four MS dissertations in *Chymische Schriften*, 2 vols. (Berlin, 1761–1767), rev. ed. of vol. I appeared in 1768. Formey's French trans. of vol. I was published by J. F. Demachy as *Opuscules chymiques*, 2 vols. (Paris, 1762). An annotated German text of three memoirs is available in Ostwalds Klassiker: *Einige neue Methoden, den Phosphor im festen Zustande sowohl leichter als bisher aus dem Urin darzustellen . . .*, G. Mielke, ed. (Leipzig, 1913), which includes "Chemische Untersuchungen eines sehr bemerkenswerten Salzes, welches die Säure des Phosphors enthält" (1746). See also *Versuche einen wahren Zucker aus verschiedenen Pflanzen, die in unseren Ländern wachsen, zu ziehen,* Edmund O. von Lippmann, ed. (Leipzig, 1907).

II. SECONDARY LITERATURE. See Condorcet, "Éloge," in *Histoire de l'Académie* for 1782 (Paris, 1785), 122–131, repr. in Condorcet's *Oeuvres,* A. Condorcet O'Connor, ed. (Paris, 1847), II, 598–610; Formey, in *Histoire de l'Académie . . . de Berlin, année 1783* (Berlin, 1785), 63–72; A. de L., in *Nouvelle biographie générale,* XXXIII (Paris, 1860), cols. 549–553; Edmund O. von Lippmann, "Andreas Sigismund Marggraf," in Eduard Farber, ed., *Great Chemists* (New York–London, 1961), 193–200; Max Speter, "Marggraf," in Gunther Bugge, ed., *Das Buch der grossen Chemiker,* I (Berlin, 1929), 231–234; and John Ferguson, *Bibliotheca chemica,* II (Glasgow, 1906), 76–77.

The best single summary is in J. R. Partington, *A History of Chemistry,* II (London, 1961), 723–729. See also Frederic L. Holmes, "Analysis by Fire and Solvent Extractions: The Metamorphosis of a Tradition," in *Isis,* **62** (1971), 129–148; Hermann Kopp, *Geschichte der Chemie,* I (Brunswick, 1843), 208–211; Max Speter, "Zur Geschichte des Marggrafschen Urin-Phosphors," in *Chemisch-technische Rundschau,* **44** (13 Aug. 1929), 1049–1051; Ferenc Szabadváry, *History of Analytical Chemistry,* G. Svehla, trans. (Oxford, 1966), 51–52, 55–59; and Mary Elvira Weeks, *Discovery of the Elements,* 7th ed. (Easton, Pa., 1968), 497–498, 560–561, 861–864.

MARTIN S. STAUM

MARGULES, MAX (*b.* Brody, Galicia [now Ukrainian S.S.R.], 23 April 1856; *d.* Perchtoldsdorf, near Vienna, Austria, 4 October 1920), *meteorology, physics.*

One of the most important meteorologists of the early twentieth century, Margules provided the first thorough, theoretical analyses of atmospheric energy processes and deeply influenced the evolution of present concepts of such processes. He studied mathematics and physics at Vienna and in 1877 joined the staff of the Zentralanstalt für Meteorologie in Vienna. From 1879 to 1880 he continued his studies at Berlin and then returned to Vienna as *Privatdozent* in physics. In 1882 he resigned from this post, thus terminating his career at the university. He was then reemployed by the Zentralanstalt until 1906. During this time Margules produced a small number of highly important papers in meteorology. In 1906, at the age of fifty, he retired and gave up meteorology, again concentrating on physical chemistry, apparently because of embitterment at the lack of recognition for his work. Lonely, unmarried and without close friends, he literally starved to death during the austere postwar period.

After returning to the Zentralanstalt in 1882, Margules continued to pursue physical and physical-chemical investigations in his free time. His publications dealt with electrodynamics, the physical chemistry of gases, and hydrodynamics. Independently of Gibbs and Duhem, he developed in 1895 a formula for the relation between the partial vapor pressures and the composition of a binary liquid mixture, now known as the Duhem-Margules equation. In 1881–1882 he furnished a theoretical analysis of the rotational oscillations of viscous fluids in a cylinder.

From 1890 to 1893 Margules produced a series of papers related to meteorology that dealt with oscillation periods of the earth's atmosphere and the solar semidiurnal barometric pressure oscillation, the universal character of which had been established by his colleague Hann. William Thomson had suggested that the magnitude of this oscillation could be explained by a resonance oscillation of the entire atmosphere. Margules substantiated this hypothesis theoretically by computing free and forced oscillations of the atmosphere on the basis of Laplace's tidal theory. He never considered his results to be a rigorous proof, however, because of several unrealistic assumptions and the lack of a physical explanation for the semidiurnal temperature variation. Margules also investigated the oscillations of a periodically heated atmosphere, using various heating models, and gave a general classification of these motions.

While these studies still tended toward theoretical physics, Margules' next investigations dealt in a novel manner with problems fundamental to meteorology. In 1901 he demonstrated that the kinetic energy displayed in storms was far too great to be derived from the potential energy of the pressure field. He reduced the pressure field, which meteorologists had previously regarded as an explanation for the genesis of atmospheric motions, to a mere "cog-wheel in the storm's machinery."

In a famous 1905 paper Margules proposed a new source for the production of kinetic energy by studying models of energy transformations in the atmosphere that involved the isentropic redistribution of warm and cold air masses from a state of instability to one of stability. He showed that the realizable kinetic energy of these closed systems was the difference between the sums of internal energy and gravitational potential energy at the beginning and the end of the redistribution process. Margules considered this quantity, which is now called available potential energy, as the source of kinetic energy in storms. His theoretical analyses supported F. H. Bigelow's view that the coexistence of warm and cold air masses is the precondition for the development of storms. The cyclone model subsequently developed by J. Bjerknes was based energetically largely on Margules' work. Margules' results formed the basis for F. M. Exner's and A. Refsdal's investigations and have continued to influence meteorological thought. The work of E. Lorenz is an example.

In his discussion of idealized situations in which there is a large store of available potential energy (1906), Margules demonstrated that on the rotating earth two air masses of different temperatures, separated by an inclined surface of discontinuity, can exist in equilibrium under certain conditions. He developed a formula for the slope of this surface, using methods developed by Helmholtz. Bjerknes and his group later applied Margules' formula to their cyclone model, in which such frontal surfaces were the salient feature.

Margules considered the study of detailed observations, distributed three-dimensionally, to be of utmost importance for progress in meteorological theory. For this reason in 1895 he began to install a small network of stations around Vienna. Observations from these and nearby mountain stations allowed him to study the progression of cold and warm air masses and sudden variations in barometric pressure and wind during the passage of storms; these observations influenced his theoretical considerations and vice versa.

Margules also attempted to determine the frictional dissipation of kinetic energy and made the first estimate of the efficiency of the general circulation of the atmosphere as a thermodynamic engine. One of his last meteorological investigations, which dealt with the development of temperature inversions by descending motion and divergence, contributed to the understanding of anticyclones.

BIBLIOGRAPHY

I. ORIGINAL WORKS. Most of Margules' papers are listed in Poggendorff, III, 870–871; IV, 960; V, 807. Many of his important publications are in *Sitzungsberichte der Akademie der Wissenschaften in Wien*, Math.-naturwiss. Kl., Abt. IIa: "Über die Bestimmung des Reibungs- und Gleitcoefficienten aus ebenen Bewegungen einer Flüssigkeit," **83** (1881), 588–602; "Die Rotationsschwingungen flüssiger Zylinder," **85** (1882), 343–368; "Über die Schwingungen periodisch erwärmter Luft," **99** (1890), 204–227, English trans. by C. Abbe in "The Mechanics of the Earth's Atmosphere," 2nd collection, *Smithsonian Miscellaneous Collection*, **34** (1893), 296–318; "Luftbewegungen in einer rotierenden Sphäroidschale," pt. 1, **101** (1892), 597–626; pt. 2, **102** (1893), 11–56; pt. 3, *ibid.*, 1369–1421; and "Über die Zusammensetzung der gesättigten Dämpfe von Mischungen," **104** (1895), 1243–1278.

See also "Über den Arbeitswert einer Luftdruckverteilung und über die Erhaltung der Druckunterschiede," in *Denkschriften der Akademie der Wissenschaften*, Math.-naturwiss. Kl., **73** (1901), 329–345, English trans. by

C. Abbe in "The Mechanics of the Earth's Atmosphere," 3rd collection, *Smithsonian Miscellaneous Collection*, **51** (1910), 501–532; "Über die Beziehung zwischen Barometerschwankungen und Kontinuitätsgleichung," in *Boltzmann-Festschrift* (Leipzig, 1904), 585–589; "Über die Energie der Stürme," in *Jahrbuch der Zentralanstalt für Meteorologie und Erdmagnetismus*, **40** (1905), 1–26, English trans. by C. Abbe in "The Mechanics of the Earth's Atmosphere," 3rd collection (see above), 533–595; "Über die Änderung des vertikalen Temperaturgefälles durch Zusammendrückung oder Ausbreitung einer Luftmasse," in *Meteorologische Zeitschrift*, **23** (1906), 241–244; "Über die Temperaturschichtung in stationär bewegter und in ruhender Luft," *ibid.* (1906), 243–254; and "Zur Sturmtheorie," *ibid.*, **23** (1906), 481–497.

II. SECONDARY LITERATURE. Some information on Margules' personal life may be found in the obituaries in *Meteorologische Zeitschrift*, **37** (1920), 322–324; and *Das Wetter*, **37** (1920), 161–165. See also F. Knoll, ed., *Österreichische Naturforscher, Ärzte und Techniker* (Vienna, 1957), 40–42.

GISELA KUTZBACH

MARIANO, JACOPO. See **Taccola, Jacopo Mariano.**

MARIE, PIERRE (*b.* Paris, France, 9 September 1853; *d.* Normandy, France, 13 April 1940), *neurology.*

Marie, the son of an upper-middle-class French family, studied law before deciding to enter medicine. After completing medical school, he was named *interne des hôpitaux* in 1878 and began his work in neurology under the tutelage of J.-M. Charcot at the Salpêtrière and Bicêtre. He soon became one of Charcot's most outstanding students and served as his laboratory and clinic chief and special assistant. Promoted to *médecin des hôpitaux* in Paris (1888), he was appointed agrégé at the Paris Faculty of Medicine (1889). As part of his work for this position, he presented to the faculty a series of lectures on diseases of the spinal cord, which were published in 1892.

From 1897 to 1907, Marie worked at the Bicêtre, where he created a neurological service that gained worldwide repute. In 1907 he successfully applied for the vacant chair of pathological anatomy in the Faculty of Medicine, and during his ten years there dedicated himself to that profession. With the aid of Gustave Roussy, his successor, Marie completely modernized the teaching of pathological anatomy in medical schools.

Marie resumed his work in clinical neurology in 1918 when he was named to the chair of clinical neurology at the Salpêtrière upon Dejerine's death. During the war, Marie and his colleagues in "Charcot's clinic" devoted most of their time to the study and treatment of neurological traumas in the wounded.

A brilliant clinician in the tradition of Charcot, Marie was an outstanding, demanding teacher. Between 1885 and 1910, the most productive period of his career, he wrote numerous articles and books and developed an international school of neurology which was to produce many distinguished pupils. He possessed a keen intuition which was sharpened by a rigorous approach to the study and practice of neurology. Capable of making shrewd clinical judgments, Marie successfully identified and described a series of disorders with which his name is linked. In one of his earliest and most significant works (1886–1891), he provided the first description and study of acromegaly. Marie's analysis of this pituitary gland disorder was a fundamental contribution to the nascent field of endocrinology. He also was the first to define muscular atrophy type Charcot-Marie (1886); pulmonary hypertrophic osteoarthropathy (1890); cerebellar heredoataxia (1893); cleidocranial dysostosis (1897); and rhizomelic spondylosis (1898).

During this early period of the neurosciences, Marie's views sometimes involved him in great controversy. After a ten-year study, his three papers on aphasia appeared in *Semaine médicale* (1906). They generated much discussion, and three special sessions of the Société Française de Neurologie de Paris convened in 1908 to compare Marie's views on language disorders, which differed from Broca's widely accepted doctrine that aphasia is caused by a lesion in the cerebral hemisphere's "speech center."

Marie led a quiet, private life with his wife and only child, André, who also became a physician. He received few visitors and avoided public appearances although he was awarded numerous honors. His abiding interests were art, the *Revue neurologique*, which he and E. Brissaud founded in 1893, and the Société Française de Neurologie, which he served as its first general secretary.

Marie resigned from his chair at the Salpêtrière and retired at the age of seventy-two, first to the Côte d'Azur and then to Normandy. Grieved by the death of his wife and son, he lived as a virtual recluse there and was increasingly troubled by ill health until his death.

BIBLIOGRAPHY

I. ORIGINAL WORKS. Marie's writings include *Essays on Acromegaly*, with bibliography and appendix of cases by other authors (London, 1891); *Exposé des titres et travaux scientifiques* (Paris, n.d.); *Leçons sur les maladies de la moelle épinière* (Paris, 1892); *Leçons de clinique médicale (Hôtel Dieu 1894–1895)* (Paris, 1896); *Lectures on Diseases of the Spinal Cord*, trans. by M. Lubbock (London, 1895); *Neurologie*, 2 vols. (Paris, 1923); and *Travaux et mémoires* (Paris, 1926).

II. SECONDARY LITERATURE. See Georges Guillain, "Nécrologie. Pierre Marie (1853–1940)," in *Bulletin de l'Académie de médecine*, **123** (1940), 524–535; "Pierre Marie (1853–1940)," in *Revue neurologique*, **72** (1940), 533–543; and Gustave Roussy, "Pierre Marie (1853–1940), Nécrologie," in *Presse médicale*, **48** (1940), 481–483.

JUDITH P. SWAZEY

MARIGNAC, JEAN CHARLES GALISSARD DE (*b.* Geneva, Switzerland, 24 April 1817; *d.* Geneva, 15 April 1894), *inorganic chemistry, physical chemistry.*

Descended from a distinguished Huguenot family, Marignac was the son of Jacob de Marignac, a judge and *conseiller d'état.* His mother was the sister of the pharmacist and physiologist Augustin Le Royer, whose house and laboratory adjoining the Marignac home was a center of Genevan scientific life. In 1835, after education at the Académie de Genève, Marignac entered the École Polytechnique in Paris, where he attended the chemistry lectures of Le Royer's former pupil J. B. Dumas. From 1837 to 1839 he studied engineering and mineralogy at the École des Mines. During 1840 Marignac traveled extensively through Europe, and for a short time he studied the derivatives of naphthalene in Liebig's laboratory at Giessen—his only research on organic chemistry. Through the influence of Dumas, he spent six months during 1841 at the porcelain factory at Sèvres; but, eager for an academic career, in the same year he succeeded Benjamin Delaplanche in the chair of chemistry at the Académie de Genève, taking on in addition the chair of mineralogy in 1845. He resigned in 1878, five years after the Academy became the University of Geneva. From 1884 on, chronic heart disease rendered Marignac a stoic but helpless invalid.

Marignac married Marie Dominicé in 1845. They had five children, one of whom, Édouard, died while a student at the École Polytechnique. Marignac worked unassisted in a damp cellar laboratory for most of his life; and this, together with his reticence and modesty, helped to create the erroneous impression that he was a recluse. From 1846 to 1857 he was a joint editor of the Swiss journal *Archives des sciences*. He commanded great respect from his students and, with the aged Berzelius' enthusiastic approval, a worldwide renown for his analytical accuracy. Always modern in outlook, he supported the work on mass action of Guldberg and Waage; he switched to two-volume formulas (such as H_2O) in 1858; and he attended the important Karlsruhe conference in 1860. In the French controversy over equivalent weights versus atomic weights in 1877, he gave statesmanlike support for the modern school.[1]

Although Marignac completed a large amount of research on mineralogy (showing, for example, that silica should be represented by the formula SiO_2, not SiO_3, because of the isomorphism between fluorstannates and fluorsilicates) and physical chemistry (where he explored the thermal effects of adding variable concentrations of different solutions together, and the alteration of the specific heats of solutions with dilution), only his contributions to inorganic chemistry will be mentioned here. In this field he accurately determined the atomic weights of nearly thirty elements and helped to unravel the tortuous chemistry of niobium and tantalum, the silicates, the tungstates, and the rare earths.

In 1842, inspired by a wave of criticism of Berzelius' atomic weights and by the plausibility of Prout's hypothesis that atomic weights were whole-number multiples of that of hydrogen, Marignac determined the atomic weights of chlorine, potassium, and silver by various methods accurate to $\pm 10^{-3}$. Although his results did not confirm "Prout's law" (as he termed it), he suggested in 1843 that the real multiple might possess only half the atomic weight of hydrogen—a suggestion not approved by Berzelius.[2]

When, in 1860, Stas dismissed Prout's law as an "illusion," Marignac cautioned that deviations from the law of definite proportions might sometimes occur —a possibility suggested by an erroneous view of the composition of acids which he then held. More speculatively, he suggested that Prout's law might be an "ideal" law (like Boyle's law) which was subject to perturbing influences such that the weights of the subatomic particles of the primordial matter (from which ordinary chemical atoms were composed) did not add up to exactly the experimentally determined "atomic" weights. This daring speculation was revived in 1915 by W. D. Harkins and E. D. Wilson, and from it the concept of the packing fraction was developed by F. W. Aston in 1920. Marignac was full of praise for Stas's reply to his challenge in 1865, but unlike Stas he was never able to accept that chance alone was the reason why atomic weights were so close to integers on the $O = 16$ scale (which he urged chemists to adopt in 1883). Unlike Crookes, Marignac did not speculate concerning the genesis of elements; although obviously sympathetic toward Crookes's hypothesis of 1887, he found it wanting for its dubious arguments drawn from rare-earth separations and spectroscopy.[3]

Marignac's groundwork with the rare earths (in which he was frequently helped spectroscopically by his physicist colleague J. L. Soret) began in the 1840's with the separation of the three cerium oxides from cerite. In 1878 he showed, after exacting fractionations based on differing solubilities, that the erbia extracted from gadolinite contained a colorless earth, ytterbia, which he correctly supposed was an oxide of a new metal, ytterbium. His ytterbia was in fact impure, for L. F. Nilson was able to extract scandia from it in 1879; and in 1907 Urbain separated it into (neo)ytterbia and lutecia (now called lutetia). In 1880 Marignac isolated white and yellow oxides from samarskite, which he uncommittedly labeled $Y\alpha$ and $Y\beta$ (samaria). In 1886, at Boisbaudran's request, he named the former gadolinia, and the element "gadolinium." Marignac is usually regarded as the discoverer of ytterbium and gadolinium. In general, his separations were a strategic and indispensable part of chemists' success in understanding the elements of the rare-earth series.

NOTES

1. For the debates at the French Academy of Sciences, see *Comptes rendus . . . de l'Académie des sciences*, **84** (1877), *passim*; and Marignac's comments, in his *Oeuvres*, II, 649–667.
2. J. J. Berzelius, in the Swedish Academy's *Jahres-Bericht*, **24** (1845), 60–62. In 1858 Marignac pointed out that a quarter-unit would preserve Prout's law. *Oeuvres*, I, 571.
3. For Crookes's reply, see *Chemical News . . .*, **56** (1887), 39–40.

BIBLIOGRAPHY

I. ORIGINAL WORKS. Marignac's 111 published papers were handsomely repr. as *Oeuvres complètes de J. C. Galissard de Marignac*, E. Ador, ed., 2 vols. (Geneva, 1902–1903), with portrait and a complete list of Marignac's atomic weights. From 1846 to 1857 Marignac was joint editor with A. de La Rive of *Archives des sciences physiques et naturelles*, **1–36** (a supp. of the *Bibliothèque universelle de Genève*). Marignac's criticism of J. S. Stas is translated in [L. Dobbin and J. Kendall,] *Prout's Hypothesis*, Alembic Club Reprints, no. 20 (Edinburgh, 1932), which also contains the relevant portions of Stas's memoir of 1860; for Stas's "answer" of 1865, see "Nouvelles recherches sur les lois des proportions chimiques, sur les poids atomiques et leurs rapports mutuels," in *Mémoires de l'Académie royale de Belgique*, **35** (1865), 3–311. For Kekulé's critique of Marignac's criticism, see his "Considérations d'un mémoire de M. Stas sur les lois des proportions chimiques," in *Bulletin de l'Académie royale de Belgique*, **19** (1865), 411–420, repr. in R. Anschütz, *August Kekulé* (Berlin, 1929), II, 357–364. For a view of Marignac's mineralogy by a lifelong friend, see A. L. O. Le Grand des Cloizeaux, *Manuel de minéralogie*, 2 vols. (Paris, 1862–1874). Two letters from the Berzelius-Marignac correspondence are in H. G. Söderbaum, ed., *Jac. Berzelius Bref*, III, pt. 7 (Uppsala, 1920), 210–216; note also 253–254.

MSS held by the Bibliothèque Publique et Universitaire de Genève include travel diaries (1839, 1840), analytical notebooks (1844 on), Swiss correspondence, and lecture notes of Marignac's students.

II. Secondary Literature. The basic life of Marignac is by Ador in the *Oeuvres*, I, i–lv, repr. from the *Archives des sciences*, **32** (1894), 183–215, and partly repr. with portrait and bibliography in *Bulletin de la Société chimique de Paris*, **17** (1894), 233–239 and *Berichte der Deutschen chemischen Gesellschaft*, **27** (1894), 979–1021. Other useful notices are P. T. Cleve, "Marignac Memorial Lecture," in *Journal of the Chemical Society*, **67** (1895), 468–489, with portrait and bibliography, repr. in *Memorial Lectures Delivered Before the Chemical Society, 1893–1900* (London, 1901); and "De Marignac," in Société des Amis de l'École, *L'École polytechnique* (Paris, 1932), pp. 194–196.

Information on Marignac's atomic weights is in *Prout's Hypothesis* (see above); I. Freund, *The Study of Chemical Composition* (Cambridge, 1904; repr. New York, 1968), pp. 599–603; and W. V. Farrar, "Nineteenth-Century Speculations on the Complexity of the Chemical Elements," in *British Journal for the History of Science*, **2** (1965), 307–308. For Marignac's contribution to rare-earth chemistry, see the obituary by Cleve (above); O. I. Deineka, "Issledovania Mariniaka po khimii redkozemelnykh elementov" ("The Research of Marignac in the Chemistry of the Rare-Earth Elements"), in *Trudy Instituta istorii estestvoznaniya i tekhniki*, **39** (1962), 87–94; and M. E. Weeks, *Discovery of the Elements*, 7th ed. (Easton, Pa., 1968), ch. 16, *passim*.

W. H. Brock

MARINUS (*b.* Neapolis [the Biblical Shechem, now Nablus], Palestine; *fl.* second half of fifth century A.D.; *d.* Athens[?]), *philosophy*.

Marinus, probably a Samaritan (perhaps also a Jew), became a convert to the Hellenic-pagan way of life.[1] He joined the Platonic Academy when Proclus, who dedicated his commentary on the Myth of Er in Plato's *Republic* to him, was its head.[2] After Proclus' death in A.D. 485 Marinus became the president of the Academy; he was evidently—and, as far as one can judge, rightly—considered the best representative of the views of Proclus, whom he praised and eulogized in an extant biography.

If one wants to assess the change in the philosophical climate since Plotinus' death in A.D. 270, it is very instructive to compare Porphyry's *Life of Plotinus* with Marinus' *Life of Proclus*, written two centuries later. Marinus, however, does not seem to have been merely a dogmatic follower of his systematizing predecessor; he did not hesitate to adopt an independent

and more realistic, down-to-earth attitude wherever he deemed it necessary. In his exegesis of Plato he rightly maintained, for instance, that Plato, when writing the *Parmenides*, had not, as Proclus' other disciples thought, been concerned with gods but with εἴδη, "Forms." Like other late Neoplatonists, he appreciated mathematics very highly: "I wished everything were mathematics."[3]

Marinus proposed a new solution to the Peripatetic-Academic problem of the Active Intelligence (νοῦς ποιητικός) by localizing it, as did the great Aristotelian Alexander of Aphrodisias (*ca.* A.D. 200), in the superlunary world but no longer identifying it with the First Cause. He placed it below the First Cause, as an "angelic, spiritual" being, making it a kind of intermediary between the highest stage of man's intellect and the unchanging superior world.[4] His view became, in due course, important for Islamic Arabic philosophers and was, with slight modifications, adopted by two of the most outstanding among them, al-Fārābī and Ibn Sīnā.[5]

NOTES

1. Damascius, *Vita Isidori*, R. Asmus, ed., ch. 141: τὸ Ἑλληνικὸν ἠγάπησεν.
2. See Proclus, *In rem publicam*, W. Kroll, ed., II, pp. 96, 200.
3. Elias, *Prolegomena*, A. Busse, ed., in *Commentaria in Aristotelem Graeca*, XVIII, pp. 28, 29: εἴθε πάντα μαθήματα ἦν.
4. See Pseudo-Philoponus, *De Anima*, M. Hayduck, ed., in *Commentaria in Aristotelem Graeca*, p. 535.
5. See R. Walzer in "Aristotle's Active Intellect (νοῦς ποιητικός) in Greek and Early Islamic Philosophy," in *Potino e il Neoplatonismo in Oriente e in Occidente* (Rome, 1974).

BIBLIOGRAPHY

I. Original Works. The only extant philosophical work by Marinus is the biography of Proclus, J. Boissonade, ed. (Leipzig, 1814), repr. in *Procli opera inedita*, V. Cousin, ed., 2nd ed. (Paris, 1864), and in Diogenes Laërtius, C. G. Cobet, ed. (Paris, 1878). There is an English trans. from the Greek in L. G. Rosan, *The Philosophy of Proclus* (New York, 1949).

II. Secondary Literature. The best account of Marinus is in F. Ueberweg and K. Praechter, *Die Philosophie des Altertums*, XIII (Tübingen, 1953), pp. 631 ff. His influence on Islamic Arabic philosophers is discussed by R. Walzer, in *Le néoplatonisme* (Paris, 1971), pp. 319 ff.

R. Walzer

MARION, ANTOINE FORTUNÉ (*b.* Aix-en-Provence, France, 10 October 1846; *d.* Marseilles, France, 22 January 1900), *zoology, geology, botany, plant paleontology*.

Marion came from a family of modest means. He attended the *lycée* at Aix, where he was a classmate of the novelist Émile Zola. His intelligence and inclination toward the natural sciences attracted the attention of Henri Coquand, professor of geology at the Faculté des Sciences of Marseilles, who had him appointed an assistant in natural history in 1862, a few days before he received his *baccalauréat ès lettres* and two years before his *baccalauréat ès sciences*. In 1868 he earned his *licence ès sciences naturelles*, and in 1869 he shared the Bordin Prize of the Academy of Sciences for his "Recherches anatomiques et zoologiques sur des nématoïdes non parasites marins." This work formed the basis of his doctoral thesis (1870).

In 1858 Marion had presented to the Marquis Gaston de Saporta a fossil leaf of *Magnolia*, a new variety, which he had discovered in the gypsum of Aix. This incident marked the beginning of a long collaboration and friendship, which remained, according to Saporta, "free from any disturbance or element of discord."

Marion's first publications (1867), which dealt with geology and paleontology, were followed by memoirs on plant paleontology published either alone or in collaboration with Saporta. Marion continued to publish works in this field until 1888, interspersing them with others on zoology, embryology, and marine biology. One of the most noteworthy is *L'évolution du règne végétal*, in three volumes (1881–1885). This synthesis, long a classic in the field, is not, properly speaking, a theoretical work but an attempt to apply transformist ideas to the history of plant life.

In 1871 as *chargé de cours* Marion gave a free course on the geology of Provence at the Faculté des Sciences of Marseilles. From October 1871 to November 1872 he took over the teaching of the natural sciences at the *lycée* of Marseilles. The Faculty of Sciences nominated him to fill the chair of geology left vacant by the death of Lespès—which, if Marion had obtained it, would have given his career a very different pattern. In 1872 he gave a course *(cours complémentaire)* on zoology for the École Pratique des Hautes Études and directed a laboratory of marine zoology established in the Allée de Meilhan, where the Faculté des Sciences had been. Numerous French and foreign researchers soon came to the laboratory, including Bobretsky, Weismann, O. Schmidt, and especially A. Kovalevsky, with whom Marion became quite friendly; he was also visited by Alexander Agassiz.

In 1876 a chair of zoology was established at Marseilles; Marion had to wait until he was thirty to assume it. This post helped him to gain acceptance for his plan to build a large marine laboratory on the coast at Endoume, a project initiated in 1878 with the support of the city of Marseilles. A source of hardship and disappointment for Marion, the project was finally accomplished after the city council of Marseilles at long last approved the necessary legislation on 16 December 1887.

In 1880 Marion succeeded the botanist E. M. Heckel as director of the Museum of Natural History of Marseilles. In 1883 he founded the museum's *Annales*, which until his death were subtitled *Travaux du Laboratoire de zoologie marine*. Indeed, Marion considered this publication to be essentially the organ of the laboratory; his wish was to coordinate the activities of the laboratory of marine zoology, the Endoume station, and the museum. In his goal of enriching the Marseilles museum with regional collections he amassed most of the material in the Salle de Provence.

A very special aspect of Marion's scientific activity was his involvement in the struggle against phylloxera from 1876 to 1878. Recognizing the inadequacy of sulfocarbonate and the efficacy of carbon disulfide, he devised methods of injecting the latter substance into the soil. These methods were of considerable value until the studies on American vines had been completed and had shown that the grafting of stock from California produced immunity. These efforts brought Marion considerable recognition as well as French and foreign honors. He received the Grande Médaille of the French National Society of Agriculture (1881) and was made Knight of the Crown of Italy (1879), Commander of Christ (Portugal, 1880), and Commander of Saint Anne (Russia, 1893). The Russian government invited him to be a member of a Commission on Vineyards, which visited Bessarabia, the Crimea, and the Caucasus. He was a member of a similar mission in the vineyards of Hungary.

Marion's extremely important zoological work dealt with marine invertebrates. After his thesis (1870), he returned to his research on the free-living nematodes (roundworms) of the Gulf of Marseilles (1870). Several of his publications are entitled "Recherches sur les animaux inférieurs du golfe de Marseille" (1873–1874). His studies in the field include memoirs on the parasitic Rotifera (1872), the nemerteans (1869–1875), the echinoderms (1873), the zoantharians (1882), the Alcyonarians (1877, 1882, 1884), the parasitic crustaceans (1882), the annelids (1874), the enteropneusts (1885, 1886), and the mollusks (1885, 1886). Two works were awarded the Grand Prize in Physical Sciences by the Academy of Sciences in 1885: "Esquisse d'une topographie zoologique du golfe de Marseille" (1883) and "Considérations sur les faunes profondes de la Méditerranée étudiées d'après les dragages opérés sur les côtes méridionales de France" (1883).

Marion constantly sought practical applications of marine zoology. He was ahead of his time in advocating the establishment of "maritime fields" where marine animals could be raised, studied, and experimented on in isolated reserves. He was a man of seductive charm and simplicity, of extremely varied talents and an open mind, and an exceptional teacher. In 1887 he was elected a corresponding member of the Institut de France. Although not cautious about his health, he dreaded long trips; before leaving for Russia, he wondered whether he would return, lamenting his "rather prematurely impaired health" (unpublished letter to Lacaze-Duthiers). Deeply grieved by the sudden death of his only daughter, he died a few months later.

BIBLIOGRAPHY

I. ORIGINAL WORKS. Marion's writings on botany and plant paleontology include "Description des plantes fossiles de Ronzon (Haute-Loire)," in *Comptes rendus . . . de l'Académie des sciences*, **74** (1872), 62–64; "Essai sur l'état de la végétation à l'époque des marnes heersiennes de Gelinden," in *Mémoires de l'Académie royale des sciences, des lettres et des beaux-arts, de Belgique*, **37**, no. 6 (1873), written with Saporta; "Recherches sur les végétaux fossiles de Meximieux (Ain), précédées d'une introduction stratigraphique par A. Falsan," in *Archives du Muséum d'histoire naturelle de Lyon*, **1** (1875–1876), 131–324, written with Saporta; "Sur les genres *Williamsonia* et *Goniolina*," in *Comptes rendus . . . de l'Académie des sciences*, **92** (1881), 1185–1188, 1268–1270, written with Saporta; *L'évolution du règne végétal. Les cryptogames*, in Bibliothèque des Sciences Internationales (Paris, 1881), written with Saporta; *L'évolution du règne végétal. Les phanérogames*, 2 vols. (Paris, 1885), written with Saporta; and "Sur les *Gomphostrobus heterophylla*, conifères prototypiques du Permien de Lodève," in *Comptes rendus . . . de l'Académie des sciences*, **110** (1890), 892–894.

On zoology and marine biology, see "Note sur l'histologie du système nerveux des némertes," in *Comptes rendus . . . de l'Académie des sciences*, **68** (1869), 1474–1475; "Recherches anatomiques et zoologiques sur des nématoïdes non parasites marins," in *Annales des sciences naturelles (zoologie)*, **13** (1870), 1–90; "Recherches sur les animaux inférieurs du golfe de Marseille. Sur un nouveau némertien hermaphrodite. Observations sur *Borlasia Kefersteini*," *ibid.*, **17** (1873), 1–21; "Recherches sur les animaux inférieurs du golfe de Marseille. II. Description de crustacés amphipodes parasites des salpes," *ibid.*, n.s. **1** (1874), 1–20; "Sur les espèces méditerranéennes du genre *Eusyllis*," in *Comptes rendus . . . de l'Académie des sciences*, **80** (1875), 498–499; "Étude des annélides du golfe de Marseille," in *Annales des sciences naturelles*, n.s. **2** (1875), 1–106, written with N. Bobretsky; "Révision des nématoïdes du golfe de Marseille," in *Comptes rendus . . . de*

l'Académie des sciences, **80** (1875), 499–501; and "Dragages au large de Marseille (juillet–septembre 1875)," in *Annales des sciences naturelles*, n.s. **8** (1879).

See also "Études sur les *Neomenia*," in *Zoologischer Anzeiger*, **5** (1882), 61–64, written with Kovalevsky; "Considérations sur les faunes profondes de la Méditerranée étudiées d'après les dragages opérés sur les côtes méridionales de la France," in *Annales du Musée d'histoire naturelle de Marseille*, **1**,2 (1883), 1–40, which won the Academy's Grand Prize in Physical Sciences in 1885; "Documents pour l'histoire embryogénique des alcyonnaires," *ibid.*, **1**,4 (1883), 1–50, written with Kovalevsky; "Esquisse d'une topographie zoologique du golfe de Marseille," *ibid.*, **1**,1 (1883), 1–120; "Organisation du *Lepidomenia hystrix*, nouveau type de solénogastre," in *Comptes rendus . . . de l'Académie des sciences*, **103** (1886), 757–759, written with Kovalevsky; "Documents ichthyologiques. Énumération des espèces rares de poissons capturés sur les côtes de Provence," in *Zoologischer Anzeiger*, **9** (1886), 375–380; "Contribution à l'histoire naturelle des solénogastres ou aplacophores," in *Annales du Musée d'histoire naturelle de Marseille*, **3** (1887), 1–76, written with Kovalevsky; and "Sur les espèces de *Proneomenia* des côtes de Provence," in *Comptes rendus . . . de l'Académie des sciences*, **106** (1888), 529–532, written with Kovalevsky.

Works on applied marine zoology appeared in the *Annales du Musée d'histoire naturelle de Marseille (Travaux du Laboratoire de zoologie marine)*. In vol. **3** (1886–1889) are articles on the anchovy and remarks on the mackerel of the Provençal coast. In vol. **4** (1890–1894) are memoirs on the fishing and the reproduction of the *Atherina hepsetus*, on floating eggs and young fish observed in the Gulf of Marseilles in 1890, and on the raising of some young fish, as well as remarks on the systematic exploitation of the shore land. Also included are climatic observations made at the zoological station at Endoume for the study of the regional fishing industry. Vol. **5** (1897–1899) contains mainly articles on climatic conditions during 1893, 1894, and 1895 designed "to aid in the statistical study of the fishing industry on the Marseilles coast."

On Marion's efforts to combat phylloxera, see "Sur l'emploi du sulfure de carbone contre le Phylloxera," in *Comptes rendus . . . de l'Académie des sciences*, **82** (1876), 1381; and "Remarques sur l'emploi du sulfure de carbone au traitement des vignes phylloxérées," *ibid.*, **112** (1891), 1113–1117, written with G. Gastine. There are various articles on agricultural techniques and applied zoology in *Revue générale d'agriculture et de viticulture méridionales* (May–Oct. 1898).

II. SECONDARY LITERATURE. See G. Gastine, "Antoine-Fortuné Marion," in *Bulletin mensuel de la Société départementale d'agriculture des Bouches-du-Rhône*, no. 2 (Feb. 1900), 33–43; Jourdan, A. Vayssière, and G. Gastine, "Notice sur la vie et les travaux de A. F. Marion," in *Annales de la Faculté des sciences de Marseille*, **11** (1901), 1–26; G. Petit, "F. A. [*sic*] Marion (1846–1900)," in *Bulletin du Muséum d'histoire naturelle de Marseille*, **1**, no. 1 (1941), 5–12; Marquis de Saporta, "Notice sur les travaux scientifiques de M. A. F. Marion," in *Mémoires de l'Aca-*

démie des sciences, agriculture, arts et belles-lettres d'Aix, **13** (1885), 241–284; and A. Vayssière, "Notice bibliographique sur A. F. Marion," in *Annales du Musée d'histoire naturelle de Marseille,* **6** (1901), 7–9.

G. Petit

MARIOTTE, EDME (*d.* Paris, France, 12 May 1684), *experimental physics, mechanics, hydraulics, optics, plant physiology, meteorology, surveying, methodology.*

Honored as the man who introduced experimental physics into France,[1] Mariotte played a central role in the work of the Paris Academy of Sciences from shortly after its formation in 1666 until his death in 1684. He became, in fact, so identified with the Academy that no trace remains of his life outside of it or before joining it. There is no documentation to support the tentative claim of most sources that he was born around 1620 in Dijon. His date of birth is entirely unknown, and his title of *seigneur de Chaseüil* makes the present-day Chazeuil in Burgundy (Côte-d'Or) the more likely site of his birth and childhood; several families of Mariottes are recorded for the immediate area in the early seventeenth century. Indirect evidence places him as titular abbot and prior of St.-Martin-de-Beaumont-sur-Vingeanne (Côte-d'Or), but his precise ecclesiastical standing is uncertain; contemporary sources generally do not refer to him by a clerical title.[2]

If not born in Dijon, Mariotte appears to have been residing there when he was named to the Academy. His letter announcing the discovery of the blind spot in the eye was sent from Dijon in 1668, as was the one extant (and perhaps only) letter he wrote to Christiaan Huygens. A letter from one Oded Louis Mathion to Huygens in 1669 suggests that Mariotte was then still in Dijon or at least that he had been there long enough to establish personal contacts.[3] By the 1670's, however, Mariotte had moved to Paris, where he spent the rest of his life.

Lack of biographical information makes it impossible to determine what drew Mariotte to the study of science and when or where he learned what he knew when named to the Academy. Seldom citing the names of others in his works, all written after joining the Academy, he left no clues about his scientific education. The circumstances surrounding his nomination, the letter to Huygens in 1668, and the nature of his scientific work combine to suggest that he was self-taught in relative isolation.[4]

It was as a plant physiologist that Mariotte first attracted the attention of the Academy shortly after its founding. Engaged in discussion of Claude Perrault's theory of the vegetation of plants, the original academicians apparently invited a contribution by Mariotte, who held the "singular doctrine"[5] that sap circulated through plants in a manner analogous to the circulation of blood in animals. Mariotte's verbal presentation of his theory and of the experiments on which it was based drew a rejoinder from Perrault, and the Academy charged the two men to return with written accounts and further experiments. At the same time, apparently, Mariotte was elected to membership as *physicien.* He carried out his charge on 27 July 1667, presenting the first draft of what was published in 1679 as *De la végétation des plantes.* His detailed argument from plant anatomy and from a series of ingenious experiments failed to resolve the controversy completely, and only in the early 1670's did accumulated evidence provided by others vindicate his position.

Whatever the advanced state of his botanical learning in 1666, Mariotte's education in other realms of science seems to have taken place in the Academy. When, for example, late in 1667 or early in 1668, he presented some of his findings on "the motion of pendulums and of heavy things that fall toward the center [of the earth]" and an account of "why the strings of the lute impress their motion on those in unison or in [the ratio of an] octave with them," he learned from Huygens that Galileo had already achieved similar results and only then, on Huygens' advice, read Galileo's *Two New Sciences.*[6] All of Mariotte's works have two basic characteristics: they treat subjects discussed at length in the Academy, and they rest in large part on fundamental results achieved by others. His own strength lay in his talent for recognizing the importance of those results, for confirming them by new and careful experiments, and for drawing out their implications.

Mariotte's career was an Academy career, which embodied the pattern of research envisaged by the founders of the institution. Although named as *physicien,* he soon shared in the work of the *mathématiciens* as well. In 1668, while continuing the debate on plant circulation, he took active part in a discussion of the comparative mechanical advantages of small and large wheels on a rocky road, read a paper containing twenty-nine propositions on the motive force of water and air (a subject to which he repeatedly returned during his career), read another paper on perspective (geometrical optics), and reviewed two recently published mathematical works.

Also in 1668 Mariotte published his first work, *Nouvelle découverte touchant la veüe,* which immediately embroiled him in a controversy that lasted until his death, although no one denied the discovery itself. Curious about what happened to light rays striking the base of the optic nerve, Mariotte devised a simple experiment: placing two small white spots on a dark background, one in the center and the other two feet to the right and slightly below the center line, he covered his left eye and focused his right eye on the center spot. When he backed away about nine feet, he found that the second spot disappeared completely, leaving a single spot on a completely dark surface; the slightest motion of his eye or head brought the second spot back into view. By experiments with black spots on a white background and with the spots reversed for the left eye, he determined that the spot disappeared when the light from it directly struck the base of the optic nerve, which therefore constituted a blind spot or, as he called it, defect of vision in the eye.

The discovery, confirmed by Mariotte's colleagues in both Paris and London, startled him into abandoning the traditional (and correct) view that images in the eye are formed on the retina (a continuous layer of tissue) and adopting the choroid coat behind the retina (discontinuous precisely where the optic nerve passes through it to attach to the retina) as the seat of vision. Mariotte's fellow academician, the anatomist Jean Pecquet, who with others had been investigating the eye since 1667, immediately disputed this conclusion and wrote to defend the traditional view. A series of experiments carried out before the Academy in August 1669 only widened the area of disagreement between the two men, as did Perrault's support of Pecquet. Mariotte published a rebuttal of Pecquet's critique in 1671, but by then the issue had become moot.[7] As the *Lettres écrites par MM. Mariotte, Pecquet et Perrault . . .* (1676) reveals, Pecquet and Perrault could not provide a convincing explanation for a blind spot on the retina (partly because they disagreed with Mariotte over the action of nerves), and Mariotte rested part of his argument on phenomena now known to be irrelevant (such as the reflection of light from the choroid of certain animals).

Although the controversy over the seat of vision dominated Mariotte's attention in 1669 and 1670, he continued his research in other areas. In 1669 he took part in discussions of the cause of weight (in which he supported some form of action at a distance against Huygens' mechanical explanation) and of the nature of coagulation of liquids. The latter issue seems to lie behind a series of experiments on freezing, carried out and presented jointly with Perrault in 1670. The experiments, which concerned the pattern of formation of ice and the trapping of air bubbles in it, enabled Mariotte to construct a burning glass of ice.

In 1671 Mariotte read a portion of his *Traité du nivellement* (published in 1672) describing a new form of level which used the surface of free-standing water as a horizontal reference and employed reflection of a mark on the sighting stick to gain greater accuracy in sighting. In the treatise itself he gave full instructions for the instrument's use in the field and a detailed analysis of its accuracy in comparison with that of traditional levels, in particular the *chorobates* of Vitruvius. In 1672 Mariotte's activities in the Academy were restricted largely to confirmation of G. D. Cassini's discoveries of a spot on Jupiter and a new satellite (Rhea) of Saturn.

As early as 1670 Mariotte had announced his intention to compose a major work on the impact of bodies. Completed and read to the Academy in 1671, it was published in 1673 as *Traité de la percussion ou choc des corps*. The first comprehensive treatment of the laws of inelastic and elastic impact and of their application to various physical problems, it long served as the standard work on the subject and went through three editions in Mariotte's lifetime.

Part I of the two-part treatise begins with the definitions of "inelastic body," "elastic body," and "relative velocity" and then makes four "suppositions": (1) the law of inertia; (2) Galileo's theorem linking the speeds of free-falling bodies to the heights from which they fall from rest ($v^2 \propto h$); (3) the independence of the speed acquired from the path taken in falling; and (4) the tautochronism of simple pendulums for small oscillations. The suppositions form the basis for an experimental apparatus consisting of two simple pendulums of equal length, the replaceable bobs (the impacting bodies) of which meet at dead center. To facilitate measurement, Mariotte makes the further assumption that for small arcs the velocity of the bob varies as the arc length (rather than the versine). Repeated experiments on inelastic clay bobs of varying sizes confirm a series of propositions, most of them termed by Mariotte "principles of experience": the additivity of motion (both directly and obliquely), the dependence of impact only on the relative velocity of the bodies (confirmed by the extent of flattening on impact), the quantity of motion (weight times speed) as the effective parameter of impact (also measured by flattening), and the laws of inelastic collision linking initial and final speeds through the conservation of quantity of motion.

The laws governing elastic collisions then follow from those of inelastic collision by means of the principle that a perfectly elastic body deformed by the impact of a "hard and inelastic" body regains its original shape and, in doing so, imparts to the impacting body its original speed. Confirmed by experiments involving the striking of a stretched string by a pendulum bob, the principle leads Mariotte to another series of experiments designed to show that such apparently "hard" bodies as ivory or glass balls in fact deform upon impact. In one test he lets ivory balls fall from varying heights onto a steel anvil coated lightly with dust; circles of varying widths show that the degree of flattening is dependent on the speed of the ball at impact, and the return of the balls to their initial heights confirms his principle of full restoration. By treating elastic bodies at the point of maximum deformation as inelastic ones, and by then distributing the added speeds acquired by restoration inversely as the weights of the bodies, Mariotte succeeds in determining the laws of elastic collision, applying them in one instance to the recoil of a cannon and directly testing the results experimentally. Part I closes with the transmission of an impulse through a chain of contiguous elastic bodies, a problem first posed by Descartes's theory of light.

Part II opens with a treatment of oblique collision, in which Mariotte employs (without citing his source) Huygens' model of impact in a moving boat. In Mariotte's use of the model, the boat is traveling at right angles to the plane of the impacting pendulums, allowing the application of the parallelogram of motion to derive the laws of oblique collision from those of direct impact. Mariotte then turns to some problems in hydrodynamics, in which he combines the hydrostatic paradox and the results thus far obtained to argue that the speed of efflux of water from a filled tank varies as the square root of the height of the surface above the opening. Examining next the force of that efflux, he determines by use of a balance beam that the force is to the weight of the water in the reservoir as the cross-sectional area of the tube (or opening) is to that of the reservoir. Combining these results, he concludes that the force of impact of a moving stream against a quiescent body varies as the cross-sectional area of the stream and the square of its speed. In a particularly suggestive passage Mariotte argues that a body moved by a steady stream of fluid striking it (that is, by a succession of uniform impulses) accelerates in the same manner as a falling body, that is, according to Galileo's law of $S \propto t^2$. In contradiction to Galileo, however, he insists that bodies falling through air must have a finite first speed of motion.

The use of a balance beam as apparatus and model leads to a treatment of the center of percussion, or center of agitation, of a compound pendulum, that is, the point on the pendulum that strikes an object with the greatest force. In one of the earliest published applications of algebraic analysis to physical problems, Mariotte obtains a solution by determining the point

that divides a rigid bar rotating about one endpoint into two segments having equal quantities of motion. His attempt then to equate the center of percussion with the center of oscillation (he calls it the center of vibration) makes his treatment of the latter far less successful than that of Huygens in his *Horologium oscillatorium*, also published in 1673. A few experiments on the fall of bodies through various media, performed in 1682 with Philippe de La Hire, bring the final version of the *Traité de la percussion* to a close.

Taken as a whole, the treatise reveals Mariotte as a gifted experimenter, learned enough in mathematics to link experiment and theory and to draw the theoretical implications of his work. He made full use of the results obtained by his predecessors and contemporaries, but his experimental mode of analysis and presentation differed markedly from their approaches to the problem. Clearly he knew of the work of Wallis, Wren, and Huygens published in the *Philosophical Transactions of the Royal Society* in 1668; and there are enough striking similarities between Mariotte's treatise and Huygens' then unpublished paper on impact (*De motu corporum ex percussione,* in *Oeuvres,* XVI) to suggest that he knew the content of the latter, perhaps verbally from Huygens himself. Certainly his colleagues in the Academy recognized Mariotte's debt to others while they praised the clarity of his presentation. And yet Galileo's name alone appears in the treatise; Huygens' in particular is conspicuously absent.

Some seventeen years later, in 1690, when Mariotte was dead, Huygens responded to this slight (whether intentional or not) by accusing Mariotte of plagiarism. "Mariotte took everything from me," he protested in a sketch of an introduction to a treatise on impact never completed,

> ... as can attest those of the Academy, M. du Hamel, M. Gallois, and the registers. [He took] the machine, the experiment on the rebound of glass balls, the experiment of one or more balls pushed together against a line of equal balls, the theorems that I had published [in the *Philosophical Transactions* (1668) and the *Journal des sçavans* (1669)]. He should have mentioned me. I told him that one day, and he could not respond.[8]

Except for the published theorems, if Mariotte took the other ideas from Huygens, he could have done so only when Huygens offered them in the course of Academy discussions. Mariotte may well have considered the content of those discussions the common property of all academicians and have felt no need to record their specific sources when publishing them under his own name.

Certainly in 1671 and 1673 Huygens made no proprietary claims in the Academy. His response to Mariotte's treatise consisted of a critique of the theory of elasticity on which parts of it were based. His commitment on cosmological grounds to the existence in nature of perfectly hard bodies that rebound from one another and transmit impulses placed him at odds with Mariotte's empirically based rejection of them.[9] Later that same theory of elastic rebound formed an integral part of Huygens' *Traité de la lumière* (1690). Huygens also later denied Mariotte any role in the determination of the center of oscillation, claiming (despite the evidence of Mariotte's treatise) that he had discussed only the center of percussion and had failed to demonstrate that it was the same as the center of oscillation.

Following the presentation of the *Traité de la percussion*, Mariotte seems largely to have withdrawn from Academy activities until 1675. Scattered evidence, particularly a striking demonstration of the hydrostatic paradox performed before a large audience at the Collège de Bourgogne in Paris in 1674 and reported in the *Journal des sçavans* in 1678 (p. 214), suggests that he was at work on the pneumatic experiments that form the basis of his *De la nature de l'air* (1679). Like his other works, *Nature de l'air* combines a review and reconfirmation of what was already known about its subject with some original contributions. Like the *Traité de la percussion*, it also omits the name of the author on whom Mariotte clearly had relied most heavily, in this case Robert Boyle, while acknowledging its debt to a more distant source, Blaise Pascal's *Équilibre des liqueurs*.

Nature de l'air focuses on three main properties of air: its weight, its elasticity, and its solubility in water. To show that air has weight, Mariotte points out the behavior of the mercury barometer and the common interpretation that the weight of the column of mercury counterbalances the weight of the column of air standing on the reservoir. Turning to the elasticity of air, he presents a series of experiments in which air is trapped in the mercury tube before it is immersed in the reservoir, thus depressing the height at which the mercury settles. He thereby establishes that

> ... the ratio of the expanded air to the volume of that left above the mercury before the experiment is the same as that of twenty-eight inches of mercury, which is the whole weight of the atmosphere, to the excess of twenty-eight inches over the height at which [the mercury] remains after the experiment. This makes known sufficiently for one to take it as a certain rule of nature that air is condensed in proportion to the weight with which it is charged.[10]

This last statement, further confirmed by experiments with a double-column barometer and extended to the expansion of air through experiments in a vacuum receiver, has gained Mariotte a share of Boyle's credit for the discovery and formulation of the volume-pressure law; indeed, it is called "Mariotte's law" in France. If, however, in the essay Mariotte gives no credit to Boyle, neither does he make any claims of originality; rather, he treats the law as one of a series of well-known properties of air.

Interested in the barometer more as a meteorological tool than as an experimental apparatus, Mariotte turns next to a discussion of the relation between barometric pressure and winds and weather (a subject to which he returned at greater length in his *Traité du mouvement des eaux*), and then to the solubility of air in water. The determination through experiment that water does absorb air in amounts dependent on pressure and temperature leads him, in one of his rare excursions into the theory of matter, to posit the existence of a *matière aérienne*, a highly condensed form of matter into which air is forced under high pressure and low temperature. His discussion, which includes the work on freezing done with Perrault in 1670, rests in part, however, on a lack of distinction between the air dissolved in water and the water vapor produced by high temperature and low pressure.

Despite his commitment to a special form of matter to explain solubility, Mariotte rejects any attempt to reduce the elasticity of air to a more fundamental mechanism. Explicitly denying the existence of an expansive subtle matter among the particles of air, he prefers to rely heuristically on Boyle's analogy (without citing the source) of a ball of cotton or wool fibers. A similar analogy of sponges piled on top of one another, together with the volume-pressure law, suggests to him a means for determining the height of the atmosphere. On the initial assumption that a given volume of air at sea level would expand some 4,000 times without the pressure of the air above it and that the pressure drops uniformly with increasing altitude, he divides the atmosphere into 4,032 strata, each stratum corresponding to a drop of 1/12 line (1/144 inch) in pressure from a height of twenty-eight inches of mercury at sea level. Comparing measurements made by Toinard and Rohault with ones made by himself with Cassini and Picard, he estimates that at sea level a rise of five feet in altitude corresponds to a drop of 1/12 line in pressure; hence the first stratum is five feet in height. By the volume-pressure law, the 2,016th stratum is ten feet thick, the 3,024th is twenty feet thick, and so on.

As a simplifying arithmetical approximation to the resulting geometric progression, Mariotte assumes an average stratum of 7.5 feet for the first 2,016 strata, an average of fifteen feet for the next 1,008, and so on. Each group of strata then has a thickness of 15,120 feet, or 5/4 league, and a total of twelve groups has a total height of fifteen leagues (sixty kilometers). As Mariotte notes, varying estimates of the full expansibility of air lead to different values for the height of the atmosphere; but even a factor of 8 million produces a height of less than thirty leagues by his method, which he further confirms by theoretical calculation of the known heights of mountains.

Mariotte's essay closes with some random observations, one of which, an assertion that air is not colorless but blue, forecasts his next major investigation. Involved in 1675–1676 in the Academy's (never completed) project to meet Louis XIV's request for a complete inventory of the machines in use in France, which was to be prefaced by a short theoretical introduction,[11] and in 1677 in a series of varied experiments, by 1678 he had begun a fairly continuous series of reports to the Academy on the rainbow and the refraction of light by lenses and small apertures. A proposal by Carcavi in March 1679 for a complete treatise on optics by Mariotte, Picard, and La Hire resulted in Mariotte's presentation in July and August of his *De la nature des couleurs*. He worked further on the essay over the next two years, reporting frequently to the Academy. In 1681 he read the final version, which then appeared as the fourth of the *Essays de physique*.

Perhaps the best example of Mariotte's experimental finesse, the essay on colors also illustrates well his scientific eclecticism, combining Cartesian epistemology, Baconian methodology, and Aristotelian modes of explanation. He begins:

> Among our sensations it is difficult not to confuse what comes from the part of objects with what comes from the part of our senses. . . . Supposing this, one clearly sees that it is not easy to say much about colors, . . . and that all one can expect in such a difficult subject is to give some general rules and to derive from them consequences that can be of some use in the arts and satisfy somewhat the natural desire we have to render account of everything that appears to us.[12]

On the basis, then, of four suppositions (the geometric structure of the light cone and of a sunbeam passing through an aperture; the gross phenomenon of refraction, including partial reflection; the refraction of light toward the normal in the denser medium, and conversely; and the focusing of the human eye),[13] Mariotte presents a comprehensive catalog of experimental results concerning refraction, emphasizing the precise order and intensity of the colors produced and the angles of the rays producing them.

Mariotte's review of various mechanisms proposed for explaining these results finds fundamental weaknesses in all of them. In particular Descartes's notion of a rotatory tendency to motion, besides being inherently unclear, implies constantly alternating patterns of color contrary to observation; and Newton's proposal of white light as a composite of monochromatic colors, although it explains much, fails on the crucial test. That is, repeating Newton's refraction of violet rays through a prism, Mariotte finds further separation in the form of red and yellow fringes about the image.[14] In the absence of an adequate theory, he retreats in the essay to eight "principles of experience," which are essentially generalizations of the behavior of refracted light as observed in the preceding experiments. Using these principles, Mariotte then undertakes to explain a long series of observed phenomena, including the chromaticism of lenses, the shape of the spectrum, and diffraction about thin objects. The explanations are merely a prelude, however, to his main concern, a complete account of the rainbow.

Reviewing previous accounts from Aristotle to Descartes, Mariotte states his basic agreement with that of Descartes but points to its lack of complete correspondence with observation. In particular Descartes failed to explain the upper and lower boundaries (40° and 44°) of the primary rainbow. Mariotte's own full account applies to Descartes's basic mechanism the precise measurements made in the preceding experiments but ends with a small divergence between the calculated and the observed height of the rainbow. The divergence, a matter of forty-six minutes of arc, led Mariotte to carry out with La Hire a protracted series of refraction experiments using water-filled glass spheres. Employing the techniques established in his essay on air, he made adjustments for the different densities of air (and hence different indexes of refraction) at sea level and at 500 feet, where the rainbow is formed, and for the heating effects of the sun on the water in the spheres. This work brings theory and observation into closer alignment. Similar but less extensive use of the "principles" offers satisfying explanations of stellar coronas (explained by refraction through water vapor in the clouds), solar and lunar coronas, and parhelia and false moons (all explained by refraction through small filaments of snow in the shape of equilateral prisms).

In contrast with the precision and clarity of the treatment of refraction in part I of the essay, Mariotte's attempt in part II to explain the color of directly observed bodies seems vague and undirected, perhaps because there was little for him to build on. A mass of undigested empirical phenomena forces him ultimately to retreat to an essentialist stance and to argue, for example, that "the weak and discontinuous light of the ignited fumes of brandy, sulfur, and other subtle and rarefied exhalations is disposed with respect to the organs of vision in a manner suited to make blue appear."[15]

For all its apparent diversity, Mariotte's research reflects a continuing concern with the motion of bodies in a resisting medium. The subject forms the core of the letter to Huygens in 1668 and of part II of the *Traité de la percussion* in 1673. Moreover, it was the subject of a full Academy investigation in 1669. A report in 1676 on the reflection and refraction of cannon balls striking water, and one in 1677 on the resistance of air to projectile motion, pursued the issue further. It was, however, only in 1678 that Mariotte broached the topic that would unite this research in a common theme: natural springs, artificial fountains, and the flow of water through pipes. His long report formed the basis for his *Traité du mouvement des eaux et des autres corps fluides*, published posthumously by La Hire in 1686. Mariotte was still working on the treatise at his death; his last two reports to the Academy dealt with the dispersion of water fired from a cannon and with the origin of the winds.

The *Mouvement des eaux*, a treatise in five parts, represents Mariotte's grand synthesis. Part I, section 1, reviews the basic properties of air and water as presented in the *Nature de l'air* and the paper on freezing. Section 2 uses these properties, together with meteorological and geological data, to argue that natural springs and rivers derive their water exclusively from rainfall (an extension of a theory first proposed by Pierre Perrault in 1674); the discussion includes an original estimate of the average total rainfall in France and of the content of its major rivers. Section 3 carries the meteorological discussion into the topic of the winds and their origin. Having met with some success in establishing a chain of weather stations across Europe,[16] Mariotte uses their reports and those of others to give a full account of the world's and Europe's major winds, basing it on the daily eastward rotation of the earth, the rarefaction and condensation of air due to heating and cooling, and changes in the distance of the moon from apogee to perigee. The account is particularly striking for the extent of detailed geographical and meteorological information from around the world.

Part II deals with the balancing forces of fluids due to weight, elasticity, and impact. Beginning with a treatment of statics based on a "universal principle of mechanics" akin to that of virtual velocities and

illustrated by the solution of what is known as "Mariotte's paradox,"[17] the treatise moves to experimental and theoretical demonstrations of the hydrostatic paradox and the Archimedean principle of floating bodies. Turning from the force of weight to that of elasticity, it recapitulates in some detail the discussion of the volume-pressure law in the *Nature de l'air*, asserting of water only that it is practically incompressible and hence has no elastic force. It does, however, have a force of impact, which Mariotte had already begun to explore in part II of the *Traité de la percussion* and which he continues to study here in the form of the speed of efflux of water through a small hole at the base of a reservoir. Five "rules of jets of water" relate the speed and force of the flow to the height of the reservoir and the cross-sectional area of the opening. The rules are then adapted to the impact of flowing water against the paddles of mills; direct measurement tends to confirm the applicability of the laws of inelastic collision to the situation.

Part III derives directly from Mariotte's work at the fountains of Chantilly in 1678 and is devoted to the experimental determination of the constants required for applying his theoretical rules to actual fountains. Of particular interest here is his report on the variation of the period of a pendulum with respect to latitude, a consideration made necessary by the use of the pendulum as a timing device for measuring rate of flow. Part IV continues in a practical vein, discussing the deviation from the ideal in real fountains. The discussion provides an opportunity to introduce Mariotte's findings on the effect of air resistance on the path and speed of projectiles, both solid and fluid, and leads directly to the subject of friction in conduit pipes, the opening topic of part V, again treated with reference to experiments performed at Chantilly.

Part V and the treatise conclude with a study of the strength of materials, in which Mariotte disputes Galileo's analysis and solution of the problem of the breaking strength of a loaded beam.[18] His solution, based largely on experimental results, is then applied to water pipes, relating the height of the reservoir to the necessary cross-sectional thickness of the pipes.

Mariotte's treatise attracted widespread attention and was the only one of his major works to be translated into English (by J. T. Desaguliers in 1718). Although eventually superseded in its theoretical portions by Daniel Bernoulli's *Hydrodynamica* (1738), it remained a standard practical guide to the construction of fountains for some time thereafter.

According to the testimony of La Hire and his colleagues, Mariotte was also the author of an unsigned *Essay de logique* that appeared in 1678. B. Rochot has shown that this work closely resembles in content and structure, and frequently quotes verbatim, an unpublished manuscript by Roberval.[19] According to Rochot, however, Mariotte did not plagiarize Roberval but, rather, succeeded him at his death in 1675 as recording secretary for an Academy project on scientific method, whence the absence of an author's name upon publication. The work bears Mariotte's unmistakable stamp, however, both in the fit between its proposed methodology and his actual research procedures and in the use of his own research as examples (in particular the argument for the choroid as the seat of vision).

Like all treatises on method in the seventeenth century, Mariotte's rests on the conviction that divergence of scientific opinion derives from faults in procedure—that is, from faulty deduction or induction, from inadequate experimentation, or from failure to observe procedure arising out of ulterior motives. Because the last is beyond his control, and because the rules of proper deduction are well known, Mariotte concentrates on the rules of induction, especially induction based on experiment. Following Descartes, he accepts both a distinction between reality and perception and also the existence of self-evident propositions that are true of reality. For the most part, however, those propositions are of a mathematical nature, abstracted from the contingencies of the natural world, which are known only through sense data. Since those data do not allow the distinction between the perception and what is perceived, they make extremely difficult, perhaps even impossible, any reliable transition from knowing the world as it appears to us to knowing the world as it really is. Hence, for Mariotte induction and analysis must stop at general principles that are directly verifiable by the senses. Although hypothetical systems that go beyond this point can have immense heuristic and organizational value (he gives as examples both the Ptolemaic and the Copernican systems), they generally can claim no epistemological status other than convenience.

In general Mariotte brought to his *Essay de logique* the precepts and procedures that characterize his actual research. Concerned more with the articulation and application of experimentally determined generalizations than with their reduction to more fundamental (and unverifiable) mechanisms or principles, Mariotte treated subjects in piecemeal fashion, relying on common sense and good judgment based on intimate familiarity with the physical situation to guide his reasoning. His essay seldom delves beyond this level of analysis, and his fifty-three general principles do little more than recapitulate what had been the common methodological stock of experimentalists since Aristotle. His distrust of theoretical systems extended to methodology itself and led him at one point in the essay to assert that there are no sure rules of method, other than constant experimentation in the study of nature.

As an active member of the Academy for over twenty-five years, Mariotte exerted influence over scientific colleagues both within and without that institution. His closest associate seems to have been La Hire, but during his tenure he carried out joint investigations with most of the other members, including Huygens. His work was known to the Royal Society and was cited by Newton in the *Principia*. Mariotte conducted an extensive correspondence (as yet unpublished) with Leibniz, for whom he was a source of information about the work of the Academy in the early and mid-1670's and who in turn cooperated with Mariotte's meteorological survey. Huygens' accusation of plagiarism in 1690 seems to have done little to dim the reputation Mariotte had earned during his career. In speaking of his death in 1684, J.-B. du Hamel summed up that career as follows:

> The mind of this man was highly capable of all learning, and the works published by him attest to the highest erudition. In 1667, on the strength of a singular doctrine, he was elected to the Academy. In him, sharp inventiveness always shone forth combined with the industry to carry through, as the works referred to in the course of this treatise will testify. His cleverness in the design of experiments was almost incredible, and he carried them out with minimal expense.[20]

NOTES

1. Condorcet, "Éloge de Mariotte," in *Éloges des académiciens de l'Académie royale des sciences, morts depuis 1666, jusqu'en 1699* (Paris, 1773), 49; *Oeuvres complètes de Christiaan Huygens*, VI (Amsterdam, 1895), 177, n. 1.
2. C. Oursel reviews previous accounts and the available documentation regarding Mariotte's birthdate and birthplace in *Annales de Bourgogne*, III (1931), 72–74.
3. Huygens, *Oeuvres*, VI, 536.
4. According to J. A. Vollgraf in Huygens, *Oeuvres*, XXII (1950), 631, Huygens knew nothing of Mariotte until the latter joined the Academy in 1666.
5. J.-B. du Hamel, *Regiae scientiarum academiae historia*, 2nd ed. (Paris, 1701), 233. For the full context of the remark, see the quotation at the end of the present article.
6. Huygens, *Oeuvres*, VI, 177–178.
7. Huygens, for one, thought Mariotte's to be the stronger argument; see Huygens, *Oeuvres*, XIII (1916), 795. William Molyneux in his *Dioptrica nova* (London, 1692) opted for the retina but acknowledged the strength of Mariotte's argument and felt the choice was immaterial.
8. Huygens, *Oeuvres*, XVI (1929), 209.

9. Mariotte's theory of elastic bodies is also a major theme of the review of the *Traité de la percussion* that appeared in the *Journal des sçavans*, **4** (1676), 122–125.
10. Mariotte, *Oeuvres* (1717), 152.
11. On this project see the full account in du Hamel, *Historia*, 150–155.
12. Mariotte, *Oeuvres*, 196–197.
13. Interestingly, Mariotte mentions the sine law of refraction only in passing and nowhere treats it as a phenomenon to be explained.
14. See in this regard the letter from Leibniz to Huygens, 26 Apr. 1694, in Huygens, *Oeuvres*, X (1905), 602. In it Leibniz refers to Mariotte's experiment and asks if Huygens has had the opportunity to investigate it further.
15. Mariotte, *Oeuvres*, 285.
16. According to Wolf, *History of Science, Technology, and Philosophy in the 16th and 17th Centuries*, 2nd ed. (New York, 1950), I, 312–314, Mariotte's effort was not the first, having been preceded most notably by that of the Accademia del Cimento in 1667.
17. The paradox is the reversal of the normal law of a bent-arm balance in the following situation: one arm of the balance is parallel to the horizon, the other inclined downward, the weights are placed at equal distances from the fulcrum along the arms, and the free-rolling weight on the inclined arm is held in place by a frictionless vertical wall. See Pierre Costabel, "Le paradox de Mariotte," in *Archives internationales de l'histoire des sciences*, **2** (1949), 864–881.
18. For details see Wolf, II, 474–477.
19. B. Rochot, "Roberval, Mariotte et la logique," in *Archives internationales de l'histoire des sciences*, **6** (1953), 38–43.
20. Du Hamel, *Historia*, p. 233.

BIBLIOGRAPHY

I. ORIGINAL WORKS. *Oeuvres de Mariotte*, 2 vols. in one (Leiden, 1717; 2nd ed., The Hague, 1740), contains all of Mariotte's published works and one unpublished paper; most of the articles that appeared under his name in vols. I, II, and X of the *Histoire de l'Académie depuis 1666 jusqu'en 1699* (Paris, 1733) are reports made to the Academy prior to their inclusion in the published works. The 1717 *Oeuvres* includes the following:

1. (I, 1–116) *Traité de la percussion ou choc des corps*, 3rd ed. (Paris, 1684)—(the *Oeuvres* dates it 1679, but the concluding experiments were not carried out until 1682); 1st ed., Paris, 1673; 2nd ed. in *Recueil de plusieurs traitez de mathématique de l'Académie royale des sciences* (Paris, 1676). Reviewed in *Journal des sçavans*, **4** (1676), 122–125.

2. (I, 117–320) *Essays de physique, pour servir à la science des choses naturelles:*

a. (I, 119–147) *De la végétation des plantes* (Paris, 1679); reviewed in *Journal des sçavans*, **7** (1679), 245–250. G. Bugler, in *Revue d'histoire des sciences . . .*, **3** (1950), 242–250, reports a 1676 version under the title *Lettre sur le sujet des plantes;* the report is confirmed by the "Avis" of the *Oeuvres* but not by other sources.

b. (I, 148–182) *De la nature de l'air* (Paris, 1679); reviewed in *Journal des sçavans*, **7** (1679), 300–304.

c. (I, 183–194) *Du chaud et du froid* (Paris, 1679); reviewed in *Journal des sçavans*, **7** (1679), 297–299.

d. (I, 195–320) *De la nature des couleurs* (Paris, 1681); reviewed in *Journal des sçavans*, **9** (1681), 369–374.

3. (II, 321–481) *Traité du mouvement des eaux et des autres corps fluides* (Paris, 1686; 2nd ed., Paris, 1690; 3rd ed., Paris, 1700), trans. into English by J. T. Desaguliers as *The Motion of Water and Other Fluids, Being a Treatise of Hydrostaticks . . .* (London, 1718).

4. (II, 482–494) "Règles pour les jets d'eau," in *Divers ouvrages de mathématique et physique par MM. de l'Académie royale des sciences* (Paris, 1693), English trans. by Desaguliers as addendum to *Traité . . . des eaux.*

5. (II, 495–534) "Lettres écrites par MM. Mariotte, Pecquet et Perrault sur le sujet d'une nouvelle découverte touchant la veüe par M. Mariotte," in *Recueil de plusieurs traitez de mathématique . . .* (Paris, 1676), and then separately (Paris, 1682). Mariotte's first letter was originally published, along with Pecquet's response, as *Nouvelle découverte touchant la veüe* (Paris, 1668) and was reviewed in *Journal des sçavans*, **2** (1668), 401–409. Mariotte's reply was published as *Seconde lettre de M. Mariotte à M. Pecquet pour montrer que la choroïde est le principal organe de la veüe* (Paris, 1671).

6. (II, 535–556) *Traité du nivellement, avec la description de quelques niveaux nouvellement inventez* (Paris, 1672), repub. in *Recueil . . .* (Paris, 1676); reviewed in *Journal des sçavans*, **3** (1672), 130–131.

7. (II, 557–566) "Traité du mouvement des pendules," from MS letter to Huygens repub. from Leiden MS in Huygens, *Oeuvres*, VI, 178–186.

[The pagination of the 1717 *Oeuvres* jumps from 566 to 600; the table of contents and the index show no work omitted.]

8. (II, 600–608) *Expériences touchant les couleurs et la congélation de l'eau*. Both taken from Academy records as later published in the *Histoire de l'Académie depuis 1666 jusqu'en 1699*, X, 507–513; the paper on freezing appeared earlier in *Journal des sçavans*, **3** (1672), 28–32.

9. (II, 609–701) *Essay de logique, contenant les principes des sciences et la manière de s'en servir pour faire des bons raisonnements* (Paris, 1678), unsigned.

Of these works only three have been republished in a modern ed.: *Discours de la nature de l'air, de la végétation des plantes. Nouvelle découverte touchant la vue*, in the series *Maîtres de la Pensée Scientifique* (Paris, 1923).

Few original papers remain, the most important of which are 28 letters from Mariotte to Leibniz and 10 from Leibniz to Mariotte. According to P. Costabel, *Archives internationales d'histoire des sciences*, **2** (1949), 882, n. 1, they are among the Leibniz papers in Hannover.

II. SECONDARY LITERATURE. There is no biography or secondary account of Mariotte's work. The above account has been culled largely from references to him in Huygens' correspondence and papers (Huygens, *Oeuvres*, passim) and in J.-B. du Hamel's *Historia*, passim. Condorcet's *éloge* gives only the broadest outline of Mariotte's career, as do other general French biographical reference works. The fullest catalog of his positive achievements remains Abraham Wolf's scattered references in the 2 vols. of his *History of Science, Technology and Philosophy in the 16th and 17th Centuries*, 2nd ed. (New York, 1950), which have the added virtue of placing those achievements in their contemporary context. For specific aspects of Mariotte's career, see Pierre Brunet, "La méthodologie de Mariotte," in *Archives internationales d'histoire des sciences*, **1** (1947), 26–59; G. Bugler, "Un précurseur de la biologie expérimentale: Edme Mariotte," in *Revue d'histoire des sciences* . . ., **3** (1950), 242–250; Pierre Costabel, "Le paradoxe de Mariotte," in *Archives internationales d'histoire des sciences*, **2** (1949), 864–881; and "Mariotte et le phénomène élastique," in *84e Congrès des sociétés savantes* (Paris, 1960), 67–69; Douglas McKie, "Boyle's Law," in *Endeavour*, **7** (1948), 148–151; Jean Pelseneer, "Petite contribution à la connaissance de Mariotte," in *Isis*, **42** (1951), 299–301; Bernard Rochot, "Roberval, Mariotte et la logique," in *Archives internationales d'histoire des sciences*, **6** (1953), 38–43; Maurice Solovine, "À propos d'un tricentenaire oublié: Edme Mariotte (1620–1920)," in *Revue scientifique* (24 Dec. 1921), 708–709; and E. Williams, "Some Observations of Leonardo, Galileo, Mariotte, and Others Relative to Size Effect," in *Annals of Science*, **13** (1957), 23–29.

On the blind-spot controversy, see John M. Hirschfield, "The Académie Royale des Sciences (1666–1683): Inauguration and Initial Problems of Method" (diss., University of Chicago, 1957), ch. 8.

MICHAEL S. MAHONEY

MARIUS, SIMON. See **Mayr, Simon.**

MARKGRAF (or **Marcgraf**), **GEORG** (*b.* Liebstadt, Meissen, Germany, 20 September 1610; *d.* Luanda, Angola, August 1644), *astronomy, botany, zoology.*

Markgraf was the son of Georg Markgraf, headmaster of the Liebstadt school, and Elisabeth Simon, daughter of the pastor there. He was educated at home, where he became proficient in Greek, Latin, music, and drawing; in 1627 he began to travel and to study with scientists throughout Germany. He matriculated at the University of Leiden on 11 September 1636, already well versed in all applications of mathematics and in both the natural sciences and medicine.

Markgraf had long been interested in observing the stars of the southern hemisphere; the opportunity to do so arose when he was invited to participate in a military and exploratory expedition to the Dutch settlements in Brazil. The expedition was under the leadership of Count Maurice of Nassau, who was at that time laying siege to the Portuguese settlement of São Salvador (now Bahía). The research staff, under the direction of Willem Pies (William Piso), left Leiden to join the count on 1 January 1638. Markgraf began his scientific work—making maps and assembling botanical and zoological collections—amid the diffi-

cult and often gravely dangerous circumstances of war.

At the end of the siege, the expedition sailed for Pernambuco. They founded the town of Mauritzstad (now part of Recife) and built the castle of Vrijburg on the island of Antonio Vaz. Markgraf drew up the plans for the new town and its fortifications, and Maurice installed an observatory for him in a tower of the castle. Markgraf determined the exact position of his site and began to make observations, including those of the planet Mercury and of the solar eclipse of 13 November 1640. He introduced into the island specimens of plants and animals that he had collected on his journeys; the park of Vrijburg Castle was subsequently made a botanical and zoological garden. With an escort of soldiers, to protect him against attack by Indians or by the Portuguese, Markgraf also mapped the region from Rio São Francisco to Ceará and Maranhão and made watercolor depictions, from nature, of flora and fauna. His methods of observing and painting were pioneering, analogous to those of Konrad Gesner in Europe.

Although Markgraf wished to return to Holland in 1644 to compile the results of his researches, he sailed to the Dutch settlements of East Africa to do further fieldwork. He died of a fever in Luanda; he had previously given all his collections and manuscripts into the keeping of Maurice of Nassau.

BIBLIOGRAPHY

I. ORIGINAL WORKS. The best evidence of Markgraf's work is in "Historiae rerum naturalium Brasiliae libri octo," in Jan De Laet, ed., *Historia naturalis Brasiliae* (Leiden–Amsterdam, 1648); Willem Pies, *De Indiae utriusque re naturali et medica libri quatuordecim*, 2nd ed. (Amsterdam, 1658), jumbles Markgraf's and Pies's observations. The printer states that he added the paper on the solar eclipse but, at the suggestion of competent astronomers, did not publish other astronomical notes. Casparis Barlaei (van Baerle), *Rerum per octennium in Brasilia et alibi gestarum sub praefectura illustrissimi comitis J. Mauritii Nassaviae &c. historia* (Amsterdam, 1657), contains Markgraf's geographical maps, which are much praised by specialists.

The dried plants introduced by Markgraf to the botanical garden of the island of Antonio Vaz and blocks for his woodcuts are at the Botanical Museum, Copenhagen. *Liber principis*, the collection of Markgraf's watercolors, was at the Preussische Staatsbibliothek, Berlin, until 1945.

II. SECONDARY LITERATURE. The first account of Markgraf's life, with important details, was written by his brother Christian and appeared in J. J. Manget, *Bibliotheca scriptorum medicorum*, XII (Geneva, 1731), 262–264. Part of it is a defense against Pies.

An early and comprehensive paper is H. Lichtenstein, "Die Werke von Marcgrave und Piso über die Naturgeschichte Brasiliens, erläutert aus den wiederaufgefundenen Originalzeichnungen," in *Abhandlungen der Preussischen Akademie der Wissenschaften*, Physikalische Abhandlungen, for 1814–1815 (1818), 201–222; for 1817 (1819), 155–178; for 1820–1821 (1823), 237–254, 267–288; and for 1826 (1829), 49–65.

An appreciation of Markgraf's significance as a pioneer of Brazilian botany is C. von Martius, "Versuch eines Commentars über die Pflanzen in den Werken von Marcgrav und Piso über Brasilien," in *Abhandlungen der Bayerischen Akademie der Wissenschaften*, Math.-phys. Kl., 7 (1853), 179–238.

Some important corrections derived from archival studies are presented by J. Moreira, "Marcgrave e Piso," in *Revista do Museu paulista*, 14 (1926), 649–673. Corresponding notes on the relationship of Markgraf and Pies are given by R. von Ihering, "George Marcgrave," *ibid.*, 9 (1914), 307–315.

The outstanding value of Markgraf's geographical maps is judged by V. Hantzsch, "Georg Marggraf," in *Berichte über die Verhandlungen der K. Sächsischen Gesellschaft der Wissenschaften zu Leipzig*, Phil.-hist. Kl., 48 (1896), 199–227.

Markgraf's plants in his books of 1648 and 1658 were identified most successfully by Dom Bento Pickel, "Piso e Marcgrave na botânica brasileira," in *Revista da flora medicinal* (Rio), 16 (1949), 155.

Markgraf's model herbarium is evaluated briefly by F. Liebmann in N. Wallich's trans. of Martius' paper in *Hooker's Journal of Botany and Kew Garden Miscellany*, 5 (1853), 167–168, note.

A recent reference of Markgraf's herbarium is given by B. MacBryde, "Rediscovery of G. Marcgrave's Brazilian Collections (1638–1644)," in *Taxon*, 19 (1970), 349.

F. MARKGRAF

MARKHAM, CLEMENTS ROBERT (*b.* Stillingfleet, Yorkshire, England, 20 July 1830; *d.* London, England, 30 January 1916), *geography.*

The second son of David F. Markham and the former Catherine Milner, Markham attended Westminster School for two years before joining the Royal Navy as a cadet in 1844. By the time he left the service in December 1851, he had acquired various technical, nautical, and geographical skills, which were the extent of his formal education. After a year in Peru (1852–1853) studying Inca ruins, he spent most of the next two decades in the service of the India Office. In 1860 Markham planned and executed a project for the acclimatization of Peruvian cinchona in India, an enterprise which had immense significance for public health and for the Indian economy. During this period he wrote extensively on topics in geography, economic botany, and technology related to the development of the British Empire. Combining these pursuits with a historical interest in the diffusion of Islamic technology, he studied the irrigation systems of southeastern Spain with an eye toward the agricultural development of the Madura district of India.

Markham left the India Office in 1877 and devoted the rest of his life to the promotion of geographical research, exploration, and education. He considered himself a comparative and historical geographer and stressed the value of historical records for the study of physical geography. Markham was secretary of the Hakluyt Society (1858–1886) and then president (1889–1909). Under his direction the society published a series of historical accounts of exploration. Markham edited twenty volumes of the series himself and was responsible for editions of several important treatises in the history of science, including a reedition of Edward Grimston's translation (1604) of José de Acosta's *Natural and Moral History of the Indies*; Robert Hues's *Tractatus de globis coelesti et terrestri ac eorem usu conscriptus* (1594); and Garcia da Orta's *Colloquies on the Simples and Drugs of India*, the latter in Markham's translation.

Markham's major work took place in the Royal Geographical Society, which he had joined in 1854. As a secretary from 1863 to 1888 and as president from 1893 to 1905, Markham helped to found a school of geography at Oxford; and it was under his aegis that the *Geographical Journal* became "the chief repository of geographical information from all parts of the world" ("Presidential Address," in *Geographical Journal*, 22 [1903], 1).

As president he enjoined both the Geographical Society and the nation to embark upon what he called an "Antarctic crusade" (*Geographical Journal*, 14 [1899], 479). He believed that the polar regions, particularly the southern one, comprised the single great geographical problem left for England to solve. Markham's program for polar exploration was carefully conceived. His chief arctic canons were that progress should always be made along the coastlines and that in order to be successful, an expedition had to remain over at least one winter in order to collect significant meteorological and magnetic data. Polar research should have two principal focuses: work conducted on the shore and that carried out aboard ship along the coasts. Research on shore would include (1) geographical exploration, (2) geology, (3) studies of glaciation, (4) magnetic observations, (5) meteorological observations, (6) pendulum observations, (7) studies of tides, and (8) inshore and land biology. Shipboard tasks would comprise (1) surveying coastlines,

(2) magnetic observations, (3) meteorological studies, (4) deep-sea soundings, and (5) marine biology (see *Geographical Journal*, **18** [1901], 13–25). As president of the Geographical Society, he established a research committee for polar exploration designed to ensure the maximum preparation and planning toward the accomplishment of these goals.

Not interested in promoting a mere race to the poles, Markham frequently stressed that the aim of such expeditions was to "secure useful scientific results" ("Presidential Address," in *Geographical Journal*, **4** [1894], 7). Although he worked closely with the Royal Society in the planning of polar expeditions, Markham always favored navy men over scientists to lead expeditions.

The last part of Markham's career was linked to the fortunes of Commander Robert F. Scott, whose first Antarctic expedition (voyage of the *Discovery*, 1901–1904) was the crowning achievement of Markham's exploration program. He continued writing on the subject and played an active role in the planning of Scott's fatal expedition on the *Terra Nova* (1910–1912). Described by a navy colleague as a "peripatetic encyclopedia," Markham's scholarship suffered from overextended interests and hasty research. His organizational and promotional talents, however, sufficed to make him the leading figure of Victorian geography.

BIBLIOGRAPHY

I. ORIGINAL WORKS. Markham's main scientific work divides into three primary categories: writings relating to Indian problems, those describing or promoting polar exploration, and essays concerning the nature and history of geographical research.

On the acclimatization of cinchona, see *Travels in Peru and India While Superintending the Collection of Chinchona Plants and Seeds in South America, and Their Introduction Into India* (London, 1862); and *Peruvian Bark* (London, 1880). See also his treatise on the diffusion of irrigation practices, *Report on the Irrigation of Eastern Spain* (London, 1867).

Of the Arctic literature, the important titles are *Franklin's Footsteps* (London, 1853); *The Threshold of the Unknown Region* (London, 1873); *Arctic and Antarctic Exploration* (Liverpool, 1895); *Antarctic Exploration: A Plea for a National Expedition* (London, 1898); "The Antarctic Expeditions," in *Geographical Journal*, **14** (1899), 473–481; "Considerations Respecting Routes for an Antarctic Expedition," *ibid.*, **18** (1901), 13–25; "The First Year's Work of the National Antarctic Expedition," *ibid.*, **22** (1903), 13–20; and his posthumous book, *Lands of Silence* (Cambridge, 1921).

For geographical history and theory, see "The Limits Between Geology and Physical Geography," in *Geographical Journal*, **2** (1893), 518–525; *Major James Rennell and*

the Rise of Modern English Geography (London, 1895); "The Field of Geography," in *Geographical Journal*, **11** (1898), 1–15; "View of the Progress of Geographical Discovery," in *Encyclopaedia Britannica*, 9th ed. (1875), under "Geography"; and "The History of the Gradual Development of the Groundwork of Geographical Science," in *Geographical Journal*, **46** (1915), 173–185.

On Markham's role as a geographical entrepreneur, see *The Fifty Years' Work of the Royal Geographical Society* (London, 1881); "The Present Standpoint of Geography," in *Geographical Journal*, **2** (1893), 481–504; *Hakluyt: His Life and Work. With a Short Account of the Aims and Achievements of the Hakluyt Society* (London, 1896); *Address on the Fiftieth Anniversary of the Foundation of the [Hakluyt] Society* (London, 1911); and his presidential addresses to the Royal Geographical Society, all pub. in *Geographical Journal*, esp. those of 1894 (**4**, 1–25), 1899 (**14**, 1–14), 1901 (**18**, 1–13), and 1903 (**22**, 1–13).

II. SECONDARY LITERATURE. Antonio Olivas, "Contribución a la bibliografía de Sir Clements Robert Markham," in *Boletín bibliográfico* (Lima), **12** (1942), 69–91, is adequate for secondary literature about Markham emanating from Latin America but is deficient otherwise. The standard biographical source is Albert H. Markham, *The Life of Sir Clements R. Markham* (London, 1917). On Markham's role in the Antarctic expeditions, see Robert F. Scott, *The Voyage of the "Discovery"* (dedicated to Markham as "the father of the expedition"), 2 vols. (London, 1905; repr. New York, 1969); Margery and James Fisher, *Shackleton and the Antarctic* (Boston, 1958); and L. B. Quartermain, *South to the Pole* (London, 1967), which cites the relevant earlier bibliography. On Markham's deficiencies as a translator, see Harry Bernstein and Bailey W. Diffie, "Sir Clements R. Markham as a Translator," in *Hispanic American Historical Review*, **17** (1937), 546–557.

THOMAS F. GLICK

MARKOV, ANDREI ANDREEVICH (*b.* Ryazan, Russia, 14 June 1856; *d.* Petrograd [now Leningrad], U.S.S.R., 20 May 1922), *mathematics*.

Markov's father, Andrei Grigorievich Markov, a member of the gentry, served in St. Petersburg in the Forestry Department and managed a private estate. His mother, Nadezhda Petrovna, was the daughter of a state employee. Markov was in poor health and used crutches until he was ten years old. He early manifested a talent for mathematics in high school but was not diligent in other courses. In 1874 Markov entered the mathematics department of St. Petersburg University and enrolled in a seminar for superior students, led by A. N. Korkin and E. I. Zolotarev. He had met them in his high school days after presenting a paper on integration of linear differential equations (which contained results already known). He also attended lectures by the head of the St. Petersburg mathematical school, P. L. Chebyshev, and afterward became a consistent follower of his ideas.

In 1878 Markov graduated from the university with a gold medal for his thesis, "Ob integrirovanii differentsialnykh uravnenii pri pomoshchi neprervnykh drobei" ("On the Integration of Differential Equations by Means of Continued Fractions") and remained at the university to prepare for a professorship. In 1880 he defended his master's thesis, "O binarnykh kvadratichnykh formakh polozhitelnogo opredelitelia" ("On the Binary Quadratic Forms With Positive Determinant"; *Izbrannye trudy*, pp. 9–83), and began teaching in the university as a docent. In 1884 he defended his doctoral dissertation, devoted to continued fractions and the problem of moments. In 1883 he married Maria Ivanovna Valvatyeva, the daughter of the proprietress of the estate managed by his father. They had been childhood friends, and Markov had helped her to learn mathematics. Later he proposed to her, but her mother agreed to the marriage only after Markov strengthened his social position.

For twenty-five years Markov combined research with intensive teaching at St. Petersburg University. In 1886 he was named extraordinary professor and in 1893, full professor. In this period he studied many questions: number theory, continued fractions, functions least deviating from zero, approximate quadrature formulas, integration in elementary functions, the problem of moments, probability theory, and differential equations. His lectures were distinguished by an irreproachable strictness of argument, and he developed in his students that mathematical cast of mind that takes nothing for granted. He included in his courses many recent results of investigations, while often omitting traditional questions. The lectures were difficult, and only serious students could understand them. He stated his opinions in a peremptory manner and was extremely exacting with his associates. During his lectures he did not bother about the order of equations on the blackboard, nor about his personal appearance. He was also a faculty adviser for a student mathematical circle. Nominated by Chebyshev, Markov was elected in 1886 an adjunct of the St. Petersburg Academy of Sciences; in 1890 he became an extraordinary academician and in 1896 an ordinary academician. In 1905, after twenty-five years of teaching, Markov retired to make room for younger mathematicians. He was named professor emeritus, but still taught the probability course at the university, by his right as an academician. At this time his scientific interests concentrated on probability theory and in particular on the chains later named for him.

A man of firm opinions, Markov participated in the liberal movement in Russia at the beginning of the twentieth century. In a series of caustic letters to academic and state authorities, he protested against the overruling, at the czar's order, of the election in 1902 of Maxim Gorky to the St. Petersburg Academy, he refused to receive decorations (1903), and he repudiated his membership in the electorate after the illegal dissolution of the Second State Duma by the government (1907). The authorities preferred not to respond to these declarations, considering them the extravagances of an academician. In 1913, when officials pompously celebrated the three-hundredth anniversary of the House of Romanov, Markov organized a celebration of the two-hundredth anniversary of the law of large numbers (in 1713 Jakob I Bernoulli's *Ars conjectandi* was posthumously published).

In September 1917 Markov asked the Academy to send him to the interior of Russia, and he spent the famine winter in Zaraisk, a little country town. There he voluntarily taught mathematics in a secondary school without pay. Soon after his return to Petrograd, his health declined sharply and he had an eye operation. In 1921 he continued lecturing, scarcely able to stand. He died after several months of intense suffering.

Markov belonged to Chebyshev's scientific school and, more than others, was faithful to the creed and the principles of his master. He inherited from Chebyshev an interest in concrete problems; a simplicity of mathematical procedures; a need to solve problems effectively, whether simple or algorithmic; and a desire to obtain exact limits for asymptotic results. These views coexisted with an underestimation of the role of some new general concepts in contemporary mathematics, namely of the axiomatic method and of the theory of functions of complex variables. Characteristic of Markov was the adherence to a chosen method of investigation and maintenance of his own view of what is valuable in science. He once said, "Mathematics is that which Gauss, Chebyshev, Lyapunov, Steklov, and I study" (N. M. Guenter, "O pedagogicheskoi deyatelnosti A. A. Markov," p. 37).

The principal aim of most of Markov's works in number theory and function theory was to evaluate the exact upper or lower bounds for various quantities (quadratic forms, integrals, derivatives). In probability theory it was at first to apply the bounds for integrals to the proof of the central limit theorem outlined by Chebyshev; later it was to discover new phenomena satisfying this theorem. Markov's work in various branches of mathematics is also united by systematic use of Chebyshev's favorite method of continued fractions, which became the principal instrument in Markov's investigations.

Markov's work in number theory was devoted mostly to the problem of arithmetical minima of indefinite quadratic forms studied previously in Russia by Korkin and Zolotarev (the topic goes back to Gauss and Hermite). These two authors had shown that if one excludes the form $f(x, y) = x^2 - xy - y^2$ (and the forms equivalent to it) for which $\min |f| = \sqrt{\frac{4}{5}d}$, then for the remaining binary forms $f(x, y) = ax^2 + 2bxy + cy^2$ with $d = b^2 - ac > 0$, one has $\min |f| \leqslant \sqrt{\frac{1}{2}d}$. By means of continued fractions Markov showed in his master's thesis (*Izbrannye trudy*, pp. 9–83) that 4/5 and 1/2 are the first two terms of an infinite decreasing sequence $\{N_k\}$ converging to 4/9, such that (1) for every N_k there exists a finite number of nonequivalent binary forms whose minimum is equal to $\sqrt{N_k d}$ and (2) if the minimum of any indefinite binary form is more than $\sqrt{\frac{4}{9}d}$, then it is equal to one of the values of $\sqrt{N_k d}$. To the limiting value $\sqrt{\frac{4}{9}d}$ there correspond infinitely many nonequivalent forms. Following the traditions of the Petersburg mathematical school, Markov also computed the first twenty numbers of $\{N_k\}$ and the forms corresponding to them. In 1901–1909 he returned to the problem of extrema of indefinite quadratic forms. He found the first four extremal forms of three variables (one of them was known to Korkin) and two extremal forms of four variables, and published a long list of ternary forms with $d \leqslant 50$. Markov's works on indefinite forms were continued both in the Soviet Union and in the West. Another problem of number theory was considered by Markov in his paper "Sur les nombres entiers dépendents d'une racine cubique d'un nombre entier ordinaire" (*Izbrannye trudy*, pp. 85–133). Following Zolotarev's ideas, Markov here obtained the final result for decomposition into ideal prime factors in the field generated by $\sqrt[3]{A}$ and calculated the units of these fields for all $A \leqslant 70$.

The next area of Markov's work concerned the evaluation of limits of functions, integrals, and derivatives. The problem of moments was the most notable among these topics. From a work of J. Bienaymé presented to the Paris Academy of Sciences in 1833 (republished in Liouville's *Journal de mathématiques pures et appliquées* in 1867), Chebyshev borrowed the problem of finding the upper and lower bounds of an integral

$$(1) \qquad \int_a^x f(x)\, dx$$

of a nonnegative function f with given values of its moments

$$(2) \qquad m_k = \int_A^B x^k f(x)\, dx \qquad (k = 0, 1, \cdots, N)$$

and the idea of applying the solution of this problem of moments to prove limit laws in probability theory. In 1874 Chebyshev published, without proofs, inequalities providing upper and lower bounds for integral (1) for some special values of a and $x(A < a < x < B)$. These bounds were expressed through the convergents of the continued fraction into which the series $\sum m_k / Z^{k+1}$ formally decomposes. The proofs of Chebyshev's inequalities appeared in 1884 in Markov's memoir "Démonstration de certaines inégalités de M. Tchebycheff" (*Izbrannye trudy po teorii nepreryvnykh drovei* . . ., pp. 15–24). The same inequalities with the same proofs were published at almost the same time by the Dutch mathematician Stieltjes. Markov claimed priority, to which Stieltjes replied that he could not have known of Markov's paper and that Chebyshev's work had indeed escaped his attention. Later Markov and Stieltjes studied the problem of moments largely side by side and sometimes one would find new proofs of the other's already published results. Both used continued fractions in their investigations and developed their theory further; but a difference in their methodological approaches manifested itself: Markov was mostly interested in the case of finite numbers of given moments and he studied the problem entirely within the limits of classical calculus; Stieltjes paid more attention to the problem of given infinite sequences of moments, and, seeking the most adequate formulation of the problem, introduced a generalization of the classical integral—the so-called Stieltjes integral.

In his doctoral dissertation Markov solved the question of the upper and lower bounds of integral (1) in the case when the first N moments are known. In subsequent papers he generalized the problem by allowing the appearance of an additional factor $\Omega(x)$ under integral (1); allowing, instead of power moments (2), moments relative to arbitrary functions $\lambda_k(x)$; and substituting the condition $c \leqslant f(x) \leqslant C$ for $f(x) \geqslant 0$. In other papers he investigated the distribution of the roots of the denominators of the convergents of the continued fraction mentioned above and the convergence of this fraction. The last question is closely related to the uniqueness of the solution of the Stieltjes problem of moments (finding a function, given its infinite sequence of power moments). In 1895, in his memoir "Deux démonstrations de la convergence de certaines fractions continues" (*Izbrannye trudy po teorii nepreryvnykh drovei* . . ., pp. 106–119), Markov obtained the following sufficient condition for the convergence, and therefore for the uniqueness of the Stieltjes problem, of functions defined on $[0, \infty)$: $\overline{\lim}_{k \to \infty} \sqrt[k]{m_k} < \infty$. Further results were obtained by O. Perron, H. Hamburger, F. Riesz, and T. G. Carleman.

Markov solved in 1889 another problem on extremal values which arose from the needs of chemistry in "Ob odnom voprose D. I. Mendeleeva" ("On a Question of D. I. Mendeleev"; *Izbrannye trudy po teorii nepreryvnykh droei* . . ., pp. 51–75). Here Markov found the maximum possible value of the derivative $f'(z)$ of a polynomial $f(z)$ of degree $\leq n$ on an interval $[a, b]$, provided that $|f(z)| \leq L$ on $[a, b]$. (This maximum value is equal to $2n^2L/(b - a)$.) Markov's result was generalized in 1892 by his younger brother Vladimir (who died five years afterward), and it was later extended for other cases by S. N. Bernstein and N. I. Akhiezer. Markov also worked on some other, practical extremal problems, namely the mapping of a part of a surface of revolution onto a plane with minimal deformations and the joining of two straight lines with a smooth curve having minimal curvature. The question of Mendeleev can be reformulated as a question about the maximum deviation of the polynomial $f(z)$ from zero, and it is therefore closely related to Chebyshev's theory of polynomials deviating least from zero and to some other topics connected with this theory, such as orthogonal polynomials (particularly Hermite and Legendre polynomials and the distribution of their roots), interpolation, and approximate quadrature formulas.

Markov obtained new results in all these areas; but unlike Chebyshev, who also studied quadrature formulas, Markov found in his formulas the expression of the remainder term. For example, in his doctoral dissertation he derived the remainder term of a quadrature formula originating with Gauss. Among other topics related to approximation calculus, Markov considered summation and improving the convergence of series. Evidence of Markov's liking for computation are his tables of the integral of probabilities calculated to eleven decimal places. Markov paid much attention to interpolation, summation, transformations of series, approximate calculation of integrals, and calculation of tables in his *Ischislenie konechnykh raznostei* ("Calculus of Finite Differences"). The difference equations themselves occupy a modest place in this book, which contains characteristic connections with the work of Briggs, Gauss, and Euler and many carefully calculated examples. Markov also obtained some results in the theory of differential equations—on Lamé's equation and the equation of the hypergeometric series—partly overlapping results of Felix Klein, and results concerning the possibility of expressing integrals in terms of elementary functions.

Markov's work in probability theory produced the greatest effect on the development of science. The basic achievements in probability theory by the middle of the nineteenth century were the law of large num-

bers, presented in its simplest version by Jakob I Bernoulli, and the central limit theorem (as it is now called) of de Moivre and Laplace. Satisfactory proofs under sufficiently wide assumptions had not been found, however, nor had the limits of their applicability. Through their closely interrelated works on these two laws Chebyshev, Markov, and Lyapunov created the foundation for the modernization of probability theory. In 1867 Chebyshev had found an elementary proof of the law of large numbers and turned to demonstrating the central limit theorem, using the solution of the problem of moments. The Bienaymé–Chebyshev problem mentioned above, translated into probability language, becomes a problem about the exact limits for the distribution function $F_\xi(x)$ of a random variable ξ with N given first moments $m_k = E\xi^k$. Let $\xi_1, \xi_2, \cdots, \xi_n, \cdots$ be a sequence of independent random variables with zero means (the case of nonzero $E\xi_n$ can be easily reduced to the considered one). According to Chebyshev's approach, one must show (a) that for every k the kth moment m_k of the normalized sum

$$\zeta_n = \frac{\xi_1 + \cdots + \xi_n}{\mathrm{var}(\xi_1 + \cdots + \xi_n)}$$

tends to the corresponding moment μ_k of the standard Gaussian distribution

$$\Phi(x) = \frac{1}{\sqrt{2\pi}} \int_{-\infty}^{x} e^{-\frac{y^2}{2}} \, dy$$

if $n \to \infty$, and (b) that if $m_k \to \mu_k$ for all k, then $F_{\zeta_n}(x) \to \Phi(x)$. When Markov published (1884) the proofs of Chebyshev's inequalities concerning the moments, Chebyshev began to work faster. In 1886 he showed that if $m_k = \mu_k$, then $F(x) = \Phi(x)$ (for him, but not for Markov, it was equivalent to assertion [b]); and in 1887 he published a demonstration of point (a) based on incorrect manipulations with divergent series.

Markov decided to turn Chebyshev's argument into a correct one and fulfilled this aim in 1898 in the paper "Sur les racines de l'équation $e^{x^2}(d^n e^{-x^2}/dx^n) = 0$" (*Izbrannye trudy po teorii nepreryvnykh droei* . . ., pp. 231–243; *Izbrannye trudy*, pp. 253–269) and in his letters to Professor A. V. Vassilyev at Kazan University, entitled "Zakon bolshikh chisel i sposob naimenshikh kvadratov" ("The Law of Large Numbers and the Method of Least Squares"; *Izbrannye trudy*, pp. 231–251). In the first of his letters Markov defined his aim thus:

The theorem which Chebyshev is proving . . . has been regarded as true for a long time, but is established by an extremely inaccurate procedure. I do not say proved because I do not recognize inaccurate proofs. . . . The known derivation of the theorem is inaccurate but simple. The derivation by Chebyshev on the contrary is very complicated, for it is based on preliminary investigations. . . . Therefore the question arises as to whether Chebyshev's derivation differs from the previous one only by its intricacy but is analogous to it in essentials, or whether one can make this derivation accurate. Your essay on Chebyshev's works strengthened my long-standing desire to simplify and at the same time to make quite accurate Chebyshev's analysis" [*Izbrannye trudy*, p. 231].

In his letters to Vassilyev, Markov established an arithmetical proof of convergence $m_k \to \mu_k$ (assertion [a]) under the following conditions: (1) for every k the sequence $E\xi_1{}^k, E\xi_2{}^k, \cdots$ is bounded and (2) $\text{var}(\xi_1 + \cdots + \xi_n) \geqslant cn$ for all n and some fixed $c > 0$. The corresponding calculation based on the expansion of the polynomial $(x_1 + \cdots + x_n)^k$ is maintained in all subsequent works by Markov on the limit theorem. In the article "Sur les racines . . ." Markov proved that $F(x) \to \Phi(x)$ if $m_k \to \mu_k$ (assertion [b]) by means of further analysis of Chebyshev's inequalities and continued fractions. He showed by examples that assumption (2), the need for which was unnoticed by Chebyshev, cannot be omitted.

In 1900 Markov published *Ischislenie veroyatnostei* ("Probability Calculus"). This book played an important role in modernizing probability theory. Characteristic features of the book are the inclusion of recent results obtained by Markov, rigorous proofs, elaborate references to classical works of the eighteenth century which for Markov had contemporary as well as historical importance, many numerical examples, and a polemical tone (Markov never missed an opportunity to mention an incorrectly solved example from another author and to correct the error).

But the triumph of the method of moments lasted only a short time. In 1901 Lyapunov, who was less influenced by their master, Chebyshev, and prized more highly the "transcendental" means (in Chebyshev's words) of the complex variable, played on Markov what he termed "a great dirty trick" (V. A. Steklov, "A. A. Markov," p. 178). Lyapunov discovered a new way to obtain and prove the limit theorems —the method of characteristic functions. The principal idea of this much more flexible method consists in assigning to the distribution of a random variable ξ not the sequence of moments $\{m_k\}$ but the characteristic function $\varphi(t) = Ee^{it\xi}$ and deducing the convergence of distributions from convergence of characteristic functions. Lyapunov proved the central limit theorem (for independent summands with zero means)

by his method under the conditions that (1) all moments $d_n = E \mid \xi_n \mid^{2+\delta}$ are finite for some $\delta > 0$, and

$$(2) \qquad \lim_{n \to \infty} \frac{[\text{var}(\xi_1 + \cdots + \xi_n)]^{2+\delta}}{(d_1 + \cdots + d_n)^2} = \infty,$$

which are near to necessary and sufficient ones. Although the second conditions of Markov and Lyapunov are of a similar character (both require rapid growth of the variance of the sum), the first condition of Lyapunov is incomparably wider than Chebyshev-Markov's, because it does not require even the existence of moments of the third and subsequent orders.

Markov struggled for eight years to rehabilitate the method of moments and was at last successful. In the memoir "Teorema o predele veroyatnosti dlya sluchaev akademika A. M. Lyapunova" ("Theorem About the Limit of Probability for the Cases of Academician A. M. Lyapunov"; *Izbrannye trudy*, pp. 319–337), included in the third edition of his *Ischislenie veroyatnostei*, Markov proved Lyapunov's result by using the new procedure of truncating the distributions, thus permitting one to reduce the general case to the case of bounded moments of every order. This procedure is still a useful device, but the method of moments could not stand the competition of the simpler and more universal method of characteristic functions. Also in the third edition of *Ischislenie veroyatnostei*, Markov showed, by means of truncating, that the law of large numbers is true for a sequence $\xi_1, \xi_2, \cdots, \xi_n, \cdots$ of independent random variables if for any $p > 1$ the moments $E \mid \xi_n \mid^p$ are bounded. (Chebyshev had proved the case $p = 2$.) Markov also deduced here the convergence of distributions from the convergence of moments for the cases when the limiting distribution is not Gaussian but has the density $Ae^{-x^2} \mid x \mid^\gamma$ or $Ae^{-x}x^\delta$ $(x \geqslant 0)$. (The theorem was demonstrated for other continuous limiting distributions in 1920 by M. Pólya.)

In his efforts to establish the limiting laws of probability in the most general situation and to enlarge the applications of the method of moments, Markov began a systematic study of sequences of mutually dependent variables, and selected from among them an important class later named for him. A sequence $\{\xi_n\}$ of random variables (or random phenomena of some other kind) is called a Markov chain if, given the value of the present variable ξ_n, the future ξ_{n+1} becomes independent of the past $\xi_1, \xi_2, \cdots, \xi_{n-1}$. If the conditional distribution of ξ_{n+1} given ξ_n (defined by transition probabilities at time n) does not depend on n, then the chain is called homogeneous. The possible values of ξ_n are the states of the chain. Such chains appeared for the first time in 1906 in Markov's paper "Rasprostranenie zakona bolshikh chisel na velichiny, zavi-syashchie drug ot druga" ("The Extension of the Law of Large Numbers on Mutually Dependent Variables"; *Izbrannye trudy*, pp. 339–361). Markov started with the statement that if the variance of the sum $(\xi_1 + \cdots + \xi_n)$ grows more slowly than n^2, then the law of large numbers is true for the sequence $\{\xi_n\}$, no matter how the random variables depend on each other. He also gave examples of dependent variables satisfying this condition, among them a homogeneous chain with a finite number of states. Markov obtained the necessary estimation of the variance from the convergence as $n \to \infty$ of the distribution of ξ_n to some final distribution independent of the values of ξ_1 (the "ergodic" property of the chain).

In his next paper, "Issledovanie zamechatelnogo sluchaya zavisimykh ispytanii" ("Investigation of a Remarkable Case of Dependent Trials"; in *Izvestiya Peterburgskoi akademii nauk*, 6th ser., **1**, no. 3 [1907], 61–80), Markov proved the central limit theorem for the sums $\xi_1 + \cdots + \xi_n$, where $\{\xi_n\}$ is a homogeneous chain with two states, 0 and 1. In 1908, in the article "Rasprostranenie predelnykh theorem ischisleniia veroyatnostei na summu velichin, svyazannykh v tsep" ("The Extension of the Limit Theorems of Probability Calculus to Sums of Variables Connected in a Chain"; *Izbrannye trudy*, pp. 365–397), he generalized this result to arbitrary homogeneous chains with finite numbers of states, whose transition probabilities satisfy some restrictions. The proof, as in all of Markov's works, was obtained by the method of moments. In "Issledovanie obshchego sluchaya ispytanii svyazannykh v tsep" ("Investigation of the General Case of Trials Connected in a Chain"; *Izbrannye trudy* [1910], pp. 467–507), Markov demonstrated the central limit theorem for nonhomogeneous chains with two states under the condition that all four transition probabilities remain in a fixed interval (c_1, c_2) $(0 < c_1 < c_2 < 1)$. In other articles, published in 1911–1912, he studied various generalizations of his chains (compound chains where ξ_n depends on several previous variables, so-called Markov-Bruns chains, partly observed chains) and deduced for them the central limit theorem under some restrictions.

Markov arrived at his chains starting from the internal needs of probability theory, and he never wrote about their applications to physical science. For him the only real examples of the chains were literary texts, where the two states denoted the vowels and the consonants (in order to illustrate his results he statistically worked up the alternation of vowels and consonants in Pushkin's *Eugene Onegin* (*Ischislenie veroyatnostei*, 4th ed., pp. 566–577). Nevertheless, the mathematical scheme offered by Markov and extended

later to families of random variables $\{\xi_t\}$ depending on continuous time t (which are called Markov processes, as suggested by Khinchin) has proved very fruitful and has found many applications. The development of molecular and statistical physics, quantum theory, and genetics showed that a deterministic approach is insufficient in natural sciences, and forced physicists to turn to probabilistic concepts. Through this evolution of scientific views, the Markov principle of statistical independence of future from past if the present is known, appeared to be the necessary probabilistic generalization of Huygens' principle of "absence of after effect." The far-reaching importance of such a generalization is shown by the fact that although Markov was the first to study the chains as a new, independent mathematical object, a number of random phenomena providing examples of Markov chains or processes were considered by other scientists before his work or concurrently with it. In 1889 the biologist Francis Galton studied the problem of survival of a family by means of a model reducing to a Markov chain with a denumerable number of states. An example of a Markov chain was considered in 1907 by Paul and T. Ehrenfest as a model of diffusion. In 1912 Poincaré, in the second edition of his *Calcul des probabilités*, in connection with the problem of card shuffling, proved the ergodic property for a chain defined on a permutation group and mentioned the possibility of an analogous approach to problems of statistical physics. An important example of a continuous Markov process was studied on a heuristic level in 1900–1901 by L. Bachelier in the theory of speculation. The same process appeared in 1905–1907 in works of Einstein and M. Smoluchowski on Brownian motion.

Markov's studies on chains were continued by S. N. Bernstein, M. Fréchet, V. I. Romanovsky, A. N. Kolmogorov, W. Doeblin, and many others. The first rigorous treatment of a continuous Markov process, the process of Brownian motion, was provided in 1923 by Wiener. The foundations of the general theory of Markov processes were laid down in the 1930's by Kolmogorov. The modern aspect of the theory of Markov processes, which became an intensively developing autonomous branch of mathematics, resulted from work by W. Feller, P. Lévy, J. Doob, E. B. Dynkin, K. Ito, and other contemporary probabilists.

Markov also studied other topics in probability: the method of least squares, the coefficient of variance, and some urn schemes.

BIBLIOGRAPHY

I. ORIGINAL WORKS. The most significant of Markov's papers are republished in *Izbrannye trudy po teorii nepre-*

ryvnykh drovei . . . and *Izbrannye trudy*, with modern commentaries; the latter contains a complete bibliography of original and secondary works to 1951. The earlier collection of his writings is *Izbrannye trudy po teorii nepreryvnykh drovei i teorii funktsii naimenee uklonyaiushchikhsya ot nulya*, N. I. Akhiezer, ed. ("Selected Works on Continued Fractions Theory and Theory of Functions Least Deviating From Zero"; Moscow, 1948), with comments by the ed. One of the memoirs was translated into English: "Functions Generated by Developing Power Series in Continued Fractions," in *Duke Mathematical Journal*, **7** (1940), 85–96. The later collection is *Izbrannye trudy. Teoria chisel. Teoria veroyatnostei*, Y. V. Linnik, ed. ("Selected Works. Number Theory. Probability Theory"; Moscow–Leningrad, 1951), which contains an essay on the papers in the volume and comments by the editor, and N. A. Sapogov, O. V. Sarmanov, and V. N. Timofeev; the most detailed biography of Markov, by his son A. A. Markov; and a full bibliography.

Individual works by Markov are *O nekotorykh prilozheniakh algebraicheskikh nepreryvnykh drobei* ("On Some Applications of Algebraic Continued Fractions"; St. Petersburg, 1884), his doctoral dissertation; *Tables des valeurs de l'intégrale $\int_x^\infty e^{-t^2}\,dt$* (St. Petersburg, 1888); *Ischislenie konechnykh raznostei* ("Differential Calculus"), 2 vols. (St. Petersburg, 1889–1891; 2nd ed., Odessa, 1910), also translated into German as *Differenzenrechnung* (Leipzig, 1896); and *Ischislenie veroyatnostei* ("Probability Calculus"; St. Petersburg, 1900; 2nd ed., 1908; 3rd ed., 1913; 4th ed., Moscow, 1924), posthumous 4th ed. with biographical note by A. S. Bezikovich, also translated into German as *Wahrscheinlichkeitsrechnung* (Leipzig–Berlin, 1912).

II. SECONDARY LITERATURE. Besides the biography in *Izbrannye trudy*, there is basic information in V. A. Steklov, "Andrei Andreevich Markov," in *Izvestiya Rossiiskoi akademii nauk*, **16** (1922), 169–184. See also *ibid.*, **17** (1923), 19–52; Y. V. Uspensky, "Ocherk nauchnoi deyatelnosti A. A. Markova" ("An Essay on the Scientific Work of A. A. Markov"); N. M. Guenter, "O pedagogicheskoi deyatelnosti A. A. Markova" ("On the Pedagogical Activity of A. A. Markov"); and A. Bezikovich, "Raboty A. A. Markova po teorii veroyatnostei" ("Markov's Works in Probability").

Various aspects of Markov's work are discussed in *Nauchnoe nasledie P. L. Chebysheva* ("The Scientific Heritage of P. L. Chebyshev"), I (Moscow–Leningrad, 1945), which compares Chebyshev's results in various fields with those of Markov, Lyapunov, and their followers in papers by N. I. Akhiezer (pp. 22–39), S. N. Bernstein (pp. 53–66), and V. L. Goncharov (pp. 154–155); and B. N. Delone, *Peterburgskaya shkola teorii chisel* ("The Petersburg School of Number Theory"; Moscow–Leningrad, 1947), which has a detailed exposition of Markov's master's thesis and a summary of further development of associated topics, pp. 141–193.

General surveys of Markov's life and work include B. V. Gnedenko, *Ocherki po istorii matematiki v Rossii* ("Essays on the History of Mathematics in Russia"; Moscow, 1946), pp. 125–133; *Istoria otechestvennoi mate-*

matiki ("History of Russian Mathematics"), II (Kiev, 1967), with an essay on Markov by I. B. Pogrebyssky, pp. 328–340; and A. P. Youschkevitch, *Istoria matematiki v Rossii do 1917 goda* ("History of Mathematics in Russia Until 1917"; Moscow, 1968), pp. 357–363, 395–403.

Post-Markov development of the theory of his chains is discussed by M. Fréchet, *Théorie des événements en chaine dans le cas d'un nombre fini d'états possible*, in *Recherches théoriques modernes sur le calcul des probabilités*, II (Paris, 1938); J. G. Kemeny and J. L. Snell, *Finite Markov Chains* (Princeton, 1960); and V. I. Romanovski, *Diskretnye tsepi Markova* ("Discrete Markov Chains"; Moscow, 1949).

An introduction to the modern theory of Markov processes is M. Loève, *Probability Theory* (Princeton, 1955), ch. 12.

ALEXANDER A. YOUSCHKEVITCH

MARKOVNIKOV, VLADIMIR VASILEVICH (*b.* Knyaginino, Nizhegorodskaya [now Gorki Region], Russia, 25 December 1837 [or 22 December 1838]; *d.* Moscow, Russia, 11 February 1904), *chemistry.*

Markovnikov was the son of an officer. In 1856 he entered Kazan University, where he was attracted to chemistry, which was taught by Butlerov. After graduating from the university in 1860 he became Butlerov's assistant in teaching inorganic and analytical chemistry. In his master's thesis, "Ob izomerii organicheskikh soedineny" ("On the Isomerism of Organic Compounds"; Kazan, 1865); in a number of other works; and especially in his doctoral thesis, "Materialy po voprosu o vzaimnom vlianii atomov v khimicheskikh soedineniakh" ("Materials on the Question of the Mutual Influence of Atoms in Chemical Compounds"; Kazan, 1869), he developed the theory of chemical structure experimentally and theoretically. From 1865 to 1867 he was in Germany, spending most of his time in the laboratories of Erlenmeyer at Heidelberg and of Kolbe at Leipzig. After his return he became assistant to, and then succeeded, Butlerov in the chair of chemistry at Kazan University. In 1871, however, protesting the arbitrariness of the administration, he resigned. Markovnikov was immediately invited to the University of Odessa, and in 1873 to Moscow University. There he improved the teaching of chemistry, set up a new chemical laboratory according to his own plans, and created his own school of chemists. His students included I. A. Kablukov, M. I. Konovalov, N. Y. Demyanov, D. N. Pryanishnikov, A. E. Chichibabin, A. A. Yakovkin, and N. M. Kizhner. In 1893, after Markovnikov had served his term in the chair of chemistry, he yielded it to N. D. Zelinsky but retained a part of the laboratory.

An important turning point in Markovnikov's

scientific career occurred in Moscow: he shifted his attention mainly to practical chemical research, and thus was reproached for betraying pure science. He devoted almost twenty-five years to the study of the hydrocarbons of Caucasian petroleum and to the chemistry of alicyclic hydrocarbons—"naphthenes," as he called them. But the range of Markovnikov's research also included the composition of the salts of the southern Russian bitter-salt lakes and the Caucasian sources of mineral waters, and methods and materials for testing railroad ties.

Markovnikov also studied the history of chemistry. He took the initiative in the publication of the Lomonosovskogo sbornika ("Lomonosov Collection"; Moscow, 1901), which included material on the history of Russian chemical laboratories; and he himself wrote a detailed sketch of the laboratories of Moscow University.

The most important results which Markovnikov obtained in his work on the theory of structure and the chemistry of petroleum and alicyclic compounds were the following. He tested certain conclusions, important in the first stage of structural theory, concerning the existence of isomers in a series of fatty acids (for example, butyric and isobutyric acid, the identity of "acetone" acid, obtained by G. Städeler in 1859 from "the mixture of acetone, hydrogen cyanide, and diluted hydrochloric acid," with isobutyric acid). Developing Butlerov's theory of the mutual influence of atoms, Markovnikov introduced certain "rules."

The rule for the substitution reaction in its general form states that with an unsymmetrical olefin, where two possible modes of addition are open, the reaction ordinarily follows a course such that the hydrogen becomes attached to the carbon atom of the olefin that already has the greater number of hydrogen atoms. Thus:

$$CH_3CH = CH_2 + HBr \rightarrow CH_3CHBrCH_3.$$

Markovnikov recognized that tertiary hydrogen atoms and hydrogen atoms in the α position to a carboxyl group are more active than hydrogen atoms in other positions. This also indicated to him at what point in a molecule destructive oxidation would take place. Markovnikov also first stated the rule that when molecules of water or hydrogen halide are obtained from alcohols or alkylhalides, the separation occurs between them and the two neighboring atoms of hydrogen.

An example from Markovnikov's doctoral thesis is:

$$\begin{matrix} CH_2OH \\ CH_2 \\ CH \\ \underbrace{CH_3CH_3} \end{matrix} \Big\rangle H_2O = \begin{matrix} CH_2{}' \\ CH' \\ CH \\ \underbrace{CH_3CH_3} \end{matrix}.$$

This rule, as Markovnikov showed, can be used to determine the structure of unsaturated compounds. The formation of compounds of the hydrogen halides or water with an unsymmetrical olefin occurs so that the halogen or hydroxyl radical adds to the carbon of the ethylenic bond with the lesser number of hydrogen atoms while the hydrogen adds to the carbon with the larger number of hydrogen atoms. This is known in textbooks as the Markovnikov rule. Markovnikov explained the mechanism of the reactions of isomerization, similar to the transition of isobutyl alcohol to tertiary isobutyl alcohol by reference to such reactions of separation and addition. He also stated rules relating to monomolecular isomerization and certain other reactions, stressing their dependence on the conditions.

In 1872, two years before van't Hoff's stereochemical representation (tetrahedral) of the carbon atom, Markovnikov showed the necessity of developing a theory of chemical structure by studying the relation between chemical interaction and the "physical position" of atoms in space; and in 1876 he stated that "the relative distribution of atoms in a molecule should be expressed by chemical formulas."

Markovnikov was one of the first to understand the importance of studying the composition of petroleum hydrocarbons for practical uses. For example, he and his colleagues showed, contrary to the current opinion, that oil from the Caucasus contained derivatives of cyclopentane and aromatic hydrocarbons along with derivatives of cyclohexane. He discovered the existence of azeotropic mixtures of both these types of hydrocarbons. Markovnikov proposed a method for obtaining aromatic nitro derivatives by means of direct nitration of petroleum fractions.

Markovnikov's contribution to the chemistry of alicyclic compounds consists, first, in his experimental refutation of the prevailing opinion that the carbocyclic compounds could have only six-atom nuclei. In 1879 he synthesized the derivative of a four-membered cycle, and ten years later (at the same time as W. H. Perkin, Jr.) the derivative of a seven-membered cycle. In 1892 Markovnikov proposed the isomerization of a seven-membered cycle into the six-membered one and laid the basis for the study of mutual transformations of alicycles. Markovnikov also made the first classifications of alicyclic compounds in 1892.

BIBLIOGRAPHY

I. ORIGINAL WORKS. Many of Markovnikov's writings are in *Izbrannye trudy* ("Selected Works"), A. F. Platé and G. V. Bykov, eds., in the series Klassiki Nauki ("Classics of Science"; Moscow, 1955), with a 448-title biblio. of Markovnikov's works in the appendix, pp. 835–889.

Among his articles are "Zur Geschichte der Lehre über die chemische Structur," in *Zeitschrift für Chemie*, n.s. **1** (1865), 280–287; "Ueber die Acetonsäure," in *Annalen der Chemie und Pharmacie*, **146** (1868), 339–352; "Ueber die Abhängigkeit der verschiedenen Vertretbarkeit des Radicalwasserstoffs in den isomeren Buttersäuren," *ibid.*, **153** (1870), 228–259; "Tetrylendicarbonsäure (Homoitakonsäure)," *ibid.*, **208** (1881), 333–349, written with G. Krestovnikov; "Recherches sur le pétrole caucase," in *Annales de chimie et de physique*, 6th ser., **2** (1884), 372–484, written with V. Ogloblin; "Die aromatischen Kohlenwasserstoffe des Kaukasischen Erdöhls," in *Justus Liebig's Annalen der Chemie*, **234** (1886), 89–115; and "Die Naphtene und deren Derivate in dem allgemeinen System der organischen Verbindungen," in *Journal für praktische Chemie*, 2nd ser., **45** (1892), 561–580, and **46** (1892), 86–106.

Some of Markovnikov's correspondence is in "Pisma V. V. Markovnikova k A. M. Butlerovu" ("Letters of V. V. Markovnikov to A. M. Butlerov"), in *Pisma russkikh khimikov k A. M. Butlerovu* ("Letters of Russian Chemists to A. M. Butlerov"), vol. IV in the series Nauchnoe Nasledstvo ("Scientific Heritage"; Moscow, 1961), pp. 212–289.

II. SECONDARY LITERATURE. There is a collection entitled *Pamyati Vladimira Vasilevicha Markovnikova* ("In Memory of Vladimir Vasilevich Markovnikov"; Moscow, 1905).

See also H. Decker, "Wladimir Wasiliewitsch Markownikow," in *Berichte der Deutschen chemischen Gesellschaft*, **38** (1906), 4249–4259; H. M. Leicester, "Vladimir Vasil'evich Markovnikov," in *Journal of Chemical Education*, **18** (1941), 53–57; and "Kekulé, Butlerov, Markovnikov. Controversies on Chemical Structure From 1860 to 1870," in *Kekulé Centennial*, no. 61 in the series Advances in Chemistry (Washington, D.C., 1966), pp. 13–23; A. F. Platé and G. V. Bykov, "Ocherk zhizni i deyatelnosti V. V. Markovnikova" ("A Sketch of the Life and Work of V. V. Markovnikov"), in Markovnikov's *Izbrannye trudy*, pp. 719–777; and A. F. Platé, G. V. Bykov, and M. S. Eventova, *Vladimir Vasilevich Markovnikov. Ocherk zhizni i deyatelnosti. 1837–1904* ("Vladimir Vasilevich Markovnikov. A Sketch of His Life and Work"; Moscow, 1962).

G. V. BYKOV

MARLIANI, GIOVANNI (*b.* Milan, Italy, early fifteenth century; *d.* Milan, late 1483), *physics, mechanics, medicine.*

There is little information on Marliani's early life. Born into a patrician family, he probably studied arts and medicine at Pavia University. In 1440 he was elected to the College of Physicians at Milan: it seems that he had received his doctorate by then, or certainly before 1442. From 1441 to 1447 Marliani taught nat-

ural philosophy and "astrologia" at Pavia, and lectured on the physics of Bradwardine and Albert of Saxony. Under the short-lived Ambrosian Republic, he left Pavia for the University of Milan (1447–1450), where he taught medicine. He was also appointed to civic office. Following the collapse of the Ambrosian Republic, Marliani returned to Pavia, where he added medicine to his previous lectureships, eventually acquiring (1469) the chair of medical theory. His salaries testify to a successful career: in 1441 Marliani earned 40 florins a year; in 1447, 200 florins; in 1463, 500 florins, plus, by secret arrangement with Duke Francesco I Sforza, an additional 150 florins. Later a special chair was set up for the Marliani family, to be held first by Giovanni's son, Paolo, in 1483. Two other sons, Girolamo and Pietro, also lectured at the university. The fortune of the family was sealed with Marliani's appointment (probably in 1472) as court physician to Galeazzo Maria Sforza. He also enjoyed the favor of the latter's successor, Gian Galeazzo.

Despite his strongly Scholastic views, Marliani was well regarded by humanists of the time. Pico della Mirandola called him the greatest mathematician of the age, and Francesco Filelfo and Marliani corresponded on medical and Scholastic matters. Giorgio Valla, who studied medicine and mathematics under Marliani at Pavia, translated the *Problemata* of Alexander of Aphrodisias at the urging of his teacher. Valla stated that Marliani owned a copy of Jacobus Cremonensis' translation of Archimedes, but the sole extant mathematical work by Marliani is on common fractions and shows little originality or knowledge of Greek mathematics. Valla's later work, however, abandoned the Scholastic physics of his teacher in favor of classical mathematics. Most Renaissance humanists and mathematicians shared Valla's preference.

Three works by Marliani deal with heat in a strongly Aristotelian fashion. The heating or cooling action is regarded as a special case of motion and thus subject to Aristotle's mistaken law of motion. In the early treatise *De reactione*, for instance, reaction (meaning the capacity of an agent to be affected by its patient) is analyzed in terms of active and resistive powers, as though it were motion. Although Marliani admits his debt to Jacopo da Forlì and others, he systematically rejects many of these predecessors' arguments in favor of some rather tortuous arguments of his own. (His criticisms soon involved Marliani in a polemic with Gaetano da Thiene.) Two main points of interest appear in the *De reactione*, although neither is original. These are Marliani's distinction between intensity of heat (temperature) and its extension (quantity of heat) and his use of a numerical scale to represent the intensity. This scale consists of eight degrees of calidity and its coextensive frigidity ($F° = 8 - C°$). The Marliani scale should not, perhaps, be taken as a forerunner of the thermometric scale, since it depends conceptually upon an Aristotelian qualitative distinction between heat and cold.

The *Disputatio cum Joanne de Arculis* discusses the reduction of hot water (that is, whether hot water is cooled by an intrinsic tendency or an extrinsic agent, such as its container). Marliani again applies his Aristotelian principle relating action and resistance to the quantity of heat and cold present. In contrast with his Avicennist opponent Giovanni Arcolani of Verona, who argues for intrinsic reduction, Marliani maintains that at any temperature, hot and cold components of the water are in equilibrium and thus cannot act upon one another. Hence, the cooling agent must be external. He also concludes that the shapes of the agent and the patient, and their distance apart, are factors in a heat action.

De caliditate corporum humanorum combines Marliani's knowledge of medicine and physics. Distinguishing between heat intensity (temperature) and its extension (the quantity of "natural" heat produced by the body), Marliani concludes that the human body, while increasing its natural heat in the winter, maintains a more or less constant temperature through most of its parts. Nevertheless, he still feels obliged (despite a youthful repudiation) to accept the notion of antiperistasis (the increase of intensity when a body is suddenly surrounded by its contrary quality), which underlay the arguments for a varying body temperature.

Marliani's writings on mechanics center on two main problems of Scholastic physics. In the *Probatio calculatoris* the kinematic mean-speed theorem of accelerated motion is outlined with some clear proofs. Of more significance is the *De proportione motuum*, designed to solve a paradox in Aristotle's law of motion. The Aristotelian law held velocity to be proportional to the ratio of the moving power to the resistance. A paradox arose when the moving power was equal to the resistance. Obviously no motion should then occur, but the ratio in the law gave a positive value to the velocity in this case. Bradwardine's law tried to eliminate the paradox by stating that the "proportions of velocities in motions follow the proportion of the power of the motor, to the power of the thing moved." Using the proportion of proportions calls for a geometrical increase in the ratio of force to resistance. Bradwardine, however, used terms which can denote either an arithmetical or a geometrical increase—for instance, "dupla," to mean "double" or "squared." Misled by this ambiguous terminology to assume that Bradwardine had fallen into the same

trap as Aristotle, Marliani severely condemned the former and advanced his own law that velocity is proportional to the excess of the motor power over the resistance. Nevertheless, Bradwardine's law remained acceptable to most Renaissance Aristotelian philosophers.

Marliani's career and works suggest that he was a competent Scholastic physicist and an adept publicist of the *calculatores* and French Scholastics in Italy. Influential among philosophers (his *De reactione* was later discussed by Pietro Pomponazzi), Marliani seems to have stayed largely outside the mainstream of Italian Renaissance mathematics. Much of his great reputation seems also to have rested on his position as physician to the Sforzas.

BIBLIOGRAPHY

I. ORIGINAL WORKS. Unless otherwise noted, Marliani's works were printed at Pavia in 2 vols. in 1482. The dates of composition are in parentheses. The writings are *Tractatus de reactione* (1448); *In defensionem Tractatus de reactione* (against Gaetano da Thiene, 1454–1456); "Annotationes in librum de instanti Petri Mantuani," unpub. work in Biblioteca Vaticana, MS Vat. Lat. 2225 (see Thorndike, below); *Probatio cuiusdam sententiae calculatoris de motu locali* (1460); *Disputatio cum Joanne de Arculis* (1461); *Difficultates missae Philippo Adiute Veneto* (before 1464); "Algorismus de minutiis" (before 1464), unpub. MSS in Bibliothèque Nationale, Paris, MS N.A.L. 761, and Biblioteca Ambrosiana, Milan, MS A.203 infra; *Questio de proportione motuum in velocitate* (1464); and *Questio de caliditate corporum humanorum* (1472).

II. SECONDARY LITERATURE. For mentions of Marliani, see Pico della Mirandola, *Disputationes adversus astrologiam divinatricem*, E. Garin, ed. (Florence, 1946), pp. 632–633; and J. L. Heiberg, "Beiträge zur Geschichte Georg Valla's und seiner Bibliothek," in *Zentralblatt für Bibliothekswesen*, supp. **16** (1896), 11–12, 85. An excellent treatment is Marshall Clagett, *Giovanni Marliani and Late Medieval Physics* (New York, 1941); see also his "Note on the *Tractatus physici* Falsely Attributed to Giovanni Marliani," in *Isis*, **34** (1942), 168, appended to D. B. Durand's review of the book. See also A. Maier, *Die Vorläufer Galileis im 14. Jahrhundert* (Rome, 1949), pp. 107–110; and the inaccurate pages in P. Duhem, *Études sur Léonard de Vinci*, III (Paris, 1913), 497–500. Two letters to Marliani are in Francesco Filelfo, *Epistolarum* (Venice, 1502), fols. 152v, 184v.

For MSS see Lynn Thorndike, "Some Medieval and Renaissance Manuscripts on Physics," in *Proceedings of the American Philosophical Society*, **104** (1960), 188–201, esp. 195; and Lynn Thorndike and Pearl Kibre, *A Catalogue of Incipits of Medieval Scientific Writings in Latin*, 2nd ed. (London, 1963), index, under "Marliani."

For Scholastic science in fifteenth-century Italy, see

Carlo Dionisotti, "Ermolao Barbaro e la fortuna di Suiseth," in *Medioevo e Rinascimento. Studi in onore di Bruno Nardi* (Florence, 1955), I, 217–253, esp. 230–231.

PAUL LAWRENCE ROSE

MARSH, OTHNIEL CHARLES (*b.* Lockport, New York, 29 October 1831; *d.* New Haven, Connecticut, 18 March 1899), *vertebrate paleontology.*

Marsh was the oldest son of a farmer and shoe manufacturer of modest means, Caleb Marsh, and Mary Gaines Peabody Marsh; his mother died before he was three years old. Aided financially by his uncle George Peabody, Marsh graduated from Phillips Academy at Andover, Massachusetts, from Yale College (1860), and from its Sheffield Scientific School (1862). After three years of study in Europe, Marsh became professor of paleontology at Yale from 1866 until his death. From 1882 to 1892 he was also the first vertebrate paleontologist of the U.S. Geological Survey. A bachelor, he lived in solitary grandeur near Yale.

Marsh was elected to the National Academy of Sciences in 1874 and served as its president from 1883 to 1895. He was the recipient of many awards and honorary degrees, among them the Bigsby Medal (1877) and the Cuvier Prize (1897).

Marsh's scientific interests began in childhood with minerals and invertebrate fossils, chiefly from formations exposed by the nearby Erie Canal. He pursued mineralogy in his education but gradually turned toward paleontology. A marked characteristic was his keen acquisitiveness, which resulted in vast collections of fossils for the Peabody Museum at Yale, a gift of his generous uncle.

Through his many scientific descriptions and his popularization of extinct animals, Marsh established the infant field of vertebrate paleontology in the United States. Accompanied by Yale students and alumni, he led four expeditions from 1870 to 1873 through the western territories, from the White River badlands of South Dakota and Nebraska, to the Bridger, Uinta, and Green River basins of Wyoming, Utah, and Colorado, to the John Day fossil fields in Oregon, and back to the Cretaceous chalk region of western Kansas. The startling fossil discoveries of these trips led Marsh into keen and bitter competition with Edward Drinker Cope for a quarter of a century. After his early collecting years, Marsh only rarely and briefly returned to the fossil fields, but he hired many amateur and professional collectors to seek specimens throughout the western United States. He urged his collectors to search out all fragments of each find, and

so was able to describe remarkably complete specimens.

In his work on fossil mammals Marsh established the evolution of the horse as North American, with a series of specimens from Eocene to Pleistocene; he presented the earliest mammals then known, from Jurassic and Cretaceous beds; in competition with Cope, he described some of the extinct horned mammals called uintatheres and some of the massive brontotheres; and he established the existence of early primates on the North American continent. On the reptiles, Marsh enlarged the classification of the dinosaurs, and described eighty new forms, both giant and tiny, and he described Cretaceous winged reptiles and marine mosasaurs. He also presented the first known toothed birds, which proved the reptile ancestry of that class. He demonstrated the gradual enlargement of the vertebrate brain from the Paleozoic era forward. Marsh's classifications and descriptions of extinct vertebrates were major contributions to knowledge of evolution.

BIBLIOGRAPHY

Marsh published about 300 papers on vertebrate fossils but left much to be completed by his successors. His work on horses was summarized in "Fossil Horses in America," in *American Naturalist*, **8** (1874), 288–294. His magnum opus was *Odontornithes: A Monograph on the Extinct Toothed Birds of North America*, vol. VII of *U.S. Geological Exploration 40th Parallel* (Washington, D.C., 1880). Dinosaurs were summarized in "The Dinosaurs of North America," in *Report of the U.S. Geological Survey*, **16**, pt. 1 (1896), 133–414; and "Vertebrate Fossils [of the Denver Basin]," in *Monographs of the U.S. Geological Survey*, **27** (1896), 473–550. Marsh's material on mammals appeared in many single papers; and much of the Mesozoic material was later synthesized by G. G. Simpson in "American Mesozoic Mammalia," in *Memoirs of the Peabody Museum of Yale University*, **3**, pt. 1 (1929).

Marsh's life, accomplishments, and bibliography are well presented in Charles Schuchert and Clara M. LeVene, *O. C. Marsh: Pioneer in Paleontology* (New Haven, 1940).

ELIZABETH NOBLE SHOR

MARSILI (or **Marsigli**), **LUIGI FERDINANDO** (*b.* Bologna, Italy, 20 July 1658; *d.* Bologna, 1 November 1730), *natural history.*

Marsili was the son of a nobleman, Carlo Marsili, and the former Margherita Ercolani. He served in the army of Emperor Leopold I from 1682 to 1704, attained high rank, and participated in the negotiations for the peace of Karlowitz—but also was wounded, imprisoned, and even suffered the humiliation of demotion. Although he did not complete his formal schooling, Marsili accumulated a vast knowledge of history, politics, geography, and the natural sciences. He traveled widely throughout Italy and the rest of Europe, particularly in the regions around the Danube, and made several long sea voyages (from Venice to Constantinople, from Leghorn to Amsterdam). While much of his work dealt with the military sciences, history, and geography, he also made a name for himself as a naturalist. He combined a love of travel with the sharp eye of an observer imbued with the Galilean method. In his scientific activity he was always guided by a prudent sagacity that once prompted this advice: "The modern method of observation is the right one, but it is still in its infancy, and we must not be so rash as to expect instantly to deduce systems from these observations; that is something which only our successors, after centuries of study by this method, will be able to do" (letter from Marsili to the astronomer F. Bianchini, 24 December 1726, in *Lettere di vari illustri italiani e stranieri* [Reggio nell'Emilia, 1841], II, 91).

As a naturalist, with that same prudent sagacity, Marsili undertook the exploration of two basic subjects: the structure of the mountains and the natural condition of the sea, lakes, and rivers. He left many local observations concerning the structure of the mountains (noteworthy among them being those on the continuity of the *linea gypsea*, the gypsum-bearing strata in the hills of the Adriatic slope of the Apennines); accurate sketches of stratigraphic profiles; and even cartographic representations of particular geologic conditions, although he was far from grasping the geologic significance of the strata. Realizing this later, he gave up systematically elaborating his many "schedae pro structura orbis terraquei."

The sea had fascinated Marsili since childhood. In 1681 he published a study of the Bosporus, the result of observations that he had made at age twenty. It contained valid findings, notably the discovery of a countercurrent with waters of different density beneath the surface current of the strait. He later traveled around the Mediterranean, doing research mainly along the coasts of the Romagna and Provence. The keenness of his mind often made up for the crudeness of his instruments during these travels. In 1724 he published the first treatise on oceanography, *Histoire physique de la mer*. In it he treated problems which until then had been veiled by error and legend. Marsili examined every aspect of the subject: the morphology of the basin and relationships between the lands under and above water; the water's properties (color, temperature, salinity) and its motion (waves, currents,

tides); and the biology of the sea, which foretold the advent of marine botany. Among the plants he numbered animals like corals, which before his time had been regarded as inorganic matter.

Finally, Marsili was the precursor of the systematic oceanographic exploration that was to begin half a century later with the famous voyage of the *Endeavour*. Using the same methods, he studied Lake Garda, the largest lake in Italy, discussing its physical and biological aspects in a very valuable report, which remained unpublished until 1930. Marsili wrote a basic work on one of Europe's greatest rivers, *Danubius . . . observationibus geographicis, historicis, physicis perlustratus . . .* (1726), in which he devoted much space to a study of the riverbed and of the waters, as well as to the flora and fauna, and the mineralogy and geology of the adjacent land.

Marsili was also a skilled organizer of scientific work. In 1712 he founded the Accademia delle Scienze dell'Istituto di Bologna, which, under his influence, immediately became an active center of scientific research, consisting mainly of natural history exploration of the area around Bologna. With Domenico Galeazzi, Marsili set an example in 1719 by climbing and studying Mount Cimone, highest peak of the northern Apennines. When Marsili went to London in 1722, to be made a member of the Royal Society, Sir Isaac Newton insisted on presenting him personally and praised him in his speech as both an already famous scientist and a founder of the new Academy of Bologna.

BIBLIOGRAPHY

I. ORIGINAL WORKS. Marsili's writings include *Osservazioni intorno al Bosforo tracio ovvero canale di Constantinopoli . . .* (Rome, 1681), a booklet repr. in *Bollettino di pesca, piscicoltura e idrobiologia,* **11** (1935), 734–758; *Histoire physique de la mer* (Amsterdam, 1725); and *Danubius-Pannonicus-Mysicus observationibus geographicis, historicis, physicis perlustratus et in sex tomos digestus* (The Hague–Amsterdam, 1726), also trans. into French (The Hague, 1744). The volume *Scritti inediti di L. F. Marsili* (Bologna, 1930) includes "Osservazioni fisiche intorno al Lago di Garda," M. Longhena and A. Forti, eds., and "Storia naturale de' gessi e solfi nelle miniere di Romagna," T. Lipparini, ed. There is also Marsili's *Autobiografia* (Bologna, 1930).

The University of Bologna library has 176 vols. of Marsili's autograph letters and cartographic sketches: see L. Frati, *Catalogo dei manoscritti di L. F. Marsili* (Florence, 1928); and M. Longhena, *L'opera cartografica di L. F. Marsili* (Rome, 1933).

II. SECONDARY LITERATURE. For information on Marsili's life and work, see G. Fantuzzi, *Memorie della vita del generale conte L. F. Marsili* (Bologna, 1770), for the period up to 1711 drawn from Marsili's *Autobiografia*. See also two works by M. Longhena: *Il conte L. F. Marsili* (Milan, 1930) and "Il conte L. F. Marsili," in *Bollettino della Società geografica italiana*, **95** (1958), 539–573, with a complete list of Marsili's published writings and an extensive bibliography on Marsili the geographer and naturalist; and *Memorie intorno a L. F. Marsili* (Bologna, 1930).

FRANCESCO RODOLICO

MARSILIUS OF INGHEN (or **Inguem** or **de Novimagio**) (*b.* near Nijmegen, Netherlands; *d.* Heidelberg, Germany, 20 August 1396), *natural philosophy.*

Almost nothing is known of Marsilius' early life. Although his name would suggest that he was born in the village of Inghen (now Lienden in de Betuwej, in the diocese of Utrecht),[1] his biographers are not all in agreement on this point; Gustav Toepke and Gerhard Ritter, for example, hold that "Inghen" is a family name, and not derived from that of a place.[2] The first document that mentions Marsilius is the register of the University of Paris, which gives 27 September 1362 as the date of his inaugural lecture as master of arts.[3]

Marsilius remained at the University of Paris for twenty years. As a *magister regens* of the arts he had a large student following. He served as rector of the university in 1367 and again in 1371; in 1362 and from 1373 until 1375 he was procurator of the English nation;[4] and in 1368 and 1376 he represented the university at the papal court in Avignon. In the latter year Marsilius accompanied the pope, Gregory XI, to Rome; he was present there in 1378 when Urban VI—whom he strongly supported—was elected pope. He was thus present at the beginning of the Great Schism.

The University of Paris was rent by the schism. It is not known exactly when Marsilius left it for the University of Heidelberg, recently founded by Urban VI, but it is certain that he was rector of that institution for the first time in 1386 (he served six other terms in the post). In 1389 he made another journey to Rome, this time to pay homage to a new pope, Boniface IX, and to ask his aid for the university as well as his personal patronage. Marsilius died in Heidelberg seven years later.

In his writings Marsilius employed the traditional medieval method of composing commentaries and questions on the works of earlier authors. He wrote on the scientific works of Aristotle—particularly the *Physica*, the *Parva naturalia*, and *De generatione et corruptione*—and on the logical works of Aristotle, including the *Posterior Analytics* and *Topics*, which he dealt with in the manner prescribed by the nominal-istic school of his time. He also wrote on theology, including a commentary on the four books of *Sentences* of Peter Lombard. His chief contributions to science lay in the field of physics, but even here his thought was always shaped by theological considerations.

Although Marsilius was born in the Netherlands and ended his life in Heidelberg, his work places him among the Parisian masters who may be considered to be the precursors of Leonardo and Galileo and the formers of the new physics of the fifteenth and sixteenth centuries. The leader of this group was Jean Buridan, of whom Marsilius declared himself a disciple, and their gift to science was the formulation of the concept of impetus—an impressed force—which they applied to the theory of gravity, acceleration, and the motion of projectiles. This was the *via moderna*, and these were the "new physicists."

Marsilius occupied a moderate and traditional position among these men, one much closer to the philosophical inquiries of Aristotle and Ibn Rushd (Averroës) than to the empiricism of Buridan. Evidence of his views may be found not only in his own writings, but also in those of his contemporaries. Blasius of Parma, for example, probably attended Marsilius' lectures during his sojourn in Paris, and in his *quaestio disputata* on the physical problem of the contact of bodies—in which he discussed whether or not the void exists and whether the action between bodies occurs directly or at a distance—he supported the traditional Aristotelian theory of the *horror vacui*, which position he attributed also to Marsilius.[5] Marsilius himself, as is clear in his works, never accepted the really novel implications for dynamics inherent in the doctrine of impetus.

Marsilius' concept of motion served him as a basis for an explanation of the entire physical world, both celestial and terrestrial. He avoided mechanical interpretation, and held that the motions of the heavens and the astronomical spheres are spiritual, eternal, incorruptible, and ungenerable; the bodies are moved in a "natural" motion that is circular and perfect and impressed upon them by angelic intelligences.[6] The initial source of this *virtus impressa* is, of course, God, the Creator and First Mover, the prime immobile mover who put into movement the whole *caelum* or the entire astrological universe. The perfect circular motion of the universe has since been maintained by the moving intelligences within the spheres.[7] Marsilius defined such motion as perfect "local" motion because it has no contrary and remains in its rightful place by nature—that is, within the order willed by God.[8]

The God-moved *caelum* in its turn moves the terrestrial world, which is not eternal but rather perpetual, beginning with time and lasting without end. The terrestrial world is not only generable but also corruptible and inferior, and moved almost exclusively by "unnatural" motions—or rather, by violent motions and "alterations" that provoke the birth and death of beings. In fact, according to Marsilius, all motion of natural, inferior, terrestrial things presupposes the action of a violent cause, whether such motion be "de loco naturale," away from the natural place, or "ad locum naturale," toward it.[9] Motion "de loco naturale" is acting against the natural order, while in the case of motion "ad locum naturale," it must already have so acted.[10] Typical violent motions are embodied in projectiles, tops, and smiths' wheels; the terrestrial world is moved by a plurality of moving causes which come from the heavens and are subordinate one to the other. These causes produce various effects by their concurrence. The destiny of an individual, for example, is shaped by concurrences among the Father, the heavens, God, and the moving intelligences;[11] God and the heavens influence human actions.

Marsilius thus appears to have accepted the doctrine of astrological determination on a philosophical level. He cited as an example the hungry dog that starves because it cannot decide which of two pieces of bread, placed at equal distances from it, to seize.[12] In reality, however, Marsilius maintained that man is free because the stars influence only the dog, but not the man, and that not everything in the world necessarily happens under absolutely determinate influences; chance or contingency can also play a part. Indeed, since he held that almost all motions of the inferior world are either violent or a mixture of local and violent movement, caused by a plurality of coincident agents, he also believed that it is possible to ascertain casual and fortuitous effects of which the primary cause is one among those that are concurrent and subordinate. Marsilius was thus able to admit the possibility of such extraordinary natural phenomena as eclipses, comets, and monsters.[13]

His system thus enabled Marsilius to explain physical mutations as instances of qualitative and violent alterations of bodies; the study of modifications of such essential physical qualities as heat and cold therefore assumed great importance.[14] Although he drew upon Buridan's doctrine of impetus to explain the special case of arrows and projectiles—or violent motion in its strictest sense—he moderated it in such a way that it might be reconciled with Aristotle's basic physical principle that every motion in the terrestrial world is caused by an agent outside the mobile. In other words, Marsilius did not fully accept Buridan's actual principle of impetus, whereby the mobile is permitted a permanent intrinsic moving force[15] which

is impeded by the medium through which it moves. In Marsilius' view, the violent motion of a projectile is not impeded by the air through which it moves; rather, it is aided by it. The impetus that moves the projectile is a disposition given it by the first mover, which sets it into motion. The impetus is initially confined to that part of the mobile that is in contact with the mover, then diffused through the whole.[16] Such impetus does not last long; it becomes corrupt in the absence of intervention from the surrounding medium. The medium may, however, receive the impetus of the mobile to reinforce the speed of the moving body.[17]

Marsilius regarded the world as a plenum. He believed that no void could exist in the physical universe and that all heavy bodies tend to the center of the earth, which is the only center as God is the only prime mover. Were the universe sustained by many prime movers, there would be many worlds and many centers, but this, too, is impossible.[18] He therefore could not admit Ockham's thesis concerning the possibility of other worlds. Here again, as in all traditional problems of physics, Marsilius stood for a reconciliation of the *via antiqua* of Aristotle with the *via moderna* of the Parisian masters.[19]

NOTES

1. J. Fruytier, *Nieuw nederlandsch biografisch woordenboek*, VIII, 908–909.
2. G. Toepke, *Die Matrikel der Universität Heidelberg von 1386 bis 1662* (Heidelberg, 1884), I, 678–685; G. Ritter, "Marsilius von Inghen und die okkamistische Schule in Deutschland," p. 210.
3. H. Denifle and E. Chatelain, *Chartularium universitatis parisiensis*, III (Paris, 1897), see index.
4. H. Denifle and E. Chatelain, *Auctarium chartularium universitatis parisiensis. Liber procuratorum nationis Anglicanae* (Paris, 1897), I, cols. 272, 559, *passim*.
5. "Quaestio magistri Blasii de Parma utrum duo corpora," MS Bologna, Biblioteca Universitaria 2567, 198 (Frati 1332), fol. 59 r-a; see G. Federici Vescovini, "Problemi di fisica," pp. 192, 209.
6. *Quaestiones subtilissimae super octo libros physicorum* (Lyons, 1518), II, qu. 4, fols. 25v-a, b ff.
7. *Quaestiones super quattuor libros Sententiarum*, II, qu. II, ad. 2, fols. 208v-a–209v-a, ff.; qu. X, ads. 2 and 3, fols. 243v-b–245r-a, b (Strasbourg, 1501).
8. . . . *Physicorum*, II, qu. 4, fols. 25v-a, b; . . . *Sententiarum*, II, qu. 10, ad. 2, fol. 244r-b; ad. 3, fols. 245r-a, b.
9. . . . *Physicorum*, VIII, qu. 1, fols. 79r-b–v-a; VIII, qu. 2, fol. 80v-a.
10. *Ibid.*, II, qu. 4, fol. 26r-a, 2a conc.
11. *Ibid.*, II, qu. 8, fol. 29r-a; II, qu. 10, fol. 30r-b.
12. *Ibid.*, II, qu. 14, fol. 33r-b.
13. *Ibid.*, II, qus. 8, 9, 10, 13, fols. 29r-b, 30r-b–v-a, 31r-b–v-a.
14. *Ibid*, VIII, qu. 4, fol. 81v-b. MS Paris, BN 16401, fols. 149v–177v; "Utrum qualitas suscipiat magis et minus," BN 6559, fol. 121r.
15. M. Clagett, *The Science of Mechanics* (Madison, Wis., 1959; 2nd ed., 1961), pp. 536–537, lines 7, 8, 9.
16. *Abbreviationes libri physicorum Aristotelis*, VIII, not. 4, qu. IV (Venice or Pavia, *ca.* 1490), fols. L 4r-a–5r-a.

17. *Ibid.*, fol. L 5r-a.
18. *Quaestiones . . . physicorum*, VIII, qu. 9, fol. 85v-b.
19. G. Ritter, Studien zur Spätscholastik, II, *Via antiqua und via moderna* . . . (Heidelberg, 1922), p. 39 and *passim*.

BIBLIOGRAPHY

I. ORIGINAL WORKS. There are no modern eds. of Marsilius' writings, but only MSS and rare eds. of collections of questions and of commentaries on the physics of Aristotle and biological problems of generation and corruption. Among the former are "Abbreviatura physicorum sive quaestiones variorum abbreviatae," MS Vienna 5437, fols. 1r–410v; Erfurt O.78, *anno* 1346, fols. 41–132; Erfurt Q.314, *anno* 1394, fol. 106; Vienna VI, 5112, fols. 181r–283v, printed at Pavia *ca.* 1480; Venice or Pavia, *ca.* 1490; Venice, 1521—see A. C. Klebs, "Incunabula scientifica et medica," in *Osiris*, **4** (1938), 667; and *Quaestiones subtilissimae super octo libros physicorum IX* (different edition from Pavia and Lyons, 1518; repr. Venice, 1617), repr. under the name of Duns Scotus and inserted in Lyons 1639 among the works of Duns Scotus (Lyons, 1639). He also wrote works in the form of questions on the psycho-biological problems of Aristotle: "Quaestiones de parvorum naturalium libris," MS Erfurt F.334, *anno* 1421, fols. 1–61; and MS Novacella [near Bressanone], Convento dei Canonici Regolari, 440, fols. 1–88, 89–268; and "Quaestiones de generatione et corruptione": MS Florence, Riccardiana 745 (N. II 26), fols. 96–137v ("per manus Nycolay Montfort"); MS Florence, Riccardiana 746 (K. II 38), fol. 76, *anno* 1407; Milan, Ambrosiana G. 102 inf., fols. 1r-a–87v-a; Modena, Estense 687 (Alpha F 5, 20) (Kristeller, *Iter italicum*, I [London–Leiden, 1963], 372); Modena, Estense Fondo Campori 1374 (Gamma T 4, 18) (Vandini, 441); Padua, Biblioteca Universitaria MS 693 (Kristeller, *Iter* . . ., II [London–Leiden, 1967], 14); Venice, Marciana, 121a (2557) (G. Valentinelli, *Bibliotheca manuscripta ad S. Marci Venetiarum*, V, Venice, 1868, pp. 50–51), *anno* 1393; Marciana, 324 (4072) (G. Valentinelli, *Bibliotheca manuscripta ad S. Marci venetiarum*, VI, Venice, 1868, p. 218); Vienna 5494, fol. 209; Munich, Staatsbibliothek, 26929, *anno* 1407, fols. 88r-a–193r-b; Erfurt Q.311, *anno* 1414, fols. 1–74v; Oxford, Bodleian cm. 238, fol. 101; Vienna 4951, *anno* 1501, fols. 164r–223v—published at Padua, 1476 (?), 1480; Venice, 1493, 1504, 1505, 1518, 1520, 1567; Strasbourg, 1501, together with works by others; Paris, 1518.

He also wrote on the mutations of qualities: "Utrum qualitas suscipiat magis et minus," MS Paris, BN 16401, fols. 149v–177v; BN 6559, fol. 121r.

Marsilius' works on logic, such as commentaries on the logic of Aristotle in the nominalists' *via moderna* form are MS Vat. lat. 3072, "Tractatus de suppositionibus"; Pistoia, Archivio Capitolare del Duomo, MS 61 (Kristeller, *Iter*, II, p. 75); Rome, Vat. lat. 3072 (Kristeller, *Iter*, II, p. 316); and Turin, Biblioteca Nazionale G.III, 12 (Pasini, lat. 449), fols. 167r-a–171r-b; "Consequentiae," MS Rome, Vat. lat. 3065; "De obligationibus, insolubilia," Rome, Pal. lat. 995 (*Arist. lat.*, II [1955], 1190–1191, n. 1777), published in *Textus dialectices de suppositionibus, ampliationibus* . . . (Cracow, n.d.); commentaries on the *Parva logicalia* (Basel, 1487; Hagenow, 1495, 1503; Vienna, 1512, 1516; Turin, 1729); and *Expositio super analitica* (Venice, 1516, 1522).

Works on theological and moral arguments are *Quaestiones super quattuor libros Sententiarum* (Hagenow, 1497; Strasbourg, 1501); and commentaries on the ethics of Aristotle, MS Rome, Pal. lat. 1022 (Kristeller, *Iter*, II, 392).

II. SECONDARY LITERATURE. There are no specific monographs on Marsilius, only partial studies. See the following, listed chronologically: P. Duhem, *Études sur Léonard de Vinci*, III (Paris, 1913), 403–405; and *Le système du monde*, IV (Paris, 1916), 164–168; G. Ritter, "Marsilius von Inghen und die okkamistische Schule in Deutschland," in *Sitzungsberichte der Heidelberger Akademie der Wissenschaften*, phil. Kl. (1921), 4, 210; J. Fruytier, *Nieuw nederlandsch biografisch woordenboek*, VIII (Leiden, 1930), 903–904; A. Maier, *Die Vorläufer Galileis in 14 Jahrhundert* (Rome, 1949), p. 3 and *passim*; *Zwei Grundprobleme der scholastischen Naturphilosophie* (Rome, 1951), pp. 275 ff.; *An der Grenze von Scholastik* . . . (Rome, 1952), pp. 118 ff.; and *Metaphysische Hintergrunde der spätscholastischen Naturphilosophie* (Rome, 1955), pp. 40, 90, 133, 146, 222, 396; G. Federici Vescovini, "Problemi di fisica aristotelica . . .," in *Rivista di filosofia*, **51** (1960), 190–193, 209; cf. Blasii qu. de gener. II, 2. MS Vat. Chig. O.IV.41, f. 37 2b; A. Birkenmajer, *Études d'histoire des sciences et de la philosophie du moyen age*, I (Wrocław, 1970), 368, 612, 654; II (Wrocław, 1972), 181, 187, 192–194; and G. Federici Vescovini, *Le questioni "De anima" de Biagio Pela cani da Parma* (Florence, 1973), p. 35 and *passim*.

GRAZIELLA FEDERICI VESCOVINI

MARTENS, ADOLF (*b.* Backendorf bei Hagenow, Mecklenburg–Schwerin, Germany, 6 March 1850; *d.* Berlin, Germany, 24 July 1914), *materials testing, metallography.*

Martens was the son of Friedrich Martens, a tenant farmer. After attending the Realschule in Schwerin and gaining two years of practical experience, he studied mechanical engineering from 1868 to 1871 at the Königliche Gewerbeakademie in Berlin (later the Technische Hochschule of Berlin-Charlottenburg). At the end of his training in 1871, he entered the service of the Prussian State Railway, where he participated in the planning of the great bridges over the Vistula near Thorn and over the Memel near Tilsit. From 1875 to 1879 he was a member of the Commission for the Berlin-Nordhausen-Wetzlar Railway. In this position he had to supervise the preliminary work done by the companies supplying the iron superstructure of the bridges. He thus became involved, early in his career, with the techniques just then being developed for testing construction materials.

Martens, in this early period, was stimulated by a short book by Eduard Schott, *Die Kunstgiesserei in Eisen* (Brunswick, 1873), to begin metallographic studies for which he built his own microscope. His first publication, "Über die mikroskopische Untersuchung des Eisens" (1878), contained his observations on freshly fractured iron surfaces, as well as drawings of etched and polished surfaces. On this topic Martens wrote:

> A careful observer of all these results cannot but come to the conclusion that in pig iron the various combinations of iron are only mechanically mixed; during the process of cooling or crystallization they arrange themselves with most surprising regularity. So the microscopical investigation of iron has a very great chance of becoming one of the most useful methods of practical analysis.

Further works followed and brought Martens into close contact with contemporary metallographers, sometimes provoking lively debates.

In 1884, after a short time as an assistant at the newly founded Königliche Technische Hochschule of Berlin-Charlottenburg, he was appointed director of the associated Mechanisch-Technische Versuchsanstalt, which in 1903 became the Königliche Materialprüfungsamt of Berlin-Dahlem. The brilliant design and organization of this institute were essentially his work.

Metallography did not at first come within the range of the Mechanisch-Technische Versuchsanstalt, and Martens could continue his very successful metallographic studies only in his free time. He could not resume them on a larger scale until 1898, when the Mechanisch-Technische Versuchsanstalt established a metallographic laboratory. His co-worker Emil Heyn greatly developed the field and made the institute a first-rate metallographic laboratory.

Martens' works from this later period were concerned with all aspects of the testing of materials, and especially the development of new measuring methods and equipment. His *Handbuch der Materialienkunde* (1899) is a comprehensive work which earned him the high regard of his colleagues as well as many honors.

Martens was chosen vice-chairman of the International Society for Testing Materials (ISTM) at its founding in 1895; in 1897 he became chairman of the German Society.

Martens combined tireless research activity and unusual talents as a designer and organizer. He made a fundamental contribution to the knowledge of the properties of materials by presenting the results of his research in clear and exhaustive reports.

BIBLIOGRAPHY

I. ORIGINAL WORKS. Martens' works on metallography include "Über die mikroskopische Untersuchung des Eisens," in *Zeitschrift des Vereins deutscher Ingenieure*, **22** (1878), 11–18; "Zur Mikrostruktur des Spiegeleisens," *ibid.*, 480–488; "Über das mikroskopische Gefüge und die Rekristallisation des Roheisens, speziell des grauen Eisens," *ibid.*, **24** (1880), 398–406; "Mikroskop für die Untersuchung der Metalle," in *Stahl und Eisen*, **2** (1882), 423–425; "Untersuchungen über das Kleingefüge des schmiedbaren Eisens, besonders des Stahles," *ibid.*, **7** (1887), 235–242; "Die Mikroskopie der Metalle auf dem Ingenieurkongress zu Chicago 1893," *ibid.*, **14** (1894), 797–809; "Ferrit und Perlit," *ibid.*, **15** (1895), 537–539, a discussion between Martens and A. Sauveur; and "F. Osmonds Methode für die mikrographische Analyse des gekohlten Eisens," *ibid.*, **15** (1895), 954–957.

For articles on materials testing, see "Die Festigkeitseigenschaften des Magnesiums," in *Mitteilungen aus den K. technischen Versuchsanstalten zu Berlin*, **5** (1887), supp. 1; "Ergebnisse der Prüfung von Apparaten zur Untersuchung der Festigkeitseigenschaften von Papier," *ibid.*, supp. 3, pt. 2; "Untersuchungen über Festigkeitseigenschaften und Leitungsfähigkeit an deutschem und schwedischem Drahtmateriale," *ibid.*, supp. 2; "Festigkeitsuntersuchungen mit Zinkblechen der schlesischen AG für Bergbau und Zinkhüttenbetrieb zu Lipine, Oberschlesien," *ibid.*, **7** (1889), supp. 4; "Untersuchungen mit Eisenbahnmaterialien," *ibid.*, **8** (1890), supp. 2; and "Untersuchungen über den Einfluss der Wärme auf die Festigkeitseigenschaften des Eisens," *ibid.*, 159–214.

With H. Sollner, Martens wrote "Verhandlungen der in Wien im Jahre 1893 abgehaltenen Conferenz zur Vereinbarung einheitlicher Prüfungsmethoden für Bau- und Konstruktionsmaterialien," in *Mitteilungen aus dem Mechanisch-technischen Laboratorium der K. technischen Hochschule in München*, no. 23 (1895). Also see "Entspricht das zur Zeit übliche Prüfungsverfahren bei der Übernahme von Stahlschienen seinem Zweck?" in *Stahl und Eisen*, **20** (1900), 302–310; "Zugversuche mit eingekerbten Probekörpern," in *Zeitschrift des Vereins deutscher Ingenieure*, **45** (1901), 805–812; *Das Königliche Materialprüfungsamt der Technischen Hochschule Berlin auf dem Gelände der Domäne Dahlem beim Bahnhof Gross-Lichterfelde-West* (Berlin, 1904), written with M. Guth.

His later works include "Prüfung der Druckfestigkeit von Portlandzement," in *Verhandlungen des Vereins zur Beförderung des Gewerbefleisses*, **88** (1909), 179–186; "Über die Grundsätze für die Organisation des öffentlichen Materialprüfungswesens," in *Dinglers polytechnisches Journal*, **93** (1912), 557–559; and "Über die in den Jahren 1892 bis 1912 im Königl. Materialprüfungsamt ausgeführten Dauerbiegeversuche mit Flusseisen," in *Mitteilungen aus dem K. Materialprüfungsamt Gross-Lichterfelde*, **32** (1914), 51–85.

Martens' handbooks are *Handbuch der Materialienkunde für den Maschinenbau*, 2 vols. (Berlin, 1898–1912); and *Das Materialprüfungswesen unter besonderer Berücksichti-*

gung der am Kgl. Materialprüfungsamt zu Berlin-Lichter-felde üblichen Verfahren im Grundriss dargestellt (Stuttgart, 1912), written with F. W. Hinrichsen.

II. Secondary Literature. Articles on Martens are "Adolf Martens," in *Metallographist*, **3** (1900), 178–181; "Adolf Martens †," in *Zeitschrift des Vereins deutscher Ingenieure*, **58** (1914), 1369–1370; and E. Heyn, "Adolf Martens †," in *Stahl und Eisen*, **34** (1914), 1393–1395.

Franz Wever

MARTÍ FRANQUÉS (or **Martí d'Ardenya**), **ANTONIO DE** (*b.* Altafulla, Tarragona, Spain, 14 June 1750; *d.* Tarragona, 19 August 1832), *biology, geology, meteorology, chemistry.*

Martí Franqués belonged to a rich and noble Catalan family. He started his studies at the University of Cervera but left because of his disgust with the Scholastic atmosphere. He continued his education himself, first learning French, and later English, German, and Italian. Initially, as a member of the Sociedad de Amigos del País, he took an interest in promoting the development of the cotton-spinning, weaving, and chinaware industries in the region of Tarragona.

Martí Franqués later became interested in scientific matters, but he was a retiring person and almost never announced his discoveries. In 1785 he started analyses of air that concluded in establishing, on 12 May 1790, that the oxygen content of the atmosphere is between 21 and 22 percent. He devised an instrument to control the air pressure and temperature in his atmospheric analyses which was a forerunner of Walter Hempel's burette. In 1791 he became absorbed in the sexual reproduction of plants, and as a result of his experiments, he understood and defended Linnaeus against Spallanzani. In all these studies he demonstrated that he was aware of the latest developments of the leading scientists of that period, notably Priestley and Cavendish. He knew and admired the work of Lavoisier, but that did not prevent him from recognizing the priority of Cavendish in the synthesis of water.

The Peninsular War (1808–1814) curtailed Martí Franqués' experiments. The bombardment of Tarragona destroyed part of his laboratory, and he himself was taken prisoner by the French. After the war he continued his reproduction experiments, but they were much less important than the ones he had done previously.

Among his disciples and friends were the botanists Mariano Lagasca y Segura, Mariano de la Paz Graells, and the physicist Juan Agell.

BIBLIOGRAPHY

A notice by Torres Amat appeared in *Diario de Barcelona*, 25–26 May 1833. The best monograph is in Catalan by Antoni Quintana i Mari, "Antoni de Martí i Franqués (1750–1832)," in *Memòries de l'Acadèmia de Ciències i Arts de Barcelona*, **24** (1935), 1–309; the first 58 pages contain Martí's preserved scientific works. A critical analysis of them was made by E. Moles in his entrance speech for the Academia de Ciencias Exactas, Físicas y Naturales de Madrid (28 March 1934).

J. Vernet

MARTIANUS CAPELLA (*b.* Carthage; *fl.* Carthage, *ca.* a.d. 365–440), *transmission of knowledge.*

Martianus may have been a secondary school teacher or a rhetorician, and he appears to have pleaded cases as a *rhetor* or advocate. He was the author of *De nuptiis philologiae et Mercurii*, the most popular textbook in the Latin West during the early Middle Ages. Cast in the form of an allegory of a heavenly marriage, in which seven bridesmaids present a compendium of each of the liberal arts, this book became the foundation of the medieval curriculum of the trivium (books III–V) and quadrivium (VI–IX). The setting (I–II) became a model of heavenly journeys as late as Dante and contributed greatly to the popularity of the book. Although Martianus understood little more of the subject matter of the disciplines than what he presented in digest form, he was a key figure in the history of rhetoric, education, and science for a thousand years.

Owing to the disappearance in the early Middle Ages of Varro's book on the mathematical disciplines (*Disciplinae*, IV–VII), Martianus' quadrivium books, inspired by Varro's archetypal work, provide the best means of reconstructing the ancient Roman mathematical disciplines. Book VI, *De geometria*, proves to be not a book on geometry but a conspectus of *terra cognita*, reduced from the geographical books of Pliny the Elder's *Natural History* (III–VI) and the *Collectanea rerum memorabilium* of Solinus. Martianus closes with a ten-page digest of Euclidean geometry, drawn from some Latin primer in the Varronian tradition. This digest assumes importance as a rare sample of pre-Boethian Latin geometry. Book VII, *De arithmetica*, was the second most important treatise on arithmetic after Boethius' *De institutione arithmetica*. Martianus' ultimate sources were Nicomachus' *Introduction to Arithmetic* and Euclid's *Elements* VII–IX, but his immediate sources were Latin primers based upon these works. A. Dick cites the original passages in the apparatus of his edition. Martianus' arithmetic proper consists of classifications and definitions of the kinds of numbers (largely Nicomachean, with some Euclidean material) and Latin translations of the enunciations of thirty-six Euclidean arithmetical propositions. Euclid developed his proofs geometrically; Martianus used numerical illustrations.

Book VIII, *De astronomia*, is the best extant ancient Latin treatise on astronomy. Because of its systematic, proportionate, and comprehensive treatment, it is the only one that bears comparison with such popular Greek handbooks as Geminus' *Introduction to Phenomena*. Its excellence indicates that Greek traditions, transmitted to the Latin world by Varro, were fairly well preserved in digest form. Martianus deals with all the conventional topics: the celestial circles; northern and southern constellations; hours of daylight at the various latitudes; anomalies of the four seasons; and a discussion of the orbits of each of the planets, including the sun and moon. Martianus was the only Latin author to give a clear exposition of Heraclides' theory of the heliocentric motions of Venus and Mercury and was commended for this by Copernicus.

Book IX, *De harmonia*, largely drawn from Aristides Quintilianus' *Peri mousikes*, book I, is important for its Latin definitions of musical terms that have long puzzled medieval musicologists. Next to Boethius, Martianus was the most important ancient Latin authority on music.

BIBLIOGRAPHY

I. ORIGINAL WORKS. The best ed. of *De nuptiis* is that of A. Dick (Leipzig, 1925). A new ed., to be published about 1976, is being prepared for the Teubner Library by J. A. Willis. A trans. of the complete work, with commentary, is W. H. Stahl, *Martianus Capella and the Seven Liberal Arts* (New York, 1971).

II. SECONDARY LITERATURE. See W. H. Stahl, *The Quadrivium of Martianus Capella; a Study of Latin Traditions in the Mathematical Sciences from 50 B.C. to A.D. 1250* (New York, 1969); and *Roman Science; Origins, Development, and Influence to the Later Middle Ages* (Madison, Wis., 1962), which contains a chapter on Martianus and places him in the stream of Latin scientific writings. C. Leonardi's book-length census of Martianus' MSS describes 243 MSS and excerpts and discusses his influence in later ages: "I codici di Marziano Capella," in *Aevum*, **33** (1959), 443–489; **34** (1960), 1–99, 411–524.

W. H. STAHL

MARTIN, BENJAMIN (*b.* Worplesdon, Surrey, England, February 1704 [?]; *d.* London, England, 9 February 1782), *experimental philosophy, scientific instrumentation*.

Benjamin Martin was the son of John Martin, gentleman, of Broadstreet, near Worplesdon. Nothing is known of his education, but it seems probable that in science he was self-taught. In 1729 he married Mary Lover of Chichester, and at the time of his marriage was described as a merchant of Guildford. The couple had two children, a daughter, Maria, and a son, Joshua Lover Martin, who joined his father in the 1770's to form the firm of B. Martin and Son.

Soon after his marriage, Martin became a teacher, running his own boarding school at Chichester, where, in 1735, he published his first work, *The Philosophical Grammar;* it ran to eight editions. The second edition includes a description of a pocket microscope, suggesting that Martin was also engaged in inventing and possibly selling optical instruments. By 1743, he had become a traveling lecturer in experimental philosophy, for in that year he first published a textbook based on his course of lectures. Martin made a curiously inept attempt during this period to secure election to the Royal Society. In letters written in 1741 to Sir Hans Sloane and the duke of Richmond, he claimed that he found it an embarrassment when lecturing not to be a fellow, and therefore requested their support in acquiring the title. This approach found no favor at all, and Martin never achieved the desired fellowship.

By 1755 Martin had settled in London, for in January he launched a monthly journal, *The General Magazine of Arts and Sciences*, publication of which continued until 1765. Between September 1755 and May 1756, he set up in business at 171 Fleet Street. His shop soon became well-known for its extensive stock and for Martin's popular lecture-demonstrations, following in the tradition of the Hauksbees and Desaguliers. Martin also stimulated business by his constant publication of catalogues of the scientific instruments that he supplied and pamphlets on a wide range of scientific subjects.

Although Martin claimed to have invented and improved numerous instruments, he is more accurately to be described as a retailer than as an instrument-maker. Instruments bearing his name are to be found in many museums, and they cover a wide range of types. Among his inventions, the best known relate to the microscope. He is credited as the first to supply, in about 1740, a microscope fitted with a micrometer. To improve the image, he produced, from 1759, an objective with two lenses set one inch apart. It has been suggested that the screw thread on this type of objective has a linear descendant in the standard thread of the Royal Microscopical Society, which was established in 1858 and continues in use today.

Martin is remarkable as one of the great popularizers of science in the mid-eighteenth century. He became known internationally, and supplied Harvard College, Massachusetts, with a large proportion of the new instruments needed after the fire of 1764. Yet Martin's industry and popularity did not bring financial stability. He was declared a bankrupt in January 1782, and died, a few weeks after a suicide attempt, on 9 February.

BIBLIOGRAPHY

Benjamin Martin's publications are too numerous to cite here. P. J. Wallis lists more than sixty works in his biobibliography of British mathematical writers.

See also R. S. Clay and T. H. Court, *The History of the Microscope* (London, 1932), ch. 9; John R. Millburn, "Benjamin Martin and the Royal Society," in *Notes and Records of the Royal Society of London*, **28** (June 1973), 15–23; "Benjamin Martin and the Development of the Orrery," in *British Journal for the History of Science*, **6** (Dec. 1973), 378–399; G. L'E. Turner, "The Apparatus of Science," in *History of Science*, **9** (1970), 129–138; John Williams, "Some Account of the Martin Microscope, Purchased for the Society at the Sale of the Late Professor Quekett's Effects," in *Transactions of the Microscopical Society of London*, n.s. **10** (1862), 31–41; and "A Few Words More on Benjamin Martin," *ibid.*, n.s. **11** (1863), 1–4, which lists forty works by Martin.

G. L'E. TURNER

MARTIN, HENRY NEWELL (*b.* Newry, County Down, Ireland, 1 July 1848; *d.* Burley-in-Wharfedale, Yorkshire, England, 27 October 1896), *physiology*.

The son of a congregational minister and sometime schoolmaster, Martin was the oldest of twelve children. Tutored at home, he entered the Medical School of University College London at sixteen, while—as was customary—apprenticing himself to a local practitioner for clinical instruction.

The youthful Martin was particularly attracted by the teaching and example of Michael Foster, then physiology instructor at the Medical School; despite his long hours as apprentice physician, Martin soon mastered the subject sufficiently to win a place as Foster's demonstrator. When Foster was called to Cambridge as praelector in physiology at Trinity College, Martin followed, receiving a sholarship at Christ's College, where he was to place first in the natural science tripos. Martin also served as assistant to T. H. Huxley in the latter's innovative biology course at the Royal College of Science, South Kensington. Under Huxley's supervision, Martin performed the "chief labour" in writing *A Course of Practical Instruction in Elementary Biology* (1875). In the same year, Martin received the D.Sc. in physiology, the first ever granted at Cambridge. Still in his twenties, Martin was clearly one of England's most promising young physiologists.

At the same time, D. C. Gilman, president of the projected Johns Hopkins University in Baltimore, was hard at work in assembling a faculty equal to his hopes of establishing a truly research-oriented university in the United States. Huxley recommended Martin, who, after some negotiation, accepted the well-paid professorship. He was only twenty-eight.

Though Martin published only fifteen research papers in his abbreviated scholarly career, he did complete a series of significant investigations based on his success in surgically isolating a mammalian heart and perfusing it so as to create experimental situations in which he could evaluate the role of such variables as temperature, alcohol, and venous and arterial pressure in cardiac function. Martin was elected a fellow of the Royal Society on the basis of this work and in 1883 delivered the Society's Croonian Lecture on the influence of temperature variation upon heart beat.

Martin's institutional role was almost certainly more significant than his scientific work. When he arrived in Baltimore, only one other course in physiology was offered in the United States (by H. P. Bowditch at Harvard). Between 1876 and 1893 the Johns Hopkins University was to play a uniquely influential role in the establishment of a research-oriented scientific community in the United States. From his strategic position at the Hopkins, Martin was to exert a significant influence in this evolution, especially in the development of physiology. Although never a magnetic lecturer, Martin was a warm and successful graduate teacher and colleague; William T. Sedwick, William Councilman, Henry Sewall, George Sternberg, W. K. Brooks, and Martin's successor at Hopkins, William H. Howells, were among his students or sometime associates. When the American physiological society was organized in 1887, six of the twenty-four founding members were Martin's students. Not only did he create and sustain an atmosphere of scholarship in his own laboratory, but Martin also consistently advocated the need for basic science excellence in the Johns Hopkins projected medical school, which opened in 1893. Although of necessity he taught general biology and animal morphology, Martin thought consistently in disciplinary terms; he never lost sight of his identity as a physiologist, and he was deeply committed to establishing the independence of physiology from the needs and attitudes of clinical medicine.

Martin was not only active in the founding and early years of the American Physiological Society, but served on the editorial board of Foster's *Journal of Physiology*—even managing to wring a small subvention for it from the Johns Hopkins administration. In addition he edited and founded *Studies from the Biological Laboratory of the Johns Hopkins University*, five volumes of which appeared between 1877 and 1893.

Martin was also a defender of the university against the attacks of antivivisectionists and spokesmen of religious orthodoxy disturbed by the encroachments of evolutionary naturalism. A fortunate marriage to Hattie Pegram, the socially prominent widow of a Confederate officer, allowed him greater access to Baltimore society, and thus to serve more effectively as an advocate of the university. The young physiologist even offered a Saturday morning course in physiology for local teachers and normal school students. In the early 1890's Martin's health began to fail and in 1893 he resigned. With a small pension from the Hopkins trustees, he returned to England in an effort to restore his health. Despite attempts to continue working at Cambridge, Martin's health did not improve and he died in 1896.

BIBLIOGRAPHY

There is no full-scale biography of Martin. The Daniel Coit Gilman Papers at the Johns Hopkins University Library contain a good many Martin letters illuminating his years at Hopkins.

For briefer sketches of Martin's work, see C. S. Breathnach, "Henry Newell Martin (1848–1893). A Pioneer Physiologist," in *Medical History*, **13** (1969), 271–279; Henry Sewall, "Henry Newell Martin, Professor of Biology in Johns Hopkins University, 1876–1893," in *Bulletin of the Johns Hopkins Hospital*, **22** (1911), 327–333; Michael Foster, *Proceedings of the Royal Society*, **60** (1897), xx–xxiii; R. H. Chittenden, "Henry Newell Martin," in *Dictionary of American Biography*, XII, 337–338. See also the sketch by William Howells, Martin's immediate successor, in *The History of the American Physiological Society. Semi-Centennial, 1887–1937* (Baltimore, 1938), 15–18.

The best evaluation of Martin's role at Hopkins is to be found in Hugh Hawkins, *Pioneer: A History of the Johns Hopkins University. 1874–1889* (Ithaca, N.Y., 1960). Supplementary information may be found in Walter J. Meek, "The Beginnings of American Physiology," in *Annals of Medical History*, **10** (1928), 122–124; Gerald B. Webb and Desmond Powell, *Henry Sewall. Physiologist and Physician* (Baltimore, 1946). Martin's *Physiological Papers* were collected and published by the Johns Hopkins University Press as volume III of the *Memoirs From the Biological Laboratory of the Johns Hopkins University* (Baltimore, 1895). His text, *The Human Body* (New York, 1881), was used widely in the United States and went through several editions. With William Moale he wrote for classroom use a *Handbook of Vertebrate Dissection* (New York, 1881).

CHARLES E. ROSENBERG

MARTIN, RUDOLF (*b*. Zurich, Switzerland, 1 July 1864; *d*. Munich, Germany, 11 July 1925), *anthropology*.

Martin, one of Germany's most important anthropologists, was born to south German parents—his father came from Württemberg and his mother from Baden. For a short time his father worked in Zurich as a mechanical engineer, but he soon established his own machine works in Offenburg in Baden. Martin began his schooling in that city, passed the final secondary school examination there in 1884, and then enrolled in the law faculty of the Baden State University in Freiburg. After two semesters he changed his field to philosophy and left Freiburg in order to continue his studies at Leipzig. He was soon drawn back to the University of Freiburg by the presence of the zoologist Weismann. The latter had developed Darwin's ideas into a theory known as neo-Darwinism; and his lectures on the theory of evolution, the theory of natural selection, and the continuity of the germ plasm as the foundation of a theory of heredity made an indelible impression on the young Martin.

Weismann emphasized the exact formulation of problems and the scientifically demonstrable axioms of the new biological theories, and this clearly satisfied Martin more than the usual, more speculative theoretical lectures on philosophy. Martin was no doubt especially attracted by the possibilities of uniting scientific conceptions with philosophic views on the origin and destiny of man, possibilities that Weismann had presented to his students in important lectures at Freiburg beginning in 1880. Martin also attended the lectures and anatomic demonstrations of Wiedersheim and enthusiastically took part in the accompanying anatomic sections. With equal interest he followed the lectures of A. Riehl on critical philosophy and positivism, and Martin's preoccupation with Kant's ideas in anthropology may have led to his decision not to become a zoologist. His doctoral dissertation was "Kants philosophische Anschauungen in den Jahren 1762–1766," which he submitted, under the supervision of Riehl, to the philosophy faculty at Freiburg in 1887. These efforts in natural science and philosophy formed the basis of all his later work.

At the conclusion of his studies Martin visited, in 1887–1890, almost all the anthropological collections in Europe. He was especially impressed by the holdings of the École d'Anthropologie in Paris, where he twice worked as a volunteer assistant. He became acquainted with leading researchers such as Duval, P. Topinard, L.-P. Manouvrier, and the Demortillet brothers; it was they who persuaded him to return to France in order to work without the obligation to teach. It was in this period that Martin decided to devote himself to anthropology. In 1890–1891 he prepared his *Habilitationsschrift*, "Zur physischen Anthropologie der Feuerländer," which was based on an exact description and comparative anatomical evaluation of five Alakaluf tribesmen from Tierra del Fuego who had died in Zurich. With this essay he qualified as privatdocent in physical anthropology on the philosophy faculty at the University of Zurich.

When Georg Ruge was appointed director of Zurich's Institute of Anatomy in 1897, he immediately provided Martin with several rooms for anthropologic work. From this a separate anthropology institute was soon formed, and recognition for Martin was not long in coming. In 1899 Martin was named extraordinary professor of anthropology at Zurich and full professor in 1905. In 1897 he had made a major research expedition to Malaysia, where he took detailed anthropological measurements of a vast number of individuals of various tribes. He presented his results in 1905 in the monograph *Die Inlandstämme der malaiischen Halbinsel*. This classic work, which has not become obsolete, dealt with tribes existing at the most primitive level of culture then known. In his investigations of the Senoi and the Semang, Martin not only recorded accurately the anatomical and physiological characteristics of these peoples but also made fundamental observations regarding their dwellings, their history, and the entire complex of their social relationships. In addition he investigated their consanguinity with other primitive populations living in Malaysia. During this project he constructed new, more exact measuring instruments. In general, these instruments, considered the best available at the time, permitted him to obtain the first exact results that could both be employed in comparative studies and be effectively submitted to statistical procedures.

Martin was able to gather around himself in Zurich many gifted students; but his health, already weak at that time, forced him to confine himself to his research. Thus in 1911 he gave up his professorship at Zurich in order to retire to Versailles. There, assisted by his French colleagues and able to draw on the rich material in Paris, he began working on a textbook of anthropology conceived on a grand scale. It appeared in 1914, just before the start of World War I. Taken by surprise by the outbreak of war, Martin managed to flee to Germany from a seaside resort in southern France; however, all his scientific collections and personal assets were impounded. In 1917 he received an appointment as professor of anthropology at the University of Munich, where he remained for the rest of his life.

Martin's textbook, which is really a sort of handbook, makes it clear that he took into account all the tendencies within the field of anthropology and that he sharply distinguished it from certain other specialties. He thereby elevated this discipline from the status of an auxiliary science and endowed it with a thoroughly autonomous character. In his introduction to the nature and tasks of anthropology he states:

> Anthropology is the natural history of the hominids throughout their temporal and spatial distribution. Hence it is established (1) that anthropology is a science of groups, and that as a result human anatomy, physiology, etc. are excluded from its domain as sciences concerned with individuals; (2) that it deals with only the nature of the hominids; and (3) that it encompasses the entire realm of forms of this zoological group without any restriction. Anthropology therefore has the task of distinguishing all the extinct and recent forms occurring among the hominids, with respect to their corporal properties, of characterizing them, and of investigating their geographical distribution. . . .

In this program Martin placed special emphasis on the technique of anthropological investigation that he had developed. He wrote repeatedly concerning "instructions for body measurements" and "anthropometry."

Furthermore, to improve the teaching of anthropology Martin created first-rate wall charts that were well made and didactically effective. He was constantly preoccupied with adapting his textbook to current developments in the young science, but the second edition was only published posthumously (1928). Prepared by his anthropological co-worker and second wife, Stephanie Oppenheim, it appeared in three volumes. In 1956–1966 a successor to Martin's chair, Karl Saller, brought out a third edition, in four volumes.

In his later years, Martin turned to the anthropology of European peoples. In particular, prompted by the years of famine in Germany during and after World War I, he undertook important investigations into the influence of hunger on the development of schoolchildren. He also studied the effect of profession and sports on the physique of certain strata of the population. His subjects were students, especially the gymnasts who had gathered for the great German gymnastic festival held at Munich in 1923. In the meantime, however, his activity was severely restricted by heart disease; his death, which was the result of a heart attack, came as a surprise to those other than his close friends.

Martin was named Geheimer Regierungsrat and was an honorary or corresponding member of many scientific societies in Germany, Italy, England, Spain, Austria, France, Holland, and Russia. Only one year before his death he founded his own journal, the *Anthropologischer Anzeiger*. In it he published sensational studies on the reduced physical development of starving Munich schoolchildren in 1921–1923. His findings helped to bring about the introduction of remedial measures financed by American institutions, which were immediately effective.

Martin was extremely tolerant and objected to the use of malicious or polemical language against his scientific opponents. In his later years his favorite field of study was Indian philosophy and art. He was warmhearted to both students and friends. By his first wife he had three sons. During the final period of his life his second wife became his trusted co-worker; she also arranged his posthumous papers. As Martin observed, "We will never be completely finished with the investigation of life, and if occasionally we seek a provisional conclusion, we know very well that even the best we can give is no more than a step towards the better."

BIBLIOGRAPHY

I. ORIGINAL WORKS. A complete bibliography of Martin's works is in *Anthropologischer Anzeiger*, **3** (1926), 15–17. His most important works and papers are "Kants philosophische Anschauungen in den Jahren 1762–1766" (Ph.D. diss., University of Freiburg im Breisgau, 1887); "Zur physischen Anthropologie der Feuerländer," in *Archiv für Anthropologie*, **22** (1893), 155–217; "Altpatagonische Schädel," in *Vierteljahrsschrift der Naturforschenden Gesellschaft in Zürich*, **41** (1896), 496–537; "Die Ureinwohner der malayischen Halbinsel," in *Correspondenzblatt der Deutschen Gesellschaft für Anthropologie Ethnologie und Urgeschichte*, **30** (1899), 125–127; "Anthropologisches Instrumentarium," *ibid.*, 130–132; *Anthropologie als Wissenschaft und Lehrfach* (Jena, 1901); *Wandtafeln für den Unterricht in Anthropologie, Ethnologie und Geographie mit Verzeichnis und Beschreibung* (Zurich, 1902); "Über einige neue Instrumente und Hilfsmittel für den anthropologischen Unterricht," in *Korrespondenzblatt der Anthropologischen Gesellschaft*, **34** (1903), 127–132; and *Die Inlandstämme der malaiischen Halbinsel* . . . (Jena, 1905). His textbook, *Lehrbuch der Anthropologie in systematischer Darstellung*, went through three eds. (Jena, 1914; 2nd ed., 1928; 3rd ed., Stuttgart, 1956–1966).

Martin's later works include "Über Domestikationsmerkmale beim Menschen," in *Naturwissenschaftliche Wochenschrift*, **30**, n.s. **14** (1915), 481–483; "Anthropologische Untersuchungen an Kriegsgefangenen," in *Umschau*, **19** (1915), 1017; "Anthropometrie," in *Münchener medizinische Wochenschrift*, **69** (1922), 383–389; *Körperverziehung* (Jena, 1922); "Anthropometrische und ärztliche Untersuchungen an Münchener Studierenden," in *Münchener medizinische Wochenschrift*, **71** (1924), 321–325, written with A. Alexander; "Die Körperbeschaffenheit der deutschen Turner," in *Monatsschrift für Turnen, Spiel und Sport*, **3** (1924), 53–61; "Körpermessungen und -wägungen an deutschen Schulkindern," in *Veröffentlichungen des K. Gesundheitsamtes*, separate supps. (pt. 1, 1922; pt. 2, 1923; pt. 3, 1924); "Die Körperentwicklung Münchener Volksschulkinder in den Jahren 1921, 1922 und 1923," in *Anthropologischer Anzeiger*, **1** (1924), 76–95; *Richtlinien für Körpermessungen und deren statistische Verarbeitung mit besonderer Berücksichtigung von Schülermessungen* (Munich, 1924); "Die Körperentwicklung Münchener Volksschulkinder im Jahre 1924," in *Anthropologischer Anzeiger*, **2** (1925), 59–78; and *Anthropometrie. Anleitung zu selbständigen anthropologischen Erhebungen und deren statistische Verarbeitung* (Berlin, 1925).

II. SECONDARY LITERATURE. See K. Saller, "Rudolf Martin†," in *Münchener medizinische Wochenschrift*, **72** (1925), 1343–1344; and E. Fischer, "Rudolf Martin†," in *Anatomischer Anzeiger*, **60** (1926), 443–448. Also see the article, "Rudolf Martin," in I. Fischer, ed., *Biographisches Lexikon der hervorragenden Ärzte der letzten fünfzig Jahre*, 2nd ed., II (Munich–Berlin, 1962), 998.

H. SCHADEWALDT

MARTÍNEZ, CRISÓSTOMO (*b*. Valencia, Spain, 1638; *d*. Flanders, 1694), *anatomy*.

Information about the early career of Martínez is almost nonexistent. He was associated with a circle of anatomists at the University of Valencia, where he began work on an anatomical atlas around 1680. In December 1686 he received a grant from Charles II which enabled him to advance his studies abroad. Martínez arrived in Paris on 19 July 1687 to continue work on his atlas, associating himself with Joseph-Guichard Duverney and other members of the recently founded Académie des Sciences. Accused of spying, he was obliged to flee France in 1690.

The anatomical work of Martínez survives in nineteen engravings and a few written descriptions. His macroscopic drawings, encompassing most of the human body, bespeak a functional interpretation of anatomy. His genius was most apparent, however, in microscopic studies of the structure of the human bone, work which placed him among the first generation of European microscopists, the only Spaniard so to qualify.

Both in his drawings and in his essay, "Generalidades acerca de los huesos" ("Generalities Concerning Bones"), Martínez sought to explain the processes of ossification from the embryo, through infantile bones without periosteum, to the mature bone structure. He studied the insertion of ligaments and muscles; the periosteum; the exterior pores of the bone; the structure of compact bone substance and of spongy bone tissue (this last was the subject of his best graphic work); and the function of bone marrow. His work rested on four concepts: the formation of fat from the blood (an iatrochemical notion accepted by most of his contemporaries); the presence of storage vesicles; the morphological and functional nature of the medulla; and the existence of adipose circulation. Martínez regarded the "adipose vessels" as his major discovery, and his work in this area reflected the great influence of Harvey among late seventeenth-century anatomists.

BIBLIOGRAPHY

I. Original Works. The extant anatomical drawings and writings of Martínez are collected in José María López Piñero, ed., *El Atlas Anatómico de Crisóstomo Martínez* (Valencia, 1964). Of the nineteen drawings published by López Piñero, only three had been published before: no. XIX in *Nouvelles figures de proportions et d'anatomie du corps humain* (Paris, 1689; repr., Frankfurt–Leipzig, 1692); and nos. XVII and XIX in *Nouvelle esposition des deux grandes planches gravées et dessinées par Chrysostome Martinez, Espagnol, représentant des figures très singulières de proportions et d'anatomie*, with revisions of Martínez's text by J. B. Winslow (Paris, 1740; repr., 1780).

II. Secondary Literature. Various aspects of Martínez' work are discussed in P. Dumaitre, "Un anatomiste espagnol à Paris au XVIIe siècle. Chrysostome Martinez et ses rarissimes planches d'anatomie," in *Médecine de France*, no. 154 (1964), 10–15; José M. López Piñero, "La repercusión en Francia de la obra anatómica de Crisóstomo Martínez," in *Cuadernos de Historia de la Medicina Española*, 6 (Salamanca, 1967), 87–100; and María Luz Terrada Ferrandis, *La anatomía microscópica en España (siglos XVII–XVIII)* (Salamanca, 1969).

Thomas F. Glick

MARTINI, FRANCESCO DI GIORGIO, also known as **Francesco di Siena** (*b.* Siena, Italy, 1439; *d.* Siena, November 1501), *architecture, sculpture, painting, technology.*

Little is known of Francesco's early life. He was apparently born to humble parents and trained as a painter, probably in Siena. He married twice, in 1467 and 1469. A document of the latter year shows that he, together with a certain Paolo d'Andrea, received a commission from the municipal authorities of Siena to improve the water supply system of that city. From this period of his career until his death it is possible to trace his activities with some facility; until 1477, when he entered the service of Federigo da Montefeltro, duke of Urbino, all records refer to Francesco as "dipintore," a painter.

In Urbino, Francesco participated not only in the decoration but also in the architectural design of the great ducal palace that Federigo was building. At a somewhat later date he became the duke's chief military engineer; he accompanied Federigo on his campaigns and was responsible for the design and maintenance of engines of war and artillery, as well as for the manufacture of gunpowder. He also built a large number of fortresses for the duke.

In 1479 Francesco went to southern Italy as artist and military engineer to Ferdinand I, king of Naples. He returned to Urbino in 1481, and remained there for some time after the death of Federigo in 1482. A number of documents suggest that he was concurrently in the employ of Sienese officials, however, and by 1485 he was again working primarily in Siena, as artist and city engineer. During this time he made frequent trips to other parts of Italy in the service of various rulers. In 1486 Francesco returned to Urbino; his skills had made him so famous that governments competed for his services.

Francesco went to Milan in 1490 at the request of Gian Galeazzo Sforza. He submitted a proposal for the completion of Milan's cathedral, which lacked its dome; his project was accepted for execution by local workmen. In the same year Francesco met Leonardo da Vinci (although they may have had some previous contact); in June the two men were at Pavia, inspecting the construction of the cathedral there. In 1490, too, Francesco may have been asked to submit a design for the façade of the cathedral of Florence.

Francesco was summoned to Naples by Alfonso, duke of Calabria (and king of Naples in 1494–1495), on three separate occasions in 1491, 1492, and 1495, respectively. On his last mission for Alfonso, Francesco used gunpowder to undermine and destroy the fortress of Castelnuovo, which was held by the forces of the invading French king, Charles VIII. Although earlier writers, among them Taccola, had discussed this technique, Francesco seems to have been the first to succeed with it. In 1499 Francesco returned to Urbino to advise on fortifications intended to stem the advance of Cesare Borgia's forces. He continued to be active as both an artist—he probably resumed painting in his later years—and an engineer until the time of his death.

Francesco's reputation lay in his work as a painter, sculptor, and architect, until the discovery that a number of important writings on technology, circulated anonymously, were in fact his. These treatises and notes mirror the full range of Francesco's concerns—among their subjects are civil and military engineering, surveying, hydraulic engineering, both offensive and defensive war machinery, mechanical technology (especially millworks, devices for raising water, and cranes), studies on proportion, and notes and relief drawings of Roman monuments (which he observed in the course of his travels). In all of these writings the influence of Vitruvius is clear. (Indeed, an autograph draft of Francesco's translation of Vitruvius' treatise is itself preserved in Florence.) Francesco's importance to the history of technology lies in these works, which were widely influential among his contemporaries and successors, although they were not printed under his name until the middle of the nineteenth century.

Although certain of Francesco's mechanical projects were borrowed from earlier authors (especially Taccola), they gain significance from the artistic and technological superiority of his rendering. His method of illustration was itself influential. One of his particular techniques, that of confining the machine illustrated within a frame, may be taken as an index to the extent of his authority in books on machinery in the sixteenth and seventeenth centuries. Many manuscript technical and military collectanea of this epoch also bear his mark, including those of Leonardo, whose inventory of books (*Codex Madrid II*) contains the entry "Francesco da Siena." Francesco's ideas appeared, without credit, in the later works of such composers of "Theaters of Machines" as Jacob de Strada, Vittorio Zonca, and Agostino Ramelli, as well as the somewhat later authors G. A. Böckler and Heinrich Zeising. The technology that Francesco represented even found its way, at quite an early date, to the Far East; the *Ch'i Ch'i T'u Shuo* ("Diagrams and Explanations of Wonderful Machines"), published in 1627 by the Jesuit missionary Johann Schreck and Wang Cheng, includes several devices that clearly derive from Francesco's work.

From such later books Francesco's ideas passed into the works of eighteenth-century writers, for example, Jacob Leupold and Stephan Switzer, and even into mechanical handbooks of the early nineteenth century. Some of the basic mechanisms described by J. A. Borgnis in his *Traité complet de mécanique appliquée* of 1818–1823, for example,

are still very similar to Francesco's prototypes. Even when the machines that Francesco described became obsolete, his renderings of them lingered as technological symbols, and as such they may be seen in the form of colophons of eighteenth- and nineteenth-century technical works.

Francesco's practical work as a military engineer was equally influential. Among his innovations was the system of fortification by bastions and curtains that replaced the medieval concept of the turreted castle. The final development of this phase of his work appeared in the great French fortresses designed by Vauban for Louis XIV.

BIBLIOGRAPHY

I. ORIGINAL WORKS. Francesco's writings are preserved in six MSS:

1. *Taccuino Vaticano*, Vatican Library, Rome, Urb. Lat. 1757, an autograph pocket encyclopedia, on vellum, with sketches of a great number of mechanical contrivances. The first entry is dated 1472; others are spread out over many years. Marginal notes in the hand of Leonardo da Vinci.

2. *British Museum Codex*, British Museum, London, 24.949, a series of drawings of machines, on vellum, dedicated to Federigo da Montefeltro, datable between 1474 and 1482. An excellent copy is Biblioteca Reale, Turin, Ser. Mil. 383.

3. *Codex Laurenziano Ashburnhamiano*, Laurentian Library, Florence, 361, datable *ca.* 1480, probably the first version of the *Treatise on Architecture*, with marginal illustrations.

4. *Codex Saluzziano*, Biblioteca Reale, Turin, 148, datable *ca.* 1485, a version of the immediately preceding, although lacking several chapters. Apparently a copy by a professional scribe, although the marginal illustrations may be in Francesco's own hand.

5. *Trattato d'architettura civile e militare*, Biblioteca Comunale, Siena, S.IV.4, datable *ca.* 1490, a copy on paper of a lost original.

6. *Codex Magliabechiano*, National Library, Florence, II.I.141, datable *ca.* 1492, an expanded and integrated version of the Siena manuscript above. Besides the beautifully illustrated text of the *Trattato*, it contains the autograph translation of Vitruvius and a collection of drawings of war machines and fortifications.

Editions of Francesco's work are Carlo Promis, ed., *Trattato di architettura civile e militare*, 2 vols. and atlas (Turin, 1841); and, more useful, Corrado Maltese, ed., *Trattati di architettura ingegniera e arte militare*, 2 vols. (Milan, 1967), which includes complete annotated transcriptions of the Turin, Siena, and Florence MSS, an index of variants and concordances, and reproductions of all illustrated pages.

II. SECONDARY LITERATURE. On Francesco and his work see Selwyn Brinton, *Francesco di Giorgio of Siena*, 2 vols.

(London, 1934–1935); P. Fontana, "I codici di Francesco di Giorgio Martini e di Mariano di Jacomo detto il Taccola," in *Actes du XIVe Congrès International d'histoire de l'art* (Brussels, 1936), p. 102, which discusses the influence of Taccola on Francesco; G. Mancini, *Giorgio Vasari: Vite cinque annotate* (Florence, 1917); Roberto Papini, *Francesco di Giorgio architetto*, 3 vols. (Florence, 1946), a superb graphic presentation, but marred by polemics and unwarranted attributions; A. E. Popham and P. Pouncey, *Italian Drawings in the Department of Prints and Drawings in the British Museum: the XIVth and XVth Centuries* (London, 1950), which gives an accurate description of the British Museum codex; Ladislao Reti, "Francesco di Giorgio Martini's Treatise on Engineering and its Plagiarists," in *Technology and Culture*, **4**, no. 3 (1963), 287–298, on the transmission of Francesco's technological drawings; E. Rostagno and T. Lodi, *Indici e cataloghi VIII. I codici Ashburnhamiani della Biblioteca Mediceo-Laurenziana*, I (Rome, 1948), 468–474, a study of the early versions of the *Trattato;* Mario Salmi, *Disegni di Francesco di Giorgio nella Collezione Chigi Saracini* (Siena, 1947); Luigi Michelini Tocci, "Disegni e appunti autografi di Francesco di Giorgio in un codice del Taccola," in *Scritti di storia dell'arte in onore di Mario Salmi*, II (Rome, 1962), 202–212, which is concerned with Taccola's influence on Francesco; and Allen Stuart Weller, *Francesco di Giorgio 1439–1501* (Chicago, 1943), the best available biography of Francesco, with a bibliography complete up to 1942.

LADISLAO RETI

MARTINOVICS, IGNÁC (*b.* Pest, Hungary, 22 July 1755; *d.* Buda, Hungary, 20 May 1795), *chemistry*.

Martinovics, the son of an army officer, entered the Franciscan order and studied philosophy and theology at the universities of Pest and Vienna, but he soon displayed an interest in natural science and technology. He left the order and served as an army chaplain at various garrisons in the Hapsburg dominions. In Galicia, he became the secretary to Count Potocki and traveled with him through western Europe. In England and France, Martinovics became acquainted with the most distinguished scientists of the age and was exposed to the progressive ideas of the French Enlightenment. Thereafter chemistry and politics determined the course of his life.

After returning home, Martinovics accepted the chair of physics at the University of Lemberg (Lvov) in Galicia, where he remained from 1783 to 1791. At Lemberg he published his *Praelectiones physicae experimentalis* (1787), which showed that he still adhered to the phlogiston theory. He continued to oppose Lavoisier's theory of combustion and sought to refute it with experiments with fulminating gold.

He asserted that although combustion is indeed determined by a substance, that substance is not oxygen; the atmosphere plays no role in the explosion of fulminating gold, which therefore takes place without oxygen. Martinovics also carried out experiments on the solubility of gases in water; he ascertained that the solubility diminishes with decreasing pressure and increasing temperature. In 1791 he undertook distillation experiments on Galician petroleum and determined the combustibility and specific gravity of various fractions.

Martinovics also published many philosophical writings that reflected the atheistic and materialistic views of d'Holbach. Inspired by the French Revolution, he fought for political reform in the Hapsburg state. At first he hoped this could be accomplished legally. He entered the service of the reform-minded Emperor Leopold II and furnished him with information concerning the political intentions of the nobility and the Jesuits. Leopold's successor, Francis, frightened by the events of the French Revolution, changed his predecessor's policies and dismissed the reformers. Martinovics then joined in a plot to proclaim a republic. The plot was discovered, and Martinovics and four other conspirators were beheaded.

BIBLIOGRAPHY

A list of Martinovics' works is in Poggendorff, I, 65. On Martinovics, see V. Fraknoi, *Martinovics* (Budapest, 1921), in Hungarian; and Z. Szökefalvi-Nagy, "Ignatius Martinovics, 18th-century Chemist and Political Agitator," in *Journal of Chemical Education*, **41** (1964), 458.

FERENC SZABADVÁRY

MARTIUS, KARL FRIEDRICH PHILIPP VON (*b.* Erlangen, Germany, 17 April 1794; *d.* Munich, Germany, 13 December 1868), *botany, ethnology*.

The Martius family, many members of which had pursued learned professions, traced its lineage back to Galeottus Martius, who was a professor at Padua in 1450. Karl was the son of Ernst Wilhelm Martius, an apothecary and honorary professor of pharmacy at Erlangen University, and Regina Weinl, a noblewoman. Martius at an early age manifested a resolve to devote himself to science. He was also much interested in classical studies and composed many of his works in a very elegant Latin. In 1810 he was admitted to Erlangen University, where he studied medicine and received his M.D. with the dissertation *Plantarum horti academici Erlangensis* (Erlangen, 1814).

In 1814 Martius became an *élève* of the Royal Bavarian Academy at Munich and was appointed assistant to the conservator of the botanic garden, F. von Schrank. In this position he published his *Flora cryptogamica Erlangensis* (Nuremberg, 1817), which he had begun while at Erlangen. In 1816 he was admitted as a member of the Leopoldine Academy, of which he became director ephemeridum in 1858.

A man of superior talents, indefatigable energy, and excellent personal qualities, Martius attracted the attention of the older members of the Royal Academy. The king of Bavaria, Maximilian Joseph I, was a lover of botany and often selected Martius as his guide when visiting the botanic garden. His plans for sending scientific explorers to South America were realized in 1817, when an Austrian expedition was sent to Brazil. Martius and several other Bavarian scientists, including his companion, the zoologist Spix, went along. They left Trieste on 2 April 1817 and returned to Munich in December 1820. The Munich herbarium received 6,500 carefully preserved species of plants, which constituted a most valuable portion of its collection, and the botanic garden received many living plants and seeds from the expedition. This voyage laid the foundation of Martius' future success, and as a result of the expedition he was appointed a member of the Royal Bavarian Academy and assistant conservator of the botanic garden. In 1826, when King Ludwig I had transferred Landshut University to Munich, Martius was appointed professor of botany, and in 1832, when Schrank retired, he was named principal conservator of the botanic garden, institute, and collections. Among his students were A. Braun, H. von Mohl, K. Schimper, and O. Sendtner. In 1840 Martius became secretary of the physicomathematical section of the academy and was charged with all correspondence and commemorative addresses. The excellent style of these eulogies is comparable to that of the *éloges* of G. Cuvier. The decision of the government to erect the glass building of the Munich industrial exhibition within the area of the botanic garden deeply disappointed Martius, who had vainly remonstrated, and it caused him to resign his professorship and superintendence of the garden. After his retirement much of his time was taken up in editorial activities and scientific labors.

In 1823 Martius married Franciska Freiin von Stengel; they had four children.

BIBLIOGRAPHY

I. ORIGINAL WORKS. Martius wrote more than 150 books, monographs, and minor works, as well as several poems. His monograph, *Historia naturalis palmarum*, 3 vols. (Munich, 1823–1853), written in cooperation with H. von Mohl, A. Braun, and O. Sendtner, was highly appreciated. In 1840 he began his great work, the *Flora Brasiliensis*, 15 vols. (Munich, 1840–1906), assisted by many collaborators and financially supported by the Brazilian government. After his death this magnificent work was continued by several others such as A. Eichler and I. Urban. Together with Spix he wrote *Reise in Brasilien . . . in den Jahren 1817–1820*, 3 vols. (Munich, 1823–1831). The eulogies are contained in *Akademische Denkreden von C. F. Ph. von Martius* (Leipzig, 1866). Those of a later date are in *Sitzungsberichte der Bayerischen Akademie der Naturwissenschaften zu München* (Munich, 1868).

Martius also wrote about the potato disease, of which he had discovered the cause, in *Die Kartoffelepidemie* (Munich, 1842). One of his more important ethnological works is *Beiträge zur Ethnographie und Sprachenkunde Amerikas zumal Brasilien*, 2 vols. (Leipzig, 1867).

II. SECONDARY LITERATURE. For works about Martius, see A. W. Eichler, "C. F. Ph. v. Martius, Nekrolog," in *Flora*, **52** (1869), 3–13, 17–24; K. Goebel, *Zur Erinnerung an K. F. Ph. von Martius* (Munich, 1905); C. F. Meissner, *Denkschrift auf Carl Friedr. Phil. von Martius* (Munich, 1869); and the article by E. Wunschmann in vol. 20, *Allgemeine Deutsche Biographie*, 517–527.

A. P. M. SANDERS

MARTONNE, EMMANUEL-LOUIS-EUGÈNE DE

(*b*. Chabris, France, 1 April 1873; *d*. Sceaux, France, 24 July 1955), *geography, geomorphology, hydrography*.

The scion of a noble Breton family, Martonne was the son of Alfred de Martonne, an archivist, and the former Caroline Cadart. He entered the École Normale Supérieure in 1892 and three years later received a degree in history and geography. He subsequently attended the courses and worked in the laboratories of Richthofen at Berlin and of Penck and Hann at Vienna. Soon after his return to Paris he began fieldwork in Rumania and became proficient in the language.

In 1899 Martonne joined the geography department of the Faculty of Letters at the University of Rennes, where in the Faculty of Sciences he established a geographical laboratory equipped with maps, geological specimens, and simple surveying instruments. He wrote several articles on the peneplain and the coastal morphology of Brittany (1904–1906); but his main publications concerned mountain glaciation, particularly in the southern Carpathians.

In 1904 Martonne went on a long excursion to the American West and Mexico with William M. Davis, professor of geology at Harvard. He became devoted to Davis' methods of teaching, presentation and landform analysis; henceforth his interests leaned increasingly toward the physical branches of geography. In 1905 he joined the Faculty of Arts at the University of Lyons.

The turning point in Martonne's life came in 1909, when he succeeded Paul Vidal de la Blache, his father-in-law, as head of the department of geography in the Faculty of Letters at the Sorbonne. He held this post for thirty-five years, and from 1927 he combined it with the directorship of the Institut de Géographie. From the death of Vidal de la Blache in 1918 to his own retirement in 1944, Martonne was the recognized leader of the French school of geography. A leading international figure, he was responsible for organizing the meetings of the International Geographical Union from 1931 to 1938, serving as its president from 1949 to his death. The Académie des Sciences elected him a member in 1942; he also held honorary membership in a dozen foreign geographical societies and honorary doctorates from Cambridge and the University of Cluj (Rumania).

Martonne was the most important influence in the development of modern French geography into an autonomous science. As a scholar he was more interested in the patient accumulation of observed facts than in deductions, but, although cautious, he was fairly open-minded toward new concepts and techniques.

Martonne's approach to the natural sciences was essentially geographical, and he took a broad view of physical geography. After describing and mapping the distribution by area of a natural phenomenon, he usually tried to associate that distribution with some general law and so to seek causes for it. Thus he tended to place more importance on comparative spatial distribution than on genetic explanation and was more inclined to determine the causes of distribution than to elucidate the scientific properties of the phenomenon itself. Among his major contributions to geographical instruction in France were his insistence on practical laboratories and on a sound knowledge of cartography and surveying. To regional geography he contributed important general descriptions of Walachia (1902) and two volumes, on central Europe, to the *Géographie universelle* (1931–1932). His smaller regional syntheses on the Alps (1926) and the major geographical regions of France (1921) are masterly summaries.

In physical geography Martonne's chief contributions were to the study of mountain glaciation, peneplains, hydrography, and climatic geomorphology. His discussions of glacial erosion and the development of Alpine valleys show keen powers of observation and are his best works. He popularized the concept that steps and over-deepenings in the floors of glaciated valleys were associated with pre-existing breaks-of-slope caused by pre-glacial and Quaternary tectonic uplifts. His elaboration of the role of snow (nivation) in sculpturing mountain landforms was one of his more original themes.

Peneplains and other erosional flattenings always interested Martonne, and he preferred Davis' peneplanation theory to the eustatic ideas of Henri Baulig. Among the investigations he helped to initiate was an international study of terraces; the findings were edited by others for the International Geographical Union.

In hydrography Martonne published (1925–1928), with the collaboration of L. Aufrère, details of areas with endoreic (interior) drainage rather than exoreic drainage (flowing to the ocean). This survey and its world map showed that 27 percent (41 million square kilometers) of the continental land area did not drain to the oceans, whereas the previously accepted measurement had been 22 percent (33 million square kilometers). Martonne then proceeded to relate the enfeebled nature of certain drainage systems to an increase of aridity; he also propounded an index of aridity and the concept of areism, or absence of stream runoff, a condition that occurs on 17 percent of the continental land area. He traveled widely in the deserts of North Africa and South America, in an attempt to determine a more precise relationship between aridity and surface runoff, or drainage. His first index of aridity was $P/(T + 10)$, P being the annual precipitation in millimeters and T the mean annual temperature on the centigrade scale, 10 being added to avoid negative values. The resulting numerical scale of values was mapped as isograms, the lower values coinciding with areism and the higher with exoreism. In 1941 Martonne improved his aridity index by adopting a scale based on the arithmetical mean of the index of aridity for the year and for the driest month. But the scheme, useful for broad correlative purposes only, was soon supplanted by the evapotranspiration concept, which allowed water deficiency and surplus to be assessed more accurately.

Most of Martonne's ideas in climatic geomorphology had already been formulated by German geographers. His account of the geomorphological problems of Brazil (1940), however, contained original observations.

As a scientific geographer and educator Martonne will probably be remembered longest for his general summaries of physical geography. His chief regional exposition of systematic physical topics was *Géographie physique de la France* for *Géographie universelle* (1942). His main global exposition, *Traité de géo-*

graphie physique, first appeared as one volume in 1909 and achieved phenomenal success. It was entirely recast in three volumes in 1925–1927 and was kept up to date by careful pruning and enlargement. Its breadth of content, wide outlook, clarity of explanation, richness in diagrams—mostly by Martonne himself—and wise choice of typical rather than exotic examples earned it a well-deserved longevity.

BIBLIOGRAPHY

I. ORIGINAL WORKS. A complete list of Martonne's more than 200 publications is in the obituary by Jean Dresch, in *Bulletin de la Société géologique de France*, **6** (1956), 623–642; and in the archives of the Académie des Sciences, Institut de France, Paris. His articles were published mainly in *Comptes rendus ... de l'Académie des sciences; Météorologie; Bulletin de la Société géologique de France;* and, above all, in *Annales de géographie*, of which he was a director from 1920 to 1940 and a chief director from 1940 to his death in 1955.

His major works are *La Valachie, essai de monographie géographique* (Paris, 1902), his dissertation for his doctorate in letters; *Recherches sur la distribution géographique de population en Valachie avec une étude critique sur les procédés de représentation de la répartition de la population* (Paris–Bucharest, 1903); "La période glaciaire dans les Karpates méridionales," in *Comptes rendus du Congrès international de géologie*, Vienne, 1903 (1904), 691–702; "La pénéplaine et les côtes bretonnes," in *Annales de géographie*, **15** (1906), 213–236, 299–328; *Recherches sur l'évolution morphologique des Alpes de Transylvanie (Karpates méridionales)* (Paris, 1907), his dissertation for his doctorate in natural sciences; "Sur l'inégale répartition de l'érosion glaciaire dans le lit des glaciers alpins," in *Comptes rendus ... de l'Académie des sciences*, **149** (1909), 1413–1415; "L'érosion glaciaire et la formation des vallées alpines," in *Annales de géographie*, **19** (1910), 289–317; **20** (1911), 1–29; "L'évolution des vallées glaciaires alpines, en particulier dans les Alpes du Dauphiné," in *Bulletin de la Société géologique de France*, **12** (1912), 516–549; *Atlas photographique des formes du relief terrestre* (Paris, 1914), compiled with J. Brunhes and E. Chaix; and "Le climat facteur du relief," in *Scientia*, **13** (1913), 339–355.

Other works are "The Carpathians," in *Geographical Review*, **3** (1917), 417–437; "Essai de carte ethnographique des pays roumains," in *Annales de géographie*, **29** (1920), 181–198; "Le rôle morphologique de la neige en montagne," in *Géographie*, **34** (1920), 255–267; *Les régions géographiques de la France* (Paris, 1921); "Le massif du Bihar (Roumanie)," in *Annales de géographie*, **31** (1922), 313–340; "Extension du drainage océanique," in *Comptes rendus ... de l'Académie des sciences*, **180** (1925), 939–942, written with L. Aufrère; "Aréisme et l'indice d'aridité," *ibid.*, **182** (1926), 1395–1398; "Une nouvelle fonction climatologique: L'indice d'aridité," in *Météorologie*, **68** (1926), 449–458; *Les Alpes. Géographie générale* (Paris, 1926); *L'extension des régions privées d'écoulement vers l'océan* (Paris, 1928), written with L. Aufrère; *Europe centrale*, 2 vols. (Paris, 1931–1932); "Les régions arides du nord argentin et chilien," in *Bulletin de l'Association de géographes français*, **79** (1934), 58–62; "Sur la formule de l'indice d'aridité," in *Comptes rendus ... de l'Académie des sciences*, **200** (1935), 166–168, written with Mme R. Fayol; "Problèmes morphologiques du Brésil tropical Atlantique," in *Annales de géographie*, **49** (1940), 16–27, 106–129; "Carte morphologique de la France," in *Atlas de France* (Paris, 1941); "Nouvelle carte mondiale de l'indice d'aridité," in *Météorologie*, **17** (1941), 3–26, and *Annales de géographie*, **51** (1942), 241–250; *Géographie physique de la France* (Paris, 1942); "Géographie zonale. La zone tropicale," in *Annales de géographie*, **55** (1946), 1–18; and *Géographie aérienne* (Paris, 1949).

His major work was *Traité de géographie physique* (Paris, 1909; 2nd ed., 1913; 3rd ed., 1921). The 4th ed. appeared in 3 vols.: I, *Notions générales. Climat. Hydrographie* (Paris, 1925); II, *Le relief du sol* (Paris, 1925); III, *Biogéographie* (Paris, 1927), written with A. Chevalier and L. Cuénot. The latest eds., as of 1970, are I, 9th ed. (Paris, 1957); II, 10th ed. (Paris, 1958); and III, 7th ed. (Paris, 1955). Abridged eds. were issued at Paris from 1922 and at London, in English, from 1927.

II. SECONDARY LITERATURE. The chief assessments of Martonne's work and influence are André Cholley, in *Annales de géographie*, **65** (1956), 1–14; Donatien Cot, in *Comptes rendus ... de l'Académie des sciences*, **241** (1955), 713–716; Jean Dresch, in *Bulletin de la Société géologique de France*, **6** (1956), 623–642, with bibliography; and André Meynier, in his *Histoire de la pensée géographique en France: 1872–1969* (Paris, 1969), *passim*.

ROBERT P. BECKINSALE

MARUM, MARTIN (MARTINUS) VAN (*b.* Delft, Netherlands, 20 March 1750; *d.* Haarlem, Netherlands, 26 December 1837), *natural philosophy, medicine, botany.*

Van Marum's father, Petrus, a construction engineer and surveyor, moved from Groningen to Delft to marry Cornelia van Oudheusden in 1744. Martin attended elementary school and grammar school at Delft; but when his father returned to Groningen in 1764, he matriculated at Groningen University. One of his teachers was Peter Camper, who greatly influenced his studies and stimulated his special interest in plant physiology, a subject hardly studied in the Netherlands in those days. On 7 August 1773, he obtained his Ph.D. for a dissertation on the circulation of plant juices. On 21 August 1773 he received his medical degree with a study of comparative animal-plant physiology. The first thesis was highly esteemed abroad. When van Marum was not appointed professor of botany, as he had been promised, he abandoned his physiological studies, although he could not refrain from experimenting in this field later on.

Van Marum subsequently studied electricity. In cooperation with Gerhard Kuyper he developed an electrical machine with shellac disks drawn through mercury. A description was published in 1776, and translations into German and French followed.

From 1776 to 1780 van Marum practiced medicine in Haarlem. He was at once elected a member of the Netherlands Society of Sciences, and was appointed a lecturer in philosophy and mathematics. In 1777 the Society appointed him director of their rapidly expanding cabinet of natural curiosities. He then lived on the museum's premises.

In 1781 van Marum married Joanna Bosch, heiress of a prosperous printer, and thus he acquired a piece of land, on which he started to cultivate plants in 1783. His new appointment as director of Teyler's Cabinet of Physical and Natural Curiosities and Library left him little time for working his own garden, but it brought him his greatest fame.

The organization of Teyler's Museum was left entirely to van Marum, and he soon obtained a large electrical machine made under his supervision by John Cuthbertson of Amsterdam. Its disks had a diameter of sixty-five inches, the largest possible at the time. Van Marum thought that results obtained with such enormous discharges were bound to bring order to the chaos of concepts about the mysterious "electrical matter." He described the experiments with this machine and great battery of Leyden jars in three volumes of *Verhandelingen uitgeven door Teyler's tweede Genootschap* (1785, 1787, 1795). These experiments were greatly admired and repeated all over Europe. From his experiments with the large machine van Marum concluded that Franklin was correct in his theory of a single electric fluid. For this support Franklin expressed his appreciation. Volta also greatly admired van Marum's work, and informed him in 1792 of his own experiments; van Marum later introduced the term "Voltaic pile." Working with C. H. Pfaff, van Marum conclusively proved static and galvanic electricity to be identical.

From 1782 van Marum regularly made trips abroad. In Paris in 1785 he met Lavoisier and saw his assistants at work. After his return home, he made his own experimental test and became convinced of the validity of Lavoisier's combustion theory. He contributed greatly to the acceptance of the "new chemistry" in the Netherlands by his *Schets der Leere van Lavoisier*, published as a supplement to the *Verhandelingen uitgeven door Teyler's tweede Genootschap* (1787). He applied himself especially to the simplification of the required instruments, thus making the experiments less costly and enabling many chemists to repeat them. His gasometer was also a very important instrument.

Van Marum, in cooperation with van Troostwijk, discovered carbon monoxide. He continued his experiments to decompose and synthesize water, as he had seen done in Paris, and he oxidized various metals and then decomposed the oxides. During his experiments he had smelled ozone, but he did not recognize it to be a form of oxygen. Van Marum was the first to observe condensation of liquid ammonia from the gas, but he failed to realize that other gases would condense under the proper conditions of temperature and pressure. For this experiment van Marum built his own convertible air pump and compressor. In 1798 the *Verhandelingen uitgeven door Teyler's tweede Genootschap* included the description of these chemical experiments.

In 1794 van Marum was appointed secretary of the Netherlands Society of Sciences. After the French occupation of the Netherlands in 1795 there was little opportunity for scientific research, and funds for acquiring new instruments steadily decreased. Van Marum then applied himself more to the study of paleontology and geology, collecting much material and information during his various travels. He had already been able to procure valuable items for Teyler's Museum. In 1784 he bought the fossil *Mosasaurus camperi*, which had been found on the St. Pietersberg hill in Limburg; it was then still called "the head of an unknown marine animal." G. Cuvier later concluded that it was a lizard's head, and thus a land animal, which it had already been assumed to be by Peter Camper's son, Adriaan.

When traveling in Switzerland in 1802, van Marum bought the *Homo diluvii testis et theoskopos*, so named by Scheuchzer (1726). Cuvier examined this fossil at Haarlem in 1811, and concluded that it was a salamander. Van Marum also rearranged the whole collection of minerals in Teyler's Museum according to the methods of Cuvier, at that time the greatest authority in paleontology, geology, and mineralogy. Van Marum always tried to procure the latest and the best scientific information.

In 1803 van Marum bought a country house with a large garden, in which he cultivated mainly South African plants, especially aloes. He contributed greatly to the publication of Prince Joseph of Salm-Dyck's descriptive catalog of aloes by supplying him with many new species and data. He also maintained correspondence with C. P. Thunberg, Banks, Jacquin, and Jacquin's son, Joseph Franz.

Van Marum was actively interested in many aspects of human welfare: the prevention of air pollution from carbon monoxide, the treatment of victims of drowning, the construction of a portable fire engine, ventilation in factories and aboard ships, improvements of Papin's digester to produce cheap food for the poor, lightning rods, especially for windmills, and steam baths for cholera patients. He also gave many public lectures on various subjects. In 1808 he and three colleagues were asked to draft a constitution for the Royal Institute of Sciences (the present Royal Netherlands Academy of Sciences). He also was a member of the committee for the organization of higher education (1814).

Although van Marum made no great scientific discoveries, he greatly influenced the dissemination of knowledge in those fields of science that made great progress during his lifetime.

BIBLIOGRAPHY

A complete bibliography by J. G. de Bruyn may be found in R. J. Forbes, ed., *Martinus van Marum, Life and Work*, I (Haarlem, 1969), 287–320. Van Marum's principal publications are numbers 1, 3, 10, 11, 35, and 40.

ALIDA M. MUNTENDAM

AL-MARWAZĪ. See Ḥabash al-Ḥāsib.

MARX, KARL HEINRICH (*b.* Trier, Germany, 5 May 1818; *d.* London, England, 14 March 1883), *philosophy*.

For a complete study of his life and works, see Supplement.

MASCAGNI, PAOLO (*b.* Pomarance, Volterra, Italy, 25 January 1755; *d.* Florence, Italy, 19 October 1815), *anatomy*.

Mascagni, whom Lalande included among the learned men of Siena, graduated at the age of twenty and, four years later was appointed professor. At that time the Academy of Sciences of Paris proposed to "determine and demonstrate the system of the lymphatic vessels." The theory of the lymphatic vessels had been all but forgotten in Italy; and although Frederik Ruysch had tried to reawaken interest in the subject in Holland, Albrecht Haller, by denying the existence of lymphatic vessels in certain parts of the body, had cast doubt on the entire lymphatic system.

By 1784 Mascagni was able to send the Academy the first part of a work on the lymphatic vessels illustrated with numerous plates. Although his work, *Vasorum lymphaticorum historia*, arrived late, he was awarded a special prize. The *Historia* paved the way for progress in anatomy, physiology, and clinical medicine, for 50 percent of the lymphatic vessels now known were discovered by Mascagni.

In studying the origin of the lymphatic vessels, Mascagni established that every vessel must in its course enter one or more lymph glands. He rearranged and completed the observations of others and overhauled their techniques. Mascagni also performed experiments, using the mercury injection method and so improving it that it surpassed all preceding techniques. In the light of his excellent results, the simplicity of the technique is truly surprising. The only instrument used was a tubular needle bent at a right angle; yet he observed, named, and described nearly all the lymph glands and vessels of the human body, checking earlier observations and carrying out new ones.

Mascagni examined the views of Boerhaave and his followers, who believed that the lymphatics arose from the tips of the arteries and were shaped like ramified conical vessels. These vessels formed canals that gradually became thinner; and these canals, by continuing into the veins, brought some sort of material to the blood and therefore were related to the lymphatic system. Mascagni demonstrated, however, that such arterial and venous lymphatics did not exist. After examining the work of Noguez, Hamberger, and Hoffmann and the results of his own researches, Mascagni concluded that the lymphatic system originates from all the cavities and surfaces of the body, both internal and external, and is related to the absorbing function. By means of colored injections he demonstrated the communication between the lymphs and the serous vessels.

Mascagni was appointed professor of anatomy at Pisa; he was also invited to hold professorships at Bologna and Padua, but accepted the vacant post at Florence. Indeed, the Tuscan government, desirous of securing his services, not only created one professorship covering anatomy, physiology, chemistry, and the teaching of art anatomy but also doubled the stipend. In appreciation of this generous offer, Mascagni began to prepare anatomical models for use by students of sculpture and painting. The various systems of the human body were to be represented on life-size figures, a grandiose concept that was carried out with the help of illustrations by Sienese artists. The huge cost of this colossal editorial venture obliged Mascagni to draw on his salary and even to mortgage the family estate. These generous efforts failed to yield the desired results, for he died before completing his lifework.

BIBLIOGRAPHY

Mascagni's first published writing was *Vasorum lymphaticorum historia* (Siena, 1784), the intro. to a work on lymphatic vessels. Of greater importance is *Vasorum lymphaticorum corporis humani historia et iconographia* (Siena, 1787). Mascagni also published commentaries on mineralogy, agriculture, chemistry, and physics.

His first posthumously published work was that on anatomy for use by students of sculpture and painting (Florence, 1816). There followed two introductions to the *Anatomia:* the Antonmarchi edition, prepared under the supervision of Antonmarchi, who was Napoleon's physician and Mascagni's pupil; and the edition prepared in Milan by Farnese. The latter is inferior because the redrawn plates are not as good as those in the Antonmarchi version. These editions were followed by the plates of the animal and vegetable organs illustrated in the intro. to the *Anatomia* (Florence, 1819).

Mascagni's monumental *Anatomia universa XLIV tabulis aeneis iuxta archetjpum hominis adulti accuratissime repraesentata* (Pisa, 1823) is so large that one of its designers—Antonio Serantoni, a designer, engraver, and modeler in wax—prepared a special personal ed. in color: *Anatomia universale descrittiva del Professor Paolo Mascagni* (Florence, 1833), reproduced with copper plates smaller than those in the Pisa ed. Mascagni's important works also include the many wax models preserved in Italian and foreign museums.

FEDERICO ALLODI

MASCART, ÉLEUTHÈRE ÉLIE NICOLAS (*b.* Quarouble, France, 20 February 1837; *d.* Paris, France, 26 August 1908), *physics.*

The son of a teacher, Mascart received his secondary education at the *collège* in Valenciennes, after which he became *maître répétiteur* at the *lycées* in Lille (1856–1857) and Douai (1857–1858). His thorough mathematical preparation at Douai enabled him in 1858 to enter the École Normale Supérieure in Paris. Three years later he became *agrégé-préparateur* at the École Normale, and in July 1864 he received his doctorate. Shortly thereafter he married a Mlle Briot.

After teaching physics at the *lycée* in Metz (1864–1866) and publishing his first book, *Éléments de mécanique* (Paris, 1866; 9th ed., 1910), his former professor of physics, Verdet, helped him secure a post in Paris at the Collège Chaptal. He soon transferred, first to the Lycée Napoléon and then to the Lycée de Versailles. In December 1868 he left secondary education for good to become Régnault's assistant at the Collège de France. In May 1872, his scientific career having been interrupted by the Franco-Prussian War, he succeeded Régnault as professor of physics, a chair that he held for the rest of his life. In 1878 he was also chosen to be the first director of the Bureau Central Météorologique. In December 1884 he was elected to J. C. Jamin's place in the French Academy of Sciences; he served as permanent secretary and, in 1904, president.

Mascart's scientific career was marked not by great discoveries but by a steady stream of first-rate experimental and theoretical work in optics, electricity, magnetism, and meteorology. The first problem he attacked—the subject of his thesis—was a systematic and precise spectroscopic exploration (using photographic detection techniques) of the ultraviolet region of the solar spectrum, in which he greatly increased the number of lines of known wavelength. At the same time he accurately determined the relative wavelengths of the principal emission lines of ten selected metals and argued that all of his data agreed well with a modified version of Cauchy's dispersion formula. A more detailed extension of these experiments, which Fizeau termed the "most thorough and most satisfying" work since Fraunhofer's, won for Mascart the 1866 Prix Bordin.

During the course of this work and in the years immediately following it, Mascart made several individually significant observations, including the existence of triplets in the spectrum of magnesium. Undoubtedly the most valuable aspect of this period was the thorough preparation in experimental optics that it gave him. When the Academy in 1870 proposed for the Grand Prix des Sciences Mathématiques the experimental determination of the modifications that light experiences in its mode of propagation and its properties as a result of the movement of the source and observer, Mascart was ready to compete for it. Several years of painstaking researches followed, at the end of which Mascart found himself forced to the purely negative conclusion that refraction and diffraction experiments, independent of whether terrestrial sources or sunlight is used, are incapable of detecting the motion of the earth through the ether. Double refraction and other optical experiments yielded the same negative result. While contemporary interpretations of these experiments had to be modified later in the light of special relativity theory, the great importance of Mascart's work was recognized by the Academy when it awarded Mascart the Grand Prix in 1874.

Optical researches of various kinds—experiments in physiological optics, determinations of the indices of refraction and dispersive powers of numerous gases, studies on metallic reflection and color, theoretical work on the rainbow and on the formation of interference fringes under various conditions—continued to occupy much of Mascart's time both before and after 1874, especially until the publication of his well-known three-volume *Traité d'optique* (Paris, 1889–1893). Concurrently, however, his interests were extending into the fields of electricity and magnetism. He studied, for example, electrical machines (1873) and the efficiencies of various motors (1877–1878), the propagation of electricity in conductors (1878), the theory of induced currents (1880, 1883), the interaction of two electrified spheres (1884), diamagnetism (1886), means for determining the positions of the poles of a bar magnet (1887), and the propagation of electromagnetic waves (1893–1894). Perhaps Mascart's most important and precise work involved the determination of the electrochemical equivalent of silver (1884) and, with F. de Nerville and R. Benoit, the absolute value of the ohm (1884–1885). Once again he incorporated his researches into textbooks: first into his two-volume *Traité d'électricité statique* (Paris, 1876) and then into his two-volume *Leçons sur l'électricité et le magnétisme*, written with J. Joubert (Paris, 1882–1886; English trans. by E. Atkinson, *A Treatise on Electricity and Magnetism* [London, 1883–1888]; 2nd French ed., 1896–1897), which was the first French textbook that attempted to treat synthetically the work of Maxwell, Kelvin, and Helmholtz. Since it concentrated on applications as well as on theory, it became a standard work for engineers as well as for physicists.

Mascart's third major area of scientific activity was meteorology, which made greater demands on his time after he became director of the Bureau Central Météorologique (1878). In succeeding years he traveled widely as a member of an international committee charged with defining pressing meteorological questions and organizing international meteorological conferences. He also helped organize an international scientific polar expedition, as well as a French expedition to Cape Horn. The major goal of the latter was to map the magnetic field of the earth, a subject to which he repeatedly returned throughout the 1880's and 1890's and on which he published an extensive textbook, *Traité de magnétisme terrestre* (Paris, 1900). Concurrently, he carried out studies on atmospheric electricity (1878–1882), on the amount of carbonic acid in the air (1882), on the terrestrial variations of gravitational attraction (1882–1883) and its possible diurnal variation (1893), on the theory of cyclone formation (1887–1888), and on the mass of the atmosphere (1892).

Beginning in the late 1870's Mascart played an increasingly prominent role in national and international scientific organizations and events. Indeed, a substantial portion of his own researches stemmed directly from issues raised at the congresses he attended or presided over. Thus, for example, at the 1881 Exposition Internationale d'Électricité in Paris, it became clear that one of the most important tasks confronting physicists was to originate a universal and coherent system of electrical units. The agreement finally reached on the definitions of the volt and the ohm prompted Mascart to carry out his determinations of the electrochemical equivalent of silver and of the absolute value of the ohm (final agreement was reached in Chicago in 1893). At a number of later congresses of electricians and meteorologists, Mascart played leading organizational and official roles. As president of the general assembly of the Société Internationale des Électriciens held in Paris in May 1887, he was instrumental in creating the Laboratoire Central d'Électricité and the École Supérieure d'Électricité. He was repeatedly called upon to serve as a consultant to the French government on matters relating to national defense, public electrical and lighting facilities, and public instruction. Only in the last year of his life did he also consent to advise two private industries. His general prestige is reflected in his election in 1892 as a foreign member of the Royal Society and in his election in 1900 as vice-president of the Institution of Electrical Engineers, the first time this post was ever held by a non-British citizen.

Of medium height and of great physical stamina, Mascart was a leader who could readily get to the essence of arguments and quietly persuade others of a course of action in which he believed. An experimental physicist with a thorough command of mathematics, he could easily step outside of his laboratory and point the way to practical results in technology or policy making. He was a *grand officier* of the Légion d'Honneur, an honor—one of his many—that he valued very highly. He died after a serious operation, at the age of seventy-one.

BIBLIOGRAPHY

I. ORIGINAL WORKS. Mascart's most important papers are "Recherches sur le spectre solaire ultra-violet et sur la détermination des longueurs d'onde," in *Annales scientifiques de l'École normale supérieure*, **1** (1864), 219–262, his doctoral thesis, also published separately (Paris, 1864); "Recherches sur la détermination des longueurs d'onde," *ibid.*, **4** (1867), 7–37, the 1866 Prix Bordin memoir; "Sur les modifications qu'éprouve la lumière par suite du mouvement de la source lumineuse et du mouvement de l'observateur," *ibid.*, 2nd ser., **1** (1872), 157–214; **3** (1874), 363–420, the 1874 Grand Prix memoir; "Sur l'équivalent électrochimique de l'argent," in *Journal de physique théorique et appliquée*, 2nd ser., **3** (1884), 283–286; and "Détermination de l'ohm et de sa valeur en colonne mercurielle," in *Annales de chimie et de physique*, 4th ser., **6** (1885), 5–86, written with F. de Nerville and R. Benoit. An interesting review article of Mascart's is "The Age of Electricity," in *Report of the Board of Regents of the Smithsonian Institution* (1894), 153–172. His many textbooks—he was one of the most prolific textbook writers of all time—are cited above.

II. SECONDARY LITERATURE. See Paul Janet, "La vie et les oeuvres de E. Mascart," in *Revue générale des sciences pures et appliquées*, **20** (1909), 574–593; the obituary notices in *Nature*, **78** (1908), 446–448; and *Journal de physique théorique et appliquée*, 4th ser., **7** (1908), 745; and the portrait in Edward D. Adams, *Niagara Power*, I (Niagara Falls, N.Y., 1927), 191.

ROGER H. STUEWER

MASCHERONI, LORENZO (*b.* Castagneta, near Bergamo, Italy, 13 May 1750; *d.* Paris, France, 14 July 1800), *mathematics.*

Mascheroni was the son of Paolo Mascheroni dell'Olmo, a prosperous landowner, and Maria Ciribelli. He was ordained a priest at seventeen and at twenty was teaching rhetoric and then, from 1778, physics and mathematics at the seminary of Bergamo. His *Nuove ricerche su l'equilibrio delle vòlte* (1785) led to his appointment as professor of algebra and geometry at the University of Pavia in 1786. In 1789 and 1793 he was rector of the university and, from 1788 to 1791, was head of the Accademia degli Affidati. Mascheroni was a member of the Academy of Padua, of the Royal Academy of Mantua, and of the Società Italiana delle Scienze. In his *Adnotationes ad calculum integrale Euleri* (1790) he calculated Euler's constant, sometimes called the Euler-Mascheroni constant, to thirty-two decimal places; the figure was corrected from the twentieth decimal place by Johann von Soldner in 1809.

In 1797 Mascheroni was appointed deputy to the legislative body in Milan. Sent to Paris by a commission to study the new system of money and of weights and measures, he published his findings in 1798 but was prevented from returning home by the Austrian occupation of Milan in 1799. Also a poet, Mascheroni dedicated his *Geometria del compasso* (1797) to Napoleon in verse; his celebrated *Invito a Lesbia Cidonia* (1793) glorifies the athenaeum of Pavia. He died after a brief illness, apparently from the complications of a cold. The poet Monti mourned his death in the *Mascheroniana*.

Mascheroni's *Nuove ricerche* is a well-composed work on statics, and the *Adnotationes* shows a profound understanding of Euler's calculus. He is best known, however, for his *Geometria del compasso*, in which he shows that all plane construction problems that can be solved with ruler and compass can also be solved with compass alone. It is understood that the given and unknowns are points; in particular, a straight line is considered known if two points of it are known.

In the preface Mascheroni recounts the genesis of his work. He was moved initially by a desire to make an original contribution to elementary geometry. It occurred to him that ruler and compass could perhaps be separated, as water can be separated into two gases; but he was also assailed by doubts and fears often attendant upon research. He then chanced to reread an article on the way Graham and Bird had divided their great astronomical quadrant, and he realized that the division had been made by compass alone, although, to be sure, by trial and error. This encouraged him, and he continued his work with two purposes in mind: to give a theoretical solution to the problem of constructions with compass alone and to offer practical constructions that might be of help in making precision instruments. The second concern is shown in the brief solutions of many specific problems and in a chapter on approximate solutions.

The theoretical solution (see especially §191) depends on the solution of the following problems: (1) to bisect a given circular arc of given center; (2) to add and subtract given segments; (3) to find the fourth proportional to three given segments; (4) to find the intersection of two given lines; and (5) to find the intersection of a given line and given circle.

In 1906 August Adler applied the theory of inversion to the Mascheroni constructions. Since this theory places lines and circles on an equal footing, it sheds light on Mascheroni's problem; but the solution via inversion is not as elegant—and certainly not as simple or as brief—as Mascheroni's.

Mascheroni's theory is but a chapter in the long history of geometrical constructions by specified means. The limitation to ruler and compass occurs in book I of Euclid's *Elements*—at least the first three postulates have been called the postulates of construction; and there are even reasons to suppose that Euclid's so-called axiomatic procedure is really only an axiomatization of the Euclidean constructions.

Euclid, of course, had inherited a tradition of restricting construction to ruler and compass. Oenopides is credited by Proclus with the construction for dropping a perpendicular (*Elements* I.12) and with the method of transferring an angle (I.23). The tradition itself appears to be of religious origin (see Seidenberg, 1959, 1962).

About 980, the Arab mathematician Abu'l-Wafā' proposed using a ruler and a compass of fixed opening, and in the sixteenth century da Vinci, Dürer, Cardano, Tartaglia, and Ferrari were also concerned with this restriction. In 1672 Georg Mohr showed that all the construction problems of the first six books of the *Elements* can be done with compass alone. Lambert in 1774 discussed the problem "Given a parallelogram, construct, with ruler only, a parallel to a given line."

Poncelet, who mentions Mascheroni in this connection, showed in 1822 that in the presence of a given circle with given center, all the Euclidean constructions can be carried out with ruler alone. This has also been credited to Jacob Steiner, although he had heard of Poncelet's result, or "conjecture," as he called it. Poncelet and others also studied constructions with ruler alone; abstractly, his result is related to the axiomatic introduction of coordinates in the projective plane. Johannes Trolle Hjelmslev and others have studied the analogue of the Mascheroni constructions in non-Euclidean geometry.

The question has recently been posed whether the notion of two points being a unit apart could serve as the sole primitive notion in Euclidean plane geometry. An affirmative answer was given, based on a device of Peaucellier's for converting circular motion into rectilineal motion.

In 1928 Mohr's *Euclides danicus*, which had fallen into obscurity, was republished with a preface by Hjelmslev, according to whom Mascheroni's result had been known and systematically expounded 125 years earlier by Mohr. The justice of this judgment and the question of the independence of Mascheroni's work will now be examined.

The term "independent invention" is used in two different but often confused senses. Anthropologists use it in reference to the appearance of identical, or similar, complex phenomena in different cultures. A controversy rages, the opposing positions of which, perhaps stripped of necessary qualifications, can be put thus: According to the "independent inventionists," the appearance of identical social phenomena in different cultures (especially in New World and Old World cultures) is evidence for the view that the human mind works similarly under similar circumstances; for the "diffusionists," it is evidence of a historical connection, but not necessarily a direct one.

The historian, dealing with a single community,

uses the term in a different sense. When he says two inventions are independent, he means that each was made without direct reliance on the other. Simultaneous and independent solutions of outstanding problems that are widely published in the scholarly press are no more surprising than the simultaneous solutions by schoolboys of an assigned problem; and the simultaneous development in similar directions of a common fund of knowledge can also be expected. Even so, examples of independent identical innovations are rare and difficult to establish.

Although five centuries separate Abu'l-Wafā' and Leonardo, presumably no one will doubt that the Italians got the compass problem of a single opening from the Arabs (or, possibly, that both got it from a third source).

When the works of Mascheroni and Mohr are compared, it is apparent that the main ideas of their solutions of individual problems are in most cases quite different. In particular, this can be said for the bisection of a given segment. Moreover, the problem of bisection plays no role in Mascheroni's general solution, whereas it is central in Mohr's constructions. Still more significantly, the general problem is not formulated in Mohr's book. Thus, any suggestion of Mascheroni's direct reliance on Mohr would be quite inappropriate. Of course, the possibility cannot be excluded that Mascheroni, who explicitly denied knowledge that anyone had previously treated the matter, had heard of a partial formulation of the problem.

It appears that Hjelmslev's judgment is not entirely accurate. Mohr's book is quite remarkable and contains the basis for a simple proof of Mascheroni's result, but there is no evidence within the book itself that Mohr formulated the problem of constructions with compass alone in complete generality.

BIBLIOGRAPHY

I. ORIGINAL WORKS. Mascheroni's mathematical works are *Nuove ricerche su l'equilibrio delle vòlte* (Bergamo, 1785); *Adnotationes ad calculum integrale Euleri* (Pavia, 1790); and *Geometria del compasso* (Pavia, 1797). A nonmathematical work is *Invito a Lesbia Cidonia* (Pavia, 1793).

II. SECONDARY LITERATURE. Biographical details are presented in A. Fiamazzo, *Nuovo contributo alla biografia di L. Mascheroni*, 2 vols. (Bergamo, 1904); J. W. L. Glaisher, "History of Euler's Constant," in *Messenger of Mathematics*, **1** (1872), 25–30; G. Loria and C. Alasia, "Bibliographie de Mascheroni," in *Intermédiaire des mathématiciens*, **19** (1912), 92–94; and G. Natali, "Mascheroni," in *Enciclopedia italiana*, XXII (Rome, 1934), 496.

Adler's application of the theory of inversion to Mascheroni's constructions is presented in his *Theorie der geometrischen Konstruktionen* (Leipzig, 1906).

Support for the view that Euclid's so-called axiomatic procedure is merely an axiomatization of Euclidean constructions may be found in A. Seidenberg, "Peg and Cord in Ancient Greek Geometry," in *Scripta mathematica*, **24** (1959), 107–122; and "The Ritual Origin of Geometry," in *Archive for History of Exact Sciences*, **1** (1962), 488–527. Opposition to the above view is presented in T. L. Heath's ed. of Euclid's *Elements*, I, 124; and A. D. Steele, "Über die Rolle von Zirkel und Lineal in der griechischen Mathematik," in *Quellen und Studien zur Geschichte der Mathematik, Astronomie und Physik*, Abt. B, **3** (1936), 287–369. W. M. Kutta, "Zur Geschichte der Geometrie mit constanter Zirkelöffnung," in *Nova acta Academiae Caesarae Leopoldino Carolinae*, **71** (1898), 71–101, discusses the use of a ruler and a compass of fixed opening.

Georg Mohr's *Euclides danicus* was translated into German by J. Pál and provided with a foreword by J. Hjelmslev (Copenhagen, 1928). Lambert's work is referred to in R. C. Archibald, "Outline of the History of Mathematics," in *American Mathematical Monthly*, **56**, no. 1, supp. (1949), note 277, 98.

Poncelet's and Steiner's contributions can be found in Poncelet's *Traité des propriétés projectives des figures*, 2nd ed. (Paris, 1865), I, 181–184, 413–414; and in Steiner's *Geometrical Constructions With a Ruler*, translated by M. E. Stark and edited by R. C. Archibald (New York, 1950), p. 10.

Analogues of Mascheroni's work in non-Euclidean geometry are discussed in J. Hjelmslev, "Om et af den danske Matematiker Georg Mohr udgivet skrift *Euclides Danicus*," in *Matematisk tidsskrift*, B (1928), 1–7; and in articles by A. S. Smogorzhevsky, V. F. Rogachenko, and K. K. Mokrishchev that are reviewed in *Mathematical Reviews*, **14** (1953), 576, 1007; **15** (1954), 148; and **17** (1956), 885, 998.

The question of whether the notion of two points being a unit apart can be the sole primitive notion in Euclidean plane geometry is discussed in R. M. Robinson, "Binary Relations as Primitive Notions in Elementary Geometry," in Leon Henkin, Patrick Suppes, and Alfred Tarski, eds., *The Axiomatic Method With Special Reference to Geometry and Physics* (Amsterdam, 1959), 68–85.

A. SEIDENBERG

MASERES, FRANCIS (*b*. London, England, 15 December 1731; *d*. Reigate, Surrey, England, 19 May 1824), *mathematics*.

Maseres was the son of a physician who was descended from a family that had been forced to flee France by the revocation of the Edict of Nantes. At Clare College, Cambridge, he obtained his B.A. degree in 1752 with highest honors in both classics and mathematics. Upon receiving the M.A. and a fellowship from his college, he moved to the Temple and was later called to the bar. After spending a few years in the practice of law with little success, he was appointed attorney general for Quebec, in which post he served until 1769. His career in the new world was distinguished "by his loyalty during the American contest and his zeal for the interests of the province." Upon his return to England he was appointed cursitor baron of the Exchequer, an office which he held until his death at the age of ninety-three. During this period of his life he was generally known as Baron Maseres. In addition he was at different times deputy recorder of London and senior judge of the sheriff's court.

Three aspects of Maseres' career are noteworthy. The first is his interest in political matters, particularly in the affairs of Canada and the American colonies. Of a considerable number of essays along these lines from Maseres' pen, the following are typical: (1) "Considerations on the expediency of admitting Representatives from the American Colonies to the House of Commons" (1770); (2) "Account of Proceedings of British and other Protestants of the Province of Quebec to establish a House of Assembly" (anon.), (1775); (3) "The Canadian Freeholder, a Dialogue shewing the Sentiments of the Bulk of the Freeholders on the late Quebeck Act" (1776–1779); (4) "Select Tracts on Civil Wars in England, in the Reign of Charles I" (1815).

A second aspect of Maseres' long career is the peculiar nature of his mathematical contributions, reflecting his complete lack of creative ability together with naive individualism. For a proper perspective, one must recall that Maseres' works were written about a century and a half after Viète and Harriot had ushered in the period of "symbolic algebra." While Viète had rejected negative roots of equations, certain immediate precursors of Maseres, notably Cotes, De Moivre, Taylor, and Maclaurin, had gone far beyond this stage, as had his contemporaries on the Continent: Lambert, Lagrange, and Laplace. Despite these advances, some quirk in the young Maseres compelled him to reject that part of algebra which was not arithmetic, probably because he could not understand it, although by his own confession others might comprehend it. Unfortunately this prejudice against "negative and impossible quantities" affected much of his later work. Thus in one of his earliest publications, *Dissertation on the Use of the Negative Sign in Algebra* (1758), he writes as follows.

If any single quantity is marked either with the sign + or the sign − without affecting some other quantity ...

the mark will have no meaning or signification; thus if it be said that the square of −5, or the product of −5 into −5, is equal to +25, such an assertion must either signify no more than 5 times 5 is equal to 25 without any regard to the signs, or it must be mere nonsense or unintelligible jargon.

Curiously enough, in addition to Maseres, two other contemporary mathematicians opposed the generalized concept of positive and negative integers: William Frend, father-in-law of De Morgan, and Robert Simson. Maseres unfortunately influenced the teaching of algebra for several decades, as may be seen from textbooks of T. Manning (1796); N. Vilant (1798); and W. Ludlam (1809).

Perhaps the many publications with which he strove to bring mathematics to a much wider public were the most notable aspect of Maseres' legacy. Some were original works; others were reprints of the works of distinguished mathematicians. His original books are characterized by extreme prolixity, occasioned by his rejection of algebra, and the consequent proliferation of particular cases. For example, in the *Dissertation* alluded to above, which is virtually a treatise on elementary algebra, the discussion of basic rules and the solution of quadratic and cubic equations occupy three hundred quarto pages.

Of the reprints that Maseres made at his own expense, the most significant is the *Scriptores logarithmici* (1791–1807), six volumes devoted to the subject of logarithms, including the works of Kepler, Napier, Snellius, and others, interspersed with original tracts on related subjects. Other republications include the following: (1) *Scriptores optici* (1823), a reprint of the optical essays of James Gregory, Descartes, Schooten, Huygens, Halley, and Barrow; (2) Jakob I Bernoulli's tract on permutations and combinations; (3) Colson's translation of Agnesi's *Analytical Institutions*; (4) Hale's Latin treatise on fluxions (1800); and (5) several tracts on English history. Presumably a number of authors were indebted to Maseres for financial assistance of this sort. There can be little doubt of his sincerity and generosity, even if somewhat misplaced.

BIBLIOGRAPHY

I. ORIGINAL WORKS.

(1) *A Dissertation on the Use of the Negative Sign in Algebra: containing a demonstration of the rules usually given concerning it; and shewing how quadratic and cubic equations may be explained, without the consideration of negative roots. To which is added, as an appendix, Mr. Machin's quadrature of the circle* (London, 1758).

(2) *Elements of Plane Trigonometry . . . with a disserta-*

tion on the nature and use of logarithms (London, 1760).

(3) *A proposal for establishing life-annuities in parishes for the benefit of the industrious poor* (London, 1772).

(4) *Principles of the Doctrine of Life Annuities explained in a familiar manner so as to be intelligible to persons not acquainted with the Doctrine of Chances, and accompanied with a variety of New Tables, accurately computed from observations* (London, 1783).

(5) *Scriptores Logarithmici, or a collection of several curious Tracts on the Nature and Construction of Logarithms, mentioned in Dr. Hutton's Historical Introduction to his New Edition of Sherwin's Mathematical Tables*, 6 vols. (London, 1791–1807).

(6) *The Doctrine of Permutations and Combinations, being an essential and fundamental part of the Doctrine of Chances; as it is delivered by Mr. James Bernoulli, in his excellent Treatise on the Doctrine of Chances, intitled, Ars Conjectandi, and by the celebrated Dr. John Wallis, of Oxford, in a tract intitled from the subject, and published at the end of his Treatise on Algebra; in the former of which tract is contained, a Demonstration of Sir Isaac Newton's famous Binomial Theorem, in the cases of integral powers, and of the reciprocals of integral powers. Together with some other useful mathematical tracts* (London, 1795).

(7) "An Appendix by F. Maseres," in William Frend, *The Principles of Algebra*, 2 vols. in 1 (London, 1796–1799), 211–456. Also "Observations on Mr. Raphson's method of resolving affected equations of all degrees by approximation," *ibid.*, vol. 2, 457–581.

(8) *Tracts on the Resolution of Affected Algebraick Equations by Dr. Halley's, Mr. Raphson's and Sir I. Newton's, Methods of Approximation [with those of W. Frend and J. Kersey]* (London, 1800).

(9) *Tracts on the Resolution of Cubick and Biquadratick Equations* (London, 1803).

II. SECONDARY LITERATURE.

(10) *The Penny Cyclopaedia of the Society for the Diffusion of Useful Knowledge*, **14** (London, 1837), 480–481.

(11) *The Gentlemen's Magazine* (June 1824); contains a list of Maseres' political writings.

(12) Moritz Cantor, *Vorlesungen über die Geschichte der Mathematik*, IV (Leipzig, 1913), 80, 86–87, 92, 149–151, 271, 302; references to some periodical articles published by Maseres.

WILLIAM L. SCHAAF

MĀSHĀ'ALLĀH (*fl.* Baghdad, 762–*ca.* 815), *astrology.*

The son of Atharī (his father's name is sometimes written Abrī or Sāriya), Māshā'allāh was a Jew from Baṣra (sometimes wrongly written Miṣr [Egypt]). His name in Hebrew was Manasse (Mīshā, according to Ibn al-Qifṭī); in Persian, Yazdān Khwāst (British Museum, MS Add. 23,400). This last form is of particular interest, since Māshā'allāh was one of those early ʿAbbāsid astrologers who introduced the Sassanian version of the predictive art to the Arabs; he was particularly indebted to the Pahlavī translation of Dorotheus of Sidon and to the *Zīk i Shahriyārān*, or *Royal Astronomical Tables*, issued under the patronage of Khusrau Anūshirwān in 556. He was also acquainted with some Greek material (perhaps through Arabic versions of Syriac texts) and would have acquired some knowledge of Indian science, both through the Pahlavī texts that he read and through such Indian scientists as the teacher of al-Fazārī and Kanaka, who visited the courts of al-Manṣūr and Hārūn al-Rashīd.

It is during al-Manṣūr's reign that Māshāʾallāh's name first appears: he participated in the astrological deliberations that led to the decision to found Baghdad on 30 July 762 (*Journal of Near Eastern Studies*, **29** [1970], 104). Several of his works contain horoscopes that can be dated between 762 and 809 and were cast during his lifetime. Ibn al-Nadīm states that Māshāʾallāh lived into the reign of al-Maʾmūn, which began in 813; but the absence of any information about his activities after 809 indicates that he probably did not live long after 813.

Māshāʾallāh wrote on virtually every aspect of astrology, as the bibliography below demonstrates. His most interesting works for the historian of astronomy are his astrological history, from which we derive almost all that we know of Anūshirwān's *Royal Tables*. His brief and rather primitive *De scientia motus orbis* combines Peripatetic physics, Ptolemaic planetary theory, and astrology in such a way that, in conjunction with its use of the Syrian names of the months, one strongly suspects that it is based on the peculiar doctrines of Ḥarrān, to which al-Kindī and Abū Maʿshar were also attracted. In fact, Māshāʾallāh's works are often echoed in Abū Maʿshar's; and in the list below references have been made to the corresponding items in the list of works given in the article on Abū Maʿshar in the *Dictionary of Scientific Biography* (I, 32–39).

The basis of this list of Māshāʾallāh's works is that given by Ibn al-Nadīm in his *Fihrist* (G. Flügel, ed. [Leipzig 1871], 273–274); this was copied by Ibn al-Qifṭī in his *Taʾrīkh al-ḥukamāʾ* (J. Lippert, ed. [Leipzig, 1903], 327), although the published text stops at book V of item 9. This bibliography is supplemented from various sources, the most important of which are F. J. Carmody, *Arabic Astronomical and Astrological Sciences in Latin Translation* (Berkeley–Los Angeles, 1956), 23–38, and L. Thorndike, "The Latin Translations of Astrological Works by Messahala," in *Osiris*, **12** (1956), 49–72.

1. Ibn al-Nadīm (hereafter N) 1. *Kitāb al-mawālīd*

al-kabīr ("Great Book of Nativities"), in fourteen books. See Abū Maʿshar 4. This apparently exists in a Latin translation made by Hugo Sanctallensis and dedicated to Michael, bishop of Tarazona from 1119 to 1151; see C. H. Haskins, *Studies in the History of Mediaeval Science*, 2nd ed. (Cambridge, Mass., 1927), 76, and Carmody, item 13.

2. N 2. *Fī al-qirānāt wa ʾl-adyān wa ʾl-milal* ("On Conjunctions and Peoples and Religions"), in twenty-one chapters. See Abū Maʿshar 8. This work, written shortly before 813, survives in an epitome by Ibn Hibintā that is published in E. S. Kennedy and D. Pingree, *The Astrological History of Māshāʾallāh* (Cambridge, Mass., 1971), 1–38. For a different interpretation of the astronomy upon which the casting of the horoscopes in this work is based, see J. J. Burckhardt and B. L. van der Waerden, "Das astronomische System der persischen Tafeln I," in *Centaurus*, **13** (1968), 1–28.

3. N 3. *Kitāb maṭraḥ al-shuʿāʿ* ("Book of the Projection of Ray[s]"). This lost work is referred to by Abū Maʿshar, as quoted by his pupil Abū Saʿīd Shādhān in his *Mudhākarāt* (*Catalogus codicum astrologorum Graecorum*, XI, pt. 1 [Brussels, 1932], 171–172); see also al-Bīrūnī, *Rasāʾil* (Hyderabad-Deccan, 1948), pt. 3, 80.

4. N 4. *Kitāb al-maʿānī* ("Book of Definitions"). Māshāʾallāh refers to this nonextant work in 16, below (Kennedy and Pingree, *The Astrological History*, p. 130).

5. N 5. *Kitāb ṣanaʿat al-asṭurlāb wa ʾl-ʿamal bihā* ("Book of the Construction of an Astrolabe and Its Use"). This survives only in a Latin translation (see Carmody, item 1). It was first published by G. Reisch, *Margarita philosophica nova* (Strasbourg, 1512; repr. Strasbourg, 1515; O. Finé, ed., Basel, 1535; repr. Basel, 1583; and trans. into Italian by G. P. Gallucci [Venice, 1599]). The third part of the treatise, on the use of the astrolabe, was edited by W. W. Skeat, *A Treatise on the Astrolabe; Addressed to His Son Lowys by Geoffrey Chaucer A.D. 1391* (London, 1872; repr. London, 1905), 88–104. The best edition now is by R. T. Gunther, *Early Science in Oxford*, V (Oxford, 1929), 195–231; for the catalog of stars see P. Kunitzsch, *Typen von Sternverzeichnissen in astronomischen Handschriften des zehnten bis vierzehnten Jahrhunderts* (Wiesbaden, 1966), 47–50. Not from Māshāʾallāh's treatise are the fragment published by J. M. Millás Vallicrosa, *Las traducciones orientales en los manuscritos de la Biblioteca catedral de Toledo* (Madrid, 1942), 313–321 (see Kunitzsch, pp. 23–30), and the first table in Gunther (see Kunitzsch, pp. 51–58, who argues that this catalog is a product of the school of al-Majrīṭī).

6. N 6. *Kitāb dhāt al-ḥalaq* ("Book of the Armillary Sphere"). Nothing more is known of this treatise.

7. N 7. *Kitāb al-amṭār wa 'l-riyāḥ* ("Book of Rains and Winds"). See Abū Ma'shar 34. The Arabic text has been published by G. Levi della Vida, "Un opuscolo astrologico di Mâsâ'allâh," in *Rivista degli studi orientali*, **14** (1933–1934), 270–281. The Latin translation by Drogon [?] was edited by M. A. Šangin, *Catalogus codicum astrologorum Graecorum*, XII (Brussels, 1936), 210–216; see Carmody, item 15, and Thorndike, pp. 67–68.

8. N 8. *Kitāb al-sahmayn* ("Book of the Two Lots [of Fortune and of the Demon]"). See Abū Ma'shar 14.

9. N 9. *Kitāb al-ma'rūf bi 'l-sābi' wa 'l-'ishrīn* ("The Book Known as the Twenty-seventh"), in six books: I, "On the Beginnings of Works"; II, "On the Overthrow of the Government"; III, "On Interrogations"; IV, "On Aspects"; V, "On Occurrences"; VI, "On the *tasyīrāt* ('astrological cycles') of the Luminaries and What They Indicate." This work is probably the source of many excerpts from Māshā'allāh found in the Arabic compendiums of al-Ṣaymarī, al-Qaṣrānī, and others, as well as in the manuscript Laleli 2122 at the Süleymaniye Library in Istanbul. It may also be the source of the following translations:

a. *De cogitationibus (De interrogationibus)*. Published by Bonetus Locatellus (Venice, 1493; repr. Venice, 1519) and by I. Heller (Nuremberg, 1549); Carmody, item 5 (who mentions a French translation of the Latin), and Thorndike, pp. 53–54 and 56–62. There is also a Hebrew translation of a work on interrogations (M. Steinschneider, *Die hebräischen Übersetzungen des Mittelalters und die Juden als Dolmetscher* [Berlin, 1893], 600–602), and a *Kitāb al-Masā'il* ("Book of Interrogations") was known to Ḥājjī Khalīfa *(Lexicon bibliographicum et encyclopaedicum*, G. Flügel, ed., VII [Leipzig, 1858], 386).

b. *De occultis*, which survives in two versions; see Carmody, items 9 and 10, and Thorndike, pp. 54–56.

c. *Liber iudiciorum;* see Carmody, item 14.

d. *De interpretationibus*, edited by I. Heller (Nuremberg, 1549). To it may also belong the *De testimoniis lune*, *De stationibus planetarum*, and *De electionibus horarum* (see Carmody, items 16, 17, and 18), if they are truly by Māshā'allāh.

10. N 10. *Kitāb al-ḥurūf* ("Book of Letters").

11. N 11. *Kitāb al-sulṭān* ("Book of Government").

12. N 12. *Kitāb al-safar* ("Book of Travel").

13. N 13. *Kitāb al-as'ār* ("Book of Prices"). This short treatise, which contains a horoscope dated 24 June 773 (Kennedy and Pingree, *The Astrological History*, p. 185), is extant in Bodleian MS Marsh 618 and Escorial MS Ar. 938. The *Liber super annona*

or *De mercibus* may be the Latin version; see Carmody, item 11, and Thorndike, pp. 68–69.

14. N 14. *Kitāb al-mawālīd* ("Book of Nativities"). This work, based largely on Dorotheus of Sidon with some additions from a Byzantine source of the sixth century, and itself the basis of the main work of Māshā'allāh's pupil al-Khayyāṭ, survives only in Latin; see Carmody, item 12. It has been edited by Kennedy and Pingree, *The Astrological History*, pp. 145–174.

15. N 15. *Kitāb taḥwīl sinī al-mawālīd* ("Book of the Revolution[s] of the Years of Nativities"). See Abū Ma'shar 19. This work is lost, but many fragments of it can be recovered from the *Majmū' aqāwīl al-ḥukamā'* of al-Dāmaghānī.

16. N 16. *Kitāb al-duwal wa 'l-milal* ("Book of Dynasties and Religions"). This is probably identical with the *Fī qiyām al-khulafā' wa ma'rifat qiyām kull malik* ("On the Installation of the Caliphs and the Knowledge of the Installation of Every King"), which was written during the reign of Hārūn al-Rashīd and is translated in Kennedy and Pingree, *The Astrological History*, pp. 129–143.

17. N 17. *Kitāb al-ḥukm 'alā 'l-ijtimā'āt wa 'l-istiqbālāt* ("Book of Judgment[s] According to the Conjunctions and Oppositions [of the Sun and Moon]").

18. N 18. *Al-Kitāb al-murdī* ("The Pleasing Book").

19. N 19. *Kitāb al-ṣuwar wa 'l-ḥukm 'alāyhā* ("Book of Constellations and Judgment[s] According to Them"). See Abū Ma'shar 17 and 18. This may be the *Kitāb al-amthāl* ("Book of Images") in Ayasofya Library, Istanbul, MS 2672.

20. *De revolutionibus annorum mundi*, in forty-six chapters. This Latin translation of a lost Arabic original was published by Bonetus Locatellus (Venice, 1493, 1519) and by I. Heller (Nuremberg, 1549); see Carmody, item 2, and Thorndike, pp. 66–67.

21. *Epistola de rebus eclipsium* or *De ratione circuli et stellarum*, in twelve chapters. This Latin translation by John of Seville was published by Bonetus Locatellus (Venice, 1493, 1519), by I. Heller (Nuremberg, 1549), and by N. Pruckner, *Iulii Firmici Materni . . . Astronomicôn libri VIII* (Basel, 1533; repr. Basel, 1551), pt. 2, 115–118; see Carmody, item 7 (who mentions a French translation of the Latin), and Thorndike, pp. 62–66. A Hebrew translation made by Ibn Ezra in 1148 was commented on by Abraham Yagel at the end of the sixteenth century (see M. Steinschneider, *Die hebräischen Übersetzungen*, pp. 602–603). Abraham's Hebrew was translated into English by B. Goldstein, "The Book on Eclipses of Masha'allah," in *Physis* (Florence), **6** (1964), 205–213.

22. *Super significationibus planetarum in nativitate,*

in twenty-six chapters. This Latin translation by John of Seville [?] of a lost Arabic original based largely on Dorotheus of Sidon was published by I. Heller (Nuremberg, 1549); see Carmody, item 4.

23. *De septem planetis*, in nine chapters, is sometimes ascribed to Jirjis; see Carmody, item 6.

24. *De receptione*, in twelve chapters, is a Latin translation by John of Seville of a lost Arabic original that contained horoscopes dated between 13 February 791 and 30 November 794 (Kennedy and Pingree, *The Astrological History*, pp. 175–178; E. S. Kennedy, "A Horoscope of Messehalla in the Chaucer Equatorium Manuscript," in *Speculum*, **34** [1959], 629–630). It was published by Bonetus Locatellus (Venice, 1493, 1519) and by I. Heller (Nuremberg, 1549); see Carmody, item 3 (who mentions a French translation of the Latin), and Thorndike, pp. 50–53.

25. *De scientia motus orbis* or *De elementis et orbibus coelestibus*, in twenty-seven chapters. This important Latin translation by Gerard of Cremona of the lost Arabic original of this exposition of apparently Harranian doctrines was published by I. Stabius (Nuremberg, 1504) and by I. Heller (Nuremberg, 1549); see Carmody, item 8. There is an Irish adaptation edited by M. Power, *An Irish Astronomical Tract* (London, 1914).

26. *De electionibus*. This text, which quotes Dorotheus, is ascribed to Māshāʾallāh and Ptolemy but is probably by neither. It was published by P. Liechtenstein (Venice, 1509) and by T. Rees (Paris, 1513).

27. *Mafātīḥ al-qaḍāʾ* ("The Keys of Judgments"). The Arabic original of this treatise is lost, but there is a manuscript of a Persian translation (see C. A. Storey, *Persian Literature*, II, pt. 1 [London, 1958], 38–39, who doubts the attribution to Māshāʾallāh) and a Latin epitome of a *Septem claves* of Māshāʾallāh, edited by M. A. Šangin, "Latinskaya parafraza iz utrachennogo sochinenia Mashallaha 'Semi Kluchey'" ("A Latin Paraphrase From Māshāʾallāh's Lost Work 'The Keys of Judgment'"), in *Zapiski kollegii vostokovedov*, **5** (1930), 235–242.

28. *Aḥkām al-qirānāt wa ʾl-mumāzajāt* ("Judgments of Conjunctions and Mixtures"). This lost work is mentioned by Ḥājjī Khalīfa, I, 175.

BIBLIOGRAPHY

Many further fragments of Māshāʾallāh's voluminous writings can be found in Arabic texts (such as those by Abū Maʿshar, al-Hāshimī, and al-Bīrūnī) and in Greek (particularly in Vaticanus Graecus 1056; see *Catalogus codicum astrologorum Graecorum*, I [Brussels, 1898], 81–82, and Kennedy and Pingree, *The Astrological History*, pp. 178–184). Until this material has been examined, it cannot be said that any complete survey of Māshāʾallāh's work exists. Apart from the present attempt to fill the gap, the only previous survey—besides the works cited in the body of this article—is H. Suter, *Die Mathematiker und Astronomen der Araber und ihre Werke* (Leipzig, 1900), 5–6.

DAVID PINGREE

MASKELYNE, NEVIL (*b.* London, England, 6 October 1732; *d.* Greenwich, England, 9 February 1811), *astronomy.*

The last male heir of an ancient Wiltshire family that probably originated in Normandy, Maskelyne was educated at Westminster School, where he received a good grounding in the classics. During his vacations, he was tutored in writing and arithmetic. He enjoyed reading and was fascinated by optics and astronomy, through which he was led to the study of mathematics as the indispensable tool for the proper understanding of these related sciences. Having mastered in a few months the elements of geometry and algebra, he then applied this knowledge to other aspects of natural philosophy, particularly mechanics, pneumatics, and hydrostatics. He furthered these studies at Trinity College, Cambridge, graduating in 1754 as seventh wrangler. After being ordained in 1755, he accepted a curacy near London; there, rather than seeking a livelihood in the Anglican Church, he devoted many of his leisure hours to assisting the astronomer royal, James Bradley, in computing tables of refraction. He was elected a fellow of Trinity College, Cambridge, in 1758 and of the Royal Society the following year.

On Bradley's recommendation, Maskelyne was sent in 1761 by the British government to the island of St. Helena to observe the transit of Venus, from which the distance of the earth from the sun can be deduced. Unfortunately, clouds prevented his observing the time of emersion of this planet; and an error in his observations of the meridian zenith distance of the bright star Sirius—due to a fault in suspending his zenith sector—prevented him from testing the supposition that it exhibited a small but measurable parallax.

Maskelyne was more successful with observations made during the voyage for the purpose of investigating the reliability of the lunar distance method of determining longitude at sea. The lunar tables that he employed were those of Tobias Mayer, transmitted to London in 1755 to support his application for a large parliamentary bounty offered to "such person or persons as shall discover the longitude at sea." The instrument used for making the necessary angular measurements of lunar distances and celestial altitudes was a reflecting quadrant of the type invented by John Hadley in 1731 and already in widespread use among seamen. In his book *The British Mariners Guide* (London, 1763) he gave detailed instructions on how to use and rectify this instrument, and examples of how to apply the lunar tables in calculating the longitude.

A prime objective of Maskelyne's second voyage, to Bridgetown in Barbados in 1764, was to assess the accuracy of the rival chronometer method of longitude determination, championed by John Harrison, before a decision could be made on its claim for a parliamentary award; this necessitated Maskelyne's making astronomical observations to establish the longitude of Barbados. He was also ordered by the Board of Longitude to investigate the comparative accuracy of two additional means of longitude determination based upon observations of the satellites of Jupiter and on occultations of stars by the moon. He was further entrusted with the testing of a marine chair designed by a certain Mr. Christopher Irwin, which he found to be quite impracticable for assisting observations made at sea.

At a memorable meeting of the Board of Longitude (9 February 1765), at which the sums to be awarded to Harrison and Mayer were specified, Maskelyne, who had just been appointed astronomer royal, arranged for four naval officers to be in attendance to testify to the general utility of the lunar-distance method for finding longitude at sea to within 1° or 60 miles. He also presented a memorial in which he proposed that the practical application of the method could be facilitated by the preparation of a nautical ephemeris with auxiliary tables and explanations. These plans crystallized less than two years later with the publication of the *Nautical Almanac* for 1767. Maskelyne also assumed the responsibility of supervising the printing and publishing of Mayer's lunar theory (1767) and his solar and lunar tables (1770), and he prepared "Requisite Tables" (1767) for eliminating the effects of astronomical refraction and parallax from the observed lunar distances. He continued to superintend the ever-increasing work of the computers and comparers of the annual *Nautical Almanac* until his death more than forty years later.

This periodical is undoubtedly Maskelyne's greatest monument to astronomical science. It is still a useful navigational aid even though the lunar distance tables themselves became obsolete by the beginning of the twentieth century, mainly as a result of the exceptionally high degree of reliability of chronometers.

Among Maskelyne's onerous duties at the Royal Observatory was to assess the performances of a considerable number of chronometers submitted for an official trial by other pioneers of watchmaking—Thomas Mudge, John Arnold, Josiah Emery, and Thomas Earnshaw. The controversial results of these comparative tests, which stemmed from an ambiguity in defining "accuracy" and "error" in the case of chronometers, had the desirable effect of establishing a consistent system of rating and the introduction in 1823 of "trial-" or "test-numbers," which were modified by George Airy in 1840 to a system that is still used.

In a famous experiment of 1774 Maskelyne attempted to determine the earth's density from measurements of the deviation of a plumb line produced by the gravitational attraction of Mt. Schiehallion, in Scotland. By observing the slight difference in the zenith distances of certain stars at two observing stations on the north and south faces of the mountain, and making due allowance for the effect of their latitude difference by means of geodetic measurements, Maskelyne identified the residual displacement of 11.7″ with the sum of the deviations in the direction of the vertical to the earth's surface on each side of this conveniently symmetrical mountain. This was the first convincing experimental demonstration of the universality of gravitation, in the sense that it operates not only between the bodies of the solar system but also between the elements of matter of which each body is composed. With the aid of his friend Charles Hutton and John Playfair, who estimated the density of the rocks and total mass of that mountain relative to the mass of the earth, Maskelyne concluded the mean density of the earth to be between 4.867 and 4.559 times that of water, a result that compares quite well with the presently accepted value of 5.52.

BIBLIOGRAPHY

No definitive biography of Maskelyne has been written, but accounts of his life and work are to be found in standard encyclopedic works such as *Encyclopaedia Britannica*, 8th ed. (1857), XIV, 334–336; and *Dictionary of National Biography*, rev. ed., XII, 1299–1301; from which references to other biographical sources may be obtained. Precise references to Maskelyne's contributions to the *Philosophical Transactions of the Royal Society* between 1760 and 1808 and to his other publications are in D. W. Dewhirst's new ed. of J. C. Houzeau and A. Lancaster, *General Bibliography of Astronomy to the Year 1880*, 2 vols. (London, 1964). No fewer than 168 batches of Maskelyne's papers are preserved in the records room of the Royal Greenwich Observatory. Photocopies of some other unpublished writings, still in the possession of one of his heirs, are in the MS department of the National Maritime Museum, London (Reference PGR/38/1). Other repositories of his correspondence include the libraries of the Royal Society of London, the British Museum, the Fitzwilliam Museum and university library in Cambridge, the Bodleian Library in Oxford, the university library in Göttingen, and the private archives of the earl of Bute at Rothesay, Scotland.

ERIC G. FORBES

MASON, CHARLES (*b.* Wherr, Gloucestershire, England, baptized 1 May 1728; *d.* Philadelphia, Pennsylvania, 25 October 1786), *astronomy, geodesy.*

Little is known of Mason's early life. He was one of at least four children of Charles Mason, and it is likely that he received his early education at Tetbury Grammar School. He may have received additional tuition from Robert Stratford, a schoolmaster and mathematician at Sapperton, a village near Wherr. At any rate, he seems to have acquired a considerable competence in plane and spherical trigonometry, practical and spherical astronomy, and geodesy. In 1756 he joined the staff of the Royal Observatory as assistant to the director, James Bradley. Mason held this post until 1760, and may be assumed to have aided Bradley in his extensive stellar, solar, planetary, and lunar astrometric work.

In 1761 Mason joined forces with an associate, Jeremiah Dixon, to observe the transit of Venus of 6 June. Their expedition was sponsored by the Royal Society as part of an international effort to establish the solar parallax (measure of the distance to the sun); they were to make observations at Bencoolen, Sumatra, but were delayed by an attack from a French frigate and actually observed the transit at the Cape of Good Hope. On their return voyage they stopped at St. Helena, where another party had observed the transit under the direction of Nevil Maskelyne, and assisted in gathering tidal and gravitational data.

In 1763 Mason and Dixon were named by Nathaniel Bliss, astronomer royal, to go to the American colonies to resolve the question of the common boundaries of Pennsylvania, Maryland, Delaware, and Virginia. (The Penn and Calvert families had disputed such boundaries for some eighty years, and it was perhaps at their request that Bliss made his recommendation.) Mason and Dixon arrived in Philadelphia on 15 November and immediately set about their task. The best-known lines of the demarcation are those that extend from east to west, mostly

along the southern border of Pennsylvania, for a combined distance of 244.483 miles westward from the west shore of the Delaware River.

While in America Mason and Dixon also undertook a commission from the Royal Society to measure a degree of latitude. They chose the latitude of $39°11'56.5''$ as the mid-point of their arc and determined the value of one degree on it as 68.7291 miles (according to the Clarke spheroid of 1866, the value is actually 68.9833 miles). Cavendish reviewed their result and concluded that the normal to the geoid had been vitiated by topographical and subsurface anomalies in mass distribution. Mason and Dixon also conducted studies of the variation of gravity with latitude before they left America for England on 11 September 1768.

Mason returned to England in time to take part in another Royal Society expedition to observe a transit of Venus. The transit occurred on 3 June 1769; Mason was in charge of the station at Cavan, County Donegal, Ireland. Despite cloud conditions, he was able to record the times of the first external contact and the first internal contact; he reported his results in a forty-three-page paper read to the Royal Society on 7 November 1770.

During this period the Royal Society also was concerned with the problem of measuring the mass and density of the earth. A procedure earlier outlined by Newton pointed out that the presence of a topographical protuberance would vitiate the direction of the vertical and therefore affect latitude observations. If a mountain of suitable shape, preferably hemispherical or conical, could be found, it would only be necessary to determine its mass by borings and topographical studies and then observe the latitude on each side of it along a meridian and finally determine the distance between the two stations by methods of plane or geodetic surveying. The mass and density of the earth would then follow from a simple trigonometrical relationship. In implementing this plan Mason, after studying the topography of northern England and Scotland, chose Mt. Schiehallion in Perthshire as most nearly answering Newton's specifications. Maskelyne then carried out the proposed research.

Mason also worked with the Commissioners of Longitude, particularly in connection with the *Nautical Almanac* of 1773, for which he prepared a catalog of the positions of 387 fixed stars from observations made by Bradley and precessed to the epoch of 1760. In 1778 he published "Lunar Tables in Longitude and Latitude According to the Newtonian Laws of Gravity"; he finished an improved set of these tables at Sapperton in 1780. Mason's

MASSA

work was cited in each annual edition of the *Nautical Almanac* for some thirty years; between 1770 and 1781 the commissioners paid him £1,317. Although there is little documentation of Mason's activities from 1781 until 1786, it is reasonable to suppose that he continued to work with the Royal Observatory and the Commissioners of Longitude.

In 1786 Mason returned to the United States, accompanied by his second wife and eight children. Although he held no particular commission, it may be surmised that he expected that considerable geodetic work would be necessary in establishing the boundaries of the states then being incorporated into the Union. During the Atlantic crossing he became ill, however; he never recovered and died in Philadelphia. He was interred in Christ Church cemetery, where Benjamin Franklin lies. He had been a member of the American Society (now the American Philosophical Society) since 1768. At his death Mason left all his manuscripts and scientific papers to John Ewing, provost of the University of Pennsylvania, who was himself a distinguished mathematician and astronomer.

BIBLIOGRAPHY

I. Original Works. The whereabouts of Mason's MSS and papers is not known today. His publications include "Observations Made at the Cape of Good Hope," in *Philosophical Transactions of the Royal Society*, **52** (1762), 378–394, written with Jeremiah Dixon; and "Astronomical Observations Made at Cavan . . .," *ibid.*, **60** (1770), 454–497.

II. Secondary Literature. On Mason and his work see T. D. Cope, "The First Scientific Expedition of Charles Mason and Jeremiah Dixon," in *Pennsylvania History*, **12**, no. 1 (1945), 3–12; "Mason and Dixon—English Men of Science," in *Delaware Notes*, 22nd ser. (1949), 13–32; T. D. Cope and H. W. Robinson, "Charles Mason, Jeremiah Dixon and the Royal Society," in *Notes and Records. Royal Society of London*, **9** (1951), 55–58, 78; A. Hughlett Mason, "The Journal of Charles Mason and Jeremiah Dixon Transcribed From the Original in the U.S. National Archives . . .," in *Memoirs of the American Philosophical Society*, **76** (1969), 25; and H. W. Robinson, "A Note on Charles Mason's Ancestry and His Family," in *Proceedings of the American Philosophical Society*, **93**, no. 2 (1949), 134–136.

A. Hughlett Mason

MASSA, NICCOLO (*b.* Venice, Italy, 1485; *d.* Venice, 1569), *medicine.*

Massa, a medical graduate of Padua, practiced medicine in Venice, where he was known chiefly as clinician and syphilologist. Owing to his belief that the physician ought to have a sound knowledge of anatomy, he undertook a program of dissection and investigation of the human body at least from 1526 to 1533, producing an unillustrated anatomical treatise entitled *Liber introductorius anatomiae* (Venice, 1536), which remained the best brief textbook of the subject for a generation.

The *Liber introductorius anatomiae* is arranged according to the medieval pattern of anatomy established by Mondino, that is, an approach to the subject derived from the necessity of dissecting the most perishable organs first. It is based partly on the work of earlier writers, especially Galen, and partly on its author's own dissections carried out in the convent of SS. John and Paul (fols. 26r, 56v) and the hospital of SS. Peter and Paul (fols. 10r, 26r) in Venice as well as others performed on the bodies of stillborn infants (fols. 7v, 43v).

Since Massa was over fifty years old when he composed his book, its text does not reflect to any large degree the alterations in favor of classical anatomical terminology that the humanists were introducing at the time. Nevertheless he was the first to employ the term *panniculus carnosus* (fol. 8r). Of his descriptions it may be said that he provided a relatively good account of the abdominal wall, noted the tendinous intersections of the rectus abdominis muscle (fol. 11v), and referred very briefly to the inguinal canal (fol. 13r). He described the intestinal canal with some accuracy (fols. 18v–23r), including an account of the appendix, which he thought tended to disappear with maturity (fols. 20v–21r). He mentioned the variation in size of the spleen in ailments involving that organ, declaring that he had observed it "very large and extending into the lower parts" (fol. 26v). He declared the liver to be divided usually into five lobes, although sometimes finding it undivided or divided into only two parts (fol. 27r), but asserted that the portal vein is always divided into five main branches (fol. 27r). In his description of the kidneys he proved by blowing through a reed that the cavity of the renal veins is not continuous with that of the sinus of the kidney (fol. 31v). This contribution was important, even though the fact had been alluded to earlier by Berengario da Carpi, since the kidneys were usually thought to be filters straining urine out of blood. Massa held that the right kidney is normally higher than the left, although he declared that he had twice seen the reverse (fol. 32r). He briefly noted the difference in the levels of origin of the spermatic vessels on the two sides (fol. 33r)—a fact known to Mondino—and made brief reference to the prostate (fol. 34r), the first reference to that organ. He denied the belief in the seven-celled uterus, declaring the uterus to contain only a single cavity (fol. 45r).

Massa gave credence to the existence of the *rete mirabile* in the human brain, but admitted that there was disagreement on this matter, and wrote "some dare to say that this *rete* is a figment of Galen . . . but I myself have often seen and demonstrated it . . . though sometimes I have found it to be very small" (fols. 89v–90r). Generally speaking, Massa's long description of the brain was traditional and unsatisfactory. On the other hand Massa was the first after Berengario da Carpi to refer to the malleus and incus, although his short statement refers to both ossicles under the word *malleolus* (fols. 93r–v). Thus he left to Vesalius the opportunity of providing better and more appropriate names.

Within its limits, and despite a considerable residue of Galenic anatomy, the *Liber introductorius anatomiae* contains shrewd observations often tempered by curious errors. For examples of the latter, it describes certain nonexistent cardiac valves (fol. 55v), and Massa asserted that he had sometimes found a third ventricle in the heart (fol. 56r). Despite its merits the book is distinctly pre-Vesalian. The reissue of the book in 1559 was unsuccessful.

BIBLIOGRAPHY

In addition to a few autobiographical notes in the *Liber introductorius anatomiae*, what further information there is on the life of Massa is in Luigi Nardo, "Dell'anatomia in Venezia," in *Ateneo Veneto*, fasc. 2–3 (1897), with additional notes by Cesare Musatti. The full title of Massa's anatomical treatise is *Liber introductorius anatomiae, sive dissectionis corporis humani, nunc primum ab ipso auctore in lucem aeditus, in quo quamplurima membra, operationes, & utilitates tam ab antiquis, quam a modernis praetermissa manifestantur. Venetiis in vico sancti Moysi, aput [sic] signum archangeli Raphaelis, in aedibus Francisci Bindoni, ac Maphei Pasini, socios [sic], accuratissimae [sic] impressum. Mense Novembri, MDXXXVI.* The reissue of 1559 was also published in Venice, *Ex officina Stellae Jordani Zilleti.*

C. D. O'MALLEY

MASSON, ANTOINE-PHILIBERT (*b.* Auxonne, France, 22 or 23 August 1806; *d.* Paris, France, 1 December 1860), *physics*.

Having completed a bachelor of arts program in Nancy, Antoine Masson subsequently received his bachelor of sciences degree from the École Normale Supérieure in Paris. In 1831, after a year of teaching mathematics at Montpellier, he moved to Caen, where he taught physical sciences at the Collège Royal until 1839. Unaware of the discoveries of Joseph Henry or William Jenkins, Masson in 1834 observed independently the self-induction of a voltaic circuit. Three years later he described his investigation of this phenomenon and, utilizing a toothed wheel as an interrupter, demonstrated the tetanic effect of a series of rapidly repeated self-induced currents. Employing a similar toothed wheel to interrupt an independent primary circuit, Masson constructed some of the earliest induction coils. In 1841, together with Louis Breguet, he described a high-tension induction coil of the type Ruhmkorff subsequently perfected. In the interim, having in 1836 successfully defended a doctoral thesis elaborating Ampère's work in electrodynamics, Masson had returned to Paris and from 1841 taught physics at the Lycée Louis-le-Grand and at the École Centrale.

Masson's subsequent researches spanned the breadth of contemporary physics and continued unabated until his death in 1860. To clarify the relations among heat, light, and electricity, between 1844 and 1854 he conducted an intensive investigation of the spark produced by electrical discharges through various media. In conjunction with L. Courtépée and J.-C. Jamin, he also examined during these years the absorption of radiant heat and light by different substances, confirming the conclusions of Melloni. In addition, he investigated aspects of electrical telegraphy, acoustics, the elasticity of solid bodies, and the discharge of induction coils through partial vacuums, as well as related chemical and physical problems. He was a member of the Académie des Sciences, Arts, et Belles-Lettres of Caen, the Société Royale of Liège, and the Société Philomatique in Paris. He received the Légion d'Honneur and, although never elected, was nominated in 1851 and 1860 to the Académie des Sciences of Paris.

BIBLIOGRAPHY

I. ORIGINAL WORKS. Accounts of Masson's researches appeared almost annually between 1835 and 1858, for the most part in the *Annales de chimie et de physique* or the *Comptes rendus . . . de l'Académie des sciences.* Although Poggendorff, II, col. 75, lists most of these papers, see also Masson's own *Notice sur les travaux scientifiques de M. A. Masson* (Paris, 1851). Several of these papers were also reprinted separately. Other published works include *Théorie physique et mathématique des phénomènes électrodynamiques et du magnétisme* (Paris, 1838); *École centrale des arts et manufactures. Cours de physique générale [1841–1842, 1843–1844]*, 2 vols. (Paris, n.d.); *Mémoire sur l'étin-*

celle électrique (Haarlem, 1854); and *Nouvelle théorie de la voix* (Paris, 1858).

II. SECONDARY LITERATURE. Although briefly mentioned in most of the standard secondary sources, the only extended discussion of Masson is Louis Jovignot, "Un grand savant bourguignon du XIXe siècle: Antoine Masson," in *Revue d'histoire des sciences et de leurs applications*, **1** (1948), 337–350.

DAVID W. CORSON

MAST, SAMUEL OTTMAR (*b*. Ann Arbor, Michigan, 3 October 1871; *d*. Baltimore, Maryland, 3 February 1947), *botany*.

Mast attended elementary school in his home state and received a teaching certificate from the State Normal College in Ypsilanti, Michigan, in 1897. He received a B.Sc. from the University of Michigan in 1899, and in 1906 he obtained his Ph.D. in zoology from Harvard. After graduation from Michigan Mast started an uninterrupted teaching career, which lasted forty-three years. From 1899 to 1908 he was professor of biology and botany at Hope College, Holland, Michigan, and then associate professor of biology and professor of botany at Goucher College, Baltimore, Maryland, where he remained until 1911. He then joined the faculty of Johns Hopkins University and became the director of the zoological laboratory in 1938, when Jennings retired. Mast published his book *Light and the Behavior of Organisms* in 1911, and most of his later research was on the reactions of lower organisms to stimuli, especially light. His contractile-hydraulic theory of amoeboid movement, proposed in 1926, continues to be basic in the explanation of this phenomenon. His careful study of the metabolism of the colorless flagellate *Chilomonas paramecium* (1933), showing its ability to synthesize organic compounds in the dark, is also a classic in the field.

During his long working life, Mast published almost 200 papers and books. Besides being a member of many scientific societies, he was awarded the Cartwright Prize by Columbia University in 1909. The State Normal College at Ypsilanti awarded him an honorary M.Pd. in 1912, and the University of Michigan the Sc.D. in 1941, a year before he retired.

BIBLIOGRAPHY

I. ORIGINAL WORKS. Among Mast's important works are "Structure, Movement, Locomotion and Stimulation in *Amoeba*," in *Journal of Morphology*, **4** (1926), 347–425, written with D. M. Pace; and "Synthesis From Inorganic

Compounds of Starch, Fats, Proteins and Protoplasm in the Colorless Animal *Chilomonas paramecium*," in *Protoplasma*, **20** (1933), 326–358.

II. SECONDARY LITERATURE. See D. H. Wenrich, "Some American Pioneers in Protozoology," in *Journal of Protozoology*, **3** (1956), 17; and the obituary notice by C. G. Wilber, "Samuel O. Mast (1871–1947)," in *Transactions of the American Microscopical Society*, **67** (1948), 82–83.

ENRIQUE BELTRÁN

MÄSTLIN, MICHAEL (*b*. Göppingen, Germany, 30 September 1550; *d*. Tübingen, Germany, 20 October 1631), *astronomy*.

Earthshine was correctly explained for the first time in print by Mästlin.[1] He matriculated at Tübingen University on 3 December 1568 and received the B.A. (30 March 1569) and M.A. (1 August 1571) before entering the theological course.[2] To his reprint of Erasmus Reinhold's *Prussian Tables*, Mästlin added a brief appendix in 1571, and his 1573 essay on the nova of 1572 was impressive enough to be incorporated in its entirety into the *Progymnasmata* of Tycho Brahe.[3] Lacking observational instruments, Mästlin stretched a thread through the nova and two pairs of previously known stars. He took the celestial longitude and latitude of these four fixed points directly from the star catalog of Copernicus, of whose *Revolutions* he had acquired a copy in 1570. (Mästlin's heavily annotated copy of the *Revolutions* is preserved at Schaffhausen, Switzerland.) The intersection of the arcs of the great circles passing through the two pairs of his reference stars gave Mästlin the position of the 1572 nova, and its nondisplacement from them convinced him that it was indeed a new star; thus, coming-into-being could occur in heaven as well as on earth, contrary to the traditional dogma.

Having served as the assistant to Philipp Apian (1531–1589), professor of mathematics at Tübingen, Mästlin replaced him when Apian went on leave in 1575. This arrangement was not renewed, however, for on 24 October 1576 Mästlin was appointed to a Lutheran pastorate in Backnang. In April 1577 he married Margaret Grüninger, who bore him three daughters and three sons.[4] Ludwig became a physician after enrolling at Tübingen on 26 February 1594 and obtaining the B.A. on 5 April 1598 and the M.A. on 13 February 1600.[5] In that year Michael, Jr., a painter, ran away from home and was later said to be hiding among the Jesuits.[6] Margaret married Tobias Olbert (M.A., Tübingen, 15 February 1598) on 7 December 1602; Anna Maria also married a Lutheran clergyman, Johann Wolfgang Mögling;

and Sabina married Burckhardt Rümelin, a Tübingen law student in 1606 and court attorney in 1624.[7]

Mästlin was designated professor of mathematics at Heidelberg University on 19 November 1580.[8] In discussing that year's comet he declared that the unsoundness of the Aristotelian cosmology had been revealed to him by three great celestial events occurring over a period of eight years: the 1572 nova and the comets of 1577 and 1580.

Having failed to detect any perceptible parallax in the comet of 1577, Mästlin concluded that it was not a sublunar but, rather, a supralunar body. Remarking that "according to Abū Ma'shar, who flourished about A.D. 844, a comet was seen above the sphere of Venus," he asked, "What would have been the physical cause of this [phenomenon], if we are to believe that comets have no place other than the region of the [four] elements?"[9] Rejecting the conventional classification of comets as meteorological phenomena, he located the comet of 1577 in the sphere of Venus.

Nevertheless, in his *Epitome of Astronomy*, an introductory textbook begun while he was still a student at Tübingen and so popular that it ran through seven editions between 1582 and 1624, Mästlin expounded the traditional view as easier for beginners to understand. He advised Protestant governments to reject the Gregorian calendar as a papal scheme to regain control over territories that had escaped from its grasp. All of his books and writings appeared on the Index of Pope Sixtus V in 1590.[10]

In a public address at Tübingen University on 22 September 1602, on the basis of his chronological researches Mästlin put Jesus' birth more than four years before the conventional date.[11] On 23 May 1584 he had replaced Apian, who had been dismissed for refusing to sign the oath of religious allegiance; he later bought Apian's library from the latter's widow.[12] Mästlin, a man of slight build, unlike the massive ancestor from whom his surname was derived,[13] was elected dean of the Tübingen Arts Faculty eight times between 1588 and 1629. He taught there for forty-seven years, until his death in 1631.

His first wife having died on 15 February 1588, on 28 January 1589 Mästlin married Margaret Burckhardt (19 March 1564–18 February 1622), who bore him nine children.[14] Sabina, the second Mästlin daughter to bear this name, was buried on 9 July 1596, before attaining the age of seven, and a third Sabina was born on 22 June 1599.[15] A second Margaret, christened on 16 December 1604, died on 31 August 1609.[16] Augustus, born on 13 January 1598, died on 16 February 1598.[17] Anna Dorothea married a Lutheran clergyman, Andrew Osiander (M.A.,

Tübingen, 1610), in 1614.[18] Gottfried, baptized on 12 October 1595, received his B.A. at Tübingen on 31 March 1612, his M.A. on 16 August 1615, and became a professor of languages there in 1627.[19] Matthew acquired his B.A. at Tübingen on 17 March 1619 and the M.A. on 20 February 1622. He married on 24 November 1622; taught school at Gerlingen; and worked as a caretaker in Knittlingen, where he was buried on 6 February 1661.[20]

In his 1578 discussion of the comet of 1577, Mästlin announced his "adoption of the cosmology of Copernicus, truly the foremost astronomer since Ptolemy." In 1632 Galileo attributed his acceptance of Copernicanism to two or three lectures, delivered shortly after he had completed his philosophy course, by a foreign professor whose identity (or very existence) is uncertain.[21] Hence in 1650 Gerhard Johann Voss (1577–1649) posthumously initiated the legend—which has been uncritically repeated by influential writers for centuries—that the foreigner responsible for Galileo's Copernicanism was Mästlin.[22] The latter, however, was really responsible for the Copernicanism of Johannes Kepler, who attended Mästlin's classroom lectures at Tübingen and heard him expound the superiority of the Copernican astronomy over the Ptolemaic.[23] The pupil-teacher relationship between Kepler and Mästlin ripened into a lifelong affectionate friendship, each sincerely acknowledging the other's valuable assistance. No finer example of the educational process at its best can be found in the entire history of science.

NOTES

1. The key passage was translated into English by Edward Rosen, *Kepler's Conversation*, 117–119, 157.
2. *Die Matrikeln der Universität Tübingen*, I (Stuttgart, 1906), 487.
3. Tycho Brahe, *Opera omnia*, III (Copenhagen, 1916), 58–62.
4. Karl Steiff, "Der Tübinger Professor . . .," 51–53.
5. *Matrikeln . . . Tübingen*, I, 708; Johannes Kepler, *Gesammelte Werke*, XIII, 211:119–121, 232:527–528.
6. Kepler, *op. cit.*, XIV, 157:73–82, 354:473–475.
7. *Diarium Martini Crusii*, III (Tübingen, 1958), 515:25–26; *Matrikeln . . . Tübingen*, I, 707; II, 7. Steiff, *op. cit.*, 53, erroneously assigned this Sabina to Mästlin's second marriage.
8. *Die Matrikel der Universität Heidelberg*, II (Heidelberg, 1886), 92.
9. Mästlin's statement about Abū Ma'shar, Islam's foremost astrologer, was undoubtedly based on Cardano's *Astronomical Aphorisms*, published as a supplement to *Hieronymi Cardani Libelli V* (Nuremberg, 1547), as cited by Willy Hartner, *Oriens-Occidens* (Hildesheim, 1968), 503.
10. Heinrich Reusch, ed., *Die Indices librorum prohibitorum des sechszehnten Jahrhunderts* (Tübingen, 1896; repr. Nieuwkoop, 1961), 504, 566.
11. Kepler, *op. cit.*, XVII, 56:100–103.
12. *Matrikeln . . . Tübingen*, I, 624; Siegmund Günther, *Peter und Philipp Apian* (Amsterdam, 1967, repr. of 1882 ed.),

107–109; Ernst Zinner, *Entstehung und Ausbreitung der Coppernicanischen Lehre*, 453.

13. Edward Rosen, *Kepler's Somnium*, 64–65; H. M. Decker, "Die Ahnen des Astronomen Mästlin," 103–104.
14. W. Bardili, "Ergänzungen zur 'Geistesmutter,' " 114.
15. *Diarium . . . Crusii*, I, 128:18–20; Kepler, *op. cit.*, XIII, 368:28; XIV, 43:5, 463:5.
16. K. E. von Marchtaler, "Ein Beitrag zur Familienforschung Mästlin," 179.
17. Kepler, *op. cit.*, XIII, 184:179–180, 209:12–14.
18. "1604" in Steiff, *op. cit.*, 53, is a misprint; *Matrikeln . . . Tübingen*, II, 41.
19. *Matrikeln . . . Tübingen*, II, 57.
20. *Ibid.*, 72; Marchtaler, *op. cit.*, 179–180.
21. Galileo Galilei, *Opere*, national ed., VII (Florence, 1897; repr. 1968), 154:5–10.
22. G. J. Voss, *De universae mathesios natura et constitutione liber, cui subjungitur chronologia mathematicorum* (Amsterdam, 1650), 192.
23. Kepler, *op. cit.*, I, 9:11–21.

BIBLIOGRAPHY

I. ORIGINAL WORKS. Writings published under Mästlin's name are *Ephemeris nova anni 1577* (Tübingen, 1576); *Observatio et demonstratio cometae aetherei, qui anno 1577 et 1578 . . . apparuit* (Tübingen, 1578); *Ephemerides novae ab anno . . . 1577 ad annum 1590* (Tübingen, 1580), preceded by Regiomontanus' brief *Commentary on the Ephemerides* and Mästlin's additions thereto, as well as a portrait of Mästlin, aged twenty-eight; *Consideratio et observatio cometae aetherei astronomica, qui anno 1580 . . . apparuit* (Heidelberg, 1581), with the same portrait as in the 1580 work; *De astronomiae principalibus et primis fundamentis disputatio* (Heidelberg, 1582), the respondent on 20 January 1582 being Jeremiah Jecklin or Jacobus of Ulm (M.A., Heidelberg, 24 July 1582)—the title is preceded by "Divino rectoris astrorum favente numine," a religious formula which has been listed as though it identified a separate publication; *De astronomiae hypothesibus sive de circulis sphaericis et orbibus theoricis disputatio* (Heidelberg, 1582), the respondent being Matthias Mener (M.A., Heidelberg, 19 Feb. 1583); *Epitome astronomiae* (Heidelberg, 1582; Tübingen, 1588, 1593, 1597, 1598, 1610, 1624); *Ausführlicher und gründtlicher Bericht von der . . . Jarrechnung* (Heidelberg, 1583), cited by Mästlin in a later calendar tract written in Latin as his "Dialexis germanica," which has sometimes been registered as though it were a separate publication, also repr., with additions, in *Nothwendige und gründtliche Bedenckhen von dem . . . Kalender* (Heidelberg, 1584); *Alterum examen novi pontificialis Gregoriani Kalendarii* (Tübingen, 1586); and *Defensio alterius sui examinis* (Tübingen, 1588).

He also wrote *Tres disputationes astronomicae et geographicae* (Tübingen, 1592): *De climatibus* (the respondent being Wolfgang Hohenfelder), *De diebus naturalibus et artificialibus* (the respondent being Ludwig Hohenfelder), *De zonis* (the respondent being George Achatius Enenckel) —each of these disputations has its own title page and separate pagination; *Disputatio de eclipsibus solis et lunae* (Tübingen, 1596), the respondent on 15 Jan. 1596 being Marcus Hohenfelder; *Geographische Landtafel, Stuttgart-*

Rome (Reutlingen, 1601); *Disputatio de multivariis motuum planetarum in coelo apparentibus irregularitatibus, seu regularibus inaequalitatibus, earumque causis astronomicis* (Tübingen, 1652), listed in Jean Graesse, *Trésor de livres rares*, IV (Dresden, 1863), 333, and Poggendorff, II, 170. Samuel Hafenreffer, ed. (Tübingen, 1641, 1646)—Hafenreffer's copy of the Mästlin MS had been approved by the author; *Synopsis chronologiae sacrae*, Johann Valentin Andreae, ed. (Lüneburg, 1642), recorded by Jacob Friderich Reimmann, *Versuch einer Einleitung in die historiam literariam*, III, pt. 2 (Halle, 1710), 369–370; and *Perpetuae dilucidationes tabularum Prutenicarum coelestium motuum* (Tübingen, 1652), listed in Jean Graesse, *Trésor de livres rares*, IV (Dresden, 1863), 333, and Poggendorff, II, 170.

Mästlin's writings published in works by other authors are "Observatio mathematica," appended to Nicodemus Frischlin, *Consideratio novae stellae, quae mense Novembri, anni . . . 1572 . . . apparuit* (Tübingen, 1573); "Demonstratio astronomica loci stellae novae," completed on 4 Mar. 1573, in Tycho Brahe, *Astronomiae instauratae progymnasmata* (Prague, 1602), pt. 3, ch. 8; "De dimensionibus orbium et sphaerarum coelestium," in Johannes Kepler, *Mysterium cosmographicum* (Tübingen, 1596), 161–181; autobiography, dated 23 Sept. 1609, in Hermann Staigmüller, "Württembergische Mathematiker," in *Württembergische Vierteljahrshefte für Landesgeschichte*, 12 (1903), 227–256, see 234–235; appendix to a proposed ed. of Copernicus written in 1621, in Christian Frisch, ed., *Joannis Kepleri astronomi opera omnia*, I (Frankfurt–Erlangen, 1858), 56–58; and "Observationes Moestlinianae," in Lucius Barrettus (Albert Curtius), *Historia coelestis* (Augsburg, 1666), esp. lxxv–lxxvi—according to the *Paralipomena* (at sig. Zzzzz2ʳ), these observations, written in Mästlin's own hand, were transferred by Wilhelm Schickard (the author of a [lost?] funeral oration for Mästlin) to a MS forming part of a collection bought for the Holy Roman Emperor Ferdinand III and preserved in the National Library, Vienna.

Original writings erroneously listed as printed, although still in MS, are "Apologia examinum suorum" (or "Examina, eorundemque apologia"), the title of a work projected by Mästlin in answer to Clavius' attack on *Ausführlicher und gründtlicher Bericht von der . . . Jarrechnung*, *Alterum examen novi pontificialis Gregoriani Kalendarii*, and *Defensio alterius sui examinis* was registered in the book fair catalog for 1593, which presence led bibliographers to list it as though it had been printed at Tübingen in 1593 (or 1597)—actually, Mästlin never finished writing this, which remains a torso in Vienna, National Library Codex 12411 (E. Zinner, *Entstehung und Ausbreitung . . .*, 435, no. 151); "Horologiorum solarium sciotericorum in superficiebus planis descriptionis et delineationis universalis informatio," which, although it has been listed as a book printed at Tübingen in 1590, is actually an unpublished MS written by the hand of Gottfried Mästlin, dated 20 July 1613, and preserved in the library of Erlangen University: *Katalog der Handschriften der Universitätsbibliothek Erlangen*, Hans Fischer, ed., II (Erlangen, 1936), 485, no. 838; and "De cometa anni 1618" (Tübingen, 1619) or "Astronomischer Discurs, von dem Cometen, so in Anno 1618 im November zu erscheinen angefangen und bis Februar dies 1619. Jars am Himmel gesehen worden" (Tübingen, 1619) is listed as printed in both Latin and German but the unpublished MS in German, ready for the printer, still lies in the Württembergische Landesbibliothek, Stuttgart, Codex math. Q. 15a–b.

Unpublished MSS, ready for the printer, are "Iudicium . . . de opere astronomico D. Frischlini," dated 18 Jan. 1586 (David Friderich Strauss, *Leben und Schriften des Dichters und Philologen Nicodemus Frischlin* [Frankfurt, 1856], 330); "Modus, ratio et fundamenta compositionis tabularum directionum Regiomontani et Reinholdi," Stuttgart Landesbibliothek, Codex math. Q. 15a–b; "Commentarius in 1. et 2. librum Euclidis cum demonstrationibus regularum algebraicarum," Vienna, National Library, Codex 12411 (E. Zinner, *Entstehung und Ausbreitung . . .*, 440, no. 260); "Emendatio sphalmatum typographicorum in Opere Palatino quodam geometrico et in Magno Canone Rhetici, nec non demonstratio, canonem tangentium et secantium in eodem Magno Canone Rhetici iuxta finem quadrantis minus exactum esse," Vienna, National Library (*Tabulae codicum manu scriptorum praeter graecos et orientales in Bibliotheca Palatina Vindobonensi asservatorum*, VI [Vienna, 1873], 253, no. 10913); and "Tractatus brevis de dimensione triangulorum rectilineorum et sphaericorum," written by the hand of Gottfried Mästlin, dated Oct. 1612, preserved at Erlangen—see *Katalog der Handschriften der Universitätsbibliothek Erlangen*, II, 485, no. 839.

Other unpublished MSS are listed in Vienna, National Library, Codex 12411; *Tabulae codicum . . . in Bibliotheca Palatina Vindobonensi; Katalog der Handschriften der Universitätsbibliothek Erlangen; Die historischen Handschriften der k. öffentlichen Bibliothek zu Stuttgart*, W. von Heyd, ed., I (Stuttgart, 1889), 257: Codex hist. fol. 603, Mästlin's correspondence with Johann Weidner; 283: Codex hist. fol. 657, biography of Mästlin by Johann Gottlieb Friedrich von Bohnenberger (1765–1831); Codex math. fol. 14b and Q. 15a–b; *Verzeichnis der Handschriften im deutschen Reich*, II, *Die Handschriften der Universitätsbibliothek Graz*, Anton Kern, ed., I (Leipzig, 1939), 82, no. 159 (15): a letter from Johann Reinhard Ziegler in Mainz to Paul Guldin in Rome, dated 5 Apr. 1611, regarding Mästlin's acquisition of a telescope; and *Kataloge der Herzog-August-Bibliothek Wolfenbüttel, Die Augusteischen Handschriften*, II (Frankfurt, 1966 [repr. of 1895 ed.]), 118–120, no. 2174: 15.3.Aug. fol.—also see Ernst Zinner, *Verzeichnis der astronomischen Handschriften des deutschen Kulturgebietes* (Munich, 1925), 217–219.

Published correspondence consists of Mästlin's letters to, from, and about Kepler, in Johannes Kepler, *Gesammelte Werke* (Munich, 1937–   ), part of which was trans. into German in *Johannes Kepler in seinen Briefen*, Max Caspar and Walther von Dyck, eds., 2 vols. (Munich–Berlin, 1930); and five letters to Mästlin from Simon Marius, in Ernst Zinner, "Zur Ehrenrettung des Simon Marius," in *Vierteljahrsschrift der Astronomischen Gesellschaft* (Leipzig), **77** (1942), 40–45.

Works edited by Mästlin are Erasmus Reinhold, *Prutenicae tabulae* (Tübingen, 1571), with an appendix, dated 1571, by Mästlin on p. 143; and George Joachim Rheticus, *Narratio prima*, in Johannes Kepler's *Mysterium cosmographicum* (Tübingen, 1596), 93–160, with Mästlin's preface, dated 1 Oct. 1596, at 86–90, and numerous notes by him in the margins.

A misattribution is *Problema astronomicum: Die Situs der Sternen, Planetarum oder Cometarum zu observiren ohne Instrumenta* (n.p., 1619). The Latin original of this trans. into German by Matthew Beger was misattributed to Mästlin by J. C. Houzeau and Albert Lancaster, *Bibliographie générale de l'astronomie*, I, pt. 1 (Brussels, 1887; repr. London, 1964), 603, where the work of Beger (miscalled Begern) is incorrectly listed as a translation of Mästlin's *Tres disputationes*. . . . Actually, Beger translated from Adriaan Metius' *Astronomiae universae institutiones* (Franeker, 1606–1608) an excerpt in which Metius discussed Mästlin's method of observing without the aid of any instruments.

II. Secondary Literature. See Peter Aufgebauer, "Die Gregorianische Kalenderreform im Urteil zeitgenössischer Astronomen," in *Sterne*, **45** (1969), 118–121; Walter Bardili, "Ergänzungen zur 'Geistesmutter,'" in *Blätter für Württembergische Familienkunde*, **8** (1939–1941), 113–119; J. G. F. von Bohnenberger, "Michael Mästlin," in *Württembergische Vierteljahrshefte für Landesgeschichte*, **12** (1903), 244–247; Erhard Cellius, *Imagines professorum Tubingensium* (Tübingen, 1596), sig. I 4v, woodcut portrait of Mästlin, aged forty-five; Hans Martin Decker, "Die Ahnen des Astronomen Mästlin," in *Blätter für Württembergische Familienkunde*, **8** (1939–1941), 102–104; Siegmund Günther, *Beiträge zur Geschichte der neueren Mathematik* (Ansbach, 1881), 15–25; and "Mästlin," in *Allgemeine deutsche Biographie*, XX, 575–580, also repr. (Berlin, 1970), XLV, 669; C. Doris Hellman, *The Comet of 1577* (New York, 1944; repr. 1971), 137–159; Johannes Kepler, *Gesammelte Werke* (Munich, 1937–   ); and Viktor Kommerell, "Michael Mästlin," in *Schwäbische Lebensbilder*, IV (Stuttgart, 1948), 86–100.

Also see Paul Löffler, "Michael Mästlin zu seinem 300. Todestag," in *Tübinger Chronik-Amtsblatt für den Oberamtsbezirk Tübingen*, **87**, no. 245 (20 Oct. 1931), 4r; Kurt Erhard von Marchtaler, "Ein Beitrag zur Familienforschung Mästlin," in *Blätter für Württembergische Familienkunde*, **8** (1939–1941), 178–180; Edward Rosen, "Kepler and the Lutheran Attitude Toward Copernicanism," in *Vistas in Astronomy* (in press); *Kepler's Somnium* (Madison, Wis., 1967), xvi, repro. of an oil portrait of Mästlin painted in 1619; Karl Steiff, "Der Tübinger Professor der Mathematik und Astronomie Michael Mästlin," in *Literarische Beilage des Staats-Anzeiger für Württemberg* (30 Apr. 1892), 49–64, 126–128; Ernst Zinner, *Entstehung und Ausbreitung der Coppernicanischen Lehre* (Erlangen, 1943); and Edward Rosen, *Kepler's Conversation with Galileo's Sidereal Messenger* (New York–London, 1965).

Edward Rosen

AL-MASʿŪDĪ, ABU ʾL-ḤASAN ʿALĪ IBN AL-ḤUSAYN IBN ʿALĪ (*b.* Baghdad; *d.* al-Fusṭāṭ [old Cairo], Egypt, September/October 956 or 957), *geography, history.*

While still relatively young, al-Masʿūdī left Baghdad about 915, spending the rest of his life traveling until he settled in Egypt toward the end of his life. He journeyed extensively in Khurāsān, Sijistān (southern Afghanistan), Kirmān, Fārs, Qūmīs, Jurjān, Ṭabaristān, Jibāl (Media), Khūzistān, Iraq (southern half of Mesopotamia), and Jazīra (northern half of Mesopotamia) until 941; and in Syria, Yemen, Ḥaḍramawt, Shaḥr, and Egypt between 941 and 956. He also visited Sind, India, and East Africa and sailed on the Caspian Sea, the Mediterranean, the Red Sea, and the Arabian Sea. His claims to have visited Java, Indochina, or China do not seem to be correct, however, since there is no internal evidence to support them in his extant works. Nor did he visit Madagascar, Ceylon, or Tibet, as is believed by some scholars. Apart from his urge to see the wonders of the world (*al-ʿajāʾib*), his travels were motivated by his conviction that true knowledge could be acquired only through personal experience and observation.

Al-Masʿūdī was a prolific writer; nearly thirty-seven works can be enumerated from his extant writings and from other sources. He wrote on a great variety of subjects: history, geography, jurisprudence, theology, genealogy, and the art of government and administration. Only two of these works have survived completely: *Murūj al-dhahab wa maʿādin al-jawhar*, completed in November/December 947 and revised in 956, and *Al-Tanbīh wa ʾl-ishrāf*, completed a year before his death. His magnum opus, *Kitāb Akhbār al-zamān wa man abādahu ʾl-ḥidthān*, a world history and geography in about thirty volumes *(funūn)*, is lost except for its first volume, which is preserved in Vienna (there is another manuscript in Berlin; see C. Brockelmann, *Encyclopaedia of Islam*, 1st ed.). A number of manuscripts and printed works bearing different titles—*Mukhtaṣar al-ʿajāʾib* (French translation by Carra de Vaux, *L'abrégé des merveilles* [Paris, 1898]); *Kitāb Akhbār al-zamān wa man abādahu ʾl-ḥidthān* (ʿAbd Allāh al-Ṣāwī, ed. [Cairo, 1938]); *Kitāb al-Ausaṭ; Kitāb ʿajāʾib al-dunyā* (or *Kitāb al-ʿajāʾib*)—are incorrectly ascribed to him. In fact, they seem to belong to the *ʿajāʾib* literature produced in abundance in the medieval period. This consisted of collections of sailors' tales about the Indian Ocean and legends about ancient Egypt, among other subjects. Similarly, *Ithbāt al-waṣiyya li ʾl-Imām ʿAlī ibn Abī Ṭālib* (al-Najaf, Iraq, 1955) does does not seem to be al-Masʿūdī's work.

Al-Masʿūdī was a Muʿtazilite thinker with Shīʿa leanings. Believing that knowledge accumulated and advanced with the passage of time, he disagreed with those scholars who uncritically accepted the "ancients" *(salaf)* as final authorities and who minimized the importance of the knowledge of the contemporary savants *(khalaf)*: "And often a latter-day writer, because of his great accumulation of experiences, and of his wariness of an uncritical imitation of his predecessors and of his caution against pitfalls, is better in his documentation and more thorough in his authorship. Again, since he discovers new things not known to former generations, the sciences steadily progress to unknown limits and ends" (*Tanbīh*, p. 76). Thus we find him openly challenging traditionalism *(taqlīd)*, which from the twelfth century exerted a deadly influence on the progress of scientific knowledge and learning and hence was the main cause of the decline of the Islamic society in the Middle Ages.

As a historian al-Masʿūdī made important contributions to Arab historiography in the Middle Ages. He believed that to obtain a true and objective picture of the history of a nation, a historian ought to consult the primary sources available in that country and not depend upon secondary sources, which are likely to distort facts. He followed this principle in the case of the history of ancient Iran (*Tanbīh*, p. 105). For his history of the world he not only utilized a large number of Arabic historical works available to him but also incorporated into it the vast amount of rich material on different countries and kingdoms that he had collected during his travels. Besides dealing with the history of Islam in the traditional manner —from the time of the Prophet Muḥammad up to his own times—he surveyed the histories of important nations and races who lived before the rise of Islam in the seventh century and also covered, as far as possible, the contemporary history of the Byzantine Empire, some European nations, India, and China. His approach to history was, therefore, both scientific and objective. He was one of the first Muslim historians to set the trend of secularism in historiography, for he included in his works the histories of a large number of non-Islamic nations. In his enthusiasm to record everything at his disposal, however, he uncritically related legends and popular beliefs side by side with history.

Al-Masʿūdī conceived of geography as an essential prerequisite of history, and hence a survey of world geography preceded his account of world history. He emphasized that geographical environment deeply influenced the character, temperament, and structure —as well as the color—of the animal and plant life found in a particular region. He was fully acquainted with ancient Greek, Indian, and Iranian geographical thought and concepts through Arabic translations of the literature and was equally well versed in contemporary Arabic geographical literature. His knowledge of geography covered almost all branches of the subject, and his system was based mainly on the Greeks. Thus, he belonged to the "secular" school of Arab geographers rather than to the Islamic Balkhī school, the followers of which took Mecca as the center of the world and made their geographical ideas conform to the concepts found in the Koran. Having had a wide experience of the οἰκουμένη and endowed with a critical mind, al-Mas'ūdī was able to challenge some of the concepts of the "theoreticians" and to rectify the confused knowledge of the Arab geographers of his time. For instance, he was not fully convinced of the Ptolemaic theory of the existence of a *terra incognita* in the southern hemisphere, according to which the Indian Ocean was believed to be surrounded by land on all sides except in the east, where it was joined with the Pacific by a sea passage. He says he was told by the sailors of the Indian Ocean *(al-baḥr al-ḥabashī)* that this sea had no limits toward the south *(Murūj,* I, 281–282; see S. M. Ahmad, "The Arabs and the Rounding of the Cape of Good Hope," in *Dr. Zakir Husain Presentation Volume* [New Delhi, 1968]).

Al-Mas'ūdī was not a philosopher, but he was deeply interested in Greek philosophy, as is evident from his writings. As a theologian he refuted the Greek materialist concept that the world is eternal *(qadīm)* and argued in favor of the Islamic belief that the world had come into existence in time *(ḥadīth).* His style is simple and direct; but he made full use of Arabic poetry, which imparts a literary touch to his writings. There is little doubt that he was one of the most original thinkers of medieval Islam.

BIBLIOGRAPHY

I. ORIGINAL WORKS. Al-Mas'ūdī's extant works are *Murūj al-dhabab wa ma'ādin al-Jawhar, Les prairies d'or,* Arabic text and French trans. by C. Barbier de Meynard and Pavet de Courteille, 9 vols. (Paris, 1861–1877), also rev. ed. of the Arabic text by Charles Pellat, 3 vols. (Beirut, 1966–1970); and *Kitāb al-Tanbīh wa 'l-ishrāf,* M. J. de Goeje, ed., vol. VIII of Bibliotheca Geographorum Arabicorum (Leiden, 1893–1894), trans. into French by Carra de Vaux as *Le livre de l'avertissement et de la revision* (Paris, 1897).

II. SECONDARY LITERATURE. See S. Maqbul Ahmad, "Al-Mas'ūdī's Contributions to Medieval Arab Geography," in *Islamic Culture* (Hyderabad), **27**, no. 2 (Apr. 1953), 61–77, and **28**, no. 1 (Jan. 1954), 275–286; and "Travels of Abu 'l-Ḥasan 'Alī ibn al-Ḥusayn al-Mas'ūdī," *ibid.,* **28**, no. 4 (Oct. 1954), 509–524; C. Brockelmann,

"Al-Mas'ūdī," in *Encyclopaedia of Islam,* 1st ed. (Leiden, 1936); I. I. Krachkovsky, *Istoria arabskoy geograficheskoy literatury* (Moscow–Leningrad, 1957), trans. into Arabic by Ṣalāḥ al-Dīn 'Uthmān Hāshim as *Ta'rīkh al-adab al-jughrāfī al-'Arabī* (Cairo, 1963), pt. 1, 177–186; *Al-Mas'ūdī Millenary Commemoration Volume,* S. Maqbul Ahmad and A. Rahman, eds. (Aligarh, 1960); and M. Reinaud, *Géographie d'Aboulféda,* I, *Introduction générale à la géographie des orientaux* (Paris, 1848).

On the *'ajā'ib* literature, see S. M. Ahmad, "The Aligarh Manuscript of al-Mas'ūdī's *'Ajāib al-Dunyā,*" in *Majalla-i 'Ulūm-i Islāmīya* (1960), 102–110; C. Brockelmann, *Geschichte der arabischen Literatur,* I (Leiden, 1943), 150–152 and supp. I (Leiden, 1937), 220–221; Carra de Vaux, in *Journal Asiatique,* 9th ser., **7;** Ḥājjī Khalīfa, *Kashf al-ẓunūn,* II (Istanbul, 1943), 1126; Quṭbuddīn Collection, no. 36/1, Maulana Azad Library, Aligarh; M. Reinaud, *Géographie d'Aboulféda,* I, *Introduction générale à la géographie des orientaux* (Paris, 1848); and P. Voorhoeve, *Handlist of Arabic Manuscripts* (Leiden, 1957), 4.

S. MAQBUL AHMAD

MATHER, WILLIAM WILLIAMS (*b.* Brooklyn, Connecticut, 24 May 1804; *d.* Columbus, Ohio, 25 February 1859), *geology.*

Mather was a member of the New England Mather family, famous in the ministry, education, and literature. A boyhood interest in chemistry led him to mineralogy. In 1823 he entered West Point Military Academy and graduated in 1828. He remained in the army for eight years, partly on regular duty and partly as assistant professor of chemistry, mineralogy, and geology. In 1833 he published *Elements of Geology for the Use of Schools,* which went through at least five editions. He was on topographic duty in 1835 as assistant geologist to G. W. Featherstonhaugh in making a geological study of the country from Green Bay, Wisconsin, to the Coteau des Prairies, in southwestern Minnesota. He resigned from the army in 1836 and Governor Marcy of New York appointed him geologist in charge of the first geological district of the New York survey. He completed his final report in 1843. During this time he organized the first geological survey of Ohio (1837–1840); was professor of natural science at Ohio University (1842–1845), and vice-president and acting president in 1845; acting professor of chemistry, mineralogy, and geology at Marietta College in 1846. He returned to Ohio University as vice-president and professor of natural science in 1847. From 1850 he was active as secretary and agricultural chemist of the Ohio State Board of Agriculture, as well as consultant on mineral resources for railroads in Kentucky and Ohio. An expert on coal geology, he had large interests in the development of Ohio coal lands.

"Not possessing the genius which dazzles, [Mather] had the intellect which achieved valuable results by patient and conscientious industry" (*Popular Science Monthly*, **49** [1896], 555). His most important geological work is embodied in his massive final report on the New York survey, which is notable not only for its vast detail of the structure and classification of the strata but also for the breadth of its coverage. His sections on the red rocks of the Catskill Mountains and his descriptions of the Quaternary deposits of the lower Hudson River valley and Long Island are still useful, although he did not accept Agassiz's glacial theory. He was one of the first to recognize that the much disputed Taconic rocks are mainly metamorphosed early Paleozoic sediments and that the rocks of the Hudson highlands correspond in age and lithology to the ancient rocks of the Adirondacks.

BIBLIOGRAPHY

I. ORIGINAL WORKS. Mather's most important work is *Geology of New York. Part I, Comprising the Geology of the First Geological District* (Albany, 1843).

II. SECONDARY LITERATURE. See the unsigned "Sketch of Williams Mather," in *Popular Science Monthly*, **49** (1896), 550–555, with portrait; I. J. Austin, "William Williams Mather," in *New England Historic-Genealogical Society, Memorial Biographies*, **3** (1883), 339–355; C. H. Hitchcock, "Sketch of William Williams Mather," in *American Geologist*, **19** (1897), 1–15, with portrait and a list of his publications; and Charles Whittlesey, "Personnel of the First Geological Survey of Ohio," in *Magazine of Western History*, **2** (1885), 73–87.

JOHN W. WELLS

MATHEWS, GEORGE BALLARD (*b*. London, England, 23 February 1861; *d*. Liverpool, England, 19 March 1922), *mathematics*.

Born of a Herefordshire family, Mathews was educated at Ludlow Grammar School; at University College, London; and at St. John's College, Cambridge. In 1883 he headed the list in the Cambridge mathematical tripos. In 1884 he was elected a fellow of St. John's, but in the same year he was appointed to the chair of mathematics at the newly established University College of North Wales at Bangor. He resigned the Bangor chair in 1896 and returned to lecture at Cambridge. He gave up this appointment in 1911, when he was appointed to a special lectureship at Bangor. He was elected to the Royal Society in 1897.

Mathews was an accomplished classical scholar; and besides Latin and Greek he was proficient in Hebrew, Sanskrit, and Arabic. He also possessed great musical knowledge and skill. His versatility led a colleague at Bangor to assert that Mathews could equally well fill four or more chairs at the college.

In mathematics Mathews' main interest was in the classical theory of numbers, and most of his research papers deal with topics in this field. His book on the theory of numbers, of which only the first of two promised volumes appeared, discusses in detail the Gaussian theory of quadratic forms and its developments by Dirichlet, Eisenstein, and H. J. S. Smith; it also contains a chapter on prime numbers that is concerned largely with describing Riemann's memoir, at that time little known in England. Since the book was published in 1892, it was not possible to mention the proofs of the prime number theorem, first given by Hadamard and Vallée Poussin in 1896. In a related field, his 1907 tract on algebraic equations gave a clear exposition of the Galois theory in relation to the theory of groups.

A collaboration with Andrew Gray, then professor of physics at Bangor, produced a book on Bessel functions, the first substantial text on this subject in English. The theory is developed carefully and rigorously, but throughout the book stress is laid on applications to electricity, hydrodynamics, and diffraction; in this respect the book retained its value even after the publication in 1922 of Watson's standard treatise on the theory of these functions.

Mathews' book on projective geometry had two main aims: first, to develop the principles of projective geometry without any appeal to the concept of distance and on the basis of a simple but not minimal set of axioms; and second, to expound Staudt's theory of complex elements as defined by real involutions. Much material on the projective properties of conics and quadrics is included. The topics were relatively novel in English texts, although the first volume of Oswald Veblen and J. W. Young's *Projective Geometry* had just become available.

Mathews' research papers advanced the study of higher arithmetic, and his books were equally valuable, since they gave English readers access to fields of study not then adequately expounded for the English-speaking world.

BIBLIOGRAPHY

I. ORIGINAL WORKS. Mathews' books are *Theory of Numbers* (Cambridge, 1892); *A Treatise on Bessel Func-*

tions and Their Applications to Physics (London, 1895; 2nd ed., rev. by T. M. MacRobert), written with A. Gray; *Algebraic Equations*, Cambridge Mathematical Tracts, no. 6 (Cambridge, 1907; 3rd ed., rev. by W. E. H. Berwick); and *Projective Geometry* (London, 1914). The 2nd ed. of R. F. Scott's *Theory of Determinants* (London, 1904) was revised by Mathews.

Most of Mathews' research papers were published in *Proceedings of the London Mathematical Society* and *Messenger of Mathematics*.

II. SECONDARY LITERATURE. See the obituary notices by W. E. H. Berwick, in *Proceedings of the London Mathematical Society*, 2nd ser., **21** (1923), xlvi–l; and A. Gray, in *Mathematical Gazette*, **11** (1922), 133–136.

T. A. A. BROADBENT

MATHIEU, ÉMILE LÉONARD (*b.* Metz, France, 15 May 1835; *d.* Nancy, France, 19 October 1890), *mathematics, mathematical physics.*

Mathieu showed an early aptitude for Latin and Greek at school in Metz; but in his teens he discovered mathematics, and while a student at the École Polytechnique in Paris he passed all the courses in eighteen months. He took his *docteur ès sciences* in March 1859, with a thesis on transitive functions, but had to work as a private tutor until 1869, when he was appointed to a chair of mathematics at Besançon. He moved to Nancy in 1874, where he remained as professor until his death.

Although Mathieu showed great promise in his early years, he never received such normal signs of approbation as a Paris chair or election to the Académie des Sciences. From his late twenties his main efforts were devoted to the then unfashionable continuation of the great French tradition of mathematical physics, and he extended in sophistication the formation and solution of partial differential equations for a wide range of physical problems. Most of his papers in these fields received their definitive form in his projected *Traité de physique mathématique*, the eighth volume of which he had just begun at the time of his death. These volumes and a treatise on analytical dynamics can be taken as the basis for assessing his achievements in applied mathematics, for they contain considered versions, and often extensions, of the results that he had first published in his research papers. Mathieu's first major investigation (in the early 1860's) was an examination of the surfaces of vibration that arise as disturbances from Fresnel waves by considering the dispersive properties of light. His later interest in the polarization of light led him to rework a number of problems in view of certain disclosed weaknesses in Cauchy's analyses.

One of Mathieu's main interests was in potential theory, in which he introduced a new distinction between first and second potential. "First potential" was the standard idea, defined, for example, at a point for a body V by an expression of the form

$$\int_V \frac{1}{r} f(x, y, z)\, dv; \tag{1}$$

but Mathieu also considered the "second potential"

$$\int_V r f(x, y, z)\, dv, \tag{2}$$

the properties of which he found especially useful in solving the fourth-order partial differential equation

$$\nabla^2 \nabla^2 w = 0. \tag{3}$$

His interest in (3) arose especially in problems of elasticity; and in relating and comparing his solutions with problems in heat diffusion (where he had investigated various special distributions in cylindrical bodies), he was led to generalized solutions for partial differential equations and to solutions for problems of the elasticity of three-dimensional bodies, especially those of anisotropic elasticity or subject to noninfinitesimal deformations. Mathieu applied these results to the especially difficult problem of the vibration of bells, and he also made general applications of his ideas of potential theory to the study of dielectrics and magnetic induction. In his treatment of electrodynamics he suggested that the traversal of a conductor by an electric current gave rise to a pair of neighboring layers of electricity, rather than just a single layer.

Mathieu introduced many new ideas in the study of capillarity, improving upon Poisson's results concerning the change of density in a fluid at its edges. His most notable achievement in this field was to analyze the capillary forces acting on an arbitrary body immersed in a liquid, but in general his results proved to be at variance with experimental findings.

In celestial mechanics Mathieu extended Poisson's results on the secular variation of the great axes of the orbits of planets and on the formulas for their perturbation; he also analyzed the motion of the axes of rotation of the earth and produced estimates of the variation in latitude of a point on the earth. Mathieu studied the three-body problem and applied his results to the calculation of the perturbations of Jupiter and Saturn. In analytical mechanics, he gave new demonstrations of the Hamiltonian systems of equations and of the principle of least action, as well as carrying out many analyses of compound motion, especially those that took into account the motion of the earth.

In all his work Mathieu built principally on solution methods introduced by Fourier and problems investigated by Poisson, Cauchy, and Lamé. The best-known of his achievements, directly linked with his name, are the "Mathieu functions," which arise in solving the two-dimensional wave equation for the motion of an elliptic membrane. After separation of the variables, both space variables satisfy an ordinary differential equation sometimes known as Mathieu's equation:

$$\frac{d^2u}{dz^2} + (a + 16b\cos 2z)u = 0, \qquad (4)$$

whose solutions are the Mathieu functions $ce_n(z, b)$, $se_n(z, b)$. These functions are usually expressed as trigonometric series in z, each of which takes an infinite power series coefficient in b; but many of their properties, including orthogonality, can be developed from (4) and from various implicit forms. Both equation (4) and the functions were an important source of problems for analysis from Mathieu's initial paper of 1868 until the second decade of the twentieth century. The functions themselves are a special case of the hypergeometric function, and Mathieu's contributions to pure mathematics included a paper on that function. He also wrote on elliptic functions and especially on various questions concerned with or involving higher algebra—the theory of substitutions and transitive functions (his earliest work, and based on extensions to the results of his thesis) and biquadratic residues. In fact, his earliest work was in pure mathematics; not until his thirties did applied mathematics assume a dominant role in his thought.

Mathieu's shy and retiring nature may have accounted to some extent for the lack of worldly success in his life and career; but among his colleagues he won only friendship and respect. Apart from a serious illness in his twenty-eighth year, which seems to have prevented him from taking over Lamé's lecture courses at the Sorbonne in 1866, he enjoyed good health until the illness that caused his death.

BIBLIOGRAPHY

I. ORIGINAL WORKS. The 7 vols. of Mathieu's *Traité de physique mathématique* were published at Paris: *Cours de physique mathématique* (1874); *Théorie de la capillarité* (1883); *Théorie du potentiel et ses applications à l'électrostatique et au magnetisme*, 2 vols. (1885–1886), also trans. into German (1890); *Théorie de l'électrodynamique* (1888); and *Théorie de l'élasticité des corps solides*, 2 vols. (1890). The 3 vols. that were still projected at his death were to

have dealt with optics, the theory of gases, and acoustics. His other book was *Dynamique analytique* (Paris, 1878). His papers were published mostly in *Journal de physique*, *Journal für die reine und angewandte Mathematik*, and especially in *Journal des mathématiques pures et appliquées*. A comprehensive list of references can be found in Poggendorff, IV, 1972.

II. SECONDARY LITERATURE. The two principal writings are P. Duhem, "Émile Mathieu, His Life and Works," in *Bulletin of the New York Mathematical Society*, **1** (1891–1892), 156–168, translated by A. Ziwet; and G. Floquet, "Émile Mathieu," in *Bulletin . . . de la Société des sciences de Nancy*, 2nd ser., **11** (1891), 1–34. A good treatment of Mathieu's equation and functions may be found in E. T. Whittaker and G. N. Watson, *A Course of Modern Analysis* (Cambridge, 1928), ch. 19.

I. GRATTAN-GUINNESS

MATHURĀNĀTHA ŚARMAN (*fl.* Bengal, India, 1609), *astronomy*.

Mathurānātha, who enjoyed the titles Vidyālaṅkāra ("Ornament of Wisdom") and Cakravartin ("Emperor"), composed the *Ravisiddhāntamañjarī* in 1609. This is an astronomical text in four chapters accompanied by extensive tables (see Supplement) based on the parameters of the *Saurapakṣa* with the admixture of some from the adjusted *Saurapakṣa*; the epoch is 29 March 1609. His is one of the primary sets of tables belonging to this school in Bengal. Another set of tables, the *Viśvahita*, is sometimes attributed to him, but its author is, rather, Rāghavānanda Śarman. Mathurānātha also wrote a *Praśnaratnāṅkura* and a *Pañcāṅgaratna*, but little is known of either.

BIBLIOGRAPHY

The *Ravisiddhāntamañjarī* was edited by Viśvambhara Jyotiṣārṇava as *Bibliotheca Indica*, no. 198 (Calcutta, 1911); the tables are analyzed by D. Pingree, *Sanskrit Astronomical Tables in England* (Madras, 1973), 128–134.

DAVID PINGREE

MATRUCHOT, LOUIS (*b.* Verrey-sous-Salmasse, near Dijon, France, 14 January 1863; *d.* Paris, France, 5 July 1921), *mycology*.

Matruchot was admitted to the École Normale Supérieure in 1885. After passing the *agrégation* he became assistant science librarian there in 1888. In 1901 he was named lecturer in botany at the Sorbonne, where he also held a professorship in mycology. He was a member of the Société Biologique and of the Société Mycologique; as president of the latter he presented an honorary membership to George Safford Torrey of Harvard. His doctoral thesis, on the Mucedinaceae, was published in 1892.

Matruchot applied Pasteur's techniques to the study of the effects of various culture media upon the polymorphism and reproduction of fungi. He was thereby able to demonstrate the facultative parasitism of *Melanosporum parasitica;* he also found three different forms of *Bulgaria sarcoides*—having solitary, coalescent, and sterile mycelia, respectively—and discovered the relationship between *Cladobotryum ternatum* and *Graphium penicilloides.* Having identified the perfect stage of *Gliocadium,* he was able to place it among the Perisporaceae; he identified *Cunninghamella africana* through the use of *Pitocephalus,* a parasite of the Mucoraceae family, as a biological indicator. Matruchot also showed symbiotic association between *Gliocephalis hyalina* and a bacterium and developed a cytological technique by which a Mortierella was inoculated with such pigmented organisms as *Bacillus violaceus* or fusaria; he was then able to demonstrate some constituents of the host in the extracted pigment.

Matruchot's research on fungi pathogenic to men and animals opened a new field of medical investigation. He showed that certain infections that had previously been treated as lymphatic tuberculosis or syphilis were in fact fungal in nature. He found the yeast stage of *Sporotrichum gougeroti* in an infected leg muscle; discovered the fungus that caused subcutaneous nodules in a forearm (naming it *Mastigocladium blochii*); described a *Trichophyton* that is pathogenic in horses and that resembles Gymnoascaceae in its conidia, although it mimics Ctenomyces in its perfect stage; and finally discovered *Microbacillus synovialis* to be the cause of a condition similar to acute arthritis.

Matruchot also obtained the conidial stage of *Cryptococcus farcimonosus* in a medium containing sugar, which he kept at 25° C., and he was first to make pure cultures of *Phytospora infestans,* the agent of potato blight, for which he was awarded the Prix Bordin in 1911. He developed new techniques for the cultivation of mushrooms and truffles, and was honored by the Academy of Agriculture for having discovered the cause of the pollution of the Étang des Suisses in the park of Versailles.

In addition to his work as a mycologist, Matruchot wrote outlines in chemistry and physics for the use of his students, and contributed to archaeological research in his native province, the Côte-d'Or. He died of appendicitis at the age of fifty-eight.

BIBLIOGRAPHY

I. ORIGINAL WORKS. A complete bibliography of Matruchot's scientific works is in Costantin (see below). His doctoral thesis was published as *Recherches sur le développement de quelques mucédinées* (Paris, 1892); and two course outlines, *Livret de chimie* (Paris, 1897) and *Livret de physique* (Paris, 1897). His papers include "Sur la culture de quelques champignons Ascomycètes," in *Bulletin de la Société mycologique de France,* **9** (1893), 246–249; "Sur la structure du protoplasma fondamental dans une espèce de *Mortierella,*" in *Comptes rendus hebdomadaires des séances de l'Académie des sciences,* **123** (1896), 1321; "Gliocéphalis hyalina," in *Bulletin de la Société mycologique de France,* **15** (1899), 254–262; "Sur la culture pure du *Phytophora infestans,* de Bary, agent de la maladie de la pomme de terre," *ibid.,* **16** (1900), 209–210, written with Marin Molliard; "Une mucorinée purement conidienne, *Cunninghamella africana,* étude éthologique et morphologique," in *Revue mycologique,* **26** (1904), 83–85; "Sur un nouveau groupe de champignons pathogènes agents des Sporotrichoses," in *Comptes rendus hebdomadaires des séances de l'Académie des sciences,* **150** (1910), 543–545; "Études sur les mauvaises odeurs de la pièce d'eau des Suisses à Versailles: nature, origine, causes et remèdes," in *Comptes rendus de la Société biologique de France,* **75** (1913), 611, written with M. Desroche; "Un microbe nouveau, *Mycobacillus synoviale* causant chez l'homme une maladie évoluant comme le rhumatisme articulaire," in *Comptes rendus hebdomadaires des séances de l'Académie des sciences,* **164** (1917), 652–655; and "Sur la forme conidienne du champignon agent de la lymphangite épizoolique," in *Bulletin de la Société mycologique de France,* **38,** supp. (1921), 76–77, written with Brocq-Rousseu.

II. SECONDARY LITERATURE. M. J. Costantin's obituary notice of Matruchot is in *Bulletin de la Société mycologique de France,* **38** (1922), 127–139. See also Paul Portier, "Louis Matruchot," in *Comptes rendus de la Société biologique de France,* **85** (1921), 322–323.

DENISE MADELEINE PLOUX, S.N.J.M.

MATTEUCCI, CARLO (*b.* Forlì, Italy, 2 June 1811; *d.* Leghorn, Italy, 24 June 1868), *physiology, physics.*

The son of Vincenzo Matteucci, a physician, and Chiara Folfi, Matteucci attended the University of Bologna from 1825 to 1828, when he graduated with a degree in physics. At the age of sixteen he prepared his first published paper, on meteorology.

In October 1829 Matteucci went to Paris, at his father's expense, and spent eight months attending scientific lectures at the Sorbonne. He returned to Forlì in June 1830. He received his first academic appointment in 1840, when he was appointed professor of physics at the University of Pisa on the advice of

Alexander von Humboldt. In the meantime he carried out electrophysiological investigations, first at his father's home and later in a small laboratory in Ravenna. His reports, presented to the Académie des Sciences by Arago and Edmond Becquerel, made his "name known through the European Continent," as Faraday wrote when Matteucci was only twenty-two. In 1842, at Pisa, he made his most famous discovery, the induced twitch, and began publishing his works in English, with a series of memoirs sent to the Royal Society.

A liberal, Matteucci was involved in the great political upheaval of 1848, which spread through most of the Continent. He was sent by the new Tuscan government to the Frankfurt Parliament, with the aim of establishing contacts with the German liberals. He did not lose his chair of physics after the restoration; in fact, the grand duke of Tuscany gave him —a political adversary—the funds necessary to build the first large institute of physics. When Italy was united, Matteucci became a senator for life and in 1862, as minister of education, reorganized the Scuola Normale Superiore of Pisa as the first Italian institute for advanced studies.

Working on torpedoes, Matteucci found that the discharges of the electric organ of the fish were due to impulses arising in the fourth lobe of the medulla, which he called the electric lobe. Mechanical or galvanic stimulation of this lobe constantly produced the electric discharge, the sign of which was not reversed when the direction of the stimulating current was reversed. This discharge could also be produced reflexly by applying pressure to the fish's eyes or by stimulating its body. Both spontaneous and induced discharges occurred after ablation of the cerebral hemispheres, of the optic lobes, and the cerebellum; but they could not be observed after bilateral ablation of the electric lobe.

The existence of the phenomenon of animal electricity had already been definitely proved with Galvani's last experiment (1797), and it is now clear that the twitch he produced was due to the currents of injury of sciatic nerve fibers, a phenomenon made possible by the polarization of their intact membranes. It remained for Matteucci to prove in 1842 that a current could constantly be detected when the electrodes of the galvanometer were placed in contact with the intact surface and with the interior (wounded portion) of a muscle. That the wounded part was always negative with respect to the normal surface was later demonstrated by Emil du Bois-Reymond, who in 1843 confirmed Matteucci's observation.

Although in 1838 Matteucci had recognized that the frog's resting potential (the difference in potential between the interior and exterior of a muscle fiber at rest) was, paradoxically, abolished by strychnine tetanus, it remained for du Bois-Reymond to demonstrate in 1848 that during the muscle contraction there was a "negative variation" of the injury currents. Just a year before the first preliminary note by du Bois-Reymond (1843), however, Matteucci had presented to the Académie des Sciences his report on the "induced twitch." In an experimental demonstration he gave at Paris in the presence of Humboldt and of several members of the Academy, he showed that when the sciatic nerve of the galvanoscopic frog leg was placed upon the leg muscle of another frog, the contraction of the latter muscle "induced" the contraction of the galvanoscopic leg. Becquerel immediately repeated and confirmed this experiment and gave the correct interpretation: that the nerve of the galvanoscopic leg was stimulated by the action currents of the contracting muscle of the other frog. Matteucci accepted this interpretation but failed to see the relation between his discovery and the phenomenon of the negative variation. This was done for the first time in Johannes Müller's *Handbuch der Physiologie des Menschen* (1844) and in du Bois-Reymond's *Untersuchungen über thierische Electricität* (1848–1849).

Although Matteucci was scientifically active until the end of his life, his most important work in physiology was carried out between 1836 and 1844 on the neural mechanisms of the electric discharge of torpedoes, on the resting potential of the frog's muscle, and especially on the action currents, which he discovered. His reluctance to admit that an electric phenomenon could disappear as a consequence of an active physiological process (the "negative variation") led Matteucci in 1845 to reject his own galvanometric findings of 1838 and Becquerel's interpretation of the induced twitch. This was the major error in the life of an outstanding scientist, whose work greatly influenced nineteenth-century electrophysiology.

BIBLIOGRAPHY

A list of 269 papers written by Matteucci is in Royal Society *Catalogue of Scientific Papers*, IV, 285–293; VIII, 354–355. His major books are *Essai sur les phénomènes électriques des animaux* (Paris, 1840); and *Traité des phénomènes électro-physiologiques des 'animaux* (Paris, 1844).

On Matteucci and his work, see N. Bianchi, *Carlo Matteucci e l'Italia del suo tempo* (Turin, 1874), with a complete bibliography of his works; and G. Moruzzi, "L'opera elettrofisiologica di Carlo Matteucci," in *Physis*, 6 (1964), 101–140, with a partial listing of his works and bibliography of secondary literature.

GIUSEPPE MORUZZI

MATTHIESSEN, AUGUSTUS (*b*. London, England, 2 January 1831; *d*. London, 6 October 1870), *chemistry*.

As a young child Matthiessen suffered a seizure that left him handicapped with a permanent twitching of the right hand. As a result he was considered unfit physically for most careers and was sent to learn farming. During his three-year stay on a farm, he nevertheless developed an interest in chemistry. At this time agricultural chemistry was attracting wide attention as a result of the writings of Liebig. It therefore was natural for Matthiessen to choose the University of Giessen to continue his education in chemistry, and he took his Ph.D. there in 1853.

From 1853 to 1857 Matthiessen worked in Bunsen's laboratory at the University of Heidelberg. Under Bunsen's direction Matthiessen prepared significant quantities of lithium, strontium, magnesium, and calcium by electrolysis of their fused salts. He then carried out a study with Kirchhoff on the electrical conductivity of these metals and of sodium and potassium.

In 1857 Matthiessen left Heidelberg and returned to London. He worked for a few months in Hofmann's laboratory at the Royal College of Chemistry. Here he studied the steps in the action of nitrous acid on aniline. Soon he moved to a small laboratory in his home and worked there for four years. At this time he began one of his most important studies, on the chemistry of narcotine and related opium alkaloids. In 1861 Matthiessen was elected a fellow of the Royal Society. He held the lectureship in chemistry at St. Mary's Hospital Medical School from 1862 to 1868, and then became a lecturer at St. Bartholomew's Medical School, both schools of the University of London. During this time he pursued research on the electrical, physical, and chemical properties of metals and their alloys. From 1862 to 1865 he served on the British Association Committee on the Standards of Electrical Resistance. Augustus Matthiessen committed suicide on 6 October 1870.

BIBLIOGRAPHY

I. ORIGINAL WORKS. A list of Matthiessen's publications can be found in the Royal Society's *Catalogue of Scientific Papers*. Among his important works are the following: "On the Electric Conducting Power of the Metals of the Alkalies and Alkaline Earths," in *Philosophical Magazine*, **13** (1857), 81–90; "On the Electric Conducting Power of Copper and its Alloys," in *Proceedings of the Royal Society*, **11** (1860–1862), 126–131; and a series of papers on the chemical constitution of the opium alkaloids.

II. SECONDARY LITERATURE. Obituaries of Matthiessen appeared in: *Nature*, **2** (1870), 517–518; *Journal of the Chemical Society*, **24** (1871), 615–617; *American Journal of Science*, **101** (1871), 73–74; and *Proceedings of the Royal Society*, **18** (1870), 111. An account of the details of his suicide appeared in *The Times* (London), 8 October 1870, p. 5, col. 5.

DANIEL P. JONES

MATTIOLI, PIETRO ANDREA GREGORIO (*b*. Siena, Italy, 12 March 1501; *d*. Trento, Italy, January/February 1577), *medicine, botany*.

Mattioli was the son of Francesco Mattioli, a physician, and Lucrezia Buoninsegni. After moving with his family to Venice, where his father practiced medicine, Mattioli was sent to Padua and began the study of Greek and Latin, rhetoric, astronomy, geometry, and philosophy. He soon developed an interest in medicine and natural history, his main concerns until he received a degree in medicine at the University of Padua in 1523.

After his father's death Mattioli returned with his mother to Siena, which was then experiencing civil unrest. He therefore left Siena and, wishing to improve his skill in surgery, moved to Perugia, where he studied under Gregorio Caravita. About 1520 he moved to Rome, where he attended the Santo Spirito Hospital and the San Giacomo Xenodochium for incurables, where he frequently dissected cadavers of syphilis victims. Mattioli also continued his interest in natural history and botany, making direct observations of herbs and plants. Following the sack of Rome in 1527, he moved to Trento.

In 1528, at Cles in Val di Non, Trentino–Alto Adige, Mattioli married a young woman named Elisabetta; in 1545 they had a son, Paolo (Pavolino), who died in childhood. During his stay in Trentino, where he began to practice medicine, Mattioli became an intimate friend, adviser, and physician to Cardinal Bernardo Clesio, bishop of Trento, who developed a great esteem for Mattioli. While in Trentino, Mattioli continued his observations of plants and wrote his first book.

After Cardinal Clesio's death in 1539, Mattioli moved to Gorizia, apparently at the request of the inhabitants of that city, to practice medicine. His medical and natural history interests had increased and had developed particularly in phytology, both through the study of books and through direct observations of plants; in 1544 he published *Di Pedacio Dioscoride anazarbeo libri cinque*, which, through revisions and expansions, made him famous. In 1554 Mattioli was called to Prague, where he served first at the court of Ferdinand I and then at that of Maximilian II.

After the death of his first wife in 1557, Mattioli married Girolama di Varmo, of a noble Friuli family; they had two sons, Ferdinando and Massimiliano. In 1570 he married his third wife, Susanna Cherubina, of Trentino, by whom he had three children—Pietro Andrea, Lucrezia, and Eufemia.

In 1570, after visiting Verona, Mattioli left Prague and returned to the Tirol. He died of the plague at Trento apparently in January or February 1577. Buried in the cathedral, he is commemorated by a monument bearing his effigy in bas-relief.

Endowed with a wide-ranging knowledge, Mattioli concerned himself with a great variety of subjects, most of them involving botany and materia medica.

De morbi gallici curandi ratione, dialogus (Bologna, 1530), written in 1528 during his stay in Trentino and reprinted several times, was Mattioli's first work. A traditional examination of the origins and treatment of syphilis, it deals with the modes of propagation and symptomatology, including buboes in the groin, which, according to some authors, Mattioli was the first to describe. The discussion centers on the efficacy of the potion obtained from guaiacum, or holy wood. In view of the large number of works dealing with holy wood, it is difficult to recognize any original contribution that Mattioli may have made.

Magno palazzo del cardinale di Trento (Venice, 1539), an elegant poem of some 450 octaves, is a description in verse of the cultured and humanistic environment of the palace of Cardinal Clesio.

Geografia di Claudio Ptolemeo Alessandrino (Venice, 1548), another nonmedical book, is an Italian translation of Ptolemy's *Geography* and one of the earliest Italian translations from Latin of classical scientific writing.

Apologia adversus Amathum Lusitanum cum censura in eiusdem enarrationes (Venice, 1558) is a short work mainly polemical in content.

Epistola de bulbocastaneo . . . (Prague, 1558) illuminates Mattioli's scientific personality through his new botanical methods and his contributions regarding the identification of plants mentioned by the ancients, their synonyms and proper names, and their spelling.

Epistolarum medicinalium libri quinque (Prague, 1561) contains the names of celebrated contemporary scientists, notably Konrad Gesner, Ulisse Aldrovandi, Francesco Calzolari, Giacomo Antonio Cortuso, and Gabriele Falloppio. It is a series of writings on various medical subjects, including alchemy, magnetism, pharmacology, and the causes and symptoms of diseases. Mattioli deals fully with plants in general

—identification, curative powers—and with medicinal plants in particular, listing properties and habitats, and outlining methods for collecting and preserving samples.

Commentarii a Dioscoride is the work with which Mattioli's name is chiefly linked. The first edition, *Di Pedacio Dioscoride anazarbeo libri cinque. Dell'historia, et materia medicinale tradotti in lingua volgare italiana da M. Pietro Andrea Mathiolo Sanese medico. Con amplissimi discorsi, et comenti, et doctissime annotationi, et censure del medesimo interprete . . .* (Venice, 1544), is an Italian version of Dioscorides' *De materia medica* ($\Pi\epsilon\rho\grave{\iota}$ $\mathring{\upsilon}\lambda\eta\varsigma$ $\mathring{\iota}\alpha\tau\rho\iota\kappa\mathring{\eta}\varsigma$). Mattioli's original purpose was relatively modest: it was to provide doctors and apothecaries with a practical treatise in Italian with a commentary that would enable them to identify the medicinal plants mentioned by Dioscorides. This first edition, without illustrations, probably had a limited circulation. The highly successful Venice edition of 1548, also in Italian and with a new commentary, was entirely rewritten and was reprinted in 1550 and 1552.

It was probably this success that induced Mattioli to publish his first Latin translation of Dioscorides' *Commentarii, in libros sex Pedacii Dioscoridis anazarbei, De medica materia. Adjectis quam plurimis plantarum et animalium imaginibus, eodem authore* (Venice, 1554), an edition of broader purpose than the previous ones. Unlike the Italian versions, this Latin edition —enriched by synonyms in various languages, provided with a special commentary, and accompanied by numerous illustrations valuable for the reader's identification of Dioscorides' simples—rendered the work accessible to scholars throughout Europe. From then Mattioli's name was linked with that of Dioscorides. Further editions and reprints of the *Commentarii* continued practically without interruption until the eighteenth century. There were versions also in German, French, and Bohemian.

A critical examination of the features and the sometimes complex vicissitudes of the *Commentarii* would constitute an interesting chapter in the history of bibliology. Fundamental to the work's success is its conception and execution as a practical scientific treatise. It was intended for daily use by physicians, herbalists, and others, who could find descriptions and notes on medicinal plants and herbs, Greek and Latin names and synonyms, and the equivalents in other languages. The work made it possible to identify and compare its plants and herbs with those mentioned by Dioscorides and also with those found in nature.

The *Commentarii* thus differed profoundly from translations by other authors, who generally insisted on lexical and grammatical aspects rather than on medical and botanical aims. Mattioli supported his work with new information, partly derived from his direct observation of plants and herbs and partly obtained from other authors. Many of the illustrations were reproductions of his own drawings or elaborations of drawings made by other authors; the rest were derived from original drawings placed at his disposal by other scholars.

From the scientific point of view, Mattioli's work did not always win approval. Sachs, for example, asserted that Mattioli's study of the medicinal effects of plants took priority to the observation of their morphological characteristics. Certainly Mattioli's interest in botany was not primary but proceeded from his interest in therapy, and it was medicine that led him back to the observation of nature. Mattioli's commentary on Dioscorides' text was aimed largely at the practical purpose of medicinal phytognosis and acquired intrinsic value both through the wealth of its descriptive details of each plant and through its accurate drawings. Mattioli may therefore be considered a member of the Vesalian school of morphological observation.

BIBLIOGRAPHY

The following works, which were consulted in the writing of this article, are also of value as sources concerning Mattioli's writings, letters, MSS, and iconography, as well as the literature on him: Vincenzo Cappelletti, "Nota sulla medicina umbra del Rinascimento: Pietro Andrea Mattioli," in *Atti del IV Convegno di studi umbri* (Perugia, 1967), pp. 513–532; Jerry Stannard, "P. A. Mattioli: Sixteenth-Century Commentator on Dioscorides," in University of Kansas Libraries, *Bibliographical Contributions*, I (Lawrence, Kans., 1969), 59–81; Giovanni Battista de Toni, "Pierandrea Mattioli," in Aldo Mieli, ed., *Gli scienziati italiani dall'inizio del medio evo ai giorni nostri . . .*, I (Rome, 1921), 382–387; and *La vita di Pietro Andrea Mattioli*, collected from his works by Giuseppe Fabiani, edited with additions and notes by Luciano Banchi (Siena, 1872).

BRUNO ZANOBIO

MATUYAMA (MATSUYAMA), MOTONORI (*b.* Uyeda [now Usa], Japan, 25 October 1884; *d.* Yamaguchi, Japan, 27 January 1958), *physics, geophysics, geology.*

Matuyama was the son of a Zen abbot, Tengai Sumiye. In 1910 he was adopted by the Matsuyama family, whose daughter, Matsuye Matsuyama, he married. He altered the romanized spelling of his adoptive name in about 1926, in conformity with a then new convention of transliteration. Matuyama received his early education in the schools of Kiyosuye and Chōfu, then entered Hiroshima Normal College (now the University of Hiroshima), where he studied physics and mathematics, graduating in 1907. After a year of teaching at a junior high school in Tomioka, Matuyama entered the Imperial University in Kyoto to further his study of physics; he graduated in 1911, then took up postgraduate study of geophysics with Toshi Shida and astronomy with Shizō Shinjō. With Shida he began work on what became one of his chief fields of research, the determination of gravity by pendulum. Matuyama's first papers were written in collaboration with his teacher and constitute the third and sixth parts of Shida's "On the Elasticity of the Earth and the Earth's Crust" of 1912.

Matuyama was appointed lecturer at the Imperial University in 1913; three years later he was promoted to assistant professor of the Geophysical Institute. His doctoral dissertation, which he published in 1918, was entitled "Determination of the Second Derivatives of the Gravitational Potential on Jaluit Atoll," and contained the results of experiments performed with the Eötvös gravity-variometer to determine the depth of the atoll. Matuyama had spent a month on Jaluit with a collaborator, H. Kaneko, in February 1915; his paper was the first to suggest that the determination of microfeatures of the gravity field of the earth could reveal geological substructure. It became the basis for the development in Japan of the torsion-balance method of prospecting for underground minerals.

In 1919 Matuyama left Japan for the United States to study geophysics with T. C. Chamberlin at the University of Chicago. While there he conducted laboratory experiments on ice designed to illuminate the mechanics of glacial movement. His results were published in 1920 as "On Some Physical Properties of Ice." Chamberlin wrote an introductory note to the paper, in which Matuyama stated the conclusion:

> These facts seem to show that gliding planes parallel to the base of each crystal are not the controlling factor in the deformation of ice and probably are not even an important factor. But instead, adjustment along the contact surfaces of adjacent crystals and perhaps the development of planes of weakness in the constituent crystals parallel to their long axis seem more effective in the process of deformation.

Matuyama returned to Japan in December 1921; the following January he was appointed professor of theoretical geology at the Imperial University. He once again took up the determination of gravity by pendulum. As early as 1911, while he was still a postgraduate student, Matuyama had participated in the national gravity survey of the Imperial Japanese Geodetic Commission; from 1927 until 1932 he extended the survey to Korea and Manchuria. In October 1934 and October 1935 he made a survey of marine gravity, using the Vening-Meinesz pendulum apparatus mounted in a navy submarine, in the Japan Trench and the area surrounding it. During the same period he also determined the gravity of nine islands in the Caroline and Mariana groups and as part of his maritime survey of 1935 landed his equipment on Chichijima in the Bonin Islands to determine its gravity. He found the free-air anomalies in these ten islands to range between $+214$ and $+357$ milligal, a result consistent in magnitude with those established for other oceanic islands.

One aspect of Matuyama's research on gravity was dictated by the request put to the Japanese delegation by the International Union of Geodesy and Geophysics, probably at the meeting of 1936, to carry out a gravity survey in the water areas surrounding the Japan Trench. The sea bottom of the landward side of the deepest line of the trench is the site of strong earthquakes that produce tsunami waves on the Pacific coast of northeast Honshu. In examining the distribution of free-air anomalies over the trench, Matuyama discovered that in the region of the northeast Pacific coasts of Honshu and the Pacific coast of Hokkaido the axis of the negative minimum of anomalies does not occur just above the axis of the maximum depths of the trench, contrary to expectations, but rather shifts landward, while the two axes are almost coincident in the southern part of the trench along the east side of the Fuji volcanic range. (Matuyama's associate Naoiti Kumagai took up these results a few years later and discovered a clear correlation between the earthquakes that occur in this area and the isostatic anomalies of great magnitude that are also apparent there.)

Matuyama was further concerned with the study of distribution of various kinds of coral reefs in the South Seas. The peculiarity of this distribution had first interested him on his visit to Jaluit atoll in 1915; he noted that in the Mariana group such reefs are elevated above sea level, while in the Marshalls they exist as atolls, or have subsided completely. He theorized that the ocean floor in this area had tilted eastward, and recommended to the Japanese Association for the Advancement of Science that a research commission be appointed to study the tilt of the sea bed in this area. A committee was appointed in 1934, and research stations were established on Saipan

and on Jaluit atoll; no reports of their observations were published, however.

The last of Matuyama's main areas of research was the remnant magnetization of rocks. The first specimen that he examined was a basalt block from Genbudō, Tazima, in western Japan; he then extended his investigation to thirty-six specimens of basalt, each taken from a different site in Japan, Korea, and Manchuria. He subjected each specimen to tests for remnant magnetization, using Gauss's analysis of the vertical component of the magnetic field of the earth originating within the earth. Matuyama published his results in a series of papers. The most important, published in 1929 and entitled "On the Direction of Magnetization of Basalt in Japan, Tyōsen [Korea] and Manchuria," offers his conclusion that "According to Mercanton the earth's magnetic field was probably in a greatly different or nearly opposite state in the Permo-Carboniferous and Tertiary Periods, as compared to the present. From my results it seems as if the earth's magnetic field in the present area has changed even reversing itself in comparatively short times in the Miocene and also Quaternary Epochs." This work was widely influential, and the term "Matuyama Epoch" was coined to indicate the paleomagnetic period—from about the late Pliocene to the middle Pleistocene—during which the direction of magnetic field of the earth is supposed to have been opposite to what it is at present.

Matuyama also wrote a series of early papers on seismology, a subject which he originally pursued with Shida, and a great number of miscellaneous papers and books addressed to laymen and students. He served the Imperial University as dean of the Faculty of Science from June 1936 until December 1937; he retired from teaching in 1944 and was made professor emeritus in 1946. In May 1949 Matuyama was appointed president of the University of Yamaguchi; the following year he was elected a fellow of the Japan Academy. He had a lifelong interest in the Noh drama, and organized Noh groups among his neighbors and colleagues in both Kyoto and Yamaguchi. He died following the onset of acute myelogenous leukemia.

BIBLIOGRAPHY

Matuyama's important papers include "Note on Hecker's Observation of Horizontal Pendulums" and "Change of Plumb Line Referred to the Axis of the Earth as Found From the Result of the International Latitude Observations," pts. 3 and 6, respectively, of Toshi Shida, "On the Elasticity of the Earth and the Earth's Crust," in *Memoirs of the College of Science and Engineering, Kyoto*

Imperial University, **4**, no. 1 (1912), both written with Shida; "Determination of the Second Derivatives of the Gravitational Potential on the Jaluit Atoll," in *Memoirs of the College of Science, Kyoto Imperial University*, **3** (1918), 17–68; "On Some Physical Properties of Ice," in *Journal of Geology*, **28** (1920), 607–631; "On the Gravitational Field at the Fushun Colliery, Manchuria," in *Japanese Journal of Astronomy and Geophysics*, **2** (1924), 91–102; "Probable Subterranean Intrusion of Magma to the North of Sakurajima Volcano," in *Proceedings of the Third Pan-Pacific Science Congress* (Tokyo, 1926), 782–783; "Torsion Balance Observation and its Value in Prospecting" [in Japanese], in *Tōyō-Gakugei-Zasshi*, **520** (1926), 476–494; "On the Subterranean Structure Around Sakurajima Volcano Considered From the State of Gravitational Fields," in *Japanese Journal of Astronomy and Geophysics*, **4** (1927), 121–138; "Gravity Measurements in Tyōsen and Manchuria," in *Proceedings of the Fourth Pacific Science Congress* (Djakarta, 1929), 745–747; "Study of the Underground Structure of Suwa Basin by Means of the Eötvös Gravity-Variometer," *ibid.*, 869–872; "On the Direction of Magnetization of Basalt in Japan, Tyōsen and Manchuria," *ibid.*, 567–569, and in *Proceedings of the Imperial Academy of Japan*, **5** (1929), 203–205; "Subterranean Structure of Takamati Oil Field Revealed by Gravitational Method," in *Japanese Journal of Astronomy and Geophysics*, **7** (1930), 47–81, written with H. Higasinaka; "Relative Measurements of Gravity in Japan, Tyōsen and Manchuria Since 1921," in *Travaux. Association internationale de géodésie*, Japan Report no. 2, **11** (1933), 1–6; "Measurements of Gravity Over Nippon Trench on Board H. I. M. Submarine Ro-57. Preliminary Report," in *Proceedings of the Imperial Academy of Japan*, **10** (1934), 625–628; "Distribution of Gravity Over the Nippon Trench and Related Areas," *ibid.*, **12** (1936), 93–95; and "Gravity Survey by the Japanese Geodetic Commission Since 1932," in *Travaux. Association internationale de géodésie*, Japan Report no. 2, **12** (1936), 1–8.

NAOITI KUMAGAI

MAUGUIN, CHARLES VICTOR (*b*. Provins, France, 19 September 1878; *d*. Paris, France, 25 April 1958), *crystallography, mineralogy*.

Mauguin was the son of a baker in Provins, a small city fifty miles southeast of Paris. He attended primary school in Provins and, from 1894 to 1897, the École Primaire of Melun, in order to become a teacher. He began his university career in Montereau, where he also taught at the elementary school, while preparing for the entrance examination for the École Normale of Saint-Cloud, which he attended from 1902 to 1904. At this school, established to train teachers for the *écoles normales primaires*, students received thorough instruction in mathematics, physics, chemistry, and the natural sciences. A brilliant student, Mauguin attracted the attention of L. J. Simon, who suggested that he work in his organic chemistry laboratory at the École Normale Supérieure in Paris. In 1910 Mauguin defended his doctoral thesis, "Les amides bromées-sodées et leur rôle dans la transposition d'Hofmann."

At the Sorbonne, Mauguin took the mathematics courses of Émile Picard, Poincaré, and Goursat and attended the lectures given there in 1905 by Pierre Curie on symmetry in physical phenomena—his first contact with crystallography. He then attended the courses of Frédéric Wallerant, the professor of mineralogy and crystallography. The latter was interested in liquid crystals, which had been discovered a few years previously by the German physicist Otto Lehmann. These organic substances, in a well-defined temperature interval between the liquid and the solid phase, present a curious state of matter that is characterized by the great fluidity of liquids and optical properties similar to those of crystals.

Attracted by crystallography, in 1910 Mauguin became Wallerant's assistant in the mineralogy laboratory of the Sorbonne in order to study liquid crystals. His memoirs on this subject published from 1910 to 1914 are fundamental to an understanding of the liquid-crystal state. He was named lecturer in mineralogy at the Faculty of Sciences of Bordeaux in 1912 and professor of mineralogy at the Faculty of Sciences of Nancy in 1913. He was mobilized in the infantry upon the declaration of war and later worked in a chemistry laboratory. At the end of 1919 he returned to the Sorbonne as a lecturer and in 1933 succeeded Wallerant in the chair of mineralogy, which he occupied until his retirement in 1948. All of his researches were concerned essentially with the diffraction of X rays by crystals.

Mauguin's marriage in 1907 to Louise Gaudebert was childless. Mme Mauguin became blind in 1930; and from then on, they led a retiring life and died within a few months of one another. Until the end of his life, Mauguin preserved an intellectual and youthful enthusiasm for science; his whole life was devoted to studies of extremely varied questions. Long interested in mathematics, especially group theory and Laplace-Fourier transforms, he later turned his interest to theoretical physics and in his last years paid special attention to biological work dealing with the genesis of life. He was a remarkable teacher who could clarify the most difficult questions. His great pleasure, before his wife's illness, was to go on excursions in the mountains and woods to search for plants; he was an excellent botanist and was president of the Mycological Society of France. He was also laureate of the Institut de France in 1922 and in 1928. In 1937 he was elected a member of the Académie des Sciences.

Mauguin's first researches in organic chemistry dealt with the amides $RCONH_2$, in which he replaced the two hydrogen atoms with one atom of bromine and one of sodium; the liberation of NaBr furnished isocyanates $O \cdot CN \cdot R$. He thus established a close link between the amides and the ureides, which play a role of great importance in the chemistry of the living cell. Throughout his life Mauguin was interested in the biological aspect of chemistry.

In the temperature region that characterizes the liquid-crystal state, the liquid crystal appears in the form of a turbid liquid, almost opaque when in a thick layer. Some scientists explained this cloudiness by the presence of an insoluble impurity. Mauguin showed that the liquid is made up of birefringent elements in random orientation so that it loses its transparency. Azoxyanisole and azoxyphenetole, which in the turbid phase are extremely fluid, become homogeneous and perfectly transparent when a uniform orientation is imposed on the birefringent elements, either by the action of a magnetic field or by suitable surface actions.

Placed between the poles of an electromagnet, the liquid-crystal phase behaves, optically, like strongly birefringent uniaxial crystal with optical axis parallel to the magnetic field. Mauguin made a complete optical study of this phenomenon, measured the indices of refraction, and reproduced all the classic experiments of optical crystallography.

Similarly, when melted between two completely clean glass plates, azoxyanisole yields a transparent homogeneous phase and behaves like a uniaxial crystal with optical axis perpendicular to the surface. If azoxyanisole is melted on a freshly cleaved flake of muscovite, the optical axis is parallel to the cleavage plane (001) in a direction (100) 30° from the plane of symmetry (010); this fact, unexpected at the time (1912), was explained by Mauguin after X rays showed that the plane of symmetry of mica is a plane of glide symmetry. If the melting takes place between the two surfaces of a cleavage, the two optical axes in contact with the two flakes form an angle of 60°; the result is a helicoidal structure, of which Mauguin made a detailed theoretical and experimental study. These results, established with remarkable rigor and clarity, excited great interest and are still valuable.

Upon returning to the laboratory after World War I, Mauguin, impressed with the importance for crystallography of the discoveries of Laue and the Braggs, devoted himself completely to X-ray crystallographic studies. In 1923 he published the atomic structure of cinnabar, followed by those of calomel and graphite. He next undertook long crystal-chemical researches on the micas and the chlorites, among which the unity of crystallographic properties is in contrast with the variety of chemical compositions. Mauguin determined the chemical composition of a great number of micas; their density; and, by means of X rays, the absolute values of their cell dimensions. He was thus able to determine the number of atoms contained in the unit cell. He established that the crystal motif for all the micas always includes twelve oxygen and fluorine atoms, although the number of cations varies within large limits and can be fractional, proving that the simple motif is not always repeated identically in the crystal (*Bulletin de la Société française de minéralogie*, **51** [1928], 285–332). In the chlorites the crystal motif always has eighteen oxygen atoms.

Mauguin worked a great deal on group theory; his memoir of 1931 represented the 230 Schönflies-Federov groups by simple symbols, describing the symmetry elements precisely and expressing directly the symmetry operations determined by X-ray diffraction. These symbols, slightly modified following a collaborative effort with C. Hermann, are universally employed and have made Mauguin's name familiar to crystallographers.

BIBLIOGRAPHY

The papers by Mauguin referred to in the text are "Étude des micas au moyen des rayons x," in *Bulletin de la Société française de minéralogie*, **51** (1928), 285–332; "La maille cristalline des chlorites," *ibid.*, **53** (1930), 279–300; and "Sur le symbolisme des groupes de répétition ou de symétrie des assemblages cristallins," in *Zeitschrift für Krystallographie*, **79** (1931), 542–558.

On Mauguin and his work see P. P. Ewald, *Fifty Years of X-Ray Diffraction* (Utrecht, 1962), pp. 335–340; and J. Wyart, "Ch. Mauguin 1878–1958," in *Bulletin de la Société française de minéralogie et de cristallographie*, **81** (1958), 171–172.

JEAN WYART

MAUNDER, EDWARD WALTER (*b.* London, England, 12 April 1851; *d.* London, 21 March 1928), *astronomy*.

Maunder attended the school attached to University College, London, and took some additional courses at King's College there. He then worked briefly in a London bank before taking the first examination ever given by the British Civil Service Commission for the post of photographic and spectroscopic assistant in the Royal Observatory at Greenwich. By appointing Maunder to this new post in 1873, the observatory—which had been concerned since its founding with positional astronomy—made a formal commitment to astrophysical observations. Maunder worked at Greenwich for forty years, largely under the direction of W. H. M. Christie; the primary task assigned him was the photographic observation of the sun and the subsequent measurement of sunspots.

Of Maunder's first wife, who died in 1888, little is known. His second wife, whom he married in 1895, was Annie S. D. Russell, a competent and active astronomer. In 1889 she graduated from Girton College, Cambridge, as Senior Optime in the Mathematical Tripos, thus earning the highest mathematical honor then available to women. In 1891 she was appointed "lady computer" at Greenwich, charged with examining and measuring the sunspot photographs taken by Maunder. Thenceforth, she worked closely with Maunder.

When the Greenwich record of sunspots was begun on 17 April 1874, the periodicity, equatorial drift, and variation of rotation with latitude of sunspots had already been established. In addition to verifying these facts, Maunder's daily photographs—taken first on wet plates, later on dry—made possible a search for other regularities. To this end Maunder tabulated various sunspot features, such as number, area, changes, position, and motion. From these data he drew conclusions concerning the relation between rotation period and latitude of sunspots, the position of the solar axis of rotation, the correlation between solar rotation and terrestrial magnetic disturbances, the variation in time of the mean spotted area of the sun, and the latitudinal distribution of sunspot centers. With a spectroscope attached to the observatory's great equatorial, Maunder observed solar prominences, the radial motion of stars, and the spectra of planets, comets, novae, and nebulae. The results were undistinguished.

Maunder traveled outside England to observe six solar eclipses. As a member of the official British party, Maunder photographed the corona from Carriacou in the West Indies in 1886, and from Mauritius in 1901. In 1905 he went to Canada as a guest of the Canadian government. The other three expeditions, under the auspices of the British Astronomical Association, were organized for the most part by Maunder himself. In 1896 they went to Vadsö, Norway, in 1898 to India, and in 1900 to Algeria. In India, Maunder and his wife made separate observations: she, with instruments of her own devising, photographed a coronal streamer extending to six solar radii—the longest ever photographed.

After his election as fellow in 1875, Maunder took an active part in the affairs of the Royal Astronomical Society, serving as council member for many years and secretary from 1892 to 1897. Despite its many advantages, this society did not satisfy the needs of many British astronomers. Therefore, in 1890, largely through the efforts of Maunder and his brother Thomas Frid Maunder, the British Astronomical Association was founded. Its purposes were twofold: "To meet the wishes and needs of those who find the subscription of the R.A.S. too high or its papers too advanced, or who are, in the case of ladies, practically excluded from becoming Fellows" and "to afford a means of direction and organization in the work of observation to amateur astronomers." For the rest of his life Maunder supported this popular organization, serving as president in 1894–1896 and as director at various times of the Mars section, the solar section, and the colored star section. He edited the association's *Journal* for about ten years; during his presidency this job was undertaken by his wife. Previously, from 1881 to 1887, he had edited *Observatory*, the journal founded by Christie.

The literary output of Maunder and his wife was prodigious. The results of their astronomical observations were communicated primarily to the *Monthly Notices of the Royal Astronomical Society* and the *Journal of the British Astronomical Association*. For many years both *Nature* and *Knowledge* contained frequent articles by the Maunders on popular astronomy, astronomical researches, and the history of astronomy—notably astronomical records in the Bible.

BIBLIOGRAPHY

I. ORIGINAL WORKS. Maunder's books include *The Royal Observatory, Greenwich, A Glance at its History and Work* (London, 1902); *Astronomy Without a Telescope* (London, 1903), derived largely from articles in *Knowledge; Astronomy of the Bible: An Elementary Commentary on the Astronomical References of Holy Scripture* (London, 1908); *The Science of the Stars* (London–Edinburgh, 1912); *Are the Planets Inhabited?* (London–New York, 1913); *Sir William Huggins and Spectroscopic Astronomy* (London–Edinburgh, 1913); and the astronomical section of *The International Standard Bible Encyclopaedia* (Chicago, 1915). See also *The Heavens and Their Story* (London, 1910), written with his wife. Articles by the Maunders are listed in the Royal Society *Catalogue of Scientific Papers*, X, 749–750; XVII, 102; and the annual volumes of the *International Catalogue of Scientific Literature*.

II. SECONDARY LITERATURE. There are obituaries of Maunder by H. P. H. in *Journal of the British Astronomical Association*, **38** (1927–1928), 229–233 (see also 165–168);

and in *Monthly Notices of the Royal Astronomical Society*, **89** (1928–1929), 313–318. The obituary of A. S. D. R. Maunder, written by M. A. Evershed, appeared in the *Journal of the British Astronomical Association*, **57** (1946–1947), 238.

DEBORAH JEAN WARNER

MAUPAS, FRANÇOIS ÉMILE (*b*. Vaudry, Calvados, France, 2 July 1842; *d*. Algiers, Algeria, 18 October 1916), *zoology, biology.*

The son of Pierre-Augustin Maupas, deputy mayor of Vaudry, and of Marie Adèle Geffroy, Maupas attended the municipal secondary school and then entered the École des Chartes. In 1867 he became archivist of the Department of Cantal, where he developed an interest in natural science and, in particular, in free protozoans. He spent his vacations in Paris in order to work in the laboratories of the Muséum d'Histoire Naturelle and of the Sorbonne. In 1870 Maupas was appointed an archivist in Algiers and then curator of the Bibliothèque Nationale. In his leisure moments he pursued research in zoological microscopy at home with rudimentary instruments. Maupas never married, and he lived the life of an isolated researcher. On 27 June 1901 he was elected a corresponding member of the Academy of Sciences, and in 1903 he received an honorary doctorate from the University of Heidelberg.

Maupas's scientific work was devoted entirely to sexuality and reproduction among the protozoans, rotifers, nematodes, and oligochaetes, which he studied with the aid of extremely ingenious breeding and culture techniques. Among the ciliates he investigated cultures bearing on hundreds of generations. He carefully examined the phenomenon of conjugation, which he discovered, and demonstrated the overall unity of the process and of its secondary features. He likewise analyzed the transformations of the nucleus. Maupas at first believed that this phenomenon effected a rejuvenation of the offspring after a long period of multiplication that produced a fatal aging—a view contradicting Weismann's theories on the "immortality" of the infusorians. His observations were later confirmed by other protistologists—who showed, however, that senescence does not have the absolute and inevitable character that Maupas had attributed to it. He also demonstrated that there are two categories of females in the rotifers: those which are parthenogenetic, producing only females, and those which can also be fertilized. If the latter are not impregnated, they yield only males; if they are fertilized they will produce special eggs which undergo later development and always produce females.

Among the nematodes, Maupas studied free and parasitic species. Among the free forms (Rhabditida) he showed that the postembryonic development consisted of five stages, of which the fifth is the adult stage (Maupas's law); and he demonstrated that in the parasitic forms the third stage, which is part of the molt of the second, is the infesting form. He provided more precise data on the phenomenon of encystment and studied parthenogenesis and hermaphroditism with self-fertilization, as well as cross-fertilization. These investigations earned him the Grand Prize in physical sciences of the Institut de France (1901). With L. G. Seurat he made observations on the strongyles, parasitic nematodes which are agents of bronchopneumonia in sheep in Algeria, and specified the etiology of this disease. Some of Maupas's investigations of the nematodes were published posthumously (1919). He also worked on freshwater oligochaetes (*Nais, Dero, Pristina, Aelosoma*), of which he obtained cultures prolonged for several months.

A complete naturalist, Maupas also published geological and botanical observations. In addition, he translated important German scientific works. Maupas was virtually unique among biologists of the second half of the nineteenth century in his ability to produce fundamental zoological work, without scientific training and through work in isolation with incredibly simple means and no real laboratory or collaborators.

BIBLIOGRAPHY

I. ORIGINAL WORKS. Maupas's principal publications are "Contribution à l'étude morphologique et anatomique des infusoires ciliés," in *Archives de zoologie expérimentale et générale*, 2nd ser., **1** (1883), 427–664; "Recherches expérimentales sur la multiplication des infusoires ciliés," *ibid.*, **6** (1888), 165–277; "Sur la multiplication agame de quelques métazoaires inférieurs," in *Comptes rendus . . . de l'Académie des sciences*, **109** (1889), 270–272; "Le rajeunissement caryogamique chez les ciliés," in *Archives de zoologie expérimentale et générale*, 2nd ser., **7** (1889), 149–517; "Sur la fécondation de l'*Hydatina scuta*," in *Comptes rendus . . . de l'Académie des sciences*, **111** (1890), 505–507; "Sur la multiplication et la fécondation de l'*Hydatina scuta*," *ibid.*, 310–312; "Sur le déterminisme de la sexualité chez l'*Hydatina scuta*," *ibid.*, **113** (1891), 388–390; "La mue et l'enkystement chez les nématodes," in *Archives de zoologie expérimentale*, 3rd ser., **7** (1900), 563–632; "Modes et formes de reproduction des nématodes," *ibid.*, **8** (1901), 463–624; "Sur le mécanisme de l'accouplement chez les nématodes," in *Comptes rendus . . . de la*

Société de biologie, **79** (1916), 614–618, written with L. G. Seurat; "Essais d'hybridation chez les nématodes," in *Bulletin biologique de la France et de la Belgique*, **52** (1919), 467–486; and "Expériences sur la reproduction asexuelle des oligochètes," *ibid.*, **53** (1919), 150–160.

Among Maupas's translations is H. Burmeister, *Histoire de la création* (Paris, 1870).

II. SECONDARY LITERATURE. See M. Caullery, *Inauguration de la plaque commémorative apposée sur la maison habitée par Émile Maupas à Alger, le mercredi 6 avril 1932* (Paris, 1932), with portrait; and E. Sergent, "Émile Maupas prince des protozoologistes," in *Archives de l'Institut Pasteur d'Algérie*, **33** (1955), 59–70.

JEAN THÉODORIDÈS

MAUPERTUIS, PIERRE LOUIS MOREAU DE (*b.* St.-Malo, France, 28 September 1698; *d.* Basel, Switzerland, 27 July 1759), *mathematics, biology, physics.*

It was said of Maupertuis, in the official eulogy by Samuel Formey, that "Madame Moreau idolized her son rather than loved him. She could not refuse him anything." It seems highly probable that the spoiled child inevitably developed some of those personality characteristics that later made him not only proud but intransigent and incapable of bearing criticism, traits that ultimately led to great unpleasantness in his life and, quite literally, to his undoing.

After private schooling Maupertuis went to Paris at the age of sixteen to study under Le Blond, but he found ordinary philosophical disciplines quite distasteful. In 1717 he began to study music; but he soon developed a strong interest in mathematics, which he pursued under the tutelage of Guisnée and, later, Nicole. Maupertuis was elected to the Academy of Sciences in 1723, at the age of twenty-five, and presented a dissertation, "Sur la forme des instruments de musique." This was soon followed by a mathematical memoir on maxima and minima, some biological observations on a species of salamander, and two mathematical works of much promise: "Sur la quadrature et rectification des figures formées par le roulement des polygones reguliers" and "Sur une nouvelle manière de développer les courbes."

In 1728 Maupertuis made a trip to London that was to exert a major influence upon his subsequent career. From a conceptual world of Cartesian vortices he was transported into the scientific milieu of Newtonian mechanics, and he was quickly converted to these views. From this time on, Maupertuis was the foremost proponent of the Newtonian movement in France and a convinced defender of Newton's ideas about the shape of the earth. After returning to France he visited Basel, where he was befriended by the Bernoullis.

While pursuing, in conjunction with Clairaut, further studies in mathematics—resulting in a steady flow of notable memoirs—Maupertuis was readying his first work on Newtonian principles, "Discours sur les différentes figures des astres" (1732). It brought him the attention of the Marquise du Châtelet and of Voltaire, both of whom he instructed in the new doctrines. His position as the leading Continental Newtonian was confirmed the following year by his "Sur la figure de la terre et sur les moyens que l'astronomie et la géographie fournissent pour la déterminer," which was accompanied by a complementary memoir by Clairaut.

Thus it came about that in 1735 France sent an expedition to Peru under the leadership of La Condamine and another to Lapland under the leadership of Maupertuis. Clairaut, Camus, and other scientists accompanied the latter. The mission of each expedition was to measure as accurately as possible the length of a degree along the meridian of longitude. If, indeed, the earth is flattened toward the poles, as Newton had predicted, the degree of latitude should be shorter in far northern latitudes than near the equator. The voyage began on 2 May 1736 and lasted over a year. The local base for the expedition's fieldwork was Torneå, in northern Sweden—then, according to Maupertuis, a town of fifty or sixty houses and wooden cabins. On the return journey the ship was wrecked in the Baltic Sea, but without loss of life, instruments, or records.

Maupertuis reached Paris on 20 August 1737, only to meet with a rather chilly reception. Envy and jealousy were already at work; he had few Newtonian supporters in France except Voltaire; and La Condamine's expedition had not yet returned from Peru. At this time Maupertuis found respite at Saint-Malo and at Cirey, where Mme du Châtelet and Voltaire made him welcome. He stayed only briefly at Cirey, however, intending to revisit Basel. There he met Samuel König, a young student of Johann I Bernoulli. He persuaded König to accompany him back to Cirey, where König behaved so arrogantly that he angered Mme du Châtelet, who through this episode became temporarily estranged from Maupertuis.

The laborious analysis of the data on the length of the arc of a meridional degree at various latitudes took much time and created much controversy. The measurements made in France had to be corrected. In December 1739 Maupertuis announced to the Academy the value found for the distance along the meridian between Paris and Amiens. The expedition to Peru having returned after an arduous three years, the degree between Quito and Cuenca was added to the comparisons. Still later (1751) measurements made by Lacaille at the Cape of Good Hope permitted a fourth comparison.

In a final revision of the reports on the "Opérations pour déterminer la figure de la terre" (*Oeuvres*, IV, 335) Maupertuis summarized the corrected measurements for a degree of longitude as follows:

	Latitude	Toises
Peru	0°30'	56,768
Cape of Good Hope	33°18'	56,994
France	49°23'	57,199
Lapland	66°10'	57,395

In 1738 Voltaire recommended Maupertuis to Frederick the Great, who was eager to rehabilitate the academy of sciences at Berlin. Frederick commenced overtures to Maupertuis, who visited Berlin after publication of his new, anonymously printed *Éléments de géographie* and his reconciliation with Mme du Châtelet. In Berlin he met Francesco Algarotti and the family of M. de Borck, whose daughter he was later to marry. After the outbreak of the War of the Austrian Succession, Maupertuis joined Frederick in Silesia, only to be captured when his horse bolted into the enemy lines. For a time he was feared dead by his friends, but Maupertuis soon emerged safely in Vienna; ominously, he took offense at the jests of Voltaire regarding his military exploit.

Maupertuis was elected to the Académie Française in 1743. In 1744 he presented the memoir "Accord de différentes lois de la nature" and published "Dissertation sur le nègre blanc." The latter was the precursor of the *Vénus physique* of 1745, which was an enlarged and more fully analyzed argument against the then-dominant biological theory of the preformation of the embryo. Maupertuis argued convincingly that the embryo could not be preformed, either in the egg or in the animalcule (spermatozoon), since hereditary characteristics could be passed down equally through the male or the female parent. He rejected the vitalistic notion that some "essence" from one of the parents could affect the preformed fetus in the other parent, or that maternal impressions could mold the characteristics of the offspring. A strict mechanist, although a believer in the epigenetic view of the origin of the embryo, he looked for some corporeal contribution from each parent as a basis of heredity.

In the middle of 1745 Maupertuis finally accepted Frederick's invitation and took up residence in Berlin. In the same year he married Mlle de Borck; and on 3 March 1746 he was installed as president of the Academy. His first contribution was the brief paper "Les lois du mouvement et du repos," in which he set forth the famous principle of least action, which he regarded as his own most significant scientific contribution. It states simply that "in all the changes that take place in the universe, the sum of the products of each body multiplied by the distance it moves and by the speed with which it moves is the least possible" (*Oeuvres*, II, 328). That is, this quantity tends to a minimum. This principle was later clarified and expounded by Euler, developed by Hamilton and Lagrange, and incorporated in modern times into quantum mechanics and the biological principle of homeostasis. As Maupertuis himself said:

The laws of movement thus deduced [from this principle], being found to be precisely the same as those observed in nature, we can admire the application of it to all phenomena, in the movement of animals, in the vegetation of plants, in the revolution of the heavenly bodies: and the spectacle of the universe becomes so much the grander, so much the more beautiful, so much worthier of its Author. . . .

These laws, so beautiful and so simple, are perhaps the only ones which the Creator and Organizer of things has established in matter in order to effect all the phenomena of the visible world [*Oeuvres*, I, 44–45].

Maupertuis clearly was successful in attracting to Berlin scientific luminaries who greatly enhanced the luster of the new Academy. Euler, one of the greatest mathematicians of the day, was already there. La Mettrie came in 1748; Mérian and Meckel in 1750; and, in the same year, after the death of Mme du Châtelet, Voltaire arrived in Berlin. With others the brusque impatience of Maupertuis rendered his efforts less successful. On the whole, however, matters were going well when the celebrated "affaire König" erupted. Samuel König, a protégé of Maupertuis, after having been elected a member of the Academy, visited Berlin, was warmly received by Maupertuis, and shortly thereafter submitted a dissertation attacking the validity of the principle of least action and then—most strangely for a devoted adherent of Leibniz—ascribed the discredited law to the latter, citing a letter from Leibniz to Hermann. Maupertuis was incensed. He demanded that the letter be produced. König produced a copy but stated that the original was in the hands of a certain Swiss named Henzi, who had been decapitated at Bern following involvement in a conspiracy. After exhaustive search no trace of the letter was found in Henzi's belongings. Maupertuis then demanded that the Academy take action against König.

At the same time Maupertuis was embroiled in a controversy between Haller and La Mettrie. The latter had dedicated to Haller, much to Haller's dismay, his *L'homme machine* (1748). La Mettrie had, in response to Haller's rejection, responded with a diatribe. Haller demanded an apology; but inasmuch as La Mettrie died at just that time, Maupertuis tried —without success—to assuage Haller with a polite letter. The episode certainly contributed to the extraordinary bitterness and tension that Maupertuis experienced in 1751.

Nevertheless, at this very time Maupertuis was able to publish one of his most significant works, later called *Système de la nature*. A sequel to the *Vénus physique*, it was a theoretical speculation on the nature of biparental heredity that included, as evidence, an account of a study of polydactyly in the family of a Berlin barber-surgeon, Jacob Ruhe, and the first careful and explicit analysis of the transmission of a dominant hereditary trait in man. Not only did Maupertuis demonstrate that polydactyly is transmitted through either the male or the female parent, but he also made a complete record of all normal as well as abnormal members of the family. He furthermore calculated the mathematical probability that the trait would occur coincidentally in the three successive generations of the Ruhe family had it not been inherited.

On the basis of this study, Maupertuis founded a theory of the formation of the fetus and the nature of heredity that was at least a century ahead of its time. He postulated the existence of hereditary particles present in the semen of the male and female parents and corresponding to the parts of the fetus to be produced. They would come together by chemical attraction, each particle from the male parent joining a corresponding particle from the female parent. Chemical affinity would also account for the proper formation of adjacent parts, since particles representing adjacent parts would be more alike than those of remote parts. At certain times the maternal character would dominate; at others the paternal character. The theory was applied to explain the nature of hybrids between species and their well-known sterility; and it was extended to account for aberrations with extra structures as well as to those characterized by a missing part. The origin of new sorts of particles, as well as the presence of those representing ancestral types, was envisaged. Finally, Maupertuis thought it possible that new species might originate through the geographical isolation of such variations.

During 1752 the König affair reached a climax and a hearing was held, from which Maupertuis absented himself. The letter cited by König was held to be unauthentic and undeserving of credence, and König resigned from the Academy—only to issue a public appeal and defense. Voltaire had already run afoul of Maupertuis, and jealousy existed between them regarding their influence with the king. Maupertuis had shown scant enthusiasm for a proposed monumental dictionary of metaphysics, to be developed by the Academy as a counterpoise to the *Encyclopédie*, for Maupertuis considered the talents of the Berlin Academy insufficient to keep such a work from being superficial. In September 1752 Voltaire attacked Maupertuis, charging him not only with plagiarism and error but also with persecution of honest opponents and with tyranny over the Academy. In the *Diatribe du Docteur Akakia*, Voltaire poured invective on the ideas that Maupertuis had expressed in his *Lettre sur le progrès des sciences* (1752) and *Lettres* (1752)—in which, among other daring speculations regarding the future course of science, Maupertuis had included his most substantial account of the investigation of polydactyly in the Ruhe family and of his own breeding experiments with Iceland dogs. In *Micromégas* Voltaire made fun of the voyage to Lapland undertaken to measure the arc of the meridian and lampooned Maupertuis's amorous adventures in the North. His mockery made a great contrast with the grandiloquent words that he had once inscribed beneath a portrait of Maupertuis. In vain Frederick supported Maupertuis and tried to restore good feeling. Maupertuis was crushed, his health gave way, and he requested a leave to recuperate at Saint-Malo. Pursued by an unceasing volley of Voltaire's most savage satires, Maupertuis withdrew. He remained at Saint-Malo until the spring of 1754, when he returned to Berlin at Frederick's insistence. Here he delivered the eulogy of his friend Montesquieu, who died at Paris early in 1755. He departed again for France, a very sick man, in May 1756. Greatly distressed by the outbreak of the Seven Years' War, he decided to return home by way of Switzerland. He went to Toulouse, whence he set out again in May 1758. At Basel, too ill to proceed, he was received warmly by his old friend Johann Bernoulli. On 27 July 1759, before his wife could reach him, he died and was buried in Dornach.

Maupertuis was a man of singular aspect. He was very short. His body was always in motion; he had numerous tics. He was careless of his apparel. Perhaps he was always endeavoring to attract attention. Perhaps he shared the Napoleonic complex of little men. Certainly he was both highly original and possessed of qualities that attracted friends, especially among the ladies; the Marquise du Châtelet and many other Frenchwomen corresponded regularly with him. He could be gay as well as fiery and violent. Above all he was proud, both of his intelligence and of his accomplishments, and to attack either was to wound him deeply. Above all, he could not understand the character of König, whom he had sponsored and who then gratuitously attacked him, or of Voltaire, whose adulation and friendship so quickly turned to malice and vituperation.

A philosopher as well as a scientist, Maupertuis proved himself a powerful and original thinker in *Essai de cosmologie* (1750). According to A. O. Lovejoy, he anticipated Beccaria and Bentham and, along with Helvétius, represents "the headwaters of the important stream of utilitarian influence which became so broad and sweeping a current through the work of the Benthamites" (*Popular Science Monthly*, **65** [1904], 340). He rejected the favorite eighteenth-century argument in favor of God—the argument from design—and instead, like Hume, he formulated a view of adaptation based on the elimination of the unfit. He recognized that Newton's laws are insufficient to explain chemistry, and even more so life, and turned to Leibniz for ideas about the properties of consciousness. In the *Système de la Nature* we may, with Ernst Cassirer (*Philosophy of the Enlightenment*, p. 86), see an attempt to "reconcile the two great opponents of the philosophy of nature of the seventeenth century," Newton and Leibniz. Yet in it must also be recognized a highly original work based on his own investigations of heredity. In his effort to introduce a calculus of pleasure and pain, in order to evaluate the good life and to measure happiness, Maupertuis proposed that the amount of pleasure or pain is a product of intensity and duration. This formulation is strictly analogous to his principle of least action in the physical world and shows how he extended his philosophy of nature into a philosophy of life.

BIBLIOGRAPHY

The works of Maupertuis are collected in *Oeuvres*, 4 vols. (Lyons, 1756). For his life see Grandjean de Fouchy, "Éloge de Maupertuis," in L. Angliviel de la Beaumelle, *Vie de Maupertuis* (Paris, 1856); Damiron, *Mémoires sur Maupertuis* (Paris, 1858); and P. Brunet, *Maupertuis*, I. *Étude Biographique* (Paris, 1929).

See also B. Glass, "Maupertuis, Pioneer of Genetics and Evolution," in B. Glass, O. Temkin, and W. Straus, Jr., eds., *Forerunners of Darwin, 1745–1859* (Baltimore, 1959); and Ernst Cassirer, *The Philosophy of the Enlightenment* (Princeton, 1951).

BENTLEY GLASS

MAURER, JULIUS MAXIMILIAN (*b.* Freiburg im Breisgau, Germany, 14 July 1857; *d.* Zurich, Switzerland, 21 January 1938), *meteorology, astronomy*.

Maurer was born in Germany but passed his whole life in Switzerland, where his parents settled when he was one year old. He attended the Eidgenössische Technische Hochschule at Zurich and studied astronomy at the University of Zurich. In 1879 he became an assistant at the Eidgenössische Sternwarte in Zurich, and in 1882 he received his doctorate for a thesis on the extinction of starlight in the atmosphere. In 1881 he was appointed adjunct at the Schweizerische Meteorologische Zentralanstalt, which was at that time connected with the observatory; soon afterwards the two institutions were separated administratively. In 1905, Maurer became director of the Zentralanstalt; he retired in 1934. He was a member of many scientific societies, and he was also an honorary citizen of Zurich from 1900.

Maurer worked mainly on radiation problems. In his early years, he dealt with the astronomical aspects of radiation. Later he became interested in questions of radiation in meteorology. His papers on total solar radiation and on nighttime heat loss by radiation made these topics as clear as was possible at the end of the nineteenth century.

Some meteorological instruments were constructed by Maurer, including barographs and instruments recording solar radiation. His position as director of the Zentralanstalt stimulated him to climatological researches. In 1909, together with R. Billwiller, Jr., and C. Hess, he published *Das Klima der Schweiz*, which contained detailed critical discussions on the climate of Switzerland. Maurer continued work on these problems and published papers on glacial variations, freezing of lakes, and general climatological problems.

BIBLIOGRAPHY

Among Maurer's works are "Die theoretische Darstellung des Temperaturgangs in der Nacht und die Wärmestrahlung der Atmosphäre," in *Meteorologische Zeitschrift*, **4** (1887), 189; "Beobachtungen über die irdische Strahlenbrechung bei typischen Formen der Luftdruckverteilung," *ibid.*, **22** (1905), 49; and *Das Klima der Schweiz*, 2 vols. (Frauenfeld, 1909–1910), written with R. Billwiller, Jr., and C. Hess. Many of Maurer's papers on detailed questions of climatology in Switzerland appeared in *Meteorologische Zeitschrift*. There is no secondary literature on Maurer.

F. SCHMEIDLER

MAURO, FRA (*d.* Murano, near Venice, Italy, 20 October 1459 [?]), *geography*.

Fra Mauro, author of the last of the great medieval world maps, was a monk of the Camaldolese order and, in all probability, head of a cartographic workshop in the Camaldolese monastery of San Michele di Murano, in the lagoon of Venice. Of his life very little is known other than that he worked at San Michele from about 1443 until his death and that his fame as a mapmaker spread as far as Portugal. There are records of payments made by the king of Portugal to the monastery in the late 1450's, and a statement on the map reads: "I have copies of maps made by the Portuguese."

The world map now in the Biblioteca Nazionale Marciana in Venice is the sole surviving work that can be positively identified as Fra Mauro's. Drawn on vellum, the circular map is 1.96 meters in diameter, and its quadrangular frame measures 2.23 meters on each side. The mapmaker's colors are well preserved, and legends both on the map and in the corners outside its circular outline are fully legible. The map is oriented to the south on the top; a legend on the back states that it was completed on 26 August 1460, which would indicate that the last touches were added after Fra Mauro's death in the previous year.

Fra Mauro's map represents an encyclopedic storehouse of contemporary geographic and cosmographic information. It is limited to Europe, Asia, and part of Africa and shows Iceland and the Canary Islands to the west. It includes the first mention of "Zimpagu" (Japan) on any European map; it extends to the Urals in the north; and its southern limit is probably at the latitude of Madagascar. The mapmaker and his assistants used a variety of sources: medieval navigation charts of European origin; the accounts of travelers, especially Marco Polo and the fifteenth-century Venetian merchant and traveler Niccolò de' Conti; Arabic sailing directions and travel accounts; and, possibly, information obtained from Ethiopian delegates to the council of Florence (1438–1445).

Fra Mauro's work is a milestone in the history of geography and cartography for the wealth of its information, both correct and distorted; its wide range of cartographic and geographic data on remote parts of Asia and, to a lesser extent, of Africa; and its clearly transitional character, from the medieval to Renaissance worldview.

BIBLIOGRAPHY

The first detailed study of Fra Mauro's world map was Placido Zurla, *Il mappamondo di Fra Mauro camaldolese* (Venice, 1808). The authoritative reference, accompanied by a complete section-by-section color reproduction of the map and transcription of all legends, is Tullia Gasparrini Leporace, *Il mappamondo di Fra Mauro* (Rome, 1956), with introductory essay by Roberto Almagià.

GEORGE KISH

MAUROLICO, FRANCESCO (*b.* Messina, Italy, 16 September 1494; *d.* near Messina, 21 or 22 July 1575), *mathematics, astronomy, optics.*

Maurolico's name is variously transcribed as Maurolyco, Marulì, Marulli, and, in Latin, Maurolicus, Maurolycus, and Maurolycius. He was the son of Antonio Maurolico, master of the Messina mint, and his wife, Penuccia or Ranuccia. The family came from Greece, from which they had fled to Sicily to escape the Turks; Maurolico learned Greek, as well as astronomy, from his father. In 1521 he was ordained priest; he later became a Benedictine. Except for short sojourns in Rome and Naples, he lived his whole life in Sicily.

Maurolico's patrons included Giovanni de Vega, Charles V's viceroy of Sicily, who entrusted him with the mathematical education of one of his sons; and Giovanni Ventimiglia and his son, Simon, both marquises of Geraci and princes of Castelbuono and governors ("stradigò") of Messina. In 1550 Simon conferred upon Maurolico the abbey of Santa Maria del Parto (today also known as the Santuario di San Guglielmo), near Castelbuono. Maurolico also held a number of civil commissions in Messina; he served as head of the mint, he was in charge (with the architect Ferramolino) of maintaining the fortifications of the city on behalf of Charles V, and he was appointed to write a history of Sicily, which, as *Sicanicarum rerum compendium*, was published in Messina in 1562. Most important, he gave public lectures on mathematics at the University of Messina, where he was appointed professor in 1569.

Although Maurolico himself referred to a vast literary production (in his *Cosmographia* and *Opuscula*), only a few of his works were printed, although these are enough to show him as an outstanding scholar. In addition to writing his own books, Maurolico translated, commented upon, reconstructed, and edited works by a number of ancient authors. His first work in this vein, published in Messina in 1558, included treatises on the sphere by Theodosius of Bythinia "ex traditione Maurolyci"; by Menelaus of Alexandria "ex traditione eiusdem"; and by Maurolico himself. The book also contained a work by Autolycus of Pitane on the moving sphere, translations of the *De habitationibus* of Theodosius and the *Phaenomena* of Euclid, trigonometric tables, a mathematical compendium, and a work entitled "Maurolyci de sphaera sermo."

This early book is especially noteworthy for two reasons. First, the Neapolitan mathematician Giuseppe d'Auria furthered the dissemination of Maurolico's work by including his annotations in later editions of Autolycus' *Sphaera* and Theodosius' *De habitationibus* (Rome, 1588), as well as of Euclid's *Phaenomena* (Rome, 1591). Second, J. B. J. Delambre, in his *Histoire de l'astronomie du moyen âge*, stated that Maurolico had been the first to make use of the trigonometric function of the secant. Maurolico did give a table of numerical values for the secants of 0° to 45° (the "tabella benefica"), but Copernicus had certainly preceded him in its use.

Maurolico's two other major books on ancient mathematics—one on Apollonius' *Conics*, and the other a collection of the works of Archimedes—were published only after his death. In *Emendatio et restitutio conicorum Apollonii Pergaei* (Messina, 1654), Maurolico attempted to reconstruct books V and VI of the *Conics* from the brief references to them that Apollonius provided in his preface to the entire work. In Maurolico's time, only the first four books were known in the Greek original; he completed his restoration in 1547, and a similar reconstruction of book V was published by Vincenzo Viviani in 1659. (Although Maurolico's work is less famous than Viviani's, both Libri and Gino Loria cite it as an example of his genius.) Maurolico's collection of Archimedes' works, *Admirandi Archimedis Syracusani monumenta omnia mathematica quae extant . . . ex traditione doctissimi viri D. Francisci Maurolici* (Palermo, 1685), was based upon an earlier partial edition by Borelli (Messina, 1670–1672), which was almost completely lost.

Among the most important of Maurolico's extant books are *Cosmographia* (Venice, 1543), written in the form of three dialogues; *Opuscula mathematica* (Venice, 1575), a collection of eight treatises; *Photismi de lumine et umbra ad perspectivam et radiorum incidentiam facientes* (possibly Venice, 1575, and certainly Naples, 1611); and *Problemata mechanica . . . et ad magnetem et ad pixidem nauticam pertinentia* (Messina, 1613). In addition to these, a number of Maurolico's manuscripts held by the Bibliothèque Nationale, Paris, were published by Federico Napoli in 1876; these include a letter of 8 August 1556, in which Maurolico reported on his mathematical studies to his patron Giovanni de Vega; a brief treatise, previously thought to be lost, entitled "Demonstratio algebrae"; books I and II of a 1555 "Geometricarum quaestionum"; and a "Brevis demonstratio centri in parabola," dated 1565.

Of the mathematical works edited by Napoli, the "Demonstratio algebrae" is elementary in its concerns, dealing with simple second-degree problems and derivations from them. "Geometricarum quaestionum" is primarily devoted to trigonometry and solid geometry, but touches upon geodesy in offering a proposal for a new method for measuring the earth, a method previously discussed in the *Cosmographia* and later taken up by Jean Picard for measuring the meridian (1669–1671). In the "Brevis demonstratio centri in parabola," Maurolico chose to deal with a problem related to mechanics—which he also treated in his edition of Archimedes—the determination of the center of gravity of a segment of a paraboloid of revolution cut off by a plane perpendicular to its axis.

The greatest number of Maurolico's mathematical writings are gathered in the *Opuscula mathematica;* indeed, the second volume of that work, "Arithmeticorum libri duo," is wholly devoted to that subject and contains, among other things, some notable research on the theory of numbers. This includes, in particular, a treatment of polygonal numbers that is more complete than that of Diophantus, to which Maurolico added a number of simple and ingenious proofs. L. E. Dickson has remarked upon Maurolico's argument that every perfect number is hexagonal, and therefore triangular, while Baldassarre Boncompagni noted his proof of a peculiarity of the succession of odd numbers. That property had been enunciated by Nicomachus of Gerasa, Iamblichus, and Boethius, among others.

Among the topics related to mathematics in the *Opuscula* are chronology (the treatise "Computus ecclesiasticus") and gnomonics (in two treatises, both entitled "De lineis horariis," one of which also discusses conics). The work also contains writing on Euclid's *Elements* (for which see also the unpublished Bibliothèque Nationale, Paris, manuscript Fonds Latin 7463). Of particular interest, too, is a passage on a correlation between regular polyhedrons, which was commented upon by J. H. T. Müller, and later by Moritz Cantor. The balance of Maurolico's known mathematical work is contained in three manuscripts, mostly on geometrical problems, in the Biblioteca Nazionale Centrale Vittorio Emanuele in Rome; they have been described by Luigi De Marchi.

Maurolico's work in astronomy includes the first treatise collected in the *Opuscula*, "De sphaera liber unus," in which he criticized Copernicus. In another item of the collection, "De instrumentis astronomicis," Maurolico described the principal astronomical instruments and discussed their theory, use, and history—a subject similar to that treated in one of his first publications, the rare and little-known tract *Quadrati fabrica et eius usus* (Venice, 1546). In practical astronomy, Maurolico observed the nova that appeared in the constellation Cassiopeia in 1572. Until recently all that was known of this observation was contained in the short extracts from an unknown work by Maurolico that were published by Clavius in his *In Sphaeram Ioannis de Sacro Bosco commentarius* (Rome, 1581). In 1960, however, C. Doris Hellman published an apograph manuscript that she had found in the Biblioteca Nazionale of Naples. This manuscript contains a full account of Maurolico's observation; it is dated 6 November 1572, and is clear evidence that Maurolico's observation preceded by at least five days the more famous one made by Tycho Brahe.

Maurolico also did important work in optics; indeed, according to Libri, "it is in his research on optics, above all, that Maurolico showed the most sagacity" (*Histoire*, III, 116). The chief record of this research is *Photismi de lumine et umbra*, in which Maurolico discussed the rainbow, the theory of vision, the effects of lenses, the principal phenomena of dioptrics and catoptrics, radiant heat, photometry, and caustics. Maurolico's work on caustics was anticipated by that of Leonardo da Vinci (as was his research on centers of gravity), but Leonardo's work was not published until long after Maurolico's. Libri further characterized the *Photismi de lumine et umbra* as "full of curious facts and ingenious research" (*Histoire*, III, 118), and Sarton suggested that it might be the most remarkable optical treatise of the sixteenth century outside the tradition of Alhazen, or even the best optical book of the Renaissance (*Six Wings*, 84, 85).

Maurolico applied his broad scientific knowledge to a number of other fields. One treatise in the *Opuscula*, "Musicae traditiones," is devoted to music. The *Problemata mechanica* published in 1613 is concerned with mechanics and magnetism, as is, to some degree, the "Brevis demonstratio." His contributions to geodesy have already been discussed; he made an additional contribution to geography with a map of Sicily, drawn about 1541 at the request of Jacopo Gastaldo (who published it in 1575)—this map was also incorporated by Abraham Ortelius in his *Theatrum orbis terrarum*. Maurolico wrote on the fish of Sicily, in a letter to Pierre Gilles d'Albi, dated 1 March 1543 and published by Domenico Sestini in 1807, and on the eruption of Mt. Etna, in a letter to Cardinal Pietro Bembo, dated 4 May 1546 and published by Giuseppe Spezi in 1862. Finally, he enjoyed some contemporary fame as a meteorologist, based upon a weather prediction that he made for John of Austria upon the latter's departure from Messina prior to the Battle of Lepanto (1571).

BIBLIOGRAPHY

I. ORIGINAL WORKS. Almost all of Maurolico's writings have been mentioned in the text. For further information see Pietro Riccardi, *Biblioteca matematica italiana* (Modena, 1870–1928; repr. Milan, 1952), articles "Archimede," "Auria," "Maurolico"; and Federico Napoli, "Intorno alla vita ed ai lavori di Francesco Maurolico," pref. to "Scritti inediti di Francesco Maurolico," in *Bullettino di bibliografia e di storia delle scienze matematiche e fisiche*, **9** (1876), 1–121, on p. 5 of which is a list of codices of the Bibliothèque Nationale, Paris, containing autographs by Maurolico (Fonds Latin, nos. 6177, 7249, 7251, 7459, 7462–7468, 7471, 7472A, 7473). On this see also Federico Napoli, "Nota intorno ad alcuni manoscritti di Maurolico della Biblioteca Parigina," in *Rivista sicula di scienze, letteratura ed arti*, **8** (1872), 185–192.

See also Luigi De Marchi, "Di tre manoscritti del Maurolicio che si trovano nella Biblioteca Vittorio Emanuele di Roma," in Eneström's *Bibliotheca mathematica* (1885), cols. 141–144, 193–195. In the codices described by De Marchi (marked 32, 33, 34; formerly S. Pantaleo 115, 116, 117), there is a letter from Maurolico to Prince Barresi di Pietraperzia, dated 11 Sept. 1571, which was published by De Marchi in "Una lettera inedita di Francesco Maurolico a proposito della battaglia di Lepanto," in *Rendiconti dell'Istituto lombardo di scienze e lettere*, **16** (1883), 464–467; this letter, De Marchi observes, may be considered as Maurolico's scientific will.

Maurolico's letter to Cardinal Bembo of 4 May 1546 is in Giuseppe Spezi, *Lettere inedite del Card. Pietro Bembo e di altri scrittori del secolo XVI, tratte da codici Vaticani e Barberiniani* (Rome, 1862), pp. 79–84. A letter from Maurolico to Cardinal Antonio Amulio dated 1 Dec. 1568 is in Baldassarre Boncompagni, "Intorno ad una proprietà de' numeri dispari," in *Bullettino di bibliografia e di storia delle scienze matematiche e fisiche*, **8** (1875), 51–62, see pp. 55–56, where a MS on arithmetic by Maurolico is cited (Codex Vat. lat. no. 3131) and the dedicatory letter which precedes it is published.

The work on the nova of 1572 is in "Maurolyco's 'Lost' Essay on the New Star of 1572," transcribed, translated, and edited by C. Doris Hellman, in *Isis*, **51** (1960), 322–336; the MS, in the Biblioteca Nazionale of Naples (cod. I E 56, fols. 2r–10r), perhaps a copy of an autograph version, is entitled "Super nova stella: Que hoc anno iuxta Cassiopes apparere cepit considerationes."

The work on Sicilian fish is in Domenico Sestini, *Viaggi e opuscoli diversi* (Berlin, 1807), 285–302, with notes on pp. 303–313 and mention of the MS used on p. xiii; see also *Tractatus per epistolam Francisci Maurolici ad Petrum Gillium de piscibus siculis Codice manu auctoris exarato, Aloisius Facciolà messanensis nunc primum edidit* (Palermo, 1893). An English translation of the *Photismi* is *The Photismi de Lumine of Maurolycus. A Chapter in Late Medieval Optics*, Henry Crew, trans. (New York, 1940).

The rare pamphlet on the quadrant is in the personal library of Dr. Carlo Viganò, Brescia, Italy. Its full title is *Quadrati fabrica et eius usus, ut hoc solo instrumento, caeteris praetermissis, uniusquisq. mathematicus, contentus esse possit, per Franciscum Maurolycum nuper edita. Illustriss. D. D. Ioanni Vigintimillio Ieraciensium Marchioni, D.* (Venice, 1546). In colophons to various parts of the text Maurolico gives the dates 6 Apr. 1541, 18 Apr. 1541, and 11 Jan. 1542. The work consists of eleven numbered pages and one unnumbered page with a table of stars.

II. SECONDARY LITERATURE. Older biographies and bibliographies on Maurolico include Francesco della Foresta, *Vita dell'abbate del Parto D. Francesco Maurolyco* (Messina, 1613), written by the nephew and namesake of the subject; Antonino Mongitore, "Maurolico," in *Biblioteca sicula*, I (Palermo, 1707), 226–227; Domenico Scinà, *Elogio di Francesco Maurolico* (Palermo, 1808); and Girolamo Tiraboschi, "Maurolico," in *Storia della letteratura italiana*, VII (Milan, 1824), 728–734.

Two valuable monographs from the late nineteenth century are Giuseppe Rossi, *Francesco Maurolico e il risorgimento filosofico e scientifico in Italia nel secolo XVI* (Messina, 1888); and Giacomo Macrì, "Francesco Maurolico nella vita e negli scritti," in R. Accademia Peloritana, *Commemorazione del IV centenario di Francesco Maurolico MDCCCXCIV* (Messina, 1896), p. iii–iv, 1–198. The latter volume also contains "Ricordi inediti di Francesco Maurolico," illustrated by Giuseppe Arenaprimo di Montechiaro, p. 199–230, with three plates reproducing handwritten items by Maurolico.

Maurolico is discussed in the following standard histories of mathematics: J. E. Montucla, *Histoire des mathématiques*, 2 vols. (Paris, 1758), I, 563, 571–572, 695–698; Guglielmo Libri, *Histoire des sciences mathématiques en Italie*, 3 vols. (Paris, 1837–1841), III, 102–118; Moritz Cantor, *Vorlesungen über Geschichte der Mathematik*, 4 vols. (Leipzig, 1880–1908), II, 558–559, 575, *passim*; and David Eugene Smith, *History of Mathematics*, 2 vols. (Boston, 1924–1925), I, 301–302, and II, 622. See also Florian Cajori, *A History of Mathematical Notations*, 2 vols. (La Salle, Ill., 1928–1929), I, 349, 362, 402, and II, 150.

Maurolico's work on mathematicians of antiquity is discussed in Vincenzo Flauti, "Sull'Archimede e l'Apollonio di Maurolico. Osservazioni storico-critiche," in *Memorie della Accademia delle scienze di Napoli*, 2 (1855–1857), lxxxiv–xciv; and Gino Loria, *Le scienze esatte nell'antica Grecia* (Milan, 1914), 219, 354, 434, 435, 502, 510, 511, 513, 515. On Maurolico as editor of Autolycus, see the following works by Joseph Mogenet: "Pierre Forcadel traducteur de Autolycus," in *Archives internationales d'histoire des sciences* (1950), 114–128; and "Autolycus de Pitane: Histoire du texte, suivie de l'édition critique des traités De la sphère en mouvement et Des levers et couchers," in Université de Louvain, *Recueil de travaux d'histoire et de philologie*, 3rd ser., fasc. 37 (1950), 23, 26, 27, 30–36, 38–42, 48–50, 176.

Arithmetic is treated in Mariano Fontana, "Osservazioni storiche sopra l'aritmetica di Francesco Maurolico," in *Memorie dell'Istituto nazionale italiano* (Bologna), Fis.-mat. cl., 2, pt. 1 (1808), 275–296; Baldassarre Boncompagni, "Intorno ad una proprietà de' numeri dispari" (see

above); and Leonard Eugene Dickson, *History of the Theory of Numbers* (Washington, D.C., 1919; repr. New York, 1952, 1966), I, 9, 20, and II, 5, 6.

On the use of the principle of mathematical induction by Maurolico, anticipated by Euclid, see the following writings by Giovanni Vacca: "Maurolycus, the First Discoverer of the Principle of Mathematical Induction," in *Bulletin of the American Mathematical Society*, 16 (1909–1910), 70–73; "Sulla storia del principio d'induzione completa," in Loria's *Bollettino di bibliografia e storia delle scienze matematiche*, 12 (1910), 33–35; and "Sur le principe d'induction mathématique," in *Revue de métaphysique et de morale*, 19 (1911), 30–33. See also W. H. Bussey, "The Origin of Mathematical Induction," in *American Mathematical Monthly*, 24 (1917), 199–207; Léon Brunschvicg, *Les étapes de la philosophie mathématique*, 3rd ed. (Paris, 1929), 481–484; and Hans Freudenthal, "Zur Geschichte der vollständigen Induktion," in *Archives internationales d'histoire des sciences* (1953), 17–37.

Maurolico's geometry is treated in Michel Chasles, *Aperçu historique sur l'origine et le développement des méthodes en géométrie*, 2nd ed. (Paris, 1875), 120, 291, 293, 345, 496, 516; J. H. T. Müller, "Zur Geschichte des Dualismus in der Geometrie," in Grunert's *Archiv der Mathematik und Physik*, 34 (1860), 1–6; and Federico Amodeo, "Il trattato sulle coniche di Francesco Maurolico," in Eneström's *Bibliotheca mathematica*, 3rd ser., 9 (1908–1909), 123–138.

On centers of gravity, see Margaret E. Baron, *The Origins of the Infinitesimal Calculus* (Oxford, 1969), 90–94.

Astronomy is discussed in J. B. J. Delambre, *Histoire de l'astronomie du moyen âge* (Paris, 1819; repr. New York, 1965), 437–441; J. L. E. Dreyer, *A History of Astronomy From Thales to Kepler* (New York, 1953), 257, 295, 356–357, formerly entitled *History of the Planetary Systems From Thales to Kepler* (Cambridge, 1906); Lynn Thorndike, *A History of Magic and Experimental Science*, V (New York, 1941), 304, 360, 421, 426, and VI (New York, 1941), 27, 74, 179–180, 382; and Edward Rosen, "Maurolico's Attitude Toward Copernicus," in *Proceedings of the American Philosophical Society*, 101 (1957), 177–194.

On Maurolico's contributions to geodesy, see Pietro Riccardi, "Cenni sulla storia della geodesia in Italia dalle prime epoche fin oltre alla metà del secolo XIX," in *Memorie della Accademia delle scienze di Bologna*, 3rd ser., 10 (1879), 431–528, see 518–519; and "Sopra un antico metodo per determinare il semidiametro della terra," *ibid.*, 4th ser., 7 (1887), 17–22; and Ottavio Zanotti-Bianco, "Sopra una vecchia e poco nota misura del semidiametro terrestre," in *Atti della Accademia delle scienze di Torino*, 19 (1883–1884), 791–794.

On optics, besides works by Libri, Crew, and Sarton already cited, see the following writings of Vasco Ronchi: *Optics, the Science of Vision*, trans. and rev. by Edward Rosen (New York, 1957), 39–40, 265; "L'optique au XVI⁰ siècle," in *La science au seizième siècle. Colloque international de Royaumont, 1–4 juillet 1957* (Paris, 1960), 47–65, and *The Nature of Light*, trans. by V. Barocas

(London, 1970), 78, 99ss., 223. See also A. C. Crombie, "The Mechanistic Hypothesis and the Scientific Study of Vision," in S. Bradbury and G. L'E. Turner, eds., *Historical Aspects of Microscopy* (Cambridge, 1967), 3–112 (see 43–46), and in *Proceedings of the Royal Microscopical Society*, 2, pt. 1 (1967).

On music, see Salvatore Pugliatti, "Le *Musicae traditiones* di Francesco Maurolico" in *Atti dell'Accademia peloritana*, **48** (1951–1967). On p. 336 is mentioned a MS, which contains three papers of Maurolico: "De divisione artium," "De quantitate," "De proportione." This MS was recently found by Monsignor Graziano Bellifemine in the Library and Museum of the Seminario Vescovile at Molfetta.

On Maurolico as a man of letters, historian, and philosopher, see G. Macrì, "Francesco Maurolico nella vita e negli scritti" (see above), pp. 48–62, 123–151; Valentino Labate, "Le fonti del *Sicanicarum rerum compendium* di Francesco Maurolico," in *Atti dell'Accademia peloritana*, **13** (1898–1899), 53–84; and L. Perroni-Grande, "F. Maurolico professore dell'Università messinese e dantista," in R. Accademia Peloritana, *CCCL anniversario della Università di Messina. Contributo storico* (Messina, 1900), 15–41, which includes the notarial act containing the nomination of Maurolico as professor at the University of Messina.

Other writings to be consulted are Luigi De Marchi, "Sull'ortografia del nome del matematico messinese Maurolicio," in Eneström's *Biblioteca mathematica* (1886), cols. 90–92; and several biobibliographical writings by Edward Rosen: "The Date of Maurolico's Death," in *Scripta mathematica*, **22** (1956), 285–286; "Maurolico Was an Abbot," in *Archives internationales d'histoire des sciences* (1956), 349–350; "De Morgan's Incorrect Description of Maurolico's Books," in *Papers of the Bibliographical Society of America*, **51** (1957), 111–118; "Was Maurolico's Essay on the Nova of 1572 Printed?," in *Isis*, **48** (1957), 171–175; "The Title of Maurolico's *Photismi*," in *American Journal of Physics*, **25** (1957), 226–228; and "The Editions of Maurolico's Mathematical Works," in *Scripta mathematica*, **24** (1959), 56–76.

ARNALDO MASOTTI

MAURY, ANTONIA CAETANA DE PAIVA PEREIRA (*b.* Cold Spring-on-Hudson, New York, 21 March 1866; *d.* Dobbs Ferry, New York, 8 January 1952), *astronomy*.

Antonia Maury was the daughter of Mytton and Virginia Draper Maury, a niece of Henry Draper, and granddaughter of John William Draper, a pioneer in the application of photography to astronomy. Her background was rich and varied, for her father was a naturalist and editor of a geographical magazine, as well as a professional minister. Although she worked chiefly as an astronomer, she was also an active ornithologist. Her sister, Carlotta Joaquina Maury (1874–1938), became a paleontologist specializing in Venezuelan and Brazilian stratigraphy.

Shortly after graduating from Vassar in 1887, Miss Maury became an assistant at the Harvard College Observatory, where during the next eight years she carried out her most perceptive and creative research. At that time the observatory director, Edward C. Pickering, was engaged in a program of stellar spectroscopy, and he had just found that the spectral lines of the star Mizar were double on one plate but single on others. Additional photographs established Mizar as the first spectroscopic binary, and Miss Maury was the first to determine its period, 104 days. In 1889 she herself discovered the second such star, Beta Aurigae, with a period of about four days.

In 1890 Williamina Fleming's *Draper Catalogue of Stellar Spectra* was published, with an initial classification of more than 10,000 stars. To Miss Maury was assigned a more detailed study of the brighter stars, made possible by placing three or four prisms in front of the eleven-inch Draper refractor (a telescope originally owned by her uncle, Henry Draper, and given to Harvard by his widow). Miss Maury soon concluded that the single spectral sequence of the *Draper Catalogue* was inadequate for representing all the observed peculiarities. In particular, for stars of the early spectral groups she assumed the existence of three collateral divisions (designated *a*, *b*, and *c*) characterized by the width and distinctness of their lines. In her system, *a* stars had normal lines, *b*, hazy lines, and *c*, sharp lines. Intermediate cases were designated *ab* or *ac*. Her catalogue, based on an examination of about 4,800 photographs, included her elaborate classification of 681 bright stars of the northern skies (*Annals of Harvard College Observatory*, **28**, pt. 1 [1896]).

Miss Maury's painstaking classifications enabled Ejnar Hertzsprung to verify his discovery of two distinct varieties of stars—dwarfs (divisions *a* and *b*) and giants (*c*). Unfortunately, Hertzsprung's interpretation was scarcely exploited at Harvard and in later work there, the *a*, *b*, *c* distinction was largely ignored in favor of the more elementary Draper system, as extended by Annie Jump Cannon. This led Hertzsprung to write to Pickering (22 July 1908):

> In my opinion the separation by Antonia C. Maury of the c- and ac-stars is the most important advancement in stellar classification since the trials by Vogel and Secchi .... To neglect the c-properties in classifying stellar spectra, I think, is nearly the same thing as if the zoologist, who has detected the deciding differences between a whale and a fish, would continue in classifying them together.

Miss Maury's temperament was little suited to the often tedious observatory routine, and as early as 1891 her connection with Harvard became an on-again off-again affair. After her paper on spectral classification was finally published, for several years she lectured on astronomy in various cities; between lectures she accepted private pupils and an occasional teaching position. In 1908 she returned to Harvard Observatory as a research associate, resuming her earlier studies of spectroscopic binaries. She spent many years investigating the complex spectrum of the binary Beta Lyrae, details of which were published in the *Annals of Harvard College Observatory*, **84**, no. 8 (1933). After retiring in 1935, at the age of sixty-nine, she served for several years as curator of the Draper Park Museum at Hastings-on-Hudson, New York, paying nearly annual visits to the Harvard Observatory to examine the current spectra of Beta Lyrae.

BIBLIOGRAPHY

Miss Maury's chief publications are noted in the text. A short obituary by Dorrit Hoffleit appears in *Sky and Telescope*, **11** (1952), 106. The best source is Bessie Zaban Jones and Lyle Gifford Boyd, *The Harvard College Observatory* (Cambridge, 1971), 395–400; see also Solon I. Bailey, *The History and Work of Harvard Observatory, 1839–1927* (New York, 1931).

OWEN GINGERICH

MAURY, MATTHEW FONTAINE (*b.* near Fredericksburg, Virginia, 14 January 1806; *d.* Lexington, Virginia, 1 February 1873), *physical geography, meteorology, oceanography.*

Maury was the seventh child of a small planter, Richard Maury, who was a descendant of a Huguenot family that had come to Virginia from Ireland about 1718, and the former Diana Minor, whose English and Dutch forebears had settled in Virginia by 1650. He grew up in Williamson County, Tennessee, where his family moved in 1810. On graduating from Harpeth Academy in 1825, he followed his deceased eldest brother into the U.S. Navy, where his career flourished until 1839, when a stagecoach accident left him too lame for further duty at sea. Maury nevertheless remained a naval officer, serving in Washington and writing vigorously in the causes of naval reform, Southern expansionism, and those branches of science that could be applied to seafaring. He resigned his commission in 1861 to join the Confederate Navy. In England for the Confederacy from 1862 until

May 1865, Maury next served Emperor Maximilian of Mexico until 1866, when he returned to England to write geography textbooks. Not until 1868 did he return to the United States, where he was professor of physics at Virginia Military Institute until his death. In 1834 Maury married Ann Herndon, a distant cousin; they had five daughters and three sons.

Maury's scientific career began with two articles and a textbook on navigation. These made him an obvious choice for the U.S. Navy exploring expedition, to which he was appointed astronomer in September 1837 by Thomas ap Catesby Jones, the expedition's first leader. Maury's role was brief; he joined Jones and the other officers in resigning in November following Jones's repeated disagreements with Secretary of the Navy Mahlon Dickerson. After a number of senior officers declined, command of the expedition went to Charles Wilkes, who as head of the Navy's Depot of Charts and Instruments had traveled to Europe to obtain the expedition's apparatus.

The depot, set up in 1830 to issue navigational supplies to the fleet, had become under Wilkes an astronomical observatory as well. His successor, James M. Gilliss, who had studied astronomy in Paris, carried the quest for a national observatory one step further when he convinced Congress in 1842 that the depot needed a new observatory building. Dispatched to Europe for ideas and instruments, Gilliss was succeeded at the depot by Maury, who was thus rewarded with an important shore billet for leading the fight to reorganize the Navy.

When the observatory building was completed in 1844, the Virginians had their revenge for Wilkes's takeover of the exploring expedition: Secretary of the Navy John Y. Mason chose Maury to head the new "National Observatory." Maury threw himself and his staff into the strenuous program of observations that his rival, Gilliss, had planned, and he began to publish the results. But Maury's poor qualifications for astronomy and his proprietary attitude toward the work done at the observatory created hostility toward him in the intense competition for scarce resources within the American scientific community. This hostility seems justified. Considering that Maury was in charge of one of the world's major observatories for almost seventeen years and that he had substantial funds at his disposal, his contributions to astronomy seem small. Between 1844 and 1861 he published fewer than twenty papers —all observational—and seven catalogs of observations.

Maury failed to accomplish more in astronomy because his main interest lay in improving the technology of navigation, for which the science of the earth was more relevant than the science of the heavens. Upon taking charge of the depot in 1842, he moved it out of Wilkes's house. The move revealed an accumulation of manuscript ships' logs. Maury's insight that the data on winds and currents in these logs could be brought together to chart the general circulation of atmosphere and ocean was the basis for his chief contribution to science.

Maury began to publish his *Wind and Current Charts*—beginning with the North Atlantic in 1847—and to issue them free to mariners in exchange for abstract logs of the winds and currents of their voyages. The result was a series of charts and (after 1850) accompanying sailing directions that presented a climatic picture of the surface winds and currents for all the oceans.

Maury's chief aim in issuing these charts was the promotion of maritime commerce. He was thus more a technologist than a scientist, interested in knowledge less for its own sake than as a means to practical achievement. Pressed by the growth of steam propulsion, the sailing fleets of the world were making great progress in improving their technology about 1850, and Maury was the leading developer of the "software" of sail. After 1849, when the rush to California's gold fields became the main stage for the conflict between sail and steam, he claimed that his charts shortened sailing passages considerably in the competition between the sailing route round Cape Horn and the steamer-railroad route across Panama.

Another commercial stimulation to Maury's scientific endeavors came from submarine telegraphy. U.S. Navy vessels sounded the North Atlantic under Maury's direction from 1849 to 1853; and from their results he prepared the first bathymetrical chart, using 1,000-fathom contours. By distributing the bottom samples collected on these cruises to J. W. Bailey at West Point and C. G. Ehrenberg at Berlin, Maury made possible their pioneering studies in marine micropaleontology.

Inspired by the example of Alexander von Humboldt, many men of science in the second quarter of the nineteenth century were devoting their efforts to collecting on a large scale the data of physical phenomena on earth. Maury's charts were important instances of this kind of empirical science. The need to standardize observations made at widely separated points required international cooperation, which led in turn to international scientific meetings. The spread of the telegraph made possible a synoptic meteorology; and in response to a tentative British proposal for American cooperation in weather observations on land, Maury organized a conference at Brussels in 1853, the first of a series of international meteoro-

logical meetings. Maury, the leader in systematizing observations at sea, hoped to extend his own efforts to land. But his plans for the conference to unify weather reporting for both land and ocean ran into opposition at home; at the insistence of the American and British governments, the Brussels conference dealt only with the sea. The final report, written largely by Maury, organized uniform weather reporting at sea. This system was extended to the land after his death.

Maury's scientific achievements were organizational and empirical; they earned him the praise of European leaders of science, including Humboldt. Maury's attempts to interpret his data, however, were largely without merit in the eyes of scientific colleagues. He expressed them in the pious language of natural theology at a time when most scientists had succeeded in purging their writings of all religious references. Beginning with an article on the Gulf Stream in 1844, Maury developed theories of the general circulation of atmosphere and ocean, first in articles and in *Explanation and Sailing Directions to Accompany the Wind and Current Charts* (1850 *et seq.*) and then in his best-known work, *Physical Geography of the Sea* (1855), a loosely organized compilation of Maury's earlier writings on meteorology and oceanography. Reprinted many times in America and England and translated into six European languages, it was received enthusiastically in general and religious journals, critically in scientific ones. Maury's response to his critics was either to ignore them or to defend vigorously his original ideas. Since he was unwilling to modify his theories in the light of criticism, Maury's importance to the history of ideas of wind and water motion lies in the stimulus he gave to others, especially William Ferrel, who were forced by the popularity of Maury's book to improve upon his unphysical interpretations.

Almost no one accepted Maury's idea—based on Faraday's demonstration that oxygen is paramagnetic—that fluctuations in the wind are due to the earth's magnetism. Maury's scheme of the general circulation of the atmosphere required that vertical air currents move across each other in the calm belts of the tropics and the equator, and that horizontal winds converge to low-pressure areas around the poles. The latter requirement was contrary to the evidence; the former, physically implausible. Maury ignored Ferrel's alternative scheme, despite its explicit challenge. Maury's ideas on ocean circulation were not much sounder. He offered a vigorous argument for "thermo-haline" forces; but unable, like most of his contemporaries, to accept multiple causes, he accompanied it with an equally vigorous argument against wind stress, based on his erroneous belief that a Gulf Stream driven by the wind would be flowing uphill. He believed (incorrectly) that variations in salinity were more important than variations in temperature in causing currents, and he was also the principal exponent of the idea that the sea around the North Pole is free of ice.

Maury's failure to revise his theories in the light of criticism and new evidence, and his aggressive promotion in Congress of his own brand of science, brought him increasingly into conflict with the leaders of the growing American scientific community. At first Maury was given the place that his position in Washington merited. He was one of the founders of the American Association for the Advancement of Science, which in 1849 formed one committee to urge more funds for Maury's charts and appointed Maury himself to several others. But in 1851, when a committee was formed to organize the land meteorology of North America, Maury was not appointed. His claim that he should organize observations on land as he had on the ocean was a threat to the network established by Joseph Henry at the Smithsonian Institution, and in the presidential address to the A.A.A.S. in that year Alexander Dallas Bache spoke of the dangers to American science "from a modified charlatanism, which makes merit in one subject an excuse for asking authority in others, or in all" (*Proceedings of the American Association for the Advancement of Science*, **6** [1852], xliv). Their opposition to Maury led to the restriction of the 1853 Brussels conference to marine meteorology. Maury continued to present papers to the annual meetings of the A.A.A.S. until his book appeared in 1855; but he ceased to be a member in 1859, after his continuing effort to take over land meteorology had again been thwarted by Henry and his allies. His hope had been to benefit agriculture as his maritime meteorology had benefited navigation.

Although his cherished theories never received the support that he believed they deserved, Maury was widely honored for his achievements, both in his lifetime and afterward. He was decorated by the sovereigns of Denmark, Portugal, and Russia; received honorary degrees from Columbian College (now George Washington University [1853]) and the universities of Cambridge (1868) and North Carolina (1852); and became a member of a number of scientific academies and societies.

BIBLIOGRAPHY

I. ORIGINAL WORKS. *The Physical Geography of the Sea* is available in a modern ed. (Cambridge, Mass., 1963).

Lists of Maury's other works are in Ralph M. Brown, "Bibliography of Commander Matthew Fontaine Maury," which is *Bulletin of the Virginia Polytechnic Institute*, **37**, no. 12 (1944); and in F. L. Williams, *Matthew Fontaine Maury* (see below), 693–710.

II. SECONDARY LITERATURE. The massive biography by Frances L. Williams, *Matthew Fontaine Maury. Scientist of the Sea* (New Brunswick, N.J., 1963), contains much more material than earlier ones, together with footnotes and a complete bibliography. Each of its three predecessors, however, is useful: Diana Corbin, *Life* (London, 1888), contains family reminiscences by Maury's daughter; Charles L. Lewis, *Maury* (Annapolis, 1927), an account of the efforts after Maury's death to keep his name alive; and John W. Wayland, *Pathfinder of the Seas* (Richmond, Va., 1930), a chronology and a number of photographs not in the other works. All four biographies take Maury's scientific achievement at his own valuation; only Williams provides the reader with the evidence for an independent judgment.

Maury's science is carefully evaluated in John Leighly's brilliant introduction to the 1963 repr. of *Physical Geography of the Sea* and his equally important "M. F. Maury in His Time," in Proceedings of the First International Congress of the History of Oceanography, *Bulletin de l'Institut océanographique de Monaco*, spec. no. 2 (1968), 147–159. Leighly's incisive treatments supersede all previous writings on Maury's work in meteorology and oceanography. For Maury's work in a broader scientific context, see Margaret Deacon, *Scientists and the Sea 1650–1900* (London, 1970), ch. 13; in the context of American science, A. Hunter Dupree, *Science in the Federal Government* (Cambridge, Mass., 1957).

HAROLD L. BURSTYN

MAWSON, SIR DOUGLAS (*b.* Bradford, Yorkshire, England, 5 May 1882; *d.* Adelaide, Australia, 14 October 1958), *geology.*

Douglas Mawson was the younger son of Robert Mawson and Margaret Ann Moore, both of long-established Yorkshire families. The family sailed to Australia in 1884, where the father eventually attained some success as a lumber merchant. The boys were educated at Rooty Hill country school and later at Fort Street Public School, Sydney. Mawson entered the University of Sydney in 1899, took a bachelor of engineering degree in 1901, and under the influence of T. W. E. David, professor of geology, decided to become a geologist. He received a bachelor of science degree in 1904, doctor of science in 1909 (South Australia), became a fellow of the Royal Society in 1923, and received a doctor of science degree from the University of Sydney in 1952. After joining the staff at the University of Adelaide in 1905 as lecturer on mineralogy and petrology, he became the first professor of geology in 1920 and remained there until his retirement in 1952.

Mawson was honorary curator of minerals at the South Australian Museum and chairman of the board of governors at the time of his death. He was also a foundation fellow of the Australian Academy of Science and an honorary member of the Geological Society of Australia. His scientific awards are too numerous to be listed here. He married Francisca Adriana ("Paquita") Deprat, daughter of a mining engineer, on 31 March 1914 and had two daughters, Patricia and Jessica.

Douglas Mawson was foremost an explorer-geologist. His first expedition to the New Hebrides in 1903 was followed by his participation in Sir Ernest Shackleton's 1907–1909 expedition to Antarctica, where Mawson ascended Mt. Erebus and mapped the position of the South Magnetic Pole, helping to man-haul sledges for some 1,300 miles. He then helped organize and commanded two Antarctic expeditions. In the first, the Australasian Antarctic Expedition (1911–1914), six parties worked in Queen Mary Land and Adélie Land. Mawson's two companions died—Ninnis falling to his death in a crevasse 315 miles from base, and Mertz dying more than a hundred miles from safety—and he struggled back alone for the last 115 miles, reaching the winter base over a month later. His epic is well told in *The Home of the Blizzard*. As commander of the British Australian New Zealand Antarctic Research Expedition (BANZARE) in 1929–1931, he traveled to subantarctic islands and to Kemp Land and Enderby Land as far west as 45°E.

Although not a first-rank scientific investigator, Mawson established an outstanding department at Adelaide and worked extensively on geological problems of South Australia, notably on the Precambrian Adelaide System. He insisted on the fusing of scientific and geographic exploration, and his highly successful pioneering expeditions inspired public and government support.

BIBLIOGRAPHY

I. ORIGINAL WORKS. Mawson's writings include "The Geology of the New Hebrides," in *Proceedings of the Linnean Society of New South Wales*, **3** (1905), 400–485; "The Australasian Antarctic Expedition, 1911–1914," in *Geographical Journal*, **44** (1914), 257–286; *The Home of the Blizzard: Being the Story of the Australasian Antarctic Expedition 1911–1914*, 2 vols. (London, 1915); *The Winning of Australian Antarctica* (Sydney, 1963), the geographical report of the BANZARE 1929–1931 research expedition;

Macquarie Island: Its Geography and Geology, Australasian Antarctic Expedition 1911–1914, Scientific Reports, ser. A, vol. V (Sydney, 1943); and "The Adelaide Series As Developed Along the Western Margin of the Flinders Ranges," in *Transactions of the Royal Society of South Australia*, **71**, pt. 2 (1947), 259–280.

II. SECONDARY LITERATURE. See Charles Francis Laseron, *South With Mawson* (London, 1947), reminiscences of the Australasian Antarctic expedition of 1911–1914; Paquita Mawson, *Mawson of the Antarctic* (London, 1961); and A. Grenfell Price, *Mawson's B.A.N.Z.A.R.E. Voyage 1929–1931, Based on the Mawson Papers* (Sydney, 1963).

J. B. WATERHOUSE

MAXIMOV. See **Maksimov.**

MAXWELL, JAMES CLERK (*b.* Edinburgh, Scotland, 13 June 1831; *d.* Cambridge, England, 5 November 1879), *physics.*

Maxwell was a descendant of the Clerks of Penicuik, a family prominent in Edinburgh from 1670 on, who had twice intermarried during the eighteenth century with the heiresses of the Maxwells of Middlebie, illegitimate offspring of the eighth Lord Maxwell. His father, John Clerk (Maxwell), younger brother of Sir George Clerk, M.P., inherited the Middlebie property and took the name Maxwell in consequence of some earlier legal manipulations which prevented the two family properties being held together. The estate, some 1,500 acres of farmland near Dalbeattie in Galloway (southwestern Scotland), descended to Maxwell; and much of his scientific writing was done there. Maxwell's mother was Frances Cay, daughter of R. Hodshon Cay, a member of a Northumbrian family residing in Edinburgh. She died when he was eight years old. On both parents' sides Maxwell inherited intellectual traditions connected with the law, as was common in cultivated Edinburgh families. John Clerk Maxwell had been trained as an advocate, but his chief interest was in practical, technical matters. He was a fellow of the Royal Society of Edinburgh and published one scientific paper, a proposal for an automatic-feed printing press. Maxwell's father was a Presbyterian and his mother an Episcopalian. Maxwell himself maintained a strong Christian faith, with a strain of mysticism which has affinities with the religious traditions of the Galloway region, where he grew up.

From 1841 Maxwell attended Edinburgh Academy, where he met his lifelong friend and biographer, the Platonic scholar Lewis Campbell, and P. G. Tait. He entered Edinburgh University in 1847 and came under the influence of the physicist and alpinist James David Forbes and the metaphysician Sir William Hamilton. In 1850 he went up to Cambridge (Peterhouse one term, then Trinity), where he studied under the great private tutor William Hopkins and was also influenced by G. G. Stokes and William Whewell. He graduated second wrangler and first Smith's prizeman (bracketed equal with E. J. Routh) in 1854. He became a fellow of Trinity in 1855. Maxwell held professorships at Marischal College, Aberdeen, and King's College, London, from 1856 to 1865, when he retired from regular academic life to write his celebrated *Treatise on Electricity and Magnetism* and to put into effect a long-cherished scheme for enlarging his house. During the four years 1866, 1867, 1869, and 1870, he also served as examiner or moderator in the Cambridge mathematical tripos, instituting some widely praised reforms in the substance and style of the examinations. In 1871 he was appointed first professor of experimental physics at Cambridge and planned and developed the Cavendish Laboratory. On 4 July 1858 he married Katherine Mary Dewar, daughter of the principal of Marischal College and seven years his senior. They had no children. He died in 1879 at the age of forty-eight from abdominal cancer.

Maxwell's place in the history of physics is fixed by his revolutionary investigations in electromagnetism and the kinetic theory of gases, along with substantial contributions in several other theoretical and experimental fields: (1) color vision, (2) the theory of Saturn's rings, (3) geometrical optics, (4) photoelasticity, (5) thermodynamics, (6) the theory of servomechanisms (governors), (7) viscoelasticity, (8) reciprocal diagrams in engineering structures, and (9) relaxation processes. He wrote four books and about one hundred papers. He was joint scientific editor with T. H. Huxley of the famous ninth edition of the *Encyclopaedia Britannica*, to which he contributed many articles. His grasp of both the history and the philosophy of science was exceptional, as may be seen from the interesting philosophical asides in his original papers and from his general writings. His *Unpublished Electrical Researches of the Hon. Henry Cavendish* (1879) is a classic of scientific editing, with a unique series of notes on investigations suggested by Cavendish's work.

It was Maxwell's habit to work on different subjects in sequence, sometimes with an interval of several years between successive papers in the same field. Six years elapsed between his first and second papers on electricity (1855, 1861), twelve years between his

second and third major papers on kinetic theory (1867, 1879). The account of his work must therefore be grouped by subject rather than in strict chronological order; a description of his juvenile papers and the studies on color vision and Saturn's rings is useful in illustrating his intellectual development up to 1859.

Juvenilia (1845–1854). Maxwell's first paper was published when he was fourteen years old. It followed the efforts of D. R. Hay, a well-known decorative artist in Edinburgh, to find a method of drawing a perfect oval similar to the string property of the ellipse. Maxwell discovered that when the string used for the ellipse is folded back on itself n times toward one focus and m times toward the other, a true oval is generated, one of the kind first studied by Descartes in connection with the refraction of light. Although Descartes had described ways of generating the curves, Maxwell's method was new. His father showed the results to J. D. Forbes, who secured publication in the *Proceedings of the Royal Society of Edinburgh*. Shortly afterward Maxwell wrote a remarkable manuscript, which is reproduced in facsimile by his biographers, on the geometrical and optical properties of ovals and related curves of higher order. It afforded a foretaste of two of his lifelong characteristics: thoroughness and a predilection for geometrical reasoning. Both qualities, traditional in Scottish education, were powerfully reinforced for Maxwell by his teacher at Edinburgh Academy, James Gloag, a man of "strenuous character and quaint originality" to whom "mathematics was a mental and moral discipline."[1]

Three of Maxwell's next four papers were on geometrical subjects. One, "On the Theory of Rolling Curves" (1848), analyzed the differential geometry of families of curves generated like the cycloid, by one figure rolling on another. Another (1853) was a brief investigation in geometrical optics, leading to the beautiful discovery of the "fish-eye" lens. The third was "Transformation of Surfaces by Bending," which extended work begun by Gauss. The only paper from this period with a strictly physical subject was "On the Equilibrium of Elastic Solids," written in 1850 shortly before Maxwell went up to Cambridge. In 1847 he had been taken by his uncle John Cay to visit the private laboratory of the experimental optician William Nicol, from whom he received a pair of polarizing prisms. With these he investigated the phenomenon of induced double refraction in strained glass, which had been discovered in 1826 by another famous Scottish experimenter, Sir David Brewster. Maxwell's studies led him to the papers of Cauchy and Stokes. He developed a simple axiomatic formulation of the general theory of elasticity, solved various problems, and offered a conjectural explanation of induced double refraction based on strain functions. The alternative interpretation based on stress functions had been given earlier by F. E. Neumann, but Maxwell's theory was independent and better. The usefulness of photoelastic techniques in studying stress distributions in engineering structures is well-known: retrospectively the paper is even more important as Maxwell's first encounter with continuum mechanics. Its significance for his researches on the electromagnetic field and (more surprisingly) gas theory will shortly appear.

Color Vision (1850–1870). Maxwell created the science of quantitative colorimetry. He proved that all colors may be matched by mixtures of three spectral stimuli, provided subtraction as well as addition of stimuli is allowed. He revived Thomas Young's three-receptor theory of color vision and demonstrated that color blindness is due to the ineffectiveness of one or more receptors. He also projected the first color photograph and made other noteworthy contributions to physiological optics.

Credit for reviving Young's theory of vision is usually given to Helmholtz. His claim cannot be sustained. The paper it is based on, published in 1852, contained useful work, but Helmholtz overlooked the essential step of putting negative quantities in the color equations and explicitly rejected the three-receptor hypothesis;[2] and although Grassmann in 1854 pointed out fallacies in his reasoning, there is no evidence that Helmholtz followed the argument through to a conclusion until after Maxwell's work appeared. Artists had indeed known centuries before Maxwell or Helmholtz that the three so-called primary pigments, red, yellow, and blue, yield any desired hue by mixture; but several things clouded interpretation of the phenomena and hindered acceptance of Young's idea. One was the weight of Newton's claim that the prismatic spectrum contains seven primary colors rather than three. Another was the cool reception given to Young's theory of light, which extended to his theory of vision. The course of speculation between Young and Maxwell has never been clearly charted. In Britain the three-receptor theory did nearly gain acceptance during the 1820's. It was favorably discussed by John Herschel and Dalton as well as by Young: Herschel in particular suggested that Dalton's red blindness might come from the absence of one of Young's three receptors.[3] A curious complication supervened, however. During the 1830's Brewster performed experiments with absorption filters by which he claimed to demonstrate the existence of three kinds of light, distributed in various proportions throughout the spectrum. Color according to him

was thus an objective property of light, not a physiological function of the human eye. Brewster's interpretations were founded on his stubborn belief in the corpuscular theory of light, but the experiments seemed good and were accepted even by Herschel until Helmholtz eventually traced the effects to imperfect focusing. During the same period from 1830 on, wide general progress was made in physiological optics throughout Europe, in which the names of Purkinje, Haidinger, Johannes Müller, and Wartmann are memorable. In Britain the first statistical survey of color deficiency was conducted by George Wilson of Edinburgh—the chemist, and biographer of Cavendish—who brought to the subject a nice touch of topical alarmism through his lurid warnings about the dangers inherent in nighttime railway signaling. It was in an appendix to Wilson's monograph *On Colour Blindness* (1855) that Maxwell's first account of his researches appeared.

Maxwell began experiments on color mixing in 1849 in Forbes's laboratory at Edinburgh. At that time Edinburgh was unusually rich in students of color: besides Forbes, Wilson, and Brewster, there were William Swan, a physician interested in the eye, and D. R. Hay, who, in addition to his work in the geometry of design, had written a book entitled *Nomenclature of Colours* (1839) and supplied Forbes and Maxwell with tinted papers and tiles for their investigations. The experiments consisted in observing hues generated by colored sectors on a rapidly spinning disk. Forbes first repeated a standard experiment in which a series of colors representing those of the spectrum combine to give gray. He then tried to produce gray from combinations of red, yellow, and blue but failed—"and the reason was found to be, that blue and yellow do not make green, but a pinkish tint, when neither prevails in the combination."[4] No addition of red to this could produce a neutral tint.

Using a top with adjustable sectors of tinted paper, Forbes and Maxwell went on to obtain quantitative color equations, employing red, blue, and *green* as primaries. Interestingly, Young, in one little-known passage, had made the same substitution.[5] The standard rules for mixing pigments were explained by Maxwell, and independently by Helmholtz, as a secondary process, with the pigments acting as filters to light reflected from the underlying surface.

In 1854, after his graduation from Cambridge, Maxwell was able to resume these researches, which Forbes had been compelled by a severe illness to abandon. He improved the top by adding a second set of adjustable sectors of smaller diameter than the first, to make accurate color comparisons, and obtained equations for several groups of observers which could be manipulated algebraically in a consistent manner. For color-deficient observers only two variables were needed. Maxwell then went on to prove that Newton's method of displaying colors on a circle with white at the center implicitly satisfies the three-receptor theory, since it is equivalent to representing each color datum by a point in a three-dimensional space. With the experimental results plotted on a triangle having red, blue, and green corners, after the method of Young and Forbes, there is a white point *w* inside and an ordered curve of spectral colors outside the triangle very similar to Newton's circle. Adapting terminology from D. R. Hay, Maxwell distinguished three new variables—hue (spectral color), tint (degree of saturation), shade (intensity of illumination)—corresponding to "angular position with respect to *w*, distance from *w*, and coefficient [of intensity]." There is an easy transformation from these variables and to the representation of colors as a sum of three primaries: hence "the relation between the two methods of reducing the elements to three becomes a matter of geometry."[6] All this is most modern. In later correspondence with Stokes (1862), Maxwell described manipulations of color coordinates to reduce data from different observers to a common white point. The advantages of this procedure were also pointed out by C. J. Monro in a letter to Maxwell, dated 3 March 1871, which was published in Campbell and Garnett's *Life of James Clerk Maxwell* (1882), although other workers in colorimetry entirely ignored the idea until Ives and Guild rediscovered it fifty years later.[7]

To go further, a new instrument less susceptible than the color top to conditions of illumination and properties of paper was called for. Accordingly Maxwell devised what he called his "colour-box," in which mixtures of spectral stimulants were directly compared with a matching white field. The original version, perfected in 1858, consisted of two wooden boxes, each about three feet long, joined at an angle, containing a pair of refracting prisms at the intersection. An eyepiece was placed at one end; at the other were three slits, adjustable in position and aperture, which could be set at positions corresponding to any three wavelengths *A*, *B*, *C* in the spectrum formed by projecting white light through the eyepiece. By the principle of reciprocity, white light entering the slits yielded mixtures of *A*, *B*, *C* at the eyepiece, with intensities determined by the widths of the slits. Light from the same source (a sheet illuminated by sunlight) entered another aperture and was reflected past the edge of the second prism to the eyepiece, where the observer saw, side by side, two fields which

he could match in hue and intensity. The spectrum locus determined by Maxwell's observer *K* (his wife) is shown in Figure 1, together with the results of König and Abney (1903, 1913) and the 1931 standard observer. The Maxwells come out of the comparison rather well. Maxwell designed two other "colour-boxes" on the same general principle. The second was made portable by the use of folded optics on the principle afterward adapted to the spectroscope by Littrow. The third gave hues of exceptional spectral purity by adopting a "double monochromator" principle, illuminating the slits with the spectrum from a second train of prisms symmetrically disposed rather than with direct sunlight. With it Maxwell studied variations of color sensitivity across the retina, a subject he had become interested in through his observations of the "Maxwell spot."

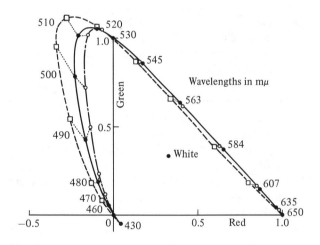

FIGURE 1. Spectrum loci determined by different experimenters.

——— ○　Maxwell 1860 (observer *K*)
— — — □　König, Abney 1903, 1913 (recalculated Weaver)
————— ●　C.I.E. standard observer 1931 (Wright, Guild)

Most people, when they look at an extended source of polarized light, intermittently perceive a curious pair of yellow structures resembling a figure eight, with purple wings at the waist. These are the "brushes" discovered by Haidinger in 1844. They may be seen especially clearly by looking at a blue surface through a Nicol prism. Maxwell studied them with the prisms he had received from Nicol; and at the British Association meeting of 1850 he proposed attributing them to a polarizing structure in the yellow spot on the retina, a hypothesis which brought him into an amusing confrontation with Brewster, who attributed them to the cornea.[8] Maxwell's explanation is now accepted. In 1855 he noticed, in the blue region of the spectrum formed by looking through a prism at a vertical slit, an elongated dark spot which moved up and down with the eye and possessed the same polarizing structure as Haidinger's brushes. This is the Maxwell spot. Later his wife discovered that she could not see the spot, there being almost no yellow pigment on her retinas. Noticing also a large discrepancy between her white point and his, Maxwell then found that his own color matches contained much less blue in the extrafoveal region and he proceeded to investigate variations of sensitivity across the retina for a large number of observers. He was able to exhibit the yellow spot—as he wrote to C. J. Monro in 1870—to "all who have it,—and all have it except Col. Strange, F. R. S., my late father-in-law and my wife,—whether they be Negroes, Jews, Parsees, Armenians, Russians, Italians, Germans, Frenchmen, Poles, etc."[9] Summaries of the work appeared in two brief papers and in a delightful correspondence with Monro, which also contains an interesting discussion on differences in color nomenclature between ancient and modern languages.

In 1861 Maxwell projected the first trichromatic color photograph at the Royal Institution before an audience which included Faraday. The subject was a tartan ribbon photographed through red, green, and blue filters by Thomas Sutton, a colleague at King's College, London, and then projected through the same filters. An odd fact which remained without explanation for many years was that the wet collodion plates used should not have given any red image, since that photographic process is completely insensitive to red. Yet contemporary descriptions make it clear that the colors were reproduced with some fidelity. In 1960 R. M. Evans and his colleagues at Kodak Research Laboratories, in a first-class piece of historical detective work, established that the red dyes in Maxwell's ribbon also reflected ultraviolet light in a region just coinciding with a pass band in the ferric thiocyanate solution used as a filter. The "red" image was really obtained with ultraviolet light! The hypothesis is confirmed by the fact that the original red plate preserved at Cambridge is slightly out of focus, although Sutton carefully refocused the camera for visible red light. A repetition of the experiment under modern conditions gave a "surprisingly colorful reproduction of the original scene."[10]

Saturn's Rings (1855–1859). In 1855 the topic of the fourth Adams prize at Cambridge was announced as an investigation of the motions and stability of the rings of Saturn. Some calculations on Saturn's rings, treated as solid bodies, had been given as early as 1787 by Laplace. He established that a uniform rigid ring would disintegrate unless (1) it is rotating at a speed where the centrifugal force balances the attraction of the planet and (2) the ratio ρ_r/ρ_s of its density to the density of Saturn exceeds a critical value 0.8, such that the attractions between inner and outer portions of the ring exceed the differences between centrifugal and gravitational forces at different radii. Also, the motions of a uniform ring are dynamically unstable: any displacement from equilibrium leads to an increased attraction in the direction of displacement, precipitating the ring against the planet. Laplace conjectured, however, that the motion is somehow stabilized by irregularities in the mass distribution; and in his dogmatic way asserted that the rings of Saturn are irregular solid bodies. That was where the theory still stood in 1855; meanwhile, a new dark ring and further divisions in the existing rings had been observed, along with some evidence for slow changes in the overall dimensions of the system during the 200 years since its discovery. The examiners, James Challis, Samuel Parkinson, and William Thomson (the future Lord Kelvin), called for explanations of each point and an investigation of dynamical stability on the hypothesis that the rings are: (1) solid, (2) fluid, (3) composed of "masses of matter not mutually coherent."[11] These were the questions on which Maxwell spent much time between 1855 and 1859 in the essay to which the prize was awarded.

Maxwell took up first the theory of the solid ring where Laplace had left it, and determined conditions for stability of a ring of arbitrary shape. Forming equations of motion in terms of the potential at the center of Saturn due to the ring, he obtained two restrictions on the first derivatives of the potential for uniform motion, and then, by a Taylor expansion, three more conditions on the second derivatives for stable motion. Maxwell next transformed these results into conditions on the first three coefficients of a Fourier series in the mass distribution. He was able to show that almost every conceivable ring was unstable except the curious special case of a uniform ring loaded at one point with a mass between 4.43 and 4.87 times the remaining mass. There the uneven distribution makes the total attraction act toward a point outside the ring in such a way that the instabilities affecting the moment of inertia are counteracted by a couple which alters the angular momentum. But such a ring would collapse under the uneven stress, and its lopsidedness would be plainly visible. The hypothesis of a solid ring is untenable.

In considering nonrigid rings Maxwell again utilized Fourier's theorem, but in a different way, examining the stability of various rings by expanding disturbances in their form into a series of waves. He took as a starting model, with which more complex structures could later be compared, a ring of solid satellites, equally spaced and all of equal mass. The motions may be resolved into four components: rotation about Saturn with constant angular velocity ω and small displacement ρ, σ, ζ in directions radial, tangential, and normal to the plane of the ring. Normal displacements of any satellite are manifestly stable, for the components of attraction to the other bodies always constitute a restoring force. Tangential disturbances might be expected to be unstable, since the attractions to neighboring satellites are in the direction of displacement; but Maxwell discovered that radial and tangential waves of a given order may be coupled together in a stable manner because the radial motions generate Coriolis forces through the rotation about the planet, which counterbalance the gravitational forces due to tangential motions. Detailed analysis revealed four kinds of waves, grouped in two pairs, all of which are stable if the mass of the central body is great enough. The motions are rather complicated; and Maxwell, "for the edification of sensible image worshippers,"[12] had a mechanical model constructed to illustrate them in a ring of thirty-six satellites. Waves of the first two kinds move in opposite directions with respect to a point on the rotating ring, with a velocity nearly equal to ω/n, where ω is the angular velocity of the ring and n the number of undulations. Thus if there are five undulations the wave velocity is 1/5 of the ring velocity. Each satellite describes an elliptical path about its mean position in a sense opposite to the rotation of the ring itself, the major axis of the ellipse being approximately twice the minor axis and lying near the tangential plane. If the number of satellites is μ, the highest-order waves, which are most likely to disrupt the ring, have $\mu/2$ undulations. The stability criterion is

$$S > 0.4352\mu^2 R, \qquad (1)$$

S and R being the masses of Saturn and the ring. Stability is determined by tangential forces; the parameter defining them must lie between 0 and $0.07\omega^2$.

For rings of finite breadth Maxwell's procedure was to examine simplified models which bracket the true situation. He began with rings whose inner and outer parts are so strongly bound together that they rotate uniformly. Such rings may be called semirigid. They are evidently subject to Laplace's criterion of cohesion $\rho_r/\rho_s > 0.8$ but, like a ring of satellites, are also subject to conditions of stability against tangential disturbances. Maxwell established that tangential forces disrupt a semirigid ring of particles unless

$\rho_r/\rho_s < 0.003$ and one of incompressible fluid unless $\rho_r/\rho_s < 0.024$. Since neither is compatible with Laplace's criterion, neither kind of semirigid ring is stable. Various arguments then disposed of other gaseous and liquid rings, leaving as the only stable structure concentric circles of small satellites, each moving at a speed appropriate to its distance from Saturn. Such rings cannot be treated independently: they attract one another. Maxwell presented a lengthy investigation of mutual perturbations between two rings. Usually they are stable; but at certain radii waves of different order may come into resonance and cause disruption, making particles fly off in all directions and collide with other rings. Maxwell estimated the rate of loss of energy and concluded that the whole system of rings would slowly spread out, as the observations indicated. He did not then study the general problem of motion among colliding bodies, but his unpublished manuscripts include one from 1863 applying the statistical methods that he later developed in the kinetic theory of gases to Saturn's rings.

In 1895 A. A. Belopolsky and C. Keeler independently confirmed the differential rotation of the rings by spectroscopic observations. Later the gaps between successive rings were attributed to resonances between orbital motions of the primary satellites of Saturn and local ring oscillations. More recently A. F. Cook and F. A. Franklin, using kinetic theory techniques, have shown that heat generated in collisions makes the rings expand in thickness unless it is removed by radiation, and have obtained closer restrictions on structure and density. Maxwell's density limit $\rho_r/\rho_s > 0.003$ has sometimes been interpreted as a limit on the actual rings; in reality it applies only to semirigid rings, where stability depends on tangential forces. With differential rotation the tangential waves are heavily damped and stability depends on radial motions. The true upper limit on ρ_r/ρ_s appears to be in the range 0.04 to 0.20. Spectroscopic evidence suggests the particles may be crystals of ice or carbon dioxide.[13]

The essay on Saturn's rings illustrates Maxwell's debt to Cambridge as sharply as the experiments on color vision reveal his debt to Edinburgh. It also established his scientific maturity. Success with a classical problem of such magnitude gave a mathematical self-assurance vitally important to his later work. Many letters testify to the concentrated effort involved, not the least interesting amongst them being one to a Cambridge friend, H. R. Droop: "I am very busy with Saturn on top of my regular work. He is all remodelled and recast, but I have more to do to him yet for I wish to redeem the character of mathematicians and make it intelligible."[14] To the graceful literary style and analytical clarity established there, two further broad qualities were added in Maxwell's work over the next two decades. The great papers of the 1860's continued at much the same level of analytical technique, with epoch-making advances in physical and philosophic insight. The books and articles of the 1870's display growing mastery of mathematical abstraction in the use of matrices, vectors and quaternions, Hamiltonian dynamics, special functions, and considerations of symmetry and topology. The contrasting ways in which these different phases of Maxwell's mature researches reflect his interaction with his contemporaries and his influence on the next scientific generation form a fascinating study which has not yet received due attention.

Electricity, Magnetism, and the Electromagnetic Theory of Light (1854–1879). Maxwell's electrical researches began a few weeks after his graduation from Cambridge in 1854, and ended just before his death twenty-five years later with a referee's report on a paper by G. F. FitzGerald. They fall into two broad cycles, with 1868 roughly the dividing point: the first a period of five major papers on the foundations of electromagnetic theory, the second a period of extension with the *Treatise on Electricity and Magnetism*, the *Elementary Treatise on Electricity*, and a dozen shorter papers on special problems. The position of the *Treatise* is peculiar. Most readers come to it expecting a systematic exposition of its author's ideas which makes further reference to earlier writings unnecessary. With many writers the expectation might be legitimate; with Maxwell it is a mistake. In a later conversation he remarked that the aim of the *Treatise* was not to expound his theory finally to the world but to educate himself by presenting a view of the stage he had reached.[15] This is a clue well worth pondering. The truth is that by 1868 Maxwell had already begun to think beyond his theory. He saw electricity not as just another branch of physics but as a subject of unique strategic importance, "as an aid to the interpretation of nature . . . and promoting the progress of science."[16] Wishing, therefore, to follow up questions with wider scientific ramifications, he gave the *Treatise* a loose-knit structure, organized on historical and experimental rather than deductive lines. Ideas are exhibited at different phases of growth in different places; different sections are developed independently, with gaps, inconsistencies, or even flat contradictions in argument. It is a studio rather than a finished work of art. The studio, being Maxwell's, is tidily arranged; and once one has grasped what is going on, it is wonderfully instructive to watch the artist at work; but anyone who finds himself there unawares is

courting bewilderment, the more so if he overlooks Maxwell's advice to read the four parts of the *Treatise* in parallel rather than in sequence. It is, for example, disconcerting to be told on reaching section 585, halfway through volume II, that Maxwell is now about to "begin again from a new foundation without any assumption except those of the dynamical theory as stated in Chapter VII." Similar difficulties occur throughout. The next fifty years endorsed Maxwell's judgment about the special importance of electricity to physics as a whole. His premature death occurred just as his ideas were gaining adherents and he was starting an extensive revision of the *Treatise*. Not the least unfortunate consequence was that the definitive exposition of his theory which he intended was never written.

Seen in retrospect, the course of physics up to about 1820 is a triumph of the Newtonian scientific program. The "forces" of nature—heat, light, electricity, magnetism, chemical action—were being progressively reduced to instantaneous attractions and repulsions between the particles of a series of fluids. Magnetism and static electricity were already known to obey inverse-square laws similar to the law of gravitation. The first forty years of the nineteenth century saw a growing reaction against such a division of phenomena in favor of some kind of "correlation of forces." Oersted's discovery of electromagnetism in 1820 was at once the first vindication and the most powerful stimulus of the new tendency, yet at the same time it was oddly disturbing. The action he observed between an electric current and a magnet differed from known phenomena in two essential ways: it was developed by electricity in motion, and the magnet was neither attracted to nor repelled by—but set transversely to—the wire carrying the current. To such a strange phenomenon widely different reactions were possible. Faraday took it as a new irreducible fact by which his other ideas were to be shaped. André Marie Ampère and his followers sought to reconcile it with existing views about instantaneous action at a distance.

Shortly after Oersted's discovery Ampère discovered that a force also exists between two electric currents and put forward the brilliant hypothesis that all magnetism is electrical in origin. In 1826 he established a formula (not to be confused with the one attached to his name in textbooks) which reduced the known magnetic and electromagnetic phenomena to an inverse-square force along the line joining two current elements idl, $i'dl'$ separated by a distance r,

$$F_{ii'} = \frac{ii'\, dl\, dl'}{r^2}\, G, \qquad (2)$$

where G is a complex geometrical factor involving the angles between r, dl, and dl'. In 1845 F. E. Neumann derived the potential function corresponding to Ampère's force and extended the theory to electromagnetic induction. Another extension developed by Wilhelm Weber was to combine Ampère's law with the law of electrostatics to form a new theory, which also accounted for electromagnetic induction, treating the electric current as the flow of two equal and opposite groups of charged particles, subject to a force whose direction was always along the line joining two particles e, e', but whose magnitude depended on their relative velocity $\dot{r}$ and relative acceleration $\ddot{r}$ along that line,

$$F_{ee'} = \frac{ee'}{r^2}\left[1 - \frac{1}{c^2}(\dot{r}^2 - 2r\ddot{r})\right], \qquad (3)$$

c being a constant with dimensions of velocity. In 1856 Kohlrausch and Weber determined c experimentally by measuring the ratio of electrostatic to electrodynamic forces. Its value in the special units of Weber's theory was about two-thirds the velocity of light. Equations (2) and (3) and Neumann's potential theory provided the starting points for almost all the work done in Europe on electromagnetic theory until the 1870's.

The determining influences on Maxwell were Faraday and William Thomson. Faraday's great discoveries—electromagnetic induction, dielectric phenomena, the laws of electrochemistry, diamagnetism, magneto-optical rotation—all sprang from the search for correlations of forces. They formed, in Maxwell's words, "the nucleus of everything electric since 1830."[17] His contributions to theory lay in the progressive extension of ideas about lines of electric and magnetic force. His early discovery of electromagnetic rotations (the first electric motor) made him skeptical about attractive and repulsive forces, and his ideas rapidly advanced after 1831 with his success in describing electromagnetic induction by the motion of lines of magnetic force through the inductive circuit. In studying dielectric and electrolytic processes he imagined (wrongly) that their transmission in curved lines could not be reconciled with the hypothesis of direct action at a distance, attributing them instead to successive actions of contiguous portions of matter in the space between charged bodies. In his work on paramagnetism and diamagnetism he conceived the notion of magnetic conductivity (permeability); and finally, in the most brilliant of of his conceptual papers, written in 1852, when he was sixty, he extended the principle of contiguous action in a general qualitative description of magnetic and electromagnetic phenomena, based on the assumption that lines of magnetic force have the physical property of shortening themselves and repelling each other sideways. A quantitative formulation of the last hypothesis was given by Maxwell in 1861.

Thomson's contribution began in 1841, while he was an undergraduate at Cambridge. His first paper established a formal analogy between the equations of electrostatics and the equations for flow of heat. Consider a point source of heat P embedded in a homogeneous conducting medium. Since the surface area of a sphere is $4\pi r^2$, the heat flux ϕ through a small area dS at a distance r from P is proportional to $1/r^2$ in analogy with Coulomb's electrostatic law; thus by appropriate substitution a problem in electricity may be transposed into one in the theory of heat. Originally Thomson used the analogy as a source of analytical technique; but in 1845 he went on to examine and dispose of Faraday's widely accepted claim that dielectric action cannot be reconciled with Coulomb's law and, conversely, to supply the first exact mathematical description of lines of electric force. Later Thomson and Maxwell between them established a general similitude among static vector fields subject to the conditions of continuity and incompressibility, proving that identical equations describe (1) streamlines of frictionless incompressible fluids through porous media, (2) lines of flow of heat, (3) current electricity, and (4) lines of force in magnetostatics and electrostatics.

Since it was Thomson's peculiar genius to generate powerful disconnected insights rather than complete theories, much of his work is best described piecemeal along with Maxwell's; but certain of his ideas from the 1840's may first be mentioned, notably the method of electric images, a second formal analogy between magnetic forces and rotational strains in an elastic solid, and, most important, the many applications of energy principles to electricity which followed his involvement with thermodynamics. Amongst other things Thomson is responsible for the standard expressions $\frac{1}{2}Li^2$ and $\frac{1}{2}CV^2$ for energy in an inductance and in a condenser. He and (independently) Helmholtz also applied energy principles to give an extraordinarily simple derivation of Neumann's induction equation. It so happened that the discussion of energy principles had a curious two-sided impact on Weber's hypothesis. In 1846 Helmholtz presented an argument which seemed to show that the hypothesis was inconsistent with the principle of conservation of energy. His conclusion was widely accepted and formed one of the grounds on which Maxwell opposed the theory, but in 1869 Weber succeeded in rebutting it. By then, however, Maxwell had developed his theory, and the implication of the Thomson-Helmholtz argument had become clearer: that any theory which is consistent with energy principles automatically predicts induction. In retrospect, therefore, although Helmholtz was wrong in his first criticism, the agreement between Weber's theory and experiment was also less compelling than Weber and his friends had supposed.

Maxwell's first paper, "On Faraday's Line of Force" (1855–1856), was divided into two parts, with supplementary examples. Its origin may be traced in a long correspondence with Thomson, edited by Larmor in 1936.[18] Part 1 was an exposition of the analogy between lines of force and streamlines in an incompressible fluid. It contained one notable extension to Thomson's treatment of the subject and also an illuminating opening discourse on the philosophical significance of analogies between different branches of physics. This was a theme to which Maxwell returned more than once. His biographers print in full an essay entitled "Analogies in Nature," which he read a few months later (February 1856) to the famous Apostles Club at Cambridge; this puts the subject in a wider setting and deserves careful reading despite its involved and cryptic style. Here, as elsewhere, Maxwell's metaphysical speculation discloses the influence of Sir William Hamilton, specifically of Hamilton's Kantian view that all human knowledge is of relations rather than of things. The use Maxwell saw in the method of analogy was twofold. It cross-fertilized technique between different fields, and it served as a golden mean between analytic abstraction and the method of hypothesis. The essence of analogy (in contrast with identity) being partial resemblance, its limits must be recognized as clearly as its existence; yet analogies may help in guarding against too facile commitment to a hypothesis. The analogy of an electric current to two phenomena as different as conduction of heat and the motion of a fluid should, Maxwell later observed, prevent physicists from hastily assuming that "electricity is either a substance like water, or a state of agitation like heat."[19] The analogy is geometrical: "a similarity between relations, not a similarity between the things related."[20]

Maxwell improved the presentation of the hydrodynamic analogy chiefly by considering the resistive medium through which the fluid moves. When an incompressible fluid goes from one medium into another of different porosity, the flow is continuous but a pressure difference develops across the boundary. Also, when one medium is replaced by another of different porosity, equivalent effects may be obtained formally by introducing appropriate sources or sinks of fluid at the boundary. These results were an impor-

tant aid to calculation and helped in explaining several processes that occur in magnetic and dielectric materials. Another step was to consider a medium in which the porosity varies with direction. The necessary equations had been supplied by Stokes in a paper on the conduction of heat in crystals. They led to the remarkable conclusion that the vector **a** which defines the direction of fluid motion is not in general parallel to α, the direction of maximum pressure gradient. The two functions are linked by the equation

$$\mathbf{a} = \mathbf{K}\alpha, \tag{4}$$

where **K** is a tensor quantity describing the porosity. Applying the analogy to magnetism, Maxwell distinguished two vectors, the magnetic induction and the magnetic force, to which he later attached the symbols **B** and **H**. The parallel quantities in current electricity were the current density **I** and the electromotive intensity **E**. The distinction between **B** and **H** provided the key to a description of "magnecrystallic induction," a force observed in crystalline magnetic materials by Faraday. Maxwell later identified the two quantities with the two definitions of magnetic force that Thomson had found to be required in developing parallel magnetostatic and electromagnetic theories of magnetism. The question of the two magnetic vectors **B** and **H** has disturbed several generations of students of electromagnetism. Maxwell's discussion gives a far clearer starting point than anything to be found in the majority of modern textbooks on the subject.

This physical distinction based on the hydrodynamic analogy led Maxwell to make an important mathematical distinction between two classes of vector functions, which he then called "quantities" and "intensities," later "fluxes" and "forces." A flux **a** is a vector subject to the continuity equation and is integrated over a surface; a force α (in Maxwell's generalized sense of the term) is a vector usually, but not always, derivable from a single-valued potential function and is integrated along a line. The functions **B** and **I** are fluxes; **H** and **E** are forces.

The close parallel that exists between electric currents and magnetic lines of force, which had been seen qualitatively by Faraday, was the concluding theme of Part 1 of Maxwell's paper. Part 2 covered electromagnetism proper. In it Maxwell developed a new formal theory of electromagnetic processes. The starting point was an identity established by Ampère and Gauss between the magnetic effects of a closed electric current and those of a uniformly magnetized iron shell of the same perimeter. In analytic method the discussion followed Thomson's "Mathematical Theory of Magnetism" (1851), as well as making extensive use of a theorem first proved by Thomson in 1847, in a letter to Stokes, and first published by Stokes as an examination question in the Smith's prize paper taken by Maxwell in February 1854. This was the well-known equality (Stokes's theorem) between the integral of a vector function around a closed curve and the integral of its curl over the enclosed surface. The original analysis given by Maxwell was Cartesian, but since in 1870 he himself introduced the terms "curl," "divergence," and "gradient" to denote the relevant vector operations, the notation may legitimately be modernized. The relationship between the flux and force vectors **a** and **α** contained in equation (4) has already been discussed. Pursuing a line of analysis started by Thomson, Maxwell now proceeded to show that any flux vector **a** may be related to a second, distinct force vector **α'** through the equation

$$\mathbf{a} = \operatorname{curl} \boldsymbol{\alpha'} + \operatorname{grad} \beta, \qquad (5)$$

where β is a scalar function. Applying (4), (5) and other equations, Maxwell obtained a complete set of equations between the four vectors **E**, **I**, **B**, **H**, which describe electric currents and magnetic lines of force. He then went on to derive another vector function, for which he afterward used the symbol **A**, such that

$$\mathbf{B} = \operatorname{curl} \mathbf{A} + \operatorname{grad} \varphi, \qquad (6)$$

where the second term on the right-hand side may, in the absence of free magnetic poles, be eliminated by appropriate change of variables. Maxwell proved that the electromotive force **E** developed during induction is $-\partial\mathbf{A}/\partial t$ and that the total energy of an electromagnetic system is $\int \mathbf{I} \cdot \mathbf{A} \, dV$. Thus the new function provided equations to represent ordinary magnetic action, electromagnetic induction, and the forces between closed currents. Maxwell called it the electrotonic function, following some speculations of Faraday's about a hypothetical state of stress in matter, the "electrotonic state." Later he identified it as a generalization of Neumann's electrodynamic potential and established other properties (to be discussed shortly).

The 1856 paper has been eclipsed by Maxwell's later work, but its originality and importance are greater than is usually thought. Besides interpreting Faraday's work and giving the electrotonic function, it contained the germ of a number of ideas which Maxwell was to revive or modify in 1868 and later: (1) an integral representation of the field equations (1868), (2) the treatment of electrical action as analogous to the motion of an incompressible fluid (1869, 1873), (3) the classification of vector functions into forces and fluxes (1870), and (4) an interesting formal symmetry in the equations connecting **A**, **B**, **E**, and **H**, different from the symmetry commonly recognized in the completed field equations. The paper ended with solutions to a series of problems, including an application of the electrotonic function to calculate the action of a magnetic field on a spinning conducting sphere.

Maxwell's next paper, "On Physical Lines of Force" (1861–1862), began as an attempt to devise a medium occupying space which would account for the stresses associated by Faraday with lines of magnetic force. It ended with the stunning discovery that vibrations of the medium have properties identical with light. The original aim was one Maxwell had considered in 1856, and although he explicitly rejected any literal interpretation of the analogy between magnetic action and fluid motion, the meaning of the analogy can be extended by picturing a magnet as a kind of suction tube which draws in fluid ether at one end and expels it from the other. That idea had been suggested by Euler in 1761;[21] it leads to a most remarkable result first published by Thomson in 1870 but probably known to Maxwell earlier.[22] Geometrically the flow between two such tubes is identical with the lines of force between two magnets, but physically the actions are reciprocal: like ends of the tubes are *attracted* according to the inverse-square law; unlike ends are repelled. The difference is that in a fluid the Bernoulli forces create a pressure minimum where the streamlines are closest, while Faraday's hypothesis requires a pressure maximum.

The clue to to a medium having a right stress distribution came from an unexpected source. During the 1840's the engineer W. J. M. Rankine (who like Maxwell had been a student of Forbes's at Edinburgh) worked out a new theory of matter with applications to thermodynamics and the properties of gases, based on the hypothesis that molecules are small nuclei in an ethereal atmosphere, fixed in space but rotating at speeds proportional to temperature. In 1851 Thomson refereed one of Rankine's papers. He was then concerning himself with thermodynamics, but five years later it dawned on him that molecular rotation was just the thing to account for the magneto-optical effect.[23] Faraday had observed a slight rotation in the plane of polarization of light passing through a block of glass between the poles of a magnet. Using an analogy with a pendulum suspended from a spinning arm, Thomson concluded that the effect could be attributed to coupling between the ether vibrations and a spinning motion of the molecules of glass about the lines of force. Maxwell's theory of physical lines of force consisted in extending this hypothesis of rotation in the magnetic field from ordinary matter to an ether. The influence of Thomson and Rankine is established by direct reference and by Maxwell's use of Rankine's term "molecular vortices" in the titles of each of the four parts of the paper. The charm of the story is that barely twelve months had passed since Maxwell had given the death blow to Rankine's theory of gases through his own work on kinetic theory.

Consider an array of vortices embedded in incompressible fluid. Normally the pressure is identical in all directions, but rotation causes centrifugal forces which make each vortex contract longitudinally and exert radial pressure. This is exactly the stress distribution proposed by Faraday for physical lines of force. By making the angular velocity of each vortex proportional to the local magnetic intensity, Maxwell obtained formulas identical with the existing theories for forces between magnets, steady currents, and diamagnetic bodies. Next came the problem of electromagnetic induction. It required some understanding of the action of electric currents on the vortex medium. That tied in with another question: how could two adjacent vortices rotate freely in the same sense,

since their surfaces move in opposite directions? Figure 2, reproduced from Part 2 of the paper, illustrates Maxwell's highly tentative solution. Each vortex is separated from its neighbors by a layer of minute particles, identified with electricity, counter-rotating like the idle wheels of a gear train.

On this view electricity, instead of being a fluid confined to conductors, becomes an entity of a new kind, disseminated through space. In conductors it is free to move (though subject to resistance); in insulators (including the ultimate insulator, space) it remains fixed. The magnetic and inductive actions of currents are then visualized as follows. When a current flows in a wire A, it makes the adjacent vortices rotate; these in turn engage the next layer of particles and so on until an infinite series of vortex rings, which constitute lines of force, fills the surrounding space. For induction consider a second wire B with finite resistance, parallel to A. A steady current in A will not affect B; but any change in A will communicate an impulse through the intervening particles and vortices, causing a reverse current in B, which is then dissipated through resistance. This is induction. Quite unex-

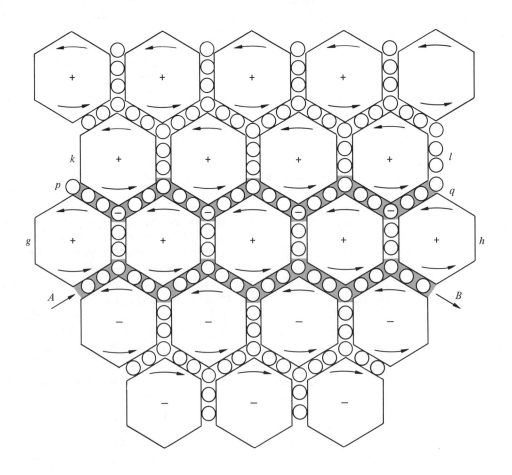

FIGURE 2.   Model of molecular vortices and electric particles (1861).

pectedly the model also suggested a physical interpretation of the electrotonic function. In analyzing machinery several engineers, including Rankine, had found it useful to add to the motion of a mechanical part terms incorporating effects of connected gears and linkages, which they called the "reduced" inertia or momentum of the system. Maxwell discovered that the electrotonic function corresponds to the reduced momentum of the vortex system at each point. The equation for induced electromotive force $\mathbf{E} = \partial\mathbf{A}/\partial t$ is the generalized electrical equivalent of Newton's equation between force and rate of change of momentum.

There is good evidence internal and external to the paper that Maxwell meant originally to end here and did not begin Part 3 until Part 2 had been printed.[24] Meanwhile, he had been considering the relation between electric currents and the induction of charge through a dielectric. In 1854 he had remarked to Thomson that a literal treatment of the analogy between streamlines and lines of electric force would make induction nothing more than an extreme case of conduction.[25] Now, with the picture of electricity as disseminated in space, Maxwell hit upon a better description, based partly on Faraday's ideas, by making the vortex medium elastic. The forces between charged bodies could be attributed to potential energy stored in the medium by elastic distortion, as magnetic forces are attributed to stored rotational energy; and the difference between conduction and static electric induction is analogous to the difference between viscous and elastic processes in matter.

Two amazing consequences swiftly followed. First, since the electric particles surrounding a conductor are now capable of elastic displacement, a varying current is no longer entirely confined like water in a pipe: it penetrates to some extent into the space surrounding the wire. Here was the first glimmering of Maxwell's "displacement current." Second, any elastic substance with density ρ and shear modulus m can transmit transverse waves with velocity $v = \sqrt{m}/\rho$. Making some ad hoc assumptions about the elastic structure of the vortex medium, Maxwell derived while he was in Scotland formulas connecting ρ and m with electromagnetic quantities, which implied a numerical relationship between v and Weber's constant c. Returning to London for the academic year, Maxwell looked up the result of Kohlrausch and Weber's experiment to determine c, and after putting their data in a form suitable for insertion into his equation he found that for a medium having a magnetic permeability μ equal to unity v was almost equal to the velocity of light. With excitement manifested in italics he wrote: "we can scarcely avoid the inference

that *light consists in the transverse undulations of the same medium which is the cause of electric and magnetic phenomena.*"[26] Thus the great discovery was made; and Maxwell, following a calculation on the dielectric properties of birefringent crystals, returned in Part 4 to his starting point, the magneto-optical effect, and replaced Thomson's spinning pendulum analogy with a more detailed theory in better accord with experiment.

In 1861 the British Association formed a committee under Thomson's chairmanship to determine a set of internationally acceptable electrical standards following the work of Weber. At Thomson's urging, a new absolute system of units was adopted, similar to Weber's, but based on energy principles rather than on a hypothetical electrodynamic force law. The first experiment was on the standard of resistance, and in 1862 Maxwell was appointed to the committee to help with that task. His third paper, "On the Elementary Relations of Electrical Quantities," written in 1863 with the assistance of Fleeming Jenkin, supplied a vital step in his development, often overlooked through its having been, most unfortunately, omitted from the *Scientific Papers*.[27] Extending a procedure begun by Fourier in the theory of heat, Maxwell set forth definitions of electric and magnetic quantities related to measures M, L, T of mass, length, and time, to provide the first—and one may also think the most lucid—exposition of that dual system of electrical units commonly but incorrectly known as the Gaussian system.[28] The paper introduced the notation, which was to become standard, expressing dimensional relations as products of powers of M, L, T enclosed in brackets, with separate dimensionless multipliers. For every quantity the ratio of the two absolute definitions, based on forces between electric charges and forces between magnetic poles, proved to be some power of a constant c with dimensions $[LT^{-1}]$ and magnitude $\sqrt{2}$ times Weber's constant, or very nearly the velocity of light. The analysis disclosed five different classes of experiments from which c might be determined. One was a direct comparison of electrostatic and electromagnetic forces carried out by Maxwell and C. Hockin in 1868, and two others were started by Maxwell at Cambridge in the 1870's.[29] The results of many experiments over the next few years progressively converged with the measured velocity of light.

By 1863, then, Maxwell had found a link of a purely phenomenological kind between electromagnetic quantities and the velocity of light. His fourth paper, "A Dynamical Theory of the Electromagnetic Field," published in 1865, clinched matters. It provided a new theoretical framework for the subject, based on experiment and a few general dynamical principles, from which the propagation of electromagnetic waves through space followed without any special assumptions about molecular vortices or the forces between electric particles. This was the work of which Maxwell, in a rare moment of unveiled exuberance, wrote to his cousin Charles Cay, the mathematics master at Clifton College: "I have also a paper afloat, containing an electromagnetic theory of light, which, till I am convinced to the contrary, I hold to be great guns."[30]

Several factors, scientific and philosophical, settled the disposition of Maxwell's artillery. From the beginning he had stressed the provisional character of the vortex model, especially its peculiar gearing of particles and vortices. Rankine was a cautionary example. In an article on thermodynamics written in 1877 Maxwell illuminated his own thought by observing that the vortex theory of matter, which at first served Rankine well, later became an encumbrance, distracting his attention from the general considerations on which thermodynamic formulas properly rest.[31] Maxwell wished to avoid that trap. Yet he did not abandon all the ground gained in 1862. The idea of treating light and electromagnetism as processes in a common medium remained sound. Furthermore, the new theory was, as the title of the paper stated, a dynamical one: the medium remained subject to the general principles of dynamics. The novelty consisted in deducing wave propagation from equations related to electrical experiments instead of from a detailed mechanism; that was why the theory became known as the electromagnetic theory of light. Again Sir William Hamilton's influence is discernible. Maxwell's decision to replace the vortex model of electromagnetic and optical processes by an analysis of the relations between the two classes of phenomena is a concretization of Hamilton's doctrine of the relativity of knowledge: all human knowledge is of relations between objects rather than of objects in themselves.

More specifically the theory rested on three main principles. Maxwell retained the idea that electric and magnetic energy are disseminated, merely avoiding commitment to hypotheses about their mechanical forms in space. Here it is worth noticing that his formal expressions $\mathbf{B} \cdot \mathbf{H}/8\pi$ and $\mathbf{D} \cdot \mathbf{E}/8\pi$ for the two energy densities simply extend and interpret physically an integral transformation of Thomson's.[32] Next Maxwell revived various ideas about the geometry of lines of force from the 1856 paper. Third, and most important, he replaced the vortex hypothesis with a new macroscopic analogy between inductive circuits and coupled dynamical systems. The analogy

seems to have germinated in Maxwell's mind in 1863, while he was working out the theory of the British Association resistance experiment.[33] In part it goes back to Thomson, especially to Thomson's use of energy principles in the theory of the electric telegraph.[34] It may be illustrated in various ways, of which the model shown in Figure 3, which Maxwell had constructed in 1874, is the most convenient.[35] Two wheels, P and Q, are geared together through a

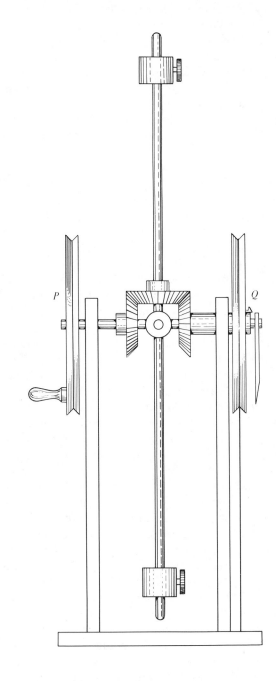

FIGURE 3. Dynamical analogy for two inductively coupled circuits (1865, 1874).

differential mechanism with adjustable flyweights. Rotations of P and Q represent currents in two circuits; the moments of inertia represent coefficients of induction; a frictional band attached to Q represents the resistance of the secondary circuit. Every feature of electromagnetic induction is seen here. So long as P rotates uniformly, Q remains stationary; but when P is started or stopped, a reverse impulse is transmitted to Q. This impulse is determined by the acceleration, the coefficient of coupling, and the inertia and resistance of Q, in exact analogy with an electrical system. Again the definitive quantity has the nature of momentum, determined in the mechanical model by the positions of the flyweights and in the electromagnetic analog by the geometry of the circuits. The total "electrokinetic momentum" $\mathbf{p}$ is $Li + \sum_j M_j i_j$, where L and i are the self-inductance and current in a particular circuit and the M_j's and i_j's are the mutual inductances and currents of neighboring circuits. Since $\mathbf{p}$ is the integral of the function $\mathbf{A}$ round the circuit, the analogy carries through at the macroscopic level Maxwell's identification of $\mathbf{A}$ with the "reduced momentum" of the field. Combined with conservation of energy, it also gives the mechanical actions between circuits. Helmholtz and Thomson had applied energy principles to deduce the law of induction from Ampère's force law; Maxwell inverted and generalized their argument to calculate forces from the induction formulas. Thus his first analytic treatment of the electrotonic function was metamorphosed into a complete dynamical theory of the field.

In the *Treatise* Maxwell extended the dynamical formalism by a more thoroughgoing application of Lagrange's equations than he had attempted in 1865. His doing so coincided with a general movement among British and European mathematicians about then toward wider use of the methods of analytical dynamics in physical problems. The course of that movement in Britain may be followed through Cayley's two British Association reports on advanced dynamics of 1857 and 1862, Routh's *Treatise on the Dynamics of a System of Rigid Bodies* (1860, 1868), and Thomson and Tait's *Treatise on Natural Philosophy* (first edition 1867). Maxwell helped Thomson and Tait with comments on many sections of their text. Then, with the freshness of outlook that makes his work so appealing, he turned the current fad to his own ends by applying it to electromagnetism. Using arguments extraordinarily modern in flavor about the symmetry and vector structure of the terms, he expressed the Lagrangian for an electromagnetic system in its most general form. Green and others had developed similar arguments in studying the

dynamics of the luminiferous ether, but the use Maxwell made of Lagrangian techniques was new to the point of being almost a new approach to physical theory—though many years were to pass before other physicists fully exploited the ground he had broken. The beauty of the Lagrangian method is that it allows new terms to be incorporated in the theory automatically as they arise, with a minimum of physical hypothesis. One that Maxwell devoted a chapter of the *Treatise* to was the magneto-optical effect. By a powerful application of symmetry considerations he put Thomson's argument of 1856 on a rigorous basis and proved that any dynamical explanation of the rotation of the plane of polarized light must depend on local rotation in the magnetic field. In later terminology, the induction $\mathbf{B}$ is an axial vector, and the electrons in matter precess about the applied field: these are the elements of truth behind the molecular vortex hypothesis. Characteristically Maxwell did not limit his thinking to the general symmetry argument: he tested it by attempting to invent counterexamples. Elsewhere he wrote, "I have also tried a great many hypotheses [to explain the magneto-optical effect] besides those which I have published, and have been astonished at the way in which conditions likely to produce rotation are exactly neutralized by others not seen at first."[36] A further instance of the power of the Lagrangian methods, covered in the *Treatise*, is Maxwell's analysis of cross-terms linking electrical and mechanical phenomena. This he did partly at the suggestion of J. W. Strutt (Lord Rayleigh).[37] He identified three possible electromechanical effects, later detected by Barnett (1908), Einstein and de Haas (1916), and Tolman and Stewart (1916). The Barnett effect is a magnetic moment induced in a rapidly spinning iron bar. Maxwell himself had looked for the inverse phenomenon in 1861 during an experiment in search of the angular momentum of molecular vortices.[38]

In 1865, and again in the *Treatise*, Maxwell's next step after completing the dynamical analogy was to develop a group of eight equations describing the electromagnetic field. They are set out in the table with subsidiary equations according to the form adopted in the *Treatise*. The principle they embody is that electromagnetic processes are transmitted by the separate and independent action of each charge (or magnetized body) on the surrounding space rather than by direct action at a distance. Formulas for the forces between moving charged bodies may indeed be derived from Maxwell's equations, but the action is not along the line joining them and can be reconciled with dynamical principles only by taking into account the exchange of momentum with the field.[39] Maxwell remarked that the equations might be condensed, but "to eliminate a quantity which expresses a useful idea would be rather a loss than a gain in this stage of our enquiry."[40] He had in fact simplified the equations in his fifth major paper, the short but important "Note on the Electromagnetic Theory of Light" (1868), writing them in an integral form without the function **A**, based on four postulates derived from electrical experiments. This may be called the electrical formulation of the theory, in contrast with the original dynamical formulation. It was later independently developed by Heaviside and Hertz and passed into the textbooks. It has the advantage of compactness and analytical symmetry, but its scope is more restricted and to some extent it concealed from the next generation of physicists ideas familiar to Maxwell which proved important later on. Two points in the table deserve comment for the modern reader. Equations (B) and (C) appear slightly unfamiliar, because (B) contains terms defined for a particular laboratory frame of reference, while (C), the so-called Lorentz force formula, contains a term in grad Ω for the force on isolated magnetic poles, should such exist. Elsewhere in the *Treatise*[41] Maxwell began the investigation of moving frames of reference, a subject which in Einstein's hands was to revolutionize physics. The second point concerns the addition of the displacement current $\dot{\mathbf{D}}$ to the

General Equations of the Electromagnetic Field (1873)

A	Magnetic Induction	$\mathbf{B} = \text{curl } \mathbf{A}$
B	Electromotive Force	$\mathbf{E} = \mathbf{v} \wedge \mathbf{B} - \dot{\mathbf{A}} - \text{grad } \psi$
C	Mechanical Force	$\mathbf{F} = \mathbf{I} \wedge \mathbf{B} + e\mathbf{E} - m \text{ grad } \Omega$
D	Magnetization	$\mathbf{B} = \mathbf{H} + 4\pi\mathbf{J}$
E	Electric Currents	$4\pi\mathbf{I} = \text{curl } \mathbf{H}$
F	Current of Conduction	$\mathbf{I}' = C\mathbf{E}$
G	Electric Displacement	$\mathbf{D} = (1/4\pi) \, \mathbf{KE}$
H	True Currents	$\mathbf{I} = \mathbf{I}' + \dot{\mathbf{D}}$
J	Induced Magnetization	$\mathbf{B} = \mu\mathbf{H}$
K	Electric Volume Density	$e = \text{div } \mathbf{D}$
L	Magnetic Volume Density	$m = \text{div } \mathbf{J}$

Note: Maxwell used **S** rather than **I** for electric current density.

current of conduction **I**'. In Maxwell's treatment (unlike later textbooks) the extra term appears almost without explanation, arising as it does from his analogy between the paired phenomena of conduction and static induction in electricity and viscous flow and elastic displacement in the theory of materials. More will be said below about the implications of Maxwell's view.

Maxwell gave three distinct proofs of the existence of electromagnetic waves in 1865, 1868, and 1873. The disturbance has dual form, consisting in waves of magnetic force and electric displacement with motions perpendicular to the propagation vector and to each other. An alternative view given in the *Treatise* is to represent it as a transverse wave of the function **A**. In either version the theory yields strictly transverse motion, automatically eliminating the longitudinal waves which had embarrassed previous theories of light.

Among later developments, the generation and detection of radio waves by Hertz in 1888 stands supreme; but there were others of nearly comparable interest. In the *Treatise* Maxwell established that light, on the electromagnetic theory, exerts a radiation pressure. Radiation pressure had been the subject of much speculation since the early eighteenth century; before Maxwell most people had assumed that its existence would be a crucial argument in favor of a corpuscular rather than a wave theory of light. When William Crookes discovered his radiometer effect in 1874, shortly after the publication of Maxwell's *Treatise*, some persons thought that he had observed radiation pressure, but the disturbance was much larger than the predicted value and in the wrong direction, and was caused, as will be explained below, by convection currents in the residual gas. Maxwell's formula was confirmed experimentally by Lebedev in 1900. The effect has implications in many branches of physics. It accounts for the repulsion of comets' tails by the sun; it is, as Boltzmann proved in 1884, critical to the theory of blackbody radiation; it may be used in deriving classically the time-dilation formula of special relativity; it fixes the mass-range of stars.

Another very fruitful new area of research started by Maxwell was on the connections between electrical and optical properties of bodies. He obtained expressions for the torque on a birefringent crystal suspended in an electric field, for the relation between refractive index and dielectric constant in transparent media, and for the relation between optical absorption and electrical conductivity in metals. In the long wavelength limit the refractive index may be expected on the simplest theory to be proportional to the square root of the dielectric constant. Measurements by Boltzmann, J. E. H. Gordon, J. Hopkinson, and others confirmed Maxwell's formula in gases and paraffin oils, but in some materials (most obviously, water) they revealed large discrepancies. These and like problems, including Maxwell's own observation of a discrepancy between the observed and predicted ratios of optical absorption to electrical conductivity in gold leaf, formed a basis for decades of research on electro-optical phenomena. Much of what was done during the 1880's and 1890's should be seen as the beginnings of modern research on solid-state physics, though a full interpretation waited on the development of the quantum theory of solids.

In classical optics Maxwell's theory worked a revolution that is now rarely perceived. A popular fiction among twentieth-century physicists is that mechanical theories of the ether were universally accepted and universally successful during the nineteenth century, until shaken by the null result of the Michelson-Morley experiment on the motion of the earth through the ether. This little piece of textbook folklore is wrong in both its positive and its negative assertions. More will be said below about the Michelson-Morley experiment, but long before that the classical ether theories were beset with grave difficulties on their own ground. The problem was to find a consistent dynamical foundation for the wave theory of light. During the 1820's Fresnel had given his well-known formulas for double refraction and for the reflection of polarized light; they were confirmed later with extraordinary experimental accuracy, but Fresnel's successors had immense trouble in reconciling them with each other on any mechanical theory of the ether. In 1862 Stokes summarized forty years of arduous research, during which a dozen different ethers had been tried and found wanting, by remarking that in his opinion the true dynamical theory of double refraction was yet to be found.[42] In 1865 Maxwell obtained Fresnel's wave surface for double refraction from the electromagnetic theory in the most straightforward way, completely avoiding the ad hoc supplementary conditions required in the mechanical theories. He did not then derive the reflection formulas, being uncertain about boundary conditions at high frequency;[43] but in 1874 H. A. Lorentz obtained them also very simply, using the static boundary condition Maxwell had given in 1856. An equivalent calculation, probably independent, appears in an undated manuscript of Maxwell's at Cambridge. The whole matter was investigated in two very powerful critical papers by Rayleigh (1881) and Gibbs (1888), and in the cycle of work begun by Thomson in his 1884 *Baltimore Lectures*. Rayleigh and Gibbs proved that Maxwell's were the only equations that give formulas for refraction, reflection, and scattering of light consistent with each other and with experiment.[44] Brief reference is appropriate here to James MacCullagh's semi-mechanical theory of 1845, in which the ether was assigned a property of rotational elasticity different from the elastic properties of any ordinary substance. After Stokes in 1862 had raised formidable objections

against the stability of MacCullagh's medium, it was taken as disproved until FitzGerald and Larmor noticed a formal resemblance between MacCullagh's and Maxwell's equations. Since then the two theories have usually been considered homologous. In truth neither Stokes's objections to, nor Larmor's claims for, MacCullagh's theory can be sustained. A dynamically stable medium with rotational elasticity supplied by gyrostatic action was invented by Thomson in 1889.[45] On the other hand, whereas MacCullagh made kinetic energy essentially linear and elastic energy rotational, Maxwell identified magnetism with rotational kinetic energy and electrification with a linear elastic displacement. Very peculiar assumptions about the action of the ether on matter are necessary to carry MacCullagh's theory through at the molecular level; Maxwell's extends naturally and immediately to the ionic theory of matter. Even as an optical hypothesis, apart from its other virtues, the position of Maxwell's theory is unique.

Maxwell's statements about the luminiferous ether have an ambiguity which needs double care in view of the intellectual confusion of much twentieth-century comment on the subject. Selective quotation can make him sound as mechanistic as Thomson became in the 1880's or as Machian as Einstein was in the early 1900's. The *Treatise* concludes flatly that "there must be a medium or substance in which . . . energy exists after it leaves one body and before it reaches [an] other";[46] a later letter dismisses the ether as a "most conjectural scientific hypothesis."[47] Some remarks simply express the ultimate skepticism behind Maxwell's working faith in science. Others hinge on the view he inherited from Whewell that reality is ordered in a series of tiers, each more or less complete in itself, each built on the one below, and that the key to discovery lies in finding "appropriate ideas"[48] to describe the tier one is concerned with. By 1865 Maxwell was convinced that magnetic and electric energy are disseminated in space. As a "very probable hypothesis" he favored identifying the two forms of energy with "the motion and the strain of one and the same medium,"[49] but definite knowledge about one tier must be distinguished from reasonable speculation about the next. That was the philosophic point of the Lagrangian method. In Hamilton's terminology the best short statement of Maxwell's position is that we may believe in the existence of the ether without direct knowledge of its properties; we know only relations between the phenomena it accounts for. In a striking passage from the article on thermodynamics mentioned above, perhaps written after seeing the famous bells at Terling near Rayleigh's estate, Maxwell compared the situation to that of a group of bellringers confronted with ropes going to invisible machinery in the bell loft. Lagrange's equations supply the "appropriate idea" expressing neither more nor less than is known about the visible motions: whether more detailed information about the machinery can be gained later remains open. In Maxwell's, as in many later applications of Lagrange's method, the energies involve electrical, not mechanical, quantities. If the "very probable hypothesis" is followed out and one term is equated with ordinary kinetic energy, then, as Thomson found in 1855, a lower limit to the density ρ of a mechanical ether can be calculated from the known energy density of sunlight.[50] The flaw in Thomson's argument lies in assuming an energy density $\frac{1}{2}\rho v^2$; it is resolved in relativistic dynamics by the mass-energy relation; the rest mass of the photon is zero. Considerations of this kind indicate the subtlety of the scientific transformation wrought by relativity theory. It eliminated the arguments for an ether of fixed position and finite density, yet it preserved intact Maxwell's equations and his fundamental idea of disseminated electrical energy. More light is thrown on Maxwell's own opinions about the problem of relative and absolute motion and the connection between dynamics and other branches of physics by the delightful monograph *Matter and Motion*, published in 1876.

Maxwell's influence in suggesting the Michelson-Morley ether-drift experiment is widely acknowleged, but the story is a curiously tangled one. It originates in the problem of the aberration of starlight. During the course of a year the apparent positions of stars, as fixed by transit measurements, vary by ± 20.5 arc-seconds. This effect was discovered in 1728 by Bradley. He attributed it to the lateral motion of the telescope traveling at velocity v with the earth about the sun. On the corpuscular theory of light the motion causes a displacement of the image, while the particles travel from the objective to the focus, through an angular range v/c just equal to the observed displacement. An explanation of aberration on the wave theory of light is harder to come by. If the ether were a gas like the earth's atmosphere (as was first supposed), it would be carried along with the telescope and one scarcely would expect any displacement. Young in 1804 therefore proposed that the ether must pass between the atoms in the telescope wall "as freely perhaps as the wind passes through a grove of trees."[51] The idea had promise, but in working it out other phenomena needed to be considered, many of which further illustrate the difficulties of classical ethers. To explain Maxwell's involvement I depart from chronology and give the facts roughly in the order in which they presented themselves to him.

In 1859 Fizeau proved experimentally that the velocity of light in a moving column of water is greater downstream than upstream. A natural supposition is that the water drags the ether along with it. This contradicts Young's hypothesis in its most primitive form; however, the modified velocity was not $c + w$ but $c + w(1 - 1/\mu^2)$, where μ is the refractive index of water, and that tallied with a more sophisticated theory of aberration due to Fresnel. Fresnel held the conviction (not actually verified until 1871) that the aberration coefficient in a telescope full of water must remain unchanged, which on Young's theory it does not. He was able to satisfy that requirement by combining Young's hypothesis with the further assumption that refraction is due to condensation of the ether in ordinary matter, the ether-density in a medium of refractive index μ being μ^2 times its value in free space. With the excess ether carried along by matter one obtains the quoted formula, which is in consequence still known as the "Fresnel drag" term, though it stands on broader foundations, as Larmor afterwards proved. Indeed Fresnel's condensation hypothesis is logically inconsistent with another principle that became accepted in the 1820's, namely, that the ether, to convey transverse but not longitudinal waves, must be an incompressible solid. A dissatisfaction with Fresnel's "startling assumptions" made Stokes in 1846 propose a radically new theory of aberration, treating the ether as a viscoelastic substance, like pitch or glass. For the rapid vibrations of light the ether acts as a solid, but for the slow motions of the solar system it resembles a viscous liquid, a portion of which is dragged along with each planetary body. A plausible circuital condition on the motion gives a deflection v/c for a beam of light approaching the earth, identical with the displacement that occurs inside the telescope in the other theories.

Some time in 1862 or 1863 Maxwell read Fizeau's paper and thought out an experiment to detect the ether wind. Since refraction is caused by differences in the velocity of light in different media, one might expect the Fresnel drag to modify the refraction of a glass prism moving through the ether. Maxwell calculated that the additional deflection in a 60° prism moving at the earth's velocity would be 17 arc-seconds. He arranged a train of three prisms with a return mirror behind them in the manner of his portable "colour-box," and set up what would now be called an autocollimator to look for the deflection, using a telescope with an illuminated eyepiece in which the image of the crosshair was refocused on itself after passing to and fro through the prisms. The displacement from ether motion could be seen by mounting the apparatus on a turntable, where the effect would reverse on rotating through 180°, giving an overall deflection after the double passage of $2\frac{1}{2}$ arc-minutes: easily measurable. Maxwell could detect nothing, so in April 1864 he sent Stokes a paper for the Royal Society concluding that "the result of the experiment is decidedly negative to the hypothesis about the motion of the ether in the form stated here."[52]

Maxwell had blundered. Though he did not then know it, the French engineer Arago had done a crude version of the same experiment in 1810 (with errors too large for his result to have real significance), and Fresnel had based his theory on Arago's negative result. Stokes knew all this, having written an article on the subject in 1845; he replied, pointing out Maxwell's error, which had been to overlook the compensating change in density that occurs because the ether satisfies a continuity equation at the boundary.[53] Maxwell withdrew the paper. He did give a description of the experiment three years later, with a corrected interpretation, in a letter to the astronomer William Huggins, who included it in his pioneering paper of 1868 on the measurement of the radial velocities of stars from the Doppler shifts of their spectral lines.[54] There the matter rested until the last year of Maxwell's life. Then in his article "Ether" for the *Encyclopaedia Britannica* he again reviewed the problem of motion through the ether. The only possible earth-based experiment was to measure variations in the velocity of light on a double journey between two mirrors. Maxwell concluded that the time differences in different directions, being of the order v^2/c^2, would be too small to detect. He proposed another method from timing the eclipses of the moons of Jupiter, which he later described in more detail in a letter to the American astronomer D. P. Todd, published after his death in the Royal Society *Proceedings* and in *Nature*.[55] His statements there about the difficulties of the earth-based experiment served as a challenge to the young Albert Michelson, who at once invented his famous interferometer to do it.

The negative result of the experiment swung Michelson and everyone else behind Stokes's theory of aberration. In 1885, however, Lorentz discovered that Stokes's circuital condition on the motion of the ether is incompatible with having the ether stationary at the earth's surface. Lorentz advanced a new theory combining some of Stokes's ideas with some of Fresnel's; he also pointed out an oversight in Michelson's (and Maxwell's) analysis of the experiment, which halved the magnitude of the predicted effect, bringing it near the limits of the observations. Michelson and Morley then repeated the experiment

with many improvements. Their conclusive results were published in 1887. In 1889 FitzGerald wrote to the American journal *Science* explaining the negative result by his contraction hypothesis.[56] The same idea was advanced independently by Lorentz in 1893. Physics texts often refer to the FitzGerald-Lorentz contraction as an ad hoc assumption dreamed up to save appearances. It was not. The force between two electric charges is a function of their motion with respect to a common frame: Maxwell had shown it (incompletely and in another context) in the *Treatise*.[57] Hence, as FitzGerald stated, all one need assume to explain the negative result of the Michelson-Morley experiment is that intermolecular forces obey the same laws as electromagnetic forces. The real (and great) merit of the special theory of relativity was pedagogical. It arranged the old confusing material in a clear deductive pattern.

Reference may be made to some more technical contributions from Maxwell's later work. A short paper of 1868, written after seeing an experiment by W. R. Grove, gave the first theoretical treatment of resonant alternating current circuits.[58] Portions of the *Treatise* applied quaternion formulas discovered by Tait to the field equations, and paved the way for Heaviside's and Gibbs's developments of vector analysis. Maxwell put these and various related matters in a wider context in a paper of 1870, "On the Mathematical Classification of Physical Quantities." He coined the terms "curl," "convergence" (negative divergence), and "gradient" for the various products of the vector operator ∇ on scalar and vector quantities, with the less familiar but instructive term "concentration" for the operation ∇^2, which gives the excess of a scalar V at a point over its average through the surrounding region.[59] He extended also his previous treatment of force and flux vectors, introduced the important distinction between what are now (after W. Voigt) known as axial and polar vectors, and in other papers gave a useful physical treatment of the two classes of tensors later distinguished mathematically as covariant and contravariant.[60] Further analytical developments in the *Treatise* include applications of reciprocal theorems to electrostatics, a general treatment of Green's functions, topological methods in field and network theory, and the beautiful polar representation of spherical harmonic functions.[61] The *Treatise* also contains important contributions to experimental technique, such as the well-known "Maxwell bridge" circuit for determining the magnitude of an inductance.[62]

A consequence of the displacement hypothesis which Maxwell himself did not truly grasp until 1869 is that all electric currents, even in apparently open circuits, are in reality closed.[63] But with that a new interpretation of electric charge became necessary. This is a subject of great difficulty, one of the most controversial in all Maxwell's writings. Many critics from Heinrich Hertz on have come to feel that a consistent view of the nature of charge and electric current, compatible with Maxwell's statements, simply does not exist. I believe these authors to be mistaken, although I admit that Maxwell gave them grounds for complaint, both by his laziness over plus and minus signs and by the fact that in parts of his work where the interpretation of charge was not the central issue he slipped back into terminology—and even ideas— not really compatible with his underlying view. The question is all the harder because the problem it touches (the relation between particles and fields) has continued as a difficulty in physics down to the present day. A full critical discussion would take many pages. I shall content myself with a short dogmatic statement, cautioning the reader that other opinions are possible.

Before Maxwell, electricity had been represented as an independent fluid (or pair of fluids), the excess or deficiency of which constitutes a charge. But if currents are invariably closed, how can charge accumulate

anywhere? Part, but only part, of the answer lies in the hypothesis, hinted at by Faraday and clearly stated by Maxwell in 1865, that electrostatic action is entirely a matter of dielectric polarization, with the conductor serving not as a receptacle for electric fluid but as a bounding surface for unbalanced polarization of the surrounding medium. The difference between the old and new interpretations of charge, illustrated in Figure 4(a) and (b), looks simple; but underneath are problems that Maxwell's followers found bafflingly obscure. One source of confusion was that the polarization in 4(b) differs from that in the theory of material dielectrics proposed earlier by Thomson and Mossotti[64] (Figure 4[c]), which made the effective charge Q at the boundary the sum of a real charge Q_0 on the conductor and an apparent charge $-Q'$ on the dielectric surface. In Maxwell's interpretation the polarization extends from material dielectrics to space itself; all charge is in a sense apparent charge, and the motion is in the opposite direction. All might have been well had Maxwell in the *Treatise* not discussed the difference between charge on a conductor and charge on a dielectric surface in language similar to Mossotti's and if he had adopted a less liberal approach to the distinction between plus and minus signs. As it was, with the

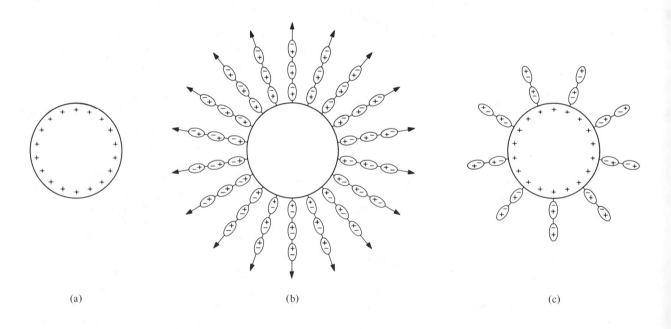

(a)

(b)

(c)

Conventional electric fluid.

Maxwell-Faraday: all charge attributed to unbalanced dielectric polarization.

Thomson-Mossotti: real charge on conductor combined with apparent charge due to unbalanced polarization of a material dielectric.

FIGURE 4.　Representations of a charged conductor.

further novelty of totally closed currents, most people from Hertz on shook their heads in despair.

Yet the two analogies on which Maxwell based his ideas—those between the motion of electricity and an incompressible fluid and between static induction and displacement—are both sound. The escape lies in recognizing the radical difference in meaning of the two charges illustrated in 4(a) and 4(b). Maxwell's current is not the motion of charge, but the motion of a continuous *uncharged* quantity (not necessarily a substance); his charge is the measure of the displacement of that quantity relative to space. To the question puzzling Hertz—whether charge is the cause of polarization or polarization the cause of charge—the answer is "neither." For Maxwell electromotive force is the fundamental quantity. It causes polarization; polarization creates stresses in the field; charge is the measure of stress. All these ideas are traceable to Maxwell; but nowhere, it must be conceded, are they fairly set out. The representation of electricity as an uncharged fluid may seem incompatible with electron theory. Actually it is not; and one of the oddities in Maxwell's development is that the clue to reconciling the two ideas rests in the treatment of charges as sources and sinks of incompressible fluid given in his 1856 paper. That essentially was the principle of the ether-electron theory worked out by Larmor in 1899.

Few things illustrate better the subtlety of physical analogy than Maxwell's developing interpretations of the function **A**. His original discussion in 1856 was purely analytic. The dynamical theory led him to its representation as a property of electricity analogous to momentum, which reached fulfillment after his death in the expression $(m\mathbf{v} + e\mathbf{A}/c)$ for the canonical momentum of the electron, $m\mathbf{v}$ being the momentum of the free particle and $e\mathbf{A}/c$ the reduced momentum contributed by sources in the surrounding field. In 1871 he perceived another, entirely different analogy for **A**. Considered in relation to electrodynamic forces it resembles a potential, as may be seen by comparing the equation $F = \text{grad}(\mathbf{i} \cdot \mathbf{A})$ for force on a conductor carrying a current with the equation $F = \text{grad}(e\Phi)$ for force on a charged body. Maxwell introduced the terms "vector" and "scalar potential" for **A** and Φ and recognized, probably for the first time, that **A** was a generalization of F. E. Neumann's electrodynamic potential, though his formulation differed in spirit and substance from Neumann's, since it started from the field equations and incorporated displacement current. The formulas were later rearranged by FitzGerald, Liénard, and Wiechert as retarded potentials of the conduction currents, thus uncovering their common ground with

L. V. Lorenz's propagated action theory of electrodynamics. Both of Maxwell's analogies may be carried through in detail: that is, equations in **A** exist analogous to every equation in dynamics involving momentum and every equation in potential theory involving Φ. The resemblance of a single function to two quantities so different as momentum and potential depends on the peculiar relation between electromotive and electrodynamic forces: the electromotive force generated by induction is proportional to the velocity of the conductor times the electrodynamic force acting on it. The momentum analogy was little appreciated until 1959, when Y. Aharonov and D. J. Bohm pointed out some unexpected effects tied to the canonical momentum in quantum mechanics.[65]

Statistical and Molecular Physics (1859–1878). The problem of determining the motions of large numbers of colliding bodies came to Maxwell's attention while he was investigating Saturn's rings. He dismissed it then as hopelessly complicated; but in April 1859, as he was finishing his essay for publication, he chanced to read a new paper by Rudolf Clausius on the kinetic theory of gases, which convinced him otherwise and made him transfer his interest to gas theory.

The idea of attributing pressure in gases to the random impacts of molecules against the walls of the containing vessel had been suggested before. Prevailing opinion, however, still favored Newton's hypothesis of static repulsion between molecules or one of its variants, such as Rankine's vortex hypothesis. Maxwell had been taught the static theory of gases as a student at Edinburgh. Behind the victory of kinetic theory led by Clausius and Maxwell lay two distinct scientific advances: the doctrine of conservation of energy, and an accumulation of enough experimental information about gases to shape a worthwhile theory. Many of the new discoveries from 1780 on, such as Dalton's law of partial pressures, the law of equivalent volumes, and measurements on the failure of the ideal gas equation near liquefaction, came as by-products of chemical investigations. Two developments especially important to Maxwell were Thomas Graham's long series of experiments on diffusion, transpiration, and allied phenomena, also begun as chemical researches, and Stokes's analysis of gas viscosity, made in 1850 as part of a study on the damping of pendulums for gravitational measurements. Maxwell had used Stokes's data in treating the hypothesis of gaseous rings for Saturn. Viscosity naturally became one of his first subjects for calculation in kinetic theory; to his astonishment the predicted coefficient was independent of the pressure of the gas. The experiments of his wife and himself between 1863 and 1865, which confirmed this seeming paradox, fixed the success of the theory.

Clausius' work appeared in two papers of 1857 and 1858, each of which contained results important to Maxwell. The first gave a greatly improved derivation of the known formula connecting pressure and volume in a system of moving molecules:

$$pV = \tfrac{1}{3}nm\overline{v^2}, \qquad (7)$$

where m is the mass of a molecule, $\overline{v^2}$ its mean square velocity, and n the total number of molecules, from which, knowing the density at a given pressure, Clausius deduced (as others had done earlier) that the average speed must be several hundred meters per second. Another matter, whose full significance only became apparent after Maxwell's work, was the exchange of energy between the translational and rotational motions of molecules. Clausius guessed that the average energies associated with the two types of motion would settle down to a constant ratio σ, and from thermodynamical reasoning he derived an equation relating σ to the ratio γ of the two specific heats of a gas.

Clausius' second paper was written to counter a criticism by the Dutch meteorologist C. H. D. Buys Ballot, who objected that gas molecules could never be going as fast as Clausius imagined, since the odor of a pungent gas takes minutes to permeate a room. Clausius replied that molecules of finite diameter must be repeatedly colliding and rebounding in new directions, and he deduced from statistical arguments that the probability W of a molecule's traveling a distance L without collision is

$$W = e^{-L/l}, \qquad (8)$$

where l is a characteristic "mean free path." Assuming for convenience that all molecules have equal velocity, Clausius found

$$\frac{1}{l} = \frac{4}{3}\,\pi s^2 N, \qquad (9)$$

where s is their diameter and N their number density. He could not determine the quantities explicitly but guessed that l/s might be about 1,000, from which l had to be a very small distance. Since by equation (8) only a minute fraction of molecules travel more than a few mean free paths without collision, Buys Ballot's objection to kinetic theory was fallacious.

Although Clausius had based his investigation on the simplifying assumption that all molecules of any one kind have the same velocity, he recognized that the velocities would in reality spread over a range of values. The first five propositions of Maxwell's "Illustrations of the Dynamical Theory of Gases"

(1860) led to a statistical formula for the distribution of velocities in a gas at uniform pressure, as follows. Let the components of molecular velocity in three axes be x, y, z. Then the number dN of molecules whose velocities lie between x and $x + dx$, y and $y + dy$, z and $z + dz$ is $Nf(x)f(y)f(z)\,dx\,dy\,dz$. But since the axes are arbitrary, dN depends only on the molecular speed v, where $v^2 = x^2 + y^2 + z^2$ and the distribution must satisfy the functional relation

$$f(x)f(y)f(z) = \phi(x^2 + y^2 + z^2), \qquad (10)$$

the solution of which is an exponential. Applying the fact that N is finite, the resolved components of velocity in a given direction may be shown to have a distribution function identical in form with Laplace's bell-shaped "normal distribution" in the theory of errors:

$$dN_x = \frac{N}{\alpha\sqrt{\pi}}\,e^{-x^2/\alpha^2}\,dx, \qquad (11)$$

where α is a quantity with dimensions of velocity. The number of particles summed over all directions with speeds between v and $v + dv$ is

$$dN_v = \frac{4N}{\alpha^3\sqrt{\pi}}\,v^2 e^{-v^2/\alpha^2}\,dv. \qquad (12)$$

Related formulas give the distributions in systems of two or more kinds of molecules. From them with (11) and (12) Maxwell was able to determine mean values of various products and powers of the velocities used in calculating gas properties.

The derivation of equations (11) and (12) marks the beginning of a new epoch in physics. Statistical methods had long been used for analyzing observations, both in physics and in the social sciences, but Maxwell's idea of describing actual physical processes by a statistical function was an extraordinary novelty. Its origin and validity deserve careful study. Intuitively equation (12) is plausible enough, since dN_v approaches zero as v approaches zero and infinity and has a maximum at $v = \alpha$, consistent with the natural physical expectation that only a few molecules will have very high or very low speeds. Empirically it was verified years later in experiments with molecular beams. Yet the assumption that the three resolved components of velocity are distributed independently is one which, as Maxwell later conceded, "may appear precarious";[66] and the whole derivation conveys a strange impression of having nothing to do with molecules or their collisions. Its roots go back to Maxwell's Edinburgh days. His interest in probability theory was aroused in 1848 by Forbes, who reexamined a statistical argument for the existence of binary stars put forward in 1767 by the Reverend John Michell. Over the next few years he read thoroughly the statistical writings of Laplace and Boole and also another item of peculiar interest, a long essay by Sir John Herschel in the *Edinburgh Review* for June 1850 on Adolphe Quetelet's *Theory of Probability as Applied to the Moral and Social Sciences.* Herschel's review ranged over many issues, social and otherwise; and a contemporary letter to Lewis Campbell leaves no doubt that Maxwell had read it.[67] One passage embodied a popular derivation of the law of least squares applied to random distributions in two dimensions, based on the supposed independence of probabilities along different axes. The family resemblance to Maxwell's derivation of equation (11) is striking. Thus early reading on statistics, study of gaseous rings for Saturn, and ideas from Clausius about probability and free path all contributed to Maxwell's development of kinetic theory.

In his second paper, published in 1867, Maxwell offered a new derivation of the distribution law tied directly to molecular encounters. To maintain equilibrium the distribution function must satisfy the relation $f(v_1)f(v_2) = f(v_1')f(v_2')$ where v_1 and v_1' are velocities of molecule 1 and v_2 and v_2' of molecule 2 before and after encounter. Combination with the energy equation yielded formulas corresponding to (11) and (12). This established the equilibrium of the exponential distribution but not its uniqueness. From considerations of cyclic collision processes Maxwell sketched an argument that any velocity distribution would ultimately converge to the same form. The proof of the theorem in full mathematical rigor is still an open problem. Boltzmann gave an interesting extended version of Maxwell's argument in his *Lectures on Gas Theory* (1892). Earlier he had formulated another approach (the *H*-theorem), which bears on the subject and is even more important as part of the development that eventually transcended gas theory and led to the separate science of statistical mechanics. One further point that has been examined by various writers is the status of Maxwell's original derivation of the exponential law. Since the result is correct the hypotheses on which it was based must in some sense be justifiable. The best proof along Maxwell's first lines appears to be one given by M. Kac in 1939.[68]

Maxwell next applied the distribution function to evaluate coefficients of viscosity, diffusion, and heat conduction, as well as other properties of gases not studied by Clausius. He interpreted viscosity as the transfer of momentum between successive layers of molecules moving, like Saturn's rings, with different transverse velocities. The probability of a molecule's starting in a layer dz and ending in dz' is found from

Clausius' equation (8) in combination with the distribution function. Integration gives the total frictional drag and an equation for the viscosity coefficient,

$$\mu = \tfrac{1}{3}\rho \bar{l}\bar{v}, \qquad (13)$$

where ρ is the density, $\bar{l}$ the mean free path, and $\bar{v}$ the mean molecular speed. Since $\bar{l}$ is inversely proportional to ρ, the viscosity is independent of pressure. The physical explanation of this result, given by Maxwell in a letter to Stokes of 30 May 1859, is that although the number of molecules increases with pressure, the average distance over which each one carries momentum decreases with pressure.[69] It holds experimentally over a wide range, only breaking down when ρ is so high that $\bar{l}$ becomes comparable with the diameter of a molecule or so low that it is comparable with the dimensions of the apparatus. Maxwell was able to calculate a numerical value for the free path by substituting into (13) a value for μ/ρ from Stokes's data and a value for $\bar{v}$ from (7). The result was 5.6×10^{-6} cm. for air at atmospheric pressure and room temperature, which is within a factor of two of the current value. The calculations for diffusion and heat conduction proceeded along similar lines by determining the number of molecules and quantity of energy transferred in the gas. Applying the diffusion formulas to Graham's experiments, Maxwell made a second, independent estimate of the free path in air as 6.3×10^{-6} cm. The good agreement between the results greatly strengthened the plausibility of the theory. There were, however, errors of principle and of arithmetic in some of the calculations, which Clausius exposed—not without a certain scholarly relish—in a new paper of 1862. The chief mistake lay in continuing to use an isotropic distribution function in the presence of density and pressure gradients. Clausius offered a corrected theory; but since he persisted in assuming constant molecular velocity, it too was unsatisfactory. Maxwell wrote out his own revised theory in 1864; but having meanwhile become dissatisfied with the whole mean free path method, he withheld the details. The true value of Clausius' criticism was to show the need for a formulation of kinetic theory consistent with known macroscopic equations. Maxwell was to produce it in 1867.

One further important topic covered in the 1860 papers was the distribution of energy among different modes of motion of the molecules. Maxwell first established an equality, which had previously been somewhat sketchily derived by both Waterston and Clausius, between the average energies of translation of two sets of colliding particles with different molecular weights. He deduced that equal volumes of gas at fixed temperature and pressure contain the same number of molecules, accounting for the law of equivalent volumes in chemistry. Later, following out Clausius' thoughts on specific heat, he studied the distribution of energy between translational and rotational motions of rough spherical particles and found that there too the average energies are equal. These two statistical equalities, between the separate translational motions of different molecular species and between the rotational and translational motions of a single species, are examples of a deep general principle in statistical mechanics, the "equipartition principle." The second was an embarrassing surprise; for if molecules are point particles incapable of rotation, Clausius' formula makes the specific heat ratio 1.666, and if they are rough spheres it makes it 1.333. The experimental mean for several gases was 1.408. Maxwell was so upset that he stated that the discrepancy "overturned the whole hypothesis."[70] His further wrestlings with equipartition in the 1870's will be discussed below.

The measurements of gaseous viscosity at different pressures and temperatures made by Maxwell and his wife[71] in 1865 were their most useful contribution to experimental physics. The "Dynamical Theory of Gases," which followed, was Maxwell's greatest single paper. The experiment consisted in observing the decay of oscillations of a stack of disks torsionally suspended in a sealed chamber. Over the ranges studied, the viscosity μ was independent of pressure, as predicted, and very nearly a linear function of the absolute temperature T. But equation (12) implies that μ should vary as $T^{1/2}$. The hypothesis that gas molecules are freely colliding spheres is therefore too simple, and Maxwell accordingly developed a new theory treating them as point centers of force subject to an inverse nth power repulsion. In a theory of this kind the mean free path ceases to be a clear-cut concept: molecules do not travel in straight lines but in complicated orbits with deflections and distances varying with velocity and initial path. Yet some quantity descriptive of the heterogeneous structure of the gas is needed. Maxwell replaced the characteristic distance l by a characteristic time, the "modulus of time of relaxation" of stresses in the gas. A second need, exposed by Clausius' critical paper of 1862, was for a systematic procedure to connect molecular motions with the known macroscopic gas laws. On both points Maxwell's thinking was influenced by Stokes's work on the general equations of viscosity and elasticity.

Elasticity may be defined as a stress developed in a body in reaction to change of form. Both solids and fluids exhibit elasticity of volume; solids alone are elastic against change of shape. A fluid resists changes of shape through its viscosity, but the resistance is evanescent: motion generates stresses proportional to velocity rather than displacement. In 1845 Stokes wrote a powerful paper giving a new treatment of the equations of motion of a viscous fluid. He noticed while doing so that if the time derivatives in the equations are replaced by spatial derivatives, they become the equations of stress for an elastic solid. Poisson also had noticed this transformation, but Stokes went further and remarked that viscosity and elasticity seem to be physically related through time. Substances like pitch and glass react as solids to rapid disturbances and as viscous liquids to slow ones. Stokes utilized this idea in the theory of aberration already described; other physicists also followed it up, among them Forbes, who, as an alpinist, applied it to the motions of glaciers. Maxwell's early letters contain several references to Forbes's opinions.[72] His youthful work on elasticity made him acquainted with Stokes's paper, and in 1861, as explained above in the section on electricity, he applied the analogy of viscosity and elasticity in another way to the processes of conduction and static induction through dielectrics.

During the experiments on gases Maxwell's attention was again directed to viscoelastic phenomena through having to correct for losses in the torsion wire from which his apparatus was suspended. His 1867 paper proposed a new method of specifying viscosity in extension of Stokes's theory. In an ideal solid free from viscosity, a distortion or strain S of any kind creates a constant stress F equal to E times S, where E is the coefficient of elasticity for that particular kind of strain. In a viscous body F is not constant but tends to disappear. Maxwell conjectured that the rate of relaxation of stress is proportional to F, in which case the process may be described formally by the differential equation

$$\frac{dF}{dt} = E\frac{dS}{dt} - \frac{F}{\tau}, \qquad (14)$$

which gives an exponential decay of stress governed by the relaxation time τ. Processes short compared with τ are elastic; processes of longer duration are viscous. The viscosity μ is equal to E_s times τ, where E_s is the instantaneous rigidity against shearing stresses. A given substance may depart from solidity either by having small rigidity or short relaxation time, or both. Maxwell seems to have arrived at (14) from a comparison with Thomson's telegraphy equations, inverting the analogy between electrical and mechanical systems that he had developed in 1865. A test that immediately occurred to him was to look for induced double refraction in a moving fluid, comparable to the double refraction in strained solids discovered by Brewster, which he himself had analyzed in his paper of 1850 on the equilibrium of elastic solids. After some difficulty Maxwell eventually demonstrated in 1873 that a solution of Canada balsam in water exhibits temporary double refraction with a relaxation time of order 10^{-2} seconds.[73] Maxwell's theory of stress relaxation formed the starting point of the science of rheology and affected indirectly every branch of physics, as may be seen from the widespread use of his term "relaxation time." Its immediate purpose lay in reaching a new formulation of the kinetic theory of gases.

Consider a group of molecules moving about in a box. Their impact on the walls exerts pressure. If the volume is changed from V to $V + dV$, the pressure will change by an amount $-p\, dV/V$. But in the theory of elasticity the differential stress due to an isotropic change of volume is $-E\, dV/V$, where E is the cubical elasticity. The elasticity of a gas is proportional to its pressure. Suppose now the pressure is reduced until the mean free path is much greater than the dimensions of the box; and let the walls be rough, so that the molecules rebound at random, and also flexible. Then in addition to the pressure there will be continued exchange of the transverse components of momentum from wall to wall, making the box, even though it is flexible, resist shearing stresses. In other words, a rarefied gas behaves like an elastic solid! Let this property be called quasi-solidity. Following the ideas expressed in equation (14), the viscosity of a gas at ordinary pressures may be conceived of as the relaxation of stresses by molecular encounters. Since elasticity varies as pressure, μ is proportional to $p\tau$ and the relaxation time of a gas at normal pressures is inversely proportional to its density. Although the concept of free path is elusive when there are forces between molecules, some link evidently exists between it and the relaxation time. Maxwell gave it in 1879 in a footnote to his last paper, added in response to a query by Thomson.[74] For a gas composed of rigid-elastic spheres, the product of τ with the mean speed $\bar{v}$ of the molecules is a characteristic distance λ, whose ratio to the mean free path $\bar{l}$ is $8/3\pi$. The free path is a special formulation of the relaxation concept applicable only to freely colliding particles of finite diameter.

To calculate the motions of a pair of molecules subject to an inverse nth power repulsion was a straightforward exercise in orbital dynamics. For the statistical specification of encounters, Maxwell wrote the number dN_1 of molecules of a particular type with molecular weight M, and velocities between ξ_1 and $\xi_1 + d\xi$, etc. as $f(\xi_1 \eta_1 \zeta_1)\, d\xi_1\, d\eta_1\, d\zeta_1$, as in the first paper, with a similar expression for molecules of another type with molecular weight M_2. The velocities of two such groups being defined, their relative velocity V_{12} is also a definite quantity; and the number of encounters between them occurring in time δt can be expressed in terms of orbit parameters. It is $V_{12} b\, db\, d\phi\, dN_1\, dN_2\, \delta t$, where b is the distance between parallel asymptotes before and after an encounter and ϕ is the angle determining the plane in which V_{12} and b lie. If Q is some quantity describing the motion of molecules in group 1, which may be any power or product of powers of the velocities or their components, and if Q' is its value after an encounter, the net rate of change in the quantity for the entire group is $(Q' - Q)$ times the number of encounters per second, or

$$\frac{\delta}{\delta t}(Q\, dN_1) = (Q' - Q)\, V_{12} b\, db\, d\phi\, dN_1\, dN_2. \quad (15)$$

Equation (13) is the fundamental equation of Maxwell's revised transfer theory, replacing the earlier equations based on Clausius' probability formula (8). With the explicit relation between V_{12} and b inserted from the orbit equation, the relative velocity enters the integral of (15) as a factor $V_{12}^{(n-5)/(n-1)}$, which means that although integration generally requires knowledge of the distribution function f_2 under nonequilibrium conditions, in the special case of molecules subject to an inverse fifth-power repulsion V_{12} drops out and the final result may be written immediately as $\bar{Q} N_2$, where $\bar{Q}$ is the mean value of the quantity and N_2 is the total number of molecules of type 2. The simplification may be understood, as Boltzmann later pointed out, by noticing that the number of deflections through a given angle is the product of two factors, one of which (the cross section for scattering) decreases with V_{12}, while the other (the number of collisions) increases with V_{12}.[75] When n is 5, the two factors are exactly balanced. Molecules subject to this law are now called Maxwellian. By a happy coincidence their viscosity is directly proportional to the absolute temperature, in agreement with Maxwell's experiment, although not with more precise measurements made later.

With this Maxwell was in a position to determine the scattering integrals and calculate physical properties of gases. Even with the simplification of inverse fifth-power forces the mathematical task remained formidable, and an impressive feature was the notation Maxwell developed to keep track of different problems. One general equation described transfer of quantities across a plane with different Q's giving the velocities, pressures, and heat fluxes in a gas. Next to be considered were variations of $\overline{Q}$ within a given element of volume. These might occur through the actions of encounters or external forces on molecules within the element or, alternatively, through the passage of molecules to or from the surrounding region. Denoting variations of the first kind by the symbol δ and variations of the second kind by ∂, Maxwell got his general equation of transfer:

$$\frac{\partial}{\partial t}\,\overline{Q}N + \left(\frac{du}{dx} + \frac{dv}{dy} + \frac{dw}{dz}\right) + \frac{d}{dx}(\overline{\xi Q}N)$$
$$+ \frac{d}{dy}(\overline{\eta Q}N) + \frac{d}{dz}(\overline{\zeta Q}N) = \frac{\delta}{\delta t}\,QN, \quad (16)$$

where u, v, w are components of the translational velocity of the gas; the differential symbol d gives total variations with respect to position and time; and subscripts are added to δ, to distinguish variations due to encounters with molecules of the same kind, molecules of a different kind, and the action of external forces. With Q equal to mass, (16) reduces to the ordinary equation of continuity in hydrodynamics. With Q equal to the momentum per unit volume, (16) in combination with the appropriate expression for $\frac{\delta}{\delta t}\,QN$ derived from (15) reduces to an equation of motion. From this, or rather from its generalization to mixtures of more than one kind of molecule, Maxwell derived Dalton's law of partial pressures, and formulas for diffusion applicable to Graham's experiments. With Q energy, (16) yields an equation giving the law of equivalent volumes and formulas for specific heats, thermal effects of diffusion, and coefficients of viscosity in simple and mixed gases. The viscosity equation replacing (12) for Maxwellian molecules is

$$\mu = 0.3416k\left(\frac{M}{K}\right)^{\frac{1}{2}} T, \quad (17)$$

where k is Boltzmann's constant, M is molecular weight, and K is the scaling constant for the forces.

The hardest area of investigation was heat conduction. That was where Maxwell had gone astray in 1860. In the exact theory effects of thermal gradients occur when Q in equation (16) is of the third order in ξ, η, ζ. Maxwell found an expression for the thermal conductivity of a gas in terms of its viscosity, density, and specific heat. The ratio of these quantities, which is known as the Prandtl number, "but which ought to be called the Maxwell number,"[76] is one of several dimensionless ratios used in applying similarity principles to the solution of problems in fluid dynamics. For a monatomic gas it is nearly a constant

over a wide range of temperatures and pressures. Another matter, in which Maxwell became interested through considering the stability of the earth's atmosphere, was the equilibrium of temperature in a vertical column of gas under gravity. The correct result was known from thermodynamics, but its derivation from gas theory gave Maxwell great trouble. It comes out right only if the ratio of the two statistical averages $\overline{\xi^4}/|\,\overline{\xi^2}\,|^2$ has the particular value 3 given by the exponential distribution law. The calculation thus supplied evidence in favor of the law. More light on the same subject came in Boltzmann's first paper on kinetic theory, written in 1868. Boltzmann investigated the distribution law by a method based on Maxwell's, but included the external forces directly in the energy equation to be combined with Maxwell's collision equation $f(v_1)\,f(v_2) = f(v_1')\,f(v_2')$. The distribution function assumed the form $e^{-E/kT}$, where E is the sum of the kinetic and potential energies of the molecule. In 1873 Maxwell gave a greatly simplified derivation of Boltzmann's result during a correspondence in *Nature* about the equilibrium of the atmosphere. He then confessed that his first calculation for the 1867 paper, which gave a temperature distribution that would have generated unending convection currents, nearly shattered his faith in kinetic theory.

Maxwell never attempted to solve the transfer equations for forces other than the inverse fifth power. In 1872 Boltzmann rearranged (16) into an integro-differential equation for f, from which the transport coefficients could in principle be calculated; but despite much effort he failed to reach any solution except for Maxwellian molecules. It was not until 1911–1917 that S. Chapman and D. Enskog developed general methods of determining the coefficients. One interesting result was Chapman's expression for viscosity of a gas made up of hard spheres, which had a form equivalent to (12) but with a numerical coefficient 50 percent higher than Maxwell's and 12 percent higher than that obtained from the mean free path method with corrections for statistical averaging and persistence of velocities derived by Tait and Jeans. So even for the hard-sphere gas the simple theory fails in quantitative accuracy.

For some years after 1867 Maxwell made only sporadic contributions to gas theory. In 1873 he gave a revised theory of diffusion for the hard-sphere gas, from which he developed estimates of the size of molecules, following the work of Loschmidt (1865), Johnstone Stoney (1868), and Thomson (1870). In 1875, following van der Waals, he applied calculations on intermolecular forces to the problem of continuity between the liquid and gaseous states of matter.

In 1876 he gave a new theory of capillarity, also based on considerations about intermolecular forces, which stimulated new research on surface phenomena. Of all the questions about molecules which Maxwell puzzled over during this period the most urgent concerned their structure. His uneasiness about the discrepancy between the measured and calculated specific heat ratios of gases has already been referred to. The uneasiness increased after 1868 when Boltzmann extended the equipartition theorem to every degree of freedom in a dynamical system composed of material particles; and it turned to alarm with the emergence of a new area of research: spectrum analysis. From 1858 onwards, following the experiments of Bunsen and Kirchhoff, several people, including Maxwell, worked out a qualitative explanation of the bright lines in chemical spectra, attributing them to resonant vibrations of molecules excited by their mutual collisions. The broad truth of the hypothesis seemed certain; but it led, as Maxwell immediately saw, to two questions, neither of which was answered until after his death. First, the identity of spectra implies that an atom in Sirius and an atom in Arcturus must be identical in all the details of their internal structure. There must be some universal dimensional constant determining vibration frequency: "each molecule . . . throughout the universe bears impressed on it the stamp of a metric system as distinctly as does the metre of the Archives of Paris, or the double royal cubit of the Temple of Karnac."[77] The royal cubit proved to be Planck's quantum of action discovered in 1900. The other question, also answered only by quantum theory, concerned the influence of molecular vibrations on the specific heat ratio. There were not three or six degrees of freedom, but dozens. There was no way of reconciling the specific heat and spectroscopic data with each other and the equipartition principle. The more Maxwell examined the problem the more baffled he became. In his last discussion, written in 1877, after summarizing and rejecting all the attempts from Boltzmann on to wriggle out of the difficulty, he concluded that nothing remained but to adopt that attitude of "thoroughly conscious ignorance that is the prelude to every real advance in knowledge."[78]

During his last two years Maxwell returned to molecular physics in earnest and produced two full-length papers, strikingly different in scope, each among the most powerful he ever wrote. The first, "On Boltzmann's Theorem on the Average Distribution of Energy in a System of Material Points," followed a line of thought started by Boltzmann, who in 1868 had offered a new conjectural derivation of the distribution law based on combinatorial theory.

A strange feature of the analysis was that it seemed to be free from restrictions on the time spent in encounters between molecules. Hence, as Maxwell was quick to point out,[79] both the distribution factor $e^{-E/kT}$ and the equipartition theorem should apply to solids and liquids as well as gases: a conclusion as fascinating and disturbing as equipartition itself.

Maxwell now gave his own investigation of the statistical problem, based partly on Boltzmann's ideas and partly on an extension of them contained in H. W. Watson's *Treatise on the Kinetic Theory of Gases*.[80] Following Watson, Maxwell used Hamilton's form of the dynamical equations, and adopted the device of representing the state of motion of a large number n of particles by the location of a single point in a "phase-space" of $2n$ dimensions, the coordinates of which are the positions and momenta of the particles. Boltzmann had applied similar methods in configuration space, but the Hamiltonian formalism has advantages in simplicity and elegance. Maxwell then postulated, as Boltzmann had done, that the system would in the course of time pass through every phase of motion consistent with the energy equation. This postulate obviously breaks down in special instances, of which Maxwell gave some examples, but he argued that it should hold approximately for large numbers of particles, where discontinuous jumps due to collisions make the particles jog off one smooth trajectory to another. The validity of this hypothesis, sometimes called the ergodic hypothesis, was afterwards much discussed, often with considerable misrepresentation of Maxwell's opinions. Maxwell next introduced a new formal device for handling the statistical averages. In place of the actual system of particles under study, many similar systems are conceived to exist simultaneously, with identical energies but different initial conditions. The statistical problem is then transformed into determining the number of systems in a given state at any instant, rather than the development in time of a single system. The method had in some degree been foreshadowed by Boltzmann in 1872. It was later very greatly extended by Gibbs, following whom it is known as the method of "ensemble averaging." Maxwell's main conclusion was that the validity of the distribution and equipartition laws in a system of material particles is not restricted to binary encounters. An important result of a more technical kind was an exact calculation of the microcanonical density of the gas, with an expression for its asymptotic form as the number n of degrees of freedom in the system goes to infinity, while the ratio E/n is held constant. According to C. Truesdell, although the hypotheses on which the theorem was based were rather special, "no better proof was given until the work of Darwin and Fowler."[76] Together with Boltzmann's articles this paper of Maxwell's marks the emergence of statistical mechanics as an independent science.

One feature of the paper "On Boltzmann's Theorem," eminently characteristic of Maxwell, is that the analysis, for all its abstraction, ends with a concrete suggestion for an experiment, based on considering the rotational degrees of freedom. Maxwell proved that the densities of the constituent components in a rotating mixture of gases would be the same as if each gas were present by itself. Hence gaseous mixtures could be separated by means of a centrifuge. The method also promised much more accurate diffusion data than was hitherto available. Maxwell's correspondence before his death discloses a plan to set up experiments at Cambridge.[81] Many years later it became a standard technique for separating gases commercially.

Maxwell's last major paper on any subject was "On Stresses in Rarefied Gases Arising From Inequalities of Temperature." Between 1873 and 1876 the scientific world had been stirred by William Crookes's experiments with the radiometer, the well-known device composed of a partially evacuated chamber containing a paddle wheel with vanes blackened on one side and silvered on the other, which spins rapidly when radiant heat impinges on it. At first many people, Maxwell included, were tempted to ascribe the motion to light pressure, but the forces were much greater than predicted from the electromagnetic theory, and in the wrong direction. The influence of the residual gas was soon established; and from 1874 on partial explanations were advanced by Osborne Reynolds, Johnstone Stoney, and others. The tenor of these explanations was that the blackened surfaces absorb radiation and, being hot, make the gas molecules rebound with higher average velocity than do the reflecting surfaces. That plausible but false notion is still perpetuated in many textbooks. A striking observation is that the stresses increase as the pressure is reduced. In 1875 Tait and James Dewar drew the significant conclusion that large stresses occur when the mean free path is comparable with the dimensions of the vanes. At higher pressures some equalizing process enters to reduce the effect.

Such was the state of affairs in 1877, when Maxwell and Reynolds independently renewed the attack. Maxwell was thoroughly familiar with the radiometer controversy, having acted as a referee for many of the original papers, as well as seeing and experimenting with radiometers himself. His work went forward in several stages, during which the comments of Thomson, who refereed his paper, and his own reaction as a referee for Reynolds' paper had important influences. He began by applying the exact transfer theory to the hypothesis that the stresses arise from the increased velocity of molecules rebounding from a heated surface, expanding the distribution function in the form

$$[1 + F(\xi, \eta, \zeta)] \, e^{-(\xi^2 + \eta^2 + \zeta^2)/kT}, \qquad (18)$$

where F is a sum of powers and products of ξ, η, ζ up to the third degree, and then calculating the effect of temperature gradients in the gas. This expansion later became the first step of Chapman's elaborate procedure for determining transport coefficients under any force law, but Maxwell kept to inverse fifth-power forces "for the sake of being able to effect the integrations."[82] The result was a stress proportional to d^2T/dn^2, the second derivative of temperature with respect to distance, correcting a formula given earlier by Stoney, where the stress was proportional to dT/dn. The stress increases when the pressure is lowered, reaching a maximum when the relaxation time τ becomes comparable with the time $d/\bar{v}$, in which a molecule traverses the dimension d of the body—that is, Tait and Dewar's conjecture in the language of the exact theory.

At this point Maxwell made an awkward discovery. Although the stresses are indeed large, when the flow of heat is uniform (as in the radiometer) they automatically distribute themselves in such a way that the forces on each element of gas are in equilibrium. The result is a very general consequence of the fact that the stresses depend on d^2T/dn^2; it is almost independent of the shape of the source; the straightforward explanation of the motions by normal stresses must, therefore, be rejected. Yet the radiometer moves. To escape the dilemma, Maxwell turned to tangential stresses at the edges of the vanes. Here the phenomenon known as "slip" proved all-important. When a viscous fluid moves past a solid body, it generates tangential stresses by sliding over the surface with a finite velocity v_s. According to experiments by Kundt and Warburg in 1875, v_s in gases is equal to SG/μ, where S is the stress and G is a coefficient expressed empirically by $G = 8/\rho$. Thus slip effects increase as the pressure is reduced; and as Maxwell pointed out in 1878,[83] convection currents due to tangential stresses should become dominant in the radiometer, completely destroying the simplicity of the original hypothesis.

The second phase of Maxwell's investigation followed a report by Thomson urging him to treat the gas–surface interaction, and his own report on Reynolds' paper. Reynolds also had decided that the effect must depend on tangential stresses, and he devised an experiment to study them under simplified conditions. When a temperature difference ΔT is set up across a porous plug between two vessels containing gas at pressure p, a pressure difference Δp develops between them proportional to $\Delta T/p$. Reynolds called this new effect "thermal transpiration." Maxwell gave a simple qualitative explanation in his report, and in an appendix added to his own paper in May 1879 he developed a semiempirical theory accounting for it and for the radiometer effect. The method was to assume that a fraction f of molecules striking any surface are temporarily absorbed and reemitted diffusely, while the remaining $(1 - f)$ are specularly reflected. Application of the transfer equations gave a formula for the velocity v_s of gas moving past an unequally heated surface, in which one term was the standard slip formula and two further terms predicted convection currents due to thermal gradients. The theory provided an explicit expression $\frac{2}{3}(2/f - 1)\bar{l}$ for the coefficient G, where $\bar{l}$ is the effective mean free path; from this, using Kundt and Warburg's data, Maxwell deduced that f is about 0.5 for air in contact with glass. Maxwell also obtained a formula for transpiration pressure, and showed that both radiometer and transpiration effects are in the correct direction and increase with reduction of pressure, in agreement with experiment.

Maxwell's paper created the science of rarefied gas dynamics. His formulas for stress and heat flux in the body of the gas were contributions of permanent value, while his investigation of surface effects started a vast body of research extending to the present day. Quantities similar to f later became known as "accommodation coefficients" and were applied to many kinds of gas–surface interaction. One other contribution of great beauty contained in notes added to the paper in May and June 1879 was an application of the methods of spherical harmonic analysis to gas theory. It exemplified the process which Maxwell elsewhere called the "cross-fertilization of the sciences."[84] He was engaged in revising the chapter on spherical harmonics for the second edition of the *Treatise on Electricity and Magnetism*, when he realized that the harmonic expansion used in potential theory could equally be applied to the expansion of the components ξ, η, ζ of molecular velocity. A standard theorem on products of surface and zonal harmonics, which is discussed in the *Treatise*, eliminates odd terms in the expansion of variations of F, greatly simplifying the calculations.[85] With this and other simplifications Maxwell carried the approximations to higher order and added an extra term to the equation of motion of a gas subject to variations in temperature.

It is a tribute to Maxwell's genius that on two occasions his papers on transfer theory stimulated fresh work long after the period at which science usually receives historical embalming. In 1910 Chapman read them, and "with the ignorant hardihood of youth,"[86] knowing nothing of the fruitless toil that had been spent on the equations during the interval, began his investigation that yielded solutions under any force law. In 1956 E. Ikenberry and C. Truesdell again returned to Maxwell. They obtained an exact representation formula for the collision integral of any spherical harmonic for inverse-fifth-power molecules, using which they explored various iterative techniques for solving the transfer equations. One technique, which they called "Maxwellian iteration" from its resemblance to Maxwell's procedure in the 1879 paper, yielded much more compact derivations than the Chapman-Enskog procedure; and with it Ikenberry and Truesdell carried solutions for pressure and energy flux in the gas one stage further than had previously been attempted. Truesdell also discovered an exact solution for steady rectilinear flow, by means of which he exposed certain shortcomings of the iterative methods. Speaking of the "magnificent genius of Maxwell" these authors concluded their appraisal by remarking that it passed over all developments in kinetic theory since 1879 and went back "for its source and inspiration to what Maxwell left us."[87]

Other Scientific Work. Maxwell's remaining work may be summarized more shortly, though not as being of small account. His early discovery of the perfect imaging properties of the "fish-eye" lens extended to a lifelong interest in the laws of optical instruments. In a medium whose refractive index varies as $\mu_0 a^2/(a^2 + r^2)$, where μ_0 and a are constants and r is the distance from the origin, all rays proceeding from any single point are focused exactly at another point. The calculation was "suggested by the contemplation of the structure of the crystalline lens in fish."[88] Real fishes' eyes of course only approximate roughly to Maxwell's medium. Not until R. K. Luneberg revived the subject in 1944 were other instances of perfect imaging devices found.[89] In 1853, shortly after discovering the "fish-eye," Maxwell came across the early eighteenth-century writings on geometrical optics by Roger Cotes and Archibald Smith, in which, as he said to his father, "I find many things far better than what is new."[90] He went on to formulate a new approach to the subject, combining the principle of perfect imaging with Cotes's neglected theorem on "apparent distance." Recent years have seen a revival of interest in Maxwell's method.[91] During the 1870's he returned to it and wrote three papers on the application of Hamilton's characteristic function to lens systems, which seems to have been about the earliest attempt to reduce Hamilton's general theory of ray optics to practice. Another striking paper was on cyclidal wave surfaces. It was illustrated with stereoscopic views of different classes of cyclide and contained a description of Maxwell's real-image stereoscope.

The most pleasing of the minor inventions was his adjustable "dynamical top" (1856), which carried a disk with four quadrants (red, blue, green, yellow) that formed gray when spinning axially, "but burst into brilliant colors when the axis is disturbed." He was led to search records at Greenwich for evidence of the earth's 10-month nutation predicted by Euler, which was detected in modified form by Chandler in 1891.

During his regular lectures at King's College, London, Maxwell was accustomed to present some of Rankine's work on the calculation of stresses in frameworks. In 1864 Rankine offered an important new theorem,[92] which Maxwell then developed into a geometrical discussion entitled "On Reciprocal Figures and Diagrams of Forces." The principle was an extension of the well-known triangle of forces in statics. Corresponding to any rectilinear figure, another figure may be drawn with lines parallel to the first, but arranged so that lines converging to a point in one figure form closed polygons in the other. The

lengths of lines in the polygon supply the ratios of forces needed to maintain the original point in equilibrium. Maxwell gave a method for developing complex figures systematically, and derived a series of general theorems on properties of reciprocal figures in two and three dimensions. He combined the method with energy principles and later, after refereeing a paper on elasticity by G. B. Airy, extended it to stresses in continuous media.[93] Figure 5 reproduces diagrams of a girder bridge and its reciprocal given by Maxwell in 1870. Reciprocal theorems and diagrams are useful in many fields of science besides elasticity. Maxwell investigated similar theorems (some of them already known) in electricity. His student Donald MacAlister, the physiologist, applied the method to bone structures. Another application from a later period is the use of reciprocal lattices to determine atomic configurations by X-ray crystallography.

In the British Association experiment on electrical resistance, Maxwell and his colleagues used a speed governor to ensure that the coil rotated uniformly. In principle it resembled James Watt's steam-engine governor: centrifugal force made weights attached to the driven shaft fly out and adjust a control valve.[94] Maxwell studied its behavior carefully; and four years later, in 1868, after reading a paper by William Siemens[95] on the practical limitations of governors, he gave an analytical treatment of the subject. He determined conditions for stability in various simple

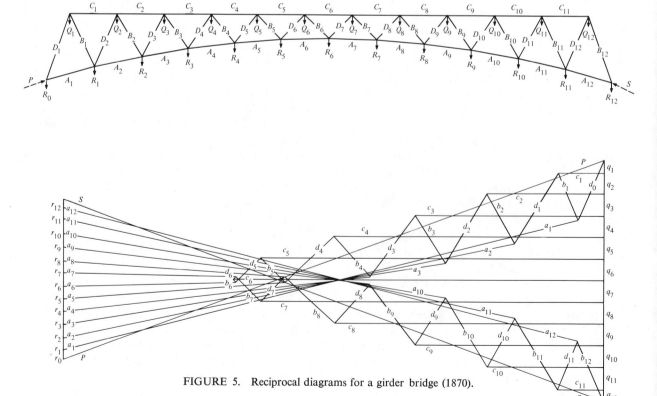

FIGURE 5. Reciprocal diagrams for a girder bridge (1870).

cases, including one fifth-order system representing a combination of two devices invented by Thomson and Fleming Jenkin, and investigated effects of natural damping and of variations in the driven load, as well as the onset of instabilities. Maxwell's paper "On Governors" is generally regarded as the foundation of control theory. Norbert Wiener coined the name "cybernetics" in its honor, from $\kappa\nu\beta\epsilon\rho\nu\eta\tau\eta\varsigma$, the Greek for "steersman," from which, via a Latin corruption, the word "governor" is etymologically descended.[96]

Maxwell's textbook *Theory of Heat* was published in 1870 and went through several editions with extensive revisions. Chiefly an exposition of standard results, it did contain one far-reaching innovation, the "Maxwell relations" between the thermodynamical variables, pressure, volume, entropy, and temperature, and their partial derivatives. In conceptual spirit they resemble Maxwell's field equations in electricity, by which they were obviously suggested; they are an ordered collection of relationships between fundamental quantities from which practically useful formulas follow. Several of the individual terms had previously been given by other writers. Maxwell's derivation was a deceptively simple geometrical argument based on the pressure-volume diagram. Applications of geometry to thermodynamics underwent an extraordinary development in 1873 through Gibbs's work on entropy-volume-temperature surfaces, of which Maxwell instantly became a powerful advocate. Maxwell's papers and correspondence contain much of related interest, including an independent development of the chemical potential and an admirable discussion of the classification of thermodynamic quantities in a little-known article, "On Gibbs' Thermodynamic Formulation for Coexistent Phases." In 1908 this paper was reprinted at the request of the energeticist W. Ostwald, with notes by Larmor.[97] One more important personage in the *Theory of Heat* was Maxwell's "sorting demon" (so named by Thomson), a member of a class of "very small BUT lively beings incapable of doing work but able to open and shut valves which move without friction and inertia"[98] and thereby defeat the second law of thermodynamics. The demon points to the statistical character of the law. His activities are related to the so-called "reversibility paradox" discussed first by Thomson in 1874—that is, the problem of reconciling the irreversible increase in entropy of the universe demanded by thermodynamics with the dynamical laws governing the motions of molecules, which are reversible with respect to time. A more formal view of the statistical basis of thermodynamics was supplied by Boltzmann in 1877 in the famous equation $S = k \overline{\log W}$, which relates entropy S to a quantity W expressing the molecular disorder of a system.

Much of Maxwell's last eight years was devoted to Cambridge and the Cavendish Laboratory. Many papers by Cambridge mathematicians of the period acknowledge suggestions by him. The design of the laboratory embodied many ingenious features: clear corridors and stairwells for experiments needing large horizontal and vertical distances, an iron-free room for magnetic measurements, built-in antivibration tables for sensitive instruments supported by piano wires from the roof brackets, and so on. The construction of the building and much of the equipment were paid for by the Duke of Devonshire, but after 1876 Maxwell had to spend substantial sums out of his own pocket to keep the laboratory going. A characteristic of the work done under his direction was an emphasis on measurements of extreme precision, in marked contrast to the "string-and-sealing-wax" tradition of research later built up by J. J. Thomson. Examples were D. MacAlister's test of the inverse-square law in electrostatics; G. Chrystal's test of the linear form of Ohm's law; J. H. Poynting's improved version (the first of many) of Cavendish's experiment to measure the gravitational constant; and R. T. Glazebrook's determination of the optical wave surface for birefringent crystals. In each instance the precision was several orders of magnitude higher than anything previously attempted. "You see," wrote Maxwell to Joule, "that the age of heroic experiments is not yet past."[99]

NOTES

1. C. G. Knott, *Life and Scientific Work of Peter Guthrie Tait* (Cambridge, 1911), 4, 5.
2. C. W. F. Everitt, in *Applied Optics*, **6** (1967), 644–645.
3. W. C. Henry, *Life of Dalton* (London, 1854), 25–27, letter of 20 May 1833; the letter was familiar to Maxwell through G. Wilson, *Researches on Colour Blindness* (Edinburgh, 1855), 60, in which his own work was first published. See also J. Herschel, "Treatise on Light," in *Encyclopaedia Metropolitana* (London, 1843), 403.
4. *Papers*, I, 146.
5. T. Young, *Lectures on Natural Philosophy*, I (London, 1807), 440; as was known to Maxwell, *Papers*, I, 150. The choice had also been suggested by C. E. Wünsch, *Versuche und Beobachtungen über die Farben des Lichtes* (Leipzig, 1792), of which an abstract is given in *Annales de chimie*, **64** (1807), 135. This rare reference is noted in one of Maxwell's memorandum books preserved at King's College, London.
6. *Papers*, I, 135.
7. J. Larmor, ed., *Memoir and Scientific Correspondence of Sir G. G. Stokes*, II (London, 1910), 22; *Life*, 376–379; W. D. Wright, *The Measurement of Colour* (London, 1944), 62 f.
8. *Life*, 489.
9. *Ibid.*, 347.
10. R. M. Evans, in *Journal of Photographic Science*, **9** (1961), 243; *Scientific American*, **205** (1961), 118.

11. *Papers*, I, 288. There is much of interest in the Challis-Thomson correspondence, Kelvin Papers, Cambridge University Library, file box 2.
12. *Life*, 295.
13. A. F. Cook and F. A. Franklin, in *Astronomical Journal*, **69** (1964), 173–200; **70** (1965), 704–720; **71** (1966), 10–19; also G. P. Kuiper, D. P. Cruikshank, and V. Fink, in *Bulletin of the American Astronomical Society*, **2** (1970), 235–236; and C. B. Pilcher, C. R. Chapman, L. A. Lebotsky, and H. H. Kieffer, *ibid.*, 239.
14. *Life*, 291.
15. Quoted by J. Larmor, in *Proceedings of the Royal Society*, **81** (1908), xix.
16. *Treatise*, preface, vi.
17. *Life*, 302.
18. *Proceedings of the Cambridge Philosophical Society. Mathematical and Physical Sciences*, **32** (1936), 695–750.
19. *Treatise*, I, sec. 72.
20. *Elementary Treatise*, sec. 64.
21. L. Euler, *Letters to a German Princess . . .*, H. Hunter, trans., II (London, 1795), 265–271; and known to Faraday, *Experimental Researches*, III, sec. 3263.
22. W. Thomson, *Papers on Electrostatics and Magnetism* (London, 1873), secs. 573 f., 733 f. See Maxwell's *Papers*, I, 453.
23. W. Thomson, in *Proceedings of the Royal Society*, **8** (1856), 150–158; repr. in *Baltimore Lectures* (London, 1890), app. F, 569–583.
24. J. Bromberg, Ph.D. thesis (Univ. of Wis., 1966); A. M. Bork, private communication.
25. *Proceedings of the Cambridge Philosophical Society*, **32** (1936), 704, letter of 13 Nov. 1854.
26. *Papers*, I, 500.
27. *Report of the British Association for the Advancement of Science*, 1st ser., **32** (1863), 130–163; repr. with interesting additions in F. Jenkin, *Reports of the Committee of Electrical Standards* (London, 1873), 59–96.
28. Gauss introduced only the definition of the magnetic pole; credit for the remaining parts of the system is shared by Weber, Thomson, and Maxwell.
29. I. B. Hopley, in *Annals of Science*, **15** (1959), 91–107.
30. *Life*, 342. Letter of 5 Jan. 1865.
31. *Papers*, II, 662–663.
32. W. Thomson, *Paper on Electrostatics and Magnetism* (London, 1873), 447–448n.
33. *Report of the British Association for the Advancement of Science*, 1st ser., **32** (1863), 163–176.
34. W. Thomson, *Mathematical and Physical Papers* (Cambridge, 1882–1911), II, 61–103.
35. *Treatise*, 3rd ed., II, 228. Other illustrations were given by Boltzmann and Rayleigh. The original MS of Maxwell's 1865 paper, preserved in the archives of the Royal Society, contains a curious canceled passage likening the action of two inductive circuits on the field to the action of two horses pulling on the swingletree of a carriage. This in essence is Rayleigh's analogy.
36. Comment on a paper by G. Forbes, in *Proceedings of the Royal Society of Edinburgh*, **9** (1878), 86.
37. *Treatise*, II, sec. 575; Rayleigh (4th Baron), *Life of Lord Rayleigh* (London, 1924), 48, letter from Maxwell to Rayleigh of 18 May 1870. *Papers*, I, 485n; *Treatise*, II, sec. 575.
39. The clearest physical treatment is by L. Page and N. I. Adams, in *American Journal of Physics*, **13** (1945), 141.
40. *Treatise*, II, sec. 615.
41. *Ibid.*, secs. 600–601.
42. G. G. Stokes, *Mathematical and Physical Papers*, IV (Cambridge, 1904), 157–202.
43. J. Larmor, ed., *Memoir and Scientific Correspondence of Sir G. G. Stokes*, II (London, 1910), 25–26. Letter to Stokes of 15 October 1864.
44. Lord Rayleigh, *Scientific Papers*, I (Cambridge, 1900), 111–

134, 518–536; J. Willard Gibbs, *Scientific Papers*, II (London, 1906), 223–246; see *Papers*, II, 772 f. for Maxwell's account.
45. Thomson, *Mathematical and Physical Papers*, III, 466, 468.
46. *Treatise*, II, sec. 866.
47. *Life*, 394, letter to Bishop Ellicott of 22 Nov. 1876.
48. W. Whewell, *Philosophy of the Inductive Sciences*, 2 vols. (London, 1840), *passim*. See *Life*, 215, letter to R. B. Litchfield of 6 June 1855.
49. *Papers*, I, 564.
50. Thomson, *Mathematical and Physical Papers*, II, 28; for Maxwell's comments, see *Papers*, II, 767–768.
51. T. Young, in *Philosophical Transactions of the Royal Society*, **94** (1804), 1.
52. Unpublished MS at Cambridge. "On an Experiment to Determine Whether the Motion of the Earth Influences the Refraction of Light."
53. The letter is lost but see Maxwell's reply dated 6 May 1864 in J. Larmor, ed., *Memoir and Scientific Correspondence of Sir G. G. Stokes*, II (London, 1910), 23–25. I am indebted to Dr. A. M. Bork for the connection, which is obscured by the first part of Larmor's footnote on p. 23.
54. W. Huggins, in *Philosophical Transactions of the Royal Society*, **158** (1868), 532.
55. *Nature*, **21** (1880), 314, 315. See Michelson's comments in *American Journal of Science*, **122** (1881), 120; also J. C. Adams to Maxwell (17 July 1879) on the feasibility of the astronomical test (Cambridge MSS).
56. Reprinted with the FitzGerald-Lorentz correspondence in S. G. Brush, *Isis*, **58** (1967), 230–232.
57. *Treatise*, II, sec. 769.
58. *Papers*, II, 121–124.
59. Cf. J. E. McDonald, in *American Journal of Physics*, **33** (1965), 706–711.
60. *Papers*, II, 329–331, 391–392.
61. In order of citation, secs. 86, 95–102, 19–21, 280–282 with app.; and 129.
62. Secs. 756–757 and app. to ch. 17.
63. The first statement is in a letter to Thomson of 5 June 1869, in *Proceedings of the Cambridge Philosophical Society*, **32** (1936), 738–739. See J. Bromberg, in *American Journal of Physics*, **36** (1968), 142–151.
64. Thomson, *Papers on Electrostatics and Magnetism*, 15–41, paper of 1845; F. O. Mossotti, in *Archives des sciences physiques et naturelles*, **6** (1847), 193.
65. Y. Aharonov and D. Bohm, in *Physical Review*, **115** (1959), 485–491; **123** (1961), 1511–1524.
66. *Papers*, II, 43.
67. *Life*, 142–143; C. C. Gillispie, *Scientific Change*, A. C. Crombie, ed. (London, 1963), 431 ff.; S. G. Brush, *Kinetic Theory*, I (Oxford, 1965), 30n.; Elizabeth Wolfe Garber, thesis (Case Institute, 1966), and in *Historical Studies in the Physical Sciences*, **2** (1970), 299; P. M. Heimann, in *Studies in History and Philosophy of Science*, **1** (1970), 189.
68. M. Kac, in *American Journal of Mathematics*, **61** (1939), 726–728. See also T. H. Gronwall, in *Acta mathematica*, **17** (1915), 1.
69. J. Larmor, ed., *Memoir and Scientific Correspondence of Sir G. G. Stokes*, II (London, 1910), 10.
70. *Report of the British Association for the Advancement of Science*, **28**, pt. 2 (1860), 16.
71. "My better ½, who did all the real work of the kinetic theory is at present engaged in other researches. When she is done I will let you know her answer to your enquiry [about experimental data]." Postcard from Maxwell to Tait, 29 Dec. 1877, Cambridge MSS.
72. *Life*, 80.
73. *Papers*, II, 379–380.
74. *Papers*, II, 681. Royal Society Archives 1878, Maxwell 70, Thomson's report marked 123 in upper right-hand corner. See S. G. Brush and C. W. F. Everitt, in *Historical Studies in the Physical Sciences*, **1** (1969), 105–125.

75. See S. G. Brush, in *American Journal of Physics*, **24** (1962), 274n.
76. Letter from C. Truesdell to C. W. F. Everitt, 16 Dec. 1971.
77. *Papers*, II, 376.
78. *Nature*, **20** (1877), 242.
79. *Life*, 570; A. Schuster, *The Progress of Physics 1875–1878* (Cambridge, 1911), 29; also *History of Cavendish Laboratory* (London, 1910), 31.
80. See the two eds. of Watson's *Treatise on the Kinetic Theory of Gases* (Oxford, 1876; 2nd ed., 1893); and the commentaries on Maxwell's paper by Boltzmann, in *Philosophical Magazine*, **14** (1882), 299–312; by Rayleigh, *ibid.*, **33** (1892), 356–359, and *Scientific Papers*, III, 554; and by J. Larmor, *Mathematical and Physical Papers*, II (Cambridge, 1929), app. III, 743–748. Rayleigh incorrectly attributes one of Watson's results to Maxwell, as Watson, in the 2nd ed. of his book (pp. 22–23), succeeds in pointing out, without appearing to do so, with the beautiful oblique courtesy to be expected from the man who was, after all, the Rector of Berkswell.
81. *Life*, 570–571.
82. *Papers*, II, 692.
83. *Proceedings of the Royal Society*, **27** (1878), 304.
84. *Papers*, II, 742.
85. *Treatise*, I, sec. 135a.
86. Letter of 11 July 1961 from S. C. Chapman to S. G. Brush quoted in S. G. Brush, in *American Journal of Physics*, **24** (1962), 276n.
87. E. Ikenberry and C. Truesdell, in *Journal of Rational Mechanics and Analysis*, **5** (1956), 4–128.
88. *Papers*, I, 79.
89. R. K. Luneberg, *Lectures on Optical Design* (Providence, R.I., 1944), mimeographed notes.
90. *Life*, 221.
91. For example, C. G. Wynne, in *Proceedings of the Physical Society of London*, **65B** (1952), 429.
92. W. J. M. Rankine, *Miscellaneous Scientific Papers* (London, 1881), 564.
93. *Papers*, II, 161–207; Royal Society Archives 1869.
94. A photograph of the governor designed by Fleeming Jenkin is given by I. B. Hopley, in *Annals of Science*, **13** (1951), 268. See also Otto Mayr, in *Isis*, **62** (1971), 425–444; and *Notes and Records. Royal Society of London*, **26** (1971), 205–228.
95. C. W. Siemens, in *Philosophical Transactions of the Royal Society*, **156** (1866), 657–670.
96. Norbert Wiener, *Cybernetics, or Control and Communication in the Animal and the Machine* (Cambridge, Mass., 1948), 11–12.
97. *Philosophical Magazine*, **16** (1908), 818.
98. C. G. Knott, *Life and Scientific Work of Peter Guthrie Tait*, 214–215.
99. *History of the Cavendish Laboratory*, 31.

BIBLIOGRAPHY

I. ORIGINAL WORKS. Most of the technical papers were reprinted in *The Scientific Papers of J. Clerk Maxwell*, W. D. Niven, ed., 2 vols. (Cambridge, 1890; repr. New York, 1952), cited in the footnotes as *Papers*. About twenty papers and short articles were omitted from the collection; most may be found in *Nature, Electrician, Reports of the British Association, Proceedings of the London Mathematical Society, Proceedings of the Royal Society of Edinburgh*, and *Cambridge Reporter*. The abstracts of longer papers printed in the *Proceedings of the Royal Society* are also often of interest. Maxwell's books are *Theory of Heat* (London, 1870; 4th ed. greatly rev., 1875; 11th ed. rev. with notes by Lord Rayleigh, 1894); *Treatise on Electricity and Magnetism*, 2 vols. (Oxford, 1873); 2nd ed., W. D. Niven, ed. (1881); 3rd ed., J. J. Thomson, ed. (1891), cited as *Treatise*—revision of the 2nd ed. was cut short by Maxwell's death; the changes in the first eight chs. are extensive and significant; references here are to the 3rd ed.; *Matter and Motion* (London, 1877), 2nd ed., with appendixes by J. Larmor (1924); *Elementary Treatise on Electricity*, W. Garnett, ed. (Oxford, 1881; 2nd ed., rev., 1888), cited as *Elementary Treatise*; and *The Unpublished Electrical Writings of Hon. Henry Cavendish* (Cambridge, 1879), 2nd ed., with further notes by J. Larmor (1924), which contains an introductory essay and extensive notes by Maxwell.

II. SECONDARY LITERATURE. The standard biography is L. Campbell and W. Garnett, *The Life of James Clerk Maxwell* (London, 1882), cited as *Life*; 2nd ed., abridged but containing letters not given in 1st ed. (1884). Extensive correspondence appears in *Memoir and Scientific Correspondence of Sir George Gabriel Stokes*, J. Larmor, ed., 2 vols. (London, 1910); C. G. Knott, *Life and Scientific Work of Peter Guthrie Tait* (Cambridge, 1911); Silvanus P. Thomson, *Life of Lord Kelvin*, 2 vols. (London, 1912); J. Larmor, "The Origin of Clerk Maxwell's Electric Ideas as described in Familiar Letters to W. Thomson," in *Proceedings of the Cambridge Philosophical Society. Mathematical and Physical Sciences*, **32** (1936), 695–750, repr. as a separate vol. (Cambridge, 1937). Other letters or personal material will be found in standard biographies of W. C. and G. P. Bond, H. M. Butler, J. D. Forbes, J. G. Fraser, F. Galton, D. Gill, F. J. A. Hort, T. H. Huxley, Fleeming Jenkin, R. B. Litchfield, by Henrietta Litchfield (London, 1903), privately printed; a copy is in the library of the Working Men's College, London; C. S. Peirce, Lord Rayleigh, H. Sidgwick, W. Robertson Smith, Sir James FitzJames Stephen, and George Wilson, and in the collected papers of T. Andrews, Sir William Huggins, J. P. Joule, J. Larmor, and H. A. Rowland.

See also C. Popham Miles, *Early Death not Premature: Memoir of Francis L. Mackenzie* (Edinburgh, 1856), 216–218; W. Garnett, *Heroes of Science* (London, 1886); R. T. Glazebrook, *James Clerk Maxwell and Modern Physics* (London, 1896); F. W. Farrar, *Men I Have Known* (London, 1897); A. Schuster, *The Progress of Physics 1875–1908* (London, 1911); *Biographical Fragments* (London, 1932); and "The Maxwell Period," in *History of the Cavendish Laboratory 1871–1910* (London, 1910), no editor identified; *Aberdeen University Quarter-Centenary Volume* (Aberdeen, 1906); D. Gill, *History of the Royal Observatory, Cape of Good Hope* (London, 1913), xi–xiv, for Maxwell at Aberdeen; F. J. C. Hearnshaw, *History of King's College, London* (London, 1929; J. J. Thomson, ed., *James Clerk Maxwell 1831–1931* (Cambridge, 1931); J. G. Crowther, *British Scientists of the Nineteenth Century* (London, 1932); K. Pearson, "Old Tripos Days at Cambridge," in *Mathematical Gazette*, **20** (1936), 27; C. Domb, ed., *Clerk Maxwell and Modern Science* (London, 1963); and R. V. Jones, "James Clerk Maxwell at Aberdeen 1856–1860," in *Notes and Records. Royal Society of London*, **28** (1973), 57–81.

Useful general bibliographies are given by W. T. Scott, in *American Journal of Physics*, **31** (1963), 819–826, for the electromagnetic field concept; and by S. G. Brush in *Kinetic Theory* (Oxford, 1965, 1966, 1972) and in *American Journal of Physics*, **39** (1971), 631–640 for kinetic theory. For thermodynamics see Martin J. Klein in *American Scientist*, **58** (1970), 84–97. For the theory of governors see two articles by Otto Mayr in *Isis*, **62** (1971), 425–444; and *Notes and Records. Royal Society of London*, **26** (1971), 205–228; references to other early papers on governors are given in the later editions of E. J. Routh, *Treatise on the Dynamics of a System of Rigid Bodies*, II (6th ed., London, 1905), sec. 107.

For reciprocal diagrams see A. S. Niles, *Engineering*, **170** (1950), 194–198, and S. Timoshenko, *History of the Strength of Materials* (New York, 1953): both authors exaggerate the neglect of Maxwell's work by his contemporaries. *Studies in History and Philosophy of Science*, **1** (1970), 189–251 contains four articles on Maxwell with lengthy bibliographies.

The two principal collections of unpublished source materials are in the Archives of the Royal Society and in Cambridge University Library, Anderson Room, where the Stokes and Kelvin MSS should also be consulted. Materials elsewhere at Cambridge are in the Cavendish Laboratory, Peterhouse and Trinity College libraries, and the Cambridge Library. Other items are at Aberdeen University; St. Andrews University (Forbes MSS); Berlin, Staatsbibliothek der Stiftung Preussischer Kultur Besitz; Bodleian Library, Oxford (Mark Pattison MSS); Burndy Library, Norwalk, Conn.; Edinburgh University (Tait MSS); Glasgow University (Kelvin MSS); Göttingen, Niedersächsische Staats- und Universitätsbibliothek; Harvard University (Bond MSS); Imperial College, Lyon Playfair Library (Huxley MSS); Institute of Electrical Engineers (Heaviside MSS); Johns Hopkins University (Rowland MSS); Manchester Institute of Science and Technology (Joule MSS); Queen's University, Belfast (Andrews, J. Thomson MSS); Royal Institution (Faraday, Tyndall MSS); University of Rochester, Rush Rhees Library; U.S. Air Force Cambridge Center (Rayleigh MSS).

A large collection of watercolor paintings of Maxwell's childhood by Jemima Wedderburn (later Mrs. Hugh Blackburn) and others are now in the possession of Brigadier J. Wedderburn-Maxwell, D.S.O., M.C.

C. W. F. EVERITT

MAYER, ALFRED MARSHALL (*b.* Baltimore, Maryland, 13 November 1836; *d.* Hoboken, New Jersey, 13 July 1897), *physics.*

Mayer invented the method of floating tiny magnets in a magnetic field, used in the early twentieth century as a key to discovering or illustrating atomic structure. He studied classics at St. Mary's College in Baltimore but left at the age of sixteen to become a machinist. A self-educated analytical chemist, he published his first research paper at the age of nineteen; it brought him to the attention of Joseph Henry, who helped him to become assistant professor of physics and chemistry at the University of Maryland when he was only twenty and professor of physical science at Westminster College, Fulton, Missouri, in 1859. From 1863 to 1865 he studied in Paris, notably under Regnault, learning advanced physics, mathematics, and physiology. On his return to America, Mayer became professor of natural science at Pennsylvania College of Gettysburg (now Gettysburg College), and then, in 1867, professor of physics and astronomy at Lehigh University; in 1871 he organized the department of physics at the newly founded Stevens Institute of Technology, with which he was associated until his death.

Mayer's only academic degree was an honorary Ph.D. from Pennsylvania College of Gettysburg in 1866; he was a member of the National Academy of Sciences (1872), the American Philosophical Society, and the American Academy of Arts and Sciences.

Mayer conducted research in sound, heat, light, and gravity; devised a number of instruments for scientific measurement; and was the author of about one hundred publications, including fifty-four research articles and three scientific books. He was selected by the U.S. Nautical Almanac Office to direct the photographing of the solar eclipse of 7 August 1869; the results were considered remarkable for those early days of photography: a set of forty-two "perfect photographs," made at exposures of 0.002 second —five of them during the eighty-three seconds of total eclipse. His major scientific work was in acoustics; Mayer's Law gives a quantitative relation between pitch and the duration of residual auditory sensation. An avid sportsman, Mayer wrote widely about fishing and invented a rod with which, in 1884, he won first prize at the Amateur Minnow-Casting Tournament of the National Rod and Reel Association.

Mayer is most remembered (and cited) for his experiments in which magnetized needles were inserted into corks, which were then floated on water with their south poles upward, under the north pole of a powerful electromagnet. Under these conditions, certain definite stable configurations were observed "which suggested the manner in which atoms of molecules may be grouped in the formation of definite compounds" (Mayer and Woodward, p. 257) and which illustrated various properties of the constitution of matter. These experiments won high praise from Kelvin (*Nature*, **18** [1878], 13–14) and were later used by J. J. Thomson (*Electricity and Matter* [New Haven, 1904], pp. 114–117, 122; *The*

Corpuscular Theory of Matter [New York, 1907], p. 110) and others as a key to the way in which a characteristic number of electrons might be arranged within the atoms of each chemical element in relation to the periodic table. Mayer thus made a small but significant contribution to the theory of atomic structure.

BIBLIOGRAPHY

I. Original Works. A full list of Mayer's scientific publications is given in Alfred G. Mayer and Robert S. Woodward, "Biographical Memoir of Alfred Marshall Mayer 1836–1897," in *Biographical Memoirs. National Academy of Sciences*, **8** (1916), 243–272. A list of his publications for the period 1871–1897 is given in his biography, in F. DeR. Furman, *Morton Memorial: A History of the Stevens Institute of Technology* (Hoboken, N.J., 1905), pp. 202–209.

Mayer's articles on the stable configurations of magnets floating freely in a magnetic field were published in *American Journal of Science*, 3rd ser., **15** (1878), 276–277, 477–478; **16** (1878), 247–256; and in *Scientific American*, supp. **5** (1878), 2045–2047, where these experiments are said "to illustrate the action of atomic forces and the molecular structure of matter . . ." (which includes allotropy, isomerism, and the kinetic theory of gases).

II. Secondary Literature. Besides the two biographies mentioned above, see F. DeR. Furman's article on Mayer in the *Dictionary of American Biography*, XII. A biography based on personal recollections was published by W. LeConte Stevens in *Science*, n.s. **6** (1897), 261–269. Obituaries appeared in *Stevens Indicator*, **14** (1897), 367; *American Journal of Science*, 4th ser., **4** (1897), 161–164; and New York *Times* (14 July 1897), p. 5.

I. B. Cohen

MAYER, CHRISTIAN (*b.* Meseritsch, Moravia [now Mederizenhi, Czechoslovakia], 20 August 1719; *d.* Heidelberg, Germany, 17 April 1783), *astronomy.*

Claims that Mayer studied Greek, Latin, philosophy, theology, and mathematics at Brno, Vienna, Turnau, Rome, and Würzburg have not been confirmed. The first authenticated fact about his life is his entering the Jesuit novitiate at Mannheim on 13 September 1745, after he had left home because his father did not approve of his decision. He subsequently taught languages and then also mathematics in a Jesuit school at Aschaffenburg, spending his evenings making astronomical observations. In 1752 he was appointed professor of mathematics and physics at Heidelberg University, and in 1753 his first physical work was printed there. He also published a series of mathe-

matical and physical works at Heidelberg, but his main interest soon turned to astronomy.

In 1762 the elector palatine, Karl Theodor, who was very interested in the arts and sciences, constructed an astronomical observatory for Mayer at Schwetzingen, his summer residence. In 1772–1774 a second and larger observatory was erected at Mannheim, then the capital of the electoral Palatinate. This observatory was well equipped for the time, with instruments from the best British workshops. A great quadrant by Bird was installed in 1775 and other instruments were made by Dollond, Troughton, and Ramsden; but Mayer did not live to see the observatory acquire all the instruments he had requested. Appointed court astronomer, Mayer was relieved of his duties as a theologian (the Jesuit order was dissolved by Pope Clement XIV in 1773, which made such a step easier).

In the 1760's Mayer participated in the measurement of a degree of the meridian, inaugurated by Cassini de Thury, visiting Paris and also executing the measurement of a geodetic base in the plain of the Rhenish Palatinate. He observed the transits of Venus across the sun in 1761 and 1769, invited to Russia for the latter by Catherine II. He also drew up a map of the Russian empire for her.

In 1776 Mayer turned to a branch of astronomy not previously investigated: the observation of double stars of all classes. Mayer could not distinguish between truly binary stars and those that are nearly on a line of sight from the earth but distantly separated in space. He made uncritical observations of all such apparent pairs and compiled a catalog of those that were near enough to be doubles in his sense of the word. The work was superseded in 1782 by W. Herschel's catalog of double stars, but his pioneer work should not be forgotten.

Mayer became involved in a polemic with his colleagues, typical of the time and caused by his name for the double stars. In his lecture at the Mannheim Academy and in his first note published in a Mannheim newspaper in 1777, Mayer said he had discovered more than 100 satellites of fixed stars. This term was misunderstood and Mayer's colleagues inferred that he was claiming to have discovered planets of other fixed stars. Today we speak of a companion of a fixed star, such as the companion of Sirius. But the contemporary astronomers, especially Hell, argued against Mayer's observations—which marked the beginning of systematic observation and an important impetus to this new branch of astronomy.

Mayer's reputation was unharmed by the quarrels. He was widely known and published his observations in various foreign journals—including the United

States, an uncommon practice among European astronomers. He was a member of scientific academies and societies in London, Philadelphia, Mannheim, Munich, Bologna, and other cities.

BIBLIOGRAPHY

Mayer's writings include *Selecta physices experimentalis elementa mathematico-physica* (Heidelberg, 1753); *Disquisitio de momento virium mechanicarum* (Heidelberg, 1756); *Basis Palatina anno 1762 ad normam Academiae Regiae Parisinae scientiarum exactam bis dimensa ...* (Mannheim, 1763); *Solis et lunae eclipseos observatio astronomica, facta Schwetzingae in specula nova electoralia* (Mannheim, 1764); "Observationes astronomicae," in *Philosophical Transactions of the Royal Society* (1764 and 1768); *Ad Augustissimam Russiarum omnium Imperatricem Catarinam II. Aliexiewnam expositio de transitu Veneris ante discum solis d. 23 Maji anno 1769* (St. Petersburg, 1769); *Nouvelle méthode pour lever ... une carte générale exacte de toute la Russie ...* (St. Petersburg, 1770); *Directio meridiani Palatini per speculam electorialem arcis aestivae Schwetzingensis ducti, observationibus et calculis definita* (Heidelberg, 1771); *Gründliche Vertheidigung neuer Beobachtungen von Fixstern Trabanten, welche zu Mannheim an der Sternwarte entdeckt worden sind* (Mannheim, 1778); *De novis in coelo sidereo phaenomenis, in miris stellarum fixarum comitibus Mannhemii detectis* (Mannheim, 1779); and "Observationes astronomicae," in *Transactions of the American Philosophical Society*, **2** (1786), 34–41.

On Mayer and his work, see J. L. Kluber, *Die Sternwarte zu Mannheim* (Mannheim, 1811), 58–59; and W. Meyer, "Geschichte der Doppelsterne," in *Vierteljahrsschrift d. naturforschenden Gesellschaft in Zurich* (1876), 695 ff.

H.-CHRIST. FREIESLEBEN

MAYER, CHRISTIAN GUSTAV ADOLPH (*b.* Leipzig, Germany, 15 February 1839; *d.* Gries bei Bozen, Austria [now Bolzano, Italy], 11 April 1908), *mathematics.*

The son of a wealthy Leipzig merchant family, Mayer studied mathematics and physics from 1857 to 1865 at Leipzig, Göttingen, Heidelberg, and chiefly at Königsberg under F. Neumann. In 1861 he received his doctorate from Heidelberg and qualified to lecture there in 1866. He became assistant professor in 1871 and full professor in 1890. In 1872 Mayer married Margerete Weigel. Poor health caused him to suspend his teaching activities early in 1908.

As a professor, Mayer enjoyed great respect from his colleagues and students. His activity as a researcher, which earned him membership in numerous learned societies, dealt essentially with the theory of differential equations, the calculus of variations, and theoretical mechanics. In his work, following Lagrange and Jacobi, he was capable of bringing out the inner relationship of these fields through emphasis on the principle of least action. Mayer achieved important individual results concerning the theory of integration of partial differential equations and the criteria for maxima and minima in variation problems. This work quickly brought him into close contact with the investigations on partial differential equations that Lie had under way at about the same time. Through subsequent works of Mayer, Lie's achievements became famous relatively quickly. Despite a great variety of methods and an outstanding mastery of calculation, Mayer was unable to develop the rigor necessary for the existence theorems of the calculus of variations; such rigor was displayed in exemplary fashion at approximately the same time by Weierstrass.

BIBLIOGRAPHY

Among Mayer's works are *Beiträge zur Theorie der Maxima und Minima einfacher Integrale* (Leipzig, 1866); *Geschichte des Prinzips der kleinsten Aktion* (Leipzig, 1877); and "Unbeschränkt integrable Systeme von linearen totalen Differentialgleichungen und die simultane Integration linearer partieller Differentialgleichungen," in *Mathematische Annalen*, **5** (1872), 448–470.

Also see the obituary notice by O. Holder, in *Berichte über die Verhandlungen der sächsischen Akademie der Wissenschaften zu Leipzig*, Math.-phys. Kl., **60** (1908), 353–373.

H. WUSSING

MAYER, JOHANN TOBIAS (*b.* Marbach, near Stuttgart, Germany, 17 February 1723; *d.* Göttingen, Germany, 20 February 1762), *cartography, astronomy.*

Mayer was the son of a cartwright, also named Johann Tobias Mayer, and his second wife, Maria Catherina Finken. In 1723 the father left his trade and went to work as the foreman of a well-digging crew in the nearby town of Esslingen, where his family joined him the following year. After his father's death in 1737 Mayer was taken into the local orphanage, while his mother found employment in St. Katharine's Hospital, where she remained until her death in 1737. It was probably through her occupation that Mayer found the opportunity to make architectural drawings of the hospital, as he did when he was barely fourteen years old. There is some evidence to indicate that he was encouraged in his draftmanship by Gottlieb David Kandler, a shoemaker who was subsequently responsible for the education of orphans in Esslingen.

Mayer's skill in architectural drawings also brought him to the attention of a certain Geiger, a non-commissioned officer in the Swabian district artillery, which was then garrisoned in Esslingen. Under Geiger's instruction, Mayer, in early 1739, prepared a book of plans and drawings of military fortifications. Later in the same year he drew a map of Esslingen and its surroundings (the oldest still extant), which was reproduced as a copper engraving by Gabriel Bodenehr of Augsburg in 1741.

Mayer's first book, written on the occasion of his eighteenth birthday, was published at about this same time. It was devoted to the application of analytic methods to the solution of geometrical problems, and in its preface Mayer acknowledged his debt to Christian von Wolff's *Anfangs-Gründe aller mathematischen Wissenschaften*, through which he had taught himself mathematics, a subject not included in the curriculum of the Esslingen Latin school, which he attended. The influence of Wolff's compendium is again apparent in the arrangement and content of Mayer's *Mathematischer Atlas* of 1745; the sixty plates of the latter work duplicate Wolff's choice of subjects—arithmetic, geometry, trigonometry, and analysis, as applied to mechanics, optics, astronomy, geography, chronology, gnomonics, pyrotechnics, and military and civil architecture. This atlas, published in Augsburg by the firm of Johann Andreas Pfeffel, for which Mayer worked during his brief stay there (from 1744 to 1746), provides a good index to the extent of his scientific and technical knowledge at that period. It was probably in Augsburg that he acquired much of his knowledge of French, Italian, and English. He also became acquainted with a local mechanic and optician, G. F. Brander.

Mayer left Augsburg to take up a post with the Homann Cartographic Bureau in Nuremberg. He spent five years there, which he devoted primarily to improving the state of cartography. To this end he collated geographical and astronomical data from the numerous printed and manuscript records to which the Homann office permitted him access. He also made personal observations of lunar occultations and other astronomical eclipse phenomena, using a nine-foot-focus telescope and a glass micrometer of his own design. Of more than thirty maps that he drew, the "mappa critica" of Germany is generally considered to be the most significant, since it established a new standard for the rigorous handling of geographical source materials and for the application of accurate astronomical methods in finding terrestrial latitude and longitude.

In order to facilitate the lunar eclipse method of longitude determination, Mayer in 1747 and 1748 made a large number of micrometric measurements of the angular diameter of the moon and of the times of its meridian transits. In his determinations of the selenographic coordinates of eighty-nine prominent lunar markings, he took account of the irregularity of the orbital and libratory motions of the moon and of the effect of its variable parallax. In addition his analysis correctly—although fortuitously—reduced twenty-seven conditional equations to three "normal" ones, a procedure that had never before been attempted, and one for which a theory had still to be developed.

Mayer was the editor of the *Kosmographische Nachrichten und Sammlungen auf das Jahr 1748*, which was published in Nuremberg in 1750, under the auspices of the newly established Cosmographical Society. The work contains Mayer's own description of his glass micrometer, his observations of the solar eclipse of 25 July 1748 and the occultations of a number of bright stars, his long treatise on the libration of the moon, and his argument as to why the moon cannot possess an atmosphere. The Cosmographical Society itself, founded by Johann Michael Franz, director of the Homann firm, was crucial in determining the nature, scope, and, to some degree, motivation of Mayer's subsequent scientific research. The aims of the mathematical class of the society, to which Mayer belonged, as set out by Franz in the preface to the *Homannisch-Haseschen Gesellschafts Atlas* (Nuremberg, 1747), define much of Mayer's later work.

In November 1750 Mayer was called to a professorship at the Georg-August Academy in Göttingen, a post that he took up after Easter of the following year. Shortly before he left Nuremberg he married Maria Victoria Gnüge; of their eight children, two, Johann Tobias and Georg Moritz, lived to maturity. Mayer's academic title, professor of economy, was purely nominal, since his actual duties were assigned, in his letter of appointment, as the teaching of practical (that is, applied) mathematics and research. His reputation as a cartographer and practical astronomer had preceded him, and was indeed the basis for his selection as professor.

Mayer's chief scientific concerns at this time were the investigation of astronomical refraction and lunar theory. In 1752 he drew up new lunar and solar tables, in which he attained an accuracy of $\pm 1'$, an achievement attributable to his skillful use of observational data, rather than to the originality of his theory or the superiority of his instruments. Mayer subsequently undertook an investigation of the celestial positions of the moon at conjunction and opposition; he compared the values that he obtained with those derivable from a systematic study of all lunar and solar eclipses reported since the invention of the astronomical telescope and the pendulum clock. His results led him to recognize that the discrepancies of up to $\pm 5'$ that he and his contemporaries had found were due largely to errors in the determination of star places and to the poor quality of their instruments.

Mayer's further astronomical researches consequently included the problem of the elimination of errors from a six-foot-radius mural quadrant made in 1755 by John Bird for installation in Mayer's newly completed observatory in Göttingen; the invention of a simple and accurate method for computing solar eclipses; the compilation of a catalogue of zodiacal stars; and the investigation of stellar proper motions. He wrote treatises on each of these topics that were published posthumously in Georg Christoph Lichtenberg's *Opera inedita Tobiae Mayeri* (Göttingen, 1775). This work also contains a treatise on the problem of accurately defining thermometric changes (an extension of Mayer's research on astronomical refraction) and another on a mathematical theory of color mixing (a topic that Mayer may have taken up in response to the need of the Homann firm, part of which had been transferred to Göttingen in 1755, to train unskilled workers in the accurate reproduction of maps). Appended to Lichtenberg's book, in accordance with one of Mayer's last wishes, is a copper engraving of Mayer's map of the moon; the original map and the forty detailed drawings from which it was constructed were also reproduced by photolithography more than a century later.

Others of Mayer's treatises, lecture notes, and correspondence have been neglected since their deposit, shortly after his death, in the Göttingen observatory archives, although abstracts of some of his lectures to the Göttingen Scientific Society were printed in the *Göttingische Anzeigen von gelehrten Sachen* between 1752 and 1762. His researches during these years included his efforts to improve the art of land measurement, for which purpose he invented a new goniometer and explored the application of the repeating principle of angle measurement, developed a new projective method for finding the areas of irregularly shaped fields, and transformed the common astrolabe into a precision instrument. He further applied the repeating principle to an instrument of his own invention, the repeating circle, which proved to be of use not only for the sea navigation for which it had been designed but also for making standard trigonometrical land surveys. (The instru-

ment used by Delambre and Méchain in their determination of the standard meter was a variant, designed by Borda, of the Mayer circle.)

Mayer also undertook to devise a method for finding geographical coordinates independently of astronomical observations. In so doing he arrived at a new theory of the magnet, based, like his lunar theory, on the principles of Newtonian mechanics. This theory represented a convincing demonstration of the validity of the inverse-square law of magnetic attraction and repulsion, and antedated Coulomb's well-known verification of that law by some twenty-five years. Mayer's manuscripts on this theory and on its application to the calculation of the variation and dip of a magnetic needle are among those that went virtually unnoticed after his death.

In 1763 Mayer's widow, acting upon another of his last requests, submitted to the British admiralty his *Theoria lunae juxta systema Newtonianum*, which contained the derivations of the equations upon which his lunar theory was based, and his *Tabulae motuum solis et lunae novae et correctae*, which were published in London in 1767 and 1770, respectively. The tables were edited by Maskelyne, and printed under his direct supervision; they were used to compute the lunar and solar ephemerides for the early editions of the *Nautical Almanac*. (They were superseded a decade later by tables employing essentially the same principles, but based upon the newer and more accurate observational data that were gradually being assembled at the Royal Observatory at Greenwich.) In 1765 the British parliament authorized Maria Mayer to receive an award of £3,000, in recognition of her husband's claim, lodged ten years before, for one of the prizes offered to "any Person or Persons as shall Discover the Longitude at Sea."

BIBLIOGRAPHY

A comprehensive list of Mayer's publications is given in Poggendorff, II, 91, the sole omission being his article "Versuch einer Erklärung des Erdbebens," in *Hannoverischen nützlichen Sammlungen* (1756), 290–296.

Mayer's scientific work is discussed by his official biographer, Siegmund Günther, in *Allgemeine deutsche Biographie*, XXI (1885), 109–116. His correspondence with Euler between 1751 and 1755, a valuable primary source of information about the former's contributions to the lunar theory, is in E. G. Forbes, ed., *The Euler-Mayer Correspondence (1751–1755)* (London, 1971).

The bulk of MS material relating to Mayer is preserved in Göttingen. The official classification of these papers is contained in the *Verzeichniss der Handschriften im Preussischen Staate I Hannover 3 Göttingen*, III (Berlin, 1894), 154–158. The title "Tobias Mayer's Nachlass, aufbewahrt in der K. Sternwarte" no longer applies, since the 70 items catalogued in this index were transferred to the Nieder-sächsische Staats- und Universitäts-Bibliothek, Göttingen, during the summer of 1965. In this same repository there is a booklet entitled "Briefe von und an J. Tobias Mayer," Cod. MS philos. 159. Cod. MS philos. 157 and Cod. MS Michaelis 320 are two other items worth consulting. *Personalakte Tobias Mayer* 4/Vb 18, and 4/Vf/1–4 are preserved in the Dekanate und Universität-Archiv, Göttingen. Some additional items of minor importance are also in the archives of the Göttingen Akademie der Wissenschaften.

The only significant MS collection outside Göttingen is "Betreffend der von Seiten des Prof. Tobias Mayer in Göttingen gelöste englische Preisfrage über die Bestimmung der Longitudo maris. 1754–1765," *Hannover Des.* 92 xxxiv no. II, 4, a′, Staatsarchiv, Hannover. A few documents relating to the payment of the parliamentary award to Mayer's widow are in vol. I of the Board of Longitude papers at the Royal Greenwich Observatory (P.R.O. Ref. 529, pp. 143–155).

E. G. Forbes, ed., *The Unpublished Writings of Tobias Mayer*, 3 vols. (Göttingen, 1972), contains Mayer's writings on astronomy and geography, his lecture notes on artillery and mechanics, and his theory of the magnet and its application to terrestrial magnetism.

Mayer's role in the development of navigation and his dealings with the British Admiralty and Board of Longitude are discussed in E. G. Forbes, *The Birth of Scientific Navigation* (London, 1973).

Eric G. Forbes

MAYER, JULIUS ROBERT (*b.* Heilbronn, Württemberg [now Baden-Württemberg], Germany, 25 November 1814; *d.* Heilbronn, 20 March 1878), *physics, physiology.*

Robert Mayer was one of the early formulators of the principle of the conservation of energy. His father, Christian Jakob Mayer, maintained a prosperous apothecary shop in Heilbronn and married Katharina Elisabeth Heermann, daughter of a Heilbronn bookbinder. The couple had three sons, of whom Robert was the youngest; both the older brothers followed their father's profession.

Mayer attended the classical Gymnasium at Heilbronn until 1829, when he transferred to the evangelical theology seminary at Schöntal. Although he was a mediocre student, he passed the *Abitur* in 1832 and enrolled in the medical faculty at the University of Tübingen. In February 1837 he was arrested and expelled from the university for participation in a secret student society. The next year Mayer was allowed to take the doctorate of medicine, and in 1838 he also passed the state medical exami-

nations with distinction. During the winter of 1839–1840 Mayer visited Paris and from February 1840 to February 1841 served as physician on a Dutch merchant ship on a voyage to the East Indies. While in Djakarta, Java, certain physiological observations convinced Mayer that motion and heat were interconvertible manifestations of a single, indestructible force in nature, and that this force was quantitatively conserved in any conversion. Mayer was inspired and occasionally obsessed by this insight. He elaborated his idea in various scientific papers which he published during the 1840's after his return to Germany.

Mayer settled in his native Heilbronn, where he took up a prosperous medical practice and held various civic posts. In 1842 he married Wilhelmine Regine Caroline Closs; the marriage produced seven children, five of whom died in infancy. Mayer maintained a conservative position during the Revolution of 1848, and this position led to his brief arrest by the insurgents and to a lasting estrangement from his brother Fritz. Depressed by these events and by his failure to obtain recognition for his scientific work, Mayer attempted suicide in May 1850. During the early 1850's he suffered recurrent fits of insanity, which necessitated several confinements in asylums at Göppingen, Kennenburg, and Winnenthal. Only after 1860 did Mayer gradually receive international recognition. He died in Heilbronn of tuberculosis in 1878.

Before his trip to Java, Mayer had shown much interest in science, but little creative ability. Flush with enthusiasm for his new idea about force, Mayer composed his "Ueber die quantitative und qualitative Bestimmung der Kräfte" immediately after his return to Heilbronn. In this paper Mayer groped toward a philosophical and mathematical expression of his new concept of force. Although he later altered the mathematical and the physical expressions of the ideas which he employed in this first paper, the philosophical and conceptual expressions remained virtually unchanged in his later work.

Mayer asserted that the task of science is to trace all phenomena back to their first causes. The laws of logic assure us that for every change there exists a first cause (*Ursache*), which is called a force (*Kraft*). In the world we observe "tension" or "difference" such as spatial separation or chemical difference existing between all matter. This tension is itself a force, and its effect is to prevent all bodies from quickly uniting themselves into a mathematical point. These tension-forces are indestructible, and their sum total in the universe is constant. Just as chemistry is the science of matter, so physics is the science of forces. Just as chemistry assumes that mass remains constant in every reaction, whatever qualitative changes the matter may undergo, so physics must also assume that forces are quantitatively conserved, no matter what conversions or qualitative changes of form they may undergo.

Although Mayer's mathematical-physical exposition of his ideas was highly original, it was also quite obscure and revealed his lack of acquaintance with the principles of mechanics. Mayer first considered a moving particle and argued that the measure of its "quantity of motion" is its mass times its speed. He then considered the special case of two particles, each having mass m and speed c and approaching each other on a straight line. The "quantitative determination" of the force of movement present is $2mc$. The "qualitative determination," however, is formally zero, since the motions are equal and opposite; this Mayer expressed by the symbolism $02mc$. Unless the particles are totally elastic, the "quantitative determination" of the force of motion present will be less after the collision than before the collision; for totally inelastic particles it will be zero after collision. The force present as motion is never lost, Mayer insisted; rather a part of it is "neutralized" in the collision and appears as heat. From this assertion Mayer generalized obscurely that all heat can be thought of as equal and opposite motions which neutralize each other, and that $02mc$ is somehow a universal mathematical expression for the force of heat. Finally Mayer showed how, in the more general case in which the colliding particles do not lie in a straight line, the parallelogram of forces may be employed to determine how much force of motion would be "neutralized" in the collision.

Upon completing "Ueber die . . . Bestimmung der Kräfte," Mayer submitted it to the *Annalen der Physik und Chemie* for publication. The editor Poggendorff ignored the paper and it was not printed. Although he was angry and disappointed, Mayer quickly became aware of the limitations of the treatise and immediately set himself to studying physics and mathematics. Between August 1841 and March 1842 Mayer discovered that mv^2, not mv, is the proper measure of the quantity of motion and that this form of force is identical to the *vis viva* of mechanics. He incorporated that discovery into his second paper, "Bemerkungen über die Kräfte der unbelebten Natur," which he had published in Liebig's *Annalen der Chemie* in May 1842.

In this second paper Mayer elaborated the conceptual basis of his theory, examining, he said, the precise meaning of the term "force." As in the previous paper, Mayer concluded that forces are first causes; hence the law *causa aequat effectum* assures us that

force is quantitatively indestructible. Like matter, forces are objects which are able to assume different forms and which are indestructible. Forces differ from matter only because they are imponderable.

Elaborating an idea mentioned in his previous paper, Mayer asserted that the spatial separation of two bodies is itself a force. This force he called "fall-force" (*Fallkraft*). Where one object is the earth and the second object is near the earth's surface, the fall-force can be written md, m being the weight of the object and d its elevation. In actual fall, fall-force is converted into force of motion. Mayer expressed this conversion as $md = mc^2$, where c is the velocity attained by an object of weight m in falling the distance d to the earth's surface.

On the basis of this concept of fall-force, Mayer concluded that gravity is not a force at all but a "characteristic of matter." Gravity cannot be a force, Mayer argued, because it is not the sufficient cause of motion; in addition to gravity, spatial separation is prerequisite to fall. If gravity were a force, then it would be a force which constantly produces an effect without itself being consumed; this, however, would violate the principle of the conservation of force. Throughout all his later papers and letters Mayer clung staunchly to this position. He continually argued that the entity "force" in its Newtonian sense is illogically and misleadingly named and that hence a different term should be introduced for it. The word "force" should be reserved for the substantial, quantitative entity conserved in conversions. Even after physics later adopted the term "energy" to describe Mayer's concept of force, Mayer continued to feel that the idea of force as a conserved entity was conceptually prior to the Newtonian entity and that hence the traditional name "force" should have been reserved for his own concept of force.

After discussing the interconvertibility of fall-force and force of motion in his 1842 paper, Mayer noted that motion is often observed to disappear without producing an equivalent amount of other motion or fall-force. In these cases motion is converted into a different form of force, namely heat. Fall-force, motion, and heat are different manifestations of one indestructible force, and hence they maintain definite quantitative relationships among themselves. This means, Mayer concluded, that there must exist in nature a constant numerical value which expresses the mechanical equivalent of heat. He stated that this value is 365 kilogram-meters per kilocalorie; that is, the fall-force in a mass of one kilogram raised 365 meters is equal to the heat-force required to raise one kilogram of water one degree centigrade.

Although Mayer's 1842 paper merely stated the

mechanical equivalent of heat without giving its derivation, later papers also gave his method. Let x be the amount of heat in calories required to raise one cubic centimeter of air from $0°$ C. to $1°$ C. at constant volume. To raise the same cubic centimeter of air one degree centigrade at constant pressure will require a larger amount of heat, say $x + y$, since, in the volume expansion, work must be done against the force which maintains constant pressure. If this latter expansion is carried out under a mercury column, then the extra heat y will go into raising that mercury column. Hence if P is the weight of the mercury column and h is the distance that it is raised in the expansion, we can write $y = Ph$; the problem is to find y. From published data Mayer knew that 3.47×10^{-4} calories are required in order to raise one cubic centimeter of air one degree centigrade under a constant pressure of 1,033 gm./cm.^{2} (that is, 76 cm. of mercury); hence $x + y = 3.47 \times 10^{-4}$ calories. He also knew from data of Dulong that the ratio of the specific heats of air at constant volume and at constant pressure is $1/1.421$; hence $x/(x + y) = 1/1.421$. Knowing the value of $x + y$, Mayer then easily found $y = 1.03 \times 10^{-4}$ calories. Since the expansion was known to raise the mercury column $1/274$ centimeters, Mayer then had for the equation $y = Ph$,

$$1.03 \times 10^{-4} \text{ cal.} = 1,033 \text{ gm.} \times 1/274 \text{ cm.}$$

The reduction of these figures yielded the equation 1 kilocalorie = 365 kilogram-meters.

Mayer's derivation of the mechanical equivalent of heat was as accurate as the value chosen for the ratio of specific heats would permit. Mayer's derivation rests upon the assumption that his cubic centimeter of air does no internal work during free expansion; that is, that all of the heat y goes to raise the mercury column. Although in 1842 Mayer already knew of an experimental result by Gay-Lussac which would substantiate this assumption, he did not invoke it publicly until three years later (1845).

The paper of 1842 set out Mayer's definitive view on the conservation of force and established his claim to priority; historically the paper also provides insight into the processes through which Mayer arrived at his theory. During the 1840's various European scientists and engineers were formulating ideas which were suggestive of the conservation of energy. Several different interests influenced these formulations. Among these interests was the growing concern with the efficiency of steam engines and with the many new conversion processes which were being discovered in electricity, magnetism, and chemistry. Mayer's early papers show little interest in these

problems but instead suggest that philosophical and conceptual considerations largely guided Mayer's theorizing. One of these considerations was his constant identification of force and cause; another was his intuitive understanding of force as a substantial, quantitative entity. The source of these ideas of Mayer's and their relationship to the larger context of German science and philosophy remain unsolved historical problems. Both concepts seem to have been unique to German science and to have led Mayer to interpret familiar phenomena in a radically new way. An example of this interpretation can be seen in the events which apparently led Mayer to his initial speculations about force conservation.

Like several other formulators of the conservation principle, Mayer was led to his theory through physiological, not physical, considerations. While letting the blood of European sailors who had recently arrived in Java in July 1840, Mayer had been impressed by the surprising redness of their venous blood. Mayer attributed this redness to the unaccustomed heat of the tropics. Since a lower rate of metabolic combustion would suffice to maintain the body heat, the body extracted less oxygen from the red arterial blood. This observation struck Mayer as a remarkable confirmation of the chemical theory of animal heat, and he quickly generalized that the oxidation of foodstuff is the only possible source of animal heat. Conceiving of the animal economy as a force-conversion process—the input and outgo of which must always balance—Mayer realized that chemical force which is latent in food is the only input and that this input could be expressed quantitatively as the heat obtained from the oxidation of the food. To this point Mayer's reasoning differed little from contemporary physiological theory, but once it was reached Mayer proceeded to a conceptual leap which was well beyond any facts at his disposal. He decided that not only the heat produced by the animal directly as body heat, but also that heat produced indirectly through friction resulting ultimately from the animal's muscular exertion must be balanced against this input of chemical force. Muscle force and also body heat must be derived from the chemical force latent in food. If the animal's intake and expenditure of force are to balance, then all these manifestations of force must be quantitatively conserved in all the force conversions which occur within the animal body. This inference, however fruitful, seemed to rest largely upon Mayer's preconceived notion of force and conversion rather than upon any empirical observations.

Immediately after his return from Java Mayer had planned a paper on physiology which would set out these ideas, but he purposely postponed the paper in order first to lay a proper physical basis for the theory. Having done so in the treatise of 1842, he published privately at Heilbronn in 1845 *Die organische Bewegung in ihrem Zusammenhang mit dem Stoffwechsel*, his most original and comprehensive paper. In this work Mayer again set out the physical basis of his theory, this time extending the ideal of force conservation to magnetic, electrical, and chemical forces. In *Die organische Bewegung* he described the basic force conversions of the organic world. Plants convert the sun's heat and light into latent chemical force; animals consume this chemical force as food; animals then convert that force to body heat and mechanical muscle force in their life processes.

Mayer intended *Die organische Bewegung* not only to establish the conservation of force as the basis of physiology, but also to refute views held by the organic chemist Liebig. In 1842 Liebig had published his influential and controversial book *Die Thierchemie oder die organische Chemie in ihrer Anwendung auf Physiologie und Pathologie*. In that work Liebig had come out as a champion of the chemical theory of animal heat, which Lavoisier and Laplace had first proposed in 1777. Reasoning much as Mayer had done, Liebig had concluded that animal heat produced from any source other than the oxidation of food was tantamount to the production of force from nothing. Hence he concluded that the oxidation of food is the sole source of animal heat. Liebig also believed that muscle force was derived ultimately from chemical force through an intermediary vital force localized in the protein substances of muscle tissue. Well aware of Liebig's acquaintance with his 1842 paper, Mayer regarded *Die organische Chemie* as possible plagiarism and as a definite threat to his priority. In his *Die organische Bewegung* Mayer joined Liebig in championing the chemical theory of animal heat, but he then proceeded to refute Liebig's other views wherever possible.

Mayer opened his attack on Liebig by criticizing Liebig's frequent recourse to vitalism. The vital force served various functions in Liebig's theory, the chief function being to prevent the living body from spontaneously beginning to putrefy, its tissues being constantly in the presence of oxygen and moisture. Mayer denied that putrefaction would occur in the tissues as spontaneously as Liebig had assumed. Mayer argued that if putrefaction did occur the putrefying parts would nevertheless be carried off in the blood as rapidly as they began to decay. Hence postulating a vital force was not merely unscientific, it was unnecessary.

Liebig had argued further that while starch and sugar are oxidized in the blood to produce heat, only the protein-bearing muscle tissue can undergo the chemical change necessary to produce mechanical muscle force. Hence those changes occur in the muscle, not in the blood; the muscle literally consumes itself in exertion. Against this argument Mayer employed his mechanical equivalent of heat to compute the amount of muscle tissue which must be consumed daily in order to support the exertions of a working animal. The high rate of assimilation necessary continuously to replace that loss, Mayer argued, made Liebig's theory improbable at best. He concluded that it seemed most reasonable to assume all oxidation to occur within the blood, whatever the form and locus of the force released. At the end of his 1845 paper Mayer finally reconciled the main observations of classical irritability theory with his own hypothesis and argued the dependence of the contractile force upon the blood supply.

Die organische Bewegung exercised little influence on German physiology, although Mayer's attack on Liebig's vital force found enthusiastic response, and the work received several favorable reviews. After 1845 Liebig's younger disciples quietly dropped his speculations about the vital force, much as Mayer had suggested. The issue of muscle decomposition remained controversial among physiologists, although by 1870 it was agreed that the oxidation of carbohydrates in addition to proteins contributed to the production of muscle energy. Mayer's writings had little direct influence on either of these developments.

Immediately after publishing his treatise on physiology, Mayer applied his theory of force conservation to a second critical problem which he had treated unsatisfactorily in 1841: the source of the heat of the sun. In 1846 he advanced an explanation of solar heat which he incorporated into a memoir submitted to the Paris Academy, "Sur la production de la lumière et de la chaleur du soleil," and into the expanded *Beiträge zur Dynamik des Himmels in populärer Darstellungen*, which was published privately at Heilbronn in 1848. After demonstrating in these papers the insufficiency of any chemical combustion to sustain the sun's enormous radiation, Mayer advanced what rapidly became known as the "meteoric hypothesis" of the sun's heat. Mayer speculated that matter, mostly in the form of meteors, daily enters the solar system in immense quantities and begins to orbit the sun. Friction with the luminiferous ether causes this matter gradually to spiral into the sun at inordinate velocities. Upon striking the sun this matter yields up its kinetic energy as light and heat. Mayer employed his mechanical equivalent of heat to show that each unit of mass striking the sun would yield four thousand to eight thousand times as much heat as would be produced by the combustion of an equivalent mass of carbon. Hence if the quantity of matter falling into the sun is assumed to be sufficiently large, this process can sustain the sun's total output of heat.

After 1850 the meteoric hypothesis received wide currency, largely on account of versions of the theory which were advanced independently of Mayer by Waterston and William Thomson. The explanation of solar heat that won general acceptance and that survived well into the twentieth century, however, was proposed by Helmholtz in a popular lecture of 1854, "Ueber die Wechselwirkung der Naturkräfte und die darauf bezüglichen Ermittlungen der Physik." According to Helmholtz the sun's heat is sustained by the gradual cooling and contraction of the sun's mass. As the sun's density increases the sun's matter yields its potential energy directly as heat. Although this was not a true meteoric hypothesis, Helmholtz' explanation of the sun's heat resembled Mayer's in many respects. Mayer's hypothesis may have influenced Helmholtz in the formulation of his own hypothesis, for by 1854 Helmholtz knew of Mayer's 1848 treatise and had discussed it in his 1854 lecture shortly before setting out his own views on the origin of solar energy.

Mayer's astronomical papers also revived another hypothesis which was to become important after 1850. In the *Dynamik des Himmels* of 1848 and in his 1851 memoir, "De l'influence des marées sur la rotation de la terre," Mayer showed that tidal friction deflects the major axis of the earth's tidal spheroid some thirty-five degrees from the earth-moon line. Hence the moon's gravitation exercises a constant retarding couple on the earth's rotation, a couple which gradually dissipates the earth's energy of rotation as heat.

Although minute, this quantity is perceptible. Citing Laplace, Mayer noted that on the basis of data from ancient eclipses, the length of the day, and hence the velocity of rotation of the earth, can be shown to have been constant to within .002 seconds over the last 2,500 years. This failure to observe the predicted retardation due to tidal friction indicated to Mayer the presence of a compensating phenomenon. He found this in geology. By 1848 many geologists believed that the earth had originally condensed as a molten mass and had since then been cooling at an undetermined rate. This theory faced a critical difficulty, for cooling should have produced a contraction of the earth, which in turn should have accelerated its rotation. No such acceleration could be observed, and Laplace had already used the apparent constancy

of the day to prove that no contraction greater than fifteen centimeters could have occurred within the last 2,500 years. At this juncture Mayer boldly hypothesized that tidal retardation of the earth's rotation is offset by the acceleration due to cooling and contraction. Mayer pointed out that this assumption rescued both hypotheses and reconciled both with the observed constancy of the day. The predicted retardation of .0625 seconds in 2,500 years, Mayer showed, would permit an offsetting contraction of the earth's radius by 4.5 meters.

The influence of Mayer's speculations is difficult to assess; the 1848 treatise was not widely read, while the memoirs to Paris had been reported upon but not printed. In 1858 Ferrel published a similar hypothesis, apparently independently of Mayer, and noted that tidal retardation and the earth's contraction might produce compensating changes in the earth's rotation. In 1865 Delaunay invoked tidal friction to explain a newly discovered inequality in the moon's motion and noted that the hypothesis of tidal friction had already been formulated in several printed works.

The *Dynamik des Himmels* marked the end of Mayer's creative career, for his numerous later articles were primarily popular or retrospective. At this point Mayer had received almost no recognition in important scientific circles, and to this disappointment was added the frustration of seeing other men independently advance ideas similar to his own. Liebig had anticipated many of Mayer's views in 1842, and in 1845 Karl Holtzmann computed a mechanical equivalent of heat without reference to Mayer. In 1847 Helmholtz set out a complete mathematical treatment of force conservation in his treatise *Ueber die Erhaltung der Kraft*. Mayer's main rival was Joule, and in 1848 Mayer became embroiled with him in a priority dispute carried out mainly through the Paris Academy. Although the dispute remained inconclusive, it later developed bitter nationalistic overtones when other scientists took up the quarrel.

After 1858 Mayer's fortunes improved. Helmholtz apparently read Mayer's early papers around 1852, and thereafter he argued Mayer's priority in his own widely read works. Clausius, too, regarded Mayer deferentially as the founder of the conservation principle and began to correspond with him in 1862. Through Clausius, Mayer was put in touch with Tyndall, who quickly became Mayer's English champion in the priority dispute with Joule, Thomson, and Tait. During the 1860's many of Mayer's early articles were translated into English, and in 1871 Mayer received the Royal Society's Copley Medal. In 1870 he was voted a corresponding member of the Paris Academy of Sciences and was awarded the Prix Poncelet.

Although the scientific world lionized Mayer before his death in 1878, in reality he exercised little influence on European science. In every field in which he worked his principal ideas were later formulated independently by others and were well established in science before his own contributions were recognized. In an age in which German science was rapidly becoming professionalized, Mayer remained a thorough dilettante. He conducted almost no experiments, and although he had an exact, numerical turn of mind, he neither fully understood mathematical analysis nor ever employed it in his papers. His scientific style, his status as an outsider to the scientific community, and his lack of institutional affiliation were all factors that limited Mayer's access to influential journals and publishers and hampered the acceptance of his ideas. Mayer was a conceptual thinker whose genius lay in the boldness of his hypotheses and in his ability to synthesize the work of others. Mayer actually possessed only one creative idea—his insight into the nature of force—but he tenaciously pursued that insight and lived to see it established in physics as the principle of the conservation of energy.

BIBLIOGRAPHY

Mayer's major scientific works were collected in Jacob J. Weyrauch, ed., *Die Mechanik der Wärme*, 3rd ed. (Stuttgart, 1893). Mayer's letters, short papers, and other documents related to his career were reprinted as Jacob J. Weyrauch, ed., *Kleinere Schriften und Briefe von Robert Mayer* (Stuttgart, 1893). In both works Weyrauch provides not only extensive nn. and commentary, but also a thorough biog. of Mayer. Other documents relating to Mayer's career and family background are included in the commemorative vol., Helmut Schmolz and Hubert Weckbach, eds., *J. Robert Mayer. Sein Leben und Werk in Dokumenten* (Weissenhorn, 1964).

Existing biographies of Mayer tend to whiggishness; one of the better ones is S. Friedländer, *Julius Robert Mayer* (Leipzig, 1905). On Mayer's place in the formulation of the principle of the conservation of energy and on the European context of his work, see Thomas S. Kuhn, "Energy Conservation as an Example of Simultaneous Discovery," in Marshall Clagett, ed., *Critical Problems in the History of Science* (Madison, Wis., 1959), 321–356. Mayer's concepts of force and causation are discussed by B. Hell in "Robert Mayer," in *Kantstudien*, **19** (1914), 222–248. Although he does not mention Mayer, Frederic L. Holmes discusses the milieu of German physiology in the 1840's in his intro. to Liebig's *Animal Chemistry*, facs. ed. (New York, 1964). On Mayer's role in astrophysical speculations see Agnes M. Clerke, *A Popular History of Astronomy During the Nineteenth Century*, 3rd ed. (London, 1893), esp. 332–334, 376–388.

R. STEVEN TURNER

MAYER-EYMAR, KARL (*b*. Marseilles, France, 29 June 1826; *d*. Zurich, Switzerland, 25 February 1907), *paleontology, stratigraphy.*

The son of a Swiss merchant, Mayer-Eymar received his early schooling in Rennes and St. Gall. He entered the University of Zurich in 1846, first studying medicine and then natural history and geology. After graduating, he worked in Paris from 1851 to 1854, principally at the Muséum d'Histoire Naturelle under d'Orbigny. In 1858 he was appointed assistant in the geology institute of the Zurich Polytechnische Hochschule. Shortly afterward, he became a *Privatdozent* and curator of the collections, and in 1875 he was named professor. He held these latter posts until his death and never married. His contemporaries described Mayer-Eymar as an original and sometimes picturesque character. Around 1865 he added the anagram Eymar to his patronymic to avert confusion with other persons of that name; thenceforth he wrote his name Mayer-Eymar.

Mayer-Eymar's chief field of interest was the biostratigraphy and paleontology of mollusks, mainly of the Tertiary and, to a lesser extent, of the Jurassic and the Cretaceous. A passionate and successful collector from his school days on, he investigated in particular the Tertiary terranes of many western European and Mediterranean countries and brought extensive collections back to Zurich. He published voluminous lists and descriptions of these materials and of others sent to him from abroad. Some of his accounts included new species and genera, but most of them were without illustrations. This omission lessened the usefulness of many of his publications.

Mayer-Eymar's stratigraphical works were of great importance and have been of some influence until the present day. In 1858 appeared his first major publication on the subdivision of the Tertiary, which he divided into stages according to lithological and faunal criteria named after typical localities in the Tertiary basins of western Europe. He introduced many new names for these stages, a number of which were never adopted or were in use for only a short time; others, however, are still employed, although with somewhat modified definitions (for instance, Bartonian, Tongrian, Aquitanian, Helvetian, Tortonian, Plaisancian). He published similar subdivisions on the Jurassic and Cretaceous. For the publication of his stratigraphic researches he employed mainly autographed synoptic tables, which he frequently republished in improved form.

His profound paleontological knowledge frequently enabled Mayer-Eymar to make definitive decisions concerning stratigraphical problems. In 1876, he clarified the much discussed question of whether the glaciers of the southern Alps were contemporaneous with the Pliocene sea in northern Italy. He demonstrated that the glacial deposits contained layers with Quaternary mollusks and therefore had to be more recent than the late Pliocene marine sediments of the Astian.

BIBLIOGRAPHY

I. ORIGINAL WORKS. Mayer-Eymar's writings include "Versuch einer neuen Klassifikation der Tertiär-Gebilde Europas," in *Verhandlungen der Schweizerischen naturforschenden Gesellschaft* (Trogen) (1858), 165–199; "Catalogue systématique et descriptif des mollusques tertiaires, qui se trouvent au Musée fédéral de Zurich," in *Vierteljahrsschrift der naturforschenden Gesellschaft in Zürich*, **11** (1866), 301–337; **12** (1867), 241–303; **13** (1868), 21–105, 163–200; **15** (1870), 31–82; "Systematisches Verzeichnis der Versteinerungen des Helvetian der Schweiz und Schwabens," in *Beiträge zur geologischen Karte der Schweiz*, **11** (1872), 475–511; "La vérité sur la mer glaciale au pied des Alpes," in *Bulletin de la Société géologique de France*, 3rd ser., **4** (1876), 199–222; the synoptic tables, published in Zurich in 1900: *Classification et terminologie des terrains tertiaires d'Europe*, *Classification et terminologie des terrains crétaciques d'Europe*, and *Classification et terminologie des terrains jurassiques d'Europe*, and many articles on fossil mollusks and stratigraphic problems in *Journal de conchyliologie*, **5–49** (1856–1902), and in *Vierteljahrsschrift der naturforschenden Gesellschaft in Zurich*, **6-49** (1861–1904).

II. SECONDARY LITERATURE. See G. F. Dollfus, "Nécrologie Ch. Mayer-Eymar," in *Journal de conchyliologie*, **56** (1908), 145–162; A. Heim and L. Rollier, "Dr. Karl Mayer-Eymar 1826–1907," in *Verhandlungen der Schweizerischen naturforschenden Gesellschaft* Vers. (Fribourg), **90** (1908), xl–xlix, with complete bibliography; and F. Sacco, "Carlo Mayer-Eymar. Cenni biografici," in *Bollettino della Società geologica italiana*, **26** (1907), 585–602, with a 127-title bibliography.

HEINZ TOBIEN

MAYO, HERBERT (*b*. London, England, 3 May 1796; *d*. Bad Weilbach, Germany, 15 May 1852), *neurology*.

Herbert was the third son of John Mayo; his father and his eldest brother, Thomas, were prominent physicians in London. He was a pupil of Charles Bell at the Windmill Street Anatomy School (1812–1815) and graduated M.D. at Leiden in 1818. He was admitted a member of the Royal College of Surgeons by examination in 1819 and elected among the first fellows in 1843. He practiced surgery and taught anatomy in London from 1819 to 1843, becoming senior surgeon to the Middlesex Hospital, where he founded the Medical School in 1836; he also wrote many successful textbooks.

Charles Bell had circulated privately in 1811 his *Idea of a New Anatomy of the Brain*, which showed that the anterior roots of the nerves alone have motor functions; ten years later, in 1821, Bell contributed to the *Philosophical Transactions* a paper "On the Nerves; Giving an Account of Some Experiments on Their Structure and Functions, Which Lead to a New Arrangement of the System." Bell here put forward the concept of the motor function of the anterior and the sensory function of the posterior roots, but as Claude Bernard later wrote, "drowned in philosophical considerations so obscure or diffuse that it is difficult to find places where his opinions are succinctly stated."

Mayo announced his independent discoveries of the physiology of the nerves in his *Anatomical and Physiological Commentaries*; the first part (August 1822) included a paper which described "Experiments to Determine the Influence of the Portio Dura of the Seventh and the Facial Branches of the Fifth Pair of Nerves"; part two (July 1823) opened with a paper "On the Cerebral Nerves With Reference to Sensation and Voluntary Motion," and it concluded with "Remarks Upon the Spinal Chord and the Nervous System Generally." Mayo attributed sensibility to the fifth nerve and motor power to the seventh nerve, and he showed that a circumscribed segment of the nervous system sufficed to produce muscular action. He wrote that "An influence may be propagated from the sentient nerves of a part to their correspondent nerves of motion through the intervention of that part alone of the nervous system to which they are mutually attached"; the term "reflex" was applied to this phenomenon by Marshall Hall in 1833. Neither Bell nor Mayo seems to have known at this time that François Magendie had achieved similar results in research which was reported in Paris (1821–1823). Bell protested his claim to priority against both Mayo and Magendie.

Mayo supplemented the fifteen plates in his *Commentaries* by an atlas of six larger plates (each plate was printed in an outline and a shaded version) entitled *A Series of Engravings Intended to Illustrate the Structure of the Brain and Spinal Chord* (1827).

The brief text is pure descriptive anatomy, except for a paragraph (p. iii) in which Mayo stated,

> The filaments of which the nerves consist have the office of conductors. We may therefore infer that the white threads which enter so largely into the composition of the spinal marrow, the medulla oblongata, and the brain, likewise serve as media for conveying impressions. The justice of this conclusion in the instance of the spinal marrow has been proved by experiments made on animals.

He added that his observations would prove useful for pathologists in explaining "interruption or impairment of functions."

Although he made no further original discoveries, Mayo retained his interest in neurology. In his *Outlines of Human Physiology* (3rd ed., 1833, p. 219) he gave "an extension of the original law respecting the place of origin of the nerves." He published a pamphlet on the *Powers of the Roots of the Nerves* in 1837 in reply to R. D. Grainger's *Observations on the Structure and Function of the Spinal Cord*; Mayo described this as a restatement of his views after conversations with Grainger. In an appendix he discussed recent demonstrations of hypnotism, "magnetic sleep," by Baron Dupotet and concluded that "persons susceptible of it may be thrown into a kind of trance by the influence of imagination excited through the senses." He also anticipated a possible value for anesthetizing a patient before "a surgical operation of little severity." In the main pamphlet Mayo also stated: "Nerves, it was discovered by the independent researches of Sir C. Bell, M. Magendie, and myself (each having contributed his separate share to the result) are of two kinds only, one *sentient*, the other *voluntary*." He did not discuss Bell's claims, but recorded (p. 8) the highest regard for Magendie's results and integrity, and he mentioned (p. 20) that "Dr Marshall Hall, who invented the term 'reflex action,' has followed out the idea with great diligence, showing fresh instances parallel to my own, which I reduced to one theory in 1823."

Mayo's final work on the subject was his monograph *The Nervous System* (1842). This was begun as a physiological introduction for a reissue of the 1827 *Engravings*, but in the event the illustrations were not reprinted. Mayo wrote that his new "survey of the nervous system and the reflections to which it gave rise did not elicit much that is new, yet display what has been discovered with new distinctness and force."

Mayo retired to Germany in 1843 for hydropathic treatment as a victim of "rheumatic gout," and while there he wrote on *The Cold Water Cure* and on *Mesmerism*. He died in Germany at the age of fifty-six, survived by his wife, a son, and two daughters.

BIBLIOGRAPHY

I. ORIGINAL WORKS. His writings on neurology are *Anatomical and Physiological Commentaries*, pt. 1 (August 1822), pt. 2 (July 1823); *A Series of Engravings Intended to Illustrate the Structure of the Brain and Spinal Chord in Man* (1827); *Powers of the Roots of the Nerves in Health and Disease, Likewise On Magnetic Sleep* (1837); and *The Nervous System and Its Functions* (1842).

Textbooks which Mayo wrote include *A Course of Dissections for Students* (1825); *Outlines of Human Physiology* (1827, 1829, 1833, 1837); *Observations on Injuries and Diseases of the Rectum* (1833; Washington, 1834); *Outlines of Human Pathology* (1836; Philadelphia, 1839; 1841), trans. into German (1838–1839); *The Philosophy of Living* (1837, 1838, 1851); *Management of the Organs of Digestion in Health and Disease* (1837, 1840); *A Treatise on Siphilis* (1840), trans. into German (1841); *The Cold Water Cure* (1845); and *Letters on the Truths in Popular Superstitions With an Account of Mesmerism* (Frankfurt, 1849), 2nd and 3rd eds. (Edinburgh, 1851).

II. SECONDARY LITERATURE. Biographical memoirs of Mayo are in the *Dictionary of National Biography*, XXXVII (1894); D'A. Power, ed., *Plarr's Lives of the Fellows of the Royal College of Surgeons* (1930); and John F. Fulton, *Selected Readings in the History of Physiology*, 2nd ed., Leonard G. Wilson, ed. (1966), 285–286.

WILLIAM LEFANU

MAYOW, JOHN (*b.* Bray, near Looe, England, December 1641; *d.* London, England, September 1679), *physiology, chemistry.*

Mayow was the second son of Phillip Mayowe, a member of the well-established, substantial, and multi-branched Mayow family of Cornwall. His grandfather Philip acquired the manor of Bray in 1564; he was one of the nine charter burgesses of East Looe when the town received its charter of incorporation from Queen Elizabeth in 1587. He appears on his altar tomb in alderman's robes with the epitaph "Phillipe Maiowe of East, Looe, Gentleman." John's father is referred to in the parish register that records John's baptism as "Mr. Phillip Mayowe, Gent."

Mayow matriculated at Wadham College, Oxford, on 2 July 1658 and was received as a commoner and admitted scholar on 23 September 1659. On 3 November 1660 he was elected to a fellowship at All Souls College, Oxford. He graduated bachelor of common law on 5 July 1670 and obtained the further privilege of studying medicine. After leaving Oxford in 1670 he entered medical practice, at least during the summer season at Bath. In the 1670's he seems to have spent considerable time during the fall and winter months in London. Robert Hooke records several meetings with Mayow in his *Diary* from 1674 through 1677. On Hooke's recommendation Mayow was elected fellow of the Royal Society on 30 November 1678.

Mayow is best known for his studies on the inter-related problems of atmospheric composition, aerial nitre, combustion, and respiration. He has occasionally been regarded, usually uncritically, as an unappreciated precursor of Lavoisier. In fact, Mayow's work was vigorously scrutinized—both in a friendly and in a hostile spirit—in his own time and again in the late eighteenth century after the discovery of oxygen. In the last several decades, the question of his originality and importance has been a subject of scholarly debate.

Mayow's first publication, a thin volume entitled *Tractatus duo*, was printed at Oxford in 1668. The two tracts, the first on respiration and the second on rickets, demonstrated his involvement in the scientific and medical issues and literature of his day. In "De respiratione," Mayow specifically cited the work of his English contemporaries Robert Boyle, Nathaniel Highmore, and Thomas Willis, and of the Italian Marcello Malpighi. He also took note of experiments on the inflation of the lungs "recently performed at the Royal Society" by Robert Hooke and Richard Lower, and textual nuances suggest that he may likewise have been familiar with such recent publications as Swammerdam's *Tractatus . . . de respiratione usuque pulmonum* (Leiden, 1667), reviewed in the *Philosophical Transactions* for October 1667. Mayow's second essay, "De rachitide," shows a familiarity with Francis Glisson's *De rachitide, sive morbo puerili . . .* (London, 1650), although Mayow departed sharply from Glisson by offering a highly abbreviated account of the symptomatology and therapeutics and a more iatromechanical version of the etiology of rickets.

The real interest and importance of the *Tractatus duo*, however, lies in the striking originality of Mayow's juxtaposition of contemporary physiological ideas in "De respiratione." Thus, after describing with some novelty the mechanics of thoracic dilatation and pulmonary inflation, Mayow argued that respiration serves principally to convey a supply of fine nitrous particles from the air to the blood. This "nitrous air" is necessary to life, for when it is missing from the mass of inspired air, respiration does not produce its usual good effect. The nitrous particles are needed to react with the "sulphureous" parts of the sanguinary stream, and this reaction causes a gentle and necessary

fermentation in the pulmonary vessels, the heart, and the arteries. Moreover, the nitrous air is also essential to the beating of the heart. Like other muscles, the heart contracts macroscopically because an "explosion" occurs microscopically within its fibres. The explosion, which inflates the muscles, results specifically from the violent interaction of the "nitro-saline" particles of the inspired air with the animal spirits fashioned from the "volatile salt" of the blood.

Mayow was here fusing in a very original way two recent Oxford physiological traditions. First, he adopted the "nitrous pabulum" theory originally advanced by Ralph Bathurst in Oxford lectures of 1654 and later remembered by Robert Hooke and improved in his *Micrographia* of 1665.[1] Second, Mayow endorsed Willis' essential ideas about the explosion mechanism for muscular contraction. But whereas Willis had attributed explosive inflation to the violent, gunpowder-like interaction of the "spirituous-saline" animal spirits with the "sulphureous" parts of the blood, Mayow contended that the blood and spirits could not possibly react explosively, for if they could they would already have done so before the spirits were distilled from the blood in the cortex of the brain. Mayow was thus able to avoid apparent contradiction and to account for the otherwise perplexing fact that death follows so suddenly upon the cessation of respiration. Failure to inspire fresh supplies of nitrous particles mixed with the larger bulk of air could now be understood to lead instantaneously to the stopping of heartbeat; and stoppage of heartbeat immediately curtails the distribution of animal spirits throughout the body. Since, Mayow asserted, the life of animals consists in the distribution of animal spirits, death quickly follows upon the cessation of respiration.

Thus, in 1668, Mayow wrote as a product of and a participant in the Oxford physiology to which he was thoroughly exposed as a student and fellow. His ideas were interesting, although his contemporaries considered them fundamentally unexceptional; they were well reviewed and apparently well received. The *Tractatus duo* was accorded the lead review in the November 1668 number of the *Philosophical Transactions* where Mayow's theories were clearly summarized in considerable detail with no suggestion of skepticism or hostility.

Between 1668 and 1674, perhaps encouraged by the initial reception of his views on respiration, Mayow attempted a clarification, expansion, and refining of his ideas, both on physiology and on chemistry. He may well have been influenced by the publication of several closely related books and essays: Richard Lower, *Tractatus de corde* (1669); Malachi Thruston,

De respirationis usus primario (1670); Thomas Willis, *De sanguinis accensione* (1670) and *De motu musculari* (1670); and Robert Boyle, *Tracts . . . Containing New Experiments, Touching the Relation Betwixt Flame and Air* (1672) and *Tracts . . . About Some Hidden Qualities of the Air* (1674). Meanwhile Hooke continued to report to the Royal Society about experiments on combustion, respiration, and the action of the air and of nitrous compounds. In 1674 Mayow referred directly to several of these efforts, and allusions suggest a familiarity with the rest. In any case, by 1674 Mayow, aware of contemporary developments, had expanded his *Tractatus duo* into *Tractatus quinque* by the addition of three new essays: "De sal-nitro et spiritu nitro-aereo"; "De respiratione foetus in utero et ovo"; and "De motu musculari et spiritibus animalibus."

In the first essay, by far the longest and most important, Mayow offered a chemical history of nitre and nitro-aerial spirit. His primary intention seems to have been to distinguish between these two distinct substances, which the vague vocabulary of his contemporaries (and his own) had previously confused. Nitre (saltpeter) is a triply complex salt. It consists of spirit of nitre (nitric acid) combined with a fixed salt of the earth; spirit of nitre is in turn derived from the "ethereal and igneous" nitro-aerial spirit of atmospheric air (oxygen?) in combination with the "Salino-metallic parts" of common "Terrestrial sulphur." It is the nitro-aerial spirit, harbored in turn in the spirit of nitre and in common nitre, that is the active and "igneous" substance in nitrous compounds. It is the chemical agent responsible for sustaining combustion and producing fermentation. Flame consists essentially in nitro-aerial spirit thrown to brisk motion by interacting with sulphureous particles, whereas fermentation is the general effervescence of nitro-aerial particles reacting with salino-sulphureous ones. Moreover, the caustic qualities of acids derive from the active and igneous nitro-aerial particles within them.

Having clarified the respective roles and chemical relations of the several nitrous substances, Mayow next explores a wide range of problems to which he makes the nitro-aerial spirit relevant. His exploration ranges from meteorological fantasies about thunder and lightning quite happily and deliberately modeled on Descartes's *Principia philosophiae*, through speculations on the role of nitro-aerial spirit in transmitting the pulse of light, to—most significantly—ingenious experimental investigations of the role of nitro-aerial spirit in combustion and respiration. In pursuit of this latter problem, Mayow dexterously employed a variety of experimental techniques that represented subtle though important improvements on contemporary practice, notably that of Boyle. For example, Mayow was able to transfer gases collected over water more neatly than did Boyle, who had to use two air pumps for this operation.[2] Mayow also experimented with animals and candles breathing or burning over water, using cupping glasses, water troughs, and bell jars. He always carefully adjusted water levels with a special siphon arrangement, and with this apparatus he was able to observe the breathing of an animal in a closed space. He was thus able to test his earlier assertion that there are nitrous particles diffused in a larger bulk of otherwise useless air, an assertion that Hooke had repeated, unverified, to the Royal Society early in the 1670's.[3] Mayow now observed the gradual rise of water into the space occupied by the breathing animal. The rise of water continued until there was a diminution of one-fourteenth of the original volume of air. Mayow explained this diminution as the result of the passing of nitro-aerial particles, which normally account for the elasticity of atmospheric air, from the air through the lungs and into the blood. There, finally, the nitro-aerial particles fermentatively interact with sulphureous particles, producing animal heat in the process and changing the blood from dark purple to light scarlet.

Mayow's two other essays in *Tractatus quinque*, "De respiratione foetus in utero et ovo" and "De motu musculari et spiritibus animalibus," primarily supplement and clarify physiological views that he had earlier expressed in *Tractatus duo*. In the first of these essays Mayow contended that embryos require nitro-aerial particles as surely as do respiring animals. The umbilical arteries convey the appropriate particles either from the maternal bloodstream or from the albuminous humor of the egg, whence they are temporarily collected from the nitro-aerial particles supplied by the heat of the incubating fowl. In the second essay Mayow somewhat revised his 1668 views on muscular contraction. Aware of Steno's recently published findings on the action of muscles, Mayow contended that muscles contract by "contortion" rather than by inflation. The active agent, however, is still nitro-aerial particles, which were now said to be the very substance of animal spirits; and these spirits produce the necessary effervescence when they come in contact with the salino-sulphureous particles of the blood in the muscular fibrils.

The immediate reception of *Tractatus quinque* was decidedly less favorable than that of *Tractatus duo*. Mayow's work again earned a lead review in the *Philosophical Transactions*, a lengthy one in the July 1674 number. Now, however, the detailed summary showed suggestions of sarcasm and disbelief. Detailed marginal notes called attention to works by Boyle and others, the impression being clearly and perhaps deliberately created that Mayow owed debts to his contemporaries that he did not fully acknowledge. The review was probably written by Henry Oldenburg, a friend of Boyle and secretary of the Royal Society. In a letter to Boyle of July 1674, Oldenburg commented that "some learned and knowing men speak very slightly of the *quinque Tractatus* of *J.M.* and a particular friend of yours and mine told me yesterday, that as far as he had read him, he would shew to any impartial and considering man more errors than one in every page."[4]

To a large extent personal loyalty and scientific partisanship lay behind Oldenburg's remark. At just this time Boyle and Hooke were conducting a polite but unmistakable debate about the nature of nitrous compounds and the role of the air in respiration and combustion. Close as Oldenburg was to Boyle, he was distant from Hooke, who several times accused Oldenburg of personal malice.[5] In striking contrast to the hostility that Hooke felt in Oldenburg, Mayow seemed eager to support Hooke's side of the debate and, indeed, to provide experimental substantiation for his long-standing but still disputed views on chemistry, combustion, and respiration. Moreover, the elaborate hypothetical excursions with which Mayow filled much of *Tractatus quinque,* and which doubtless ran counter to the skeptical mood of Boyle and Oldenburg, were consistent with Hooke's increasing enthusiasm for explicit Cartesian hypothesizing in the 1670's and 1680's.[6] It seems no accident that Hooke and Mayow were friendly after the publication of *Tractatus quinque*, and that Hooke proposed Mayow for membership in the Royal Society after Oldenburg's death in 1677.

Other reactions to Mayow's views were mixed. In one work Boyle seemed to flirt briefly with Mayow's ideas, but by and large he remained aloof.[7] Other contemporaries cited Mayow's theories and experiments, and some continued to do so for several decades. In the eighteenth century Mayow still had a following which included the chemist and physiologist Stephen Hales. The bitter critics who also appeared were usually, like Archibald Pitcairne, of iatromechanical persuasion. There were also judicious conciliators. Albrecht von Haller, for example, a magisterial figure in mid-eighteenth-century physiology, devoted significant attention to Mayow's views in his much studied *Elementa physiologiae.* Yet with the passage of time, Mayow's special originality faded, even in the minds of his supporters; and he became generally but vaguely identified with Boyle, Hooke,

Lower, and those other seventeenth-century virtuosi who speculated on the role of the air in respiration and combustion.

With the discovery of oxygen the ground shifted considerably. Lavoisier himself had a copy of Mayow's book in his library, and several of his ideas and experiments seem to show important traces of Mayow's techniques and perhaps even his theories.[8] Lavoisier's contemporary Fourcroy discussed Mayow explicitly and remarked that his experiments had been more ingenious than those of his much noted countrymen Boyle and Hales.[9] In England, no doubt due in part to the hunger for national priority, a small Mayow revival began. Participating with various degrees of enthusiasm were Thomas Thomson, Thomas Beddoes, and G. D. Yeats. Beddoes and Yeats published extracts and analyses of *Tractatus quinque*, Beddoes' being *Chemical Experiments and Opinions. Extracted From a Work Published in the Last Century* (1790) and Yeats's *Observations on the Claims of the Moderns, to Some Discoveries in Chemistry and Physiology* (1798). Among the claims made for Mayow were that in 1674 he already knew the true cause of increased weight in metallic calcination (fixation of nitro-aerial particles = oxygen) and clearly recognized that certain bases are made acid by the addition of nitro-aerial particles (= oxygen, the acidifying principle).[10]

Throughout the nineteenth century and into the early twentieth there was a steady flow of commentary on Mayow and his originality *vis-à-vis* both immediate contemporaries and late-eighteenth-century successors. Several editions of his works were also published. More recently, especially since the publication of T. S. Patterson's long and biting "John Mayow in Contemporary Setting" (1931), Mayow's originality and importance have been considerably debated. Patterson contended that Mayow had been elevated to scientific preeminence by confusion and poor scholarship. In fact, Mayow was a derivative thinker who owed his important ideas to Boyle, Hooke, and Lower, among others. If Mayow contributed anything of his own to the *Tractatus quinque*, it was a penchant for fanciful hypothesizing and a general confusion. Further, according to Patterson, the later resurrection of Mayow as a precursor of Lavoisier was based on misreading and special pleading; Mayow's ideas, closely studied, were fundamentally different from Lavoisier's. Not the least significant difference was that Mayow and Lavoisier wrote in completely different ways about the gaseous state.

A steadily growing body of scholarship has developed since Patterson's article appeared. Working from one perspective, Henry Guerlac and Allen Debus have attempted to trace the alchemical and meteorological roots of Mayow's theories on aerial nitre and the nitrous compounds, and have thus begun to situate his work in a more cogent contemporary context. Partington has taken up Mayow's cause by working from another angle. Both in separately published articles and in long chapters in *A History of Chemistry*, he has vigorously defended Mayow's special talents and insights against Patterson's assault. Other scholars have also contributed important information.

It seems probable that both pro- and anti-Mayow scholars will be vindicated to a certain extent. It is already clear that several of Mayow's ideas had a long history or at least prehistory and that he acquired many notions from the contemporary intellectual environment. It seems equally to be correct that Mayow gave common and contemporary views— especially those popular at Oxford and the Royal Society—unique twists and imaginative interpretations. He was also an ingenious and talented experimenter, whatever his passion for Cartesian hypotheses. With regard to his putative anticipation of Lavoisier, it is well established that Mayow covered many of the basic chemical and physiological phenomena that Lavoisier later interpreted with new theories and improved data. Nonetheless, as Patterson has argued, a certain unbridgeable conceptual gulf separated Mayow's seventeenth-century formulations from Lavoisier's in the eighteenth century, even when they used similar or approximately similar experimental apparatus and materials. But quantumlike theoretical discontinuities within a continuum of experimental and theoretical concern with certain problems are a commonplace in the history of science.

NOTES

1. Bathurst's lectures were published by T. Warton in *The Life and Literary Remains of Ralph Bathurst* (London, 1761). Hooke's discussion of the "dissolution" of combustible bodies by "a substance inherent, and mixt with the Air, that is like, if not the very same, with that which is fixt in *Salt-peter*" is found on pp. 103–105 of the *Micrographia*.
2. J. R. Partington, *A History of Chemistry*, II (London, 1961), 604.
3. Douglas McKie, "Fire and the Flamma Vitalis: Boyle, Hooke and Mayow," in E. Ashworth Underwood, ed., *Science, Medicine and History*, I (Oxford, 1953), 482.
4. Robert Boyle, *Works*, VI (London, 1772), 285.
5. Margaret 'Espinasse, *Robert Hooke* (Berkeley, 1962), 63–65.
6. Theodore M. Brown, "Introduction," *The Posthumous Works of Robert Hooke* (London, 1971), 4–7.
7. McKie, *op. cit.*, pp. 483–484.
8. Partington, *op. cit.*, II, 592, 595.
9. *Ibid.*, p. 595.
10. T. S. Patterson, "John Mayow in Contemporary Setting," in *Isis*, 15 (1931), 49.

BIBLIOGRAPHY

I. ORIGINAL WORKS. Mayow's two major publications were both originally published at Oxford but enjoyed several seventeenth-century reissues and special editions. *Tractatus duo* has never been translated as such. *Tractatus quinque*, which includes modified versions of the two 1668 essays, has been translated several times: in a Dutch trans. by Steven Blankaart (Amsterdam, 1683); a German trans. by J. Koellner (Jena, 1799); and an English trans. by A. Crum Brown and L. Dobbin (Edinburgh, 1907). A partial German trans. was published by F. G. Donnan, in Ostwalds Klassiker der exacten Wissenschaften no. 125 (Leipzig, 1901); and a partial French trans. by L. Ledru and H. C. Gaubert was published at Paris in 1840.

Contemporary English reactions to Mayow, with references, have been cited above. A favorable contemporaneous review of *Tractatus quinque* was published in *Journal des sçavans* (3 Feb. 1676).

II. SECONDARY LITERATURE. There is as yet no detailed, full-length study of Mayow, although it has recently been called for. A number of recent studies are, however, of considerable utility. Principal among them are Walter Böhm, "John Mayow und Descartes," in *Sudhoffs Archiv für Geschichte der Medizin und der Wissenschaften*, 46 (1962), 45–68; "John Mayow and His Contemporaries," in *Ambix*, 11 (1963), 105–120; and "John Mayow und die Geschichte des Verbrennungsexperiments," in *Centaurus*, 11 (1967), 241–258; Allen G. Debus, "The Paracelsian Aerial Niter," in *Isis*, 55 (1964), 43–61; Henry Guerlac, "John Mayow and the Aerial Nitre," in *Actes du septième congrès international d'histoire des sciences* (Jerusalem, 1953), 332–349; and "The Poets' Nitre: Studies in the Chemistry of John Mayow—II," in *Isis*, 45 (1954), 243–255; Diana Long Hall, *From Mayow to Haller: A History of Respiratory Physiology in the Early Eighteenth Century*, unpublished diss. (Yale, 1966); Douglas McKie, "Fire and the Flamma Vitalis: Boyle, Hooke and Mayow," in E. Ashworth Underwood, ed., *Science, Medicine and History*, I (Oxford, 1953), 469–488; J. R. Partington, "The Life and Work of John Mayow," in *Isis*, 47 (1956), 217–230, 405–417; "Some Early Appraisals of the Work of John Mayow," *ibid.*, 50 (1959), 211–226; and *A History of Chemistry*, II (London, 1961), *passim*, but esp. ch. 16; and T. S. Patterson, "John Mayow in Contemporary Setting," in *Isis*, 15 (1931), 47–96, 504–546.

Of considerable help also are John F. Fulton, *A Bibliography of Two Oxford Physiologists: Richard Lower (1631–1691) and John Mayow (1643–1679)* (Oxford, 1935); and Douglas McKie, "John Mayow, 1641–1679," in *Nature*, 148 (1941), 728; and "The Birth and Descent of John Mayow," in *Philosophical Magazine and Journal of Science*, 33 (1942), 51–60.

Additional studies that throw light on Mayow's contemporary context are Hansruedi Isler, *Thomas Willis 1621–1675* (New York, 1968); D. J. Lysaght, "Hooke's Theory of Combustion," in *Ambix*, 1 (1937), 93–108; Alfred Myer and Raymond Hierons, "On Thomas Willis' Concepts of Neurophysiology," in *Medical History*, 9

(1965), 1–15, 142–155; and H. D. Turner, "Robert Hooke and Theories of Combustion," in *Centaurus*, **4** (1955–1956), 297–310.

Theodore M. Brown

MAYR (MARIUS), SIMON (*b*. Gunzenhausen, Germany, 20 January 1573; *d*. Ansbach, Germany, 26 December 1624), *astronomy*.

Mayr was the first to mention in print the nebula in Andromeda, to publish tables of the mean periodic motions of the four satellites of Jupiter then known, to direct attention to the variation in their brightness, and to identify the brightest of the four, which are still called by the names he bestowed on them.[1]

After studying at Gunzenhausen and later at the Margrave's School in Heilbronn from 1589 to 1601, he was appointed mathematician of the margrave of Ansbach and was sent to Prague, with a recommendation dated 22 May 1601, to join the staff of Tycho Brahe, the mathematician of Emperor Rudolph II.[2] Arriving toward the end of May, Mayr learned how to use Tycho's observational instruments.[3] He remained less than four months, however, since on 25 September he passed through Znojmo in Moravia, and then Vienna, on his way south.[4]

Deciding to study medicine in Padua, Mayr was admitted to the Association of German Students of the Arts in the University of Padua on 18 December 1601.[5] On that occasion he donated six Venetian lire to the association and in each of the next four years he contributed ten lire. Because the association's proctor was unable to complete his term of office, Mayr replaced him on 5 March 1604. On 27 July Mayr was elected librarian or second counselor, to serve until 14 April 1605. On 1 July 1605 he announced that he had to return home.[6] During this journey he spent the night of 25 July near Donauwörth.[7] The association's official minutes for 24 July record that he had misinformed it about the German law students' attitude toward an impending election.[8] In 1606 Mayr married Felicitas Lauer, daughter of his publisher in Nuremberg. In October 1613 he met Johannes Kepler, Brahe's successor as imperial mathematician, at the Diet held in Regensburg.[9]

Shortly after his arrival in Padua, on 24 December 1601, Mayr had observed a solar eclipse; and on 10 October 1604 he noticed that year's nova while with a pupil who a few years later, on 4 May 1607, was convicted of plagiarizing a work by Galileo.[10] In his published denunciation of the plagiarist, Galileo refrained from mentioning Mayr by name, perhaps to avoid arousing the powerful Association of German Students, but referred to him as an "old adversary," poisonous reptile, and "enemy . . . of all mankind."[11] Sixteen years later, after Galileo had left Padua, in his *Assayer* of 1623 he condemned Mayr's *World of Jupiter* (Nuremberg, 1614, two editions) as itself an outright plagiarism. But Galileo spoiled his case by suggesting that Jupiter's satellites had never been seen by Mayr, whose tables of their mean motions preceded and surpassed Galileo's.[12]

NOTES

1. Pierre Humberd, "Le baptême des satellites de Jupiter," in *Revue des questions scientifiques*, **117** (1940), 171, 175.
2. J. Klug, "Simon Marius . . .," p. 397; E. Zinner, "Zur Ehrenrettung . . .," pp. 25, 66, 70; Johannes Kepler, *Gesammelte Werke*, XIV (Munich, 1946), 168, 170.
3. Zinner, *op. cit.*, pp. 49, 54, 66.
4. *Ibid.*, pp. 59, 60, 61, 70.
5. Antonio Favaro, *Galileo Galilei a Padova*, Contributi alla Storia dell'Università di Padova, V (Padua, 1968), 218, with Mayr's coat of arms on p. 219; repr. of Favaro's *Stemmi ed inscrizioni concernenti personaggi galileiani nella Università di Padova* (Padua, 1893).
6. *Atti della nazione germanica artista nello studio di Padova*, Antonio Favaro, ed., Monumenti Storici Pubblicati dalla R. Deputazione Veneta di Storia Patria, no. 19–20, 1st ser., Documenti, no. 13–14 (Venice, 1911–1912), II, 189, 195, 211, 214, 220, 225, 231, 236.
7. Zinner, *op. cit.*, pp. 60, 72.
8. *Atti*, II, 238, 239.
9. Zinner, *op. cit.*, pp. 63, 71.
10. *Ibid.*, pp. 48, 51; *Le opere di Galileo Galilei*, II (Florence, 1891; repr. 1968), 293, 560.
11. *Ibid.*, 519.
12. *Ibid.*, VI (1896), 215, 217.

BIBLIOGRAPHY

I. Original Works. Mayr's writings were listed by Ernst Zinner, in "Zur Ehrenrettung des Simon Marius," in *Vierteljahrsschrift der Astronomischen Gesellschaft*, **77** (Leipzig, 1942), 27–32. Mayr's *Mundus Jovialis* was trans. into English by Arthur Octavius Prickard in *Observatory*, **39** (1916), 367–381, 403–412, 443–452, 498–503.

II. Secondary Literature. The Dutch Academy of Sciences announced a contest for the best essay submitted by 1 Jan. 1900 on the question whether Galileo was justified in condemning Mayr as a plagiarist. The only entry was submitted by Josef Klug, who published a revised version as "Simon Marius aus Gunzenhausen und Galileo Galilei," in *Abhandlungen der Bayerischen Akademie der Wissenschaften*, Math.-phys. Kl., **22** (1906), 385–526. Klug's attack on Mayr stimulated one of the judges, J. A. C. Oudemans, together with Johannes Bosscha, to defend Mayr in "Galilée et Marius," in *Archives néerlandaises des sciences exactes et naturelles*, 2nd ser., **8** (1903), 115–189. After the publication of Klug's article, the reasons why its original version had been rejected were explained by Bosscha in "Simon Marius, réhabilitation d'un astronome calomnié," in *Archives*

AL-MĀZINĪ

MAZURKIEWICZ

néerlandaises . . ., 2nd ser., **12** (1907), 258–307, 490–528; G. S. Braddy, "Simon Marius (1570–1624)," in *Journal of the British Astronomical Association,* **81** (1970), 64–65.

See also J. B. J. Delambre, *Histoire de l'astronomie moderne* (New York–London, 1969 [repr. of 1821 ed.]), I, 634, 693–703; Antonio Favaro, "Galileo Galilei e Simone Mayr," in *Bibliotheca mathematica,* 3rd ser., **2** (1901), 220–223; "Galileo and Marius," in *Observatory,* **27** (1904), 199–200; "A proposito di Simone Mayr," in *Atti e memorie dell'Accademia di scienze, lettere ed arti* (Padua), n.s. **34** (1917–1918), 17–19; and *Galileo Galilei e lo studio di Padova,* 2 vols., Contributi alla Storia dell'Università di Padova, nos. 3–4 (Padua, 1966 [repr. of 1883 ed.]), I, 184, 192, 234, 340–347; Siegmund Günther, "Mayr," in *Allgemeine deutsche Biographie,* XXI (Leipzig, 1885; repr. Berlin, 1970), 141–146; J. H. Johnson, "The Discovery of the First Four Satellites of Jupiter," in *Journal of the British Astronomical Association,* **41** (1930–1931), 164–171; William Thynne Lynn, "Simon Marius and the Satellites of Jupiter," in *Observatory,* **26** (1903), 254–256; "Galilée et Marius," *ibid.,* 389–390; "Galileo and Marius," *ibid.,* **27** (1904), 63–64, 200–201; and "Simon Mayr," *ibid.,* **32** (1909), 355–356; Julius Meyer, "Osiander und Marius," in *Jahresbericht des historischen Vereins für Mittelfranken,* **44** (1892), 59–71; Pietro Pagnini, "Galileo and Simon Mayer," trans. by W. P. Henderson, in *Journal of the British Astronomical Association,* **41** (1930–1931), 415–422; and Emil Wohlwill, *Galilei und sein Kampf für die Copernicanische Lehre* (Wiesbaden, 1969 [repr. of 1909–1926 ed.]), II, 343–426.

EDWARD ROSEN

AL-MĀZINĪ. See **Abū Ḥāmid.**

MAZURKIEWICZ, STEFAN (*b.* Warsaw, Poland, 25 September 1888; *d.* Grodżisk Mazowiecki, near Warsaw, 19 June 1945), *mathematics.*

Mazurkiewicz was, with Zygmunt Janiszewski and Wacław Sierpiński, a founder of the contemporary Polish mathematical school and, in 1920, of its journal *Fundamenta mathematicae,* which is devoted to set theory and to related fields, including topology and foundations of mathematics.

The son of a noted lawyer, Mazurkiewicz received his secondary education at the lyceum in Warsaw. He passed his baccalaureate in 1907, studied mathematics at the universities of Cracow, Lvov, Munich, and Göttingen, and was awarded a Ph.D. in 1913 by the University of Lvov for his thesis, done under Sierpiński, on curves filling the square ("O krzywych wypełniajacych kwadrat"). Named professor of mathematics in 1915 at the University of Warsaw, he held

this chair until his death. He was several times elected dean of the Faculty of Mathematical and Natural Sciences and, in 1937, prorector of the University of Warsaw. He was a member of the Polish Academy of Sciences and Letters; of the Warsaw Society of Sciences and Letters, which elected him its secretary-general in 1935; of the Polish Mathematical Society, which elected him its president for the years 1933–1935; and member of the editorial boards of *Fundamenta mathematicae* and the *Monografie matematyczne* from their beginnings. His book on the theory of probability was written in Warsaw during the German occupation of Poland. The manuscript was destroyed in 1944 when the Germans burned and destroyed Warsaw before their retreat; it was partly rewritten by Mazurkiewicz and published in Polish eleven years after his death. Gravely ill, Mazurkiewicz shared the lot of the people of Warsaw. He died in the outskirts of the city during an operation for gastric ulcer.

Mazurkiewicz' scientific activity was in two principal areas: topology with its applications to the theory of functions, and the theory of probability. The topology seminar given by him and Janiszewski, beginning in 1916, was probably the world's first in this discipline. He exerted a great influence on the scientific work of his students and collaborators by the range of the ideas and problems in which he was interested, by the inventive spirit with which he treated them, and by the diversity of the methods that he applied to them.

As early as 1913 Mazurkiewicz gave to topology an ingenious characterization of the continuous images of the segment of the straight line, known today as locally connected continua. He based it on the notions of the oscillation of a continuum at a point and on that of relative distance; the latter concept, which he introduced, was shown to be valuable for other purposes. This characterization therefore differs from those established at about the same time by Hans Hahn and by Sierpiński, which were based on other ideas. It is also this characterization that is linked with the Mazurkiewicz-Moore theorem on the arcwise connectedness of continua.

Mazurkiewicz' theorems, according to which every continuous function that transforms a compact linear set into a plane set with interior points takes the same value in at least three distinct points (a theorem established independently by Hahn), while every compact plane set that is devoid of interior point is a binary continuous image, enabled him to define the notion of dimension of compact sets as follows: the dimension of such a set C is at most n when n is the smallest whole number for which there exists a

248

continuous function transforming onto C a nondense compact linear set and taking the same value in at most $n + 1$ distinct points of this set. This definition preceded by more than seven years that of Karl Menger and Pavel Uryson, to which it is equivalent for compact sets.

In a series of later publications Mazurkiewicz contributed considerably to the development of topology by means of solutions to several fundamental problems posed by Sierpiński, Karl Menger, Paul Alexandroff, Pavel Uryson, and others, through which he singularly deepened our knowledge, especially of the topological structure of the Euclidean plane. In solving the problem published by Sierpiński (in *Fundamenta mathematicae*, **2** [1921], 286), he constructed on the plane a closed connected set which is the sum of a denumerable infinity of disjoint closed sets (1924) and which, in addition, has the property that all these summands except one are connected; at the same time he showed (independently of R. L. Moore) that on the plane the connectedness of all the summands in question is impossible, although, according to a result of Sierpiński's, it ought to be possible in space. Mazurkiewicz also solved, affirmatively, Alexandroff's problem (1935) on the existence of an indecomposable continuum (that is, one which is the sum of not fewer than $2^{\aleph_0}$ subcontinua different from itself) in every continuum of more than one dimension; that of Menger (1929) on the existence, for every positive integer n, of weakly n-dimensional sets; and that of Uryson (1927) on the existence, for every integer $n > 1$, of separable complete n-dimensional spaces devoid of connected subsets containing more than one point. He also showed (1929) that if R is a region in n-dimensional Euclidean space and E is a set of $n - 2$ dimensions, then the difference $R - E$ is always connected and is even a semicontinuum.

Mazurkiewicz also contributed important results concerning the topological structure of curves, in particular concerning that of indecomposable continua, as well as an ingenious demonstration, by use of the Baire category method, that the family of hereditarily indecomposable continua of the plane, and therefore that the continua of less paradoxical structure occur in it only exceptionally (1930).

By applying the same method to the problems of the theory of functions, Mazurkiewicz showed (1931) that the set of periodic continuous functions f, for which the integral $\int_0^1 t^{-1} f(x + t) + f(x - t) - 2f(x) dt$ diverges everywhere, is of the second Baire category in the space of all continuous real functions, and that the same is true with the set of continuous functions which are nowhere differentiable. In addition he provided the quite remarkable result that the set of continuous functions transforming the segment of the straight line into plane sets which contains Sierpiński's universal plane curve (universality here designating the presence of homeomorphic images of every plane curve) is also of the second Baire category. Among Mazurkiewicz' other results on functions are those concerning functional spaces and the sets in those spaces that are called projective (1936, 1937), as well as those regarding the set of singular points of an analytic function and the classical theorems of Eugène Roché, Julius Pál, and Michael Fekete.

In the theory of probability, Mazurkiewicz formulated and demonstrated, in a work published in Polish (1922), the strong law of large numbers (independently of Francesco Cantelli); established several axiom systems of this theory (1933, 1934); and constructed a universal separable space of random variables by suitably enlarging that of the random variables of the game of heads or tails to a complete space (1935). These results and many others were included and developed in his book on the theory of probability.

BIBLIOGRAPHY

I. Original Works. Among the 130 of Mazurkiewicz' mathematical publications listed in *Fundamenta mathematicae*, **34** (1947), 326–331, the most important are "Sur les points multiples des courbes qui remplissent une aire plane," in *Prace matematyczno-fizyczne*, **26** (1915), 113–120; "Teoria zbiorów G_δ" ("Theory of G_δ Sets"), in *Wektor*, **7** (1918), 1–57; "O pewnej nowej formie uogólnienia twierdzenia Bernoulli'ego" ("On a New Generalization of Bernoulli's Theorem"), in *Wiadomości aktuarjalne*, **1** (1922), 1–8; "Sur les continus homogènes," in *Fundamenta mathematicae*, **5** (1924), 137–146; "Sur les continus plans non bornés," *ibid.*, 188–205; "Sur les continus absolument indécomposables," *ibid.*, **16** (1930), 151–159; "Sur le théorème de Rouché," in *Comptes rendus de la Société des sciences et des lettres de Varsovie*, **28** (1936), 78, 79; and "Sur les transformations continues des courbes," in *Fundamenta mathematicae*, **31** (1938), 247–258. See also the posthumous works *Podstawy rachunku prawdopodobieństwa* ("Foundations of the Calculus of Probability"), J. Łoś, ed., *Monografie Matematyczne*, no. 32 (Warsaw, 1956); and *Travaux de topologie et ses applications* (Warsaw, 1969), with a complete bibliography of Mazurkiewicz' 141 scientific publications.

II. Secondary Literature. See P. S. Alexandroff, "Sur quelques manifestations de la collaboration entre les écoles mathématiques polonaise et soviétique dans le domaine de topologie et théorie des ensembles," in *Roczniki Polskiego towarzystwa matematycznego*, 2nd ser., *Wiadomości matematyczne*, **6** (1963), 175–180, a lecture delivered at the Polish Mathematical Society, Warsaw, 18 May 1962; and

C. Kuratowski, "Stefan Mazurkiewicz et son oeuvre scientifique," in *Fundamenta mathematicae*, **34** (1947), 316–331, repr. in S. Mazurkiewicz, *Travaux de topologie et ses applications* (Warsaw, 1969), pp. 9–26.

B. KNASTER

MÉCHAIN, PIERRE-FRANÇOIS-ANDRÉ (*b*. Laon, Aisne, France, 16 August 1744; *d*. Castellón de la Plana, Spain, 20 September 1804), *geodesy, astronomy*.

Méchain was the son of Pierre-François Méchain, a master ceiling plasterer of modest means, and Marie-Marguerite Roze. Young Méchain's mathematical ability attracted the attention of various local notables, who advised sending him to the École des Ponts et Chaussées in Paris. Lack of financial resources interrupted his studies there and he accepted a tutorship for two young noblemen about thirty miles from Paris.

In some fashion Méchain came into communication with Lalande, who sent him the proofs of the new second edition of his *Astronomie*. Filled with enthusiasm, Méchain made such rapid progress in this study that in 1772 Lalande procured for him a position as hydrographer at the naval map archives (Depôt de la Marine) in Versailles. The archives were then a seat of political patronage and intrigue, and, caught in the political crosscurrents, Méchain twice lost his job; but each time he was reinstated because of his competence as a map-maker. The archives were soon transferred to Paris, and there he drew up the maps for the shoreline from Nieuwpoort in Flanders to Saint-Malo. In 1777 Méchain married Thérèse Marjou, whom he met while working in Versailles; they had a daughter and two sons.

Beginning in 1780 he determined the network of fundamental points for large military maps of Germany and northern Italy.

Meanwhile Méchain was also active as an astronomical observer, his early efforts being crowned in 1781 with the discovery of not one but two comets. He calculated the orbits for both, and in the following year he calculated orbits for the comets of 1532 and 1661, proving, contrary to general expectation, that they were not the same. This research won both the 1782 prize of the Académie Royale des Sciences and admission to its ranks. Encouraged by these successes, Méchain threw himself into observing with still greater zeal, and ultimately discovered nine more comets, including the remarkable short-period one now named after Encke. He calculated the orbits for all of these, as well as thirteen found by other observers. In addition he found many nebulae, which were incorporated by Charles Messier into his famous catalogue of clusters and nebulae. In 1785 he became editor of the French national almanac, *Connoissance des temps*, and he prepared the seven volumes for 1788 to 1794.

In 1787 a joint Anglo-French project undertook the triangulation between the Greenwich and Paris observatories. Méchain was chosen as one of the French commissioners, along with Legendre and J.-D. Cassini. Both countries engaged in a friendly rivalry to produce new and more accurate measuring instruments. In France, Borda developed the principle of the *cercle répétiteur*, or repeating transit, in which after the first set of readings, the circle was clamped with the telescope and moved back to the original line of sight. In this way the angles could be measured against different segments of the circle, thus averaging out graduation errors. The commissioners systematically tested Borda's device, with Méchain using the older equipment; the tests demonstrated the great superiority of this new circle.

In 1790 the National Assembly approved an Academy proposal to establish a decimal system of measures, and Méchain and Delambre were designated to carry out the fundamental geodetic measurements for a new unit of length. This unit, the meter, was intended to be one ten-millionth part of the distance from the terrestrial pole to the equator, and it was to be based on an extended survey from Dunkerque to Barcelona. Méchain was assigned the shorter but more difficult southern zone, the previously unsurveyed region across the Pyrenees.

The new repeating transit became the fundamental instrument of the survey, but not until June 1792 was the new equipment, including parabolic mirrors for reflecting signals, ready. By this time the Revolution was engulfing France and the monarchy was tottering; Méchain with his suspicious array of instruments was arrested at Essonnes just south of Paris as a potential counterrevolutionary. Only with much difficulty was he located and released two months later, so that he could continue his journey to Spain. In September and October he swiftly carried out the triangulation between Perpignan and Barcelona. During the winter of 1792–1793 he undertook the astronomical observations to establish the latitude of Barcelona, almost at the southernmost limit of the meridian. At the same time he investigated the possibility of extending the meridian $2\frac{1}{2}°$ southward to the Balearic Islands. Otherwise, only a few weeks of work remained to complete the network across the frontier.

His plans were abruptly interrupted at the beginning of spring in 1793 when, invited by a friend to inspect a new hydraulic pump in the outskirts of Barcelona, he was involved in an accident. While trying to start the machine, the friend and an assistant were caught in the mechanism. Méchain, rushing to aid them, was struck by a lever that knocked him violently against the wall, breaking some ribs and a collarbone. He was unconscious for three days and afterward was forced to remain completely immobile for two months. By June he still did not have the use of his right arm; but, undeterred, he used his left hand to make the solar observations at the summer solstice.

During Méchain's convalescence, open war had broken out between Spain and France, and he was denied a passport to return home. Profiting from his captivity, he determined the latitude of Montjouy, just south of Barcelona, and surveyed the triangle connecting these points. He then noticed a 3″ discrepancy in the latitude previously obtained for Barcelona. Anguished by his failure to find the cause, and blaming himself for the error, he kept the discrepancy a carefully guarded secret. In the remaining years of his life he became a driven and tormented man, whose behavior was mysterious and inexplicable to his colleagues. Delambre, who found out the secret only when he inherited the notes, intimated that Méchain simply put too much trust in the precision of the repeating transit.

Eventually Méchain obtained a passport for Italy, and he managed to reach Genoa in September 1794. Saddened by the guillotining of several of his colleagues and in poor health, he delayed his return to France, not embarking for Marseilles until the following year. After additional hesitation he journeyed to the vicinity of Perpignan and in September 1795 resumed the triangulation. Méchain slowly continued his work through 1796 and 1797. Meanwhile, after a fifteen-month suspension for political reasons, Delambre proceeded with measurements of the northern part of the network and in April 1798 he invited Méchain to join him in linking the sections. Méchain remained incommunicado, and Delambre finally sought him out in Carcassonne. Méchain expressed a stubborn desire, inexplicable to Delambre, to return to Spain for further latitude determinations. Faced with the choice of returning to Paris and the warm welcome of his colleagues, or remaining forever an expatriate, Méchain reluctantly came back to Paris. There he was less than cooperative in presenting his observations to the commissioners charged with setting up the decimal metric system.

In Paris he was made director of the observatory, considered a just and tranquil reward for an astronomer who had labored so faithfully without a real astronomical position. But to Méchain nothing was

right, and in a remarkable letter to Franz von Zach he aired his complaints publicly (*Monatliche Correspondenz*, **2** [1800], 290–302). Always he yearned to return to Spain; and eventually the Bureau of Longitudes approved the extension of the meridian to the Balearic Islands, a project that would render his imperfect latitude of Barcelona unnecessary. The Bureau, believing that Méchain's abilities were best employed as director of the observatory, appointed another astronomer to extend the meridian, but to their surprise Méchain insisted on doing it himself.

The expedition left Paris on 26 April 1803, but encountered unexpected delays in Spain. When the ship at last departed for the islands, an epidemic of yellow fever broke out on board. Méchain eventually reached Ibiza, but he discovered that his mainland station at Montsia could not be sighted from the island. Thus he was obliged to change the pattern of triangles and survey a greater distance southward in the mountains along the Spanish coast. Exhausted by the work and further weakened by fever and a poor diet, he collapsed and died on 20 September 1804. Several years later the extension of the meridian was completed by Biot and Arago.

BIBLIOGRAPHY

I. ORIGINAL WORKS. Méchain's bibliography is found in J. M. Quérard, *La France littéraire*, V (Paris, 1830), 10–11, which along with several biographical encyclopedias, gives his death erroneously as 1805. Méchain's most important work was edited in three volumes by J. B. Delambre, *Base du système métrique décimal, ou Mesure de l'arc du méridien compris entre les parallèles de Dunkerque et Barcelone, exécutée en 1792 et années suivantes, par MM. Méchain et Delambre* (Paris, 1806–1810).

Many of Méchain's MSS are preserved at the Paris observatory; they are catalogued as C6.6–7 and E2.1–21 in G. Bigourdan, "Inventaire des manuscrits," in *Annales de l'observatoire de Paris. Mémoires*, **21** (Paris, 1895), 1–60. See also his "La prolongation de la méridienne de Paris, de Barcelone aux Baléares, d'après les correspondances inédites de Méchain, de Biot et d'Arago," in *Bulletin astronomique*, **17** (Paris, 1900), 348–368, 390–400, 467–480.

II. SECONDARY LITERATURE. J. B. Delambre's florid "Notice," in *Mémoires de l'Institut des sciences, lettres et arts, sciences mathématiques et physiques*, **6** (Paris, 1806), 1–28, was written before he knew why Méchain desired so ardently to return to Spain; for the earlier biography he refers the reader to Franz von Zach, "Pierre-François-André Méchain," in *Monatliche Correspondenz*, **2** (Gotha, 1800), 96–120. More balanced accounts by Delambre are in *Histoire de l'astronomie au dix-huitième siècle* (Paris, 1827), 755–767, and in Michaud's *Biographie universelle ancienne et moderne*, XXVII (Paris, after 1815), 454–458.

See also Delambre's *Grandeur et figure de la terre*, G. Bigourdan, ed. (Paris, 1912).

An excellent modern account with new material is by Joseph Laissus, "Un astronome Français en Espagne: Pierre-François-André Méchain (1744–1804)," in *Comptes rendus 94ᵉ Congrès national des sociétés savantes, Pau, 1969, sciences*, **1** (Paris, 1970), 37–59.

OWEN GINGERICH

MECHNIKOV. See **Metchnikoff.**

MECKEL, JOHANN FRIEDRICH (*b.* Halle, Germany, 13 October 1781; *d.* Halle, 31 October 1833), *anatomy, embryology, comparative anatomy.*

Belonging to the third generation of an illustrious family of physicians, Meckel was one of the greatest anatomists of his time. His painstaking observations in comparative and pathological anatomy furnished a wealth of new knowledge, which Meckel attempted to organize along certain evolutionary schemes popular in his day.

Although influenced by the contemporary ideas of *Naturphilosophie*, Meckel rejected pure speculation and stressed instead the acquisition of empirical data from which certain useful conclusions could be derived. Among his most lasting and impressive contributions was the study of the abnormalities occurring during the embryological development. Hence, Meckel's teratology was the first comprehensive description of birth defects, a detailed and sober analysis of a topic which had hitherto been approached with a great deal of fantasy and moral bias.

Meckel's grandfather, Johann Friedrich the Elder (1714–1774), had been one of Haller's most brilliant disciples, an anatomist endowed with great powers of observation and notable skill in the preparation of anatomical specimens. In turn Meckel's father, Phillip Friedrich (1755–1803), was a famous surgeon and obstetrician in Halle, where he taught for twenty-six years.

Young Meckel spent his student years between 1798 and 1801 in Halle, then a bastion of academic freedom and objective scientific inquiry. Among his teachers were Kurt Sprengel, famous for his botanical and historical studies, and, above all, his mentor Johann C. Reil, who inspired Meckel's studies in cerebral anatomy and was the true leader of the local medical school.

After studying anatomy under the direction of his father—he apparently detested the discipline in the beginning—Meckel transferred in 1801 to the University of Göttingen. There he studied comparative anatomy with the famous physician and anthropologist Blumenbach. His subsequent doctoral dissertation, defended at Halle in 1802, dealt with the subject of cardiac malfunctions. Meckel expanded the topic and eventually published it as an article in Reil's journal, the *Archiv für die Physiologie.*

Following his graduation Meckel visited Würzburg, then a stronghold of Schelling's philosophy of nature, and Vienna, where he met Johann P. Frank. In 1803 Meckel temporarily interrupted his travels, returning to Halle at his father's death. Thereafter he went to Paris where he met and worked with Georges Cuvier, Étienne Geoffroy Saint-Hilaire, and Alexander von Humboldt. Together with Cuvier, Meckel systematically analyzed the immense anatomical collection located at the Jardin des Plantes. The available material, sent back from Napoleon's campaigns abroad, was described by Cuvier in his *Leçons d'anatomie comparée*. Meckel translated Cuvier's five-volume work into German, a task which he completed in 1810.

Meckel returned to his native Halle in 1806 under tragic circumstances. The Napoleonic forces had occupied the city and dissolved the local university. Napoleon used Meckel's own home as temporary headquarters, an intrusion which may have aided in preserving the valuable anatomical collection of the Meckel family.

When the newly organized University of Halle opened its doors in May 1808, Meckel was appointed professor of normal and pathological anatomy, surgery, and obstetrics. He was at Halle until his death and set a harsh working schedule for himself. He gradually withdrew from social activities and grew bitter in the face of the academic mediocrity surrounding him. Furthermore, Meckel was impatient with the bureaucratic fetters imposed by the Prussian government, which treated Halle as a secondary and provincial city compared to the capital, Berlin.

Meckel's scientific aim was to arrive at an understanding of the great variety of organic forms. Such knowledge would, he hoped, reveal the uniformity of nature and expose its general laws. Meckel sought fundamental types amid the multiplicity of organisms, and accepted the idea that each higher evolutionary product must have traversed all the lower stages of development before achieving its position.

Moreover, Meckel adopted an Aristotelian position by clearly distinguishing between matter and form, the latter being provided by the *Lebenskraft*. His morphological studies were, therefore, geared to discovering the fundamental laws regulating the formation of the various organic categories. Meckel was interested in malformations because structural aberrations were the result of normal actions attributable to the vital force, which he tried to understand.

Meckel achieved such insights through exhaustive observations rather than arm-chair philosophy—witness his three-volume *Handbuch der pathologischen Anatomie* (1812–1816) and six-volume *System der vergleichenden Anatomie* (1821–1831). Among Meckel's discoveries was the diverticulum—which now carries his name—in the distal small bowel, a vestige between the intestinal tract and the yolk sac.

In 1815 Meckel became the editor of Reil's journal, then known as *Deutsches Archiv für die Physiologie*, which listed among its distinguished collaborators Autenrieth, Blumenbach, Döllinger, Kielmeyer, Sprengel, and others. Meckel wrote a preface to the first volume stressing that only articles based on observations and experiments would be printed. He hoped that such an approach would gradually prevail in German science in order to obviate the ridicule incurred by speculation. But Meckel also decried mindless experimentation.

Meckel emphasized the need for work in comparative anatomy and embryology. An early article concerned the development of the central nervous system in mammals, a study followed by new observations related to the evolution of the gut, heart, and lungs. Interspersed with these careful monographs were numerous shorter articles. They covered subject matters as varied as the generation of earthworms, bleeding diatheses, development of human teeth, and the cerebral anatomy of birds.

From 1826 until his death, Meckel was the editor of the *Archiv für Anatomie und Physiologie*, a continuation of the previous publication. His last articles dealt to a considerable degree with malformations as well as with vascular and pulmonary development.

Meckel's adherence to a *Lebenskraft* or vital force and denial of mechanical factors in embryological development were strongly disputed by successors who viewed life in strict physicochemical terms. Although his interpretations rapidly became obsolete, Meckel's material remained an extremely valuable source for those interested in comparative anatomy and in congenital malformations.

BIBLIOGRAPHY

I. ORIGINAL WORKS. A complete bibliography of Meckel's works, prepared in a chronological order, is in Beneke (see below), pp. 155–159. Meckel's work on comparative anatomy is contained in his *Beyträge zur vergleichenden Anatomie*, 2 vols. (Leipzig, 1808–1812), and the more extensive publication, *System der vergleichenden Anatomie*, 6 vols. (Halle, 1821–1831). He also published a large number of articles which can be found in the *Deutsches Archiv für die Physiologie*, 1–8 (1815–1823) and the *Archiv für Anatomie und Physiologie*, 1–6 (1826–1832).

Among those works of Meckel available in English is the *Manual of General Anatomy*, translated from a French version by A. S. Doane (London, 1837), and the *Manual of Descriptive and Pathological Anatomy*, also from the French by the same English translator, 2 vols. (London, 1838).

II. SECONDARY LITERATURE. The only extensive biography of Meckel is Rudolf Beneke, *Johann Friedrich Meckel der Jüngere* (Halle, 1934). Shorter notices appeared in the *Medicinische Wochenschrift für Hamburg*, **1** (1833), as well as in August Hirsch's *Biographisches Lexikon*, 2nd ed., IV (Munich, 1932), 145–146. A recent article reviewing Meckel's interest in birth defects is Owen E. Clark, "The Contributions of J. F. Meckel, the Younger, to the Science of Teratology," in *Journal of the History of Medicine and Allied Sciences*, **24** (1969), 310–322. A short description of the genealogy of the Meckel family and the collection of their skulls is contained in H. Schierhorn and R. Schmidt, "Beitrag zur Genealogie und Kraniologie der Familie Meckel," in *Verhandlungen der anatomischen Gesellschaft Jena*, **63** (1969), 591–599. The conceptual background for the contemporary German interest in embryology is given by Owsei Temkin, "German Concepts of Ontogeny and History Around 1800," in *Bulletin of the History of Medicine*, **24** (1950), 227–246.

GUENTER B. RISSE

MEDICUS, FRIEDRICH CASIMIR (Medikus, Friedrich Kasimir) (*b.* Grumbach, Rhineland, Germany, 6 January 1736; *d.* Mannheim, Germany, 15 July 1808), *botany.*

Medicus was the most prolific, bitter, witty, and sarcastic of those contemporary opponents of Linnaeus who ignored or assailed his innovations. Apart from invective and references to Linnaeus' shortcomings, errors, and inconsistencies, Medicus' numerous works contain many firsthand botanical observations, particularly on *Leguminosae, Cruciferae, Malvaceae,* and *Rosaceae,* and are of both historical and nomenclatural importance.

After studying medicine at the universities of Tübingen, Strasbourg, and Heidelberg, in 1759 he became garrison doctor at Mannheim, in the Palatinate (Pfalz), then ruled by the Elector Carl Theodor. The latter founded in 1763 the Academia Theodoro-Palatina of Mannheim and in 1766 an associated botanic garden at Medicus' instigation. In 1766 Medicus spent five months on sick leave in Paris, becoming friendly with the botanists Duhamel du

Monceau, Bernard de Jussieu, and Adanson, all of whom favored the generic concepts and nomenclature of Tournefort rather than those of Linnaeus, his successor. Returning to Mannheim, Medicus abandoned medicine for botany and became director of the Mannheim botanic garden. In his many publications based on the study of living plants, he thereafter never lost an opportunity to criticize the works and character of Linnaeus. He angrily called attention to a basic practical weakness of Linnaeus' generic descriptions: being originally based on the study of one or two species, they often failed to cover adequately other species later added by Linnaeus to the genus by virtue of similarity in habit rather than technical details. He restored many Tournefortian genera and rejected many Linnaean generic names.

It is uncertain how widely his works, written mostly in German rather than Latin, were read at the time. He undoubtedly greatly influenced Conrad Moench at Marburg, whose relatively well-known *Methodus plantas horti botanici et agri Marburgensis* (1794–1802) brought the same Tournefortian concepts and names to more general notice, although their valid post-Linnaean publication dates from Miller's *Gardeners Dictionary Abridged* of 1754. Unfortunately, both the Mannheim academy and garden, to which Medicus had given so much attention, suffered almost irreparable damage when Mannheim was heavily bombarded in 1795 and 1799. The academy was closed, its books sold, and the garden did not long outlast the death of Medicus in 1808.

BIBLIOGRAPHY

I. ORIGINAL WORKS. Medicus was a prolific and combative writer. G. Pritzel, *Thesaurus litterature botanicae* (Leipzig, 1872), 211, lists 16 publications, of which the most notable are *Theodora speciosa* (Mannheim, 1786); "Versuch einer neuen Lehrart der Pflanzen," in *Vorlesungen der churpfälzischen physikalisch-ökonomischen Gesellschaft*, **2** (1787), 327–460; *Philosophische Botanik*, 2 vols. (Mannheim, 1789–1791); *Pflanzengattungen* (Mannheim, 1792); *Geschichte der Botanik unserer Zeiten* (Mannheim, 1793); and *Beyträge zur Pflanzen-Anatomie* (Leipzig, 1799–1801).

II. SECONDARY LITERATURE. See (listed in chronological order) A. Kistner, *Die Pflege der Naturwissenschaften in Mannheim zur Zeit Karl Theodors* (Mannheim, 1930); G. Schmid, "Linné im Urteil Johann Beckmanns, mit besonderer Bezeihung auf F. C. Medicus," in *Svenska linnésällskapets årsskrift*, **20** (1937), 47–70; W. T. Stearn, "Botanical gardens and botanical literature in the eighteenth century," in Rachel Hunt, *Catalogue of Botanical Books . . .*, II (Pittsburgh, 1961), xli–cxl; and "Early Marburg Botany," in Conrad Moench, *Methodus plantas*

horti . . ., Otto Koeltz, ed. (Koenigstein–Taunus, 1966), xi–xv; and F. A. Stafleu, *Linnaeus and the Linnaeans* (Utrecht, 1971), 260–265.

WILLIAM T. STEARN

MEDINA, PEDRO DE (*b.* Seville [?], Spain, 1493; *d.* Seville, 1576), *cosmography, navigation.*

Little is known about the life of Medina, who wrote both literary and navigational works during Spain's golden age. He was a cleric, and may have graduated from the University of Seville. It is certain that for part of his career he was librarian to the duke of Medina-Sidonia, for whom he composed a family chronicle that was published in 1561. His other philosophical and historical works date from the same period of his life.

Medina was also a teacher of mathematics and a founder of marine science. King Charles I gave him a warrant (dated Toledo, 20 December 1538) to draw charts and prepare pilot books and other devices necessary to navigation to the Indies. In 1549 he was named "cosmógrafo de honor." In addition to charts and sailing directions, Medina made astrolabes, quadrants, mariner's compasses, forestaffs, and other navigational instruments; it is possible that he himself performed some actual practical navigation.

Medina's first navigational book was the *Arte de navegar*, which is supposed to have inspired Bernardino Baldi's great didactic poem "Nautica." Medina's work contains two fundamental errors, however, in that he failed to recognize the exactness of the plane chart (Mercator's projection of the world was made in 1569) and posited the constancy of the magnetic declination (although his sailing directions do take account of the variation of the magnetic pole).

Medina further became active in the professional argument between cosmographers and pilots, maintaining that the existing instructions for the determination of altitude and use of the mariner's compass prepared by the Seville school of cosmographers contained a number of serious errors. His "Suma de Cosmografia" was never published; it is preserved in the Biblioteca Capitular Colombina in Seville, and contains material on astrology, philosophy, and navigation.

BIBLIOGRAPHY

I. ORIGINAL WORKS. Medina's literary works include *Libro de verdad* (Valladolid, 1545; repr. Málaga, 1620); *Libro de las grandezas y cosas memorables de España* (Seville, 1548 and 1549; repr. Alcalá de Henares, 1566);

and *Crónica breve de España* (Seville, 1548), which was commissioned by Queen Isabella I before her death in 1504.

His navigational writings comprise *El arte de navegar* (Seville–Valladolid, 1545; Seville, 1548), with trans. into German, French, English, and Italian; *Tabulae Hispaniae geographica* (Seville, 1560); "Suma de Cosmografía," an unpublished work of 1561 (MS in Biblioteca Capitular Colombina, Seville); and *Regimiento de navegación* (Seville, 1563).

II. SECONDARY LITERATURE. On Medina and his work, see *Diccionario Enciclopédico Hispano-Americano*, XII (Barcelona, 1893), 692-693. See also Angel Gonzalez Palencia, *La primera guia de la España Imperial* (Madrid, 1940); Rafael Pardo de Figueroa, *Pedro de Medina y su "Libro de las Grandezas"* (Madrid, 1927); *Regimiento de navegación de Pedro de Medina 1563* (Cadiz, 1867); Martin Fernandez de Navarrete, *Biblioteca marítima española* (Madrid, 1851); and *Dissertación sobre la historia de la naútica* (Madrid, 1846).

J. M. LÓPEZ DE AZCONA

MEEK, FIELDING BRADFORD (*b.* Madison, Indiana, 10 December 1817; *d.* Washington, D. C., 21 December 1876), *paleontology, geology.*

Meek was an exceptionally able paleontologist who made substantial pioneering contributions to knowledge of extinct faunas and stratigraphic geology. Little is known of his early life in Indiana and Kentucky, except that frail health and the early death of his father rendered his childhood and young adulthood difficult. As a youth he devoted most of his time to the study of natural history. He failed in business, partly because of his preoccupation with fossils, and supported himself as a portrait painter. His first work in geology was as an assistant to D. D. Owen in 1848 and 1849.

From 1852 until 1858 he served as an assistant to James Hall, paleontologist of New York, at Albany. During this interval he worked two summers for the Geological Survey of Missouri and spent the summer of 1853 exploring, with F. V. Hayden, the Badlands of South Dakota and surrounding areas. Accounts and letters indicate that throughout this time Hall tyrannized and exploited his modest and retiring assistant.

In 1857 Meek first recognized the occurrence of Permian fossils in North America. Unfortunately, he became involved in a bitter controversy concerning the priority of this discovery. He also felt, probably justifiably, that Hall was claiming credit for other significant age determinations that he had made.

In 1858 Meek left Albany and became the first full-time paleontologist associated with the Smithsonian Institution. Although he received no salary, Joseph

Henry permitted him to live in the south tower of the Smithsonian building. Progressive deafness and continued poor health combined to limit his professional contacts.

Despite his physical handicaps, Meek completed a prodigious quantity of descriptions of invertebrate fossils and probably ranks only behind Hall and C. D. Walcott in sheer volume of published pages. Although many works were published jointly with his associate Hayden or with A. H. Worthen of the Illinois Geological Survey, it is likely that almost all of this work, including many plates of illustrations, was done entirely by Meek. Meek also published for the Ohio and California Geological Surveys.

Meek's principal contributions may be divided into three parts. The first, begun while he was still at Albany, included descriptions and interpretations of fossils collected by Hayden before the Civil War and laid the groundwork for stratigraphic and age interpretations of rocks of the Great Plains. The second part comprised his descriptions of the Paleozoic fossils of Illinois, especially those of Mississippian and Pennsylvanian age. His third great body of work relates to the U. S. Geological and Geographical Survey of the Territories, with which his name is closely associated, although he was never formally employed by it. He is particularly noted for investigations of freshwater faunas at the Mesozoic-Cenozoic boundary.

Meek described invertebrate fossils of almost every phylum from all geologic periods, from Cambrian to Tertiary and over a wide area. His descriptions and observations on fossils are still valid; and his geologic interpretations, based on these fossils, have contributed materially to a better understanding of the geology of about half of the United States.

BIBLIOGRAPHY

I. ORIGINAL WORKS. Meek's bibliography (see below) contains 106 titles, some of which were written with F. V. Hayden or A. H. Worthen. Following the conventional practice of the times, a number of these are merely preliminary accounts of fossil descriptions, but they are more than balanced by his massive contributions to the Geological Survey of Illinois and to the U.S. Geological and Geographical Survey.

II. SECONDARY LITERATURE. C. A. White, in *American Geologist*, **18** (1896), 337–350, gives a brief account of Meek's life based in part on a memorial of 1877. This work and a similar shorter version for *Biographical Memoirs. National Academy of Sciences* (1902), 77–91, include a bibliography of Meek's works. G. P. Merrill, *Contributions to the History of American Geology*, U.S.

National Museum Annual Report for 1904 (Washington, 1906), provides some additional material on his relations with Hall and priority in scientific discovery.

ELLIS L. YOCHELSON

MEGGERS, WILLIAM FREDERICK (*b.* Clintonville, Wisconsin, 13 July 1888; *d.* Washington, D. C., 19 November 1966), *physics.*

Meggers was reared on a farm near Clintonville, Wisconsin, to which his parents, John Meggers and the former Bertha Bork, had emigrated from Pomerania in 1872. A self-made man who placed a high value on education, he supported himself while attending Ripon College (1910), largely by playing a trombone in a dance orchestra, and earned his doctorate at Johns Hopkins University (1917).

Inspired by Bohr's classical work "On the Constitution of Atoms and Molecules," he chose spectroscopy as his lifework. After serving as assistant in the new spectroscopy section at the U.S. Bureau of Standards, he became chief of the section in 1919, a post he held until 1958.

His achievements provide a lasting contribution to our knowledge of atomic structure. He was an expert in observing, measuring, and interpreting optical spectra. From intricate spectra he deciphered the quantum structure of many atoms and ions: energy levels, terms, quantum numbers, and configurations. He worked on some fifty spectra of about thirty elements—a unique record. His generosity in sharing his splendid line lists with co-workers was outstanding.

Meggers' name will always be associated with standard wavelengths of light. For almost a decade he was president of the Commission of the International Astronomical Union, which was responsible for recommending international standards. The first such standards, dating from 1910, were interferometric measurements of lines observed with an iron arc as source.

He developed an electrodeless ^{198}Hg lamp, which he hoped would be accepted as the source for the primary standard of length, but it was not adopted. In 1958, however, he reported the first interferometric wavelength measurement, by using radiation emitted from an electrodeless thorium halide lamp. Some five hundred thorium wavelengths now supersede the early iron standards, with at least a tenfold increase in accuracy.

A pioneer in spectrochemistry, Meggers fully realized the value of utilizing laboratory spectra for identification and quantitative analysis of commercial substances. To this end he instituted an extensive program in which the spectra of seventy metallic elements were observed under standardized conditions, and the relative intensities of some 39,000 spectral lines were determined to provide requisite data for quantitative work.

He was an excellent photographer and was a member of the Bureau orchestra in the early days. He had valuable coin and stamp collections, which he left to the American Institute of Physics to establish a foundation for training students in science.

He served also on many important committees and received many high honors. It has been truly said that "all of these are trivial compared to the esteem and admiration of his colleagues" [P. D. Foote]. Meggers was content to spend hours at a comparator measuring wavelengths of spectral lines. From these patiently accumulated, reliable data he solved many intricate problems on atomic structure, which to him was an abundant reward for his years of painstaking effort.

His zest for knowledge took other turns. He invested heavily in many collections, ranging from buttons to light bulbs to phonographs, with the belief that they had educational value. They were housed in the "Meggers Museum of Science and Technology," located over his garage, and he and his friends spent pleasant evenings in this museum enjoying travelogues illustrated by the splendid color slides he accumulated on his extensive travels, or listening to favorite records played on the large "Regina" music box. On 13 July 1920 Meggers married Edith (Marie) Raddant; they had two sons and a daughter.

BIBLIOGRAPHY

A complete bibliography of Meggers' writings is included in Paul D. Foote's biographical memoir; see *Biographical Memoirs. National Academy of Sciences*, **41** (1970), 319–340. His most important writings include "Measurements on the Index of Refraction of Air for Wavelengths From 2218 Å to 9000 Å," in *Scientific Papers of the United States Bureau of Standards*, **14** (1918), 697–740, with C. G. Peters; "Solar and Terrestrial Absorption in the Sun's Spectrum From 6500 Å to 9000 Å," in *Publications of the Allegheny Observatory, University of Pittsburgh*, **6**, no. 3 (1919), 13–44; "Interference Measurements in the Spectra of Argon, Krypton, and Xenon," in *Scientific Papers of the National Bureau of Standards*, **17** (1921), 193–202; "Standard Solar Wavelengths (3592–7148 Å)," in *Journal of Research of the National Bureau of Standards*, **1** (1928), 297–317, with K. Burns and C. C. Kiess; "Infrared Arc Spectra Photographed with Xenocyanine," in *Journal of Research of the National Bureau of Standards*, **9** (1932), 309–326, with C. C. Kiess; "Term Analysis of the First Spectrum of Vanadium," in *Journal of Research of the National Bureau of Standards*, **17** (1936), 125–192, with H. N. Russell;

Index to the Literature on Spectrochemical Analysis 1920–1939, (Philadelphia, 1941), with B. F. Scribner; "Spectroscopy, Past, Present, and Future," in *Journal of the Optical Society of America*, **36** (1946), 431–448; "Dr. W. F. Meggers, Ives Medalist for 1947," in *Journal of the Optical Society of America*, **38**, no. 1 (1948), 1–6; "Zeeman Effect," in *Encyclopaedia Britannica*, XXIII; "Wavelengths From Thorium-Halide Lamps," in *Journal of Research of the National Bureau of Standards*, **61**, no. 2 (1958), 95–103, with Robert W. Stanley; "Table of Wavenumbers," in *National Bureau of Standards Monographs*, **3** (1960), with Charles DeWitt Coleman and William R. Bozman; "Tables of Spectral-Line Intensities," *ibid.*, **32** (1961), with Charles H. Corliss and Bourdon F. Scribner; "Spectra Inform Us About Atoms," in *Physics Teacher*, **2**, no. 7 (1964), 303–311; *Key to the Welch Periodic Chart of the Atoms* (Chicago, 1965); "More Wavelengths From Thorium Lamps," in *Journal of Research of the National Bureau of Standards*, **69A**, no. 2 (1965), 109–118, with Robert W. Stanley; "Mees Medal Ceremony," in *Journal of the Optical Society of America*, **55**, no. 4 (1965), 341–345; "Dr. William F. Meggers," in *Arcs and Sparks*, **12**, no. 1 (1967), 3–4; and "The Second Spectrum of Ytterbium (Yb II)," in *Journal of Research of the National Bureau of Standards*, **71A**, no. 6 (1967), 396–544, edited by Charlotte E. Moore.

CHARLOTTE E. MOORE

MEINESZ, F. A. VENING. See **Vening Meinesz, Felix A.**

MEINZER, OSCAR EDWARD (*b.* near Davis, Illinois, 28 November 1876; *d.* Washington, D. C., 14 June 1948), *ground-water hydrology.*

Meinzer was the son of William and Mary Julia Meinzer. He graduated *magna cum laude* from Beloit College, Wisconsin, in 1901. He was a graduate student in geology at the University of Chicago (1906–1907) and received the Ph.D., *magna cum laude*, in 1922. His career in the United States Geological Survey began as geologic aide in June 1906. He married Alice Breckenridge Crawford in October 1906. Meinzer became junior geologist on ground-water investigations (1907), acting chief (1912), and chief, ground-water division (1913), a post which he held until retirement on 30 November 1946. In that same year he received an honorary doctorate from Beloit College.

During his thirty-four years as the chief, ground-water division (now branch) of the United States Geological Survey, Meinzer became the main architect in development of the modern science of ground-water

hydrology. He organized and trained a large number of scientists and engineers, many of whom became recognized international authorities in this vastly expanded field. When he began, the study of underground water was an insignificant and poorly appreciated art.

During his early years as chief, he initiated the development of the science of ground-water hydrology. He realized that in addition to locating and defining ground-water basins, as had been the earlier practice, the principles governing occurrence, movement, and discharge of ground water must be determined, and methods had to be devised and tested for determining the quantity and quality of available ground water. In order to standardize terms and describe principles, he prepared *Outline of Ground-Water Hydrology, With Definitions*, and *The Occurrence of Ground Water in the United States, With a Discussion of Principles*. Among his definitions he proposed the term "phreatophyte" taken from Greek roots meaning a "well plant," which like a water well taps the ground-water supply especially in arid regions in contrast to most plants which derive their water from soil moisture in humid regions. That term, together with many of his logical definitions, continues to be used. In his definitions he explained the significant difference between "porosity" and "effective porosity" and the relation of these terms to specific yield, which many hydrologists failed to recognize.

The need for more precise and comprehensive methods to determine the perennial yield of aquifers led him to devise a quantitative approach. In a report, *Outline of Methods for Estimating Ground-Water Supplies*, Meinzer described twenty-six approaches, eleven of which are applicable, though not exclusively, to aquifers and parts of aquifers under water table conditions. Five of the methods are applicable to aquifers in which water moves considerable distances from intake to discharge areas.

As part of the study of ground-water hydrology, Meinzer established a laboratory, where, along with other experiments and tests, he was able to prove that as long as the flow of water through granular material is laminar, the velocity is directly proportional to the hydraulic gradient—that is, the flow conforms to Darcy's law. For field investigations Meinzer proposed and encouraged development of geophysical methods and such instrumentation as automatic water-stage recorders on wells. He was in the vanguard of those pioneers who urged pumping tests and other analytical tests on wells to obtain quantitative information on the water-bearing properties of aquifers. Among these was the method of Gunter Thiem, which was tested in the field and described by L. K. Wenzel.

The quantitative methods described by Wenzel, and those developed by C. V. Theis, and later C. E. Jacobs and others under Meinzer's supervision, provided additional means for determining the perennial yield of aquifers.

Meinzer also emphasized the need for studying the chemical quality and geochemistry of water, as well as salt-water encroachment in aquifers. Among the research studies on geochemistry were investigations of natural softening of water and the source of some elements, such as fluoride. One of the early reports, prepared by John S. Brown under Meinzer's supervision, introduced to this country the Ghyben-Herzberg formula to estimate the extent of salt-water encroachment in aquifers in which fresh water is in dynamic equilibrium with sea water.

Meinzer recognized that aquifers are functional components of the hydrologic cycle and that ground-water investigations require special skills of the geologist, engineer, physicist, chemist, and others. He pioneered in the teaming of men of these disciplines, in particular geologists and engineers. Beginning about 1930, as the demand for ground-water investigations began to increase rapidly, Meinzer and his assistants trained and supervised dozens of geologists and engineers, many of whom, with that fundamental training, were able to develop more sophisticated tools and techniques.

BIBLIOGRAPHY

I. ORIGINAL WORKS. Meinzer was author or coauthor of more than 100 reports and papers dealing with ground water, as listed on pages 202–206 of a memorial to Oscar Edward Meinzer by A. Nelson Sayre, published in *Proceedings Volume (1948) of the Geological Society of America* (April 1949), 197–206. "The Occurrence of Ground Water in the United States, With a Discussion of Principles," was published in 1923 as *U.S. Geological Survey Water-Supply Paper 489*. The report served as his dissertation for his Ph.D. at the University of Chicago.

His "Outline of Ground-Water Hydrology, With Definitions" was also published in 1923 as *U.S. Geological Survey Water-Supply Paper 494*. His definition of the coefficient of hydraulic permeability as used in hydrologic work of the U.S. Geological Survey, defined and illustrated in his paper "Movements of Ground Water," in *Bulletin of the American Association of Petroleum Geologists*, **20**, no. 6 (1936), is an example of Meinzer's ability to express technical terms so clearly that they can be understood by the layman.

His report "Large Springs in the United States," published in 1927 as *U.S. Geological Survey Water-Supply Paper 557*, continues to serve as a model for later reports on large springs. Meinzer's *Water-Supply Paper 577*, "Plants as Indicators of Ground Water," was published

in the same year. *Water-Supply Paper 640*, prepared under Meinzer's direction and close supervision, is cited in the fourth edition of *Suggestions to Authors of the Reports of the United States Geological Survey* as a model for the preparation of ground-water reports. "Outline of Methods for Estimating Ground-Water Supplies" was published as *U.S. Geological Survey Water-Supply Paper 638-C*, 99–144.

Meinzer's "Ground-Water in the United States, a Summary of Ground-Water Conditions and Resources, Utilization of Water from Wells and Springs, Methods of Scientific Investigation, and Literature Relating to the Subject" was published in 1939 as *U.S. Geological Survey Water-Supply Paper 836-D*, 157–232.

Meinzer was editor and coauthor with twenty-three associates of a book, *Hydrology*, published in the *Physics of the Earth Series*, vol. 9 (New York, 1942). The part of the book written by Meinzer, as with all of his reports, stands out as an example of his excellent, plain, terse, readable style of writing.

Three of his latest papers were (1) "Problems of the Perennial Yield of Artesian Aquifers," in *Economic Geology*, **40** (1945), 159–163; (2) "General Principles of Artificial Ground-Water Recharge," in *Economic Geology*, **41** (1946), 191–201; and (3) "Hydrology in Relation to Economic Geology," *ibid.*, 1–12. His last paper, *Suggestions as to Future Research in Ground-Water Hydrology*, serves as a guide for continuing research.

II. SECONDARY LITERATURE. A biography and a complete bibliography are given by Sayre, in *Proceedings Volume (1948) of the Geological Society of America*, pp. 197–206, cited above. A brief biography by Sayre is published in *Transactions, American Geophysical Union*, **29**, no. 4 (1948), 455–456. A longer discussion of Meinzer is given by O. M. Hackett, "The Father of Modern Ground-Water Hydrology," in *Ground Water*, **3**, no. 2 (April 1965). A brief biography is given in *American Men of Science*. Thirty-one of his U.S. Geological Survey Water-Supply Papers are listed in *Publications of the Geological Survey, U.S. Department of the Interior, 1879–1961*. *U.S. Geological Survey Water-Supply Paper 992* lists his reports up to 1946, and *Water-Supply Paper 1492* gives an annotated bibliography for his later reports. The following U.S. Geological Survey bulletins in *Bibliography of North American Geology* list most of his reports for years indicated: Bull. 746–747 for the period through 1918; Bull. 823, pts. 1 and 2 for years 1919–1928; Bull. 937, pts. 1 and 2 for years 1929–1939; and Bull. 1049, pts. 1 and 2, 1940–1949.

V. T. STRINGFIELD

MEISSNER, GEORG (*b*. Hannover, Germany, 19 November 1829; *d*. Göttingen, Germany, 30 March 1905), *anatomy, physiology*.

Meissner was the son of a senior law-court official, Adolf Meissner. As a student he displayed only average talents, his wish to study medicine becoming strong only during his last years at school. After passing the final examination, he left school in the spring of 1849. He began to study medicine at Göttingen in the summer term of 1849. There he had a fatherly friend and patron in Rudolph Wagner, who was professor of physiology, comparative anatomy, and zoology. While still a student Meissner took an active part in Wagner's investigations in anatomy and, especially, in microscopy. At Göttingen he became a lifelong friend of Theodor Billroth. The two were united by their great love for music and by their interest in microscopic anatomy. In the autumn of 1851 Meissner and Billroth accompanied Wagner on a research expedition to Trieste, in order to investigate the origins and endings of the nerves in the torpedo. Meissner provided the drawings, which were printed by Wagner in his *Icones physiologicae.* The expedition was also concerned with analyzing the electrical organ of the torpedo. At Trieste, Meissner became acquainted with Johannes Müller, whom he esteemed highly. It was also in 1851 that Meissner conducted intensive comparative microscopic investigations on the cells and fibers of the *nervus acusticus.* In 1852 he studied the tactile corpuscles of the skin which today bear his name. The results were first published under the names of Wagner and Meissner; but in Meissner's doctoral thesis the same results were again published, this time under his name alone, as *Beiträge zur Anatomie und Physiologie der Haut* (Leipzig, 1853), and a fierce controversy over priority ensued between Wagner and Meissner.

After finishing his studies in the spring of 1853, Meissner went to Berlin to attend the lectures of Johannes Müller and Lukas Schönlein. In April 1853 he left for Munich to attend the lectures of Siebold, Emil Harless, and Liebig. In August he there received a letter from Wagner, who claimed the discovery of the tactile cells for his own and demanded a public resolution of the matter. Meissner rejected this proposal politely but firmly, and bad feeling between teacher and pupil persisted until 1859.

In 1855, at the age of twenty-six, Meissner was appointed full professor of anatomy and physiology at the University of Basel. Two years later he accepted a professorship of physiology, zoology, and histology at the University of Freiburg im Breisgau. At Freiburg he married the daughter of the mineralogist and poet Franz Ritter von Kobell; they had two sons.

In 1859 Wagner and Meissner were reconciled. Wagner, who until then had held the joint chair of physiology, comparative anatomy, and zoology, turned over his duties in physiology to Meissner, who thus became the first occupant of the separate chair of

physiology at Göttingen. He took office after Easter of 1860 and held the chair until 1901, when he retired because of asthma.

He also was not very sociable. His lectures on physiology, which were illustrated by many experiments, were always well prepared and vivid. Here his talent for drawing, especially for microscopic drawings, served him well.

The number of Meissner's publications is not great. Those composed between 1853 and 1858 dealt chiefly with problems of microscopy, especially as related to the skin (Meissner's tactile cells). In 1857 he described the submucosal nerve plexus of the intestinal wall, which is now called Meissner's plexus. After 1858 he wrote largely on physiological-chemical problems. He was mainly concerned with the nature and the decomposition of protein compounds in the digestive system. The results of his investigations, undertaken alone as well as with collaborators, were published in *Zeitschrift für rationelle Medizin,* edited by his friend Jakob Henle. In 1861 Meissner constructed a new electrometer, a mirror galvanometer. The ensuing electrophysiological investigations led him to propose a new theory concerning the generation of electric potentials through the deformation of biological tissues. This suggestion provoked a devastating critique in 1867 by Emil du Bois-Reymond, the Nestor of electrophysiology. Meissner's experiments on protein also failed to meet with recognition. He was so offended that after 1869 he published nothing more under his own name. His collaborators included Carl Büttner, Friedrich Jolly, Heinrich Boruttau, Otto Weiss, and Karl Flügge. The bacteriologist Robert Koch was among his pupils.

BIBLIOGRAPHY

I. ORIGINAL WORKS. Meissner's memoirs are listed in the Royal Society *Catalogue of Scientific Papers,* IV, 326–327; and VIII, 375. His writings include *Über das Vorhandensein bisher unbekannter eigentümlicher Tastkörperchen (Corpuscula tactus) in den Gefühlswärzchen der menschlichen Haut und über die Endausbreitung sensitiver Nerven* (Göttingen, 1852), written with R. Wagner; *Beiträge zur Anatomie und Physiologie der Haut* (Leipzig, 1853); "Über die Nerven der Darmwand," in *Zeitschrift für rationelle Medizin,* n.s. **8** (1857), 364–366; "Über die Verdauung der Eiweisskörper," *ibid.,* 3rd ser. **7** (1859), 1–26; **8** (1859), 280–303; **10** (1860), 1–32; **12** (1861), 46–67, written with C. Büttner; **14** (1862), 78–96, 303–319, written with L. Thiry; "Zur Kenntnis des elektrischen Verhaltens des Muskels," *ibid.,* **12** (1861), 344–353; "Über das Entstehen der Bernsteinsäure im tierischen Stoffwechsel," *ibid.,* **24** (1865), 97–112, written with F. Jolly; and *Untersuchungen über das Entstehen der Hippursäure* *im tierischen Organismus* (Hannover, 1866), written with C. K. Shepard.

II. SECONDARY LITERATURE. See Heinrich Boruttau, "Zum Andenken an Georg Meissner," in *Pflügers Archiv für die gesamte Physiologie* . . ., **110** (1905), 351–399, with portrait and bibliography; Otto Damsch, "Georg Meissner†," in *Deutsche medizinische Wochenschrift*, **31** (1905), 758–759; Gottfried Müller, *Georg Meissner, sein Leben und seine Werke* (Düsseldorf, 1935), especially for letters to Wagner; Gernot Rath, "Georg Meissners Tagebuch seiner Triestreise (1851)," in *Sudhoffs Archiv für Geschichte der Medizin und der Naturwissenschaften*, **38** (1954), 129–164; and Otto Weiss, "Georg Meissner," in *Münchener medizinische Wochenschrift*, **52** (1905), 1206–1207.

See also Walter von Brunn, ed., *Jugendbriefe Theodor Billroths an Georg Meissner* (Leipzig, 1941).

K. E. ROTHSCHUH

MEITNER, LISE (*b.* Vienna, Austria, 7 November 1878; *d.* Cambridge, England, 27 October 1968), *physics.*

Meitner was the third of eight children of Hedwig Skovran and Philipp Meitner, a lawyer. Although both parents were of Jewish background (and the father was a freethinker), all the children were baptized and Meitner was raised as a Protestant. Her interest in physics apparently began very early, but her parents encouraged her to study for the state examination in French, so that she could support herself as a teacher should the need arise. Meitner passed the examination, then studied privately for the test that permitted her to enter the University of Vienna in 1901. At the university she met with some rudeness from her fellows (a female student then being something of a freak) but was inspired by her teachers, particularly Boltzmann. She was the second woman to receive a doctorate in science from the university; her dissertation, in 1905, was on heat conduction in non-homogeneous materials.

After graduation Meitner remained in Vienna for a time, during which she was introduced to the new subject of radioactivity by Stephan Meyer. Although she then had no notion of making the study of radioactivity her life's work, she did design and perform one of the first experiments to demonstrate that alpha rays are slightly deflected in passing through matter. But she was also interested in theoretical physics; she requested and obtained her father's (very modest) financial support to go to Berlin to study with Planck—for a year or two, as she thought.

Meitner enrolled for Planck's lectures, but had some difficulty finding a place to do experimental work until she met Otto Hahn, who was looking for a physicist to help him in his work on the chemistry of radioactivity. Hahn himself was working at the Chemical Institute, under Emil Fischer, who did not allow women in his laboratory (although two years later, after women's education had been regularized in Berlin, he welcomed Meitner). Hahn and Meitner equipped a carpenter's workshop for radiation measurement and set to work. Hahn, the chemist, was primarily interested in the discovery of new elements and the examination of their properties, while Meitner was more concerned with disentangling their radiations. They were pioneers, and a great deal of their first work was based on false assumptions—such as H. W. Schmidt's idea that beta rays of defined energy follow an exponential absorption law—so that most of their early papers are of largely historical interest.

In 1912 Meitner joined the Kaiser-Wilhelm Institut für Chemie, newly opened in Berlin-Dahlem. World War I interrupted her work; Hahn was called to military service, and Meitner volunteered as a roentgenographic nurse with the Austrian army. On her leaves she went back to Berlin to measure radioactive substances; Hahn's leaves sometimes coincided with hers, so that they could occasionally continue their collaboration. Since in the study of radioactive substances measurements made at fairly long intervals may be desirable to allow some activities to build up and others decay, Hahn and Meitner were able to make a virtue of necessity. By this time they were searching for the still unknown precursor of actinium; they reported their success at the end of the war, naming the new element protactinium.

In 1918 Meitner was appointed head of the physics department of the Kaiser-Wilhelm Institut. She also maintained her rather tenuous connection with the University of Berlin—from 1912 to 1915 she had been Planck's assistant, and after the war she became a docent. Her inaugural lecture was given in 1922; it concerned cosmic physics (reported in the press as "cosmetic physics"). Meitner was appointed extraordinary professor in 1926; she never gave any courses, although she did contribute to the weekly physics colloquia, in which her colleagues included Planck, Einstein, Nernst, Gustav Hertz, and Schroedinger.

Meitner continued her work toward clarifying the relationships between beta and gamma rays. It had by then become clear that while some radioactive substances emitted an electron from the nucleus, others did not. The electrons that these latter substances, the alpha emitters, released must therefore come from the outer shell (as was presumably the case for some of those issued by the true electron emitters), and must therefore be regarded as secondary. There remained the determination of which of the many electron lines then identified were primary electrons from the nucleus. Ellis, at Cambridge, thought that none of them were, but rather that the primary electrons constituted the continuous spectrum that Chadwick had discovered as early as 1914. Meitner disagreed, in the belief that Chadwick's method was inadequate for the discrimination of such a spectrum. Chadwick had counted electrons deflected by a fixed angle in a variable magnetic field; Meitner had always put her faith in photographing electrons. In 1922 she published the measurements that she had made using Danysz' method of focusing the electrons by deflection through 180°. This technique emphasized the narrow electron lines, while the continuous spectrum appeared to be very faint, and Meitner attributed the latter to secondary effects.

Meitner's skepticism was, moreover, a product of her belief that, like alpha particles, primary electrons must form a group of well-defined energy. Her conviction was in the spirit of the quantum theory, which was then being applied to nuclei (largely by Gamow).

If the primary electrons display a continuous spectrum it must be, she thought, because they lose varying amounts of energy in the form of X rays on passing through the strong electric field that surrounds the nucleus, or perhaps in collisions with atomic electrons. The primary energy would then correspond to the highest energy found in the continuous spectrum. In 1927, however, Ellis and Wooster measured the heat generated by electron-emitting nuclei and thus found that each electron gives to its surrounding material an energy equal to the mean energy of the continuous spectrum, not its top energy as Meitner's view demanded. With Wilhelm Orthmann, Meitner immediately set out to check Ellis and Wooster's result; she reported good agreement with their data in the paper that she and Orthmann published jointly in 1929.

The growing evidence for the continuous distribution of energy of primary electrons emitted in beta decay led Pauli to write to Meitner and Geiger a letter in which he proposed the existence of a new neutral particle—later called the neutrino—that should be emitted together with the electron and would at random share the energy available to it. Pauli had to assume that this particle was too elusive to be detected by means then available, and, indeed, effects due to free neutrinos were not actually found until 1956.

Although Meitner's belief in the simplicity of nature had led her astray in regard to the distribution of energy of primary electrons, she was correct in her theory that electron lines were generated from the outer electron shell. She measured the electron lines of actinium to demonstrate that they were produced from the shells of the newly formed—rather than the decaying—nucleus. She thus showed that gamma rays follow upon radioactive transformation, rather than acting as the triggering mechanism for it (as Ellis had suggested). She further observed and correctly interpreted those radiationless transitions in which an electron, on dropping into a vacancy in an inner shell, ejects another electron from the atom, a phenomenon usually named for Auger, who independently described it about two years later in a different context.

Although Meitner never invented a laboratory instrument or technique of her own, she rapidly adopted any new methods developed by others that seemed to her to be useful in her work. For example, she encouraged her student Gerhard Schmidt to make use of Millikan's droplet technique to study the ionization density of alpha particles, and introduced C. T. R. Wilson's cloud chamber—which had been neglected since its invention in 1911—into her Berlin laboratory and applied it in innovative researches (as, for instance, the study of slow electrons, for which she employed the device at greatly lowered pressure). Among her own investigations, she was one of the first (with Phillip, in a paper dated March 1933) to observe and report on positrons formed from gamma rays.

Meitner had accurately measured the attenuation of hard gamma rays in their passage through matter even earlier, when she realized the potential of the new Geiger-Mueller counter for measuring the attenuation of well-collinated, narrow beams of gamma rays. The main purpose of that measurement was to test the Klein-Nishina formula for the Compton effect, and she found good agreement for light elements, up to magnesium. She discovered, however, that attenuation increased with atomic number; she suspected an effect of nuclear structure, perhaps a resonance, and therefore searched for the scattering of gamma rays with unchanged wavelength. (In 1933 it was discovered that the excess attenuation was due to the formation of electron-positron pairs, rather than to scattering.)

In the early 1930's nuclear physics advanced dramatically: the neutron was discovered in 1932, the positron in 1933, and artificial radioactivity in 1934. Meitner and her colleagues published a number of short papers in the light of these rapid developments. In 1934 Meitner resumed work with Hahn to follow up results obtained by Fermi, who had bombarded uranium with neutrons and had found several radioactive products which he thought must be due to a transuranic element since neutron bombardment had invariably led to the formation of a heavier, usually beta-radioactive, isotope of the bombarded element (except for the lightest elements, where a nucleus of lower atomic number might result from the ejection of a charged particle such as a proton or a helium nucleus).

In his investigation of this phenomenon, Hahn discovered several decay products for uranium, some of which might be presumed to be transuranic, with atomic numbers greater than 92. He and Meitner set out to isolate such elements by precipitating an irradiated and acidified uranium salt solution with hydrogen sulfide in order to eliminate all elements between polonium (84) and uranium (92); they assumed that the remaining precipitate must contain only transuranic elements. To be sure, Ida Noddack had suggested that the formation of transuranic elements could not be regarded as proven until it could be established that such elements were not, in fact, identical with any elements between hydrogen and uranium, but her paper was little read and uninfluential. Meitner and Hahn were thus considerably surprised when Irene Curie and Savitch reported irradiating uranium to find a product with penetrating beta rays and a half-life of three-and-one-half hours. Curie further noted that this substance behaved chemically somewhat like thorium. (Hahn and Meitner, using sulfide precipitation, would have missed that.) By implication, then, a uranium nucleus upon being hit by a neutron might emit an alpha particle—a helium nucleus—which seemed unlikely. Later Curie changed her view and pointed to the similarity of her three-and-one-half-hour substance with lanthanum, foreshadowing, but not formulating, the concept of nuclear fission.

Meitner set one of her students, Gottfried von Droste, to look for such alpha particles, but he failed to find them. By studying substances not precipitated as sulfide, Hahn and Strassmann found yet more products, with actinium-like properties and, startlingly, three others with the properties of radium, four places below uranium on the atomic scale.

These results puzzled Meitner, who was unable to reconcile them with nuclear theory. It was at this point, however, that she was forced to interrupt her researches and leave Germany, where the Nazi racial laws had made it increasingly difficult for her to work. Although the Kaiser-Wilhelm Institut was to some degree autonomous, Nazi policies were being enforced even there, and Meitner's situation became critical when the occupation of Austria robbed her of the protection of her foreign nationality. She had never concealed her Jewish origin; her Austrian passport was invalid, and her dismissal from the institute certain. The Dutch physicist Peter Debye communicated (through Scherrer in Zürich) with Dirk Coster at the University of Groningen, and Coster arranged that Meitner be allowed to enter Holland, despite her

lack of papers. No one except Hahn knew that she was leaving Germany for good.

Meitner remained in Holland for only a short time, then went to Denmark, where she was the guest of Niels Bohr and his wife. Although Copenhagen offered her good facilities for research, and although there were a number of younger nuclear physicists working there (including her nephew, O. R. Frisch), Meitner soon chose to accept an invitation from Manne Siegbahn to work in the new Nobel Institute in Stockholm, where a cyclotron was being constructed. Meitner was sixty years old when she went to Sweden; she nonetheless acquired a good command of the language, built up a small research group, and eventually published a number of short papers, most of them on the properties of new radioactive species formed with the cyclotron.

Meitner made her most famous contribution to science shortly after she arrived in Stockholm, however. Worried by Hahn's statement that neutron bombardment of uranium leads to isotopes of radium, she had written to Hahn to ask for irrefutable data concerning the properties of these substances. Her request led Hahn and Strassmann to undertake a series of tests designed to demonstrate that these products were chemically identical to radium, as their earlier investigations had suggested. Hahn wrote to inform her that in these tests, he and Strassmann had found that, like radium, these substances could be precipitated with barium but, surprisingly, were then inseparable from it. They therefore reluctantly concluded that the decay products were isotopes of barium, rather than radium. The evidence for transuranic elements was thus placed in doubt, since sulfide precipitation did not eliminate elements lighter than polonium.

Meitner discussed this news with Frisch. It soon became clear that Bohr's droplet model of the nucleus must provide the clue to understanding how barium nuclei could be formed from uranium nuclei, which are almost twice as heavy. Frisch suggested that the division into two smaller nuclei was made possible through the mutual repulsion of the many protons of the uranium nucleus, making it behave like a droplet in which the surface tension has been greatly reduced by its electric charge. Meitner estimated the difference between the mass of the uranium nucleus (plus the extra neutron with which it had been bombarded) and the slightly smaller total mass of the two fragment nuclei; from this she worked out (by Einstein's mass-energy equivalence) the large amount of energy that was bound to be released. The two mutually repulsed fragments would, indeed, be driven apart with an energy that agreed with her value, so it all fitted.

Meitner and Frisch reported these findings in a joint

paper that described this "nuclear fission" (composed over the telephone, since she was in Stockholm and he had returned to Copenhagen). A few months later they jointly demonstrated experimentally that radioactive fission fragments could be collected on a water surface close to a uranium layer undergoing neutron irradiation. They further showed that the sulfide precipitated from the material so obtained had a decay curve of the same shape as the precipitate derived directly from the irradiated uranium. They concluded that no observable amounts of transuranic elements were produced, capable of affecting their counters.

Meitner had, however, previously demonstrated that one of the products of slow-neutron irradiation of uranium was a uranium isotope of twenty-four minutes half-life; by measuring the resonance cross section she concluded that it was U-239, formed by the capture of a neutron in U-238. Although she realized that its observed beta decay must lead to the formation of a transuranic element, she was not able to observe the very soft radiation of that daughter substance. (Macmillan later found this substance, neptunium; the next generation, plutonium, was the explosive of the first atomic bomb.) Meitner was invited to join the team at work on the development of the nuclear-fission bomb; she refused, and hoped until the very end that the project would prove impossible. Except for a brief note on the asymmetry of fission fragments, she did no more work in nuclear fission.

In 1946 Meitner spent half a year in Washington, D. C., as a visiting professor at Catholic University. In 1947 she retired from the Nobel Institute and went to work in the small laboratory that the Swedish Atomic Energy Commission had established for her at the Royal Institute of Technology. She later moved to the laboratory of the Royal Academy for Engineering Sciences, where an experimental nuclear reactor was being built. In 1960, having spent twenty-two years in Sweden, Meitner retired to Cambridge (England). She continued to travel, lecture, and attend concerts (her love of music was lifelong), but she gradually gave up these activities as her strength ebbed. She died a few days before her ninetieth birthday.

BIBLIOGRAPHY

Meitner's earlier writings include "Über die Zerstreuung der α-Strahlen," in *Berichte der Deutschen physikalischen Gesellschaft*, **8** (1907), 489; "Eine neue Methode zur Herstellung radioaktiver Zerfallsprodukte; Thorium D, ein kurzlebiges Produkt des Thoriums," in *Verhandlungen der Deutschen physikalischen Gesellschaft*, **11** (1909), 55, written with O. Hahn; "Vorträge aus dem Gebiet der Radioaktivität," in *Physikalische Zeitschrift*, **12** (1911), 147; "Magnetische Spektren der β-Strahlen des Radiums," *ibid.*, 1099, written with O. von Baeyer and O. Hahn; "Die Muttersubstanz des Actiniums, ein neues radioaktives Element von langer Lebensdauer," *ibid.*, **19** (1918), 208, written with O. Hahn; "Über die verschiedenen Arten des radioaktiven Zerfalls und die Möglichkeit ihrer Deutung aus der Kernstruktur," in *Zeitschrift für Physik*, **4** (1921), 146; "Die γ-Strahlung der Actiniumreihe und der Nachweis, dass die γ-Strahlen erst nach erfolgtem Atomzerfall emittiert werden," *ibid.*, **34** (1925), 807; "Einige Bemerkungen zur Isotopie der Elemente," in *Naturwissenschaften*, **14** (1926), 719; "Experimentelle Bestimmung der Reichweite homogener β-Strahlen," *ibid.*, 1199; "Über eine absolute Bestimmung der Energie der primären β-Strahlen von Radium E," in *Zeitschrift für Physik*, **60** (1930), 143, written with W. Orthmann; and "Über das Absorptionsgesetz für kurzwellige γ-Strahlung," *ibid.*, **67** (1931), 147, written with H. H. Hupfeld.

Later works are "Die Anregung positiver Elektronen durch γ-Strahlen von Th C″," in *Naturwissenschaften*, **21** (1933), 468, written with K. Philipp; *Kernphysikalische Vorträge am Physikalischen Institut der Eidgenössischen technischen Hochschule* (Berlin, 1936); "Über die Umwandlungsreihen des Urans, die durch Neutronenbestrahlung erzeugt werden," in *Zeitschrift für Physik*, **106** (1937), 249, written with O. Hahn and F. Strassmann; "Künstliche Umwandlungsprozesse bei Bestrahlung des Thoriums mit Neutronen; Auftreten isomerer Reihen durch Abspaltung von α-Strahlen," *ibid.*, **109** (1938), 538, written with O. Hahn and F. Strassmann; "Trans-Urane als künstliche radioaktive Umwandlungsprodukte des Urans," in *Scientia* (Jan. 1938), written with O. Hahn; "Disintegration of Uranium by Neutrons; a New Type of Nuclear Reaction," in *Nature*, **143** (1939), 239, written with O. R. Frisch; "Resonance Energy of the Th Capture Process," in *Physical Review*, **60** (1941), 58; "Spaltung und Schalenmodell des Atomkernes," in *Arkiv för fysik*, **4** (1950), 383—see *Nature*, **165** (1950), 561; "Die Anwendung des Rückstosses bei Atomkernprozessen," in *Zeitschrift für Physik*, **133** (1952), 141; "Einige Erinnerungen an das Kaiser-Wilhelm-Institut für Chemie in Berlin-Dahlem," in *Naturwissenschaften*, **41** (1954), 97; and "Looking Back," in *Bulletin of the Atomic Scientists* (Nov. 1964), 2.

A biography with extensive bibliography is by O. R. Frisch, in *Biographical Memoirs of Fellows of the Royal Society*, **16** (Nov. 1970), 405–420.

O. R. Frisch

MELA, POMPONIUS. See **Pomponius Mela.**

MELLANBY, EDWARD (*b.* West Hartlepool, Durham, England, 8 April 1884; *d.* Mill Hill, London, England, 30 January 1955), *pharmacology*.

For a complete study of his life and work see Supplement.

MELLO, FRANCISCO DE (*b.* Lisbon, Portugal, 1490; *d.* Évora, Portugal, 27 April 1536), *mathematics.*

The son of a nobleman, Manuel de Mello, and Beatriz de Silva, Mello was a protégé of the Portuguese king Manuel I, who sent him to Paris to study. Mello graduated in theology and mathematics; his teacher was Pierre Brissot, who gave him a thorough grounding in the works of Euclid and Archimedes. On his return to Portugal, Mello was appointed tutor to the king's children. He may have served in an official capacity in navigating the Atlantic in order to determine the boundaries of the Spanish and Portuguese territories as defined by the Holy See. He was also to some degree involved in Portuguese politics, and shortly before his death was rewarded with the bishopric of Goa (it is not known whether he actually accepted this post, although it is certain that he never went there).

Mello enjoyed considerable fame as a scientist; as such, Gil Vicente dedicated to him some verses in the introduction to the *Auto da feira.* He was also firmly within the humanistic tradition of his time; his friends included Nicolás Clenard, Juan Luis Vives, and his fellow mathematician Gaspar de Lax. Many of his own works were destroyed by the fire that followed the Lisbon earthquake of 1755. Among his mathematical writings are "De videndi ratione atque oculorum forma," a commentary on Euclid's *Optica*; "De incidentibus in humidis," a commentary on Archimedes' hydrostatics; and an "Elements of Geometry," which would seem to be derived from Jābir ibn Aflaḥ. His nonscientific writings included translations from Latin authors and funerary poems.

Mello should not be confused with the great historian Francisco Manuel de Mello (1611–1667), who also wrote on mathematics.

BIBLIOGRAPHY

On Mello and his work see M. Bataillon, "Erasme et la cour de Portugal," in *Études sur le Portugal au temps de l'humanisme* (Coimbra, 1952), 49–100; Diego Barbosa Machado, *Bibliotheca Lusitana* (Lisbon, 1747), 197–198; Felipe Picatoste y Rodríguez, *Apuntes para una biblioteca científica española del siglo XVI* (Madrid, 1891), 167; Antonio Ribeiro dos Santos, *Memoria da vida e escritos de Don Francisco de Mello*, Memórias de literatura portuguesa publicadas pela Academia Real das sciencias de Lisboa, VII (Lisbon, 1806), 237–249; and Inocencio Francisco da Silva, *Diccionario bibliografico portugues*, III (Lisbon, 1859), 8–10.

JUAN VERNET

MELLONI, MACEDONIO (*b.* Parma, Italy, 11 April 1798; *d.* Portici, Italy, 11 August 1854), *physics.*

Melloni spent his early professional life as a professor of physics at the University of Parma between 1824 and 1831. Political difficulties arising out of the 1830 rebellions forced him to flee to Paris, where he lived without a position until 1839. He returned to Italy to become director of a conservatory of physics in Naples, where, until 1848, he also directed the meteorological observatory on Vesuvius.

As a physicist Melloni was concerned mainly with the properties of radiant heat, or calorific radiation as it was then called. Since 1820 it had been thought, after a suggestion by Ampère, that the rays of light and those of heat are different manifestations of the same process. According to Young and Fresnel light is a transverse disturbance or wave propagated through a ubiquitous medium called the "luminiferous ether." Rays of heat were thought to be modifications of these waves, not very, if at all, different from the rays of light. There were, however, difficulties with this view which centered about the propagation of heat and light through matter. If, it was thought, heat rays and light rays are the same things, then there should be little difference in the ways they are transmitted through matter. Thus an optically transparent body should transmit the heat it intercepts in the same manner it transmits the light that impinges upon it. The problem was that in many instances the effects of radiant heat on matter are very unlike those of light.

Early in his career Melloni became convinced that, while radiant heat is a wave in ether as is light, it is not the same kind of a disturbance as a ray of light. His first experiments were designed to pinpoint the differences between these two kinds of propagation. In 1832, for example, Melloni attempted to show that, contrary to general belief, radiant heat and light are not transmitted in the same amounts through a given transparent body. In his experiment he demonstrated that the quantity of heat a body transmits is not related in any degree to its transparency, that is, to the amount of light which it can transmit. In fact the only relationship he could find was between the body's permeability to heat and its index of refraction (1). This strengthened him in his belief that heat rays and light are both ethereal vibrations but not the same kind of vibration.

Between 1833 and 1840 Melloni continued in his attempts to find the differences between heat and light. His experimental accounts, if read singly, often seem to be showing that heat and light are the same, not that they differ. But Melloni expected to find many similarities—since both are ethereal vibrations. What he sought were the differences in their effects. He showed in 1833 that the same body has different effects on heat and light; radiant heat, for example, is unaffected by the same polarization arrangement that extinguishes light (3). By 1835 Melloni was certain ". . . that light and radiant heat are effects directly produced by two different causes" (4). These causes are both molecular vibrations that set the ether in motion, but the molecules move differently when producing the vibrations of heat than they do when producing those of light. As he continued in his investigations he found that the rays of heat can actually be polarized, but in a way bearing no relation to the effects of the same polarization apparatus on light—further evidence, he thought, of the different origins of these two kinds of ethereal motions (2).

Between 1834 and 1840 Melloni began to concentrate on the behavior of bodies transmitting heat radiation. He was by then certain that heat and light are distinct modes of the same process, ethereal propagation, and he looked for those details of the transfer of heat that distinguished it from that of light. His theoretical distinction between the two kinds of waves aided him because he was always looking for differences and not similarities. For example, it was held that the amount of light that is transmitted by a body depends somehow upon the state of its surface, upon its degree of smoothness or lack of irregularities. Melloni, because he distinguished between heat and light, did not believe that this would be strictly true for heat. In 1839 he tried to show that, if a smooth-surfaced body transmits heat more readily than one with a rough surface, it is not because of some special surface regularity, as it is with light, but because the polishing necessary to make the surface smooth had so altered its elasticity that the rate at which it could transmit the vibrations of heat was also changed (6, 7). The importance of Melloni's work for later generations lies in the detailed investigations he made into the behavior of heat and the new knowledge about it that resulted.

BIBLIOGRAPHY

Melloni's writings include (1) "Expériences relatives à la transmission du calorique rayonnant par divers liquides," in *Bibliothèque universelle des sciences et arts . . .,* **49** (1832), 337–340, written with P. Prevost; (2) "Mémoire sur la transmission libre de la chaleur rayonnante par différens corps solides et liquides," in *Annales de chimie,* **53** (1833), 5–73; (3) "Nouvelles recherches sur la transmission immédiate de la chaleur rayonnante par différens corps solides et liquides," *ibid.,* **55** (1833), 337–397; (4) "Note sur la réflexion de la chaleur rayonnante," *ibid.,* **60** (1835), 402–426; (5) "Mémoire sur la polarisation de la chaleur," *ibid.,* **61** (1836), 375–410; (6) "De la prétendue influence que les aspérités et le poli des surfaces exercent sur le pouvoir émissif des corps," *ibid.,* **70** (1839), 435–444; and (7) "Sur la constance de l'absorption . . .," *ibid.,* **75** (1840), 337–388.

JED Z. BUCHWALD

MELTZER, SAMUEL JAMES (*b.* Ponevyezh, Russia [now Panevezhis, Lithuanian S.S.R.], 22 March 1851; *d.* New York, N.Y., 7 November 1920), *physiology, pharmacology.*

Meltzer was the son of Simon Meltzer, a teacher, and Taube Kowars. The family were orthodox Jews and Samuel's early education was obtained at a rabbinical seminary, but he decided against a religious vocation. After his marriage he studied at the Realgymnasium in Königsberg. He also attempted unsuccessfully to operate a soap-manufacturing business. In 1876 Meltzer entered the University of Berlin, where he studied philosophy and medicine. Under the direction of the physiologist Hugo Kronecker, he pursued experimental studies on the mechanism of swallowing. He received his medical degree in 1882.

Meltzer soon immigrated to the United States, settling in New York City, where he practiced medicine. In order to continue research he made arrangements to have access to laboratory facilities, particularly William Henry Welch's laboratory at Bellevue Hospital. In 1904 he joined the staff of the recently created Rockefeller Institute for Medical Research; he was head of the department of physiology and pharmacology until his retirement in 1919.

Meltzer's intimate acquaintance with both medical practice and research allowed him to serve as a liaison between practitioners and scientific investigators. He played an important role in the founding and early development of several scientific or medical societies, including the Society for Experimental Biology and Medicine—familiarly called the Meltzer Verein for many years. His strong belief in the cosmopolitanism of science led him to organize the Fraternitas Medicorum, an international medical brotherhood, during World War I, and thousands of American medical men joined the organization before its activities were suspended when the United States entered the conflict.

Meltzer also made contributions to pharmacology, pathology, and clinical medicine, as well as to physiology, his major field of interest. He was too inclined to speculate about his experimental results and to seek the general principles that govern physiological phenomena. In his 1882 medical dissertation on the swallowing reflex, he first outlined the theory of inhibition, which influenced much of his subsequent work. During the course of his studies on the act of swallowing, he noted that reflex stimulation of inspiratory muscles is accompanied by reflex inhibition of expiratory muscles, and vice versa. He postulated that this reciprocal arrangement must exist for other antagonistic muscles in the body for the purpose of efficient motor action. In 1893 Charles Sherrington, apparently unaware of Meltzer's work, showed that the contraction of an extensor in the limb is accompanied by a relaxation of its opposing flexor, and vice versa (specifically predicted by Meltzer). Sherrington called this relationship reciprocal innervation.

Meltzer developed the idea of combined action of opposing processes into a general theory. He believed that every excitation or stimulation of a tissue was accompanied by a corresponding inhibitory impulse. Physiological phenomena are a result of the compromise between these two fundamental, antagonistic life forces—excitation and inhibition. Although his dualistic conception of life processes did not gain wide acceptance, it was an important stimulus to his own experimental work.

One of Meltzer's most important experimental studies dealt with the pharmacological effect of magnesium salts. These compounds were shown to produce a state of unconsciousness and muscle relaxation in animals which was readily reversed by the injection of calcium chloride. This work added magnesium to the elements known to play a part in the activity of the cell, and Meltzer believed he had found the element in the body that is especially concerned with inhibition.

Another important series of researches dealt with artificial respiration. Meltzer and John Auer developed the technique of intratracheal insufflation, whereby the lungs are kept inflated by blowing a stream of air through a tube inserted into the trachea. By including an anesthetic vapor in the air stream, anesthetization could be produced at the same time as artificial respiration. The technique was valuable in thoracic surgery as a simple means of keeping the lungs inflated after the chest had been opened.

Meltzer's other significant contributions included the hypothesis that bronchial asthma is a phenomenon of anaphylaxis, the introduction of the engineering term "factors of safety" to describe the reserve powers of organisms, and researches with his daughter Clara on the effects of adrenaline on the blood vessels and on the muscles of the iris.

BIBLIOGRAPHY

I. ORIGINAL WORKS. For a bibliography of Meltzer's publications, see William Howell, "Biographical Memoir, Samuel James Meltzer, 1851–1920," in *Memoirs of the National Academy of Sciences*, Scientific Memoir Series, **21** (1926), 15–23. Meltzer discussed his theory of inhibition in detail in "Inhibition," in *New York Medical Journal*, **69** (1899), 661–666, 699–703, 739–743. The enduring influence of this theory on his work is illustrated in one of his last papers, "The Dualistic Conception of the Processes of Animal Life," in *Transactions of the Association of American Physicians*, **35** (1920), 247–257. His main experimental work on magnesium salts is described in a series of four papers written with John Auer under the general title "Physiological and Pharmacological Studies of Magnesium Salts," in *American Journal of Physiology*, **14-17** (1905–1906). For a description of the technique of intratracheal insufflation, see "The Method of Respiration by Intratracheal Insufflation: Its Scientific Principle and Its Practical Availability in Medicine and Surgery," in *Medical Record*, **77** (1910), 477–483, which is followed by several articles by other authors on the surgical use of this technique. Other important papers include "Bronchial Asthma as a Phenomenon of Anaphylaxis," in *Journal of the American Medical Association*, **55** (1910), 1021–1024; and "The Factors of Safety in Animal Structure and Animal Economy," in *Science*, **25** (1907), 481–498.

II. SECONDARY LITERATURE. The best biography is the memoir by William Howell cited above. See also Howell's memoir in *Science*, n.s. **53** (1921), 99–106; and R. H. Chittenden's article in the *Dictionary of American Biography*, XII (1933), 519–520. George Corner, *A History of The Rockefeller Institute, 1901–1953, Origins and Growth* (New York, 1964), discusses Meltzer and his work—see esp. pp. 117–120. A special supp. to vol. **18** (1921) of *Proceedings of the Society for Experimental Biology and Medicine*, entitled "Memorial Number for Samuel James Meltzer, Founder and First President of the Society for Experimental Biology and Medicine," contains several biographical sketches by colleagues.

JOHN PARASCANDOLA

MELVILL, THOMAS (*b.* Glasgow [?], Scotland, 1726; *d.* Geneva, Switzerland, December 1753), *astronomy, physics.*

Thomas Melvill noted the yellow spectrum of sodium and considered a means of testing a suggested relation between the velocity of light and its color.

Melvill's origins are obscure, but at a relatively late age, in 1748 and 1749, he studied divinity at the University of Glasgow, where he acquired a taste for experimental philosophy from Alexander Wilson. Together they used kites to investigate the change in atmospheric temperature with altitude. Melvill studied Newton's *Opticks* closely. In a paper "Observations on Light and Colours," given to the Medical Society of Edinburgh on 3 January and 7 February 1752, he wrote of his use of a prism for examining color in flames (*Edinburgh Physical and Literary Essays*, II [Edinburgh, 1752], 35). The property of common salt, whereby it turns a flame yellow, was probably well recognized, but Melvill was seemingly the first to treat the coloration in any way quantitatively. He studied the spectrum of burning alcohol into which he introduced in turn sal ammoniac, potash, alum, niter, and sea salt, noting the persistence of the yellow component of the spectrum, and he remarked that this yellow color was of a definite degree of refrangibility. His work appears to have had little influence, and the origins of spectrum analysis are not usually traced back before W. H. Wollaston's discovery of dark solar lines (1802).

As an explanation of the different refrangibilities of light of different colors, in terms of the corpuscular theory, Melvill suggested that the several colored rays were projected with various velocities from the luminous body—the violet with the least. A letter to Bradley, written from Geneva, dated 2 February 1753, and read to the Royal Society on 8 March 1753, pointed out an interesting consequence regarding aberration. Depending on velocity, the aberration would be different for different colors, and the satellites of Jupiter would gradually change color in one way (white to violet) on entering the planet's shadow, and another way (red to white) on leaving the shadow. This effect was not observed, and the suggestion was soon forgotten. Its originality is in some doubt, since Courtivron's *Traité d'optique*, published in 1752 and readily available to Melvill in Geneva, contained not only the fundamental hypothesis but its consequences for the appearance of Jupiter.

Melvill developed the idea further in a letter of 2 June 1753, and suggested that Bradley's observations of aberration revealed the ratio of the velocities of light, not in space and air, but in space and in the humors of the eye. He believed that it would be necessary to reject what he called "Sir Isaac Newton's whole doctrine of refraction by an accelerating or retarding power," if the consequences of his new hypothesis were not confirmed. Dying at the age of twenty-seven, he scarcely lived long enough to be disappointed at the neglect of his letter, which contains essentially the same idea as that adopted by Alexander Wilson's son, Patrick, who long afterwards discussed the consequences for an observer with a water-filled telescope. Conclusions drawn by Wilson and Robison (*Philosophical Transactions of the Royal Society*, **74** [1784], 35) were put to the test by Arago and communicated to the Institut de France in 1810 (*Comptes-rendus hebdomadaires des séances de l'Académie des sciences*, **8** [1839], 326, and **36** [1853], 38).

At the close of his second letter to Bradley, Melvill stated that he had designed and had made in Geneva a timepiece with a conical pendulum, the virtues of which he extolled. His early death deprived physics of a gifted and ingenious experimenter.

BIBLIOGRAPHY

I. ORIGINAL WORKS. Melvill published only the papers cited above. The letter of 2 June 1753 was unpublished until it was printed by S. P. Rigaud in *Miscellaneous Works and Correspondence of the Rev. James Bradley* (Oxford, 1832), 483–487. The letter is now at Oxford, Bodleian Library, MS Bradley 44, f. 112.

II. SECONDARY LITERATURE. On Melvill and his work see Brewster's *Edinburgh Journal of Science, Technology and Photographic Art*, **10** (1829), 5; A. M. Clerke, *History of Astronomy During the Nineteenth Century*, 3rd ed. (London, 1893), 165; and E. T. Whittaker, *History of the Theories of Aether and Electricity*, 2nd ed., I (Edinburgh, 1951), 99, 367.

J. D. NORTH

MENABREA, LUIGI FEDERICO (*b.* Chambéry, Savoy, 4 September 1809; *d.* St. Cassin [near Chambéry], France, 24 May 1896), *structural and military engineering, mathematics.*

Menabrea is known to scientists as one of the most important men in the development of energy methods in the theory of elasticity and structures, and to others as a distinguished general and statesman, each group being generally little aware of Menabrea's accomplishments in the other fields. Indeed, it is remarkable that he was able to make significant contributions in both types of activities.

Menabrea first studied engineering and then mathematics at the University of Turin. Upon graduation he entered the army corps of engineers. When Charles Albert acceded to the throne in 1831, Cavour resigned his army commission, and Menabrea replaced Cavour at the Alpine fortress of Bardo. Menabrea soon left to become professor of mechanics and construction at the Military Academy of the Kingdom of Sardinia at Turin and at the University of Turin. To this early period belongs his exposition and extension of Babbage's invention of a mechanical calculator to be published in 1842.

His political career started at this time. Between the years 1848 and 1859 King Charles Albert entrusted Menabrea with diplomatic missions to Modena and Parma. Menabrea then entered Parliament (where he championed proposals for Alpine tunneling) and was attached successively to the ministries of war and foreign affairs. At the same time he attained the rank of major general and was commander in chief of the army engineers in the Lombard campaigns of 1859. He directed siege and fortification works and also the artifical flooding of the plains between the Dora Baltea and the Sesia rivers to obstruct the Austrian advance.

During this time (1857–1858) Menabrea's early scientific papers were published, in which he gave the first precise formulation of the methods of structural analysis based on the "virtual work principle" earlier examined by A. Dorna. He studied an elastic truss in these papers and enunciated his "principle of elasticity," calling it also "principle of least work." He stated that when an elastic system attains equilibrium under external forces, the work done by the tensions and compressions in the internal members of the system is a minimum.

Menabrea's political and military advance continued. In 1860 he became lieutenant-general, conducted sieges at Ancona, Capua, and Gaeta, was appointed senator, and was granted the title of count. He was minister of the navy under Ricasoli from June 1861 to May 1862 and from January to April 1863 and minister of public works from December 1862 to September 1864 (under Farini and Minghetti). He was named Italian plenipotentiary for the peace negotiations with Austria in 1866. In October 1867 he succeeded Rattazzi as premier, holding simultaneously the portfolio of foreign minister, and remained in these posts in three cabinets until December 1869. During this turbulent period he was faced with the difficult situation created by Garibaldi's invasion of the Papal States. Menabrea issued the famous proclamation of 27 October 1868, in which he disavowed Garibaldi, against whom he instituted judicial proceedings. He protested against the pope's temporal power, insisted on the Italian prerogative of interference in Rome, and contended against infringement of Italian rights in repeated negotiations with Napoleon III and the pope.

In 1868 Menabrea published a new demonstration of his principle of least work, which, although superior to the preceding one, still failed to note the independence of the variations of the internal forces and of the elongations of the members of the structure. This oversight was criticized by Sabbia, Genocchi, and Castigliano, giving rise to a controversy lasting until 1875, which is described in the article on Castigliano. In 1870 Menabrea published jointly with the French mathematician J. L. F. Bertrand (1822–1900) a note that advanced the first valid proof of his principle.

In order to deprive Menabrea of influence as aide-de-camp to King Victor Emmanuel II, and to get him out of the country, Giovanni Lanza, his successor as premier, appointed him ambassador to London, and

in 1882 to Paris. In 1875 he was made marquis of Valdora; he retired from public life in 1892.

Menabrea's place in the history of Italy is assured; his role in the introduction of concepts of work and energy into analytical mechanics and engineering has been overshadowed by the greater fame of Castigliano. In the United States, for example, Menabrea is hardly mentioned, although in Continental and particularly Italian textbooks the correct distinction between Menabrea's and Castigliano's theorems is generally made. Menabrea's methods placed these concepts for the first time very clearly before the engineering profession and thus started the essential work of education which was completed by Castigliano.

BIBLIOGRAPHY

I. Original Works. Menabrea's principal scientific works consist of seven papers, as follows: "Notions sur la machine analytique de Charles Babbage," in *Bibliothèque Universelle de Genève*, n.s. **41** (1842), 352–376; "Principio generale per determinare le tensioni e le pressioni in un sistema elastico," a seminar presented to the Reale Accademia delle Scienze di Torino in 1857, which was then printed as "Nouveau principe sur la distribution des tensions dans les systèmes élastiques," in *Comptes rendus hebdomadaires des séances de l'Académie des sciences*, **46** (1858), 1056. Then followed "Étude de Statique Physique —Principe général pour déterminer les pressions et les tensions dans un système élastique," in *Memorie della Reale Accademia delle scienze di Torino*, 2nd ser., **25** (1868), 141. An abstract of Bertrand's letter to General Menabrea was published jointly by Menabrea and Bertrand in *Atti della Reale Accademia delle scienze*, **5** (1 May 1870), 702.

The last two contributions are the reply to criticism in "Un'ultima lettera sulle peripezie della serie di Lagrange in risposta al Prof. Angelo Genocchi per L. F. Menabrea, A. D. B. Boncompagni," in *Bullettino di bibliografia e di storia delle scienze matematiche e fisiche*, **6** (October 1873), 435, and the memoir which raised the dispute with Castigliano, i.e., "Sulla determinazione delle tensioni e delle pressioni ne sistemi elastici," in *Atti della Reale Accademia dei Lincei*, 2nd ser., **2** (1875), 201.

II. Secondary Literature. The reader is referred to the article on Alberto Castigliano for a listing of pertinent works, and to Menabrea's autobiography, covering the years up to 1871, published as *Memorie*, L. Briguglio and L. Bulferetti, eds. (Florence, 1971).

Bruno A. Boley

MENAECHMUS (*fl.* Athens and Cyzicus, middle of fourth century B.C.), *mathematics.*

In the summary of the history of Greek geometry given by Proclus, derived at this point from Eudemus, it is stated that "Amyclas of Heraclea, one of the friends of Plato, and Menaechmus, a pupil of Eudoxus and associate of Plato, and his brother Dinostratus made the whole of geometry still more perfect."[1] There is no reason to doubt that this Menaechmus is to be identified with the Manaechmus who is described in the *Suda Lexicon* as "a Platonic philosopher of Alopeconnesus, or according to some of Proconnesus, who wrote works of philosophy and three books on Plato's *Republic*."[2] Alopeconnesus was in the Thracian Chersonese, and Proconnesus (the Island of Marmara) was in the Propontis (the Sea of Marmara), no great distance from it; and both were near Cyzicus (Kapidaği Yarimadasi, Turkey), where Eudoxus took up his abode and where Helicon, another pupil, was born.[3] This dating of Menaechmus, about the middle of the fourth century B.C., accords with an agreeable anecdote reproduced by Stobaeus from the grammarian Serenus; when Alexander the Great requested Menaechmus to teach him geometry by an easy method, Menaechmus replied: "O king, for traveling through the country there are private roads and royal roads, but in geometry there is one road for all."[4] A similar story is told of Euclid and Ptolemy I;[5] but it would be natural to transfer it to the more famous geometer, and the attribution to Menaechmus is to be preferred. If true, it would suggest that Menaechmus was the mathematical tutor of Alexander. He could have been introduced to Alexander by Aristotle, who had close relations with the mathematicians of Cyzicus.[6] A phrase used by Proclus in two places—οἱ περὶ Μέναιχμον μαθηματικοί—implies that Menaechmus had a school;[7] and Allman has argued cogently that this was the mathematical school of Cyzicus, of which Eudoxus and Helicon (probably) were heads before him and Polemarchus and Callippus after him.[8]

According to Proclus, Menaechmus differentiated between two senses in which the word στοιχεῖον, "element," is used.[9] In one sense it means any proposition leading to another proposition, as Euclid I.1 is an element in the proof of I.2, or I.4 is in that of I.5; and in this sense propositions may be said to be elements of each other if they can be established reciprocally—for example, the relation between the sum of the interior angles of a rectilineal figure and the sum of the exterior angles. In the second sense an element is a simple constituent of a composite entity, and in this sense not every proposition is an element but only those having a primordial relation to the conclusion, as the postulates have to the theorems. As Proclus notes, this is the sense in which "element" is used by Euclid, and Menaechmus may have helped to fix this terminology.

In another passage Proclus shows that many so-called conversions of propositions are false and are not properly called conversions, that is, not every converse of a proposition is true.[10] As an example he notes that every hexagonal number is triangular but not every triangular number is hexagonal, and he adds that these matters have not escaped the notice of the mathematicians in the circle of Menaechmus and Amphinomus.

In yet another passage Proclus discusses the division of propositions into problems and theorems.[11] While the followers of Speusippus and Amphinomus held that all propositions were theorems, the school of Menaechmus maintained that they were all problems but that there were two types of problems: at one time the aim is to find the thing sought, at another to see what some definite thing is, or to what kind it belongs, or what change it has undergone, or what relation it has to something else. Proclus considers that both schools were right; it might be argued with equal justice that both were wrong and that the distinction between theorem and problem is valid.

It is clear from these references that Menaechmus gave much attention to the philosophy and technology of mathematics. He must also have applied himself to mathematical astronomy, for Theon of Smyrna records that Menaechmus and Callippus introduced the system of "deferent" and "counteracting" spheres into the explanation of the movements of the heavenly bodies (οἱ τὰς μὲν φερούσας, τὰς δὲ ἀνελιττούσας εἰσηγήσαντο).[12] The terms mean that one of the spheres bears the heavenly body; the other corrects its motion so as to account for the apparent irregularities of their paths. Eudoxus was the first to devise a mathematical model to explain the motions of the sun and planets, and he did so by a highly ingenious system of concentric spheres, the common center being the center of the earth. The sun, moon, and planets were each regarded as fixed on the equator of a moving sphere; the poles of that sphere were themselves borne round on a larger concentric sphere moving about two different poles with a different speed; and so on. For the sun and moon Eudoxus postulated three spheres; for the planets, four. The modifications in this system made by Callippus are known in some detail. For example, he added one sphere for each planet except Jupiter and Saturn and two spheres for the sun and the moon—five in all. Nothing more is known of Menaechmus' contribution than what Theon relates, but he would appear to have been working on the same lines as Callippus. T. H. Martin conjectured that Menaechmus made his contribution in his commentary on Plato's *Republic* when dealing with the passage on the distaff of the Fates.[13]

It is not, however, on these achievements but on the discovery of the conic sections that the fame of Menaechmus chiefly rests. Democritus had speculated on plane sections of a cone parallel to the base and very near to each other,[14] and other geometers must have cut the cone (and cylinder) by sections not parallel to the base; but Menaechmus is the first who is known to have identified the resulting sections as curves with definite properties.

The discovery was a by-product of the search for a method of duplicating the cube. Hippocrates had shown that this could be reduced to the problem of finding two mean proportionals between two lines, and Menaechmus showed that the two means could be obtained by the intersection of a parabola and a hyperbola. His solution is given in a collection of such solutions preserved by Eutocius in his commentary on Archimedes' *On the Sphere and the Cylinder*.[15] Another of the solutions, by Eratosthenes, is introduced by a letter purporting to be from Eratosthenes to Ptolemy Euergetes.[16] The letter is spurious, but it quotes a genuine epigram by Eratosthenes written on a votive pillar to which was attached a device for effecting the solution mechanically. The epigram included the lines:[17]

> Try not to do the difficult task of the cylinders of Archytas, or to cut the cones in the triads of Menaechmus or to draw such a pattern of lines as is described by the god-fearing Eudoxus.

Proclus, in a passage derived from Geminus, also attributes the discovery of the conic sections to Menaechmus and cites a line from the verses of Eratosthenes in the form $M\grave{\eta}$ $\delta\grave{\epsilon}$ $M\epsilon\nu\alpha\iota\chi\mu\acute{\iota}ous$ $\kappa\omega\nu\sigma\tau\sigma$-$\mu\epsilon\hat{\iota}\nu$ $\tau\rho\iota\acute{\alpha}\delta\alpha s$.[18] He notes again in his commentary on Plato's *Timaeus* that Menaechmus solved the problem of finding two means by "conic lines" but says that he prefers to transcribe Archytas' solution.[19]

Eratosthenes' epigram implies not only that Menaechmus was aware of the conic sections but that he was aware of all three types and saw them as sections of a cone—that is, not as plane curves that he later identified with sections of a cone. The proof itself shows also that he knew the properties of the asymptotes of a hyperbola,[20] at least of a rectangular hyperbola, which is astonishing when it is remembered that Apollonius does not introduce the asymptotes until his second book, after the properties of the diameter and ordinates have been proved. There are no signs of any knowledge of the conic sections before Menaechmus, but with him it suddenly blossomed forth into full flower.[21]

The proof as we have it cannot reproduce the words of Menaechmus himself and no doubt has been recast

by Eutocius in his own language, or by someone earlier.[22] It uses the terms $\pi\alpha\rho\alpha\beta o\lambda\acute{\eta}$ and $\acute{\upsilon}\pi\epsilon\rho\beta o\lambda\acute{\eta}$, although these words were first coined by Apollonius; and we have the evidence of Geminus, as transmitted by Eutocius, that "the ancients" ($o\acute{\iota}$ $\pi\alpha\lambda\alpha\iota o\acute{\iota}$) used the names "section of a right-angled cone" for the parabola, "section of an obtuse-angled cone" for the hyperbola, and "section of an acute-angled cone" for the ellipse.[23] This is undeniable evidence that at the time of "the ancients" the three curves were conceived as sections of three types of cone. But how ancient were "the ancients"? Pappus gives a similar account to that of Geminus but says these names were given by Aristaeus;[24] and there is some reason to believe that the name used by Menaechmus for the ellipse was $\theta\upsilon\rho\epsilon\acute{o}s$, because its oval shape resembled a shield.[25] The question of name is not so important as the question behind it: whether Menaechmus discovered his curves as sections of cones or whether he investigated them as plane curves, which were only later (by Aristaeus?) identified with the curves obtained by plane sections of a cone. It will be necessary to return to this question later.

The term $\dot{\alpha}\sigma\acute{\upsilon}\mu\pi\tau\omega\tau o\iota$, employed by Eutocius, would also not have been used by Menaechmus, who probably used the expression $\alpha\acute{\iota}$ $\check{\epsilon}\gamma\gamma\iota\sigma\tau\alpha$ $\epsilon\grave{\upsilon}\theta\epsilon\hat{\iota}\alpha\iota$ $\tau\hat{\eta}s$ $\tau o\hat{\upsilon}$ $\dot{\alpha}\mu\beta\lambda\upsilon\gamma\omega\nu\acute{\iota}o\upsilon$ $\kappa\acute{\omega}\nu\sigma\upsilon$ $\tau o\mu\hat{\eta}s$, or simply $\alpha\acute{\iota}$ $\check{\epsilon}\gamma\gamma\iota\sigma\tau\alpha$, which is found in Archimedes, who also employed the old names for the sections. Other terms that Menaechmus would not have used are $\check{\alpha}\xi\omega\nu$, "axis," and $\dot{o}\rho\theta\acute{\iota}\alpha$ $\pi\lambda\epsilon\upsilon\rho\acute{\alpha}$, or *latus rectum*.

By way of introduction to Menaechmus' proof it may be pointed out that if x, y are two mean proportionals between a, b, so that

$$a : x = x : y = y : b,$$

then

$$x^2 = ay$$

and

$$xy = ab.$$

These are easily recognized today as the equations of a parabola referred to a diameter and a tangent at its extremity as axes and the equation of a hyperbola referred to its asymptotes as axes; the means may therefore be obtained as the intercepts on the axes of a point of intersection of a parabola and hyperbola, but Menaechmus had to discover *ab initio* that there were such curves and to ascertain their properties.

He proceeded by way of analysis and synthesis.

Suppose the problem solved. Let a, b be the given straight lines and x, y the mean proportionals—where the letters both indicate the lines and are a measure of

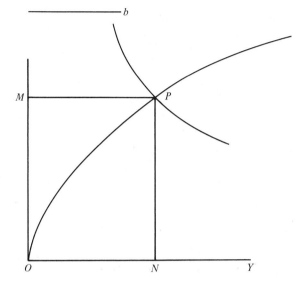

FIGURE 1

and let the perpendiculars PM, PN be drawn. Then by the property of the parabola

$$PN^2 = a \cdot ON,$$

that is,

$$a : PN = PN : ON,$$

and by the property of the hyperbola

$$ab = PN \cdot PM$$
$$= PN \cdot ON.$$

Therefore

$$a : PN = ON : b,$$

and

$$a : PN = PN : ON = ON : b.$$

Let a straight line x be drawn equal to PN and a straight line y equal to ON. Then a, x, y, b are in continuous proportion.

This solution is followed in the manuscripts of Eutocius by another solution introduced with the word ″$A\lambda\lambda\omega s$, "Otherwise," in which the two means are obtained by the intersection of two parabolas.

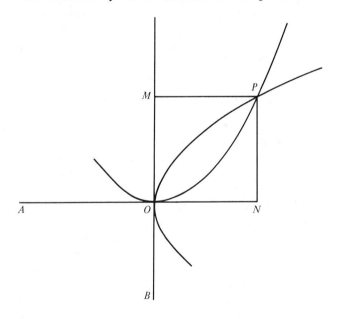

FIGURE 2

In the figure, AO, BO are the two given straight lines, the two parabolas through O intersect at P, and it is easily shown that

$$BO : ON = ON : OM = OM : OA,$$

or

$$a : x = x : y = y : b.$$

their length—so that $a : x = x : y = y : b$. On a straight line OY given in position and terminating at O, let $ON = y$ be cut off, and let there be drawn perpendicular to it at N the straight line $PN = x$. Because $a : x = x : y$, it follows that $ay = x^2$ or $a \cdot ON = x^2$, that is, $a \cdot ON = PN^2$, so that P lies on a parabola through O. Let the parallels PM, OM be drawn. Since xy is given, being equal to ab, $PM \cdot PN$ is also given; and therefore P lies on a hyperbola in the asymptotes OM, ON. P is therefore determined as the intersection of the parabola and hyperbola.

In the synthesis the straight lines a, b are given, and OY is given in position with O as an end point. Through O let there be drawn a parabola having OY as its axis and *latus rectum* a. Then the squares on the ordinates drawn at right angles to OY are equal to the rectangle contained by the *latus rectum* and the abscissa. Let OP be the parabola, let OM be drawn perpendicular to OY, and in the asymptotes OM, OY let there be drawn a hyperbola such that the rectangle contained by the straight lines drawn parallel to OM, ON is equal to the rectangle contained by a, b (that is, $PM \cdot PN = ab$). Let it cut the parabola at P,

271

The proof is established by analysis and synthesis as in the first proof, and it corresponds to the equations

$$x^2 = ay$$
$$y^2 = bx.$$

It has hitherto been assumed by all writers on the subject that this second proof is also by Menaechmus, but G. J. Toomer has discovered as proposition 10 of the Arabic text of Diocles' *On Burning Mirrors* a solution of the problem of two mean proportionals by the intersection of two parabolas with axes at right angles to each other, and with *latera recta* equal to the two extremes, which looks remarkably like the second solution;[26] and it is followed as proposition 11 by another solution which is identical in its mathematical content with that attributed to Diocles by Eutocius. Toomer believes that the second solution should therefore be attributed to Diocles, not to Menaechmus. A final judgment must await publication of his edition of Diocles, but it may at once be noted that there are differences as well as resemblances. In particular, in the Arabic text Diocles starts from the focus-directrix property of the parabola—of which Menaechmus shows no awareness—and in order to get his means deduces from it the property that the ordinate at any point is a mean proportional between the abscissa and the *latus rectum*. It could be that Diocles found his solution independently, or he may have made a conscious adaptation of Menaechmus' solution in order to start from the focus-directrix property.

C. A. Bretschneider first showed how Menaechmus could have investigated the curves, and his suggestion has been generally followed.[27] In a semicircle the perpendicular from any point on the circumference to the diameter is a mean proportional between the segments of the diameter. This property would have been familiar before Menaechmus, and Bretschneider thinks it probable that he would have sought some similar property for the conic sections. We know from Geminus, as transmitted by Eutocius, that "the ancients" generated the conic sections by a plane section at right angles to one side (generator) of the cone, getting different curves according to whether the cone was right-angled, obtuse-angled, or acute-angled.[28] If *ABC* is a right-angled cone and *DEF* is a plane section at right angles at *D* to the generator *AC*, the resulting curve where the plane intersects the cone is a parabola. Let *J* be any point in *DE*, and through *J* let there be drawn a plane parallel to the base of the cone. It will cut the cone in a circle. Let it meet the parabola at *K*. The planes *DEF* and *HKG* are both perpendicular to the plane *BAC*, and their line of

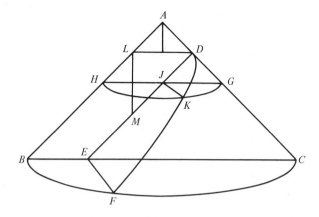

FIGURE 3

intersection *JK* is thus perpendicular to the diameter *HG*. Therefore,

$$JK^2 = HJ \cdot JG$$
$$= LD \cdot JG$$
$$= DJ \cdot DM,$$

because *JDG* and *DLM* are similar triangles. That is to say, the square on the ordinate of the parabola is equal to the rectangle contained by the abscissa and a given straight line *(latus rectum)*, which is the fundamental property of the curve. Bretschneider demonstrates in similar manner the corresponding properties for the ellipse and hyperbola.

Despite Eratosthenes' epigram, the clear statement of Geminus, and the evidence of the early names, it has been doubted whether Menaechmus first obtained the curves as sections of a cone. Charles Taylor suggests that they were discovered as plane loci in investigations of the problem of doubling the cube.[29] In support he argues that Menaechmus used a machine for drawing conics, that in his solutions he uses only the parabola and hyperbola, and that the ellipse—the most obvious of the sections of a cone— is treated last by Apollonius; but he agrees that the conception of a conic as a plane locus was immediately lost. If it be the case that such names as "section of a right-angled cone" were introduced by Aristaeus after the time of Menaechmus, this raises a slight presumption that Menaechmus did not obtain the curves as sections of a cone; but it can hardly outweigh the evidence of Eratosthenes and Geminus.[30]

Allman believes that Menaechmus was led to his discovery by a study of Archytas' solution of the problem of doubling the cube. "In the solution of Archytas the same conceptions are made use of and the same course of reasoning is pursued, which, in the hands of his successor and contemporary Menaechmus, led to the discovery of the three conic sections."[31] This is more than likely. The brilliant solution of Archytas must have made a tremendous splash in the mathematical pool of ancient Greece.

If it be granted that Menaechmus knew how to obtain a hyperbola by a section of an obtuse-angled cone perpendicular to a generator, how did he obtain the rectangular hyperbola required for his proof? H. G. Zeuthen showed how this could be done.[32] In Figure 4, TKC is a plane section through the axis

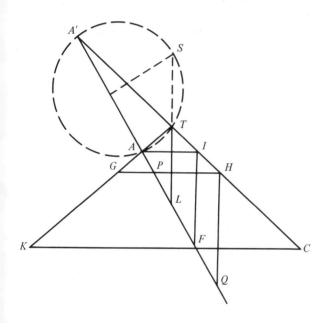

FIGURE 4

of an obtuse-angled cone, AP is a perpendicular to the generator TK, and a plane section through A parallel to the base meets TC at I. If P is the foot of an ordinate to the hyperbola with value y, then

$$y^2 = GP \cdot PH$$
$$= AP \cdot PQ$$
$$= AP \cdot \frac{AF}{A'A} \cdot A'P$$
$$= \frac{2AL}{A'A} \cdot x \cdot x',$$

where $AP = x$ and $A'P = x'$.

The hyperbola will be rectangular if $A'A = 2AL$. The problem is therefore as follows: Given a straight line $A'A$, and AL along $A'A$ produced equal to

$A'A/2$ to find a cone such that L is on its axis and the section through AL perpendicular to the generator through A is a rectangular hyperbola with $A'A$ as transverse axis. That is to say, the problem is to find a point T on the straight line through A perpendicular to $A'A$ such that TL bisects the angle that is the supplement of $A'TA$. Suppose that T has been found. The circle circumscribing the triangle $A'AT$ will meet LT produced in some point S; and because the angle $A'AT$ is right, $A'T$ is its diameter. Therefore $A'SL$ is right and S lies on the circle having $A'L$ as its diameter. But

$$\angle AA'S = \text{supplement of } \angle ATS$$
$$= \angle ATL$$
$$= \angle LTC$$
$$= \angle A'TS,$$

whence it follows that the segments AS, $A'S$ are equal and S lies on the perpendicular to the midpoint of $A'A$. Therefore S is determined as the intersection of the perpendicular to the midpoint of $A'A$ with the circle drawn on $A'L$ as diameter; and if SL is drawn, T, the vertex of the cone, is obtained as the intersection of SL with the perpendicular to $A'A$ at A.

Some writers, such as Allman, have doubted whether Menaechmus could have been aware of the asymptotes of a hyperbola;[33] but unless it is held that Eutocius rewrote Menaechmus' proof so completely that it really ceased to be Menaechmus, the evidence is compelling. It is easy to see (again following

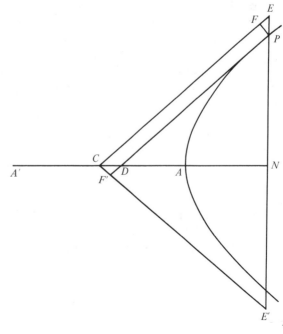

FIGURE 5

Zeuthen) how in the case of a rectangular hyperbola Menaechmus could have deduced the asymptote property from the axial property without difficulty.[34]

Let AA' be the transverse axis of a rectangular hyperbola, and CE, CE' its asymptotes meeting at right angles at C. Let P be any point on the curve and N the foot of the perpendicular to AA' (the principal ordinate). Let PF, PF' be drawn perpendicular to the asymptotes. Then

$$CA^2 = CN^2 - PN^2, \text{ by the axial property}$$
$$= CN \cdot NE - PN \cdot ND$$
$$= 2(\triangle CNE - \triangle PND)$$
$$= 2 \text{ quadrilateral } CDPE$$
$$= 2 \text{ rectangle } CF'PF, \text{ since } \triangle PEF = \triangle CDF',$$
$$= 2PF \cdot PF'.$$

$\therefore PF \cdot PF' = \frac{1}{2}CA^2$, which is the asymptote property.

Alternatively,

$$CA^2 = CN^2 - PN^2$$
$$= EN^2 - PN^2$$
$$= (EN - PN)(EN + PN)$$
$$= (EN - PN)(PN + NE')$$
$$= EP \cdot PE'$$
$$= \sqrt{2}PF \cdot \sqrt{2}PF', \text{ because } \angle PEF \text{ is } 45°.$$

$\therefore PF \cdot PF' = \frac{1}{2}CA^2.$

The letter of the pseudo-Eratosthenes to Ptolemy Euergetes says that certain Delians, having been commanded by an oracle to double one of their altars, sent a mission to the geometers with Plato in the Academy. Archytas solved the problem by means of half-cylinders, and Eudoxus by means of the so-called curved lines. Although they were able to solve the problem theoretically, none of them except Menaechmus was able to apply his solution in practice—and Menaechmus only to a small extent and with difficulty.[35] According to Plutarch, Plato censured Eudoxus, Archytas, Menaechmus, and their circle for trying to reduce the doubling of the cube to mechanical devices, for in this way geometry was made to slip back from the incorporeal world to the things of sense.[36]

Despite this emphatic evidence, Bretschneider considers it doubtful whether Menaechmus had an instrument for drawing his curves.[37] He notes that it is possible to find a series of points on each curve but agrees that this is a troublesome method of obtaining a curve without some mechanical device. Allman develops this hint and believes that by the familiar Pythagorean process of the "application" ($\pi\alpha\rho\alpha\beta\omega\lambda\acute{\eta}$) of areas, which later gave its name to the parabola,

Menaechmus could have found as many points as he pleased—"with the greatest facility"—on the parabola $y^2 = px$; that, having solved the Delian problem by the intersection of two parabolas, he later found it easier to employ one parabola and the hyperbola $xy = a^2$, "the construction of which by points is even easier than that of the parabola"; and that this was the way by which in practice he drew the curves.[38] He also implies that this was what the pseudo-Eratosthenes and Plutarch had in mind. The evidence, however, seems inescapable that Menaechmus attempted to find some mechanical device for tracing the curves. Bretschneider's objection that no trace of any such instrument has survived is not substantial. Centuries later, Isidorus of Miletus is said to have invented a compass, $\delta\iota\alpha\beta\acute{\eta}\tau\eta\varsigma$, for drawing the parabola in Menaechmus' first solution.[39] Every schoolchild knows, of course, how to draw the conic sections with a ruler, string, and pins;[40] but this easy method was not open to Menaechmus, since it depends upon the focus-directrix property.

There is a possible solution to this dilemma, so simple that apparently it has not hitherto been propounded, although Heath came near to doing so. In Eutocius' collection of solutions to the problem of doubling the cube is a mechanical solution attributed to Plato.[41] It is now universally agreed that it cannot be by Plato because of his censure of mechanical solutions, which fits in with his whole philosophy. M. Cantor, however, thought it possible that he worked it out in a spirit of contempt, just to show how

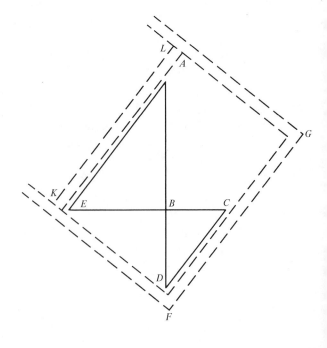

FIGURE 6

easy such things were in comparison with the real business of the philosopher.[42] The lines between which it is desired to find two means are placed at right angles, as *AB*, *BC*. The dotted figure *FGLK* is an instrument in which a ruler *KL* moves in slots in the two vertical sides so as to be always parallel to the base *FG*. The instrument is moved so that *FG* is made to pass through *C*, and *F* lies on *AB* produced. The ruler is then moved so that *KL* passes through *A*. If *K* does not then lie on *CB* produced, the instrument is manipulated until the four following conditions are all fulfilled: *FG* passes through *C*; *F* lies on *AB* produced; *KL* passes through *A*; *K* lies on *CB* produced. The conditions can be satisfied with difficulty—$\delta\upsilon\sigma\chi\epsilon\rho\hat{\omega}\varsigma$, as the pseudo-Eratosthenes says —and when it is done,

$$AB : BE = BE : BD = BD : BC,$$

so that *EB*, *BD* are the required means.

The arrangement of the extremes and the means in Figure 6 is exactly the same as in the second solution attributed to Menaechmus. "Hence," says Heath, "it seems probable that someone who had Menaechmus' second solution before him worked to show how the same representation of the four straight lines could be got by a mechanical construction as an alternative to the use of conics."[43] But why not Menaechmus himself? If he was the author, it would be easy for the tradition to refer it to his master, Plato. This cannot be proved or disproved, but it would be the simplest explanation of all the facts.

NOTES

1. Proclus, *In primum Euclidis*, G. Friedlein, ed. (Leipzig, 1873; repr. Hildesheim, 1967), 67.8–12. An English trans. is Glenn R. Morrow, *Proclus: A Commentary on the First Book of Euclid's Elements* (Princeton, 1970), 55–56. For Dinostratus see *Dictionary of Scientific Biography*, IV, 103–105.
2. *Suda Lexicon*, A. Adler, ed., *M*. No. 140, I, pt. 3 (Leipzig, 1933), 317–318. It is entirely in character that the *Suda* not only misspells Menaechmus' name but omits his most important achievement. The *Suda* is followed by Eudocia, *Violarium*, No. 665, H. Flach, ed. (Leipzig, 1880), p. 494.3–5.
3. Also Athenaeus of Cyzicus, if that is the correct interpretation of the name in Proclus, *op. cit.*, 67.16, as seems probable, but it could possibly be understood as Cyzicinus of Athens.
4. Stobaeus, *Anthologium*, C. Wachsmuth, ed., II (Leipzig, 1884), 228.30–33.
5. Proclus, *op. cit.*, 68.13–17.
6. G. J. Allman, *Greek Geometry From Thales to Euclid* (Dublin–London, 1889), 154, n. 2, 179 and n. 42.
7. *Op. cit.*, 78.9, 254.4.
8. *Op. cit.*, 171–172.
9. *Op. cit.*, 72.23–73.14. This passage is subjected to an elaborate analysis by Malcolm Brown in "A Pre-Aristotelian Mathematician on Deductive Order," in *Philosophy and Humanism: Essays in Honor of Paul Oskar Kristeller* (New

York, in press). Brown sees Menaechmus as the champion of the relativity of mathematical principles and Aristotle as the champion of their absolute character.
10. *Ibid.*, 253.16–254.5.
11. *Ibid.*, 77.6–79.2.
12. *Liber de astronomia*, T. H. Martin, ed. (Paris, 1849; repr. Groningen, 1971), 330.19–332.3; *Expositio rerum mathematicarum ad legendum Platonem utilium*, E. Hiller, ed. (Leipzig, 1878), 201.22–202.2.
13. *Liber de astronomia*, "Dissertatio," 59–60; *Republic*, X, 616–617.
14. Plutarch, *De communibus notitiis contra Stoicos* 39.3, 1079E, M. Pohlenz and R. Westman, eds., in *Plutarchi Moralia*, VI, fasc. 2 (Leipzig, 1959), 72.3–11. Plutarch writes on the authority of Chrysippus.
15. *Commentarii in libros De sphaera et cylindro*, in *Archimedis opera omnia*, J. L. Heiberg, ed., 2nd ed., III (Leipzig, 1915; repr. Stuttgart, 1972), 54.26–106.24.
16. For Eratosthenes there is now available P. M. Fraser, *Eratosthenes of Cyrene* (London, 1971), which was the 1970 "Lecture on a Master Mind" of the British Academy.
17. Eutocius, *op. cit.*, 96.16–19; $\delta\iota\zeta\acute{\eta}\sigma\eta$ is the conjecture of Ulrich von Wilamowitz-Moellendorff for the solecism $\delta\iota\zeta\eta\alpha\iota$ of the MS.
18. *Op. cit.*, 111.20–23.
19. *In Platonis Timaeum ad 32A, B*, E. Diehl, ed. (Leipzig, 1914), pp. 33.29–34.4. The promise to transcribe Archytas is not redeemed in this work.
20. Allman, *op. cit.*, is skeptical on this point. "Menaechmus may have discovered the asymptotes; but, in my judgement, we are not justified in making this assertion, on account of the fact, which is undoubted, that the solutions of Menaechmus have not come down to us in his own words" (p. 170). "There is no evidence, however, for the inference that Menaechmus . . . knew of the existence of the asymptotes of the hyperbola, and its equation in relation to them" (p. 177).
21. The first historian of Greek mathematics, J. E. Montucla, was deeply impressed by this fact. Writing of the proof by means of the parabola and hyperbola between asymptotes he notes, "Cette dernière montre même qu'on avoit fait à cette époque quelque chose de plus que les premiers pas dans cette théorie" (*Histoire des mathématiques*, I [Paris, 1758], 178). And again, "On ne peut y méconnoître une théorie déjà assez sçavante de ces courbes" (p. 183).
22. This appears to have been first recognized by N. T. Reimer, *Historia problematis de cubi duplicatione* (Göttingen, 1798), 68, n.
23. *Commentaria in Conica*, in *Apollonii Pergaei quae . . . exstant . . .*, J. L. Heiberg, ed., II (Leipzig, 1893), 168.17–170.27.
24. *Collectio*, F. Hultsch, ed., II (Berlin, 1877), VII.30–31, pp. 672.20–24, 674.16–19; Hultsch attributes the second passage to an interpolator.
25. In the following passages of Greek authors $\theta\upsilon\rho\epsilon\acute{o}\varsigma$ and $\acute{\epsilon}\lambda\lambda\epsilon\iota\psi\iota\varsigma$ are used interchangeably: Eutocius, *Commentaria in Conica*, Heiberg, ed., II, 176.6; Proclus, *In primum Euclidis*, Friedlein, ed., 103.6, 9, 10, 111.6 (citing Geminus), 126.19, 20–21, 22. The name appears also to have been familiar to Euclid, for in the preface to the *Phaenomena*, in *Euclidis opera omnia*, J. L. Heiberg and H. Menge, eds., VIII (Leipzig, 1916), 6.5–7, he says: "If a cone or cylinder be cut by a plane not parallel to the base, the section is a section of an acute-angled cone which is like a shield ($\theta\upsilon\rho\epsilon\acute{o}\varsigma$)." From such passages Heiberg concluded that $\theta\upsilon\rho\epsilon\acute{o}\varsigma$ was the term used for the ellipse by Menaechmus ("Nogle Bidrag til de graeske Mathematikeres Terminologi," in *Philologisk-historiske Samfunds Mindeskrift*, XXVI [Copenhagen, 1879], 7; *Litterärgeschichtliche Studien über Euklid* [Leipzig, 1882], 88). The primary meaning of $\theta\upsilon\rho\epsilon\acute{o}\varsigma$ is "stone put against a door" (to keep it shut)—so H. G. Liddell and R. Scott, *A Greek-English Lexicon*, new ed.,

H. Stuart Jones (Oxford, 1940)—whence it comes to mean "oblong shield" (shaped like a door).

26. Dublin, Chester Beatty Library, Chester Beatty MS Arabic no. 5255, fols. 1–26.

27. *Die Geometrie und die Geometer vor Euklides* (Leipzig, 1870), 157–158.

28. *Commentaria in conica*, Heiberg, ed., II, 168.17–170.18.

29. *Introduction to the Ancient and Modern Geometry of Conics* (Cambridge, 1881), xxxi, xxxiii, xliii.

30. There is a similar uncertainty about the term "solid loci" (στερεοὶ τόποι). According to Pappus (*Collectio*, VII.30, Hultsch, ed., II, 672.21), Aristaeus wrote five books of *Solid loci* connected with (or continuous with) the *Conics*. This implies that "solid loci" were conics; and the name suggests that when it was given, the curves were regarded as sections of a solid, in contrast with "plane loci" such as straight lines and "linear loci," which were higher curves. But there can be no certainty that the name is older than Aristaeus. T. L. Heath, *A History of Greek Mathematics*, II (Oxford, 1921), 117–118, gives an alternative explanation, deriving plane, solid, and linear loci from plane, solid, and linear problems; but he concedes that it would be natural to speak of the conic sections as solid loci, "especially as they were in fact produced from sections of a solid figure, the cone."

31. *Op. cit.*, 115. In detail he writes:

In each investigation two planes are perpendicular to an underlying plane; and the intersection of the two planes is a common ordinate to two curves lying one in each plane. In one of the intersecting planes the curve is in each case a semi-circle, and the common ordinate is, therefore, a mean proportional between the segments of its diameter. So far the investigation is the same for all. Now, from the consideration of the figure in the underlying plane—which is different in each case—it follows that:—in the first case—the solution of Archytas—the ordinate in the second intersecting plane is a mean proportional between the segments of its base, whence it is inferred that the extremity of the ordinate in this plane also lies on a semi-circle; in the second case—the section of the right-angled cone—the ordinate is a mean proportional between a given straight line and the abscissa; and, lastly, in the third case—the section of an acute-angled cone—the ordinate is proportional to the geometric mean between the segments of the base [p. 169].

32. *Die Lehre von den Kegelschnitten im Altertum*, R. von Fischer-Benzon, ed. (Copenhagen, 1886), repr. with foreword and index by J. E. Hofmann (Hildesheim, 1966), 464–465. T. L. Heath, who followed Zeuthen's method in *Apollonius of Perga* (Cambridge, 1896), xxvi–xxviii, gives a different method in *A History of Greek Mathematics*, II, 113–114, for determining *T*. He shows that *T* is on the circle which is the locus of all points such that their distances from *A′*, *A* are in the ratio 3:1, and *T* is determined as the intersection of the perpendicular to *A′A* at *A* with this circle.

33. See n. 20.

34. *Op. cit.*, 463–464.

35. Eutocius, *Commentarii in libros De sphaera et cylindro*, in *Archimedis opera omnia*, J. L. Heiberg, ed., III, 88.23–90.11. There are similar accounts of the Delian mission in other authors. Plutarch, *De genio Socratis*, 7, 579A–D, P. H. De Lacy and B. Einarson, eds., Loeb Classical Library (London–Cambridge, Mass., 1959), 396.17–398.22, says that Plato referred the Delians to Eudoxus of Cnidus and Helicon of Cyzicus; John Philoponus, *Commentary on the Posterior Analytics of Aristotle*, 1.vii, 75b12, M. Wallies, ed., *Commentaria in Aristotelem Graeca*, XIII, pt. 3 (Berlin, 1909), p. 102.7–18, is in general agreement but omits the references to the geometers. Theon of Smyrna, *Expositio*, E. Hiller, ed., 2.3–12, quoting a lost work of Eratosthenes entitled *Platonicus*, says the god gave this oracle to the

Delians, not because he wanted his altar doubled but because he wished to reproach the Greeks for their neglect of mathematics and contempt for geometry. Plutarch also in another work, *De E apud Delphos*, c. 6, 386E, F. C. Babbitt, ed., Loeb Classical Library, Plutarch's *Moralia*, V (London–Cambridge, Mass., 1936), p. 210.6–11 agrees that the god was trying to get the Greeks to pursue geometry rather than to have his altar doubled.

36. Plutarch, *Quaestiones conviviales*, viii.2.1, 718E–F, E. L. Minar, W. C. Helmbold, and F. H. Sandbach, eds., Loeb Classical Library, Plutarch's *Moralia*, IX, trans. as *Table Talk* (London–Cambridge, Mass., 1961), pp. 120.20–122.7. The same censure of Eudoxus and Archytas is repeated in Plutarch, "Vita Marcelli," xiv.5–6, *Plutarch's Lives*, B. Perrin, ed., V, Loeb Classical Library (London–Cambridge, Mass., 1917; repr. 1961), pp. 470.17–472.6, but here there is no mention of Menaechmus.

37. *Op. cit.*, 162.

38. *Op. cit.*, 176–177.

39. [Eutocius], *Commentarii in libros De sphaera et cylindro*, in *Archimedis opera omnia*, Heiberg, ed., III, 84.7–11. The words are bracketed by Heiberg and are no doubt an interpolation made by one of the pupils of Isidorus, who revised Eudocius' text.

40. Charles Smith, *Geometrical Conics* (London, 1894), 32, 84, 125.

41. Eutocius, *Commentarii in libros De sphaera et cylindro*, in *Archimedis opera omnia*, Heiberg, ed., III, 56.13–58.14.

42. *Vorlesungen über Geschichte der Mathematik*, 3rd ed., I (Leipzig, 1907), 234.

43. T. L. Heath, *A History of Greek Mathematics*, I (Oxford, 1921), 256.

BIBLIOGRAPHY

Menaechmus is known to have written a commentary on Plato's *Republic* in three books and other philosophical works, and he must have written at least one work in which he described his discovery of the conic sections. (Whether he wrote a separate book on the subject has been doubted, since Pappus, *Collectio*, F. Hultsch, ed., II [Berlin, 1877], VII 30, p. 672, does not mention any treatise on conics before those of Euclid and Aristaeus.) None of his works has survived. The fragments relating to his life and work are collected in Max C. P. Schmidt, "Die Fragmente des Mathematikers Menaechmus," in *Philologus*, **42** (1884), 77–81. Malcolm Brown (see below) believes that a passage in Proclus, *op. cit.*, 72.23–73.9, may be a quotation from Menaechmus.

The most complete account of Menaechmus is still that of G. J. Allman, *Greek Geometry From Thales to Euclid* (Dublin–London, 1889), 153–179, reproducing an article which appeared in *Hermathena*, no. 12 (July 1886), 105–130. Other accounts to which reference may profitably be made are C. A. Bretschneider, *Die Geometrie und die Geometer vor Eukleides* (Leipzig, 1870), 155–163; H. G. Zeuthen, *Keglesnitslaeren i Oldtiden* (Copenhagen, 1885), German trans. *Die Lehre von den Kegelschnitten im Altertum*, R. von Fischer-Benzon, ed. (Copenhagen, 1886), repr. with foreword and index by J. E. Hofmann (Hildesheim, 1966), 457–467; T. L. Heath, *Apollonius of Perga* (Cambridge, 1896), xvii–xxx; and *A History of Greek Mathematics* (Oxford, 1921), I, 251–255, II, 110–116; J. L. Coolidge, *A History of the Conic Sections and Quadric*

Surfaces (Oxford, 1945), 1–5; Malcolm Brown, "A Pre-Aristotelian Mathematician on Deductive Order," in *Philosophy and Humanism: Essays in Honor of Paul Oskar Kristeller* (New York, in press).

IVOR BULMER-THOMAS

MENDEL, JOHANN GREGOR (*b.* Heinzendorf, Austria [now Hynčice, Czechoslovakia], 22 July 1822; *d.* Brno, Austria [now Czechoslovakia], 6 January 1884), *genetics, meteorology.*

At the February and March 1865 meetings of the Natural Sciences Society of Brno, J. G. Mendel first presented an account of his eight years of experimental work on artificial plant hybridization. His paper was published in the Society's *Verhandlungen* in 1866 but went unnoticed. In 1900, within a two-month period, there appeared three preliminary reports by Hugo de Vries, Carl Correns, and Erich von Tschermak, who, working independently in Amsterdam, Tübingen, and Vienna respectively, attained the same results almost simultaneously. Each of them stated that just before completing his work, he learned that he had been preceded, by several decades, by a virtually unknown monk. Mendel's experimental work, designed after long contemplation of the problem, painstakingly executed on an extensive scale, intelligently analyzed and interpreted, and presented straightforwardly and clearly, yielded results of such general and far-reaching significance that his paper became the basis of the science of genetics.

Mendel's father, Anton, was a peasant. He served in the army during the Napoleonic Wars and later was able to turn his experience acquired in other regions to the improvement of his farm. Mendel's mother was the daughter of a village gardener, and other of his ancestors were professional gardeners in service at the local manor. Heinzendorf was on the border between the Czech- and German-speaking areas; the Mendel family spoke German, but about one-fourth of their ancestors were of Czech extraction. Mendel himself later lived on excellent terms with both. After the death of two infant girls, a daughter, Veronica, was born to the couple (1820), followed by Johann (1822), and another daughter, Theresia (1829). In the primary school the enlightened Reverend Schreiber, vicar of the neighboring village, taught the children natural science in addition to elementary subjects, and encouraged the cultivation of fruit trees both at the school and by the parishioners. Mendel also helped his father in grafting fruit trees in their orchard.

Since Mendel showed exceptional abilities, his parents, on the advice of the vicar and the village schoolmaster, sent the boy in 1833 to the Piarist secondary school in nearby Leipnik (Lipník), and a year later to the Gymnasium in Troppau (Opava), where he spent six years. He left with a certificate *primae classis cum eminentia*, the designation referring to his industriousness, knowledge, and ability. In 1838 Mendel's father suffered serious injuries during his statutory labor, and had to retire and turn over his farm to his son-in-law. Mendel had to earn his living by private tutoring. The physical and mental strain affected his health so much that in his fifth year at the Gymnasium he had to interrupt his studies for several months. He recovered, completed his secondary studies, and in 1840 enrolled in the philosophy course, as preparation for higher studies, at the University of Olmütz (Olomouc). His efforts to find private pupils were in vain, however, because he had no references. This new distress and frustration brought on further illness, this time for a longer period, so that Mendel had to spend a year with his parents to recover. But he refused to give up his studies and become a farmer. His younger sister, Theresia, offered him a part of her dowry to help him return to Olmütz and complete the two-year philosophy course. He accepted and thus became acquainted with the elements of philosophy, physics, and mathematics, including—in a course given by J. Fux—the principles of combinatorial operations, which he later employed with great success in his research. His physics professor, F. Franz, in 1843 recommended the admission of this "young man of very solid character, almost the best in his own branch," to the Augustinian monastery in Brno where Franz himself had stayed for nearly twenty years.

Mendel entered the monastery on 9 October 1843 with the name Gregor. He did so out of necessity and without feeling in himself a vocation for holy orders. But he soon realized that now he was free of all financial worries and that he had found the best possible conditions for pursuing his studies. The monastery was a center of learning and scientific endeavor, and many of its members were teachers at the Gymnasium or at the Philosophical Institute. The monastery was supported mainly by the income from its estates. F. C. Napp (1792–1867), who was the abbot from 1824, devoted much energy to the improvement of agriculture; he was a member of the Central Board of the Moravian Agricultural Society and later its president. He wanted one of his monks to teach natural sciences and agriculture at the Philosophical Institute, and he established the tradition of experimenting with plants in the monastery garden. When Mendel entered the order, Matthew Klácel, a teacher of philosophy (1808–1882), was directing the experimental garden and investigating variation, heredity, and evolution in plants. He enjoyed a high reputation among the Brno botanists, and drew upon his experience in natural history to formulate ideas on the Hegelian philosophy of gradual development that ultimately led to his dismissal and emigration to America. Klácel guided Mendel in his first studies in science and later put him in charge of the experimental garden.

During his theological studies (1844–1848) Mendel, in accordance with the abbot's interests, also attended courses at the Philosophical Institute in agriculture, pomology, and viticulture given by F. Diebl (1770–1859), who, in his textbook of plant production, *Abhandlungen aus der Landwirtschaftskunde für Landwirthe . . .* (2 vols., Brno, 1835), had described artificial pollination as the main method of plant improvement. In these lectures Mendel also learned of the methods of sheep breeding introduced by F. Geisslern (1751–1824). Diebl was, with Napp, among the main organizers of the Congress of German Agriculturists at Brno in 1840, where hybridization as a method of fruit-tree breeding was discussed.

After he finished his theological studies, Mendel was appointed chaplain to the parish served by the monastery, his duty being to see to the spiritual welfare of the patients in the neighboring hospital. But Mendel was extremely sensitive, and could not bear to witness suffering; he was overcome by fear and shyness and again became very depressed, almost to the point of illness. The sympathetic abbot relieved him of this duty and sent him as a substitute teacher to the grammar school at Znojmo in southern Moravia.

Mendel enjoyed teaching and so impressed both his pupils and his colleagues that at the end of his first year the headmaster recommended him for the university examination for teachers of natural sciences, which would allow him a regular appointment. Mendel passed well in physics and meteorology, but failed in geology and zoology. Since the failure seemed to be due largely to Mendel's lack of a university education, Andreas Baumgartner, the professor of physics, advised Napp to send Mendel to the University of Vienna.

At the university Mendel attended lectures on experimental physics by Doppler and on the construction and use of physical apparatus by Andreas von Ettinghausen (1796–1878), who had earlier published a book on combinatorial analysis, *Die combinatorische Analysis* (Vienna, 1826). It is clear that some of the methods described there later influenced Mendel's derivation of series in the hybrid progeny. Mendel also attended lectures on paleontology, botany, zoology, and chemistry, and was especially influenced by the professor of plant physiology, Franz Unger, who also gave a practical course on organizing botanical experiments. In his research Unger turned from the investigation of forms of fossil plants, to the influence of soil upon plants, to the causes of variation. He was known for his views on evolution, and in his lectures emphasized that sexual generation was the basis of the origin of the great variety in cultured plants. He sought to explain the evolution of new plant forms by the combination of the simplest elements in the cell, surmising their existence but unable to prove it. In Vienna, Mendel also thoroughly studied Gaertner's *Versuche und Beobachtungen über die Bastardzeugung im Pflanzenreich* (Stuttgart, 1849), in which nearly 10,000 separate experiments with 700 plant species yielding hybrids were described. In his copy, preserved in the Mendelianum, Mendel marked pages where pea hybrids are mentioned and also made notes on pairs of *Pisum* characteristics, some of which later appeared in his experimental program.

During his studies Mendel became a member of the Zoologisch-botanischer Verein in Vienna and published his first two short communications in its *Verhandlungen* (1853, 1854). They concerned damage to plant cultures by some insects.

After his return to Brno, Mendel was appointed substitute teacher of physics and natural history at the Brno Technical School. His superior was A. Zawadski (1798–1868), previously professor of physics and applied mathematics and dean of the Philosophical Faculty at the University of Lvov. He had come to Brno in 1854, after being dismissed for alleged responsibility for student uprisings in 1848. Zawadski was a man of wide scientific interests, ranging from botany, zoology, and paleontology to evolution. On his nomination Mendel became a member of the natural science section of the Agricultural Society in Brno.

As a teacher Mendel was highly appreciated both by students and colleagues. His task was not easy, for the classes were large, some with over 100 students. After his first year his headmaster reported that Mendel was a good experimentalist and, with rather scanty equipment, was able to give excellent demonstrations in both physics and natural history.

After his return from Vienna, Mendel also began experimenting with peas. The most arduous aspect was artificial pollination. He began the work in 1856, when he was also preparing for his second university examination for teachers of natural science, which took place in May 1856. Once again the instability of his psychological constitution betrayed him. He broke down during the written examination, withdrew from the other parts, and returned to Brno. There he became so seriously ill that his father and uncle came all the way from Silesia to see him. It was their only visit to him in Brno. Indirect evidence suggests that his indisposition derived from the stress of his studies and preoccupation with his research problems, compounded by the memory of his experience with the previous examination. After this failure he attempted no further degrees, and remained a substitute teacher until 1868, when he gave up teaching.

In 1868 Mendel was elected abbot of the monastery, which involved many official duties. He also became a member of the Central Board of the Agricultural Society and was entrusted with distribution of subsidies for promoting farming; from 1870 he was frequently elected to its executive committee. He took an active part in the organization of the first statistical service for agriculture. Later he also reported on scientific literature and cooperated with the editorial board of the society's journal, *Mittheilungen der K. K. Mährisch-schlesischen Gesellschaft zur Beförderung des Ackerbaues, der Natur- und Landeskunde.* Near the end of his life he was considered for the society's presidency, but he refused to accept this honor because of his poor health. From 1863 Mendel was also a member of the Brno Horticultural Society and after 1870 he belonged to the Society of Apiculturists, and influenced the development of both these fields.

Mendel soon became known for his liberal views, which he demonstrated by public support for the nominees of the Liberal Party in the general election of 1871. When the victorious Liberal government issued a law requiring a large contribution by the monastery to the religious fund, Mendel refused to pay the new taxes. After 1875 he was thus involved in a lengthy conflict with the authorities, which led to the sequestration of monastery land. In a last attempt to regain Mendel's support for the Liberal party, he was offered a place on the board of directors of the Moravian Mortgage Bank in 1876, and was even proposed for the office of its governor in 1881. Mendel persisted in his opposition, however, convinced that he was fighting for the rights of the monastery. The tension eventually had a deleterious effect on Mendel's health; he died of chronic inflammation of the kidneys with edema, uremia, and cardiac hypertrophy. At his funeral nobody was aware that an outstanding scientist was being buried, even though Mendel's experiments on *Pisum* were remembered by the fruit growers of Brno and in the obituary notices of local newspapers.

Work. *Meteorological Studies.* Mendel began his meteorological studies in 1856 and was soon recognized as the only authority on this subject in Moravia. In his first meteorological paper, published in 1863, he summarized graphically the results of observations at Brno, using the statistical principle to compare the data for a given year with average conditions of the previous fifteen years. Between 1863 and 1869, the paper was followed by five similar communications concerned with the whole of Moravia. Later Mendel published three meteorological reports describing exceptional storm phenomena. He also devoted much time to the observation of sunspots, assuming that they had some relation to the weather. In 1877, with his support, weather forecasts for farmers in Moravia were issued, the first in central Europe.

As always, there was a practical aspect to Mendel's pursuits and interests, but the primary motivation was scientific. The best example is his paper on the tornado that he observed at Brno in 1870. He began with a careful description and followed it with a new interpretation, which was that the observed phenomena were vortices engendered by encounters between conflicting air currents. This interpretation was overlooked, as were his hybridization experiments, even by those who advanced similar explanations many years later. These studies, although remote from his main work and far less important, have much in common methodologically with his studies of hybridization. They grew out of his habit of scrupulously collecting and recording data, thinking in quantitative terms, and subjecting observational data to statistical treatment.

Plant Hybridization. Mendel's principal work was the outcome of ten years of tedious experiments in plant growing and crossing; seed gathering and careful labeling; and observing, sorting, and counting almost 30,000 plants. It is hardly conceivable that it could have been accomplished without a precise plan and a preconceived idea of the results to be expected. There is evidence that Mendel began with an inductive hypothesis carefully framed so as to be testable in his experimental program.

From his previous experience Mendel was familiar with methods for improvement of cultivated plants. Both his teachers, Diebel and Unger, had pointed to hybridization as the source for this improvement. But hybridization was an empirical procedure, and Unger therefore stressed the necessity of studying the nature of variability. He suggested that it might be possible to find an explanation in the combination of hypothetical elements within the cells. Mendel's approach was the common one of reducing the problem to an elementary level and formulating a hypothesis that could be proved or disproved by experiments.

Between 1856 and 1863 Mendel cultivated and tested at least 28,000 plants, carefully analyzing seven pairs of seed and plant characteristics. This was his main experimental program. His original idea was that heredity is particulate, contrary to the model of "blending inheritance" generally accepted at that time. In the pea plants hereditary particles to be investigated are in pairs. Mendel called them "elements" and attributed them to the respective parents. From one parent plant comes an element determining, for instance, round seed shape; from the other parent, an element governing the development of the angular shape. In the first generation all hybrids are alike, exhibiting one of the parental characteristics (round seed shape) in unchanged form. Mendel called such a characteristic "dominant"; the other (angular shape), which remains latent and appears in the next generation, he called "recessive." The "elements" determining each paired character pass in the germ cells of the hybrids, without influencing each other, so that one of each pair of "elements" passes in every pollen (sperm) and in every egg (ovule) cell. In fertilization, the element marked by Mendel A, denoting dominant round seed shape, and the element a, denoting the recessive angular shape, meet at random, the resulting combination of "elements" being

$$\tfrac{1}{4}AA + \tfrac{1}{4}Aa + \tfrac{1}{4}aA + \tfrac{1}{4}aa.$$

In hybrid progeny both parental forms appear again; and Mendel's explanation of this segregation of parental traits was called, after 1900, Mendel's law (or principle) of segregation.

In his simplest experiments with crossing pea plants that differed in only one trait pair, Mendel cultivated nearly 14,000 plants and explained the progeny of the hybrid in terms of the series $A + 2 Aa + a$. At the same time he conformed to the view of K. F. von Gaertner and J. G. Koelreuter that hybrids have a tendency to revert to the parental forms. Mendel then called his explanation of hybrid progeny "the law of development thus found," which he tested further in a case "when several different traits are united in the hybrid through fertilization." Hereditary elements belonging to different pairs of traits, for example A and a for the round and angular seed shapes and B and b for the yellow and green seed colors, recombine the individual series $A + 2 Aa + a$ and $B + 2 Bb + b$, resulting in terms of a combination series:

$$AB + Ab + aB + ab + 2 ABb + 2 aBb$$
$$+ 2 AaB + 2 Aab + AaBb.$$

In his paper Mendel also illustrated a recombination of three trait pairs, showing every expected combination of characteristics and relevant elements in actual counts of offspring. He also observed that 128 constant associations of seven alternative and mutually exclusive characteristics were actually obtained—that being the expansion of 2^7, and the maximum number theoretically possible. His conclusion was that the "behavior of each of different traits in a hybrid association is independent of all other differences in the two parental plants," which principle was later called Mendel's law of independent assortment.

The generalization of Mendel's explanation was that "if n denotes the number of characteristic differences in two parental plants, then 3^n is the number of terms in the combination series, 4^n the number of individuals that belong to the series, and 2^n the number of combinations that remain constant."

In his second lecture to the Natural Sciences Society of Brno (March 1865) Mendel presented his hypothesis explaining "the development of hybrids in separate generations," and furnished both a theoretical and experimental proof of his assumption by crossing hybrids with constant dominant and constant recessive forms. His explanation was "that hybrids form germinal and pollen cells that in their composition correspond in equal numbers to all the constant forms resulting from the combination of traits united through fertilization."

At this meeting Mendel also described briefly his experiments with other plant species, the object of which was to determine "whether the law of development discovered for *Pisum* is also valid for hybrids of other plants." He predicted that "through this approach we can learn to understand the extraordinary diversity in the coloration of our ornamental flowers." In his concluding remarks he compared the observations made on *Pisum* with the results obtained by Koelreuter and Gaertner, especially those on hybrid characteristics. Mendel could not agree with the assumption that in some cases hybrid offspring remain "exactly like the hybrid and propagate unchanged," as Gaertner believed.

Mendel also touched on the experimental transformation of one species into another by artificial fertilization. In a simple experiment with *Pisum* he exhibited transformation from the viewpoint of his theory, explaining why some transformations took longer than others. He noted that Koelreuter's transformation experiments proceeded "in a manner similar to that in *Pisum*," and that in this way "the entire process of transformation would have a rather simple explanation." In this connection Mendel opposed Gaertner's view "that a species has fixed limits beyond which it cannot change," and he thus asserted his conviction that his theory favored the assumption of continuous evolution.

In the opening paragraph of his paper Mendel stated

that his experiments were initiated by "artificial fertilization undertaken on ornamental plants to obtain new color variants," and he emphasized the significance of such investigations in establishing "the evolutionary history of organic forms." Subsequently he often mentioned that "law of development" but without using the terms "inheritance" or "hereditary." By the "law of development" he certainly meant the law governing the evolution of cultured plants. He also assumed that his theory was valid in generality because "no basic difference could exist in important matters, since unity in the plan of development of organic life is beyond doubt."

In comparison with his predecessors, Mendel was original in his approach, in his method, and in his interpretation of experimental results. He reduced the hitherto extremely complex problem of crossing and heredity to an elementary level appropriate to exact analysis. He left nothing to chance. The choice of *Pisum* as his main experimental material resulted both from his study of the literature and from numerous preliminary experiments. He very carefully selected varieties whose purity had been assured by several years of cultivation under strictly controlled conditions. The hybridized varieties differed in only a few characteristics, and those that did not allow a clear distinction were discarded. Limiting the characteristics to a small number enabled Mendel to distinguish all possible combinations. His introduction of simple symbols that permitted comparing the experimental results definitively with the theory was very important. Altogether new was his use of large populations of experimental plants, which allowed him to express his experimental results in numbers and subject them to mathematical treatment. By the statistical analysis of large numbers Mendel succeeded in extracting "laws" from seemingly random phenomena. This method, quite common today, was then entirely novel. Mendel, inspired by physical sciences, was the first to apply it to the solution of a basic biological problem and to explain the significance of a numerical ratio. His great powers of abstraction enabled him to synthesize the raw experimental data and to reveal the basic principles operating in nature.

Mendel's manuscript, as read at the 1865 meetings, was published without change in the Natural Sciences Society's *Verhandlungen* in 1866. The other members, however, could hardly have grasped either the main idea or the great significance of his discoveries. The *Verhandlungen* was distributed to 134 scientific institutions in various countries, including those in New York, Chicago, and Washington. Mendel commissioned forty reprints, two of which have been found in Brno and five others elsewhere; one was sent to

Naegeli and another to Anton Kerner, two contemporary authorities on hybridization.

Further Research. The main results of Mendel's experiments and their interpretation, which constituted his whole theory, were reported in "Versuche über Pflanzenhybriden" (1866). This memoir was his magnum opus, one of the most important papers in the history of biology, and the foundation of genetic studies. Mendel had confidence in his experimental work and its rational interpretation; but he did observe "that the results I obtained were not easily compatible with our contemporary scientific knowledge and that under [such] circumstances publication of one such isolated experiment was doubly dangerous, dangerous for the experimenter and for the cause he represented. Thus I made every effort to verify the results obtained with *Pisum*." He was aware that there would be difficulty in "finding plants suitable for another extended series of experiments and that under unfavorable circumstances years might elapse without my obtaining the desired information." Nevertheless he tried very hard to confirm, as far as possible, the general validity of the experiments, first with other genera of plants (some experiments extending to four to six generations) and then with animals. After 1866, however, he published only a single short paper on *Hieracium* hybrids (1869). But the great efforts he devoted to this goal are evident from his letters written from 1866 to 1873 to Naegeli. They amount to scientific reports containing many details of his work, of the problems he encountered, and of the results he obtained. The discussion is extremely sober, objective, and scientific—a patient reaction to Naegeli's "mistrustful caution" regarding his experiments. The letters also convey Mendel's personality: his sincere endeavor to reach the truth, his truly scientific spirit, his modesty in the calm defense of his viewpoint. This correspondence remained unknown until 1905, when it was published by Naegeli's pupil, Carl Correns.

Mendel's experiments demonstrated that hybrids of *Matthiola*, *Zea*, and *Mirabilis* (like those of *Phaseolus* reported in the first paper) "behave exactly like those of *Pisum*." There still remained the question "whether variable hybrids of other plant species show complete agreement in their behavior with hybrids of *Pisum*." In the relevant contemporary literature it was reported that some hybrids (such as *Aa*) remain constant (*A*), which contradicted the generalization of Mendel's results. The genus *Hieracium* (hawkweed) seemed to Mendel most suitable for solving this question.

Mendel's *Hieracium* research project was also connected with some taxonomical questions, since the transitional forms of a highly polymorphic genus like *Hieracium* were very difficult to classify. The results of his four years' work, reported at the meeting of the Natural Sciences Society in Brno on 9 June 1869 and published in the society's *Verhandlungen* in 1870, were disappointing. He had to admit that in his *Hieracium* experiments "the exactly opposite phenomenon seems to be exhibited" as compared with *Pisum*. Subsequently, however, he carefully added that the whole matter "is still an open question, which may well be raised but not as yet answered." These experiments were extremely laborious and delicate because of the minuteness of the flowers and their particular structure; Mendel succeeded in obtaining only six hybrids, and only one to three specimens of each. Another obstacle was to be explained only in 1903: that *Hieracium* reproduces partly by apogamy, so that in many instances offspring are not formed by cross-pollination and are all alike, as though derived from cuttings.

Mendel discussed these experiments and the problems involved in more detail in his letters to Naegeli. The small number of *Hieracium* hybrids he obtained did not allow any definite conclusion, and it is surprising that eventually he found the theoretical explanation even in this case. Notes in Mendel's handwriting brought to light only recently in the Mendelianum indicate that he insisted on his idea of variable hybrids and, assuming polygene action, he tried to explain that in *Salix*, as in *Hieracium*, a multifactor crossing takes place and that the segregation of their hybrids follows the same principle as in *Pisum*. According to this assumption, the reported constancy exhibited by the extremely variable *Hieracium* hybrids would be only apparent.

Mendel centered his efforts on proving that a certain system operates in nature and that its laws could be formulated. It required a great capacity for abstraction and simplification of the extremely complex set of observed phenomena. He had to focus his attention on the main issues; otherwise he would have become lost in the complexities of nature, as had all his predecessors who found many isolated phenomena but did not synthesize them into a coherent system. Many potentially interesting observations had to be left out of consideration in the first phase. Thus, it has been often overlooked that besides the main findings, Mendel noted several phenomena attributed, after 1900, to other scientists: the intermediate forms of heredity, the additive action of his "elements," like that of genes, and complete linkage. He also described, in principle, the frequencies of "elements" in the population. Later, as his extant fragmentary notes show, he imagined the existence of the interaction of the "elements" and of an action like

that later called polygenic. He also suggested that egg cells and pollen cells contain different hereditary units for the development of sex.

After 1871 Mendel conducted hybridization experiments on bees, hoping to prove his theory in the animal kingdom. He kept about fifty bee varieties, which he attempted to cross in order to obtain "a new synthetic race." He was not successful, however, because of the complex problem of the controlled mating of queen bees. In these experiments he also proved the hybrid effect on fertility of bees.

Mendel must have been greatly disappointed that there was no recognition of his scientific work and that even Naegeli missed its essential feature and did not grasp the historical significance of his theory. Naegeli's attention was focused on other problems, and Mendel's findings did not fit into his manner of thinking. Thus he raised objections—in fact not relevant—to Mendel's experiments and rejected his rational conclusions. Mendel was not understood in his time. Only in the following decades did the discoveries of the material basis of what was later called Mendelian—behavior of the nucleus in cell division, constancy in each species of the number of chromosomes, the longitudinal splitting of chromosomes, the reduction division during the maturation of germ cells, and the restitution of the number of chromosomes in fertilization—prepare the way for understanding the cytological basis of Mendelian inheritance and for its general acceptance.

The absence of response and recognition was one of the reasons that Mendel stopped publishing the results of his later experiments and observations. He did, nonetheless, take satisfaction and pleasure in the application of his theory in the breeding of new varieties of fruit trees and in propagating the idea of hybridization among local gardeners and horticulturists.

Mendel and Darwin. After Mendel's rediscovery in 1900, Mendelism was often opposed to the Darwinian theory of natural selection, and unfortunately so, for the apparent opposition was based on misunderstanding. In fact, the modern theory of descent and heredity has two foundations, one laid by Darwin and the other by Mendel, and both indispensable. It was in 1926, however, that Chetwerikov attempted a synthesis of Darwinian and Mendelian theories, a move completed in 1930–1932 by R. A. Fisher, Sewall Wright, and J. B. S. Haldane. The importance of Mendel for the theory of evolution rests in his demonstration of the mechanism that is the primary source of variability in plant and animal populations, on which natural selection subsequently operates.

When he wrote his paper, Mendel was already acquainted with Darwin's *On the Origin of Species*.

A copy of its German translation with Mendel's marginalia, preserved in the Mendelianum in Brno, shows Mendel's deep interest in this work. Similar marginalia are in the Brno copies of other Darwin books. Mendel's notes show his readiness to accept the theory of natural selection. He rejected the Darwinian provisional hypothesis of pangenesis, however, as contradicting in principle his own interpretation of the formation and development of hybrids. Like Darwin, Mendel was convinced that "it is impossible to draw a sharp line between species and variations."

On the other hand, Darwin, looking for the causes of variations, seems never to have realized that the clue to this problem was in the hybridization experiments. He never learned about Mendel's work, although almost the only book in which it was cited, S. O. Focke's *Pflanzenmischlinge . . .* (1881), is known to have passed through his hands.

Mendel was a lonely, unrecognized genius. Yet the rediscovery of his work brought to a close an era of speculation on heredity, which then became a subject of scientific analysis. He opened a new path to 'he study of heredity and revealed a new mechan sm operating in the process of evolution. Every generation of biologists has found something new in his fundamental experiments. The science of genetics, which had both its origins and a powerful impetus in Mendel's work, has advanced with prodigious speed, linking many branches of biology (cytology in particular) with mathematics, physics, and chemistry. This development has led to a deeper understanding of man and nature with far-reaching theoretical implications and practical consequences.

BIBLIOGRAPHY

I. ORIGINAL WORKS. Mendel published thirteen papers, two of which were on plant-damaging insects, nine on meteorology, and two—the most important—on plant hybrids. Most of them, including seven on meteorology and the two on plant hybrids, were published in *Verhandlungen des Naturforschenden Vereins in Brünn*, **1** (1863) to **9** (1871). A list of them was published by J. Kříženecký in *Gregor Johann Mendel 1822–1884. Texte und Quellen zu seinem Wirken und Leben* (Leipzig, 1965), along with the text of the 1865 paper, revised according to the original MS. Another critical ed. of the 1865 paper was edited, with the text of the 1869 paper on *Hieracium* hybrids, an introduction, and commentaries, by F. Weiling as *Ostwalds Klassiker der Exakten Wissenschaften*, n.s. VI (Brunswick, 1970). There are numerous trans. in many languages of Mendel's magnum opus, "Versuche über Pflanzenhybriden." Most are listed in M. Jakubíček and J. Kubíček, *Bibliographia Mendeliana* (Brno, 1965); and M. Jakubíček, *Bibliographia Mendeliana. Supplementum 1965–1969* (Brno, 1970). Trans. into English have been published several times since C. T. Druery's trans. was modified and corrected by W. Bateson in *Mendel's Principles of Heredity. A Defence* (Cambridge, 1902); this was republished by J. H. Bennett as *Experiments in Plant Hybridization—Gregor Mendel* (Edinburgh, 1965). A new English trans. was edited by C. Stern and E. R. Sherwood, *The Origins of Genetics. A Mendel Source Book* (San Francisco–London, 1966), which includes the paper on *Hieracium* hybrids and ten of Mendel's letters to Naegeli. The letters were first published by Carl Correns, "Gregor Mendel's Briefe an Carl Nägeli 1866–1873," in *Abhandlungen der Königlichen sächsischen Gesellschaft der Wissenschaften*, Math.-phys. Kl., **29** (1905), 189–265, and their English trans. in *Genetics*, **35** (1950), 1–29.

Besides the thirteen full-length papers by Mendel, over twenty minor publications, mostly book reviews from 1870–1882 signed with the initial "M" or "m," have been identified.

II. SECONDARY LITERATURE. The first—and still the best—detailed biography is H. Iltis, *Gregor Johann Mendel. Leben, Werk und Wirkung* (Berlin, 1924), translated into English as *Life of Mendel* (New York, 1932; 2nd ed., 1966). Additional information was published by O. Richter, *Johann Gregor Mendel wie er wirklich war* (Brno, 1943). Literature on Mendel's work and his importance in the history of biology is extremely plentiful. Over 800 titles are listed in Jakubíček and Kubíček's *Bibliographia Mendeliana* and in its supp. (see above). The most important recent literature includes "Commemoration of the Publication of Gregor Mendel's Pioneer Experiments in Genetics," in *Proceedings of the American Philosophical Society*, **109**, no. 4 (1965), 189–248; F. A. E. Crew, *The Foundations of Genetics* (Oxford, 1966); L. C. Dunn, *Genetics in the 20th Century. Essays on the Progress of Genetics During Its First 50 Years* (New York, 1951); and *A Short History of Genetics* (New York, 1965); R. A. Fisher, "Has Mendel's Work Been Rediscovered?" in *Annals of Science*, **1** (1936), 115–137; A. E. Gaissinovich, *Zarozhdenie genetiky* (Moscow, 1967); J. Kříženecký, ed., *Fundamenta genetica* (Brno, 1965); M. Sosna, ed., *G. Mendel Memorial Symposium (Brno, 1965)* (Prague, 1966); R. C. Olby, *Origins of Mendelism* (London, 1966); H. Stubbe, *Kurze Geschichte der Genetik bis zur Wiederentdeckung der Vererbungsregeln Gregor Mendels* (Jena, 1963); and A. H. Sturtevant, *A History of Genetics* (New York, 1965).

Over 700 original documents relating to Mendel have been preserved in the Mendelianum, established in 1964 by the Moravian Museum in the former Augustinian monastery at Brno. Since 1966 it has published annually the series *Folia Mendeliana;* no. 6 (1971) contains important papers, including an elucidation of the problem of the triple rediscovery of Mendel's work, presented at the international Gregor Mendel Colloquium, held at Brno, 29 June–3 July 1970. The rediscovery of Mendel's work is further discussed in V. Orel, *The Secret of Mendel's Discovery* (Tokyo, 1973), in Japanese.

V. KRUTA
V. OREL

MENDEL, LAFAYETTE BENEDICT (*b.* Delhi, New York, 5 February 1872; *d.* New Haven, Connecticut, 9 December 1935), *physiological chemistry.*

Lafayette Benedict Mendel was the elder of two sons of Benedict Mendel, a merchant in Delhi, New York, and Pauline Ullman. Both parents migrated to the United States from Germany. Mendel prepared for Yale College in the local schools and in 1887, at the age of fifteen, obtained a New York State scholarship by competitive examination. At Yale he studied mainly the classics, economics, and the humanities, but on graduation in 1891 was awarded a fellowship that enabled him to enter the graduate course in physiological chemistry under Russell H. Chittenden of the Sheffield Scientific School.

Although poorly prepared in chemistry and physics and without experience in laboratory work, Mendel completed the requirements for the Ph.D. in two years with a thesis on the proteolysis of the crystalline hempseed protein edestin. This investigation was of no lasting significance, but the work aroused his interest in protein chemistry, and especially in the properties of the proteins of plant seeds. This was to have great influence upon his later career.

Mendel joined the teaching staff of the Sheffield Laboratory of Physiological Chemistry in the fall of 1893. He took a year's leave of absence in 1895 and went to Germany, where he studied physiology at Breslau with R. Heidenhain and chemistry at Freiburg im Breisgau with E. Baumann. He quickly established his position in Heidenhain's laboratory by demonstrating the preparation of crystalline edestin from hempseed, something that the professor had never seen before.

In 1897 Mendel was appointed assistant professor and in 1903 full professor of physiological chemistry in the Sheffield Scientific School. In 1921 he was made Sterling professor of physiological chemistry at Yale University, one of the first of these distinguished appointments, with membership in the faculties of the graduate and medical schools. Mendel was a most effective teacher. His lectures were clear and forceful, scholarly and stimulating. In his seminars, where weekly discussions of current advances in biochemistry were held, he taught his students to teach, always insisting on dignity in presentation, good English style, and accuracy of statement, qualities that he illustrated in his own discussions of the matter at hand. He communicated his own enthusiasm to his students and took a close personal interest in all of them, winning their confidence and respect and caring for their interests long after they had left Yale for positions elsewhere.

Mendel's early scientific work illustrates the in-fluence of his teachers: from Chittenden the interest in proteins, from Heidenhain that in experimental physiology, and from Baumann the biochemistry of such compounds as creatine, choline, taurine, the purines, and especially the iodine-containing substances of the thyroid gland. With the collaboration of colleagues on the faculty and of a steadily increasing number of graduate students, papers were published on such subjects as the nitrogen metabolism of the cat, the formation of uric acid, the paths of excretion of a number of inorganic ions, the metabolism of iodine, of allantoin, of kynurenic acid, and of several pyrimidines and purines. His interest in nutrition was aroused early. The first paper on the subject was published in 1898 and dealt with the nutritive value of various edible fungi. As early as 1906 he began his studies of growth, one or another aspect of which was dealt with in about seventy subsequent papers.

Mendel is remembered chiefly as coauthor, with Thomas B. Osborne of the Connecticut Agricultural Experiment Station in New Haven, of more than one hundred papers on the subject of nutrition. This collaboration began in 1905 with a study of the proteins of the castor bean, especially the highly toxic albumin ricin. Osborne, with his assistant Isaac F. Harris, prepared the proteins, and Mendel did the physiological tests. Their most active preparation of ricin was fatal to rabbits at a dose of two one-thousandths of a milligram per kilogram, and they thoroughly established the protein nature of this extraordinary poison, then a matter of debate.

The collaboration with Osborne on the nutritive effect of proteins began in 1909. Osborne had become an internationally known authority on the preparation and properties of the proteins of plant seeds and upon their analysis by the then current methods. The question had arisen whether substances of such widely varied amino acid composition were equally effective for the protein nutrition of animals, a matter upon which divergent opinions were held, and he invited Mendel to join him in an investigation of the subject. Together they devised improved experimental methods for feeding white rats, including accurate measurements of food intake, and adopted the general principle that in a successful experiment the food must maintain the weight of the rat for a substantial fraction of its life span. Rats had been used previously for studies of nutrition by a number of European investigators, such as V. Henriques in Copenhagen and W. Falta in Basel. They had also been used in this country by Henry H. Donaldson of the Wistar Institute in Philadelphia and by E. V. McCollum of the University of Wisconsin.

Aided by annual grants from the Carnegie Institution of Washington, they soon established that the diet must contain adequate amounts of lysine and tryptophan, supplied either as such or combined in the protein of the diet. This was the first convincing proof that certain amino acids are essential components of the diet and cannot be synthesized by the animal organism. Despite every care nutritive failure ultimately supervened. A preparation obtained by evaporating milk serum to dryness, the so-called protein-free milk, greatly postponed such failures when included in the food; but, even so, sudden and dramatic losses of weight and ultimately death occurred. The animals could be saved by furnishing a diet which contained dried whole milk, but not by one containing dried skim milk. This difference led to an examination of the effect of supplying butter for a part of the lard in the food. The result was a demonstration that butter contains an organic substance, obviously present only in trace amounts, which is essential in the nutrition of the rat. This finding was the discovery of vitamin A. Unfortunately for Osborne and Mendel, McCollum at the University of Wisconsin had made a similar observation and had submitted his results for publication a few weeks before Osborne and Mendel's paper was received. McCollum is thus regarded in the history of biochemistry as the discoverer of vitamin A.

Nevertheless, the independent and almost simultaneous publication of so important a conclusion from two laboratories greatly strengthened the position of both groups of investigators, and the doctrine that an adequate diet must supply trace amounts of hitherto unrecognized organic substances was at once widely accepted. It led within a few years to a complete revolution in the science of nutrition. The "protein-free milk" was recognized to contain a second such essential, which was soon designated water-soluble vitamin B, and Osborne and Mendel for some ten years devoted much study to the distribution of these essential substances in natural food products and to the elucidation of the complexities of the water-soluble vitamins.

Once they had established that diets of purified food materials could be compounded upon which rats would grow well and live indefinitely, Osborne and Mendel carried out many studies of growth. From the beginning of the period of collaboration with Osborne, Mendel's bibliography contains 289 titles. Of these, 24 percent contain the word "growth." Perhaps the most striking result was the demonstration that the growth of a young animal could be indefinitely suppressed and then resumed by manipulation of the diet with respect to the supply of lysine. Growth could occur at any age.

Notwithstanding the time and effort expended upon the collaboration with Osborne, Mendel also directed the activities of many graduate students. Especial attention was given to the chemistry and metabolism of fats and to the regulation of blood volume by the supply of salts. He was frequently invited to give lectures and also served as a consultant not only to his medical colleagues but also to the food industry. Withal, he was a professor who is remembered by his students with deep affection and respect. Always approachable and kindly, ever ready with suggestions for the solution of practical problems or with helpful reference to the literature of his subject, of which he was a complete master, his influence upon science at Yale and especially upon the medical applications of nutrition was great. Largely through his own efforts nutrition was transformed during his lifetime from empiricism to a clearly recognized branch of biochemistry founded upon scientific principles. When it is realized that the modern poultry and meat industries rely entirely upon the proper supply to the animals of vitamins and of proteins of correct and adequate amino acid composition, the benefit to practical agriculture of Mendel's work is almost incalculable.

BIBLIOGRAPHY

I. Original Works. Mendel's bibliography in Russell H. Chittenden's memoir, *Biographical Memoirs of the National Academy of Sciences*, XVIII (1937), lists titles of 326 papers, nearly all being records of scientific research published between 1894 and 1935. It is incomplete since Mendel, who prepared it, did not include numerous editorials in the *Journal of the American Medical Association* and many book reviews and similar less important writings. Of these papers 111 were written in collaboration with Thomas B. Osborne of the Connecticut Agricultural Experiment Station and record joint experiments made between 1910 and 1928 in Osborne's laboratory with financial support by the Carnegie Institution of Washington and the Connecticut Station. Most of them were published in the *Journal of Biological Chemistry*. After Osborne's retirement in 1928 and death in 1929, the work at the station was continued until Mendel's death in 1935, some fifteen papers being published with several junior collaborators. During this same period Mendel published ninety-seven papers in collaboration with graduate students or colleagues on the Yale faculty, and forty-one, mainly reviews of various aspects of nutrition, of which he was sole author. Mendel gave lectures subsequently published before the Harvey Society of New York in 1906 and 1914 and also gave the Herter Lectures at University and Bellevue Hospital Medical College in New York in 1914. In 1923 he gave the Hitchcock Lectures at the University of California and in 1930 was Cutler Lecturer at the Harvard Medical School.

Publications in book form include *Feeding Experiments with Isolated Food-substances*, 2 pts., Carnegie Institution of Washington Publication No. 156 (1911); *Changes in the Food Supply and Their Relation to Nutrition* (New Haven, 1916); and *Nutrition the Chemistry of Life* (New Haven, 1923).

II. Secondary Literature. In addition to the memoir by Chittenden mentioned above, there is a tribute by A. H. Smith, in *Journal of Nutrition*, **60** (1956), 3.

H. B. Vickery

MENDELEEV, DMITRY IVANOVICH (*b.* Tobolsk, Siberia [now Tyumen Oblast, R.S.F.S.R.], Russia, 8 February 1834; *d.* St. Petersburg [now Leningrad], Russia, 2 February 1907), *chemistry.*

Mendeleev was the fourteenth and last child of Ivan Pavlovich Mendeleev, a teacher of Russian literature, and Maria Dmitrievna Kornileva, who came of an old merchant family (she herself owned a glass factory near Tobolsk). His mother, who died when he was fifteen, played a large part in Mendeleev's early education and was strongly influential in shaping the views that he held throughout his life. Mendeleev entered the Tobolsk Gymnasium when he was seven, and graduated from it in 1849; while there he learned to dislike ancient languages and theology and to enjoy history, mathematics, and physics. In Tobolsk Mendeleev, who lived with his family near the glassworks, also acquired an interest in industrial affairs and, through the group of Decembrists exiled there, a love of liberty.

Shortly before her death, Mendeleev's mother took him to St. Petersburg, where in 1850 he enrolled in the faculty of physics and mathematics of the Main Pedagogical Institute, a progressive institution in which the revolutionary democrat Nikolai Dobrolyubov was a fellow student. Among his teachers were the chemist A. A. Voskresensky, who gave his pupils a taste for chemical experiment (and of whose lectures Mendeleev wrote down detailed descriptions that are preserved in the Mendeleev Museum in Leningrad); the zoologist Brandt, who interested Mendeleev in the classification of animals (his notes on this subject are also preserved); the geologist and mineralogist Kutorga, who immediately assigned him the chemical analysis of orthosilicate and pyroxene, and thus introduced him to research techniques; and the pedagogue Vyshnegradsky, who influenced his ideas on education. Mendeleev graduated from the Institute in 1855 with a brilliant record, but his hot temper led him into a quarrel with an important official of the Ministry of

Education, and his first teaching assignment was therefore to the Simferopol Gymnasium, which was closed because of the Crimean War.

After two months in the Crimea, where he was unable to work, Mendeleev went to Odessa as a teacher in the lyceum, and there took up the continuation of his early scientific work. He had already begun to investigate the relationships between the crystal forms and chemical composition of substances. On graduating from the Institute, he had written a dissertation entitled "Izomorfizm v svyazi s drugimi otno sheniami formy k sostavu" ("Isomorphism in Connection With Other Relations of Form to Composition"). It was published in *Gorny zhurnal* ("Mining Journal") in 1856. The writing of this work in itself led Mendeleev still further into the comparative study of the chemical properties of substances; his master's dissertation, prepared while he was in Odessa, was entitled "Udelnye obemy" ("Specific Volumes") and was a direct extension of the earlier articles, in which he had raised the question of whether the chemical and crystallographic properties of substances have any relation to their specific volumes.

During this same period, Mendeleev, in order to support himself, also wrote articles for *Novosti estestvenykh nauk* ("News of Natural Sciences") and reviews for *Zhurnal ministerstva narodnogo prosveshchenia* ("Journal of the Ministry of Public Education"). At a slightly later date he wrote an article on gas fuel and the Bessemer process for *Promyshlenny listik* ("Industrial Notes"). From this time, the application of science to industry and economics was a pronounced and recurrent preoccupation in his work.

In September 1856 Mendeleev defended a master's thesis at the University of St. Petersburg, expressing his adherence to the chemical ideas of Gerhardt, to which he remained loyal throughout his life. Among other topics, he made known his agreement with unitary and type theories and his opposition to Berzelius' electrolytic theory of the formation of chemical compounds. Mendeleev adhered to Gerhardt's ideas all his life, and in consequence later years found him resisting Arrhenius' electrolytic theory, rejecting the concept of the ion as an electrically charged molecular fragment, and refusing to recognize the reality of the electron. He was opposed in general to linking chemistry with electricity and preferred associating it with physics as the science of mass. His predilection found its most brilliant vindication in the correlation he achieved between the chemical properties and the atomic weights of elements. Nor was he a chemical mechanist in the methodological sense then fashionable in certain quarters. Chemistry in his view was an independent science, albeit a physical one.

In October 1856 Mendeleev defended a thesis *pro venia legendi* to obtain the status of privatdocent in the university. His subject was the structure of silicon compounds. In January 1857 he began to give lectures in chemistry and to conduct research at the university's laboratories. In 1859–1860 Mendeleev worked at the University of Heidelberg, where he first collaborated with Bunsen, and then established his own laboratory. He studied capillary phenomena and the deviations of gases and vapors from the laws of perfect gases. In 1860, he discovered the phenomenon of critical temperature—the temperature at which a gas or vapor may be liquefied by the application of pressure alone—which he called the "absolute temperature of boiling." He was thereby led to consider once again the relationship between the physical and chemical properties of particles and their mass. He was convinced that the force of chemical affinity was identical to the force of cohesion; he looked upon his work, then, as falling within the realm of physical chemistry, the ground upon which chemistry, physics, and mathematics met.

Mendeleev took part in the first International Chemical Congress, held at Karlsruhe in 1860. The idea of the congress had been Kekulé's; its purpose was the standardization of such basic concepts of chemistry as atomic, molecular, and equivalent weights, since the prevailing use of a variety of atomic and other weights had considerably impeded the development of the discipline. At the congress Mendeleev met a number of prominent chemists, including Dumas, Wurtz, Zinin, and Cannizzaro, whose championship of Gerhardt's notions impressed him deeply. His account of the congress, in a letter to his teacher Voskresensky, was published in the St. Petersburg *Vedomosti* ("Record") in the same year.

In February 1861, on his return to St. Petersburg, Mendeleev published *Opyt teorii predelov organicheskikh soedineny* ("Attempt at a Theory of Limits of Organic Compounds"), in which he stated that the percentage of such elements as oxygen, hydrogen, and nitrogen could not exceed a certain maximum value when combined with carbon—a theory that brought him into direct opposition to the structural theories of organic chemistry. On this theory he based his text *Organicheskaya khimia* ("Organic Chemistry"), which was published in the same year (a second edition was brought out in 1863) and which won the Demidov Prize.

From January 1864 to December 1866 Mendeleev was professor of chemistry at the St. Petersburg Technological Institute and a docent on the staff of the university. In addition, he traveled abroad on scientific assignments for three or four months of each year, wrote books, edited translations, and participated in the compilation of a technical encyclopedia, for which he wrote articles on the production of chemicals and technical chemistry, including the production of alcohol and alcoholometry. In 1865 he defended a thesis for the doctorate in chemistry, "O soedinenii spirta s vodoyu" ("On the Compounds of Alcohol With Water"). In it he first developed the characteristic view that solutions are chemical compounds and that dissolving one substance in another is not to be distinguished from other forms of chemical combination. In this thesis, he also adhered to the principles of chemical atomism.

At this same time Mendeleev, stimulated by the stormy social conditions that followed on the abolition of serfdom in Russia, became more strongly attracted to the practical problems facing the national economy. He began to study petroleum, traveling to Baku for that purpose in 1863; he attended the great industrial expositions held in Moscow in 1865 and in Paris in 1867; he purchased the estate of Boblovo, near Klin, in 1865 to demonstrate how agriculture could be put on a rational scientific basis; and he joined the Free Economic Society, where he lectured on such subjects as experimental agriculture, cooperative cheesemaking, and experiments with fertilizers. In 1862 he also married Feozva Nikitichna Leshchevaya; they had a son and a daughter.

It is appropriate here to consider how Mendeleev's work was situated in the context of the major scientific advances of the earlier nineteenth century. He had learned of the cell theory in the lectures on botany given at the Pedagogical Institute. Out of a belief in the unity of forces of nature had developed the law of the conservation of energy, which was fundamental to the study of chemical transformations in their relation to physical properties. The Darwinian theory of evolution had come to dominate the study of living nature, and considering the way in which the periodic law in its turn followed out of these fundamental discoveries, William Crookes could later refer to Mendeleev's theory as "inorganic Darwinism."

Mendeleev was also active in the growth of Russian chemical organizations during the 1860's. He participated in drawing up the bylaws of the Russian Chemical Society, which was founded in 1868, and he systematically presented the results of his researches at its meetings. The journal of the society was also important to the growth of the discipline within Russia, and Mendeleev chose to publish many of his findings in it. He further communicated his work to the chemical section of the Russian congresses for natural sciences, of which the first was held in December 1867.

A turning point in Mendeleev's career occurred in October 1867, when he was appointed to the chair of chemistry at the University of St. Petersburg. In preparing for his lectures he found nothing which he could recommend to his students as a text, so he set out to write his own. He derived his basic plan for his book from Gerhardt's theory of types, whereby elements were grouped by valence in relation to hydrogen. The typical elements hydrogen (1), oxygen (2), nitrogen (3), and carbon (4) were listed first, followed, in the same order, by the halogens (1) and the alkali metals (1).

Mendeleev entitled his book *Osnovy khimii* ("Principles of Chemistry"); he finished the first part of it, ending with the halogens, at the end of 1868. During the first two months of 1869 he wrote the first two chapters—on alkali metals and specific heat—of the second part. In spite of their common univalency, he organized the halogens and alkali metals so as to point up their contrary chemical relationships. It then remained to organize them according to another, more basic quantitative variable (or system of ordering), namely, their atomic weight. It may be noted that all Mendeleev's early work—his studies of the chemical properties of substances, his work on specific weights and their relationships to atomic and molecular weights, his investigations of the limits of compounds, and his study of atomic weights and their correlation with elements—had fitted him to undertake such a task, which was to culminate in the grand synthesis of the periodic law.

On 1 March 1869 Mendeleev was making preparations to leave St. Petersburg for a trip to Tver (now Kalinin) and then to other provincial towns. The Free Economic Society had given him a commission to investigate the methods in use for making cheeses in artels. It was on the very day of his departure that he realized the answer to the question of what group of elements should be placed next after the alkali metals in his *Osnovy khimii*. The principle of atomicity required treating copper and silver as a transitional group, since they gave compounds of both the $CuCl_2$ and $AgCl$ types; it therefore seemed logical to place them next to the alkali metals, which they most closely resemble in chemical properties. In seeking a quantitative basis to justify such a transition, Mendeleev had the crucial idea of arranging the several groups of elements in the order of atomic weights, a sequence which gave him the following tabulation:

		Ca = 40	Sr = 87.6	Ba = 137
Li = 7	Na = 23	K = 39	Rb = 85.4	Cs = 133
	F = 19	Cl = 35.5	Br = 80	Te = 127

Clearly there was a regular progression in the differences between the atomic weights of the elements in the vertical columns (the future periods), and this arrangement made it possible to place other elements of intermediate atomic weight in the gaps in the table. In working out the final stages of his discovery, Mendeleev used the method of "chemical solitaire," writing out the names or symbols of the elements, together with their atomic weights and other properties, on cards. The procedure was an adaptation of the game of patience, which he liked to play for relaxation.

Mendeleev's work toward the *Osnovy khimii* thus led him to the periodic law, which he formulated in March 1869: "Elements placed according to the value of their atomic weights present a clear periodicity of properties." The work of the Karlsruhe congress had contributed to its discovery; clearly, it would have been impossible to find any relationship between the elements using the old atomic weights—Ca = 20, Sr = 43.8, and Ba = 63.5, for example. The necessity to establish correct atomic weights was indeed what first led Mendeleev to investigate the connections among the elements; from this investigation he proceeded inductively to the periodic law, upon which he was then able to construct a system of elements. He used deduction, however, to predict consequences from his still incomplete discovery, moving from the general to the particular to test the validity of the law. For example, immediately following his discovery of the periodic law, Mendeleev proposed changing the generally accepted weight for beryllium—14—to 9.4, ascribing to its oxide (after I. Avdeev) the formula BeO (by analogy with magnesia, MgO) and not Be_2O_3 (by analogy with alumina, Al_2O_3). He thus correctly determined the place of beryllium in his system of elements. He also predicted three undiscovered elements in the future groups III and IV of his system, which he called eka-aluminum, ekasilicon, and ekazirconium.

Mendeleev's first report of his discovery was "Opyt sistemy elementov, osnovannoy na ikh atomnom vese i khimicheskom skhodstve" ("Attempt at a System of Elements Based on Their Atomic Weight and Chemical Affinity"); he presented it in more detail in "Sootnoshenie svoystv s atomnym vesom elementov" ("Relation of the Properties to the Atomic Weights of the Elements"), which was read to the Russian Chemical Society in March 1869 by N. A. Menshutkin (since Mendeleev himself was away visiting cheesemaking cooperatives). In preparing the latter report, Mendeleev developed several variant tables of elements, including one in which even- and odd-valenced elements were placed in two separate columns. He discovered gaps at three points—between hydrogen and lithium, between fluorine and sodium, and between chlorine and potassium—and predicted that these lacunae would be filled by then-unknown elements having atomic weights of approximately 2, 20, and 36—that is, by helium, neon, and argon.

At first Mendeleev could subsume under the periodic law only isomorphism and atomic weight; in each of these early papers, too, he presented only the quantitative argument for the analytical expression of the law in the form of the increase of atomic weights. The first paper in particular contained many ambiguities and imprecisions; lead, for example, was placed in the same group as calcium and barium, while thallium occupied the same group as sodium and potassium, and uranium was grouped with boron and aluminum. Having been occupied with studies leading up to the law for fifteen years—since 1854—Mendeleev then formulated it in a single day. He spent the next three years in further perfecting it, and continued to be concerned with its finer points until 1907.

In the work that immediately followed his statement of the law, Mendeleev returned to his earlier investigation of specific volumes, studying the physical function of the rule that showed that uranium should be ascribed a doubled atomic weight. Many elements, including tellurium and lead, had therefore to be assigned new places on the table. Mendeleev presented this new result to the Second Congress of Russian Natural Scientists in August 1869, in a report entitled "Ob atomnom vese prostykh tel" ("On the Atomic Weight of Simple Bodies"). He then proceeded to use the same argument to determine chemical functions; having recognized the importance of the simplicity of oxides as compound types, he proceeded to clarify the seven fundamental groups that extend from the alkali metal oxides of form R_2O to the halogen oxides of form R_2O_7. Mendeleev communicated this finding to the October meeting of the Russian Chemical Society in the memoir "O kolichestve kisloroda v solyanykh okislakh i ob atomnosti elementov" ("On the Quantity of Oxygen in Salt Oxides and on the Valence of Elements"). By 1870, he had taken into account the presence of compounds of the type RO_4 for osmium and ruthenium, and had therefore introduced an eighth group into his classification.

Mendeleev himself summarized the studies that had brought him to the periodic law in a later edition of Osnovy khimii, in which he commented on "four aspects of matter," representing the measurable properties of elements and their compounds: "(a) isomorphism, or the similarity of crystal forms and their ability to form isomorphic mixtures; (b) the relation of specific volumes of similar compounds or elements; (c) the composition of their compound salts; and (d) the relations of the atomic weights of elements." He concluded that these "four aspects" are important because "when a certain property is measured, it ceases to have an arbitrary and subjective character and gives objectivity to the equation."

Since the periodic law was dependent upon the quantitative relation between atomic weight, as an independent variable, and its physical and chemical properties, Mendeleev in 1870 took up the problem of developing an entire "natural system of elements." He employed deduction to reach the boldest and most far-reaching logical consequences of the law that he had discovered, so that he might, by verification of these consequences, confirm the law itself.

Mendeleev simultaneously described various groups of elements for inclusion in the Osnovy khimii and made them the subject of extended laboratory research. He examined molybdenum, tungsten, titanium, uranium, and the rare metals, and in November 1870 he wrote two articles. In the first, "O meste tseria v sisteme elementov" ("On the Place of Cerium in the System of Elements"), he introduced a theoretically corrected value for the atomic weight of cerium—138, instead of the previously accepted 92—and determined its new place within his system. In the second article, "Estestvennaya sistema elementov i primenenie ee k ukazaniyu svoystv neotkrytykh elementov" ("The Natural System of Elements and Its Application to Indicate the Properties of Undiscovered Elements"), Mendeleev predicted that because of the volatility of its salts, eka-aluminum would be discovered by spectroscopic means.

The Osnovy khimii was finished in February 1871. Among the important ideas that the work embodied was Mendeleev's notion of the complexity of the chemical elements and their formation from "ultimates." He stated that the bivalence (II) of magnesium and calcium could be explained as a result of the close blending of monovalent (I) sodium and potassium with monovalent (I) hydrogen:

$$Na_{23}^I + H_1^I = Mg_{24}^{II}; \qquad K_{39}^I + H_1^I = Ca_{40}^{II},$$

a formulation that may be seen as a confused premonition of the later rule of displacement.

In March 1871, two years after his discovery of the law, Mendeleev first named it "periodic." That summer he published in Justus Liebigs Annalen der Chemie his article "Die periodische Gesetzmässigkeit der chemischen Elemente," which he later characterized as "the best summary of my views and ideas on the periodicity of the elements and the original after which so much was written later about this system. This was the main reason for my scientific fame, because much was confirmed—much later." In the fall of that year Mendeleev turned to conducting research on rare earth metals to determine their place among the elements of group IV. One of his goals was to find the ekasilicon (later called germanium) that he had predicted. He also conducted research on hydrates and complex compounds, especially those of ammonia, and gave public lectures in which he combined chemical topics with philosophical ones.

The reception of the periodic law caused Mendeleev considerable mental anguish. In the sharp and prolonged battle that was soon joined, the law at first had few advocates, even among Russian chemists. Its opponents, who were especially vocal in Germany and England, included those chemists who thought in exclusively empirical terms and who were unable to acknowledge the validity of theoretical thinking; Bunsen, Zinin, Lars Nilson, and Carl Petersen were prominent among them. Petersen not only doubted the generality of the periodic law but also defended the contradictory view of the trivalence of beryllium. In Germany, Rammelsberg also took issue with a particular point, attempting in 1872 to refute Mendeleev's proposed correction for the atomic weights of cerium and its close neighbors. Mendeleev answered this charge the following year in an article entitled "O primenimosti periodicheskogo zakona k tseritovym metallam" ("On the Application of the Periodic Law to Cerite Metals"), in which he demonstrated that the facts introduced by Rammelsberg "strengthen, not refute, my proposed changes in the atomic weight of cerium."

A number of other chemists specializing in the system of the elements either attacked Mendeleev's law or disputed his priority. Lothar Meyer, for example, proposed in 1870 a representation for the atomic volumes of the elements in the form of a broken zigzag line. Blomstrand and E. H. von Blomhauer developed a spiral system, also in 1870. Mendeleev answered these and other claims to the periodic law— and also claims against it—in the article "K voprosu o sisteme elementov" ("Toward the Question of a System of Elements"), published in March 1871. Basically, however, he had no patience with disputes over priority, and although by taste an internationalist in science, he engaged in such disputes only when others denigrated Russian achievements.

The years 1871 to 1874 saw the acceptance by a number of chemists of Mendeleev's corrected atomic weights for several elements. Bunsen consented to Mendeleev's value for indium; Rammelsberg and

Roscoe, to that for uranium; Cleve, to that for the rare earth metals (which for yttrium confirmed the values found earlier by Marc Delafontaine, Bunsen, and J. F. Bahr); and Chidenius and Delafontaine, to that for thorium. Nevertheless, the majority of scientists did not accept Mendeleev's discovery for some time; the first textbook on organic chemistry to be based on the law was published in St. Petersburg by Richter only in 1874. Wurtz's *Théorie atomique* further helped to propagate Mendeleev's ideas. At about the same time, Brauner spoke in favor of Mendeleev's corrected weight for chlorine and set out to determine the density of the vapors of beryllium chloride, as Nilson and Petersen had also done— Brauner's determination of the weight of beryllium was a major confirmation of the generality of Mendeleev's law.

The discovery of the three elements predicted by Mendeleev was, however, of decisive importance in the acceptance of his law. In 1875 Lecoq de Boisbaudran, knowing nothing of Mendeleev's work, discovered by spectroscopic methods a new metal, which he named gallium. Both in the nature of its discovery and in a number of its properties gallium coincided with Mendeleev's prediction for eka-aluminum, but its specific weight at first seemed to be less than predicted. Hearing of the discovery, Mendeleev sent to France "Zametka po povodu otkrytia gallia" ("Note on the Occasion of the Discovery of Gallium"), in which he insisted that gallium was in fact his eka-aluminum. Although Lecoq de Boisbaudran objected to this interpretation, he made a second determination of the specific weight of gallium and confirmed that such was indeed the case. From that moment the periodic law was no longer a mere hypothesis, and the scientific world was astounded to note that Mendeleev, the theorist, had seen the properties of a new element more clearly than the chemist who had empirically discovered it. From this time, too, Mendeleev's work came to be more widely known; in 1877 Crookes placed in the *Quarterly Journal of Science* an abstract, entitled "The Chemistry of the Future," of Mendeleev's summarizing article of 1871, while in 1879 a French translation of the full article, with a new introduction, was published by G. G. Quesneville in the *Moniteur scientifique*.

The discovery of gallium was incorporated into the third edition of *Osnovy khimii* in 1877. The fourth edition, of 1881–1882, mentioned the discovery of scandium—the ekaboron predicted by Mendeleev—by Nilson, in 1879. Winkler discovered germanium in 1886; its properties matched precisely those of Mendeleev's ekasilicon, and the discovery of germanium figured in the fifth edition of Mendeleev's book in 1889. This edition also contained, within a single frame, reproductions of portraits of Lecoq de Boisbaudran, Nilson, Winkler, and Brauner. The composite bore the caption "Reinforcers of the Periodic Law."

The periodic law might now be considered proven, and Mendeleev presented a summary of the research leading to it in his Faraday lecture, "The Periodic Law of the Chemical Elements," which he delivered at the invitation of the Chemical Society of London on the occasion of the twentieth anniversary of his discovery. He spoke of the scientists who preceded him in his work as well as those who later contributed to the development of the law, and dealt with both the history and what might be called the prehistory of it. During the same visit to London Mendeleev was also invited to lecture before the Royal Institution of Great Britain. In this speech, "An Attempt at the Application to Chemistry of One of Newton's Principles," he sought to oppose the concept of chemical structure to the hypothesis that the mutual influence of atoms within the molecule is in concord with Newton's third law of motion. In both these lectures (which he published in 1889 as *Dva Londonskikh chtenia* ["Two London Lectures"]), Mendeleev did not confine himself to chemistry, but went on to draw philosophical generalities that embraced the whole of the natural sciences.

Even while he was working toward the periodic law and its proofs, Mendeleev was also concerned with the problem of the liquefaction of gases. As early as 1870 he discussed the necessity of intensive cooling in the process, while in December 1871 he suddenly turned to purely physical research on permanent gases and their compressibility. In initiating this investigation, he hoped to find the hypothetical "universal ether," which he believed to be an extraordinarily thin gas that must, in his system, occupy the place above hydrogen. Although his primary goal was unreachable, in the course of his studies Mendeleev discovered a number of deviations of gases from the Boyle-Mariotte law, and gave a more precise equation for the state of real gases. His work then assumed a more practical slant, and he turned to aeronautical research; giving a general form to his experiments on the temperature of the upper layers of the atmosphere (1875) in the report "Ob opytakh nad uprugostyu gazov" ("On Experiments on the Elasticity of Gases"), published in 1881. In 1887 Mendeleev made a solo balloon ascension from Klin, for the purpose of observing a solar eclipse.

After 1884, Mendeleev concerned himself with the expansion of liquids and in particular with the specific weights of aqueous solutions of various substances. He was able to conclude that in such solutions discontinuous relationships exist between the solvent and the solute, attesting to the existence of determinate chemical relationships—a necessary condition, according to chemical atomic theory. Mendeleev thus arrived at a chemical theory of solutions, which he opposed to the theory of electrolytic dissociation of dilute aqueous solutions set forth by Arrhenius. Mendeleev stated his theory both in his *Issledovanie vodnykh rastvorov po udelnomu vesu* ("Research on Aqueous Solutions According to Their Specific Weight") of 1887 and in the fifth edition of *Osnovy khimii.*

From the last years of the 1870's Mendeleev was also concerned with the production of petroleum. In 1876 he visited the United States; in the resultant book, *Neftyanaya promyshlennost v Severo-amerikanskom shtate Pensilvanii i na Kavkaze* ("Petroleum Production in the North American State of Pennsylvania and in the Caucasus"), he advanced a theory of the inorganic origin of petroleum. Mendeleev traveled to Baku, too, to study oilfields. He did further research on the uses of petroleum, including its medical applications; in 1878 he employed petroleum as a self-treatment for pleurisy. In 1880–1881, Mendeleev wrote a series of reports of the results of his Caucasian journeys, and thus became engaged in a dispute with Nobel over the proper location of petroleum refineries. In 1883, with *Po voprosy o nefti* ("On a Question of Petroleum"), he entered into a discussion with Markovnikov; in the same year and the one following he wrote a series of works on the refining of both Baku and American oil. In an article of 1889 he denied recurrent rumors of the exhaustion of the Baku fields.

By the end of the 1880's Mendeleev had added an investigation of the coal industry to his practical concerns; he visited the Donets Basin to study mining and wrote *Budushchaya sila, pokoyushchayasya na beregakh Dontsa* ("Future Power Lying on the Banks of the Donets"; 1888). None of his efforts toward the development of domestic industry was successful, however; the czarist government chose to dismiss his remarkable ideas and projects as "professorial dreams."

The decades of the 1870's and 1880's marked a major transitional period in Mendeleev's life. The law that he had discovered was developed and confirmed, and he had turned to more commercial matters in the interest of the national economy; he had also left his family and entered into a second marriage, in 1882, with a young artist, Anna Ivanova Popova, by whom he had two sons and two daughters. He became increasingly concerned with philosophical matters, of which he wrote, "Much in me was changing; at that time I read much on religion, on sects, and philosophy, economic articles." His writings on philosophical themes included the articles "Ob ediniyse" ("On Unity"; 1870), "Pered kartinoyu A. I. Kuindzhi" ("Before a Picture of A. I. Kuindzhi"; 1880), "O edinstve veshchestva" ("On the Unity of Matter"; 1886), and, most importantly, the book *Materialy dlya suzhdenia o spiritizma* ("Material for an Opinion on Spiritism"; 1876), which embodied the results of the work of a special commission of the Physical Society of St. Petersburg. Of the last, Mendeleev later wrote, "I tried to fight against superstition . . . it took professors to act against the authority of professors. The result was right, spiritism was rejected."

The same period also saw a change in Mendeleev's academic status. In 1876 he was elected a corresponding member of the St. Petersburg Academy of Sciences; in 1880 he was defeated in an election for extraordinary membership by the reactionary majority of members of the physics and mathematics section, who had come to fear his democratic tendencies. In the course of the protests that followed this event, Butlerov published an article entitled "Russkaya ili tolko Imperatorskaya Akademia Nauk v S.-Peterburge" ("The Russian or Only an Imperial Academy of Sciences in St. Petersburg") and some twenty other scientific institutions elected Mendeleev an honorary member. In 1890 disorders broke out among the students at the University of St. Petersburg and Mendeleev undertook to deliver a student petition to the ministry of education. He was given a rude and insulting answer, tantamount to a demand for his personal resignation; he thus left the university, where he had taught for more than thirty years. On 3 April 1890 Mendeleev gave his last lecture to the students of the general chemistry course.

His teaching career at an end, Mendeleev decided to publish a newspaper in support of the protectionist policies that his investigations of petroleum and coal production had convinced him were the "sole means of saving Russia." Before he could begin this project, however, he received a commission from the naval ministry to conduct large-scale laboratory research on the production of smokeless powder—a secret project that took high priority. From 1890 to 1892, he also participated in a study of the tariff structure, at the invitation of the ministry of finance; this resulted in his *Tolkovy tarif* ("Comprehensive Tariff"), which was published in 1891–1892. The government was appreciative of his services, and Mendeleev rose rapidly in the bureaucracy, being appointed privy councillor in 1891.

From 1892 on, Mendeleev was concerned in the regulation of the system of weights and measures in Russia, a task that he discharged "with enthusiasm, since here the purely scientific was closely interwoven with the practical." In 1893 he was named director of the newly created Central Board of Weights and Measures, a post that he held until his death, and in connection with which he frequently traveled abroad. In the 1890's Mendeleev was also actively involved in problems of shipbuilding and the development of shipping routes. He participated in the design of the icebreaker *Ermak* (launched in 1899) and wrote on the progress of research in the northern Arctic Ocean (1901). He simultaneously studied the development of heavy industry in Russia, traveled to the Urals and to Siberia to observe the production of iron, began to publish a series entitled Biblioteka Promyshlennykh Znany ("Library of Industrial Knowledge")—for which he also compiled a curriculum—and wrote, in addition to several related books and articles, *Uchenie o promyshlennosti* ("Theory of Industry"; 1901), which contained a number of ideas that he later developed in his *Zavetnye mysli* ("Private Thoughts"; 1903–1905).

Nor was Mendeleev unconcerned with theoretical chemistry during these years. A sixth edition of *Osnovy khimii* was published in 1895; in it he expressed some skepticism about the discoveries (in 1894 and 1895, respectively) of the first inert gases, argon and helium. After Ramsay's 1898 discovery of their three analogues (which Ramsay had himself predicted), and after the determination of the place of the whole group as a zero valence group within the periodic system (1900), Mendeleev reconsidered his position and not only accepted the new elements but also grouped Ramsay among the "reinforcers of the periodic law."

Mendeleev denied the discovery of the electron, however, and in particular the explanation of radioactivity as the disintegration of atoms and the transformation of elements, thinking that these discoveries destroyed the very foundations of the periodic law. He disputed the transmutation of elements in his article "Zoloto iz serebra" ("Gold From Silver") of 1898; while in *Popytka khimicheskogo ponimania mirovogo efira* ("An Attempt at a Chemical Conception of the Universal Ether") of 1902, he introduced the erroneous notion that the universal ether is similar in nature to a very light inert gas and that it takes part in radioactive processes. In 1902 he visited the Paris laboratories of the Curies and Becquerel to study radioactivity further. He dealt with these questions in the seventh and eighth editions (1903 and 1906) of his textbook.

The fifth and later editions of the *Osnovy khimii*

were translated into the western European languages. In addition, from 1892 Mendeleev took an active part in the preparation of the great Brockhaus encyclopedia, which provided another vehicle for the dissemination of his ideas in western Europe. He introduced a section on chemistry and the production of chemicals and wrote the articles on matter, the periodic regularity of the chemical elements, and technology, among a number of other topics. His work as a whole amounted to more than 400 books and articles, as well as a large number of manuscripts, which are preserved in the D. I. Mendeleev Museum-Archive, Leningrad State University.

In 1894 Mendeleev was awarded the doctorate by both Oxford and Cambridge; his seventieth birthday was widely observed in 1904, as was the fiftieth anniversary of his scientific career the following year. In 1905 he attended the commemorative session of the Royal Society of London, and was awarded the Copley Medal; he was also a member of many Russian and foreign scientific societies. He held several czarist orders and the French government made him a member of the Legion of Honor. Following his death from heart failure, students followed his funeral to the Volkov Cemetery in St. Petersburg. They carried the periodic table of the elements high above the procession as the fitting emblem of Mendeleev's career.

BIBLIOGRAPHY

I. Original Works. Mendeleev's works were published in Russian during his lifetime in the form of monographs and magazine and encyclopedia articles, many of which were translated into English, French, and German. For Mendeleev's own annotated bibliography of his works (1899) see *Sochinenia,* XXV, 686–776. After his death *Osnovy khimii* was reprinted many times, collections of his work appeared in the series Klassiki Nauki ("Classics of Science"), his archival material was published, and his complete works were published as *Sochinenia* ("Works"), 25 vols. (Leningrad, 1934–1952); a supp. vol. contains a detailed index.

Osnovy khimii ("Principles of Chemistry"), Mendeleev's main work, went through eight eds. during his lifetime: 1868–1871; 1872–1873; 1877; 1881–1882; 1889; 1895; 1903; 1906. The 1st ed., in 4 pts., was published in two vols. Starting with the 5th ed. the book was no longer divided into parts. In the 8th ed. all notes were placed at the end of the book as special appendixes. The periodic law is the focus of the work; beginning with the 3rd ed. (after the discovery of gallium in 1875) it is more prominent because it had been verified experimentally. Mendeleev rewrote each ed., including all new scientific data—particularly confirmations of the periodic law—and reanalyzing difficulties that had arisen to hinder its confirmation (inert

gases, radioactivity, radioactive and rare-earth elements). He also expanded the sections on the chemical industry and on philosophy and methodology. To update the posthumous 9th (1928) and 10th (1931) eds. a section by G. V. Wulff *et al.* was added on new trends in the construction of chemical principles. This new section was omitted from the 11th to 13th eds. (1932–1947).

The first English trans., *The Principles of Chemistry* (London, 1891), was made from the 5th ed. The 2nd (London, 1897) and 3rd (London, 1905; repr. New York, 1969) eds. were based on the 6th and 7th Russian eds., respectively. A German version, *Grundlagen der Chemie* (St. Petersburg, 1890), was based on the 5th ed. and a French trans., *Principes de chimie* (Paris, 1895), on the 6th ed.

Several articles from *Periodichesky zakon* ("Periodic Law"), B. M. Kedrov, ed., in the series Klassiki Nauki (Moscow, 1958), were published in other languages. Those in English are "The Periodic Law of the Chemical Elements" (written in 1871), in *Chemical News and Journal of Physical (Industrial) Science,* **40,** nos. 1042–1048 (1879); **41,** nos. 1049–1060 (1880), reviewed by W. Crookes, "The Chemistry of the Future," in *Quarterly Journal of Science,* no. 55 (July 1877), also published separately, George Kamensky, trans., as *An Attempt Towards a Chemical Conception of the Ether* (London, 1904); "An Attempt to Apply to Chemistry One of the Principles of Newton's Natural Philosophy," in *Chemical News,* **60,** no. 1545 (1889), 1–4; no. 1546, pp. 15–17; no. 1547, pp. 30–32, also published separately under the same title (London, 1889); and "The Periodic Law of the Chemical Elements (Faraday Lecture)," in *Journal of the Chemical Society,* **55** (1889), 634–656, repr. in *Faraday Lectures 1869–1928* (London, 1928). See also "The Relations Between the Properties and Atomic Weights of the Elements," in Henry M. Leicester and Herbert S. Klickstein, eds., *A Source Book in Chemistry 1400–1900* (New York, 1952), 438–444.

French translations include "Remarques à propos de la découverte du gallium," in *Comptes rendus . . . de l'Académie des sciences,* **81** (1875), 969–972; "La loi périodique des éléments chimiques (Faraday Lecture)," in *Moniteur scientifique,* 3rd ser., **9** (1879), 691 ff.; "La loi périodique des éléments chimiques," *ibid.,* 4th ser., **3,** pt. 2, no. 572 (1889), 899–904; and "Comment j'ai trouvé le système périodique des éléments," in *Revue générale de chimie pure et appliquée,* **1** (1899), 210 ff., 510 ff.—here, repr. under another title, is "Periodicheskaya zakonnost khimicheskikh elementov" ("Periodic Law of Chemical Elements"), from Brockhaus and Efron's *Entsiklopedichesky slovar* ("Encyclopedic Dictionary").

For German translations, see "Zur Frage über das System der chemischen Elemente," in *Berichte der Deutschen chemischen Gesellschaft,* **4** (1871), 348–352; "Die periodische Gesetzmässigkeit der chemischen Elemente," in *Justus Liebigs Annalen der Chemie,* supp. **8,** no. 2 (1871), 133–229, repr. in Ostwald's Klassiker, no. 68 (Leipzig, 1913), pp. 41–118; "Ueber die Stellung des Ceriums im System der Elemente," in *Bulletin de l'Académie des sciences de St. Petersbourg,* **16** (1871), 45–50; "Ueber die Anwendbarkeit des periodischen Gesetzes bei die Ceritmetallen," in *Justus Liebigs Annalen der Chemie*, **168**, no. 1 (1873), 45–63; "Zur Geschichte des periodischen Gesetzes," in *Berichte der Deutschen chemischen Gesellschaft*, **8** (1875), 1796–1804; and "Das natürliche System der chemischen Elemente," in Ostwald's Klassiker, no. 68 (Leipzig, 1913), pp. 20–40.

Publications from Mendeleev's scientific archives include *Novye materialy po istorii otkrytia periodicheskogo zakona* ("New Material on the History of the Discovery of the Periodic Law"), B. M. Kedrov, ed. (Moscow, 1950); *Nauchny arkhiv* ("Scientific Archive"), I, *Periodichesky zakon* ("The Periodic Law"), compiled and edited by B. M. Kedrov (Moscow, 1953), which includes theoretical material from 1869–1871. Notes on experiments related to the periodic law during this period will appear in vol. II, *Eksperimentalnye raboty* ("Experimental Works"). Material preceding the periodic law and directed toward it will be in vol. III, *Podgotovlenie otkrytia* ("Preparation for the Discovery"), and that which followed the discovery will be in vol. IV, *Razrabotka otkrytia* ("Development of the Discovery"); *Rastvory* ("Solutions"), K. P. Mishchenko, ed., in the series Klassiki Nauki (Leningrad, 1959); *Nauchny arkhiv. Osvoenie kraynego Severa* ("Scientific Archive. The Conquest of the Far North"), I, *Vysokie shiroty Severnogo Ledovitogo okeana* ("High Latitudes of the Northern Arctic Ocean"), A. I. Dubravin, ed. (Moscow–Leningrad, 1960); *Nauchny arkhiv. Rastvory* ("Scientific Archive. Solutions"), compiled by R. B. Dobrotin (Moscow–Leningrad, 1960); and *Izbrannye lektsii po khimii* ("Selected Lectures in Chemistry"), compiled by A. A. Makarenya *et al.* (Moscow, 1968), Mendeleev's chemistry lectures from 1864, 1870–1871, and 1889–1890.

For autobiographical source material, see *D. I. Mendeleev. Literaturnoe nasledstvo*, I, *Zametki i materialy D. I. Mendeleeva biograficheskogo kharaktera* (Leningrad, 1938), a collection of biographical notes and material; *Arkhiv D. I. Mendeleeva*, I, *Avtobiograficheskie materialy. Sbornik dokumentov* (Leningrad, 1951), with Mendeleev's bibliography of his works, pp. 39–130, as well as diary notes, chronology, a catalog of Mendeleev's personal library, and a list of the contents of his scientific archives; and M. D. Mendeleeva, ed., *Nauchnoe nasledstvo* ("Scientific Heritage"), Natural Science Ser., II (Moscow–Leningrad, 1951): see pp. 111–256 for Mendeleev's diaries for 1861–1862, and pp. 257–294 for his letters concerning his work on smokeless powder, P. M. Lukyanov, ed.

II. SECONDARY LITERATURE. Biographical works on Mendeleev include N. A. Figurovsky, *Dmitry Ivanovich Mendeleev* (Moscow, 1961); B. Kedrov and T. Chentsova, *Brauner—spodvizhnik Mendeleeva* (Moscow, 1960), written by Mendeleev's associate, with their correspondence and material on their elaboration of the periodic law; A. A. Makarenya *et al.*, eds., *D. I. Mendeleev v vospominaniakh sovremennikov* ("Mendeleev Recalled by His Contemporaries"; Moscow, 1969), with recollections of his friends, students, acquaintances, and relatives; A. I. Mendeleeva, *Mendeleev v zhizni* ("Mendeleev in Life"; Moscow, 1928), written by Mendeleev's second wife; M. N. Mladentsev and V. E. Tishchenko, *Mendeleev, ego zhizn i deyatelnost*

("Mendeleev, His Life and Work"), I (Moscow–Leningrad, 1938)—Tishchenko was Mendeleev's laboratory assistant and the biography goes as far as 1861; O. N. Pisarzhevsky, *Dmitry Ivanovich Mendeleev* (Moscow, 1959); and *Semeynaya khronika* ("Family Chronicle"; St. Petersburg, 1908), letters from his relatives and recollections by his niece, N. Y. Gubkinaya.

On the history of the discovery and development of the periodic law, see L. A. Chugaev, *Periodicheskaya sistema khimicheskikh elementov* ("The Periodic System of the Chemical Elements"; St. Petersburg, 1913); K. Danzer, *Dmitri I. Mendelejew und Lothar Meyer. Die Schöpfer des Periodensystems der chemischen Elemente* (Leipzig, 1971); B. M. Kedrov, *Razvitie ponyatia elementa ot Mendeleeva do nashikh dney* ("The Development of the Concept of the Element From Mendeleev to Our Times"; Moscow–Leningrad, 1948), pp. 24–71, 220–239; *Evolyutsia ponyatia elementa v khimii* ("The Evolution of the Concept of the Element in Chemistry"; Moscow, 1956), pp. 137–161, 188–294; *Den odnogo velikogo otkrytia* ("The Day of One Great Discovery"; Moscow, 1958), also translated into French as "Le 1er mars 1869; jour de la découverte de la loi périodique par D. I. Mendeléev," in *Cahiers d'histoire mondiale*, VI, 3, 644–656, gives the history of the discovery of the periodic law on 1 Mar. 1869 and includes many archival documents; *Filosofsky analiz pervykh trudov D. I. Mendeleeva o periodicheskom zakone (1869–1871)* ("A Philosophical Analysis of the First Works of D. I. Mendeleev on the Periodic Law [1869–1871]"; Moscow, 1959), a continuation of the preceding work; *Tri aspekta atomistiki* ("Three Aspects of Atomic Theory"), III, *Zakon Mendeleeva. Logiko-istorichesky aspekt* ("Mendeleev's Law. Logical-Historical Aspect"; Moscow, 1969); *Mikroanatomia velikogo otkrytia. K 100-letiyu zakona Mendeleeva* ("Microanatomy of the Great Discovery. For the 100th Anniversary of Mendeleev's Discovery"; Moscow, 1970); B. M. Kedrov and D. N. Trifonov, *Zakon periodichnosti i khimicheskie elementy. Otkrytia i khronologia* ("The Law of Periodicity and the Chemical Elements. Discoveries and Chronology"; Moscow, 1969); Paul Kolodkine, *Dmitri Mendeleiv et la loi périodique* (Paris, 1963); V. A. Krotikov, "The Mendeleev Archives and Museum of the Leningrad University," in *Journal of Chemical Education*, **37** (1960), 625–628; V. Y. Kurbatov, *Zakon D. I. Mendeleeva* ("D. I. Mendeleev's Law"; Leningrad, 1925); Henry M. Leicester, "Dmitrii Ivanovich Mendeleev," in Eduard Farber, ed., *Great Chemists* (New York, 1961), 719–732; A. A. Makarenya, *D. I. Mendeleev o radioaktivnosti i slozhnosti elementov* ("D. I. Mendeleev on Radioactivity and the Complexity of the Elements"), 2nd ed. (Moscow, 1965); and *D. I. Mendeleev i fiziko-khimicheskie nauki* ("D. I. Mendeleev and the Physicochemical Sciences"; Moscow, 1972), an attempt at a scientific biography; F. A. Paneth, "Radioactivity and the Completion of the Periodic System," in *Nature*, **149** (23 May 1942), 565–568; *Periodichesky zakon D. I. Mendeleeva i ego filosofskoe znachenie* ("Periodic Law of D. I. Mendeleev and Its Philosophical Significance"; Moscow, 1947), a collection of articles by A. N. Bakh, A. F. Joffe, A. E. Fersmann,

A. V. Rakovsky, B. M. Kedrov, G. S. Vasetsky, and B. N. Vyropaev; *Periodichesky zakon i stroenie atoma* ("The Periodic Law and Atomic Structure"; Moscow, 1971), a collection of articles; O. N. Pisarzhevsky, *Dmitrii Ivanovich Mendeleev. His Life and Work* (Moscow, 1959); Daniel Posin, *Mendeleyev, the Story of a Great Scientist* (New York, 1948); E. Rabinowitsch and E. Thilo, *Periodisches System. Geschichte und Theorie* (Stuttgart, 1930), trans. into Russian as *Periodicheskaya sistema elementov. Istoria i teoria* (Moscow–Leningrad, 1933), pt. 1, esp. ch. 5; N. N. Semenov, ed., *Sto let periodicheskogo zakona khimicheskikh elementov* ("A Hundred Years of the Periodic Law of the Chemical Elements") (Moscow, 1969); *75 let periodicheskogo zakona D. I. Mendeleeva i Russkogo khimicheskogo obshchestva* ("Seventy-five Years of Mendeleev's Periodic Law and the Russian Chemical Society"; Moscow–Leningrad, 1947), a collection of articles with a bibliography on pp. 261–265; T. E. Thorpe, *Essays in Historical Chemistry* (London, 1911), 483–499; W. A. Tilden, *Famous Chemists* (London, 1921), 241–258; D. N. Trifonov, *O kolichestvennoy interpretatsii periodichnosti* ("On the Quantitative Interpretation of Periodicity"; Moscow, 1971), pp. 16–31; *Voprosy estestvoznania i tekhniki* ("Questions of Natural Science and Technology"), no. 4 (29) (Moscow, 1969), a collection commemorating the centenary of the periodic law; Alexander Vucinich, "Mendeleev's Views on Science and Society," in *Isis*, **58** (1967), 342–351; and *Science in Russian Culture* (Stanford, Calif., 1970), 147–165; and *Yubileynomu Mendeleevskomu sezdu v oznamenovanie 100-letney godovshchiny so dnya rozhdenia D. I. Mendeleeva. Varianty periodicheskoy sistemy* ("Mendeleev Anniversary Congress in Recognition of the Centenary of D. I. Mendeleev's Birth. Variants of the Periodic System"), collected by M. A. Blokh (Leningrad, 1934).

On other aspects of Mendeleev's life and work, see V. P. Barzakovsky and R. B. Dobrotin, *Trudy D. I. Mendeleeva v oblasti khimii silikatov i stekloobraznogo sostoyania* ("Mendeleev's Works on the Chemistry of Silicates and Glass-Making"; Moscow, 1960); T. S. Kudryavtseva and M. E. Shekhter, *Mendeleev i ugolnaya promyshlennost Rossii* ("Mendeleev and the Russian Coal Industry"; Moscow, 1952); A. A. Makarenya and I. N. Filimonova, *D. I. Mendeleev i Peterburgsky universitet* ("Mendeleev and St. Petersburg University"; Leningrad, 1969); V. E. Parkhomenko, *D. I. Mendeleev i russkoe neftyanoe delo* ("Mendeleev and the Russian Petroleum Industry"; Moscow, 1957); S. I. Volfkovich *et al.*, eds., *D. I. Mendeleev. Raboty po selskomu khozyaystvu i lesovodstvu* ("Mendeleev. Works on Agriculture and Forestry; Moscow, 1954); and G. A. Zabrodsky, *Mirovozzrenie D. I. Mendeleeva* ("Mendeleev's World View"; Moscow, 1957), issued to commemorate the fiftieth anniversary of his death.

B. M. KEDROV

MENEGHETTI, EGIDIO (*b.* Verona, Italy, 14 November 1892; *d.* Padua, Italy, 4 March 1961), *experimental pharmacology*.

Meneghetti was the son of Umberto Meneghetti and Clorinda Stegagno; both his father and grandfather were physicians. He attended school in Verona, then entered the University of Padua, where from 1913 to 1914 he worked under Luigi Sabbatani, director of the Institute of Pharmacology. He graduated *cum laude* in 1916. Meneghetti was named chief of the pharmacology laboratory in 1919; in 1922 he became teacher of experimental pharmacology. In 1926 he left Padua to take up a similar post at Camerino. Two years later he accepted an appointment at the University of Palermo, then, in 1932, returned to Padua as professor of pharmacology. He was chosen rector of the university in 1945 after the Liberation, and in 1951 founded the Centro di Studio per la Chemoterapia there.

Meneghetti's first important scientific work, in 1921, concerned the relationship of the hemolytic and fixative actions of metals to their place in the table of atomic weights. He used the techniques of quantitative biophysics to demonstrate that such actions increase in intensity as the ionic tension of the metallic solution decreases. He was also concerned with technology; in 1925 he improved the apparatus by which artifical circulation could be maintained in the isolated heart of a frog. By 1928, however, he had formulated the basis for his subsequent researches; in his inaugural lecture at the University of Palermo, given on 5 March of that year, Meneghetti stated that experimental biochemistry and cell physiology must constitute the basis of modern pharmacology.

The greatest part of Meneghetti's pharmacological contributions concern the occasionally overlapping fields of colloids and toxicology. His first work on colloids was his article *Über die pharmakologischen Wirkung des kolloidalen Arsensulfids*, published in 1921. He developed this line of inquiry in a series of works, written between 1924 and 1934, on the trivalent and pentavalent compounds of antimony and in articles (in 1930 and 1937) on the salts of silver, gold, and copper. He demonstrated that substances that are of limited solubility in water may be introduced into the circulatory system (or injected locally) in the colloidal state.

In a series of researches conducted between 1924 and 1926 Meneghetti showed a specific effect of colloidal sulfur of antimony when it is injected into a vein. He was able to demonstrate that it lodged in the histiocytes of the bone marrow, thereby disrupting erythropoiesis, as is evidenced by the appearance of immature erythrocytes in the blood. (Two conditions are necessary to the observation of this erythroneocytosis—the granules of the colloidal preparation must be extremely fine, and the injection must be made slowly; otherwise the substance is exhibited more strongly in the macrophagic reticuloendothelial cells of the lungs and liver.) Meneghetti's proof that erythroblasts can be produced in the blood by the fixation of a toxic substance in the reticuloendothelial cells of the bone marrow was a major contribution toward the understanding of the pathogenesis of blood diseases (see his very important work *Emopatia primitiva da solfuro di antimonio colloidale* [1926]).

Meneghetti devoted ten works to toxicology between 1928 and 1936. In them he examined the efficacy of the therapeutic use of sodium thiosulfate and tetrathionate to counter mercury, lead, and cyanide poisoning. In 1936 he also conducted research on the thiazinic dyes, including methylene blue and toluidin blue, two highly dispersed electropositive colloids, and on the action, similar to that of digitalis, of alkaloid substances derived from *Erythrophloeum* (work which was developed in 1939 by Meneghetti's student Renato Santi). In the same year, too, he returned to the study of the reticuloendothelial system to demonstrate the presence of histiocytes in the pulmonary alveoli. In his article "Il polmone come sede di azione e come via di assorbimento di farmaci," Meneghetti posited a relationship between the activity of these alveolar cells and the absorption of drugs into the lung, a new route for the chemotherapeutic treatment of infectious diseases. He further took up the question of the special conditions that might macrophagically fix such electropositive colloids as the cupric oxide in the lung: in particular he noted the relation of blood coagulability to cell permeability, and therefore to granulopexis.

Meneghetti conducted a long series of investigations of the factors that can modify the relation between the degree of dispersion of colloids and the intensity of their action. In 1939 he drew upon his results to state that such highly stable colloids as the electropositive ones have a rapid and intense action that recommends them for pharmacological use. He thus completed some of his earlier researches; a decade before, he had pronounced the colloidal state to be the "pharmacological" one. In later work (1943 and 1954) he went on to investigate the relationship between the molecular structure and pharmacological action of the sulfa drugs and antibiotics.

Meneghetti wrote more than a hundred scientific works. He was also concerned with the history of medicine and with social problems—his book *Biologia rivoluzionaria* was published posthumously in 1962. He was a humanist and a patriot; he was decorated four times for valor in World War I, and in World War II, in which he lost both his wife and his daughter in an air raid, he showed equal courage as a member of the Resistance.

BIBLIOGRAPHY

I. ORIGINAL WORKS. Among Meneghetti's earliest publications is a brief *Curriculum vitae* (Padua, 1925). His inaugural lecture at the University of Palermo is "Chimismo, forma, funzione e fenomeni colloidali," in *Biochimica e terapia sperimentale*, **15** (1928), 77–98. His *Elementi di Farmacologia* (Padua, 1934) went through many eds. and revisions; the 9th ed. was entitled *Farmacologia*, 2 vols. (Padua, 1958). Meneghetti wrote more than 100 articles on his research, including reports on the toxicology of arsenic, erythrocytes, antimony, sulfur, and the histiocytic system. For a complete bibliography, see Renato Santi, "Commemorazione . . ." (below).

Meneghetti published two reports on his chemotherapy study center, "Centro di Studio per la Chemoterapia. Attività svolta nel quinquennio 1951–1955," in *Ricerca scientifica*, **26** (1956), 72–99; and "Centro di Studio . . . 1956–1960," *ibid.*, **30** (1960), 2228–2240. He also contributed many articles to *Enciclopedia medica italiana* (Rome, 1952). Two works were published posthumously, *Biologia rivoluzionaria* (Padua, 1962); and *Poesie e prose* (Vicenza, 1963), with a preface by E. Opocher and D. Valeri, which is a collection of works showing his civil and social concerns.

II. SECONDARY LITERATURE. See M. Aloisi, "Ricordo di Egidio Meneghetti nell'anniversario della sua morte," in *Atti dell'Accademia nazionale dei Lincei. Rendiconti*, fasc. 2 (1962), 78–83; A. Cestari, "In memoria di Egidio Meneghetti," in *Ricerca scientifica*, 2nd ser., **1**, pt. 1 (1961), 123–128; M. Messini, "Egidio Meneghetti a un anno dalla morte," in *Clinica terapeutica*, **22** (1962), 461–465; and R. Santi, "Egidio Meneghetti," in *Archivio italiano di scienze farmacologiche*, 3rd ser., **11** (1961), 183–186; and "Commemorazione di Egidio Meneghetti," in *Annuario dell'Università di Padova* (Padua, 1963), which contains a complete bibliography of Meneghetti's writings.

PIETRO FRANCESCHINI

MENELAUS OF ALEXANDRIA (*fl.* Alexandria and Rome, A.D. 100), *geometry, trigonometry, astronomy.*

Ptolemy records that Menelaus made two astronomical observations at Rome in the first year of the reign of Trajan, that is, A.D. 98.[1] This dating accords with Plutarch's choice of him as a character in a dialogue supposed to have taken place at or near Rome some time after A.D. 75.[2] He is called "Menelaus of Alexandria" by Pappus and Proclus.[3] Nothing more is known of his life.

The first of the observations that Ptolemy records was the occultation of the star Spica by the moon at the tenth hour in the night (that is, 4 A.M. in seasonal hours or 5 A.M. in standard hours) of the fifteenth-sixteenth of the Egyptian month Mechir and its emergence at the eleventh hour.[4] In the second observation Menelaus noticed that at the eleventh hour in the night of 18–19 Mechir the southern horn of the moon appeared to fall in line with the middle and southern stars in the brow of Scorpio, while its center fell to the east of this straight line and was as distant from the star in the middle of Scorpio as the middle star was from the southern, and the northern star of the brow was occulted.[5] Both these observations took place in year 845 of the era of Nabonassar (reigned 747–734 B.C.). By comparing the position of the stars as observed by Timocharis in year 454 of the era of Nabonassar, Ptolemy (and presumably Menelaus before him) concluded that the stars had advanced to the east by 3°55′ in 391 years, from which he confirmed the discovery originally made by Hipparchus that the equinox was moving westward at the rate of 1° a century. (The true figure is 1° in about seventy-two years.) It was by comparing the position of Spica in his day with that recorded by Timocharis that Hipparchus had been led to postulate the precession of the equinoxes.

A list of works attributed to Menelaus is given in the register of mathematicians in the *Fihrist* ("Index") of Ibn al-Nadīm (second half of tenth century). His entry reads:[6]

> He lived before Ptolemy, since the latter makes mention of him. He composed: *The Book on Spherical Propositions. On the Knowledge of the Weights and Distribution of Different Bodies*, composed at the commission of Domitian.[7] Three books on the *Elements of Geometry*, edited by Thābit ibn Qurra. The *Book on the Triangle*. Some of these have been translated into Arabic.

From references by the Arabic writers al-Battānī (*d.* 929), al-Ṣūfī (*d.* 986), and Ḥajjī-Khalīfa it has been deduced that Menelaus composed a catalog of the fixed stars, but there is some uncertainty whether the observations that he undoubtedly made were part of a full catalog.[8]

According to Pappus, Menelaus wrote a treatise on the settings of the signs of the zodiac.[9] Hipparchus had shown "by numbers" that the signs of the zodiac take unequal times to rise, but he had not dealt with their settings. Menelaus appears to have remedied the omission.[10] The work has not survived, nor did Pappus redeem his promise to examine it later, not at least in any surviving writings.

The problem can be solved rigorously only by the use of trigonometry,[11] and it is on his contributions to trigonometry that the fame of Menelaus chiefly rests. Theon of Alexandria noted that Hipparchus had treated chords in a circle in twelve books and Menelaus in six.[12] Almost certainly this means that Menelaus,

like Hipparchus before him, compiled a table of sines similar to that found in Ptolemy. For the Greeks, if *AB* is a chord of a circle, sin *AB* is half the chord subtended by double of the arc *AB* and a table of chords is, in effect, a table of sines. Menelaus' work has not survived.

Menelaus' major contribution to the rising science of trigonometry was contained in his *Sphaerica*, in three books. It is this work which entitles him to be regarded as the founder of spherical trigonometry and the first to have disengaged trigonometry from spherics and astronomy and to have made it a separate science. The work has not survived in Greek; but it was translated into Arabic, probably through a lost Syriac rendering, and from Arabic into Latin and Hebrew. There have been three printed Latin versions; and although it is debatable how much of them is due to Menelaus and how much to their editors, a modern study in German by A. A. Björnbo and a critical edition of the Arabic text with German translation by Max Krause make the content of Menelaus' work tolerably clear.[13]

Book I opens with the definition "A spherical triangle is the space included by arcs of great circles on the surface of a sphere," subject to the limitation that "these arcs are always less than a semicircle." This is the earliest known mention of a spherical triangle. Since the Arabic tradition makes Menelaus address a prince with the words, "O prince, I have discovered a splendid form of demonstrative reasoning," it would appear that he was claiming originality. This is, indeed, implied in a reference by Pappus, who, after describing how a spherical triangle is drawn, says, "Menelaus in his *Sphaerica* calls such a figure a *tripleuron* [τρίπλευρον]."[14] Euclid (in *Elements* I, defs. 19, 20) had used τρίπλευρον for plane rectilinear figures having three sides—that is, triangles —but in the body of his work, beginning with proposition 1, he regularly employed the term τρίγωνον, "triangle." Menelaus' deliberate choice of *tripleuron* for a spherical triangle shows a consciousness of innovation.

In book I Menelaus appears to make it his aim to prove for a spherical triangle propositions analogous to those of Euclid for a plane triangle in *Elements* I. In proposition 11 it is proved that the three angles of a spherical triangle are together greater than two right angles. Menelaus did not always use Euclid's form of proof even where it can be adapted to the sphere, and he avoided the use of indirect proofs by *reductio ad absurdum*. Sometimes his treatment, as of the "ambiguous case" in the congruence of triangles (prop. 13), is more complete than Euclid's.

Book I is an exercise in spherics in the old sense of

that term—the geometry of the surface of the sphere—and book II consists only of generalizations or extensions of Theodosius' *Sphaerica* needed in astronomy; the proofs, however, are quite different from those of Theodosius. It is in book III that spherical trigonometry is developed. It opens (prop. 1) with the proposition long since known as "Menelaus' theorem." This is best known from the proof in Ptolemy's *Syntaxis mathematica*, along with preliminary lemmas, but it is not there attributed by name to Menelaus.[15] According to the Arabic of Mansūr ibn 'Irāq as contained in a Leiden manuscript, the proof runs:[16]

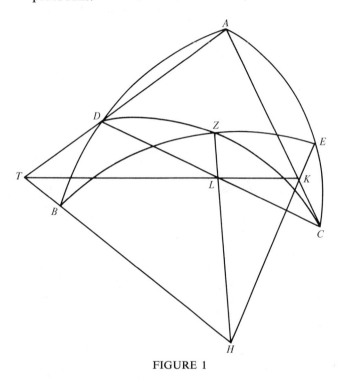

FIGURE 1

Between two arcs of great circles *ADB* and *AEC* let two other arcs of great circles intersect in *Z*. All four arcs are less than a semicircle. It is required to prove

$$\frac{\sin CE}{\sin EA} = \frac{\sin CZ}{\sin ZD} \cdot \frac{\sin DB}{\sin BA}.$$

Let *H* be the center of the circle and let *HZ*, *HB*, *HE* be drawn. *AD* and *BH* lie in a plane and, if they are not parallel, let *AD* meet *BH* in the direction of *D* at *T*. Draw the straight lines *AKC*, *DLC*, meeting *HE* in *K* and *HZ* in *L*, respectively. Because the arc *EZB* is in one plane and the triangle *ACD* is in another plane, the points *K*, *L*, *T* lie on the straight line which is the line of their intersection. (More clearly, because *HB*, *HZ*, *HE*, which are in one plane, respectively intersect the straight lines *AD*, *DC*, *CA*, which are also in one

plane, in the points *T*, *L*, *K*, these three points of intersection must lie on the straight line in which the two planes intersect.) Therefore, by what has become known as Menelaus' theorem in plane geometry (which is proved by Ptolemy, although not here),

$$\frac{CK}{KA} = \frac{CL}{LD} \cdot \frac{DT}{TA}.$$

But, as Ptolemy also shows,

$$\frac{CK}{KA} = \frac{\sin CE}{\sin EA}, \quad \frac{CL}{LD} = \frac{\sin CZ}{\sin ZD}, \quad \frac{DT}{TA} = \frac{\sin DB}{\sin BA},$$

and the conclusion follows.

Menelaus proceeds to prove the theorem for the cases where *AD* meets *HB* in the direction of *A* and where *AD* is parallel to *HB*. He also proves that

$$\frac{\sin CA}{\sin AE} = \frac{\sin CD}{\sin DZ} \cdot \frac{\sin ZB}{\sin BE}.$$

Björnbo observed that Menelaus proved the theorem in its most general and most concise form; Ptolemy proved only what he needed, and Theon loaded his pages with superfluous cases. But A. Rome challenged this view.[17] He considered that Ptolemy really covered all cases, that the completeness of Menelaus' treatment may have been due to subsequent amplification, and that Theon's prolixity was justified by the fact that he was lecturing to beginners.

In Ptolemy's *Syntaxis*, Menelaus' theorem is fundamental. For Menelaus himself it led to several interesting propositions, of which the most important is book III, proposition 5; it is important not so much in itself as in what it assumes. The proposition

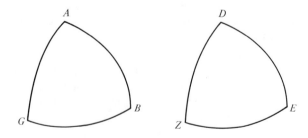

FIGURE 2

is that if in two spherical triangles *ABG*, *DEZ*, the angles *A*, *D* are both right, and the arcs *AG*, *DZ* are each less than a quarter of the circumference,

$$\frac{\sin(BG + GA)}{\sin(BG - GA)} = \frac{\sin(EZ + ZD)}{\sin(EZ - ZD)},$$

from which may be deduced the modern formula

$$\frac{\sin(a + b)}{\sin(a - b)} = \frac{1 + \cos C}{1 - \cos C},$$

or

$$\tan b = \tan a \cos C.$$

In the proof Menelaus casually assumes (to use modern lettering) that if four great circles drawn through any point O on a sphere are intersected in

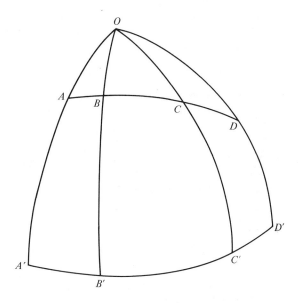

FIGURE 3

A, B, C, D and A', B', C', D' by two other great circles (transversals), then

$$\frac{\sin AD}{\sin DC} \cdot \frac{\sin BC}{\sin AB} = \frac{\sin A'D'}{\sin D'C'} \cdot \frac{\sin B'C'}{\sin A'B'}.$$

This is the anharmonic property, the property that the cross ratio or double ratio of the range $(A, D : B, C)$ is unaltered by projection on to another great circle. There is, of course, a corresponding property for four concurrent lines in a plane cut by a transversal.

It is possible that Menelaus did not prove this property and the preliminary lemmas needed for book III, proposition 1, because he had done so in another work; but the balance of probability is that they were well known in his time and had been discovered by some earlier mathematician. The fact that Menelaus' theorem is proved, not as a proposition about a spherical triangle, but as a proposition about four arcs of great circles, suggests that this also was taken over from someone else. It would not be the

first time that credit has been given to the publicist of a discovery rather than to the discoverer. If this is so, it is tempting to think that both Menelaus' theorem and the anharmonic property go back to Hipparchus. This conjecture is reinforced by the fact that the corresponding plane theorems were included by Pappus as lemmas to Euclid's *Porisms* and therefore presumably were assumed by Euclid as known.[18]

When Ptolemy in the former of his two references to Menelaus called him "Menelaus the geometer,"[19] he may have had his trigonometrical work in mind, but Menelaus also contributed to geometry in the narrower sense. According to the *Fihrist*, he composed an *Elements of Geometry* which was edited by Thābit ibn Qurra (d. 901) and a *Book on the Triangle*. None of the former has survived, even in Arabic, and only a small part of the latter in Arabic;[20] but it was probably in one of these works that Menelaus gave the elegant alternative proof of Euclid, book I, proposition 25, which is preserved by Proclus.[21]

Euclid's enunciation is as follows: "If two triangles have the two sides equal to two sides respectively, but have [one] base greater than the base [of the other], they will also have [one of] the angle[s] contained by the equal straight lines greater [than the other]." He proved the theorem by *reductio ad absurdum*. Menelaus' proof was direct and is perhaps further evidence of his distaste for indirect proofs already manifested in the *Sphaerica*. Let the two triangles be ABC, DEF, with $AB = DE$, $AC = DF$, and $BC > EF$. From BC cut off BG equal to EF. At B make the angle GBH on the side of BC remote from A equal to angle DEF. Draw BH equal to DE. Join HG and produce HG to meet AC at K. Then the triangles BGH, DEF are congruent and $HG = DF = AC$. Now HK is greater than HG or AC, and therefore greater than AK. Thus angle KAH is greater than angle KHA. And since $AB = BH$, angle BAH = angle BHA. Therefore, by addition, angle BAC is greater than angle BHG, that is, greater than angle EDF.

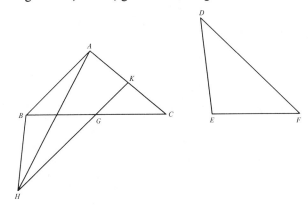

FIGURE 4

The *Liber trium fratrum de geometria*, written by Muḥammad, Aḥmad, and al-Ḥasan, the three sons of Mūsā ibn Shākir (Banū Mūsā) in the first half of the ninth century,[22] states that Menelaus' *Elements of Geometry* contained a solution of the problem of doubling the cube, which turns out to be Archytas' solution.

This bears on a statement by Pappus that Menelaus invented a curve which he called "the paradoxical curve" (γραμμὴ παράδοξος).[23] Pappus, writing of the so-called "surface loci," says that many even more complicated curves having very remarkable properties were discovered by Demetrius of Alexandria in his *Notes on Curves* and by Philo of Tyana as a result of weaving together plektoids[24] and other surfaces of all kinds. Several of the curves, he continues, were considered by more recent writers to be worthy of a longer treatment, in particular the curve called "paradoxical" by Menelaus.

If Menelaus really did reproduce Archytas' solution, which relies on the intersection of a tore and a cylinder, this lends support to a conjecture by Paul Tannery that the curve was none other than Viviani's curve of double curvature.[25] In 1692 Viviani set the learned men of Europe the problem "how to construct in a hemispherical cupola four equal windows such that when these areas are taken away, the remaining part of the curved surface shall be exactly capable of being geometrically squared." His own solution was to take through O, the center of the sphere, a diameter BC and to erect at O a perpendicular OA to the plane $BDCO$. In the plane $BACO$ semicircles are described on the radii BO, CO, and on each a right half-cylinder is described. Each half-cylinder will, of course, touch the sphere internally; and the two half-cylinders will cut out of the hemispherical surface the openings BDE, CDF, with corresponding openings on the other side. The curve in which the half-cylinders

intersect the hemisphere is classified as a curve of the fourth order and first species, and it is a particular case of the *hippopede* used by Eudoxus to describe the motion of a planet. The portion left on the hemispherical surface is equal to the square on the diameter of the hemisphere, and Tannery conjectures that the property of this area being squarable was considered at that time, when the squaring of the circle was much in the air, to be a paradox. It is an attractive conjecture but incapable of proof on present evidence.

According to several Arabic sources[26] Menelaus wrote a book on mechanics, the title of which was something like *On the Nature of Mixed Bodies*.[27] This is presumably to be identified with the unnamed work by Menelaus on which al-Khāzinī draws in his *Kitāb mīzān al-ḥikma* ("Book of the Balance of Wisdom," 1121/1122). The fourth chapter of the first book quotes theorems by Menelaus respecting weight and lightness; the first chapter of the fourth book describes Archimedes' balance on the evidence of Menelaus; and the second and third chapters of the same book describe the balance devised by Menelaus himself and his use of it to analyze alloys, with a summary of the values he found for specific gravities.[28]

NOTES

1. *Syntaxis mathematica*, VII, 3, in *Claudii Ptolemaei opera quae exstant omnia*, J. L. Heiberg, ed., I, pt. 2 (Leipzig, 1903), pp. 30.18–19, 33.3–4.
2. Plutarch, *De facie quae in orbe lunae apparet*, 17, 930A, H. Cherniss and William C. Helmbold, eds., in *Moralia*, Loeb Classical Library, XII (London–Cambridge, Mass., 1957), 106.7–15. Lucius is the speaker and says, "In your presence, my dear Menelaus, I am ashamed to confute a mathematical proposition, the foundation, as it were, on which rests the subject of catoptrics. Yet it must be said that the proposition, 'All reflection occurs at equal angles,' is neither self-evident nor an admitted fact." Menelaus is not allowed by Plutarch to speak for himself, and it would be rash to assume from this reference that he made any contribution to optics. Cherniss thinks that "the conversation was meant to have taken place in or about Rome some time—and perhaps quite a long time—after A.D. 75" (p. 12).
3. Pappus, *Collectio*, VI.110, F. Hultsch, ed., II (Berlin, 1877), p. 102.1; Proclus, *In primum Euclidis*, G. Friedlein, ed. (Leipzig, 1873; repr. Hildesheim, 1967), 345.14; English trans., G. R. Morrow (Princeton, 1970).
4. Ptolemy, *op. cit.*, 30.18–32.3.
5. *Ibid.*, 33.3–34.8.
6. Heinrich Suter, "Das Mathematiker Verzeichniss im *Fihrist* des Ibn Abî Ja'kûb an-Nadim (Muhammad Ibn Ishāk)," in *Abhandlungen zur Geschichte der Mathematik*, no. 6 (Leipzig, 1892), 19.
7. This is unlikely to be correct and is probably an embroidering of the reference to Trajan in Ptolemy.
8. A. A. Björnbo, "Hat Menelaos einen Fixsternkatalog verfasst?" in *Bibliotheca mathematica*, 3rd ser., **2** (1901), 196–212.
9. Pappus, *op. cit.*, VI.110, vol. II, 600.25–602.1.
10. This at least is what the text of Pappus as we have it implies, but there is some reason to doubt whether the text can be correct. See Hultsch's note at the point.

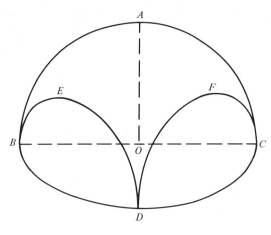

FIGURE 5

11. The inequality of the times was already known to Euclid, *Phaenomena, Euclidis opera omnia*, J. L. Heiberg and H. Menge, eds., VIII (Leipzig, 1916), props. 9, 12, 13, pp. 44, 62, 78; and Hypsicles (q.v.) attempted to calculate the times by an arithmetical progression. When Hipparchus is said to have solved the problem "by numbers," it presumably means that he was the first to have given a correct solution by trigonometrical methods.

12. *Commentary on the Syntaxis mathematica of Ptolemy*, A. Rome, ed., in the series Studi e Testi, LXXII (Vatican City, 1936), I.10, p. 451.4–5. For further discussion see A. Rome, "Premiers essais de trigonométrie rectiligne chez les Grecs," in *L'antiquité classique*, **2**, fasc. 1 (1933), 177–192; and a brief earlier note by the same author with the same title in *Annales de la Société scientifique de Bruxelles*, ser. A, **52**, pt. 1 (1932), 271–274.

13. The trans. and eds. are summarized by George Sarton, *Introduction to the History of Science*, I (Baltimore, 1927; repr. 1968), 253–254; and are more fully examined by A. A. Björnbo, *Studien über Menelaos' Sphärik* (Leipzig, 1902), 10–22, and Max Krause, *Die Sphärik von Menelaos aus Alexandrien* (Berlin, 1936), 1–116. See also the bibliography at the end of this article.

14. Pappus, *op. cit.*, VI.1, p. 476.16–17. This is part of the evidence for the genuineness of the definitions even though they do not appear in Gerard's Latin trans.

15. Ptolemy, *Syntaxis mathematica*, I.13, J. L. Heiberg, ed., I, pt. 1 (Leipzig, 1898), pp. 68.14–76.9. See also the commentary of Theon of Alexandria with the valuable notes of A. Rome, ed., *Commentaires de Pappus et de Théon d'Alexandrie sur l'Almageste*, II, *Théon d'Alexandrie*, which is Studi e Testi, LXXII (Vatican City, 1936), 535–570.

16. A. A. Björnbo, *Studien*, 88–92. Menelaus omits a general enunciation (πρότασις) and goes straight to the particular enunciation (ἔκθεσις). Björnbo (p. 92) regards this as partial evidence that the proposition was taken from some other work; but Rome, "Les explications de Théon d'Alexandrie sur le théorème de Ménélas," in *Annales de la Société scientifique de Bruxelles*, ser. A, **53**, pt. 1 (1933), 45, justly says that the length and complexity of a general enunciation, as given by Theon writing for his pupils, is a sufficient reason for the omission.

17. Björnbo, *Studien*, 92. A. Rome, "Les explications de Théon d'Alexandrie sur la théorème de Ménélas" (see n. 16), 39–50; and *Commentaires de Pappus et de Théon d'Alexandrie sur l'Almageste*, II, 554, n. 1 ("L'on est tenté de conclure que le complément de preuve établissant le théorème de Ménélas pour tous les cas, a été inventé à une date située entre Théon et les auteurs arabes qui nous font connaître les Sphériques.")

18. Pappus, *op. cit.*, VII.3–19, props. 129, 136, 137, 140, 142, 145, Hultsch ed., vol. II, pp. 870.3–872.22, 880.13–882.16, 882.17–884.9, 886.23–888.8, 890.3–892.2, 894.14–28. M. Chasles, "Aperçu historique sur l'origine et le développement des méthodes en géométrie," in *Mémoires couronnés par l'Académie royale des sciences et des belles-lettres de Bruxelles*, **2** (1837), 33, 39; and *Les trois livres de Porismes d'Euclide* (Paris, 1860), 11, 75–77, was the first to recognize the anharmonic property in the lemmas of Pappus and to see that "les propositions d'Euclide étaient de celles auxquelles conduisent naturellement les développements et les applications de la notion du rapport anharmonique, devenu fondamentale dans la géométrie moderne." Actually, in prop. 129 Pappus does not use four concurrent lines cut by two transversals but three concurrent lines cut by two transversals issuing from the same point. (The generality is not affected.) Props. 136 and 142 are the converse; prop. 137 is a particular case and prop. 140 its converse; prop. 145 is another case of prop. 129.

19. Ptolemy, *op. cit.*, 30.18.

20. M. Steinschneider, *Die arabischen Uebersetzungen aus dem Griechischen*, 2. Abschnitt, Mathematik §111–112, in *Zeit-*

schrift der Deutschen morgenländischen Gesellschaft, **50** (1896), 199.

21. Proclus, *op. cit.*, 345.9–346.13.

22. M. Curtze first edited Gerard of Cremona's trans. in *Nova acta Academiae Caesareae Leopoldino Carolinae germanicae naturae curiosorum*, **49** (1885), 105–167. This is now superseded by the later and better ed. of M. Clagett, *Archimedes in the Middle Ages*, I (Madison, Wis., 1964), 223–367, see particularly 334–341, 365–366.

23. Pappus, *op. cit.*, IV.36, vol. I, p. 270.17–26.

24. A plektoid (πλεκτοειδὴς ἐπιφάνεια) is a twisted surface; the only other example of the word, also in Pappus, suggests that it may mean a conoid.

25. Paul Tannery, "Pour l'histoire des lignes et surfaces courbes dans l'antiquité," in *Bulletin des sciences mathématiques*, 2nd ser., **7** (1883), 289–291, repr. in his *Mémoires scientifiques*, II (Toulouse–Paris, 1912), 16–18. On Viviani's curve see *Acta eruditorum* (Leipzig, 1692), "Aenigma geometricum de miro opificio testudinis quadrabilis hemispherica a D. Pio Lisci Posillo geometra propositum die 4 April. A. 1692," pp. 274–275, also pp. 275–279, 370–371; Moritz Cantor, *Vorlesungen über Geschichte der Mathematik*, III (Leipzig, 1898), 205.

26. Among them the *Fihrist*, see n. 7.

27. In *Codex Escurialensis* 905 the title is given as *Liber de quantitate et distinctione corporum mixtorum* and in *Codex Escurialensis* 955 as *De corporum mistorum quantitate et pondere*; but J. G. Wenrich, *De auctorum graecorum versionibus et commentariis Syriacis, Arabicis, Persicisque* (Leipzig, 1842), 211, gives *De cognitione quantitatis discretae corporum permixtorum*.

28. N. Khanikoff, "Analysis and Extracts of the Book of the Balance of Wisdom," in *Journal of the American Oriental Society*, **6** (1859), 1–128, especially pp. 34, 85. Unfortunately Khanikoff does not translate the passage referring to Menelaus, but the whole Arabic text has since been published—*Kitāb mīzān al-ḥikma* (Hyderabad, 1940). For further information see *Dictionary of Scientific Biography*, VII.

BIBLIOGRAPHY

I. ORIGINAL WORKS. Menelaus wrote a work on spherics (the geometry of the surface of a sphere) in three books (the third treating spherical trigonometry); a work on chords in the circle, which would have included what is now called a table of sines; an elements of geometry, probably in three books; a book on the triangle, which may or may not have been a publication separate from the last-mentioned one; possibly a work on transcendental curves, including one called "paradoxical" that he discovered himself; a work on hydrostatics, dealing probably with the specific gravities of mixtures; a treatise on the setting of the signs of the zodiac; and a series of astronomical observations which may or may not have amounted to a catalog of the fixed stars.

None of these has survived in Greek, but after earlier efforts the *Sphaerica* was translated into Arabic by Isḥāq ibn Ḥunayn (*d.* 910/911), or possibly by his father, Ḥunayn ibn Isḥāq (*d.* 877), and the translation was revised by several eds., notably by Manṣūr ibn ʿIrāq (1007/1008), whose redaction survives in the University library at Leiden as *Codex Leidensis* 930, and by Nasir al Dīn al-Ṭūsī (1265), whose work exists in many manuscripts. From Arabic the work was translated into Latin by Gerard of Cremona (*d.* 1187), and his trans. survives to varying extents in some

17 MSS; in many of them the author is called Mileus. The work was rendered into Hebrew by Jacob ben Māḥir ibn Tibbon (ca. 1273). The first printed ed. was a Latin version by Maurolico (Messina, 1558) from the Arabic; based on a poor MS, it is replete with interpolations. Nor was the Latin version of Mersenne (Paris, 1644) much better. Halley produced a Latin version which was published posthumously (Oxford, 1758) with a preface by G. Costard. Halley made some use of Arabic MSS, but in the main he has given a free rendering of the Hebrew version, with some mathematical treatment of his own. It held the field until Axel Anthon Björnbo produced his "Studien über Menelaos' Sphärik. Beiträge zur Geschichte der Sphärik und Trigonometrie der Griechen," in *Abhandlungen zur Geschichte der Mathematischen Wissenschaften*, **40** (1902), 1–154. After the introductory matter this amounts to a free German rendering of the *Sphaerica* based mainly on Halley's ed. and *Codex Leidensis* 930. It was the best work on Menelaus that existed for many years, but as a doctoral thesis, the work of a young man who had to rely on secondhand information, it had many deficiencies. The need for a satisfactory ed. of the Arabic text with a German trans. and notes on the history of the text was finally met when Max Krause, basing his work on the same Leiden MS, published "Die Sphärik von Menelaos aus Alexandrien in der Verbesserung von Abū Naṣr Manṣūr b. 'Alī b. 'Irāq mit Untersuchungen zur Geschichte des Textes bei den islamischen Mathematikern," in *Abhandlungen der Gesellschaft der Wissenschaften zu Göttingen*, Phil.-Hist. Klasse, 3rd ser., no. 17 (1936).

None of Menelaus' other works survives even in trans. except for a small part of his *Book on the Triangle* (if this is different from his *Elements of Geometry*). For notes on the Arabic translations, see M. Steinschneider, "Die arabischen Uebersetzungen aus dem Griechischen, 2. Abschnitt, Mathematik 111–112," in *Zeitschrift der Deutschen Morgenländischen Gesellschaft*, **50** (1896), 196–199.

It is possible that the proof of Menelaus' theorem given by Ptolemy, *Syntaxis mathematica*, in *Claudii Ptolemaei opera quae exstant omnia*, J. L. Heiberg, ed., I, pt. 1 (Leipzig, 1898), 74.9–76.9, reproduces, at least to some extent, the language of Menelaus; but in the absence of direct attribution there can be no certainty.

II. SECONDARY LITERATURE. The various references to Menelaus by Plutarch, Pappus, Proclus, and Arabic authors are given in the notes above. The chief modern literature is A. A. Björnbo, "Studien über Menelaos' Sphärik," mentioned above; and his "Hat Menelaos einen Fixsternkatalog verfasst?" in *Bibliotheca mathematica*, 3rd ser., **2** (1901), 196–212; Thomas Heath, *A History of Greek Mathematics*, II (Oxford, 1921), 260–273; A. Rome, "Premiers essais de trigonométrie rectiligne chez les Grecs," in *Annales de la Société scientifique de Bruxelles*, ser. A, **52**, pt. 2 (1932), 271–274; an expanded version with the same title is in *L'antiquité classique*, II, fasc. 1 (Louvain, 1933), 177–192; "Les explications de Théon d'Alexandrie sur le théorème de Ménélas," in *Annales de la Société scientifique de Bruxelles*, ser. A, **53**, pt. 1 (1933), 39–50; and *Commentaires de Pappus et de Théon d'Alexan-*

drie sur l'Almageste, II, *Théon d'Alexandrie*, Studi e Testi, LXXII (Vatican City, 1936), 535–570; and Max Krause, *Die Sphärik von Menelaos aus Alexandrien* (mentioned above).

IVOR BULMER-THOMAS

MENGHINI, VINCENZO ANTONIO (*b.* Budrio, Italy, 15 February 1704; *d.* Bologna, Italy, 27 January 1759), *medicine, chemistry.*

Menghini was the son of Domenico Menghini and Bartolomea Benelli. He graduated from the University of Bologna in philosophy and medicine on 18 June 1726. Ten years later he lectured there on logic. In the following year he lectured on theoretical medicine, and then, from 1738 until his death, he taught practical medicine at the same university. He was a member of the Academy of Sciences of the Bologna Institute, of which he was also president in 1748.

Menghini developed Galeazzi's research on the presence of iron in the blood, identifying the red corpuscles as the chief site of iron within the organism. He experimentally reduced an organ to ashes, then used a magnetized knife to extract iron particles from it. Since he suspected that the iron particles might be contained in the blood within the organ so treated (especially if it were muscle) he carefully washed out the blood before incinerating it, and found that the number of iron particles did perceptibly decrease. Having thus demonstrated that the iron in an organism is located primarily in the blood, Menghini continued his investigation in an effort to determine which of the three components of the blood—as described by Malpighi in *De polypo cordis*—actually contained the iron. He used Malpighi's method of repeatedly washing coagulum to separate the blood into its three parts, then examined each, thus determining that while iron was absent from both the serum and the bleached coagulum, it was abundant in the red corpuscles that remained in the washing liquid.

BIBLIOGRAPHY

I. ORIGINAL WORKS. Menghini's writings include "De ferrearum particularum sede in sanguine," in *De Bononiensi scientiarum et artium instituto atque academia commentarii*, II, pt. 2 (1746), 244–266; and "De ferrearum particularum progressu in sanguinem," *ibid.*, pt. 3 (1747), 475–488.

II. SECONDARY LITERATURE. On Menghini and his work see L. Belloni, "La scoperta del ferro nell'organismo," in *Atti dell'Accademia medica lombarda*, **20** (1965), 1809–1815; "Dal polipo del cuore (Malpighi 1666) al ferro dei glo-*

buli rossi," in *Simposi clinici*, **3** (1966), xvii–xxiv; A. Brighetti, "Il Menghini e la scoperta del ferro nel sangue," in *Atti del XXIII Congresso Nazionale di storia della medicina, Modena, 1967* (Rome, n.d.), 63–80; "Una lettera inedita di Vincenzo Menghini ad Ercole Lelli," in *Scritti in onore di Adalberto Pazzini* (Rome, 1968), 212–218; and "Una importante memoria inedita del Menghini sul ferro nel sangue," in *La clinica*, **28** (1968), 46–56; M. Medici, "Elogio di Vincenzo Menghini," in *Memorie dell'Accademia delle scienze dell'Istituto di Bologna. Memorie*, **9** (1858), 455–479; and V. Busacchi, "Vincenzo Menghini e la scoperta del ferro nel sangue," in *Bullettino delle scienze mediche*, **130** (1958), 202–205.

LUIGI BELLONI

MENGOLI, PIETRO (*b.* Bologna, Italy, 1625; *d.* Bologna, 1686), *mathematics.*

Mengoli's name appears in the register of the University of Bologna for the years between 1648 and 1686. He studied with Cavalieri, whom he succeeded in the chair of mathematics, and also took a degree in philosophy in 1650 and another in both civil and canon law in 1653. He was in addition ordained to the priesthood and from 1660 until his death served the parish of Santa Maria Maddalena, also in Bologna.

Mengoli's mathematics were superficially conservative. He did not subscribe to the innovations of Torricelli, and his own discoveries were set out in an abstruse Latin that made his works laborious to read. His books were nevertheless widely distributed in the seventeenth century, and were known to Collins, Wallis, and Leibniz; they were then almost forgotten, so that Mengoli's work has been studied again only recently. His significance to the history of science lies in the transitional position of his mathematics, midway between Cavalieri's method of indivisibles and Newton's fluxions and Leibniz' differentials.

In *Novae quadraturae arithmeticae* (Bologna, 1650), Mengoli took up Cataldi's work on infinite algorithms. As Eneström (1912) and Vacca (1915) have pointed out, he was the first to sum infinite series that were not geometric progressions and to demonstrate the existence of a series which, although its general term tends to zero, has a sum that can be greater than any number. In particular, he showed the divergence of the harmonic series

$$\sum_{n=1}^{\infty} \frac{1}{n} = 1 + \frac{1}{2} + \frac{1}{3} + \frac{1}{4} + \cdots,$$

preceding Jakob Bernoulli's demonstration of it by nearly forty years (it was known to Oresme in the fourteenth century). From this, Mengoli made the general deduction that any series formed from the reciprocals of the terms of an arithmetic progression must diverge.

Mengoli also considered the series of the reciprocals of the triangular numbers

$$\frac{1}{3} + \frac{1}{6} + \frac{1}{10} + \cdots \frac{1}{\frac{n(n+1)}{2}} + \cdots,$$

and said that the sum is 1, because the sum of the first *n* terms is $n/(n + 2)$, which (for suitably large *n*) differs from 1 by less than any given quantity. He then demonstrated the convergence of the series of the reciprocals of the numbers $n(n + r)$ to the result that

$$\sum_{n} \frac{1}{n(n+r)} = \frac{1}{n}\left(1 + \frac{1}{2} + \frac{1}{3} + \cdots + \frac{1}{r}\right),$$

and summed the reciprocals of the solid numbers,

$$\sum \frac{1}{n(n+1)(n+2)} = \frac{1}{4}.$$

In the *Geometriae speciosae elementa* (1659), Mengoli set out a logical arrangement of the concepts of limit and definite integral that anticipated the work of nineteenth-century mathematicians. In establishing a rigorous theory of limits, he considered a variable quantity as a ratio of magnitudes and hence needed to consider only positive limits. He then made the following definitions: a variable quantity that can be greater than any assignable number is called "quasi-infinite"; a variable quantity that can be smaller than any positive number is "quasi-nil"; and a variable quantity that can be both smaller than any number larger than a given positive number *a* and greater than any number smaller than *a* is "quasi-*a*."

Using these precise concepts of the infinite, the infinitesimal, and the limit, and working from simple inequalities valid between numerical ratios, he demonstrated (as Agostini recognized by translating his obscure exposition into modern symbols and terminology) the properties of the limit of the sum and the product, and showed that the properties of proportions are conserved also at the limit. The proofs obtain when such limits are neither 0 nor ∞; for this case Mengoli set out the properties of the infinitesimal calculus and the calculus of infinites some thirty years before Newton published them in his *Principia*.

Mengoli's predecessors (among them Archimedes, Kepler, Valerio, and Cavalieri) had assumed as intuitively evident that a plane figure has an area. By contrast, he proved the existence of the area by dividing an interval of the continuous figure $f(x)$ into *n* parts and considering, alongside the figure to be squared (which he called the "form"), the figures formed by parallelograms constructed on each segment of the interval and having the areas (in modern notation):

$$s_n = \sum_{i=1}^{n} l_i(x_{i+1} - x_i), \quad \text{(inscribed figure)}$$

$$S_n = \sum_{i=1}^{n} L_i(x_{i+1} - x_i), \quad \text{(circumscribed figure)}$$

$$\left.\begin{array}{l} \sigma_n = \sum_{i=1}^{n} f(x_i)(x_{i+1} - x_i), \text{ or} \\[2ex] \sigma'_n = \sum_{i=1}^{n} f(x_{i+1})(x_{i+1} - x_i), \end{array}\right\} \text{(adscribed figure)}$$

where l_i and L_i denote, respectively, the minimum and maximum of $f(x)$ on the interval (x_i, x_{i+1}). Drawing upon the theory of limits that had worked so well in the study of series, Mengoli demonstrated that the sequences of the s_n and S_n tend to the same limit to which the sequences of the σ_n and σ'_n, compressed between them, also tend. Hence, since the figure to be squared is always compressed between the s_n and the S_n, it follows that this common limit is the area of the figure itself.

Mengoli also used this method to integrate the binomial differentials $Z^s(a - x)^r \, dx$ with whole and positive exponents. (He had, preceding Wallis, already integrated these some time before by the method of indivisibles.) Before publishing his results, however, he wished to give a rigorous basis to the method of indivisibles or to develop in its stead another method that would be immune to criticism. He therefore set out a purely arithmetic theory of logarithms; having given a definition of the logarithmic ratio similar to Euclid's definition of ratio between magnitudes, he then extended Euclid's book V to encompass his own logarithmic ratio. Mengoli also did significant work in logarithmic series (thirteen years before N. Mercator published his *Logarithmotecnia*).

In a short work of 1672, entitled *Circolo*, Mengoli calculated the integrals of the form

$$\int_0^1 x^{\frac{m}{2}} (1 - x)^{\frac{n}{2}} \, dx,$$

finding for $n/2$ the same infinite product that had already been given by Wallis. Mengoli published other, minor mathematical writings; in addition he was interested in astronomy, and wrote a short vernacular book on music, published in 1670.

BIBLIOGRAPHY

I. ORIGINAL WORKS. Mengoli's writings include *Novae quadraturae arithmeticae* (Bologna, 1650); *Geometriae speciosae elementa* (Bologna, 1659); *Speculazioni di musica* (Bologna, 1670); and *Circolo* (Bologna, 1672).

II. SECONDARY LITERATURE. On Mengoli and his work, see A. Agostini, "La teoria dei limiti in Pietro Mengoli," in *Periodico di matematiche*, 4th ser., **5** (1925), 18–30; "Il concetto di integrale definito in Pietro Mengoli," *ibid.*, 137–146; and "Pietro Mengoli," in *Enciclopedia italiana*, XXII (Milan, 1934), 585; E. Bortolotti, *La storia della matematica nella università di Bologna* (Bologna, 1947), 98–101, 137–138; G. Eneström, "Zur Geschichte der unendlichen Reihen in die Mitte des siebzehnten Jahrhunderte," in *Bibliotheca mathematica* (1912), 135–148; and G. Vacca, "Sulle scoperte di Pietro Mengoli," in *Atti dell'Accademia nazionale dei Lincei. Rendiconti* (Dec. 1915), 512.

A. NATUCCI

MENSHUTKIN, NIKOLAY ALEKSANDROVICH

(*b.* St. Petersburg, Russia, 24 October 1842; *d.* St. Petersburg, 5 February 1907), *chemistry.*

Menshutkin graduated from St. Petersburg University in 1862, having studied under A. A. Voskresensky and N. N. Sokolov. In 1866 he defended his master's thesis, "O vodorode fosforistoy kisloty, ne sposobnom k metallicheskomu zameshcheniyu pri obyknovennykh usloviyakh dlya kislot" ("On the Hydrogen of Phosphorous Acid, Which Is Not Replaceable by Metal Under the Usual Conditions for Acids"). After the defense of his doctoral dissertation, "Sintez i svoystva ureidov" ("The Synthesis and Properties of the Ureides"), in 1869, Menshutkin was appointed professor at St. Petersburg. From 1902 he was professor at the Petersburg Polytechnic Institute. His basic research involved the rate of chemical transformation of organic compounds in relation to the composition and structure of the reacting substances. This work appeared in the early stages of physical chemistry and played an important role in the development of chemical kinetics.

Menshutkin began his research in 1877 with the study of the reversibility of complex ester formation from alcohols and organic acids. He first showed that the reactivity of monovalent, bivalent, and multivalent alcohols depends on the structure of this carbon chain. He found that in the formation of esters the secondary alcohols react less rapidly than the primary ones. The kinetic method of determining the isomers of alcohols, which Menshutkin developed, was used to determine the structure of newly synthesized alcohols.

Studying the esterification of monobasic and polybasic organic acids, and oxy acids (1882), Menshutkin explained the influence of the molecular weights of alcohols and acids, and of temperature, on the formation of their complex esters.

After van't Hoff introduced the concept of the reaction-rate constant and gave examples of its application to the solution of structural questions (1884), Menshutkin compared the structures of alcohols and acids with the constants of the corresponding rates of reaction. The results confirmed his earlier conclusions. From 1887, Menshutkin investigated the influence of solvents on the rate of reaction. Defining the rate of reaction of triethylamine with ethyl iodide,

$$N(C_2H_5)_3 + C_2H_5I \rightarrow N(C_2H_5)_4I$$

in twenty-three different solvents, he found (1890) that this reaction in benzyl alcohol was 742 times faster than in hexane. Menshutkin's discovery of the influence of the medium on the rate of reaction was one of the most important achievements in the field of chemical kinetics.

Explaining the quantitative relation between the reactivity of organic compounds and their structure, Menshutkin found that any branching of the chain leads to a decrease in the value of the constant of velocity; substitution in the ortho position causes a decrease in velocity, but in meta- and para-positions it causes an increase in velocity. The accumulation of methyl groups in the α position decreases the velocity of esterification of aliphatic acids and alcohols.

In Menshutkin's study of the kinetics of esterification reactions he discovered that the reaction proceeds more readily with straight-chain than with cyclic compounds. For his work in chemical kinetics the Petersburg Academy of Sciences awarded Menshutkin the Lomonosov Prize in 1904.

Menshutkin was an outstanding teacher. He wrote a textbook, *Analiticheskaya khimia* ("Analytic Chemistry," 1871; 16th ed., 1931), which was translated into German and English and for several decades served as a reference book. In 1883 and 1884 he published the most detailed course of organic chemistry in Russian at that time (4th ed., 1901). He also wrote the first original work on the history of chemistry in the Russian language.

Menshutkin was one of the founders and leaders of the Russian Physical-Chemical Society, and from 1869 to 1900 he was editor of its journal. On his initiative and under his leadership were established the chemical laboratory of St. Petersburg University (1894) and the Petersburg Polytechnical Institute (1902).

BIBLIOGRAPHY

I. ORIGINAL WORKS. Menshutkin's works, apart from his textbook *Analiticheskaya khimia*, include "Issledovanie obrazovania uksusnykh efirov pervichnykh spirtov" ("Research on the Formation of Acetic Esters of Primary Alcohols"), in *Zhurnal Russkago fisiko-khimicheskago obshchestva*, **9** (1877), 318–319; "Rukovodstvo k opredeleniyu izomerii spirtov i kislot pri pomoshchi eterifikatsionnykh dannykh" ("A Guide to the Determination of Isomers of Alcohols and Acids With the Aid of Esterification Data"), *ibid.*, **13** (1881), 572; "O metode opredelenia khimicheskogo znachenia sostavlyayushchikh organicheskikh soedineny" ("On the Method of Determining the Chemical Significance of the Components of Organic Compounds"), *ibid.*, 67; "Issledovania raspadenia uksusnogo tretichnogo amida pri nagrevanii" ("Research on the Decomposition of Acetic Tertiary Amide During Heating"), *ibid.*, **14** (1882), 292–300; and "Issledovanie obrazovania amidov kislot" ("Research on the Formation of Amides of Acids"), *ibid.*, **16** (1884), 191–206.

See also *Lektsii organicheskoy khimii* ("Lectures in Organic Chemistry"), 2 vols. (St. Petersburg, 1883–1884); *Ocherk razvitia khimicheskikh vozzreny* ("A Sketch of the Development of Chemical Views"; St. Petersburg, 1888); "O vlianii khimicheski nedeyatelnoy zhidkoy sredy na skorost soedinenia trietilamina s iodgidrinami" ("On the Influence of Chemically Inactive Liquid Mediums on the Speed of Compound Formation of Triethylamine With Iodohydrins"), in *Zhurnal Russkago fisiko-khimicheskago obshchestva*, **22** (1890), 393–409; "Vlianie chisla tsepey na skorost obrazovania aminov" ("Influence of the Number of Chains on the Rate of Formation of Amines"), *ibid.*, **27** (1895), 96–118, 137–157; "O vlianii khimicheski nedeyatelnykh rastvoriteley na izmenenie raspredelenia skorostey reaktsii v ryadakh izomernykh aromaticheskikh soedineny" ("On the Influence of Chemically Inactive Solvents on the Change in the Determination of the Rate of Reaction in a Series of Isomeric Aromatic Compounds"), *ibid.*, **32** (1900), 46–60; and "Vlianie katalizatorov na obrazovanie anilidov i amidov" ("The Influence of Catalysts on the Formation of Anilides and Amides"), *ibid.*, **35** (1903), 343.

II. SECONDARY LITERATURE. See B. N. Menshutkin, *Zhizn i deyatelnost N. A. Menshutkina* ("Life and Work of N. A. Menshutkin"; St. Petersburg, 1908); and P. I. Staroselsky and Y. I. Soloviev, *Nikolay Aleksandrovich Menshutkin. Zhizn i deyatelnost* ("... Life and Work"; Moscow, 1968).

Y. I. SOLOVIEV

MENURET DE CHAMBAUD, JEAN JACQUES (*b.* Montélimar, France, 1733; *d.* Paris, France, 15 December 1815), *physiology, medicine.*

Having completed preliminary philosophical studies, presumably in Montélimar, Menuret received his medical degree from Montpellier and journeyed to

Paris to seek his fortune. His activity in the capital is evidenced by several notable contributions to the final series of volumes (1765) of the Diderot-d'Alembert *Encyclopédie*. Medical practice appears to have been his principal occupation; Menuret held minor appointments in the royal household (physician to the staff of the royal stables and consulting physician to the Comtesse d'Artois, wife of the future Charles X). Active in military service during the revolutionary wars (he joined General C. F. Dumouriez in his disgrace and flight from France), Menuret passed his final years in providing medical care for the poor of his section of Paris.

Menuret applied considerable learning and subtlety to consideration of the principal phenomena of life and to the methods dccmcd suitablc for the study of vital activities. In his views he echoed those expressed by Théophile de Bordeu and others of similar conviction, and thus expounded further the central tenets of Montpellier vitalism. Observation assumed methodological preeminence; experimentation was virtually rejected. Observation respects the autonomy, the naturalness of the object or process under scrutiny; experiment "dismembers and combines and thereby produces phenomena quite different from those which nature presents."[1] This approach confirmed the faith upon which the Montpellier physiologists acted—the uniqueness of vital structure and processes necessarily excludes the possibility of an easy or even legitimate application of the methods or substance of other sciences to matters physiological.

Menuret regarded mechanistic explanation and its associated manipulative, experimental method as the principal enemy. One must begin with the characteristic manifestations of life (motion and sensory impressions), recognize in their ceaseless interaction the very essence of life, and then concede that these processes, while grounded in the physical constitution of the body (Menuret speaks of *molécules organiques*), will be no better understood even if translated into the language of contemporary physics or chemistry.

Such views were fully in accord with Menuret's central preoccupation, which was man and his relation to the medical art. Menuret was a physician with broad clinical experience; he was also a physician who claimed to regard practice as being far more important than theory. In medicine observation must absolutely displace experiment; the nature of man and the art demand no less.[2] For this reason the prospective physician required more the clinical experience won through apprenticeship or in the hospital and less a rigorous introduction to the sciences ancillary to medicine. "No profession," Menuret remarked of medicine, "demands more imperiously the harmonious

cooperation of *science* and *virtue*."[3] The physician's virtue referred not to scientific learning but to the recognized and willingly accepted responsibility of the physician to the healing art and to its primary concern, the patient.

Montpellier medicine was noted for its clinical emphasis. It venerated Hippocrates and Sydenham, condemned (but often yielded to) speculative excesses, and sought, as part of its ambition to embrace all that affected the ailing patient, to determine with comprehensiveness and precision the factors which dictated the well-being or afflictions of man. Among these factors were climate and the general physical and social conditions under which men lived. Menuret devoted considerable attention to these matters. His medical surveys of Montélimar and Hamburg are worthy examples of medicogeographical interest; that of Paris is an invaluable guide to conditions influencing health and disease in the French capital during the closing years of the *ancien régime*.

NOTES

1. "Observation," in *Encyclopédie*, XI (1765), 313 (cited in Roger, *Sciences de la vie*, 632).
2. *Ibid.*, 315.
3. *Essai . . . de former des bons médecins*, 4.

BIBLIOGRAPHY

I. Original Works. Menuret contributed some forty articles (signed either "M" or by his name: see Roger, *Sciences de la vie*, p. 631) to the Diderot-d'Alembert *Encyclopédie*. Among these "Oeconomie animale" and "Observation" are of general interest; Albert Von Haller replied, under the same heading and in the *Supplément* to the *Encyclopédie*, to Menuret's views on the animal economy. Other works by Menuret include *Nouveau traité du pouls* (Amsterdam, 1767); *Avis aux mères sur la petite vérole et la rougeole* (Lyons, 1770); *Essai sur l'action de l'air dans les maladies contagieuses* (Paris, 1781); *Essai sur l'histoire médico-topographique de Paris* (Paris, 1786; 2nd ed., rev. and enl., 1804); *Essai sur les moyens de former des bons médecins* (Paris, 1791; 2nd ed., 1814); and *Discours sur la réunion de l'utile à l'agréable, même en médecine* (Paris, 1809). Menuret also published reflections on the utilization of fallow lands and the implications for health of the salt tax *(gabelle)*. He prepared two obituary notices: G. F. Venel (1777) and P. Chappon (1810).

II. Secondary Literature. Henri Zeiller provides the unique but, regrettably, muddled catalog of medical contributors to the *Encyclopédie: Les collaborateurs médicaux de l'Encyclopédie de Diderot et d'Alembert* (Paris, 1934); Zeiller wrongly reassigns the signature "M" from Menuret to the Dijon practitioner Hugues Maret. Biographical notices concerning Menuret are brief and provide few details; see F. L. Chaumeton, "Notice sur Jean Jacques Menuret de Chambaud," in *Journal universel des sciences médicales*, **1** (1816), 384–390; M. de Cubières–Palmézeaux, "Lettre sur Jean Jacques Menuret de Chambaud," in *Journal général de médecine*, **54** (1816), 415–429.

Reference to Menuret has virtually disappeared from modern historical literature. Exceptions are the invaluable accounts by Jacques Roger, *Les sciences de la vie dans la pensée française du XVIIIᵉ siècle* (Paris, 1963), 631–634, and "Méthodes et modèles dans la préhistoire du vitalisme française," in *Actes du XIIᵉ Congrès international d'histoire des sciences* (Paris, 1971), IIIB, 101–108. A roster of Menuret's contributions to the *Encyclopédie* is given in R. N. Schwab and W. E. Rex, "Inventory of Diderot's Encyclopédie," in *Studies on Voltaire and the Eighteenth Century*, **93** (1972), 216–217. For the context of Menuret's theoretical work see Frédéric Bérard, *Doctrine médicale de l'école médicale de Montpellier, et comparaison de ses principes avec ceux d'autres écoles d'Europe* (Montpellier, 1819), 3–77, and François Granel, "Théophile de Bordeu (1722–1776)," in *Pages médico-historiques montpelliéraines* (Montpellier, 1964), 87–97.

WILLIAM COLEMAN

MÉRAY, HUGUES CHARLES ROBERT (*b.* Chalon-sur-Saône, France, 12 November 1835; *d.* Dijon, France, 2 February 1911), *mathematics*.

Méray entered the École Normale Supérieure in 1854. After teaching at the lycée of St. Quentin from 1857 to 1859 he retired for seven years to a small village near Chalon-sur-Saône. In 1866 he became a lecturer at the University of Lyons and, in 1867, professor at the University of Dijon, where he spent the remainder of his career.

In his time he was a respected but not a leading mathematician. Méray is remembered for having anticipated, clearly and with only minor differences of style, Cantor's theory of irrational numbers, one of the main steps in the arithmetization of analysis.

Méray first expounded his theory in an article entitled "Remarques sur la nature des quantités définies par la condition de servir de limites à des variables données" (1869). His precise formulation in the framework of the terminology of the time and the place is of considerable historical interest.

I shall now reserve the name number or quantity to the integers and fractions; I shall call *progressive variable* any quantity v which takes its several values successively in unlimited numbers.

Let v_n be the value of v of rank n: if as n increases to infinity there exists a number V such that beginning with a suitable value of n, $V - v_n$ remains smaller than any quantity as small as might be supposed, one says that V is the limit of v and one sees immediately that

$v_{n+p} - v_n$ has zero for limit whatever the simultaneous laws of variation imposed on n and p.

If there is no such number it is no longer legitimate, analytically speaking, to claim that v has a limit; but if, in this case, the difference $v_{n+p} - v_n$ still converges to zero then the nature of v shows an extraordinary similarity with that of the variables which really possess limits. We need a special term in order to express the remarkable differentiation with which we are concerned: I shall say that the progressive variable is *convergent*, whether or not a numerical limit can be assigned to it.

The existence of a limit to a convergent variable permits greater ease in stating certain of its properties which do not depend on this particular question [i.e., whether or not there exists a numerical limit] and which frequently can be formulated directly only with much greater difficulty. One sees therefore that it is advantageous, in cases where there is no limit, to retain the same abbreviated language which is used properly when a limit exists, and in order to express the convergence of the variable one may say simply that *it possesses a fictitious limit*.

Here is a first example of the usefulness of this convention: if, when m and n both increase to infinity, the difference $u_m - v_n$ between two convergent variables tends to zero for a certain mutual dependence between the subscripts, then one proves easily that it remains infinitely small also for any other law [i.e., law of dependence between m and n]: I shall then say that the variables u and v are *equivalent*, and one sees immediately that two variables which are equivalent to a third variable are equivalent to each other [*loc. cit.* p. 284, in translation].

In this paper Méray also discussed the question of how to assign values to a given function for irrational values of the argument or arguments, and he suggested that this problem could always be solved by a passage to the limit. In this connection, as well as elsewhere in his writings, he did in fact assume a somewhat constructive point of view of the notion of a function, taking it for granted that a function can always be obtained constructively either by rational operations or by limiting processes.

The paper marked the first appearance in print of an "arithmetical" theory of irrational numbers. Some years earlier Weierstrass had, in his lectures, introduced the real numbers as sums of sequences or, more precisely, indexed sets, of rational numbers; but he had not published his theory and there is no trace of any influence of Weierstrass' thinking on Méray's. Dedekind also seems to have developed his theory of irrationals at an earlier date, but he did not publish it until after the appearance of Cantor's relevant paper in 1872. In that year Méray's *Nouveau précis d'analyse infinitésimale* was published in Paris. In the first chapter the author sketches again his theory of irrationals and remarks that however peculiar it might appear to be, compared with the classical traditions, he considers it more in agreement with the nature of the problem than the physical examples required in other approaches.

The *Nouveau précis* had as its principal aim the development of a theory of functions of complex variables based on the notion of a power series. Thus here again, Méray followed unconsciously in the footsteps of Weierstrass; consciously, he was developing the subject in the spirit of Lagrange but felt—rightly—that he could firmly establish what Lagrange had only conjectured. The book is in fact written with far greater attention to rigor than was customary in Méray's time.

Little regard was paid to Méray's main achievement until long after it was first produced, partly because of the obscurity of the journal in which it was published. But even in his review (1873) of the *Nouveau précis,* H. Laurent pays no attention to the theory, while gently chiding the author for using too narrow a notion of a function and for being too rigorous in a supposed textbook. At that time there was not in France—as there was in Germany—a sufficient appreciation of the kind of problem considered by Méray, and not until much later was it realized that he had produced a theory of a kind that had added luster to the names of some of the greatest mathematicians of the period.

Although Méray's theory of irrationals stands out above the remainder of his work, his development of it may be regarded as more than an accident. For elsewhere he also showed the same critical spirit, the same regard to detail, and the same independence of thought that led him to his greatest discovery.

BIBLIOGRAPHY

I. Original Works. Méray's theory was published as "Remarques sur la nature des quantités définies par la condition de servir de limites à des variables données," in *Revue des sociétés savantes des départements*, Section sciences mathématiques, physiques et naturelles, 4th ser., **10** (1869), 280–289. Among his many treatises and textbooks are *Nouveaux éléments de géométrie* (Paris, 1874); *Exposition nouvelle de la théorie des formes linéaires et des déterminants* (Paris, 1884); and *Sur la convergence des développements des intégrales ordinaires d'un système d'équations différentielles totales ou partielles*, 2 vols. (Paris, 1890), written with Charles Riquier.

II. Secondary Literature. Laurent's review of the *Nouveau précis* was published in *Bulletin des sciences mathématiques*, **4** (1873), 24–28. See also the biography of Méray in *La grande encyclopédie* XXIII (Paris, 1886), 692; and J. Molk, "Nombres irrationels et la notion de limite," in *Encyclopédie des sciences mathématiques pures et appliquées*, French ed., I (Paris, 1904), 133–160, after the German article by A. Pringsheim.

Abraham Robinson

MERCATI, MICHELE (*b.* San Miniato, Italy, 13 April 1541; *d.* Rome, Italy, 25 June 1593), *medicine, natural sciences.*

Mercati was the elder son of Pietro Mercati, a doctor, and Alfonsina Fiaminga. He received his early education from his father and later enrolled at the University of Pisa, where he studied under Cesalpino. Possibly as a result of the specialized knowledge he received from Cesalpino, Mercati was called by Pope Pius V to direct the Vatican botanical garden, a post he retained under Gregory XIII and Sixtus V. He was very active in botany and contributed greatly to the development of the simples section of the botanical garden. He was early famous for his scientific achievements; when he was only twenty-seven, the future Grand Duke Ferdinand I of Tuscany honored him by elevating his family into the ranks of the Florentine aristocracy, while the same privilege was bestowed on him the following year by the Roman Senate. Pope Gregory XIII named him a member of the "pontifical family," and Mercati showed his gratitude by caring for the pope during his final illness. For Gregory XIII, Mercati wrote *Istruzione sopra la peste* (Rome, 1576), with the addition "Tre altre istruzioni sopra i veleni occultamente ministrati, podagra e paralisi"—which is, however, of limited scientific value. Pope Sixtus V held Mercati in great esteem and created him apostolic protonotary. He also sent him to Poland with Cardinal Aldobrandini (later Clement VIII) on a mission to King Sigismund III. It was during that journey that Mercati began writing from memory a book on the obelisks of Rome; it was published at Rome in 1589 under the title *Degli obelischi di Roma*. Clement VIII made Mercati chief physician and knight of the Order of Santo Spirito in Sassia.

Mercati suffered from bladder and kidney stones and gout. In 1582 he recovered from an attack of renal colic; he had another four years later. On the advice of his friend Filippo Neri he retired to the Oratorian monastery at Santa Maria in Vallicella, where he could receive better care. Again he recovered; but following a third attack in 1593 he died, at the age of fifty-two. He was treated by Filippo Neri and by his teacher Cesalpino, whom he had called to Rome years previously and for whom he had procured appointments as papal physician and professor at the

University of Rome. An autopsy on Mercati, possibly carried out by Cesalpino himself, showed the existence of two stones in the ureters, about sixty in the kidneys, and thirty-six in the gallbladder. Mercati was buried in the church of Santa Maria in Vallicella, in the tomb of the Mediobarba family. He had been a friend and correspondent not only of Filippo Neri and of famous cardinals, but also of persons renowned in the arts and sciences, including Marsilio Cognati, Pier Angelo Bargeo, Latino Latini, Girolamo Mercuriale, Aldrovandi, and Melchior Wieland (better known as Guilandinus). A portrait of Mercati engraved by Benoit Fariat (after Tintoretto) is the frontispiece of his *Metallotheca*.

As a naturalist Mercati's greatest interest lay in collecting minerals and fossils; this collection later formed the basis of the work that has made him famous: *Metallotheca. Opus posthumum, auctoritate et munificentia Clementis undecimi pontificis maximi e tenebris in lucem eductum*; *opera autem et studio Joannis Mariae Lancisii archiatri pontificii illustratum* (Rome, 1717).

Mercati was a good mineralogist and one of the founders of paleontology. He understood the true origin of stone implements, which in his day were considered to be the product of lightning. In his book he described, besides the Vatican collection of minerals, some stones of animal origin and the bladder stones that had been found by Lancisi during the autopsy of Pope Innocent XI.

Mercati's book is illustrated by beautiful copper engravings which, with the manuscript of the work, were rediscovered by Carlo Roberto Dati in the eighteenth century.

BIBLIOGRAPHY

In addition to Mercati's writings mentioned above, on Mercati and his work see P. Capparoni, "Michele Mercati (1541–1593)," in *Profili bio-bibliografici di medici e naturalisti celebri italiani dal secolo XV al secolo XVIII* (Rome, 1932), pp. 48–50; E. Gurlt, *Geschichte der Chirurgie und ihrer Ausübung*, II (repr. Hildesheim, 1964), 482–483; W. Haberling and J. L. Pagel, "Mercati, Michele," in *Lexicon der hervorragen den Ärzte aller Zeiten und Völker*, W. Haberling, F. Hübotter, and H. Vierordt, eds., 2nd ed., IV (Berlin–Vienna, 1932), 169–170; and G. Montalenti, *Storia delle scienze*, N. Abbagnano, ed., III, pt. 1 (Turin, 1962), 353.

Loris Premuda

MERCATOR, GERARDUS (or **Gerhard Kremer**) (*b.* Rupelmonde, Flanders, 5 March 1512; *d.* Duisburg, Germany, 2 December 1594), *geography.*

Mercator's family name was Kremer, but he latinized it on entering the University of Louvain in 1530. Philosophy and theology were his principal subjects at Louvain, and he retained a concern with these matters throughout his life. Soon after his graduation he became concerned with mathematics and astronomy, studied these subjects informally under the guidance of Gemma Frisius, and acquired considerable skills as an engraver. His first known work was a globe, made in 1536; the following year he published his first map—of Palestine. Mercator was a man of many talents, well versed in mathematics, astronomy, geography, and theology, and was also a great artist whose contributions to calligraphy and engraving influenced several generations of artisans. His lasting fame rests on his contributions to mapmaking: he was undoubtedly the most influential of cartographers.

Mercator's maps cover a variety of subjects. During his sojourn at Louvain (1530–1552), besides his map of Palestine, he made maps of the world, globes, and scientific instruments and also established a reputation as a surveyor. Accused of heresy in 1544, and imprisoned for several months, he was released for lack of evidence, and in 1552 moved to Duisburg, where he became cosmographer to the duke of Cleves. His years at Duisburg were most fruitful: he published the first modern maps of Europe and of Britain, prepared an excellent edition of Ptolemy, and in 1569 published a world map on a new projection that still bears his name.

The 1569 world map of Mercator was designed for seamen. In order to lay out his course easily, the navigator needed a map where a line of constant bearing would cross all meridians at the same angle. Mercator designed a cylindrical projection, tangent at the equator; on it meridians and parallels are straight lines, intersecting at right angles, and distortion gradually increases toward the poles. Such a map shows loxodromes as straight lines, and for small areas it conforms to shapes, but tends to distort large areas, especially at high latitudes. Nonetheless, the Mercator projection, as modified at the end of the sixteenth century by Wright and Molyneux, remains the most important tool of the navigator.

Mercator's second great contribution to geography and cartography was the collection of maps he designed, engraved, and published during the last years of his life. It consisted of detailed and remarkably accurate maps of western and southern Europe. In 1595, the year after Mercator's death, his son, Rumold, published the entire collection under the title "Atlas— or Cosmographic Meditations on the Structure of the World," the first time the word "atlas" was used to designate a collection of maps.

BIBLIOGRAPHY

The most detailed and authoritative biography of Mercator is the work of H. Averdunk and J. Müller-Reinhard, *Gerhard Mercator und die Geographen unter seinen Nachkommen*, which is *Petermanns Mitteilungen, Ergänzungsheft*, no. 182 (1914). His correspondence was published by M. Van Durme, *Correspondance Mercatorienne* (Anvers, 1959). Among the many studies dealing with Mercator's life and works, a special publication, on the occasion of the 450th anniversary of his birth, is "Gerhard Mercator— 1512–1594: zum 450. Geburtstag," in *Duisburger Forschungen*, **6** (1962).

GEORGE KISH

MERCATOR, NICOLAUS (Kauffman, Niklaus) (*b.* Eutin [?], Schleswig-Holstein, Denmark [now Germany], *ca.* 1619; *d.* Paris, France, 14 January [?] 1687), *mathematics, astronomy.*

His father was probably the Martin Kauffman who taught school at Oldenburg in Holstein from 1623 and died there in 1638. Doubtless educated in boyhood at his father's school, Nicolaus graduated from the University of Rostock and was appointed to the Faculty of Philosophy in 1642. At Copenhagen University in 1648 he superintended a "Disputatio physica de spiritibus et innato calido" and over the next five years produced several short textbooks on elementary astronomy and spherical trigonometry; one of his title pages at this time describes him as "mathematician and writer on travels to the Indies."

His tract on calendar improvement (1653) caught Cromwell's eye in England and, whether invited or not, he subsequently left Denmark for London. There he resided for almost thirty years and came universally to be known by his latinized name, an "anglicization" which he himself soon adopted. Unable to find a position in a university, Mercator earned a living as a mathematical tutor, but soon he made the acquaintance of Oughtred, Pell, Collins, and other practitioners. In November 1666, on the strength of his newly invented marine chronometer, he was elected a fellow of the Royal Society; earlier, in Oldenburg's *Philosophical Transactions of the Royal Society*, he had wagered the profits (seemingly nonexistent) from his invention against anyone who could match his expertise in the theory of Gerard Mercator's map. Through his Latin version (1669) of Kinckhuysen's Dutch *Algebra*, commissioned by Collins at Seth Ward's suggestion, he came into contact with Newton, and the two men later exchanged letters on lunar theory. Aubrey portrays Mercator at this time as "of little stature, perfect; black haire, . . . darke eie,

but of great vivacity of spirit . . . of a soft temper . . . *(amat Venerem aliquantum)*: of a prodigious invention, and will be acquainted (familiarly) with nobody." In September 1676 Hooke unsuccessfully proposed Mercator as Mathematical Master at Christ's Hospital. In 1683 he accepted Colbert's commission to plan the waterworks at Versailles, but died soon afterward, having fallen out with his patron.

Mercator's early scientific work is known only through the university textbooks which he wrote in the early 1650's; if not markedly original, they show his firm grasp of essentials. His *Trigonometria sphaericorum logarithmica* (1651) gives neat logarithmic solutions of the standard cases of right and oblique triangles and tabulates the logarithms of sine, cosine, tangent, and cotangent functions (his "Logarithmus," "Antilogarithmus," "Hapsologarithmus," and "Anthapsologarithmus") at 1′ intervals. His *Cosmographia* (1651) and *Astronomia* (1651) deal respectively with the physical geography of the earth and the elements of spherical astronomy. In his *Rationes mathematicae* (1653) he insists on drawing a basic distinction between rational and irrational numbers: the difference in music is that between harmony and dissonance; in astronomy that between a Keplerian "harmonice mundi" and the observable solar, lunar, and planetary motions. In the tract *De emendatione annua* (1653[?]) he urges the reform of the 365-day year into months of (in sequence) 29, 29, 30, 30, 31, 31, 32, 31, 31, 31, 30, and 30 days.

Mercator's first published book in England, *Hypothesis astronomica nova* (1664), in effect combines Kepler's hypothesis (that planets travel in elliptical orbits round the sun, with the sun at one focus) with his vicarious hypothesis (in which the equant circle is centered in the line of apsides at a distance from the sun roughly 5/8 times the doubled eccentricity): Mercator sets this ratio exactly equal to the "divine section" $(\sqrt{5}-1)/2$, with an error even in the case of Mars of less than 2′. (Here a mystical streak in his personality gleams through, for he compares his hypothesis to a knock-kneed man standing with arms outstretched, a "living image of Eternity and the Trinity." He later expounded similar insights in an unpublished manuscript on *Astrologia rationalis*.) Subsequently, in 1670, he showed his skill in theoretical astronomy by demolishing G. D. Cassini's 1669 method for determining the lines of apsides of a planetary orbit, given three solar sightings. He showed that it reduced to the Boulliau-Ward hypothesis of mean motion round an upper-focus equant and pointed out its observational inaccuracy. (His enunciation of the "true" Keplerian hypothesis, that time in orbit is proportional to the focal sector swept out by the planet's radius vector, may well have been the source of Newton's knowledge of this basic law.) The two books of his *Institutiones astronomicae* (1676) offered the student an excellent grounding in contemporary theory, and Newton used them to fill gaps in his rather shaky knowledge of planetary and lunar theory. Some slight hint of the practical scientist is afforded by the barometric measurements made during the previous half year, which Mercator registered at the Royal Society in July 1667. No working drawings are extant of the Huygenian pendulum watch—which he designed in 1666—or of its marine mounting (by gimbal suspension), but an example "of a foote diameter" was made.

Mercator is remembered above all as a mathematician. In 1666 he claimed to be able to prove the identity of "the Logarithmical Tangent-line beginning at 45 deg." with the "true Meridian-line of the Sea-Charte" (Mercator map). This declaration is not authenticated but not necessarily empty. It is difficult to determine how far his researches into finite differences—which were restricted to the advancing-differences formula—were independent of Harriot's unpublished manuscripts on the topic, to which Mercator perhaps had access. In his best-known work, *Logarithmotechnia* (1668), he constructed logarithms from first principles (if $a^b = c$, then $b = \log_a c$), making ingenious use of the inequality

$$\left(\frac{a + px}{a - px}\right) < \left(\frac{a + x}{a - x}\right)^p, \qquad p = 1/2, 1/3, \cdots,$$

while in supplement (a late addition to the manuscript submitted in 1667) he used the St. Vincent-Sarasa hyperbola-area model to establish, independently of Hudde and Newton, the series expansion

$$\text{lognat}(1 + x) = \int_0^x 1/(1 + x) \cdot dx$$
$$= x - \tfrac{1}{2}x^2 + \tfrac{1}{3}x^3 \cdots.$$

The circulation by Collins of the "De analysi," composed hurriedly by Newton as a riposte, seems to have effectively blocked Mercator's plans to publish a complementary *Cyclomathia* with allied expansions (on Newtonian lines) of circle integrals. The "Introductio brevis" which he added in 1678 to Martyn's second edition of the anonymous *Euclidis elementa geometrica* commendably sought to simplify the Euclidean definitions for the beginner by introducing motion proofs: a circle is generated as the ripple on the surface of a stagnant pool when a stone is dropped at its center, a line as the instantaneous meet of two such congruent wave fronts. His

Hypothesis astronomica nova contains the first publication of the polar equation of an ellipse referred to a focus.

BIBLIOGRAPHY

I. ORIGINAL WORKS. The trio of textbooks put out by Mercator at Danzig in 1651 appeared under the titles *Cosmographia, sive descriptio coeli et terrae in circulos . . .; Trigonometria sphaericorum logarithmica, . . . cum canone triangulorum emendatissimo . . .;* and *Astronomia sphaerica decem problematis omnis ex fundamento tradita.* They were reissued shortly afterward at Leipzig . . . *Conformatae ad exactissimas docendi leges pro tironibus, . . . privatis hactenus experimentis comprobatae.* At Copenhagen Mercator published in 1653 his study on mathematical rationality, *Rationes mathematicae subductae,* and also his propagandist tract on calendar improvement, *De emendatione annua diatribae duae. . . .*

During the next ten years he apparently wrote nothing for the press, but at length produced his *Hypothesis astronomica nova, et consensus ejus cum observationibus* (London, 1664). His wager regarding the logarithmic nature of the Mercator map was announced in "Certain Problems Touching Some Points of Navigation," in *Philosophical Transactions of the Royal Society,* **1,** no. 13 (4 June 1666), 215–218. In 1668 appeared his major work, *Logarithmotechnia: sive methodus construendi logarithmos nova, accurata, & facilis; scripto antehàc communicata, anno sc. 1667 nonis Augusti: cui nunc accedit vera quadratura hyperbolae, & inventio summae logarithmorum* (London, 1668), later reprinted in F. Maseres, *Scriptores logarithmici,* I (London, 1791), 169–196; to Wallis' "account" of it in *Philosophical Transactions of the Royal Society,* **3,** no. 38 (17 Aug. 1668), 753–759, Mercator added "Some further Illustration," *ibid.,* 759–764. For a page-by-page analysis of the bk. see J. E. Hofmann, "Nicolaus Mercators *Logarithmotechnia* (1668)," in *Deutsche Mathematik,* **3** (1938), 446–466. His "Some Considerations . . . Concerning the Geometrick and direct Method of Signior Cassini for finding the Apogees, Excentricities and Anomalies of the Planets; as that was printed in the *Journal des sçavans* of *Septemb. 2. 1669"* appeared in *Philosophical Transactions of the Royal Society,* **5,** no. 57 (25 Mar. 1670), 1168–1175.

His astronomical compendium, *Institutionum astronomicarum libri duo, de motu astrorum communi & proprio, secundum hypotheses veterum & recentiorum praecipuas . . .* came out at London in 1676 (reissued Padua, 1685): Newton's lightly annotated copy is now at Trinity College, Cambridge, NQ.10.152. In an app. to the compendium Mercator reprinted his earlier *Hypothesis nova* (the preface excluded). His "Introductio brevis, qua magnitudinum ortus ex genuinis principiis, & ortarum affectiones ex ipsa genesi derivantur" was adjoined in 1678 to John Martyn's repr. of the "Jesuit's Euclid," in *Euclidis elementa geometrica, novo ordine ac methodo fere, demonstrata* (London, 1666).

None of Mercator's correspondence with his contemporaries seems to have survived, although that with

Newton (1675–1676) on lunar vibration is digested in the *Institutiones astronomicae*, 286–287; compare the remark added by Newton to the third bk. of the third ed. of his *Principia*, Propositio XVII (London, 1726), 412. The MS (Bodleian, Oxford, Savile G.20⁴) of Mercator's Latin rendering of Kinckhuysen's *Algebra ofte Stelkonst* (Haarlem, 1661) is reproduced in *The Mathematical Papers of Isaac Newton*, II (Cambridge, 1968), 295–364, followed by Newton's "Observations" upon it, *ibid.*, 364–446.

Thomas Birch in his biography in *A General Dictionary Historical and Critical*, VII (London, 1738), 537–539 [= J. G. de Chaufepié, *Nouveau Dictionnaire historique et critique*, III (Amsterdam, 1753), 79], records the existence in Shirburn Castle of "a manuscript containing Theorems relating to the Resolution of Equations, the Method of Differences, and the Construction of Tables; and another, intitled, *Problema arithmeticum ad doctrinam de differentialium progressionibus pertinens*"; these are now in private possession. Birch also lists (*ibid.*, 539) the section titles of Mercator's unpublished "Astrologia rationalis, argumentis solidis explorata" (now Shirburn 180.F.34). Details of his chronometer are given by Birch in his *History of the Royal Society*, II (London, 1756), 110–114, 187, and in Oldenburg's letter to Leibniz of 18 December 1670 (C. I. Gerhardt, *Die Briefwechsel von G. W. Leibniz*, I [Berlin, 1899], 48). References to the lost treatise on circle quadrature, *Cyclomathia*, occur in John Collins' contemporary correspondence with James Gregory (see the *Gregory Memorial Volume* [London, 1939], 56, 60, 153).

II. SECONDARY LITERATURE. J. E. Hofmann's *Nicolaus Mercator (Kauffman), sein Leben und Wirken, vorzugsweise als Mathematiker* (in *Akademie der Wissenschaften und der Literatur in Mainz [Abh. der Math.-Nat. Kl.]*, no. 3 [1950]) is the best recent survey of Mercator's life and mathematical achievement. On his personality and habits see John Aubrey, *Letters . . . and Lives of Eminent Men*, II (London, 1813), 450–451, 473, or *Brief Lives* (London, 1949), 135, 142, 153–154. J.-B. Delambre gives a partially erroneous estimate of Mercator's equant hypothesis of planetary motion in his *Histoire de l'astronomie moderne*, II (Paris, 1821), 539–546; see also Curtis Wilson, "Kepler's Derivation of the Elliptical Path," in *Isis*, **59** (1968), 5–25, esp. 23.

D. T. WHITESIDE

MERICA, PAUL DYER (*b.* Warsaw, Indiana, 17 March 1889; *d.* Tarrytown, New York, 20 October 1957), *metallurgy.*

After receiving his A.B. from the University of Wisconsin in 1908, Merica taught physics in Wisconsin and "Western Subjects" in Hangchow, China, for two years before going to the University of Berlin, from which he obtained his Ph.D. in 1914. From 1914 to 1919 he was a physical metallurgist at the U.S. National Bureau of Standards, where he was involved with R. G. Waltenberg and H. Scott in research on the new alloy Duralumin, the properties of which

depended on precipitation hardening—the only method of hardening metals and alloys that was not known in or before classical times.

Merica was the first to show that such hardening resulted from a change of solubility with temperature (almost unsuspected in solids at the time), and he postulated that hardening required a critical dispersion of submicroscopic particles of $CuAl_2$ within the matrix crystals.

This theory inspired many studies of hardening in other alloy systems, especially when it was combined with the slip-interference theory of hardening advanced by Jeffries and R. S. Archer in 1921. X-ray diffraction later showed that the greatest hardness occurred before there were detectable particles of compound, and in 1932 Merica advanced his theory of "knots," that is, clusters with high concentration of solute atoms forming without breaking coherence with the parent lattice. The idea soon received quite independent experimental proof in the X-ray studies of Guinier and Preston. From 1919 until his death, Merica was with the International Nickel Company (director of research, 1919–1949; president, 1949–1952). He developed several precipitation-hardening alloys of nickel, nickel-bearing cast irons, and sheet alloys for corrosion-resistant and high temperature applications.

BIBLIOGRAPHY

I. ORIGINAL WORKS. Merica's most influential papers are "Heat Treatment and Constitution of Duralumin," in *Scientific Papers of the United States Bureau of Standards*, **347** (1919), with R. G. Waltenberg and H. Scott; repr. in *Transactions of the American Institute of Mining and Metallurgical Engineers*, **64** (1920), 41–79; and "The Age-Hardening of Metals," *ibid.*, **99** (1932), 13–54.

II. SECONDARY LITERATURE. For a complete list of Merica's publications and further biographical information, see Z. Jeffries, "Paul Dyer Merica," in *Biographical Memoirs. National Academy of Sciences*, **33** (1959), 226–239.

For a review of the history of precipitation-hardening, see H. Y. Hunsicker and H. C. Stumpf, "History of Precipitation Hardening," in C. S. Smith, ed., *The Sorby Centennial Symposium on the History of Metallurgy* (New York, 1965), 271–311.

CYRIL STANLEY SMITH

MERRETT, CHRISTOPHER (*b.* Winchcomb, England, 16 February 1614; *d.* London, England, 19 August 1695), *natural history, glassmaking.*

Merrett, who like his father and younger son was baptized Christopher, was educated at Oxford, where he obtained the M.B. in 1636 and the M.D. in 1643. He had married Ann Jenour of Kempsford by the time he commenced practice in London about 1640. Admitted as a fellow of the College of Physicians in 1651, he became the first keeper of their new library and museum given by William Harvey and compiled its first printed catalog (1660). The founding of the Royal Society (1660), of which he was an original member, provided an outlet for Merrett's varied interests.

In 1662, at the suggestion of Robert Boyle and the instigation of the Royal Society, Merrett translated Antonio Neri's *L'arte vetraria* (1612), a pioneer work. By adding his own extensive observations on the construction of glassmaking furnaces, the types of glass being manufactured in England, and the raw materials used, Merrett gave considerable impetus to glassmaking in England and other European countries.

William How's *Phytologia* (1650) was still in demand when it went out of print. At the publisher's request Merrett wrote *Pinax rerum naturalium Britannicarum* (1666) to replace it. Since he was no fieldworker but a sedentary and inexpert naturalist, he enlisted all the help possible and revealed a wide knowledge of the relevant literature by giving more precise references than his predecessors had. By his own admission the list was imperfect ("inchoatus"). Although the large botanical section, with over 1,400 species and synonyms from Gerard and John Parkinson, was soon superseded, the section on mammals and birds is important as the first attempt to construct a British fauna. The name *Merrettia* was given to a group of unicellular algae by S. F. Gray.

When the College of Physicians was destroyed by fire in 1666, Merrett saved and looked after 150 books; but the College argued that since they now had no library, they had no need of a keeper. The last years of Merrett's life were clouded by the consequent dispute, which cost him his fellowship (1681), allegedly for nonattendance. His only medical publications were those that contributed to the war of mutual denigration between the physicians and the apothecaries.

BIBLIOGRAPHY

I. ORIGINAL WORKS. Merrett's main work is *Pinax rerum naturalium Britannicarum* (London, 1666; another ed., 1667). He translated A. Neri, *L'arte vetraria* as *The Art of Glass, . . . With Some Observations on the Author, . . .* (London, 1662); it also appeared in Latin (London, 1668), in German (Amsterdam, 1679; Frankfurt, 1689), and in French (Paris, 1752).

II. SECONDARY LITERATURE. See C. E. Raven, "William How and Christopher Merrett," in his *English Naturalists From Neckham to Ray* (Cambridge, 1947), 298–338; and W. E. S. Turner, "A Notable British Seventeenth Century Contribution to the Literature of Glass-Making," in *Glass Technology*, **6** (1962), 201–213.

LEONARD M. PAYNE

MERRIAM, CLINTON HART (*b.* New York, N. Y., 5 December 1855; *d.* Berkeley, California, 19 March 1942), *biology.*

Merriam was raised in Locust Grove, New York, on a farm bought by an ancestor in 1800; his father, Clinton L. Merriam, had retired there early from a brokerage firm. The boy's mother, Caroline Hart Merriam, and father both encouraged his interest in collecting birds. Through Spencer F. Baird, to whom his father introduced him, the boy was appointed, when only sixteen, to collect bird skins and eggs on Hayden's geological and geographical survey of the territories.

Merriam studied for three years at Yale, then received an M.D. from the College of Physicians and Surgeons in 1879. For six years he practiced medicine in Locust Grove, then accepted the new post of ornithologist in the entomological division of the U.S. Department of Agriculture. In 1905 this unit became the Bureau of Biological Survey, which Merriam continued to direct until 1910, when a trust fund established by Mrs. E. H. Harriman provided him with independent research money through the Smithsonian Institution.

Although most widely known for his definition of life zones of faunal distribution, Merriam also did significant groundwork in mammalian studies. Many of the 600 species he proposed have proved invalid, for Merriam was indeed a "splitter"; but he established the significance of cranial characters in mammalian classification, and he perfected preservation techniques. Under his direction the Biological Survey accumulated a vast collection of mammals, on which Merriam published extensively, from shrews to grizzly bears. Under the sponsorship of the Harriman fund, he devoted his later years to gathering data on the Indians of California.

An early conservationist, Merriam was the most active founder of the American Ornithologists' Union and a founder of the Washington Academy of Sciences, the American Society of Mammalogists, and the National Geographic Society. He served on the American-British fur-seal commission in 1891, and he was the scientific director and editor of all reports of the Harriman Alaska Expedition of 1899.

BIBLIOGRAPHY

I. ORIGINAL WORKS. Merriam's report on life zones was "Results of a Biological Survey of the San Francisco Mountain Region and Desert of the Little Colorado, Arizona," in *North American Fauna*, **3** (1890), 119–136. He was especially proud of his "Monographic Revision of the Pocket Gophers Family Geomyidae (Exclusive of the Species of *Thomomys*)," in *North American Fauna*, **8** (1895), 1–258. His publications, totaling nearly 500 titles, are included in Osgood (see below).

II. SECONDARY LITERATURE. Wilfred H. Osgood presented a thorough account of Merriam's life in "Biographical Memoir of Clinton Hart Merriam," in *Biographical Memoirs. National Academy of Sciences*, **24** (1944), 1–57, which includes a bibliography. A. L. Kroeber, "C. Hart Merriam as Anthropologist," in C. Hart Merriam, *Studies of California Indians* (Berkeley, 1955), vii–xiv, covers Merriam's work on California Indians; and Peter Matthiesen, *Wildlife in America* (New York, 1959), touches on Merriam as a biologist and conservationist.

ELIZABETH NOBLE SHOR

MERRIAM, JOHN CAMPBELL (*b*. Hopkinton, Iowa, 20 October 1869; *d*. Oakland, California, 30 October 1945), *paleontology*.

Merriam was the son of Charles and Margaret Merriam. He took a B.S. degree at Lenox College, Iowa, then in 1887 moved with his family to Berkeley, California, where he attended the University of California, studing botany under E. L. Green and geology with Joseph Le Conte. In 1893 he took his doctorate in vertebrate paleontology at Munich. The following year he became an instructor at the University of California and in 1912 chairman of the new department of paleontology. In 1896 he married Ada Gertrude Little, who bore him three sons. She died in 1940, and in 1941 he married Margaret Webb.

Merriam helped pioneer the study of paleontology on the West Coast. The early explorations were followed by a period of stagnation, broken in the early 1890's by workers from both Stanford and the University of California. Between 1896 and 1908 Merriam published papers on Tertiary molluscan faunas, Tertiary echinoids, and the Triassic Ichthyosauria. In 1901 he published an important work on the John Day Basin in Oregon, and after 1905 he published many descriptions of the Tertiary mammalian faunas of the Rancho La Brea tar pits in Los Angeles.

In 1917 Merriam began two additional activities that led him away from research in paleontology: he helped found and became president of the pioneer conservation organization, the Save-the-Redwoods League, and he was elected chairman of the Committee

on Scientific Research of the California State Council of Defense. The latter post led him to Washington, D.C., to aid in the war effort, and eventually to his becoming president (1920–1938) of the Carnegie Institution of Washington. During these two decades he was very active in the National Academy of Sciences and was a leader in science administration and conservation. Conservative in his social beliefs, he was considered more serious than jovial by his colleagues.

BIBLIOGRAPHY

I. ORIGINAL WORKS. Merriam's writings are collected in *Published Papers and Addresses of John Campbell Merriam*, 4 vols. (Washington, D.C., 1938); vol. IV contains his nonscientific writings. His MSS are in the Manuscript Division, Library of Congress and the Bancroft Library, Univ. of California, Berkeley.

II. SECONDARY LITERATURE. A biographical sketch and extensive bibliography of Merriam's work, by Chester Stock, appear in *Biographical Memoirs. National Academy of Sciences*, **26** (1951), 208–232; see also Chester Stock, "Memorial to John Campbell Merriam," in *Proceedings. Geological Society of America* for 1946 (1947), 182–197; and Ralph W. Chaney, "John Campbell Merriam," in *Yearbook. American Philosophical Society* for 1945 (1946), 381–387. A sketch of his scientific and philosophical thought appears in Chester Stock, "John Campbell Merriam as Scientist and Philosopher," in Carnegie Institution of Washington, *Cooperation in Research* (Washington, D.C., 1938), 765–778.

CARROLL PURSELL

MERRILL, GEORGE PERKINS (*b*. Auburn, Maine, 31 May 1854; *d*. Auburn, 15 August 1929), *geology, meteoritics*.

Merrill was a descendant of Nathaniel Merrill, who had settled in the Massachusetts Bay Colony in the 1630's. His father, Lucius Merrill, was a carpenter and cabinetmaker. His mother, Anne Elizabeth Jones, was the daughter of the Reverend Elijah Jones of Minot, Maine, whose scholarship had a profound influence upon the intellectual life of his grandson. One of seven children, Merrill began to earn his own living at an early age, finding employment as a farmhand and in shoe factories. He was twenty-two years old when he entered the Maine State College of Agriculture and the Mechanic Arts (now the University of Maine), where he majored in chemistry and graduated with a B.Sc. degree in 1879. He then became a laboratory assistant, working on the chemistry of foods, at Wesleyan University, Middletown,

Connecticut. There he became acquainted with G. Brown Goode, formerly curator of Wesleyan's museum collections and at that time in charge of the U.S. National Museum in the Smithsonian Institution and director of the survey of fisheries for the tenth census. Goode appointed Merrill to the census staff in 1880 and to the Museum staff in 1881 as aid to George W. Hawes, who had just become curator of the geological collections. After Hawes's death in 1882, Merrill was put in charge of petrology and physical geology; and in 1897 he became head curator of the department of geology in the U.S. National Museum, a position that he held until his death. He also served as part-time professor of geology and mineralogy at Columbian College (renamed George Washington University in 1904) from 1893 to 1915.

Merrill was married in 1883 to Sarah Farington, of Portland, Maine. She died in 1894, leaving one son and three daughters. He was married again in 1900 to Katherine L. Vancey of Virginia, by whom he had one daughter. He was of sturdy build, alert and active, accustomed to long hours in office and laboratory; an avid reader, he was fond of poetry and music. Austere and reserved on first acquaintance, to his many friends Merrill displayed a warm and generous heart. He was a member of the National Academy of Sciences; the American Philosophical Society; the Geological Society of America, of which he was vice-president in 1920; the Washington, the Maryland, and the Philadelphia academies of science; and the Geological Society of Washington, of which he was president in 1906–1907; and a corresponding member of the American Institute of Architects.

Merrill made scientific contributions in at least five distinct areas. As a museum administrator he built up the department of geology in the U.S. National Museum and made it one of the world's greatest and best-organized geological collections. Introduced by Hawes to the new petrologic technique of microscopic study of thin sections of rocks, he applied that procedure to the large collection of building stones assembled in connection with the tenth census and enlarged in succeeding years. This not only led to the publication of his most widely read book, *Stones for Building and Decoration*, but also established such a reputation for him that he was influential in the selection of stones for many governmental buildings, most notably the Lincoln Memorial. These studies naturally turned Merrill's attention to the processes of rock weathering. Here his early training in chemistry proved valuable and his *Treatise on Rocks, Rock Weathering, and Soils* was immediately hailed by European as well as American geologists, and led to his recognition in agricultural circles as the out-

standing authority of his time on soils and their origin.

A fourth area in which Merrill's scientific contributions were especially notable involved the study of meteorites. He was one of the first to "regard the meteorites as world matter"; and in 1906 he correctly identified Coon Butte, near Canyon Diablo, Arizona (now known as Meteor Crater), as an impact crater. Last among his areas of accomplishment was that of the history of geological science; his three works on this subject (1906, 1920, 1924) are indispensable source material.

Among Merrill's many honors were an honorary Ph.D. from what is now the University of Maine, in 1889, the Sc.D. from George Washington University in 1917, and the J. L. Smith gold medal of the National Academy of Sciences in 1922.

BIBLIOGRAPHY

I. ORIGINAL WORKS. Merrill's writings include "On the Collection of Maine Building Stones in the United States National Museum," in *Proceedings of the United States National Museum*, **6** (1883), 165–183; "Report on the Building Stones of the United States and Statistics of the Quarry Industry for 1880," in *Tenth Census United States*, X (Washington, D.C., 1884), bound as part of vol. X, but with separate pagination; "The Collection of Building and Ornamental Stones in the United States National Museum," in *Report of the Board of Regents of the Smithsonian Institution* for 1886 (Washington, D.C., 1889), pt. 2, 277–648; *Stones for Building and Decoration* (New York, 1891; 2nd ed., 1897; 3rd ed., 1903); "Disintegration of the Granitic Rocks of the District of Columbia," in *Bulletin of the Geological Society of America*, **6** (1895), 321–332; "The Principles of Rock Weathering," in *Journal of Geology*, **4** (1896), 704–724, 850–871; and *A Treatise on Rocks, Rock Weathering, and Soils* (New York, 1897; new ed., 1906).

Later works are *The Non-metallic Minerals, Their Occurrence and Uses* (New York, 1904); "Contributions to the History of American Geology," in *Report of the United States National Museum* for 1904 (1906), 189–733; "The Meteor Crater of Canyon Diablo, Arizona," in *Smithsonian Miscellaneous Collections*, **50** (1908), 461–498; "The Composition of Stony Meteorites, Compared With That of Terrestrial Igneous Rocks and Considered With Reference to Their Efficacy in World Making," in *American Journal of Science*, 4th ser., **27** (1909), 469–474; "Handbook and Descriptive Catalogue of the Meteorite Collections in the United States National Museum," *Bulletin. United States National Museum*, no. 94 (1916); "Contributions to a History of American State Geological and Natural History Surveys," *ibid.*, no. 100 (1920); "On Chondrules and Chondritic Structure in Meteorites," in *Proceedings of the National Academy of Sciences . . .*, **6** (1920), 449–472; *The First One Hundred Years of American Geology* (New Haven, Conn., 1924); "The Present Condition of

Knowledge on the Composition of Meteorites," in *Proceedings of the American Philosophical Society*, **65** (1926), 119–130; and "Composition and Structure of Meteorites," *Bulletin. United States National Museum*, no. 149 (1930).

II. SECONDARY LITERATURE. Biographies of Merrill are Waldemar Lindgren, in *Biographical Memoirs. National Academy of Sciences*, **17** (1935), 31–53, which includes a bibliography of 196 titles; and Charles Schuchert, in *Bulletin of the Geological Society of America*, **42** (1931), 95–122, with a bibliography of 196 titles.

KIRTLEY F. MATHER

MERSENNE, MARIN (*b*. Oizé, Maine, France, 8 September 1588; *d*. Paris, France, 1 September 1648), *natural philosophy, acoustics, music, mechanics, optics, scientific communication.*

> The sciences have sworn among themselves an inviolable partnership; it is almost impossible to separate them, for they would rather suffer than be torn apart; and if anyone persists in doing so, he gets for his trouble only imperfect and confused fragments. Yet they do not arrive all together, but they hold each other by the hand so that they follow one another in a natural order which it is dangerous to change, because they refuse to enter in any other way where they are called. . . .[1]

Mersenne's most general contribution to European culture was this vision of the developing community of the sciences. It could be achieved only by the cultivation of the particular:

> Philosophy would long ago have reached a high level if our predecessors and fathers had put this into practice; and we would not waste time on the primary difficulties, which appear now as severe as in the first centuries which noticed them. We would have the experience of assured phenomena, which would serve as principles for a solid reasoning; truth would not be so deeply sunken; nature would have taken off most of her envelopes; one would see the marvels she contains in all her individuals. . . .[1]

These complaints had long been heard, yet "most men are glad to find work done, but few want to apply themselves to it, and many think that this search is useless or ridiculous."[1] He offered his scientific study of music as a particular reparation of a general fault.

Born into a family of laborers, Mersenne entered the new Jesuit *collège* at La Flèche in 1604 and remained there until 1609. After two years of theology at the Sorbonne, in 1611 he joined the Order of Minims and in 1619 returned to Paris to the Minim Convent de l'Annonciade near Place Royale, now

Place des Vosges. There he remained, except for brief journeys, until his death in 1648.[2] The Minims recognized that Mersenne could best serve their interests through an apostolate of the intellect. He made his entry upon the European intellectual scene in his earliest publications, with a discussion of ancient and modern science in support of a characteristic theological argument. He aimed to use the certifiable successes of natural science as a demonstration of truth against contemporary errors dangerous to religion and the morals of youth. In his vast and diffuse *Quaestiones in Genesim* (1623) he defended orthodox theology against "atheists, magicians, deists and suchlike,"[3] especially Francesco Giorgio, Telesio, Bruno, Francesco Patrizzi, Campanella, and above all his contemporary Robert Fludd, by attacking atomism and the whole range of Hermetic, Cabalist, and "naturalist" doctrines of occult powers and harmonies and of the Creation. In the same volume he included a special refutation of Giorgio,[4] and he continued his attack on this group in *L'impiété des déistes, athées, et libertins de ce temps* (1624). This attack on magic and the occult in defense of the rationality of nature attracted the attention of Pierre Gassendi, whom he met in 1624 and who became his closest friend.[5]

Mersenne's next work, the *Synopsis mathematica* (1626), was a collection of classical and recent texts on mathematics and mechanics. After that came *La vérité des sciences, contre les sceptiques ou Pyrrhoniens*, a long defense of the possibility of true human knowledge against the Pyrrhonic skepticism developed especially by Montaigne. Thus religion and morality had some rational basis. Yet while he stood with Aristotle in arguing that nature was both rational and knowable, he denied that theologians had to be tied to Aristotle.[6] Against the qualitative, verbal Aristotelian physics he came to argue that nature was rational, its actions limited by quantitative laws, because it was a mechanism.[7]

From about 1623 Mersenne began to make the careful selection of *savants* who met at his convent in Paris or corresponded with him from all over Europe and as far afield as Tunisia, Syria, and Constantinople. His regular visitors or correspondents came to include Peiresc, Gassendi, Descartes,[8] the Roman musicologist Giovanni Battista Doni, Roberval, Beeckman, J. B. van Helmont, Fermat, Hobbes, and the Pascals. It was in Mersenne's quarters that in 1647 the young Blaise Pascal first met Descartes.[9] Mersenne's role as secretary of the republic of scientific letters, with a strong point of view of his own, became institutionalized in the Academia Parisiensis, which he organized in 1635.[10] His monument as an architect of the European scientific community is the rich edition of his *Correspondance* published in Paris in the present century.

Mersenne developed his mature natural philosophy in relation to two fundamental questions. The first was the validity in physics of the axiomatic theory of truly scientific demonstration described in Aristotle's *Posterior Analytics* and exemplified in contemporary discussions especially by Euclid's geometry. Mersenne entered in the wake of the sixteenth-century debate on skepticism. The second question was the acceptability of a strictly mechanistic conception of nature. Opinions about these two questions decided what was believed to be discoverable in nature and what any particular inquiry had discovered. Opinions about the second also decided how to deal with the relationship of perceiver to world perceived, and so with the information communicated, especially through vision and hearing.

Mersenne's approach to these problems represents a persistent style in science. He took up his characteristic position on the first in the course of the debate over the new astronomy. He treated the decree of 1616 against Copernicus with Northern independence and moved in his early writings from rejection of the hypothesis of the earth's motions because sufficient evidence was lacking,[11] to preference for it as the most plausible. Copernicus' hypothesis, he said, had been neither refuted nor demonstrated: "I have never liked the attitude of people who want to look for, or feign, or imagine reasons or demonstrations where there are none; it is better to confess our ignorance than abuse the world."[12] But Mersenne reacted strongly against theologically sensitive extensions of the new cosmology, especially the doctrines of a plurality of worlds and of the infinity of the universe.[13] He took particular exception to Giordano Bruno: "one of the wickedest men whom the earth has ever supported . . . who seems to have invented a new manner of philosophizing only in order to make underhand attacks on the Christian religion."[14] He maintained that ecclesiastics had the right to condemn opinions likely to scandalize their flocks and merely advised moderation in censorship, because in the end "the true philosophy never conflicts with the belief of the Church."[15]

Through the Christian philosopher defending true knowledge against the skeptic in *La vérité des sciences*, and in later essays, Mersenne defined the kind of rational knowledge he held to be available. He found in Francis Bacon a program for real scientific knowledge, but he reproached him for failing to keep abreast of the "progress of the sciences" and for proposing the impossible goal of penetrating "the nature of things."[16] Only God knew the essences of things.

God's inscrutable omnipotence, which denied men independent rational knowledge of his reasons, and the logical impossibility of demonstrating causes uniquely determined by effects reduced the order of nature for men simply to an order of contingent fact. Mersenne concluded that the only knowledge of the physical world available to men was that of the quantitative externals of effects, and that the only hope of science was to explore these externals by means of experiment and the most probable hypotheses. But this was true knowledge, able to guide men's actions, even though theology and logic showed it to be less than that claimed to be possible by Aristotle.[17]

In 1629, after some earlier approaches, Mersenne wrote to Galileo, offering his services in publishing "the new system of the motion of the earth which you have perfected, but which you cannot publish because of the prohibition of the Inquisition."[18] Galileo did not reply to this generous offer—nor, indeed, to any of Mersenne's later letters to him. But Mersenne was not put off. He had come to see in Galileo's work a supreme illustration of the rationality of nature governed by mechanical laws and, so far as these laws went, of the true program for natural science.[19] In 1633 he published his first critique of Galileo's *Dialogo* (1632) in his *Traité des mouvemens et de la cheute des corps pesans et de la proportion de leurs différentes vitesses, dans lequel l'on verra plusieurs expériences très exactes.*[20] His first response to hearing of Galileo's condemnation in that year was to agree with the need for the Church to preserve Scripture from error;[21] yet he came forward at once with a French version (with additions of his own) of Galileo's unpublished early treatise on mechanics under the title *Les méchaniques de Galilée* (1634), and with a summary account of the first two days of the *Dialogo* and of the trial in *Les questions théologiques, physiques, morales, et mathématiques* (1634).

Mersenne's mature natural philosophy appeared in *Les questions* and three other works in the same year: *Questions inouyës, Questions harmoniques,* and *Les préludes de l'harmonie universelle.*[22] He made it plain that Galileo had not been condemned for heresy; and although he wrote later that he would not be prepared to risk schism for the new astronomy,[23] in 1634 he planned to write a defense of Galileo.[24] He gave this up. Mersenne disagreed with Galileo's claim to "necessary demonstrations" on the general ground that no physical science had "the force of perfect demonstration;"[25] and like most of his contemporaries he was unconvinced by the dynamical arguments so far produced by Galileo or anyone else. Yet while he saw the question of the earth's motion as undecided, he encouraged the search for fresh quantitative evidence which alone would make it possible "to distinguish the way nature acts in these movements, and to make a decision about it."[26]

Mersenne's conclusion that an inescapable "ignorance of true causes"[27] was imposed by the human situation gave him a scientific style interestingly different from that of Galileo and of Descartes. They aimed at certainty in physical science; Mersenne, disbelieving in the possibility of certainty, aimed at precision. Galileo's lack of precision in his first published mention of his experiments on acceleration down an inclined plane in the *Dialogo* led Mersenne to doubt whether he had really performed them. His own carefully repeated experiments, using a seconds pendulum to measure time, confirmed the "duplicate proportion" between distance and time deduced by Galileo but gave values nearly twice as great for the actual distances fallen. He commented that "one should not rely too much only on reasoning."[28] On many occasions Mersenne's too close attention to the untidy facts of observation may have deprived him of theoretical insight; but his insistence on the careful specification of experimental procedures, repetition of experiments, publication of the numerical results of actual measurements as distinct from those calculated from theory, and recognition of approximations marked a notable step in the organization of experimental science in the seventeenth century. Amid many words and some credulity, the works of his maturity, especially on acoustics and optics, contain models of "expériences bien reglées et bien faites"[29] and of rational appreciation of the limits of measurement and of discovery.

While strict demonstration was beyond natural science, Mersenne maintained that the imitation of God's works in nature by means of technological artifacts gave experimental natural philosophy an opening into possible explanations of phenomena. In this way he linked his experimental method with the second fundamental question for his natural philosophy—the conception of nature as a mechanism—and with the method of the hypothetical model. Characteristically it was through theological issues that he developed the central idea that living things were automatons. He used it as a weapon in his campaign for the uniqueness of human reason and of its power to grasp true knowledge and moral responsibility, against the false doctrines both of "les naturalistes,"[30] who asserted human participation in a world soul, and of the skeptics, who threw doubt on human superiority over the animals. After his visit to Beeckman, Descartes, and J. B. van Helmont in the Netherlands in 1630,[31] Mersenne came to hold that, on the analogy of sound, light was a form of purely corporeal propagation. Although he remained unconvinced by the evidence for any of the current theories of light and sound and changed his views several times, his restriction of the choice to physical motions gave him (like Descartes) a method of asking how these motions affected a sentient being.[32] He disposed finally of the arguments against the uniqueness of man by declaring animals to be simply automatons, explicitly first in *Les préludes de l'harmonie universelle* (1634):

> ... for the animals, which we resemble and which would be our equals if we did not have reason, do not reflect upon the actions or the passions of their external or internal senses, and do not know what is color, odor or sound, or if there is any difference between these objects, to which they are moved rather than moving themselves there. This comes about by the force of the impression that the different objects make on their organs and on their senses, for they cannot discern if it is more appropriate to go and drink or eat or do something else, and they do not eat or drink or do anything else except when the presence of objects, or the animal imagination *[l'imagination brutalle]*, necessitates them and transports them to their objects, without their knowing what they do, whether good or bad; which would happen to us just as to them if we were destitute of reason, for they have no enlightenment except what they must have to take their nourishment and to serve us for the uses to which God has destined them.[33]

So one could say of the animals that they knew nothing of the world impinging upon them, "that they do not so much act as be put into action, and that objects make an impression on their senses such that it is necessary for them to follow it just as it is necessary for the wheels of a clock to follow the weights and the spring that pulls them."[34] Yet Mersenne did not say, like Descartes, that animals were machines identical in kind with the artificial machines made by men. He wrote that the movements of the heart would be understood without mystery if one could discover its mechanism,[35] but men could imitate God's productions in nature only externally and quantitatively. The essence remained hidden. Nevertheless, men's artificial imitations could become testable hypotheses or models for explaining natural phenomena.[36] The quantitative relations within natural phenomena represented the rational and stable *harmonie universelle* that God had chosen to exhibit, both within the structure of his physical creation and in the information about it that men were in a position to discover and communicate.

Mersenne selected for his own particular field of positive inquiry, and for the elimination of magic and the irrational, the mode of operation of vision and of heard sound, and of the languages of men and animals. His first original contributions to acoustics (on vibrating strings), as well as analyses of ancient and modern musical theory and optics, appeared in *Quaestiones in Genesim* (1623). In the same year he announced in his *Observationes*[37] on Francesco Giorgio's plans for a systematic science of sound, "le grand oeuvre de la musique,"[38] which henceforth became his chief intellectual preoccupation. The first sketches appeared in the *Traité de l'harmonie universelle* (1627),[39] *Questions harmoniques* (1634), and *Les préludes de l'harmonie universelle* (1634). Meanwhile, by 1629 Mersenne had planned and soon afterward began writing simultaneously two sets of treatises, in French and in Latin, which together form his great systematic work and were published as the two parts of *Harmonie universelle, contenant la théorie et la pratique de la musique* (1636, 1637), and the eight books of *Harmonicorum libri* with *Harmonicorum instrumentorum libri IV* (1636).[40] Before the final sections of *Harmonie universelle* were in print, he read in Paris, in the winter of 1636–1637, a manuscript of the first day of Galileo's *Discorsi* (1638) containing an account of conclusions about acoustics and the pendulum similar to his own.[41] Mersenne's next work on these subjects was his French summary and critical discussion of Galileo's book in *Les nouvelles pensées de Galilée* (1639). Later he published the results of further acoustical researches in three related works, *Cogitata physico-mathematica* (1644), *Universae geometriae mixtaeque mathematicae synopsis* (1644), and *Novarum observationum physico-mathematicarum tomus III* (1647). The last contains a summary of his contributions to the science of sound.

Parallel discussions of light and vision, beginning in *Quaestiones in Genesim* and Mersenne's correspondence from this time, run especially through *Harmonie universelle* and *Harmonicorum libri*, the *Cogitata*, and *Universae geometriae synopsis*. The inclusion in the optical section of *Universae geometriae synopsis* of unpublished work by Walter Warner, and of a version of Hobbes's treatise on optics with its mechanistic psychology, reflects Mersenne's close English connections at this time. His final contributions to optics, including experimental studies of visual acuity and binocular vision and a critical discussion of current hypotheses on the nature of light, appeared posthumously in *L'optique et la catoptrique* (1651).

Mersenne's scientific analysis of sound and of its effects on the ear and the soul began with the fundamental demonstration that pitch is proportional to frequency and hence that the musical intervals (octave, fifth, fourth, and so on) are ratios of frequencies of vibrations, whatever instrument produces them.

The essential propositions were established by G. B. Benedetti (*ca.* 1563), Galileo's father, Vincenzio Galilei (1589–1590), Beeckman (1614–1615), and, finally, Mersenne (1623–1634). Mersenne gave an experimental proof by counting the slow vibrations of very long strings against time measured by pulse beats or a seconds pendulum. He then used the laws he had completed (now bearing his name), relating frequency to the length, tension, and specific gravity of strings, to calculate frequencies too rapid to count. Similar relations were established for wind and percussion instruments. The demonstration of these propositions made it possible to offer quantitative physical explanations of consonance, dissonance, and resonance.[42]

An allied outstanding discovery apparently made first by Mersenne was the law that the frequency of a pendulum is inversely proportional to the square root of the length. His first statement of this was printed by 30 June 1634, about a year before Galileo's was written.[43] Exploring further acoustical quantities, Mersenne pioneered the scientific study of the upper and lower limits of audible frequencies, of harmonics, and of the measurement of the speed of sound, which he showed to be independent of pitch and loudness. He established that the intensity of sound, like that of light, is inversely proportional to the distance from its source.[44] Mersenne's discussions, after his visit to Italy in 1644, of the Italian and later French experiments with a Torricellian vacuum helped to make a live issue of this whole subject and its bearing on the true medium of sound and on the existence of atmospheric pressure.[45] Besides these contributions to science, collaboration with Doni on an ambitious plan for a comprehensive historical work on the theory and practice of ancient and modern music[46] yielded a rich collection of descriptions and illustrations of instruments, making *Harmonie universelle* and its Latin counterpart essential sources for musicology.

In keeping with his empirical philosophy, Mersenne looked for purely rational explanations of the motions and dispositions of the soul brought about by music. He aimed to put an end to all ideas of magical and occult powers of words and sounds.[47] At the same time he offered a rational analysis of language, arguing that if it was language that chiefly distinguished men from animals, this was a fundamental distinction, for language meant conscious understanding of meaning. The speech and jargon of animals was a kind of communication, but not language, for they mindlessly emitted and responded to messages simply as automatons.[48] Mersenne soon rejected any idea that there were natural names revealing the natures of things and firmly proposed a purely rational theory of language that made words simply conventional physical signs. Because all men possessed reason, they had developed languages in which spoken or written words signified meanings. But just as the effects of music varied with temperament, race, period, and culture, so different groups of men had come to express their common understanding of meaning in a variety of languages diversified by their different historical experiences, environments, needs, temperaments, and customs.[49] In this analysis of common elements Mersenne saw a means of inventing a perfect universal language that could convey information without error. Basing his linguistic experiments on a calculus of permutations and combinations, he proposed a system that would convey the only knowledge of things available to men, that of their quantitative externals. Such a language of quantities "could be called natural and universal"[50] and would be a perfect means of philosophical communication.

Descartes's famous comment that this perfect language could be achieved only in an earthly paradise[51] was true in a way perhaps not intended, for "le bon Père Mersenne" seems to have lived mentally in just such a paradise. "A man of simple, innocent, pure heart, without guile," Gassendi wrote three days after his friend had died in his arms. "A man than whom none was more painstaking, inquiring, experienced. A man whom all the arts and sciences to whose advance he tirelessly devoted himself, by investigating or by deliberating or by stimulating others, will justly mourn."[52] With almost his last breath Mersenne asked for an autopsy to discover the cause of his death. *Maxime de Minimis.*[53] He illustrates the creativeness of gifts of personality distinct from those of sheer originality in the scientific movement.

NOTES

1. Mersenne, *Les préludes de l'harmonie universelle* (Paris, 1634), 135–139.
2. Lenoble, *Mersenne*, 15 ff.
3. Mersenne, *Quaestiones celeberrime in Genesim, cum accurata textus explicatione. In hoc volumine athei, et deistae impugnantur, et expurgantur, et Vulgata editio ab haereticorum calumniis vindicatur. Graecorum et Hebraeorum musica instauratur. . . . Opus theologis, philosophis, medicis, iuriconsultis, mathematicis, musicis vero, et catoptricis praesertim utile . . .* (Paris, 1623), preface.
4. Mersenne, *Observationes et emendationes ad Francisci Georgii Veneti problemata* (Paris, 1623).
5. Mersenne, *Correspondance*, I, 190–193; Lenoble, *Mersenne*, xviii, 28.
6. *Quaestiones in Genesim*, preface.
7. Mersenne, *Les méchaniques de Galilée* (1634), "Épistre dédicatoire," ch. 1.
8. It seems likely that he met Descartes in either 1623 or 1625, before or after the latter's journey to Italy: see *Correspon-*

dance, I, 149; and Lenoble, *Mersenne*, 1, 17, 31, 314–316, for the improbability of their friendship at La Flèche as boys separated by seven and a half years in age, and other misconceptions of their relationship promulgated by Descartes's biographer Adrien Baillet.
9. A. Baillet, *La vie de Monsieur Des-Cartes*, II (1691), 327–328; Jacqueline Pascal's letter of 25 Sept. 1647, in Blaise Pascal, *Oeuvres*, L. Brunschvicg, P. Boutroux, and F. Gazier, eds., II (Paris, 1908), 39–48. Pascal in his "Histoire de la roulette" gave Mersenne the credit for being the first to consider, about 1615, the curve produced by "le roulement des roues": *Oeuvres*, VIII (1914), 195; cf. Mersenne, *Correspondance*, I, 13, 183–184, and II, 598–599.
10. *Correspondance*, I, xliii–xliv, V, 209–211, 371; Lenoble, *Mersenne*, 1, 35–36, 48, 233–234, 586–594. Mersenne had for more than a decade been a member of the Cabinet des Frères Dupuy; for this and the various proposals he made beginning in 1623 for national and international cooperation through academies of theology in which scientific and other experts assisted, of science and mathematics, and of music, see the *Correspondance*, I, 45, 106–107, 129, 136–137, 169–172, V, 301–302; *Quaestiones in Genesim*, preface, dedication, and cols. 1510–1511, 1683–1687; *La vérité des sciences*, 206–224, 751–752, 913–914; *Traité de l'harmonie universelle*, 50, 255–256.
11. *Quaestiones in Genesim*, preface and cols. 841–850, 879–920. He gave considerable attention to Kepler, whom he supported against Fludd: cf. cols. 1016, 1556–1562; *Correspondance*, I, 131–132, 147–148; Lenoble, *Mersenne*, 224–225, 367–370, 394–413.
12. *L'impiété des déistes*, II, 200–201; cf. 198.
13. *Quaestiones in Genesim*, cols. 57, 85, 892–893, 903–904, 1081–1096, 1164; cf. *Correspondance*, I, 130–135.
14. *L'impiété*, I, 230–231; cf. II, 326–342, 363–364, 475.
15. *La vérité*, 111; cf. *L'impiété*, II, 479, 494–495.
16. *La vérité*, 109, 212–213; cf. 913–914.
17. *Ibid.*, 13–15, 226; *Les questions théologiques* (1634), "Épistre" and pp. 9–11, 16–19, 116–117, 123–124, 178–183, 229; *Questions inouyës* (1634), 69–78, 130–131, 153–154; see notes 25–27.
18. *Correspondance*, II, 175.
19. Cf. note 7.
20. *Correspondance*, III, 437–439, 561–568, 630–633.
21. Letter of 8 Feb. 1634, *ibid.*, IV, 37–38.
22. *Ibid.*, IV, 76–78, 156–157; cf. III, 570–572.
23. Mersenne to Martinus Ruarus, 1 Apr. 1644, in Ruarus' *Epistolarum selectarum centuria* (Paris, 1677), 269; Lenoble, *Mersenne*, 413.
24. *Correspondance*, IV, 226, 232, 267–268, 406–407, 411–412, V, 106, 127, 214; note 22. Cf. Descartes's letters to Mersenne during 1633–1635 on Galileo: *ibid.*, III, 557–560; IV, 26–29, 50–52, 97–99, 297–300; V, 127.
25. *Les questions théologiques*, 116–117; cf. 18–19, 43–44, 164.
26. *Harmonie universelle* (1636), "Traitez de la nature des sons et des mouvemens," I, prop. xxxiii, p. 76; cf. II, props. xix, xxi, pp. 149–150, 154–155. The same attitude appears in the *Cogitata physico-mathematica* (1644), "Hydraulica," 251, 260, and "Ballistica," 81–82; in the *Universae geometriae synopsis* (1644), "Cosmographia," preface, 258; and in Roberval's dedication to his "Aristarchus," printed in Mersenne's *Novarum observationum tomus III* (1647).
27. *Les questions théologiques*, 18–19; cf. *Harmonie universelle*, "Première preface générale"; see notes 20, 29.
28. *Harmonie universelle*, "Traitez . . . des sons," II, prop. vii, coroll. 1, p. 112; cf. prop. i, pp. 85–88, and prop vii, pp. 108–112; and for his seconds pendulum, prop. xv, pp. 135–137, prop. xxii, coroll. 9, p. 220; *Correspondance*, IV, 409–411; A. Koyré, "An Experiment in Measurement." These criticisms may have provoked Galileo to describe his experiment in more detail in the *Discorsi* (1638); Mersenne again repeated the experiment and wrote in *Les nouvelles pensées de Galilée* (1639) that, with a ball

heavy enough not to be significantly affected by air resistance, he found "les mesmes proportions" (pp. 188–189).

29. *Harmonie universelle*, "Traitez . . . des sons," III, prop. v, p. 167; cf. *Novarum observationum . . . tomus III*, 113, on reason guiding the senses.

30. *Les préludes*, 118; cf. *Quaestiones in Genesim*, cols. 130, 937–948, 1262–1272; *L'impiété*, II, 360–378, 390–391, 401–437; *La vérité*, 15–20, 25–36, 179–189; *Les questions théologiques*, 229–232.

31. See for this visit, *Correspondance*, II, 486, 506–507, 522–525.

32. See for discussions about light and sound, *ibid.*, I, 329–330, 333–335, II, 107–108, 116–124, 248–249, 282–283, 293–296, 353, 456–459, 467–477, 669–670, III, 35–42, 48–49; *Quaestiones in Genesim*, cols. 742, 1561, 1892; *La vérité*, 69–72; *Traité de l'harmonie universelle*, preface and p. 7; *Les questions théologiques*, 67–69, 105–106, and 164 of the expurgated ed. (Lenoble, *Mersenne*, xx, 399–401, 518; *Correspondance*, IV, 74–76, 203–206, 267–271); *Harmonie universelle*, "Traitez de la nature des sons," I, props. i–ii, viii–x, xxv, xxxii, pp. 1–6, 14–19, 44–48, 73–74; *Harmonicorum libri*, I, props. ii–vi, pp. 1–3; *Cogitata physico-mathematica*, "Harmonia," 261–271, "Ballistica," preface and pp. 74–82 (on Hobbes, etc.); *Universae geometriae . . . synopsis*, "Praefatio utilis in synopsim mathematicam," sec. x, and pp. 471 bis–487, 548, 567–571 (by Hobbes); *L'optique et la catoptrique* (1651), 1–3, 49–54, 77–92; Lenoble, *Mersenne*, 107–108, 317–318, 370–371, 414–418, 421–424, 478–486; note 44.

33. *Les préludes*, 156–159. Their correspondence and Mersenne's publications leave uncertain what Mersenne knew at this time of the earlier ideas developed by Descartes in the *Regulae*, left unfinished in 1629, and *Le monde* and *L'homme*, begun in the same year.

34. *Harmonie universelle* (1637), "Traitez de la voix," I, prop. lii, p. 79; cf. note 47.

35. *Les questions théologiques*, 76–81; cf. 183. Mersenne sent a copy of William Harvey's *De motu cordis* (1628) together with a set of Fludd's works to Gassendi in Dec. 1628 and discussed the circulation of the blood with Descartes: *Correspondance*, II, 181–182, 189, 268; III, 346, 349–350; VIII, 296.

36. *La vérité*, preface; *Harmonie universelle*, "Traitez de la voix," II, prop. xxii, pp. 159–160, "Nouvelles observations physiques et mathématiques," I, coroll. 5, pp. 7–8.

37. Cols. 439–440; see note 4.

38. *La vérité*, 567; cf. preface and 370–371, 579, 981.

39. "Sommaire"; cf. *Correspondance*, I, 195–196, 204.

40. Mersenne created a major bibliographical problem by writing these treatises simultaneously with numerous revisions and repetitions, and by having the different sections printed separately: scarcely any two of the extant copies have the same contents in the same order; cf. Lenoble, *Mersenne*, xxi–xxvi; see note 41.

41. Galileo Galilei, *Opere*, A. Favaro, ed., XVI (Florence, 1905), 524, XVII (1907), 63–64, 80–81; Mersenne, *Correspondance*, VI, 83–84, 241–243, cf. 216, 237; *Harmonie universelle* (1637), "Seconde observation"; cf. *Les nouvelles pensées de Galilée* (1639), preface and 66–67, 72, 92, 96–99, 104–105, 109–110; cf. *Correspondance*, VII, 107–109, 317–320; see note 42.

42. *Quaestiones in Genesim*, cols. 1556–1562, 1699, 1710; *La vérité*, 370–371, 567, 614–620; *Traité de l'harmonie universelle*, 147–148, 447; *Harmonicorum libri*, I, prop. ii, II, props. vi–viii, xvii–xxi, xxxiii–xxxv, IV, prop. xxvii; *Harmonie universelle*: "Traité des instrumens," I, props. v, xii, xvi, III, props. vii, xvii; "Traitez . . . des sons," I, props. i, vii, xiii, III, props. i, v, vi, xv; "Traitez de la voix," I, prop. lii; "Traitez des consonances," I, props. vi, x, xiii, xvii, xviii, xix, xxii, II, prop. x. Mersenne wrote from Paris on 20 Mar. 1634 to Peiresc in Aix-en-Provence that after more than ten years of work he had finished his "grand

oeuvre de l'*Harmonie universelle*," of which he sent "le premier cayer" (*Correspondance*, IV, 81–82). The earliest section in which he gave an extensive analysis of the physical quantities determining the notes and intervals produced by vibrating strings, bells, and pipes, and used this to explain resonance, consonance, and dissonance seems to have been the "Traitez des consonances," I, "Des consonances." This was in print by 2 Feb. 1635 (Mersenne to Doni, *Correspondance*, V, 40–41). Internal references and the *Correspondance*, IV–V, indicate that he was writing at the same time, during 1634, the "Traité des instrumens" (I–III) and the *Harmonicorum libri* (I–IV); see Crombie, *Galileo and Mersenne* (forthcoming).

43. Mersenne published this law first in one of his original additions to *Les méchaniques de Galilée*, 7th addition, p. 77. The "privilège du roy" gives 30 June 1634 as the date on which the printing was completed: cf. Mersenne, *Correspondance*, IV, 76–77, 207–212, and the new ed. of *Les méchaniques* by Rochot (1966). The work was bound with Mersenne's *Les questions théologiques* and presumably sent with that to Doni by way of Peiresc in 1634 (Mersenne to Peiresc, 28 July 1634; Doni to Mersenne, 8 Nov. 1634; *Correspondance*, IV, 267, 384–385, appendix III, 444–455). Élie Diodati sent a copy of *Les méchaniques* from Paris to Galileo on 10 Apr. 1635 (*ibid.*, V, 132; cf. VI, 242). For Mersenne's use of this pendulum law, and his possible derivation of it from the law of falling bodies, see also *Harmonicorum libri*, II, props. xxvi–xxix; *Harmonie universelle*, "Traitez des instrumens," I, props. xix–xx; "Traitez . . . des sons," III, "Du mouvement," props. xxi, xxiii. Galileo's correspondence with Fulgenzio Micanzio in Venice between 19 Nov. 1634 and 7 Apr. 1635 (*Opere*, XVI, 163, 177, 193, 200–201, 203, 208–210, 214, 217–233, 236–237, 239–244, 254) indicates that he had not written the last part of the first day of the *Discorsi* (in which he discussed the pendulum and acoustics) by the latter date. His letter of 9 June 1635 to Diodati, saying that he had sent a copy to Giovanni Pieroni, and subsequent correspondence (*Opere*, XVI, 272–274, 300–304, 359–361) establishes this as the latest date of composition. This copy survives in Biblioteca Nazionale Centrale, Florence, MS Banco Raro 31; cf. note 40.

44. For these subjects see, respectively, *Harmonie universelle*, "Traitez des instrumens," I, prop. xix, III, prop. xvii, "Traitez . . . des sons," III, prop. vi, "Traitez de la voix," I, prop. lii; *Harmonicorum libri*, II, props. xviii, xxxiii; *Harmonie universelle*, "Traitez des instrumens," IV, prop. ix, VI, prop. xlii, VII, prop. xviii, "Nouvelles observations," IV; *Harmonicorum instrumentorum libri IV*, I, prop. xxxiii, III, prop. xxvii; cf. *Quaestiones in Genesim*, col. 1560; *Harmonie universelle*, "Traitez . . . des sons," I, props. vii, viii, xiii, xvii, xxi, III, prop. xxii, "De l'utilité de l'harmonie," prop. ix; *Novarum observationum . . . tomus III*, "Reflectiones physico-mathematicae," ch. 20; *Harmonie universelle*, "Traitez . . . des sons," I, props. xii, xv, cf. props. iii, iv (coroll. 30), and III, prop. xxi, coroll. 4; *Harmonicorum libri*, II, prop. xxxix.

45. *Novarum observationum . . . tomus III*, "Praefatio ad lectorem," "Praefatio secunda," and pp. 84–96, 216–218; cf. de Waard, *L'expérience barométrique*, 117–131; Lenoble, *Mersenne*, xxx, 431–436; Middleton, *The History of the Barometer*, 33–54.

46. Cf. *Correspondance*, III, 395, 512–513, IV, 80, 345, 368, VII, 393–394; G. B. Doni, *Annotazioni sopra il compendio de' generi, e de' modi della musica* (Rome, 1640), 277–280.

47. *Quaestiones in Genesim*, cols. 1619–1624; *La vérité*, 16–17, 32, 69–72; *Les préludes*, 212, 219–222; *Questions harmoniques*, 91–99; *Harmonie universelle*, "Préface générale au lecteur," "Traitez . . . des sons," I, props. i–ii, "Traitez de la voix," I, "Traitez des consonances," I, prop. xxxiii; *Harmonicorum libri*, I, prop. ii; Lenoble, *Mersenne*, 522–531.

48. *Harmonie universelle*, "Traitez de la voix," I, prop. xxxix, pp. 49–52; cf. props. vii–xii, xxxviii; cf. note 33.
49. *Quaestiones in Genesim*, cols. 23–24, 470–471, 702–704, 1197–1202, 1217, 1383–1398, 1692; MS continuation, Bibliothèque Nationale, Paris, MS lat. 17, 262, pp. 511, 536 (Lenoble, *Mersenne*, xiii–xiv, 514–517); *L'impiété*, 167; *La vérité*, 67–76, 544–580; *Traité de l'harmonie universelle*, "Sommaire," item 9; *Questions inouyës*, 95–101, 120–122; *Harmonie universelle*, "Préface générale au lecteur" and "Traitez de la voix," preface, I, II, props. vii–xii; *Harmonicorum libri*, VII; Mersenne's discussions from 1621 to 1640 with Guillaume Bredeau, Descartes, Jean Beaugrand, Peiresc, Gassendi, Comenius, and others are in *Correspondance*, I, 61–63, 102–103; II, 323–329, 374–375; III, 254–262; IV, 329; V, 136–140; VI, 4–6; VII, 447–448; X, 264–274; Lenoble, *Mersenne*, 96–109, 514–521.
50. *Les questions théologiques*, quest. xxxiv, "Peut-on inventer une nouvelle science des sons, qui se nomme psophologie?" p. 158 (expurgated ed.); *Harmonie universelle*, "Traitez . . . des sons," I, prop. xxiv (language played on a lute), "Traitez de la voix," I, props. xii, xlvii–l (artificial rational languages), "De l'utilité de l'harmonie," prop. ix (symbolic language, acoustical telegraph). Cf. his proposals for methods of imitating human speech with instruments and for teaching deaf-mutes to speak and communicate: *Harmonie universelle*, "Traitez de la voix," I, props. x–xi, li, "Traitez des instrumens," II, prop. ix; cf. *Correspondance*, III, 354, 358–359, 375, 378, IV, 258–259, 262–263, 280, 289, 294 (1633–1634). On instruments for imitating human speech see "Traitez des instrumens," VI, props. xxxi–xxxii, xxxvi, VII, prop. xxx; *Correspondance*, III, 2–9, 538–553, 578–597, V, 269–272, 293–294, 299–300, 410–415, 478–482 (1631–1635).
51. Descartes to Mersenne, 20 Nov. 1629, in Mersenne, *Correspondance*, II, 323–329; cf. 374–375, IV, 329, 332, V, 134–140, VI, 4, 6.
52. Gassendi to Louis de Valois, 4 Sept. 1648, *Opera*, VI (Lyons, 1658), 291; Lenoble, *Mersenne*, 596, cf. 58; cf. Coste, *Vie*, 13, 99–101; Mersenne, *Correspondance*, I, xxx.
53. Constantijn Huygens, in a poem cited by Thuillier, *Diarium . . .*, II, 104; cf. Lenoble, *Mersenne*, 597, who also quotes a poem by Hobbes on Mersenne.

BIBLIOGRAPHY

I. ORIGINAL WORKS. A list of Mersenne's published and unpublished writings is in R. Lenoble, *Mersenne ou la naissance du mécanisme* (Paris, 1943), "Bibliographie," which also contains a list of publications on Mersenne from the seventeenth century. His main books are named in the text; all were published at Paris. There is a recent edition of *Les méchaniques de Galilée* by B. Rochot (Paris, 1966); and Mersenne's own copy of *Harmonie universelle*, with his annotations made during 1637–1648, has been reprinted in facsimile by the Centre National de la Recherche Scientifique (Paris, 1965). Above all there is Mersenne's *Correspondance*, C. de Waard, R. Pintard, and B. Rochot, eds. (Paris, 1932–   ), which includes information about his publications and MSS.

II. SECONDARY LITERATURE. The first biography was the valuable study written immediately after Mersenne's death by a fellow Minim, Hilarion de Coste, *La vie du R. P. Marin Mersenne, théologien, philosophe et mathématicien, de l'Ordre des Pères Minim* (Paris, 1649). A second main source for his life is René Thuillier, *Diarium patrum, fratrum et sororum Ordinis Minimorum Provinciae Franciae*

sive Parisiensis qui religiose obierunt ab anno 1506 ad annum 1700 (Paris, 1709). The critical problems are discussed in the *Correspondance*, I (1932), xix–lv; in this his career, publications, and relations with his contemporaries can be followed in detail from 1617.

The major study of Mersenne's life and thought is Lenoble's *Mersenne*. A valuable monograph is H. Ludwig, *Marin Mersenne und seine Musiklehre* (Halle–Berlin, 1935). For particular aspects there are C. de Waard, *L'expérience barométrique* (Thouars, 1936), and W. E. K. Middleton, *The History of the Barometer* (Baltimore, 1964), on the Torricellian vacuum; Mario M. Rossi, *Alle fonti del deismo e del materialismo moderni* (Florence, 1942), on his relation to deism; A. Koyré, "An Experiment in Measurement," in *Proceedings of the American Philosophical Society*, **97** (1953), 222–237, repr. in his *Metaphysics and Measurement* (London, 1968), on his critique of Galileo's experiments on acceleration; R. H. Popkin, *The History of Scepticism From Erasmus to Descartes* (Assen, Netherlands, 1964), on his relation to contemporary skepticism; F. A. Yates, *Giordano Bruno and the Hermetic Tradition* (London, 1964), on his relation to Hermeticism; A. C. Crombie, "Mathematics, Music and Medical Science," in *Actes du XIIe Congrès international d'histoire des sciences: Paris 1968* (Paris, 1971), 295–310, on his science of sound; and W. L. Hine, "Mersenne and Copernicanism," in *Isis*, **64** (1973), 18–32. A further substantial discussion of his natural philosophy, with special reference to vision, heard sound and language, is included in A. C. Crombie, with the collaboration of A. Carugo, *Galileo and Mersenne: Science, Nature and the Senses in the Sixteenth and Early Seventeenth Centuries*, 2 vols. (forthcoming).

A. C. CROMBIE

MÉRY, JEAN (*b.* Vatan, France, 6 January 1645; *d.* Paris, France, 3 November 1722), *anatomy, surgery, pathology.*

Intent on following in his father's profession, Méry traveled to Paris at the age of eighteen to study surgery at the Hôtel-Dieu, then the best place to learn surgical practice. In addition to his regular studies, Méry undertook clandestine dissections whenever fresh human material became available to him. After completing his preparations he set up a private surgical practice, becoming well known, particularly in lithotomy. Much of his career was centered at the Hôtel-Dieu, where he was surgeon from 1681 and chief surgeon from 1700. He was appointed a senior surgeon at Les Invalides, Paris, in 1683. In 1684 Méry was elected to the Academy of Sciences. He also had connections with the French court. In 1681 he was appointed surgeon to the queen and later was sent by the court on at least two medical missions. Méry traveled to England in 1692 for the court, but the purpose of this trip is unknown.

Méry tended to be taciturn, has been described as argumentative, and often saw his family only at meals. He did a thorough job at the Hôtel-Dieu, both in his hospital practice and in training young surgeons. The balance of his time was divided between the Academy and his anatomical research.

Most of Méry's researches were comparative-anatomical and pathological. The pathological researches were mostly descriptive in character and covered a wide range of situations, although most of them were concerned with human developmental malformations. Of greater interest are his researches in comparative anatomy, including his physiological investigations. In the latter his methods were comparative and were based on preserved and dried anatomical specimens. Because of the limited preservation techniques available, this approach could be deceptive.

After his election to the Academy in 1684, Méry became closely associated with the comparative-anatomical work led by Claude Perrault and J.-G. Duverney. As a member of this group, Méry made contributions to their joint publications, in which each man's specific contributions usually cannot be determined. Méry worked closely with Duverney until about 1693, when their differing interpretations of mammalian fetal circulation estranged them. The coolness that resulted was apparent to Martin Lister, when he visited Paris in 1698. Méry probably did more to retard than to aid the understanding of this problem. Méry claimed that the blood flowed from the left to the right through the foramen ovale in the interatrial septum. This view was prevalent enough that Haller took time to refute it. Méry initially formulated his theory from a false analogy between a tortoise heart and a fetal mammalian heart. Ultimately he based his theory of fetal circulation on a comparison of the cross sections of the pulmonary artery and the aorta, concluding that not all of the blood passing through the pulmonary artery and returning to the heart by the pulmonary vein could pass into the aorta. Instead, he thought, a portion of that blood passed through the foramen ovale from the left to the right side of the heart.

Méry erred in assuming that the cross section of an artery is the only factor determining the amount of blood that can flow through it. He compounded this error by his method of measuring the relative cross sections of the arteries. He may have used fresh preparations for his measurements on cows and sheep. For those on human beings, he probably used preserved specimens, dried ones as a rule. The results were inconsistent at best. For example, Martin Lister described a fetal heart that he saw in Méry's collection

which had no valve for the foramen ovale, and which was open in both directions and had a diameter nearly equal to that of the aorta. For two decades numerous arguments were presented on both sides of the controversy between Méry's views and the traditional views dating back to Harvey and Lower. Méry held his views against all opposition to the end.

In other areas of anatomy Méry demonstrated that he was a capable and careful worker, making a number of valuable contributions to the anatomy of a wide range of animals. He described the urethral glands named after Cowper some years before Cowper's description, and he preceded Winslow in a description of the eustachian valve, although he misinterpreted its function as part of his concept of the mammalian fetal circulation.

BIBLIOGRAPHY

I. ORIGINAL WORKS. For a list of Méry's anatomical studies in *Mémoires de l'Académie royale des sciences,* see either the 1734 index volumes or J. D. Reuss, *Repertorium commentationum ... a societatibus litterariis editorum,* 16 vols. (Gottingen, 1801–1820). Much of the controversy on fetal circulation is contained in *Nouveau système de la circulation du sang par le trou ovale dans le foetus humain; avec les réponses aux objections de Messieurs Duverney, Tauvri, Verheyen, Silvestre & Buissiere contre cette hypothèse* (Paris, 1700), as well as scattered papers. He made unidentified contributions which were incorporated in *Mémoires pour servir à l'histoire naturelle des animaux* (Paris, 1732–1734).

II. SECONDARY LITERATURE. Principal biographical sources are the article in *Biographie universelle ancienne et moderne,* XXVIII (Paris); Bernard Le Bouyer de Fontenelle, "Éloge," in *Oeuvres de M. de Fontenelle,* VI (Amsterdam, 1754); and Martin Lister, *A Journey to Paris in the Year 1698* (London, 1699), *passim.* There are numerous references to Méry throughout the appropriate years of the *Histoire de l'Académie royale des sciences.* For Méry's theory of fetal circulation see Kenneth J. Franklin, "Jean Méry (1645–1722) and His Ideas on the Foetal Blood Flow," in *Annals of Science,* **5** (1945), 203–338; and for some of his other anatomical work see F. J. Cole, *A History of Comparative Anatomy from Aristotle to the Eighteenth Century* (London, 1944).

WESLEY C. WILLIAMS

MESHCHERSKY, IVAN VSEVOLODOVICH (*b.* Arkhangelsk, Russia, 10 August 1859; *d.* Leningrad, U.S.S.R., 7 January 1935), *mechanics, mathematics.*

Meshchersky was born into a family of modest means, but succeeded in obtaining a good education. He was enrolled in the Arkhangelsk Gymnasium in

1871, and graduated from it with a gold medal after seven years. He entered St. Petersburg University in 1878, and undertook the study of mathematics, attending the lectures of Chebyshev, A. N. Korkin, and A. Possé; he simultaneously studied mechanics. He graduated in 1882 but remained at the university to begin his own academic career. He passed the examinations for the master's degree in applied mathematics in 1889 and became a *Privatdozent* the following year.

In 1891 Meshchersky was appointed to the chair of mechanics at the St. Petersburg Women's College, a post that he retained until 1919, when the college was incorporated into the university. In 1897 he defended a dissertation entitled *Dinamika tochki peremennoy massy* ("The Dynamics of a Point of Variable Mass") before the Physics and Mathematics Faculty of St. Petersburg University and was awarded a doctorate in applied mathematics. In 1902 Meshchersky was invited to head the department of applied mathematics at the newly founded St. Petersburg Polytechnic Institute (now the Leningrad M. I. Kalin Polytechnic Institute), for which he had helped to develop a curriculum.

Meshchersky taught at St. Petersburg University for twenty-five years and at the Polytechnic Institute for thirty-three. He was a conscientious and innovative pedagogue. Among other things, he was concerned with drafting a scientific-methodological guide to the teaching of mathematics and mechanics; his *Prepodavanie mekhaniki i mekhanicheskie kollektsii v nekotorykh vysshikh uchebnykh zavedeniakh Italii, Frantsii, Shveytsarii i Germanii* ("The Teaching of Mechanics and Mechanics Collections in Certain Institutions of Higher Education in Italy, France, Switzerland, and Germany"; 1895) contributed significantly toward raising the standards of the teaching of mechanics in Russia. Meshchersky's own course in theoretical mechanics became famous, while his textbook on that subject, *Sbornik zadach po teoreticheskoy mekhanike* ("A Collection of Problems in Theoretical Mechanics"), published in 1914, went through twenty-four editions and became a standard work.

Meshchersky's purely scientific work was devoted to the motion of bodies of variable mass. He reported the results of his first investigations of the problem at a meeting of the St. Petersburg Mathematical Society held on 27 January 1893, then made it the subject of the doctoral dissertation that he presented four years later. He began the thesis *Dinamika tochki peremennoy massy* with a discussion of the many instances in which the mass of a moving body changes, citing as examples the increase of the mass

of the earth occasioned by meteorites falling on it; the increase of the mass of an iceberg with freezing and its decrease with thawing; the increase of the mass of the sun through its gathering of cosmic dust and its decrease with radiation; the decrease of the mass of a rocket as its fuel is consumed; the decrease of the mass of a balloon as its ballast is discarded; and the increase of the mass of a captive balloon as it draws its tether with it in rising.

Having defined the problem, Meshchersky considered it physically. He established that if the mass of a point changes during motion, then Newton's second law of motion must be replaced by an equation of the motion of a point of variable mass wherein $m \dfrac{d\bar{v}}{dt} = \bar{F} \mid \bar{R}$ ($\bar{F}$ and $\bar{R}$ being the given and the reactive forces, respectively), where $\bar{F}$ and $\bar{R} = \dfrac{dm}{dt}\bar{U}_r$.

This natural generalization of the equation of motion of classical mechanics is now called Meshchersky's equation. In his second important work, "Uravnenia dvizhenia tochki peremennoy massy v obshchem sluchae" ("Equations of the Motion of a Point of Variable Mass in the General Case," 1904), Meshchersky gave his theory a definitive and elegant expression, establishing the general equation of motion of a point of which the mass is changing by the simultaneous incorporation and elimination of particles.

In developing the theoretical foundations of the dynamics of a point of variable mass, Meshchersky opened a new area of theoretical mechanics. He also examined a number of specific problems, including the ascending motion of a rocket and the vertical motion of a balloon. His exceptionally thorough general investigation of the motion of a point of variable mass under the influence of a central force led to a new celestial mechanics; he was further concerned with the motions of comets. He was, moreover, the first to formulate, from given external forces and given trajectories, the so-called inverse problems in determining the law for the change of mass.

Meshchersky published a number of papers on general mechanics. In "Differentsialnye svyazi v sluchae odnoy materialnoy tochki" ("Differential Ties in the Case of One Material Point"; 1887), he examined the motion of a point subjected to a nonholonomic tie, which is neither ideal nor linear. In "O teoreme Puassona pri sushchestvovanii uslovnykh uravneny" ("On Poisson's Theorem on the Existence of Conditional Arbitrary Equations"; 1890), he took up the integration of dynamical equations, while in "Sur un problème de Jacobi" (1894), he gave a generalization of Jacobi's results. A paper of 1919, "Gidrodinamicheskaya analogia prokatki" ("A Hydrodynamic Analogue of Rolling"), is of particular interest because it contains Meshchersky's ingenious attempt to elucidate the equations of motion rolling bodies in terms of those for a viscous fluid.

Meshchersky's work on the motion of bodies of variable mass remains his most important contribution to science. His pioneering studies formed the basis for much of the rocket technology and dynamics that was developed rapidly following World War II.

BIBLIOGRAPHY

I. ORIGINAL WORKS. Meshchersky's writings include "Davlenie na klin v potoke neogranichennoy shiriny dvukh izmereny" ("The Pressure on a Wedge in a Two-Dimensional Stream of Unbounded Width"), in *Zhurnal Russkago fiziko-khimicheskago obshchestva pri Imperatorskago St.-Peterburskago universitete*, **18** (1886); "Differentsialnye svyazi v sluchae odnoy materialnoy tochki" ("Differential Bonds in the Case of One Material Point"), in *Soobshchenie Kharkovskogo matematicheskogo obshchestva* (1887), 68–79; *Prepodavanie mekhaniki i mekhanicheskie kollektsii v nekotorykh vysshikh uchebnykh zavedeniakh Italii, Frantsii, Shveytsarii i Germanii* ("The Teaching of Mechanics and Mechanics Collections in Certain Institutions of Higher Education in Italy, France, Switzerland, and Germany"; St. Petersburg, 1895); *Dinamika tochki peremennoy massy* ("The Dynamics of a Point of Variable Mass"; St. Petersburg, 1897); and *O vrashchenii tyazhelogo tverdogo tela s razvertyvayushcheysya tyazheloy nityu okolo gorizontalnoy osi* ("On the Rotation of a Heavy Solid Body Having an Unwinding Heavy Thread About Its Horizontal Axis"; St. Petersburg, 1899).

Later writings include "Über die Integration der Bewegungsgleichungen im Probleme zweier Körper von veränderlicher Masse," in *Astronomische Nachrichten*, **159**, no. 3807 (1902); "Uravnenia dvizhenia tochki peremennoy massy v obshchem sluchae" ("Equations of the Motion of a Point of Variable Mass in the General Case"), in *Izvestiya S-Peterburgskago politekhnicheskago instituta Imperatora Petra Velikago*, **1** (1904); *Sbornik zadach po teoreticheskoy mekhanike* ("Collection of Problems in Theoretical Mechanics"; St. Petersburg, 1914); "Zadachi iz dinamiki peremennoy massy" ("Problems From the Dynamics of a Variable Mass"), in *Izvestiya S-Peterburgskago politekhnicheskago instituta Imperatora Petra Velikago*, **27** (1918); and "Gidrodinamicheskaya analogia prokatki" ("A Hydrodynamic Analogue of Rolling"), *ibid.*, **28** (1919).

II. SECONDARY LITERATURE. See Y. L. Geronimus, "Ivan Vsevolodovich Meshchersky (1859–1935)," in *Ocherki o rabotakh korifeev russkoy mekhaniki* ("Essays on the Works of the Leading Figures of Russian Mechanics"; Moscow, 1952); A. T. Grigorian, "Ivan Vsevolodovich Meshchersky (k 100-letiyu so dnya rozhdenia)" ("Ivan Vsevolodovich Meshchersky [on the Centenary of His Birth]"), in *Voprosy istorii estestvoznaniya i tekhniki* (1959), no. 7; and "Mekhanika tel peremennoy massy I. V. Meshcherskogo" ("I. V. Meshchersky's Mechanics of Bodies of Variable Mass"), in *Evolyutsia mekhaniki v Rossii* ("The Evolution of Mechanics in Russia"; Moscow, 1967); A. A. Kosmodemyansky, "Ivan Vsevolodovich Meshchersky (1859–1935)," in *Lyudi russkoy nauki* ("People of Russian Science"; Moscow, 1961), 216–222; and E. L. Nikolai, "Prof. I. V. Meshchersky [Nekrolog]," in *Prikladnaya matematika i mekhanika*, **3**, no. 1 (1936).

A. T. GRIGORIAN

MESMER, FRANZ ANTON (*b.* Iznang, Germany, 23 May 1734; *d.* Meersburg, Germany, 5 March 1815), *medicine, origins of hypnosis.*

Mesmer was born and raised in the Swabian village of Iznang near the Lake of Constance. His father was a forester employed by the archbishop of Constance; his mother, the daughter of a locksmith; and his family, large (Franz Anton was the third of nine children), Catholic, and not particularly prosperous. By the time he began to propound his theory of animal magnetism or mesmerism, Mesmer had risen through the educational systems of Bavaria and Austria and had advanced to a position of some prominence in Viennese society through his marriage to a wealthy widow, Maria Anna von Posch, on 16 January 1768. Mesmerism therefore may have been the product of an ambitious *arriviste* but not of a mountebank. The man and the "ism" represent a period when medicine was attempting to assimilate advances in the physical and biological sciences and when scientists often indulged in cosmological speculations that read like science fiction today but passed as respectable varieties of Newtonianism in the eighteenth century.

After preliminary studies in a local monastic school, Mesmer spent four years at the Jesuit University of Dillingen (Bavaria), presumably as a scholarship student preparing for the priesthood. He then attended the University of Ingolstadt for a brief period and in 1759 entered the University of Vienna as a law student. Having changed to medicine and completed the standard course of studies, he received his doctorate in 1766. A year later he began practice as a member of the faculty of medicine in what was one of Europe's most advanced medical centers; for the Vienna school was then in its prime, owing to the patronage of Maria Theresa and the leadership of Gerhard van Swieten and Jan Ingenhousz.

Mesmer later traced his theory of animal magnetism to his doctoral thesis, *Dissertatio physico-medica de planetarum influxu.* At the time of its defense, however, the thesis did not strike the Viennese authorities as a revolutionary new theory of medicine. On the contrary, it showed a common tendency to speculate about invisible fluids, which derived both from Cartesianism and from the later queries in Newton's *Opticks* as well as from Newton's remarks about the "most subtle spirit which pervades and lies hid in all gross bodies" in the last paragraph of his *Principia.* The immediate source of Mesmer's fluid was Richard Mead's *De imperio solis ac lunae in corpora humana et morbis inde oriundis* (London, 1704), a work upon which Mesmer's thesis drew heavily. Mead had argued that gravity produced "tides" in the atmosphere as well as in water and that the planets could therefore affect the fluidal balance of the human body. Mesmer associated this "animal gravitation" with health: physical soundness resulted from the "harmony" between the organs of the body and the planets—a proposition, he emphasized, that had nothing to do with the fictions of astrology.

The proposition took on new life for Mesmer when he began treating his own patients. Inspired by the experiments of Maximilian Hell, a court astronomer and Jesuit priest, who used magnets in the treatment of disease, Mesmer applied magnets to his patients' bodies and produced remarkable results, especially in the case of a young woman suffering from hysteria. Unlike Hell, Mesmer did not attribute his cures to any power in the magnets themselves. Instead, he argued that the body was analogous to a magnet and that the fluid ebbed and flowed according to the laws of magnetic attraction. Having moved from "animal gravitation" to "animal magnetism," he announced his new theory in *Sendschreiben an einen auswärtigen Arzt . . .* (Vienna, 1775).

By this time Mesmer had moved into a comfortable town house in Vienna, which he used as a clinic. His marriage brought him enough wealth to pursue his experiments at his leisure and enough leisure to indulge his passion for music. Mesmer knew Gluck, seems to have been acquainted with Haydn, and saw a great deal of the Mozarts. The first production of a Mozart opera, *Bastien und Bastienne,* took place in Mesmer's garden, and Mozart later made room for mesmerism in a scene in *Così fan tutte.* In general, the ten years between Mesmer's marriage in 1768 and his departure from Vienna in 1778 seem to have been a time of prosperity and some prominence. He built up a repertoire of techniques and cures; he gave lectures and demonstrations; and he traveled through Hungary, Switzerland, and Bavaria, where he was made a member of the Bavarian Academy of Sciences at Munich in 1775. Mesmer also developed a taste for

publicity. He staged and announced his cures in a manner that offended some of Vienna's most influential doctors. Offense developed into open hostility in 1777 during a dispute over Mesmer's treatment of Maria-Theresa von Paradies, a celebrated blind pianist who was eventually removed from Mesmer's care by her parents. In these circumstances Mesmer decided to leave Vienna and perhaps also to leave his wife, who did not accompany him through the later episodes of his career.

The next and most spectacular episode began with Mesmer's arrival in Paris in February 1778. He set up a clinic in the Place Vendôme and the nearby village of Créteil and then began an elaborate campaign to win recognition of his "discovery" from France's leading scientific bodies. Helped by some influential converts and an ever-increasing throng of patients, who testified that they had been cured of everything from paralysis to what the French then called "vapeurs," Mesmer seized the public's imagination and alienated the Faculty of Medicine of the University of Paris, the Royal Society of Medicine, and the Academy of Sciences. The defenders of orthodox medicine took offense at what the public found most appealing about mesmerism—not its theory but its extravagant practices. Instead of bleeding and applying purgatives, the mesmerists ran their fingers over their patients' bodies, searching out "poles" through which they infused mesmeric fluid. By the 1780's Mesmer had given up the use of magnets; but he had perfected other devices, notably his famous "tub," a mesmeric version of the Leyden jar, which stored fluid and dispensed it through iron bars that patients applied to their sick areas. Mesmer transmitted his invisible fluid through all sorts of media—ropes, trees, "chains" of patients holding hands—and he usually sent it coursing through the air by gestures with his hands. He reasoned that his own body acted as an animal type of magnet, reinforcing the fluid in the bodies of his patients. Disease resulted from an "obstacle" to the flow of the fluid. Mesmerizing broke through the obstacle by producing a "crisis," often signaled by convulsions, and then restoring "harmony," a state in which the body responded to the salubrious flow of fluid through all of nature.

Mesmerism presented itself to the French as a "natural" medicine at a time when the cult of nature and the popular enthusiasm for science had reached a peak. Mesmer did not produce any proof of his theory or any rigorous description of experiments that could be repeated and verified by others; but like contemporary chemists and physicists, he seemed able to put his invisible fluid to work. Scores of Parisians fell into "crises" at the touch of Mesmer's hand and recovered with a new sense of being at harmony with the world. The mesmerists published hundreds of carefully documented and even notarized case histories. And they produced an enormous barrage of propaganda—at least 200 books and pamphlets, more than were written on any other single subject during the decade before the opening phase of the Revolution in 1787.

Thus mesmerism became a *cause célèbre*, a movement, which eventually even eclipsed Mesmer himself. He limited his part in the polemics to two pamphlets, written by or for him: *Mémoire sur la découverte du magnétisme animal* (1779) and *Précis historique des faits relatifs au magnétisme animal* (1781). The first contained twenty-seven rather vague propositions, which is as close as Mesmer came to systematizing his ideas. He left the system-building to his disciples, notably Nicolas Bergasse, who produced many of the articles and letters issued in Mesmer's name as well as his own mesmeric treatise, *Considérations sur le magnétisme animal* (1784). The disciples also formed a sort of Masonic secret society, the Société de l'Harmonie Universelle, which developed affiliates in most of France's major cities. The spread of the new medicine alarmed not only the old doctors but also the government. A royal commission composed of distinguished doctors and academicians, including Bailly, Lavoisier, and Franklin, reported in 1784 that, far from being able to cure disease, Mesmer's fluid did not exist. The report badly damaged the movement, which later dissolved into schisms and heresies. Mesmer finally left his followers to their quarrels and, after a period of traveling through England, Austria, Germany, and Italy, settled in Switzerland, where he spent most of the last thirty years of his life in relative seclusion.

Considered as a movement, mesmerism suggests some of the varieties of pre-Romanticism and popular science in the late eighteenth century. It did not spend itself as an intellectual force for almost a hundred years, as the mesmerist passages in the works of Hoffmann, Hugo, and Poe testify. But as a scientific theory mesmerism offered only a thin and unoriginal assortment of ideas. Although Mesmer's own writings contained little sustained theorizing, they provided enough for his enemies to detect all manner of occultist and vitalistic influences and to align him with William Maxwell, the Scottish physician, author of *De Medicina Magnetica* (1769), Robert Fludd, J. B. van Helmont, and Paracelsus—when they did not categorize him with Cagliostro. This version of his intellectual ancestry seems convincing enough, if one adds Newton and Mead to the list. But nothing proves that Mesmer was a charlatan. He seems to have believed sincerely in his theory, although he also showed a fierce determination to convert it into cash: he charged ten louis a month for the use of his "tubs"; and he made a fortune from the Société de l'Harmonie Universelle, which, in return, claimed exclusive proprietorship of his deepest "secrets."

In terms of the development of medicine, the techniques of mesmerizing proved more influential than its theory. By concentrating on the "rapport" of patient and doctor, Mesmer seems to have dealt effectively with nervous disorders. He certainly had, to put it mildly, a forceful bedside manner; and in 1784 his followers, led by the Chastenet de Puységur brothers, extended mesmeric "rapport" into something new: mesmerically induced hypnosis. Later groups of hypnotists, particularly in the mesmerist sects of Lyons and Strasbourg, abandoned the hypothesis of a cosmic fluid. In the nineteenth century hypnosis, shorn of Mesmer's cosmology and perfected by James Braid and J. M. Charcot, became an accepted medical practice. And finally, through Charcot's impact on Freud, mesmerism exerted some influence on the development of psychoanalysis, another unorthodox product of the Viennese school.

BIBLIOGRAPHY

I. ORIGINAL WORKS. Mesmer's own works contain only a sketchy version of his system. The most important are *Dissertatio physico-medica de planetarum influxu* (Vienna, 1766); *Schreiben über die Magnetkur* (n.p., 1766); *Mémoire sur la découverte du magnétisme animal* (Geneva, 1779); *Précis historique des faits relatifs au magnétisme animal* (London, 1781); and *Mesmerismus oder System der Wechselwirkungen, Theorie und Anwendung des thierischen Magnetismus als die allgemeine Heilkunde zur Erhaltung des Menschen*, K. C. Wolfart, ed. (Berlin, 1814). There are some unpublished letters by Mesmer and his followers in the Bibliothèque Nationale, fonds français, 1690.

II. SECONDARY LITERATURE. For a thorough but incomplete bibliography of early works on Mesmer and mesmerism, see Alexis Dureau, *Notes bibliographiques pour servir à l'histoire du magnétisme animal* (Paris, 1869). Most of the important source material is contained in the fourteen enormous volumes of the mesmerist collection in the Bibliothèque Nationale, 4° Tb 62.1.

Biographies of Mesmer tend to treat him as a forgotten pioneer of hypnosis and Freudianism: Margaret Goldsmith, *Franz Anton Mesmer: The History of an Idea* (London, 1934); E. V. M. Louis, *Les origines de la doctrine du magnétisme animal: Mesmer et la Société de l'harmonie* (Paris, 1898); Rudolf Tischner, *Franz Anton Mesmer, Leben, Werk und Wirkungen* (Munich, 1928); Jean Vinchon, *Mesmer et son secret* (Paris, 1936); and Stefan Zweig, *Mental Healers: Franz Anton Mesmer, Mary Baker Eddy, Sigmund Freud* (London, 1933).

For a more scholarly treatment of aspects of Mesmer's life and thought, see R. Lenoir, "Le mesmérisme et le système du monde," in *Revue d'histoire de la philosophie*, **1** (1927), 192–219, 294–321; Bernhard Milt, *Franz Anton Mesmer und Seine Beziehungen zur Schweiz: Magie und Heilkunde zu Lavaters Zeit* (Zurich, 1953); and Frank Pattie, "Mesmer's Medical Dissertation and Its Debt to Mead's *De Imperio Solis ac Lunae*," in *Journal of the History of Medicine and Allied Sciences*, **11** (1956), 275–287.

Works concentrating on mesmerism as a movement rather than as a philosophy are Robert Darnton, *Mesmerism and the End of the Enlightenment in France* (Cambridge, Mass., 1968); and Louis Figuier, *Histoire du merveilleux dans les temps modernes*, 2nd ed., III (Paris, 1860).

ROBERT DARNTON

MESNIL, FÉLIX (*b.* Ormonville-la-Petite, Manche, France, 12 December 1868; *d.* Paris, France, 15 February 1938), *zoology, general biology, tropical medicine*.

Mesnil, whose family had been farmers in Normandy for several generations, attended the school in his village. One of his uncles, a physician in the navy, recognized Mesnil's exceptional ability and arranged for him to enter the lycée in Cherbourg and then the Lycée Saint-Louis in Paris. At the age of eighteen he was accepted by both the École Polytechnique and the École Normale Supérieure; he chose the latter because of the interest in natural history he had developed during his boyhood. He passed the *agrégation* in the natural sciences in 1891 and obtained his doctorate in 1895 with a work on the resistance of lower vertebrates to microbial invasions. After passing the *agrégation* he spent several months at universities in central Europe. Upon returning, he entered the Institut Pasteur and remained there throughout his career. While serving as assistant and secretary to Pasteur, he began to work in Metchnikoff's laboratory, where he acquired experimental technique. Mesnil became *agrégé préparateur* in 1892, laboratory director in 1898, and professor in 1910.

Mesnil's work was varied, much of it oriented toward general biology; important memoirs dealt with systematic, ecological, and ethological zoology. For more than thirty years, during summer vacations Mesnil had the opportunity to study—first alone and then, beginning in 1914, with his brother-in-law Caullery—the fauna of St. Martin Cove, near the Cap de la Hague, and of the neighboring coasts. This research resulted in the description of many new genera and species of annelids, crustaceans, enteropneusts, turbellarians, Orthonectida, and protozoans. A great number of investigations were devoted to the annelid polychaetes—to their morphology, in order to establish their phylogenetic relationships; to their sexual maturity (epitokous forms); and to their asexual reproduction (schizogenesis, regeneration). Mesnil, who was interested in parasitism, discovered that condition in the Monstrillidae. With Caullery, he described *Xenocoeloma brumpti*, a parasite of *Polycirrus arenivorus*; the two scientists furnished a precise analysis of its morphology, of the penetration of the larva into the annelid, and of its complex development. They also studied isopod parasites of sea acorns and spheromes; *Fecampia* (turbellarian rhabdocoeles that are internal parasites of crustaceans); and the Orthonectida and their life cycle.

Alone or with Caullery and A. Laveran (the latter discovered the hematozoon of malaria), Mesnil examined the parasitic protozoans: gregarines, coccidia, Myxosporidia, Microsporidia, infusoria, and flagellates. From 1900 to 1916 Mesnil was concerned especially with the trypanosomes and trypanosomiases: chemotherapy, determination of the species, experimental constitution of heritable strains, infectious power and virulence, reactions of the organism, and the resistance of certain strains to medicines and serums. He was also interested in natural and acquired immunity. He and Laveran devised the test that bears their names for detecting the specific identity of the trypanosomes.

Mesnil reported on many works in microbiology and general biology for various French journals. With G. Bertrand, A. Besredka, Amédée Borrel, C. Delezenne, and A. C. Marie he founded the *Bulletin de l'Institut Pasteur* and he was also its editor. In 1907 he participated in founding the Société de Pathologie Exotique, of which he was secretary-general (1908–1920), then vice-president and president (1924–1927).

Mesnil belonged to the Académie des Sciences (1921), the Académie de Médecine, and (as founding member) the Académie des Sciences Coloniales. He was a commander of the Légion d'Honneur. In 1920 he received the Mary Kingsley Medal of the Liverpool School of Tropical Medicine, and in 1926 C. M. Wenyon dedicated his *Textbook of Protistology* to Mesnil. Among Mesnil's students were E. Roubaud, E. Chalton, A. Lwof, and S. Volkonsky.

Mesnil's learning was prodigious and his memory legendary. Kind and easily approachable, he gave advice and support to everyone. The archives of the Académie des Sciences contain his portrait and autograph manuscripts.

BIBLIOGRAPHY

I. ORIGINAL WORKS. A complete bibliography is in *Titres et travaux scientifiques (1893–1920)* (Laval, 1921). Mesnil's major book was *Trypanosomes et trypanosomiases* (Paris, 1904; 2nd ed., enl., 1912), written with Laveran.

His early articles include "Sur la résistance des vertébrés inférieurs aux infections microbiennes artificielles," his doctoral diss., published in *Annales de l'Institut Pasteur*, **9** (1895), 301–351; "Études de morphologie externe chez les annélides. I. Les spionidiens des côtes de la Manche," in *Bulletin scientifique de la France et de la Belgique*, **29** (1896), 110–268; ". . . II. Remarques complémentaires sur les spionidiens. La nouvelle famille des disomidiens. La place des aonides" and ". . . III. Formes intermédiaires entre les maldaniens et les arénicoliens," *ibid.*, **30** (1897), 83–101 and 144–168; ". . . IV. La famille nouvelle des levinséniens. Révision des ariciens. Affinités des deux familles. Les apistobranchiens," *ibid.*, **31** (1898), 126–149, written with Caullery.

Between 1900 and 1910 he wrote "Recherches sur l'*Hemioniscus balani* épicaride parasite des balanes," in *Bulletin scientifique de la France et de la Belgique*, **34** (1901), 316–362, written with Caullery; "Recherches sur les orthonectides," in *Archives d'anatomie microscopique*, **4** (1901), 381–470, written with Caullery; "Les trypanosomes des poissons," in *Archiv für Protistenkunde*, **1** (1902), 475–498, written with Laveran; "Recherches sur les *Fecampia*, turbellariés rhabdocèles parasites internes des crustacés," in *Annales de la Faculté des sciences de Marseille*, **13** (1903), 131–167, written with Caullery; "Contribution à l'étude des entéropneustes," in *Zoologische Jahrbuch Abteilung für Anatomie*, **20** (1904), 227–256, written with Caullery; "Recherches sur les haplosporidies," in *Archives de zoologie expérimentale et générale*, 4th ser., **4** (1905), 101–181, written with Caullery; and "Sur les propriétés préventives du sérum des animaux trypanosomiés. Races résistantes à ces sérums," in *Annales de l'Institut Pasteur*, **23** (1909), 129–154, written with E. Brimont.

After 1910 he published "Sur deux monstrilides parasites d'annélides," in *Bulletin scientifique de la France et de la Belgique*, **48** (1914), 15–29, written with Caullery; "Notes biologiques sur les mares à *Lithothamnion* de la Hague," in *Bulletin de la Société zoologique de France*, **40** (1915), 160–161, 176–178, 198–200, written with Caullery; "*Xenocoeloma brumpti*, copépode parasite de *Polycirrus arenivorus*," in *Bulletin biologique de la France et de la Belgique*, **53** (1919), 161–233, written with Caullery; and "*Ancyroniscus bonnieri*, épicaride parasite d'un sphéromide *(Dynamene bidentulata)*," *ibid.*, **44** (1920), 1–36, written with Caullery.

II. SECONDARY LITERATURE. Obituaries include M. Caullery, in *Presse médicale*, no. 21 (12 Mar. 1938), 401–402; and in *Bulletin biologique de la France et de la Belgique*, **77** (1938); and G. Ramon, in *Bulletin de l'Académie de médecine*, **119** (1938), 241–247. Unsigned obituaries are in *Bulletin de la Société de pathologie exotique*, **31** (1938), 173–177; *Bulletin de l'Institut Pasteur*, **36** (1938), 177–179; *Annales de l'Institut Pasteur*, **60** (1938), 221–226; and *Archives de l'Institut Pasteur*, **16** (1938), 1–2.

ANDRÉE TÉTRY

MESSAHALA. See **Māshāllāh.**

MESSIER, CHARLES (*b.* Badonviller, Lorraine, France, 26 June 1730; *d.* Paris, France, 11 or 12 April 1817), *astronomy.*

Messier was the tenth of twelve children; his father died when the boy was eleven years old. In October 1751 he arrived in Paris, where (according to J. B. Delambre, virtually the sole biographical source) he had only a neat, legible hand and some practice in drawing to recommend him. The astronomer Joseph-Nicolas Delisle hired him to record observations and to copy maps of Peking and of the great wall of China. In 1755 Delisle, by trading his large collection of books and maps to the French government, received for himself an annuity and for Messier an appointment as clerk with a salary of 500 francs plus room and board at the observatory in the Hôtel de Cluny. There Messier undertook the series of observations that gradually secured his fame.

In 1759 the comet predicted by Halley reached perihelion, and in anticipation of the return Delisle set Messier on a systematic search for the object. Unfortunately the perturbations from Jupiter were underestimated and Messier surveyed too restricted an area. Finally he recovered the comet on 21 January 1759, but Delisle demanded strict secrecy. Unknown to French astronomers, Halley's comet had already been observed in Saxony; and only after this news reached Paris did Delisle reveal Messier's discovery. The incorrigible Delisle followed the same procedure with a comet that Messier discovered on 21 January 1760.

Soon thereafter Delisle, who was in his seventies, retired and left Messier to carry out comet searches. For the next fifteen years Messier claimed a virtual monopoly on comet discoveries. According to Lalande, Messier observed a total of forty-one comets, claiming twenty-one as his own (*Bibliographie astronomique* [Paris, 1803], 796). By stricter modern standards twelve or thirteen initial discoveries from Comet 1759 III to Comet 1798 I, and three additional independent ones, can be attributed to him. J.-F. La Harpe records that Messier's having to tend his wife on her deathbed cost him the discovery of yet another comet, which was identified instead by a certain Montagne of Limoges. When friends consoled him for the loss he had suffered, he wept for the comet and barely remembered to sigh, "Ah, cette pauvre femme."

As a result of these discoveries Messier became a member of the Royal Society of London in 1764 and of the academies at Berlin and St. Petersburg, and his title was changed from clerk to *astronome*. The French savants were reluctant to admit a mere observer to their academy, but finally in 1770, two years after Delisle's death, he gained entry. Ultimately he also became a member of several academies in Sweden, of the Netherlands Society of Sciences, and of the Institute of Bologna; and in 1806 he received the cross of the Legion of Honor.

Immediately after his election to the Academy, Messier began publication of a long series of memoirs, invariably devoted to observations and often accompanied by elegant maps of his own design. His first memoir, "Catalogue des nébuleuses et des amas d'étoiles, que l'on découvre parmi les étoiles fixes" (*Mémoires de mathématiques et physique de l'Académie des sciences* for 1771 [1774], 435–461), remains his most enduring contribution. In it he describes forty-five of what are today the most celebrated nebulae and clusters, including M1, the Crab Nebula; M13, the globular star cluster in Hercules; and M31, the Andromeda Galaxy.

In 1780 Messier added twenty-three new objects, publishing a list of sixty-eight objects in the *Connoissance des temps* for 1783. A year later he again augmented his list, to a total of 103, for the *Connoissance des temps* for 1784; many of the new nebulae were first found by his colleague P. F. A. Méchain. The two supplements revealed for the first time the remarkable abundance of faint nebulae in the constellations Virgo and Coma Berenices, now recognized as the Virgo cluster of galaxies.

At various times Messier used over a dozen telescopes for his observations, but none larger than his favorite 7.5-inch Gregorian. His contemporary Jean-Sylvain Bailly carried out some experimental comparisons showing that the inefficient speculum metal surfaces of the reflector gave it a light-gathering power equivalent to a 3.5-inch refractor. Messier also undertook observations with one of the new Dollond achromatic refractors, which had an aperture of 3.5 inches and a magnification of 120. These small telescopes stand in marked contrast to the giant reflectors constructed by William Herschel during Messier's lifetime. Herschel quickly outstripped Messier's brief list, finding literally thousands of faint nebulae. Looking back on his work, Messier wrote in the *Connaissance des temps* for 1800/1801:

> What caused me to undertake the catalog was the nebula I discovered above the southern horn of Taurus on September 12, 1758, while observing the comet of that year. . . . This nebula had such a resemblance to a comet, in its form and brightness, that I endeavored to find others, so that astronomers would not confuse these same nebulae with comets just beginning to shine. I

observed further with the proper refractors for the search of comets, and this is the purpose I had in forming the catalog. After me, the celebrated Herschel published a catalog of 2,000 that he had observed. This unveiling of the sky, made with instruments of great aperture, does not help in a perusal of the sky for faint comets. Thus my object is different from his, as I only need nebulae visible in a telescope of two feet [length].

Besides the comets and nebulae Messier observed eclipses, occultations, sunspots, the new planet Uranus, and the transits of Mercury and Venus. In 1767 he sailed aboard the *Aurore* for nearly four months, testing instruments for longitude determinations at the request of the Academy. A man of single-minded purpose, he pressed his observational abilities to the utmost but, unskilled in mathematics, left the calculations to others.

On 6 November 1781 a severe accident interrupted Messier's observing for an entire year. Walking in a park with his friend Bochart de Saron, a presiding judge of the Parlement of Paris, Messier entered what he assumed to be a grotto. Instead it was an icehouse; and he fell nearly twenty-five feet onto the ice, breaking an arm, thigh, wrist, and two ribs. Although he was attended by the leading Academy surgeons, Messier sustained a permanent limp.

In 1802, when Messier was seventy-two, Herschel visited Paris and wrote in his diary:

> A few days ago I saw Mr. Messier at his lodgings. He complained of having suffered much from his accident of falling into an ice cellar. He is still very assiduous in observing, and regretted that he had not interest enough to get the windows mended in a kind of tower where his instruments are; but keeps up his spirits. He appeared to be a very sensible man in conversation. Merit is not always rewarded as it ought to be.

As Harlow Shapley wrote in *Star Clusters* (New York, 1930), ". . . the systematic listing by Messier in 1784 marked an epoch in the recording of observations. . . . He is remembered for his catalogue; forgotten as the applause-seeking discoverer of comets."

BIBLIOGRAPHY

I. Original Works. Messier's bibliography is found in J. M. Quérard, *La France littéraire*, V (Paris, 1830), 90–91. Messier published detailed accounts of his lifetime of observations, with incidental biographical material, in *Connaissance des tems* for 1798–1799, 1799–1800, 1800–1801, 1807–1808, 1809, and 1810. His catalog of nebulae and clusters, cited in the text, found its final form in *Connoissance des temps* for 1784 (Paris, 1781), 227–269. For an English trans. of this paper, see Kenneth Glyn

Jones, "The Search for the Nebulae—VIII," in *Journal of the British Astronomical Association*, **79** (1969), 357–370.

II. SECONDARY LITERATURE. The principal sources are J. B. Delambre's "Notice," in *Histoire de l'Académie royale des sciences de l'Institut de France* for 1817 (1819), 83–92; and his somewhat rewritten biography in *Histoire de l'astronomie au dix-huitième siècle* (Paris, 1827), 767–774; see also J.-F. La Harpe, *Correspondance littéraire*, 6 vols. in 5 (Paris, 1801–1807), I, 97–98.

Recent material includes Owen Gingerich, "Messier and His Catalogue," in *Sky and Telescope*, **12** (1953), 255–258, 288–291; and Kenneth Glyn Jones, *Messier's Nebulae and Star Clusters* (London, 1968), 376–410. See also C. Flammarion, "Nébuleuse et amas d'étoiles de Messier," in *Bulletin de la Société astronomique de France*, **31** (1917), 385–400.

OWEN GINGERICH

MESYATSEV, IVAN ILLARIONOVICH (*b.* 1885; *d.* Moscow, U.S.S.R., 7 May 1940), *earth science, oceanography.*

In 1908 Mesyatsev entered the natural sciences section of the department of physics and mathematics of Moscow University. During his student years he displayed great ability in teaching and research; and in 1912, after he graduated from the university, he remained in the department of invertebrate zoology to prepare for a professorship.

Although Mesyatsev's first works were in embryology, histology, and protozoology, his scientific interests later became more involved with marine ichthyology and its application to the fishing industry. In 1920 he headed a pioneering group of Soviet zoologists who worked on a plan to set up the first special scientific oceanographic institute in the country, the Plavmornin.

Established in 1921, Plavmornin was oriented to research on the country's northern seas, especially the Barents Sea. Mesyatsev was placed in charge of all its expeditionary activity, and in 1928 was appointed its director. He organized the construction of the institute's special scientific ship, the *Perseus*, and its systematic research expeditions on the Barents Sea. He himself participated in many of these trips. Mesyatsev succeeded, through careful study of the biology and environment of fish in that sea, in establishing its rich potential for the fishing industry.

Mesyatsev was director of the State Oceanographic Institute from 1929 and manager of the oceanography laboratory of the All-Union Scientific-Research Institute of Ocean Fishing Economy and Oceanography from 1933.

In 1934, as president of the Government Commission for the Determination of Fish Resources of the Caspian Sea, Mesyatsev developed a special method for measuring the supply of fish, which was later used by other specialists in the Sea of Azov. In 1937–1939 Mesyatsev came to conclusions, important for the fishing industry, about the behavior of schools of fish and introduced into ichthyology a clear definition of schooling.

Despite his great load of scientific and administrative work, Mesyatsev found time for teaching as professor of zoology at Moscow University.

BIBLIOGRAPHY

I. ORIGINAL WORKS. Mesyatsev's most important works are "K embriologii mollyuskov" ("Toward an Embryology of Mollusks"), in *Dnevnik zoologicheskogo otdelenia Obshchestva lyubiteley estestvoznania, antropologii i ethografii*, n.s., **1**, no. 4 (1913); *Plavychy morskoy nauchny institut i ego eksepditsia 1921 god. Otchet nachalnika polyarnoy ekspeditsii* ("The Floating Ocean Scientific Institute and Its Expedition of 1921: An Account by the Chief of the Polar Expedition"; Moscow, 1922); *Materialy k zoogeografii russkikh severnykh morey* ("Material for a Zoogeography of the Russian Northern Oceans"; Moscow, 1923); "Stroenie kosyakov rybnykh stad" ("The Structure of Schools of Fish"), in *Izvestia Akademii nauk SSSR, seria biologicheskaya*, no. 3 (1937); "O strukture kosyakov treski" ("On the Structure of Schools of Cod"), in *Trudy VNIRO*, IV (Moscow, 1939).

II. SECONDARY LITERATURE. See the foreword to *Trudy VNIRO*, LX (Moscow, 1966); "Kratkaya biografia I. I. Mesyatsev" ("A Short Biography of I. I. Mesyatsev"), in *Bolshaya sovetskaya entsiklopedia* ("Great Soviet Encyclopedia"), 2nd ed., XXVII (1954), 209; and A. D. Starostin, "Zhizn i nauchnaya deyatelnost I. I. Mesyatseva" ("Life and Scientific Career of I. I. Mesyatsev"), in *Trudy VNIRO*, LX (Moscow, 1966), 11–18.

A. F. PLAKHOTNIK

METCHNIKOFF, ELIE (*b.* Ivanovka, Kharkov Province, Russia, 16 May 1845; *d.* Paris, France, 15 July 1916), *embryology, comparative anatomy, pathology, bacteriology, immunology.*

Elie was the youngest of the five children of Ilia Ivanovitch Metchnikoff and Emilia Nevahovna, the daughter of the Jewish writer Leo Nevahovna. His mother played an important role in the boy's education and encouraged his scientific career. A tutor to the family stimulated Elie to become interested in the wonders of natural history at an early age. In 1856 he enrolled in the Kharkov Lycée, where he made a splendid academic record, his main passion being biology. At this time he read Buckle's *History of*

Civilization in England, and throughout his life strongly adhered to one of Buckle's main tenets, that through science would come man's advancement.

Elie's mother dissuaded him from the study of medicine because she believed that he was too sensitive for such a career. He did win her approval to study physiology and zoology, to which he increasingly devoted his life. The seventeen-year-old student was especially interested in the subject of protoplasm and decided to go to Würzburg to study with Koelliker. The German term did not begin in September. Disappointed, lonely, and bewildered in the strange city, Metchnikoff hurried back home, content to study for two years at the university in Kharkov. In 1864 he studied the sea fauna on the North Sea island of Heligoland, a naturalist's paradise. Here the botanist Ferdinand Cohn gave Metchnikoff friendly guidance and advised him to continue his work with Rudolf Leuckart at Giessen. Metchnikoff made his first real scientific discovery in Leuckart's laboratory when he found an interesting example of alternation of generations (sexual and asexual) in nematodes. In Giessen, Metchnikoff also read Fritz Müller's *Für Darwin*. The German enthusiasm for the theory of evolution greatly influenced him. He worked feverishly and began to suffer from severe eyestrain. This malady prevented him for a time from using his chief research tool, the microscope.

In 1865 Metchnikoff went to Naples, where he began a systematic study of the development of germ layers in invertebrate embryos, a subject less well understood at the time than the similar development in vertebrate embryos. Metchnikoff devoted many years to studying the comparative development of the embryonic layers of lower animals. Like many zoologists of the immediate post-*Origin of Species* period, Metchnikoff's constant aim was to show that in their development the lower animals follow a plan similar to that of the higher animals. He thus attempted to establish a definite link between the two divisions and to add to the theory of evolution. In Naples he befriended another young Russian zoologist, Aleksandr Kovalevsky, with whom he collaborated on several embryological studies.

Because cholera was epidemic in Naples in the autumn of 1865, Metchnikoff decided to continue his studies in Germany. He went to Göttingen, where he briefly worked with W. M. Keferstein and then with Henle. In the following summer he went to Munich to study with Siebold. After again doing research together in Naples, Kovalevsky and Metchnikoff returned to Russia in 1867 to obtain their doctoral degrees in St. Petersburg. For their work on the development of germ layers in invertebrate embryos,

they shared the prestigious Karl Ernst von Baer prize, presented by the discoverer of the human ovum. Metchnikoff also received a faculty position at the new University of Odessa. At age twenty-two the instructor was younger than some of his pupils. He soon was embroiled in a controversy with a senior colleague over attendance at a scientific meeting. The conflict was resolved, but Metchnikoff thought the atmosphere at the university in St. Petersburg would be more conducive to work and teaching and when he was offered a job there in 1868 he gladly accepted. The move proved a disappointment, for the working conditions were, if anything, worse than in Odessa. Metchnikoff was barely able to make ends meet, and he led a lonely existence.

He did meet Ludmilla Federovna, who on one occasion nursed him during an illness. They were married in 1869. Trouble was already on the horizon. The bride was disabled by severe "bronchitis," and she had to be carried to the church in a chair. For the next five years Metchnikoff devoted himself to caring for his wife, who subsequently died of the tuberculous disease already present on her wedding day. To enable him to take Ludmilla to a warmer climate, he did translations besides his teaching and researches. His eyesight again weakened, and he became extremely distraught. In the winter of 1873 he hurried to Madeira to see Ludmilla, who by now was extremely sick. She died in April 1873, and Metchnikoff collapsed. He did not attend the funeral and on his way back to Russia attempted suicide. He swallowed a large dose of morphine, which caused him to vomit, thereby sparing his life.

After this period of tragedy and exhaustion, Metchnikoff slowly returned to his scientific work, but his eyesight was not sufficiently restored to allow microscopic work. Instead he planned an anthropological trip to the Kalmuk steppes, where he observed the natives and carried out comparative physical measurements. He concluded that the development of Mongol natives was arrested in comparison with that of the Caucasian race, although relative bodily proportions were the same. He ascribed the growth lag of the Kalmuks to a state of slight but chronic intoxication, which was the effect of the habitual drinking of fermented milk.

The trip helped Metchnikoff to recover from the hardships of the previous five years and restored his eyesight. He again returned to his job in Odessa, to which he had been recalled in 1872. Metchnikoff was already well established in the scientific world by this time. He had published twenty-five papers, most of which dealt with the development and characteristics of invertebrates, and Odessa afforded him ample material for collecting sea fauna. Moreover, he was a successful and popular lecturer. In 1875 he married Olga Belokopitova, a young student who lived with her large family in the apartment directly over Metchnikoff's. It was a happy marriage, and his wife was a devoted companion and co-worker for the remainder of his life.

Political pressures, student unrest, and Olga's severe bout of typhoid fever in 1880 led Metchnikoff to a second suicide attempt. He injected himself with the spirochete of relapsing fever. A long illness resulted, but he recovered with a renewed zest for life. Cardiac disturbances, from which he suffered in his last years, seem to have begun with his bout of relapsing fever, but the eyestrain, a great cause of worry and inconvenience in earlier years, never did return.

In 1880 the Metchnikoffs spent the summer on Mme Metchnikoff's family farm. A beetle infestation was destroying the grainfields, and Metchnikoff studied the insects and found that some had died from a fungus infection. He conceived the idea of starting an epidemic among the beetles. After experimenting with the idea in the laboratory, he had some success in its implementation in the fields. This study was the starting point for his interest in the infectious diseases. A remarkably similar chain of events occurred in the career of Pasteur, who would in future years play a significant role in Metchnikoff's life.

By 1882 the unrest in Russia, and at the University of Odessa in particular, was so great that the nonpolitical Metchnikoff wished to leave for the quiet atmosphere of Messina, where he could better devote himself to science. In Messina he made his greatest scientific discovery, the role of phagocytes in the defense of the animal body; but the related strands of this concept of the cellular mechanism of immunity had begun to take shape somewhat earlier.

While working in Giessen in 1865, Metchnikoff had studied and observed intracellular digestion in roundworm (*Fabricia*). He compared this type of digestion to that found in some protozoans and saw in the similarity one more proof of a genetic connection between a lower and somewhat higher animal form. A dozen years later he published another paper that dealt with the digestive process and in 1880 "Über die intracelluläre Verdauung bei Coelenteraten." Here he showed that endodermal and mesodermal cells take up carmine granules suspended in water. He did not discover the exact mode of uptake of dye by the cell.

This phenomenon was not an original discovery by Metchnikoff. In 1862 Ernst Haeckel had described in his monograph on *Radiolaria* white blood cells ingesting dye particles. Several other investigators reported similar results, but it was Metchnikoff who made the proper interpretation and who realized the significance of the link between phagocytic digestion and the body's defense.

In Messina in 1882 Metchnikoff observed that the mobile cells in a transparent starfish larva surrounded intruding foreign bodies, a phenomenon similar to the inflammatory response in animals with a vascular system. These mobile cells were derived not from the endoderm, the layer that gives rise to the digestive system, but from the mesoderm. Metchnikoff reasoned correctly that these mesodermal cells might serve in the defense of the animal against intruders and that this observation had very wide implications. He devoted the next twenty-five years to the development and popularization of his theory. As he later explained, "Thus it was in Messina that the great event of my scientific life took place. A zoologist until then, I suddenly became a pathologist."

Both Kleinenberg and Virchow, who were in Messina that summer of 1882, encouraged Metchnikoff. Carl Claus in Vienna urged Metchnikoff to publish his findings, and in 1883 the first of many papers appeared in which Metchnikoff explored the newly developing field of immunology. In Claus's *Arbeiten*, Metchnikoff first used the term phagocyte, derived from the Greek, instead of *Fresszellen* (eating cells). Metchnikoff had been studying the evolution of the alimentary tract. One question that had arisen was whether the lower metazoa retained the power of using mesodermal and also endodermal cells for digestion. He observed that in starfish larvae the wandering or mobile cells of mesodermal origin were active in the metamorphosis of the larva. These cells resorbed the parts of the larva that were no longer used. By simple experiment Metchnikoff showed that it was but a short step from resorption of useless parts to a similar role when a foreign particle was introduced into the organism.

In the next years Metchnikoff showed that the mobile cells (the white blood corpuscles) of the higher animals and man also developed from the mesodermal layer of the embryo and were responsible for ridding the body of foreign invaders, especially bacteria. Although Virchow supported him and published Metchnikoff's papers in his *Archiv für pathologische Anatomie und Physiologie und für klinische Medizin*, the phagocyte theory ran counter to many commonly held theories of the time.

For instance, Julius Cohnheim, a pupil of Virchow's, had shown that the pus cells of the inflammatory process were derivatives of the bloodstream, and not of the surrounding connective tissue, as Virchow had claimed. Cohnheim further maintained that without blood vessels to bring the white blood cells, there could be no inflammation. Metchnikoff claimed that the action of mobile cells in clearing an organism of foreign material or no-longer-useful parts was a form of inflammation. According to Metchnikoff, furthermore, one could observe this action in starfish larva altogether lacking a vascular system.

A serious objection to this new theory of bodily defense was the currently held idea that the white blood cells took up invading particles or bacteria and spread them throughout the body. These phagocytes of Metchnikoff were then far from salutary and were believed to be helpful to the invader rather than to the host. There was also the usual resistance to major innovations in thought or approach.

Metchnikoff had been in a number of scientific and personal fights in his early career, and it was natural that he now became a staunch defender of his new theory. He devised new experiments and new arguments and warded off one attack after another upon his brainchild. Much of his voluminous writing in the years 1883 to 1910 was dedicated to elaboration or modification of the role of phagocytes in inflammation and immunity; but he always held tenaciously to the underlying idea of the central role of the phagocytes.

By 1886 Metchnikoff was well known as a biologist and also as a microbiologist and pathologist, and was invited back to Odessa, where he had taught from 1873 to 1882. The city had established a bacteriological institute similar to the Pasteur Institute of Paris. In Odessa there was to be a combination of basic research and the production of antirabies vaccine.

Metchnikoff headed the Institute in 1886 and part of 1887, but found that the internal strife among the members and his inability to carry out immunizations himself, because he was not a physician, combined to make life and work there unpleasant. He and his wife traveled to various centers in Europe in search of a congenial place to settle. It was Pasteur in Paris who made them most welcome and who gave Metchnikoff a laboratory in which to work. In 1888 the Metchnikoffs moved to Paris, where Elie worked for the last twenty-eight years of his life. This was an honorary position because Metchnikoff had sufficient income from his parents-in-law's estate to live without salary.

Metchnikoff quickly became a revered member of the small circle of the Institute, where friendships and working relationships were close. He began to attract students to his laboratory and set most of them to work answering the various objections to the theory of phagocytosis, elucidating ways in which the white blood cells were attracted to and ingested bacteria, or determining how, in general, the mechanism of immunity worked. Among his many talented students

was Bordet, who in 1919 received the Nobel Prize for his work on complement fixation.

Metchnikoff also gave public lectures, for he believed the popularization of science to be important. In 1891 he delivered a series of talks on inflammation. In these talks Metchnikoff dealt with the history of the various theories of inflammation and their investigation, and chiefly with the role of phagocytes in the animal kingdom. The lectures were well-attended and Pasteur himself came. The series was published as *Leçons sur la pathologie comparée de l'inflammation* in 1892 and in English translation in the following year.

Metchnikoff felt that the decade 1895–1905 was the happiest period of his life. He and his wife lived outside of Paris in Sèvres, and he came to the Institute each morning by train. He continued his research in immunity and also into the problem of fever and the mechanisms of infection. While attending the International Medical Congress in Paris in 1900, he realized that there should be a summary of his and his antagonists' different theories. He began to write a large and comprehensive book, *L'immunité dans les maladies infectieuses* (1901). This book was a magnificent review of the entire field of both comparative and human immunology. The work was also, of course, a defense of the theory of phagocytosis, which the humoral theory of immunity seriously challenged. The work of the German bacteriologists, especially Emil Behring, Paul Ehrlich, and Robert Koch, which led to discovery of many new bacteria, toxins, and antitoxins, strengthened the beliefs of those who held to a noncellular theory of immunity. Even before the English edition of *Immunity in Infectious Diseases* was issued in 1905, two British investigators, A. E. Wright and S. R. Douglas, put forth their theory of opsonins, which postulated that something in the fluid portions of the blood helped the white blood cells to digest bacteria. Hence a compromise was beginning to take shape. In 1908 Metchnikoff and Ehrlich shared the Nobel Prize for their researches illuminating the understanding of immunity.

After the *Immunity* was finished, Metchnikoff turned his attention to the problems of aging and the idea of death. With his friend and co-worker Émile Roux, he began to study syphilis, one disease that was known to be implicated in cardiovascular pathology. In 1903 Metchnikoff and Roux discovered that syphilis was transmissible to monkeys, thereby destroying the old theory that the disease was exclusively human and inaccessible to experiment. They also showed the importance and efficacy of early treatment of the primary lesion with mercurial ointment.

In a series of books and lectures between 1903 and 1910 Metchnikoff developed his thoughts on the prolongation of life. He stressed proper hygienic and dietary rules. His idea of orthobiosis, or right living, included careful attention to the flora of the intestinal canal. He believed that intestinal putrefaction was harmful and that the introduction of lactic-acid bacilli, as in yogurt, accounted for the longevity of the Bulgars. He introduced sour milk into his own diet and thought that his health improved. Although his name became associated with a commercial yogurt preparation, he had not endorsed it and realized no profit from it.

In his *Nature of Man* Metchnikoff argued that when diseases have been suppressed and life has been hygienically regulated, death would come only with extreme old age. Death would then be natural, accepted gratefully, and robbed of its terrors.

The outbreak of World War I in 1914 was a profound shock to Metchnikoff. Not only was there an interruption of the work of the Pasteur Institute, but Metchnikoff was forced to acknowledge that science had not yet brought man to that stage of civilization which he had envisioned. When he became ill and weaker in the summer of 1916, he faced death placidly, according to the tenets of his own philosophy. He was moved from his country house to the rooms at the Pasteur Institute that had been occupied by Pasteur. There he died of cardiac failure on 15 July 1916.

BIBLIOGRAPHY

I. ORIGINAL WORKS. A complete list of Metchnikoff's arts. and bks. is available in the Zeiss trans. of Olga Metchnikoff's biog. (see below). A less complete list may be found in the English trans. and in the Dover repr. of the *Lectures on Inflammation*. "Metchnikoff" is the preferred spelling. The name appears that way on the French original eds. of his work. American catalogs usually list it under "Mechnikov."

The major bks. by Metchnikoff that have been translated into English are included here. The French original and the German translations often predated the English by one to four years: *Lectures on the Comparative Pathology of Inflammation*, delivered at the Pasteur Institute in 1891, trans. by F. A. and E. H. Starling (London, 1893), repr. with a new intro. by Arthur M. Silverstein (New York, 1968); *The Nature of Man; Studies in Optimistic Philosophy*, trans. by P. C. Mitchell (New York, 1903); *Immunity in Infectious Diseases*, trans. by F. G. Binnie (Cambridge, 1905), repr. with a new intro. by Gert H. Brieger (New York, 1968); *The New Hygiene. Three Lectures on the Prevention of Infectious Diseases* (London, 1906); *The Prolongation of Life: Optimistic Studies*, trans. by P. C. Mitchell (New York, 1908); and *The Founders of Modern Medicine; Pasteur, Koch, Lister*, trans. by D. Berger (New York, 1939), which was originally published in 1933.

II. SECONDARY LITERATURE. The most important source for details of Metchnikoff's life and work is the memoir written by his wife, Olga Metchnikoff, *Life of Elie Metchnikoff 1845–1916*, trans. by E. Ray Lankester (Boston, 1921). Heinz Zeiss translated the original French into a German ed., *Elias Metschnikow, Leben und Werk* (Jena, 1932), in which he included many additional letters, excellent nn., and the most complete bibliog. of Metchnikoff's writings that I have seen. A. Besredka, a devoted student and co-worker, wrote *Histoire d'une idée, l'oeuvre de E. Metchnikoff* (Paris, 1921). Pierre Lépine, *Elie Metchnikoff et l'immunologie* (Vichy, 1966), is helpful for personal details and for its many photographs.

Useful arts. include Alice G. Elftman, "Metchnikoff as a Zoologist," in *Victor Robinson Memorial Volume* (New York, 1948), 49–60; R. B. Vaughn, "The Romantic Rationalist, a Study of Elie Metchnikoff," in *Medical History*, **10** (1965), 201–215; and Denise Wrotnowska, "Elie Metchnikoff quelques documents inédits conservés au Musée Pasteur," in *Archives internationales d'histoire des sciences*, **21** (1968), 115–136.

GERT H. BRIEGER

METIUS, ADRIAEN (*b.* Alkmaar, Netherlands, 3 December 1571; *d.* Franeker, Frisia [now Netherlands], 1635), *mathematics, instrument making;* **[METIUS], ADRIAEN ANTHONISZ** (*b.* Alkmaar [?], Netherlands, *ca.* 1543; *d.* Alkmaar [?], 20 November 1620), *military engineering, cartography;* **METIUS, JACOB** (*b.* Alkmaar, Netherlands; *d.* Alkmaar, June 1628), *mathematics, instrument making.*

The father, Adriaen Anthonisz, was a cartographer and military engineer for the States of Holland, and between 1582 and 1601 he was burgomaster of Alkmaar several times. In an unpublished pamphlet *Tegens de quadrature des circkels van Mr. Simon van Eycke* (1584), he gave, according to his son Adriaen (1625), the value of 355/113 for what we now denote by π, stating that it differs from the true value by less than 1/100,000. He obtained it by averaging numerators and denominators of the values 377/120 and 333/106. (This value had already been obtained by Tsu Chung-chih in the fifth century.) Anthonisz built fortifications in the war against Spain, drew charts of cities and military works, and wrote on sundials and astronomical problems. In the receipt for his burial the name Metius, adopted by some of his sons, is mentioned. The origin of the name is uncertain: some derive it from Metz, others from the family name Schelven (*schelf* = *rick* = Latin *meta*), it may also simply be related to *metiri* (to measure). Anthonisz and his wife Suida Dircksd. had one daughter and six sons, of whom two, Adriaen and Jacob, became widely known.

The second son, Adriaen, educated at the Latin school in Alkmaar, entered the recently founded University of Franeker in Frisia in 1589, and in 1594 continued his studies at the University of Leiden. Among his teachers in Leiden were the mathematicians Rudolf Snellius and Van Ceulen. Like his townsman Blaeu, Adriaen worked under Tycho Brahe at his observatory on the island of Hven; he then went to Rostock and Jena, where in 1595 he gave his first lectures. He returned to the Netherlands where he assisted his father in his military engineering until, in 1598, he was appointed professor extraordinarius at Franeker; in the same year he published his first book, *Doctrina spherica*.

Adriaen became professor ordinarius of mathematics, surveying, navigation, military engineering, and astronomy at Franeker in 1600, a position he held until his death. He bought mathematical and astronomical instruments, observed sunspots, and showed familiarity with the telescope, of which his brother Jacob was a coinventor. He especially appreciated its use for measuring instruments. In his *Geometria practica* (Franeker, 1625) he described a triangulation of part of Frisia, made shortly after Rudolf Snellius' son Willebrord had published his triangulation of the west Netherlands in *Eratosthenes batavus* (1617). Adriaen was a popular and efficient teacher who stressed the training of Frisian surveyors. His lectures were well attended by an international audience including, in 1629, Descartes. In 1625 Adriaen received an honorary doctorate in medicine from Franeker. He was married twice, first to Jetske Andreae, and then to Cecelia Vertest. He left no children. His motto was "Simpliciter et sine strepitu."

Adriaen's books cover all fields that he taught, and although they show little originality, they were widely used in his time. He followed Tycho Brahe's theory of the solar system, but also showed respect for the Copernican system. While not accepting astrology, he did believe in alchemy, and spent money in the search for the transmutation of metals.

His brother Jacob was as shy as Adriaen was sociable. He became an instrument maker in Alkmaar, specializing in the grinding of lenses. He made several inventions but rarely showed them to others, even to his brother. He was one of the claimants to the invention of the telescope, and is mentioned as such by Descartes in his *Dioptrique* (1637). Jacob was indeed one of the first to bring a concave and a convex lens together in a tube, thus constructing a telescope. In 1608 he applied for a patent on such an instrument but unfortunately a similar request had been made

a few weeks earlier by H. Lippershey of Middelburg. This disappointment may have intensified Jacob's shyness. Adriaen, in several of his books after 1614, refers to his brother's "perspicilla" (telescope). He expresses the hope that he would allow others to share in his discoveries, but Jacob remained secretive. Before his death he destroyed his instruments so that, as a contemporary said, "the perfection of his art has died and been buried with him."

BIBLIOGRAPHY

I. ORIGINAL WORKS. A satisfactory bibliography of Adriaen Metius' works does not exist. Boeles lists seventeen titles, de Waard eighteen, and Bierens De Haan thirty-three, but some are reprints, trans., or collections. Boeles also lists a map of Frisia and a celestial globe from J. Janssonius' cartographic workshop (1648). Some titles are *Institutiones astronomiae et geographicae*, found together with *Geographische Onderwysinghe, waer in ghehandeld wordt die Beschryvinghe ende Afmetinghe des Aertsche Globe* (Franeker, 1614; Amsterdam, 1621); *Arithmetica et geometrica nova* (Franeker, 1625); *Arithmeticae libri II et geometriae libri VI. Hic adiungitur trigonometriae planorum methodus succincta* (Leiden, 1626); *Geometria practica* (Franeker, 1625), which states that "Parens P. M. illustrium D. D. Ordinum Confoederatarum Belgiae Provinciarum Geometra" found $\pi = 355/113$ (pp. 88–89; "P. M." is clearly "pia memoria"—Anthonisz died in 1620—and not P. Metius, as has occasionally been claimed to justify the term "ratio of Metius"); *Maet-constigh Lineael ... alsmede de Sterckten-Bouwinghe ofte Fortificatie* (Franeker, 1626), which is a trans. of part of *Arithmeticae libri II ...*, in which is described an early form of a calculating mechanism; *Eeuwighe Handt-calendrier* (Amsterdam, 1627; Rotterdam, 1628); *Tafelen van de Declinatie des Sons* (Franeker, 1627); *Astronomische ende Geographische Onderwysinghe* (Amsterdam, 1632); *Manuale arithmeticae et geometriae practica* (Franeker, 1633; 1646); *Opera omnia astronomia* (Amsterdam, 1632–1633), which contains the *canon sinuum, tangentium et secantium ad radium* 10,000,000.

II. SECONDARY LITERATURE. On the father and sons see C. de Waard, "Anthonisz" and "Metius," in *Nieuw Nederlandsch Biographisch Woordenboek*, I (Leiden, 1911), 155–158, 1325–1329 (in Dutch). In "Anthonisz," he gives an account of Anthonisz' MSS and published material. On Anthonisz' value of π see D. Bierens De Haan, "Adriaan Metius," in *Bouwstoffen voor de geschiedenis der wis- en natuurkundige wetenschappen in de Nederlanden*, XII, repr. from *Verslagen en Mededeelingen K. Akademie van Wetenschappen Amsterdam, Afdeling Natuurkunde*, 2nd ser., **12** (1878), 1–35. The same author's *Bibliographie neerlandaise historique et scientifique sur les sciences mathématiques et physiques* (Rome, 1883; Nieuwkoop, 1960) lists thirty-three works of Adriaen; this is a reprint of articles in *Bullettino di bibliografia e di storia delle scienze matematiche*, **14**

(1881), and **15** (1882), esp. 258–259. Also see his "Notice sur quelques quadrateurs du cercle dans les Pays-Bas," *ibid.*, **7** (1874), 99–104; and "Notice sur un pamphlet mathématique hollandais," *ibid.*, **11** (1878), 383–452. On Adriaen also see W. B. S. Boeles, *Frieslands Hoogeschool*, II (Leeuwarden, 1879), 70–75; and H. K. Schippers, "Fuotprinten fan in mannich Fryske stjerrekundigen," in *Beaken*, **24** (1962), 77–104 (in Frisian). On Jacob, see C. de Waard, *De uitvinding der verrekijkers* (The Hague, 1906).

D. J. STRUIK

METON (*fl.* Athens, second half of fifth century B.C.), *astronomy.*

Meton was the son of Pausanias, an Athenian from the deme (local subdivision of Attica) Leuconoe. He is dated by his observation of the summer solstice on 27 June 432 B.C.[1] He was still active nearly eighteen years later, for the story is told of a ruse whereby he avoided military service either for himself or for his son (the more probable version) on the Athenian expedition to Sicily that set out in 415, by pretending to be mad and setting fire to his house.[2] Furthermore, he is introduced as a character in Aristophanes' comedy *The Birds*, produced early in 414, and mentioned in Phrynichus' *Monotropos*, produced at the same festival as *The Birds*.[3] That is all we know about his life.

No written work by Meton survives; and we have to reconstruct what he did, and its purpose, from a few scattered references in ancient literature. Any such reconstruction involves some guesswork, and there is considerable disagreement on this subject among modern scholars. The following account seems to me to represent the evidence best.

Meton was famous in antiquity for his introduction of a nineteen-year lunisolar calendaric cycle, in ancient times usually called ἐννεακαιδεκετηρίς ("nineteen-year period") and sometimes the "great year" or "year of Meton," in modern times often called the "Metonic cycle." In Meton's time all Greek civil calendars, including the Athenian, were lunisolar. That is, the months were theoretically (although often not in practice) true lunar months, with the new moon occurring on the first of the month. Since the mean synodic month is slightly longer than 29.5 days, calendar months were normally either 30 days long ("full" months) or 29 days long ("hollow" months). The years, on the other hand, were supposed to be solar. Now twelve true lunar months make up only about 354 days. Therefore it was necessary to intercalate a thirteenth month in some years in order to keep the calendar roughly in step with the seasons (the year at Athens, as in most Greek states, began near the summer solstice). All the extant primary evidence (in the form of dated inscriptions and coins) from Athens and elsewhere suggests that at no period was a fixed rule for intercalation of a thirteenth month adopted by any Greek state. Instead, each intercalation was determined by the decision of a magistrate (probably the eponymous archon at Athens) or an official body.

In constructing his nineteen-year cycle Meton used the fact (known before his time) that nineteen solar years correspond very well, on the average, to 235 true synodic months. This means that during one nineteen-year period a thirteenth month has to be intercalated seven times. Meton must have prescribed rules for the places of those seven intercalations. In addition, we are told that his cycle contained precisely 6,940 days.[4] This means that 110 of the 235 months in the cycle were hollow and the other 125 full. Thus Meton must also have prescribed rules for the sequence of full and hollow months throughout the cycle. Since it is probable that he derived the equation of nineteen years with 235 months from Babylonian practice (see below), it might seem plausible to assume that, like the Babylonians, he intercalated in the third, sixth, eighth, eleventh, fourteenth, seventeenth, and nineteenth years of his cycle. There is no evidence whatever for his having adopted this or any other scheme, however, and all modern attempts to reconstruct his intercalation system are futile. The same is true for reconstructions of his sequence of full and hollow months (in this case we cannot even refer to Babylonian practice, since their nineteen-year cycle did not contain a fixed number of days: instead, the length of each month was determined by observation of the new crescent). The only other information we have on the Metonic cycle is that it used the month names of the Athenian civil calendar[5] and that the first Metonic cycle began on 27 June 432 B.C., in the archonship of Apseudes at Athens—which was, according to Meton, the day of the summer solstice and the thirteenth day of Skirophorion (the twelfth month) in his calendar.

The evidence for the last statement needs examination. The chief authority is Diodorus, who says: "In the archonship of Apseudes at Athens . . . Meton the son of Pausanias, who has a reputation in astronomy, set out the so-called nineteen-year period, taking the beginning from the thirteenth of Skirophorion at Athens . . ." (XII, 36). One might assume that Meton would start his cycle from the beginning of the first month—Hekatombaion 1—and it would be possible to interpret Diodorus as meaning that the first day of Meton's cycle (his Hekatombaion 1) coincided with

Skirophorion 13 of the Athenian civil calendar of that year. That this is not so is shown by a fragment of an inscription found at Miletus, probably part of a calendar, which reads: "from the summer solstice in the archonship of Apseudes on Skirophorion 13, which was Phamenoth 21 according to the Egyptians, until the [solstice] in the archonship of Polykleitos on Skirophorion 14, or Pauni 1 according to the Egyptians."[6]

The equations with the Egyptian calendar plus the archon years enable us to determine these dates as 27 June 432 B.C. and 26 June 109 B.C., respectively. They are exactly seventeen nineteen-year periods apart; and one can deduce, first, that the second date is not an observed solstice but one computed from the first by means of a fixed calendrical scheme and, second, that the Skirophorion dates must have been taken not from the Athenian civil calendar but from an artificial astronomical calendar. The first is the starting date of Meton's first cycle. The second must be the date in the current Callippic cycle, for the interval corresponds to 323 years of $365\frac{1}{4}$ days.[7] The transition from Metonic to Callippic reckoning may also explain the shift of one day (from 13 to 14 Skirophorion) of the date of the solstice after an integer number of nineteen-year cycles. It may seem strange that Meton began his cycle elsewhere than at the beginning of a year (in contrast, when Callippus introduced his "improved" cycle, he began it at a solstice that coincided roughly with a new moon and thus was able to begin the year there too). But one can begin a cycle at an arbitrary point within it, and we know that Meton himself had observed the solstice of 432. No doubt he wished to begin at a point well-established astronomically.

Enormous confusion has arisen over the Metonic cycle because many scholars since Scaliger have assumed that Meton's purpose was to reform the Athenian civil calendar and that he succeeded (at least in part). Not only is there no evidence for the latter belief, but the former too is not supported by our texts. Instead, we may state confidently that his purpose in publishing the "nineteen-year cycle" was to provide a fixed calendrical scheme for recording astronomical data. Thus, if one was told that event A occurred on day 6 of the month Metageitnion in the second year of the cycle and event B on day 21 of the month Anthesterion of the fourth year, it was possible to determine the exact interval between the two—this was, in general, not possible between dates in the Athenian or any other Greek civil calendar. We can see from references in Ptolemy's *Almagest* that the cycle, as reformed by Callippus, was used for astronomical dating as late as Hipparchus (128 B.C.).

Ptolemy preferred to use the Egyptian calendar, with a fixed year length of 365 days, for the same purpose.

It is likely that what Meton published was in fact an astronomical calendar for nineteen years. If this assumption is correct, the calendar listed for each year the dates of the solstices and equinoxes, of the morning and evening risings and settings of certain prominent stars and constellations (that is, their first and last appearances just before dawn and just after sunset), and weather predictions associated with the various astronomical phenomena. All of this, except the equinoxes, was traditional Greek "astronomy,"[8] but Meton's observations of the intervals between phenomena may have been more accurate than his predecessors'. There exist a number of astronomical calendars of the type described, both in manuscript and (fragmentarily) on stone,[9] but all cover just one year. In these the months are either "zodiacal" (the time taken by the sun to traverse one sign of the zodiac), Egyptian, or Julian.[10] All three types are impossible or unlikely for Meton's time; and it is preferable to assume that he used synodic months and hence, necessarily, a nineteen-year calendar.[11] It is certain, at least, that Meton did publish an astronomical calendar. It is one of the authorities listed by Ptolemy,[12] and it is referred to by others.

Meton's chief claim to fame, apart from his cycle, is that he is the first Greek of whom we can say with certainty that he undertook serious astronomical observations. His solstice observations are the earliest that Ptolemy thought worth attention, even if inaccurate. (The only recorded observation is more than a day too early.) We are told that Meton erected an instrument for observing solstices (a ἡλιοτρόπιον) on the hill of the Pnyx in Athens.[13] The form of this instrument is entirely conjectural; but any upright gnomon would serve the purpose, provided one could observe its longest and shortest midday shadow.

It is commonly supposed that Meton also observed equinoxes, since he assumed unequal lengths for the seasons. One source quotes figures for the lengths of the seasons, beginning with the summer solstice, of ninety, ninety, and ninety-two days (and hence, by inference, ninety-three for the fourth).[14] But these figures (which are very inaccurate) can equally well be explained by a crude schematic distribution of the times spent by the sun in each of the twelve zodiacal signs into intervals of thirty and thirty-one days.[15] This would imply that the equinoxes were not observed. The first Greek of whom we can say with certainty that he determined the lengths of the seasons by observation, and drew the conclusion that the sun has an anomalistic motion, is Callippus (*ca.* 330 B.C.).

The question of Babylonian influence on Meton is

relevant. A standard nineteen-year intercalation cycle was in regular use in the civil calendar of Babylonia from 367 B.C. and seems to have been known there, although not uniformly used, from the early fifth century. We are informed by an early source that Meton derived his nineteen-year cycle from a certain Phaeinos who was a resident alien at Athens.[16] It is possible that Phaeinos was an Asian Greek who acted as transmitter of Babylonian astronomical knowledge. A further connection of Meton with Babylonian astronomy is that he put the equinoxes and solstices at 8° of their respective zodiacal signs.[17] This is characteristic of "System B" in Babylonian astronomical texts.

Meton is also called a "geometer" in some ancient sources. In Aristophanes' *The Birds* he comes on stage equipped with the geometer's traditional rule and compasses and proceeds to "square the circle" in an absurd manner. It would be hazardous to draw any inference about the real Meton's mathematical interests from this burlesque. Similarly, we cannot conclude that he engaged in town planning or hydrography from the representations of him in contemporary comedies drawing plans for Cloudcuckooland or drilling wells. He remains an obscure figure; but he and his associate Euctemon were probably of importance in giving an initial impetus to astronomical observation, however crude, in Greece. His cycle, although later superseded by other reference systems, provided the first adequate framework for recording astronomical data.

NOTES

1. Reported by Ptolemy, *Almagest*, III, 1.
2. Differing versions in Plutarch, "Nicias," 13.5; Plutarch, "Alcibiades," 17.4–5; Aelian, *Varia Historia*, 10.7.
3. Aristophanes, *Birds*, 992–1020, with the scholion on 997.
4. Ptolemy, *Almagest*, III, 1; Geminus, VIII, 51; Censorinus, 18.8.
5. The best evidence for this is that Callippus used them in his cycle, which was a slight modification of Meton's. See, for instance, Ptolemy, *Almagest*, VII, 3 (Heiberg, ed., II, 28).
6. *Sitzungsberichte der K. Preussischen Akademie der Wissenschaften zu Berlin* (1904), no. 1, 96.
7. The Callippic cycle differed from the Metonic in assuming a year length of $365\frac{1}{4}$ days. Hence to 76 years it assigned 27,759 days, a day less than the corresponding 4 Metonic cycles (Geminus, VIII, 59–60). Otherwise the Callippic cycle of 76 years was, presumably, identical to 4 consecutive Metonic cycles. The first Callippic cycle began at the summer solstice (probably 28 June) 330 B.C.
8. Found, for example, in Hesiod's *Works and Days*, which is some 300 years earlier.
9. Such a calendar, when inscribed on permanent material, was often provided with a hole at each entry to receive a peg to mark the current date. Hence it was called παράπηγμα ("that which has a peg beside it").
10. Examples of the first are Geminus' calendar (Geminus,

Elementa astronomiae, Manitius, ed., 210–232) and the Miletus parapegma (*Sitzungsberichte der K. Preussischen Akademie der Wissenschaften zu Berlin* [1904], no. 1, 102–111); of the second, Ptolemy's *Phaseis* (*Calendaria Graeca*, Wachsmuth, ed., 211–274); and of the third, the calendar of Clodius Tuscus (*Sitzungsberichte der Heidelberger Akademie der Wissenschaften*, Phil.-hist. Kl. [1914], no. 3).
11. A nineteen-year calendar is suggested by the scholion on Aratus, 753 (Maass, ed., *Commentariorum in Aratum reliquiae*, 478), and by Diodorus, XII, 36.3.
12. *Phaseis*, in *Calendaria Graeca*, Wachsmuth, ed., 275.
13. Philochorus, fr. 122, Jacoby, ed.
14. *Eudoxi Ars astronomica*, Blass, ed., 25. The figures quoted are attributed to Euctemon, but the two are so often coupled in the context of astronomical observations (as by Ptolemy in the passages referred to in notes 1 and 10) that it is plausible to associate Meton with these season lengths too. This is confirmed by Simplicius, *In De caelo*, Heiberg, ed., 497.
15. See Albert Rehm, "Das Parapegma des Euktemon" ("Griechische Kalender," F. Boll, ed., III), in *Sitzungsberichte der Heidelberger Akademie der Wissenschaften*, Phil.-hist. Kl. (1913), no. 3, 9.
16. Theophrastus, *De signis*, 4.
17. Columella, *De re rustica*, IX, 14.12.

BIBLIOGRAPHY

The principal ancient passages concerning Meton are Ptolemy, *Almagest*, III, 1 (*Claudii Ptolemaei Opera quae exstant omnia*, I, *Syntaxis mathematica*, J. L. Heiberg, ed., pt. 1 [Leipzig, 1898], 205–207); Diodorus, *Bibliotheca historica*, XII, 36 (F. Vogel, ed. [Leipzig, 1890], II, 395); Geminus, *Elementa astronomiae*, VIII, 50–56 (K. Manitius, ed. [Leipzig, 1898], 120–122); Censorinus, *De die natali*, 18.8 (Otto Jahn, ed. [Berlin, 1845], 54); Aristophanes, *The Birds*, ll. 992–1020, with the scholion on l. 997 (printed as fr. 122 of Philochorus by F. Jacoby, *Die Fragmente der Griechischen Historiker*, III B [Leiden, 1950], 135); Theophrastus, *De signis*, 4 (*Theophrasti Opera*, F. Wimmer, ed. [Leipzig, 1862], III, 116); the scholion on Aratus, 752–753, in *Commentariorum in Aratum reliquiae*, E. Maass, ed. (Berlin, 1898), 478; Ptolemy, *Phaseis*, 93D, in *Calendaria Graeca*, C. Wachsmuth, ed. (with Ioannes Lydus, *Liber de Ostentis*) (Leipzig, 1897), 275 (see also index, 360, under Μέτων); Columella, *De re rustica*, IX, 14.12 (*ibid.*, 303); *Eudoxi Ars astronomica*, F. Blass, ed. (Kiel, 1887), 25; Simplicius, *In Aristotelis De caelo commentaria*, J. L. Heiberg, ed. (*Commentaria in Aristotelem Graeca*, VII) (Berlin, 1894), 497; Plutarch, "Life of Nicias," 13.5, and "Life of Alcibiades," 17.4–5 (*Plutarch's Lives*, B. Perrin, ed. [Cambridge, Mass., 1916], III, 254–256, and IV, 44–46); and Aelian, *Varia Historia*, 10.7 and 13.12 (R. Hercher, ed. [Leipzig, 1866], 109, 149).

There is no satisfactory modern account of Meton. Most are vitiated by the belief that the "Metonic cycle" is somehow reflected in the Athenian civil calendar. For a refutation of this belief and a history of scholarly discussion of the question, see W. Kendrick Pritchett, "The Choiseul Marble," in *University of California Publications. Classical Studies*, 5 (1970), 39–97. The fragments of the Miletus parapegma(ta) were published by H. Diels and

A. Rehm, "Parapegmenfragmente aus Milet," in *Sitzungsberichte der K. Preussischen Akademie der Wissenschaften zu Berlin* (1904), no. 1, 92–111; and by A. Rehm, "Weiteres zu den milesischen Parapegmen," *ibid.*, 752–759. For references to other partially preserved ancient parapegmata, see A. Rehm, "Parapegma," in Pauly-Wissowa, *Realencyclopädie der classischen Altertumswissenschaft*, 1st ser., XVIII, pt. 2, cols. 1299–1302; for eds. of the Italian ones, with excellent photographs, see *Inscriptiones Italiae*, XIII, pt. 2, A. Degrassi, ed. (Rome, 1963), 299–313.

Many of the ancient astronomical calendars preserved in MS were collected by C. Wachsmuth in his *Calendaria Graeca* (see above). See also the series "Griechische Kalender," F. Boll, ed., in *Sitzungsberichte der Heidelberger Akademie der Wissenschaften*, Phil.-hist. Kl. (1910), no. 16, (1911), no. 1, (1913), no. 3, (1914), no. 3 and (1920), no. 15. On the nineteen-year cycle in Babylonia, see R. A. Parker and W. H. Dubberstein, *Babylonian Chronology 625 B.C.–A.D. 75*, Brown University Studies, XIX (Providence, 1956), esp. 1–6. On System B in Babylonian astronomy see O. Neugebauer, *Astronomical Cuneiform Texts*, I (London, 1955), 69–85 (for the vernal point in Aries 8° see 72). A well-informed conjecture about the site of Meton's instrument for observing solstices on the Pnyx at Athens was advanced by K. Kouroniotes and Homer A. Thompson, "The Pnyx in Athens," in *Hesperia*, **1** (1932), 207–211.

G. J. Toomer

METTENIUS, GEORG HEINRICH (*b.* Frankfurt am Main, Germany, 24 November 1823; *d.* Leipzig, Germany, 19 August 1866), *botany.*

Mettenius was the son of a Frankfurt merchant. He studied medicine at Heidelberg from 1841 to 1845, defending a doctoral dissertation, *De Salvinia*, in 1845. After periods of further study in Heligoland, Berlin, Vienna, and Fiume, he became a *Privatdozent* at Heidelberg. He moved to Freiburg im Breisgau as professor extraordinarius in 1851. In 1852, following the death of Gustav Kunze, he became professor of botany and director of the botanic garden at Leipzig. Kunze had assembled a rich collection of living plants and herbarium specimens of ferns. This abundant material led Mettenius to concentrate his attention thereafter on pteridological studies.

In June 1859 Mettenius married Cecile Braun, second daughter of the Berlin botanist Alexander Braun; her sister married the Königsberg botanist Robert Caspary on the same day. Mettenius' life, according to Caspary, was the most regular possible:

> At five o'clock he began the work of the day and finished it punctually at ten in the evening. His whole mind was turned towards the study of plants and especially of ferns. . . . Mettenius generally took the whole

management of the garden upon himself, being out by six o'clock in the morning and directing the operations of each of the labourers.

By such incessant toil, he accomplished much in his relatively short life. An athletic man of great bodily strength, he was in splendid health when he contracted cholera during the epidemic of 1866, and died suddenly at the age of forty-two. His years of intensive study of ferns were leading to comprehensive monographs on the species of individual genera. He had already published, in the *Abhandlungen der Senckenbergischen Naturforschenden Gesellschaft*, on *Polypodium* in 1856; *Plagiogyra*, *Phegopteris*, and *Aspidium* in 1858; and *Cheilanthes* and *Asplenium* in 1859. These writings revised existing classifications, and, together with his *Filices Lechlerianae Chilenses et Peruanae* (1856–1859), *Filices horti botanici Lipsiensis* (1856), and *Ueber die Hymenophyllaceae* (1864), place him among the leading pteridologists of the nineteenth century.

Like J. W. Hooker, Mettenius preferred a few large genera divided into sections instead of a multitude of smaller genera as proposed by his older contemporaries C. B. Presl, John Smith, and Fée (see *Webbia*, **17** [1962], 207–222). His work remains important for its detailed and precise descriptions of species.

BIBLIOGRAPHY

I. Original Works. Some of Mettenius' pteridological publications were issued both as parts of the *Abhandlungen herausgegeben von der Senckenbergischen Naturforschenden Gesellschaft*, **2–3** (Frankfurt am Main, 1856–1860), and as separates; for their precise dates of publication, see W. T. Stearn, "Pteridological Publications of G. H. Mettinius," in *Journal of the Society for the Bibliography of Natural History*, **4** (1967), 287–289.

II. Secondary Literature. The main source of biographical information is in the obituary by his brother-in-law, Robert Caspary, in *Gardeners' Chronicle* (1866), 1018; repr. in *Journal of Botany*, **4** (1866), 388–391, with a list of publications. Various references occur in the biography of his father-in-law by Cecile Braun Mettenius, *Alexander Braun's Leben nach seinem handschriftlichen Nachlass dargestellt* (Berlin, 1882).

William T. Stearn

METZGER, HÉLÈNE (*b.* Chatou, near Paris, France, 26 August 1889; *d.* on the way to Auschwitz, Poland, after February 1944), *chemistry, history of science, philosophy.*

The daughter of Paul Bruhl and Jenny Adler, and the niece of Lucien Lévy-Bruhl, Hélène Metzger earned the *brevet supérieur* at a time when girls did not go to lycée to prepare for the *baccalauréat* and rarely attended university. She studied mineralogy with Frédéric Wallerant in his laboratory at the Sorbonne and in May 1912 received a *diplôme d'études supérieures* in physics under him for her memoir "Étude cristallographique du chlorate de lithium."

On 13 May 1913 she married Paul Metzger, a professor of history and geography at Lyons. In September 1914 he was reported missing in one of the first battles of World War I. A widow at twenty-five, she henceforth devoted herself entirely to research, commencing with her work on crystallography. Animated by a wide-ranging curiosity and an eminently philosophic cast of mind, she became interested in a historical approach to science.

Hélène Metzger's abilities as historian of science were evident as early as 1918 in her doctoral thesis, *La genèse de la science des cristaux*, which she defended in Paris. In this work she showed how crystallography slowly became differentiated from mineralogy, physics, and chemistry, until at the end of the eighteenth century it had developed into an independent science. Convinced that scientific revolutions are the visible effect of a previous underlying current, she took chemistry as an example and conceived a vast plan that was to lead her research from the beginning of the seventeenth century to Lavoisier. The general title was *Les doctrines chimiques en France, du début du XVIIe à la fin du XVIIIe siècle*. The first part appeared in 1923, and in 1924 it won for its author the Prix Binoux of the Académie des Sciences. Nicolas Lemery was the central character, but importance was accorded to authors little known until then.

Newton, Stahl, Boerhaave et la doctrine chimique, published in 1930, may be considered the second installment of the projected work. Sections of what would have been the third part can be found in lectures delivered in 1932–1933 at the Institut d'Histoire des Sciences of the University of Paris; they were published in 1935 under the title *La philosophie de la matière chez Lavoisier*.

A synthetic view of the history of chemistry, as Hélène Metzger conceived of it, can be found in the little volume *La chimie*, written in 1926 (published in 1930) for the series Histoire du Monde, edited by Eugène Cavaignac.

Drawn to philosophical reflection and preoccupied by epistemological problems, Hélène Metzger submitted an essay to a contest held by the Académie des Sciences Morales et Politiques that won the Prix Bordin in philosophy in 1925 and was published in 1926 under the title *Les concepts scientifiques*. This study deals with both psychology and logic and takes examples from the history of science to show how concepts arise and become transformed and how they can be classified.

A disciple of both Émile Meyerson and Lucien Lévy-Bruhl, Hélène Metzger was not satisfied with a strictly positivist position. She followed Meyerson in seeking out the philosophical bases of science, and Lévy-Bruhl in extending her investigations to the nonrational aspects of thought, which are as prevalent in the civilized as in the primitive mind. It was the activity of the entire human intellect that she wanted to uncover by following the scientists in their groping; she was as interested in "false" ideas as in those currently considered "true." For her, religious, metaphysical, and scientific ideas formed a unified whole in a given historical period, and she believed that one group could not properly be studied by artificially separating it from others. If that viewpoint is widely accepted today, it is partly because she helped to establish it.

The best example of Hélène Metzger's approach is the thesis she presented to the fifth section of the École Pratique des Hautes Études in 1938, published as *Attraction universelle et religion naturelle chez quelques commentateurs anglais de Newton*. Encouraged by Léon Brunschvicg, she wished to pursue this synthetic study of the development of scientific and philosophical thought by a thorough examination of the work of Condillac in relation to that of Lavoisier and the chemists of the end of the eighteenth century. Unfortunately, this project was never carried out; after the beginning of the German occupation of Paris, she moved to Lyons to work, at the Bureau d'Études Israélites, on a study of Jewish monotheism. Her conclusions appeared in 1947 in *Revue philosophique*; the preamble was published in 1954 in a volume prepared by her brother, Adrien Bruhl: *La science, l'appel de la religion et la volonté humaine*. Hélène Metzger was arrested in the Rue Vaubécour in Lyons in February 1944, deported to Drancy, and then sent to Auschwitz; it has proved impossible to establish the circumstances and the date of her death.

During the interwar years Hélène Metzger's works—which included many articles in *Isis*, *Archeion*, and *Scientia* and her contributions to the *Vocabulaire historique* (prepared by the Centre International de Synthèse), which appeared in *Revue de synthèse* and *Revue d'histoire des sciences et de leurs applications*—had a considerable impact on the history of science and on epistemology. Her personal influence among historians of science was still more decisive. She participated in the first four international congresses of the history of science and was a charter member of the Comité International d'Histoire des Sciences (converted into the Académie Internationale d'Histoire des Sciences in 1929). She served as its administrator-treasurer from 5 June 1931 until her arrest. She organized the Academy's library in the Rue Colbert in Paris, and her philosophical concerns are clearly reflected in its holdings.

In 1939 Hélène Metzger was placed in charge of the history of science library of the Centre International de Synthèse. An ardent participant in all the meetings of its history of science section—as is attested by the issues of *Archeion*—and secretary of the Groupe Français d'Historiens des Sciences, she enlivened the discussions with her subtle and often ironic remarks, which were always pertinent and erudite, if somewhat disconcerting in their impulsiveness. Cordial to young scholars and to French and foreign colleagues, she was an inceptive influence for many studies. Aldo Mieli, Pierre Brunet, Federigo Enriques, Alexandre Koyré, George Sarton, Paul Mouy, and Robert Lenoble all derived inspiration for their work from their contact with Hélène Metzger.

BIBLIOGRAPHY

I. Original Works. Metzger's books are *La genèse de la science des cristaux* (Paris, 1918; repr. 1970); *Les doctrines chimiques en France, du début du XVIIe à la fin du XVIIIe siècle* (Paris, 1923; repr. 1970); *Les concepts scientifiques* (Paris, 1926); *La civilisation européenne*, pt. 4, *La chimie* (Paris, 1930); *Newton, Stahl, Boerhaave et la doctrine chimique* (Paris, 1930); *La philosophie de la matière chez Lavoisier* (Paris, 1935); *Attraction universelle et religion naturelle chez quelques commentateurs anglais de Newton* (Paris, 1938); and *La science, l'appel de la religion et la volonté humaine* (Paris, 1954).

Her principal articles are "L'évolution du règne métallique d'après les alchimistes du XVIIe siècle," in *Isis*, **4** (1922), 464–483; "La philosophie de la matière chez Stahl et ses disciples," *ibid.*, **8** (1925), 427–464; "Newton et l'évolution de la théorie chimique," in *Archeion*, **9** (1928), 243–256, 433–461, and **10** (1929; incorrectly numbered **11**), 13–25—the three parts were also brought together in a booklet (Rome, n.d.); "La philosophie de Lucien Lévy-Bruhl et l'histoire des sciences," in *Archeion*, **12** (1930), 15–24; "Eugène Chevreul historien de la chimie," *ibid.*, **14** (1932), 6–11; "Introduction à l'étude du rôle de Lavoisier dans l'histoire des sciences," *ibid.*, 31–50; "L'historien des sciences doit-il se faire le contemporain des savants dont il parle?" *ibid.*, **15** (1933), 34–44; "Tribunal de l'histoire et théorie de la connaissance scientifique," *ibid.*, **17** (1935), 1–18; "La signification de l'histoire de la pensée scientifique," in *Scientia*, **57** (June 1935), 449–453; "L'*a priori* dans la doctrine scientifique et l'histoire des sciences," in *Archeion*, **18** (1936), 29–79; "La méthode philosophique dans l'histoire des sciences," *ibid.*, **19** (1937), 204–216; "Alchimie. Communication pour servir au Vocabulaire historique," in *Revue de synthèse*, **16**, no. 1 (Apr. 1938), 43–53; and "Atomisme. Communication pour servir au Vocabulaire historique," in *Revue d'histoire des sciences et de leurs applications*, **1**, no. 1 (July 1947), 51–62.

II. Secondary Literature. See Marie Boas, in *Archives internationales d'histoire des sciences*, **12** (1959), 432–435; Pierre Brunet, in *Revue d'histoire des sciences et de leurs applications*, **1**, no. 1 (July 1947), 68–70; and Suzanne Delorme, in *Archives internationales d'histoire des sciences*, **1** (1948), 326–327.

Suzanne Delorme

MEUSNIER DE LA PLACE, JEAN-BAPTISTE-MARIE-CHARLES (*b.* Tours, France, 19 June 1754; *d.* Mainz, Germany, 17 June 1793), *mathematics, physics, engineering.*

Meusnier was the son of Jean-Baptiste Meusnier and Anne le Normand Delaplace. The family was for generations engaged in law and administration; the father was a counsel attached to a court (*présidial*) at Tours. He tutored his son, and only during his last years at Tours did Meusnier go to school.

From 1771 to 1773 Meusnier was privately tutored at Paris for entrance into the military academy at Mézières, where he studied in 1774–1775 and graduated as second lieutenant in the Engineering Corps. His mathematics teacher was Gaspard Monge, under whom Meusnier did his only published mathematical work, on the theory of surfaces.

His paper, read at the Paris Academy of Sciences in 1776, supposedly led d'Alembert to state: "Meusnier commence comme je finis." It also led to Meusnier's election, at twenty-one, as a corresponding member of the Academy. He was placed in charge of continuing the descriptions of machines approved by the Academy, and in February 1777 he presented to the Academy the seventh volume of the *Recueil des machines approuvées par l'Académie*. During 1777, now a first lieutenant, he was sent to Verdun to study mining and sapping. From 1779 to 1788 he worked as a military engineer on the harborworks of Cherbourg, where he displayed great ingenuity and perseverance, despite red tape and intrigues, in the building of the breakwater and the fortification of Île Pelée. To provide drinking water for this island he spent much time on experiments on the desalinization of seawater. In March 1783 Meusnier, sent into debt by his work, presented his machine to the Academy.

During 1783 the first balloon ascensions took place. Meusnier, on leaves of absence from Cherbourg, began to study the theory of this new field, aerostation. In December 1783 he read before the Academy

342

his "Mémoire sur l'équilibre des machines aéro-statiques." The next month he was elected a full member of the Academy and was immediately appointed to a committee on aerostation, other members of which were Lavoisier, Berthollet, and Condorcet. The results of his work on this committee were presented in November 1784 in "Précis des travaux faits à l'Académie des sciences pour la perfection des machines aérostatiques," with a theory and detailed construction plans for dirigible balloons. It led to no practical results at the time.

During this period Meusnier began a collaboration with Lavoisier on the synthesis and analysis of water; Meusnier was especially interested in the production of hydrogen in quantity from water. On 21 April 1784 they presented to the Academy a continuation of the paper presented in June 1783 by Lavoisier and Laplace on the synthesis of water from oxygen and hydrogen: "Mémoire où l'on prouve par la décomposition de l'eau que ce fluide n'est point une substance simple . . . ." It was also a heavy blow against the phlogiston theory, which Berthollet and others soon abandoned in favor of Lavoisier's "théorie française." Meusnier also collaborated with Lavoisier on the improvement of oil lamps for city street illumination. Their ideas were contemporary with those of Aimé Argand and perhaps inspired the construction of his lamp.

In May 1787 Meusnier became a captain; in July 1788, he was promoted to *aide-maréchal général des logis au corps de l'État Major* and major.

From then on his career was with the army, and in July 1789 he became a lieutenant colonel. With his friends Monge and Berthollet he joined the Jacobins in 1790. With many other academicians he was appointed to the Bureau de Consultation Pour les Arts et Métiers to study inventions useful to the state. Meusnier invented a machine for engraving assignats that greatly reduced the possibility of producing counterfeit notes. In February 1792 he was appointed colonel, then *adjutant général colonel*, and in September 1792 field marshal. Sent in February 1793 to the armies of the Rhine commanded by Custine, he participated in the defense of the fortress of Kassel during the siege of Mainz by the Prussians. He was wounded on 5 June and died twelve days later. His remains were brought to Paris (Goethe witnessed the procession leaving Mainz; see his *Kampagne in Frankreich*), and were later transferred to Tours, where in 1888 a bust was erected on a pedestal containing his ashes.

The "Mémoire sur la courbure des surfaces," read in 1776 and published in 1785, was written after Monge had shown him Euler's paper on this subject

(*Mémoires de l'Académie des Sciences* [Berlin, 1760]). In the "Mémoire" Meusnier derived "Meusnier's theorem" on the curvature, at a point of a surface, of plane sections with a common tangent and also found, as special solutions of Lagrange's differential equation of the minimal surfaces (1760), the catenoid and the right helicoid. His results can be found in any book on differential geometry. In the "Mémoire" on aerostation (1783) Meusnier presented a theory of the equilibrium of a balloon, the dynamics of ascension, and the rules for maneuvering a balloon. To maintain appropriate altitude even with the disposal of ballast he proposed a balloon filled with hydrogen containing a smaller balloon filled with air (known as *ballonet d'air*); he also suggested a model with air in the larger balloon and hydrogen in the smaller. In the "Précis" of 1784, the result of a great many test experiments, Meusnier gave a detailed plan for the construction of a dirigible balloon in the form of an elongated ellipsoid with another balloon inside. For propulsion he suggested revolving air screws worked by a crew. He described two possibilities: a small dirigible 130 feet long carrying six men and one 260 feet long (130 feet minor axis) with a crew of thirty and food for sixty days, able to fly around the earth. In his formula for the stability of the balloon,

$$n = \left(\frac{P+E}{P}\right) \times \frac{3}{2}\left(\frac{l^2 - h^2}{h^2}\right) \times \frac{(h-x)^2}{3h - 2x},$$

n is the distance from the metacenter to the center of the balloon, P the weight of the objects collected at the center of the gondola, E the weight of the balloon as concentrated at the center, l and h the major and minor axes of the balloon, and x the height of the hydrogen when the balloon is on earth, the hydrogen rising above the air in the balloon.

The principle of the revolving screw had also occurred to David Bushnell of Connecticut in the construction of his submarine (1776–1777). Meusnier knew of Bushnell's invention.

After Cavendish had shown nonquantitatively in 1781 that the combination of oxygen and hydrogen yields water, Lavoisier and Laplace in 1783 presented to the Academy an account of their work on the synthesis of water; Monge had also performed this experiment. Meusnier suggested more exact measurements to Lavoisier and constructed precision instruments for this purpose. Their "Mémoire" of April 1784 showed how they had decomposed water into its components; the hydrogen was obtained as a gas and the oxygen in the form of an iron oxide. For many this famous experiment carried convincing evidence against the phlogiston theory.

BIBLIOGRAPHY

I. ORIGINAL WORKS. "Mémoire sur la courbure des surfaces" appeared in *Mémoires de mathématique et de physique présentés par divers sçavans*, **10** (1785), pt. 2, 477–510. The "Mémoire" and the "Précis" on aerostation were published, with other material, by G. Darboux in *Mémoires de l'Académie des sciences*, 2nd ser., **51** (1910), 1–128. This includes the "Atlas de dessins relatifs à un projet de machine aérostatique" of 1784, presented in a photographic reproduction to the Academy in 1886 by General Perrier. The "Mémoire où l'on prouve par la décomposition de l'eau . . .," written with Lavoisier, is in *Mémoires de l'Académie royale des sciences pour 1781* (1784), 269–283. See also "Description d'un appareil propre à manoeuvrer différentes espèces d'air dans les expériences qui exigent des volumes considérables," *ibid.*, *1782* (1785), 466; "Sur les moyens d'opérer l'entière combustion de l'huile et d'augmenter la lumière des lampes," *ibid.*, *1784* (1787), 390–398. There is MS material in the Archives of the Académie des Sciences, the Institut de France, the Archives Historiques de la Guerre, and the Bibliothèque du Génie, all in Paris. Details are given by J. Laissus (see below).

II. SECONDARY LITERATURE. "Notice sur le général Meusnier," in *Revue rétrospective*, 2nd ser., **4** (1835), 77–99, contains biographical notes on Meusnier by Monge and others, the originals of which have not been found. Partly based on these is Darboux's "Notice historique sur le général Meusnier," in his *Éloges académiques et discours* (Paris, 1912), 218–262, also in *Mémoires de l'Académie des Sciences*, 2nd ser., **51** (see above). In it are many particulars on Meusnier's work in Cherbourg and in the army of the Revolution. See also L. Louvet, in *Nouvelle biographie générale*, XXXV (1865), cols. 264–267. Bibliographical details based on independent research in the printed and MS materials are in J. Laissus, "Le général Meusnier de la Place, membre de l'Académie royale des sciences," in *Comptes rendus du 93ᵉ Congrès national des sociétés savantes, Tours, 1968*, Section des Sciences, II (Paris, 1971), 75–101. Meusnier's work on decomposition of water can be studied in books on Lavoisier. His works on aerostation have been analyzed by F. Letonné, "Le général Meusnier et ses idées sur la navigation aérienne," in *Revue du génie militaire*, **2** (1888), 247–258; and by Voyer, "Les lois de Meusnier," *ibid.*, **23** (1902), 421–430; "Le ballonet de Meusnier," *ibid.*, 521–532; and "Le général Meusnier et les ballons dirigibles," *ibid.*, **24** (1902), 135–156—German trans. in *Illustrierte aeronautische Mitteilungen*, **9** (1905), 137–144, 353–361, 373–387. The third of these papers gives a proof of Meusnier's stability formula. See also G. Béthuys, *Les aérostations militaires* (Paris, 1894), 137–146. On the Argand lamp see S. T. McCoy, *French Inventions of the Eighteenth Century* (Lexington, Ky., 1952), 52–56.

On the papers relating to the collaboration between Lavoisier and Meusnier, see also D. I. Duveen and H. S. Klickstein, *Bibliography of the Works of Antoine Laurent Lavoisier 1743–1794* (London, 1954), index, p. 462.

On Bushnell see D. J. Struik, *Yankee Science in the Making* (New York, 1962), 83, 453.

D. J. STRUIK

MEYEN, FRANZ JULIUS FERDINAND (*b.* Tilsit, Prussia [now Sovetsk, U.S.S.R.], 28 June 1804; *d.* Berlin, Germany, 2 September 1840), *botany*.

Meyen's father, who was president of the commercial court in Tilsit, died in 1811. Meyen attended the Gymnasium in Tilsit until 1819, when he had to begin an apprenticeship to an apothecary in Memel, Prussia. In 1821 his brother in Berlin offered him the chance to continue his schooling so that he could enter a university. From 1823 to 1826 Meyen studied medicine at the Friedrich Wilhelms Institut, where military physicians were trained. At the same time, however, he also attended the zoology lectures of H. Lichtenstein and K. A. Rudolphi and the botany lectures of Johann Horkel, K. H. Schultz, and H. F. Link at the University of Berlin. He received his medical degree in October 1826 with a dissertation entitled "De primis vitae phaenomenis in fluidis formativis et de circulatione sanguinis in parenchymate." Until 1830 he was a military physician in Berlin, Cologne, Bonn, and Potsdam. In this period he published three monographs and eleven journal articles.

In 1830, through the influence of Alexander von Humboldt, Meyen obtained the post of doctor on the royal cargo ship *Prinzess Louise*. His assignment, during a world cruise lasting nearly two years, was to collect natural history specimens and make scientific observations. He made long excursions in the western part of South America, climbing the Andes in Chile and Peru up to the snow line. Later in the voyage he spent some time in the Sandwich Islands (Hawaii), the Philippines, and China. Wherever he went, he collected plants, and in China took an interest in Chinese gardening.

Upon returning to Prussia, Meyen began to prepare his material; he published a general account of the voyage in 1834–1835. The scientific presentation of the collections appeared in the *Nova Acta Leopoldina* (XVI, XVII, and XIX [1832–1834, 1835, 1843]). At first Meyen published only the articles on zoology and ethnography; of the plants he described only the lichens (with J. von Flotow).

In August 1834 Meyen was named extraordinary professor in the Philosophy Faculty of the University of Berlin. He received this appointment—which followed his being granted an honorary doctorate by the University of Bonn—on the basis of the description of his voyage and of his earlier works. He then continued his study of phytotomy and plant physiology. Meyen's most important scientific publication was *Phytotomie*. Written when he was twenty-five, it presented the new field of microscopic plant anatomy. The book appeared at the beginning of about ten years of intensive microscopic investigations of plants and animals. At the end of this period the Schleiden-Schwann cell theory had fully emerged. Meyen's *Phytotomie* did not in every respect represent progress, but its comprehensive summary of the subject provided a strong impetus to further research. In response to the much-discussed question of the type and number of the elementary plant organs, Meyen described the cells, spiral tubes, and sap vessels. For the various forms of cellular tissue he introduced new designations —mesenchyma and pleurenchyma—to be added to those already used by Link—parenchyma and prosenchyma.

Before Meyen's research, only the structure of the cellular reticulum was considered important, but he investigated the contents of the cell as well. Most notably, he described in detail the movements that could be observed within it. As early as 1827 he published a paper entitled "Über die eigentümliche Säftebewegung in den Zellen der Pflanzen." This movement had first been observed in 1774 by Bonaventura Corti in the cells of *Chara*. For this reason Meyen's cognomen as a member of the Imperial Leopoldine-Caroline Academy of Science was Corti. In *Phytotomie* he also treated movements of fluids throughout the plant. He viewed the lactiferous tubes as circulatory organs and "as the highest thing that the plant produces." The fluid circulating within them corresponded, he thought, to the blood of animals.

In 1837 there appeared the first volume of Meyen's other major work: *Neues System der Pflanzen-Physiologie*. Meyen expressed the wish that the book be considered a continuation and improvement of his *Phytotomie*. This first volume is in fact a reworking of *Phytotomie*; once again the content of plant cells is examined and described more fully than in the writings of other contemporary students of microscopic plant anatomy. The amalgamation of physiological and morphological problems, more evident here than in *Phytotomie*, corresponded to the conception of the relationship between anatomy and physiology held in zoology since the beginning of the nineteenth century. In the preface to the *Neues System* Meyen stated that the time had arrived "when one could attempt to study plant physiology in just the same way as animal physiology."

For his essay "Ueber die neusten Fortschritte der Anatomie und Physiologie der Gewächse" Meyen received, among other honors, the prize offered by the Teyler Society of Haarlem for the best paper on that subject. For his "Ueber die Secretions-Organe der Pflanzen" he was awarded the prize of the Royal Society of Science of Göttingen. Both these awards were presented in 1836.

Meyen's importance for botany lies much less in discovery than in the intensive and wide-ranging study of microscopic anatomy in connection with physiology. The breadth of his interests can be seen from the fact that he published the "Jahresberichte über die Resultate der Arbeiten im Felde der physiologischen Botanik" for the years 1834–1839 for A. F. A. Wiegmann's *Archiv für Naturgeschichte*. He also wrote a work on plant geography and one on plant pathology.

BIBLIOGRAPHY

Meyen's most important works are *Phytotomie* (Berlin, 1830); *Ueber die Bewegung der Säfte in den Pflanzen. Ein Schreiben an die Königliche Akademie der Wissenschaften zu Paris* (Berlin, 1834); and *Neues System der Pflanzen-Physiologie*, 3 vols. (Berlin, 1837–1839).

A biography with a complete bibliography is J. T. C. Ratzeburg, "Meyen's Lebenslauf," in *Nova acta Academiae Caesareae Leopoldina Carolinae germanicae naturae curiosorum*, **19** (1843), xiii–xxxii.

HANS QUERNER

MEYER, CHRISTIAN ERICH HERMANN VON (*b.* Frankfurt, Germany, 3 September 1801; *d.* Frankfurt, 2 April 1869), *paleontology*.

Meyer came from on old Frankfurt family; his father was a lawyer and later mayor of Frankfurt. Meyer was born with clubfeet, which handicapped his movement. He was educated in Frankfurt, then he worked for a year in a glasswork and for three years as an apprentice in a banking house. From 1822 to 1827 he studied finance and natural science, especially geology and mineralogy, at Heidelberg, Munich, and Berlin, where he met Hegel and Humboldt. On returning to Frankfurt he devoted all his time and energy to paleontology, publishing numerous works in rapid succession, visiting museums and collections, and attending professional congresses. He soon became a known and respected paleontologist, who received material for study and publication from all of Germany and neighboring countries.

In 1837, Meyer entered the financial administration of the Bundestag—the parliament of the German Confederation, which was then under Austrian leadership; in 1863 he became its director of finances. Consequently, from 1837 he could carry out his paleontological studies only during his spare time. Nevertheless, in 1860 he rejected an appointment as professor at the University of Göttingen in order to maintain his scientific independence. Meyer never married. He was sociable, had charming manners, and was respected and loved by his fellow citizens for his sincerity and civic service.

The main subject of Meyer's scientific studies was the fossil vertebrates. His chief work in this field is the four-volume *Fauna der Vorwelt*, which contains 132 plates of outstanding drawings done by Meyer himself. In these books Meyer described vertebrates—chiefly from Germany—of the Miocene, Jurassic, Triassic, Permian, and Carboniferous. He also wrote articles for journals; 103 of his paleontological writings were published in *Palaeontographica*, which he founded in 1846 with Wilhelm Dunker.

Meyer considered all classes of vertebrates—fishes, amphibians, reptiles, birds, and mammals—and was, in fact, one of the most distinguished vertebrate paleontologists of his time in Europe. He also published studies on the crustaceans, the crinoids, the Asterozoans, and the cephalopods. His descriptions are characterized by great accuracy, by clarity of expression, and by first-rate drawings. Meyer produced no original theories, no ingenious hypotheses; yet he did not lose sight of the broader connections among his detailed studies. He repeatedly criticized Cuvier's law of correlation and, as early as 1832, wrote a survey of the vertebrates then known, their stratigraphic distribution, emergence, and evolution.

BIBLIOGRAPHY

I. ORIGINAL WORKS. Meyer's books include *Palaeologica, zur Geschichte der Erde und ihrer Geschöpfe* (Frankfurt, 1832); *Die fossilen Zähne und Knochen und ihre Ablagerung in der Gegend von Georgensgmünd in Bayern* (Frankfurt, 1834); *Beiträge zur Paläontologie Württemberg's, enthaltend die fossilen Wirbelthierreste aus den Triasgebilden mit besonderer Rücksicht auf die Labyrinthodonten des Keupers* (Stuttgart, 1844), written with T. Plieninger; *Zur Fauna der Vorwelt: Erste Abtheilung. Fossile Säugetiere, Vögel und Reptilien aus dem Molasse-Mergel von Oeningen* (Frankfurt, 1845); *Zweite Abtheilung. Die Saurier des Muschelkalks mit Rücksicht auf die Saurier aus Buntem Sandstein und Keuper* (Frankfurt, 1847–1855); *Dritte Abtheilung. Saurier aus dem Kupferschiefer der Zechsteinformation* (Frankfurt, 1856); and *Vierte Abtheilung. Reptilien aus dem lithographischen Schiefer in Deutschland und Frankreich* (Frankfurt, 1860).

Among his articles are "Reptilien aus der Steinkohlenformation in Deutschland," in *Palaeontographica*, **6** (1856–1858), 59–219; "Reptilien aus dem Stubensandstein des obern Keupers," *ibid.*, **7** (1861), 253–346; and "Studien über das Genus Mastodon," *ibid.*, **17** (1867–1870), 1–72.

II. SECONDARY LITERATURE. On Meyer or his work, see T. H. Huxley, "The Life of Hermann Christian Erich von Meyer. The Anniversary Address of the President," in *Quarterly Journal of the Geological Society of London*, **26** (1870), xxxiv–xxxvi; F. von Kobell, "Nekrolog auf Hermann v. Meyer," in *Sitzungsberichte der Bayerischen Akademie der Wissenschaften zu München*, **1** (1871), 403–407; M. Pfannenstiel, "Unbekannte Briefe von Sir Charles Lyell an Hermann von Meyer," in *Bulletin of the Geological Institution of the University of Uppsala*, **40** (1961), 1–15; J. J. Rein, "Dr. Christian Erich Hermann von Meyer. Eine biographische Skizze," in *Bericht der Senckenbergischen naturforschenden Gesellschaft in Frankfurt a. M.* (1868–1869), report of the anniversary of 30 May 1869, pp. 13–17; W. Struve, "H. von Meyer und die Senckenbergische Paläontologie," in "Zur Geschichte der paläozoologisch-geologischen Abteilung des Natur-Museums und Forschungs-Instituts Senckenberg," in *Senckenbergiana lethaea*, **48** (1967), 64–75; and C. A. Zittel, *Denkschrift auf Christ. Erich Hermann von Meyer* (Munich, 1870), with complete bibliography.

HEINZ TOBIEN

MEYER, JOHANN FRIEDRICH (*b.* Osnabrück, Hannover [now German Federal Republic], 24 October 1705; *d.* Osnabrück, 2 November 1765), *chemistry.*

Meyer is best known for having been wrong. Just a few years after Joseph Black explained that the difference between the mild and the caustic alkalies lies in the presence or absence of "fixed air" (1756), Meyer published his *Chymische Versuche zur näheren Erkenntniss des ungelöschten Kalchs* (1764). In this work he argued that causticity in alkalies arose from a substance that entered the mild alkalies from the fire. He called this substance *acidum pingue* and characterized it as a combination of a previously unknown acid substance with the matter of fire or light. It was, Meyer said, responsible for "sharp" properties and thus was found in acids, caustic alkalies, and fire. It was not to be confused with phlogiston, which turned calxes into metals; *acidum pingue* calcined metals, causing the famous—or notorious—weight gain. Meyer's peculiar theory of the *acidum pingue* combined features from the Paracelsian sulfur of metals and Lemery's "matter of fire." The theory avoided one set of errors by attributing the "augmented calx" to a gain of matter, but it incurred others by claiming causticity to be a result of an accession of *acidum pingue*.

Meyer's father, who died in 1714, was a physician; his mother, the daughter of an apothecary. Meyer was intended for the clergy but, he said, "Providence made me a pharmacist." He went at age fifteen into his grandmother's apothecary shop, where he served six years as an apprentice. After working as a journeyman in Leipzig, Nordhausen (where he studied mining and metallurgy), Frankfurt am Main, Trier, and Halle, he returned to Osnabrück and in 1737 inherited his grandmother's shop. The following year he married a clergyman's daughter, who died in 1759; they had no children.

Meyer's work was highly respected on the Continent in the 1760's and early 1770's; and his theory of causticity was accepted by a number of chemists, including Baumé, Pörner, and Wiegleb. Black took special care to answer point by point this challenge to his own findings. Lavoisier and Guyton de Morveau avowed at different times that Meyer's writings had considerable merit. But with the explication over the next fifteen years of the role of oxygen in combustion and acidification, and with the recognition that Black's work had inaugurated this great train of discoveries in pneumatic chemistry, Meyer's claim to a place among the builders of eighteenth-century chemistry suffeʹed a blow from which it has never recovered.

BIBLIOGRAPHY

I. ORIGINAL WORKS. *Chymische Versuche zur näheren Erkenntniss des ungelöschten Kalchs* ... (Hannover–Leipzig, 1764; 2nd ed., 1770) was translated into French by F. F. Dreux (Paris, 1765). *Alchymistische Briefe* ... (Hannover, 1767) is available in a French trans. by Dreux (Paris, 1767). Johann Christian Wiegleb, *Kleine chymische Abhandlungen von dem grossen Nutzen der Erkenntniss des Acidi pinguis bey der Erklärung vieler chymischen Erscheinungen* (Langensalza, 1767), draws freely and expands upon Meyer's *Versuche*, and contains a short autobiographical sketch by Meyer (on which the present article is based), edited by E. G. Baldinger.

II. SECONDARY LITERATURE. See Henry Guerlac, *Lavoisier: The Crucial Year* (Ithaca, N.Y., 1961), 48–49, and the literature cited there; and J. R. Partington, *A History of Chemistry*, III (London, 1962), 145–146, 152–153, 388–389, 519–520.

STUART PIERSON

MEYER, JULIUS LOTHAR (*b.* Varel, Oldenburg, Germany, 19 August 1830; *d.* Tübingen, Germany, 11 April 1895), *chemistry.*

(Julius) Lothar Meyer was the fourth of seven children of Heinrich Friedrich August Jacob Meyer, a prominent physician in Varel. His mother, the former Anna Sophie Wilhelmine Biermann, was the daughter of another physician of that town. Both Lothar and his brother, Oskar Emil, later a physicist, began their studies with the intention of entering medicine. Brought up as a Lutheran, Meyer first attended a private school, then the newly founded Bürgerschule in Varel, supplementing this education with private instruction in Latin and Greek. Delicate in his early years, he suffered such severe headaches at age fourteen that his father advised complete discontinuance of academic studies and placed him as an assistant to the chief gardener at the summer palace of the grand duke of Oldenburg, at Rastede. After a year his health was sufficiently restored for him to enter the Gymnasium at Oldenburg, from which he graduated in 1851. In the summer of that year Meyer began to study medicine at the University of Zurich, and in 1853 he moved to Würzburg, where Virchow was lecturing on pathology. He received the M.D. the following year. Encouraged by Carl Ludwig, his former physiology professor at Zurich, Meyer turned from medicine to physiological chemistry and went to Heidelberg to study under Bunsen. The latter's work on gas analysis particularly attracted him, and in 1856 Meyer completed his investigation *Ueber die Gase des Blutes*, which was accepted by the Würzburg Faculty of Medicine as his doctoral dissertation. F. Beilstein, H. H. Landolt, H. E. Roscoe, A. von Baeyer, and F. A. Kekulé were in Heidelberg at the same time. Lectures by Kirchhoff moved Meyer further toward physical chemistry.

At the suggestion of his brother, Meyer moved to Königsberg in the fall of 1856, to attend Franz Neumann's lectures on mathematical physics. He also pursued there his earlier physiological interests by studying the effect of carbon monoxide on the blood. When he moved to Breslau in 1858, this investigation was accepted by the Philosophy Faculty as his dissertation for the Ph.D. In February 1859 Meyer established himself as *Privatdozent* in physics and chemistry at Breslau with a critical historical work, "Über die chemischen Lehren von Berthollet und Berzelius." That same spring he took over the direction of the chemical laboratory in the physiological institute and lectured on organic, inorganic, physiological, and biological chemistry. During his stay at Breslau the first edition of his *Die modernen Theorien der Chemie und ihre Bedeutung für die chemische Statik* appeared (1864). It went through five editions and was translated into English, French, and Russian.

Meyer had attended the 1860 Karlsruhe Congress, where he heard Cannizzaro and read his paper on the use of Avogadro's hypothesis and the law of Dulong and Petit in establishing atomic weights and formulas. Meyer edited Cannizzaro's paper for Ostwald's

Klassiker der Exacten Wissenschaften and describes in that work how "the scales fell from my eyes and my doubts disappeared and were replaced by a feeling of quiet certainty." Meyer's *Moderne Theorien* was a direct outcome of that experience. In a brief obituary in 1895 the book was described as "not especially well received at first, but as years passed it exerted a more and more powerful influence on the thoughts of chemists. From a flimsy pamphlet it grew to a stately volume, and it has generally been recognized as the best presentation of the fundamental principles of chemistry until the physicochemical movement began."[1]

Meyer was called to the School of Forestry at Neustadt-Eberswalde in 1866 for his first independent position. The same year he married Johanna Volkmann; they had four children. In 1868 Meyer succeeded C. Weltzien as professor of chemistry and director of the chemical laboratories at the Karlsruhe Polytechnic Institute. His final move, in 1876, was to Tübingen, where he taught until his death.

Two major events occurred in the early years of Meyer's stay at Karlsruhe. Mendeleev's 1869 paper on the periodic table led him to submit his own matured ideas for publication in December of that year. The paper was published in March 1870. In the summer of 1870 the Franco-Prussian War broke out; and Meyer made use of his medical abilities, helping to organize an emergency hospital in the buildings of the Polytechnic.

Meyer's Tübingen years at last offered an opportunity for intensive pursuit of his major interests. In excellent health until his sudden death, he guided the work of over sixty doctoral candidates; and with his associate Karl Seubert he published a careful analysis of the best atomic weight determinations available until then. In 1890 Meyer published *Grundzüge der theoretischen Chemie*, a less technical account of the theoretical foundations of chemistry than the later editions of his *Moderne Theorien* had become.

Outside his work in chemistry, Meyer read Greek and Latin classics and retained his love for gardening, learned in his youth. He was concerned with higher education and gave a number of lectures—later published—on that subject. For the year 1894–1895 he was elected rector of Tübingen University.

Meyer received the Davy Medal of the Royal Society jointly with Mendeleev in 1882. In 1883 he became a foreign honorary member of the Chemical Society (London) and in 1888 and 1891 corresponding member of the Prussian and St. Petersburg Academies of Sciences, respectively. He was given a title of nobility by decree of the Württemberg crown in 1892.

Meyer's earliest research dealt with physiological aspects of the uptake of gases by the blood. Building on previous studies by G. Magnus, he was able to demonstrate in 1856 that oxygen absorption by blood in the lungs occurs independently of pressure. This suggested to him that some possibly loose chemical linkage occurred. When he turned his attention to carbon monoxide poisoning, Meyer demonstrated a similar chemical linkage between that gas and a constituent of the blood. Further, he found that the amounts of oxygen and carbon monoxide taken up by the blood were in a simple molecular ratio, the carbon monoxide being able to expel volume for volume the oxygen already in the blood. This suggested to him that the same constituent of blood reacted with both gases. His preliminary searches for this constituent were unsuccessful. Hemoglobin was discovered by Hoppe-Seyler in 1864.[2]

Although these physiological studies were of considerable importance, Meyer's greatest achievement is no doubt tied to his work on the periodic classification of the elements. Meyer and Mendeleev both received their major stimulus for these considerations at the 1860 Karlsruhe Congress through Cannizzaro's paper on atomic weights. By 1862, Meyer had completed the manuscript of *Moderne Theorien*, including a table of twenty-eight elements in order of increasing atomic weight. Meyer felt that by the early 1860's considerable unity had finally been achieved regarding the fundamental principles of chemistry; and it was the purpose of his book to present these theoretical foundations.

Meyer saw J. W. Döbereiner and M. von Pettenkofer as his direct precursors and later edited their key papers. In 1816–1817, and more fully in 1829, Döbereiner had drawn attention to the fact that similar chemical elements often occurred in groups of three and that the arithmetic mean of the atomic weights of the lightest and heaviest elements often corresponded closely to the atomic weight of the third member of the group. His "triads" included calcium, strontium, and barium; lithium, sodium, and potassium; chlorine, bromine, and iodine; sulfur, selenium, and tellurium. Such a quantitative relationship suggested to some the likelihood that atoms were not the ultimate building blocks of nature—that they were composite, with the differences in weight of successive members of triads representing weights of more fundamental units.

Pettenkofer, pursuing Döbereiner's ideas, pointed to the parallelism between regular increases in equivalent weights of similar elements and increases in molecular weights of successive members of homologous series in organic chemistry.[3] Thus $CH_3 = 15$, $C_2H_5 = 29$, $C_3H_7 = 43$, $C_4H_9 = 57$,

$C_5H_{11} = 71$. The common increment (of 14) in these weights suggested that organic radicals may well hold the clue to the nature of the internal structure of inorganic atoms. Similar ideas were independently developed by Dumas, who spoke about them to the British Association for the Advancement of Science in 1851 but did not publish them until 1857.[4] Further attempts at systematizing the elements known to Meyer were made by J. H. Gladstone (1853), J. P. Cooke (1854), W. Odling (1857), and E. Lenssen (1857).

No progress beyond the arithmetic comparisons of weights of similar elements was likely as long as no clear distinction was made between equivalent and atomic weights, and no path to the values of the latter was generally accepted. That clarification was achieved by Cannizzaro at Karlsruhe in 1860, and almost immediately further relations between the elements became apparent. In 1862 A. E. Béguyer de Chancourtois plotted atomic weights of elements on a "telluric screw," on which similar elements would fall directly below each other. J. A. R. Newlands, beginning in 1863, organized the elements by their atomic weights, as computed by Cannizzaro's methods, into ten families (later reduced to eight). In an early table blanks were left for undiscovered elements; but these later disappeared in the eight-family version of 1865, which Newlands claimed as illustrating a "law of octaves."

Near the end of the first edition of Meyer's *Moderne Theorien*, the author points to the evidences for the composite nature of atoms, emphasizing the parallelism between series of related elements and organic compounds. He then appends a tabulation (see Figure 1) of twenty-eight elements, arranged according to increasing atomic weight, in six families that have valences of 4, 3, 2, 1, 1, and 2, respectively. Thus the integral stepwise change in valence as atomic weight increases was in print by 1864. A relation between families, and hence between dissimilar yet neighboring elements, was clearly established. Meyer remained interested also in constant increments within families and left a space for an as yet undiscovered element between silicon and tin, clearly indicating its probable atomic weight to be $28.5 + 44.55$, or 73.1. His next publication on the subject appeared after Mendeleev's historic 1869 paper, which Meyer had seen only in its abbreviated German form.[5]

Meyer's independent establishment of the central principles underlying the periodic table of the elements was demonstrated in 1893, when Adolf Remelé, his successor at Neustadt-Eberswalde, showed him a handwritten draft periodic table (Figure 2) designed by Meyer for the second edition of *Moderne Theorien* and given to Remelé in July 1868. Its notation "§91" makes clear its intended use for the second edition. It differs from the 1864 table mainly by the addition of twenty-four elements and nine families. These were the B-subgroups, the characteristics of which Meyer later claimed to have discovered independently. Hydrogen, boron, and indium are not in the table, and aluminum appears in both column 3

	4 werthig	3 werthig	2 werthig	1 werthig	1 werthig	2 werthig
	— —	— —	— —	— —	Li = 7.03	(Be = 9.3?)
Differenz =	— —	— —	— —	— —	16.02	(14.7)
	C = 12.0	N = 14.04	O = 16.00	Fl = 19.0	Na = 23.05	Mg = 24.0
Differenz =	16.5	16.96	16.07	16.46	16.08	16.0
	Si = 28.5	P = 31.0	S = 32.07	Cl = 35.46	K = 39.13	Ca = 40.0
Differenz =	$\frac{89.1}{2} = 44.55$	44.0	46.7	44.51	46.3	47.6
	— —	As = 75.0	Se = 78.8	Br = 79.97	Rb = 85.4	Sr = 87.6
Differenz =	$\frac{89.1}{2} = 44.55$	45.6	49.5	46.8	47.6	49.5
	Sn = 117.6	Sb = 120.6	Te = 128.3	I = 126.8	Cs = 133.0	Ba = 137.1
Differenz =	89.4 = 2 x 44.7	87.4 = 2 x 43.7	— —	— —	(71 = 2 x 35.5)	— —
	Pb = 207.0	Bi = 208.0	— —	— —	(Tl = 204?)	— —

FIGURE 1.   Meyer's periodic table of 1864.

1	2	3	4	5	6	7	8	9	10	11	12	13	14	15
§ 91														
											Li = 7.03	Be = 9.3		
											16.02	14.7		
		Al = 27.3	Al = 27.3				C = 12.00	N = 14.04	O = 16.00	Fl = 19.0	Na = 23.05	Mg = 24.0		
		$\frac{28.7}{2} = 14.3$					16.5	16.96	16.07	16.46	16.08			
							Si = 28.5	P = 31.0	S = 32.07	Cl = 35.46	K = 39.13	Ca = 40.0	Ti = 48	Mo = 92
							$\frac{89.1}{2} = 44.55$	44.0	46.7	44.51	46.3	47.6	42	45
Cr = 52.6	Mn = 55.1	Fe = 56.0	Co = 58.7	Ni = 58.7	Cu = 63.5	Zn = 65.0	—	As = 75.0	Se = 78.8	Br = 79.97	Rb = 85.4	Sr = 87.6	Zr = 90	Vd = 137
	49.2	48.3	47.3		44.4	46.9	$\frac{89.1}{2} = 44.55$	45.6	49.5	46.8	47.6	49.5	47.6	47
	Ru = 104.3	Rh = 104.3	Pd = 106.0		Ag = 107.94	Cd = 111.9	Sn = 117.6	Sb = 120.6	Te = 128.3	I = 126.8	Cs = 133.0	Ba = 137.1	Ta = 137.6	W = 184
	92.8 = 2 × 46.4	92.8 = 2 × 46.4	93 = 2 × 46.5		88.8 = 2 × 44.4	88.3 = 2 × 44.15	89.4 = 2 × 44.7	87.4 = 2 × 43.7			71 = 2 × 35.5			
	Pt = 197.1	Ir = 197.1	Os = 199.0		Au = 196.7	Hg = 200.2	Pb = 207.0	Bi = 208.0			?Tl = 204?			

FIGURE 2. Meyer's 1868 table, published in 1895.

and column 4. Boron, indium, and aluminum properly belong in a family between columns 7 and 8. Meyer placed lead (Pb) correctly in column 8, while Mendeleev put it with calcium, strontium, and barium. Remelé's disclosure was published by Seubert after Meyer's death.[6]

In Meyer's classic paper of 1870, he adopted Mendeleev's use of a vertical form for the periodic table, publishing a table (Figure 3) in which the relation of the A- and B-subgroups of the chemical families is for the first time clearly indicated.[7] He also attached his graphical representation of the variation of atomic volume of the solid elements (volume divided by atomic weight) when plotted against atomic weight (Figure 4), for which he is most generally known. Both Meyer and Mendeleev emphasized that there is a periodic variation, a succession of maxima and minima, in several physical and chemical properties when they are examined as functions of atomic weight. Meyer began this paper with the assertion that it is most improbable that the chemical elements are absolutely undecomposable and referred to the ideas of Prout, Pettenkofer, and Dumas. As for the gaps in the table, he suggested that they would be filled through careful redeterminations of the atomic weights of known elements or through the discovery of new ones.

The significance of atomic weights in the demonstration of chemical periodicity, and the suspicion that some atomic weights were not accurate, led Meyer and Seubert to examine critically and to recalculate all atomic weights then considered important. Their study was published in 1883. All atomic weights were referred to the standard of unity for the atomic weight of hydrogen, a standard Meyer championed. Wilhelm Ostwald, on the other hand, strongly urged the adoption of O = 16.000 as standard, a view accepted in 1898 by a special committee of the German Chemical Society consisting of Landolt, Ostwald, and Seubert. In 1903 the newly created International

I	II	III	IV	V	VI	VII	VIII	IX
	B = 11.0	Al = 27.3		—		?In = 113.4		Tl = 202.7
		—		—			—	
	C = 11.97	Si = 28		—		Sn = 117.8		Pb = 206.4
			Ti = 48		Zr = 89.7		—	
	N = 14.01	P = 30.9		As = 74.9		Sb = 122.1		Bi = 207.5
			V = 51.2		Nb = 93.7		Ta = 182.2	
	O = 15.96	S = 31.98		Sc = 78		Te = 128?		—
			Cr = 52.4		Mo = 95.6		W = 183.5	
—	F = 19.1	Cl = 35.38		Br = 79.75		I = 126.5		—
			Mn = 54.8		Ru = 103.5		Os = 198.6?	
			Fe = 55.9		Rh = 104.1		Ir = 196.7	
			Co = Ni = 58.6		Pd = 106.2		Pt = 196.7	
Li = 7.01	Na = 22.99	K = 39.04		Rb = 85.2		Cs = 132.7		—
			Cu = 63.3		Ag = 107.66		Au = 196.2	
?Be = 9.3	Mg = 23.9	Ca = 39.9		Sr = 87.0		Ba = 136.8		—
			Zn = 64.9		Cd = 111.6		Hg = 199.8	

Difference from I to II and from II to III about = 16.
Difference from III to V, Iv to VI, V to VII fluctuating around 46.
Difference from VI to VIII, from VII to IX = 88 to 92.

FIGURE 3.   Meyer's 1870 table showing subgroups A (columns V, VII, IX) and B (IV, VI, VIII).

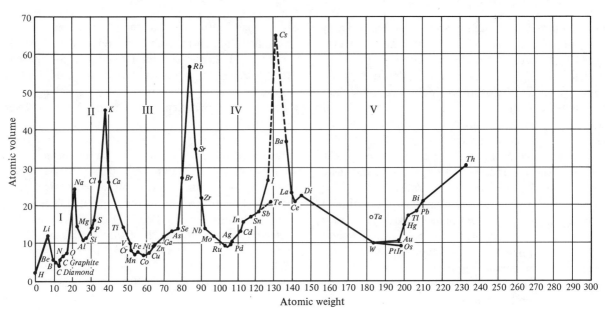

FIGURE 4. Meyer's 1870 graph of atomic volume plotted against atomic weight. (A redraft of Bailey's version which appeared in *Philosophical Magazine*, 5th ser., **13** [1882], 31.)

Commission on Atomic Weights decided to publish parallel tables based on H = 1 and O = 16, a practice followed for many years. The arguments for the oxygen standard were that the O:H ratio was for many years in doubt and that far more elements formed stable compounds with oxygen than with hydrogen.

In organic structural theory Meyer became involved in discussions of the structure of benzene. In 1865 Kekulé had proposed ring formula I; but this predicted two substances $C_6H_4X_2$, each having the two substituents, X, on adjacent carbons (II and III). Only one

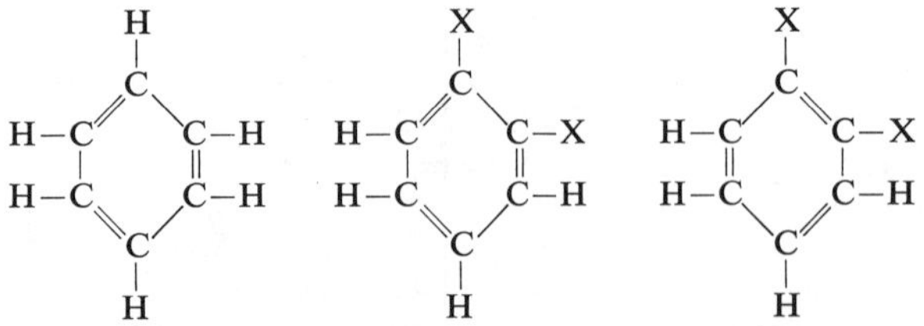

FIGURE 5

was ever found, and in 1872 Kekulé proposed a complex atomic oscillation mechanism in order to make all carbon atoms equivalent.[8] In the same year, in the second edition of *Moderne Theorien*, Meyer proposed a much simpler solution. He suggested that each carbon used only three of its four affinities, leaving one valence unsatisfied. His formula was the

first of a series of "centric" formulas proposed by a number of chemists. The unused valences point to the center of the ring.

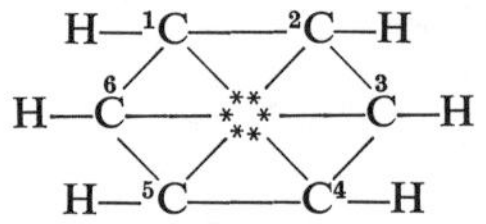

FIGURE 6

Meyer studied a number of benzene substitution reactions, particularly the nitration of benzene and its derivatives. He examined the effects of time, temperature, solvent, and concentration of reagents, feeling that chemists must go beyond a mere interest in the nature and quantity of products and must subject chemical reactions themselves to quantitative study. He examined reagents that facilitated chlorination and oxidation, the so-called chlorine and oxygen carriers, thus laying some of the groundwork for Ostwald's extensive revision of the concept of catalysis (1894). Meyer's studies of the effects of reagent concentration on chemical reactions served to confirm the law of mass action enunciated by C. M. Guldberg and P. Waage. In the fourth edition of *Moderne Theorien* (1883), he included a major new section, constituting more than a third of the book, entitled "Dynamik der Atome."

Meyer and his students investigated a number of physical properties, such as the boiling points, of structurally related organic compounds, seeking relations between these properties and molecular structure. His wide-ranging interests and mechanical skill led Meyer to devise or improve many pieces of apparatus, often adopted by other chemists. He pleaded with chemists to systematize inorganic chemistry on the basis of the periodic table, in order to approach the organization of subject matter achieved in organic chemistry.

NOTES

1. I. R., "Lothar Meyer," in *Journal of the American Chemical Society*, **17** (1895), 471–472.
2. F. Hoppe-Seyler, "Ueber die optischen und chemischen Eigenschaften des Blutfarbstoffs," in *Virchows Archiv für pathologische Anatomie und Physiologie und für klinische Medizin*, **29** (1864), 233–235.
3. M. Pettenkofer, "Ueber die regelmässigen Abstände der Aequivalentzahlen der sogenannten einfachen Radicalen," in *Münchener Gelehrten Anzeigen*, **30** (1850), 261–272; repr. with new intro. by the author in *Annalen der Chemie*, **105** (1858), 187–202.
4. J. B. A. Dumas, "Mémoire sur les équivalents des corps simples," in *Comptes rendus . . . de l'Académie des sciences*, **45** (1857), 709–731; **46** (1858), 951–953; **47** (1858), 1026–1034.
5. D. Mendeleev, "Sootnoshenie svoistv s atomnym vesom elementov" ("The Correlation Between the Properties and the Atomic Weights of the Elements"), in *Zhurnal Russkago fiziko-khimicheskago obshchestva pri Imperatorskago St-Peterburgskago universitete*, **1** (1869), 60–77; *Zeitschrift für Chemie*, **12** (1869), 405–406.
6. K. Seubert, "Zur Geschichte des periodischen Systems," in *Zeitschrift für anorganische . . . Chemie*, **9** (1895), 334–338.
7. L. Meyer, "Die Natur der chemischen Elemente als Function ihrer Atomgewichte," in *Justus Liebigs Annalen der Chemie*, supp. **7** (1870), 354–364.
8. F. A. Kekulé, "Ueber einige Condensationsproducte des Aldehyds," in *Annalen der Chemie*, **162** (1872), 77–124.

BIBLIOGRAPHY

I. Original Works. Meyer's publications and those of students under his direction are listed in the extensive obituaries by K. Seubert, in *Berichte der Deutschen chemischen Gesellschaft*, **28R** (1895), 1109–1146; and P. P. Bedson, in *Journal of the Chemical Society*, **69** (1896), 1402–1439, repr. in *Memorial Lectures Delivered Before the Chemical Society, 1893–1900* (London, 1901). Bedson's bibliography is copied from Seubert's. Unfortunately the listing of doctoral publications is given under the year of the dissertation and not the year of publication of the journal article. The volume of each journal article is, however, given. The bibliography includes, in addition to Meyer's technical articles, a number of his obituaries and more general lectures and papers, particularly on education and the nature of the university.

Meyer's major work is *Die modernen Theorien der Chemie und ihre Bedeutung für die chemische Statik* (Breslau, 1864, 1872, 1876, 1883, 1884). The 1st ed. was translated into Russian as *Novieishie teorii khimii ikh znacherie dlya khimicheskoy statiki* (St. Petersburg, 1866). The 5th ed. was translated into English by P. Phillips Bedson and W. Carleton Williams, *Modern Theories of Chemistry* (London, 1888), and French by A. Bloch and J. Meunier, *Les théories modernes de la chimie et leur application à la mécanique chimique*, 2 vols. (Paris, 1887–1889). His less technical account of the same subject appeared as *Grundzüge der theoretischen Chemie* (Leipzig, 1890; 2nd ed., 1893); there is an English trans. by P. Phillips Bedson and W. C. Williams, *Outlines of Theoretical Chemistry* (London, 1892).

Lothar Meyer and Karl Seubert published *Die Atomgewichte der Elemente aus den Originalzahlen neu berechnet* (Leipzig, 1883).

Meyer edited two works in Ostwald's Klassiker der Exacten Wissenschaften: no. 30, *Abriss eines Lehrganges der theoretischen Chemie, vorgetragen von Prof. S. Cannizzaro* (Leipzig, 1891); and no. 66, *Die Anfänge des natürlichen Systemes der chemischen Elemente. Abhandlungen von J. W. Döbereiner und Max Pettenkofer* (Leipzig, 1895), which contains a historical survey by Meyer of the further development of the doctrine of the triads of the elements. Meyer's major contributions on the periodic law (1864, 1870) were published with those of Mendeleev in Ostwald's Klassiker, no. 68, edited with commentary by Karl Seubert: *Das natürliche System der chemischen Elemente, Abhandlungen von L. Meyer und D. Mendeleeff* (Leipzig, 1895).

The paper that established Meyer as codiscoverer of the periodic law is "Die Natur der chemischen Elemente als Function ihrer Atomgewichte," in *Justus Liebigs Annalen der Chemie*, supp. **7** (1870), 354–364.

II. Secondary Literature. In addition to the major obituaries by Seubert and Bedson (see above), there is P. Walden, "Meyer, Mendelejeff, Ramsay, und das periodische System der Elemente," in G. Bugge, ed., *Das Buch der grossen Chemiker*, II (Berlin, 1930), 229–287, with further bibliographic sources on p. 508. Brief biographical sketches were written by J. H. Long, in *Journal of the American Chemical Society*, **17** (1895), 664–666; and R. Winderlich, in *Journal of Chemical Education*, **27** (1950), 365–368.

Meyer's work and its context are discussed in some detail in P. Venable, *The Development of the Periodic Law* (Easton, Pa., 1896), 96–108; and in Ida Freund, *The Study of Chemical Composition* (Cambridge, 1904; repr. New York, 1968), esp. ch. 16.

The question of Mendeleev's priority in enunciating the periodic law was discussed under the title "Zur Geschichte der periodischen Atomistik" by L. Meyer, in *Berichte der Deutschen chemischen Gesellschaft*, **13** (1880), 259–265, 2043–2044; by D. Mendeleev, *ibid.*, 1796–1804; and K. Seubert, in *Zeitschrift für anorganische . . . Chemie*, **9** (1895), 334–338. See also J. W. van Spronsen, "The Priority Conflict Between Mendeleev and Meyer," in *Journal of Chemical Education*, **46** (1969), 136–139; J. W. van Spronsen, *The Periodic System of Chemical Elements: A History of the First Hundred Years* (Amsterdam–London–New York, 1969), 124–132; and H. Cassebaum and G. B. Kauffman, "The Periodic System of the Chemical Elements: The Search for Its Discoverer," in *Isis*, **62** (1971), 314–327.

Otto Theodor Benfey

MEYER, KURT HEINRICH (*b.* Dorpat, Russia, 29 September 1883; *d.* Menton, France, 14 April 1952), *organic chemistry.*

Meyer's father, Hans Horst Meyer, was a pharmacologist who also taught at the German-speaking University of Dorpat. Meyer was educated mainly in Germany and studied chemistry at the universities of Marburg, Freiburg, and Leipzig; among his teachers were Hantzsch and Ostwald. After receiving the doctorate in 1907, he traveled for a year to America, to Rutherford's department of physics at Manchester, and to Ramsay's department of chemistry at London; he then settled in Munich, where the school of organic chemistry was led by Adolf von Baeyer. There Meyer carried out the studies on keto-enol tautomerism that first made his reputation, including the determination by a simple titration of the amount of enol in samples of ethyl acetoacetate. He also discovered some new coupling reactions of diazonium salts and worked on a possible industrial synthesis of formamide from carbon monoxide and ammonia and, with F. Bergius, on the large-scale hydrolysis of chlorobenzene to phenol.

Meyer spent three years on war service as an artillery officer but was recalled in 1917 to work with Haber on chemical warfare. When peace came, he returned to Munich, where organic chemistry was under Willstätter's direction, and returned to his previous interests. Notably, with H. Hopff he isolated the pure enol form of ethyl acetoacetate by "aseptic distillation," avoiding the presence of any impurities that might catalyze the conversion to the keto form.

In 1921 Meyer left academic life to become director of the headquarters laboratories of the firm of Badische Anilin- und Sodafabrik (BASF) at Ludwigshafen. He organized a large and active research group whose interests, although wide, were concerned mainly with dyeing and dyestuffs. His own interests became increasingly centered on the chemistry of natural high polymers, a study to which he recruited the young physical chemist Herman Francis Mark. Their results, published in 1930, were a landmark in the development of the subject. In a lucid survey of naturally occurring organic polymers (cellulose, starch, proteins, rubber) the authors, although giving due weight to the then fashionable theory of "small building blocks," found themselves more in sympathy with the concept of true macromolecules, which was being vigorously promoted by Staudinger.

Especially after the incorporation of BASF into the huge I. G. Farbenindustrie complex in 1926, Meyer found that his research work was increasingly hindered by the cares of administration. The political situation in Germany also caused him justifiable anxiety, and in 1932 he left the country to take the chair of inorganic and organic chemistry at the University of Geneva. Although he had to accustom himself to lecturing in French, a language with which he was unfamiliar, he was a successful teacher and again built up a fine research school. With his collaborators (notably A. J. A. van der Wyk) Meyer continued his studies of cellulose and chitin, the permeability of synthetic membranes, and the thermodynamics of large molecules in solution; and developed a theory of muscle contraction by analogy with the contraction of rubber. Investigation of the structure of amylopectin, the branched-chain constituent of starch, led to extensive work on the crystallization, characterization, and specificity of enzymes, a subject that occupied his last years. Meyer died suddenly while on holiday.

BIBLIOGRAPHY

I. ORIGINAL WORKS. Meyer's main work is *Der Aufbau der hochpolymeren organischen Naturstoffe* (Leipzig, 1930), written with H. F. Mark. The 2nd ed. (Leipzig, 1940) was in 2 vols., one by Meyer, dealing with the chemical aspects of the subject, and the other by Mark, concerned with physics. Meyer's volume was trans. into English by L. E. R. Picken and published as *Natural and Synthetic High Polymers* (New York, 1942; 2nd ed., 1950). Meyer wrote many scientific papers, usually with collaborators. A complete list of works is in Poggendorff, V, 843–844, VI, 1717, and VIIa, 3, 283–285.

II. SECONDARY LITERATURE. There is a memorial article by R. Jeanloz in *Advances in Carbohydrate Chemistry*, **11** (1956), xiii–xviii, with portrait. The more important obituary notices include H. Mark, in *Angewandte Chemie*, **64** (1952), 521–523; L. E. R. Picken, in *Nature*, **169** (1952), 820; and A. J. A. van der Wyk, in *Helvetica chimica acta*, **35** (1952), 1418–1422.

W. V. FARRAR

MEYER, VICTOR (*b.* Berlin, Germany, 8 September 1848; *d.* Heidelberg, Germany, 8 August 1897), *chemistry.*

Victor Meyer was the second son of Jacques and Bertha Meyer. The elder Meyer, a prosperous Jewish merchant in calico printing and dyeing, wanted his sons to become chemists, but Victor's foremost desire was to be an actor. Hoping that his interests would change, the family persuaded him to attend some lectures at Heidelberg, where his brother Richard was a student. He then enrolled at Berlin and transferred to Heidelberg in 1865. He was suddenly converted to chemistry on encountering the renowned Bunsen. His dramatic gifts were to be employed as a striking and effective teacher.

At Heidelberg, Meyer studied under Bunsen, Erlenmeyer, Kopp, Kirchhoff, and Helmholtz. He headed the lists in all his courses and progressed so rapidly that he was awarded the Ph.D., *summa cum laude*, at the age of eighteen.

Bunsen was so impressed with Meyer that he immediately selected him to be his assistant. Meyer worked with Bunsen for one year, performing analyses of the mineral waters of Baden for the government. Bunsen then recommended him to Baeyer, whose laboratory at the Gewerbeakademie in Berlin was one of the most famous in Europe. Meyer spent three years with Baeyer (1868–1871), beginning his publications in organic chemistry during this period. His first professorship was at the Stuttgart Polytechnic in 1871. At the age of twenty-four he became a full professor and director of the chemical laboratories at the Eidgenössische Technische Hochschule at Zurich. Meyer celebrated this appointment with his engagement to Hedwig Davidson. They were married in 1873 and had five daughters.

Meyer had rejected offers from several institutions during his thirteen years in Zurich, when in 1885 the University of Göttingen asked him to take charge of the construction of its new chemical laboratories. No sooner were the laboratories completed in 1888 than Meyer received the offer of Bunsen's chair at Heidelberg. He felt obliged to remain at Göttingen and declined the offer. It was only Bunsen himself, intervening with the Prussian ministry for the release of Meyer from Göttingen, who enabled Meyer to become his successor at Heidelberg, a position Meyer held until his death in 1897.

An extremely attractive personality, Meyer was also a brilliant lecturer and attracted many students from both Europe and North America. He was a member of the academies of Berlin, Uppsala, and Göttingen, and president of the German Chemical Society.

Meyer's health declined during the 1880's. He suffered several breakdowns and was so frequently ill that he resorted to drugs in order to sleep. He became conscious that his long suffering was affecting his thinking and suddenly in the summer of 1897, suffering from continuous neuralgic pains, his nervous system shattered, he ended his life by taking prussic acid.

There is an unusually large number of important contributions in the more than 300 papers that Meyer published. His first significant paper appeared in 1870, when he described a new method for introducing the carboxyl group into an aromatic substance by heating the potassium salts of aromatic sulfonic acids with sodium formate. This method has been used ever since by organic chemists for the synthesis of aromatic acids. Meyer's primary purpose in this paper was to ascertain the constitution of benzene derivatives. The determination of the ring position of substituents in isomeric aromatic compounds was an unsettled problem, and Meyer established that salicylic acid and other compounds which had been assigned to the *meta* series were *ortho* derivatives. In a paper one year later Meyer listed many disubstituted benzoic acids in columns according to whether they were *ortho*, *meta*, or *para* compounds.

Meyer first attracted wide attention in 1872 with his work on the nitroparaffins. Aromatic nitro compounds were well known and easily prepared from aromatic hydrocarbons, but aliphatic nitro compounds existed only in scattered examples, obtained mostly by accident. Meyer proposed the existence of two series of isomeric organic nitrogen compounds: the nitrite and nitro compounds. During his year at Stuttgart, he discovered a general method for the preparation of nitroparaffins and made the subject his main research problem for his first four years at Zurich. Meyer found that alkyl iodides combined with silver nitrite to form true nitro compounds, the nitrogen atom being bound directly to a carbon atom, whereas the isomeric nitrites were esters. He explored this area so thoroughly that at the time of his death almost all that was known about nitro compounds was due to Meyer and his students.

Meyer disclosed the existence of two new classes of organic nitrogen substances from the surprising reactions of nitrous acid with the primary and secondary nitroparaffins. Primary nitro compounds formed acidic products which dissolved in alkali to form red salts. Secondary nitro compounds formed blue nonacidic products which did not dissolve in alkali. Meyer named these products "nitrolic acids" and "pseudonitroles" respectively. Since tertiary nitroparaffins did not react with nitrous acid, these color reactions served as a test to differentiate between primary, secondary, and tertiary radicals. Meyer and his students established the structural formulas for the nitrolic acids and pseudonitroles.

Meyer then explored a variety of organic nitrogen compounds and discovered several new types. His most important compound was his preparation of the first oximes by means of the reaction between hydroxylamine and an aldehyde or ketone. He proved that this was a general reaction with the carbonyl group, and he established the structure of oximes.

Victor Meyer's name is most closely associated with his vapor density method. Devised in three stages from 1876, the method was a product of his researches in organic chemistry, since it was necessary for him to determine the molecular formulas of the substances with which he was working. There were several methods available for determining the density of the vapors of liquids or solids, each having particular advantages and disadvantages. Meyer wanted a method that (1) utilized small amounts of a substance (he was working with new substances usually obtainable only in small quantities) and that (2) could be used at high temperatures (his substances often possessed high boiling points).

In 1876 he measured for the first time the vapor density of diphenyl, anthracene, anthraquinone, triphenylamine, *p*-dibromobenzene, and *p*-diphenylbenzene by volatilizing them at the temperature of boiling sulfur. The following year, in order to make his method more flexible, he used the vapors of a variety of liquids, depending on the temperature required, instead of sulfur. Finally, in 1878 he presented his third and best-known modification. The vapor of a weighed substance displaced an equal volume of air, which in turn was measured by means of a burette. This method is more commonly used than any other, and Meyer's apparatus is found in most chemical laboratories.

These vapor density studies led to his endless series of pyrochemical researches, which he investigated until his last days at Heidelberg. Meyer hoped to get vapor density estimations at ever higher temperatures. He employed molten lead baths and platinum, platinum-iridium, and porcelain bulbs, which enabled him to study vapors at temperatures up to 3000° C. Very little was known about the molecular constitution of vapors at high temperatures. Meyer's method made possible the determination of the molecular state of many elements and inorganic substances. In 1879 he showed that the halogens dissociated at high temperatures. His pyrochemical investigations included vapor density determinations, the effect of temperature on the dissociation of substances, and the study of the ignition temperatures of explosive gas mixtures. In 1885 he published *Pyrochemische Untersuchungen*, a monograph on the subject.

A new area for investigation came about through a lecture demonstration that failed. In 1882 Meyer gave a series of lectures on benzene and its derivatives. His lectures were brilliant as usual, and the experiments performed before the class were well prepared. At one of these lectures Meyer wanted to show the students Baeyer's indophenine test for benzene, in which the addition of isatin and sulfuric acid to benzene produces the deep blue indophenine. The results were negative. Sandmeyer, Meyer's assistant, reminded him afterward that the benzene sample that had been used

was not commercial benzene from coal tar but synthetic benzene prepared by the decarboxylation of benzoic acid.

Meyer's investigation of the indophenine reaction began the same day. He found that the purest samples of benzene from coal tar invariably gave the blue color reaction, but the color could be eliminated by first extracting the benzene with sulfuric acid. The sulfonated product on distillation gave Meyer an active "benzene," which again showed the indophenine reaction. Meyer proposed several hypotheses, one of which was that coal-tar benzene was a mixture of two substances with similar properties and that only one of these substances combined with isatin. In 1883 he isolated this substance and named it "thiophene" because it contains sulfur and is similar to phenyl compounds. He then rapidly developed the subject and was able in five years to publish a 300-page monograph, *Die Thiophengruppe* (1888), which contains a list of 106 papers by Meyer and his students. His main interest was in demonstrating the similarity between the chemistry of thiophene compounds and the chemistry of benzene. By 1885 he proved that thiophene has a ring structure and suggested that pyrrole and furan were analogous ring compounds.

Meyer contributed many papers on the negative nature of radicals, a topic which had interested him ever since he detected the acidic properties of the nitroparaffins. He could replace hydrogen in a nitroparaffin by an alkali metal and thereby form a salt. He explained that the acidic character is due to the influence of the nitro group on the hydrogen atoms bound to the same carbon atom. He noted that acidity can be induced in an inert hydrocarbon by the introduction of certain substituents, such as nitro, cyano, or phenyl radicals. In 1887 Meyer defined those groups which possessed acid-inducing properties as "negative," later termed "electrophilic" by Ingold in the context of the electronic theory of valence.

Continuing research on the oximes led Meyer into the realm of stereochemistry and the spatial effects of radicals. Meyer and his students noted that benzil forms more than one dioxime. In 1888 he and Karl von Auwers explained that the isomerism is due to lack of free rotation about the carbon-carbon single bond, an explanation at variance with van't Hoff's assumption of free rotation about such bonds. In so doing, they proposed the term "stereochemistry" in place of van't Hoff's "chemistry in space" as a more suitable name for phenomena involving spatial effects. Their explanation never appealed to chemists, and Hantzsch and Werner in 1890 presented an explanation based on the stereochemistry of nitrogen. This explanation proved to be more satisfactory, although

Meyer was critical to the end of his life of their spatial formulas.

Meyer's interest in spatial aspects of organic reactions continued during the 1890's, and in 1894 he identified the inhibiting effect known as "steric hindrance." Benzoic acid and most of its substitution products readily formed esters with alcohol at room temperature, but Meyer observed that trisubstituted benzoic acids do not form esters unless the carboxyl group is extended well beyond the ring by the interposition of a chain of carbon atoms. Further study showed that *meta* and *para* derivatives of benzoic acid esterify almost completely while their isomeric *diortho* compounds yield little or no ester. Meyer explained this as a spatial effect, the *ortho* substituent exerting a blocking action, which suppresses the esterification.

Concurrently with his stereochemical researches, Meyer published many papers exploring new types of aromatic iodine compounds. He revealed that iodine can exist in higher oxidation states in aromatic compounds. He first prepared an "iodoso" compound in 1892 by the oxidation of *o*-iodobenzoic acid; further oxidation produced an "iodoxy" compound. What was remarkable to him was his discovery in 1894 of a class of free organic bases, the "iodonium" compounds. He obtained the first member of this class by the interaction of iodoso- and iodoxyhydrocarbons. Iodosobenzene and iodoxybenzene yielded diphenyliodonium hydroxide:

$$C_6H_5—IO + C_6H_5—IO_2 + AgOH$$
$$\rightarrow (C_6H_5)_2I—OH + AgIO_3$$

Iodonium hydroxides are strong bases resembling the quaternary ammonium and ternary sulfonium bases.

Victor Meyer's concern for excellence in teaching found expression in a project with which he was occupied at the time of his death. With Paul Jacobson, his assistant at Heidelberg, he wrote a comprehensive treatise on organic chemistry, the *Lehrbuch der organischen Chemie*. This two-volume work, the second volume of which was incomplete when he died, remains the best extended treatment of the subject. The book was meant to be fresh and comprehensive. It included the most recent developments in theory, the authors being the first to use stereochemistry as a background for the subject. Written in an attractive style, it remains a rich source of information about both the principles of organic chemistry and of the chemistry of the classes of organic compounds and their individual members.

BIBLIOGRAPHY

I. ORIGINAL WORKS. Meyer wrote three major chemical treatises: *Pyrochemische Untersuchungen* (Brunswick,

1885), written with Carl Langer; *Die Thiophengruppe* (Brunswick, 1888); and *Lehrbuch der organischen Chemie*, 2 vols. (Leipzig, 1893–1903), written with Paul Jacobson.

Among his important papers are "Untersuchungen über die Constitution der zweifach-substituirten Benzole," in *Justus Liebigs Annalen der Chemie*, 156 (1870), 265–301, and 159 (1871), 1–27; "Über die Nitroverbindungen der Fettreihe," in *Berichte der Deutschen chemischen Gesellschaft*, 5 (1872), 399–406, 514–518, written with O. Stüber; "Über die Nitroverbindungen der Fettreihe," in *Justus Liebigs Annalen der Chemie*, 171 (1874), 1–56, and 175 (1875), 88–140; "Über die Pseudonitrole, die Isomeren der Nitrolsäuren," *ibid.*, 180 (1876), 133–155, written with J. Locher; "Zur Dampfdichtebestimmung," in *Berichte der Deutschen chemischen Gesellschaft*, 11 (1878), 1867–1870; "Über das Verhalten des Chlors bei höher Temperatur," *ibid.*, 12 (1879), 1426–1431, written with Carl Meyer; "Über stickstoffhaltige Acetonderivate," *ibid.*, 15 (1882), 1164–1167, written with Alois Janny; "Untersuchungen über die Strukturformel des Thiophens," *ibid.*, 18 (1885), 3005–3012, written with L. Gattermann and A. Kaiser; "Untersuchungen über die zweite van't Hoffsche Hypothese," *ibid.*, 21 (1888), 784–817, and "Über die isomeren Oxime unsymmetrischer Ketone und die Configuration der Hydroxylamins," *ibid.*, 23 (1890), 2403–2409, written with Karl Auwers; "Über Jodosobenzoësäure," *ibid.*, 25 (1892), 2632–2635, written with Wilhelm Wachter; "Über ein seltsames Gesetz bei der Esterbildung aromatischer Säuren," *ibid.*, 27 (1894), 510–512; "Das Gesetz der Esterbildung aromatischer Säuren," *ibid.*, 1580–1592, and "Weiteres über die Esterbildung aromatischer Säuren," *ibid.*, 3146–3156, written with J. Sudborough.

II. SECONDARY LITERATURE. The principal source on the life and work of Victor Meyer was composed by his brother Richard Meyer: *Victor Meyer. Leben und Wirken eines deutschen Chemikers und Naturforschers 1848–1897* (Leipzig, 1917).

Other important studies are Heinrich Biltz, in *Zeitschrift für anorganische Chemie*, 16 (1898), 1–14; Margaret Davis Cameron, "Victor Meyer and the Thiophene Compounds," in *Journal of Chemical Education*, 26 (1949), 521–524; Friedrich Challenger, "Victor Meyer's and Paul Jacobson's 'Lehrbuch der organischen Chemie': the Authors and Their Work," in *Journal of the Royal Institute of Chemistry*, 82 (1958), 164–169; Benjamin Harrow, *Eminent Chemists of Our Time*, 2nd ed. (New York, 1927), 177–195, 407–422; F. Henrich, in G. Bugge, ed., *Das Buch der grossen Chemiker*, II (Berlin, 1930), 374–390; B. Horowitz, in *Journal of the Franklin Institute*, 182 (1916), 363–394; Paul Jacobson, in *Allgemeine Deutsche Biographie*, LV (Leipzig, 1910), 833–841; and C. Liebermann, in *Berichte der Deutschen chemischen Gesellschaft*, 30 (1897), 2157–2168.

See also G. Lunge, in *Vierteljahrsschrift der Naturforschenden Gesellschaft in Zürich*, 42 (1897), 347–361; J. McCrae, "Recollections of Heidelberg and Victor Meyer: 1893–1895," in *Journal of the Royal Institute of Chemistry*, 82 (1958), 77–82; Richard Meyer, in *Berichte der Deutschen chemischen Gesellschaft*, 41 (1909), 4505–

4718; Gustav Schmidt, "The Discovery of the Nitroparaffins by Victor Meyer," in *Journal of Chemical Education*, **27** (1950), 557–559; J. Sudborough, "Victor Meyer," in *Proceedings of the Chemical Society* (1959), 137–141; and Edward Thorpe, *Essays in Historical Chemistry*, 3rd ed. (London, 1911), 422–482, which originally appeared as the "Victor Meyer Memorial Lecture," in *Journal of the Chemical Society*, **77** (1900), 169–206.

ALBERT B. COSTA

MEYER, WILHELM FRANZ (*b*. Magdeburg, Germany, 2 September 1856; *d*. Königsberg, Germany [now Kaliningrad, U.S.S.R.], 11 June 1934), *mathematics*.

Meyer studied in Leipzig and Munich, where he received his doctorate in 1878. He studied further in Berlin, where at that time Weierstrass, Kummer, and Kronecker were active. In 1880 he qualified for lecturing at the University of Tübingen, and in 1888 he became full professor at the Bergakademie of Clausthal–Zellerfeld. From October 1897 until October 1924, when he retired, he taught at the University of Königsberg.

Meyer was a many-sided and very knowledgeable mathematician, whose list of writings includes 136 titles. His principal field of interest, however, was geometry, especially algebraic geometry and the related projective invariant theory. His *Habilitationsschrift*, which was published in 1883 as *Apolarität und rationale Kurven*, shows this direction of his research. In this work he extended the apolarity theory, created by Reye, to a multidimensional projective geometry based on the theory of rational curves. At the time such considerations were not completely obvious.

Other of Meyer's works from this period deal with algebraic curves and their production, and with related algebraic questions. He early showed himself to be one of the leading experts on invariant theory. In 1892 he composed for the Deutsche Mathematiker-vereinigung a long report on this subject, which was translated into French, Italian, and Polish. In this work he presented the development of invariant theory from its beginning in the middle of the nineteenth century to the end of the century and the appearance of the decisive finiteness theorems of Gordan and Hilbert. Meyer also made many individual contributions to invariant theory. This area of research went somewhat out of fashion during his lifetime, however, chiefly as a result of Hilbert's work.

Meyer was one of the founders of the *Encyklopädie der mathematischen Wissenschaften*. He, H. Weber, and F. Klein were responsible for planning this project.

The *Encyklopädie*, which was conceived on a large scale, was supported from 1895 by a syndicate of German academies. From the turn of the century until the 1930's some twenty volumes appeared; they treated all fields of mathematics and their applications. Meyer wrote the articles on potential theory (with H. Burkhardt), invariant theory, the new geometry of the triangle (with G. Berkhan), third-order surfaces, and surfaces of the fourth and higher orders.

The editing of such a vast work required great effort and presupposed considerable knowledge. In this regard Meyer benefited from his extensive familiarity with the literature, gained in large measure through the 2,000 reviews that he wrote for *Fortschritte der Mathematik*; his knowledge of foreign languages was also very useful to him. Of special note are the articles on third- and fourth-degree surfaces, which he composed at an advanced age. At that period, around 1930, Meyer was the only German mathematician who still possessed a comprehensive view of the abundant material, produced mainly in the nineteenth century, on special algebraic curves and surfaces. Meyer conducted investigations in geometry of the triangle, handled in the spirit of Klein's Erlangen program, and gave lectures discussing the essential aspects of mathematical research in the spirit of the time and emphasizing the importance of simple algebraic identities, the symmetries of group theory, and transformation principles as a source of geometric theorems.

Meyer was an excellent teacher who had many students. Most East Prussian mathematics teachers at the beginning of the twentieth century were trained by him.

BIBLIOGRAPHY

An extensive listing of Meyer's writings can be found in Poggendorff, IV, 1001–1002; V, 841; and VI, 1714. They include *Apolarität und rationale Kurven, eine systematische Voruntersuchung zu einer allgemeinen Theorie der linearen Räume* (Tübingen, 1883); "Bericht über den gegenwärtigen Stand der Invariantentheorie," in *Jahresberichte der Deutschen Mathematiker-vereinigung*, **1** (1892), 79–292; and the following articles in *Encyklopädie der mathematischen Wissenschaften:* "Invariantentheorie," I, pt. 1, 320–403; "Potentialtheorie," II-A, pt.7-b, 464–503, written with H. Burkhardt; "Neuere Dreiecksgeometrie," III, pt.1-b, 1173–1276, written with G. Berkhan; "Flächen 3. Ordnung," III-C, pt. 10-a, 1437–1532; and "Flächen 4. und höherer Ordnung," III-C, 1533–1779.

An article on Meyer is B. Arndt, "W. F. Meyer zum Gedächtnis," in *Jahresberichte der Deutschen Mathematiker-vereinigung*, **45** (1935), 99–113.

W. BURAU

MEYERHOF, OTTO (*b.* Hannover, Germany, 12 April 1884; *d.* Philadelphia, Pennsylvania, 6 October 1951), *biochemistry.*

The son of Felix Meyerhof, merchant, and Bettina May, Meyerhof received the M.D. from the University of Heidelberg in 1909. While at the medical clinic of Ludolf von Krehl, he came under the influence of Otto Warburg, who turned Meyerhof's interest from psychology and philosophy to cellular physiology. He worked at the Institute of Physiology at the University of Kiel from 1913 to 1924 and then at the Kaiser Wilhelm Institute for Biology in Berlin–Dahlem until 1929, when he became head of the department of physiology at the Kaiser Wilhelm Institute for Medical Research in Heidelberg. Like other scientists of Jewish extraction, Meyerhof left Germany (in 1938) after the Nazi rise to power; he joined the Institute of Physicochemical Biology in Paris but was forced to flee in 1940, when the Germans invaded France. He came to the United States late in 1940 and was research professor of physiological chemistry at the School of Medicine of the University of Pennsylvania until his death.

Meyerhof's work on the chemical processes in muscle laid the basis for the elucidation of the chemical pathway in the intracellular breakdown of glucose to provide energy for biological processes. In 1919 he demonstrated that during muscle contraction in the absence of oxygen, muscle glycogen is converted to lactic acid. In the presence of oxygen, about one-fifth of the lactic acid is oxidized to carbon dioxide and water; and the energy yielded by this oxidation is used to regenerate glycogen from the remaining lactic acid. This discovery provided a chemical basis for the interpretation of the heat changes during muscle contraction and subsequent recovery, studied by A. V. Hill in 1913–1914. Hill and Meyerhof shared the 1922 Nobel Prize in physiology or medicine, awarded in 1923.

Meyerhof's choice of muscle as the experimental material was prompted by his philosophical commitment to the idea that the dynamics of biological processes can be described in the language of chemistry and physics. From his early article "Zur Energetik der Zellvorgänge" (1913) to his last writings, this affirmation of the antivitalist position is evident; outside his scientific work he retained an attachment to the transcendental idealism of Kant and J. F. Fries.

In 1925 Meyerhof succeeded in extracting from muscle the group of enzymes responsible for the conversion of glycogen to lactic acid. This preparation of a cell-free glycolytic system was a counterpart of the earlier successful extraction from yeast of the enzyme system (zymase) that converts glucose to alcohol and carbon dioxide during fermentation (Buchner, 1897). In 1917–1918 Meyerhof had shown the presence in animal tissues of the "cozymase" of yeast fermentation, discovered by Harden and W. J. Young in 1906. After 1925 the study of the chemical pathway in the breakdown of glucose by muscle and by yeast was found to be very similar, and this pathway was shown to be operative in many other biological systems. The development pioneered by Meyerhof thus provided striking evidence of the unity of biochemical processes amid the manifold diversity of the forms of life.

The discovery of phosphocreatine (in 1926) and of adenosine triphosphate (in 1929) as constituents of mammalian muscle was followed by Einar Lundsgaard's demonstration in 1930 that muscle contraction depends more directly on the enzymic cleavage of these two substances than on the production of lactic acid. These advances provided essential links in Meyerhof's later analysis of the energy relations between chemical change and the physical events in muscle contraction, and his studies led to the recognition of the central role in muscle contraction (and in other energy-requiring processes of biological systems) of adenosine triphosphate, the resynthesis of which in muscle is driven by the breakdown of phosphocreatine and the production of lactic acid. His measurements of the heat released in the hydrolysis of adenosine triphosphate and related compounds, although subsequently refined, permitted the first quantitative estimates to be made of the efficiency of muscle operating as a chemical machine.

Meyerhof's influence on the development of biochemistry was profound and continued past the middle of the twentieth century through the work of his former students; noteworthy among them were Fritz Lipmann and Severo Ochoa, both Nobel laureates.

BIBLIOGRAPHY

I. ORIGINAL WORKS. Meyerhof's books include *Chemical Dynamics of Life Phenomena* (Philadelphia, 1925); *Die chemischen Vorgänge im Muskel und ihr Zusammenhang mit Arbeitsleistung und Wärmebildung* (Berlin, 1930); and *Chimie de la contraction musculaire* (Paris, 1933). He published some 400 scientific articles; a list of his publications is given in *Biographical Memoirs. National Academy of Sciences,* **34** (1960), 164–182.

II. SECONDARY LITERATURE. See D. Nachmansohn, S. Ochoa, and F. A. Lipmann, in *Science,* **115** (1952), 365–369; and C. L. Gemmill, *Medical College of Virginia Quarterly,* **2** (1966), 141–142.

JOSEPH S. FRUTON

MEYERSON, ÉMILE (*b*. Lyublin, Russia [now Lublin, Poland], 1859; *d*. Paris, France, 1933), *philosophy.*

For a complete study of his life and work, see Supplement.

MICHAEL, ARTHUR (*b*. Buffalo, New York, 7 August 1853; *d*. Orlando, Florida, 8 February 1942), *chemistry.*

Michael studied chemistry under Hofmann at Berlin (1871, 1875–1878), under Bunsen at Heidelberg (1872–1874), and under Wurtz at Paris (1879). He was professor of chemistry at Tufts College (1881–1889, 1894–1907) and Harvard University (1912–1936). A severe critic of mechanical interpretations of chemical phenomena, he introduced thermodynamic conceptions into organic chemical theory.

Michael's earliest studies included the discovery of several synthetic reactions. He was the first to synthesize a natural glucoside (helicin, 1879), and the method that he introduced became the standard synthetic route to this class of organic substances. His best-known synthetic method is the direct addition of the sodium derivatives of malonic, acetoacetic, or cyanoacetic esters to α, β-unsaturated esters, ketones, nitriles, amides, and sulfones (the Michael reaction, 1887). There had been no general method available for the conversion of unsaturated compounds into saturated compounds of a higher carbon series until Michael described his method of additive condensation in his article "Über die Addition von Natriumacetessig- und Natriummalonsäureäthern zu den Aethern ungesättigter Säuren."

Michael's primary concern was organic theory. From 1888 he developed a novel theory of organic reactions based on the thermodynamic concepts of free energy and entropy. He maintained that organic structural theory was too qualitative, pictorial, and mechanical; and he hoped to overcome these deficiencies with energetic conceptions. Michael made the second law of thermodynamics the fundamental principle governing organic reactions. He related the course of reactions to energy conversions, including addition and substitution reactions, molecular rearrangements, tautomerism, and stereochemistry within his theory. He concentrated much of his research on the theoretical aspects of addition to the double bond and the behavior of active methylene compounds.

In his role as critic of accepted views, Michael refused to accept Wislicenus' assumption that addition to unsaturated compounds always proceeded in the *cis* manner and that elimination reactions occurred more easily with *cis* isomers than with *trans* isomers. By carefully planned experiments he proved that *trans* additions and eliminations did occur and that all of the then accepted configurations of geometric isomers were erroneous (1895–1918).

Michael was a critic of all purely mechanical interpretations of organic reactions, such as steric hindrance and the strain hypothesis. He attempted over many years to show experimentally that these conceptions were extremely limited and inadequate as explanations of chemical phenomena and that they needed to be modified by considerations of chemical affinity and energy.

BIBLIOGRAPHY

I. ORIGINAL WORKS. Important papers by Michael include "On the Synthesis of Helicin and Phenolglucoside," in *American Chemical Journal*, **1** (1879), 305–312: "Über die Addition von Natriumacetessig- und Natriummalonsäureäthern zu den Aethern ungesättigter Säuren," in *Journal für praktische Chemie*, **35** (1887), 349–356; **43** (1891), 390–395; **45** (1892), 55–63; and **49** (1894), 20–25; "Untersuchungen über Alloisomerie," *ibid.*, **52** (1895), 289–325; "Über die Gesetze der Alloisomerie und Anwendung derselben zur Classificirung ungesättigter organischer Verbindungen," *ibid.*, 344–372; "Über einige Gesetze und deren Anwendung in der organischen Chemie," *ibid.*, **60** (1899), 286–384, 409–486; "Valenzhypothesen und der Verlauf chemischer Vorgänge," *ibid.*, **68** (1903), 487–520; "Stereoisomerism and the Law of Entropy," in *American Chemical Journal*, **39** (1908), 1–16; "Outline of a Theory of Organic Chemistry Founded on the Law of Entropy," in *Journal of the American Chemical Society*, **32** (1910), 990–1007; and "The Configurations of Organic Compounds and Their Relation to Chemical and Physical Properties," *ibid.*, **40** (1918), 704–723, 1674–1707.

II. SECONDARY LITERATURE. Brief accounts of Michael's life and work are Louis F. Fieser, Edward W. Forbes, and Arthur B. Lamb, in *Harvard University Gazette* (22 May 1943), 246–248; *Dictionary of American Biography*, supp. 3 (New York, 1973), 520–521; *National Cyclopaedia of American Biography*, XV (New York, 1916), 172; and W. T. Read, in *Industrial and Engineering Chemistry*, **22** (1930), 1137–1138. His thermodynamic conceptions are discussed by Ferdinand Henrich, in *Theories of Organic Chemistry*, T. B. Johnson and D. A. Hahn, trans. (New York, 1922), 569–584, and Albert B. Costa, "Arthur Michael (1853–1942): The Meeting of Thermodynamics and Organic Chemistry," in *Journal of Chemical Education*, **48** (1971), 243–246.

ALBERT B. COSTA

MICHAEL PSELLUS. See **Psellus, Michael.**

MICHAEL SCOT (*b.* before 1200; *d. ca.* 1235), *astrology, popularization of science, translation of scientific and philosophical works from the Arabic.*

Almost all information about Michael's life and work is uncertain; his posthumous fame as a wise or wicked magician bred legends and was increased by them. Although imaginative scholars have established undocumented traditions, no satisfactory analysis—linguistic, stylistic, or doctrinal—of writings ascribed to him has been carried out. It is thus impossible to determine the accuracy of many attributions.

Life. Michael's place of birth and details of his family are unknown. His appointment to an Irish archbishopric suggests that "Scot" meant, as it often did, "Irish"; his refusal on the ground of ignorance of the vernacular does not support this suggestion but does not exclude it. He was given benefices in England and Scotland with no explicit residence requirements; there is no sign that he lived in either place at any time. Nowhere in his geographical and meteorological works does he show any special interest in those countries or knowledge of them, and nothing points to English being his mother tongue. The ten Anglo-Saxon names of months that he mentioned are found in Bede; there might just be a nostalgic element in his words referring to the "Anglici" as beginning the year on Christmas night and flocking to their main churches on the first day of every month with offerings while the bells ring festively. Otherwise his examples, topical anecdotes, and descriptions give a more prominent position to several other countries than to Britain.

The date of Michael's birth can be guessed with a vague approximation by considering that he produced a translation about 1217 and that by early 1236 he was dead. The insistent recommendations for benefices in the mid-1220's suggest that he was then still young with a precocious interest in learning. It may well be that he was not born before 1195. In 1217, or not long after, Michael was in Toledo and had perhaps acquired some knowledge of Arabic. It was there that, with the help of Abuteus (or Andreas) Levita, a Jew later converted to Christianity, he translated a work of al-Biṭrūjī and—with or without help—some Aristotle from the Arabic. He may also have learned some Hebrew. In 1220 or 1221 he was staying in Bologna in the house of the widow of one Albertus Gallus; while there he had the opportunity to examine and describe in detail a tumor of the womb. By 1224 Michael was a priest and was addressed as "magister" by Pope Honorius III, who obtained for him benefices in Britain and appointed him to be archbishop of Cashel, Ireland; for the latter see he was also recommended by the pope to King Henry III. He renounced

the appointment as archbishop and was given further benefices by Stephen Langton, archbishop of Canterbury, at the instance of Honorius III and of Gregory IX (1225 and 1227). Nothing is known of Michael's university studies: his references to Paris and to some teaching he had done may indicate that he had been there, as either a student or teacher or both. In 1228 Leonardo Fibonacci of Pisa sent a revised copy of the *Liber abaci*, which Michael had solicited. The date 1231 occurs in one manuscript of a poetical prophecy on the future of many towns in northern Italy, requested by Bolognese dignitaries and ascribed to Michael in the later part of the century by the Parma chronicler Salimbene and by many manuscripts. It is probably to this prophecy that the poet Henry of Avranches referred when writing around the beginning of 1236 to Emperor Frederick II; Henry noted that Michael, who had predicted the fate of others, had himself succumbed to fate.

In his writings Michael presented himself as a highly regarded scientific companion and consultant to Frederick II, the most faithful among astrologers. It is as "astrologer to the emperor" that he is often referred to by Salimbene and other writers. No state or other documents survive to confirm or disprove the truth of this title; nor is it clear whether it would imply a regular attachment to the court. No doubt Frederick would have welcomed Michael's contribution to his scientific knowledge. Michael's dedications to him probably represent more than pleas or thanks for moral and social support; his skill in astrological forecasts may have been very welcome to an intelligent ruler who was not above putting some trust in this kind of "science" and to a shrewd politician for whom favorable prophecies had an undeniable propaganda value. Before the middle of 1232 Michael had dedicated to Frederick II his translation of Ibn Sīnā's treatise on animals. This is the only definite date concerning his relationship with the emperor. The dedications, introductions, and contents of his astrological works—taken together with Henry of Avranches's remarks—suggest that the relationship had been neither trivial nor brief in the last years before Michael's death. But it is reasonable to assume that Michael was not depending on Frederick II when, from 1224 to 1227, he relied on the papal curia's support for an income from British benefices or, perhaps, when he was dedicating to Stephen of Provins a translation of Aristotle's *De caelo* with commentary of Ibn Rushd (Averroës); in 1231 Stephen was in a key position to decide on the introduction and study of approved texts of Aristotle in the University of Paris.

There is not much evidence for assessing Michael's standing with his learned contemporaries, apart from

Frederick's interest, Leonardo Fibonacci's complimentary words, and the formal praises contained in the papal recommendations. The two first philosopher-scientists to express views on him after his death, Albertus Magnus and Roger Bacon, were scathing about his scientific knowledge and honesty, although Bacon recognized his merit in having introduced some Aristotle and Ibn Rushd into the Latin West in the early 1230's. What fame, praise, and blame he was accorded later were the result of his reputed magic powers, and the variety of scientific and astrological information presented in his treatises—especially, perhaps, the systematic section on the generation of human beings (book I of his *Physionomia*), which gained great popularity in the late fifteenth and the sixteenth centuries.

It may be that Roger Bacon hit on the one activity for which Michael deserves a place in the history of serious philosophical and scientific speculation. It was through his efforts that some of Ibn Rushd's commentaries on Aristotle came into circulation in Latin and led the way to a penetrating analysis of fundamental problems, such as those of the eternity of the world and the immortality of the soul; they also provided models of methodical interpretation of Aristotle's texts on sound, objective bases. It may also be that Bacon was right when he minimized Michael's linguistic achievements. There is no clear evidence of what his share was in producing the translations ascribed wholly or partly to him. The collaboration in this work by other interpreters, such as Abuteus, may have been much greater than just occasional help. The only research—and that very limited—made on Michael's method of translating suggests considerable inaccuracy, systematic changes in style between one work and another, and occasional recourse to existing Latin translations for increased ease and reliability.

Works. *Translations.* Al-Biṭrūjī's *In astrologia* (as given in a manuscript) or *De sphaera* (both these titles correspond to the Arabic *Fi 'l-hay'at*) or *De motibus celorum circularibus* (the title given by Roger Bacon) is preserved completely or incompletely in eleven manuscripts, all of the fourteenth or fifteenth century. The translation was made with the collaboration of Abuteus and was finished in Toledo on 18 August, probably in 1217 (other, less likely, years are 1207 and 1221). Michael took over long passages of Ptolemy, included by al-Biṭrūjī in his work, from Gerard of Cremona's translation of the *Almagest*. The *In astrologia* made accessible in Latin some recent Spanish-Arabic learning (the original Arabic text was finished about 1190). In it an attempt was made with the use of new mathematical methods to revive Aristotle's cosmology of concentric spheres as against

Ptolemy's system of epicycles and eccentrics. The translation had a certain success, as use of it by Roger Bacon, Grosseteste, Pseudo-Grosseteste, and Albertus Magnus in the thirteenth century testifies, and as the several manuscripts of the next centuries confirm.

Aristotle's *De animalibus* (*Historia animalium*, *De partibus animalium*, *De generatione animalium*), a literal translation made at Toledo, possibly before 1220, of the ninth-century Arabic version by Ibn al-Bitriq, is preserved in more than sixty manuscripts of the thirteenth and early fourteenth centuries. It exerted a considerable influence, mainly through Albertus Magnus' exposition and elaboration. It may have been used by Frederick II, but was soon superseded by William of Moerbeke's translation from the Greek (*ca.* 1260) and later—at the time of printing—by Theodore of Gaza's version (*ca.* 1475). It has been suggested, on the basis of two short references by Michael and an attribution in a manuscript, that he produced a complete translation of the *Nicomachean Ethics* from the Greek, of which book I (*Ethica nova*) was widely circulated and other sections or fragments still survive.

Ibn Sīnā's *De animalibus* or *Abbreviatio de animalibus* was the relevant part of the philosophical encyclopedia *Shifa*. The translation was dedicated to Frederick II, who used it in the preparation of his *De arte venandi*; it is preserved in thirty or more manuscripts of the thirteenth and early fourteenth centuries, some of which derive from the 1232 copy of the volume presented to the emperor.

Of Ibn Rushd's commentaries on and expositions of Aristotle's works, only the first one listed here was certainly edited by Michael, the second quite probably, the others with a smaller degree of probability. *Great Commentary on the De caelo*, with Aristotle's full text; a preface addressed to Stephen of Provins, mentioning the translation of al-Biṭrūjī's *In astrologia*, was written after 1217. Thirty or forty manuscripts still exist; and the work was often quoted by the Latin commentators and philosophers from the thirteenth to the sixteenth centuries. *Great Commentary on the De anima*, with the full text, is preserved in more than fifty manuscripts, three or four of which give Michael as the translator. This version was also translated into Hebrew in the fifteenth century. *Great Commentary on the Physics* is available with full text. At least the prologue, and possibly the whole work, was translated by Theodore of Antioch; only a few manuscripts of the more than fifty extant suggest Michael as the author of the translation. Suggestions of Michael's authorship of the following have no documentary support: *Great Commentary on Metaphysics* with the full text, I.1–4, II–X, XII, extant in fifty manuscripts (in another sixty there is only Aristotle's text, under the name of *Metaphysica nova*, by which title it is quoted in the commentary on Sacrobosco's *Sphaera* ascribed to Michael), *Expositions (Middle Commentaries) of Meteorologica* IV (twenty manuscripts), *De generatione et corruptione* (forty manuscripts), and the *Epitome of Parva naturalia* (fifty manuscripts).

Original Writings. A trilogy, consisting of the *Liber introductorius, Liber particularis,* and *Physionomia* (*De secretis nature,* including a section *De urinis*), is presented by Michael as a unit in a general foreword addressed to Frederick II after the middle of 1228; but the unity does not go further than that of a collection of independent treatises, each of which seems also to lack a definite unity of its own, even though each has an introduction. It appears that Michael collected those of his writings that he thought might interest the emperor, whatever their state of elaboration, and started preparing a volume to be presented to him. This work seems never to have been completed, since the epilogue mentioned in the foreword is nowhere to be found and the manuscript tradition—all later than about 1270—is too varied in form and content to suggest its dependence on a properly edited text.

The *Liber introductorius* is preserved in four manuscripts, each of them differing in many respects from the others and some of them containing later interpolations. It is divided into four sections and is said to have been written at Frederick's request. It is meant to be a compendium of astrological, scientific, and general lore extracted from the works of many authors and enlarged with some personal observation. It is directed to beginners and written in a simple style. The main matters dealt with are astronomy, partly mixed with and partly distinguishable from what may be more properly called astrology, including a systematic treatment of the individual heavenly bodies, their spheres, and their movements; general geography and meteorology (the five main zones of the earth, the climes of the northern temperate zone, the seven regions of the air—that is, dew, snow, hail, rain, honey, laudanum, manna); the tides, including a "new" theory based on the mixture of cold influx from the moon and hot influx from the sun; and some descriptive geography, unsystematic and poorly informed. Other matters discussed at some length concern medicine (advice on food, on how to cure mental states with the help of enchantresses and divines); music; the calendar; important numbers (especially the number seven); and some theology. Altogether there is little more than a collection of secondhand, blindly accepted, information; occa-

sionally there is an assessment of the views of other authorities (for instance, on the distance of the heavens from earth) or an exposition of contradictory doctrines that suggest a critical approach to a problem, indications that scientific inquiry must be applied to research on the terrestrial paradise, hell, and purgatory, some incidental information on things seen, habits of and differences between people of different races, reports of simple experiments made by him and the emperor, and some of his predictions. His sources range from the Bible and Ptolemy to al-Farghānī, Abū Ma'shar, and the *Toledan Tables.* From what has been published so far it is not clear whether Aristotle and Ibn Rushd have been put directly to use, and if so, to what extent.

The *Liber particularis,* a shorter book, is intended to supplement with a fuller and more advanced treatment some of the things expounded for "novices" in the *Liber introductorius.* All that this second book includes is to be "new" but necessary for a better acquaintance with the grand science: "He who has assimilated both books will have qualified for the title of new astrologer." According to Haskins, these additions concern mainly the reckoning of time; sun, moon, and stars; winds and tides; and various meteorological questions. Compared with the *Liber introductorius,* it is based more extensively on an Italian background and on Latin authors like Isidore of Seville, and on Aristotle's meteorological theories. The last part of the *Liber particularis* is the most interesting, for it contains a large number of questions purporting to have been put to Michael by Frederick II, together with his answers. Frederick had heard enough about the sun, moon, fixed stars, elements, world soul, pagan and Christian peoples, creatures moving on earth, plants, and metals; he now wanted to hear about the more inaccessible things leading to spiritual enjoyment and wisdom: paradise, purgatory, and hell; what supports the earth, the abyss, the heavens; the relationships and relative distances between the heavens; where God dwells and what angels and saints do in front of Him; where fresh and salt waters come from; how volcanic eruptions and sulfur springs come about. The answers are less interesting than the questions, and provide little more than known facts or pseudo facts, apart from some information on specific volcanic phenomena in Italy and some attempts to explain them; one chapter, alchemical in character, deals with metals and would seem not to belong among the answers to Frederick's questions.

The third part of the trilogy is preserved in three or four manuscripts. The title *Physionomia* fits little more than half of it; *De secretis nature,* in its vagueness, is more appropriate. The contents of this part are at least threefold. Most of what appears as book I in the printed editions contains a detailed treatise on generation of human beings, with anatomical and physiological descriptions, information on the best time for conception, on sexual behavior, and on the state of the fetus during each of the nine months after conception. The rest of book I deals with differences between genera and species of animals. Books II and III contain the *Physionomia* proper (apart from some chapters on dreams and auguries from sneezes). In these, a systematic survey of the different parts of the body, in connection with the basic or other qualities affecting them, is meant to show how souls are intrinsically dependent for their natures on the bodies that they inhabit: "animae sequuntur corpus." Book III is particularly concerned with showing that such parts of the body as hair, forehead, eyes, nails, and heels, if properly studied, can inform one of the virtues and vices of men and women. A section not included in the printed editions of the *Physionomia,* but published by itself, contains the short treatise *De urinis.*

A *Commentary on Sacrobosco's Sphaera* is ascribed to Michael Scot in the two old printed editions and—with some doubts—in the recent one, but it is anonymous in the two manuscripts containing some parts of it. Thorndike suggested that its twenty-eight "lectiones" somehow reflect a course of lectures. Whether authentic or not, this work is an important document belonging most probably to Michael's time. The following authors and works are the only ones mentioned or quoted, and none of them was unknown in Latin before Michael's death: Aristotle (*Physica, De caelo, Metaphysica* [*Prima philosophia*], *De generatione et corruptione, Meteorologica, De anima, De sensu, Analytica posteriora*), pseudo-Aristotle (*De plantis, De proprietatibus elementorum*), Plato, al-Farghānī, Euclid, Boethius, Ibn Sīnā, Ibn Rushd (*De substantia orbis, Commentary on Metaphysics*), Theodosius of Bythinia (*Spherica*), and Mercurius (*De vita Deorum*); the method followed in the work is also consistent with the habits of that time.

The *Questiones Nicolai peripatetici* contains a few discussions on physical, chemical, and physiological topics, similar to those of the trilogy. They were ascribed to Michael by Albertus Magnus, who condemned them as rubbish, and are preserved, without the author's name, in several manuscripts. Six fragments from a *Divisio philosophie,* quoted as coming from a work by Michael in Vincent of Beauvais's *Speculum doctrinale,* contain a definition of philosophy, the basic classification into theoretical and practical sciences, and some account of two of the former sciences: mathematics and metaphysics. One of the sources of these fragments is Dominicus Gundissalinus; others are found in Arabic texts.

The *Ars alchemie,* preserved in three manuscripts (two of them containing additional material, perhaps spurious), was designed to reveal the "secret of philosophers." In it metals are assimilated to planets, both classes being studied in their special natures; the transformation of Venus into the sun, of mercury into silver, and of lead into gold, and the nature of salts are the other main topics discussed in this treatise. The *Lumen luminum* may be one of the forms, perhaps the basic one, in which the *Dedalus grecus*—a work translated from the Arabic—has been preserved. It contains an alchemical and descriptive study of salts. A *Geomantia,* ascribed to Michael in one manuscript, has never been studied. Together with a short text, of *Experimenta necromantica,* which appears under his name in another manuscript, it completes what is known of Scot's or pseudo-Scot's more fanciful writings on the margins of science and magic. The few lines of the *Description of a Tumor* and two *Recipes* ascribed to him are all we have of his medical texts; and the *Vaticinium* (the prophecy in verse of 1231, mentioned above) is the only "prophetic" text, apart from the few passages of this kind in the *Liber introductorius.*

A *Theorica planetarum,* the *Ten Categories in Theology,* and the *Mensa philosophica* (the last printed under Michael's name and constituting a handbook of dietetics and of characterization of people, with many references to Latin authors known only after Michael's death) are probably to be ascribed, respectively, to Gerard of Cremona, John Scot Eriugena or one of his followers (being perhaps extracts from the *De divisione naturae*), and an anonymous author of the sixteenth century.

BIBLIOGRAPHY

I. ORIGINAL WORKS. The fifteenth-century eds. of the translations of Ibn Rushd are listed in *Gesamtkatalog der Wiegendrucke* under "Aristoteles." Other translations are al-Biṭrūjī, *In astrologia* (*De motibus celorum*), critical text by F. J. Carmody (Berkeley–Los Angeles, 1952); Aristotle, *De animalibus,* bk. X only in Gunnar Rudberg, *Zum sogenannten zehnten Buche der aristotelischen Tiergeschichte,* which is Skrifter Utgivna af Humanistika Vetenskapssampfundet i Uppsala, XIII, pt. 6 (Uppsala–Leipzig, 1911), 109–120, cf. 64–70; Ibn Sīnā, *Abbreviatio de animalibus* (Venice, *ca.* 1500, 1508); and Ibn Rushd, *Great Commentaries on De caelo, De anima, Physica, and Metaphysica,* expositions of *Meteorologica* and *De generatione et corruptione,* and *Epitome of Parva naturalia,* several eds. between 1472 and 1575, also critical eds. of *Great Commentary on De anima* and *Epitome of Parva naturalia* in Corpus

Commentariorum Averrois in Aristotelem, VI, pt. 1 (1953), and VII (1949), by F. S. Crawford and A. L. Shields, respectively. What remains of the Greco-Latin translation of the *Nicomachean Ethics* tentatively ascribed to Michael is available in a critical ed. by R. A. Gauthier in *Aristoteles Latinus*, XXVI, pts. 1–3, fasc. 2 (Leiden–Brussels, 1972), 63–165.

Michael's own works are *Liber introductorius*, unpub. except for a few passages (see Haskins and Thorndike, below); *Liber particularis*, unpub. except for an important section containing Frederick II's questions and Scot's answers in Haskins, *Studies in the History of Mediaeval Science*, pp. 292–298, and a few quotations, in Haskins, *op. cit.*, and Thorndike, *Michael Scot; Physionomia* or *De secretis naturae*, about 20 eds. in the fifteenth century and 20 in the sixteenth (the earliest, Bologna, 1477), all without *De urinis*—which is pub. in A. H. Querfeld, *Michael Scottus und seine Schrift De secretis naturae* (Leipzig, 1919) —as well as sixteenth- and seventeenth-century eds. (without *De urinis*) in Spanish, Italian, and French; *Commentary on Sacrobosco's Sphaera* (Bologna, 1495; Venice, 1531), new ed. by Lynn Thorndike in *The Sphere of Sacrobosco and Its Commentators* (Chicago, 1949), pp. 248–342; *Questiones Nicolai peripatetici*, unpub. except for some quotations and an English summary in Thorndike, *Michael Scot*, pp. 127–131; *Ars alchimie*, or *Magisterium*, available in complete ed. by S. H. Thomson in *Osiris*, **5** (1938), 523–559, and partially, from MSS not used by Thomson, in Haskins, "The Alchemy . . .," in *Isis*, **10** (1928), 350–356, and *Studies in Mediaeval Culture*, pp. 148–159.

Also available are *Lumen luminum* (and/or *Dedalus grecus*, perhaps trans. from the Arabic), edited by W. J. Brown in *An Enquire . . .*, pp. 240–265; The *Experimenta necromantica, ibid.*, pp. 231–234; *Division of Philosophy*, extant fragments pub. in Ludwig Baur, "Dominicus Gundissalinus, De divisione philosophie," in Beiträge zur Geschichte der Philosophie des Mittelalters, **4**, nos. 2–3 (1903), 365–368; *Vaticinium*, or *Futura praesagia*, critical ed. by O. Holder-Egger in "Italienische Prophetieen des 13. Jahrhunderts, II," in *Neues Archiv der Gesellschaft für ältere deutsche Geschichtskunde*, **30** (1905), 321–386, see 349–377, also in *Cronica Fratris Salimbene*, Monumenta Germaniae Historica: Scriptores, XXXII (1905–1913), pp. 361–362; some recipes, in "Notes on Medical Texts," in *Janus*, **48** (1959), 148; *Description of a Tumor*, in M. R. James, *Catalogue of Manuscripts in . . . Gonville and Caius College*, I (Cambridge, 1907), 112, and facs. from MS in *Edinburgh Medical Journal* (1920), 56; and *Mensa philosophica*, several eds. in the sixteenth and seventeenth centuries, such as Leipzig, 1603.

The main MSS of most works and references to printed lists of MSS are in Lynn Thorndike and Pearl Kibre, *A Catalog of Incipits of Mediaeval Scientific Works in Latin*, rev. ed. (Cambridge, Mass., 1963), see index, col. 1864. For the MSS of translations of Aristotle and Ibn Rushd, and of the *Questiones Nicolai peripatetici*, see G. Lacombe, L. Minio-Paluello *et al.*, *Aristoteles latinus*, codices I–II with suppl., suppl. alt. (Rome–Cambridge–Bruges, 1939–

1961). For those of Ibn Sīnā, see M. T. d'Alverny, "Avicenna latinus," in *Archives d'histoire littéraire et doctrinale du moyen-âge*, 1961 (1962)–1970 (1971). For texts of which there are critical eds., see the relevant introductions or appendixes. For the rest, see Haskins, Thorndike, and Thomson (below).

II. SECONDARY LITERATURE. The most precise study of Michael's life and some of his works is still C. H. Haskins, *Studies in the History of Mediaeval Science*, 2nd ed. (Cambridge, Mass., 1927; repr. ch. 12). A more ample account of his life and of more of his works is Lynn Thorndike, *Michael Scot* (London, 1965). There is still much of value in J. W. Brown, *An Enquire Into the Life and Legend of Michael Scot* (Edinburgh, 1897); and in Lynn Thorndike, *A History of Magic and Experimental Science*, II (London, 1923), 307–337. See also John Ferguson, "A Short Biography and Bibliography of Michael Scot," in *Records of the Glasgow Bibliographical Society*, **9** (1931), 75–100; and George Sarton, *Introduction to the History of Science*, II, pt. 2 (Baltimore, 1931), 579–582.

For special aspects of Michael's life or works, see the introductions to the critical eds. and C. H. Haskins, *Studies in Mediaeval Culture* (Oxford, 1929), pp. 148–159 (on *Liber particularis*, Alchemy, Abuteus/Andreas of Palencia); "Two Roman Formularies in Philadelphia," in *Miscellanea Ehrle*, **4** (*Studi e testi*, 40) (1924), 275–286, on Michael's appointment as an archbishop and on Stephen of Provins; Petrus Pressutti, *Regesta Honorii Papae III*, II (Rome, 1895), 194, 227, 254, 334, on the appointments to the benefices; H. A. Wolfson, "Revised Plan of a Corpus Commentariorum Averrois in Aristotelem," in *Speculum*, **38** (1963), 88–104, on attribution of translations of Ibn Rushd to Michael; G. Rudberg, "Die Tiergeschichte des Michael Scotus und ihre mittelbare Quelle," in *Eranos*, **9** (1909), 92–128, on the relationship between the Greek original and Michael's trans. from the Arabic; and J. D. Comrie, "Michael Scot, a 13th Century Scientist and Physician," in *Edinburgh Medical Journal*, n.s. **25** (1920), 50–60.

On the *Physionomia*, its eds., contents, and sources, see A. H. Querfeld, *Michael Scottus und seine Schrift De secretis naturae* (Leipzig, 1919); C. Klebs, "Incunabula scientifica et medica," in *Osiris*, **4** (1938), 297–299; and R. Foerster, *De translatione latina physiognomonicorum quae feruntur Aristotelis* (Kiel, 1884), p. 22; and *De Aristotelis quae feruntur secretis secretorum commentarium* (Kiel, 1888), p. 29.

LORENZO MINIO-PALUELLO

MICHAUX, ANDRÉ (*b.* Satory, near Versailles, France, 7 March 1746; *d.* Madagascar [now Malagasy Republic], 13 November 1802), *botany.*

Michaux compiled the first flora for eastern America, mainly from his own specimens. He introduced many American plants into French horticulture and disseminated the camellia, silk tree, and tea olive in the Carolinas.

Michaux was the eldest son of the manager of the 500-acre royal farm at Satory. His father died when the boy was seventeen, and after four years at a boarding school Michaux became manager of the farm. At twenty-three he married Cécile Claye, daughter of a wealthy Beauce farmer. She died following the birth of their son, François André, and Michaux then began to concentrate on botany and horticulture.

He first studied under Le Monnier at Montreuil, near Versailles. Michaux pursued his botanical studies with Bernard de Jussieu at the Trianon (1777) and then at the Jardin des Plantes. He returned from a visit to England with sccds for Le Monnier and the influential Louis, duc de Noailles of Perpignan. In 1780 he botanized with Lamarck and André Thouin in the Auvergne and the Pyrenees, and he then spent three years traveling through Persia.

Although he wished to explore Tibet, the French government commissioned him to visit North America, particularly to investigate potential sources of ship timbers. With his son and a journeyman gardener, Pierre Paul Saunier, he embarked for New York on 1 September 1785. After two years of collecting and cultivating at Hackensack, New Jersey, Michaux sent 5,000 trees and twelve parcels of seed to France. In September 1787 he set out for Charleston, South Carolina, where he established a second garden. The reputed quinine substitute *Pinckneya* (Rubiaceae) especially interested him. William Bartram's discoveries early attracted Michaux's attention, and he botanized along the St. Johns River and its savannahs in Spanish Florida. Mark Catesby's *Natural History* prompted him to visit the Bahamas.

The last trip for which Michaux received support from the pre-Revolutionary French government was a difficult journey into the mountains of North Carolina. In 1792 Michaux made his longest trip— eight months—to Hudson Bay via the Saguenay River and Lake Mistassini. Then followed a three-month trip to Kentucky. Sponsored by the French revolutionary minister to the United States, Edmond Charles Genet, it was a political as well as botanical journey. In 1794 he returned to the coniferous forests of the southern Appalachians and then journeyed to the Illinois prairies. Thus ended ten years of exploration in America.

His personal funds exhausted, Michaux sailed for France from Charleston on 13 August 1796. He was shipwrecked off the Netherlands coast, and his specimens were only tediously salvaged. Arriving in France, he found the Republic unwilling to indemnify his private expenses or to support future explorations in America. He then set about writing on the American oaks, but before the engravings were completed, he accepted a position with Nicolas Baudin's government expedition to Australia. He stipulated that he be permitted to disembark at Mauritius for exploration and to establish a garden there. He then sailed for Madagascar to found a garden at Tamatave. Before he was able to explore the interior, he contracted a fever and died. Three different death dates have been published although Deleuze offers none; most, though weak, evidence supports that of 13 November 1802.

Michaux's success no doubt encouraged the French naturalists Louis Bosc and Palisot de Beauvois to visit Amcrica. His *Histoire des chênes de l'Amérique*, illustrated with twenty plates, mostly by Redouté, inspired an interest in forestry that was furthered by his son.

BIBLIOGRAPHY

I. ORIGINAL WORKS. Michaux's writings are *Histoire des chênes de l'Amérique* (Paris, 1801); "Mémoire sur les dattiers, avec des observations sur quelques moyens utiles aux progrès de l'agriculture dans les colonies occidentales," in *Journal de physique, de chemie, d'histoire naturelle et des arts*, **52** (1801), 325–335; and *Flora boreali-americana* (Paris, 1803; repr. with extensive introduction by J. Ewan, New York, 1973); none of them was seen in print by Michaux. G. A. Pritzel and others attributed the *Flora* to L. C. M. Richard, who assisted Michaux's son François in bringing out the work, but there is no peremptory reason for doubting its authorship. The type specimens of the *Flora* are preserved in a separate herbarium at the Muséum d'Histoire Naturelle, Paris.

II. SECONDARY LITERATURE. The primary source is J. P. F. Deleuze, "Notice historique sur André Michaux," in *Annales du Muséum national d'histoire naturelle*, **3** (1804), 191–227. Additional references will be found in *Dictionary of American Biography*, VI, 591–592; and Frans A. Stafleu, *Taxonomic Literature* (Utrecht, 1967), 309–310. Useful papers not mentioned by Stafleu are W. J. Robbins and Mary C. Howson, "André Michaux's New Jersey Garden and Pierre Paul Saunier, Journeyman Gardener," in *Proceedings of the American Philosophical Society*, **102** (1958), 351–370; and C. V. Morton, "Fern Herbarium of André Michaux," in *American Fern Journal*, **57** (1967), 166–182.

JOSEPH EWAN

MICHEL-LÉVY, AUGUSTE (*b.* Paris, France, 7 August 1844; *d.* Paris, 27 September 1911), *geology, mineralogy.*

The greater part of Michel-Lévy's scientific work was accomplished in collaboration with Ferdinand Fouqué. Jointly they introduced into France the study of rocks by microscopical petrography and artificially synthesized many igneous rocks in order to determine the conditions necessary for the production of their mineral constituents. In addition Michel-Lévy, working with feldspars, founded the method of statistical research on the constituents of rocks. He was the first to demonstrate the importance of birefringence in petrographic studies, and he determined this optical constant for a large number of minerals. In the field Michel-Lévy devoted twenty years to the study of the eastern part of the Massif Central, of the Morvan Massif, and of the western Alps.

Michel-Lévy's father was a noted military hygienist and president of the Académie de Médecine. Independently wealthy, Michel-Lévy was raised in an intellectual atmosphere devoted to both science and literature. As a result he won numerous prizes in the general *concours* from 1859 to 1861 and entered the École Polytechnique in 1862. He ranked first there and at the École des Mines, from which he graduated in 1867. Michel-Lévy entered the service of the Carte Géologique in 1870 and was its director from 1887 until his death. In 1879 he was appointed engineer of mines of the first class; he became chief engineer of this division in 1883; and he attained the highest rank, inspector general, in 1907. He was elected to the Académie des Sciences in 1896. In addition Michel-Lévy was a member of the administrative council of the Conservatoire National des Arts et Métiers, of the national council of hygiene of France, and of the council of public hygiene for the department of the Seine.

Together with Fouqué and Alfred Lacroix, Michel-Lévy pioneered the science of microscopical petrography in France. His two-volume work, *Minéralogie micrographique: Roches éruptives françaises* (1879), written with Fouqué, demonstrated the results of this method to French scientists. In it they also employed a new classificatory system for volcanic rocks, using as criteria mineralogical composition, structure, and chemical composition. In 1880, with Lacroix, Michel-Lévy published *Tableaux des minéraux des roches*; and in 1889 they produced *Les minéraux des roches*, which described the optical and chemical methods of studying minerals in thin sections and the microscopical features of rock-forming minerals. Michel-Lévy's *Étude sur la détermination des feldspaths dans les plaques minces* (1894) was a significant contribution to the microscopic study of feldspars in thin sections.

From 1878 to 1882 Michel-Lévy and Fouqué worked to synthesize igneous rocks artificially, believing that if they could determine the peculiar conditions surrounding the rocks' genesis, they might arrive at important geological conclusions. Despite the meager equipment of the laboratory at the Collège de France in which they carried out their experiments, they produced rocks having the mineralogical composition and structure identical with most of the volcanic rocks found in nature. They published jointly twenty-two articles and a book, *Synthèse des minéraux et des roches* (1882), which incorporated the results of their work. Their most important conclusions were that the degree of crystallinity depended largely upon the rate of cooling and that rocks of distinctly different mineralogical compositions would be formed from the same magma, depending upon the conditions of crystallization. Their failure to reproduce the trachytes and rhyolites demonstrated that in order to obtain the characteristic elements of these rocks, the presence of mineralizers was necessary to lessen the viscosity of the magma and to allow crystallization.

The principal result of Michel-Lévy's fieldwork in geology was in distinguishing the two successive phases of folding and dislocation of the Massif Central toward the end of the late Paleozoic. His analysis of the dislocations superimposed on the eastern portion of the Massif Central caused him to trace them to the east and to search for their influence on the tectonic movements and volcanic phenomena of the Tertiary. In conjunction with this research he mapped Clermont-Ferrand, studied the lavas of the neighboring regions, and explored and mapped portions of the western Alps.

BIBLIOGRAPHY

Michel-Lévy's chief publications treating the application of optical methods to the study of minerals in thin sections are *Minéralogie micrographique: Roches éruptives françaises* 2 vols. (Paris, 1879), written with F. Fouqué; *Tableaux des minéraux des roches* (Paris, 1880), written with A. Lacroix; *Les minéraux des roches* (Paris, 1889), written with A. Lacroix; and *Étude sur la détermination des feldspaths dans les plaques minces* (Paris, 1894). In petrography he collaborated with F. Fouqué in *Synthèse des minéraux et des roches* (Paris, 1882) and also published *Structures et classifications des roches éruptives* (Paris, 1889). In addition he published either alone or jointly some 150 articles on geological or mineralogical subjects, and he contributed 10 maps to the Carte Géologique.

A biography is Alfred Lacroix, *Notice historique sur Auguste Michel-Lévy* (Paris, 1914), which includes a complete bibliography.

JOHN G. BURKE

MICHELI, PIER ANTONIO (*b.* Florence, Italy, 11 December 1679; *d.* Florence, 1 January 1737), *botany.*

Micheli was the son of Pier Francesco Micheli, a laborer, and Maria Salvucci. The boy had only the most elementary schooling (Haller, in 1772, described him as "illiteratus et pauper"). He was, however, interested in plants from childhood, and his native talent won him the respect of, and eventually a prominent position among, the botanists of his time. He obtained the patronage of both the Grand Duke Cosimo III de' Medici and his successor Gian Gastone de' Medici; the generosity of these two men permitted him to devote himself completely to his studies. Micheli was nonetheless hampered by the lack of an academic degree and never held a post worthy of his talents. He was obliged to content himself with modest positions in the botanical gardens of Pisa and Florence, although he enjoyed considerable contemporary fame among both Italian and foreign botanists and conducted an extensive correspondence with them. He was further influential in founding, with a group of friends, the Società Botanica Fiorentina in 1716 and in the tutelage of a student of great ability, Giovanni Targioni-Tozzetti.

Micheli was a lifelong and tireless collector of plants. His travels for this purpose took him to the provinces of Venetia, Emilia-Romagna, Lazio, Abruzzi e Molise, the Marches, Campania, and Puglia; he was also extremely active in his native Tuscany. In 1708 and 1709 he made collecting expeditions to the Tirol, Austria, Bohemia, Thuringia, and Prussia. He was occasionally accompanied on these trips by Targioni-Tozzetti, who wrote of his skills as a collector:

> He was perspicacious and possessed a talent made expressly for natural history, and particularly for botany; his eye was so keen that as soon as he reached a meadow or other place full of grasses, he could immediately distinguish the rarest or most worthy of observation. He was also gifted with an acute critical capacity . . ., so that he could tell in an instant why other illustrious botanists had been in error, confusing one species with another, or multiplying them [*Notizie della vita e delle opere di P. A. Micheli* (Florence, 1858), 330].

In the seventeenth and eighteenth centuries the concept of species was crucial to the great botanical task of classification. Micheli's views on species were in large part derived from those of Joseph de Tournefort, but even more than Tournefort, Micheli realized the need for great caution in the problem of definition. Micheli's attitude was, in fact, quite close to that of Linnaeus, who expressed his admiration for

him. His concern is evident in the first part of his *Nova plantarum genera* of 1729. In this work Micheli considered some 1,900 species, of which nearly 1,400 were new. The greater number of these new species were thallophytes—fungi, lichens, liverworts, and mosses—which Micheli classified for the first time. Using two primitive microscopes, he was able to observe, again for the first time, such notable anatomical details as the antheridia and the archegones of mosses and the spores of fungi. He thus discovered, too, the generative function and the anatomy of the mycelium; for this discovery, among others, he may properly be considered the founder of mycology.

The *Nova plantarum genera* remained unfinished at the time of Micheli's death, and a considerable amount of the data that he had gathered—particularly material relating to algae, which attracted him as much as did fungi—was therefore never incorporated into it.

In addition to his botanical studies, Micheli was also concerned with zoology, paleontology, and geology. In 1710, while he was botanizing in Campania, he noticed the similarity of the rocks on the islands of Ischia and Procida to those of Vesuvius, and realized that the islands were, in fact, extinct volcanoes. In 1722, recalling this earlier observation, Micheli concluded that the hill of Radicofani in Tuscany and a number of outcroppings in nearby Lazio might also be extinct volcanoes; in 1734 he reached the same conclusion about Monte Amiata, also in Tuscany. His intuition proved to be correct; Micheli's suggestion represented the first recognition of an extinct volcano far from regions still active volcanically.

BIBLIOGRAPHY

I. Original Works. Micheli's major work was *Nova plantarum genera* . . . (Florence, 1729). Some of his reports of his journey in 1708–1709 were published posthumously in G. Targioni-Tozzetti, *Relazioni di alcuni viaggi fatti in diverse parti della Toscana* . . . (Florence, 1768–1779), IX (1776), 338; X (1777), 134, 159, 177. Several unpublished manuscripts are in the library of the Istituto di Botanica of the University of Florence.

II. Secondary Literature. G. Targioni-Tozzetti, *Notizie della vita e delle opere di P. A. Micheli* (Florence, 1858), with copious historical notes and extracts from Micheli's correspondence, edited by Antonio and Adolfo Targioni-Tozzetti, is an excellent source on the life of Micheli and his relations with contemporary scientists.

See also G. Negri, "P. A. Micheli (1679–1737)," in *Nuovo giornale botanico italiano*, n.s. **45** (1938), lxxxi–cvii; and "P. A. Micheli botanico," in *Atti della Società Colombaria fiorentina*, meeting of 27 Dec. 1937, 47–67, and with an extensive bibliography on Micheli; and F. Rodolico, "P. A. Micheli e le prime ricerche sui vulcani spenti," in *Atti dell'Accademia toscana di scienze e lettere "La Colombaria,"* **27** (1962–1963), 353–360.

FRANCESCO RODOLICO

MICHELINI, FAMIANO (*b.* Rome, Italy, 31 August 1604; *d.* Florence, Italy, 20 January 1665), *hydraulics, medicine.*

In 1619, when he was only fifteen, Michelini entered the Piarists as a lay brother, under the name of Francesco di San Giuseppe. In 1621 the congregation was raised to the status of a religious order with the task of running free schools open to all, so that poor boys from their earliest years could be given both religious and secular education. The program of the order included a number of interesting teaching principles: religious instruction was not to take priority over other subjects; not only literature but also mathematics was to be taught; and all teachers had received a thorough training.

Michelini was therefore sent to continue his education in Genoa, where he studied mathematics with Somasco Antonio Santini. In 1629 he went to Florence, where the first schools of the order were to be opened, and took a letter (still extant) of introduction and recommendation to Galileo (from Giovanni Battista Baliani). One can follow his activities and movements until September 1641 through Galileo's manuscripts, which contain several letters from Michelini; and he is frequently mentioned in other letters of the time, especially those of Benedetto Castelli, to whom Michelini was introduced by Galileo when he went to Rome in 1634. In April 1634 Castelli wrote to Galileo, who had already been obliged to retire to Arcetri: "I am amazed by his [Michelini's] knowledge, surprised by the subtlety of his mind, delighted by the sincere love that he bears for you, and fascinated by his goodness."

In 1635 Michelini was called to teach mathematics at the Florentine court, and a little later he was asked to give instruction to the brothers of Ferdinand II de' Medici, Gian Carlo and Leopoldo—to whom he also gave lectures in physics and astronomy, which apparently were attended by the grand duke. In November 1636 Michelini was ordained a priest, and in 1648 he obtained the chair of mathematics at Pisa—vacant after the death of Vincenzo Renieri— because negotiations with Ismael Boulliau and William Oughtred had failed. In his inaugural lecture, now lost, Michelini declared that all knowledge is derived from the exact sciences. Following this principle, he applied the experimental method even to medicine, in which he was much interested although he was not a doctor; he is generally credited with paving the way for Redi's experiments and Borelli's theories. Among other things he recommended the use in many illnesses of abundant quantities of orange and lemon juice, and advised people to control their weight. But being generally misunderstood, he was made the object of much derision.

In 1655 Michelini left the chair at Pisa and sought the appointment of Borelli, then at Messina, to replace him. In 1657, already in poor health and afflicted by gout, he received permission to leave his religious order and, remaining a simple priest under his old name, went to Sicily as a pro-vicar to the new bishop of Patti. The latter soon died, however, and Michelini had to return to Florence, in even worse health and with no financial resources. He had the good fortune, though, to attract the patronage of Prince Leopold de' Medici, through whom, at the end of 1664, he was able to publish his book *Della direzione de' fiumi.* A few weeks after its publication he became seriously ill and died within a few days.

In the following years books and papers by Michelini appeared in the possession of Vincenzo Santini—certainly not through inheritance, as was once believed—and in 1671 Santini copied from one of the books the marginal notes written by the young Galileo on Archimedes' *De sphaera et cylindro,* comments that otherwise would have been lost.

Michelini was always reluctant to publish, and he left unpublished a number of "Discourses on Health," now lost, which however were known at least to Redi, his direct follower in the field. As for the book he published just before his death, Michelini's contemporary fame as an expert in hydraulics obviously did not depend on it. The relevant authorities had, many years before, sought his advice on important problems concerning water, such as the course of the Chiana River, the threat of silting in the Lagoon of Venice, and the control of the Arno for the protection of Pisa against floods. But now one can judge him only from his book, which has been much criticized because it includes several serious mistakes—for instance, the belief that stagnant water exerts pressure only on the bottom of its bed and not on the sides, even though Pascal's basic principle of hydrostatics had been known for sixteen years. But since Michelini was dealing exclusively with running water, this error does not invalidate the rest of the work, which contains many good suggestions. One of them was that it should be possible to protect and repair riverbeds with boxes— or, rather, bulkheads—full of stones. Unlike Torricelli, he recognized the theory that one of the factors determining the velocity of current is the gradient of the riverbed. Most significantly, he attributed to the viscosity of water the fact that the current is faster in the middle of a stream than near the banks. This idea was not accepted by his contemporaries.

For these and for other reasons, Michelini was referred to as his partial source by Domenico Guglielmini when, in his fundamental treatise *Della natura de' fiumi* (1697), he dealt with the control of riverbeds. This resulted in a reprint of Michelini's book (Bologna, 1700) and its inclusion in all the editions of the *Raccolta d'autori che trattano dell'acque*, beginning with the Florence edition of 1723.

BIBLIOGRAPHY

I. ORIGINAL WORKS. Michelini's surviving works are *Trattato della direzione de' Fiumi* (Florence, 1664; 2nd ed., Bologna, 1700); and "Risposta alla scrittura del Sig. Torricelli," in *Raccolta d'autori che trattano del moto dell'acque*, 2nd ed. (Florence, 1768), 121. His letters to Galileo are in Galileo, *Opere*, National Ed. (Florence, 1890–1909), XVI, 76, 139–140; XVII, 234–235, 316–317, 321–322, 399–400, 407, 411–412; XVIII, 35–36, 39–40, 128.

II. SECONDARY LITERATURE. See the following, listed chronologically: G. Targioni-Tozzetti, *Notizie degli aggrandimenti della scienze fisiche accaduti in Toscana nel corso di anni 60 del secolo XVII* (Florence, 1780; repr. Bologna, 1967), I, 188–204, 365; P. Riccardi, *Biblioteca matematica italiana*, II (Modena, 1870), 156–157; and A. Neri, "Il Padre Staderone," in *Rivista europea*, n.s. **23** (1881), 756–764; and G. Giovannozzi, *Scolopi galileiani*, which is *Pubblicazioni dell'Osservatorio Ximeniano dei PP. Scolopi*, no. 124 (Florence, 1917); and "Un capitolo inedito della storia del metodo sperimentale in Italia di R. Caverni," in *Atti della Pontificia Accademia Romana dei Nuovi Lincei*, **71** (1918), 171–189.

GIORGIO TABARRONI

MICHELL, JOHN (*b.* Nottinghamshire [?], England, 1724[?]; *d.* Thornhill, near Leeds, England, 21 April 1793), *astronomy*.

Michell earned a permanent place in the history of stellar astronomy for two signal accomplishments: he was the first to make a realistic estimate of the distance to the stars, and he discovered the existence of physical double stars. He was educated at Cambridge. After graduating from Queens' College with the M.A. (1752) and the B.D. (1761), he held the Woodwardian chair of geology at Cambridge (1762–1764). In 1767 he was appointed rector of St. Michael's Church in Thornhill, near Leeds—a post he held for the rest of his life. He is buried at Thornhill, where the parish register describes him as aged sixty-eight (hence the surmise that he was born in 1724).

Michell's published scientific work, which earned him election to the Royal Society in 1760, covered many subjects, including the cause of earthquakes (1760), observations of the comet of January 1760, a method for measuring degrees of longitude "upon parallels of the Equator" (1766), and an independent discovery with Coulomb of the torsion balance (1784). His greatest accomplishments were two investigations published in the *Philosophical Transactions of the Royal Society*: "An Inquiry Into the Probable Parallax and Magnitude of the Fixed Stars From the Quantity of Light Which They Afford Us, and the Particular Circumstances of Their Situation" (1767) and "On the Means of Discovering the Distance, Magnitude, etc. of the Fixed Stars" (1784).

In the first of these papers, Michell pointed out that the frequency of the angular separation of close pairs of stars known at that time deviated grossly from what one could expect for chance projection of stars uniformly distributed in space—there appeared to be an excessive number of close pairs—and, according to Michell: ". . . The natural conclusion from hence is, that it is highly probable, and next to a certainty in general, that such double stars as appear to consist of two or more stars placed very near together, do really consist of stars placed nearly together, and under the influence of some general law . . . to whatever cause this may be owing, whether to their mutual gravitation, or to some other law or appointment of the Creator." The directness of Michell's language perhaps leaves something to be desired; but the unimpeachable logic of his arguments gave a convincing theoretical proof of the existence of physical binary stars in the sky long before Herschel (1803) provided a compelling observational proof.

Michell's second great achievement was a realistic estimate of the distance to the stars, and he made it more than half a century before the first parallax of any fixed star had been measured. His argument was very neat and can be regarded as the precursor of the "photometric" parallaxes of the twentieth century. Michell noticed that Saturn at opposition appears in the sky as bright as the star Vega and exhibits an apparent disk about twenty seconds in diameter, one which from the sun would be seen as seventeen seconds across. Therefore Saturn's illuminated hemisphere clearly intercepts $(17/3600)^2(\pi/720)^2$ of the light sent out by the sun.

Now—and this is essential—if the sun and Vega were of equal intrinsic brightness, and Vega's apparent brightness is equal to that of Saturn, it follows (from the inverse-square law of the attenuation of brightness, already established by Bouguer) that Vega must be $(3600/17)(720/\pi)$, or 48,500, times as far from the sun as Saturn is. Moreover, since Saturn is known to be 9.5 times as far from the sun as the earth is, it follows that the distance to Vega should amount to $9.5 \times 48,500$, or some 460,000 astronomical units.

Although this value represents only about a quarter of the actual distance of Vega, first measured trigonometrically by F. G. W. Struve in 1837 (the underestimate resulting from Vega's being intrinsically much brighter than the sun), Michell's value was the first realistic estimate of the distance to any star.

Michell was apparently a man of wide interests, including music. Tradition has it that William Herschel was a frequent guest at Thornhill during his years as a young musician in Yorkshire, and he is even said to have received his introduction to mirror grinding from Michell. There is, however, no real evidence that Herschel turned to astronomical observation before his move to Bath some years later; and the story of his apprenticeship with Michell may, therefore, be apocryphal.

BIBLIOGRAPHY

I. ORIGINAL WORKS. Michell's papers appeared mainly in the *Philosophical Transactions of the Royal Society* and include "Conjectures Concerning the Cause and Observations Upon the Phenomena of Earthquakes," **51**, pt. 2 (1760), 566–634, also published separately (London, 1760); "Observations on the Same Comet [January 1760]," *ibid.*, 466–467; "A Recommendation of Hadley's Quadrant for Surveying," *ibid.*, **55** (1765), 70–78, also published separately (London, 1765); "Proposal of a Method for Measuring Degrees of Longitude Upon Parallels of the Equator," **56** (1766), 119–125, also published separately (London, 1767); "An Inquiry Into the Probable Parallax and Magnitude of the Fixed Stars From the Quantity of Light Which They Afford Us," *ibid.*, **57** (1767), 234–264, also published separately (London, 1768); and "On the Means of Discovering the Distance, Magnitude, etc. of the Fixed Stars," *ibid.*, **74** (1784), 35–57.

Michell was also author of *A Treatise of Artificial Magnets* (Cambridge, 1750; 2nd ed., 1751), translated into French as *Traité sur les aimans artificiels* (Paris, 1752); and *De arte medendi apud priscos musices* (London, 1766; 1783).

II. SECONDARY LITERATURE. See Archibald Geikie, *Memoir of John Michell* (Cambridge, 1918); and *Dictionary of National Biography*, XIII, 333–334.

ZDENĔK KOPAL

MICHELSON, ALBERT ABRAHAM (*b.* Strelno, Prussia [now Poland], 19 December 1852; *d.* Pasadena, California, 9 May 1931), *physics, optics, metrology.*

Precision measurement in experimental physics was Michelson's lifelong passion. In 1907 he became the first American citizen to win a Nobel Prize in one of the sciences, being so honored "for his precision optical instruments and the spectroscopic and metrological investigations conducted therewith." Michelson measured the speed of light in 1878 as his first venture into scientific research, and he repeatedly returned to the experimental determination of this fundamental constant over the next half century. Never fully satisfied with the precision of former measurements, he developed and took advantage of more advanced techniques and tools to increase the accuracy of his observations. He died, after several strokes, during an elaborate test of the velocity of light in a true partial vacuum over a mile-long course at Irvine, California; but the value later published by his colleagues (299,774 ± 11 km./sec.) was probably less precise than Michelson's own optical determination over a twenty-two-mile course between mountains in southern California during 1924–1926 (299,796 ± 4 km./sec.).

Born to parents of modest means in disputed territory between Prussia and Poland, Michelson at the age of four emigrated with his parents, Samuel and Rosalie Michelson, to San Francisco via New York and Panama. The elder Michelson became a merchant to gold-rush miners in California and later in Virginia City, Nevada, while his son was sent after the sixth grade to board first with relatives in San Francisco and then with Theodore Bradley, the headmaster of Boys' High School there. Bradley seems to have aroused young Michelson's interest in science and to have recognized and rewarded his talents in the laboratory. At Bradley's suggestion Michelson competed for a state appointment to the U.S. Naval Academy; but when three boys tied for first place in the scholastic examination and another was appointed, young Michelson decided to take his case, with a letter of recommendation from his congressman, to the White House. In 1869 he traveled to Washington, saw President Grant, and gained his appointment to Annapolis.

Graduating with the class of 1873, Michelson went to sea for several cruises before being reassigned to the academy as instructor in physical sciences. On 10 April 1877 Michelson married Margaret Heminway from a prosperous New York family; this marriage lasted twenty years and produced two sons and a daughter.

While teaching physics in 1878, Michelson became interested in improving upon Foucault's method for measuring the speed of light terrestrially. In July 1878, with a $2,000 gift from his father-in-law, Michelson was able to improve the revolving-mirror apparatus and to perfect his experiment—the fourth terrestrial measurement of the speed of light. He was preceded by Fizeau, Foucault, and Cornu. Simon Newcomb, superintendent of the Nautical Almanac Office, became interested in his work. In consequence, his first scientific notices and papers were published in 1878–1879, and he began to collaborate with Newcomb on a government-sponsored project to refine further the determination of the velocity of light. He obtained a leave of absence to do postgraduate study in Europe during 1880–1882. He studied with Helmholtz in Berlin, with Quincke in Heidelberg, and with Cornu, Mascart, and Lippman in Paris.

In the winter of 1880–1881, while working in Helmholtz' laboratory, Michelson thought of a means to try a second-order measurement of Maxwell's suggestion for testing the relative motion of the earth against the ubiquitous, if hypothetical, luminiferous ether. Drawing on the credit that Alexander Graham Bell maintained in his account with the Berlin instrument makers Schmidt and Haensch, Michelson designed an apparatus called an interferential refractometer, which he then used to test for relative motion, or an "aether-wind," by comparing the speed of two pencils of light split from a single beam and caused to traverse paths at right angles to each other upon a base that could be rotated between observations. At different azimuths it was expected that the recombined pencils forming interference fringes would shift past a fiducial mark and thereby give data from which could be calculated the "absolute motion" of the earth, with respect to the ether or the "fixed" stars, as it hurtles through space. This first ether-drift experiment was tried in Berlin, then at the Astrophysicalisches Observatorium at Potsdam, with disappointingly null results. The instrument itself was amazingly sensitive and versatile; but errors in experimental design, pointed out by A. Potier and later by H. A. Lorentz, together with the null results themselves and the theoretical difficulties with regard to what was meant by "absolute velocity," later led Michelson to consider the experiment a failure. The hypotheses of A. J. Fresnel concerning a universal stationary ether and of G. G. Stokes concerning astronomical aberration were thus called into question.

The undulatory theory of light as generally accepted in the 1880's simply assumed a luminiferous medium. This "aether" must pervade intermolecular spaces, of both transparent and opaque materials, as well as interstellar space. Hence, it should be at rest or stationary in the universe and therefore provide a reference frame against which to measure the earth's velocity. Michelson boldly denied the validity of this hypothesis of a *stationary* ether, but he always maintained the need for some kind of ether to explain the phenomena of the propagation of light. Ad hoc hypotheses soon seemed necessary to explain why no relative ether-wind or relative motion appeared to be detectable in Michelson's interferometer at the surface of the earth. This curious puzzle piqued the interest of Lorentz, W. Thomson (later Lord Kelvin), and FitzGerald, among others.

In 1881 Michelson resigned from active duty, and the next year he joined the faculty of the new Case School of Applied Science in Cleveland, Ohio. There he set up improved apparatus, helping to check Simon Newcomb's velocity-of-light measurements and testing various colored lights for indexes of refraction in various media. In 1885 Michelson began a collaborative project with Edward W. Morley of Western Reserve, a senior experimentalist (and primarily a chemist) with an elaborate laboratory. Their first effort, undertaken at the suggestion of W. Thomson, and of Rayleigh and Gibbs, was to verify the Fizeau experiment, reported in 1859, that supposedly had confirmed Fresnel's drag coefficient by comparing the apparent velocities of light moving with and against a current of water. This "ether-drag" experiment worked out well and corroborated the suppositions of Fresnel, Maxwell, Stokes, and Rayleigh concerning astronomical aberration and an all-pervasive immaterial luminiferous medium.

Michelson and Morley next redesigned the 1881 ether-drift experiment to increase the path length almost tenfold and to reduce friction of rotation by floating a sandstone slab on a mercury bearing. During five days in July 1887 Michelson and Morley performed their test for the relative motion of the earth in orbit against a stationary ether. Their results were null and so discouraging that they abandoned any effort to continue with the tests they intended in the following autumn, winter, and spring. The sensitivity they had achieved with this new interferometer, about one-fourth part in one billion, was its own reward, however; and both innovators began to think of other uses for such instruments. Although the experimenters quickly forgot their disappointment, theorists, and notably FitzGerald, Larmor, Lorentz, and Poincaré, made much of their failure to find fringe shifts and to corroborate Fresnel and Stokes's wave theory of light.

Michelson accepted an offer in 1889 to move to the new Clark University at Worcester, Massachusetts. Concurrently he began to carry out a monumental metrological project that he and Morley had envisioned to determine experimentally the length of the international meter bar at Sèvres in terms of wavelengths of cadmium light. Adapting his refractometer as a comparator for lengths that could be reduced through spectroscopy and interferometric techniques to nonmaterial standards of length, Michelson found in 1892–1893 that the Paris meter bar was equal to 1,553,163.5 wavelengths of the red cadmium line. So elegant were the success and precision of this project that Michelson became internationally famous.

In 1893 Michelson moved to the new University of Chicago to head its department of physics. There he began to develop his interests in astrophysical spectroscopy. Diffraction gratings, a new harmonic analyzer, and the echelon spectroscope, as well as a large-scale vertical interferometer, were designed by and built for Michelson around the turn of the century. He was clearly recognized as one of the foremost experimental physicists of the nation and was invited to give the Lowell lectures at Harvard in 1899, later published as *Light Waves and Their Uses* (Chicago, 1903). Also in 1899, Michelson remarried, having been divorced, and took as his second wife Edna Stanton, who bore him three daughters.

When Einstein's three famous papers of 1905 appeared, one of which inaugurated the special theory of relativity by dispensing with the idea of an ether and by elevating the velocity of light into an absolute constant, Michelson was much too busy with prior commitments and with receiving honors to pay much heed.

The relation between Michelson's experimental work and Einstein's theories of relativity is complex and historically indirect. But the influence of his ether-drift tests on Lorentz, FitzGerald, Poincaré, W. Thomson, Lodge, Larmor, and other theoreticians around 1900 is less problematic and quite direct. Although scholars continue to debate the role of his classic ether-drift experiment, Michelson himself in his last years still spoke of "the beloved old ether (which is now abandoned, though I personally still cling a little to it)." He advised in 1927 in his last book that relativity theory be accorded a "generous acceptance," although he remained personally skeptical.

From 1901 to 1903 he had served as president of the American Physical Society, and in 1907 he received the Copley Medal from the Royal Society (London) in addition to the Nobel Prize. In all, during his half-century as an active scientist he was elected to honorary membership in more than twenty-five societies, was awarded eleven honorary degrees, and received seventeen medals. In 1910–1911 he served as president of the American Association for the Advancement of Science, and from 1923 to 1927 he presided over the National Academy of Sciences.

During World War I, Michelson returned to the navy as a sixty-five-year-old reserve officer. He helped perfect an optical range finder and demonstrated tolerances for imperfections in striated optical glasses. After the war the Eddington eclipse expedition of 1919 made Einstein and relativity theory almost synonymous with esoteric modern science. Although legend has much inflated the role of the Michelson-Morley experiment in supposedly providing the basis for Einstein's first work on the principle of relativity applied to electrodynamics, Michelson's corroborations of the speed of light as a virtual constant did in fact prove significant equally for the special and for the general theories of relativity.

Early in the 1920's Michelson began to spend more time in California at Mt. Wilson, in Pasadena, and at the California Institute of Technology. Besides teaching, his main work for almost a decade had been to perfect ruling engines for the production of better diffraction gratings. But administrative duties at the University of Chicago also weighed heavily upon him. In southern California, he could work and play in several well-equipped laboratories and also indulge his interest in tennis, billiards, chess, and watercolor painting. Tests for the rigidity of the earth (or earth-tide experiments) were followed by work with H. G. Gale toward an elaborate test near Chicago for the effect of the earth's rotation on the velocity of light. Other studies of the application of interference methods to astronomical problems led to the construction in 1920 of the celebrated stellar interferometer on the Hooker 100-inch telescope that measured the amazing angular diameter of α Orionis (Betelgeuse), which was found to have a disk subtending 0.047″ arc, or approximately 240 million miles in diameter. Still other tests and a geodetic survey under Michelson's supervision in southern California prepared the way for a measurement of the velocity of light between mountain peaks. The Mt. Wilson to the San Jacinto Mountains measurement (eighty-two miles) was scuttled because of smog in 1925; the Mt. Wilson to Mt. San Antonio measurement (twenty-two miles) was completed in 1926, and the value remains one of the best optical determinations ever made.

Meanwhile, George Ellery Hale, director of the Mt. Wilson Observatory, had invited to southern California Michelson's friend and successor at Case, Dayton C. Miller, who had worked with Morley on other ether-drift tests in 1900–1906 and had achieved eminence in acoustics. Miller was supposed to perfect the original Michelson-Morley experiment for all seasons and at a 6,000-foot altitude. After many vicissitudes he did so in 1925–1926 and, to the consternation or delight of a divided profession, Miller announced in his retiring address as president of the American Physical Society that he had finally found the absolute velocity of the solar system: about 200 km./sec. toward the head of the constellation Draco! This challenge spurred Michelson to take up ether-drift tests once again. In conjunction with F. G. Pease and F. Pearson, several very elaborate interferometers were built and operated briefly from 1926 through 1928 but to little avail. Neither Michelson nor his team—nor any other experimentalists later in the 1920's—were able to corroborate Miller's slight but positive results; and so Einstein stood verified largely on the authority of Michelson's reiterated word.

Michelson's second book, *Studies in Optics*, was published in 1927, the year before the Optical Society of America dedicated its annual meeting to him on the fiftieth anniversary of his scientific career. Michelson had used "Light Waves as Measuring Rods for Sounding the Infinite and the Infinitesimal," as the title of one of his last papers. When he died in 1931, he was hardly less a believer in the wave theory of light and its concomitant ether. Although he supported Einstein with few reservations, he was secure in the knowledge that he had indeed sounded the nature of light and found its field both infinite and infinitesimal.

BIBLIOGRAPHY

I. ORIGINAL WORKS. Michelson's books are *Light Waves and Their Uses* (Chicago, 1903); and *Studies in Optics* (Chicago, 1927). Translations and 78 articles are listed in Harvey B. Lemon, "Albert Abraham Michelson: The Man and the Man of Science," in *American Physics Teacher*, **4** (Feb. 1936), 1–11.

MS and memorabilia material are widely scattered, but the best collection is held by the Michelson Laboratory, Naval Weapons Center, China Lake, California. See D. Theodore McAllister, "Collecting Archives for the History of Science," in *American Archivist*, **32** (Oct. 1969), 327–332; and *Albert Abraham Michelson: The Man Who Taught a World to Measure*, Publication of the Michelson Museum, no. 3 (China Lake, Calif., 1970). See also holdings of the Bohr Library, American Institute of Physics, Center for History and Philosophy of Physics, 335 East 45th Street, New York, N.Y. 10017.

II. SECONDARY LITERATURE. See Bernard Jaffe, *Michelson and the Speed of Light*, Science Study series (Garden City, N.Y., 1960); Dorothy Michelson Livingston, "Michelson in the Navy; the Navy in Michelson," in *Proceedings of the United States Naval Institute*, **95**, no. 6 (June 1969), 72–79, a collection of papers and memorabilia that forms the basis for a biography of her father, *The Master of Light* (New York, 1973); Robert A. Millikan, "Albert A. Michelson," in *Biographical Memoirs. National Academy of Sciences*, **19**, no. 4 (1938), 120–147; "Pro-

ceedings of the Michelson Meeting of the Optical Society of America," in *Journal of the Optical Society of America*, **18**, no. 3 (Mar. 1929), 143–286; Robert S. Shankland, "Albert A. Michelson at Case," in *American Journal of Physics*, **17** (Nov. 1949), 487–490; and Loyd S. Swenson, Jr., *The Ethereal Aether: A History of the Michelson-Morley-Miller Aether-Drift Experiments 1880–1930* (Austin, Tex., 1972); Gerald Holton, "Einstein, Michelson, and the 'Crucial Experiment,'" in *Isis*, **60**, no. 202 (Summer 1969), 133–197; Jean M. Bennett, *et al.*, "Albert Michelson, Dean of American Optics–Life, Contributions to Science, and Influence on Modern-Day Physics," together with Robert S. Shankland, "Michelson's Role in the Development of Relativity," in *Applied Optics*, **12**, no. 10 (Oct. 1973), 2287 and 2253; Loyd S. Swenson, Jr., "The Michelson-Morley-Miller Experiments Before and After 1905," in *Journal for the History of Astronomy*, **1**, no. 1 (1970), 56–78.

LOYD S. SWENSON, JR.

MICHURIN, IVAN VLADIMIROVICH (*b.* Dolgoye, Russia [now Michurovka, U.S.S.R.], 28 October 1855; *d.* Michurinsk, U.S.S.R., 7 June 1935), *plant breeding*.

For a complete study of his life and work, see Supplement.

MIDDENDORF, ALEKSANDR FEDOROVICH (*b.* St. Petersburg, Russia, 6 August 1815; *d.* Khellenurme [now Estonian S.S.R.], 16 January 1894), *biogeography*.

Middendorf graduated from the Third Petersburg Gymnasium, of which his father was director, and, in 1837, from Dorpat University with an M.D. For two years he studied zoology, botany, and geognosy at universities in Germany and Austria. In 1839 and 1840 he taught zoology at Kiev University. During the summer of 1839, he traveled to the Kola Peninsula with Karl Ernst von Baer.

In 1844 Middendorf completed a two-year journey to northern and eastern Siberia commissioned by the St. Petersburg Academy of Sciences. In 1845 he was elected to membership in the Academy, and in 1852 he became its permanent secretary. A sharp decline in his health obliged Middendorf in 1865 to relinquish his post as academician, but he was retained as an honorary academician. Middendorf subsequently resided at his estate, Khellenurme, where he completed a multivolume account of his Siberian journey and also journeyed to the Baraba Steppe in Western Siberia, and to the Fergana Valley in Central Asia.

Middendorf gave a brilliant geographical description and an ecological and geographical analysis of the fauna of Siberia, in which he examined in detail the concept of species, the causes of changes of species, the adaptation of animals to their environment, and laws of the geographical distribution of animals, including the distribution of boreal species in a zone surrounding the pole. No less valuable is his description of the geographical distribution and ecological peculiarities of Siberia's vegetation.

Two tasks had been assigned to the expedition to Siberia: to study the quality and quantity of organic life and to verify the presence and distribution of the permafrost discovered in many Siberian locations, especially in Yakutsk.

Middendorf twice crossed the Taymyr Peninsula and in Yakutsk revealed the mysterious phenomenon of permafrost and laid the scientific bases of the study of frozen soil. He calculated the geothermal gradient in the Fedor Shergin well and, on the basis of this calculation, determined the depth of the frozen layer under Yakutsk to be 204 meters (10 meters less than the current value). In the third stage of the expedition Middendorf crossed the Dzhugdzhur Range and investigated the flora and fauna of the Okhotsk Sea coastal areas and of the Shantar Islands.

Middendorf's Siberian journey led to the establishment of the Russian Geographical Society.

BIBLIOGRAPHY

I. ORIGINAL WORKS. Middendorf's main work, published originally in German, was *Reise in den aussersten Norden und Osten Sibiriens während der Jahre 1843 und 1844*, 4 vols. (St. Petersburg, 1848–1875), dealing with the climatology, geognosy, botany, zoology, ethnography, and the flora and fauna of the region; the Russian ed. was *Puteshestvie na sever i vostok Sibiri*, 2 pts. (St. Petersburg, 1860–1878).

His other writings include "Medved bury" ("The Brown Bear"), in Y. Simashko, *Russkaya fauna ili opisanie i izobrazhenie zhivotnykh, vodyashchikhsya v Imperii Rossyskoy* ("Russian Fauna or a Description and Depiction of the Animals Found in the Russian Empire"), 2 pts. (St. Petersburg, 1850–1851), 187–295; "O sibirskikh mamontakh" ("On Siberian Mammoths"), in *Vestnik estestvennykh nauk*, nos. 26–27 (1860), 843–868, with additions by N. Lyaskovsky; "Golfstrim na vostoke ot Nordkapa" ("The Gulfstream East of North Cape"), in *Zapiski Imperatorskoi akademii nauk*, **19**, no. 1 (1871), 73–101; "Baraba," *ibid.*, **19**, supp. (1871); and *Ocherki Ferganskoy doliny* ("Essays on the Fergana Valley"; St. Petersburg, 1882), with app. by F. Schmidt.

II. SECONDARY LITERATURE. There are biographies of Middendorf by K. Kirt (Tartu, 1963); N. I. Leonov (Moscow, 1967); and S. P. Naumov (Moscow, 1959), 323–331.

G. NAUMOV

MIDGLEY, THOMAS, JR. (*b.* Beaver Falls, Pennsylvania, 18 May 1889; *d.* Worthington, Ohio, 2 November 1944), *chemistry.*

Midgley was the son of Thomas Midgley, a successful inventor, and Hattie Lena Emerson. Following graduation from Betts Academy, he entered Cornell University where he took the course in mechanical engineering; he graduated in 1911. His subsequent work, however, lay in industrial chemistry, to which he brought a mastery of scientific fundamentals and a talent for ingenious experimentation.

In 1916, shortly after he had joined Charles F. Kettering's Dayton Engineering Laboratories, Midgley was assigned the problem of reducing internal-combustion knock. Gaseous detonation, or knock, was obstructing the development of Kettering's Delco engine; the phenomenon was only imperfectly understood, and was initially attributed to the ignition of the battery employed in Kettering's self-starting device for automobiles. From 1917 until 1921, Midgley used a variety of experimental techniques directed toward finding a chemical antiknock agent.

Midgley and Kettering assumed that knock was an inverse function of volatility. They first tried iodine as a fuel additive, supposing that a red dye might cause low-volatility fuel to absorb radiant heat and citing the example of the red-backed trailing arbutus that blooms under snow. They found that knocking did decrease greatly, although further tests demonstrated that the red color was inconsequential. They had nonetheless shown that a chemical antiknock agent does exist; since iodine proved impractical as a fuel additive, Midgley undertook to test at least one compound of each chemical element in an attempt to find something better. After months of research he established that aniline and its homologues—as well as other nitrogenous compounds—are effective chemical agents, but that they also give off an unbearable smell.

Midgley then began to make use of Robert E. Wilson's arrangement of the periodic table, which was based on Langmuir's theory of atomic structure and chemical valence. He employed a bouncing-pin indicator to measure knock quantitatively, then correlated these measurements with the table to establish trends; he thus discovered the antiknock properties of lead and, on 9 December 1921, singled out the remarkable effectiveness of tetraethyl lead. Added to gasoline, tetraethyl lead improved the engine compression ratio, thereby economizing fuel. The substance was put into large-scale production after a number of difficulties had been overcome, including a moratorium placed upon its use to allow a U. S. surgeon general's committee to investigate any possible danger of widespread lead poisoning. Midgley was then appointed general manager of the Ethyl Gasoline Corporation.

In 1930 Kettering, who was in charge of research at General Motors, asked Midgley to find a nontoxic, nonflammable, and cheap refrigerant for use in household appliances. Again drawing upon the periodic table, Midgley discovered dichlorodifluoromethane within three days, a compound that possessed all the desired qualities. He was less successful in his investigations of natural and synthetic rubber, however. These constitute the most purely scientific of his works.

Midgley received the four principal American medals for achievement in chemistry: the Nichols Medal (1922), the Perkin Medal (1937), the Priestley Medal (1941), and the Willard Gibbs Medal (1942). He was president of the American Chemical Society; in his presidential address to that body, delivered only a few months before he died, he suggested that scientists older than forty should remove themselves from positions requiring a high order of creativity, since he thought that most of the great discoveries and inventions had been made by workers between the ages of twenty-five and forty-five.

BIBLIOGRAPHY

I. ORIGINAL WORKS. Two articles indicative of Midgley's formal and informal communications are, respectively, "The Chemical Control of Gaseous Detonation With Particular Reference to the Internal-Combustion Engine," in *Journal of Industrial and Engineering Chemistry*, **14** (1922), 894–898, written with T. A. Boyd; and "From the Periodic Table to Production," *ibid.*, **29** (1937), 241–244.

II. SECONDARY LITERATURE. Biographical essays on Midgley include one by his mentor, professional associate, and friend, Charles F. Kettering: "Biographical Memoir of Thomas Midgley, Jr.," in *Biographical Memoirs. National Academy of Sciences*, **24** (1947), 361–380, which includes a list of 57 articles by Midgley but does not list his patents. An extremely useful essay is by Midgley's close research associate, T. A. Boyd: "Thomas Midgley, Jr.," in *Journal of the American Chemical Society*, **75** (1953), 2791–2795, also with a list of articles. Williams Haynes, the historian of the American chemical industry, has written an essay on Midgley in *Great Chemists*, Eduard Farber, ed. (New York, 1961), 1589–1597. On Midgley's work, especially the history of tetraethyl lead, see T. A. Boyd, "Pathfinding in Fuels and Engines," in *S.A.E. Quarterly Transactions*, **4** (1950), 182–185; and *Professional Amateur: The Biography of Charles Franklin Kettering* (New York, 1957); and Williams Haynes, *American Chemical Industry*, IV (New York, 1948).

THOMAS PARKE HUGHES

MIE, GUSTAV (*b*. Rostock, Germany, 29 September 1868; *d*. Freiburg im Breisgau, German Federal Republic, 13 February 1957), *physics*.

Mie, the son of a pastor, spent his childhood and went to high school in Rostock. He studied mathematics and the physical sciences at the University of Heidelberg, completing his doctorate in 1891 with a dissertation on a mathematical problem in partial differential equations. He then went to Dresden as a teacher of mathematics and physical sciences in a private school but did not stay there long. He accepted an assistantship in the Physics Institute of the Technische Hochschule in Karlsruhe. Mie's interests had turned from mathematics to physics, and in Karlsruhe, where Heinrich Hertz had done his famous experiments on electrical oscillations, Mie reassembled Hertz's apparatus and repeated the experiments. On completing his *Habilitation*, Mie became a *Privatdozent* at Karlsruhe in 1897.

Mie married in the spring of 1901. He was appointed extraordinary professor of experimental physics at the University of Greifswald in 1902 and was made an ordinary professor and director of its Physics Institute in 1905. He remained at Greifswald for fifteen years, a happy and scientifically productive period. From Greifswald, Mie moved to the University of Halle in 1917 as professor of experimental physics and stayed there until 1924. Mie then became the director of the Institute of Physics of the University of Freiburg im Breisgau, where he remained until his retirement in 1935.

In 1908 Mie published the rigorous electrodynamic calculation of light diffraction from spherical dielectric and conducting particles. This, together with the explanation of color effects, led to the discovery of the asymmetry in the intensity distribution and the precise determination of the optical constants of suspended particles. Called the Mie effect, it has had increasing importance in the determination of molecular clusters in solutions and the investigation of interstellar matter.

Before 1914, encouraged by Russian researches in the field, Mie solved the problem of the anomalous dispersion of water by using his quenched-spark oscillator as emitter and thermal elements connected with a spherically coated galvanometer as receiver. These experiments led to the determination of an invariant dielectric constant for water. They also brought understanding of the free rotation of polar groups in molecules and the frictional dispersion of dipole molecules in highly viscous solutions.

The other main direction of Mie's experimental work was a series of X-ray analyses of the crystal structure of organic compounds, especially anthracene and naphthalene, which he began at Halle soon after the work of the Braggs. Mie continued these studies in collaboration with Staudinger at Freiburg, and the investigation of different polyoxymethylenes (as the model substances for cellulose) led to the verification of the molecular lattice.

Mie's greatest personal involvement was in his effort to understand the fundamental and general principles of physical phenomena, and to state them suitably. At Greifswald, during 1912–1913, he made the first attempt to construct a complete theory of matter in the twentieth century. In an imaginative extension of Maxwell's theory in the framework of special relativity, the elementary particles known at that time (electrons and protons) appear in Mie's work as offspring of a universal electromagnetic field. His goal was to overcome the traditional opposition between "field" and "matter," thereby seeking to obtain a "unity of the physical world-view." In particular, he wanted to explain "the existence of an indivisible electron and to relate the phenomenon of gravitation to the existence of matter."

Three assumptions formed the basis of Mie's theory:

1. Electric and magnetic fields exist both inside and outside the electrons.

2. The principle of special relativity is valid throughout.

3. "The hitherto known states of the ether, namely the electric field, the magnetic field, the electric charge, and the charge current are entirely sufficient to describe all phenomena in the material world."

From these three assumptions, Mie was led to a generalization of the equations for the ether. This extension of the Maxwell-Lorentz theory was determined on the basis of the validity of the principle of conservation of energy and the existence of a localizable energy. Mie did not realize how many unnecessary assumptions were built into his derivation. His theory ran into serious difficulties, some of which stem from the fact that no one has succeeded in deriving solutions for static electrons in which the charge is "quantized."

Mie was the first to recognize the necessity of "quantizing" the field variables of the electromagnetic field, long before Heisenberg and Pauli developed the first fundamentals of a rational quantum field theory. This insight aroused the admiration of David Hilbert, who was inspired by the "deep ideas and original concepts on which Mie had built his electrodynamics." This theory, together with Einstein's ideas on gravitation and relativity, led Hilbert to develop an axiomatic theory of the foundations of physics, from which he derived the field equations of gravitation (together with their auxiliary conditions as given by the Bianchi identities) and the equations of the electromagnetic field.

Although Mie did not discover the appropriate "world function" that could account for the existence, asymmetry, and stability of the proton and the electron, his investigations later inspired the work of Max Born and Leopold Infeld on "nonlinear electrodynamics," which corresponded entirely with Mie's program. Mie's theory of matter would probably be regarded as his greatest contribution to physics.

The originality of Mie's ideas lay in his treatment of electromagnetic phenomena. His *Textbook on Electricity and Magnetism*, with the subtitle *An Experimental Physics of World-Ether*, was constructed on the fundamental distinction between the "quantities of intensity" and "quantities of magnitude." The "quantities of magnitude" are the length of a path or the duration of an event, the inertial mass of a body, and the electric charge. The "intensive" quantities are, for instance, "force" in mechanics, and the electric and magnetic field strengths E and B in the expression for Lorentz' ponderomotive force.

BIBLIOGRAPHY

I. ORIGINAL WORKS. Mie's articles include "Grundlagen einer Theorie der Materie," in *Annalen der Physik*, **37** (1912), 511–534; **39** (1912), 1–40; and **40** (1913), 1–66; and the autobiographical sketch "Aus meinem Leben," in *Zeitwende*, **19** (1948), 733–743.

Among his books are *Textbook of Electricity and Magnetism* (1910; 2nd ed., 1941; 3rd ed., 1948); *Die Grundlage der Quantentheorie* (Freiburg im Breisgau, 1926); *Elektrodynamik*, XI, pt. 1 of the series Handbuch der Experimentalphysik, W. Wien and F. Harms, eds. (Leipzig, 1932); *Molecules, Atoms and Ether; Die Einsteinsche Gravitationstheorie; Die Denkweise der Physik; Naturwissenschaft und Theologie;* and *Die Grundlagen der Mechanik* (1950).

II. SECONDARY LITERATURE. Articles on Mie are H. Hönl, "Intensitäts- und Quantitätsgrössen," in *Physikalische Blätter*, **24** (1968), 498–502, commemorating the centenary of Mie's birth; and W. Kast, "Gustav Mie," *ibid.*, **13** (1957), 129–131.

Max Born's review article on the Born-Infeld nonlinear electrodynamics, in *Annales de l'Institut Henri Poincaré*, **7** (1937), 155, gives an explicit discussion of the relation of this work to Mie's. For Mie's work on the field theory of matter see J. Mehra, "Einstein, Hilbert, and the Theory of Gravitation," in *The Physicist's Conception of Nature* (Dordrecht, 1973).

JAGDISH MEHRA

MIELI, ALDO (*b.* Leghorn, Italy, 4 December 1879; *d.* Florida, Argentina, 16 February 1950), *chemistry, history of science.*

Mieli graduated in chemistry from the University of Pisa, then went to Leipzig to attend Ostwald's lectures on physical chemistry. He next studied mathematics with Ulisse Dini and chemistry with Stanislas Cannizzaro and Emanuele Paternò. He was Paternò's assistant at the University of Rome from 1905 to 1912. He became a docent at that university in 1908; in the same year he also published two articles on chemistry. Mieli's interests at this time were not confined to chemistry, however. During the same period he wrote a number of articles of general cultural interest, as well as works on the philosophy of science and the relationships between science and art.

From 1912 on, Mieli began to devote himself to the history of science. He was one of the first to consider this study as an autonomous discipline, and his position was consolidated when he became the Italian bibliographic editor for *Isis*, which had just been founded. He also collaborated in editing the journal *Scientia* and edited the *Rivista di storia delle scienze mediche e naturali*, in whose pages he initiated a campaign to have the history of science taught in the universities. In 1913–1914, with E. Trollo, he began the series *Classici della Scienza e della Filosofia*, which was inspired by Ostwald's collection. In 1919 the publisher Nardecchia suggested that he take over the bibliographical work *Gli scienziati italiani*; while only one full volume (and part of another) appeared, it included studies by A. Boffito, A. Corsini, A. Favaro, G. Loria, and Mieli himself. The journal *Archivio di storia della scienza* was founded in the same year; Mieli became its editor in 1921 (after 1925, the journal was called *Archeion*).

At this time, too, Mieli invested his own money in founding the Leonardo da Vinci publishing house. Among its first publications were the *Universitas scriptorum* and the *Rivista di studi sessuali e di eugenetica* (Mieli was secretary of the Italian Society for Sexual Studies, and editor of its journal until 1928). The most notable of Mieli's own works that were published by his house are *Pagine di storia della chimica* (1922) and *Manuale di storia della scienza: Antichità* (1925).

In 1928 political considerations forced Mieli to leave Italy. He went to Paris, where he became director of the section for the history of science of the Centre International de Synthèse, to which he gave his large history of science library. He also continued to edit *Archeion* and, at a meeting of the International Congress of the Historical Sciences held in Oslo, proposed the formation of an International Committee for the History of Sciences. The members of the committee included Abel Rey, Sarton, Sigerist, Charles Singer, Sudhoff, and Lynn Thorndike; Mieli

served as secretary of the group and organized the First International Congress of the History of Science, held in Paris in May 1929. During this congress, the committee transformed itself into the International Academy for the History of Science, and *Archeion*, under Mieli's editorship, became its official journal.

In 1939, on the eve of World War II, Mieli again exiled himself. He went to Argentina, where from 1940 until 1943 he taught the history of science at the Universidad Nacional del Litoral in Santa Fé. He created an Institute for the History and Philosophy of Science, continued to edit *Archeion*, and published a summary of his lectures. In 1943, however, the political situation in Argentina, and especially the intervention of the government into university affairs, forced Mieli to leave Santa Fé, and he retired to Florida, near Buenos Aires, sadly spent in both health and finances. To recoup the latter, he began to write his *Panorama general de historia de la ciencia*. Of the eight volumes that he planned, two were published before his death; he saw only the proofs of volumes III, IV and V. (The work was finished by Desiderio Papp and José Babini; it was eventually published in twelve volumes.) Mieli was gravely ill for the last three years of his life. He gave up the editorship of *Archeion*, which became the *Archives internationales d'histoire des sciences*. In the first issue of the newly renamed journal Mieli wrote, "Je puis mourir tranquille en sachant qu'une partie, au moins, des multiples efforts que j'ai amplement déployés pendant ma vie, pour la réalisation de maints idéaux, va continuer à exercer son action bienfaisante."

BIBLIOGRAPHY

I. ORIGINAL WORKS. Mieli contributed significantly to the journals *Archeion, Archives internationales d'histoire des sciences, Archivio di storia della scienza, Gazzetta chimica italiana, Isis, Miniera italiana, Rendiconti della R. Accademia dei Lincei, Rendiconti della Società chimica italiana, Rivista di biologia, Rivista di filosofia, Rivista di storia critica delle scienze mediche e naturali*, and *Scientia* (Bologna).

His separate publications include *Influenza che esercita un sale in varie concentrazioni sulla velocità di decolorazione di soluzioni acquose di sostanze organiche sotto l'influenza della luce* (Rome, 1906), written with G. Bargellini; *Catalogo ragionato per una biblioteca di cultura generale. Storia delle scienze* (Milan, 1914); *Programma del corso di storia della chimica tenuto nella Università di Roma durante l'anno scolastico 1913–1914* (Chiusi, 1914); *La scienza greca: I prearistotelici. I. La scuola ionica, la scuola pitagorica, la scuola eleata, Herakleitos* (Florence, 1915); *Programma del corso di storia della chimica tenuto nell'Università di Roma durante l'anno scolastico 1914–1915* (Florence, 1915); *La storia della scienza in Italia* (Florence, 1916; Rome, 1926); *Per una cattedra di storia della scienza* (Florence, 1916); *Il libro dell'amore* (Florence, 1916), which he considered his spiritual testament; *Lavoisier* (Genoa, 1916; 2nd ed., Rome, 1926); *Lavori e scritti di Aldo Mieli. I. (1906–1916)* (Florence, 1917); *Pagine di storia della chimica* (Rome, 1922); *Manuale di storia della scienza: Antichità, storia, antologia, bibliografia* (Rome, 1925), trans. into French as *Histoire des sciences. Antiquité* (Paris, 1935); *Un viaggio in Germania. Impressioni ed appunti di uno storico della scienza* (Rome, 1927); *La science arabe et son rôle dans l'évolution scientifique mondiale* (Leiden, 1938), with additional material by H.-P.-J. Renaud, Max Mayerhof, and Julius Ruska; *El desarollo histórico de la historia de la ciencia y la función actual de los institutos de historia de la ciencia* (Santa Fé, 1939); *Sumario de un curso de historia de la ciencia en ciento veinte números* (Santa Fé, 1943); *Lavoisier y la formación de la teoría química moderna* (Buenos Aires, 1944); *Volta y el desarollo de la electricidad hasta el descubrimiento de la pila y de la corriente eléctrica* (Buenos Aires, 1945); *Panorama general de historia de la ciencia. I. El mundo antiguo: Griegos y Romanos. II. La época medieval, mundo islamico y occidente cristiano* (Buenos Aires, 1945); and *La teoría atómica química moderna desde sus orígenes con J. B. Richter, John D lton y Gay-Lussac, hasta su definitivo desarollo con Sta islao Cannizzaro, el sistema periódico de los elementos y el número atómico* (Buenos Aires, 1947).

II. SECONDARY LITERATURE. On Mieli and his work see Andrea Corsini, "Aldo Mieli," in *Rivista di storia delle scienze mediche e naturali*, **41**, no. 1 (1950), 111–113; and P. Sergescu, "Aldo Mieli," in *Actes du VIe Congrès International d'histoire des sciences* (Amsterdam, 1951), 79–95.

MARIA LUISA RIGHINI BONELLI

MIERS, HENRY ALEXANDER (*b.* Rio de Janeiro, Brazil, 25 May 1858; *d.* London, England, 10 December 1942), *mineralogy*.

Miers was the third son of Francis Charles Miers, a civil engineer, and Susan Fry Miers. He won scholarships at Eton and Oxford, where he studied classics and science, and gained second-class honors in mathematics upon graduation in 1881. He was small in stature, and the handsome appearance and trim build that he retained for most of his life made him look younger than his years. His manner was that of a pleasant diplomat who endeavored to get things done by friendly negotiation, without controversy. Tempering this quality was an adventurous streak, which was nearly his undoing in an early balloon attempt and led him to travel extensively, to the Klondike, to South Africa, and to Russia.

Although he had no formal training directly related to crystallography, Miers prepared within a year for an opening as assistant at the British Museum by short stays with N. Story-Maskelyne at Oxford, W. J. Lewis at Cambridge, and Paul von Groth at Strasbourg. His work at the Museum, under Lazarus Fletcher, was largely concerned with descriptions of crystal forms. He first detected the merohedrism in cooperite and other minerals, thus helping to complete the recognition of the classes of naturally occurring crystal symmetry. He described a number of sulfosalt minerals in detail, and the complex morphology of these crystals led to his interest in crystal growth. Miers constructed an ingenious inverted goniometer for measuring crystal faces while they were growing in solution. This led to direct observation of the ubiquitous but variable presence of slightly divergent (vicinal) faces during crystal growth. He realized the role of growth in the matching of a low reticular density on such faces with the lower density of matter in the solution, but the real underlying reason did not become apparent for another fifty years.

After his appointment as Wayneflete professor of mineralogy at Oxford in 1895, Miers refined his apparatus so that it could also measure the concentration of solution at the growing crystal surface through the index of refraction as determined by total reflection. This ingenious approach enabled him and a group of students to observe directly the conditions of various styles of crystallization, from slow, regular growth of large crystals at low supersaturations to the shower of microscopic crystals at a critical high degree of supersaturation, which were correlative with Ostwald's metastable and labile conditions, respectively.

Miers began his teaching in 1886, when H. E. Armstrong asked him to give a course in crystallography at the recently opened City and Guilds of London Institute (later part of Imperial College). His most famous student was William J. Pope, later professor of chemistry at Manchester and then at Cambridge. William Barlow is said to have learned his crystallography from Miers, apparently without formally registering as a student. Students at Oxford who later gained prominence in the field were Thomas V. Barker and Harold Hartley. Miers completed his textbook on mineralogy in 1902; it went through a second edition in 1929 and was translated into French. It drew heavily on Dana's *Mineralogy* and, for its time, contained excellent discussions of, for example, crystal optics as treated by Lazarus Fletcher.[1] The description of internal structures of crystals, which was then only a theory, was relegated to an appendix, although Miers had certainly followed closely developments in the field.[2]

In 1908 Miers was appointed principal of the University of London, the first of a long series of administrative and committee posts that occupied most of his time and energy for the remainder of his life. Although his next appointment, as vice-chancellor of the University of Manchester (1915), also created a special chair of crystallography, he did not publish further scientific work. That his teaching there was influential is amply shown by one student, H. E. Buckley, who later became head of the crystallography department at Manchester and who included much from Miers in his book *Crystal Growth*.[3] Published in 1951, just as F. C. Frank's revolutionary theory of crystal growth by dislocations had appeared, it was the most important inspiration and source of information on earlier work for the new school of crystal growth in the 1950's.[4]

Miers was elected to the Royal Society in 1896 and knighted in 1912. He received half a dozen honorary degrees, the Wollaston Medal of the Geological Society of London, and numerous other honors.

NOTES

1. J. D. Dana, *System of Mineralogy*, 6th ed. by E. S. Dana (New York, 1892); E. S. Dana, *Textbook of Mineralogy* (New York, 1877). On crystal optics see Miers, *Mineralogy*, pp. 118–165; and Lazarus Fletcher, *The Optical Indicatrix and the Transmission of Light in Crystals* (London, 1892).
2. *Mineralogy*, pp. 283–287; Miers, "Some Recent Advances in the Theory of Crystal Structures," in *Nature*, **39** (1889), 277–283; "Homogeneity of Structure the Source of Crystal Symmetry," *ibid.*, **51** (1894), 79–142; "The Arrangement of Molecules in a Crystal," in *Science Progress*, **1** (1894), 483–500; "The Arrangement of the Atoms in a Crystal," *ibid.*, **3** (1895), 129–142; William Barlow, H. A. Miers, and G. F. Smith, "Report of Committee on the Structure of Crystals. Part I. Report on the Development of the Geometrical Theories of Crystal Structure," in *Report of the British Association for the Advancement of Science* (1901), 297–337.
3. H. E. Buckley, *Crystal Growth* (New York, 1951).
4. F. C. Frank, *Discussions of the Faraday Society*, **5** (1948), 48–54.

BIBLIOGRAPHY

I. ORIGINAL WORKS. A complete bibliography is provided by L. J. Spencer in *Mineralogical Magazine*, **27** (1944), 23–28, including works on descriptive mineralogy and crystallography and on the growth of crystals; Miers's extensive reports on museums, libraries, and education; and the known MSS. In addition to his textbook, *Mineralogy, an Introduction to the Scientific Study of Minerals* (London, 1902; 2nd ed., revised by H. L. Bowman, 1929), the most important publications are perhaps "Contributions to the Study of Pyrargyrite and Proustite," in *Mineralogical Magazine*, **8** (1888), 27–102; "Xanthoconite and Ritteringite, With Remarks on the Red Silvers," *ibid.*, **10** (1893), 185–216; "An Enquiry Into the Variation of Angles Observed in Crystals; Especially of Potassium-Alum and

Ammonium-Alum," in *Philosophical Transactions of the Royal Society*, **202A** (1903), 459–523; and "The Refractive Indices of Crystallising Solutions With Especial Reference to the Passage From the Metastable to the Labile Condition," in *Journal of the Chemical Society*, **89** (1906), 413–454, written with F. Isaac.

II. SECONDARY LITERATURE. A number of memorials were published, the most important being those of L. J. Spencer, in *Mineralogical Magazine*, **27** (1944), 17–23, with photograph; H. T. Tizard, in *Dictionary of National Biography*, *1941–1950* (London, 1959), 588–590; and Sir Thomas Holland, in *Obituary Notices of Fellows of the Royal Society of London*, **12** (1943).

WILLIAM T. HOLSER

MIESCHER, JOHANN FRIEDRICH II (*b.* Basel, Switzerland, 13 August 1844; *d.* Davos, Switzerland, 26 August 1895), *physiology, physiological chemistry.*

Shortly after his birth Miescher (who was known as Fritz because his father bore the same names) was taken with his family to the Bernese Emmental, where his father had just been appointed professor of pathological anatomy at the University of Bern. When local canton politics caused the elder Miescher's resignation in 1850, the family returned to Basel, where Miescher excelled at the Gymnasium and in the musical circle to which his father and his uncle, Wilhelm His, Sr., belonged, together with the famous chemist C. F. Schönbein.

Apart from a semester spent at Göttingen Miescher remained in Basel, where he qualified in 1868. He then did research on the composition of pus cells, working in Hoppe-Seyler's laboratory at Tübingen (from 1868 to 1869). From there he went to Carl Ludwig's physiological institute in Leipzig (from 1869 to 1870), and finally returned to Basel for the *Habilitation*. In 1871 he was appointed to the chair of physiology at the University of Basel, which had been vacated by his uncle. There he worked to isolate nuclein from the sperm heads of Rhine salmon and to solve the mystery of the fasting-reproductive stage of the male. His resources were meager, and working conditions poor. When the effects of tuberculosis necessitated treatment in a sanatorium, Miescher used the opportunity to study the effects of altitude on the constitution of the blood and observed the increase in the blood count of erythrocytes with increasing altitude. In 1885 the university built an institute for him, the Vesalianum, where he was joined by Bunge.

Miescher was an unimpressive teacher and an often obscure writer. His subtle mind demanded caution and qualifications; his indecision over the rival claims of physical and of chemical reductionism, coupled with his distrust of cytology, prevented him from unifying that subject with chemistry, as he had sought to do. Miescher nonetheless had an eye for a good problem and for appropriate research material; he was observant and painstaking, and if he sought to achieve more than the best techniques of his day could allow, he established a method for physiological chemistry that his successors eagerly exploited.

Miescher's first and most important discovery was a new class of compounds rich in organic phosphorus and forming the major constituent of cell nuclei. He rightly concluded that these "nucleins," as he called them, were as important a center of metabolic activity as the proteins. The product he obtained from the pepsin digestion of pus cells in 1869 was nucleohistone; five years later he isolated a purer form of nuclein from salmon spermatozoa and demonstrated the saltlike union between its two major constituents, an acid fraction ("pure nuclein," or DNA) and a basic fraction (which he called "protamine" and regarded as an alkaloid, rather than a protein). He left to others the task of establishing the detailed chemical constitution of these compounds, and his own knowledge of them was limited to their solubility characteristics, elementary components, and reactions with histochemical tests.

Instead Miescher preferred investigating the formation of large amounts of nuclein by the male salmon during the fasting period. He concluded that this activity is achieved only at the expense of the trunk muscles of the fish, and suggested that these muscles are progressively decomposed or "liquidated" because of reduced oxygen supply. He further attempted to trace these changes at the chemical level, thereby providing a remarkable early example of the "dynamic biochemistry" that F. G. Hopkins later advocated. (Miescher's effort was, however, unfortunate in that it encouraged others in persistent attempts to show that nucleic acids are derived from proteins; these researches in turn fostered a false view of nucleic acids as compounds formed between proteins and phosphoric acid.)

In histochemistry Miescher established a clear chemical distinction between the nucleus and cytoplasm, based on the presence of nuclein in the nucleus, a distinction that supporters of the chemical theory of staining gladly embraced. Miescher's own parallel studies of the staining reactions of the spermatozoa were in conflict with the results achieved by Walther Flemming and P. Schweigger-Seidel, however, and thus lent credence to those who, like Albert Fischer, sought to discredit the chemical theory of staining altogether.

Miescher's interpretation of fertilization vacillated between the extreme physicalist reductionism of his uncle and the chemical theory that his own work suggested. He was never able to accept the notion that the structures revealed by cytological staining are themselves carried by the sperm into the egg to contribute to the structure of the embryo. Miescher saw this morphological theory, as Hertwig called it, as flying in the face of the reductionist program to which he was committed.

Miescher died of tuberculosis before he had completed his last paper on nuclein. A full account of his studies on this was compiled and published by Ostwald Schmiedeberg.

BIBLIOGRAPHY

I. ORIGINAL WORKS. All of Miescher's papers and a selection of his letters will be found in *Die histochemischen und physiologischen Arbeiten von Friedrich Miescher. Gesammelt und herausgegeben von seinen Freunden* (Leipzig, 1897). His paper on the isolation of nuclein from pus is "Ueber die chemische Zusammensetzung der Eiterzellen," in F. Hoppe-Seyler's *Medisch-chemische Untersuchungen*, IV (Berlin, 1871), 441–460. His first publication on salmon nuclein is "Die Spermatozoen einiger Wirbelthiere. Ein Beitrag zur Histochemie," in *Verhandlungen der Naturforschenden Gesellschaft in Basel*, 6 (1874), 138–208; his last and posthumous paper was "Physiologisch-chemische Untersuchungen über die Lachsmilch von F. Miescher, nach den hinterlassenen Aufzeichnungen und Versuchsprotokollen des Autors und herausgegeben von O. Schmiedeberg," in *Archiv für experimentelle Pathologie und Pharmakologie*, 37 (1896), 100–155.

II. SECONDARY LITERATURE. The clearest account of Miescher's work on the chemistry of the nucleus is by A. Mirsky, "The Discovery of DNA," in *Scientific American*, 218 (1967), 78–88. Biographical information is in M. de Meuron-Landot, "Friedrich Miescher, l'homme qui a découvert les acides nucléiques," in *Histoire de la médecine*, 15 (1965), 2–25. Earlier studies include the following: K. Spiro, *Zur Erinnerung an Schönbein, Miescher und Bunge* (Basel, 1922), originally published in *Basler Nachrichten*, 12 and 19 Feb. 1922; J. P. Greenstein, "Friedrich Miescher, 1844–1895. Founder of Nuclear Chemistry," in *Scientific Monthly*, 57 (1943), 523–532. The best obituary notice is A. Jaquet, "Professor Friedrich Miescher Nachruf," in *Verhandlungen der Naturforschenden Gesellschaft in Basel*, 11 (1897), 399–417. Further information is contained in F. Suter et al., *Friedrich Miescher, 1844–1895. Vorträge gehalten anlässlich der Feier zum hunderten Geburtstag von Professor Friedrich Miescher in der Aula der Universität Basel am 15 Juni 1944* (Basel, 1944); and in *Helvetica physiologica et pharmacologica acta*, supp. 2 (1944).

ROBERT OLBY

MILHAUD, GASTON (*b.* Nîmes, France, 10 August 1858; *d.* Paris, France, 1 October 1918), *mathematics, philosophy of science.*

Milhaud, a village near Nîmes, once belonged to the bishop of Nîmes and thus was able to shelter a Marrano community. Gaston Milhaud's ancestors came from this locality. He was the third of the famous trio with the same Christian name who brought fame to Nîmes during the nineteenth century; the other two were the historian Gaston Boissier and the mathematician Gaston Darboux, whose student he was at the École Normale Supérieure.

In 1878 Milhaud qualified for both the École Normale Supérieure and the École Polytechnique; he chose the former. *Agrégé* in mathematics in 1881, he then taught mathematics at Le Havre for ten years. His meeting with Pierre Janet and the fruitful collaboration that followed during this period induced a shift in his interests. He translated du Bois-Reymond's *Théorie générale des fonctions*; wrote a number of articles for such journals as *Revue scientifique, Revue des études grecques,* and *Revue philosophique de la France et de l'étranger*; and was henceforth concerned with the philosophy of mathematics.

Appointed professor of mathematics at Montpellier in 1891, Milhaud gave a series of lectures on the origins of Greek science (published in 1893). In 1894, at Paris, he defended a Ph.D. dissertation on the conditions and limits of logical certainty. This remarkable work was decisive for his career. He was appointed to the chair of philosophy at the Faculty of Letters of Montpellier in 1895 and rapidly became, through his lectures and publications, a respected authority in a field that was then quite new. He also arranged meetings between investigators in various disciplines. In 1909 a chair was created for Milhaud at the Sorbonne in the history of philosophy in relation to science. Despite the decline in his health, which had always been delicate, he continued to be active and held this chair with distinction until his death.

It has been observed that the end of the nineteenth century witnessed two complementary movements in response to the crisis in the foundations of science: that of philosophers becoming scientists and that of scientists becoming philosophers. Milhaud is one of the best representatives of the latter trend. He modestly presented himself as a teacher who wished to do useful work in the history of science, which he conceived of as "inseparable from a critical examination of fundamental notions and inseparable from philosophical views that, underneath the precise data that are constantly accumulating, attempt to appear and to evaluate the progressive and continuous work being accomplished" (quoted in Pierre Janet, "Notice," p. 57).

Acutely aware of the effort required to amass and criticize data, Milhaud declared that he was not learned in this respect. Nevertheless, his many works on Greek science show that he accepted the burdens of scholarship; and his study of the arguments of Zeno of Elea is important and still worth consulting. He was also responsible for renewing knowledge of Descartes as a scientist, and his writings on this subject remain a reliable source. It was Milhaud's second son, Gérard, who with Charles Adam produced an improved edition of Descartes's correspondence.

Milhaud oriented the study of the history of science more toward philosophy. Certain of his views, although representative of his time, are now outmoded, notably those of continuous progress and the analysis of the conditions, role, and scope of demonstration in mathematics and physics. But his writings on logical contradiction, the limits of the affirmations that it appears to permit, and the critique of scientifically inspired deterministic metaphysical systems are still of interest and justify the considerable influence he has exerted. Milhaud also illustrated his contention that "science progresses in proportion to the disinterestedness with which it is pursued." Émile Boutroux said in proposing Milhaud's election to the Académie des Sciences Morales et Politiques in 1918: "By the soundness and originality of his findings in both the theoretical and the historical domains regarding a question of paramount importance, that of the relation between certainty and truth, this conscientious, modest, and penetrating investigator has performed a lasting service to science and to philosophy" (Pierre Janet, "Notice," p. 58).

BIBLIOGRAPHY

I. ORIGINAL WORKS. Milhaud's books include *Leçons sur les origines de la science grecque* (Paris, 1893); *Essai sur les conditions et les limites de la certitude logique* (Paris, 1894; 4th ed., 1924); *Le rationnel* (Paris, 1898); *Les philosophes géomètres de la Grèce: Platon et ses prédécesseurs* (Paris, 1900; 2nd ed., 1934); *Le positivisme et le progrès de l'esprit (Études critiques sur Auguste Comte)* (Paris, 1902); *Études sur la pensée scientifique chez les Grecs et chez les modernes* (Paris, 1906); *Nouvelles études sur l'histoire de la pensée scientifique* (Paris, 1911); *Descartes savant* (Paris, 1921); *Études sur Cournot* (Paris, 1927); and *La philosophie de Charles Renouvier* (Paris, 1927).

Among Milhaud's many articles, the following appeared in *Revue de métaphysique et de morale:* "Le concept du nombre chez les Pythagoriciens" (1893), 140–156; "Réponse à Brochard" (1893), 400–404, concerning Zeno of Elea; "L'idée d'ordre chez Auguste Comte" (1901), 385–406; "Le hasard chez Aristote et chez Cournot" (1902), 667–681; and "*La science et l'hypothèse* par H. Poincaré"

(1903), 773–781. See also "Science et religion chez Cournot," in *Bulletin de la Société française de philosophie* (Apr. 1911), 83–104.

II. SECONDARY LITERATURE. See André Bridoux, "Souvenirs concernant Gaston Milhaud," in *Bulletin de la Société française de philosophie*, **55**, no. 2 (1960), 109–112; Edmond Goblot, "Gaston Milhaud (1858–1918)," in *Isis*, **3** (1921), 391–395; Pierre Janet, "Notice sur Gaston Milhaud," in *Annuaire des anciens élèves de l'École normale supérieure* (1919), pp. 56–60; André Nadal, "Gaston Milhaud (1858–1918)," in *Revue d'histoire des sciences . . .* (Paris), **12**, no. 2 (1959), 97–110; Dominique Parodi, *La philosophie contemporaine en France* (Paris, 1919), pp. 211–216; and René Poirier, *Philosophes et savants français du XXᵉ siècle*, II, *La philosophie de la science* (Paris, 1926), 55–80; and "Meyerson, Milhaud et le problème de l'épistémologie," in *Bulletin de la Société française de philosophie*, **55**, no. 2 (1960), 65–94.

PIERRE COSTABEL

MILL, JOHN STUART (*b.* London, England, 20 May 1806; *d.* Avignon, France, 8 May 1873), *philosophy, economics.*

Mill was the son of James Mill, a London Scot who had risen from humble origins to become a prominent intellectual, a collaborator of Jeremy Bentham, and a leading exponent of utilitarianism. Mill's childhood was a singular one. He was educated at home by his father, learning both Greek and Latin before he was nine years old. All religion was excluded from his upbringing. James Mill, an even more rigid adherent than Bentham to the rationalism of the Enlightenment, was determined to educate his son to be another philosopher in the same mold. At sixteen the younger Mill started to earn his living as a clerk in the East India Office, where his father was a senior official. At seventeen he published his first article, in the *Westminster Review*, and in the same year he made his debut as a radical reformer, spending a day or two in the police cells for distributing pamphlets recommending contraception as a solution to the population problem.

In his posthumously published *Autobiography*, Mill recalls that at the age of twenty he went through a period of acute depression, from which he was delivered by reading the poetry of Wordsworth. Through Wordsworth he met a romanticism that challenged the whole rationalistic ethos in which he had been so carefully bred. After this experience, wrote Mill, "I did not lose sight of . . . that part of the truth I had learned before . . . but I thought that it had consequences which required to be corrected, by joining other kinds of cultivation with it" (*Autobiog-*

raphy, p. 34). Mill's aim thenceforth became to produce a philosophy that combined the virtues of rationalism with those of romanticism; but the contradictions between them proved to be too fundamental for even the ablest mind to reconcile, and Mill's philosophy is marred by a certain incoherence that even his most fervent admirers cannot deny.

Mill's most important work in pure philosophy was his *System of Logic*, which he began at the age of twenty-four and completed thirteen years later. Soon after he had started work on it, Mill met a beautiful, intelligent, and imperious young woman named Harriet Taylor. He fell in love with her, and she with him; but she was already the wife of a wholesale druggist and the mother of two children. In the nineteen years before the druggist's death enabled them to marry, Mill and Harriet Taylor were constantly in each other's company—"Seelenfreunden" ("soul friends"), as they put it, but not lovers. Victorian society's frowns (and his own sense of guilt) drove Mill to lead a lonely life, and Mrs. Taylor's hold over his thinking was immense. She was not a Wordsworthian but a rationalist of the left—and, paradoxically, she reinforced the influence of James Mill's training rather than that of romanticism.

Mill's marriage to Harriet Taylor took place in 1851; but seven years later she died at Avignon, and Mill bought a house there to live near her tomb. But by this time Mill's books had made him famous, and in 1865 he was persuaded by the controversial and progressive Viscount Amberley to stand for election to Parliament in Westminster. Mill was elected, and he sat until 1868 as an independent Liberal M.P. He died at the age of sixty-six, having just become the agnostic's equivalent of a godfather to Amberley's son, Bertrand Russell.

Mill's central endeavor as a philosopher was to provide science with a better claim to truth than that afforded by the skeptical philosophers of the seventeenth and eighteenth centuries. Locke had written: "As to a *perfect* science of natural bodies (not to mention spiritual beings) we are, I think, so far from being capable of any such thing that I conclude it lost labour to look after it." Mill disagreed. Indeed, he wrote his *Logic* precisely to formulate "a *perfect* science of natural bodies"—or, in other words, a demonstrative theory of induction—by which he hoped to reduce the conditions of scientific proof "to strict rules and scientific tests, such as the syllogism is for ratiocination."

Mill called himself a "philosopher of experience"; he believed that all knowledge of the universe is derived from sensory observation, and he opposed those who claimed that some knowledge of synthetic truth is either innate or acquired by rational insight.

He was what has come to be known as an empiricist, although that word did not then have the commonly accepted usage it has today and Mill rejected it. But he tried to give what we should call empiricism a form that would satisfy the nineteenth century's demand for certainty.

Mill's *Logic* seeks to diminish the value of knowledge achieved deductively—that is, by deriving particulars from universals—and to vindicate the importance of knowledge derived inductively, by the accumulation of evidence from particulars. Our "universal" knowledge, Mill argued, comes from particulars. We begin with particulars and end with particulars, and it is the method of science that enables us to formulate the "universals" or "generalities" that the mind knows.

In book II of his *Logic*, Mill claimed that even mathematics is, in a way, inductive. In the manner of Kant he said that mathematical propositions are synthetic propositions about the world of measurable things, but he denied the Kantian view that the mind imposes categories on experience. He argued instead that mathematical propositions are experimental truths of a highly general kind. Mill did not even admit that they are necessarily true, except in the sense that it is psychologically impossible for us to doubt them. His mathematical theory has not had much support from theorists of later generations. He is generally considered to have failed to solve logical problems by proposing psychological answers, and the reform of deductive logic that was begun by Boole and completed by Mill's "godson" Russell has suggested that Mill was mistaken about what could be done with inductive logic.

Book III of Mill's *Logic* has been more influential. Here Mill explained what he means by induction. He said it depends on the "assumption" that nature is uniform and that its future course will be like that of the past. Elementary induction is based on the enumeration of like instances: "All the crows we have seen are black, therefore all crows are black." Mill next distinguished between uniformity of "togetherness" and uniformity of sequence. In the first class he put properties that exist at the same time and can be measured or counted so as to give our knowledge a formal order. The second class, uniformities of sequence, he called "causal"; and here, instead of mere enumerations, he believed we can establish laws. These laws are discovered with the aid of Mill's famous "eliminative methods of induction."

These methods are (1) the canon of agreement, which asserts that if those instances in which a phenomenon occurs have only one feature in common, then that feature contains the cause of the phenom-enon; (2) the canon of difference, which asserts that if those instances in which the phenomenon occurs differ from instances in which it does not occur in only one feature, then that one feature contains the cause of the phenomenon; (3) the canon of residues (a variant of the canon of difference), which asserts that if we take away from a phenomenon all the effects we know to be caused by certain antecedents, then the remainder is the effect of the remaining antecedents; (4) the canon of concomitant variations, which asserts that when one phenomenon varies only when another varies, there is either a causal relation between them or they are both causally related to a third factor.

Although Mill's "eliminative methods of induction" have figured prominently in subsequent controversies about scientific method, their value has been criticized on several grounds. First, they cannot be used to vindicate the assumptions on which they are grounded: the uniformity of nature and the ubiquity of causality. Second, no method of elimination can yield demonstrably certain conclusions about the candidates that remain, although it may well yield high probabilities. Third, science is not primarily interested in the kind of "common sense" causal relations that Mill's methods can be used to discover. Fourth, science is not properly understood as an inductive enterprise; it does not proceed by the observation of regularities in nature but by the use of conjecture and "experimental refutation."

Some of these objections to Mill's inductivism can be met by a more sophisticated reformulation of his thesis, but the consensus among twentieth-century specialists in scientific method is that the more skeptical approach of Mill's predecessors, including perhaps Kant as well as Locke and Hume, comes closer both to the realities of scientific discovery and to the exigencies of logic.

In 1848, shortly after the publication of his *Logic*, Mill brought out another of his most influential books, *The Principles of Political Economy*. This is a curious mixture of orthodox economic theories and arresting, original ideas. Some of the new ideas are expressed in the language of classical economics, so that the shock of them is softened. Mill maintained that "the economic man" is a fiction, a way of registering the tendency of men to pursue wealth. He suggested that economic principles should be tested by their stability in a particular era and by their ability to promote transition to another era.

In his review of political economy as a static science, Mill did little but repeat the principles laid down by Adam Smith and others about production and exchange, the dependence of wealth on production and of profit on the cost of labor, the token nature of money, the need to balance imports with exports, and so forth. It was when he turned from the static to the dynamic side of the subject, to economics as related to social progress, that Mill propelled economic thought into new channels.

He remained true to Malthus on the subject of the population problem. There was no remedy for poverty, he thought, unless excessive numbers could be reduced, although, unlike Malthus, Mill favored contraception as well as "moral restraint." But Mill differed from his predecessors in his understanding of the concept of property and on the distribution of wealth. Property rights were conventional; and although private property was a useful institution, the only basis of a sound entitlement was a man's own labor. There was no natural right to inheritance or to the ownership of land. In the first edition of his *Political Economy*, Mill criticized the socialist theories put forward by Louis Blanc and others as unrealistic, but in later editions he withdrew these strictures and wrote sympathetically of socialism. It is probable that he made these changes under pressure from Harriet Taylor, a convert to socialism.

The intrusion of socialist sentiments into a book that was substantially based on the principles of classical economics has seemed to some readers to be yet another mark of Mill's inconsistency. A similar criticism might be addressed to his writings on politics and ethics. In ethics Mill affirmed his adherence to his father's utilitarianism, the doctrine that the rightness of an act is to be measured by the extent to which it promotes pleasure. But Mill rejected his father's belief that pleasure has only quantitative differences. Ever since he had read Wordsworth, Mill had believed in the superiority of the "pleasures of the mind" over the brutish pleasures of the uncultivated: "Better Socrates dissatisfied than a pig satisfied." But Mill was never able to produce any utilitarian or empirical reason to justify this preference.

The same ardent belief in the values of culture influenced Mill's political theorizing. He was an eloquent champion of freedom; but although he sometimes defined freedom as the absence of constraint, he went on to speak of it as "self-perfection" and said that men should be free in order to improve themselves. Although Mill came out (as his father had done) in favor of democratic government, he proposed that democratic institutions should be carefully designed to prevent government by the majority: he wanted a form of government by a cultured elite that would rest upon the assent of a progressively more educated populace. Like many another intellectual of the Victorian period, Mill made something of a religion of the culture of the sensi-

bilities, notwithstanding his general belief, as an empiricist, that all real knowledge is scientific knowledge.

BIBLIOGRAPHY

I. ORIGINAL WORKS. A bibliography is N. MacMinn, J. R. Hainds, and J. M. McCrimmon, *The Bibliography of Published Works . . .* (Evanston, Ill., 1945). A more up-to-date bibliography is being published in serial form in the *Mill News Letter* (Toronto, 1965–   ).

Publication is in progress on the 13-vol. *Collected Works of John Stuart Mill* (Toronto, 1963–   ). Among his earlier writings are *A System of Logic*, 2 vols. (London, 1843); *Essays on Some Unsettled Questions of Political Economy* (London, 1844); *Principles of Political Economy*, 2 vols. (London, 1848); *On Liberty* (London, 1859), also repr. with *Representative Government* and an intro. by R. B. MacCallum (Oxford, 1946); *Thoughts on Parliamentary Reform* (London, 1859); *Dissertations and Discussions*, 2 vols. (London, 1859), articles repr. from periodicals, chiefly *Edinburgh Review* and *Westminster Review; Considerations on Representative Government* (London, 1861); *Utilitarianism* (London, 1863), also edited by J. Plamenatz (Oxford, 1949); *Auguste Comte and Positivism* (London, 1865); *An Examination of Sir William Hamilton's Philosophy* (London, 1865); *Inaugural Address to the University of St. Andrews* (London, 1867); *England and Ireland* (London, 1869); and *The Subjection of Women* (London, 1869), also edited with an intro. by S. Coit (London, 1906).

Works published later are *Autobiography*, Helen Taylor, ed. (London, 1873), repr., with an appendix of unpublished papers, and an intro. by H. J. Laski (London, 1924); *Three Essays on Religion* (London, 1874); *Chapters on Socialism*, W. D. F. Bliss, ed. (New York, 1891); *Early Essays*, J. W. M. Gibbs, ed. (London, 1897); *On Education*, F. A. Cavanagh, ed. (London, 1931), printed with writings on the same subject by James Mill; *The Spirit of the Age*, edited, with an intro., by F. A. von Hayek (Chicago, 1942), articles contributed to *Examiner* in 1831; *Four Dialogues of Plato*, edited, with an intro., by R. Borchardt (London, 1946); *On Bentham and Coleridge*, edited, with an intro., by F. R. Leavis (London, 1950), repr. from *Dissertations and Discussions; An Early Draft of John Stuart Mill's Autobiography*, J. Stillinger, ed. (Urbana, Ill., 1961); *Mill's Essays on Politics and Culture*, G. Himmelfarb, ed. (New York, 1962); *Mill's Essays on Literature and Society*, J. B. Schneewind, ed. (New York–London, 1965); and *Mill's Ethical Writings*, J. B. Schneewind, ed. (New York–London, 1965).

II. SECONDARY LITERATURE. Biographical studies are A. Bain, *John Stuart Mill* (London, 1882), a study of Mill and his philosophy, with personal recollections by a close friend of his later years; W. D. Christie, *J. S. Mill and Mr. Abraham Cowley, Q.C.* (London, 1873); G. J. Holyoak, *John Stuart Mill as the Working Class Knew Him* (London, 1873), a short personal memoir; and M. St. J. Packe, *The

Life of John Stuart Mill (London, 1954), the standard biography.

Critical and expository studies include R. P. Anschutz, *The Philosophy of J. S. Mill* (Oxford, 1953), and K. Britton, *John Stuart Mill* (London, 1953), both written from the point of view of modern analytic philosophy; Britton stresses the lasting value of Mill's achievement and is less sharply critical of Mill's shortcomings than is the Anschutz work; M. Cowling, *Mill and Liberalism* (Cambridge, 1963), a brisk conservative criticism of Mill; R. Jackson, *Examination of the Deductive Logic of J. S. Mill* (Oxford, 1941); O. A. Kubitz, *The Development of John Stuart Mill's System of Logic* (Urbana, Ill., 1932); E. Nagel, ed., *Mill's Philosophy of Scientific Method* (New York, 1950)—this and the works of Jackson and Kubitz are up-to-date books concerned with Mill as a logician; E. Neff, *Carlyle and Mill* (New York, 1926), an instructive comparison of rival philosophies; B. Russell, *John Stuart Mill* (London, 1955); J. B. Schneewind, ed., *Mill: A Collection of Critical Essays* (New York, 1968), short studies of Mill's thought from a philosophical perspective; and C. L. Street, *Individualism and Individuality in the Philosophy of John Stuart Mill* (Milwaukee, Wis., 1926), a study of Mill's ethical and political theory.

MAURICE CRANSTON

MILLER, DAYTON CLARENCE (*b.* Strongsville, Ohio, 13 March 1866; *d.* Cleveland, Ohio, 22 February 1941), *physics.*

Miller was the son of Charles Webster Dewey Miller and the former Vienna Pomeroy. In 1886 he was graduated from Baldwin-Wallace College and in 1890 was awarded a doctorate in science from Princeton University, having studied under the astrophysicist Charles A. Young. Miller then joined the faculty of the Case School of Applied Science in Cleveland as an instructor of mathematics. He transferred to the physics department in 1893, the year he married Edith Easton. He remained at Case as professor of physics until his death.

Miller was an effective teacher, a captivating public lecturer, and a respected research scientist. In 1914 he was elected to the American Academy of Arts and Sciences, in 1919 to the American Philosophical Society, and in 1921 to the National Academy of Sciences. From 1918 until his death he held various offices in the American Physical Society, including the presidency for the 1925–1926 term. From 1927 to 1930 he was chairman of the National Research Council's Division of Physical Sciences, and from 1913 to 1933 he was president of the Acoustical Society of America.

Miller's work in acoustics grew out of a love for music that dated from childhood. His mother had been the church organist, his father had sung in the choir, and Miller himself was an accomplished flutist. Keenly interested in the physics of musical tones, he invented what he called the phonodeik in 1908, a mechanical device that recorded sound patterns photographically. During World War I he used the apparatus to analyze the nature of gun wave-forms for the National Research Council, which was developing improved techniques to locate enemy artillery by sonic means. After the war Miller became an expert in architectural acoustics, consulting on the interior design of a number of college chapels as well as Severance Hall, the home of the Cleveland Orchestra.

As a research physicist Miller was best known for his elaborate repetitions of the experiment that Albert A. Michelson and Edward W. Morley had performed with an interferometer in 1887 to detect the stationary luminiferous ether postulated by Maxwell. Miller did the experiment in collaboration with Morley between 1902 and 1904. Repeating it by himself on Mt. Wilson, California, between 1921 and 1926, he found a positive effect corresponding to an apparent relative motion of the earth and the ether of some ten kilometers per second in the plane of the interferometer. Though this velocity was about 70 percent less than expected, Miller fastened on his result as a refutation of Einstein's theory of relativity, which he was unwilling to accept on principle to the end of his life.

When Miller presented his data in 1925, he provoked considerable interest among physicists and was awarded a $1,000 annual prize by the American Association for the Advancement of Science. Anti-relativists hailed his findings; relativists believed that they probably rested on experimental error. In the 1950's a group of physicists subjected Miller's Mt. Wilson data to statistical analysis. They found that only part of his positive readings could be attributed to random fluctuations. The rest seemed to result from an appreciable systematic effect whose magnitude varied with the conditions of observation. The cause of this effect appeared to be the large temperature changes which undoubtedly occurred in the poorly insulated shack that housed Miller's apparatus atop Mt. Wilson.

BIBLIOGRAPHY

I. ORIGINAL WORKS. Miller's personal papers and research notebooks are at Case Western Reserve University but are as yet unavailable to scholars. Miller's "The Ether Drift Experiment and the Determination of the Absolute Motion of the Earth," in *Reviews of Modern Physics*, **5** (July 1933), 203–242, is a comprehensive résumé of all Miller's repetitions of the Michelson-Morley experiment.

II. SECONDARY LITERATURE. Complete bibliographies of Miller's writings and introductions to his career are Harvey Fletcher, "Dayton Clarence Miller," in *Biographical Memoirs. National Academy of Sciences*, **23** (1945), 61–74; and Robert S. Shankland, "Dayton Clarence Miller: Physics Across Fifty Years," in *American Journal of Physics*, **9** (Oct. 1941), 273–283. Miller's ether-drift data is assessed in Shankland, *et al.*, "New Analysis of the Interferometer Observations of Dayton C. Miller," in *Reviews of Modern Physics*, **27** (Apr. 1955), 167–178; and his research is set in context in Loyd S. Swenson, Jr., *The Ethereal Aether: A History of the Michelson-Morley-Miller Aether-Drift Experiments, 1880–1930* (Austin, 1972).

DANIEL J. KEVLES

MILLER, GEORGE ABRAM (*b.* Lynnville, Pennsylvania, 31 July 1863; *d.* Urbana, Illinois, 10 February 1951), *mathematics.*

The description of Miller's rise to prominence in the world of mathematics is one of those Horatio Alger stories with which American intellectual history of the late nineteenth century is studded. He was the son of Nathan and Mary Sittler Miller and a descendant of one Christian Miller who had emigrated from Switzerland around 1720. Unable to continue his education without self-support, he began to teach school at the age of seventeen. He studied at Franklin and Marshall Academy in Lancaster during 1882–1883, then enrolled at Muhlenberg College, where he received the baccalaureate with honorable mention in 1887, the master of arts in 1890, and an honorary doctor of letters in 1936. Miller served as principal of the schools in Greeley, Kansas, during the year 1887–1888 and as professor of mathematics at Eureka College (Illinois) from 1888 to 1893. Cumberland University in Lebanon, Tennessee, granted him a doctorate in 1892; it was then possible to do course work by correspondence, and examinations in advanced courses were an acceptable substitute for thesis requirements. Miller was offering the same courses toward a doctorate to his students at Eureka. He spent the summers of 1889 and 1890 at the Johns Hopkins University and the University of Michigan but probably did not come under Frank Nelson Cole's influence until 1893, when he became an instructor at Michigan for three years and lived in Cole's home during the first two years of that period. It was Cole who inspired him to pursue the research in group theory that was to engage his talents for the rest of his life. Cole, incidentally, had been a pupil of Felix Klein, who had made groups basic in his "Erlanger Programm."

Miller spent the years 1895–1897 in Europe, attending the lectures of Sophus Lie at Leipzig and Camille Jordan in Paris. He soon was publishing papers independent of their specializations, although Lie had become instrumental in Miller's study of commutators and commutator subgroups and Jordan's interest in questions of primitivity and imprimitivity was reflected in Miller's investigations of those problems throughout his career.

Upon Miller's return to the United States, his European experience and mathematical productivity gained him an assistant professorship at Cornell (1897–1901). This was followed by an associate professorship at Stanford University (1901–1906), and in 1906 an appointment at the University of Illinois, an affiliation that lasted for the rest of his life—first as associate professor, then as professor, and finally as professor emeritus. His retirement in 1931 was from classroom responsibilities only, for he continued his research and writing in his office at the university. The university undertook, as "a fitting memorial of his contributions to mathematical scholarship and to the renown of the University," the collection and reprinting of Miller's studies in the theory of finite groups as well as other studies. It is said that of the more than 800 titles that appeared in some twenty periodicals over forty years approximately 400 made direct scientific contributions to that theory. Other papers were written in the hope that teachers of elementary and secondary mathematics might be inspired to study advanced mathematics. Miller himself aided in the preparation of these memorial volumes.

This was not the only legacy that Miller left to the University of Illinois. To the great surprise of the colleagues who knew him well, the university found itself after his death the beneficiary of a bequest valued at just under one million dollars, the accumulation of judicious investments. His wife, the former Cassandra Boggs of Urbana, had predeceased him in 1949 and there were no children.

Miller's interest in the history of mathematics was second only to that in the theory of finite groups. His articles on the history of his own subject were of particular significance and value. He became a severe critic of historical methodology in mathematics and was zealous in rooting out error in conjecture or assumed fact. His letters in the David Eugene Smith collection at Columbia University offer ample evidence of this missionary fervor. His "History of Elementary Mathematics" remained unpublished, although there was originally the intention to include it in a volume of the *Collected Papers*.

Miller was elected to the National Academy of Sciences in 1921 and was a fellow of the American

Academy of Arts and Sciences. In 1900, for his work in group theory, the Academy of Sciences of Cracow awarded him a prize that had not been given for fourteen years. This is said to have been the Academy's first award to an American for work in pure mathematics. He was a member of the London Mathematical Society, the Société Mathématique de France, and the Deutsche Mathematiker-Vereinigung, an honorary life member of the Indian Mathematical Society, and a corresponding member of the Real Sociedad Matemática Española. He was an active member and served in various high offices of the American Mathematical Society and the Mathematical Association of America, serving also as an editor of the latter's *American Mathematical Monthly* (1909–1915).

BIBLIOGRAPHY

I. ORIGINAL WORKS. Miller's writings were brought together in *The Collected Works of George Abram Miller*, 5 vols. (Urbana, Ill., 1935–1959). Two of his books are *Determinants* (1892) and *Historical Introduction to Mathematical Literature* (New York, 1916). A more detailed bibliography is in Poggendorff, IV, 1013; V, 855–857; and VI, 1737–1738. *Theory and Application of Finite Groups* (New York, 1916; repr. 1961) was written in collaboration with H. F. Blichfeldt and L. E. Dickson.

II. SECONDARY LITERATURE. See H. R. Brahana, "George Abram Miller (1863–1951)," in *Biographical Memoirs. National Academy of Sciences*, **30** (1957), 257–312, with a complete bibliography of Miller's writings (1892–1947) on 277–312. E. T. Bell makes occasional reference to Miller's work in "Fifty Years of Algebra in America, 1888–1938," in *American Mathematical Society Semicentennial Publications*, II, as does Florian Cajori in his *History of Mathematics*. See also J. W. A. Young, *Monographs on Topics of Modern Mathematics Relevant to the Elementary Field* (New York, 1911, 1915, 1927; repr. 1955); *American Men of Science* (New York, 1906), 219–220; and *National Cyclopedia of American Biography*, XVI (New York, 1918), 388.

CAROLYN EISELE

MILLER, HUGH (*b.* Cromarty, Scotland, 10 October 1802; *d.* Portobello, Scotland, 24 December 1856), *geology.*

Miller was the elder son of Hugh Miller by his second wife, Harriet. His father, the master of a fishing sloop, was drowned when Miller was five. At school the boy was unruly and independent; and instead of following the conventional education that was open to one of his intelligence and social position, he apprenticed himself to a stonemason at the age of seventeen and thereafter used his leisure to educate himself in natural history and literature. His geological studies arose directly from his work as a mason and from his interest in the history, scenery, and folklore of the Highlands. Miller discovered that the Old Red Sandstone was not (as was commonly believed) virtually devoid of fossils but contained in certain strata an abundant fauna of spectacular bony fish that constituted one of the earliest vertebrate faunas then known. His *Scenes and Legends of the North of Scotland* (1835) brought him recognition as a descriptive writer of striking power; in addition its chapter "The Antiquary of the World," describing geology as "the most poetical of all the sciences," led to correspondence with Roderick Murchison and thus to contact with the scientific community at large.

In 1834, after some twelve years as a journeyman mason, Miller exchanged an outdoor life for that of an accountant in a Cromarty bank; and in 1837 he married Lydia Fraser, an author of children's books. In 1839 he entered the patronage controversy in the Church of Scotland by publishing a powerful open letter to Lord Brougham; his abilities were immediately recognized by the "nonintrusion" party, and he was invited to Edinburgh to edit their newspaper, *The Witness*. Miller's leading articles (from 1840) made him at once one of the most prominent and influential figures in public life in Scotland. His eloquent style, passionate commitment, and independent position were deployed with great effectiveness in the protracted struggle for the right of Scottish people to control the appointment of ministers in the national church. When at the Disruption (1843) the Free Kirk seceded on this issue, Miller used his influence to try to prevent the new body from retreating into a "sectarian" position and to keep alive the ideal of a truly national but non-Erastian church. As an integral part of this ideal, he pleaded for public education to be undenominational and fully grounded in modern science: he believed that this would defend Christian faith not only against the "infidelity" of materialism but also against the "Puseyite" anti-intellectualism of the Oxford Movement and the literalistic obscurantism of the scriptural "antigeologists."

Soon after Miller's arrival in Edinburgh, the meeting there of the British Association for the Advancement of Science gave him an opportunity to meet many of the leading British scientists and also Louis Agassiz, then the greatest authority on fossil fish. His subsequent articles in *The Witness* on his own research and its implications were amplified into his first scientific book, *The Old Red Sandstone* (1841). Like all his books this was not a conventional scientific monograph but a series of discursive essays, leading the ordinary reader from a starting point in everyday experience, through the details of the anatomy of the most ancient fossil fish and the reconstruction of their environment, toward the broader implications that geology held for the place of man in nature and his relation to God. Miller was no naïve literalist (he had, for example, a most vivid sense of the vast antiquity of the earth), but he did believe that the fossil record confirmed in broad outline the cosmic drama depicted symbolically in the Bible. More particularly, his strong sense of man as a moral being, ultimately responsible to God, led him to attack vehemently any attempt to diminish that responsibility by blurring the distinction between man and the lower animals. Hence Lamarck's "theory of progression" by transmutation was abhorrent to him, and its revival in Robert Chambers' anonymous *Vestiges of Creation* (1844) disturbed him particularly because he saw its "infidel" tendencies spreading to the artisan classes.

Miller's reply to Chambers was delayed by a breakdown of health brought on by overwork and by silicosis contracted during his years as a mason; and in 1845 he left Scotland for the first time, subsequently publishing his *First Impressions of England* (1846) with perceptive comments on the new industrial society as well as frequent digressions on geology. On his return to Scotland he wrote *Foot-Prints of the Creator* (1847) in answer to the *Vestiges*. It was explicitly an attack on the metaphysical and theological implications of Chambers' work, but Miller used his own scientific research to focus his attack on one of the weakest points in the "development hypothesis." The fish of the Old Red Sandstone (and the few that had been found in still earlier strata) were not, he argued, the rudimentary quasi-embryonic forms that Chambers' theory required; on the contrary, these "Ganoids"—earliest vertebrates then known—"enter large in their stature and high in their organisation." The geological history of the fish suggested that they had been created already perfect, clearly distinct from other animals, and that a better case could be made out for their subsequent "degradation" than for their "progress," since the earliest representatives were in some ways the most complex. Miller's interpretation could be extended to the rest of the fossil record; and he thus derived (like Agassiz) a picture of overall "progress" achieved by distinct creative steps, each initiating a new and higher form of organization, culminating in man.

More accurately, however, Miller saw the final culmination of this vast history in an eschatological future kingdom of Christ. This Christological focus to Miller's interpretation of science distinguishes his work sharply from that of most of his contemporaries, who were concerned to "reconcile" geology with religion. Miller was not interested in defending natural theology except as a prelude to existential commitment to God as revealed in Christ: dissociated from distinctively Christian beliefs, "a belief in the existence of a God is," he asserted, "of as little *ethical* value as a belief in the existence of the great sea-serpent." But by stating his opposition to evolutionary theory in terms of a characteristically stark antithesis—"the *law* of development *versus* the *miracle* of creation"—he placed his theology and his science in a vulnerable position during a period in which scientific plausibility was seen increasingly in terms of a metaphysical "principle of uniformity" that excluded the category of miracle altogether.

Miller published an attractive account of his life up to 1840 in *My Schools and Schoolmasters* (1854), which romanticizes his early life and exaggerates his humble origins. In the last years of his life he suffered increasingly from mental illness, and he finally committed suicide at his home near Edinburgh while seeing his last collection of essays, *The Testimony of the Rocks* (1857), through the press.

Miller's strictly original scientific work was of limited scope; his studies of early fossil fish did little more than amplify and correct some details of Murchison's stratigraphy and Agassiz's paleontology. His importance for nineteenth-century science lies, rather, in his use of outstanding literary abilities to broaden the taste for science in general and for geology in particular, and to encourage a humane concern for the fundamental significance of such studies: in the words of his biographer Mackenzie, "probably no single man since has so powerfully moved the common mind of Scotland, or dealt with it on more familiar and decisive terms."

BIBLIOGRAPHY

I. ORIGINAL WORKS. Miller's principal publications are *Scenes and Legends of the North of Scotland or the Traditional History of Cromarty* (Edinburgh, 1835); *The Old Red Sandstone: Or New Walks in an Old Field* (Edinburgh, 1841); *First Impressions of England and Its People* (London, 1846); *Foot-Prints of the Creator: Or, the Asterolepis of Stromness* (Edinburgh, 1847); *My Schools and Schoolmasters; Or, the Story of My Education* (Edinburgh, 1854); and *The Testimony of the Rocks; Or, Geology in Its Bearings on the Two Theologies, Natural and Revealed* (Edinburgh, 1857). Some of his correspondence is published in Peter Bayne, ed., *The Life and Letters of Hugh Miller*, 2 vols. (London, 1871).

II. SECONDARY LITERATURE. A short biography is W. Keith Leask, *Hugh Miller* (Edinburgh, 1896). W. M.

Mackenzie, *Hugh Miller. A Critical Study* (London, 1905) is penetrating although somewhat unsympathetic; Mackenzie's *Selections From the Writings of Hugh Miller* (Paisley, 1908) contains a well-chosen and balanced sample of Miller's work.

M. J. S. RUDWICK

MILLER, PHILIP (*b.* Bromley, Greenwich, or Deptford, London, England, 1691; *d.* Chelsea, London, 18 December 1771), *botany, horticulture.*

Miller, the most important horticultural writer of the eighteenth century, was curator of the Chelsea Physic Garden from 1722 to 1770. During this period, and largely through his skill as a grower and propagator and his extensive correspondence, the Chelsea botanic garden belonging to the Society of Apothecaries of London became famous throughout Europe and the North American colonies for its wealth of plants, which was continuously enriched by new introductions, notably from the West Indies, Mexico, eastern North America, and Europe. These plants Miller recorded in successive editions of his *Gardeners Dictionary* from 1731 to 1768. As Richard Pulteney stated in 1790, "He added to the theory and practice of gardening, that of the structure and characters of plants, and was early and practically versed in the methods of Ray and Tournefort." He derived his concept of genus and his generic names from the *Institutiones rei herbariae* (1700) of Tournefort, who recognized genera of first rank based on floral and fruiting characters and genera of second rank based on vegetative characters, whereas Linnaeus, Tournefort's successor and Miller's contemporary, in his *Species plantarum* (1753) recognized only first-rank genera and united Tournefort's second-rank genera with them— thus including, for example, *Abies* and *Larix* in *Pinus*. Miller accepted Linnaeus' classification and nomenclature reluctantly and never wholeheartedly. Thus in 1754, by continuing to use, in the fourth abridged edition of his *Gardeners Dictionary*, the Tournefortian names he had always used but which Linnaeus had suppressed in 1753, Miller brought these names back into post-Linnaean botanical literature and so became an inadvertent innovator; he is now cited as the authority for some eighty generic names, among them *Abies* and *Larix*, really derived from his predecessors. In the eighth edition (16 April 1768) of his *Gardeners Dictionary*, Miller at last adopted Linnaean binomial nomenclature for species; he also published some 400 specific names for plants imperfectly known or unknown to Linnaeus, or otherwise classified by him. These works of 1754 and 1768 earned Miller his lasting place in systematic botany.

Miller's father, a gardener of Scots origin, gave him a good schooling, and Miller early set up in business in the London area as a florist, grower of ornamental shrubs, and planter and designer of gardens. Thus he came to the notice of Sir Hans Sloane, who had bought the manor of Chelsea in 1712 and had become the ground landlord of the Chelsea site which the Society of Apothecaries had leased since 1673 for their physic garden, or *hortus medicus*. In 1722 Sloane transferred it in perpetuity to the Society of Apothecaries for use as a botanic garden. A condition in the deed of conveyance was that every year the Apothecaries should give the Royal Society fifty good herbarium specimens of distinct plants grown that year in the garden "and no one offered twice until the compleat number of two thousand plants have been delivered." This necessitated the continual introduction of new plants. On Sloane's recommendation, Miller was appointed head gardener in 1722. The Chelsea botanic garden then became possibly the most richly stocked of any garden of the mid-eighteenth century, and Miller recorded his firsthand experience in his publications. In 1724 he published his two-volume octavo *Gardeners and Florists Dictionary*, replaced in 1731 by the one-volume folio *Gardeners Dictionary*, of which eight editions appeared in his lifetime. They provide not only cultural but also descriptive botanical information, including the characters of each genus and diagnoses of the species. From them there arose after Miller's death a series of encyclopedic works on cultivated plants culminating in the Royal Horticultural Society's *Dictionary of Gardening* (1951).

Miller remained in charge of the Chelsea Physic Garden until 1770, when most reluctantly he retired with a pension. The mainspring of his life gone, he died the year after.

BIBLIOGRAPHY

Miller's main work is his *Gardeners Dictionary* (London, 1731; 2nd ed., 1733; 3rd ed., 1737–1739; 4th ed., 1743; 5th ed., 1747 [apparently no copy extant]; 6th ed., 1752; 7th ed., 1756–1759; 8th ed., 1768); the 8th is the most important ed., for it contains the binomial specific names attributed to Miller. The concise ed. "abridged" from the 1731 ed. was first published in 1733–1740; 2nd ed., 1741; 3rd ed., 1748; 4th ed., 1754; 5th ed., 1763; 6th ed., 1771— of these only the 4th ed. (1754) in 3 vols. is nomenclaturally important; it has been reprinted in facsimile (Lehre, German Federal Republic, 1969), with an introduction by W. T. Stearn (see below).

The main biographical sources are R. Pulteney, *Historical and Biographical Sketches of the Progress of Botany in England*, II (London, 1790), 241–242; J. Rogers, *The Vegetable Cultivator* (London, 1839), pp. 335–343, reprinted by W. T. Stearn in 1969 (see below); and C. Wall and H. C. Cameron, *A History of the Worshipful Society of Apothecaries of London* (London, 1963). Some further information is given in W. T. Stearn, "The Abridgement of Miller's Gardeners Dictionary," prefixed to the Historiae Naturalis Classica facs. of the 4th ed. (1754) of *Gardeners Dictionary Abridged* (Lehre, 1969); Hazel Le Rougetel, "Gardener Extraordinary: Philip Miller of Chelsea, 1691–1771," in *Journal of the Royal Horticultural Society*, **96** (1971), 556–563; and W. T. Stearn, "Philip Miller and the Plants from the Chelsea Physic Garden Presented to the Royal Society of London, 1723–1796," in *Botanical Society of Edinburgh Transactions*, **41** (1972), 293–307, in which is printed the deed of conveyance of the Chelsea Physic Garden.

WILLIAM T. STEARN

MILLER, WILLIAM ALLEN (*b.* Ipswich, England, 17 December 1817; *d.* Liverpool, England, 30 September 1870), *chemistry, spectroscopy, astronomy.*

A writer of widely used textbooks on chemistry, Miller pioneered the use of spectroscopic analysis in chemistry and the application to it of photography. With Huggins he subsequently extended his studies to planetary and stellar spectra, making comparisons with terrestrial sources, and was one of the first to produce reliable information as to the chemical constitution of the stars.

Miller was the son of Frances (née Bowyer) and William Miller, who after being secretary to the Birmingham General Hospital became a brewer in the Borough, London. After a year at Merchant Taylors' School, Northwood, Middlesex, Miller was sent to a Quaker school at Ackworth, Yorkshire. There he was taught by William Allen, after whom he had been named and who introduced him to chemistry and astronomy.

In 1832 Miller was apprenticed to his uncle, Bowyer Vaux, a surgeon at the Birmingham General Hospital; he left in 1837 to read medicine at King's College, London. In 1839 Miller received the Warneford Prize in theology and in 1840 was made demonstrator of chemistry after having spent some months in Liebig's laboratory at Giessen. He took his M.B. and M.D. in 1841–1842 and in 1841 became assistant lecturer under J. F. Daniell, whom he succeeded as professor of chemistry in 1845. He was made a fellow of the Royal Society in the latter year.

Miller's lecture notes formed the basis of his well-known textbooks of inorganic and organic chemistry, which went into many editions. His first important and original paper, "Additional Researches on the Electrolysis of Secondary Compounds," was written jointly with Daniell (*Philosophical Transactions of the Royal Society*, **134** [1844], 1–19; *Philosophical Magazine*, **25** [1844], 175–188). It was well known that the passage of a current through an electrolyte yielded decomposition products at the electrodes and also resulted in changes of concentration of the solution at different points in relation to the cathodes. Faraday had investigated the subject in 1835, but it was left to Daniell and Miller to propose an explanation in terms of a discrepancy between the mobilities of cation and anion, movement of the latter often being much the greater. Their hypothesis was more fully exploited by Hittorf (from 1853) and later by F. W. Kohlrausch (from 1876), whose theory of ionic movement is substantially that taught at an elementary level today.

During the 1840's it was slowly becoming appreciated that the spectra of flames were far from simple; Miller was perhaps the first to publish drawings of flame and absorption spectra, which he observed in a makeshift laboratory under the King's College lecture theater. He presented his findings to the Cambridge meeting of the British Association for the Advancement of Science in 1845, and they were printed in *Philosophical Magazine* (**27** [1845], 81–91). He drew the spectra of calcium, copper, and barium chlorides, boric acid, and strontium nitrate, each having bright lines and bands, and the common yellow D-line of sodium. The ubiquity of the D-line, and the newly found complexity in spectral structure, led some chemists almost to despair of ever laying down rigid rules of chemical spectrum analysis.

In 1861, before the Manchester meeting of the British Association, and 1862, before the Pharmaceutical Society of Great Britain, Miller gave another address with far-reaching consequences for spectroscopy. Rather than use a fluorescent screen for the study of the ultraviolet spectra of metals, he photographed them. This made it possible to record accurately the extraordinary complexity of the spectra. He was now able to find similarities between the characteristic spectra of certain metals, such as the cadmium, zinc, and magnesium group. His photographic plates were wet collodion, and his prisms were quartz. In all, he published spectra of twenty-five metals (*Philosophical Transactions of the Royal Society*, **152** [1862], 861–887).

In 1862 Miller joined forces with William Huggins, his neighbor at Tulse Hill, London, and together they arranged a telescope to give the spectra of celestial objects side by side with a comparison spectrum from a laboratory source. In Italy, Pietro Secchi was making a very extensive survey and in due course classified the spectra of more than 4,000 stars. Miller and Huggins, however, aimed at a smaller survey but one made with much greater precision. In March 1863, Miller was able to show a fine spectrophotograph of Sirius to an audience at the Royal Institution. He and Huggins paid especial attention to the spectra of the moon, Jupiter, and Mars. In 1867 they were jointly awarded the gold medal of the Royal Astronomical Society for their work (for a report of which see *Proceedings of the Royal Society*, **12** [1862–1863], 444–445; *Philosophical Transactions of the Royal Society*, **154** [1864], 413–436). Other examples of their work on fixed stars can be found in *Monthly Notices of the Royal Astronomical Society* (**26** [1866], 215–218) and *Proceedings of the Royal Society* (**15** [1867], 146–149). Their general conclusion was that although stars differ considerably one from another, they all have much chemically in common with the sun.

Miller was a deeply religious man and was indefatigable in applying his scientific talents to social ends, whether advising on the chemistry of the Metropolitan water supply, on the uniformity of weights and measures, on the establishment of regular meteorological observations (under the Board of Trade), or on the affairs of the Royal Mint, where he was Assayer. He worked hard for the Royal Society, of which he was treasurer for nine years, and helped to found the Chemical Society, of which he was twice president. He married Eliza Forrest in 1842. She died a year before him, and two daughters and a son survived them both.

BIBLIOGRAPHY

I. ORIGINAL WORKS. For papers in addition to those mentioned in the text, see the Royal Society's *Catalogue of Scientific Papers*, IV (1870), 390, and VIII (1879), 406–407. Miller's longer works include *On the Importance of Chemistry to Medicine* (London, 1845); *Report . . . in Reference to the Composition of the Lambeth Stone Ware and Aylesford Pottery Pipes* (London, 1855); *Elements of Chemistry Theoretical and Practical*, 3 vols. (London, 1855; 6th ed., 1877–1878); *Practical Hints to the Medical Student* (London, 1867); and *Introduction to the Study of Inorganic Chemistry* (London, 1871).

II. SECONDARY LITERATURE. For further details of Miller's public life, see Agnes Clerke's notice on him in the *Dictionary of National Biography*, XIII, 429–430 and the first three obituaries listed there. See especially *Proceedings of the Royal Society*, **19** (1871), xix–xxvi. In most histories of spectroscopy Miller is given only slight attention. For the background to his work, see W. McGucken, *Nineteenth-Century Spectroscopy* (Baltimore, 1969).

J. D. NORTH

MILLER, WILLIAM HALLOWES (*b.* Llandovery, Carmarthenshire, Wales, 6 April 1801; *d.* Cambridge, England, 20 May 1880), *crystallography, mineralogy.*

His father, Captain Francis Miller, who served in the American war, had a long military ancestry. By his first wife Captain Miller had three sons, all of whom entered the army, and two daughters. After losing his estate near Boston, Massachusetts, he retired to Wales to the small estate of Velindre near Llandovery and in 1800 married Ann Davies, the daughter of a Welsh vicar. William was the only child of this second marriage; his mother died a few days after his birth, but his father lived to the age of eighty-six, dying in 1820.

William was educated privately until he entered St. John's College, Cambridge, where he graduated B.A. as fifth wrangler in mathematics in 1826. In 1829 he became a fellow of St. John's and in 1831 published his first book—written in his characteristically lucid but terse style—*The Elements of Hydrostatics and Hydrodynamics*, which survived as a standard, though difficult, textbook into the fifties. Another mathematical textbook, *An Elementary Treatise on the Differential Calculus*, appeared in 1833 and passed through several editions. By then Miller had, in 1832, succeeded William Whewell as professor of mineralogy. He was elected F.R.S. in 1838. In 1841 came a curious diversion: the statutes of St. John's College required all fellows to proceed in time to holy orders except for four who should be doctors of medicine, and in order to retain his fellowship, Miller prepared himself for and took the M.D. He was obliged to vacate his fellowship on his marriage (5 November 1844) to Harriet Susan Minty, the daughter of R. V. Minty, a retired civil servant. They had two sons and four daughters. In 1875 he became a fellow of St. John's again under new statutes. In 1876 he suffered a stroke, which effectively brought his scientific life to a close four years before his death.

Miller's significant contribution to crystallography was made in *A Treatise on Crystallography* published in 1839 (translated into French by H. de Senarmont [1842], and into German with two new chapters by J. Grailich [1856], and again into German in abbreviated form by P. Joerres [1864]). Miller started with the fundamental assertion that crystallographic reference axes should be parallel to possible crystal edges; his system of indexing, a derivative from Whewell (*Philosophical Transactions of the Royal Society*, 1825), was based on a parametral plane (111) making intercepts a, b, c on such reference axes and was such that indices (hkl) were assigned to a plane making intercepts on the reference axes in the ratio $a/h : b/k : c/l$, where h, k, l are integers. The established German school of C. F. Naumann and C. S. Weiss had, to use the same nomenclature, assigned indices (*hkl*) to a plane making intercepts in the ratio *ah* : *bk* : *cl* on reference axes not restricted to parallelism with possible crystal edges. The algebraic advantages of "Millerian indices" were immediately apparent; the crystallographic superiority of Miller's reciprocal indices over Weiss's direct indices did not become apparent until Bravais's development of Haüy's rudimentary lattice concept in 1848, and not fully appreciated until Bragg's interpretation of the diffraction of X rays by crystals in 1912. But Miller's notation had quickly found favor with his contemporaries on grounds of convenience and had already served to codify an immense corpus of morphological observations in a thoroughly well-understood manner.

In the *Treatise* Miller had little to say about symmetry, but he explored crystal geometry to the full. The zone law of Weiss was simplified by the new notation, and zone symbols were defined in familiar form; the equations to the normal and the cos θ formula were developed; and the rational sine ratio, which was to be further developed in *A Tract on Crystallography* (1863), made its first appearance here. For the representation of three-dimensional angular relationships Miller followed F. E. Neumann in using spherical projection, but the stereographic projection, which subsequently acquired greater currency, and the gnomonic projection were discussed in the final chapter.

The new edition (1852) of William Phillips' *Elementary Introduction to Mineralogy* by H. J. Brooke and W. H. Miller was an entirely new book largely written, as Brooke states in the preface, by Miller, and it represents his principal contribution to mineralogy. It incorporated a vast amount of accurate goniometric data provided by Miller himself; it followed the *Treatise* in using spherical projection; and it made a tentative start in the use of polarized light for the characterization of transparent minerals. The *Introduction*, like the *Treatise*, soon eclipsed its contemporaries; it inspired Des Cloizeaux to produce his more elaborate *Manuel de minéralogie* (Paris, 1862–1893), and determined the form of all subsequent texts on descriptive mineralogy.

In 1843 Miller branched out into a new field on appointment to the parliamentary committee concerned with the preparation of new standards of length and weight consequent on the destruction of the old standards in the burning of the Houses of Parliament in 1834. His exceptionally accurate work was responsible for the construction of the new standard of weight (*Philosophical Transactions of the Royal Society*, 1856). In 1870 he was appointed to the Commission Internationale du Mètre.

Many honors fell to Miller in his lifetime. He was president of the Cambridge Philosophical Society (1857–1859) and foreign secretary of the Royal Society (1856–1873), being awarded a Royal Medal in 1870.

The exceptional breadth of Miller's scientific knowledge was recognized by his contemporaries. He was generous and hospitable to a point, yet remarkably spartan in his way of life. His ingenuity in constructing surprisingly accurate apparatus from simple, often homely, materials was notable. While no great traveler, he obviously enjoyed his trips to Paris for meetings of the Meter Commission, and he regularly holidayed in the Italian Tirol, where he simply enjoyed the scenery while his wife sketched it.

BIBLIOGRAPHY

I. ORIGINAL WORKS. Miller's works include *The Elements of Hydrostatics and Hydrodynamics* (Cambridge, 1831); *An Elementary Treatise on the Differential Calculus* (Cambridge, 1833); *A Table of Mineralogical Species* (Cambridge, 1833); *A Treatise on Crystallography* (Cambridge, 1839); and William Phillips, *An Elementary Introduction to Mineralogy*, new ed. by H. J. Brooke and W. H. Miller (London, 1852).

II. SECONDARY LITERATURE. On Miller or his work see N. Storey Maskelyne, *Nature*, **22** (1880), 247–249; T. G. Bonney, *Proceedings of the Royal Society*, **31** (1881), ii–vii; and J. P. Cooke, *Proceedings of the American Academy of Arts and Sciences*, **16** (1881), 460–468. Memorial of William Hallowes Miller by his wife (privately printed, Cambridge, 1881[?]).

DUNCAN MCKIE

MILLER, WILLIAM LASH (*b*. Galt, Canada, 10 September 1866; *d*. Toronto, Canada, 1 September 1940), *chemistry*.

During the last decade of the nineteenth century and the early part of the twentieth century Miller played a leading role in the development of chemistry in Canada. In accordance with the tradition of the time, classical studies formed a significant part of his undergraduate training at the University of Toronto, although this was leavened with courses in mathematics, physics, chemistry, biology, mineralogy, and geology. After receiving his B.A. degree in natural philosophy in 1887, Miller went to Germany. After spending some time in Berlin and Göttingen he proceeded to Munich, where he took his Ph.D. under Baeyer in 1890. He then moved to Leipzig to work in Ostwald's laboratory. This was a turning point in Miller's life, and he continued to spend his summers with Ostwald after he had become a member of the staff of the department of chemistry at Toronto and had received his second Ph.D. at Leipzig in 1892. It was at Leipzig that he first became acquainted with the elegance and applicability to chemistry of the thermodynamic approach of Josiah Willard Gibbs.

In a series of papers published largely in the *Transactions of the Connecticut Academy of Arts and Sciences* (1873–1883), Gibbs had laid the foundation for chemical thermodynamics; but it was left to Ostwald and Miller to translate Gibbs's highly theoretical treatment into laboratory terms. Most of Miller's academic life was devoted to this activity. In one of his first scientific papers (1892), he showed that the electromotive force of an electrochemical cell having a metallic electrode of mercury, lead, or tin was independent, at the melting point, of whether the metal was in the liquid or the solid state, and hence the factor determining the electromotive force at constant temperature and pressure must therefore be the chemical potential, or Gibbs free energy.

Miller was particularly adroit in applying and extending Gibbs's theories to polycomponent systems. When one of his students found that the addition of salt to an aqueous solution of alcohol raised the partial pressure of alcohol, an effect opposite to that predicted by then-current theories, he was able to show that this was a logical consequence of thermodynamic reasoning and proved it in the paper "On The Second Differential Coefficients of Gibbs Function ζ" (1897). His appreciation of the importance of having quantitative values for these second differential coefficients dominated his approach to both teaching and research. It also led, in later years, to some controversy between the Miller school of thermodynamics and that of G. N. Lewis at Berkeley. In retrospect, this controversy was seen to relate to a choice of formalism rather than logic. Lewis chose to use two quantities, the standard state and the activity coefficient, which are related to the first derivative of the Gibbs free energy; Miller used the second differential coefficient of the Gibbs free energy. Whereas Miller's choice was mathematically more elegant, and possibly a better method for the beginning student, the Lewis approach was more practical insofar as getting accurate data for real systems is concerned. Miller realized this difference, and saw to it that his students were exposed to the two approaches.

During his long career at Toronto, Miller had a record of outstanding research in several areas of physical chemistry: chemical equilibriums, rates of reaction, electrochemistry, transference numbers, overvoltage, high-current electric arcs, and diffusion. In his later years he was attracted by problems of a biochemical nature, particularly by growth factors for simple cells like the yeasts.

In 1894, when Miller was promoted to lecturer in chemistry, he introduced research as a regular part of the fourth year of the undergraduate honors program, a step that has been followed in many other disciplines and institutions. He was made associate professor in 1900 and professor in 1908. When World War I broke out, the head of Miller's department took leave to enter active military service and Miller became *de facto* head. He remained in that capacity until his retirement in 1937.

Miller's leadership in research was paralleled by his active role in many professional societies and academies in Canada, Great Britain, and the United States. As early as 1910 he was chairman of the Canadian Section of the Society of Chemical Industry. He was one of the chief organizers of the Canadian Institute of Chemistry (its president in 1926) and of the Canadian Chemical Association. All three societies were eventually amalgamated to form the present Chemical Institute of Canada. In 1926 Miller was the first Canadian to be made an honorary member of the American Chemical Society and served for many years as an associate editor of its *Journal*. He was active in the establishment of, and as an associate editor of, *Journal of Physical Chemistry*. His distinction as a scientist and his leadership in his profession were recognized when he was made a commander of the Order of the British Empire.

BIBLIOGRAPHY

I. ORIGINAL WORKS. Miller's works include "Über die Umwandlung Chemischer Energie in Elektrische," in *Zeitschrift für physikalische Chemie*, **10** (1892), 459–466; "On the Second Differential Coefficients of Gibbs Function ζ. The Vapour Tensions, Freezing and Boiling Points of Ternary Mixtures," in *Journal of Physical Chemistry*, **1** (1896–1897), 633–642; "The Theory of the Direct Method of Determining Transport Numbers," in *Zeitschrift für physikalische Chemie*, **69** (1910), 436–441; "Mathematical Theory of the Changes in Concentration at the Electrode Brought About by Diffusion and by Chemical Reactions," in *Journal of Physical Chemistry*, **14** (1910), 816–885, written with T. R. Rosebrugh; "The Influence of Diffusion on Electromotive Force Produced in Solutions by Centrifugal Action," in *Transactions of the Electrochemical Society*, **21** (1912), 209–217; "Toxicity and Chemical Potential," in *Journal of Physical Chemistry*, **24** (1920), 562–569; "The Method of Willard Gibbs in Chemical Thermodynamics," in *Chemical Reviews*, **1** (1924–1925), 293–344; "Numerical Evaluation of Infinite Series and Integrals Which Arise in Certain Problems of Linear Heat Flow, Electrochemical Diffusion, etc.," in *Journal of Physical Chemistry*, **35** (1931), 2785–2884, written with A. R. Gordon.

II. SECONDARY LITERATURE. See C. J. S. Warrington and R. V. V. Nicholls, *A History of Chemistry in Canada* (New

York, 1949), and the obituaries by Frank B. Kenrick, in *Proceedings of the Royal Society of Canada*, 3rd ser., **35** (1941), 131–134, and by Wilder D. Bancroft, in *Journal of the American Chemical Society*, **63** (1941), 1–2.

DONALD J. LE ROY

MILLIKAN, ROBERT ANDREWS (*b.* Morrison, Illinois, 22 March 1868; *d.* Pasadena, California, 19 December 1953), *physics.*

Millikan was the son of Silas Franklin Millikan, a Congregational preacher, and Mary Jane Andrews, a graduate of Oberlin who had been dean of women at a small college in Michigan. Raised in Maquoketa, Iowa, where his family moved in 1875, young Millikan enjoyed a storybook Midwestern American boyhood, fishing, farming, fooling, and learning next to nothing about science. In 1886 he enrolled in the preparatory department of Oberlin College and, in 1887, in the classical course of the college itself. Mainly because he did quite well in Greek, at the end of his sophomore year he was asked to teach an introductory physics class. Glad to have the job, Millikan plunged into the subject, liked it, and soon decided to make it his career.

Millikan graduated from Oberlin in 1891 and continued to teach physics to the preparatory students while successfully pursuing a course of self-instruction in Silvanus P. Thomson's *Dynamic Electric Machinery.* Awarded an M.A. for this achievement, in 1893 Millikan entered Columbia University on a fellowship as the sole graduate student in physics. He was impressed by the lectures of Michael I. Pupin, who emphasized the importance of mathematical techniques, and by the experimental deftness of Michelson, under whom he studied at the University of Chicago in the summer of 1894. Receiving his Ph.D. in 1895, Millikan went to Europe for postgraduate study, financed by a loan from Pupin. He heard Poincaré lecture at Paris, took a course from Planck at Berlin, and did research with Nernst at Göttingen. In 1896, the excitement of the discovery of X rays still fresh in his mind, Millikan joined the faculty of the University of Chicago as an assistant in physics.

There he soon met Greta Irvin Blanchard, the daughter of a successful manufacturer from Oak Park, Illinois. By the time the young couple was married in 1902, Millikan was pouring a large fraction of his considerable energies into the development of the physics curriculum, especially the introductory courses. In conjunction with this work, he wrote or coauthored a variety of textbooks and laboratory manuals which, like his *First Course in Physics* (1906), written with

Henry Gale, quickly became standards and sold steadily through the years. In 1907, largely because of his outstanding pedagogical achievements, Millikan was promoted to an associate professorship.

But Millikan was acutely aware that at the University of Chicago the major rewards went to those who contributed to the advancement of knowledge. Although he had consistently done research, even his most recent investigation, on the photoelectric effect, had failed to yield significant results. Unaware of Einstein's explanation of the effect, Millikan used a spark source of ultraviolet light to determine conclusively whether the photocurrent from various metals varied with temperature; as he found, it did not. Eager to earn a reputation in research, about 1908 he decided to shelve the writing of textbooks and concentrate on his work in the laboratory.

By 1909 Millikan was deeply involved in an attempt to measure the electronic charge. No one had yet obtained a reliable value for this fundamental constant, and some antiatomistic Continental physicists were insisting that it was not the constant of a unique particle but a statistical average of diverse electrical energies. Millikan launched his investigation with a technique developed by the British-born physicist H. A. Wilson; it consisted essentially of measuring, first, the rate at which a charged cloud of water vapor fell under the influence of gravity and then the modified rate under the counterforce of an electric field. Using Stokes's law of fall to determine the mass of the cloud, one could in principle compute the ionic charge. Millikan quickly recognized the numerous uncertainties in this technique, including the fact that evaporation at the surface of the cloud confused the measure of its rate of fall. Hoping to correct for this effect, he decided to study the evaporation history of the cloud while a strong electric field held it in a stationary position.

But when Millikan switched on the powerful field, the cloud disappeared; in its place were a few charged water drops moving slowly in response to the imposed electrical force. He quickly realized that it would be a good deal more accurate to determine the electronic charge by working with a single drop than with the swarm of particles in a cloud. Finding that he could make measurements on water drops for up to forty-five seconds before they evaporated, Millikan arrived at a value for *e* in 1909 which he considered accurate to within 2 percent. More important, he observed that the charge on any given water drop was always an integral multiple of an irreducible value. This result provided the most persuasive evidence yet that electrons were fundamental particles of identical charge and mass.

Late in 1909 Millikan greatly improved the drop method by substituting oil for water. Because of the relatively low volatility of this liquid, he could measure the rise and fall of the drops for up to four and a half hours. Spraying the chamber with radium radiation, he could change the charge on a single drop at will. His overall results decisively confirmed the integral-multiple values of the total charge. As for the determination of e itself, Millikan found that Stokes's law was inadequate for his experimental circumstances because the size of the drops was comparable with the mean free path of the air. Using the so-called Stokes-Cunningham version of the law, which took this condition into account, by late 1910 he had computed a charge for e of 4.891×10^{-10} e.s.u. Realizing that the accuracy of this figure was no better than that of the key constants involved in the computation, Millikan painstakingly reevaluated the coefficient of viscosity of air and the mean-free-path term in the Stokes-Cunningham law. In 1913 he published the value for the electronic charge, $4.774 \pm .009 \times 10^{-10}$ e.s.u., which would serve the world of science for a generation.

Off and on all the while, Millikan had continued his exploration of the photoelectric effect; about 1912, now aware of Einstein's interpretation of it, he began an intensive experimental study of the phenomenon, with the aim of testing the formula relating the frequency of the incident light to the retarding potential which cut off the photocurrent. No experimentalist had yet succeeded in proving or disproving the validity of the equation. Millikan took great care to avoid the mistakes that he and other physicists had previously made. Since a spark source of ultraviolet light induced spurious voltages in the apparatus, he used a high-pressure mercury-quartz lamp arranged to suppress stray light, especially on the short wavelength side. To extend the range of test well into the visible region, he made targets of alkali metals which were photosensitive up to 6,000 Å. Where others had adulterated their results by using photosensitive materials as the reference for the cutoff voltage, Millikan employed a Faraday cage of well-oxidized copper netting which was not photosensitive in the range of his incident radiation. Finally, he sought to reduce the inaccuracies introduced when the photocurrent near the cutoff point was too low to measure with precision. Having noticed that this current was highest when the metal was fresh, he fashioned his targets into thick cylinders and rigged up an electromagnetically operated knife to shave off the ends of the blocks.

By 1915, as the result of these meticulous investigations, Millikan had confirmed the validity of Einstein's equation in every detail. He not only demonstrated the linear relationship between the cutoff potential and the frequency of the incident light but also showed that the intercept of the graphed data on the voltage axis equaled the contact electromotive force, or work potential, of the target metal, a quantity which he had measured independently, to within 0.5 percent. In addition Millikan proved that the slope of the line equaled the ratio of Planck's constant to the electronic charge, and his work provided the best measure of h then available. Despite the conclusiveness of these results, Millikan did not believe that he had confirmed Einstein's theory of light quanta but only his equation for the photoeffect. In the face of all the evidence for the wave nature of light, he was convinced, as were most other physicists of the day, that the equation had to be based on a false, albeit evidently quite fruitful, hypothesis.

By 1916, when Millikan completed his major work on the photoeffect, he had already assumed more than a mere professor's role in the world of science. In 1913 he became a consultant to the research department of Western Electric, primarily to advise the company on vacuum tube problems. In 1914 he was elected to the American Philosophical Society and the American Academy of Arts and Sciences, in 1915 to the National Academy of Sciences, and in 1916 to the presidency of the American Physical Society, an office which he held for two years. Millikan also served as an associate editor of *Physical Review* from 1903 to 1916; and he was made an editor of *Proceedings of the National Academy of Sciences* . . ., which was started in the year of his election.

Early in 1917, after the United States broke diplomatic relations with Germany, Millikan went to Washington as a vice-chairman and director of research for the National Research Council, the organization which the National Academy of Sciences had recently created to help mobilize science for defense. Commissioned a lieutenant colonel in the Army Signal Corps, he served in his military capacity as the director of the Signal Corps Division of Science and Research and, in his National Research Council identity, as a member of the U.S. Navy's Special Board on Antisubmarine Devices. After a brief postwar period back at Chicago, in 1921 Millikan accepted appointment as chairman of the executive council and director of the Norman Bridge Laboratory at the newly renamed California Institute of Technology in Pasadena. In effect the president of the school, he was an able fund raiser and its enthusiastic spokesman; and under his leadership it quickly developed into one of the most distinguished scientific centers in the world.

Managing all the while to supervise many doctoral and postdoctoral fellows, Millikan maintained an active research career throughout the interwar years. One of the important subjects he investigated was the ability of electric fields on the order of a few hundred thousand volts per centimeter to draw electrons out of cold metals. By 1926, working in collaboration with Carl F. Eyring, a Caltech graduate student, Millikan had completed a thorough study of the phenomenon, using tungsten wires threaded along the axis of a hollow cylinder in high vacuum. The two men found that the field current, to use the term they introduced, depended only on the field gradient, not on the potential difference, between the wire and the walls of the cylinder. More important, within wide limitations the current was also entirely independent of temperature. Pointing out that these results violated Owen W. Richardson's theory of thermionic emission, Millikan and Eyring speculated that at relatively low temperatures some metallic electrons must not obey the law of equipartition. But in 1928 Oppenheimer, R. H. Fowler, and their co-workers showed independently that cold emission was a quantum mechanical result of the leakage of electrons through a potential barrier. In 1929 Charles C. Lauritsen, who was completing his doctoral research under Millikan, derived an empirical formula from their data which related the field current to the field gradient; and this equation was ultimately found to be experimentally indistinguishable from the quantum mechanical expression.

During the 1920's Millikan also did significant research in the "hot spark" spectra. As he knew, a high potential difference would maintain a spark source of ultraviolet radiation across two electrodes in a vacuum. The relative ease with which such radiation was absorbed had made its study difficult. In 1915 Millikan proposed that one could get around the problem of absorption by enclosing the path between the spark and a photographic plate entirely in a vacuum. To maximize the intensity of the spectrum, he had a grating ruled that would throw most of the light into the first order. Shortly after the war, with the apparatus now working reliably, Millikan and Ira S. Bowen, another Caltech graduate student, embarked upon a thorough study of the ultraviolet spectra of the lighter elements up to copper. By 1924 they had found and identified some 1,000 new lines. They had also extended the observable spectrum down to 136.6 Å and had helped to close the last gap between the optical and the X-ray frequencies.

In the course of this work, Millikan and Bowen found that the strongest lines were produced by atoms which had been stripped of their valence electrons.

Since the spectra of such hydrogen-like atoms ought to contain multiplets, they began, about the end of 1923, to study the fine spectra in the ultraviolet. By early 1924 they had found that the $2s$, $2p_1$, and $2p_2$ terms of the ultraviolet doublets corresponded precisely to the L_I, L_{II}, and L_{III} levels associated with the X-ray spectra of the heavier elements. Moreover, exactly the same relationship existed between the M and N X-ray levels and the higher ultraviolet multiplet terms. Millikan and Bowen concluded that the X-ray doublet laws based on Sommerfeld's relativistic orbital analysis could account for the doublets in the whole field of optics.

Yet, as they also pointed out in 1924, independently of Alfred Landé, this conclusion raised a serious difficulty for the theory of spectra. On the one hand, Sommerfeld's relativistic analysis of the X-ray doublets assigned a different azimuthal quantum number to the L_{II} and L_{III} terms. On the other hand, Bohr's spectral scheme accounted for the optical doublets by assuming different orientations for the same orbit; by definition, the p_1 and p_2 terms of the optical doublets possessed the same azimuthal quantum number. Since the results of Millikan and Bowen identified the L_{II} and L_{III} terms with the p_1 and p_2 levels, it seemed that one had to give up either Sommerfeld's relativistic explanation or the Bohr scheme of spectra. Millikan and Bowen could find no way out of the dilemma; but their forceful statement of it in 1924, coupled with Landé's, contributed to the ultimate resolution of the difficulty through G. E. Uhlenbeck and S. A. Goudsmit's postulation of electron spin in 1925.

In the 1920's Millikan also began an increasingly intensive program of research into the penetrating radiation which in mid-decade he would name "cosmic rays." In 1912 the Austrian-born physicist Victor Hess had found that atmospheric ionization increased with altitude up to 12,000 feet. But although Hess had argued that some kind of radiation was coming from the heavens, most physicists still attributed the phenomenon to some terrestrial cause, such as electrical discharges from thunderstorms or radioactivity. Millikan's initial experiments in the field, done with an unmanned sounding balloon in 1922 to a height of fifteen kilometers and with lead-shielded electroscopes atop Pike's Peak in 1923, failed to decide in favor of either interpretation. In the summer of 1925 Millikan proposed to settle the question by measuring the variation of ionization with depth in Muir Lake and Lake Arrowhead in the mountains of California. Snow-fed and separated by many miles, as well as 6,675 feet of atmosphere, each was likely to be free of both local radioactive disturbances and whatever atmospheric peculiarities might affect the ionization in the other.

Millikan's electroscopic measurements showed that the intensity of ionization at any given depth in Lake Arrowhead was the same as the intensity six feet lower in Muir Lake. Since the layer of atmosphere between the surfaces of the two lakes had precisely the absorptive power of six feet of water, the results decisively confirmed that the radiation was coming from the cosmos. Moreover, since the intensity of the ionization showed no diurnal variation, the radiation was uniformly distributed over all directions in space. And, finally, since Millikan detected ionization as far below the top of the atmosphere as the combined air and water equivalent of six feet of lead, it was evident that cosmic rays were a good deal more energetic than even the hardest known gamma rays.

To penetrate six feet of lead, charged particles would have to possess stores of energy then considered impossibly large; accordingly, Millikan assumed that cosmic rays must consist of photons. In 1926 he tested this assumption experimentally with what he considered confirmatory results. If cosmic rays were charged particles, their trajectories would be affected by the earth's magnetic field, so that more of them would strike the earth at higher than at lower latitudes. But Millikan could detect virtually no difference in cosmic ray flux at Lake Titicaca in South America from that at Muir Lake. And, although he ran his electroscope while sailing back from Mollendo, Peru, to Los Angeles, he found no variation of intensity with latitude at sea level.

Employing the photonic interpretation of cosmic rays, Millikan developed a theory of their origin in 1928. Combining the data from the balloon flight of 1922 with that of his terrestrial surveys, he graphed a curve of ionization versus depth which covered the range from sea level up to virtually the top of the atmosphere. Because no single coefficient of absorption could account for the curve, he inferred that cosmic rays were spread across a spectrum of energies. Going further, he argued that the experimental curve could be constructed from three different curves, each representing a different coefficient of absorption. According to this analysis, cosmic ray energies were not generally distributed but were clustered in three distinct bands.

To account for these bands, Millikan introduced what he called the "atom-building hypothesis." Using Dirac's formula for absorption through Compton scattering, Millikan computed the energy of the three bands from their absorption coefficients and found them equal to 26, 110, and 220 MEV. These figures equaled the mass defects of hydrogen, oxygen,

and silicon, which were known to be three of the most abundant elements of the universe. Millikan concluded that the photons striking the earth must be produced when four atoms of hydrogen somehow fused to form helium, sixteen to form oxygen, and twenty-eight to form silicon. In his summary of the argument, cosmic rays were the "birth cries" of atoms, a phrase which quickly achieved a good deal of notoriety among both the scientific and the lay publics.

Although in the late 1920's most physicists agreed with Millikan that cosmic rays were photons, few accepted his atom-building hypothesis. He had no proof of the uniqueness of his three absorption coefficients and could not convincingly explain away the kinetic difficulties involved in the spontaneous union of sixteen hydrogen atoms into oxygen, let alone twenty-eight into silicon. Moreover, some of his own experimental evidence cast doubt on the validity of using the Dirac formula to compute cosmic ray energies. Then, at the beginning of the 1930's, Millikan's assumption that the primary radiation consisted of photons was refuted by the work of other experimentalists, especially by Arthur Compton's conclusive detection of a latitude effect in 1932.

Millikan hotly contested Compton's findings. He had repeated his search for a latitude effect in the late 1920's, and in late 1932 he did so once more, again without success. But Millikan was the victim of experimental circumstance. In the longitudinal region of California, the dip in cosmic ray intensity began quite suddenly in the neighborhood of Los Angeles and quickly reached its maximum fall of some 7 percent less than two days' sail south of the city. In Millikan's initial search for the latitude effect—the voyage from Mollendo, Peru, to Los Angeles—his estimated error had been 6 percent. In most of his later searches, he went to the north of Pasadena, where the rise in intensity was too small to detect easily. In 1932 he sent H. Victor Neher, a young collaborator at Caltech, on a voyage to the south; but Neher did not get his electroscope working before he had passed the region of the dip.

By 1933, with Neher having now found a latitude effect, Millikan had admitted that some percentage of cosmic radiation must consist of charged particles. By 1935 he had also rejected the atom-building hypothesis, mainly because it was now clear that the bulk of cosmic radiation possessed energies much higher than the mass defects of the abundant elements. All the same, despite a vast array of contrary evidence and the overwhelming body of professional opinion, Millikan clung tenaciously to the assumption that some fraction of the primary cosmic radiation could be photons. In the late 1930's and early 1940's he searched for evidence in support of this view, measuring cosmic ray intensities around the world at sea level, in airplanes at high altitudes, and with unmanned sounding balloons up to the top of the atmosphere. On the basis of this data, he also developed a theory that cosmic ray photons originated in the spontaneous annihilation of atoms in interstellar space. No more convincing than its predecessor, this hypothesis became completely untenable, as Millikan himself admitted a few years before his death, after the detection of the π-meson in 1947 made it clear that the primary cosmic radiation consisted almost entirely of protons.

But however wrongheaded Millikan had been, his cosmic ray research yielded a valuable fund of experimental data. Moreover, in 1934, independently of Jacob Clay, he detected the variation of the latitude effect with longitude because of the dissymmetry of the earth's magnetic field. In a roundabout way even the atom-building hypothesis strikingly benefited the progress of science. In the late 1920's, troubled by the discrepancy between his experimental data and the predictions of both the Dirac absorption formula and its successor, the Klein-Nishina formula, Millikan recognized that he needed a measure of cosmic ray energies that was not based on absorption coefficients. To obtain a direct determination, he put Carl Anderson, a young research fellow at Caltech, to work with a cloud chamber set in a powerful magnetic field. In 1931 Anderson's studies of the trajectories of charged particles showed conclusively that the absorption of cosmic rays resulted from nuclear encounters as well as from Compton scattering. They also led to his detection of the positron in 1932.

Between the wars Millikan played a prominent role in the affairs of his profession. The president of the American Association for the Advancement of Science in 1929 and the holder of various offices in the National Academy of Sciences and the National Research Council, he was especially active as a member of the NRC fellowship board and as foreign secretary of the Academy. From 1922 to 1932 Millikan served as the American representative to the Committee on Intellectual Cooperation of the League of Nations. Throughout the interwar period he participated in the International Research Council; its successor, the International Council of Scientific Unions; and the affiliate of both, the International Union of Pure and Applied Physics. In 1933 Millikan was appointed by President Franklin D. Roosevelt to the Science Advisory Board, a joint venture of the Academy and the federal government to find ways to use science for economic recovery.

Millikan was an able popularizer and lecturer, and

after he won the Nobel Prize in 1923 he became perhaps the most famous American scientist of his day. An outspoken religious modernist, he was a leading exponent of the reconcilability of science and religion in the 1920's, the decade of the Scopes trial. Politically, Millikan was a conservative Republican. During the 1930's he vigorously opposed the New Deal, repeatedly denounced governmental intervention in the economy, and argued that the promotion of science, because it led to new industries and new jobs, was a much sounder way to achieve economic recovery. Always an internationalist, Millikan believed firmly in collective security. In the late 1930's, unlike many conservative Republicans at the time, including his good friend Herbert Hoover, he helped propagandize in favor of aid to the Allies; by early 1941 he was encouraging the conversion of Caltech from academic to military purposes.

During the war Millikan turned over an increasing fraction of his administrative responsibilities at Caltech to the younger staff members who were running the various defense projects. In 1946 he retired from his professorship and the chairmanship of the executive council. He remained active as a public lecturer and spoke frequently on the subject of science and religion. He was cool to the creation of the National Science Foundation and spoke often against federal aid to education. By the time of his death, Millikan had been awarded numerous medals, even more honorary degrees, and membership in twenty-one foreign scientific societies, including the Royal Society of London and the Institut de France.

BIBLIOGRAPHY

I. ORIGINAL WORKS. A complete bibliography of Millikan's published work, which includes close to 300 scientific papers, is in Lee A. DuBridge and Paul S. Epstein, "Robert Andrews Millikan," in *Biographical Memoirs. National Academy of Sciences*, **33** (1959), 241–282. In *The Autobiography of Robert A. Millikan* (New York, 1950) Millikan provided valuable accounts of his childhood and education, work on the electronic charge and the photoeffect, and involvement in the mobilization of science during World War I; curiously, he devoted little space to his research in hot spark spectra or cosmic rays, and his account of the development of the California Institute of Technology must be used with special care. Millikan left a voluminous body of correspondence, which is now in the Caltech archives. Dating in the main from 1921, the collection contains substantial materials on the National Academy of Sciences–National Research Council and the administration of the California Institute of Technology, as well as a sizable amount of family and scientific letters. Another important batch of Millikan's letters is in the papers of

George Ellery Hale in the Caltech archives, which were also published in a microfilm edition (Pasadena, 1968) under the editorship of Daniel J. Kevles. The locations of other letters to and from Millikan are given in Thomas S. Kuhn, *et al.*, *Sources for the History of Quantum Mechanics* (Philadelphia, 1967), 68.

II. Secondary Literature. Paul S. Epstein wrote an excellent résumé of Millikan's scientific work in "Robert A. Millikan as Physicist and Teacher," in *Reviews of Modern Physics*, **20** (Jan. 1948), 10–25, a volume published in honor of Millikan's eightieth birthday. A condensed version of Epstein's essay occupies part of the memoir written with DuBridge (see above), which is on the whole a useful introduction to Millikan's life. Millikan the famous scientist is treated in Daniel J. Kevles, "Millikan: Spokesman for Science in the Twenties," in *Engineering and Science*, **32** (Apr. 1969), 17–22.

Daniel J. Kevles

MILLINGTON, THOMAS (*b.* Newbury, Berkshire, England, 1628; *d.* London, England, 5 January 1704), *medicine.*

The son of Thomas Millington, Esquire, of Newbury, Millington was educated at Westminster School during the headmastership of Richard Busby. From Westminster he was elected a scholar to Trinity College, Cambridge, on 31 May 1645 and matriculated the following Easter. At Cambridge he was under the tutorship of James Duport and graduated B.A. in 1649. Shortly afterward he moved to Oxford, where he became fellow of All Souls and proceeded M.A. on 30 May 1651 (incorporated Cambridge 1657), B.D. on 8 July 1659, and M.D. on 9 July 1659. Why Millington moved is unclear, but Carter states that he was invited.[1] Possibly, in view of his close connections with the groups engaged in natural philosophy at Wadham, All Souls, and elsewhere, his interest in these activities was the cause.[2] He was particularly involved in the research on the brain conducted by the Willis-Wren-Lower circle.[3]

Candidate of the Royal College of Physicians on 30 September 1659, Millington became a fellow on 2 April 1672, although he still seems to have been living at Oxford. Appointed Sedleian professor of natural philosophy as successor to Willis in 1675, he gave his inaugural lecture on 12 April 1676; according to Wood it was "much commended."[4] Shortly afterward, however, he seems to have moved to London, although he retained his post at Oxford until his death, discharging his duties by deputy. (Among these deputies was John Keill, in 1700.) Physician in ordinary to William and Mary, as he was also to Queen Anne, Millington was knighted in March 1680.

Censor of the Royal College of Physicians in 1678, 1680, 1681, and 1684; Harveian orator in 1679; and treasurer from 1686 to 1689, he was consiliarius in 1691 and 1695, and served as president from 1696 until his death. Millington was licensed to marry Hannah King, widow of Henry King, on 23 February 1680; they had a son, Thomas (who seems to have gained notoriety as a beau), and a daughter, Anne.[5] In 1680 Bishop John Fell put forward Millington's name to William Sancroft, archbishop of Canterbury, for the vacant chair of medicine at Oxford.[6]

Millington's claim upon the attention of posterity is threefold: as a physician, as a man of wide-ranging intellectual activities, and as the reputed discoverer of sexuality in plants. This last claim rests entirely on a remark by Nehemiah Grew (to whom the discovery would otherwise be credited) in a discourse upon the anatomy of flowers read to the Royal Society on 9 November 1676. While discussing the role of the attire (stamens) in flowers, he said, "In Discorse hereof with our Learned *Savilian* [*sic*] Professor Sir *Thomas Millington*, he told me he conceived, That the Attire doth serve in the *Male*, for the *Generation* of the *Seed*. I immediately reply'd, That I was of the Opinion; and gave him some reasons for it, and answered some *Objections*, which might oppose them."[7] The emphasis here seems to be on the immediacy of Grew's reply and the reasons he provided in support of Millington's idea. This suggests that Grew had been considering the problem for some time, while for Millington it was still little more than an undeveloped insight. In the lack of any other evidence that Millington had been working on this particular subject, there seems no reason to ascribe priority to him over Grew in the enunciation and explanation of this phenomenon.

Millington impressed his contemporaries as a learned man, as indeed he still does, despite the scarcity of materials relating to him. Thanks partly to his learning and partly to a notably amiable disposition, he was much respected as a physician, being praised in Samuel Garth's "The Dispensary" under the name of Machaon and also by Sydenham. He had a highly fashionable practice—amassing, says Carter, a fortune of £60,000.[8] Millington was called to the deathbed of Charles II and was one of the physicians to perform the dissection of William III's body. As an officer and as president of the Royal College of Physicians, he was active and zealous during a somewhat disturbed period of the College's history but was unable to restore it from its moribund state. He was, however, instrumental in liquidating the College's £7,000 debt to the executors of Sir John Cutler in 1701, providing £2,000 himself. Linnaeus named the genus *Millingtonia* among the Bignoniaceae after him.

NOTES

1. Edmund Carter, *History of Cambridge* (London, 1753), 329.
2. Anthony Wood, *Life and Times*, A. Clarke, ed., I (Oxford, 1891), 201.
3. Thomas Willis, *Cerebri anatomici* (Oxford, 1664), dedicatory epistle.
4. *Op. cit.*, II, 343.
5. William Musgrave, ed., *Obituaries Prior to 1800 (as far as Relates to England, Scotland and Ireland)*, which is *Harleian Society*, XLVII (London, 1900), 201–202. Bodleian Library MS Rawlinson D 1160, fol. 34a.
6. Bodleian MS Tanner 36*, fol. 51.
7. Nehemiah Grew, *The Anatomy of Plants With an Idea of a Philosophical History of Plants* (London, 1682), bk. IV, ch. 5, 171.
8. *Op. cit.*, p. 350.

BIBLIOGRAPHY

I. ORIGINAL WORKS. Millington's only published work, prepared jointly with Sir Richard Blackmore and Sir Edward Hannes, was *The Report of the Physicians and Surgeons, Commanded to Assist at the Dissecting the Body of His Late Majesty at Kensington, March the Tenth MDCCII. From the Original Delivered to the Right Honourable the Privy Council* (London, 1702). Bodleian MS Rawlinson D 1041, "Sententiae collectae a Tho. Millington Cantabrigiensi anno 1648," and British Museum, Sloane MS 3565, "Celebriorum distinctionum synopsis Thomas Millington April 7 1646," are undergraduate notebooks. Sloane 2148, "De morbis in specie eorumque remedius: Tum logice tum empirice," is more a recipe book than a treatise.

II. SECONDARY LITERATURE. In addition to the references given in the notes, further discussions of Millington and the Royal College of Physicians can be found in W. Munk, *The Roll of the Royal College of Physicians of London* (London, 1878), 363–365; and Sir George Clark, *A History of the Royal College of Physicians*, 2 vols. (London, 1964–1966), I, 323, 358, 370; II, 469, 472, 474–475, 483, 487. For most purposes, the entry in *Dictionary of National Biography*, XIII, 442, is perfectly adequate.

A. J. TURNER

MILLON, AUGUSTE-NICOLAS-EUGÈNE (*b.* Châlons-sur-Marne, France, 24 April 1812; *d.* St.-Seine-l'Abbaye, France, 22 October 1867), *chemistry, agronomy, pharmacy.*

After completing his secondary education in Châlons-sur-Marne and working as a teaching assistant at the Collège Rollin in Paris, Millon decided on a medical career in the army. In 1832 he was admitted as a student to the Val-de-Grâce military teaching hospital in Paris, and two years later he qualified for active duty as a surgeon. He served successively in Bitche, Lyons, Algeria, and Metz; in 1836 he received the M.D. from the Paris Faculty of Medicine.

Millon's long interest in chemistry prompted him, however, to take up military pharmacy rather than surgery and medicine. A brief appointment as *préparateur* and tutor at the Val-de-Grâce was followed by a tour of duty at several military installations, and finally by a professorship of chemistry at the Val-de-Grâce in 1841. During the next six years Millon established himself as an outstanding chemist and teacher. Probably because of his unorthodox views, he was abruptly transferred in 1847 as professor to the military teaching hospital in Lille, which not only separated him from his students and friends but also seriously disrupted his scientific work. From 1850 until his retirement in 1865, he served as the top-ranking pharmacist for the French army in Algeria.

The years from 1837 to 1847, scientifically the most important period of Millon's life, were devoted largely to basic chemistry. His circle of friends and occasional collaborators included Pelouze, J. Reiset, F. Hoefer, Regnault, Louis Laveran, and F.-J.-J. Nicklès. Particularly noteworthy at this time were his studies of the nitrides of bromine, iodine, and cyanogen; of oxides of chlorine and iodine; of reactions of nitric acid on metals and of mercury salts with ammonia; and the investigation with J. Reiset of the nature of catalytic reactions. It was also during this decade that Millon discovered iodine dioxide, chlorites, ethyl nitrate, and the production of potassium iodate. In 1845 he launched the *Annuaire de chimie* with Reiset (in collaboration with Hoefer and Nicklès), seven volumes of which appeared before it was discontinued in 1851. Millon's lectures at the Val-de-Grâce formed the basis for his two-volume treatise, *Éléments de chimie organique* (Paris, 1845–1848).

After 1847 the direction and emphasis of Millon's scientific work changed, becoming more applied and diffuse. He devoted considerable time to studying wheat, especially its classification, constituents, conservation, and processing. Millon showed in 1848 that urea could be quantitatively analyzed by decomposing it with nitrous acid and determining the amount of carbon dioxide released. In 1849 he published his discovery of a sensitive reagent for detecting proteins made by dissolving mercury in concentrated nitric acid and diluting with water, which proved effective in the presence of tyrosine. Among the varied projects he pursued in Algeria were the extraction of perfumes from Algerian flowers, the chemistry of nitrification, the raising and commerce of leeches, quality control of milk, the study of alcoholic fermentation, and the analysis of mineral water.

BIBLIOGRAPHY

1. ORIGINAL WORKS. Most of Millon's work was published in entirety or as extracts in the *Comptes rendus . . . de l'Académie des sciences*, often appearing at the same time in the *Annales de chimie et de physique*. For a chronological list of Millon's published and unpublished material, see J. Reiset *et al.*, *E. Millon, sa vie, ses travaux de chimie et ses études économiques et agricoles sur l'Algérie* (Paris, 1870), 321–327. A bibliography of Millon's articles is in the Royal Society's *Catalogue of Scientific Papers*, IV (London, 1870), 393–395; VIII (London, 1879), 407–408.

II. SECONDARY LITERATURE. See A. Balland, *Les travaux de Millon sur les blés* (Paris, 1905); H.-P. Faure, *E. Millon, notice biographique, lue à la Société d'agriculture, commerce, sciences et arts du département de la Marne, dans la séance publique du 26 août 1868* (Châlons-sur-Marne, 1868); J. R. Partington, *A History of Chemistry*, IV (London–New York, 1964), 57, 84, 90, 342, 364, 427–429, 603; and J. Reiset *et al.*, *E. Millon . . .* (see above).

The centenary of Millon's death was the occasion for three articles: Jean Delga, "La carrière militaire d'Eugène Millon," in *Revue d'histoire de la pharmacie*, **19** (1968), 69–72; Pierre Malangeau, "L'oeuvre scientifique d'Eugène Millon," *ibid.*, 73–82; and André Quevauviller, "À propos du centenaire de la mort de Millon," *ibid.*, 83–86.

ALEX BERMAN

MILLS, WILLIAM HOBSON (*b.* London, England, 6 July 1873; *d.* Cambridge, England, 22 February 1959), *organic chemistry*.

Although Mills was born in London, his father, William Henry Mills, an architect, and his mother, Emily Wiles Quincey Hobson, came from Lincolnshire and returned there in the autumn of 1873; thus he always regarded himself as a Lincolnshire man. He was educated first at Spalding Grammar School and then at Uppingham School, where an accident in the snow caused the severing of an Achilles tendon and limited his outdoor activities, although in his mature years he could walk and cycle with considerable vigor. He entered Jesus College, Cambridge, in October 1892 and in due course obtained a first class in the natural sciences tripos, part I, in 1896 and in part II (chemistry) in 1897.

Mills then began research in the Cambridge University Chemical Laboratory under T. H. Easterfield (later Sir Thomas Easterfield); when the latter accepted the professorship of chemistry at Wellington, New Zealand, in 1899, Mills continued the work alone and was elected to a fellowship (tenable for six years) at Jesus College in 1899.

In October 1899 Mills went to Tübingen to work under Hans von Pechmann for two years, during which period he met N. V. Sidgwick of Oxford

University; the two chemists, so similar in their interests, became friends for life. In 1902 Mills was appointed head of the chemical department of Northern Polytechnic Institute in London. In 1912 he returned to Cambridge, having been appointed to the demonstratorship to the Jacksonian professorship of natural philosophy and to a fellowship and lectureship at Jesus College. In 1919 he was appointed university lecturer, and in 1931 the university recognized the high quality of his work by creating a personal readership in stereochemistry, which he held until his retirement in 1938.

The major part of Mills's scientific work was devoted to stereochemistry and the cyanine dyes. Only brief mention of some of the highlights in each of these divisions will be made.

Stereochemistry. Certain types of oximes were known to exist in two or more isomeric forms, for which an explanation had been suggested by Hantzsch and Werner. This explanation was not accepted by many chemists, and Mills sought decisive experimental evidence for its accuracy. After investigating several compounds, Mills and B. C. Saunders (*Journal of the Chemical Society* [1931], 537) prepared the *o*-carboxyphenylhydrazone of β-methyl-trimethylene-dithiolcarbonate, which they resolved into optically active forms. Optical activity could arise in this compound only if the Hantzsch-Werner theory were correct.

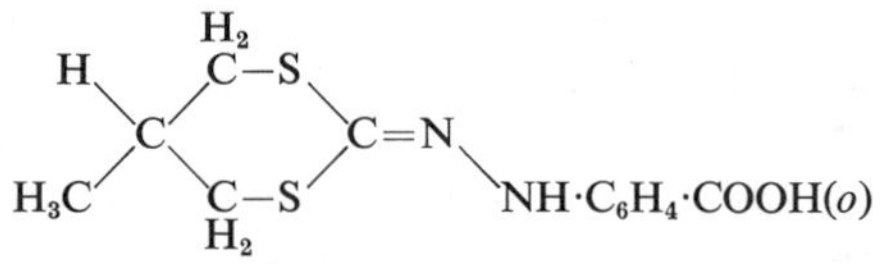

FIGURE 1. *o*-Carboxyphenylhydrazone of β-methyl-trimethylene-dithiolcarbonate.

It had been recognized that a spirocyclic compound consisting of two carbon rings linked together by a common carbon atom might show optical activity if the rings possessed appropriate substituents to ensure molecular dissymmetry. Mills and C. R. Nodder (*ibid.*, **119** [1921], 2094) synthesized and resolved into optically active forms the first such compound, the ketodilactone of benzophenone-2,4,2′,4′-tetracarboxylic acid.

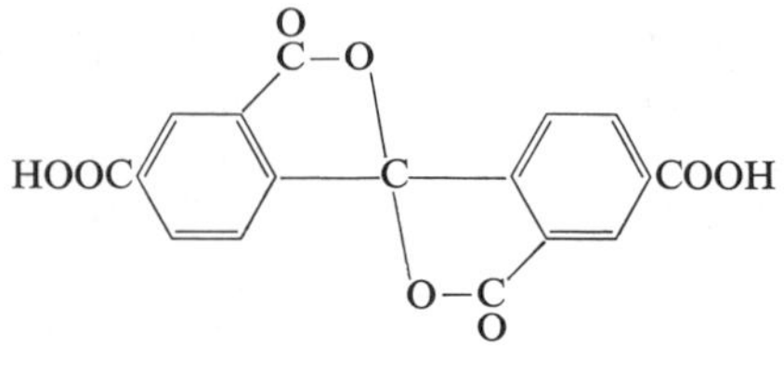

FIGURE 2. Ketodilactone of benzophenone-2,4,2′,4′-tetracarboxylic acid.

It had also long been recognized that a suitably substituted allene compound (3) would be dissym-

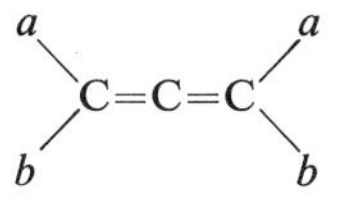

FIGURE 3. Example of an unsymmetrically substituted allene.

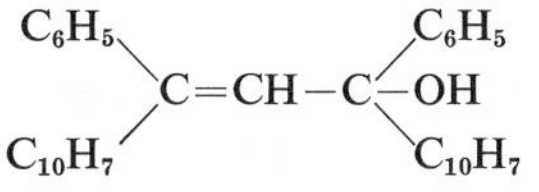

metric; but the synthesis of such a compound, bearing acidic or basic groups for resolution, had defied synthesis. P. Maitland and Mills, after about six years of persistent work, synthesized $\alpha\gamma$-biphenyl-$\alpha\gamma$-di-α-naphthylallyl alcohol, which by a stereospecific dehy-

FIGURE 4. $\alpha\gamma$-Biphenyl-$\alpha\gamma$-di-α-naphthylallyl alcohol.

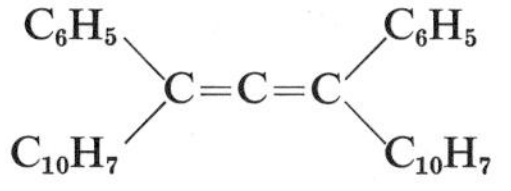

FIGURE 5. $\alpha\gamma$-Biphenyl-$\alpha\gamma$-di-α-naphthyl allene.

dration using dextro and levo camphorsulfonic acid, was converted into the optically active forms of $\alpha\gamma$-biphenyl-$\alpha\gamma$-di-α-naphthyl allene (5).

By extensions of these general methods, Mills and E. H. Warren (*ibid.*, **127** [1925], 2507) showed that the nitrogen atom of a quaternary ammonium salt had the tetrahedral configuration and was not situated in the center of a square-based pyramid. Furthermore, Mills and T. H. H. Quibell (*ibid.* [1935], 839) produced stereochemical evidence that the four-coordinated platinum atom had the planar, as distinct from the tetrahedral, configuration.

The fact that a biphenyl molecule, having suitable substituents in the 2,2'6,6'-positions, could show optical activity at first puzzled chemists. Mills was the first to point out in a simple diagram (*Chemistry and Industry*, **45** [1926], 884) that the size of these substituents could obstruct the free rotation of the two phenyl groups about their common axis, C_6H_5—C_6H_5, and such molecules could thus show optical activity. He became greatly interested in this subject of "restricted rotation" and later applied it to suitably substituted derivatives of naphthalene, quinoline, and benzene (with K. A. C. Elliott, *et al.*, in *Journal of the Chemical Society* [1928], 1291; [1932], 2209; [1939], 460).

Cyanine Dyes. In 1914 photographic plates and films were normally prepared with a silver bromide–silver iodide emulsion, which was sensitive only in the ultraviolet, violet, and blue regions. In 1905, however,

a German firm had synthesized a "photographic sensitizer" which, when incorporated into the emulsion, extended the sensitivity well into the red region. When in 1914–1915 the Western Front became essentially two parallel bands of heavily entrenched positions, it became imperative to detect as early as possible each day any work on these positions which the enemy had carried out during the previous night. The photographic reconnaissance of the British Royal Flying Corps (later the Royal Air Force) was under a great disadvantage, for their silver bromide–silver iodide plates were at their least sensitive in the red light of the early morning. The British authorities sent an urgent request to W. J. Pope, the head of the Cambridge University chemical department, to investigate the structure and the synthesis of Pinacyanol, which the Germans were using.

Pope enlisted the help of Mills and other workers, notably F. M. Hamer. This small team showed that Pinacyanol had the structure shown in Figure 6 and

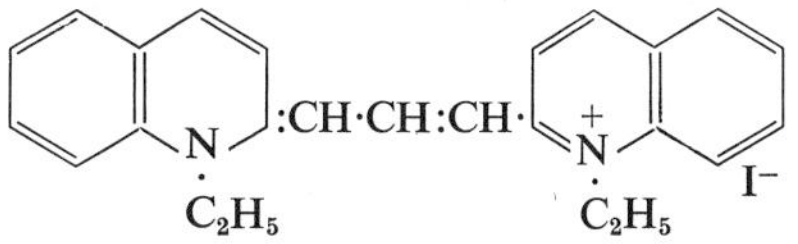

FIGURE 6. Pinacyanol (systematic name: 1,1'-Diethyl-2,2'-carbocyanine iodide).

developed a rapid synthesis of this compound and of other novel sensitizers such as the isocyanines. After the war Pope and Mills stated: "Throughout the war practically all the sensitizing dyestuffs used by the Allies in the manufacture of panchromatic plates were produced in this (i.e. the *Cambridge*) Laboratory" (*Photographic Journal*, **60** [1920], 183, 253). Mills and his co-workers subsequently continued the investigations of the various new types of sensitizers.

Mills was elected a fellow of the Royal Society in 1923 and received its Davy Medal in 1935. He was president of the Chemical Society for the years 1942–1943 and 1943–1944; his presidential addresses, entitled "The Stereochemistry of Labile Compounds" and "Old and New Views on Some Chemical Problems," respectively form the last of his chemical publications.

Retirement allowed Mills to devote himself to the study of natural history, in particular to the many subspecies of British bramble. His collection of *Rubi* is housed in the botany department of Cambridge University and is composed of about 2,200 specimens mounted in sheets and arranged in systematic order: he had specimens of 320 of the 389 "microspecies" of *Rubus fructicosus*.

BIBLIOGRAPHY

In addition to the works cited in the text see F. G. Mann's much fuller account of Mills and his work (with a photograph and a bibliography containing 73 entries) in *Biographical Memoirs of Fellows of the Royal Society*, **6** (Nov. 1960), 201.

FREDERICK G. MANN

MILNE, EDWARD ARTHUR (*b*. Hull, England, 14 February 1896; *d*. Dublin, Ireland, 21 September 1950), *astrophysics, cosmology*.

Milne was one of the foremost pioneers of theoretical astrophysics and modern cosmology, and his name was often linked with those of Eddington and Jeans, although, unlike them, he wrote no books on astronomy for the general public. He was the eldest of three brothers who all entered on scientific careers. His father, Sydney Arthur Milne, was headmaster of a Church of England school at Hessle, near Hull, in Yorkshire. His mother, born Edith Cockcroft, lived to an advanced age and survived her famous son.

Milne went to school at Hymers College, Hull, and entered Trinity College, Cambridge, in 1914 with an open scholarship in mathematics and natural science, having gained a record number of marks in the examination. The first World War had already begun when Milne went to Cambridge, but his defective eyesight prevented him from undertaking active military duties. Early in 1916 he accepted the invitation of the biophysicist A. V. Hill to join the Anti-Aircraft Experimental Section (as it was later called) of the Munitions Inventions Department. The work was largely concerned with ballistics and sound ranging and involved many problems relating to the atmosphere of the earth. This was the beginning of Milne's deep interest in atmospheric theory. For his war services he was awarded the M.B.E.

In 1919 Milne returned to Cambridge and soon afterward was elected a fellow of Trinity College. In 1920 H. F. Newall appointed him assistant director of the solar physics observatory. In 1924 he succeeded Sydney Chapman as Beyer professor of applied mathematics at the University of Manchester, a post which he held until January 1929, when he became first Rouse Ball professor of mathematics at Oxford and fellow of Wadham College. He held both these posts for the rest of his life but was granted leave of absence during the Second World War, from 1939 to 1944, in order to work at the Ordnance Board, at Chislehurst in Kent. There he dealt with a wide variety of problems, including ballistics, rockets, sound ranging, and the optimum distribution of guns.

Important though his researches were to military science, it is by his original contributions to astrophysics and cosmology that Milne's reputation must be judged. These fall into three clearly defined parts, which can be associated with three consecutive stages in his career. From 1920 to 1929 his researches centered on problems of radiative equilibrium and the theory of stellar atmospheres. From 1929 to 1932 he was mainly concerned with the theory of stellar structure, and from 1932 onward with relativity and cosmology.

Milne's interest in stellar atmospheres was first aroused by H. F. Newall, who sensed his special fitness for the task of tackling theoretical problems associated with the outermost layers of stars, particularly those that concern the transfer of radiation through an atmosphere and those relating to the ionization of the material. These problems have to be combined, and this combination leads to subtle considerations of the detailed interaction of matter and radiation. It was in this field that Milne first made his name in astronomical circles, and it was primarily through his work that by the end of the 1920's astrophysicists were provided with the theoretical techniques appropriate for the study of stellar atmospheres.

The idealized problem of radiative transfer through a scattering atmosphere without absorption was first studied by A. Schuster in a fundamental paper in 1905, and the theory of radiative equilibrium in an absorbing atmosphere was examined by K. Schwarzschild in 1906. Milne combined and extended these pioneer investigations. He soon realized that the concept of radiative equilibrium implies the constancy of the net flux of radiation through the atmosphere. He obtained an integral equation for this net flux and used it to derive a useful approximation for the dependence of temperature on optical depth. This work was not only of great scientific importance but also yielded a result of mathematical interest, an integral equation now known as Milne's equation.

In his Smith's Prize essay of 1922 (*Philosophical Transactions of the Royal Society*, **223A** [1922], 201–255), Milne extended his theory to the law of darkening of a stellar disk toward the limb and its relation to the distribution of energy in the star's continuous spectrum, assuming radiative equilibrium with a "gray" coefficient of absorption (that is, independent of wavelength). He demonstrated how closely this prediction was obeyed by the sun. Milne was the first to investigate the inverse problem of obtaining the temperature distribution from the observed darkening toward the limb. He also showed how both this and the observed continuous spectrum could be used to infer the dependence of opacity on frequency. (Milne's method was later used to show that the sun's opacity is in fact due to the negative hydrogen ion.) An admirable account of his theory and of related investigations was given by Milne in his "Thermodynamics of the Stars," published as a part of the *Handbuch der Astrophysik* (1930). It was a milestone in the history of the subject and was the starting point for more elaborate investigations by E. Hopf, S. Chandrasekhar, and others.

In 1923 Milne began a fruitful collaboration with R. H. Fowler on the intensities and widths of absorption lines in stellar spectra. This work was based on M. N. Saha's theory of high-temperature ionization. By modifying Saha's technique, Milne and Fowler developed a theory of the maximum intensity, instead of the marginal appearance, of any given absorption line in the spectral sequence. They deduced that the pressures of the levels in stellar atmospheres at which absorption lines are formed are of the order of 10^{-4} atmosphere, a value considerably lower than had previously been assumed. They were also able to determine a reliable temperature scale for the sequence of stellar spectra, one of the greatest advances in modern astrophysics. Although much of the work was Fowler's, the original key idea was Milne's.

Another problem to which Milne made a classic contribution was that of the escape of molecules from stellar and planetary atmospheres. In particular he investigated the equilibrium of the calcium chromosphere under the balance of gravitational forces and radiation pressure. He discovered that, for varying radiation from below, this equilibrium is unstable, so that in certain circumstances atoms can ultimately be ejected from the sun with velocities of the order of 1,000 kilometers a second.

Milne's numerous papers on stellar atmospheres led to his election as a fellow of the Royal Society in 1926, at the age of thirty, and culminated in his Bakerian lecture of 1929, "The Structure and Opacity of a Stellar Atmosphere." It was for his researches in this field that he was awarded the gold medal of the Royal Astronomical Society in 1935.

The second phase of Milne's research career in astrophysics began with his move to Oxford in 1929. During the following three years he devoted his main energies to elaborating a theory of stellar structure based on a constructive mathematical criticism of the pioneer researches of Sir Arthur Eddington. Although much of Milne's criticism of Eddington's work has not been generally accepted, his methods led to important developments, notably T. G. Cowling's fundamental study of the stability of gaseous stars and Chandrasekhar's standard theory of white dwarf stars. In particular Milne seems to have been the first to

suggest an association between the nova phenomenon and the collapse of a star from one configuration to another as a result of decreasing luminosity.

In May 1932 Milne turned his attention to the problem of the expansion of the universe. It occurred to him that this phenomenon might not be essentially different from the inevitable dispersion of a gas cloud liberated in empty space. Whatever its ultimate value, this simple idea fired his imagination and led him to develop an entirely new approach to theoretical cosmology. He abandoned the mathematically recondite method of general relativity and worked as far as possible in terms of special relativity and Euclidean space. He regarded Hubble's empirical law of simple proportionality of the distances and recessional speeds of the galaxies as immediately explicable in terms of uniform motion. This meant that the common ratio of the respective distances and speeds of the galaxies provided a direct measure of the age of the universe, that is, of the time that has elapsed since the initial pointlike singularity when the universe was "created." According to the scale of extragalactic distances determined by Hubble, this age was only some 2,000 million years, which is less than the ages now assigned to the sun and earth. According to the latest data bearing on Hubble's law, based on revised estimates of extragalactic distances, the age of the universe, if it is of Milne's type, would be nearer 10,000 million years, which is about twice the age now assigned to the sun and comparable with that currently assigned to the galaxy.

Milne's world model is not in accord with general relativity, according to which a homogeneous isotropic world model that expands uniformly must be devoid of matter. Milne was not dismayed at this discrepancy, but instead made his kinematic approach to cosmology the basis of a new deductive system of theoretical physics, which came to be called kinematic relativity. He introduced the useful term "cosmological principle" to signify that observers associated with galaxies in his model and in many others, including those based on general relativity, would see similar "world pictures."

Milne went on to derive from his model many properties analogous to the laws of dynamics, gravitation, and electromagnetic theory. These developments of his theory were not generally accepted, and it is now thought that the most important effect of his work was that it led to fresh attempts to analyze the concepts of time and space-time. In particular A. G. Walker proved, as a generalization of Milne's work, that the general metric of "orthodox" relativistic cosmology could be derived without appeal to general relativity, and G. J. Whitrow showed that special relativity can be based on determinations of distance in terms of time measurement by what is now called the radar technique. Milne himself was led to the conclusion that there may be different uniform scales of time operating in nature and that some of the fundamental constants of physics may vary with the cosmic epoch. He also showed, partly in collaboration with W. H. McCrea, that there exist useful Newtonian analogues of the expanding world models of relativistic cosmology and so was the founder of modern "Newtonian" cosmology.

Milne's philosophical outlook was best expressed in his inaugural lecture at Oxford in 1929, in which he claimed that the primary aim of mathematical physics is to build up a system of theory rather than to seek the solution of particular problems.

Small in stature, Milne had outstanding qualities of mind and was a continual fount of inspiration to others as well as to himself. Although as a young man he was stricken with epidemic encephalitis ("sleepy sickness"), he made a remarkable recovery. For about twenty years after he recovered from the initial attack in 1923 he remained a man who radiated energy and vitality. Nevertheless, he did not escape the usual long-delayed aftereffects of encephalitis; and in the last five years of his life he suffered from rigidity of muscles and tremor of the left arm. Milne had the humility and simplicity of character that often goes with scientific genius, and he bore personal misfortunes with courage, dignity, and religious conviction. Both his wives predeceased him in tragic circumstances, and he was left to bring up three young daughters and a son. In his later years his heart became affected, and he died suddenly in Dublin, where he had gone to attend a scientific meeting.

BIBLIOGRAPHY

I. ORIGINAL WORKS. "Thermodynamics of the Stars" and "Theory of Pulsating Stars" are in *Handbuch der Astrophysik*, III (Berlin, 1930), pt. 1, 65–255, and pt. 2, 804–821, respectively. Two of his books are *Relativity, Gravitation and World-Structure* (Oxford, 1935); and *Kinematic Relativity* (Oxford, 1948).

Milne contributed many original papers to scientific journals, notably *Monthly Notices of the Royal Astronomical Society*, *Philosophical Transactions of the Royal Society*, *Proceedings of the Royal Society* (sec. A), *Philosophical Magazine*, and *Zeitschrift für Astrophysik*. A complete list will be found at the end of the obituary notice by McCrea (see below).

Milne also wrote a valuable and highly individual textbook, *Vectorial Mechanics* (London, 1948), and left two other books in MS which were published posthumously: *Modern Cosmology and the Christian Idea of God the Creator* (Oxford, 1952) and *Life of James Hopwood Jeans* (Cambridge, 1952).

II. SECONDARY LITERATURE. Biographical notices are W. H. McCrea, in *Obituary Notices of Fellows of the Royal Society of London*, **7** (1951), 421–443; and in *Monthly Notices of the Royal Astronomical Society*, **111** (1951), III, 160–170. The latter is followed by a short notice, pp. 170–172, by H. H. Plaskett. There is a short biographical notice by G. J. Whitrow in *Nature*, **166** (1950), 715–716. McCrea also wrote the article on Milne in the *Dictionary of National Biography, 1941–1950*, pp. 594–595.

G. J. WHITROW

MILNE, JOHN (*b*. Liverpool, England, 30 December 1850; *d*. Shide, Isle of Wight, England, 31 July 1913), *seismology*.

The son of John Milne and Emma Twycross, Milne was educated at Liverpool and King's College, London, and later studied geology and mineralogy at the Royal School of Mines. He was an ardent and adventurous traveler, starting, as a schoolboy and without parental leave, with a dangerous exploration of the Vatnajökull in Iceland. After early experience as a mining engineer in Great Britain and Germany he spent two years investigating the mineral resources of Newfoundland and Labrador, and later wrote geological notes on his observations in Egypt, Arabia, and Siberia. He visited Funk Island, off the coast of Newfoundland, where he made a large collection of skeletons of the great auk. In 1874 he served as geologist in an expedition that sought to fix the site of Mt. Sinai.

In 1875 Milne was appointed professor of geology and mining at the Imperial College of Engineering, Tokyo. The journey to Japan took eleven months, part of it crossing Mongolia by camel in subzero weather. In Japan he turned to the study of earthquakes, the field in which he became world famous. He married Tone Noritsune, daughter of Horikawa Noritsune, the high priest of Hakodate. Milne retired from Japan in 1895 and went with his wife to Shide, on the Isle of Wight, where he continued in active seismological work until his death after a short illness in 1913. Throughout his work at Shide he was assisted by the British Association for the Advancement of Science, which had established a seismological committee and appointed Milne its secretary. The work was a labor of love in which he had the devoted services of a Japanese assistant, Shinobu Hirota; many of the expenses were defrayed by Milne himself. He became a fellow of the Royal Society in 1887 and was awarded the Lyell Medal of the Geological Society of London in 1894 and a Royal Medal in 1908. The emperor of Japan conferred upon him the Order of the Rising Sun.

Milne was the most noted of a group of British scientists in Japan who pioneered modern seismology. An earthquake at Yokohama on 22 February 1880 led them, on Milne's initiative, to form with their Japanese colleagues the Seismological Society of Japan, the first organization devoted exclusively to the study of earthquakes and volcanoes. Its work was crucial at a time when seismology was developing from a qualitative science, resting largely on geological observations and concerned with such matters as cataloging earthquake effects, into a science in which precise physical measurements are brought to bear. By 1892 Milne, in association with his colleagues J. A. Ewing and T. Gray, had developed a seismograph for recording horizontal components of the ground motion. It was reliable, compact, and simple enough to be installed on a worldwide basis and to provide a global coverage of ground movements due to large earthquakes. From that date the science of seismology as a branch of geophysics advanced apace, and seismological data began to be applied to unraveling the internal structure of the earth.

Milne's researches touched on nearly all aspects of seismology. From his Tokyo records he deduced that in large earthquakes the ground accelerations can be comparable with the vertical acceleration of gravity. He showed that earthquake accelerations are in general greater—and therefore more dangerous—on soft ground than on hard rock. He initiated experiments to study properties of earthquake waves by generating artificial shocks by explosives and other means and by examining records of the ensuing ground motions. In this way he obtained records showing groups of waves corresponding to the P and S (primary and secondary) waves of modern seismology. He devised methods of locating distant earthquake sources from his records and evolved early travel-time curves of earthquake waves in terms of the distances from the source. He compiled important earthquake catalogs, including one covering the seismic history of Japan from 295 B.C. and another on destructive world earthquakes. (It is estimated that he examined about one hundred thousand documents in the course of this work.)

Starting in 1881, Milne produced the seismological *General Reports* of the British Association for the Advancement of Science. Subsequently, on the Isle of Wight, he produced the "Shide circulars," which summarized the data gathered by a worldwide network of seismological stations set up by Milne and using his instruments. The Shide circulars were the forerunners of the *International Seismological Summary*, which, after Milne's death, became the basic source of instrument-gathered data on earthquakes.

With its recent successor, the *Bulletin of the International Seismological Center*, it has long been centrally important in world research on earthquakes.

Milne's success was due to the combination of scientific brilliance and adaptability with a genial disposition and capacity to interest others in his enthusiasms. He was modest, notably hospitable, gifted with a sense of humor, and generous to others in his scientific and pecuniary help.

BIBLIOGRAPHY

I. ORIGINAL WORKS. Milne was a prolific writer who contributed nearly 2,000 pages (about two-thirds of the entire content) of the *Transactions* and *Journal* of the Seismological Society of Japan during his editorship (1880–1895). He also published papers in the *Proceedings of the Royal Society*, *Geological Magazine*, and *Bulletin of the Seismological Society of America*. He wrote *Earthquakes and Other Earth Movements* (London, 1886); two eds. of a supp. volume, *Seismology*, appeared in 1898 and 1908. Among his noted publications are the Shide circulars and *A Catalogue of Destructive Earthquakes, A.D. 7 to A.D. 1899* (London, 1912), published by the British Association.

II. SECONDARY LITERATURE. A list of Milne's publications is given by H. Woodward, in *Geological Magazine*, **9** (Aug. 1912), 337–346. For details of Milne's life, see J.W.J., "Prof. John Milne, F.R.S.," in *Nature*, **91** (Aug. 1913), 587–588; J.P., "John Milne, 1850–1913," in *Proceedings of the Royal Society*, **84A** (Mar. 1914), xxii–xxv; **91** (Aug. 1913), 587–588; and C. Davison, *The Founders of Seismology* (Cambridge, 1927), ch. 10. Milne's *Earthquakes and Other Earth Movements* was revised by A. W. Lee (Philadelphia, 1939). For an account of Milne's work at Shide, see Mrs. Lou Henry Hoover, "John Milne, Seismologist," in *Bulletin of the Seismological Society of America*, **2**, no. 1 (Mar. 1912), 2–7.

K. E. BULLEN

MILNE-EDWARDS, HENRI (*b.* Bruges, Belgium, 23 October 1800; *d.* Paris, France, 29 July 1885), *zoology.*

The son of William Edwards, an English planter and militia colonel in Jamaica, and Elisabeth Vaux, Milne-Edwards was born in Bruges, where his parents had retired. (Milne, which he added to his father's name, was the married name of his godmother and half-sister by a previous marriage of his father.) When Belgium became independent, he chose French citizenship. After medical studies in Paris he acquired a solid background in zoology and in 1832 accepted a post as professor of hygiene and natural history at the École Centrale des Arts et Manufactures. Despite his delicate health, in addition to his teaching he undertook a vast program of research on the invertebrates. His success in this field earned him the Academy of Sciences' prize in experimental physiology in 1828 and his election to the zoology section of the Academy in 1838. Three years later he was appointed to the chair of entomology of the Museum of Natural History, where he had long had a laboratory. At the time the holder of this chair was responsible for the crustaceans, the myriapods, and the arachnids as well as the insects. Twenty years later the chair of mammalogy became vacant on the death of Isidore Geoffroy Saint-Hilaire, and Milne-Edwards was transferred to it at his own request. In the meantime he had been named professor, and then dean, of the Faculty of Sciences. He was a member of most of the scientific societies of his time and a commander of the Legion of Honor.

In contrast with the tendencies of his contemporaries, Milne-Edwards had been attracted since his youth to the study of the invertebrates, especially those inhabiting the coastal regions. With his friends from the Museum and later with his students, he organized scientific excursions along the shores of the English Channel. Not content with collecting and classifying the animals, he insisted on examining them in their habitat and observed their behavior, their movements, their localization according to the level of the tides, and their modes of obtaining food and of reproducing. Milne-Edwards recorded a wealth of observations in which physiological data were joined with data from comparative morphology. This method, essentially that of ecology, appeared to afford a novel approach to the marine invertebrates, although it was inspired by one that Georges Cuvier had applied to other groups. It led Milne-Edwards to brilliant discoveries and started the creation of maritime laboratories in France and abroad.

Milne-Edwards' first investigations were primarily concerned with crustaceans. He published a series of memoirs on most of their systems, including circulation, respiration, nerve, and muscle. He began this work with his friend Jean Audouin, who preceded him in the chair of entomology at the Museum, and who accompanied him on his expeditions to the Chausey Isles; he then continued it alone.

These anatomicophysiological investigations served as the basis for the comprehensive three-volume synthesis to which Milne-Edwards dedicated many years—the classic *Histoire naturelle des crustacés* (1834–1840). In this work he developed some highly original ideas. He reported that the Crustaceae are made up of some twenty homologous metameric segments, the "zoonites," which are variously fashioned according to the functions they fulfill and the mode of life (free, fixed, or parasitic) of the species. The variety of possible natural combinations, within the limits of a basic framework, is thus virtually infinite. Among Milne-Edwards' other works are *Histoire naturelle des coralliaires* (1858–1860), *Monographie des polypes des terrains paléozoïques*, and the two-volume *Recherches pour servir à l'histoire des mammifères* (1868–1874).

As an adjunct to his teaching duties at the Faculty of Sciences Milne-Edwards gathered his lectures into a fourteen-volume publication, *Leçons sur la physiologie et l'anatomie comparée de l'homme et des animaux*, the composition of which was spread over more than twenty years (1857–1881). At the same time he provided a valuable development of his ideas on animal organization in *Introduction à la zoologie générale, ou considérations sur les tendances de la nature dans la constitution du règne animal* (1858). In this book Milne-Edwards set forth his principal discoveries. These concern the variations that obtain between animal groups, variations which in the final analysis display a great fundamental principle, the law of the division of labor within organisms. Milne-Edwards suspected the existence of this law with his first studies of crustaceans, and he verified it subsequently among the other groups. In the lower animals the same tissue can adapt to different functions. He observed this phenomenon, for example, in the coelenterates, where a single fragment was seen to be capable of regenerating the entire animal. But in animals of higher zoological order, this ability tends to disappear and is progressively replaced by a specialization of the tissues. Systems, or groups of related organs, become individualized in order to carry out precise and exclusive functions: a digestive system, a respiratory system, a reproductive system, and so on.

Within each system each organ has a well-defined role. Therefore the digestive system is divided into a digestive tube and the attached glands; and the digestive tube itself consists of a first region into which food is introduced, a second in which the nutriments undergo the action of the digestive juices, and a third where substances that are useful to the organism are absorbed and where waste products are eliminated. One could reconsider each of these regions and ascertain further subdivisions within them, varying according to diet and other factors. Such specializations, which become more and more precise, determine the rank of an organism in the animal series. It is in large part through the discovery, analysis, and application of these fundamental principles that

Milne-Edwards was for years the leader of the French naturalists and that his work remained famous long after his death.

BIBLIOGRAPHY

I. ORIGINAL WORKS. In addition to those described above, Milne-Edwards' works include *Manuel de matière médicale* (Paris, 1825), written with Vavasseur; *Manuel d'anatomie chirurgicale* (Paris, 1826); *Anatomie des crustacés* (Paris, 1832); *Recherches pour servir à l'histoire naturelle du littoral de la France* (Paris, 1832); *Éléments de zoologie* (Paris, 1834); *Discours sur les progrès des sciences dans les départements* (Paris, 1861); and *Rapport sur les progrès récents des sciences zoologiques en France* (Paris, 1867).

II. SECONDARY LITERATURE. See M. Berthelot, *Notice historique sur Henri Milne-Edwards, membre de l'Académie des sciences* (Paris, 1891); *Médaille d'honneur offerte à M. H. Milne Edwards. Allocutions de MM. de Quatrefages, Blanchard et J. B. Dumas* (Paris, 1881); and G. Pennetier, *Discours sur l'évolution des connaissances en histoire naturelle*, IV, *XVIIIᵉ–XIXᵉ siècles—zoologie* (Rouen, 1920), 497–501.

JEAN ANTHONY

MINDING, ERNST FERDINAND ADOLF (or **Ferdinand Gotlibovich**) (*b.* Kalisz, Poland, 23 January 1806; *d.* Dorpat, Russia [now Tartu, Estonian S.S.R.], 13 May 1885), *mathematics*.

Minding was a son of the town lawyer in Kalisz. After graduation in 1824 from the Gymnasium at Hirschberg (now Jelenia Góra, Poland), where the family had moved in 1807, Minding studied philology, philosophy, and physics at the universities of Halle and Berlin. In mathematics he was a self-taught amateur. After graduating from Berlin University in 1827, Minding taught mathematics in Gymnasiums for several years. In 1829 he received at Halle the doctorate in philosophy for his thesis on approximating the values of double integrals; from 1831 to 1843 he lectured on mathematics at Berlin University and from 1834 also at the Berlin Bauschule. At the university he lectured in 1831 and 1834 on the history of mathematics and gave a general introduction to the foundations and goals of the mathematical sciences. During these years he published thirty works, including several textbooks. Despite intensive pedagogical and scientific activity, Minding's position at Berlin was unsatisfactory; and he eagerly accepted an invitation to the University of Dorpat, where in 1842 the chair of mathematics of the Faculty of Philosophy was divided between one of pure mathematics, which was occupied by K. E. Senff, and one of applied mathematics, which was vacant. From 1843 to 1883 Minding was at the University of Dorpat as a full professor, giving both general and special courses in algebra, analysis, geometry, theory of probability, mechanics, and physics. In 1850 the Faculty of Philosophy was divided into that of physicomathematics and that of history-philology, and in 1851 Minding was elected to a four-year term as dean of the former division. In 1864 Minding and his family became Russian citizens. (In 1838 he had married Augusta Regler, and they had several children.) In the same year he was elected a corresponding member, and in 1879 an honorary member, of the St. Petersburg Academy of Sciences.

Minding's most important discoveries were in the differential geometry of surfaces; in these works he brilliantly continued the researches of Gauss, which had been published in 1828. In his first paper (1830), which dealt with the isoperimetric problem of determining on a given surface the shortest closed curve surrounding a given area (on the plane it is the circumference of a circle), he introduced the concept of geodesic curvature. It was independently discovered in 1848 by O. Bonnet, and it was he who named it geodesic curvature. Minding soon proved, as did Bonnet after him, the invariance of the geodesic curvature under bending of the surface. Neither of them knew that the same results had been presented in an earlier, unpublished paper of Gauss's (1825).

Minding's studies on the bending or the applicability of surfaces were especially remarkable. He first examined the bending of a particular class of surfaces (1838); incidentally, in the case of surfaces of revolution, he studied an example of the "applicability on a principal basis," which later became a preferred research topic for his disciple K. M. Peterson and for Peterson's followers in Moscow. He then proceeded to the general problem of determining the conditions for applicability of surfaces. Gauss had discovered (1828) that if one surface can be isometrically applied to another (so that the bending does not alter the lengths of curves), then the total curvature will be the same at all corresponding points.

In his article "Wie sich entscheiden lässt, ob zwei gegebene krumme Flächen auf einander abwickelbar sind oder nicht . . ." (1839), Minding stated the following sufficient condition for applicability: Two given surfaces of equal constant total curvature are applicable to one another isometrically, and this can be done in infinitely many different ways. He also investigated the corresponding problem for surfaces with a variable total curvature. Today "Minding's theorem" is found in all textbooks of differential geometry. Minding's papers, as well as Gauss's work of 1828, were great influences on the development of this branch of mathematics. In the article "Beiträge zur Theorie der kürzesten Linien auf krummen Flächen," which was published in Crelle's *Journal für die reine und angewandte Mathematik* (1840), Minding pointed out that when the trigonometric functions are replaced by corresponding hyperbolic ones, the trigonometric formulas in spherical trigonometry for the geodesic triangles on the surfaces with constant positive curvature are converted into the hyperbolic formulas for the surfaces with negative curvature. In 1837, Lobachevski showed (in an article that also appeared in Crelle's *Journal*) that the same relation exists between the trigonometric formulas for the sphere and the formulas in his "imaginary" (hyperbolic) geometry. The confrontation of these results might have led to the conclusion that two-dimensional hyperbolic geometry can be (partly) interpreted as the geometry of geodesics on a surface of constant negative curvature; but it was not until 1868 that Beltrami established this connection.

Starting from Euler's ideas, Minding proposed the method of solving the differential equation $M(x, y) \, dx + N(x, y) \, dy = 0$, where M and N are polynomials of some degree, based on determining the integrating factor by means of particular integrals of the equation. Minding's method, expounded in the paper "Beiträge zur Integration der Differentialgleichungen erster Ordnung zwischen zwei Veränderlichen," for which he received in 1861 the Demidov Prize of the St. Petersburg Academy of Sciences, was developed further by A. N. Korkin and others. Darboux (1878) worked independently, followed by E. Picard and others, in the same direction. Minding also published works on algebra (the elimination problem), the theory of continued fractions, the theory of algebraic functions, and analytic mechanics.

BIBLIOGRAPHY

I. Original Works. Minding's writings include "Ueber die Curven kürzesten Perimeters auf krummen Flächen," in *Journal für die reine und angewandte Mathematik*, **5** (1830), 297–304; "Bemerkung über die Abwickelung krummer Linien von Flächen," *ibid.*, **6** (1830), 159–161; "De valore integralium duplicum quam proxime inveniendo" (his doctoral diss., in the archives of the University of Halle), pub. with minor modifications as "Ueber die Berechnung des Näherungswertes doppelter Integrale," *ibid.*, 91–95; *Anfangsgründe der höheren Arithmetik* (Berlin, 1832); *Handbuch der Differential- und Integralrechnung nebst Anwendung auf die Geometrie* (Berlin, 1836); *Handbuch der Differential- und Integralrechnung und ihrer Anwendungen auf Geometrie und Mechanik. Zweiter Teil, enthaltend die Mechanik* (Berlin, 1838); "Ueber die Biegung gewisser Flächen," in *Journal für die reine und angewandte Mathematik*, **18** (1838), 297–302; "Wie sich entscheiden lässt, ob zwei gegebene krumme Flächen auf einander abwickelbar sind oder nicht; nebst Bemerkungen über die Flächen von unveränderlichem Krümmungsmasse," *ibid.*, **19** (1839), 370–387; "Beiträge zur Theorie der kürzesten Linien auf krummen Flächen," *ibid.*, **20** (1840), 323–327; and "Beiträge zur Integration der Differentialgleichungen erster Ordnung zwischen zwei Veränderlichen," in *Mémoires de l'Académie des sciences de St. Petersbourg*, 7th ser., **5**, no. 1 (1863), 1–95, also pub. separately in Russian trans. (St. Petersburg, 1862).

II. Secondary Literature. See R. I. Galchenkova *et al.*, *Ferdinand Minding. 1806–1885* (Leningrad, 1970), which includes a complete list of Minding's works, pp. 205–210 (nos. 1–72), and extensive secondary literature, pp. 210–220 (nos. 73–289); A. Kneser, "Übersicht der wissenschaftlichen Arbeiten Ferdinand Minding's nebst biographischen Notizen," in *Zeitschrift für Mathematik und Physik*, Hist.-lit. Abt., **45** (1900), 113–128; I. Z. Shtokalo, ed., *Istoria otechestvennoy matematiki*, II (Kiev, 1967); A. Voss, "Abbildung und Abwickelung zweier Flächen auf einender abwickelbarer Flächen," in *Encyklopädie der mathematischen Wissenschaften*, III, pt. 6a (Leipzig, 1903), 355–440; and A. P. Youschkevitch, *Istoria matematiki v Rossii do 1917 g.* (Moscow, 1968).

A. Youschkevitch

MINEUR, HENRI (*b.* Lille, France, 7 March 1899; *d.* Paris, France, 7 May 1954), *astronomy, astrophysics, mathematics.*

Although he was first on the admissions list of the École Normale Supérieure in 1917, Mineur enlisted in the army and did not enter the school until after the end of World War I. After passing the *agrégation* in mathematics in 1921, he taught at the French *lycée* in Düsseldorf while pursuing the mathematical research he had begun in 1920. He received his doctorate in science in 1924 for his work on functional equations, in which he established an addition theorem for Fuchsian functions.

Mineur had been interested in astronomy from his youth, and in 1925 he left his teaching post to become astronomer at the Paris observatory. He made important contributions to several fields related to mathematical astronomy: celestial mechanics, analytic mechanics, statistics, and numerical calculus. His treatise on the method of least squares has become a classic.

It was in stellar astronomy, however, that Mineur's work was most sustained and fruitful. In particular, he detected the variation in the speed of near stars according to the distance from the galactic plane and the retrograde rotation of the system of globular clusters. He also corrected the coordinates of the galactic center and studied interstellar absorption. As early as 1944 he showed that an important correction had to be made in the zero of the period-luminosity relation of the Cepheids; this change led to a doubling of the scale of distances in the universe. All these results have been confirmed by recent investigations.

Mineur was a brilliant and unusual person who became thoroughly involved in many areas. Between 1940 and 1944 he was a member of the Resistance. He was a founder of the Centre National de la Recherche Scientifique and of the observatory of Saint-Michel in Haute-Provence. At his initiative the Institute d'Astrophysique was created at Paris in 1936; he was its director until his death, which occurred after five years of a serious heart and liver ailment. Mineur twice won prizes of the Académie des Sciences.

BIBLIOGRAPHY

I. ORIGINAL WORKS. In kinematics and stellar dynamics Mineur's principal writings are "Rotation de la gala⋅ ⋅e," in *Bulletin astronomique*, **5** (1925), 505–543; "Étud⋅ de mouvements propres moyens d'étoiles," *ibid.*, **6** (1930), 281–304; "Recherches sur les vitesses radiales résiduelles . . .," *ibid.*, **7** (1931), 321–352, written with his wife; *Éléments de statistique . . . stellaire*, Actualités Scientifiques et Industrielles, no. 116 (Paris, 1934); *Photographie stellaire . . .*, *ibid.*, no. 141 (Paris, 1934); *Dénombrement d'étoiles . . .*, *ibid.*, no. 225 (Paris, 1935); "Recherches sur la distribution de la matière absorbante . . .," in *Annales d'astrophysique*, **1** (1938), 97–128; "Sur la rotation galactique . . .," *ibid.*, 269–281; "Équilibre des nuages galactiques . . .," *ibid.*, **2** (1939), 1–244; "Zéro de la relation période-luminosité . . .," *ibid.*, **7** (1944), 160–186; and "Recherches théoriques sur les accélérations stellaires," *ibid.*, **13** (1950), 219–242. Some of these investigations are summarized in *L'espace interstellaire* (Paris, 1947).

In celestial and analytical mechanics, see especially "La mécanique des masses variables . . .," in *Annales scientifiques de l'École normale supérieure*, 3rd ser., **50** (1933), 1–69; "Étude théorique du mouvement séculaire de l'axe terrestre," in *Bulletin astronomique*, **13** (1947), 197–252; "Quelques propriétés . . . équations de la mécanique," *ibid.*, **13** (1948), 309–328; and "Recherche . . . dans le groupe canonique linéaire," *ibid.*, **15** (1950), 107–141.

In statistics and numerical calculus, see *Technique de la méthode des moindres carrés* (Paris, 1938); "Nouvelle méthode de lissage . . . période d'un phénomène," in *Annales d'astrophysique*, **6** (1943), 137–158; "Sur la meilleure représentation d'une variable aléatoire . . .," *ibid.*, **7** (1944), 17–30; and *Techniques de calcul numérique* (Paris, 1952).

His works in analysis include his doctoral dissertation, "Théorie analytique des groupes continus finis," in *Journal de mathématiques pures et appliquées*, 9th ser., **4** (1925),

23–108; and "Calcul différentiel absolu," in *Bulletin des sciences mathématiques*, 2nd ser., **52** (1928), 63–76.

II. SECONDARY LITERATURE. See D. Barbier, "Henri Mineur," in *Annales d'astrophysique*, **17** (1954), 239–242; and J. Dufay, "Henri Mineur," in *Astronomie*, **70** (1956), 235–238.

JACQUES R. LÉVY

MINKOWSKI, HERMANN (*b.* Alexotas, Russia [now Lithuanian S.S.R.], 22 June 1864; *d.* Göttingen, Germany, 12 January 1909), *mathematics*.

Minkowski was born of German parents who returned to Germany and settled in Königsberg [now Kaliningrad, R.S.F.S.R.] when the boy was eight years old. His older brother Oskar became a famous pathologist. Except for three semesters at the University of Berlin, he received his higher education at Königsberg, where he became a lifelong friend of both Hilbert, who was a fellow student, and the slightly older Hurwitz, who was beginning his professorial career. In 1881 the Paris Academy of Sciences had announced a competition for the Grand Prix des Sciences Mathématiques to be awarded in 1883, the subject being the number of representations of an integer as a sum of five squares of integers; Eisenstein had given formulas for that number but without proof. The Academy was unaware that in 1867 H. J. Smith had published an outline of such a proof, and Smith now sent a detailed memoir developing his methods. Without knowledge of Smith's paper, the eighteen-year-old Minkowski, in a masterly manuscript of 140 pages, reconstructed the entire theory of quadratic forms in n variables with integral coefficients from Eisenstein's sparse indications. He gave an even better formulation than Smith's because he used a more natural and more general definition of the genus of a form. The Academy, unable to decide between two equally excellent, and substantially equivalent, works, awarded the Grand Prix to both Smith and Minkowski.

Minkowski received his doctorate in 1885 at Königsberg; he taught at Bonn until 1894, then returned to Königsberg for two years. In 1896 he went to Zurich, where he was Hurwitz' colleague until 1902; Hilbert then obtained the creation of a new professorship for him at Göttingen, where Minkowski taught until his death.

From his Grand Prix paper to his last work Minkowski never ceased to return to the arithmetic of quadratic forms in n variables ("n-ary forms"). Ever since Gauss's pioneering work on binary quadratic forms at the beginning of the nineteenth century, the generalization of his results to n-ary forms had been the goal of many mathematicians, including Eisenstein, Hermite, Smith, Jordan, and Poincaré. Minkowski's most important contributions to the theory were (1) for quadratic forms with rational coefficients, a characterization of equivalence of such forms under a linear transformation with rational coefficients, through a system of three invariants of the form and (2) in a paper of 1905, the completion of the theory of reduction for positive definite n-ary quadratic forms with real coefficients, begun by Hermite. The latter had defined a process yielding in each equivalence class (for transformations with integral coefficients) a finite set of "reduced" forms; but it was still possible for this set to consist in more than one form. Minkowski presented a new process of "reduction" giving a unique reduced form in each class. In the space of n-ary quadratic forms (of dimension $n(n + 1)/2$), the "fundamental domain" of all reduced forms proves to be a polyhedron; Minkowski made a detailed investigation of the relation of this domain to its neighbors and computed its volume, which enabled him to obtain asymptotic formulas for the number of equivalence classes of a given determinant, when the value of that determinant tends to infinity.

This 1905 paper was greatly influenced by the geometric outlook that Minkowski had developed fifteen years earlier—the "geometry of numbers," as he called it, his most original achievement. He was led to it by the theory of ternary quadratic forms. Following brief indications given by Gauss, Dirichlet had developed a geometrical method of reduction of positive definite ternary forms; Minkowski's brilliant idea was to use the concept of volume in conjunction with this geometric method, thus obtaining far better estimates than had been possible before. To make matters simpler, consider a binary positive definite quadratic form $F(x, y) = ax^2 + 2bxy + cy^2$. To say that F takes a value m when $x = p, y = q$ are integers, means, geometrically, that the ellipse E_m of equation $F(x, y) = m$ passes through the point (p, q). To find the minimum M of all such values m, obtained for p, q not both 0, Minkowski observed that for small α, certainly the ellipse E_α will not contain any such points; if one considers the ellipse $\frac{1}{2}E_\alpha$ and translates it by sending its center to every point (p, q) with integral coordinates, one obtains an infinite pattern of ellipses which do not touch each other. When α increases and reaches the value M, some of the corresponding ellipses will touch each other but no two will overlap. Now, if $A = ac - b^2$ is the area of the ellipse E_1, the ellipse $\frac{1}{2}E_M$ has area $AM^2/4$, $AM^2/4$ and the total area of the nonoverlapping ellipses which are translations of $\frac{1}{2}E_M$ and which have centers

at the points (p, q) with $|p| \leqslant n$ and $|q| \leqslant n$ is $(2n + 1)^2(AM^2/4)$. It is easy to see, however, that there is a constant $c > 0$ independent of n, such that all these ellipses are contained in a square of center 0 and of side $2n + 1 + c$, so that

$$(2n + 1)^2(AM^2/4) \leqslant (2n + 1 + c)^2;$$

letting n grow to infinity gives the inequality

$$\frac{AM^2}{4} \leqslant 1 \quad \text{or} \quad M \leqslant \frac{2}{\sqrt{A}}.$$

Not only can this argument be at once extended to spaces of arbitrary finite dimension, but Minkowski had a second highly original idea: He observed that in the preceding geometric argument, ellipses could be replaced by arbitrary convex symmetric curves (and, in higher-dimensional spaces, by symmetric convex bodies). By varying the nature of these convex bodies with extreme ingenuity (polyhedrons, cylinders), he immediately obtained far-reaching discoveries in many domains of number theory. For instance, by associating to an algebraic integer x in a field of algebraic numbers K of degree n over the rationals, the point in n dimensions having as coordinates the rational integers which are the coefficients of x with respect to a fixed basis, Minkowski gave lower bounds for the discriminant of K, which in particular proved that when $n > 1$, the discriminant may never be equal to 1 and that there are only a finite number of fields of discriminants bounded by a given number.

Minkowski's geometric methods also enabled him to reach a far better understanding of the theory of continued fractions and to generalize it into an algorithm which, at least theoretically, gives a criterion for a number to be algebraic. It was similar in principle to Lagrange's well-known criterion that quadratic irrationals are characterized by periodic continued fractions; but Minkowski also showed that, for his criterion, periodicity occurs in only a small number of cases, which he characterized completely. Finally, if, for instance, one considers (as above, but in three dimensions) an ellipsoid $F(x, y, z) = 1$ in relation to the lattice L of points with integral coordinates, the largest possible number M will be obtained when the translated ellipsoids are "packed together" as closely as possible. If one makes a linear transformation of the space transforming the ellipsoid in a sphere, L is transformed into another lattice consisting of linear combinations with integral coefficients of three vectors. The problem of finding the largest M, then, is equivalent to the "closest packing of spheres" in space, when the centers are at the vertices of a

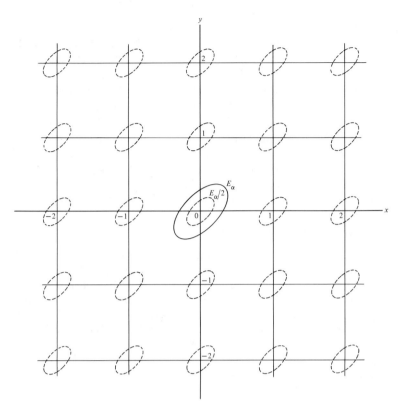

FIGURE 1

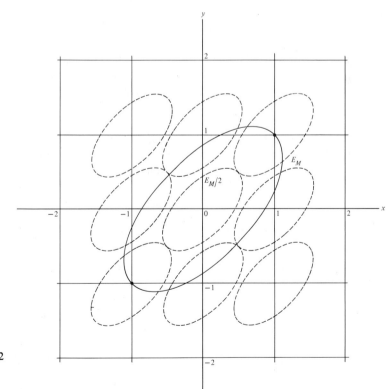

FIGURE 2

lattice L'; one has to find the lattice L' that gives this closest packing. Minkowski began the study of that difficult problem (which extends to any n-dimensional space) and of corresponding problems when spheres are replaced by some other type of convex set (particularly polyhedrons); they have been the subject of fruitful research ever since.

The intensive use of the concept of convexity in his "geometry of numbers" led Minkowski to investigate systematically the geometrical properties of convex sets in n-dimensional space, a subject that had barely been considered before. He was the first to understand the importance of the notion of hyperplane of support (both geometrically and analytically), and he proved the existence of such hyperplanes at each point of the boundary of a convex body. Long before the modern conception of a metric space was invented, Minkowski realized that a symmetric convex body in an n-dimensional space defines a new notion of "distance" on that space and, hence, a corresponding "geometry." His ideas thus paved the way for the founders of the theory of normed spaces in the 1920's and became the basis for modern functional analysis.

The evaluation of volumes of convex bodies led Minkowski to the very original concept of "mixed volume" of several convex bodies: when K_1, K_2, K_3 are three convex bodies in ordinary space and t_1, t_2, t_3 are three real numbers $\geqslant 0$, the points $t_1 x_1 + t_2 x_2 + t_3 x_3$, when x_j varies in K_j for $j = 1, 2, 3$, fill a new convex body, written $t_1 K_1 + t_2 K_2 + t_3 K_3$. When the volume of this new convex body is computed, it is seen to be a homogeneous polynomial in t_1, t_2, t_3 and the mixed volume $V(K_1, K_2, K_3)$ is the coefficient of $t_1 t_2 t_3$ in that polynomial. Minkowski discovered remarkable relations between these new quantities and more classical notions: if K_1 is a sphere of radius 1, then $V(K_1, K, K)$ is one third of the area of the convex surface bounding K; and $V(K_1, K_1, K)$ is one third of the mean value of the mean curvature of that surface. He also proved the inequality between mixed volumes

$$V(K_1, K_2, K_3)^2 \geqslant V(K_1, K_1, K_3)\, V(K_2, K_2, K_3),$$

from which he derived a new and simple proof of the isoperimetric property of the sphere. As a beautiful application of his concepts of hyperplane of support and of mixed volumes, Minkowski showed that a convex polyhedron having a given number m of faces is determined entirely by the areas and directions of the faces, a theorem that he generalized to convex surfaces by a passage to the limit. He also determined all convex bodies having constant width.

Minkowski was always interested in mathematical physics but did not work in that field until the last years of his life, when he participated in the movement of ideas that led to the theory of relativity. He was the first to conceive that the relativity principle formulated by Lorentz and Einstein led to the abandonment of the concept of space and time as separate entities and to their replacement by a four-dimensional "space-time," of which he gave a precise definition and initiated the mathematical study; it became the frame of all later developments of the theory and led Einstein to his bolder conception of generalized relativity.

BIBLIOGRAPHY

Minkowski's writings were collected in *Gesammelte Abhandlungen*, D. Hilbert, ed., 2 vols. (Leipzig–Berlin, 1911). Among his books are *Geometrie der Zahlen* (Leipzig, 1896; 2nd ed., 1910); and *Diophantische Approximationen* (Leipzig, 1907; repr. New York, 1957).

On Minkowski's work, see Harris Hancock, *Development of the Minkowski Geometry of Numbers* (New York, 1939); and Frederick W. Lanchester, *Relativity. An Elementary Explanation of the Space-Time Relations As Established by Minkowski* (London, 1935).

J. DIEUDONNÉ

MINNAERT, MARCEL GILLES JOZEF (*b.* Bruges, Belgium, 12 February 1893; *d.* Utrecht, Netherlands, 26 October 1970), *astronomy.*

Minnaert, one of the pioneers of solar research in the first half of the twentieth century, was professor at the University of Utrecht and director of its observatory from 1937 to 1963. His parents were teachers at normal schools and many of his other relatives were involved in teaching, which background undoubtedly determined his later interest in science and education. He studied biology at the University of Ghent and in 1914 defended—with the highest distinction—his doctoral thesis, "Contributions à la photobiologie quantitative."

At that time the University of Ghent, although situated in the heart of the Dutch-speaking part of Belgium, used French as the language of instruction, as did the other universities and most of the secondary schools in Flemish Belgium. Minnaert, who gradually realized that the linguistic problem was also a social problem related to the underdeveloped status of Flanders, joined associations of Flemish students and intellectuals who sought political equality and, later, relative independence (federalism) for both parts of Belgium. They also wished to convert the University of Ghent to the Dutch language. During

414

the German occupation of Belgium in World War I, the latter goal was attained.

The urgent need for teachers at the new Flemish university induced Minnaert to go to Leiden in 1915–1916 to study physics. After his return to Ghent he was named associate professor of physics and remained in that post until 1918. At the end of the war those who had cooperated in the linguistic reform of the University of Ghent were accused of collaboration with the Germans, and many received long prison sentences. In order to escape that fate Minnaert moved to Utrecht, to which place he was attracted by the technique of objective photometry, then being developed at its physics laboratory by W. H. Julius, Ornstein, and Moll. He readily understood the importance of the technique because of his previous experience in photobiology, in which specialty the lack of quantitative measures was deeply felt. The director, W. H. Julius, had just set up a solar spectrograph—at that time the third in the world—intending to apply spectrophotometric techniques to the solar spectrum. Minnaert became interested in this work, and after Julius' death in 1924 he assumed the main responsibility for solar research at Utrecht. In 1925 he defended—*cum laude*—another thesis, this time in physics:"Onregelmatige straalkromming"("Irregular Refraction of Light").

At that time the basic requirements were available for quantitative research in solar physics: Bohr's atomic model, Saha's ionization law, the developments of quantum theory, and the new technique of quantitative spectrophotometry made possible the quantitative interpretation of the solar spectrum. Minnaert developed the concepts "equivalent width" and "curve of growth"; the theory of weak lines was carried further; and the intensity measurements of sunspots made possible the physical interpretation of these phenomena. This work, performed in the physics laboratory at Utrecht, culminated in 1940 in *Photometric Atlas of the Solar Spectrum* (in collaboration with Houtgast and Mulders), which is still a standard reference.

In 1937 Minnaert was named director of the University of Utrecht observatory, which he transformed into an astrophysical institute' devoted mainly to the investigation of solar and stellar spectra. Yet his interests were wider than the sun: he studied comets and gaseous nebulae and was involved in lunar photometry; and during the last few years of his life he was a member of a working group of the International Astronomical Union concerned with naming the newly discovered formations on the hemisphere of the moon that is not visible from the earth. In 1970, two months before his death, Minnaert was elected president of the Commission for the Moon of the International Astronomical Union. His broad interest in science and nature led him to write these books: *De Natuurkunde van het vrije veld* ("Physics of the Open Field"), translated into many languages; *Dichters over Sterren* ("Poets on Stars"); and *De sterrekunde en de mensheid* ("Astronomy and Mankind").

In 1928 Minnaert married Maria Boergonje Coelingh, who defended her thesis in physics in 1938; they had two sons. Philosophically Minnaert defended determinism; politically he had strong left-wing sympathies but was too committed to science to link himself to any political party. Yet his political ideas were sufficiently known to the Germans for him to be imprisoned in 1942–1944.

Minnaert spoke ten languages fluently and could read in even more. Even so he was an enthusiastic defender of Esperanto—attracted by the simplicity and regularity of this artificial language which, he felt, could be of great importance for both scientific and social communication. He loved music and painting and cultivated both actively. Above all, Minnaert had a strong interest in humanity and its problems. He was admired and loved by his friends, students and co-workers, and respected by those who did not agree with his social or political ideas. In the last few years of his life he was very active in an international group purchasing books for the University of Hanoi, North Vietnam.

Minnaert was a member of the Royal Netherlands Academy of Arts and Sciences, of the Royal Belgian Academy of Science, Letters and Fine Arts, of the Kungl. Vetenskapsamhället of Uppsala, and of the Instituto de Coimbra; and associate of the Royal Astronomical Society of London. In 1947 he received the gold medal of the Royal Astronomical Society and in 1951 that of the Astronomical Society of the Pacific (Bruce Medal). He held honorary doctorates from the universities of Heidelberg, Moscow, and Nice.

BIBLIOGRAPHY

Minnaert's main publications were his diss., *Onregelmatige straalkromming* (Utrecht, 1925); *De Natuurkunde van de Zon* (The Hague, 1936); *De betekenis der zonnephysica voor de astrophysica* (Utrecht, 1937); *De Natuurkunde van het vrije veld* (Zutphen, 1937); *Photometric Atlas of the Solar Spectrum* (Amsterdam, 1940), with G. F. Mulders and J. Houtgast; *De sterrekunde en de mensheid* (The Hague, 1946); *Dichters over Sterren* (Arnhem, 1949); and *Practical Works in Elementary Astronomy* (Dordrecht, 1969). He was the author of many papers, of which a review is given in M. Minnaert, "Forty Years of Solar Spectroscopy," in C. de Jager, ed., *The Solar Spectrum*, the proceedings of a symposium held at the occasion of the seventieth anniversary of the birth of M. G. J. Minnaert (Dordrecht, 1965).

Minnaert also prepared the articles on Hoek, Kaiser, Hortensius, W. Julius, Pannekoek, and Stevin for this *Dictionary*.

C. DE JAGER

MINOT, CHARLES SEDGWICK (*b.* Roxbury, Massachusetts, 23 December 1852; *d.* Milton, Massachusetts, 19 November 1914), *anatomy, embryology.*

Minot was the second of three sons of William Minot and Katharine Sedgwick. A paternal ancestor was Jonathan Edwards; and on both sides, there were several distinguished lawyers and public figures. Growing up on his wealthy father's country estate, he early became interested in natural history and at the age of seventeen published articles on insect and bird life. He graduated from the Massachusetts Institute of Technology in 1872 and then entered the graduate school of Harvard College, where he worked under Henry P. Bowditch, spending a summer with Louis Agassiz at Penikese, Massachusetts. In 1873 he went to Leipzig to work with Karl Ludwig and Rudolf Leuckart. He was also at Paris for a few months with Louis-Antoine Ranvier and at Würzburg.

After his return to America in 1876, Minot completed in 1878 the requirements for his Harvard doctorate in science. After two years of private biological research, in 1880 he joined the Harvard faculty, at first in the dental school and, after 1883, in the department of histology and embryology of the school of medicine. There he began what became an outstanding collection of vertebrate embryos. To facilitate the work of sectioning them, he invented in 1886 the automatic rotary microtome, ever since in worldwide use. In 1892 Minot published his chief work, *Human Embryology*, a masterly summation of an unwieldy literature and a highly original presentation of the major problems of that branch of science. Among his many research accomplishments were an account of the microscopic structure of the human placenta and a description of the blood channels in the liver since known by his term "sinusoids."

Minot's wide-ranging intellect led him into very broad fields of thought. For a few years he was active in the American Society for Psychical Research, from which he withdrew when finally convinced of its unscientific outlook. Deep reflection about the nature of life, its origin, course, and termination guided his protracted studies of the growth of animals and the progressive changes in cell structure from birth to death.

Minot exerted wide influence on American biology of his time in his books, numerous papers in scientific journals, and lectures, all presented with clarity and stylistic elegance. Reserved in professional manner and sometimes sharply critical of other workers in matters of scientific judgment, he was a genial participant in the professional societies of natural history, anatomy, and physiology. He was one of a small group of biologists and medical scientists who broadened the study and teaching of anatomy in the United States to include not only gross morphology but also embryology, histology, and physical anthropology, and transformed the American Association of Anatomists from a small society with limited interests to its present breadth and strength.

Minot was elected to the National Academy of Sciences in 1897 and served as president of the American Society of Naturalists, the American Association of Anatomists, and the American Association for the Advancement of Science. His eminence in human and comparative embryology was recognized by honorary degrees from Yale, Toronto, St. Andrews, and Oxford universities and by a visiting professorship at Berlin.

BIBLIOGRAPHY

I. ORIGINAL WORKS. Minot's primary publications are *Human Embryology* (New York, 1892); "A Bibliography of Vertebrate Embryology," in *Memoirs of the Boston Society of Natural History*, **4**, no. 11 (1893), 487–614; *Laboratory Textbook of Embryology* (Philadelphia, 1903); and *The Problem of Life, Growth, and Death* (New York, 1908), in addition to about 180 scientific papers and lectures.

II. SECONDARY LITERATURE. On Minot and his work, see (listed chronologically) Henry H. Donaldson, "Charles Sedgwick Minot," in *Science*, n.s. **40** (1914), 926–927, a character study; Charles W. Eliot, "Charles Sedgwick Minot," *ibid.*, **41** (1915), 701–704; Frederic T. Lewis, "Charles Sedgwick Minot," in *Anatomical Record*, **10** (1915–1916), 133–164, with portrait and bibliography; Edward S. Morse, "Charles Sedgwick Minot, 1852–1914," in *Biographical Memoirs. National Academy of Sciences*, **14** (1920), 263–285, with portrait and complete bibliography.

GEORGE W. CORNER

MINOT, GEORGE RICHARDS (*b.* Boston, Massachusetts, 2 December 1885; *d.* Brookline, Massachusetts, 25 February 1950), *medicine.*

Minot was the eldest son of James Jackson Minot, a physician, and Elizabeth Whitney, from whom he inherited the inquisitiveness and industry of cultured forebears successful in Boston's business and professional life. In 1915 Minot married Marian Linzee

Weld of Milton, Massachusetts. He was an amateur naturalist, cultivator of irises, and summer sailor of the coast of Maine.

Minot received the A.B. from Harvard College in 1908 and the M.D. from Harvard Medical School in 1912. He was professor of medicine at Harvard University and director of the Harvard Medical Unit at the Boston City Hospital from 1928 to 1948. His outstanding contribution to medical science was the discovery in 1926, with William P. Murphy, of the successful treatment of pernicious anemia by liver feeding, for which they shared the 1934 Nobel Prize in physiology or medicine, with George H. Whipple of Rochester, N. Y.

While in the private practice of medicine in Boston with an appointment as associate in medicine at the Massachusetts General Hospital (1918–1923), Minot had become convinced of the inadequacy of the diets of many patients with pernicious anemia. Consequently, he was prepared to make a thorough trial of the effects of liver feeding when reported by Whipple as especially potent in preventing an experimental anemia due to chronic, periodic blood removal in dogs. In Minot's patients a prompt increase in the number of reticulocytes (young red blood cells) objectified the repeated association observed between daily liver feeding, clinical improvement, and progressive lessening of their hitherto fatally progressive anemia. This observation led shortly to the development of therapeutic liver extracts and to research by others eventually identifying their active principle as vitamin B_{12} in 1948.

Minot's work and that of numerous pupils during the decade after 1926 initiated a new era in clinical hematology by replacing the largely morphologic studies of the blood and of the blood-forming and blood-destroying organs with dynamic measurements of their functions. Today the use of radioisotopic labeling of the formed elements of the blood, together with biochemical and biophysical analyses, are extending this revolution in depth. Among the many significant contributions of Minot and his associates were early work on blood transfusion, blood coagulation, and blood platelets, and classical studies of the hematological effects of irradiation in chronic leukemias and lymphoid tumors. Later came successful treatment of hypochromic anemia with sufficient iron; and demonstration that hemophilia is due to lack of a globulin substance present in normal plasma.

BIBLIOGRAPHY

Minot's principal writings include "The Development of Liver Therapy in Pernicious Anaemia: A Nobel Lecture," in *Lancet* (1935), **1**, 361–364.

On Minot and his work, see W. B. Castle, "The Contributions of George Richards Minot to Experimental Medicine," in *New England Journal of Medicine*, **247** (1952), 585–592; and F. M. Rackemann, *The Inquisitive Physician: The Life and Times of George Richards Minot, A.B., M.D., D.Sc.* (Cambridge, Mass., 1956).

W. B. CASTLE

MIQUEL, FRIEDRICH ANTON WILHELM (*b.* Neuenhaus, Germany, 24 October 1811; *d.* Utrecht, Netherlands, 23 January 1871), *botany.*

Miquel was the son of a country physician. His university studies and subsequent academic career took place in the Netherlands. Trained as a physician at the University of Groningen, Miquel specialized in botany and was director of the Rotterdam botanic garden (1835–1846), professor of botany at Amsterdam (1846–1859) and at Utrecht (1859–1871), and director of the Rijksherbarium at Leiden (1862–1871). His numerous (296 items in his bibliography) botanical publications deal mainly with the floras of the former Netherlands East Indies, Surinam, and Japan, and with the Cycadaceae, Moraceae, and Piperaceae. He collaborated with C. F. P. von Martius on the *Flora brasiliensis*; was the first to publish a comprehensive flora of the Netherlands East Indies; and played an important background role in the development of the East Indian quinine industry. Miquel was also the founder of the University of Utrecht herbarium.

BIBLIOGRAPHY

I. ORIGINAL WORKS. Miquel's main writings are *Commentarii phytographici, quibus varia rei herbariae capita illustrantur* (Leiden, 1839); *Monographia Cycadearum* (Utrecht, 1842); *Sertum exoticum contenant des figures et descriptions de plantes nouvelles ou peu connues* (Rotterdam, 1843); *Systema Piperacearum* (Rotterdam, 1843); "Symbolae ad floram surinamensem," a series of 12 articles in *Linnaea*, **18**–**22** (1844–1849); "Illustrationes Piperacearum," *Nova acta Academiae Caesareae Leopoldino Carolinae germanicae naturae curiosorum*, **21**, supp. 1 (1846); *Stirpes surinamenses selectae* (Leiden, 1850); *Plantae junghuhnianae. Enumeratio plantarum, quas in insulis Java et Sumatra detexit Fr. Junghuhn* (Leiden, 1851[–1857]); *Flora van Nederlandsch Indië*, 3 vols. (Amsterdam–Leipzig, 1855–1859); and *Annales Musei botanici Lugduno Batavi*, 4 vols. (Amsterdam, 1863–1869).

II. SECONDARY LITERATURE. See F. A. Stafleu, "F. A. W. Miquel, Netherlands Botanist," in *Wentia*, **16** (1966), 1–95, a biography with complete bibliography and secondary references to published and unpublished sources; and *Taxonomic Literature* (Utrecht, 1967), pp. 315–324, for further bibliographical details.

FRANS A. STAFLEU

MIRBEL, CHARLES FRANÇOIS BRISSEAU DE (*b.* Paris, France, 27 March 1776; *d.* Paris, 12 September 1854), *botany.*

The son of a jurist, Mirbel began his studies in Paris at a private boarding school run by the Congregation of the Picpus; the Revolution forced him to interrupt his education and to seek refuge with his parents at Versailles. When he was conscripted during the Terror, he hid in Toulouse, waiting for conditions to become more settled. Through the aid of a friend Mirbel returned to Paris, where, in the office of Lazare Carnot, then member of the Committee of Public Safety, he did work in topography and military history.

In 1796, through his position as secretary to General Henri Clarke, Mirbel learned of a proscription list that included the name of a friend's relative. He warned the man immediately so that he could escape. Clarke ordered Mirbel's arrest, but he fled to the Pyrenees, profiting from his enforced exile by studying physics, mineralogy, and botany. His first research, carried out in collaboration with Louis Ramond de Carbonnières, professor of natural history at the École Centrale of Tarbes, dealt with the geological configuration of the Pyrenees.

In 1798, after two years of exile—and in order to escape military conscription to serve in the Egyptian campaign—Mirbel returned to Paris and obtained a post in the Museum of Natural History.

The following year Mirbel presented a memoir on ferns to the Academy of Sciences, and in 1801 and 1802 he submitted a series of articles on the structure of plants and on the seed and embryo. Thus he inaugurated the study of microscopical plant anatomy in France. He showed that the characteristics of the seed and of the embryo are identical for all plants of the same natural family, thereby laying the foundations of embryogenic classifications, which are still used.

In 1802 Mirbel married a young woman who was related through her mother to the Dandolo family of Venice. His wife's influence helped him to obtain the post of head gardener at Malmaison, the country palace of Napoleon and Josephine. On 9 May 1806 Mirbel presented to the Academy remarks on a system of comparative plant anatomy based on the organization of the flower and a memoir on plant fluids. At the end of 1806, seeking the financial security he had always lacked, Mirbel entered the service of Louis Bonaparte, king of Holland. On 17 March 1807 he was named a correspondent of the Academy of Sciences.

Mirbel's new duties afforded him sufficient time to pursue his research on plant organography and physiology. He was elected a member of the botany section of the Academy of Sciences on 31 October 1808. He was appointed supplementary professor of botany at the Faculty of Sciences in Paris, then the center of a renaissance in scientific education; Desfontaines, Gay-Lussac, and Poisson were teaching there. Mirbel's work benefited greatly from his association with the Faculty. Between 1809 and 1815, in a series of notes and memoirs and in his lectures, he challenged orthodox opinion by asserting the independence of plant cells in the different tissues. His *Éléments de botanique* (1815), which revealed his talents as a draftsman, contributed to the acceptance of his ideas.

In 1816, after Napoleon's abdication, Mirbel accepted the posts of secretary-general of the Ministry of the Interior and *maître des requêtes* of the Council of State, using his positions to obtain passage of measures that would advance science. He resigned on 20 February 1820 when, following a cabinet reorganization, his friend Élie Decazes, who had helped him to secure those posts, was named ambassador to England. Mirbel joined Decazes in England, where he met several English scientists.

On his return to France, Mirbel, whose wife had died, married Lezinka Rue, curator of Louis XVIII's art collection. He was appointed professor-administrator of the Jardin des Plantes in 1829, replacing Bosc, who had died. Mirbel's research activity was at its height from 1825 to 1846. Using the microscope he followed the first stages of the formation of the tissues and studied the origin, development, and organization of the phloem and wood. In his papers on the development of the ovule, on the structure of *Marchantia*, on the formation of the embryo, on the disposition of the tissues in the stems and roots of the monocotyledons, and on the cambium the rigor of the draftmanship strengthened the presentation of the material.

Between 1843 and 1845 Mirbel demonstrated that the cambium contains ternary substances and various nitrogenous materials, which include the most active parts of the plant, those that secrete cellulose and produce all the organic and mineral substances. In this pioneering work, therefore, living protoplasm was differentiated from the cell wall. The formation of the cuticle and the lignification of the vessels were described by means of simple chemical reactions in which Mirbel anticipated plant cytochemistry. Through the study of vascularization in date trees and other monocotyledons, he demonstrated that the plant is formed of two parts that begin at the collar: one descends into the soil, and the other rises above the substratum.

Mirbel was continuing his investigations when his second wife died of cholera in 1849. Exhausted and bereaved, Mirbel gradually lost his memory. He spent his last years in peaceful retirement, cared for by his daughter. He died in 1854, at the age of seventy-eight.

BIBLIOGRAPHY

I. ORIGINAL WORKS. A list of 67 papers written or coauthored by Mirbel is in Royal Society *Catalogue of Scientific Papers*, IV, 405–407. His principal works include *Traité d'anatomie et de physiologie végétales*, 2 vols. (Paris, 1802); *Histoire naturelle générale et particulière des plantes*, 18 vols. (Paris, 1802–1806), to which he contributed; *Éléments de physiologie végétale et de botanique*, 3 vols. (Paris, 1815); and *Physique végétale, ou Traité élémentaire de botanique* (Paris, 1832).

II. SECONDARY LITERATURE. See Élie Margollé, *Vie et travaux de M. de Mirbel d'après sa correspondance et des documents inédits* (St. Germain, 1863); and M. Payen's biography of Mirbel, in Michaud, *Biographie universelle*, new ed., XXVIII, 382–387; based largely on his *Éloge historique de M. de Mirbel* (Paris, 1858).

A. NOUGARÈDE

MISES, RICHARD VON (*b*. Lemberg, Austria [now Lvov, U.S.S.R.], 19 April 1883; *d*. Boston, Massachusetts, 14 July 1953), *mathematics, mechanics, probability.*

Von Mises was the second son of Arthur Edler von Mises, a technical expert with the Austrian state railways, and Adele von Landau. His elder brother, Ludwig, became a prominent economist; the younger brother died in infancy. After earning his doctorate in Vienna in 1907, Richard taught at universities in Europe and Turkey and then, from 1939, in the United States. In 1944 he became Gordon McKay professor of aerodynamics and applied mathematics at Harvard. During his European period, he married Hilda Pollaczek-Geiringer, one of his pupils. Proud to call himself an applied mathematician, he was the founder and editor, from 1921 to 1933, of the well-known *Zeitschrift für angewandte Mathematik und Mechanik*. He was a scholar with wide interests, who wrote perceptively on the philosophy of science from a positivist point of view, and who was also an authority on the poet Rilke.

Von Mises' early preoccupation with fluid mechanics led him into aerodynamics and aeronautics, subjects that in the years immediately before 1914 had received a major fillip from the success of heavier-than-air flying machines. He himself learned to fly and in the summer of 1913 gave what is believed to be the first university course on the mechanics of powered flight. After the outbreak of World War I he helped develop an Austrian air arm, and in 1915 the team he led produced a giant 600-horsepower military plane with an original wing profile of his own design (wing theory was perhaps his specialty).

In 1916 he published a booklet on flight, under the auspices of the Luftfahrarsenal in Vienna. It went into many enlarged editions and is the basis of *Theory of Flight*, published with collaborators in English toward the end of World War II. Other, allied topics to which he contributed were elasticity, plasticity, and turbulence. He also worked in various branches of pure mathematics, particularly numerical analysis.

Von Mises' concern with the border areas of mathematics and the experimental sciences was reflected in his giving much thought to probability and statistics. In 1919 he published two papers that, although little noticed at the time, inaugurated a new look at probability that was destined to become famous. The background to this contribution was the slow buildup, during the nineteenth century, of a frequency theory of probability, in contrast to the received classical theory of Laplace. The fathers of the frequency theory, Poisson in France and Ellis in England, had identified the probability of a given event in specified circumstances with the proportion of such events in a set of exactly similar circumstances, or trials. The weakness of this position is the necessary finiteness of the set, and there is no obvious way of extending the idea to those very large or infinite sets that in practice must be sampled for probabilistic information. The Cambridge logician John Venn improved the theory in 1866 by equating probability with the relative frequency of the event "in the long run," thereby introducing a mathematical limit and the infinite set. Nevertheless, even this reformation failed to make the theory compelling enough to tempt mathematicians to put it into rigorous terms; and Keynes in his *Treatise on Probability* (1921) expressed his inability to assess the frequency theory adequately because it had never been unambiguously formulated. This was the deficiency that Von Mises attempted to correct.

What Von Mises did was to splice two familiar notions, that of the Venn limit and that of a random sequence of events. Let us consider the matter in terms of a binary trial, the outcome of which is either a "success" or otherwise. Given an endless sequence of such trials, in the sense of Bernoulli binomial sampling, what can we say about it probabilistically? A meaningful answer, said Von Mises, is possible only if we postulate (1) the mathematical existence of a limiting value to the fraction successes/trials, and (2) the invariance of this limit for all possible infinite subsequences formed by any rule of place selection of trials that is independent of their outcomes. Then the limit can be called the probability of a success in the particular system. It then follows that the probability of a single event is formally meaningless; random sampling is a *sine qua non*; and the sequence (otherwise collective or sample space) must be clearly defined before any discussion of probability—in this strictly operational sense—can be undertaken.

The intuitive appeal of Von Mises' limiting-frequency theory is strong, and its spirit has influenced all modern statisticians. Remarkably, however, the mathematics of the theory, even after sophistication by leading probabilists, has never been rendered widely acceptable, and some authorities today do not mention Von Mises. In advanced work, the measure-theoretic approach initiated by Kolmogorov in 1933 is most favored. On the practical side, his statistical writings suffered from a foible: he denied the importance of small-sample theory. Von Mises' *Probability, Statistics, and Truth*, published in German in 1928 and in English in 1939, is not a pedagogic text but a semipopular account, very subjective in tone, good on the historic side, and in general notably stimulating.

BIBLIOGRAPHY

I. ORIGINAL WORKS. The core of Von Mises' work is to be found in the following six books: *Probability, Statistics and Truth* (New York, 1939; 2nd ed., 1957); *Theory of Flight* (New York, 1945); *Positivism, a Study in Human Understanding* (Cambridge, Mass., 1951); *Mathematical Theory of Compressible Fluid Flow* (New York, 1958), completed by Hilda Geiringer and G. S. S. Ludford; *Selected Papers of Richard von Mises*, Philipp P. Frank *et al.*, eds. (Providence, R.I., 1963); and *Mathematical Theory of Probability and Statistics* (New York, 1964), edited and complemented by Hilda Geiringer. His first papers on probability are "Fundamentalsätze der Wahrscheinlich-keitsrechnung," in *Mathematische Zeitschrift*, **4** (1919), 1–97; and "Grundlagen der Wahrscheinlichkeitsrech-nung," *ibid.*, **5** (1919), 52–99. A good bibliography of 143 works is in Garrett Birkhoff, Gustav Kuerti, and Gabor Szego, eds., *Studies in Mathematics and Mechanics Presented to Richard von Mises* (New York, 1954), which contains a portrait.

II. SECONDARY LITERATURE. The opening chapters of *Mathematical Theory of Probability and Statistics* (see above) contain a survey by Hilda Geiringer of other workers' developments of Von Mises' controversial theory, as well as a synopsis of Kolmogorov's rival theory and its relation to that of Von Mises. A critical essay review of this book by D. V. Lindley is in *Annals of Mathematical Statistics*, **37** (1966), 747–754. W. Kneale, *Probability and*

Induction (London, 1949), marshals some logical arguments against limiting-frequency theories. On the other hand, H. Reichenbach, *The Theory of Probability* (Berkeley, 1949); Rudolf Carnap, *Logical Foundations of Probability* (Chicago, 1950); and Karl Popper, *The Logic of Scientific Discovery* (New York, 1958), are all, in different ways and with various emphases, derivative and sympathetic.

NORMAN T. GRIDGEMAN

MITCHELL, ELISHA (*b.* Washington, Connecticut, 19 August 1793; *d.* on Mount Mitchell, North Carolina, 27 June 1857), *natural history.*

Elisha Mitchell was the eldest son of Abner Mitchell, a farmer with substantial property, and Phoebe Eliot, a direct descendant of John Eliot, the "apostle to the Indians," who translated the Bible into Algonkian. A precocious child, he learned to read at an early age and possessed a nearly photographic memory. He attended school in Litchfield County, Connecticut, where he showed an interest in scholarship of all kinds, especially in natural science. Mitchell graduated from Yale in 1813 and in 1819 married Maria Sybil North, the educated daughter of a physician, by whom he had three sons (two of whom died in infancy) and four daughters. All his children received an excellent education, much of it imparted at home by their parents.

In 1816 Mitchell became a tutor at Yale, and in 1817 he was appointed professor of mathematics and natural philosophy at the University of North Carolina, in Chapel Hill. In the same year he received a license to preach and was ordained a minister of the Presbyterian church in 1821. In 1825 he transferred to the chair of chemistry, geology, and mineralogy at the university, a post that he held until his death. He received the honorary degree of doctor of divinity from the University of Alabama in 1838.

Mitchell accepted the scriptural account of the creation, and yet his *Elements of Geology*, a work published when he had been a minister for over twenty years, contained a sentence that showed him to be ahead of most of his contemporaries in accepting the principle of uniformitarianism: "A knowledge of the present will assist in explaining the past" (p. 64).

Mitchell was endowed not only with culture and learning but with humor, and he employed all these gifts in doing work in various fields. He earned respect as an authority on botany, geography, and geology, particularly of North Carolina. In his many cross-country trips on horseback and on foot, he not only took abundant scientific notes but made numerous friends. In 1842 he published the first geologic map of North Carolina and was the first to explain the origin of its gold deposits. He also wrote on meteorology and was a pioneer in applied soil science and in conservation.

Mitchell's lectures were all the more effective for being enlivened with humor. He was state geologist in 1826 and, during two brief periods, was acting president of the University of North Carolina. In addition he served as justice of the peace and town commissioner for Chapel Hill, preached on Sundays in the university chapel and the village church, prepared student manuals on natural history, botany, chemistry, geology, mineralogy, and certain religious topics, and contributed several scientific papers to the *American Journal of Science*.

Mitchell's richly varied career came to a tragic end in his sixty-fourth year. In 1839 he determined the altitude of the North Carolina peak now called Mount Mitchell, the highest mountain in the eastern United States. On a return to the summit in 1857 to verify his measurements and to attempt to settle a priority dispute, he lost his life in a fall from a cliff into a pool, in which he drowned. He is buried at the top of the mountain.

BIBLIOGRAPHY

I. ORIGINAL WORKS. Mitchell's writings include *Agricultural Speculations*, North Carolina Board of Agriculture (Raleigh, N.C., 1825), 49–58; *Report on the Geology of North Carolina*, pt. 3 (Raleigh, 1827), 1–27; "On the Character and Origin of the Low Country of North Carolina," in *American Journal of Science*, **13** (1828), 336–347; "On the Geology of the Gold Region of North Carolina," *ibid.*, **16** (1829), 1–19, **17** (1829), 400; "On the Effect of Quantity of Matter in Modifying the Force of Chemical Attraction," *ibid.*, **16** (1829), 234–242; "On a Substitute for Welther's Tube of Safety, with Notices on other Subjects," *ibid.*, **17** (1830), 345–350; "On the Proximate Causes of Certain Winds and Storms," *ibid.*, **19** (1831), 248–292; "Analysis of the Protogaea of Leibnitz," *ibid.*, **20** (1831), 56–64; "On Storms and Meteorological Observations," *ibid.*, **20** (1831), 361–369; "Notice of the Height of Mountains in North Carolina," *ibid.*, **35** (1839), 377–380; *Elements of Geology With an Outline of the Geology of North Carolina* (n.p., 1842), with map; and *Diary of a Geological Tour*, James Sprunt Historical Monograph, no. 6 (Chapel Hill, N.C., 1905), with introduction and notes by K. P. Battle.

II. SECONDARY LITERATURE. See K. P. Battle, *History of the University of North Carolina*, I (Raleigh, N.C., 1907), II (1912); H. S. Chamberlain, "Life Story of Elisha Mitchell, D.D., 1793–1857, Professor in the University of North Carolina From 1818 Until His Death in 1857," unpublished MS, University of North Carolina Library, Chapel Hill (1951); F. B. Dexter, *Biographic Sketches of the Graduates of Yale College*, VI (New York, 1912), 586–589; G. P. Merrill, *Contributions to the History of American Geology*, U.S. National Museum Annual Report for 1904 (Washington, 1906), 285–286, 706; and *The First One Hundred Years of American Geology* (New York, 1964), 114–116; and Charles Phillips, "A Sketch of Elisha Mitchell," in *Journal of the Elisha Mitchell Scientific Society*, **1** (1883–1884), 9–18, with portrait on frontispiece.

ELLEN J. MOORE

MITCHELL, MARIA (*b.* Nantucket, Massachusetts, 1 August 1818; *d.* Lynn, Massachusetts, 28 June 1889), *astronomy.*

Maria Mitchell, the first woman astronomer in America, was the third of ten children of William and Lydia Coleman Mitchell. She was educated chiefly by her father, a man of wide culture. As a small child she helped him with the observations that he made for the purpose of checking the chronometers of whaling ships, while in 1831, during an annular eclipse of the sun, she assisted him in timing the contacts that he used to determine the longitude of Nantucket.

When she was eighteen, Maria Mitchell became librarian of the Nantucket Atheneum, a post that she held for twenty-four years. At the same time she conducted astronomical observations, sweeping the skies on clear evenings. It was thus that, on 1 October 1847, she discovered a new telescopic comet, for which discovery she was awarded a gold medal by the king of Denmark and became world famous.

From 1849 until 1868 Maria Mitchell was employed by the U.S. Nautical Almanac Office to compute the ephemerides of the planet Venus. She resigned her post there reluctantly when her academic duties at Vassar Female College, of which she had been a faculty member since its founding in 1865, demanded her full attention. At Vassar she was both professor of astronomy and director of the college observatory, positions that she fulfilled with great distinction until her death.

Maria Mitchell was the first woman to be elected to the American Academy of Arts and Sciences; she also belonged to the American Philosophical Society and to the American Association for the Advancement of Women (of which she was president in 1870 and subsequently chairman of its Committee on Women's Work in Science). Of the many memorials established in her honor after her death, the Nantucket Maria Mitchell Association (founded in 1902, with headquarters at her birthplace that incorporate an observatory, a science library, and a natural science museum) is of particular interest.

BIBLIOGRAPHY

Maria Mitchell's published scientific papers were short and few. Apart from the notes listed in Poggendorff, she wrote a few less technical articles for *Atlantic Monthly*, *Hours at Home*, and *Century*.

On Maria Mitchell and her work, see Phoebe Mitchell Kendall, *Maria Mitchell. Life, Letters, and Journals* (Boston, 1896); Helen Wright, *Sweeper in the Sky* (New York, 1950), and "Mitchell, Maria," in *Notable American Women*, II (Cambridge, Mass., 1971), 554–556; and *Annual Reports. Nantucket Maria Mitchell Association*.

DORRIT HOFFLEIT

MITCHELL, SILAS WEIR (*b*. Philadelphia, Pennsylvania, 15 February 1829; *d*. Philadelphia, 4 January 1914), *medicine, neurology.*

Mitchell's father, John Kearsley Mitchell, was professor of medicine at Jefferson Medical College, where the younger Mitchell graduated in 1850. He then spent a year in Europe attending the lectures of Claude Bernard and the microscopist Charles Philippe Robin. In 1851 he joined his father's Philadelphia medical practice, a practice that S. W. Mitchell continued until his own death in 1914.

Between 1852 and 1863 Mitchell published more than thirty papers on a variety of topics ranging from the toxic effects of rattlesnake venom to the crystalline forms of uric acid. The physiological bent of these early papers reflected his Paris experience; and though Mitchell always maintained an interest in toxicology, pharmacology, and physiology, his later writings were generally more clinically oriented. The nature of his clinical work was deeply affected by his medical experiences in the Civil War. He treated many patients with nerve injuries, post-traumatic epilepsy, and other neurological conditions at the Turner's Lane Hospital, the 400-bed army neurological hospital in Philadelphia where Mitchell was assigned.

In collaboration with William Keen, Jr., and George Morehouse, he published in 1864 the important study *Gunshot Wounds and Other Injuries of Nerves*. Mitchell later extended this treatise into a definitive monograph, *Injuries of Nerves and Their Consequences* (1872). By this time he was widely recognized as the outstanding American neurologist. His additional contributions to clinical neurology included papers on posthemiplegic chorea,[1] causalgia and traumatic neuralgia,[2] the effects of weather on painful amputation stumps,[3] various forms of headache,[4] and a rare condition he called erythromelalgia.[5] Mitchell also investigated the physiology of the cerebellum[6] and the cutaneous distribution of nerves,[7] described the cremasteric reflex,[8] and (with Morris J. Lewis) gave an early account of the phenomenon of sensory reinforcement of the deep tendon reflexes.[9]

Mitchell published two general neurological works, *Lectures on Diseases of the Nervous System—Especially in Women* (1881) and *Clinical Lessons on Nervous Diseases* (1897). His preeminence as a neurologist brought him many patients with functional and neurotic complaints. He was especially interested in hysteria, and a large portion of the two general treatises is devoted to the description and treatment of hysteria and related disorders. Mitchell popularized the "rest cure" in the management of many kinds of nervous diseases, both functional and organic.[10] His concern with therapeutics also resulted in a number of papers on the pharmacology of the bromides, lithium, and chloral hydrate.

In addition, Mitchell wrote novels, short stories, and poetry; indeed, in his later years his fame as a man of letters equaled his reputation as a physician. His home on Walnut Street was a longtime center of Philadelphia's intellectual life. Mitchell's intimates included William Osler, William Henry Welch, and John Shaw Billings.

NOTES

1. "Post-Paralytic Chorea," in *American Journal of the Medical Sciences*, **61** (1874), 342–352.
2. "Clinical Lecture on Certain Painful Affections of the Feet," in *Philadelphia Medical Times*, **3** (1872), 81–82, 113–115.
3. "The Relations of Pain to Weather," in *American Journal of the Medical Sciences*, **73** (1877), 305–329.
4. "Headaches, From Heat Stroke, From Fevers, After Meningitis, From Over Use of the Brain, From Eye Strain," in *Medical and Surgical Reporter*, **31** (1874), 67–70.
5. "On a Rare Vaso-motor Neurosis of the Extremities, and on the Maladies With Which It May Be Confounded," in *American Journal of the Medical Sciences*, **76** (1878), 17–36.
6. "Researches on the Physiology of the Cerebellum," *ibid.*, **57** (1869), 320–338.
7. "The Supply of Nerves to the Skin," in *Philadelphia Medical Times*, **4** (1874), 401–403.
8. "The Cremaster-Reflex," in *Journal of Nervous and Mental Diseases*, **6** (1879), 577–586.
9. "Physiological Studies of the Knee-jerk, and of the Reactions of Muscles Under Mechanical and Other Excitants," in *Medical News*, **48** (1886), 169–173, 198–203.
10. "Rest in Nervous Disease: Its Use and Abuse," in *A Series of American Clinical Lectures*, E. C. Seguin, ed., I (New York, 1875), 83–102; also in the semipopular book by Mitchell, *Fat and Blood* (Philadelphia, 1877).

BIBLIOGRAPHY

I. ORIGINAL WORKS. Mitchell wrote more than 170 medical and scientific papers. To the books and articles mentioned in the text and notes should be added the following: *Wear and Tear* (Philadelphia, 1871); *Nurse and Patient*

(Philadelphia, 1877); and *Doctor and Patient* (Philadelphia, 1888), all popular works.

Mitchell's lifelong interest in the physiological effects of snake venom led to a number of papers, including "Researches Upon the Venoms of Poisonous Snakes," in *Smithsonian Contributions to Knowledge*, **26** (1886), 1–186, written with Edward Reichert.

A complete bibliography of Mitchell's medical, scientific, and literary works may be found in Richard D. Walter, *S. Weir Mitchell, M.D., Neurologist: A Medical Biography* (Springfield, Ill., 1970), 207–222.

II. SECONDARY LITERATURE. Anna Robeson Burr, *Weir Mitchell—His Life and Letters* (New York, 1929), is an important source of letters and includes an autobiographical fragment. Ernest Earnest, *S. Weir Mitchell—Novelist and Physician* (Philadelphia, 1950), emphasizes his literary achievements. Mitchell's scientific and medical work is extensively considered in the volume by Richard Walter (see above). This book also contains a full bibliography of additional secondary literature.

WILLIAM F. BYNUM

MITSCHERLICH, EILHARD (*b*. Neuende, Oldenburg, Germany, 7 January 1794; *d*. Berlin, Germany, 28 February 1863), *chemistry, mineralogy*.

Mitscherlich was the son of a minister, also named Eilhard Mitscherlich, and Laura Meier. He received his early education at Jever, in the school directed by the historian F. C. Schlosser, who encouraged him to apply himself to the liberal arts. In 1811 Mitscherlich entered the University of Heidelberg, where he studied Oriental languages; he continued this pursuit at the University of Paris, which he entered in 1813. He learned Persian with particular enthusiasm and hoped to be a member of the legation that Napoleon intended to send there. When Napoleon's fall ended that prospect, Mitscherlich returned to Germany, where in 1817 he enrolled in the University of Göttingen to read science and medicine—a choice dictated by his determination to reach the Orient, as a ship's doctor if not as a diplomat.

Simultaneously with his medical studies, Mitscherlich completed the research on ancient Persian texts for which he was awarded the doctorate. At the same time, his interest increasingly turned toward chemistry, which was taught at Göttingen by F. Strohmeyer, who, in addition to his lectures, gave his students the opportunity to carry out certain laboratory experiments.

In 1818 Mitscherlich went to Berlin to work in the laboratory of the botanist Heinrich Link. There he began to study crystallography. He observed that the crystals of potassium phosphate and potassium arsenate appeared to be nearly identical in form and, his curiosity spurred, asked Gustav Rose to instruct him in exact crystallographic methods so that he could make precise measurements. He then applied spherical trigonometry to the data that he obtained, and was thereby able to confirm his first impression. He reported this finding in an article entitled "Ueber die Krystallisation der Saltze, in denen das Metall der Basis mit zwei Proportionen Sauerstoff verbunden ist," published in the *Abhandlungen der Preussischen Akademie der Wissenschaften* for 1818–1819 and translated into French for publication in the *Annales de chimie* in the following year.

In this important article Mitscherlich discussed the crystals of the sulfates of various metals. He demonstrated that these sulfates—as well as the double sulfates of potassium and ammonium—crystallize in like forms, provided that they bind the same quantity of water of crystallization. Thus, for sulfates of copper and manganese, he found the ratio between the oxygen of the oxide and that of the water of crystallization to be 1:5; while for zinc, nickel, and magnesium, the ratio is 1:7. He further stated his hope "that through crystallographic examination the composition of bodies will be determined with the same certainty and exactness as through chemical analysis."

Mitscherlich met Berzelius in 1819, when the latter was passing through Berlin. Berzelius had heard of Mitscherlich's work and recognized the significance of his findings. When the Prussian Ministry of Education offered Berzelius the chair of chemistry at the University of Berlin, left vacant on the death of Klaproth, Berzelius suggested appointing Mitscherlich in his stead. Mitscherlich was thought to be too young to fill the post, however, and a compromise was arranged whereby he would be sent to work with Berzelius in Stockholm for two years, in order to enlarge his knowledge of chemistry. In the course of this fruitful partnership Mitscherlich worked in Berzelius' laboratory, visited and studied the mines and metallurgical works at Falun, and acquired further experience in chemical analysis and inorganic chemistry. Most important, he continued his work on isomorphism.

In his second article on his crystallographic researches, "Om Förhållandet einellan Chemiska Sammansättningen och Krystallformen hos Arseniksyrade och Phosphorsyrade Salter" ("On the Relation Between the Chemical Composition and the Crystal Form of Salts of Arsenic and Phosphoric Acids"), published in *Kungliga Svenska vetenskapsakademiens handlingar* in 1822, Mitscherlich reported on new observations that he had made with Berzelius. Among his findings were that

... each arsenate has its corresponding phosphate, composed according to the same proportions, combined with the same amount of water and having nearly equal solubilities in water and acids; in fact the two series of salts differ in no respect except that the radical of the acid in one series is arsenic, while in the other it is phosphorus. ... Certain elements have the property of producing the same crystal form when in combination with an equal number of atoms of one or more common elements, and the elements, from this point of view, can be arranged in certain groups. For convenience I have called the elements belonging to the same group ... *isomorphous.*

He then stated his conclusion that

... an equal number of atoms, combined in the same way produce the same crystal forms and the crystal form does not depend on the nature of the atoms, but only on their number and mode of combination.

Mitscherlich further noted that the hydrate crystals of $NaO^2 + 2 PO^5$ (written today as $NaH_2PO_4 \cdot H_2O$) and $NaO^2 + 2 AsO^5$ ($NaH_2AsO_4 \cdot H_2O$) ordinarily exist in two different forms; but since the phosphate crystal also exists in another form identical to the usual form of the arsenate crystal, the criterion for isomorphism is met. He was thus the first to recognize the phenomenon now called dimorphism. In his next paper, "Ueber die Körper, welche in zwei verschiedene Formen krystallisieren," published in *Abhandlungen der Preussischen Akademie der Wissenschaften* for 1822–1823, he investigated this phenomenon in greater detail and presented a number of examples, including the rhombic and monoclinic forms of sulfur. (He thus refuted Haüy's crystallographic axiom, whereby crystal angles, particularly the angles of cleavage, are characteristic of a given substance.)

The statement of the law of isomorphism, made early in his career, marks Mitscherlich's most important contribution to chemistry—indeed, Berzelius considered Mitscherlich's discovery to be the most significant since that of chemical proportions. Berzelius himself found Mitscherlich's work to be of great use; he was at this time concerned with the determination of the atomic weights of the elements and the law of isomorphism provided him with a valuable tool. Since the relative atomic weight of an element could be determined only through a knowledge of how many atoms are contained in the molecule, Berzelius' task was simplified by the application of Mitscherlich's law—once he had established the atomic composition of one of the isomorphic compounds, those of the others could be assumed to correspond to it. He was thus able to check the atomic weights that he had set out in his *Lärbok i kemien*

of 1814 and presented corrected values for twenty-one elements in the second edition, which was published in 1826.

Mitscherlich refined his work on isomorphism from time to time throughout his scientific life. When it became clear that his original formulation of the law was too broad, he modified it (in 1832) to state more precisely that only certain elements can substitute for each other in crystal form. During the following years, too, Mitscherlich established the isomorphism that exists between a number of specific compounds, including sulfates, metallic selenates, potassium chromate and potassium manganate, and potassium perchlorate and potassium permanganate. All of his later work was conducted in Berlin, where he returned in 1822 to take up the post of assistant professor of chemistry at the university. He became full professor three years later. He was also a member of the Berlin Academy of Sciences and director of its laboratory, located in the observatory. He made extensive use of this installation for teaching as well as for research, since the university offered no facilities for practical instruction in chemistry.

Besides his sojourn in Sweden, Mitscherlich made other trips abroad to work with foreign scientists. In 1823–1824 he was in Paris, where he collaborated with Fresnel in investigating the alteration of the double refraction of crystals as a function of temperature; he also met Thenard and Gay-Lussac. In 1824 he visited Humphry Davy, Faraday, Wollaston, and Dalton in England, where he inspected a number of factories. Back in Berlin, he worked in a number of areas of both organic and inorganic chemistry, in addition to his studies of isomorphism.

In inorganic chemistry Mitscherlich investigated the higher compounds of manganese, including the mixture of manganate and permanganate that Glauber, in the seventeenth century, had called the "chameleon mineral." Mitscherlich offered an explanation for the transformations of this substance, establishing that its red and green salts are the derivatives of two different (manganic and permanganic) acids; he determined their chemical composition in 1830. Aschoff produced the anhydride of permanganic acid in Mitscherlich's laboratory, and Mitscherlich himself was the first to obtain iodine azide and selenic acid.

During the same period Mitscherlich was also concerned with vapor-density determinations. He modified Dumas's apparatus by employing a metal bath for measuring higher temperatures; he was thus able to determine the vapor densities of bromine, sulfur phosphorus, arsenic, mercury, sulfur trioxide, phosphorus pentachloride, calomel, and arsenic oxide.

His results were highly accurate in most instances. He further measured the pressure of water vapor over Glauber's salt, in response to a suggestion of Berzelius, who had hoped—erroneously—that a numerical indication of the affinity of water for various substances might be determined from the differences between the pressure of water vapor over those substances.

In organic chemistry, Mitscherlich in 1834 obtained benzene by the dry distillation of the calcium salt of benzoic acid. He found the product of the distillation to be identical with the "bicarburet of hydrogen" that Faraday had isolated from compressed oil-gas five years earlier. From his observation that benzoic acid might be a compound of benzene and carbon dioxide, Mitscherlich concluded that all organic acids must consist of hydrocarbons plus carbonic acid—a misconception that was long perpetuated. By vapor-density measurements, he reached the formula C_3H_3 (the present C_6H_6) for the composition of benzene, a quantity that corresponds in volume to one atom of hydrogen.

Mitscherlich went on to conduct experiments on various benzene derivatives. He obtained nitrobenzene from the reaction of benzene with fuming nitric acid (ordinary nitric acid does not react with benzene) and benzenesulfonic acid from the reaction of benzene with fuming sulfuric acid. He also obtained azobenzene, trichlorobenzene, hexachlorobenzene, and their corresponding bromine derivatives.

In 1834 Mitscherlich also showed that a mixture of ether and water distills out of a mixture of alcohol and diluted sulfuric acid; he suggested that in this case the sulfuric acid acts as a dehydrating agent. From this observation he developed his contact theory, whereby certain chemical reactions can take place only in the presence of certain other substances. Mitscherlich's theory was a direct predecessor of Berzelius' catalyst theory, which was, in fact, a refinement of it.

Mitscherlich further sought to explain fermentation by this theory, the "contact" in this process being yeast, which is necessary for the conversion of sugar into alcohol. He observed that if a test tube filled with yeast is dipped into a sugar solution, no fermentation occurs, whereas if the sugar is introduced directly into the tube that contains the yeast—or is brought into contact with it—fermentation does take place. Since it is not necessary that a contact agent be a chemical substance in Mitscherlich's theory, he was able to accept Cagniard de La Tour's assertion (of 1842) that yeast is a microorganism; indeed, Mitscherlich was the first chemist to do so.

In his experiments on fermentation Mitscherlich further discovered that yeast does not act directly on cane sugar; instead, an invert sugar, a kind of levorotatory "modified cane sugar" identical to the sugar formed by the action of acids on cane sugar, is formed first. He also established that 0.001 percent acid is sufficient to invert sugar solutions. He gave impetus to the sugar industry both by developing the first practical polarization apparatus and by devising a method to control polarization through polarimetric analysis.

Mitscherlich worked to improve the methods and accuracy of both organic and inorganic analytical chemistry. In 1855 he developed a toxicological detection index for white phosphorus, by which the substance to be tested was distilled with steam and the presence of phosphorus determined by luminescence in the condenser of the distilling apparatus. He was also the first to employ a mixture of potassium carbonate and sodium carbonate to produce fusion. For analyzing organic compounds, Mitscherlich constructed a combustion apparatus that differed from those of Berzelius and Liebig in that the combustion tube was heated by a spirit lamp, rather than by burning charcoal. The oxygen produced by the potassium chlorate was used to regenerate cupric oxide. Liebig, who was never on very good terms with Mitscherlich, pronounced the apparatus to be of little value.

Mitscherlich's early interest in geology and mineralogy continued throughout his life. He was particularly concerned with the production of artificial minerals through the fusion of silica with various metallic oxides, and achieved some valuable results in such experiments. In his last years he made a number of journeys to the most important European volcanoes to gather data toward a general theory of volcanoes, the subject of his last, posthumously published, articles. (It must be noted, however, that his work in volcanology produced little of significant value.)

Mitscherlich was perhaps most successful as a writer of textbooks. His *Lehrbuch der Chemie* was first published in 1829; by 1847 it had had four new editions in German, as well as two editions in French and one in English. The work contained Mitscherlich's lectures on all aspects of pure and applied chemistry, as well as a considerable amount of material on physics, all illustrated with a number of beautiful woodcuts. The lectures themselves are characterized by their exemplary clarity and ingenious experiments; the book was highly praised by Mitscherlich's contemporaries, including Berzelius and Liebig. As a teacher, Mitscherlich was aware that his students needed practical instruction; although his efforts to this end were in fact little more than perfunctory, he did take them on visits to factories.

Mitscherlich married and had five children, of whom the youngest, Alexander, also became a chemist. It was he, rather than his father, who discovered the Mitscherlich process for extracting cellulose from wood through boiling with calcium bisulfite, upon which discovery the German cellulose industry was based.

BIBLIOGRAPHY

I. ORIGINAL WORKS. A more complete list of Mitscherlich's writings can be found in Poggendorff, II, cols. 160–162. His major book is *Lehrbuch der Chemie*, 2 vols. (1829; 4th ed., 1847), also trans. into French (1835) and into English by S. L. Hammick as *Practical Experimental Chemistry Adapted to Arts and Manufactures* (1838). Many of his shorter writings were brought together as *Gesammelte Schriften von Eilhard Mitscherlich. Lebensbild, Briefwechsel und Abhandlungen* (1896).

II. SECONDARY LITERATURE. See G. Bugge, "Mitscherlich," in *Das Buch der Grossen Chemiker*, I (Berlin, 1929); F. Heinrich, "Zur Erinnerung an Eilhard Mitscherlich," in *Chemische Zeitung*, **37** (1913), 1369, 1398; H. Kopp, *Geschichte der Chemie*, I (Brunswick, 1843), 414; and *Die Entwicklung der Chemie in der neueren Zeit* (Munich, 1873), p. 417; A. Mitscherlich, in *E. Mitscherlichs Gesammelte Schriften* (Berlin, 1896); J. R. Partington, *A History of Chemistry*, IV (London, 1964); W. Prandtl, *Deutsche Chemiker* (Weinheim, 1956); G. Rose, "Eilhard Mitscherlich," in *Zeitschrift der Deutschen geologischen Gesellschaft*, **16** (1864), 21; and Williamson, "Eilhard Mitscherlich," in *Journal of the Chemical Society,* **17** (1864), 440.

F. SZABADVÁRY

MITTAG-LEFFLER, MAGNUS GUSTAF (GÖSTA) (*b*. Stockholm, Sweden, 16 March 1846; *d*. Stockholm, 7 July 1927), *mathematics.*

Mittag-Leffler was the eldest son of John Olaf Leffler and Gustava Wilhelmina Mittag. His father was a school principal and from 1867 to 1870 a deputy in the lower house of the Swedish parliament. The atmosphere at home was intellectually stimulating, and Mittag-Leffler's aptitude for mathematics was recognized and encouraged at an early age.

Mittag-Leffler entered the University of Uppsala in 1865 and obtained the doctorate in 1872. He remained at the university as a lecturer for a year, but in 1873 left on a traveling scholarship for Paris, Göttingen, and Berlin. On the advice of Hermite, whom he met in Paris, he went to study under Weierstrass in Berlin. Weierstrass exerted a decisive influence on his subsequent development.

In 1877 Mittag-Leffler wrote his *Habilitationsschrift* on the theory of elliptic function and, in consequence, was appointed professor of mathematics at the University of Helsinki. In 1881 he left Helsinki for Stockholm, where he became professor of mathematics at the newly established Högskola (later the University of Stockholm). He twice served as rector. Among his colleagues there were Sonya Kovalewsky and E. Phragmén. Mittag-Leffler was an excellent lecturer. Among his students were I. O. Bendixson, Helge von Koch, and E. I. Fredholm.

Mittag-Leffler was not among the mathematical giants of his time, but he did contribute several methods and results that have found a lasting place in the mathematical literature. His most important contributions clearly reflect Weierstrass' influence. Thus, where Weierstrass had given formulas for the representation of entire functions and of elliptic functions, Mittag-Leffler set himself the task of finding a representation for an arbitrary meromorphic function $f(z)$ which would display its behavior at its poles. The answer is of classical simplicity. Let $\{z_n\}$ be the set of poles of $f(z)$ so that $\{z_n\}$ is either finite or is infinite and possesses a limit point at infinity. In the former case the answer is trivial since $f(z)$ differs from an entire function only by the sum of its principal parts, $\sum h_n(z - z_n)$. In the latter case, this sum is infinite and may diverge. Convergence is reestablished by adding to the individual terms of the sums certain suitable polynomials.

A generalization of this result, also due to Mittag-Leffler, is concerned with the case where the set $\{z_n\}$ while still consisting of isolated points may have limit points also in the finite plane. Another field that was pioneered by Mittag-Leffler is that of the representation of an analytic function $f(z)$ beyond the circle of convergence of its power series round a given point. Taking this point, without loss of generality, as the origin, one defines the (principal) Mittag-Leffler star of the function with respect to the origin as the union of the straight segments extending from the origin to the first singularity of the function in that direction (or to infinity). Mittag-Leffler developed analytic expressions that represent $f(z)$ in the entire Mittag-Leffler star. The later evolutions of this subject led to its being subsumed under the heading of the theory of summability, where certain infinite matrices are now known as Mittag-Leffler matrices.

Mittag-Leffler was a prolific writer, and the list of his publications includes 119 items. But his importance as a research worker is overshadowed by his prominence as an organizer in many spheres of scientific activity. He was the founder and, for many years, the chief editor of the highly influential *Acta mathe-*

matica, to which he attracted important contributions by men such as E. Borel, G. Cantor, J. Hadamard, D. Hilbert, J. Jensen, V. Volterra, H. Weber, and above all H. Poincaré.

Mittag-Leffler's relationship with Cantor is of particular interest. Mittag-Leffler himself said of his work on meromorphic functions and of its generalizations that it had been his endeavor to subsume Weierstrass' and Cantor's approaches to analysis under a single point of view. And Cantor regarded Mittag-Leffler as one of his most influential friends and supporters in a hostile world.

Mittag-Leffler was very conscious of the importance of maintaining a record of the contemporary history of mathematics for posterity, and the pages of *Acta mathematica* contains reprints of many exchanges of letters between notable mathematicians of the period. He wrote a moving account of the relationship between Weierstrass and Sonya Kovalewsky.

He was one of the organizers of the first and of subsequent international congresses of mathematicians and was the recipient of many honors, including doctorates from the universities of Bologna, Oxford, Cambridge, Christiania (Oslo), Aberdeen, and St. Andrews.

Mittag-Leffler was married to Signe af Lindfors in 1882.

BIBLIOGRAPHY

Articles by Mittag-Leffler include "Sur la représentation analytique des fonctions monogènes uniformes d'une variable indépendante," in *Acta mathematica*, **4** (1884), 1–79, and "Weierstrass et Sonja Kowalewsky," *ibid.*, **39** (1923), 133–198. A complete bibliography is in N. E. G. Nörlund, "G. Mittag-Leffler," *ibid.*, **50** (1927), I-XXIII. See also A. Schoenflies, "Die Krisis in Cantor's mathematischem Schaffen," *ibid.*, 1–23, with "Zusätzliche Bemerkungen," by Mittag-Leffler, *ibid.*, 25–26; and E. Hille, "In Retrospect," 1962 Yale Mathematical Colloquium (mimeographed).

ABRAHAM ROBINSON

MITTASCH, ALWIN (*b.* Grossdehsa, Germany, 27 December 1869; *d.* Heidelberg, Germany, 4 June 1953), *physical chemistry*.

Mittasch was a leading authority on contact catalysis. As head of catalytic research for the Badische Anilin- und Soda-Fabrik (BASF) he guided the research that led to inexpensive, durable compound catalysts for the Haber-Bosch synthetic ammonia process, the Ostwald process for oxidation of ammonia to nitric acid, the water gas reaction, and various hydrogenations in the gas phase. His career and private life were singularly untroubled, touched only marginally by the military and political upheavals through which he lived.

The fourth of the six children of a village schoolmaster, Mittasch grew up happily in Wendish Saxony. For lack of money the boy was sent to a teacher-training school instead of a university. At nineteen Mittasch began teaching in a rural grade school. Three years later, in 1892, he secured an appointment to a city school in Leipzig. Here he soon attended public lectures at the university, being particularly drawn to those of Ostwald on energy relations in chemical systems. He resolved to become a middle school science teacher, but as his undergraduate studies progressed, and with Ostwald's encouragement, he determined to become a physical chemist. After seven years of university studies (in addition to full-time teaching), Mittasch advanced to doctoral candidacy. His thesis under Max Bodenstein, on the kinetics and catalytic aspects of nickel carbonyl formation and decomposition, led directly to a career in catalytic chemistry.

After short interludes as Ostwald's assistant and then as analyst in a lead and zinc fabricating company, he was hired by the BASF in 1904. Here Mittasch assisted Carl Bosch in seeking an industrial process for fixing nitrogen via cyanides or nitrides. This work was abandoned in 1909 in favor of the commercially more promising Haber ammonia synthesis directly from nitrogen and hydrogen. However, the experience that he had just gained helped Mittasch in seeking a cheaper catalyst than Haber's osmium. Assuming that in the Haber process the metal catalyst briefly forms a nitride intermediate and remembering that nitride formation occurs best in the presence of certain stable oxides, Mittasch directed an exhaustive search that led not only to an optimal, cheap catalyst of iron, aluminum, and potassium oxides, but also to much knowledge about catalyst poisons and activators. His discovery of the utility of compounded catalysts formed the basis of a massive research program he directed at the BASF for the next two decades. Besides the catalysts for important industrial processes his research yielded much data on high pressure and temperature reactions in the gaseous phase. He became particularly impressed by the manner in which the selection of a specific catalytic mixture can favor the yield of a desired compound while inhibiting the formation of other possible products.

After his retirement in 1934, he made this last observation the basis of an elaborate philosophy of causality, about which he wrote two books and several articles. Much better received by critics than these often abstruse writings were his scholarly and extensive publications on the history of catalysis.

BIBLIOGRAPHY

I. Original Works. Mittasch wrote some twenty-one books, which include *Von Davy und Döbereiner bis Deacon, ein Halbes Jahrhundert grenzfläschenkatalyse* (Berlin, 1932), with Erich Theis; *Julius Robert Mayer's Kausalbegriff* (Berlin, 1940); *Geschichte der Amoniaksynthese* (Weinheim, 1951); *Wilhelm Ostwalds Auflösungstheorie* (Heidelberg, 1951); and *Friedrich Nietzsche als Naturphilosoph* (Stuttgart, 1952).

Mittasch's numerous patents and technical articles and much else that he published can be located through the *Chemisches Zentralblatt* and *Chemical Abstracts*. His published speeches, books, book reviews, and articles of a historical, philosophical, and broader scientific nature are listed in an "Autobibliography" on pp. 747–759 of his book *Von der Chemie zur Philosophie* (Ulm, 1948); a supplement covering 1948–1953 is provided at the close of the obituary written by Karl Holdermann, "Alwin Mittasch in Memoriam," in *Chemische Berichte*, **90** (1957), LIV The archive of the BASF reports that it holds some of Mittasch's correspondence and other papers of minor historical significance. There is also an unpublished autobiography, *Chronik meines Lebens*, the location of which is unknown.

II. Secondary Literature. Several obituaries and tributes are listed in the bibliography by Karl Holdermann. There are others at the BASF archives. In addition there is a chapter on "Alwin Mittasch," by Alfred von Nagel, in *Ludwigshafener Chemiker*, I (Dusseldorf, 1958), 137–170.

See also Eduard Farber, "From Chemistry to Philosophy: The Way of Alwin Mittasch (1869–1953)," in *Chymia*, **11** (1966), 157–178.

JOHN J. BEER

MIVART, ST. GEORGE JACKSON (*b.* London, England, 30 November 1827; *d.* London, 1 April 1900), *biology, natural history.*

Mivart was born of well-to-do parents who were members of the rising nonprofessional middle class. His father's associations in natural history encouraged him to develop his own interests in that field, which he did through reading and collecting. An expected enrollment at either Oxford or Cambridge was prevented by his conversion at seventeen to Roman Catholicism. He prepared instead for a career in law and was admitted to the bar in 1851.

Mivart's primary interests in natural history persisted, and he came to know many of the naturalists of his day, particularly Owen and Huxley. The latter demonstrated to Mivart the excitement of natural history as a discipline in its own right. It was undoubtedly through Huxley's influence that Mivart worked in the 1860's and 1870's on his series of papers on Primate comparative anatomy. Huxley viewed the development of a precise body of knowledge about the Primates as significant to the elaboration and documentation of Darwinian evolution. The prosimians themselves had not been systematically studied as a group; and Mivart's work, which culminated in "On *Lepilemur* and *Cheirogaleus* and the Zoological Rank of the Lemuroidea" (1873), was a major contribution to an understanding of this enigmatic Primate group and their systematic relationship to the rest of the order.

Meanwhile Mivart attained a modest reputation in comparative anatomy; he published a series of descriptive studies, lectured to lay audiences, and from 1862 to 1884 taught anatomy at St. Mary's Hospital Medical School in London. Mivart had been a member of the Royal Institution since 1849, and he was elected fellow of the Zoological Society in 1858, of the Linnean Society in 1862, and of the Royal Society in 1869.

Although he was initially an adherent of the new biology for which Huxley was the most articulate spokesman, and for which Darwinism was the most influential method, Mivart regarded the tendency to universalize and to reify organic evolution as a threat both to the truths of his own Catholicism and to his more restricted definition of the canons of science. The conflict led to the publication of *On the Genesis of Species* (1871) and *Man and Apes* (1873); in both works Mivart criticized Darwinism as insufficient to explain anomalies in the data of observation or to answer the more general questions which dealt with the initiation of specific forms which must precede the action of natural selection. Such attacks on Darwinism—which were coupled with what were defined as insulting personal allusions—precipitated a formal break with Huxley and the Darwinians and through them Mivart's removal from the main current of natural science, so that after 1873, although he continued to publish, his work appeared more and more dated.

Mivart's attempts to reconcile his Catholicism with his science were equally destructive to his position as a prominent Catholic layman. In a series of articles and books which began with *Contemporary Evolution* (1873–1876), he sought to inject a modernist spirit born of the new science into the still conservative theology, structure, and practice of the Catholic Church. His arguments were attacked with increasing bitterness and finally rejected. Six weeks before his death he was excommunicated. Mivart stands as an important symbol and victim of the deep conflicts in science and in the intellectual milieu of the nineteenth century.

BIBLIOGRAPHY

For descriptions of Mivart's life and works see Jacob W. Gruber, *A Conscience in Conflict: The Life of St. George Mivart* (New York, 1960), and Peter Vorzimmer, *Charles Darwin: The Years of Controversy* (Philadelphia, 1970).

JACOB W. GRUBER

MÖBIUS, AUGUST FERDINAND (*b*. Schulpforta, near Naumburg, Germany, 17 November 1790; *d*. Leipzig, Germany, 26 September 1868), *mathematics, astronomy*.

Möbius was the only child of Johann Heinrich Möbius, a dancing teacher in Schulpforta until his death in 1793, and the former Johanne Catharine Christiane Keil, a descendant of Luther. His father's unmarried brother succeeded him as dancing teacher and as provider for the family until his own death in 1804. Möbius was taught at home until his thirteenth year, by which time he had already shown an interest in mathematics. He pursued formal education from 1803 to 1809 in Schulpforta, where he studied mathematics under Johann Gottlieb Schmidt. In 1809 he entered Leipzig University with the intention of studying law, but his early love for mathematics soon came to dominance. Consequently he studied mathematics under Moritz von Prasse, physics with Ludwig Wilhelm Gilbert, and astronomy with Mollweide, whose assistant he became.

Having been selected for a traveling fellowship, he left Leipzig in May 1813, a few months before the Battle of Leipzig, and went to Göttingen, where he spent two semesters studying theoretical astronomy with Gauss. He then proceeded to Halle for studies in mathematics with Johann Friedrich Pfaff. When in 1814 Prasse died, Mollweide succeeded him as mathematics professor, thereby opening up the position in astronomy at Leipzig. The position was given to Möbius, who received his doctorate from Leipzig in 1814 and qualified for instruction in early 1815 with his *De peculiaribus quibusdam aequationum trigonometricarum affectionibus*. In the same year he published his doctoral thesis entitled *De computandis occultationibus fixarum per planetas*. In spring 1816 he became extraordinary professor of astronomy at Leipzig and also observer at the observatory. In preparation for these duties he visited a number of

the leading German observatories and eventually made recommendations for the refurbishing and reconstruction of the observatory at Leipzig; these were carried out by 1821. Other instruments were added later, including a six-foot Fraunhofer refractor.

In 1820 Möbius' mother, who had come to live with him, died. Shortly thereafter he married Dorothea Christiane Johanna Rothe, whose subsequent blindness did not prevent her from raising a daughter, Emilie, and two sons, Theodor and Paul Heinrich, both of whom became distinguished literary scholars. The former is best known for his research on Scandinavian and Icelandic literature; the latter is sometimes confused with Paul Julius Möbius the neurologist, who was Möbius' grandson.

Although Möbius was offered attractive positions as an astronomer at Greifswald in 1816 and as a mathematician at Dorpat in 1819, he refused them both to remain at Leipzig. In 1829 he became a corresponding member of the Berlin Academy of Sciences, but it was not until 1844, after he had been invited to succeed J. F. Fries at Jena, that Leipzig promoted him to ordinary professor of astronomy and higher mechanics. The slowness of his promotion and his modest salary have been attributed to his quiet and reserved manner, while his refusal to leave Leipzig stemmed from his love for his native Saxony and the quality of Leipzig University. In 1848 Möbius became director of the observatory, and d'Arrest became the observer and eventually his son-in-law. Möbius rarely traveled, and in general his life centered around his study, the observatory, and his family. His writings were fully developed and original; he was not widely read in the mathematical literature of his day and consequently found at times that others had previously discovered ideas presented in his writings. Also his investigations were frequently aimed not so much at finding new results, but rather at developing more effective and simpler means for treating existing areas. In 1868, not long after having celebrated his fiftieth year of teaching at Leipzig, Möbius died; his wife's death had come nine years earlier.

Möbius' scientific contributions may be divided into two areas—astronomy and mathematics. Like his contemporaries Gauss and W. R. Hamilton, Möbius was employed as an astronomer but made his most important contributions to mathematics.

His early publications were in astronomy; two short papers on Juno and Pallas were followed by the separate publication in 1815 of his doctoral dissertation on occultation phenomena (see above) and in 1816 by his *De minima variatione azimuthi stellarum circulos parallelos uniformiter describentium com-*

mentatio. By 1823 his observational activities had borne fruit to the extent that he published his only work of that sort, his *Beobachtungen auf der Königlichen Universitäts-Sternwarte zu Leipzig*. He published a few observational papers in later decades and in the 1830's made measurements on terrestrial magnetism. He also published two popular treatises on the path of Halley's comet (1835) and on the fundamental laws of astronomy (1836), the latter having gone through many editions. His greatest contribution to astronomy was his *Die Elemente der Mechanik des Himmels* (1843), wherein he gave a thorough mathematical treatment of celestial mechanics without the use of higher mathematics. Although astronomical amateurs could therefore read the book, it nevertheless contained results important to professionals. Moreover he introduced (for the first time, he thought) the use of vectorial addition and subtraction to represent velocities and forces and effectively showed the computational usefulness of that very ancient mathematical device, the epicycle.

When Mollweide died in 1825, Möbius hoped to follow his example by exchanging his own position in astronomy for that in mathematics, but in 1826 M. W. Drobisch was selected. In the following year Möbius published his greatest work, which later became a mathematical classic. Möbius' *Der barycentrische Calcul: Ein neues Hülfsmittel zur analytischen Behandlung der Geometrie* (1827) was not only his most important mathematical publication, but also the source of much of his later work. He had come upon the fundamental ideas for his barycentric calculus in 1818 and by 1821 decided that they merited book-length treatment. In an appendix to his 1823 astronomical treatise, he had given a first discussion of his new method. As he stated in the foreword to his 1827 treatise, the concept of the centroid had been recognized by Archimedes as a useful tool for geometrical investigations.

Möbius proceeded from the well-known law of mechanics, that a combination of weights positioned at various points can be replaced by a single weight of magnitude equal to the sum of the individual weights and positioned at the center of gravity of the combination. Thus Möbius constructed a mathematical system, the fundamental entities of which were points, to each of which a weight or numerical coefficient was assigned. The position of any point could be expressed in this system by varying the numerical coefficients of any four or more noncoplaner points. Thus Möbius used an equation such as $aA + bB + cC \equiv D$, where a, b, c are numerical coefficients (positive or negative), and A, B, C, D are points, to express the fact that if A, B, C are not

collinear, then D must lie in the plane of A, B, C. Möbius went on in his treatise to apply this method with noteworthy success to many important geometrical problems. Since barycentric coordinates are a form of homogeneous coordinates, their creator is recognized with Feuerbach and Plücker, whose publications were independent and nearly simultaneous, as a discoverer of homogeneous coordinates.

Moreover Möbius developed important results in projective and affine geometry and also was among the first fully to appreciate the principle of duality and to give a thorough treatment of the cross ratio. He was the first mathematician to make use of a system wherein geometrical entities, such as lines, plane figures, and solids, were consistently treated as spatially oriented and to which a positive or negative sign could be affixed. Moreover he presented in this work the construction now known as the Möbius net. Finally at one point in the treatise he commented that two equal and similar solid figures, which are however mirror images of each other, could be made to coincide, if one were "able to let one system make a half revolution in a space of four dimensions. But since such a space cannot be conceived, this coincidence is impossible in this case" (*Werke*, I, 172).

Nearly all of Möbius' subsequent mathematical publications appeared in Crelle's *Journal für die reine und angewandte Mathematik* and from 1846 in either the *Abhandlungen* . . . or the *Berichte der Königlichen Sächsischen Gesellschaft der Wissenschaften zu Leipzig*. Some of these merit special attention. An 1828 paper discussed two tetrahedrons which mutually circumscribe and inscribe each other; such tetrahedrons are now known as Möbius tetrahedrons. Two dioptrical papers appeared in 1830 wherein Möbius used continued fractions to develop his results; another optical paper appeared in 1855 based on the concept of collineation. The Möbius function in number theory was presented in an 1832 paper, but most of his energies during the 1830's went into a series of papers on statics, which culminated in his 1837 two-volume *Lehrbuch der Statik*, wherein he treated the subject, following Poinsot, through combining individual forces with couples of forces and introduced the concept of a null system.

It is frequently stated that in 1840 Möbius posed for the first time the four-color conjecture, that is, that four colors are sufficient for the unambiguous construction of any map, no matter how complex, on a plane surface. This attribution is, however, incorrect; its source lies in the correct statement that in 1840 Möbius presented a lecture in which he posed the problem of how a kingdom might be divided into five regions in such a way that every region would border on each of the four other regions. In 1846 Möbius published a treatment of spherical trigonometry based on his barycentric calculus and in 1852 a paper on lines of the third order. His 1855 "Theorie der Kreisverwandschaft in rein geometrischer Darstellung" is the culmination of a number of studies on circular transformations, which are now frequently called Möbius transformations.

Möbius had been visited in 1844 by a high school teacher, Hermann Grassmann, whose now famous *Ausdehnungslehre* of 1844 contained among other things results similar to Möbius' point system of analysis. Grassmann requested Möbius to review the book, but Möbius failed to appreciate it, as did many others. When Grassmann in 1846 won the prize in a mathematical contest, which he had entered at Möbius' suggestion, Möbius did agree to write a commentary on the prize-winning essay. This 1847 work was the only significant published analysis of Grassmann's ideas until the late 1860's, when their significance was realized. Möbius was stimulated in the early 1860's to write his own treatise, "Ueber geometrische Addition und Multiplication," but this was not published until nineteen years after his death.

Möbius is now most frequently remembered for his discovery of the one-sided surface called the Möbius strip, which is formed by taking a rectangular strip of paper and connecting its ends after giving it a half twist (*Werke*, II, 484–485). The Paris Academy had offered a prize for research on the geometrical theory of polyhedrons, and in 1858 Möbius began to prepare an essay on this subject. The results of his essay were for the most part given in two important papers: his "Theorie der elementaren Verwandtschaft" of 1863 and his "Ueber die Bestimmung des Inhaltes eines Polyëders" of 1865. The latter contains his discovery of the "Möbius strip" and proof that there are polyhedrons to which no volume can be assigned. Curt Reinhardt has shown from an examination of Möbius' notebooks that he discovered this surface around September 1858 (*Werke*, II, 517–521); this date is significant, since it is now known that Johann Benedict Listing discovered the same surface in July 1858 and published his discovery in 1861. Listing and Möbius, who worked independently of each other, should thus share the credit for this discovery.

BIBLIOGRAPHY

I. ORIGINAL WORKS. Möbius' main publications, including his three long books, are collected in R. Baltzer, F. Klein, and W. Scheibner, eds., *Gesammelte Werke*, 4 vols. (Leipzig, 1885–1887). The second and fourth volumes contain previously unpublished writings, and the fourth contains a useful discussion of Möbius' manuscripts by C. Reinhardt.

II. SECONDARY LITERATURE. The best discussion of Möbius' life and astronomical activities is contained in C. Bruhns, *Die Astronomen auf der Pleissenburg* (Leipzig, 1879). The best treatment of his mathematical work is by R. Baltzer in Möbius, *Werke*, I, v–xx. See also H. Gretschel, "August Ferdinand Möbius," in *Archiv der Mathematik und Physik*, **49**, *Literarischer Bericht*, CLXXXXV (1869), 1–9; and M. Cantor, "Möbius," in *Allgemeine deutsche Biographie*, XX (1885), 38–43. A useful discussion of his barycentric calculus is R. E. Allardice, "The Barycentric Calculus of Möbius," in *Proceedings of the Edinburgh Mathematical Society*, **10** (1892), 2–21, and selections from his treatise on this subject are given in English in D. E. Smith, ed., *A Source Book in Mathematics*, II (New York, 1959), 525.–526, 670–676. K. O. May in his "The Origin of the Four-Color Conjecture," in *Mathematics Teacher*, **60** (1967), 516–519, clarifies Möbius' relationship to this conjecture, and M. Crowe, in *A History of Vector Analysis* (Notre Dame, Ind., 1967), treats Möbius' relationship to vectorial analysis and to Grassmann. See also E. Kötter, "Die Entwickelung der synthetischen Geometrie," in *Jahresbericht der Deutschen Mathematikervereinigung*, **5**, pt. 2 (1901), 1–486.

MICHAEL J. CROWE

MÖBIUS, KARL AUGUST (*b.* Eilenburg, Germany, 7 February 1825; *d.* Berlin, Germany, 26 April 1908), *zoology.*

The son of Gottlob Möbius, a wheelwright, and the former Sophie Kaps, Möbius was trained as an elementary school teacher at the private training college in Eilenburg, and from 1844 to 1849 he taught at Seesen, in the Harz Mountains. His strong interest in science led him, despite difficulties, to Berlin, where he had to pass the *Reifeprüfung* (certificate examination) to enter the university. He studied natural science until 1853 under Johannes Müller, C. G. Ehrenberg, Eilhard Mitscherlich, E. H. Beyrich, and the zoologist A. A. H. Lichtenstein; for a time he was the latter's assistant. Inspired by Humboldt's writings, Möbius hoped to join scientific expeditions to the tropics. Since it appeared that the first opportunity would arise in Hamburg, he took a position there in 1853 as a teacher of natural science at the Johanneum grammar school. In 1855 he married Helene Meyer, sister of the zoologist and philosopher Jürgen-Bona Meyer. He soon joined the administration of the Hamburg Museum of Natural History and in 1863 was a cofounder of the Hamburg zoo. Möbius was responsible for the construction of Germany's first public aquarium.

From 1860 Möbius carried out regular investigations of the fauna of the Kieler Bucht; the first volume of his *Fauna der Kieler Bucht* appeared in 1865. In the introduction to this work he set forth a program and methodology for modern ecology. The topography and variations in depth, the plant and animal life of the Kieler Bucht were characterized. The concept of "life community" ("Lebensgemeinschaft" or "Biocönose") was introduced, although Möbius did not define it more precisely until 1877. Through his scientific connections, Möbius was appointed to the chair of zoology at the University of Kiel in 1868. The following year, on a commission from the Prussian government, he traveled along the coasts of France and England to investigate the possibility of promoting mussel and oyster breeding on the German coasts. During the succeeding years he took part in further marine biological research expeditions to the North Sea and Baltic Sea. In 1874 Möbius' wish to visit the tropics was fulfilled; he joined an expedition to Mauritius and the Seychelles, during which he studied chiefly marine fauna and coral reefs. His *Die Auster und die Austernwirtschaft* (1877) contains a clear definition of the concept of "Biocönose."

In 1881 a zoology institute built according to Möbius' plans was opened at Kiel. Its museum was for decades considered a model for such establishments. Möbius left Kiel in 1887 to become director of the new natural history museum in Berlin. He considered his chief task there to be the creation of a large and impressive collection. In 1901 he presided over the International Congress of Zoology at Berlin.

Möbius' scientific work was extensive and very broad in scope. The major portion was devoted to marine biology, including applied research on invertebrates and fishery biology. He also studied the formation of pearls and investigated the biology and anatomy of the whale. Of special scientific importance are his studies on the Foraminifera and the related discovery that the *Eozoon canadense*, which had been considered a living creature, is actually a mineral aggregate. Also of value are Möbius' works on species and the theory of evolution, on animal geography and nomenclature, on animal psychology, on the administration of museums, and on ornithology.

BIBLIOGRAPHY

A complete bibliography is included in Friedrich Dahl, "Karl August Möbius. Ein kurzes Lebensbild, nach authentischen Quellen entworfen," in *Zoologische Jahrbücher*, supp. no. 8 (1905), 1–22, with four potraits. See also obituaries by R. von Hanstein, in *Naturwissenschaft-*liche Rundschau, **23** (1908), 361–373; H. Conwentz, in *Schriften der Naturforschenden Gesellschaft in Danzig*, n.s. **12** (1909), xviii–xx; and C. Matzdorff, in *Monatshefte für den Naturwissenschaftlichen Unterricht*, **2** (1909), 433–448; and L. Gebhardt, *Die Ornithologen Mitteleuropas* (Giessen, 1964).

HANS QUERNER

MOCIÑO, JOSÉ MARIANO (*b.* Temascaltepec, Mexico, 24 [?] September 1757; *d.* Barcelona, Spain, 19 May 1820), *botany.*

In botanical literature Mociño is the generally accepted spelling. The owner of the name always signed it Moziño. On the title page of *Noticias de Nutka* it is written Moziño Suarez de Figueroa. His mother's name was Manuela Losada, and the name under which his degree of bachelor of medicine was conferred was José Mariano Moziño Suares Losada. Nineteenth-century authors wrote the name Mocinno, Moçino, Mozino, or Mozinno.

Mociño studied for a career in theology, philosophy, and history, then about 1784 turned to the natural sciences. After medical training at the University of Mexico, he became committed to botany, and in March 1790 he joined the Royal Botanical Expedition to New Spain, which under the direction of Martín Sessé had been exploring in Mexico since 1787. He continued as a member of the expedition until its effective termination in 1804, traveling to western Mexico (1790–1791), to the coast of California and Nutka Island (1792–1793), the Atlantic slope of Mexico (1793–1794), and Central America (1795–1799).

When the period of exploration came to an end (1803), Mociño and Sessé went to Spain to complete their work and to get support for a new *Flora Mexicana*, to be based on their collections and the approximately 1,400 paintings made by the expedition's artists, Athanasio Echeverría and Vicente de la Cerda. The Napoleonic government then in power in Spain did not support the *Flora*; Sessé died in 1808; Mociño assumed responsibility for the manuscripts and paintings, and when he was forced to leave Madrid with the retreating French (1812), he carried a part of the material with him to Montpellier, where he worked with the botanist Augustin-Pyramus de Candolle. Most of the manuscripts of the *Flora Mexicana* were lost before they came into Candolle's hands, but most of the paintings were saved and some of them formed the bases for almost three hundred new species of plants described by Candolle. Mociño returned to Spain, probably in 1820, and died the same year.

The *Plantae Novae Hispaniae* (1887–1891) and the *Flora Mexicana* (1891–1897), two posthumous volumes, together comprise almost the sum of the original publication which resulted from the Royal Botanical Expedition. The names of Sessé and Mociño are commonly linked (and in that order) in any mention of the botanical work of the expedition, but their contributions seem to have been quite different. Both were competent and active botanists as shown by their existing analyses and descriptions of plants according to the Linnaean method. Sessé was the more competent administrator, with numerous responsibilities and an enormous amount of paper work, and he seems to have delegated much of the purely botanical work to Mociño. The latter was charged, for example, with the preparation of *Plantae Novae Hispaniae*—the entire manuscript is in his handwriting—which he completed in a little over a year after joining the group. The archives at the Instituto Botánico in Madrid contain various inventories of paintings and specimens summarizing the botanical activities of the expedition; which inventories are also for the most part in Mociño's hand. The herbarium of Sessé and Mociño, which is also at Madrid, contains much internal evidence that Mociño began and attempted to carry on some final organization of the specimens leading to publication of a flora of New Spain. That Mociño was a scholar—neither merely a collector nor a menial assistant—is attested by the opinions of his contemporaries and by his surviving reports on his expeditions to Nutka and to the Volcán de Tuxtla in Veracruz. He had some facility with languages; he wrote Latin well, and when in Nutka he soon learned the language of the aborigines well enough to serve as the interpreter for the Spanish party.

The surviving remains of the abortive Botanical Expedition include more than 10,000 herbarium specimens, some 1,300 paintings, and a mass of sorted and unsorted manuscript material. Through Mociño's efforts the paintings came to play a part in the development of nineteenth-century botany, and twentieth-century interpretation of the collections and other documents, including the paintings, has been made possible largely through the manuscripts that Mociño compiled systematically while he was in Mexico.

BIBLIOGRAPHY

I. Original Works. Two vols., attributed to Sessé and Mociño jointly, were published in Mexico between 1887 and 1897. These appeared first in pts., as supps. to the periodical *La Naturaleza*. *Plantae Novae Hispaniae* (1887–1891), 1–184, I–XIII, was based on a MS written by Mociño, completed at Guadalajara, Jalisco, forwarded from there to the Viceroy, the Conde de Revilla-Gigedo, in July 1791, and now in the archives of the Instituto Botánico "A. J. Cavanilles," Madrid. It is a complete flora, including the species of flowering plants studied by the Botanical Expedition up to about the beginning of 1791. A 2nd ed. was published in bk. form in 1893.

Flora Mexicana (1891–1897, pp. I–XI [intro.], 1–263, and I–XV [index]) was based on a very heterogeneous sers. of nn. on individual plant species from many parts of Spanish America. These nn. comprised a part, but by no means all, of those prepared by the members of the Botanical Expedition. The nn. were found in no particular order in the archives in Madrid; they were organized by the editor into the Linnaean classes and published without careful study or collation. A 2nd ed. was published in bk. form in 1894, before the later pts. of the 1st ed. appeared in *La Naturaleza*.

Mociño's own writings are listed by Rickett (v. inf.). The most important are the *Noticias de Nutka*, first published in *Gazeta de Guatemala* (1803–1804), then in bk. form by Alberto M. Carreño, ed. (Mexico, 1913), I–CIX, 1–117. This account of Nutka includes descriptions of the island itself and its inhabitants, comments on the then prevailing political situation, and a vocabulary of about five hundred words. Included in the 1913 vol. is Mociño's report on his ascent of the Mexican volcano of Tuxtla "Descripción del Volcán de Tuxtla," in *Noticias de Nutka* (Mexico, 1913), 103–117.

Original letters, memoranda, and other documents relative to the Botanical Expedition to New Spain and the Royal Botanical Garden in Mexico, are to be found in the Mexican National Archives, the *Archivo General de la Nación*, in the section *Historia*, vols. 460–466, 527. A few documents apparently of similar origin are in the William L. Clements Library, University of Michigan, Ann Arbor. The richest source of MS material in Spain is the archive of the Instituto Botánico "A. J. Cavanilles," Madrid; here are most of the existing MSS having to do with strictly botanical matters, for example, the MS of *Plantae Novae Hispaniae*, various botanical descriptions, fragments of unpublished floras including a *Flora guatemalensis* by Mociño, inventories of paintings and collections from the different excursions which were carried out in Mexico. Descriptions or copies of most of these inventories have been published by Arias Divito or in the papers cited by him. Arias Divito also lists (p. 307) the other major sources of MS material in Madrid and Seville.

II. Secondary Literature. An extensively documented account of Sessé, Mociño, and their co-workers, based primarily on materials in the *Archivo General de la Nación*, Mexico, is H. W. Rickett, "The Royal Botanical Expedition to New Spain," in *Chronica botanica*, **11** (1947), 1–86. Juan Carlos Arias Divito, *Las expediciones científicas españolas durante el siglo XVIII* (Madrid, 1968), is based primarily on Spanish archival sources; it includes copies of many previously unpublished inventories of plants, animals, and paintings, and a considerable bibliog. that supplements the references cited by Rickett. Additional information, especially relative to the members of the Malaspina Expedition who were in Mexico at the same time as the Royal Botanical Expedition, may be found in Iris Higbie Wilson, "Scientific Aspects of Spanish Exploration in New Spain During the Late Eighteenth Century," (Ph.D. diss., Univ. of Southern California, 1962).

The original account of the intercourse between Mociño and Candolle and the story of the copying of some 1,200 paintings by 120 artists in ten days is told in *Mémoires et souvenirs de Augustin-Pyramus de Candolle* (Geneva, 1862), 219–221, 288–290.

ROGERS MCVAUGH

MOENCH, CONRAD (*b*. Kassel, Germany, 20 August 1744; *d*. Marburg, Germany, 6 January 1805), *botany*.

Moench, the son of a pharmacist, worked for six and a half years in pharmacies in Hannover, Bern, and Strasbourg before returning to take charge of his family's pharmacy, the Apotheke zum Einhorn, in Kassel. He had studied pharmacy, botany, chemistry, and mineralogy; in 1781 he became professor of botany at the Collegium Medicum Carolinianum attached to the court of the landgrave of Hesse-Kassel at Kassel. In 1785 the college and its professors were transferred to Marburg and incorporated into the Philipps-Universität, which had been founded there in 1527.

Moench initiated a new period in the teaching of botany at Marburg. Under his supervision, a site for a botanic garden was prepared in 1786; the following year it was planted with material brought from the Kassel botanic garden. Strongly influenced by the Mannheim botanist C. F. Medicus, a bitter opponent of Linnaeus, Moench likewise frequently rejected Linnaean classification, generic concepts, and nomenclature, restoring the names and genera of Tournefort wherever possible, and often subdividing Linnaean genera. His rebellion against the dominant Linnaean taxonomy of the period found expression in the *Methodus plantas horti botanici et agri Marburgensis a staminum situ describendi*, published at Marburg in 1794. The *Methodus* deals with 674 species, including both those cultivated in the Marburg botanic garden and those growing wild in the Marburg district. In it Moench used names at variance with those of most floristic works of the time. A *Supplementum ad Methodum plantas* of 1802 added 634 more flowering plants.

Moench's works remain nomenclaturally important because, although they include many generic names that have never been adopted elsewhere, they also comprise a number of generally accepted ones, among them *Bergenia, Cedronella, Froelichia, Kniphofia,*

Myosoton, Olearia, and *Sorghum*. Moench also provided various binomial specific names needed to avoid tautonyms—*Fagopyrum esculentum* and *Omphalodes verna*, for example—although most of his new names are now considered technically illegitimate. In dealing with the *Leguminosae*, Moench, a follower of Medicus, overlooked an important paper published by the latter in 1787 and thus renamed plants already named by him.

BIBLIOGRAPHY

I. ORIGINAL WORKS. Moench's most important works, *Methodus plantas horti botanici et agri Marburgensis a staminum situ describendi* (Marburg, 1794) and *Supplementum ad Methodum plantas* (Marburg, 1802), are repr. in facsimile with an intro. by W. T. Stearn, *Early Marburg Botany* (Königstein, 1966).

II. SECONDARY LITERATURE. On Moench and his work, see F. Grundlach, *Catalogus professorum academiae Marburgensis* (Marburg, 1927); C. Rommel, *Memoriam Conradi Moench* (Marburg, 1805); R. Schmitz, "Naturwissenschaft an der Universität Marburg," in *Sitzungsberichte der Gesellschaft zu Beförderung der gesamten Naturwissenschaften zu Marburg*, **83–84** (1963), 1–33, with portrait; and W. T. Stearn, *Early Marburg Botany* (cited above), for the nomenclature and concepts of Medicus and Moench in opposition to those of Linnaeus, as well as for their nomenclatural conflict with each other.

WILLIAM T. STEARN

MOERBEKE, WILLIAM OF, also known as **Guillelmus de Moerbeka** (*b*. Moerbeke, Belgium [?], *ca*. 1220–1235; *d*. before 26 October 1286), *philosophy, geometry, biology*.

Moerbeke, a Dominican, was one of the most productive and eminent translators from Greek into Latin of philosophical and scientific works written between the fourth century B.C. and the sixth century A.D. A spectacular widening and increase of the Greek sources for study and speculation in the second half of the thirteenth century and later times were due to Moerbeke's insatiable desire to pass on to Latin-reading students the yet undiscovered or rediscovered treasures of Greek civilization, his extensive linguistic knowledge, his indefatigable search for first-class works, and his philosophical vision.

There is little evidence concerning Moerbeke's life apart from some names of places and dates at which he produced a particular translation: they are enough, however, to suggest reasons why he did not have much time to write original works. He was at Nicaea, Asia Minor, in the spring of 1260, and in Thebes—where Dominicans had been present at least since 1253—in December of that same year. He was at

Viterbo, then a papal residence, in November 1267, May 1268, and June 1271. From 1272, at the latest, until April 1278 Moerbeke held the office of chaplain and penitentiary to the pope: in this capacity he visited the courts of Savoy and France pleading for help with the Ninth Crusade (March 1272), absolved an Augustinian prior from excommunication (1272, from Orvieto, seat of the papal curia), and authorized Albertus Magnus to absolve two abbeys in Cologne from censures (November 1274). In the same period (May–July 1274) he took part in the Second Council of Lyons, which was meant to bring about the reunion with the Greek Church; there, with Greek dignitaries, he sang the Creed in Greek in a pontifical mass. In October 1277 he was active at Viterbo. From April 1278 until his death he was archbishop of Corinth.

The three or four contemporaries with whom it is definitely known that Moerbeke had some contact were all scientists. The Silesian Witelo, who was in Viterbo toward the end of 1268, dedicated his *Perspectiva* to Moerbeke. In the introduction Witelo sheds some light on Moerbeke's philosophical doctrines and explains that they were never put in writing because Moerbeke was kept too busy by his ecclesiastical and pastoral duties and by his work as a translator. Henry Bate of Malines, a distinguished astronomer, was asked by Moerbeke, whom he met at Lyons in 1274, to write a treatise on the astrolabe; Henry immediately obliged and dedicated his *Magistralis compositio astrolabii* to his compatriot (October 1274). The physician Rosellus of Arezzo, who may have attended Pope Gregory X at his deathbed in Arezzo (1276), is the addressee of Moerbeke's dedication of his version of Galen's *De alimentis* (Viterbo, October 1277). Finally, some evidence seems to suggest that Moerbeke met the mathematician and astronomer Campanus of Novara at the papal curia.

Moerbeke may well have been in touch with Aquinas at or near Rome before 1269 or between 1271 and 1274, but there is no reliable direct evidence of any personal relationship. It is a commonplace, repeated *ad nauseam* by almost all historians and scholars concerned with either Aquinas or Moerbeke, that the latter was prompted by the former to undertake his work as a translator, especially as a translator of Aristotle. This is most probably nothing more than a legend originating in hagiography, when "evidence" was offered by William of Tocco, a confrère of Aquinas, for the latter's canonization, about forty years after his death. What remains true is that Aquinas, like other philosophers of his time, used some—by no means all—of Moerbeke's translations soon after they were made.

Works. Only one original work by Moerbeke is preserved, under the title *Geomantia* ("Divination From Earth"). It was dedicated to his nephew Arnulphus and seems to have been quite popular: several manuscripts in Latin and one manuscript of a French translation made in 1347 by Walter of Brittany are still extant, but the treatise does not seem to have been studied by modern scholars. The authenticity of the attribution has been doubted on the ground that a "faithful follower of Aquinas" could not have written a treatise on matters condemned by the master, but the premise is unfounded. There is no reason to believe that the *Geomantia* is a translation from the Greek or Arabic. Witelo's evidence strongly supports the evidence for the attribution found in the manuscripts. Addressing himself to Moerbeke in the introduction to his *Perspectiva*, he says:

> As an assiduous investigator of the whole of reality, you saw that the intelligible being which proceeds from the first principles is connected in a causal way with individual beings; and when you were inquiring into the individual causes of these individual beings, it occurred to you that there is something wonderful in the way in which the influence of divine power flows into things of the lower world passing through the powers of the higher world . . .; you saw that what is acted upon varies not only in accordance with the variety of the acting powers, but also in accordance with the variety of the modes of action; consequently you decided to dedicate yourself to the "occult" inquiry of this state of affairs.

Preliminary studies of the *Geomantia* have shown a vocabulary consistent with Moerbeke's translations; explicit mentions of the causal chain from God through the heavens to events on earth, of the occult nature of at least some part of geomancy, as indicated by Witelo; and attribution to "Frater Guillelmus de Moerbeka domini pape penitentiarius." MCCCLXXXVII in some manuscripts should be MCCLXXVI because of Moerbeke's death date, Witelo's statements, and the description "penitentiarius" and not "archiepiscopus Corinthiensis."

We do not know whether Moerbeke wrote any other original works; but we still possess many, if not all, of the translations which he made from the Greek. He undertook this activity, he says, "in spite of the hard work and tediousness which it involves, in order to provide Latin scholars with new material for study" and "in order that my efforts should add to the light to which Latins have access." His knowledge of the Greek language, perhaps scanty when he first embarked on translations, improved greatly in the course of the more than twenty years which he devoted partly to them. In this field Moerbeke was a very exacting scholar and philologist, comparable only, in the thirteenth century, with Robert Grosseteste. The unfavorable criticism brought against his versions—even against his latest and best—does not take into account two facts: first, that Greek scholarship among Latins in the thirteenth century was not the product of a long tradition and well-organized schools but the hard-won possession of isolated individuals; and second, that a very sound philosophy of language, accompanied by the need for detailed, literal interpretation of authoritative texts—biblical, legal, scientific, philosophical—required that translations should be strictly faithful, word by word, to the original. Within these limits Moerbeke was often excellent, although, like all translators, he made mistakes. He was meticulous in his quest for exactitude; he would search the Latin vocabulary with a sound critical sense and great knowledge, in order to find words which could convey to the intelligent reader the meaning of the Greek terms. If his search failed to produce the necessary results, he would form new Latin words by compounding two terms, or adding prefixes and suffixes on the Greek pattern, or even combining Greek and Latin elements: a typical example would be his rendering of αὐτοκίνητον by "automobile." In extreme cases he would resort to that great source of enrichment of a language, the transliteration, with slight adaptations, of foreign—in his case Greek—words. A test of Moerbeke's care in trying to pass on as much as possible of Aristotle's "light" can be found in his revisions, based on Greek manuscripts, of translations produced by such scholars as Boethius and James of Venice: in most cases where he introduced a change, a misinterpretation was put right, a serious mistake corrected, or a more appropriate shade of meaning introduced if his predecessor had missed a finer point. His scientific attitude toward language is also revealed by his attempt to reproduce the exact Greek sounds (mainly those of Byzantine Greek) in his transliteration of names or of newly introduced technical terms, for instance, by using "kh" for the Greek χ.

Moerbeke applied his interest and gifts as a translator to four aims: (a) completing and improving the Latins' knowledge of Aristotle's works in all their encyclopedic extent; (b) making available to Latin readers some of the most valuable and comprehensive elaborations of Aristotle's treatises on logic, philosophy of nature, and psychology which had been written between about A.D. 200 and 550; (c) propagating the doctrines of Proclus, the greatest systematizer of Neoplatonic philosophy, on which Moerbeke's own philosophy so much depended; (d) introducing into the Latin West a more exact and extensive knowledge

of Archimedes' achievements in mathematics and physics. He also contributed in a smaller measure to the knowledge of Greek medical literature and of works by Ptolemy, Hero, Alexander of Aphrodisias, and Plato.

Evidence for Moerbeke's authorship of the versions ascribed to him varies in strength. In a number of instances—all concerning works by Aristotle—he only revised, more or less thoroughly, versions made by earlier scholars. In the following survey two asterisks indicate the titles of translations for which the evidence of authorship is direct (the translator's name accompanying the text itself); a single asterisk indicates the titles of versions for which evidence is elicited from a linguistic analysis. Dates and places where the translations were made are given only in the relatively few instances for which the evidence is found at the end of the translation itself. Within square brackets is a short indication of the best existing edition, whenever this has been ascertained, or of one old edition (original or a reprint). Some additional information on editions will be found in the bibliography.

The translations or revisions so far identified and ascribed with some degree of probability to Moerbeke are the following:

I. Plato (see Proclus).

II. Aristotle.

(1) Works never before translated into Latin:
　　**Politica*, *one version of bks. I–II.11 [P. Michaud-Quantin, Bruges, 1961] and **one complete [F. Susemihl, Leipzig, 1872; bks. I–III.8, with Aquinas' commentary, H. F. Dondaine and L. Bataillon, Rome, 1972]. .
　　Poetica, 1278 [L. Minio-Paluello, Brussels, 1968].
　　**Metaphysica*, bk. XI [Venice, 1562, and with Aquinas' commentary].
　　De motu animalium [L. Torraca, Naples, 1958].
　　De progressu animalium [unpublished].

(2) Works never before translated into Latin from the Greek:
　　Historia animalium [bk. I, G. Rudberg, Uppsala, 1908; bk. X.6, Rudberg, Uppsala, 1911; bks. II–IX unpublished].
　　**De partibus animalium*, Thebes, 1260 [unpublished].
　　De generatione animalium, two recensions [H. J. Drossaart Lulofs, Bruges, 1966].
　　**Meteorologica*, bks. I–III [with Aquinas].
　　De caelo, bks. III–IV [with Aquinas].

(3) Works of which Latin translations from the Greek already existed and which were translated anew by Moerbeke:

Categoriae, 1266 [L. Minio-Paluello, Bruges, 1961; also (first half) A. Pattin, Louvain, 1971].
De interpretatione, 1268 [G. Verbeke and L. Minio-Paluello, Bruges, 1965].
**Meteorologica*, bk. IV [with Aquinas' commentary].
De caelo, bks. I–II [with Aquinas' commentary].
**Rhetorica*, *first recension [unpublished]; **second recension [L. Spengel, Leipzig, 1867].

(4) Works translated from the Greek by other scholars and revised by Moerbeke:

Analytica posteriora, translated by James of Venice [L. Minio-Paluello and B. G. Dod, Bruges, 1968].
De sophisticis elenchis, translated by Boethius [B. G. Dod and L. Minio-Paluello, in press].
Physica, translated by James of Venice [with Aquinas' commentary].
De generatione et corruptione, translated by an unknown scholar [with Aquinas].
De anima, translated by James of Venice [with Aquinas].
Parva naturalia, translated by various scholars: *De sensu* and *De memoria* [with Aquinas]; *De somno et vigilia* [H. J. Drossaart Lulofs, n.p., 1943]; *De insomniis et De divinatione* [H. J. Drossaart Lulofs, Leiden, 1947]; *De longitudine*, *De iuventute*, *De morte*, *De respiratione* [Venice, 1496].
De coloribus, incomplete, translated by an unknown scholar [E. Franceschini, Louvain, 1955].
**Metaphysica*, bks. I–X and XII–XIV [Venice, 1562] and I–X, XII [with Aquinas].
Ethica Nicomachea, translated by Robert Grosseteste [R.-A. Gauthier, Leiden–Brussels, 1973].

III. Commentators on Aristotle.

(1) Alexander of Aphrodisias:

**In meteorologica*, Nicaea, 1260 [A. J. Smet, Louvain, 1968]; *In De sensu* [C. Thurot, Paris, 1875].

(2) Themistius:

In De anima, Viterbo, 1267 [G. Verbeke, Louvain, 1957].

(3) Ammonius:

In De interpretatione, 1268 [G. Verbeke, Louvain, 1961].

(4) Philoponus:

In De anima, bks. I.3 and **III.4–9, 1268 [G. Verbeke, Louvain, 1966].

(5) Simplicius:

In categorias, 1266 [Venice, 1516; also (first half) A. Pattin, Louvain, 1971]; **In De caelo*, Viterbo, 1271 [Venice, 1540].

IV. Proclus.

**Elementatio theologica*, Viterbo, 1268 [C. Vansteenkiste, 1951].
**De decem dubitationibus*, **De providentia et fato*, **De malorum subsistentia*, Corinth, 1280 [H. Boese, Berlin, 1960].
In Platonis Parmenidis priorem partem commentarium, including Plato's *Parmenides* as far as 142A (shortly before 1286; authenticated by Henry Bate) [extensive sections edited by V. Cousin, Paris, 1820; last section, lost in Greek, and Plato's text edited by R. Klibansky and L. Labowsky, London, 1953].
In Platonis Timaeum commentarium, extracts, containing also a few passages from Plato's *Timaeus* [G. Verbeke, Louvain, 1953].

V. Alexander of Aphrodisias (see also above, under Commentators on Aristotle).

De fato ad imperatores and *De fato*, which is *De anima*, bk. II (authorship authenticated by the surviving Greek manuscript owned by Moerbeke and carrying his autograph title of possession) [P. Thillet, Paris, 1963].

VI. Archimedes.

De quam pluribus theorematibus, which is *De lineis spiralibus*, 1269 [J. L. Heiberg, Leipzig, 1890].
De centris gravium, which is *De planis aeque repentibus*, 1269 [N. Tartaglia, Venice, 1543].
Quadratura parabolae, 1269 [L. Gauricus, Venice, 1503].
Dimensio circuli, 1269 [L. Gauricus, Venice, 1503; partly, J. L. Heiberg, Leipzig, 1890].
De sphaera et cylindro, 1269 [introduction to bks. I and II, J. L. Heiberg, Copenhagen, 1887, and Leiden, 1890, respectively].
De conoidalibus et sphaeroidalibus, 1269 [unpublished].
De insidentibus aquae, 1269 [bk. I, N. Tartaglia, Venice, 1543; both books, Curtius Troianus, Venice, 1565; collations by J. L. Heiberg, Leipzig, 1890; several sections edited by Heiberg, Leipzig, 1913].

VII. Commentator on Archimedes and Eutocius.

On the De sphaera et cylindro, 1269 [a small section edited by J. L. Heiberg, Leipzig, 1890]; *On the De centris gravium*, 1269 [unpublished].

VIII. Hero of Alexandria.

Catoptrica, which is Pseudo-Ptolemy, *De speculis*, 1269 [W. (G.) Schmidt, Leipzig, 1901].

IX. Ptolemy.

De Analemmate, 1269 [J. L. Heiberg, Leipzig, 1907].

X. Galen.

**De alimentis*, Viterbo, 1277 [Venice, 1490].

XI. Pseudo-Hippocrates.

**De prognosticationibus aegritudinum secundum motum lunae*, which is (?) *Astronomia* [Padua, 1483].

Some Latin translations attributed at different times to Moerbeke have been proved or can be proved not to be by him. Among them are the Pseudo-Aristotelian *Rhetorica ad Alexandrum* (by Anaximenes of Lampsacus) and *Oeconomica* and Hero of Alexandria's [?] *Pneumatica* or *De aquarum conductibus*.

Influence. Moerbeke's influence can be assessed from different points of view.

1. The popularity of many of his Aristotelian translations is evidenced by the surviving manuscripts of the thirteenth to the fifteenth centuries; printed editions of the fifteenth, sixteenth, and later centuries; and versions or adaptations into French, English, Greek, and Spanish made in the fourteenth, sixteenth, nineteenth, and twentieth centuries. From about 100 to nearly 300 manuscripts and up to a dozen printed editions exist of works which had never before been translated into Greek, or older translations of which had been superseded or revised. This means that these works became accessible—and to a large extent comprehensible—to most Latin-reading students of philosophy, and that the philosophical language adopted by Moerbeke (and in some cases his interpretations) has influenced philosophical culture since the thirteenth century.

2. The introduction into the Western Latin world—and the consequent extensive study in universities and ecclesiastical and monastic schools, or the less extensive but still very influential study by specialists—of works of Aristotle, Proclus, and Archimedes which had been practically lost sight of for several centuries. An extreme example is provided by Aristotle's *Politica*, which had never been the object of more than exceptional study—and that only among Greeks before the fifth and in the eleventh century—and was almost discovered for the world at large, and introduced as one of the basic classics of

political thought, by Moerbeke through his translation. Again, it was through Moerbeke's translations that some of Proclus' works were taken up by eminent philosophers and became essential ingredients of philosophical outlooks which affected the background and contents of some of the great schools of thought of the later Middle Ages, Renaissance, and more recent times. This is particularly true for the *Elementatio theologica*, in which Moerbeke discovered the original text of those propositions which formed the nucleus of the *De causis*, possibly the most influential carrier of Neoplatonic doctrines, transmitted via the Arabic to the Latin schools of the late twelfth and following centuries; it is also true of Proclus' *In Platonis Parmenidis . . . commentarium* and *De providentia et fato*. A similar influence was exerted by the translations of Themistius' and Philoponus' commentaries on the *De anima*, of Simplicius' on the *Categoriae* and *De caelo*, and of Ammonius' on the *De interpretatione*. The importance of Moerbeke's translations of Archimedes has been sketched in a masterly way by M. Clagett in his article on Archimedes in vol. I of this Dictionary.

3. A better knowledge of the actual Greek texts of several works came about through Moerbeke's versions. In a few cases they are the only evidence for lost Greek texts (the whole of Hero's *Catoptrica* and Pseudo-Hippocrates' *De prognosticationibus secundum motum lunae*; an important section of Proclus' *Commentary on the Parmenides*; and some sections of Proclus' smaller treatises and of Archimedes' *De insidentibus aquae*). Apart from two instances, Moerbeke used for his translations manuscripts now lost or not yet identified; on many points some of them provide us with better evidence of the Greek originals than the known Greek manuscripts: this is especially the case for Aristotle's *Politica*. For every single work Moerbeke's translations add to our knowledge of the tradition and history of the Greek texts.

BIBLIOGRAPHY

I. ORIGINAL WORKS. Abundant bibliographical information can be found in G. Lacombe, L. Minio-Paluello *et al.*, *Aristoteles Latinus, Codices: Pars prior* (Rome, 1939; Bruges, 1957), 21–38; *Pars posterior et supplementa* (Cambridge, 1955), 773–782, 1277; and *Supplementa altera* (Bruges, 1961), 7–17. See also M. Grabmann, *Guglielmo di Moerbeke O.P., il traduttore delle opere di Aristotele*, vol. II of *I papi del duecento e l' Aristotelismo*, Miscellanea Historiae Pontificiae, XI (no. 20) (Rome, 1946), *passim*; and in the relevant sections of *Bulletin thomiste* and *Bulletin de théologie ancienne et médiévale*.

The existing MSS of translations from Aristotle and his commentators are listed and described in the 3 vols. cited above of *Aristoteles Latinus, Codices*; those of Archimedes in V. Rose, *Deutsche Literaturzeitung* (1884), 210–213, and J. L. Heiberg, "Neue Studien zu Archimedes," in *Abhandlungen zur Geschichte der Mathematik*, **5** (1890), 1–84; for Pseudo-Hippocrates, see H. Diels, "Die Handschriften der antiken Aerzte I," in *Abhandlungen der Preussischen Akademie der Wissenschaften* (1905). Most scholarly eds. contain additional information on the MSS. No survey was made of the printed eds. of Latin versions of Aristotle, accompanied or not accompanied by commentaries by Ibn Rushd, Aquinas, or others. The *Gesamtkatalog der Wiegendrucke* contains, under "Aristoteles" and the names of commentators, short descriptions of most eds. printed, or thought to have been printed, before 1501 but does not give sufficient identifications of the authors of the translations. The oldest eds. of some of Moerbeke's Aristotelian translations are found in fifteenth- and sixteenth-century printed texts. The eds. of Aristotelian writings mentioned in the text and carrying dates later than 1960 are contained in the series Aristoteles Latinus (part of the Corpus Philosophorum Medii Aevi); those of Aristotelian commentaries cited in the text dated 1957 and later are part of the Corpus Latinum Commentariorum in Aristotelem Graecorum; and most of those mentioned as being "with Aquinas" are found in the modern critical ed. of Aquinas' works, the Leonine ed. (Rome, 1882–   ). A critical ed. of all the translations from Archimedes and Eutocius based on Moerbeke's autograph is in vol. II of M. Clagett, *Archimedes in the Middle Ages*. The MSS of the *Geomantia* known to date are listed in L. Thorndike and P. Kibre, *A Catalog of Incipits of Mediaeval Scientific Writings in Latin* (1963).

II. SECONDARY LITERATURE. By far the most exhaustive study of Moerbeke's life and the best collection of evidence, information on the works which he translated and on the opinions expressed on them through the centuries, and references to modern scholarly studies is Grabmann's *Guglielmo di Moerbeke, il traduttore . . .* (cited above). This work suffers, however, from the wartime circumstances in which the material was being assembled and from the fact that it was left to be edited and translated into "Germitalian" by rather incompetent hands; the misprints affecting essential data are far too numerous. Its extreme bias in favor of Aquinas' and the popes' share in providing Moerbeke with the initiatives which were in fact his own is all-pervasive and misleading. Among the older works mention should be made of the article on Moerbeke in I. Quétif and I. Échard, *Scriptores ordinis praedicatorum*, I (Paris, 1791), 388–391. The best modern, concise, and critical survey listing Moerbeke's translations and their more important eds. is in P. Thillet's version of Alexander of Aphrodisias' *De fato* (cited in text); unfortunately, he ascribes to Moerbeke more recent Latin translations of Archimedes and Eutocius. For Moerbeke's early stay in Greece, see O. van der Vat, *Die Anfänge der Franziskaner Mission . . . im nahen Orient . . .* (Werl, 1934); B. Altaner's two long articles on the missionaries'

linguistic knowledge in the Middle East, in *Zeitschrift für Kirchengeschichte*, **53** (1934) and **55** (1936); V. Laurent, "Le Pape Alexandre IV et l'empire de Nicée," in *Échos d'Orient*, **38** (1935), 26–55; K. M. Setton, "The Byzantine Background to the Italian Renaissance," in *Proceedings of the American Philosophical Society*, **100** (1956), esp. 31–35.

On Witelo and Bate, see C. Baeumker, *Witelo, ein Philosoph . . .*, III, pt. 2 of *Beiträge zur Philosophie des Mittelalters* (Münster, 1908); G. Wallerand, "Henri Bate de Malines et Thomas d'Aquin," in *Revue néoscolastique de philosophie*, **36** (1934), 387–410, and his ed. of the first part of Bate's *Speculum, Les philosophes belges*, XI, pt. 1 (Louvain, 1931). On the question of Aquinas' influence on Moerbeke, the best critical assessment is in R.-A. Gauthier's intro. to *Sententia libri ethicorum*, I, which is vol. XLVIII of *S. Thomae de Aquino opera omnia* (Rome, 1969), 232*–235*, 264*–265*. A page from Moerbeke's holograph of his Archimedes is reproduced in B. Kattenbach *et al.*, *Exempla scripturarum*, II (Rome, 1929), pl. 20; and his autograph inscription of property of the Greek MS of Alexander's *De fato* is reproduced in L. Labowsky, "William of Moerbeke's Manuscript of Alexander of Aphrodisias," in *Mediaeval and Renaissance Studies*, n.s. **5** (1961), 155–162.

Studies on Moerbeke's works and on their influence have been directed mainly to aspects of his method as a translator, particularly to his vocabulary, often for the purpose of ascertaining or suggesting his authorship or of distinguishing his version from those by other scholars. Apart from the extensive "indices verborum" which accompany most modern eds. of his texts (esp. the vols. in Aristoteles Latinus and Corpus Latinum Commentariorum Graecorum and the eds. by Drossaart Lulofs and Thillet cited in text as well as the forthcoming vol. II of Clagett's *Archimedes*) and linguistic analyses in some of the introductions (again by Clagett, Lulofs, Thillet, Verbeke, Vansteenkiste, and Rudberg), there are many special inquiries: F. H. Fobes, "Mediaeval Versions of Aristotle's *Meteorology*," in *Classical Philology*, **10** (1915), 297–314; F. Pelster, "Die griechisch-lateinischen Metaphysikuebersetzungen des Mittelalters," in *Beiträge zur Geschichte der Philosophie des Mittelalters*, supp. **2** (1923), 89–118; L. Minio-Paluello, "Guglielmo di Moerbeke traduttore della 'Poetica' d'Aristotele, 1278," in *Rivista di filosofia neoscolastica*, **39** (1947), 1–17; "Henri Aristippe, Guillaume de Moerbeke et les traductions latines médiévales des *Météorologiques* et du *De generatione et corruptione*," in *Revue philosophique de Louvain*, **45** (1947), 206–235; D. J. Allan, "Mediaeval Versions of Aristotle *De caelo* and the *Commentary* of Simplicius," in *Mediaeval and Renaissance Studies*, **2** (1950), 82–120.

Various points and aspects of Moerbeke's influence, the dates of his translations in relation to the dates of works in which use was made of them, and commentaries based on his versions have been the object of scholarly study in many books and articles devoted to wider issues. To those already mentioned one may add D. A. Callus, "Les sources de Saint Thomas," in *Aristote et Saint-Thomas d'Aquin: Journées d'études* (Louvain–Paris, 1957), 93–174;

A. Dondaine, reviews in *Bulletin Thomiste* (1924 ff.); R. A. Gauthier's intro. to Thomas Aquinas, *Contra gentiles, livre premier* (Paris, 1961); B. Geyer, "Die Uebersetzungen der aristotelischen Metaphysik bei Albertus Magnus und Thomas . . .," in *Philosophisches Jahrbuch*, **30** (1917), 392–415; J. Isaac, *Le Peri Hermeneias en Occident de Boèce à Saint Thomas* (Paris, 1953); R. Klibansky, "Ein Proklosfund und seine Bedeutung," in *Sitzungsberichte der Heidelberger Akademie der Wissenschaften*, Phil.-hist. Kl., **5** (1928–1929); and "Plato's *Parmenides* in the Middle Ages and the Renaissance," in *Mediaeval and Renaissance Studies*, **1** (1943), 281–330; H. Lohr, "Mediaeval Latin Aristotle Commentaries," in *Traditio*, **23** (1967 ff.); A. Mansion, "Le commentaire de Saint Thomas sur le *De sensu et sensato* d'Aristote," in *Mélanges Mandonnet* (Paris, 1930), 83–102; C. Martin, "The Commentaries on the *Politics* of Aristotle in the Late Thirteenth and Early Fourteenth Centuries" (D. Phil. thesis, Oxford University, 1949; copy at the Bodleian Library, Oxford); and B. Schneider, *Die mittelalterlichen griechisch-lateinischen Uebersetzungen der Aristotelischen Rhetorik* (Berlin, 1971).

The study of the Greek tradition through Moerbeke's texts has been carried out extensively both in the process of editing the Greek texts—see, for instance, Heiberg's ed. of Archimedes and Eutocius (Leipzig, 1910–1915) and W. L. Newman's ed. of Aristotle's *Politica* (Oxford, 1887) —and as part of the Aristoteles Latinus (every vol. contains the results of this study). Many separate studies were devoted to problems in this field, including E. Lobel, "The Medieval Latin Poetics," in *Proceedings of the British Academy*, **17** (1931), 309–334, and the work of B. Schneider cited above.

LORENZO MINIO-PALUELLO

MOFFETT (MOUFET, MUFFET), THOMAS (*b.* London, England, 1553; *d.* Bulbridge, Wiltshire, England, 5 June 1604), *medicine, entomology.*

Moffett was the second son of a London haberdasher, Thomas Moffett, and Alice Ashley. He was educated at the Merchant Taylors' School and at Cambridge, where he studied medicine under John Caius, with Thomas Penny as a fellow student. After graduating in 1573 he read medicine in Basel, where he received the M.D. in 1578 and published several of his medical theses. He accepted the Paracelsian system of treating disease with drugs and advocated it on his return to England. In 1579 he visited Italy and Spain; he there studied the silkworm, which he later described in his entomology. A poem on the silkworm by "T.M." is usually attributed to him. In 1580 Moffett married his first wife, Jane Wheeler, who died in 1600, and traveled again in Germany before returning to Cambridge in 1582. On a visit to Denmark in 1582 he met Severinus, to whom he dedicated his book *De jure et praestantia chemicorum medicamentorum dialogus apologeticus*, published in Frankfurt in 1584 and widely read on the Continent.

Settling in England, Moffett practiced as a physician in Ipswich for a time, then established himself in London. He became a fellow of the College of Physicians in 1588. His professional attendance on the nobility brought him to the court of Elizabeth I, but the Earl of Pembroke persuaded him to move to Wiltshire, and secured for him a seat as Member of Parliament for Wilton in 1597, which he held until his death in 1604. His second wife, the widow Catherine Brown, survived him with their daughter.

Moffett is remembered today mainly for two works, both published posthumously. The *Theatrum insectorum* which bears his name has a complex history. When Konrad Gesner died in 1565 he left an unfinished book on entomology: this was eventually sold to his friend Thomas Penny, who had already done some work of his own on Gesner's collection of insects. Penny also acquired the notes on insects made by Edward Wotton of Oxford, and made some progress in amalgamating the information before his death in 1589. The work was then rescued from Penny's heir by his Cambridge friend Moffett, who added a number of descriptions and drawings from his own observations in England and on the Continent— including a number of "lesser living creatures," spiders, crustacea, and worms. Moffett prepared a manuscript with a fine title page engraved by William Rogers, with portraits of Moffett himself, Gesner, Wotton, and Penny. The title page was dated 1589; by 1590 Moffett was negotiating for publication in The Hague. That fell through, however, and he was unable to find a printer in England, probably because there was little demand in England for a book on natural history. After Moffett's death his apothecary Darnell sold the manuscript to Sir Theodore Mayerne, who, after having found a printer only with great difficulty, eventually published it in 1634. Since illustrations for the mass market had to be done cheaply, the book appeared with woodcuts and without the original title page. It was translated into English and issued as part of Edward Topsell's *History of Four-Footed Beasts and Serpents* in 1658.

The *Theatrum insectorum* itself is a systematic treatise dealing with the habits, habitat, breeding, and economic importance of insects, beginning with bees, which are accorded the most detailed treatment. The observations and illustrations, although usually thought inferior to those published by Aldrovandi in 1602, are of considerable interest. Moffett usually described the larval and adult forms separately, and was aware that "There are so many kindes of Butterflies as there are of the Cankerwormes." He further discussed the emergence of either normal butterflies or "ordinaries Flyes" (now known to be parasitic wasps) from similar pupae and described both the depredations wrought by locusts in Europe and their use as food. He also recorded observations of the movement of the tongue of the chameleon in feeding.

Moffett's other work was *Health's Improvement*, which was edited by Christopher Bennet and published in 1655. Designed for the layman to a greater degree than were the medical works published in Frankfurt and Nuremberg, the book is concerned mainly with food and diet, but includes descriptions of animals and fishes used for food, and is of particular interest for two chapters of observations about wild birds. While few of these observations were original, together they constitute the first printed list of British birds. Moffett was one of the first to recognize migration in birds and referred to woodcock and snipe "when they have rested themselves after their long flight from beyond the Seas, and are fat"; he also mentioned crane breeding in the fens, a practice that ceased shortly after Moffett's own time. In his *Panzoologia* (1661), Robert Lovell used Moffett's descriptions of both birds and insects extensively.

BIBLIOGRAPHY

I. ORIGINAL WORKS. Of Moffett's two scientific works, the manuscript of *Theatrum insectorum* is in the British Museum, Sloane MS 4014. It was first published as *Insectorum, sive minimorum animalium* (London, 1634); the English trans. by John Rowland was issued as the third vol. of Edward Topsell, *The History of Four-Footed Beasts and Serpents. Whereunto is now added The Theater of Insects; or, lesser Living Creatures: as Bees, Flies, Caterpillars, to Spiders, Worms &c. by T. Muffet* (London, 1658); facs. repr. has a new intro. by Willy Ley (London, 1967). For *Health's Improvement: or, Rules Comprizing and Discovering the Nature, Method and Manner of Preparing all Sorts of Foods Used in this Nation* (London, 1655), a second ed. was "corrected and enlarged by Christopher Bennet . . . to which is now prefix'd, a short view of the author's life and writings by Mr. Oldys, and an introduction by R. James" (London, 1746). Other of Moffett's works are listed in the *Athenae Cantabrigiensis* (below).

II. SECONDARY LITERATURE. The first biography of Moffett is by W. Oldys in the 1746 edition of *Health's Improvement*, pp. vii–xxxii, and includes the text of his will. There is an early evaluation of Moffett's medical ability by John Aikin in *Biographical Memoirs of Medicine in Great Britain* (London, 1780), 168–175. The article by Sidney Lee in *Dictionary of National Biography* includes a comprehensive survey of the manuscript sources for Moffett's life, and gives a short bibliography of early sources. C. H. Cooper and T. Cooper, *Athenae Cantabri-* *giensis, II, 1568–1609* (Cambridge, 1861), 400–402, 554, is a good account of Moffett's life with the most comprehensive bibliography of his works and sources of biographical material, mainly nineteenth century. The chapter on Moffett in C. E. Raven, *English Naturalists from Neckham to Ray* (Cambridge, 1947), ch. 10, 172–191, is critical of Moffett's originality and accuracy, and collates his information with that given by other naturalists.

The history of the *Theatrum insectorum* is in M. Burr, "Unpublished for Over Three Hundred Years: the Original Title-Page for the First Book on Natural History Printed in England," in *Field*, 27 August (1938), 495; H. M. Fraser, "Moufet's Theatrum Insectorum," in *Gesnerus*, **3** (1946), 131; and B. Milt, "Some Explanatory Notes to Mr. H. M. Fraser's Article about Moufet's Theatrum Insectorum," *ibid.*, 132–134. The text of Moffett's notes on ornithology is included in W. H. Mullens, *Thomas Muffett*, Hastings and St. Leonards Natural History Society, Occasional Publication no. 5 (1911).

DIANA M. SIMPKINS

MOHL, HUGO VON (*b.* Stuttgart, Germany, 8 April 1805; *d.* Tübingen, Germany, 1 April 1872), *biology*.

Born into a respected bourgeois family, Mohl had three brothers who gained reputations in scholarship, economics, and politics; his brother Jules, a naturalized Frenchman, was professor of Persian at the Collège de France from 1847 to 1876 (on the Mohl brothers see Vapereau [1870]). Mohl had a classical education; but from an early age he demonstrated a predilection for science, especially for botany and optics, thus early revealing his vocation: it was in the field of microscopic botany that he made his most remarkable contributions. He studied medicine at Tübingen and in 1827 presented a work on the structure and movement of the climbing plants, a problem which concerned botanists throughout the nineteenth century. Mohl's doctoral thesis (1828) was devoted to the constitution of the pores of plants. In 1832 he was appointed professor of physiology at Bern and in 1835 professor of botany at Tübingen, a chair he occupied until his death.

Mohl's scientific work deals with extremely diverse areas of botany. Among them he devoted himself to the technical problems of the microscope that he himself was able to construct and recorded his findings in a manual on microscopy (1846).

After a century Mohl remains famous for his works on the microscopic anatomy of plants and for his contributions to knowledge of the plant cell. The publication which gives the best picture of his way of working and thinking is the comprehensive memoir "Die vegetabilische Zelle" (1853), in Wagner's *Handwörterbuch der Physiologie*, a classic reference work of the period. In this study Mohl sketched a veritable panorama of botany, taking as a base the cell, which he viewed as an "elementary organ." He summarized his own work, claiming priority in certain cases, and subjected the publications of his predecessors and contemporaries to a critical examination. He recalled that he was the first to demonstrate the fusion of aligned cells in the formation of ducts and to observe intracellular movements. He examined the structure of the cell and its derivatives, its generation by division or free formation, and its physiology as an organ of nutrition, of reproduction, and of movement. For Mohl the cell is composed of the membrane, the primordial utricle, the protoplasm, the nucleus, and the cellular fluid. He arrived at this conception after meticulous studies, which were the first efforts in cytochemistry. An impartial examination makes it appear that Mohl did not go beyond the discoveries of Raspail, which he did not know of and nowhere cites (cf. Klein [1936]).

The history of biology credits Mohl with the invention of the term "protoplasm" (independently of Purkyně, who had already used it in a different sense). A careful reading of the texts reveals that Mohl saw in this substance a preliminary material in cellular generation. This position is all the more surprising because he was one of the first to describe the generation of cells in plants by division starting from preexisting cells. The notion of protoplasm was integrated into the knowledge of the period; but it is a derivative sense of this word, defined by Max Schultze in 1861, which has survived in contemporary biology (cf. Robin [1872] and Klein [1936]). Mohl always limited himself to descriptions of concrete facts and carefully avoided drawing general conclusions from them; moreover, he did not write a synthetic exposition clearly summarizing his stand on the cell theory, which was then in full development.

Mohl had a happy childhood and adolescence, and a university career and personal life without difficulties. He remained unmarried and never attempted to surround himself with a circle of pupils. He was known as a meticulous worker, very clever with his hands, who brought a great number of precise details to bear on very circumscribed problems. He was one of the founders of the *Botanische Zeitung* (1843), one of the most famous periodicals of modern botany. He was also one of the promoters of the creation of the Faculty of Sciences at Tübingen, the first of its kind in Germany. From the time of his inauguration (1863) he proclaimed his hostility toward speculative thought and, in particular, to German *Naturphilosophie*, which was then in its dying stages (cf. Bünning [1963]).

Mohl's great ability was recognized very early in his life. He was awarded many decorations and honorific titles, a fact to which he drew attention. Among numerous academies and learned societies, he was a corresponding member of the Institut de France at a very early age (1838). The Order of the Crown of Württemberg, bestowed in 1843, conferred upon him a title of nobility. He died in his sleep on Easter day 1872. The laudatory obituary of De Bary (1872) and the biography of Sachs (1875) have perpetuated his memory in contemporary biology.

BIBLIOGRAPHY

I. ORIGINAL WORKS. Among Mohl's writings are *Mikrographie, oder Anleitung zur Kenntniss und zum Gebrauche des Mikroskops* (Tübingen, 1846); "Sur le mouvement du suc à l'intérieur des cellules," in *Annales des sciences naturelles*. Botanique, 3rd ser., **6** (1846), 84–96; "Saftbewegung im Inneren der Zellen," in *Botanische Zeitung*, **4** (1846), 73–78; 89–94; and "Die vegetabilische Zelle," in Rudolph Wagner, *Handwörterbuch der Physiologie*, IV (Brunswick, 1853), 167–309.

II. SECONDARY LITERATURE. See A. de Bary, "Hugo von Mohl," in *Botanische Zeitung*, **30** (1872), 561–579, with complete bibliography; E. Bünning, "Hugo von Mohl (1805–1872)," in H. Freund and A. Berg, *Geschichte der Mikroskopie*, I (Frankfurt, 1963), 273–280, with portrait; M. Klein, *Histoire des origines de la théorie cellulaire* (Paris, 1936), 60; C. Robin, *Anatomie et physiologie cellulaires* (Paris, 1873), 7, 249; J. Sachs, *Geschichte der Botanik vom 16. Jahrhundert bis 1860* (Munich, 1875), 315–335; and G. Vapereau, *Dictionnaire universel des contemporains*, 4th ed. (Paris, 1870), 1285–1286.

MARC KLEIN

MOHN, HENRIK (*b.* Bergen, Norway, 15 May 1835; *d.* Christiania [now Oslo], Norway, 12 September 1916), *meteorology, oceanography.*

Mohn became interested in science during his first year at the University of Christiania, where in 1858 he received his master's degree in mineralogy. Soon afterward he wrote a prize essay on the position of cometary orbits and became assistant professor of astronomy. In 1866 he was appointed director of the new Norwegian Meteorological Institute, which under his guidance grew into an important organization with 450 stations all over the country. From 1866 to 1913 he occupied the first chair of meteorology at the University of Christiania.

Mohn's many endeavors in practical meteorology resulted in a very wide range of publications. A series of annual bulletins and other reports and climatic tables for Norway were written or edited by him. Particular works dealt with thunderstorms, fog signals, and other subjects. He also perfected the hypsometer for measuring altitude by the boiling point of water.

With G. O. Sars, Mohn planned the Norwegian North Atlantic expeditions, participating in the three voyages that occupied the summers of 1876, 1877, and 1878. He edited the general report and wrote the sections on meteorology and oceanography. He also edited the meteorological data from Nansen's *Fram* expedition through the Northwest Passage of 1893–1896, from the Arctic expedition of 1898–1902, and from Amundsen's South Pole expedition of 1912–1913. Beginning in 1870, Mohn became very active in international organizations. At his death he was recognized as the grand old man of European meteorology.

Mohn was not only a brilliant empiricist, but also deeply interested in theory. His studies, with C. M. Guldberg, on the motions of the atmosphere (1876–1880) utilized the Coriolis law and also took into account friction between the atmosphere and the earth. The Mohn-Guldberg equations meant a breakthrough for dynamical meteorology and gave the first (although incomplete) theoretical foundation of the work begun by Buys Ballot, William Ferrel, and others. The new ideas were propagated in a book which, from a small and rather popular Norwegian version, developed into a comprehensive manual of meteorology that was translated into several languages.

BIBLIOGRAPHY

I. ORIGINAL WORKS. Mohn's writings include *Om Kometbaners indbyrdes Beliggenhed* (Christiania, 1861), trans. into French as *Mémoire sur la situation réciproque des comètes* (Christiania, 1861); *Norsk meteorologisk Aarbog* (Christiania, 1867 ff.); *Det Norske Meteorologiske Instituts Storm Atlas, Atlas des tempètes* (Christiania, 1870); "Om Vind og Vejr," in *Folkevennen*, (1872), ; *Grundzüge der Meteorologie* (Berlin, 1875, 1879, 1883), trans. into Finnish (Helsinki, 1880) and into French (Paris, 1884); *Études sur les mouvements de l'atmosphère*, 2 vols. (Christiania, 1876–1880), written with C. M. Guldberg; *Den Norske Nordhavs-Expedition 1876–1878. The Norwegian North Atlantic Expedition. General Report* (Christiania, 1880); "Studien über Nebelsignale," in *Annalen der Hydrographie und maritime Meteorologie* (1882, 1893, 1895); *Den Norske Nordhavs-Expedition 1876–1878. . . . Meteorology* (Christiania, 1883); *Nordhavets Dybder, Temperatur og Strømninger* (Christiania, 1887); *Tordenbygernes Hyppighed i Norge 1867–83* (Christiania, 1887); *Les orages dans la peninsule scandinave* (Uppsala, 1888), written with H. H. Hildebrandsson; *Om Taage-*

signaler (Christiania, 1897); "Das Hypsometer als Luftdruckmesser und seine Anwendung zur Bestimmung der Schwerekorrektion," in *Skrifter . . . det Norske videnskapsakademi i Oslo*, Math.-naturvis. Kl. (1899); and "Neue Studien über das Hypsometer," in *Meteorologische Zeitschrift* (1908).

II. SECONDARY LITERATURE. See V. Bjerknes, *Meteorologien in Norge* (Christiania, 1917); the unsigned *Christian Joachim Mohn, hans forfaedres liv og efterkommere* (Oslo, 1928), 318 ff.; and T. Hesselberg, "Mohn, Henrik," in *Norsk Biografisk Leksikon*, IX (Oslo, 1939), 290 ff.

OLAF PEDERSEN

MOHOROVIČIĆ, ANDRIJA (*b.* Volosko, Istria, Croatia, 23 January 1857; *d.* Zagreb, Yugoslavia, 18 December 1936), *meteorology, seismology.*

Mohorovičić's father was a shipwright; his mother died shortly after his birth. A brilliant pupil at the grammar school in Rijeka, he entered the University of Prague in 1875, graduating in mathematics and physics. One of his teachers at Prague was Ernst Mach. There followed a period as a secondary school teacher, and in 1882 he was appointed to the Royal Nautical School in Bakar (Buccari), near Rijeka, where he taught, among other things, meteorology and oceanography. His interests turned strongly toward meteorology and in 1887 he founded the Meteorological Station of Bakar. In 1891 he was appointed professor at the Main Technical School in Zagreb, where in 1892 he became director of the meteorological observatory. In 1897 he received his doctorate from the University of Zagreb, becoming an unsalaried lecturer there the same year and reader in 1910.

Most meteorological centers in Croatia had been set up under control from Budapest. Mohorovičić campaigned actively to remove this control and in 1900 he succeeded in having the Zagreb observatory established as meteorological center for all Croatia and Slavonia, completely independent of Budapest, as the Royal Regional Center for Meteorology and Geodynamics. Soon after the turn of the century, seismology became his dominant interest and the field in which he became famous. He retired in 1921 when the reorganized Royal Regional Center was renamed the Geophysical Institute. Although troubled by weak eyesight since 1916, he continued in active seismological research until 1926.

Mohorovičić's fame rests nearly entirely on the results of his very thorough investigation of the destructive earthquake of 8 October 1909, which occurred about thirty miles south of Zagreb, in the Kulpa valley of Croatia. Previously the velocities of *P* and S (primary and secondary) seismic waves had been treated as varying continuously with the depth, apart from minor discontinuities connected with geographic features such as sedimentary deposits, mountain ranges, and ocean floors. In his study of records of the Kulpa valley earthquake, however, Mohorovičić detected the presence of two distinct pairs of P and S phases recorded by seismographs at distances between 125 and 450 miles from the epicenter of the earthquake. His careful analysis showed that one of the pairs was associated with markedly slower speeds of wave travel. Mohorovičić correctly interpreted the observations as showing that the focus of the earthquake was inside a distinct outer layer of the earth; that the slower P and S waves had traveled directly to observing stations through this layer; and that the paths of the faster waves had been mostly below the layer, having been refracted when passing down from the focus through the separating boundary and later refracted upward through this boundary to the surface. He calculated the thickness of the layer as 30–35 miles. The work of other seismologists later confirmed this discovery, though reducing his estimate of the layer thickness, and showed that the layer was worldwide.

This outer layer of the earth is now conventionally called the crust, and the region between the crust and the central core (at a depth of 1,800 miles) is known as the mantle (originally the shell). The boundary separating the crust and the mantle is now called the Mohorovičić discontinuity (sometimes the M discontinuity). Its depth below the earth's surface is now known to be about 20 miles in continental shield areas, somewhat greater under large mountain ranges, but only 3–6 miles below ocean floors. Across the Mohorovičić discontinuity, the seismic P velocity changes fairly rapidly from about 4 to 5 miles per second and the S velocity from about $2\frac{1}{2}$ to 3 miles per second. The Mohorovičić discontinuity was the second major boundary to be discovered below the earth's surface by seismological means; in 1906 R. D. Oldham had established the existence of the central core. The depth of the boundary separating mantle and core was accurately determined by Beno Gutenberg in 1914.

One outcome of Mohorovičić's work on the 1909 earthquake was a classification of earthquakes for investigative purposes into near and distant earthquakes. Near earthquakes are well recorded within about 600 miles of the epicenter and within this range show low-velocity phases of the type that Mohorovičić had found. In distant earthquakes the dominant interest is on records taken beyond this distance.

The name of Mohorovičić caught the public fancy in 1957 when the International Association of Seismology and Physics of the Earth's Interior sponsored a proposal, the Mohole project, to drill through the earth's crust to just below the Mohorovičić discontinuity. In the hands of journalists the discontinuity soon came to be referred to as the Moho.

As a meteorologist Mohorovičić was noted for his great organizational ability, his insistence on high standards of precision wherever he had responsibilities, and his success in circumventing bureaucracy. His research papers dealt with such subjects as cloud movements, the variation of atmospheric temperature with height, rainfall in Zagreb, and a tornado. Mohorovičić was also noted for his deep appreciation of what was needed to produce an up-to-date seismological observatory. In 1901, soon after the Royal Regional Center had been set up, a strong earthquake was felt in Zagreb, as a consequence of which he and a colleague were able to secure for the center a reliable seismograph constructed by the noted Italian seismologist G. Agamennone. With the addition of further equipment, including a Wiechert seismograph and equipment for ensuring precise timing, the center had become by 1908 one of the leading seismological observatories in central Europe. The thoroughness of this preparation was a major factor in the success of his study of the 1909 Kulpa valley earthquake.

Mohorovičić contributed research papers on a variety of other topics in seismology. He evolved a method of determining earthquake epicenters and constructed curves giving the travel times of seismic waves over distances of up to 10,000 miles from the source. He also carried out macroseismic studies—the investigation of the salient features of an earthquake from reports of observations of surface effects (geological effects, effects on buildings, bridges, among others) taken over a wide area surrounding the epicenter, and he investigated the subject of constructing earthquake-proof buildings. His ideas on improving the construction of seismographs were unrealized due to insufficient financial support.

Mohorovičić was most punctilious, tenacious, and meticulous in all his scientific work, and was known for always talking to the point. He was keen and enthusiastic in his undertakings and intolerant of slipshod work. Even after his eyesight began to fail, he worked far into the night reading specialized papers on seismology. An extremely good-humored man, he was idolized by his colleagues. His success was partly due to his unusual linguistic ability. At the age of fifteen, he spoke Croatian, Italian, English, and French, and later spoke German, Czech, Latin, and classical Greek as well. His son Stjepan is also a distinguished seismologist. Mohorovičić was elected to the Yugoslav Academy of Sciences in 1898 and was secretary of the mathematics and science section from 1918 to 1922.

BIBLIOGRAPHY

I. ORIGINAL WORKS. Mohorovičić's best-known publication is "Das Beben vom 8.X.1909," in *Jahrbuch des meteorologischen Observatoriums in Zagreb* for 1909, pt. 4, par. 1 (1910), 1–67. The essential content of this paper was published in French by E. Rothé, "Sur la propagation des ondes sismiques au voisinage de l'épicentre. Préliminaires continues et trajets à réfraction," in *Publications du Bureau central séismologique international*, ser. A, Travaux scientifiques, fasc. **1** (1924), 17–59. Mohorovičić published a total of 21 papers on meteorology and seismology from 1888 to 1926, mostly in journals published in Zagreb; a full list is held by the University of Zagreb.

II. SECONDARY LITERATURE. The proceedings of a symposium held in Zagreb in March 1968 in honor of Mohorovičić include an article by his student and collaborator Josip Mokrović, "Andrija Mohorovičić—sein Leben und Wirken."

K. E. BULLEN

MOHR, CARL FRIEDRICH (*b.* Koblenz, then France [now Germany], 4 November 1806; *d.* Bonn, Germany, 28 September 1879), *analytical chemistry, physical chemistry, agricultural chemistry, geology.*

Mohr's father, Karl, was an apothecary and city councillor. After completing secondary school, Mohr attended the University of Bonn, where he studied botany, chemistry, and mineralogy. After gaining practical experience with his father, he attended the chemistry lectures of Leopold Gmelin at Heidelberg and those of Heinrich Rose in analytical chemistry at Berlin. He then obtained his degree in pharmacy and took over his father's business. Besides attending to the business, he was interested in various areas of science. In 1833 he married Jacobine Derichs; they had three sons and two daughters.

In 1837 Mohr published an essay, "Ansichten über die Natur der Wärme," in which he wrote: "Apart from the known chemical elements, there exists in nature only one agent, and that is force; it can show itself in appropriate relationships as motion, chemical affinity, cohesion, electricity, light, heat or magnetism. And out of each of these kinds of phenomena all the others can be produced" (*Zeitschrift für Physik, Mathematik und verwandte Wissenschaften*, **5** [1837], 419). On the basis of this statement, he later claimed priority regarding the law of the conservation of energy. In 1847 Mohr

wrote a commentary on the Prussian pharmacopoeia and *Lehrbuch der pharmazeutischen Technik*. In this period he also carried out titrimetric experiments, the results of which are in *Lehrbuch der chemisch-analytischen Titriermethode* (1855). In the meantime Mohr had established a vinegar factory with his son-in-law and purchased an estate at Metternich, to which he moved after selling his apothecary's shop. He now concerned himself with fermentation and the cultivation of grapes, and wrote popular books on these subjects. He also experimented with artificial fertilizers, following the ideas of his close friend Liebig. In 1863 his factory failed, and Mohr found himself and his family in financial difficulties. Through Liebig's help he qualified as a *Privatdozent* at the University of Berlin, at the age of fifty-nine. A short time later he moved to Bonn and was *Dozent* at the university there. He soon turned to new areas of research and studied geology, on which science he developed original but wholly incorrect opinions.

Mohr's attention next turned to thermodynamics. He recalled his statement on force and had it reprinted in a book. His commentary on it is chracteristic: "This passage was written by me thirty-three years ago and contains, as I now see, the main features of the mechanical theory of heat." Mohr published several works in this field; but because he lacked the necessary mathematical knowledge, his works were failures.

Mohr had a passionate, critical, and combative nature and was therefore unpopular with his colleagues. He was active in many areas, yet much of his activity was marked by dilettantism. He was a skillful author and wrote many books.

Mohr's most lasting contribution was in titration. His *Lehrbuch . . . Titriermethode* was the first successful compendium in this new field of analytical chemistry; It went through eight new editions between 1856 and 1913 and was translated into several languages. Mohr invented many new titration procedures and examined and often improved most of the older ones. Many methods and designs of apparatus bear his name: the Mohr test for iron and chloride determination, the Mohr pinchcock burette, the Mohr balance for the determination of specific gravity, and Mohr's salt (ferrous ammonium sulfate) are evidence of his skill in the laboratory. The cooling device generally called the Liebig reflux and the useful cork borer were also invented by Mohr.

BIBLIOGRAPHY

I. ORIGINAL WORKS. Mohr's writings include *Commentar zur preussischen Pharmacopöe*, 2 vols. (Brunswick,

1847; 2nd ed., 1853); *Lehrbuch der pharmazeutischen Technik* (Brunswick, 1847; 2nd ed., 1853), also trans. into English as *Practical Pharmacy* (London, 1848) and as *Practice of Pharmacy* (Philadelphia, 1849); *Taschenbuch der chemischen Receptirkunst* (Hamburg, 1854); *Lehrbuch der chemisch-analytischen Titriermethode* (Brunswick, 1855; 6th and 7th eds., with A. Classen, 1880 and 1896; 8th ed., with H. Beckurts, 1912), trans. into French as *Traité d'analyse chimique par la méthode des liqueurs titrées* (Paris, 1888); *Der Weinstock und der Wein* (Koblenz, 1864); *Geschichte der Erde* (Bonn, 1866); *Mechanische Theorie der chemischen Affinität und die neuere Chemie* (Brunswick, 1868); and *Allgemeine Theorie der Bewegung und der Kraft als Grundlage der Physik und Chemie* (Brunswick, 1874). Many of his articles appeared in Poggendorff's *Annalen der Physik* and *Justus Liebigs Annalen der Chemie*.

II. Secondary Literature. See E. E. Aynsley and W. A. Campbell, "Karl Friedrich Mohr's Contributions to Chemical Apparatus," in *School Science Review* (1959), 312; R. Hasenclever, "Erinnerungen an Friedrich Mohr," in *Berichte der Deutschen chemischen Gesellschaft*, **33** (1900), 3827; G. Kahlbaum, *Justus von Liebig und Friedrich Mohr in ihren Briefen* (Leipzig, 1904); Ralph E. Oesper, "Karl Friedrich Mohr," in *Journal of Chemical Education*, **4** (1927), 1357; J. R. Partington, *History of Chemistry*, IV (London, 1964), 317–318; J. M. Scott, "Karl Friedrich Mohr, Father of Volumetric Analysis," in *Chymia*, **3** (1950), 191; and F. Szabadváry, *History of Analytical Chemistry* (Oxford–New York, 1966), pp. 241–250, also available in German, *Geschichte der analytischen Chemie* (Brunswick, 1966), pp. 245–257.

F. Szabadváry

MOHR, CHRISTIAN OTTO (*b.* Wesselburen, Holstein, 8 October 1835; *d.* Dresden, Germany, 2 October 1918), *civil engineering*.

A descendent of Holstein landowners, Mohr studied engineering at the Polytechnic Institute of Hannover, which he entered in 1851. As an engineer for the state railroads of Hannover and Oldenburg, he not only built some notable bridges but also began to publish original research papers. He published continuously into his eighties. In 1867 Mohr became professor of mechanics and civil engineering at the Technische Hochschule in Stuttgart and, from 1873, at Dresden, where he remained after his retirement in 1900. Commensurate with his steadily growing fame, he received the usual honors in generous measure. Many of his students, including Föppl, described Mohr as their most remarkable teacher. He was of imposing height, proud and taciturn; his ideals in lecturing as well as in writing were simplicity, clarity, and conciseness.

With the exception of one textbook, Mohr published only original research papers. In his first publication

(1860), on the theory of continuous beams, he presented the three-moments equation (derived earlier by Clapeyron and Bertot) for the first time in general form by adding terms to account for vertical variations of the supports. In 1868, recognizing that the differential equation of the elastic line has the same form as that of the funicular curve, he developed the method of influence lines, which makes it possible to determine the deflections of a loaded beam, even of varying cross section, without requiring the integration of its differential equation. In 1874 he independently rediscovered a method of determining the stresses in statically indeterminate frameworks that had been published, somewhat obscurely, by Maxwell ten years earlier. "Mohr's stress circle," his most widely known contribution, was described in a paper of 1882. Following a suggestion (1866) of Karl Culmann he devised a simple graphic representation of the stresses at one point. Mohr's theory of failure, based upon the concept of the stress circle, has been widely accepted in engineering practice.

BIBLIOGRAPHY

I. Original Works. Apart from the textbook *Technische Mechanik* (Stuttgart, 1877), Mohr wrote only research papers, which were collected as *Abhandlungen aus dem Gebiete der technischen Mechanik* (Berlin, 1906).

II. Secondary Literature. Biographical information on Mohr can be found in three articles by W. Gehler: "Christian Otto Mohr," in *Festschrift Otto Mohr zum 80. Geburtstag* (Berlin, 1916), v–vii; "Otto Mohr," in *Zentralblatt der Bauverwaltung*, **38** (1918), 425; and "Otto Mohr," in *Zeitschrift des Vereins deutscher Ingenieure*, **62** (1918), 114. See also Poggendorff, V, 868; VI, 1761. Mohr's technical work is discussed extensively in Stephen P. Timoshenko, *History of Strength of Materials* (New York, 1953), *passim*.

Otto Mayr

MOHR, GEORG (*b.* Copenhagen, Denmark, 1 April 1640; *d.* Kieslingswalde, near Görlitz, Germany, 26 January 1697), *mathematics*.

Mohr was the son of David Mohrendal (or Mohrenthal), a hospital inspector and tradesman. His parents taught him reading, writing, and basic arithmetic, but his love for mathematics could not be satisfied in Denmark, and in 1662 he went to Holland, where Huygens was teaching, and later to England and France. He returned to Denmark, but about 1687 he went again to Holland, this time because of a difference with King Christian V. Wishing to be scientifically independent, he remained aloof from official positions; but Tschirnhausen finally persuaded him to come to Kieslingswalde to participate in his mathematical projects. Mohr went there in 1695, accompanied by his wife, whom he had married in 1687, and by his three-year-old son. Only one of his works, the *Euclides danicus* (1672), a valuable short work, is known today; but his son claimed that he wrote three books on mathematics and philosophy that were well received by scholars.

Mohr is often mentioned in the intellectual correspondence of the day. He corresponded with Leibniz, with Pieter van Gent, and with Ameldonck Bloeck, a member of Spinoza's circle. In 1675 Oldenburg sent Leibniz a work of Mohr's on the root extraction of $A + \sqrt{B}$. Leibniz, in a letter of 1676 to Oldenburg in which he refers to "Georgius Mohr Danus, in geometria et analysi versatissimus," mentions that he learned from Mohr that Collins had the expansions for sin x and arcsin x. Unfortunately, little else of Mohr's scientific activity is known.

In 1928 Mohr's *Euclides danicus*, which had fallen into obscurity, was republished with a preface by J. Hjelmslev. Hjelmslev recognized that in 1672 Mohr had been dealing with a problem made famous 125 years later by Mascheroni, namely, that of making constructions with compass alone.

The book has two parts: the first consists of the constructions of the first six books of Euclid; the second, of various constructions. The problem of finding the intersection of two lines, which is of some theoretical importance, is solved incidentally in the second part in connection with the construction of a circle through two given points and tangent to a given line.

Hjelmslev made the acute observation that a minor variant of Mohr's constructions enables one to add and subtract segments on the sphere and in the hyperbolic plane.

The obscurity that befell Mohr and his book can be attributed, in some degree, to the presentation of the material. In the body of the book, Mohr does not state the issue until the very last paragraph, although the lines are referred to as "imagined" *(gedachte)*. In the dedication to Christian V, he does say that he believes he has done something new, and on the title page the issue is explicitly stated. Still, it would be easy for an inattentive reader to misjudge the value of the book.

According to Hjelmslev, Mascheroni's result—that all ruler and compass constructions can be done by compass alone—was already known and systematically expounded by Mohr. (The justice of this judgment and the question of the independence of Mascheroni's work are examined in the article on Mascheroni.)

The laconic Mohr tells us nothing about the genesis of his ideas. A guess is that the fundamental problem stems from a similar problem, that of the compass of a single opening, which was posed in the contests of the great Renaissance mathematicians. This conjecture might be supported by a historical study of the problems in the second part of the book: νεύσεις (inclinations) problems; maxima-minima problems; the problem of Pothenot, solved in 1617 by Snellius in his *Eratosthenes batavus;* and problems in perspective.

BIBLIOGRAPHY

Mohr's *Euclides danicus* (Amsterdam, 1672) was translated into German by J. Pál, with a foreword by J. Hjelmslev (Copenhagen, 1928).

Hjelmslev has written two articles on Mohr: "Om et af den danske matematiker Georg Mohr udgivet skrift *Euclides Danicus*," in *Matematisk Tidsskrift*, B (1928), 1–7; and "Beiträge zur Lebenabschreibung von Georg Mohr (1640–1697)," in *Kongelige Danske Videnskabernes Selskabs Skrifter, Math.-fysiske Meddelelser*, **11** (1931), 3–23.

A. SEIDENBERG

MOHS, FRIEDRICH (*b.* Gernrode, Anhalt-Bernburg, Germany, 29 January 1773; *d.* Agordo, Tirol, Italy, 29 September 1839), *mineralogy, geology.*

One of Abraham Werner's outstanding students, Mohs made his primary scientific contribution in systematic mineralogy. He also proposed the scale of hardness for minerals, which is named for him and which is still in use.

Mohs displayed a marked interest in science at an early age and received a private education before entering the University of Halle in 1797. In 1798 he matriculated at the mining academy at Freiberg, where in addition to physics and mathematics he studied mineralogy under Werner. In 1802 Mohs was invited to Great Britain by his fellow students George Mitchell and Robert Jameson to participate in the planning of a mining academy at Dublin. Although the proposed academy was never established, the journey enabled Mohs to study the geology and mineralogy of Ireland and Scotland and to make lasting friends among Scottish geologists.

On Mitchell's recommendation, Mohs in 1802 was commissioned by J. F. von der Null, a Viennese banker, to prepare a systematic description of his important mineral collection. In 1804 Mohs published a two-volume description of this collection and two other works, *Beschreibung des Gruben gebäudes*

Himmelfürst ohnweit Freiberg and *Über die oryk-tognostische Klassifikation, nebst Versuch eines auf blosse äussere Kennzeichen gegründeten Mineralsystems*, in which he first expressed his misgivings with Werner's approach to mineralogy. In 1810, while on one of his frequent mining and mineralogical explorations, he encountered Werner at Carlsbad. Failing to convince Werner of the inadequacy of his mineralogical method, Mohs determined to establish systematic mineralogy on a completely new basis.

In 1811 Archduke Johann established the Johanneum in Graz, and Mohs was appointed curator of the mineral collection and charged with adding to it the minerals and rocks of Styria. Mohs enlisted the young scholar Wilhelm Haidinger to help in this work, and Haidinger remained with Mohs until 1822. In 1812 Mohs became professor of mineralogy at the Johanneum and in the same year revealed the basis of his new classificatory system in his *Versuch einer Elementar-Methode zur naturhistorischen Bestimmung und Erkennung der Fossilien*, in which he first proposed his hardness scale for minerals. Miners and mineralogists had long been accustomed to scratch a mineral to aid in determining the species. In an attempt to make this method more certain, Mohs proposed a scale of increasing hardness from one to ten as follows: talc, 1; gypsum, 2; calcite, 3; fluorite, 4; apatite, 5; feldspar, 6; quartz, 7; topaz, 8; corundum, 9; and diamond, 10. Intermediate degrees of hardness were subsequently added to the Mohs scale. Mineralogists did not commonly employ the scale until the 1820's, after the publication of the English translation of Mohs's *Die Charaktere der Klassen, Ordnungen, Geschlecter, und Arten der naturhistorischen Mineral-Systems*, in which the scale was prominently featured.

Mohs remained at Graz until 1817. During this period he worked at perfecting his method of mineral classification, giving particular attention to the possible arrangements of minerals in crystal systems based on external symmetry. In 1817 he again traveled extensively in Great Britain, impressing Scottish mineralogists in particular with his novel ideas. Following Werner's death in 1817 Mohs was called to Freiberg as professor of mineralogy, and he assumed this post in the autumn of 1818.

In 1822 and 1824 Mohs published his two-volume *Grund-Riss der Mineralogie*, the first volume of which was largely devoted to the explanation of his ideas concerning crystallography and the second to a systematic description of minerals. Mohs postulated four crystal systems based on external symmetry: rhombohedral (hexagonal), pyramidal (tetragonal), prismatic (orthorhombic), and tessular (cubic). These divisions were similar to those proposed in 1816–1817

by Christian Samuel Weiss, who had approached the problem in much the same manner. Mohs, however, did not refer to Weiss's prior publication, and Weiss publicly accused him of plagiarism. Mohs defended himself in a letter to the *Edinburgh Philosophical Journal* (**8** [1823], 275–290), explaining that his dissatisfaction with Haüy's crystallographic concepts had led him to develop his own ideas.

Mohs, however, had surpassed Weiss in his analysis. In the first volume of *Grund-Riss* (pp. 56 ff.) he mentioned the possible existence of symmetry systems in which the crystallographic axes were not mutually perpendicular; and in the second volume (pp. vi–viii) he affirmed their existence. These new systems, the monoclinic and the triclinic, were described by K. F. Naumann in 1824 and were fully developed by Mohs in 1832. *Grund-Riss*, substantially amended and revised, was translated into English by Wilhelm Haidinger as *Treatise on Mineralogy* (1825). Mohs's classificatory system, based primarily on crystal form, hardness, and specific gravity, was not received favorably by most mineralogists.

In 1826 Mohs resigned his professorship at Freiberg to accept a position in Vienna, first to reorganize the imperial collection, which he augmented by the acquisition of the von der Null collection, and then as professor of mineralogy at the university in 1828. In 1835 he resigned to become imperial counselor of the exchequer in charge of mining and monetary affairs. This position required him to travel frequently to all parts of the Austro-Hungarian empire. He died while on a journey to inspect the volcanic areas of southern Italy.

BIBLIOGRAPHY

I. ORIGINAL WORKS. Mohs's chief works are *Beschreibung des Gruben gebäudes Himmelfürst ohnweit Freiberg* (Vienna, 1804); *Des Herrn J. F. von der Null Mineralien-Kabinet*, 2 vols. (Vienna, 1804); *Über die oryktognostische Klassifikation, nebst Versuch eines auf blosse äussere Kennzeichen gegründeten Mineralsystems* (Vienna, 1804); *Versuch einer Elementar-Methode zur naturhistorischen Bestimmung und Erkennung der Fossilien* (Vienna, 1812); *Die Charaktere der Klassen, Ordnungen, Geschlecter, und Arten der naturhistorischen Mineral-Systems* (Dresden, 1820), translated as *The Characters of the Classes, Orders, Genera, and Species, or the Characteristics of the Natural-History System of Mineralogy* (Edinburgh, 1820); *Grund-Riss der Mineralogie*, 2 vols. (Dresden, 1822–1824), translated, revised, and expanded by Wilhelm Haidinger as *Treatise on Mineralogy, or the Natural History of the Mineral Kingdom*, 3 vols. (Edinburgh, 1825); *Leichtfässliche Anfangsgründe der Naturgeschichte des Mineralreiches*, 2 vols. (Vienna, 1832–1839), vol. II completed by F. X. M.

Zippe; *Anleitung zum Schürfen* (Vienna, 1838); and the posthumous *Die ersten Begriffe der Mineralogie und Geognosie für engehende Bergbeamte*, 2 vols. (Vienna, 1842).

II. SECONDARY WORKS. The principal biographical source for Mohs, which includes an autobiography to 1830, is Wilhelm Fuchs, G. Haltmeyer, and F. Leydolt, eds., *Friedrich Mohs und sein Wirken in wissenschaftlicher Hinsicht: ein biographischer Versuch entworfen und zur Enthüllingsfeier seines Monumentes im st. st. Johanneums-Garten zu Grätz* (Vienna, 1843). Other sources are *Festschrift zur hundertjährigen Jubiläum der Bergakademie zu Freiberg* (Dresden, 1866), 24–28; Franz von Kobell, *Geschichte der Mineralogie* (Munich, 1864), 216–222; Paul Groth, *Entwicklungsgeschichte der mineralogischen Wissenschaften* (Berlin, 1926), 249–250; and *Allgemeine deutsche Biographie*, XXII (Leipzig, 1885), 76–79.

JOHN G. BURKE

MOISEEV, NIKOLAY DMITRIEVICH (*b*. Perm, Russia, 16 December 1902; *d*. Moscow, U.S.S.R., 6 December 1955), *celestial mechanics, astronomy, mathematics.*

After graduating from the Perm Gymnasium in 1919, Moiseev entered the Faculty of Physics and Mathematics of Perm State University, where he also worked as a laboratory assistant. In 1922 he transferred to Moscow University, from which he graduated in 1923, having specialized in astronomy. In 1922 he became a junior scientific co-worker at the State Astrophysics Institute (since 1931 part of the P. K. Sternberg Astronomical Institute). After completing his graduate work there, in 1929 Moiseev defended his dissertation "O nekotorykh osnovnykh voprosakh teorii proiskhozhdenia komet, meteorov i kosmicheskoy pyli" ("On Certain Basic Questions of the Theory of the Origin of Comets, Meteors, and Cosmic Dust"). From 1929 to 1947 he taught mathematics at the N. E. Zhukovsky Military Air Academy. In 1935 he was awarded the degree of doctor of physics and mathematical sciences and the title of professor. He was director of the department of celestial mechanics at the University of Moscow from 1938 to 1955 and was head of the P. K. Sternberg Astronomical Institute from 1939 to 1943.

The recognized leader of the Moscow school of celestial mechanics, Moiseev published more than 120 works on the mechanical theory of cometary forms; the cosmogony of comets, meteors, and cosmic dust; theoretical gravimetry, including an original method (the "nonregularized earth") used in the theory of determining the forms of geoids from gravimetric observations; and dynamic cosmogony. To the study of the general characteristics of the

trajectories of celestial bodies he applied qualitative methods based on the use of differential equations of movement and certain known integrals. He introduced qualitative regional characteristics of the trajectory, such as its contacts with certain given curves and surfaces and its longitudinal and transversal stability.

Moiseev applied qualitative methods to problems of certain specific celestial bodies, and his investigations of the characteristics of stability of orbital motion found many applications in problems of airplane and missile dynamics. His 1949 monograph *Ocherki razvitia teorii ustoychivosti* ("Essays of the Development of the Theory of Stability") presented a historical analysis of the subject from antiquity to the twentieth century.

From 1940 to 1955 Moiseev published the results of his investigations on secular and periodic perturbations and the motions of celestial bodies. He developed concepts of the internal and external environments and twofold averaging. Moiseev established his own, interpolational-average scheme to supplement those of Gauss and Delaunay. Approximate empirical integrals of motion, deduced from observations of celestial bodies, were used for averaging the force function. This method allows the integration of the averaged differential equations of motion and the computation of the ephemerides of perturbation of motion.

Moiseev's chief contributions to mathematics were his two new methods of solving systems of linear differential equations: the method of determinant integrals and the interational method.

BIBLIOGRAPHY

I. ORIGINAL WORKS. Moiseev's early writings are "O vychislenii kometotsentricheskikh koordinat chastitsy kometnogo khvosta" ("On the Computation of the Comet-centered Coordinates of the Particles of the Comet Tail"), in *Russkii astronomicheskii zhurnal*, **1**, pt. 2 (1924), 79–86; "O khvoste komety 1901 I" ("On the Tail of the Comet of 1901 I"), *ibid.*, **2**, pt. 1 (1925), 73–84; "O vychislenii effektivnoy sily i momenta izverzhenia chastitsy kometnogo khvosta" ("On the Computation of the Effective Force and Moment of Ejection of the Particles of a Comet Tail"), *ibid.*, **2**, pt. 2 (1925), 54–60; and "O stroenii sinkhronnykh konoidov" ("On the Structure of Synchronic Conoids"), *ibid.*, **4**, pt. 3 (1927), 184–190.

Subsequent works are "Über einige Grundfragen der Theorie des Ursprungs der Kometen, Meteoren und des kosmischen Staubes (Kosmogonische Studien)" in *Trudy Gosudarstvennogo astrofizicheskogo instituta*, **5**, no. 1 (1930), 1–87; *Trudy Gosudarstvennogo astronomicheskogo instituta im P. K. Sternberga*, **5**, no. 2 (1933), 1–63; *Astronomicheskii Zhurnal*, **9**, nos. 1–2 (1932), 30–52; and *Trudy*

Gosudarstvennogo astronomicheskogo instituta im P. K. Sternberga, **6**, no. 1 (1935), 5–28, 50–58; "Intorno alla legge di resistenza al moto dei corpi in un mezzo pulviscolare," in *Atti dell'Accademia nazionale dei Lincei. Rendiconti*, **15** (1932), 135–139, 377–381, 443–447; "Sulle curve definite da un sistema di equazioni differenziali di secondo ordine," *ibid.*, Ser. 6a, **20** (1934), 178–182, 256–265, 321–327; "O nekotorykh obshchikh metodakh kachestvennogo izuchenia form dvizhenia v problemakh nebesnoy mekhaniki" ("On Certain General Methods of Qualitative Study of the Forms of Motion in Problems of Celestial Mechanics"), in *Trudy Gosudarstvennogo astronomicheskogo instituta im P. K. Sternberga*, **7**, pt. 1 (1936), 5–127; **9**, pt. 2 (1939), 5–45, 47–81, 165–166; **14**, pt. 1 (1940), 7–68; **15**, pt. 1 (1945), 7–26; and "O nekotorykh svoystvakh traektory v ogranichennoy probleme trekh tel" ("On Certain Properties of Trajectories in a Limited Three-Body Problem"), *ibid.*, **7**, pt. 1 (1936), 129–225; **9**, pt. 1 (1936), 44–71; **9**, pt. 2 (1939), 82–114, 116–131, 167–170; **15**, pt. 1 (1945), 27–74.

Also published in the 1930's were "Über die Relativkrummung der zwei benachbarten Trajektorien. Zum Frage über die Stabilität nach Jacobi," in *Astronomicheskii Zhurnal*, **13**, no. 1 (1936), 78–83; "Su alcune proposizioni di morfologia dei movimenti nei problemi dinamichi analoghi a quello del tre corpi," in *Revista de ciencias* (Lima), **39** (1937), 45–50; "Über Stabilität Wahrscheinlichkeitsstrehnung," in *Mathematische Zeitschrift*, **42**, no. 4 (1937), 513–537; "O postroenii oblastey sploshnoy ustoychivosti i neustoychivosti v smysle Lyapunova" ("On the Construction of Areas of Continuous Stability and Instability in Lyapunov's Sense"), in *Doklady Akademii nauk SSSR*, **20**, no. 6 (1938), 419–422; and "O fazovykh oblastyakh sploshnoy ustoychivosti i neustoychivosti" ("On Phase Areas of Continuous Stability and Instability"), *ibid.*, 423–425.

Moiseev's later works are "O nekotorykh osnovnykh uproshchennykh skhemakh nebesnoy mekhaniki, poluchaemykh pri pomoshchi osredenenia raznykh variantov problemy trekh tel" ("On Certain Basic Simplified Schemes of Celestial Mechanics Obtained With the Aid of Averaging Different Variants of the Problem of Three Bodies"), in *Trudy Gosudarstvennogo astronomicheskogo instituta im Sternberga*, **15**, pt. 1 (1945), 75–117; **20** (1951), 147–176; **21** (1952), 3–18; **24** (1954), 3–16; and in *Vestnik Moskovskogo gosudarstvennogo universiteta*, no. 2 (1950), 29–37; "A. M. Lyapunov i ego trudy po teorii ustoychivosti" ("A. M. Lyapunov and His Works on the Theory of Stability"), in *Uchenya zapiski Moskovskogo gosudarstvennogo universiteta*, no. 91 (1947), 129–147; "Kosmogonia" ("Cosmogony"), in the collection of papers, *Astronomia v SSSR za 30 let* ("Astronomy in the U.S.S.R. for Thirty Years"; Moscow–Leningrad, 1948), 184–191; *Ocherki razvitia teorii ustoychivosti* ("Essays of the Development of the Theory of Stability"; Moscow–Leningrad, 1949); and "Obshchii ocherk razvitia mekhaniki vo Rossii i v SSSR" ("A General Sketch of the Development of Mechanics in Russia and in the U.S.S.R."), in the collection of papers, *Mekhanika v SSSR za 30 let*

("Mechanics in the U.S.S.R. for Thirty Years"; Moscow–Leningrad, 1950), 11–57.

The following were published posthumously: "Ob ortointerpolyatsionnom osrednennom variante ogranichennoy zadachi trekh tochek" ("On the Orthointerpolational Averaging Variant of the Limited Problem of Three Points"), in *Trudy Gosudarstvennogo astronomicheskogo instituta im P. K. Sternberga*, **28** (1960), 9–24; and *Ocherk razvitia mekhaniki* ("Essay of the Development of Mechanics"; Moscow, 1961).

II. SECONDARY LITERATURE. On Moiseev and his work, see (listed in chronological order) the obituary in *Astronomicheskii tsirkulyar Akademii nauk SSSR*, no. 166 (1956), 24–25; and the biographies in *Trudy Gosudarstvennogo astronomicheskogo instituta im P. K. Shternberga*, **28** (1960), 5–9, with a list of 15 of Moiseev's works; *Ocherk razvitia mekhaniki* (cited above), 4–11, with bibliography of 21 works; and E. N. Rakcheev, in *Vestnik Moskovskogo gosudarstvennogo universiteta*, Seria matematika, mekhanika, no. 4 (1961), 71–77, published on the fifth anniversary of his death, with bibliography of 49 works. See also M. S. Yarov-Yarovoy, "Raboty v oblasti nebesnoy mekhaniki v MGU za 50 let (1917–1967 gg.)" ("Works in the Area of Celestial Mechanics at Moscow State University for Fifty Years"), in *Trudy Gosudarstvennogo astronomicheskogo instituta im P. K. Sternberga*, **41** (1968), 86–103.

P. G. KULIKOVSKY

MOISSAN, FERDINAND-FRÉDÉRIC-HENRI (*b.* Paris, France, 28 September 1852; *d.* Paris, 20 February 1907), *chemistry*.

Born into a family of modest means, Moissan lived in Paris until 1864, when his parents moved to Meaux. He attended the municipal college there but did not complete his studies; instead he returned to Paris to work for two years as a pharmacy apprentice. In 1872 he went to work in the laboratory of Edmond Frémy at the Muséum d'Histoire Naturelle but shortly after transferred to Pierre-Paul Dehérain's laboratory, also at the Muséum, where he began research in plant physiology under Dehérain's direction. Conscious of his need for more formal academic training, Moissan studied in Paris, earned the *baccalauréat* (1874) and the *licence* (1877), qualified as first-class pharmacist at the École Supérieure de Pharmacie (1879), and received the *docteur ès sciences physiques* from the Faculté des Sciences (1880).

Impressed with Moissan's ability, Dehérain collaborated with his young protégé in a study of plant respiration which was published in 1874. By this time Moissan had definitely decided on inorganic chemistry as his main interest. His early investigation of the oxides of iron and related metals, and particularly the compounds of chromium, attracted the attention of

Henri Sainte-Claire Deville and H. J. Debray, who encouraged him. This work formed the basis of Moissan's doctoral thesis of 1880 and preoccupied him to a large extent during the next three years. For some time he directed a private analytical laboratory and also served as *maître de conférences* and *chef des travaux pratiques* at the École Supérieure de Pharmacie (1879–1883). A happy marriage with Léonie Lugan of Meaux in 1882 and the financial and moral support of his father-in-law enabled Moissan to pursue his scientific objectives with a minimum of distraction. That same year he also competed successfully for an *agrégation* at the École Supérieure de Pharmacie.

In 1884 Moissan began his remarkable research on the compounds of fluorine, which was to lead him to the isolation of this element. Previous attempts by others to obtain fluorine had not been successful because of the toxicity of fluorine compounds and the difficulty in designing suitable apparatus. Efforts by Davy, Gay-Lussac, and Thenard had not only been fruitless but injurious to their health. George J. and Thomas Knox of Ireland were seriously affected; and for the Belgian chemist Paulin Louyet and the French chemist Jérôme Nicklès these investigations proved fatal. Frémy was equally unsuccessful in preparing fluorine, as was George Gore of England. Although Moissan's initial experiments to isolate fluorine, including the electrolytic decomposition of phosphorus trifluoride and arsenic trifluoride, had also failed and proved injurious to his health, he persisted and on 26 June 1886 finally succeeded. This difficult feat was accomplished by using an electrolyte of dry potassium acid fluoride dissolved in anhydrous hydrofluoric acid. For the reaction Moissan employed a platinum U-tube containing two platinum-iridium electrodes, closed by fluorite caps and cooled by methyl chloride. At the anode an electric current yielded a gas which by its strong reaction with silicon was shown to be fluorine. Moissan's continuing investigation of the chemistry of fluorine subsequently resulted in the discovery of a number of fluorides such as carbon tetrafluoride, ethyl fluoride, methyl and isobutyl fluorides (with M. Meslans), and sulfuryl fluoride (with P. Lebeau). In collaboration with James Dewar he both liquefied (1897, 1903) and solidified fluorine (1903).

Meanwhile, Moissan had turned his attention to the production of artificial diamonds and in the process constructed his famous electric furnace, which, although simple in design, proved to be a technological tool of the first order. The original model, which he subsequently improved, was demonstrated to the Academy of Sciences in December 1892. It consisted of two blocks of lime, one laid on the other, with a hollow space in the center for a crucible, and a longitudinal groove for two carbon electrodes which produced a high-temperature electric arc. In one experiment Moissan heated iron and carbonized sugar in his electric furnace, causing the carbon to dissolve in the molten iron. He then subjected the mixture to rapid cooling in cold water, causing the iron to solidify with enormous pressure, producing carbon particles of microscopic size that appeared to have the physical characteristics of diamond. Moissan and his contemporaries believed that diamonds had finally been synthesized by this method, a conclusion that has been rejected in recent years. Nevertheless, Moissan's electric furnace provided great impetus to the development of high-temperature chemistry. With this apparatus he prepared and studied refractory oxides, silicides, borides, and carbides; he succeeded in volatilizing many metals; and, by reducing metallic oxides with carbon, he obtained such metals as manganese, chromium, uranium, tungsten, vanadium, molybdenum, titanium, and zirconium. The electrochemical and metallurgical applications to industry of Moissan's work became immediately apparent, for example in the large-scale production of acetylene from calcium carbide.

Academic recognition came to Moissan in December 1886 with his appointment to a professorship in toxicology at the École Supérieure de Pharmacie. In 1899 he became professor of inorganic chemistry at this same institution and in 1900 he succeeded Troost in the chair of inorganic chemistry at the Faculty of Sciences. Moissan received the Nobel Prize for chemistry in 1906 and was elected to membership in numerous learned societies both in France and abroad. Through the originality of his research and the effectiveness of his teaching, Moissan attracted an increasing number of students and exerted a remarkable influence on the progress of inorganic chemistry.

BIBLIOGRAPHY

I. ORIGINAL WORKS. Moissan was a prolific writer and his papers and monographs (written either by himself or in collaboration with others) number more than three hundred—including such major works as *Le four électrique* (Paris, 1897), *Le fluor et ses composés* (Paris, 1900), and the five-volume collaborative work which he edited, *Traité de chimie minérale* (Paris, 1904–1906).

Comprehensive listings of Moissan's publications were compiled by Alexander Gutbier, *Zur Erinnerung an Moissan* (Erlangen, 1908), 268–285; and Paul Lebeau, "Notice sur la vie et les travaux de Henri Moissan," in *Bulletin. Société chimique de France*, 4th ser., **3** (1908), xxv–xxxviii.

II. SECONDARY LITERATURE. For accounts of Moissan's life and work, see *Centenaire de l'École supérieure de phar-*

macie de l'Université de Paris, 1803–1903 (Paris, 1904), 249–257; Alexander Gutbier, *Zur Erinnerung an Moissan* (Erlangen, 1908); Benjamin Harrow, *Eminent Chemists of Our Time*, 2nd ed. (New York, 1927), 135–154, 374–388; A. J. Ihde, *The Development of Modern Chemistry* (New York, 1964), 367–369; Paul Lebeau, "Notice sur la vie et les travaux de Henri Moissan," in *Bulletin. Société chimique de France*, 4th ser., **3** (1908), i–xxxviii; J. R. Partington, *A History of Chemistry*, IV (London–New York, 1964), 911–914; Sir William Ramsay, "Moissan Memorial Lecture," in *Journal of the Chemical Society*, **101** (1912), 477–488; and Alfred Stock, "Henri Moissan," in *Berichte der Deutschen chemischen Gesellschaft*, **40** (1907), 5099–5130.

Evidence disputing Moissan's claim to the production of diamonds has been presented by F. P. Bundy, *et al.*, "Man-Made Diamonds," in *Nature*, **176** (9 July 1955), 51–55.

For a discussion of the background and discovery of fluorine, see Louis Domange, "Les débuts de la chimie du fluor," in *Proceedings of the Chemical Society* (June/July 1959), 172–176; and M. E. Weeks, *Discovery of the Elements*, 6th ed. (Easton, Pa., 1956), 755–770.

ALEX BERMAN

MOIVRE, ABRAHAM DE (*b.* Vitry-le-François, France, 26 May 1667; *d.* London, England, 27 November 1754), *probability*.

De Moivre was one of the many gifted Protestants who emigrated from France to England following the revocation of the Edict of Nantes in 1685. His formal education was French, but his contributions were made within the Royal Society of London. His father, a provincial surgeon of modest means, assured him of a competent but undistinguished classical education. It began at the tolerant Catholic village school and continued at the Protestant Academy at Sedan. After the latter was suppressed for its profession of faith, De Moivre had to study at Saumur. It is said that he read mathematics on the side, almost in secret, and that Christiaan Huygens' work on the mathematics of games of chance, *De ratiociniis in ludo aleae* (Leiden, 1657), formed part of this clandestine study. He received no thorough instruction in mathematics until he went to Paris in 1684 to read the later books of Euclid and other texts under the supervision of Jacques Ozanam.

His Protestant biographers say that De Moivre, like so many of his coreligionists, was imprisoned during the religious tumult of 1685 and not released until 1688. Other, nearly contemporary sources report him in England by 1686. There he took up his lifelong, unprofitable occupation as a tutor in mathematics. On arrival in London, De Moivre knew many of the classic texts, but a chance encounter with

Newton's *Principia* showed him how much he had to learn. He mastered the book quickly; later he told how he cut out the huge pages and read them while walking from pupil to pupil. Edmond Halley, then assistant secretary of the Royal Society, was sufficiently impressed to take him up after meeting him in 1692; it was he who communicated De Moivre's first paper, on Newton's doctrine of fluxions, to the Royal Society in 1695 and saw to his election by 1697. (In 1735 De Moivre was elected fellow of the Berlin Academy of Sciences, but not until 1754 did the Paris Academy follow suit.)

Once Halley had made him known, De Moivre's talents became esteemed. He was able to dedicate his first book, *The Doctrine of Chances*, to Newton; and the aging Newton would, it is said, turn students away with "Go to Mr. De Moivre; he knows these things better than I do." He was admired in the verse of Alexander Pope ("Essay on Man" II, 104) and was appointed to the grand commission of 1710, by means of which the Royal Society sought to settle the Leibniz-Newton dispute over the origin of the calculus. Yet throughout his life De Moivre had to eke out a living as tutor, author, and expert on practical applications of probability in gambling and annuities. Despite his powerful friends he found little patronage. He canvassed support in England and even begged Johann I Bernoulli to get Leibniz to intercede on his behalf for a chair of mathematics at Cambridge, but to no avail. He was left complaining of the waste of his time spent walking between the homes of his pupils. At the age of eighty-seven De Moivre succumbed to lethargy. He was sleeping twenty hours a day, and it became a joke that he slept a quarter of an hour more every day and would die when he slept the whole day through.

De Moivre's masterpiece is *The Doctrine of Chances*. A Latin version appeared as "De mensura sortis" in *Philosophical Transactions of the Royal Society* (1711). Successively expanded versions under the English title were published in 1718, 1738, and 1756. The only systematic treatises on probability printed before 1711 were Huygens' *De ratiociniis in ludo aleae* and Pierre Rémond de Montmort's *Essay d'analyse sur les jeux de hazard* (Paris, 1708). Problems which had been posed in these two books prompted De Moivre's earliest work and, incidentally, caused a feud between Montmort and De Moivre on the subject of originality and priority.

The most memorable of De Moivre's discoveries emerged only slowly. This is his approximation to the binomial probability distribution, which, as the normal or Gaussian distribution, became the most fruitful single instrument of discovery used in probability theory and statistics for the next two centuries. In De Moivre's own time his discovery enormously clarified the concept of probability. At least since the fifteenth century there had been substantial work on games of chance that recognized the existence of stable frequencies in nature. But in the classic work of Huygens and even in that of Montmort, the reader was usually given, in the context of a game or lottery, a set of events of equal probability—a set of what were often called "chances"—and he was asked to derive further probabilities or expectations from this fundamental set. No one had a clear mathematical formulation of how "chances" and stable frequencies are related. Jakob I Bernoulli provided a first answer in part IV of his *Ars conjectandi* (Basel, 1713), where he proved what is now called the weak law of large numbers; De Moivre's approximation to the binomial distribution was conceived as an attempt to improve on Bernoulli.

In some experiment, let the ratio of favorable to unfavorable "chances" be p. In n repeated trials of the experiment, let m be the number of successes. Consider any interval around p, bounded by two limits. Bernoulli proved that the probability that m/n should lie between these limits increases with increasing n and approaches 1 as n grows without bound. But although he could establish the fact of convergence, Bernoulli could not tell at what rate the probability converges. He did obtain some idea of this rate by computing numerical examples for particular values of n and p, but he was unable to state the principles that underlie his discovery. That was left for De Moivre.

De Moivre's solution was published as a Latin pamphlet dated 13 November 1733. Introducing his translation of, and comments on, this work at the end of the last edition of *The Doctrine of Chances*, he took "the liberty to say, that this is the hardest Problem that can be proposed on the Subject of Chance" (p. 242). In this problem the probability of getting exactly m successes in n trials is expressed by the mth term in the expansion of $(a + b)^n$—that is, $\binom{n}{m} a^m b^{n-m}$, where a is the given ratio of chances and $b = 1 - a$. Hence the probability of obtaining a proportion of successes lying between the two limits is a problem in "approximating the Sum of the Terms of the Binomial $(a + b)^n$ expanded into a Series" (p. 243).

Working first with the binomial expansion of $(1 + 1)^n$, De Moivre obtained what is now recognized as $n!$ approximated by Stirling's formula—that is, $cn^{n+\frac{1}{2}}e^{-n}$. He knew the constant c only as the limiting sum of an infinite series: "I desisted in proceeding

farther till my worthy and learned Friend Mr. James Stirling, who had applied after me to that inquiry," discovered that $c = \sqrt{2\pi}$ (p. 244). Hence what is now called Stirling's formula is at least as much the work of De Moivre as of Stirling.

With his approximation of $n!$ De Moivre was able, for example, to sum the terms of the binomial from any point up to the central term. This summation is equivalent to the modern normal approximation and is, indeed, the first occurrence of the normal probability integral. He even appears to have perceived, although he did not name, the parameter now called the standard deviation σ. It was left for Laplace and Gauss to construct the equation of the normal curve in its form

$$\int \frac{1}{\sigma \sqrt{2\pi}} e^{-\frac{1}{2}\left(\frac{x-\mu}{\sigma}\right)^2} dx;$$

but De Moivre obtained, in a series of examples, expressions that are logically equivalent to this. He understood the rate of the convergence that Bernoulli had discovered and saw that the "error" —that is, the likely difference of the observed frequency from the true ratio of "chances"—decreases in inverse proportion to the square of the number of trials.

De Moivre's approximation is a theorem in probability theory: given the initial law about the distribution of chances, he could approximate the probability that observed frequencies should lie within any two assigned limits. Unlike some later workers, he did not imagine that his result would solve the converse statistical problem—namely, given the observed frequencies, to approximate the probability that the initial law about the ratio of chances lies within any two limits. But he did think his theorem bore on statistics. After summarizing his theorem, he reasoned:

> *Conversely*, if from numberless Observations we find the Ratio of the Events to converge to a determinate quantity, as to the Ratio of P to Q; then we conclude that this Ratio expresses the determinate Law according to which the Event is to happen. For let that Law be expressed not by the ratio P : Q, but by some other, as R : S; then would the Ratio of the Events converge to this last, and not to the former: which contradicts our *Hypothesis* [p. 251].

Nowhere in *The Doctrine of Chances* is this converse reasoning put to any serious mathematical use, yet its conceptual value is great. For De Moivre, it seemed to resolve the philosophical paradox of finding regularities within events postulated to be random. As he expressed it in the third edition, "altho' Chance produces Irregularities, still the Odds will be infinitely great, that in process of Time, those Irregularities will bear no proportion to the recurrency of that Order which naturally results from ORIGINAL DESIGN" (p. 251).

All the mathematical problems treated by De Moivre before setting out his approximation to the binomial distribution are closely related to earlier work by Huygens and Montmort. They include the first intimation of another approximation to the binomial distribution, now usually named for Poisson. In the normal approximation, the given ratio of chances is constant at p; and as n increases, so does np. In the Poisson approximation, np is constant, so that as n grows, p tends to zero. It is useful in studying the probabilities of rather infrequent events. Although De Moivre worked out a particular case of the Poisson approximation, he does not appear to have guessed its subsequent uses in probability theory.

Also included in *The Doctrine of Chances* are great advances in problems concerning the duration of play; a clearer formulation of combinatorial problems about chances; the use of difference equations and their solutions using recurring series; and, as illustrated by the work on the normal approximation, the use of generating functions, which, by the time of Laplace, came to play a fundamental role in probability mathematics.

Although no statistics are found in *The Doctrine of Chances*, De Moivre did have a great interest in the analysis of mortality statistics and the foundation of the theory of annuities. Perhaps this originated from his friendship with Halley, who in 1693 had written on annuities for the Royal Society, partly in protest at the inane life annuities still being sold by the British government, in which the age of the annuitant was not considered relevant. Halley had very meager mortality data from which to work; but his article, together with the earlier "political arithmetic" of John Graunt and William Petty, prompted the keeping of more accurate and more relevant records. By 1724, when De Moivre published the first edition of *Annuities on Lives*, he could base his computations on many more facts. Even so, he found it convenient to base most of his computations on Halley's data, derived from only five years of observation in the city of Breslau; he claimed that other results confirmed the substantial accuracy of those data. In his tables De Moivre found it convenient to suppose that the death rate is uniform after the age of twelve. He did not pretend that the rate is absolutely uniform, as a matter of objective fact, but argued for uniformity partly because of its mathematical simplicity and partly because the mortality

records were still so erratically collected that precise curve fitting was unwarranted.

De Moivre's contribution to annuities lies not in his evaluation of the demographic facts then known but in his derivation of formulas for annuities based on a postulated law of mortality and constant rates of interest on money. Here one finds the treatment of joint annuities on several lives, the inheritance of annuities, problems about the fair division of the costs of a tontine, and other contracts in which both age and interest on capital are relevant. This mathematics became a standard part of all subsequent commercial applications in England. Yet the authorship of this work was a matter of controversy. De Moivre's first edition appeared in 1725; in 1742 Thomas Simpson published *The Doctrine of Annuities and Reversions Deduced From General and Evident Principles*. De Moivre republished in the next year, bitter at what, with some justice, he claimed to be the plagiarization of his work. Since the sale of his books was a real part of his small income, money must have played as great a part as pride in this dispute.

Throughout his life De Moivre published occasional papers on other branches of mathematics. Most of them offered solutions to fairly ephemeral problems in Newton's calculus; in his youth some of this work led him into yet another imbroglio about authorship, involving some minor figures from Scotland, especially George Cheyne. In these lesser works, however, there is one trigonometric equation the discovery of which is sufficiently undisputed that it is still often called De Moivre's theorem:

$$(\cos \varphi + i \sin \varphi)^n = \cos n\varphi + i \sin n\varphi.$$

This result was first stated in 1722 but had been anticipated by a related formula in 1707. It entails or suggests a great many valuable identities and thus became one of the most useful steps in the early development of complex number theory.

BIBLIOGRAPHY

I. ORIGINAL WORKS. De Moivre's two books are *The Doctrine of Chances* (London, 1718; 2nd ed., 1738; 3rd ed., 1756; photo. repr. of 2nd ed., London, 1967; photo. repr. of 3rd ed., together with the biography by Helen M. Walker, New York, 1967); and *A Treatise of Annuities on Lives* (London, 1725), repr. in the 3rd ed. of *The Doctrine of Chances*. Mathematical papers are in *Philosophical Transactions of the Royal Society* between 1695 and 1744 (nos. 216, 230, 240, 265, 278, 309, 329, 341, 345, 352, 360, 373, 374, 451, 473). "De mensura sortis" is no. 329; the trigonometric equation called De Moivre's formula is in 373 and is anticipated in 309. *Approximatio ad summam terminorum binomii* $(a + b)^n$ *in seriem expansi* is reprinted by R. C. Archibald, "A Rare Pamphlet of De Moivre and Some of His Discoveries," in *Isis,* **8** (1926), 671–684. Correspondence with Johann I Bernoulli is published in K. Wollenshläger, "Der mathematische Briefwechsel zwischen Johann I Bernoulli und Abraham de Moivre," in *Verhandlungen der Naturforschenden Gesellschaft in Basel,* **43** (1933), 151–317. I. Schneider (below) lists all known publications and correspondence of De Moivre.

II. SECONDARY LITERATURE. Ivo Schneider, "Der Mathematiker Abraham de Moivre," in *Archive for History of Exact Sciences,* **5** (1968–1969), 177–317, is the definitive study of De Moivre's life and work. For other biography, see Helen M. Walker, "Abraham de Moivre," in *Scripta mathematica,* **2** (1934), 316–333, reprinted in 1967 (see above), and Mathew Maty, *Mémoire sur la vie et sur les écrits de Mr. Abraham de Moivre* (The Hague, 1760).

For other surveys of the work on probability, see Isaac Todhunter, *A History of Probability From the Time of Pascal to That of Laplace* (London, 1865; photo. repr. New York, 1949), 135–193; and F. N. David, *Gods, Games and Gambling* (London, 1962), 161–180, 254–267.

IAN HACKING

MOLDENHAWER, JOHANN JACOB PAUL (*b.* Hamburg, Germany, 11 February 1766; *d.* Kiel, Germany, 22 August 1827), *plant anatomy.*

Moldenhawer was one of the principal founders of plant anatomy. His chief work, published in 1812, reflects substantially the extensive knowledge acquired in this field during the period 1800–1830. During these years, which were characterized by a wealth of polemical tracts on the structure of the basic plant organs, he went his own way in his research and in his studies.

Moldenhawer was the son of the theologian and preacher Johann Heinrich Daniel Moldenhawer and his third wife. He studied theology, following the example of his elder brother, Daniel Gotthilf, a distinguished scholar of Greek and oriental languages and of dogmatics. Moldenhawer lived with his brother both as a student in Kiel until 1783 and in Copenhagen, where he was a candidate in theology. It is not known when he turned to the study of science, a change concurrent with his interest in literature, but it was probably in the mid-1780's. He was especially attracted to botany. Evidence of Moldenhawer's interest in these two areas is provided by his first publication, *Tentamen in historiam plantarum Theophrasti* (Hamburg, 1791), a philological study of Peripatetic botany based on ancient sources. On 13 April 1792 Moldenhawer was appointed extraordinary professor of botany and fruit-tree culture at the Faculty of Philosophy of the University of Kiel. He also lectured regularly on classical Greek literature, especially Pindar. He was able to do this because of his philological training while a theology student.

The scene of Moldenhawer's most important work was the fruit-tree nursery in Düsternbrook, near Kiel, which was associated with his professorship and which was run by Moldenhawer with great conscientiousness. He botanized only occasionally and directed work at the nursery toward applied botany, especially phytotomy, his major interest. Through use of the nursery, its library, and its five microscopes, Moldenhawer was able to produce his *Beiträge zur Anatomie der Pflanzen* (Kiel, 1812). This lifework, prepared over eighteen years of unremitting research, is notable for its critical insights and methodical observations.

The *Beiträge* contains important findings concerning plant anatomy that were made possible by a preparation method of Moldenhawer's own devising. He allowed the cells and vessels, which he recognized as structural elements, to macerate by decaying in water, then separated out the parts to be examined. His success was attributable to his use of the monocotyledonous corn plant as a subject of investigation and demonstration because of its simple structure and quick growth. The illustrations surpassed all earlier representations of plant anatomy. They included the first accurate depictions of the structure of the disputed fissured openings of the epidermis. By completely isolating the cells and vessels in his preparations Moldenhawer demonstrated that the cell wall is closed on all sides. This discovery clarified a long-contested question, since the membrane was seen to be doubled between two closely packed cellular spaces in intact tissue. Moldenhawer's later reputation was diminished primarily because he assumed, incorrectly, that cells and vessels were held together by a fibrous network. This assumption, all the more misleading because of his mistaken nomenclature, was in accord with Grew's hypothesis. On the other hand, Moldenhawer devised the concept of the vascular tissue (*Gefässbündel*), opposing it to that of the parenchyma. Herein lay his greatest achievement; with this radical new histological orientation he created the foundation of the theory of secondary thickening of woody stems, thereby separating himself most strikingly from the ideas of his predecessors (Grew, Malpighi) and contemporaries (Mirbel). Unfortunately, he never carried out his intention, expressed in 1812, of publishing a detailed work on the structure and development of the spiral vessels, one of his favorite objects of study.

In 1795 Moldenhawer married Catherina Dorothea Gädechens. They had one daughter, Pauline Mathilde, born in 1803. The family lived in Brunswick and later in Düsternbrook. Widely known and honored as a botanist, Moldenhawer was awarded the Danebrog Order in 1813 and was named king's counsel in 1824. He received a further honor in 1821, when H. A. Schrader named a legume genus *Moldenhawera*. Through a bequest of his daughter (1845) the botanical gardens of the University of Kiel received Moldenhawer's herbarium, which encompassed 120 files of plants arranged according to the systems of Forskål, Förster, and Linnaeus.

BIBLIOGRAPHY

Moldenhawer's two major works are mentioned in the text. See the obituary in *Neuer Nekrolog der Deutschen*, V, pt. 2 (Ilmenau, 1829); and the article in *Allgemeine deutsche Biographie*, XXII (Leipzig, 1885). For the report on his daughter's bequest to the University of Kiel, see *Botanische Zeitung*, **3** (1845), 262; and Prahl (cited below), II, 38.

For works about Moldenhawer see J. H. Barnhart, *The New York Botanical Garden. Biographical Notes Upon Botanists* (Mschr.), II (Boston, 1965); C. Harms, *Lebenbeschreibung* (Kiel, 1851); E. Hofmann, "Philologie," in *Geschichte der Philosophischen Fakultät*, pt. 2, K. Jordan and E. Hofmann, eds., *Geschichte der Christian-Albrechts-Universität Kiel 1665–1965*, V (Neumünster, 1969); P. Knuth, *Geschichte der Botanik in Schleswig-Holstein* (Kiel–Leipzig, 1890–1892); M. Möbius, *Geschichte der Botanik* (Jena, 1937); and F. Overbeck, "Botanik," in *Geschichte der Mathematik, der Naturwissenschaften und der Landwirtschaftswissenschaften*, K. Jordan, ed., *Geschichte der Christian-Albrechts-Universität Kiel 1665–1965*, VI (Neumünster, 1968). See also P. Prahl, ed., *Kritische Flora der Provinz Schleswig-Holstein, des angrenzenden Gebiets der Hansastädte Hamburg und Lübeck und des Fürstentums Lübeck*, 2 vols. (Kiel, 1888–1890); H. Röhrich, "Memoria horti medici Academiae Kiliensis III," in *Schleswig-Holsteinisches Ärzteblatt*, **18** (1965), 376–382; J. Sachs, "Geschichte der Botanik vom 16. Jahrhundert bis 1860," in *Geschichte der Wissenschaften in Deutschland*, XV (Munich, 1875); F. Volbehr and R. Weyl, in R. Bülck and H. J. Newiger, eds., *Professoren und Dozenten der Christian-Albrechts-Universität zu Kiel 1665–1954*, 4th ed. (Kiel, 1956); and O. F. Wiegand, *Bibliographie zur Geschichte der Christian-Albrechts-Universität Kiel* (Kiel, 1964).

JÖRN HENNING WOLF

MOLESCHOTT, JACOB (*b.* 's Hertogenbosch, Netherlands, 9 August 1822; *d.* Rome, Italy, 20 May 1893), *medicine, physiology.*

Moleschott's father, Johannes Franciscus Gabriel Moleschott, was a physician; his mother was the former Elisabeth Antonia van der Monde. He attended the Gymnasium at Cleves, Germany, then studied medicine at Heidelberg (1842–1845) under Tiedemann, Naegele, and Henle, while also pursuing his interest in the philosophy of Ludwig Feuerbach, Karl Vogt, and Hegel. On 22 January 1845 he received his medical degree with the thesis *De Malpighianis pulmonum vesiculis*. It had been written under the direction of Jakob Henle, who had instructed him in the use of the microscope. Moleschott then settled in Utrecht as a general practitioner. He was a pupil of G. J. Mulder from 1845 to 1847; in Mulder's laboratory he met the physiologists F. C. Donders (who did not share his enthusiasm for materialistic monism) and I. van Deen. With them he conducted an extensive scientific and private correspondence, much of which has been preserved.

While still a student Moleschott entered a prize competition sponsored by the Teyler's Society in Haarlem and received an award. His growing interest in physiology led him to publish, with Donders and van Deen, the first Dutch journal (in German) for anatomy and physiology: *Holländische Beiträge zu den anatomischen und physiologischen Wissenschaften* (1846–1848). In 1848 Moleschott violently opposed the appointment of H. J. Halbertsma as professor of anatomy and physiology at Leiden because his friend van Deen had been passed over. Moleschott was made *Privatdozent* in physiology and anthropology at Heidelberg in 1847 and resigned in 1854 after having been sharply reprimanded by the rector and senate because they felt that he was misleading the students. (Among other things, he had spoken out in favor of cremation.) He was appointed professor of physiology at Zurich in 1856, giving an oration, "Licht und Leben," based upon his own observations. In the following year he started publication of the journal *Untersuchungen zur Naturlehre des Menschen und der Thiere*, which he edited and which was continued until 1894 under G. Colasanti and S. Fubini.

While in Zurich, Moleschott was especially concerned with research on the cardiac nervous system, the respiratory system, the smooth muscles, and embryology, the last being closely connected with his teaching. At the urging of Cavour he was appointed professor of experimental physiology and physiological chemistry at Turin in 1861. He became professor at the "Sapienza" in Rome in 1879 and later a senator. (For many years he had studied the Italian language and literature; he became an Italian citizen.)

Moleschott's special interests were in the metabolism of plants and animals, and in the effect of light on it in nutrition. His most important work, *Kreislauf des Lebens* (1852), concerned the structure and function of the brain and contained arguments against Liebig's *Chemische Briefe* and strong statements favoring Moleschott's own materialistic view of life. The work was highly praised by Humboldt and was translated into French, Italian, and Russian.

It cannot be said that Moleschott possessed great creativity; but he did have a strong love for science, especially experimental physiology. He died of erysipelas at the age of seventy. His library was donated to the University of Turin.

BIBLIOGRAPHY

I. ORIGINAL WORKS. Moleschott's writings include *Kritische Betrachtung von Liebig's Theorie der Ernährung der Pflanzen, im Jahre 1844 gekrönte Preisschrift* (Haarlem, 1845); *De Malpighianis pulmonum vesiculis dissertatio* (Heidelberg, 1845); *Holländische Beiträge zu den anatomischen und physiologischen Wissenschaften* (Dusseldorf–Utrecht, 1846–1848), written with F. C. Donders and I. van Deen; *Lehre der Nahrungsmittel* (Erlangen, 1850; 2nd ed., 1858); *Physiologie der Nahrungsmittel* (Darmstadt, 1850; 2nd ed., Giessen, 1859); *Physiologie des Stoffwechsels in Pflanzen und Thieren* (Erlangen, 1851); *Der Kreislauf des Lebens* (Mainz, 1852; 5th ed., 2 vols., Mainz–Giessen, 1877, 1887), also trans. into French (Paris, 1866); *Untersuchungen zur Naturlehre des Menschen und der Thiere*, 15 vols. (Frankfurt–Giessen, 1857–1892), of which the last contains a list of his works; and *Für meine Freunde. Lebenserinnerungen* (Giessen, 1894; 2nd ed., 1901).

With G. E. V. Schneevoogt, Moleschott translated Carl Rokitansky's *Handboek der bijzondere ziektekundige ontleedkunde*, 2 vols. (Haarlem, 1849).

II. SECONDARY LITERATURE. See A. Cantani and W. Haberling, in *Biographisches Lexicon hervorragender Aerzte*, IV (Berlin–Vienna, 1932), 232; I. van Esso, "Jacob Moleschott," in *Nederlands tijdschrift voor geneeskunde*, **93** (1949), 1; A. A. Guye, "Jacob Moleschott," *ibid.*, **28** (1892), 325; R. E. de Haan, *Jacob Moleschott* (Haarlem, 1883); M. A. van Herwerden, "Eine Freundschaft von drei Physiologen," in *Gids*, **7** (1914), 448, also in *Janus*, **20** (1915), 174–201, 409–436; and the article in *Nieuw nederlandsch biografisch woordenboek*, III (Leiden, 1914), 874; C. A. Pekelharing, "Jakob Moleschott," in *Nederlands tijdschrift voor geneeskunde*, **29** (1893), 1741; and B. J. Stokvis, "Jacob Moleschott," in *Gids*, **5** (1892), 339.

See also *In memoria di Jacopo Moleschott* (Rome, 1894), which includes a bibliography of his works.

A. M. GEIST-HOFMAN

MOLIÈRES, JOSEPH PRIVAT DE. See **Privat de Molières, Joseph.**

MOLIN, FEDOR EDUARDOVICH (*b.* Riga, Russia, 10 September 1861; *d.* Tomsk, U.S.S.R., 25 December 1941), *mathematics.*

Molin graduated from the same Gymnasium in Riga at which his father was a teacher. He then entered the Faculty of Physics and Mathematics at Dorpat University (now Tartu University), from which he graduated in 1883 with the rank of candidate and remained in the department of astronomy to prepare for a teaching career. In the same year he was sent to Leipzig University, where he attended the lectures of Felix Klein and Carl Neumann. Under the guidance of Klein he wrote his master's thesis ("Über die lineare Transformation der elliptischen Functionen"), which he defended in 1885 at Dorpat, where he then became *Dozent.*

During this period Molin became acquainted with the works of Sophus Lie and began to study hypercomplex systems. His most profound results in this field were presented in his doctoral dissertation, which he defended in 1892. Despite his outstanding work, Molin was unable to obtain a professorship at Dorpat and in 1900 moved to Tomsk, in west-central Siberia, where he found himself cut off from centers of scientific activity. He occupied the chair of mathematics at Tomsk Technological Institute and from 1918 was professor at Tomsk University. In 1934 he received the title Honored Worker of Science.

Molin obtained fundamental results in the theory of algebras and the theory of representation of groups. In his doctoral dissertation, which concerned the structure of an arbitrary algebra of finite rank over a field of complex numbers C, he showed that a simple algebra over C is isomorphic to a complete ring of matrices. He also introduced the concept of a radical (the term was introduced by Frobenius) and showed that the structure of an arbitrary algebra is reduced essentially into the case where factor algebra by a radical decomposes into a direct sum of simple algebras. Cartan later obtained the same results, which he introduced into the case of an algebra over a field of real numbers. In 1907 Wedderburn extended Molin's and Cartan's results into the case of an algebra over an arbitrary field.

Studying the theory of representation of groups, Molin explicitly introduced a group ring and showed that it is a semisimple algebra broken into the direct sum of S simple algebras, where S is the order of the center. This proved the decomposability of the regular representation into irreducible parts. Molin showed that every irreducible representation of the group is contained in the regular representation. He also demonstrated that representations of groups up to equivalence are determined by their traces. At the same time analogous results were obtained in a different way by Frobenius, who later became acquainted with Molin's research and valued it highly.

BIBLIOGRAPHY

I. Original Works. Molin's writings include "Über die lineare Transformation der elliptischen Functionen" (Dorpat, 1885), his master's thesis; "Über Systeme höherer complexer Zahlen," in *Mathematische Annalen*, **41** (1893), 83–156, his doctoral dissertation; "Eine Bemerkung über endlichen linearen Substitutionsgruppen," in *Sitzungsberichte der Naturforscher-Gesellschaft bei der Universität Jurjew*, no. 11 (1896–1898), 259–276; "Über die Anzahl der Variabelen einer irreductibelen Substitutionsgruppen," *ibid.*, 277–288; and "Über die Invarianten der linearen Substitutionsgruppen," in *Sitzungsberichte der Preussischen Akademie der Wissenschaften zu Berlin*, **52** (1897), 1152–1156.

II. Secondary Literature. See N. Bourbaki, *Éléments d'histoire des mathématiques* (Paris, 1969), 152, 154; and N. F. Kanunov, *O rabotakh F. E. Molina po teorii predstavlenia grupp* ("On the Works of F. E. Molin on the Theory of the Representation of Groups"), no. 17 in the series Istoriko-Matematicheskie Issledovania ("Historical–Mathematical Research"), G. F. Rybkin and A. P. Youschkevitch, eds. (Moscow, 1966), 57–88.

J. G. Bashmakova

MOLINA, JUAN IGNACIO (*b*. Guaraculen, Talca, Chile, 24 June 1740; *d*. Bologna, Italy, 12 September 1829), *natural history*.

Molina received his early education at Talca; when he was sixteen, he entered the Jesuit college at Concepción, where he studied languages and the natural sciences. He entered the Jesuit order and was made librarian of the college, but in 1768 he had to leave Chile because of the expulsion of the Jesuits from the Spanish dominions. Molina received holy orders upon arrival at Imola, Italy; and in 1774 he was appointed professor of natural sciences at the Institute of Bologna, where he wrote most of his works. Some of his lectures maintained the analogy of the matter of living organisms and of minerals and the idea of the evolution of human beings, and he was censured by his superiors. Molina, who remains the classic author on the natural history of Chile, incorporated the observations of A. F. Frézier and Feuillée in the 1776 revised edition of his *Compendio*.

BIBLIOGRAPHY

I. Original Works. Molina first published *Compendio della storia geografica, naturale, e civile del regno del Chile* (Bologna, 1776) anonymously; it was greatly improved in its 2nd ed., *Saggio sulla storia naturale del Chile* (Bologna, 1782). *Storia civile* (Bologna, 1786) was trans. into German, Spanish, French, and English. Molina's pupils published his 14 major essays on natural history under the title *Memorie di storia naturale lette in Bologna nelle adunaza dell'Istituto*, 2 vols. (Bologna, 1821).

II. Secondary Literature. See Rodolfo Jaramillo Barriga, *El abate Juan Ignacio Molina, primer evolucionista y precursor de Teilhard de Chardin* (Santiago de Chile, 1963); Enrique Laval, "La medicina en el abate Molina," in *Anales chilenos de historia de la medicina* (1965); and Miguel Rojas Mix, in *Anales de la Universidad de Chile* (1965).

Francisco Guerra

MOLL, FRIEDRICH RUDOLF HEINRICH CARL (*b*. Culm, Germany, 31 January 1882; *d*. Berlin, Germany, 8 May 1951), *naval engineering, wood technology*.

After working on the docks and as a shipwright, Moll studied shipbuilding at the Technische Hochschule in Berlin-Charlottenburg from 1902 to 1907. Following his graduation he worked as an engine operator on English trawlers, then, in 1909, received a doctorate in engineering for a work on the possible causes of disappearance of long-missing trawlers.

From 1907 Moll was chiefly concerned with the preservation of wood. He obtained contracts to construct plants for impregnating telephone poles with mercuric chloride and studied this process (kyanizing) from a scientific, as well as technical, point of view. He published a large number of papers on both the biological and chemical aspects of preserving wood. From 1911 Moll privately built wood treatment works in a number of countries and his operations acquired an international reputation.

In 1920 Moll was awarded the doctor of philosophy degree by the University of Berlin for a study of the toxic effects of salts on fungi. Without giving up his profession of wood technologist, he qualified as a university lecturer and was *Privatdozent* at the Technische Hochschule in Berlin-Charlottenburg from 1922 until 1936, lecturing on the preservation of wood. He was thus led to prepare a comprehensive course of lectures on wood technology. Moll was a member of several national and international wood preservation societies and his authoritative papers appeared in many technical journals.

All of Moll's publications endorse the kyanization process that he had helped to develop. Although he rejected in principle the use of arsenic as a preservative agent, he fully recognized that other, more sophisticated agents—including the bifluorides—must be the future means of preserving wood. After World War II Moll did work in Berlin on a number of topics, including the geographic distribution of the Teredinidae in Africa. He also worked closely with American wood experts on developing techniques for protection against shipworms.

BIBLIOGRAPHY

I. ORIGINAL WORKS. Moll's writings include *Über die Ursachen des Unterganges der verschollenen Fisch-dampfer* (Berlin, 1909); "Schutz des Bauholzes in den Tropen gegen die Zerstörung durch die Termiten," in *Tropenpflanzer*, **18** (1915), 591–605; "Untersuchungen über Gesetzmässigkeiten in der Holzkonservierung. Die Giftwirkung anorganischer Verbindungen (Salze) auf Pilze," in *Zentralblatt für Bakteriologie, Parasitenkunde, Infektionskrankheiten und Hygiene*, **51** (1920), 257–279; *Das Schiff in der bildenden Kunst* (Bonn, 1929); *Der Schiffbauer in der bildenden Kunst* (Berlin, 1930); *Künstliche Holztrocknung* (Berlin, 1930); "Teredinidae of the British Museum," in *Proceedings of the Malacological Society of London*, **19** (1931), 201–218; *Der Schutz des Bauholzes und die Schädlingsbekämpfung mit chemischen Mitteln* (Karlsruhe, 1939); *Die Terediniden im königlichen Museum für Naturkunde zu Brüssel* (Brussels, 1940); "Übersicht über die Terediniden des Museums für Naturkunde Berlin," in *Sitzungsberichte der Gesellschaft naturforschender Freunde zu Berlin 1940* (1941), 152–219; *Zeitgemässe Verwendung von Holz in Bauwesen* (Berlin, 1942); and *Geographical Distribution of the Teredinidae of Africa* (London, 1949).

II. SECONDARY LITERATURE. Obituaries include that by Max Seidel, in *Norddeutsche Holzwirtschaft*, **5** (1951), 148–149; and those in *Chemiker Zeitung*, **75** (1951), 313–314; and *Holz-Zentralblatt*, **77** (1951), 794.

KURT MAUEL

MOLL, GERARD (*b.* Amsterdam, Netherlands, 18 January 1785; *d.* Amsterdam, 17 January 1838), *astronomy, physics.*

Moll had an enthusiastic interest in many of the physical sciences of his day. His contributions ranged from observing a transit of Mercury to determining the speed of sound.

Moll's father, a well-to-do businessman, was also named Gerard; his mother was the former Anna Diersen. Although destined for a commercial career, Moll met and talked to sea captains while serving his apprenticeship in Amsterdam, and became intrigued with the art of celestial navigation—to such an extent that he decided to change to astronomy as his life's work.

Moll studied at the University of Amsterdam, receiving his Ph.D. in 1809, and then continued his studies for some months in Paris. Returning to Holland in 1812, he was appointed director of the observatory in Utrecht. When that university was reorganized in 1815, Moll became professor of physics as well, and continued in both these positions until his death.

With little financial support and a crumbling observatory building, Moll contributed to astronomy rather more by personal contacts with scientists in other countries—especially in Great Britain—than by observing the heavens. His main astronomical accomplishment seems to have been his observation of the transit of Mercury of 5 May 1832.

In physics Moll made several contributions. With Albert van Beek he measured the speed of sound; an artillery battalion was placed at the experimenters' disposal, cannon were fired simultaneously—at night —from hills about nine miles apart, and the interval between light flash and sound was recorded at either end and then averaged. The value obtained was 332.05 m./sec. (the currently accepted value is 331.45).

Moll also extended the pioneering observations of H. C. Oersted, published in 1820, on the magnetic field that surrounds a wire carrying an electric current. He also investigated the lifting capacities of the electromagnets based on this phenomenon.

In recognition of his services on a commission dealing with weights and measures, the Kingdom of the Netherlands in 1815 appointed Moll *chevalier* of the Order of the Belgian Lion. In 1835 the University of Edinburgh gave him an honorary LL.D., and in 1836 the University of Dublin followed suit. Moll was buried beside his mother in Amerongen, some fifteen miles east of Utrecht.

BIBLIOGRAPHY

I. ORIGINAL WORKS. The experimental work referred to above is described in "An Account of Experiments on the Velocity of Sound, Made in Holland," in *Philosophical Transactions of the Royal Society*, **114** (1824), 425–456, written with A. van Beek; *Electro-magnetische Proeven* (Amsterdam, 1830); "Ueber die Bildung künstlicher Magnete mittelst der Voltaschen Kette," in *Annalen der Physik und Chemie*, 2nd ser., **29** (1833), 468–479; and "On the Transit of Mercury of May 5, 1832," in *Memoirs of the Royal Astronomical Society*, **6** (1833), 111–117.

A list of fifty articles by Moll appears in the Royal Society *Catalogue of Scientific Papers*, IV, 433–434. There is some duplication of subject matter, as Moll's work tended to appear simultaneously in at least two countries.

II. SECONDARY LITERATURE. An unsigned obituary of Moll appeared in *Annual Register* for 1838 (London, 1839), app., p. 198. A more extensive notice, by A. Quetelet, in *Annuaire de l'Académie des sciences, des lettres, et des beaux-arts de Belgique*, **5** (1839), 63–79, refers to other sources of biographical information in Dutch and Latin.

Note: The so-called Moll's thermopile was invented by Willem Jan Henri Moll in 1913.

SALLY H. DIEKE

MÖLLER, DIDRIK MAGNUS AXEL (*b.* Sjörup, Sweden, 16 February 1830; *d.* Lund, Sweden, 26 October 1896), *astronomy.*

Möller studied at the University of Lund, where in 1853 he became a docent, in 1855 associate professor, and in 1863 full professor. He was also director of the observatory, which he founded, until his resignation in 1895.

Möller's predecessor, John Mortimer Agardh, had sought to establish an observatory at Lund, but his efforts bore fruit only after his death. The government granted money for an observatory in 1863, and during the following years Möller devoted much of his time to its completion. The main building, still in use, was dedicated in 1867. Among the instruments Möller ordered were a refractor with a nine-inch objective and a thirteen-foot focal length, installed in 1867, and a meridian circle with a six-inch aperture and a seven-foot focal length, mounted in 1874. He intended these two instruments to be used simultaneously. Differential measurements of moving objects (comets and planets) and stars were made with the refractor, and accurate positions of the stars were determined with the meridian circle. Positions of about 11,000 stars in the declination zone +35° to +40° were measured with the meridian circle, and observations of planets, comets, and double stars were made with the refractor. The recently modernized meridian circle is still in use.

Möller's most important contributions to astronomy concern the motion of the comet discovered in 1843 by the French astronomer H. Faye and that of the asteroid Pandora. His interest in Faye's comet was aroused by J. F. Encke, who, on the basis of Newton's law, had computed the orbit of a comet of very short period (3.3 years) first seen in 1786 (later called Encke's comet) and found that the comet showed a retardation in relation to the computed positions. His explanation was that in interplanetary space there is a low-density medium which slows the motion of the comet. Möller started to test Encke's hypothesis by using Faye's comet, which has a period of 7.5 years.

He first concluded that the observations of this comet indeed indicated a retardation in comparison with theory. But through new, laborious, and careful calculations he was able to show that full agreement between theory and observations was obtained on the basis of Newton's theory: when the comet was observed in 1865, 1873, and 1880, the agreement was perfect. Encke's hypothesis could be rejected. For this brilliant work Möller was awarded the gold medal of the Royal Astronomical Society in 1881.

In several papers Möller studied the motion of the asteroid Pandora. He first calculated the special perturbations and later, according to Hansen's method, the general perturbations, including certain second-order perturbations depending on the masses of Jupiter and Saturn. In this case too Möller, through his skillful and accurate calculations, reached extremely good agreement with the observations. Besides his theoretical work Möller performed extensive series of observations of planets and comets.

Möller was an exceptionally able person, and was frequently called on for special commissions by the university and other agencies. He served as rector of the university in 1874–1875 and 1891–1895.

BIBLIOGRAPHY

Möller contributed many articles on Faye's comet and Planet 55 Pandora to *Astronomische Nachrichten* and *Vierteljahrsschrift der Astronomischen Gesellschaft*. Poggendorff, III, 924, gives a list of these and other publications.

In addition see Erik Holmberg, "Lundensisk astronomi under ett sekel," in *Cassiopeia* (1949), 21–27; Anders Lindstedt, "Didrik Magnus Axel Möller," in *Minnesteckning, Kungliga Svenska Vetenskapakademiens lefnadsteckningar*, **4** (1899–1912), no. 79; and Carl Schalén, Nils Hansson, and Arvid Leide, "Astronomiska Observatoriet vid Lunds Universitet," in *Lunds Universitets historia*, **4** (1968), 52–72.

C. SCHALÉN

MOLLIARD, MARIN (*b.* Châtillon-Colligny, Loiret, France, 8 June 1866; *d.* Paris, France, 24 July 1944), *plant physiology.*

For fifty years after his graduation from the École Normale Supérieure in 1894, Molliard taught and worked at the Faculty of Sciences of the Sorbonne; he was its dean for six years and had the first chair of plant physiology in France created for him. Deeply imbued with Lamarck's ideas, he devoted all his writings to the influence of the environment on plants. He stated:

The most general idea which emerges from my studies is that plants, even the most differentiated, are extremely plastic, much more so than has been admitted until now, that their structure is closely dependent on their chemistry, the latter being influenced by external conditions; it is therefore an experimental confirmation that my researches contribute to Lamarck's theory, insofar as its essential features are concerned [*Oeuvres scientifiques*, p. 6].

Molliard began by investigating the morphological transformations that certain parasites produce in plants and that lead to the formation of galls; his last works were concerned with the conditions of tuberization in the potato *(Solanum tuberosum)*. In order to carry out precise studies, he controlled all the nutrients. He also eliminated all possible parasites by cultivating his plants, particularly radishes, in an aseptic environment from germination to fructification.

Molliard systematically investigated the mineral nourishment of the mold *Sterigmatocystis nigra*. Normally it lives on sucrose, which it transforms into carbon dioxide and water by respiration; only a very small amount of organic acids appears in the medium. If too little nitrogen is furnished, a large quantity of citric acid is produced in the medium. If too little phosphorus is supplied, the medium abounds in both citric acid and oxalic acid. If potassium is lacking, only oxalic acid is abundant. If all of the mineral elements are reduced, gluconic acid appears. Hence, well before Hans Krebs discovered the acid cycle named for him, Molliard drew attention to the importance of organic acids in intermediate metabolism.

Molliard's studies on the radish are famous. Cultivated aseptically, provided with light but in an atmosphere without any carbon dioxide—and consequently without the assimilation of chlorophyll—the plant absorbs glucides through its roots. If the supply of glucides is abundant, the form of the radish is altered; the glucides no longer accumulate in the tissues as sucrose or monosaccharides but as starch. The reserves, instead of remaining in the root, move into the stem, which swells and acquires the characteristics and appearance of a subterranean stem. The organic nutrition totally transforms the physiology of the plant and thus modifies its microscopic appearance and morphology.

Through multiple experiments of the same kind Molliard showed that it is possible, by varying only the nutrition, to transform ordinary leaves into cotyledons. In addition, certain leaves can be changed into thorns and certain thorns into leaves. He also showed that parasitic plants like the dodder *(Cuscuta)*,

which ordinarily feeds on clover and alfalfa to which it attaches itself by its suckers, are able, if they receive suitable nutrition, to live independently without suckers but with abundant chlorophyll.

Molliard was the leading authority on plant physiology in France. He spent the whole of his professional life in the same laboratory, on the second floor of the old Sorbonne building. From 1894 to 1940 all plant physiologists in France were more or less his direct pupils. His writings consist primarily of short notes to the Academy of Sciences, of which he was a very influential member.

BIBLIOGRAPHY

Molliard's most important book is *Nutrition de la plante*, 4 vols. (Paris, 1923). Among his many papers are the series with the general title "Recherches physiologiques sur les galles," in *Revue générale de botanique*, **25** (1913), 225–252, 285–307, 341–370; most appear in *Oeuvres scientifiques* (Paris, 1936), which consists of works republished or abridged under the supervision of a group of his students and friends. See also the obituary by Charles Maurain in *Comptes rendus hebdomadaires des séances de l'Académie des sciences*, **219** (1944), 144–147.

JULES CARLES

MOLLIER, RICHARD (*b.* Trieste, 30 November 1863; *d.* Dresden, Germany, 13 March 1935), *thermodynamics.*

Mollier was the eldest son of German parents. His father, Eduard Mollier, a Rhinelander, was a naval engineer at, and later director of, a Trieste machine factory; his mother (née von Dyck) was a native of Munich. After graduating *summa cum laude* (1882) from the local German Gymnasium, Mollier studied mathematics and physics at the universities of Graz and Munich. He soon transferred to the Technische Hochschule of Munich, where Moritz Schröter and Carl von Linde became his most influential teachers. He graduated in 1888 and, after a brief engineering practice at his father's factory in Trieste, became Schröter's assistant in 1890. His first scientific investigations were his *Habilitation* thesis on thermal diagrams in the theory of machines (1892) and his doctoral dissertation at the University of Munich on the entropy of vapors (1895). In 1896 Felix Klein, who was conducting a wide-ranging campaign to reunite science and technology, called Mollier to the University of Göttingen to introduce "technical physics" into the curriculum. Mollier's stay was brief. Feeling isolated in a purely scientific atmosphere, he was delighted to answer an invitation in 1897 to succeed Gustav Zeuner at the Technische Hochschule of Dresden. In this post, as professor of the theory of machines and director of the machine laboratory, Mollier spent his working life. He subsequently received international recognition for his contributions to thermodynamics, as well as the concomitant honors.

Mollier was unassuming and kindly, if somewhat retiring, and he took his teaching duties seriously. His lectures, prepared by a unique method, were much praised. Using no notes, he would compose and memorize them, so that clarity of organization and simplicity of style were combined with spontaneity. Several of his pupils became notable contributors to thermodynamics, including F. Bošnjaković, F. Merkel, Wilhelm Nusselt, and Rudolf Plank—as well as his own sister Hilde Mollier, who later married the electronics pioneer H. G. Barkhausen.

Although his engineering colleagues considered Mollier a pure theoretician—instead of experimenting himself, he based his findings upon the empirical data of others—his role was actually that of a mediator between the theoretical work of Clausius and J. W. Gibbs (whose work he knew through Ostwald's 1892 German translation) and the realm of practical engineering. From the beginning his interest centered on the properties of thermodynamic media and their effective presentation in the form of charts and diagrams. It was here that he made his crucial contribution. Engineers had traditionally visualized thermodynamic processes in terms of the pressure-volume (P-V) diagram with which they were familiar from practical experience with the steam engine indicator. This diagram, however, obscured the significance of the second law of thermodynamics. In 1873 Gibbs had suggested an alternative in the temperature-entropy (T-S) diagram where Carnot processes stand out as simple rectangles, and the degree of approximation of actual thermodynamic processes to ideal ones can be easily judged. It was at this point that Mollier introduced the concept of *enthalpy,* a property of state that was then little known (1902). This property had been defined in 1875 by Gibbs, under the name "heat function for constant pressure," as the sum of internal energy and of the product of pressure and volume (the term "enthalpy" was coined later by Kamerlingh Onnes). Like Clausius' entropy, enthalpy is an abstract property that cannot be measured directly. Its great advantage is that it describes energy changes in thermodynamic systems without requiring a distinction between heat and work. Employing this new property of state, in 1904 Mollier devised an enthalpy-entropy (H-S) diagram, which

retained most of the advantages of the T-S diagram, while acquiring some additional ones. While vertical lines signified, as before, reversible processes, horizontal lines in it described processes of constant energy; the diagram thus demonstrated in strikingly simple fashion the essence of both the first and the second law of thermodynamics. Quantities of work, which in the P-V and the T-S diagrams had appeared as an area, as well as discharge velocities through adiabatic nozzles, were represented here simply as vertical distances. Although the H-S diagram quickly became a principal tool of power and refrigeration engineers, to Mollier it was merely an element in a broad reorganization of thermodynamic practice. He also developed a new system of thermodynamic computation in which enthalpy played an important role, and as a basis for such calculations he published charts and diagrams of the properties of steam and of various refrigerants (his steam tables, first published in 1906, quickly went through seven editions). Besides the H-S diagram he proposed a number of other enthalpy diagrams, which have all become known, upon recommendation of the U.S. Bureau of Standards in 1923, as Mollier diagrams.

Mollier also contributed to other areas of thermodynamics. In 1897 he published an important study on heat transfer, before turning this subject over to his pupil Wilhelm Nusselt, who soon made fundamental contributions to it. His presentation of the first mathematical analysis of the process of combustion (1921) has proven of lasting utility.

BIBLIOGRAPHY

I. ORIGINAL WORKS. Except for the chapter "Wärme," in Akademischer Verein Hütte, *Hütte: Des Ingenieurs Taschenbuch*, 18th ed. (Berlin, 1902), and *Neue Tabellen und Diagramme für Wasserdampf* (Berlin, 1906; 7th ed., 1932), Mollier's publications were confined to journals. His most important research papers are "Über die kalorischen Eigenschaften der Kohlensäure und anderer technisch wichtiger Dämpfe," in *Zeitschrift für die gesamte Kälteindustrie*, **2** (1895), 66–70, 85–91; "Über die kalorischen Eigenschaften der Kohlensäure ausserhalb des Sättigungsgebietes," *ibid.*, **3** (1896), 65–69, 90–92; "Über den Wärmedurchgang und die darauf bezüglichen Versuchsergebnisse," in *Zeitschrift des Vereins deutscher Ingenieure*, **41** (1897), 153–162, 197–202; "Über die Beurteilung der Dampfmaschinen," *ibid.*, **42** (1898), 685–689; "Dampftafel für schweflige Säure," in *Zeitschrift für die gesamte Kälteindustrie*, **10** (1903), 125–127; "Neue Diagramme zur technischen Wärmelehre," in *Zeitschrift des Vereins deutscher Ingenieure*, **48** (1904), 271–275; "Gleichungen und Diagramme zu den Vorgängen im Gasgenerator," *ibid.*, **51** (1907), 532–536; "Die physikalischen

Grundlagen der Kältetechnik," in *Zeitschrift für die gesamte Kälteindustrie*, **16** (1909), 186–190; "Die technische Darstellung der Zustandsgleichungen," in *Physikalische Zeitschrift*, **21** (1920), 457–463; "Die Gleichungen des Verbrennungsvorganges," in *Zeitschrift des Vereins deutscher Ingenieure*, **65** (1921), 1095–1096; "Ein neues Diagramm für Gasluftgemische," *ibid.*, **67** (1923), 869–872; and "Das i/x-Diagramm für Dampfluftgemische," *ibid.*, **73** (1929), 1009–1013.

II. SECONDARY LITERATURE. The following are particularly useful for biographical data: N. Elsner, "Richard Mollier als Mensch und Wissenschaftler," in *Wissenschaftliche Zeitschrift der Technischen Universität Dresden*, **13** (1964), 1101–1103; Heinz Jungnickel, "Kältetechnik—Stand und Entwicklung," *ibid.*, 1105–1106; Walter Pauer, "Erinnerungen an Richard Mollier," *ibid.*, 1103–1104; and Rudolf Plank, "Richard Mollier zum 70. Geburtstag," in *Zeitschrift für die gesamte Kälteindustrie*, **40** (1933), 165–167; and "Richard Mollier," in *Kältetechnik*, **15** (1963), 342–344. Poggendorff, VI, 1766; and VIIa, pt. 3, 342; gives a number of further references.

OTTO MAYR

MOLLWEIDE, KARL BRANDAN (*b*. Wolfenbüttel, Germany, 3 February 1774; *d*. Leipzig, Germany, 10 March 1825), *astronomy, mathematics.*

Mollweide graduated from the University of Halle, then became a teacher of mathematics in the Pädagogium of the Franckesche Stiftung there. In 1811 he was appointed to a position at Leipzig University, where he worked in the astronomical observatory that had been established in the old castle of Pleissenburg; the post carried with it the title of professor, and the following year he was made full professor of astronomy. In 1814 Mollweide was appointed to the chair of mathematics, one of the old and privileged university posts that carried with it the right to become dean or rector; he was twice dean during his eleven-year tenure at Leipzig.

Mollweide's two professorships left him little time to make astronomical observations—during term he usually gave four courses that met for fourteen to sixteen hours weekly. In his astronomy courses he emphasized the fixing of stellar positions, although he also treated the other branches of the subject; his mathematical courses comprised arithmetic, algebra, analysis, stereometry, trigonometry, analytical geometry, conics, and the theory of probability. He nevertheless was able to publish a number of scientific works; some of them represented his own researches, others were editions of standard authors and logarithmic tables.

Certain trigonometrical formulas and a conformal map projection are named for Mollweide. He is also known for his youthful dispute with Goethe over the latter's *Farbenlehre*, in which he defended the Newtonian theory of colors that Goethe was never able to accept.

BIBLIOGRAPHY

A more complete list of Mollweide's writings is in Poggendorff, II, cols. 180–181. They include "Beweis dass die Bonne'sche Entwerfungsart die Länder ihrem Flächeninhalt auf der Kugel gemäss darstellt," in *Monatliche Correspondenz zur Beförderung der Erd- und Himmelskunde*, **13** (1806); "Analytische Theorie der stereographische Projektion," *ibid.*, **14** (1806); "Einige Projektionsarten der sphäroidischen Erde," *ibid.*, **16** (1807); *Prüfung der Farbenlehre des Herrn von Göthe und Verteidigung des Newtonschen Systems gegen dieselbe* (Halle, 1810); *Darstellungen der optischen Irrtümer in Herrn von Göthes Farbenlehre* (Halle, 1811); *Commentatio mathematico-philologica* (Leipzig, 1813); *Kurzgefasste Beschreibung der künstliche Erd- und Himmelskugel ...* (Leipzig, 1818); *Multiplex et continuus seriorum transformatio exemplo quodem illustratur* (Leipzig, 1820); and *Formula valorem praesentem pensionum annuarum comptandi recognitio et disputatio* (Leipzig, 1823).

A short biography of Mollweide by Siegmund Günther is in *Allgemeine deutsche Biographie*, XXII (Leipzig, 1885), 151–154.

H.-CHRIST. FREIESLEBEN

MOLYNEUX, SAMUEL (*b*. Chester, England, 18 July 1689; *d*. Kew, England, 13 April 1728), *astronomy, optics.*

Samuel Molyneux was the only son of William Molyneux to survive infancy. His mother died in 1691 and he was raised by his father, who zealously undertook his education on Lockean principles. His father died in 1698, leaving him to the care of his uncle, Thomas Molyneux. He entered Trinity College, Dublin, when he was sixteen, and there formed a friendship with George Berkeley, who dedicated his *Miscellanea Mathematica* to him in 1707. Molyneux received the B.A. in 1708, and the M.A. in 1710. In 1717 he married Lady Elizabeth Capel, who inherited a large sum of money and a residence outside London, Kew House, in 1721. Molyneux was thus able to devote himself to the study of astronomy and optics. He was elected fellow of the Royal Society in 1712.

Molyneux's most important astronomical investigations were undertaken in collaboration with his close friend James Bradley. In 1725 the two scientists decided to examine for themselves the validity of Robert Hooke's supposed detection of a large parallax for γ Draconis. To this end they ordered a large zenith sector with a radius of twenty-four feet from George Graham, the distinguished London instrument maker. The sector was set up on 26 November 1725 at Molyneux's residence, passing through holes in the ceilings and roof. Observations of γ Draconis on 3, 5, 11, and 12 December 1725 did not, however, reveal any change in the apparent position of the star.

Bradley observed the star again on 17 December "chiefly through curiosity," and to his great surprise found that it had moved southward, in the opposite direction to that which would arise from the projected parallax. Observations performed throughout the next twelve months revealed that the star exhibited an annual circular movement. Anxious to ascertain the laws of this phenomenon and to discover its physical cause, Bradley had Graham construct a more versatile sector than Molyneux's—one having a larger angular range. Molyneux helped set up this instrument at Bradley's aunt's residence at Wanstead on 19 August 1727. By 29 December 1727, Bradley had completed the observations necessary for his discovery of the aberration of light. The two scientists further worked together from 1723 to 1725 to improve methods of constructing reflecting telescopes; their efforts here did much to help bring reflecting telescopes into more general use.

In addition to his scientific activities, Molyneux pursued an active and noteworthy career in politics. He was a member of the English parliaments of 1715, 1726, and 1727, and a member of the Irish parliament of 1727. On 29 July 1727 he was appointed a lord of the admiralty, in which office he devised several schemes for the improvement of the navy. It was probably because of the pressure of public business arising from this appointment that he was unable to continue his astronomical observations after helping Bradley to set up his sector. Kew House was demolished in 1804 and a sundial, erected by William IV in 1834, now commemorates the observations made there.

BIBLIOGRAPHY

I. Original Works. Molyneux's writings include "A Relation of the Strange Effects of Thunder and Lightning, which Happened at Mrs. Close's House at New-Forge, in the County of Down in Ireland, on the 9th of August, 1707," in *Philosophical Transactions of the Royal Society*, **26** (1708), 36–40; "Sectio Oculorum Duorum Cataractâ Affectorum," *ibid.*, **33** (1724), 149–150; "The Method of Grinding and Polishing Glasses for Telescopes, Extracted from Mr. Huygens and Other Authors," in Robert Smith, *A Compleat System of Optics* (Cambridge, 1738), 281–301; "The Method of Casting, Grinding and Polishing Metals for Reflecting Telescopes, Begun by the Honourable Samuel Molyneux Esquire, and Continued by John Hadley Esquire, Vice-President of the Royal Society," *ibid.*, 301–312; "Sir Isaac Newton's Reflecting Telescope Made and Described by the Honourable Samuel Molyneux Esquire, and Presented by Him to His Majesty John V. King of Portugal: with Other Kinds of Mechanisms for This and for Mr. Gregory's Reflecting Telescope," *ibid.*, 363–368; "A Description of an Instrument Set up at Kew, in Surrey, for Investigating the Annual Parallax of the Fixed Stars, with an Account of the Observations Made Therewith," in James Bradley, *Miscellaneous Works and Correspondence of the Rev. James Bradley*, S. P. Rigaud, ed. (Oxford, 1832), 93–115; and "Observations Made at Kew," *ibid.*, 116–193, which includes observations by James Bradley after 22 April 1726.

II. Secondary Literature. An excellent account of Molyneux's education can be gleaned from the extensive exchange of letters between William and Thomas Molyneux and John Locke in *The Works of John Locke*, IX (London, 1823), 289–472. See also the article on Molyneux by Agnes M. Clerke in *Dictionary of National Biography*.

Howard Plotkin

MOLYNEUX, WILLIAM (*b.* Dublin, Ireland, 17 April 1656; *d.* Dublin, 11 October 1698), *astronomy, physics.*

Molyneux was the son of Samuel and Margaret Dowdall Molyneux. He was born at his father's house in New Row near Ormond-Gate. The father was of an old family and, although trained in law, took up a military career during the turbulent 1640's. He was proficient in mathematics and as master gunner of Ireland, performed numerous gunnery experiments.

A delicate child, Molyneux was educated at a Dublin grammar school and entered Trinity College, Dublin, on 10 April 1671, under the tutelage of William Palliser (later archbishop of Cashel). After taking his bachelor of arts degree in 1675 he was sent by his father to prepare for the legal profession at the Middle Temple. He had little zeal for the law and, expecting an independent income, preferred to follow his own interests in natural philosophy. In 1678 he returned to Dublin to marry Lucy Domvile, daughter of the attorney general Sir William Domvile. Her ill health and subsequent blindness imposed a tragic family burden until her death on 9 May 1691.

After a vain attempt to secure a cure for his wife's ills, Molyneux returned to his studies in natural philosophy. While at Trinity he had already turned from Aristotelianism and began the study of Descartes, Gassendi, Bacon, and Digby. He also studied the *Philosophical Transactions of the Royal Society*. In April 1680 Molyneux published his first work, a translation of Descartes's *Six Metaphysical Meditations*, for which he wrote a brief introduction and a short sketch of Descartes's life. This book, published in London, appears to be among the first English translations of Descartes. In the summer of 1682 he undertook to publish some queries concerning a description of Ireland in connection with Moses Pitt's *English Atlas*, an abortive attempt that left Molyneux with vast heaps of uncorrelated materials.

In October 1683 Molyneux formed a Dublin scientific society in an attempt to emulate the Royal Society of London. He brought together at a coffee-house on Cock Hill about a dozen men to discourse on philosophy and mathematics. They were soon invited by Robert Huntington, provost of Trinity College, to meet at his home, where in January 1684 they adopted the name Dublin Philosophical Society and elected Molyneux their first secretary. Despite his initial pessimism, the society flourished, many scientific papers were read, and correspondence was initiated with the Royal Society and with the Oxford Philosophical Society. The Dublin group was dispersed under the government of Tyrconnell in 1687, but resumed its activities for a brief period in 1693.

During this time Molyneux began a lengthy correspondence with John Flamsteed, astronomer royal, and strengthened his connections with the Royal Society, of which he was elected a fellow in 1685. The most important of his numerous articles published in the *Philosophical Transactions* include papers on the hygroscope, optics, and astronomy. In the short work *Sciothericum telescopium* (1686), he described a telescopic sundial constructed for him in London by Richard Whitehead.

Fearing for their lives under Tyrconnell's rule, Molyneux and his family left Ireland in January 1689 to settle in Chester, England. There Molyneux wrote his best-known scientific work, the *Dioptrica nova*, the first treatise on optics published in English. Printed at London in 1692, it was intended as a complete and clear treatise of current optical knowledge independent of any hypothesis concerning the nature of light. Appended to it was Halley's famous theorem for finding the foci of lenses. A popular text, it was reprinted in 1709 and provided a scientific base for Berkeley's *Essay Towards a New Theory of Vision*. The book was widely distributed, and Molyneux personally sent copies to Newton, Halley, Locke, Hooke, Boyle, Flamsteed, and Huygens. Its publication ended his friendship with Flamsteed, who, according to Molyneux, took umbrage at the lack of prominence accorded his work.

In the dedicatory epistle, Molyneux lavishly praised Locke's *Essay Concerning Human Understanding*;

Locke's letter thanking him initiated a lengthy correspondence that was ended only by Molyneux's death in 1698. It was during the course of this exchange that Molyneux first posed the famous problem known by his name: Would a blind man, suddenly granted his vision, be able to distinguish by sight alone between a sphere and a cube that he had touched when sightless? Both Molyneux and Locke, as well as Berkeley, decided in the negative.

Through the influence of the duke of Ormonde, in 1684 Molyneux shared the post of surveyor general with William Robinson, but he was removed from the position by Tyrconnell in 1688. In 1691 the family returned from Chester to Dublin, where his wife died. Their son, Samuel, later became a noted astronomer. In 1692 Molyneux was chosen to represent the University of Dublin in Parliament and served for a short time. His services pleased the government and the university, and he was nominated a commissioner of forfeited estates in Ireland (a post he declined) and was awarded an honorary doctorate of laws in 1693.

Molyneux is best remembered for *The Case of Ireland's Being Bound by Acts of Parliament in England Stated* (1698), in which he argued for Ireland's autonomy and against the English Parliament's right to legislate for it. He died of a lifelong affliction, kidney stones, and was interred in St. Audoen's Church, Dublin, in the tomb of his grandfather, Sir William Usher.

BIBLIOGRAPHY

I. ORIGINAL WORKS. Molyneux's major published works are his translation of Descartes's *Six Metaphysical Meditations* (London, 1680); *Sciothericum telescopium* (Dublin, 1686); *Dioptrica nova* (London, 1692; 2nd ed., 1709); and *The Case of Ireland's Being Bound by Acts of Parliament in England Stated* (Dublin, 1698). His published articles appeared mainly in the *Philosophical Transactions of the Royal Society*, **14–19** (1684–1697).

The main repositories of his MSS are the Civic Centre Archives, Southampton, which possesses the bulk of his correspondence with Flamsteed and his translations of Galileo and Torricelli on mechanics, now being edited for publication by R. Kargon; and Trinity College, Dublin. The British Museum has Molyneux's own copy of the *Dioptrica nova* with MS notes; letters of Molyneux to Hans Sloane on Newton's *Principia*, a 2nd ed. of which Molyneux offered to underwrite (Sloane MS 4036); and the minute and register book of the Dublin Philosophical Society (Add. MS 4811).

Much of the society's correspondence with the Oxford Philosophical Society is in R. T. Gunther, *Early Science in Oxford*, IV (Oxford, 1925), 129–208; and its correspondence with the Royal Society, in T. Birch, *History of the*

Royal Society, IV (London, 1757), *passim. Dublin University Magazine*, **18** (1841), 305–327, 470–490, 604–619, 744–764, contains four articles with long extracts from Molyneux's correspondence with his brother Thomas. Molyneux's correspondence with John Locke, first published in *Some Familiar Letters Between Mr. Locke and Several of His Friends* (London, 1708), is reprinted in *The Works of John Locke*, 11th ed., IX (London, 1812).

II. SECONDARY LITERATURE. The major biographical source is still Molyneux's autobiographical sketch (1694)· in Capel Molyneux, *An Account of the Family and Descendants of Sir Thomas Molyneux, Kt.* (Evesham, 1820). *Biographia Britannica* (London, 1760) has a lengthy account, as does Pierre Bayle, *A General Dictionary Historical and Critical*, J. P. Bernard, T. Birch, and J. Lockman, eds., 10 vols. (London, 1734–1741), which also contains part of the Molyneux-Flamsteed correspondence. There is a short MS biography, probably by Birch, in the British Museum, Add. MS 4223. See also Robert Dunlop's article in *Dictionary of National Biography*.

More recent works are Colin Turbayne, "Berkeley and Molyneux on Retinal Images," in *Journal of the History of Ideas*, **16** (1955); I. Ehrenpreis, *Swift: The Man, His Works and the Age*, I (Cambridge, Mass., 1962), 43–88; K. T. Hoppen, "The Royal Society and Ireland: William Molyneux, F.R.S.," in *Notes and Records. Royal Society of London*, **18** (1963), 125–135; and *The Common Scientist in the Seventeenth Century* (Charlottesville, Va., 1970), 90–190, which contains a good account of Molyneux's work and an excellent bibliography.

ROBERT H. KARGON

MONARDES, NICOLÁS BAUTISTA (*b.* Seville, Spain, *ca.* 1493; *d.* Seville, 10 October 1588), *medicine, natural history.*

Monardes was the son of Nicoloso de Monardis, an Italian bookseller, and Ana de Alfaro, daughter of a physician. He received a bachelor's degree in arts in 1530 and in medicine in 1533, both from the University of Alcalá de Henares, and the doctorate in medicine at Seville in 1547. In 1537 he married Catalina Morales, daughter of García Perez Morales, professor of medicine at Seville. They had seven children, some of whom went to America; their father, however, had to learn about American drugs at Seville's docks. Monardes had a good medical practice as well as considerable investments and businesses, which included the importation of drugs and the slave trade, the latter involving him in bankruptcy. After the death of his wife in 1577 Monardes took holy orders; he died eleven years later of a cerebral hemorrhage.

Monardes was the best-known and most widely read Spanish physician in Europe in the sixteenth century: his books were translated into Latin, English, Italian, French, German, and Dutch; and through his writings the American materia medica began to be known. He also published works on pharmacology, toxicology, medicine, therapeutics, phlebotomy, iron, and snow. He was an expert botanist; and because of his careful descriptions of drugs and the tests he carried out in animals to ascertain their medicinal properties, he is considered one of the founders of pharmacognosy and experimental pharmacology.

BIBLIOGRAPHY

I. ORIGINAL WORKS. The earliest book by Monardes, a survey of materia medica prior to the introduction of American drugs, was *Pharmacodilosis* (Seville, 1536). The study on venesection, *De secanda vena in pleuritii* (Seville, 1539), was reprinted at Antwerp in 1551, 1564, and 1943. The booklet on the medicinal properties of the rose, *De rosa et partibus eius* (Seville, *ca.* 1540), was also reprinted in *Archaeion* (Santa Fé, Argentina, 1941–1942). Monardes' fame grew after the publication of his first book on American drugs, *Dos libros. El uno que trata de todas las cosas que traen de nuestras Indias Occidentales . . .* (Seville, 1565; repr. 1569). The *Segunda parte del libro de todas las cosas . . .* (Seville, 1571) contains the description of tobacco, among other drugs. He also published a book on snow, *Libro que trata de la nieve* (Seville, 1571). Some of these works were translated and published abroad. A book containing all of Monardes' printed works on the American drugs plus those on the bezoar, viper's-grass, iron, and snow, *Primera, y segunda y tercera partes de la historia medicinal de las cosas que se traen de nuestras Indias Occidentales que se sirven en medicina . . .* (Seville, 1574), was soon translated into Italian, English, Latin, and French, and reprinted in Spanish (1580); up to 50 eds. of his works have been recorded. Monardes also edited Jean d'Avignon's *Sevillana medicina* (Seville, 1545; repr. 1885).

II. SECONDARY LITERATURE. There are several biographies on Monardes. Joaquín Olmedilla y Puig, *Estudio histórico de . . . Monardes* (Madrid, 1897); and Carlos Pereyra, *Monardes y el exotismo médico en el siglo XVI* (Madrid, 1936), were superseded by the data found in Seville's archives by Francisco Rodríguez Marín and presented in *La verdadera biografía del doctor Nicolás Monardes* (Madrid, 1925). Corrected biographical information, a study of Monardes' pharmacological work, and a bibliographical survey are in Francisco Guerra, *Nicolás Bautista Monardes, su vida y su obra* (Mexico City, 1961).

FRANCISCO GUERRA

MOND, LUDWIG (*b.* Kassel, Germany, 7 March 1839; *d.* London, England, 11 December 1909), *industrial chemistry.*

Mond is remembered for three contributions to the chemical industry: the establishment of the ammonia soda process in England; the development of an efficient power gas plant; and the discovery of nickel carbonyl, which led to a new process for extracting nickel from its ores.

Born into a wealthy and cultured Jewish family, Mond began his chemical education in 1855 under Kolbe at Marburg; from 1856 to 1859 he worked with Bunsen at Heidelberg. The next eight years were spent in acquiring experience in chemical manufacturing, especially of soda, ammonia, and acetic acid, in Germany, England, and Holland. In 1867 he settled in Widnes, one of the centers of the Leblanc soda trade in England.

Many unsuccessful attempts had been made to develop a simpler alternative to the Leblanc process by treating salt solutions with ammonia and carbon dioxide. By 1865 Ernest Solvay in Belgium had brought the process to some measure of efficiency, and a meeting between Mond and Solvay led to Mond's acquisition in 1872 of a license to use the process in England. Seven years of unceasing effort (during which time he often slept at the plant) enabled Mond to solve the chemical engineering problems posed by the handling of large volumes of liquids and gases, and by 1880 the success of the venture was assured. The corporation of Brunner and Mond (1881) was the first real threat to the survival of the Leblanc soda trade.

The search for a cheap source of ammonia for his soda works led Mond to examine ways of obtaining ammonia from coal. He devised in 1889 a system that burned coal in gas producers using a mixture of air and steam. In addition to ammonia the system yielded a cheap gas suitable for most industrial heating purposes. To promote its local use, the South Staffordshire Mond Gas Company was formed, and to develop the process overseas Mond founded the Power Gas Corporation.

From 1884 Mond and his assistant Carl Langer were concerned with recovering chlorine from waste ammonium chloride by distilling over heated metal oxides. Nickel valves in the plant became corroded, although this did not happen in the laboratory apparatus. Carbon monoxide in the kiln gases used to sweep ammonia out of the plant proved to be the reason. Experiments showed that nickel combined with carbon monoxide under gentle heat to form nickel carbonyl $Ni(CO)_4$, which on thermal decomposition yielded pure nickel. Mond's last industrial enterprise was the creation in 1900 of the Mond Nickel Company to link mines in Canada with extraction works in Wales.

Mond believed that the study of pure science is the best preparation for a career in industry. He used his great wealth wisely; particularly notable gifts were the Davy-Faraday Laboratory at the Royal Institution and financial support to the Royal Society for the *Catalogue of Scientific Papers*. He also bequeathed his collection of Italian paintings to his adopted country.

BIBLIOGRAPHY

I. ORIGINAL WORKS. Seventeen papers are listed in the Royal Society *Catalogue of Scientific Papers*, XVII, 318. The developments outlined in the text were all described by Mond in their historical setting. For ammonia-soda see "On the Origin of the Ammonia-soda Process," in *Journal of the Society of Chemical Industry*, **4** (1885), 527–529; on power gas, "The Commercial Production of Ammonium Salts," *ibid.*, **8** (1889), 505–510; on nickel carbonyl and nickel extraction, "On Nickel Carbon Oxide and Its Application in Arts and Manufactures," in *Report of the British Association for the Advancement of Science* (1891), 602–607; and "The History of the Process of Nickel Extraction," in *Journal of the Society of Chemical Industry*, **14** (1895), 945–946. There is also a valuable historical survey of chlorine manufacture in *Report of the British Association for the Advancement of Science* (1896), 734–745.

II. SECONDARY LITERATURE. The most useful obituary is in *Journal of the Chemical Society*, **113** (1918), 318–334. Not well known but very valuable is F. G. Donnan's published lecture to the (Royal) Institute of Chemistry, *Ludwig Mond F.R.S., 1839–1909* (London, 1939). More general is J. M. Cohen, *Life of Ludwig Mond* (London, 1956).

W. A. CAMPBELL

MONDEVILLE. See **Henry of Mondeville.**

MONDINO DE' LUZZI (also **Liucci** or **Liuzzi**) (*b.* Bologna, Italy, *ca.* 1275; *d.* Bologna, 1326), *anatomy.*

The name Mondino was probably an endearing form of Raimondo. The Luzzi family was prominent in Florence, but Mondino's father, Nerino Frazoli de' Luzzi, and his uncle, Liuccio, had moved to Bologna by 1270, where Mondino was born about 1275. Little is known of his youth, but since his father was an apothecary and his uncle, who made him his heir, taught medicine, it seems probable that he early became interested in the subject of medicine. Mondino attended the University of Bologna, where he studied under Alderotti (Thaddeus of Florence), and received his doctorate in 1300. He probably joined the faculty of the college of medicine and philosophy shortly after his graduation, but the earliest inscription of his name that has been found there is 1321.

Mondino's chief work is his compendium of anatomy, *Anatomia Mundini*, completed in 1316, which made him, in Castiglione's words, "the first outstanding anatomist worthy of the name."[1] Mondino's book dominated anatomy for over two hundred years. The major reason for Mondino's great popularity was the simplicity, conciseness, and systematic arrangement of his book, which is divided into six parts: (1) an introduction to the whole body and a discussion of authorities; (2) the natural members including the liver, spleen, and other organs in the abdominal cavity; (3) the generative members; (4) the spiritual members, the heart, lungs, trachea, esophagus, and other organs of the thoracic cavity up to the mouth; (5) the animal members of the skull, brain, eyes, ears; and (6) the peripheral parts, bones, spinal column, extremities. This organization was not the result of any philosophical approach to the subject but rather derived from the necessity of dissecting the most perishable organs first.

There is some scholarly discussion over whether Mondino dissected human cadavers himself, even though he spoke of a female cadaver that he anatomized in January 1316, who had a womb "double as big as her." George Sarton[2] and Charles Singer[3] felt that Mondino must have done his own dissection, but Moritz Roth[4] was convinced that he utilized a dissector to perform the manual operations. Regardless of what Mondino himself did, it seems likely from the way in which his book was written that he intended it to be read aloud while others were doing the actual dissection. For example, in describing the chest, Mondino stated, "After the muscle, the bones. Now the bones of the chest are many and are not continuous in order that it may be expanded and contracted, since it has to be ever in motion. . . . The bones are of two kinds, namely the ribs and the bones of the thorax. . . ."[5]

Illustrations from the last part of the fourteenth century usually indicate a professor on an elevated platform reading from a book (probably that of Mondino), while an *ostensore* points to the part and a dissector, a barber or surgeon, performs the actual manual operation. Guy de Chauliac described the same sort of method in his *Chirurgia* when he talked about his master Bertruce, a student of Mondino.[6] The subjects for Mondino's dissections were apparently criminals, since he stated that anatomization begins by placing the body of "one who has died from beheading or hanging" in a supine position.

Mondino was not a particularly accurate observer

of the actual results of his anatomies, perhaps because his purpose was not so much to enlarge knowledge through dissection as to memorize the works of the Arabic authorities. His book added very little to knowledge and instead repeated many old errors, thereby giving them new currency. He described the five-lobed liver (derived from dog anatomy), although he did say these were not always separate in man. Mondino reported black bile as coming from the spleen and being conducted to the stomach by a vein; his description of the heart was crude and also erroneous as was most of his physiology, which was that of Galen modified by Aristotelian or pseudo-Aristotelian notions. Surprisingly, his descriptions of the bones, muscles, nerves, veins, and arteries were also very inadequate, perhaps because physicians held that medicine should be concerned primarily with curing internal afflictions; consequently they gave their greatest attention to the viscera. Even though he performed anatomies on at least two women, his female anatomy seems to have been based almost entirely upon either that of animals or the erroneous notions of his predecessors; he described the womb as having seven chambers. He did give an interesting account of the sexual organs and tried to establish analogies between the male and female organs. He was also at some pains to emphasize the differences between the anatomy of the pig (as in Copho's *Anatomia porci*) and that of human beings.

Although Mondino regarded Galen as an almost infallible authority, he made errors that Galen did not. The trouble may have been that Mondino relied upon an abbreviated Latin translation of the *De juvamentis membrorum*, an incomplete Arabic version of the first nine books (of a total of seventeen) of Galen's *De usu partium*. This mixture of Arabic and Greek sources also helped create the confusion in terminology evident in Mondino's work. The sacrum, for example, is variously identified as *alchatim*, *allanis*, and *alhavius*. The pubic bone is called *os femoris* and *pecten*. On the other hand the same terms are used for different parts; *pomum granatum* can refer to either the thyroid cartilage or the xiphoid process, and *anchae* can mean the hips in general, the pelvic skeleton, the acetabulum, or the corpora quadrigemina of the brain. Much of Mondino's difficulty over terms was caused by the lack of standardization of anatomical nomenclature. Mondino himself seems to have introduced the words "matrix" and "mesentery" into anatomy.

In spite of the above criticism, Mondino should be regarded as the restorer of anatomy if only because his popular textbook and his experimental teaching were instrumental in preparing the revival of the subject. His text was the first book written on anatomy during the Middle Ages that was based on the dissection of the human cadaver; his efforts consolidated anatomy as a part of the medical program at Bologna and encouraged further study. His book also dominated the teaching of anatomy, and no real improvements were made upon it until 1521, when Berengario da Carpi wrote his famous commentary on Mondino.

Although he is best known for his *Anatomia*, Mondino wrote at least nine *consilia* dealing with such ailments as catarrh, fevers, stone, melancholic humors, and so forth. He also wrote a number of commentaries on the collection of classical writings known as the *Ars medicinae* including *Super libro prognosticorum Hippocratis, Super Hippocratis de regimine acutorum, Annotata in Galeni de morbo et accidenti*, and perhaps others. His commentary *Lectura super primo, secundo et quarto de juvamentis* is on part of Galen's *De usu partium*. Another commentary on the *Canones* of Mesue the Younger includes material from his *Anatomia*. Mondino also wrote treatises on weights and measures, human viscera, prescriptions and drugs, medical practice, and fevers.

NOTES

1. Arturo Castiglioni, *A History of Medicine*, trans. by E. B. Krumbhaar (New York, 1947), 341.
2. Sarton, *Introduction to the History of Science*, III, 1, 842.
3. Singer, *The Evolution of Anatomy*, 75–76.
4. Moritz Roth, *Andreas Vesalius Bruxellensis*, 20.
5. Mondino dei Luzzi, *Anothomia*, in Joannes Ketham, *Fasciculo di medicina*, trans. by Charles Singer, I, 80–81.
6. Guy de Chauliac, *La grande chirurgie*, trans. by E. Nicaise (Paris, 1890), 30–31.

BIBLIOGRAPHY

The first printed edition of the *Anatomia* appeared at Padua in 1476. Other editions in Latin appeared at Pavia, Bologna, Leipzig, Venice, Strasbourg, Paris, Milan, Geneva, Rostock, Lyons, and Marburg. All told there are approximately forty printed editions in Latin and other languages. Only a few of the editions include woodcuts or illustrations, but one or more appear in the Leipzig (1493), Venice (1494), Strasbourg (1513), Rostock (1514), Bologna (1521), and Marburg (1541) editions. A modern facsimile of the 1478 Pavia edition was edited by Ernest Wickersheimer, *Anatomies de Mondino dei Luzzi et de Guido de Vigevano* (Paris, 1926).

There were at least two early French translations, one by Richard Roussat (Paris, 1532), and another by an unknown translator (Paris, 1541). This last was erroneously labeled as the first French translation by LeRoy Crummer, who discovered it and published several interesting woodcuts from it. See Crummer, "La première traduction française de l'*Anatomie* de Mondini," in *Aesculape*, **20** (1930), 204–207. An Italian translation by Sebastian M. Romano was included in Joannes Ketham, *Fasciculo di medicina* (1493), and this was translated into English by Charles Singer in the reprinting of *Fasciculo di medicina* (Florence, 1924 and 1925). Another fifteenth-century Italian translation, with a photographic reproduction of a fourteenth-century MS of the *Anatomia*, was printed in Lino Sighinolfi, ed., *Mondino de Liucci Anatomia, Riprodotta da un Codice Bolognese del secolo XIVᵉ; volgarizzata nel secolo XV* (Bologna, 1930).

Seven of Mondino's *consilia* were printed by Balduin Vonderlage in his dissertation, *Consilien des Mondino dei Luzzi aus Bologna* (Leipzig, 1922). Mondino's commentary on the *Canones* of Mesue, *Mesuë cum expositione Mondini super canones universales*, was printed at Venice in 1490, 1495, 1497, 1570, 1638, and Lyons in 1525. The following works have not been printed: *Practica de accidentibus morborum secundum Magistrum Mundinum de Liucius de Bononis; Tractatus de ponderibus secundum Magistrum Mundinum; De visceribus humani corporis; Super libro prognosticorum Hippocratis; Mundinus super Hippocratis de regimine acutorum; Annotata in Galeni de morbo et accidente; Super libro prognosticorum Hippocratis; Mundinus super Hippocratis de regimine acutorum; Annotata in Galeni de morbo et accidente; Super libro de pulsibus; Tractatus de dosis medicinae; De medicinis simplicibus; Practicae medicinae libri X; Consilia medicinalia; Consilium ad retentionem menstruorum*, and *De accidentibus febrium*.

For further information see Howard B. Adelmann, *Marcello Malpighi and the Evolution of Embryology*, 5 vols., I (Ithaca, 1966), 74–84; Vern L. Bullough, *The Development of Medicine as a Profession* (Basel–New York, 1966); Giovanni Fantuzzi, *Notizie degli Scrittori Bolognese*, 9 vols., VI (Bologna, 1782–1790), 41–46; Giovanni Martinotti, "L'insegnamento del'Anatomia in Bologna prima del secolo XIX," in *Studi e memorie per la storia dell'università di Bologna*, II (Bologna, 1911), 1–146; Moritz Roth, *Andreas Vesalius Bruxellensis* (Berlin, 1892); George Sarton, *Introduction to the History of Science*, 3 vols. in 5, III (Baltimore, 1927–1942), 1, 842–845; and Charles Singer, *The Evolution of Anatomy* (London, 1925), 75.

VERN L. BULLOUGH

MONGE, GASPARD (*b.* Beaune, France, 9 May 1746; *d.* Paris, France, 28 July 1818), *geometry, calculus, chemistry, theory of machines.*

Monge revived the study of certain branches of geometry, and his work was the starting point for the remarkable flowering of that subject during the nineteenth century. Beyond that, his investigations extended to other fields of mathematical analysis, in particular to the theory of partial differential equations, and to problems of physics, chemistry, and technology. A celebrated professor and peerless *chef d'école*, Monge assumed important administrative and political responsibilities during the Revolution and the Empire. He was thus one of the most original mathematicians of his age, while his civic activities represented the main concerns of the Revolution more fully than did those of any other among contemporary French scientists of comparable stature.

The elder son of Jacques Monge, a merchant originally of Haute-Savoie, and the former Jeanne Rousseaux, of Burgundian origin, Monge was a brilliant student at the Oratorian *collège* in Beaune. From 1762 to 1764 he completed his education at the Collège de la Trinité in Lyons, where he was placed in charge of a course in physics. After returning to Beaune in the summer of 1764, he sketched a plan of his native city. The high quality of his work attracted the attention of an officer at the École Royale du Génie at Mézières, and this event determined the course of his career.

Created in 1748, the École Royale du Génie at Mézières had great prestige, merited by the quality of the scientific and practical training that it offered. Admitted to the school at the beginning of 1765 in the very modest position of draftsman and technician, Monge was limited to preparing plans of fortifications and to making architectural models, tasks he found somewhat disappointing. But barely a year after his arrival he had an opportunity to display his mathematical abilities. The result was the start of a career worthy of his talents.

Monge was requested to solve a practical exercise in defilading—specifically, to establish a plan for a fortification capable of shielding a position from both the view and the firepower of the enemy no matter what his location. For the very complicated method previously employed he substituted a rapid graphical procedure inspired by the methods of what was soon to become descriptive geometry. This success led to his becoming *répétiteur* to the professor of mathematics, Charles Bossut. In January 1769 Monge succeeded the latter, even though he did not hold the rank of professor. The following year he succeeded the Abbé Nollet as instructor of experimental physics at the school. In this double assignment, devoted partially to practical ends, Monge showed himself to be an able mathematician and physicist, a talented draftsman, a skilled experimenter, and a first-class teacher. The influence he exerted until he left the school at the end of 1784 helped to initiate several brilliant careers of future engineering officers and to give the engineering corps as a whole a solid technical training and a marked appreciation for science. The administrators of the school recognized his ability and, after obtaining for him the

official title of "royal professor of mathematics and physics" (1775), steadily increased his salary.

Parallel to this brilliant professional career, Monge very early commenced his personal work. His youthful investigations (1766–1772) were quite varied but exhibit several characteristics that marked his entire output: an acute sense of geometric reality; an interest in practical problems; great analytical ability; and the simultaneous examination of several aspects of a single problem: analytic, geometric, and practical.

This was the period in which Monge developed descriptive geometry. He systematized its basic principles and applied it to various graphical problems studied at the École du Génie—problems taken, for example, from fortification, architecture, and scaffolding. That Monge left only a few documents bearing on this work is not surprising, since he was essentially coordinating and rationalizing earlier knowledge, rather than producing really original material. Elements of descriptive geometry appeared very early in his teaching—to the degree that his familiarity with the graphical procedures currently in use and with the various branches of geometry allowed him to make the necessary synthesis. The documents from this period record the many investigations inspired by his readings in the rich collections of the library of the École du Génie. This research dealt with topics in infinitesimal calculus, infinitesimal geometry, analytic geometry, and the calculus of variations. His first important original work was "Mémoire sur les développées, les rayons de courbure et différents genres d'inflexions des courbes à double courbure." He published an extract from it in June 1769 in the *Journal encyclopédique*, and in October 1770 he finished a more complete version that he read before the Académie des Sciences in August 1771; the latter, however, was not published until 1785 (*Mémoires de mathématiques et de physique présentés à l'Académie . . . par divers sçavans . . .*, **10**, 511–550). By then some of the most important ideas in the memoir no longer seemed so original, because Monge had employed them in other works published in the intervening years. Nevertheless, this memoir is of exceptional interest, for it presents most of the new conceptions that Monge developed in his later works, as well as his very personal method of exposition, which combined pure geometry, analytic geometry, and infinitesimal calculus.

Wishing to make himself known and to have his work discussed, Monge sought out d'Alembert and Condorcet at the beginning of 1771. On the latter's advice, he later in the same year presented before the Paris Academy four memoirs corresponding to the main areas of his research. The first, which was not published, dealt with a problem to which he never returned: the extension of the calculus of variations to the study of extrema of double integrals. The second was the memoir on infinitesimal geometry mentioned above. The fourth treated a problem in combinatorial analysis related to a card trick.

In the third memoir Monge entered a field of study that was to hold his interest for many years: the theory of partial differential equations. In particular he undertook the parallel examination of certain equations of this type and of the families of corresponding surfaces. The geometric construction of a particular solution of the equations under consideration allowed him to determine the general nature of the arbitrary function involved in the solutions of a partial differential equation. Moreover, this finding enabled him to take a position on a question then being disputed by d'Alembert, Euler, and Daniel Bernoulli. Monge developed the ideas set forth in this memoir in two others sent to the Academy in 1772. The work presented in these papers was extended in four publications dating from 1776; two of these appeared in the *Mémoires* of the Academy of Turin and two in the *Mémoires* of the Paris Academy. In another paper (1774) Monge discussed the nature of the arbitrary functions involved in the integrals of finite difference equations. He also considered the equation of vibrating strings, a topic he later investigated more fully.

In May 1772 the Academy of Sciences elected Monge to be Bossut's correspondent. At this time he became friendly with Condorcet and Vandermonde. The latter's influence was probably responsible for two unpublished memoirs Monge wrote during this period, on the theory of determinants and on the knight's moves on a chessboard.

In 1775 Monge returned to infinitesimal geometry. Working on the theory of developable surfaces outlined by Euler in 1772, he applied it to the problem of shadows and penumbrae and treated several problems concerning ruled surfaces. A memoir composed in 1776 on Condorcet's prompting (and reworked in 1781 on the basis of a more thorough understanding) is of major importance, although not for its contributions to the practical problem of cuts and fills that served as its point of departure. Its great interest lies in its introduction of lines of curvature and congruences of straight lines.

Although in 1776 Monge was still interested in Lagrange's memoir on singular integrals, his predilection for mathematics was meanwhile slowly yielding to a preference for physics and chemistry. In 1774, while traveling in the Pyrenees, he had collaborated with the chemist Jean d'Arcet in making altitude measurements with the aid of a barometer. Having some instruments at his disposal in Mézières and working with Vandermonde and Lavoisier during his stays in Paris, Monge carried out experiments on expansion, solution, the effects of a vacuum, and other phenomena; acquired an extensive knowledge of contemporary physics; and participated in the elaboration of certain theories, including the theory of caloric and triboelectricity.

In 1777 Monge married Catherine Huart. They had three daughters, the two elder of whom married two former members of the National Convention, N.-J. Marey and J. Eschassériaux: the two present branches of Monge's descendants are their issue.

During the period 1777–1780 Monge was interested primarily in physics and chemistry and arranged for a well-equipped chemistry laboratory to be set up at the École du Génie. Moreover, having for some time been responsible for supervising the operation of a forge belonging to his wife, he had become interested in metallurgy.

His election to the Academy of Sciences as *adjoint géomètre* in June 1780 altered Monge's life, obliging him to stay in Paris on a regular basis. Thus for some years he divided his time between the capital and Mézières. In Paris he participated in the Academy's projects and presented memoirs on physics, chemistry, and mathematics. He also substituted for Bossut in the latter's course in hydrodynamics (created by A.-R.-J. Turgot in 1775) and in this capacity trained young disciples such as S. F. Lacroix and M. R. de Prony. At Mézières, where he arranged for a substitute to give some of his courses—although he kept his title and salary—Monge conducted research in chemistry. In June–July 1783 he synthesized water. He then turned his attention to collecting stores of hydrogen and to the outer coverings of balloons. Finally, with J. F. Clouet he succeeded in liquefying sulfur dioxide.

In October 1783 Monge was named examiner of naval cadets, replacing Bézout. He attempted to reconcile his existing obligations with the long absences required by this new post, but it proved to be impossible. In December 1784 he had to give up his professorship at Mézières, thus leaving the school at which he had spent twenty of the most fruitful years of his career. From 1784 to 1792 Monge divided his time between his tours of inspection of naval schools and his stays in Paris, where he continued to participate in the activities of the Academy and to conduct research in mathematics, physics, and chemistry. A list of the subjects of his communications to the Academy attests to their variety: the composition of nitrous acid, the generation of curved surfaces, finite difference equations, and partial differential equations (1785); double refraction and the structure of Iceland spar, the composition of iron, steel, and cast iron, and the action of electric sparks on carbon dioxide gas (1786); capillary phenomena (1787); and the causes of certain meteorological phenomena; and a study in physiological optics (1789).

Meanwhile, with other members of the Academy, Monge assisted Lavoisier in certain experiments. For example, in February 1785 he participated in the analysis and synthesis of water. In fact, he was one of the first to accept Lavoisier's new chemical theory. After having collaborated with Vandermonde and Berthollet on a memoir on "iron considered in its different metallurgical states" (1786), he participated in several investigations of metallurgy in France. In 1788 he joined in the refutation, instigated by Lavoisier, of a treatise by the Irish chemist Kirwan, who was a partisan of the phlogiston theory. That Monge was among the founders of the *Annales de chimie* testifies to his standing in chemistry. During this period Monge's position as naval examiner obliged him to write a course in mathematics to replace Bézout's. Only one volume was published, *Traité élémentaire de statique* (1788).

When the Revolution began in 1789, Monge was among the most widely known of French scientists. A very active member of the Academy of Sciences, he had established a reputation in mathematics, physics, and chemistry. As an examiner of naval cadets he directed a branch of France's military schools, which were then virtually the only institutions offering a scientific education of any merit. This position also placed him in contact, in each port he visited, with bureaucracy that was soon to come under his administration. It also enabled him to visit iron mines, foundries, and factories, and thus to become an expert on metallurgical and technological questions. Furthermore, the important reform of teaching in the naval schools that he had effected in 1786 prepared him for the efforts to renew scientific and technical education that he undertook during the Revolution.

Although Monge was a resolute supporter of the Revolution from the outset, his political role remained discreet until August 1792. He joined several revolutionary societies and clubs but devoted most of his time to tours of inspection as examiner of naval cadets and to his functions as a member of the Academy, particularly to the work of the Academy's Commission on Weights and Measures.

After the fall of the monarchy on 10 August 1792, a government was created to carry on the very diffi-

cult struggle imposed on the young republic by adherents of the *ancien régime*. On the designation of the Legislative Assembly, Monge accepted the post of minister of the navy, which he held for eight months. Although not outstanding, his work showed his desire to coordinate all efforts to assure the nation's survival and independence. His politics, however, were judged by some to be too moderate; and attacked from several sides and exhausted by the incessant struggle he had to wage, he resigned on 10 April 1793. Henceforth he never played more than a minor political role. A confirmed republican, he associated with Jacobins such as Pache and Hassenfratz; but he never allied himself with any faction or participated in any concrete political action. On the other hand, he was an ardent patriot, who placed all his energy, talent, and experience in the service of the nation, and he played a very important role in developing the manufacture of arms and munitions, and in establishing a new system of scientific and technical education.

Monge resumed his former activities for a short time; but after the suppression of the Academy of Sciences on 8 August 1793 his work came under the direct control of the political authorities, especially of the Committee of Public Safety. From the beginning of September 1793 until October 1794, he took part in the work of the Committee on Arms. He wrote, with Vandermonde and Berthollet, a work on the manufacture of forge and case-hardened steels, drew up numerous orders concerning arms manufacture for Lazare Carnot and C. L. Prieur, supervised Paris arms workshops, assembled technical literature on the making of cannons, gave "revolutionary courses" on this latter subject (February–March 1794), and wrote an important work on it. He also was involved in the extracting and refining of saltpeter and the construction and operation of the great powderworks of Paris. In addition, he participated in the development of military balloons.

Monge also engaged in tasks of a different sort. After the suppression of the Academy he joined the Société Philomatique; participated in the work of the Temporary Commission on Weights and Measures, which continued the projects of the Academy's commission; and took part in the activities of the Commission on the Arts, which was responsible for preserving the nation's artistic and cultural heritage. He was also active in the projects for educational reform then under discussion. His experience at the École de Mézières and in the naval schools explains the special interest that the renewal of scientific and technical instruction held for him. At the elementary level, he prepared for the department of Paris a plan for schools for artisans and workers that the Convention adopted on 15 September 1793 but rejected the next day. At a more advanced level, he was convinced of the value of creating a single national school for training civil and military engineers. Consequently, when he was appointed by the Convention (11 March 1794) to the commission responsible for establishing an École Centrale des Travaux Publics, he played an active role in its work. The memoir that Fourcroy prepared in September 1794 to guide the first steps of the future establishment ("Développements sur l'enseignement . . .") shows the influence of Monge's thinking, which derived from his experience at Mézières. Appointed instructor of descriptive geometry on 9 November 1794, Monge supervised the operation of the training school of the future *chefs de brigade*, or foremen, taught descriptive geometry in "revolutionary courses" designed to complete the training of the future students, and was one of the most active members of the governing council. After a two-month delay caused by political difficulties, the school—soon to be called the École Polytechnique—began to function normally in June 1795. Monge's lectures, devoted to the principles and applications of infinitesimal geometry, were printed on unbound sheets; these constituted a preliminary edition of his *Application de l'analyse à la géométrie*.

Monge was also one of the professors at the ephemeral École Normale de l'An III. From 20 January to 20 May 1795 this school brought together in Paris 1,200 students, who were to be trained to teach in the secondary schools then being planned. The lectures he gave, assisted by his former student S.-F. Lacroix and by J. Fournier, constituted the first public course in descriptive geometry. Like those of the other professors, the lectures were taken down by stenographers and published in installments in the *Journal des séances des écoles normales*.

Monge, who regretted the suppression of the Academy of Sciences, actively participated in the meetings held from December 1795 to March 1796 to prepare its rebirth as the first section of the Institut National, created by the Convention on 26 October 1795. But just when Monge's activities seemed to be returning to normal, events intervened that prevented this from happening.

Monge was named, along with his friend Berthollet, one of the six members of the Commission des Sciences et des Arts en Italie, set up by the Directory to select the paintings, sculptures, manuscripts, and valuable objects that the victorious army was to bring back. He left Paris on 23 May 1796. His mission took him to many cities in northern and central Italy, including Rome, and allowed him to become

friendly with Bonaparte. At the end of October 1797 Monge returned to Paris, officially designated, with General Louis Berthier, to transmit to the Directory the text of the Treaty of Campoformio.

Immediately after returning, Monge resumed his former posts, as well as a new one, that of director of the École Polytechnique. But his stay in Paris was brief; at the beginning of February 1798 the Directory sent him back to Rome to conduct a political inquiry. While there, Monge took an active interest in the organization of the short-lived Republic of Rome. The following month, at the request of Bonaparte, he took part in the preparations for the Egyptian expedition. Although reluctant at first, he finally agreed to join the expedition. His boat left Italy on 26 May 1798, joining Bonaparte's squadron two weeks later. Monge arrived in Cairo on 21 July and was assigned various administrative and technical tasks. As president of the Institut d'Égypte, created on 21 August, he played an important role in the many scientific and technical projects undertaken by this body. He accompanied Bonaparte on a brief trip in the Suez region, on the disastrous Syrian expedition (February–June 1799), and, after another brief stay in Cairo, on his return voyage to France (17 August–16 October). During this period of three and a half years, in which he was for almost the whole time away from France, Monge's correspondence and communications to the Institut d'Égypte show that he was working on new chapters of his *Application de l'analyse à la géométrie*. Moreover, the observation of certain natural phenomena, such as mirages, and the study of certain techniques, including metallurgy and the cultivation of the vine, provided him with fruitful sources for thought. Meanwhile, at the request of his wife and without his knowledge, his *Géométrie descriptive* was published in 1799 by his friend and disciple J. N. Hachette, who limited himself to collecting Monge's École Normale lectures previously published in the *Séances*.

On his return to Paris, Monge resumed his duties as director of the École Polytechnique but relinquished them two months later when, following the *coup d'état* of 18 Brumaire, Bonaparte named him senator for life. By accepting this position Monge publicly attached himself to the Consulate. Although this decision may seem to contradict his republican convictions and revolutionary faith, it can be explained by his esteem for and admiration of Bonaparte and by his dissatisfaction with the defects and incompetence of the preceding regime. Dazzled by Napoleon, Monge later rallied to the Empire with the same facility and accepted all the honors and gifts the emperor bestowed upon him: grand officer of the

Legion of Honor in 1804, president of the Senate in 1806, count of Péluse in 1808, among others.

Monge had to divide his time among his family, his teaching of infinitesimal geometry at the École Polytechnique, and his obligations as a member of the Academy of Sciences and of the Conseil de Perfectionnement of the École Polytechnique, and his duties as a senator. Further tasks were soon added. He was founder of the Société d'Encouragement pour l'Industrie Nationale and vice-president of the commission responsible for supervising the preparation and publication of the material gathered on the Egyptian expedition, *Description de l'Égypte*. Even though his duties as senator took him away on several occasions from his courses at the École Polytechnique, he maintained his intense concern for the school. He kept careful watch over the progress of the students, followed their research, and paid close attention to the curriculum and the teaching.

Most of Monge's publications in this period were written for the students of the École Polytechnique. The wide success of the *Géométrie descriptive* was responsible for the rapid spread of this new branch of geometry both in France and abroad. It was reprinted several times; the edition of 1811 contained a supplement by Hachette; and the fourth, posthumous edition, published in 1820 by Barnabé Brisson, included four previously unpublished lectures on perspective and the theory of shadows.

In 1801 Monge published *Feuilles d'analyse appliquée à la géométrie*, an expanded version of his lectures on infinitesimal geometry of 1795. In 1802, working with Hachette, he prepared a brief exposition of analytic geometry that was designed to replace the few remarks on the subject contained in the *Feuilles*. Entitled *Application de l'algèbre à l'analyse*, it was published separately in 1805; in 1807 it became the first part of the final version of *Feuilles d'analyse*, now entitled *Application de l'analyse à la géométrie*. This larger work was republished in 1809 and again in 1850 by J. Liouville, who appended important supplements.

Aside from new editions of the *Traité élémentaire de statique*, revised by Hachette beginning with the fifth edition (1810), and some physical and technical observations made in Italy and Egypt and published in 1799, Monge's other publications during this period dealt almost exclusively with infinitesimal and analytic geometry. For the most part they were gradually incorporated into successive editions of his books. His production of original scientific work began to decline in 1805.

A decline likewise occurred in Monge's other activities. Suffering from arthritis, he stopped teaching

at the École Polytechnique in 1809, arranging for Arago to substitute for him and then to replace him. Although he wrote a few more notes on mathematics and several official technical reports, his creative period had virtually come to an end. In November 1812, overwhelmed by the defeat of the Grande Armée, he suffered a first attack of apoplexy, from which he slowly recovered. At the end of 1813 he was sent to his senatorial district of Liège to organize its defenses but fled a few weeks later before the advancing allied armies. Absent from Paris at the moment of surrender, he did not participate in the session of 3 April 1814, in which the Senate voted the emperor's dethronement. He returned shortly afterward and resumed a more or less normal life. In 1815, during the Hundred Days, he renewed his contacts with Napoleon and even saw him several times after Waterloo and the abdication. In October 1815, fearing for his freedom, Monge left France for several months. A few days after his return to Paris, in March 1816, he was expelled from the Institut de France and harassed politically in other ways. Increasingly exhausted physically, spiritually, and intellectually, he found his last two years especially painful. Upon his death, despite government opposition, many current and former students at the École Polytechnique paid him tribute. Throughout the nineteenth century mathematicians acknowledged themselves as his disciples or heirs.

Scientific Work. Monge's scientific work encompasses mathematics (various branches of geometry and mathematical analysis), physics, mechanics, and the theory of machines. His principal contributions to these different fields will be discussed in succession, even though his mathematical work constitutes a coherent ensemble in which analytic developments were closely joined with material drawn from pure, descriptive, analytic, and infinitesimal geometry, and even though his investigations in physics, mechanics, and the theory of machines were also intimately linked.

Descriptive and Modern Geometry. Elaborated during the period 1766–1775, Monge's important contribution is known from his *Géométrie descriptive*, the text of his courses at the École Normale de l'An III (1795), and from the manuscript of his lectures given that year at the École Polytechnique. Before him various practitioners, artists, and geometers, including Albrecht Dürer, had applied certain aspects of this technique. Yet Monge should be considered the true creator of descriptive geometry, for it was he who elegantly and methodically converted the group of graphical procedures used by practitioners into a general uniform technique based on simple and rigorous geometric reasoning and methods. Within a few years this new discipline was being taught in French scientific and technical schools and had spread to several other Continental countries.

Monge viewed descriptive geometry as a powerful tool for discovery and demonstration in various branches of pure and infinitesimal geometry. His persuasive example rehabilitated the study and use of pure geometry, which had been partially abandoned because of the success of Cartesian geometry. Monge's systematic use of cylindrical projection and, more discreetly, that of central projection, opened the way to the parallel creation of projective and modern geometry, which was to be the work of his disciples, particularly J.-V. Poncelet. The definition of the orientation of plane areas and volumes, the use of the transformation by reciprocal polars, and the discreet introduction in certain of his writings of imaginary elements and of elements at infinity confirms the importance of his role in the genesis of modern geometry.

Analytic and Infinitesimal Geometry. Analytic and infinitesimal geometry overlap so closely in Monge's work that it is sometimes difficult to separate them. Whereas from 1771 to 1809 he wrote numerous memoirs on the infinitesimal geometry of space, it was not until 1795, in his lectures at the École Polytechnique, that he specifically developed analytic geometry.

Nevertheless, even in his earliest works, Monge sought to remedy the chief weaknesses of analytic geometry, although this discipline was then for him only an auxiliary of infinitesimal geometry. Rejecting the restrictive Cartesian point of view that was still dominant, he considered analytic geometry as an autonomous branch of mathematics, parallel to pure geometry and independent of it. Consonant with this approach, as early as 1772 and at the same time as Lagrange, Monge systematically introduced into the subject the elements defined by first-degree equations (straight lines and planes) that had previously not been part of it. He also solved the basic problems posed by this extension. Parallel with this endeavor, he sought, following Clairaut and Euler, to make up for the long delay in the development of three-dimensional analytic geometry. In addition Monge introduced an absolute symmetry into the use of the coordinate axes. He showed great analytic virtuosity in his calculations, some of which display, except for the symbolism, a skillful handling of determinants and of certain algorithms of vector calculus. His ability in this regard very early allowed him to establish the foundations of the geometry of the straight line (in Plücker's sense), which he systematized in 1795.

The first two editions (1795 and 1801) of Monge's course in "analysis applied to geometry" at the École Polytechnique contain as an introduction a brief statement of the principles and fundamental problems of this renewed analytic geometry, which was soon taught in upper-level French schools. With his disciple J. N. Hachette, Monge published in 1802 an important memoir, "Application de l'algèbre à la géométrie," which completed the preceding study, notably regarding the theory of change of coordinates and the theory of quadrics. In 1805 Monge collected these various contributions to analytic geometry in a booklet entitled *Application de l'algèbre à la géométrie,* which in 1807 became the first part of his great treatise *Application de l'analyse à la géométrie.* The many articles that Monge and his students devoted to individual problems of analytic geometry (change of coordinates, theory of conics and quadratics, among others) in the *Journal de l'École polytechnique* and in the *Correspondance sur l'École polytechnique* attest to the interest stimulated by the discipline's new orientation.

Throughout his career infinitesimal geometry remained Monge's favorite subject. Here his investigations were directed toward two main topics: families of surfaces defined by their mode of generation, which he examined in connection with the corresponding partial differential equations, and the direct study of the properties of surfaces and space curves. Since the first topic is discussed below, only the principal research relating to the second topic will be presented here. In 1769 Monge defined the evolutes of a space curve and showed that these curves are the geodesics of the developable envelope of the family of planes normal to the given curve. In 1774, after having returned to this question in a memoir presented in 1771, Monge completed the study of developable surfaces outlined by Euler. Concurrently utilizing geometric considerations and analytic arguments, he established the distinction between ruled surfaces and developable surfaces; gave simple criteria for judging, from its equation, whether a given surface is developable; applied these results to the theory of shadows and penumbrae; and solved various problems concerning surfaces. In particular, he determined by means of descriptive geometry the ruled surface passing through three given space curves. Still more important is the memoir on cuts and fills, of which Monge made two drafts (1776 and 1781). The point of departure was a technical problem: to move a certain quantity of earth, determining the trajectory of each molecule in such a way that the total work done is a minimum. Through repeated schematizations he derived the formulation of a question concerning the theory of surfaces that he examined very generally, introducing such important notions as the congruence of straight lines, line of curvature, normal, and focal surface. This memoir served as a starting point for several of Monge's later works, as well as for important investigations by Malus in geometrical optics and by Dupin in infinitesimal geometry.

Several memoirs written between 1783 and 1787 contain numerous studies of families of surfaces and some new results relating to the general theory of surfaces and to the properties of certain space curves.

In *Feuilles d'analyse appliquée à la géométrie* (1795 and 1801) Monge assembled, along with general considerations regarding the theory of surfaces and the geometric interpretation of partial differential equations, monographs on about twenty families of surfaces defined by their mode of generation. *Application de l'analyse à la géométrie* (1807) includes some supplementary material, notably attempts to find families of surfaces when one of the nappes of their focal surface is known. The manuscript of Monge's course for 1805–1806 also contains important additional findings (transformations by reciprocal polars, conoids, etc.). The richness and originality in Monge's lectures, qualities evident in this manuscript and confirmed by the testimony of former students, explain why so many French mathematicians can be considered his direct followers. Among them we may cite Tinseau and Meusnier at the École de Mézières, Lacroix, Fourier, and Hachette at the École Normale, and Lancret, Dupin, Livet, Brianchon, Malus, Poncelet, Chasles, Lamé, and still others at the École Polytechnique. Certain aspects of their writings show the direct influence of Monge, who thus emerges as a true *chef d'école*.

Mathematical Analysis. The theory of partial differential equations and that of ordinary differential equations occupies—often in close connection with infinitesimal geometry—an important place in Monge's work. Yet, despite his great mastery of the techniques of analysis and the importance and originality of certain of the new methods he introduced, his writings in this area are sometimes burdened by an excessive number of examples and are blemished by insufficiently rigorous argumentation.

As early as 1771 the memoirs presented to the Academy and the letters to Condorcet reflect two of the guiding ideas of Monge's work: the geometric determination of the arbitrary function involved in the general solution of a partial differential equation, and the equivalence established between the classification of families of surfaces according to their mode of generation and according to their partial differential equation. He returned to these questions several times between 1771 and 1774, developing many examples and extending his study to finite difference equations. Also, in the memoir of 1775 on developable surfaces he discussed the partial differential equation of developable surfaces and that of ruled surfaces.

From 1773 to 1786 Monge carried out new research in this area. In seven memoirs of varying importance he presented flawlessly demonstrated results, and a progressively elaborated outline of very fruitful new methods. His essentially geometric inspiration drew upon the ideas of his earliest papers and on the division, introduced by Lagrange, of the integral surfaces of a first-order partial differential equation into a complete integral, a general integral, and a singular integral. By means of his theory of characteristics Monge gave a geometric interpretation of the method of the variation of parameters. In addition he introduced such basic notions as characteristic curve, integral curve, characteristic developable, trajectory of characteristics, and characteristic cone. Monge was also interested in second-order partial differential equations.

In particular he created the theory of "Monge equations"—equations of the type

$$Ar + Bs + Ct + D = 0,$$

where A, B, C, D are functions of x, y, z, p, q, and where p, q, r, s, and t have the classical meanings—and solved the equation of minimal surfaces. Investigating the theory of partial differential equations from various points of view, Monge—despite some errors and a somewhat disorganized and insufficiently rigorous presentation—contributed exceptionally fruitful methods of approaching this topic. For example, he demonstrated the geometric significance of the total differential equations that do not satisfy the condition of integrability, thus anticipating J. F. Pfaff's treatment of the question in 1814–1815. Monge also introduced contact transformations, the use of which was generalized by Lie a century later. In addition he determined the partial differential equations of many families of surfaces and perfected methods of solving and studying various types of partial differential equations.

Monge resumed his research in this area in 1795–1796 and in 1803–1807, when he completed his courses in infinitesimal geometry at the École Polytechnique, with a view toward their publication. He perfected the theories sketched in 1783–1786, corrected or made certain arguments more precise, and studied the area of their application.

Mechanics, Theory of Machines, and Technology. From the time he came to Mézières, Monge was interested in the structure, functioning, and effects of machines; in the technical and industrial problems of fortification and construction; and in local industry, particularly metallurgy. He held that technical progress is a key factor governing the happiness of humanity and depends essentially on the rational application of theoretical science. His interest in physics, mechanics, and the theory of machines derived in part from his view that they are the principal factors of industrial progress and, therefore, of social progress.

Monge discussed the theory of machines in his course in descriptive geometry at the École Polytechnique (end of 1794). His ideas, employed by Hachette in *Traité élémentaire des machines* (1809), were derived from the principle that the function of every machine is to transform a motion of a given type into a motion of another type. Although this overly restrictive conception has been abandoned, it played an important role in the creation of the theory of machines in the nineteenth century.

Monge's *Traité élémentaire de statique* (1788) was a useful textbook, and its successive editions recorded the latest developments in the subject, for example the theory of couples introduced by Poinsot. The fifth edition (1810) included important material on the reduction of an arbitrary system of forces to two rectangular forces.

The unusual experience that Monge had acquired in metallurgy was frequently drawn upon by the revolutionary government and then by Napoleon.

Physics and Chemistry. Although the details regarding Monge's contributions to physics are poorly known, because he never published a major work in this field, his reputation among his contemporaries was solid. His main contributions concerned caloric theory, acoustics (theory of tones), electrostatics, and optics (theory of mirages).

In 1781 Monge was selected to be editor of the *Dictionnaire de physique* of the *Encyclopédie méthodique.* He did not complete this task, but he did write certain articles.

His most important research in chemistry dealt with the composition of water. As early as 1781 he effected the combination of oxygen and hydrogen in the eudiometer, and in June–July and October 1783 he achieved the synthesis of water—at the same time as Lavoisier and independently of him. Although Monge's apparatus was much simpler, the results of his measurements were more precise. On the other hand, his initial conclusions remained tied to the phlogiston theory, whereas Lavoisier's conclusions

signaled the triumph of his new chemistry and the overthrow of the traditional conception of the elementary nature of water. Monge soon adhered to the new doctrine. In February 1785 he took part in the great experiment on the synthesis and analysis of water; he was subsequently an ardent propagandist for the new chemistry and actively participated in its development.

In the experimental realm, in 1784 Monge achieved, in collaboration with Clouet, the first liquefaction of a gas, sulfurous anhydride (sulfur dioxide). Finally, between 1786 and 1788 Monge investigated with Berthollet and Vandermonde the principles of metallurgy and the composition of irons, cast metals, and steels. This research enabled them to unite previous findings in these areas, to obtain precise theoretical knowledge by means of painstaking analyses, and to apply this knowledge to the improvement of various techniques.

BIBLIOGRAPHY

I. ORIGINAL WORKS. A partial list of Monge's works is given in Poggendorff, II, 184–186. More precise and more complete bibliographies are in L. de Launay, *Un grand français: Monge* . . . (Paris, 1933), pp. 263–276, which includes a list of MSS and portraits; and in R. Taton, *L'oeuvre scientifique de Gaspard Monge* (Paris, 1951), pp. 377–393, which contains lists of MSS, scientific correspondence, and memoirs presented to the Académie des Sciences.

Works that were published separately are *Traité élémentaire de statique* (Paris, 1788; 8th ed., 1846), trans. into Russian, German, and English; *Avis aux ouvriers en fer sur la fabrication de l'acier* (Paris, 1794), written with Berthollet and Vandermonde; *Description de l'art de fabriquer les canons* . . . (Paris, 1794); *Géométrie descriptive* . . . (Paris, 1799), a collection in 1 vol. of the lectures given in 1795 at the École Normale de l'An III and published in the *Séances des écoles normales* . . . (7th ed., Paris, 1847; repr. 1922), trans. into German, Italian, English, Spanish, and Russian; *Feuilles d'analyse appliquée à la géométrie* . . . (Paris, 1801), a collection, with various additions, of lectures given at the École Polytechnique, which were published on separate sheets in 1795; *Application de l'algèbre à la géométrie* (Paris, 1805), written with Hachette; and *Application de l'analyse à la géométrie* (Paris, 1807), a new ed. of the *Feuilles d'analyse appliquée à la géométrie,* preceded, with special pagination, by *Application de l'algèbre à la géométrie* (new ed., 1809; 5th ed., 1850), J. Liouville, ed., with several appendixes, including Gauss's *Disquisitiones circa superficies curvas;* trans. into Russian, with commentary (Moscow, 1936).

II. SECONDARY LITERATURE. The most recent biography of Monge is P.-V. Aubry, *Monge, le savant ami de Napoléon: 1746–1818* (Paris, 1954). The most complete study of his scientific work is R. Taton, *L'oeuvre scientifique de Gaspard Monge* (Paris, 1951).

A few older monographs are still worth consulting, in particular the following, listed chronologically: C. Dupin, *Essai historique sur les services et les travaux scientifiques de Gaspard Monge* (Paris, 1819; repr. 1964); F. Arago, "Biographie de Gaspard Monge ...," in *Oeuvres de François Arago, Notices biographiques*, II (Paris, 1853; repr. 1964), 426–592, trans. into German, English, and Russian; E. F. Jomard, *Souvenirs sur Gaspard Monge et ses rapports avec Napoléon* ... (Paris, 1853); L. de Launay, *Un grand français: Monge, fondateur de l'École polytechnique* (Paris, 1933); and E. Cartan, *Gaspard Monge, sa vie, son oeuvre* (Paris, 1948).

A very complete bibliography of other works dealing with Monge is given in R. Taton, *L'oeuvre scientifique*, pp. 396–425. This list should be completed by some more recent studies, listed chronologically: C. Bronne, "La sénatorerie de Monge," in *Bulletin de la Société belge d'études napoléoniennes*, no. 9 (1953), 14–19; Y. Laissus, "Gaspard Monge et l'expédition d'Égypte (1798–1799)," in *Revue de synthèse*, **81** (1960), 309–336; R. Taton, "Quelques lettres scientifiques de Monge," in *84ᵉ Congrès des sociétés savantes. Dijon, Section des sciences* (Paris, 1960), pp. 81–86; J. Duray, "Le sénateur Monge au château de Seraing (près de Liège)," in *Bulletin de la Société belge d'études napoléoniennes*, no. 36 (1961), 5–17; A. Birembaut, "Deux lettres de Watt, père et fils, à Monge," in *Annales historiques de la Révolution française*, **35** (1963), 356–358; J. Booker, "Gaspard Monge and His Effect on Engineering Drawing and Technical Education," in *Transactions of the Newcomen Society*, **34** (1961–1962), 15–36; and R. Taton, "La première note mathématique de Gaspard Monge," in *Revue d'histoire des sciences*, **19** (1966), 143–149.

RENÉ TATON

MONIZ, EGAS. See **Egas Moniz, A. A. F.**

MONNET, ANTOINE-GRIMOALD (*b.* Champeix, Puy-de-Dôme, France, 1734; *d.* Paris, France, 23 May 1817), *chemistry, mineralogy.*

Little is known about Monnet's early life and education. He attended the chemistry lectures of G.-F. Rouelle at the Jardin du Roi in Paris (*ca.* 1754) and was for a time a pharmacist's assistant in Nantes. By 1767 papers on the analysis of mineral springs had attracted the attention of some scientists and of Malesherbes, who became Monnet's patron; and Monnet was able to secure a post with the Bureau du Commerce, then under the direction of Daniel Trudaine. Beginning in 1772, he also worked for Henri

Bertin, minister and secretary of state in charge of mining; in 1776 he was named France's first *inspecteur général des mines et minières du royaume*. Although his title and duties varied somewhat in later years, he survived many changes in the organization of the government corps of mining engineers and was finally retired in 1802.

His employment took Monnet to Alsace and the German states to study mining and metallurgy, and after 1772 his principal duty was to inspect and to suggest improvements in the French mining industry. Many of his published works were the result of these activities, and his post as mineralogist traveling at government expense was partly responsible for his appointment, in 1777, to direct the national geological survey earlier begun by Guettard and Lavoisier.

Monnet incorporated into his writings some of the findings of contemporary German and Swedish scientists, often before their treatises were available in French. Although French scientists considered his works useful, their judgments varied when they tried to assess Monnet's talents. Early in his career, he was pronounced a chemist of genuine ability by Macquer; but despite influential patronage, he failed in his attempts to become a member of the Académie Royale des Sciences. (He belonged to learned societies in Clermont-Ferrand, Rouen, Stockholm, and Turin.) After 1790 his persistent and violent adherence to the phlogiston theory and his personal eccentricities isolated him increasingly from the scientific community.

Monnet's first wife, by whom he had a son and a daughter, died in 1779. He married the writer Mariette Moreau in 1781. Monnet's brother was a mineralogist active in the Société Littéraire de Clermont-Ferrand.

BIBLIOGRAPHY

I. ORIGINAL WORKS. Monnet's publications are *Traité des eaux minérales* (Paris, 1768); *Traité de la vitriolisation & de l'alunation* (Amsterdam, 1769); *Exposition des mines, ... à laquelle on a joint ... une dissertation pratique sur le traitement des mines de cuivre, traduite de l'allemand, de M. Cancrinus* (London, 1772)—the major part of this work is a free rendering of A. F. Cronstedt, *Försök til mineralogie* (Stockholm, 1758), but is based on a German trans. of Cronstedt; *Traité de l'exploitation des mines* (Paris, 1773), based on an unidentified work published by the Council of Mines of Freiberg, Saxony; *Dissertation sur l'arsenic, qui a remporté le prix proposé par l'Académie Royale [de Berlin] pour l'année 1773* (Berlin, 1774); *Traité de la dissolution des métaux* (Paris, 1775); *Nouveau système de minéralogie* (Bouillon, 1779); *Mémoire historique et*

478

politique sur les mines de France (Paris, 1790); and *Démonstration de la fausseté des principes des nouveaux chymistes* (Paris, 1798).

Monnet published articles in the *Journal de médecine, chirurgie, pharmacie, &c.; Observations sur la physique, sur l'histoire naturelle et sur les arts; Journal des mines; Mélanges de philosophie et de mathématiques de la Société royale de Turin pour les années 1766–1769* (*Miscellanea Taurinensia*, vol. IV); and *Mémoires de l'Académie royale des sciences* (Turin). He contributed maps to and was author of the text of *Atlas et description minéralogiques de la France, entrepris par ordre du roi ... première partie* (Paris, 1780); and 2nd ed., *Collection complette de toutes les parties de l'atlas minéralogique de la France, qui ont été faites jusqu'aujourd'hui* ([Paris, *ca.* 1799]); for an analysis, see the Lavoisier bibliographies cited below. Monnet was also translator of Ignaz von Born, *Voyage minéralogique fait en Hongrie et en Transilvanie* (Paris, 1780).

Approximately 20 vols. of papers are at the École des Mines, Paris; many were written after Monnet's retirement and are of varying reliability, showing evidence of increasing paranoia. Extracts from MS 4672, a volume of inaccurate copies of letters, have been published in *Nouvelle revue rétrospective*, **19** (1903), 289–298, 361–384; and **20** (1904), 1–24, 100–120, 169–192, 245–264, 445–446. Two travel journals, MSS 4688 and 8286, respectively, have been edited and published by Henry Mosnier, *Voyage de Monnet, inspecteur général des mines, dans la Haute-Loire et le Puy-de-Dôme, 1793–1794* (Le Puy, 1875); and "Les bains du Mont-Dore en 1786. Voyage en Auvergne de Monnet," in *Mémoires de l'Académie des sciences, belles-lettres et arts de Clermont-Ferrand*, **29** (1887), 71–174. Important papers and letters are at the Archives Nationales, Paris, F[14]1313–1314; Bibliothèque Centrale du Muséum National d'Histoire Naturelle, Paris, MS 283; Bibliothèque Nationale, Paris, MSS fr. 11881, 12306; and Bibliothèque Municipale de Clermont-Ferrand, MSS 1339, 1390, 1400. One letter has been published by R. Rappaport, "The Early Disputes Between Lavoisier and Monnet, 1777–1781," in *British Journal for the History of Science*, **4** (1969), 233–244.

Letters by and about Monnet are in *Torbern Bergman's Foreign Correspondence*, G. Carlid and J. Nordström, eds., I (Stockholm, 1965).

II. SECONDARY LITERATURE. There is an anonymous eulogy of Monnet in *Annales des mines*, **2** (1817), 483–485. See also Louis Aguillon, "L'École des mines de Paris: notice historique," in *Annales des mines*, 8th ser., **15** (1889), 433–686; Denis I. Duveen and Herbert S. Klickstein, *A Bibliography of the Works of Antoine Laurent Lavoisier 1743–1794* (London, 1954); Denis I. Duveen, *Supplement to a Bibliography of the Works of Antoine Laurent Lavoisier 1743–1794* (London, 1965); and R. Rappaport, "The Geological Atlas of Guettard, Lavoisier, and Monnet: Conflicting Views of the Nature of Geology," in Cecil J. Schneer, ed., *Toward a History of Geology* (Cambridge, Mass., 1969).

RHODA RAPPAPORT

MONRO, ALEXANDER (Primus) (*b.* London, England, 8 September 1697; *d.* Edinburgh, Scotland, 10 July 1767), *anatomy.*

Monro was the only child of John Monro, military surgeon, and Jean Forbes, granddaughter of Duncan Forbes of Culloden. John Monro—who was the youngest son of Sir Alexander Monro, advocate, of Bearcroft, Stirlingshire—retired from the army in 1700 and took up private practice in Edinburgh. Alexander entered Edinburgh University in 1710, where he remained for three years, studying Latin, Greek, and philosophy. He also learned French, arithmetic, and bookkeeping under private teachers and received instruction in fencing, dancing, music, and painting. He did not graduate in arts, but, having decided on a medical career, was formally apprenticed to his father in 1713. He also attended such medical courses as were available locally, but these did not amount to much. He says "the dissection of a human body was shewed once in two or three years by Mr. Robert Elliot, and afterwards by Messrs. Adam Drummond and John Macgill, Surgeon-Apothecaries," who, he adds pointedly, "had the Title of Professors of Anatomy."

John Monro had studied at Leiden University under Archibald Pitcairne, whose idea of founding a medical school of repute in Edinburgh seems to have fired his imagination, and once his son's aptitude became apparent, he spared no efforts in preparing him to play a major role in the scheme. In 1717 Alexander was sent to London, where he studied physics under Whiston and Hauksbee and attended demonstrations by the great anatomist William Cheselden. With the encouragement of their master, Cheselden's students had formed a scientific society; and a paper read by Monro on "the bones in general" was a forerunner of his own important work on that subject. He also made a number of anatomical preparations, which he sent home and which were so admired by Adam Drummond, one of the professors of anatomy at Edinburgh, that he offered to resign in Monro's favor when he should return to Scotland. In the spring of 1718 he went to Paris, where he attended a course in anatomy by Bouquet and frequented the hospitals. He performed operations under the direction of Thibaut, was instructed in midwifery by Grégoire, in bandaging by Cesau, and in botany by Chomel. In the autumn of 1718 he went to Leiden, where he won the favorable attention of Boerhaave, his father's old fellow student. He returned to Edinburgh in 1719.

Monro had come to realize the value of the history of anatomy in the academic teaching of the subject, and with his customary thoroughness he enrolled as a student in Charles Mackie's newly inaugurated class of universal history. On 20 November 1719 he was admitted a fellow of the Royal College of Surgeons of Edinburgh, after passing the usual tests. On 29 January 1720, even though he was still only twenty-two years of age, the town council appointed him professor of anatomy. On Cheselden's recommendation he was elected a fellow of the Royal Society in 1723. In 1724 and 1725 there was a popular outcry against grave robbing in Edinburgh. Surgeons' Hall was beset, and there were threats to demolish it. In 1725 the town council accordingly provided Monro with an anatomy theatre and museum for his preparations within the comparative safety of the university precinct, and thereafter he undertook all the duties of a professor. One of these was to take his turn in delivering the public oration that inaugurated each session, and the subject of his first, delivered on 3 November 1725, was "De origine et utilitate anatomes," which he later incorporated into his course on the history of anatomy.

On 3 January 1725 Monro married Isabella MacDonald, daughter of Sir Donald MacDonald of Sleat (Isle of Skye). They had three sons and five daughters. Only one of his daughters survived infancy, and for her Monro wrote an "Essay on Female Conduct," which included a section on "The Laws of Nature, the Mosaical Institution and the Christian System." John, the eldest son, became an advocate; Donald, his second son, graduated M.D. (1753) and was physician to St. George's Hospital, London; and Alexander, the youngest, succeeded his father in the chair of anatomy at the University of Edinburgh. In 1726 Monro published his major work, *The Anatomy of the Humane Bones*. It had no illustrations, being intended as a commentary on actual demonstrations and dissections. Moreover, Monro knew that his old master Cheselden was preparing a set of accurate plates for his *Osteographia* (1733), made with the help of the improved camera obscura. The work is enlivened by Monro's acute and original comments based on close observation: for example, that different nationalities are distinguishable by the form of the cranium, that the nasal sinuses improve the power and tone of the voice, that a man's stature decreases as evening approaches, and that the bone at a healed fracture is stronger than before. In the second (1732) and later editions there is added "An Anatomical Treatise of the Nerves, an Account of the Reciprocal Motions of the Heart and a Description of the Human Lacteal Sac and Duct." Here he observes that the nerves consist of "a great many threads lying parallel to each other," and seems to anticipate Müller's law of specific nerve

energies, noting that "when all light is excluded from the eyes an idea of light and colour may be excited in us by coughing, sneezing, rubbing or striking the eyeball." The work continued to be reprinted as late as 1828, by which time it had gone through nineteen English editions and appeared in several translations, the most notable being the large, illustrated French edition (1759) by Jean-Joseph Sue.

The Edinburgh Medical School had now a nucleus of medical professors, but there was still no hospital for clinical teaching. As early as 1721 John Monro had agitated for the establishment of a regular hospital in Edinburgh, and Alexander himself had published appeals for funds for the purpose, but it was not until 1725 that the matter was seriously pursued with the help of George Drummond, lord provost of Edinburgh. In 1729 a small hospital for the sick poor was opened, and it was from its case register that much of the material was derived for the *Medical Essays and Observations*, 6 vols. (1732–1744), edited by Monro for the Society for the Improvement of Medical Knowledge, of which he was secretary. The series owed much to the individual efforts of Monro, who contributed many of the papers, his most important being an "Essay on the Nutrition of Foetuses," in which he showed "that there is no Anastomosis, Inosculation or Continuation between the vessels of the Womb and those of the Secundines and that the Liquors are not carried from the Mother to the Foetus or from the Foetus to the Mother by continued Canals." The *Medical Essays* became a standard work of reference, went through five editions, and was translated into several languages.

The scope of this society was widened in 1737 at the suggestion of Monro's friend Colin Maclaurin, professor of mathematics, and it was renamed the Society for Improving Philosophy and Natural Knowledge, or the Philosophical Society, but Maclaurin's death and the rebellion of 1745 caused its decline. In 1752 it was revived and Monro was elected joint secretary with David Hume the philosopher, contributing six medical papers to their *Essays and Observations, Physical and Literary* (1754, 1756). This society became the Royal Society of Edinburgh in 1783. Monro belonged to several other societies: the Honorable Society of Improvers of the Knowledge of Agriculture in Scotland (disbanded in 1745); the Select Society, founded by Allan Ramsay the Younger; and the Edinburgh Society for Encouraging Arts, Sciences, Manufactures and Agriculture in Scotland, an offshoot of the Select Society. He was also a manager of the Royal Infirmary and a director of the Bank of Scotland. In addition he was a justice of the peace, a manager of the Orphans

Hospital and of the Scheme for the Widows of Ministers and Professors, although he was less active in these roles.

In politics Monro was a staunch Hanoverian but no bigot. After the battle of Prestonpans in 1745, which went against his cause, he impartially assisted the wounded of both armies. Upon the death of his friend Maclaurin (1745), he delivered before the university a memorial lecture that formed the basis of the memoir prefixed to Maclaurin's posthumously published *Account of Sir Isaac Newton's Philosophical Discoveries* (London, 1748). Monro actively fostered the career of his gifted youngest son, Alexander, with the parental concern characteristic of the family. For his benefit he wrote a "commentary" on his *Anatomy of the Human Bones* and in 1754 persuaded the town council to admit him as joint professor of anatomy with himself, although he had not yet graduated. After his son Alexander, Secundus as he was thenceforth designated, had taken his M.D. (1755), Monro Primus—to use the father's new epithet—was granted the degree of M.D., *honoris causa* (1 January 1756). The system of joint professorships was to provide emoluments for the retiring professor, but Monro Primus, having secured the succession for his son, continued to share the duties of the chair until 1758, after which he confined himself to his favorite clinical lectures in the new Royal Infirmary, which had been completed in 1741. The infirmary was designed by William Adam under the supervision of Monro and Lord Provost Drummond.

Monro Primus was of medium height, strongly built, and energetic, but subject to periodical inflammatory fevers. He continued to take an active part in university business until the end of 1765, although by 1762 he was beginning to feel the symptoms of cancer of the rectum, which caused his death on 10 July 1767 at his home in Covenant Close, Edinburgh. He was buried in Greyfriars Churchyard, Edinburgh. He had bought an estate at Auchenbowie, Stirlingshire, but his plan to retire there was thwarted by circumstances, although he often visited it and took a close interest in its management. He was a commissioner of supply and highroads for Stirlingshire and a benefactor of the local parish church. Earlier he had provided a country home at Carolside, Berwickshire, for his father in his declining years.

Monro was not ambitious as an author. His great work on the human bones was published rather as a teaching aid, and many of his important contributions to *Medical Essays and Observations* were anonymous. His lectures that exist in manuscript reveal his wide reading in their references to past and contemporary anatomical works. A section of

them was published without his authority in *An Essay on Comparative Anatomy* (1744). In 1762 he published *An Expostulatory Epistle to William Hunter*, in which he rebuked his old pupil for some criticisms of himself included in Hunter's *Medical Commentaries* (1762), a work primarily directed against Monro Secundus, but there is little doubt that it was parental concern rather than personal pique that stirred him to the attack. His last publication, *An Account of the Inoculation of Smallpox in Scotland* (1765), was also due to external prompting. It contains the answers conscientiously gathered by Monro to a questionnaire sent to him by the Faculty of Medicine in Paris about the efficacy of inoculation, of which Monro himself was a strong advocate. After his death his course of lectures on the history of anatomy, which included his remarks on the usefulness of the study of the subject and the best method of teaching it, were plagiarized by William Northcote in *A Concise History of Anatomy* (1772).

Monro was not a great innovating genius (eighteenth-century anatomy indeed was marked more by advances in the field of description than by new discoveries), but his extraordinary industry, his wide reading, his accuracy of observation, and his open, original mind sometimes led him to correct conclusions that could only be verified by the more refined equipment of later times. He was a supreme teacher and demonstrator. A gifted technician, Monro improved methods of injecting minute vessels and preserving anatomical preparations. He had the manual dexterity of a master craftsman and was a cool and expert surgeon, in spite of a strong natural abhorrence of inflicting pain. His practice of lecturing informally in English was then a novelty, Latin being still the academic language, and he spoke from only the briefest notes. Oliver Goldsmith, who was a medical student at the University of Edinburgh (1752–1754), said he was "an able orator," explaining "things in their nature obscure in so easy a manner that the most unlearned might understand him." In 1720 his class numbered fifty-seven, but by 1749 he had 182 students, and by 1751 it had outgrown the anatomy theater and had to be taught at two separate meetings daily. His reputation attracted students from all parts of Europe, so that his father's dream of Edinburgh as a medical center rivaling Leiden began to come true. The advance guard of students from America also began to appear, and the influence of the Edinburgh Medical School was carried to the New World. The inspiration of Monro's teaching was frequently acknowledged in grateful dedications in the M.D. theses of his students, among whom were such distinguished names as William Hunter,

Robert Whytt, John Fothergill, Andrew Duncan, and, of course, his own son, Alexander Monro (Secundus).

BIBLIOGRAPHY

I. ORIGINAL WORKS. Note references in the text. Also Monro Secundus, ed., *The Works of Alexander Monro* (Edinburgh, 1781), with a life of A. Monro Primus by Donald Monro. This book contains the published works, including contributions to *Medical Essays and Observations*. The largest collection of his manuscript lectures and other unpublished material is in the Otago University Library, New Zealand; see W. J. Mullin, "The Monro Family and the Monro Collection of Books and MSS," in *New Zealand Medical Journal*, **35** (1936), 221. See also for MSS of his lectures *The Index Catalogue of the Library of the Surgeon General's Office*, IX (1888), 384. Monro's own carefully kept account book for his students' fees, 1720–1749, is in Edinburgh University Library (Dc.5.95). The short biography, "Alexander Monro, Primus," in *University of Edinburgh Journal*, **17** (1953), 77–105, although from an apparently holograph MS, may be wholly or partly the work of Monro's young friend, William Smellie, the printer, who published verbatim extracts from it in his "Life of the Celebrated Dr. Monro," in *Edinburgh Magazine and Review*, **1** (1744), 302–306, 337–344.

II. SECONDARY LITERATURE. On Monro and his work see A. Duncan, Sr., *An Account of the Life and Writings of Alexander Monro, Senr.* (Edinburgh, 1780); D. J. Guthrie, "The Three Alexander Monros," in *Journal of the Royal College of Surgeons of Edinburgh*, **2** (1956), 24–34; J. A. Inglis, *The Monros of Auchenbowie* (Edinburgh, 1911); K. F. Russell, *British Anatomy, 1525–1800: a Bibliography* (Melbourne, 1963); S. W. Simon, "The Influence of the Three Monros on the Practice of Medicine and Surgery," in *Annals of Medical History*, **9** (1927), 244–266; and R. E. Wright-St. Clair, *Doctors Monro, a Medical Saga* (London, 1964).

C. P. FINLAYSON

MONRO, ALEXANDER (Secundus) (*b.* Edinburgh, Scotland, 10 March 1733; *d.* Edinburgh, 2 October 1817), *anatomy*.

The third and youngest son of Alexander Monro (Primus), Monro was educated first at James Mundell's private school, Edinburgh, and then at the University of Edinburgh. His name appears in his father's account book for his anatomy class in 1744, when he was only eleven years of age. In the following year he matriculated in the Faculty of Arts and studied Latin, Greek, philosophy, mathematics, physics, and history. Like the majority of arts students in the university at that time, he did not graduate, individual professors' certificates being then more highly valued than the official diploma. In 1750 he

began the serious study of medicine under Andrew Plummer (chemistry), Charles Alston (botany), John Rutherford (practice of physic), Robert Whytt (institutes of medicine), and Robert Smith (midwifery).

His father encouraged his natural bent for medicine, making for him in 1750 a manuscript commentary on his *Anatomy of the Human Bones*, and entrusting him in 1753 with the teaching of the evening anatomy class necessitated by the growing numbers of students. After only one session of this arrangement Monro Primus petitioned the town council, the patrons of the university, to appoint his son joint professor of anatomy, and his request was backed by a certificate from the students of his son's evening class (they included Joseph Black) testifying to their satisfaction with his teaching. On 10 June 1754 the desired appointment was ratified, although Monro was still only twenty-one years of age. On 25 October 1755 he graduated M.D. with the thesis *De testibus et semine in variis animalibus*. Edinburgh M.D. theses were printed at this period, but most were essays based on secondary sources. Monro's thesis extended the knowledge of the seminiferous tubules by some original research. He injected the tubules with mercury and showed their connection with the epididymis, observing that semen has a close relationship with blood and lymph, although his later lectures show that his notions about the real nature of the substance were quite fanciful. Whereas his father considered that the spermatozoa alone formed the embryo, Monro Secundus taught that "these animalculae are no more essential to generation than the animals found in vinegar are to its acidity."

Soon after graduating he went to London, where he attended the lectures of William Hunter, an old student of Monro Primus. He then went on to Paris but had to return hastily to Edinburgh in 1757 to deputize for his father during an illness. He returned to the Continent later in the same year, spending several months in the home of the famous Berlin anatomist Meckel, with whom he performed the operation of paracentesis of the thorax. While there, he published his treatise *De venis lymphaticis valvulosis* (Berlin, 1757), in which he showed that the lymphatics were absorbents and distinct from the circulatory system. There was a counterclaim for priority in this discovery from William Hunter, which sparked off an acrimonious exchange of pamphlets. Monro Secundus replied to Hunter's claim in his *Observations, Anatomical and Physiological, Wherein Dr. Hunter's Claim to Some Discoveries Is Examined* (1758). Hunter retorted in *Medical Commentaries, Part I: Containing a Plain and Direct Answer to Professor*

Monro, Jun., Interspersed with Remarks on the Structure, Functions and Diseases of Several Parts of the Human Body (London, 1762–1764). Monro seems to have been ahead of Hunter in the matter of the lymphatics, but their mutual jealousy blinded them to the earlier discoveries of Friedrich Hoffman in this field.

Monro extended his attacks to include Hewson, his own former pupil and a colleague of Hunter, who in 1767 had recommended the operation of paracentesis of the thorax in traumatic pneumothorax and at the same time had published his own discovery of the existence of lacteals and lymphatics in non-mammalians. Monro asserted his own priority in both fields in *A State of Facts Concerning the First Proposal of Performing the Paracentesis of the Thorax and the Discovery of the Lymphatic Valvular Absorbent System of Oviparous Animals. In Answer to Mr. Hewson* (Edinburgh, 1770).There is no doubt that Monro had preceded Hewson in performing the operation of paracentesis of the thorax. Although he had earlier shown injections of the lymphatics and described them to his class, Hewson was the first to publish a full and accurate account of them in nonmammalian animals.

From Berlin, Monro went on to Leiden, where he met the anatomist B. S. Albinus, once a fellow student of Monro Primus and Peter Camper, professor of anatomy at Amsterdam. In January 1758, his father being again taken ill, Monro, now in his twenty-fifth year, had to cut short his European tour in order to conduct the anatomy class at Edinburgh. His father recovered and delivered the opening lecture of the session (1758–1759), but thereafter Monro Secundus undertook the main work of the chair and continued to do so for the next fifty years. On 1 May 1759 he became a fellow of the Royal College of Physicians of Edinburgh. His course started with a detailed history of anatomy and proceeded to anatomy itself, beginning with the bones; then came physiology, and finally the operations of surgery. His clear informal style of lecturing was even more effective than his father's. The official records of the Faculty of Medicine give him 228 students in 1808.

His earlier publications were largely polemical, and it was not until he had been teaching for twenty-five years that his three main contributions to medical literature appeared:

His *Observations on the Structure and Functions of the Nervous System* (Edinburgh, 1783; German ed., Leipzig, 1787) advanced the study of the subject by making several original discoveries, the most famous being of the foramen connecting the lateral and third ventricles of the brain, thereafter known as the "foramen of Monro."

The Structure and Physiology of Fishes Explained and Compared With Those of Man and Other Animals (Edinburgh, 1785) was the first important Edinburgh textbook on comparative anatomy, a subject that had been recently introduced to their London students by the Hunters.

A Description of All the Bursae Mucosae of the Human Body; Their Structure Explained and Compared With That of the Capsular Ligaments of the Joints, and of Those Sacs Which Line the Cavities of the Thorax and Abdomen: With Remarks on the Accidents and Diseases Which Affect Those Several Sacs, and on the Operations Necessary for Their Cure (London, 1788), trans. into German by J. C. Rosenmüller (Leipzig, 1799), was a practical manual for direct use in surgery. Although next to nothing was known of germ life at that time, Monro's acute observation and independent empirical judgment led him to the conclusion that the chief danger of infection in surgery of the joints lay in exposure to the air.

Monro published three lesser but original works:

In *Experiments on the Nervous System, With Opium and Metalline Substances; Made Chiefly With the View of Determining the Nature and Effects of Animal Electricity* (Edinburgh, 1793), he showed that stimulation of a nerve by Galvani's couple (tinfoil and silver) produced muscle contraction, but he failed to deduce the true nature of nervous energy, clinging to the old theory of nervous fluid. Still he did at least conclude that the nerves conducted "that matter by which the muscle is influenced more readily than the skin, flesh or blood vessels."

In *Observations on the Muscles and Particularly on the Effects of Their Oblique Fibres: With an Appendix, in Which the Pretension of Dr. Gilbert Blane, That He First Demonstrated the Same Effect to Be Produced by Oblique Muscles as by Straight Ones, With a Less Proportional Decurtation of Fibres is Proved to Be Quite Unfounded* (Edinburgh, 1794), his old combative spirit is shown not to be quite dead.

The third work was his *Three Treatises on the Brain, the Eye and the Ear* (Edinburgh, 1797).

Like his father, Monro Secundus was a sociable man. He was a member of the Harveian Society of Edinburgh, which cultivated conviviality as well as oratory, in both of which fields Monro shone brilliantly. He was joint secretary of the Philosophical Society of Edinburgh along with David Hume (1760–1763) and sole secretary (1763–1783) when it became the Royal Society of Edinburgh. He was also a district commissioner for cleansing, lighting, and watching the streets, a manager of the Royal Infirmary, and a member of the committee of defense for

Midlothian during the French invasion scare of 1794.

On 25 September 1762 Monro married Katherine Inglis, daughter of David Inglis, treasurer of the Bank of Scotland, and by her had three sons and two daughters. He lived first in a flat in Carmichael's Land in the Lawnmarket, Edinburgh. In 1766 he moved to a house with a garden in Nicolson Street, near the university, where he stayed until 1801, when he took up residence in the New Town, in St. Andrew Square. In 1773 he bought a property of 271 acres at Craiglockhart on the outskirts of the town, not as a residence but purely to indulge his passion for gardening.

In 1798 he persuaded the town council to appoint his elder son, Alexander, thereafter known as Monro Tertius, to be joint professor of anatomy with him. He himself continued to share the duties of the chair until 1808, when he retired at age seventy-five. He died of apoplexy on 2 October 1817, at age eighty-four. He had bequeathed his fine collection of anatomical and pathological specimens for the use of his son and his successors in the chair of anatomy.

Monro Secundus was a kindly man in family and social life but perhaps overjealous of his professional reputation. He used his powerful influence, for instance, to prevent until almost the end of his teaching career the establishment of a separate chair of surgery—a clear necessity as Monro, although officially professor of anatomy and surgery, was not himself a practicing surgeon. His medical ability had been proved in the most testing of situations, having to follow a great father and work with such colleagues as William Cullen, Joseph Black, Daniel Rutherford, James Gregory, and Andrew Duncan.

BIBLIOGRAPHY

I. ORIGINAL WORKS. Most are referred to in the text. Read also *Essays and Heads of Lectures on Anatomy, Physiology, and Surgery. With a Memoir of His Life ... by His Son* (Edinburgh, 1840).

II. SECONDARY LITERATURE. On Monro and his work see A. Duncan, Senior, *An Account of the Life, Writings and Character of Alexander Monro, Secundus* (Edinburgh, 1818). Other relevant works are in the bibliography under Alexander Monro, Primus. See especially R. E. Wright-St. Clair, *Doctors Monro: a Medical Saga* (London, 1964).

C. P. FINLAYSON

MONTANARI, GEMINIANO (*b.* Modena, Italy, 1 June 1633; *d.* Padua, Italy, 13 October 1687), *astronomy, geophysics, biology, ballistics.*

When Montanari was ten, his father, Giovanni, died; and he and his brothers, who died very young, were educated by his mother, Margherita Zanasi. His adolescence may therefore have been somewhat unrestrained and turbulent. One of his last works, *L'astrologia convinta di falso*, contains many autobiographical notes which show that besides suffering several serious illnesses and severe falls, he was involved in brawls in which he both sustained and inflicted injuries. At the age of twenty he was sent to Florence to study law, a profession which would have enabled him to ease his family's financial problems. Montanari remained in Tuscany for three years, absorbed by many interests—but above all by a passion for a woman prominent in Tuscan society. The latter involvement led to trouble; and he was obliged to spend the last few months of this period at Grosseto, which was then in the middle of the swamps of the Maremma.

Fortunately Montanari was invited to go to Vienna and at Salzburg, he received a degree in both church and civil law. The epigraph on his tomb in the church of San Benedetto in Padua indicates that he also, probably at a different time, obtained degrees in philosophy and in medicine. In Vienna he practiced law and formed a friendship with Paolo del Buono (1625–1659), a young Florentine who had studied under Michelini at Pisa, where he had become imbued with the ideas and principles of Galileo. Del Buono was the director of the Imperial Mint and a correspondent of the Accademia del Cimento; and from him Montanari rapidly acquired a proficiency in mathematics and natural science, which until then he had considered merely a hobby. At the end of 1657 he accompanied Del Buono on a long trip to the mines which supplied the mint, visiting Styria, Bohemia, and Bergstetten, in the Carpathian Mountains of Upper Hungary (now Horni Mesto, Czechoslovakia). It appears that their research and inquiries aroused suspicions and accusations from which Del Buono fled to Poland, where he died at the age of thirty-four.

Montanari began the long and perilous journey back to Modena, where he entered the service of Duke Alfonso IV d'Este, and married a woman named Elisabetta. They had no children. She was an active and skillful collaborator in his work, including the construction of instruments and the polishing of lenses. After a few months Montanari tired of the ducal court and moved to Florence, where he became legal adviser to Prince Leopoldo de' Medici, who soon discovered his scientific abilities.

But in Florence too, Montanari's fiery character stirred up trouble; and when, at the beginning of

1661, the duke of Modena invited him to return to that city as court philosopher and mathematician, he accepted. The appointment was a brief one —Alfonso IV died in July 1662—but during this time he met Cornelio Malvasia, a Bolognese nobleman who commanded the duke's militia and who was passionately interested in astronomy. An active patron of talented scientists, in 1650 he had recommended to the Bolognese Senate G. D. Cassini, who had worked for him in the observatory that Malvasia had built in his house at Panzano, near Castelfranco Emilia. Now Malvasia became interested in Montanari, who had helped him to compile his volume of ephemerides covering 1661–1666 (Modena, 1662). Montanari left the court of Modena with him and went to Bologna and Panzano, where Malvasia died in March 1664 after having obtained the chair of mathematics at Bologna for his protégé.

Montanari began teaching at Bologna the following December, and the fourteen years that he spent there were the most productive years of his life. A. Fabroni, in the preface to his biography of Montanari, states that the extraordinary flowering of science in Bologna at the beginning of the eighteenth century had its beginning in Montanari's work. This flowering was of course attributable also to others, such as Cassini and Malpighi; but Montanari's influence must have been important, for he taught not only at the Archiginnasio but also at the many academies of natural philosophy. Soon after he arrived in Bologna, he founded such a school, which was modeled on the Florentine Accademia del Cimento and was called the Accademia della Traccia, or Accademia dei Filosofi (this was the precursor of the Accademia degli Inquieti, founded in 1690 by Eustachio Manfredi, which in 1712 became the Accademia delle Scienze dell'Istituto di Bologna).

Montanari also edited a volume of ephemerides and astronomical tables (1665). From 1669 he was concerned with Cassini's sundial in the church of San Petronio, and from the same year he published an annual almanac, in which he poked fun at judicial astrology because its predictions, rather than being deduced from the appearance of the heavens, were picked at random in the presence of friends.

The University of Bologna suffered a financial crisis in the late 1670's, during which the professors' salaries were greatly reduced and paid after long delays; Montanari decided to go to Padua, where a new chair of astronomy and meteorology was created for him, carrying a very high salary. But the Republic of Venice, not content to have him merely teaching, expected his advice and assistance on the control of rivers and the protection of the Venetian

Lagoon, military fortifications and the training of the artillery, and especially the organization of the mint and all problems having to do with currency. This last, heavy duty occupied Montanari for the rest of his life and was detrimental to his health—he was obese and inclined to apoplexy. He gradually became almost blind and died suddenly of apoplexy in 1687.

The volume of Montanari's work was enormous. G. Venturi summarizes his achievements by saying that he was an astronomer in Modena, a physicist in Bologna, and an engineer in Venice. It could be said that in a relatively short life he continually added to his interests but he never abandoned old ones when he took up a new ones.

Montanari's major contribution to Malvasia's ephemerides (1662) consisted of a map of the moon thirty-eight centimeters in diameter, the largest at the time and one of the most exact and detailed. Its precision resulted from his use of a reticle, which he described in this work as a network of silver wires; it must certainly have been more sophisticated than those used, but not described, by Divini and Grimaldi. As for the richness of detail, Montanari probably engraved the map himself, thus saving it from the arbitrary simplification that often accompanied the transition from drawing to engraving, a fate that ten years earlier had befallen Grimaldi's similar map. The ephemerides also contains the description of an attempt to work a clock by means of a pendulum, a project with which Montanari was in all probability concerned.

It was in Bologna that Montanari showed his exceptional skill in inventing and making precision instruments. He constructed enormous objective lenses, that were greatly praised by Cassini; one of them, dated 1666, is preserved in Bologna. In 1674 he published a description of the "dioptric level," an instrument that gave extremely accurate levelings because the level was fitted onto a telescope. This telescope was also equipped with a distance-measuring reticle made from hairs arranged on the focal plane of the eyepiece.

In physics Montanari conducted experiments to obtain drops of tempered glass and to observe the curious way in which they shattered. He also made studies, much admired by Huygens, of the behavior of liquids in capillary tubes (1672–1678), which suggested a similarity of the ascent of water in capillary tubes and that of the sap in plant stems. Yet Montanari had already done some experimental biology; at Vienna in 1657 he had artificially incubated chicks, and at Udine in 1668 he had performed a blood transfusion between animals. It is likely that he had also taken part in similar experiments conducted in 1667 at Cassini's house in Bologna.

In 1673, in a note on the "tromba parlante," Montanari demonstrated that the principle of the megaphone, invented two years previously by Morland, could be reversed and used as an ear trumpet. With a pair of such instruments he was able to send and receive signals over distances of up to four miles.

Montanari was also interested in meteorological phenomena and was the first to use the term "atmospheric precipitation." In a work of 1675, published by C. Bonacini in 1934, he speaks of the barometer as a "meteoroscope," an instrument the variations of which can forecast weather conditions; and in 1671 he had used a barometer as an altimeter, first on the Asinelli tower in Bologna and then on Monte Cimone, the highest mountain in the Tuscan Apennines.

Montanari's greatest achievements, however, were in astronomy, particularly in his observations of the star Algol, which contributed to one of the earliest and most important chapters in the history of astrophysics, the study of the variable stars. He sent the results of his observations, which struck a fresh blow at the Aristotelian concept of the heavens' immutability, to the Royal Society in London and gave the first report on them in the paper "Sopra la sparizione d'alcune stelle et altre novità celesti," published in *Prose de' signori accademici Gelati* (1671; French version, 1672). In this paper he catalogued many stars of variable brightness, again drawing particular attention to Algol; having observed it when it was fairly bright, in 1667 he noticed that it was only of the fourth magnitude, in 1669 it was of the second magnitude, and in 1670 again fourth magnitude. Montanari seems not to have noticed the regularity of the phenomenon, but he was reasonably accurate in indicating the extremes of the variations. In fact Algol (β Persei) has a period, determined by Goodrike in 1782, of less than three days; but its magnitude varies from approximately 3.4 to 2.1.

Montanari failed to perceive either the regularity or the period of variation because the deterioration of his sight prevented him from making regular observations, as he stated in the same paper. But his considerations of these phenomena are extremely interesting, expressed as they were against the prevailing opinions of the time. He mentioned Boulliau's fairly accurate calculation, made in 1638, that it took 332 days for Mira Ceti to complete its cycle of appearances and disappearances (this strange behavior, but not the periodicity, had been noted in 1596 by David Fabricius), then stated that nothing was known about the causes of the appearance of new stars and of variations in the brightness of known stars, but offered the hypothesis that they might be phenomena analogous to sunspots.

At Padua, although his sight continued to fail, Montanari did not abandon astronomy—indeed, he made instruments for new observatories in Padua and in the Palazzo Corner in Venice.

Montanari contributed to the martial arts through his *Manualetto dei bombisti . . . con le tavole delle inclinazioni . . . secondo la dottrina di Galileo* (1680, 1682, 1690), a manual for gunners, which contains tables for firing based on the hypothesis that it is possible to ignore the resistance of air. His works on fortifications have never been published; and very little has been published of his valuable research in hydraulics, the results of which he passed on to his pupil D. Guglielmini. Perhaps influenced by Michelini, Montanari declared that to keep the lagoon surrounding Venice unpolluted and to prevent its silting up, it was necessary to divert directly into the sea the rivers that emptied into the lagoon. Fortunately his advice was heeded. A posthumous paper on the same topic, "Il mare Adriatico e sua corrente, et la naturalezza dei fiumi . . .," appeared in 1696 and was reprinted several times. Another, on civil engineering, has almost certainly been lost.

Montanari's final project, undertaken after he had become almost blind, was the compilation of two important works on money, which are still considered the precursors of modern ideas in this field: *Trattato del valore ed abuso delle monete* and "La zecca in consulta di stato."

His battles against astrology, in which he was passionately engaged all his life, are summarized in *L'astrologia convinta di falso . . .* (Venice, 1685), which aroused great interest and brought about the banning of this pseudoscience from the universities. He left unfinished a dialogue on a tornado which had devastated the Venetian hinterland in 1686; and it was completed and published in 1694 with the title *La forza d'Eolo . . .* by one of his students, Francesco Bianchini (1662–1729), who included a biography of his teacher in the introduction.

Montanari observed comets in 1664, 1665, 1680, 1681, and 1682; a solar eclipse on 2 July 1666; and several lunar eclipses: 29 September 1670, 18 September 1671, an unknown date in 1674, and in September 1681.

BIBLIOGRAPHY

I. ORIGINAL WORKS. Montanari's map of the moon and some of his poems are in Cornelio Malvasia, *Ephemerides novissimae motuum coelestium . . . ad longitudinem urbis Mutinae . . .* (Modena, 1662). His other works include *Cometes . . . observatus anno 1664 et 1665. Astronomicophysica dissertatio . . .* (Bologna, 1665); *Ephemeris Lans-*

bergiana ad longitudinem . . . Bononiae, ad annum 1666 . . . (Bologna, 1665); *Intorno diversi effetti de' liquidi in cannucce di vetro . . .* (Bologna, 1667); *Speculazioni fisiche . . . sopra gli effetti di que' vetri temprati che rotti in parte si risolvono tutti in polvere . . .* (Bologna, 1671); and "Sopra la sparizione di alcune stelle ed altre novità celesti discorso astronomico," in *Prose de' signori accademici gelati*, V. Zani, ed. (Bologna, 1671), 369–392.

He also wrote *La livella diottrica . . . per livellare col cannocchiale . . .* (Bologna, 1674; Venice, 1680); *Discorso sopra la tromba parlante . . . con dotte osservazioni della natura dell'eco e del suono* (Guastalla, 1678); *Manualetto dei bombisti . . . per ben maneggiare i mortari . . .* (Venice, 1680; Verona, 1682); *L'Astrologia convinta di falso col mezzo di nuove esperienze e ragioni fisico-astronomiche . . .* (Venice, 1685); *Le forze d'Eolo, dialogo fisico-matematico . . .*, F. Bianchini, ed. (Parma, 1694); and "Il mare Adriatico e sua corrente . . . et la naturalezza dei fiumi . . .," in *La Galleria di Minerva . . .* (Venice, 1696), 320. His tract on money, "La Zecca in Consulta di Stato, trattato mercantile ove si mostrano . . . le vere ragioni dell'aumentare giornalmente di valuta delle monete . . . co' modi di preservarne gli Stati," appeared first in C. Casanova, ed., *In Philippi Argelati tractatus de monetis Italiae appendix (seu pars VI)* (Milan, 1759), 3–70, and in A. Graziani, ed., *Economisti del Cinque e Seicento* (Bari, 1913), pp. 252 ff.

II. Secondary Literature. See G. Albenga and F. Porro, "Montanari," in *Enciclopedia italiana*, XXIII (Rome, 1934), 720; C. Bonacini, "Una carta lunare di Geminiano Montanari," in *Nel primo centenario della fondazione dell'osservatorio geofisico dell'Università* (Modena, 1927), 1–14; "Sull'opera scientifica svolta a Modena da Geminiano Montanari," in *Annuario della R. Università di Modena*, 1933, Appendice (1935), 17–24; and "Nel terzo centenario della nascita di Geminiano Montanari," in *Atti e memorie. Accademia di scienze, lettere ed arti* (Modena), 4th ser., **4** (1934), 63–76; G. Campori, "Notizie e lettere inedite di Geminiano Montanari," in *Atti e memorie della Deputazione di storia patria di Modena e Parma*, **8** (1876), 65–96; P. Dore, "Origini e funzione dell'Istituto e dell'Accademia delle scienze di Bologna," in *Archiginnasio*, XXXV (Bologna, 1940), 192–214; A. Fabroni, *Vitae Italorum*, III (Pisa, 1779), 64–119; G. Horn-D'Arturo, "Montanari," in *Piccola enciclopedia astronomica*, II (Bologna, 1938; 2nd ed., 1960), 304–306; P. di Pietro, "Modena e la trasfusione del sangue," in *Bollettino dell'Ordine dei medici* (Modena) (1969), 123–128; P. Riccardi, *Biblioteca matematica italiana*, II (Modena, 1870; repr. Milan, 1952), col. 170–177; G. Targioni-Tozzetti, *Notizie degli aggrandimenti delle scienze fisiche accaduti in Toscana nel corso di anni 60 del secolo XVII* (Florence, 1780; repr. Bologna, 1967), I, 303–304; G. Tiraboschi, *Biblioteca modenese*, III (Modena, 1783; repr. Bologna, 1969), 254–279; G. Venturi, *Elogio di Geminiano Montanari recitato nel solenne aprimento delle scuole* (Modena, 1790); and Count Valerio Zani, ed., *Le memorie, imprese, ritratti e notizie dei signori accademici Gelati* (Bologna, 1672).

GIORGIO TABARRONI

MONTE, GUIDOBALDO, MARCHESE DEL (*b.* Pesaro, Italy, 11 January 1545; *d.* Montebaroccio, 6 January 1607), *mechanics, mathematics, astronomy.*

[He is known as Guidobaldo del Monte, although his signature reads Guidobaldo dal Monte. The form Guido Ubaldo (from the Latinized version) is often used, Ubaldo being taken incorrectly as the family name.]

Guidobaldo was born into a noble family in the territory of the dukes of Urbino. While at the University of Padua in 1564 he studied mathematics and befriended the poet Torquato Tasso. Later Guidobaldo served in campaigns against the Turks and in 1588 was appointed visitor general of the fortresses and cities of the grand duke of Tuscany. Soon afterward Guidobaldo retired to the family castle of Montebaroccio near Urbino, where he pursued his scientific studies until his death.

Guidobaldo was a prominent figure in the renaissance of the mathematical sciences. At Urbino he was a friend and pupil of Federico Commandino and an intimate of Bernardino Baldi, the mathematical historian. In 1588 Guidobaldo saw Commandino's Latin translation of Pappus through the press at Pesaro. The autograph transcript had initially been sent to the Venetian mathematician Barocius for publication; but Barocius, having refused to edit the work without making extensive changes, sent the manuscript to Guidobaldo, who published the text exactly as he found it. Concerning Pappus, Guidobaldo also corresponded with the Venetian senator Jacomo Contarini, who helped Guidobaldo secure an appointment at Padua for Galileo. Guidobaldo's correspondence with these and other friends is an important source for the history of the mathematics of the period.

Guidobaldo's first book, the *Liber mechanicorum* (1577), was regarded by contemporaries as the greatest work on statics since the Greeks. It was intended as a return to classical Archimedean models of rigorous mathematical proof and as a rejection of the "barbaric" medieval proofs of Jordanus de Nemore (revived by Tartaglia in his *Quesiti* of 1546), which mixed dynamic principles with mathematical analysis.

The *Liber* may be seen as a forceful argument that statics and dynamics are entirely separate sciences; hence no unified science of mechanics is possible. This attitude is evident in Guidobaldo's treatment of equilibrium in the simple machines, which he terms the case where the power sustains the weight. He stresses that a greater power is needed to move the weight than to sustain it and that the power which moves has a greater ratio to the weight moved than does the power which sustains to the weight sustained. Consequently, the same principle and proportions cannot hold good for both moving and sustaining.

Galileo overcame this objection to a unified mechanics by positing that an insensibly greater amount of power was needed to move, than to sustain, a given weight. Guidobaldo had scorned the use of *insensibilia* in mechanics, probably because they were not susceptible of precise mathematical definition. Like his contemporary Benedetti, Guidobaldo attacked Jordanus, Cardano, and Tartaglia for assuming that the lines of descent of heavy bodies were parallel rather than convergent to the center of the earth. The answer of both Tartaglia and Galileo to this demand for unreasonable exactitude in mechanics was that, at a great distance from the center, the difference between the parallel and convergent descents was insensible.

This extreme concern for precision led Guidobaldo to reject the valid inclined-plane theorem of Jordanus in favor of the erroneous theorem of Pappus. Pappus' premise that a definite amount of force was needed to move a body horizontally was in accord with the view of Guidobaldo that more power was required to move than to sustain the body. Moreover, Jordanus' theorem seemed vitiated by its neglect of the angle of convergence of the descents. By supposing against Pappus (whom he named) and Guidobaldo (whom he did not name) that an insensible amount of power was required to move a body horizontally, Galileo was able to apply the principle of virtual displacements to both static and dynamic cases and was able to frame useful principles of virtual work and inertia. Guidobaldo's quest for mathematical rigor may have barred such imaginative concepts from his mind.

The most fruitful section of the *Liber mechanicorum* deals with pulleys, reducing them to the lever. This analysis—which is far superior to that of Benedetti—was adopted by Galileo. In two subsequent mechanical works Guidobaldo developed other ideas of this first book. These works were the *Paraphrase of Archimedes: Equilibrium of Planes* (1588), a copy of which was sent to Galileo, and the posthumous *De cochlea* (1615).

Guidobaldo was Galileo's patron and friend for twenty years and was possibly the greatest single influence on the mechanics of Galileo. In addition to giving Galileo advice on statics, Guidobaldo discussed projectile motion with him, and both scientists reportedly conducted experiments together on the trajectories of cannonballs. In Guidobaldo's notebook (Paris MS 10246), written before 1607, it is asserted that projectiles follow parabolic paths; that this path is similar to the inverted parabola (actually a catenary) which is formed by the slack of a rope held horizon-tally; and that an inked ball that is rolled sideways over a near perpendicular plane will mark out such a parabola. Remarkably the same two examples are cited by Galileo at the end of the *Two New Sciences*, although only as postscripts to his main proof—which is based on the law of free fall—of the parabolic trajectory.

Among Guidobaldo's nonmechanical works are three manuscript treatises on proportion and Euclid; two astronomical books, the *Planisphaeriorum* (1579) and the posthumous *Problematum astronomicorum* (1609); and the best Renaissance study of perspective (1600).

Guidobaldo helped to develop a number of mathematical instruments, including the proportional compass, the elliptical compass, and a device for dividing the circle into degrees, minutes, and seconds.

BIBLIOGRAPHY

I. Original Works. Guidobaldo's published works are *Liber mechanicorum* (Pesaro, 1577; repr. Venice, 1615); Italian trans. by Filippo Pigafetta, *Le mechanice* (Venice, 1581; repr. Venice, 1615); *Planisphaeriorum universalium theorica* (Pesaro, 1579; repr. Cologne, 1581); *De ecclesiastici kalendarii restitutione opusculum* (Pesaro, 1580); *In duos Archimedis aequeponderantium libros paraphrasis* (Pesaro, 1588); *Perspectivae libri sex* (Pesaro, 1600); *Problematum astronomicorum libri septem* (Venice, 1609); and *De cochlea libri quatuor* (Venice, 1615).

MS works of Guidobaldo are the *Meditatiunculae*, Bibliothèque Nationale (Paris), MS Lat. 10246; *In quintum Euclidis elementorum commentarius* and *De proportione composita opusculum*, Biblioteca Oliveriana (Pesaro), respectively MSS 630 and 631; and a treatise on the reform of the calendar, Biblioteca Vaticana, MS Vat. Lat. 7058. A collection of drawings of machines by Francesco di Giorgio Martini in the Biblioteca Marciana (Venice), MS Lat. VIII 87(3048), was formerly owned by Guidobaldo. The present location of the MS *In nonnulla Euclidis elementorum expositiones* (item 194 bis in the Boncompagni Sale Catalogue of 1898) is not known.

Guidobaldo's letters (some are copies) are scattered: Biblioteca Nazionale Centrale (Florence), MSS Galileo 15, 16, 88; Biblioteca Comunale "A. Saffi" (Forlì), MSS Autografi Piancastelli Nos. 755, 1508; Archivio di Stato (Mantua), Corrispondenza Estera, E.XXVIII, 3; Biblioteca Ambrosiana (Milan), MSS D.34 inf., J.231 inf., R.121 sup.; Bodleian Library (Oxford), MS Canon. Ital. 145; Bibliothèque Nationale (Paris), MS 7218 Lat.; Biblioteca Oliveriana (Pesaro), MSS 193 Ter.; 211/ii; 426; 1580 (MS 1538 = Tasso to Guidobaldo); Archivum Pontificiae Universitatis Gregorianae (Rome), Cassetta 1, MSS 529-530; Biblioteca Comunale degli Intronati (Siena), MS K.XI.52; Biblioteca Universitaria (Urbino), MS Carità Busta 47, Fasc. 6; and Biblioteca Nazionale Marciana (Venice), MS Ital. IV, 63 (Rari V.259).

Favaro has printed the Galileo correspondence in the *Opere* of Galileo, vol. X; and the two Marciana letters in *Due Lettere*. Rose, *Origins*, prints Ambrosiana MS J.231 inf., and Arrighi, *Un grande*, has six letters from Oliveriana MS 426, with the prefaces of MSS 630 and 631. Most of *Le mechanice* is translated in Drake and Drabkin, *Mechanics*. Important pages from Paris MS 10246 are in Libri, *Histoire*, IV, 369–398.

II. SECONDARY LITERATURE. A bibliog. is in Paul Lawrence Rose, "Materials for a Scientific Biography of Guidobaldo del Monte," in *Actes du XIIe congrès international d'histoire des sciences, Paris, 1968*, **12** (1971), 69–72. The earliest biography is the short note by Guidobaldo's friend Bernardino Baldi, *Cronica de' matematici* (Urbino, 1707), 145–147. Baldi's full *Vita* has disappeared. Giuseppe Mamiani, *Elogi storici di Federico Commandino, G. Ubaldo del Monte . . .* (Pesaro, 1828), is informative, although few references are given. The Guidobaldo section was earlier published in the *Giornale arcadico*, vols. IX, X (Senigallia, 1821). The 1828 ed. is reprinted in Mamiani, *Opuscoli scientifici* (Florence, 1845).

On Guidobaldo's mechanics see Antonio Favaro, "Due lettere inedite di Guidobaldo del Monte a Giacomo Contarini," in *Atti del Istituto veneto di scienze, lettere ed arti*, **59** (1899–1900), 303–312. Pierre Duhem, *Les origines de la statique*, I (Paris, 1905), 209–226, was very critical of Guidobaldo. Stillman Drake and I. E. Drabkin, *Mechanics in Sixteenth Century Italy* (Madison, Wis., 1969), 44–52 and *passim*, are more favorably disposed.

Guidobaldo's astronomical interests are illustrated in Gino Arrighi, "Un grande scienziato italiano; Guidobaldo dal Monte . . .," in *Atti dell' Accademia lucchese di scienze, lettere ed arti*, n.s. **12** (1965), 183–199.

For mathematical instruments see Paul Lawrence Rose, "The Origins of the Proportional Compass," in *Physis*, **10** (1968), 54–69, and "Renaissance Italian Methods of Drawing the Ellipse and Related Curves," in *Physis*, **12** (1970), 371–404.

See also Guillaume Libri, *Histoire des sciences mathématiques en Italie*, IV (Paris, 1841), 79–84, 369–398; and Antonio Favaro, "Galileo e Guidobaldo del Monte," (*Scampoli Galileani 146*), in *Atti dell' Accademia di scienze, lettere ed arti di Padova*, **30** (1914), 54–61.

PAUL LAWRENCE ROSE

MONTELIUS, GUSTAV OSCAR (*b.* Stockholm, Sweden, 9 September 1843; *d.* Stockholm, 4 November 1921), *archaeology*.

Montelius was strongly influenced by the great Scandinavian archaeologists Thomsen, Worsaae, and Nilsson, the creators of the scientific method of archaeology and the authors of the theory of three successive ages of Stone, Bronze, and Iron, of which they had demonstrated the stratigraphical validity in their work in the peat bogs and barrows of Denmark.

At the age of twenty Montelius joined the Swedish Archaeological Service and began working in the Swedish National Museum, where he remained for fifty years. In 1913, when he was seventy, he retired from the museum as director and state antiquary. Montelius traveled extensively; his reputation in his maturity and old age was international.

Montelius adopted the three-age system of Thomsen and Worsaae and expanded it into a four-age system comprising the Paleolithic, Neolithic, Bronze, and Iron. He was particularly interested in the Neolithic and Bronze ages, into which he introduced further subdivisions. He divided the Scandinavian Neolithic, for example, into four phases: the premegalithic (Neolithic I); the dolmen period (Montelius II); the passage grave period (Montelius III); and the long-stone cist period (Montelius IV). Montelius' work in subdividing epochs of the prehistoric past paralleled that of G. de Mortillet in France and his Neolithic subdivisions, although never widely adopted, were a model for the subdivisions of the Bronze age made by Déchelette in France and Fox in Britain.

Montelius believed firmly in the exact description and classification of prehistoric artifacts—indeed, he may be considered the founder of prehistoric taxonomy. He distinguished between open and closed finds and carefully classified prehistoric artifacts according to form, design, and ornament. He further taught the importance of studying the associations among these properly described and classified artifacts, and began to arrange them in sequences based upon changes in form, design, and ornament. This notion of typological sequence was developed by Worsaae; Montelius' refinement of it allowed him to establish a relative time sequence for Scandinavian artifacts.

Montelius next addressed himself to the problem of translating this relative chronology into an absolute one, and to the question of how new forms of implements and new customs came into existence in Scandinavia. Drawing upon the man-made chronologies of Egypt and Mesopotamia—going back to 3000 B.C.—he devoted himself to establishing links between the ancient East and Barbarian Europe; his *Bronze Age Chronology in Europe* was published in 1889. Between 1889 and 1891 Flinders Petrie was able, by cross-dating, to establish the absolute dates of Mycenaean Greece; Montelius then set out to develop these dated connections, although he realized that the opportunity to establish cross-dating diminished in proportion to geographical distance from the eastern Mediterranean. In 1892 Montelius published his own account of the relationship between Greek and Oriental chronology; in 1897 he extended this from Greece to Italy; in 1898 he developed the connections between Mediterranean chronology and that of Germany and Scandinavia; in 1900 he dealt with France and the Netherlands; and in 1908 with England and Scotland. By 1910 he had done all that it was possible to do with the methods of his time in correlating the undated prehistoric sequences of Barbarian Europe with the dated sequences of the eastern Mediterranean and Egypt. He thus established a historical chronology of prehistoric Europe that, however modified, served prehistory until the advent of geochronology and carbon-14 dating.

Montelius' popularizing work, *The Civilization of Sweden in Heathen Times* (1888), is a model of early *haute vulgarisation.* In the study of cultural origins he was, in the end, an advocate of the sort of modified diffusionism which was taken up after his death by Gordon Childe. Montelius applied his theories and tested them in relation to megalithic monuments; his *The Orient and Europe* (1894) is a classic statement of the theory of megalithic origins in the eastern Mediterranean.

BIBLIOGRAPHY

Montelius' chief works are discussed in the text; a *Festschrift* presented to him on the occasion of his seventieth birthday, *Opuscula archaeologica Oscari Montelio dicata* (Stockholm, 1913), contains a list of 346 of his writings.

GLYN DANIEL

MONTESQUIEU, CHARLES-LOUIS DE SE-CONDAT, BARON DE LA BRÈDE ET DE (*b.* La Brède, near Bordeaux, France, January 1689; *d.* Paris, France, 10 February 1755), *philosophy, political theory.*

Montesquieu was born into a noble family traditionally in the service of the king of Navarre. Since the seventeenth century a member of the family had been *président à mortier* of the Parlement of Guyenne, at Bordeaux.

After attending the Oratorian *collège* in Juilly (1700–1705) and studying law at Bordeaux (1705–1708), Charles-Louis de Secondat inherited from an uncle the name of Montesquieu and the office of *président à mortier*, which he held without enthusiasm and sold in 1726. Montesquieu was often received in Bordeaux society and became a member of the Académie de Bordeaux in 1716; traveled frequently to Paris, where he moved easily in high society. Famous for his *Lettres persanes* (1721) and elected to the Académie

Française (1728), he took a long trip through central Europe, Italy, and Germany and stayed for more than fifteen months in England (November 1729–May 1730), where he became a fellow of the Royal Society and a Mason. After his return to France he wrote his most important work, *L'esprit des lois,* which provoked a vigorous debate upon its publication in 1748.

Attracted to science in his youth, Montesquieu presented to the Académie de Bordeaux several reports on scientific memoirs submitted to it (1718–1720). In 1719 he commenced the compilation of a physical history of the earth in ancient and modern times, requesting scientists from all over Europe to send him papers on the subject. The project was never completed, but the topic long concerned Montesquieu. Several passages from *Mes pensées* and various memoirs on mines written during the period 1731–1751 are evidence of his continuing interest. On 20 November 1721 Montesquieu read before the Académie de Bordeaux his "Observations sur l'histoire naturelle." His remarks on insects, parasitic plants, and the anatomy of the frog show that he was well informed on the work of the Paris Académie des Sciences and accepted the primacy of observation, but that he remained much closer to the integral mechanism of Descartes than to the limited mechanism of Malebranche. Montesquieu denied the preexistence theory and interpreted the phenomena of vegetation purely mechanistically—including the formation of new tissues and of parasitic plants, such as mistletoe and mosses, which he considered to be vegetable excrescences rather than plants of a definite species. He was less interested in the notion of species than in the activity of nature, a position that led him to reject the intervention of Providence and to see the living world in a state of perpetual change, thus foreshadowing Diderot and recalling Lucretius. Similarly, the *Essai sur les causes qui peuvent affecter les esprits* outlines a psychophysiology with a clearly materialistic cast.

Montesquieu's most significant work was in the social and political sciences, of which he has been considered a founder. In *Considérations sur les causes de la grandeur des Romains, et de leur décadence* (1734) he rejected the moral and religious point of view in order to establish a historical science capable of discovering the real causes of major historical events. The goal of *L'esprit des lois* was to discover the scientific law of social institutions and phenomena, which, according to Montesquieu, depend neither on Providence nor on chance. Montesquieu's intention, in short, was to extend to human events his own mechanistic view of nature. This extension was effected without an unwarranted reduction of ethical, political, and social phenomena to physical factors. The analysis of political institutions led Montesquieu to link them with the *esprit général* of the societies that they govern, and analysis of this *esprit général* brought out the diversity of the factors that act on it. In addition to purely physical factors, such as climate and geography, there are also those pertaining to economics, demography, and ethical and religious traditions. (It may be remarked that Montesquieu studied religion as a social fact, without considering its truth or falsity.) The *esprit général* sustained the psychological principle proper to each type of government. Thus republican government rested on civic virtue, monarchy on honor, and despotism on fear. These moral foundations, without which governments could not remain what they are, can endure only so long as the various factors that determine the *esprit général* of the nation are appropriate. A healthy government should seek to maintain the vitality of its proper principle because the balance of physical, ethical, social, and political factors is never stable, and the government is always in danger of degenerating. That is why there is history. But knowledge of all the mediating factors and of their different combinations is bound to permit the rational explanation of even the most bizarre institutions and the most unexpected events.

This attempt to establish a science of social and political facts was based on the conviction that a rational order exists in seemingly diverse phenomena—a conviction especially evident in the writings of Malebranche, who influenced Montesquieu, and one shared by most scientists of the age. God, creator of this order and guarantor of its constancy and intelligibility, is also guarantor of the success of science. Montesquieu's originality consists in having applied this approach to human societies and institutions.

Not everything in *L'esprit des lois*, however, is original; the theory of climates, in particular, is very old. Montesquieu's writing often lacks scientific rigor: he selected the facts that suited his argument and rejected or misinterpreted those that did not. Moreover, he was far from possessing the knowledge necessary for the realization of his immense project. His purpose was not purely scientific: he wished to turn the French monarchy away from its despotic tendency and to introduce more humanity and reason into the laws. Although excessively moralistic and too involved in contemporary philosophical and political struggles to be a pure scientist, Montesquieu nevertheless offered an example of a scientific approach to political and social problems.

BIBLIOGRAPHY

I. ORIGINAL WORKS. Montesquieu's writings were collected as *Oeuvres complètes . . .*, Roger Caillois, ed., 2 vols. (Paris, 1949–1951). His major work is *L'esprit des lois*, J. Brethe de la Gressaye, ed., 4 vols. (Paris, 1950–1961).

II. SECONDARY LITERATURE. See L. Althusser, *Montesquieu, la politique et l'histoire* (Paris, 1969); S. Cotta, *Montesquieu e la scienza della società* (Turin, 1953); J. Ehrard, *L'idée de nature en France dans la première moitié du XVIIIe siècle*, 2 vols. (Paris, 1963), II, 493–515, 718–786, and *passim*; and R. Shackleton, *Montesquieu. A Critical Biography* (Oxford, 1961).

JACQUES ROGER

MONTGÉRY, JACQUES-PHILIPPE MÉRIGON DE (*b.* Paris, France, 25 July 1781; *d.* Paris, 9 September 1839), *military technology.*

Montgéry began his career as a seaman in the French navy in 1794. Commissioned midshipman second class in 1798, he rose to the rank of captain (1828). His commands included the gunboat *Enflammée* (1803) and the corvettes *Émulation* (1816–1818) and *Prudente* (1819–1820). In 1820 he made a military and naval tour of America. It was Montgéry's long-standing position as a member of the Conseil des Travaux de la Marine, however, that afforded him the opportunity to undertake his extensive scientific and military studies.

Montgéry was an analytical scientific chronicler and a prolific writer. He suggested the adoption of new weapons, including a flamethrower, the use of military railroads, and a rocket-firing submarine called *L'invisible*. A strong advocate of steamships, ironclads, mines, torpedoes, and rockets, he examined the *Steam Battery* and submarines of Robert Fulton and subsequently wrote extensively on the historical development of all phases of underwater warfare and exploration. He also wrote on Fulton's life and made elaborate critiques of his experiments.

Montgéry's investigations of pyrotechnics led him to do research into the war rockets of William Congreve and to the production of what may be the first documented history of rocketry, *Traité des fusées de guerre* (1825). Known throughout Europe, it became the standard work on the subject, appeared in serial form in several official journals, and was republished in 1841. Montgéry's coverage was exhaustive and analytical, particularly in his treatment of rocket physics.

Montgéry also wrote essays on the aeolipile designed by Hero of Alexandria (*circa* A.D. 60) and discussed in his *Pneumatica*; the origin of cannon shells; the development of whaling implements; and the rise of industrial education in England.

Montgéry's *Traité des fusées* earned him membership in the Royal Swedish Academy of Sciences (1825) and in other learned societies. He was also an officer of the Légion d'Honneur, Knight of Saint-Louis, and Knight of the Sword of Sweden. He never married.

BIBLIOGRAPHY

I. ORIGINAL WORKS. A bibliography of Montgéry's work is found in the *Catalogue général des Livres Imprimés de la Bibliothèque nationale*, **118** (1933), 417–419. *Traité des fusées de guerre* (Paris, 1825) was also printed in part or in whole in *Journal des sciences militaires*, **1** (1825), 260–286; *Bulletin des sciences militaires*, **1** (1824), 368–380; and *Annales maritimes et coloniales*, **26**, pt. 2 (1825), 565–741; and was republished in J. Corréard, ed., *Histoire des fusées de guerre*, I (Paris, 1841), 77–288. The work was reviewed at length in *Revue encyclopédique*, **28** (Dec. 1825), 699–711; and *Allgemeine Militär-Zeitung*, **1** (12 July 1826), 25–28; (5 July 1826), 20–23; (7 Mar. 1827), 148–151.

Other works include *Règles de pointage à bord des vaisseaux* (Paris, 1816, 1828); *Mémoire sur les mines flottantes* (Paris, 1819); *Mémoire sur les navires en fer* (Paris, 1824); *Notice sur la navigation et sous-marines* (Paris, n.d.); *Notice sur la vie et les travaux de Robert Fulton* (Paris, 1825); *Observations relatives aux ouvrages de M. Paixhans* (Paris, n.d.); and *Réflexions sur quelques institutions . . . sur . . . les progrès de l'industrie* (Paris, n.d.). Montgéry was also a regular contributor to the journals cited above.

II. SECONDARY LITERATURE. See Howard I. Chapelle, *Fulton's "Steam Battery": Blockship and Catamaran*, Museum of History and Technology Paper no. 39 (Washington, D.C., 1964), 147, 149, 150–152, 159; F. Forest and H. Noalhat, *Les bateaux sous-marins* (Paris, 1900), 21–23; H. J. Paixhans, *Nouvelle force maritime et application de cette force à quelques parties du service de l'armée de terre* (Paris, 1822), 6–7, 41, 136, 294; G. L. Pesce, *La navigation sous-marine* (Paris, 1906), 7, 15, 242–254; A. Pralon, *Les fusées de guerre* (Paris, 1883), 46–48; and [Louis Auguste Victor Vincent] Susane, *Les fusées de guerre* (Metz, 1863), 48–49.

FRANK H. WINTER

MONTGOLFIER, ÉTIENNE JACQUES DE[1] (*b.* Vidalon-les-Annonay, France, 6 January 1745; *d.* Serrières, France, 1 August 1799); **MONTGOLFIER, MICHEL JOSEPH DE**[2] (*b.* Vidalon-les-Annonay, 26 August 1740; *d.* Balaruc-les-Bains, France, 26 June 1810), *technology, aeronautics.*

The Montgolfier brothers were two of the sixteen children of Pierre Montgolfier, a paper manufacturer near Annonay, south of Lyons, and Anne Duret. Joseph de Montgolfier traveled widely in his youth, married in 1771, and settled in Vidalon, having founded his own paper factory fifty miles away at Voiron. He was a skillful and imaginative technologist, self-taught in mathematics and science. Étienne Jacques de Montgolfier excelled in mathematics at school in Paris and studied architecture under J. G. Soufflot; he practiced architecture until 1772, then returned to Annonay to direct his father's factory.

It is not known why Joseph and Étienne de Montgolfier first became interested in the problem of flight. At any rate, their early experiments were based upon the belief that a man could be raised by a balloon filled with a light gas. In 1782 they made small paper and silk model balloons filled with hydrogen; the balloons rose but the gas quickly escaped. They then found that air heated to about 80° R. (100° C.) became sufficiently rarefied to lift a balloon and did not diffuse. In November 1782 they made a balloon of forty-cubic-foot capacity, which reached a height of seventy feet; and on 5 June 1783 a paper and cloth globe thirty-five feet in diameter rose 6,000 feet above Annonay.

An incomplete account of the Montgolfiers' experiment convinced scientists in Paris that hydrogen had been used, and J. A. C. Charles began to develop a hydrogen balloon. Before Charles launched it on 27 August Étienne de Montgolfier himself arrived in Paris, where he constructed several hot-air balloons. The first human flight was made on 20 November 1783 by J. F. Pilatre de Rozier and the Marquis d'Arlandes in one of these "Montgolfières."

The Montgolfier brothers were elected as corresponding members of the Paris Académie des Sciences, and at its meeting on 15 November 1783 Étienne de Montgolfier discussed in mathematical terms the problem of navigating balloons. Joseph de Montgolfier was then in Lyons, where he described the brothers' discovery to the Lyons Academy and constructed a balloon of more than 100 feet in diameter, in which he and six others flew on 19 January 1784.

Joseph de Montgolfier may have witnessed Le Normand's parachute trials at Montpellier. He made a parachute of his own design and in March 1784 dropped a sheep from a tower at Avignon. After that the brothers withdrew from aeronautics. They had been helped in their efforts by F. P. A. Argand, and in October 1785 Joseph de Montgolfier visited London to support him in a patent case concerning his oil lamp.

After spending the winter of 1783–1784 in Paris, Étienne de Montgolfier returned to his father's factory, which produced high-quality paper and which in 1784 was given the appellation *manufacture royale*. He became the proprietor in 1787, and the factory remained his principal interest, apart from a brief excursion into local politics in 1790–1791.

Joseph de Montgolfier subsequently made several inventions, the most important of which was the hydraulic ram, a simple device for raising water, which was widely adopted. The machine consisted of two valves in an iron box (the "ram's head") which were automatically operated by the changing pressure of water flowing into it from a reservoir. When the valve leading to the waste pipe closed suddenly, that leading to the outlet pipe opened and a small volume of water was driven by its own momentum to a considerable height. The valves then recovered their original positions and the action was repeated. Argand helped to develop the ram, but in the *Journal des mines* for 1802–1803 Joseph de Montgolfier claimed that the invention was his own.

Never a very successful businessman, Joseph de Montgolfier retired from paper making after the French Revolution and moved to Paris, where in 1800 he was appointed demonstrator at the Conservatoire des Arts et Métiers. In 1801 he helped to found the Société d'Encouragement pour l'Industrie Nationale; he was elected to the Institut de France in 1807.

NOTES

1. He was named Étienne Jacques in the baptismal register, but Jacques Étienne in his death certificate. These documents, cited by Rostaing (see bibliography), are now in the Archives Départementales de l'Ardèche, Privas. The family was ennobled in December 1783 and only then acquired the right to use the prefix "de."
2. He was baptized Michel Joseph but was generally known as Joseph Michel.

BIBLIOGRAPHY

I. ORIGINAL WORKS. Joseph de Montgolfier's *Discours prononcé à l'Académie des Sciences de Lyon* (Paris, 1784) was also printed by Saint-Fond (see below). Different versions of the hydraulic ram are described in several papers by Montgolfier: "Note sur le bélier hydraulique, et sur la manière d'en calculer les effets," in *Journal des mines*, **13** (1802–1803), 42–51; "Sur le bélier hydraulique," *ibid.*, **15** (1803–1804), 23–37; "Du bélier hydraulique et de son utilité," in *Bulletin de la Société d'encouragement pour l'industrie nationale*, **4** (1805), 170–181; "Mémoire sur la possibilité de substituer le bélier hydraulique à l'ancienne machine de Marly," *ibid.*, **7** (1808), 117–124, 136–152; and "Sur quelques perfectionnemens du bélier hydraulique," *ibid.*, **8** (1809), 215–220. Also by Joseph de Montgolfier is "Description et usage d'un calorimetre, ou appareil propre

à déterminer le degré de chaleur ainsi que l'économie qui résulte de l'emploi du combustible," *ibid.*, **4** (1805), 43–46; repr. in *Journal des mines*, **19** (1806), 67–72.

Many letters and other manuscripts concerning the Montgolfier family, formerly in the archives of the Château de Colombier le Cardinal (Ardèche), are now in the *Fonds Montgolfier* of the Musée de l'Air, Paris.

II. SECONDARY LITERATURE. The earliest account of Joseph de Montgolfier is J. B. J. Delambre, "Notice sur la vie et les ouvrages de M. Montgolfier," in *Mémoires de la classe des sciences mathématiques et physiques de l'Institut de France ... Histoire* for 1810 (1814), xxvii–xliv; more detail is given by J. M. de Gerando, "Notice sur M. Joseph Montgolfier," in *Bulletin de la Société d'encouragement pour l'industrie nationale*, **13** (1814), 91–108; some information about his work in Paris is in R. Tresse, "La Conservatoire des Arts et Métiers et la Société d'encouragement pour l'industrie nationale au début du XIXe siècle," in *Revue d'histoire des sciences*, **5** (1952), 246–264.

There is a short account of Étienne Jacques de Montgolfier (with incorrect dates of birth and death) in Michaud's *Biographie universelle*, **29** (Paris, 1821), 570–571. A valuable study of the entire family is L. Rostaing, *La famille de Montgolfier, ses alliances, ses descendants* (Lyons, 1910).

The early balloon flights are described in B. Faujas de Saint-Fond, *Description des expériences de la machine aérostatique de MM. de Montgolfier, et de celles auxquelles cette découverte a donné lieu* (Paris, 1783; 2nd ed., 1784) and in his second volume, *Première suite de la description des expériences ...* (Paris, 1784). The *Première suite* includes Joseph de Montgolfier's "Mémoire lu à l'Académie de Lyon" (pp. 98–111) and Étienne de Montgolfier's "Mémoire sur les moyens mécaniques appliqués à la direction des machines aérostatiques, lu à l'Académie royale des sciences" (pp. 287–295).

Joseph de Montgolfier's parachute is described by C. A. Prieur, "Note historique sur l'invention et les premiers essais des parachutes," in *Annales de chimie*, **31** (1799), 269–273, with an extract from a letter by him. Le Normand's claim to priority is published by C. A. Prieur, "Réclamation relative à l'invention des parachutes," *ibid.*, **36** (1800), 94–99. For a guide to the extensive early literature of aeronautics, see G. Tissandier, *Bibliographie aéronautique* (Paris, 1887; repr., Amsterdam, 1971).

The relations between Argand and the Montgolfier brothers are discussed by M. Schrøder, *The Argand Burner, Its Origin and Development in France and England, 1780–1800* (Odense, 1969), see index. Argand is named as one of the inventors of the hydraulic ram by Schrøder (p. 57); the first published account of the ram, by "L. C." (probably Lazare Carnot), is "Sur une nouvelle espèce de machine hydraulique, par les CC. [Citoyens] Montgolfier et Argant," in *Bulletin des sciences par la Société philomathique*, no. 8 (1797), 58–60 with plate facing p. 72.

A final anonymous description of the ram is "Note sur le bélier hydraulique de feu M. Joseph Montgolfier," in *Bulletin de la Société d'encouragement pour l'industrie nationale*, **12** (1813), 10–11. Two other inventions are described posthumously by Desormes and Clément: "Description d'un procédé économique pour l'évaporation, imaginé par feu Joseph Montgolfier," in *Annales de chimie*, **76** (1810), 34–53, and "Fabrication du blanc de plomb (procédé de Montgolfier)," *ibid.*, **80** (1811), 326–329, repr. in *Bulletin de la Société d'encouragement pour l'industrie nationale*, **11** (1812), 16–17. A previously unpublished memoir by Joseph de Montgolfier, describing a device for raising water by the expansion of hot air, is printed, with a useful commentary and a misleading title, by C. Cabanes, "Joseph de Montgolfier: Inventeur du moteur à combustion interne," in *Nature* (Paris), **64**, pt. 1 (1936), 364–368, and *ibid.*, pt. 2, 252–255.

W. A. SMEATON

MONTGOMERY, EDMUND DUNCAN (*b*. Edinburgh, Scotland, 19 March 1835; *d*. Hempstead, Texas, 17 April 1911), *cell biology, philosophy.*

Montgomery was the illegitimate son of Duncan MacNeill, a famous Scottish jurist, and Isabella Montgomery. He received his early education in Paris and later in Frankfurt am Main. In 1852 Montgomery entered the University of Heidelberg as a medical student and in the same year met the sculptor Elisabet Ney. Her desire for intellectual and artistic success and her indifference toward the normal social standards mirrored Montgomery's philosophy of life, and they became close friends. The relationship in time was to be restrictive for Montgomery because it isolated him spatially, intellectually, and socially from fellow scientists. But he did continue his studies in Berlin (1855), Bonn (1856), and Würzburg (1857), and he later observed clinical practices at Prague (1858) and Vienna (1859). Although fully trained in the medical arts, there is doubt as to whether he received an official degree.

In 1860 Montgomery became a resident physician at the German Hospital in London. The following year he served as an attendant physician at Bermondsey Dispensary. In 1861–1862 he also served as demonstrator of morbid anatomy at St. Thomas' Hospital and in 1863 became a lecturer on that subject. Mostly for reasons of health Montgomery left London in 1863 for Madeira and set up private practice there. He was joined in November by Elisabet and was married to her by the British consul, although she later denied the legality of their relationship.

From 1864 to 1867 the couple worked and traveled in Italy, but to facilitate Elisabet's work as a sculptor, they became permanent residents of Munich in 1867. Although Munich society was liberal, the Montgomerys were socially ostracized and they moved

to the United States. Their two sons were born during a two-year stay in Thomasville, Georgia. Georgia was even less tolerant of their nonconformist behavior, and they left for Texas, arriving in Hempstead in March of 1873. Soon after his purchase of the Liendo plantation, their firstborn son died of diphtheria. They remained in Texas for the rest of their lives and Montgomery became a United States citizen in 1886. They both continued to travel widely in the United States and Europe.

During his student years (1852–1859), Montgomery encountered the divergent philosophies of materialism and idealism and participated in the blossoming of German experimental physiology. His work at St. Thomas' (1861–1863), at the London Zoological Gardens (summer of 1867), and at Munich (1869) culminated in a research publication, *On the Formation of So-Called Cells in Animal Bodies* (London, 1867), and a philosophical treatise, *Die Kant'sche Erkenntnisslehre widerlegt vom Standpunkt der Empirie* (Munich, 1871).

From 1873 to 1879 Montgomery performed intensive microscopical investigations of protozoans and multicellular organisms at his laboratory on his Texas plantation. From 1879 to 1892 his activities centered increasingly on synthesizing the results of his biological investigations and his philosophical viewpoints. He maintained an active correspondence with scientists and published occasional articles in American and European scientific and philosophic journals.

The years 1892–1911 were a time of great emotional stress aggravated by financial difficulties and the aberrant behavior of their surviving son. In 1907 Elisabet died and Montgomery suffered a paralytic stroke that obliged him to spend the rest of his life on his ranch. Yet he wrote three books that summarized his intellectual beliefs.

The Vitality and Organization of Protoplasm (Austin, 1904) was a statement of his biological researches. Believing mechanistic and vitalistic explanations of life to be in error, Montgomery felt that the vital properties of life resided in the protoplasm of living organisms. The vitality of this substance was due to the interdependencies of the chemical constituents of which it was composed, not to an aggregate of qualities of the atoms or to a vital spirit with which protoplasm might be imbued. This vitality was demonstrated by the ability of protoplasm to reconstitute itself from its elements, not by the activities carried out by living organisms.

Montgomery also disagreed with many cell theorists. He thought that cellular specialization was the result of "ontogenetic" differentiation of the protoplasm of the germ cell and not simply a division of labor in an

494

aggregate of cells. For him this was an evolutionary development in which the resultant cells of the mature organism were not coequals of the original germ cell but lineal descendants with increased specialization. The inheritance of characteristics was effected through the protoplast of the germ cell, not through the nucleus, which structure he thought to be involved in oxidation. He disapproved of any theories of inheritance in which characteristics were carried to the germ cell or were attributed to vital properties of the process.

His biological researches as summarized in his book and articles show him to be a careful observer with original thoughts. Yet, at the same time, isolation from the scientific community partially explains misconceptions that appear in his writings and the lack of a broader acceptance of his work.

Montgomery's philosophy is characterized by an attempt to utilize biological observations as a basis for philosophical generalizations. He was neither a materialist nor an idealist but a monist, who strongly defended the concept of the unity and indivisibility of the living substance and thus of life as a whole. His book *Philosophical Problems in the Light of Vital Experience* (New York, 1907) presented these views in greater detail. His last book, *The Revelation of Present Experience* (Boston, 1910), was a philosophic treatise concerned with the function of the mind. Montgomery believed perception to be not reality but the subjective appearance of things in the mind of the viewer. The only reality was the substance and not its activities. Because of his unique approach to philosophy, Montgomery was not always understood. He gained a larger following among philosophers than biologists.

BIBLIOGRAPHY

I. ORIGINAL WORKS. Among Montgomery's significant works are *On the Formation of So-Called Cells in Animal Bodies* (London, 1867); *Die Kant'sche Erkenntnisslehre widerlegt vom Standpunkt der Empirie* (Munich, 1871); *The Vitality and Organization of Protoplasm* (Austin, Tex., 1904); *Philosophical Problems in the Light of Vital Experience* (New York, 1907); and *The Revelation of Present Experience* (Boston, 1910).

II. SECONDARY LITERATURE. See Morris Keeton, *The Philosophy of Edmund Montgomery* (Dallas, 1950), with bibliography, pp. 319–338; Vernon Loggins, *Two Romantics and Their Ideal Life* (New York, 1946); Ira Stephens, *The Hermit Philosopher of Liendo* (Dallas, 1951); and Bride Taylor, *Elisabet Ney, Sculptor* (New York, 1916).

LARRY T. SPENCER

MONTGOMERY, THOMAS HARRISON, JR. (*b.* New York, N.Y., 5 March 1873; *d.* Philadelphia, Pennsylvania, 19 March 1912), *zoology.*

In his brief life span of thirty-nine years Montgomery became one of the leaders in American zoology; he made substantial contributions in several fields and ranks as a major figure in one of them, cytology. He rose steadily in his profession and at his death he was professor and chairman of the department of zoology at the University of Pennsylvania.

Montgomery's background and early life were favorable for a scholarly career. His family, which settled in New Jersey in 1701, included paternal ancestors distinguished in religion, law, and business; one of them was "the first bishop of English consecration in the United States." Montgomery's maternal forebears included prominent physicians and scientists: his grandfather, Samuel George Morton, was one of the founders of anthropology and served a term as president of the Academy of Natural Sciences of Philadelphia. Montgomery's father, an insurance executive of scholarly bent, published a voluminous early history of the University of Pennsylvania, which made him an honorary doctor of letters. When Montgomery was nine years old, the family moved to a country home near West Chester, Pennsylvania; and there he developed the strong interest in natural history, particularly birds, that he maintained throughout his life.

After graduation from a private secondary school, Montgomery attended the University of Pennsylvania (1889–1891) and completed his studies at Berlin, taking his Ph.D. in 1894. At Berlin he met a group of scholars whose interest in the maturing science of cytology turned him toward the area in which he was to make his greatest contributions: the histologist F. E. Schultze; H. W. G. Waldeyer, who had published extensively on the structure of spermatozoa, on the differentiation of germ cells, and on cell division in the fertilized egg; and Oscar Hertwig, who had carried out pioneer investigations of fertilization and early development. Montgomery's doctoral thesis, however, dealt not with cytology but with a variety of lesser problems in phylogeny, taxonomy, and anatomy. The most extensive essay was on one of the nemerteans, a subject that interested him enough to continue work on the group for many years and to publish a series of ten papers.

After returning to the United States in 1895, Montgomery served for three years as an investigator at the Wistar Institute of Anatomy in Philadelphia. He spent the summer at marine laboratories, notably at Woods Hole, Massachusetts (1897), to which he returned nearly every summer. From 1897 to 1903 he taught zoology at Pennsylvania, serving for the last three years as assistant professor. After five years as professor at the University of Texas, Montgomery was recalled to Pennsylvania in 1908 to become professor and chairman of zoology, the post he held until his death. An arduous executive achievement during this tenure was the construction of a new laboratory building. Montgomery was a coeditor of the *Journal of Morphology* from 1908 until his death and was president of the American Society of Zoologists in 1910.

An unusually diligent researcher who published promptly, Montgomery produced more than eighty papers between 1894 and 1912. Like many biologists of his time, he was interested in certain older problems, such as animal behavior and taxonomy, while remaining active in cytology, a more modern, laboratory-oriented field. As classified by Conklin, sixteen of his papers were devoted mainly to taxonomy, five to animal distribution, eleven to ecology and behavior, sixteen to morphology, and eight to phylogeny—in addition to twenty-five papers in cytology, on which his reputation is based.

Montgomery made a fortunate choice of animals for cell studies, working mainly on the males of the Hemiptera-Heteroptera, or true bugs, which are represented in the United States by many common, easily collected species. They are peculiarly suitable for investigating the processes of meiosis and differentiation of the spermatozoon, because the successive stages are arranged in a series of follicles along the cylindrical lobes of the testis and culminate in the mature spermatozoa, in the follicle nearest the efferent duct. In 1898 Montgomery noted that in the premeiotic, spermatogonial metaphases in the bug testis, the chromosomes consist of a definite number of pairs, many of which are individually recognizable; for example, there are often exceptionally large or small pairs. In the later maturation divisions he found that the half number of chromosomes show the same size differences. Thus, for example, if there formerly was a single large pair there now would be one large chromosome. Although not all the individual chromosomes could be followed throughout the prophase of the first maturation division, certain exceptional chromosomes did remain condensed and hence continuously recognizable. These latter he called heterochromosomes, resulting from the property of heteropycnosis.

From these observations Montgomery concluded in 1901, "Through the germinal cycle the chromosomes preserve their individuality from generation to generation—that is, a particular chromosome of one generation is represented by a particular one of the preceding, so that the chromosomes are not produced anew in each generation." From this general conclusion he suggested that (1) the members of each chromosome pair are homologous and are of maternal and paternal origin respectively; (2) in each case the two are synaptic mates and conjugate to form a "bivalent" chromosome of the same size relative to the rest of the complement; (3) this pairing may be regarded as "the final step in the process of conjugation of the germ cells"; (4) the homologues separate at the reduction division to form the reduced number of "univalent" chromosomes; and (5) finally each spermatid receives a set of "semivalent" chromosomes, resulting from division of each univalent chromosome.

These conclusions, which contain the essentials of the chromosomal basis of biparental inheritance, were announced just prior to the rediscovery of Mendel's laws of segregation and recombination. Montgomery himself failed to see such a possible relationship to inheritance, concluding only that the broader significance of the basic process (synapsis and subsequent separation) was that it might lead to a "rejuvenation of the chromosomes." Despite what seems in retrospect an unimaginative and narrow interpretation, it appears undeniable that the speed with which Sutton, Wilson, and others subsequently established the correlation with the rediscovered laws of inheritance was due in large measure to Montgomery's masterly analysis—not the least aspect of which was its clarifying terminology.

Another facet of Montgomery's work was basic to the theory of sex determination. In certain species of bugs he noted that one chromosome, which he called the X chromosome, is single, as distinct from the paired "autosomes." Consequently the somatic, or diploid, chromosome number is odd, and during meiosis only half the spermatids receive the X chromosome. It remained for others to point out the obvious possibility that this chromosomal mechanism could be the basis of determination of sex. Instead, Montgomery assumed that the females of these species, like the males, had the uneven chromosome number; and he became engrossed with the puzzle of how the odd number could be preserved to the next generation. Realizing that random fertilization in such a situation should also produce two even-numbered chromosomal complements with respectively one more and one less than the X type, he was led to postulate a sort of selective fertilization to maintain the odd number.

Montgomery's other cytological contributions were not inconsiderable. In his 1901 paper he brought order to the rather chaotic views of the nucleolus by clearly distinguishing "chromatin nucleoli" (among which were the heterochromosomes) from the plasmosomes, or true nucleoli. His careful seriation of steps in the differentiation of the hemipteran spermatozoon (1911) was a model for many later studies, including R. H. Bowen's (1920). Noteworthy in the 1911 paper was Montgomery's confirmation of observations on fixed material by examining cells teased out in a physiological fluid. This degree of sophistication was not attained by other cytologists for nearly two decades.

BIBLIOGRAPHY

Montgomery's writings include "Comparative Cytological Studies With Especial Reference to the Morphology of the Nucleolus," in *Journal of Morphology*, **15** (1898), 204–265; "A Study of the Chromosomes of Germ Cells," in *Transactions of the American Philosophical Society*, **20** (1901), 154–236; and "The Spermatogenesis of an Hemipteron, *Euschistus*," in *Journal of Morphology*, **22** (1911), 731–799.

On his life and work, see R. H. Bowen, "Studies on Insect Spermatogenesis. I.," in *Biological Bulletin. Marine Biological Laboratory, Woods Hole, Mass.*, **39** (1920), 316–362; and E. G. Conklin, "Professor Thomas Harrison Montgomery, Jr.," in *Science*, n.s. **38** (1913), 207–214, an excellent obituary; the complete bibliography prepared to accompany it, however, seems to have been omitted.

ARTHUR W. POLLISTER

MONTMOR, HENRI LOUIS HABERT DE (*b.* Paris [?], France, *ca.* 1600; *d.* Paris, 21 January 1679), *scientific patronage.*

Montmor's family, which originally came from Artois, moved to Paris in the sixteenth century. Its leading members were high government officials who grew rich in the king's service. Related to the greatest families in the kingdom, including the Lamoignons, the Bethunes, and the Phélypeaux, and a grandnephew of Guillaume Budé, he received an excellent education. When Montmor was twenty-five, his father obtained for him a position as *conseiller* in the Parlement of Paris; and on 6 April 1632 he was appointed *maître des requêtes*. His connections with two of his cousins, the brothers Philippe Habert, artillery commissioner and poet, and Germain Habert, *abbé* of Cerisy and likewise a poet, undoubtedly account for his having been well enough known to Valentin Conrart to have been included in the small group forming the Académie Française. He was elected to the latter in December 1634 and formally welcomed by his cousin Germain Habert on 2 January 1635. On 30 April 1635 the group met at the handsome

town house on the Rue Sainte-Avoye (now 79 Rue du Temple) that his father had constructed about 1623.

On 29 March 1637 Montmor married a cousin, Marie Henriette de Buade de Frontenac, whose brother Louis later became governor of New France. Between 1638 and 1659 they had fifteen children, most of whom died young. The eldest son, who also became *maître des requêtes*, suffered a bankruptcy of 600,000 *livres*—which, if Jean Chapelain's letter to François Bernier is to be believed, was the cause of the "fatal melancholy" that overtook Montmor beginning in 1669. On 7 September 1671 Chapelain wrote to Nikolaas Heinsius: "M. de Montmor's fate is deplorable. For a year he has lived only on milk and to his distress he is unable to leave this life . . ." (*Lettres de Jean Chapelain*, II, 752). He was obliged to sell his post and as a result "suffered such a great mental disturbance that he became almost insane" (*ibid.*, to Régnier de Graff, 28 August 1671). Yet he lived until the beginning of 1679, having survived his wife by more than two years.

Very few of Montmor's writings are extant. In the *Histoire de l'Académie françoise* Pellisson states that Montmor delivered an address to the Academy on 3 March 1635, "De l'utilité des conférences." Today he is of interest for his role as patron of the scientists and philosophers of his age. A fine scholar, Montmor assembled a very rich library, in which the correspondence of important contemporaries (for example, Gui Patin and Chapelain) had a major place, and he attracted to his residence on Rue Saint-Avoye both men of letters and scientists. A Cartesian throughout his life, he offered Descartes "the full use of a country house [Mesnil-Saint-Denis] worth 3,000 to 4,000 *livres* in rent" (A. Baillet, *La vie de Monsieur Descartes*, II, 462), which the latter declined.

No document proves conclusively that regular scientific meetings took place at Montmor's residence before 1653. Toward the middle of the century political agitation attracted more attention than did scientific activity, and Montmor did not escape this preoccupation. Although not really a rebel, he stood with the Parlement and princes against the king and court. When the disorders of the Fronde had died down, and after Descartes was dead, Montmor, while remaining a Cartesian, offered Gassendi, Descartes's great adversary, lodgings in his house. Gassendi moved into the second floor of the house in the Rue du Temple on 9 May 1653. He spent the month of August 1654 at Mesnil, where he made astronomical observations. Montmor encouraged him to write *La vie de Tycho Brahe*, and Gassendi dedicated it to his patron, whom he also made the executor of his will and to whom he left all his books, manuscripts, and the telescope Galileo had given him. When Gassendi died, on 24 October 1655, Montmor returned in haste from Mesnil to arrange his friend's funeral; Gassendi was buried in the Montmor chapel in the church of St.-Nicolas-des-Champs. Montmor collected his writings—with the help of François Henri, Samuel Sorbière, and Antoine de La Poterie—and wrote a preface to the six-volume Latin edition published at Lyons in 1658.

Gassendi's presence in Montmor's household certainly contributed to the development of the meetings held there by the cultivated men who had previously gathered around Mersenne, the brothers Pierre and Jacques Du Puy, the Abbé Picot, and François Le Pailleur and who, with several newcomers, now assembled on the Rue du Temple: Boulliau, Pascal, Roberval, Gérard Desargues, Carcavi, Jean Segrais, Gui Patin, Michel de Marolles, Balthazar de Monconys, and others. In a letter to his friend Regnault of Lyons, dated 4 August 1656, Monconys described a meeting in Montmor's house in which experiments on glass drops were conducted. Although Monconys spoke of an "assemblée" (of which he was not then a member), it is only from the end of 1657 that the weekly gatherings of what came to be called the Académie Montmor can be dated.

At Montmor's request, Sorbière prepared a plan for the organization of meetings in the form of nine articles. The goals of the meetings "will not be the vain exercise of the mind on useless subtleties; rather, one should always propose the clearest knowledge of the works of God and the advancement of the conveniences of life, in the arts and sciences that best serve to establish them." Sorbière was also charged with preparing the *Mémoires* of the assembly, but unfortunately they have been lost.

Among the members of the Académie Montmor were Chapelain, Sorbière, Montmor (named the "Modérateur"), Clerselier, Rohault, Pierre Huet, Roberval (until he was "uncivil," boorish, and rude to Montmor, who supported the opinions of Descartes), and Huygens (when he was in Paris). The latter's journal provides information on the weekly sessions and on those he met there; on 9 November 1660 he was introduced to Auzout, Frenicle de Bessy, Desargues, Pecquet, Rohault, La Poterie, Sorbière, and Boulliau. Oldenburg also visited the house in the Rue du Temple when he stayed in Paris.

The activities of the Académie Montmor during its first years included Chapelain's announcement of Huygens' discoveries (the pendulum clock, the first known satellite of Saturn, a diagram of his system of Saturn—planet and ring), Rohault's experiments on the magnet, Pecquet's dissections, and Thévenot's

presentation of his tubes "made expressly to examine the ascension of water that mounts by itself beyond its level."

Two currents soon appeared within the Académie Montmor: the first, a tendency to seek natural causes, was associated with the philosophers, both Cartesians and Gassendists; the second, a preference for observation and experiment, was emphasized by Auzout, Petit, and Rohault, who complained of sterile discussions and prating that explained nothing.

The problem worsened in the following years, and on 3 April 1663 Sorbière delivered "Discours à l'ouverture de l'Académie des Physiciens qui s'assemblent tous les mardis chez Monsieur de Montmor," which he sent to Colbert. Although he began by honoring the "illustrious moderator" who "first aroused interest in Paris in the studies we cultivate" and by praising the early meetings and experiments, he soon turned to severe criticism of the disputes and interruptions; "people who have come here only to waste time and to acquire esteem"; and the mutual intolerance of the partisans of experiment and philosophy. Even though Montmor had provided his guests with "an infinity of machines and instruments with which he has stimulated his curiosity for thirty years," he could not furnish them with a forge and a laboratory or an observatory. That was not within the power of an individual but, rather, of a sovereign. This implicit appeal to the king for the creation of an institution under royal patronage explains why the Académie Montmor has been seen as a forerunner of the Académie Royale des Sciences.

In response to all the criticism, the Académie Montmor attempted to reform itself. Experiments were tried there with an air pump constructed according to Huygens' plans. Nevertheless, so Huygens wrote to Moray in March 1664, a widespread desire was felt to establish the academy on a new basis. On 12 June, he wrote to Moray that "the academy has ended forever *chez* M. de Montmor" but that another was being born from its ruins. Montmor, meanwhile, continued to receive scientists and to take an interest in philosophers. Experiments on blood transfusion were carried out by Jean-Baptiste Denis at his home in 1668; and when the human subject died, Montmor exerted his influence to save the experimenters from legal penalties. It was to Montmor that Henri Justel sent Hooke's *Micrographia*. He also received the first copy of the *Saggi dell'esperienze naturali fatte nell'Accademia del Cimento*, and on the advice of Chapelain, Louis de La Forge dedicated his *Traitté de l'esprit de l'homme*, which was inspired by Descartes's philosophy, to Montmor (as had Mersenne his *Harmonie universelle*). Montmor's continued attachment to Descartes is proved by the fact that he undertook the writing of a Latin poem on Cartesian physics with the Lucretian title *De rerum natura* and by the fact that he was among the faithful Cartesians who followed Descartes's bier to the church of Ste.-Geneviève-du-Mont on 25 June 1667.

BIBLIOGRAPHY

Information on Montmor's life may be found in Adrien Baillet, *La vie de Monsieur Descartes* (Paris, 1691), 266–267, 346–347, 462; Faustin Foiret, "L'hôtel de Montmor," in *La Cité, Bulletin trimestriel de la Société historique et archéologique du IVᵉ arrondissement de Paris*, **13** (1914), 309–339; "Habert, Henri Louis, de Montmor," in Moreri's *Dictionnaire . . .* (1759); René Kerviler, "Henri-Louis Habert de Montmor, de l'Académie française et bibliophile (1600–1679)," in *Le bibliophile français*, VI (Paris, 1872), 198–208; Frédéric Lachèvre, *Bibliographie des recueils collectifs de poésies publiés de 1597 à 1700*, III (Paris, 1903), 455; Pellisson and d'Olivet, *Histoire de l'Académie françoise depuis son établissement jusqu'en 1652* (Paris, 1729), 81, 175, 276, 344; and Tallemant des Réaux, *Historiettes*, Antoine Adam, ed., 2 vols. (Paris, 1960–1961).

On Montmor's relations with Gassendi see, in addition to the classic works on Gassendi, Georges Bailhache and Marie-Antoinette Fleury, "Le testament, l'inventaire après décès, la sépulture et le monument funéraire de Gassendi," in *Tricentenaire de Pierre Gassendi, 1655–1955. Actes du Congrès Gassendi, 4–7 août 1955* (Paris, 1957), 19–68.

Information on the Académie Montmor is in *The Correspondence of Henry Oldenburg*, A. Rupert Hall and Marie Boas-Hall, eds., I–IV (Madison–Milwaukee, Wis., 1965–1971); F. Graverol, ed., *Sorberiana* (Toulouse, 1691), 28–29; *Lettres et Discours de Monsieur de Sorbière sur diverses Matières Curieuses* (Paris, 1660), 60, 181, 190, 193, 369, 631, 694, 701; *Lettres de Gui Patin*, J.-H. Réveillé-Parise, ed., II (Paris, 1846), 107, 211, 317–318; *Lettres de Jean Chapelain*, P. Tamizey de Larroque, ed., II (Paris, 1883), *passim;* Balthazar de Monconys, *Journal des voyages* (Lyons, 1666), 162–169; and *Oeuvres de Christiaan Huygens*, J. Volgraf, ed., I–V, XXII (The Hague, 1888–1950).

The following studies have made extensive use of the above sources: M. G. Bigourdan, *Les premières sociétés savantes de Paris au XVIIᵉ siècle et les origines de l'Académie des sciences* (Paris, 1919); Harcourt Brown, *Scientific Organizations in Seventeenth Century France (1620–1680)* (Baltimore, 1934), 64–134; and René Taton, *Les origines de l'Académie royale des sciences* (Paris, 1966), 47–54.

Suzanne Delorme

MONTMORT, PIERRE RÉMOND DE

MONTMORT, PIERRE RÉMOND DE (*b.* Paris, France, 27 October 1678; *d.* Paris, 7 October 1719), *probability.*

Montmort was the second of the three sons of François Rémond and Marguerite Ralle. On the advice of his father he studied law, but tired of it and ran away to England. He toured extensively there and in Germany, returning to France only in 1699, just before his father's death. He had a substantial inheritance, which he did not exploit frivolously.

Having recently read, and been much impressed by, the work of Nicolas Malebranche, Montmort began study under that philosopher. With Malebranche he mastered Cartesian physics and philosophy, and he and a young mathematician, François Nicole, taught themselves the new mathematics. When Montmort visited London again in 1700, it was to meet English scientists; he duly presented himself to Newton. On his return to Paris, his brother persuaded him to become a canon at Notre Dame de Paris. He was a good ecclesiastic until he bought an estate at Montmort and went to call on the grand lady of the neighborhood, the duchess of Angoulême. He fell in love with her niece, and in due course gave up his clerical office and married. It is said to have been an exceptionally happy household.

Montmort's book on probability, *Essay d'analyse sur les jeux de hazard*, which came out in 1708, made his reputation among scientists and led to a fruitful collaboration with Nikolaus I Bernoulli. The Royal Society of London elected Montmort fellow when he was visiting England in 1715 to watch the total eclipse of the sun in the company of the astronomer royal, Edmond Halley. The Académie Royale des Sciences made him an associate member the following year—he could not be granted full membership because he did not reside in Paris. He died during a smallpox epidemic in 1719.

It is not clear why Montmort undertook a systematic exposition of the theory of games of chance. Gaming was a common pastime among the lesser nobility whom he frequented, but it had not been treated mathematically since Christiaan Huygens' monograph of 1657. Although there had been isolated publications about individual games, and occasional attempts to come to grips with annuities, Jakob I Bernoulli's major work on probability, the *Ars conjectandi*, had not yet been published. Bernoulli's work was nearly complete at his death in 1705; two obituary notices give brief accounts of it. Montmort set out to follow what he took to be Bernoulli's plan.

One obituary gave a fair idea of Bernoulli's proof of the first limit theorem in probability, but Montmort, a lesser mathematician, was not able to reach a comparable result unaided. He therefore continued along the lines laid down by Huygens and made analyses of fashionable games of chance in order to solve problems in combinations and the summation of series. For example, he drew upon the game that he calls "treize," in which the thirteen cards of one suit are shuffled and then drawn one after the other. The player who is drawing wins the round if and only if a card is drawn in its own place, that is, if the nth card to be drawn is itself the card n. In the generalized game, the pack consists of m cards marked in serial order. The chance of winning is shown to be

$$\sum_{i=1}^{i=m} \frac{(-1)^{i-1}}{i!}.$$

A 1793 paper by Leibniz provided Montmort with a rough idea of the limit to which this tends as m increases, but Euler was the first to state it as $1 - e^{-1}$.

The greatest value of Montmort's book lay perhaps not in its solutions but in its systematic setting out of problems about games, which are shown to have important mathematical properties worthy of further work. The book aroused Nikolaus I Bernoulli's interest in particular and the 1713 edition includes the mathematical correspondence of the two men. This correspondence in turn provided an incentive for Nikolaus to publish the *Ars conjectandi* of his uncle Jakob I Bernoulli, thereby providing mathematics with a first step beyond mere combinatorial problems in probability.

The work of De Moivre is, to say the least, a continuation of the inquiries of Montmort. Montmort put the case more strongly—he accused De Moivre of stealing his ideas without acknowledgment. De Moivre's *De mensura sortis* appeared in 1711 and Montmort attacked it scathingly in the 1713 edition of his own *Essay*. Montmort's friends tried to soothe him, and largely succeeded. He tried to correspond with De Moivre, but the latter seldom replied. In 1717 Montmort told Brook Taylor that two years earlier he had sent ten theorems to De Moivre; he implied that De Moivre could be expected to publish them.

Taylor was doing his best work at this time. He and Montmort had struck up a close friendship in 1715, and corresponded about not only mathematics but also general questions of philosophy, Montmort mildly defending Cartesian principles against the sturdy Newtonian doctrines of Taylor. Montmort's only other mathematical publication, an essay on summing infinite series, has an appendix by Taylor. It is notable that in this period of vigorous strife between followers of Newton and Leibniz, Montmort was able to remain on the best of terms with both the Bernoullis and the Englishmen.

BIBLIOGRAPHY

I. ORIGINAL WORKS. Montmort's mathematical writings are *Essay d'analyse sur les jeux de hazard* (Paris, 1708),

2nd ed. revised and augmented with correspondence between Montmort and N. Bernoulli (Paris, 1713; 1714); and "De seriebus infinitis tractatus," in *Philosophical Transactions of the Royal Society*, **30** (1720), 633–675. Part of Montmort's correspondence with Taylor is in William Young, ed., Brook Taylor, *Contemplatio philosophica* (London, 1793).

II. SECONDARY LITERATURE. See "Éloge de M. de Montmort," in *Histoire de l'Académie royale des sciences pour l'année 1719* (1721), 83–93.

IAN HACKING

MONTUCLA, JEAN ÉTIENNE (*b.* Lyons, France, 5 September 1725; *d.* Versailles, France, 19 December 1799), *mathematics, history of mathematics.*

Montucla, the son of a merchant, attended the Jesuit *collège* in Lyons, where he received a thorough education in mathematics and ancient languages. Following the death of his father in 1741 and of his grandmother, who was caring for him, in 1745, he began legal studies at Toulouse. On their completion he went to Paris, drawn by the many opportunities for further training. Soon after his arrival there he undertook the study of the history of mathematics. His work on the quadrature of the circle (1754) brought him a corresponding membership in the Berlin Academy. In the same year he announced the forthcoming publication of what was to be his masterpiece, *Histoire des mathématiques.* The exchange of ideas in the literary circle that had formed around the bookseller and publisher Charles Antoine Jombert (1712–1784), which included Diderot, d'Alembert, and Lalande, was very valuable to him. Before the appearance of *Histoire des mathématiques*, Montucla published, in collaboration with the physician Pierre Joseph Morisot-Deslandes, a collection of sources on smallpox vaccination (1756).

From 1761 Montucla held several government posts. His first appointment was as secretary of the intendance of Dauphiné in Grenoble, where in 1763 he married Marie Françoise Romand. In 1764–1765 he was made royal astronomer and secretary to Turgot on a mission to Cayenne. After his return, Montucla became inspector of royal buildings (1766–1789) and, later, royal censor (1775). From this period date his new edition of Ozanam's *Récréations mathématiques* (1778) and his translation of Jonathan Carver's account of travels in North America (1784).

As a result of the Revolution, Montucla lost his posts and most of his wealth. He was again given public office in 1795—examination of the treaties deposited in the archives of the Ministry of Foreign

Affairs—but the salary was not sufficient to meet his expenses, so he also worked in an office of the national lottery. During these years Montucla, at the insistence of his friends, began to prepare an improved and much enlarged edition of *Histoire des mathématiques*. The first two volumes appeared in August 1799, four months before his death, just when he had been promised a pension of 2,400 francs.

Montucla's major work, the first classical history of mathematics, was a comprehensive and, relative to the state of contemporary scholarship, accurate description of the development of the subject in various countries. The account also included mechanics, astronomy, optics, and music, which were then considered subdivisions of mathematics; these branches *(mathématiques mixtes)* receive a thorough treatment in both editions, and only a third of the space is devoted to pure mathematics. The first volume of the two-volume edition of 1758 covers the beginnings, the Greeks (including the Byzantines), and the West until the start of the seventeenth century; the second volume is devoted entirely to the latter century. Montucla originally planned to take his work up to the middle of the eighteenth century in a third volume but could not do so, principally because of the abundance of material. In the second edition, extended to cover the whole of the eighteenth century, he was able to reach this goal. Much remained unfinished, however, since Montucla died during the printing of the third volume. Lalande, his friend from childhood, assisted by others, completed volumes III (pure mathematics, optics, mechanics) and IV (astronomy, mathematical geography, navigation) and published them in 1802.

Many authors before Montucla—beginning with Proclus and al-Nadīm—had written on the history of mathematics. Their accounts can be found in the citations of ancient authors, in the prefaces to many mathematical works of the sixteenth through eighteenth centuries, in university addresses (for example, that of Regiomontanus at Padua in 1464), and in two earlier books that, as their titles indicate, were devoted to the history of mathematics: G. I. Vossius' *De universae mathesios natura et constitutione* (1650) and J. C. Heilbronner's *Historia matheseos universae* (1742). All these early efforts constituted only a modest beginning, containing many errors and legends, and the latter two works give only a jumble of names, dates, and titles. Montucla was familiar with all this material and saw what was required: a comprehensive history of the development of mathematical ideas, such as had been called for by Bacon and Montmor. Inspired by them Montucla undertook the immense labor, the difficulty of which he recognized and which

he carried out with his own research in and mastery of the original texts.

Montucla had no successor until Moritz Cantor. The *Histoire des mathématiques* is, of course, obsolete—as, in many respects, is Cantor's *Vorlesungen*. Yet even today the expert can, with the requisite caution, go back to Montucla, especially with regard to the mathematics of the seventeenth century.

BIBLIOGRAPHY

I. ORIGINAL WORKS. Montucla's writings include *Histoire des recherches sur la quadrature du cercle* (Paris, 1754); *Recueil de pièces concernant l'inoculation de la petite vérole et propres à en prouver la sécurité et l'utilité* (Paris, 1756), written with Morisot-Deslandes; *Histoire des mathématiques*, 2 vols. (Paris, 1758; 2nd ed., 4 vols., Paris, 1799–1802); a new ed. of Ozanam's *Récréations mathématiques et physiques* (Paris, 1778); and a translation, from the 3rd English ed., of Jonathan Carver, *Voyages dans les parties intérieures de l'Amérique septentrionale, pendant les années 1766, 1767, et 1768* (Paris, 1784).

II. SECONDARY LITERATURE. See the following, listed chronologically: Auguste Savinien Le Blond, "Sur la vie et les ouvrages de Montucla. Extrait de la notice historique lue à la Société de Versailles, le 15 janvier 1800. Avec des additions par Jérôme de Lalande," in Montucla's *Histoire des mathématiques*, IV, 662–672; G. Sarton, "Montucla (1725–1799). His Life and Works," in *Osiris*, **1** (1936), 519–567, with a portrait, the title page of each of his works, two previously unpublished letters, and further bibliographical information; and Kurt Vogel, "L'historiographie mathématique avant Montucla," in *Actes du XIᵉ Congrès international d'histoire des sciences*. III, 179–184.

KURT VOGEL

MOORE, ELIAKIM HASTINGS (*b*. Marietta, Ohio, 26 January 1862; *d*. Chicago, Illinois, 30 December 1932), *mathematics*.

Moore was prominent among the small circle of men who greatly influenced the rapid development of American mathematics at the turn of the twentieth century. The son of David Hastings Moore, a Methodist minister, and Julia Sophia Carpenter, he had an impressive preparation for his future career. While still in high school he served one summer as an assistant to Ormond Stone, the director of the Cincinnati Observatory, who aroused his interest in mathematics. He later attended Yale University, from which he received the A.B. in 1883 as class valedictorian and the Ph.D. in 1885. The mathematician Hubert Anson Newton, his guiding spirit at Yale, then financed a year's study abroad for him at the universities of Göttingen and Berlin. He spent the summer of 1885 in Göttingen, where he studied the German language; and the winter of 1885–1886 in Berlin, where Kronecker and Weierstrass were lecturing. The work of Kronecker impressed him, as did the rigorous methods of Weierstrass and Klein, who was then at Leipzig.

In 1886 Moore returned to the United States to begin his career in mathematics. He accepted an instructorship at the academy of Northwestern University for 1886–1887. During the next two years he was a tutor at Yale. In 1889 he returned to Northwestern as an assistant professor and in 1891 was promoted to associate professor. When the University of Chicago first opened in the autumn of 1892, Moore was appointed professor and acting head of the mathematics department. In 1896, after successfully organizing the new department, he became its permanent chairman, a post he held until his partial retirement in 1931. Shortly before assuming his post at Chicago, he married a childhood playmate, Martha Morris Young, on 21 June 1892, in Columbus, Ohio. They had two sons, David and Eliakim.

During his career Moore became a leader at the University of Chicago and in mathematical associations. He helped shape the character of the university and gave it great distinction. With his faculty colleagues Oskar Bolza and Heinrich Maschke, he modified the methods of undergraduate instruction in mathematics. Casting aside textbooks, he stressed fundamentals and their graphical interpretations in his "laboratory courses." Although a gentle man, he sometimes displayed impatience as he strove for excellence in his classes. He became a teacher of teachers. Among his supervised Ph.D.'s were L. E. Dickson, O. Veblen, and G. D. Birkhoff.

Moore also advanced his profession outside the classroom. In 1894 he helped transform the New York Mathematical Society into the American Mathematical Society, of which he was vice-president from 1898 to 1900 and president from 1900 to 1902. A founder of the society's *Transactions* in 1899, he was chief editor until 1907. He served on the editorial boards of the *Rendiconti del Circolo matematico di Palermo* (1908–1932), the University of Chicago Science Series (chairman, 1914–1929), and the *Proceedings of the National Academy of Sciences* (1915–1920). With his encouragement in, 1916 H. E. Slaught saw through the formation of the Mathematical Association of America. In 1921 Moore was president of the American Association for the Advancement of Science.

Rigor and generalization characterized the mathematical research of Moore. His research fell principally into the areas of (1) geometry; (2) algebra, groups, and number theory; (3) the theory of functions; and (4) integral equations and general analysis. Among these he emphasized the second and fourth areas. In geometry he examined the postulational foundations of Hilbert, as well as the earlier works of Pasch and Peano. He skillfully analyzed the independence of the axioms of Hilbert and formulated a system of axioms for n-dimensional geometry, using points only as undefined elements instead of the points, lines, and planes of Hilbert in the three-dimensional case. During his investigation of the theory of abstract groups, he stated and proved for the first time the important theorem that every finite field is a Galois field (1893). He also discovered that every finite group G of linear transformations on n variables has a Hermitian invariant (1896–1898). His probe of the theory of functions produced a clarified treatment of transcendentally transcendental functions and a proof of Goursat's extension of the Cauchy integral theorem for a function $f(z)$ without the assumption of the continuity of the derivative $f'(z)$.

His work in the area of integral equations and general analysis sparkled most. He brought to culmination the study of improper definite integrals before the appearance of the more effective integration theories of Borel and Lebesgue. He diligently advanced general analysis, which for him meant the development of a theory of classes of functions on a general range. The contributions of Cantor, Russell, and Zermelo underlay his research here. While inventing a mathematical notation for his analytical system, he urged Florian Cajori to prepare his two-volume *History of Mathematical Notations* (1928–1929). Throughout his work in general analysis, Moore stressed fundamentals, as he sought to strengthen the foundations of mathematics. His research set a trend for precision in American mathematical literature at a time when vagueness and uncertainty were common.

Honors were bestowed upon Moore for his distinguished contributions to mathematics and education. The University of Göttingen awarded him an honorary Ph.D. in 1899, and the University of Wisconsin an LL.D. in 1904. Yale, Clark, Toronto, Kansas, and Northwestern subsequently granted him honorary doctorates in science or mathematics. In 1929 the University of Chicago established the Eliakim Hastings Moore distinguished service professorship, while he was still an active member of the faculty. Besides belonging to American, English, German, and Italian mathematical societies, he was a member of the American Academy of Arts and Sciences, the American Philosophical Society, and the National Academy of Sciences.

BIBLIOGRAPHY

I. ORIGINAL WORKS. Moore wrote *Introduction to a Form of General Analysis* (New Haven, 1910); and *General Analysis*, published posthumously in *Memoirs of American Philosophical Society*, **1** (1935).

His articles include "Extensions of Certain Theorems of Clifford and Cayley in the Geometry of n Dimensions," in *Transactions of the Connecticut Academy of Arts and Sciences*, **7** (1885), 1–18; "Note Concerning a Fundamental Theorem of Elliptic Functions, As Treated in Halphen's Traité," **1**, 39–41, in *Rendiconti del Circolo matematico di Palermo*, **4** (1890), 186–194; "A Doubly-Infinite System of Simple Groups," in *Bulletin of the New York Mathematical Society*, **3** (1893), 73–78; "A Doubly-Infinite System of Simple Groups," in *Mathematical Papers Read at the International Mathematical Congress in Chicago 1893* (New York, 1896), 208–242; "Concerning Transcendentally Transcendental Functions," in *Mathematische Annalen*, **48** (1897), 49–74; "On Certain Crinkly Curves," in *Transactions of the American Mathematical Society*, **1** (1900), 72–90; "A Simple Proof of the Fundamental Cauchy-Goursat Theorem," *ibid.*, 499–506; "The Undergraduate Curriculum," in *Bulletin of the American Mathematical Society*, **7** (1900), 14–24; "Concerning Harnack's Theory of Improper Definite Integrals," in *Transactions of the American Mathematical Society*, **2** (1901), 296–330; and "On the Theory of Improper Definite Integrals," *ibid.*, 459–475.

Subsequent articles are "Concerning Du Bois-Reymond's Two Relative Integrability Theorems," in *Annals of Mathematics*, 2nd ser., **2** (1901), 153–158; "A Definition of Abstract Groups," in *Transactions of the American Mathematical Society*, **3** (1902), 485–492; "On the Foundations of Mathematics," in *Bulletin of the American Mathematical Society*, **9** (1903), 402–424; also in *Science*, 2nd ser., **17** (1903), 401–416, his retiring address as president of the American Mathematical Society; "The Subgroups of the Generalized Finite Modular Group," in *Decennial Publications of the University of Chicago* (1903), 141–190; "On a Form of General Analysis with Application to Linear Differential and Integral Equations," in *Atti del IV Congresso internazionale dei matematici* (Rome, 6–11 Apr. 1908), II (1909), 98–114; "The Role of Postulational Methods in Mathematics" (address at Clark University, 20th Anniversary), in *Bulletin of the American Mathematical Society*, **16** (1909), 41; "On the Foundations of the Theory of Linear Integral Equations," *ibid.*, **18** (1912), 334–362; "On the Fundamental Functional Operation of a General Theory of Linear Integral Equations," in *Proceedings of the Fifth International Congress of Mathematicians* (Cambridge, 1912), I (1913), 230–255; "Definition of Limit in General Integral Analysis," in *Proceedings of the National Academy of Sciences of the United States of America*, **1** (1915), 628–632; "On Power Series in General Analysis," in *Mathematische Annalen*, **86** (1922), 30–39; and "A General Theory of Limits," in *American Journal of Mathematics*, **44** (1922), 102–121, written with H. L. Smith.

II. SECONDARY LITERATURE. Articles on Moore are G. A. Bliss, "Eliakim Hastings Moore," in *Bulletin of the American Mathematical Society*, **39** (1933), 831–838; and "The Scientific Work of Eliakim Hastings Moore," *ibid.*, **40** (1934), 501–514, with bibliography of Moore's publications; and G. A. Bliss and L. E. Dickson, "Eliakim Hastings Moore (1862–1932)," in *Biographical Memoirs. National Academy of Sciences*, **17** (1937), 83–102.

RONALD S. CALINGER

MOORE, JOSEPH HAINES (*b.* Wilmington, Ohio, 7 September 1878; *d.* Oakland, California, 15 March 1949), *astronomy.*

The only child of John Haines Moore and Mary Ann Haines, Moore graduated from Wilmington College in 1897. His field was classics, but during his senior year he became interested in astronomy. He then entered Johns Hopkins University, concentrating in physics, with minors in mathematics and astronomy, and received his Ph.D. in 1903. He immediately took a position as assistant to W. W. Campbell, director of the Lick Observatory; he passed through all the grades to astronomer (1923), was appointed assistant director in 1936, and director in 1942. Poor health forced him to relinquish the directorship in 1945 and move to a lower altitude, but he was engaged in teaching and research at the University of California at Berkeley until six months before his death. He was a member of the National Academy of Sciences, served as chairman of the astronomical section of the American Association for the Advancement of Science, and was twice president of the Astronomical Society of the Pacific. Moore was married in 1907 to Fredrica Chase, a computing assistant at the Lick Observatory. Modest and unassuming, he held to the Quaker philosophy all his life and was always regarded with deep affection by his colleagues and students.

Moore's principal scientific work was concerned with astronomical spectroscopy, in particular with the measurement of radial velocities of stars. From 1909 to 1913 he was in charge of the observatory's southern station at Santiago, Chile, and during this time some 2,700 stellar spectrograms were obtained there. As progressively more of Campbell's time was expended on other duties, Moore took over more of the spectrographic work of the observatory. In collaboration with Campbell he published (1918) the radial velocities of 125 bright-line nebulae, and he played the major role in producing (1928) the exhaustive discussion of the Mount Hamilton and Santiago radial-velocity measurements of all stars brighter than visual magnitude 5.51. Moore also made observations of fainter stars, and his "A General Catalogue of the Radial Velocities of Stars, Nebulae and Clusters" (1932) remained the standard work for two decades.

A by-product of radial-velocity studies was the discovery of spectroscopic binary stars. Moore discovered and calculated the orbits of many of these objects, and the results of this work were included in the third, fourth, and fifth catalogues of spectroscopic binaries, which he published in 1924, 1939, and 1948, respectively.

Moore took part in five eclipse expeditions, obtaining important spectroscopic information on the structure and composition of the solar corona. He led the Lick eclipse expeditions to Camptonville, California (1930) and to Fryeburg, Maine (1932).

Among Moore's other contributions were spectroscopic observations of novae, the companion of Sirius, the comet Pons-Winnecke, and the eclipsed moon. He also made spectroscopic determinations of the rotation periods of Saturn, Uranus, and Neptune (the last two in collaboration with D. H. Menzel); the result for Neptune, published in 1928, is still the accepted value.

BIBLIOGRAPHY

I. Original Works. Moore's most important writings are "Methods of Measurement and Reduction of Spectrograms for the Determination of Radial Velocities," in *Publications of the Astronomical Society of the Pacific*, **19** (1907), 13–26; "The Spectrographic Velocities of the Bright-Line Nebulae," in *Publications of the Lick Observatory*, **13** (1918), 75–186, written with W. W. Campbell; "Third Catalogue of Spectroscopic Binary Stars," in *Lick Observatory Bulletin*, **11** (1924), 141–185; "Radial Velocities of Stars Brighter than Visual Magnitude 5.51 As Determined at Mount Hamilton and Santiago," in *Publications of the Lick Observatory*, **16** (1928), written with W. W. Campbell; "The Crocker Eclipse Expedition of the Lick Observatory to Camptonville, California, April 28, 1930," in *Publications of the Astronomical Society of the Pacific*, **42** (1930), 131–144; "The Lick Observatory Crocker Eclipse Expedition to Fryeburg, Maine, August 31, 1932," *ibid.*, **44** (1932), 341–352; "A General Catalogue of the Radial Velocities of Stars, Nebulae and Clusters," in *Publications of the Lick Observatory*, **18** (1932); "Fourth Catalogue of Spectroscopic Binary Stars," in *Lick Observatory Bulletin*, **18** (1936), 1–37; and "Fifth Catalogue of Spectroscopic Binary Stars," *ibid.*, **20** (1948), 1–31, written with F. J. Neubauer. Moore also wrote numerous shorter items and survey papers for the *Lick Observatory Bulletin* and the *Publications of the Astronomical Society of the Pacific*.

II. Secondary Literature. For biographical information see R. G. Aitken, "Joseph Haines Moore: 1878–1949 —A Tribute," in *Publications of the Astronomical Society of the Pacific*, **61** (1949), 125–128; R. G. Aitken, C. D. Shane, R. J. Trumpler and W. H. Wright, "Joseph Haines Moore, 1878–1949," in *Popular Astronomy*, **57** (1949), 372–375; and F. J. Neubauer, "J. H. Moore—A Good Neighbor," in *Sky and Telescope*, **8** (1949), 197–198.

Brian G. Marsden

MOORE, WILLIAM (*fl. ca.* 1806–1823), *rocketry.*

Moore's origin and education are unknown. His writings and position as a mathematical instructor at the Royal Military Academy at Woolwich, England, suggest that he was influenced by Charles Hutton, professor of mathematics at the academy from 1773 to 1807. Moore was chosen as an assistant mathematical master in October 1806, Hutton being one of the three examiners on the selection board. By August 1807, Moore had advanced to the post of mathematical master. In July 1823 he left the academy because of a staff reduction.

Moore's first published writing was "Observations on the Problem Respecting the Radius of Curvature," in Nicholson's *Journal of Natural Philosophy* (1808). In 1810 he was prompted to investigate the ballistics of rockets when the Royal Danish Academy of Sciences offered a prize for the best paper describing the motion of rockets. Moore examined hypothetical rocket motion, both with and without air resistance, but did not submit a paper.

His theories on rockets first appeared in Nicholson's *Journal* for 1810 and 1811. In 1813 Moore published his collected findings as *A Treatise on the Motion of Rockets*. The world's first mathematical treatise on rocket dynamics, it had many shortcomings; and Moore admitted that lack of data had hindered his calculations. Nonetheless, he correctly recognized and demonstrated that Newton's third law of motion explained the principle of rocket motion. Moore was the first to consider rocket performance in terms other than range and altitude, and he arrived at calculations for thrust and specific impulse. He also suggested the use of the ballistic pendulum for a more accurate determination of performance.

BIBLIOGRAPHY

I. Original Works. Moore's principal writings are "Observations on the Problem Respecting the Radius of Curvature," in *Journal of Natural Philosophy, Chemistry, and the Arts*, 2nd ser., **21** (1808), 256–259; "On the Penetration of Balls Into Uniform Resisting Substances," in

Tilloch's *Philosophical Magazine*, **36** (1810), 325–334, and in *Emporium of Arts and Sciences*, **1** (July 1812), 277–289; "On the Destruction of An Enemy's Fleet at Sea by Artillery," in *Journal of Natural Philosophy*, **28** (1811), 81–93, also published in French in *Bibliothèque britannique*, **48** (1811), 365–379; *Treatise on the Doctrine of Fluxations* (London, 1811); and *A Treatise on the Motion of Rockets: To Which Is Added An Essay on Naval Gunnery* (London, 1813).

II. SECONDARY LITERATURE. A note on Moore appears in *Journal of Natural Philosophy*, **27** (1810), 318. See also Harry Harper, *Dawn of the Space Age* (London, 1946), p. 19; J. G. von Hoyer, *System der Brandraketen* (Leipzig, 1827), pp. 55–58; B. Allerslev Jensen, "Fra Leipzig til London," in *Dansk Artilleri-Tidsskrift*, **49** (June 1959), 61–79; Jacques-Philippe Mérigon de Montgéry, "Traité des fusées de guerre," in *Annales maritimes et coloniales*, **26**, pt. 2 (1825), 576–580; and H. D. Turner, "Sir William Congreve and the Development of the War Rocket," in *Research* (London), **19** (Aug. 1961), 326–328.

FRANK H. WINTER

MORAT, JEAN-PIERRE (*b.* St.-Sorlin, Saône-et-Loire, France, 18 April 1846; *d.* La Roche-Vineuse, near St.-Sorlin, 25 July 1920), *physiology.*

Morat contributed to physiological knowledge primarily by his studies of what is now called the autonomic nervous system; in particular, he and Albert Dastre showed in 1880 that stimulation of the cervical portion of the sympathetic nerve led to vasodilation in the gums and hard palate of the dog. His subsequent research emphasized the general significance of vasodilator nerves in the regulation of organic function.

Morat studied at Lyons in the early 1870's and then in Paris, where he worked in Claude Bernard's laboratory at the Museum of Natural History and in 1873 received his medical degree from the Faculty of Medicine.

From 1873 to 1876 Morat served as the *chef des travaux anatomiques* for the medical school at Lyons. A year after the founding of a new Faculty of Medicine at Lille in 1875, he became the *chargé de cours* for physiology and assistant professor of the subject in 1878. He returned to Lyons in 1882 as professor of physiology at the Faculty of Medicine and Pharmacy, which had been established in 1877. He became an associate member of the Société de Biologie in 1906 and a correspondent of the medical and surgical section of the Academy of Sciences in 1916.

In 1899 Morat received the Academy's Lacaze Prize for his research career, which was dominated especially

by the ideas and techniques of Bichat, Bernard, and Chauveau. In 1877 he had won the Academy's Montyon Prize in experimental physiology for his first significant research, which concerned muscle physiology.[1]

This work was performed with J. J. H. Toussaint in Chauveau's laboratory and depended on both the graphical recording techniques developed by Chauveau and Marey and the use of Chauveau's unipolar electrode. It clarified the analogy which had been drawn between induced tetanus and the voluntary contraction of skeletal muscle by showing that the two were comparable only when the tremors composing tetanus were perfectly fused.

Morat and his later, more renowned, collaborator, Dastre, began their joint study of the vasodilator nerves (1876–1882) in Chauveau's laboratory. They were among the first to employ simultaneous graphical recordings of arterial and venous pressures to study vasomotor action. Chauveau and Marey had previously used a similar method to study other aspects of circulation.

At the beginning of their study of the vasomotor nerves, Morat and Dastre assumed the inverse of what they are best known for having discovered: the vasodilatory effect of excitation of the cervical portion of the sympathetic nerve. In their attempt (1878) to resolve the question of whether the sciatic nerve had vasodilator properties, they used the cervical sympathetic nerve as an exemplar of a vasoconstrictor. They argued that since excitation of the sciatic had the same effect on vascular pressures as excitation of the cervical sympathetic, then the sciatic, like the cervical sympathetic, must be a vasoconstrictor.

Two years later, however, Morat and Dastre announced that the cervical sympathetic was a vasodilator.[2] The reason for the change is unclear; but it seems to have resulted from interpreting the recent studies of other researchers in the light of views they already held, which were based on those of Bernard. In Bernard's view, vasodilation resulted from nervous inhibition or paralysis of vasoconstrictor nerves. On this basis Dastre and Morat had supposed that if vasodilator nerves were present, they were most likely to be found related to ganglia (where the inhibitory action would be localized), especially those of the sympathetic. At about the same time that they were studying the sciatic nerve, work by others had shown that branches of the submaxillary nerve caused dilation in vessels of certain parts of the face and that the submaxillary nerve's dilator fibers did not originate in the brain. Their expectation that vasodilators were related to ganglia probably led Dastre and Morat to suppose that the submaxillary dilator fibers arose at the cervical ganglia of the sympathetic and, hence, to test the cervical sympathetic to see whether it caused dilation in the specified regions of the face, which it did.

Morat's research frequently returned to topics related to vasodilation or to other functions mediated by the sympathetic system. For example, he collaborated with Maurice Doyon in demonstrating (1891) that sympathetic nerve fibers have an effect opposite to that of the ciliary nerve: they accommodate the lens of the eye for nearby objects.

In their study of the consumption of sugar by resting muscle (1892), Morat and E. Dufourt used Bernard's chemical test for blood sugar to treat a problem that Bernard's work had suggested and that Chauveau had explicitly raised: the role of glycogen in the various organs of the body. This work led to a study of nervous control over liver glycogenesis (1894), which showed that liver glycolysis can be stimulated without changing the flow of blood through hepatic vessels and, thus, independently of its vasomotor nerves.

In Morat's later studies of vasodilation he rejected Bichat's view that had dominated his early thought about the sympathetic system: that the sympathetic system was independent of the cerebrospinal system and was the sole regulator of visceral organ function. In 1894 Morat expressed the view that vasodilation was localized neither in the spinal nor in the sympathetic trunk but was a property of certain nervous elements which compose both these trunks.

In rejecting Bichat's distinction Morat made no reference to the histological results which had led Gaskell to a similar conclusion a decade earlier. Rather, Morat had come to regard metabolic and functional behaviors as concurrent activities, elicited at the same time by the same fiber. For this reason there was no longer any motive for ascribing all "organic" vasomotor functions to a distinct nervous structure, the sympathetic trunk.

Morat's last significant research, performed with M. Petzetakis just before the beginning of World War I, showed that cardiac fibrillation can result from a disequilibration between cardiac excitor and inhibitor nerves, and that the rhythms of auricles and ventricles are mutually related but not in a totally dependent fashion.

NOTES

1. The report of the commission awarding the Lacaze Prize appears in *Comptes rendus de l'Académie des sciences*, **129** (1899), 1140–1144. The Montyon commission report is *ibid.*, **84** (1877), 848–851.

2. Dastre and Morat claimed that this result of work begun in 1876 was first announced to the Société de Biologie in 1878; but there are no citations to Morat in the Society's 1878 *Mémoires et comptes rendus*, and the note he alluded to seems to be one of a series submitted in 1880 and published in 1881. The error may have been part of an effort to avoid a priority dispute with Jolyet and Laffont, whose study of dilation in the facial region, published in 1878, may in fact have led immediately to the finding by Dastre and Morat.

BIBLIOGRAPHY

I. ORIGINAL WORKS. Morat published about 75 scientific papers, more than half of which were written in collaboration with other physiologists. Papers published prior to 1900 are indexed in the Royal Society's *Catalogue of Scientific Papers*, X, 843–844; XII, 184, 519; XVII, 342–343; and XVIII, 278.

Among the most important are "Variations de l'état électrique des muscles dans différents modes de contraction . . .," in *Archives de physiologie normale*, 2nd ser., **4** (1877), 156–182, written with H. Toussaint; "Sur l'expérience du grand sympathique cervical," in *Comptes rendus . . . de l'Académie des sciences*, **91** (1880), 393–395, written with A. Dastre; "Sur la fonction vasodilatatrice du nerf grand sympathique," in *Archives de physiologie*, **9** (1882), 177–236, 337–382, written with A. Dastre; "Le grand sympathique nerf de l'accommodation pour la vision des objets éloignés," in *Comptes rendus . . . de l'Académie des sciences*, **112** (1891), 1327–1329, written with M. Doyon; "Les fonctions vaso-motrices des racines postérieures," in *Archives de physiologie*, 5th ser., **4** (1892), 689–698; "Nerfs et centres inhibiteurs," *ibid.*, **6** (1894), 7–18; "Les nerfs glyco-sécréteurs," *ibid.*, 371–380, written with E. Dufort; and "Le système nerveux et la nutrition. Les nerfs thermiques. [Les nerfs trophiques.]," in *Revue scientifique*, 4th ser., **4** (1895), 487–495; **5** (1896), 193–199, 234–241.

In addition to papers indexed in the Royal Society's *Catalogue of Scientific Papers*, Morat published the following: "Réserve adipeuse de nature hivernale dans les ganglions spinaux de la grenouille," in *Mémoires de la Société de biologie*, **53** (1901), 473–474; "La réforme des études médicales," in *Revue scientifique*, 5th ser., **5** (1906), 524–526; "Les racines du système nerveux: Le mot et la chose," in *Archives internationales de physiologie . . .*, **8** (1909), 75–103; "Les variations de la formule sanguine chez les morphinomanes et les héroïnomanes au cours de désintoxication rapide par la méthode de Sollier," in *Mémoires de la Société de biologie*, **66** (1909), 1025–1027, written with Chartier; and three articles written with M. Petzetakis: "Production de la fibrillation des oreillettes par voie nerveuse, au moyen de l'excitation du pneumogastrique," *ibid.*, **77** (1914), 222–224; "Fibrillation auriculaire et ventriculaire produite par voie nerveuse," *ibid.*, 377–379; and "Production expérimentale d'extrasystoles ventriculaires retrogrades, et de rythme inverse, par inversion de la conduction des excitations dans le coeur," in *Comptes rendus . . . de l'Académie des sciences*, **163** (1916), 969–971.

Morat also published with A. Dastre a book-length account of their study of the vasomotor system: *Recherches expérimentales sur le système nerveux vasomoteur* (Paris, 1884). With Maurice Doyon he wrote *Traité de physiologie*, 5 vols. (Paris, 1899–1918). Vol. II, Morat's treatment of the nervous system, was translated into English and published under the title *Physiology of the Nervous System* (London, 1906).

II. SECONDARY LITERATURE. The longest published account which I have been able to locate of Morat's life is a paragraph in *Biographisches Lexikon der hervorragenden Ärzte, 1800–1930*, II (Munich, 1962), 1065.

Background for Morat's work on vasodilation is in Donal Sheehan, "Discovery of the Autonomic Nervous System," in *Archives of Neurology and Psychiatry*, **35** (1936), 1081–1115, esp. 1102–1105; and E. A. Schafer, ed., *Textbook of Physiology*, II (London, 1898), 71, 130–136, 618, 626, 659–661.

MICHAEL GROSS

MORAY (or **MURREY** or **MURRAY**), **SIR ROBERT** (*b*. Scotland, 1608 [?]; *d*. London, England, 4 July 1673), *chemistry, metallurgy, mineralogy, natral history*.

Moray was the first president of the Royal Society of London and contributed significantly to its survival and growth during the early years. Little is known of his early life. He was born between 10 March 1608 and 10 March 1609 and was the elder of two sons. His grandfather was Robert Moray of Abercairney. His father, Sir Mungo Moray of Craigie, in Perthshire, married a daughter of George Halket of Pitfirran.

As a soldier, statesman, and diplomat, Moray played an important part in the politics of England, France, and Scotland. He served in the Scottish regiment which joined the French army under Colonel Hepburn in 1633, was Richelieu's agent in England during the Puritan Revolution, was knighted by Charles I on 10 January 1643, and became a colonel in the Scottish Foot Guards in April 1645. In the autumn of 1645 Moray went to London as political mediary for the Scots. He later served as royal secretary to Charles I and as confidential agent to the duke of Hamilton. While engaged by John Maitland (1616–1682), duke of Lauderdale, in 1648 to negotiate with the Prince of Wales, Moray befriended the future Charles II. In 1651 he was a justice clerk and privy councillor in Scotland.

Moray's political life has an indirect relationship to his importance in the history of science. His title, which lent prestige, and his close friendship with Charles II, which aided in obtaining a charter for the Royal Society, may have occasioned his election as first president of the Society. He was also religious,

a gentleman of high character, and was well-liked and respected by his colleagues. Furthermore, he was very enthusiastic about the group's scientific interests. Besides his terms as president, Moray served frequently on the council and on numerous committees and carried on a vast correspondence to procure scientific information. He was fluent in French, German, Dutch, and Italian and strengthened the international character of the Society through his powerful connections and many foreign correspondents. He often served as a liaison for men of science in England, Scotland, America, and on the Continent.

Moray's versatility and utilitarian view of science are reflected in his knowledge of trades and industrial processes, such as fishing, lumbering, mining, shipbuilding, windmills, watermills, magnetism, and mineralogy. Chronometry was a major interest; at times he offered advice to Huygens. When Alexander Bruce and Huygens argued over patent rights to the pendulum clock for use at sea, Moray, often a peacemaker when disputes arose, proposed that the patent be taken in the name of the Royal Society.

It appears that Moray earned some renown in chemistry, although no lasting influence can be detected. Anthony Wood recorded that Thomas Vaughan, who became eminent in medical chemistry, served in London under the patronage of "that noted Chymist Sir *Rob. Murrey* or *Moray* Kt." Metallurgy and pharmaceutical preparations were enduring interests throughout Moray's life; and although nothing is known of his formal education, he wrote to Bruce in 1657 that he was "so far advanced towards the gown as to be already about half an Apothecary" and "I was as long at the Anatomy school as the Chimicall."

Moray's direct contributions to science are difficult to assess, but his unabated interest in and work on behalf of the Royal Society and its scientific developments from 1660 until his death in 1673 were perhaps unequaled.

BIBLIOGRAPHY

I. ORIGINAL WORKS. According to Royal Society records (*Letter Books*, 16 and 26 Sept. 1665), Moray worked on a history of masonry. Sprat (see below), pp. 257–258, recorded the title, but the work is not preserved in the Archives of the Royal Society. Some of Moray's papers and letters are listed in Sir Arthur Church, *The Royal Society. Some Account of the "Classified Papers" in the Archives. With an Index of Authors* (Oxford, 1907). An MS catalog, also compiled by Church, "The Royal Society 'Classified Papers' in the Archives Titles and Authors" (1907) (Church MS) includes several references to Moray not in the printed volume. Other Royal Society archival sources include *The Letter Books* (LBC), *The Register Books* (RBC), *Boyle's Letters* (BL), and *Miscellaneous Manuscripts* (MM).

Among his writings are "Relation Concerning Barnacles" (RBC.1.19); "A Copie of the Letter Sent to Paris to Mr. de Monmort" (LBC.1.1); "An Account of Glass Dropps" (RBC.1.57); "Clarke, Dr. Observations on ye Humble & Sensible Plants in Mr. Chaffin's Garden in St. James' Parke, Made August 9, 1661. Present the Lord Brouncker, Sr. *Robert Murrey*, Dr. Wilkins, W(?) Evelin, Dr. Goddard, Dr. Henshaw, Dr. Clarke" [Read 21 Aug. 1661] (Church MS), "The Description of the Island Hirta" (RBC.1.97); "Account of the Sounding of the Depth of the Sea Between Portsmouth & the Isle of Wight," written with Brouncker [FRS] (Church MS); "Sr R, Moray's Letter to Sr Phil. Vernatti in Java, by Order of the R. Society" (LBC.1.79); "An Account of an Echo" (RBC. 1.263); "The Way How Malt is Made in Scotland" (RBC. 1.306); "Sr Robert Moray's Story of Persons Killed With Subterraneous Damps" (RBC.1.319).

Additional writings are "Of the Minerall of Liege Yeilding Both Brimstone and Vitrioll and the Way of Extracting Them Out of it Used at Liege" (RBC.2.35); "The Measure of the Parts of a Gyant Child Borne in Scotland" (RBC.2.50); "An Account Englished Out of French of an Unusuall Way of Cutting the Stone of the Bladder Practiced by a Frenchman" (RBC.2.72); "Of a Spring Near Chertsey" (Church MS); "Of the Way Used Upon the Coast of Coromandell of Cooling of Drincks by Exposing Them to the Heat of the Sun" (RBC.2.50); Murray to Sir P. Vernatti, 21 Sept. 1664 (LBC.1.237); "On a Tumulus in Lord Seamore's Garden at Marlborough" (Church MS); "Directions for Observing the Conjunction of Mercury With the Sun" (Church MS); "Observations Made in Their Late Excursion Into ye Country," written with Sir P. Neil and Dr. Wren [FFRS] (Church MS); "Description in German With Drawing, of a Comet of January, 1663/4" [Read 9 March 1663 (1664); comm. by R. M.].

See also "An Extract of a Letter of Mr. Moray" (LBC. 1.280); "Eclipse as Observed by ——, on June 22, 1666," written with F. Willughby *et al.* [FFRS] (Church MS); "A Fair Copy of No. 13, With Diagram on a Folded Leaf" [Read 27 June 1666]; "Letter of 30 Jan 1667/8 to Him . . . From the Counsell of the Royal Society . . . Solliciting Contribution in Scotland for Building a Colledge" (LBC.2.160); "Calendarium ecclesiasticum et astronomicum, cum tabulis et figuris zodiaci." In the folding form, of the first half of the fifteenth century, on vellum. The *homo signorum* and the *homo venarum* appear to be deficient. Presented by Moray (Church MS); Murray to R. Boyle, 15 Jul. 1672 (BL4.75); "Experiments Propounded by Sir Robert Moray and Recommended to Dr. Power" (RBC.1.167); no origin, undated, to the grand duke of Tuscany (MM.3.119), credited to Moray; "Sounding Between Portsmouth & the Isle of Wight With the Wooden Globe & Lead," written with Brouncker (Church MS).

Some of the papers were published in the *Philosophical Transactions of the Royal Society* without clear authorship; e.g., Moray suggested several experiments to Newton in "Some Experiments Propos'd in Relation to Mr. Newton's Theory of Light, Printed in Numb. 80; Together With the Observations Made Thereupon by the Author of That Theory; Communicated in a Letter of His From Cambridge, April 13. 1672," **7** (1672), 4059–4062. Although Moray is not cited as the author of the paper, Newton in a letter to Oldenburg indicates that he was (Rigaud, II, 324).

II. SECONDARY LITERATURE. The major source of information about Moray is Alexander Robertson, *The Life of Sir Robert Moray, Soldier, Statesman and Man of Science* (London, 1922); it does not include a bibliography of Moray's writings, but some of his work and scientific interests are discussed. Other biographical sketches, which add little, include D. C. Martin, "Sir Robert Moray, F.R.S. (1608?–1673)," in *Notes and Records. Royal Society of London*, tercentenary no., **15** (1960), 239–250; Agnes Mary Clerke, *Dictionary of National Biography*, XIII (1967–1968), 1298–1299, under "Murray." An obituary appeared in Thomas Birch, *History of the Royal Society of London*, 4 vols. (London, 1756–1757), III, 113–114. See also John Aubrey, *Brief Lives*, A. Clark, ed., 2 vols. (Oxford, 1898); *Bishop Burnet's History of His Own Time*, Osmund Airy, ed., 2 vols. (London, 1840), *passim*; Patrick Gordon, *A Shorte Abridgement of Britane's Distemper* (Aberdeen, 1844), pp. 5–6; L. C. Martin, ed., *The Works of Henry Vaughan*, 2 vols. (Oxford, 1914), *passim*; W. Shaw, *Knights of England* (London, 1906), *passim*; Thomas Sprat, *History of the Royal Society of London*, 2nd ed. (London, 1702), *passim*; T. Thomson, *History of the Royal Society From Its Institution to the End of the Eighteenth Century* (London, 1812), *passim*; C. R. Weld, *A History of the Royal Society With Memoirs of the Presidents*, 2 vols. (London, 1848), *passim*; Anthony Wood, *Athenae Oxonienses*, 3rd ed., 5 vols. (London, 1813–1820), *passim*; and Dudley Wright, *England's Masonic Pioneers* (London, 1925), *passim*.

Two helpful general articles are Marie Boas Hall, "Sources for the History of the Royal Society in the Seventeenth Century," in *History of Science*, **5** (1966), 62–76; and R. K. Bluhm, "A Guide to the Archives of the Royal Society and to Other Manuscripts in Its Possession," in *Notes and Records. Royal Society of London*, **12** (Aug. 1956), 21–39.

For additional information relating to Moray's scientific interests and his activities in the Royal Society, see the *Journal Books, Council Minutes, Minutes of Meetings of Committees*, and the *Kincardine Papers* (transcript, MS246) in the archives; *Oeuvres complètes de Christiaan Huygens*, 22 vols. (The Hague, 1888–1950), *passim*; S. P. Rigaud and S. J. Rigaud, eds., *Correspondence of Scientific Men of the Seventeenth Century*, 2 vols. (Oxford, 1841–1842), *passim*; and A. R. Hall and M. B. Hall, eds., *The Correspondence of Henry Oldenburg*, 8 vols. (Madison, Wis., 1965–   ), *passim*.

Most sources relating to his career in the French army are in the archives at Paris. For MS collections see

Robertson (above), pp. 201–203; sources of a political nature include Osmund Airy, ed., *Lauderdale Papers*, 3 vols. (Westminster, 1884–1885), *passim*; J. G. Fotheringham, ed., *The Diplomatic Correspondence of Jean de Montereul and the Brothers de Bellièvre*, 2 vols. (Edinburgh, 1898–1899), *passim*; and Samuel R. Gardiner, ed., *The Hamilton Papers; Being Selections From Original Letters in the Possession of His Grace the Duke of Hamilton and Brandon, Relating to the Years 1638–1650* (Westminster, 1880).

BARBARA ROSS

MORE, HENRY (*b.* Grantham, Lincolnshire, England, October 1614; *d.* Cambridge, England, 1 September 1687), *philosophy, theology.*

The youngest child of Alexander More, a fairly prosperous gentleman and several times mayor of Grantham, Henry More was educated at Grantham School, Eton, and Christ's College, Cambridge, from which he graduated B.A. in 1636. In 1639 he received the M.A., took orders, and was appointed a fellow of his college—which position he held, refusing preferment, all his life. More became doctor of divinity in 1660 and was elected fellow of the Royal Society on 25 May 1664. (He had been among the original fellows under the first charter but was omitted when the Society was refounded.)

In theology More was a moderate latitudinarian, known for piety and an almost saintly nature. He wrote extensively against sectarians and enthusiasts, for their uncharitable doctrinal wrangling and their depreciation of reason in religion, and against the Roman Catholic Church, on the usual contemporary grounds. He concerned himself particularly with the interpretation of prophetic and apocalyptic Scriptures.

In the history of philosophy More is counted among the Cambridge Platonists. His "Platonism" was rather vague and highly eclectic; its basic themes were those of the middle Platonists and Neoplatonists, and he found them in a great variety of ancient thinkers, including Democritus, Hermes Trismegistus, and Moses. The central point is the primacy of spirit over matter. Dissatisfaction with the scholastic fare of his undergraduate studies led More to turn briefly to the ascetic-mystical side of Neoplatonism: true knowledge requires spiritual purification, and devotion is more important than learning. Both doctrines were soon greatly moderated, as his bent for philosophy (including natural philosophy) reasserted itself. Under the influence of the *Theologia Germanica*, More came to emphasize moral goodness over asceticism; and the "spiritual purification" idea had little real effect on his mature writings, unless in a certain tendency

to overrate the rational perspicuity of arguments which have edifying conclusions.

A factor in More's return to philosophy was his discovery, sometime before 1647, of Descartes, whose writings seemed to show how to combine a scientific interest in nature with a primary concern for vindicating the reality of God and immortal human souls. This suited More admirably: his interest in the new experimental philosophy was genuine (he was the only fellow of the Royal Society among the Cambridge Platonists), but he conceived his main philosophical mission to be the refutation of mechanistic materialism.

Appropriately, More's first major work was *An Antidote Against Atheisme* (1652), one of the most prominent early responses to Thomas Hobbes. The first part of this three-part work is primarily an elaboration of the ontological argument as found in Descartes. The second part enumerates a great range of natural phenomena that can be understood only as showing a divine providence. This section provided the structure and core of John Ray's *Wisdom of God Manifested in the Works of Creation*, and thus considerably influenced the subsequent tradition of scientifically elaborated teleological arguments. Two points should be noted, however. First, relatively little of More's argumentation really depends on contemporary science; the majority of his examples had been, or could have been, used in antiquity. Second, the comparison with machinery (such as the watch) is not made. The emphasis is, rather, on the usefulness to man or other creatures of various features of nature, and on phenomena which show the working of immaterial substances, such as an unintelligent "spirit of nature" which can be invoked to account for botches in nature as well as for phenomena (such as gravity and the formation of animals) which cannot be explained mechanically. The relation between this "spirit of nature" and the intelligent Designer remains unclear, but just showing the reality of spiritual agents is what More really cares about. Thus it is perfectly in accord with his design when he devotes the third part of his treatise to stories of witches, hauntings, and so on. These direct empirical evidences of the activities of spirits should convince those on whom the arguments of the first two sections are lost.

More's opposition to mechanism eventually led him to a repudiation (in large part) of Descartes and a sad skirmish with Robert Boyle. In his early enthusiasm he had been instrumental in introducing Cartesian philosophy to England; but an unsatisfactory correspondence with Descartes, further reflection on his metaphysical principles, and observation of the path taken by Spinoza and other Cartesians convinced More that there were great dangers in Cartesianism. More was persuaded that to *be,* a thing must be *somewhere;* Descartes's identification of matter with extension thus seemed to exclude spirits (including God) from reality. Therefore, in *The Immortality of the Soul* (1659) and *Enchiridion metaphysicum* (1671) More argued at length that spirits are extended. The defining characteristic of body is not extension but impenetrability and physical divisibility ("discerpibility"); spirits, More deduced, are by definition "indiscerpible" and capable of penetrating themselves, other spirits, and matter. He adds that bodies are passive and spirits are capable of initiating activity. If spirits are extended, God in particular is (infinitely) extended. More does not flinch from this consequence but, listing a long series of properties predicable both of God and of space, concludes that absolute space is an attribute of the substance, God; it is the medium in which God acts upon bodies.

The Immortality of the Soul is actually an elaborate treatise on the nature, kinds, and habits of spirits—by far More's most systematic work—in which many doctrines of Descartes and others are criticized. It defies summary.

More consistently argued that gravity, magnetism, and various of Boyle's experimental results in hydrostatics could not be accounted for mechanistically. In the *Enchiridion metaphysicum* he treated the latter point in detail, attempting with physical as well as metaphysical arguments to refute Boyle's interpretation of his own experiments. Boyle found it necessary to demolish More's efforts, carefully adding that one could be a great scholar without being a good hydrostatician. He patiently corrected More's mistakes, pointed out that a mechanical explanation is one based on the laws of mechanics and need not (for instance) specify the cause of gravity, and suggested that the watchmaker version of the design argument is more effective than any that resort to such dubious entities as the spirit of nature. More was rather hurt but eager to maintain their friendship. Unlike Descartes, the Royal Society and its virtuosos were never even partially repudiated by More, who distinguished sharply between their "experimental philosophy" and the "mechanical philosophy" he combated.

The exchange with Boyle shows that More's grasp of the new natural philosophy was limited. His interest was genuine, but he was himself no virtuoso. His main contributions lay in introducing generations of students to Descartes, in lending to the Royal Society the prestige of his great reputation for learning

and piety, and (arguably) in his influence upon Newton. The nature and extent of that influence are hard to assess. It appears that More and Newton were well acquainted and perhaps close. More left Newton a funeral ring; a letter survives in which he good-humoredly reports to a friend that Newton stubbornly clings to a misinterpretation of a passage in the Apocalypse, which More thought he had corrected; Newton informs a correspondent that he had "engaged Dr More to be of" a "Philosophick Meeting" then proposed at Cambridge. E. A. Burtt and A. Koyré have argued powerfully that More influenced Newton's views on space and on such matters as the (immaterial) cause of gravity. Certainly there are interesting parallels; other evidence for or against direct influence is, unfortunately, scarce.

BIBLIOGRAPHY

I. ORIGINAL WORKS. *Philosophical Writings of Henry More,* Flora I. MacKinnon, ed. (New York, 1925), contains a useful selection, with intro., extensive notes, and a bibliography of works by and about More. For more recent bibliographical information see Aharon Lichtenstein, *Henry More: The Rational Theology of a Cambridge Platonist* (Cambridge, Mass., 1962).

II. SECONDARY LITERATURE. Marjorie Nicolson, ed., *Conway Letters* (New Haven, 1930), has biographical information as well as letters. More's views and their relation to those of Descartes and Newton are discussed by Edwin A. Burtt, *The Metaphysical Foundations of Modern Physical Science* (Garden City, N.Y., 1955), esp. 135–148; and Alexandre Koyré, *From the Closed World to the Infinite Universe* (Baltimore, 1957), esp. 110–154, 190. On More's relation to Hobbes, see Samuel I. Mintz, *The Hunting of Leviathan* (Cambridge, 1962), esp. 80–95; and on his relation to Boyle, see Robert A. Greene, "Henry More and Robert Boyle on the Spirit of Nature," in *Journal of the History of Ideas,* **23** (1962), 451–474. Also of interest are C. A. Staudenbaur, "Galileo, Ficino, and Henry More's *Psychathanasia,*" in *Journal of the History of Ideas,* **29** (1968), 565–578; and C. Webster, "Henry More and Descartes: Some New Sources," in *British Journal for the History of Science,* **4** (1969), 359–377.

WILLIAM H. AUSTIN

MORGAGNI, GIOVANNI BATTISTA (*b.* Forlì, Italy, 25 February 1682; *d.* Padua, Italy, 5 December 1771), *medicine, anatomy, pathological anatomy.*

Morgagni was the son of Fabrizio Morgagni and Maria Tornielli. After completing his early studies at Forlì, in 1698 he went to Bologna, where he attended the university, taking the degree in philosophy and medicine in 1701. His principal university teachers were Antonio Maria Valsalva and Ippolito Francesco Albertini, both former pupils of Malpighi, who trained him in Malpighi's methods and in the rational medicine that follows from them. Having received his degree, Morgagni remained in Bologna to work in the three hospitals of that city and carry out further anatomical studies with Valsalva.

Morgagni was admitted to the Accademia degli Inquieti in 1699 and became its head in 1704. He reformed the academy on the model of the Paris Académie Royale des Sciences and accepted an invitation to hold meetings in the mansion belonging to Luigi Ferdinandino Marsili, thus paving the way for its incorporation into the Istituto delle Scienze that was founded by Marsili in 1714. It was to the Inquieti that Morgagni in 1705 communicated his *Adversaria anatomica prima*, which he also dedicated to them. The *Adversaria* was published in Bologna in 1706 and earned Morgagni international fame as an anatomist.

At the beginning of 1707 Morgagni moved to Venice, where he stayed through May 1709. Venice offered him the opportunity to study chemistry with Gian Girolamo Zanichelli, to investigate the anatomical structure of the great fishes, and to secure a number of rare and choice books. He also conducted a number of dissections of human cadavers with Gian Domenico Santorini, who was at that time dissector and lector in anatomy at the Venetian medical college. In June 1709 Morgagni returned to Forlì, where he practiced medicine with great success. In September 1711 he was called to the second chair of theoretical medicine at Padua University; the chair had become vacant when Antonio Vallisnieri was promoted to the first chair, following the death of Domenico Guglielmini. Morgagni delivered his inaugural lecture, *Nova institutionum medicarum idea*, on 17 March 1712. He was appointed to the first chair of anatomy at Padua in September 1715 and began teaching that subject on 21 January 1716. He held this post until his death. Morgagni's teaching was always clear and gave the impression of a perpetually fresh mind.

The *Adversaria anatomica prima* is a series of researches on fine anatomy conducted according to the tradition established by Malpighi, although Morgagni showed greater caution in the use of the microscope and in making anatomical preparations. Morgagni's profoundly inquiring intellect is apparent in even this early work. Despite the modesty of its title—"Notes on Anatomy"—Morgagni's book actually records a whole succession of discoveries regarding minute organic mechanisms, including the glands of the trachea, of the male urethra, and of the female genitals. These represent new contributions to the mechanical interpretation of the structure of the organism, as do the descriptions contained in Morgagni's five subsequent *Adversaria* (1717–1719), *Epistolae anatomicae duae* (published in Leiden by Boerhaave in 1728), and *Epistolae anatomicae duodeviginti* on Valsalva's writings (1740).

Morgagni's most important work, however, is his *De sedibus et causis morborum per anatomen indagatis* of 1761. This book grew out of a concept of Malpighi, which Morgagni then developed into a major work. The concept may be stated simply as the notion that the organism can be considered as a mechanical complex. Life therefore represents the sum of the harmonious operation of organic machines, of which many of the most delicate and minute are discernible, hidden within the recesses of the organs, only through microscopic examination.

Like inorganic machines, organic machines are subject to deterioration and breakdowns that impair their operation. Such failures occur at the most minute levels, but, given the limits of technique and instrumentation, it is possible to investigate them only at the macroscopic level, by examining organic lesions on the dissecting table. These breakdowns give rise to functional impairments that produce disharmony in the economy of the organism; their clinical manifestations are proportional to their location and nature.

This thesis is implicit in the very title *De sedibus et causis morborum per anatomen indagatis*. In this book Morgagni reasons that a breakdown at some point of the mechanical complex of the organism must be both the seat and cause of a disease or, rather, of its clinical manifestations, which may be conceived of as functional impairments and investigated anatomically. Morgagni's conception of etiology also takes into account what he called "external" causes, including environmental and psychological factors, among them the occupational ones suggested to Morgagni by Ramazzini.

The parallels that exist between anatomical lesion and clinical symptom served Morgagni as the basis for his "historiae anatomico-medicae," the case studies from which he constructed the *De sedibus*. There had, to be sure, been earlier collections of case histories, in particular Théophile Bonet's *Sepulchretum* (1679), but Bonet's work was, as René Laënnec wrote of it, an "undigested and incoherent compilation," while the special merit of Morgagni's work lies in its synthesis of case materials with the insights provided by his own anatomical investigations. In his book Morgagni made careful evaluations of anatomic-medical histories drawn exhaustively from the existing literature. In addition, he describes a great number of previously unpublished cases, including both those that he had himself observed in sixty years of anatomical investigation and those collected by his immediate predecessors, especially Valsalva, whose posthumous papers Morgagni meticulously edited and commented upon. The case histories collected in the *De sedibus* therefore represent the work of an entire school of anatomists, beginning with Malpighi, then extending through his pupils Valsalva and Albertini to Morgagni himself.

Morgagni may thus be considered to be the founder of pathological anatomy. This work was, in turn, developed by Baillie, who classified organic lesions as types (1793); Auenbrugger and Laënnec, who recognized organic lesions in the living subject (1761 and 1819, respectively); Bichat, who found the pathological site to be in the tissue, rather than the organ (1800); and Virchow, who traced the pathology from the tissue to the cell (1858).

BIBLIOGRAPHY

I. ORIGINAL WORKS. Morgagni's writings include *Adversaria anatomica prima* (Bologna, 1706); *Nova institutionum medicarum idea* (Padua, 1712); *Adversaria anatomica altera et tertia* (Padua, 1717); *Adversaria anatomica quarta, quinta et sexta* (Padua, 1719); *Epistolae anatomicae duae* (Leiden, 1728); *Epistolae anatomicae duodeviginti ad scripta pertinentes celeberrimi viri A. M. Valsalvae* (Venice, 1740); *De sedibus et causis morborum per anatomen indagatis* (Venice, 1761); *Opuscula miscellanea* (Venice, 1763); *Opera omnia* (Venice, 1764); and *Opera postuma* (Rome, 1964–1969), vol. I, *Le autobiografie*, and vols. II–IV, *Lezioni di medicina teorica.*

A bibliography is Renato Zanelli, "Catalogo ragionato delle edizioni Morgagnane in ordine cronologico," in *Le onoranze a G. B. Morgagni, Forlì, 24 maggio 1931–IX* (Siena, 1931), 137–147.

II. SECONDARY LITERATURE. Bibliographies are Carlo Fiorentini, *Giovanni Battista Morgagni: Primo saggio di bibliografia sintetica* (Bologna, 1930); and Loris Premuda, "Versuch einer Bibliographie mit Anmerkungen über das Leben und die Werke von G. B. Morgagni," in Markwart Michler, ed. and trans., *Sitz und Ursachen der Krankheiten* (Bern–Stuttgart, 1967), 163–195.

More recent works include Luigi Belloni, "Aus dem Briefwechsel von G. B. Morgagni mit L. Schröck und J. F. Baier," in *Nova acta leopoldina,* **36** (1970), 107–139; "Lettere del 1761 fra D. Cotugno e G. B. Morgagni," in *Physis,* **12** (1970), 415–423; "Contributo all'epistolario Boerhaave–Morgagni. L'edizione delle Epistolae anatomicae duae, Leida 1728," *ibid.,* **13** (1971), 81–109; "L'epistolario Morgagni–Réaumur alla Biblioteca Civica di Forlì," in *Gesnerus,* **29** (1972), 225–254; "L'opera di Giambattista Morgagni: dalla strutturazione meccanica dell'or-

ganismo vivente all'anatomia patologica," in *Simposi clinici,* **9** (1972), I–VIII; and in *Morgagni,* **4** (1971), 71–80; and "G. B. Morgagni und die Bedeutung seines 'De sedibus et causis morborum per anatomen indagatis,' " in Erna Lesky and Adam Wandruzka, eds., *Gerard van Swieten und seine Zeit* (Vienna–Cologne–Graz, 1973), 128–136. See also Giuseppe Ongaro, "La biblioteca di Giambattista Morgagni," in *Quaderni per la storia dell'Università di Padova,* **3** (1970), 113–129.

LUIGI BELLONI

MORGAN, CONWY LLOYD (*b.* London, England, 6 February 1852; *d.* Hastings, England, 6 March 1936), *comparative psychology, philosophy.*

Lloyd Morgan, as he was usually called, was a pioneer of animal psychology and an outstanding contributor to the evolutionary understanding of animal behavior. He was the second son of James Arthur Morgan, a solicitor, and received his early education at the Royal Grammar School, Guildford. When he was seventeen he entered the School of Mines at the Royal College of Science in London, intending to become a mining engineer. His progress was brilliant and at the same time he studied philosophy and biology. After a spell of traveling in the Americas he worked under T. H. Huxley, who influenced him profoundly. From 1878 to 1883 he taught physical sciences, English literature, and constitutional history at the Diocesan College of Rondebosch, South Africa. On his return to England Lloyd Morgan took the chair of geology and zoology at University College, Bristol, and stayed there for the rest of his professional career. In 1887 he was elected principal of the college and when a university charter was granted in 1909 he became the first vice-chancellor, although he held the position for only a few months. On resigning from it he returned to his studies as professor of psychology and ethics. He retired in 1919. In 1899 he became the first fellow of the Royal Society to be elected for work in psychology. He was also the first president of the psychological section of the British Association (Edinburgh, 1921); in 1910 he received the honorary D.Sc. from Bristol University. He married Emily Charlotte Maddock, the daughter of a vicar, and had two sons.

Lloyd Morgan's academic activity comprised work in geology and general science, comparative psychology, and philosophy. His geological writings include *Water and Its Teachings* (1882) and *Facts Around Us* (1884). He also wrote introductions to books on the geology of the Bristol region. His chief accomplishments, however, lie in the area of comparative psychology. Lloyd Morgan extended the work of G. J. Romanes and, together with E. L. Thorndike of the United States, helped to establish modern animal psychology. He was one of the first psychologists to recognize the need for an experimental as well as an observational approach to learning. Instead of using casual, recorded observations (the "anecdotal method" of Romanes), Lloyd Morgan resorted to rigorously controlled experiments.

Like Romanes, Lloyd Morgan relied on the concept of continuity in evolution as a justification for comparative psychology. He argued that because mind evolved from a lower to a higher mental state, the existence of the latter means that all others below it in the evolutionary scale also exist. To fathom the minds of animals, therefore, it is necessary to proceed from the lowest and simplest to the highest and most complex forms, rather than assuming human mental processes for all animals. A dictum embodying this basic prerequisite, "a law of parsimony," is now known as "Lloyd Morgan's canon." It states that "in no case is an animal activity to be interpreted as the outcome of the exercise of a higher psychical faculty, if it can be fairly interpreted as the outcome of his exercise of one which stands lower in the psychological scale" (*An Introduction to Comparative Psychology*, p. 59). This was a salutary warning; like his insistence that new levels of adaptive response are not necessarily the sum of simpler processes, it is still useful to recall.

Lloyd Morgan's literary output was astonishing. His experimental work, although not extensive, was nonetheless characterized by precise observations and vivid accounts of behavior. He advanced extremely cautious interpretations concerning instinctive behavior and its relationship to intelligence, and these appeared in *Animal Life and Intelligence* (1890–1891), *Animal Sketches* (1891), *An Introduction to Comparative Psychology* (1895), and *Animal Behavior* (1900). A more detailed consideration is in *Instinct and Experience* (1912). No one has written with more sense about the animal mind than Lloyd Morgan and although there is some disharmony and ambivalence in his writings, his contribution to psychology, especially in the area of methodology, is nevertheless important.

During this same period Lloyd Morgan published books on general biology and psychology; his influence spread to the United States, where he lectured in the 1890's.

Following his retirement Lloyd Morgan became primarily concerned with general philosophy and metaphysical speculation. He developed the theory of "emergent evolution," which maintained that evolution is not a steady, continuous process and that during it new properties suddenly emerge at certain levels of complexity. He developed this theory in a number of works—*Emergent Evolution* (1923), *Life, Mind and Spirit* (1926), *Mind at the Crossways* (1929), *The Animal Mind* (1930), and *The Emergence of Novelty* (1933).

BIBLIOGRAPHY

I. ORIGINAL WORKS. Lloyd Morgan published a great number of articles in journals of psychology and philosophy and numerous books based upon them. His works include *Water and Its Teachings in Chemistry, Physics and Physiography. A Suggestive Handbook* (London, 1882); *Facts Around Us: Simple Readings in Inorganic Science; with Experiments* (London, 1884); *Springs of Conduct; an Essay in Evolution* (London, 1885); *Animal Biology. An Elementary Textbook* (London, 1887; 2nd ed., 1889); *Animal Life and Intelligence* (London, 1890–1891); *Animal Sketches* (London, n.d. [1891], 1893); *Psychology for Teachers* (London, [1894], new ed., 1906); *An Introduction to Comparative Psychology* (London, 1895; 2nd ed., 1904); *Habit and Instinct* (London, 1896); *Animal Behavior* [rev. version of *Animal Life and Intelligence*] (London, 1900); *The Interpretation of Nature* (Bristol, 1905); *Instinct and Experience* (London, 1912); *Eugenics and Environment* (London, 1919); *Emergent Evolution* (London, 1923); *Life, Mind, and Spirit* (London, 1926); *Mind at the Crossways* (London, 1929); and *The Emergence of Novelty* (London, 1933).

II. SECONDARY LITERATURE. On the development of Lloyd Morgan's thought, especially concerning philosophic topics, see C. Murchison, ed., *A History of Psychology in Autobiography*, II (Worcester, Mass., 1932), 237–264. The best obituary notices are G. C. G., "Professor C. Lloyd Morgan 1852–1936," in *British Journal of Psychology*, **27** (1936), 1–3, with portrait; J. H. Parsons, "Conwy Lloyd Morgan 1852–1936," in *Obituary Notices of Fellows of the Royal Society of London*, **2** (1936–1938), 25–27, with portrait; and *Dictionary of National Biography 1931–1940*. There is an excellent account of Lloyd Morgan's contributions to psychology and philosophy in L. S. Hearnshaw, *A Short History of British Psychology 1840–1940* (London, 1964), 96–100. Briefer assessments are E. G. Boring, *A History of Experimental Psychology*, 2nd ed. (New York, 1957), 472–476 and 497–498; R. Watson, *The Great Psychologists* (Philadelphia, 1963), 296–298; and R. J. Herrnstein and E. G. Boring, eds., *A Source Book in the History of Psychology* (Cambridge, Mass., 1965), 462–468, which incorporates pp. 47–59 of Lloyd Morgan's *An Introduction to Comparative Psychology*.

EDWIN CLARKE

MORGAN, HERBERT ROLLO (*b.* Medford, Minnesota, 21 March 1875; *d.* Washington, D.C., 11 June 1957), *astronomy.*

Morgan was the son of Henry D. and Olive Sabre Smith Morgan. He received the B.A. from the

University of Virginia in 1899 and the Ph.D. in 1901. On 25 May 1904 he married Fannie Evelyn Wallis; they had one daughter. Morgan was a member of the American Astronomical Society (vice-president 1940–1942), American Geophysical Union, International Astronomical Union, American Association for the Advancement of Science (vice-president 1935–1936), and the Washington Academy of Science. He was president of the Commission on Meridian Astronomy of the International Astronomical Union from 1938 to 1948 and associate editor of *Astronomical Journal* from 1942 to 1948. He received the Watson Medal of the National Academy of Sciences in 1952 for his achievements in fundamental astronomy.

As a child Morgan suffered from an asthmatic condition, and at the age of nine to avoid the rigors of Minnesota winters his mother took him to Tennessee. He obtained his early education in a country school there whenever his bouts with asthma allowed him to attend. This intermittent schooling was supplemented by instruction at home under the guidance of his mother. When Morgan entered the University of Virginia, his primary interests were mathematics and astronomy. After one year, however, he had to withdraw in order to support his mother, who was growing old. At the university he had met Ormond Stone, director of the Leander McCormick Observatory. Stone took an interest in Morgan and helped him to obtain a Vanderbilt fellowship at the observatory. Stone's was probably the most important single influence in directing Morgan's interest toward classical astronomy.

With the aid of the fellowship, Morgan resumed his studies in 1896 and went on to receive the Ph.D. in 1901. During his last year of graduate study he taught mathematics at Pantops Academy, Charlottesville, Virginia, until receiving an appointment as a calculator at the U.S. Naval Observatory in Washington from 1901 to 1905. In 1905 he accepted an appointment as professor of astronomy and mathematics at Pritchett College, Glasgow, Missouri, and director of the Morrison Observatory. In 1907 he returned to the U.S. Naval Observatory, where he remained for the rest of his career. He started as an assistant astronomer on the staff of the nine-inch transit circle and by 1913 was in charge. For the next thirty-one years Morgan carried out a series of fundamental observations of the sun, moon, planets, and selected stars with this instrument. The resulting catalogs are milestones in fundamental astronomy and demonstrate Morgan's outstanding ability to analyze observations.

Morgan's earliest scientific papers dealt with the orbits of comets and asteroids. As the precise observations obtained with the nine-inch transit circle ac-

cumulated, Morgan turned his attention to the analysis of these observations and those from other observatories to obtain information on some of the fundamental constants on which astronomy is built. These analyses led to an extensive series of papers dealing with the position and motion of the equinox, the elements of the principal planets and their variations, and the constants of nutation and aberration. Although he formally retired in 1944, Morgan continued to work voluntarily at the U.S. Naval Observatory on his research, from 1947 to 1950 as a research associate under the auspices of Yale University.

It was during this period that Morgan produced what may be considered his most important work, "Catalog of 5,268 Standard Stars, 1950.0 Based on the Normal System N30." The N30 catalog is probably the most accurate source of positions and proper motions available today. It proved so useful in the interpretation of problems involving both astrophysical and astrometric data that Morgan was besieged with requests to extend the N30 proper motion system to a larger group of stars. Until several months before his death, he was engaged in deriving the proper motions on the N30 system of several hundred O- and B-type stars.

BIBLIOGRAPHY

Morgan's writings include "Results of Observations with the Nine-Inch Transit Circle, 1903–1911," in *Publications of the United States Naval Observatory*, 2nd ser., **9**, pt. 1 (1920), 1–452; "Observations made with the Nine-Inch Transit Circle, 1912–1913," *ibid.*, pt. 4 (1918), 1–116; "Results of Observations with the Nine-Inch Transit Circle, 1913–1926: Observations of the Sun, Moon, and Planets: Catalogue of 9,989 Standard and Intermediary Stars: Miscellaneous Stars," *ibid.*, **13** (1933), 1–228; "Results of Observations on the Nine-Inch Transit Circle, 1932–1934: Positions and Proper Motions of 1117 Reference Stars in Declination $-10°$ to $-20°$: Miscellaneous Stars," *ibid.*, **14**, pt. 2 (1938), 81–125; "Vertical Circle Observations made with the Five-Inch Alt-Azimuth Instrument, 1916–1933: Catalog of Declinations of Standard Stars: Declinations of the Sun, Mercury and Venus," *ibid.*, pt. 3 (1938), 127–216; "Proper Motions of 2916 Intermediary Stars, Mostly in Declination $-5°$ to $-30°$," *ibid.*, pt. 4 (1938), 217–283; "Results of Observations made with the Nine-Inch Transit Circle, 1935–1945: Observations of the Sun and Planets: Catalog of 5446 Stars: Corrections to GC and FK3," *ibid.*, **15**, pt. 5 (1948), 115–390; and "Catalog of 5,268 Standard Stars, 1950.0 based on the Normal System N30," which is *Astronomical Papers of the American Ephemeris*, **13**, pt. 3 (1952).

RAYNOR L. DUNCOMBE

MORGAN, THOMAS HUNT (*b.* Lexington, Kentucky, 25 September 1866; *d.* Pasadena, California, 4 December 1945), *embryology, genetics.*

Although known best for his studies in heredity with the small vinegar fly *Drosophila melanogaster* (often called fruit fly), Morgan contributed significantly to descriptive and experimental embryology, cytology, and, to a lesser extent, evolutionary theory. In recognition of his work in establishing the chromosome theory of heredity (the idea that genes are located in a linear array on chromosomes), Morgan was awarded the Nobel Prize in medicine or physiology for 1933.

The son of Charlton Hunt Morgan and the former Ellen Key Howard, Morgan came from two prominent family lines. His father had been American consul at Messina, Sicily, in the early 1860's and had given assistance to Giuseppe Garibaldi and his Red Shirts. John Hunt Morgan, Charlton's brother, was a colonel and later general in the Confederate Army and leader of his own guerrilla band, "Morgan's Raiders." His mother's maternal grandfather was Francis Scott Key, composer of the national anthem.

As a boy Morgan spent much time roaming the hills and countryside of rural Kentucky. His visits to his mother's family in western Maryland, provided the opportunity for further explorations during summers, and particularly for collecting fossils. He also worked for two summers in the Kentucky mountains with the U.S. Geological Survey. All of these activities gave Morgan an ease and familiarity with natural history which he retained throughout his life.

Morgan entered the preparatory department of the State College of Kentucky in 1880 and, after two years, the college itself (now the University of Kentucky). In 1886 he received a B.S., *summa cum laude,* in zoology. While an undergraduate Morgan was particularly influenced toward science by one of his teachers, A. R. Crandall, a geologist, and an undergraduate friend, Joseph H. Kastle. Kastle graduated two years ahead of Morgan and went to Johns Hopkins University in 1884 to do graduate work in chemistry. Perhaps through Kastle's influence, and because his mother's family lived in and around Baltimore, Morgan was attracted to Hopkins for graduate work. The summer before he entered graduate school (1886), Morgan went to the Boston Society of Natural History's marine biological station at Annisquam, Massachusetts. This was his first experience in working with marine organisms, an interest he was to continue throughout his life, primarily in association with the Marine Biological Laboratory, Woods Hole, Massachusetts.

At Hopkins, Morgan took courses in general biology, anatomy, and physiology with H. Newell Martin, a former student of Michael Foster and assistant to T. H. Huxley; anatomy with William N. Howard; and morphology and embryology with William Keith Brooks. He concentrated on morphology with Brooks. In 1890 he completed his doctoral work, on sea spiders, and received his Ph.D. He stayed on at Hopkins for a postdoctoral year on a Bruce fellowship; and in the fall of 1891 he went to Bryn Mawr College, where he remained until 1904, when E. B. Wilson offered him the chair of experimental zoology at Columbia. He was a member of the Columbia zoology department from 1904 until 1928, when he resigned to found the division of biological sciences at California Institute of Technology. He remained at Cal Tech and was active in scientific and administrative work until his death, after a short illness, in 1945.

During his academic life Morgan was involved not only in research and teaching but also in numerous professional organizations and activities. He was a member of the Genetics Society of America, the American Morphological Society (president, 1900), the American Society of Naturalists (president, 1909), the Society for Experimental Biology and Medicine (president, 1910–1912), and the American Association for the Advancement of Science (president, 1930). He also served as president of the Sixth International Conference on Genetics held in Ithaca, New York, in 1932. He was a member of the American Philosophical Society and the National Academy of Sciences (president, 1927–1931); through the National Academy he was intimately involved with the function of the National Research Council, especially in its formative years between 1921 and 1940. In addition to the Nobel Prize, Morgan received numerous scientific honors, including the Darwin Medal (1924) and the Copley Medal (1939) of the Royal Society.

In 1904 Morgan married Lilian Vaughan Sampson, one of his former students at Bryn Mawr. Lilian Morgan was a cytologist of considerable skill who always maintained an active interest in her husband's work. After the four Morgan children were in school, she returned to the laboratory and made important contributions to the *Drosophila* work.

Morgan was known to his friends, colleagues, and students as a man of quick mind, incisive judgment, and sparkling humor. While rarely showing his inner feelings, he nonetheless enjoyed people immensely and was a personal friend as well as teacher to many of his students. Frequently he paid the salaries of laboratory assistants from his own funds; and he shared his Nobel Prize money with his lifelong assistants and co-workers C. B. Bridges and A. H.

Sturtevant, to provide for the education of their children.

Morgan has been described as down-to-earth, practical, and sensitive. He retained an alert inquisitiveness and excitement for new ideas throughout his life. An extremely hard worker, Morgan pursued his scientific interests enthusiastically and relentlessly. He seldom took vacations, and used only one sabbatical during his twenty-four years at Columbia (in the year 1920–1921, when he went to Stanford University, where he continued his work in heredity and embryology). Despite his busy schedule and concentration on work, however, Morgan always found a small part of every day to spend with his family.

Early Scientific Work. As a student of W. K. Brooks, Morgan was trained as a morphologist—one who sought to discover evolutionary (phylogenetic) relationships among organisms by studying their comparative anatomy, embryology, cytology, and, to some extent, physiology. Morphologists relied heavily on descriptive and comparative methods, drawing their conclusions by analogy and inference. Such conclusions necessarily were highly speculative, because they could not be tested in any direct way. Brooks had been a student of Louis Agassiz and later Alexander Agassiz, and was thoroughly grounded in comparative anatomy and embryology, two of the hallmarks of late nineteenth-century descriptive biology. Through detailed studies of early and later embryonic stages of various groups of marine organisms, Brooks sought to elucidate phylogenetic relationships which were not apparent simply from examining the adult forms. Marine organisms seemed particularly important to Brooks, because he felt they were the oldest and most basic types of animals, and thus demonstrated most clearly the fundamental principles of animal organization. Like most morphologists, he viewed his own special subdiscipline, embryology, less as a field in its own right than as a tool for studying evolutionary relationships.

According to Bateson, Brooks taught his students to see subjects such as heredity not as completed axioms but rather as unsolved problems for further investigation. Brooks elucidated for them the interrelationships among such disparate areas of biology as heredity, anatomy, embryology, cytology, and evolution. He had, according to Morgan, a wide-ranging and philosophical mind, which, if not always rigorous, was at least provocative.

Morgan's doctoral dissertation under Brooks involved a study of four species of marine invertebrates, the Pycnogonida (sea spiders), focusing largely, but not exclusively, on their comparative embryology. The purpose of this study was to determine whether the Pycnogonida belonged to the Arachnida (a group including spiders and scorpions) or to the Crustacea (including crabs, lobsters, and crayfish). Observing both large anatomical and smaller cellular changes during embryogenesis, Morgan found that the pattern of development more closely resembled that of the Arachnida than that of the Crustacea. He continued this line of work during the first several years after leaving Hopkins, extending his investigations of early embryology to other forms such as *Balanoglossus* and the ascidians (both primitive chordates). Morgan had, however, become increasingly dissatisfied with morphology during his graduate days; he objected to the subordination of disciplines such as embryology almost exclusively to phylogenetic and evolutionary problems. Increasingly, he saw embryology and other disciplines as having their own sets of problems for study; moreover, he felt that an experimental approach to problem-solving would make it possible to draw more firm and rigorous conclusions than the inferences and speculations that characterized morphology.

Several factors contributed to Morgan's growing disaffection with the morphological tradition. The first was perhaps his association with the physiologist H. Newell Martin (head of the biology department at Hopkins),[1] an emphatic and vocal exponent of the experimental method. Following Michael Foster's lead, he had introduced experimental teaching laboratories when he came to Hopkins; and he made it clear from the outset that he regarded physiology as the queen of the sciences, with morphology as its servant.[2] A second factor was Morgan's early acquaintance with Jacques Loeb. Both joined the faculty of Bryn Mawr in the same year (1891) and maintained a lifelong friendship. Loeb was a strong proponent of the mechanistic conception of life. He believed that (1) organisms function in accordance with the laws of physics and chemistry, so that to understand living phenomena, it is necessary to approach them from a physicochemical standpoint; and (2) only quantitative and experimental methods would allow biologists to get at the fundamental chemical and physical processes involved with life. These methods, in contrast with those of descriptive biologists, would yield rigorous and testable conclusions. Loeb believed that biologists should emulate the methods used in the physical sciences. Loeb's views no doubt strongly influenced Morgan at a time when the latter was beginning to turn away, on his own accord, from descriptive methods.

A third, and perhaps crucial, factor which may have caused Morgan to embrace the experimental approach was his association with Hans Driesch, his colleague in 1894–1895 at the zoological station in Naples. Driesch was at the time an enthusiastic proponent of experimental embryology (the school of *Entwicklungsmechanik*) and had performed some highly controversial experiments on sea urchin eggs. Morgan and Driesch collaborated on experimental studies of development in Ctenophora (published in 1895).[3] Not only was Driesch's influence important, but so was that of the zoological station itself. Morgan had visited the station first in 1890 and had become intrigued with the many possibilities that the institute offered for research on marine forms. During his ten months at Naples, he was excited by the work, the constant stream of visitors, the exchange of ideas, and the emphasis on new modes of thought, such as performing experiments in areas of biology, like embryology, previously approached only descriptively. He wrote in 1896: "No one can fail to be impressed [at the Naples Station] and to learn much in the clash of thought and criticism that must be present where such diverse elements come together."[4] By contrast, Morgan found the situation in America more parochial and less exciting: "Isolated as we are in America, from much of the newer current feeling, we are able at Naples as in no other laboratory in the world to get in touch with the best modern work."[5]

After he returned to the United States in 1895, Morgan's biological interests expanded in scope; and his research methods became largely experimental for the remainder of his life. Between 1895 and 1902 he focused on experimental embryology; between 1903 and 1910, on evolution, especially heredity and cytology in relation to sex determination; between 1910 and 1925, on problems of heredity in *Drosophila*; and from 1925 to 1945, on embryology and its relations to heredity and evolution. Yet in none of these periods was Morgan exclusively concerned with a single subject. The breadth of his interests was such that he always worked simultaneously on several problems, often of a divergent nature. At almost any point in his career he moved back and forth between the broad areas of evolution, heredity, and development with considerable ease and grasp of fundamental concepts.

Embryological Studies (1895–1902). Morgan's earliest work in experimental embryology largely concerned the factors influencing normal embryonic development. These studies were motivated by the controversy raging in the early 1890's between Driesch and the founder of the *Entwicklungsmechanik* school, Wilhelm Roux, on the question of whether the differentiation of embryonic cells is directed by internal (hereditary) or external (environmental) forces. Morgan studied

fertilization of egg fragments, both nucleated and nonnucleated, in the sea urchin and in amphioxus. Both types of fragments were able to undergo varying degrees of normal development and even to produce partial larvae. Morgan carried out other studies in which he removed cells from normally fertilized blastulae to produce embryos which, although modified, still developed along the major outlines of their normal course. Other experiments during the same period involved the effects of various salt solutions and of the force of gravity (or lack of it) on the course of development in the eggs of sea urchins, mollusks, and teleost fishes. Beginning in 1902 he published an extensive series of papers on normal and abnormal development in the frog's egg. Here, Morgan tested the effects of such factors as injury to the egg yolk; varying concentrations of lithium chloride; and injuries to the embryo at various stages, including repetition of Roux's experiment involving injury to the first blastomere. The results of all these experiments showed Morgan that despite the alterations in development which could be brought about by various physical constraints, the embryo still displayed a tendency to reach its prescribed goal. It became clear to him that environmental influences might shape the embryo's development within certain limits, but that the overriding factors determining the sequence of events in development must lie within the embryo itself: the interaction of embryonic tissues and of specific embryonic regions with each other.

Coupled with Morgan's interest in early embryonic development was a corresponding interest in the regeneration of lost or injured tissues (or organs) in adults. While still a student at Hopkins, he had studied regeneration in the earthworm; and in the late 1890's he pursued these studies in flatworms (*Planaria* and *Bipalium*); jellyfish (*Gonionemus*); bony fishes (teleosts); and ciliate protozoa (*Stentor*). In 1901 he published his first major book, *Regeneration*, a compendium of contemporary information on this subject. More than simply a review of the literature, *Regeneration* provided a foretaste of Morgan's writing and analytical skill. He saw that the events in regeneration (regrouping of cells in the wound area, despecialization, and renewed differentiation) were the other side of the coin from those of early embryonic development. In regeneration there was a return to the embryonic state. The same essential questions lay behind both processes: How could different components of a cell's hereditary information be signaled to turn on or off at different periods in its life? Morgan emphasized the relationship between the two processes (he was not alone in making the connection); he saw that any explanation for one must be able to account for the other.

As in most of his later writings, Morgan presented the problem of regeneration as one composed largely of questions—of unknowns—rather than of knowns. He made clear the gaps in contemporary knowledge, in terms of specific experiments or broad interpretations. Morgan sought an understanding of problems such as regeneration (or embryonic differentiation) in terms of underlying (and continuing) processes. He was not content to "explain" one event simply by describing the events or organizational relationships which might lead to it. For example, he felt that those who saw embryonic differentiation as the result of "formative stuffs" already organized in the cytoplasm of the unfertilized egg, or regeneration as simply the work of special cells which congregate at a wound site, really explained nothing. They simply pushed the causal factors back to a further point in the organism's life history.

It was important to Morgan to view such phenomena less as series of events than as processes. These processes were chemical and physical, and they followed regular laws—if only one could discover what they were. Development and regeneration were to some extent programmed events, but they were not simply the unfolding of preexisting structures. Specific interactions were programmed between structures that gave rise to new and qualitatively different structures. The job of the developmental biologist, he argued, was to seek the general laws governing these interactions. This discovery could not come about simply by describing anatomy—it required experimental analysis as well.

Study of Sex Determination (1903–1910). In the latter part of the nineteenth and the first years of the twentieth centuries there were two schools of thought on the problem of sex determination. One maintained that the causal factors were environmental: temperature, or amount of food available to the embryo or to the mother during development. This argument derived from the observation that changes in various environmental factors affected the sex ratio in many species, particularly insects. Another school, however, felt that sex was by and large determined at the moment of fertilization, or perhaps even before, by factors internal to the egg or sperm or both. This school emphasized the hereditary, as opposed to environmental, factors in determining sex differentiation.

After 1900 there were several attempts by those favoring the hereditary view to understand sex in terms of the newly discovered Mendelian principles. In 1903 Morgan published a review of the sex deter-

mination problem, criticizing all of the existing theories, including those based on Mendel's laws. His major argument was that there was relatively little evidence substantiating the claims of either the environmentalists or the hereditarians. Most of the current theories of sex determination tried to explain only the customary 1:1 sex ratio found in most species. Any theory of sex determination, however, had to account for a number of other phenomena, such as the process of parthenogenesis, either natural or artificially induced; the appearance of gynandromorphs, often observed in insects (in gynandromorphs, one half of the organism has male characteristics and the other half female characteristics); and sex reversals, as observed in fowl and other species, especially under the influence of hormonal changes.

In his analysis of the sex determination problem, Morgan displayed his deep-rooted embryological bias. He was unwilling to see sex as a primarily hereditary phenomenon, determined at the moment of fertilization, but, rather, he analyzed it as a developmental process, guided by natural laws; he was clearly an epigenesist. He found most of the environmentalists' experiments inconclusive, but this did not mean that sex could be explained by postulating hereditary units, such as Mendel's "factors," or by reference to visible cell structures, such as chromosomes. To Morgan, structures such as chromosomes were only indicators of underlying processes—they were not causal factors themselves. For this reason he was not initially sympathetic to C. E. McClung's suggestion in 1901 and 1902 that sex was determined by the disposition of the accessory (or X) chromosome. Before 1910 Morgan admitted only that the fertilized egg might inherit a predisposition toward maleness or femaleness. The realization of that sexual potential, however, was largely a result of the same developmental forces involved in differentiation, organogenesis, and regeneration.

Through his interest in sex determination, Morgan carried out important cytological studies on the movement and disposition of chromosomes during the formation of eggs of naturally parthenogenetic forms. Studying in detail two kinds of insects, the aphids and phylloxerans, Morgan was able to demonstrate conclusively that the production of parthenogenetic males was associated with the loss of a chromosome during development from a diploid egg. His papers on this subject, published in 1909 and 1910, show the beginning of Morgan's realization that chromosomes might actually be related to sex determination.[6] He did not conclude at the time, however, that the accessory chromosome (X) was a sex determiner. Morgan maintained that the real sex-determining process occurred *before* the actual loss of the chromosome; the latter was only an indication of this process, not the cause. He wrote in 1909: "The preliminaries of the sex determination for both sexes go on in the presence of all chromosomes . . . clearly I think the results show that changes of profound importance may take place without change in the number of chromosomes."[7]

Evidence had been accumulating since the 1870's that the chromosomes were somehow intimately involved with general hereditary processes. Morgan had remained skeptical of such conclusions, however, not only because the idea had been inferred from circumstantial evidence but also because of his bias against explaining phenomena in terms of preexisting structures. Yet shortly after 1910 increasing experimental evidence led him to change his mind and to accept the chromosomes as important hereditary structures. It was largely work on sex determination that brought Morgan to accept these new ideas. His own studies on chromosomes in aphids and phylloxerans suggested that more attention ought to be paid to the possible role of chromosomes in determining sex. At the same time, between 1901 and 1905, E. B. Wilson, Morgan's colleague at Columbia, and Nettie M. Stevens, at Bryn Mawr, amassed considerable evidence suggesting that the accessory (X) chromosome was responsible for sex determination. Although Morgan did not accept these findings unequivocally, Wilson's concern for the hereditary implications of these chromosome studies strongly influenced Morgan's ideas about sex determination. Morgan and Wilson had been close friends and colleagues for many years and Morgan had great respect for Wilson's judgment.

Evolution and Heredity (1903–1910). Morgan had become interested in the Darwinian theory of natural selection, through the influence of W. K. Brooks and through his own studies on regeneration. He reported that he constantly wondered how the regenerative power in higher organisms could have evolved by a mechanism such as natural selection. In 1903 Morgan published *Evolution and Adaptation* (dedicated to W. K. Brooks), a lengthy attack on the Darwinian theory of natural selection as it was interpreted around the turn of the century by the neo-Darwinians. Morgan believed that Darwin himself was an outstanding naturalist who approached his conclusions with caution, reasoning only within narrow bounds from the data itself. He felt, however, that many of Darwin's followers had become "ultra selectionists," investing natural selection with more powers than was legitimate. While maintaining that evolution was a fact, Morgan argued that the theory of the mechanism

by which evolution was brought about—natural selection—had many loopholes.

Morgan's many objections to natural selection have been discussed in detail in the secondary literature; but one major criticism deserves mention here. Morgan shared with many prominent biologists (especially embryologists) the view that the Darwinian theory (as stated by Darwin or modified in the 1890's by his followers) was incomplete because it lacked a concept of heredity. Although Darwin had emphasized that selection acts on slight individual variations (what some people at the time came to call "continuous variations"), more recent evidence had suggested that such variations were not usually heritable. It was a cardinal principle to Darwin and the neo-Darwinians that the only variations upon which selection could act were hereditary ones. Thus, Morgan and many of the less orthodox Darwinians came to believe that variations of evolutionary significance must be large-scale, or discontinuous, because these were the only ones which appeared to be inherited. Morgan maintained that in the face of this dilemma, the neo-Darwinians, rather than abandoning the idea of small, individual variations as the raw material for evolution, interpreted selection itself as the creative agent. Morgan believed that selection was only a negative factor, however, which sorted out the favorable from the unfavorable variations already present. It could not, as some neo-Darwinians believed, create new variations in the germ plasm.

Morgan's view of evolution was like his view of heredity and development, in that it was fashioned by a skepticism about single answers or mechanisms for which experimental proof was inconclusive. From his graduate days on, he felt that heredity was in some ways central to an understanding of all biological phenomena, especially development and evolution. Recognizing the lack of what seemed to be any coherent theory of heredity in the period before 1910, he was skeptical of any attempts to explain processes such as cell differentiation or the origin of species by analogies, inferences, or speculative hypotheses. A change in Morgan's ideas led him to the dramatic discoveries with *Drosophila*. A brief examination of his ideas on heredity, especially in relation to cytology and evolution between 1900 and 1910, will be useful in understanding this change.

Three concepts of heredity, representing several lines of reasoning and experimentation, had become well-known to most biologists by the first decade of the twentieth century. The first of these was the newly discovered Mendelian laws, based on data from plant-breeding experiments. The second was the chromosome theory of heredity, based on cytological studies of chromosome movement during gametogenesis in both animals and plants. Third was the publication of *The Mutation Theory*, a monumental treatise on heredity, variation, and evolution by the Dutch botanist Hugo de Vries.

Although by 1902 several workers had suggested the possible relationship between chromosome movements and the segregation of Mendel's alternate factors, there was no agreement that this relationship was anything more than coincidental. Morgan's objections to the Mendelian scheme can be summarized as follows:

1. If the Mendelian theory were correct, and if Mendelian "factors" (what Mendel more commonly called *Anlagen*, and what later became known as genes) were actually associated with chromosomes, then breeding results ought to show large groups of characteristics inherited together (as many groups as there were chromosome pairs). Because few "linkage groups" had been observed in the period before 1910 (Bateson and Punett in England had shown some in 1905 and 1906), the identification of Mendel's factors with chromosomes seemed less than likely.

2. The results of animal- and plant-breeding experiments showed that many characteristics in an offspring were a mixture of parental types, and not simple dominance or recessiveness. Thus, Mendel's "laws" might apply only to special, exceptional cases.

3. The Mendelian theory of dominance and recessiveness could not explain the normal 1:1 sex ratio. According to Mendel's scheme, the sex ratio would be 3:1 (if one sex factor were dominant over the other) or 1:2:1 (if incomplete dominance were involved). Since neither sex ratio occurred in nature, Mendel's laws provided no clear way to account for the important phenomenon of sex inheritance.

4. On methodological grounds, Mendel's laws called for too neat a set of categories among the offspring of any cross. Since such categories seldom occurred in nature, Morgan claimed that Mendelians often placed borderline organisms into whichever category was necessary to give the expected ratios.

5. On a more philosophical level, both the Mendelian and the chromosome theories seemed to be preformationist in character; they referred basic hereditary characteristics to preexisting particles or units in the cell. Morgan felt that, like all preformationist theories of the past, the Mendelian and chromosome doctrines simply pushed a basic problem back further in the life history of the organism.

6. In addition, the Mendelian and chromosome theories seemed to Morgan to be based too much on speculation, and too little on sound experimental evidence. They reminded him of the speculative theories—especially those of Ernst Haeckel and August Weismann—that attempted to explain all of biology that abounded during his student years. Morgan was inalterably opposed to speculation that could not be subjected to experimental tests.

Skeptical of both the Mendelian and the chromosome theories, Morgan was, however, an outspoken advocate of de Vries's mutation theory (published in a two-volume work between 1901 and 1903). De Vries proposed that large-scale heritable variations occurring in one generation could produce offspring that were of species different from their parents. De Vries's evidence was based largely on experiments with the evening primrose *(Oenothera lamarckiana)*. What he called "mutations" are now known to be the result not of actual changes in genetic material, but complex chromosome arrangements which are peculiar to *Oenothera*. Thus they did not produce species-level changes in a single generation, as de Vries claimed. Nevertheless, the mutation theory is historically important, for Morgan and others saw in it, as did de Vries himself, an answer to the perplexities of Mendelian heredity and Darwinian selection. It accounted for the origin of new variations which were definite enough to be of evolutionary significance (that is, would not be lost by swamping), and yet were also heritable. Furthermore, Morgan's acceptance of the mutation theory was influenced by the sound experimental evidence behind de Vries's work. De Vries had a large experimental garden where he grew his plants and made crosses under carefully controlled conditions. New mutants could be isolated and shown to breed true. Thus de Vries not only provided a new concept that made evolution conceivable; he also provided an experimental approach by which his conclusions could be tested.

Morgan's Work With Drosophila. Morgan appears to have begun breeding the fruit fly *Drosophila melanogaster* somewhere around 1908 or 1909. It is not clear how he came to use this organism, or where he obtained his original cultures. *Drosophila* seems to have been an organism favorable for laboratory studies, however, between 1900–1910. It was used in Castle's laboratory at the Bussey Institution (Harvard) as early as 1900–1901; by W. J. Moenkhaus at Indiana in 1903; by F. E. Lutz at the Carnegie Institution Laboratory (Cold Spring Harbor, New York, and after 1909, when Lutz was at the American Museum of Natural History); by Nettie Stevens at Bryn Mawr in 1906; and by Fernandus Payne and L. S. Quackenbush in the Columbia laboratory itself prior to 1909. Morgan's original purpose had been to test de Vries's mutation theory in animals. He exposed *Drosophila* cultures to radium in an attempt to induce the formation of new mutants, but he never obtained mutations of the magnitude which de Vries claimed for *Oenothera*.

In 1910 Morgan discovered a small, distinct variation in one male fly in one of his culture bottles. This fly had white, as opposed to the normal (wild-type) red, eye color. This variation did not make a new species, but Morgan thought he would try to breed the fly with its red-eyed sisters to see what would happen. All of the offspring (F_1) were red-eyed. Brother-sister matings among the F_1 generation produced a second generation (F_2) with some white-eyed flies—all of which, Morgan noticed with astonishment, were males. Further matings showed that while the white-eye condition almost always occurred in males, occasionally a white-eyed female would appear. Morgan noted that the white-eye and red-eye conditions behaved as typical Mendelian factors, with red being dominant over white.

The limitation of the white-eye condition largely, but not exclusively, to males presented a very curious problem. Morgan found that the only way he could explain this phenomenon was to assume that the red- and white-eye conditions were determined by Mendelian factors, and that these somehow associated with the element which determined sex in the cell. In his first paper on heredity in *Drosophila*, Morgan refrained from identifying the eye color with chromosomes in general, or the accessory chromosomes in particular.[8] Within a year, however, he concluded that such caution was unwarranted. The cytological studies on chromosomes and sex determination by Wilson and others, and his own work with *Drosophila*, convinced Morgan that chromosomes could in fact be the real bearers of Mendelian factors. Much to his credit, he rejected his skepticism about both the Mendelian and the chromosome theories when he saw from two independent lines of evidence (breeding experiments and cytology) that one could be treated in terms of the other.

Morgan called the white-eye condition sex-limited (later sex-linked), meaning that the genes for this character were carried on (linked to) the X chromosome. Sex-linked genes, if recessive to their wild-type alleles, will show up almost exclusively in males, who do not have a second X chromosome to mask genes on the first. Sex linkage was found to hold for all sexually reproducing organisms and accounted for many other perplexing hereditary patterns, including red-green color blindness and hemophilia in man. Morgan's *Drosophila* work showed for the first time the clear association of one or more hereditary characters with a specific chromosome.

Early in 1910 Morgan had taken into his laboratory several enthusiastic Columbia undergraduates: A. H. Sturtevant and Calvin B. Bridges, both juniors in the college and Hermann J. Muller, a graduate student of E. B. Wilson's. With Morgan these men quickly developed the *Drosophila* work into an intensively active project. As more breeding experiments were initiated, new mutants began to appear. Careful records were kept of the mutants, and their hereditary patterns were studied through various crosses and backcrosses. It would be impossible to describe or list all of the new findings which emerged from the *Drosophila* studies. A few major developments will illustrate the enormous breakthroughs which Morgan and his colleagues were able to make with this new experimental organism.

At first the relationship between Mendelian genes and chromosomes was purely inferential. While it was not possible to make that relationship more concrete (no one could "see" a gene on a chromosome), a means appeared by which the inference could be tested. In 1909 the Belgian cytologist F. A. Janssens had published a careful series of cytological observations of what he called chiasmatype formation (intertwining of chromosomes during meiosis).[9] Janssens believed he could show that occasionally homologous chromosome strands exchanged parts during chiasma. Morgan was familiar with Janssens' concept and applied it to the conception of genes as parts of chromosomes. He reasoned that the strength of linkage between any two factors must be related in some way to their distances apart on the chromosome. The farther apart any two genes, the more likely that a break could occur somewhere between them, and hence the more likely that the linkage relationship would be disturbed. During a conversation with Morgan in 1911, Sturtevant, then still an undergraduate, suddenly realized that the variations in strength of linkage could be used as a means of determining the relative spatial distances of genes on a chromosome. According to Sturtevant's own report, he went home that night and produced the first genetic map in *Drosophila* for the sex-linked genes y, w, z, m, and r. The order and relative spacing which Sturtevant determined at that time are essentially the same as those appearing on the recent standard map of *Drosophila*'s X chromosome.

Following the initial success of this technique, positions were determined for many other genes. The *Drosophila* group depended upon the appearance of mutants to determine the existence and chromosomal location of specific genes. Thus the initial work of the group took two directions: the location of mutants and the maintenance of a stock for each mutant (or group of mutants), and the mapping of these mutant gene positions on the appropriate chromosomes. The success of the mapping technique added further weight to the inferred relationship between genes and chromosomes and at the same time provided an increasingly clear picture of the architecture of the germ plasm. The major outcome of the mapping work was the idea that genes are arranged in a linear fashion and occupy specific positions, or loci, on the chromosomes. While the direct and final proof of this relationship had to wait until proper cytological materials (the giant salivary glands of *Drosophila*) and techniques were developed by T. S. Painter and others in the 1930's, the mapping work firmly established the inference in the years between 1912 and 1915.

As the work progressed, other problems arose. Genes were discovered which, when combined in the homozygous condition, caused the embryo to die before birth (so-called lethal genes). Various traits proved to be determined by a number of alternative genes (alleles) at the same locus, which could be combined in various forms to give a series of phenotypes (multiple alleles). Because crossover frequencies did not always turn out as predicted, they arrived at the idea of crossover interference, in which segments of a homologous chromosome pair showed little or no crossing over, often as the result of alterations in chromosome structure which prevented normal intertwining during chiasma. A furor among orthodox Mendelians was aroused by Sturtevant's suggestion that the expression of a given gene was affected by its position on the chromosome (the "position effect"). Position effect became the target of one of the most persistent attacks on the Mendelian and chromosome theories to be launched in the twentieth century, by Richard Goldschmidt, for many years director of the Kaiser Wilhelm Institute for Biology in Berlin–Dahlem. Goldschmidt argued that the suggestion that a gene's effect could be modified by a change in its position along the chromosome (that is, by what genes were on either side of it) violated the basic Mendelian conception of the purity of the gametes. The necessity of invoking a hypothesis such as position effect was, to Goldschmidt, tantamount to an admission that the Mendelian and chromosome theories were not compatible, and that a new conception had to be substituted. Yet position effect and its cytological basis, as worked out in the 1930's by Muller, Prokofieva, Bridges, and others, proved to be a valid conception—a modification, if not a contradiction, of orthodox Mendelian theory.

Among the most important ideas to emerge from the *Drosophila* work was the balance concept of sex, developed largely by Bridges between 1913–1925

through an analysis of the cytological phenomenon of nondisjunction. Nondisjunction is a condition occurring during oogenesis, in which the X chromosomes fail to segregate, so that a haploid egg may end up with two X chromosomes. Bridges' work of 1916, in particular, showed clearly that sex was determined not simply by the inheritance of one or two X chromosomes but, rather, by the ratio of X chromosomes to autosomes (the other, nonsex chromosomes in the nucleus). According to this idea, organisms could inherit various degrees of sexuality based upon variations in this ratio. The genes governing male and female characteristics (such as production of testes or ovaries) are found in both sexes and apparently are not located exclusively on the sex chromosomes but throughout the genome. Which of these sets of genes express themselves is a result not simply of their presence or absence but, rather, of some complex and little-understood relationship between the sex chromosomes and autosomes.

The major early findings of the *Drosophila* group were summarized in an epoch-making book, *The Mechanism of Mendelian Heredity*, published by Morgan, Bridges, Sturtevant, and Muller in 1915. They presented evidence to suggest that genes were linearly arranged on chromosomes and that it was possible to regard the Mendelian laws as based on observable events taking place in cells. Most important, however, they demonstrated that heredity could be treated quantitatively and rigorously. For almost the first time since the advent of experimental embryology in the 1880's, a previously descriptive area of biology had proved itself accessible to quantitative and experimental methods. Through *The Mechanism of Mendelian Heredity*, the new science of genetics reached many teachers, students, and specialists in other areas.

All of the early work on *Drosophila* between 1910 and 1925 was carried out in the winter in Morgan's small laboratory space, called the "fly room," at Columbia, and during the summers at the Marine Biological Laboratory in Woods Hole, Massachusetts. Although Morgan was considerably older than his co-workers, there was a give-and-take atmosphere in the "fly room" that precluded formal barriers and rigid teacher-student distinctions. There was little consideration of priority in new ideas or discoveries at the time (although some did emerge in later years); and each was free to criticize anyone else openly, and sometimes vehemently. Sturtevant has described the relationship among the workers in the "fly room" as follows:

> As each new result or new idea came along, it was discussed freely by the group. The published accounts

do not always indicate the source of ideas. It was often not only impossible to say but was felt to be unimportant, who first had an idea. A few examples come to mind. The original chromosome map made use of a value represented by the number of recombinations divided by the number of parental types as a measure of distances; it was Muller who suggested the simpler and more convenient percentage, the recombinance formed of the whole population. The idea that "crossover reducers" might be due to inversions of sections was first suggested by Morgan, and this does not appear in my published account of the hypothesis. I first suggested to Muller that lethals might be used to give an objective measure of the frequency of mutation. These are isolated examples, but they represent what was going on all the time. I think we came out somewhere near even in this give and take, and it certainly accelerated the work.[10]

However, all was not idyllic within the *Drosophila* group. H. J. Muller, perhaps Morgan's most independent and brilliant student, felt that Morgan had a tendency to use his students' ideas without fully acknowledging them. While recognizing Morgan's unsurpassed abilities as a leader, his fiery and quick imagination, and his frequently penetrating insights, Muller claimed that Morgan was frequently confused about rather fundamental issues involved in the work—such as the theory of modifier genes, or the supposed swamping effect of dominant genes in a population. According to Muller, Morgan frequently had to be "straightened out" on such issues by hardheaded arguments with his students—mostly Muller and Sturtevant, with occasional help from E. B. Wilson. Sturtevant concurs with this evaluation at least with regard to the idea of natural selection, which he claims Morgan persisted in misunderstanding until as late as 1914 or 1915.

What is clear from an analysis of the reports of many people who worked in the "fly room" during the years 1911–1915, was that Morgan's primary role was that of leader and stimulator. He was constantly coming up with ideas—some wrong, others right—and throwing these out to the eager and brilliant group of young people whom he had working with him. That many of the most far-reaching ideas (such as a quantitative method of making chromosome maps, crossover interference, modifier genes) were first proposed by his students, not directly by Morgan, is also clear. His genius in the development of the *Drosophila* work may have rested more in bringing together the right group of people, in working together with them in a democratic and informal way, and in letting them alone, than in producing all the major ideas himself. In fact, it is clear from an analysis of Morgan's published work that he frequently proposed ideas "off the top of his head" and was not always careful to work out their details or implications.

Morgan's laboratory became the training ground for a school of Mendelian genetics—one generation of which emphasized particularly the relationship between genes and chromosomes. Besides Bridges, Sturtevant, and Muller, Morgan's students or postdoctoral associates at Columbia included Alexander Weinstein, E. G. Anderson, H. H. Plough, Theodosius Dobzhansky, L. C. Dunn, Donald Lancefield, Curt Stern, and Otto Mohr. These workers, and many others, developed what has come to be called "classical genetics"—that is, genetics at the chromosome level.

Morgan's mind ranged freely over the broad areas of genetics, embryology, cytology, and evolution. Soon after the *Drosophila* work had gotten under way, he saw that the Mendelian concept could throw considerable light on the problem of natural selection. In 1916 Morgan published his second major work on evolution, *A Critique of the Theory of Evolution* (revised in 1925 as *Evolution and Genetics*), showing clearly his altered views about Darwinian selection. Although he had previously regarded de Vries's mutation theory as an alternative to natural selection, Mendelism now provided a mechanism for understanding the Darwinian theory itself. Mendelian variations (called also "mutations" by Morgan) were not as large or as drastic as those postulated by de Vries. Yet they were more distinct and discontinuous than the slight individual variations which Darwin had emphasized. Most important, they could be shown to be inherited in a definite pattern and were therefore subject to the effects of selection. The Mendelian theory filled the gap which Darwin had left open so long before.

Morgan found it more difficult to make explicit the relationships which he instinctively knew existed between the new science of heredity and the old problems of development (such as cell differentiation or regeneration). In 1934 Morgan attempted to make these connections in a book titled *Embryology and Genetics*. The work proved to be less an analysis of interrelated mechanisms and more a summary of efforts in the two separate fields. Morgan knew well that the time was not ripe for understanding such problems as how gene action could be controlled during development. Yet *Embryology and Genetics* served an important function of keeping before biologists the idea that ultimately any theory of heredity had to account for the problem of embryonic differentiation. Morgan wisely refrained from drawing conclusions or proposing hypotheses which could not be experimentally verified. One of the most important characteristics of his genius was the ability to restrict the number and kinds of questions which he asked at any one time. For example, by focusing primarily upon the relationships between the Mendelian theory and chromosome structure, he was able to work out the chromosome theory of heredity in great detail. In contrast, other workers, such as Richard Goldschmidt, tried to make those relationships more explicit than the evidence at the time would allow. Consequently, they were often drawn into realms of speculation where no concrete advances could be made.

Later Work (1925–1945). After the mid-1920's Morgan's interest shifted away from the specific *Drosophila* work. His new concerns took two forms. One was the attempt to summarize the conclusions deriving from his genetic studies. To this category belong those broader works relating heredity to development and evolution. His other interest turned him to some of the original problems of development and regeneration which had launched his career thirty-five years previously. During the summers at Woods Hole, and especially after his move to California in 1928, Morgan returned to studies of early embryonic development. The cleavage of eggs; the effects of centrifuging eggs before and after fertilization; the behavior of spindles in cell division; preorganization in the egg; self-sterility in ascidians; and the factors affecting normal and abnormal development were some of the problems in experimental embryology. They represented the type of biological work that Morgan was most interested in. Although he approached the *Drosophila* studies enthusiastically, the mathematics of mapping and many other highly technical problems were less interesting to him than working directly with living organisms. Morgan had the naturalist's love of whole organisms and of studying organisms in their natural environment. He was a good naturalist with a knowledge of many species.[11] His strong interest in laboratory and experimental work in no way detracted from his interest in whole systems. He was not in spirit a mechanist, although he recognized the value of studying systems in isolation to obtain rigorous and useful data.

Methodology in Science. Being a thorough experimentalist, Morgan saw that unbounded speculation was detrimental to the development of sound scientific ideas. He did not object to the formulation of hypotheses for he saw them as essential to developing new concepts and experimental ideas. For Morgan, however, the only acceptable hypotheses were those which suggested experimental tests.

Yet Morgan was not a mere empiricist—that is,

one who simply tries to amass large amounts of basically similar kinds of evidence before drawing a conclusion. As an experimentalist he drew conclusions most readily when several different types of data sources were available (for example, determining the existence of a chromosomal deletion from breeding data and from cytological examination of chromosome preparations). By 1909 considerable breeding data suggested that Mendel's laws had wide application. Yet Morgan remained skeptical because there was no evidence (in his mind) that Mendel's "factors" had any reality. What began to change his mind was not the finding that he could apply the Mendelian theory to yet another organism (*Drosophila*), but that he could test the Mendelian theory (studied by breeding experiments) with evidence from a wholly different area—cytology—in the observed behavior of chromosomes during gametogenesis. As soon as he saw that the white-eye mutation acted as if it were part of the X chromosome, he began to view the Mendelian theory in a completely different light.

That Morgan saw Mendel's "factors" as having a possible material basis in the chromosome does not imply that he automatically accepted the idea that genes were physical entities; nor was he primarily concerned with determining how much of the chromosome a mutant gene occupied whenever a new mutation was discovered. The physical existence of genes was unnecessary for the validity of the original Mendelian theory and for much of the *Drosophila* work. The Mendelian-chromosome theory was largely a formalism: it stood on its own as a consistent scheme without necessarily being tied to observable physical structures. Until cytological techniques materials were developed by Painter and others in the late 1920's, it was impossible to determine a point-by-point correspondence between genetic maps (determined by crossover frequencies) and chromosome structure (determined cytologically). Nevertheless, from the outset Morgan was never content to deal with a purely formalistic theory. In the preface to *The Mechanism of Mendelian Heredity*, he and his coauthors admitted that Mendel's theories could be viewed independently of chromosomes. But they hastened to point out that this was not the course they were going to follow:

> Why then, we are often asked, do you drag in the chromosomes? Our answer is that since the chromosomes furnish exactly the kind of mechanism that the Mendelian laws call for; and since there is an ever-increasing body of information that points clearly to the chromosomes as the bearers of the Mendelian factors, it would be folly to close one's eyes to so patent a relation.[12]

Preliminary evidence suggested that genes were real entities on chromosomes, even though it could not be proved conclusively.

As an experimentalist Morgan urged other biologists to employ the quantitative and rigorous methodology which had been so successful in experimental embryology and in his own work on heredity. For biology to attain the same level of development as the physical sciences, it was necessary to adopt the same standards. Yet Morgan did not believe that biology should be reduced simply to expressions of physical and chemical interactions. He believed too much in the naturalist's view of living systems. Reductionism was too simplistic for Morgan; he could never follow Loeb to the logical conclusions of the mechanistic conception of life. What Morgan did believe, however, was that biology should be placed on the same footing as the physical sciences: that is, that the criteria for evaluating ideas in biology should be the same as those in physics and chemistry (quantitative measurement, experimentation, and rigorous analysis).

The California Institute of Technology. In 1927 George Ellery Hale invited Morgan to come to the California Institute of Technology to establish its first division of biology. After weighing the matter for a short time, Morgan accepted with enthusiasm. Although he had doubts about his abilities as an administrator (he wrote to Hale that he was a "laboratory animal, who has tried most of his life to keep away from such entanglements"),[13] the opportunity of heading a new department seemed to far outweigh the possible administrative problems. This move offered several advantages to Morgan, who was then sixty-two. Because the Kerckhoff Laboratory had a generous endowment (from the Kerckhoff family) as well as assistance from the Rockefeller Foundation, Morgan was able from the start to attract a first-rate staff. At Caltech, Morgan developed a modern department based on the concept of biology as he thought it should be studied and taught, where the new experimentalism could play a predominant role. Moving to Caltech also provided Morgan with the opportunity of achieving on a permanent basis the kind of scientific interaction and cooperation which he found so productive first at Naples and later during summers at the Marine Biological Laboratory, Woods Hole. As he wrote to Hale: "The participation of a group of scientific men united in a common venture for the advancement of research fires my imagination to the kindling point."[14]

In the Caltech period Morgan's influence in genetics extended beyond the *Drosophila* work and the classical chromosome theory. Although he did not

pioneer in the newer biochemical and molecular genetics that began to emerge in the 1940's, he nourished that trend. Both George Beadle, as a National Research Council fellow in 1935, and Max Delbrück, as an international research fellow in biology of the Rockefeller Foundation in 1939, worked with Morgan's group at Caltech; both saw that the next logical questions arising out of the *Drosophila* work were those of gene function. It was their work on the relationships between genes and proteins in simple organisms, such as yeasts and bacteriophages, that prepared the way for the revolution in molecular genetics during the 1950's and 1960's.

Morgan's influence was central to the transformation of biology in general, and heredity and embryology in particular, from descriptive and highly speculative sciences arising from a morphological tradition, into ones based on quantitative and analytical methods. Beginning with embryology, and later moving into heredity, he brought first the experimental, and then the quantitative and analytical, approach to biological problems. Morgan's work on the chromosome theory of heredity alone would have earned him an important place in the history of modern biology. Yet in combination with his fundamental contributions to embryology, and his enthusiasm for a new methodology, he can be ranked as one of the most important biologists in the twentieth century.

NOTES

1. A. H. Sturtevant, "Thomas Hunt Morgan," p. 285.
2. D. M. McCullough, "W. K. Brooks' Role in the History of American Biology," in *Journal of the History of Biology,* **2** (1969), 411–438, esp. p. 420.
3. T. H. Morgan and Hans Driesch, "Zur Analyse der ersten Entwickelungsstadien des Ctenophoreneies. I. Von der Entwickelung einzelner Ctenophorenblastomeren. II. Von der Entwickelung ungefurchter Eier mit Protoplasmadefekten," in *Archiv für Entwicklungsmechanik der Organismen,* **2** (1895), 204–215, 216–224.
4. T. H. Morgan, "Impressions of the Naples Zoological Station," in *Science,* **3** (1896), 16–18.
5. *Ibid.*
6. T. H. Morgan, "A Biological and Cytological Study of Sex Determination in Phylloxerans and Aphids," in *Journal of Experimental Zoology,* **7** (1909), 239–352; "Chromosomes and Heredity," in *American Naturalist,* **44** (1910), 449–496.
7. T. H. Morgan, "A Biological and Cytological Study . . .," p. 263.
8. T. H. Morgan, "Sex-Limited Inheritance in *Drosophila,*" in *Science,* **32** (1910), 120–122.
9. F. A. Janssens, "La théorie de la chiasmatypie," in *La Cellule,* **25** (1909), 389–411.
10. A. H. Sturtevant, *A History of Genetics* (New York, 1965), pp. 49–50.
11. A. H. Sturtevant, "Thomas Hunt Morgan," p. 297.

12. T. H. Morgan, A. H. Sturtevant, H. J. Muller, and C. B. Bridges, *The Mechanism of Mendelian Heredity*, p. viii.
13. Morgan to George Ellery Hale, 9 May 1927, G. E. Hale papers, California Institute of Technology Archives, microfilm roll 26, frame 29.
14. *Ibid.*

BIBLIOGRAPHY

I. ORIGINAL WORKS. A complete bibliography of Morgan's published writings can be found in Sturtevant's "Thomas Hunt Morgan" (see below). Among the more important books and articles are "The Relationships of the Sea-Spiders," in *Biological Lectures Delivered at the Marine Biological Laboratory of Woods Hole in the Summer Session of 1890* (Boston, 1891), pp. 142–167; "Regeneration: Old and New Interpretations," in *Biological Lectures Delivered . . . Summer Session of 1899* (Boston, 1900), pp. 185–208; *Regeneration* (New York, 1901); *Evolution and Adaptation* (New York, 1903); "Recent Theories in Regard to the Determination of Sex," in *Popular Science Monthly*, **64** (1903), 97–116; "The Assumed Purity of the Germ Cells in Mendelian Results," in *Science*, **22** (1905), 877–879; *Experimental Zoology* (New York, 1907); "A Biological and Cytological Study of Sex Determination in Phylloxerans and Aphids," in *Journal of Experimental Zoology*, **7** (1909), 293–352; "What Are 'Factors' in Mendelian Explanations?," in *American Breeders' Association Report*, **5** (1909), 365–368; "Chromosomes and Heredity," in *American Naturalist*, **44** (1910), 449–496; and "Sex-Limited Inheritance in Drosophila," in *Science*, **32** (1910), 120–122.

After 1910 there appeared "An Attempt to Analyze the Constitution of the Chromosomes on the Basis of Sex-Limited Inheritance in Drosophila," in *Journal of Experimental Zoology*, **11** (1911), 365–412; "Random Segregation Versus Coupling in Mendelian Inheritance," in *Science*, **34** (1911), 384; "The Explanation of a New Sex Ratio in Drosophila," *ibid.*, **36** (1912), 718–719; *Heredity and Sex* (New York, 1913); "Multiple Allelomorphs in Mice," in *American Naturalist*, **48** (1914), 449–458; *The Mechanism of Mendelian Heredity* (New York, 1915; reiss., New York, 1972), written with A. H. Sturtevant, H. J. Muller, and C. B. Bridges; *A Critique of the Theory of Evolution* (Princeton, 1916), rev. as *Evolution and Genetics* (Princeton, 1925); *Sex Linked Inheritance in Drosophila*, Carnegie Institution Publication no. 237 (Washington, D.C., 1916), written with C. B. Bridges; "The Theory of the Gene," in *American Naturalist*, **51** (1917), 513–544; "The Origin of Gynandromorphs," in *Contributions to the Genetics of Drosophila Melanogaster*, Carnegie Institution Publication no. 278 (Washington, D.C., 1919), 3–124, written with C. B. Bridges; *The Physical Basis of Heredity* (Philadelphia, 1919); "Chiasmatype and Crossing Over," in *American Naturalist*, **54** (1920), 193–219, written with E. B. Wilson; "The Evidence for the Linear Order of the Genes," in *Proceedings of the National Academy of Sciences of the United States of America*, **6** (1920), 162–164, written with A. H. Sturtevant and C. B. Bridges; "The Bearing of Mendelism on the Origin of Species," in *Scientific Monthly*, **16** (1923), 237–247; "The Modern Theory of Genetics and the Problem of Embryonic Development," in *Physiological Reviews*, **3** (1923), 603–627; *The Theory of the Gene* (New Haven, 1926); "The Relation of Physics to Biology," in *Science*, **65** (1927); *The Scientific Basis of Evolution* (New York, 1932); *Embryology and Genetics* (New York, 1934); and "The Conditions That Lead to Normal or Abnormal Development of *Ciona*," in *Biological Bulletin*, **88** (1945), 50–52.

There is no single collection of Morgan's letters, notebooks, or other unpub. materials. Numerous Morgan letters can be found, however, in the papers of Ross G. Harrison (Yale University), Edwin Grant Conklin (Princeton University), William Bateson (American Philosophical Society), and George Ellery Hale (Mount Wilson and Palomar Observatories Library, Pasadena). The American Philosophical Society Library, Philadelphia, is collecting the papers of important American geneticists; Morgan letters appear prominently in many of these collections.

II. SECONDARY LITERATURE. The fullest account of Morgan's life to date remains A. H. Sturtevant, "Thomas Hunt Morgan," in *Biographical Memoirs. National Academy of Sciences*, **33** (1959), 283–325. Selected writings about Morgan and his work include G. E. Allen, "Thomas Hunt Morgan and the Problem of Natural Selection," in *Journal of the History of Biology*, **1** (1968), 113–139; "T. H. Morgan and the Emergence of a New American Biology," in *Quarterly Review of Biology*, **44** (1969), 168–188; "T. H. Morgan and the Problem of Sex Determination," in *Proceedings of the American Philosophical Society*, **110** (1966), 48–57; "T. H. Morgan, Richard Goldschmidt and the Opposition to Mendelian Theory 1900–1940," in *Biological Bulletin*, **139** (1970), 412–413; and a slightly fuller treatment of this same material, "Richard Goldschmidt's Opposition to the Mendelian-Chromosome Theory," in *Folia Mendeliana*, **6** (1971), 299–303. See also Edward Manier, "The Experimental Method in Biology. T. H. Morgan and the Theory of the Gene," in *Synthese*, **20** (1969), 185–205; and A. H. Sturtevant, "The Fly Room," ch. 6 of *A History of Genetics* (New York, 1965). An analysis of the work of the *Drosophila* group from Muller's point of view is given in E. A. Carlson, "The Drosophila Group: the Transition From the Mendelian Unit to the Individual Gene," in *Journal of the History of Biology* (in press).

Background material on much of the development of Mendelian genetics after 1900 can be found in three general historical studies: E. A. Carlson, *The Gene, a Critical History* (Philadelphia, 1966); L. C. Dunn, *A Short History of Genetics* (New York, 1965); and Sturtevant's *History of Genetics*.

GARLAND E. ALLEN

MORICHINI, DOMENICO LINO (*b*. Civitantino, Aquila, Italy, 23 September 1773; *d*. Rome, Italy, 19 November 1836), *medicine, chemistry*.

The son of Anselmo Morichini and Domitilla Moratti, Morichini went to Rome for his university studies and remained there. In 1792 he graduated in philosophy and in medicine. The following year he was appointed assistant physician, and later head physician, at the Arcispedale di Santo Spirito. While holding important executive positions in public hygiene and health organizations and continuing to practice medicine throughout his lifetime, Morichini lectured in chemistry at the University of Rome from 1800. In fact, he introduced this new subject of instruction by presenting Lavoisier's doctrines instead of the phlogiston theory and by setting up an experimental chemistry laboratory.

In 1802 Morichini was appointed to make the chemical examination of fossil elephant teeth found in Rome. Treating them with concentrated sulfuric acid, he noted a lively effervescence due to the discharge of gas, which he recognized to consist of both carbon dioxide and fluorine. The latter, released especially from the tooth enamel, was recognizable because it corroded the glass of the vessel used in the experiment and because, when brought into contact with limewater, it formed lime fluoride "endowed with all the properties of natural fluorspar" (that is, CaF_2). Morichini subsequently found elemental fluorine in human teeth.

BIBLIOGRAPHY

I. ORIGINAL WORKS. Morichini's "Analisi chimica del dente fossile" is included in Carlo Lodovico Morozzo, "Sopra un dente fossile trovato nelle vicinanze di Roma," in *Memorie di matematica e di fisica della Società italiana delle scienze*, **10**, pt. 1 (1803), 166–170. Morichini returned to the subject in "Analisi dello smalto di un dente fossile di elefante e dei denti umani," *ibid.*, **12**, pt. 2 (1805), 73–88, 268–269. See also *Raccolta degli scritti editi ed inediti del Cav. Dott. Domenico Morichini* (Rome, 1852).

II. SECONDARY LITERATURE. See the anonymous "Memorie storiche del Cavalier Domenico Morichini," in *Raccolta degli scritti* (see above); and Luigi Belloni, "Il fluoro dentario scoperto a Roma nel 1802 da Domenico Morichini," in *Scritti in onore di Adalberto Pazzini* (Rome, 1968), 199–205.

LUIGI BELLONI

MORIN, JEAN-BAPTISTE (*b.* Villefranche, Beaujolais, France, 23 February 1583; *d.* Paris, France, 6 November 1656), *medicine, astronomy, astrology.*

Morin was a strange person but typical of his age in that he undertook very varied activities. He was sufficiently successful and intelligent to acquire a reputation in his own time, but he failed to demand of himself the thorough discipline that would have enabled him to produce truly scientific work. A medical doctor at first, he then took an interest in all the topics associated with hermetic literature. In order to penetrate the mysteries of nature he studied mining and astrology (he was later to draw up an astrological chart for the infant Louis XIV). His talents won him the support of influential people, and in 1630 he was professor of mathematics at the Collège Royal (now Collège de France), a position he held until his death.

A polemicist by disposition, he quickly sought to distinguish himself in the major debates most likely to bring him widest attention. In 1624 he published a defense of Aristotle in conjunction with the refutation of the theses of Antoine de Villon and of Étienne de Claves, both of which had been condemned by the Sorbonne. He opposed Galileo before and after the trial of 1633. He attacked Descartes in 1638, flattering himself that he had detected how bad his philosophy was from the moment that they had met, before Descartes's departure for Holland.

If Morin suffered injustice in the judgments reached by his contemporaries, especially Boulliau, he owed his poor reputation to the way in which he conducted his disputes. This fault is best illustrated in the matter of the determination of longitudes. During the very period when he was presenting himself as the champion of the immobility of the earth, Morin simultaneously wished to prove that he was capable of drawing inspiration from Kepler, of correcting the Rudolphine Tables, and of proposing a method for finding longitudes that would always be usable at sea and sufficiently precise for navigation. The only original thing about this method, which was based on the observation of the moon, was its claim of utilizing the movements of the moon relative to the stars as a universal clock and of generalizing this phenomenon to calculate the difference in hours between two positions on the earth. The method required new observational instruments, which could be used with sufficient precision on ships, the improvement of the mathematical solution of spherical triangles, and the possibility of a systematic checking of tables of lunar motion established for a given position. Morin glimpsed these three facets of the problem and made an important contribution to instrumental technique by utilizing telescopes for the sights and verniers for the measurement of angles; but he was incapable of mastering the complex problem of precision in a process involving both observation and computation. Ambition and the desire to obtain a pension from Richelieu made him deaf to all objections.

From 1626 to 1628 Morin undertook research in optics with the engineer Ferrier, in whom Descartes had placed his hopes. Shortly afterward his friendships with Peiresc and Gassendi helped him in observational astronomy. But the affair of the longitudes, with the prolonged debate (1634–1639) that put him in opposition to Étienne Pascal, Mydorge, and Beaugrand, alienated the scientific community, and he continued his work largely in isolation.

Morin's posthumously published *Astrologia gallica* reveals that he had interesting ideas concerning the theory of heat and the temperature of mixtures. Moreover, in the correspondence of Mersenne and Descartes references to Morin are not entirely negative. Despite his undoubted talents, Morin's philosophical and scientific choices were too often political ones and prevented him from producing the caliber of work of which we now see he was capable.

BIBLIOGRAPHY

I. ORIGINAL WORKS. For works of Morin see *Astronomicarum domorum cabala detecta* (Paris, 1623); *Refutation des thèses ... d'A. Villon et E. de Claves ... contre les doctrines d'Aristote* (Paris, 1624); *Famosi et antiqui problematis de telluris motu vel quiete hactenus optata solutio* (Paris, 1631); *Trigonometriae canonicae libri tres quibus planorum et sphaericorum triangulorum theoria ... adjungitur liber quartus pro calculi tabulis logarithmorum* (Paris, 1633); *Pro telluris quiete* (Paris, 1634); *Astronomia jam a fundamentis integre et exacte restituta ...* (Paris, 1640); *La science des longitudes ...* (Paris, 1647); and *Astrologia gallica ...* (The Hague, 1661), published posthumously under the auspices of Marie Louise of Gonzaga, queen of Poland, which also contains a Latin trans. of the anonymous *Vie de Morin* (1660), the French original of which is lost.

II. SECONDARY LITERATURE. For information about Morin see P. Bayle, *Dictionnaire historique et critique*, II, pt. 1 (Rotterdam, 1697), 602–612; M. Delambre, *Histoire de l'astronomie moderne*, II (Paris, 1821), 236–273; G. de Fouchy, "Sur la date de l'application des lunettes aux instruments d'observation ...," in *Mémoires de l'Académie royale des sciences pour l'année 1783* (1787), 385–392; L. Moreri, *Le grand dictionnaire historique*, VII (Paris, 1759), 786–788; and J. Montucla, *Histoire des mathématiques* (Paris, 1799), II, 336–1802; IV, 543–545.

PIERRE COSTABEL

MORISON, ROBERT (*b.* Dundee, Scotland, 1620; *d.* London, England, 11 November 1683), *botany.*

Morison was the son of John Morison and his wife, Anna Gray. He was educated at Aberdeen University, where he obtained the M.A. in 1638. He taught at that university until his career was interrupted by the Civil War. In 1644, after fighting against the Covenanters, he fled to France as a Royalist. Morison studied medicine at Paris and obtained an M.D. at Angers in 1648. Botany quickly became his main interest, an interest that led to his appointment as gardener to Gaston d'Orléans at Blois. He undertook extensive journeys to collect material for the gardens and probably contributed to the catalog of the Blois garden and was regarded as a suitable authority to revise the first English plant list, *Phytologia Britannica* (1650).

Charles II brought Morison back to England at the Restoration in 1660 as royal physician and botanist. In 1669 he became the first professor of botany at Oxford. His activities were closely related to the Oxford botanical garden, founded forty years earlier. Its gardener, Jacob Bobart the younger, became Morison's closest colleague. At Oxford, Morison's duties were not onerous, allowing him ample leisure for the compilation of his major botanical works.

Morison's first publication, *Praeludia botanica* (1669), was a composite volume containing an augmented list of the plants at Blois arranged according to the orthodox classification; critical animadversions on the taxonomic work of Jean and Gaspard Bauhin; and an intriguing dialogue announcing his dissatisfaction with current approaches to a plant classification. The latter stressed the need for a single, key criterion for determining the *nota generica*, or natural relationships, of plants. He revived Cesalpino's suggestion that classification should be based on fruit and seed characteristics.

This principle was first applied in Morison's monograph on umbelliferous plants (1672), which successfully isolated the Umbelliferae from other plants with similar inflorescence forms. The family was then subdivided into a series of genera that closely resembled later categories. Vegetative characteristics were consulted only for subsidiary taxonomic affinities.

Morison next endeavored to apply his taxonomic principles to the entire plant kingdom. Like so many similar enterprises this was destined for a fragmentary conclusion. He pursued the work with great enthusiasm, however, even obtaining substantial financial assistance from Oxford and private patrons to meet the cost of publishing his voluminous illustrated *Historia plantarum*. Part I, on trees, was thought to exist by contemporaries but was never published. Morison himself published part II (1680), and his colleague Bobart completed part III (1699). Bobart also assembled an herbarium of 5,000 plants organized to illustrate Morison's system. The published sections of the *Historia* include about 6,000 plants. Those in part II are described in considerable detail, and the illustrations are of higher quality than has usually been appreciated by modern commentators. The crucial system of classification was a poor application of Morison's original idea. His taxonomic principle was not followed consistently, vegetative criteria frequently being invoked to establish major divisions. Hence Morison's system had many of the defects which it was designed to counteract.

In reputation Morison was quickly eclipsed by his gifted contemporary John Ray. This decline in fortune was undoubtedly reinforced by the excessively critical tone of his writings and his false claims to originality. Thus even during his lifetime Morison was a relatively isolated figure, avoiding association with such major organizations as the Royal Society and the London College of Physicians. Certain major taxonomists, however, recognized the importance of his declared principles. Both Tournefort and Linnaeus regarded Morison as the "instaurator" of taxonomy. His pioneer attempt to apply a single, clear taxonomic criterion to the whole plant kingdom had the potential to rescue taxonomy from a state of strangled confusion resulting from the rapid accumulation of data.

BIBLIOGRAPHY

I. ORIGINAL WORKS. Morison's writings are *Hortus regius Blesensis auctus, praeludium botanicorum* (London, 1669), known as *Praeludia botanica; Plantarum umbelliferarum distributio nova* (Oxford, 1672); *Plantarum historiae universalis Oxoniensis pars secunda* (Oxford, 1680); and *Plantarum historiae universalis Oxoniensis pars tertia,* Jacob Bobart, ed. (Oxford, 1699), introduced with a biographical sketch of Morison by Archibald Pitcairne.

Morison also edited Paulo Boccone's *Icones et descriptiones rariorum plantarum . . .* (Oxford, 1674).

II. SECONDARY LITERATURE. See G. S. Boulger, "Robert Morison," in *Dictionary of National Biography*, XIII (1967–1968), 958–960; J. Reynolds Green, *A History of Botany in the United Kingdom* (London, 1914), 98–110; R. Pulteney, *Historical and Biographical Sketches of the Progress of Botany*, I (London, 1790), 289–327; and S. H. Vines, "Robert Morison and John Ray," in *Makers of British Botany*, F. W. Oliver, ed. (Cambridge, 1913), 8–43; and *An Account of the Morisonian Herbarium* (Oxford, 1914).

CHARLES WEBSTER

MORLAND, SAMUEL (*b.* Sulhamstead Bannister, Berkshire, England, 1625; *d.* Hammersmith, Middlesex [now London], England, 30 December 1695), *mathematics, technology.*

Morland was the son of Thomas Morland, rector of Sulhamstead Bannister, from which village Samuel

Morland took his title when he was created baronet. He was educated at Winchester College from 1639 and at Magdalene College, Cambridge, from 1644. Elected a fellow of the college on 30 November 1649, he continued his studies of mathematics there until 1653.

From 1653 to the Restoration in 1660, Morland was deeply involved in politics. He was a supporter and associate of Oliver Cromwell, who employed him on two foreign embassies, to Sweden in 1653 and to the duke of Savoy in 1655, with the object of persuading him to grant amnesty to the Waldenses. Morland was finally successful and in 1658 he published a history of the Waldensian church.

Close association with the intrigues of Cromwell and John Thurloe and, in particular, knowledge of a plot to murder Charles II and his brother disgusted Morland with the Commonwealth cause; and he began working as an agent to promote the restoration of the monarchy. Despite serious charges brought against him, he was granted a full pardon by Charles II in 1660, knighted, and later in the same year was created a baronet. He was also appointed gentleman of the privy chamber, but he did not receive the financial help he had hoped for. From 1660 he devoted himself to experimental work with occasional support from the king, who named him "Master of Mechanicks" in 1681.

Morland was married five times and was survived by only one son, Samuel, who became the second and last baronet of the family. Morland became blind three years before his death and retired to Hammersmith, where he died on 30 December 1695.

Morland's studies of mathematics and his inventiveness led him to make two "arithmetick instruments," or hand calculators, with gear wheels operated by a stylus, for pedagogic use. His perpetual almanac was a concise form of pocket calendar, adapted for use on coin-sized disks, sundials and other instruments, and snuffboxes. A speaking trumpet, described in his treatise on the subject as a "*tuba stentoro-phonica,*" was another of his inventions; with it he estimated that a conversation could be carried on at a distance of three-quarters of a mile. It has recently been established that Morland also invented the balance barometer and the diagonal barometer.

Morland's most important work was in the field of hydrostatics. There was much interest in the mid-seventeenth century in mechanical methods of raising water. Morland invented an apparatus using an airtight cistern from which air was expelled by a charge of gunpowder, the water below rising to fill the vacuum thus produced. The *London Gazette* for

30 July 1681 describes how at Windsor Castle, "Sir Samuel Morland, with the strength of eight men, forced the water (mingled with a Vessel of Red Wine to make it more visible) in a continuous stream, at the rate of above sixty Barrels an hour, from the Engine below at the Parkpale, up to the top of the Castle, and from thence into the Air above sixty Foot high."

Morland's efforts to raise water at Versailles led to the publication of *Élévation des eaux* in 1685, but in the manuscript version in the British Museum, written in 1683, there is an account of the use of steam power to raise water. Although he did not develop the steam engine, his experiment is one of the first to show the practical possibilities of steam power.

BIBLIOGRAPHY

I. ORIGINAL WORKS. A list of Samuel Morland's published works is given in app. 2 to the biography by H. W. Dickinson cited below. The main scientific works are *Tuba stentoro-phonica, an Instrument of Excellent use as Well at sea as at Land; Invented, and Variously Experimented, in the year 1670* ... (London, 1671); *The Description and Use of two Arithmetick Instruments. Together With a Short Treatise Explaining and Demonstrating the Ordinary Operations of Arithmetick. As Likewise A Perpetual Almanack, and Several Useful tables* (London, 1673); *Élévation des eaux par toute sorte de machines réduite à la mesure, au poids, à la balance par le moyen d'un nouveau piston & Corps de pompe d'un nouveau mouvement cyclo-elliptique* ... (Paris, 1685); *The Poor Man's Dyal With an Instrument to Set It. Made Applicable to Any Place in England, Scotland, Ireland, &c.* (London, 1689); and *Hydrostaticks: or Instructions Concerning Water-works, Collected out of the Papers of Sir Samuel Morland. Containing the Method Which he Made use of in This Curious art*, Joseph Morland, ed. (London, 1697). There is MS material at the British Museum, Lambeth Palace Library, and Cambridge University Library, all noted by Dickinson. Examples of Morland's calculating machines are at the Science Museum, London; Museum of the History of Science, Oxford; and the Museo di Storia della Scienza, Florence; a speaking trumpet is at Trinity College, Cambridge.

II. SECONDARY LITERATURE. See J. O. Halliwell, *A Brief Account of the Life, Writings, and Inventions of Sir Samuel Morland* (Cambridge, 1838); W. E. Knowles Middleton, "Sir Samuel Morland's Barometers," in *Archives internationales d'histoire des sciences*, **5** (1962), 343–351. H. W. Dickinson, *Sir Samuel Morland, Diplomat and Inventor 1625–1695* (Cambridge, 1970), published eighteen years after the author's death by the Newcomen Society, includes an iconography and a bibliography; and D. J. Bryden, "A Didactic Introduction to Arithmetic, Sir Charles Cotterell's 'Instrument for Arithmeticke' of 1667," in *History of Education*, **2** (1973), 5–18.

G. L'E. TURNER

MORLEY, EDWARD WILLIAMS (*b.* Newark, New Jersey, 29 January 1838; *d.* West Hartford, Connecticut, 24 February 1923), *chemistry, physics.*

Morley was the eldest of four children of Sardis Brewster Morley and Anna Clarissa Treat; his father was a Congregational minister. Morley's education was conducted under his parents' tutelage until he was nineteen, when the family moved to Williamstown, Massachusetts, so that the three boys could attend Williams College, the father's alma mater. With the intention of entering the ministry, Morley entered Andover Theological Seminary in 1861, after receiving the B.A. from Williams College in 1860. While pursuing his theological studies he completed work for the master's degree at Williams College in 1863. Having finished his theological studies in 1864, he served for a year on the Sanitary Commission at Fort Monroe, Virginia.

Morley resumed studies at Andover for another year before receiving an offer to teach at South Berkshire Academy, Marlboro, Massachusetts. There he became acquainted with Isabella Ashley Birdsall, whom he married on Christmas Eve 1868. In September of that year he had accepted an invitation to become minister at the Congregational Church in Twinsburg, Ohio, and was then invited to teach at Western Reserve College in nearby Hudson. In 1882, when the college was transferred to Cleveland and became Adelbert College of Western Reserve University, he was chosen to fill the chair of chemistry and natural history, a position he held until his retirement to West Hartford, Connecticut, in 1906. From 1873 to 1888 Morley also held the professorship of chemistry and toxicology at the Cleveland Medical School, a position which required him to travel back and forth between Cleveland and Hudson for many years. He died just three months after his wife and was buried in the family lot in Pittsfield, Massachusetts. The couple had no children.

Morley's scientific achievements fall into three rather well-defined periods, which are linked by a passionate concern for precise and accurate quantitative measurements, in the tradition of Berzelius and Stas, and by a keen interest in the theories of the structure of matter, especially as reflected in Prout's hypothesis. During his first period, while at Hudson, he employed and refined eudiometric methods for analyzing the oxygen content of the atmosphere to within .0025 percent. His purpose was to test the correctness of Loomis' hypothesis that cold waves were caused by the descent of air from high elevations at times of high barometric pressure rather than by horizontal currents moving from north to south. He reasoned that if Loomis' theory were correct, the oxygen content of air collected during a cold wave should be lower due to gravitational separation at high altitudes of molecules of different mass, which had been predicted by Dalton on the basis of his atomic theory. A careful correlation of meteorological records with precise determinations of the oxygen content of air revealed good agreement with Loomis' theory.

After spending three years constructing apparatus and equipping his laboratory at Western Reserve University in Cleveland, Morley undertook a painstakingly planned program to determine the atomic weight of oxygen relative to hydrogen taken as unity. His aim was to test the validity of Prout's hypothesis that the elements of the periodic table were built up from hydrogen or hydrogenlike units. First, he determined quantitatively the extent to which moisture could be removed from gases by drying with sulfuric acid and phosphorus pentoxide. He then made a careful study of the volume proportions in which oxygen and hydrogen unite. Finally, by two independent methods, one based on direct weighings of components and product and the other based on density determinations of oxygen and hydrogen, he found the atomic weight of oxygen to be 15.879, with an uncertainty of the order of one part in ten thousand. With this result Morley felt that he had laid Prout's hypothesis to rest. It was only during the last two decades of his life, after his retirement, that the full significance of exact atomic weight determinations by chemical methods was fully grasped: that they measure the weighted average of all the stable isotopes of the elements in question.

The final period of Morley's career overlaps the previous one and is characterized by extensive collaborative studies with A. A. Michelson, H. T. Eddy, W. A. Rogers, D. C. Miller, C. F. Brush, and J. P. Iddings. His famous ether-drift experiments with Michelson, and much of his work with the others, reflect a fascination with the sensitivity of the interferometer and measuring techniques based upon the wavelengths of light. They also suggest an interest in the rapidly developing field of spectroscopy because it seemed to hold out promise for a deeper understanding of the structure of matter, a promise which was being realized during the last decade and a half of his life. Even in retirement Morley kept active professionally. He constructed an analytical laboratory of his own and carried out an extensive series of exact analyses of rocks collected by J. P. Iddings in Java and the Celebes. In the latter years of his life Morley was honored with the Davy Medal of the Royal Society, the Elliot Cresson Medal of the Franklin Institute, and the Willard Gibbs Medal of the Chicago section of the American Chemical Society, principally for his careful atomic weight determinations.

BIBLIOGRAPHY

A bibliography of 55 papers was collected by F. W. Clarke in *Biographical Memoirs. National Academy of Sciences*, **21** (1927), 1–8. This bibliography was extended by Morley's biographer, H. R. Williams (see below). Morley's letters are preserved at the Library of Congress, and photostat copies are kept in the archives of Case Western Reserve University. Morley's notebooks are in the custody of Frank Hovorka, Hurlbut professor of chemistry at Case Western Reserve University.

A biography is Howard R. Williams, *Edward Williams Morley* (Easton, Pa., 1957).

ERNEST G. SPITTLER

MORO, ANTONIO-LAZZARO (*b.* San Vito del Friuli, Italy, 16 March 1687; *d.* San Vito del Friuli, 12 April 1764), *geology.*

Moro was the son of Bernardino and Felicita Mauro. Although his early education was marked by frequent changes of instructors whose academic preparation and instruction were poor, he distinguished himself in mathematics, music, languages, literature, natural sciences, and ecclesiastical studies. Upon completion of the latter and his ordination into the ministry, he was offered a post as professor of philosophy and rhetoric at the seminary in Feltre and shortly thereafter became its director. Following the death of the bishop of Feltre and because of his own poor health, Moro returned to Friuli, where he became chapelmaster of the cathedral at Portogruaro. Of his varied interests Moro's involvement was greatest in scientific studies, which he pursued with a Galilean conviction that the proper research methodology would inevitably, if gradually, reveal the secrets of nature.

In 1721 Vallisnieri had concluded a study of fossils with a categorical rejection of all theories on the subject, on the ground that none of them could hold up under analytical scrutiny. Moro agreed with this polemical judgment and decided to accept the challenge of the question left open by Vallisnieri. In 1740 he published his best-known work, *Dei crostacei e degli altri corpi marini che si trovano sui monti*, a study of the origin and development of fossiliferous deposits.

The logical order of Moro's work reflects an empirical spirit characteristic of the enlightened intellectual climate of eighteenth-century Europe. Despite its archaic language the book is a model of cogent reasoning. Moro first provides a survey of fossil occurrence, with regional and global distribution, based on personal and reported observations. Elaborating upon earlier indications of L. Marsili and Vallisnieri, he gives a stratigraphic compilation of various fossilized marine flora and fauna, with sequential indications that preceded by almost a century the chronological intuitions of William Smith and his French contemporaries, G. Cuvier and Alexandre Brongniart.

Moro divided prevailing opinions into two groups—neptunist and nonneptunist. After singling out the most popular current opinions—the theory of total submersion and the diluvial views of Burnet and J. Woodward—for reexamination, he finally rejected them as scientifically invalid. Moro's main objection to Burnet was that the British theologian, in *Sacred Theory of the Earth*, had contrived an elaborate antediluvian system in order to force proofs of conjectures established a priori as scientific conclusions.

The examination of Woodward's theories presented a greater challenge, since the noted British naturalist had made considerable and valid contributions to the fossil debate. Moro asserted that when Woodward attempted to construct a scientific theory upon two antithetical principles, one factual (direct observation of fossils) and one hypothetical (miraculous, divine causes of geological phenomena), he violated his stated objectives and thereby undermined his preliminary observations.

Like Burnet before him, Woodward espoused the notion of an aqueous abyss, to which he ascribed, among other functions, the process of supplying water to the rivers and streams of the earth. Moro denied such hydrologic function to an "imaginary abyss," pointing out that, among the scholars (P. Perrault, P. de la Hire, Mariotte, Purchot, J. Cassini, and, more recently, Vallisnieri, D. Corradi, and Giorgi) concerned with the origin and sources of the water on earth, Woodward alone disregarded or rejected the theory of natural precipitation as the initial source of groundwater. The most questionable theory, according to Moro, was Woodward's theory of global disintegration, which attributed to divine intervention the loss of gravity during diluvial submersion.

Because of his plutonism Moro categorically rejected the view that stratification was caused by aqueous agents, arguing that such a view was contradicted by the evidence of both chromatic and density differentiation among layers. Unlike the Plutonist views, which were characteristic largely of the eighteenth century, the theory of total submersion had persisted from ancient times (Plutarch, Strabo) and was espoused among moderns by Fracastoro and Leibniz; Vallisnieri had also been tempted by the notion until he found it unprovable. Moro's main objection to the theory was that it did not take into account the structural or dynamic aspects of mountain formation or land building, which were crucial to an understanding of the fossil problem.

Moro began his contribution to the fossil debate with a clear statement of his scientific credo: It is within nature that one must search for laws governing physical reality. But since nature seldom reveals efficient causes to man, one may use its constancy and uniformity and the stability of natural law to deduce from certain effects their dynamic causes, which are similar if not identical. The guiding principle and unifying theme of Moro's theory and proof was thus the Newtonian axiom that affirms this concept: "Effectuum naturalium ejusdem generis eadem sunt causae."

Moro noted that in order to understand his fossil theory, it was first necessary to determine the dynamic forces and physical laws involved in the formation of fossiliferous deposits. The refuted theories had been exclusively neptunist and Moro's attitudes were exclusively plutonic, attributing geomorphological development of the globe to igneous agents. On the basis of observations made by contemporary and classical scientists of volcanic mountains and islands, Moro established a chronological framework within which he synthesized and historicized the two aspects of his theory: the formation and development of the earth and fossil phenomena. Proceeding retrospectively, he quoted from a detailed report, made to Vallisnieri by a student, G. C. Condilli, of the volcanic island of Mea Kaumen, which surfaced in the Greek archipelago near Santorini in 1707 and which continued to rise, shift, and settle until it stabilized in 1711. This, Moro noted, was but the latest incident in a long history of volcanic island formation in the archipelago, accounts of which had been left by Strabo, Pliny, and Justinian.

Historical incidents of volcanological mountain formation similarly offered proof that these land masses were the result of igneous forces within the earth. Referring to reports by N. Madrisio and Agricola, Moro cited the example of the volcanic birth in 1538 of Monte Nuovo in the Bay of Naples. On the basis of these and other historical incidents, Moro concluded that (1) mountains and islands are volcanic in origin and (2) the presence of marine fossils on the surface of these landmasses justifies the belief that the newly formed surfaces were once submerged and, in the process of rising, brought marine organisms to the surface with other materials.

Although Vallisnieri, Steno, Woodward, and F. Colonna (Hooke and J. Ray are not mentioned) had already indicated that a necessary and essential relationship exists between mountain fossils and marine organisms, Moro defended the originality of his work on the ground that he had integrated the incidental, particular observations of his predecessors into a systematic and generalized theory that correctly placed the problem of fossil deposits in the broader framework of mountain formation and tectonics. He also indicated that the dynamic processes involved are igneous rather than aqueous.

Moro then passed from dynamic to structural geology, describing insular mountains as massive gneisses folded in numerous directions—concave, convex, perpendicular, oblique—all caused by the pressure of intense subterranean heat. Generalizing his plutonic theory to include mainland masses, Moro differentiated these masses into two types: primary mountains—massive orthogneisses pushed up from the center of the earth when that part of the surface of the earth was submerged by water, and secondary (or stratified) mountains—composed largely of paragneisses and formed on the surface of the earth. To corroborate his distinction between volcanic and sedimentary mountains, Moro quoted Marsili, who had similarly distinguished two types of ocean floor: essential (or original) crustal rock and accidental rock, composed of sands and mineral deposits carried back to sea by returning lava flows. If the essential bedrock of the sea is similar to the massive gneisses of primary mountains, Marsili noted, it is reasonable to conclude that the latter were once a part of the ocean floor and that they were thrust above the water when that portion of the crust was still uncovered by secondary deposits.

The divergent views of mountain formation held by Moro and Vallisnieri reflect, to a large extent, the basic tensions between plutonists and neptunists, who often agreed upon structural effects but disagreed upon the dynamic principles involved. While Vallisnieri considered mountain building to be the result of successive aqueous "inundations," Moro insisted that these "inundations" actually consisted of igneous materials, of which each successive crust is composed.

Moro attributed the striated and undulating patterns of marble to seismic action and ground shifts during the cooling and solidifying of the magma. The process of crystallogenesis, which Vallisnieri was unable to define to his satisfaction, also was interpreted by Moro as dependent upon intense heat.

The effects of volcanic action upon biological development were indicated in Moro's consideration of the problem of skeletal remains of extinct or exotic animals in areas that are no longer a natural habitat for such species. He refuted Woodward's diluvial explanations in favor of a surprisingly modern, naturalistic theory: that the areas of occurrence of extinct animal fossils were once their life-supporting environment. He insisted that the chronological aspect of the plutonic principle must always be kept in mind. Between volcanic eruptions, vast periods of time elapse. If the volcanic deposit is organically sterile, animal life will become extinct. As subsequent volcanic activity deposits additional strata, in which organic matter fosters the development of vegetation, the chain of being is reestablished.

Moro terminated his work with a chronological résumé of his system that is in effect a miniature composite biogeological history. Briefly, his summary is as follows: On the first day God created, among other things, the terraqueous globe surrounded by fresh water. There followed a division of the waters, without mountainous protrusions to disturb the uniform spheroid form. On the second day Moro's concept of an active Creator as the dynamic principle is abruptly replaced by natural actualism. The shift is so sudden that one suspects that the scriptural inclusion was but a token gesture made by Moro to insure permission from religious and civic authorities to publish his work. Moro makes it quite clear that the natural potential combustibility of the core of the earth was set in motion according to divine plan or will, not according to a divine act.

Once activated, igneous pressures push up the rocky surface of the submerged lithosphere, forming primary mountains. Volcanic matter may return to the sea, depositing salts, minerals, and bitumen into the fresh water and changing its chemical composition. Secondary mountains and landmasses are built up by the same volcanic activity, with the seabed rising also as a result of the deposition of strata.

Biology also was determined by physical conditions. Marine flora and fauna were not yet formed. As the accidental seabed was transformed by mineral deposits into life-supporting systems, marine vegetation appeared, followed by marine animals, the latter with their origin and habitat in the soft earth, sand, and clay of redeposited lava flows. As landmasses built up, marine biological processes and patterns were repeated on land, with the formation of vegetation preceding that of animal life. Crowning this chain of being is man, viewed, as all else in Moro's system, as a product of natural evolution rather than of divine creation. This audacious view of anthropogenesis by a clergyman was vehemently condemned by contemporary scripturalists. For Moro subsequent volcanic activity involved the uplift of part of the ocean floor with its flora and fauna. This in turn caused the formation of fossiliferous deposits in stratified mountains which preserved the fossils as a museum preserves evidences of human life, arts, and crafts. Superposed strata were for Moro indicative of chronological as well as environmental and cultural data.

The work provoked vehement reactions among European scientists, polarizing them into neptunists and plutonists. Among the former were Zolmann, who published his views in the *Philosophical Transactions of the Royal Society* in 1745, and Giuseppe Costantini, whose impassioned defense of the diluvial theory, *La verità del diluvio dimostrata* (1747), was so indiscriminate in its attack on Moro that it had the effects of confirming rather than disproving Moro's views. In 1749 G. C. Generelli added his support to Moro's theory in a paper read before the Academy of Sciences at Cremona. A German translation of the work, *Neue Untersuchung über die Abanderungen der Erde* (1751), was followed by a lengthy review by C. Delius in his *Anleitung zu der Bergbaukunst*. Moro was one of the authorities cited by Knorr (1755) and Desmarest. His scheme of periodization (primary, secondary) was amplified in Italy by Arduino and in Sweden by Bergman, who referred to Moro; by J. G. Lehmann and Pallas, in Germany and Russia, respectively; and the ultimate development was the establishment of the geologic column by A. G. Werner. Oddly enough, this work, so influential on German neptunism, made little impression on British volcanism. In 1767 E. King, in an article in the *Philosophical Transactions*, admitted that his geological theories had been anticipated by those of Moro, whose work he allegedly had discovered only after the publication of his own study. King's cursory reference to Moro disappeared altogether in J. Hutton's *Theory of the Earth*, which contains observations and conclusions similar to those made by Moro half a century earlier. Hutton's ideas were influenced by the still earlier plutonic geodynamics of Hooke, and his neglect of Moro was emphasized by J. Playfair in his *Illustration of the Huttonian Theory of the Earth* (1802).

Moro's work was most thoroughly analyzed and its importance emphasized by Hutton's principal successor, Lyell.

The modernity of Moro's views, the broad scope of knowledge that he brought to his investigations, and the breadth of the scientific fields that he examined and illuminated place him at the center of the intense intellectual activity of Italy's "second Renaissance."

BIBLIOGRAPHY

I. ORIGINAL WORKS. Moro's works include *Dei crostacei e degli altri corpi marini che si trovano sui monti*, 2 vols. (Venice, 1740); *Lettera, ossia dissertazione sopra la calata de' fulmini dalle nuvole* (Venice, 1750); and his MS "Due lettere latine sul sistema dei crostacei" (see below).

II. SECONDARY LITERATURE. See A. Altan, "Memorie biografiche della terra di Sanvito," in *Memorie storiche della terra di Sanvito al Tagliamento* (Venice, 1832), pp. 87–89, which also locates a number of Moro's MSS; G. Dandolo, *La caduta della Repubblica di Venezia, Appendice* (Venice, 1857), p. 70; G. Generelli, "Dissertazione de' crostacei, e dell'altre produzioni marine che sono ne'monti," in *Raccolta Milanese dell'anno 1757* (Milan, 1757), pp. 1–22; C. Lyell, *Principles of Geology* (London, 1830), pp. 42–47; F. di Manzano, *Cenni biografici dei letterati ed artisti friulani* (Udine, 1887), p. 135; Saccardo, "La botanica in Italia, materiali per la storia di questa scienza," in *Memorie del R. Istituto veneto di scienze, lettere ed arti*, **25**, no. 4 (1895), 76; **26**, no. 6 (1901), 76, 114; and P. Zecchini, *Vita di A. L. Moro* (Padua, 1865).

See also the article on Moro in *Biografia degli italiani illustri nelle scienze, lettere ed arti del secolo XVIII e dei contemporanei* (Venice, 1834–1845), pp. 304–305.

ROSE THOMASIAN

MOROZOV, GEORGY FEDOROVICH (*b.* St. Petersburg, Russia, 7 January 1867; *d.* Simferopol, U.S.S.R., 9 May 1920), *biogeography, ecology.*

Morozov's father, a cutter in a linen draper's shop who later became commissar of the administration of city property in St. Petersburg, planned a military career for his son. Morozov accordingly entered the Pavlovsk military academy in 1884, graduating as a second lieutenant of artillery two years later. He was then sent to the fortress of Daugavpils, in Latvia, where he became acquainted with a group of youthful students and began to broaden his own education. In particular he was influenced by a young woman revolutionary, O. N. Zandrok, who had been sent into exile for participating in the People's Will movement. Zandrok, for whom he clearly felt affection, fostered Morozov's sympathy for the peasant classes; when he decided to devote himself to studying science, it was therefore natural for him to select the agricultural sciences as being closest and most necessary to the people. He chose forestry as his specialty and spent his free time reading the works of Timiryazev, Gustavson, and the other professors at the Petrovskaya (now Timiryazev) Agricultural Academy and in attending lectures on Russian village economy.

When Zandrok's term of exile was up she returned to St. Petersburg. Morozov gave up his commission and accompanied her there, entering the St. Petersburg Forestry Institute in 1889. Morozov's father thought that he was mad and refused him financial aid; he broke with his family and was forced to live on the meager earnings that he made giving lessons. At the Forestry Institute Morozov's teachers included the botanist I. P. Borodin, the soil scientist P. A. Kostichev, and the zoologist N. A. Kholodkovsky. His scientific views were, however, more strongly influenced by the anatomist and social activist P. F. Lesgaft, whom he had met in the Zandrok household, Lesgaft having been introduced into the family circle by O. N. Zandrok's sister, Lidia Nikolaevna. When Lesgaft, who had been dismissed from the University of St. Petersburg for his radical social opinions, began to give an anatomy course in his own home, Morozov was his most eager auditor. From Lesgaft, a convinced evolutionist, Morozov learned to consider the mutual relationship of the form and function of an organism and to view the animal within its environment; he learned to think in broad terms and to see in each biological phenomenon the outcome of development under the influence of a complex chain of interrelationships in nature.

Morozov's student years were saddened by the death from diphtheria of O. N. Zandrok. He remained on good terms with her family, however, and after a period of shared mourning married her sister. He graduated from the Forestry Institute in 1894 and became an assistant forester and teacher at the school in the Khrenovk forest preserve in Voronezh gubernia, where he was faced with the complex problems of managing a forest on sandy soil in a dry climate. His first article, published in 1896, was "O borbe s zasukhoy pri kulture sosny" ("On the Struggle Against Drought in the Culture of Pine Trees").

In May 1896 Morozov received a commission to go abroad to study forest management in Germany and Switzerland for two years. He attended lectures on forestry at Munich University and at the Eberswalde Academy and met the leading German specialists in forest economy. Shortly after his return to Russia he was appointed director of the Kammeno-Steppe experimental forestry preserve in Voronezh gubernia, one of the preserves created by Dokuchaev in an attempt to prevent drought. (The drought of 1891 had led to a terrible famine in the steppe chernozem zone.)

It was Dokuchaev who had founded soil science in Russia, and in 1899 Morozov, who had completed a zealous study of Dokuchaev's works, published a paper of his own, "Pochvovedenie i lesovodstvo" ("Soil Science and Forestry"). In this, and in a series of other works, Morozov proposed the establishment

of forest management as a specific discipline with a theoretical basis in forestry and forest economy. Morozov's work became widely known, and in 1901 he was appointed professor of forestry at St. Petersburg University; from 1904 until 1919 he also edited the *Lesnoy zhurnal* ("Forest Journal"). He had always been an ardent advocate of women's education, and he participated actively in the creation of the first women's institution of higher learning devoted to agricultural sciences—the Stebutovsky Higher Women's Agricultural Courses—of which he was the director from 1905. He took part in congresses on forestry and lectured to foresters throughout Russia; he was head of the Forestry Institute until 1917.

Morozov began to suffer from a serious nervous disorder in 1904, and his health, undermined by the deprivations of his student years and by overwork, rapidly grew worse until in 1917 he was forced to leave St. Petersburg for the milder climate of the Crimea. His condition was aggravated by professional frustration; all about him he observed the senseless, irremediable despolation of Russia's forests. Morozov's health did not improve in the Crimea; isolated from his work, his friends, and from a scientific environment, he had nothing to live for. He therefore joyfully accepted the offer to give a course in forest management when a new university was opened in Simferopol in 1918. Those who knew him at this time were struck by the contrast between his physical disability and the cheerfulness and enthusiasm that he brought to his work. He died two years later, at the age of fifty-three.

Morozov's work laid the theoretical bases for rational forest management. His theory of the forest as "a single complex organism with regular interconnections among its parts and, like every other organism, distinguished by a definite stability" had a further general biological importance. Morozov showed that "a forest is not simply an accumulation of trees, but is itself a society, a community of trees that mutually influence each other, thus giving rise to a whole series of new phenomena that are not the properties of trees alone." He further stated that trees in a forest display an influence "not only on each other, but also on the soil and atmosphere . . .." In one of his last articles, in 1920, he wrote, "The forest is not only a community of trees; it is also a community of a broader order, in which . . . plants are adapted to each other, as well as animals to plants and plants to animals; all this is influenced by the external environment . . .." Morozov also showed the forest to be a geographical as well as a biological phenomenon. He made detailed investigations of the interrelationships of plants, animals, soils, and their geographical distribution; he may thus be considered to have been one of the founders of the modern studies of phytosociology, phytobiology, phytogeography, biogeography, and ecology.

Morozov was a firm Darwinist, maintaining that "the forest is not some sort of homogeneous thing in space, unchanging in time," but that "for every forest community, as for every living substance, there is a tendency toward development." He drew upon a wealth of ecological data to reach a theory of the transformation of species, showing that the replacement of species can be understood only in the dynamical context of climate, geography, soil, plant communities, the activities of man as they affect natural processes, and the complex interrelationships among all these factors. His theory of the types of forest plantation had considerable practical significance. He distinguished among types of plantations according to soil composition, climate, geological considerations, topography, the processes by which the soil had been formed, and distribution of plants. Morozov considered this concept of plantation type to be analogous to the botanical-geographical classification of vegetational zones.

BIBLIOGRAPHY

I. ORIGINAL WORKS. Morozov's major writings are "Pochvovedenie i lesovodstvo" ("Soil Management and Forest Management"), in *Pochvovedenie* (1899), no. 1; *Biologia nashikh lesnykh porod* ("Biology of Our Forest Species"; St. Petersburg, 1912); *Les kak rastitelnoe soobshchestvo* ("The Forest as a Plant Society"; St. Petersburg, 1913); "O biogeograficheskikh osnovaniakh lesovodstva" ("On the Biogeographical Bases of Forest Management"), in *Lesnoy zhurnal* (1914), no. 1; *Les kak yavlenie geograficheskoe* ("The Forest as a Geographical Phenomenon"; St. Petersburg, 1914); *Smena porod* ("The Replacement of Species"; St. Petersburg, 1914); *Uchenie o tipakh nasazhdeny* ("The Theory of Types of Plantation"; Petrograd, 1917; Moscow–Leningrad, 1931); *Uchenie o lese* ("Theory of the Forest"), 7th ed. (Moscow, 1949); and *Ocherki po lesokulturnomu delu* ("Sketches on Forest Matters"), 2nd ed. (Moscow–Leningrad, 1950).

II. SECONDARY LITERATURE. See I. G. Beylin, *G. F. Morozov—vydayushchysya lesovod i geograf* ("G. F. Morozov—Distinguished Forest Manager and Geographer"; Moscow, 1954); *Istoria estestvoznania v Rossii* ("History of the Natural Sciences in Russia"), III (Moscow, 1962); V. G. Nesterov, "G. F. Morozov," in *Vydayushchiesya deyateli otechestvennogo lesovodstva* ("Distinguished Workers in Forestry of Our Country"), no. 2 (Moscow–Leningrad, 1950); and V. N. Sukachev and S. I. Vanin, *G. F. Morozov kak ucheny i pedagog* ("G. F. Morozov as Scientist and Teacher"; Leningrad, 1947).

S. R. MIKULINSKY

MORSE

MORSE, EDWARD SYLVESTER (*b.* Portland, Maine, 18 June 1838; *d.* Salem, Massachusetts, 20 December 1925), *zoology.*

Morse was the son of Jonathan Kimball Morse, a businessman, and Jane Seymour Beckett Morse, who claimed descent from Thomas à Becket and who was much interested in science. He collected shells when very young, even corresponding with such experts as A. A. Gould and Amos Binney, and showed a remarkable drawing ability.

Morse attended Bethel Academy and Bridgeton Academy, both in Maine, before going to Harvard as one of Louis Agassiz's special students at Lawrence Scientific School, where he had an assistantship from 1859 to 1862. He then went to the Essex Institute in Salem, Massachusetts; and when the Peabody Academy of Science (later Peabody Museum) was founded there in 1867, Morse became curator of its Radiata and Mollusca. From 1871 to 1874 he was professor of zoology at Bowdoin College, and from 1877 to 1880 was a professor at Tokyo University, before returning to the Peabody Academy as a highly effective director (emeritus from 1916).

Among many other honors Morse received an honorary Ph.D. from Bowdoin (1871) and a D.Sc. from Yale (1918); he was elected to the National Academy of Sciences in 1876. With Alpheus Hyatt, Alpheus Spring Packard, and Frederic Ward Putnam he founded the *American Naturalist* in 1867.

From his childhood interest in shells, Morse went on to considerable work with mollusks, including pioneering studies on land snails of Maine. He soon specialized in brachiopods, which were then considered mollusks. Through detailed studies on the anatomy and the larval development of this relict group, he proved them to be more closely affiliated with worms.

Having gone to Japan in 1877 to enlarge his brachiopod studies, Morse established one of the earliest marine stations, at Enoshima. Charmed by the country, he delved deeply into its architecture, archaeology, and midden pottery. The Boston Museum of Fine Arts bought his great collection.

Morse illustrated effectively all his own articles and some for others, most notably Gould's *Invertebrata of Massachusetts.*

BIBLIOGRAPHY

I. ORIGINAL WORKS. Morse's most significant publications on mollusks and brachiopods were "Observations on the Terrestrial Pulmonifera of Maine," in *Journal of the Portland Society of Natural History*, **1**, no. 1 (1864), 1–63; and "Systematic Position of the Brachiopoda," in *Memoirs of the Boston Society of Natural History*, **15** (1873), 315–

372. He also wrote many valuable shorter papers in these fields. The Peabody Museum at Salem has a full bibliography of his publications, in both science and Japanese studies.

II. Secondary Literature. An account of Morse's life and interests, by L. O. Howard, is in *Biographical Memoirs. National Academy of Sciences*, **17**, 1–29; it includes a selected bibliography and citations to shorter memorials on Morse. A fine tribute by J. S. Kingsley appeared in *Proceedings of the American Academy of Arts and Sciences*, **61** (1925–1926), 549–555.

Elizabeth Noble Shor

MORSE, JEDIDIAH (*b.* Woodstock, Connecticut, 23 August 1761; *d.* New Haven, Connecticut, 9 June 1826), *geography.*

The son of Jedidiah Morse, holder of local offices and deacon of the Congregational Church, and Sarah Child, Morse was educated in the Academy of Woodstock. He entered Yale College in 1779, graduating in 1783. He remained in New Haven to study theology and supported himself by conducting a school for young girls. In 1785 he became pastor of the church at Norwich, Connecticut, but returned to Yale as tutor in 1786. During part of 1787 Morse was pastor of a church in Midway, Georgia, after which he returned to Yale. In 1789 he was installed as minister of the First Parish Church of Charlestown, Massachusetts, where he remained until 1819. In 1789 he married Elizabeth Ann Breese of New Jersey. Of their eleven children only three sons survived to maturity, one of whom was Samuel Finley Breese Morse, painter and inventor of the telegraph. Throughout his career at Charlestown Morse took a leading (and often inflexible) part in upholding orthodox Calvinist views in opposition to the growing liberal Unitarianism in New England. Morse was one of the founders of Andover Theological Seminary (1808) to train orthodox ministers, and of the Park Street Church in Boston (1809) to provide an orthodox center in that city. He moved to New Haven in 1820. He visited Indian tribes in the Northwest as a representative of the government and prepared a report of his investigation in 1822.

For use in his school in New Haven in 1784, Morse wrote *Geography Made Easy*, the first geography to be published in the United States. This was so well received that he extensively revised and extended the work and published in 1789 the famous *American Geography; or A View of the Present Situation of the United States of America*. It was an immediate success and the edition of 3,000 copies was soon sold. The book was quickly reprinted in Edinburgh, Dublin, and London, and translated into German and Dutch, but no benefit except scholarly recognition came to the author from the foreign editions. A second edition, published in 1793, *The American Universal Geography*, was much enlarged and a second volume added on "The Eastern Continent." Later editions included even more on foreign countries. Morse was immediately established as the "American Geographer" and as such he commanded the field for the next twenty-five years.

In the preparation of his books Morse sought the aid by consultation and correspondence of any who would contribute information. Elaborate questionnaires were circulated far and wide and produced information of varying reliability and importance. Many well-known people provided data. He submitted sections for criticism to men like Jeremy Belknap, who reviewed and corrected his work on New England. Morse's limited travels to the several states provided some firsthand information, but he was in no sense a field geographer.

The first edition of the *American Geographer*, and also later editions, contained much historical and political material as well as sections on the "Face of the Country" for the various states and regions. The latter contained paraphrases or direct quotations (not always acknowledged) from the work of Lewis Evans, Thomas Pownall, Robert Rogers, Jonathan Carver, John and William Bartram, Thomas Jefferson, Thomas Hutchins, Mark Catesby, Noah Webster, John Filson, Gilbert Imlay, Samuel Mitchell, and many others. They form an instructive summary of geomorphology and geology of the United States in 1789. Although his own interests were historical and political geography, Morse paid enough attention to topography in his travels to understand physical geography and geology as described by others. One of the greatest criticisms of his geographies was that their maps were small and inadequate.

BIBLIOGRAPHY

I. Original Works. Morse's earliest geography was *Geography Made Easy* (1784). This went through 20 editions by 1819. His great work was *American Geography* (Elizabethtown, 1789; London, 1792, also Edinburgh and Dublin); the 2nd ed., in two vols. (now called *The American Universal Geography*), with the 2nd vol. on the eastern hemisphere (Boston, 1793), was followed by successive two-volume eds. to the 7th in 1819. *The American Gazetteer* (Boston, 1796) was followed by successive editions and abridgments; *The New Gazetteer of the Eastern Continent* (Charlestown, 1802), with Elijah Parish, was followed by several editions to 1823. His historical works, sermons, editorials and other theological works are listed in the

50-page typed bibliography in the Yale University Library; a copy is in the Clements Library, University of Michigan. The locations of the extraordinarily voluminous Morse papers and correspondence are given in detail by J. K. Morse, R. H. Brown, and W. R. Waterman.

II. Secondary Literature. A biography by his son, R. C. Morse, remains unpublished. The only published biography is W. B. Sprague, *The Life of Jedidiah Morse, D.D.* (New York, 1874). It contains excerpts from letters and a long chapter of personal notes from Morse's three sons and from his other associates. The short biography by W. R. Waterman in the *Dictionary of American Biography* is an excellent summary. J. K. Morse, *Jedidiah Morse, a Champion of New England Orthodoxy* (New York, 1939), with elaborate bibliographies, is devoted entirely to Morse's career as a minister and strenuous advocate of orthodox Calvinism, without any mention of his geographical work. R. H. Brown, "The American Geographies of Jedidiah Morse," in *Annals of the Association of American Geographers*, **31** (1941), 145–217, is the definitive work on Morse as a geographer, his geographical associations, and sources. It contains Winfield Shires' "Geographical Works of Jedidiah Morse, a Brief Bibliography, . . . adapted and simplified from a list of the works of Jedidiah Morse with notes, from the fifty typewritten pages in the Yale Library," 214–217.

George W. White

MORTILLET, LOUIS-LAURENT GABRIEL DE (*b.* Meylan, Isère, France, 29 August 1821; *d.* St.-Germain-en-Laye, France, 25 September 1898), *archaeology, anthropology.*

Mortillet was educated at a Jesuit seminary in Chambéry and in Paris, where he became a revolutionary freethinker. He took part in the revolution of 1848 and was a fervent disciple of Ledru-Rollin. He was found guilty of violating the laws governing the press, and left France in 1849 for Switzerland and Savoy, where he engaged in scientific and archaeological work. His scientific work was concerned mainly with zoology (particularly the study of mollusks) and geology. He organized and classified the material in the museums of Geneva and Annecy. In addition to many scientific papers he was author of the *Guide de l'étranger en Savoie* (Chambéry, 1856), generally accepted as a model of its kind and an early demonstration of his flair for popularization. In Italy he directed works exploiting hydraulic lime and collaborated on the construction of railroads in Lombardy. Together with Stoppiani and Édouard Desor he explored the lakes of Lombardy and discovered the first Italian Neolithic settlement, at Isolino (Isola di San Giovanni), on Lake Varese, in 1863.

He edited the *Revue scientifique italienne*, published in French at Turin (1862–1863) as part of the political daily *Italie*; under his direction it became a complete and masterly summary of current scientific progress.

Mortillet returned to Paris in 1864, at a period when the study of prehistoric man through the interpretation of archaeological remains was in its infancy. Indeed, the word "prehistory" itself was not widely known until John Lubbock published his *Prehistoric Times* (1865): in Italy, Mortillet and his colleagues had used the term "antéhistoire." Lyell's *Antiquity of Man* and Huxley's *Man's Place in Nature*, both published in 1863, greatly affected Mortillet, and he decided to devote himself to the new and growing science of early man. In September 1864 he founded in Paris a new journal, *Matériaux pour l'histoire positive et philosophique de l'homme*, to discuss and summarize all new prehistoric discoveries. Salomon Reinach has described the founding and editing of *Matériaux* as one of the greatest services ever rendered to the development of prehistoric science in France. Mortillet edited it with vigor and distinction and in his first year exposed the forgeries of M. Meillet in the rock shelters of Poitou.

The first public recognition of prehistory at a scientific congress occurred at the meeting of the Italian Scientific Congress at La Spezia in 1865. Mortillet was invited by the president to give a survey of prehistory, and as a result it was decided to found an international congress of anthropology and prehistoric archaeology. Mortillet was secretary of the first two congresses, held in Neuchâtel in 1866 and in Paris the following year.

In 1867 Mortillet was appointed secretary of the committee charged with setting up an *Exposition des oeuvres caractérisant les diverses époques de l'histoire du travail* at the Paris universal exposition, and he prepared a guide, *Promenades préhistoriques à l'Exposition universelle* (Paris, 1867), for it. In his enthusiastic appraisal of prehistory as a new discipline Mortillet declared that prehistorians had already discovered three important facts, which he termed the *loi de progrès de l'humanité*, the *loi de développement similaire*, and the *haute antiquité de l'homme*.

In 1867 Mortillet joined the staff of the newly created Musée des Antiquités Nationales at St.-Germain-en-Laye and subsequently became its director. Largely responsible for saving the museum during the Franco-Prussian war, he directed it until 1885, when he became deputy to the national assembly from Seine-et-Oise. He held this post for four years, always voting with the extreme left.

The pressure of work at St.-Germain forced him to give up the editorship of *Matériaux* in 1872, but his keen journalistic sense and devotion to the need for publishing archaeological results in an easily accessible form led him in September 1872 to found a new journal, *Indicateur de l'archéologue*, which lasted only two years, and then *Homme*, published every two weeks from 1884 to 1887. In a spate of learned papers, published beginning in 1869, he dealt with such varied subjects as Paleolithic art, megalithic monuments and bronze axes, lake dwellings, and the archaeology of the Celts. His views on prehistory were set out in *Musée préhistorique* (Paris, 1881), written with his son Adrien, and in *Formation de la nation française* (Paris, 1897).

By the early 1870's Lubbock had expanded the basic three-age system of Thomsen and Worsaae into the four-age system of Paleolithic, Neolithic, Bronze, and Iron ages. Mortillet subdivided these ages into periods, and the periods into epochs. His "Tableau de la classification," published in *Musée préhistorique*, proposed a succession of fourteen epochs. This idea of successive epochs as well as its underlying theoretical structure represented an extension of the idea of stratigraphic geology to prehistory. It predominated until the mid-1920's, when the concept of culture was borrowed from anthropologists and anthropogeographers, and the prehistorian's role was seen to be the definition and description of cultures. Certain French archaeologists had already realized that Mortillet's fourteen epochs did not represent a true and universal succession. If the Thomsen-Worsaae three-age system was the foundation stone of modern prehistory, then Mortillet's idea of epochs was an important stepping-stone between the beginnings of classification in prehistory and the ideas of the second and third quarters of the twentieth century.

From the time of his decision to devote himself to prehistory Mortillet remained in the forefront of French archaeology. He became professor of prehistory at the school of anthropology in the École des Hautes-Études in Paris. In 1878 the French government established a commission charged with listing and classifying the megalithic monuments of France and Algeria. Henri-Martin was appointed the first president and Mortillet vice-president; on Henri-Martin's death, Mortillet became president. He retained his interest in the International Congress of Anthropology and Prehistoric Archaeology, which he had helped to found, and participated in subsequent meetings until 1880. Following the discovery in 1879 of the Altamira cave paintings and the controversy over the authenticity of Upper Paleolithic art, Mortillet wisely accepted their true nature. "This is not the art of a child," he declared, "It is the childhood of art."

BIBLIOGRAPHY

A listing of nearly 100 of Mortillet's papers may be found in Royal Society *Catalogue of Scientific Papers*, IV, 487; VI, 730; VIII, 443–444; X, 858; XII, 521; and XVII, 367–368. On his life and work, see E. Cartailhac's obituary in *Anthropologie*, **9** (1898), 601–612.

GLYN DANIEL

MORTON, JOHN (*b.* England, 18 July 1670–18 July 1671; *d.* Great Oxendon, England, 18 July 1726), *natural history.*

Morton, rector of Great Oxendon, wrote *The Natural History of Northamptonshire* (1712). In it he discusses the general geography, topography, natural history, and prehistory of the county. The work shows careful observation and the descriptions are generally good. Morton chose popular theoretical assumptions as bases for his commentary.

His observations on geology and paleontology are of interest. Although Morton knew John Ray, Martin Lister, and others who were especially concerned with geology and with fossils, he chose to follow the ideas of John Woodward. The latter believed that the biblical Deluge was responsible for geological features and for the presence of fossils, and Morton applied this assumption to his regional study. As Woodward had done, Morton interpreted the strata as having originated in water and as having settled out of the Flood waters according to the specific gravity of the matter of which they were composed. The remains of invertebrate sealife and the teeth and bones of land vertebrates destroyed by the Deluge settled out concurrently, also according to their specific gravity, and were entombed in the strata as fossils.

The botanical section of Morton's *Natural History* received significant attention from his contemporaries and is notable for its attempt to arrange the flora of Northamptonshire systematically. The arrangement is principally that of John Ray's *Synopsis methodica stirpium Britannicarum* (1690).

Morton was elected a member of the Royal Society in 1703. In 1705 a letter from him on fossils was published in the *Philosophical Transactions*. It is Morton's only other publication.

BIBLIOGRAPHY

Morton's two works are "Letter From the Reverend Mr. Morton Containing a Relation of River and Other Shells Digg'd up, Together With Various Vegetable Bodies, in a Bituminous Marshy Earth, Near Mears Ashby in Northamptonshire: With Some Reflections Thereupon:

As Also an Account of the Progress He Has Made in the Natural History of Northamptonshire," in *Philosophical Transactions of the Royal Society*, **25** (1705), 2210–2214; and *The Natural History of Northamptonshire; With Some Account of the Antiquities. To Which Is Annexed a Transcript of Doomsday-Book, so Far as It Relates to That County* (London, 1712).

A biographical notice is George Simonds Boulger, "John Morton," in *Dictionary of National Biography*, XIII, 1050–1051.

PATSY A. GERSTNER

MORTON, SAMUEL GEORGE (*b.* Philadelphia, Pennsylvania, 26 January 1799; *d.* Philadelphia, 15 May 1851), *anthropology*.

Morton was the last of nine children of George and Jane Cummings Morton (he a native of Ireland, she a birthright Friend). He was educated at the Quakers' Westtown School in Pennsylvania and at the Friends' School in Burlington, New Jersey, kept by John Gummere, mathematician and astronomer. An early interest in science was encouraged by his stepfather, Thomas Rogers, an amateur mineralogist. After a period as a merchant's clerk, he became a student of Dr. Joseph Parrish of Philadelphia and attended the University of Pennsylvania Medical School, graduating in 1800. With support from a wealthy Irish uncle, Morton studied medicine at Edinburgh University, where he also attended Robert Jameson's geology lectures, and received a second M.D. degree in 1823; his thesis was *De corporis dolore*. In 1827 he married Rebecca Pearsall of Philadelphia; they had eight children, seven of whom survived their father.

Morton was inclined to study and had time as well because, without influential connections, he acquired a practice slowly. His first medical paper, on the use of cornine to treat intermittent fevers, appeared in 1826; in another he recommended fresh air for pulmonary consumption. He was a physician to the Almshouse in 1829; taught after 1830 in the Philadelphia Association for Medical Instruction, which grew out of Parrish's private lectures; prepared an American edition of John Mackintosh's *Principles of Pathology* in 1836; was professor of anatomy in the Medical Department of Pennsylvania College (1839–1843); and in 1849 published *An Illustrated System of Human Anatomy*. He achieved this sound medical reputation while engaged in scientific research.

Richard Harlan, one of Parrish's assistants, directed Morton's scientific interests and saw that he was elected to the Academy of Natural Sciences of Philadelphia in 1820. The Academy was Morton's scientific home until his death. His papers were chiefly on geology and fossils; his description of fossil remains collected by Lewis and Clark, *Synopsis of the Organic Remains of the Cretaceous Group of the United States* (Philadelphia, 1834), marks him as a founder of invertebrate paleontology in the United States. As secretary of the Academy for many years, Morton maintained a wide scholarly correspondence. In 1849 he was elected president.

In 1830 Morton began collecting human craniums and eventually owned over 1,000 specimens. Never a field anthropologist, he depended on military and naval officers, consuls, physicians, missionaries, and naturalists for both skulls and identifications. He devised ingenious ways to measure and calculate the capacity of craniums and concluded that races are distinguished by their skulls as well as by color. His *Crania Americana* (1839), by its use of physical measurements, the classification and comparison of data, and its accurate drawings, was a landmark in anthropology. In an "Introductory Essay" of ninety-five pages Morton asserted that the American Indians are a separate race, not descendants of migrants from Asia. "We are left to the reasonable conclusion," he continued, "that each Race was adapted from the beginning to its peculiar local destination. In other words, it is assumed, that the physical characteristics which distinguish the different Races, are independent of external causes." The work was hailed in the *American Journal of Science* as "the most extensive and valuable contribution to the natural history of man, which has yet appeared on the American continent." In *Crania Aegyptiaca* (1844), based on 137 skulls collected by George R. Gliddon, United States consul at Cairo, Morton concluded that the ancient Egyptians were not Negroes and indicated that the races are of great antiquity. Morton's last important work on the subject, "Hybridity in Animals . . ." (*American Journal of Science*, 2nd ser., **3** [1847], 39–50, 203–212), rejected the argument of other anthropologists that the capacity of individuals of different races to produce fertile progeny is a proof of the unity of the human species.

Morton's conclusions were in conflict with the biblical version of creation, James Ussher's chronology, and widely accepted ideas, reinforced in the United States by ideals of the Enlightenment, that all men are descended from a single pair and that racial differences are the effects of environment. Morton preferred not to press the issue with his critics. His views received support in the scientific community from Louis Agassiz; approval of another kind came from Josiah C. Nott, of Mobile, Alabama, who used them enthusiastically to attack the orthodox clergy and to support notions of white superiority congenial to the American South. At the time of his death Morton was said to have been planning a comprehensive work to be called *Elements of Ethnology*.

BIBLIOGRAPHY

I. ORIGINAL WORKS. Most of Morton's scientific writings, in addition to works cited in the text, appeared in *American Journal of Science*, *Journal of the Academy of Natural Sciences of Philadelphia*, *Philadelphia Journal of the Medical and Physical Sciences*, and *Transactions of the American Philosophical Society*. A fairly complete bibliography is in Charles D. Meigs, *A Memoir of Samuel George Morton, M.D. . . .* (Philadelphia, 1851), 45–47. His books include *Crania Americana; or, a Comparative View of the Skulls of Various Aboriginal Nations of North and South America* (Philadelphia–London, 1839); and *Crania Aegyptiaca; or, Observations on Egyptian Ethnography* (Philadelphia–London, 1844), which first appeared as "Observations on Egyptian Ethnography, Derived from Anatomy, History, and the Monuments," in *Transactions of the American Philosophical Society*, n.s., **9** (1844), 93–159.

II. SECONDARY LITERATURE. Biographical data are recorded in memoirs prepared by medical colleagues after Morton's death: Charles D. Meigs, *A Memoir of Samuel George Morton, M.D.* (Philadelphia, 1851); William R. Grant, *Lecture Introductory to a Course on Anatomy and Physiology in the Medical Department of Pennsylvania College* (Philadelphia, 1852); and George B. Wood, "A Biographical Memoir of Samuel George Morton, M.D.," in *Transactions of the College of Physicians of Philadelphia*, n.s., **1** (1853), 372–388. A fuller sketch, with emphasis on Morton's ethnographic work, is by Henry S. Patterson, in Josiah C. Nott and George R. Gliddon, eds., *Types of Mankind* (Philadelphia, 1854), xvii–lvii. William Stanton, *The Leopard's Spots: Scientific Attitudes Toward Race in America, 1815–1859* (Chicago, 1960), 25–35, 39, 40–44, and *passim*, places Morton's work in its scientific and social background.

WHITFIELD J. BELL, JR.

MOSANDER, CARL GUSTAF (*b.* Kalmar, Sweden, 10 September 1797; *d.* Ångsholm, Sweden, 15 October 1858), *chemistry, mineralogy.*

Mosander began his career as a pharmacist, at the age of fifteen becoming an apprentice at a Stockholm apothecary. Seven years later he began medical studies and, after serving as an army surgeon, received the M.D. in 1825. In 1824 he was appointed teacher of chemistry at the Caroline Institute and shortly thereafter was put in charge of the chemistry laboratory. He succeeded Berzelius as professor of chemistry and pharmacy in 1832, when the latter retired. He held this position until his death. For a time Mosander and his wife lived with Berzelius, who had befriended the couple. Berzelius was taught Dutch by Mrs. Mosander. A close friendship also developed between Friedrich Wöhler and Mosander, who was affectionately referred to as "Father Moses" in correspondence between Wöhler and Berzelius.

Among Mosander's responsibilities was the mineral collection at the Royal Swedish Academy of Sciences, which was helpful to his research on minerals containing rare earth elements. In 1839 he showed that ceria (isolated independently by Berzelius and by Martin Klaproth in 1803) was really a mixture of earths. When he treated the substance with dilute nitric acid, a yellow residue, true ceria (cerium oxide), remained; he named the dissolved component lanthana (from Greek *lanthanein*, "to escape notice").

Two years later Mosander showed that lanthana was really a mixture of a white substance, lanthana, and a brown earth named didymia (Greek, "two") because of its similarity to lanthana. Didymia was regarded as a pure earth until 1885, when Auer von Welsbach decomposed it into praseodymia and neodymia. Mosander also separated yttria (discovered by Gadolin in 1794) into yttria, erbia, and terbia (the oxides of yttrium, erbium, and terbium) in 1843. These names were derived from the Swedish town of Ytterby, where the mineral was discovered. In 1839 Axel Erdmann discovered lanthana in a new Norwegian mineral, which he named mosandrite.

BIBLIOGRAPHY

I. ORIGINAL WORKS. Many of Mosander's discoveries were communicated orally at meetings of chemists; Wöhler and Berzelius have commented on his reluctance to publish. A summary of his important work is his article, "On the New Metals, Lanthanum and Didymium, Which Are Associated With Cerium; and on Erbium and Terbium, New Metals Associated With Yttria," in *Philosophical Magazine*, **23** (1843), 241–254. A short portion of this paper was published as "Lanthanum and Didymium, Erbium and Terbium: A Classic of Chemistry," in *Chemistry*, **20** (Sept. 1946), 53–59.

II. SECONDARY LITERATURE. Biographical material on Mosander is rather scanty. A short article describing his work is J. Erik Jorpes, "Carl Gustaf Mosander," in *Acta chemica scandinavica*, **14** (1960), 1681–1683. A more informal treatment and a photograph can be found in Mary Elvira Weeks and Henry M. Leicester, *Discovery of the Elements*, 7th ed. (Easton, Pa., 1968), 671–678.

SHELDON J. KOPPERL

MOSELEY, HENRY GWYN JEFFREYS (*b.* Weymouth, Dorsetshire, England, 23 November 1887; *d.* Gelibolu, Gallipoli Peninsula, Turkey, 10 August 1915), *physics.*

Like his friends Julian Huxley and Charles Galton Darwin, Harry Moseley came from a family long distinguished for its contributions to science. His paternal grandfather, Canon Henry Moseley (1801–1872), F.R.S., the first professor of natural philosophy at Kings College, London, was an international authority on naval architecture. His maternal grandfather, John Gwyn Jeffreys (1809–1885), F.R.S., was the dean of England's conchologists. His father, Henry Nottidge Moseley (1844–1891), F.R.S., a protégé of Charles Darwin's, was the founder of a strong school of zoology at Oxford. Harry inherited a full share of his ancestors' energy, intelligence, and singleness of purpose.

H. N. Moseley died just before his son's fourth birthday. The family then moved to Chilworth, near Guildford, in Surrey, where Harry and his two older sisters received their elementary education. The elder sister died in childhood; the other, Margery, four years Harry's senior, was his constant companion, confidante, and collaborator in neighborhood bird's-nesting expeditions. Harry's mother and such old family friends as E. R. Lankester encouraged the children in their natural-historical interests, which they never discarded. Margery later published a paper about protozoa in the *Journal of the Royal Microscopical Society*, and Harry, who became an enthusiastic gardener, was to fill his last letters from Gallipoli with observations of the flora near his campsites. At the age of nine Harry was sent to Summer Fields near Oxford, a preparatory school that specialized in training King's Scholars, winners of scholarships to Eton. He and six of his fellow students won seven of the dozen scholarships available for 1901.

Although Edwardian Eton tended to produce over-cocky, undereducated sportsmen, those who preferred not to conform, to be "saps" like most King's Scholars, could obtain a thorough grounding in almost any respectable branch of learning. After passing through the required curriculum with a prize for Latin prose, Harry specialized under T. C. Porter, the discoverer of Porter's law of flicker and one of the first Englishmen to experiment with X rays. Harry fought with most of the stodgy science masters, but found Porter's unconventional approach—a minimum of books and a maximum of challenging experiments—exactly to his taste. Before leaving Eton he had begun quantitative analysis, had measured the freezing point of saline solutions of diverse concentrations, and had discovered, to his "astonishment," that helium, despite its lower boiling point, is denser than hydrogen. For the rest Harry shared quietly in Eton life. He won a place in the college boats, participated in college debates, and found nourishment for his natural patrician aloofness.

In 1906 Harry entered Trinity College, Oxford, on one of the few science scholarships then available in the university. He was the only important early contributor to atomic physics who attended Oxford, which he chose over Cambridge in order to be near his mother, who had returned there in 1902. He read for honors in physics, rowed for his college, and worked in the Balliol-Trinity laboratory where, among other things, he built a primitive Wilson cloud chamber. The official curriculum provided no opportunity for the kind of independent work to which Porter had exposed him; and partly from tension over his course of study he obtained a second-class degree in the bookish Oxford honors schools. Nonetheless with the help of testimonials gathered before the examinations and a visit to Rutherford, he obtained a demonstratorship in the physics department at the University of Manchester for the fall of 1910.

Harry found the students dense and provincial and taught them without enthusiasm. But he worked happily as much as sixteen hours a day on the research problem Rutherford had given him—to determine how many β particles are expelled in the disintegration of one atom of radium B (Pb^{214}) or radium C (Bi^{214}). No one at Manchester doubted that the answer would be one; but it was an important fact to establish, and gave the neophyte opportunity to master the techniques of high vacuums and radioactivity. Difficulty with the apparatus prevented completion of the work until 1912. Meanwhile Harry also collaborated on small projects that trained him to handle α particles and γ rays. These exercises, typical of Rutherford's laboratory, came to an end in the spring of 1912, along with Harry's teaching duties. In the fall he was to enjoy a fellowship established by a Manchester industrialist and the congenial obligation of devoting himself entirely to research.

As Moseley passed from being Rutherford's apprentice to independent journeyman, news of the success of Laue's experiment on the diffraction of X rays by crystals reached Manchester. Moseley decided to exercise his new independence in investigating how the wavelike properties of the rays, established by Laue, could be reconciled with the phenomena from which W. H. Bragg had persuasively deduced their particulate nature. Anticipating that he would need the help of a mathematical physicist, Moseley asked C. G. Darwin, then also between researches, to collaborate. They approached Rutherford for permission. The master refused, apparently on the ground that he knew nothing about the subject and consequently could not offer his usual paternal guidance. Moseley insisted, won, and went off to Leeds to W. L. Bragg, who had generously agreed to instruct him. It appears that Moseley's habit of having his own way and his Etonian experience with circumventing inconvenient masters made possible the researches that made his reputation.

To explain his diffraction patterns Laue assumed that the radiation striking the crystal contained precisely six groups of effective monochromatic rays. Darwin and Moseley rejected the assumption, which indeed was scarcely plausible, and concluded instead that only certain planes through the crystal—namely, those "rich" in atoms—gave rise to sensible interference. W. L. Bragg reached the same conclusion earlier and undertook to confirm it by reflecting X rays from the atom-rich cleavage surface of mica. He found, as he had anticipated, that in reflection the crystal did behave like a pile of semitransparent mirrors, causing interference of reflected radiation of wavelength λ incident upon the surface at glancing angle θ in accordance with the formula $n\lambda = 2d \sin \theta$, where n is the order of the interference and d the separation of the atom-rich planes. Moseley adopted Bragg's method with the important exception, typical of Rutherford's school, of substituting an ionization chamber for a photographic plate as detector of the reflected radiation. In their first research, in which the Braggs again anticipated them, Moseley and Darwin determined that the reflected radiation had the same penetrating power as the incident. They next undertook to measure the intensity of reflection I as a function of θ; their results, obtained with the sensitive ionization chamber, quickly became standard. The narrowness of their collimation, however, caused them to miss places where $I(\theta)$ had sharp peaks. The Braggs found the peaks, which Darwin and Moseley then examined carefully. All agreed in referring the maxima to monochromatic radiation characteristic of their platinum anticathodes and identical to the L rays earlier identified by Barkla.

On the publication of their paper in July 1913, the collaborators separated, Darwin to consider theoretical problems relating to $I(\theta)$ and Moseley to map the characteristic K and L spectra of the elements. It appears that Moseley aimed to test the doctrine of atomic number, then recently suggested by van den Broek of Amsterdam, and doubtless also independently invented at Manchester where, besides Moseley, Bohr, Russell, and Hevesy also worked in 1912. According to measurements made at the Cavendish Laboratory in 1911 by R. Whiddington,

the minimum velocity w that a cathode ray requires to stimulate the emission of a K ray from a target of atomic weight A is $w = A \cdot 10^8$ cm/sec. The doctrine of atomic number, however, would suggest a dependence on Z, rather than on A; and apparently Moseley's initial intent was to discover, by examining those places in the periodic table where the sequence of weights inverts the chemical order, whether the frequency ν_K followed A or Z. Very likely he expected ν_K to grow roughly as Z^2; for assuming that the entire energy of Whiddington's electron goes into the K ray,

$$\nu_K \approx (m/2h) A^2 \cdot 10^{16} = 2.7 Z^2 \cdot 10^{15} \text{ sec}^{-1},$$

where the approximation $Z = A/2$, derived from Rutherford's famous analysis of α scattering, has been used. By no means did Moseley expect the coefficient of Z^2 to be exactly constant independent of Z. Among other considerations, the atomic model of his friend Niels Bohr, which assumed that the population of the innermost ring changed with Z, ruled out any such regularity.

Moseley began his experiments in the fall of 1913, using a clever device that enabled him to bring different anticathodes into position without interrupting the vacuum in his X-ray tube. He photographically recorded the position of constructive interference and found the K rays to consist of a soft, intense line, which he called K_α, and a harder, weaker satellite K_β. The L rays appeared to be more complicated, a soft intense line L_α and several weaker companions. Measurements of Co and Ni immediately showed, as expected, that ν_{K_α} followed Z. In addition, and most unexpectedly, the frequencies for ten elements from Ca to Zn satisfied to a precision of 0.5 percent the astonishingly simple relation

$$\nu_{K_\alpha}/R = (3/4)(Z-1)^2,$$

where R stands for the Rydberg frequency. Emboldened by this formula, and especially by the appearance of the Bohr-Balmer coefficient $3R/4$, Moseley guessed on very little evidence that

$$\nu_{L_\alpha}/R = (5/36)(Z-7.4)^2.$$

These results were published in the *Philosophical Magazine* for December 1913, about one month after the photographic measurements began.

Moseley quickly perceived that if these formulas held exactly, they could be used to test the periodic table for completeness; he need only obtain ν_{K_α} and/or ν_{L_α} for the known elements and determine which values of Z (if any) were not represented. The task offered no difficulties of principle. The procurement of rare substances, however, presented a severe problem which could not be confronted with any particular advantage in Manchester. Since at this juncture (December 1913) Moseley considered that Rutherford had no more to teach him, he decided to resign his fellowship and to migrate to Oxford. There he could be near home and might build up local support for a possible candidacy for the professorship of experimental physics, then still held by its first incumbent, R. B. Clifton, whose overdue resignation was expected momentarily. Moseley obtained permission to work in a private capacity in the new electrical laboratory of Townsend, where he carried his examination of the elements into the fuzzy family of the rare earths.

Chemists then officially recognized thirteen "lanthanides" stretching from cerium to lutetium, the most recently accepted elemental earth, which had been independently isolated in 1906/1907 by Auer von Welsbach and Urbain. There were a great many other putative earths, including Auer's thulium II and Urbain's celtium; and it was this confusion that Moseley proposed to resolve with his X-ray machine. After hasty examination he announced with his usual confidence that three elements remained undiscovered between aluminum and gold. Two of these, numbers 43 and 75, corresponded to spaces the chemists had long recognized. The other, 61, lay in the middle of the lanthanides which, according to Moseley, began at cerium 58 and terminated with Auer's two thuliums, 69 and 70, and Urbain's ytterbium 71 and lutetium 72.

This assignment, which Moseley deduced without examining any samples of the heaviest earths, annoyed Urbain, as it left no room for celtium and established the second thulium of his rival Auer. In May 1914 Urbain brought his best specimens to Oxford, where the X rays revealed (to his mixed satisfaction) neither celtium nor thulium II nor any trace of element 72 and demonstrated that the known lanthanides terminated at lutetium, which Moseley had given a Z one unit too high. Elements 43 and 61 do not appear to exist naturally; number 75 was found among manganese ores as expected; and hafnium, which is not a rare earth, came to light in Copenhagen in the fall of 1922 after Urbain, in violation of one of Bohr's theories, had again tried to install celtium as element 72.

Moseley's formulas not only terminated what Urbain called the "romantic age of chemical discovery"; they also seemed to lift the veil guarding the atom's innermost recesses. Moseley himself argued that the factors

$$3R/4 = (1/1^2 - 1/2^2)R$$

and

$$5R/36 = (1/2^2 - 1/3^2)R$$

showed that Bohr's rules of behavior for the single electron of hydrogen also governed the deep-lying electrons of the heavy elements; in particular, according to Moseley, his formulas confirmed the principle of quantization of the angular momentum. The factor $(Z - 1)^2$ gave him more trouble; and, in the event, neither he nor Bohr was able to design a plausible model of X-ray emission that satisfied Bohr's principles and agreed even qualitatively with the K_α formula. Nonetheless Moseley expressed his confidence in a letter to *Nature*, for which he was roundly and unfairly cudgeled by F. A. Lindemann. In the fall of 1914 W. Kossel supplied a successful qualitative model of X-ray emission. No one, however, has managed to derive Moseley's formulas from Bohr's initial principles of atomic structure.

In June 1914 Moseley and his mother left England for a leisurely trip to Australia, where the British Association for the Advancement of Science was to meet in September. When news of the outbreak of war reached him, he determined to return home immediately upon concluding his obligations to the association—participation in a discussion on atomic structure and delivery of a paper on the rare earths. He practiced semaphore and Morse code aboard ship and upon arrival rushed to procure a commission in the Royal Engineers. Neither the solicitations of his family nor the initial refusal of the engineers deterred him from what he and his former classmates at Summer Fields and Eton considered their duty. His obstinacy eventually secured the commission, and after eight months' training he was shipped to the Dardanelles as signal officer of the thirty-eighth brigade of the new army. He participated in some minor skirmishes near Cape Helles before his unit went to reinforce the last desperate effort to reach the critical ridge of Sari Bair. He was killed in a furious counterattack led by Kemal Ataturk.

BIBLIOGRAPHY

I. ORIGINAL WORKS. Moseley's important papers are "The Number of β Particles Emitted in the Transformations of Radium," in *Proceedings of the Royal Society of London*, **87A** (1912), 230–255; "The Attainment of High Potentials by the Use of Radium," *ibid.*, **88A** (1913), 471–476; "The Reflexion of X Rays," in *Philosophical Magazine*, **26** (1913), 210–232, written with C. G. Darwin; "The High-Frequency Spectra of the Elements," *ibid.*, 1025–1034, and **27** (1914), 703–713. A full list of Moseley's published writings is in Poggendorff and in J. B. Birks, ed., *Rutherford at Manchester* (London, 1962), 349. Moseley's correspondence, scientific and personal, has been published in J. L. Heilbron, *H. G. J. Moseley. The Life and Letters of an English Physicist, 1887–1915* (Berkeley, 1974).

II. SECONDARY LITERATURE. For Moseley's career and work see the *Life and Letters;* Rutherford in *Proceedings of the Royal Society of London*, **93A** (1916), xxii–xxviii; C. G. Darwin, "Moseley and the Atomic Number of the Elements," 17–26 in the volume edited by Birks; and (for high-schoolers) B. Jaffe, *Moseley and the Numbering of the Elements* (New York, 1971).

J. L. HEILBRON

MOSS, WILLIAM LORENZO (*b.* Athens, Georgia, 23 August 1876; *d.* Athens, 12 August 1957), *medicine, pathology.*

Moss was the son of Elizabeth Luckie Moss and Rufus Lafayette Moss. After being educated privately and in local public schools, he enrolled in the University of Georgia and earned his B.S. in 1902. He proceeded to The Johns Hopkins University Medical School and received his M.D. three years later. Several years later he undertook further study on the continent. Moss spent his career teaching and doing research at Johns Hopkins, at the State Institute for the Study of Malignant Diseases in Buffalo, New York, and at Yale and Harvard Universities. In 1926 he became acting dean of the School of Public Health at Harvard, where he served for one year. Moss served as dean of the University of Georgia School of Medicine from 1931 to 1934, when he retired and returned to Athens.

Moss' most renowned contribution to medicine was a classification system of the four blood groups, which he designated by the roman numerals I through IV. He took as the basis of these blood groups the content of the serum, because in a small sample experiment he found that Landsteiner's classification, which was based on the agglutination (clumping) properties of serum and red blood cells, did not hold in all cases. Landsteiner's blood groups were labeled A, B, O, and AB. A study published in 1929 by Moss' colleague, James Kennedy, revealed that prior to 1921, ninety percent of the hospitals that were surveyed used the Moss system and that five years later seventy-eight percent continued to use it, although a committee of immunologists recommended use of a rival system that was proposed by Jansky in which the blood groups were labeled in reverse to Moss' system, although the system had the same basis. The three systems relate in this way:

Moss	Jansky	Landsteiner
I	IV	AB
II	III	A
III	II	B
IV	I	O

With the advent of World War II and the need to systematize large quantities of blood for use in transfusions, the Landsteiner system was adopted on a worldwide basis and remains in use at present.

Moss gathered information on blood groups among a wide variety of the population of Santo Domingo during scientific expeditions in 1920 and 1925. His other scientific trips were to Peru (1916), the Pacific Islands (1928), and New Guinea (1937).

Moss was a member of the small group who in the early twentieth century studied the immunization properties of the blood components, especially in relation to tuberculosis, diphtheria, and allergy reactions leading to the extreme effects in anaphylactic shock.

His honors include decoration by the French government for service in World War I in the French Medical Corps and membership in the Cosmos Club and Phi Beta Kappa.

In 1925 he married Marguerite E. Widle and they had three children: Marguerite, Elizabeth, and William Lorenzo II.

BIBLIOGRAPHY

I. ORIGINAL WORKS. Moss' widow in Athens and the library of the University of Georgia retain the personal papers and correspondence. Moss' works include "Studies in Opsonins," in *Johns Hopkins Hospital Bulletin*, **18** (1907), 237–245; "Traumatic Pneumothorax," in *Journal of the American Medical Association* (1908), 1971; "A Recent Visit to Some of the Medical Laboratories Abroad," in *Johns Hopkins Hospital Bulletin*, **19** (1908), 188–192; "Studies on Iso-Agglutinins and Isohemolysins," in *Transactions of the Association of American Physicians*, **19** (1909), 419–437; "The Relationship of Bovine to Human Tuberculosis," in *Johns Hopkins Hospital Bulletin*, **20** (1909), 39–49; and "Tuberculosis: A Plan of Study," *ibid.*, p. 87.

See also "Studien über Isoaglutinine und Isohämolysine," in *Folia serologica*, **5** (1910), 267–276; "A Cutaneous Anaphylactic Reaction as a Contra-Indication to the Administration of Antitoxin," in *Journal of the American Medical Association*, **55** (1910), 776–777; "Studies on Isoagglutinins and Isohemolysins," in *Johns Hopkins Hospital Bulletin*, **21** (1910), 63–70; "Subcutaneous Reaction of Rabbits to Horse Serum," in *Journal of Experimental Medicine*, **12** (1910), 562–574, written with J. W. Mason Knox and G. L. Brown; "Paroxysmal Hemoglobinuria: Blood Studies in Three Cases," in *Johns Hopkins Hospital Bulletin*, **22** (1911), 238–247; "Paroxysmale Hamoglobinurie. Blutstudien in drei Fallen," in *Folia serologica*, **7** (1911), 1117–1142; "Concerning the Much–Holzmann Reaction," in *Johns Hopkins Hospital Bulletin*, **22** (1911), 278–282, written with F. M. Barnes, Jr.; "Variations in the Leucocyte Count in Normal Rabbits, in Rabbits Following the Injection of Normal Horse Serum, and During a Cutaneous Anaphylactic Reaction," *ibid.*, 258–268, written with G. L. Brown; "Serum Treatment of Hemorrhagic Diseases," *ibid.*, 272–278, written with J. Gelien; "Diphtheria Bacillus Carriers," in *Transactions of the XV International Congress on Hygiene and Demography*, IV (1913), 156–170; "Diphtheria Bacillus-Carriers," in *Transactions of the International Congress of Medicine*, pt. 2, sec. 4 (1914), 75–79, written with G. Guthrie and J. Gelien; and "A Simple Method for the Treatment for the Indirect Transfusion of Blood," in *American Journal of the Medical Sciences*, **147** (1914), 698–703.

Other works are "An Attempt to Immunize Calves Against Tuberculosis by Feeding the Milk of Vaccinated Cows," in *Johns Hopkins Hospital Bulletin*, **26** (1915), 241–245; "A Simplified Method for Determining the Iso-Agglutinin Group in the Selection of Donors for Blood Transfusion," in *Journal of the American Medical Association*, **68** (1917), 1905–1906; "Diphtheria Bacillus Carriers; Second Communication," in *Johns Hopkins Hospital Bulletin*, **31** (1920), 388–403, and "The Effect of Diphtheria Antitoxin in Preventing Lodgement and Growth of Diphtheria Bacillus in the Nasal Passages Of Animals," *ibid.*, 381–388, both written with C. G. Guthrie and J. Gelien; "Experimental Inoculation of Human Throats with Virulent Diphtheria Bacilli," *ibid.*, **32** (1921), 369–378, written with C. G. Guthrie and B. C. Marshall; "Diphtheria Bacillus Carriers. A Report on Conditions Found in an Orphan Asylum," *ibid.*, 109–113, written with C. G. Guthrie and J. Gelien; "Experimental Inoculation of Human Throats with Virulent Diphtheria Bacilli," *ibid.*, 37–44, written with C. G. Guthrie and B. C. Marshall; "Yaws; An Analysis of 1,046 Cases in the Dominican Republic," *ibid.*, **33** (1922), 43–55, written with G. H. Bigelow; and "Hospitalization of Pneumonia Cases. Criticisms of the Recommendation of the Chicago Pneumonia Commission," in *Modern Hospital*, **25** (1926), 425–430.

Also see "Yaws; Results of Neosalvarsan Therapy After Five Years," in *American Journal of Tropical Medicine*, **20** (1926–1927), 365–384; "Malta Fever; Laboratory Infection in Humans," in *Transactions of the Association of American Physicians*, **43** (1928), 272–284, written with M. Castenada; "Blood Groups in Peru, Santo Domingo, Yucatan and Among Mexicans at Blue Ridge Prison Farm in Texas," in *Journal of Immunology*, **16** (1921), 159–174, written with J. A. Kennedy; and "From the South Seas," in *Harvard Alumni Bulletin* (1930), 532.

II. SECONDARY LITERATURE. Work done with data gathered by Moss includes William W. Howells, "Anthropometry and Blood Types in Fiji and the Solomon Islands Based Upon Data of Dr. William L. Moss," in *Anthropological Papers of the American Museum of Natural History*, **33**, pt. 4 (1933).

AUDREY B. DAVIS

MOSSO, ANGELO (*b.* Turin, Italy, 30 May 1846; *d.* Turin, 24 November 1910), *physiology, archaeology.*

Mosso was taken at a very early age to Chieri, where his father had a carpenter's shop and where he completed his elementary and part of his secondary schooling. He then went to Cuneo and Asti to attend the *liceo*, aided by a small scholarship given by the town of Chieri. In 1865 he enrolled in the medical school of the University of Turin. Two of Mosso's professors, the zoologist Filippo de Filippi and the botanist Moris, procured him a post as teacher of natural sciences in the *liceo* of Turin, so that he could support himself. His financial situation improved somewhat during his last two years at the university, when he became an intern at the Mauriziano Hospital in Turin and was given free board and lodging.

Mosso graduated *summa cum laude* on 25 July 1870. The board of examiners was so impressed by his experimental thesis on the growth of bones that it decided to have it printed. His work at the hospital decided his future, for it was there that he met Luigi Pagliari, who introduced him to Jacob Moleschott, then teaching physiology at Turin. At the end of his military service (which had prevented him from accepting an assistantship offered by Moleschott), Mosso obtained a scholarship, on Moleschott's recommendation, to the University of Florence, where for two years he worked in the laboratory of Moritz Schiff and wrote his first scientific papers. Next he went to Leipzig, where he studied under Ludwig (1873–1874), and learned to use machines for the graphic registration of physiological phenomena, an approach he later used extensively in his work. It was also at this time that Mosso first proposed his plethysmograph.

Refusing a number of assistantships offered him by German universities, Mosso returned to Italy after visiting Paris, where he met Bernard, Brown-Séquard, and Marey. (Marey stirred his interest in graphic registration machines yet again.) At Turin, Mosso entered Moleschott's institute and began fundamental studies on blood circulation; in 1875 he became professor of pharmacology (materia medica, as it was then called), and in 1879 he succeeded Moleschott, who had moved to Rome, as professor of physiology.

Under Mosso's leadership the physiology institute of the University of Turin became an extremely active center of research, especially in experimental physiology and biology, and attracted many foreign researchers. During this period Mosso founded *Archivio italiano di biologia* (1882), and established the Institute of Physiology in the Parco del Valentino (1893) and a station in the Alps (1895) for the study of human physiology at high altitudes.

In recognition of his scientific achievements Mosso was named a senator in 1904 but almost immediately contracted locomotor ataxia, which forced him to give up his physiological studies. Because of his illness, however, he dedicated himself to archaeology and conducted studies in the Roman Forum, in Crete, and in southern Italy. His last publications were all in this field, in which he acquired as great fame as he had in physiology. He died in 1910, following a more serious attack of his illness.

The importance of Mosso's physiological research lies in his emphasis on experimenting directly on man whenever possible, as well as on animals, so that his research was truly in human physiology. His scientific experiments were carried out with special equipment, which he devised to suit the requirements of the studies. He pursued two main lines of research, the analysis of motor functions and the relationship between physiological and psychic phenomena. On the first topic, Mosso carried out highly accurate studies on movements of both smooth and striated muscles, in relation to a great variety of physiological and pathological conditions, such as heat, cold, sleeping, waking, and hibernation. He also considered movement from a mechanical point of view and from that of heat production. On the second topic he studied the variations in the frequency and energy of cardiac systoles, the increase of blood flow into the brain, and the increase of blood pressure in situations that today would be described as those of intellectual or emotional stress.

Outstanding among the many machines that Mosso perfected for his physiological research is the plethysmograph, with which he measured slow changes in the volume of the blood vessels; as a result he was able to determine which part of the movement of the pulse was due to cardiac pulsation and which to contraction of the vessels' walls. With the ergograph and the ponograph, he completed very accurate studies on fatigue.

Mosso also studied respiration during sleep, pointing out the inversion of thoracic respiration; and human physiology at high altitude, demonstrating the phenomenon of acapnia, the difficulty of breathing due to absence or scarcity of carbon dioxide in the organism.

BIBLIOGRAPHY

I. ORIGINAL WORKS. Mosso wrote about 200 articles and books. His most important works are *Saggio di alcune ricerche fatte intorno all'accrescimento delle ossa* (Naples, 1870), his thesis for the M.D.; "Sopra alcuni sperimenti di trasfusione del sangue," in *Sperimentale*, **30** (1872), 369–375; "Sull'irritazione chimica dei nervi cardiaci," *ibid.*, 358–368; "Sopra un nuovo metodo per scrivere i movimenti dei vasi sanguigni dell'uomo," in *Atti dell'Accademia delle scienze (Turin)*, **11** (1875), 21–81; "Introduzione ad una serie di esperienze sui movimenti del cervello

nell'uomo," in *Archivio per le scienze mediche*, **1** (1876), 216–244; "Sul polso negativo e sui rapporti della respirazione addominale e toracica nell'uomo," *ibid.*, **2** (1878), 401–464; "Sulla circolazione del sangue nel cervello dell'uomo," in *Atti dell'Accademia nazionale dei Lincei. Memorie*, 3rd ser., **5** (1879–1880), 237–358; "Ricerche sulla fisiologia della fatica," in *Rendiconti dell'Accademia de medicina Torino*, **31** (1883), 667; *La paura* (Milan, 1884); *La respirazione dell'uomo sulle alte montagne* (Turin, 1884), a volume in honor of C. Sperino; "Le leggi della fatica studiate nei muscoli dell'uomo," in *Atti dell'Accademia nazionale dei Lincei. Memorie*, 4th ser., **5** (1888), 410–426; *La fatica* (Milan, 1891); *La temperatura del cervello* (Milan, 1894); "Descrizione di un miometro per studiare la tonicità dei muscoli nell'uomo," in *Memorie della Accademia delle scienze di Torino*, **46** (1896), 93–120; and *Fisiologia dell'uomo sulle Alpi* (Milan, 1897).

Major twentieth-century publications are "La fisiologia dell'apnea studiata nell'uomo," in *Memorie della Accademia delle scienze di Torino*, **53** (1902), 367–386; "Crani preistorici trovati nel foro romano," in *Notizie degli scavi* (Rome, 1906), fasc. 1, 46–54; *Escursioni nel Mediterraneo e gli scavi di Creta* (Milan, 1908); and *Le origini della civiltà mediterranea* (Milan, 1910).

II. SECONDARY LITERATURE. See *Angelo Mosso, la sua vita e le sue opere* (Milan, 1912), with a detailed bibliography of his writings; A. Botto Micca, "A. Mosso archeologo," in *Atti del XV Congresso italiano di storia della medicina* (Turin, 1957), 20; L. Ferretti, *A. Mosso, apostolo dello sport* (Milan, 1951); and A. Gallassi, "Angelo Mosso e la medicina sportiva," in *Atti del XV Congresso italiano di storia della medicina* (Turin, 1957), 22–25.

CARLO CASTELLANI

MOSSOTTI, OTTAVIANO FABRIZIO (*b.* Novara, Italy, 18 April 1791; *d.* Pisa, Italy, 20 March 1863), *physics.*

Mossotti came from a moderately well-to-do family. Little is known directly of his life. He attended the University of Milan, and spent ten years as an assistant at the Milan observatory before leaving to seek a position in England. Following four years without an appointment, Mossotti accepted the post of astronomer at the topographical bureau in Buenos Aires and also served as professor of physics at the university. He remained in Argentina for several years, returning to become a professor of mathematics at the Ionian University at Corfu, founded by Frederick North in 1824. In 1841 he became professor of mathematics, theoretical astronomy, and geodesy at the University of Pisa, where he remained until his death.

As is evident in his first work (1815), Mossotti's outlook and methods derived from that French group of analysts best exemplified by Laplace, Poisson, and Ampère. Like them, he believed that the proper way to explain all physical phenomena was by means of forces acting centrally at a distance between various fluids. Given such a force and its subject, the correct application of mathematics to the equilibrium situations of the fluids in all circumstances should, they believed, lead one to the observed phenomena. Poisson had been the first to subject Coulomb's magnetic and electrical fluids to extensive analysis, while Ampère had postulated a central force acting between the elements of galvanic circuits. While most scientists granted the general applicability of these forces and fluids, they were utilized fully only by the Continental "action-at-a-distance" group. Indeed, after Faraday's work of the 1830's and 1840's many refused to grant even the existence of the requisite fluids. Mossotti, however, held firmly to the outlook of Poisson and others, and much of his work derived from this belief.

Between 1815 and 1832 Mossotti concentrated on the question of which forces were responsible for cohesion and aggregation in liquids and solids. In accordance with the French tradition, Mossotti thought that these forces were best explained by means of a fluid distributed in an atmosphere about the particles of the ordinary matter that constituted all bodies. Mossotti's "ether" was subject to the two central forces of self-repulsion and attraction to "natural matter"—these two were the only electrical forces he imagined. In the absence of an external concentration of the electrical ether the molecules of a body were evenly surrounded by the electrical fluid, and the whole existed in stable equilibrium under the action of the two forces, with the self-repulsion of the ether atmospheres balancing the mutual attraction of the ether and the matter. Mossotti thought these two forces were sufficient to explain cohesion as well as a number of other phenomena, including the propagation of light in transparent bodies.

In the mid-1840's Mossotti read of Faraday's investigation of dielectric, or nonconducting, bodies. Until then it was believed that nonconductors were unlike conductors only in opposing the motion of electrical fluid. Faraday sought an explanation predicated upon the ability of an intervening medium to propagate electrical force from point to point between "charged" bodies. Faraday assumed that all dielectrics—solid and liquid as well as gaseous—were constituted so that their smallest parts were somehow "polarized" under electrical influence, each part transmitting the action to its neighbor by its "polarity." Although Faraday rejected the notion of an absolute electrical fluid, Mossotti ignored this rejection and accepted Faraday's assumption of

dielectric polarity, preferring to explain polarity by means of ether-bearing molecules.

Mossotti thought that all bodies were built of ether-matter molecules in which the ether acted as an electrical fluid. The difference between conductors and nonconductors resulted from a variation in the abilities of bodies to retain the ethereal atmosphere about a single molecule under electrical action. A conductor had no retentive strength, while a dielectric retained the ether, but in a condition of varying density about the molecule. Mossotti believed this conception to be in the tradition of Franklin's single electrical fluid.

When Mossotti's dielectric was placed near an element of ether, its molecules became polarized in the sense that the ether density about each particle varied from point to point. The variation resulted in a net force on the molecule because of the change in distance of the ether and matter from the external fluid. In Mossotti's notation, if ρ is the distance from a volume element $d\psi \, d\xi \, ds$ of the ether to the electricity, then $\mu \, d\psi \, d\xi \, ds/\rho^2$ represents the force produced on the electricity by $d\psi \, d\xi \, ds$. The function μ was positive if the density in $d\psi \, d\xi \, ds$ was greater than the equilibrium value, and negative if it was less, since a decrease in ether density yielded an attraction. Mossotti was thus able to show that, if k' represents the ratio of ether volume to total volume, then the force on a unit of electrical fluid is given by the negative coordinate derivatives of

$$Q = \iiint \left(\frac{d^1/\rho}{dx'} \alpha' + \frac{d^1/\rho}{dy'} \beta' + \frac{d^1/\rho}{dz'} \gamma' \right) k' \, dx' \, dy' \, dz'$$

integrating over the dielectric. The functions α', β', γ' are the "dipole moments" of a molecule, their form being $\alpha' = \iiint \mu\psi \, d\psi \, d\xi \, ds$, integrating over a molecule.

Mossotti made an extensive analysis of the internal conditions in a dielectric subject to electrical action. Employing the mathematics derived by Poisson in 1826 for the actions of magnetic molecules, he obtained expressions like $\frac{4}{3}\pi k' \alpha'$ for the force in a small element of the dielectric resulting from the distribution of the molecules therein. This result is the Clausius-Mossotti relation in its original form. In addition Mossotti showed that the action of a polarized dielectric can be fully represented by an imaginary distribution of ether on its boundary surfaces. By means of this equivalent surface distribution Mossotti demonstrated in 1846 that "[because of] the polarization of the atmospheres of its molecules the dielectric simply transmits the action between conducting bodies . . ." ("Sull'influenza . . ." [1850], p. 73).

Mossotti reached this last result—which he considered his most important—only after an extensive analysis. In Faraday's view such a transmission was an elementary proposition. The difference between Mossotti's and Faraday's work—and, eventually, between the proponents of mediated action and those of electrical matter—was a deep one, hinging on the acceptance or rejection of fluids acting directly at a distance. Thus, Mossotti felt called upon to explain why electrical fluid does not fly off the surface of a conductor under its self-repulsion; while Faraday considered the question unnecessary because he did not employ an electrical fluid like Mossotti's ether. Mossotti's success in accounting for dielectric behavior may be considered together with the impact of Faraday's work to illustrate the conceptual flux that characterized the study of electricity and magnetism from 1840 to 1870. Mossotti's work was not very influential theoretically; it remained in the Continental action-at-a-distance tradition and used none of Faraday's newer ideas on the distribution of force in space. However it was an important formal development in that it showed that one could explain dielectric behavior without abandoning the scheme of central forces and subtle fluids as Faraday had. During the late 1840's and early 1850's, electrical fluids came to be viewed with increasing distrust both because of Faraday's work and because they seemed to imply unacceptable behavior for the fluid *qua* fluid.

BIBLIOGRAPHY

I. ORIGINAL WORKS. Mossotti's writings include "Del movimento di un fluido elastico che sorte da un vase e della pressione che fa sulle pareti dello stesso," in *Memorie di matematica e di fisica della Società italiana delle scienze*, **17** (1815), 16–72; *Sur les forces qui régissent la constitution intérieure des corps, aperçu pour servir à la détermination de la cause et des lois de l'action moléculaire* (Turin, 1836), repr. in Richard Taylor, ed., *Scientific Memoirs*, I (London, 1837), 448; *Dell'azione delle forze molecolari nella produzione dei fenomeni della capillarità* (Milan, 1840), repr. in *Scientific Memoirs* (London, 1841); *Lezioni elementari di fisica matematica*, 2 vols. (Florence, 1843–1845); and "Sull'influenza che l'azione di un mezzo dielettrico ha sulla distribuzione dell'elettricità alla superficie di più corpi elettrici disseminati in esso," in *Memorie di matematica e di fisica della Società italiana delle scienze*, **24**, pt. 2 (1850), 49–74.

II. SECONDARY LITERATURE. On Mossotti and his work, see Salvatore de Benedetti, *Ottaviano Fabrizio Mossotti. Elogio pronunziato nella inaugurazione del monumento all'illustre scienziato li di 16 giugno 1867, e le interpretazioni del Mossotti ai versi astronomici della Divina Commedia* (Pisa, 1867).

Related works are Samuel Earnshaw, "On the Nature of the Molecular Forces Which Regulate the Constitution of the Luminiferous Ether," in *Transactions of the Cambridge Philosophical Society*, **7** (1839), 97–112; Michael Faraday, *Experimental Researches in Electricity*, I (London, 1839); George Green, *An Essay on the Application of Mathematical Analysis to the Theories of Electricity and Magnetism* (Nottingham, 1828), repr. in his *Mathematical Papers* (New York, 1970), 3–115; James Clerk Maxwell, *A Treatise on Electricity and Magnetism* (Cambridge, 1891); and William Thomson, *Reprint of Papers on Electrostatics and Magnetism* (London, 1872).

JED ZACHARY BUCHWALD

MOTTRAM, JAMES CECIL (*b.* Slody, Norfolk, England, 12 December 1879; *d.* London, England, 4 October 1945), *medicine, natural history.*

Mottram was the only son of James Alfred Mottram and Clara Ellen Swanzy. He qualified as a doctor in 1903 at University College Hospital, London. After research in Cambridge, he joined the Cancer Research Laboratories of the Middlesex Hospital Medical School in 1908. He remained there until 1919, except for service in the Royal Navy from 1916 to 1918. Mottram was director of the research department of the Radium Institute from 1919 to 1937 and then director of the research laboratories of Mount Vernon Hospital until his death in 1945. He married Rhoda Pritchard, and they had two sons and one daughter.

Mottram's first published work (1909) was on spectroscopic analysis of tissues for sodium and potassium, and he did some work on nutrition; but his professional life was mainly devoted to the study of cancer. This work can be broadly divided into three phases: (1) the effects of X rays and radium on the cells of normal and malignant tissues; (2) carcinogenesis; and (3) the development of methods of treating cancer. He was a competent experimenter who usually planned and executed his work alone, but he also collaborated easily. He was always particularly concerned with the practical applications of his discoveries, and much of his work in all fields appears to have been conceived as exploring the basic concepts likely to improve some known practical problem.

Mottram's most important discovery came early and was published in 1913. He showed that in both plant and animal tissues (the tips of bean shoots and ova of *Ascaris megalocephala*) cells are more vulnerable to damage by beta and gamma radiation when they are in process of division than in the resting stage, and that the metaphase is the most vulnerable stage. This damage results in profound nuclear changes affecting chromatin. Further work on radiation damage (published in 1926 in collaboration with G. M. Scott and S. Russ) showed that although there were no immediate changes apparent in cells subjected to beta rays from radium, subsequent examination of the tissue showed an absence of cells in active division. The final changes, which were profound enough to prevent the growth of a tumor, were interpreted as due to the incapacity of daughter cells to divide normally. In 1934 Mottram showed that if cells from bean roots were treated with X rays the chromosomes were fragmented and migration to the poles of the spindle was delayed, preventing normal cell division.

The practical applications of this work were not only in the treatment of tumors by exposure to radium but also careful measures to protect those working with X rays and radium. Mottram and Russ studied dosage in radium therapy (1916–1917) and Mottram and Clarke the leucocyte blood content of those handling radium for therapeutic purposes. Mottram served on the X-ray and radium protection committee, and he reorganized safety programs at the Radium Institute.

The most important paper on the part played by lymphocytes in carcinogenesis and immunity was published by Mottram and Russ in 1917–1918. Rats immune to Jensen's sarcoma showed a high content of lymphocytes in the spleen and accumulation of lymphocytes around a graft of sarcoma cells; if this accumulation were delayed, growth of the sarcoma occurred. Rats could be made immune by inoculation of sarcoma cells previously exposed to beta and gamma rays from radium, and immune rats could be made tumor-bearing by exposure to X rays. This work was used to test for radiation hazard by lymphocyte counts.

Mottram's only book on tumors (1942) examined the effect of blastogenic agents on populations of *Paramecium* and showed that fission time was prolonged and that abnormal individuals (often polyploids) were produced and spread through the population. These changes were related to an increase in viscosity of the protoplasm, which inhibited normal fission.

Mottram's contribution in World War I was his fundamental and applied work on the principles of camouflage. He was already interested in the coloration of animals, and the most useful part of his book *Controlled Natural Selection and Value Marking* (1914) was his discussion of the function of color and pattern in natural selection. The main theme of the book is that individuals differ in their value to society, as for example in age and sex, and this influences natural selection, which may result in the destruction of the less valuable; alternatively, natural selection may act not upon the individual but upon a group such as the family.

Zoologists had been aware for some time that a color pattern that broke the outline of an animal was protective. But Mottram, using plain and patterned objects against plain and patterned backgrounds for human vision, showed experimentally that if the pattern interrupts the margin of the object blurring of the outline occurs. Also the near presence or contact with the object of an area of tone similar to the object makes it less visible. Blending of patterns near the margin also masks an outline and even small details of pattern may be sufficiently important in concealment to have a survival value. This work (published 1915–1917) was of obvious value in Mottram's service at the Camouflage School during the war.

Mottram's research on fish was related to his hobby of fly-fishing. He published numerous articles in sporting journals—particularly *The Field, Salmon and Trout Magazine, Game and Gun Magazine*, and *Flyfisher's Journal*—ranging from personal anecdotes to papers on cultivating weed beds, breeding trout, breeding food for trout, pollution, and disease. The three books *Fly-fishing* (1915), *Sea Trout* (1925), and *Trout Fisheries* (1928) were based on these articles and included chapters designed to encourage the recreational fisherman to take a greater interest in ecology and in special techniques such as reading the age of fish from their scales. He served on the furunculosis committee of the Ministry of Agriculture and Fisheries, which studied the susceptibility of trout to this epizootic and the influence of temperature on its spread. The committee recommended legislation to aid in the control of the disease.

BIBLIOGRAPHY

I. ORIGINAL WORKS. Mottram's books are *Controlled Natural Selection and Value Marking* (London, 1914); *Fly-fishing: Some New Arts and Mysteries* (London, 1915; 2nd ed., actually only a new impression, London, 1921); *Sea Trout and Other Fishing Studies* (London, 1922); *Trout Fisheries: Their Care and Preservation* (London, 1928); *The Problem of Tumours. The Application of Blastogenic Agents to Ciliates. A Cytoplasmic Hypothesis* (London, 1942).

The medical papers are "A Method of Quantitative Analysis of the Tissues for Potassium and Sodium by Means of the Spectroscope," in *Archives of the Middlesex Hospital*, **15** (1909), 106–117; "On the Action of Beta and Gamma Rays of Radium on the Cell in Different States of Nuclear Division," *ibid.*, **30** (1913), 98–119; with G. M. Scott and S. Russ, "On the Effects of Beta Rays

From Radium Upon the Division and Growth of Cancer Cells," in *Proceedings of the Royal Society of London*, **100B** (1926), 326–335; "Some Effects of Cancer-Producing Agents on Chromosomes," in *British Journal of Experimental Pathology*, **15** (1934), 71–73; with S. Russ, "A Contribution to the Study of Dosage in Radium Therapy," in *Proceedings of the Royal Society of Medicine*, **10** (1916/ 1917), section electrotherapy, 121–140; with J. R. Clarke, "The Leucocytic Blood Content of Those Handling Radium for Therapeutic Purposes," *ibid.*, **13** (1919/1920), 25–32; with S. Russ, "Observations and Experiments on the Susceptibility and Immunity of Rats Towards Jensen's Rat Sarcoma," in *Proceedings of the Royal Society of London*, **90B** (1917/1918), 1–33; "A Diurnal Variation in the Production of Tumours," in *Journal of Pathology and Bacteriology*, **57** (1945), 265–267.

Papers on camouflage are "Some Observations on Pattern-Blending With Reference to Obliterative Shading and Concealment of Outline," in *Proceedings of the Zoological Society of London* (1915), 679–692; "An Experimental Determination of the Factors Which Cause Patterns to Appear Conspicuous in Nature," *ibid.* (1916), 383–419; "Some Observations Upon Concealment by the Apparent Disruption of Surface in a Plane at Right-Angles to the Surface," *ibid.* (1917), 253–257.

II. SECONDARY LITERATURE. There is no bibliography of Mottram's papers, but they may be traced through *Index Medicus, Zoological Record*, and individual indices of periodicals mentioned in the text.

Three obituaries are worth noting: one by "S.R." in *Lancet* (1945), **2**, 581, with a photograph; R. J. Ludford, in *Nature*, **157** (1946), 399–400; and an anonymous notice in *Salmon and Trout Magazine*, no. 116 (1946), 16.

The furunculosis committee issued three reports: interim report (Edinburgh, 1930); second interim report (Edinburgh, 1933); final report (Edinburgh, 1935).

DIANA M. SIMPKINS

MOUCHEZ, ERNEST BARTHÉLÉMY (*b.* Madrid, Spain, 24 August 1821; *d.* Wissous, Seine-et-Oise, France, 29 June 1892), *cartography, astronomy.*

Mouchez studied at Versailles and then entered the École Navale to prepare for a career in the navy. He became an ensign in 1843, a post captain in 1868, and a rear admiral in 1878. In the meantime he was elected a member of the Bureau des Longitudes (1873) and of the astronomy section of the Académie des Sciences (1875, replacing L. Mathieu), and became director of the Paris observatory (26 June 1878).

As soon as he began his voyages, Mouchez started making important hydrographical studies along the coasts of Korea, China, and South America (sailing 320 kilometers up the Paraguay River). In 1862 he was sent by the minister of the navy to explore the Abrolhos Islands of Brazil, and he then explored

4,000 kilometers of coastline between the Amazon and the Río de la Plata. From 1867 to 1873 Mouchez charted the coast of Algeria during several expeditions. In all, he published about 140 maps and determined many geographical positions.

From the beginning Mouchez worked to improve surveying techniques and promoted the use by the navy of suitably adapted stationary observational instruments. In particular he employed a meridian telescope designed to determine lunar culminations and generalized the use of the theodolite in topographical surveying. By means of these modifications, he reduced the margin of error in the determination of longitudes from 30″ to 3″ or 4″.

In 1874 Mouchez was sent by the Académie des Sciences to St. Paul Island in the Indian Ocean to observe the transit of Venus on 9 December; he succeeded in making more than 400 extremely sharp exposures. In 1878 he succeeded Le Verrier as director of the Paris observatory. Dissatisfied both with its state of repair and with the quality of the work being done there, he tried in vain to persuade the authorities to build a branch of the observatory outside of Paris, in Versailles or in the forest of Verrières, and to construct lodgings for the staff nearby. Although he failed in this project, Mouchez improved the observatory of the Bureau des Longitudes in the park of Montsouris (at the southern edge of Paris), created a school of practical astronomy at the Paris observatory (1879), and founded the *Bulletin astronomique* (1884).

Arago had embarked on a program of improving Lalande's catalog of 50,000 stars with the aid of new, more precise measurements. Mouchez published the part observed up to 1875. Most notably, however, he enlisted the support of Sir David Gill, director of the Cape observatory, to bring about an international astronomical congress at Paris in 1887. It was there decided to produce photographically a large-scale general map of the heavens and to establish a catalog giving the position and brightness of all stars up to the eleventh magnitude. Two young astronomers at the Paris observatory, the brothers Prosper and Paul Henry, both of whom were also talented opticians, had just completed an astrograph and Mouchez had it adopted for this gigantic undertaking, which took more than fifty years. Four French observatories covered nearly half of the northern hemisphere, and eighteen other observatories participated.

BIBLIOGRAPHY

I. ORIGINAL WORKS. A list of Mouchez's books and other scientific publications can be found in the Royal

Society's *Catalogue of Scientific Papers*, IV, 498; VIII, 448; and X, 864; in Poggendorff, III, 940; and IV, 1034–1035; in *Catalogue général des livres imprimés de la Bibliothèque nationale*, CXX, cols. 533–538; and in *Notice sur les travaux scientifiques de M. Mouchez* (Paris, 1875).

On hydrography he wrote *Recherches sur la longitude de la côte orientale de l'Amérique du Sud* (Paris, 1866); *Río de la Plata. Description et instructions nautiques* (Paris, 1873); *Instructions nautiques sur les côtes d'Algérie* (Paris, 1879); and *Instructions nautiques sur les côtes du Brésil* (Paris, 1890).

In astronomy, see *La photographie astronomique à l'Observatoire de Paris et la Carte du ciel* (Paris, 1887); and *Rapport annuel de l'Observatoire de Paris* for 1884–1891 (Paris, 1885–1892). He was also responsible for the publication of vols. **24–39** of *Annales de l'Observatoire de Paris*.

II. Secondary Literature. Besides the *Notice sur les travaux scientifiques de M. Mouchez* (Paris, 1875), written at the time of his candidacy for the Académie des Sciences, there are only a few brief biographies: that in the *Dictionnaire universel des contemporains*, G. Vapereau, ed., 5th ed. (Paris, 1880), 1323; that in *Polybiblion*, 2nd ser., **36** (July–Dec. 1892), 78; the pamphlet containing the speeches given at his funeral (Paris, 1892); and an address delivered by B. Baillaud at the unveiling of his statue at Le Havre in 1921 (Paris, 1921).

Juliette Taton

MOULTON, FOREST RAY (*b.* Osceola County, Michigan, 29 April 1872; *d.* Wilmette, Illinois, 7 December 1952), *astronomy, mathematics.*

Forest Ray Moulton was the eldest of eight children born to Belah and Mary Smith Moulton. He was named Forest Ray because his poetic mother thought him a "perfect ray of light and happiness in that dense forest." He received his early education in a typical frontier school and at home. At the age of sixteen he taught in this same school, where one of the students was his brother Harold, who was later to become the first head of the Brookings Institute. At the age of eighteen he enrolled in Albion College, where he received his B.A. in 1894. Moulton received his Ph.D. in astronomy, *summa cum laude*, from the University of Chicago (1899). He also received honorary degrees from Albion College (1922), Drake University (1939), and the Case School of Applied Science (1940).

Moulton had a variety of careers. His academic life began at the University of Chicago with his appointment, while a graduate student, in 1896 as an assistant in astronomy, and it continued through his appointment as professor in 1912 until retirement in 1926. From 1927 to 1936 he was a director of the

Utilities Power and Light Corporation of Chicago. He was also a trustee of the Exposition Committee of Chicago's Century of Progress from 1920 to 1936, and its director of concessions from 1931 to 1933. From 1936 to 1940 he was executive secretary of the American Association for the Advancement of Science. During his tenure in this position, he edited more than twenty symposium volumes.

In 1898, while still a graduate student at Chicago, Moulton was invited by Thomas Crowder Chamberlin, then chairman of the geology department, to participate in an investigation of the earth's origin. Chamberlin's investigations on glacial movements had raised doubts that were relevant to the then existing theories. Kant had originally proposed his nebular hypothesis in 1755. Half a century later, and with no knowledge of Kant's theory, Laplace developed a similar theory. He suggested that the earth had originated in a vast mass of blazing gas, which had been thrown off by the sun and had liquefied into a molten sphere. According to this theory the earth was steadily cooling off from its original molten state. After the primordial sun, which was five and a half billion miles in diameter, had cast away the planets, it reached its present diameter and a rotational velocity of 270 miles per second. When Moulton imagined that all planets were returned to the sun, his calculations indicated that the sun would not have enough momentum to hurl off any rings of matter. Chamberlin and his group investigated everything that was written on the origin of the solar system. The most promising prospect was offered by that of nebular "knots," revealed in photographs of spiral nebulae, which could have served as collecting centers.

On 28 May 1900 there was an eclipse of the sun. Chamberlin and Moulton meticulously studied the photographs that illustrated the sun's eruptive nature. Their observation of great clouds of gaseous matter flaring out and away from the sun's surface led to the planetesimal hypothesis proposed in 1904. They proposed that the nebula quickly cooled and solidified, creating small chunks of matter, the planetesimals. Although today neither the Laplace-Kant nor the Moulton-Chamberlin hypothesis stands by itself, both provide a basis for current theories.

During World War I Moulton was assigned to do ballistics research at Fort Sill. Here he is said to have effectively doubled the range of artillery. His work was the forerunner of the efforts in World War II to get improved ballistics tables faster and more accurately, which was one of the links giving impetus to contemporary high speed electronic computing equipment.

In 1920, Moulton, one of the founders of the Society for Visual Education, gave the first radio address broadcast from the University of Chicago. As one of the pioneers of educational broadcasting, Moulton was heard weekly in Chicago from 1934 to 1936, and in Washington from 1938 to 1940.

Moulton was a fellow of the American Academy of Arts and Sciences, the American Physical Society, and the Royal Astronomical Society; President of Sigma Xi, and Honorary Foreign Associate of the British Association for the Advancement of Science. He was also an active member of many other professional societies.

In 1897 Moulton married Estelle Gillete. They had four children and were divorced in 1938. In 1939 he married Alicia Pratt of Winnetka, Illinois. They were divorced in 1951.

BIBLIOGRAPHY

I. ORIGINAL WORKS. No complete bibliog. of Moulton's publications has been published. His major bks. include *An Introduction to Celestial Mechanics* (London–New York, 1902, 1914, 1935); *Descriptive Astronomy* (Chicago, 1912, 1921, 1923); *Periodic Orbits* (Washington, 1920); *Differential Equations* (New York, 1930), written with D. Buchanan, T. Buck, F. Griffin, W. Longley, and W. MacMillan; and *New Methods in Exterior Ballistics* (Chicago, 1926). The last work is the beginning of a contemporary mathematical approach to the science of ballistics. For interesting and opposing reviews of this work see J. E. Rowe, *Bulletin of the American Mathematical Society*, **34** (1928), 229–332, and L. S. Dederick, *ibid.*, 667.

The first twenty vols. of the *Carnegie Institution of Washington Yearbooks* (1902–1921) provide a clear picture of the work of Moulton and Chamberlin on the planetesimal hypothesis. *Yearbook*, no. 2 (1903), 261–270, contains a report by T. C. Chamberlin, entitled "Fundamental Problems of Geology," in which he describes the progress of the investigators and collaborators, and their roles and status to date. *Yearbook*, no. 3 (1904), 255–256, contains a letter from Moulton to Chamberlin describing different hypotheses and their applications, pertinent observational data, and the laws derived from the data. Other vols. of the *Yearbook* and their relevant p. nos. are no. 1, 25–43; no. 3, 195–258; no. 4, 171–190; no. 5, 166–172; no. 6, 195; no. 7, 204–205; no. 8, 28–52, 224–225; no. 9, 48–222; no. 10, 45, 222–225; no. 11, 13–44, 264–266; no. 12, 52, 292, 297; no. 13, 45–46, 356–357, 376; no. 14, 36–37, 289, 368; no. 15, 358–362; no. 16, 307–319; no. 17, 297; no. 18, 39, 343–345, 349–351; no. 19, 21, 366–382, 386; and no. 20, 412–425. See also "The Development of the Planetesimal Hypothesis," in *Science*, n.s. **30** (1909), 642–645, written with T. C. Chamberlin; "An Attempt to Test the Nebular Hypothesis by an Appeal to the Laws of Dynamics," in *Astrophysical Journal*, **11** (1900), 103; and "Evolution of the Solar System," *ibid.*, **22** (1905), 166.

One of Moulton's more important papers from the standpoint of current research is "The Straight Line Solutions of the Problem of n Bodies," in *Annals of Mathematics*, **12** (1910–1911), 1–17. In this paper the number of straight line solutions is found for n arbitrary masses. This is the generalization of the problem solved by Lagrange for three bodies. Moulton also attacks what is a sort of converse of this problem by determining, when possible, n masses such that if they are placed at n arbitrarily collinear points, they will, under proper initial projection, always remain in a straight line. This paper was originally presented to the Chicago Section of the American Mathematical Society on 28 December 1900 (see *Bulletin of the American Mathematical Society*, **7** [1900–1901], 249–250).

II. SECONDARY LITERATURE. The anonymous "The Washington Moultons, Forest Ray, '94, and Harold Glenn, 1907," in *Io Triumphe* (March, 1947) (Albion College Alumni Magazine), is an excellent art. with portraits of both Forest and Harold. This art. also contains a photograph of the seven Moulton brothers and their sister on the occasion of the awarding of an M.A. to Mary Moulton by Wayne University in 1945. Other biographical arts. appear in *Current Biography* (1946), 421–423; *The National Cyclopaedia of American Biography*, XLIII (1946), 314–315; and A. J. Carlson, "Forest Ray Moulton: 1872–1952," in *Science*, **117** (1953), 545–546.

HENRY S. TROPP

MOUTARD, THÉODORE FLORENTIN (*b.* Soultz, Haut-Rhin, France, 27 July 1827; *d.* Paris, France, 13 March 1901), *geometry, engineering.*

Moutard was educated at the École Polytechnique from 1844 to 1846. Like many of his fellow students and alumni, he was both an engineer and a geometer. He was graduated from the École des Mines in 1849 and entered the engineering corps. He was discharged in 1852, because as a republican he refused to take the required loyalty oath after the coup d'état by Napoleon III. He was reinstated in 1870. The majority of his mathematical publications date from these years. In 1875 Moutard was appointed professor of mechanics at the École des Mines, but he retained his army rank and was named *ingénieur en chef* in 1878 and *inspecteur général* in 1886. He retired with the latter rank in 1897 but retained his position at the École des Mines. From 1883 he also served as an outside examiner at the École Polytechnique. He was one of the collaborators on *La grande encyclopédie*.

Moutard's mathematical work was primarily in the theory of algebraic surfaces, particularly anallagmatic surfaces, differential geometry, and partial differential equations. His broadest work was a memoir on elliptic functions, which was published as an appendix in Victor Poncelet's *Applications d'analyse et de géométrie.*

BIBLIOGRAPHY

I. ORIGINAL WORKS. The works by Moutard in Victor Poncelet, *Applications d'analyse et de géométrie*, 2 vols. (Paris, 1862–1864), are "Rapprochements divers entre les principales méthodes de la géométrie pure et celles de l'analyse algébrique" (I, 509–535); the work on elliptic functions, "Recherches analytiques sur les polygons simultanément inscrits et circonscrits à deux coniques" (I, 535–560), and a short note "Addition au IVe cahier" (II, 363–364), on the principle of continuity. A bibliography of Moutard's papers in various journals can be found in Poggendorff, IV, 1037.

II. SECONDARY LITERATURE. An account of Moutard's life is in *La grande encyclopédie*, XXIV, 504. His work is also mentioned in Michel Chasles, *Rapport sur les progrès de la géométrie* (Paris, 1870).

ELAINE KOPPELMAN

MOUTON, GABRIEL (*b.* Lyons, France, 1618; *d.* Lyons, 28 September 1694), *mathematics, astronomy.*

Mouton became *vicaire perpétuel* of St. Paul's Church in Lyons in 1646, after taking holy orders and obtaining a doctorate in theology. He spent his whole life in his native city, fulfilling his clerical responsibilities and untroubled by any extraordinary events. During his leisure time he studied mathematics and astronomy and rapidly acquired a certain renown in the city. Jean Picard, who also was an *abbé*, held Mouton in high esteem and always visited him when in Lyons to work on the determination of the city's geographic position.

The book that made Mouton famous, *Observationes diametrorum solis et lunae apparentium* (1670), was the fruit of his astronomical observations and certain computational procedures he had developed. Lalande later stated: "This volume contains interesting memoirs on interpolations and on the project of a universal standard of measurement based on the pendulum."

Mouton was a pioneer in research on natural and practicable units of measurement. He had been struck by the difficulties and disagreements resulting from the great number of units of length, for example, which varied from province to province and from country to country. First he studied how the length of a pendulum with a frequency of one beat per second varies with latitude. He then proposed to deduce from these variations the length of the terrestrial meridian, a fraction of which was to be taken as the universal unit of length. Mouton selected the minute of the degree, which he called the *mille*. The divisions and subdivisions of this principal unit, all in decimal fractions, were called *centuria, decuria, virga, virgula, decima, centesima,* and *millesima*—or alternatively,

in the same order, *stadium, funiculus, virga, virgula, digitus, granum,* and *punctum.*

The *virgula geometrica* (geometric foot), for example, was 1/600,000 of the degree of meridian. In order to be able to determine the true length of this foot at any time, Mouton counted the number of oscillations of a simple pendulum of the same length over a span of thirty minutes and found it to be 3,959.2. These ideas were espoused by Picard shortly after the book appeared and a little later, in 1673, by Huygens. They were also favorably received by members of the Royal Society.

Although Mouton's proposals were seriously considered in theoretical terms in his own time, they led to no immediate practical results. Contemporary measuring procedures were too unsatisfactory to assure their valid and definitive application. It was not until 1790 that projects like Mouton's were taken up again. At a session of the Academy of Sciences on 14 April of that year, M. J. Brisson proposed that a new system be based on a natural standard. The Academy preferred to press for a geodesic survey, however, and decided to adopt one ten-millionth of the quadrant of the meridian of Paris as the standard for the meter.

In the *Observationes diametrorum* Mouton presented a very practical computational device for completing ordered tables of numbers when their law of formation is known. He used successive numerical differences, an idea previously employed by Briggs to establish his logarithmic tables.

When Leibniz went to London in January 1673, he took with him his *Dissertatio de arte combinatoria.* He summarized its contents to John Pell and, in particular, explained what he called "différences génératrices." Pell remarked that he had read something very similar in Mouton's book, which had appeared three years earlier. Leibniz had learned, during his stay in Paris, that the book was in preparation but did not know that it had been published. While visiting Oldenburg, Leibniz found Mouton's book and observed that Pell had been right; but he was able to prove that his own, more theoretical and general ideas and results had been reached independently of Mouton's.

A skillful calculator, Mouton produced ten-place tables of logarithmic sines and tangents for the first four degrees, with intervals of one second. He also determined, with astonishing accuracy, the apparent diameter of the sun at its apogee. A skilled experimentalist, he constructed an astronomical pendulum remarkable for its precision and the variety of its movements. It was long preserved at Lyons but was ultimately lost.

BIBLIOGRAPHY

I. ORIGINAL WORKS. Mouton's major work is *Observationes diametrorum solis et lunae apparentium, meridianarumque aliquot altitudinum, cum tabula declinationum solis; Dissertatio de dierum naturalium inaequalitate,* . . . (Lyons, 1670).

His trigonometric tables, which remained in MS, are now in the library of the Academy of Sciences. Esprit Pézénas, director of the observatory at Avignon, consulted this MS and used Mouton's method in preparing the new ed. of William Gardiner's *Tables of Logarithms* (to seven decimals) (Avignon, 1770).

II. SECONDARY LITERATURE. See the following (listed chronologically): J. B. Delambre, *Base du système métrique décimal ou mesure de l'arc de méridien compris entre les parallèles de Dunkerque et Barcelone, exécutée en 1792 et années suivantes par MM. Méchain et Delambre,* I (Paris, 1806), 11; *Biographie universelle ancienne et moderne,* XXIX (Paris, 1861), 485; Rudolf Wolf, *Geschichte der Astronomie* (Munich, 1877), 623; and Moritz Cantor, *Vorlesungen über Geschichte der Mathematik,* III (Leipzig, 1901), 76–77, 310, 389, and IV (Leipzig, 1908), 362, 440.

PIERRE SPEZIALI

MUḤYI 'L-DĪN AL-MAGHRIBĪ (Muḥyi 'l-Milla wa 'l-Dīn Yaḥyā ibn Muḥammad ibn Abi 'l-Shukr al-Maghribī al-Andalusī) (*fl.* Syria, and later Marāgha, *ca.* 1260–1265), *trigonometry, astronomy, astrology.*

Al-Maghribī was a Hispano-Muslim mathematician and astronomer, whose time and place of birth and death cannot be determined. Little is known about his life except that he was born in the Islamic West and flourished for a time in Syria and later in Marāgha, where he joined the astronomers of the Marāgha directed by Naṣīr al-Dīn al-Ṭūsī. He made observations in 1264–1265. It has been said that he was a guest of Hūlāgū Khān (Īl-khān of Persia, 1256–1265) and met Abu 'l Faraj (Bar Hebraeus, 1226–1286).

Suter and Brockelmann ascribe quite a long list of writings to al-Maghribī.

Trigonometry

1. *Kitāb shakl al-qaṭṭāʿ* ("Book on the Theorem of Menelaus").

2. *Ma yanfariʿu ʿan shakl al-qaṭṭāʿ* ("Consequences Deduced From *shakl al-qaṭṭāʿ*").

3. *Risāla fī kayfiyyat istikhrāj al-juyūb al-wāqiʿa fi 'l-dāʾira* ("Treatise on the Calculation of Sines").

Astronomy

4. *Khulāṣat al-Majisṭī* ("Essence of the *Almagest*"). It contains a new determination of the obliquity of the ecliptic made at Marāgha in 1264, 23; 30° (the real value in 1250 was 23; 32, 19°).

5. *Maqāla fī istikhrāj ta ʿdīl al-nahār wa sa ʿat al-mashriq wa 'l-dāʾir min al-falak* ("Treatise on Finding the Meridian, Ortive Amplitude, and Revolution of the Sphere").

6. *Muqaddamāt tataʿallaqu bī-ḥarakāt al-kawākib* ("Premises on the Motions of the Stars").

7. *Tasṭīḥ al-asṭurlāb* ("The Flattening of the Astrolabe").

Editions of the Greek classics; they are called recensions (sing. *taḥrīr*).

8. Euclid's *Elements.*

9. Apollonius' *Conics.*

10. Theodosius' *Spherics.*

11. Menelaus' *Spherics.*

He also wrote more than six books on astrology and a memoir on chronology.

Al-Maghribī's writings on trigonometry contain original developments. For example, two proofs are given of the sine theory for right-angled spherical triangles, and one of them is different from those given by Naṣir al-Dīn al-Ṭūsī; this theorem is generalized for other triangles. He also worked in several other branches of trigonometry.

Ptolemy (A.D 150) used an ingenious method of interpolation in the calculation of chord 1°. This is of course approximately equivalent to chord 1°. The same method was used for sines in Islam. To find the exact value, one must solve a cubic equation. This was done later by the Persian astronomer al-Kāshī (*d.* 1429/1430). Al-Maghribī, and before him Abu 'l-Wafāʾ (940–997/998), tried to find the value of the sine of one-third of an arc. For that purpose Abu 'l-Wafāʾ laid down a preliminary theorem that the differences of sines of arcs having the same origin and equal differences become smaller as the arcs become larger.

Using this preliminary theorem, al-Maghribī calculated sin 1° in the following way (see Fig. 1):

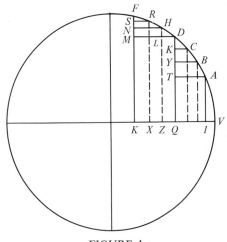

FIGURE 1

$VF = 1; 7, 30°$ and sin $VF = FK = 1; 10, 40, 12, 34^{p}$

$AV = 0; 45°$ and sin $AV = AI = 0; 44, 8, 21, 8, 38^{p}$.

The arc AF is divided into six equal parts and each part $= 0; 3, 45°$; therefore,

arc DV + arc $DH = 1°$ and sin $HV(=1°) = HZ$.

The perpendiculars AT, BY, and CK divide DT into three unequal parts: $TY > YK > DK$; $TD/3 > HL$; $DQ + TD/3 (=1; 2, 49, 43, 36, 9^{p}) > HZ(=\sin 1°)$. FM is divided into three unequal parts: $MN > NS > SF$; $DQ + FM/3 (=1; 2, 49, 42, 50, 40, 40^{p}) < NK = HZ(=\sin 1°)$. Then he found sin $1° = 1; 2, 49, 43, 24, 55^{p}$.

Al-Maghribī calculated sin $1°$ by using another method of interpolation based on the ratio of arcs greater than the ratios of sines. He found sin $1° = 1; 2, 49, 42, 17, 15, 12^{p}$ and said that the difference between two values of sines found by using different methods is $0; 0, 0, 0, 56^{p}$, which is correct to four places.

Using these methods, al-Maghribī calculated the ratio of the circumference to its diameter (that is, π).

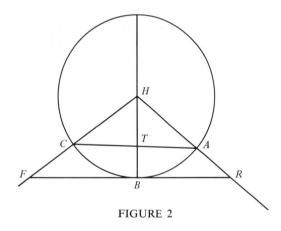

FIGURE 2

$$AC(=2AT) < \text{arc } ABC < RF$$
$$\sin AB(=3/4°) = AT = 0; 47, 7, 21, 7, 37^{p}$$
$$\Delta RFH \sim \Delta AHC, \qquad RF/AC = BH/TH.$$
$$RF = 1; 34, 15, 11, 19, 25^{p}$$
$$\text{arc } ABC = \frac{AC + RF}{2} = 1; 34, 14, 16, 47, 19, 30^{p}.$$

The circumference $= 240$. Arc $AB = 6; 16, 59, 47, 18^{p}$, the diameter being 2^{p}. The diameter being 1^{p}, the circumference $= 3; 8, 29, 53, 34, 39^{p} < 3R + 1/7$, since $1/7 = 0; 8, 34, 17, 8, 34, 17^{p}$.

Al-Maghribī compared the latter and Archimedes' value, $3R + 1/7 <$ the circumference $< 3R + 10/71$, found by computing the lengths of inscribed and circumscribed regular polygons of ninety-six sides. Half of the difference between $10/71$ and $10/70$ is equal to $0; 8, 30, 40^{p}$.

Al-Maghribī determined two mean proportionals between two lines, that is, the duplication of the cube (the problem of Delos). In antiquity many solutions were produced for this problem. It was thought that in terms of solving this problem the mathematicians of Islam stood strangely apart from those of antiquity; but recently many examples have been discovered, thus altering this opinion. The following example of al-Maghribī's is of interest in this respect. He finds two values (see Fig. 3):

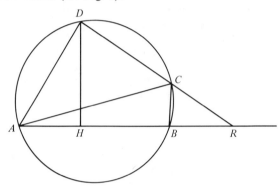

FIGURE 3

AB and BC are given and $AB > BC$, and $AB \perp BC$. AC are joined. Triangle ABC is circumscribed by a circle. The perpendicular DH is drawn so that DC must pass through point R.

$HR = AB$, $RH/DH = BA/DH$
$AH = BR$, $RH/HD = DH/HA$, since angle $D = 90°$
$BA/DH = DH/HA$

But

$$RH/DH = RB(=HA)/BC$$
$$BA/DH = DH/HA = HA/BC.$$

BIBLIOGRAPHY

The following works should be consulted: A. Aaboe, "Al-Kāshī's Iteration Method for Determination of sin 1°," in *Scripta mathematica*, **20** (1954), 24–29; and *Episodes From the Early History of Mathematics* (New Haven, 1964), 120; C. Brockelmann, *Geschichte der arabischen Literatur*, I (Leiden, 1943), 626, and supp. I (Leiden, 1937), 868–869; P. Brunet and A. Mieli, *Histoire des sciences d'antiquité* (Paris, 1935), 333–415; H. Bürger and K. Kohl, "Zur Geschichte des Transversalenzatzes," in *Abhandlungen zur Geschichte der Naturwissenschaften und der Medizin*, no. 7 (1924), 55–57, 67, 70, 71, 73–75, 89; Carra de Vaux, "Remaniement des sphériques de Théodose par Iahia ibn Muhammed ibn Abī Schukr Almaghrabī Alandalusī," in *Journal asiatique*, **17** (1891), 287–295; T. Heath, *A History of Greek Mathematics*, I (Oxford, 1921), 244–270; P. Luckey, "Der Lehrbrief über Kreis

Umfang (ar-Risāla al-Muhītīya) von Čamšid b. Mas'ūd al-Kāšī," in *Abhandlungen der Deutschen Akademie der Wissenschaften zu Berlin*, Math.-naturwiss. Kl., no. 5 (1950); and G. Sarton, *Introduction to the History of Science*, II, pt. 2 (Baltimore, 1931), 1015–1017.

H. Suter, *Die Mathematiker und Astronomen der Araber und ihre Werke* (Leipzig, 1900), 155; S. Tekeli, "Taqī al-Din's Work on Extracting the Chord 2° and sin 1°," in *Araştirma*, **3** (1965), 123–131; and "The Works on the Duplication of the Cube in the Islamic World," *ibid.*, **4** (1966), 87–105; F. Woepcke, "Sur une mesure de la circonférence du cercle due aux astronomes arabes et fondée sur un calcul d'Aboul Wafā," in *Journal asiatique*, **15** (1860), 281–320; and S. Zeki, *Asari bakiye*, I (Istanbul, 1913), 106–120.

S. TEKELI

MUIR, MATTHEW MONCRIEFF PATTISON (*b.* Glasgow, Scotland, 1 November 1848; *d.* Epsom, England, 2 September 1931), *chemistry.*

Muir was the son of a Glasgow merchant and received his elementary education in his native city. He began the study of chemistry at the University of Glasgow. In 1870 he entered the University of Tübingen in order to continue his studies, but the Franco-Prussian War forced his return home in 1871. He then served for two years as demonstrator in chemistry at Anderson College in Glasgow, and in 1873 he accepted a similar position at Owens College in Manchester.

In 1877 he was appointed to the praelectorship in chemistry at Gonville and Caius College at Cambridge, where he remained for the rest of his scientific career. He received an honorary M.A. in 1880, and in 1881 he became a fellow of the College. At the time of his death he was its senior fellow. By his 1873 marriage to Florence Haslam he had two sons, both of whom became clergymen. He retired from active teaching in 1908 and spent the remainder of his life at his home in Epsom, devoting himself to writing. Throughout his life he took an active part in politics.

Muir's laboratory investigations, which were carried out between 1876 and 1888, chiefly related to compounds of bismuth. He and his students published eighteen papers in this field. It is said that for a time his students called him Mr. Bismuth. He was not essentially interested in laboratory work, but rather preferred teaching and writing. His courses, especially those for medical students, were considered outstanding, and he excelled in encouraging weaker students, to whom his home was always open.

He closely followed the scientific literature and was therefore able to write a number of textbooks for the use of his students. His text *Principles of Chemistry* (1889) went through two editions, and his *Elements of Thermal Chemistry* (1885) was highly successful. He translated Ostwald's book on solutions (1891) and took part in a major revision of Watts' *Dictionary of Chemistry*. From the beginning of his career, his chief interest lay in the philosophical aspects of chemistry. In the 1880's he turned to historical studies, and in this field he made his greatest contributions. His historical writings began when he was asked to prepare a biographical work on famous chemists for a series on heroes of science. The book appeared in 1883 and so aroused his interest in the subject that for the rest of his life he devoted himself with increasing frequency to historical studies. In his later years Muir worked entirely in this field. Besides his biographical studies of famous chemists, he wrote on the chemical elements and alchemy. His chief historical work, which was published in 1907, was his *History of Chemical Theories and Laws*. He said that in this book he was trying to picture the steps in the development of major advances in chemistry without obscuring them by details. The book remains a classic in the history of chemistry.

BIBLIOGRAPHY

I. ORIGINAL WORKS. Muir's original studies lay in the field of historical works. His books included *Heroes of Science: Chemists* (London, 1883); *The Alchemical Essence and the Chemical Element* (London, 1894); *History of Chemical Theories and Laws* (London and New York, 1907); *The Story of the Chemical Elements* (London, 1908); and *The Story of Alchemy and the Beginnings of Chemistry* (London, 1914).

II. SECONDARY LITERATURE. The only substantial biography is R. S. Morrell, "M. M. Pattison Muir 1848–1931," in *Journal of the Chemical Society* (1932), 1330–1334.

HENRY M. LEICESTER

MULDER, GERARDUS JOHANNES (*b.* Utrecht, Netherlands, 27 December 1802; *d.* Bennekom, Netherlands, 18 April 1880), *chemistry.*

Mulder studied medicine at the University of Utrecht (1819–1825), from which he graduated with a dissertation on the action of alkaloids of opium, *De opio ejusque principiis, actione inter se comparatis* (1825). He practiced medicine in Amsterdam and then in Rotterdam, where he also lectured at the Bataafsch Genootschap der Proefondervindelijke Wijsbegeerte and taught botany to student apothecaries. At the foundation of a medical school at

Rotterdam (1828), Mulder became lecturer in botany, chemistry, mathematics, and pharmacy. His attention was directed primarily to the practical training of his students. In 1840 Mulder succeeded N. C. de Fremery as professor of chemistry at the University of Utrecht. He applied for his retirement in 1868 and spent the rest of his life in Bennekom. Besides publishing on scientific subjects, Mulder took an active part in education, politics, and public health. The works of Faraday and Berzelius exerted a great influence on him; his *Leerboek voor Scheikundige Werktuigkunde* (1832–1835) was written in the spirit of Faraday's *Chemical Manipulation*. Mulder edited a Dutch translation by three of his students of Berzelius' textbook of chemistry as *Leerboek der Scheikunde* (6 vols., 1834–1845). His difficult character caused problems with some of his pupils and with other chemists.

From 1826 to 1865 Mulder edited five Dutch chemical journals (see bibliography), in which most of his work was published. He worked in physics and in both general and physical chemistry, the latter in combination with medicine, physiology, agriculture, and technology. Most of his work had a polemic character. His most important contributions are in the field of physiological chemistry and soil chemistry, in which he published two extensive works that attracted much attention in translation despite their many mistakes and erroneous speculations.

Studies on proteins led Mulder to his protein theory (1838): he supposed that all albuminous substances consist of a radical compound (protein) of carbon, hydrogen, nitrogen, and oxygen, in combination with varying amounts of sulfur and phosphorus. The differences among proteins resulted from multiplication of the protein units in conjunction with the two other elements. Thus, casein was formulated as

$$10 \text{ protein units} + S,$$

and serum albumin as

$$10 \text{ protein units} + SP_2.$$

In 1843 Mulder published the first volume of a treatise on physiological chemistry, which was translated into English as *The Chemistry of Vegetable and Animal Physiology* (1845–1849). At first both Liebig and Berzelius accepted Mulder's analysis of proteins; but Liebig soon opposed the theory vigorously, and a deep conflict with Mulder ensued. In 1839–1840 Mulder investigated humic and ulmic acids and humus substances and determined the amounts of geic acid (acidum geïcum), apocrenic acid (acidum apocrenicum or Quellsatzsäure), crenic acid

(acidum crenicum or Quellsäure), and humic acids in fertile soils (1844). The structure of these various brown or black substances is unknown. They are a group of aromatic acids of high molecular weight, which can be extracted from peat, turf, and decaying vegetable matter in the soil. The difference between these acids is the oxygen content. In the decay of vegetable matter ulmic acid is formed. According to Mulder, this has the formula $C_{20}H_{14}O_6$ (in modern equivalents). In contact with air and water more oxygen is absorbed, which results in the successive formation of humic acid ($C_{20}H_{12}O_6$), geic acid ($C_{20}H_{12}O_7$), apocrenic acid ($C_{24}H_{12}O_{12}$), and crenic acid ($C_{12}H_{12}O_8$).

His studies on agricultural chemistry led to the treatise *De scheikunde der bouwbare aarde* (1860). Mulder confirmed Berzelius' suggestion that theine and caffeine are identical (1838) and was the first to analyze phytol correctly in his researches on chlorophyll. Among his other works are technical chemical publications on indigo (1833), wine (1855), and beer (1857), detailed research on the assaying method for analyzing silver in relation to the volumetric silver determination of Gay-Lussac (1857), and a study on drying oils (1865).

BIBLIOGRAPHY

I. Original Works. Mulder's writings include *Dissertatio de opio ejusque principiis, actione inter se comparatis* (Utrecht, 1825); *Leerboek voor Scheikundige Werktuigkunde*, 2 vols. (Rotterdam, 1832–1835); *Proeve eener algemeene physiologische scheikunde*, 2 vols. (Rotterdam, 1843–1850), translated as *The Chemistry of Vegetable and Animal Physiology* (Edinburgh, 1845–1849); *De vraag van Liebig aan de zedelijkheid en de wetenschap getoetst* (Rotterdam, 1846), also in *Scheikundige onderzoekingen*, **3** (1846), 357–487, and translated as *Liebig's Question to Mulder Tested by Morality and Science* (London, 1846); "De essayeermethode van het zilver scheikundig onderzocht," which is *Scheikundige verhandelingen en onderzoekingen*, **1**, pt. 1 (1857); *De scheikunde der bouwbare aarde*, 4 vols. (Rotterdam, 1860); and "De scheikunde der droogende oliën en hare toepassing," which is *Scheikundige verhandelingen en onderzoekingen*, **4**, pt. 1 (1865). Journals edited by Mulder are *Bijdragen tot de natuurkundige wetenschappen*, 7 vols. (Amsterdam, 1826–1832), with H. C. van Hall and W. Vrolik; *Natuur- en scheikundig archief*, 6 vols. (Rotterdam, 1833–1838); *Bulletin des sciences physiques et naturelles en Néerlande*, 3 vols. (Leiden, 1838–1840), with F. A. W. Miquel and W. Wenckebach; *Scheikundige onderzoekingen, gedaan in het laboratorium der Utrechtsche hoogeschool*, 6 vols. (Utrecht, 1842–1851); and *Scheikundige verhandelingen en onderzoekingen*, 4 vols. (Utrecht, 1857–1865).

II. Secondary Literature. See the biography by W. Labruyère, *G. J. Mulder (1802–1880)* (Leiden, 1938), with bibliography, pp. 113–130. See also E. Cohen, "Wat leeren ons de archieven omtrent Gerrit Jan Mulder?" which is *Verhandeling der K. akademie van wetenschappen. Afdeling Natuurkunde*, **19**, pt. 2 (1948). An autobiographical sketch of Mulder was published posthumously as *Levensschets. Door hemzelven geschreven en door drie zijner vrienden uitgegeven* (Rotterdam, 1881; 2nd ed. Utrecht, 1883). Mulder's correspondence with Berzelius was published as *Jac. Berzelius Bref*, H. G. Söderbaum, ed., V (Uppsala, 1916), *Briefväxling mellan Berselius och G. J. Mulder (1834–1837)*.

H. A. M. SNELDERS

MÜLLER, FRANZ (FERENC), BARON DE REICHENSTEIN (*b*. Nagyszeben, Transylvania [now Sibiu, Rumania], 1 July 1740; *d*. Vienna, Austria, 12 October 1825), *chemistry*.

Müller, the son of a treasury official, was educated in Nagyszeben and then studied law in Vienna. By then he had become interested in chemistry and mineralogy and went to Selmecbánya (Schemnitz, in Hungary), where a short time earlier one of the world's first mining academies had been opened. There he studied mining and metallurgy under N. J. Jacquin. After completing his studies in 1768, Müller entered the service of the state saltworks in Transylvania; later he was active in mining in southern Hungary. From 1775 to 1778 he was director of the state mines in the Tirol; and from 1778 to 1802 he directed all mining operations in Transylvania from his office in Nagyszeben. In 1802 Müller moved to Vienna to head the council that had jurisdiction over minting and mining in Austria and Hungary. He held this position until 1818. On his retirement he received the Order of St. Stephen and the title of baron.

Müller discovered the chemical element tellurium in 1784 at Nagyszeben. For several years sylvanite, an auriferous mineral from Transylvania, had been causing problems because its processing always yielded less gold than expected. Anton von Ruprecht, a former schoolmate of Müller's and professor of chemistry at the Selmecbánya Mining Academy, analyzed the ore in 1782 and published his finding that it contained antimony as well as gold. Müller did not share this view, asserting in print that the substance involved was bismuth. Ruprecht responded by stating the reasons it could not be bismuth. In his next publication Müller admitted his error and announced that a new, previously unrecognized semimetal was present in the ore; he also enumerated its characteristic chemical reactions (*Physikalische Arbeiten der einträchtigen Freunde in Wien*, **1**, no. 2 [1783], 63).

Müller, however, did not name the new element.

Instead, he sent a sample of the ore to Torbern Bergman at Uppsala, wishing to confirm his conclusion by submitting the substance for examination to the most famous analyst of the century. Bergman reported in a letter that he was starting to work on the matter, but he died soon after. Ten years later the Berlin chemist Martin Klaproth asked Müller to send him a sample. He carried out an analysis, confirmed Müller's finding, and gave a lecture on the subject at the Berlin Academy, where he proposed that the previously unnamed element be called tellurium.

Müller also contributed to mineralogy. He discovered a variety of tourmaline and a variety of opal that is also called Müller glass.

BIBLIOGRAPHY

Müller's publications are listed in Poggendorff, II, 231.

Secondary sources include R. Jagnaux, *Histoire de la chimie*, I (Paris, 1891), 500–504; F. Szabadváry, *Az elemek nyomában* ("In the Traces of Elements"; Budapest, 1961), 142–148; and M. E. Weeks, *Discovery of the Elements* (1956), 303–304, also in *Journal of Chemical Education*, **12** (1935), 403.

F. SZABADVÁRY

MÜLLER, FRITZ (JOHANN FRIEDRICH THEODOR) (*b*. Windischholzhausen, Thuringia, Germany, 31 March 1822; *d*. Blumenau, Brazil, 21 May 1897), *natural history*.

Although described by Blandford in *Nature* (1897) as "one of the greatest and most original naturalists" of the nineteenth century, Müller's reputation has always been overshadowed by those of his illustrious scientific contemporaries. His innate modesty, complete indifference to fame, and physical isolation in southern Brazil further contributed to obscure the significance of his work. His book, *Für Darwin* (1864), however, was a fundamental contribution to evolutionary biology at a critical moment during its infancy; and his name has been immortalized in scientific literature with the term "Müllerian mimicry."

Müller was born in a small village outside Erfurt. His father, Johann Friedrich Müller, was a minister, and his mother was the daughter of the distinguished Erfurt chemist and pharmacist J. B. Trommsdorff. Both parents had a strong interest in natural history, and his father in particular greatly influenced Fritz and his younger brother Hermann, who became a well-known botanist at Lippstadt.

Müller's early formal education began at the village school of Mühlberg and continued at the Erfurt

Gymnasium (1835–1840), where his extraordinary linguistic ability—he learned Italian, Russian, Syriac, Arabic, English, and later Portuguese—became evident. After studying pharmacy for one year at Naumburg, he began advanced work at the University of Berlin, where he studied mathematics and the natural sciences. His anatomy professor was Johannes Müller. He spent the following academic year (1842–1843) at the University of Greifswald, then returned to Berlin, where he completed the Ph.D. on 14 December 1844. His dissertation, "De hirudinibus circa Berolinum hucusque observatis," dealt with the leeches found near Berlin. He continued work at Berlin for his advanced teaching certificate (Oberlehrerexamen) before returning to the Gymnasium at Erfurt for his teaching period as a probationary candidate (1845). Later that year, however, he decided to study medicine, with the intention of becoming a ship's surgeon and seeing the world, especially the tropics, where he hoped to study zoology.

Returning to the University of Greifswald (1845–1849), Müller completed all the work for his medical degree except for the state certification examination (Staatsexamen). The Ministry of Education would not allow him to take the examination because he had sided with the democrats in the Revolution of 1848 and had refused to take a religious oath recognizing the established church and orthodox religious views. (Müller believed in free love, and Katherine Töllner bore three of their ten children out of wedlock.) In 1849, unable to obtain his degree, Müller in October became a private tutor at Roloffshagen (near Grimmen). He eventually received honorary medical degrees from Bonn in 1868 and from Tübingen in 1877. Prussian religious intolerance finally led Müller to abandon his homeland in 1852 and sail to Blumenau, Brazil, on the Itajai River near the coast between Río de Janeiro and Buenos Aires. Most of his important scientific work was done in South America, where he spent the rest of his life.

Despite his superb education, Müller lived there as a farmer until 1856, when he was appointed mathematics teacher at the provincial lyceum at Desterro (now Florianópolis), Santa Catarina Island. Various conflicts—particularly with the Jesuits—led to the termination of his employment in 1867, and he returned to Blumenau, where he worked as a civil servant for the provincial government until 1876. Müller was then appointed traveling naturalist for the National Museum in Río de Janeiro, a post which he lost (including the pension) in 1891 when he refused to move to Río de Janeiro. His last years in Blumenau were marred by a variety of misfortunes—imprisonment and trial by rebels, and the death of both his

wife and his daughter—although he resumed his work before his death.

Müller's scientific contributions ranged from anatomical work on Coelenterata, Annelida, and especially Crustacea to entomology, emphasizing mimicry, and to botany, particularly in his later years. After moving to Desterro, he began to study the marine invertebrates of the Brazilian coastal waters. Darwin's Origin of Species (1859) led Müller to test those evolutionary ideas by applying them to the Crustacea. He traced the genealogies of various groups, hoping to uncover affinities and the origins of fundamental (primitive) forms. While Darwin offered general propositions, Müller provided a specific test case in the development of the Crustacea. His verdict was rendered in favor of Darwin's views: "In one thing, I hope, I have succeeded,—in convincing unprejudiced readers, that Darwin's theory furnishes the key of intelligibility for the developmental history of the Crustacea" and "many other facts [are] inexplicable without it" (Für Darwin, 1869 translation, p. 141). Publication of such enthusiastic, sympathetic views led to a lengthy correspondence with Darwin, who provided the financial backing for the English translation of Für Darwin in 1869, frequently sent Müller's letters to journals for publication, and quoted him extensively in his own work.

During the 1870's and 1880's Müller published many articles on entomology, the most famous of which discussed mimetic phenomena. In 1862 the English naturalist Henry Walter Bates had first published his own observations concerning examples of relatively scarce, palatable, and defenseless species of insects (primarily) which closely resembled other species which were plentiful and relatively unpalatable or were protected in some other manner. He thought such situations arose through the process of evolution by means of natural selection; that is, those mimics which most closely resembled the protected species would be rejected by predators and therefore survive, but those which varied greatly in appearance from the protected species would be eliminated in the struggle for existence. Bates, however, did not explain why two or more distasteful but unrelated species resembled one another.

In a series of articles beginning in 1878, Müller explained that predators must learn through warning characteristics which species are palatable, and that in this process some of the prey population must be sacrificed. If there are two or more similar, unpalatable species, then predators will be educated faster by the warning characteristics, the similar species will be better protected, fewer deaths will result, and the losses will be absorbed by a larger group. These views were quickly adopted and expanded by other evolutionists, including A. R. Wallace and E. B. Poulton, and form an important part of contemporary literature in evolutionary biology.

Müller's botanical work dealt mainly with the fertilization of plants, with discussions of hybridization and sterility, including self-sterility. Darwin cited this work frequently in his book *The Effects of Cross and Self Fertilisation in the Vegetable Kingdom* (1876).

Altogether, Müller published almost 250 articles in which he demonstrated extraordinary powers of observation, while his book *Für Darwin* and his articles on mimicry reflect his considerable ability to formulate perceptive conclusions. In general, however, he was content to allow others to build upon his smaller, albeit valuable, contributions.

BIBLIOGRAPHY

I. ORIGINAL WORKS. All of Müller's works were conveniently collected by his nephew, Dr. Alfred Möller, *Fritz Müller. Werke, Briefe und Leben*, 3 vols. in 4 plus atlas (Jena, 1915–1921). His works are in vol. I (in two pts., plus an atlas of plates [1915]). *Für Darwin* (Leipzig, 1864) was trans. into English as *Facts and Arguments for Darwin* (London, 1869; repr., 1968) and into French. The rich and extensive correspondence, including letters to Charles Darwin and many German biologists, is in vol. II (1921); and his biography is in vol. III (1920), with a valuable map of his excursions. MS letters and additions to *Für Darwin* are in the Darwin Papers, University Library, Cambridge.

Important early articles on mimicry are "Ueber die Vortheile der Mimicry bei Schmetterlingen," in *Zoologischer Anzeiger*, **1** (1878), 54–55; a note on a remarkable case of mimicry of *Eueides pavana* with *Acraea thalia* referred to in *Proceedings of the Entomological Society of London* (1879), ii; and particularly "Ituna and Thyridia; a Remarkable Case of Mimicry in Butterflies," *ibid.*, pp. xx–xxviii, discussion pp. xxviii–xxix—the article first appeared in *Kosmos*, **5** (1879), 100–108. The latter two articles were trans. by R. Meldola. English trans. of some other entomological articles appear in George B. Longstaff, *Butterfly-Hunting in Many Lands. Notes of a Field Naturalist. To Which Are Added Translations of Papers by Fritz Müller on the Scentorgans of Butterflies and Moths: With a Note by E. B. Poulton, D.Sc., F.R.S.* (London–Bombay–Calcutta, 1912). Numerous letters exchanged by Darwin and Müller appear in both Francis Darwin, ed., *The Life and Letters of Charles Darwin*, 3 vols. (London, 1887); and Francis Darwin and A. C. Seward, eds., *More Letters of Charles Darwin*, 2 vols. (New York, 1903).

II. SECONDARY LITERATURE. The best biography of Müller is Alfred Möller's *Leben* (vol. III of the *Works*). He also lists some obituaries on p. 163, two of which are F. Ludwig, "Ueber das Leben und die botanische Thätigkeit Dr. Fritz Müller's," in *Botanisches Centralblatt*, **71** (1897), 291–302, 347–363, 401–408, plus 4 plates, (100

articles are listed on pp. 404–408); and W. F. H. B. [Walter F. H. Blandford], "Fritz Müller," in *Nature*, **56** (1897), 546–548. Blandford's observations on mimicry are quite interesting, as are those of Roland Trimen, "President's Address. Mimicry in Insects," in *Proceedings of the Entomological Society of London* (1897), lxxiv–xcvii. Also on mimicry, see Mary Alice Evans, "Mimicry and the Darwinian Heritage," in *Journal of the History of Ideas*, **26** (1965), 211–220. For additional references on mimicry, see H. Lewis McKinney, "Henry Walter Bates," *Dictionary of Scientific Biography*, I, 504. For Müller's work on botany, see the article by F. Ludwig cited above.

H. LEWIS MCKINNEY

MÜLLER, GEORG ELIAS (*b.* Grimma, Germany, 20 July 1850; *d.* Göttingen, Germany, 23 December 1934), *psychology.*

Müller was the son of Oberpfarrer Müller and Rosalie Zehme. At the *Fürstenschule* in Grimma, where his father was a professor of religion, Müller became intellectually awakened by Goethe's *Faust* and by romantic poetry, an enthusiasm from which he was rescued by studying Lessing. He briefly attended the Gymnasium at Leipzig, and on leaving there he determined to accept in philosophy only what could be proved by strict logic, a resolve from which he never wavered.

He spent the next two years, successively, at the universities of Leipzig and Berlin. During this time he was much concerned with whether science or history should be the propaedeutic to philosophy. During a year in the Franco-Prussian War, he decided in favor of science. Müller then returned to Leipzig but went on to Göttingen to study psychology and philosophy under Lotze. Lotze's influence is shown in Müller's 1873 doctoral dissertation, *Zur Theorie der sinnlichen Aufmerksamkeit*, a nonexperimental but exhaustive study of sensory attention which was soon cited extensively.

While at Leipzig, Müller had heard Fechner lecture, and he continued a somewhat belligerent scientific correspondence with him from Rötha, where he had a job as a tutor. This correspondence resulted in his *Grundlegung der Psychophysik*, which Müller presented as his *Habilitationsschrift* at Göttingen. He became *Dozent* there in 1876. The work was published in 1878, and its meticulous discussion of Weber's law and many innovations in psychophysical method became the chief reason for Fechner's own *Revision der Hauptpunkte der Psychophysik* (1882).

Müller remained at Göttingen as *Dozent* for four years, then spent a year as professor of philosophy at Czernowitz. He succeeded Lotze at Göttingen in 1881.

Remaining at Göttingen for the rest of his life, he became an institution, much like Wundt at Leipzig. His laboratory vied with Wundt's as the best in Germany for psychological research. Müller was methodological, austere, and had a mania for impartiality. His work was characterized by a fierce insistence on order, so much so that he refused to partake in seminars, regarding them as too improvised to be of value.

Müller's work can be classified as being in three areas: psychophysics, learning, and vision.

Psychophysics. Beyond the work already mentioned, Müller's contributions in psychophysics comprised articles on method and the muscle basis of weight judgments. He and L. J. Martin published jointly in 1899 *Zur Analyse der Unterschiedsempfindlichkeit*, which, after Fechner's *Elemente der Psychophysik*, is the classical study of the psychophysics of lifted weights, the most thoroughly investigated psychophysical function. In 1903 there appeared the definitive handbook on psychophysics, the *Gesichtspunkte und Tatsachen der psychophysischen Methodik*. This book did not present anything very new, but it was a thorough summing-up of the entire field and what, in Müller's view, psychophysics had accomplished up to that time.

Learning. In 1885 Ebbinghaus published his classic work on the learning and memory of nonsense syllables. In 1887 Müller made the problem his own. Where Ebbinghaus was original, Müller was thorough. He carefully extended Ebbinghaus' findings, and he and Schumann invented the memory drum for more accurate presentations; most interestingly, they recorded introspections while the learning was going on. This innovation contradicted the feeling one gets from Ebbinghaus that learning is a mechanical and automatic process occurring through mere contiguity. During learning, subjects are active, not passive, using groupings and rhythms, finding meanings even in nonsense materials, and, in general, consciously organizing material. And the preparatory set or *Anlage* of the subject is a determining factor in memory.

These emphases are entirely absent from Ebbinghaus' work and anticipated those of the Würzburg laboratory under Külpe, once Müller's student. Another student at this time was Adolph Jost, whose work with Müller led to Jost's law—when two associations are of equal strength, a repetition is more strengthening to that which occurred first. This work led to the theory of the advantage of distributed practice; and with another student, Alfons Pilzecker, Müller published a joint monograph in 1900 showing the significance of reaction times as indicators of the strength of associations.

Müller summarized his work in learning in the encyclopedic, three-volume *Zur Analyse der Gedächtnistätigkeit und des Vorstellungsverlaufes* (1911–1917).

Vision. Müller's first work dealing with vision presented the hypothesis that cortical gray is the zero point from which all color sensations diverge. This work was an attempt to solve the paradox of color mixture inherent in Hering's theory of three reversible photochemical substances which in equilibrium should, but do not, result in visual silence. After occasional publications in this area over the years, Müller published his lengthy *Über die Farbenempfindungen: Psychophysische Untersuchungen* in 1930, reviewing and evaluating the entire field.

From 1907 to 1918, David Katz worked with Müller, and he published his important work in surface and volumic colors, a landmark in experimental phenomenology, during this time. Edgar Rubin, working in Müller's laboratory, printed his own phenomenological analysis of visual perceptions into figure, ground, and contour, instead of into the more conventional sensory ultimates.

Müller's position at the beginnings of phenomenology and Gestalt psychology was extremely important, if complex. One of his last works was his 1923 *Komplextheorie und Gestalttheorie*, a work on methodology in perception that criticized the more strident claims to newness on the part of Gestalt psychology.

Much more than Wundt, Müller's approach and work set the ideal pattern for experimental psychology. It was Müller who established the precedent that psychology had to be separated from philosophy if it was to become a rigorous science. Titchener, although Wundt's student, always turned to Müller for criticism and guidance. Among German psychologists of the period, Müller is usually ranked second to Wundt partly because of the large numbers of American students whom Wundt managed to attract, partly because of Müller's austerity and occasional absentminded ungraciousness to visitors, and partly because Müller, unlike Wundt, never wrote popular or systematic books. But at the present day Wundt's influence is difficult to find because it is so diffuse; whereas the emphases, problems, and particularly the tough-minded experimentalism represented by Müller are at the heart of contemporary experimental psychology.

BIBLIOGRAPHY

I. ORIGINAL WORKS. Müller's major works are *Zur Theorie der sinnlichen Aufmerksamkeit* (Leipzig, 1873); *Zur Grundlegung der Psychophysik* (Berlin, 1878); "Ueber die psychologischen Grundlagen für die Vergleichung der gehobenen Gewichte," in *Pflüger's Archiv für die gesamte Physiologie des Menschen und der Tiere*, **47**, 37-112, written with F. Schumann; "Theorie der Muskel-contraction," in *Nachtrichten von der Gesellschaft der Wissenschaften zu Göttingen* (1889); "Experimentelle Beiträge zu Untersuchungen des Gedächtnisses," in *Zeitschrift für Psychologie und Physiologie der Sinnesorgane*, **6** (1893), 81–190, 257–339, written with F. Schumann; with L. J. Martin, *Zur Analyse der Unterschiedsempfindlichkeit* (Leipzig, 1899); with A. Pilzecker, *Experimentelle Beiträge zur Lehre vom Gedächtnis* (Leipzig, 1900); *Zur Analyse der Gedächtnistätigkeit und des Vorstellungsverlaufes* (Leipzig, 1911–1917), also published in *Zeitschrift für Psychologie*, vols. 5, 8, 9; *Komplextheorie und Gestalttheorie: ein Beitrag zur Wahrnehmungspsychologie* (Göttingen, 1923); and *Über die Farbenempfindungen: psychophysische Untersuchungen* (Göttingen, 1930).

II. Secondary Literature. The only biographical source for Müller's early life is two letters from him to E. G. Boring, which are now in the Boring Papers of the Harvard archives. These letters are excerpted in Boring's obituary of Müller, in *American Journal of Psychology*, **47** (1935), 344–348; see also E. G. Boring, *A History of Experimental Psychology* (New York, 1929), 361–373, which contains a more complete bibliog.; for his personality see the obituary by his student D. Katz, in *Psychological Bulletin*, **32** (1935), 377–380; and for a description of the Göttingen Laboratory in 1892, see O. Krohn, in *American Journal of Psychology*, **5** (1893), 282–284.

E. B. Titchener discusses Müller's work throughout his own publications, but see particularly his *Lectures on the Elementary Psychology of Feeling and Attention* (New York, 1908), esp. 188–206, 356–359; *Experimental Psychology*, II (New York, 1905), 2, esp. 300–313; and for Müller on introspection, see "Prolegomena to a Study of Introspection," in *American Journal of Psychology*, **23** (1912), 490–494.

See also O. Klemm, *A History of Psychology* (New York, (1914), esp. 257–262, 296; H. Münsterberg, *Professor G. E. Müller's "Berichtigung"* (Boston, 1893); and reviews of Müller's books by: J. A. Bergstrom, in *American Journal of Psychology*, **6** (1894), 301–303; F. Angell, *ibid.* **11** (1899), 266–271; J. W. Baird, in *Psychological Bulletin*, **13** (1916), 373–375; and K. Koffka, *ibid.* **19** (1922), 572–576.

W. Köhler's reply to Müller's 1923 criticism of Gestalt psychology can be found in *Psychologische Forschung*, **6** (1925), 358–416; and Müller's rejoinder, in *Zeitschrift für Psychologie*, **99** (1926), 1–15.

JULIAN JAYNES

MÜLLER, GUSTAV (*b.* Schweidnitz, Germany [now Świdnica, Poland], 7 May 1851; *d.* Potsdam, Germany, 7 July 1925), *astronomy, astrophysics.*

Müller's father, a merchant, died when Gustav was only six years old. Müller was educated at a private school and then at the Gymnasium of his native city. After passing the final secondary school examination in 1870, he began the study of mathematics and natural science at Leipzig. From 1872 he continued his studies in Berlin, where the lectures of Wilhelm Foerster induced him to give up his original plans of becoming a teacher and to dedicate himself to astronomy. His other professors included Helmholtz, Weierstrass, and Ernst Kummer. Even before completing his studies Müller took part in the calculations made for the Berlin *Astronomisches Jahrbuch* and in Auwers' new reduction of Bradley's observations. He also assisted Hermann Vogel in his spectroscopic work and in 1877 followed him to the newly created astrophysical observatory at Potsdam. The work done there in spectrophotometry provided the decisive impetus for Müller's enduring interest in photometry. In 1877 he earned his doctorate with *Untersuchungen über Mikrometerschrauben*. He became an observer in 1882, chief observer in 1888, and professor in 1891.

Müller's contributions to the development of astrophysics, especially to the gathering of primary data, were distinguished less by bold innovations than by a clear grasp of the needs of an organically growing science and by the persevering and precise execution of the vast programs of research required by such growth. His photometric studies began in 1877 with investigations of the luminosities of the planets and of the absorption of starlight in the earth's atmosphere. He published extensive results in 1883 and 1893. His absorption tables for Potsdam were used for decades.

Müller's most important photometric project was the *Photometrische Durchmusterung des nördlichen Himmels*—the *Potsdamer Durchmusterung*—which he planned and, for the most part, carried out himself. Observations were begun in 1886, in collaboration with Paul Kempf. Utilizing an astrometer that he had constructed in accordance with the principle described in 1861 by Zöllner, he ascertained the luminosities of more than 14,000 stars in the Northern sky listed in the *Bonner Durchmusterung* (*BD*) to a magnitude of 7.5. The work devoted to this undertaking lasted for several decades. Partial results appeared in 1894, 1899, 1903, and 1906; the general catalog appeared in 1907. The *Potsdamer Durchmusterung* furnished, with the *Harvard Photometry* of Pickering and his co-workers, the most exact photometric information on stars then available; and it is still an indispensable standard work. Moreover, through its consistent use of Pogson's scale it played a decisive role in the general adoption of this scale.

In 1909 Müller began with E. Kron a further series of zonal observations of the luminosities and colors of the *BD* stars from magnitude 7.6 to 9.5. However,

World War I hindered the progress of this undertaking, and Müller was unable to publish a compendium of the results.

Müller's proposal at the 1900 meeting of the Astronomische Gesellschaft that a complete catalog of the variable stars be produced led to the *Geschichte und Literatur des Lichtwechsels der bis Ende 1915 als sicher veränderlich anerkannten Sterne (GuL)*. This three-volume work (1918–1922), which exercised an extremely positive influence on variable-star research, was continued by R. Prager, who published two volumes of a new edition in 1934 and 1936; the third volume, prepared by H. Schneller, appeared in 1952.

Müller's work in the field of spectroscopy includes his contribution to Vogel's *Spektroskopische Beobachtungen der Sterne* (1883) and his determination with Kempf (1886) of the absolute wavelengths of 300 lines of the solar spectrum.

Müller also produced a series of works that contributed to increasing the precision of observations and to improving reduction elements. This series included an investigation of the influence of temperature on the refraction of light in various types of glass (1885).

Müller participated in many of the great scientific expeditions sponsored by the Potsdam observatory. He led the 1882 expedition to the United States to observe the transit of Venus and participated in the expeditions to Russia to observe the total solar eclipses of 1887 and 1900. He also assisted in absorption studies conducted in Tenerife.

Müller exercised important functions in the organization of scientific research; from 1896 to 1924 he was secretary of the Astronomische Gesellschaft. Following the death of the second director of the Potsdam observatory, Karl Schwarzschild, he directed the institution from 1917 to 1921. During this time the Einstein Foundation was created and the Einstein Tower for solar physics was constructed. Admitted to the Prussian Academy of Sciences in 1918, Müller became chairman of the commission on the *Geschichte des Fixsternhimmels* ("history of the fixed stars"). Although not active in popularizing astronomy, he wrote the well-known monograph *Die Photometrie der Gestirne* (1897).

BIBLIOGRAPHY

A complete bibliography of Müller's 107 published works, compiled by his son Rolf, is in *Vierteljahrsschrift der Astronomischen Gesellschaft*, **60** (1925), 174–177. See also H. Ludendorff's obituary, *ibid.*, 158–174; and his notice on Müller, in *Astronomische Nachrichten*, **225** (1925), cols. 199–200.

DIETER B. HERRMANN

MULLER, HERMANN JOSEPH (*b*. New York, N.Y., 21 December 1890; *d*. Indianapolis, Indiana, 5 April 1967), *genetics, evolution, eugenics.*

Muller's grandfather came to the United States from Germany following the revolution of 1848. His father, for whom he was named, was to train in the law, but instead had to take over the family business of manufacturing bronze artworks. He died when Muller was nine years old, leaving his widow, Frances Lyons, a modest income. Even as a child, Muller was interested in evolution and the sciences; as a student at Morris High School, he founded a science club. Upon graduation, Muller entered Columbia College, where in his sophomore year he decided to make genetics his major study, after reading R. H. Lock's *Heredity, Variation, and Evolution.*

At Columbia, Muller attended Edmund B. Wilson's course, which, through its emphasis on the chromosome theory of heredity, shaped his genetic view of biological problems. He received the B.A. in 1910, then enrolled in Cornell Medical School and the Columbia University department of physiology. His master's thesis (1912) concerned the transmission of nerve impulses. His interest in genetics was unabated, however, and he remained in daily contact with two of his classmates, Alfred H. Sturtevant and Calvin B. Bridges, who were working at Columbia with Thomas Hunt Morgan. In 1912 Muller himself was accepted by Morgan as a graduate student; he rapidly established a reputation for imaginative theorizing and ingenious experimental design. His dissertation, on crossing-over, contributed the new concepts of coincidence and interference in the resolution of genetic maps and established the law of linear linkage. After taking the Ph.D. in 1916, Muller accepted an invitation from Julian Huxley to teach at Rice Institute. His own research at this time comprised studies of the complex relationship of gene and character in which he isolated and mapped the modifier genes that control the quantitative expression of inherited characteristics. From these investigations Muller was led to recognize the significance of the individual gene.

Muller analyzed mutations to conclude that the concept should be confined to variations arising in the individual gene. Upon his return to Columbia (1918–1920), he produced his most important theoretical work. Since genes, unlike all other cellular components, can reproduce the alterations arising in them (indeed, this property of self-replication is their unique feature), Muller argued that all other cellular components must be ultimately produced by genes; he theorized that life must have begun with the appearance of self-replicating molecules, or "naked genes," which he thought similar to viruses.

From 1921 until 1932 Muller worked at the University of Texas, where he became full professor. During this period he studied mutation frequency and designed complex genetic stocks to detect the most commonly occurring lethal mutations (which kill unless protected by a normal allele). In 1926 Muller induced mutations by exposing *Drosophila* to X rays. He reported his findings in an article entitled "Artificial Transmutation of the Gene," which was published in *Science* in 1927. This work won Muller an international reputation, stimulated a number of other workers to take up the subject, and became the basis for the study of radiation genetics.

In 1932 Muller went to Berlin, where he spent a year as a Guggenheim fellow working in Oskar Vogt's Brain Research Institute. He did research with N. W. Timofeev-Ressovsky on mutation, evaluating and criticizing physical models (among them the "target theory"), and exploring the structure of the gene. As Hitler achieved increasing power, Muller, a staunch supporter of socialist causes, decided to leave Germany. He accepted an invitation from N. I. Vavilov to do research in the Soviet Union, which he regarded as an experimental society that would support genetics and eugenics.

Muller worked in Leningrad and Moscow at the Academy of Sciences from 1933 until 1937. He was chiefly concerned with radiation genetics, cytogenetics, and gene structure. By 1935 he was embroiled in the growing controversy about the work of Lysenko, which he could not support. Muller himself hoped to win Soviet sponsorship for basic genetics and for the program of positive eugenics that he presented in his book *Out of the Night* (New York, 1935). Lysenko won Stalin's backing, and Muller left the Soviet Union, after volunteering to serve in the Spanish Civil War.

In 1938 Muller received an appointment to the University of Edinburgh, where he analyzed the chromosomal basis of embryonic death from radiation damage. With the outbreak of World War II, he returned to the United States. He thought that it would be difficult to continue scientific research in Great Britain and since he and his second wife, Dorothea Kantorowicz, were part Jewish, he was also concerned with their own safety. He first went to Amherst College, where he continued his genetic studies, and then, in 1945, he secured a permanent post at Indiana University, where he remained until his death.

In 1946 Muller was awarded the Nobel Prize in physiology or medicine. He took advantage of the concomitant fame to publicize his campaign against the medical, industrial, and military abuse of radiation.

He also publicly criticized the doctrines of Lysenko and resigned from the Soviet Academy of Sciences in 1947. In his later years Muller also worked for the reform of the teaching of biology in secondary schools (he advocated a strong genetic and evolutionary viewpoint) and set out a positive eugenic program, based on what he called "germinal choice," whereby the semen of unusually healthy and gifted men would be frozen for use by later generations. Although his social views were the subject of considerable public controversy, they were prompted by his genuine scientific concern about the accumulating load of human spontaneous mutations produced by the relaxation of natural selection through modern culture and technology.

Muller was a member of a number of scientific societies, including the National Academy of Sciences. He was a fellow of the Royal Society, and served as president of several genetic societies and congresses.

BIBLIOGRAPHY

I. ORIGINAL WORKS. Muller published 372 works, of which a complete bibliography is given by Pontecorvo (below). His published and manuscript articles, correspondence, notebooks, and other scholarly documents are in the Lilly Library of Indiana University in Bloomington. His most famous publication is "Artificial Transmutation of the Gene," in *Science*, **66** (1927), 84–87.

II. SECONDARY LITERATURE. On Muller and his work see E. A. Carlson, "The Legacy of Hermann Joseph Muller: 1890–1967," in *Canadian Journal of Genetics and Cytology*, **9**, no. 3 (1967), 437–448; "H. J. Muller," in *Genetics*, **70** (1972), 1–30; G. Pontecorvo, "Hermann Joseph Muller 1890–1967," in *Biographical Memoirs of Fellows of the Royal Society*, **14** (1968), 349–389, with complete bibliography; and T. M. Sonneborn, "H. J. Muller, Crusader for Human Betterment," in *Science*, **162** (1968), 772–776.

ELOF AXEL CARLSON

MÜLLER, JOHANN. See **Regiomontanus.**

MÜLLER, JOHANN HEINRICH JACOB (*b.* Kassel, Germany, 30 April 1809; *d.* Freiburg im Breisgau, Germany, 3 October 1875), *physics.*

Müller's father originally studied law, but gave it up to become, in 1807, a painter to the court of the prince of Waldeck. Johann Müller—as he unfortunately called himself in publications, thus creating confusion in bibliographies—inherited his father's talent and excelled in illustrating his books. He spent his early years in Frankfurt am Main and Darmstadt, then, in 1829, enrolled in the University of Bonn to study mathematics and physics. His teachers included Karl von Münchow and Heinrich Plücker. In 1832 he entered Giessen University, where he attended lectures on chemistry, physics, and mathematics given by Heinrich Buff, Liebig, and Hermann Umpfenbach. He received the Ph.D. in 1833 with a thesis on the optics of crystals. In 1837 he became a teacher at the Giessen Realschule and was eventually, in 1844, appointed professor of physics and technology at the University of Freiburg.

Early in his career Müller developed a systematic concept of physics which was partially determined by his didactic talent. He wrote a number of synopses (see Bibliography) before his appointment at Freiburg and from 1842 he edited his work *Lehrbuch der Physik und Meteorologie*. The subsequent enlargements of this work, which remained well-known until the 1930's, constituted Müller's principal activity for the rest of his life.

When the University of Freiburg ended the mandatory study of physics in 1836 Müller attempted to attract students by improving the instruction. He purchased and built many improved instruments for the laboratory and asked Lerch to draw large-scale illustrations for his lectures. During the university's summer term Müller lectured on statics of solids, fluids, and aeriform bodies; acoustics; heat; electricity; and magnetism. His winter lectures comprised wave mechanics and optics, his preferred subjects. He also introduced experimental practice in physics and in 1850 the first doctorate was awarded.

Müller conducted research into optics, magnetism, and light and heat radiation. Applying George Airy's theory to crystal optics, Müller calculated the isochromatic curves of plates of crystals with one parallel optical axis; he also calculated the black hyperbolic beams in the system of lemniscates of crystals with two axes. In 1849 he experimentally found, independently of Joule, the limits of intensity of the magnetism (m) of iron, which according to Siméon Poisson, Heinrich Lenz, and Carl Jacobi should have been proportional to the magnetizing force (p). Müller's findings were more precise than Joule's results, and he calculated $p \propto$ an m for a constant diameter of the rod, a relation subsequently used with success.

From 1846 Müller studied Fraunhofer lines, an investigation that led him to explore ultraviolet radiation by fluorescence; with the chemist Lambert Babo, he was able to measure, in 1855, the first Fraunhofer lines beyond the violet end of the spectrum. He then examined the *Thermische Wirkungen des Sonnenspectrums* ("infrared spectrum") by drawing upon the work of Rudolph Franz, who first measured the infrared spectrum quantitatively by heat transmission (1856). This process, known as "diathermancy," plus the analysis of the spectrum with rocksalt and the use of a linear thermopile to determine the spectral energy, permitted Müller to draw a curve of the intensity of radiation by heat effect. This was the first such curve ever to be drawn. Müller thus proved, contrary to Antoine Masson and Jules Jamin, that the maximum of heat energy lies in the dark region. He also found the wavelength of the rays on the extreme end of the spectrum to be 0.0048 mm. Although it was impossible to take a curve in the diffraction of heat because of the energy distribution by diffraction, Müller construed the curve geometrically from the prismatic spectrum and concluded that the heat spectrum is three times as extensive as the visible one.

Müller's most significant textbook, the *Lehrbuch*, first appeared as *Pouillet's Lehrbuch der Physik und Meteorologie*, a "free adaptation" of the 1837 edition of C. S. Pouillet's *Éléments de physique expérimentale et de météorologie*. Müller's innovations included numerous woodcuts inserted directly into the text, whereas Pouillet had inserted copper-plate engravings after the text. The illustrations of the apparatus were particularly useful for the mechanician. The book was initially styled for the nonphysics major. He supplied the derivations of mathematical formulas and stressed mechanical theorems. Müller incorporated Gauss's works on magnetism for the first time and recast the chapters on galvanism, light, and meteorology. Each of the seven editions that were published during his lifetime underwent considerable emendation. A third volume, *Lehrbuch der Kosmischen Physik*, based upon Müller's own observations was added in 1856. A supplement, *Die medizinische Physik*, edited by A. Fick, also appeared in the same year. The rapid evolution in physics was characteristically reflected by the "Reports on the Most Recent Developments in Physics," edited by Müller in 1849.

Liebig's letters of 1844 induced Müller to edit *Physikalische Briefe für Gebildete aller Stände*, a new edition of Euler's *Letters to a German Princess*. Müller subsequently added a third and fourth part; there, in his own letters on physics, he was emphatic that natural sciences taught objective truth, the limits of man's intellectual power, and tolerance. Müller encouraged the criticism of traditional preoccupations. He read many papers before the Naturforschende

Gesellschaft zu Freiburg im Breisgau and helped to extend the society's activity into physics.

BIBLIOGRAPHY

I. Original Works. MS documents on Müller's employment at Freiburg are in the university archive. A letter to an unknown person, 3 April 1843, is in the Germanisches Nationalmuseum, Nuremberg, Germany; 6 letters to Karl Mohr, 30 December 1838, 1 January 1848, 30 April 1848, 4 August 1862, 1 April 1868, and 17 April 1868, are at the University of Bonn. The Staatsbibliothek Preussischer Kulturbesitz, Berlin, contains one letter to an unknown person, 15 October 1843, one letter to Peter Riess, 30 September 1856, and one letter to an unknown person (perhaps Steeg), 9 September 1873.

Müller's dissertation was *Erklärung der isochromatischen Curven, welche einaxige parallel mit der Axe geschnittene Krystalle im homogenen polarisirten Lichte zeigen* (Darmstadt, 1833); his main synopses are *Kurze Darstellung des Galvanismus* (Darmstadt, 1836); *Elemente der ebenen Geometrie* (Darmstadt, 1838); *Elemente der ebenen Trigonometrie* (Darmstadt, 1838); *Elemente der sphärischen Trigonometrie* (Darmstadt, 1840); *Pouillet's Lehrbuch der Physik und Meteorologie, für deutsche Verhältnisse frei bearbeitet*, 2 vols. (Brunswick, 1842–1844); a second ed. (Brunswick, 1844–1845), with subtitle *Lehrbuch der Physik und Meteorologie*, includes about 1,000 woodcuts; the eleventh and final ed., 13 pts. in 5 vols., was published in Brunswick in 1926–1935; extended by the third volume, *Lehrbuch der Kosmischen Physik* (Brunswick, 1856), and by a supplement edited by A. Fick, *Die medizinische Physik* (Brunswick, 1856).

Subsequent writings are *Grundzüge der Krystallographie* (Brunswick, 1845); *Grundriss der Physik und Meteorologie. Für Lyceen, Gymnasien, Gewerbe- und Realschulen, sowie zum Selbstunterrichte* (Brunswick, 1846); *Mathematischer Supplementband zum Grundriss der Physik und Meteorologie* (Brunswick, 1860); *Physikalische Briefe für Gebildete aller Stände von Leonhard Euler und Dr. Johann Müller*, edited by Müller, contains a third part, *Die neuesten Ergebnisse und Bereicherungen der Physik in Briefform behandelnd* (Stuttgart, 1847–1848), and a fourth part (Stuttgart, 1854) edited by Müller; *Bericht über die neuesten Fortschritte der Physik, in ihrem Zusammenhange dargestellt*, 2 vols. (Brunswick, 1849–1851), with an English trans. in *Report of the Board of Regents of the Smithsonian Institution*, (1856), 311–423; (1857), 357–456; (1858), 333–431; (1859), 372–415; *Die constructive Zeichnungslehre, oder die Lehre vom Grund- und Aufriss*, 2 parts (Brunswick, 1865); *Die Schule der Physik. Eine Anleitung zum ersten Unterricht in der Naturlehre* (Brunswick, 1874); Müller was an editor of *Berichte über die Verhandlungen der Gesellschaft für Beförderung der Naturwissenschaften zu Freiburg im Breisgau* (from vol. 2 on entitled *Berichte über die Verhandlungen der naturforschenden Gesellschaft . . .*, ed. by Müller et al.), of which vol. 1 (1855) contains a number of his papers.

Papers cited in the article are "Berechnung der hyperbolischen dunkeln Büschel, welche die farbigen Ringe zweiaxiger Krystalle durchschneiden," in *Annalen der Physik* **120** (1838), 273–291; "Fraunhofer'sche Linien auf einem Papierschirm," *ibid.*, **145** (1846), 93–115; "Anwendung der stroboskopischen Scheibe zur Versinnlichung der Grundgesetze der Wellenlehre," *ibid.*, **143** (1846), 271–272; "Entwickelung der Gesetze des Elektromagnetismus," repr. from *Bericht über die neuesten Fortschritte*, **1**, 494–538, with a defense of his statement on saturation of magnetism (Brunswick, 1850); "Photographirte Spectren," in *Annalen der Physik* **173** (1856), 135–138; "Die Photographie des Spectrums," *ibid.*, **185** (1860), 151–157; *Programm, wodurch zur Feier des Geburtsfestes Seiner Königlichen Hoheit . . . Grossherzogs Friedrich . . . einladet der gegenwärtige Prorector Dr. J. Müller* (Freiburg im Breisgau, 1858), which contains a sketch of the *Geschichte . . . des physikalischen Kabinets* of Freiburg University; "Untersuchungen über die thermischen Wirkungen des Sonnenspectrums," repr. in *Annalen der Physik* **181** (1858), 337–359, 543–547; and "Rutherfurd's Photographie des Spectrums," *ibid.*, **202** (1865), 435–440. For a bibliography of Müller's papers see *Annalen der Physik, Namenregister und Sachregister 1875*; and Poggendorff (not always exact), II, 228–229; III, 944; VI, 1799. A bibliography of his books is given in W. Heinsius, **10** (1849), III, 124; British Museum, Catalogue of Printed Books; Catalogue Général de la Bibliothèque Nationale, Paris.

II. Secondary Literature. The only biographical sketch is Emil Warburg, *Gedächtnisrede auf Johann Heinrich Jacob Müller bei dessen academischer Todtenfeier am 16. Juli 1877* (Freiburg im Breisgau, 1877), repr. in large part in Friedrich von Weech, ed., *Badische Biographien*, III (Karlsruhe, 1881), 114–121; on Müller's activities at Freiburg see Hans Kangro, "Die Geschichte der Physik an der Universität Freiburg," MS copy (1954) deposited at Freiburg University, pp. 77–89.

Hans Kangro

MÜLLER, JOHANNES PETER (*b.* Coblenz, Germany, 14 July 1801; *d.* Berlin, Germany, 28 April 1858), *physiology, anatomy, zoology.*

Müller introduced a new era of biological research in Germany and pioneered the use of experimental methods in medicine. He overcame the inclination to natural-philosophical speculation widespread in German universities during his youth, and inculcated respect for careful observation and physiological experimentation. He required of empirical research that it be carried out "with seriousness of purpose and thoughtfulness, with incorruptible love of truth and perseverance." Anatomy and physiology, pathological anatomy and histology, embryology and zoology—in all these fields he made numerous fundamental discoveries. Almost all German scientists who achieved fame after the middle of the nineteenth century considered themselves his students or adopted his methods or views. Their remarks reveal his preeminent position in medical and biological research. Helmholtz, one of his most brilliant students, termed Müller a "man of the first rank" and stated that his acquaintance with him had "definitively altered his intellectual standards."

Life. Müller came from a family of winegrowers in the Moselle Valley. His father learned shoemaking and moved to Coblenz, where he became fairly well-to-do. Müller distinguished himself as a student through his unusual gifts, methodical and assiduous work habits, and craving for knowledge. From the works of Aristotle to Goethe's scientific writings, he devoured all the books he could obtain. His most striking trait was evident from childhood: a powerful ambition that drove him to be first, on the playing field as in the classroom. In the winter semester of 1819–1820 he entered the University of Bonn, which, founded in 1818, prided itself on being open to the latest intellectual currents. The Ministry of Education in Berlin had selected the professors chiefly according to their opinions of Schelling's *Naturphilosophie*. Almost all the members of the Faculty of Medicine were adherents of the latter. Some embraced the Romantic belief in supernatural cures; others endorsed Mesmer's animal magnetism.

During his second year at Bonn, Müller's father died; and henceforth he had to appeal to his family, friends, native city, and the state for support. He then attracted the attention of Philipp Jakob Rehfues, curator of the University of Bonn, who assisted his academic career. Rehfues applied, on Müller's behalf, to the authorities in Berlin—he could only give the grounds for such requests, not grant them himself—for stipends, travel allowances, and printing subsidies; later, money for vacations, remission of a loan, and a new microscope; and, finally, an increase in salary. Nevertheless Müller had financial problems even after he had become famous, since he unhesitatingly spent money for scientific purposes, for printing, and for books and instruments.

Müller received his medical degree in December 1822 and went to Berlin to continue his studies for another year and a half. In Berlin he came under the influence of Carl Rudolphi, Germany's most distinguished anatomist. Rudolphi sought to lead scientific research out of the "turbid mire of mysticism" and endow it with an exact method. Müller credited Rudolphi with having enabled him to escape the dangers of natural-philosophical speculation and to cease adorning his writings with the fashionable

vocabulary of electrical, magnetic, and chemical polarities, and positive and negative forces. In 1824 Müller passed the state medical examination in Berlin and then returned to Bonn, where in the same year he qualified as lecturer in physiology and comparative anatomy. A year later the Faculty of Medicine assigned him to lecture on general pathology as well. In 1826, not yet twenty-five years old, he became extraordinary professor; and in 1830 he was appointed full professor at an annual salary of 1,000 talers. In April 1827, after a long engagement, Müller married Nanny Zeiller of Coblenz; they had a daughter, Maria, and a son, Max.

The recognition that Müller enjoyed in the scientific world resulted in an offer from the University of Freiburg in 1832. He turned it down, even though it would have considerably improved his financial situation. He knew that Rudolphi was deathly sick and probably assumed that he would succeed him. Rudolphi died at the end of 1832. Since the Berlin Faculty of Medicine first offered the post to the Heidelberg anatomist Friedrich Tiedemann—who, however, did not accept it—Müller decided on a most unusual step for a German professor. In a letter to the Prussian minister of education, he proposed himself for the position and described the tasks that the holder of the Berlin chair of anatomy ought to fulfill. He must survey human, comparative, and pathological anatomy and must have done distinguished work in physiology, the foundation of all medicine. He must be familiar with microscopic observation, with experimental techniques in physiology, and with studying problems from an embryological perspective. Furthermore, he must be able to attract and encourage talented students. All this was necessary if Berlin were to take its rightful place in the international competition for scientific predominance. He convinced the minister of the brilliance of his own attainments, and during the Easter season of 1833 he assumed the Berlin professorship of anatomy and physiology.

At Berlin, Müller lost interest in experimental physiology. The appearance of the last section of his two-volume *Handbuch der Physiologie des Menschen* in 1840 more or less marks this shift. Henceforth he devoted himself almost exclusively to comparative anatomy and zoology, especially to research on marine animals—down to the protozoans. In the last years of his life he also contributed to paleontology through his publications on fossil fish and echinoderms. He gathered the material for these studies on numerous expeditions in the North Sea and in the Mediterranean.

By 1834 Müller had become a member of the Prussian Academy of Sciences. In 1841 and in 1853 he received offers from the University of Munich, both of which he declined. He was elected rector for the years 1838–1839 and 1847–1848. This position burdened him with many difficulties during the Revolution of 1848. As rector he was caught between those students who longed for a new German Reich and those who demanded reforms in the university and in the state. At the same time, he stood between the government and the student body. Considering himself a loyal servant of the state, he did not contemplate a revolutionary change in the form of government and could not grasp the thinking of the students, who were filled with new ideals. Furthermore, he was dominated by the fear that the university, and with it the irreplaceable treasures of his anatomical collection, might go up in flames.

At the end of 1857 Müller complained of insomnia, with which he had been afflicted for years. He is reported to have wandered through backstreets in Berlin, driven by inexplicable anxiety. On the morning of 28 April 1858 his wife found Müller dead in his bed. Since he had forbidden an autopsy, the cause of death remains unknown.

In his lifetime Müller experienced several periods of depression, which, like his periods of intense productivity, are traceable to a manic-depressive condition. The first depression, lasting for about five months, occurred in the summer of 1827. He was unable to work and ceased lecturing. The minister of education, when informed of Müller's condition, granted him a leave of absence and provided him with money for a trip. In 1840 Müller again became depressed, but he did not suffer as much as the previous time. This depression may have been precipitated by his realization that he was no longer the leader in physiological research. He lacked a thorough knowledge of chemistry and physics, which was becoming a necessity in an age that had set out to investigate causal relationships among vital phenomena.

The depression that Müller experienced at the end of his year as rector in 1848 was far more serious. Incapable of working, he obtained an indefinite leave of absence, gave up his residence in Berlin, and fled with his family to Coblenz; but, unable to find peace in his native city, he traveled on to Bonn. In a state of extreme despair, he finally sought refuge in Belgium, at Ostend. He did not lecture during the winter semester of 1848–1849, returning to Berlin only at the end of March 1849. The end of his life was also marked by depression. He was obsessed by the fear that his field of research was exhausted and that his productivity was ended. Many of his contemporaries suspected that his sudden death was a suicide, a hypothesis that accorded with the clinical record of his depression. Ernst Haeckel, his last close student, had no doubt that he had ended his life with an overdose of morphine.

Physiology. Müller's first publication and the doctoral dissertation based on it, *De phoronomia animalium* (1822), dealt with locomotion in animals. The investigation began with the movements of arthropods but continued in a comparative physiological manner, involving other classes of animals. He couched his excellent observations in terms of the doctrines of *Naturphilosophie*, which he had fully assimilated.

In 1820 the Bonn Faculty of Medicine posed its first prize question: Does the fetus breathe in the mother's womb? Müller entered the competition and sought to clarify the problem through experiments on live animals. Experiments on pregnant cats were not successful. He was first able to demonstrate that the fetus breathes in an experiment on a ewe. Observing the umbilical cord, he ascertained that bright red blood flows to the fetus through the umbilical vein and that dark blood flows back to the placenta through the umbilical artery. Fetal respiration is one of the few problems that Müller solved through vivisection on warm-blooded animals. Later he had harsh words for this crude, "knife-happy" type of experimentation.

In 1826 Müller published an extensive work that attracted the attention of the scientific world: *Zur vergleichenden Physiologie des Gesichtssinnes des Menschen und der Tiere nebst einem Versuch über die Bewegungen der Augen und über den menschlichen Blick*. The book, in nine parts, reported on Müller's various studies and interests. It opened with his inaugural lecture, "Von dem Bedürfnis der Physiologie nach einer philosophischen Naturbetrachtung," in which he outlined his views on science at the time of his habilitation. The succeeding sections offered a wealth of new findings on human and animal vision, brilliant investigations into the compound eyes of insects and crabs, and truly perceptive analyses of human sight. Moreover, the book recorded the young physiologist's most important achievement, the discovery that each sensory system responds to various stimuli only in a fixed, characteristic way—or, as Müller stated, with the energy specific to itself: the eye always with a sensation of light, the ear always with a sensation of sound, and so forth. This "law of specific nerve energies" led to the insight that man does not perceive the processes of the external world but only the alterations they produce in his sensory systems: "In intercourse with the external world we continually sense ourselves." This statement had important implications for epistemology.

Later, in the *Handbuch der Physiologie* (4th ed., I, 534), Müller maintained that all stimuli acting on the nerves have the same effect, whether they be mechanical, chemical, thermal, or galvanoelectric; each nerve can react only with its "specific energy." The reaction is determined by the properties of the stimulated organic substance, not by the quality of the stimulus. Müller conducted many experiments on isolated nerve-muscle preparations. Those on frog legs, employing galvanic electricity, were designed to determine the general conditions under which muscle contraction occurs. He utilized both changes in the galvanic stimulus and the effect of closing and opening the electric circuit. These experiments, along with those of Humboldt and Johann Wilhelm Ritter, constituted the first advances in electrophysiology. Matteucci's research in this field led Müller to encourage his student Emil du Bois-Reymond to enter it.

Müller's second book, *Über die phantastischen Gesichtserscheinungen* (1826), is still of interest. In it he showed that the sensory system of the eye not only reacts to external optical stimuli but also can be excited by interior stimuli arising from organic malfunction, lingering mental images, or the play of the imagination. He himself found it easy to make luminous images of people and things appear suddenly, move about, and disappear whenever he closed his eyes and concentrated on his darkened field of vision. With such self-observation and self-experimentation, supplemented with reports of earlier and contemporary authors—including Goethe, who in his scientific research likewise commenced from subjective experience—Müller demonstrated that optical perceptions can arise without an adequate external stimulus. When the stimulus is mistakenly assumed to have originated outside the body, the result—depending on the situation—is the reporting of religious or magical visions, or the seeing of ghosts.

Embryology. In the following years Müller's research made him the most celebrated member of the Bonn Faculty of Medicine. During a period of extraordinary productivity he investigated problems of physiology, embryology, and comparative anatomy and was usually busy with several investigations at once. In 1830, under the title *De glandularum secernentium structura penitiori earumque prima formatione*, he published his studies on the emergence and structure of the glands; in the course of this research he employed anatomical preparations, injections, and especially the microscope. The book considerably fostered the advance of embryology and histology. In it he demonstrated that glands are invaginations of the covering membranes that are closed at one end and that blood vessels do not open into the glandular

ducts but lie like capillaries in the walls of these ducts.

Simultaneously, Müller studied the origin of the omentum and its relation to the peritoneal sac (1830) in the human embryo and the embryonic development of the sexual organs in man and other vertebrates. In his *Bildungsgeschichte der Genitalien* he clarified the very complicated relationships between the initial form of the kidneys and their ducts, on the one hand, and the sexual organs, on the other. He discovered that the embryonic duct (described by Heinrich Rathke) now called "Müller's duct" forms the Fallopian tubes, uterus, and vagina: only rudiments of it are found in the male.

Neurophysiology and Neurology. Müller was responsible for a remarkable advance in neurophysiology: confirmation of the Bell-Magendie law by means of a simple experiment performed on the frog. In 1822 Magendie had reported experiments indicating that the anterior roots of the spinal nerves conduct motor impulses outward, while posterior roots transmit sensations from the periphery to the central nervous system. Bell thereupon interpreted, in this sense, experiments that he had published in 1811, claiming priority of discovery. Unfortunately, in remarks made shortly afterward, Magendie partially retracted his findings. Since the research of several others had further complicated the situation, in 1831 Müller took up this important subject. At first he experimented on rabbits, but the work was difficult and yielded ambiguous results. Consequently he continued his investigation on frogs, the use of which in the laboratory had almost entirely ceased. In the frog the spinal cord was far easier to remove, the relationships between the nerve roots were much more apparent, and the results unambiguous and always reproducible. Cutting through the posterior roots leading to a hind leg, he found that the limb became insensible but was not paralyzed. When he cut through the anterior roots, however, he observed that the limb was paralyzed but not rendered insensible. The simplicity, conclusiveness, and memorableness of the experiment— which has been repeated countless times in physiology courses—made a marked impression on Müller's contemporaries. To be sure, Müller, who was driven by ambition throughout his life, did not hesitate to make his experiment widely known. In that very year (1831) he reproduced it in Paris for Cuvier and Humboldt and, in Heidelberg, for Tiedemann. He also had his friend Anders Retzius perform the experiment in Stockholm.

Closely related to the demonstration of the Bell-Magendie law were Müller's efforts to determine the sensory and motor portions of the cranial nerves. He established experimentally that the first and second branches of the trigeminal nerve are sensory and that the third branch contains, in addition to sensory fibers, motor fibers for the jaw muscles. He also asserted, again on the basis of his own research, that the glossopharyngeal and vagus nerves are of the mixed type. In opposition to Magendie, however, Müller held that the hypoglossal nerve is of the motor type. This research led to the first comprehensive scientific conception of the nervous system as a unit. At the same time he postulated motor and sensory fibers for the autonomic nervous system; otherwise, it would be incapable of governing the intestinal functions under its control. Its motor function was demonstrated, he contended, by the fact that when a caustic is lightly applied to the celiac ganglion in the opened abdominal cavity of a rabbit the peristaltic movements of the intestine become far stronger.

In 1833 Müller studied the phenomenon of reflection, by which he meant the involuntary transition—occurring in the spinal cord or brain—of excitation from the centripetally conducting nerves to the centrifugally conducting ones. He took up this subject independently of Marshall Hall, who, shortly before Müller presented his results, had published *On the Reflex Function of the Medulla Oblongata and Medulla Spinalis. . . .* That the stimulation of an afferent or sensory nerve can provoke involuntary movements had already been asserted by Descartes. Müller provided the experimental proof that no communication exists between the afferent and efferent fibers, even though they are in the same nerve. Only the spinal cord or brain, he held, can mediate between the site of stimulation and the effector organ. According to Müller, the sensory organ, which receives the stimulus, can lie in the skin, the mucous membrane, or a muscle. As soon as the stimulus has reached the spinal cord, the impulse goes directly to the appropriate motor nerves—it does not pass up through the entire spinal cord: "The easiest path for the current or vibration of the nervous principle is from the posterior root of a nerve to its anterior root or to the anterior roots of several neighboring nerves." He thought that the simplest type of reflex movement was the quick jerk with which one withdraws an injured limb when the skin has been burned. He also described coughing, sneezing, hiccuping, vomiting, and ejaculation as reflex arcs located along the spinal cord and medulla oblongata, thereby contributing a fundamental new insight into the study of such phenomena.

To elucidate reflex action, physiology no longer had to fall back upon the old concept of "sympathies," as Boerhaave had been constrained to do. Even Procháska, professor of anatomy and physiology at Prague and later at Vienna, who had observed the reflex mechanism in 1784, still spoke of the *consensus nervorum* or the "polar interaction of the organs."

Hall preceded Müller by some months in the publication of his *Reflex Function*, but he diminished the practical value of his observations by assuming the existence of a special "excitomotor" nervous system—leading only to the spinal cord—that supposedly carried stimuli to the central nervous system. With his reflex theory Müller was able to explain many processes in the human organism and was also able to demonstrate his ideas on animals—an achievement that his era, so fond of experimentation, considered of no less importance.

Handbuch der Physiologie. While at Bonn, Müller had planned to write a handbook of human physiology. The first section appeared in 1833, shortly before he moved to Berlin; the last section of the second volume was published in 1840. The work became a milestone in the history of European medicine. In Germany it established a fruitful interaction between physiology and clinical practice. Beyond a critical examination of the established knowledge of the subject, it furnished a wealth of new findings derived from his own work. Among the many topics in which Müller was able to draw upon the results of his own research, or upon his verification of the work of others, were the composition and coagulation of the blood, the origin of fibrin, the nature of lymph, the occurrence of the retinal image, the origin of the voice in the human and animal larynx, the propagation of sound in the tympanic cavity, the process of secretion, the nerves of the erectile sexual organs, and the function of the sympathetic nerve and other elements of the nervous system.

Müller's *Handbuch der Physiologie* was a powerful stimulus to physiological research and one of the sources of the mechanistic conception of the life processes that prevailed in the second half of the nineteenth century. Yet the book's fundamental tendency was vitalistic. For Müller, the cause and supreme organizer of life phenomena—the vital force (*Lebenskraft*)—acted in accord with the law of rational adaptation to function and had nothing in common with the forces of physics and chemistry. He attributed to this vital force the peculiar nature of the physiology experiment, which sets it apart from every other type. In the chemical experiment, both the reagents and the substance under consideration enter into the final product; in the result of a physiology experiment, however, the applied stimulus is by no means a more important component than is living nature: the result is determined solely by the vital energies of the organism. In whatever manner a muscle is stimulated, it always reacts with a contraction. A stimulus, however, can only provoke something fundamentally different from itself. This vitalistic tendency explains why Müller took a very critical view of the validity of physiological experimentation. Consequently, he underestimated the importance of Magendie, who must be considered the true creator of the techniques of experimental physiology.

The Soul. Müller's *Handbuch der Physiologie* contains an extensive section on the soul (II, 505–588). This emphasis is understandable, in the light of his initial adherence to the ideas of *Naturphilosophie*. Starting from the philosophy and psychology of Aristotle, Giordano Bruno, Spinoza, Schelling, Hegel, and especially Herbart, he approached the questions of the identity of the psychic principle and the vital principle, the divisibility of the soul, and the seat of the soul. He arrived at the following alternative: The soul, which utilizes the organization of the brain in its activity, is either foreign to the physical body, not a force of organic nature, and only temporarily united with the body; or else it is inherent in all matter, a force of matter itself. Müller appears to have inclined more to the panpsychic conception when he wrote:

> The relationship of the psychic forces to matter differs from that of other physical forces to matter solely because the spiritual forces appear only in organic and especially animal bodies, [whereas] the general physical forces, which are also called imponderables [light, electricity], are much more commonly active and widespread in nature. Since, however, the organic bodies take root in inorganic nature and draw their nourishment from it, . . . it remains uncertain whether or not the rudiments [*Anlage*] of psychic activities, like the common physical forces, is present in all matter and attains expression in a definite manner through the existing structures [brain and nervous system] [*Handbuch*, II, 553].

This panpsychism accounts for the fact that Müller considered the brain to be the seat of the soul but still suspected that it "might perhaps be more widespread in the organism." To support this supposition, he pointed out that lower animals like polyps and worms are divisible; and that therefore among such lower creatures, and thus in organic matter in general, the life principle and psychic principle can be separated.

Pathological Anatomy. At Berlin, Müller displayed a new interest in pathological anatomy as a result of his access to the holdings of the Anatomisch-zootomische Museum, which had come under his direction when he assumed his professorship. The surgeons of Berlin had contributed many operation preparations, tumors, and deformities. Through studying these specimens Müller realized that the traditional description of the external form of tumors could lead to no further advance. It had to be supplanted by the chemical analysis and microscopic examination of the pathological elements and by the study of their development. He saw the desirability of establishing a system able to distinguish between benign and malignant growths.

At the end of 1837 Müller's student Theodor Schwann began working on his new cell theory, according to which the cells were the ultimate constituents of the animal body. On this basis Müller investigated pathological tumors, observing the similarity between the development of embryonic tissue and the formation of tumors from cells and showing that elements of normal tissue could be detected in the tumors. In 1838 he published the first, and only, part of his *Über den feineren Bau und die Formen der krankhaften Geschwülste*. The publication of this work fostered the use of the microscope in the study of pathological formations. Müller thus founded pathological histology as an independent field and provided physicians with diagnostic procedures that are now used in daily clinical work. Several decades later his brilliant student Rudolf Virchow—to whom, in 1856, Müller entrusted the lectures on pathological anatomy at Berlin—greatly expanded research on pathological growths.

Zoology. After 1840 Müller devoted himself primarily to comparative anatomy and zoology, and his research in these fields made him the most respected scientist of his day. Collecting, describing, and classifying were now virtually the only methods he employed. He accomplished little of interest in physiological experimentation, which was by then drawing increasingly on the methods of physics and chemistry.

As early as 1832 Müller published a systematic classification of the amphibians and reptiles. In 1834 he turned his attention to the Cyclostomata, members of the most primitive class of vertebrates. In broadly conceived comparative anatomical studies, completed in 1842, he examined their skeleton and musculature, sensory organs and nervous system, and vessels and intestines. His considerations ranged from the muscles of the Cyclostomata to human trunk musculature and to the homology of the cranial and spinal nerves.

Müller placed the Cyclostomata among the fishes. He was thus led to study the sharks (which he called Plagiostomi) and the rays; he published the results of his research jointly with his student Jakob Henle as *Systematische Beschreibung der Plagiostomen* (1841). A further product of this investigation was "Über den glatten Hai des Aristoteles" (1842). In *Historia animalium*, Aristotle had reported that the embryos of the "so-called smooth shark" are attached to the uterus of the mother by a placenta, as is the case among mammals. Rondelet had described such a shark in 1555 and Steno had observed one in 1673 off the coast of Tuscany, but it had not been referred to in more recent times. Müller was the first who was able to corroborate the earlier testimony.

In conjunction with the study of the shark, Müller constructed a natural system of the fishes based on work as painstaking as it was perceptive. He also devoted attention to the systematics of the songbirds, employing the vocal apparatus as his chief criterion of classification.

In the introduction to his account of the Cyclostomata Müller emphasized that the animals most apt to provoke the curiosity of the scientist are those "standing on the border of a class." This was doubly true of the Cyclostomata because they stood at the border of the fishes and at that "of the vertebrates in general." This curiosity led him to study the lancelet, especially the *Branchiostoma lanceolatum*. The latter could serve, Müller stated, as the simplest model of the basic plan of the vertebrate subphylum. A flexible member which serves as an axial skeleton, the *Chorda dorsalis*, extends through the animal's body. This same skeleton is also found in the embryos of birds and mammals. The lancelet had been described in 1774, but Müller was the first to recognize its great systematic importance. In 1841, with Retzius at Stockholm—who had been his close friend since the Berlin scientific congress of 1828—he studied this primitive animal in Bohuslan, Sweden, and on the Felsen Islands near Göteborg and in the same year sent a description of it to the Prussian Academy of Sciences. This research, to which later work could add but little, became of great importance for knowledge of general vertebrate structure.

In his zoological research, Müller at first was satisfied to rely on the material in the collections available to him or that had been sent to him, but beginning in 1845 he traveled to the seashore to examine its animal life, especially the microscopic forms, *in situ*. During vacations he made many trips to the coasts of the North Sea and the Baltic, Adriatic, and Ligurian seas. He was rewarded with many unexpected results. In carefully executed studies, some of them lasting for years, Müller explored the echinoderms and sea slugs. Through observation and comparison he was able to elucidate their complex metamorphoses. He recognized the connection, bordering on the fantastic, between the double-ray larvae of the echinoderms, which are microscopic,

transparent plankton animals, and the squat, five-ray, sexually mature animals of the same phylum. Thus, from the bilaterally symmetric larva that he named *Pluteus paradoxus* he was able to derive the radially symmetric sea urchin. In short, Müller not only opened this field, with its wealth of forms, to research but also penetrated its secrets conceptually. It is characteristic of the way he worked that he never attempted to clarify the development of the echinoderms through experimentation, which might have led him to his goal more quickly.

These studies, which revealed the extraordinary creative powers of nature in a unique manner, received greater recognition than any of Müller's previous achievements. He was awarded the Copley Medal of the Royal Society and the Prix Cuvier of the Académie des Sciences. His last research was devoted to the single-celled marine animals the Radiolaria and the Foraminifera. Haeckel continued this work.

Research Methods. Müller began his scientific career with vivisection in order to demonstrate that the fetus breathes in the uterus. He also utilized vivisection later when he was convinced that it could elucidate a question. Yet in setting forth his views on the study of living nature in his Bonn inaugural lecture, he stressed the technical difficulties of physiological experimentation and dissociated himself from the fondness for experiment characteristic of Magendie's school. Müller's scientific ideal was not Magendie or brilliant experimentalists like Haller or Spallanzani, but Cuvier, who devoted himself to the description and comparison of the forms and species of living creatures.

In the same inaugural lecture Müller spoke against the speculative interpretation of biological processes. Accordingly, he criticized the ideas of *Naturphilosophie*, the temptations of which he had escaped through Rudolphi's influence. He advocated a physiology that united "exact empirical training" in all the methods suited to the investigation of living nature with a philosophical penetration of the data. Such a union, he claimed, would uncover the *Urphänomene*, the ideas underlying everything in the universe.

As he grew older, Müller expressed his methodology in much more modest terms in the *Handbuch der Physiologie*. He remained convinced that a strictly empirical approach to physiology could not solve the ultimate questions of life, but he conceded that philosophy could not yield results usable in an empirical science. Only a union of the two paths to knowledge, which he termed "critically evaluated experience" ("denkende Erfahrung"), could lead to scientific truth, although he added that there would always be something unsolved. He no longer expressed the hope of penetrating to the *Urphänomene*. At Berlin, holding that the researcher's principal instrument was "conceptual empiricism," not experiment, Müller increasingly shifted his attention from functional to morphological issues. While his students sought to elucidate vital phenomena with the methods of physics and chemistry, he was satisfied to describe them. In opposition to this new generation of scientists, he remained a vitalist throughout his life, never doubting the existence of the "vital force."

BIBLIOGRAPHY

I. ORIGINAL WORKS. A complete list of Müller's 267 writings is in *Abhandlungen der K. Preussischen Akademie der Wissenschaften* for 1859 (1860), 157–175; it was reprinted in Koller's biography (see below), pp. 241–260.

His works include *De phoronomia animalium* (Bonn, 1822); *De respiratione foetus* (Leipzig, 1823); *Von dem Bedürfnis der Physiologie nach einer philosophischen Naturbetrachtung* (Bonn, 1825), his inaugural lecture, repr. in Müller's *Zur vergleichenden Physiologie* and in Adolf Meyer-Abich, *Biologie der Goethezeit* (Stuttgart, 1949), 256–281; *Über die phantastischen Gesichtserscheinungen* (Coblenz, 1826), repr. in vol. XXXII of Klassiker der Medizin (Leipzig, 1927) and in Ulrich Ebbecke, *Johannes Müller* (see below), 77–187; *Zur vergleichenden Physiologie des Gesichtssinnes des Menschen und der Tiere nebst einem Versuch über die Bewegungen der Augen und über den menschlichen Blick* (Leipzig, 1826); *Bildungsgeschichte der Genitalien aus anatomischen Untersuchungen an Embryonen des Menschen und der Tiere* (Düsseldorf, 1830); *De glandularum secernentium structura penitiori earumque prima formatione in homine atque animalibus* (Leipzig, 1830); and "Über den Ursprung der Netze und ihr Verhältnis zum Peritonealsacke beim Menschen, aus anatomischen Untersuchungen an Embryonen," in *Archiv für Anatomie und Physiologie* (1840), 395–411.

See also "Bestätigung des Bell'schen Lehrsatzes, dass die doppelten Wurzeln der Rückenmarksnerven verschiedene Funktionen haben, durch neue und entscheidende Experimente," in *Notizen aus dem Gebiete der Natur- und Heilkunde*, **30** (1831), 113–117; *Handbuch der Physiologie des Menschen für Vorlesungen*, 2 vols. (I, Coblenz, 1833–1834; 4th ed., 1841–1844; II, Coblenz, 1837–1840), also trans. into English by Baly, 2 vols. (London, 1840–1843) and into French by Jourdan, 2 vols. (Paris, 1845; 2nd ed., 1851); "Vergleichenden Anatomie der Myxinoiden (Cyclostomen)," in *Abhandlungen der K. Preussischen Akademie der Wissenschaften*, Phys. Kl., for 1834, 1837–1839, 1843 (1836–1845); *Über den feineren Bau und die Formen der krankhaften Geschwülste* (Berlin, 1838); *Systematische Beschreibung der Plagiostomen* (Berlin, 1841), written with Jakob Henle; "Über den glatten Hai des Aristoteles und über die Verschiedenheiten unter den

Haifischen und Rochen in der Entwicklung des Eies," in *Abhandlungen der K. Preussischen Akademie der Wissenschaften*, Phys. Kl., for 1840 (1842), 187–257; and "Über den Bau und die Lebenerscheinungen des *Branchiostoma lubricum* Costa, *Amphioxus lanceolatus* Yarrell," *ibid.*, for 1842 (1844), 79–116.

Müller's many publications on echinoderms began with his description of a sea lily—"Über den Bau des *Pentacrinus caput* Medusae," in *Abhandlungen der K. Preussischen Akademie der Wissenschaften*, Phys. Kl., for 1841 (1843), 177–248—and was followed by eight papers on larvae, metamorphoses, and the structure of echinoderms, *ibid.*, for 1846, 1848, 1850–1854 (1848–1855).

II. SECONDARY LITERATURE. Although the writings of du Bois-Reymond and Virchow on their teacher are limited by the thinking of their generation and do not always do justice to Müller, they are nevertheless indispensable: Emil du Bois-Reymond, "Gedächtnisrede auf Johannes Müller," in *Abhandlungen der K. Preussischen Akademie der Wissenschaften*, for 1859 (1860), 25–191, repr. with additional material in du Bois-Reymond's *Reden*, 2nd ed., I (Leipzig, 1912), 135–317; and Rudolf Virchow, *Johannes Müller. Gedächtnisrede* (Berlin, 1858).

Among the more recent literature, see Wulf Emmo Ankel, "Branchiostoma ist ein Wirbeltier—eine 130 Jahre alte Erkenntnis," in *Natur und Museum*, no. 101 (1971), 321–339; Ulrich Ebbecke, *Johannes Müller, der grosse rheinische Physiologe* (Hannover, 1951), which includes a reprint of Müller's *Über die phantastischen Gesichtserscheinungen*; Wilhelm Haberling, *Johannes Müller. Das Leben des rheinischen Naturforschers* (Leipzig, 1924), a detailed biography with letters from Müller to his family, friends, and colleagues; Gottfried Koller, *Das Leben des Biologen Johannes Müller* (Stuttgart, 1958), with a bibliography of secondary literature, pp. 261–263; Walther Riese and George E. Arrington, Jr., "The History of Johannes Müller's Doctrine of the Specific Energies of the Senses," in *Bulletin of the History of Medicine*, **37** (1963), 179–183; Robert Rössle, "Die pathologische Anatomie des Johannes Müller. Nach einem aufgefundenen Kollegheft aus dem Jahre 1834," in *Sudhoffs Archiv für Geschichte der Medizin und Naturwissenschaften*, **22** (1919), 24–47; Johannes Steudel, *Le physiologiste Johannes Müller*, Conférences du Palais de la Découverte, D85 (Paris, 1963); and Manfred Stürzbecher, "Auf dem Briefwechsel des Physiologen Johannes Müller mit dem preussischen Kulturministerium," in *Janus*, **49** (1960), 273–284.

JOHANNES STEUDEL

MÜLLER, OTTO FREDERIK (*b.* Copenhagen, Denmark, 2 March 1730; *d.* Copenhagen, 26 December 1784), *botany, zoology.*

The son of a court trumpeter, Müller was educated from the age of ten by his mother's family at the grammar school of Ribe, Jutland. Five years later he was sent to the University of Copenhagen, where he earned his living by teaching music while studying theology and law. In 1750 he enrolled in Borch's College and wrote two short theological theses (1751, 1753) while there. Three years later he was appointed tutor to Sigismund Schulin, the son of an influential noble family; for nearly twenty years he lived with this family on their estate, Frederiksdal (northern Zealand), and traveled with them to Germany, Switzerland, Italy, France, and the Netherlands. In this way Müller met many outstanding scientists and became a member of the Academia Caesarea Leopoldina (1764), the Royal Swedish Academy of Sciences (1769), and the Norwegian Society of Sciences (1770). In 1771–1773 he was secretary to the Danish Ministry for Norway; but his marriage on 26 May 1773 to a wealthy widow, Anna Carlsen, née Paludan (1735–1787), daughter of a Norwegian bishop, made Müller financially independent, so that he was able to devote the last thirteen or fourteen years of his life to science. In 1774 he became a corresponding member of the Paris Academy of Sciences and the Berlin Society of Friends of Natural Science, and two years later he became a member of the Royal Danish Academy of Sciences and of the Academy of Sciences of the Institute of Bologna.

During Müller's years as a tutor, his pupil's mother interested him in the flora and fauna of the estate, procuring microscopes and other equipment for his growing absorption in biological studies. Because of his theological training Müller examined nature in terms of natural theology, considering it his task to investigate and point out the wisdom of the Creator everywhere in nature. The mainsprings of his scientific research were his love of beauty and his discovery that previous systematists had largely neglected to take microorganisms into account. Even the smallest puddle was full of organisms, and his unique powers of observation enabled Müller to demonstrate their adaptations. Thus he stands as the foremost representative of the Linnaean period in Danish natural history. He established the classification of several groups of animals—including Hydrachnellae, Entomostraca and Infusoria—completely disregarded by Linnaeus. He pursued his zoological studies and became one of the first field naturalists, using surprisingly modern methods long before the development of experimental biology.

At the age of thirty-four Müller published his first zoological work, a systematic study written entirely in the Linnaean spirit, *Fauna insectorum Fridrichsdalina* (1764). The description of 858 "insects" found at the estate of Frederiksdal—including spiders, wood lice

and centipedes—contains several errors. Two years later there appeared *Flora Fridrichsdalina* (1766), a description of 1,100 species intended to represent most of the Danish flora. But Müller's main work in systematics consisted of studying animal groups that were very little known before he identified them: the Hydrachnida, the Tardigrada, the Entomostraca, and the Infusoria. Among his works on these groups he lived to see only one published, *Hydrachnae, quas in aquis Daniae palustribus detexit* (1781). The other fundamental writings—"Von dem Bärthierchen" (1785), *Entomostraca, seu Insecta testacea* (1785), and *Animalcula infusoria* (1786)—were printed after his death. Müller applied the term "Entomostraca" to the small, often minute crustaceans: many genera belonging to this group—including *Daphnia, Cyclops caligus,* and *Argulus*—were first defined by him, and he formulated a systematic classification and created Danish names for the various groups. His classic paper was for many years accepted as the best study of the Infusoria, which he placed near the order Acarina, where it is still placed by most authors. Illustrated with fifty plates, his work on Infusoria describes algae, bacteria, and many other microorganisms as well as some protozoans.

It was not only on microscopic animals that Müller did significant and fundamental studies. His main work concerning mollusks and worms, *Vermium terrestrium et fluviatilium* (1773–1774), in which he first described a large number of new species of freshwater and terrestrial mollusks, also presented his primary system on the Infusoria. Neither did Müller limit his interests to invertebrate taxonomy; in *Von Würmern des süssen und salzigen Wassers* (1771) he clearly demonstrated the propagation of naiads. He wrote on annelids (1771), on the moth (1779), and on helminths (1779). In this field of studies his contemporaries rightly called him the Danish Linnaeus.

Besides his systematic works Müller wrote several entomological papers in Réaumur's style, for instance, on the propagation of the daphnids. For the collection of sea animals he invented a special dredge; and with his brother, Christian Frederik Müller (1744–1814), who was responsible for some of the illustrations in Müller's books, he published an anonymous account of a journey from Norway (1778).

In 1776 Müller published *Zoologiae Danicae prodromus,* an excellent survey of the fauna of Norway and Denmark. It was the first manual on this topic and was for many years the most comprehensive. It was planned as the beginning of a large illustrated fauna, but only one volume appeared before his death; the following volumes—the last published in 1806—prepared by Abildgaard, Rathke, and others, never reached the standard of the *Flora Danica* begun by Georg Christian Oeder.

Müller also wrote a number of botanical studies dealing especially with fungi and other groups, but they were little known during his lifetime. They presented findings and views pointing far beyond what was known then—for instance, in his "Ueber die Feld-Lilie" (1766).

Besides his systematic and exact studies Müller was also occupied with the philosophy of biology, advancing his "monadic" theory, a view that all living things are composed of minute elements—monads—that are set free by putrefaction and reunited in propagation. This belief indicates that he was an adherent of the preformation hypothesis.

BIBLIOGRAPHY

I. ORIGINAL WORKS. A full list of Müller's writings is in H. Ehrencron-Müller, *Forfatterlexikon*, VI (Copenhagen, 1929), 17–22.

Among his works are *Fauna insectorum Fridrichsdalina* . . . (Copenhagen–Leipzig, 1764); *Flora Fridrichsdalina* . . . (Strasbourg, 1766; 1767); *Von Würmern des süssen und salzigen Wassers* (Copenhagen, 1771); *Vermium terrestrium et fluviatilium* . . ., 2 vols. (Copenhagen–Leipzig, 1773–1774); *Zoologiae Danicae prodromus* (Copenhagen, 1776); *Zoologiae Danicae seu animalium Daniae et Norvegiae rariorum ac minus notorum icones*, 2 vols. (Copenhagen, 1777–1780); *Rejse igiennem Övre-Tillemarken til Christianssand og tilbage 1775* (Copenhagen, 1778), published anonymously; *Hydrachnae, quas in aquis . . . palustribus detexit* . . . (Leipzig, 1781); *Zoologia Danica eller Danmarks og Norges sieldne og ubekiente Dyrs Historie* (Copenhagen, 1781), also trans. into German (Leipzig–Dessau, 1782); *Entomostraca seu insecta testacea, quae in aquis Daniae et Norvegiae reperit* . . . (Copenhagen, 1785); *Animalcula infusoria, fluviatilia et marina* . . . (Copenhagen, 1786); and *Zoologia Danica seu animalium Daniae et Norvegiae rariorum ac minus notorum descriptiones et historia*, 4 vols. (Copenhagen, 1788–1806).

His diary, which remains unpublished, is in the Royal Library, Copenhagen, Add. 4° no. 710.

II. SECONDARY LITERATURE. See Jean Anker, *Otto Friderich Müller* (Copenhagen, 1943), of which only the 1st vol., covering 1730–1767, has been published (because of the author's death); V. Meisen, *Prominent Danish Scientists* (Copenhagen, 1932), 60–64; and Jens Worm, *Lexicon over laerde Maend*, II (Copenhagen, 1773), 88–91, and III (Copenhagen, 1784), 548–549.

E. SNORRASON

MÜLLER, PAUL (*b.* Olten, Solothurn, Switzerland, 12 January 1899; *d.* Basel, Switzerland, 13 October 1965), *chemistry.*

Müller's earliest years were spent in Lenzburg, in the canton of Aargau; but when he was nearly five, his father, an employee of the Swiss Federal Railroads, was transferred to Basel, where the boy received his elementary education. His interest in chemistry was stimulated in 1916, when he was employed as a laboratory assistant in the chemical factory of Dreyfus and Company. In 1917 he became a chemical assistant in the research laboratories of Lonza A. G., where he remained for a year. The experience thus gained convinced Müller that his future lay in chemistry, and he therefore entered the University of Basel in 1919 to work with F. Fichter and Hans Rupe. He received his doctorate in April 1925, with a thesis on the chemical and electrochemical oxidation of *m*-xylidine and its derivatives. His minor subjects were physical chemistry and botany. In May 1925 he became a research chemist in the dye factory of J. R. Geigy A.G., where he remained for the rest of his active career, rising to the post of deputy head of pest control research.

Because of his interest in botany Müller began a study of plant pigments and natural tanning agents. His work on the preservation and disinfection of animal skins led him to utilize biological studies, and he soon turned his attention to pesticides. He discovered a mercury-free seed dressing, which was of value for Swiss farmers.

In 1935 Müller began to study contact insecticides and drew up a set of criteria to guide him in his search for an ideal agent. These included great toxicity for insects, rapid toxic action, no or slight toxicity for plants and warm-blooded animals, no odor, long-lasting action, and low price. He did not consider stability and decomposability.

Müller first studied the action of a great many different types of compounds on a variety of insects. He believed that besides its practical value, this work was of philosophical interest in bringing together biologists and chemists. He soon concluded that it would be possible to discover safe insecticides, since the absorption of toxic substances by insects was entirely different physiologically from their absorption by warm-blooded animals. His attention was drawn to the group of chlorinated derivatives of phenyl ethane. In September 1939 he observed that in its action on flies 4,4′-dichloro-diphenyl-trichloro-ethane satisfied nearly all of his criteria, and he soon found that it was equally effective against other types of insects. The compound had first been prepared in 1873 by Othmar Zeidler, who had published a brief description of the substance in *Berichte der Deutschen chemischen Gesellschaft* (**7** [1874], 1181), but with no indication that it had any physiological action.

Müller called the substance DDT, and the basic Swiss patent was secured in March 1940. He described his work on this and other insecticides in papers published in the *Helvetica chimica acta* in 1944 and 1946. In the second of these papers he reported the investigation of a large number of compounds related to DDT, none of which proved as effective as the original. Müller was unable to find any general relations between structure and insecticidal action and noted, "A chemical compound is an individual whose characteristic action can be understood only from its totality, and the molecule means more in this case than the sum of its atoms."

The first commercial preparation of DDT appeared on the market at the beginning of 1942, and its value was immediately recognized. Müller noted rather proudly that in spite of many attempts by the combatants in World War II to obtain a better insecticide, none had improved on his DDT. Since much of the war was fought in tropical areas, the value of DDT in destroying disease-bearing insects was obvious. The early work seemed to indicate that the substance was completely safe for humans. After the war it was widely used in the Mediterranean area to eradicate malaria-bearing mosquitoes, and the incidence of this disease was greatly reduced. Honors quickly came to Müller, culminating in his receipt of the Nobel Prize in physiology or medicine in 1948. In 1963 he was awarded an honorary doctorate by the University of Thessaloniki because of the value of DDT in the Mediterranean region. In 1961 Müller retired from the Geigy Company and established a private laboratory at his home in Oberswil, where he continued his investigations until his death.

In establishing his criteria for the properties of an ideal insecticide, Müller had not considered the possibility of the accumulation of the agent in various biological species until it reached dangerous proportions. He was not unaware of this possibility, however, for in concluding his 1946 article he wrote, "Pyrethrum and rotenone, like all natural insecticides, are completely destroyed in a short time by light and oxidation, as opposed to the synthetic contact insecticides which have been shown to be very stable. Nature must and will behave in this way, for what a catastrophe would result if the natural insecticide poisons were stable. Nature plans for life and not for death!" Such considerations, disregarded at first, were gradually recognized in the years immediately following his death. As the use of DDT increased, it became apparent that the stability of the compound caused it to accumulate to a harmful degree in some animal species. Far-reaching ecological changes could be foreseen, and numerous controversies arose as doubts

were expressed concerning the long-range safety of DDT. It was apparent that the contact insecticides of the DDT type were not the complete answer in the search for an ideal agent of this sort.

BIBLIOGRAPHY

The basic papers describing Müller's investigations are "Über Konstitution und toxische Wirkung von natürlichen und neuen synthetischen insektentötenden Stoffen," in *Helvetica chimica acta*, **27** (1944), 892–928, written with P. Läuger and H. Martin; and "Über Zusammenhänge zwischen Konstitution und insektizider Wirkung. I," *ibid.*, **29** (1946), 1560–1580. An account of his discovery of DDT is given by Müller in "Dichlorodiphenyltrichloroäthan und neuere Inzekticide," in *Les Prix Nobel en 1948* (Stockholm, 1949), 122–132. His autobiography is on pp. 75–76 of this work.

An obituary in *Nature*, **208** (1965), 1043–1044, gives information on the latter part of Müller's life.

HENRY LEICESTER

MÜLLER-BRESLAU, HEINRICH (FRANZ BERNHARD) (*b*. Breslau, Germany [now Wrocław, Poland], 13 May 1851; *d*. Berlin, Germany, 23 April 1925), *theory of structures.*

The son of a merchant, Müller (after the 1870's he styled himself Müller-Breslau) grew up in Breslau. In 1869, upon graduation from the Realgymnasium, he joined the Prussian army engineers and saw action in the Franco-Prussian War. After the war he gave up plans for a military career in order to become a civil engineer. Müller then moved to Berlin and embarked upon an informal program of study consisting of courses in engineering at the Gewerbeakademie and of lectures in mathematics at the university. Without graduating he began practicing as an independent consulting engineer at Berlin in 1875, specializing in the design of iron structures, chiefly bridges. At the same time he prepared his first major book, *Theorie und Berechnung der eisernen Bogenbrücken* (1880) and a series of articles on problems of statically indeterminate structures. In 1883 Müller-Breslau was appointed professor of bridge design at the Polytechnic Institute of Hannover. The steady flow of his publications and some notable designs rapidly established his reputation, and in 1888 he was appointed to the chair of structural engineering at the Berlin-Charlottenburg Institute of Technology. Here he worked as a teacher, researcher, and consultant for the rest of his life, serving in 1895–1896 and 1910–1911 as rector.

Müller-Breslau has been termed the founder of modern structural engineering in Germany. Among the more significant designs credited to him are the Volga bridge at Kazan, Russia, and the new cathedral in Berlin. He also participated in the construction of Count Ferdinand von Zeppelin's airships, designed large aircraft hangars, and contributed to the introduction of cantilever wings on airplanes.

Müller-Breslau exerted even greater influence through his publications, notably his books. His numerous monographs deal with specific problems in the theory of structures, such as cantilevers, arches, lattice structures, earth pressure on retaining walls, and buckling of straight bars. He did not present methods or theories of fundamental novelty; his strength was the refinement and elaboration of earlier methods, presented in systematic and unified form. In *Die neueren Methoden der Festigkeitslehre und der Statik der Baukonstruktionen* (1886) he consistently based the solution of statically indeterminate structural systems upon the strain energy methods of L. F. Menabrea and C. A. Castigliano. The last three decades of Müller-Breslau's life were devoted chiefly to his magnum opus, *Die graphische Statik der Baukonstruktionen*. A three-volume handbook (1887, 1892, 1908), it came to be internationally regarded as the definitive presentation of the graphical methods of structural design.

BIBLIOGRAPHY

I. ORIGINAL WORKS. A bibliography of Müller-Breslau's most important publications is given by H. Reissner in *Zeitschrift für angewandte Mathematik und Mechanik*, **5** (1925), 277–278. Besides some 35 research papers it lists the following books: *Theorie und Berechnung der eisernen Bogenbrücken* (Berlin, 1880); *Die neueren Methoden der Festigkeitslehre und der Statik der Baukonstruktionen* (Leipzig, 1886); *Die graphische Statik der Baukonstruktionen*, 3 vols. (Leipzig, 1887–1908); and *Erddruck auf Stützmauern* (Stuttgart, 1906).

II. SECONDARY LITERATURE. Biographical information on Müller-Breslau can be found in Karl Bernhard, "Müller-Breslau," in *Bautechnik*, **3** (1925), 261–262; A. Hertwig, "Rede, gehalten bei der Gedenkfeier für Müller-Breslau am 25. Juni 1925," in *Stahlbau*, **20** (1951), 53–54; H. Müller-Breslau, Jr., "Heinrich Müller-Breslau," in H. Boost *et al.*, *Festschrift Heinrich Müller-Breslau* (Leipzig, 1912), v–viii; Poggendorff, VI, 1802; H. Reissner, "H. Müller-Breslau," in *Zeitschrift für angewandte Mathematik und Mechanik*, **1** (1921), 159–160, and **5** (1925), 277–278; and Stephen P. Timoshenko, *History of Strength of Materials* (New York, 1953), *passim*.

OTTO MAYR

MUNCKE, GEORG WILHELM (*b*. Hillingsfeld, near Hameln, Germany, 28 November 1772; *d*. Grosskmehlen, Germany, 17 October 1847), *physics*.

Muncke served as an overseer at the Georgianum in Hannover. In 1810 he became professor of physics at the University of Marburg, where he stayed until 1817, when he went to the University of Heidelberg to take up the professorship that he held until his death.

Muncke's chief importance lies in his critical attitude toward much of the scientific speculation of his time, and in particular in his opposition to Kant's dynamical theory of matter. He attributed the wide influence of *Naturphilosophie* in German science to Kant's "mystical play with unknown forces," and to the elaboration of Kant's ideas by Fichte, Schelling, Ritter, Steffens, Oken, and Hegel. Muncke was himself an advocate of the atomic theory of matter and his views are explicitly stated in his books *System der atomistischen Physik* (1809) and *Handbuch der Naturlehre* (1829–1830).

Muncke also collaborated with Brandes, Gmelin, Horner, Pfaff, and Littrow in preparing the edition of Johann Gehler's *Physikalisches Wörterbuch* published in Leipzig in eleven volumes (1825–1845). This work constitutes one of the best records of the state of the natural sciences in the first quarter of the nineteenth century; Muncke's contribution consists of descriptions of individual physical subjects and an excellent, objective general discussion of current physical theories and knowledge.

Muncke's own experimental results were trivial. He published a series of observations on the expansion and boiling of water, and he determined the densities of water, alcohol, and ether (1816). He adhered to the theory that the earth has four magnetic poles, and he explained Brownian movement as the passage of light and heat rays through the liquid medium.

BIBLIOGRAPHY

Muncke's works are *System der atomistischen Physik* (Hannover, 1809); *Anfangsgründe der Naturlehre*, 2 vols. (Heidelberg, 1819–1820); *Handbuch der Naturlehre*, 2 vols. (Heidelberg, 1829–1830); *Physicalische Abhandlungen. Ein Versuch zur Erweiterung der Naturkunde* (Giessen, 1816); "Hypothesen zur Erklärung einiger räthselhafter Naturphänomene," in *Journal für Chemie und Physik*, **25** (1819), 17–28; "Versuche über den Elektromagnetismus zur Begründung einer genügenden Erklärung desselben," in *Annalen der Physik*, **70** (1822), 141–174, **71** (1822), 20–38, 411–435; and "Ueber Robert Brown's microscopische Beobachtungen," *ibid.*, **17** (1829), 159–176.

Bibliographies of Muncke are given in the Royal Society *Catalogue of Scientific Papers*, IV, 543–544; and Poggendorff, II, 238–239.

H. A. M. SNELDERS

MUNIER-CHALMAS, ERNEST CHARLES PHILIPPE AUGUSTE (*b.* Tournus, France, 7 April 1843; *d.* Saint-Simon, near Aix-les-Bains, France, 8 August 1903), *paleontology, stratigraphy.*

Although he received only an inferior early education, Munier-Chalmas was able, through a combination of intelligence and willpower, to teach himself geology so successfully that he eventually became a professor of that subject at the Sorbonne. After holding a number of menial jobs, Munier-Chalmas in 1863 was made an assistant in the Sorbonne's geology laboratory. He worked under the supervision of Edmond Hébert and, as a result of his diligence in his task of self-education, became Hébert's closest collaborator, accompanying him on geological trips throughout France, northern Italy, Austria, and Hungary. Having attained the master's degree, Munier-Chalmas was able to begin teaching at the École Normale in 1882, but his doctoral thesis, on the Jurassic, Cretaceous, and Cenozoic deposits of the Vicentin, presented in 1891, was never published. He succeeded Hébert in the chair of geology at the Sorbonne in 1891 and was elected to the Académie des Sciences in 1903.

Munier-Chalmas's paleontological skills complemented Hébert's stratigraphical ones, and their collaboration was mutually profitable. His own paleontological work included his investigations of the Brachiopoda, Cephalopoda, Gastropoda, Foraminifera, and Calcareous Algae. He was also concerned with classification and nomenclature, and established a number of new genera, comprising *Toucasia, Matheronia*, and *Heterodiceras.* One of his findings—that the shapes of the shells of some groups of ammonites clearly indicate sexual dimorphism, while those of other groups do not—led to a number of taxonomic changes and reclassifications.

In stratigraphy, Munier-Chalmas's chief contribution concerned the Cenozoic of the Paris Basin. "Note sur la nomenclature des terrains sédimentaires," published in 1893 and written with Lapparent (who had earlier supervised the surveying of that region) is his chief work. He also collaborated with Paul Henri Fischer, to whom he supplied much data for his *Manuel de conchyliologie* (1880–1887).

Munier-Chalmas was not particularly successful as a teacher; he was unable to communicate his ideas to large audiences, and he was continually hampered by

an awareness of his own educational deficiencies. His influence, however, is clear in the works of his students—the more so that he was reluctant to publish his findings himself. Nevertheless, he had published more than sixty papers before 1900, including nine in the single year of 1892.

BIBLIOGRAPHY

I. ORIGINAL WORKS. Munier-Chalmas's publications include "Prodrome d'une classification des Rudistes," in *Journal de conchyliologie*, 3rd ser., **13** (1873), 71–75; "Matériaux pour servir à la description du terrain Crétacé supérieur en France," "Description du bassin d'Uchaux," and "Appendice paléontologique (Fossiles du bassin d'Uchaux)," in *Annales de la Société géologique*, **6**, pt. 2 (1875), 1–132, all written with E. Hébert; "Mollusques nouveaux des terrains paléozoïques des environs de Rennes," in *Journal de conchyliologie*, 3rd ser., **16** (1876), 102–109; "Diagnosis generis novi Molluscorum Cephalopodorum fossilium," *ibid.*, 3rd ser., **20** (1880), 183–184; *Étude du Tithonique, du Crétacé et du Tertiaire du Vicentin* (Paris, 1891); "Note sur la nomenclature des terrains sédimentaires," in *Bulletin de la Société géologique de France*, 3rd ser., **21** (1893), 438–488, written with A. de Lapparent; and "Note préliminaire sur les assises montiennes du bassin de Paris," *ibid.*, 3rd ser., **25** (1897), 82–91.

II. SECONDARY LITERATURE. Munier-Chalmas's *Étude du Tithonique . . .* is reviewed by A. Andreae in *Neues Jahrbuch für Mineralogie, Geologie und Paläontologie*, pt. 1 (1894), 156–160. See also G. F. Dollfus, "Nécrologie de E. Munier-Chalmas," in *Journal de conchyliologie*, **52** (1904), 100–106.

ALBERT V. CAROZZI

MUNĪŚVARA VIŚVARŪPA (*b.* Benares, India, 17 March 1603), *astronomy, mathematics.*

The member of a noted family of astronomers who originated at Dadhigrāma on the Payoṣṇī River in Vidarbha with Cintāmaṇi, a Brahmana of the Devarātragotra, in the middle of the fifteenth century, and continued with successive generations represented by Rāma (who was patronized by a king of Vidarbha), Trimalla, and Ballāla, Munīśvara was a grandson of Ballāla, born after the latter had moved the family to Benares. Ballāla had had five sons: Rāma, who wrote a commentary on the *Sudhārasasāraṇī* of Ananta (*fl.* 1525); Kṛṣṇa (*fl.* 1600–1625); Govinda, whose son Nārāyaṇa wrote commentaries on the *Grahalāghava* of Gaṇeśa (*b.* 1507) and, in 1678, on the *Jātakapaddhati* of Keśava (*fl.* 1496); Raṅganātha, who finished his commentary on the *Sūryasiddhānta*, the *Gūḍhārthaprakāśa*, in 1603; and Mahādeva. Munīśvara was the son of Raṅganātha and the pupil of Kṛṣṇa, who traces his *guruparamparā*, or lineage of teachers, back through Viṣṇu (*fl. ca.* 1575–1600) and Nṛsiṃha (*b.* 1548) to the great Gaṇeśa himself.

Although thus tracing his intellectual genealogy back to the school of Gaṇeśa and Keśava (see essay in Supplement), Munīśvara followed his uncle's example of studying the works of Bhāskara II (*b.* 1115); as Kṛṣṇa had written a commentary, the *Bījāṅkura*, on Bhāskara's *Bījagaṇita*, Munīśvara continued the task by commenting on the *Līlāvatī* in the *Nisṛṣṭārthadūtī* and on the two parts of the *Siddhāntaśiromaṇi* in the immense *Marīci*, begun in 1635 and finished in 1638.

In the 1640's and 1650's Munīśvara's family entered into a scientific controversy with another Benares family of astronomers whose intellectual genealogy was traced back to Gaṇeśa. This second family had originated in Golagrāma in Mahārāṣṭra at about the same time that Cintāmaṇi appeared in Dadhigrāma; its representatives contemporary with Munīśvara were the three brothers Divākara (*b.* 1606), Kamalākara (*fl.* 1658), and Raṅganātha. They generally favored the *Saurapakṣa* (see essay in Supplement). And, in this connection, it should be noted that Munīśvara's greatest work, the *Siddhāntasārvabhauma*, which was completed in 1646 and on which he wrote a commentary, the *Āśayaprakāśinī*, in 1650, is fundamentally *Saura* in character; there is, however, a strong admixture of material from the *Brāhmapakṣa* (see essay in Supplement), reflecting his intense study of Bhāskara II's *Siddhāntaśiromaṇi* and of the *Siddhāntasundara* of Jñānarāja (*fl.* 1503). He also demonstrates some knowledge of Islamic astronomy, although much less acceptance of it than is shown by Kamalākara. It is around their respective attitudes toward Islamic astronomy that the controversy between the two families principally turned. Despite his negative attitude, however, the author of the *Siddhāntasārvabhauma* seems to have enjoyed the patronage of Shāh Jahān (reigned 1628–1658).

Munīśvara also composed a *Pāṭīsāra* on mathematics, of which the earliest manuscript, still in Benares, was copied in 1654.

BIBLIOGRAPHY

I. Original Works. Only one of Munīśvara's works has been published in full. Of the *Marīci* the part relating to the *Golādhyāya* was edited by Dattātreya Āpṭe as Ānandāśrama Sanskrit Series 122, 2 vols. (Poona, 1943–1952). Of the part relating to the *Gaṇitādhyāya*, the first chapter only was edited by Muralīdhara Jhā (Benares, 1917) and the rest by Kedāradatta Jośī in vols. II and III of his ed. of the *Grahagaṇitādhyāya* (Benares, 1964).

Muralīdhara Ṭhakkura edited 2 vols. containing the first two chs. and a part of the third of the *Siddhāntasārvabhauma* with the *Āśayaprakāśinī* as Saraswati Bhavana Texts 41 (Benares, 1932–1935); no more has appeared.

II. Secondary Literature. There are notices on Munīśvara in S. Dvivedin, *Gaṇakataraṅgiṇī* (Benares, 1933), repr. from *The Pandit*, n.s. **14** (1892), 91–94; Ś. B. Dīkṣita, *Bhāratīya Jyotiḥśāstra* (Poona, 1896, 1931), 286–287; and M. M. Patkar in *Poona Orientalist*, **3** (1938), 170–171.

David Pingree

MUÑJĀLA (*fl.* India, 932), *astronomy.*

A Brahmana of the Bhāradvājagotra, in 932 Muñjāla composed a *Bṛhanmānasa* which was known to al-Bīrūnī (*India*, translated by E. C. Sachau, 2 vols. [London, 1910], I, 157), who claims that a commentary on it was written by Utpala (*fl.* 966–968). We now have only fragments of the *Bṛhanmānasa*; but the second treatise of Muñjāla mentioned by al-Bīrūnī, the *Laghumānasa*, is extant. It was composed after the lost work at a place called Prakāśa; the earliest commentary on it, by Praśastidhara of Kashmir, was written in 958.

The *Laghumānasa* is a rather eclectic work (see essay in Supplement), although it possesses elements derived from the two schools of Āryabhaṭa I (*b.* 476) and some original insights into lunar theory. Besides the commentary of Praśastidhara mentioned above there are others by Sūryadeva Yajvan of Kerala (*b.* 1191), Parameśvara (*ca.* 1380–1460), and Yallaya (*fl.* 1482). All three of these later commentators lived in southern India, where Muñjāla's work maintained some influence although it had been forgotten in the northwest since the eleventh century.

BIBLIOGRAPHY

The *Laghumānasa* was edited with Parameśvara's *vyākhyā* by Dattātreya Āpṭe as Ānandāśrama Sanskrit Series 123 (Poona, 1952), and with an English trans. and commentary by N. K. Majumdar (Calcutta, 1951). Majumdar had previously published a short note on it, "Laghumānasam of Muñjāla," in *Journal of the Department of Letters, University of Calcutta*, **14** (1927), art. 8.

David Pingree

MÜNSTER, SEBASTIAN (*b.* Nieder-Ingelheim, Germany, 1489; *d.* Basel, Switzerland, 26 May 1552), *geography.*

Münster began his studies at Heidelberg and entered the Minorite order at the age of sixteen.

Early in his career he became fascinated with Hebrew and Greek and mastered both. His first printed work was a Hebrew edition of the Psalms (Basel, 1516). During the first part of his life his primary concern was with the publication of Hebrew texts, dictionaries, and grammars; and on the strength of his important contributions he was elected to the chair of Hebrew at Basel in 1527. Münster moved to Basel in 1529, having become a Protestant in the same year, married, and spent most of the rest of his life there, except for extensive travels in Germany and Switzerland.

Münster's first major contribution to geography dates from 1540, the year of the publication of his Latin translation of Ptolemy's *Geography*, illustrated with maps of his own design. Having addressed an appeal in 1528 "to all lovers of the joyful art of geography to help him in a true and correct description of the German nation," he spent fifteen years collecting up-to-date information on Germany and adjacent lands and in 1544 published his most important work, *Cosmographei*, "a description of the whole world and everything in it." This book set a new standard in the field, diverging widely from such earlier works as Gregor Reisch's *Margarita philosophica* (1496) and following both a regional and an encyclopedic approach. The work ran to 660 pages in the first edition and to nearly twice as many in later editions; its most valuable parts are those dealing with Germany and Central Europe, as well as the illustrations and maps, the latter drawn by Münster himself. Besides the *Cosmographei* Münster is noted for his common-sense approach to geography: when he asked his German colleagues for information about their districts, he provided them with detailed directions, including a simple plane-table survey, the first of its kind. The *Cosmographei* was among the most popular treatises of the sixteenth and seventeenth centuries: forty-six editions, in six languages, were published prior to 1650.

Although Münster was celebrated in his lifetime as a Hebraic scholar, his influence was most widely felt through his understanding of the interests of the reading public: he was not at all reluctant to include some choice miraculous happenings in his otherwise sober and factual narrative. *Cosmographei* may still be consulted with profit by those interested in the humanist world view in the Reformation.

BIBLIOGRAPHY

An outstanding facs. of the 1550 Basel ed. of the *Cosmographei* was published by Theatrum Orbis Terrarum (Amsterdam, 1967), with 910 woodcuts and intro. by Ruthardt Oehme.

For many years the standard source on the life and works of Münster was Victor Hantzsch, "Sebastian Münster—Leben, Werk, wissenschaftliche Bedeutung," which is *Abhandlungen der K. Sächsischen Gesellschaft der Wissenschaften*, Phil.-hist. kl., **18**, no. 3 (1898). A more recent biography is Karl Heinz Burmeister, *Sebastian Münster: Versuch eines biographischen Gesamtbildes*, Basler Beiträge zur Geschichtswissenschaft no. 91 (Basel, 1963). The most up-to-date bibliography is Karl Heinz Burmeister, *Sebastian Münster—eine Bibliographie* (Wiesbaden, 1964).

GEORGE KISH

MURALT, JOHANNES VON (*b.* Zürich, Switzerland, 18 February 1645; *d.* Zürich, 12 January 1733), *surgery, medicine, anatomy.*

Muralt was a member of the old noble de Muralto family, which had been driven from its seat in Locarno in 1555 upon its conversion to Protestantism. The refugees were eventually invested with citizenship in the Reformed Swiss cities of Bern and Zürich, and found new prosperity. Some of Muralt's ancestors were physicians and diplomats; his father, Johann Melchior Muralt, was a merchant.

Muralt was educated at the Zürich Carolinum. When he was twenty he published his *Schola mutorum ac surdorum*, then set out on his academic travels, which took him to Basel, Leiden, London, Oxford, Paris, and Montpellier. He studied anatomy, surgery, and obstetrics with a number of famous teachers, among them Franciscus Sylvius. He returned to Switzerland to take the M.D. at the University of Basel in 1671 with a dissertation "De morbis parturientium et accidentibus, quae partum insequuntur." The following year he settled in Zürich, where he married Regula Escher; they had many children, including the distinguished physician Johann Conrad Muralt.

The Zürich surgeons' guild challenged Muralt's right to practice in that city, and he encountered widespread disapproval for conducting public animal dissections. His success as a physician overcame all opposition, however, and after five years of argument the Zürich Bürgerrat authorized him to dissect the bodies of executed criminals and of hospital patients who had died of rare diseases. Muralt was admitted to the Academia Caesario-Leopoldina Naturae Curiosorum (with the name "Aretaeus") in 1681; forgetting their old feud, the surgeons also made him an honorary member of their guild.

In 1686 Muralt gave a course of lectures at the surgeons' guildhall, "Zum Schwarzen Garten." His audience was composed of surgeons, their apprentices, medical students, and laymen; the lectures themselves were the first on anatomical subjects to be given in the vernacular. Once a week, for an entire year, Muralt displayed dissected bodies (chiefly animal) and discussed the anatomy, physiology, and pathology of the organs. He expounded the theory of diseases and outlined medical and surgical treatment, including precise directions for the use of medicinal plants and detailed instructions for military surgeons.

In 1688 Muralt was named archiater of Zürich, with duties that comprised devising sanitary measures to protect the city against infectious diseases, advising the municipal marriage court, inspecting apothecaries, supervising the training of midwives, and treating internal diseases in the city's hospital. Ex officio, Muralt also performed all operations for fractures, the stone, and cataracts. In 1691 he was appointed professor of natural sciences at the cathedral school and also became canon of its chapter. He made use of this multitude of offices to transform Zürich into an important center for the study of anatomy and surgery.

Muralt's considerable achievement was largely based upon his surgical skill. He developed new procedures and set them forth systematically in his writings. His work is, however, more notable for the quantity and range of his material than for the depth of his knowledge. His twenty-one titles on anatomy, medicine, and physiology, as well as his thirteen separate publications on mineralogy, zoology, and botany, are marred by repetitiousness. Many of his printed works represent a collection of what are, in effect, his laboratory notes on experiments, natural objects, or the course of a disease (for example, the 174 "Observationes" that he published in *Miscellanea curiosa medico-physica Academiae naturae curiosorum*); others, among them the *Anatomisches Collegium* of 1687, record his lectures more or less verbatim. His principle work on natural history was *Systema physicae experimentalis* . . . (1705–1714); a manuscript regional pharmacopoeia has also been preserved. The last of his writings, *Kurtze und Grundlich Beschreibung der ansteckenden Pest* (1721) remains of interest for its suggestion of the "animal" nature of the plague contagium.

In general, Muralt was a keen observer and a poor critic. He was occasionally prey to superstition, and elements of popular medical beliefs are apparent in his theory of disease. But if some of his therapeutic measures derive from the operations of magic, Muralt was nevertheless an effective physician and a tireless popularizer and communicator of genuinely scientific knowledge.

BIBLIOGRAPHY

I. ORIGINAL WORKS. Muralt's writings include *Vademecum anatomicum sive clavis medicinae* (Zurich, 1677); *Anatomisches Collegium* (Nuremberg, 1687); *Curationes medicae observationibus et experimentis anatomicis mixtae* (Amsterdam, 1688); *Kinder- und Hebammenbüchlein* (Zurich, 1689; Basel, 1697); *Chirurgische Schriften* (Basel, 1691); *Hippocrates Helveticus oder der Eydgenössische Stadt- Land- und Hauss-Artzt* (Basel, 1692); *Systema physicae experimentalis*, 4 vols. (Zurich, 1705–1714), of which the fourth part, *Botanologia seu Helvetiae paradisus*, was trans. into German as *Eydgenössischer Lust-Garte* (Zurich, 1715); *Schriften von der Wund-Artzney* (Basel, 1711); *Kriegs- und Soldaten-Diaet* (Zurich, 1712); and *Sichere Anleitung wider den dissmal grassirenden Rothen Schaden* (Zurich, 1712).

II. SECONDARY LITERATURE. On Muralt and his work, see C. Brunner, *Die Verwundeten in den Kriegen der alten Eidgenossenschaft* (Tübingen, 1903); and *Aus den Briefen hervorragender Schweizer Ärzte des 17. Jahrhunderts* (Basel, 1919), written with W. von Muralt; E. Eidenbenz, "Dr. Leonhard von Muralts 'Pharmocopoea domestica,' " in *Schweizerische Apothekerzeitung*, **60** (1922), 393–399; J. Finsler, *Bemerkungen aus dem Leben des Johannes von Muralt* (Zurich, 1833); H. Koller, "Das anatomische Institut der Universität Zürich in seiner geschichtlichen Entwicklung," in *Zürcher medizingeschichtliche Abhandlungen*, **11** (1926); K. Meyer-Ahrens, "Die Arztfamilie von Muralt, insbesondere Joh. v. Muralt, Arzt in Zürich," in *Schweizerische Zeitschrift für Heilkunde*, **1** (1862), 268–289, 423–436, and **2** (1863), 25–47; O. Obschlager, "Der Zürcher Stadtarzt Joh. von Muralt und der medizinische Aberglaube seiner Zeit," M.D. dissertation, University of Zurich (1926); G. Sticker, *Abhandlungen aus der Seuchengeschichte und Seuchenlehre*, vol. I *Die Pest* (Giessen, 1910); and G. A. Wehrli, "Die Bader, Barbiere und Wundärzte im alten Zürich," in *Mitteilungen der Antiquarischen Gesellschaft Zürich*, **30**, pt. 3 (1927), 99.

JÖRN HENNING WOLF

MURCHISON, RODERICK IMPEY (*b.* Tarradale, Ross and Cromarty, Scotland, 19 February 1792; *d.* London, England, 22 October 1871), *geology*.

Murchison was born into a long-established family of Highland landowners. His father, Kenneth Murchison, died when the boy was only four; and after his childhood he never lived in Scotland. He was educated at the military college at Great Marlow and in 1808 saw active service briefly in the Peninsular War. In 1815 he married Charlotte Hugonin and soon afterward resigned his commission. From 1816 to 1818 Murchison traveled in Italy and under his wife's influence showed signs of artistic interests, but on his return he sold his family estate and for several years devoted himself chiefly to fox hunting. A chance acquaintance with Humphry Davy turned his attention toward science, however, and in 1824 he settled in London and attended lectures at the Royal Institution. Encouraged by his wife, Murchison soon focused his interests on geology, chiefly through the influence of William Buckland; he was elected a fellow of the Geological Society of London in 1825 and of the Royal Society in 1826. With the advantages of a private income, he was able thereafter to devote himself entirely to science.

Taking as a model the stratigraphical handbook of W. D. Conybeare and W. Phillips (1822), Murchison began the long series of geological studies which brought him worldwide fame and recognition. Almost every summer, for over twenty years, he undertook long and often arduous journeys in search of new successions of strata which would help to bring order to the reconstruction of the history of the earth. He entered geology during the first great period of stratigraphical research, and stratigraphy remained his chief area of interest. He was not a theoretician and generally delegated the paleontological parts of his work to others, but he was an excellent observer with a flair for grasping the major features of an area from a few rapid traverses.

Some of his earliest work convinced Murchison of the superiority of fossils over lithology as criteria of geological age: in 1826 he showed that the fauna and flora of the isolated coalfield of Brora in northeastern Scotland indicated it to be of the same age as the English Oolites (that is, Jurassic), although the rock types resembled the Coal Measures (that is, Carboniferous). In 1828 he accompanied Charles Lyell through the celebrated volcanic districts of the Massif Central into northern Italy, and their joint papers suggest that Murchison was at this time much influenced by Lyell's theoretical views. His subsequent work in the Alpine region, some of it in the company of Adam Sedgwick, included an attempt to show the continuity of the Secondary and Tertiary strata; but, at the same time, firsthand experience of the vast scale of folding and faulting in the Alps led Murchison toward an increasing catastrophist emphasis on the role of occasional episodes of drastic disturbance in the crust of the earth.

During these first years of research Murchison's travels brought him into contact with most of the leading geologists on the Continent, and his position as secretary (from 1827) of the Geological Society made him equally well known in Britain. In 1831 he was elected president of the Geological Society (he held office until 1833, and again from 1841 to 1843), and in the same year began his most important research.

At this time the major features of the stratigraphical succession had been clarified down to the Old Red Sandstone underlying the Carboniferous rocks, but below that was what Murchison called "interminable grauwacke"—rocks containing few fossils, in which no uniform sequence had been detected. It was widely doubted whether the method of correlation by fossils would even be applicable to these ancient Transition strata, yet in them—if anywhere—lay the possibility of finding evidence for the origin of life itself. Acting on a hint of Buckland's, Murchison was fortunate to find in the Welsh borderland an area in which there was a conformable sequence downward from the Old Red Sandstone into Transition strata with abundant fossils. He gave a preliminary report of his work at the first meeting (1831) of the British Association for the Advancement of Science; and in 1835, after further fieldwork, he named the strata Silurian after the Silures, a Romano-British tribe that had lived in the region.

The Silurian constituted a major system of strata with a highly distinctive fauna, notable for an abundance of invertebrates and for the complete absence, except in the youngest strata, of any remains of vertebrates or land plants. It thus seemed to Murchison to mark a major period in the progressive history of life on earth. Even before he had completed his great monographic account The Silurian System (1839), its validity had been rapidly recognized by geologists in many other parts of the world. The striking uniformity of the Silurian fauna, in contrast with the highly differentiated faunal provinces of the present day, was taken by Murchison to underline the limitations of Lyell's uniformitarian approach, and was attributed by him to the greater climatic uniformity of the globe in Silurian times, a result of the greater influence of conducted heat from the still incandescent interior of the earth.

Murchison was well aware of the vast economic implications of his delineation of a Silurian system. If the Silurian period had truly predated the establishment of terrestrial vegetation, the recognition of Silurian fossils in any part of the world would reliably indicate a base line beneath which it was pointless to search for coal: this would save much useless expenditure and also help to assess more accurately the possible reserves of undiscovered coal. A report by Henry de la Beche of coal plants in the "grauwacke" of Devonshire (1834) therefore seemed to Murchison to be a very serious anomaly, and he devoted several years to an attempt to explain it away. He and Sedgwick discovered first that the fossil plants were in fact in strata of Coal Measure age overlying the true "grauwacke"; and later, in 1839, following a suggestion of William Lonsdale's, they concluded that even these older strata were not pre-Silurian, as they had originally thought, but were the lateral equivalents of the Old Red Sandstone. This definition of a Devonian system was at first criticized as being based purely on paleontological criteria and not on any plain evidence of superposition; but Murchison and Sedgwick soon showed that the distinctive Devonian invertebrate fauna occurred in Westphalia in the expected position immediately below the Carboniferous strata. The following year (1840) Murchison resolved the matter by discovering in European Russia a sequence of undisturbed strata in which the Devonian was clearly underlain by Silurian and overlain by Carboniferous, and in which Devonian invertebrates were interbedded with Old Red Sandstone fish. This established the temporal equivalence of the Devonian and Old Red Sandstone despite their contrasting lithology and fauna.

A second expedition to Russia in 1841 took Murchison as far as the Urals and confirmed this Paleozoic sequence. At the same time it showed him how undisturbed and unaltered sediments could change their appearance radically when traced laterally into a region of mountain-building, and this convinced him of the validity of Lyell's hypothesis of metamorphism. He also found a vast development of Paleozoic strata overlying the Carboniferous and named them Permian after the Perm region near the Urals.

In 1839 Murchison's financial position had greatly improved, and he had moved into a grander house, which thereafter became a fashionable salon of the London intelligentsia. His enhanced social position, coupled with the many distinctions conferred on him for his work in Russia, unfortunately made him increasingly conscious of social prestige and increasingly arrogant and intolerant of opposition in scientific matters.

Murchison's capacity for transforming scientific controversies into paramilitary "campaigns" against opponents had already been evident in his treatment of de la Beche over the Devonian problem. It was now shown much more seriously in his controversy with Sedgwick over the base of the Silurian. In the same year that Murchison had first investigated the Transition strata, Sedgwick had begun to unravel still older strata in Wales; and when Murchison first established the Silurian, Sedgwick had suggested the name Cambrian system (after the Latin name for Wales) for the older rocks. During their only joint fieldwork in Wales (in 1834) Murchison had assured Sedgwick that the latter's Upper Cambrian lay below his own Lower Silurian strata, although, as expected,

there was a faunal gradation between the two. But when Murchison later realized that the fossils of the Upper Cambrian Bala series were indistinguishable from his own Lower Silurian Caradoc series, he boldly proclaimed their identity and annexed the Upper Cambrian into his Silurian system.

Sedgwick protested that the Cambrian had been clearly defined by reference to an undisputed succession of strata in northern Wales and that it was wrong to alter the meaning of the term just because its upper part contained Silurian fossils. But Murchison continued to annex more and more of the Cambrian into his Lower Silurian, until the two terms were virtually synonymous. Sedgwick claimed that this unjustified annexation was designed to cover two major mistakes of Murchison's. He had misinterpreted the Lower Silurian succession in its type area and had therefore believed that these strata were younger than the Upper Cambrian when in fact they were of the same age; and—an even more serious mistake—he had wrongly incorporated some Upper Silurian strata (May Hill sandstone) into the Lower Silurian Caradoc series, despite their very different faunas, thus giving the Silurian fauna a spurious uniformity down into Sedgwick's Cambrian.

But there was even more to the controversy than technical mistakes and a priority dispute over stratigraphical nomenclature. Each geologist, as a firm believer in a progressionist interpretation of the fossil record, ardently desired the distinction of showing that his own system contained the evidence for the origin of life on earth. Thus, when Murchison wrote his *Geology of Russia* (1845), he asserted that the "unequivocal base-line of palaeozoic existence" was to be seen in the Lower Silurian strata, within which there was a "gradual decrement and disappearance of fossils" toward the base. Furthermore, in Scandinavia (where he had traveled in 1844) these strata were immediately underlain by "Azoic" crystalline schists, in which Murchison believed that it was "hopeless to expect" to find fossils. This was not because they had been metamorphosed (although he agreed that they resembled the metamorphic rocks of later periods) but because they had been formed under conditions too hot to support life. He therefore argued, against Lyell, that geology provided "undeniable proofs of a beginning" to life on earth. His desire to have sole credit for providing these "proofs" is shown by his obstinate insistence that the Silurian fauna was the earliest. Thus when Joachim Barrande first described a distinctive "primordial" fauna (the Cambrian of modern geology) below the previously known Lower Silurian faunas, Murchison did not allow it as a possible paleontological basis for Sedgwick's

Cambrian but incorporated it too into his Silurian.

In 1846 Murchison was knighted and served as president of the British Association for the Advancement of Science; and in 1849 his work was recognized by the award of the Royal Society's Copley Medal. He later published an updated and more popular version of his work as *Siluria* (1854), expressly in order to deliver a "knock-down blow" (the aggressive metaphor is characteristic) to those, like Lyell, who still denied the reality of organic progression. The book also contained an assessment of the world's probable resources of gold, designed to reassure those who feared that the recent Australian gold rush presaged a slump in that metal's monetary value. In 1855 he succeeded de la Beche as director general of the Geological Survey of Great Britain (thus becoming a professional scientist for the first time), and in 1856 he was appointed to a royal commission to report on the nation's coal reserves.

From the 1840's Murchison became increasingly rigid and intolerant of scientific innovation. He opposed the glacial theory of Louis Agassiz and continued to assert that icebergs alone had been responsible for the transport of erratic blocks ("drift") long after most other geologists had accepted at least a modified glacialism: under his influence the Geological Survey's maps long continued to use the term "drift" for glacial and postglacial deposits. Murchison's last major fieldwork, in 1858–1860, was devoted to arguing that the Moine schists of the northwestern Highlands were Silurian sediments, although he had always favored relatively catastrophist interpretations of mountain tectonics and had been convinced a decade earlier of the reality of large-scale thrusting in the Alps. He was totally opposed to Darwin's evolutionary theory.

Murchison retired temporarily from the council of the Geological Society in 1863 and was therefore eligible to be awarded the Wollaston Medal the following year. He was created a baronet in 1866. He had earlier been one of the founders of the Royal Geographical Society and was for many years its president. Indeed, despite his post with the Geological Survey, he was better known as a geographer than as a geologist in his later years, being prominent in the support of David Livingstone's and other expeditions. The Murchison Falls of the Nile in Uganda are named after him.

BIBLIOGRAPHY

The following are the more important of Murchison's published works: "On the Coal-Field of Brora in Sutherlandshire, and on Some Other Stratified Deposits in the North of Scotland," in *Transactions of the Geological Society of London*, 2nd ser., **2**, pt. 2 (1829), 293–326; "A Sketch of the Structure of the Eastern Alps . . .," *ibid.*, **3**, pt. 2 (1832), 301–420, written with Adam Sedgwick; *The Silurian System, Founded on Geological Researches in the Counties of Salop, Hereford, Radnor, Montgomery, Caermarthen, Brecon, Pembroke, Monmouth, Gloucester, Worcester, and Stafford; With Descriptions of the Coal-Fields and Overlying Formations* (London, 1839); "Classification of the Older Rocks of Devonshire and Cornwall," in *Philosophical Magazine*, **14** (1839), 242–260, written with Adam Sedgwick; "On the Classification and Distribution of the Older or Palaeozoic Rocks of the North of Germany and of Belgium, as Compared With Formations of the Same Age in the British Isle," in *Transactions of the Geological Society of London*, 2nd ser., **6**, pt. 2 (1842), 221–302, written with Adam Sedgwick; and *The Geology of Russia in Europe and the Ural Mountains*, 2 vols. (London–Paris, 1845), written with Édouard de Verneuil and Alexander von Keyserling—Murchison wrote the stratigraphy in vol. I.

See also "On the Palaeozoic Deposits of Scandinavia and the Baltic Provinces of Russia, and Their Relations to Azoic or More Ancient Crystalline Rocks; With an Account of Some Great Features of Dislocation and Metamorphism Along Their Northern Frontiers," in *Quarterly Journal of the Geological Society of London*, **1** (1845), 467–494; "On the Meaning Originally Attached to the Term 'Cambrian System,' and on the Evidences Since Obtained of Its Being Geologically Synonymous With the Previously Established Term 'Lower Silurian,'" *ibid.*, **3** (1847), 165–179; "On the Geological Structure of the Alps, Apennines and Carpathians, More Especially to Prove a Transition From Secondary to Tertiary Rocks, and the Development of Eocene Deposits in Southern Europe," *ibid.*, **5** (1849), 157–312; *Siluria. The History of the Oldest Known Rocks Containing Organic Remains, With a Brief Sketch of the Distribution of Gold Over the Earth* (London, 1854); and "On the Succession of the Older Rocks in the Northernmost Counties of Scotland; With Some Observations on the Orkney and Shetland Islands," in *Quarterly Journal of the Geological Society of London*, **15** (1859), 353–418.

Murchison's field notebooks and a collection of his scientific correspondence are in the library of the Geological Society of London. Material in the Institute of Geological Sciences, London (formerly Geological Survey), is described by John C. Thackray, "Essential Source-Material of Roderick Murchison," in *Journal of the Society for the Bibliography of Natural History*, **6**, pt. 3 (1972), 162–170.

Some excerpts from Murchison's journals and letters are published in the only full-length biography, Archibald Geikie, *Life of Sir Roderick Murchison . . . Based on His Journals and Letters With Notices of His Scientific Contemporaries and a Sketch of the Rise and Growth of Palaeozoic Geology in Britain*, 2 vols. (London, 1875), which also includes a fairly full list of Murchison's publications.

M. J. S. RUDWICK

MURPHY, JAMES BUMGARDNER (*b.* Morganton, North Carolina, 4 August 1884; *d.* Bar Harbor, Maine, 24 August 1950), *biology.*

Murphy was the son of Patrick Livingston Murphy, a pioneer in modern psychiatric therapy and director of the state mental hospital at Morganton. He received the B.S. from the University of North Carolina in 1905 and the M.D. in 1909 from the Johns Hopkins University, where his surgical finesse was appreciated by Harvey Cushing, who became his good friend. From 1910 to 1950, the year of his retirement and of his death, he pursued research on cancer and related physiological problems as a scientist and later as administrator at the Rockefeller Institute in New York City.

Brought to the Institute in 1910 by the noted cancer researcher and his first collaborator, Peyton Rous, Murphy soon demonstrated a skill in developing methods to answer the ill-defined and unlimited questions concerning the origin and growth of malignant tissues. In showing that the frozen and dried tissue extract of the spindle-cell chicken sarcoma (Chicken Tumor I) could be used to transmit this form of cancer, he produced one of the earliest successful applications of the process known as lyophilization, now commonly used in biological research. Later he perfected the technique of growing a chicken tumor virus in fertilized eggs, a method of fundamental importance to virus research.

In 1923 Murphy was placed in charge of the department of cancer research, succeeding Rous, who had turned to other research interests. Thus Murphy began to play a significant role in determining the direction of cancer research for over a quarter of a century at the Rockefeller Institute.

Two lectures presented by Murphy as Thayer lecturer at Johns Hopkins in 1935 summarized the four main lines of cancer research, for which he had helped to lay the foundation, in the first half of the twentieth century. These areas, which were explored independently, included the discovery that certain tumors in mice could be transplanted, that specific chemical substances produced malignancies after an animal was repeatedly exposed to them, that certain cancers occurred more frequently in individuals whose ancestors had expressed the same disease, and that chicken tumors were transplantable and were equivalent to cancer in mammals. He emphasized that the data gathered to test the inheritance factor were useful in examining the possibility that cancer was an infectious disease transmitted by a parasite; he later became more skeptical of this mode of transmission as an explanation of the origin of cancer. By 1942 Murphy had reduced the main lines of research to the first three of these areas and pointed out how study of chemical carcinogens was on the rise.

A skilled administrator, Murphy wisely marshaled public support for cancer research and stressed the need for increasing public awareness of the early signs of treatable cancers, especially breast and uterine tumors. He encouraged the formation of the Woman's Field Army, which campaigned for women to seek medical aid when suspicious symptoms in these areas first developed. As a member of the board of the American Society for Control of Cancer, which became the American Cancer Society in 1929, he sought to change public opinion from one of shame toward cancer victims to sympathy for them and interest in their care.

Murphy contributed his knowledge and talents to a broad range of activities in the field of cancer research. He lectured extensively and was a member of the National Academy of Sciences, National Research Council, American Society for Cancer Research, of which he was president from 1921 to 1922, American Society of Experimental Pathology, and a number of foreign scientific societies. He was on the editorial board of the journal *Cancer Research*, and served as a delegate to several international congresses devoted to cancer studies. He received numerous medals and awards, and honorary doctorates from the University of Louvain, the University of North Carolina, and Oglethorpe University.

At his death, Murphy was survived by his wife, Ray Slater Murphy, and his two sons.

BIBLIOGRAPHY

Murphy published over 130 papers between 1907 and 1950. Over three-quarters of these were collaborative papers published in association with visitors and staff at the Rockefeller Institute, including F. Duran-Reynals, Arthur W. M. Ellis, Fred Gates, R. G. Hussey, Karl Landsteiner, Douglas A. MacFayden, J. Maisin, John J. Morton, Waro Nakahara, Peyton Rous, H. D. Taylor, W. H. Tytler, and especially Ernest Sturm, his last assistant and colleague for 31 years.

His more significant papers include a series published between 1911 and 1914 describing transplanted chicken tumors. Most appear in the *Journal of Experimental Medicine;* "The Lymphocyte in Resistance to Tissue Grafting, Malignant Disease, and Tuberculosis Infection. An Experimental Study," in Rockefeller Institute *Monographs*, no. 21 (1926); "Experimental Approach to the Cancer Problem. I. Four Important Phases of Cancer Research. II. Avian Tumors in Relation to the General Problem of Malignancy," in *Bulletin of the Johns Hopkins Hospital*, **56** (1935), 1–31, two lectures of the Thayer lectureship; "An Analysis of the Trends in Cancer Research," in *Journal of the American Medical Association*,

120 (1942), 107–111, Barnard Hospital lecture; and "The Cancer Control Movement," in *North Carolina Medical Journal*, **5** (Apr. 1944), 121–125.

A series of papers produced in the last decade of his life on the development of experimental leukemia and its relationship to basic physiological processes, written with Ernest Sturm, include "The Transmission of an Induced Lymphatic Leukemia and Lymphosarcoma in the Rat," in *Cancer Research*, **1** (1941), 379–383; "The Effect of Sodium Pentobarbital, Paradichlorbenzene, Amyl Acetate, and Sovasol on Induced Resistance to a Transplanted Leukemia of the Rat," *ibid.*, **3** (1943), 173–175; "The Adrenals and Susceptibility to Transplanted Leukemia of Rats," in *Science*, **98** (1943), 568–569; "The Effect of Adrenalectomy on the Susceptibility of Rats to a Transplantable Leukemia," in *Cancer Research*, **4** (1944), 384–388; "Effect of Adrenal Cortical and Pituitary Adrenotropic Hormones on Transplanted Leukemia in Rats," in *Science*, **99** (1944), 303; "The Inhibiting Effect of Ethyl Urethane on the Development of Lymphatic Leukemia in Rats," in *Cancer Research*, **7** (1947), 417–420; "The Effect of Diethylstilbestrol on the Incidence of Leukemia in Male Mice of the Rockefeller Institute Leukemia Strain (R.I.L.)," *ibid.*, **9** (1949), 88–89; and "The Effect of Adrenal Grafting on Transplanted Lymphatic Leukemia in Rats," *ibid.*, **10** (1950), 191–193.

For brief biographies of Murphy see *National Cyclopedia of American Biography*, XXXVIII, 69; an obituary in *Journal of the American Medical Association*, **144** (14 Oct. 1950), 562, and a longer biography by C. C. Little in *Biographical Memoirs of the National Academy of Sciences*, **34** (1960), 183–203, which contains a bibliography arranged chronologically.

For a discussion of his work at the Rockefeller Institute see George W. Corner, *A History of the Rockefeller Institute 1901–1953. Origins and Growth* (New York, 1964), *passim.* Personal business papers and correspondence are held by the Rockefeller University and Murphy's family.

AUDREY B. DAVIS

MURRAY, GEORGE ROBERT MILNE (*b.* Arbroath, Scotland, 11 November 1858; *d.* Stonehaven, Scotland, 16 December 1911), *botany.*

Murray was one of eight children born to George and Helen Margaret Murray. He was educated in Arbroath until 1875, when he spent a year in Strasbourg studying under Anton de Bary. In 1876 he became an assistant in the botany department of the British Museum, where he was put in charge of the cryptogamic collections. He spent the rest of his career in this department, becoming Keeper in 1895. His early research was in mycology, and was of sufficient taxonomic interest to result in his election as fellow of the Linnean Society in 1878, before he was twenty-one, and an invitation to write an article on fungi for the *Encyclopaedia Britannica* in 1879. The natural history departments of the British Museum moved to South Kensington in 1881, and Murray was responsible for the transfer of the cryptogams, reorganization of the herbarium, and later development of the section.

Murray always maintained his links with Scotland and worked there during vacations, investigating salmon disease and collecting diatoms and pelagic algae from the sea and the lochs while on board the Fishery Board's vessel *Garland*. New techniques in trawling for phytoplankton by pumping water through fine silk nets allowed him to study seasonal variations in forms; he taught these methods to captains of trawlers, who then collected for him in the course of their normal business. His work on the reproduction of diatoms by asexual spore formation was published in 1897. Working with *Biddulphia* spp., *Chaetoceros* spp., and *Coscinodiscus concinnus*, he showed that small specimens growing inside the shells of adult forms were not only a means of rejuvenating those individuals, but might divide into two, four, or eight new individuals, which would eventually be released and grow to full size.

Murray was associated with several expeditions, generally sponsored by the museum. In 1886 he visited the West Indies as a naturalist attached to the solar eclipse expedition, and in 1888 he sorted the algae and fungi from the expedition to Fernando de Noronha, and wrote those sections of the report. He was secretary to the West Indies exploration committee from 1891, and in 1897 he returned there on the *Para* to visit Barbados, Haiti, Jamaica, and Panama, collecting particularly *Coccosphaera*, a hitherto little-known unicellular alga. He differentiated the species, and showed how the cover of overlapping calcareous scales, arranged in a definite order, provide defensive armor while still allowing for growth, an evolutionary advance on the structure of diatoms.

In 1898 he organized an expedition under the Royal Geographic Society in the *Oceana* to collect material in an area off the coast of Ireland where the sea bed dropped steeply. He was also scientific director of the *Discovery* expedition of 1901, but sailed only as far as Cape Town. He organized the ship's stores and apparatus, and edited *The Antarctic Manual* for the expedition, writing a brief section, "Notes on Botany and How to Collect Specimens."

In 1884 he married Helen Welsh; they had one son and one daughter. He was elected fellow of the Royal Society in 1897. In 1905 he retired because of ill health and returned to Scotland.

Massee named the new fungal species *Schizophyllum murrayi* after Murray.

BIBLIOGRAPHY

I. ORIGINAL WORKS. Murray published approximately forty papers on cryptogams and oceanography, in which he described new species, surveyed distribution, and listed specimens in the British Museum collections. Many of these papers appeared in *Journal of Botany, British and Foreign*. His reports on the work of the botany department from 1895 to 1903 appeared in *Journal of Botany*. His other works include the section on fungi in A. Henfrey, ed., *An Elementary Course of Botany*, 3rd ed. (London, 1878), 455–472; 4th ed. (London, 1884), 428–449; *A Handbook of Cryptogamic Botany* (London, 1889), written with A. W. Bennett; two articles, "Algae" and "Fungi," in *The natural history of the Island of Ferdinand de Noronha, based on the Collections made by the . . . Expedition of 1887* (London, 1890), 75–81, extracted from *Journal of the Linnean Society*, Botany, **26** (1888), 1–95, and *ibid.*, Zoology, **20** (1888), 473–570; *Phycological Memoirs*, pts. I–III (London, 1892–1895), edited by Murray; *Introduction to the Study of Seaweeds* (London, 1895); "Report of Observations on Plant Plankton," in *Edinburgh Fisheries Board Report*, **15** (1897), pt. 3, 212–218; *Report of the Lords Commissioners of H.M. Treasury of the Departmental Committee on the Botanical Work and Collections at the British Museum and at Kew . . . 1901*, questions 1–198 (London, 1901), 1–13; and *The Antarctic Manual* (London, 1901), edited by Murray.

His scientific papers may be traced through the Royal Society *Catalogue of Scientific Papers*, XVII, 429–430. Papers mentioned in the text are "On the Reproduction of Some Marine Diatoms," in *Proceedings of the Royal Society of Edinburgh*, **21** (1897), 207–218; and "On the Nature of the Coccospheres and Rhabdospheres," in *Philosophical Transactions of the Royal Society*, **190B** (1898), 427–441, plus 2 plates, written with V. H. Blackman.

II. SECONDARY LITERATURE. The most useful obituary is James Britten, in *Journal of Botany*, **50** (1912), 73–75. There is also an obituary by K.F. and W.C., in *Proceedings of the Royal Society*, **B86** (1913), xxi–xxiii; and the entry by G. S. Boulger, in Sidney Lee, ed., *Dictionary of National Biography*, supp. 1901–1911 (Oxford, 1912), 667–668.

DIANA M. SIMPKINS

MURRAY, JOHN (*b.* Cobourg, Ontario, Canada, 3 March 1841; *d.* Kirkliston, Scotland, 16 March 1914), *oceanography, marine geology.*

As editor (after C. Wyville Thomson's death) of the fifty-volume *Report on the Scientific Results of the Voyage of H.M.S. Challenger During the Years 1872–1876* (London, 1880–1895; reprinted New York, 1966) and coauthor (with Johan Hjort) of *The Depths of the Ocean* (London, 1912), Murray presided over the organization of oceanography as a separate science. His most significant personal contribution was the mapping and classification of the sediments on the ocean bottom.

Raised "on the plains of Canada which lie between the great lakes of Erie, Huron, and Ontario" (J. L. Graham, p. 173), to which his parents, Robert and Elizabeth (Macfarlane) Murray, migrated in 1834, John Murray first saw the ocean when he sailed to their native Scotland at age seventeen. This voyage and his first glimpse of the rise and fall of the tide along the Scottish coast was the beginning of his lifelong interest in the ocean. His maternal grandfather, for whom he founded a natural history museum at Bridge of Allan, Stirlingshire, sent him to Stirling High School and the University of Edinburgh, where he studied medicine with John Goodsir and his successor, William Turner. In 1868 he sailed on the whaler *Jan Mayen* from Peterhead to Spitzbergen and the Arctic, returning with a large collection of marine organisms and observations on currents, temperatures, and sea ice. Murray then returned to Edinburgh to enter the physical laboratory of P. G. Tait. Marine biology was left for vacations, when Murray, Laurence Pullar, and the anatomist Morison Watson would hire a fishing boat from which to dredge along the rugged Scottish coast.

Murray's days as a gentleman-student (he took no examination or degree) ended in 1872. Tait and Sir William Thomson (who in a chance encounter with Murray aboard his yacht, was impressed by Murray's knowledge of the sea) recommended him to C. Wyville Thomson, regius professor of natural history at Edinburgh and scientific director of the voyage of circumnavigation which the Royal Society was organizing for the British navy. Murray spent the next three and a half years aboard H.M.S. *Challenger*. At thirty-one he was the oldest of Thomson's four scientific assistants. Less clearly bent on a scientific career than the two younger naturalists, Henry N. Moseley and Rudolph von Willemoes-Suhm (the fourth assistant, John Y. Buchanan, was a chemist), Murray took over, perhaps by default, the newest of the *Challenger* expedition's scientific quests: investigating the deposits on the sea bottom.

Of the major sedimentary types, only globigerina ooze had been named prior to Murray's work aboard *Challenger*, and there was no agreement whether the organisms whose calcareous skeletons made up this deposit lived at the surface or the bottom. By his careful towing of fine nets to catch living specimens, Murray proved that they were surface dwellers, confirming the earlier view of J. W. Bailey. Murray also collected the surface-living forms whose skeletons made up the other major organic sediment types, which he named radiolarian, diatom, and pteropod oozes. The most widespread deposit was a brownish, largely inorganic mud. Murray named it red clay, showed that it originated mainly from volcanic dust, and deduced that it covered those parts of the deep ocean where calcareous skeletons were so few that they had almost all dissolved, an explanation that still stands. Murray's work with the sediments he collected demonstrated conclusively the surprising slowness of deposition over much of the ocean. Murray increased considerably the collection of pelagic animals by towing at depths the nets previously used only at the surface. He also took charge of the small collection of vertebrates.

Murray combined exceptional organizing skill with a strong desire to stake out new scientific territory. When Thomson's scientific staff was disbanded in 1877 shortly after the return of the *Challenger*, Murray stayed on to help Thomson set up, in Edinburgh, the Challenger Expedition Commission, charged with preparing a report in five years. With Thomson and Alexander Agassiz, Murray sorted into groups the contents of 600 cases of specimens, each group to be the subject of a specialist's monograph. Thomson's health soon gave way under the combined pressures of teaching, public lecturing, and accounting personally to the British Treasury for the *Challenger Report*. He died early in 1882 as the grant expired; printing had barely begun.

Pressed by the Royal Society, the Treasury appointed Murray as Thomson's successor with another five-year grant, increased by 20 percent. Murray distributed those specimens which remained in Edinburgh, hounded his dilatory authors, and saw their contributions through the press. When the Treasury tried to halt the project in 1889, with the *Report* still unfinished after a year's extension to his original five, Murray fought back. He saved the *Report*, but the Treasury paid him only a small lump sum for the editorial work from 1889 to 1895; he thus must have been put to considerable personal expense. Among the final volumes of the *Challenger Report*, which might otherwise have failed of publication, were Murray's masterly two-volume *Summary* (1895) and the volume on *Deep-Sea Deposits* (1891), written with Alphonse Renard of the University of Ghent.

Murray did not neglect his own researches, even during the years he traveled around Europe to prod his authors and edited the thousands of manuscript pages they sent him. Cruises in *Knight Errant* (1880) and *Triton* (1882) enabled him and his *Challenger* shipmate, Commander Thomas Tizard, to confirm, by their discovery of the Wyville Thomson ridge, Thomson's proposed solution to the problem of faunal distribution at the bottom of the Faroe-Shetland channel.

Murray's study of coral reefs led him to challenge Charles Darwin's widely accepted notion that they were universally built up on subsiding island bases. Murray suggested instead that reef building could begin when deposition brought a submerged base close enough to the surface for corals to grow, so that uplift rather than subsidence could be the dominant mechanism in some localities. In Murray's view the reef grew seaward on a talus of dead shells, while the retardation of coral growth away from the sea and the solvent action of seawater on the dead coral accounted for the formation of lagoons. Although his views have not stood up, Murray stimulated much contemporary debate, especially after the eighth duke of Argyll charged that only Darwin's great name had prevented the replacement of his theory by Murray's. Murray stimulated the long series of reef explorations by his friend Alexander Agassiz, and with another friend, Robert Irvine, Murray studied the deposition of carbonate and silicate by organisms and the composition of manganese nodules. In 1886, working from a few samples of rocks and deep-ocean sediments, Murray deduced that the Antarctic ice sheet must be underlain by continental rocks. He was one of the strongest advocates of the renewal of polar exploration which began about 1900. His 1888 estimates of the proportion of the ocean floor at different depths, based on rope and wire soundings, have not been much altered by the incalculably greater number of soundings provided since the 1920's by the sonic fathometer.

Murray and Renard's 1891 volume, *Deep Sea Deposits*, in the *Challenger Report*, was the first treatment of its subject for the entire ocean. Murray and Renard classified and named the major sediment types, delineated their provinces, and provided their successors with most of their subsequent research problems, including the origin of glauconite and manganese nodules. From his comparison of marine sediments with the sedimentary rocks found on land, Murray came to a firm belief that the ocean basins have been a persistent feature of the surface of the earth throughout geologic time. To the International Congress of Zoology meeting in Leiden in 1895, Murray gave the classic statement of the relations between the physical conditions of life and the faunal and floral provinces of the ocean.

In spite of his commitments to the *Challenger Report* and his own researches for it, Murray also found time to use his organizational skills in other areas. From 1883 to 1894 he dredged on the east and west coasts of Scotland in *Medusa*, a specially equipped steam yacht. He founded marine stations at Granton and Millport; the latter is still in operation. Murray was also a founder of the short-lived meteorological observatory atop Ben Nevis. He served as a scientific member of the Fishery Board for Scotland, and he represented the British government at the 1899 Stockholm conference, which founded the International Council for the Exploration of the Sea. After the *Challenger Report* was completed in 1895, Murray organized a survey of the freshwater lochs of Scotland with Frederick Pullar, carrying it to completion (6 vols., 1910) after the latter's drowning.

Murray was able to continue his scientific career after the dissolution of the Challenger Expedition Commission in 1889 because of the independence provided by his marriage (1889) to Isabel Henderson, the only daughter of a Glasgow shipowner. His fortune was further increased by his development of phosphate mining on Christmas Island, in the Indian Ocean. Murray discovered the island's rich deposits when a small specimen was sent to him by a *Challenger* shipmate. He persuaded the British government to annex the island in 1887 and to grant him a lease in 1891. Phosphate exploitation began in earnest about 1900; Murray used his substantial profits to support both a new "Challenger Office" at his home outside Edinburgh and the four-month cruise of the Norwegian fisheries' vessel *Michael Sars* in 1910. The general account by Murray and Johan Hjort of his voyage became the leading textbook of oceanography for three decades after its publication in 1912, and the small volume *The Ocean*, written by Murray himself and published in 1913, served as a popular introduction to the subject. Murray died in an automobile accident in 1914, leaving most of his mining fortune to subsidize oceanographic research.

BIBLIOGRAPHY

I. Original Works. A complete list of Sir John Murray's scientific writings is given in *Proceedings of the Royal Society of Edinburgh*, 35 (1914–1915), 313–317. Murray's major works are his coral-reef theory, "On the Structure and Origin of Coral Reefs and Islands," *ibid.*, 10 (1880), 505–518, and the volume on *Deep Sea Deposits* (Edinburgh, 1891), in *Challenger Reports*, written with A. Renard. Of primary biographical material, the most important items are the bound volume of outgoing letters and the corrected copy of Murray's autobiography (in the form of an obituary booklet); both are in the Mineralogical Library of the British Museum (Natural History). The 126 letters to Alexander Agassiz in the Library of the Museum of Comparative Zoology, Harvard University, and the typescript narrative of the Christmas Island phosphate industry in the possession of the Murray family, are also important. Additional material is in the Public Record Office and the Royal Society.

II. Secondary Literature. There is no biography of Murray. In its absence one must turn to the general histories of oceanography: Margaret Deacon, *Scientists and the Sea 1650–1900* (London, 1971), and Susan Schlee, *To the Edge of an Unfamiliar World* (New York, 1973), and to the obituary articles: G. R. Agassiz, in *Proceedings of the American Academy of Arts and Sciences*, 52 (1917), 853–859; J. Graham Kerr, in *Proceedings of the Royal Society of Edinburgh*, 35 (1914–1915), 305–317; Robert C. Mossman, in *Symons's Meteorological Magazine*, 49 (1914), 45–47; and Sir Arthur Shipley, in *Proceedings of the Royal Society*, 89B (1915–1916), vi–xv, and in *Cornhill Magazine*, 34 (1914), 627–636. The latter is reprinted in Shipley's *Studies in Insect Life and Other Essays* (London, 1917). There are reminiscences by Murray himself in J. Lascelles Graham, "*Old Boys*" *and Their Stories of the High School of Stirling* (Stirling, 1900); other reminiscences are included in Hugh R. Mill, *An Autobiography* (London, 1951), 43–44; Laurence Pullar, *Lengthening Shadows* (privately printed, 1910), *passim*; and A. L. Turner, *Sir William Turner* (Edinburgh, 1919), 496. A recent summary is William N. Boog Watson, "Sir John Murray—A Chronic Student," in *University of Edinburgh Journal*, 23 (1967), 123–138.

Harold L. Burstyn

MŪSĀ IBN MUḤAMMAD IBN MAḤMŪD AL-RŪMĪ QĀḌĪZĀDE. See **Qāḍī Zāda al-Rūmī.**

MŪSĀ IBN SHĀKIR, SONS OF. See **Banū Mūsā.**

MUSHET, DAVID (*b.* Dalkeith, Scotland, 2 October 1772; *d.* Monmouth, Wales, 7 June 1847), *metallurgy.*

Mushet was the son of William Muschet, an iron founder, and Margaret Cochrane. The family name, the origins of which have been traced to the Norman period, appears in various forms, including Mushett. Educated at Dalkeith Grammar School, Mushet frequented his father's and other foundries in the Glasgow area, although his first job (1792) was as an accountant at the Clyde ironworks at Tollcross, near Glasgow, where he began experiments with iron in 1793. Working after business hours, he used the firm's reverberatory furnace and other facilities until he was summarily denied access to them in 1798. His first three papers, published in 1798 in Tilloch's *Philosophical Magazine,* were the product of this period.

Mushet stayed with Clyde until 1800, continuing his research in his own laboratory and reporting the results in thirteen more papers. From 1801 to 1805 he was associated in partnership with William Dixon and Walter Neilson (the father of J. B. Neilson, the inventor of the hot-blast stove) at the Calder Ironworks. During this period he discovered, in the parish of Old Monkland, some ten miles east of Glasgow, the blackband ironstone formation (1801). This discovery, later to put Scotland in a favorable competitive position vis-à-vis England and Wales, brought no advantage to Mushet because its full utilization had to await the introduction of the hot blast (1828–1830), first tried out at the Clyde and Calder Ironworks. Because of the "speculative habits of one partner and the constitutional nervousness of another," Mushet abandoned his interests in Calder and in the blackband leases and left Scotland. Another group of ten papers was published during this period.

From 1805 to 1810 Mushet was associated with the ironworks at Alfreton, Derbyshire. Not much is known of his activities there, apart from his publication of six papers and, apparently, the writing of articles on the blast furnace and blowing engine for Rees's *Cyclopaedia* and on iron for the 1824 supplement to the fourth, fifth, and sixth editions of the *Encyclopaedia Britannica*.

In 1810 Mushet moved to Coleford, in the Forest of Dean. He published nothing during his first six years there, being occupied with a partnership in the Whitecliff Ironworks until he became "dissatisfied with his partners" and withdrew. Between 1816 and 1823 he published eight more papers in *Philosophical Magazine*; but subsequently—apart from three studies on the alloying of copper with iron (1835)—seems to have confined himself to his experiments, to consultation with neighboring ironmasters, and, presumably, to the training of his son Robert. He was also active during the early 1830's in a controversy over the right of the Free Miners of the Forest of Dean to transfer their rights to "foreigners" like himself, that is, those who had not worked a year and a day in the iron mines.

Mushet was granted five patents: one (2,447 of 1800) in the Clyde period, three (3,944 of 1815; 4,248 of 1818, and 4,697 of 1822) from the middle period at Coleford, and one in 1835 (6,908). The technical value of the processes described, especially in the patent of 1800, was the subject of controversy in the trade press of the 1870's, and Robert Forester Mushet proved an aggressive defender of his father.

The publication of Mushet's collected papers in 1840 was initiated by his son David, and Mushet appended considerable material in the form of illustrative notes to the originals. He notes, for example, that in 1798, as is claimed in the 1800 patent, he had asserted that carbon in a gaseous state passes into iron by the mouth or through the pores of the crucible to form steel. His later note states: "This opinion I have long considered the effort of a young mind eager to account for the whole phenomena before it, without that knowledge of the subject which long experience and observation confer" (*Papers*, p. 33). This disclaimer was overlooked in the later arguments, particularly by J. S. Jeans.

In a field in which scientific research had to wage a long battle with the empiricism of the ironmaster, it is not surprising that Mushet's acknowledged contributions to the development of the iron and steel industry were less spectacular than those of his son. It is, indeed, strange that eight years after the publication of the *Papers*, the Institution of Civil Engineers, of which Mushet had been an associate, expressed the hope that his family would collate the papers of one "whose researches were carried on with such indefatigable industry and perseverance and yet of whose labor so little is really known" (*Proceedings, Institution of Civil Engineers*, **7** [1848], 12).

Samuel Smiles credits Mushet with the successful application of the hot-blast stove to anthracite in iron smelting and states that Heath developed his patent cast steel from Mushet's experiments on the "beneficial effects of oxide of manganese on steel." Mushet's work on ferromanganese (1817) may have given Robert Forester Mushet the hint that led to his involvement with the Bessemer process.

BIBLIOGRAPHY

I. Original Works. Mushet's papers were collected by his son David as *Papers on Iron and Steel* (London, 1840). Three other papers are "Blast" and "Blowing Machine," in Abraham Rees, ed., *The Cyclopaedia or Universal Dictionary of the Arts, Sciences, and Literature* (London, 1819) (the *Papers*, p. xix, states that the volume was in the hands of a committee appointed by the iron trade in 1807 "on occasion of the proposed tax on iron"), and "Ironmaking," in *Supplement to the Fourth, Fifth and Sixth Editions* of the *Encyclopaedia Britannica* (London, 1824), 114–127.

II. Secondary Literature. See F. W. Baty, *Forest of Dean* (London, 1952), 98; *Dictionary of National Biography*, repr. ed., XIII, 1326–1327; Henry Hamilton, *The Industrial Revolution in Scotland* (London, 1932; repr. 1966), 179; C. E. Hart, *The Free Miners* (Gloucester, 1953), 136, 272, 290, 506; J. S. Jeans, *Steel* (London, 1880), 23; Fred M. Osborn, *The Story of the Mushets* (London, 1952); H. S. Osborn, *The Metallurgy of Iron and Steel* (Philadelphia, 1869), esp. 124–142, which frequently confuses the work of

Mushet and his son Robert; John Percy, *Metallurgy (Iron and Steel)* (London, 1875), 424–425; and Samuel Smiles, *Industrial Biography* (London, 1863), 141–148.

PHILIP W. BISHOP

MUSHKETOV, IVAN VASILIEVICH (*b.* Alekseevskaya, Voronezh, Russia, 21 January 1850; *d.* St. Petersburg, Russia, 23 January 1902), *geology, geography.*

Mushketov was born to a family of modest means. After the death of his father, Vasily Kuzmich, in 1864, he continued his education at the Gymnasium at Novocherkassk, supporting himself by tutoring children of wealthy parents. While attending the Gymnasium he acquired an interest in natural history, inspired by his teacher, S. F. Nomikosov, that determined his life as a scientist. Recommended by the Gymnasium authorities on the basis of his progress in ancient languages, in 1867 he entered the department of history and philology of St. Petersburg University. He quickly realized his mistake, however, and transferred to the St. Petersburg Mining Institute, where he studied mineralogy and petrography under P. V. Eremeev.

While still a student Mushketov published his first scientific work, a description of volynite. Immediately after graduating from the Mining Institute in 1872 he was sent to the Urals to study deposits of precious stones. Here he continued his work in mineralogy, discovered several arsenical minerals, and journeyed along the Chusovaya River. In 1873 he was assigned to Turkistan and began his many years of research in central Asia. The following year Mushketov worked in the Karatau Mountains and the western spur of the Tien Shan, also investigating the Badamsky Mountains and the plain of the Syr Darya River between Tashkent and Samarkand. In 1875 he traveled from Tashkent through the central Tien Shan to Kuldja; climbed the Talass Ala-Tau, Terskei Ala-Tau, Kungei Ala-Tau, and Zailissky Ala-Tau ranges; traversed the Kirghiz Range, went around the high mountain lakes Son Kul and Issyk Kul; and visited the Dzungarian Ala-Tau. In 1876 he published *Kratky otchet o geologicheskom puteshestvii po Turkestanu v 1875 g.* ("A Short Account of a Geological Trip Through Turkistan in 1875").

In 1876 Mushketov reported to the St. Petersburg Mineralogical Society and the Russian Geographical Society on the scientific results of his trips. For these communications the two societies elected Mushketov a member, and the Geographical Society gave him a silver medal. The Mining Institute invited Mushketov to teach; but in order to do so he had to present a dissertation, and the material on central Asia was too extensive and required more work. In December 1877 he defended his dissertation, "Materialy dlya izuchenia geognosticheskogo stroenia i rudnykh bogatstv Zlatoustovskogo gornogo okruga v yuzhnom Urale" ("Material for the Study of the Geognostic Structure and Ore Resources of the Zlatoust Mountain Region in the Southern Urals"), which contained extensive material on mineralogy and descriptions of ore deposits. He was then appointed adjunct professor, and from 1896 professor, of geology, geognosy, and ore deposits.

In the summer of 1877 Mushketov returned to central Asia, completing a trip from Fergana across the Alai and Trans-Alai ranges to the Pamir, studying the relations between the Tien Shan and the Pamir. In 1878 he investigated the region where the Fergana and the Alai ranges meet, and in 1879 he participated in an expedition to the Turan lowlands.

In 1880 Mushketov made his last trip to central Asia, to study glaciers, and climbed Zeravshan glacier, previously considered inaccessible. Part of the results of these investigations appeared in publications of the Russian Geographical Society and the St. Petersburg Mineralogical Society. These articles brought him a gold medal from the Russian Geographical Society and a prize from the Academy of Sciences. He was also elected an honorary member of the Vienna Geographical Society. In 1884 Mushketov and G. D. Romanovsky published *Geologicheskaya karta Turkestanskogo kraya* ("A Geological Map of the Turkistan Region"), the first summarizing work on the geology of the region.

In 1886 there appeared the first volume of Mushketov's *Turkestan*, which contained a description of the geological structure of that territory. This was widely recognized as a basic work and was awarded a prize by the Academy of Sciences and the St. Petersburg Mineralogical Society. Unfortunately, Mushketov did not finish working out all the material; and the second volume, containing journals of the trip, was published posthumously. The first volume was reprinted in 1915 by his students and for many years served as the basic source on the geology of central Asia.

After completing his expeditions in central Asia, Mushketov undertook diverse projects in various regions of Russia. In 1881 he traveled to the Caucasus to study mineral sources and ore deposits, as well as the Elbrus glacier. During this trip he participated in the fifth Congress of Archaeologists in Tiflis, where he presented a report on nephrite from the tombs at Samarkand.

In 1882 Mushketov became senior geologist of the Geological Committee of Russia. The following year, on instructions from the Committee, he studied the Lipetsk mineral waters and suggested measures to increase their flow. In 1884–1885 he investigated the geological structure of the Kalmuck steppe region along the lower Volga and inspected the Caucasian mineral waters and salt lakes of the Crimea. In 1886 he organized a geological study of the Transcaspian region for his students V. A. Obruchev and K. I. Bogdanovich and established a research program for them.

Mushketov published the results of this work in 1891—"Kratky ocherk geologicheskogo stroenia Zakaspyskoy oblasti" ("A Short Account of the Geological Structure of the Transcaspian Region")—in *Zapiski Imperatorskogo mineralogicheskogo obshchestva*; a supplementary geological map was based on the data of Obruchev and Bogdanovich.

In 1887 a government commission headed by Mushketov was organized to study the causes and consequences of the powerful earthquake that had struck the town of Verny (now Alma-Ata) that year. This work led to his lasting interest in earthquakes. The preliminary data from the investigation were published by Mushketov in a series of articles as early as 1888; and in "Vernenskoe zemletryasenie 28 maya 1887 g." ("The Verny Earthquake of 28 May 1887"), which appeared in 1890, he analyzed the causes of the earthquake and of seismic phenomena in general.

In 1888, at Mushketov's initiative, a seismic commission was organized in the Russian Geographical Society. To collect information on earthquakes it compiled a list of questions, which it sent to all seismically active regions. In a supplement to the list Mushketov wrote an explanatory note on earthquakes, methods of observing them, and the reasons for them. On the basis of the material thus compiled and processed by Mushketov, *Materialy dlya izuchenia zemletryaseny Rossii* ("Materials for the Study of Earthquakes in Russia") was published (1891, 1899).

During this time A. P. Orlov began to compile "Katalog zemletryaseny Rossyskoy imperii" ("Catalog of Earthquakes in the Russian Empire") but died without finishing the work. Mushketov expanded and published this catalog (1893), which became the most valuable source of information on earthquakes in Russia.

In 1900 Mushketov became a member of the Permanent Central Seismic Commission of the Academy of the Sciences, as a representative of the Geological Committee. In the same year he investigated the severe earthquake at Akhalkaliki in the Caucasus.

In 1888–1891 Mushketov published his two-volume *Fizicheskaya geologia* ("Physical Geology"). In 1892 he studied the upper reaches of the Don, with the aim of organizing hydrological research. He worked in the lower Volga region and in the Kirghiz steppe in 1894 and investigated the salt lakes of the Crimea. He revisited the Crimean lakes in 1895 and also studied the plain of the Teberda and Chkhalta rivers and the glaciers of the Caucasus. In his account of this trip (1896), Mushketov described the rocks he had seen and suggested that the gradual formation of the main Caucasus range had been accompanied by dislocations caused by horizontal pressure.

In the region where he worked, Mushketov discovered fifteen previously unknown glaciers. In 1895 Mushketov was elected to the International Commission for the Study of Glaciers. As director of glaciological studies in Russia he attracted a large group of young scientists. Also in 1895 Mushketov published *Kratky kurs petrografii* ("A Short Course in Petrography").

In 1900, having investigated the aftereffects of the Akhalkaliki earthquake, Mushketov traveled to Transbaikalia as consultant on a projected new railroad and then to Paris to take part in the eighth session of the International Geological Congress.

To explain the complex orography and tectonics of central Asia, Mushketov considered it necessary to study its geological history, believing that "every contemporary phenomenon can be fully explained and understood only by the study of its history . . ." (*Kratky otchet o geologicheskom puteshestvii po Turkestanu . . .*, p. 24). Such an approach to the study of the geological structure of central Asia distinguishes Mushketov from his predecessors, who understood its tectonics only through purely external, morphological signs.

On the basis of his own observation and study of its geological history, Mushketov offered a scheme for the geological structure of central Asia which showed that the Tien Shan and Pamir-Alai consist of folded arcs that extend to the northeast and northwest but are bent toward the south by tangential pressure from the north.

Mushketov also distinguished the stratigraphic relations of the formations in the region and described many deposits of useful minerals. He worked out the particular details of the glaciers of central Asia, arguing that the mountain glaciers were retreating; he described the central Asiatic loess; and he provided a classification of quicksands.

Mushketov believed the earth to be so complex that it can be studied only through the aggregate efforts of many sciences. Physical geology, to which his major work is devoted, examines tectonic and erosional processes. He felt, however, that to study processes it is necessary first to understand the position of the earth in space, the hypotheses concerning its origins, and its physical properties, and then to grasp the interplay of tectonic processes, volcanic and seismic phenomena, the record of surface features, and development of the phenomena of denudation. The book is organized according to this plan. The second volume is devoted to a description of the geological activity of the atmosphere, water, and ice.

Mushketov was an adherent of the Kant-Laplace nebular hypothesis, the generally accepted cosmogony of the time. He considered the internal heat of the earth a remnant of the previous molten state. But the idea of a molten state and thin crust was contradicted, as Mushketov stressed, by the phenomena of precession and nutation; yet in the assertion of a solid earth or thick crust, volcanic phenomena remained inexplicable. He saw the solidification of the earth as proceeding both from the center to the periphery and from the periphery to the center. Thus he considered the present structure of the earth to comprise a hard crust and nucleus, with an intermediate belt, possibly of olivine composition.

On the causes of tectonic processes, Mushketov started from the then widely accepted contraction hypothesis. "The main reasons for dislocation and plasticity," he asserted, "are found in the gradual tightening or shrinking of the crust as a consequence of the decreasing volume of the nucleus due to cooling and the loss of volcanic material" (*Fizicheskaya geologia,* vol. 1, p. 599). The contraction of the crust as a consequence of the cooling of the earth was, in Mushketov's opinion, the main cause of seismic phenomena. The statistics of earthquakes and data on the geological structure of various areas led him to distinguish five seismically active regions in Russia: the Caucasus, Turkistan, Transbaikalia, Altai, and Kamchatka.

In the Mining Institute Mushketov taught geology and physical geography for twenty-five years. He also taught in the Institute of Communications Engineers, in the Higher Courses for Women, and in the Historical-Philological Institute, as well as giving many public lectures. Among his students were V. A. Obruchev, K. I. Bogdanovich, and L. I. Lutugin.

Mushketov was president of the physical geography section of the Russian Geographical Society, a member of the St. Petersburg Mineralogical Society, a member of the Council of the St. Petersburg Biological Laboratory, and a representative of Russia at the International Commission for Research on Glaciers.

His work was especially influential in the study of the geology of central Asia, tectonics, seismology, and glaciology.

BIBLIOGRAPHY

I. ORIGINAL WORKS. His works include "Volynit" ("Volynite"), in *Zapiski Imperatorskogo mineralogicheskogo obshchestva,* 2nd ser., **7** (1872), 320–329; *Kratky otchet o geologicheskom puteshestvii po Turkestanu v 1875 g.* ("A Short Account of a Geological Journey Through Turkistan in 1875"; St. Petersburg, 1876); "Geologicheskie issledovania v Kalmytskoy stepi v 1885 g." ("Geological Research in the Kalmuck Steppe in 1885"), in *Izvestiya Geologicheskogo komiteta,* **5** (1886), 203–233; *Turkestan,* 2 vols. (St. Petersburg, 1886–1906); *Fizicheskaya geologia* ("Physical Geology"), 2 vols. (St. Petersburg, 1888–1891); "Vernenskoe zemletryasenie 28 maya 1887 g." ("The Verny Earthquake of 28 May 1887"), in *Trudy Geologicheskogo komiteta,* **10,** no. 1 (1890), 1–154; *Zemletryasenia, ikh kharakter i sposoby nablyudenia. . . .* ("Earthquakes, Their Character and Methods of Observing Them. . . ."; St. Petersburg, 1890); "Kratky ocherk geologicheskogo stroenia Zakaspyskoy oblasti" ("A Short Sketch of the Geological Structure of the Transcaspian Region"), in *Zapiski Imperatorskogo mineralogicheskogo obshchestva,* **28** (1891), 391–429; and "Katalog zemletryaseny Rossyskoy imperii" ("Catalog of Earthquakes in the Russian Empire"), in *Zapiski Russkogo geograficheskogo obshchestva,* **26** (1893), a completion of the work begun by A. P. Orlov.

II. SECONDARY LITERATURE. See D. N. Anuchin, "I. V. Mushketov i ego nauchnye trudy" ("I. V. Mushketov and His Scientific Work"), in *Zemlevedenie,* **1,** 9, no. 1 (1902), 113–133; B. A. Fedorovich, "I. V. Mushketov kak geograf" ("I. V. Mushketov as a Geographer"), in *Izvestiya Akademii nauk SSSR,* Geog. ser., no. 1 (1952), 63–67; A. P. Karpinsky, "Pamyati I. V. Mushketova" ("Memories of I. V. Mushketov"), in *Gornyi zhurnal,* **1,** no. 2 (1902), 203–207; V. A. Obruchez, "Ivan Vasilievich Mushketov," in *Lyudi russkoy nauki* ("People of Russian Science"; Moscow, 1962), 54–62; and L. A. Vayner, *Ivan Vasilevich Mushketov i ego rol v poznanii geologii Sredney Azii* ("Ivan Vasilievich Mushketov and His Role in the Knowledge of the Geology of Central Asia"; Tashkent, 1954).

IRINA V. BATYUSHKOVA

MUSSCHENBROEK, PETRUS VAN (*b.* Leiden, Netherlands, 14 March 1692; *d.* Leiden, 19 September 1761), *physics.*

Musschenbroek belonged to a well-known family of brass founders and instrument makers who were originally from near Tournai in Hainaut and who settled at Leiden in the latter part of the sixteenth century. His grandfather, Joost Adriaensz (1614–1693), manufactured lamps, especially church lamps, and was also a gauger of weights and measures. He was succeeded in his craft by his sons Samuel (1639–1681) and Johan (1660–1707). In accordance with the spirit of the times, they turned their skill to the making of scientific apparatus such as air pumps, microscopes, and telescopes. Christiaan Huygens mentions one of them as a maker of microscopes (letters of 1678 and 1683, *Oeuvres complètes,* VIII, 64, 422; see also *ibid.,* XXII, 762). Swammerdam used a microscope made by Samuel (Boerhaave, in his preface to the *Biblia naturae,* 1737); and the anatomist Regnier de Graaf employed anatomical injection spouts also made by him. Leeuwenhoek's aquatic microscope (letter to the Royal Society of 12 January 1689) was made by Johan. Many of Johan's instruments are still preserved; the extant instrument made by Samuel is an air pump of 1675.

In 1685 Johan married Maria van der Straeten; they had two sons, Jan (1687–1748) and Petrus. Jan, who succeeded his father in the workshop, obtained a good education at the Latin school and studied under Boerhaave. He was offered teaching positions but preferred to remain an instrument maker. This may well have been due to his friendship and collaboration with 'sGravesande, who based his physics lectures on experiments and had many of his instruments constructed by Jan. They can be studied in 'sGravesande's *Physices elementa mathematica experimentis confirmata* (Leiden, 1720–1721; 2nd ed., 1742); some seventy-five of them still exist. The popularity of this book brought numerous orders to Jan from universities and amateurs; many of his instruments were imitated, for instance, by George Adams for the cabinet of George III of England. His workshop on the Rapenburg (now no. 66) was famous. Albrecht von Haller, who visited him between 1725 and 1727, especially admired Jan's magic lantern. Apart from catalogs of his works—as many as 200 items—he published only a description of new air pumps and of "agreeable and instructive" experiments to be performed with them (1736).

Petrus, not yet fifteen when his father died, owed his further education to his brother. He studied at the University of Leiden and in 1715 received his doctorate in medicine. He then made a study trip to London, where he met Desaguliers, then famous as lecturer and demonstrator of scientific experiments, who visited Holland in 1730. Back in Leiden he practiced medicine and shared with his brother both the friendship and the philosophy of 'sGravesande. In 1719 Petrus received his doctorate in philosophy and accepted a professorate in mathematics and philosophy at Duisburg, where in 1721 he also became extraordinary

professor of medicine. From 1723 to 1740 he occupied the chair of natural philosophy and mathematics at Utrecht, in 1732 also holding the chair of astronomy. Here he lectured on experimental philosophy, presenting views like those of 'sGravesande and Newton and often using apparatus made by his brother.

Musschenbroek became increasingly famous, especially because of his lecture notes collected in ever larger volumes; the *Epitome elementorum physico-mathematicorum conscripta in usus academico* (Leiden, 1726), *Elementa physicae* (Leiden, 1734), *Institutiones physicae* (Leiden, 1748), and the posthumous two-volume *Introductio ad philosophiam naturalem,* edited by J. Lulofs (Leiden, 1762), were widely used and translated into Dutch, English, French, and German. He refused offers of academic chairs at Copenhagen in 1731 and at Göttingen in 1737; but at the end of 1739 he accepted a chair at Leiden, where he taught from 1740 until his death. In 1742, after the death of 'sGravesande, Musschenbroek became his logical successor in the teaching of experimental physics. The excellence of his lectures maintained the reputation that Leiden had acquired under Boerhaave and 'sGravesande, and students interested in experimentation came from all parts of Europe. One of them was Jean-Antoine Nollet (in 1736), who became the leading exponent of this school in France.

Primarily a lecturer and author, Musschenbroek tended more to supervise than to become involved in the construction of apparatus. He devised many of his experiments, in the process consulting records of other experimenters, among them those of the Accademia del Cimento. Musschenbroek translated their accounts into Latin, adding reports concerning his own work (1731). Many of his instruments were made by his brother Jan; but those of other craftsmen —for instance, Jan Paauw—were also employed. The Musschenbroeks never made barometers or thermometers; these were supplied by Gabriel Daniel Fahrenheit in Amsterdam and by others.

The experiments can be studied in Musschenbroek's books, which contain many fine illustrations; they deal with the mechanics of rigid bodies, air pressure, heat, cohesion, capillarity, phosphorescence, magnetism, and electricity. Many of these experiments have become classics in elementary instruction. One of the better-known apparatuses is the pyrometer (the name was given by Musschenbroek), first described in the *Tentamina* of 1731; it consists of a horizontal metal bar fixed at one end and connected at the other end to wheelwork that shows the expansion of the bar when it is heated. It was originally used without a thermometer, which was not mentioned until the

Introductio of 1762. Musschenbroek's best-known experiment is that with the Leyden jar, discussed below.

Underlying Musschenbroek's lectures demonstrated with experiments was the experimental philosophy. This philosophy, which he proclaimed in Holland along with Boerhaave and 'sGravesande, was set forth in their books and in academic lectures such as Musschenbroek's inaugural address at Utrecht, *Oratio de certo methodo philosophiae* (Leiden, 1723). The principal source of inspiration was Newton; but Galileo, Torricelli, Huygens, Réaumur, and others were important to this school. Since the mind, Musschenbroek states in his *Elementa physicae* (1734), has no innate idea of what bodies and their qualities are, we can obtain knowledge about them only by observation and experiments. But we must be extremely careful, use good instruments, and take into consideration all circumstances—atmospheric pressure, temperature, locality, and weather. Thus we can discover the laws that govern the behavior of bodies, provided the results of experiments, repeated over and over again, are the same, and specific causes are admitted only when the phenomena investigated leave no doubt. The stress is therefore on induction; but deduction, for example, by means of mathematics, is admissible, as Newton had shown, provided such deductions are constantly tested by experiment. The success of such reasoning on the basis of careful experimentation finds its guarantee in the infinite wisdom of the Supreme Being. This philosophy inspired the founding of many amateur societies in Holland and abroad for experimental study.

Musschenbroek is generally credited with originating the Leyden jar. He knew that a charged conductor surrounded by air loses its charge very rapidly, especially in a rainy climate like that of Holland. He had a gun barrel suspended by two silk lines and the barrel charged by means of a rapidly rotating glass globe rubbed by hand. A brass wire from the barrel led a few inches through a cork into a bottle and extended into water in the bottle. Thus the water was charged. Musschenbroek's assistant, Andreas Cunaeus, accidentally took hold of the bottle, thus giving it the necessary outer coating. Then he touched the wire with his other hand—and received a fearful shock. He had unintentionally experienced the effect of a true capacitor.

This accident occurred in January 1746. Musschenbroek reported the experiment to Réaumur, who showed the letter to Nollet. Musschenbroek's other collaborator, Jean Nicolas Sebastien Allamand (who later wrote a biography of 'sGravesande), wrote directly to Nollet. The latter, quite excited about this "Leiden experiment," reported to the Académie des Sciences at its April meeting. Nollet continued to write on and repeat the "expérience nouvelle mais terrible" in a sensational way. Thus the *bouteille de Leyde* became widely known during 1746. Musschenbroek first described it in the *Institutiones physicae* of 1748.

As to the priority of the experiment, in 1744 Georg Matthias Bose in Wittenberg had published the drawing of "fire" from electrified water in a glass, an experiment that Musschenbroek knew and wanted to repeat. Early in 1745 Allamand had received a terrible shock in the same way that Cunaeus did the following year, and he had reported on it in the *Philosophical Transactions* of 1746. And on 4 October 1745 J. G. von Kleist, dean of the cathedral at Kammin (now Kamień Pomorski), made a similar experiment. This was reported to other Germans interested in electricity and was published by J. G. Krüger in his *Geschichte der Erde* (Halle, 1746), but it passed unnoticed for a long time. Yet it was Musschenbroek's communication which, through Nollet, made the capacitor known, so that there were soon improvements in its construction, and Benjamin Franklin analyzed the experiments on "M. Musschenbroek's wonderful bottle" in his third and fourth letters to Peter Collinson (1747, 1748). In them he established that the charge is not in the wire or the water, but in the glass. He also corresponded with Musschenbroek and in 1761 visited him in Leiden.

Musschenbroek married Adriana van de Water (d. 1732) in 1724 and Helena Alstorphius in 1738. A son, Jan Willem (1729–1807), wrote the family history.

BIBLIOGRAPHY

I. ORIGINAL WORKS. There is no modern critical bibliography of the Musschenbroeks' works. For a preliminary listing see D. Bierens De Haan, *Bibliographie néerlandaise historique-scientifique* (Rome, 1883; repr. Nieuwkoop, 1960), 202–204. Petrus' main literary production consists of a gradual extension of his Utrecht lecture notes published first in the already mentioned *Epitome* of 1726. These include *Beginsels der natuurkunde* (Leiden, 1736; 2nd ed., 1739); *Essai de physique . . . avec une description de nouvelles sortes de machines pneumatiques et un recueil d'expériences par Mr. J. Musschenbroek*, translated by P. Massuet, 2 vols. (Leiden, 1736–1739; 2nd ed., 1751); *The Elements of Natural Philosophy*, translated by J. Colson, 2 vols. (London, 1744); *Grundlehren der Naturwissenschaft nach der zweiten lateinischen Ausgabe*, translated by J. C. Gottsched (Leipzig, 1747); and *Cours de physique expérimentale et mathématique*, translated by Sigault de la Fond, 3 vols. (Paris, 1769), with a preface by J. Lulofs, a colleague of Musschenbroek's at Leiden, and a description of the relation of these different books to each other.

See also *Disputatio medica inauguralis de aeris praesentia in humoribus animalibus* (Leiden, 1715; 2nd ed., 1749); *Oratio de certo methodo philosophiae* (Leiden, 1723), his inaugural address at Utrecht—on its influence on the spread of Newtonianism in France see Brunet, below; *Physicae experimentales et geometricae, de magnete, tuborum capillarium vitreorumque speculorum attractione magnitudine terrae, cohaerentia corporum firmorum dissertationes ut et ephemerides meteorologicae ultrajectinae (anni 1728)* (Leiden, 1729; Vienna–Prague–Trieste, 1754); *Tentamina experimentorum naturalium captorum in Accademia del Cimento . . . quibus commentarios, nova experientia, et orationem de methodo instituendi experimenta physica additit P. v. M.*, 2 vols. (Leiden, 1731; Vienna–Prague–Trieste, 1756), French version in *Collection académique* (Dijon–Auxerre, 1755); *Oratio de mente humana semet ignorante* (Leiden, 1740), his inaugural address at Leiden; *Oratio de sapientia divina* (Leiden–Vienna, 1744); *Institutiones logicae, praecipue comprehentes artem argumentandi* (Leiden, 1746; Venice, 1763); and *Compendium physicae experimentales conscripta in usus academicos*, J. Lulofs, ed. (Leiden, 1762). Jan Musschenbroek's writings are *Liste de diverses machines de physique, de mathématique, d'anatomie et de chirurgie* (Leiden, 1736); and *Description de nouvelles sortes de machines pneumatiques tant doubles que simples* (Leiden, 1738), also published in Dutch.

The only extant instrument of Samuel Musschenbroek, the air pump of 1675, was made for Professor Burchard de Volder and is in the Leiden Museum of Science, which also has a considerable number of Johan's instruments, identifiable by the trademark of Samuel and Johan, an oriental lamp. The collection of Johan's instruments includes the aquatic and other microscopes, as well as air pumps. The instruments gathered by Jan for 'sGravesande have been to a great extent preserved at Leiden. Many other instruments made by Jan may still exist, but they cannot be identified with certainty because he seldom used the family trademark (see the books by Rooseboom, Crommelin, and van der Star). Many of Jan's models are illustrated, however, in the *Physices elementa* of 1720–1721 and 1742. The university museum at Utrecht has an air pump and two microscopes by Jan, the first pyrometer made by Petrus, and three cylinders of Petrus' friction meter. Instruments made by the Musschenbroeks and their imitations exist elsewhere—for instance, in the cabinet of George III. The *Catalogus van Mathematische, Physische, Astronomische, Chirurgische, en andere Instrumenten te Bekomen in de Fabricq van Mr. J. H. Onderdewyngaart Canzius te Delft* (1804) lists a number of Jan's apparatuses, with reference to the pictures in 'sGravesande.

Another catalog that mentions instruments of Musschenbroek is *Collectio exquisitissima Instrumentorum in Primis ad Physicam experimentalem Pertinentium, quibus, dom vivebat, Usus fuit Celeberrimus Petrus van Musschenbroek . . . quorum Auctio fiet . . . ad Diem 15 Martii et Seqq. 1762* (Leiden, 1762).

In the university library at Leiden there is much MS material by Petrus and some by Jan. They are listed in J. Geel, *Catalogus librorum manuscriptorum qui inde ab 1741 bibliotheca Lugduno Batavae accesserunt* (Leiden, 1852), 221–223. The municipal archives of Utrecht contain Petrus' handwritten copy of the list of instruments he was authorized to buy for the Theatrum Physicum and the observatory at Utrecht.

II. SECONDARY LITERATURE. No full-length modern biography of the Musschenbroeks exists. A concise sketch of their lives, based partially on the MS history by Jan Willem van Musschenbroek (Petrus' son), together with a description of their extant instruments, can be found in M. Rooseboom, *Bydrage tot de geschiedenis der instrumentmakerskunst in de Noordelyke Nederlanden tot omstreekts 1840* (Leiden, 1950); supplemented by C. A. Crommelin, *Descriptive Catalogue of the Physical Instruments in the National Museum of the History of Science at Leyden* (Leiden, 1951); "Leidsche leden van het geslacht Musschenbroek," in *Leidsch Jaarboekje* (1939), 135–149; and "Huizen der Leidsche van Musschenbroeks," *ibid.* (1945), 127–133; and P. van der Star, *Descriptive Catalogue of the Simple Microscopes in the National Museum of the History of Science at Leyden* (Leiden, 1953). See also H. J. M. Bos, *Mechanical Instruments in the Utrecht University Museum* (Utrecht, 1968), where 22 of the *ca.* 100 mentioned instruments are copies of instruments described in Musschenbroek's books; and E. J. Dijksterhuis, "Uit het Utrechts verleden der fysica," in *Nederlands tijdschrift voor natuurkunde*, **22** (1956), 163–180; and A. Savérien, *Histoire des philosophes*, VI (Paris, 1768); A. N. Condorcet, "Éloge de Musschenbroek," in *Oeuvres*, II (1847), 125–127; and F. Boerne, *Nachrichten von den vornehmsten Leben und Schriften jetz lebender berühmter Aertze und Naturforscher*, I (Wolfenbüttel, 1749), 529–541.

On Musschenbroek's influence on the spread of Newtonianism see P. Brunet, *Les physiciens hollandais et la méthode expérimentale en France au XVIII siècle* (Paris, 1928), 68–100; and his *L'introduction des théories de Newton en France au XVIIIe siècle* (Paris, 1931), 124, 326. The invention of the Leyden jar is discussed in C. Dorsman and C. A. Crommelin, "The Invention of the Leyden Jar," in *Janus*, **46** (1957), 275–280; F. M. Feldhaus, *Die Erfindung der elektrischen Verstärkungsflasche durch E. J. von Kleist* (Heidelberg, 1903); and J. L. Heilbron, "G. M. Bose: The Prime Mover in the Invention of the Leyden Jar?" *Isis*, **57** (1966), 264–267. The original publications on this subject are G. M. Bose, *Tentamina electrica in academiis regiis Londensi et Parisina primum habita omni studio repitata . . .* (Wittenberg, 1744), 64; *Tentamina electrica tandem aliquando hydraulicae chymiae et vegetabilibus utilia* (Wittenberg, 1747), 36–37; J. A. Nollet, "Observations sur quelques nouveaux phénomènes d'électricité," in *Histoire de l'Académie des sciences* for 1746, 1–33; and "Recherches sur la communication de l'électricité," *ibid.*, 447. See also J. Priestley, *The History and Present State of Electricity* (London, 1767); P. F. Mottelay, *Bibliographical History of Electricity and Magnetism* (London, 1922); and I. B. Cohen, *Benjamin Franklin's Experiments* (Cambridge, Mass., 1941).

D. J. STRUIK

MYDORGE, CLAUDE (*b.* Paris, France, 1585; *d.* Paris, July 1647), *mathematics, physics.*

Mydorge belonged to one of France's richest and most illustrious families. His father, Jean Mydorge, *seigneur* of Maillarde, was *conseiller* at the Parlement of Paris and judge of the Grande Chambre; his mother's maiden name was Lamoignon. He decided to pursue a legal career and was, first, *conseiller* at the Châtelet, then treasurer of the *généralité* of Amiens. In 1613 he married the sister of M. de la Haye, the French ambassador at Constantinople. His duties as treasurer left him sufficient time to devote himself to his passion, mathematics.

About 1625 Mydorge met Descartes and became one of his most faithful friends. In 1627, to aid Descartes in his search for an explanation of vision, Mydorge had parabolic, hyperbolic, oval, and elliptic lenses made for him. He also determined and drew their shapes with great precision. He subsequently had many lenses and burning glasses made. It was said that altogether he spent more than 100,000 écus for this purpose.

After a thorough study of Descartes's *Dioptrique*, Mydorge at first criticized the book on various points but later completely adopted his friend's theories. Fermat, however, in 1638, wrote to Mersenne to refute the *Dioptrique*. On 1 March 1638 (see *Oeuvres de Descartes*, C. Adam and P. Tannery, eds., II, *Correspondance*, 15–23) Descartes sent a long letter to Mydorge—he knew that the latter had openly taken his side in the dispute—in which he provided him with the seven documents relating to the case and asked him to be judge and intermediary. He also asked Mydorge to make a copy of the letter and send the original to Fermat's friends Étienne Pascal and Roberval. (It should be noted that Fermat's correspondence indicates that Pascal and Roberval were in no way his friends.) Through the good offices of Mydorge and Mersenne, Descartes and Fermat were reconciled.

Mydorge was held in high regard by other famous contemporaries; for instance, on 2 March 1633 Peiresc wrote from Aix to Gassendi, who was then at Digne: "If you have any special observations by M. Mydorge, you would do me a great favor by communicating them to me" (see Galileo Galilei, *Opere* [Edizione nazionale], XVIII [Florence, 1966], 430).

Mydorge's work in geometry was directed to the study of conic sections. In 1631 he published a two-volume work on the subject, which was enlarged to four volumes in 1639. The four volumes were reprinted several times under the title *De sectionibus conicis*. A further portion of the work, in manuscript, is lost. It seems that two English friends of the

Mydorge family, William Cavendish, duke of Newcastle, and Thomas Wriothesley, earl of Southampton, took it to England, where apparently it disappeared.

In his study of conic sections Mydorge continued the work of Apollonius, whose methods of proof he refined and simplified. Among the ways of describing an ellipse, for example, two from volume II may be cited. According to the first definition, an ellipse is the geometric locus of a point of a straight line the extremities of which move along two fixed straight lines. (This definition had already been demonstrated by Stevin, who attributed it to Ubaldi; actually, it goes back to antiquity, as Proclus indicates in his commentaries on Euclid.) According to the second definition, the ellipse can be deduced from a circle by extending all its ordinates in a constant relationship. In the same book Mydorge asserts that if from a given point in the plane of a conic section radii to the points of the curve are drawn and extended in a given relationship, then their extremities will be on a new conic section similar to the first. This statement constitutes the beginnings of an extremely fruitful method of deforming figures; it was successfully used by La Hire and Newton, and later by Poncelet and, especially, by Chasles, who named it *déformation homographique*.

Mydorge posed and solved the following problem in volume III: "On a given cone place a given conic section"—a problem that Apollonius had solved only for a right cone. Mydorge was also interested in geometric methods used in approximate construction, such as that of the regular heptagon. Another problem that Mydorge solved by approximation—although he did not clearly indicate his method—was that of transforming a square into an equivalent regular polygon possessing an arbitrary number of sides.

Mydorge's works on conic sections contain hundreds of problems published for the first time, as well as a multitude of ingenious and original methods that later geometers frequently used, usually without citing their source. The collection of Mydorge's manuscripts held by the Académie des Sciences contains more than 1,000 geometric problems. Finally, it should be noted that the term "parameter" of a conic section was introduced by Mydorge.

A friend of Descartes and an eminent geometer, Mydorge was also well versed in optics. He possessed a lively curiosity and was open to all the new ideas of his age. Like Fermat, he belonged to that elite group of seventeenth-century scientists who pursued science as amateurs but nevertheless made contributions of the greatest importance to one or more fields of knowledge.

BIBLIOGRAPHY

I. ORIGINAL WORKS. Mydorge's first major writing, *Examen du livre des Récréations mathématiques* (Paris, 1630; repr. 1643), with notes by D. Henrion, is a commentary on *Récréations mathématiques* (Pont-à-Mousson, 1624), published under the pseudonym H. Van Etten (actually Leurechon).

The second was *Prodromi catoptricorum et dioptricorum, sive conicorum operis . . . libri duo* (Paris, 1631), enlarged to *Conicorum operis . . . libri quattuor* (Paris, 1639, 1641, 1660), also issued as *De sectionibus conicis, libri quattuor* (Paris, 1644), which Mersenne inserted in his *Universae geometriae, mixtaeque mathematicae synopsis . . .* (Paris, 1644).

A selection of the geometry problems preserved in Paris was published by C. Henry in *Bullettino di bibliografia e di storia delle scienze matematiche e fisiche*, **14** and **16**. Mydorge's son assembled three short treatises from his father's MSS—*De la lumière*, *De l'ombre*, and *De la sciotérique*—but all trace of them has been lost.

II. SECONDARY LITERATURE. See the following, listed chronologically: C. G. Jöcher, *Allgemeines Gelehrten-Lexicon*, III (Leipzig, 1751), 787; *Biographie universelle*, XXIX (Paris, 1860), 666; *La grande encyclopédie*, XXIV (Paris, 1899), 657; M. Chasles, *Aperçu historique sur l'origine et le développement des méthodes en géométrie* (Paris, 1889), 88–89; and M. Cantor, *Vorlesungen über Geschichte der Mathematik*, II (Leipzig, 1913), 673–674, 768–769.

PIERRE SPEZIALI

MYLON, CLAUDE (*b.* Paris, France, *ca.* 1618; *d.* Paris, *ca.* 1660), *mathematics*.

Mylon's place in the history of science derives from the service he provided in facilitating communication among more learned men in the decade from 1650 to 1660. He was the third son of Benoist Mylon, counselor to Louis XIII and Controller-General of Finance; he himself was admitted to the bar as an advocate before Parlement in 1641, even though he lacked two years of being twenty-five, the legal age of majority.

As early as 1645 Mylon had become concerned with mathematics, making written notes of new Cartesian mathematical problems. He was also in contact with Mersenne, Debeaune, and Roberval, and when Schooten passed through Paris he was able to transmit a considerable amount of new information to him. Mylon also served as secretary to the "Académie Parisienne," a continuation of the Mersenne group, under the direction of F. le Pailleur, which in 1654 received Pascal's famous "Adresse." Mylon achieved a certain importance when the death of Pailleur, in November 1654, left the papers of the society at his disposal; it was thus he who told Schooten (who told

Huygens) of Fermat's and Pascal's problems and solutions concerning games of chance. He also forwarded to Holland Fermat's and Frenicle's problems in number theory. In 1655 Huygens, who was making his first trip to France, visited Mylon; the following year he suggested the "commerce scientifique" that provides the chief documentation of Mylon's career.

Mylon maintained a number of rather delicate relationships with other mathematicians. He had access to Pascal in his retirement (although to a lesser degree than did Carcavi), and while his affection for Conrart threatened his friendship with Roberval, the latter continued to make use of him as an intermediary. He was less happy in his two attempts at personal achievement: in 1658 he hazarded his own solution to the quadrature of the cubic curves known as the "perles de M. Sluse" and in January 1659, in the wake of the debate provoked by Pascal, he proposed to prove Wren's solution of the length of the cycloid. These efforts stand as a monument to his inadequacies as a mathematician, and it is with them that all mention of Mylon by Huygens stops. No publication by him is known.

BIBLIOGRAPHY

On Mylon and his work, see J.-B. du Hamel, *Astronomia physica*, . . . *Accessere P. Petiti observationes*. . . . (Paris, 1660), 12, which includes an account of Pierre Petit's pamphlet on the observation made by Mylon and Roberval of the solar eclipse of 8 Apr. 1652.

See also C. Adam and P. Tannery, eds., *Oeuvres de Descartes*, IV (Paris, 1901), 232, 397, which deals with the problem of the "trois bâtons" and Roberval's "Aristarchus."

See L. Brunschvicg, P. Boutroux, and F. Gazier, eds., *Oeuvres de Blaise Pascal*, IX (Paris, 1914), 151–156; the letter referred to here (Mylon to Pascal, 27 Dec. 1658) is at the Bibliothèque Nationale, Paris, Res. V 859, with a demonstration by Mylon of "the equality of the cycloid and its partner."

There are numerous references to Mylon in Huygens' correspondence, as well as letters from him, in *Oeuvres complètes de Christiaan Huygens*, 22 vols. (The Hague, 1888–1950); see esp. I, 517, for Roberval's demonstration on the surface of spherical triangles; II, 8–25, for Frenicle's results on compatible numbers; "Propositio Domini Wren Angli. Demonstrata a Claudio Mylon die 26 Januarii 1659," II, 335; and "La quadrature des perles de M. Sluse par Claude Mylon. En juin 1658," II, 337. Mylon's role in the problem of games of chance is discussed in "Avertissement," XIV, 4–9. See also *The Correspondence of H. Oldenburg*, I (London, 1965), 225.

PIERRE COSTABEL

NAEGELI, CARL WILHELM VON (*b.* Kilchberg, near Zurich, Switzerland, 27 March 1817; *d.* Munich, Germany, 10 May 1891), *botany, microscopy.*

The son of a physician, Naegeli was educated at a private school, the Zurich Gymnasium, and Zurich University. His enthusiasm for science was stimulated by Oken's lectures on zoology, and in 1839 he gave up medicine at Zurich to study botany under Alphonse de Candolle at Geneva. In 1840 he received the doctorate for his study of Swiss *Circia*, a work marked by the same precision and detail as his later studies. There followed a summer semester in Berlin when he studied Hegel's philosophy. Hegel had been dead eleven years, but his writings were still much admired. Although in retrospect Naegeli claimed that he had found nothing useful in Hegelianism, his work is characterized by a Hegelian search for universal concepts which at times seems pedantic and misdirected.

In the autumn of 1842 Naegeli left Berlin for Jena, where he worked with Schleiden. Together they published the new, and short-lived, journal *Zeitschrift für wissenschaftliche Botanik.* Naegeli's eighteen months in Jena were highly productive. From 1845 to 1852 he worked in Zurich, first as *Privatdozent,* then as assistant professor. There his collaboration with Carl Cramer in plant physiology research began in 1850. This work was continued when he became full professor at Freiburg im Breisgau in 1852. Finally in 1857 he accepted the chair of botany in the University of Munich. There, in 1890, he celebrated the fiftieth anniversary of his degree.

When Naegeli arrived in Jena, Schleiden had just published his famous *Grundzüge der wissenschaftlichen Botanik*, which begins with a lengthy critique of the philosophy of science and goes on to enunciate the Schleiden-Schwann theory of free cell formation, the analogy of cryptogamous spores with phanerogamous pollen, and the assertion that the embryo in phanerogams is the transformed tip of the pollen tube. Like Schleiden, Naegeli began with a philosophical essay, "Über die gegenwärtige Aufgabe der Naturgeschichte, insbesondere der Botanik," in which he eschewed compilations of empirical data, since science is concerned not with the changing characteristics of individuals but with the unchanging laws relevant to all individuals. When he sought to practice science in harmony with this definition he ran into difficulties. His early studies of cell division (1844, 1846) appeared to show two types of cell formation—free cell formation and division of preexisting cells. At first he found the latter process in all cells of algae and diatoms and in all spore mother and pollen mother cells of lower and higher plants. Two years later he altered this

decision, making a simple distinction between reproductive tissues, in which free cell formation rules, and vegetative tissues, in which cell division rules. Meanwhile a more decisive stand in favor of cell division had been taken by Unger.

These studies of cell formation illustrate Naegeli's striving for general laws, the strong influence of Schleiden on him, and his eye for detail. Thus he realized that in cell division the wall formed between the two daughter cells is the result, not the cause, of cell division. The latter he recognized as the function of the whole protoplast. These studies also gave valuable support to Robert Brown's assertion of the invariable presence of a single nucleus in every cell, and it is to his and Mohl's credit that the protoplasmic lining of the cell (Naegeli's *Schleimschicht*) was recognized as the living substance.

Naegeli's failure in 1846 to limit correctly the application of Schleiden's theory of cell formation must be balanced against his brilliant achievement in 1845, when he studied apical growth. This work culminated thirteen years later in his researches into the formation of tissues in the stems and roots of vascular plants, which constituted a major contribution to plant anatomy. For his study of apical growth he began with simple cases—from the Bryophyta—and in his thorough manner he traced back the various tissues and organs in a cell lineage to the apical cell. The regular way in which this cell cuts off daughter cells in either one, two, or three rows gave Naegeli an example of the operation of laws which he could represent mathematically and which for him pointed the way to absolute concepts of the sort characteristic of science proper. It was no accident that he used the phrase *wissenschaftliche Botanik* in the title of two of his series of papers, nor was it uncharacteristic for him to represent apical cell division in terms of equations. This was the realization of Schleiden's hope that the development of plants would one day be expressed by mathematical laws. Naegeli's success in thus tracing cell lineages had a profound impact on the botanists of his time.

Extending these studies to the vascular cryptogams and the angiosperms, Naegeli arrived at the important distinction between formative tissues (*Bildungsgewebe*), which he divided into cambia and meristems, and structural tissues (*Dauergewebe*) no longer actively multiplying. In the stems and roots of plants was a strain of cells (cambial and meristematic) which remained untouched by differentiation and whose origin could be traced back to the original "foundation cell" or zygote. Unfortunately he did not draw the same conclusion from these findings as did Weismann from his study of the Coelenterata.

Naegeli's conception of an hereditary and a nutritive component in every cell derived instead from the facts of sexual reproduction.

In 1844 Naegeli discovered the antherozoids of ferns and in 1850 those of the Rhizocarps. He also discovered the protonema and archegonia in *Ricciocarpus*, but it was left for Hofmeister to arrive at the correct analogies between these organs and those of the phanerogams. It seems that Naegeli was too much under Schleiden's influence. How else could he have rejected the discovery of antherozoids in *Fucus* by Decaisne and Thuret in 1849?

Naegeli made a major contribution to the field of cell ultrastructure when he published his detailed study of starch grains in 1858. Here he arrived at his micellar theory, according to which such amorphous substances as starch and cellulose consist of building blocks, which he later termed "micelles," packed in crystalline array. Each micelle was an aggregate of up to nine thousand molecules ("atoms" in Naegeli's terminology) of starch. Water could penetrate between the micelles, and new micelles could form in the interstices between old micelles. The swelling property of starch grains and their growth by intussusception were thus based on a molecular-aggregate model, which he also applied to the cellulose of the cell wall. Three years later (1861) he reported on the anisotropy of starch grains and of cell walls from observations with the polarimeter, which he took as supporting his assumption of crystalline ultrastructure. Other botanists, notably Strasburger, put a different interpretation upon this anisotropy.

Nevertheless, Naegeli's micellar theory stimulated studies of ultrastructure and initiated a tradition of the study of botanical ultrastructure in Germany and Switzerland, a tradition continued by Hermann Ambronn in Jena and Alfred Frey-Wyssling at the Polytechnic in Zurich. Naegeli's work also fostered a belief in micellar aggregates at the expense of the macromolecular concept; a lengthy debate ensued in the 1920's and 1930's between the concept of a long chain polymer and an aggregate or micell of several shorter chains.

In his search for general laws, Naegeli used his micellar theory, which was based on carbohydrate products, to arrive at a molecular-aggregate model of the hereditary substance, its expression, growth, and modification. This inspired piece of deductive thinking appeared in his famous *Mechanisch-physiologische Theorie der Abstammungslehre* (1884), where the important distinction is made between the nutritive trophoplasm and the hereditary idioplasm—the egg being rich in trophoplasm, the spermatozoon almost completely without it. Since paternal and maternal characteristics are transmitted approximately equally, they must be carried by the idioplasm and not by the trophoplasm. Other biologists, notably Weismann and Nussbaum, developed this idea in relation to current work in cytology. Whereas Naegeli made his idioplasm a continuous web of fibers which penetrated cell walls, Weismann limited it to the chromosomes in each cell. Oscar Hertwig, on the other hand, who was much influenced by Naegeli, did not restrict the idioplasm to the chromosomes but to the nuclear substance as a whole.

Naegeli's micellar theory can be seen as the fulfillment of his aim to put Schwann's crystal model of cell growth on a sound footing. The studies he published on the cell wall of *Caulerpa* in 1844 mark the beginning of this work which culminated in his grand synthesis of 1884.

Despite Naegeli's creation of molecular models, he never made a complete break with the vitalistic and teleological ideas so popular among German-speaking biologists of his youth. Consequently natural selection was for him only a pruning device, evolution being the result of an internal perfecting principle. To the end of his days he believed in the spontaneous generation of cells and that, in view of the time required for complexity to be achieved, simple organisms must be younger than complex ones. His search for discontinuities between species and between the plant and animal kingdoms was consistent with his desire for absolute concepts. It was to Naegeli—who had denied the existence of antherozoa in *Fucus*, of genuine species of microorganisms responsible for infectious diseases, and of Darwin's role for natural selection—that Gregor Mendel sent his "Versuche über Pflanzenhybriden." Naegeli, who believed he himself knew how hybrids behaved from his study of crosses in the genus *Hieracium*, regarded Mendel's hybrid ratios and demonstration of complete reversion as of purely empirical significance, irrelevant to genuine species.

As one of the nineteenth century's foremost botanists and influential theoreticians, Naegeli deserves sympathetic evaluation as both an innovator and a victim of the biological thinking to which he contributed so much. Where he failed so conspicuously his famous pupil Carl Correns succeeded. Correns was one of the three rediscoverers of Mendel's laws.

BIBLIOGRAPHY

I. ORIGINAL WORKS. A complete list of Naegeli's publications will be found in S. Schwendener's obituary notice (see below). With Schwendener he wrote the very popular *Das Mikroskop; Theorie und Anwendung desselben*,

2 vols. (Leipzig, 1867), English trans. by F. Crisp (London, 1887; 2nd ed., London, 1892). Naegeli introduced the term *Micell* in the 2nd German ed. of 1877. With A. Peters he wrote *Die Hieracien Mittel Europas. Monographische Bearbeitung der Piloselloiden mit besonderer Berücksichtigung der mitteleuropaischen Sippen*, 2 vols. (Munich, 1885–1889). Naegeli's final statements on heredity, growth, and ultrastructure will be found in *Mechanisch-physiologische Theorie der Abstammungslehre* (Munich–Leipzig, 1884).

The majority of his earlier cytological papers appeared in the short-lived journal which he and Schleiden edited, *Zeitschrift für wissenschaftliche Botanik* (Jena, 1844–1847). The most important papers from this journal were translated into English by Arthur Henfrey and published in the Ray Society's *Reports and Papers on Botany* (London, 1846, 1849). Naegeli's studies of starch grains, his micellar theory, and his work with C. Cramer were published in the series *Pflanzenphysiologische Untersuchungen von C. Naegeli und C. Cramer*, nos. 1–4 (Zurich, 1855–1858). A selection from Naegeli's contributions was published by Albert Frey in *Die Micellartheorie . . . Auszüge aus den grundlegenden Originalarbeiten Nägelis, Zusammenfassung und kurze Geschichte der Micellartheorie*, in Ostwald's *Klassiker der exakten Wissenschaften*, no. 227 (Leipzig, 1908).

Forty-two papers presented by Naegeli to the Bavarian Academy are in *Botanische Mitteilungen aus den Sitzungsberichten der k. b. Akademie der Wissenschaft in München*, III (Munich, 1863–1881). Extracts from Naegeli's letters to Mendel were published by Hugo Iltis in his *Life of Mendel* (London, 1932; repr. 1966).

II. SECONDARY LITERATURE. A long list of obituary notices is given in the Royal Society *Catalogue of Scientific Papers*, 17 (1891), 443. Readily available is D. H. Scott's notice in *Nature*, 44 (1891), 580–583.

The only biographical notice which includes a full bibliography is that by S. Schwendener in *Bericht der deutschen botanischen Gesellschaft*, 9 (1891), (26)–(42). Most accounts of Naegeli's life rely on C. Cramer, *Leben und Wirken von Carl Wilhelm Nägeli* (Zurich, 1896; first published in the *Neue Zürcher Zeitung*, 16 May 1891). For a critical account of Naegeli's botanical work see Sidney Vines's obituary notice in the *Proceedings of the Royal Society*, 51 (1892), 27–36. The work of Naegeli and Schwendener is included in A. Frey-Wyssling's paper, "Frühgeschichte und Ergebnisse der submikroskopischen Morphologie," in *Mikroskopie*, 19 (1964), 2–12. Naegeli's micellar theory has been analyzed in depth by J. S. Wilkie. His summary of this work appeared in *Nature*, 209 (1961), 1145–1150, and his detailed papers are "Nägeli's Work on the Fine Structure of Living Matter," nos. I, II, IIIa, IIIb, in *Annals of Science*, 16 (1960), 11–42, 171–207, 209–239, and *ibid.*, 17 (1961), 27–62.

Naegeli's philosophical position and his attitude to Mendel are discussed in J. S. Wilkie's commentary to the paper by Bentley Glass, "The Establishment of Modern Genetical Theory as an Example of the Interaction of Different Models, Techniques, and Inferences," in A. C. Crombie, ed., *Scientific Change, Symposium on the History*

of Science, . . . Oxford (London, 1963), 521–541, commentary on 597–603. Naegeli's attitude to Mendel has also been discussed by A. Weinstein, "The Reception of Mendel's Paper by His Contemporaries," in *Proceedings of the Tenth International Congress of the History of Science* (Ithaca, 1962; Paris, 1964), 997–1001, and in R. C. Olby and P. Gautrey, "Eleven References to Mendel Before 1900," in *Annals of Science*, 24 (1968), 7–20. C. C. Gillispie has compared the speculative ideas of Naegeli and Weismann in *The Edge of Objectivity: An Essay in the History of Scientific Ideas* (Princeton–London, 1960), 322–328.

ROBERT OLBY

IBN AL-NAFĪS, ʿALĀʾ AL-DĪN ABU ʾL-ḤASAN ʿALĪ IBN ABI ʾL-ḤAZM AL-QURASHĪ (or AL-QARASHĪ) (*b.* al-Qurashiyya, near Damascus, thirteenth century; *d.* Cairo, 17 December 1288), *medicine.*

Ibn al-Nafīs' *nisba*, al-Qurashī, is from his birthplace or, according to other authorities, from Qarash, a village beyond the River Oxus from which his family originally came. He studied medicine in Damascus, at the great Nūrī Hospital (al-Bīmāristān al-Nūrī al-Kabīr) founded by the Turkish prince Nūr al-Dīn Maḥmūd ibn Zankī (Nureddin) in the twelfth century. Among his teachers was Muhadhdhab al-Dīn ʿAbd al-Raḥīm ibn ʿAlī al-Dakhwār (*d.* 1230), founder of al-Dakhwāriyya Medical School at Damascus, and among his students at Damascus was Abu ʾl-Faraj ibn Yaʿqūb ibn Isḥāq al-Masīḥī ibn al-Quff Amīn al-Dawla al-Karakī (1233–1286), who at one time was Ibn Abī Uṣaybiʿa's student.

The hospital in which Ibn al-Nafīs practiced and taught in Egypt is not known with certainty. Eventually he became *raʾīs al-aṭibbāʾ* (chief of physicians), possibly appointed by the Mamlūk ruler al-Ẓāhir Baybars al-Bunduqdārī (reigned 1260–1277), for whom Ibn al-Nafīs worked in the capacity of personal physician; this post was not merely honorific but conferred disciplinary powers over medical practitioners. His name does not anywhere appear in connection with the al-Bīmāristān al-Nāṣirī, founded in 1171 by Ṣalāḥ al-Dīn al-Ayyūbī (or Saladin, who reigned from 1169 to 1193), where Ibn Abī Uṣaybiʿa (*d.* 1270) was an oculist during the one year (1236–1237) he spent in Egypt. Toward the end of his life Ibn al-Nafīs bequeathed his house and library to the newly founded Dār al-Shifāʾ (House of Recovery), also called Qalāwūn Hospital or al-Manṣūrī Hospital, founded in 1284 by the Mamlūk al-Manṣūr Sayf al-Dīn Qalāwūn al-Alfī (reigned from 1279 to 1290), during whose time Ibn al-Nafīs died in

Cairo—he had then reached the age of about eighty lunar years—on 21 Dhu 'l-Qaʿda 687 or 17 December 1288.

In addition to being a physician Ibn al-Nafīs lectured on *fiqh* (jurisprudence) at al-Masrūriyya School in Cairo. The inclusion of his name in the *Ṭabaqāt al-Shāfiʿiyyīn al-Kubrā* ("Great Classes of Shāfiʿī Scholars") of Tāj al-Dīn al-Subkī (*d.* 1370) indicates his eminence in religious law. He wrote his *Kitāb al-Shāmil fi 'l-Ṣināʿa al-Ṭibbiyya* ("Comprehensive Book on the Art of Medicine") when he was in his thirties. It was said to consist of 300 volumes of notes, of which he published only eighty. This voluminous work was thought to have been lost until 1952, when one large but fragmentary volume was cataloged among the Cambridge University Library Islamic manuscripts. Much earlier, the Bodleian Library cataloged four manuscripts of this work, without identifying the author. In 1960 three autograph manuscripts (MS Z276) were found in Lane Medical Library, Stanford University, of which one is referred to by the author as the thirty-third *mujallad* (volume). The two other manuscripts are its forty-second and forty-third volumes, the latter dated 641/1243–1244. Another manuscript of the same book is extant in al-Muthaf al-ʿIrāqī, Baghdad; and al-Ziriklī mentions one manuscript in Damascus (not in the Ẓāhiriyya collection) without specifying any particular library.

The *Kitāb al-Shāmil*, so far unpublished, contains an interesting section on surgical technique and throws new light on Ibn al-Nafīs as a surgeon. In it he defines three stages for each operation—al-iʿṭāʾ (the presentation for diagnosis, upon which a patient entrusts a surgeon with his body and life), al-ʿamal (the operative procedure), and al-ḥifẓ (preservation, that is, postoperative care)—and gives detailed descriptions of the duties of surgeons and the relationships among patients, surgeons, and nurses. He discusses each stage in detail, touching upon such subjects as the decubitus of the patient and the posture, bodily movement, and manipulation of instruments of the surgeon in the course of carrying out his duties. Ibn al-Nafīs illustrates his points with examples of specific operations.

Ibn al-Nafīs' book *Sharḥ Ṭabīʿat al-Insān li-Buqrāṭ* ("Commentary on Hippocrates' *Nature of Man*") was housed in a private library at Damascus owned by Aḥmad ʿUbayd and in 1933 was owned by professor A. S. Yahuda in London. (The medical manuscripts that were in the Yahuda collection are now in the National Library of Medicine, Bethesda, Maryland; the *Sharḥ Ṭabīʿat al-Insān li-Buqrāṭ* is MS A69.) It has an *ijāza* (license) written and signed by Ibn al-Nafīs stating that a physician named Shams al-Dawla Abu 'l-Faḍl ibn Abi 'l-Ḥasan al-Masīḥī had studied the entire book under him. Perhaps one of Ibn al-Nafīs' earliest books is *Sharḥ Tashrīḥ al-Qānūn* ("Commentary on Anatomy in Books I and III of Ibn Sīnā's *Kitāb al-Qānūn*"), of which a copy was written forty-seven lunar years before his death, and is presently at the University of California, Los Angeles (MS Ar. 80). In this book he gives the earliest known account of the pulmonary blood circulation. His major work, *Sharḥ al-Qānūn* ("Commentary on *Kitāb al-Qānūn*") is in four books: "A Commentary on Generalities"; "A Commentary on Materia Medica and Compound Drugs"; "A Commentary on Head-to-Toe Diseases"; and "A Commentary on Diseases Which Are Not Specific to Certain Organs." In the first of these books, the "Commentary on Generalities," Ibn al-Nafīs repeats his account of the lesser circulations of the blood:

> . . . This is the right cavity of the two cavities of the heart. When the blood in this cavity has become thin, it must be transferred into the left cavity, where the pneuma is generated. But there is no passage between these two cavities, the substance of the heart there being impermeable. It neither contains a visible passage, as some people have thought, nor does it contain an invisible passage which would permit the passage of blood, as Galen thought. The pores of the heart there are compact and the substance of the heart is thick. It must, therefore, be that when the blood has become thin, it is passed into the arterial vein [pulmonary artery] to the lung, in order to be dispersed inside the substance of the lung, and to mix with the air. The finest parts of the blood are then strained, passing into the venous artery [pulmonary vein] reaching the left of the two cavities of the heart, after mixing with the air and becoming fit for the generation of pneuma . . . .

According to one manuscript of *Sharḥ Tashrīḥ al-Qānūn* (MS Ar. 80), the *terminus ante quem* of Ibn al-Nafīs' discovery of the lesser circulation can be fixed at 1242, three centuries before those published by Servetus (1553) and Colombo (1559). The determination by Iskandar of discussions of the lesser circulation in commentaries on book I of the *Kitāb al-Qānūn* of Sadīd al-Dīn Muḥammad ibn Masʿūd al-Kāzarūnī (completed in 1344) and ʿAlī ibn ʿAbdallāh Zayn al-ʿArab al-Miṣrī (written in 1350), who used Ibn al-Nafīs' *Sharḥ Tashrīḥ al-Qānūn* and his *Sharḥ al-Qānūn,* may serve to reopen the widely debated question of whether the Latin West had access to Ibn al-Nafīs' description of the lesser circulation. It is believed that Andrea Alpago of Belluno (*d.* 1520) may have transmitted Ibn al-Nafīs' work orally or in hitherto unpublished writings.

Alpago lived in the Middle East (mainly in Syria) for thirty years, collecting, translating, and editing the writings of Arab physicians. He made a Latin translation (Venice, 1547) of the commentary on compound drugs that is a part of Ibn al-Nafīs' *Sharḥ al-Qānūn*. In a section (fds. 24v–30r) entitled "Consideratio sexta de pulsibus ex libro Sirasi arabico," Alpago gives some interesting statements on the Galenic doctrine related to the heart and arterial system, together with Ibn al-Nafīs' criticism.

Ibn al-Nafīs' *Kitāb al-Mūjiz* or *Mūjiz al-Qānūn* ("Epitome of *Kitāb al-Qānūn*") is a concise book divided into four sections corresponding to the four books of the *Sharḥ al-Qānūn*, except that in *Kitāb al-Mūjiz* he does not deal with anatomy or with the lesser circulation. The popularity of *Kitāb al-Mūjiz* led many physicians to write commentaries on it and to translate it into other languages. Two Turkish translations are known, one by Muṣliḥ al-Dīn Muṣṭafā ibn Shaʻbān al-Surūrī (d. 1464) and the other by Aḥmad Kamāl, a physician in Adrianople. There is a Hebrew translation entitled *Sefer-ha-Mūjiz*. The author of *Kitāb Tadhkirat al-Suwaydī*, ʻIzz al-Dīn Abū Isḥāq Ibrāhīm ibn Muḥammad ibn Ṭarkhān al-Suwaydī (d. 1291), also wrote a commentary on *Kitāb al-Mūjiz*. Other commentaries, still preserved in manuscript, were written by Jalāl al-Dīn Muḥammad ibn ʻAbd al-Raḥmān al-Qazwīnī (d. 1308), Muzaffar al-Dīn Abu ʼl-Thanāʼ Maḥmūd ibn Aḥmad al-ʻAyntābī ibn al-Amshāṭī (d. 1496), and Shihāb al-Dīn Muḥammad al-Ījī al-Bulbulī. Three major commentaries, widely used until recently, are *Kitāb al-Mughnī fī Sharḥ al-Mūjiz*, by Sadīd al-Dīn al-Kāzarūnī; *Kitāb Ḥall al-Mūjiz* ("Key to *Kitāb al-Mūjiz*") by Jamāl al-Dīn Muḥammad ibn Muḥammad al-Āqṣarāʼī (d. 1378); and *Kitāb al-Nafīsī*, also known as *Sharḥ Mūjiz Ibn al-Nafīs*, by Burhān al-Dīn Nafīs ibn ʻAwaḍ al-Kirmānī (written in 1437). Among many marginal commentaries to the *Kitāb al-Nafīsī* are *Ḥāshiya ʻAlā Sharḥ Nafīs Ibn ʻAwaḍ al-Kirmānī ʻAlā Mūjiz Ibn al-Nafīs* ("Marginal Commentaries on the Commentary of Nafīs ibn ʻAwaḍ al-Kirmānī on *Kitāb al-Mūjiz* of Ibn al-Nafīs") by Ghars al-Dīn Ibrāhīm al-Ḥalabī (d. 1563) and *Ḥall al-Nafīsī* ("Key to *Kitāb al-Nafīsī*"), which was begun by Muḥammad ʻAbd al-Ḥalīm and posthumously completed by his son, Muḥammad ʻAbd al-Ḥayy, who published the whole work in 1872. Ibn al-Nafīs wrote out his *Sharḥ Fuṣūl Buqrāṭ* ("Commentary on Hippocrates' *Aphorisms*") more than once, each time to meet certain requests made to him by physicians. An introductory note to a lithographed edition of this book (dated 1892) repeats the statement that he made in his *Sharḥ Tashrīḥ al-Qānūn* and *Sharḥ al-Qānūn*—that he decided to ". . . throw light on and stand by true opinions, and forsake those which are false and erase their traces. . . ." This statement seems to suggest that he rebelled against the authority of books, a view substantiated by his rejection of Galen's concept of invisible pores in the interventricular septum, his notion of blood flow, and his belief that arterial blood was produced in the left ventricle.

Other books written by Ibn al-Nafīs are: *Sharḥ Abīdhīmyā li-Buqrāṭ* ("Commentary on Hippocrates' *Epidemics*"); *Sharḥ Masāʼil Ḥunayn* ("Commentary on Ḥunayn [ibn Isḥāq's] *Questions*"); *al-Muhadhdhab fi ʼl-Kuḥl* ("Polished Book on Ophthalmology"); and *Bughyat al-Ṭālibīn wa Ḥujjat al-Mutaṭabbibīn* ("Reference Book for Physicians"). He also wrote on logic and theology, including such books as his commentary on Ibn Sīnā's *Kitāb al-Hidāya* ("Guidance"), and *Fāḍil Ibn Nāṭiq* (also entitled *al-Risāla al-Kāmiliyya fi ʼl-Sīra al-Nabawiyya*), a counterpart to Ibn Ṭufayl's (d. 1185) *Ḥayy Ibn Yaqẓān*. Ibn Ṭufayl's purpose was to show the discovery of philosophical truths by an individual who had been created by spontaneous generation on a desert island, while that of Ibn al-Nafīs was to show the discovery by independent reasoning (under similar conditions) of the main principles of Islamic religion and natural sciences.

Ibn al-Nafīs was reputed to have recorded his own experiences, observations, and deductions rather than using reference books. His religion (Islam) and his mercy toward animals, he tells us, prevented him from practicing anatomy. His major contribution—the discovery of the lesser circulation—was nonetheless a physiological one and would probably have been more adequately documented had he resorted to animal dissection. His experimental approach to physiology is evident in his *Sharḥ Tashrīḥ al-Qānūn*: ". . . In determining the use of each organ we shall rely necessarily on verified examinations and straightforward research, disregarding whether our opinions will agree or disagree with those of our predecessors."

BIBLIOGRAPHY

I. ORIGINAL WORKS. Ibn al-Nafīs' books are *Manāfiʻ al-Aʻḍāʼ al-Insāniyya* (Dār al-Kutub al-Miṣriyya, MS 209, III, *majāmīʻ*); *al-Muhadhdhab fiʼl-Kuḥl* (Vatican, MS Arabo 1307); *Mūjiz al-Qānūn* (Calcutta, 1244/1828, 1261/1845; Lucknow, 1288/1871, 1302/1884, 1324/1906); *Kitāb al-Shāmil fiʼl-Ṣināʻa al-Ṭibbiyya* (incomplete autograph copy, Lane Medical Library, Stanford University, MS Z276; al-Mutḥaf al-ʻIrāqī, Baghdad, MS 1271; Cambridge

University Library, MS Or. 1546 (10); Bodleian Library, MSS Pocock 248 and 290–292); *Sharḥ Abīdhīmyā li-Buqrāṭ* (Aya Sofya, MS 3642, fols. 1-200a; Dār al-Kutub al-Miṣriyya, MS 583 Ṭibb Ṭalʿat); *Sharḥ Fuṣūl Buqrāṭ* (Teheran [?], 1310/1892; Aya Sofya, MSS 3554, fols. 35b–137b, 3644, fols. 1–109b; Dār al-Kutub al-Miṣriyya, MS 1448 Ṭibb; Forschungsbibliothek, Gotha, MSS 1897–1898; Deutsche Staatsbibliothek, Berlin, MS 6224); *Sharḥ Masāʾil Ḥunayn* (Leiden University Library, MS Or. 49, II, fols. 101b–174a); *Sharḥ Ṭabīʿat al-Insān li-Buqrāṭ* (National Library of Medicine, Bethesda, Md., MS A69; MS owned by Aḥmad ʿUbayd, Damascus and later by A. S. Yahuda, London); *Sharḥ Taqdimat al-Maʿrifa li-Buqrāṭ* (Forschungsbibliothek, Gotha, MS 1899; Leiden University Library, MS Or. 49, I, fols. 1–98; Bodleian Library, MS Marsh 81); *Sharḥ al-Qānūn* (Wellcome Historical Medical Library, London, WMS. Or. 51; and incomplete in WMS. Or. 154); and *Sharḥ Tashrīḥ al-Qānūn* (University of California, Los Angeles, MSS Ar. 80, and Ar. 102, I, pp. 1–298; Bibliothèque Nationale, Paris, MS 2939; Al-Ẓāhiriyya Library, Damascus, MS 3145 Ṭibb XX). See also M. Meyerhof and J. Schacht, *The Theologus Autodidactus of Ibn al-Nafīs*, ed. with intro., trans., and notes (Oxford, 1968).

Other works directly related to Ibn al-Nafīs' writings are al-Āqṣarāʾī, *Ḥall al-Mūjiz* (Lucknow, 1325/1907; Urdu trans., 2 vols., Lucknow, 1325–1326/1907–1908); al-Kāzarūnī, *Kitāb al-Mughnī fī Sharḥ al-Mūjiz* (Calcutta, 1244/1828, 1832; Lucknow, 1295/1878, 1307/1890, 1894); al-Kirmānī's *Kitāb al-Nafīsī* (Lucknow, 1282/1865); ʿAbd al-Ḥalīm and ʿAbd al-Ḥayy, *Ḥall al-Nafīsī* (Cawnpore, 1288/1872; Lucknow, 1302/1885); al-Kāzarūnī, *Sharḥ al-Qānūn (al-Kulliyyāt)* (Wellcome Historical Medical Library, WMS. Or. 89); Zayn al-ʿArab al-Miṣrī, *Sharḥ al-Qānūn* (Wellcome Historical Medical Library, WMS. Or. 119); and Ibn Rushd, . . . *Avicenna . . . libellus de removendes nocumentis quae accidunt in regimine sanitatis . . .*, A. Alpago, trans. (Venice, 1547).

II. SECONDARY LITERATURE. General works are Ibn Abī Uṣaybiʿa, *ʿUyūn al-Anbāʾ* (al-Ẓāhiriyya Library, MS 4883, I, ʿāmm, fol. 104)—the concise account at the end of this manuscript seems to have been written by a later author, not by Ibn Abī Uṣaybiʿa himself, and does not appear in the Būlāq ed., 2 vols. (1882–1884); Ibn Faḍlallāh al-ʿUmarī, *Masālik al-Abṣār . . .* (Dār al-Kutub al-Miṣriyya, MS 8 mīm, Maʿārif ʿāmma, VIII, 119a); al-Ṣafadī, *Kitāb al-Wāfī biʾl-Wafayāt* (British Museum, MS Or. 6587, fols. 20v–21v); Abū ʿAbdallāh Muḥammad ibn Aḥmad al-Dhahabī, *Tārīkh al-Islām* (Bodleian Library, MS Laud Or. 279, fol. 170a); ʿAbdallāh ibn Asʿad al-Yāfiʿī, *Mirʾāt al-Janān*, IV (Hyderabad, 1920–1921), 207; Tāj al-Dīn al-Subkī, *Ṭabaqāt al-Shāfiʿiyyīn al-Kubrā*, V (Cairo, 1906–1907), 129; J. Uri, *Bibliothecae Bodleianae codicum manuscriptorum orientalium* (Oxford, 1787), pt. 1, 130; A. Nicoll and E. B. Pusey, *Bibliothecae Bodleianae . . .*, (Oxford, 1821–1835), pt. 2, 586; L. Leclerc, *Histoire de la médecine arabe*, II (Paris, 1876), 207–209; W. Pertsch, *Die arabischen Handschriften der herzoglichen Bibliothek zu Gotha* (Gotha, 1878–1892), III, 444–446; W. Ahlwardt, *Verzeichniss der*

arabischen Handschriften (Berlin, 1887–1899), V, 496; *The Encyclopaedia of Islam* (Leiden–London, 1913–1938), supp., 94–95; *ibid.*, new ed. (Leiden–London, 1960–1971), III, 897–898; G. Sarton, *Introduction to the History of Science* (Baltimore, 1927–1948), II, 1099–1101; A. Issa, *Histoire des bimaristans (hôpitaux) à l'époque islamique* (Cairo, 1928); and *Tārīkh al-Bīmāristānāt fī ʾl-Islām* (Damascus, 1939).

See also C. A. Wood, "The Lost Manuscript on Ophthalmology by the Thirteenth-Century Surgeon Ibn al-Nafīs," in *Journal of the American Medical Association*, **104** (1935), 2122–2123; C. Brockelmann, *Geschichte der arabischen Litteratur* (Leiden, 1943–1949), I, 649, and supp. (Leiden, 1937–1942), I, 899; I. al-Baghdādī, *Īḍāḥ al-Maknūn . . .* (Istanbul, 1945), I, 188; and *Hadiyyat al-ʿĀrifīn . . .* (Istanbul, 1951), I, 714; J. ʿAwwād, *Jawla fī Dūr al-Kutub al-Amrīkiyya* (Baghdad, 1951), 46; A. J. Arberry, *A Second Supplementary Hand-List of the Muḥammadan Manuscripts in the University and Colleges of Cambridge* (Cambridge, 1952), 57; Kh. al-Ziriklī, *al-Aʿlām . . .*, 2nd ed. (Cairo, 1954–1959), V, 78, and pl. 740; J. Schacht, "Ibn al-Nafīs et son *Theologus Autodidactus*," in *Homenaje a Millás-Vallicrosa*, II (Barcelona, 1956), 325–345; ʿU. R. Kaḥḥāla, *Muʿjam al-Muʾallifīn . . .* (Damascus, 1957–1961), VII, 58; Ṣ. al-Munajjid, "Maṣādir Jadīda ʿAn Tārīkh al-Ṭibb ʿInd al-ʿArab," in *Majallat Maʿhad al-Makhṭūṭāt al-ʿArabiyyaʾ*, 5, no. 2 (1959), 270; M. J. L. Young, "Some Observations on the Use of Arabic as a Scientific Language as Exemplified in the *Mūjiz al-Qānūn* of Ibn al-Nafīs (d. 1288)," in *Abr-Nahrain*, **1** (1959–1960), 68–72; N. Heer, "Thalāthat Mujalladāt Min Kitāb al-Shāmil lʾ Ibn al-Nafīs," in *Majallat Maʿhad al-Makhṭūṭāt al-ʿArabiyya*, **6** (1960), 203–210; S. K. Hamarneh, *Index of Manuscripts on Medicine, Pharmacy, and Allied Sciences in the Ẓāhiriyah Library* (Damascus, 1969), 476–481, and pl. 7; M. Ullmann, *Die Medizin im Islam*, which is pt. 1, supp. VI, of *Handbuch der Orientalistik* (Leiden–Cologne, 1970), 172–176.

On blood circulation see M. Tatawi, *Der Lungenkreislauf nach el-Koraschi*, inaugural diss. (Freiburg, 1924); M. Meyerhof, "M. El-Tatawi: Der lungenkreislauf nach el-Koraschi," in *Mitteilungen zur Geschichte der Medizin und Naturwissenschaften*, **30** (1931), 55–57; "La découverte de la circulation pulmonaire par Ibn an-Nafīs, médecin arabe du Caire (xiiie siècle)," in *Bulletin de l'Institut d'Égypte*, **16** (1934), 33–46; "Ibn an-Nafīs und seine Theorie des Lungenkreislaufs," in *Quellen und Studien zur Geschichte der Naturwissenschaften und Medizin*, **4** (1935), 37–88, and 1–22 (Arabic text); and "Ibn An-Nafīs (XIIIth cent.) and His Theory of the Lesser Circulation," in *Isis*, **23** (1935), 100–120; S. Ḥaddād and A. Khairallah, "A Forgotten Chapter in the History of the Circulation of the Blood," in *Annals of Surgery*, **104** (1936), 1–8; S. Ḥaddād, "Who Is the Discoverer of the Lesser Circulation?" in *al-Muqtaṭaf*, **89** (1936), 264–271; and "Arabian Contributions to Medicine," in *Annals of Medical History*, **3** (1941), 60–72; O. Temkin, "Was Servetus Influenced by Ibn an-Nafīs?" in *Bulletin of the History of Medicine*, **8** (1940), 731–734; T. Bannurah, "Enthüllungen in der

Geschichte der Medizin, Ibn al-Nafīs oder Serveto?" in *Münchener medizinische Wochenschrift*, **88** (1941), 1088 ff.; L. Binet and A. Herpin, "Sur la découverte de la circulation pulmonaire," in *Bulletin de l'Académie nationale de médecine*, 3rd ser., **132**, nos. 31–32 (1948), 542–549.

See also A. Chéhadé, *Ibn al-Nafīs et la circulation pulmonaire*, M.D. dissertation (Faculté de Médecine, Paris, 1951), no. 1143; *Ibn al-Nafīs et la découverte de la circulation pulmonaire* (Damascus, 1955); and "Ibn al-Nafīs et la découverte de la circulation pulmonaire," in *Maroc médical*, **35** (1956), 1013–1016; C. D. O'Malley, *Michael Servetus, A Translation of His Geographical, Medical, and Astrological Writings With Introductions and Notes* (Philadelphia, 1953), 195–200; and "A Latin Translation of Ibn Nafis (1547) Related to the Problem of the Circulation of the Blood," in *Journal of the History of Medicine and Allied Sciences*, **12**, no. 2 (1957), 248–253; E. E. Bittar, "A Study of Ibn Nafis," in *Bulletin of the History of Medicine*, **29** (1955), 352–368, 429–447; and "The Influence of Ibn Nafis: A Linkage in Medical History," in *University of Michigan Medical Bulletin*, **22** (1956), 274–278; G. Wiet, "Ibn al-Nafīs et la circulation pulmonaire," in *Journal Asiatique*, **244** (1956), 95–100; E. D. Coppola, "The Discovery of the Pulmonary Circulation: A New Approach," in *Bulletin of the History of Medicine*, **31** (1957), 44–77; and J. Schacht, "Ibn an-Nafīs, Servetus and Colombo," in *al-Andalus*, **22** (1957), 317–336.

Also of value are L. G. Wilson, "The Problem of the Discovery of the Pulmonary Circulation," in *Journal of the History of Medicine*, **17** (1962), 229–244; R. E. Siegel, "The Influence of Galen's Doctrine of Pulmonary Bloodflow on the Development of Modern Concepts of Circulation," in *Sudhoffs Archiv für Geschichte der Medizin und der Naturwissenschaften*, **46** (1962), 311–332; A. Z. Iskandar, *A Catalogue of Arabic Manuscripts on Medicine and Science in the Wellcome Historical Medical Library* (London, 1967), 38–42, 47–50; and E. Lagrange, "Réflexions sur l'historique de la découverte de la circulation sanguine," in *Episteme*, **3** (1969), 31–44.

ALBERT Z. ISKANDAR

NAGAOKA, HANTARO (*b.* Nagasaki, Japan, 15 August 1865; *d.* Tokyo, Japan, 11 December 1950), *physics.*

Nagaoka graduated from the department of physics of the University of Tokyo in 1887 and entered the graduate school, where he began experimental research in magnetostriction under the British physicist C. G. Knott, who was in Japan between 1883 and 1891. After receiving a doctorate, Nagaoka studied at the universities of Berlin, Munich, and Vienna from 1893 to 1896. He was especially impressed by Boltzmann's course on the kinetic theory of gases at the University of Munich. In 1900 Nagaoka was stimulated to study atomic structure to explain radioactivity by the lecture of the Curies at the first international congress of physics in Paris, where he had been invited to deliver a paper on magnetostriction.

From 1901 to 1925 Nagaoka, a leading professor of physics at the University of Tokyo, was primarily responsible for promoting the advancement of physics in Japan. In addition to studying magnetostriction, he did work in atomic structure, geophysics, mathematical physics, spectroscopy, and radio waves. The present Japanese tradition of experimental and theoretical physics has been formed almost entirely by Nagaoka and his successors. These include his pupils Kotaro Honda, Jun Ishiwara, Shoji Nishikawa, and Yoshio Nishina and his protégé Hideki Yukawa. For his efforts the Japanese government awarded him the National Cultural Prize in 1937.

Nagaoka is known for his Saturnian atomic model, published in 1904. His criticism of Lord Kelvin's Aepinus atom, proposed in the paper "Aepinus Atomized" (*Philosophical Magazine*, 6th ser., **3** [1902], 257–283), was essential for the formation of his model. Rejecting the interpenetrability of two kinds of electricity, which had been supposed by Kelvin, Nagaoka arranged electrons outside the central positive charge. Thus his model consists of a number of electrons of equal mass, arranged uniformly in a ring, and a positively charged sphere of large mass at the center of the ring. This material view of electricity played the most important role in Nagaoka's theory of the structure of matter, which was based partly on Boltzmann's atomistic influence and partly on the reflection of unsophisticated scientific thought in Japan during the early Meiji period. He obtained the equations of motion of the ring in his model according to Maxwell's work on the stability of the motion of Saturn's rings, which he had read in Germany. In 1904–1905 Nagaoka dealt with band spectra, dispersion of light, and mutual action of atoms, based on the assumption of a Saturnian atom. Having renounced his model, he started spectroscopic experiments to investigate the actual arrangement of electrons in the atom in 1908.

BIBLIOGRAPHY

Nagaoka's paper on the Saturnian atomic model is in *Proceedings of the Tokyo Mathematico-Physical Society*, 2nd ser., **2** (1904), 92–107; and *Philosophical Magazine*, 6th ser., **7** (1904), 445–455.

Anniversary Volume Dedicated to Professor Nagaoka by His Friends and Pupils on the Completion of Twenty-Five Years of His Professorship (Tokyo, 1925) contains a bibliography of his works. The origin of his model is discussed in Eri Yagi, "On Nagaoka's Saturnian Atomic Model," in *Japanese Studies in the History of Science* (1964), no. 3, 29–47. Nagaoka's life and work are fully discussed in connection with the development of physics in Japan in Kiyomobu Itakura, Tosaka Kimura, and Eri Yagi, *A Biography of Hantaro Nagaoka* (Tokyo, 1973), written in Japanese.

ERI YAGI

NĀGEŚA (*fl.* Gujarat, India, *ca.* 1630), *astronomy.*

Nāgeśa was born into a family of learned Brāhmaṇas of the Gārgyagotra, who resided at Khecaramaṇḍala in Gujarat. His father, Śiva, and his grandfather, Keśava, are otherwise unknown to us; but his son, Śiva, was the author of a *Sankrāntipaṭala* on the entry of the sun into the signs of the zodiac. Nāgeśa's principal astronomical work is an unpublished *Grahaprabodha* in thirty-seven verses, which gives instructions for computing the true longitudes of the sun, the moon, and the planets according to the parameters of the *Gaṇeśapakṣa* founded by Gaṇeśa in 1520 (see essay in Supplement); its epoch is 5 March 1619. In 1663 Nāgeśa's pupil, Yādava, wrote a set of tables based on the *Grahaprabodha* (see D. Pingree, *Sanskrit Astronomical Tables in the United States* [Philadelphia, 1968], 63a–64b, and *Sanskrit Astronomical Tables in England* [Madras, 1973], 149).

Nāgeśa also wrote a *Nirṇayatattva* in 102 verses which describe the computation of the *tithis* in a synodic month. Based on the *Nirṇayasindhu*, which was composed by Kamalākara Bhaṭṭa in 1612, the *Nirṇayatattva* uses as an example the year 1629–1630. A *Parvādhikāra* on the syzygies is also attributed to Nāgeśa.

BIBLIOGRAPHY

Aside from the articles mentioned above, see Ś. B. Dikṣita, *Bhāratiya Jyotiḥśāstra* (Poona, 1896; repr. Poona, 1931), 285–286, and D. Pingree, "On the Classification of Indian Planetary Tables," in *Journal for the History of Astronomy*, **1** (1970), 95–108, esp. 99–100.

DAVID PINGREE

AL-NAIRĪZĪ. See **al-Nayrīzī.**

NAIRNE, EDWARD (*b.* Sandwich [?], England, 1726; *d.* London, England, 1 September 1806), *mathematics, optics, physics.*

Nairne achieved an international reputation as one of the foremost makers of mathematical, optical, and philosophical instruments of the eighteenth century. He became free of the Spectaclemakers Company in 1748 and established his business in London at 20 Cornhill, not far from the shop of Matthew Loft, to whom he had been apprenticed in 1741. Nairne took Thomas Blunt, his own former apprentice, into partnership in 1774, and the firm, which in 1791 was moved to 22 Cornhill, continued as Nairne and Blunt until the latter's death in 1822.

In 1771 Nairne contributed to the *Philosophical Transactions of the Royal Society* the first of many papers on experiments in optics, pneumatics, and, most notably, electricity. He was elected a fellow of the Royal Society in 1776.

In 1772 Nairne invented an improved form of electrostatic machine using a cylindrical glass vessel as the generator. Its quick acceptance in England and on the Continent did much to enhance his reputation. The regular production from Nairne's shop included microscopes, telescopes, navigating and surveying instruments, electrical machines, vacuum pumps, and measuring equipment required by the new philosophical laboratories.

Franklin seems to have had a long acquaintance with Nairne and his work. In 1758 Nairne made a set of artificial magnets for him, and the swelling and shrinking of the mahogany case led to a later correspondence between them on a possible design for a hygrometer. After the Harvard College fire of 1764, Nairne was one of the makers commissioned, on Franklin's recommendation, to replace the lost instruments.

Nairne also reported on his experiments on the specific gravity and freezing point of seawater, desiccation by means of a vacuum, and the adaptation of the mercury barometer for use at sea.

BIBLIOGRAPHY

I. ORIGINAL WORKS. Nairne's papers published in the *Philosophical Transactions of the Royal Society* include "Description of a New Constructed Equatorial Telescope," **61** (1771), 223–225; "Water From Sea Ice," **66** (1776), 249–256; "Experiments With the Air-pump," **67** (1777), 614–648; and "Experiments on Electricity," **68** (1778), 823–860. Other works are *Description of a Pocket Microscope* (n.p., 1771); *Directions for Using the Electrical Machine as Made and Sold by E. Nairne* (London, 1773); *Directions for the Use of the Octant* (n.d.).

Many of Nairne's instruments survive and some may be seen in the collections of the Adler Planetarium, Chicago; Conservatoire National des Arts et Métiers, Paris; Harvard University; the museums of the history of science at Oxford and Florence; National Maritime Museum, Greenwich; Naval Museum, Madrid; Science

Museum, London; and the Smithsonian Institution, Washington.

II. Secondary Literature. See Maria Luisa Bonelli, *Catalogo degli strumenti dei Museo di storia della scienza* (Florence, 1954), 92, 131, 194, 200, 208, 210, 251–252, 254, 256; I. Bernard Cohen, *Some Early Tools of American Science* (Cambridge, Mass., 1950), 166, 169; Maurice Daumas, *Les instruments scientifiques au XVII et XVIII siècles* (Paris, 1953), 316–317; Nicholas Goodison, *English Barometers, 1680–1860* (New York, 1968), 52–53, 123, 168–170, 257; W. E. Knowles Middleton, *The History of the Barometer* (Baltimore, 1964), 163; Leslie Stephen and Sidney Lee, eds., *Dictionary of National Biography*, XIV, 25–26; E. G. R. Taylor, *The Mathematical Practitioners of Hanoverian England* (London, 1966), 50, 53, 62–63, 66, 214; Carl Van Doren, ed., *Benjamin Franklin's Autobiographical Writings* (New York, 1945), 490–494; and David P. Wheatland, *The Apparatus of Science at Harvard, 1765–1800* (Cambridge, Mass., 1968), 22–23, 79, 155–161.

RODERICK S. WEBSTER

NAJĪB AL-DĪN. See **al-Samarqandī, Najīb al-Dīn.**

NAMETKIN, SERGEY SEMENOVICH (*b.* Kazan, Russia, 3 July 1876; *d.* Moscow, U.S.S.R., 5 August 1950), *chemistry.*

Nametkin's parents died when he was ten; he graduated from a Gymnasium in Moscow, earning his living as a private tutor. From 1896 to 1902 he studied and then taught at Moscow University. From 1910 he was assistant professor and, from 1911, when he was awarded his M.Sc., professor of organic chemistry at the Higher Women's Courses, which in 1917 became the Second Moscow University (of which he was rector from 1919 to 1924), and then the Moscow Institute of Fine Chemical Technology. In 1938 he returned to Moscow University, where he was head of the organic chemistry department until 1950. From 1927 to 1950 he was professor of organic chemistry and petrochemistry in the petroleum department of the Moscow Mining Academy; the department was reorganized as the Moscow Petroleum Institute.

In 1917 Nametkin was awarded the degree of D.Sc.; from 1932 he was a corresponding member, and from 1939 a member, of the Academy of Sciences of the U.S.S.R. From 1939 to 1950 he was director of the Institute of Fossil Fuels, and then of the Petroleum Institute of the Soviet Academy of Sciences, which was created from the former.

Nametkin's scientific work concentrated on the nitration of saturated hydrocarbons, the chemistry of terpenes, stereochemistry, the chemistry and technology of petroleum, and the synthesis of growth stimulators and perfumes.

Having examined in his thesis the effect of dilute nitric acid on monocyclic and bicyclic hydrocarbons, Nametkin developed a logical scheme of the resulting transformations: if hydrogen at a tertiary carbon atom reacts, then a tertiary nitric compound is immediately formed; in other cases intermediate isonitroso compounds are formed, which subsequently either isomerize into stable nitric compounds or are decomposed into nitrous oxide and a ketone or an aldehyde. The latter is easily oxidized into carboxylic acid.

Nametkin's many investigations in the chemistry of terpenes, particularly the study of the hydration of α-methylcamphene with the formation of 4-methylisoborneol, led him to an important generalization—the discovery of the camphene rearrangement of the second type; this has been called the Nametkin rearrangement and has led to the clarification of phenomena in the chemistry of camphor and camphene that were not explained by the Wagner-Meerwein rearrangement.

Nametkin studied the composition of petroleum and natural gases of the Soviet Union; investigated the chemical nature of petroleum paraffins and ceresins; showed the presence in them of a significant quantity of hydrocarbons with branched structure; discovered a new type of transformation of ethylene hydrocarbons—the hydro-dehydropolymerization reaction—as a result of which hydrogenated lower-polymers and highly unsaturated higher-polymers are formed in the presence of acid catalysts; showed that thiophanes constitute a significant part of the sulfur compounds of petroleum; conducted systematic research on the desulfurization of shale gasolines and on the chemistry of the cracking process with simultaneous partial oxidation, developed a method for the analysis of cracking gasolines. Nametkin published more than 300 works on organic and petroleum chemistry.

BIBLIOGRAPHY

I. Original Works. Many of Nametkin's writings were brought together in *Izbrannye trudy* ("Selected Works"; Moscow–Leningrad, 1949); and almost all are in *Sobranie trudov* ("Collected Works"), 3 vols. (Moscow–Leningrad, 1954–1956), with a sketch of his life and works and a bibliography in vol. I. Among his works are *K voprosu o deystvii azotnoy kisloty na uglevodorody predelnogo kharaktera* ("On the Action of Nitric Acid on Saturated Hydrocarbons"; Moscow, 1911), his master's thesis;

Issledovania iz oblasti bitsiklicheskikh soedineny ("Research on Bicyclic Compounds"; Moscow, 1916), his doctoral thesis; and *Khimia nefti* ("The Chemistry of Petroleum"; Moscow–Leningrad, 1955).

II. SECONDARY LITERATURE. See V. M. Rodionov, A. K. Ruzhentseva, A. S. Nekrasov, and N. N. Melnikov, "Akademik Sergey Semenovich Nametkin," in *Zhurnal obshchey khimii*, **21** (1951), 2101–2146; P. I. Sanin, "Sergey Semenovich Nametkin (k devyanostoletiyu so dnya rozhdenia)" ("Sergey Semenovich Nametikin [on the Ninetieth Anniversary of His Birth]"), in *Neftekhimia*, **6** (1966), 649–658; *Sergey Semenovich Nametkin*, in the series *Materialy k bibliografii uchenykh SSSR* ("Material for a Bibliography of Scientists of the U.S.S.R."; Moscow–Leningrad, 1946); A. V. Topchiev, S. R. Sergienko, and P. I. Sanin, "Trudy vydayushchegosya sovetskogo uchenogo S. S. Nametkina v oblasti khimicheskoy nauki i neftyanoy promyshlennosti" ("Works of the Distinguished Soviet Scientist S. S. Nametkin in Chemistry and the Petroleum Industry"), in *Izvestia Akademii nauk SSSR*, Otd. tekhn. nauk (1951), no. 1, 3–21; and G. D. Vovchenko and A. F. Platé, "Akademik Sergey Semenovich Nametkin (k godovshchine so dnya smerti)" ("Academician S. S. Nametkin [on the Anniversary of His Death]"), in *Vestnik Moskovskogo . . . universiteta* (1951), no. 10, 89–94.

A. PLATÉ

NANSEN, FRIDTJOF (*b.* Fröen, Norway, 10 October 1861; *d.* Oslo, Norway, 13 May 1930), *anatomy*.

For a detailed study of his life and work, see Supplement.

NAPIER, JOHN (*b.* Edinburgh, Scotland, 1550; *d.* Edinburgh, 4 April 1617), *mathematics*.

The eighth laird of Merchiston, John Napier was the son of Sir Archibald Napier by his first wife, Janet Bothwell, daughter of an Edinburgh burgess. At the age of thirteen he went to St. Salvator's College, St. Andrews, where he lodged with John Rutherford, the college principal. Little is known of his life at this time save that he gained some impetus toward theological studies during the brief period at St. Andrews. His mother's brother, Adam Bothwell, bishop of Orkney, recommended that he continue his studies abroad and it seems likely that he did so, although no explicit evidence exists as to his domicile, or the nature of his studies. At all events, by 1571 he had returned to Scotland and, in 1572, he married Elizabeth, daughter of Sir James Stirling, and took up residence in a castle at Gartnes (completed in 1574). On the death of his father in 1608, he moved to

Merchiston Castle, near Edinburgh, where he lived for the rest of his life. In 1579 his wife died and he subsequently married Agnes Chisholm of Cromlix, Perthshire. There were two children by the first marriage, a son, Archibald, who in 1627 was raised to the peerage by the title of Lord Napier, and a daughter, Joanne. By the second marriage there were ten children; the best known of these is the second son, Robert, his father's literary executor.

Napier lived the full and energetic life of a sixteenth-century Scottish landowner, participating vigorously in local and national affairs. He embraced with great fervor the opinions of the Protestant party, and the political activities of his papist father-in-law, Sir James Chisholm, involved him in continuous embarrassment. There were quarrels with his half brothers over the inheritance and disputes with tenants and neighboring landlords over land tenure and rights. In all these matters, Napier seems to have shown himself forthright and determined in the pursuit of his aims, but nonetheless just and reasonable in his demands and willing to accept a fair settlement. As a landowner, Napier gave more than the usual attention to agriculture and to the improvement of his crops and his cattle. He seems to have experimented with the use of manures and to have discovered the value of common salt for this purpose, a monopoly for this mode of tillage being granted to his eldest son, Archibald, in 1698. A monopoly was granted to Napier also for the invention of a hydraulic screw and revolving axle to keep the level of water down in coal pits (1597). In 1599 Sir John Skene mentioned that he had consulted Napier, "a gentleman of singular judgement and learning, especially in mathematic sciences," with reference to the proper methods to be used in measuring lands.

In sixteenth-century Scotland, intellectual interest centered on religion, theology, and politics rather than on science and mathematics and Napier's first literary work arose out of the fears entertained in Scotland of an invasion by Philip II of Spain. *A Plaine Discovery of the Whole Revelation of Saint John* occupied him for about five years before its publication in 1593. In this tract Napier urged the Scottish king, James VI (the future James I of England), to see that "justice be done against the enemies of Gods church" and implored him to "purge his house, family and court of all Papists, Atheists and Newtrals." Through this publication, Napier gained a considerable reputation as a scholar and theologian and it was translated into Dutch, French, and German, going through several editions in each language. It is possible that, in later life, his authority as a divine saved him from persecution as a warlock, for there are many stories told suggesting that, locally, he was suspected of being in league with the powers of darkness. Not content with opposing popery by the pen, Napier also invented various engines of war for the defense of his faith and his country. In a document preserved in the Bacon Collection at Lambeth Palace, Napier outlines four inventions, two varieties of burning mirrors for setting fire to enemy ships at a distance, a piece of artillery for destroying everything round the arc of a circle, and an armored chariot so constructed that its occupants could fire in all directions. It is not known whether any of these machines were ever constructed.

Although documentary evidence exists to substantiate the active part Napier played in public affairs in this tumultuous age, it is more difficult to trace the development of his mathematical work, which seems to have begun in early life and persisted, through solitary and indefatigable labors, to the very end, when he made contact with Henry Briggs. Some material was, apparently, assembled soon after his first marriage in 1572 and may have been prompted by knowledge he had gleaned during his travels abroad. This treatise, dealing mainly with arithmetic and algebra, survived in manuscript form and was transcribed, after Napier's death, by his son Robert for the benefit of Briggs. It was published in 1839 by a descendant, Mark Napier, who gave to it the title *De arte logistica*. From this work, it appears that Napier had investigated imaginary roots of equations, a subject he refers to as a great algebraic secret.

There is evidence that Napier began to work on logarithms about 1590; the work culminated in the publication of two Latin treatises, known respectively as the *Descriptio* (1614) and the *Constructio* (1619). The *Descriptio* bears evidence of having been written all at one time and contains, besides the tables, a brief general account of their nature and use. An English translation of this work was made by Edward Wright but was published only after Wright's death by his son, Samuel Wright (1616). Napier approved the translation, both in substance and in form. The *Constructio* was brought out by Robert Napier, after the death of his father, and consists of material which Napier had written many years before. The object of the *Constructio* was to explain fully the way in which the tables had been calculated and the reasoning on which they were based. In the *Constructio* the phrase "artificial numbers" is used instead of "logarithms," the word "logarithm" being apparently of later invention. Napier offered no explanation for the choice but Briggs, in the *Arithmetica logarithmica* (1624), explains that the name came from their inventor because they exhibit numbers which preserve always the same ratio to one another.

Although it is as the inventor of logarithms that Napier is known in the history of mathematics, the two works mentioned above contain other material of lesser importance but nonetheless noteworthy. In the course of illustrating the use and application of logarithms Napier made frequent use of trigonometric theorems and the contribution he made to the development and systematization of spherical trigonometry has been rated highly. Napier's rules (called the Napier analogies) for the right-angled spherical triangle were published in the *Descriptio* (Bk. II, Ch. 4). He expressed them in logarithmic form and exhibited their character in relation to the star pentagon with five right angles. Another achievement was the effective use he made of decimal notation (which he had learnt of from Stevin) in conjunction with the decimal point. Although he was not the first to use a decimal separatrix in this way, the publicity that he gave to it and to the new notation helped to establish its use as standard practice. In 1617 Napier's intense concern for the practicalities of computation led him to publish another book, the *Rabdologiae*, which contains a number of elementary calculating devices, including the rods known as "Napier's bones." These rods, which in essence constitute a mechanical multiplication table, had a considerable vogue for many years after his death. Each rod is engraved with a table of multiples of a particular digit, the tens and units being separated by an oblique stroke. To obtain the product 267 × 8, the rods 2, 6, 7 are assembled and the result is read off from the entries in the eighth row; thus ⬚⬚⬚ gives 2,136. Book II is a practical treatment of mensuration formulas. Book III, the method of the promptuary, deals with a more complicated system of multiplication by engraved rods and strips, which has been called the first attempt at the invention of a calculating machine. The concluding section deals with a mechanical method of multiplication that was based on an "areal abacus" consisting of a checkerboard with counters, in which numbers were expressed in the binary scale.

Until recently the historical background of the invention of logarithms has remained something of an enigma. At the Napier tercentenary celebrations, Lord Moulton referred to Napier's invention as a "bolt from the blue" and suggested that nothing had led up to it, foreshadowed it, or heralded its arrival. Notwithstanding, Joost Bürgi, a maker of watches and astronomical instruments, had turned his attention to the problem about the same time and developed a system of logarithms entirely independently. Many Continental historians have accorded him priority in the actual invention, although he certainly did not have it in the publication of his *Arithmetische und geometrische Progress-Tabulen* (1620).

After the revival of learning in western Europe some of the first advances made were in trigonometry, which was developed as an independent field of study, largely in the interests of astronomy but also for surveying, mapmaking, and navigation. Much time was spent in calculating extensive tables of sines and tangents. Trigonometric tables were appearing in all parts of Europe, and stress was laid on the development of formulas, analogous to

$$\sin A \sin B = \tfrac{1}{2}(\cos \overline{A - B} - \cos \overline{A + B}),$$

which could, by converting the product of sines into sums and differences, reduce the computational difficulties. This conversion process was known as prosthaphaeresis. Formulas generated in this way were much used in astronomical calculations and were linked with the names of Longomontanus and Wittich, who both worked as assistants to Tycho Brahe. It is said that word of these developments came to Napier through a fellow countryman, John Craig, who accompanied James VI to Norway in 1590 to meet his bride, Anne of Denmark. The party landed near Tycho Brahe's observatory at Hven and was entertained by the astronomer. Although the construction of Napier's logarithms clearly owes nothing to prosthaphaeresis, the aim—that of substituting addition and subtraction for multiplication and division in trigonometrical calculations—was the same, and if Napier was already working on the problem, he may well have been stimulated to further efforts by the information he received through Craig. There is evidence in a letter written by Kepler in 1624 that he had received an intimation of Napier's work as early as 1594. This information presumably came through Tycho Brahe and Craig.

Napier's own account of his purpose in undertaking the work is printed in the author's preface to the *Descriptio* and is reprinted with slight modification in Wright's translation. Napier says that there is nothing more troublesome to mathematical practice than the "multiplications, divisions, square and cubical extractions of great numbers" and that these operations involve a tedious expenditure of time, as well as being subject to "slippery errors." By means of the tables all these operations could be replaced by simple addition and subtraction.

As presented, Napier's canon is specifically associated with trigonometric usage, in the sense that it gives logarithms of natural sines (from the tables of Erasmus Reinhold). The sine of an arc was not, at that time, given as a ratio but as the length of the semichord of a circle of given radius, subtending a specified angle at the center. In tabulating such sines, it was customary to choose a large number for the radius of the circle (or whole sine); Napier's choice of 10^7 gave him seven significant figures before introducing fractions.

The theory of arithmetic and geometric progressions, which played a central role in Napier's constructions, was of course available from ancient times (Napier quotes Euclid). The correspondence between the terms of an arithmetic and a geometric progression had been explored in detail by many sixteenth-century mathematicians; and Stifel in *Arithmetica integra* (1544) had enunciated clearly the basic laws—but without the index notation—corresponding to

$$a^m a^n = a^{m+n}, \qquad (a^m)^n = a^{mn}.$$

But, in all this work, only the relation between discrete sets of numbers was implied. In Napier's geometric model the correspondence between the terms of an arithmetic and a geometric progression was founded on the idea of continuously moving points and involved concepts of time, motion, and instantaneous speed. Although such notions had played a prominent part in the discussions of the fourteenth-century philosophers of the Merton school (most notably Swineshead in his *Liber calculationum*), there is nothing to suggest that any of this work directly influenced Napier.

Most historical accounts of Napier's logarithms have suffered considerably through translation into modern symbolism. Napier himself used virtually no notation, and his explanatory detail is almost wholly verbal. Without any of the tools of modern analysis for handling continuous functions, his propositions inevitably remained on an intuitive basis. He had, nonetheless, a remarkably clear idea of a functional relation between two continuous variables.

Briefly, two points move along parallel straight lines, the first moving arithmetically through equal distances in equal times and the second moving geometrically toward a fixed point, cutting off proportional parts of the whole line and then of subsequent remainders, also in equal times.

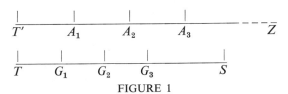

FIGURE 1

If the first point moves through the spaces $T'A_1$, A_1A_2, A_2A_3, $\cdots$, in equal times, then

$$T'A_1 = A_1A_2 = A_2A_3 = \cdots.$$

If the second point moves toward a fixed point S and is at T, G_1, G_2, G_3, $\cdots$, when the first point is at T', A_1, A_2, A_3, $\cdots$, then the spaces TG_1, G_1G_2, G_2G_3, $\cdots$, are also covered in equal times. But since the second point moves geometrically,

$$TG_1/TS = G_1G_2/G_1S = G_2G_3/G_2S = \cdots.$$

It follows that the velocity of the second point is everywhere proportional to its distance from S.

The definition of the logarithm follows: Two points start from T' and T respectively, at the same instant and with the same velocities, the first point moving uniformly and the second point moving so that its velocity is everywhere proportional to its distance from S; if the points reach L and G respectively, at the same instant, the number that measures the line $T'L$ is defined as the logarithm of GS (GS is the sine and TS, the whole sine, or radius).

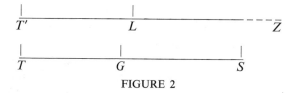

FIGURE 2

From the definition, it follows that the logarithm of the whole sine (10^7) is 0 and that the logarithm of n, where $n > 10^7$, is less than 0. In modern notation, if $T'L = y$, $y_0 = 0$, $GS = x$, $TS = x_0 = r = 10^7$, $dx/dt = -kx$, $dy/dt = kr$, $dy/dx = -r/x$, $\log_e(x/r) = -y/r$, or $\log_{1/e}(x/r) = y/r$. It remained to apply this structure in the calculation of the canon. Without any machinery for handling continuous functions it was necessary for Napier to calculate bounds, between which the logarithm must lie. His entire method depends upon these bounds, together with the corresponding bounds for the difference of the logarithms of two sines.

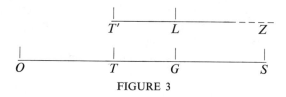

FIGURE 3

If the point O lies on ST produced such that $OS/TS = TS/SG$, then the spaces OT and TG are covered in equal times. But, since $OS > TS > GS$, the velocity at $O >$ the velocity at $T >$ the velocity at G. It follows that $OT > T'L > TG$, and $OS - TS > \log SG > TS - GS$. If $TS = r$, $GS = x$, we have

$$\frac{r - x}{x} > \frac{\log x}{r} > \frac{r - x}{r};$$

the corresponding bounds for the difference between two logarithms are given by

$$\frac{x_1 - x_2}{x_2} > \frac{\log x_2 - \log x_1}{r} > \frac{x_1 - x_2}{x_1}.$$

Napier then calculates in a series of tables the values of

$$10^7 \left(1 - \frac{1}{10^7}\right)^n, \quad n = 0, 1, 2, 3, \cdots, 100;$$

$$10^7 \left(1 - \frac{1}{10^5}\right)^n, \quad n = 0, 1, 2, \cdots, 50;$$

and finally,

$$10^7 \left(1 - \frac{5}{10^4}\right)^n \left(1 - \frac{1}{10^2}\right)^m, \quad n = 0, 1, 2, \cdots, 20;$$

$$m = 0, 1, 2, \cdots, 68.$$

The terms in each progression were obtained by successive subtraction, the last figure in the first table giving the starting point for the second. The final figure in the last table gave a value very little less than $10^7/2$, so that Napier had available a very large number of geometric means distributed over the interval 10^7, $10^7/2$. Using his inequalities, he was able to derive bounds for the logarithms of these numbers and, by taking an arithmetic mean between the bounds, to obtain an accuracy of seven significant figures. By interpolation, he tabulated the values of the logarithms of the sines (and tangents) of angles, taken at one-minute intervals, extending the tables to cover angles between 0 and 90 degrees.

Napier did not think in terms of a base, in the modern sense of the word, although since $\left(1 - \frac{1}{10^7}\right)^{10^7}$ is very nearly $\lim_{n\to\infty} \left(1 - \frac{1}{n}\right)^n$, it is clear that we have virtually a system of logarithms to base $1/e$. In Napier's system, the familiar rules for the logarithms of products, quotients, and exponents did not hold because of the choice of the whole sine (10^7), rather than 1, as the logarithm whose number was zero. Napier's tables were also awkward to use in working with ordinary numbers, rather than sines or tangents.

The calculation of the canon was a tremendous task and occupied Napier personally for over twenty years. Although not entirely free from error the calculations were essentially sound and formed the basis for all subsequent logarithm tables for nearly a century. The publication in 1614 received immediate recognition. Henry Briggs, then Gresham professor of geometry in the City of London, was enthusiastic and visited Napier at Merchiston in the summers of 1615 and 1616. During discussions that took place there the idea emerged of changing the system so that 0 should become the logarithm of unity and 10^{10} that of the whole sine. Briggs in the preface to *Arithmetica logarithmica* (1624) clearly attributes this suggestion to Napier and apparently believed that Napier had become convinced of the desirability of making this change, even before the publication of the *Descriptio*. Because of failing health, however, Napier did not have the energy to embark on this task, and it was left to Briggs to recalculate the tables, adapting them to use with a decimal base. The first 1,000 logarithms of the new canon were published after Napier's death by Briggs, without place or date (but at London before 6 December 1617), as *Logarithmorum chilias prima*. The earliest publication of Napier's logarithms on the Continent was in 1618, when Benjamin Ursinus included an excerpt from the canon, shortened by two places, in his *Cursus mathematici practici*. Through this work Kepler became aware of the importance of Napier's discovery and expressed his enthusiasm in a letter to Napier dated 28 July 1619, printed in the dedication of his *Ephemerides* (1620).

In matters of priority in the invention of logarithms the only serious claims have been made on behalf of Joost Bürgi. Many German historians have accorded him priority in the actual invention on the grounds that his tables had been computed about 1600, although they were not published until 1620. Since Napier's own work extended over a long period of time, both must be accorded full credit as independent inventors. The tables were quite differently conceived, and neither author owed anything to the other. Napier enjoyed the right of priority in publication.

BIBLIOGRAPHY

I. ORIGINAL WORKS. Napier's works are *A Plaine Discovery of the Whole Revelation of Saint John* (Edinburgh, 1593); *Mirifici logarithmorum canonis descriptio, ejusque usus, . . .* (Edinburgh, 1614); *Rabdologiae, seu numerationis per virgulas libri duo* (Edinburgh, 1617); *Mirifici logarithmorum canonis constructio; et eorum ad naturales ipsorum numeros habitudines* (Edinburgh, 1619); *De arte logistica*, Mark Napier, ed. (Edinburgh, 1839); *A Description of the Admirable Table of Logarithmes*: . . ., translated by Edward Wright, published by Samuel Wright (London, 1616). *The Construction of the Wonderful Canon of Logarithms* (Edinburgh, 1889), W. R. Macdonald's trans. of the *Constructio*, contains an excellent catalog of all the editions of Napier's works and their translations into French, Dutch, Italian, and German. Details are also included of the location of these works at that date. Further details and descriptions are included in R. A. Sampson, ed., "Bibliography of Books Exhibited at the Napier Tercentenary Celebrations, July 1914," in C. G.

Knott, ed., *Napier Tercentenary Memorial Volume* (London, 1915).

II. SECONDARY LITERATURE. Such information as is available about Napier's life and work has been fairly well documented by his descendants. Mark Napier, *Memoirs of John Napier of Merchiston; His Lineage, Life and Times* (Edinburgh, 1834), based on careful research of the private papers of the Napier family, is the source of most modern accounts. The tercentenary of the publication of the *Descriptio* was celebrated by an international congress, organized by the Royal Society of Edinburgh. The papers communicated to this congress were published in the *Napier Tercentenary Memorial Volume* (see above) and supply much detail on the historical background to Napier's work. E. M. Horsburgh, ed., *Modern Instruments and Methods of Calculation: A Handbook of the Napier Tercentenary Exhibition* (London, 1914), is also useful. Of the various reconstructions of Napier's work, Lord Moulton's, in the *Tercentenary Memorial Volume*, pp. 1–24, is the most imaginative; E. W. Hobson, *John Napier and the Invention of Logarithms* (Cambridge, 1914), is the most useful.

Still valuable on the early history of logarithms are J. W. L. Glaisher's articles, "Logarithms," in *Encyclopaedia Britannica*, 11th ed. (1910), XVI, 868–877; and "On Early Tables of Logarithms and Early History of Logarithms," in *Quarterly Journal of Pure and Applied Mathematics*, **48** (1920), 151–192. Florian Cajori, "History of the Exponential and Logarithmic Concepts," in *American Mathematical Monthly*, **20** (1913), 5–14, 35–47, 75–84, 107–117, 148–151, 173–182, 205–210, is also useful. A more recent discussion of the development of the concept of logarithm is that of D. T. Whiteside, "Patterns of Mathematical Thought in the Later Seventeenth Century," in *Archive for History of Exact Sciences*, **1** (1961), 214–231.

MARGARET E. BARON

NĀRĀYAṆA (*fl.* India, 1356), *mathematics*.

Nārāyaṇa, the son of Nṛsiṃha (or Narasiṃha), was one of the most renowned Indian mathematicians of the medieval period. His *Gaṇitakaumudī*, on arithmetic and geometry, was composed in 1356; in it he refers to his *Bījagaṇitāvataṃsa*, on algebra (see Supplement). The *Karmapradīpikā*, a commentary on the *Līlāvatī* of Bhāskara II (*b.* 1115), is found in several south Indian libraries attributed to Nārāyaṇa; but the author, a follower of Āryabhaṭa I (*b.* 476), may be the Kerala astronomer and mathematician Mādhava of Saṅgamagrāma (*ca.* 1340–1425).

The *Gaṇitakaumudī* consists of rules (*sūtras*) and examples (*udāharaṇas*), which in the only edition, the two-volume one of P. Dvivedi (Benares, 1936–1942), are given separate numberings that do not coincide with the division of the work into chapters *(vyavahāras)*. In fact, the edition is based on a single manuscript which was evidently corrupt and perhaps incomplete. We do not really know in detail the contents of the *Gaṇitakaumudī*. The *Bījagaṇitāvataṃsa* is preserved in a unique and incomplete manuscript at Benares; only the first part has been edited, by K. S. Shukla as a supplement to *Ṛtam* (**1**, pt. 2 [1969–1970]).

BIBLIOGRAPHY

Various rules from the *Gaṇitakaumudī* are discussed by B. Datta and A. N. Singh, *History of Hindu Mathematics*, 2 vols. (Lahore, 1935–1938), *passim;* and the section of that work devoted to magic squares is analyzed by S. Cammann, "Islamic and Indian Magic Squares," in *History of Religions*, **8** (1968–1969), 181–209, 271–299, esp. 274 ff. The algebra of the *Bījagaṇitāvataṃsa* has been commented on by B. Datta, "Nārāyaṇa's Method for Finding Approximate Value of a Surd," in *Bulletin of the Calcutta Mathematical Society*, **23** (1931), 187–194. See also R. Garver, "Concerning Two Square-Root Methods," *ibid.*, **23** (1932), 99–102; and "The Algebra of Nārāyaṇa," in *Isis*, **19** (1933), 472–485.

DAVID PINGREE

AL-NASAWĪ, ABU 'L-ḤASAN, ʿALĪ IBN AḤMAD (*fl.* Baghdad, 1029–1044), *arithmetic, geometry.*

Arabic biographers do not mention al-Nasawī, who has been known to the scholarly world since 1863, when F. Woepcke made a brief study of his *al-Muqniʿ fi 'l-Ḥisāb al-Hindī* (Leiden, MS 1021). The introduction to this text shows that al-Nasawī wrote, in Persian, a book on Indian arithmetic for presentation to Magd al-Dawla, the Buwayhid ruler in Khurasan who was dethroned in 1029 or 1030. The book was presented to Sharaf al-Mulūk, vizier of Jalāl al-Dawla, ruler in Baghdad. The vizier ordered al-Nasawī to write in Arabic in order to be more precise and concise, and the result was *al-Muqniʿ*. Al-Nasawī seems to have settled in Baghdad; another book by him, *Tajrīd Uqlīdis* (Salar-Jang, MS 3142) was dedicated in highly flattering words to al-Murtadā (965–1044), an influential Shīʿite leader in Baghdad. Nothing else can be said about his life except that al-Nasawī refers to Nasā, in Khurasan, where he probably was born.

Al-Nasawī has been considered a forerunner in the use of the decimal concept because he used the rules $\sqrt{n} = \sqrt{nk^2}/k$ and $\sqrt[3]{n} = \sqrt[3]{nk^3}/k$, where k is taken as a power of 10. If k is taken as 10 or 100, the root is found correct to one or two decimal places.

There is now reason to believe that al-Nasawī cannot be credited with priority in this respect. The two rules were known to earlier writers on Hindu-Arabic arithmetic. The first appeared in the *Paṭīgaṇita* of Śrīdhārācārya (750–850). Like others, al-Nasawī rather mechanically converted the decimal part of the root thus obtained to the sexagesimal scale and suggested taking k as a power of sixty, without showing signs of understanding the decimal value of the fraction. Their concern was simply to transform the fractional part of the root to minutes, seconds, and thirds. Only al-Uqlīdisī (tenth century), the discoverer of decimal fractions, retained some roots in the decimal form.

In *al-Muqniʿ*, al-Nasawī presents Indian arithmetic of integers and common fractions and applies its schemes to the sexagesimal scale. In the introduction he criticizes earlier works as too brief or too long. He states that Kūshyār ibn Labbān (*ca.* 971–1029) had written an arithmetic for astronomers, and Abū Ḥanīfa al-Dīnawārī (*d.* 895) had written one for businessmen; but Kūshyār's proved to be rather like a business arithmetic and Abū Ḥanīfa's more like a book for astronomers. Kūshyār's work, *Uṣūl Ḥisāb al-Hind*, which is extant, shows that al-Nasawī's remark was unfair. He adopted Kūshyār's schemes on integers and, like him, failed to understand the principle of "borrowing" in subtraction. To subtract 4,859 from 53,536, the Indian scheme goes as follows:

Arrange the two numbers as 53536

4859.

Subtract 4 from the digit above it; since 3 is less than 4, borrow 1 from 5, to turn 3 into 13, and subtract. And so on. Both Kūshyār and al-Nasawī would subtract 4 from 53, obtain 49, subtract 8 from 95, and so on. Only finger-reckoners agree with them in this.

In discussing subtraction of fractional quantities, al-Nasawī enunciated the rule $(n_1 + f_1) - (n_2 + f_2) = (n_1 - n_2) + (f_1 - f_2)$, where n_1 and n_2 are integers and f_1 and f_2 are fractions. He did not notice the case when $f_2 > f_1$ and the principle of "borrowing" should be used.

Al-Nasawī gave Kūshyār's method of extracting the cube root and, like him, used the approximation $\sqrt[3]{n} = p + \dfrac{r}{3p^2 + 1}$, where p^3 is the greatest cube in n and $r = n - p^3$. Arabic works of about the same period used the better rule

$$\sqrt[3]{n} = p + \frac{r}{3p^2 + 3p + 1}.$$

Later works called $3p^2 + 3p + 1$ the conventional denominator.

Al-Muqniʿ differs from Kūshyār's *Uṣūl* in that it explains the Indian system of common fractions, expresses the sexagesimal scale in Indian numerals, and applies the Indian schemes of operation to numbers expressed in this scale. But al-Nasawī could claim no priority for these features, since others, such as al-Uqlīdisī, had already done the same thing.

Three other works by al-Nasawī, all geometrical, are extant. One of them is *al-Ishbāʿ*, in which he discusses the theorem of Menelaus. One is a corrected version of Archimedes' *Lemmata* as translated into Arabic by Thābit ibn Qurra, which was later revised by Naṣīr al-Dīn al-Ṭūsī. The last is *Tajrīd Uqlīdis* ("An Abstract From Euclid"). In the introduction, al-Nasawī points out that Euclid's *Elements* is necessary for one who wants to study geometry for its own sake, but his *Tajrīd* is written to serve two purposes: it will be enough for those who want to learn geometry in order to be able to understand Ptolemy's *Almagest*, and it will serve as an introduction to Euclid's *Elements*. A comparison of the *Tajrīd* with the *Elements*, however, shows that al-Nasawī's work is a copy of books I–VI, on plane geometry and geometrical algebra, and book XI, on solid geometry, with some constructions omitted and some proofs altered.

BIBLIOGRAPHY

I. ORIGINAL WORKS. Al-Nasawī's writings include "On the Construction of a Circle That Bears a Given Ratio to Another Given Circle, and on the Construction of All Rectilinear Figures and the Way in Which Artisans Use Them," cited by al-Ṭūsī in *Maʾkhūdhāt Arshimīdis*, no. 10 of his *Rasāʾil*, II (Hyderabad-Deccan, 1940); *al-Ishbāʿ*, trans. by E. Wiedemann in his *Studien zur Astronomie der Araber* (Erlangen, 1926), 80–85—see also H. Burger and K. Kohl, *Geschichte des Transversalensätze* (Erlangen, 1924), 53–55; *Kitāb al-lāmiʿ fī amthilat al-Zīj al-jāmiʿ* ("Illustrative Examples of the Twenty-Five Chapters of the *Zīj al-jāmiʿ* of Kūshyār"), in Ḥājji Khalīfa, *Kashf* (Istanbul, 1941), col. 970; and *Risāla fī maʿrifat al-taqwīm waʾl-asṭurlāb* ("A Treatise on Chronology and the Astrolabe"), Columbia University Library, MS Or. 45, op. 7.

II. SECONDARY LITERATURE. See H. Suter, "Über des Rechenbuch des Ali ben Ahmed el-Nasawi," in *Bibliotheca mathematica*, 2nd ser., **7** (1906), 113–119; and F. Woepcke, "Mémoires sur la propagation des chiffres indiens," in *Journal asiatique*, 6th ser., **1** (1863), 492 ff.

See also Kūshyār ibn Labbān, *Uṣūl Ḥisāb al-Hind*, in M. Levey and M. Petruck, *Principles of Hindu Reckoning* (Madison, Wis., 1965), 55–83.

A. S. SAIDAN

NAṢĪR AL-DĪN AL-ṬŪSĪ. See **al-Ṭūsī, Naṣīr al-Dīn.**

NASMYTH, JAMES (*b.* Edinburgh, Scotland, 19 August 1808; *d.* London, England, 7 May 1890), *engineering, astronomy.*

Although best known for his steam hammer, Nasmyth also did much to improve the design of machine tools in general. His mechanical skills were used to help William Lassell build a very fine Newtonian reflector, and he published astronomical observations of some interest.

Nasmyth was the youngest son of Alexander Nasmyth, a portrait and landscape painter, and Barbara Foulis. Leaving the high school in Edinburgh, in 1820, he was educated privately—and not at all well—but he acquired some knowledge of chemistry, mathematics, and natural philosophy. Through his skill in model engineering, he met John Leslie, who gave him admittance to his classes at Edinburgh University. In 1821 he attended the Edinburgh School of Arts. In 1829 he found employment as personal assistant to Henry Maudslay in his works in London, and in 1834 he established his own business in Manchester. So successful was he there, and in many later enterprises, that by 1856 he was able to retire with a large fortune.

The steam hammer was Nasmyth's most successful invention. Power hammers had previously been worked by steam, but in an imprecise and relatively uncontrolled manner, actuated by levers or cams. Nasmyth produced his design in November 1839 with a view to its being used for the forging of a thirty-inch-diameter paddle shaft in prospect for the steamship *Great Britain*, then on the stocks at Bristol. (The shaft was forged, but not actually used, since the ship was eventually screw-driven.) His first solution was a single-acting hammer, operating by gravity, the steam merely lifting the hammer for each successive drop. Acceptance of his design was at first slow; and the first steam hammer to be built, in 1842 at the ironworks founded by Adolphe and Eugène Schneider at Creuzot, was copied without Nasmyth's knowledge from his private "scheme-book" design. James Watt is said to have anticipated the idea. The steam hammer, with many improvements, soon became perhaps the most dramatic symbol of steam power, particularly as it made possible the forging of very large guns for the British navy.

Nasmyth also designed a milling machine, a planing or shaping machine, and a steam pile driver. After his early retirement, he took up astronomy, creating something of a stir when he announced that the solar surface was patterned like willow leaves (1862). A book on the moon, written jointly with James Carpenter, was well illustrated.

BIBLIOGRAPHY

I. Original Works. Nasmyth's book on the moon is *The Moon Considered as a Planet, a World, and a Satellite* (London, 1874; 4th ed., London, 1903), written with James Carpenter; his paper on the sun is "On the Structure of the Luminous Envelope of the Sun," in *Manchester Philosophical Society Memoirs*, **1** (1862), 407–411. For his astronomical work, all of minor importance, see Agnes Clerke, *History of Astronomy During the Nineteenth Century*, 3rd ed. (London, 1893), 103, 204, 313, 326, 352. Apart from his autobiography, Nasmyth published nothing on the steam hammer except his patent (No. 9382, 9 June 1842).

II. Secondary Literature. The literature is very extensive, but the main biographical source is Nasmyth's autobiography, edited by Samuel Smiles (London, 1883; Cambridge, 1931). See also Thomas Baker, *Elements of Mechanism. With Remarks on Tools and Machines by J. Nasmyth*, 2nd ed. (London, 1858).

J. D. North

NATALIS, STEPHANUS. See **Nöel, Étienne.**

NATANSON, WŁADYSŁAW (*b.* Warsaw, Poland, 18 June 1864; *d.* Cracow, Poland, 26 February 1937), *physics.*

Natanson, a son of Ludwik Natanson and Natalia Epstein, came from a distinguished literary and scientific family. His father, a physician, wrote papers on medicine and edited a medical journal. While attending school in Warsaw from 1874 to 1882, he wrote his first memoirs, collaborating with his older brother Edward. In 1882 he began to study mathematics and physics at the University of St. Petersburg. During vacations, in a laboratory in the family home in Warsaw, Władysław and Edward carried out experiments on the dissociation of nitrogen tetroxide— one of the first experimental confirmations of the law of mass action. In 1886 Natanson completed his studies at St. Petersburg and went to England, where he worked for a time in the Cavendish Laboratory, then directed by J. J. Thomson. In 1887 he received a master's degree in science after presenting "Über die kinetische Theorie unvollkommener Gase" at the University of Dorpat, where the physics department was directed by A. von Oettingen. Here too in 1888 he presented his thesis, "Über die kinetische Theorie der Jouleschen Erscheinung," and was awarded a doctorate. In 1891 Natanson received the *veniam legendi* at the Jagiellonian University in Cracow and became a professor there, occupying the chair of theoretical physics until his retirement in 1935. In 1911 he married Elżbieta Baranowska.

A member of the physical societies of London, Berlin, and Paris, Natanson was a founder and first president (1920) of the Polish Physical Society. From 1893 he was a corresponding member, and from 1900 a member, of the Academy of Sciences at Cracow (later the Polish Academy of Sciences). In 1925 he was elected vice-president of the International Union of Pure and Applied Physics. Rector of the Jagiellonian University in 1922, Natanson received an honorary Ph.D. from that university in 1930.

After the experiments on the dissociation of nitrogen tetroxide that he had carried out with his brother, Natanson worked on the kinetic theory of gases. His "Über die kinetische Theorie unvollkommener Gase" contains, for example, a proof that the aggregate of molecules of a gas, however great, underlies both Maxwell's law of distribution and the law of the equipartition of energy—a statement important for the theory of Brownian movement, although the paper was not seen then in this light. Beginning in 1891 Natanson published several papers on thermodynamics. The most important are "On the Laws of Irreversible Phenomena" (1896) and "Sur les propriétés thermocinétiques des potentiels thermodynamiques" (1897). Considering the thermodynamics of his time to be mere thermostatics, Natanson sought a way to achieve genuine thermodynamics, examining the function of energy dissipation and introducing a generalization of Hamilton's principle. For postulating this principle of many applications and general scope (sometimes called Natanson's thermokinetic principle), Natanson is considered a pioneer in the thermodynamics of irreversible processes.

Natanson later worked on the hydrodynamics of viscous liquids and on such related phenomena as the double refraction of light in moving viscous liquids. Inspired by Lorentz' works, he published several papers on the optical properties of matter. In "On the Elliptic Polarization of Light Transmitted Through an Absorbing Naturally-Active Medium" (1908) Natanson gave a theory of Cotton's phenomenon and a rule governing it (known in the French literature as the *règle de Natanson*). In 1929–1933 he worked on Fermat's principle in relation to wave mechanics.

Natanson published five volumes of lectures and essays on scientists, writers, and philosophers, as well as on intellectual and religious trends. He was also an author and a coauthor of university and secondary school textbooks.

BIBLIOGRAPHY

I. ORIGINAL WORKS. German, English, and French versions of Natanson's works are in Wiedemann's *Annalen der Physik und Chemie* (1885–1891); *Philosophical Magazine* (1890–1933); and *Bulletin international de l'Académie polonaise des sciences et des lettres de Cracovie* (1891–1933). Lists of his publications are in the articles by Weyssenhoff and Klecki (see below).

His articles include "Über die Dissoziation des Untersalpetersäuredampfes," in *Annalen der Physik und Chemie*, **24** (1885), 454, and **27** (1886), 606, written with E. Natanson; "Über die kinetische Theorie unvollkommener Gase," *ibid.*, **33** (1888), 683; "On the Laws of Irreversible Phenomena," in *Philosophical Magazine*, **41** (1896), 385; "Sur les propriétés thermocinétiques des potentiels thermodynamiques," in *Bulletin de l'Académie des sciences de Cracovie* (1897), 247; "On the Elliptic Polarization of Light Transmitted Through an Absorbing Naturally-Active Medium," *ibid.* (1908), 764; and "Fermat's Principle," in *Philosophical Magazine*, **16** (1933), 178. Natanson's autobiography was published in *Postępy fizyki*, **9** (1958), 115.

II. SECONDARY LITERATURE. See the following, listed chronologically: J. W. Weyssenhoff, in *Acta physica polonica*, **6** (1937), 295, an obituary in English and Polish with a bibliography of Natanson's works; L. Klecki, in *Prace matematyczno-fizyczne*, **46** (1939), 1, an obituary in French with a bibliography without titles; and articles by J. W. Weyssenhoff, A. Piekara, and L. Infeld, in *Postępy fizyki*, **9** (1958).

See also Armin Teske, "Sur un travail de Ladislas Natanson de 1888," in *Actes du VIIIe Congrès international d'histoire des sciences*, II (Florence, 1958), 123; and K. Gumiński, "O pracach termodynamicznych Władysława Natansona" ("Władysław Natanson's Works on Thermodynamics"), in *Postępy fizyki*, **17** (1966), 101.

ANDRZEJ A. TESKE

NATHORST, ALFRED GABRIEL (*b.* Väderbrunn, Södermanland, Sweden, 7 November 1850; *d.* Stockholm, Sweden, 20 January 1921), *paleobotany, geology, exploration.*

His parents, Hjalmar Otto Nathorst and Maria Charlotta af Georgii, moved in 1861 to Alnarp, in Skåne, where his father had been appointed professor at the Institute of Agriculture. Nathorst was educated at Malmö and entered the University of Lund in 1868. In 1871 he enrolled at the University of Uppsala but returned to Lund, where he took his doctorate in 1874. He was docent of geology from 1874 to 1879. In 1873 he became a member of the staff of the Geological Survey of Sweden serving until 1884. He was then given the post of professor and director of the newly created Department of Archegoniates and Fossil Plants at the Swedish Museum of Natural History in Stockholm. Nathorst held this position until he resigned in 1917.

Nathorst showed an early inclination for the outdoors and natural science, particularly botany. In Lund, he turned to geology under the influence of N. P. Angelin, a pioneer in Swedish paleontology and stratigraphic geology. Nathorst's first published paper (1869) was a detailed study of a Cambrian sequence in Skåne. As an officer in the Geological Survey, Nathorst made many important contributions to the knowledge of the geology of south Sweden. He discovered and described (1871) the remains of glacial plants in a freshwater clay in Skåne. This paper was the first in a series of contributions to the study of vegetational history in Sweden in postglacial times.

Nathorst's paleobotanical investigations of the Rhaeto-Liassic flora of Skåne resulted in a number of publications beginning in 1875. Since all but one of these papers are in Swedish, they failed to win the recognition they deserved. His international reputation as a paleobotanist is based instead on his monographs of the Tertiary floras of Japan (1882, 1888) and of the Paleozoic and Mesozoic floras of the Arctic. He was not speculative, and he treated his material from a strictly morphological and taxonomic point of view. Nathorst himself assembled part of the collections that he studied.

In 1871 Nathorst took part in an expedition to Spitsbergen. He led several expeditions to Svalbard and Greenland to study the geography, geology, and biology of the Arctic. Nathorst also attempted to elucidate experimentally the origin of fossil trails and tracks (1881, 1886). Although temperamental he won friendship and respect in international circles. His enormous capacity for work explains in part the great extent and importance of his scientific output.

BIBLIOGRAPHY

Nathorst's most important works were "Zur fossilen Flora Japans," in *Palaeontologische Abhandlungen*, **4**, pt. 3 (1888), 195–250; *Zur fossilen Flora der Polarländer*, 2 vols. in 5 pts. (Stockholm, 1894–1920); and "Beiträge zur Geologie der Bären-Insel, Spitzbergens und des König-Karl-Landes," in *Bulletin of the Geological Institute of Uppsala*, **10** (1910), 257–416. The most exhaustive biography of Nathorst is by T. G. Halle, in *Geologiska föreningens i Stockholm förhandlingar*, **43** (1921), 241–280, with bibliography, 281–311. See also the obituary by A. C. Seward in *Botanical Gazette* (Chicago), **71** (1921), 464–465.

GERHARD REGNÉLL

NAUDIN, CHARLES (*b.* Autun, France, 14 August 1815; *d.* Villa Thuret, near Antibes, France, 19 March 1899), *horticulture, experimental botany.*

Naudin was the son of a petty entrepreneur whose financial successes were rare. His childhood and youth were thus marked by frequent moves and numerous schools but also by an extraordinary determination to prepare for a medical, and then a scientific, career. After receiving the baccalaureate in science at Montpellier in 1837, he moved on to Paris. Working as bookkeeper, tutor, private secretary, and gardener, he earned his doctorate in 1842 and awaited an opening in the French educational system. He occupied minor posts until 1846, when, recommended by his lifelong friend and supporter, the botanist Joseph Decaisne, he joined the herbarium staff at the Muséum d'Histoire Naturelle and became professor of zoology at the Collège Chaptal. Almost immediately Naudin was obliged to resign these posts and to seek his livelihood elsewhere in Paris and in the provinces. He had been struck by a severe nervous disorder which left him totally deaf and in constant pain; his public career, so arduously earned, had to be abandoned. He finally settled at Collioure, in 1869, and established a private experimental garden devoted especially to problems of acclimatization and earned his living by the sale of seeds and specimens. In 1878 he became the first director of the experimental garden at Antibes given to the state by the family of the horti-culturist Gustave Thuret. Thus, not until age sixty-three did Naudin find financial security and suitable institutional support. His plaint rings true: "Happy is the professor who enjoys an assured income and whom the government provides with assistance and collaborators."

Throughout these years of insecurity and frequent isolation, Naudin pursued a remarkably varied program of research and horticultural promotion. His primary interests focused on acclimatization and economic botany and on the relation of hybridization to the formation of new biological species. The gardens at Collioure and Antibes were directed toward the introduction into France and her colonies, notably Algeria, of foreign plants of potential eco-nomic value. Naudin demonstrated exceptional skill as horticulturist and arboriculturist, and the garden at Antibes soon became a primary means of communi-cation among French botanists and agronomists and their foreign colleagues. Naudin himself paid particular attention to the economic potential of the Australian import *Eucalyptus* for dry and saline areas of southern France.

By far Naudin's most celebrated scientific work was done on problems of plant hybridization. His research began in 1854 and continued for two decades. Decaisne had suggested hybridization as a seemingly fruitful approach to the issue of species stability; Linnaeus' famous experiments (1759) with speedwell and goats-beard had suggested that man might indeed modify nature's creations. Working primarily with *Datura* species, Naudin pursued this suggestion and arrived at results of interest to the history of both the study of inheritance and of evolution theory. He ascertained that the first generation of hybrids was relatively homogeneous in appearance and that reciprocal crosses produced identical results. From this first generation of hybrids he then produced a second and thereby established that second-generation hybrids display extraordinary diversity; "disjunction" of all the species' characters seems to occur, with new and unexpected combinations appearing in the offspring. His contemporary Gregor Mendel also recognized these phenomena, but, unlike Naudin, he marshaled the data from the second generation and sought its explanation in the statistical distribution of hereditary factors. Naudin overlooked this crucial step and could only emphasize the seemingly chaotic distribu-tion of characters in second-generation hybrids, a phenomenon now called segregation.

Hybridization proved effective for Naudin in the limited production of new species. His faith in evolution was real but constrained. He held that the present diversity of specific forms had been produced from a reduced number of aboriginal forms. Hybridization was the primary agency of change, not natural selection or environmental action. The ancestral or primary forms were of basic importance; all other species were secondary productions and might or might not exhibit permanence. Naudin's scheme, remarkably consonant with the century-old conclusions of Linnaeus, thus reveal his belief in the reality of species transformation as well as his res-ervations regarding proposed evolutionary mech-anisms. Hybridization, the object of Naudin's most prolonged and assiduous investigations, thus provided a seemingly plausible alternative mechanism. At the same time it ensured the creation of but one more explanation of evolutionary change in those confused years between 1859 and 1900, when the phenomena of inheritance were brought by Darwin to the center of attention of natural history and left there unresolved.

BIBLIOGRAPHY

I. ORIGINAL WORKS. Naudin published voluminously and widely. The principal listing of his scattered writings is the Royal Society *Catalogue of Scientific Papers*, IV, 575–576; VIII, 483; X, 901; XVII, 459. These lists are nonetheless incomplete. His major papers on hybridization are "Réflexions sur l'hybridation dans les végétaux," in *Revue horticole*, 4th ser., **4** (1855), 351–354; "Sur les plantes hybrides," *ibid.*, **10** (1861), 396–399; "Nouvelles recherches sur l'hybridité dans les végétaux," in *Annales des sciences naturelles*, Botanique, 4th ser., **19** (1863), 180–203, which offers only the "Conclusions" to Naudin's foremost contribution to the study of heredity, published under the same title in *Nouvelles archives du Muséum d'histoire naturelle*, **1** (1865), 25–176; and "De l'hybridité considérée comme cause de la variabilité dans les végétaux," in *Comptes rendus . . . de l'Académie des sciences*, **59** (1864), 837–845.

Other publications include "Les espèces affines et la théorie de l'évolution," in *Bulletin. Société botanique de France*, **21** (1874), 240–272; *Le jardin du cultivateur* (Paris, 1857); *Manuel de l'amateur des jardins, traité général d'horticulture*, 4 vols. (Paris, 1862–1871), written with Joseph Decaisne; *Mémoire sur les eucalyptus introduits dans la région méditerranéenne* (Paris, 1883), also published in *Annales des sciences naturelles*, Botanique, 5th ser., **16** (1883), 337–430.

II. SECONDARY LITERATURE. The principal account of Naudin's life is Marcelin Berthelot, *Notice historique sur la vie et les travaux de M. Naudin, lue à l'Académie des sciences le 17 décembre 1900* (Paris, 1900); see also E. Bornet's brief notice in *Comptes rendus . . . de l'Académie des sciences*, **128** (1899), 127–128. The only comprehensive study of Naudin's scientific inquiries, especially those dealing with plant hybridization, is Louis Blaringham, "La notion de l'espèce et la disjonction des hybrides d'après Charles Naudin (1852–1875)," in *Progressus rei botanicae*, **4** (1913), 27–108. Shorter accounts are H. F. Roberts, *Plant Hybridization Before Mendel* (Princeton, 1929), 129–136; R. C. Olby, *Origins of Mendelism* (London, 1966), 62–66; Jean F. Leroy, "Naudin, Spencer et Darwin dans l'histoire des théories de l'hérédité," *Actes du XIe Congrès international d'histoire des sciences*, V (Warsaw–Krakow, 1968), 64–69; and A. E. Gaisinovich, *Zarozhdenie genetiki* ("The Origin of Genetics"; Moscow, 1967), 54–71.

WILLIAM COLEMAN

NAUMANN, ALEXANDER (*b.* Eudorf, near Alsfeld, Prussia, 31 July 1837; *d.* Giessen, Germany, 16 March 1922), *chemistry*.

The son of a Protestant minister, Naumann attended the Gymnasium in Darmstadt and then studied chemistry and mathematics at the University of Giessen. After graduating in 1858 he became an assistant at the technical school in Darmstadt and in 1860–1861 was an assistant in the chemistry institute of the University of Tübingen. He moved to Giessen in 1862 and taught mathematics in the Gymnasium there. Naumann qualified as a lecturer in chemistry in 1864 at the University of Giessen. While continuing to teach at the Gymnasium, he lectured and conducted research at the university. He became associate professor in 1869 and, in 1882, full professor and director of the chemistry laboratory, where he remained active until his retirement in 1913.

Naumann began his scientific work in organic chemistry with an investigation of the chlorination of butyric acid and studies of the esters of benzoic acid. In his *Habilitationsschrift*, which dealt with the bromination of acetyl chloride, his interest in the study of the reaction mechanism was already evident. From this time on, he dedicated himself to physical chemistry, especially thermochemistry. During the 1860's the thermodynamic knowledge recently acquired in physics was slowly penetrating the field of chemistry, and the results of Naumann's tireless work contributed significantly to preparing the way for later important discoveries in chemical thermodynamics.

In an essay (1867) that can be considered a contribution toward Guldberg and Waage's law of mass action, Naumann expressed the view that only those molecules which possess energy higher than the critical energy can react with each other. At a constant temperature molecules form and disintegrate, thus producing an equilibrium. With increasing temperature the number of molecular collisions increases while the reaction velocity increases at an ever greater rate.

In his investigation in 1878 of the dissociation process $N_2O_4 \rightleftharpoons 2NO_2$ Naumann demonstrated the validity of the law of mass action, which had already been formulated. Many of his papers dealt with the equilibrium ratios between water vapor and various crystal hydrates, as well as determinations of vapor densities and heats of decomposition.

Naumann's scientific activity diminished drastically after his appointment as full professor at Giessen. Perhaps the legacy of his predecessors at the chemistry laboratory, Liebig and Heinrich Will, both great organic chemists, proved too heavy a burden. Moreover, his responsibility for teaching primarily organic chemistry diverted him from the field in which he had originally done creative work—without providing a substitute. As a result he devoted himself to university administration and took an interest in politics.

BIBLIOGRAPHY

Naumann's most important books are *Grundriss der Thermochemie* (Brunswick, 1869); and *Lehr- und Handbuch der Thermochemie* (Brunswick, 1882). Many of his other publications are listed in Poggendorff, III, 958–959; IV, 1059; V, 895. An obituary is *Akademische Rede zur Jahresfeier der Hessischen Ludwigs Universität* (Giessen, 1922), 43.

F. SZABADVÁRY

NAUMANN, KARL FRIEDRICH (*b.* Dresden, Germany, 30 May 1797; *d.* Dresden, 26 November 1873), *mineralogy, geology.*

Naumann discovered tetartohedrism in the isometric, tetragonal, and hexagonal crystal systems and was the first to observe hemimorphism. His *Lehrbuch der Geognosie* (1850–1854) was the most authoritative work on petrography in the mid-nineteenth century and served as a standard textbook for decades.

Naumann's father, Johann Gottlieb, was a noted composer of church music. In 1816 Naumann went to the mining academy at Freiberg to study mineralogy under Werner. After Werner's death in 1817, Naumann continued his education at Leipzig and at Jena, where he received his doctorate in 1819. During 1821 and 1822 he traveled in Norway observing its geology and collecting minerals. In 1823 he became a *Privatdozent* at Jena, in 1824 a *Dozent* at Leipzig, and in 1826 he was named professor of crystallography at Freiberg. His first book, *Beiträge zur Kenntnis Norwegens* (1824), described his observations in Norway. In 1825 he published his *Grundriss der Krystallographie,* in which he introduced the concept of a "crystal series," that is, the aggregate of all crystal forms that can be developed from a basic form in accordance with Weiss's law of zones (the law of rational intercepts). In this work also Naumann examined Mohs's 1822 suggestion that crystal systems might exist in which the crystallographic axes are not mutually perpendicular and successfully identified the present monoclinic system. His *Lehrbuch der reinen und angewandten Krystallographie* (1830) was even more important in that Naumann introduced a novel method for the designation and treatment of crystal forms, which greatly simplified and coordinated those of Weiss and Mohs and which was adopted almost immediately by German crystallographers. Naumann also analyzed the tetragonal system in this work and commenced an examination of the incomplete symmetry of some crystals, which led to many published descriptions of tetartohedrism and hemimorphism.

Although Naumann continued to write on crystallography, his central interest turned to geognosy. In 1835 he was named professor of geognosy at Freiberg, and in 1842 he became professor of mineralogy and geognosy at Leipzig. His 1846 *Elemente der Mineralogie,* which successfully coordinated the systems of Mohs and Berzelius, went to fifteen editions, seven of them posthumous.

Naumann's most important work was his *Lehrbuch der Geognosie.* In it he differentiated rocks primarily according to their origin, which he determined from their texture, for example, crystalline, clastic, and hyaline. He supported the theory that most gneisses and schists had been formed from sedimentary rocks but admitted that some gneisses had been produced by the deformation of igneous rocks. The first text to devote considerable space to tectonics, his *Lehrbuch* contained all of the scientific information known about earthquakes at that time. Naumann held that certain earthquakes occurred independently of any volcanic activity and might therefore be termed "plutonic." This view was in opposition to that of Humboldt, who believed earthquakes and volcanoes to be merely different manifestations of the same causes.

Late in life, Naumann's proficiency in mathematics led him to the study of symmetry in plants and conch shells. In 1872 he retired from his chair at Leipzig and returned to Dresden, where he died the following year. He was a corresponding member of the academies of Berlin, Munich, St. Petersburg, and Paris, of the Royal Society of London and of the American Philosophical Society, and he received the Wollaston Medal of the Geological Society of London in 1865.

BIBLIOGRAPHY

I. ORIGINAL WORKS. Naumann's principal works are *Beiträge zur Kenntnis Norwegens,* 2 vols. (Leipzig, 1824); *De granite juxta calcem transitorium posito* (Jena, 1823); *De hexagonali crystallinarium formarum systemate* (Leipzig, 1824); *Grundriss der Krystallographie* (Leipzig, 1825); *Entwurf der Lithurgik oder ökonomischen Mineralogie* (Leipzig, 1826); *Lehrbuch der Mineralogie* (Leipzig, 1828); *Lehrbuch der reinen und angewandten Krystallographie,* 2 vols. (Leipzig, 1830); *Geognostische Beschreibung des Königreiches Sachsen,* 5 vols. (Dresden–Leipzig, 1834–1844), written with B. von Cotta; *Anfangsgründe der Krystallographie* (Dresden–Leipzig, 1841); *Elemente der Mineralogie* (Leipzig, 1846); *Lehrbuch der Geognosie,* 2 vols. (Leipzig, 1850–1854); *Elemente der theoretischen Krystallographie* (Leipzig, 1856); *Geognostische Beschreibung des Kohlenbassins von Flöha* (Leipzig, 1864); and *Geognostische Karte des erzgebirgschen Bassins* (Leipzig, 1866). He published about 100 scientific articles; see Royal Society *Catalogue of Scientific Papers,* IV, 576–578; VIII, 484.

II. SECONDARY LITERATURE. See *Allgemeine Deutsche Biographie,* XXIII, 316–319; H. B. Geinetz, "Zur Erinnerung an Dr. Carl Friedrich Naumann," in *Neues Jahrbuch für Mineralogie, Geologie und Paläontologie* for 1874, p. 147; and Franz von Kobell, "Nekrolog auf Dr. K. F. Naumann," in *Sitzungsberichte der Bayerischen Akademie der Wissenschaften zu München,* Math.-phys. Kl., for 1874, pp. 81–84.

JOHN G. BURKE

DICTIONARY
OF
SCIENTIFIC BIOGRAPHY

PUBLISHED UNDER THE AUSPICES OF
THE AMERICAN COUNCIL OF LEARNED SOCIETIES

The American Council of Learned Societies, organized in 1919 for the purpose of advancing the study of the humanities and of the humanistic aspects of the social sciences, is a nonprofit federation comprising forty-five national scholarly groups. The Council represents the humanities in the United States in the International Union of Academies, provides fellowships and grants-in-aid, supports research-and-planning conferences and symposia, and sponsors special projects and scholarly publications.

MEMBER ORGANIZATIONS
AMERICAN PHILOSOPHICAL SOCIETY, 1743
AMERICAN ACADEMY OF ARTS AND SCIENCES, 1780
AMERICAN ANTIQUARIAN SOCIETY, 1812
AMERICAN ORIENTAL SOCIETY, 1842
AMERICAN NUMISMATIC SOCIETY, 1858
AMERICAN PHILOLOGICAL ASSOCIATION, 1869
ARCHAEOLOGICAL INSTITUTE OF AMERICA, 1879
SOCIETY OF BIBLICAL LITERATURE, 1880
MODERN LANGUAGE ASSOCIATION OF AMERICA, 1883
AMERICAN HISTORICAL ASSOCIATION, 1884
AMERICAN ECONOMIC ASSOCIATION, 1885
AMERICAN FOLKLORE SOCIETY, 1888
AMERICAN DIALECT SOCIETY, 1889
AMERICAN PSYCHOLOGICAL ASSOCIATION, 1892
ASSOCIATION OF AMERICAN LAW SCHOOLS, 1900
AMERICAN PHILOSOPHICAL ASSOCIATION, 1901
AMERICAN ANTHROPOLOGICAL ASSOCIATION, 1902
AMERICAN POLITICAL SCIENCE ASSOCIATION, 1903
BIBLIOGRAPHICAL SOCIETY OF AMERICA, 1904
ASSOCIATION OF AMERICAN GEOGRAPHERS, 1904
HISPANIC SOCIETY OF AMERICA, 1904
AMERICAN SOCIOLOGICAL ASSOCIATION, 1905
AMERICAN SOCIETY OF INTERNATIONAL LAW, 1906
ORGANIZATION OF AMERICAN HISTORIANS, 1907
AMERICAN ACADEMY OF RELIGION, 1909
COLLEGE ART ASSOCIATION OF AMERICA, 1912
HISTORY OF SCIENCE SOCIETY, 1924
LINGUISTIC SOCIETY OF AMERICA, 1924
MEDIAEVAL ACADEMY OF AMERICA, 1925
AMERICAN MUSICOLOGICAL SOCIETY, 1934
SOCIETY OF ARCHITECTURAL HISTORIANS, 1940
ECONOMIC HISTORY ASSOCIATION, 1940
ASSOCIATION FOR ASIAN STUDIES, 1941
AMERICAN SOCIETY FOR AESTHETICS, 1942
AMERICAN ASSOCIATION FOR THE ADVANCEMENT OF SLAVIC STUDIES, 1948
METAPHYSICAL SOCIETY OF AMERICA, 1950
AMERICAN STUDIES ASSOCIATION, 1950
RENAISSANCE SOCIETY OF AMERICA, 1954
SOCIETY FOR ETHNOMUSICOLOGY, 1955
AMERICAN SOCIETY FOR LEGAL HISTORY, 1956
AMERICAN SOCIETY FOR THEATRE RESEARCH, 1956
SOCIETY FOR THE HISTORY OF TECHNOLOGY, 1958
AMERICAN COMPARATIVE LITERATURE ASSOCIATION, 1960
AMERICAN SOCIETY FOR EIGHTEENTH-CENTURY STUDIES, 1969
ASSOCIATION FOR JEWISH STUDIES, 1969

DICTIONARY

OF

SCIENTIFIC BIOGRAPHY

CHARLES COULSTON GILLISPIE

Princeton University

EDITOR IN CHIEF

Volume 10

S. G. NAVASHIN – W. PISO

CHARLES SCRIBNER'S SONS · NEW YORK

Panel of Consultants

Contributors to Volume 10

The following are the contributors to Volume 10. Each author's name is followed by the institutional affiliation at the time of publication and the names of the articles written for this volume. The symbol † means that an author is deceased.

CONTRIBUTORS TO VOLUME 10

FRANK N. EGERTON III
University of Wisconsin-Parkside
PETTY

CAROLYN EISELE
Hunter College, City University of New York
B. PEIRCE; B. O. PEIRCE II; C. S. PEIRCE

CHURCHILL EISENHART
National Bureau of Standards
K. PEARSON

ANN MARIE ERDMAN
Florida State University
PEKELHARING

VASILIY A. ESAKOV
Academy of Sciences of the U.S.S.R.
PALLAS

JOSEPH EWAN
Tulane University
ORTON

JOAN M. EYLES
T. OLDHAM

W. V. FARRAR
University of Manchester
NERI; NIEUWLAND; PAYEN; W. H. PERKIN, JR.

LUCIENNE FÉLIX
PAINLEVÉ; C. É. PICARD

KONRADIN FERRARI D'OCCHIEPPO
University of Vienna
OPPENHEIM; OPPOLZER; PALISA

MARTIN FICHMAN
York University
P. PETIT

M. FIERZ
Federal Institute of Technology, Zurich
PAULI

KARIN FIGALA
Deutsches Museum
PAULLI

BERNARD S. FINN
Smithsonian Institution
PELTIER

PAUL FORMAN
Smithsonian Institution
ORNSTEIN; PASCHEN

ROBERT FOX
University of Lancaster
A. T. PETIT; M.-A. PICTET

PIETRO FRANCESCHINI
University of Florence
PACINI

EUGENE FRANKEL
NICOL

H.-CHRIST. FREIESLEBEN
C. A. F. PETERS; C. F. W. PETERS

RICHARD D. FRENCH
Privy Council Office, Government of Canada
OLIVER

HANS FREUDENTHAL
Rijksuniversiteit, Utrecht
NIEUWENTIJT

KURT VON FRITZ
University of Munich
PHILOLAUS OF CROTONA

GERALD L. GEISON
Princeton University
PASTEUR

PATSY A. GERSTNER
Howard Dittrick Museum of Historical Medicine
J. PARKINSON

PAUL GLEES
University of Göttingen
NISSL

A. GOUGENHEIM
Académie des Sciences de l'Institut de France
G. PERRIER

EDWARD GRANT
Indiana University, Bloomington
PETER PEREGRINUS

FRANK GREENAWAY
Science Museum, London
PERCY

A. T. GRIGORIAN
Academy of Sciences of the U.S.S.R.
NEKRASOV; PETERSON; N. P. PETROV

N. A. GRIGORIAN
Academy of Sciences of the U.S.S.R.
ORBELI; I. P. PAVLOV

A. RUPERT HALL
Imperial College of Science and Technology
OLDENBURG

WALLACE B. HAMBY, M.D.
PARÉ

RICHARD HART
Boston University
PEASE

HAROLD HARTLEY †
PARTINGTON

JOHN L. HEILBRON
University of California, Berkeley
NOLLET

DIETER B. HERRMANN
Archenhold Observatory, Berlin
NICOLAI

PIERRE HUARD
René Descartes University
PECQUET; PEYER

KARL HUFBAUER
University of California, Irvine
C. NEUMANN

MARIE-JOSÉ IMBAULT-HUARD
René Descartes University
PECQUET; PEYER

JEAN ITARD
Lycée Henri IV
OCAGNE

S. A. JAYAWARDENE
Science Museum, London
PACIOLI

BØRGE JESSEN
University of Copenhagen
PETERSEN

PAUL JOVET
Centre National de Floristique
PÉRON

GEORGE B. KAUFFMAN
California State University, Fresno
PFEIFFER

ALEX G. KELLER
University of Leicester
D'ORTA; C. PERRAULT

SUZANNE KELLY, O.S.B.
Stonehill College
NORMAN

HUBERT C. KENNEDY
Providence College
PADOA; PARSEVAL DES CHÊNES; PEANO; PIERI

MILTON KERKER
Clarkson College of Technology
PAMBOUR

DANIEL J. KEVLES
California Institute of Technology
NICHOLS

PEARL KIBRE
Hunter College, City University of New York
PETRUS BONUS

GEORGE KISH
University of Michigan
A. E. NORDENSKIÖLD

MARC KLEIN
Louis Pasteur University
OKEN

ZDENĚK KOPAL
University of Manchester
E. AND N. PIGOTT

ELAINE KOPPELMAN
Goucher College
PEACOCK

HANS GÜNTHER KÖRBER
Zentralbibliothek des Meteorologischen Dienstes der DDR, Potsdam
OLSZEWSKI; C. W. W. OSTWALD

T. W. KORZYBSKI
Polish Academy of Sciences
PARNAS

EDNA E. KRAMER
Polytechnic Institute of New York
A. E. NOETHER; M. NOETHER

FRIDOLF KUDLIEN
University of Kiel
ORIBASIUS; PHILINUS OF COS

P. G. KULIKOVSKY
Moscow University
NEUYMIN; NUMEROV; A. Y. ORLOV;
S. V. ORLOV; PARENAGO

LOUIS I. KUSLAN
Southern Connecticut State College
NORTON

BENGT-OLOF LANDIN
University of Lund
PAYKULL

GEORGE H. M. LAWRENCE
*Hunt Institute for Botanical Documen-
tation*
F. NYLANDER; W. NYLANDER

JACQUES R. LÉVY
Paris Observatory
PERROTIN

O. A. LEZHNEVA
Academy of Sciences of the U.S.S.R.
V. V. PETROV

JOHN H. LIENHARD
University of Kentucky
NUSSELT

C. LIMOGES
University of Montreal
E. PERRIER

DAVID C. LINDBERG
University of Wisconsin
PECHAM

STEN LINDROTH
University of Uppsala
N. E. NORDENSKIÖLD; OLAUS MAGNUS

JAMES LONGRIGG
University of Newcastle Upon Tyne
NICOLAUS OF DAMASCUS

J. M. LÓPEZ DE AZCONA
Comisión Nacional de Geología, Madrid
NUÑEZ SALACIENSE

AVERIL M. LYSAGHT
S. PARKINSON

RUSSELL McCORMMACH
Johns Hopkins University
J. W. NICHOLSON

MARVIN W. McFARLAND
Library of Congress
PICCARD

ROBERT McKEON
Babson College
NAVIER

PATRICIA P. MacLACHLAN
College of DuPage
PAPIN

ROGERS McVAUGH
University of Michigan
PALMER

EDWARD P. MAHONEY
Duke University
NIFO

MICHAEL S. MAHONEY
Princeton University
NICERON

J. C. MALLET
Centre National de Floristique
PÉRON

BRIAN G. MARSDEN
Smithsonian Astrophysical Observatory
NEWCOMB; W. H. PICKERING

OTTO MAYR
Smithsonian Institution
PERRONET; R.-P. PICTET

A. J. MEADOWS
University of Leicester
NEWALL

JEAN MESNARD
University of Paris
NOEL

S. R. MIKULINSKY
Academy of Sciences of the U.S.S.R.
PIROGOV

MARCEL MINNAERT †
PANNEKOEK

A. M. MONNIER
University of Paris
PEZARD

ERNEST A. MOODY
OCKHAM

LETTIE S. MULTHAUF
NIESTEN; NORWOOD; OLBERS

ARNE MÜNTZING
University of Lund
NILSSON-EHLE

G. NAUMOV
Academy of Sciences of the U.S.S.R.
OBRUCHEV

AXEL V. NIELSEN †
OLUFSEN

WŁODZIMIERZ NIEMIERKO
Nencki Institute of Experimental Biology
NENCKI

CALVERT E. NOLAND
San Diego State University
PACKARD

J. D. NORTH
University of Oxford
PARSONS; T. E. R. PHILLIPS

H. OETTEL
NIELSEN

C. D. O'MALLEY †
NEMESIUS

JANE M. OPPENHEIMER
Bryn Mawr College
NICHOLAS

WALTER PAGEL
*Wellcome Institute of the History of
Medicine*
PARACELSUS

JOHN PARASCANDOLA
University of Wisconsin-Madison
NOVY

FRANKLIN PARKER
West Virginia University
PEARL

LINUS PAULING
Institute of Orthomolecular Medicine
A. A. NOYES

OLAF PEDERSEN
University of Aarhus
ORTELIUS; PETER PHILOMENA OF DACIA

RUDOLF PEIERLS
University of Oxford
OPPENHEIMER

J. PELSENEER
University of Brussels
NEUBERG

VICENTE R. PILAPIL
California State University, Los Angeles
PÉREZ DE VARGAS

DAVID PINGREE
Brown University
NĪLAKAṆṬHA; PARAMEŚVARA; PAULIŚA;
PAUL OF ALEXANDRIA; PETOSIRIS

LUCIEN PLANTEFOL
University of Paris
NICOT; PEYSSONNEL

HOWARD PLOTKIN
University of Western Ontario
E. C. PICKERING

JESSIE POESCH
Newcomb College, Tulane University
T. R. PEALE

JOHANNES PROSKAUER †
NEES VON ESENBECK

SAMUEL X. RADBILL
College of Physicians of Philadelphia
OTT

GLORIA ROBINSON
Yale University
PAULY; PFEFFER

JOEL M. RODNEY
Widener College
PALEY

COLIN A. RONAN
*Journal of the British Astronomical Asso-
ciation*
PARSONS; PINGRÉ

PAUL LAWRENCE ROSE
New York University
B. PEREIRA

EDWARD ROSEN
City College, City University of New York
NOSTRADAMUS; NOVARA; OSIANDER

K. E. ROTHSCHUH
Universität Münster/Westphalia
PFLÜGER

A. I. SABRA
Harvard University
AL-NAYRĪZĪ

WILLIAM L. SCHAAF
Brooklyn College
OZANAM

H. SCHADEWALDT
University of Düsseldorf
NEISSER

CHARLES B. SCHMITT
Warburg Institute
PATRIZI

IVO SCHNEIDER
University of Munich
NEANDER

E. L. SCOTT
Stamford High School, Lincolnshire
NEWLANDS; G. PEARSON

J. F. SCOTT †
OUGHTRED

T. K. SCOTT, JR.
Purdue University
PAUL OF VENICE

DANIEL SEELEY
Boston University
PERRINE

A. SEIDENBERG
University of California, Berkeley
PASCH

E. M. SENCHENKOVA
Academy of Sciences of the U.S.S.R.
NAVASHIN; PALLADIN

W. A. SMEATON
University College London
B. PELLETIER; PILATRE DE ROZIER

ROBERT SOULARD
Musée du Conservatoire National des Arts et Métiers
NIEPCE

HAROLD SPEERT
Columbia University
PAPANICOLAOU

ERNEST G. SPITTLER, S.J.
John Carroll University
PANETH

NILS SPJELDNAES
University of Aarhus
NIEBUHR

ROGER H. STUEWER
University of Minnesota
PERRIN

CHARLES SÜSSKIND
University of California, Berkeley
PIERCE

FERENC SZABADVÁRY
Technical University, Budapest
NODDACK; PÉAN DE SAINT-GILLES

LEONARDO TARÁN
Columbia University
NICOMACHUS OF GERASA; PARMENIDES OF ELEA

JULIETTE TATON
PEZENAS; J. PICARD

RENÉ TATON
École Pratique des Hautes Études
B. PASCAL; É. PASCAL; J. PICARD

KENNETH L. TAYLOR
University of Oklahoma
NECKER

SEVIM TEKELI
Ankara University
PIRĪ RAIS

ARNOLD THACKRAY
University of Pennsylvania
W. NICHOLSON

PHILLIP DRENNON THOMAS
Wichita State University
NUTTALL; PAUL OF AEGINA

V. V. TIKHOMIROV
Academy of Sciences of the U.S.S.R.
OZERSKY

RONALD TOBEY
University of California, Riverside
OMALIUS D'HALLOY

HEINZ TOBIEN
University of Mainz
NEHRING; NEUMAYR; OPPEL; D'ORBIGNY

G. J. TOOMER
Brown University
NICOMEDES

F. G. TRICOMI
Academia delle Scienze de Torino
PINCHERLE

G. L'E. TURNER
University of Oxford
NOBERT

CAROL URNESS
University of Minnesota
PENNANT

PETER W. VAN DER PAS
South Pasadena, Calif.
OUDEMANS; PISO

GERALD R. VAN HECKE
Harvey Mudd College
W. A. NOYES

J. J. VERDONK
PELETIER

HUBERT BRADFORD VICKERY
Connecticut Agricultural Experiment Station
OSBORNE

P. J. WALLIS
University of Newcastle Upon Tyne
PELL

J. L. WALSH †
OSGOOD

DEBORAH JEAN WARNER
Smithsonian Institution
S. B. NICHOLSON; OUTHIER; C. H. F. PETERS

RACHEL HORWITZ WESTBROOK
NEEDHAM

RICHARD S. WESTFALL
Indiana University
PEMBERTON

GEORGE W. WHITE
University of Illinois
D. D. OWEN

L. PEARCE WILLIAMS
Cornell University
OERSTED

WESLEY C. WILLIAMS
Case Western Reserve University
R. OWEN

M. L. WOLFROM †
NEF

H. WUSSING
Karl Marx University
C. G. NEUMANN; PFAFF

A. P. YOUSCHKEVITCH
Academy of Sciences of the U.S.S.R.
OSTROGRADSKY; PETERSON; NEWTON

BRUNO ZANOBIO
University of Pavia
NEGRI

DICTIONARY
OF
SCIENTIFIC BIOGRAPHY

DICTIONARY OF SCIENTIFIC BIOGRAPHY

NAVASHIN— PISO

NAVASHIN, SERGEY GAVRILOVICH (*b.* Tsarevshin, Saratov guberniya, Russia, 14 December 1857; *d.* Detskoye Selo [now Pushkin], U.S.S.R., 10 December 1930), *biology, plant cytology, plant embryology.*

The son of a physician, Navashin graduated from the Saratov Gymnasium in 1874 and entered the St. Petersburg Academy of Medicine and Surgery although medicine did not especially interest him. His courses in chemistry with Borodin led to a strong interest in the subject. After four years Navashin transferred to the University of Moscow and entered second-year courses in the natural sciences section of the department of physics and mathematics. He was especially enthusiastic about his courses in chemistry with Markovnikov and in botany with Timiryazev. After his graduation from the university, Markovnikov offered Navashin an assistantship, first at Moscow University (1881) and then at the Petrov Academy (1884). After he passed his master's examination in 1887 Navashin began to teach courses: "Introduction to the Taxonomy of Fungi" at the university and plant pathology at the Petrov Academy. After the dissolution of the Petrov Academy in 1888, he became Borodin's assistant at the University of St. Petersburg. His interest in mycology drew Navashin to the mycologist Voronin, who suggested a study of the fungus *Sclerotinia betulae* (*Woroninaceae*), a parasite of the birch tree, for the subject of his master's thesis, which he defended in 1894 at St. Petersburg. After receiving the master's degree in botany, Navashin became professor of botany at the University of Kiev, where he accomplished his most fruitful scientific research and teaching. In 1915 serious illness obliged him to leave Kiev for the warmer climate of Tbilisi, where he devoted much energy to the university. Invited to Moscow in 1923 to organize the K. A. Timiryazev Institute of Plant Physiology, he was its director until 1929, the year before his death.

Navashin's basic research was devoted to the morphology and taxonomy of mosses and parasitic fungi. The study of the development of *Sclerotinia* in the ovaries of birches (1894) led him in 1895 to the discovery of chalazogamy in birches and, in 1899, in alders, elms, and other trees. Chalazogamy is a process of fertilization in which the pollen tubes penetrate to the embryo sac not through the micropyle but through its base, the chalaza. The phenomenon had previously been observed in the beefwood (*Casuarina*), only by Melchior Treub (1891), who considered it a distinguishing feature of these flowering plants. Navashin's observations introduced a correction into Treub's division of angiosperms into chalaziferous and porogamic.

Navashin's embryological research led him to the important discovery in 1898 of double fertilization in angiosperms. Observing fertilization in the Turk's-cap lily (*Lilium martagon*) and *Fritillaria tenella*, he was the first to note that this process involves not one but two sperm, which form in the pollen tube. One of them merges with the ovicell; the other, with the nucleus of the embryo sac, so that both the embryo and the endosperm develop as a result of the sexual process. On 24 August 1898 Navashin communicated this discovery to the Tenth Congress of Natural Scientists and Physicians, held in Kiev; later that year a description of the phenomenon appeared in print. The important discovery of double fertilization immediately attracted international attention; several scientists had already observed this phenomenon but had not given it proper attention. The presence of double fertilization made possible discovery of the fact that the endosperms of angiosperms and gymnosperms are not homologous formations but are completely distinct in nature and origin, despite the external similarity caused by their identical functions.

The last period of Navashin's life was dedicated to research in karyology. His work contributed to the comparative karyological trend in cytology, which was especially intensively developed in the Soviet Union. The success of Navashin's research was determined in substantial measure by his outstanding abilities as a microscope technician and observer.

For his research in embryology and plant cytology Navashin was made a corresponding member (1901) and an academician (1918) of the Russian Academy of Sciences, and an active member of the Ukrainian S.S.R. Academy of Sciences (1924). In 1929 he was awarded the title of Honored Scientist of the R.S.F.S.R.

BIBLIOGRAPHY

I. ORIGINAL WORKS. Navashin's collected works were published as *Izbrannye trudy* (Moscow–Leningrad, 1951). His writings include "Sklerotinia berezy (*Sclerotinia betulae Woroninaceae*)" ("Sclerotinia of Birches"), in *Trudy Sankt-Peterburskago obshchestva estestvoispytatelei*, Otd. botanichesky, **23** (1893), 56–64; "Ein neues Beispiel der Chalazogamie," in *Botanisches Zentralblatt*, **63**, no. 12 (1895), 353–357; "Resultate einer Revision der Befruchtungsvorgänge bei *Lilium martagon* und *Fritillaria tenella*," in *Mélanges biologiques tirés du Bulletin de l'Académie des sciences de Pétersbourg*, **9**, no. 9 (Nov. 1898), 377–382; "Neue Beobachtungen über Befruchtung bei *Fritillaria tenella* und *Lilium martagon*," in *Botanisches Zentralblatt*, **77** (1899), 62; and "Getero- i idiokhromozomy rastitelnogo yadra kak prichina dimorfizma nekotorykh vidov rastenii" ("Hetero- and Idiochromosomes in the Plant Nucleus as the Reason for Dimorphism in Certain Plant Species"), in *Izvestiya Akademii nauk*, no. 17 (1915), 1812–1834.

II. SECONDARY LITERATURE. On Navashin and his work, see his autobiography, in *Izbrannye trudy*, 13–20, with illustrations and portrait; V. V. Finn, "K 50-letiyu otkrytia S. G. Navashinym dvoynogo oplodotvorenia y pokrytosemennykh rastenii" ("On the Fiftieth Anniversary of Navashin's Discovery of Double Fertilization in Angiosperms"), in *Priroda*, no. 9 (1948), 80–81; and D. A. Granovsky, *Sergey Gavrilovich Navashin* (Moscow, 1947), with illustrations, portrait, and bibliography of nineteen works on Navashin.

E. M. SENCHENKOVA

NAVIER, CLAUDE-LOUIS-MARIE-HENRI (*b.* Dijon, France, 10 February 1785; *d.* Paris, France, 21 August 1836), *engineering, mechanics.*

During the French Revolution, Navier's father was a lawyer to the Legislative Assembly at Paris and his mother's uncle, the engineer Emiland Gauthey, worked in the head office of the Corps des Ponts et Chaussées at Paris. After her husband's death in 1793, Navier's mother moved back to Chalon-sur-Saône and left her son in Paris, under the tutelage of her uncle. In 1802, after receiving preparation from his granduncle, Navier entered the École Polytechnique near the bottom of the list; but he did so

well during his first year that he was one of ten students sent to work in the field at Boulogne instead of spending their second year in Paris. Navier's first year at the École Polytechnique had critical significance for the formation of his scientific style, which reflects that of Fourier because the latter was briefly his professor of analysis. He subsequently became Fourier's protégé and friend.

In 1804 Navier entered the École des Ponts et Chaussées, from which he graduated in 1806 near the top of his class. After spending a few months in the field, he was brought to Paris to edit the works of his granduncle, who had just died and who had become France's leading engineer. Navier, who seems to have been insecure financially, lived for the rest of his life in the St.-Germain-des-Prés quarter of Paris. His wife, Marie Charlot, whom he married around 1812, came from a family of small landowners in Burgundy.

Navier was a member of the Société Philomatique (1819) and of the Académie des Sciences (1824). In 1831 he became *Chevalier* of the Legion of Honor. From 1819 he taught and had complete charge of the courses in applied mechanics at the École des Ponts et Chaussées but did not become titular professor until 1830, when A.-J. Eisenmann died. In 1831 Navier replaced Cauchy at the École Polytechnique. Navier participated in Saint-Simonianism and the positivist movements. He had Auguste Comte appointed to be one of his assistants at the École Polytechnique and participated actively in Raucourt de Charleville's Institut de la Morale Universelle.

Navier sought to complete the publishing project of his granduncle Gauthey. The administration of the Corps des Ponts et Chaussées, which looked with favor on this project, had him brought back to Paris in 1807 to publish Gauthey's manuscripts. This convergence of interests turned Navier into a theoretician who wrote textbooks for practicing engineers. His taste for scholarship and his background of higher analysis at the École Polytechnique and of practical engineering learned from his granduncle gave him the ideal preparation to make significant contributions to engineering science. During the period 1807–1820 he made mathematical analysis a fundamental tool of the civil engineer and codified the nascent concept of mechanical work for the science of machines.

Navier contributed only a few notes, of little scientific interest, to the first volume of Gauthey's works, which appeared in 1809. But during the next three years (1809–1812) he did a great deal of research in analytical mechanics and its application to the strength of materials as preparation for the second volume of Gauthey's works and for the revised edition of Bélidor's *Science des ingénieurs*, both of which appeared in 1813. The traditional engineering approach as exemplified by Gauthey studied experimentally the materials used in construction. These materials—primarily stone and wood—possess poor resistance to bending and were used rigidly. They also have widely varying properties that depend on their type and origin. In the traditional approach the engineer designed to avoid rupture and gave no thought to bending. He used large safety factors to compensate for the widely varying properties of the materials, which he viewed as rigid bodies subjected only to extension and to compression. On the other hand, the analytical tradition, which belonged to mathematical physics and did not form part of an engineer's training until after the creation of the École Polytechnique, studied idealized flexible bodies that can vibrate, such as strings, thin bars, and thin columns. In the derivation of analytical expressions these bodies were assumed to be subjected uniquely to pure bending; compression and tension were, therefore, ignored.

Navier, who had received training in both traditions, united them when he considered iron, which was just beginning to be used for bridges. He used a sort of principle of superposition: two sets of independent forces developed when bodies were bent—those resisting compression and extension and those resisting bending. He drew on the traditional engineering approach for the study of the first set of forces and on the analytical one for the second set. The first set follows Hooke's law that stress is proportional to strain. For the second set Navier used the relationship that the resistance varies as the angle of contingency (one divided by the radius of curvature), for which he referred to Euler's *Methodus inveniendi lineas curvas maximi minimive proprietate gaudentes* (1744).

Navier found an expression for the static moment of the resistance of any given fiber and then integrated it to find the total resisting moment, which he equated with the total moment of the applied forces. He concluded that for simple cases the moment of elasticity varies as the thickness squared, which quantity measures the resistance to extension, and as the thickness cubed times the length, which measures the resistance to bending. In his later study on the bending of an elastic plane (1820), Navier used the same general approach, which, however, led to fourth-order partial differential equations that had already been set down by Lagrange and Poisson. He showed how these can be solved in certain cases by applying methods that Fourier had used in an unpublished study.

In the 1813 editions of Bélidor and of Gauthey, Navier added notes which drew on the research of Coulomb and on the experimental tradition of eighteenth-century physics that had given him data for tables of the strength of stone and of wood. He appealed for further experiments on the strength of materials so that they could be used well in construction.

Navier's success as editor of Bélidor's *Science des ingénieurs* and of Gauthey's works led their publisher, Firmin Didot, to invite him to prepare a revised edition of Bélidor's *Architecture hydraulique*. Navier sought to correct the errors found in this work and to give it a mathematical sophistication that would make it useful to the graduates of the École Polytechnique. One item needed particular attention—the study of machines, for which Navier sought a quantitative criterion that would facilitate the selection of the best machines and motors. Research on this topic, conducted during 1814–1818, led him to the concept of quantity of action, which Coriolis shortly afterward transformed into that of mechanical work. A body animated by a force, Navier argued, can produce an effect observable to our senses only if this body covers a distance and at the same time exerts a pressure against an obstacle. Thus, for any interval of time, we can measure the effect of an acting force by the integral of $F\,dx$, where F designates the acting force and x the distance through which it acts. Navier, who drew on Lazare Carnot, then equated this to half the *vis viva* (mv^2) acquired by the moving body during the same interval, less that lost through sudden changes of speed. Because he thought that the above relation applied only to cases for which the various parts of a system are linked by expressions independent of time, he did not achieve a full concept of the conservation of mechanical work as did G.-G. Coriolis. Navier called the action of a force over a distance "quantity of action," an expression taken from Coulomb, and related this to the quantity of work (in a nontechnical sense) used to run the machine. Citing Montgolfier, who said that it was the quantity of action which pays, Navier called the quantity of action a mechanical form of money. In Navier's writings the march of the argument leads to the concept of work, whereas in those of J.-V. Poncelet and of Coriolis it flows from this concept. It took an embryonic form in the writings of Lazare Carnot, found its birth in those of A.-T. Petit, Poncelet, and Navier, and achieved the status of a general principle of applied mechanics in those of Coriolis. In Navier's revision of *Architecture hydraulique*, the engineer found this concept so defined as to give a measure of the usefulness of motors and a criterion that permitted rational design of motors and machines.

Editing the works of Gauthey and Bélidor in the years between 1809 and 1819 led Navier to make significant contributions to engineering science and placed him in a position to institute a new era in the teaching of engineering. His courses, the style of which was influential for well over a half a century, built on the creative physics of his generation, liberally applied analytical mechanics, and thus gave to the civil engineer tools adapted to an industrializing age.

During the years 1820–1829 Navier's research moved in two directions. In practical engineering he designed a suspension bridge that spanned the Seine in front of the Invalides, where today the Pont Alexandre III stands. Just as his bridge was in the final stages of construction, a sewer broke and the resultant flooding caused the bridge to list. This accident, which, in the view of the Corps des Ponts et Chaussées, could have been easily repaired, gave to the Municipal Council of Paris, which had opposed Navier's project, the opportunity to put pressure on the government to order the bridge torn down, to Navier's great chagrin.

In theoretical science Navier studied the motion of solid and liquid bodies, deriving partial differential equations to which he applied Fourier's methods to find particular solutions. This theoretical research led him to formulate the well-known equation identified with his name and that of Stokes. Navier viewed bodies as made up of particles which are close to each other and which act on each other by means of two opposing forces—one of attraction and one of repulsion—which, when in a state of equilibrium, cancel each other out. The repelling force resulted from the caloric that a body possessed. When equilibrium is disturbed in a solid, a restoring force acts which is proportional to the change in distance between the particles. In a liquid this force becomes proportional to the difference in speed of the particles. For both cases Navier derived equations that proved to have the same mathematical form. Although he had no concept of shear and used a concept of inter-molecular forces that is unacceptable today, he achieved results of which the expressions remain valid because he carefully summed moments of forces about orthogonal axes. This guaranteed that he did not overlook any forces that were acting even though he did not possess an efficient formulation for them.

Following the July Revolution, Navier became an active technical consultant to the state. He reported on the policy that should be adopted for policing the road transportation of heavy loads, for bidding to obtain government contracts, for constructing roads, and for laying out a national railway system. His reports exhibit Navier's high engineering ability and his continuing commitment to the Saint-Simonian and positivist movements.

BIBLIOGRAPHY

I. Original Works. A. Barré de Saint-Venant, in C. Navier, *Résumé des leçons données à l'École des ponts et chaussées . . . première partie . . . première section. De la résistance des corps solides*, 3rd ed. (Paris, 1864), I, lv–lxxxiii, lists Navier's major works. The *Catalogue des manuscrits de la Bibliothèque de l'École des ponts et chaussées* (Paris, 1886) details Navier's extant scientific MSS. Autograph copies of Navier's "Mémoire sur la flexion des plans élastiques" (1820) are at the New York City Public Library and at the Archives Nationales, Paris, in F[14]2289[1], dossier Navier. Navier's debate with Poisson concerning the molecular structure of matter is recorded in *Annales de chimie et de physique*, 2nd ser., **38** (1828), 304–314, 435–440; **39** (1828), 145–151, 204–211; **40** (1829), 99–110. Navier wrote a preface to J. B. Fourier, *Analyse des équations déterminées* (Paris, 1831). Navier's views on determinism are in *Comptes rendus . . . de l'Académie des sciences*, **2** (1836), 382. At the Archives Nationales, Paris, the following are of interest: F[14]2289[1], dossier Navier, contains many autograph letters; F[14]11057 contains reports by Navier on the École des Ponts et Chaussées and on the teaching of applied mechanics; F[14]11139 documents Navier's bridge-building activities. The library of the Institut de France, Paris, has a few letters by Navier.

II. Secondary Literature. A. Barré de Saint-Venant, *op cit.*, pp. xxxix–liv, lists obituaries by P. S. Girard, by C. H. Emmery de Sept-Fontaines, and by G. C. Prony, who errs in Navier's date of birth; in the library of the École des Ponts et Chaussées a MS biography of Navier contains obituaries by Coriolis and by Raucourt, the latter from *Éducateur, journal de l'Institut de la morale universelle . . .*, **1**, no. 5 (Sept.–Oct. 1836), 38–39; there is also a notice by Fayolle in *Biographie universelle*, LXXV (Paris, 1844), 314–317. Navier's standing as a student at the École des Ponts et Chaussées is detailed in Archives nationales, Paris, F[14]2148, F[14]11054, F[14]11055 (which also contains remarks about Navier's course).

A. Barré de Saint-Venant, *op. cit.*, pp. xc–cccxi, gives a comprehensive history of the strength of materials and locates Navier's work within it; S. Timoshenko, in *History of Strength of Materials* (New York, 1953), 70–80; and I. Todhunter, in *A History of the Theory of Elasticity and of the Strength of Materials*, I (repr. New York, 1960), 133–146, cover the same ground. F. Stüssi, "Baustatik vor 100 Jahren—die Baustatik Naviers," in *Schweizerische Bauzeitung*, **116**, no. 18 (2 Nov. 1940), 201–205, discusses Navier's ill-fated bridge and the correctness of his research on the strength of materials.

For Navier's contribution to mechanics consult M. R. Rühlmann, *Geschichte der technischen Mechanik* (Leipzig, 1885); and R. Dugas, *Histoire de la mécanique* (Neuchâtel, 1950), 393–401. Navier's views on *vis viva* and on the composition of matter are discussed in W. L. Scott, *The Conflict Between Atomism and Conservation Theory 1644 to 1860* (London, 1970), 104–135, 155–182; and C. C. Gillispie, *Lazare Carnot Savant* (Princeton, 1971), 111–115. Navier's contribution to thermodynamics is mentioned in D. S. L. Cardwell, *From Watt to Clausius . . .* (Ithaca, N.Y., 1971), 167–169. F. Klemm, "Die Rolle der Mathematik in der Technik des 19. Jahrhunderts," in *Technikgeschichte*, **33** (1966), 72–90, devotes a few pages to Navier.

The official *Moniteur universel* (24 Feb. 1828), 251–252, defends Navier's bridge, which Balzac ridicules in his *Oeuvres*, M. Boutreron and H. Longnon, eds., XXV (Paris, 1922), 196, and notes on 299–302. G. Le Franc, "The French Railroads 1823–1842," in *Journal of Economic and Business History*, **2** (1929–1930), 299–333, gives the background that enables one to situate Navier's writings on railway systems.

Navier's association with positivism may be traced through *Journal du génie civil . . .*, *Éducateur, journal de l'Institut de la morale universelle* (Bibliothèque Nationale, Paris, cote R7368), and through the archives of the École Polytechnique, Paris, "Registres du Conseil d'instruction," **7bis**, 30 Oct. 1833 and 31 Oct. 1834.

At the archives of the École Polytechnique, the "Registre matricule des élèves" describes Navier; the cartons for 1802 and for 1803 give Navier's standing; that for 1831 contains the letter appointing Navier professor; the "Registre du Conseil d'instruction," **7bis**, 18 Dec. 1830 and 29 Jan., 18 Feb., 4 Mar., and 18 Mar. 1831, details Navier's election to his professorship at the École Polytechnique; the same vol. reveals the conflict between Navier and Poisson concerning the teaching of Fourier's theory of heat (see 20 July 1831, 29 May 1832); on this conflict also see the "Registre du Conseil de perfectionnement," **6**, 54.

Also consult the following works by R. McKeon: "A Study in the History of Nineteenth Century Science and Technology: Engineering Science in the Works of Navier," in *Proceedings of the XIII*[th] *International Congress of the History of Science*, section 11, history of technology; "Profile chronologique de Navier," deposited at the Centre Alexandre Koyré, Paris, on 23 Nov. 1970; and "Navier éditeur de *l'Architecture hydraulique* de Bélidor," unpublished report read at the Congrès de l'Association française pour l'avancement des sciences, Chambéry, France, 8 July 1971, of which an extract is in *Sciences*, **3** (1972), 256–257.

Robert M. McKeon

AL-NAYRĪZĪ, ABU'L-ʿABBĀS AL-FAḌL IBN ḤĀTIM (*fl.* Baghdad, *ca.* 897; *d. ca.* 922), *geometry, astronomy.*

As his name indicates, al-Nayrīzī's origins were in Nayrīz, a small town southeast of Shīrāz, Fārs, Iran. For at least part of his active life he lived in Baghdad, where he probably served the ʿAbbāsid caliph al-Muʿtaḍid (892–902), for whom he wrote an extant treatise on meteorological phenomena (*Risāla fī aḥdāth al-jaww*) and a surviving work on instruments for determining the distances of objects.

The tenth-century bibliographer Ibn al-Nadīm refers to al-Nayrīzī as a distinguished astronomer; Ibn al-Qifṭī (*d.* 1248) states that he excelled in geometry and astronomy; and the Egyptian astronomer Ibn Yūnus (*d.* 1009) takes exception to some of al-Nayrīzī's astronomical views but shows respect for him as an accomplished geometer.

Of the eight titles attributed to al-Nayrīzī by Ibn al-Nadīm and Ibn al-Qifṭī, two are commentaries on Ptolemy's *Almagest* and *Tetrabiblos* and two are astronomical handbooks (*zījes*). Ibn al-Qifṭī indicates that the larger handbook (*Kitāb al-zīj al-kabīr*) was based on the *Sindhind*. None of these works has survived, but the commentary on the *Almagest* and one (or both?) of the handbooks were known to al-Bīrūnī. Ibn Yūnus cites, critically, a certain *zīj* in which, he states, al-Nayrīzī adopted the mean motion of the sun as determined in the *Mumtaḥan zīj*, which was prepared under the direction of Yaḥyā ibn Abī Manṣūr in the time of al-Maʾmūn (813–833). Ibn Yūnus wonders at al-Nayrīzī's adoption of this "erroneous" determination without further examination and, continuing his criticism of the "excellent geometer," refers further to oversights and errors, particularly in connection with the theory of Mercury, the eclipse of the moon, and parallax.

Al-Nayrīzī has been known mainly as the author of a commentary on Euclid's *Elements* that was based on the second of two Arabic translations of Euclid's text, both of which were prepared by al-Ḥajjāj ibn Yūsuf ibn Maṭar (see *Dictionary of Scientific Biography*, IV, 438–439). The commentary survives in a unique Arabic manuscript at Leiden (bks. I–VI) and in a Latin version (bks. I–X), made in the twelfth century by Gerard of Cremona. (The Arabic manuscript lacks the comments on definitions 1–23 of book I, but these are preserved in the Latin translation.) In the course of his own comments al-Nayrīzī quotes extensively from two commentaries on the *Elements* by Hero of Alexandria and Simplicius, neither of which has survived in the original Greek.

The first of these must have covered at least the first eight books (Hero's last comment cited by al-Nayrīzī deals with Euclid VIII.27), whereas the second, entitled "A Commentary on the Premises [*ṣadr, muṣādara, muṣādarāt*] of Euclid's *Elements*," was concerned solely with the definitions, postulates, and axioms at the beginning of book I of the *Elements.*

Simplicius' *Commentary*, almost entirely reproduced by al-Nayrīzī, played a significant part in arousing the interest of Islamic mathematicians in methodological problems. It further quotes verbatim a full proof of Euclid's postulate 5, the parallels postulate, by "the philosopher Aghānīs." The proof, which is based on the definition of parallel lines as equidistant lines and which makes use of the "Eudoxus-Archimedes" axiom, has left its mark on many subsequent attempts to prove the postulate, particularly in Islam.

Aghānīs is no longer identified with Geminus, as Heiberg and others once thought because of a similarity between their views on parallels. He almost certainly lived in the same period as Simplicius; and Simplicius' reference to him in the *Commentary* as "our associate [or colleague] Aghānīs," or, simply, "our Aghānīs" (*Aghānīsu, ṣāḥibunā*, rendered by Gerard as *socius noster Aganis*) strongly suggests that the two philosophers belonged to the same school. There is an anonymous fifteenth-century Arabic manuscript that aims to prove Euclid's parallels postulate and refers in this connection to Simplicius and Aghānīs, but spells the latter's name "Aghānyūs," thus supplying a vowel that can only be conjectured in the form "Aghānīs." Given that the Arabic "gh" undoubtedly stood for the letter γ, "Aghānyūs" may very easily have been a mistranscription of the recognizable Greek name "Agapius." Reading "Aghānyūs" for "Aghābyūs" (Arabic has no "p") may well have resulted from misplacing a single diacritical point, thereby transforming the "b" (that is, "p") into an "n." This hypothesis is the more plausible since we know that diacritical points were often omitted in Arabic manuscripts. It therefore seems reasonable to assume that Aghānīs-Aghānyūs was no other than the Athenian philosopher Agapius, a pupil of Proclus and Marinus who lectured on the philosophy of Plato and Aristotle about A.D. 511 and whose versatility was praised by Simplicius' teacher, Damascius. Agapius' name, place, date, affiliation, and interests agree remarkably with the reference in Simplicius' *Commentary.*

In his commentary on the *Elements*, al-Nayrīzī followed a conception of ratio and proportion that had previously been adopted by al-Māhānī (see *Dictionary of Scientific Biography*, IX, 21–22). Al-Nayrīzī's treatise "On the Direction of the *qibla*" (*Risāla fī samt al-qibla*) shows that he knew and utilized the equivalent of the tangent function. But in this, too, he is now known to have been preceded, for example, by Ḥabash (see *Dictionary of Scientific Biography*, V, 612).

Again, his unpublished treatise "On the Demonstration of the Well-Known Postulate of Euclid" (Paris,

Bibliothèque Nationale, arabe 2467, fols. 89r–90r) clearly depends on Aghānīs. In it al-Nayrīzī argues that, because equality is "naturally prior" to inequality, it follows that straight lines that maintain the same distance between them are prior to those that do not, since the former are the standard for estimating the latter. From this reasoning he concludes the existence of equidistant lines, accepting as a "primary proposition" that equidistant lines do not meet, however extended. His proof consists of four propositions, of which the first three state that: (1) the distance (that is, shortest line) between any two equidistant lines is perpendicular to both lines; (2) if a straight line drawn across two straight lines is perpendicular to both of them, then the two lines are equidistant; and (3) a line falling on two equidistant lines makes the interior angles on one side together equal to two right angles. These three propositions correspond to Aghānīs's propositions 1–3, while the fourth is the same as Euclid's postulate 5: If a straight line falling on two straight lines makes the interior angles on one side together less than two right angles, then the two lines will meet on that side. The proof closely follows Aghānīs.

Al-Nayrīzī, however, claims originality for the theorems that he proves in the extant but unpublished treatise for al-Muʿtaḍid—"On the Knowledge of Instruments by Means of Which We May Know the Distances of Objects Raised in the Air or Set Up on the Ground and the Depths of Valleys and Wells, and the Widths of Rivers." Al-Bīrūnī also states that al-Nayrīzī, in his commentary on the *Almagest*, was the only writer known to him who had provided a method for computing "a date for a certain time, the known parts of which are various *species* that do not belong to one and the same *genus*. There is, *e.g.*, a day the date of which within a Greek, Arabic, or Persian month is known; but the name of this month is unknown, whilst you know the name of another month that corresponds with it. Further, you know an era, to which, however, these two months do *not* belong, or such an era, of which the name of the month in question is not known" (*Chronology*, p. 139).

Al-Nayrīzī's work on the construction and use of the spherical astrolabe (*Fi 'l-asṭurlāb al-kurī*), in four *maqālas*, is considered the most complete treatment of the subject in Arabic.

BIBLIOGRAPHY

I. ORIGINAL WORKS. The Arabic text of al-Nayrīzī's commentary on the *Elements* (bks. I–VI and a few lines from bk. VII) was published as *Codex Leidensis 399,*

I. *Euclidis Elementa ex interpretatione al-Hadschdschadschii cum commentariis al-Narizii*, R. O. Besthorn and J. L. Heiberg, eds. (Copenhagen, 1893–1932). This ed. is in three pts., each comprising two fascicules, of which pt. III, fasc. II (bks. V–VI), is edited by G. Junge, J. Raeder and W. Thomson. Gerard of Cremona's Latin trans. is *Anaritii in decem libros priores Elementorum Euclidis commentarii . . . in codice Cracoviensi 569 servata*, Maximilianus Curtze, ed. (Leipzig, 1899), in Euclid's *Opera omnia*, J. L. Heiberg and H. Menge, eds., supp. (Suter mentions the probable existence of another MS of Gerard's trans. in "Nachträge," p. 164).

A German trans. and discussion of al–Nayrīzī's treatise on the direction of the *qibla* (*Risāla fī samt al-qibla*) is C. Schoy, "Abhandlung von al-Faḍl b. Ḥātim an Nairîzî: Über die Rechtung der Qibla," in *Sitzungsberichte der Bayerischen Akademie der Wissenschaften zu München*, Mathematisch-physikalische Klasse (1922), 55–68.

A short "chapter" (perhaps drawn from a longer work by al Nayrīzī) on the hemispherical sundial was published as *Faṣl fī takhṭīṭ al-sāʿāt al-zamāniyya fī kull qubba aw fī qubba tustaʿmal lahā* ("On Drawing the Lines of Temporal [that is, unequal] Hours in Any Hemisphere or in a Hemisphere Used for That Purpose"); see *al-Rasāʾil al-mutafarriqa fi ʾl-hayʾa l ʾl-mutaqaddimīn wa-muʿāṣiri ʾl-Bīrūnī* (Hyderabad, 1947).

II. Secondary Literature. MSS of al-Nayrīzī's works are in C. Brockelmann, *Geschichte der arabischen Literatur*, supp. vol. I (Leiden, 1937), 386–387; 2nd ed., I, (Leiden, 1943), 245; H. Suter, "Die Mathematiker und Astronomen der Araber und ihre Werke," in *Abhandlungen zur Geschichte der mathematischen Wissenschaften mit Einschluss ihrer Anwendungen*, 10 (1900), no. 88, 45; and "Nachträge und Berichtigungen zu 'Die Mathematiker . . . ,'" *ibid.*, 14 (1902), 164; and H. P. J. Renaud, "Additions et corrections à Suter 'Die Mathematiker . . . ,'" in *Isis*, 18 (1932), 171.

The little information that we have of al-Nayrīzī's activities and a list of his works are in Ibn al-Nadīm, *al-Fihrist*, G. Flügel, ed., I (Leipzig, 1871), 265, 268, 279; and Ibn al-Qifṭī, *Taʾrikh al-ḥukamāʾ*, J. Lippert, ed. (Leipzig, 1930), 64, 97, 98, 254.

For the references to al-Nayrīzī's *zīj* in Ibn Yūnus' Ḥākimite *zīj*, see *Notices et extraits des manuscrits de la Bibliothèque nationale . . .*, VII (Paris, 1803), 61, 65, 69, 71, 73, 121, 161, 165. Al-Bīrūnī refers to al-Nayrīzī in *Rasāʾil*, 2 (Hyderabad, 1948), 39, 51, and in *The Chronology of Ancient Nations*, C. E. Sachau, trans. (London, 1879), 139. See also E. S. Kennedy, "A Survey of Islamic Astronomical Tables," in *Transactions of the American Philosophical Society*, n.s. 46, pt. 2 (1956), nos. 46, 63, 75.

For a description of the contents and character of Hero's commentary on the *Elements* as preserved by al-Nayrīzī, see T. L. Heath, *The Thirteen Books of Euclid's Elements*, 2nd ed. (Cambridge–New York, 1956), 21–24.

Simplicius' commentary on the *Elements*, including a proof of Euclid's parallels postulate that seems to have been omitted from the text quoted by al-Nayrīzī, is discussed by A. I. Sabra in "Simplicius's Proof of Euclid's Parallels Postulate," in *Journal of the Warburg and Courtauld Institutes*, 32 (1969), 1–24.

For a detailed description of al-Nayrīzī's work on the spherical astrolabe, see Hugo Seemann and T. Mittelberger, "Das kugelförmige Astrolab nach den Mitteilungen von Alfonso X. von Kastilien und den vorhandenen arabischen Quellen," in *Abhandlungen zur Geschichte der Naturwissenschaften und der Medizin*, 8 (1925), 32–40.

For a discussion of al-Nayrīzī's concept of ratio, see E. B. Plooij, *Euclid's Conception of Ratio and His Definition of Proportional Magnitudes as Criticized by Arabian Commentators* (Rotterdam, 1950), 51–52, 61; and J. E. Murdoch, "The Medieval Language of Proportions," in A. C. Crombie, ed., *Scientific Change* (London, 1963), 237–271, esp. 240–242, 253–255.

The identity of Aghānīs is discussed in Paul Tannery, "Le philosophe Aganis est-il identique à Géminus?" in *Bibliotheca mathematica*, 3rd ser., 2 (1901), 9–11, reprinted in *Mémoires scientifiques*, III (Toulouse–Paris, 1915), 37–41; Sir Thomas Heath, *A History of Greek Mathematics*, II (Oxford, 1921), 224; *The Thirteen Books of Euclid's Elements*, I (Cambridge–New York, 1956), 27–28; A. I. Sabra, "Thābit ibn Qurra on Euclid's Parallels Postulate," in *Journal of the Warburg and Courtauld Institutes*, 31 (1968), 13. The information on Agapius is summarized in Pauly-Wissowa, *Real-Encyclopädie der classischen Altertumswissenschaft*, 1st ser., I (Stuttgart, 1894), 735.

See also the notice on al-Nayrīzī in Sarton's *Introduction*, I (Baltimore, 1927), 598–599.

A. I. Sabra

NEANDER, MICHAEL (*b.* Joachimsthal, Bohemia, 3 April 1529; *d.* Jena, Germany, 23 October 1581), *mathematics, medicine.*

The assessment of Neander and his work is complicated by confusion with another Michael Neander (1525–1595), who came from Sorau and was a school principal in Ilfeld. The achievements of each have been credited to the other, and to date no library has correctly cataloged their respective writings. Neander from Joachimsthal, like his namesake, studied at the Protestant university in Wittenberg, where he earned his baccalaureate degree in 1549 and his master's degree in 1550; he was eighth among fifty candidates. Beginning in 1551, he taught mathematics and Greek at the Hohe Schule in Jena. In 1558, when this school became a new Protestant university, Neander obtained the doctor of medicine degree with a work on baths, *De thermis*. In 1560 he advanced from professor at the faculty of arts to the more lucrative position of professor of medicine at Jena, which post he held until his death.

Neander's scholarly reputation was based on textbooks written primarily for students at the faculty of arts. He considered the writings of the ancients, especially Galen, absolutely authoritative. In the introduction to his *Methodorum in omni genere artium . . .* (1556), he based his exposition on Galen's opinion that the best kind of demonstration is mathematical. Neander distinguished the analytic and synthetic methods and introduced proof by contradiction as a third independent possibility.

In opposition to his contemporary Petrus Ramus, Neander contended that, even from a pedagogical point of view, Euclid's *Elements* contained the essence of a satisfactory synthetic demonstration. Neander's account of the metrology of the Greeks and Romans seems to have served for a time as a sort of reference work. His *Elementa sphaericae doctrinae* (1561), which includes an appendix on calendrical computation, endorsed Melanchthon's rejection of the Copernican view of the universe. The *Elementa* influenced one of Neander's colleagues at Jena, Victorinus Strigelius, whose *Epitome doctrinae de primo motu* (1564) also placed the earth at rest in the center of the universe.

Although Neander typified the close connection between mathematics and medicine frequently seen in the sixteenth century, this link appears only indirectly in his writings.

BIBLIOGRAPHY

I. ORIGINAL WORKS. Neander's major works are Σύνοψις *mensurarum et ponderum, ponderationisque mensurabilium secundum Romanos, Athenienses . . . Accesserunt etiam quae apud Galenum hactenus extabant de ponderum et mensurarum ratione* (Basel, 1555); *Methodorum in omni genere artium brevis et succincta* ὑφήγησις (Basel, 1556); *Gnomologia graecolatina, hoc est . . . Sententiae . . . ex magno anthologio Joannis Stobaei excerptae . . . Accessit praeterea* Ὄνειρος *vel* Ἀλεκτρυὼν, *id est somnium vel Gallus, dialogus Luciani . . . graece et latine . . .* (Basel, 1557); and *Elementa sphaericae doctrinae, seu de primo motu: in usum studiosae iuventutis methodicé et perspicué conscripta. Accessit praecipua computi astronomici materia, ubi temporis pleraeque differentiae explicantur* (Basel, 1561).

Biographisches Lexikon hervorragender Ärzte, IV (Berlin–Vienna, 1932), 331–332, lists a work entitled *De thermis* (Jena, 1558), but the author has been unable to verify this title in any library.

II. SECONDARY LITERATURE. Works on Neander and his work (in chronological order) are Heinrich Pantaleon, *Prosopographiae heroum atque illustrium virorum totius Germaniae* (Basel, 1566), 553; also in *Teutscher Nation Heldenbuch . . .* (Basel, 1578), 515; Paul Freher, *Theatrum virorum eruditione clarorum* (Nuremberg, 1688), 1279; Johann Caspar Zeumer, *Vitae professorum theologiae omnium Jenensium* (Jena, 1711), 14; *Hamburgische vermischte Bibliothek*, pt. 1 (Hamburg, 1743), 695–701;

Christian Gottlieb Jöcher, ed., *Allgemeines Gelehrten-Lexicon*, III (Leipzig, 1751), 840; Johannes Günther, *Lebensskizzen der Professoren der Universität Jena von 1558 bis 1858* (Jena, 1858); *Allgemeine deutsche Biographie*, XXIII (Leipzig, 1886), 340; and Otto Knopf, *Die Astronomie an der Universität Jena von der Gründung der Universität im Jahre 1558 bis zur Entpflichtung des Verfassers im Jahre 1927* (Jena, 1937), 1–6.

IVO SCHNEIDER

NECKER, LOUIS-ALBERT, known as **NECKER DE SAUSSURE** (*b.* Geneva, Switzerland, 10 April 1786; *d.* Portree, Skye, Scotland, 20 November 1861), *geology, mineralogy, zoology.*

The name Necker de Saussure represents the union of two illustrious Swiss families. Louis-Albert's father, Jacques Necker, was professor of botany and a magistrate at Geneva, and the nephew of Louis XVI's director general of finance (and thus first cousin of Mme de Staël). His mother, Albertine de Saussure, was the daughter of Horace-Bénédict de Saussure, the eminent geologist and naturalist.

The eldest of four children, Necker studied at the Academy of Geneva, then went to Edinburgh in 1806 to pursue university studies. Already versed in mineralogy and geology, in Scotland he was exposed to both Huttonian and Wernerian geological doctrines and became personally acquainted with Playfair, Hall, and other Edinburgh intellectuals. After visiting many parts of Scotland, with special attention to geological features, Necker returned to Geneva, where he became a professor of mineralogy and geology at the Academy in 1810. He retained a chair there for over two decades. During these years he traveled widely, sometimes conducting excursions with his students, and undertook geological investigations, especially in the Alps, concentrating particularly on the eastern and western extremities of the Alpine ranges.

During the 1830's Necker lived restlessly in Edinburgh, London, and Paris, as well as Geneva. He suffered increasingly from depressions that may have stemmed from declining health. Following his mother's death in 1841, he settled at Portree, Skye. He passed most of his remaining years there as a recluse.

Necker's scientific work was marked by a deep concern with the special methods and procedures that distinguished geology and mineralogy from other sciences. He emphasized the dependence of mineralogists and geologists upon real characteristics of actual objects, as opposed to abstractions. In mineralogical classification, one of his most serious concerns, he opposed the use of chemical composition as a major taxonomic criterion, viewing chemical entities as fundamentally abstract. A suitable organizational scheme for minerals, he believed, ought to depend on the characteristics that the observer perceives directly in mineral objects. The integrant molecule, in his opinion an abstract conception without real existence, could not define the "individual" that gives meaning to mineral species, whereas the crystal could. Together with zoology and botany, the other branches of natural history, mineralogy was "positive and descriptive," not speculative. As a member of the tradition of the natural method of classification, Necker claimed inspiration from Augustin-Pyramus de Candolle and Cuvier.

A strong advocate of field observation in geology, Necker presented the first geological map of the whole of Scotland to the Geological Society of London in 1808. Although he resisted committing himself completely to the theoretical schemes of either the Huttonians or the Wernerians, in Scotland Necker did become convinced of the igneous origin of granite. His fieldwork included studies of the volcanoes of Italy; the geological features of parts of Savoy, Carniola, Carinthia, Istria, and Illyria; and the Arran dike swarms. He also investigated the origins of mineral deposits and concluded that metalliferous veins are formed by sublimation from igneous intrusions. In 1832 he gave the Royal Society of Edinburgh an improved "clinometrical compass" of his own design for rapid determination of the positions of strata. As his geological outlook matured, he showed an increasing tendency to favor a uniformitarian approach.

Necker disdained artificial boundaries of scientific specialization. Notable among his publications are studies of birds and of meteorological optical phenomena, including the aurora borealis and parhelia. He was inclined to see links between phenomena conventionally regarded as unrelated, as in his endeavor to establish relationships among temperature, stratigraphic configuration, and magnetic intensity in various geographic locations.

BIBLIOGRAPHY

I. ORIGINAL WORKS. Necker's major works include *Voyage en Écosse et aux Îles Hébrides*, 3 vols. (Geneva, 1821), an appreciative account of Scotland and the character and accomplishments of its inhabitants, with some geological observations; *Le règne minéral ramené aux méthodes de l'histoire naturelle*, 2 vols. (Paris–Strasbourg, 1835), an expansion of the ideas found in "On Mineralogy Considered as a Branch of Natural History, and Outlines of an Arrangement of Minerals Founded on the Principles

of the Natural Method of Classification," in *Edinburgh New Philosophical Journal*, **12** (1832), 209–265; and *Études géologiques dans les Alpes* (Paris–Strasbourg, 1841), dealing with the geology of the environs of Geneva. Later volumes, intended to present results of Necker's investigations in the eastern Alps and along the southern flank of the Alps, were never published. Much of his research, in fact, was never published, and what did appear in print was sometimes delayed. Partial listings of Necker's works are given in the Candolle obituary notice, *Mémoires*, pp. 455–456, or *Verhandlungen*, pp. 276–278; by Eyles, pp. 125–126; and in the Royal Society's *Catalogue of Scientific Papers*, IV (1870), 581–582, and X (1894), 904.

II. SECONDARY LITERATURE. A useful recent account is V. A. Eyles, "Louis Albert Necker, of Geneva, and His Geological Map of Scotland," in *Transactions of the Edinburgh Geological Society*, **14** (1952), 93–127. Among contemporary biographical sketches the fullest is James David Forbes, "Biographical Account of Professor Louis Albert Necker," in *Proceedings of the Royal Society of Edinburgh*, **5** (1862–1866), 53–76. Others include Henri de Saussure, "Nécrologie de M. Louis Necker," in *Revue et magasin de zoologie . . .*, 2nd ser., **13** (1861), 553–555; and an obituary notice by Alphonse de Candolle, in *Mémoires de la Société de physique et d'histoire naturelle de Genève*, **16** (1862), 452–456, also in *Verhandlungen der Schweizerischen naturforschenden Gesellschaft*, **46** (1862), 272–278. Forbes's account served as the basis for a "notice biographique" in a republication of Necker's *Mémoire sur les oiseaux des environs de Genève* (Geneva–Paris, 1864), pp. 5–45.

KENNETH L. TAYLOR

NEEDHAM, JOHN TURBERVILLE (*b.* London, England, 10 September 1713; *d.* Brussels, Belgium, 30 December 1781), *biology, microscopy.*

Needham's most important contributions to science were early observations of plant pollen and the milt vessels of the squid, a forward-looking theory of reproduction (1750), and a classic experiment for determining whether spontaneous generation occurs on the microscopic level (1748).

The son of recusants, John Needham and Margaret Lucas, Needham received a religious education in French Flanders, which prepared him for the intellectual life of the Continent. Ordained a secular priest in 1738, he supported himself first by teaching, and then by accompanying young English Catholic noblemen on the grand tour, until he settled in Brussels in 1768 as director of what was to become the Royal Academy of Belgium. His scientific interests were motivated largely by a desire to defend religion in an age when biological question had serious theological and philosophical meanings for many. Needham's extrascientific activities made him equally well known throughout educated Europe; in these he also defended the faith. He was particularly notable for his dispute with Voltaire over miracles and for a linguistic theory of the biblical chronology based on a supposedly Egyptian statue.

Needham was elected a fellow of the Royal Society (1747) and of the Society of Antiquaries of London (1761), as Buffon's correspondent for the Académie des Sciences (1768), a member of the Royal Basque Society of Amis de la Patrie, and first director of the Royal Academy of Belgium (1773), where he did much to disseminate advanced laboratory techniques. A genus of Australian plants, *Needhama*, was named for him.

Until cell theory reconciled both aspects of the problem of reproduction, explanations emphasized either the preformed nature of the primordia out of which new organisms came into being (were generated) or the gradual differentiation of growing tissue apparent in the embryo. During Needham's lifetime iatromechanists insisted on preformation, since known mechanical principles could not account for extensive differentiation; vitalists, led by Buffon, accounted for extensive differentiation through chance combining of genetic factors brought together by hypothetical natural principles that were peculiar to living things but that contemporary science had yet to discover.

In 1748, at Buffon's invitation, Needham examined fluids extracted from the reproductive organs of animals and infusions of plant and animal tissue. Given the weak, indistinct magnifying power of instruments then available, it is not surprising that the two men observed globules under their microscopes. For Buffon these were genetic factors, which he termed "organic molecules."

The second volume of Buffon's *Histoire naturelle* (1749) based proof of the "organic molecules" largely on these experiments, which in turn rested on Needham's skill and reknown as an empirical scientist. Thus Needham found himself at the focal point of the controversy over generation.

Buffon never claimed to have observed the microscopic joining of molecules that he speculated took place, but Needham thought he actually did see new organisms taking shape out of disorganized material. This was his famous experiment with boiled mutton gravy (1748). Foreshadowing recapitulation theory, he "saw" certain species of microscopic creatures giving birth to other species of animalcules and imagined that in embryonic development of higher organisms a similar phenomenon must occur. Needham's own theory of generation (1750) placed him in the vitalist camp through its reliance on principles peculiar to living things and its assignment of self-patterning powers to matter. It differed from Buffon's in its denial of chance combinations of mathematically countable genetic traits.

In Needham's view God would not allow chance to play a role in reproduction. The embryo was not preformed but predetermined. Two kinds of physical force were the building blocks of all matter. In each embryo a specific combination of these elements was contributed by each parent. This combination produced a unique vibratory motion which simultaneously molded the growing embryonic tissue into new shapes and changed their chemistry. Thus Needham considered the organism on physical, chemical, and biological levels, an approach through which the mechanist-vitalist controversy was later transcended. In his correspondence Needham, in the tradition of Aristotle and Descartes, referred to "my system of spontaneous generation and epigenesis."

The many attempts to refute Needham's claim were based either on logic or on inconclusive experiments until 1765, when Spallanzani boiled hermetically sealed mutton gravy and, upon opening the flasks, found nothing there where Needham claimed to have found animalcules. For Needham's sterilization techniques had in fact been faulty. While the iatromechanists sided with Spallanzani, the matter was not settled until Pasteur replied to Needham's contention (in footnotes to Spallanzani's *Nouvelles recherches*, 1769) that through using a longer boiling period Spallanzani had destroyed something in the air responsible for sustaining life.

BIBLIOGRAPHY

I. Original Works. For further references and a bibliography of Needham's works consult *Dictionary of National Biography*, XIV (1967–1968), 157–159; *Bibliographical Dictionary of the English Catholics*.

On early observations and the spontaneous generation controversy, see Needham's *An Account of Some New Microscopical Discoveries Founded on an Examination of the Calamary and Its Wonderful Milt-Vessels* (London, 1745); "A Summary of Some Late Observations Upon the Generation, Composition, and Decomposition of Animal and Vegetable Substances," in *Philosophical Transactions of the Royal Society*, **45**, no. 490 (1748), 615–666; and *Nouvelles observations microscopiques, avec des découvertes intéressantes sur la composition et la décomposition des corps organisés* (Paris, 1750).

Correspondence between Needham and Charles Bonnet and a rare pamphlet, *Idées républicaines, par un membre d'un corps, M.D.V.*, published anonymously (Geneva, 1766), are in the Bibliothèque Publique et Universitaire, Geneva; *Mémoire sur la maladie contagieuse des bêtes à cornes* (Brussels, 1770) is in the Belgian National Library (Bibliothèque Royale Albert 1er); and a portrait of

Needham (by Henry Edridge after Reynolds) is in the Holburne of Menstrie Museum, Great Pulteney St., Bath, England.

II. SECONDARY LITERATURE. Lazzaro Spallanzani's works on spontaneous generation are *Saggio de observazione microscopiche concernante il systema della generazione de Needham et Buffon* (Modena, 1765); *Nouvelles recherches sur les découvertes microscopiques et la génération des corps organisés* (London–Paris, 1769); and *Opuscoli de fisica animale et vegetabile* (Modena, 1776).

See also Silvio Curto, "Storia di un falso celebre," in *Bollettino della Società piemontese d'archeologia e belle arti* (1962–1963), 5–15, with four plates; Elizabeth Gasking, *Investigations Into Generation 1651–1828* (Baltimore, 1967); Stephen F. Milliken, "Buffon and the British" (doctoral diss., Columbia University, 1965); Jacques Roger, *Les sciences de la vie dans la pensée française du XVIIIᵉ siècle* (Paris, 1963); Jean Rostand, *La genèse de la vie* (Paris, 1943) and *Les origines de la biologie expérimentale et l'abbé Spallanzani* (Paris, 1951); and Rachel Westbrook, "John Turberville Needham and His Impact on the French Enlightenment" (diss. in progress, Columbia University).

RACHEL HORWITZ WESTBROOK

NEES VON ESENBECK, CHRISTIAN GOTT-FRIED (DANIEL) (b. Reichenberg Castle, near Erbach, Hesse, 14 February 1776; d. Breslau, Silesia [now Wrocław, Poland], 16 March 1858), *botany*.

Nees's father, estate administrator for the count of Erbach, had as official residence a castle in the Odenwald. There Nees was born and raised. He received a highly liberal education at home, and in 1792 he entered the humanistic high school in Darmstadt. From 1796 to 1799 he attended the University of Jena, studying medicine and natural history under Batsch, and philosophy under Schelling. In this period Nees was drawn into the nearby Weimar circle. His personal friendship with Goethe, which led to years of correspondence, greatly influenced his career. After receiving a doctorate from Giessen in 1800, Nees practiced medicine in the Odenwald (according to his autobiography; some others state Frankfurt) and experimented with Mesmerian magnetic techniques. Financially his practice was a failure—he cared only for sick patients. His wife of one year died in childbirth, and in 1802 Nees retired to a small estate—Sickershausen, near Kitzingen, Bavaria—which she had left him. A move to the University of Jena, encouraged by Goethe, was prevented by war in 1806. At Sickershausen he lived the life of a country gentleman, happily married to his second wife, née von Mettingh. Nees acquired a working knowledge of the major European languages except the Slavic ones. (He was coauthor of a book on modern Greek history and poetry in translation, pub-

lished in 1825.) He also assembled natural history collections, some of them with his younger brother, Friedrich, and wrote scientific papers and reviews. By 1818 the estate was in such bad shape that he had to get a job.

Thus Nees's career proper started when he was forty-two. He assumed the professorship of botany at Erlangen, as Scheber's belated successor; but in the same year the Prussian minister of education, Karl von Stein zum Altenstein, Nees's protector, appointed him to the chair of botany at the newly founded University of Bonn.[1] With zest he established a botanical garden, aided by his brother, whom he appointed inspector. Goethe helped even here, by contributing seeds.[2] In 1830, ostensibly at his request, Nees was allowed to exchange professorships with Treviranus at Breslau, where he started by reorganizing the botanical garden. As professor of botany he was progressively less successful. Presumably his lectures were too full of obscurantist *Naturphilosophie*, and the students turned to his colleague Goeppert. But Nees made a name for himself in courses on speculative philosophy and social ethics. Having, at Altenstein's request, drafted the requirements for the high school teaching certificate in natural science, he became the first examiner in the field at Breslau (1839). He was promptly relieved of this post when Altenstein died and was replaced, in 1840, by the reactionary Eichhorn. In 1852 Nees provided a textbook for the teachers of natural science, dealing with the study of form in nature. But natural history teaching in the schools faltered after charges that it was conducive to agnosticism.

At Breslau, Nees played an active role in civic affairs. It started with the organization of public scientific lectures and gradually broadened. He was cofounder of a successful health insurance scheme. In 1845 he became the beloved leader ("Father Nees") of a community of "Christian Catholics" following J. Ronge. This radical movement soon aroused the active opposition of the state. Nees freely expressed his deeply devout views in articles and tracts, such as his 1845 publication on matrimony in an intelligent society, and its relation to state and church. He practiced what he preached. His third marriage, to a weaver's daughter from Warmbrunn, in the Riesengebirge, was without state or church sanction but was reputedly a model marriage.

As a boy Nees had been deeply impressed by the French Revolution. The Napoleonic era brought disillusionment. He became a fervent and conspiratorial supporter of German unity, placing his hope in enlightened Prussian leadership, but was in turn disillusioned by the conservatism of the crown. Thus the way was opened to his becoming an active radical democrat, a liberal intellectual. (Other professors at Breslau were also politically active, but as conservatives or moderates.) Nees stood alone then—a German botanist, a member of a group not renowned for liberalism. In 1848 he helped found the Breslau Workers' Club. He was elected a deputy from Breslau to the Prussian National Assembly meeting at Berlin in 1848. Even the other members of Waldeck's left group —striving, like Nees, for an English-style constitutional monarchy—shuddered at his speeches and at the "extremism" of the draft constitution he presented. Apparently the most hair-raising article of that document was "The people is sovereign, and the concept 'subject' is struck from the life of the state for all time." Nees presided over the Berlin Workers' Congress and also founded (in Berlin) the German Workers' Brotherhood. In January 1849 he was banished from Berlin for life "because of dangerous socialistic tendencies."

Back in Breslau, Nees's house was subject to constant police search. In 1851 he was suspended from his professorship; in June 1852, at the age of seventy-six, he was dismissed without pension. The official charge was moral turpitude; specifically, concubinage. (In the Roman law of Germany that was the term for the "common law" marriage of Germanic Anglo-Saxon law; it was a "bad thing" if it caused "public annoyance.") The real reason, his political activity, apparently was sufficiently protected. Nees had no money; it had all been spent subsidizing scientific publications and on charity.

Pitiful advertisements announced the sale of his herbarium, containing some 80,000 specimens or 40,000 species.[3] His library also was sold. He moved to a garret. When he died, at the age of eighty-two, an immense crowd (reputedly 10,000) of mourners, mainly artisans and students, accompanied him to his grave.

Nees had many children. I have found information on only one: Carl Nees von Esenbeck was inspector of the Breslau Botanical Garden from 1853 to 1880. Friedrich Nees von Esenbeck, a Christian Catholic writer, was probably another. He even left some minor children; the Leopoldina contributed 100 copies of an engraved portrait of him for sale by their guardian.[4]

Nees's greatest contribution to science has not yet been mentioned. The venerable *Academia Caesarea Leopoldina-Carolina Naturae Curiosorum*, the Imperial German Academy of Natural Science (or Leopoldina), in Erlangen, had fallen on bad times during the Napoleonic era. Nees was elected a member in 1816, with the cognomen Aristoteles, and in the same year he became an officer. On 3 August 1818 he was elected president. To German science he would remain "Herr President" until his death. The Leopoldina, however,

was not viable. When he was invited to move from Erlangen to Bonn, Nees suggested to the Prussian chancellor K. A. von Hardenberg and to Altenstein that he should bring the Leopoldina with him from Bavaria to Prussia. The two gentlemen were not only enlightened, but as politicians they were quick to recognize a splendid propaganda stroke. They gladly accepted and guaranteed financial support.[5] Outlasting even the privileges of the house of Thurn und Taxis (the last hereditary postmaster generalship was lost in 1918), the Leopoldina is the only surviving institution of the Holy Roman Empire. The proposal by Nees and D. G. Kieser, addressed to the abortive Frankfurt Parliament in 1848, to make an expanded Leopoldina the center of German cultural life, was bound to fail. The officers of the Leopoldina insisted on Nees's remaining in his unpaid presidency even when he had been stripped of his professorship; and the Prussian government continued its subsidies, for the Hapsburgs were only too anxious to regain the Leopoldina.[6] Nees commented that he was "dead to the Prussian state, but lives still for the Academy."[7] Nees personally edited forty-seven volumes of the *Nova acta Academiae Caesareae Leopoldino Carolinae germanicae naturae curiosorum*, the high quality of which was partly the result of his bringing the artist Aimé Henry to Bonn. His last direct contribution was the preface to *Nova acta*, **26**, Abt. 2, dated 1 February 1858. Nees spared no money, including the last of his own, on publications. He apologized for the overdraft he had incurred.[8] His successor was not amused, but he was left a flourishing and proud academy.[9]

A logical outgrowth of the rebirth of the Leopoldina was the invention of the "annual meeting" by L. Oken in 1822. This type of gathering, which almost immediately assumed the detailed form so familiar to the present-day American scientist, was a major force in the blossoming of German science. It was copied by the British Association (1831) and the American Association for the Advancement of Science (1848). It is incomprehensible that the visionary Nees (perhaps in uncharacteristic jealousy) bungled the request to have it made an official practice of the Leopoldina,[10] although its adoption became inevitable.

Nees's primary services to German science were, in order of importance, as an organizer; as an editor, not only of the *Nova acta* but also of a vast body of additional material; and as a transmitter of outstanding foreign publications. In the last field the item of greatest impact was the five-volume edition of Robert Brown's botanical works, . . . *vermischte botanische Schriften* (Nuremberg, 1825–1834). Volume III is the original publication of the second edition of the *Prodromus florae Novae hollandiae* in Latin; the bulk of the remaining volumes consists of German translations, largely by Nees himself, with his occasional notes and additions. Nees was highly conscious of the antiquated Linnaean flavor persisting in much of German botany; and one of the avowed aims of his two-volume *Handbuch der Botanik* (1821–1822) was to stress the modern structural work, including that done in France (he leaned heavily on Mirbel) and by Robert Brown in England, and even the inclusion of what now are called physiological aspects. The work seems to be Nees's most dismal failure, although Cohn (1858) reported that it acted as a major stimulus. It is permeated not only by *Naturphilosophie* but also by an attempt to apply Goethe's theory of metamorphosis (the work is dedicated to him). The books consist of a series of aphorisms that to an unsympathetic reviewer appear to be outpourings from a strange dream world. If his lectures were similar, that would explain Nees's failure as a botany professor. But he did keep trying. As late as 1850 he asked Cohn to give a demonstration of microscopy to his students, but the only decent microscope available was, shamefully, Cohn's private one. In his later writings Nees managed to disentangle his botany and *Naturphilosophie*, and they both became "pure"—the latter in his book *Naturphilosophie* (1841).

Finally there are Nees's direct scientific contributions; they are far-reaching, even if the medical ones are excluded. In zoology there are major contributions to the taxonomy of ichneumon flies. His earliest botanical monograph (1814) dealt with freshwater algae. This was followed by a systematic treatment of the fungi, *System der Pilze und Schwämme* (1816).

A digression is required here. Nees's younger brother, Theodor Friedrich Ludwig Nees von Esenbeck (1787–1837), progressed from being his admirer to coauthor of numerous botanical works, especially during their joint period at Bonn. In 1805 he became an apprentice in the Martius pharmacy in Erlangen. (This led to his connection with Martius and Brazilian plants, for he interested Carl von Martius in botany.) In 1817 he was appointed inspector at the botanical garden in Leiden (hence the link to C. F. Blume and Javanese plants). Friedrich's brother brought him to the new Bonn garden in 1819, and at Bonn he became professor of pharmacy. Some confusion in the literature needs disentangling: It was Friedrich who made the contributions to the development of the mosses and to the discovery of spermatozoids in plants (in *Sphagnum*). The genus *Neesia* Blume is dedicated to him, while the genus *Esenbeckia* Humboldt et Bonpland ex Kunth commemorates C. G. Nees. The most important joint works by the brothers deal with cinnamon and with mycological subjects.

Most of Nees's botanical work is in the form of taxonomic monographs, initially dealing with the German flora: his treatment of the genus *Rubus* and of the Astereae, and the moss flora *Bryologica germanica*, written with Hornschuch and Sturm. (Volume I of the latter [1823] contains a superb Neesian history of the field.) In keeping with the age of discovery, he became a world expert on certain groups; their diversity is spectacular. In the flowering plants the main ones are the Lauraceae, Acanthaceae, Solanaceae, Restionaceae, Juncaceae, Cyperaceae, and Gramineae, which Nees treated variously for the Brazilian (Martius), Indian, Australian, and South African floras, or for worldwide works (Acanthaceae in A. de Candolle's *Prodromus, XI*).

Of major interest are Nees's contributions to the study of liverworts. He wrote monographs on two tropical floras, those of Java (1830) and Brazil (1833, for Martius). His four-volume account of the European liverworts, *Naturgeschichte der europäischen Lebermoose* (1833–1838), is regarded as his botanical masterpiece. It constitutes the only published part of a projected series of reminiscences from the Riesengebirge, the mountains in which Nees spent his spare time on excursions, accompanied by a local amateur botanist, J. von Flotow. The work begins with a superb introduction to the subject. The main body presents a completely new level of detail, with the recognition of innumerable subspecific variation and growth forms and a wealth of information on substrate and habitat. The final volume ends, characteristically, with a German translation, by Flotow, of Mirbel's studies on *Marchantia*, annotated by Nees. In 1841 Nees published an annotated reprint of G. Raddi's important but overlooked *Jungermanniografia etrusca* (1818). Nees's system of the world's liverworts was originally contributed to the second edition of J. Lindley's *Natural System of Botany* in 1836—Nees used Lindley's system in arranging his own general herbarium.[11] It was refined for the *Synopsis hepaticarum*. This work, still a standard reference volume, deals with all the liverworts then known. It was organized by Lehmann, principal of the Hamburg classical high school. Nees's coauthors were C. M. Gottsche, who practiced medicine in Altona and became the greatest hepaticologist of the century, and J. B. W. Lindenberg, Lübeck-Hamburgian administrator of Bergedorf. Nees's talent is perhaps most concisely displayed in his review of Corda's *Deutschlands Jungermannien* and in some notes on liverworts published in 1833.[12] The latter range from a pungent attack on chauvinism to the publication of a revolutionary report by Flotow on the culturing of liverworts in his room and the results obtained.

NOTES

1. *Flora*, **1** (1818), 137, 411, 518.
2. *Ibid.*, **2** (1819), 406.
3. *Ibid.*, **34** (1851), 559; **35** (1852), 347; *Bonplandia*, **2** (1854), 161–162.
4. *Leopoldina*, **2** (1861), 75.
5. For documents see *Nova acta*, **10** (1820), vii–xii; **11** (1823), lx–x.
6. Cf. *Bonplandia*, **1** (1852), 24–26; **6** (1858), 1–2; *Nova acta*, **24**, Abt. 1 (1854), lii–lviii, lxxxviii–lxxxix.
7. *Ibid.*, **23**, Abt. 1 (1851), xxiii.
8. *Bonplandia*, **6** (1858), 152.
9. *Nova acta*, **27** (1860), xciv.
10. Cf. *Verhandlungen Gesellschaft deutscher Naturforscher und Ärzte*, **10** (1832), 4, 6; *Nova acta*, **24**, Abt. 1 (1854), xi ff.
11. See *Flora*, **35** (1852), 347.
12. *Ibid.*, **18** (1835), "Literaturberichte," 145–165; **14** (1833), 385–412.

BIBLIOGRAPHY

I. Original Works. No proper Nees bibliography has been published, but it would comprise some 1,500 entries. Listings of his major works are available in the following standard volumes: *British Museum General Catalogue of Printed Books*, CLXIX (1963), 477–478, including nonscientific publications; *Catalogue of the Library of the British Museum* (*Natural History*), III (1910), 1407–1408; and the Royal Society *Catalogue of Scientific Papers*, IV, 583–585, which lists 72 of his major papers. The magnitude becomes clear from the general index to *Flora*, **1** (1818)–**25** (1842), which lists more than 300 items by Nees.

Nees's insect collection is at the University of Bonn, his remaining private papers in the municipal archives at Wrocław, and the liverwort section of his herbarium at the University of Strasbourg.

II. Secondary Literature. Abundant contemporary information is found in the journals *Bonplandia* and *Flora*, and in the introductory material to volumes of the *Nova acta Academiae Caesareae Leopoldino Carolinae germanicae naturae curiosorum*. Some of this has been cited in the notes. A one-page leaflet issued by Nees on 1 Feb. 1851, entitled *Erklärung* and dealing with his suspension from his professorship, is tipped into the copy of his *Handbuch der Botanik*, I, in the library of the German Society of Philadelphia. The MS of a full biography by H. Winkler, obviously unpublishable during Nazi times, came into the hands of an unidentified West German free church organization (*Nova acta*, 2nd ser., **15** [1952], 41). There is also a personal communication from the late Professor R. Zaunick of the Leopoldina, 17 Nov. 1956. For a charming photograph of Nees's head taken from the Weigelt photograph, a copy of which Nees presented to Humboldt, see *Bonplandia*, **6** (1858), 144 (an unsatisfactory engraving from this photograph is in F. Cohn [1858]).

Other works of value are the unsigned "Nees von Esenbeck, 1," in *Der grosse Brockhaus*, XIII (Leipzig, 1932), 250; F. T. Bratranek, *Neue Mittheilungen aus Johann Wolfgang von Goethe's handschriftlichem Nachlasse*, II (Leipzig, 1874), Nees-Goethe correspondence, pp. 13–180; F. Cohn, "Christian Gottfried Daniel Nees von Esenbeck," in *Illustrirte Zeitung* (Leipzig), **30**, no. 778 (29 May 1858), 345–347, published anonymously; P. Cohn, *Ferdinand Cohn*, 2nd ed. (Breslau, 1901); G. Kaufmann, ed., *Festschrift zur Feier des hundertjährigen Bestehens der Universität Breslau*, 2 vols. (Breslau, 1911); D. G. Kieser, "Lebensbeschreibung des . . . Dr. Christian Gottfried Daniel Nees von Esenbeck," in *Nova acta Academiae Caesareae Leopoldino Carolinae germanicae naturae curiosorum*, **27** (1860), lxxxv–xcii, which includes Nees's autobiography, written in 1836; C. Nissen, *Die botanische Buchillustration*, I (Stuttgart, 1966), 217–218, for the relationship of A. Henry to Nees and the Leopoldina; G. Schmid, *Goethe und die Naturwissenschaften* (Halle, 1940); B. Seemann and W. E. G. Seemann, eds., "Christian Gottfried Daniel Nees von Esenbeck," in *Bonplandia*, **6** (1858), 145–152, which includes a list of the more than 70 scientific societies to which Nees belonged, an account of the funeral, an obituary by a Dr. M. Elsner originally published in a Breslau newspaper, a description of Nees's final illness by his physician, and his official "testament" explaining the overdraft for which he was responsible to the officers of the Leopoldina; H. Winkler, "Christian Gottfried Nees von Esenbeck als Naturforscher und Mensch," in *Naturwissenschaftliche Wochenschrift*, **36** (1921), 337–346; and "Christian Gottfried Nees von Esenbeck," in Historische Kommission für Schlesien, *Schlesische Lebensbilder*, II (Breslau, 1926), 203–208; and E. Wunschmann, "Christian Gottfried Daniel Nees von Esenbeck," in *Allgemeine deutsche Biographie*, XXIII (Leipzig, 1886), 368–376; and "Theodor Friedrich Ludwig Nees von Esenbeck," *ibid.*, pp. 376–380.

Johannes Proskauer

NEF, JOHN ULRIC (*b.* Herisau, Switzerland, 14 June 1862; *d.* Carmel, California, 13 August 1915), *chemistry.*

Nef, a pioneer in the transfer of the German university traditions in organic chemistry to the United States, immigrated with his parents to Housatonic, Massachusetts, in 1866. He graduated from Harvard University with honors, in 1884, and received from the university a traveling fellowship that enabled him to obtain the Ph.D. under Adolf von Baeyer at Munich in 1886. He remained as a post-doctoral student for one year and later was instrumental in establishing postdoctoral fellowship study in the United States. He held academic positions at Purdue (1887–1889), Clark (1889–1892), and Chicago (1892–1915) universities.

Nef was a great experimentalist and, as a pioneer in theoretical organic chemistry, contributed new methods to synthetic organic chemistry, in which three separate reactions are termed "Nef reactions." He studied the apparently bivalent carbon compounds and their dissociation. His theoretical work clearly contains the germs of the present concepts of free radicals, transition states, and polymerization. He was concerned with all products formed in an organic reaction and not just with the desired end product.

The later work of Nef and his students at Chicago was concerned with the action of alkali and alkaline oxidizing agents on the sugars. They isolated and characterized various types of saccharinic acids and used enolization and subsequent carbonyl migrations to interpret the alkaline transformations and degradations of the sugars. They discovered the two types of aldonolactones.

Nef's students helped to establish graduate research in organic chemistry in the universities of the American Middle West. An intense individual, Nef impressed his personality and aims on his students, who strove to continue and extend his work.

BIBLIOGRAPHY

A listing of Nef's writings is in Poggendorff, IV, 1060–1061, and V, 896; see also the biography by Wolfrom (1960).

Two biographies, both by M. L. Wolfrom, are in Eduard Farber, ed., *Great Chemists* (New York, 1931); and *Biographical Memoirs. National Academy of Sciences*, **34** (1960), 204–227.

M. L. WOLFROM

NEGRI, ADELCHI (*b*. Perugia, Italy, 2 August 1876; *d*. Pavia, Italy, 19 February 1912), *pathology*.

Negri studied medicine and surgery at Pavia University, where, as a resident student, he worked in the pathology laboratory directed by Camillo Golgi. After graduating with honors in 1900, he became Golgi's assistant. He was named lecturer in general pathology in 1905 and in 1909 was appointed to teach bacteriology, thus becoming the first official teacher of that subject at Pavia. In 1906 he married his colleague Lina Luzzani and six years later, at the age of thirty-five, died of tuberculosis.

Trained in Golgi's school, Negri conducted research in histology, hematology, cytology, protozoology, and hygiene. His fundamental scientific contribution was the discovery, announced to the Pavia Medical Society on 27 March 1903, of the rabies corpuscles, now known as "Negri bodies." During histological research undertaken to clarify the etiology of rabies and performed on Golgi's advice, Negri found that in animals suffering from rabies, certain cells of the nervous system, especially the pyramidal cells of the horn of Ammon, contain endocellular bodies with

an internal structure so evident and regular as to constitute a characteristic feature. These bodies consist of single or multiple eosinophile, spherical, ovoid, or pyriform endocytoplasmic (never endonuclear) formations with a well-defined outline, varying in size from two to more than twenty microns (apparently in proportion to the size of the animal) and containing minute basophil granules having a diameter of 0.2–0.5 micron.

This cytological phenomenon proved to be almost constant and was found typically and abundantly in the histological material from living victims of advanced spontaneous rabies (street virus) or from those who had died of it. On the other hand, it was absent or very rare in cases of infection following inoculation of fixed virus. Rabbits and dogs infected experimentally with the street virus, dogs dead from spontaneous rabies, a cat infected experimentally by subdural injection, and one human case (a woman of sixty-four who had died of rabies after being bitten by a rabid dog) furnished the material on which Negri gave the first demonstrations of his discovery.

From the beginning Negri believed that the endocellular bodies he had observed in the nerve cells were the pathogenic agents of rabies and that they were forms belonging to the developmental cycle of a protozoan, the systematic position of which he could not define. This opinion, which Negri never abandoned, immediately became the object of scientific discussion. Some months after Negri's discovery, Alfonso Di Vestea in Naples, and Paul Remlinger and Riffat Bey in Constantinople, showed that the etiological agent of rabies is a filterable virus; and the argument about the significance of Negri's bodies became wider and more intense, with eminent parasitologists taking conflicting positions. Even today, despite research with the electron microscope, the significance of Negri's bodies has not been definitively clarified. Thus, as Luigi Bianchi wrote, it is still possible to accept Emilio Veratti's opinion that Negri's bodies are to be interpreted as specific formations closely linked to the virus and not as products of the cell containing it, without thereby assuming that they constitute the sole, infallible manifestation of the virus.

The specificity of Negri's bodies and their importance for diagnosis are universally recognized; the search for them, however, has absolute probative value in diagnosis only when there is a positive result. Negri himself indicated the rules to be observed in identifying the bodies for diagnostic purposes in animals suspected of rabies. The bodies, in material simply fixed for eighteen to twenty-four hours in Zenker's fluid and delicately pulped between cover glass and slide in a drop of glycinerinated water, appear under small enlargement as light yellow, glassy formations in the cytoplasm of the pyramidal cells; at enlargements of 400–600 diameters they reveal their characteristic internal structure.

BIBLIOGRAPHY

I. ORIGINAL WORKS. Negri produced some thirty publications, some of them joint works, which appeared in Italian and foreign journals between 1899 and 1911. They are listed in Veratti's article (see below) and in *Archives de parasitologie*, **16** (1913), 166. Documentary material concerning Negri is kept in the Museum of University History, Pavia.

II. SECONDARY LITERATURE. See Luigi Bianchi, "Rabbia," in Paolo Introzzi, ed., *Trattato italiano di medicina interna*, pt. 4, *Malattie infettive e parassitarie*, II (Bologna, 1965), 1351–1364; and *I corpi del Negri nello sviluppo della microbiologia all'Università di Pavia* (Pavia, 1967); and Emilio Veratti, "Adelchi Negri. La vita e l'opera scientifica," in *Rivista di biologia*, **16**, no. 3 (1934), 577–601.

BRUNO ZANOBIO

NEHRING, ALFRED (*b*. Gandersheim, Germany, 29 January 1845; *d*. Berlin, Germany, 29 September 1904), *paleontology, zoology*.

After graduating from the Gymnasium, Nehring studied natural sciences, especially zoology, from 1863 to 1867 at the universities of Göttingen and Halle, receiving his doctorate from Halle in 1867. He then taught biology at the Gymnasiums in Wesel and Wolfenbüttel. In 1881, on the strength of his scientific achievements, he was appointed professor of zoology at the Agricultural College in Berlin, where he later also became curator of the zoological collections. He held these posts until his death. Nehring never recovered from the psychological effects of a gas explosion in 1902 under the museum of the Agricultural College; this accident, which destroyed or damaged one portion of the collections and disrupted another, also weakened his health during his last years.

Nehring's scientific works covered Recent, postglacial, and Pleistocene vertebrates, particularly domestic animals; their domestication; their history; their relations in the wild; and the zoology of untamed game animals. His publications on Pleistocene mammals are of major importance. Nehring's works on Recent zoology concern the distribution of *Mus rattus* and *Mus decumanus*; canine teeth in horses, wild swine, Saiga antelopes, and various species of deer; the skeleton and systematic position of the seal *Halichoerus*; dog skulls with abnormal dentition; the craniological differences between the lion and the tiger; the origin, descent, and hybridizations of the South American rodent *Cavia cobaya*; the distribution and agricultural importance of the hamster *Cricetus cricetus* in Germany; various genera and species of Cricetidae; the origin of the duck *Anas moschata*; the distribution of the snake *Coronella austriaca* and of the freshwater fish *Pelecus*; and the presence of the snail *Helix candicans* in Pomerania.

Nehring's investigations of postglacial and domesticated animals concerned dwarf swine from Pomerania; primitive domesticated dogs near Berlin; the influence of domestication on the size of an animal's body; lake-dwelling fauna of East Prussia; the remains of *Bos primigenius* and *Alces* in the regions around Berlin; Herberstain's woodcuts (1557) of *Bison priscus* and *Bos primigenius*; Inca dogs from Peru; ancient Egyptian animal mummies; and the descent of domestic sheep.

The largest portion of Nehring's work consists of writings on Pleistocene mammals and birds, especially small mammals. In them he treated the morphology, taxonomy, and biogeographic distribution of many Pleistocene species; he also compared them in these respects with living representatives and relatives. He covered such rodents as lemmings and other Arvicolidae, including the genus *Dolomys*, which Nehring himself established, and members of other rodent families. He also dealt with the dog and cat families; the bear; several species of deer, ox, goat, sheep, and antelope; *Bison priscus*, camels, wild asses, horses, and the mammoth; and such birds as *Tetrao* (wood grouse), *Lyrurus* (black grouse), *Lagopus* (ptarmigan), *Nyctea scandiaca* (snowy owl), and *Scolopax rusticola* (woodcock). Most of Nehring's material was of German origin; but it also came from Europe (as far east as Russia and as far west as Portugal and England), Lebanon, and China.

Besides these individual descriptions Nehring wrote many accounts of the Pleistocene fauna: a survey of twenty-four central European Quaternary local faunas; the Quaternary local faunas of Thiede and Westeregeln near Brunswick; micromammals from the caves of Upper Franconia; diluvial vertebrates from Pösneck, Thuringia; the local fauna of a Pleistocene cave near Schaffhausen, Switzerland; diluvial animal remains from the Seveckenberg near Quedlinburg; and mammal remains from a Pleistocene peatbog near Cottbus.

Nehring not only studied the morphological, taxonomic, and phylogenetic relationships of the Pleistocene mammals but also treated some of the ecological problems involved. He wrote at length on these matters in *Über Tundren und Steppen der Jetzt- und Vorzeit* (1890), after having already treated them in several shorter papers. Following a description of the tundra (arctic steppes) in northern Russia and Siberia, as well as of the steppes of southern Russia and southwest Siberia, and their characteristic mammals, Nehring showed that regions of tundra and steppe, with their corresponding fauna, had existed in the later Pleistocene in central and western Europe. This study, filled with numerous and exact data, is still among the most important foundations of the paleoecology and paleobiogeography of the later Pleistocene in central and western Europe. Nehring also was interested in the fossil remains of man and his implements, and he participated vigorously in the discussions on the *Pithecanthropus erectus* of Java at the end of the nineteenth century.

Nehring's painstaking and thorough works still provide useful and much-employed data for research on Pleistocene mammals.

BIBLIOGRAPHY

I. ORIGINAL WORKS. A nearly complete bibliography is in A. S. Romer, N. E. Wright, T. Edinger, and R. van Frank, *Bibliography of Fossil Vertebrates Exclusive of North America, 1509–1927*, II, *L–Z*, Geological Society of America Memoir no. 87 (New York, 1962), 978–985. Nehring's writings include "Länge und Lage der Schneidezahnalveolen bei den wichtigsten Nagethieren," in *Zeitschrift für Naturwissenschaften*, **45** (1875), 217–239; "Beiträge zur Kenntniss der Diluvialfauna," *ibid.*, **47** (1876), 1–68; "Beiträge zur Kenntniss der Diluvialfauna (Fortsetzung)," *ibid.*, **48** (1876), 177–236; "Die quaternären Faunen von Thiede und Westeregeln nebst Spuren des vorgeschichtlichen Menschen," in *Archiv für Anthropologie*, **10** (1878), 359–398; "Fortsetzung und Schluss," *ibid.*, **11** (1879), 1–24; "Übersicht über vierundzwanzig mitteleuropäische Quartärfaunen," in *Zeitschrift der Deutschen geologischen Gesellschaft*, **32** (1880), 478–509; "Über die Abstammung unserer Hausthiere," in *Jahresberichte und Abhandlungen des Naturwissenschaftlichen Vereins in Magdeburg* for 1885–1886 (1886), 129–144; *Über Tundren und Steppen der Jetzt- und Vorzeit, mit besonderer Berücksichtigung ihrer Faunen* (Berlin, 1890); "Die geographische Verbreitung der Säugethiere in dem Tschernosem-Gebiete des rechten Wolga-Ufers," in *Zeitschrift der Gesellschaft für Erdkunde zu Berlin*, **26** (1891), 297–351; "Über einen Molar aus dem Diluvium von Taubach," in *Zeitschrift für Ethnologie*, **27** (1895), 573–577; "Die kleineren Wirbeltiere aus dem Schweizersbild bei Schaffhausen," in J. Nüesch *et al.*, "Das Schweizersbild, eine Niederlassung aus paläolithischer und neolithischer Zeit," in *Neue Denkschriften der Allgemeinen schweizerischen Gesellschaft für die gesamten Naturwissenschaften*, **35** (1902), 159–198.

II. SECONDARY LITERATURE. See the following, listed chronologically: J. V. Zelizko, "Alfred Nehring. Črta Životopisná," in *Pravěk*, **2** (1904), 150–155, with portrait; E. Friedel, "Alfred Nehring als Erforscher unserer Heimat," in *Brandenburgia*, **13** (1905), 289–301, with partial bibliography; and G. Tornier, "Rückblick auf Anatomie und Zoologie," in *Sitzungsberichte der Gesellschaft naturforschender Freunde zu Berlin* (1923), 12–71, see 43–44; and "Rückblick auf die Paläontologie," *ibid.* (1925), 72–106, titles of many of Nehring's paleontological papers, 101–103.

HEINZ TOBIEN

NEISSER, ALBERT LUDWIG SIGESMUND (*b.* Schweidnitz, Germany [now Swidnica, Poland], 22 January 1855; *d.* Breslau, Germany [now Wrocław, Poland], 30 July 1916), *dermatology.*

Neisser's father was a highly respected physician; his mother died before he was a year old, and he was raised by his stepmother. Neisser attended the Volksschule in Münsterberg, then entered the St. Maria Magdalena Humanistic Gymnasium in Breslau, where Paul Ehrlich was a classmate. In 1872 he began his medical studies, which, with the exception of one semester of clinical work in Erlangen, were carried out entirely in Breslau. His studies were not outstanding—in fact, he had to repeat the chemistry test—but he passed the state examination and received the medical degree in 1877 with a thesis on echinococosis, prepared under the direction of the internist Anton Biermer. His other teachers included Rudolf Heidenhain, Julius Cohnheim, Carl Weigert, and C. J. Salomonsen. Neisser originally planned to become a specialist in internal medicine, but there were no openings for assistants in Biermer's clinic. It was therefore purely by chance that he turned to dermatology, becoming an assistant in Oskar Simon's clinic, where he worked for two years. It was there that in 1879 Neisser discovered the gonococcus.

Neisser's discovery occurred in the wake of the rapid development of the new field of bacteriology. It was made possible in large part by his close association with the botanist Ferdinand Cohn, who taught him Koch's smear tests for the identification of bacteria, and with Cohnheim and Weigert, who taught him staining techniques, including those with methylene blue. Neisser was further able to make use of a new Zeiss microscope that incorporated Abbe's innovative condenser and oil-immersion system. He at first called the microorganisms that he thus observed "micrococcus"; they were then given the name "gonococcus" by Ehrlich. Neisser's paper "Über eine der Gonorrhoe eigenthümliche Micrococcenform," published in 1879, was a milestone in elucidating the etiology of venereal diseases.

Neisser made a research trip to Norway in the same year. He was able to examine more than 100 patients with leprosy in Trondheim, Molde, and Bergen, and to take secretion smears back to Germany to study. In examining the smears he found, in almost all cases, "bacilli as small, thin rods, whose length amounts to about half the diameter of a human red blood corpuscle and whose width I estimate at one-fourth the length." These results embroiled him in a priority dispute with the Norwegian bacteriologist G. H. A. Hansen, who had found similar microorganisms in leprosy secretions as early as 1873; when Neisser published his findings in 1880, Hansen responded with a paper, published in four languages, in which he stated his earlier claim. It is clear, however, that while Hansen first discovered the leprosy bacillus, Neisser was the first to identify it as the etiological agent of the disease. The etiology, diagnosis, and prophylaxis of leprosy occupied him for much of his subsequent career.

His early publications made Neisser's name well known. On his return to Breslau, he was able to qualify as a lecturer in dermatology on the university medical faculty, and he was named *Privatdozent* on 6 August 1880. In 1882 Simon died suddenly, and Neisser was appointed his successor in the chair of dermatology and as director of the clinic. His promotion at the age of twenty-seven was sponsored by Friedrich Althoff, the Prussian councillor for education and cultural affairs. In the following year Neisser married Toni Kauffmann, who assisted him in his investigations and accompanied him on research trips. At about the same time he became involved in planning a new dermatological clinic, which, built to his design, was opened in 1892 and became an internationally famous research center.

Neisser's work with leprosy led him to study another infectious skin disease, lupus. He early suspected a connection between lupus and tuberculosis and went on to distinguish non-tubercular forms of the disease, including lupus erythematosus, lupus pernio, and sarcoidosis of the skin. His attempts to cure lupus with tuberculin came to nothing, however; he remained particularly concerned in alleviating the lot of those scarred by lupus. His servant, Hein, was so afflicted, and Neisser often used him as an object lesson in what might be done toward rehabilitation.

Neisser also devoted intensive study to syphilis, although his therapeutic suggestions are of little significance. His attempt to discover the cause of the disease through a series of inoculation experiments were unfortunate; he was accused of having "maliciously inoculated innocent children with syphilis poison," and a scandal resulted. Neisser was misled by drawing an analogy with the serum therapy that Behring had used against diphtheria and tetanus; the supposed serum with which he inoculated young prostitutes was probably highly infectious in itself.

In 1903 Metchnikoff and Roux demonstrated that syphilis could be communicated to apes, and Neisser immediately repeated their experiments and confirmed their findings. He made two trips to Java, in 1905 and 1906, to obtain ape specimens and to continue his research toward determining the cause of the disease. On 16 May 1905, however, he heard of Schaudinn and Hoffmann's discovery of the syphilis spirochete; the news must have been disappointing to him, and he at first was disinclined to accept that the spirochete was actually the causative agent. In a letter of June 1905, he wrote, "We are still toiling with the syphilis spirilla. Here and there we find something positive, but on the whole we are more convinced than ever that these spirilla . . . are not really the syphilis spirilla." By the time of his second Java trip he was convinced, however, and he turned to the investigation of the transmission of syphilis among both apes and men—the temporary stationing of Dutch sailors in Java supplied him with the human syphilis patients formerly lacking on that island. His observations yielded valuable data concerning reinfection and superinfection.

Neisser encouraged Wassermann to study seroreaction in syphilis in 1906. With him and with Carl Bruck, he developed the serological test, now named for Wassermann. He also worked in testing therapeutically the arsenic preparations, especially arsenophenylglycine, with which Ehrlich provided him. He found these to be effective but dangerous as remedies for syphilis. He also contributed to Ehrlich's introduction of Salvarsan (1910).

Neisser's work with venereal diseases brought him into the field of public health. He propagandized widely for better prophylactic measures and for more public education about these diseases; he was active in founding the Deutschen Gesellschaft zur Bekämpfung der Geschlechtskrankheiten and served as its president in 1901. He strongly supported stricter regulation of prostitution, and favored increased sanitary measures rather than police action. He also supported the establishment of a central board of health but objected to the obligation to inform the police and advocated the confidentiality of the doctor-patient relationship.

In 1907 Neisser was named full-time professor of dermatology at Breslau. He trained a number of eminent dermatologists, and conducted research on lichen infestations and urticaria. He experimentally explored the emergence of weals in the latter condition, and contributed substantially to Heidenhain's conclusion that the weal is a vasodilatory edema. As early as 1894 he described vitiligo with lichenoid eruptions, and he published a number of other findings about skin tumors, infectious diseases (including anthrax, actinomycosis, glanders, blastomycosis, and skin diphtheria), psoriasis, mycosis fungoids, and various forms of pemphigus. His work received wide official recognition.

The death of his wife, in 1913, affected Neisser deeply. His own health began to fail rapidly in 1916, and he died shortly after he was named a member of the Imperial Health Council. In 1920 his house was made a museum; in 1933 it was confiscated by the Nazis and turned into a guesthouse. Neisser's papers were salvaged by a Schweinfurt physician named Brock and form the basis for recent works about him.

BIBLIOGRAPHY

I. ORIGINAL WORKS. A full list of Neisser's publications may be found in the biography by Sigrid Schmitz, cited below. His most important works include "Über eine der Gonorrhoe eigenthümliche Micrococcenform," in *Centralblatt für die medizinischen Wissenschaften*, **28** (1879), 497–500; "Über die Aetiologie des Aussatzes," in *Jahresbericht der Schlesischen Gesellschaft für vaterländische Kultur*, **57** (1880), 65–72; "Weitere Beiträge zur Aetiologie der Lepra," in *Archiv für pathologische Anatomie und Physiologie*, **84** (1881), 514–542; "Die Mikrokokken der Gonorrhoe," in *Deutsche medizinische Wochenschrift*, **8** (1882), 279–283; "Die chronischen Infektionskrankheiten der Haut," in H. W. von Ziemssen, ed., *Handbuch der speciellen Pathologie und Therapie*, XIV (Leipzig, 1883), 560–723; "Über das Leukoderma syphiliticum," in *Vierteljahrsschrift für Dermatologie und Syphilis*, **15** (1883), 491–508; "Über die Mängel der zur Zeit üblichen Prostituiertenuntersuchungen," in *Deutsche medizinische Wochenschrift*, **16** (1890), 834–837; "Pathologie des Ekzems," in *Archiv für Dermatologie und Syphilis*, **1** (1892), suppl., 116–161; "Über den gegenwärtigen Stand der Lichenfrage," *ibid.*, **28** (1894), 75–99; and "Über Vitiligo mit lichenoiden Eruptionen," in *Verhandlungen der Deutschen Dermatologischen Gesellschaft. IV. Kongress zu Breslau* (Vienna-Leipzig, 1894), 435–439.

See also "Syphilis maligne," in *Journal des maladies cutanées et syphilitiques*, **9** (1896), 210–213; "Was wissen wir von einer Serumtherapie der Syphilis und was haben wir von ihr zu hoffen?," in *Archiv für Dermatologie und Syphilis*, **44** (1898), 431–439; "Über Versuche, Syphilis auf Schweine zu übertragen," *ibid.*, **59** (1902), 163–170; "Meine Versuche zur Übertragung der Syphilis auf Affen," in *Deutsche medizinische Wochenschrift*, **30** (1904), 1369–1373, 1431–1434; "Weitere Mitteilungen über den Nachweis spezifischer luetischer Substanzen durch Komplement-

bindung," in *Zeitschrift für Hygiene und Infektionskrankheiten*, **55** (1906), 451–477, written with A. Wassermann, C. Bruck, and A. Schucht; *Über die Bedeutung der Lupuskrankheit und die Notwendigkeit ihrer Bekämpfung* (Leipzig, 1908); "Über das neue Ehrlich'sche Mittel," in *Deutsche medizinische Wochenschrift*, **36** (1910), 1212–1213; "Beiträge zur Pathologie und Therapie der Syphilis," in *Arbeiten aus dem Kaiserlichen Gesundheitsamt*, **37** (1911), 1–624; *Syphilis und Salvarsan* (Berlin, 1913); "Ist es wirklich ganz unmöglich, die Prostitution gesundheitlich unschädlich zu machen?," in *Deutsche medizinische Wochenschrift*, **41** (1915), 1385–1388; "Über das urtikarielle Ekzem," in *Archiv für Dermatologie und Syphilis*, **121** (1916), 579–612; and *Die Geschlechtskrankheiten und ihre Bekämpfung* (Berlin, 1916).

II. SECONDARY LITERATURE. In addition to obituary notices in a number of medical journals, see K. Bochmann, "Albert Neisser," in *Heilberufe*, **7** (1955), 179; E. Czaplewski, "Albert Neisser und die Entdeckung des Leprabazillus," in *Archiv für Dermatologie und Syphilis*, **124** (1917), 513–530; G. L. Flite and H. W. Wade, "The Contribution of Neisser to the Establishment of the Hansen Bacillus as the Etiologic Agent of Leprosy and the So-called Hansen-Neisser Controversy," in *International Journal of Leprosy*, **23** (1955), 418–428; J. Jadassohn, "Albert Neisser," in F. Andreae, ed., *Schlesische Lebensbilder*, I (Breslau, 1922), 111–115; J. Schäffer, *Albert Neisser* (Berlin-Vienna, 1917); W. Schönfeld, "In Memoriam Albert Neisser zum 100. Geburtstag," in *Hautarzt*, **6** (1955), 94–96; A. Stühmer, "Albert Neisser," in *Dermatologische Wochenschrift*, **131** (1955), 214–216; and T. M. Vogelsang, "The Hansen-Neisser Controversy, 1879–1880," in *International Journal of Leprosy*, **31** (1963), 74–80, and **32** (1964), 330–331.

The best and most comprehensive biography, which draws upon Neisser's posthumous papers and other previously unpublished sources, is Sigrid Schmitz, "Albert Neisser. Leben und Werk auf Grund neuer, unveröffentlicher Quellen," in H. Schadewaldt, ed., *Düsseldorfer Arbeiten zur Geschichte der Medizin*, XXIX (Düsseldorf, 1968).

H. SCHADEWALDT

NEKRASOV, ALEKSANDR IVANOVICH (*b.* Moscow, Russia, 9 December 1883; *d.* Moscow, 21 May 1957), *mechanics, mathematics.*

Nekrasov graduated from the Fifth Moscow Gymnasium in 1901 with a gold medal and entered the mathematical section of the Faculty of Physics and Mathematics at Moscow University. In 1906 he graduated with a first-class diploma and received a gold medal for "Teoria sputnikov Yupitera" ("Theory of the Satellites of Jupiter"). Nekrasov remained at the university to prepare for a professorship. At the same time he taught in several secondary schools in Moscow. In 1909–1911 Nekrasov passed his master's examinations in two specialties, astronomy and mechanics. In 1912 he became assistant professor in the department of astronomy and geodesy of the Faculty of Physics and Mathematics at the university, and in 1913 he was appointed to the same post in the department of applied mathematics (theoretical mechanics) of the same faculty. From 1917 until his death Nekrasov taught and conducted research at Moscow University, the Higher Technical School, the Central Aerohydrodynamics Institute, the Sergo Orjonikidze Aviation Institute, and the Institute of Mechanics of the Academy of Sciences of the U.S.S.R.

In 1922 Nekrasov was awarded the N. E. Zhukovsky Prize for "O volnakh ustanovivshegosya vida na poverkhnosti tyazheloy zhidkosti" ("On Smooth-Form Waves on the Surface of a Heavy Liquid"). For his distinguished scientific services he was elected corresponding member of the Academy of Sciences of the U.S.S.R. in 1932 and an active member in 1946. He was awarded the title Honored Worker in Science and Technology in 1947 for his services in the development of aviation technology. Nekrasov was a brilliant representative of the trend in the development of precise mathematical methods in hydromechanics and aeromechanics that is associated with Zhukovsky and S. A. Chaplygin. He published basic works on the theory of waves, the theory of whirlpools, the theory of jet streams, and gas dynamics.

Nekrasov's *Tochnaya teoria voln ustanovivshegosya vida na poverkhnosti tyazheloy zhidkosti* ("A Precise Theory of Smooth-Form Waves on the Surface of a Heavy Liquid"), on classical problems of hydromechanics, was awarded the State Prize of the U.S.S.R. in 1951. In an extensive monograph on aerodynamics, *Teoria kryla v nestatsionarnom potoke* ("Theory of the Wing in a Nonstationary Current"; 1947), he presented a systematic and detailed account of all the basic scientific works dealing with the theory of the unsmooth motion of a wing in the air without allowing for its compressibility. He not only systematized material published earlier but also analyzed and compared it, in a number of cases providing a new mathematical treatment of the subject. Other important works in aerodynamics are *Primenenie teorii integralnykh uravneny k opredeleniyu kriticheskoy skorosti flattera kryla samoleta* ("Application of the Theory of Integral Equations to the Determination of the Critical Velocity of the Flutter of an Airplane Wing"; 1947) and *Obtekanie profilya Zhukovskogo pri nalichii na profile istochnika i stoka* ("Flow on a Zhukovsky Cross Section in the Presence of a Cross Section of the Source and Outflow"). Besides his work on aerohydrodynamics Nekrasov published an excellent two-volume textbook on theoretical vector mechanics (1945–1946).

Nekrasov's works also enriched mathematics. Among his contributions are the first fruitful investigations of nonlinear integral equations with symmetrical nuclei, the books *O nelineynikh integralnykh uravneniakh s postoyannymi predelami* ("On Nonlinear Integral Equations With Constant Limits"; 1922) and *Ob odnom klasse lineynykh integro-differentsialnykh uravneny* ("On One Class of Linear Integral-Differential Equations"; 1934), and many investigations in an important area of aerohydrodynamics. The extremely varied mathematical apparatus that he used contains many original details developed by Nekrasov himself.

Nekrasov translated into Russian É. Goursat's *Cours d'analyse mathématique* as *Kurs matematicheskogo analiza*. To a substantial degree this project made possible Nekrasov's assimilation of the mathematical methods that he later applied so skillfully to the solution of concrete problems in aerodynamics.

A fully worthy disciple of and successor to Zhukovsky, Nekrasov enriched Soviet science with his scientific works and, through his work in education, aided the development of many scientists and engineers.

BIBLIOGRAPHY

Many of Nekrasov's writings are in his *Sobranie sochineny* ("Collected Works"), 2 vols. (Moscow, 1961–1962).

Secondary literature includes *Aleksandr Ivanovich Nekrasov* (Moscow–Leningrad, 1950); and Y. I. Sekerzh-Zenkovich, "Aleksandr Ivanovich Nekrasov," in *Uspekhi matematicheskikh nauk*, **15**, no. 1 (1960).

A. T. GRIGORIAN

NEMESIUS (*fl.* Emesa [now Homs], Syria, A.D. 390–400), *medicine.*

Nemesius, possibly although not certainly a provincial governor of Cappadocia, is believed to have been converted to Christianity about 390, and sometime thereafter he became bishop of Emesa. During these final years of the fourth century he composed his treatise Περὶ φύσεως ἀνθρώπου ("On the Nature of Man"), which is essentially concerned with the reconciliation of Platonic doctrines on the soul with Christian philosophy and also, importantly, with the interpretation of Greek scientific knowledge of the human body from the standpoint of Christian doctrine. For a time the work was attributed to Gregory of Nyssa; and it was not until the seventh century that there was any ascription of it to Nemesius, of whom almost nothing is known except for such self-revelations as are to be found in his text. From these it is apparent that he was well-read in the writings of Galen and may even have had some medical training.

Although Nemesius' book contains many passages dealing with Galenic anatomy and physiology, the most important contribution of the work was to establish the idea that the mental faculties were localized in the ventricles of the brain, a belief that was generally accepted and retained as late as the sixteenth century. Actually the belief in such localization had been advanced even earlier in the fourth century by the Greek physician Posidonius, to whom Nemesius referred; but because only fragments of Posidonius' writings survived, the doctrine of ventricular localization gained prominence only through the later treatise.

According to Nemesius' doctrine, all sensory perceptions were received in the anterior—now called lateral—ventricles of the brain. Later this area came to be designated the "sensus communis," that is, the region where all the sensory perceptions were held in common by a force known as the faculty of imagination. The middle or, as it is now called, third ventricle was the region of the faculty of intellect, which controlled the "judging, approving, refuting, and assaying" of the sensory perceptions gathered in the lateral ventricles. The third faculty was that of memory, the storehouse of sensory perceptions after they had been judged by the faculty of intellect. Memory was located by Nemesius in the cerebellum but, according to succeeding interpretations, in the fourth ventricle. Moreover, later writers extended Nemesius' doctrine by causing the intellectual or rational faculty to draw upon memory in the making of decisions. The faculties operated through the agency of the animal spirit, the very refined spirit which, according to Galen, was produced from vital spirit after it had been carried through the supposititious network of arteries, called the *rete mirabile*, at the base of the brain. Nemesius was convinced of the correctness of his doctrine of the ventricular localization of the mental faculties, since in his opinion injury to those areas of the brain caused the loss of the faculties.

After its composition Nemesius' book seems to have gone through a period of disregard, to be rediscovered only after the passage of several centuries. It was cited by John of Damascus in the eighth century, by Timothy I, the Nestorian *catholicos*, and by the Phrygian monk and physician Meletius. Possibly it was from one of these sources that ventricular localization came to be accepted and described as early as the late ninth and early tenth centuries by Qusṭā ibn Lūqā and by al-Rāzī, the latter of whom was important in the diffusion of the doctrine.

The first extensive and medically important treatment of Nemesius' book was that by Alphanus, a monk of the Benedictine abbey of Monte Cassino and later archbishop (1058–1085) of Salerno. Alphanus translated it into Latin under the title of *Premnon physicon* ("Tree of Nature") and thus made the medical portions available to the Salernian medical school, although there is no clear evidence of their influence upon Salernian medicine. The work was again translated in 1155 by Burgundio of Pisa; neither translator appears to have been aware of the identity of the author. These Latin translations were definitely effective in promotion of the idea of the localization of the mental faculties in the ventricles—the former, for example, on Albertus Magnus and the latter on Thomas Aquinas. Localization began to be described with some frequency in the twelfth and thirteenth centuries. It was also illustrated by drawings of the head in which the lateral ventricles were often identified by circular figures, frequently called "cellulae" and specified as the area of "imaginatio," "phantasia," or "sensus communis"; a further circle representing the third ventricle was commonly designated as the area of "aestimativa" or "cogitativa," and the circle of the fourth ventricle was most often labeled "memoria."

The first translation of Nemesius' work to be printed (Strasbourg, 1512) was that by John Cono of Nuremberg, and the first printed edition of the Greek text was published under the editorship of Nicasius Ellebodius by the Plantin Press of Antwerp in 1565.

The idea of ventricular localization of the mental faculties in the form presented by Nemesius was first attacked in 1521 by Berengario da Carpi, who grouped the three faculties in three separate areas of the lateral ventricles. Vesalius delivered the coup de grace to the entire theory in 1543, when he denied any role to the ventricles except the collection of fluid and declared that in some manner the mind was in the brain at large. Although Vesalius did not elaborate upon this point, his theme was picked up by Costanzo Varolio, who in 1573 asserted more clearly that there was a single mental faculty in the brain as a whole and that the ventricles served merely to collect and drain off superfluous fluid.

BIBLIOGRAPHY

There is an English trans. of Περὶ φύσεως ἀνθρώπου by William Telfer in *Cyril of Jerusalem and Nemesius of Emesa*, W. Telfer, ed., vol. IV in Library of Christian Classics (London, 1955). It includes references to all the pertinent literature on Nemesius and eds. of his work.

C. D. O'MALLEY

NEMORE, JORDANUS DE. See **Jordanus de Nemore.**

NENCKI, MARCELI (*b.* Boczki, near Kielce, Russia [now Poland], 15 January 1847; *d.* St. Petersburg, Russia, 14 December 1901), *biochemistry.*

Nencki was the son of Wilhelm Nencki, a landowner, and the former Katarzyna Serwaczyńska. In 1863 he graduated from a classical secondary school in Piotrków Trybunalski; active participation in the Polish uprising made his situation uncertain and forced him to emigrate. For about a year Nencki studied philosophy and ancient languages at the University of Berlin before transferring to the Medical Faculty of the same university, where he studied the chemistry of living organisms. To increase his knowledge of inorganic and organic chemistry, Nencki worked during two years of his medical studies under the direction of Baeyer in his laboratory at the Gewerbeinstitut.

While still a university student Nencki published, with his friend O. Schultzen, a paper dealing with precursors of urea in mammals. He received the M.D. in 1870 with a dissertation on the oxidation of aromatic compounds in the body. In 1871 Nencki published a paper related to the chemistry of uric acid and similar compounds. These three subjects were the center of his interest throughout his life.

Nencki's initial publications appeared to be of such value that shortly after receiving the M.D. he was engaged as an assistant at the University of Bern, where his scientific career developed swiftly. Within a few years he was appointed to a professorship and became head of the department of biochemistry. A pioneer of a chemical approach to microorganisms, he also lectured on pharmacology and bacteriology. Nencki became an internationally recognized authority on biochemistry and theoretical medicine, attracting students from all parts of Europe and America.

In 1890 Nencki left Switzerland for St. Petersburg, to help organize a new institute of experimental medicine. As head of the department of chemistry and biochemistry he began work in a new building specially constructed and equipped according to his plans. Partly in collaboration with Pavlov, who was head of the department of physiology, Nencki started to reinvestigate the method and site of formation of urea in the body. These fundamental and beautiful experiments showed that urea is formed chiefly, if not exclusively, in the liver. According to Nencki, urea was synthesized from amino groups of amino acids and from carbon dioxide and did not preexist in the protein molecule, as was then quite generally believed. This important idea anticipated modern views of the utilization of carbon dioxide in certain synthetic processes that occur in the animal body. Speculating on the various possibilities of biosynthesis, especially of fatty acids, Nencki proposed a hypothesis according to which a gradual condensation of some active two-carbon-atom fragments takes place; a splitting-off of these fragments may occur during oxidation. This hypothesis, together with the results of Nencki's earlier investigation on the oxidation of aromatic compounds in the animal body, formed a basis for Knoop's β oxidation theory. Nencki supposed that the active two-carbon-unit compound was acetaldehyde and hence, in principle, he was not far from the modern view of the role of acetyl coenzyme A.

Nencki's best-known investigations concerned hemoglobin. Using original methods he systematically studied for many years the degradation products of this blood pigment. Somewhat later Marchlewski, working first in England and later in Cracow, performed similar studies on chlorophyll. It soon appeared that some of the degradation products of these two pigments resembled each other. Nencki and Marchlewski thus initiated a peculiar art of collaboration at a distance. They exchanged and carefully investigated the particular degradation products of the pigments and finally obtained, both from hemoglobin and from chlorophyll, the same substance, hemopyrrole.

On the basis of these results Nencki put forward a hypothesis concerning the chemical relationship between the animal and plant kingdoms. He planned vast investigations in the field, which now would be called evolutionary biochemistry. Unfortunately these plans were not realized; at the end of 1901 Nencki died of stomach cancer.

Nencki's work is unusually impressive for its magnitude, as well as for the variety of the problems he investigated, the ingenuity and the precision of his experiments, and his perseverance in achieving his aims. His chief interest was biochemistry; but some of his papers deal with analytical and organic chemistry, bacteriology, pharmacology, pharmacy, hygiene, and practical medicine.

All of Nencki's scientific life took place outside Poland, but he was always in contact with his motherland: he published many of his papers in Polish scientific journals, held an honarary doctorate from the University of Cracow, was a member of several Polish scientific societies, and often participated in Polish congresses. In accordance with his wishes, he was buried in Warsaw.

Shortly after Nencki's death it was suggested that a scientific institute be built and named for him. This idea was not realized until 1918, with the creation of the Nencki Intstiute of Experimental Biology, devoted chiefly to biochemistry and physiology, in Warsaw.

BIBLIOGRAPHY

Marceli Nencki. Opera omnia. Gesammelte Arbeiten von Prof. M. Nencki, Nadine Sieber and J. Zaleski, eds., (Brunswick, 1904), contains Nencki's 150 papers and 450 papers by his colleagues at Bern and St. Petersburg.

Fifty Years of Activity of the M. Nencki Institute of Experimental Biology (1918–1968) (Warsaw, 1968), in Polish with English summaries, contains articles on the history of the Nencki Institute (W. Niemierko), the investigations of the physiology of the brain (J. Konorski), biochemistry (W. Niemierko), neurochemistry (Stella Niemierko), biology (S. Dryl), hydrobiology (R. Klekowski), data concerning the library of the Institute (H. Adler), and a complete list of workers there (1918–1968). *Marceli Nencki, Materiały biograficzne i bibliograficzne*, Aniela Szwejcerowa and Jadwiga Groszyńska, eds. (Warsaw, 1956), contains a full bibliography of Nencki's works and of works on him, his correspondence with Marchlewski, some of his letters to his family, a biographical article by W. Niemierko, some other biographical articles, photographs, and documents. See also M. H. Bickel, *Marceli Nencki, 1847–1901* (Bern–Stuttgart–Vienna, 1972).

WŁODZIMIERZ NIEMIERKO

NERI, ANTONIO (*b*. Florence, Italy, 29 February 1576; *d*. Pisa or Florence, *ca*. 1614), *chemical technology*.

Almost nothing is known with certainty about Neri, except that his father, Jacopo, was a physician; that Neri was ordained a priest before 1601; and that he led a wandering life. He appears to have learned the art of glassmaking at Murano, near Venice, and to have continued his studies of this and other chemical arts in the Low Countries. From about 1604 to 1611 he was at Antwerp, lodging in the house of Emanuel Ximenes, a Portuguese; he published his book and spent the last years of his life in northern Italy. The evidence for the date of his death is very scanty.

Neri is remembered only for *L'arte vetraria* (1612), a little book in which many, although by no means all, of the closely guarded secrets of glassmaking were printed for the first time. He recommended that glass be made from *rocchetta* (a fairly pure sodium sesquicarbonate from the Near East) and *tarso*, which he described as a kind of marble but which must have been some form of silica. He did not indicate the source of the necessary proportion of lime. The main part of the text deals with the coloring of glass with metallic oxides to give not only clear and uniform colors but also various veined effects. There are chapters on making lead glass of high refractive index and enamel (opaque) glass by the addition of tin oxide.

There are no illustrations, and the operations are not described in much detail. The proportions of ingredients are often left to the experience of the operator. It is difficult to believe that the book could have been of great value to a practical glassmaker, but it served as a nucleus for the observations of later writers.

BIBLIOGRAPHY

I. ORIGINAL WORKS. The full title of Neri's book is *L'arte vetraria distinta in libri sette, ne quali si scoprone, effetti maravigliosi, & insegnano segreti bellissimi del vetro nel fuoco & altre cose curiose* (Florence, 1612). Later Italian eds. appeared at Florence (1661), Venice (1663, 1678), and Milan (1817). The book and its various eds. and trans. are discussed by Luigi Zecchin in "Il libro di prete Neri," in *Vetro e silicati*, **7** (1963), 17–20.

An English version was prepared by C. M. (Christopher Merrett) for the Royal Society as part of its plan for "histories" of trades, and published as *The Art of Glass . . . With Observations on the Author* (London, 1662). The "observations" are a collection of explanations, additions, and emendations which double the length of the text. The British Museum copy is heavily annotated, perhaps by Merrett, clearly in preparation for a drastically rev. 2nd ed., which never appeared.

Merrett's version and notes were trans. into Latin as *Ars vitraria . . .* by Andreas Frisius (Amsterdam, 1668). It was also trans. into German by Friedrich Geissler (1678) and appeared as part of Johann Kunckel's *Ars vitraria experimentalis* (Frankfurt–Leipzig, 1679, 1689). A French version by M. D. (Baron d'Holbach), entitled *Art de la verrerie* (Paris, 1752), incorporated the additions of Merrett, Kunckel, and d'Holbach himself. *De l'Art de la Verrerie*, by Haudicquer de Blancourt (Paris, 1697), is a French version, without acknowledgment, of the Neri-Merrett text, expanded by redundant verbiage to nearly twice the original length.

An "alchemical" MS by Neri is mentioned by G. F. Rodwell in "On the Theory of Phlogiston," in *Philosophical Magazine*, **35** (1868), 10, without any indication of its location.

II. SECONDARY LITERATURE. The lack of information about Neri has resulted in his omission from most works of reference, but he is mentioned in J. Ferguson, *Bibliotheca chemica*, II (Glasgow, 1906), 135. Luigi Zecchin, in "Lettere a prete Neri," in *Vetro e silicati*, **8** (1964), 17–20, discovered the record of Neri's baptism in Florence and 28 letters from Ximenes to Neri, most of them dated between 1601 and 1603. The Neri-Merrett text is discussed by

W. E. S. Turner, "A Notable British Seventeenth-Century Contribution to the Literature of Glassmaking," in *Glass Technology*, **3** (1962), 201–213.

W. V. FARRAR

NERNST, HERMANN WALTHER (*b*. Briesen, Prussia [now Wąbrzeźno, Poland], 25 June 1864; *d*. Bad Muskau, Prussia [now German Democratic Republic], 18 November 1941), *chemistry*.

For a detailed study of his life and work, see Supplement.

NETTESHEIM. See **Agrippa, Heinrich Cornelius.**

NETTO, EUGEN (*b*. Halle, Germany, 30 June 1848; *d*. Giessen, Germany, 13 May 1919), *mathematics*.

Netto was the grandson of a Protestant clergyman and the son of an official of the "Franckeschen Stiftungen," Heinrich Netto, and his wife, Sophie Neumann. He attended elementary school in Halle and at the age of ten entered the Gymnasium in Berlin. There he was a pupil of Karl Heinrich Schellbach, who had been Eisenstein's teacher; this famous educator aroused his interest in mathematics. In 1866, following his graduation from the Gymnasium, Netto enrolled at the University of Berlin, where he was influenced mainly by Kronecker, Kummer, and Weierstrass. In 1870 he graduated with honors from Berlin with the dissertation *De transformatione aequationis* $y^n = R(x)$, *designante* $R(x)$ *functionem integram rationalem variabilis* x, *in aequationem* $\eta^2 = R_1(\xi)$ (Weierstrass was chief referee). After teaching at a Gymnasium in Berlin, he became an associate professor at the University of Strasbourg in 1879.

In 1882, on Weierstrass' recommendation, Netto was appointed associate professor at the University of Berlin. Besides the introductory lectures for first-semester students, he gave those on higher algebra, the calculus of variations, Fourier series, and theoretical mechanics; he also lectured on synthetic geometry. His textbook *Substitutionentheorie und ihre Anwendung auf die Algebra* (Berlin, 1882) is a milestone in the development of abstract group theory. In it two historical roots of abstract group theory are united— the theory of permutation groups and that of implicit group-theoretical thinking in number theory. Even though Netto did not yet include transformation groups in his concept of groups, he nevertheless clearly recognized the far-reaching importance of the theory of composition in a group and its significance for future developments.

In 1888 Netto became professor at the University of Giessen, where he remained until his retirement in 1913. He contributed to the dissemination of group theory in further papers; and in *Lehrbuch der Combinatorik* (Leipzig, 1901; 2nd ed., enlarged by T. Skolem and Viggo Brun, 1927) he skillfully gathered the scattered literature in this area. His *Die Determinanten* (Leipzig, 1910) was translated into Russian in 1911. Netto was a clever, persuasive, and witty teacher who demonstrated his educational abilities and productivity through additional textbooks and other publications on algebra.

BIBLIOGRAPHY

Netto's works are listed in Poggendorff, III, 962; IV, 1064; and V, 897–898.

On Netto or his work see Wilhelm Lorey, "Die Mathematiker an der Universität Giessen vom Beginn des 19. Jahrhunderts bis 1914," in *Nachrichten der Giessener Hochschulgesellschaft*, **11** (1937), 54–97; Egon Ullrich, "Die Naturwissenschaftliche Fakultät," in *Ludwigs-Universität–Justus-Liebig-Hochschule. 1607–1957. Festschrift zur 350-Jahrfeier* (Giessen, 1957), 267–287; Hans Wussing, "Zum historischen Verhältnis von Intension und Extension des Begriffes Gruppe im Herausbildungsprozess des abstrakten Gruppenbegriffes," in *NTM—Schriftenreihe für Geschichte der Naturwissenschaften, Technik und Medizin*, **4** (1967), 23–34; and Kurt-R. Biermann, "Die Mathematik und ihre Dozenten an der Berliner Universität 1810–1920" (Berlin, 1973).

KURT-R. BIERMANN

NEUBERG, JOSEPH (*b*. Luxembourg City, Luxembourg, 30 October 1840; *d*. Liège, Belgium, 22 March 1926), *geometry*.

Neuberg was one of the founders of the modern geometry of the triangle. The considerable body of his work is scattered among a large number of articles for journals; in it the influence of A. Möbius is clear. In general, his contribution to mathematics lies in the discovery of new details, rather than in any large contribution to the development of his subject.

Neuberg was educated at the Athénée de Luxembourg, and later at the Normal School of Sciences, which was then a part of the Faculty of Sciences of the University of Ghent. From 1884 to 1910 he was a professor at the University of Liège. He was a naturalized citizen of Belgium and was a member of the sciences section (which he headed in 1911) of the Belgian Royal Academy. From 1874 to 1880 Neuberg, with Catalán and Mansion, published the *Nouvelle correspondance mathématique*; subsequently he collaborated with Mansion in publishing *Mathesis.*

BIBLIOGRAPHY

A portrait of Neuberg and a notice with a complete bibliography of his work by A. Mineur may be found in *Annuaire de l'Académie royale de Belgique*, **98** (1932), 135–192; see also L. Godeaux, in *Biographie nationale publiée par l'Académie royale de Belgique*, XXX (1958), cols. 635–637; and in *Liber Memorialis. L'Université de Liège de 1867 à 1935*, II (Liège, 1936), 162–175.

J. PELSENEER

NEUMANN, CARL GOTTFRIED (*b.* Königsberg, Prussia [now Kaliningrad, R.S.F.S.R.], 7 May 1832; *d.* Leipzig, Germany, 27 March 1925), *mathematics, theoretical physics.*

Neumann's father, Franz Ernst Neumann, was professor of physics and mineralogy at Königsberg; his mother, Luise Florentine Hagen, was a sister-in-law of the astronomer F. W. Bessel. Neumann received his primary and secondary education in Königsberg, attended the university, and formed particularly close friendships with the analyst F. J. Richelot and the geometer L. O. Hesse. After passing the examination for secondary school teaching he obtained his doctorate in 1855; in 1858 he qualified for lecturing in mathematics at Halle, where he became *Privatdozent* and, in 1863, assistant professor. In the latter year he was called to Basel, and in 1865 to Tübingen. From the autumn of 1868 until his retirement in 1911 he was at the University of Leipzig. In 1864 he married Hermine Mathilde Elise Kloss; she died in 1875.

Neumann, who led a quiet life, was a successful university teacher and a productive researcher. More than two generations of future Gymnasium teachers received their basic mathematical education from him. As a researcher he was especially prominent in the field of potential theory. His investigations into boundary value problems resulted in pioneering achievements; in 1870 he began to develop the method of the arithmetical mean for their solution. He also coined the term "logarithmic potential." The second boundary value problem of potential theory still bears his name; a generalization of it was later provided by H. Poincaré.

Neumann was a member of the Berlin Academy, and the Societies of Göttingen, Munich, and Leipzig. He performed a valuable service in founding and editing the important German mathematics periodical *Mathematische Annalen.*

BIBLIOGRAPHY

I. ORIGINAL WORKS. Neumann's writings include *Vorlesungen über Riemanns Theorie der Abelschen Integrale* (Leipzig, 1865); *Untersuchungen über das logarithmische und Newtonsche Potential* (Leipzig, 1877); and *Über die nach Kreis-, Kugel- und Zylinderfunktionen fortschreitenden Entwicklungen* (Leipzig, 1881).

II. SECONDARY LITERATURE. See H. Liebmann, "Zur Erinnerung an Carl Neumann," in *Jahresberichte der Deutschen Mathematikervereinigung*, **36** (1927), 175–178; and H. Salié, "Carl Neumann," in *Bedeutende Gelehrte in Leipzig*, II, G. Harig, ed. (Leipzig, 1965), 13–23.

H. WUSSING

NEUMANN, CASPAR (*b.* Züllichau, Germany [now Sulechów, Poland], 11 July 1683; *d.* Berlin, Germany, 20 October 1737), *chemistry.*

The first child of a merchant-musician, Caspar Neumann was intended for the clergy. He learned music from his father and studied at the local Latin school. But, orphaned at the age of twelve, he had to go into pharmacy as an apprentice to his godfather. He showed such aptitude that three years later his guardian put him in charge of an apothecary shop, brewery, and distillery in nearby Unruhstadt. Neumann remained there until 1704, when the Great Northern War forced him to flee to Berlin. In the Prussian capital he soon became an assistant to the traveling pharmacist of Frederick I. As part of the royal entourage, he played the clavier for the king, he traveled throughout Germany and Holland, and he pursued a growing interest in science and medicine.

Neumann's serious scientific education began in 1711, when, apparently at the urging of the renowned royal physician F. Hoffmann, he was sent abroad to study chemistry. He first visited the Harz mining towns, where he learned assaying and smelting, and then went to Holland, where he inspected large chemical works and studied with Boerhaave. In 1713 he went to London, where he was stranded because of the recent death of his royal patron. He found employment as a laboratory assistant to the wealthy Dutch surgeon A. Cyprian, who spent £1,000 annually on chemical experiments. In his free time, Neumann gave private courses on chemistry and participated in the scientific life of London. After three years there he returned to Berlin to collect his belongings. Stahl, who had recently been made royal physician, persuaded Neumann to reenter Prussian service by obtaining a continuation of his travel stipend and promising him a position in the court apothecary shop. On his second tour Neumann first visited his friends in London. Then he proceeded to Paris, where he attended courses on chemistry and botany; taught a course of his own on chemistry; experimented two afternoons weekly with C. J. and E. F. Geoffroy; and made the acquaintance of all the leading scientists. In 1719 he returned to Berlin by way of Rome.

Upon his return, Neumann as court apothecary took on the demanding job of running one of Europe's busiest pharmacies. Nevertheless, he managed to find time for other activities. In 1721 he began active membership in Berlin's Society of Sciences, and in 1724 he became a member of the chief Prussian medical board and began teaching in the new Medical-Surgical College as professor of practical (experimental) chemistry. He remained in all these positions until his death in 1737 at the age of fifty-four.

In the mid-1720's, after more than a decade of serious work in chemistry, Neumann began his short yet prolific career as an author with a series of articles in the *Philosophical Transactions*. After his death his collected lectures appeared in two German versions and, partially at least, in English, Dutch, and French translations.

Though Neumann was not a highly original chemist, he did influence the development of chemistry in a variety of ways. First, during his *Wanderjahre*, he conveyed knowledge of German techniques and theories to chemists in London and Paris. Second, as master pharmacist and as professor, he gave the young Marggraf his initial instruction in chemistry. It may well have been Neumann's exhortations that inspired Marggraf to develop "wet" analysis. Third, as an author he contributed significantly to the establishment of Stahlian chemistry, especially in Germany but also abroad. Like Stahl's other main disciples—Pott, Henckel, and Juncker—Neumann distinguished clearly between pure and applied chemistry and insisted that the chemical approach to nature was vastly superior to the mechanical philosophy. He envisioned many levels of chemical aggregation and invoked the phlogiston theory to explain combustion and calcination-reduction. Unlike Stahl's other main disciples, Neumann concentrated on pharmaceutical chemistry, thereby reaching and inspiring a generation of pharmacists which included Scheele and Klaproth.

BIBLIOGRAPHY

I. ORIGINAL WORKS. See *Praelectiones chemicae seu chemia medico-pharmaceutica experimentalis & rationalis, oder gründlicher Unterricht der Chemie . . .* (Berlin, 1740), edited on the basis of student notes by J. C. Zimmermann with the assistance of J. H. Pott, with portrait. Zimmermann republished this ed. with minor changes under his own name in 1755. Neumann's nephew C. H. Kessel put out a different ed. which was based on Neumann's own notes, *Chymiae medicae dogmatico-experimentalis . . . oder der gründlichen und mit Experimenten erwiesenen Medicinischen Chymie . . .*, 4 vols. (Züllichau, 1749–1755; partial reprint, 1755–1756). W. Lewis' English ed. appeared in 1759 and 1773, the Dutch ed. in 1766, and Roux's French ed. in 1781. A complete bibliography of Neumann's works which mentions many reviews appears in Exner's biography, cited below, and an annotated partial bibliography in J. R. Partington, *A History of Chemistry*, II (London–New York, 1961), 702–706.

II. SECONDARY LITERATURE. The best biography is Alfred Exner, *Der Hofapotheker Caspar Neumann (1683–1737)* (Berlin, 1938). Exner begins with an annotated trans. of A. P. Queitsch's Latin biography (1737) and then, relying heavily on Kessel's ed. of the lectures, he assesses Neumann's role in chemistry and pharmacy. Some additional materials are in Herbert Lehmann, *Das Collegium medico-chirurgicum in Berlin als Lehrstätte der Botanik und der Pharmazie* (Berlin, 1936).

KARL HUFBAUER

NEUMANN, FRANZ ERNST (*b.* Joachimsthal, Germany [now Jachymov, Czechoslovakia], 11 September 1798; *d.* Königsberg, Germany [now Kaliningrad, R.S.F.S.R.], 23 May 1895), *mineralogy, physics, mathematics.*

Neumann extended the Dulong-Petit law—that the specific heats of the elements vary inversely as their atomic weights—to include compounds having similar chemical constitutions. His work in optics contributed to the establishment of the dynamical theory of light, and he formulated mathematically the laws of induction of electric currents. He also aided in developing the theory of spherical harmonics. Neumann was a highly influential teacher; many of his students became outstanding scientists, and he inaugurated the mathematical science seminar at German universities.

Neumann's mother was a divorced countess whose family prevented her marrying his father, a farmer who later became an estate agent, because he was not of noble birth. Neumann was therefore raised by his paternal grandparents. He attended the Berlin Gymnasium, where he displayed an early talent for mathematics. His education was interrupted in 1814, when he became a volunteer in the Prussian army to fight against Napoleon. He was seriously wounded on 16 June 1815 at the battle of Ligny, the prelude to Waterloo. After recovering in a Düsseldorf hospital, he rejoined his company and was mustered out of the army in February 1816.

Because his father had lost all of his resources in a fire, Neumann pursued his education under severe financial difficulties. He completed his studies at the Gymnasium and in 1817 entered the University of Berlin, studying theology in accordance with his father's wishes. In April 1818 he left Berlin for Jena, where he began his scientific studies and was particularly attracted to mineralogy. In 1819 Neumann returned to Berlin to study mineralogy and crystallography under Christian S. Weiss, who became his close friend as well as his mentor. Weiss made the financial arrangements for Neumann to take a three-month geological field trip in Silesia during the summer of 1820, and Neumann was planning other trips for 1822 and 1823 when his father died. Thereafter Neumann and his mother became very close; his concern for her health and financial independence caused him to leave the university during 1822–1823 and manage her farm. Nevertheless, in 1823 he published his first work, *Beiträge zur Kristallonomie*, which was highly regarded in Germany; and on Weiss's recommendation he was appointed curator of the mineral cabinet at the University of Berlin in November 1823.

Neumann received the doctorate at Berlin in November 1825; and in May 1826, together with Jacobi and Dove, he became a *Privatdozent* at the University of Königsberg. Dove and Neumann were destined to assume the physics and mineralogy courses, respectively, of Karl G. Hagen, who had been teaching botany, zoology, mineralogy, chemistry, and physics. In 1828 Neumann was advanced to the rank of lecturer, and in 1829 he was named professor of mineralogy and physics. He married Hagen's daughter, Luise Florentine, in 1830; they had five children before her death in 1838. He married Wilhelmina Hagen, her first cousin, in 1843.

Neumann's early scientific works, published between 1823 and 1830, concerned crystallography; in these he introduced the method of spherical projection and extended Weiss's work on the law of zones (law of rational intercepts). At Königsberg, however, he was influenced by Bessel, Dove, and Jacobi; and he began to concentrate on mathematical physics. His first two important papers were published in Poggendorff's *Annalen der Physik und Chemie* (**23** [1831], 1–39 and 40–53); the first was entitled "Untersuchung über die specifische Wärme der Mineralien" and the second "Bestimmung der specifischen Wärme des Wassers in der Nähe des Siedpunctes gegen Wasser von niedriger Temperatur." In the first article Neumann investigated the specific heats of minerals and extended the Dulong-Petit law to include compound substances having similar chemical constitutions. He arrived at what has been termed Neumann's law, that the molecular heat

of a compound is equal to the sum of the atomic heats of its constituents. In the second paper Neumann considered the specific heat of water. In earlier investigations physicists had noticed that when equal quantities of hot and cold water are mixed the temperature of the mixture is lower than the arithmetic mean of the temperatures of the original quantities. This result was generally interpreted as being due to a progressive decrease in the specific heat of water from the point of fusion to that of vaporization, a conclusion that appears to be validated by a number of experiments. Neumann disclosed errors in these experiments and concluded instead that the specific heat of water increases as its temperature increases. He failed to determine, however, that an increase occurs over only a portion of the temperature range from fusion to vaporization.

In 1832 Neumann published another important paper, again in Poggendorff's *Annalen*, "Theorie der doppelten Strahlenbrechung abgeleitet aus der Gleichungen der Mechanik." Many physicists and mathematicians of the period were concerned with determining the conditions under which waves are propagated in ordinary elastic bodies so that they might develop a model which could serve as the optical medium; that is, they wished to evolve an elastic-solid theory of the ether in order to promote the undulatory theory of light. In his article Neumann reported obtaining a wave surface identical with that determined earlier by Augustin Cauchy, and he succeeded in deducing laws of double refraction agreeing with those of Fresnel except in the case of biaxial crystals.

Neumann encountered difficulty in explaining the passage of light from one medium to another. He attempted to overcome this obstacle in an article entitled "Theoretische Untersuchungen der Gesetze, nach welchen das Licht an der Grenze zweier vollkommen durchsichtigen Medien reflectirt und gebrochen wird," published in *Abhandlungen der Preussischen Akademie der Wissenschaften*, mathematische Klasse ([1835], 1–160). In this paper Neumann raised the question of the mathematical expression of the conditions which must hold at the surface separating the two crystalline media, and he adopted the view that the density of the ether must be identical in all media.

Neumann and his contemporary Wilhelm Weber were the founders of the electrodynamic school in Germany, which later included, among others, Riemann, Betti, Carl Neumann, and Lorenz. The investigations and analyses of this group were guided by the assumption, held originally by Ampère, that electromagnetic phenomena resulted from direct action at a distance rather than through the mediation of a field. Neumann's major contributions were contained in two papers published in 1845 and 1848, in which he established mathematically the laws of induction of electric currents. The papers, transmitted to the Berlin Academy, were entitled "Allgemeine Gesetze der inducirten elektrischen Ströme" and "Über ein allgemeines Princip der mathematischen Theorie inducirter elektrischer Ströme."

As a starting point Neumann took the proposition, formulated in 1834 by F. E. Lenz after Faraday's discovery of induction, that the current induced in a conductor moving in the vicinity of a galvanic current or a magnet will flow in the direction that tends to oppose the motion. In his mathematical analysis Neumann arrived at the formula $E \cdot Ds = -\epsilon v\, C \cdot Ds$, where Ds is an element of the moving conductor, $E \cdot Ds$ is the elementary induced electromotive force, v is the velocity of the motion, $C \cdot Ds$ is the component of the inducing current, and ϵ is a constant coefficient. With this formula Neumann was able to calculate the induced current in numerous particular instances. At present a common formulation is $E = -\, dN/dt$, where E is the electromotive force generated in the circuit through which the number of magnetic lines of force is changing at the rate of dN/dt.

Continuing his analysis Neumann noticed a way in which the treatment of currents induced in closed circuits moving in what is now termed a magnetic field might be generalized. He saw that the induced current depends only on the alteration, caused by the motion, in the value of a particular function. Considering Ampère's equations for a closed circuit, Neumann arrived at what is known as the mutual potential of two circuits, that is, the amount of mechanical work that must be performed against the electromagnetic forces in order to separate the two circuits to an infinite distance apart, when the current strengths are maintained unchanged. In modern notation the potential function, Vii', is written:

$$Vii' = -ii' \iint \frac{\mathbf{ds} \cdot \mathbf{ds'}}{r},$$

$\mathbf{ds} \cdot \mathbf{ds'}$ is the scalar product of the two vectors $\mathbf{ds}$ and $\mathbf{ds'}$, and r their distance apart. If a fixed element $\mathbf{ds'}$ is taken and integrated with respect to $\mathbf{ds}$, the vector potential of the first circuit at the point occupied by $\mathbf{ds}$ is obtained. Maxwell arrived at the concept of vector potentials by another method and interpreted them as analytical measures of Faraday's electrotonic state.

According to his contemporaries, only a small portion of Neumann's original scientific work was published. But he was an extremely effective teacher, and he made known many of his discoveries in heat, optics, electrodynamics, and capillarity during his lectures, thinking that priority of discovery extended equally to lectures and publications. Thus he made numerous contributions to the theory of heat without receiving credit; on occasion he thought about raising questions concerning priority but never did.

In 1833, with Jacobi, Neumann inaugurated the German *mathematisch-physikalische* seminar, employing such sessions to supplement his lectures and to introduce his students to research methodology. Gustav Kirchhoff attended these seminars from 1843 to 1846; his first papers on the distribution of electrical conductors, and H. Weld's development of the photometer and polarimeter, were among the direct results of Neumann's seminars. Neumann pleaded continually for the construction of a physics laboratory at Königsberg, but his hopes were thwarted during his tenure as professor; a physics institute was not completed at Königsberg until 1885. In 1847, however, the inheritance from the estate of the parents of his second wife enabled Neumann to build a physics laboratory next to his home, the facilities of which he shared with his students. He retired as professor in 1873, although he continued his seminar for the next three years. He maintained his good health by making frequent walking tours throughout Germany and Austria, and he was still climbing mountains at the age of eighty.

Throughout his life Neumann was an ardent Prussian patriot. He aided in keeping peace in Königsberg during the uprisings of 1848. He pleaded continually for the unification of Germany under the leadership of Prussia, and in the early 1860's he made numerous political speeches supporting Bismarck and the war against Austria. At the fiftieth anniversary of his doctorate in 1876, he was congratulated by the crown prince, later Wilhelm II; and he received honors from Bismarck in 1892 as a veteran of the campaign of 1815. Neumann was a corresponding member of every major European academy of science; he received the Copley Medal of the Royal Society in 1887.

BIBLIOGRAPHY

I. Original Works. Three of Neumann's most important works were published in Ostwalds Klassiker der Exakten Wissenschaften: *Die mathematischen Gesetze der inducirten elektrischen Ströme*, no. 10 (Leipzig, 1889); *Über ein allgemeines Princip der mathematischen Theorie inducirter elektrischer Ströme*, no. 36 (Leipzig, 1892); and *Theorie der doppelten Strahlenbrechung*, no. 76 (Leipzig, 1896). Other books are *Beiträge zur Kristallonomie* (Berlin–Posen, 1823); *Über den Einfluss der Krystallflächen bei der Reflexion des Lichtes und über die Intensität des gewöhnlichen und ungewöhnlichen Strahls* (Berlin, 1837); and *Beiträge zur Theorie der Kugelfunktionen* (Leipzig, 1878). Some of his lectures were published in *Vorlesung über mathematischen Physik gehalten an der Universität Königsberg von Franz Neumann*, Carl Neumann, C. Pape, Carl Vondermühll, and E. Dorn, eds., 5 vols. (Leipzig, 1881–1887). His collected works were published as *Franz Neumanns Gesammelte Werke*, 3 vols. (Leipzig, 1906–1928).

II. SECONDARY LITERATURE. See C. Voit, "Nekrolog auf Franz Ernst Neumann," in *Sitzungsberichte der Akademie München*, **26** (1896), 338–343; Luise Neumann, *Franz Neumann: Erinnerungsblätter* (Tübingen–Leipzig, 1904); W. Voigt, "Gedächtnissrede auf Franz Neumann," in *Franz Neumanns Gesammelte Werke*, I (Leipzig, 1906), 1–19; and Paul Volkmann, *Franz Neumann . . . den Andenken an dem Altmeister der mathematischen Physik gewidmete Blätter* (Leipzig, 1896).

See also James Clerk Maxwell, *A Treatise on Electricity and Magnetism*, 3rd ed. (Oxford, 1891), art. 542; and Sir Edmund Whittaker, *A History of the Theories of Aether and Electricity*, I (London, 1951), 137–138, 166–167, 198–200.

JOHN G. BURKE

NEUMAYR, MELCHIOR (*b.* Munich, Germany, 24 October 1845; *d.* Vienna, Austria, 29 January 1890), *paleontology, geology.*

Neumayr was the son of Max von Neumayr, a Bavarian government minister. He attended secondary schools in Stuttgart and in Munich, where he also studied law. Under the influence of Oppel, he soon turned to paleontology and geology and received the doctorate in Munich in 1867. The following year he moved to Vienna and joined the Imperial Austrian Geological Survey. In 1872 he became *Privatdozent* in paleontology and stratigraphy at the University of Heidelberg. He returned to Vienna in 1873 to fill the new professorship of paleontology that had been created for him at the university and held this post until his death from a heart ailment. On 2 April 1879 he married Paula Suess, the daughter of his colleague Eduard Suess; they had three daughters.

Neumayr's first scientific work was his geological mapping of southern Germany, the Carpathians, and the eastern Alps. His subsequent paleontological and stratigraphical investigations of the Jurassic period (1870–1871, 1874) soon established him as an expert on this period and its fauna. In studies of the Upper Tertiary freshwater mollusks of Yugoslavia, work that he began in 1869, he showed the gradual transformation of the shell morphology in the various horizons. A follower of Darwin from his student days, Neumayr

was the first to give a concise demonstration of the Darwinian theory of variation and evolution of species in invertebrate fossils (1875).

Neumayr's geological and paleontological investigations in Greece and in the Aegean Islands (1874–1876) made fundamental contributions to the knowledge o the geological structure of this region. In the Upper Tertiary of the island of Cos he found an even finer example of the evolution of freshwater snails (1880). A result of his studies in the Aegean was the first geological history of the eastern Mediterranean, in which methodological principles of paleogeography were demonstrated (1882).

Continuing his studies of the ammonites, Neumayr then turned his attention to the Cretaceous species, dividing them into new genera, as Suess had done for the Triassic and Jurassic forms. He clarified the relationships of the straightened forms to the parent curled genera and discussed the Cretaceous ammonites in relation to those of the Jurassic and Triassic periods. By extending his paleobiogeographic research to the Jurassic and Cretaceous periods he created the foundations of the present conceptions of the faunal regions in the Jurassic seas (1885). He also studied climatic differentiation during the Jurassic and Cretaceous periods (1883).

During his last years, Neumayr was occupied with writing comprehensive surveys and in his *Erdgeschichte* he produced a popular synthesis with strictly scientific methods. His unfinished *Die Stämme des Thierreiches* was Darwinian in approach and showed the close relationship between zoology and paleontology. He thereby raised paleontology—previously considered simply a study of index fossils—to the level of a basic biological science.

BIBLIOGRAPHY

I. ORIGINAL WORKS. Neumayr's writings include "Jurastudien," in *Jahrbuch der Geologischen Bundesanstalt*, **20** (1870), 549–558; **21** (1871), 297–379; "Die Fauna der Schichten mit Aspidoceras acanthicum," in *Abhandlungen der Geologischen Bundesanstalt*, **5** (1874), 141–257; "Die Congerien- und Paludinenschichten Slavoniens und deren Fauna," *ibid.*, **7**, no. 3 (1875), written with C. M. Paul; "Über den geologischen Bau der Insel Kos und die Gliederung der jungtertiären Binnenablagerungen im Archipel," in *Denkschriften der Akademie der Wissenschaften*, **40** (1880), 213–240; *Zur Geschichte des östlichen Mittelmeerbeckens* (Berlin, 1882); "Ueber klimatische Zonen während der Kreide- und Jurazeit," in *Denkschriften der Akademie der Wissenschaften*, **47** (1883), 277–310; "Die geographische Verbreitung der Juraformation," *ibid.*, **50** (1885), 57–145; *Erdgeschichte,*

2 vols. (Leipzig, 1886–1887); and the unfinished, posthumously published *Die Stämme des Thierreiches. Wirbellose Thiere* (Vienna–Prague, 1889).

II. SECONDARY LITERATURE. See W. T. Blanford's obituary in *Quarterly Journal of the Geological Society of London*, **46** (1890), 54–56; F. Toula, "Zur Erinnerung an Melchior Neumayr," in *Annales géologiques de la Péninsule balkanique*, **3** (1891), 1–9, with bibliography of 30 titles covering Neumayr's work in the Balkans; V. Uhlig, "Melchior Neumayr. Sein Leben und Wirken," in *Jahrbuch der Geologischen Bundesanstalt*, **40** (1891), 1–20, with complete bibliography of 133 works; K. Lambrecht and W. and A. Quenstedt, "Palaeontologi. Catalogus bio-bibliographicus," in *Fossilium Catalogus I: Animalia,* **72**(1938), 311; and F. Steininger and E. Thenius: "Die Ära Melchior Neumayr (1873–1890)," in *100 Jahre Paläontologisches Institut der Universität Wien, 1873–1973* (Vienna, 1973), 14–17, with portrait.

HEINZ TOBIEN

NEUYMIN, GRIGORY NIKOLAEVICH (*b.* Tiflis, Georgia [now Tbilisi, Georgian S.S.R.], 3 January 1886; *d.* Leningrad, U.S.S.R., 17 December 1946), *astronomy.*

Neuymin was the son of a military oculist-physician. In 1904 he graduated with a gold medal from the Second Tiflis Gymnasium and entered the Faculty of Physics and Mathematics at St. Petersburg University. Among his teachers were the astronomers A. A. Ivanov and S. P. Glazenap, and the mathematician V. A. Steklov. He graduated with a first-class diploma in 1910 and remained in the department of astronomy to prepare for a scientific career. From 1908 he was an assistant at the Pulkovo Observatory, where he was directed by F. F. Renz (astrometry) and A. A. Belopolsky (astrospectroscopy). Under their direction Neuymin conducted his first scientific research and mastered the techniques of astronomical observation. His first published works dealt with the determination of the radial velocity of the star α Cygni (Deneb) and the photographic observations of the annular eclipse of 12 April 1912. In June 1910 Neuymin became supernumerary astronomer at Pulkovo Observatory. After working on stellar spectroscopy in the astrophysical laboratory he participated in the processing of observations with the great transit instrument and began to observe comets and double stars on the thirty-eight-centimeter refractor.

In December 1912 Neuymin was sent as an adjunct astronomer to the recently created southern section of Pulkovo Observatory, at Simeiz, in the Crimea. Almost all his subsequent scientific work was associated with the Simeiz Observatory. In 1922 he returned for three years to Pulkovo, where he made observations with the seventy-six-centimeter refractor and made extensive computations of the final orbit of the comet Neuymin II, discovered by him in 1916. In 1924 the Scientific Council of Pulkovo Observatory elected him senior astronomer. The following year he returned to Simeiz as director of the observatory. In 1935 Neuymin was awarded a doctorate in the physical and mathematical sciences.

At Simeiz, Neuymin developed a broad program for the systematic search and photographic observation of comets and asteroids. With the help of a very modest 125-millimeter double astrograph the observatory soon held second place for the number of asteroids discovered there. Neuymin discovered sixty-three of the 110 numbered asteroids (those for which enough observations had been collected for the orbit to be calculated). About 400 others discovered at Simeiz were not numbered at that time. Widely known as the "comet hunter," Neuymin discovered six comets, five of which were periodic, with periods from 5.4 to 17.9 years. The comet Neuymin II was especially interesting. Having computed its orbit and calculated the planetary perturbations, Neuymin obtained very precise ephemerides, with which the comet was rediscovered in 1927. Neuymin developed a special method of calculating higher-order terms for use in computing perturbations.

Neuymin discovered thirteen variable stars, including the bright variable X Trianguli, and developed a method for discovering short-period variables on photographic plates.

Neuymin's work at Pulkovo included his measurements of double stars, micrometric measurements of the satellites of Neptune, and the determination of the proper motions of seventeen stars.

In connection with compiling a catalog of faint stars he selected and tested galaxies in order to attach the fundamental stars of the catalog to them.

In the fall of 1941, when the Simeiz Observatory was evacuated, Neuymin saved some of the valuable equipment and the archive of astronegatives. At Kitab, to which some of the observatory workers were sent, he continued his work on asteroids and his study of the comet Neuymin II. One of the asteroids he discovered was named Uzbekistan.

In 1944 Neuymin was named director of the Pulkovo Observatory, then in ruins. Charged with the difficult task of restoring and reorganizing the institution, he did not live to complete it. On 17 December 1946, exhausted by the evacuation and by the hard conditions of Central Asia, he died after a brief illness. Asteroid 1129 and a crater on the moon were named after him. In 1945 Neuymin was awarded the order of the Red Banner of Labor. His discoveries of comets were recognized by three prizes of the Russian Astronomical Society and six medals from the Astronomical Society of the Pacific.

BIBLIOGRAPHY

I. ORIGINAL WORKS. Neuymin's writings include "Sur les éléments et le prochain retour de la comète Neujmin (1916 a)," in *Astronomie*, **35** (1921), 160–162; "Mikrometrennye izmerenia dvoynykh zvezd v Pulkove" ("Micrometric Measurements of Double Stars at Pulkovo"), in *Izvestiya Glavnoi astronomicheskoi observatorii v Pulkove*, **9**, pt. 1, no. 88 (1923), 1–84; and "Vyvod sobstvennykh dvizhenii 17 zvezd" ("Definition of the Proper Motions of Seventeen Stars"), *ibid.*, **10**, pt. 3, no. 96 (1925), 305–314.

On his research on the orbit of the comet Neuymin II see "Definitive Bahnbestimmung des periodischen Kometen 1916 II (Neujmin) aus der Erscheinung in Jahre 1916," in *Izvestiya Glavnoi astronomicheskoi observatorii v Pulkove*, **10**, pt. 6, no. 99 (1927), 531–584; "Svyaz poyavlenia komety v 1916 i 1926 gg. (1916 II–1927 I)" ("Relations of the Appearance of the Comet in 1916 and 1926 [1916 II–1927 I]"), in *Tsirkulyar Glavnoi astronomicheskoi observatorii v Pulkove* (1941), no. 32, 25–61; and "Issledovanie orbity komety Neuymina II" ("Research on the Orbit of the Comet Neuymin II"), in *Izvestiya Glavnoi astronomicheskoi observatorii v Pulkove*, **17**, pt. 6, no. 141 (1948), 6–23.

See also "On a Method of Discovering Short-Period Variables With Rapid-Changes in Brightness," in *Tsirkulyar Glavnoi astronomicheskoi observatorii v Pulkove* (1932), no. 4, 22–24; "Rabochii katalog vnegalakticheskikh tumannostey dlya privyazki Kataloga slabykh zvezd" ("Working Catalog of Extragalactic Nebulae for Attachment to the Catalog of Faint Stars"), in *Uchyenye zapiski Kazanskogo gosudarstvennogo universiteta*, **100**, bk. 4 (1940), 116–127; "Simeizskoe otdelenie Pulkovskoy observatorii za 25 let (1908–1933)" ("Simeiz Section of Pulkovo Observatory for 25 Years"), in *Astronomicheskii Kalendar na 1934 god* ("Astronomical Calendar for 1934"; Nizhny Novgorod, 1934), 115–137; "Ob uchete vozmushcheny vysshikh poryadkov pri vychislenii spetsialnykh vozmushcheny" ("On Taking Into Account Perturbations of Higher Orders in Calculating Special Perturbations"), in *Astronomicheskii zhurnal*, **11**, pt. 2 (1934), 140–143; "Prostoy obiektivny mikrofotometr" ("Simple Objective Microphotometer"), in *Optiko-mekhanicheskaya promyshlennost* (1936), no. 9, 22–23; and "Periodicheskaya kometa Neuymina II i ee predstoyashchee vozvrashchenie k perigeliyu v 1943 godu" ("Periodic Comet Neuymin II and Its Forthcoming Return to Perihelion in 1943"), in *Astronomicheskii zhurnal*, **20**, pt. 1 (1943), 34–40.

II. SECONDARY LITERATURE. See the unsigned obituary, "Grigory Nikolaevich Neuymin," in *Izvestiya Glavnoi astronomicheskoi observatorii v Pulkove*, **17**, pt. 6, no. 141 (1948), 1–3; N. I. Idelson, "Pamyati Grigoria Nikolaevicha Neuymina" ("Memories of . . . Neuymin"), in *Astronomicheskii Kalendar na 1948 god* ("Astronomical Calendar for 1948"; Gorky, 1947), 138–142; B. Yu. Levin, "G. N. Neuymin," in *Priroda* (1948), no. 3, 86–87; and G. A. Shayn, "G. N. Neuymin," in *Izvestiya Krymskoi astrofizicheskoi observatorii*, **2** (1948), 136–138, with portrait.

P. G. KULIKOVSKY

NEWALL, HUGH FRANK (*b.* Gateshead, England, 21 June 1857; *d.* Cambridge, England, 22 February 1944), *astrophysics.*

Newall was the son of R. S. Newall, a wealthy manufacturer and a fellow of the Royal Society. During the 1860's Thomas Cooke of York constructed a twenty-five-inch refracting telescope—for a short time the largest in the world—for the father, who installed it at his home. Newall was not then interested in astronomy; he read mathematics as an undergraduate at Cambridge and subsequently worked under J. J. Thomson in the Cavendish Laboratory. In 1889, only weeks before his death, Newall's father offered his refractor to Cambridge University, with the request that it be used primarily for work in stellar physics. Certain financial problems arose, however, which were resolved when Newall offered the university a sum of money in addition to his own services as an unpaid observer. Throughout the early 1890's he concentrated on putting the telescope into service and providing it with appropriate instrumentation. He continued to be interested in the design of instrumentation throughout his life.

From the mid-1890's Newall's interest focused increasingly on the sun. He took part in four eclipse expeditions between 1898 and 1905, studying the flash and coronal spectra and the polarization of the corona. As the result of a substantial bequest of money by Frank McClean in 1905, Newall was able to construct a horizontal solar telescope at Cambridge and to begin a program of solar observations there. He concerned himself primarily with sunspot spectra and the rotation of the sun. His time was much occupied, however, by the transfer of Sir William Huggins' instrumentation to Cambridge in 1908, followed by the transfer of the Solar Physics Observatory (formerly under the direction of Sir Norman Lockyer) from South Kensington in 1911.

Newall was appointed professor of astrophysics in 1909 and held this post until his retirement in 1928. He published less than many of his contemporaries; partly, perhaps, because he was a perfectionist and partly because, being financially independent, he was under no pressure. His influence in the astronomical community was felt mainly through his local, national, and international organizational work.

BIBLIOGRAPHY

Some of Newall's MS diaries are preserved at the Cambridge Observatories. The majority of his published astronomical work appeared in the *Monthly Notices of the Royal Astronomical Society* between 1892 and 1927.

There is a detailed obituary of Newall by E. A. Milne in *Obituary Notices of Fellows of the Royal Society of London*, **4** (1944), 717–732. Additional details, especially concerning instrumentation, can be found in *Annals of the Solar Physics Observatory* (Cambridge), **1** (1949).

A. J. MEADOWS

NEWBERRY, JOHN STRONG (*b.* Windsor, Connecticut, 22 December 1822; *d.* New Haven, Connecticut, 7 December 1892), *paleontology, geology.*

Newberry was the son of Elizabeth Strong Newberry and Henry Newberry, an entrepreneur who prospered in the development of the Western Reserve lands in Ohio. When James Hall studied the geology of Ohio in 1841, he met young Newberry and encouraged his interest in the fossils of nearby coal fields. Newberry graduated from Western Reserve College in 1846 and from Cleveland Medical School as an M.D. in 1848. In 1849 and 1850 he attended scientific lectures by Adolphe Brongniart, Charles Robin, and Louis Cordier at the Jardin des Plantes in Paris. He returned to practice medicine in Cleveland from 1851 to 1855. Newberry married Sarah Brownell Gaylord of Cleveland; they had five sons and a daughter. From 1861 to 1865 he served as a doctor and executive with the United States Sanitary Commission.

Newberry served as physician-naturalist for several important army exploring expeditions in the trans-Mississippi West. He was with the Pacific Railroad Survey group led by Lieut. R. S. Williamson, which explored the northern Pacific coast in 1855 and 1856. He then joined the party under Lieut. Joseph C. Ives, which surveyed the Colorado River in 1857 and 1858. In 1859 he accompanied Capt. John N. Macomb on the survey of the area around Santa Fe. Newberry was professor of geology at the Columbia University School of Mines from 1866 to his death, and he is credited with making that part of the university a first-rate scientific institution. He worked as a paleobotanist for the Hayden and Powell Surveys in the 1870's and directed the Ohio State Geological Survey from 1869 to 1874. Newberry was a charter member of the National Academy of Sciences (1863) and of the Geological Society of America (1888), and in 1867 he presided over the American Association for the Advancement of Science. He helped revitalize the Lyceum of Natural History of New York City, which became, with his guidance, the New York Academy of Sciences.

As his appointments suggest, Newberry was a field scientist; when he needed a petrographic sample studied under the microscope or a chemical analysis done, he usually asked a colleague or student to do it. Although he contributed to nearly every branch of geology, he concentrated on paleobotany, especially on the stratigraphic relations and the fossil flora of American coal beds. Beginning in 1859, he argued for a Cretaceous age for the Western lignites, opposing Lesquereux, who thought they were Tertiary, and Marcou, who said they were Jurassic.[1] Lester Frank Ward's work (1885) on the Laramie flora convinced Newberry that there were several distinct beds of both Tertiary and Cretaceous age.[2] Newberry also wrote on glacial phenomena in the Great Lakes and Midwest area, but he was unaware that more than one stage of glaciation affected the region. Newberry was a staunch uniformitarian in geological philosophy. For example, his theory of cycles of deposition (1873), which fits American rocks into sequence by texture and by the nature of organic contents, was based on an analogy to shores, continental shelves, and ocean bottoms.[3] He is best known for his accurate description (1861) of the Grand Canyon as erosion on a large scale, an explanation he buttressed with analogies to present-day erosion patterns in the Colorado River Basin.[4]

NOTES

1. See Newberry's letter in Ferdinand Hayden and Fielding Meek, "On the So-Called Triassic Rocks of Kansas and Nebraska," in *American Journal of Science*, 2nd ser., **27** (1859), 33; and Newberry, "Explorations in New Mexico," *ibid.*, **28** (1859), 298–299.
2. Ward, "Synopsis of the Flora of the Laramie Group," in *Report of the United States Geological Survey* (Washington, 1885), 399–557. Newberry's rather grudging admission appeared in his article, "The Laramie Group," in *Bulletin of the Geological Society of America*, **1** (1890), 524–541.
3. "Circles of Deposition in American Sedimentary Rocks," in *Proceedings of the American Association for the Advancement of Science*, **22** (1873), 185–196.
4. "Geological Report," in Joseph C. Ives, *Report Upon the Colorado River of the West, Explored in 1857 and 1858*, U.S., Congress, Senate, Executive Document (1861), pp. 25, 32, 41–48, 103.

BIBLIOGRAPHY

I. ORIGINAL WORKS. For a bibliography of Newberry's writings see Charles A. White, "Biographical Memoir of John Strong Newberry," in *Memoirs of the National Academy of Sciences*, **6** (1909), 1–24. White's list is full but not exhaustive. For other publications by Newberry, see Max Meisel, *A Bibliography of American Natural History:*

The Pioneer Century, 1769–1865, 3 vols. (Brooklyn, N. Y., 1924–1929); and Lawrence Schmeckebier, *Catalogue and Index of the Hayden, King, Powell, and Wheeler Surveys*, in *Bulletin of the United States Geological Survey*, **222** (Washington, 1904). The citations in White are casual and must be checked against the sources. White occasionally paraphrased titles, omitted page numbers, or failed to indicate whether the item was an abstract rather than the full piece. White's list is especially inaccurate for government documents.

Newberry's frequent articles (1880–1889) in the *Columbia University School of Mines Quarterly* appear to be written versions of his classroom lectures. Newberry's MS notes from the lectures at Paris are at the New York Botanical Garden.

II. SECONDARY LITERATURE. White's memoir is a convenient and adequate account of Newberry's life. For citations to other biographies, see George P. Merrill's article on Newberry in *Dictionary of American Biography;* and Meisel, I, 214. Merrill has a useful ch. on the lignite controversy in *The First One Hundred Years of American Geology* (New Haven, 1924), 579–593. William H. Goetzmann reevaluates Newberry's work in the American West, esp. his Grand Canyon monograph, in *Army Exploration in the American West 1803–1863* (New Haven, 1959), 317 ff., and in *Exploration and Empire: The Explorer and the Scientist in the Winning of the American West* (New York, 1966), 307 ff.

MICHELE L. ALDRICH

NEWCOMB, SIMON (*b*. Wallace, Nova Scotia, Canada, 12 March 1835; *d*. Washington, D.C., 11 July 1909), *astronomy*.

Simon Newcomb was the most honored American scientist of his time. During his lifetime his influence on professional astronomers and laymen was unparalleled, and it is still widely felt today. Having revolutionized the observational methods of the United States Naval Observatory, he reformed the entire theoretical and computational basis of the *American Ephemeris*. The planetary theories and astronomical constants that he derived are either still in official use or have been superseded only recently. Newcomb's discovery of the departure of the moon from its predicted position led to the investigations on the variations in the rate of rotation of the earth. These inquiries dominated dynamical astronomy during the first half of the twentieth century.

Though almost wholly of New England ancestry, Newcomb was born in Canada, the elder son of John Burton Newcomb, an itinerant country schoolteacher, and Emily Prince, daughter of a New Brunswick magistrate. Newcomb's early years were spent in various villages in Nova Scotia and Prince Edward Island. At the age of sixteen he was apprenticed to one

Dr. Foshay, on the understanding that in return for schooling in "medical botany" he would serve as general assistant for five years. Dr. Foshay was a quack, and Newcomb ran away empty-handed, after serving two years. He walked most of the 120 miles to Calais, Maine, where he was befriended by a sea captain who agreed to let him work his passage to Salem, Massachusetts. There he was met by his father and they journeyed together to Maryland.

Newcomb obtained a teaching post at a country school at Massey's Cross Roads, Kent County, and a year later he moved to a school in nearby Sudlersville. In his spare time he taught himself mathematics, studying in particular Newton's *Principia*. In 1856 Newcomb became a private tutor nearer Washington and frequently traveled to the capital; he visited the library of the Smithsonian Institution and secured secretary Joseph Henry's permission to borrow the first volume of Bowditch's translation of Laplace's *Mécanique céleste*—a work that proved then to be somewhat beyond his mathematical powers. Soon afterward he met Henry, who suggested he seek employment at the Coast Survey. He was in turn recommended to the Nautical Almanac Office, then located in Cambridge, Massachusetts. Newcomb arrived there at the beginning of 1857 and a few weeks later was given a trial appointment as an astronomical computer. He also took the opportunity of studying mathematics under Benjamin Peirce at the Lawrence Scientific School of Harvard University and graduated the following year.

The outbreak of the Civil War in 1861 brought the resignations of several of the professors of mathematics attached to the United States Navy, and Newcomb was invited to fill a vacancy at the Naval Observatory. He was assigned to assist in observing the right ascensions of stars with the transit circle. He deplored the random observation of stars, as was customary, and was dismayed that there was no concerted action with the person observing declinations with the mural circle. In 1863 he was placed in charge of the mural circle, and he proposed to Superintendent Gilliss a plan, based largely on the practice at European observatories, whereby the right ascension and declination observations would be conducted more systematically. When a new transit circle was acquired in 1865 Newcomb initiated a four-year program of fundamental observations of stellar positions, involving both day and night measurements.

Newcomb had great respect, but no particular love, for observational work. While in Cambridge he had put the principles of the *Mécanique céleste* to good use and studied the secular variations in the motions of some of the minor planets. He showed that their orbits did not intersect and that there was no reason for accepting the then prevalent hypothesis that the minor planets were fragments of a larger planet that had exploded or been shattered by a collision.

After moving to Washington, Newcomb became especially interested in the motion of the moon and in the accuracy of Hansen's lunar tables. It soon became clear that the moon was starting to deviate from its predicted position. Hansen had fitted his theory to observations back to 1750, and in order to study the deviation it was desirable to make use of even earlier observations. Surmising that older records of occultations of stars by the moon existed in the archives of the Paris Observatory, Newcomb visited Paris during the siege of 1871 (departing only three weeks before the observatory found itself in the line of retreat of the Commune) and located a wealth of high-quality observations extending back to 1672. His analysis of these and other observations revealed that Hansen's tables were considerably in error prior to 1750. He suspected that the discrepancy was due to variations in the rate of rotation of the earth—and thus in the astronomical reckoning of time—but his attempt to verify this from observations of transits of Mercury was inconclusive (1882). Newcomb again took up the problem of the "fluctuation" in the motion of the moon during the final years of his life, and his exhaustive discussion of lunar observations from 720 B.C. to A.D. 1908 was completed only a month before his death. It remained for Brown, Innes, Spencer Jones, de Sitter, and others to prove that the cause of the fluctuation is indeed the irregular rotation of the earth.

In 1875 Newcomb was offered the directorship of the Harvard College Observatory, which he declined. In 1877 he was appointed superintendent of the Nautical Almanac Office, which had by then been transferred to Washington. After improving the efficiency with which the calculations for the *American Ephemeris* were made, he embarked on two ambitious projects: discussing the observations of the sun, moon, and planets obtained since 1750 at thirteen of the leading observatories throughout the world, and developing new theories and tables for the motions of these bodies. (He had published preliminary theories and tables for Uranus and Neptune several years earlier.) The project was clearly too much for one individual; and Newcomb thus went to considerable pains to obtain the best possible assistance. The most difficult part of the work, that of constructing the theories of Jupiter and Saturn, was entrusted to G. W. Hill. For these, Hansen's method was employed, and Newcomb subsequently regretted that he had not

used the same method for the other planets; the use of Encke's method, although much more straightforward, introduced problems into the determination of the orbital constants that Newcomb was not able to solve. Most of the work was completed by 1895, although it was left for E. W. Brown to construct the lunar theory.

In the course of his work on planetary theory Newcomb devised a useful procedure for developing the "disturbing function" that gives the perturbative action of one planet on another. In the case of circular orbits it is usual to develop the reciprocal of the distance between the planets as a cosine series in multiples of the longitude difference between the planets, each term being multiplied by a "Laplace coefficient." Newcomb showed that the process could easily be extended to elliptical orbits by the introduction of quantities dependent upon the multiple of the mean longitude difference and differential operators that act on the Laplace coefficients. He tabulated these quantities, now commonly known as "Newcomb operators," out to those corresponding to the eighth power of the orbital eccentricities, although some of the final ones have been found to be incorrect.

During his early years at the Naval Observatory, Newcomb made an investigation of the solar parallax, principally from observations of Mars at its 1862 opposition. In 1870 he proposed the establishment of a committee to plan observations of the 1874 and 1882 transits of Venus, with a view to obtaining a more precise value of the solar parallax. The committee became the Transit of Venus Commission, and Newcomb was appointed secretary. The results from the 1874 transit were disappointing; and although he was very much in the minority, Newcomb seriously questioned the wisdom of dispatching expeditions to observe the 1882 transit. (He did, however, conduct an expedition to South Africa in 1882.) He felt that a better value of the parallax could be obtained from the velocity of light and the constant of aberration. Newcomb's investigation of the velocity of light, using mirrors at the Naval Observatory, the Washington Monument, and Fort Myer, Virginia, was essentially a refinement of Foucault's method. The value obtained was long the astronomical standard.

Newcomb's study of the transits of Mercury confirmed Leverrier's conclusion that the perihelion of Mercury is subject to an anomalous advance (now known to be due to relativity), and he sought vainly for an explanation. In the course of his work on the transits of Venus of 1761 and 1769 he resolved the doubts surrounding the 1769 observations of Maximilian Hell. The value for the mass of Jupiter which he determined from the observations of Polyhymnia has still not been significantly improved. Newcomb also established that the retrograde motion of the line of apsides of Saturn's satellite Hyperion is due to the resonant influence of Titan. He was able to show that the fourteen-month period found by Chandler in the variation of latitude is due to some lack of rigidity of the earth. He studied the zodiacal light, the distribution and motions of the stars, and solar radiation.

Around 1880 Newcomb founded the *Astronomical Papers Prepared for the Use of the American Ephemeris and Nautical Almanac*, and the greater part of the above-mentioned researches was printed in the first seven volumes of this series. He also published a short account of his work on astronomical constants under the title *The Elements of the Four Inner Planets and the Fundamental Constants of Astronomy* (1895). At an international conference in Paris in 1896, it was agreed that from 1901 onwards, these constants (with only minor modifications) should be used in all the national ephemerides of the world. Newcomb was also charged with completing a catalogue of the positions and motions of the brighter stars and with making a new determination of the constant of precession. Completion of this work was complicated by his automatic retirement on his sixty-second birthday (1897), but arrangements were made for him to continue on a consulting basis.

Newcomb was instrumental in securing from Alvan Clark and Sons a twenty-six-inch refractor for the Naval Observatory, and with it he made measurements of the satellites of Uranus and Neptune. He was also prominently involved in negotiations with the Clarks for a thirty-inch refractor for the Pulkovo Observatory and in the establishment of the Lick Observatory.

In addition to his many scientific papers Newcomb wrote *A Compendium of Spherical Astronomy* (1906). It was intended to be the first of a series of texts, and it is regrettable that he never produced any further volumes. He wrote popular works on astronomy as well as three novels, some mathematical texts, several papers on economics, psychical research, and rain-making, and one on the "flying machine" (in which his gift of foresight completely failed him: his view that man would never fly brought him into direct conflict with the astrophysicist Samuel Pierpont Langley).

Newcomb was a member or foreign associate of the national academies or astronomical societies of seventeen countries, and he received honorary degrees from as many universities. He was one of the first lecturers at the Johns Hopkins University and became a professor there in 1884; he was awarded the Sylvester

prize in 1901. Among his other awards were the Copley Medal of the Royal Society, the Gold Medal of the Royal Astronomical Society, and the (first) Bruce Medal of the Astronomical Society of the Pacific. In 1863 he married Mary Caroline Hassler. He retired from the navy with the rank of captain and was promoted to rear admiral (retired) in 1906. Newcomb was buried with military honors in Arlington National Cemetery; President Taft and the representatives of several foreign governments attended the funeral.

BIBLIOGRAPHY

I. ORIGINAL WORKS. An exhaustive bibliography of Newcomb, compiled by R. C. Archibald, is contained in *Biographical Memoirs. National Academy of Sciences*, **17** (1924), 19–69. The best single source of biographical information is Newcomb's autobiography, *The Reminiscences of an Astronomer* (Boston–New York, 1903). Most of Newcomb's important writings are contained in *Astronomical Papers Prepared for the Use of the American Ephemeris and Nautical Almanac*, **1–9** (1879–1913).

Among other astronomical writings, in addition to those cited in the text, are "On the Secular Variations and Mutual Relations of the Orbits of the Asteroids," in *Memoirs of the American Academy of Arts and Sciences*, n.s. **5** (1860), 123–152; "An Investigation of the Distance of the Sun and of the Elements Which Depend Upon It," in *Washington Observations for 1865* (1867), app. 2; "Researches on the Motion of the Moon. Part I: Reduction and Discussion of Observations of the Moon Before 1750," in *Washington Observations for 1875* (1878), app. 2; *Popular Astronomy* (New York, 1878); *The Stars* (New York, 1901); *Astronomy for Everybody* (New York, 1902); "On the Position of the Galactic and Other Principal Planes Toward Which the Stars Tend to Crowd," which is *Carnegie Institute of Washington Contributions to Stellar Statistics*, no. 10 (1904); "An Observation of the Zodiacal Light to the North of the Sun," in *Astrophysical Journal*, **22** (1905), 209–212; *Sidelights on Astronomy* (New York–London, 1906); "A Search for Fluctuations in the Sun's Thermal Radiation Through Their Influence on Terrestrial Temperature," in *Transactions of the American Philosophical Society*, n.s. **21** (1908), 309–387.

Among his mathematical works are "A Generalized Theory of the Combination of Observations so as to Obtain the Best Result," in *American Journal of Mathematics*, **8** (1886), 343–366; "The Philosophy of Hyperspace," in *Science*, **7** (1898), 1–7.

Newcomb's works on economics include *The ABC of Finance* (New York, 1877); *Principles of Political Economy* (New York, 1886); *A Plain Man's Talk on the Labor Question* (New York, 1886).

II. SECONDARY LITERATURE. Among the many accounts of Newcomb's life and work are G. W. Hill, "Simon Newcomb as an Astronomer," in *Science*, **30** (1909), 353–357; T. J. J. See, "An Outline of the Career of Professor

Newcomb," in *Popular Astronomy*, **17** (1909), 465–481; E. W. Brown, "Simon Newcomb," in *Bulletin of the American Mathematical Society*, **16** (1910), 341–355; an obituary notice by H. H. Turner in *Monthly Notices of the Royal Astronomical Society*, **70** (1910), 304–310; W. W. Campbell, "Simon Newcomb," in *Biographical Memoirs. National Academy of Sciences*, **17** (1916), 1–18.

For recent work on astronomical constants, see W. de Sitter (and D. Brouwer), "On the System of Astronomical Constants," in *Bulletin of the Astronomical Institutes of the Netherlands*, **8** (1938), 213–231; G. M. Clemence, "On the System of Astronomical Constants," in *Astronomical Journal*, **53** (1948), 169–179; "Colloque International sur les Constants Fondamentales de l'Astronomie," in A. Danjon, ed., *Bulletin astronomique*, **15** (1950), 163–292; *International Astronomical Union Symposium No. 21: On the System of Astronomical Constants*, in J. Kovalevsky, ed., *Bulletin astronomique*, **25** (1965), 1–324; *International Astronomical Union Colloquium No. 9: The IAU System of Astronomical Constants*, in B. Emerson and G. A. Wilkins, eds., *Celestial Mechanics*, IV (1971), 128–280.

For material on the rotation of the earth, see W. de Sitter, "On the Secular Accelerations and the Fluctuations of the Longitudes of the Moon, Sun, Mercury, and Venus," in *Bulletin of the Astronomical Institutes of the Netherlands*, **4** (1927), 21–38; H. Spencer Jones, "The Rotation of the Earth, and the Secular Accelerations of the Sun, Moon, and Planets," in *Monthly Notices of the Royal Astronomical Society*, **99** (1939), 541–558. For further calculations of the Newcomb operators, see Sh. G. Sharaf, "Teoriya dvizheniya Plutona" ("Theory of the Motion of Pluto"), in *Trudy Instituta Teoreticheskoi astronomii. Akademiya nauk SSSR*, **4** (1955); I. G. Izsak *et al.*, "Construction of Newcomb Operators on a Digital Computer," which is *Smithsonian Astrophysical Observatory Special Report*, no. 140 (1964).

BRIAN G. MARSDEN

NEWCOMEN, THOMAS (*b.* Dartmouth, England; christened 24 February 1663; *d.* London, England, 5 August 1729), *steam technology*.

Newcomen is renowned as the inventor of the steam engine. He was descended from an aristocratic family that had lost its property during the reign of Henry VIII. His grandfather and father were merchants and nonconformists, and Newcomen followed them in both respects. During the 1680's he became an ironmonger in partnership with John Calley, an artisan and fellow Baptist who later collaborated with him on the development of the steam engine. Newcomen became a leader of the local Baptists and often preached to their congregations. His formal education appears to have been rudimentary, and he published nothing. Few details are known of his personal life or of the circumstances that surrounded his invention.

Newcomen's first successful engine, which was erected in the Midlands in 1712, was the reward of years of trials and tinkering. The increasingly troublesome problem of removing water from mines had already provided the stimulus for attempts by Newcomen and others to design an improved machine to serve either as a pump or as an engine to drive a pump. In 1698 Thomas Savery (also of Devon) invented a steam pump which he protected with a broad patent that covered all "vessells or engines for raiseing water or occasioning motion to any sort of millworks by the impellent force of fire." Because of the scope of Savery's patent, Newcomen was later prevented from patenting his own engine and was required to build his engines under license from Savery, although his work was entirely independent of Savery's and his engine was totally different from Savery's pump.

Newcomen's engine was an ingenious combination of familiar elements: piston and cylinder, pumps, levers, valves, and the process of producing low pressure by the condensation of steam in a vessel. The key invention, which was the injection of cold water directly into the cylinder, was hit upon accidentally in the course of experiments that used cold water jackets to produce condensation. Later James Watt significantly increased the efficiency of the engine through his invention of the separate condenser (1765), which avoided the necessity of alternately heating and cooling the cylinder. Nevertheless, unmodified Newcomen engines continued to be used long after Watt's improvement, but because of their low efficiency, they were confined largely to collieries, where coal was cheap.

At the end of the eighteenth century John Robison propagated the belief that Newcomen's achievement somehow depended upon the application of scientific principles gained through an alleged correspondence between Newcomen and Robert Hooke. (Robison advanced a similar claim for the derivation of Watt's separate condenser from Joseph Black's theory of latent heat.) Robison's allegation has been discredited; the records reveal no contact whatever between Newcomen and his contemporaries in science. His invention was the product of a familiarity with technical operations and needs in the mining industry, a close knowledge of contemporary craftsmanship, repeated trials and improvements, and a stroke of luck.

BIBLIOGRAPHY

For a full biography, see L. T. C. Rolt, *Thomas Newcomen: The Prehistory of the Steam Engine* (London, 1963); this work modifies some of the views presented in

H. W. Dickinson, *A Short History of the Steam Engine* (London, 1938, 1963), ch. 3. An important contemporary account of Newcomen's work, which is based apparently on firsthand knowledge, is Mårten Triewald, *Beskrifning om eld- och luftmachin vid Dannemora grufvor* (Stockholm, 1734), trans. as *A Short Description of the Fire- and Air-Machine at the Dannemora Mines*, and published by the Newcomen Society as *Mårten Triewald's Short Description of the Atmospheric Engine*, Extra Publication no. 1 (London, 1928). On the question of the influence of science on Newcomen's work, see Rhys Jenkins, "The Heat Engine Idea in the Seventeenth Century," in *Transactions of the Newcomen Society*, **17** (1936–1937), 1–11.

HAROLD DORN

NEWLANDS, JOHN ALEXANDER REINA (*b.* London, England, 26 November 1837; *d.* London, 29 July 1898), *chemistry.*

Newlands was one of the precursors of Mendeleev in the formulation of the concept of periodicity in the properties of the chemical elements. He was the second son of a Presbyterian minister, William Newlands, from whom he received his general education. In 1856 he entered the Royal College of Chemistry, where he studied for a year under A. W. Hofmann. He then became assistant to J. T. Way, chemist to the Royal Agricultural Society. He stayed with Way until 1864, except for a short interlude in 1860, when he served as a volunteer with Garibaldi in Italy. Newlands' mother, Mary Sarah Reina, was of Italian descent.

In 1864 he set up practice as an analytical chemist and supplemented his income by teaching chemistry. He seems to have made a special study of sugar chemistry and in 1868 became chief chemist in a refinery belonging to James Duncan, with whom he developed a new system of cleaning sugar and introduced a number of improvements in processing. The business declined as a result of foreign competition, and in 1886 he left the refinery and again set up as an analyst, this time in partnership with his brother, B. E. R. Newlands. The brothers collaborated with C. G. W. Lock, one of the previous authors, in the revision of an established treatise on sugar growing and refining. Newlands died of influenza in 1898; he was survived by his wife, a daughter, and a son.

Newlands' early papers on organic compounds, the first suggesting a new nomenclature, the second proposing the drawing up of tables to show the relationships between compounds, were vitiated by the absence at that time of clear ideas regarding structure and valency; but they are interesting because they show the cast of his mind toward systematization. His first communication (*Chemical News*, 7 February 1863) on the numerical relationships existing between the atomic weights of similar elements was a summing-up, with some of his own observations, of what had been pointed out by others (of whom he mentioned only Dumas). Two main phenomena had been observed: (*a*) there existed "triads" (first noticed by Döbereiner), groups of three elements of similar properties, the atomic weight of one being the numerical mean of the others, and (*b*) it was also found that the difference between the atomic weights of analogous elements seemed often to be a multiple of eight.

Like many of his contemporaries, Newlands at first used the terms "equivalent" and "atomic weight" without distinction of meaning, and in this first paper he employed the values accepted by his predecessors. In a July 1864 letter he used A. W. Williamson's values,[1] which were based on Cannizzaro's system. The letter contains a table of the sixty-one known elements in the order of their "new" atomic weights. In a second table he grouped thirty-seven elements into ten classes, most of which contained one or more triads. The incompleteness of the table was attributed to uncertainty regarding the properties of some of the more recently discovered elements and also to the possible existence of undiscovered elements. He considered silicon (atomic weight 28) and tin (atomic weight 118) to be the extremities of a triad, the middle term of which was unknown; thus his later claim to having predicted the existence of germanium (atomic weight 73) before Mendeleev is valid.

About a month later he said that if the elements were numbered in the order of their atomic weights (giving the same number to any two with the same weight) it was observed "that elements having consecutive numbers frequently either belong to the same group or occupy similar positions in other groups." The following table[2] was given in illustration:

TABLE I

Group	No.		No.		No.		No.		No.	
a	N	6	P	13	As	26	Sb	40	Bi	54
b	O	7	S	14	Se	27	Te	42	Os	50
c	F	8	Cl	15	Br	28	I	41	—	—
d	Na	9	K	16	Rb	29	Cs	43	Tl	52
e	Mg	10	Ca	17	Sr	30	Ba	44	Pb	53

The difference between the number of the lowest member of a group and that immediately above it was seven: "in other words, the eighth element starting from a given one is a kind of repetition of the first, like the eighth note in an octave of music." One or two transpositions had been made to give an acceptable grouping; the element omitted (no. 51) would have been mercury, which clearly could not be grouped with the halogens.

Newlands was groping toward an important discovery, although it excited little comment. A year later (August 1865) he again drew attention to the difference of seven (or a multiple thereof) between the ordinal numbers of elements in the same horizontal group: "This peculiar relationship I propose to provisionally term the 'Law of Octaves.'" This time he put all sixty-two elements (he included the newly discovered indium) in his table[3]:

TABLE II

No.	No.	No.	No.	No.	No.	No.	No.
H 1	F 8	Cl 15	Co, Ni 22	Br 29	Pd 36	I 42	Pt, Ir 50
Li 2	Na 9	K 16	Cu 23	Rb 30	Ag 37	Cs 44	Tl 53
Be 3	Mg 10	Ca 17	Zn 25	Sr 31	Cd 38	Ba, V 45	Pb 54
B 4	Al 11	Cr 19	Y 24	Ce, La 33	U 40	Ta 46	Th 56
C 5	Si 12	Ti 18	In 26	Zr 32	Sn 39	W 47	Hg 52
N 6	P 13	Mn 20	As 27	Di, Mo 34	Sb 41	Nb 48	Bi 55
O 7	S 14	Fe 21	Se 28	Rh, Ru 35	Te 43	Au 49	Os 51

But this forcing of the elements into too rigid a framework weakened his case. It seemed to preclude (a conclusion that he subsequently denied) the possibility of gaps in the sequence which, when filled, would lead to a more acceptable grouping. The resulting anomalies were seized upon by his critics, when on 1 March 1866 he read a paper to the Chemical Society presenting the same table—except that the elements in the last column now appeared in numerical order. The facetious inquiry of G. C. Foster, professor of physics at University College, London, as to whether Newlands had ever examined the elements when placed in alphabetical order, has often been quoted; but Foster also made the cogent criticism that no system of classification could be accepted which separated chromium from manganese and iron from cobalt and nickel.[4]

The hostile reception of his paper and the disinclination of the Society to publish it (on the grounds of its purely theoretical nature) seem to have discouraged Newlands from following up his ideas until after the publication of Mendeleev's table in 1869. After that table appeared, Newlands continued to seek numerical relationships among atomic weights, while attempting, in a series of letters to *Chemical News*, to establish his priority. He set out his claims more specifically in December 1882, on hearing of the award of the Davy Medal of the Royal Society to Mendeleev and Lothar Meyer. His persistence was eventually rewarded in 1887, when the medal was awarded to him.

NOTES

1. A. W. Williamson, "On the Classification of the Elements in Relation to Their Atomicities," in *Journal of the Chemical Society*, **17** (1864), 211–222.
2. *Chemical News* (20 Aug. 1864), 94; *On the Discovery of the Periodic Law*, 11.
3. *Chemical News* (18 Aug. 1865), 83; *On the Discovery of the Periodic Law*, 14. A few symbols have been altered to conform with modern usage (Di = "didymium," shown in 1885 to be a mixture of neodymium and praseodymium).
4. For a report of the meeting see *Chemical News*, **13** (1866), 113–114. There was no hint of both a "vertical" and a "horizontal" relationship between elements prior to the publication of Mendeleev's table.

BIBLIOGRAPHY

I. ORIGINAL WORKS. Newlands' writings on periodicity were republished in a small book, *On the Discovery of the Periodic Law* (London, 1884); those published after the appearance of Mendeleev's table, with some additional notes, are in the appendix. *Sugar: A Handbook for Planters and Refiners* (London–New York, 1888), written with C. G. W. Lock and B. E. R. Newlands, was based on C. G. W. Lock, G. W. Wigner, and R. H. Harland, *Sugar Growing and Refining* (London–New York, 1882); a further rev. ed. by "the late J. A. R. Newlands and B. E. R. Newlands," with no mention of a third author, was published in 1909.

An incomplete list of Newlands' papers is in the Royal Society *Catalogue of Scientific Papers*, IV (London, 1870), 600; VIII (London, 1879), 494; X (London, 1894), 916–917; and XVII (Cambridge, 1921), 506. Those particularly mentioned in the text are "On Relations Among the Equivalents," in *Chemical News*, **7** (1863), 70–72; "Relations Between Equivalents," *ibid.*, **10** (1864), 59–60, 94–95; "On the Law of Octaves," *ibid.*, **12** (1865), 83; a reply to his critics at the Chemical Society meeting is *ibid.*, **13** (1866), 130; "On the Discovery of the Periodic Law," *ibid.*, **46** (1882), 278–279, is the most detailed of Newlands' claims for priority over Mendeleev. All of the above papers, except the last, are reprinted in *On the Discovery of the Periodic Law*.

II. SECONDARY LITERATURE. Most of the biographical details stem from the obituary by W. A. Tilden, in *Nature*, **58** (1898), 395–396; another obituary is W. Smith, in *Journal of the Society of Chemical Industry*, **17** (1898), 743.

W. A. Smeaton, "Centenary of the Law of Octaves," in *Journal of the Royal Institute of Chemistry*, **88** (1964), 271–274, reproduces the more important of Newlands' tables and gives a useful summary of the relevant work of others, particularly W. Odling.

See also J. A. Cameron, "J. A. R. Newlands (1837–1898), A Pioneer Whom the Chemists Ridiculed," in *Chemical Age*, **59** (1948), 354–356; W. H. Taylor, "J. A. R. Newlands: A Pioneer in Atomic Numbers," in *Journal of*

Chemical Education, **26** (1949), 491–496; J. W. van Spronsen, "One Hundred Years of the 'Law of Octaves,'" in *Chymia*, **11** (1966), 125–137.

For a detailed history of the periodic table see J. W. van Spronsen, *The Periodic System of Chemical Elements* (Amsterdam–London–New York, 1969). An earlier and less comprehensive work, but giving a good summary of Newlands' work, is A. E. Garrett, *The Periodic Law* (London, 1909). See also H. Cassebaum and G. B. Kaufman, "The Periodic System of the Chemical Elements: the Search for Its Discoverer," in *Isis*, **62** (1971), 314–317.

E. L. SCOTT

NEWPORT, GEORGE (*b.* 4 July 1803, Canterbury, England; *d.* 7 April 1854, London, England), *entomology, natural history.*

Newport was the son of a wheelwright and after receiving a simple schooling, he became an apprentice to his father's trade at age fourteen. During the next nine years he read widely in many subjects and by dint of tireless application extended his scanty education. From an early age he had been interested in insect life, and now he began serious entomological studies that were to continue throughout his life. He took advantage of the Canterbury Philosophical and Literary Institution and made liberal use of its library, lectures, and natural history collections. In 1825 and 1826 he gave lectures there on mechanics, and in 1826 he became general exhibitor of the museum when the institution's new building was opened. Among his various activities were lectures and demonstrations on entomology, and he donated many specimens of British insects, which he himself had preserved.

During the two-year tenure of this post, Newport became acquainted with William Henry Weekes, a surgeon of Sandwich, and in 1828 he began an apprenticeship with him. Throughout his early life he suffered great privations and was at times dependent upon friends for financial support, debts which he in later life honorably liquidated. After his apprenticeship Newport enrolled in the University of London (now University College, London), on 16 January 1832. In 1835 he was admitted a licentiate of the Society of Apothecaries of London and a member of the Royal College of Surgeons of England, which at that time was the usual combination of diplomas for medical practice. Newport held the post of house surgeon to the Chichester Infirmary until January 1837, when he established himself in practice at 30 Southwick Street, London. He was more interested in scientific pursuits so that his practice gradually declined; and when in 1847 he was awarded a pension from the civil list of £100 per annum for his contributions to natural history, he was able to devote all his time to research.

Newport never married, and as his habits were of the most frugal kind he was able to subsist on this limited income. His extensive researches were rewarded with several honors. On 11 December 1843 he was elected a fellow of the Royal College of Surgeons of England, of which he was one of the original 300 fellows, and from 1844 to 1845 he was president of the Entomological Society. On 26 March 1846 he became a fellow of the Royal Society, and at the time of his death he was a Member of Council. He was also a fellow of the Linnean Society and of several foreign natural history societies. He contracted an illness—from which he died—in the marshy ground west of London while collecting research material.

Newport was a man of the strictest honesty, both in his scientific studies and in his dealings with the world. He had a nervous temperament and a morbid sensitivity to criticism which caused him to make enemies readily. He possessed unwearied patience and remarkable digital dexterity, evidenced in his dissections, demonstrations, and insect preparations; he could draw equally well with either hand; and his powers of observation were acute. He was exceedingly zealous and industrious and was interested only in the advancement of science. His services were commemorated in a public monument in Kensal Green Cemetery, erected by fellows of the Royal Society and of the Linnean Society.

Newport's contributions to biology lay mostly within the field of entomology and the embryology of the Insecta and Amphibia. His first papers—sufficiently important and original to appear in the *Philosophical Transactions*—were on the bumblebee, butterflies, and moths; and he investigated the nervous system, respiration, and temperature of these and other insects. He also published many subsequent papers on insect structure, which included an important survey of Insecta (1839). For his essay on the turnip fly (1838) he was awarded a medal by the Agricultural Society of Saffron Walden. Newport's most outstanding contribution to biology was his discovery that during fertilization in higher animals impregnation of the ovum by the spermatozoon is by penetration and not just by contact as previously thought. For this work on the frog (1851) he was awarded the Royal Medal of the Royal Society. He was also the first to observe the coincidence between the first plane of cleavage in the egg made by the spermatozoon at its place of entry and the median plane of the body of the embryo and thus of the adult body (1854).

BIBLIOGRAPHY

I. ORIGINAL WORKS. A list of Newport's writings (thirty-five items produced during a period of twenty-two years) is in *Proceedings of the Royal Society*, **7** (1855), 281–283. They were published mainly in periodicals; his excellent article on Insecta appeared in Robert B. Todd, ed., *The Cyclopaedia of Anatomy and Physiology*, II (London, 1836–1839), 835–994, and his prize essay on the turnip fly was a monograph, *Observations on the Anatomy, Habits, and Economy of "Athalia centrifoliae," the Saw-fly of the Turnip, and on the Means Adopted for the Prevention of Its Ravages* (London, 1838). His *Catalogue of the Myriapoda in the British Museum* (London, 1856) appeared posthumously. Of Newport's earlier papers those on *Sphinx* are outstanding; "On the Nervous System of the *Sphinx Ligustri*," in *Philosophical Transactions of the Royal Society*, pt. 2 (1832), 383–398; and "On the Nervous System of the *Sphinx* During the Latter Stages of Its Pupa and Imago States," *ibid.*, pt. 2 (1834), 389–423.

Newport's classic papers on embryology are "On the Impregnation of the Ovum in Amphibia," in *Philosophical Transactions of the Royal Society*, 1st ser., **141** (1851), 169–242; "On the Impregnation of the Ovum in Amphibia (2nd Series Revised), and on the Direct Agency of the Spermatozoon," *ibid.*, **143** (1853), 233–290; and "Researches on the Impregnation of the Ovum in the Amphibia," *ibid.*, **144** (1854), 229–244; this article contains material selected and arranged by G. V. Ellis from the author's MSS after his death.

II. SECONDARY LITERATURE. There are only a few brief biographical notices on Newport; the best are *Proceedings of the Linnean Society of London*, **2** (1855), 309–312; *Dictionary of National Biography*, **14** (1844), 357–358; *Gentleman's Magazine* (June 1854), 660–661; *Medical Times and Gazette* (London), n.s. **8** (1854), 392–393; *Proceedings of the Royal Society*, **7** (1855), 278–285; and *Plarr's Lives of the Fellows of the Royal College of Surgeons of England*, **2** (1930), 95–96. An account of his epitaph is in *Lancet* (1855), **2**, 554.

Newport's embryological investigations are discussed in F. J. Cole, *Early Theories of Sexual Generation* (Oxford, 1930), 193–196, and in A. W. Meyer, *The Rise of Embryology* (Stanford, 1939), 188–190.

EDWIN CLARKE

NEWTON, EDWIN TULLEY (*b*. Islington, London, England, May 1840; *d*. Canonbury, London, 28 January 1930), *paleontology*.

Newton began his scientific career as a student at the Royal School of Mines. In 1865 he became assistant naturalist under T. H. Huxley at the Geological Survey Museum. He was appointed paleontologist and curator of fossils in 1882; he retired in 1905. Newton took an active part in the work of the Geological Society: he was vice-president from 1903 to 1905 and again from 1916 to 1918, and was awarded the Lyell Medal in 1893. In the same year (1893) he was elected a fellow of the Royal Society. He was president of the Paleontographical Society from 1921 to 1928 and president of the Geologists' Association in 1896–1898. He also served on committees of the Zoological Society.

Newton's work was always thorough, and he showed remarkable patience and skill in performing delicate manipulations. His beautiful model of the brain of a cockroach, constructed by means of serial sections, is displayed in the museum of the Royal College of Surgeons (London). He began his research by inventing a new method for making microsections of coal. A number of these sections, as well as many skeletons prepared by him, are in the British Museum. In his official work Newton studied a wide variety of fossils, but he is noted for his original investigations in vertebrate paleontology. He first studied the Cretaceous fishes, and later made observations on the bones of birds and the remains of man, but he devoted his chief energies to studying the vertebrate fragments from the Pleistocene and Pliocene deposits in England. His most important monographs are probably those on the brain of the Jurassic flying reptile, the pterodactyl (1888), and on the reptilian remains from the Permotriassic rocks at Elgin, Scotland (1893 and 1894). In the latter research Newton lacked the actual bones to work with; he had only their cavities in sandstone, from which, after great labor, he obtained and fitted together casts of gutta-percha. He discovered dicynodonts and pareiasaurs for the first time in Europe, showing how closely they resembled the descriptions of those found in the Karroo Formation of South Africa.

All of Newton's published work was descriptive. He was a realist and thought that although the search back in evolutionary forms might end in so-called fish forms, such retrospection should not lead away "from scientific facts to a slough of unscientific imagination."

BIBLIOGRAPHY

I. ORIGINAL WORKS. Among Newton's important works are *The Chimaeroid Fishes of the British Cretaceous Rocks* (London, 1878); *The Vertebrata of the Forest-Bed Series of Norfolk and Suffolk* (London, 1882); "On the Skull, Brain, and Auditory Organ of a New Species of Pterosaurian From the Upper Lias Near Whitby, Yorkshire," in *Philosophical Transactions of the Royal Society*, ser. B, **179** (1888), 503–537; "Notes on Pterodactyls," in *Proceedings of the Geologists' Association*, **10** (1888), 406–424; *The Vertebrata of the Pliocene Deposits of Britain* (London,

1891); "Reptiles From the Elgin Sandstone," in *Philosophical Transactions of the Royal Society*, ser. B, **184** (1893), 431–503, and **185** (1894), 573–607; and "The [Pleistocene] Vertebrate Fauna Collected by Mr. Lewis Abbott From the Fissure Near Ightham, Kent," in *Quarterly Journal of the Geological Society of London*, **50** (1894), 188–211.

II. SECONDARY LITERATURE. For an article on Newton's work, see *Journal of Microscopical Science* (1879), p. 340. Obituary notices include *Geological Magazine*, **67** (1930), 286–287; A. S. Woodward, in *Quarterly Journal of the Geological Society of London*, **86** (1930), lix–lxii; and A. S. W[oodward], in *Obituary Notices of the Fellows of the Royal Society*, **1** (1932), 5–7.

                                            JOHN CHALLINOR

NEWTON, HUBERT ANSON (*b.* Sherburne, New York, 19 March 1830; *d.* New Haven, Connecticut, 12 August 1896), *astronomy, mathematics.*

Hubert was one of eleven children of William and Lois Butler Newton, both of whom were descendants of the first Puritan settlers in New England. After attending public schools in Sherburne, Newton entered Yale at age sixteen. He was an outstanding student; he won election to the Phi Beta Kappa Society and first prize for the solution of mathematics problems.

Following his graduation in 1850, Newton studied mathematics for two and a half years at his home and in New Haven. He became tutor at Yale in 1853, and almost immediately thereafter, on the death of A. D. Stanley, he was asked to chair the mathematics department. Two years later Newton was elected professor and at age twenty-five was one of the youngest persons ever to have reached that rank at Yale.

The professorship included a year's leave of absence, which he took at the Sorbonne with the geometer Chasles. That experience clearly influenced Newton, who subsequently published several important papers on mathematics.

Even though mathematics constituted his education and vocation, his principal efforts began to shift to astronomy and meteorology. His interest in those subjects was sparked by the spectacular meteor shower of 13 November 1833. Although Newton was too young to remember it, others in New Haven, like Edward C. Herrick, Alexander C. Twining, and Denison Olmsted (his undergraduate teacher in astronomy), had written about the event and had checked the records of earlier showers. Thus by 1860 rudimentary data on meteors existed and tentative hypotheses about their orbits were being proffered.

Newton's first papers on the subject (1860–1862) dealt primarily with the orbits and velocities of fireballs. In 1861 the Connecticut Academy of Arts and Sciences established a committee to obtain systematic sightings from diverse observers of the meteor showers of August and November. As one of the leaders of that group, Newton soon accumulated vast amounts of information.

From a careful study of all extant records of the shower of November 1861 Newton in 1864 published his important finding that the shower had occurred thirteen times since A.D. 902, in a cycle of 33.25 years. He reasoned that the phenomenon was caused by a swarm of meteoroids orbiting the sun and concluded that the number of revolutions they must make in one year would be $2 \pm 1/33.25$ or $1 \pm 1/33.25$ or $1/33.25$. These frequencies correspond to periods of 180.0, 185.4, 354.6, and 375.5 days and 33.25 years. Using these five values, the position of the radiant point, and the knowledge that the meteoroids' heliocentric motion is retrograde, Newton calculated five possible orbits.

He noted that the real orbit could be distinguished from the others by calculating the secular motion of the node that was due to planetary perturbations for each of the hypothetical orbits. J. C. Adams, who undertook those calculations, found that the four short periods were not compatible with the observations; the period of 33.25 years, however, corresponds to an elliptical orbit, which extends past Uranus and is subject to perturbations by Uranus and Saturn. Since Adams' determination of the effect of perturbations agreed with Newton's data for the Leonids, these meteoroids were proved to be in such an orbit with a period of 33.25 years.

The Leonids' dramatic reappearance in 1866 spurred meteoroid research and added credence to Newton's calculations; moreover, the reappearance led to the positive identification of the swarm with a comet. By 1865 Newton in the United States and Schiaparelli in Italy had independently concluded that the mean velocities of meteoroids are nearly parabolic and resemble those of comets. When it was found in 1866 that a comet and the Leonids had virtually identical orbits, their relationship was firmly established.

From about 1863 to 1866 Newton amassed and published extensive statistics from observations of sporadic meteors. From this information, he derived the paths and the numbers of meteors, plus the spatial density of meteoroids near the earth's orbit and their velocity about the sun.

Newton's next major contribution to meteor studies came in the mid-1870's when he compared the statistical distribution of known cometary orbits with the hypothetical distributions that would result from two currently leading theories for the origin of the solar system—those of Kant and Laplace. According to Kant, comets formed as part of the primeval solar nebula, while according to Laplace they originated independently from the solar system. Newton found that the distribution of comets' aphelia and inclinations agrees better with the latter theory, although he noted that the problem was unsettled.

These calculations included considerations of the effect of large planetary perturbations on the distribution of cometary orbits; such studies culminated in 1891 in his most famous paper on perturbations. During the 1870's and 1880's Newton accumulated statistical data that indicated that long period comets could be captured by Jupiter, shortening their periods.

Newton devoted the last decade of his research to Biela's comet and meteor shower, to fireballs, and to meteorites. At his death he was probably the foremost American pioneer in the study of meteors.

Besides his scientific research, Newton was active in teaching and educational reform, especially about the metric system. He was a founder of the American Metrological Society, and he persuaded many manufacturers of scientific instruments and publishers of school arithmetic texts to adopt the system.

In 1868 the University of Michigan awarded Newton an honorary LL.D. After joining the American Association for the Advancement of Science in 1850, he served as the vice-president of its Section A in 1875, and as president of the Association in 1885. He was a president of the Connecticut Academy of Arts and Sciences, a member of the American Philosophical Society, and one of the original members of the National Academy of Sciences. In 1888 the National Academy awarded him its J. Lawrence Smith Gold Medal in recognition of his research on meteoroids. At his death he was the vice-president of the American Mathematical Society and an associate editor of the *American Journal of Science*.

Aside from societies in the United States, he was elected in 1860 corresponding member of the British Association for the Advancement of Science, in 1872 associate of the Royal Astronomical Society of London, in 1886 foreign honorary fellow of the Royal Philosophical Society of Edinburgh, and in 1892 foreign member of the Royal Society of London.

Newton's association with Yale and New Haven was long and rich. He directed the Yale mathematics department and also the observatory, which he helped organize in 1882, and he helped build the extensive collection of meteorites in the Peabody Museum. He also provided considerable assistance to poor students who wanted to attend Yale. For a time he was the only Democrat on the Yale faculty and became alderman in the strongly Republican first ward of New Haven.

BIBLIOGRAPHY

I. ORIGINAL WORKS. Newton published approximately seventy papers, an extensive bibliography of which is included in the memoir by Gibbs that is cited below. Newton's most significant writings included the following: "Explanation of the Motion of the Gyroscope," in *American Journal of Science*, **24** (1857), 253–254; "On the Geometrical Construction of Certain Curves by Points," in *Mathematics Monthly*, **3** (1861), 235–244, 268–279; "On November Star-Showers," in *American Journal of Science*, **37** (1864), 377–389; **38** (1864), 53–61; "On Shooting Stars," in *Memoirs of the National Academy of Sciences*, **1** (1866), 291–312; *The Metric System of Weights and Measures* (Washington, 1868); "On the Transcendental Curves Whose Equation Is $\sin y \sin my = a \sin x \sin nx + b$," in *Transactions of the Connecticut Academy of Arts and Sciences*, **3** (1875), 97–107, written with A. W. Phillips; "On the Origin of Comets," in *American Journal of Science*, **16** (1878), 165–179; "The Story of Biela's Comet," *ibid.*, **31** (1886), 81–94; and "On the Capture of Comets by Planets, Especially Their Capture by Jupiter," in *Memoirs of the National Academy of Sciences*, **6** (1891), 7–23.

II. SECONDARY LITERATURE. An article on meteors that gives a critique of Newton's work is M. Faye, in *Comptes rendus hebdomadaires des séances de l'Académie des sciences*, **64** (1867), 550. Biographical sketches, which were written about the time of Newton's becoming president of the American Association for the Advancement of Science, are in *Science*, **6** (1885), 161–162; in *Popular Science Monthly*, **27** (1885), 840–843; and in James Grant Wilson and John Fisk, eds., *Appleton's Cyclopedia of American Biography*, IV (New York, 1888), 506–507.

Obituaries on Newton are William L. Elkin, in *Astronomische Nachrichten*, **141** (1896), 407; unsigned writers, in *Popular Astronomy*, **4** (1896), 236–240; in *Monthly Notices of the Royal Astronomical Society*, **57** (1897), 227–231; and in *New York Times* (13 Aug. 1896), 5. Biographical articles that were written after his death were J. Willard Gibbs, in *Biographical Memoirs. National Academy of Sciences*, **4** (1902), 99–124, which includes a bibliography; Anson Phelps Stokes, in *Memorials of Eminent Yale Men* (New Haven, 1914), 48–54; and David Eugene Smith, in Dumas Malone, ed., *Dictionary of American Biography*, XIII (New York, 1934), 470–471.

RICHARD BERENDZEN

NEWTON, ISAAC (*b.* Woolsthorpe, England, 25 December 1642; *d.* London, England, 20 March 1727), *mathematics, dynamics, celestial mechanics, astronomy, optics, natural philosophy*.

Isaac Newton was born a posthumous child, his father having been buried the preceding 6 October.

Newton was descended from yeomen on both sides: there is no record of any notable ancestor. He was born prematurely, and there was considerable concern for his survival. He later said that he could have fitted into a quart mug at birth. He grew up in his father's house, which still stands in the hamlet of Woolsthorpe, near Grantham in Lincolnshire.

Newton's mother, Hannah (née Ayscough), remarried, and left her three-year-old son in the care of his aged maternal grandmother. His stepfather, the Reverend Barnabas Smith, died in 1653; and Newton's mother returned to Woolsthorpe with her three younger children, a son and two daughters. Their surviving children, Newton's four nephews and four nieces, were his heirs. One niece, Catherine, kept house for Newton in the London years and married John Conduitt, who succeeded Newton as master of the Mint.

Newton's personality was no doubt influenced by his never having known his father. That he was, moreover, resentful of his mother's second marriage and jealous of her second husband may be documented by at least one entry in a youthful catalogue of sins, written in shorthand in 1662, which records "Threatning my father and mother Smith to burne them and the house over them."[1]

In his youth Newton was interested in mechanical contrivances. He is reported to have constructed a model of a mill (powered by a mouse), clocks, "lanthorns," and fiery kites, which he sent aloft to the fright of his neighbors, being inspired by John Bate's *Mysteries of Nature and Art*.[2] He scratched diagrams and an architectural drawing (now revealed and preserved) on the walls and window edges of the Woolsthorpe house, and made many other drawings of birds, animals, men, ships, and plants. His early education was in the dame schools at Skillington and Stoke, beginning perhaps when he was five. He then attended the King's School in Grantham, but his mother withdrew him from school upon her return to Woolsthorpe, intending to make him a farmer. He was, however, uninterested in farm chores, and absent-minded and lackadaisical. With the encouragement of John Stokes, master of the Grantham school, and William Ayscough, Newton's uncle and rector of Burton Coggles, it was therefore decided to prepare the youth for the university. He was admitted a member of Trinity College, Cambridge, on 5 June 1661 as a subsizar, and became scholar in 1664 and Bachelor of Arts in 1665.

Among the books that Newton studied while an undergraduate was Kepler's "optics" (presumably the *Dioptrice*, reprinted in London in 1653). He also began Euclid, which he reportedly found "trifling,"

throwing it aside for Schooten's second Latin edition of Descartes's *Géométrie*.[3] Somewhat later, on the occasion of his election as scholar, Newton was reportedly found deficient in Euclid when examined by Barrow.[4] He read Descartes's *Géométrie* in a borrowed copy of the Latin version (Amsterdam, 1659–1661) with commentary by Frans van Schooten, in which there were also letters and tracts by de Beaune, Hudde, Heuraet, de Witt, and Schooten himself. Other books that he studied at this time included Oughtred's *Clavis*, Wallis' *Arithmetica infinitorum*, Walter Charleton's compendium of Epicurus and Gassendi, Digby's *Two Essays*, Descartes's *Principia philosophiae* (as well as the Latin edition of his letters), Galileo's *Dialogo* (in Salusbury's English version)—but not, apparently, the *Discorsi*—Magirus' compendium of Scholastic philosophy, Wing and Streete on astronomy, and some writings of Henry More (himself a native of Grantham), with whom Newton became acquainted in Cambridge. Somewhat later, Newton read and annotated Sprat's *History of the Royal Society*, the early *Philosophical Transactions*, and Hooke's *Micrographia*.

Notebooks that survive from Newton's years at Trinity include an early one[5] containing notes in Greek on Aristotle's *Organon* and *Ethics*, with a supplement based on the commentaries by Daniel Stahl, Eustachius, and Gerard Vossius. This, together with his reading of Magirus and others, gives evidence of Newton's grounding in Scholastic rhetoric and syllogistic logic. His own reading in the moderns was organized into a collection of "Questiones quaedam philosophicae,"[6] which further indicate that he had also read Charleton and Digby. He was familiar with the works of Glanville and Boyle, and no doubt studied Gassendi's epitome of Copernican astronomy, which was then published together with Galileo's *Sidereus nuncius* and Kepler's *Dioptrice*.[7]

Little is known of Newton's friends during his college days other than his roommate and onetime amanuensis Wickins. The rooms he occupied are not known for certain; and we have no knowledge as to the subject of his thesis for the B.A., or where he stood academically among the group who were graduated with him. He himself did record what were no doubt unusual events in his undergraduate career: "Lost at cards twice" and "At the Taverne twice."

For eighteen months, after June 1665, Newton is supposed to have been in Lincolnshire, while the University was closed because of the plague. During this time he laid the foundations of his work in mathematics, optics, and astronomy or celestial mechanics. It was formerly believed that all of these discoveries were made while Newton remained in seclusion at Woolsthorpe, with only an occasional excursion into nearby Boothby. During these "two plague years of 1665 & 1666," Newton later said, "I was in the prime of my age for invention & minded Mathematicks & Philosophy more then at any time since." In fact, however, Newton was back in Cambridge on at least one visit between March and June 1666.[8] He appears to have written out his mathematical discoveries at Trinity, where he had access to the college and University libraries, and then to have returned to Lincolnshire to revise and polish these results. It is possible that even the prism experiments on refraction and dispersion were made in his rooms at Trinity, rather than in the country, although while at Woolsthorpe he may have made pendulum experiments to determine the gravitational pull of the earth. The episode of the falling of the apple, which Newton himself said "occasioned" the "notion of gravitation," must have occurred at either Boothby or Woolsthorpe.[9]

Lucasian Professor. On 1 October 1667, some two years after his graduation, Newton was elected minor fellow of Trinity, and on 16 March 1668 he was admitted major fellow. He was created M.A. on 7 July 1668 and on 29 October 1669, at the age of twenty-six, he was appointed Lucasian professor. He succeeded Isaac Barrow, first incumbent of the chair, and it is generally believed that Barrow resigned his professorship so that Newton might have it.[10]

University statutes required that the Lucasian professor give at least one lecture a week in every term. He was then ordered to put in finished form his ten (or more) annual lectures for deposit in the University Library. During Newton's tenure of the professorship, he accordingly deposited manuscripts of his lectures on optics (1670–1672), arithmetic and algebra (1673–1683), most of book I of the *Principia* (1684–1685), and "The System of the World" (1687). There is, however, no record of what lectures, if any, he gave in 1686, or from 1688 until he removed to London early in 1696. In the 1670's Newton attempted unsuccessfully to publish his annotations on Kinckhuysen's algebra and his own treatise on fluxions. In 1672 he did succeed in publishing an improved or corrected edition of Varenius' *Geographia generalis*, apparently intended for the use of his students.

During the years in which Newton was writing the *Principia*, according to Humphrey Newton's recollection,[11] "he seldom left his chamber except at term time, when he read in the schools as being Lucasianus Professor, where so few went to hear him, and fewer that understood him, that ofttimes he did in a manner,

for want of hearers, read to the walls." When he lectured he "usually staid about half an hour; when he had no auditors, he commonly returned in a 4th part of that time or less." He occasionally received foreigners "with a great deal of freedom, candour, and respect." He "ate sparingly," and often "forgot to eat at all," rarely dining "in the hall, except on some public days," when he was apt to appear "with shoes down at heels, stockings untied, surplice on, and his head scarcely combed." He "seldom went to the chapel," but very often "went to St Mary's church, especially in the forenoon."[12]

From time to time Newton went to London, where he attended meetings of the Royal Society (of which he had been a fellow since 1672). He contributed £40 toward the building of the new college library (1676), as well as giving it various books. He corresponded, both directly and indirectly (often through Henry Oldenburg as intermediary), with scientists in England and on the Continent, including Boyle, Collins, Flamsteed, David Gregory, Halley, Hooke, Huygens, Leibniz, and Wallis. He was often busy with chemical experiments, both before and after writing the *Principia*, and in the mid-1670's he contemplated a publication on optics.[13] During the 1690's, Newton was further engaged in revising the *Principia* for a second edition; he then contemplated introducing into book III some selections from Lucretius and references to an ancient tradition of wisdom. A major research at this time was the effect of solar perturbations on the motions of the moon. He also worked on mathematical problems more or less continually throughout these years.

Among the students with whom Newton had friendly relations, the most significant for his life and career was Charles Montague, a fellow-commoner of Trinity and grandson of the Earl of Manchester; he "was one of the small band of students who assisted Newton in forming the Philosophical Society of Cambridge"[14] (the attempt to create this society was unsuccessful). Newton was also on familiar terms with Henry More, Edward Paget (whom he recommended for a post in mathematics at Christ's Hospital), Francis Aston, John Ellis (later master of Caius), and J. F. Vigani, first professor of chemistry at Cambridge, who is said to have eventually been banished from Newton's presence for having told him "a loose story about a nun." Newton was active in defending the rights of the university when the Catholic monarch James II tried to mandate the admission of the Benedictine monk Alban Francis. In 1689, he was elected by the university constituency to serve as Member of the Convention Parliament.

While in London as M.P., Newton renewed contact

with Montague and with the Royal Society, and met Huygens and others, including Locke, with whom he thereafter corresponded on theological and biblical questions. Richard Bentley sought Newton's advice and assistance in preparing the inaugural Boyle Lectures (or sermons), entitled "The Confutation of Atheism" and based in part on the Newtonian system of the world.

Newton also came to know two other scientists, each of whom wanted to prepare a second edition of the *Principia*. One was David Gregory, a professor at Edinburgh, whom Newton helped to obtain a chair at Oxford, and who recorded his conversations with Newton while Newton was revising the *Principia* in the 1690's. The other was a refugee from Switzerland, Nicolas Fatio de Duillier, advocate of a mechanical explanation of gravitation which was at one time viewed kindly by Newton. Fatio soon became perhaps the most intimate of any of Newton's friends. In the early autumn of 1693, Newton apparently suffered a severe attack of depression and made fantastic accusations against Locke and Pepys and was said to have lost his reason.[15]

In the post-*Principia* years of the 1690's, Newton apparently became bored with Cambridge and his scientific professorship. He hoped to get a post that would take him elsewhere. An attempt to make him master of the Charterhouse "did not appeal to him"[16] but eventually Montague (whose star had risen with the Whigs' return to power in Parliament) was successful in obtaining for Newton (in March 1696) the post of warden of the mint. Newton appointed William Whiston as his deputy in the professorship. He did not resign officially until 10 December 1701, shortly after his second election as M.P. for the university.[17]

Mathematics. Any summary of Newton's contributions to mathematics must take account not only of his fundamental work in the calculus and other aspects of analysis—including infinite series (and most notably the general binomial expansion)—but also his activity in algebra and number theory, classical and analytic geometry, finite differences, the classification of curves, methods of computation and approximation, and even probability.

For three centuries, many of Newton's writings on mathematics have lain buried, chiefly in the Portsmouth Collection of his manuscripts. The major parts are now being published and scholars will shortly be able to trace the evolution of Newton's mathematics in detail.[18] It will be possible here only to indicate highlights, while maintaining a distinction among four levels of dissemination of his work: (1) writings printed in his lifetime, (2) writings

circulated in manuscript, (3) writings hinted at or summarized in correspondence, and (4) writings that were published only much later. In his own day and afterward, Newton influenced mathematics "following his own wish," by "his creation of the fluxional calculus and the theory of infinite series," the "two strands of mathematical technique which he bound inseparably together in his 'analytick' method."[19] The following account therefore emphasizes these two topics.

Newton appears to have had no contact with higher mathematics until 1664 when—at the age of twenty-one—his dormant mathematical genius was awakened by Schooten's "Miscellanies" and his edition of Descartes's *Géométrie*, and by Wallis' *Arithmetica infinitorum* (and possibly others of his works). Schooten's edition introduced him to the mathematical contributions of Heuraet, de Witt, Hudde, De Beaune, and others; Newton also read in Viète, Oughtred, and Huygens. He had further compensated for his early neglect of Euclid by careful study of both the *Elements* and *Data* in Barrow's edition.

In recent years[20] scholars have come to recognize Descartes and Wallis as the two "great formative influences" on Newton in the two major areas of his mathematical achievement: the calculus, and analytic geometry and algebra. Newton's own copy of the *Géométrie* has lately turned up in the Trinity College Library; and his marginal comments are now seen to be something quite different from the general devaluation of Descartes's book previously supposed. Rather than the all-inclusive "Error. Error. Non est geom." reported by Conduitt and Brewster, Newton merely indicated an "Error" here and there, while the occasional marginal entry "non geom." was used to note such things as that the Cartesian classification of curves is not really geometry so much as it is algebra. Other of Newton's youthful annotations document what he learned from Wallis, chiefly the method of "indivisibles."[21]

In addition to studying the works cited, Newton encountered the concepts and methods of Fermat and James Gregory. Although Newton was apparently present when Barrow "read his Lectures about motion," and noted[22] that they "might put me upon taking these things into consideration," Barrow's influence on Newton's mathematical thought was probably not of such importance as is often supposed.

A major first step in Newton's creative mathematical life was his discovery of the general binomial theorem, or expansion of $(a + b)^n$, concerning which he wrote, "In the beginning of the year 1665 I found the Method of approximating series & the Rule for reducing any dignity [power] of any Binomial into such a series. . . ."[23] He further stated that:

> In the winter between the years 1664 & 1665 upon reading Dr Wallis's *Arithmetica Infinitorum* & trying to interpole his progressions for squaring the circle [that is, finding the area or evaluating $_0\int^1 (1 - x^2)^{\frac{1}{2}} dx$], I found out another infinite series for squaring the circle & then another infinite series for squaring the Hyperbola. . . .[24]

On 13 June 1676, Newton sent Oldenburg the "Epistola prior" for transmission to Leibniz. In this communication he wrote that fractions "are reduced to infinite series by division; and radical quantities by extraction of roots," the latter

. . . much shortened by this theorem,

$$\overline{P + PQ}^{\frac{m}{n}} = P^{\frac{m}{n}} + \frac{m}{n} AQ + \frac{m-n}{2n} BQ$$
$$+ \frac{m-2n}{3n} CQ + \frac{m-3n}{4n} DQ + \cdots \&c.$$

where $P + PQ$ signifies the quantity whose root or even any power, or the root of a power, is to be found; P signifies the first term of that quantity, Q the remaining terms divided by the first, and m/n the numerical index of the power of $P + PQ$, whether that power is integral or (so to speak) fractional, whether positive or negative.[25]

A sample given by Newton is the expansion

$$\sqrt{(c^2 + x^2)} \quad \text{or} \quad (c^2 + x^2)^{\frac{1}{2}} = c + \frac{x^2}{2c} - \frac{x^4}{8c^3}$$
$$+ \frac{x^6}{16c^5} - \frac{5x^8}{128c^7} + \frac{7x^{10}}{256c^9} + \text{etc.}$$

where

$$P = c^2, \quad Q = x^2/c^2, \quad m = 1, \quad n = 2, \quad \text{and}$$
$$A = P^{\frac{m}{n}} = (c^2)^{\frac{1}{2}} = c, \quad B = (m/n) AQ = x^2/2c,$$
$$C = \frac{m-n}{2n} BQ = -x^4/8c^3,$$

and so on.

Other examples include

$$(y^3 - a^2 y)^{-\frac{1}{3}}$$
$$(c^5 + c^4 x - x^5)^{\frac{1}{5}},$$
$$(d + e)^{-\frac{3}{5}}.$$

What is perhaps the most important general statement made by Newton in this letter is that in dealing with infinite series all operations are carried out "in the symbols just as they are commonly carried out in decimal numbers."

Wallis had obtained the quadratures of certain curves (that is, the areas under the curves), by a technique of indivisibles yielding $_0\int^1 (1 - x^2)^n dx$ for certain positive integral values of n (0, 1, 2, 3); in attempting to find the quadrature of a circle of unit radius, he had sought to evaluate the integral $_0\int^1 (1 - x^2)^{\frac{1}{2}} dx$ by interpolation. He showed that

$$\frac{4}{\pi} = \frac{1}{_0\int^1 (1 - x^2)^{\frac{1}{2}} dx} = \frac{3 \cdot 3 \cdot 5 \cdot 5 \cdot 7 \cdot 7 \cdots}{2 \cdot 4 \cdot 4 \cdot 6 \cdot 6 \cdot 8 \cdots}.$$

Newton read Wallis and was stimulated to go considerably further, freeing the upper bound and then deriving the infinite series expressing the area of a quadrant of a circle of radius x:

$$x - \frac{\frac{1}{2}x^3}{3} - \frac{\frac{1}{8}x^5}{5} - \frac{\frac{1}{16}x^7}{7} - \frac{\frac{5}{128}x^9}{9} - \cdots.$$

In so freeing the upper bound, he was led to recognize that the terms, identified by their powers of x, displayed the binomial coefficients. Thus, the factors $\frac{1}{2}, \frac{1}{8}, \frac{1}{16}, \frac{5}{128}, \ldots$ stand out plainly as $\binom{q}{1}, \binom{q}{2}, \binom{q}{3}, \binom{q}{4}, \ldots$, in the special case $q = \frac{1}{2}$ in the generalization

$$\int_0^x (1 - x^2)^q dx = X - \binom{q}{1} \cdot \frac{1}{3} X^3 + \binom{q}{2} \cdot \frac{1}{5} X^5$$
$$- \binom{q}{3} \cdot \frac{1}{7} X^7 + \frac{q}{5} \cdot \frac{1}{9} X^9 + \cdots,$$

where

$$\binom{q}{n} = \frac{q(q-1) \cdots (q - n + 1)}{n!}.$$

In this way, according to D. T. Whiteside, Newton could begin with the indefinite integral and, "by differentiation in a Wallisian manner," proceed to a straightforward derivation of the "series-expansion of the binomial $(1 - x^p)^q$ . . . virtually in its modern form," with "$| x^p |$ implicitly less than unity for convergence." As a check on the validity of this general series expansion, he "compared its particular expansions with the results of algebraic division and square-root extraction $(q = \frac{1}{2})$." This work, which was done in the winter of 1664–1665, was later presented in modified form at the beginning of Newton's *De analysi.*

He correctly summarized the stages of development of his method in the "Epistola posterior" of 24 October 1676, which—as before—he wrote for Oldenburg to transmit to Leibniz:

At the beginning of my mathematical studies, when I had met with the works of our celebrated Wallis, on considering the series, by the intercalation of which he himself exhibits the area of the circle and the hyperbola, the fact that in the series of curves whose common base or axis is x and the ordinates

$$(1-x^2)^{\frac{0}{2}},\ (1-x^2)^{\frac{1}{2}},\ (1-x^2)^{\frac{2}{2}},\ (1-x^2)^{\frac{3}{2}},\ (1-x^2)^{\frac{4}{2}},\ (1-x^2)^{\frac{5}{2}},$$

etc., if the areas of every other of them, namely

$$x,\ x-\tfrac{1}{3}x^3,\ x-\tfrac{2}{3}x^3+\tfrac{1}{5}x^5,\ x-\tfrac{3}{3}x^3+\tfrac{3}{5}x^5-\tfrac{1}{7}x^7,\quad \text{etc.}$$

could be interpolated, we would have the areas of the intermediate ones, of which the first $(1-x^2)^{\frac{1}{2}}$ is the circle. . . .[26]

The importance of changing Wallis' fixed upper boundary to a free variable x has been called "the crux of Newton's breakthrough," since the "various powers of x order the numerical coefficients and reveal for the first time the binomial character of the sequence."[27]

In about 1665, Newton found the power series (that is, actually determined the sequence of the coefficients) for

$$\sin^{-1}x = x + \tfrac{1}{6}x^3 + \tfrac{3}{40}x^5 + \cdots,$$

and—most important of all—the logarithmic series. He also squared the hyperbola $y(1+x) = 1$, by tabulating

$$\int_0^x (1+t)^r\, dt$$

for $r = 0, 1, 2, \cdots$ in powers of x and then interpolating

$$\int_0^x (1+t)^{-1}\, dt.[28]$$

From his table, he found the square of the hyperbola in the series

$$x - \frac{x^2}{2} + \frac{x^3}{3} - \frac{x^4}{4} + \frac{x^5}{5} - \frac{x^6}{6} + \frac{x^7}{7}$$
$$- \frac{x^8}{8} + \frac{x^9}{9} - \frac{x^{10}}{10} + \cdots,$$

which is the series for the natural logarithm of $1 + x$. Newton wrote that having "found the method of infinite series," in the winter of 1664–1665, "in summer 1665 being forced from Cambridge by the Plague I computed the area of the Hyperbola at Boothby . . . to two & fifty figures by the same method."[29]

At about the same time Newton devised "a completely general differentiation procedure founded on the concept of an indefinitely small and ultimately vanishing element o of a variable, say, x." He first used the notation of a "little zero" in September 1664, in notes based on Descartes's *Géométrie*, then extended it to various kinds of mathematical investigations. From the derivative of an algebraic function $f(x)$ conceived ("essentially") as

$$\operatorname*{Lim.}_{o\to\text{zero}} \frac{1}{0}\left[f(x+0) - f(x)\right]$$

he developed general rules of differentiation.

The next year, in Lincolnshire and separated from books, Newton developed a new theoretical basis for his techniques of the calculus. Whiteside has summarized this stage as follows:

[Newton rejected] as his foundation the concept of the indefinitely small, discrete increment in favor of that of the "fluxion" of a variable, a finite instantaneous speed defined with respect to an independent, conventional dimension of time and on the geometrical model of the line-segment: in modern language, the fluxion of the variable x with regard to independent time-variable t is the "speed" dx/dt.[30]

Prior to 1691, when he introduced the more familiar dot notation ($\dot{x}$ for dx/dt, $\dot{y}$ for dy/dt, $\dot{z}$ for dz/dt; then $\ddot{x}$ for d^2x/dt^2, $\ddot{y}$ for d^2y/dt^2, $\ddot{z}$ for d^2z/dt^2), Newton generally used the letters p, q, r for the first derivatives (Leibnizian dx/dt, dy/dt, dz/dt) of variable quantities x, y, z, with respect to some independent variable t. In this scheme, the "little zero" o was "an arbitrary increment of time,"[31] and op, oq, or were the corresponding "moments," or increments of the variables x, y, z (later these would, of course, become $o\dot{x}$, $o\dot{y}$, $o\dot{z}$).[32] Hence, in the limit ($o \to$ zero), in the modern Leibnizian terminology

$$q/p = dy/dx \qquad r/p = dz/dx,$$

where "we may think of the increment o as absorbed into the limit ratios." When, as was often done for the sake of simplicity, x itself was taken for the independent time variable, since $x = t$, then $p = \dot{x} = dx/dx = 1$, $q = dy/dx$, and $r = dz/dx$.

In May 1665, Newton invented a "true partial-derivative symbolism," and he "widely used the notation $\ddot{p}$ and $\dot{p}$ for the respective homogenized derivatives $x(dp/dx)$ and $x^2(d^2p/dx^2)$," in particular to express the total derivative of the function

$$\sum_i (p_i y^i) = 0$$

before "breaking through . . . to the first recorded use of a true partial-derivative symbolism." Armed with this tool, he constructed "the five first and second order partial derivatives of a two-valued function" and composed the fluxional tract of October 1666.[33]

Extracts were published by James Wilson in 1761, although the work as a whole remained in manuscript until recently.[34] Whiteside epitomizes Newton's work during this period as follows:

> In two short years (summer 1664–October 1666) Newton the mathematician was born, and in a sense the rest of his creative life was largely the working out, in calculus as in his mathematical thought in general, of the mass of burgeoning ideas which sprouted in his mind on the threshold of intellectual maturity. There followed two mathematically dull years.[35]

From 1664 to 1669, Newton advanced to "more general considerations," namely that the derivatives and integrals of functions might themselves be expressed as expansions in infinite series, specifically power series. But he had no general method for determining the "limits of convergence of individual series," nor had he found any "valid tests for such convergence."[36] Then, in mid-1669, he came upon Nicolaus Mercator's *Logarithmotechnica*, published in September 1668, of which "Mr Collins a few months after sent a copy . . . to Dr Barrow," as Newton later recorded.[37] Barrow, according to Newton, "replied that the Method of Series was invented & made general by me about two years before the publication of" the *Logarithmotechnica* and "at the same time," July 1669, Barrow sent back to Collins Newton's tract *De analysi.*

We may easily imagine Newton's concern for his priority on reading Mercator's book, for here he found in print "for all the world to read . . . his [own] reduction of $\log(1 + a)$ to an infinite series by continued division of $1 + a$ into 1 and successive integration of the quotient term by term."[38] Mercator had presented, among other numerical examples, that of $\log(1.1)$ calculated to forty-four decimal places, and he had no doubt calculated other logarithms over which Newton had spent untold hours. Newton might privately have been satisfied that Mercator's exposition was "cumbrous and inadequate" when compared to his own, but he must have been immeasurably anxious lest Mercator generalize a particular case (if indeed he had not already done so) and come upon Newton's discovery of "the extraction of roots in such series and indeed upon his cherished binomial expansion."[39] To make matters worse, Newton may have heard the depressing news (as Collins wrote to James Gregory, on 2 February 1668/1669) that "the Lord Brouncker asserts he can turne the square roote into an infinite Series."

To protect his priority, Newton hastily set to work to write up the results of his early researches into the properties of the binomial expansion and his methods for resolving "affected" equations, revising and amplifying his results in the course of composition. He submitted the tract, *De analysi per aequationes infinitas*, to Barrow, who sent it, as previously mentioned, to Collins.

Collins communicated Newton's results to James Gregory, Sluse, Bertet, Borelli, Vernon, and Strode, among others.[40] Newton was at that time unwilling to commit the tract to print; a year later, he incorporated its main parts into another manuscript, the *Methodus fluxionum et serierum infinitarum.* The original Latin text of the tract was not printed until long afterward.[41] Among those who saw the manuscript of *De analysi* was Leibniz, while on his second visit to London in October 1676; he read Collins' copy, and transcribed portions. Whiteside concurs with "the previously expressed opinions of the two eminent Leibniz scholars, Gerhardt and Hofmann," that Leibniz did not then "annex for his own purposes the fluxional method briefly exposed there," but "was interested only in Newton's series expansions."[42]

The *Methodus fluxionum* provides a better display of Newton's methods for the fluxional calculus in its generality than does the *De analysi.* In the preface to his English version of the *Methodus fluxionum*, John Colson wrote:

> The chief Principle, upon which the Method of Fluxions is here built, is this very simple one, taken from the Rational Mechanicks; which is, That Mathematical Quantity, particularly Extension, may be conceived as generated by continued local Motion; and that all Quantities whatever, at least by analogy and accommodation, may be conceived as generated after a like manner. Consequently there must be comparative Velocities of increase and decrease, during such generations, whose Relations are fixt and determinable, and may therefore (problematically) be proposed to be found.[43]

Among the problems solved are the differentiation of any algebraic function $f(x)$; the "method of quadratures," or the integration of such a function by the inverse process; and, more generally, the "inverse method of tangents," or the solution of a first-order differential equation.

As an example, the "moments" $\dot{x}o$ and $\dot{y}o$ are "the infinitely little accessions of the flowing quantities [variables] x and y": that is, their increase in "infinitely small portions of time." Hence, after "any infinitely small interval of time" (designated by o), x and y become $x + \dot{x}o$ and $y + \dot{y}o$. If one substitutes these for x and y in any given equation, for instance

$$x^3 - ax^2 + axy - y^3 = 0,$$

"there will arise"

$$x^3 + 3\dot{x}ox^2 + 3\dot{x}^2oox + \dot{x}^3o^3$$

$$- ax^2 - 2a\dot{x}ox - a\dot{x}^2oo$$

$$+ axy + a\dot{x}oy + a\dot{y}ox + a\dot{x}\dot{y}oo$$

$$- y^3 - 3\dot{y}oy^2 - 3\dot{y}^2ooy - \dot{y}^3o^3 = 0.$$

The terms $x^3 - ax^2 + axy - y^3$ (of which "by supposition" the sum $= 0$) may be cast out; the remaining terms are divided by o, to get

$$3\dot{x}x^2 + 3\dot{x}^2ox + \dot{x}^3oo - 2ax\dot{x} - a\dot{x}^2o + a\dot{x}y$$

$$+ a\dot{y}x + a\dot{x}\dot{y}o - 3\dot{y}y^2 - 3\dot{y}^2oy - \dot{y}^3oo = 0.$$

"But whereas o is suppos'd to be infinitely little, that it may represent the moments of quantities, consequently the terms that are multiplied by it will be nothing in respect of the rest."[44] These terms are therefore "rejected," and there remains

$$3x^2\dot{x} - 2a\dot{x}x + a\dot{x}y + a\dot{y}x - 3\dot{y}y^2 = 0.$$

It is then easy to group by $\dot{x}$ and $\dot{y}$ to get

$$\dot{x}(3x^2 - 2ax + ay) + \dot{y}(ax - 3y^2) = 0$$

or

$$\frac{\dot{y}}{\dot{x}} = - \frac{3x^2 - 2ax + ay}{ax - 3y^2},$$

which is the same result as finding dy/dx after differentiating

$$x^3 - ax^2 + axy - y^3 = 0.^{[45]}$$

Problem II then reverses the process, with

$$3\dot{x}x^2 - 2a\dot{x}x + a\dot{x}y + a\dot{y}x - 3\dot{y}y^2 = 0$$

being given. Newton then integrates term by term to get $x^3 - ax^2 + axy - y^3 = 0$, the validity of which he may then test by differentiation.

In an example given, o is an "infinitely small quantity" representing an increment in "time," whereas, in the earlier *De analysi*, o was an increment x (although again infinitely small). In the manuscript, as Whiteside points out, Newton canceled "the less precise equivalent 'indefinitè' (indefinitely)" in favor of "infinitely."[46] Certainly the most significant feature is Newton's general and detailed treatment of "the converse operations of differentiation and integration (in Newton's terminology, constructing the 'fluxions' of given 'fluent' quantities, and vice versa)," and "the novelty of Newton's . . . reformulation of the calculus of continuous increase."[47]

Other illustrations given by Newton of his method are determining maxima and minima and drawing tangents to curves at any point. In dealing with maxima and minima, as applied to the foregoing equation, Newton invoked the rule (Problem III):

When a quantity is the greatest or the least that it can be, at that moment it neither flows backwards nor forwards: for if it flows forwards or increases it was less, and will presently be greater than it is; and on the contrary if it flows backwards or decreases, then it was greater, and will presently be less than it is.

In an example Newton sought the "greatest value of x" in the equation

$$x^3 - ax^2 + axy - y^3 = 0.$$

Having already found "the relation of the fluxions of x and y," he set $\dot{x} = o$. Thus, $\dot{y}(ax - 3y^2) = 0$, or $3y^2 = ax$, gives the desired result since this relation may be used to "exterminate either x or y out of the primary equation; and by the resulting equation you may determine the other, and then both of them by $-3y^2 + ax = 0$." Newton showed how "that famous Rule of *Huddenius*" may be derived from his own general method, but he did not refer to Fermat's earlier method of maxima and minima. Newton also found the greatest value of y in the equation

$$x^3 - ay^2 + \frac{by^3}{a + y} - xx\sqrt{ay + xx} = 0$$

and then indicated that his method led to the solution of a number of specified maximum-minimum problems.

Newton's shift from a "loosely justified conceptual model of the 'velocity' of a 'moveing body' . . ." to the postulation of "a basic, uniformly 'fluent' variable of 'time' as a measure of the 'fluxions' (instantaneous 'speeds' of flow) of a set of dependent variables which continuously alter their magnitude" may have been due, in part, to Barrow.[48] This concept of a uniformly flowing time long remained a favorite of Newton's; it was to appear again in the *Principia*, in the scholium following the definitions, as "mathematical time" (which "of itself, and from its own nature, flows equably without relation to anything external"), and in lemma 2, book II (see below), in which he introduced quantities "variable and indetermined, and increasing or decreasing, as it were, by a continual motion or flux." He later explained his position in a draft review of the *Commercium epistolicum* (1712),

I consider time as flowing or increasing by continual flux & other quantities as increasing continually in time & from the fluxion of time I give the name of fluxions to the velocitys with which all other quantities increase. Also from the moments of time I give the name of moments to the parts of any other quantities generated in moments of time. I expose time by any quantity flowing uniformly & represent its fluxion by an unit, & the fluxions of other quantities I represent by any other fit symbols & the fluxions of their fluxions by other fit symbols & the fluxions of those fluxions by others, & their moments generated by those fluxions I represent by the symbols of the fluxions drawn into the letter o & its powers o^2, o^3, &c: vizt their first moments by their first fluxions drawn into the letter o, their second moments by their second fluxions into o^2, & so on. And when I am investigating a truth or the solution of a Probleme I use all sorts of approximations & neglect to write down the letter o, but when I am demonstrating a Proposition I always write down the letter o & proceed exactly by the rules of Geometry without admitting any approximations. And I found the method not upon summs & differences, but upon the solution of this probleme: *By knowing the Quantities generated in time to find their fluxions*. And this is done by finding not prima momenta but primas momentorum nascentium rationes.

In an addendum (published only in 1969) to the 1671 *Methodus fluxionum*,[49] Newton developed an alternative geometrical theory of "first and last" ratios of lines and curves. This was later partially subsumed into the 1687 edition of the *Principia*, section 1, book I, and in the introduction to the *Tractatus de quadratura curvarum* (published by Newton in 1704 as one of the two mathematical appendixes to the *Opticks*). Newton had intended to issue a version of his *De quadratura* with the *Principia* on several occasions, both before and after the 1713 second edition, because, as he once wrote, "by the help of this method of Quadratures I found the Demonstration of Kepler's Propositions that the Planets revolve in Ellipses describing . . . areas proportional to the times," and again, "By the inverse Method of fluxions I found in the year 1677 the demonstration of Kepler's Astronomical Proposition. . . ."[50]

Newton began *De quadratura* with the statement that he did not use infinitesimals, "in this Place," considering "mathematical Quantities . . . not as consisting of very small Parts; but as describ'd by a continued Motion."[51] Thus lines are generated "not by the Apposition of Parts, but by the continued Motion of Points," areas by the motion of lines, solids by the motion of surfaces, angles by the rotation of the sides, and "Portions of Time by a continual Flux." Recognizing that there are different rates of increase and decrease, he called the "Velocities of the Motions or Increments *Fluxions*, and the generated

Quantities *Fluents*," adding that "Fluxions are very nearly as the Augments of the Fluents generated in equal but very small Particles of Time, and, to speak accurately, they are in the *first Ratio* of the nascent Augments; but they may be expounded in any Lines which are proportional to them."

As an example, consider that (as in Fig. 1) areas *ABC*, *ABDG* are described by the uniform motion of

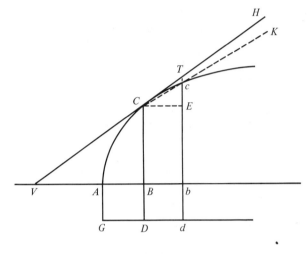

FIGURE 1

the ordinates *BC*, *BD* moving along the base in the direction *AB*. Suppose *BC* to advance to any new position *bc*, complete the parallelogram *BCEb*, draw the straight line *VTH* "touching the Curve in *C*, and meeting the two lines *bc* and *BA* [produced] in *T* and *V*." The "augments" generated will be: *Bb*, by *AB*; *Ec*, by *BC*; and *Cc*, by "the Curve Line *ACc*." Hence, "the Sides of the Triangle *CET* are in the *first Ratio* of these Augments considered as nascent." The "Fluxions of *AB*, *BC* and *AC*" are therefore "as the Sides *CE*, *ET* and *CT* of that Triangle *CET*" and "may be expounded" by those sides, or by the sides of the triangle *VBC*, which is similar to the triangle *CET*.

Contrariwise, one can "take the Fluxions in the *ultimate Ratio* of the evanescent Parts." Draw the straight line *Cc*; produce it to *K*. Now let *bc* return to its original position *BC*; when "*C* and *c* coalesce," the line *CK* will coincide with the tangent *CH*; then, "the evanescent Triangle *CEc* in its ultimate Form will become similar to the Triangle *CET*, and its evanescent Sides *CE*, *Ec*, and *Cc* will be *ultimately* among themselves as the sides *CE*, *ET* and *CT* of the other Triangle *CET*, are, and therefore the Fluxions of the Lines *AB*, *BC* and *AC* are in this same Ratio."

Newton concluded with an admonition that for the line *CK* not to be "distant from the Tangent *CH* by a small Distance," it is necessary that the points *C* and c not be separated "by any small Distance." If the points C and c do not "coalesce and exactly coincide," the lines CK and CH will not coincide, and "the ultimate Ratios in the Lines CE, Ec, and Cc" cannot be found. In short, "The very smallest Errors in mathematical Matters are not to be neglected."[52]

This same topic appears in the mathematical introduction (section 1, book I) to the *Principia*, in which Newton stated a set of lemmas on limits of geometrical ratios, making a distinction between the limit of a ratio and the ratio of limits (for example, as $x \rightarrow 0$, lim. $x^n/x \rightarrow 0$; but lim. x^n/lim. $x \rightarrow 0/0$, which is indeterminate).

The connection of fluxions with infinite series was first publicly stated in a scholium to proposition 11 of *De quadratura*, which Newton added for the 1704 printing, "We said formerly that there were first, second, third, fourth, *&c.* Fluxions of flowing Quantities. These Fluxions are as the Terms of an infinite converging series." As an example, he considered z^n to "be the flowing Quantity" and "by flowing" to become $(z + o)^n$; he then demonstrated that the successive terms of the expansion are the successive fluxions: "The first Term of this Series z^n will be that flowing Quantity; the second will be the first Increment or Difference, to which consider'd as nascent, its first Fluxion is proportional . . . and so on *in infinitum*." This clearly exemplifies the theorem formally stated by Brook Taylor in 1715; Newton himself explicitly derived it in an unpublished first version of *De quadratura* in 1691.[53] It should be noted that Newton here showed himself to be aware of the importance of convergence as a necessary condition for expansion in an infinite series.

In describing his method of quadrature by "first and last ratios," Newton said:

> Now to institute an Analysis after this manner in finite Quantities and investigate the *prime* or *ultimate* Ratios of these finite Quantities when in their nascent or evanescent State, is consonant to the Geometry of the Ancients: and I was willing [that is, desirous] to show that, in the Method of Fluxions, there is no necessity of introducing Figures infinitely small into Geometry.[54]

Newton's statement on the geometry of the ancients is typical of his lifelong philosophy. In mathematics and in mathematical physics, he believed that the results of analysis—the way in which things were discovered—should ideally be presented synthetically, in the form of a demonstration. Thus, in his review of the *Commercium epistolicum* (published anonymously), he wrote of the methods he had developed in *De quadratura* and other works as follows:

> By the help of the new *Analysis* Mr. *Newton* found out most of the Propositions in his *Principia Philosophiae*: but because the Ancients for making things certain admitted nothing into Geometry before it was demonstrated synthetically, he demonstrated the Propositions synthetically, that the Systeme of the Heavens might be founded upon good Geometry. And this makes it now difficult for unskilful Men to see the Analysis by which those Propositions were found out.[55]

As to analysis itself, David Gregory recorded that Newton once said "Algebra is the Analysis of the Bunglers in Mathematicks."[56] No doubt! Newton did, nevertheless, devote his main professorial lectures of 1673–1683 to algebra,[57] and these lectures were printed a number of times both during his lifetime and after.[58] This algebraical work includes, among other things, what H. W. Turnbull has described as a general method (given without proof) for discovering "the rational factors, if any, of a polynomial in one unknown and with integral coefficients"; he adds that the "most remarkable passage in the book" is Newton's rule for discovering the imaginary roots of such a polynomial.[59] (There is also developed a set of formulas for "the sums of the powers of the roots of a polynomial equation.")[60]

Newton's preference for geometric methods over purely analytical ones is further evident in his statement that "Equations are Expressions of Arithmetical Computation and properly have no place in Geometry." But such assertions must not be read out of context, as if they were pronouncements about algebra in general, since Newton was actually discussing various points of view or standards concerning what was proper to geometry. He included the positions of Pappus and Archimedes on whether to admit into geometry the conchoid for the problem of trisection and those of the "new generation of geometers" who "welcome" into geometry many curves, conics among them.[61]

Newton's concern was with the limits to be set in geometry, and in particular he took up the question of the legitimacy of the conic sections in solid geometry (that is, as solid constructions) as opposed to their illegitimacy in plane geometry (since they cannot be generated in a plane by a purely geometric construction). He wished to divorce synthetic geometric considerations from their "analytic" algebraic counterparts. Synthesis would make the ellipse the simplest of conic sections other than the circle; analysis would award this place to the parabola. "Simplicity in figures," he wrote, "is dependent on the simplicity of their genesis and conception, and it is not its equation but its description (whether geometrical or mechanical) by which a figure is generated and rendered easy to conceive."[62]

The "written record of [Newton's] first researches in the interlocking structures of Cartesian co-ordinate geometry and infinitesimal analysis"[63] shows him to have been establishing "the foundations of his mature work in mathematics" and reveals "for the first time the true magnitude of his genius."[64] And in fact Newton did contribute significantly to analytic geometry. In his 1671 *Methodis fluxionum*, he devoted "Prob. 4: To draw tangents to curves" to a study of the different ways in which tangents may be drawn "according to the various relationships of curves to straight lines," that is, according to the "modes" or coordinate systems in which the curve is specified.[65]

Newton proceeded "by considering the ratios of limit-increments of the co-ordinate variables (which are those of their fluxions)."[66] His "Mode 3" consists of using what are now known as standard bipolar coordinates, which Newton applied to Cartesian ovals as follows: Let x, y be the distances from a pair of fixed points (two "poles"); the equation $a \pm (e/d)x - y = 0$ for Descartes's "second-order ovals" will then yield the fluxional relation $\pm(e/d)\dot{x} - \dot{y} = 0$ (in dot notation) or $\pm em/d - n = 0$ (in the notation of the original manuscript, in which m, n are used for the fluxions $\dot{x}$, $\dot{y}$ of x, y). When $d = e$, "the curve turns out to be a conic." In "Mode 7," Newton introduced polar coordinates for the construction of spirals; "the equation of an Archimedean spiral" in these coordinates becomes $(a/b)x = y$, where y is the radius vector (now usually designated r or ρ) and x the angle (ϑ or ϕ).

Newton constructed equations for the transformation of coordinates (as, for example, from polar to Cartesian), and found formulas in both polar and rectangular coordinates for the curvature of a variety of curves, including conics and spirals. On the basis of these results Boyer has quite properly referred to Newton as "an originator of polar coordinates."[67]

Further geometrical results may be found in *Enumeratio linearum tertii ordinis*, first written in 1667 or 1668, and then redone and published, together with *De quadratura*, as an appendix to the *Opticks* (1704).[68] Newton devoted the bulk of the tract to classifying cubic curves into seventy-two "*Classes*, *Genders*, or *Orders*, according to the Number of the Dimensions of an Equation, expressing the relation between the *Ordinates* and the *Abscissae*; or which is much at one [that is, the same thing], according to the Number of Points in which they may be cut by a Right Line."

In a brief fifth section, Newton dealt with "The Generation of Curves by Shadows," or the theory of projections, by which he considered the shadows produced "by a luminous point" as projections "on an infinite plane." He showed that the "shadows" (or projections) of conic sections are themselves conic sections, while "those of curves of the second genus will always be curves of the second genus; those of the third genus will always be curves of the third genus; and so on *ad infinitum*." Furthermore, "in the same manner as the circle, projecting its shadow, generates all the conic sections, so the five divergent parabolae, by their shadows, generate all the other curves of the second genus." As C. R. M. Talbot observed, this presentation is "substantially the same as that which is discussed at greater length in the twenty-second lemma [book III, section 5] of the *Principia*, in which it is proposed to 'transmute' any rectilinear or curvilinear figure into another of the same analytical order by means of the method of projections."[69]

The work ends with a brief supplement on "The Organical Description of Curves," leading to the "Description of the Conick-Section by Five Given Points" and including the clear statement, "*The Use of Curves in Geometry is, that by their Intersections Problems may be solved*" (with an example of an equation of the ninth degree). Newton in this tract laid "the foundation for the study of Higher Plane Curves, bringing out the importance of asymptotes, nodes, cusps," according to Turnbull, while Boyer has asserted that it "is the earliest instance of a work devoted solely to graphs of higher plane curves in algebra," and has called attention to the systematic use of two axes and the lack of "hesitation about negative coordinates."[70]

Newton's major mathematical activity had come to a halt by 1696, when he left Cambridge for London. The *Principia*, composed in the 1680's, marked the last great exertion of his mathematical genius, although in the early 1690's he worked on porisms and began a "Liber geometriae," never completed, of which David Gregory gave a good description of the planned whole.[71] For the most part, Newton spent the rest of his mathematical life revising earlier works.

Newton's other chief mathematical activity during the London years lay in furthering his own position against Leibniz in the dispute over priority and originality in the invention of the calculus. But he did respond elegantly to a pair of challenge problems set by Johann [I] Bernoulli in June 1696. The first of these problems was "mechanico-geometrical," to find the curve of swiftest descent. Newton's answer was brief: the "brachistochrone" is a cycloid. The second problem was to find a curve with the following property, "that the two segments [of a right line drawn from a given point through the curve], being

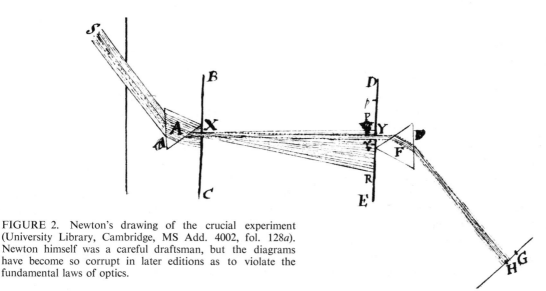

FIGURE 2. Newton's drawing of the crucial experiment (University Library, Cambridge, MS Add. 4002, fol. 128a). Newton himself was a careful draftsman, but the diagrams have become so corrupt in later editions as to violate the fundamental laws of optics.

raised to any given power, and taken together, may make everywhere the same sum."[72]

Newton's analytic solution of the curve of least descent is of particular interest as an early example of what became the calculus of variations. Newton had long been concerned with such problems, and in the *Principia* had included (without proof) his findings concerning the solid of least resistance. When David Gregory asked him how he had found such a solid, Newton sent him an analytic demonstration (using dotted fluxions), of which a version was published as an appendix to the second volume of Motte's English translation of the *Principia*.[73]

Optics. The study of Newton's work in optics has to date generally been limited to his published letters relating to light and color (in *Philosophical Transactions*, beginning in February 1672), his invention of a reflecting telescope and "sextant," and his published *Opticks* of 1704 and later editions (in Latin and English). There has never been an adequate edition or a full translation of the *Lectiones opticae*. Nor, indeed, have Newton's optical manuscripts as yet been thoroughly studied.[74]

Newton's optical work first came to the attention of the Royal Society when a telescope made by him was exhibited there. Newton was elected a fellow shortly thereafter, on 11 January 1672, and responded by offering the Society an account of the discovery that had led him to his invention. It was, he proudly alleged, "the oddest if not the most considerable detection yet made in the operations of nature": the analysis of dispersion and the composition of white light.

In the published account Newton related that in 1666 ("at which time I applied myself to the grinding of Optick glasses of other figures than *Spherical*") he procured a triangular glass prism, "to try therewith

the celebrated *Phaenomena of Colours*." Light from a tiny hole in a shutter passed through the prism; the multicolored image—to Newton's purported surprise—was of "an *oblong* form," whereas "according to the received laws of Refraction, I expected [it] should have been *circular*." To account for this unexpected appearance, Newton looked into a number of possibilities, among them that "the Rays, after their trajection through the Prisme did not move in curve lines," and was thereby led to the famous "experimentum crucis."[75] In this experiment Newton used two prisms: the first was employed to produce a spectrum on an opaque board (*BC*) into which a small hole had been drilled; a beam of light could thus pass through the hole to a second board (*DE*) with a similar aperture; in this way a narrow beam of light of a single color would be directed to a second prism, and the beam emerging from the second prism would project an image on another board (Fig. 2). Thus, all light reaching the final board had been twice subjected to prismatic dispersion. By rotating the first prism "to and fro slowly about its Axis," Newton allowed different portions of the dispersed light to reach the second prism.

Newton found that the second prism did not produce any further dispersion of the "homogeneal" light (that is, of light of about the same color); he therefore concluded that "Light it self is a *Heterogeneous mixture of differently refrangible Rays*"; and asserted an exact correspondence between color and "degree of Refrangibility" (the least refrangible rays being "disposed to exhibit a *Red* colour," while those of greatest refrangibility are a deep violet). Hence, colors "are not *Qualifications* of Light, derived from Refractions, or Reflections of natural Bodies," as commonly believed, but "*Original* and *connate properties*," differing in the different sorts of rays.[76]

The same experiment led Newton to two further conclusions, both of real consequence. First, he gave up any hope of "the perfection of Telescopes" based on combinations of lenses and turned to the principle of the reflector; second, he held it to be no longer a subject of dispute "whether Light be a Body." Observing, however, that it "is not so easie" to determine specifically "what Light is," he concluded, "I shall not mingle conjectures with certainties."[77]

Newton's letter was, as promised, read at the Royal Society on 6 February 1672. A week later Hooke delivered a report in which he criticized Newton for asserting a conclusion that did not seem to Hooke to follow necessarily from the experiments described, which—in any event—Hooke thought too few. Hooke had his own theory which, he claimed, could equally well explain Newton's experimental results.

In the controversy that followed with Hooke, Huygens, and others, Newton quickly discovered that he had not produced a convincing demonstration of the validity and significance of the conclusions he had drawn from his experiments. The objection was made that Newton had not explored the possibility that theories of color other than the one he had proposed might explain the phenomena. He was further criticized for having favored a corporeal hypothesis of light, and it was even said that his experimental results could not be reproduced.

In reply, Newton attacked the arguments about the "hypothesis" that he was said to have advanced about the nature of light, since he did not consider this issue to be fundamental to his interpretation of the "experimentum crucis." As he explained in reply to Pardies[78] he was not proposing "an hypothesis," but rather "properties of light" which could easily "be proved" and which, had he not held them to be true, he would "rather have . . . rejected as vain and empty speculation, than acknowledged even as an hypothesis." Hooke, however, persisted in the argument. Newton was led to state that he had deliberately declined all hypotheses so as "to speak of *Light* in *general* terms, considering it abstractly, as something or other propagated every way in straight lines from luminous bodies, without determining what that Thing is." But Newton's original communication did assert, "These things being so, it can be no longer disputed, whether there be colours in the dark, nor . . . perhaps, whether Light be a Body." In response to his critics, he emphasized his use of the word "perhaps" as evidence that he was not committed to one or another hypothesis on the nature of light itself.[79]

One consequence of the debate, which was carried on over a period of four years in the pages of the

Philosophical Transactions and at meetings of the Royal Society, was that Newton wrote out a lengthy "Hypothesis Explaining the Properties of Light Discoursed of in my Several Papers,"[80] in which he supposed that light "is something or other capable of exciting vibrations in the aether," assuming that "there is an aetherial medium much of the same constitution with air, but far rarer, subtler, and more strongly elastic." He suggested the possibility that "muscles are contracted and dilated to cause animal motion," by the action of an "aethereal animal spirit," then went on to offer ether vibration as an explanation of refraction and reflection, of transparency and opacity, of the production of colors, and of diffraction phenomena (including Newton's rings). Even "the gravitating attraction of the earth," he supposed, might "be caused by the continual condensation of some other such like aethereal spirit," which need not be "the main body of phlegmatic aether, but . . . something very thinly and subtilly diffused through it."[81]

The "Hypothesis" was one of two enclosures that Newton sent to Oldenburg, in his capacity of secretary of the Royal Society, together with a letter dated 7 December 1675. The other was a "Discourse of Observations," in which Newton set out "such observations as conduce to further discoveries for completing his theory of light and colours, especially as to the constitution of natural bodies, on which their colours or transparency depend." It also contained Newton's account of his discovery of the "rings" produced by light passing through a thin wedge or layer of air between two pieces of glass. He had based his experiments on earlier ones of a similar kind that had been recorded by Hooke in his *Micrographia* (observation 9). In particular Hooke had described the phenomena occurring when the "lamina," or space between the two glasses, was "*double concave*, that is, thinner in the middle then at the edge"; he had observed "various coloured rings or lines, with differing consecutions or orders of Colours."

When Newton's "Discourse" was read at the Royal Society on 20 January 1676, it contained a paragraph (proposition 3) in which Newton referred to Hooke and the *Micrographia*, "in which book he hath also largely discoursed of this . . . and delivered many other excellent things concerning the colours of thin plates, and other natural bodies, which I have not scrupled to make use of so far as they were for my purpose."[82] In recasting the "Discourse" as parts 1, 2, and 3 of book II of the *Opticks*, however, Newton omitted this statement. It may be assumed that he had carried these experiments so much further than Hooke, introducing careful measurements and quantitative analysis, that he believed them to be his own. Hooke, on the other hand, understandably thought that he deserved more credit for his own contributions —including hypothesis-based explanations—than Newton was willing to allow him.[83] Newton ended the resulting correspondence on a conciliatory note when he wrote in a letter of 5 February 1676, "What Des-Cartes did was a good step. You have added much in several ways, and especially in taking the colours of thin plates into philosophical consideration. If I have seen further it is by standing on the shoulders of Giants."[84]

The opening of Newton's original letter on optics suggests that he began his prism experiments in 1666, presumably in his rooms in Trinity, but was interrupted by the plague at Cambridge, returning to this topic only two years later. Thus the famous eighteen months supposedly spent in Lincolnshire would mark a hiatus in his optical researches, rather than being the period in which he made his major discoveries concerning light and color. As noted earlier, the many pages of optical material in Newton's manuscripts[85] and notebooks have not yet been sufficiently analyzed to provide a precise record of the development of his experiments, concepts, and theories.

The lectures on optics that Newton gave on the assumption of the Lucasian chair likewise remain only incompletely studied. These exist as two complete, but very different, treatises, each with carefully drawn figures. One was deposited in the University Library, as required by the statutes of his professorship, and was almost certainly written out by his roommate, John Wickins,[86] while the other is in Newton's own hand and remained in his possession.[87] These two versions differ notably in their textual content, and also in their division into "lectures," allegedly given on specified dates. A Latin and an English version, both based on the deposited manuscript although differing in textual detail and completeness, were published after Newton's death. The English version, called *Optical Lectures*, was published in 1728, a year before the Latin. The second part of Newton's Latin text was not translated, since, according to the preface, it was "imperfect" and "has since been published in the *Opticks* by Sir Isaac himself with great improvements." The preface further states that the final two sections of this part are composed "in a manner purely Geometrical," and as such they differ markedly from the *Opticks*. The opening lecture (or section 1) pays tribute to Barrow and mentions telescopes, before getting down to the hard business of Newton's discovery "that . . . Rays [of light] in respect to the Quantity of Refraction differ from one another." To show the reader that he had not set forth "Fables instead of Truth," Newton at once gave

"the Reasons and Experiments on which these things are founded." This account, unlike the later letter in the *Philosophical Transactions*, is not autobiographical; nor does it proceed by definitions, axioms, and propositions (proved "by Experiment"), as does the still later *Opticks*.[88]

R. S. Westfall has discussed the two versions of the later of the *Lectiones opticae*, which were first published in 1729;[89] he suggests that Newton eliminated from the *Lectiones* those "parts not immediately relevant to the central concern, the experimental demonstration of his theory of colors." Mathematical portions of the *Lectiones* have been analyzed by D. T. Whiteside, in Newton's *Mathematical Papers*, while J. A. Lohne and Zev Bechler have made major studies of Newton's manuscripts on optics. The formation of Newton's optical concepts and theories has been ably presented by A. I. Sabra; an edition of the *Opticks* is presently being prepared by Henry Guerlac.

Lohne finds great difficulty in repeating Newton's "experimentum crucis,"[90] but more important, he has traced the influence of Descartes, Hooke, and Boyle on Newton's work in optics.[91] He has further found that Newton used a prism in optical experiments much earlier than hitherto suspected—certainly before 1666, and probably before 1665—and has shown that very early in his optical research Newton was explaining his experiments by "the corpuscular hypothesis." In "Questiones philosophicae," Newton wrote: "Blue rays are reflected more than red rays, because they are slower. Each colour is caused by uniformly moving globuli. The uniform motion which gives the sensation of one colour is different from the motion which gives the sensation of any other colour."[92]

Accordingly, Lohne shows how difficult it is to accept the historical narrative proposed by Newton at the beginning of the letter read to the Royal Society on 8 February 1672 and published in the *Philosophical Transactions*. He asks why Newton should have been surprised to find the spectrum oblong, since his "note-books represent the sunbeam as a stream of slower and faster globules occasioning different refrangibility of the different colours?" Newton must, according to Lohne, have "found it opportune to let his theory of colours appear as a Baconian induction from experiments, although it primarily was deduced from speculations." Sabra, in his analysis of Newton's narrative, concludes that not even "the 'fortunate Newton' could have been fortunate enough to have achieved this result in such a smooth manner." Thus one of the most famous examples of the scientific method in operation now seems to have been devised as a sort of scenario by which Newton attempted to convey the impression of a logical train of discovery based on deductions from experiment. The historical record, however, shows that Newton's great leap forward was actually a consequence of implications drawn from profound scientific speculation and insight.[93]

In any event, Newton himself did not publish the *Lectiones opticae*, nor did he produce his planned annotated edition of at least some (and maybe all) of his letters on light and color published in the *Philosophical Transactions*.[94] He completed his English *Opticks*, however, and after repeated requests that he do so, allowed it to be printed in 1704, although he withheld his name, save on the title page of one known copy. It has often been alleged that Newton released the *Opticks* for publication only after Hooke —the last of the original objectors to his theory of light and colors—had died. David Gregory, however, recorded another reason for the publication of the *Opticks* in 1704: Newton, Gregory wrote, had been "provoked" by the appearance, in 1703, of George Cheyne's *Fluxionum methoda inversa* "to publish his [own tract on] Quadratures, and with it, his Light & Colours, &c."[95]

In the *Opticks*, Newton presented his main discoveries and theories concerning light and color in logical order, beginning with eight definitions and eight axioms.[96] Definition 1 of book I reads: "By the Rays of Light I understand its least Parts, and those as well Successive in the same Lines, as Contemporary in several Lines." Eight propositions follow, the first stating that "Lights which differ in Colour, differ also in Degrees of Refrangibility." In appended experiments Newton discussed the appearance of a paper colored half red and half blue when viewed through a prism and showed that a given lens produces red and blue images, respectively, at different distances. The second proposition incorporates a variety of prism experiments as proof that "The Light of the Sun consists of Rays differently refrangible."

The figure given with experiment 10 of this series illustrates "two Prisms tied together in the form of a Parallelopiped" (Fig. 3). Under specified conditions, sunlight entering a darkened room through a small hole *F* in the shutter would not be refracted by the parallelopiped and would emerge parallel to the incident beam *FM*, from which it would pass by refraction through a third prism *IKH*, which would by refraction "cast the usual Colours of the Prism upon the opposite Wall." Turning the parallelopiped about its axis, Newton found that the rays producing the several colors were successively "taken out of the transmitted Light" by "total Reflexion"; first "the

Rays which in the third Prism had suffered the greatest Refraction and painted [the wall] with violet and blew were . . . taken out of the transmitted Light, the rest remaining," then the rays producing green, yellow, orange, and red were "taken out" as the parallelopiped was rotated yet further. Newton thus experimentally confirmed the "experimentum crucis," showing that the light emerging from the two prisms "is compounded of Rays differently Refrangible, seeing [that] the more Refrangible Rays may be taken out while the less Refrangible remain." The arrangement of prisms is the basis of the important discovery reported in book II, part 1, observation 1.

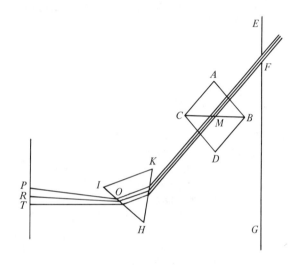

FIGURE 3

In proposition 6 Newton showed that, contrary to the opinions of previous writers, the sine law actually holds for each single color. The first part of book I ends with Newton's remarks on the impossibility of improving telescopes by the use of color-corrected lenses and his discussion of his consequent invention of the reflecting telescope (Fig. 4).

In the second part of book I, Newton dealt with colors produced by reflection and refraction (or transmission), and with the appearance of colored objects in relation to the color of the light illuminating them. He discussed colored pigments and their mixture and geometrically constructed a color wheel, drawing an analogy between the primary colors in a compound color and the "seven Musical Tones or Intervals of the eight Sounds, *Sol, la, fa, sol, la, mi, fa, sol.* . . ."[97]

Proposition 9, "Prob. IV. By the discovered Properties of Light to explain the Colours of the Rain-bow," is devoted to the theory of the rainbow. Descartes had developed a geometrical theory, but had

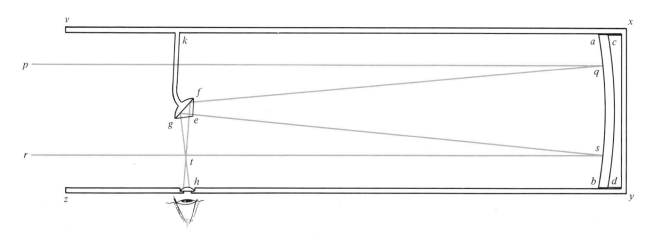

FIGURE 4. Newton's method "To shorten Telescopes": *efg* represents the prism, *abcd* the speculum, and *h* the lens.

used a single index of refraction (250:187) in his computation of the path of light through each raindrop.[98] Newton's discovery of the difference in refrangibility of the different colors composing white light, and their separation or dispersion as a consequence of refraction, on the other hand, permitted him to compute the radii of the bows for the separate colors. He used 108:81 as the index of refraction for red and 109:81 for violet, and further took into consideration that the light of the sun does not proceed from a single point. He determined the widths of the primary and secondary bows to be 2°15′ and 3°40′, respectively, and gave a formula for computing the radii of bows of any order *n* (and hence for orders of the rainbow greater than 2) for any given index of refraction.[99] Significant as Newton's achievement was, however, he gave only what can be considered a "first approximation to the solution of the problem," since a full explanation, particularly of the supernumerary or spurious bows, must require the general principle of interference and the "rigorous application of the wave theory."

Book II, which constitutes approximately one third of the *Opticks*, is devoted largely to what would later be called interference effects, growing out of the topics Newton first published in his 1675 letter to the Royal Society. Newton's discoveries in this regard would seem to have had their origin in the first experiment that he describes (book II, part 1, observation 1); he had, he reported, compressed "two Prisms hard together that their sides (which by chance were a very little convex) might somewhere touch one another" (as in the figure provided for experiment 10 of book I, part 1). He found "the place in which they touched" to be "absolutely transparent," as if there had been one "continued piece of Glass," even though there was

total reflection from the rest of the surface; but "it appeared like a black or dark spot, by reason that little or no sensible light was reflected from thence, as from other places." When "looked through," it seemed like "a hole in that Air which was formed into a thin Plate, by being compress'd between the Glasses." Newton also found that this transparent spot "would become much broader than otherwise" when he pressed the two prisms "very hard together."

Rotating the two prisms around their common axis (observation 2) produced "many slender Arcs of Colours" which, the prisms being rotated further, "were compleated into Circles or Rings." In observation 4 Newton wrote that

> To observe more nicely the order of the Colours . . . I took two Object-glasses, the one a Plano-convex for a fourteen Foot Telescope, and the other a large double Convex for one of about fifty Foot; and upon this, laying the other with its plane side downwards, I pressed them slowly together, to make the Colours successively emerge in the middle of the Circles, and then slowly lifted the upper Glass from the lower to make them successively vanish again in the same place.

It was thus evident that there was a direct correlation between particular colors of rings and the thickness of the layer of the entrapped air. In this way, as Mach observed, "Newton acquired a complete insight into the whole phenomenon, and at the same time the possibility of determining the thickness of the air gap from the known radius of curvature of the glass."[100]

Newton varied the experiment by using different lenses, and by wetting them, so that the gap or layer was composed of water rather than air. He also studied the rings that were produced by light of a single color, separated out of a prismatic spectrum; he found that in a darkened room the rings from a single color extended to the very edge of the lens. Furthermore, as he noted in observation 13, "the Circles which the red Light made" were "manifestly bigger than those which were made by the blue and violet"; he found it "very pleasant to see them gradually swell or contract accordingly as the Colour of the Light was changed." He concluded that the rings visible in white light represented a superimposition of the rings of the several colors, and that the alternation of light and dark rings for each color must indicate a succession of regions of reflection and transmission of light, produced by the thin layer of air between the two glasses. He set down the latter conclusion in observation 15: "And from thence the origin of these Rings is manifest; namely that the Air between the Glasses, according to its various thickness, is disposed in some places to reflect, and in others to transmit the Light

of any one Colour (as you may see represented . . .) and in the same place to reflect that of one Colour where it transmits that of another" (Fig. 5).

Book II, part 2, of the *Opticks* has a nomogram in which Newton summarized his measures and computations and demonstrated the agreement of his analysis of the ring phenomenon with his earlier conclusions drawn from his prism experiments— "that whiteness is a dissimilar mixture of all Colours, and that Light is a mixture of Rays endued with all those Colours." The experiments of book II further confirmed Newton's earlier findings "that every Ray have its proper and constant degree of Refrangibility connate with it, according to which its refraction is ever justly and regularly perform'd," from which he argued that "it follows, that the colorifick Dispositions of Rays are also connate with them, and immutable." The colors of the physical universe are thus derived "only from the various Mixtures or Separations of

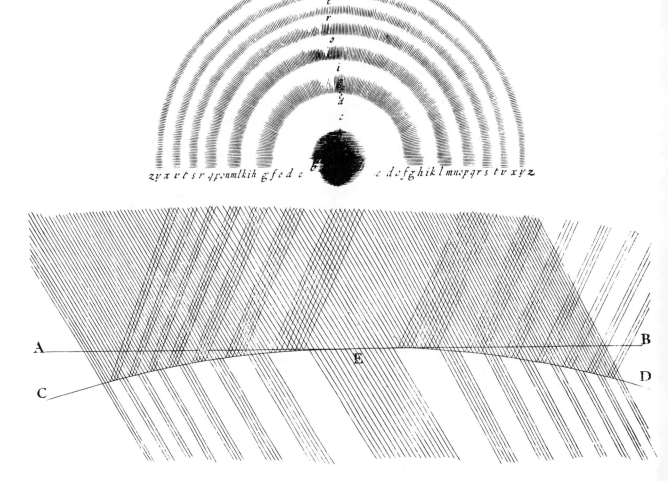

FIGURE 5. Two drawings from book II, part 1, plate 1 of the 1704 edition of the *Opticks*, illustrating Newton's studies of what are now called Newton's rings.

Rays, by virtue of their different Refrangibility or Reflexibility"; the study of color thus becomes "a Speculation as truly mathematical as any other part of Opticks."[101]

In part 3 of book II, Newton analyzed "the permanent Colours of natural Bodies, and the Analogy between them and the Colours of thin transparent Plates." He concluded that the smallest possible subdivisions of matter must be transparent, and their dimensions optically determinable. A table accompanying proposition 10 gives the refractive powers of a variety of substances "in respect of . . . Densities." Proposition 12 contains Newton's conception of "fits":

> Every Ray of Light in its passage through any refracting Surface is put into a certain transient Constitution or State, which in the progress of the Ray returns at equal Intervals, and disposes the Ray at every return to be easily transmitted through the next refracting Surface, and between the returns to be easily reflected by it.

The succeeding definition is more specific: "The returns of the disposition of any Ray to be reflected I will call its *Fits of easy Reflection*, and those of its disposition to be transmitted its *Fits of easy Transmission*, and the space it passes between every return and the next return, the *Interval of its Fits*."

The "fits" of easy reflection and of easy refraction could thus be described as a numerical sequence; if reflection occurs at distances 0, 2, 4, 6, 8, $\cdots$, from some central point, then refraction (or transmission) must occur at distances 1, 3, 5, 7, 9, $\cdots$. Newton did not attempt to explain this periodicity, stating that "I do not here enquire" into the question of "what kind of action or disposition this is." He declined to speculate "whether it consists in a circulating or a vibrating motion of the Ray, or of the Medium, or something else," contenting himself "with the bare Discovery, that the Rays of Light are by some cause or other alternately disposed to be reflected or refracted for many vicissitudes."

Newton thus integrated the periodicity of light into his theoretical work (it had played only a marginal part in Hooke's theory). His work was, moreover, based upon extraordinarily accurate measurements— so much so that when Thomas Young devised an explanation of Newton's rings based on the revived wave theory of light and the new principle of interference, he used Newton's own data to compute the wavelengths and wave numbers of the principal colors in the visible spectrum and attained results that are in close agreement with those generally accepted today.

In part 4 of book II, Newton addressed himself to "the Reflexions and Colours of thick transparent polish'd Plates." This book ends with an analysis of halos around the sun and moon and the computation of their size, based on the assumption that they are produced by clouds of water or by hail. This led him to the series of eleven observations that begin the third and final book, "concerning the Inflexions of the Rays of Light, and the Colours made thereby," in which Newton took up the class of optical phenomena previously studied by Grimaldi,[102] in which "fringes" are produced at the edges of the shadows of objects illuminated by light "let into a dark Room through a very small hole." Newton discussed such fringes surrounding the projected shadows of a hair, the edge of a knife, and a narrow slit.

Newton concluded the first edition of the *Opticks* (1704) with a set of sixteen queries, introduced "in order to a further search to be made by others." He had at one time hoped he might carry the investigations further, but was "interrupted," and wrote that he could not "now think of taking these things into farther Consideration." In the eighteenth century and after, these queries were considered the most important feature of the *Opticks*—particularly the later ones, which were added in two stages, in the Latin *Optice* of 1706 and in the second English edition of 1717–1718.

The original sixteen queries at once go beyond mere experiments on diffraction phenomena. In query 1, Newton suggested that bodies act on light at a distance to bend the rays; and in queries 2 and 3, he attempted to link differences in refrangibility with differences in "flexibility" and the bending that may produce color fringes. In query 4, he inquired into a single principle that, by "acting variously in various Circumstances," may produce reflection, refraction, and inflection, suggesting that the bending (in reflection and refraction) begins before the rays "arrive at the Bodies." Query 5 concerns the mutual interaction of bodies and light, the heat of bodies being said to consist of having "their parts [put] into a vibrating motion"; while in query 6 Newton proposed a reason why black bodies "conceive heat more easily from Light than those of other Colours." He then discussed the action between light and "sulphureous" bodies, the causes of heat in friction, percussion, putrefaction, and so forth, and defined fire (in query 9) and flame (in query 10), discussing various chemical operations. In query 11, he extended his speculations on heat and vapors to sun and stars. The last four queries (12 to 16) of the original set deal with vision, associated with "Vibrations" (excited by "the Rays of Light") which cause sight by "being propagated along the solid Fibres of the optick Nerves into the Brain." In query 13 specific wavelengths are associated with each of several colors. In query 15 Newton discussed binocular vision, along with other aspects of seeing, while in query 16 he took up the phenomenon of persistence of vision.

Newton has been much criticized for believing dispersion to be independent of the material of the prism and for positing a constant relation between deviation and dispersion in all refractive substances. He thus dismissed the possibility of correcting for chromatic aberration in lenses, and directed attention from refraction to reflecting telescopes.[103]

Newton is often considered to be the chief advocate of the corpuscular or emission theory of light. Lohne has shown that Newton originally did believe in a simple corpuscular theory, an aspect of Newton's science also forcibly brought out by Sabra. Challenged by Hooke, Newton proposed a hypothesis of ether waves associated with (or caused by) these corpuscles, one of the strongest arguments for waves probably being his own discovery of periodicity in "Newton's rings." Unlike either Hooke or Huygens, who is usually held to be the founder of the wave theory but who denied periodicity to waves of light, Newton postulated periodicity as a fundamental property of waves of (or associated with) light, at the same time that he suggested that a particular wavelength characterizes the light producing each color. Indeed, in the queries, he even suggested that vision might be the result of the propagation of waves in the optic nerves. But despite this dual theory, Newton always preferred the corpuscle concept, whereby he might easily explain both rectilinear propagation and polarization, or "sides." The corpuscle concept lent itself further to an analysis by forces (as in section 14 of book I of the *Principia*), thus establishing a universal analogy between the action of gross bodies (of the atoms or corpuscles composing such bodies), and of light. These latter topics are discussed below in connection with the later queries of the *Opticks*.

Dynamics, Astronomy, and the Birth of the "Principia." Newton recorded his early thoughts on motion in various student notebooks and documents.[104] While still an undergraduate, he would certainly have studied the Aristotelian (or neo-Aristotelian) theory of motion and he is known to have read Magirus' *Physiologiae peripateticae libri sex*; his notes include a "Cap:4. De Motu" (wherein "Motus" is said to be the Aristotelian ἐντελέχεια). Extracts from Magirus occur in a notebook begun by Newton in 1661;[105] it is a repository of jottings from his student years on a variety of physical and non-physical topics. In it Newton recorded, among other extracts, Kepler's third law, "that the mean distances of the primary Planets from the Sunne are in

sesquialter proportion to the periods of their revolutions in time."[106] This and other astronomical material, including a method of finding planetary positions by approximation, comes from Thomas Streete's *Astronomia Carolina*.

Here, too, Newton set down a note on Horrox' observations, and an expression of concern about the vacuum and the gravity of bodies; he recorded, from "Galilaeus," that "an iron ball" falls freely through "100 braces Florentine or cubits [or 49.01 ells, perhaps 66 yards] in 5″ of an hower." Notes of a later date—on matter, motion, gravity, and levity—give evidence of Newton's having read Charleton (on Gassendi), Digby (on Galileo), Descartes, and Henry More.

In addition to acquiring this miscellany of information, making tables of various kinds of observations, and supplementing his reading in Streete by Wing (and, probably, by Galileo's *Sidereus nuncius* and Gassendi's epitome of Copernican astronomy), Newton was developing his own revisions of the principles of motion. Here the major influence on his thought was Descartes (especially the *Principia philosophiae* and the Latin edition of the correspondence, both of which Newton cited in early writings), and Galileo (whose *Dialogue* he knew in the Salusbury version, and whose ideas he would have encountered in works by Henry More, by Charleton and Wallis, and in Digby's *Two Essays*).

An entry in Newton's Waste Book,[107] dated 20 January 1664, shows a quantitative approach to problems of inelastic collision. It was not long before Newton went beyond Descartes's law of conservation, correcting it by algebraically taking into account direction of motion rather than numerical products of size and speed of bodies. In a series of axioms he declared a principle of inertia (in "Axiomes" 1 and 2); he then asserted a relation between "force" and change of motion; and he gave a set of rules for elastic collision.[108] In "Axiome" 22, he had begun to approach the idea of centrifugal force by considering the pressure exerted by a sphere rolling around the inside surface of a cylinder. On the first page of the Waste Book, Newton had quantitated the centrifugal force by conceiving of a body moving along a square inscribed in a circle, and then adding up the shocks at each "reflection." As the number of sides were increased, the body in the limiting case would be "reflected by the sides of an equilateral circumscribed polygon of an infinite number of sides (i.e. by the circle it selfe)." Herivel has pointed out the near equivalence of such results to the early proof mentioned by Newton at the end of the scholium to proposition 4, book I, of the

Principia. Evidently Newton learned the law of centrifugal force almost a decade before Huygens, who published a similar result in 1673. One early passage of the Waste Book also contains an entry on Newton's theory of conical pendulums.[109]

According to Newton himself, the "notion of gravitation" came to his mind "as he sat in a contemplative mood," and "was occasioned by the fall of an apple."[110] He postulated that, since the moon is sixty times as far away from the center of the earth as the apple, by an inverse-square relation it would accordingly have an acceleration of free fall $1/(60)^2 = 1/3600$ that of the apple. This "moon test" proved the inverse-square law of force which Newton said he "deduced" from combining "Kepler's Rule of the periodical times of the Planets being in a sesquialterate proportion of their distances from the Centers of the Orbs"—that is, by Kepler's third law, that $R^3/T^2 = $ constant, combined with the law of central (centrifugal) force. Clearly if $F \propto V^2/R$ for a force F acting on a body moving with speed V in a circle of radius R (with period T), it follows simply and at once that

$$F \propto V^2/R = 4\pi^2 R^2/T^2 R = 4\pi^2/R^2 \times (R^3/T^2).$$

Since R^3/T^2 is a constant, $F \propto 1/R^2$.

An account by Whiston states that Newton took an incorrect value for the radius of the earth and so got a poor agreement between theory and observation, "which made Sir *Isaac* suspect that this Power was partly that of Gravity, and partly that of *Cartesius*'s Vortices," whereupon "he threw aside the Paper of his Calculation, and went to other Studies." Pemberton's narration is in agreement as to the poor value taken for the radius of the earth, but omits the reference to Cartesian vortices. Newton himself said (later) only that he made the two calculations and "found them [to] answer pretty nearly."[111] In other words, he calculated the falling of the moon and the falling of a terrestrial object, and found the two to be (only) approximately equal.

A whole tradition has grown up (originated by Adams and Glaisher, and most fully expounded by Cajori)[112] that Newton was put off not so much by taking a poor value for the radius of the earth as by his inability then to prove that a sphere made up of uniform concentric shells acts gravitationally on an external point mass as if all its mass were concentrated at its center (proposition 71, book I, book III, of the *Principia*). No firm evidence has ever been found that would support Cajori's conclusion that the lack of this theorem was responsible for the supposed twenty-year delay in Newton's announcement of his "discovery"

of the inverse-square law of gravitation. Nor is there evidence that Newton ever attempted to compute the attraction of a sphere until summer 1685, when he was actually writing the *Principia*.

An existing document does suggest that Newton may have made just such calculations as Whiston and Pemberton described, calculations in which Newton appears to have used a figure for the radius of the Earth that he found in Salusbury's version of Galileo's *Dialogue*, 3,500 Italian miles *(milliaria)*, in which one mile equals 5,000, rather than 5,280, feet.[113] Here, some time before 1669, Newton stated, to quote him in translation, "Finally, among the primary planets, since the cubes of their distances from the Sun are reciprocally as the squared numbers of their periods in a given time, their endeavours of recess from the Sun will be reciprocally as the squares of their distances from the Sun," and he then gave numerical examples from each of the six primary planets. A. R. Hall has shown that this manuscript is the paper referred to by Newton in his letter to Halley of 20 June 1686, defending his claim to priority of discovery of the inverse-square law against Hooke's claims. It would have been this paper, too, that David Gregory saw and described in 1694, when Newton let him glance over a manuscript earlier than "the year 1669."

This document, however important it may be in enabling us to define Newton's values for the size of the earth, does not contain an actual calculation of the moon test, nor does it refer anywhere to other than centrifugal "endeavours" from the sun. But it does show that when Newton wrote it he had not found firm and convincing grounds on which to assert what Whiteside has called a perfect "balance between (apparent) planetary centrifugal force and that of solar gravity."[114]

By the end of the 1660's Newton had studied the Cartesian principles of motion and had taken a critical stand with regard to them. His comments occur in an essay of the 1670's or late 1660's, beginning "De gravitatione et aequipondio fluidorum,"[115] in which he discussed extensively Descartes's *Principia* and also referred to a letter that formed part of the correspondence with Mersenne. Newton further set up a series of definitions and axioms, then ventured "to dispose of his [Descartes's] fictions." A large part of the essay deals with space and extension; for example, Newton criticized Descartes's view "that extension is not infinite but rather indefinite." In this essay Newton also defined force ("the causal principle of motion and rest"), conatus (or "endeavour"), impetus, inertia, and gravity. Then, in the traditional manner, he reckoned "the quantity of these powers" in "a double way: that is, according to intension or extension." He defined bodies, in the later medieval language of the intension and remission of forms, as "denser when their inertia is more intense, and rarer when it is more remiss."

In a final set of "Propositions on Non-Elastic Fluids" (in which there are two axioms and two propositions), axiom 2, "Bodies in contact press each other equally," suggests that the eventual third law of motion (*Principia*, axiom 3: "To every action is always opposed an equal and opposite reaction") may have arisen in application to fluids as well as to the impact of bodies. The latter topic occurs in another early manuscript, "The Lawes of Motion," written about 1666 and almost certainly antedating the essay on Descartes and his *Principia*.[116] Here Newton developed some rules for the impact of "bodyes which are absolutely hard," and then tempered them for application to "bodyes here amongst us," characterized by "a relenting softnesse & springynesse," which "makes their contact be for some time in more points than one."

Newton's attention to the problems of elastic and inelastic impact is manifest throughout his early writings on dynamics. In the *Principia* it is demonstrated by the emphasis he there gave the concept of force as an "impulse," and by a second law of motion (Lex II, in all editions of the *Principia*) in which he set forth the proportionality of such an impulse (acting instantaneously) to the change in momentum it produces.[117] In the scholium to the laws of motion Newton further discussed elastic and inelastic impact, referring to papers of the late 1660's by Wallis, Wren, and Huygens. He meanwhile developed his concept of a continuously acting force as the limit of a series of impulses occurring at briefer and briefer intervals *in infinitum*.[118]

Indeed, it was not until 1679, or some time between 1680 and 1684, following an exchange with Hooke, that Newton achieved his mature grasp of dynamical principles, recognizing the significance of Kepler's area law, which he had apparently just encountered. Only during the years 1684–1686, when, stimulated by Halley, he wrote out the various versions of the tract *De motu* and its successors and went on to compose the *Principia*, did Newton achieve full command of his insight into mathematical dynamics and celestial mechanics. At that time he clarified the distinction between mass and weight, and saw how these two quantities were related under a variety of circumstances.

Newton's exchange with Hooke occurred when the latter, newly appointed secretary of the Royal Society, wrote to Newton to suggest a private philosophical correspondence. In particular, Hooke asked Newton for his "objections against any hypothesis or opinion of mine," particularly "that of compounding the celestiall motions of the planetts of a direct motion by the tangent & an attractive motion towards the centrall body. . . ." Newton received the letter in November, some months after the death of his mother, and evidently did not wish to take up the problem. He introduced, instead, "a fancy of my own about discovering the Earth's diurnal motion, a spiral path that a freely falling body would follow as it supposedly fell to Earth, moved through the Earth's surface into the interior without material resistance, and eventually spiralled to (or very near to) the Earth's centre, after a few revolutions."[119]

Hooke responded that such a path would not be a spiral. He said that, according to "my theory of circular motion," in the absence of resistance, the body would not move in a spiral but in "a kind [of] Elleptueid," and its path would "resemble an Ellipse." This conclusion was based, said Hooke, on "my Theory of Circular Motions [being] compounded by a Direct [that is, tangential] motion and an attractive one to a Centre." Newton could not ignore this direct contradiction of his own expressed opinion. Accordingly, on 13 December 1679, he wrote Hooke that "I agree with you that . . . if its gravity be supposed uniform [the body would] not descend in a spiral to the very centre but circulate with an alternate descent & ascent." The cause was "its *vis centrifuga* & gravity alternately overballancing one another." This conception was very like Borelli's, and Newton imagined that "the body will not describe an Ellipsoeid," but a quite different figure. Newton here refused to accept the notion of an ellipse produced by gravitation decreasing as some power of the distance—although he had long before proved that for circular motion a combination of Kepler's third law and the rule for centrifugal force would yield a law of centrifugal force in the inverse square of the distance. There is no record of whether his reluctance was due to the poor agreement of the earlier moon test or to some other cause.

Fortunately for the advancement of science, Hooke kept pressing Newton. In a letter of 6 January 1680 he wrote ". . . But my supposition is that the Attraction always is in a duplicate proportion to the Distance from the Centre Reciprocall, and Consequently that the Velocity will be in a subduplicate proportion to the Attraction, and Consequently as Kepler Supposes Reciprocall to the Distance." We shall see below that this statement, often cited to support Hooke's claim to priority over Newton in the discovery of the inverse-square law, actually shows that Hooke was not a very good mathematician. As Newton proved, the force law here proposed contradicts the alleged velocity relation.

Hooke also claimed that this conception "doth very Intelligibly and truly make out all the Appearances of the Heavens," and that "the finding out the proprietys of a Curve made by two principles will be of great Concerne to Mankind, because the Invention of the Longitude by the Heavens is a necessary Consequence of it." After a few days, Hooke went on to challenge Newton directly:

> . . . It now remaines to know the proprietys of a curve Line (not circular nor concentricall) made by a centrall attractive power which makes the velocitys of Descent from the tangent Line or equall straight motion at all Distances in a Duplicate proportion to the Distances Reciprocally taken. I doubt not but that by your excellent method you will easily find out what that Curve must be, and its proprietys, and suggest a physicall Reason of this proportion.[120]

Newton did not reply, but he later recorded his next steps:

> I found now that whatsoever was the law of the forces which kept the Planets in their Orbs, the areas described by a Radius drawn from them to the Sun would be proportional to the times in which they were described. And . . . that their Orbs would be such Ellipses as Kepler had described [when] the forces which kept them in their Orbs about the Sun were as the squares of their . . . distances from the Sun reciprocally.[121]

Newton's account seems to be reliable; the proof he devised must have been that written out by him later in his "De motu corporum in gyrum."[122]

Newton's solution is based on his method of limits, and on the use of infinitesimals.[123] He considered the motion along an ellipse from one point to another during an indefinitely small interval of time, and evaluated the deflection from the tangent during that interval, assuming the deflection to be proportional to the inverse square of the distance from a focus. As one of the two points on the ellipse approaches the other, Newton found that the area law supplies the essential condition in the limit.[124] In short, Newton showed that if the area law holds, then the elliptical shape of an orbit implies that any force directed to a focus must vary inversely as the square of the distance.

But it was also incumbent upon Newton to show the significance of the area law itself; he therefore proved that the area law is a necessary and sufficient condition that the force on a moving body be directed to a center. Thus, for the first time, the true significance of Kepler's first two laws of planetary

motion was revealed: that the area condition was equivalent to the action of a central force, and that the occurrence of the ellipse under this condition demonstrates that the force is as the inverse square of the distance. Newton further showed the law of areas to be only another aspect of the law of inertia, since in linear inertial motion, in the absence of external forces, equal areas are swept out in equal times by a line from the moving body directed toward any point not on the line of motion.[125]

Newton was thus quite correct in comparing Hooke's claim and Kepler's, as he wrote to Halley on 20 June 1686:

> But grant I received it [the hypothesis of the inverse-square relation] afterwards [that is, after he had come upon it by himself, and independently of Hooke] from Mr Hook, yet have I as great a right to it as to the Ellipsis. For as Kepler knew the Orb to be not circular but oval & guest it to be Elliptical, so Mr Hook without knowing what I have found out since his letters to me, can know no more but that the proportion was duplicate *quam proximè* at great distances from the center, & only guest it to be so accurately & guest amiss in extending that proportion down to the very center, whereas Kepler guest right at the Ellipsis. And so Mr Hook found less of the Proportion than Kepler of the Ellipsis.[126]

What Newton "found out" after his correspondence with Hooke in 1679 was the proof that a homogeneous sphere (or a sphere composed of homogeneous spherical shells) will gravitate as if all its mass were concentrated at its geometric center.

Newton refrained from pointing out that Hooke's lack of mathematical ability prevented him (and many of those who have supported his claim) from seeing that the "approximate" law of speed ($v \propto 1/r$) is

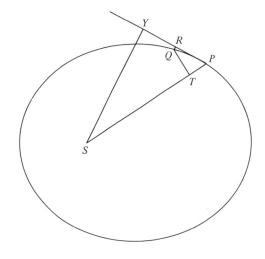

FIGURE 6

inconsistent with the true area law and does not accord with a force law of the form $f \propto 1/r^2$. Newton proved (Fig. 6: *Principia*, book I, proposition 16), that the speed at any point in an elliptical orbit is inversely proportional to the perpendicular dropped from the sun (focus) to the tangent drawn to the ellipse at that point, rather than being inversely proportional to the simple distance as Hooke and others had supposed; these two quantities being, of course, the same at the apsides. In the second edition of the *Principia* (1713) Newton shifted the corollaries to propositions 1 and 2, introducing a new set of corollaries to proposition 1, with the result that a prominent place was given to the true speed law.

Newton therefore deserves sole credit for recognizing the significance of the area law, a matter of some importance between 1679 and 1684. Following the exchange with Hooke in the earlier year, however, Newton did not at once go on to complete his work in celestial mechanics, although he did become interested in comets, corresponding with Flamsteed about their motion. He was converted from a belief in the straight-line motion of comets to a belief in parabolic paths, and thereafter attributed the motions of comets (in conic sections) to the action of the inverse-square law of the gravitation of the sun. He was particularly concerned with the comet of 1680, and in book III of the *Principia* devoted much space to its path.

In 1684, Halley visited Newton to ask about the path a planet would follow under the action of an inverse-square force: Wren, Hooke, and he had all been unsuccessful in satisfactorily resolving the matter, although Hooke had asserted (vainly) that he could do it. When Newton said to Hooke that he himself had "calculated" the result and that it was "an Ellipsis," Halley pressed him "for his calculation," but Newton could not find it among his papers and had to send it to Halley at a later date, in November. Halley then went back to Cambridge, where he saw "a curious treatise, *De Motu*." He obtained Newton's promise to send it "to the [Royal] Society to be entered upon their Register,"[127] and Newton, thus encouraged, wrote out a *De motu corporum*, of which the first section largely corresponds to book I of the *Principia* (together with an earnest of what was to become book II), while the second represents a popular account of what was later presented in book III.

Texts of both parts were deposited in the University Library, as if they were Newton's professorial lectures for 1684, 1685, and 1687; the second was published posthumously in both Latin and English, with the introduction of a new and misleading title of *De mundi*

systemate, or *The System of the World*. (This misnomer has ever since caused the second part of *De motu* to be confused with book III of the *Principia*, which is subtitled "De mundi systemate.")

Newton composed the *Principia* in a surprisingly short time.[128] The manuscript of book I was presented on 28 April 1686 to the Royal Society, which ordered it to be printed, although in the event Halley paid the costs and saw the work through the press. Halley's job was not an easy one; when Hooke demanded credit in print for his share in the inverse-square law, Newton demurred and even threatened to suppress book III. Halley fortunately dissuaded Newton from so mutilating his great treatise.

On 1 March 1687 Newton wrote to Halley that book II had been sent to him "by the Coach." The following 5 April Halley reported to Newton that he had received book III, "the last part of your divine Treatise." The printing was completed on 5 July 1687. The first edition included a short preface by Newton and an introductory ode to Newton by Halley—but book III ended abruptly, in the midst of a discussion of comets. Newton had originally drafted a "Conclusio" dealing with general aspects of natural philosophy and the theory of matter,[129] but he suppressed it. The famous conclusion, the "Scholium Generale," was first published some twenty-six years later, in 1713, in the second edition.

The development of Newton's views on comets may be traced through his correspondence with Flamsteed[130] and with Halley, and by comparing the first and second editions of the *Principia*. From Flamsteed he obtained information not only on comets, but also on the distances and periods of the satellites of Jupiter (which data appear in the beginning of book III of the *Principia* as a primary instance of Kepler's third law), and on the possible influence of Jupiter on the motion of Saturn. When Newton at first believed the great comet observed November 1680–March 1681 to be a pair of comets moving (as Kepler proposed) in straight lines, although in opposite directions, it was Flamsteed who convinced him that there was only one, observed coming and going, and that it must have turned about the sun.[131] Newton worked out a parabolic path for the comet of 1680 that was consistent with the observations of Flamsteed and others, the details of which occupy a great part of book III of the *Principia*. Such a parabolic path had been shown in book I to result from the inverse-square law under certain initial conditions, differing from those producing ellipses and hyperbolas.

In 1695, Halley postulated that the path of the comet of 1680 was an elongated ellipse—a path not very distinguishable from a parabola in the region of the sun, but significantly different in that the ellipse implies periodic returns of the comet—and worked out the details with Newton. In the second and third editions of the *Principia*, Newton gave tables for both the parabolic and elliptical orbits; he asserted unequivocally that Halley had found "a remarkable comet" appearing every seventy-five years or so, and added that Halley had "computed the motions of the comet in this elliptic orbit." Nevertheless, Newton himself remained primarily concerned with parabolic orbits. In the conclusion to the example following proposition 41 (on the comet of 1680), Newton said that "comets are a sort of planets revolved in very eccentric orbits about the sun." Even so, the proposition itself states (in all editions): "From three given observations to determine the orbit of a comet moving in a parabola."

Mathematics in the "Principia." The *Philosophiae naturalis principia mathematica* is, as its title suggests, an exposition of a natural philosophy conceived in terms of new principles based on Newton's own innovations in mathematics. It is too often described as a treatise in the style of Greek geometry, since on superficial examination it appears to have been written in a synthetic geometrical style.[132] But a close examination shows that this external Euclidean form masks the true and novel mathematical character of Newton's treatise, which was recognized even in his own day. (L'Hospital, for example—to Newton's delight—observed in the preface to his 1696 *Analyse des infiniment petits*, the first textbook on the infinitesimal calculus, that Newton's "excellent Livre intitulé *Philosophiae Naturalis principia Mathematica* . . . est presque tout de ce calcul.") Indeed, the most superficial reading of the *Principia* must show that, proposition by proposition and lemma by lemma, Newton usually proceeded by establishing geometrical conditions and their corresponding ratios and then at once introducing some carefully defined limiting process. This manner of proof or "invention," in marked distinction to the style of the classical Greek geometers, is based on a set of general principles of limits, or of prime and ultimate ratios, posited by Newton so as to deal with nascent or evanescent quantities or ratios of such quantities.

The doctrine of limits occurs in the *Principia* in a set of eleven lemmas that constitute section 1 of book I. These lemmas justify Newton in dealing with areas as limits of sums of inscribed or circumscribed rectangles (whose breadth → 0, or whose number → ∞), and in assuming the equality, in the limit, of arc, chord, and tangent (lemma 7), based on the proportionality of "homologous sides of similar figures, whether curvilinear or rectilinear" (lemma 5),

whose "areas are as the squares of the homologous sides." Newton's mathematical principles are founded on a concept of limit disclosed at the very beginning of lemma 1, "Quantities, and the ratios of quantities, which in any finite time converge continually to equality, and before the end of that time approach nearer to each other than by any given difference, become ultimately equal."

Newton further devoted the concluding scholium of section 1 to his concept of limit, and his method of taking limits, stating the guiding principle thus: "These lemmas are premised to avoid the tediousness of deducing involved demonstrations *ad absurdum*, according to the method of the ancient geometers." While he could have produced shorter ("more contracted") demonstrations by the "method of indivisibles," he judged the "hypothesis of indivisibles "to be "somewhat harsh" and not geometrical:

> I chose rather to reduce the demonstrations of the following propositions to the first and last sums and ratios of nascent and evanescent quantities, that is, to the limits of those sums and ratios; and so to premise, as short as I could, the demonstrations of those limits. For hereby the same thing is performed as by the method of indivisibles; and now those principles being demonstrated, we may use them with greater safety. Therefore if hereafter I should happen to consider quantities as made up of particles, or should use little curved lines for right ones, I would not be understood to mean indivisibles, but evanescent divisible quantities; not the sums and ratios of determinate parts, but always the limits of sums and ratios; and that the force of such demonstrations always depends on the method laid down in the foregoing Lemmas.

Newton was aware that his principles were open to criticism on the ground "that there is no ultimate proportion of evanescent quantities; because the proportion, before the quantities have vanished, is not the ultimate, and when they are vanished, is none"; and he anticipated any possible unfavorable reaction by insisting that "the ultimate ratio of evanescent quantities" is to be understood to mean "the ratio of the quantities not before they vanish, nor afterwards, but [that] with which they vanish." In a "like manner, the first ratio of nascent quantities is that with which they begin to be," and "the first or last sum is that with which they begin and cease to be (or to be augmented or diminished)." Comparing such ratios and sums to velocities (for "it may be alleged, that a body arriving at a certain place, and there stopping, has no ultimate velocity; because the velocity, before the body comes to the place, is not its ultimate velocity; when it has arrived, there is none"), he imagined the existence of "a limit which the velocity at the end of the motion may attain, but not exceed," which limit is "the ultimate velocity," or "that velocity with which the body arrives at its last place, and with which the motion ceases." By analogy, he argued, "there is the like limit in all quantities and proportions that begin and cease to be," and "such limits are certain and definite." Hence, "to determine the same is a problem strictly geometrical," and thus may be used legitimately "in determining and demonstrating any other thing that is also geometrical."

In short, Newton wished to make a clear distinction between the ratios of ultimate quantities and "those ultimate ratios with which quantities vanish," the latter being "limits towards which the ratios of quantities decreasing without limit do always converge. . . ." He pointed out that this distinction may be seen most clearly in the case in which two quantities become infinitely great; then their "ultimate ratio" may be "given, namely, the ratio of equality," even though "it does not from thence follow, that the ultimate or greatest quantities themselves, whose ratio that is, will be given."

Section 1 of book I is unambiguous in its statement that the treatise to follow is based on theorems of which the truth and demonstration almost always depend on the taking of limits. Of course, the occasional analytical intrusions in book I and the explicit use of the fluxional method in book II (notably in section 2) show the mathematical character of the book as a whole, as does the occasional but characteristic introduction of the methods of expansion in infinite series. A careful reading of almost any proof in book I will, moreover, demonstrate the truly limital or infinitesimal character of the work as a whole. But nowhere in the *Principia* (or in any other generally accessible manuscript) did Newton write any of the equations of dynamics as fluxions, as Maclaurin did later on. This continuous form is effectively that published by Varignon in the *Mémoires* of the Paris Academy in 1700; Newton's second law was written as a differential equation in J. Hermann's *Phoronomia* (1716).

The similarity of section 1, book I, to the introductory portion of the later *De quadratura* should not be taken to mean that in the *Principia* Newton developed his principles of natural philosophy on the basis of first and last ratios exclusively, since in the *Principia* Newton presented not one, but rather three modes of presentation of his fluxional or infinitesimal calculus. A second approach to the calculus occurs in section 2, book II, notably in lemma 2, in which Newton introduced the concept and method of moments. This represents the first printed statement (in the first edition of 1687) by Newton himself of

his new mathematics, apart from its application to physics (with which the opening discussion of limits in section 1, book I is concerned). In a scholium to lemma 2, Newton wrote that this lemma contains the "foundation" of "a general method," one

> . . . which extends itself, without any troublesome calculation, not only to the drawing of tangents to any curve lines . . ., but also to the resolving other abstruser kinds of problems about the crookedness, areas, lengths, centres of gravity of curves, &c.; nor is it . . . limited to equations which are free from surd quantities. This method I have interwoven with that other of working in equations, by reducing them to infinite series.

He added that the "last words relate to a treatise I composed on that subject in the year 1671,"[133] and that the paragraph quoted above came from a letter he had written to Collins on 10 December 1672, describing "a method of tangents."

The lemma itself reads: "The moment of any *genitum* is equal to the mome. 's of each of the generating sides multiplied by the indices of the powers of those sides, and by their coefficients continually."[134] It may be illustrated by Newton's first example: Let AB be a rectangle with sides A, B, diminished by $\frac{1}{2}a$, $\frac{1}{2}b$, respectively. The diminished area is $(A - \frac{1}{2}a)(B - \frac{1}{2}b) = AB - \frac{1}{2}aB - \frac{1}{2}bA + \frac{1}{4}ab$. Now, by a "continual flux," let the sides be augmented by $\frac{1}{2}a$, $\frac{1}{2}b$, respectively; the area ("rectangle") will then become $(A + \frac{1}{2}a)(B + \frac{1}{2}b) = AB + \frac{1}{2}aB + \frac{1}{2}bA + \frac{1}{4}ab$ (Fig. 7). Subtract one from the other, "and there will

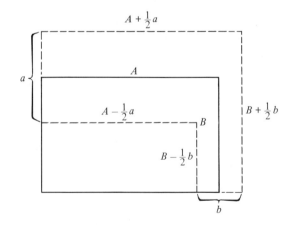

FIGURE 7

remain the excess $aB + bA$." Newton concluded, "Therefore with the whole increments a and b of the sides, the increment $aB + bA$ of the rectangle is generated." Here a and b are the moments of A and B, respectively, and Newton has shown that the moment of AB, corresponding to the moments a and b of A and B, respectively, is $aB + bA$. And, for the special case of $A = B$, the moment of A^2 is determined as $2aA$.

In order to extend the result from "area" to "content" or ("bulk"), from AB to ABC, Newton set $AB = G$ and then used the prior result for AB twice, once for AB, and again for GC, so as to get the moment of ABC to be $cAB + bCA + aBC$; whence, by setting $A = B = C$, the moment of A^3 is determined as $3aA^2$. And, in general, the moment of A^n is shown to be naA^{n-1} for n as a positive integer.

The result is readily extended to negative integral powers and even to all products $A^m B^n$, "whether the indices m and n of the powers be whole numbers or fractions, affirmative or negative." Whiteside has pointed out that by using the decrements $\frac{1}{2}a$, $\frac{1}{2}b$ and the increments $\frac{1}{2}a$, $\frac{1}{2}b$, rather than the increments a, b, "Newton . . . deluded himself into believing" he had "contrived an approach which avoids the comparatively messy appeal to the limit-value of $(A + a)/(B + b) - AB$ as the increments a, b vanish." The result is what is now seen as a "celebrated *non-sequitur*."[135]

In discussing lemma 2, Newton defined moments as the "momentary increments or decrements" of "variable and indetermined" quantities, which might be "products, quotients, roots, rectangles, squares, cubes, square and cubic sides, and the like." He called these "quantities" *genitae*, because he conceived them to be "generated or produced in arithmetic by the multiplication, division, or extraction of the root of any terms whatsoever; in geometry by the finding of contents and sides, or of the extremes and means of proportionals." So much is clear. But Newton warned his readers not "to look upon finite particles as such [moments]," for finite particles "are not moments, but the very quantities generated by the moments. We are to conceive them as the just nascent principles of finite magnitudes." And, in fact, it is not "the magnitude of the moments, but their first proportion [which is to be regarded] as nascent."

Boyer has called attention to the difficulty of conceiving "the limit of a ratio in determining the moment of AB."[136] The moment of AB is not really a product of two independent variables A and B, implying a problem in partial differentiation, but rather a product of two functions of the single independent variable time. Newton himself said, "It will be the same thing, if, instead of moments, we use either the velocities of the increments and decrements (which may also be called the motions, mutations, and fluxions of quantities), or any finite quantities proportional to those velocities."

Newton thus shifted the conceptual base of his procedure from infinitely small quantities or moments —which are not finite, and clearly not zero—to the "first proportion," or ratio of moments (rather than "the magnitude of the moments") "as nascent." This nascent ratio is generally not infinitesimal but finite, and Newton thus suggested that the ratio of finite quantities may be substituted for the ratio of infinitesimals, with the same result, using in fact the velocities of the increments or decrements instead of the moments, or "any finite quantities proportional to those velocities," which are also the "fluxions of the quantities." Boyer summarized this succinctly:

> Newton thus offered in the *Principia* three modes of interpretation of the new analysis: that in terms of infinitesimals (used in his *De analysi* . . .); that in terms of prime and ultimate ratios or limits (given particularly in *De quadratura*, and the view which he seems to have considered most rigorous); and that in terms of fluxions (given in his *Methodus fluxionum*, and one which appears to have appealed most strongly to his imagination).[137]

From the point of view of mathematics, proposition 10, book II, may particularly attract our attention. Here Newton boldly displayed his methods of using the terms of a converging series to solve problems and his method of second differences. Expansions are given with respect to "the indefinite quantity o," but there are no references to (nor uses of) moments, as in the preceding lemma 2, and, of course, there is no use made of dotted or "pricked" letters.

The proposition is of particular interest for at least two reasons. First, its proof and exposition (or exemplification) are highly analytic and not geometric (or synthetic), as are most proofs in the *Principia*. Second, an error in the first edition and in the original printed pages of the second edition was discovered by Johann [I] Bernoulli and called to Newton's attention by Nikolaus [I] Bernoulli, who visited England in September or October 1712. As a result, Newton had Cotes reprint a whole signature and an additional leaf of the already printed text of the second edition; these pages thus appear as cancels in every copy of this edition of the *Principia* that has been recorded. The corrected proposition, analyzed by Whiteside, illustrates "the power of Newton's infinitesimal techniques in the *Principia*," and may thus confute the opinion that "Newton did not (at least in principle, and in his own algorithm) know how 'to formulate and resolve problems through the integration of differential equations.' "[138]

From at least 1712 onward, Newton attempted to impose upon the *Principia* a mode of composition that could lend support to his position in the priority dispute with Leibniz: he wished to demonstrate that he had actually composed the *Principia* by analysis and had rewritten the work synthetically. He affirmed this claim, in and after 1713, in several manuscript versions of prefaces to planned new editions of the *Principia* (both with or without *De quadratura* as a supplement). It is indeed plausible to argue that much of the *Principia* was based upon an infinitesimal analysis, veiled by the traditional form of Greek synthetic geometry, but the question remains whether Newton drew upon working papers in which (in extreme form) he gave solutions in dotted fluxions to problems that he later presented geometrically. But, additionally, there is no evidence that Newton used an analytic method of ordinary fluxional form to discover the propositions he presented synthetically.

All evidence indicates that Newton had actually found the propositions in the *Principia* in essentially the way in which he there presented them to his readers. He did, however, use algebraic methods to determine the solid of least resistance. But in this case, he did not make the discovery by analysis and then recast it as an example of synthesis; he simply stated his result without proof.[139]

It has already been mentioned that Newton did make explicit use of the infinitesimal calculus in section 2, book II, of the *Principia*, and that in that work he often employed his favored method of infinite series.[140] But this claim is very different indeed from such a statement of Newton's as: ". . . At length in 1685 and part of 1686 by the aid of this method and the help of the book on Quadratures I wrote the first two books of the mathematical Principles of Philosophy. And therefore I have subjoined a Book on Quadratures to the Book of Principles."[141] This "method" refers to fluxions, or the method of differential calculus. But it is true, as mentioned earlier, that Newton stated in the *Principia* that certain theorems depended upon the "quadrature" (or integration) of "certain curves"; he did need, for this purpose, the inverse method of fluxions, or the integral calculus. And proposition 41 of book I is, moreover, an obvious exercise in the calculus.

Newton himself never did bring out an edition of the *Principia* together with a version of *De quadratura*.[142] In the review that he published of the *Commercium epistolicum*,[143] Newton did announce in print, although anonymously, that he had "found out most of the Propositions in his *Principia*" by using "the new *Analysis*," and had then reworked the material and had "demonstrated the Propositions synthetically." (This claim cannot, however, be substantiated by documentary evidence.)

Apart from questions of the priority of Newton's

method, the *Principia* contains some problems of notable mathematical interest. Sections 4 and 5 of book I deal with conic sections, and section 6 with Kepler's problem; Newton here introduced the method of solution by successive iteration. Lemma 5 of book III treats of a locus through a given number of points, an example of Newton's widely used method of interpolating a function. Proposition 71, book I, contains Newton's important solution to a major problem of integration, the attraction of a sphere, called by Turnbull "the crown of all." Newton's proof that two spheres will mutually attract each other as if the whole of their masses were concentrated at their respective centers is posited on the condition that, however the mass or density may vary within each sphere as a function of that radius, the density at any given radius is everywhere the same (or is constant throughout any concentric shell).

The "Principia": General Plan. Newton's masterwork was worked up and put into its final form in an incredibly short time. His strategy was to develop the subject of general dynamics from a mathematical point of view in book I, then to apply his most important results to solving astronomical and physical problems in book III. Book II, introduced at some point between Newton's first conception of the treatise and the completion of the printer's manuscript, is almost independent, and appears extraneous.

Book I opens with a series of definitions and axioms, followed by a set of mathematical principles and procedural rules for the use of limits; book III begins with general precepts concerning empirical science and a presentation of the phenomenological bases of celestial mechanics, based on observation.

It is clear to any careful reader that Newton was, in book I, developing mathematical principles of motion chiefly so that he might apply them to the physical conditions of experiment and observation in book III, on the system of the world. Newton maintained that even though he had, in book I, used such apparently physical concepts as "force" and "attraction," he did so in a purely mathematical sense. In fact, in book I (as in book II), he tended to follow his inspiration to whatever aspect of any topic might prove of mathematical interest, often going far beyond any possible physical application. Only in an occasional scholium in books I and II did he raise the question of whether the mathematical propositions might indeed be properly applied to the physical circumstances that the use of such words as "force" and "attraction" would seem to imply.

Newton's method of composition led to a certain amount of repetition, since many topics are discussed twice—in book I, with mathematical proofs, to illustrate the general principles of the motions of bodies, then again in book III, in application to the motions of planets and their satellites or of comets. While this mode of presentation makes the *Principia* more difficult for the reader, it does have the decided advantage of separating the Newtonian principles as they apply to the physical universe from the details of the mathematics from which they derive.

As an example of this separation, proposition 1 of book III states that the satellites of Jupiter are "continually drawn off from rectilinear motions, and are retained in their proper orbits" by forces that "tend to Jupiter's centre" and that these forces vary inversely as the square of their distances from that center. The proof given in this proposition is short and direct; the centripetal force itself follows from "Phen. I [of book III], and Prop. II or III, Book I." The phenomenon cited is a statement, based upon "astronomical observations," that a radius drawn from the center of Jupiter to any satellite sweeps out areas "proportional to the times of descriptions"; propositions 2 and 3 of book I prove by mathematics that under these circumstances the force about which such areas are described must be centripetal and proportional to the times. The inverse-square property of this force is derived from the second part of the phenomenon, which states that the distances from Jupiter's center are as the $\frac{3}{2}$th power of their periods of revolution, and from corollary 6 to proposition 4 of book I, in which it is proved that centripetal force in uniform circular motion must be as the inverse square of the distance from the center.

Newton's practice of introducing a particular instance repeatedly, with what may seem to be only minor variations, may render the *Principia* difficult for the modern reader. But the main hurdle for any would-be student of the treatise lies elsewhere, in the essential mathematical difficulty of the main subject matter, celestial mechanics, however presented. A further obstacle is that Newton's mathematical vocabulary became archaic soon after the *Principia* was published, as dynamics in general and celestial mechanics in particular came to be written in the language of differentials and integrals still used today. The reader is thus required almost to translate for himself Newton's geometrical-limit mode of proof and statement into the characters of the analytic algorithms of the calculus. Even so, dynamics was taught directly from the *Principia* at Cambridge until well into the twentieth century.

In his "Mathematical Principles" Whiteside describes the *Principia* as "slipshod, its level of verbal fluency none too high, its arguments unnecessarily diffuse and repetitive, and its content on occasion markedly irrelevant to its professed theme: the theory of bodies moving under impressed forces." This view is somewhat extreme. Nevertheless, the work might have been easier to read today had Newton chosen to rely to a greater extent on general algorithms.

The *Principia* is often described as if it were a "synthesis," notably of Kepler's three laws of planetary motion and Galileo's laws of falling bodies and projectile motion; but in fact it denies the validity of both these sets of basic laws unless they be modified. For instance, Newton showed for the first time the dynamical significance of Kepler's so-called laws of planetary motion; but in so doing he proved that in the form originally stated by Kepler they apply exactly only to the highly artificial condition of a point mass moving about a mathematical center of force, unaffected by any other stationary or moving masses. In the real universe, these laws or planetary "hypotheses" are true only to the limits of ordinary observation, which may very well have been the reason that Newton called them "Hypotheses" in the first edition. Later, in the second and third editions, he referred to these relations as "Phaenomena," by which it may be assumed that he now meant that they were not simply true as stated (that is, not strictly deducible from the definitions and axioms), but were rather valid only to the limit of (or within the limits of) observation, or were phenomenologically true. In other words, these statements were to be regarded as not necessarily true, but only contingently (phenomenologically) so.

In the *Principia*, Newton proved that Kepler's planetary hypotheses must be modified by at least two factors: (1) the mutual attraction of each of any pair of bodies, and (2) the perturbation of a moving body by any and all neighboring bodies. He also showed that the rate of free fall of bodies is not constant, as Galileo had supposed, but varies with distance from the center of the earth and with latitude along the surface of the earth.[144] In a scholium at the end of section 2, book I, Newton further pointed out that it is only in a limiting case, not really achieved on earth, that projectiles (even *in vacuo*) move in Galilean parabolic trajectories, as Galileo himself knew full well. Thus, as Karl Popper has pointed out, although "Newton's dynamics achieved a unification of Galileo's terrestrial and Kepler's celestial physics," it appears that "from a logical point of view, Newton's theory, strictly speaking, contradicts both Galileo's and Kepler's."[145]

The "Principia": Definitions and Axioms. The *Principia* opens with two preliminary presentations: the "Definitions" and the "Axioms, or Laws of

Motion." The first two entities defined are "quantity of matter," or "mass," and "quantity of motion." The former is said to be the measure of matter proportional to bulk and density conjunctively. "Mass" is, in addition, given as being generally known by its weight, to which it is proportional at any given place, as shown by Newton's experiments with pendulums, of which the results are more exact than Galileo's for freely falling bodies. Newton's "quantity of motion" is the entity now known as momentum; it is said to be measured by the velocity and mass of a body, conjunctively.

Definition 3 introduces *vis insita* (probably best translated as "inherent force"), a concept of which the actual definition and explanation are both so difficult to understand that much scholarly debate has been expended on them.[146] Newton wrote that the *vis insita* may be known by "a most significant name, *vis inertiae*." But this "force" is not like the "impressed forces" of definition 4, which change the state of rest or uniform rectilinear motion of a body; the *vis inertiae* merely maintains any new state acquired by a body, and it may cause a body to "resist" any change in state.[147]

Newton then defined "centripetal force" (*vis centripeta*), a concept he had invented and named to complement the *vis centrifuga* of Christiaan Huygens.[148] In definitions 6 through 8, Newton gave three "measures" of centripetal force, of which the most important for the purposes of the *Principia* is that one "proportional to the velocity which it generates in a given time" (for point masses, unit masses, or for comparing equal masses). There follows the famous scholium on space and time, in which Newton opted for concepts of absolute space and absolute time, although recognizing that both are usually reckoned by "sensible measures"; time, especially, is usually "relative, apparent, and common." Newton's belief in absolute space led him to hold that absolute motion is sensible or detectable, notably in rotation, although contemporaries as different in their outlooks as Huygens and Berkeley demurred from this view.

The "Axioms" or "Laws of Motion" are three in number: the law of inertia, a form of what is today known as the second law, and finally the law that "To every action there is always opposed an equal and opposite reaction." There is much puzzlement over the second law, which Newton stated as a proportionality between "change in motion" (in momentum) and "the motive force impressed" (a change "made in the direction . . ., in which that force is impressed"); he did not specify "per unit time" or "in some given time." The second law thus seems clearly to be stated for

an impulse, but throughout the *Principia* (and, in a special case, in the antecedent definition 8), Newton used the law for continuous forces, including gravitation, taking account of time. For Newton, in fact, the concepts of impulse and continuous force were infinitesimally equivalent, and represented conditions of action "altogether and at once" or "by degrees and successively."[149] There are thus two conditions of "force" in the second law; accordingly, this Newtonian law may be written in the two forms $f \propto d(mv)$ and $f \propto d(mv)/dt$, in which both concepts of force are taken account of by means of two different constants of proportionality. The two forms of the law can be considered equivalent through Newton's concept of a uniformly flowing time, which makes dt a kind of secondary constant, which can arbitrarily be absorbed in the constant of proportionality.

There may be some doubt as to whether or not Newton himself was unclear in his own mind about these matters. His use of such expressions as "vis impressa" shows an abiding influence of older physics, while his continued reference to a "vis" or a "force" needed to maintain bodies in a state of motion raises the question of whether such usage is one of a number of possibly misleading "artifacts left behind in the historical development of his [Newton's] dynamics."[150] It must be remembered, of course, that throughout the seventeenth and much of the eighteenth century the word "force" could be used in a number of ways. Most notably, it served to indicate the concept now called "momentum," although it could also even mean energy. In Newton's time there were no categories of strict formalistic logic that required a unitary one-to-one correspondence between names and concepts, and neither Newton nor his contemporaries (or, for that matter, his successors) were always precise in making such distinctions.

The careful reader of books I–III should not be confused by such language, however, nor by the preliminary intrusion of such concepts. Even the idea of force as a measure of motion or of change of motion (or of change *per se*, or rate of change) is not troublesome in practice, once Newton's own formulation is accepted and the infinitesimal level of his discourse (which is not always explicitly stated) understood. In short, Newton's dynamical and mathematical elaboration of the three books of the *Principia* is free of the errors and ambiguities implicit in his less successful attempt to give a logically simple and coherent set of definitions and axioms for dynamics. (It is even possible that the definitions and axioms may represent an independent later exercise, since there are, for example, varying sets of definitions and axioms for the same system of dynamics.) One of the most important consequences of Newton's analysis is that it must be one and the same law of force that operates in the centrally directed acceleration of the planetary bodies (toward the sun) and of satellites (toward planets), and that controls the linear downward acceleration of freely falling bodies. This force of universal gravitation is also shown to be the cause of the tides, through the action of the sun and the moon on the seas.

Book I of the "Principia." Book I of the *Principia* contains the first of the two parts of *De motu corporum.* It is a mathematical treatment of motion under the action of impressed forces in free spaces—that is, spaces devoid of resistance. (Although Newton discussed elastic and inelastic impact in the scholium to the laws, he did not reintroduce this topic in book I.) For the most part, the subject of Newton's inquiries is the motion of unit or point masses, usually having some initial inertial motion and being acted upon by a centripetal force. Newton thus tended to use the change in velocity produced in a given time (the "accelerative measure") of such forces, rather than the change in momentum produced in a given time (their "motive" measure).[151] He generally compared the effects of different forces or conditions of force on one and the same body, rather than on different bodies, preferring to consider a mass point or unit mass to computing actual magnitudes. Eventually, however, when the properties and actions of force had been displayed by an investigation of their "accelerative" and "motive" measures, Newton was able to approach the problem of their "absolute" measure. Later in the book he considered the attraction of spherical shells and spheres and of nonsymmetrical bodies.

Sections 2 and 3 are devoted to aspects of motion according to Kepler's laws. In proposition 1 Newton proceeded by four stages. He first showed that in a purely uniform linear (or purely inertial) motion, a radius vector drawn from the moving body to any point not in the line of motion sweeps out equal areas in equal times. The reason for this is clearly shown in

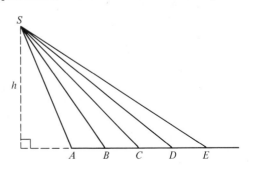

FIGURE 8

Figure 8, in which in equal times the body will move through the equal distances *AB, BC, CD, DE,* · · ·. If a radius vector is drawn from a point *PS*, then triangles *ABS, BCS, CDS, DES,* · · · have equal bases and a common altitude *h*, and their areas are equal. In the second stage, Newton assumed the moving object to receive an impulsive force when it reaches point *B*. A component of motion toward *S* is thereby added to its motion toward *C*; its actual path is thus along the diagonal *Bc* of a parallelogram (Figure 9).

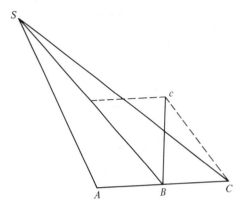

FIGURE 9

Newton then showed by simple geometry that the area of the triangle *SBc* is the same as the area of the triangle *SBC*, so that area is still conserved. He repeated the procedure in the third stage, with the body receiving a new impetus toward *S* at point *C*, and so on. In this way, the path is converted from a straight line into a series of joined line segments, traversed in equal intervals of time, which determine triangles of equal areas, with *S* as a common vertex.

In Newton's final development of the problem, the number of triangles is increased "and their breadth diminished *in infinitum*"; in the limit the "ultimate perimeter" will be a curve, the centripetal force "will act continually," and "any described areas" will be proportional to the times. Newton thus showed that inertial motion of and by itself implies an area-conservation law, and that if a centripetal force is directed to "an immovable centre" when a body has such inertial motion initially, area is still conserved as determined by a radius vector drawn from the moving body to the immovable center of force. (A critical examination of Newton's proof reveals the use of second-order infinitesimals.)[152] The most significant aspect of this proposition (and its converse, proposition 2) may be its demonstration of the hitherto wholly unsuspected logical connection, in the case of planetary motion, between Descartes's law of inertia and Kepler's law of areas (generalized to hold for an arbitrary central orbit).

Combining proposition 1 and proposition 2, Newton showed the physical significance of the law of areas as a necessary and sufficient condition for a central force (supposing that such forces exist; the "reality" of accelerative and motive forces of attraction is discussed in book III). In proposition 3, Newton dealt with the case of a body moving around a moving, rather than a stationary, center. Proposition 4 is concerned with uniform circular motion, in which the forces (F, f) are shown not only to be directed to the centers of the circles, but also to be to each other "as the squares of the arcs $[S, s]$ described in equal times divided respectively by the radii $[R, r]$ of the circles" $(F : f = S/R^2 : s/r^2)$. A series of corollaries demonstrate that $F : f = V^2/R : v^2r = R/T^2 : r/t^2$, where V, v are the tangential velocities, and so on; and that, universally, T being the period of revolution, if $T \propto R^n$, $V \propto 1/R^{n-1}$, then $F \propto 1/R^{2n-1}$, and conversely. A special case of the last condition (corollary 6) is $T \propto R^{3/2}$, yielding $F \propto 1/R^2$, a condition (according to a scholium) obtaining "in the celestial bodies," as Wren, Hooke, and Halley "have severally observed." Newton further referred to Huygens' derivation, in *De horologio oscillatorio*, of the magnitude of "the centrifugal force of revolving bodies" and introduced his own independent method for determining the centrifugal force in uniform circular motion. In proposition 6 he went on to a general concept of instantaneous measure of a force, for a body revolving in any curve about a fixed center of force. He then applied this measure, developed as a limit in several forms, in a number of major examples, among them proposition 11.

The last propositions of section 2 were altered in successive editions. In them Newton discussed the laws of force related to motion in a given circle and equiangular (logarithmic) spiral. In proposition 10 Newton took up elliptical motion in which the force tends toward the center of the ellipse. A necessary and sufficient cause of this motion is that "the force is as the distance." Hence if the center is "removed to an infinite distance," the ellipse "degenerates into a parabola," and the force will be constant, yielding "Galileo's theorem" concerning projectile motion.

Section 3 of book I opens with proposition 11, "If a body revolves in an ellipse; it is required to find the law of the centripetal force tending to the focus of the ellipse." The law is: "the centripetal force is inversely . . . as the square of the distance." Propositions 12 and 13 show that a hyperbolic and a parabolic orbit imply the same law of force to a focus. It is obvious that the converse condition, that the centripetal force varies inversely as the square of the distance, does not by itself specify which conic section will constitute the orbit. Proposition 15 demonstrates that in ellipses "the periodic times are as the 3/2th power of their greater axes" (Kepler's third law). Hence the periodic times in all ellipses with equal major axes are equal to one another, and equal to the periodic time in a circle of which the diameter is equal to the greater axis of each ellipse. In proposition 17, Newton supposed a centripetal force "inversely proportional to the squares of the distances" and exhibited the conditions for an orbit in the shape of an ellipse, parabola, or hyperbola. Sections 4 and 5, on conic sections, are purely mathematical.

In section 6, Newton discussed Kepler's problem, introducing methods of approximation to find the future position of a body on an ellipse, according to the law of areas; it is here that one finds the method of successive iteration. In section 7, Newton found the rectilinear distance through which a body falls freely in any given time under the action of a "centripetal force . . . inversely proportional to the square of the distance . . . from the centre." Having found the times of descent of such a body, he then applied his results to the problem of parabolic motion and the motion of "a body projected upwards or downwards," under conditions in which "the centripetal force is proportional to the . . . distance." Eventually, in proposition 39, Newton postulated "a centripetal force of any kind" and found both the velocity at any point to which any body may ascend or descend in a straight line and the time it would take the body to get there. In this proposition, as in many in section 8, he added the condition of "granting the quadratures of curvilinear figures," referring to his then unpublished methods of integration (printed for the first time in the *De quadratura* of 1704).

In section 8, Newton often assumed such quadrature. In proposition 41 he postulated "a centripetal force of any kind"; that is, as he added in proposition 42, he supposed "the centripetal force to vary in its recess from the center according to some law, which anyone may imagine at pleasure, but [which] at equal distances from the centre [is taken] to be everywhere the same." Under these general conditions, Newton determined both "the curves in which bodies will move" and "the times of their motions in the curves found." In other words, Newton presented to his readers a truly general resolution of the inverse problem of finding the orbit from a given law of force. He extended this problem into a dynamics far beyond that commonly associated with the *Principia*. In the ancillary proposition 40, for example, Newton (again under the most general conditions of force) had sought the velocity at a point on an orbit, finding a result that is the equivalent of an integral, which (in E. J. Aiton's words) in "modern terms . . . expresses the invariance of the sum of the kinetic and gravitational potential energies in an orbit."[153]

In section 11, Newton reached a level of mathematical analysis of celestial motions that fully distinguishes the *Principia* from any of its predecessors. Until this point, he there explained, he had been "treating of the attractions of bodies towards an immovable centre; though very probably there is no such thing existent in nature." He then outlined a plan to deal with nature herself, although in a "purely mathematical" way, "laying aside all physical considerations"—such as the nature of the gravitating force. "Attractions" are to be treated here as originating in bodies and acting toward other bodies; in a two-body system, therefore, "neither the attracted nor the attracting body is truly at rest, but both . . . being as it were mutually attracted, revolve about a common centre of gravity." In general, for any system of bodies that mutually attract one another, "their common centre of gravity will either be at rest, or move uniformly" in a straight line. Under these conditions, both members of a pair of mutually attractive bodies will describe "similar figures about their common centre of gravity, and about each other mutually" (proposition 57).

By studying such systems, rather than a single body attracted toward a point-center of force, Newton proved that Kepler's laws (or "planetary hypotheses") cannot be true within this context, and hence need modification when applied to the real system of the world. Thus, in proposition 59, Newton stated that Kepler's third law should not be written $T_1{}^2 : T_2{}^2 = a_1{}^3 : a_2{}^3$, as Kepler, Hooke, and everybody else had supposed, but must be modified.

A corollary that may be drawn from the proposition is that the law might be written as $(M + m_1)T_1{}^2 : (M + m_2)T_2{}^2 : a_1{}^3 : a_2{}^3$, where m_1, m_2 are any two planetary masses and M is the mass of the sun. (Newton's expression of this new relation may be reduced at once to the more familiar form in which we use this law today.) Clearly, it follows from Newton's analysis and formulation that Kepler's own third law may safely be used as an approximation in most astronomical calculations only because m_1, m_2 are very small in relation to M. Newton's modification of Kepler's third law fails to take account of any possible interplanetary perturbations. The chief function of proposition 59 thus appears to be not to reach the utmost generalization of that law, but rather to reach a result that will be useful in the problems that follow, most notably proposition 60 (on the orbits described when each of two bodies attracts the other with a force proportional to the square of the distance, each body "revolving about the common centre of gravity").

From proposition 59 onward, Newton almost at once advanced to various motions of mutually attractive bodies "let fall from given places" (in proposition 62), "going off from given places in given directions with given velocities" (proposition 63), or even when the attractive forces "increase in a simple ratio of their [that is, the bodies'] distances from the centres" (proposition 64). This led him to examine Kepler's first two laws for real "bodies," those "whose forces decrease as the square of their distances from their centres." Newton demonstrated in proposition 65 that in general it is not "possible that bodies attracting each other according to the law supposed in this proposition should move exactly in ellipses," because of interplanetary perturbations, and discussed cases in astronomy in which "the orbits will not much differ from ellipses." He added that the areas described will be only "very nearly proportional to the times."

Proposition 66 presents the restricted three-body problem, developed in a series of twenty-two corollaries. Here Newton attempted to apply the law of mutual gravitational attraction to a body like the sun to determine how it might perturb the motion of a moonlike body around an earthlike body. Newton examined the motion in longitude and in latitude, the annual equation, the evection, the change of the inclination of the orbit of the body resembling the moon, and the motion on the line of apsides. He considered the tides and explained, in corollary 22, that the internal "constitution of the globe" (of the earth) can be known "from the motion of the nodes." He further demonstrated that the shape of the globe can be derived from the precession constant (precession being caused, in the case of the earth, by the pull of the moon on the equatorial bulge of the spinning earth). He thus established, for the first time, a physical theory, elaborated in mathematical expression, from which some of the "inequalities" of the motion of the moon could be deduced; and he added some hitherto unknown "inequalities" that he had found. Previous to Newton's work, the study of the irregularities in the motion of the moon had been posited on the elaboration of geometric models, in an attempt to make predicted positions agree with actual observations.[154]

Section 12 of book I contains Newton's results on the attractions of spheres, or of spherical shells. He dealt first with homogeneous, then nonhomogeneous spheres, the latter being composed of uniform and concentric spherical shells so that the density is the same at any single given distance from the center. In proposition 71 he proved that a "corpuscle" situated outside such a nonhomogeneous sphere is "attracted towards the centre of the sphere with a force inversely proportional to the square of its distance from the centre." In proposition 75, he reached the general conclusion that any two such spheres will gravitationally attract one another as if their masses were concentrated at their respective centers—or, in other words, that the distance required for the inverse-square law is measured from their centers. A series of elegant and purely mathematical theorems follow, including one designed to find the force with which a corpuscle placed inside a sphere may be "attracted toward any segment of that sphere whatsoever." In section 13, Newton, with a brilliant display of mathematics (which he did not fully reveal for the benefit of the reader) discussed the "attractive forces" of nonspherical solids of revolution, concluding with a solution in the form of an infinite series for the attraction of a body "towards a given plane."[155]

Book I concludes with section 14, on the "motion of very small bodies" acted on by "centripetal forces tending to the several parts of any very great body." Here Newton used the concept of "centripetal forces" that act under very special conditions to produce motions of corpuscles that simulate the phenomena of light—including reflection and refraction (according to the laws of Snell and Descartes), the inflection of light (as discovered by Grimaldi), and even the action of lenses. In a scholium, Newton noted that these "attractions bear a great resemblance to the reflections and refractions of light," and so

> ... because of the analogy there is between the propagation of the rays of light and the motion of bodies, I thought it not amiss to add the following Propositions for optical uses; not at all considering the nature of the rays of light, or inquiring whether they are bodies or not; but only determining the curves of [the paths of] bodies which are extremely like the curves of the rays.

A similar viewpoint with respect to mathematical analyses (or models and analogies) and physical phenomena is generally sustained throughout books I and II of the *Principia*.

Newton's general plan in book I may thus be seen as one in which he began with the simplest conditions and added complexities step by step. In sections 2 and 3, for example, he dealt with a mass-point moving under the action of a centripetal force directed toward a stationary or moving point, by which the dynamical significance of each of Kepler's three laws of planetary motion is demonstrated. In section 6, Newton developed methods to compute Keplerian motion (along an ellipse, according to the law of areas), which leads to "regular ascent and descent" of bodies when the force is not uniform (as in Galilean free fall) but varies, primarily as the inverse square of the distance, as in Keplerian orbital motion. In section 8 Newton considered the general case of "orbits in which bodies will revolve, being acted upon by any sort of centripetal force." From stationary orbits he went on, in section 9, to "movable orbits; and the motion of the apsides" and to a mathematical treatment of two (and then three) mutually attractive bodies. In section 10 he dealt with motion along surfaces of bodies acted upon by centripetal force; in section 12, the problems of bodies that are not mere points or point-masses and the question of the "attractive forces of spherical bodies"; and in section 13, "the attractive forces of bodies that are not spherical."

Book II of the "Principia." Book II, on the motion of bodies in resisting mediums, is very different from book I. It was an afterthought to the original treatise, which was conceived as consisting of only two books, of which one underwent more or less serious modifications to become book I as it exists today, while the other, a more popular version of the "system of the world," was wholly transformed so as to become what is now book III. At first the question of motion in resisting mediums had been relegated to some theorems at the end of the original book I; Newton had also dealt with this topic in a somewhat similar manner at the end of his earlier tract *De motu*. The latter parts of the published book II were added only at the final redaction of the *Principia*.

Book II is perhaps of greater mathematical than physical interest. To the extent that Newton proceeded by setting up a sequence of mathematical conditions and then exploring their consequences, book II resembles book I. But there is a world of difference between the style of the two books. In book I Newton made it plain that the gravitational force exists in the universe, varying inversely as the square of the distance, and that this force accordingly merits our particular attention. In book II, however, the reader is never certain as to which of the many conditions of resistance that Newton considers may actually occur in nature.[156]

Book II enabled Newton to display his mathematical ingenuity and some of his new discoveries. Occasionally, as in the static model that he proposed to explain the elasticity and compressibility of gases according to Boyle's law, he could explore what he believed might be actual physical reality. But he nonetheless reminded his readers (as in the scholium at the end of section 1) that the condition of resistance that he was discussing was "more a mathematical hypothesis than a physical one." Even in his final argument against Cartesian vortices (section 9), he admitted the implausibility of the proposed hypothesis that "the resistance . . . is, other things being equal, proportional to the velocity." Although a scholium to proposition 52 states that "it is in truth probable that the resistance is in a less ratio than that of the velocity," Newton in fact never explored the consequences of this probable assumption in detail. Such a procedure is in marked contrast to book I, in which Newton examined a variety of conditions of attractive and centripetal forces, but so concentrated on the inverse-square force as to leave the reader in no doubt that this is the chief force acting (insofar as weight is concerned) on the sun, the planets, the satellites, the seas, and all terrestrial objects.

Book II differs further from book I in having a separate section devoted to each of the imagined conditions of resistance. In section 1, resistance to the motions of bodies is said to be as "the ratio of the velocity"; in section 2, it is as "the square of their velocities"; and in section 3, it is given as "partly in the ratio of the velocities and partly as the square of the same ratio." Then, in section 4, Newton introduced the orbital "motion of bodies in resisting mediums," under the mathematical condition that "the density of a medium" may vary inversely as the distance from "an immovable centre"; the "centripetal force" is said in proposition 15 to be as the square of the said density, but is thereafter arbitrary. In a very short scholium, Newton added that these conditions of varying density apply only to the motions of very small bodies. He supposed the resistance of a medium, "other things being equal," to be proportional to its "density."

In section 5, Newton went on to discuss some general principles of hydrostatics, including properties of the density and compression of fluids. Historically, the most significant proposition of section 5 is proposition 23, in which Newton supposed "a fluid [to] be composed of particles fleeing from each other," and then showed that Boyle's law ("the density" of a gas varying directly as "the compression") is a necessary and a sufficient condition for the centrifugal forces to "be inversely proportional to distances of their [that is, the particles'] centers."

Then, in the scholium to this proposition, Newton generalized the results, showing that for the com-

pressing forces to "be as the cube roots of the power E^{n+2}," where E is "the density of the compressed fluid," it is both a necessary and sufficient condition that the centrifugal forces be "inversely as any power D^n of the distance [between particles]." He made it explicit that the "centrifugal forces" of particles must "terminate in those particles that are next [to] them, or are diffused not much farther," and drew upon the example of magnetic bodies. Having set such a model, however, Newton concluded that it would be "a physical question" as to "whether elastic fluids [gases] do really consist of particles so repelling each other," and stated that he had limited himself to demonstrating "mathematically the property of fluids consisting of particles of this kind, that hence philosophers may take occasion to discuss that question."[157]

Section 6 introduces the "motion and resistance of pendulous bodies." The opening proposition (24) relates the quantity of matter in the bob to its weight, the length of the pendulum, and the time of oscillation in a vacuum. Because, as corollary 5 states, "in general, the quantity of matter in the pendulous body is directly as the weight and the square of the time, and inversely as the length of the pendulum," a method is at hand for using pendulum experiments to compare directly "the quantity of matter" in bodies, and to prove that the mass of bodies is proportional "to their weight." Newton added that he had tested this proposition experimentally, then further stated, in corollary 7, that the same experiment may be used for "comparing the weights of the same body in different places, to know the variation of its gravity."[158] This is the first clear recognition that "mass" determines both weight (the amount of gravitational action) and inertia (the measure of resistance to acceleration)—the two properties of which the "equivalence" can, in classical physics, be determined only by experiment.

In section 6 Newton also considered the motion of pendulums in resisting mediums, especially oscillations in a cycloid, and gave methods for finding "the resistance of mediums by pendulums oscillating therein." An account of such experiments makes up the "General Scholium" with which section 6 concludes.[159] Among them is an experiment Newton described from memory, designed to confute "the opinion of some that there is a certain aethereal medium, extremely rare and subtile, which freely pervades the pores of all bodies."

Section 7 introduces the "motion of fluids," and "the resistance made to projected bodies," and section 8 deals with wave motion. Proposition 42 asserts that "All motion propagated through a fluid diverges from a rectilinear progress into the unmoved spaces"; while proposition 50 gives a method of finding "the distances of the pulses," or the wavelength. In a scholium, Newton stated that the previous propositions "respect the motions of light and sound" and asserted that "since light is propagated in right lines, it is certain that it cannot consist in action alone (by Prop. XLI and XLII)"; there can be no doubt that sounds are "nothing else but pulses of the air" which "arise from tremulous bodies." This section concludes with various mathematical theorems concerning the velocity of waves or pulses, and their relation to the "density and elastic force of a medium."

In section 9, Newton showed that in wave motion a disturbance moves forward, but the parts (particles) of the medium in which the disturbance occurs only vibrate about a fixed position; he thereby established the relation between wavelength, frequency, and velocity of undulations. Proposition 47 (proposition 48 in the first edition) analyzes undulatory motion in a fluid; Newton disclosed that the parts (or particles) of an undulating fluid have the same oscillation as the bob of a simple pendulum. Proposition 48 (proposition 47 in the first edition) exhibits the proportionality of the velocity of waves to the square root of the elastic force divided by the density of an elastic fluid (one whose pressure is proportional to the density). The final scholium (much rewritten for the second edition) shows that Newton's propositions yield a velocity of sound in air of 979 feet per second, whereas experiment gives a value of 1,142 feet per second under the same conditions. Newton offered an ingenious explanation (including the supposition, in the interest of simplicity, that air particles might be rigid spheres separated from one another by a distance of some nine times their diameter), but it remained for Laplace to resolve the problem in 1816.[160]

Section 9, the last of book II, is on vortices, or "the circular motion of fluids." In all editions of the *Principia*, this section begins with a clearly labeled "hypothesis" concerning the "resistance arising from the want of lubricity in the parts of a fluid . . . other things being equal, [being] proportional to the velocity with which the parts of the fluid are separated from one another." Newton used this hypothesis as the basis for investigating the physics of vortices and their mathematical properties, culminating in a lengthy proposition 52 and eleven corollaries, followed by a scholium in which he said that he has attempted "to investigate the properties of vortices" so that he might find out "whether the celestial phenomena can be explained by them." The chief "phenomenon" with which Newton was here concerned is Kepler's third (or harmonic) law for the motion of the satellites

of Jupiter about that planet, and for the primary "planets that revolve about the Sun"—although Newton did not refer to Kepler by name. He found "the periodic times of the parts of the vortex" to be "as the squares of their distances." Hence, he concluded, "Let philosophers then see how that phenomenon of the 3/2th power can be accounted for by vortices."

Newton ended book II with proposition 53, also on vortices, and a scholium, in which he showed that "it is manifest that the planets are not carried round in corporeal vortices." He was there dealing with Kepler's second or area law (although again without naming Kepler), in application to elliptic orbits. He concluded "that the hypothesis of vortices is utterly irreconcilable with astronomical phenomena, and rather serves to perplex than to explain the heavenly motions." Newton himself noted that his demonstration was based on "an hypothesis," proposed "for the sake of demonstration . . . at the beginning of this Section," but went on to add that "it is in truth probable that the resistance is in a less ratio than that of the velocity." Hence "the periodic times of the parts of the vortex will be in a greater ratio than the square of the distances from its centre." But it must be noted that it is in fact probable that the resistance would be in a greater "ratio than that of the velocity," not a lesser, since almost all fluids give rise to a resistance proportional to the square (or higher powers) of the velocity.[161]

Book III, "The System of the World." In the Newtonian system of the world, the motions of planets and their satellites, the motions of comets, and the phenomena of tides are all comprehended under a single mode of explanation. Newton stated that the force that causes the observed celestial motions and the tides and the force that causes weight are one and the same; for this reason he gave the name "gravity" to the centripetal force of universal attraction. In book III he showed that the earth must be an oblate spheroid, and he computed the magnitude of the equatorial bulge in relation to the pull of the moon so as to produce the long-known constant of precession; he also gave an explanation of variation in weight (as shown by the change in the period of a seconds pendulum) as a function of latitude on such a rotating non-spherical earth. But above all, in book III Newton stated the law of universal gravitation. He showed that planetary motion must be subject to interplanetary perturbation —most apparent in the most massive planets, Jupiter and Saturn, when they are in near conjunction—and he explored the perturbing action of the sun on the motion of the moon.

Book III opens with a preface in which Newton

stated that in books I and II he had set forth principles of mathematical philosophy, which he would now apply to the system of the world. The preface refers to an earlier, more popular version,[162] of which Newton had recast the substance "into the form of Propositions (in the mathematical way)."

A set of four "rules of reasoning in [natural] philosophy" follows the preface. Rule 1 is to admit no more causes than are "true and sufficient to explain" phenomena, while rule 2 is to "assign the same causes" insofar as possible to "the same natural effects." In the first edition, rules 1 and 2 were called "hypotheses," and they were followed by hypothesis 3, on the possibility of the transformation of every body "into a body of any other kind," in the course of which it "can take on successively all the intermediate grades of qualities." This "hypothesis" was deleted by the time of the second edition.[163]

A second group of the original "hypotheses" (5 through 9) were transformed into "phenomena" 1 and 3 through 6. The first states (with phenomenological evidence) the area law and Kepler's third law for the system of Jupiter's satellites (again Kepler is not named as the discoverer of the law). Phenomenon 2, which was introduced in the second edition, does the same for the satellites of Saturn (just discovered as the *Principia* was being written, and not mentioned in the first edition, where reference is made only to the first [Huygenian] satellite discovered). Phenomena 3 through 6 (originally hypotheses 6 through 9) assert, within the limits of observation: the validity of the Copernican system (phenomenon 3); the third law of Kepler for the five primary planets and the earth— here for the first time in the *Principia* mentioning Kepler by name and thus providing the only reference to him in relation to the laws or hypotheses of planetary motion (phenomenon 4); the area law for the "primary planets," although without significant evidence (phenomenon 5); and the area law for the moon, again with only weak evidence and coupled with the statement that the law does not apply exactly since "the motion of the moon is a little disturbed by the action of the sun" (phenomenon 6).

It has been mentioned that Newton probably called these statements "phenomena" because he knew that they are valid only to the limits of observation. In this sense, Newton had originally conceived Kepler's laws as planetary "hypotheses," as he had also done for the phenomena and laws of planetary satellites.[164]

The first six propositions given in book III display deductions from these "phenomena," using the mathematical results that Newton had set out in book I. Thus, in proposition 1, the forces "by which the circumjovial planets are continually drawn off from rectilinear motions, and retained in their proper orbits" are shown (on the basis of the area law discussed in propositions 2 and 3, book I, and in phenomenon 1) to be directed toward Jupiter's center. On the basis of Kepler's third law (and corollary 6, proposition 4, book I) these forces must vary inversely as the square of the distance; propositions 2 and 3 deal similarly with the primary planets and our moon.

By proposition 5, Newton was able to conclude (in corollary 1) that there "is . . . a power of gravity tending to all the planets" and that the planets "gravitate" toward their satellites, and the sun "towards all the primary planets." This "force of gravity" varies (corollary 2) as the inverse square of the distance; corollary 3 states that "all the planets do mutually gravitate towards one another." Hence, "near their conjunction," Jupiter and Saturn, since their masses are so great, "sensibly disturb each other's motions," while the sun "disturbs" the motion of the moon and together both sun and moon "disturb our sea, as we shall hereafter explain."

In a scholium, Newton said that the force keeping celestial bodies in their orbits "has been hitherto called centripetal force"; since it is now "plain" that it is "a gravitating force" he will "hereafter call it gravity." In proposition 6 he asserted that "all bodies gravitate towards every planet"; while at equal distances from the center of any planet "the weight" of any body toward that planet is proportional to its "quantity of matter." He provided experimental proof, using a pair of eleven-foot pendulums, each weighted with a round wooden box (for equal air resistance), into the center of which he placed seriatim equal weights of wood and gold, having experimented as well with silver, lead, glass, sand, common salt, water, and wheat. According to proposition 24, corollaries 1 and 6, book II, any variation in the ratio of mass to weight would have appeared as a variation in the period; Newton reported that through these experiments he could have discovered a difference as small as less than one part in a thousand in this ratio, had there been any.[165]

Newton was thus led to the law of universal gravitation, proposition 7: "That there is a power of gravity tending to all bodies, proportional to the several quantities of matter which they contain." He had shown this power to vary inversely as the square of the distance; it is by this law that bodies (according to the third law of motion) act mutually upon one another.

From these general results, Newton turned to practical problems of astronomy. Proposition 8 deals with gravitating spheres and the relative masses and densities of the planets (the numerical calculations in this proposition were much altered for the second edition). In proposition 9, Newton estimated the force of gravity within a planet and, in proposition 10, demonstrated the long-term stability of the solar system. A general "Hypothesis I" (in the second and third editions; "Hypothesis IV" in the first) holds the "centre of the system of the world" to be "immovable," which center is given as the center of gravity of the solar system in proposition 11; the sun is in constant motion, but never "recedes" far from that center of gravity (proposition 12).

It is often asserted that Newton attained his results by neglecting the interplanetary attractions, and dealing exclusively with the mutual gravitational attractions of the planets and our sun. But this is not the case, since the most fully explored example of perturbation in the *Principia* is indeed that of the sun-earth-moon system. Thus Newton determined (proposition 25) the "forces with which the sun disturbs the motions of the moon," and (proposition 26) the action of those forces in producing an inequality ("horary increment") of the area described by the moon (although "in a circular orbit").

The stated intention of proposition 29 is to "find the variation of the moon," the inequality thus being sought being due "partly to the elliptic figure of the Moon's orbit, partly to the inequality of the moments of the area which the Moon by a radius drawn to the Earth describes." (Newton dealt with this topic more fully in the second edition.) Then Newton studied the "horary motion of the nodes of the moon," first (proposition 30) "in a circular orbit," and then (proposition 31) "in an elliptic orbit." In proposition 32, he found "the mean motion of the nodes," and, in proposition 33, their "true motion." (In the third edition, following proposition 33, Newton inserted two propositions and a scholium on the motion of the nodes, written by John Machin.) Propositions 34 and 35, on the inclination of the orbit of the moon to the ecliptic plane, are followed by a scholium, considerably expanded and rewritten for the second edition, in which Newton discussed yet other "inequalities" in the motion of the moon and developed the practical aspects of computing the elements of that body's motion and position.

Propositions 36 and 37 deal at length and in a quantitative fashion with the tide-producing forces of the sun and of the moon, yielding, in proposition 38, an explanation of the spheroidal shape of the moon and the reason that (librations apart) the same face of it is always visible. A series of three lemmas introduces the subject of precession and a fourth lemma (transformed into hypothesis 2 in the second

and third editions) treats the precession of a ring. Proposition 39 represents an outstanding example of the high level of mathematical natural science that Newton reached in the *Principia*. In it he showed the manner in which the shape of the earth, in relation to the pull of the moon, acts on its axis of rotation so as to produce the observed precession, a presentation that he augmented and improved for the second edition. Newton here employed the result he had previously obtained (in propositions 20 and 21, book III) concerning the shape of the earth, and joined it to both the facts and theory of precession and yet another aspect of the perturbing force of the moon on the motion of the earth. He thus inaugurated a major aspect of celestial mechanics, the study of a three-body system.

Lemma 4, book III initiates a section on comets, proving that comets are "higher" than the moon, move through the solar system, and (corollary 1) shine by reflecting sunlight; their motion shows (corollary 3) that "the celestial spaces are void of resistance." Comets move in conic sections (proposition 40) having the sun as a focus, according to the law of areas. Those comets that return move in elliptic orbits (corollary 1) and follow Kepler's third law, but (corollary 2) "their orbits will be so near to parabolas, that parabolas may be used for them without sensible error."

Almost immediately following publication of the *Principia*, Halley, in a letter of 5 July 1687, urged Newton to go on with his work on lunar theory.[166] Newton later remarked that his head so ached from studying this problem that it often "kept him awake" and "he would think of it no more." But he also said that if he lived long enough for Halley to complete enough additional observations, he "would have another stroke at the moon." In the 1690's Newton had depended on Flamsteed for observations of the moon, promising Flamsteed (in a letter of 16 February 1695) not to communicate any of his observations, "much less publish them, without your consent." But Newton and Flamsteed disagreed on the value of theory, which Newton held to be useful as "a demonstration" of the "exactness" of observations, while Flamsteed believed that "theories do not command observations; but are to be tried by them," since "theories are . . . only probable" (even "when they agree with exact and indubitable observations"). At about this same time Newton was drawing up a set of propositions on the motion of the moon for a proposed new edition of the *Principia*, for which he requested from Flamsteed such planetary observations "as tend to [be useful for] perfecting the theory of the planets," to serve Newton in the preparation of a second edition of his book.

Revision of the "Opticks" (the Later Queries); Chemistry and Theory of Matter. Newton's *Opticks*, published in 1704, concluded with a Third Book, consisting of eleven "Observations" and sixteen queries, occupying a bare five pages of print. A Latin translation, undertaken at Newton's behest by Samuel Clarke, appeared in 1706, and included as its most notable feature the expansion of the original sixteen queries into twenty-three. The new queries 17 through 23 correspond to the final queries 25–31 of the later editions. In a series of "Errata, Corrigenda, & Addenda," at the beginning of the Latin volume, lengthy additions are provided to be inserted at the end of query 8 and of query 11; there is also a short insertion for query 14.

In a second English edition (London, 1717) the number of queries was increased to thirty-one. The queries appearing for the first time are numbered 17 to 24, and they have no counterparts in the 1706 Latin version. Newton's own copy of the 1717 English edition, in the Babson Institute Library, contains a number of emendations and corrections in Newton's hand, some of which were incorporated into the third edition (London 1721), as was a postscript to the end of the last sentence, referring to Noah and his sons.

The queries new to the 1717 edition cover a wide range of topics. Query 17 introduces the possibility that waves or vibrations may be excited in the eye by light and that vibrations of this sort may occur in the medium in which light travels. Query 18 suggests that radiant heat may be transmitted by vibrations of a medium subtler than air that pervades all bodies and expands by its elastic force throughout the heavenly spaces—the same medium by which light is put into "fits" of "easy" reflection and refraction, thus producing "Newton's rings." In queries 19 and 20, variations in the density of this medium are given as the possible cause of refraction and of the "inflection" (diffraction) of light rays. Query 21 would have the medium be rarer within celestial bodies than in empty celestial spaces, which may "impel Bodies from the denser parts of the Medium towards the rarer"; its elasticity may be estimated by the ratio of the speed of light to the speed of sound. Although he referred in this query to the mutually repulsive "particles" of ether as being "exceedingly smaller than those of Air, or even those of Light," Newton confessed that he does "not know what this *Aether* is."

In query 22, the resistance of the ether is said to be inconsiderable; the exhalations emitted by "electrick" bodies and magnetic "effluvia" are offered as other instances of such rareness. The subject of vision is introduced in query 23. Here vision is again said to be chiefly the effect of vibrations of the medium, propagated through the "optick Nerves"; an analogy is made to hearing and the other senses. Animal motion (query 24) is considered as a result of vibrations in the medium propagated from the brain through the nerves to the muscles.

Queries 25 to 31 are the English recasting of queries 17 to 23 of the Latin edition. Query 25 contains a discussion of double refraction in calcite (Iceland spar) and a geometrical construction of both the ordinary ray and (fallaciously) the extraordinary ray; query 26 concludes that double refraction may be caused by the two "sides" of rays of light. Then, in query 27, Newton attacked as erroneous all hypotheses explaining optical phenomena by new modifications of rays, since such phenomena depend upon original unalterable properties.

Query 28 questions "all Hypotheses" in which light is supposed to be a "Pression or Motion, propagated through a fluid Medium." Newton showed that Huygens' wave theory of double refraction would fail to account for the heating of bodies and the rectilinear propagation of light. Those who would fill "the Heavens with fluid Mediums" come under attack, while Newton praised the ancient philosophers who "made a *Vacuum*, and Atoms, and the Gravity of Atoms, the first Principles of their Philosophy." He added that "the main Business of natural Philosophy is to argue from Phaenomena without feigning Hypotheses"; we are to "deduce Causes from Effects, till we come to the very first Cause, which certainly is not mechanical," since nature exhibits design and purpose.

In query 29, Newton suggested that rays of light are composed of "very small Bodies emitted from shining Substances," since rays could not have a permanent virtue in two of their sides (as demonstrated by the double refraction of Iceland spar) unless they be bodies. This query also contains Newton's famous theory that rays of light could be put into "Fits of easy Reflexion and easy Transmission" if they were "small Bodies which by their attractive Powers, or some other Force, stir up Vibrations in what they act upon." These vibrations would move more swiftly than the rays themselves, would "overtake them successively," and by agitating them "so as by turns to increase and decrease their Velocities" would put them into those "fits."[167] Newton further argued that if light were to consist of waves in an ethereal medium, then in order to have the fits of easy reflection and easy transmission, a second ether would be required, in which there would be waves (of higher velocity) to put the waves of the first ether into the necessary fits. He had, however, already argued in query 28 that it would be inconceivable for two ethers to be "diffused through all

Space, one of which acts upon the other, and by consequence is re-acted upon, without retarding, shattering, dispersing and compounding one another's Motions."

In query 30, Newton discussed the convertibility of gross bodies and light, with examples showing that nature delights in transmutations. In illustration, he cited Boyle's assertion that frequent distillations had turned water into earth. In query 31, he discussed questions ranging from the forces that hold particles of matter together to the impact of bodies on one another; also causes of motion, fermentation, the circulation of the blood and animal heat, putrefaction, the force of inertia, and occult qualities. He stated a general philosophy and concluded with the pious hope that the perfection of natural philosophy will enlarge the "Bounds of Moral Philosophy."

Newton's queries, particularly the later ones, thus go far beyond any simple questions of physical or geometrical optics. In them he even proposed tentative explanations of phenomena, although explanations that are perhaps not as fully worked out, or as fully supported by experimental evidence, as he might have wished. (Some queries even propose what is, by Newton's own definition, a hypothesis.) In each case, Newton's own position is made clear; and especially in the queries added in the Latin version of 1706 (and presented again in the English version of 1717/1718), his supporting evidence is apt to be a short essay.

One notable development of the later queries is the emphasis on an "Aethereal Medium" as an explanation for phenomena. In his first papers on optics, in the 1670's, Newton had combined his cherished conception of corpuscular or globular light with the possibly Cartesian notion of a space-filling ether, elastic and varying in density. Although Newton had introduced this ether to permit wave phenomena to exist as concomitants of the rays of light, he also suggested other possible functions for it—including causing sensation and animal motion, transmitting radiant heat, and even causing gravitation. His speculations on the ether were incorporated in the "Hypothesis" that he sent to the Royal Society (read at their meetings in 1675 and 1676) and in a letter to Boyle of 28 February 1679.[168]

In the second English edition of the Opticks (1717/1718) Newton made additions which "embodied arguments for the existence of an elastic, tenuous, aetherial medium." The new queries in the Latin version of 1706 did not deal with an ether, however, and by the time of the Principia, Newton may have "rejected the Cartesian dense aether" as well as "his own youthful aetherial speculations."[169]

Newton thus did not propose a new version of the ether until possibly the 1710's; he then suggested, in the general scholium at the conclusion of the second edition of the Principia (1713), that a most subtle "spiritus" ("which pervades and lies hid in all gross bodies") might produce just such effects as his earlier ether (or the later ethereal medium of queries 18 through 24). In the general scholium of the Principia, however, Newton omitted gravitation from the list of effects that the "spiritus" may produce. There is evidence that Newton conceived of this "spiritus" as electrical, and may well have been a precursor of the ether or ethereal medium of the 1717/1718 queries.[170] In a manuscript intended for the revised second English edition of the Opticks,[171] Newton wrote the heading, "The Third Book of Opticks. Part II. Observations concerning the Medium through which Light passes, & the Agent which emits it," a title that would thus seem to link the ethereal medium with the emission of electrical effluvia. It would further appear that Newton used both the earlier and later concepts of the ether to explain, however hypothetically, results he had already obtained; and that the concept of the ether was never the basis for significant new experiments or theoretical results. In a general scholium to book II, Newton described from memory an experiment that he had performed which seemed to him to prove the nonexistence of an ether; since Newton's original notes have never been found, this experiment, which was presumably an important element in the decline of his belief in an ether, cannot be dated.

The later queries also develop a concept of matter, further expounded by Newton in his often reprinted De natura acidorum (of which there appear to have been several versions in circulation).[172] Newton here, as a true disciple of Boyle, began with the traditional "mechanical philosophy" but added "the assumption that particles move mainly under the influence of what he at first called sociability and later called attraction."[173] Although Newton also considered a principle of repulsion, especially in gases, in discussing chemical reactions he seems to have preferred to use a concept of "sociability" (as, for example, to explain how substances dissolve).

He was equally concerned with the "aggregation" of particles (in queries 28 and 31 as well as at the end of De natura acidorum) and even suggested a means of "differentiating between reaction and transmutation."[174] Another major concern was the way in which aqua regia dissolves gold but not silver, while aqua fortis dissolves silver but not gold,[175] a phenomenon Newton explained by a combination of the attraction of particles and the relation between the size of the acid particles and the "pores" between the particles of metal. He did not, however, have a sound operational definition of acid, but referred to acids theoretically, in *De natura acidorum*, as those substances "endued with a great Attractive Force; in which Force their Activity consists." He maintained this definition in query 31, in which he further called attention to the way in which metals may replace one another in acid solutions and even "went so far as to list the six common metals in the order in which they would displace one another from a solution of aqua fortis (strong nitric acid)."[176]

Alchemy, Prophecy, and Theology. Chronology and History. Newton is often alleged to have been a mystic. That he was highly interested in alchemy has been embarrassing to many students of his life and work, while others delight in finding traces of hermeticism in the father of the "age of reason." The entries in the *Catalogue of the Portsmouth Collection* give no idea of the extent of the documents in Newton's hand dealing with alchemy; these were listed in the catalogue, but not then presented to Cambridge University. Such information became generally available only when the alchemical writings were dispersed in 1936, in the Sotheby sale. The catalogue of that sale gives the only full printed guide to these materials, and estimates their bulk at some 650,000 words, almost all in Newton's hand.

A major problem in assessing Newton's alchemical "writings" is that they are not, for the most part, original compositions, nor even critical essays on his readings (in the sense that the early "De gravitatione et aequipondio fluidorum" is an essay based on his reading in Descartes's *Principia*). It would be necessary to know the whole corpus of the alchemical literature to be able to declare that any paper in Newton's hand is an original composition, rather than a series of extracts or summaries.[177]

In a famous letter to Oldenburg (26 April 1676), Newton offered an explanation of Boyle's presentations of the "incalescence" of gold and mercury (*Philosophical Transactions*, **9**, no. 122 [1675], 515–533), and presented an explanation based on the size of the particles of matter and their mechanical action. Newton particularly commended Boyle for having concealed some major steps, since here was possibly "an inlet into something more noble, and not to be communicated without immense dammage to the world if there be any verity in the Hermetick writers." He also gave some cautionary advice about alchemists, even referring to a "true Hermetic Philosopher, whose judgment (if there be any such)" might be of interest and highly regarded, "there being other things beside the transmutation of metalls (if those pretenders bragg not) which none but they understand." The apparently positive declarations in Newton's letter thus conflict with the doubts expressed in the two parenthetical expressions.

Newton's studies of prophecy may possibly provide a key to the method of his alchemical studies. His major work on the subject is *Observations upon the Prophecies of Daniel, and the Apocalypse of St. John* (London, 1733). Here Newton was concerned with "a figurative language" used by the prophets, which he sought to decipher. Newton's text is a historical exegesis, unmarked by any mystical short-circuiting of the rational process or direct communication from the godhead. He assumed an "analogy between the world natural, and an empire or kingdom considered as a world politic," and concluded, for example, that Daniel's prophecy of an "image composed of four metals" and a stone that broke "the four metals into pieces" referred to the four nations successively ruling the earth ("*viz.* the peoples of Babylonia, the Persians, the Greeks, and the Romans"). The four nations are represented again in the "four beasts."

"The folly of interpreters," Newton wrote, has been "to foretell times and things by this Prophecy, as if God designed to make them Prophets." This is, however, far from God's intent, for God meant the prophecies "not to gratify men's curiosities by enabling them to foreknow things" but rather to stand as witnesses to His providence when "after they were fulfilled, they might be interpreted by events." Surely, Newton added, "the event of things predicted many ages before, will then be a convincing argument that the world is governed by providence." (It may be noted that this book also provided Newton with occasion to refer to his favorite themes of "the corruption of scripture" and the "corruption of Christianity.")

The catalogue of the Sotheby sale states that Newton's manuscript remains include some 1,300,000 words on biblical and theological subjects. These are not particularly relevant to his scientific work and— for the most part—might have been written by any ordinary divinity student of that period, save for the extent to which they show Newton's convinced anti-Trinitarian monotheism or Unitarian Arianism. (His tract *Two Notable Corruptions of Scripture*, for example, uses historical analysis to attack Trinitarian doctrine.) "It is the temper of the hot and superstitious part of mankind in matters of religion," Newton wrote, "ever to be fond of mysteries, and for that reason to like best what they understand least."[178]

Typical of Newton's theological exercises is his "Queries regarding the word *homoousios*." The first query asks "Whether Christ sent his apostles to preach metaphysics to the unlearned common people, and to their wives and children?" Other queries in this set are also historical; in the seventh Newton marshaled his historico-philological acumen in the matter of the Latin rendering *unius substantiae*, which he considered to have been imposed on the Western churches instead of *consubstantialis* by "Hosius (or whoever translated that [Nicene] Creed into Latin)." Another manuscript entitled "Paradoxical Questions" turns out to be less a theological inquiry than a carefully reasoned proof of what Lord Keynes called "the dishonesty and falsification of records for which St Athanasius [and his followers] were responsible." In it Newton cited, as an example, the spreading of the story that Arius died in a house of prostitution.

In a Keynes manuscript (in King's College, Cambridge), "The First Book Concerning the Language of the Prophets," Newton explained his method:

> He that would understand a book written in a strange language must first learn the language. . . . Such a language was that wherein the Prophets wrote, and the want of sufficient skill in that language is the reason why they are so little understood. John . . ., Daniel . . ., Isaiah . . . all write in one and the same mystical language . . . [which] so far as I can find, was as certain and definite in its signification as is the vulgar language of any nation. . . .

Having established this basic premise, Newton went on: "It is only through want of skill therein that Interpreters so frequently turn the Prophetic types and phrases to signify whatever their fancies and hypotheses lead them to." Then, in a manner reminiscent of the rules at the beginning of book III of the *Principia*, he added:

> The rule I have followed has been to compare the several mystical places of scripture where the same prophetic phrase or type is used, and to fix such a signification to that phrase as agrees best with all the places: . . . and when I had found the necessary significations, to reject all others as the offspring of luxuriant fancy, for no more significations are to be admitted for true ones than can be proved.

Newton's alchemical manuscripts show that he sometimes used a similar method, drawing up comparative tables of symbols and of symbolic names used by alchemists, no doubt in the conviction that a key to their common language might be found thereby. His careful discrimination among the alchemical writers may be seen in two manuscripts in the Keynes Collection, one a three-page classified list of alchemical writers and the other a two-page

selection of "authores optimi," by whom Newton perhaps meant authorities who described processes that might be repeated and verified. The Babson Collection of Newtoniana contains a two-page autograph manuscript listing 113 writers on alchemy arranged by nationalities and another seven-page manuscript of "chemical authors and their writings" in which Newton commented on the more important ones. At least two other such bibliographical works by Newton are known. An "Index Chemicus," an elaborate subject index to the literature of alchemy with page references to a number of different works (described as containing more than 20,000 words on 113 pages), is one of at least five such indexes, all in autograph manuscripts.[179]

It must be emphasized that Newton's study of alchemy was not a wholly rational pursuit, guided by a strict code of linguistic and historical investigative procedures. To so consider it would be to put it on the same plane as his chronological inquiries.[180] The chronological studies are, to a considerable degree, the result of the application of sound principles of astronomical dating to poor historical evidence—for which his *Chronology of Ancient Kingdoms Amended* was quite properly criticized by the French antiquarians of his day—while his alchemical works show that he drew upon esoterical and even mystical authors, far beyond the confines of an ordinary rational science.

It is difficult to determine whether to consider Newton's alchemy as an irrational vagary of an otherwise rational mind, or whether to give his hermeticism a significant role as a developmental force in his rational science. It is tempting, furthermore, to link his concern for alchemy with his belief in a secret tradition of ancient learning. He believed that he had traced this *prisca sapientia* to the ancient Greeks (notably Pythagoras) and to the Chaldean philosophers or magicians; he concluded that these ancients had known even the inverse-square law of gravitation. Cohen, McGuire, and Rattansi have shown that in the 1690's, when Newton was preparing a revised edition of the *Principia*, he thought of including references to such an ancient tradition in a series of new scholia for the propositions at the beginning of book III of the *Principia*, along with a considerable selection of verses from Lucretius' *De natura rerum*. All of this was to be an addendum to an already created *Principia*, which Newton was revising for a new edition.

There is not a shred of real evidence, however, that Newton ever had such concerns primarily in mind in those earlier years when he was writing the *Principia* or initially developing the principles of dynamics and of mathematics on which the *Principia* was ultimately to be based. In Newton's record of alchemical experiments (University Library, Cambridge, MS Add. 3975), the experiments dated 23 May [1684] are immediately followed by an entry dated 26 April 1686. The former ends in the middle of a page, and the latter starts on the very next line; there is no lacuna, and no possibility that a page—which chronologically might concern experiments made while the *Principia* was being written—might be missing from the notebook.[181]

The overtones of alchemy are on occasion discernible in Newton's purely scientific writings. In query 30 of the *Opticks* (first published in the Latin version, then in the second English edition), Newton said that "Nature . . . seems delighted with Transmutations," although he was not referring specifically to changing metals from one to another. (It must be remembered in fact that "transmutation" would not necessarily hold an exclusively chemical or alchemical meaning for Newton; it might, rather, signify not only transformations in general, but also particular transformations of a purely mathematical sort, as in lemma 22 of book I of the *Principia*.) This is a far cry, indeed, from Newton's extracts from the mystical Count Michael Maier and kindred authors. P. M. Rattansi particularly calls attention to the alchemist's "universal spirit," and observes: "It is difficult to understand how, without a conviction of deep and hidden truths concealed in alchemy, Newton should have attached much significance to such ideas."[182]

Notable instances of the conflation of alchemical inspiration and science occur in Newton's letter to Boyle (1679) and in the hypothesis he presented to explain those properties of light of which he wrote in his papers in the *Philosophical Transactions*. While it is not difficult to discover alchemical images in Newton's presentation, and to find even specific alchemical doctrines in undisguised form and language, the problem of evaluating the influence of alchemy on Newton's true science is only thereby compounded, since there is no firm indication of the role of such speculations in the development of Newton's physical science. The result is, at best, one mystery explained by another, like the alchemist's confusing doctrine of *ignotum per ignotius*. Rattansi further suggests that alchemy may have served as a guiding principle in the formulation of Newton's views on fermentation and the nourishment of the vegetation of the earth by fluids attracted from the tails of comets. He would even have us believe that alchemical influences may have influenced "the revival of aetherical notions in the last period of Newton's life."[183] This may be so; but what, if any, creative effect such "aetherical notions" then had on Newton's thought would seem to be a matter of pure hypothesis.

Scholars do not agree whether Newton's association with some "Hermetic tradition" may have been a creative force in his science, or whether it is legitimate to separate his alleged hermeticism from his positive science. Apart from the level of general inspiration, it must be concluded that, excluding some aspects of the theory of matter and chemistry, notably fermentation, and possibly the ether hypotheses, the real creative influence of alchemy or hermeticism on Newton's mathematics and his work in optics, dynamics, and astronomy (save for the role of the tails of comets in the economy of nature) must today be evaluated in terms of the Scottish verdict, "not proven." Investigations of this topic may provide valuable insights into the whole man, Newton, and into the complexities of his scientific inspiration. His concern for alchemy and theology should not be cast aside as irrelevant aberrations of senility or the product of a mental breakdown. Yet it remains a fact beyond dispute that such early manuscripts as the Waste Book—in which Newton worked out and recorded his purely scientific discoveries and innovations—are free from the tinges of alchemy and hermeticism.

The London Years: the Mint, the Royal Society, Quarrels with Flamsteed and with Leibniz. On 19 March 1696, Newton received a letter from Charles Montagu informing him that he had been appointed warden of the mint. He set up William Whiston as his deputy in the Lucasian professorship, to receive "the full profits of the place." On 10 December 1701 he resigned his professorship, and soon afterward his fellowship. He was designated an *associé étranger* of the Paris Académie des Sciences in February 1699, chosen a member of the Council of the Royal Society on the following 30 November, and on 30 November 1703 was made president of the Royal Society, an office he held until his death. He was elected M.P. for Cambridge University, for the second time, on 26 November 1701, Parliament being prorogued on 25 May 1702. Queen Anne knighted Newton at Trinity College on 16 April 1705; on the following 17 May he was defeated in his third contest for the university's seat in Parliament.

At the mint, Newton applied his knowledge of chemistry and of laboratory technique to assaying, but he apparently did not introduce any innovations in the art of coinage. His role was administrative and his duties were largely the supervision of the recoinage and (curious to contemplate) the capture, interrogation, and prosecution of counterfeiters. Newton used the patronage of the mint to benefit fellow scientists. Halley entered the service in 1696 as comptroller of the Chester mint, and in 1707 David

Gregory was appointed (at a fee of £250) as general supervisor of the conversion of the Scottish coinage to British.

Newton ruled over the Royal Society with an iron hand. When Whiston was proposed as a fellow in 1720, Newton said that if Whiston were chosen, he "would not be president." At Newton's urging, the council brought the society from the verge of bankruptcy to solvency by obtaining regular contributions from fellows. When a dispute arose between Woodward and Sloane, Newton had Woodward ejected from the council. Of Newton's chairmanship of meetings, Stukeley reported, "Everything was transacted with great attention and solemnity and dignity," for "his presence created a natural awe in the assembly"; there was never a sign of "levity or indecorum." As England's foremost scientist, president of the Royal Society, and civil servant, Newton appeared before Parliament in Spring 1714, to give advice about a prize for a method of finding longitude.

When Newton moved from Cambridge to London in the 1690's to take up the wardenship of the mint, he continued to work on the motion of the moon. He became impatient for Flamsteed's latest observations and they soon had a falling-out, no doubt aggravated by the strong enmity which had grown up between Halley and Flamsteed. Newton fanned the flames by the growing arrogance of his letters: "I want not your calculations but your observations only." And when in 1699 Flamsteed let it be known that Newton was working to perfect lunar theory, Newton sent Flamsteed a letter insisting that on this occasion he not "be brought upon the stage," since "I do not love to be printed upon every occasion much less to be dunned & teezed by foreigners about Mathematical things or to be thought by our own people to be trifling away my time about them when I should be about the King's business." Newton and Halley published Flamsteed's observations in an unauthorized printing in 1712, probably in the conviction that his work had been supported by the government and was therefore public property. Flamsteed had the bitter joy of burning most of the spurious edition; and he then started printing his own *Historia coelestis Brittanica*.

A more intense quarrel arose with Leibniz. This took two forms: a disagreement over philosophy or theology in relation to science (carried out through Samuel Clarke as intermediary), and an attempt on Newton's part to prove that Leibniz had no claim to originality in the calculus. The initial charge of plagiarism against Leibniz came from Fatio de Duillier, but before long Keill and other Newtonians were involved and Leibniz began to rally his own supporters. Newton held that not only had Leibniz stolen the calculus from him, but that he had also composed three tracts for publication in the *Acta eruditorum* claiming some of the main truths of the *Principia* as independent discoveries, with the sole original addition of some mistakes. Today it appears that Newton was wrong; no doubt Leibniz had (as he said) seen the "epitome" or lengthy review of the *Principia* in the *Acta eruditorum* of June 1688, and not the book, when (to use his own words) "Newton's work stimulated me" to write out some earlier thoughts on "the causes of the motions of the heavenly bodies" as well as on the "resistance of a medium" and motion in a medium.[184] Newton stated, however, that even if Leibniz "had not seen the book itself, he ought nevertheless to have seen it before he published his own thoughts concerning these matters."[185]

That Newton should have connived at declaring Leibniz a plagiarist gives witness to his intense possessiveness concerning his discoveries or inventions; hence his consequent feeling of violation or robbery when Leibniz seemed to be publishing them. Newton was also aware that Leibniz must have seen one or more of his manuscript tracts then in circulation; and Leibniz had actually done so on one of his visits, when, however, he copied out some material on series expansions, not on fluxions.[186]

No one today seriously questions Leibniz' originality and true mathematical genius, nor his independence—to the degree that any two creative mathematicians living in the same world of mathematical thought can be independent—in the formulation of the calculus. Moreover, the algorithm in general use nowadays is the Leibnizian rather than the Newtonian. But by any normal standards, the behavior of both men was astonishing. When Leibniz appealed to the Royal Society for a fair hearing, Newton appointed a committee of good Newtonians. It has only recently become known that Newton himself wrote the committee's report, the famous *Commercium epistolicum*,[187] which he presented as if it were a set of impartial findings in his own favor.

Newton was not, however, content to stop there; following publication of the report there appeared an anonymous review, or summary, of it in the *Philosophical Transactions*. This, too, was Newton's work. When the *Commercium epistolicum* was reprinted, this review was included, in Latin translation, as a kind of introduction, together with an anonymous new preface "To the Reader," which was also written by Newton. This episode must be an incomparable display of thoroughness in destroying an enemy, and Whiston reported that he had heard directly that Newton had "once pleasantly" said to Samuel Clarke that "He had broke Leibnitz's Heart with his Reply to him."

Newton's later London years were marked by creative scientific efforts. During this time he published the *Opticks*, with the two mathematical tracts, and added new queries for its later editions. He also produced, with Roger Cotes's aid, a second edition of the *Principia*, including the noteworthy general scholium, and, with assistance from Henry Pemberton, a third edition. In the last, however, Newton altered the scholium to lemma 2, book II, to prevent its being read as if Leibniz were entitled to a share of credit for the calculus—although Leibniz had been dead for nearly twelve years.

Newton died on Monday, 20 March 1727,[188] at the age of eighty-five, having been ill with gout and inflamed lungs for some time. He was buried in Westminster Abbey.

Newton's Philosophy: The Rules of Philosophizing, the General Scholium, the Queries of the "Opticks." Like others of his day, Newton believed that the study of natural philosophy would provide evidence for the existence of God the Creator in the regularities of the solar system. In the general scholium at the end of book III of the *Principia*, he said "it is not to be conceived that mere mechanical causes could give birth to so many regular motions," then concluded his discussion with observations about God, "to discourse of whom from phenomena does certainly belong to Natural Philosophy" ("Experimental Philosophy" in the second edition). He then went on to point out that he had "explained the phenomena of the heavens and of our sea, by the power of Gravity" but had not yet "assigned the cause of this power," alleging that "it is enough that Gravity does really exist, and act according to the laws which we have explained" and that its action "abundantly serves to account for all the motions of the celestial bodies, and of our sea." The reader was thus to accept the facts of the *Principia*, even though Newton had not "been able to discover the cause of those properties of gravity from phenomena." Newton here stated his philosophy, "Hypotheses non fingo."[189]

Clearly, Newton was referring here only to "feigning" a hypothesis about the cause of gravitation, and never intended that his statement should be applied on all levels of scientific discourse, or to all meanings of the word "hypothesis." Indeed, in each of the three editions of the *Principia*, there is a "hypothesis" stated in book II. In the second and third editions there are a "Hypothesis I" and a "Hypothesis II" in book III. The "phaenomena" at the beginning of book III, in the second and third editions, were largely the "hypotheses" of the first

edition. It may be that Newton used these two designations to imply that these particular statements concerning planetary motions are not mathematically true (as he proved), but could be only approximately "true," on the level of (or to the limits of) phenomena.

Newton believed that his science was based upon a philosophy of induction. In the third edition of the *Principia*, he introduced rule 4, so that "the argument of induction may not be evaded by hypotheses." Here he said that one may look upon the results of "general induction from phenomena as accurately or very nearly true," even though many contrary hypotheses might be imagined, until such time as the inductive result may "either be made more accurate or liable to exceptions" by new phenomena. In rule 3, in the second and third editions, he stated his philosophical basis for establishing general properties of matter by means of phenomena.

Newton's philosophical ideas are even more fully developed in query 31, the final query of the later editions of the *Opticks*, in which he argued for both the philosophy of induction and the method of analysis and composition (or synthesis). In both mathematics and natural philosophy, he said, the "Investigation of difficult Things by the method of Analysis, ought ever to precede the Method of Composition." Such "Analysis consists in making Experiments and Observations, and in drawing general Conclusions from them by Induction, and admitting of no Objections against the Conclusions, but such as are taken from Experiments, or other certain Truths."

In both the *Principia* and the *Opticks*, Newton tried to maintain a distinction among his speculations, his experimental results (and the inductions based upon them), and his mathematical derivations from certain assumed conditions. In the *Principia* in particular, he was always careful to separate any mathematical hypotheses or assumed conditions from those results that were "derived" in some way from experiments and observations. Often, too, when he suggested, as in various scholiums, the applicability of mathematical or hypothetical conditions to physical nature, he stated that he had not proved whether his result really so applies. His treatment of the motion of small corpuscles, in book I, section 14, and his static model of a gas composed of mutually repulsive particles, in book II, proposition 23, exemplify Newton's use of mathematical models of physical reality for which he lacked experimental evidence sufficient for an unequivocal statement.

Perhaps the best expression of Newton's general philosophy of nature occurs in a letter to Cotes (28 March 1713), written during the preparation of the second edition of the *Principia*, in which he referred to the laws of motion as "the first Principles or Axiomes" and said that they "are deduced from Phaenomena & made general by Induction"; this "is the highest evidence that a Proposition can have in this philosophy." Declaring that "the mutual & mutually equal attraction of bodies is a branch of the third Law of motion," Newton pointed out to Cotes "how this branch is deduced from Phaenomena," referring him to the "end of the Corollaries of the Laws of Motion." Shortly thereafter, in a manuscript bearing upon the Leibniz controversy, he wrote, "To make an exception upon a mere Hypothesis is to feign an exception. It is to reject the argument from Induction, & turn Philosophy into a heap of Hypotheses, which are no other than a chimerical Romance."[190] That is a statement with which few would disagree.

NOTES

1. See R. S. Westfall, "Short-writing and the State of Newton's Conscience, 1662," in *Notes and Records. Royal Society of London*, **18** (1963), 10–16. L. T. More, in *Isaac Newton* (New York, 1934), p. 16, drew attention to the necessary "mental suffering" of a boy of Newton's physical weakness, living in a lonely "farmhouse situated in a countryside only slowly recovering from the terrors of a protracted and bitter civil war," with "no protection from the frights of his imagination except that of his grandmother and such unreliable labourers as could be hired."

F. E. Manuel, in *A Portrait of Isaac Newton* (Cambridge, Mass., 1968), has subjected Newton's life to a kind of psychoanalytic scrutiny. He draws the conclusion (pp. 54–59) that the "scrupulosity, punitiveness, austerity, discipline, industriousness, and fear associated with a repressive morality" were apparent in Newton's character at an early age, and finds that notebooks bear witness to "the fear, anxiety, distrust, sadness, withdrawal, self-belittlement, and generally depressive state of the young Newton."

For an examination of Manuel's portrait of Newton, see J. E. McGuire, "Newton and the Demonic Furies: Some Current Problems and Approaches in the History of Science," in *History of Science*, **11** (1973), 36–46; see also the review in *Times Literary Supplement* (1 June 1973), 615–616, with letters by Manuel (8 June 1973), 644–645; D. T. Whiteside (15 June 1973), 692, and (6 July 1973), 779; and G. S. Rousseau (29 June 1973), 749.

2. See E. N. da C. Andrade, "Newton's Early Notebook," in *Nature*, **135** (1935), 360; and G. L. Huxley, "Two Newtonian Studies: I. Newton's Boyhood Interests," in *Harvard Library Bulletin*, **13** (1959), 348–354, in which Andrade has first called attention to the importance of Bate's collection, an argument amplified by Huxley.

3. Newton apparently came to realize that he had been hasty in discarding Euclid, since Pemberton later heard him "even censure himself for not following them [that is, 'the ancients' in their 'taste, and form of demonstration'] yet more closely than he did; and speak with regret of his mistake at the beginning of his mathematical studies, in applying himself to the works of Des Cartes and other algebraic writers, before he had considered the elements of Euclide with that attention, which so excellent a writer

deserves" (*View of Sir Isaac Newton's Philosophy* [London, 1728], preface).

4. Newton's college tutor was not (and indeed by statute could not have been) the Lucasian professor, Barrow, but was Benjamin Pulleyn.

5. University Library, Cambridge, MS Add. 3996, discussed by A. R. Hall in "Sir Isaac Newton's Notebook, 1661–1665," in *Cambridge Historical Journal*, **9** (1948), 239–250.

6. *Ibid.*; also partially analyzed by R. S. Westfall, in "The Foundations of Newton's Philosophy of Nature," in *British Journal for the History of Science*, **1** (1962), 171–182. Westfall has attempted a reconstruction of Newton's philosophy of nature, and his growing allegiance to the "mechanical philosophy," in ch. 7 of his *Force in Newton's Physics* (London, 1971).

7. On Newton's entrance into the domains of mathematics higher than arithmetic, see the account by A. De Moivre (in the Newton MSS presented by the late J. H. Schaffner to the University of Chicago) and the recollections of Newton assembled by John Conduitt, now mainly in the Keynes Collection, King's College, Cambridge.

8. See D. T. Whiteside, "Newton's Marvellous Year. 1666 and All That," in *Notes and Records. Royal Society of London*, **21** (1966), 37–38.

9. See A. H. White, ed., William Stukeley, *Memoirs of Sir Isaac Newton's Life* (London, 1936). Written in 1752, this records a conversation with Newton about his discovery of universal gravitation (the apple story), pp. 19–20.

10. In November 1669 John Collins wrote to James Gregory that "Mr Barrow hath resigned his Lecturers place to one Mr Newton of Cambridge" (in the Royal Society ed. of Newton's *Correspondence*, I, 15). Newton himself may have been referring to Barrow in an autobiographical note (*ca.* 1716) that stated, "Upon account of my progress in these matters he procured for me a fellowship ... in the year 1667 & the Mathematick Professorship two years later"—see University Library, Cambridge, MS Add. 3968, §41, fol. 117, and I. B. Cohen, *Introduction to Newton's Principia*, supp. III, p. 303, n. 14.

11. Among the biographical memoirs assembled by Conduitt (Keynes Collection, King's College, Cambridge). Humphrey Newton's memoir is in L. T. More, *Isaac Newton*, pp. 246, 381, and 389.

12. According to J. Edleston (p. xlv in his ed. of *Correspondence of Sir Isaac Newton and Professor Cotes ...*; see also pp. xlix–1), in 1675 (or March 1674, OS), "Newton obtained a Royal Patent allowing the Professor to remain Fellow of a College without being obliged to go into orders." See also L. T. More, *Isaac Newton*, p. 169.

13. This work might have been an early version of the *Lectiones opticae*, his professorial lectures of 1670–1672; or perhaps an annotated version of his letters and communications to Oldenburg, which were read at the Royal Society and published in major part in its *Philosophical Transactions* from 1672 onward.

14. Quoted in L. T. More, *Isaac Newton*, p. 217.

15. It has been erroneously thought that Newton's "breakdown" may in part have been caused by the death of his mother. But her death occurred in 1679, and she was buried on 4 June. "Her will was proved 11 June 1679 by Isaac Newton, the executor, who was the residuary legatee"; see *Correspondence*, II, 303. n. 2. David Brewster, in *Memoirs ...*, II, 123, suggested that Newton's "ailment may have arisen from the disappointment he experienced in the application of his friends for a permanent situation for him." On these events and on contemporary discussion and gossip about Newton's state of mind, see L. T. More, *Isaac Newton*, pp. 387–388, and F. E. Manuel, *A Portrait of Isaac Newton*, pp. 220–223. Newton himself, in a letter to Locke of 5 October 1693, blamed his "distemper" and insomnia on "sleeping too often by my fire."

16. L. T. More, *Isaac Newton*, p. 368.
17. See J. Edleston, ed., *Correspondence . . . Newton and . . . Cotes*, pp. xxxvi, esp. n. 142.
18. *Mathematical Papers of Isaac Newton*, D. T. Whiteside, ed., in progress, to be completed in 8 vols. (Cambridge, 1967–   ); these will contain edited versions of Newton's mathematical writings with translations and explanatory notes, as well as introductions and commentaries that constitute a guide to Newton's mathematics and scientific life, and to the main currents in the mathematics of the seventeenth century. Five volumes have been published (1973).
19. See D. T. Whiteside, "Newton's Discovery of the General Binomial Theorem," in *Mathematical Gazette*, **45** (1961), 175.
20. Especially because of Whiteside's researches.
21. Whiteside, ed., *Mathematical Papers*, I, 1–142. Whiteside concludes: "By and large Newton took his arithmetical symbolisms from Oughtred and his algebraical from Descartes, and onto them . . . he grafted new modifications of his own" (I, 11).
22. *Ca.* 1714; see University Library, Cambridge, MS Add. 3968, fol. 21. On this often debated point, see D. T. Whiteside, "Isaac Newton: Birth of a Mathematician," in *Notes and Records. Royal Society of London*, **19** (1964), n. 25; but compare n. 48, below.
23. University Library, Cambridge, MS Add. 3968. 41, fol. 85. This sentence occurs in a passage canceled by Newton.
24. *Ibid.*, fol. 72. This accords with De Moivre's later statement (in the Newton manuscripts recently bequeathed the University of Chicago by J. H. Schaffner) that after reading Wallis' book, Newton "on the occasion of a certain interpolation for the quadrature of the circle, found that admirable theorem for raising a Binomial to a power given."
25. Translated from the Latin in the Royal Society ed. of the *Correspondence*, II, 20 ff. and 32 ff.; see the comments by Whiteside in *Mathematical Papers*, IV, 666 ff. In the second term, A stands for $P^{m/n}$ (the first term), while in the third term B stands for $(m/n) AQ$ (the second term), and so on. This letter and its sequel came into Wallis' hands and he twice published summaries of them, the second time with Newton's own emendations and grudging approval. Newton listed some results of series expansion—coupled with quadratures as needed—for $z = r \sin^{-1} [x/r]$ and the inverse $x = r \sin[z/r]$; the versed sine $r(1 - \cos[z/r])$; and $x = e^{z/b} - 1$, the inverse of $z = b \log(1 + x)$, the Mercator series (see Whiteside, ed., *Mathematical Papers*, IV, 668).
26. Translated from the Latin in the Royal Society ed. of the *Correspondence*, II, 110 ff., 130 ff.; see the comments by Whiteside in *Mathematical Papers*, IV, 672 ff.
27. See Whiteside, *Mathematical Papers*, I, 106.
28. *Ibid.*, I, 112 and n. 81.
29. The Boothby referred to may be presumed to be Boothby Pagnell (about three miles northeast of Woolsthorpe), whose rector, H. Babington, was senior fellow of Trinity and had a good library. See further Whiteside, *Mathematical Papers*, I, 8, n. 21; and n. 8, above.
30. *The Mathematical Works of Isaac Newton*, I, x.
31. *Ibid.*, I, xi.
32. Here the "little zero" o is not, as formerly, the "indefinitely small" increment in the variable t, which "ultimately vanishes." In the *Principia*, bk. II, sec. 2, Newton used an alternative system of notation in which $a, b, c, \cdots$ are the "moments of any quantities $A, B, C, \&c.$," increasing by a continual flux or "the velocities of the mutations which are proportional" to those moments, that is, their fluxions.
33. See Whiteside, *Mathematical Works*, I, x.
34. See A. R. and M. B. Hall, eds., *Unpublished Scientific Papers of Isaac Newton* (Cambridge, 1962).
35. *Mathematical Works*, I, xi.
36. *Ibid.*, xii.
37. University Library, Cambridge, MS Add. 3968.41, fol. 86, v.
38. Whiteside, *Mathematical Papers*, II, 166.
39. *Ibid.*, 166–167.
40. *Ibid.*, I, 11, n. 27. where Whiteside lists those "known to have seen substantial portions of Newton's mathematical papers during his lifetime" as including Collins, John Craig, Fatio de Duillier, Raphson, Halley, De Moivre, David Gregory, and William Jones, "but not, significantly, John Wallis," who did, however, see the "Epistola prior" and "Epistola posterior" (see n. 25, above); and II, 168. Isaac Barrow "probably saw only the *De analysi.*"
41. The *Methodus fluxionum* also contained an amplified version of the tract of October 1666; it was published in English in 1736, translated by John Colson, but was not properly printed in its original Latin until 1779, when Horsley brought out *Analysis per quantitatum series, fluxiones, ac differentias*, incorporating William Jones's transcript, which he collated with an autograph manuscript by Newton. Various MS copies of the *Methodus fluxionum* had, however, been in circulation many years before 1693, when David Gregory wrote out an abridged version. Buffon translated it into French (1740) and Castillon used Colson's English version as the basis of a retranslation into Latin (*Opuscula mathematica*, I, 295 ff.). In all these versions, Newton's equivalent notation was transcribed into dotted letters. Horsley (*Opera*, I) entitled his version *Artis analyticae specimina vel geometria analytica*. The full text was first printed by Whiteside in *Mathematical Papers*, vol. III.
42. *Mathematical Papers*, II, 170.
43. P. xi; and see n. 41, above.
44. The reader may observe the confusion inherent in using both "indefinitely small portions of time" and "infinitely little" in relation to o; the use of index notation for powers (x^3, x^2, o^2) together with the doubling of letters (oo) in the same equation occurs in the original. These quotations are from the anonymous English version of 1737, reproduced in facsimile in Whiteside, ed., *Mathematical Works*. See n. 46.
45. In this example, I have (following the tradition of more than two centuries) introduced $\dot{x}$ and $\dot{y}$ where Newton in his MS used m and n. In his notation, too, r stood for the later $\dot{z}$.
46. *Mathematical Papers*, III, 80, n. 96. In the anonymous English version of 1737, as in Colson's translation of 1736, the word "indefinitely" appears; Castillon followed these (see n. 41). Horsley first introduced "*infinité.*"
47. *Ibid.*, pp. 16–17.
48. See Whiteside, *ibid.*, p. 17; on Barrow's influence, see further pp. 71–74, notes 81, 82, 84.
49. *Ibid.*, pp. 328–352. On p. 329, n. 1, Whiteside agrees with a brief note by Alexander Witting (1911), in which the "source of the celebrated 'fluxional' Lemma II of the second Book of Newton's *Principia*" was accurately found in the first theorem of this addendum; see also p. 331, n. 11, and p. 334, n. 16.
50. On this topic, see the collection of statements by Newton assembled in supp. I to I. B. Cohen, *Introduction to Newton's Principia*.
51. This and the following quotations of the *De quadratura* are from John Stewart's translation of 1745.
52. As C. B. Boyer points out, in *Concepts of the Calculus*, p. 201, Newton was thus showing that one should not reach the conclusion "by simply neglecting infinitely small terms, but by finding the ultimate ratio as these terms become evanescent." Newton unfortunately compounded the confusion, however, by not wholly abjuring infinitesimals thereafter; in bk. II, lemma 2, of the *Principia* he warned the reader that his "moments" were not finite

quantities. In the eighteenth century, many English mathematicians, according to Boyer, "began to associate fluxions with the infinitely small differentials of Leibniz."

53. University Library, Cambridge, MS Add. 3960, fol. 177. Newton, however, was not the first mathematician to anticipate the Taylor series.

54. Introduction to *De quadratura*, in John Stewart, trans., *Two Treatises of the Quadrature of Curves, and Analysis by Equations of an Infinite Number of Terms . . .* (London, 1745), p. 4.

55. *Philosophical Transactions*, no. 342 (1715), 206.

56. Attributed to Newton, May 1708, in W. G. Hiscock, ed., *David Gregory, Isaac Newton and Their Circle* (Oxford, 1937), p. 42.

57. Henry Pemberton recorded, in his preface to his *View of . . . Newton's Philosophy* (London, 1728), that "I have often heard him censure the handling [of] geometrical subjects by algebraic calculations; and his book of Algebra he called by the name of Universal Arithmetic, in opposition to the injudicious title of Geometry, which Des Cartes had given to the treatise wherein he shews, how the geometer may assist his invention by such kind of computations."

58. There were five Latin eds. between 1707 and 1761, of which one was supervised by Newton, and three English eds. between 1720 and 1769.

59. For details, see Turnbull, *The Mathematical Discoveries of Newton*, pp. 49-50.

60. See C. B. Boyer, *History of Mathematics*, p. 450.

61. *Arithmetica universalis*, English ed. (London, 1728), p. 247; see Whiteside, *Mathematical Papers*, V, 428–429, 470–471.

62. *Arithmetica universalis*, in Whiteside's translation, *Mathematical Papers*, V, 477.

63. Published by Whiteside, *Mathematical Papers*, I, pp. 145 ff.

64. See especially *ibid.*, pp. 298 ff., pt. 2, sec. 5, "The Calculus Becomes an Algorithm."

65. *Ibid.*, III, pp. 120 ff.

66. *Ibid.*

67. In "Newton as an Originator of Polar Coördinates," in *American Mathematical Monthly*, **56** (1949), 73–78.

68. Made available in English translation (perhaps supervised by Newton himself) in John Harris, *Lexicon technicum*, vol. II (London, 1710); reprinted in facsimile (New York, 1966). The essay entitled "Curves" is reprinted in Whiteside, *Mathematical Papers*, II.

69. C. R. M. Talbot, ed. and trans., *Enumeration of Lines of the Third Order* (London, 1860), p. 72.

70. On other aspects of Newton's mathematics see Whiteside, *Mathematical Papers*, specifically III, 50–52, on the development of infinite series; II, 218–232, on an iterative procedure for finding approximate solutions to equations; and I, 519, and V, 360, on "Newton's identities" for finding the sums of the powers of the roots in any polynomial equation. See, additionally, for Newton's contributions in porisms, solid loci, number theory, trigonometry, and interpolation, among other topics, Whiteside, *Mathematical Papers*, passim, and Turnbull, *Mathematical Discoveries*.

71. See Whiteside, *Mathematical Works*, I, XV, and Boyer, *History of Mathematics*, p. 448. Drafts of the "Liber geometria" are University Library, Cambridge, MS Add. 3963 passim and MS Add. 4004, fols. 129–159. Gregory's comprehensive statement of Newton's plans as of summer 1694 is in Edinburgh University Library, David Gregory MS C42; an English version in Newton's *Correspondence*, III, 384–386, is not entirely satisfactory.

72. Newton's laconic statement of his solution, published anonymously in *Philosophical Transactions*, no. 224 (1697), p. 384, elicited from Bernoulli the reply "Ex ungue, Leonem" (the claw was sufficient to reveal the lion); see *Histoire des ouvrages des savans* (1697), 454–455.

73. See I. B. Cohen, "Isaac Newton, John Craig, and the Design of Ships," in *Boston Studies for the Philosophy of Science* (in press).

74. Even the variants in the eds. of the *Opticks* have never been fully documented in print (although Horsley's ed. gives such information for the Queries), nor have the differences between the Latin and English versions been fully analyzed. Zev Bechler is in the process of publishing four studies based on a perceptive and extensive examination of Newton's optical MSS. Henry Guerlac is presently engaged in preparing a new ed. of the *Opticks* itself.

75. The expression "experimentum crucis" is often attributed to Bacon, but Newton in fact encountered it in Hooke's account of his optical experiments as given in *Micrographia* (observation 9), where Hooke referred to an experiment that "will prove such a one as our *thrice excellent Verulam* [that is, Francis Bacon] calls *Experimentum crucis*." While many investigators before Newton— Dietrich von Freiberg, Marci, Descartes, and Grimaldi among them—had observed the oval dispersion of a circular beam of light passing through a prism, they all tended to assign the cause of the phenomenon to the consideration that the light source was not a point, but a physical object, so that light from opposite limbs of the sun would differ in angle of inclination by as much as half a degree. Newton's measurements led him from this initial supposition to the conclusion that the effect—a spectrum some five times longer than its width—was too great for the given cause, and therefore the prism must refract some rays to a considerable degree more than others.

76. This account of the experiment is greatly simplified, as was Newton's own account, presented in his letter to Oldenburg and published in *Philosophical Transactions*. See J. A. Lohne, "Experimentum Crucis," in *Notes and Records. Royal Society of London*, **23** (1968), 169-199; Lohne has traced the variations introduced into both the later diagrams and descriptions of the experiment. Newton's doctrine of the separation of white light into its component colors, each corresponding to a unique and fixed index of refraction, had been anticipated by Johannes Marcus Marci de Kronland in his *Thaumantias, liber de arcu coelesti* (Prague, 1648). An important analysis of Newton's experiment is in A. I. Sabra, *Theories of Light*.

77. See R. S. Westfall, "The Development of Newton's Theory of Color," in *Isis*, **53** (1962), 339–358; and A. R. Hall, "Newton's Notebook," pp. 245–250.

78. Dated 13 April 1672, in *Philosophical Transactions*, no. 84.

79. See R. S. Westfall, "Newton's Reply to Hooke and the Theory of Colors," in *Isis*, **54** (1963), 82–96; an edited text of the "Hypothesis" is in *Correspondence*, I, 362–386.

80. Published in Birch's *History of the Royal Society* and in I. B. Cohen, ed., *Newton's Papers and Letters*.

81. R. S. Westfall has further sketched Newton's changing views in relation to corpuscles and the ether, and, in "Isaac Newton's Coloured Circles Twixt Two Contiguous Glasses," in *Archive for History of Exact Sciences*, **2** (1965), 190, has concluded that "When Newton composed the *Opticks*, he had ceased to believe in an aether; the pulses of earlier years became 'fits of easy reflection and transmission,' offered as observed phenomena without explanation." Westfall discusses Newton's abandonment of the ether in "Uneasily Fitful Reflections on Fits of Easy Transmission [and of Easy Reflection]," in Robert Palter, ed., *The Annus Mirabilis of Sir Isaac Newton 1666–1966*, pp. 88–104; he emphasizes the pendulum experiment that Newton reported from memory in the *Principia* (bk. II, scholium at the end of sec. 7, in the first ed., or of sec. 6, in the 2nd and 3rd eds.). Henry Guerlac has discussed Newton's return to a modified concept of the ether in a series of studies (see Bibliography, sec. 8).

82. Birch, *History of the Royal Society*, III, 299; the early text of the "Discourse" is III, 247–305, but Newton had

already published it, with major revisions, as book II of the *Opticks*. Both the "Hypothesis" and the "Discourse" are reprinted in Newton's *Papers and Letters*, 177–235. Newton's original notes on Hooke's *Micrographia* have been published by A. R. and M. B. Hall, *Unpublished Scientific Papers of Isaac Newton*, 400 ff., especially sec. 48, in which he refers to "coloured rings" of "8 or 9 such circuits" in this "order (white perhaps in the midst) blew, purple, scarlet, yellow, greene, blew. . . ."

83. Newton's notes on Hooke were first published by Geoffrey Keynes in *Bibliography of Robert Hooke* (Oxford, 1960), pp. 97–108. Hooke claimed in particular that Newton's "Hypothesis" was largely taken from the *Micrographia*; see Newton's letters to Oldenburg, 21 December 1675 and 10 January 1676, in *Correspondence*, I, 404 ff. Hooke then wrote to Newton in a more kindly vein on 20 January 1676, provoking Newton's famous reply.

84. In this presentation, attention has been directed only to certain gross differences that exist between the texts of Newton's "Discourse of Observations" of 1675 and bk. II of the *Opticks*. The elaboration of Newton's view may be traced through certain notebooks and an early essay "On Colours" to his optical lectures and communications to the Royal Society. In particular, R. S. Westfall has explored certain relations between the essay and the later *Opticks*. See also his discussion on Newton's experiments cited in n. 81, above.

85. Chiefly in University Library, Cambridge, MS Add. 3970; but see n. 76.

86. University Library, Cambridge, MS Dd. 9.67.

87. Now part of the Portsmouth Collection, University Library, Cambridge, MS Add. 4002. This MS has been reproduced in facsimile, with an introduction by Whiteside, as *The Unpublished First Version of Isaac Newton's Cambridge Lectures on Optics* (Cambridge, 1973).

88. The development of the *Opticks* can be traced to some degree through a study of Newton's correspondence, notebooks, and optical MSS, chiefly University Library, Cambridge, MS Add. 3970, of which the first 233 pages contain the autograph MS used for printing the 1704 ed., although the final query 16 is lacking. An early draft, without the preliminary definitions and axioms, begins on fol. 304; the first version of prop. 1, book I, here reads, "The light of one natural body is more refrangible than that of another." There are many drafts and versions of the later queries, and a number of miscellaneous items, including the explanation of animal motion and sensation by the action of an "electric" and "elastic" spirit and the attribution of an "electric force" to all living bodies. A draft of a proposed "fourth Book" contains, on fol. 336, a "Conclusion" altered to "Hypoth. 1. The particles of bodies have certain spheres of activity with in which they attract or shun one another . . ."; in a subsequent version, a form of this is inserted between props. 16 and 17, while a later prop. 18 is converted into "Hypoth. 2," which is followed shortly by hypotheses 3 to 5. It may thus be seen that Newton did not, in the 1690's, fully disdain speculative hypotheses. On fol. 409 there begins a tract, written before the *Opticks*, entitled "Fundamentum Opticae," which is similar to the *Opticks* in form and content. The three major notebooks in which Newton entered notes on his optical reading and his early thoughts and experiments on light, color, vision, the rainbow, and astronomical refraction are MSS Add. 3975, 3996, and 4000.

89. In "Newton's Reply to Hooke and the Theory of Colors," in *Isis*, **54** (1963), 82–96; an analysis of the two versions of Newton's lectures on optics is given in I. B. Cohen, *Introduction to Newton's 'Principia*,' supp. III.

90. See "Experimentum Crucis," in *Notes and Records. Royal Society of London*, **23** (1968), 169–199.

91. See, notably, "Isaac Newton: The Rise of a Scientist 1661–1671," in *Notes and Records. Royal Society of London*, **20** (1965), 125–139.

92. University Library, Cambridge, MS Add. 3996.

93. See Sabra, *Theories of Light*; also Westfall, "The Development of Newton's Theory of Color," in *Isis*, **53** (1962), 339–358. A major source for the development of Newton's optical concepts is, of course, the series of articles by Lohne, esp. those cited in nn. 90 and 91.

94. The surviving pages of this abortive ed. are reproduced in I. B. Cohen, "Versions of Isaac Newton's First Published Paper, With Remarks on the Question of Whether Newton Planned to Publish an Edition of His Early Papers on Light and Color," in *Archives internationales d'histoire des sciences*, **11** (1958), 357–375, 8 plates. See also A. R. Hall, "Newton's First Book," in *Archives internationales d'histoire des sciences*, **13** (1960), 39–61.

95. In W. C. Hiscock, ed., *David Gregory*, p. 15. The preface to the first ed. of the *Opticks* is signed "I.N."

96. See the "Analytical Table of Contents" prepared by Duane H. D. Roller for the Dover ed. of the *Opticks* (New York, 1952) for the contents of the entire work.

97. *Opticks*, book I, part 2, proposition 6. Newton's first statement of a musical analogy to color occurs in his "Hypothesis" of 1675; for an analysis of Newton's musical theory, see *Correspondence*, I, 388, n. 14, which includes a significant contribution by J. E. Bullard.

98. As Boyer has pointed out, "In the Cartesian geometrical theory [of the rainbow] it matters little what light is, or how it is transmitted, so long as propagation is rectilinear and the laws of reflection and refraction are satisfied"; see *The Rainbow from Myth to Mathematics* (New York, 1959), ch. 9.

99. Although Newton had worked out the formula at the time of his optical lectures of 1669-1671, he published no statement of it until the *Opticks*. In the meantime Halley and Johann [I] Bernoulli had reached this formula independently and had published it; see Boyer, *The Rainbow*, pp. 247 ff. In the *Opticks*, Newton offered the formula without proof, observing merely that "The Truth of all this Mathematicians will easily examine." His analysis is, however, given in detail in the *Lectiones opticae*, part 1, section 4, propositions 35 and 36, as a note informs the reader of the 1730 ed. of the *Opticks*.

For a detailed analysis of the topic, see Whiteside, *Mathematical Papers*, III, 500–509.

100. Ernst Mach, *The Principles of Physical Optics*, John S. Anderson and A. F. A. Young, trans. (London, 1926), 139.

101. This final sentence of book II, part 2, is a variant of a sentiment expressed a few paragraphs earlier: "Now as all these things follow from properties of Light by a mathematical way of reasoning, so the truth of them may be manifested by Experiments."

102. The word "diffraction" appears to have been introduced into optical discourse by Grimaldi, in his *Physico-mathesis de lumine, coloribus, et iride* (Bologna, 1665), in which the opening proposition reads: "Lumen propagatur seu diffunditur non solùm Directè, Refractè, ac Reflexè, sed etiam alio quodam Quarto modo, DIFFRACTÈ." Although Newton mentioned Grimaldi by name (calling him "Grimaldo") and referred to his experiments, he did not use the term "diffraction," but rather "inflexion," a usage the more curious in that it had been introduced into optics by none other than Hooke (*Micrographia*, "Obs. LVIII. Of a new Property in the Air and several other transparent *Mediums* nam'd *Inflection* . . ."). Newton may thus have been making a public acknowledgment of his debt to Hooke; see n. 83.

103. Newton's alleged denial of the possibility of correcting chromatic aberration has been greatly misunderstood. See the analysis of Newton's essay "Of Refractions" in Whiteside, *Mathematical Papers*, I, 549–550 and 559–576,

esp. the notes on the theory of compound lenses, pp. 575–576, and notes 60 and 61. This topic has also been studied by Zev Bechler; see " 'A Less Agreeable Matter'— Newton and Achromatic Refraction" (in press).

104. Many of these are available in two collections: A. R. and M. B. Hall, eds., *Unpublished Scientific Papers;* and John Herivel, *The Background to Newton's Principia.* See also the Royal Society's ed. of the *Correspondence.*

105. University Library, Cambridge, MS Add. 3996, first analyzed by A. R. Hall in 1948.

106. *Ibid.,* fol. 29. See also R. S. Westfall, *Force in Newton's Physics.* Newton's entry concerning the third law was first published by Whiteside in 1964; see n. 114.

107. University Library, Cambridge, MS Add. 4004; Herivel also gives the dynamical portions, with commentaries.

108. Def. 4; see Herivel, *Background,* p. 137.

109. *Ibid.,* p. 141.

110. See William Stukeley, *Memoirs of Sir Isaac Newton's Life,* p. 20; see also Douglas McKie and G. R. de Beer, "Newton's Apple," in *Notes and Records. Royal Society of London,* **9** (1952), 46–54, 333–335.

111. Various nearly contemporary accounts are given by W. W. Rouse Ball, *An Essay on Newton's "Principia,"* ch. 1.

112. See F. Cajori, "Newton's Twenty Years' Delay in Announcing the Law of Gravitation," in F. E. Brasch, ed., *Sir Isaac Newton,* pp. 127–188.

113. This document, a tract on "circular motion," University Library, Cambridge, MS Add. 3958.5, fol. 87, was in major part published for the first time by A. R. Hall in 1957. It has since been republished, with translation, in *Correspondence,* I, 297–300, and by Herivel in *Background,* pp. 192 ff.

114. In "Newton's Early Thoughts on Planetary Motion: A Fresh Look," in *British Journal for the History of Science,* **2** (1964), 120, n. 13.

115. In A. R. and M. B. Hall, *Unpublished Papers,* pp. 89 ff.

116. University Library, Cambridge, MS Add. 3958, fols. 81–83; also in Turnbull, *Correspondence,* III, 60–64.

117. Newton's concept of force has been traced, in its historical context, by Westfall, *Force in Newton's Physics;* see also Herivel, *Background,* and see I. B. Cohen, "Newton's Second Law and the Concept of Force in the *Principia,*" in R. Palter, ed., *Annus Mirabilis,* pp. 143–185.

118. In the scholium to the Laws of Motion, Newton mentioned that Wren, Wallis, and Huygens at "about the same time" communicated their "discoveries to the Royal Society"; they agreed "exactly among themselves" as to "the rules of the congress and reflexion of hard bodies."

119. Almost all discussions of Newton's spiral are based on a poor version of Newton's diagram; see J. A. Lohne, "The Increasing Corruption of Newton's Diagrams," in *History of Science,* **6** (1967), 69–89, esp. pp. 72–76.

120. Whiteside, "Newton's Early Thoughts," p. 135, has paraphrased Hooke's challenge as "Does the central force which, directed to a focus, deflects a body uniformly travelling in a straight line into an elliptical path vary as the inverse-square of its instantaneous distance from that focus?"

121. University Library, Cambridge, MS Add. 3968.41, fol. 85r, first printed in *Catalogue of the Portsmouth Collection,* p. xviii; it is in fact part of a draft of a letter to Des Maizeaux, written in summer 1718, when Des Maizeaux was composing his *Recueil.* In a famous MS memorandum (University Library, Cambridge, MS Add. 3968, fol. 101), Newton recalled the occasion of his correspondence with Hooke concerning his use of Kepler's area law in relation to elliptic orbits; see I. B. Cohen, *Introduction to Newton's Principia,* supp. I, sec. 2.

122. University Library, Cambridge, MS Add. 3965.7, fols. 55r–62(bis)r; printed versions appear in A. R. and M. B. Hall, *Unpublished Papers;* J. Herivel, *Background;* and W. W. Rouse Ball, *Essay.*

123. See Whiteside, "Newton's Early Thoughts," pp. 135–136; and see I. B. Cohen, "Newton's Second Law and the Concept of Force in the *Principia,*" in R. Palter, ed., *Annus Mirabilis,* pp. 143–185.

124. Analysis shows that great care is necessary in dealing with the limit process in even the simplest of Newton's examples, as in his early derivation of the Huygenian rule for centrifugal force (in the Waste Book, and referred to in the scholium to prop. 4, bk. I, in the *Principia*), or in the proof (props. 1–2, bk. I) that the law of areas is a necessary and sufficient condition for a central force. Whiteside has analyzed these and other propositions in "Newtonian Dynamics," pp. 109–111, and "Mathematical Principles," pp. 11 ff., and has shown the logical pitfalls that await the credulous reader, most notably the implied use by Newton of infinitesimals of an order higher than one (chiefly those of the second, and occasionally those of the third, order).

125. See the *Principia,* props. 1–3, bk. I, and the various versions of *De motu* printed by A. R. and M. B. Hall, J. Herivel, and W. W. Rouse Ball.

126. In *Correspondence,* II, 436–437. This letter unambiguously shows that Newton did not have the solution to the problem of the attraction of a sphere until considerably later than 1679, and declaredly not "until last summer [1685]."

127. There is considerable uncertainty about what "curious treatise, *De Motu*" Halley saw; see I. B. Cohen, *Introduction,* ch. 3, sec. 2.

128. *Ibid.,* sec. 6.

129. First published by A. R. and M. B. Hall, *Unpublished Papers.*

130. Newton at first corresponded with Flamsteed indirectly, beginning in December 1680, through the agency of James Crompton.

131. In 1681, Newton still thought that the "comets" seen in November and December 1680 were "two different ones" (Newton to Crompton for Flamsteed, 28 February 1681, in *Correspondence,* II, 342); in a letter to Flamsteed of 16 April 1681 (*ibid.,* p. 364), Newton restated his doubts that "the Comets of November & December [were] but one." In a letter of 5 January 1685 (*ibid.,* p. 408), Flamsteed hazarded a "guess" at Newton's "designe": to define the curve that the comet of 1680 "described in the aether" from a general "Theory of motion," while on 19 September 1685 (*ibid.,* p. 419), Newton at last admitted to Flamsteed that "it seems very probable that those of November & December were the same comet." Flamsteed noted in the margin of the last letter that Newton "would not grant it before," adding, "see his letter of 1681." In the *Arithmetica universalis* of 1707, Newton, in problem 52, explored the "uniform rectilinear motion" of a comet, "supposing the 'Copernican hypothesis' "; see Whiteside, *Mathematical Papers,* V, 299, n. 400, and esp. pp. 524 ff.

132. As far as actual Greek geometry goes, Newton barely makes use of Archimedes, Apollonius, or even Pappus (mentioned in passing in the preface to the 1st ed. of the *Principia*); see Whiteside, "Mathematical Principles," p. 7.

133. This is the tract "De methodis serierum et fluxionum," printed with translation in Whiteside, ed., *Mathematical Papers,* III, 32 ff.

134. Motte has standardized the use of the neuter *genitum* in his English translation, although Newton actually wrote: "Momentum Genitae aequatur . . .," and then said "Genitam voco quantitatem omnem quae . . .," where *quantitas genita* (or "generated quantity") is, of course, feminine.

135. Whiteside, *Mathematical Papers,* IV, 523, note 6.

136. *Concepts,* p. 200.

137. *Ibid.;* on Newton's use of infinitesimals in the *Principia,* see also A. De Morgan, "On the Early History of Infinitesimals in England," in *Philosophical Magazine,* **4** (1852), 321–330, in which he notes especially some changes in

138. Newton's usage from the 1687 to the 1713 eds. See further F. Cajori, *A History of the Conceptions of Limits*, pp. 2–32.

138. Whiteside, "Mathematical Principles," pp. 20 ff.

139. Newton's method, contained in University Library, Cambridge, MS Add. 3965.10, fols. 107v and 134v, will be published for the first time in Whiteside, *Mathematical Papers*, VI.

140. Halley refers to this specifically in the first paragraph of his review of the *Principia*, in *Philosophical Transactions of the Royal Society*, no. 186 (1687), p. 291.

141. Translated from University Library, Cambridge, MS Add. 3968, fol. 112.

142. *De quadratura* was printed, together with the other tracts in the collection published by W. Jones in 1711, as a supp. to the second reprint of the 2nd ed. of the *Principia* (1723).

143. In *Philosophical Transactions of the Royal Society* (1715), p. 206.

144. Newton was aware that a shift in latitude causes a variation in rotational speed, since $v = 2r/T \times \cos \varphi$, where v is the linear tangential speed at latitude φ; r, T being the average values of the radius of the earth and the period of rotation. The distance from the center of the earth is also affected by latitude, since the earth is an oblate spheroid. These two factors appear in the variation with latitude in the length of a seconds pendulum.

145. "The Aim of Science," in *Ratio*, **1** (1957), 24–35; repr. in Karl Popper, *Objective Knowledge* (Oxford, 1972), 191–205.

146. See, for example, R. S. Westfall, *Force in Newton's Physics*. See also Alan Gabbey, "Force and Inertia in 17th-century Dynamics," in *Studies in History and Philosophy of Science*, **2** (1971), 1–67; Gabbey contests Westfall's point of view concerning the *vis insita*, in *Science*, **176** (1972), 157-159.

147. This would no longer even be called a force; some present translations, among them F. Cajori's version of Motte, anachronistically render Newton's *vis inertiae* as simple "inertia."

148. University Library, Cambridge, MS Add. 3968, fol. 415; published in A. Koyré and I. B. Cohen, "Newton and the Leibniz-Clarke Correspondence," in *Archives internationales d'histoire des sciences*, **15** (1962), 122–141.

149. See I. B. Cohen, "Newton's Second Law and the Concept of Force in the *Principia*," in R. Palter, ed., *Annus Mirabilis*, pp. 143–185.

150. R. S. Westfall, *Force*, p. 490. It is with this point of view in particular that Gabbey takes issue; see n. 146. See further E. J. Aiton, "The Concept of Force," in A. C. Crombie and M. A. Hoskin, eds., *History of Science*, X (Cambridge, 1971), 88–102.

151. In prop. 7, bk. III (referring to prop. 69, bk. I, and its corollaries), Newton argued from "accelerative" measures of forces to "absolute" forces, in specific cases of attraction.

152. See D. T. Whiteside, in *History of Science*, V (Cambridge, 1966), 110.

153. E. J. Aiton, "The Inverse Problem of Central Forces," in *Annals of Science*, **20** (1964), 82.

154. This position of the *Principia* was greatly altered between the 1st and 2nd eds.; Newton's intermediate results were summarized in a set of procedural rules for making up lunar tables and were published in a Latin version in David Gregory's treatise on astronomy (1702). Several separate English versions were later published; these are reprinted in facsimile in I. B. Cohen, *Newton's Theory of the Moon* (London, 1974).

155. W. W. Rouse Ball gives a useful paraphrase in *Essay*, p. 92.

156. See the analyses by Clifford Truesdell, listed in the bibliography to this article.

157. In his review of the *Principia*, in *Philosophical Transactions* (1687), p. 295, Halley referred specifically to this proposition, "which being rather a Physical than Mathematical Inquiry, our Author forbears to discuss."

158. This problem had gained prominence through the independent discovery by Halley and Richer that the length of a pendulum clock must be adjusted for changes in latitude.

159. This "General Scholium" should not be confused with the general scholium that ends the *Principia*. It was revised and expanded for the 2nd ed., where it appears at the end of sec. 6; in the 1st ed. it appears at the end of sec. 7.

160. In *Mécanique céleste*, V, bk. XII, ch. 3, sec. 7. Newton failed to take into account the changes in elasticity due to the "heat of compression and cold of rarefaction"; Laplace corrected Newton's formula ($v = k\sqrt{p/d}$), replacing it with his own ($v = k\sqrt{1.41\,p/d}$, where p is the air pressure and d the density of the air).

Laplace, who had first published his own results in 1816, later said that Newton's studies on the velocity of sound in the atmosphere were the most important application yet made of the equations of motion in elastic fluids: "sa théorie, quoique imparfaite, est un monument de son génie" (*Méchanique céleste*, V, bk. XII, ch. 1, pp. 95–96). Lord Rayleigh pointed out that Newton's investigations "established that the velocity of sound should be independent of the amplitude of the vibration, and also of the pitch."

161. The confutation of Descartes's vortex theory was thought by men of Newton's century to be one of the major aims of bk. II. Huygens, for one, accepted Newton's conclusion that the Cartesian vortices must be cast out of physics, and wrote to Leibniz to find out whether he would be able to continue to believe in them after reading the *Principia*. In "my view," Huygens wrote, "these vortices are superfluous if one accepts the system of Mr. Newton."

162. On the earlier tract in relation to bk. III of the *Principia*, see the preface to the repr. (London, 1969) and I. B. Cohen, *Introduction*, supp. VI.

163. At one time, according to a manuscript note, Newton was unequivocal that hypothesis 3 expressed the belief of Aristotle, Descartes, and unspecified "others." It was originally followed by a hypothesis 4, which in the 2nd and 3rd eds. was moved to a later part of bk. III. For details, see I. B. Cohen, "Hypotheses in Newton's Philosophy," in *Physis*, **8** (1966), 163–184.

164. See *De motu* in A. R. and M. B. Hall, *Unpublished Papers*, and J. Herivel, *Background*.

165. Newton apparently never made the experiment of comparing mass and weight of different quantities of the same material.

166. There has been little research on the general subject of Newton's lunar theory; even the methods he used to obtain the results given in a short scholium to prop. 35, bk. I, in the 1st ed., are not known. W. W. Rouse Ball, in *Essay*, p. 109, discusses Newton's formula for "the mean hourly motion of the moon's apogee," and says, "The investigation on this point is not entirely satisfactory, and from the alterations made in the MS. Newton evidently felt doubts about the correctness of the coefficient $\frac{11}{2}$ which occurs in this formula. From this, however, he deduces quite correctly that the mean annual motion of the apogee resulting would amount to $38°51'51''$, whereas the annual motion" is known to be $40°41'30''$. His discussion is based upon the statement, presumably by J. C. Adams, in the preface to the *Catalogue of the Portsmouth Collection* (Cambridge, 1888), pp. xii–xiii. Newton's MSS on the motion of the moon—chiefly University Library, Cambridge, MS Add. 3966—are one of the major unanalyzed collections of his work. For further documents concerning this topic, and a scholarly analysis by A. R. Hall of some aspects of Newton's researches on the motion of the moon, see *Correspondence*, V (in press), and I. B. Cohen, intro. to a facsimile repr. of Newton's pamphlet on the motion of the moon (London, in press).

167. Although Newton had suspected the association of color with wavelength of vibration as early as his "Hypothesis" of 1675, he did not go on from his experiments on rings, which suggested a periodicity in optical phenomena, to a true wave theory—no doubt because, as A. I. Sabra has suggested, his a priori "conception of the rays as discrete entities or corpuscles" effectively "prevented him from envisaging the possibility of an undulatory interpretation in which the ray, as something distinguished from the waves, would be redundant" (*Theories of Light*, p. 341).

168. Both printed in facsimile in I. B. Cohen, ed., *Isaac Newton's Papers and Letters on Natural Philosophy*. They were published and studied in the eighteenth century and had a significant influence on the development of the concept of electric fluid (or fluids) and caloric. This topic is explored in some detail in I. B. Cohen, *Franklin and Newton* (Philadelphia, 1956; Cambridge, 1966; rev. ed. in press), esp. chs. 6 and 7.

169. Henry Guerlac has studied the development of the queries themselves, and in particular the decline of Newton's use of the ether until its reappearance in a new form in the queries of the 2nd English ed. He has also noted that the concept of the ether is conspicuously absent from the Latin ed. of 1706. See especially his "Newton's Optical Aether," in *Notes and Records. Royal Society of London*, **22** (1967), 45–57. See, further, Joan L. Hawes, "Newton's Revival of the Aether Hypothesis . . .," *ibid.*, **23** (1968), 200–212.

170. A. R. and M. B. Hall have found evidence that Newton thought of this "spiritus" as electrical in nature; see *Unpublished Papers*, pp. 231 ff., 348 ff. Guerlac has shown that Newton was fascinated by Hauksbee's electrical experiments and by certain experiments of Desaguliers; see bibliography for this series of articles.

171. University Library, Cambridge, MS Add. 3970, sec. 9, fols. 623 ff.

172. These works, especially queries 28 and 31, have been studied in conjunction with Newton's MSS (particularly his notebooks) by A. R. and M. B. Hall, D. McKie, J. R. Partington, R. Kargon, J. E. McGuire, A. Thackray, and others, in their elucidations of a Newtonian doctrine of chemistry or theory of matter. *De natura acidorum* has been printed from an autograph MS, with notes by Pitcairne and transcripts by David Gregory, in *Correspondence*, III, 205–214. The first printing, in both Latin and English, is reproduced in I. B. Cohen, ed., *Newton's Papers and Letters*, pp. 255–258.

173. According to M. B. Hall, "Newton's Chemical Papers," in *Newton's Papers and Letters*, p. 244.

174. *Ibid.*, p. 245.

175. Discussed by T. S. Kuhn, "Newton's '31st Query' and the Degradation of Gold," in *Isis*, **42** (1951), 296–298.

176. M. B. Hall, "Newton's Chemical Papers," p. 245; she continues that there we may find a "forerunner of the tables of affinity" developed in the eighteenth century, by means of which "chemists tried to predict the course of a reaction."

177. In "Newton's Chemical Experiments," in *Archives internationales d'histoire des sciences*, **11** (1958), 113–152—a study of Newton's chemical notes and papers—A. R. and M. B. Hall have tried to show that Newton's primary concern in these matters was the chemistry of metals, and that the writings of alchemists were a major source of information on every aspect of metals. Humphrey Newton wrote up a confusing account of Newton's alchemical experiments, in which he said that Newton's guide was the *De re metallica* of Agricola; this work, however, is largely free of alchemical overtones and concentrates on mining and metallurgy.

178. R. S. Westfall, in *Science and Religion in Seventeenth-Century England*, ch. 8, draws upon such expressions by Newton to prove that "Newton was a religious rationalist who remained blind to the mystic's spiritual communion with the divine."

179. These MSS are described in the Sotheby sale catalog and by F. Sherwood Taylor, in "An Alchemical Work of Sir Isaac Newton," in *Ambix*, **5** (1956), 59–84.

180. These have been the subject of a considerable study by Frank E. Manuel, *Isaac Newton, Historian* (Cambridge, Mass., 1964).

181. Newton's interest in alchemy mirrors all the bewildering aspects of that subject, ranging from the manipulative chemistry of metals, mineral acids, and salts, to esoteric and symbolic (often sexual) illustrations and mysticism of a religious or philosophical kind. His interest in alchemy persisted through his days at the mint, although there is no indication that he at that time still seriously believed that pure metallic gold might be produced from baser metals—if, indeed, he had ever so believed. The extent of his notes on his reading indicate the seriousness of Newton's interest in the general subject, but it is impossible to ascertain to what degree, if any, his alchemical concerns may have influenced his science, beyond his vague and general commitment to "transmutations" as a mode for the operations of nature. But even this belief would not imply a commitment to the entire hermetic tradition, and it is not necessary to seek a unity of the diverse interests and intellectual concerns in a mind as complex as Newton's.

182. P. M. Rattansi, "Newton's Alchemical Studies," in Allen Debus, ed., *Science, Medicine and Society in the Renaissance*, II (New York, 1972), 174.

183. The first suggestion that Newton's concept of the ether might be linked to his alchemical concerns was made by Taylor; see n. 179, above.

184. Leibniz, *Tentamen . . .* ("An Essay on the Cause of the Motions of the Heavenly Bodies"), in *Acta eruditorum* (Feb. 1689), 82–96, English trans. by E. J. Collins. Leibniz' marked copy of the 1st ed. of the *Principia*, presumably the one sent to him by Fatio de Duillier at Newton's direction, is now in the possession of E. A. Fellmann of Basel, who has discussed Leibniz' annotations in "Die Marginalnoten von Leibniz in Newtons Principia Mathematica 1687," in *Humanismus und Technik*, **2** (1972), 110–129; Fellmann's critical ed., G. W. Leibniz, *Marginalia in Newtoni Principia Mathematica 1687* (Paris, 1973), includes facsimiles of the annotated pages.

185. Translated from some MS comments on Leibniz' essay, first printed in Edleston, *Correspondence*, pp. 307–314.

186. Leibniz' excepts from Newton's *De analysi*, made in 1676 from a transcript by John Collins, have been published from the Hannover MS by Whiteside, in *Mathematical Papers*, II, 248–258. Whiteside thus demonstrates that Leibniz was "clearly interested only in its algebraic portions: fluxional sections are ignored."

187. Several MS versions in his hand survive in University Library, Cambridge, MS Add. 3968.

188. At this period the year in England officially began on Lady Day, 25 March. Hence Newton died on 20 March 1726 old style, or in 1726/7 (to use the form then current for dates in January, February, and the first part of March).

189. In the 2-vol. ed. of the *Principia* with variant readings edited by A. Koyré, I. B. Cohen, and Anne Whitman; Koyré has shown that in the English *Opticks* Newton used the word "feign" in relation to hypotheses, in the sense of "fingo" in the slogan, a usage confirmed by example in Newton's MSS. Motte renders the phrase as "I frame no hypotheses." Newton himself in MSS used both "feign" and "frame" in relation to hypotheses in this regard; see I. B. Cohen, "The First English Version of Newton's *Hypotheses non fingo*," in *Isis*, **53** (1962), 379–388.

190. University Library, Cambridge, MS Add. 3968, fol. 437.

BIBLIOGRAPHY

This bibliography is divided into four major sections. The last, by A. P. Youschkevitch, is concerned with Soviet studies on Newton and is independent of the text.

ORIGINAL WORKS (numbered I–IV): Newton's major writings, together with collected works and editions, bibliographies, manuscript collections, and catalogues.

SECONDARY LITERATURE (numbered V–VI): including general works and specific writings about Newton and his life.

SOURCES (numbered 1–11): the chief works used in the preparation of this biography; the subdivisions of this section are correlated to the subdivisions of the biography itself.

SOVIET LITERATURE: a special section devoted to Newtonian scholarship in the Soviet Union.

The first three sections of the bibliography contain a number of cross-references; a parenthetical number refers the reader to the section of the bibliography in which a complete citation may be found.

ORIGINAL WORKS

I. MAJOR WORKS. Newton's first publications were on optics and appeared in the *Philosophical Transactions of the Royal Society* (1672–1676); repr. in facs., with intro. by T. S. Kuhn, in I. B. Cohen, ed., *Isaac Newton's Papers & Letters on Natural Philosophy* (Cambridge, Mass., 1958; 2nd ed., in press). His *Opticks* (London, 1704; enl. versions in Latin [London, 1706], and in English [London, 1717 or 1718]) contained two supps.: his *Enumeratio linearum tertii ordinis* and *Tractatus de quadratura curvarum*, his first published works in pure mathematics. The 1704 ed. has been repr. in facs. (Brussels, 1966) and (optical part only) in type (London, 1931); also repr. with an analytical table of contents prepared by D. H. D. Roller (New York, 1952). French trans. are by P. Coste (Amsterdam, 1720; rev. ed. 1722; facs. repr., with intro. by M. Solovine, Paris, 1955); a German ed. is W. Abendroth, 2 vols. (Leipzig, 1898); and a Rumanian trans. is Victor Marian (Bucharest, 1970). A new ed. is currently being prepared by Henry Guerlac.

The *Philosophiae naturalis principia mathematica* (London, 1687; rev. eds., Cambridge, 1713 [repr. Amsterdam, 1714, 1723], and London, 1726) is available in an ed. with variant readings (based on the three printed eds., the MS for the 1st ed. and Newton's annotations in his own copies of the 1st and 2nd eds.) prepared by A. Koyré, I. B. Cohen, and Anne Whitman: *Isaac Newton's Philosophiae naturalis principia mathematica, the Third Edition (1726) With Variant Readings*, 2 vols. (Cambridge, Mass.–Cambridge, England, 1972). Translations and excerpts have appeared in Dutch, English, French, German, Italian, Japanese, Rumanian, Russian, and Swedish, and are listed in app. VIII, vol. II, of the Koyré, Cohen, and Whitman ed., together with an account of reprs. of the whole treatise. The 1st ed. has been printed twice in facs. (London, 1954[?]; Brussels, 1965).

William Jones published Newton's *De analysi* in his ed. of *Analysis per quantitatum series, fluxiones, ac differentias* . . . (London, 1711), repr. in the Royal Society's *Commercium epistolicum D. Johannis Collins, et aliorum de analysi promota* . . . (London, 1712–1713; enl. version, 1722; "variorum" ed. by J.-B. Biot and F. Lefort, Paris, 1856), and as an appendix to the 1723 Amsterdam printing of the *Principia*. Newton's *Arithmetica universalis* was published from the MS of Newton's lectures by W. Whiston (Cambridge, 1707); an amended ed. followed, supervised by Newton himself (London, 1722). For bibliographical notes on these and some other mathematical writings (and indications of other eds. and translations), see the introductions by D. T. Whiteside to the facs. repr. of *The Mathematical Works of Isaac Newton*, 2 vols. (New York–London, 1964–1967). Newton's *Arithmetica universalis* was translated into Russian with notes and commentaries by A. P. Youschkevitch (Moscow, 1948); English eds. were published in London in 1720, 1728, and 1769.

After Newton's death the early version of what became bk. III of the *Principia* was published in English as *A Treatise of the System of the World* (London, 1728; rev. London, 1731, facs. repr., with intro. by I. B. Cohen, London, 1969) and in Latin as *De mundi systemate liber* (London, 1728). An Italian trans. is by Marcella Renzoni (Turin, 1959; 1969). The first part of the *Lectiones opticae* was translated and published as *Optical Lectures* (London, 1728) before the full Latin ed. was printed (1729); both are imperfect and incomplete. The only modern ed. is in Russian, *Lektsii po optike* (Leningrad, 1946), with commentary by S. I. Vavilov.

For Newton's nonscientific works (theology, biblical studies, chronology), and for other scientific writings, see the various sections below.

II. COLLECTED WORKS OR EDITIONS. The only attempt ever made to produce a general ed. of Newton was S. Horsley, *Isaaci Newtoni opera quae exstant omnia*, 5 vols. (London, 1779–1785; photo repr. Stuttgart–Bad Cannstatt, 1964), which barely takes account of Newton's available MS writings but has the virtue of including (vol. I) the published mathematical tracts; (vols. II–III) the *Principia* and *De mundi systemate*, *Theoria lunae*, and *Lectiones opticae*; (vol. IV) letters from the *Philosophical Transactions* on light and color, the letter to Boyle on the ether, *De problematis Bernoullianis*, the letters to Bentley, and the *Commercium epistolicum*; (vol. V) the *Chronology*, the *Prophecies*, and the *Corruptions of Scripture*. An earlier and more modest collection was the 3-vol. *Opuscula mathematica, philosophica, et philologica*, Giovanni Francesco Salvemini (known as Johann Castillon), ed. (Lausanne–Geneva, 1744); it contains only works then in print.

A major collection of letters and documents, edited in the most exemplary manner, is Edleston (1); Rigaud's *Essay* (5) is also valuable. S. P. Rigaud's *Correspondence of Scientific Men of the Seventeenth Century . . . in the collection of . . . the Earl of Macclesfield*, 2 vols. (Oxford, 1841; rev., with table of contents and index, 1862) is of special importance because the Macclesfield collection is not at present open to scholars.

Four vols. of the Royal Society's ed. of Newton's *Correspondence* (Cambridge, 1959–   ) have (as of 1974) been published, vols. I–III edited by H. W. Turnbull, vol. IV by J. F. Scott; A. R. Hall has been appointed editor of the succeeding volumes. The *Correspondence* is not limited to letters but contains scientific documents of primary importance. A recent major collection is A. R. and M. B. Hall, eds., *Unpublished Scientific Papers of Isaac Newton, a Selection From the Portsmouth Collection in the University Library, Cambridge* (Cambridge, 1964). Other presentations of MSS are given in the ed. of the *Principia* with variant readings (1972, cited above), Herivel's *Background* (5), and in D. T. Whiteside's ed. of Newton's *Mathematical Papers* (3).

III. BIBLIOGRAPHIES. There are three bibliographies of Newton's writings, none complete or free of major error. One is George J. Gray, *A Bibliography of the Works of Sir Isaac Newton, Together With a List of Books Illustrating His Works*, 2nd ed., rev. and enl. (Cambridge, 1907; repr. London, 1966); H. Zeitlinger, "A Newton Bibliography," pp. 148–170 of the volume ed. by W. J. Greenstreet (VI); and *A Descriptive Catalogue of the Grace K. Babson Collection of the Works of Sir Isaac Newton . . .* (New York, 1950), plus *A Supplement . . .* compiled by Henry P. Macomber (Babson Park, Mass., 1955), which lists some secondary materials from journals as well as books.

IV. MANUSCRIPT COLLECTIONS AND CATALOGUES. The Portsmouth Collection (University Library, Cambridge) was roughly catalogued by a syndicate consisting of H. R. Luard, G. G. Stokes, J. C. Adams, and G. D. Liveing, who produced *A Catalogue of the Portsmouth Collection of Books and Papers Written by or Belonging to Sir Isaac Newton . . .* (Cambridge, 1888); the bare descriptions do not always identify the major MSS or give the catalogue numbers (*e.g.*, the Waste Book, U.L.C. MS Add. 4004, the major repository of Newton's early work in dynamics and in mathematics, appears as "A common-place book, written originally by B. Smith, D.D., with calculations by Newton written in the blank spaces. This contains Newton's first idea of Fluxions"). There is no adequate catalogue or printed guide to the Newton MSS in the libraries of Trinity College (Cambridge), the Royal Society of London, or the British Museum. The Keynes Collection (in the library of King's College, Cambridge) is almost entirely based on the Sotheby sale and is inventoried in the form of a marked copy of the sale catalogue, available in the library; see A. N. L. Munby, "The Keynes Collection of the Works of Sir Isaac Newton at King's College, Cambridge," in *Notes and Records. Royal Society of London*, **10** (1952), 40–50. The "scientific portion" of the Portsmouth Collection was given to Cambridge University in the 1870's; the remainder was dispersed at public auction in 1936. See Sotheby's *Catalogue of the Newton Papers, Sold by Order of the Viscount Lymington, to Whom They Have Descended From Catherine Conduitt, Viscountess Lymington, Great-niece of Sir Isaac Newton* (London, 1936). No catalogue has ever been made available of the Macclesfield Collection (rich in Newton MSS), based originally on the papers of John Collins and William Jones,

for which see S. P. Rigaud's 2-vol. *Correspondence . . .* (I). Further information concerning MS sources is given in Whiteside, *Mathematical Papers*, I, xxiv–xxxiii (3).

Many books from Newton's library are in the Trinity College Library (Cambridge); others are in public and private collections all over the world. R. de Villamil, *Newton: The Man* (London, 1931[?]; repr., with intro. by I. B. Cohen, New York, 1972), contains a catalogue (imperfect and incomplete) of books in Newton's library at the time of his death; an inventory with present locations of Newton's books is greatly to be desired. See P. E. Spargo, "Newton's Library," in *Endeavour*, **31** (1972), 29–33, with short but valuable list of references. See also *Library of Sir Isaac Newton. Presentation by the Pilgrim Trust to Trinity College Cambridge 30 October 1943* (Cambridge, 1944), described on pp. 5–7 of *Thirteenth Annual Report of the Pilgrim Trust* (Harlech, 1943).

SECONDARY LITERATURE

V. GUIDES TO THE SECONDARY LITERATURE. For guides to the literature concerning Newton, see . . . *Catalogue . . . Babson Collection . . .* (III); and scholarly eds., such as *Mathematical Papers* (3), *Principia* (I), and *Correspondence* (II). A most valuable year-by-year list of articles and books has been prepared and published by Clelia Pighetti: "Cinquant'anni di studi newtoniani (1908–1959)," in *Rivista critica di storia della filosofia*, **20** (1960), 181–203, 295–318. See also Magda Whitrow, ed., *ISIS Cumulative Bibliography . . . 1913–65*, II (London, 1971), 221–232. Two fairly recent surveys of the literature are I. B. Cohen, "Newton in the Light of Recent Scholarship," in *Isis*, **51** (1960), 489–514; and D. T. Whiteside, "The Expanding World of Newtonian Research," in *History of Science*, **1** (1962), 16–29.

VI. GENERAL WORKS. Biographies (*e.g.*, by Stukeley, Brewster, More, Manuel) are listed below (1). Some major interpretive works and collections of studies on Newton are Ferd. Rosenberger, *Isaac Newton und seine physikalischen Principien* (Leipzig, 1895); Léon Bloch, *La philosophie de Newton* (Paris, 1908); S. I. Vavilov, *Isaak Nyuton; nauchnaya biografia i stati*, 3rd ed. (Moscow, 1961), German trans. by Josef Grün as *Isaac Newton* (Vienna, 1948), 2nd ed., rev., German trans. by Franz Boncourt (Berlin, 1951); Alexandre Koyré, *Newtonian Studies* (London–Cambridge, Mass., 1965) which, posthumously published, contains a number of errors—a more correct version is the French trans., *Études newtoniennes* (Paris, 1968), with an *avertissement* by Yvon Belaval; and Alberto Pala, *Isaac Newton, scienza e filosofia* (Turin, 1969).

Major collections of Newtonian studies include W. J. Greenstreet, ed., *Isaac Newton 1642–1727* (London, 1927); F. E. Brasch, ed., *Sir Isaac Newton 1727–1927* (Baltimore, 1928); S. I. Vavilov, ed., *Isaak Nyuton 1643[n.s.]–1727*, a symposium in Russian (Moscow–Leningrad, 1943); Royal Society, *Newton Tercentenary Celebrations, 15–19 July 1946* (Cambridge, 1947); and Robert Palter, ed., *The Annus Mirabilis of Sir Isaac Newton 1666–1966* (Cambridge, Mass., 1970), based on an earlier version in *The Texas Quarterly*, **10**, no. 3 (autumn 1967).

On Newton's reputation and influence (notably in the eighteenth century), see Hélène Metzger, *Newton, Stahl, Boerhaave et la doctrine chimique* (Paris, 1930), and *Attraction universelle et religion naturelle chez quelques commentateurs anglais de Newton* (Paris, 1938); Pierre Brunet, *L'introduction des théories de Newton en France au XVIIIe siècle*, I, *Avant 1738* (Paris, 1931); Marjorie Hope Nicolson, *Newton Demands the Muse, Newton's Opticks and the Eighteenth Century Poets* (Princeton, 1946); I. B. Cohen, *Franklin and Newton, an Inquiry Into Speculative Newtonian Experimental Science . . .* (Philadelphia, 1956; Cambridge, Mass., 1966; rev. repr. 1974); Henry Guerlac, "Where the Statue Stood: Divergent Loyalties to Newton in the Eighteenth Century," in Earl R. Wasserman, ed., *Aspects of the Eighteenth Century* (Baltimore, 1965), pp. 317–334; R. E. Schofield, *Mechanism and Materialism, British Natural Philosophy in an Age of Reason* (Princeton, 1970); Paolo Casini, *L'universo-macchina, origini della filosofia newtoniana* (Bari, 1969); and Arnold Thackray, *Atoms and Powers, an Essay in Newtonian Matter-Theory and the Development of Chemistry* (Cambridge, Mass., 1970). Still of value today are three major eighteenth-century expositions of the Newtonian natural philosophy, by Henry Pemberton, Voltaire, and Colin Maclaurin.

Whoever studies any of Newton's mathematical or scientific writings would be well advised to consult J. A. Lohne, "The Increasing Corruption of Newton's Diagrams," in *History of Science*, 6 (1967), 69–89.

Newton's MSS comprise some 20–25 million words; most of them have never been studied fully, and some are currently "lost," having been dispersed at the Sotheby sale in 1936. Among the areas in which there is a great need for editing of MSS and research are Newton's studies of lunar motions (chiefly U.L.C. MS Add. 3966); his work in optics (chiefly U.L.C. MS Add. 3970; plus other MSS such as notebooks, etc.); and the technical innovations he proposed for the *Principia* in the 1690's (chiefly U.L.C. MS Add. 3965); see (4), (7). It would be further valuable to have full annotated editions of his early notebooks and of some major alchemical notes and writings.

Some recent Newtonian publications include Valentin Boss, *Newton and Russia, the Early Influence 1698–1796* (Cambridge, Mass., 1972); Klaus-Dietwardt Buchholtz, *Isaac Newton als Theologe* (Wittenburg, 1965); Mary S. Churchill, "The Seven Chapters With Explanatory Notes," in *Chymia*, 12 (1967), 27–57, the first publication of one of Newton's complete alchemical MS; J. E. Hofmann, "Neue Newtoniana," in *Studia Leibnitiana*, 2 (1970), 140–145, a review of recent literature; D. Kubrin, "Newton and the Cyclical Cosmos," in *Journal of the History of Ideas*, 28 (1967), 325–346; J. E. McGuire, "The Origin of Newton's Doctrine of Essential Qualities," in *Centaurus*, 12 (1968), 233–260; and L. Trengrove, "Newton's Theological Views," in *Annals of Science*, 22 (1966), 277–294.

SOURCES

1. *Early Life and Education.* The major biographies of Newton are David Brewster, *Memoirs of the Life, Writings,* *and Discoveries of Isaac Newton*, 2 vols. (Edinburgh, 1855; 2nd ed., 1860; repr. New York, 1965), the best biography of Newton, despite its stuffiness; for a corrective, see Augustus De Morgan, *Essays on the Life and Work of Newton* (Chicago–London, 1914); Louis Trenchard More, *Isaac Newton* (New York–London, 1934; repr. New York, 1962); and Frank E. Manuel, *A Portrait of Isaac Newton* (Cambridge, Mass., 1968). Of the greatest value is the "synoptical view" of Newton's life, pp. xxi–lxxxi, with supplementary documents, in J. Edleston, ed., *Correspondence of Sir Isaac Newton and Professor Cotes . . .* (London, 1850; repr. London, 1969). Supplementary information concerning Newton's youthful studies is given in D. T. Whiteside, "Isaac Newton: Birth of a Mathematician," in *Notes and Records. Royal Society of London*, 19 (1964), 53–62, and "Newton's Marvellous Year: 1666 and All That," *ibid.*, 21 (1966), 32–41.

John Conduitt assembled recollections of Newton by Humphrey Newton, William Stukeley, William Derham, A. De Moivre, and others, which are now mainly in the Keynes Collection, King's College, Cambridge. Many of these documents have been printed in Edmund Turnor, *Collections for the History of the Town and Soke of Grantham* (London, 1806). William Stukeley's *Memoirs of Sir Isaac Newton's Life* (1752) was edited by A. Hastings White (London, 1936).

On Newton's family and origins, see C. W. Foster, "Sir Isaac Newton's Family," in *Reports and Papers of the Architectural Societies of the County of Lincoln, County of York, Archdeaconries of Northampton and Oakham, and County of Leicester*, 39 (1928–1929), 1–62. Newton's early notebooks are in Cambridge in the University Library, the Fitzwilliam Museum, and Trinity College Library; and in New York City in the Morgan Library. For the latter, see David Eugene Smith, "Two Unpublished Documents of Sir Isaac Newton," in W. J. Greenstreet, ed., *Isaac Newton 1642–1727* (London, 1927), pp. 16 ff. Also, E. N. da C. Andrade, "Newton's Early Notebook," in *Nature*, 135 (1935), 360; George L. Huxley: "Two Newtonian Studies: I. Newton's Boyhood Interests," in *Harvard Library Bulletin*, 13 (1959), 348–354; and A. R. Hall, "Sir Isaac Newton's Notebook, 1661–1665," in *Cambridge Historical Journal*, 9 (1948), 239–250. Elsewhere, Andrade has shown that Newton did not write the poem, attributed to him, concerning Charles II, a conclusion supported by William Stukeley's 1752 *Memoirs of Sir Isaac Newton's Life*, A. Hastings White, ed. (London, 1936).

On Newton's early diagrams and his sundial, see Charles Turnor, "An Account of the Newtonian Dial Presented to the Royal Society," in *Proceedings of the Royal Society*, 5 (1851), 513 (13 June 1844); and H. W. Robinson, "Note on Some Recently Discovered Geometrical Drawings in the Stonework of Woolsthorpe Manor House," in *Notes and Records. Royal Society of London*, 5 (1947), 35–36. For Newton's catalogue of "sins," see R. S. Westfall, "Short-writing and the State of Newton's Conscience, 1662," in *Notes and Records. Royal Society of London*, 18 (1963), 10–16.

On Newton's early reading, see R. S. Westfall, "The

Foundations of Newton's Philosophy of Nature," *British Journal for the History of Science*, **1** (1962), 171–182, which is repr. in somewhat amplified form in his *Force in Newton's Physics*. On Newton's reading, see further I. B. Cohen, *Introduction to Newton's Principia* (7) and vol. I of Whiteside's ed. of Newton's *Mathematical Papers* (3). And, of course, a major source of biographical information is the Royal Society's edition of Newton's *Correspondence* (II).

2. *Lucasian Professor.* For the major sources concerning this period of Newton's life, see (1) above, notably Brewster, Cohen (*Introduction*), Edleston, Manuel, More, Whiteside (*Mathematical Papers*), and *Correspondence*.

Edleston (pp. xci–xcviii) gives a "Table of Newton's Lectures as Lucasian Professor," with the dates and corresponding pages of the deposited MSS and the published ed. for the lectures on optics (U.L.C. MS Dd. 9.67, deposited 1674; printed London, 1729); lectures on arithmetic and algebra (U.L.C. MS Dd. 9.68; first published by Whiston, Cambridge, 1707); lectures *De motu corporum* (U.L.C. MS Dd. 9.46), corresponding *grosso modo* to bk. I of the *Principia* through prop. 54; and finally *De motu corporum liber secundus* (U.L.C. MS Dd. 9.67); of which a more complete version was printed as *De mundi systemate liber* (London, 1728)—see below.

Except for the last two, the deposited lectures are final copies, complete with numbered illustrations, as if ready for the press or for any reader who might have access to these MSS. The *Lectiones opticae* exist in two MS versions, an earlier one, which Newton kept (U.L.C. MS Add. 4002, in Newton's hand), having a division by dates quite different from that of the deposited lectures; this has been printed in facs., with an intro. by D. T. Whiteside as *The Unpublished First Version of Isaac Newton's Cambridge Lectures on Optics 1670–1672* (Cambridge, 1973). See I. B. Cohen, *Introduction*, supp. III, "Newton's Professorial Lectures," esp. pp. 303–306.

The deposited MS *De motu corporum* consists of leaves corresponding to different states of composition of bk. I of the *Principia*; the second state (in the hand of Humphrey Newton, with additions and emendations by Isaac Newton) is all but the equivalent of the corresponding part of the MS of the *Principia* sent to the printer, but the earlier state is notably different and more primitive. See I. B. Cohen, *Introduction*, supp. IV, pp. 310–321.

Edleston did not list the deposited copy of the lectures for 1687, a fair copy of only the first portion of *De motu corporum liber secundus* (corresponding to the first 27 sections, roughly half of Newton's own copy of the whole work, U.L.C. MS Add. 3990); he referred to a copy of the deposited lectures made by Cotes (Trinity College Library, MS R.19.39), in which the remainder of the text was added from a copy of the whole MS belonging to Charles Morgan. See I. B. Cohen, *Introduction*, supp. III, pp. 306–308, and supp. VI, pp. 327–335. This MS, an early version of what was to be rewritten as *Liber tertius: De mundi systemate* of the *Principia*, was published in English (London, 1728) and in Latin (London, 1728); see I. B. Cohen, "Newton's *System of the World*," in *Physis*,

11 (1969), 152–166; and intro. to repr. of the English *System of the World* (London, 1969).

The statutes of the Lucasian professorship (dated 19 Dec. 1663) are printed in the appendix to William Whiston's *An Account of . . . [His] Prosecution at, and Banishment From, the University of Cambridge* (London, 1718) and are printed again by D. T. Whiteside in Newton's *Mathematical Papers*, III, xx–xxvii.

It is often supposed, probably mistakenly, that Newton actually read the lectures that he deposited, or that the deposited lectures are evidence of the state of his knowledge or his formulation of a given subject at the time of giving a particular lecture, because the deposited MSS may be divided into dated lectures; but the statutes required that the lectures be rewritten after they had been read.

The MSS of Humphrey Newton's memoranda are in the Keynes Collection, King's College, Cambridge (K. MS 135) and are printed in David Brewster, *Memoirs*, II, 91–98, and again in L. T. More, *Isaac Newton*, pp. 246–251.

The evidence for Newton's plan to publish an ed. of his early optical papers, including the letters in the *Philosophical Transactions*, is in a set of printed pages (possibly printed proofs) forming part of such an annotated printing of these letters, discovered by D. J. de S. Price. See I. B. Cohen, "Versions of Isaac Newton's First Published Paper With Remarks on . . . an Edition of His Early Papers on Light and Color," in *Archives internationales d'histoire des sciences*, **11** (1958), 357–375; D. J. de S. Price, "Newton in a Church Tower: The Discovery of an Unknown Book by Isaac Newton," in *Yale University Library Gazette*, **34** (1960), 124–126; A. R. Hall, "Newton's First Book," in *Archives internationales d'histoire des sciences*, **13** (1960), 39–61. On 5 Mar. 1677, Collins wrote to Newton that David Loggan "informs me that he hath drawn your effigies in order to [produce] a sculpture thereof to be prefixed to a book of Light [&] Colours [&] Dioptricks which you intend to publish."

The most recent and detailed analysis of the Newton-Fatio relationship is given in Frank E. Manuel, *A Portrait of Isaac Newton*, ch. 9, "The Ape of Newton: Fatio de Duillier," and ch. 10, "The Black Year 1693." For factual details, see Newton, *Correspondence*, III. The late Charles A. Domson completed a doctoral dissertation, "Nicolas Fatio de Duillier and the Prophets of London: An Essay in the Historical Interaction of Natural Philosophy and Millennial Belief in the Age of Newton" (Yale, 1972).

Newton's gifts to the Trinity College Library are listed in an old MS catalogue of the library; see I. B. Cohen: "Newton's Attribution of the First Two Laws of Motion to Galileo," in *Atti del Symposium internazionale di storia, metodologia, logica e filosofia della scienza: "Galileo nella storia e nella filosofia della scienza"* (Florence, 1967), pp. xxii–xlii, esp. pp. xxvii–xxviii and n. 22.

3. *Mathematics.* The primary work for the study of Newton's mathematics is the ed. (to be completed in 8 vols.) by D. T. Whiteside: *Mathematical Papers of Isaac Newton* (Cambridge, 1967–   ). Whiteside has also provided a valuable pair of introductions to a facs. repr. of early translations of a number of Newton's tracts, *The Mathe-*

matical Works of Isaac Newton, 2 vols. (New York–London, 1964–1967); these introductions give an admirable and concise summary of the development of Newton's mathematical thought and contain bibliographical notes on the printings and translations of the tracts reprinted, embracing *De analysi*; *De quadratura*; *Methodus fluxionum et serierum infinitarum*; *Arithmetica universalis* (based on his professorial lectures, deposited in the University Library); *Enumeratio linearum tertii ordinis*; and *Methodus differentialis* ("Newton's Interpolation Formulas"). Attention may also be directed to several other of Whiteside's publications: "Isaac Newton: Birth of a Mathematician," in *Notes and Records. Royal Society of London*, **19** (1964), 53–62; "Newton's Marvellous Year: 1666 and All That," *ibid.*, **21** (1966), 32–41; "Newton's Discovery of the General Binomial Theorem," in *Mathematical Gazette*, **45** (1961), 175–180. (See other articles of his cited in (6), (7), (8) below.)

Further information concerning the eds. and translations of Newton's mathematical writings may be gleaned from the bibliographies (Gray, Zeitlinger, Babson) cited above (III). Various Newtonian tracts appeared in Johann Castillon's *Opuscula* . . . (II), I, supplemented by a two-volume ed. (Amsterdam, 1761) of *Arithmetica universalis*. The naturalist Buffon translated the *Methodus fluxionum . . .* (Paris, 1740), and James Wilson replied to Buffon's preface in an appendix to vol. II (1761) of his own ed. of Benjamin Robins' *Mathematical Tracts*; these two works give a real insight into "what an interested student could then know of Newton's private thoughts." See also Pierre Brunet, "La notion d'infini mathématique chez Buffon," in *Archeion*, **13** (1931), 24–39; and Lesley Hanks, *Buffon avant l'"Histoire naturelle"* (Paris, 1966), pt. 2, ch. 4 and app. 4. Horsley's ed. of Newton's *Opera* (II) contains some of Newton's mathematical tracts. A modern version of the *Arithmetica universalis*, with extended notes and commentary, has been published by A. P. Youschkevitch (Moscow, 1948). A. Rupert Hall and Marie Boas Hall have published Newton's October 1666 tract, "to resolve problems by motion" (U.L.C. MS Add. 3458, fols. 49–63) in their *Unpublished Scientific Papers* (II); see also H. W. Turnbull, "The Discovery of the Infinitesimal Calculus," in *Nature*, **167** (1951), 1048–1050.

Newton's *Correspondence* (II) contains letters and other documents relating to mathematics, with valuable annotations by H. W. Turnbull and J. F. Scott. See, further, Turnbull's *The Mathematical Discoveries of Newton* (London–Glasgow, 1945), produced before he started to edit the *Correspondence* and thus presenting a view not wholly borne out by later research. Carl B. Boyer has dealt with Newton in *Concepts of the Calculus* (New York, 1939; repr. 1949, 1959), ch. 5; "Newton as an Originator of Polar Coordinates," in *American Mathematical Monthly*, **56** (1949), 73–78; *History of Analytic Geometry* (New York, 1956), ch. 7; and *A History of Mathematics* (New York, 1968), ch. 19.

Other secondary works are W. W. Rouse Ball, *A Short Account of the History of Mathematics*, 4th ed. (London, 1908), ch. 16—even more useful is his *A History of the Study of Mathematics at Cambridge* (Cambridge, 1889), chs. 4–6; J. F. Scott, *A History of Mathematics* (London, 1958), chs. 10, 11; and Margaret E. Baron, *The Origins of the Infinitesimal Calculus* (Oxford–London–New York, 1969).

Some specialized studies of value are D. T. Whiteside, "Patterns of Mathematical Thought in the Later Seventeenth Century," in *Archive for History of Exact Sciences*, **1** (1961), 179–388; W. W. Rouse Ball, "On Newton's Classification of Cubic Curves," in *Proceedings of the London Mathematical Society*, **22** (1891), 104–143, summarized in *Bibliotheca mathematica*, n.s. **5** (1891), 35–40; Florian Cajori, "Fourier's Improvement of the Newton-Raphson Method of Approximation Anticipated by Mourraile," in *Bibliotheca mathematica*, **11** (1910–1911), 132–137; "Historical Note on the Newton-Raphson Method of Approximation," in *American Mathematical Monthly*, **18** (1911), 29–32; and *A History of the Conceptions of Limits and Fluxions in Great Britain From Newton to Woodhouse* (Chicago–London, 1919); W. J. Greenstreet, ed., *Isaac Newton 1642–1727* (London, 1927), including D. C. Fraser, "Newton and Interpolation"; A. R. Forsyth, "Newton's Problem of the Solid of Least Resistance"; J. J. Milne, "Newton's Contribution to the Geometry of Conics"; H. Hilton, "Newton on Plane Cubic Curves"; and J. M. Child, "Newton and the Art of Discovery"; Duncan C. Fraser, *Newton's Interpolation Formulas* (London, 1927), repr. from *Journal of the Institute of Actuaries*, **51** (1918–1919), 77–106, 211–232, and **58** (1927), 53–95; C. R. M. Talbot, *Sir Isaac Newton's Enumeration of Lines of the Third Order, Generation of Curves by Shadows, Organic Description of Curves, and Construction of Equations by Curves*, trans. from the Latin, with notes and examples (London, 1860); Florence N. David, "Mr. Newton, Mr. Pepys and Dyse," in *Annals of Science*, **13** (1957), 137–147, on dice-throwing and probability; Jean Pelseneer, "Une lettre inédite de Newton à Pepys (23 décembre 1693)," in *Osiris*, **1** (1936), 497–499, on probabilities; J. M. Keynes, "A Mathematical Analysis by Newton of a Problem in College Administration," in *Isis*, **49** (1958), 174–176; Maximilian Miller, "Newton, Aufzahlung der Linien dritter Ordnung," in *Wissenschaftliche Zeitschrift der Hochschule für Verkehrswesen, Dresden*, **1**, no. 1 (1953), 5–32; "Newtons Differenzmethode," *ibid.*, **2**, no. 1 (1954), 1–13; and "Über die Analysis mit Hilfe unendlicher Reihen," *ibid.*, no. 2 (1954), 1–16; Oskar Bolza, "Bemerkungen zu Newtons Beweis seines Satzes über den Rotationskörper kleinsten Widerstandes," in *Bibliotheca mathematica*, 3rd ser., **13** (1912–1913), 146–149.

Other works relating to Newton's mathematics are cited in (6) and (for the quarrel with Leibniz over priority in the calculus) (10).

4. *Optics*. The eds. of the *Opticks* and *Lectiones opticae* are mentioned above (I); the two MS versions of the latter are U.L.C. MS Add. 4002, MS Dd.9.67. An annotated copy of the 1st ed. of the *Opticks*, used by the printer for the composition of the 2nd ed. still exists (U.L.C. MS Adv.b.39.3—formerly MS Add. 4001). For information Cohen, *Introduction to Newton's Principia* (7), p. 34;

and R. S. Westfall, "Newton's Reply," pp. 83–84—extracts are printed with commentary in D. T. Whiteside's ed. of Newton's *Mathematical Papers* (3). At one time Newton began to write a *Fundamentum opticae*, the text of which is readily reconstructible from the MSS and which is a necessary tool for a complete analysis of bk. I of the *Opticks*, into which its contents were later incorporated; for pagination, see *Mathematical Papers* (3), III, 552. This work is barely known to Newton scholars. Most of Newton's optical MSS are assembled in the University Library, Cambridge, as MS Add. 3970, but other MS writings appear in the Waste Book, correspondence, and various notebooks.

Among the older literature, F. Rosenberger's book (VI) may still be studied with profit, and there is much to be learned from Joseph Priestley's 18th-century presentation of the development and current state of concepts and theories of light and vision. See also Ernst Mach, *The Principles of Physical Optics: An Historical and Philosophical Treatment*, trans. by John S. Anderson and A. F. A. Young (London, 1926; repr. New York, 1953); and Vasco Ronchi, *The Nature of Light: An Historical Survey*, trans. by V. Barocas (Cambridge, Mass., 1970)—also 2 eds. in Italian and a French translation by Juliette Taton.

Newton's MSS have been used in A. R. Hall, "Newton's Notebook" (1), pp. 239–250; and in J. A. Lohne, "Newton's 'Proof' of the Sine Law," in *Archive for History of Exact Sciences*, 1 (1961), 389–405; "Isaac Newton: The Rise of a Scientist 1661–1671," in *Notes and Records. Royal Society of London*, 20 (1965), 125–139; and "Experimentum crucis," *ibid.*, 23 (1968), 169–199. See also J. A. Lohne and Bernhard Sticker, *Newtons Theorie der Prismenfarben, mit Übersetzung und Erläuterung der Abhandlung von 1672* (Munich, 1969); and R. S. Westfall, "The Development of Newton's Theory of Color," in *Isis*, 53 (1962), 339–358; "Newton and his Critics on the Nature of Colors," in *Archives internationales d'histoire des sciences*, 15 (1962), 47–58; "Newton's Reply to Hooke and the Theory of Colors," in *Isis*, 54 (1963), 82–96; "Isaac Newton's Coloured Circles Twixt Two Contiguous Glasses," in *Archive for History of Exact Sciences*, 2 (1965), 181–196; and "Uneasily Fitful Reflections on Fits of Easy Transmission [and of easy reflection]," in Robert Palter, ed., *The Annus Mirabilis* (VI), pp. 88–104.

Newton's optical papers (from the *Philosophical Transactions* and T. Birch's *History of the Royal Society*) are repr. in facs. in *Newton's Papers and Letters* (I), with an intro. by T. S. Kuhn. See also I. B. Cohen, "I prismi del Newton e i prismi dell'Algarotti," in *Atti della Fondazione "Giorgio Ronchi"* (Florence), 12 (1957), 1–11; Vasco Ronchi, "I 'prismi del Newton' del Museo Civico di Treviso," *ibid.*, 12–28; and N. R. Hanson, "Waves, Particles, and Newton's 'Fits,'" in *Journal of the History of Ideas*, 21 (1960), 370–391. On Newton's work on color, see George Biernson, "Why did Newton see Indigo in the Spectrum?," in *American Journal of Physics*, 40 (1972), 526–533; and Torger Holtzmark, "Newton's *Experimentum Crucis* Reconsidered," *ibid.*, 38 (1970), 1229–1235.

An able account of Newton's work in optics, set against the background of his century, is A. I. Sabra, *Theories of Light From Descartes to Newton* (London, 1967), ch. 9–13. An important series of studies, based on extensive examination of the MSS, are Zev Bechler, "Newton's 1672 Optical Controversies: A Study in the Grammar of Scientific Dissent," in Y. Elkana, ed., *Some Aspects of the Interaction Between Science and Philosophy* (New York, in press); "Newton's Search for a Mechanistic Model of Color Dispersion: A Suggested Interpretation," in *Archive for History of Exact Sciences*, 11 (1973), 1–37; and an analysis of Newton's work on chromatic aberration in lenses (in press). On the last topic, see also D. T. Whiteside, *Mathematical Papers*, III, pt. 3, esp. pp. 442–443, 512–513 (n. 61), 533 (n. 13), and 555–556 (nn. 5–6).

5. Dynamics, Astronomy, and the Birth of the "Principia." The primary documents for the study of Newton's dynamics have been assembled by A. R. and M. B. Hall (II) and by J. Herivel, *The Background to Newton's Principia* (Oxford, 1965); other major documents are printed (with historical and critical essays) in the Royal Society's ed. of Newton's *Correspondence* (II); S. P. Rigaud, *Historical Essay on the First Publication of Sir Isaac Newton's Principia* (Oxford, 1838; repr., with intro. by I. B. Cohen, New York, 1972); W. W. Rouse Ball, *An Essay on Newton's Principia* (London, 1893; repr. with intro. by I. B. Cohen, New York, 1972); and I. B. Cohen, *Introduction* (7).

The development of Newton's concepts of dynamics is discussed by Herivel (in *Background*, and in a series of articles summarized in that work), in Rouse Ball's *Essay*, I. B. Cohen's *Introduction*, and in R. S. Westfall's *Force in Newton's Physics* (London–New York, 1971). On the concept of inertia and the laws of motion, see I. B. Cohen, *Transformations of Scientific Ideas: Variations on Newtonian Themes in the History of Science*, the Wiles Lectures (Cambridge, in press), ch. 2; and "Newton's Second Law and the Concept of Force in the *Principia*," in R. Palter ed., *Annus mirabilis* (VI), pp. 143–185; Alan Gabbey, "Force and Inertia in Seventeenth-Century Dynamics," in *Studies in History and Philosophy of Science*, 2 (1971),1–68; E. J. Aiton, *The Vortex Theory of Planetary Motions* (London–New York, 1972); and A. R. Hall, "Newton on the Calculation of Central Forces," in *Annals of Science*, 13 (1957), 62–71. Newton's encounter with Hooke in 1679 and his progress from the Ward-Bullialdus approximation to the area law are studied in J. A. Lohne, "Hooke Versus Newton, an Analysis of the Documents in the Case of Free Fall and Planetary Motion," in *Centaurus*, 7 (1960), 6–52; D. T. Whiteside, "Newton's Early Thoughts on Planetary Motion: A Fresh Look," in *British Journal for the History of Science*, 2 (1964), 117–137, "Newtonian Dynamics," in *History of Science*, 5 (1966), 104–117, and "Before the *Principia*: The Maturing of Newton's Thoughts on Dynamical Astronomy, 1664–84," in *Journal for the History of Astronomy*, 1 (1970), 5–19; A. Koyré, "An Unpublished Letter of Robert Hooke to Isaac Newton," in *Isis*, 43 (1952), 312–337, repr. in Koyré's *Newtonian Studies* (VI); and R. S. Westfall, "Hooke and the Law of Universal Gravitation," in *British Journal for the History of Science*, **3** (1967), 245–261. "The Background and Early Development of Newton's Theory of Comets" is the title of a Ph.D. thesis by James Alan Ruffner (Indiana Univ., May 1966).

6. *Mathematics in the Principia.* The references for this section will be few, since works dealing with Newton's preparation for the *Principia* are listed under (5), and additional sources for the *Principia* itself are given under (7). See, further, Yasukatsu Maeyama, *Hypothesen zur Planetentheorie des 17. Jahrhunderts* (Frankfurt, 1971), and Curtis A. Wilson, "From Kepler's Laws, So-called, to Universal Gravitation: Empirical Factors," in *Archive for History of Exact Sciences*, **6** (1970), 89–170.

Two scholarly studies may especially commend our attention: H. W. Turnbull, *Mathematical Discoveries* (3), of which chs. 7 and 12 deal specifically with the *Principia*; D. T. Whiteside, "The Mathematical Principles Underlying Newton's *Principia Mathematica*," in *Journal for the History of Astronomy*, **1** (1970), 116–138, of which a version with less annotation was published in pamphlet form by the University of Glasgow (1970). See also C. B. Boyer, *Concepts of Calculus* and *History* (3), and J. F. Scott, *History* (3), ch. 11. Valuable documents and commentaries also appear in the Royal Society's ed. of Newton's *Correspondence*, J. Herivel's *Background* (5) and various articles, and D. T. Whiteside, *Mathematical Papers* (3). Especially valuable are three commentaries: J. M. F. Wright, *A Commentary on Newton's Principia*, 2 vols. (London, 1833; repr., with intro. by I. B. Cohen, New York, 1972); Henry Lord Brougham and E. J. Routh, *Analytical View of Sir Isaac Newton's Principia* (London, 1855; repr., with intro. by I. B. Cohen, New York, 1972); and Percival Frost, *Newton's Principia, First Book, Sections I., II., III., With Notes and Illustrations* (Cambridge, 1854; 5th ed., London–New York, 1900). On a post-*Principia* MS on dynamics, using fluxions, see W. W. Rouse Ball, "A Newtonian Fragment Relating to Centripetal Forces," in *Proceedings of the London Mathematical Society*, **23** (1892), 226–231; A. R. and M. B. Hall, *Unpublished Papers* (II), pp. 65–68; and commentary by D. T. Whiteside, in *History of Science*, **2** (1963), 129, n. 4.

7. *The Principia.* Many of the major sources for studying the *Principia* have already been given, in (5), (6), including works by A. R. Hall and M. B. Hall, J. Herivel, R. S. Westfall, and D. T. Whiteside. Information on the writing of the *Principia* and the evolution of the text is given in I. B. Cohen, *Introduction to Newton's Principia* (Cambridge, 1971) and the 2-vol. ed. of the *Principia* with variant readings, ed. by A. Koyré, I. B. Cohen, and Anne Whitman (I). Some additional works are R. S. Westfall, "Newton and Absolute Space," in *Archives internationales d'histoire des sciences*, **17** (1964), 121–132; Clifford Truesdell, "A Program Toward Rediscovering the Rational Mechanics of the Age of Reason," in *Archive for History of Exact Sciences*, **1** (1960), 3–36, and "Reactions of Late Baroque Mechanics to Success, Conjecture, Error, and Failure in Newton's *Principia*," in Robert Palter, ed., *The Annus Mirabilis* (VI), pp. 192–232—both articles by

Truesdell are repr. in his *Essays in the History of Mechanics* (New York–Berlin, 1968); E. J. Aiton, "The Inverse Problem of Central Forces," in *Annals of Science*, **20** (1964), 81–99; J. A. Lohne, "The Increasing Corruption" (VI), esp. "5. The Planetary Ellipse of the *Principia*"; and Thomas L. Hankins, "The Reception of Newton's Second Law of Motion in the Eighteenth Century," in *Archives internationales d'histoire des sciences*, **20** (1967), 43–65. Highly recommended is L. Rosenfeld, "Newton and the Law of Gravitation," in *Archive for History of Exact Sciences*, **2** (1965), 365–386: see also E. J. Aiton, "Newton's Aether-Stream Hypothesis and the Inverse-Square Law of Gravitation," in *Annals of Science*, **25** (1969), 255–260; and L. Rosenfeld, "Newton's Views on Aether and Gravitation," in *Archive for History of Exact Sciences*, **6** (1969), 29–37.

I. B. Cohen has discussed some further aspects of *Principia* questions in the Wiles Lectures (5) and a study of "Newton's Second Law" (5); and in "Isaac Newton's *Principia*, the Scriptures and the Divine Providence", in S. Morgenbesser, P. Suppes, and M. White, eds., *Essays in Honor of Ernest Nagel* (New York, 1969), pp. 523–548, esp. pp. 537 ff.; and "New Light on the Form of Definitions I–II–VI–VIII," where Newton's concept of "measure" is explored. On the incompatibility of Newton's dynamics and Galileo's and Kepler's laws, see Karl R. Popper, "The Aim of Science," in *Ratio*, **1** (1957), 24–35; and I. B. Cohen, "Newton's Theories vs. Kepler's Theory," in Y. Elkana, ed., *Some Aspects of the Interaction Between Science and Philosophy* (New York, in press).

8. *Revision of the Opticks (The Later Queries); Chemistry, and Theory of Matter.* The doctrine of the later queries has been studied by F. Rosenberger, *Newton und seine physikalischen Principien* (VI), and by Philip E. B. Jourdain, in a series of articles entitled "Newton's Hypothesis of Ether and of Gravitation. . . ," in *The Monist*, **25** (1915), 79–106, 233–254, 418–440; and by I. B. Cohen in *Franklin and Newton* (VI).

In addition to his studies of the queries, Henry Guerlac has analyzed Newton's philosophy of matter, suggesting an influence of Hauksbee's electrical experiments on the formation of Newton's later concept of ether. See his *Newton et Epicure* (Paris, 1963); "Francis Hauksbee: Expérimentateur au profit de Newton," in *Archives internationales d'histoire des sciences*, **17** (1963), 113–128; "Sir Isaac and the Ingenious Mr. Hauksbee," in *Mélanges Alexandre Koyré: L'aventure de la science* (Paris, 1964), pp. 228–253; and "Newton's Optical Aether," in *Notes and Records. Royal Society of London*, **22** (1967), 45–57. See also Joan L. Hawes, "Newton and the 'Electrical Attraction Unexcited,' " in *Annals of Science*, **24** (1968), 121–130; "Newton's Revival of the Aether Hypothesis and the Explanation of Gravitational Attraction," in *Notes and Records. Royal Society of London*, **23** (1968), 200–212; and the studies by Bechler listed above (4).

The electrical character of Newton's concept of "spiritus" in the final paragraph of the General Scholium has been disclosed by A. R. and M. B. Hall, in *Unpublished Papers* (II). On Newton's theory of matter, see Marie Boas [Hall],

"Newton's Chemical Papers," in *Newton's Papers and Letters* (I), pp. 241–248; and A. R. Hall and M. B. Hall, "Newton's Chemical Experiments," in *Archives internationales d'histoire des sciences*, **11** (1958), 113–152; "Newton's Mechanical Principles," in *Journal of the History of Ideas*, **20** (1959), 167–178; "Newton's Theory of Matter," in *Isis*, **51** (1960), 131–144; and "Newton and the Theory of Matter," in Robert Palter, ed., *The Annus Mirabilis* (VI), pp. 54–68.

On Newton's chemistry and theory of matter, see additionally R. Kargon, *Atomism in England From Hariot to Newton* (Oxford, 1966); A. Koyré, "Les Queries de l'Optique," in *Archives internationales d'histoire des sciences*, **13** (1960), 15–29; T. S. Kuhn, "Newton's 31st Query and the Degradation of Gold," in *Isis*, **42** (1951), 296–298, with discussion *ibid.*, **43** (1952), 123–124; J. E. McGuire, "Body and Void . . .," in *Archive for History of Exact Sciences*, **3** (1966), 206–248; "Transmutation and Immutability," in *Ambix*, **14** (1967), 69–95; and other papers; D. McKie, "Some Notes on Newton's Chemical Philosophy," in *Philosophical Magazine*, **33** (1942), 847–870; and J. R. Partington, *A History of Chemistry*, II (London, 1961), 468–477, 482–485.

For Newton's theories of chemistry and matter, and their influence, see the books by Hélène Metzger (VI), R. E. Schofield (VI), and A. Thackray (VI).

Geoffroy's summary ("extrait") of the *Opticks*, presented at meetings of the Paris Academy of Sciences, is discussed in I. B. Cohen, "Isaac Newton, Hans Sloane, and the Académie Royale des Sciences," in *Mélanges Alexandre Koyré*, I, *L'aventure de la science* (Paris, 1964), 61–116; on the general agreement by Newtonians that the queries were not so much asking questions as stating answers to such questions (and on the rhetorical form of the queries), see I. B. Cohen, *Franklin and Newton* (VI), ch. 6.

9. *Alchemy, Theology, and Prophecy. Chronology and History.* Newton published no essays or books on alchemy. His *Chronology of Ancient Kingdoms Amended* (London, 1728) also appeared in an abridged version (London, 1728). His major study of prophecy is *Observations Upon the Prophecies of Daniel, and the Apocalypse of St. John* (London, 1733). A selection of *Theological Manuscripts* was edited by H. McLachlan (Liverpool, 1950).

For details concerning Newton's theological MSS, and MSS relating to chronology, see secs. VII–VIII of the catalogue of the Sotheby sale of the Newton papers (IV); for other eds. of the *Chronology* and the *Observations*, see the Gray bibliography and the catalogue of the Babson Collection (III). There is no analysis of Newton's theological writings based on a thorough analysis of the MSS; see R. S. Westfall, *Science and Religion in Seventeenth-Century England* (New Haven, 1958), ch. 8; F. E. Manuel, *The Eighteenth Century Confronts the Gods* (Cambridge, 1959), ch. 3; and George S. Brett, "Newton's Place in the History of Religious Thought," in F. E. Brasch, ed., *Sir Isaac Newton* (VI), pp. 259–273. For Newton's chronological and allied studies, see F. E. Manuel, *Isaac Newton, Historian* (Cambridge, 1963).

On alchemy, the catalogue of the Sotheby sale is most illuminating. Important MSS and annotated alchemical books are to be found in the Keynes Collection (King's College, Cambridge) and in the Burndy Library and the University of Wisconsin, M.I.T., and the Babson Institute. A major scholarly study of Newton's alchemy and hermeticism, based on an extensive study of Newton's MSS, is P. M. Rattansi, "Newton's Alchemical Studies," in Allen G. Debus, ed., *Science, Medicine and Society in the Renaissance: Essays to Honor Walter Pagel*, II (New York, 1972), 167–182; see also R. S. Westfall, "Newton and the Hermetic Tradition," *ibid.*, pp. 183–198.

On Newton and the tradition of the ancients, and the intended inclusion in the *Principia* of references to an ancient tradition of wisdom, see I. B. Cohen, " 'Quantum in se est': Newton's Concept of Inertia in Relation to Descartes and Lucretius," in *Notes and Records. Royal Society of London*, **19** (1964), 131–155; and esp. J. E. McGuire and P. M. Rattansi, "Newton and the 'Pipes of Pan'," *ibid.*, **21** (1966), 108–143; also J. E. McGuire, "Transmutation and Immutability," in *Ambix*, **14** (1967), 69–95. On alchemy, see R. J. Forbes, "Was Newton an Alchemist?," in *Chymia*, **2** (1949), 27–36; F. Sherwood Taylor, "An Alchemical Work of Sir Isaac Newton," in *Ambix*, **5** (1956), 59–84; E. D. Geoghegan, "Some Indications of Newton's Attitude Towards Alchemy," *ibid.*, **6** (1957), 102–106; and A. R. and M. B. Hall, "Newton's Chemical Experiments," in *Archives internationales d'histoire des sciences*, **11** (1958), 113–152.

A salutary point of view is expressed by Mary Hesse, "Hermeticism and Historiography: An Apology for the Internal History of Science," in Roger H. Stuewer, ed., *Historical and Philosophical Perspectives of Science*, vol. V of Minnesota Studies in the Philosophy of Science (Minneapolis, 1970), 134–162. But see also P. M. Rattansi, "Some Evaluations of Reason in Sixteenth- and Seventeenth-Century Natural Philosophy," in Mikuláš Teich and Robert Young, eds., *Changing Perspectives in the History of Science, Essays in Honour of Joseph Needham* (London, 1973), pp. 148–166.

10. *The London Years: the Mint, the Royal Society, Quarrels With Flamsteed and With Leibniz.* On Newton's life in London and the affairs of the mint, see the biographies by More and Brewster (1), supplemented by Manuel's *Portrait* (1). Of special interest are Augustus De Morgan, *Newton: His Friend: and His Niece* (London, 1885); and Sir John Craig, *Newton at the Mint* (Cambridge, 1946). On the quarrel with Flamsteed, see Francis Baily, *An Account of the Rev^d. John Flamsteed* (London, 1835; supp., 1837; repr. London, 1966); the above-mentioned biographies of Newton; and Newton's *Correspondence* (II). On the controversy with Leibniz, see the *Commercium epistolicum* (I). Newton's MSS on this controversy (U.L.C. MS Add. 3968) have never been fully analyzed; but see Augustus De Morgan, "On the Additions Made to the Second Edition of the *Commercium epistolicum*," in *Philosophical Magazine*, 3rd ser., **32** (1848), 446–456; and "On the Authorship of the Account of the *Commercium epistolicum*, Published in the *Philosophical Transactions*," *ibid.*, 4th ser., **3** (1852), 440–444. The most recent ed. of

The Leibniz-Clarke Correspondence was edited by H. G. Alexander (Manchester, 1956).

11. *Newton's Philosophy: The Rules of Philosophizing, the General Scholium, the Queries of the Opticks.* Among the many books and articles on Newton's philosophy, those of Rosenberger, Bloch, and Koyré (VI) are highly recommended. On the evolution of the General Scholium, see A. R. and M. B. Hall, *Unpublished Papers* (II), pt. IV, intro. and sec. 8; and I. B. Cohen, *Transformations of Scientific Ideas* (the Wiles Lectures, in press) (5) and "Hypotheses in Newton's Philosophy," in *Physis*, **8** (1966), 163–184.

The other studies of Newton's philosophy are far too numerous to list here; authors include Gerd Buchdahl, Ernst Cassirer, A. C. Crombie, N. R. Hanson, Ernst Mach, Jürgen Mittelstrass, John Herman Randall, Jr., Dudley Shapere, Howard Stein, and E. W. Strong.

I. B. COHEN

SOVIET LITERATURE ON NEWTON

A profound and manifold study of Newton's life and work began in Russia at the beginning of the twentieth century; for earlier works see the article by T. P. Kravets, cited below.

The foundation of Soviet studies on Newton was laid by A. N. Krylov, who in 1915–1916 published the complete *Principia* in Russian, with more than 200 notes and supplements of a historical, philological, and mathematical nature. More than a third of the volume is devoted to supplements that present a complete, modern analytic exposition of various theorems and proofs of the original text, the clear understanding of which is often too difficult for the modern reader: "Matematicheskie nachala naturalnoy estestvennoy filosofii" ("The Mathematical Principles of Natural Philosophy"), in *Izvestiya Nikolaevskoi morskoi akademii*, **4–5** (1915–1916); 2nd ed. in *Sobranie trudov akademika A. N. Krylova* ("Collected Works of Academician A. N. Krylov"), VII (Moscow–Leningrad, 1936). Krylov devoted special attention to certain of Newton's methods and demonstrated that after suitable modification and development they could still be of use. Works on this subject include "Besedy o sposobakh opredelenia orbit komet i planet po malomu chislu nabludenii" ("Discourse on Methods of Determining Planetary and Cometary Orbits Based on a Limited Number of Observations"), *ibid.*, VI, 1–149; a series of papers, *ibid.*, V, 227–298; and "Nyutonova teoria astronomicheskoy refraktsii" ("Newton's Theory of Astronomical Refraction"), *ibid.*, V, 151–225; see also his "On a Theorem of Sir Isaac Newton," in *Monthly Notices of the Royal Astronomical Society*, **84** (1924), 392–395. On Krylov's work, see A. T. Grigorian, "Les études Newtoniennes de A. N. Krylov," in I. B. Cohen and R. Taton, eds., *Mélanges Alexandre Koyré*, II (Paris, 1964), 198–207.

A Russian translation of Newton's *Observations on the Prophecies . . . of Daniel and the Apocalypse of St. John* was published simultaneously with the first Russian edition of *Principia* as *Zamechania na knigu Prorok Daniil i*

Apokalipsis sv. Ioanna (Petrograd, 1916); the translator's name is not given.

An elaborately annotated translation of Newton's works on optics is S. I. Vavilov, ed., *Optika ili traktat ob otrazheniakh, prelomleniakh, izgibaniakh i tsvetakh sveta* ("Optics"; Moscow–Leningrad, 1927; 2nd ed., Moscow, 1954). Vavilov also published Russian translations of two of Newton's essays, "Novaya teoria sveta i tsvetov" ("A New Theory of Light and Colors") and "Odna gipoteza, obyasnyayushchaya svoystva sveta, izlozhennaya v neskolkikh moikh statyakh" ("A Hypothesis Explaining the Properties of Light Presented in Several of My Papers"), in *Uspekhi fizicheskikh nauk*, **2** (1927), 121–163; and *Lektsii po optike* ("Lectiones opticae"; Leningrad, 1946). Vavilov was the first to study thoroughly the significance of the last work in the development of physics.

Newton's mathematical works published by Castillon in vol. I of *Opuscula mathematica* (1744) were translated by D. D. Mordukhay-Boltovskoy as *Matematicheskie raboty* ("Mathematical Works"; Moscow–Leningrad, 1937); the editor's 336 notes constitute nearly a third of the volume. *Arithmetica universalis* was translated by A. P. Youschkevitch with commentary as *Vseobshchaya arifmetika ili kniga ob arifmeticheskikh sintese i analise* (Moscow, 1948).

Many works dedicated to various aspects of Newton's scientific activity and to his role in the development of science were included in the tercentenary volumes *Isaak Nyuton. 1643–1727. Sbornik statey k trekhsotletiyu so dnya rozhdenia*, S. I. Vavilov, ed. (Moscow–Leningrad, 1943); and *Moskovsky universitet—pamyati Nyutona—sbornik statey* (Moscow, 1946). These works are cited below as *Symposium I* and *Symposium II*, respectively.

Z. A. Zeitlin, in *Nauka i gipotesa* ("Science and Hypothesis"; Moscow–Leningrad, 1926), studied the problem of Newton's methodology, particularly the roles of Bentley and Cotes in preparing the 2nd ed. of the *Principia*, and emphasized that both scientists had falsified Newtonian methods; the majority of other authors did not share his viewpoint. In "Efir, svet i veshchestvo v fisike Nyutona" ("Ether, Light, and Matter in Newton's Physics"), in *Symposium I*, 33–52, S. I. Vavilov traced the evolution of Newton's views on the hypothesis of the ether, the theory of light, and the structure of matter. Vavilov also dealt with Newton's methods and the role of hypothesis in ch. 10 of his biography *Isaak Nyuton* (Moscow–Leningrad, 1943; 2nd ed., rev. and enl., 1945; 3rd ed., 1961). The 3rd ed. of this work appeared in vol. III of Vavilov's *Sobranie sochinenii* ("Selected Works"; Moscow, 1956), which contains all of Vavilov's papers on Newton. The biography also appeared in German trans. (Vienna, 1948; Berlin, 1951).

B. M. Hessen in *Sotsialno-ekonomicheskie korni mekhaniki Nyutona* ("The Socioeconomic Roots of Newton's Mechanics"), presented to the Second International Congress of the History of Science and Technology held in London in 1931 (Moscow–Leningrad, 1933), attempted to analyze the origin and development of Newton's work in Marxist terms. Hessen examined the *Principia* in the light of contemporary economic and technological problems and in the context of the political, philosophical, and religious views which reflected the social conflict occurring during the period of revolution in England. His essay appeared in English as *Science at the Crossroads* (London, 1931), which is reprinted in facsimile with a foreword by Joseph Needham and an introduction by P. G. Werskey (London, 1971) and with a foreword by Robert S. Cohen (New York, 1971).

In his report on Newton's atomism, "Newton on the Atomic Theory," in Royal Society, *Newton Tercentenary Celebrations: 15–19 July, 1946* (Cambridge, 1947), Vavilov compared Newtonian chemical ideas with the development of chemistry in the nineteenth and twentieth centuries and, in particular, with the work of Mendeleev. The latter topic was also discussed in T. I. Raynov, "Nyuton i russkoe estestvoznanie" ("Newton and Russian Natural Science"), in *Symposium I*, 329–344, which also examined Lomonosov's attitude toward Newton. See also P. S. Kudriavtsev, "Lomonosov i Nyuton," in *Trudy Instituta istorii estestvoznaniya i tekhniki. Akademiya nauk SSSR*, **5** (1955), 33–51. On Newton's role in the development of chemistry see also N. I. Flerov, "Vlianie Nyutona na razvitie khimii" ("Newton's Influence on the Development of Chemistry"), in *Symposium II*, 101–106.

For detailed comments on some important problems of the *Principia*, see L. N. Sretensky, "Nyutonova teoria prilivov i figury zemli" ("Newton's Theory of Tides and of the Figure of the Earth"), in *Symposium I*, 211–234; and A. D. Dubyago, "Komety i ikh znachenie v obshchey sisteme Nyutonovykh Nachal ("Comets and Their Significance in the General System of Newton's *Principia*"), *ibid.*, 235–263. N. I. Idelson dealt with the history of the theory of lunar motion and presented a detailed study of the St. Petersburg competition of 1751, through which the theory of universal gravitation received lasting recognition, in "Zakon vsemirnogo tyagotenia i teoria dvizhenia luny" ("The Law of Universal Gravitation and the Theory of Lunar Motion"), *ibid.*, 161–210. See also Idelson's paper "Volter i Nyuton," in *Volter 1694–1778. Stati i materialy* (Moscow–Leningrad, 1948), 215–241; and A. D. Lyublinskaya's paper on the discussions between the Newtonians and the Cartesians, "K voprosu o vlianii Nyutona na frantsuzkuyu nauku" ("On the Problem of Newton's Influence on French Science"), in *Symposium I*, 361–391. On Newton's physics, see V. G. Fridman, "Ob uchenii Nyutona o masse" ("Newton's Doctrine of Mass"), in *Uspekhi fizicheskikh nauk*, **61**, no. 3 (1957), 451–460.

On Newton's optics, apart from the fundamental studies of Krylov and Vavilov, see G. G. Slyusarev, "Raboty Nyutona po geometricheskoy optike" ("Newton's Works in Geometrical Optics"), in *Symposium I*, 127–141; I. A. Khvostikov, "Nyuton i razvitie uchenia o refraktsii sveta v zemnoy atmosfere" ("Newton and the Development of Studies of the Refraction of Light in the Earth's Atmosphere"), *ibid.*, 142–160; and L. I. Mandelshtam, "Opticheskie raboty Nyutona" ("Newton's Works in Optics"), in *Uspekhi fizicheskikh nauk*, **28**, no. 1 (1946), 103–129.

P. S. Kudriavtsev treated Newtonian mechanics and physics in his *Istoria fiziki* ("History of Physics"), 2nd ed. (Moscow, 1956), I, 200–258; and also published a biography, *Isaak Nyuton* (Moscow, 1943; 2nd ed., 1955). The basic ideas of Newton's mechanics are described in A. T. Grigorian and I. B. Pogrebyssky, eds., *Istoria mekhaniki s drevneyshikh vremen do kontsa 18 veka* ("The History of Mechanics from Antiquity to the End of the 18th Century"; Moscow, 1971).

Many works on Newton as mathematician were devoted to an analysis of his views on the foundations of infinitesimal calculus and, in particular, of his conceptions of the limiting process and of moment. S. Gouriev dealt with this question in "Kratkoe izlozhenie razlichnykh sposobov izyasnyat differentsialnoe ischislenie" ("A Brief Account of Various Methods of Explaining the Differential Calculus"), in *Umozritelnye issledovanie SPb. Akademii nauk*, **4** (1815), 159–212. Gouriev's conception was subsequently reinterpreted—occasionally with disagreement—in the commentaries of Krylov and Mordukhay-Boltovskoy (see above); and in the papers of S. A. Yanovskaya related to the publication of the mathematical MSS of Karl Marx, "O matematicheskikh rukopisyakh Marksa" ("On Marx's Mathematical Manuscripts"), in *Marksism i estestvoznanie* (Moscow, 1933), 136–180. See also K. Marx, *Matematicheskie rukopisi* ("Mathematical Manuscripts"; Moscow, 1968), 573–576; S. A. Bogomolov, *Aktualnaya beskonechnost* ("Actual Infinity"; Leningrad–Moscow, 1934); N. N. Luzin, "Nyutonova teoria predelov" ("Newton's Theory of Limits"), in *Symposium I*, 53–74; S. Y. Lurie, "Predshestvenniki Nyutona v filosofii beskonechno malykh" ("Newton's Predecessors in the Philosophy of Infinitesimal Calculus"), *ibid.*, 75–98; A. N. Kolmogorov, "Nyuton i sovremennoe matematicheskoe myshlenie" ("Newton and Modern Mathematical Thought"), *ibid.*, II, 27–42; and F. D. Kramar, "Voprosy obosnovania analisa v trudakh Vallisa i Nyutona" ("The Problems of the Foundation of the Calculus in the Works of Wallis and Newton"), in *Istoriko-matematicheskie issledovaniya*, **3** (1950), 486–508.

K. A. Rybnikov studied the role of infinite series as a universal algorithm in Newton's method of fluxions in "O roli algoritmov v istorii obosnovania matematicheskogo analisa" ("On the Role of Algorithms in the History of the Origin of the Calculus"), in *Trudy Instituta istorii estestvoznaniya i tekhniki. Akademiya nauk SSSR*, **17** (1957), 267–299. The history of Newton's parallelogram and its applications was discussed in N. G. Chebotaryov, "Mnogougolnik Nyutona i ego rol v sovremennom razvitii matematiki" ("Newton's Polygon and his Role in the Modern Development of Mathematics"), in *Symposium I*, 99–126. I. G. Bashmakova examined the research of Newton and Waring on the problem of reducibility of algebraic equations in "Ob odnom voprose teorii algebraicheskikh uravneny v trudakh I. Nyutona i E. Varinga" ("On a Problem of the Theory of Algebraic Equations in the Works of I. Newton and E. Waring"), in *Istoriko-matematicheskie issledovaniya*, **12** (1959), 431–456. Newton's use of asymptotic series was discussed in M. V.

Chirikov, "Iz istorii asimptoticheskikh ryadov" ("On the History of Asymptotic Series"), *ibid.*, **13** (1960), 441–472. On Newton's calculations equivalent to the use of multiple integrals, see V. I. Antropova, "O geometricheskom metode 'Matematicheskikh nachal naturalnoy filosofii' I. Nyutona" ("On the Geometrical Method in Newton's *Philosophiae naturalis mathematica principia*"), *ibid.*, **17** (1966), 208–228; and "O roli Isaaka Nyutona v razvitii teorii potentsiala" ("On Isaac Newton's Role in the Development of Potential Theory"), in *Uchenye zapiski Tulskogo gosudarstvennogo pedagogicheskogo instituta, Mat. kafedr,* **3** (1970), 3–56. N. I. Glagolev described Newton's geometrical ideas in "Nyuton kak geometr" ("Newton as Geometer"), in *Symposium II,* 71–80; and his mathematical discoveries were summarized in vols. II and III of A. P. Youschkevitch, ed., *Istoria matematiki s drevneyshikh vremen do nachala XIX stoletia* ("A History of Mathematics From Antiquity to the Beginning of the Nineteenth Century"; Moscow, 1970–1972).

See also two papers on Newton as historian of antiquity: S. Y. Lurie, "Nyuton—istorik drevnosti" ("Newton—Historian of Antiquity"), in *Symposium I,* 271–311; and E. C. Skrzhinskaya, "Kembridgsky universitet i Nyuton" ("Cambridge University and Newton"), *ibid.*, 392–421.

On Soviet studies of Newton, see T. P. Kravets, "Nyuton i izuchenie ego trudov v Rossii" ("Newton and the Study of His Works in Russia"), *ibid.*, 312–328; A. P. Youschkevitch, "Sovetskaya yubileynaya literatura o Nyutone" ("Soviet Jubilee Literature on Newton"), in *Trudy Instituta istorii estestvoznaniya. Akademiya nauk SSSR,* **1,** 440–455; and *Istoria estestvoznaniya. Bibliografichesky ukazatel. Literatura, opublikovannaya v SSSR* (*1917–1948*) ("History of Natural Science. Bibliography. Literature Published in the U.S.S.R. 1917–1948"; Moscow–Leningrad, 1949).

A. P. YOUSCHKEVITCH

NICERON, JEAN-FRANÇOIS (*b.* Paris, France, 1613; *d.* Aix-en-Provence, France, 22 September 1646), *geometrical optics.*

Niceron was the eldest child of Claude Niceron and Renée Barbière. He studied under Mersenne at the Collège de Nevers in Paris and then entered the Order of Minims, where he took his second name to distinguish him from a paternal uncle, also named Jean. In 1639 Niceron was appointed professor of mathematics at Trinità dei Monti, the order's convent in Rome. From 1640 he also served as auxiliary visitor for Minim monasteries. The frequent travels required by the latter post weakened his already frail health, and he died at the age of thirty-three while visiting Aix.

Having been a student of Mersenne, Niceron shared his mentor's broad interest in natural philosophy as well as his penchant for gathering and disseminating news of the latest developments. Niceron's journeys to Rome brought him into contact with many Italian scientists, to whom he communicated the results of French investigations and whose work he in turn forwarded to Paris. In 1639 Niceron informed Cavalieri of the work of Fermat, Descartes, and Roberval on the quadrature and cubature of curves of the form $y = x^n$ and on the properties of the cycloid. Niceron's revelations concerning the cycloid angered Roberval, who apparently wished to keep his results secret until he could publish them or use them in the triennial defense of his chair at the Collège Royal. Not knowing the true source of Cavalieri's information, Roberval accused Beaugrand of having betrayed confidences. The affair seems to have become something of a *cause célèbre* until Cavalieri clarified matters in 1643 (see Cavalieri's letters in *Correspondance de Mersenne,* C. de Waard *et al.,* eds., XII [Paris, 1972], *passim*). In 1640 Niceron returned to Paris with the first copies of Cavalieri's *Geometria indivisibilibus . . . promota.*

While in Italy in 1639–1640, Niceron measured the declination of the magnetic compass in Ligurno, Rome, and Florence. From 1643 to 1645 he collaborated with a group of scientists in Rome (including Magiotti, Baliani, Kircher, Ricci, and Maignan) in conducting experiments suggested by the work of Galileo. It was from Niceron that Mersenne first heard of Galileo's death (see Niceron to Mersenne, 2 Feb. 1642, *Correspondance de Mersenne,* XI, 30–34).

Niceron's major work, however, dealt with perspective and geometrical optics. His *Perspective curieuse* (1638) defines the range and nature of the problems he addressed; later editions of the work simply provide more detail. Although aware of the latest theoretical developments, Niceron concentrated primarily on the practical applications of perspective, catoptrics, and dioptrics, and on the illusory effects of optics then traditionally associated with natural magic. The work is divided into four books, of which the first presents briefly the fundamental geometrical theorems that are necessary for what follows; it then develops a general method of perspective collineation, borrowing heavily from Alberti and Dürer. Book II, which is addressed to the problem of establishing perspective for paintings executed on curved or irregular surfaces (for example, vaults and niches), presents a general technique of anamorphosis; that is, the determination of the surface distortions necessary to bring a picture into perspective when viewed from a given point. Niceron showed, for example, how to construct on the interior surface of a cone a distorted image which, when viewed end on through the base, appears in proper proportion.

Book III discusses the anamorphosis of figures that are viewed by reflection from plane, cylindrical, and conical mirrors. He explained how to draw on a plane surface a distorted figure which, when viewed by means of a cylindrical mirror standing perpendicular to the plane, appears in normal proportion. Book IV deals with the distortions created by refraction. Here Niceron abandoned any effort at general treatment and concentrated instead on constructing an optical device consisting of a polyhedral lens that gathers elements of one figure and unites them into another, totally different figure. The discussion contains perhaps the first published reference to Descartes's derivation of the law of refraction (1638) and thus gains some historical significance.

Later editions of Niceron's work, particularly the Latin version of 1646, do not differ from the 1638 edition in their basic content. Although clearly a capable mathematician, Niceron was interested more in practice than in theory. Sympathetic to the natural magic still current in his time, he tended to view optics as the art of illusion rather than the science of light.

BIBLIOGRAPHY

I. ORIGINAL WORKS. Niceron's major works are *La perspective curieuse, ou magie artificielle des effets merveilleux de l'optique, par la vision directe; la catoptrique, par la réflexion des miroirs plats, cylindriques et coniques; la dioptrique, par la réfraction des crystaux . . .* (Paris, 1638; expanded Latin version of I and II under title *Thaumaturgus opticus, seu admiranda optics . . . catoptrices . . . dioptrices . . .* , Paris, 1646; 3rd ed., heavily edited by Roberval, together with Mersenne's *L'optique et la catoptrique,* Paris, 1651; 4th ed., in Latin and French, Paris, 1663); and *L'interprétation des chiffres, ou règle pour bien entendre et expliquer facilement toutes sortes de chiffres simples, tirée de l'italien d'Antonio Maria Cospi, augmentée et accommodée particulièrement à l'usage des langues française et espagnole* (Paris, 1641). Of Niceron's correspondence, only two letters survive, both written to Mersenne and published in *Correspondance de Mersenne,* Cornelis de Waard, *et al.,* eds. (Paris, 1932–   ), X, 811–814 (8 Dec. 1641); XI, 30–34 (2 Feb. 1642).

II. SECONDARY LITERATURE. See Maria Luisa Bonelli, "Una lettera di Evangelista Torricelli a Jean François Niceron," in *Convegno di studi Torricelliani* (Faenza, 1959), 37–41; Robert Lenoble, "Roberval 'éditeur' de Mersenne et du P. Niceron," in *Revue d'histoire des sciences et de leurs applications,* **10** (1957), 235–254, and *Mersenne, ou la naissance du méchanisme* (Paris, 1943), *passim*; and *Correspondance de Mersenne,* VIII–XII, *passim*, with a short biography in X, 811.

MICHAEL S. MAHONEY

NICHOLAS CHUQUET. See **Chuquet, Nicolas.**

NICHOLAS OF CUSA. See **Cusa, Nicholas.**

NICHOLAS OF DAMASCUS. See **Nicolaus of Damascus.**

NICHOLAS, JOHN SPANGLER (*b.* Allegheny, Pennsylvania, 10 March 1895; *d.* New Haven, Connecticut, 11 September 1963), *biology.*

Nicholas was a descendant of two old Pennsylvania families. His father, Samuel Trauger Nicholas, was a Lutheran minister; his mother, formerly Elizabeth Spangler, had been trained as a teacher before her marriage. An only child, Nicholas was educated at Pennsylvania (now Gettysburg) College (B.S., 1916; M.S., 1917). He entered Yale as a graduate student in the autumn of 1917 but in 1918 interrupted his work there to enlist in the Army Medical Corps. Nicholas was assigned to the vaccine department of the Army Medical School in Washington, D.C., where he worked on methods of improving typhoid vaccine until his discharge in 1919. He then returned to Yale, where he received a Ph.D. in zoology in 1921. In the same year he married Helen Benton Brown.

After teaching for six years in the department of anatomy at the University of Pittsburgh, Nicholas returned to Yale in 1926 to teach zoology and remained there until his retirement in 1963. He became Sterling professor of zoology in 1939 and was chairman of the department of zoology from 1946 to 1956 and Master of Trumbull College from 1945 until his retirement. Nicholas performed many other administrative duties, at Yale and elsewhere. He served as editor of a number of biological journals, most notably the *Journal of Experimental Zoology.* He held office in a number of professional societies, including the American Philosophical Society and the National Academy of Sciences. As adviser to government agencies he was particularly influential as a consultant to the National Research Council on the effective use of scientific manpower during World War II.

Nicholas' primary interest, however, was in embryology; and within this area his activities were quite varied. He studied experimentally the development of fishes, amphibians, and mammals; worked in endocrinology, reproductive physiology, and neurology; and conducted original and pioneering investigations so numerous and diverse that only the most important will be mentioned here.

Nicholas began his experimental work by studying various aspects of the development of asymmetry in the amphibian limb. Ross Harrison had previously shown, through grafting experiments on salamander larvae, that whether a limb will become a left or a right limb depends on the orientation of the limb bud with respect to its surroundings in the embryo at certain specified periods in development. That is, by rotating only a narrow ring of tissue surrounding the limb bud, Nicholas demonstrated that this ring contains the factors that interact with the limb bud to determine its asymmetry. Nicholas also performed experiments on amphibian embryos and larvae to elucidate the development of the nervous system. His vital staining experiments on amphibian eggs showed that extensive movements take place in the endoderm before gastrulation; previously it had been believed that such movements begin only at gastrulation.

Nicholas made an important contribution to the study of teleost development by being the first to improvise a method for removing the horny covering of the egg, thereby making it possible to apply the modern methods of experimental embryology to the eggs of these fishes. He also applied the new methods of experimental embryology to the study of mammalian eggs—this was his most important scientific contribution. Although not the first to attempt mammalian experimental embryology, Nicholas was the first to carry out an intensive program in this field. He studied young rat eggs, or young rat embryos or their parts, both in tissue culture and in grafts implanted at a number of sites in adult rats—and even on the chick chorioallantois. His most noteworthy experiments along this line demonstrated that single blastomeres, isolated at the two-cell stage and then transplanted into the uterus of foster mothers, could develop to the egg cylinder stage. This was the first experiment to demonstrate the flexible nature of mammalian development and to prove that the important embryological principles of induction and progressive differentiation are applicable to higher as well as to lower vertebrates.

BIBLIOGRAPHY

The most detailed biography of Nicholas, with a complete bibliography of his articles, is Jane M. Oppenheimer, "John Spangler Nicholas 1895–1963," in *Biographical Memoirs. National Academy of Sciences*, **40** (1969), 239–289.

Nicholas' professional correspondence is in the Archives Collection of the Sterling Library, Yale University.

JANE OPPENHEIMER

NICHOLAS ORESME. See **Oresme, Nicole.**

NICHOLS, ERNEST FOX (*b*. Leavenworth, Kansas, 1 June 1869; *d*. Washington, D.C., 29 April 1924), *physics.*

The son of Alonzo Curtis Nichols, a photographer, and Sophronia Fox, Nichols was orphaned at an early age and raised in Manhattan, Kansas, by his aunt and uncle, General and Mrs. S. M. Fox. Having inherited some money for his education, Nichols graduated from Kansas State College of Agriculture in 1888 and did graduate work at Cornell University from 1888 to 1892. While holding an associate professorship at Colgate University from 1892 to 1898, he married Katharine W. West, the daughter of a prominent family in Hamilton, New York (1894); did further graduate work in the laboratory of Emil Warburg at Berlin (1894 to 1896); and received his D.Sc. from Cornell (1897).

A professor of physics at Dartmouth College from 1898 to 1903 and at Columbia University from 1903 to 1909, Nichols was president of Dartmouth from 1909 to 1916. While serving on the physics faculty at Yale from 1916 to 1920 he did war work for the National Research Council and the ordnance department of the U.S. Navy. The president of M.I.T. for some months in 1920, Nichols was director of research in pure science at the laboratories of the National Electric Light Association in Cleveland, Ohio, from 1921 until his death. Nichols was awarded the Rumford Medal in 1904, was coeditor of the *Physical Review* from 1913 to 1916, and was elected to the National Academy of Sciences, the American Academy of Arts and Sciences, and the American Philosophical Society.

As a research scientist Nichols' reputation rested on his development and use of the Nichols radiometer. While in Berlin, with the help of Ernst Pringsheim he constructed a radiometer far more sensitive than any other then in existence. In his new model he made the moving parts as light as possible, reduced the torsional moment, employed a delicate suspension fiber, and set the gas pressure for maximum operational effectiveness. Free of the chief disturbances suffered by thermoelements and the bolometer, Nichols' device was superior to those instruments for measurements in the infrared range. With his radiometer Nichols successfully explored the reflection and transmission of infrared rays, measured the relative heats of fixed stars, and, independently of Peter Lebedev, quantitatively confirmed the existence of the pressure of light predicted by Maxwell's laws. Near the end of his life Nichols used a resonant form of the radiometer to close the final gap between the radiations produced by thermal and electric means. He was reporting his results at a meeting of the National Academy when, in mid-sentence, he collapsed and died of heart failure.

BIBLIOGRAPHY

Leigh Page, "Ernest Fox Nichols," in *Dictionary of American Biography*, XIII, 491–494, is an authoritative introduction to Nichols' life and work. It may be supplemented by Edward L. Nichols, "Ernest Fox Nichols," in *Biographical Memoirs. National Academy of Sciences*, **12** (1929), 97–131, which contains a complete bibliography of Nichols' scientific writings. In the archives of Colgate University are eight Nichols items: six letters from Joseph Larmor and two books of notes on Larmor's lectures at Cambridge University, all from 1904–1905, when Nichols spent a sabbatical year in England.

DANIEL J. KEVLES

NICHOLSON, JOHN WILLIAM (*b.* Darlington, England, 1 November 1881; *d.* Oxford, England, 10 October 1955), *mathematical physics, astrophysics.*

The eldest son of John William Nicholson and Alice Emily Kirton, Nicholson received his early education at Middlesbrough High School. He studied mathematics and physical science at the University of Manchester from 1898 to 1901. He went on to Trinity College, Cambridge, where he took the mathematical tripos in 1904. At Cambridge he was Isaac Newton student in 1906, Smith's prizeman in 1907, and Adam's prizeman in 1913 and again in 1917. He lectured at the Cavendish Laboratory, Cambridge, and later at the Queen's University, Belfast, before being appointed professor of mathematics at King's College, London, in 1912. In 1921 he became fellow and director of studies in mathematics at Balliol College, Oxford, retiring in 1930 because of bad health. In 1922 he married Dorothy Wrinch, fellow of Girton College, Cambridge; they had one daughter. Their marriage was dissolved in 1938.

Nicholson became fellow of the Royal Astronomical Society in 1911 and fellow of the Royal Society in 1917. He was vice-president of the London Physical Society, president of the Röntgen Society, and member of the London Mathematical Society and the Société de Physique. He received the M.A. from the universities of Oxford and Cambridge, the D.Sc. from the University of London, and the M.Sc. from the University of Manchester.

Nicholson's most original work was his atomic theory of coronal and nebular spectra, which he published in a series of papers, beginning in November 1911, in the *Monthly Notices of the Royal Astronomical Society*. The spectra of the solar corona and galactic nebulae contained lines of unknown origin, which Nicholson, following a common astrophysical speculation at the time, supposed were produced by elements that were primary in an evolutionary sense to terrestrial elements. The presumed simplicity of the primary elements opened the possibility of their exact dynamical treatment. Adapting an atomic model of J. J. Thomson, Nicholson viewed an atom of a primary element as a single, planetary ring of electrons rotating about a small, massive, positively charged nucleus. Associating the frequencies of the unidentified spectral lines with those of the transverse modes of oscillation of the electrons about their equilibrium path, he accounted for most coronal and nebular lines with impressive numerical accuracy, even predicting a new nebular line that was soon observed.

In his first two papers on celestial spectra, Nicholson had no theoretical means for fully specifying his atomic systems, having to fix empirically the radius and angular velocity of the electron rings from observed spectral frequencies. In his third paper (June 1912) he rectified the incompleteness of his theory by introducing the Planck constant, h. He did so by observing that the angular momentum of primary atoms was a multiple of $h/2\pi$. Niels Bohr read Nicholson's papers in the *Monthly Notices* in late 1912, at the time he was working out his own early thoughts on the relation of the Planck constant to the structure of atoms and molecules. Impressed by the unprecedented spectral capability of Nicholson's theory, Bohr sought its relation to his own theory. In so doing Bohr came to a deeper understanding of his own atomic model, in particular of his need to attribute excited states to it. After Bohr published his theory in 1913, Nicholson challenged it and extended his own theory. But it was Bohr's theory and not his that led to a full understanding of spectra and beyond that to a new quantum atomic physics. The significance of Nicholson's theory for the development of twentieth-century atomic physics lies chiefly in the early impetus it gave Bohr for exploring the spectral implications of his very different quantum theory.

BIBLIOGRAPHY

I. ORIGINAL WORKS. Nicholson's active career spanned the years 1905–1925, during which he published roughly seventy-five papers. His most important papers on coronal and nebular spectra are "The Spectrum of Nebulium," in *Monthly Notices of the Royal Astronomical Society*, **72** (1911), 49–64; and "The Constitution of the Solar Corona," *ibid.*, 139–150; *ibid.* (1912), 677–692, 729–739. In addition to his astrophysical papers, Nicholson published on a wide range of topics that included electric and elastic vibrations, electron theory of metals, electron structure, atomic structure and spectra of terrestrial elements, relativity principle, and special mathematical functions.

Although Nicholson wrote no books, he contributed to Arthur Dendy, ed., *Problems of Modern Science. A Series of Lectures Delivered at King's College—University of London* (London, 1922). He collaborated with Arthur Schuster in revising and enlarging the third ed. of the latter's *An Introduction to the Theory of Optics* (London, 1924); and with Joseph Larmor *et al.* he edited the *Scientific Papers of S. B. McLaren* (Cambridge, 1925).

II. SECONDARY LITERATURE. Nicholson's contribution to modern atomic theory has been recently assessed by John L. Heilbron and Thomas S. Kuhn, "The Genesis of the Bohr Atom," in *Historical Studies in the Physical Sciences*, **1** (1969), 211–290; T. Hirosige and S. Nisio, "Formation of Bohr's Theory of Atomic Constitution," in *Japanese Studies in the History of Science*, no. 3 (1964), 6–28; and Russell McCormmach, "The Atomic Theory of John William Nicholson," in *Archives for History of Exact Sciences*, **3** (1966), 160–184.

In his introduction to Niels Bohr, *On the Constitution of Atoms and Molecules* (Copenhagen, 1963), xi–liii, Léon Rosenfeld has, in addition to discussing Nicholson's theory, published and analyzed letters by Bohr in 1912 and 1913 that bear on his reading of the theory. An older historical discussion of Nicholson's theory is Edmund Whittaker, *A History of the Theories of Aether and Electricity. The Modern Theories 1900–1926* (London, 1953), 107. A contemporary scientific account of Nicholson's theory is W. D. Harkens and E. D. Wilson, "Recent Work on the Structure of the Atom," in *Journal of the American Chemical Society*, **37** (1915), 1396–1421.

For biographical information, see William Wilson, "John William Nicholson 1881–1955," in *Biographical Memoirs of Fellows of the Royal Society*, **2** (1956), 209–214.

RUSSELL MCCORMMACH

NICHOLSON, SETH BARNES (*b.* Springfield, Illinois, 12 November 1891; *d.* Los Angeles, California, 2 July 1963), *observational astronomy.*

Nicholson spent his youth in rural communities in Illinois, where his father, a somewhat trained geologist, alternated between farming and teaching in elementary and high schools. At Drake University, Nicholson's career choice was influenced by the professor of astronomy, D. W. Morehouse, who was well-known for his discovery in 1908 of a particularly bright comet. In 1912 Nicholson and Alma Stotts, a classmate at Drake whom he soon married, enrolled as graduate students in astronomy at Berkeley; from then on, they were part of the West Coast astronomical community.

Nicholson's most noted astronomical work was his discovery of four faint moons of Jupiter. In 1914, while at Lick Observatory photographing Jupiter VIII, which had been found by P. Melotte a few years before, Nicholson discovered Jupiter IX. This small satellite was at the limit of detectability of the thirty-six-inch telescope and the then available photographic plates, and for many years other astronomers had to take Nicholson's word for its existence. Nicholson's Ph.D. dissertation concerned the discovery of Jupiter IX and calculations of its orbit. After receiving his doctorate Nicholson was appointed to the staff at the Mt. Wilson Observatory, where he found Jupiter X and Jupiter XI in 1938, and Jupiter XII in 1951.

In addition to his work on the minor bodies of the solar system, Nicholson tackled several astrophysical problems. He made long-term and detailed observations of the surface features and spectrum of the sun. With Edison Pettit he used a vacuum thermocouple to measure the temperatures of stars, planets, and the eclipsed moon. He charted the profiles of spectral lines in Cepheid stars, and from spectrograms of Venus he derived a value for the solar parallax and a verification of the absence of oxygen and water vapor from the Venusian atmosphere.

A mentor to numerous young astronomers, Nicholson was an active member of the Astronomical Society of the Pacific and several other astronomical and civic organizations. He received the Catherine Bruce Gold Medal of the Astronomical Society of the Pacific (1963) and was elected to membership in the National Academy of Sciences (1937).

BIBLIOGRAPHY

For biography and bibliography see Paul Herget, "Seth Barnes Nicholson," in *Biographical Memoirs. National Academy of Sciences*, **42** (1971), 201–227. See also R. M. Petrie, "Award of the Bruce Gold Medal to Seth B. Nicholson," in *Publications of the Astronomical Society of the Pacific*, **75** (1963), 305–307.

DEBORAH JEAN WARNER

NICHOLSON, WILLIAM (*b.* London, England, 1753; *d.* London, 21 May 1815), *chemistry, technology.*

As is characteristic of minor scientific figures of the British industrial revolution, only fragmentary information survives on William Nicholson's variegated activities. Many of these endeavors were of considerable significance within the rapidly developing and changing scientific world of his day. Nicholson was successively a servant of the East India Company, a European commercial agent of Josiah Wedgwood, the potter, master of a London mathematical school, patent agent, and water engineer. He also found time to translate foreign scientific works, compile a chemical dictionary, perfect a number of inventions, devise new instruments, act as secretary of the General Chamber of Manufacturers of Great Britain, undertake significant original research and, for sixteen years, edit and promote the monthly scientific journal for which he is most often remembered. Despite—or possibly because of—advanced scientific knowledge and practical ingenuity, "he lived in trouble and died poor."

The son of a London solicitor, Nicholson was educated in North Yorkshire, before entering the service of the East India Company in 1769. In 1776 he returned home from India. He then spent time in Amsterdam, as Dutch sales agent for Wedgwood. By 1780 he apparently had settled in London with the proceeds of his foreign ventures and had begun to find his métier as inventor, translator, and scientific projector. It was presumably about this time that he married Catherine, daughter of Peter Boullie of London and remote descendant of Edward III. Nothing is known of their family life, save that at least one son reached maturity.

On arriving in London, Nicholson seems to have intended an assault on its literary world. Initially he lodged with the dramatist Thomas Holcroft, with whom he collaborated on at least one novel. The burgeoning scientific life of the capital soon captured his fancy, although a taste for literary and historical works remained with him. Nicholson appears to have run a mathematical school for some years, until other pursuits crowded it out. Reestablished in 1799, the school again experienced its earlier fate. Pedagogic concerns were certainly paramount in Nicholson's first scientific publication, *An Introduction to Natural Philosophy* (1781), which enjoyed some success as a Newtonian text.

In December 1783 Nicholson's serious scientific interests were recognized in his election to the Chapter Coffee House Society, or Philosophical Society. This ephemeral research club flourished throughout the 1780's and Nicholson soon became its secretary. Among its twenty-five participants the club numbered J. H. de Magellan, Richard Kirwan, and Tiberius Cavallo; Joseph Priestley and Thomas Percival figured among its provincial honorary members. Association with the group no doubt prompted Nicholson's interest in the intellectual and commercial possibilities of the new French chemistry, then generating intense debate. His translation of A. F. de Fourcroy's *Élémens d'histoire naturelle et de chimie* appeared in 1788; that of the French rebuttal of Richard Kirwan's *Essay on Phlogiston*, in 1789; and

that of J. A. C. Chaptal's *Élémens de chimie*, in 1791. A natural consequence of this activity was the publication of Nicholson's *First Principles of Chemistry* in 1790 and of a weighty, competent, but pedestrian *Dictionary of Chemistry* in 1795. At a slightly later date he translated Fourcroy's *Tableaux synoptiques de chimie* and also his authoritative eleven-volume *Système des connaissances chimiques*, as well as Chaptal's four-volume *Chimie appliquée aux arts*.

Just as this cluster of works is indicative of growing British concern with chemistry, so the success accorded Nicholson's decision to found a monthly journal of scientific news and commentary reflects the quickening of British interest across a wider range of natural knowledge. The *Journal of Natural Philosophy, Chemistry and the Arts* began publication in April 1797. Its success invited emulation. Alexander Tilloch's *Philosophical Magazine* appeared in June 1798 and offered a continuing threat to the less worldly Nicholson. When in 1813 the field was crowded still further by Thomas Thomson's *Annals of Philosophy*, Nicholson, already ill, withdrew. His *Journal* was merged with Tilloch's, which throughout had shown greater commercial if less scientific acumen, and which continues to flourish (having also ingested Thomson's *Annals*).

The reception accorded Nicholson's *Journal* reveals the growing number of cultivators of science to be found in the urbanizing and industrializing culture of late-Georgian Britain. Reliable news of scientific discoveries, technical processes, instruments, books, translations, and meetings met an evident demand. The medium itself also created a fresh audience and new possibilities for scientific controversy and intellectual fashion. In July 1800 Nicholson's *Journal* enjoyed its greatest coup, when it gave the first report of its proprietor's sensational electrolysis of water, in collaboration with Anthony Carlisle. The *Journal* immediately became the accepted vehicle and the powerful reinforcer of the resulting scientific fashion for electrolysis, a fashion which Humphry Davy effectively exploited in his own brilliant demonstration of the newly possible art of scientific careerism. Another illustration of the changes wrought by this fresh medium of scientific communication may be seen in the work of John Dalton. He used the monthly journals to engage critics of his theory of mixed gases and thereby was encouraged to persevere in the work which finally led to his chemical atomic theory.

Nicholson's real genius was that of a projector. As a researcher he was competent but uninspired; as an entrepreneur, persistent but empty-handed. The range of his inventions was wide, running from hydrometers to machinery for manufacturing files. All were as commercially unrewarding as they were technically excellent. His plans for a new Middlesex waterworks, for supplying Southwark, and for piping water to Portsmouth were important and practical pieces of urban engineering from which he drew little reward. Indeed, Nicholson's financial problems were such that he spent time in debtors' prison, deliberately sold his name to the proprietors of the six-volume *British Encyclopaedia* in 1809, and died in poverty after a lingering illness. He neither became a fellow of the Royal Society nor did he enjoy other public recognition. If his activities illustrate the widening scientific opportunities of a new age, they also show that energy, imagination, and expert knowledge provided no infallible route to personal fortune or social reward.

BIBLIOGRAPHY

I. ORIGINAL WORKS. There is no bibliography of Nicholson's works, many of which are now very rare. The following list is necessarily tentative, not definitive. Scientific books are *An Introduction to Natural Philosophy*, 2 vols. (London, 1781; 5th ed., 1805); *First Principles of Chemistry* (London, 1790; 3rd ed., 1796); and *A Dictionary of Chemistry*, 2 vols. (London, 1795), rev. as *A Dictionary of Practical and Theoretical Chemistry* (London, 1808). Other books are *The History of Ayder Ali Khan, Nabob Buhader; or New Memoirs Concerning the East Indies, With Historical Notes*, 2 vols. (London, 1783); *The Navigator's Assistant* (London, 1784); and *Abstract of Such Acts of Parliament as Are Now in Force for Preventing the Exportation of Wool* (London, 1786).

Nicholson's editions and translations of works by others are *Ralph's Critical Review of the Public Buildings, Statues and Ornaments in and About London and Westminster . . . With Additions* (London, 1783); Fourcroy's *Elements of Natural History and Chemistry*, 4 vols. (London, 1788) plus *Supplement* (London, 1789); the French reply to Kirwan's *Essay on Phlogiston, and the Constitution of Acids . . . With Additional Remarks . . .* (London, 1789); *Memoirs and Travels of the Count de Benyowsky*, 2 vols. (London, 1789); Chaptal's *Elements of Chemistry*, 3 vols. (London, 1791; 4th ed., 1803); Pajot des Charmes's *The Art of Bleaching Piece Goods, Cottons, and Threads . . . by . . . Oxygenated Muriatic Acid* (London, 1799); G. B. Venturi's *Experimental Enquiries Concerning . . . Motion in Fluids* (London, 1799); Fourcroy's *Synoptic Tables of Chemistry* (London, 1801); Fourcroy's *General System of Chemical Knowledge*, 11 vols. (London, 1804); and Chaptal's *Chemistry Applied to Arts and Manufactures*, 4 vols. (London, 1807).

Scientific papers by Nicholson include "Description of a New Instrument for Measuring the Specific Gravity of Bodies," in *Memoirs of the Manchester Literary and Philosophical Society*, **2** (1785), 386–396; "The Principles

and Illustration of an Advantageous Method of Arranging the Differences of Logarithms, on Lines Graduated for the Purpose of Computation," in *Philosophical Transactions of the Royal Society*, **77** (1787), 246–252; "Experiments and Observations on Electricity," ibid., **79** (1789), 265–287; "Account of the New Electrical or Galvanic Apparatus of Sig. Alex. Volta, and Experiments Performed With the Same," in *Journal of Natural Philosophy, Chemistry and the Arts*, **4** (1800), 179–187, written with A. Carlisle; and numerous other contributions (many anonymous) to his own journal. A list of 62 papers is in the Royal Society *Catalogue of Scientific Papers*, IV, 610–612.

II. SECONDARY LITERATURE. The best obituary of Nicholson is that in *New Monthly Magazine*, **3** (1815), 569; **4** (1816), 76–77, on which the *Dictionary of National Biography* leans heavily. There is some additional information in *Gentlemen's Magazine*, **85** (1815), 570. His mechanical inventions are treated briefly in Samuel Smiles, *Men of Invention and Industry* (London, 1884), 164, 177, 194, 202; his chemical work is mentioned in J. R. Partington, *A History of Chemistry*, IV (London, 1964), 19–20. S. Lilley, "Nicholson's Journal (1797–1813)," in *Annals of Science*, **6**, (1948), 78–101, discusses the content and significance of the *Journal*. Nicholson's other publications are exhaustively examined in R. S. Woolner, "Life and Scientific Work of William Nicholson" (M.Sc. diss., University College, London, 1959). Some further information, and reference to a manuscript biography by his son, are in R. W. Corlass, "A Philosophical Society of a Century Ago," in *Reliquary*, **18** (1878), 209–211. The MS minute book of the Chapter Coffee House Society, to which Corlass refers, is now in the Museum of the History of Science, Oxford (MS Gunter 4).

ARNOLD THACKRAY

NICOL, WILLIAM (*b.* 1768; *d.* Edinburgh, Scotland, 2 September 1851), *optics, petrology, paleontology*.

Although he achieved fame in physics as the inventor of the first polarizer, the Nicol prism, Nicol was primarily a geologist who made important but unappreciated contributions to petrology and paleontology.

Little is known about Nicol's early career at the University of Edinburgh, where he lectured in natural philosophy, for he did not publish until he was fifty-eight. His first known scientific research dealt with the structure of crystals, and it was undoubtedly in connection with this work that he invented the prism. Nicol constructed his device by splitting a parallelepiped of calcite spar along its shorter diagonal and then cementing the halves together with Canada balsam, a substance with an index of refraction intermediate to the two indices of the doubly refracting calcite spar. The balsam allows the extraordinary ray to pass almost undeviated through the prism while it reflects the ordinary beam (see Figure 1). The two beams emerging from the prism are so widely separated they can be used independently. When it was invented (1828) Nicol's device was the most convenient means of producing polarized light, and it became an important tool in physical optics and petrography.

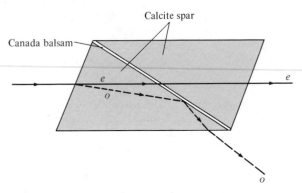

Extraordinary ray, *e*, passes through undeviated. Ordinary ray, *o*, reflected by first Canada balsam surface.

FIGURE 1

Nicol's inventive talents were equally well displayed in geology, although he received less credit than he deserved for his work in this field. To aid his early studies of crystals and rocks, he developed a technique for preparing transparent slivers for viewing directly through a microscope. Previous microscopic studies of minerals had been done with reflected light, which could reveal only surface qualities. Nicol's technique was to cement the mineral in question to a glass plate and then grind it down to extreme thinness, thereby making possible for the first time direct microscopic investigation of the innermost structure of rocks and crystals.

Unfortunately, the potential of Nicol's new method was not realized in petrology for more than forty years after its invention, in or around 1815. Nicol himself was partly to blame for this, since he never published any structural studies of his slide specimens (indeed he may never have made any such studies), except for two papers on fluid cavities in crystals. Furthermore, the first printed account of his technique did not appear until 1831, and then in a book on fossil woods (Witham's *Observations on Fossil Vegetables*), which few petrologists were likely to read. Thus, it is not altogether surprising that this promising new method was not incorporated into the science of petrology until 1853, when Henry Sorby obtained Nicol's slides and showed how they could divulge the secrets of mineral structure.

Nicol had far more success in the field of paleontology. He found that the same slidemaking technique could be used in the study of fossil woods in order to obtain a view of the minute cell structure with a microscope. Knowledge of the cell pattern thus obtained could be used as a basis for classifying and identifying the specimens being examined. Nicol made these identifications for a large number of fossil woods and displayed the arrangement of the cells in Witham's *Observations*. Yet he seems not to have been accorded full recognition for this important work either.

BIBLIOGRAPHY

Nicol's articles include "Observations on the Fluids Contained in Crystallized Minerals," in *Edinburgh New Philosophical Journal*, **5** (1828), 94–96; "On a Method of So Far Increasing the Divergency of the Two Rays in Calcareous Spar That Only One Image May Be Seen at a Time," *ibid.*, **6** (1829), 83–84; "On the Cavities Containing Fluids in Rock Salt," *ibid.*, **7** (1829), 111–113; *ibid.*, **10** (1831), 361–364; *ibid.*, **14** (1833), 153–158; "On the Anatomical Structure of Recent and Fossil Woods," in *British Association for the Advancement of Science Report* (1834), 660–666; *Edinburgh New Philosophical Journal*, **18** (1835), 335–339; and "Observations on the Structure of Recent and Fossil Coniferae," *ibid.*, **29** (1840), 175.

The only biographical material on Nicol is in Poggendorff, II, 151. The account of his mounting technique is in H. T. M. Witham, *Observations on Fossil Vegetables* (Edinburgh, 1831).

EUGENE FRANKEL

NICOLAI, FRIEDRICH BERNHARD GOTTFRIED (*b.* Brunswick, Germany, 25 October 1793; *d.* Mannheim, Germany, 4 July 1846), *astronomy.*

After Nicolai had begun to study theology in Göttingen, he started to attend Gauss's mathematics lectures. In 1813 he became an assistant at the observatory at Seeberg near Gotha, which was then under the direction of Lindenau. When Schumacher left the observatory at Mannheim, Nicolai succeeded him as director, a position he held until his death.

Nicolai spent most of his career observing comets and planets. He made preliminary calculations of lunar occultations which were important in astronomical geography. In particular he pointed out the distorting influence of the profile of the moon.

Of greater influence were Nicolai's works on the determination of differences of longitude from lunar observations. He modified a method devised in the first third of the eighteenth century, in which the right ascension of the moon could be determined by culmination observations. From this information, and with the aid of the ephemerides, the true time at the place of observation could be derived. Nicolai proposed that instead of measuring the culmination of the moon, what should be measured is the time between the transit of the rim of the moon through the local meridian and that of several nearby fixed stars of similar declination. The defective reduction to the center of the moon thereby became unnecessary, while at the same time the lesser error introduced by setting up the meridian telescope was not of great importance.

In pure mathematics, Nicolai worked on series expansions and integral functions. He improved the uncertain values of the mass of Jupiter by employing Gauss's perturbation equations of the planetoid Juno.

BIBLIOGRAPHY

I. ORIGINAL LITERATURE. See "Berechnung der Meridiandifferenz zweier Orte, aus correspondirenden Mondsculminationen," in *Astronomische Nachrichten*, **2** (1824), cols. 17–24. About eighty additional works that Nicolai published in *Astronomischen Nachrichten* are listed in H. Kobold, ed., *Generalregister der Bände 1–40 der Astronomischen Nachrichten Nr. 1–960 (1821–1855)* (Kiel, 1936), cols. 78–79.

II. SECONDARY LITERATURE. See the article by S. Gunther, in *Allgemeine Deutsche Biographie*, XXIII (Leipzig, 1886), 590–591. On the Mannheim Observatory and Nicolai's activity there see G. Klare, "Ein Jahrhundert wechselvoller Geschichte der Mannheimer Sternwarte 1783–1883," in *Sterne und Weltraum*, **9** (1970), 148–150. On his correspondence with Gauss, see W. Valentiner, ed., *Briefe von C. F. Gauss an B. Nicolai* (Karlsruhe, 1877).

DIETER B. HERRMANN

NICOLAUS OF DAMASCUS (*b.* Damascus, 64 B.C.), *botany.*

Nicolaus was the son of wealthy parents, whose names, Antipater and Stratonice, suggest that they were of Macedonian origin. He received an expensive liberal education, probably from Greek tutors, and became so distinguished a scholar that he attracted the attention of Herod the Great, king of Judaea. He subsequently spent his life in the service of Herod, accompanying him twice to Rome during the last ten years of his rule (14–4 B.C.). Nicolaus served the king as secretary, adviser, and court historian and, in Rome, endeavored to explain Herod's anti-Nabataean politics to the Roman Senate. After Herod's death he sought to retire but was obliged to represent Herod's son, Archelaus, and to travel again to Rome to undertake the latter's defense against complaints

by the Jews. In spite of Nicolaus' efforts, Archelaus was banished by Augustus to Vienne and died there. It is not known what happened subsequently to Nicolaus.

Besides dramatic compositions, an autobiography, a panegyrical biography of Augustus' youth, a *Universal History* in 144 books from the earliest times to the death of Herod, and a collection of writings on the manners and customs of some fifty nations (Παραδόξων ἐθῶν συναγωγή), Nicolaus wrote commentaries on Aristotle, now largely lost, and also an extant treatise on plants in two books; the latter were written in Peripatetic style and dealt with the generalities of plant life. Indeed, so Peripatetic in style and structure are these books that they were believed to have been the work of Aristotle himself.

The first book is divided into seven chapters, in which are discussed the nature of plant life; sex in plants; the parts, structure, classification, composition, and products of plants; their methods of propagation and fertilization; and their changes and variations. Book II contains ten chapters, which describe the origins of plant life; the material of plants; the effects of external conditions and climate; water and rock plants; effects of topography upon plants; parasitism; the production of fruits and leaves; the colors and shapes of plants; and fruits and their flavors.

The original Greek text of the *De plantis* has been lost. It was, however, translated into Syriac in the ninth century; and a few scattered fragments have survived in Cambridge MS. Gg. 2.14 (fifteenth-sixteenth century), together with the translation of Nicolaus' Περὶ τῆς τοῦ 'Αριστοτέλους φιλοσοφίας. It has been suggested by Hemmerdinger, but denied by Drossaart Lulofs, that this Syriac translation was made by Ḥunayn ibn Isḥāq, court physician at Baghdad. The fragments of the Syriac translation of the *De plantis* consist of a series of dislocated sentences from the first book. Bar-Hebraeus possessed a copy of it and preserved a brief but valuable excerpt of book I in Syriac in his *Candelabrum Sanctorum*. The Syriac version was subsequently translated into Arabic by Isḥāq ibn Ḥunayn about 900. This Arabic translation is badly preserved and four pages toward the end are missing. In 1893 Steinschneider discovered a Hebrew translation made verbatim from the Arabic by the Provençal Kalonymus ben Kalonymus in 1314. The Arabic text was also translated into Latin by Alfred of Sareshel (first half of the thirteenth century), and during the Middle Ages it exercised a wide influence, as is attested by numerous manuscripts and several commentaries. The Latin translation, however, was superseded by the clumsy thirteenth-century translation of the Latin into Greek by Maximus

Planudes, which has been printed in Bekker's edition of Aristotle (815A–830B).

The *De plantis*, apart from the herbals deriving from Dioscorides and pseudo-Apuleius, became the most important single source for later medieval botany. As has been seen above, its two volumes were long credited to Aristotle himself and were included in his *Opera*. Scaliger actually devoted a commentary to these books, entitled "In libros duos qui inscribuntur *De plantis*, Aristotele autore" (Paris, 1556). He subsequently corrected his mistake in the heading of his Preface to "In libros De plantis falso Aristoteli attributos," but it may be assumed that the majority of Renaissance botanists were ready to accept uncritically Aristotle's authority upon the basis of these two incorrectly attributed volumes.

It was not only in botany that Nicolaus' work was influential. So great was his prestige as an Aristotelian commentator that Porphyry and even Simplicius used to appeal to his authority. The following titles of treatises written by him on Aristotelian philosophy have survived: Περὶ τῆς τοῦ Ἀριστοτέλους φιλοσοφίας, Περὶ θεῶν, Περὶ τῶν ἐν τοῖς πρακτικοῖς καλῶν and, possibly, Περὶ τοῦ παντός. The first of these works, and the sole survivor, is preserved only in the Syriac abridgment described above (Cantab. MS Gg. 2.14). Although Nicolaus was eclipsed by other commentators in Greek, notably Alexander and Simplicius, it is clear from the Islamic bibliographers that his commentaries were read and studied in the East.

BIBLIOGRAPHY

For the Latin medieval translation see E. H. F. Meyer, *Nicolai Damasceni "De plantis" libri duo Aristoteli vulgo adscripti* (Leipzig, 1841). For the Greek text (which is a retranslation from the Latin) see I. Bekker, *Aristote is Opera*, II (Berlin, 1831), 815A–830B. There is an English translation by E. S. Foster in *The Works of Aristotle Translated into English*, VI (Oxford, 1913), 815A–830B, and by W. S. Hett in his Loeb volume, *Aristotle, Minor Works* (London–Cambridge, Mass., 1936). Secondary literature includes A. J. Arberry, "An Early Arabic Translation from the Greek," in *Bulletin of the Faculty of Arts* (Cairo University), **1** (1933), 48 ff., and **2** (1934), 72 ff.; R. P. Bouyges, "Sur le *De plantis* d'Aristote-Nicolas à propos d'un manuscrit arabe de Constantinople," in *Mélanges de la Faculté orientale, Université St.-Joseph* (Beirut), **9**, no. 2 (1932), 71–89; H. J. Drossaart Lulofs, "Aristotle's ΠΕΡΙ ΦΥΤΩΝ," in *Journal of Hellenic Studies*, **77**, 1 (1957), 75–80; and *Nicolaus Damascenus on the Philosophy of Aristotle*, Philosophia Antiqua, XIII (Leiden, 1965); B. Hemmerdinger, "Le *De Plantis*, de Nicolas de Damas à Planude," in *Philologus*, **111** (1967), 56–65; E. H. F. Meyer, *Geschichte der Botanik* (Königsberg, 1854); and G. Sarton, *The Appreciation of Ancient and*

Medieval Science During the Renaissance (Philadelphia, 1955), 63 ff.

JAMES LONGRIGG

NICOLLE, CHARLES JULES HENRI (*b.* Rouen, France, 21 September 1866; *d.* Tunis, 28 February 1936), *bacteriology*.

For a detailed study of his life and work, see Supplement.

NICOMACHUS OF GERASA (*fl. ca.* A.D. 100), *mathematics, harmonics*.

That Nicomachus was from Gerasa, probably the city in Palestine, is known from Lucian (*Philopatris*, 12), from scholia to his commentator Philoponus, and from some manuscripts that contain Nicomachus' works. The period of his activity is determined by inference. In his *Manual of Harmonics* Nicomachus mentions Thrasyllus, who died in A.D. 36; Apuleius, born about A.D. 125, is said to have translated the *Introduction to Arithmetic* into Latin; and a character in Lucian's *Philopatris* says, "You calculate like Nicomachus," which shows that Lucian, born about A.D. 120, considered Nicomachus a famous man.[1] Porphyry mentions him, together with Moderatus and others, as a prominent member of the Pythagorean school, and this connection may also be seen in his writings.[2] Only two of his works are extant, *Manual of Harmonics* and *Introduction to Arithmetic*. He also wrote a *Theologumena arithmeticae*, dealing with the mystic properties of numbers, and a larger work on music, some extracts of which have survived.[3] Other works are ascribed to him, but it is not certain that he wrote any of them.[4]

In the *Manual of Harmonics*, after an introductory chapter, Nicomachus deals with the musical note in chapters 2–4 and devotes the next five chapters to the octave. Chapter 10 deals with tuning principles based on the stretched string; chapter 11, with the extension of the octave to the two-octave range of the Greater Perfect System in the diatonic genus; and the work ends with a chapter in which, after restating the definitions of note, interval, and system, Nicomachus gives a survey of the Immutable System in the three genera: diatonic, chromatic, and enharmonic. He deals with notes, intervals, systems, and genera, the first four of the seven subdivisions of harmonics recognized by the ancients, but not with keys, modulation, or melodic composition. The treatise exhibits characteristics of both the Aristoxenian and the Pythagorean schools of music. To the influence of the latter must be ascribed Nicomachus' assignment of number and numerical ratios to notes and intervals, his recognition of the indivisibility of the octave and the whole tone, and his notion that the musical consonances are in either multiple or superparticular ratios. But unlike Euclid, who attempts to prove musical propositions through mathematical theorems, Nicomachus seeks to show their validity by measurement of the lengths of strings. Hence his treatment of consonances and of musical genera, as well as his definition of the note, are Aristoxenian.

The *Introduction to Arithmetic* is in two books. After six preliminary chapters devoted to the philosophical importance of mathematics, Nicomachus deals with number per se, relative number, plane and solid numbers, and proportions. He enunciates several definitions of number and then discusses its division into even and odd. He states the theorem that any integer is equal to half the sum of the two integers on each side of it and proceeds to give the classification of even numbers (even times even, odd times even, and even times odd), followed by that of odd numbers (prime, composite, and relative prime).[5] The fundamental relations of number are equality and inequality, and the latter is divided into the greater and the less. The ratios of the greater are multiples, superparticulars, superpartients, multiple superparticulars, and multiple superpartients; those of the less are the reciprocal ratios of these. Book I concludes with a general principle whereby all forms of inequality of ratio may be generated from a series of three equal terms.[6] At the beginning of the second book the reverse principle is given. It is followed by detailed treatments of squares, cubes, and polygonal numbers. Nicomachus divides proportions into disjunct and continuous, and describes ten types. He presents no abstract proofs (as are found in Euclid's *Elements*, VII–IX), and he limits himself for the most part to the enunciation of principles followed by examples with specific numbers.[7] On one occasion this method leads to a serious mistake,[8] but there are many other mistakes which are independent of the method of exposition—for example, his inclusion of composite numbers, a class which belongs to all numbers, as a species of the odd. Yet despite its notorious shortcomings, the treatise was influential until the sixteenth century and gave its author the undeserved reputation of being a great mathematician.

NOTES

1. For references to modern discussions, see Tarán, *Asclepius on Nicomachus*, p. 5, n. 3. J. M. Dillon, "A Date for the Death of Nicomachus of Gerasa?" in *Classical Review*, n.s. **19** (1969), 274–275, conjectures that Nicomachus died in A.D. 196, because Proclus, who was born in A.D. 412, is said by Marinus, *Vita Procli*, 28, to have believed that he was a reincarnation of Nicomachus, and because some

Pythagoreans believed that reincarnations occur at intervals of 216 years. But Dillon fails to cite any passage in which Proclus would attach particular importance to the number 216 and, significantly enough, this number is not mentioned in Proclus' commentary on the creation of the soul in Plato's *Timaeus*, a passage where one would have expected this number to occur had Dillon's conjecture been a probable one.

2. In Eusebius of Caesarea, *Historia ecclesiastica*, VI, xix, 8.

3. Some of the contents of the *Theologumena* can be recovered from the summary of it given by Photius, *Bibliotheca*, codex 187, and from the quotations from it in the extant *Theologumena arithmeticae* ascribed to Iamblichus.

   In his *Manual of Harmonics*, I, 2, Nicomachus promises to write a longer and complete work on the subject; and the extracts in some MSS, published by Jan in *Musici scriptores Graeci*, pp. 266–282, probably are from this work. They can hardly belong to a second book of the *Manual*, because Nicomachus' words at the end of this work indicate that it concluded with chapter 12. Eutocius seems to refer to the first book of the larger work on music; see *Eutocii Commentarii in libros De sphaera et cylindro*, in *Archimedis Opera omnia*, J. L. Heiberg, ed., III (Leipzig, 1915), 120, ll. 20–21.

4. In his *Introduction to Arithmetic*, II, 6, 1, Nicomachus refers to an *Introduction to Geometry*. Some scholars attribute to him a *Life of Pythagoras* on the grounds that Nicomachus is quoted by both Porphyry and Iamblichus in their biographies of Pythagoras. It is also conjectured that he wrote a work on astronomy because Simplicius, *In Aristotelis De caelo*, Heiberg ed., p. 507, ll. 12–14, says that Nicomachus, followed by Iamblichus, attributed the hypothesis of eccentric circles to the Pythagoreans. A work by Nicomachus with the title *On Egyptian Festivals* is cited by Athenaeus and by Lydus, but the identity of this Nicomachus with Nicomachus of Gerasa is not established. Finally, the "Nicomachus the Elder" said by Apollinaris Sidonius to have written a life of Apollonius of Tyana in which he drew from that of Philostratus cannot be the author of the *Manual*, since Philostratus was born ca. A.D. 170.

5. Nicomachus considers prime numbers a class of the odd, because for him 1 and 2 are not really numbers. For a criticism of this and of Nicomachus' classifications of even and odd numbers, see Heath, *A History of Greek Mathematics*, I, 70–74. In I, 13, Nicomachus describes Eratosthenes' "sieve," a device for finding prime numbers.

6. This principle is designed to show that equality is the root and mother of all forms of inequality.

7. Euclid represents numbers by lines with letters attached, a system that makes it possible for him to deal with numbers in general, whereas Nicomachus represents numbers by letters having specific values.

8. See *Introduction to Arithmetic*, II, 28, 3, where he infers a characteristic of the subcontrary proportion from what is true only of the particular example (3, 5, 6) that he chose to illustrate this proportion. See Tarán, *Asclepius on Nicomachus*, p. 81 with references.

BIBLIOGRAPHY

I. ORIGINAL WORKS. The best, but not critical, ed. of the *Introduction to Arithmetic* is *Nicomachi Geraseni Pythagorei Introductionis arithmeticae libri II*, R. Hoche, ed. (Leipzig, 1866), also in English with notes and excellent introductory essays as *Nicomachus of Gerasa, Introduction to Arithmetic*, trans. by M. L. D'Ooge, with studies in Greek arithmetic by F. E. Robbins and L. C. Karpinski (New York, 1926); Boethius' Latin trans. and adaptation is *Anicii Manlii Torquati Severini Boetii De institutione arithmeticae libri duo*, G. Friedlein, ed. (Leipzig, 1867).

The *Manual of Harmonics* is in Carolus Jan, *Musici scriptores Graeci* (Leipzig, 1895), 235–265; an English trans. and commentary is F. R. Levin, "Nicomachus of Gerasa, Manual of Harmonics: Translation and Commentary" (diss., Columbia University, 1967).

II. Secondary Literature. Ancient commentaries are an anonymous "Prolegomena" in P. Tannery, ed., *Diophanti Opera omnia*, II (Leipzig, 1895), 73–76; Iamblichus' commentary, *Iamblichi in Nicomachi Arithmeticam introductionem liber*, H. Pistelli, ed. (Leipzig, 1894); Philoponus' commentary, R. Hoche, ed., 3 fascs. (Wesel, 1864, 1865; Berlin, 1867); another recension of this commentary in Hoche (Wesel, 1865), pp. ii–xiv, for the variants corresponding to the first book, and in A. Delatte, *Anecdota Atheniensia et alia*, II (Paris, 1939), 129–187, for those corresponding to the second book; Asclepius' commentary, "Asclepius of Tralles, Commentary to Nicomachus' Introduction to Arithmetic," edited with an intro. and notes by L. Tarán, *Transactions of the American Philosophical Society*, n.s., **59**, pt. 4 (1969); there is an anonymous commentary, still unpublished, probably by a Byzantine scholar—see Tarán, *op. cit.*, pp. 6, 7–8, 18–20.

For an exposition of the mathematical contents of Nicomachus' treatise and a criticism of it, see T. Heath, *A History of Greek Mathematics*, I (Oxford, 1921), 97–112.

Leonardo Tarán

NICOMEDES (*fl. ca.* 250 B.C. [?]), *mathematics.*

Nothing is known of the life of Nicomedes. His period of activity can be only approximately inferred

from the facts that he criticized the solution of Eratosthenes (*fl.* 250 B.C.) to the problem of doubling the cube and that Apollonius (*fl.* 200 B.C.) named a curve "sister of the cochlioid," presumably as a compliment to Nicomedes, who had discovered the curve known as cochlioid, cochloid, or conchoid.[1] The second inference is far from secure, but what we know of Nicomedes' mathematical investigations fits well into the period of Archimedes (d. 212 B.C.).

The work for which Nicomedes became famous was called *On Conchoid Lines* (Περὶ κογχοειδῶν γραμμῶν).[2] We know it only through secondhand references. In it Nicomedes described the generation of a curve, which he called the "first conchoid," as follows (see Figure 1): Given a fixed straight line *AB* (the "canon") and a fixed point *E* (the "pole"), draw *EDG* perpendicular to *AB*, cutting it at *D*, and make *DG* a fixed length (the "interval"); then let *GDE* move about *E* in such a way that *D* is always on *AB* (thus when *D* reaches *H*, *G* will have reached *T*). *G* will then describe a curve, *LGTM*, the first conchoid. The advantages of this curve are that it is very easy to construct (Nicomedes described a mechanical instrument for drawing it)[3] and that it can be used to solve a variety of problems, including the "classical" problems of doubling the cube and trisecting the angle. These are all soluble by means of the auxiliary construction which we may call the "lemma of Nicomedes": Given two straight lines, *X*, *Y*, meeting in a given angle and a point *P* outside the angle, it is possible to draw a line through *P* cutting *X* and *Y* so

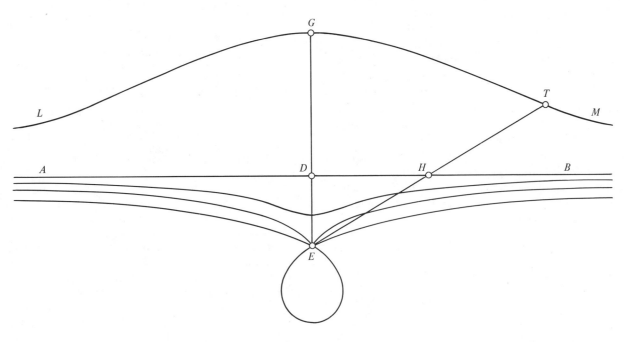

FIGURE 1

that a given length, *l*, is intercepted between *X* and *Y*. This is done by constructing a conchoid with "canon" *X*, "pole" *P*, and "interval" *l*; the intersection of this conchoid and *Y* gives the solution.

Nicomedes solved the problem of finding two mean proportionals (to which earlier Greek mathematicians had reduced the problem of doubling the cube) as shown in Figure 2.

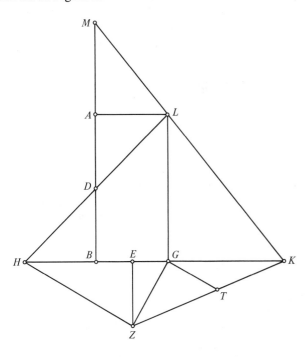

FIGURE 2

Given two straight lines *AB*, *BG*, between which it is required to find two mean proportionals. Complete the rectangle *ABGL*. Bisect *AB*, *BG* in *D* and *E*, respectively. Join *LD* and produce it to meet *GB* produced in *H*. Draw *EZ* perpendicular to *EG*, of length such that *GZ = AD*. Join *HZ*, and draw *GT* parallel to it. Then draw *ZTK*, meeting *BG* produced in *K*, so that *TK = AD* (this is possible by the "lemma of Nicomedes"). Join *KL* and produce it to meet *BA* produced in *M*. Then

$$\frac{AB}{GK} = \frac{GK}{MA} = \frac{MA}{BG},$$

and *GK*, *MA* are the required mean proportionals.[4]

Nicomedes also showed how to trisect the angle by means of his lemma and proved that the "first conchoid" is asymptotic to its "canon."[5] In addition he described what he called the "second, third, and fourth conchoids" and their uses. The ancient sources tell us nothing about them beyond their names, but it has been plausibly conjectured that they are to be

identified with the other branch of the curve in its three possible forms. In modern terms the curve is a quartic whose equation is, in polar coordinates,

$$\rho = \frac{a}{\cos \theta} \pm l,$$

or, in Cartesian coordinates,

$$(x - a)^2 (x^2 + y^2) - l^2 x^2 = 0.$$

This curve has two branches, both asymptotic to the line *x = a*. The lower branch (see Figure 1, in the lower part of which the three forms are depicted) has a double point at the pole *E*, which is either a node, a cusp, or an isolated point according as $l \gtreqless a$ (here *l* corresponds to the interval, *a* to the distance from the pole to the canon). This second branch can be constructed in the same way that Nicomedes constructed the first, with the sole difference that the interval is taken on the same side of the canon *AB* as the pole *E*.[6]

As far as is known, all applications of the conchoid made in antiquity were developed by Nicomedes himself. It was not until the late sixteenth century, when the works of Pappus and Eutocius describing the curve became generally known, that interest in it revived and new applications and properties were discovered. Viète used the "lemma of Nicomedes" as a postulate in his *Supplementum geometriae* (1593) to solve a number of problems leading to equations of the third and fourth degrees, including the construction of the regular heptagon. Johann Molther, in his little-known but remarkable *Problema deliacum* (1619), used the same lemma for an elegant reworking of the old problem of finding two mean proportionals. The conchoid attracted the attention of some of the best mathematicians of the seventeenth century. Descartes discussed the construction of tangents to it (in his *Géométrie* of 1637); Fermat and Roberval treated the same problem with respect to both branches. Huygens discovered a neat construction of the point of inflection (1653). Newton discussed the curve more than once, and in his *Arithmetica universalis* he recommended its use as an auxiliary in geometry because of the ease of its construction (it is in fact sufficient to solve any problem involving equations of the third and fourth degrees). In the appendix to the *Arithmetica*, on the linear construction of equations, he makes extensive use of the "lemma of Nicomedes." The seventeenth century also saw the first generalization of the conchoid, produced by taking a circle instead of a straight line for canon: this generates Pascal's limaçon.

NOTES

1. Eutocius, *Commentary on Archimedes' Sphere and Cylinder*, in *Archimedis Opera omnia*, J. L. Heiberg, ed., III, 98; Simplicius, *Commentary on Aristotle's Categories*, Kalbfleisch, ed., p. 192.
2. Eutocius, *loc. cit.* It is uncertain whether Nicomedes called the curve κοχλοειδής, κοχλιοειδής, or κογχοειδής (all three are found in our sources). The first two mean "snail-shaped"; the third, "mussel-shaped."
3. *Ibid.*
4. For a proof see Pappus, *Synagoge*, bk. 4, sec. 43, Hultsch, ed., I, 248–250, repro. in Heath, *History of Greek Mathematics*, I, 260–262. In fact all of the construction after the determination of point K is superfluous, for $ZT = MA$ and therefore GK, ZT are the required mean proportionals. As D. T. Whiteside pointed out to me, this was realized by Molther (*Problema deliacum*, pp. 55–58), and by Newton (*Mathematical Papers*, II, prob. 15, pp. 460–461).
5. On trisection of the angle see Pappus, *Synagoge*, bk. 4, sec. 62, Hultsch, ed., I, 274–276; see also Proclus, *Commentary on Euclid I*, Friedlein, ed., p. 272. On the "first conchoid" being asymptotic to its "canon," see Eutocius, *op. cit.*, in *Archimedis Opera omnia*, J. L. Heiberg, ed., III, 100–102.
6. I do not know who first proposed this identification of Nicomedes' "second, third, and fourth conchoids," but a probable guess is either Fermat or Roberval. In a letter to Roberval of 1636, Fermat (*Oeuvres*, II, 94) mentions the "second conchoid of Nicomedes." Roberval refers to the two branches as "conchoide de dessus" and "conchoide de dessous," respectively ("Composition des mouvemens," in *Ouvrages de mathématique* [1731], p. 28).

BIBLIOGRAPHY

The principal ancient passages concerning Nicomedes are Pappus, *Synagoge*, F. Hultsch, ed., I (Berlin, 1875), bk. 3, sec. 21, p. 56; bk. 4, secs. 39–45, pp. 242–252, and secs. 62–64, pp. 274–276; Eutocius, *Commentary on Archimedes' Sphere and Cylinder*, in *Archimedis Opera omnia*, J. L. Heiberg, ed., 2nd ed., III (Leipzig, 1915), 98–106; Proclus, *Commentary on Euclid I*, G. Friedlein, ed. (Leipzig, 1873), 272; and Simplicius, *Commentary on Aristotle's Categories*, K. Kalbfleisch, ed., which is Commentaria in Aristotelem Graeca, VIII (Berlin, 1907), 192. The best modern account of Nicomedes is Gino Loria, *Le scienze esatte nell'antica Grecia*, 2nd ed. (Milan, 1914), 404–410. See also T. L. Heath, *A History of Greek Mathematics* (Oxford, 1921), I, 238–240, 260–262, and II, 199. There is no adequate account of the treatment of the conchoid in the sixteenth and seventeenth centuries. The best available is Gino Loria, *Spezielle algebraische und transzendente ebene Kurven*, 2nd ed. (Leipzig–Berlin, 1910), I, 136–142, which also gives a good description of the mathematical properties of the curve; see also F. Gomes Teixeira, *Traité des courbes spéciales remarquables*, I, which is vol. IV of his *Obras sobre matematica* (Coimbra, 1908), 259–268.

On generalizations of the conchoid see Loria, *Spezielle . . . Kurven*, pp. 143–152. Viète's *Supplementum geometriae* is printed in his *Opera mathematica*, F. van Schooten, ed. (Leiden, 1646; repr. Hildesheim, 1970), 240–257. Johann Molther's extremely rare opuscule, *Problema deliacum de cubi duplicatione*, was printed at Frankfurt in 1619. For Descartes's treatment of the conchoid see *The Geometry of René Descartes*, trans. by D. E. Smith and M. L. Latham (Chicago–London, 1925), 113–114. The discussions of Fermat and Roberval are printed in Pierre de Fermat, *Oeuvres*, P. Tannery and C. H. Henry, eds., II (Paris, 1894), 72, 82, 86–87. See also Roberval, *Ouvrages de mathématique* (The Hague, 1731), 28–32 (on Pascal's limaçon, see p. 35). Huygens' solution is printed in *Oeuvres complètes de Christiaan Huygens*, XII (The Hague, 1910), 83–86. For Newton's treatment see *The Mathematical Papers of Isaac Newton*, D. T. Whiteside, ed., II (Cambridge, 1968), prob. 15, pp. 460–461, and especially the app. to his "Universal Arithmetick," printed in *The Mathematical Works of Isaac Newton*, D. T. Whiteside, ed., II (New York–London, 1967), 118–134.

G. J. Toomer

NICOT, JEAN (*b.* Nîmes, France, *ca.* 1530; *d.* Paris, France, 10 May 1604), *philology, botany.*

Any dictionary of scientific biography would be incomplete without an entry for Jean Nicot. His name designates in French "the nicotian plant, admirably suited to curing all wounds, sores, cankers, scurfs, and other such misfortunes of the human body" (*Thresor*, p. 429); it is preserved in the Linnaean designation *Nicotiana tabacum*, which to a certain extent renders Nicot a botanist.

The son of a court clerk, he studied letters at Nîmes, his native city, then at Paris, where he became a friend of the poet Ronsard. He was admitted to the king's household and took charge of charters. As councillor to the king he was sent on a diplomatic mission to Portugal, from which in 1560 he sent to Queen Mother Marie de' Medici seeds and leaves of "petun" (the Indian name for tobacco), pointing out the therapeutic value of the plant. Its cultivation later spread from France.

After two years of diplomatic service, Nicot began to dedicate himself to historical and literary study in his vast library at Brie-Comte Robert, near Paris. In 1568 he published *Historiae francorum lib. IV* of Aimonius (960–1010); in 1573 he supervised the publication of a new edition of the *Dictionnaire francois latin* of Robert Estienne, then began his own magnum opus, *Thresor de la langue francoyse*, published posthumously in 1606. This new French-Latin dictionary was enriched with a commentary in French that facilitated the compilation of subsequent French dictionaries. Although not strictly a scientist, Nicot was concerned in this work with animals and plants—the above citation is the prime example. He also left an unpublished treatise on nautical subjects.

BIBLIOGRAPHY

Nicot's works are *Aimonii monachi . . . Historiae francorum libri IV* (Paris, 1568) and *Thresor de la langue francoyse tant ancienne que moderne* (Paris, 1606).

On Nicot and his work, see Maxime Lanusse, *De Joanne Nicotio philologo* (Paris, 1893).

L. PLANTEFOL

NIEBUHR, CARSTEN (*b.* Altendorf, Holstein, 17 March 1733; *d.* Meldorf, Holstein, 26 April 1815), *cartography, exploration.*

The son of a farmer, Niebuhr did not attend school until he was eighteen. After inheriting some money he began training to be a surveyor by studying mathematics and astronomy at the University of Göttingen, but he never obtained a degree. In 1758 he was hired as a cartographer for a Danish expedition to Arabia that lasted from 1761 to 1767. During the expedition he made very exact determinations of longitude and latitude of localities in the eastern Mediterranean, made maps of cities, and mapped the Middle East, especially Arabia and Yemen. These maps were the best available for a long time. All the other members of the expedition died, and after having returned most of the scientific collections by ship from Bombay to Denmark, Niebuhr returned overland through Persia, Palestine, and Constantinople. During this trip he continued his geographic observations and made exact copies of the cuneiform inscriptions at Persepolis. The interpretation of the cuneiform alphabet by R. C. Rask and others was based on these copies, which were the best and most complete available. After the expedition Niebuhr declined several offers of high positions and became registrar at Meldorf, near his birthplace. The real value of his contributions was discovered later, partly because of the advanced and not always accepted mathematical methods used in his calculations and partly because he shunned publicity. Niebuhr's success as an explorer was based on his ability to make exact observations under highly adverse conditions and his ability to win the acceptance and cooperation of the local population. He preferred to write and speak in low German, published his papers in German, and regarded himself as Danish. (Holstein was then under the Danish crown.)

BIBLIOGRAPHY

I. ORIGINAL WORKS. Niebuhr's writings are *Beschreibung von Arabien* (Copenhagen, 1772), the first, preliminary account of the results of the expedition; *Reisebeschreibung nach Arabien und andern umliegenden Ländern*, 2 vols. (Copenhagen, 1774–1778), vol. III, edited posthumously by J. N. Gloyer and J. Olshausen (Hamburg, 1837).

II. SECONDARY LITERATURE. There are few biographical papers on Niebuhr, the most important being a short biography by his son, B. G. Niebuhr, "Carsten Niebuhrs Leben," in *Kieler Blätter*, **3** (1816), 1–86. A semipopular narrative of the expedition and Niebuhr's life, which became a best seller in Denmark, is Thorkild Hansen, *Det lykkelige Arabien* (Copenhagen, 1962).

NILS SPJELDNAES

NIELSEN, NIELS (*b.* Ørslev, Denmark, 2 December 1865; *d.* Copenhagen, Denmark, 16 September 1931), *mathematics.*

Nielsen's father was a small farmer, and his family lived in modest circumstances. He originally wished to attend the polytechnical institute, but he was early attracted to pure science. In 1885 he began his studies at the University of Copenhagen, where he passed the government examination in 1891 and received his doctorate in 1895. He had been teaching in the secondary schools since 1887, and in 1900 he began to give preparatory courses for the polytechnic institute. From 1903 to 1906 he belonged to the University Inspectorate for secondary schools. In 1905 he became *Dozent* and in 1909 he succeeded Julius Petersen as full professor of mathematics at the University of Copenhagen.

He became a member of the Leopoldina of Halle in 1906 and an honorary member of the Wiskundig Genootschap of Amsterdam in 1907. Nielsen's principal achievements were his many textbooks, which dealt with various classes of special functions. Before he prepared these books he wrote numerous papers. His textbooks on cylindrical functions (1904) and on the gamma function (1906) were widely used.

Nielsen developed no new ideas and did not even present any fundamental theorems, but he possessed great knowledge and the ability to generalize existing formalisms. Moreover, he did make an important contribution to the theory of gamma function and factorial series. The theory originated with W. V. Jensen; Nielsen gave it further impetus, and Nörlund provided its definitive clarification. Nielsen's abilities were thus very restricted. He was a master in the treatment of unmethodical calculations and came up with a multitude of particular points. He playfully conceived new things that were not always in a completed form, and he was a significant influence on his students.

In 1917 Nielsen suffered a breakdown. He never fully recovered but his powers were not perceptibly diminished. He turned his attention to number theory (Bernoulli's numbers, Fermat's equation), which he treated unsystematically. In the history of mathematics he occupied himself primarily with accounts of personalities and the historical development of specific mathematical problems. Two books on Danish mathematicians and two on French mathematicians are the fruits of his work in this area.

BIBLIOGRAPHY

I. ORIGINAL WORKS. Nielsen's works include the following: *Handbuch der Theorie der Zylinderfunktionen* (Leipzig, 1904), which contains sixteen pp. of bibliography; *Theorie der Integrallogarithmus und verwandter Transzendenten* (Leipzig, 1906), which has ten pp. of bibliography, tables, and applications; *Handbuch der Theorie der Gammafunktion* (Leipzig, 1906), which represents twenty years of work and is the first comprehensive treatment of the gamma function since Legendre's *Traité;* and *Lehrbuch der unendlichen Reihen; Vorlesung gehalten an der Universität Kopenhagen* (Leipzig, 1908), an elementary treatment without the use of calculus.

See also *Laeren on Graensvaerdier som indledning til analysen* (Copenhagen, 1910); *Mathematiken i Danmark, 1528–1800,* I; *1801–1908,* II (Copenhagen–Oslo, 1910), which contains data on his life and a compilation of his published works; *Elemente der Funktionentheorie Vorlesung gehalten an der Universität Kopenhagen* (Leipzig, 1911); *Géomètres français sous la révolution* (Copenhagen, 1929), treats of seventy-six mathematicians; and *Géomètres français du dix-huitième siècle,* Niels Nörlund, ed. (Copenhagen–Paris, 1929), which is a posthumous work and treats of 153 mathematicians.

II. SECONDARY LITERATURE. Nielsen also published about 100 articles in twenty-one different Danish and foreign periodicals. For further information see Harald Bohr, "Niels Nielsen 2 December 1865–16 September 1931," in *Matematisk Tidsskrift,* 41–45; and Poggendorff, IV, 1073; V, 905; VI, 1855.

H. OETTEL

NIEPCE, JOSEPH (later **NICÉPHORE**) (*b.* Chalon-sur-Saône, France, 1765; *d.* St. Loup de Varenne, France, 5 July 1833), *photography.*

In 1789 Niepce was a professor at an Oratorian *collège* in Angers; he then took up a military career, eventually becoming a staff officer with the French army in Italy. In 1794 he left the army to settle in Nice, where he married. He was appointed administrator of the district of Nice at the beginning of 1795, but resigned after a few months; his elder brother Claude, who had also retired from the army, came to join him, and together they pursued their common interest in research.

By 3 August 1807, the date upon which they patented their "pyréolophore," the brothers had left Nice for their paternal home at Chalon-sur-Saône and their country estate of St. Loup de Varenne. The "pyréolophore" was an internal combustion engine fueled by lycopodium powder. It was sufficiently powerful to move a boat, and trials of it were conducted on a pond at St. Loup de Varenne and on the Saône. The tests were reported favorably to the Académie des Sciences by Berthollet and Lazare Carnot, but the invention never became practical, although the Niepce brothers attempted to render it more economical by substituting first pulverized coal, then petroleum, for the expensive lycopodium powder. A further invention of the same year, a sort of hydraulic ram, devised in response to a government competition to replace the apparatus formerly used to supply Versailles with water from the Seine, won them only an encouraging letter from Carnot.

The brothers next turned to agricultural research. In 1813 the government offered a prize for a woad that could replace indigo, which was then totally unavailable because of the Continental Blockade. The Niepces investigated a variety of materials, but did not succeed in extracting a suitable dye. They also tried to derive sugar from beets and starch from pumpkins, and examined plants that might yield fibers for textiles. Their only rewards were flattering letters from the government.

By 1813 Nicéphore Niepce had taken up the then fashionable occupation of lithography. The development of his researches during the next few years is not known, but it would seem probable that having himself tried to sketch some simple subjects for lithographic reproduction, he next tried to copy engravings automatically by rendering them transparent for transfer to the stone. He thus may have reached the idea of reproducing nature itself. Claude Niepce assisted him in this work. At the same time, the brothers tried to recoup their finances, depleted since the failure of the "pyréolophore," by searching their neighborhood for a supply of stone suitable for being made into lithographic tablets. Although they were unsuccessful in this effort, Nicéphore Niepce was in 1817 recompensed for his attempt by the Société d'Encouragement pour l'Industrie Nationale.

In the meantime, Claude Niepce had, in March 1816, gone first to Paris and then to England to conduct further tests and solicit support for the "pyréolophore." The brothers remained in constant correspondence, and their letters document Nicéphore Niepce's subsequent invention of photography. The latter continued in his efforts to reproduce nature directly on a specially prepared surface; a letter to

Claude of 5 May 1816 refers to a photographic apparatus that produced a negative image. A letter of the following 28 May adds, "I am hurrying to get these four new prints to you . . ." It is thus clear that Nicéphore had succeeded in fixing the images.

Nicéphore Niepce then began work on the chambers, diaphragms, and shutters that constituted his camera. In his 1816 experiments he used paper impregnated with silver chloride fixed with nitric acid; in March 1817 he began to use the Judean bitumen process of reproduction, making use of a light-sensitive lithographer's asphalt. By 1821, he was using this method to produce images on both glass and metal, notably tin. In 1822 he recorded the first fixed positive image, which he called a "point de vue," to distinguish it from transparency-copied engravings. In January 1826 Nicéphore Niepce received from Daguerre, then unknown to him, a letter of inquiry about his work, which he answered courteously but uninformatively. A year later, Daguerre wrote a second letter, which prompted Nicéphore Niepce to make inquiries about him; in June 1827 he relented, and offered Daguerre a heliograph. The two men met when Nicéphore Niepce, alarmed by news about his brother, passed through Paris on his way to London.

In London Nicéphore Niepce discovered that his brother had been out of his senses for several years, and that the inventions about which Claude had written him—and for which he had in fact ruined himself financially—were mere follies. Disappointed, he returned to France at the beginning of 1828. His brother died a few days after his departure. Although Nicéphore Niepce was old, tired, and in debt, he still hesitated to reveal his secrets to Daguerre; it was only in October 1829 that he offered to "cooperate" with the latter, and an agreement was signed on 14 December of the same year. Little is known of the development of this association, save that in 1831 Daguerre suggested to Niepce that he experiment with silver iodide. Niepce was struck by apoplexy on 3 July 1833, and died two days later.

BIBLIOGRAPHY

I. ORIGINAL WORKS. Niepce's only writings, apart from the patent for the pyréolophore (No. 405 at the Institut National de la Propriété Industrielle), are his letters, some of which are preserved together with some of his devices at the Conservatoire National des Arts et Métiers. Others are at the Musée de Chalons and at the Academy of Sciences of the U.S.S.R.

II. SECONDARY LITERATURE. See Raymond Lécuyer, *Histoire de la photographie* (Paris, 1945); Georges Potonniée, *Histoire de la découverte de la photographie* (Paris, 1925); and B. Newhall, *Image* (New York, 1967).

The Berthollet and Carnot "Rapports sur une nouvelle machine inventée par MM. Niepce et nommée par eux pyréolophore," are in *Mémoires de la classe de sciences mathématiques et physiques de l'Institut,* **8** (1807), 146–153.

ROBERT SOULARD

NIESTEN, JEAN LOUIS NICOLAS (*b.* Visé, Liège, Belgium, 4 July 1844; *d.* Laeken, Brussels, Belgium, 27 December 1920), *astronomy.*

Niesten initially made a career in the military, where he served as a captain in the artillery. He wrote two textbooks on military science: *Artillerie. Passage des rivières* (Brussels, 1876) and *Précis des connaissances exigées des officiers sortis des cadres et des sous-officiers de l'artillerie par les programmes de 1876.* In 1877 he resigned from the service, and Jean Charles Houzeau, director of the Brussels observatory, appointed him assistant astronomer. In 1878 he was promoted to full astronomer, and he became *chef de service* in 1884.

Niesten was particularly interested in planetary astronomy, and in 1882 he was appointed chief of the Belgian mission to Santiago, Chile, to observe the transit of Venus. Houzeau himself directed another mission to Texas. (This was the first time Belgium officially participated in an international astronomical expedition.) In contrast with the mission to Texas, Niesten had favorable weather and was able to make useful observations. His report on the expedition was published in 1884. In 1887 he traveled to Yurievets, Russia, to observe the total solar eclipse. Niesten was a systematic observer, and many of his articles were concerned with the physical aspects of the planets: the "canals" on Mars (he claimed also to have seen rivers) and the red spot on Jupiter. He also subscribed to the now discredited theory of the daily nutation of the earth. Most of his astronomical observations appeared in the various publications of the Brussels observatory between 1878 and 1900. Niesten was one of the founders of the journal *Ciel et terre,* which first appeared in 1880; he contributed more than seventy articles, all of a semipopular nature, between 1880 and 1899.

BIBLIOGRAPHY

A complete bibliography of Niesten's works has been compiled by A. Collard, in *Ciel et terre,* **38** (1922), 330–338, 400–406; **39** (1923), 21–23, 41–44, 62–68, 86–89.

Niesten's report on the 1882 expedition is "Passage de Venus du 6 décembre 1882," in *Annuaire de l'observatoire de Bruxelles*, **51** (1884).

LETTIE S. MULTHAUF

NIEUWENTIJT, BERNARD (*b.* Westgraftdijk, North Holland, 10 August 1654; *d.* Purmerend, North Holland, 30 May 1718), *mathematics, philosophy.*

Bernard was the son of Emmanuel Nieuwentijt, minister at Westgraftdijk, and Sara d'Imbleville. Although he was expected to enter the ministry, he chose instead to study natural sciences. On 28 February 1675 he was enrolled as a student in medicine at Leiden University; later in the year he was also enrolled at Utrecht University, where he studied law and defended his medical thesis in 1676 [1]. He then settled as a medical practitioner in Purmerend. On 12 November 1684 he married Eva Moens, the widow of Philips Munnik, a naval captain in the service of the Dutch States-General. He was elected a member of the city council and became a burgomaster of Purmerend. As a youth Nieuwentijt was influenced by Cartesianism and he acquired a thorough knowledge of mathematics and natural philosophy. In 1695–1700 he was engaged in a controversy with Leibniz and his school on the foundations of calculus. On 12 March 1699 he married his second wife, Elisabeth Lams, the daughter of Willem Lams, burgomaster of Wormer.

Nieuwentijt became famous in his home country and abroad because of the publication of two lengthy works. One [6] was originally published in Dutch in 1714; according to [12] it was reedited in 1717, 1720, 1725, 1741, and 1759; editions with other dates are incidentally found in various libraries (1715, 1718, 1730; see [13]). The work was translated into English by J. Chamberlayne in 1718 [6a]; a fourth edition (1730) is mentioned. It was also translated into French by P. Noguez [6b]; this translation was published in 1725 (Paris), 1727 (Amsterdam), and 1760 (Amsterdam–Leipzig). A German translation by W. C. Baumann appeared in 1732 and another by J. A. von Segner in 1747 [6c]. The second of his two works [8] was posthumously published in Dutch in 1720; it was published again in 1741 and 1754; and was translated into French (1725) and English (1760). Nieuwentijt's portrait, painted by D. Valkenburg, is in the University of Amsterdam; the portrait in his 1714 publication [6] was engraved by P. van Gunst.

The title of Nieuwentijt's *Analysis infinitorum* [3] reminds the historian of the title of Leibniz' article of 1686, "De geometria recondita et analysi indivisibilium atque infinitorum." Nieuwentijt's [3] was the first comprehensive book on "analysis infinitorum." By L. Euler's *Introductio in analysin infinitorum*, analysis became the name of a mathematical discipline. To this field Nieuwentijt contributed little more than the name. What is surprising, however, is the erudite scholarship of a small-town physician who, except for limited university study, does not seem to have cultivated many learned colleagues. Nieuwentijt's work reveals his full acquaintance with the mathematics of his period and a remarkable self-reliance.

Nieuwentijt rejected Leibniz' approach to analysis. He did not admit infinitesimals of higher order. Nieuwentijt's method consists, in modern terms, in adjoining to the real field an element e with $e^2 = 0$. Leibniz' answer [9] to Nieuwentijt's objections [2] (see also [4]) was not convincing. Nieuwentijt's objections, however, may have contributed to improving the insight into higher-order differentials. It is disappointing that he did not sufficiently appreciate Leibniz' integral calculus.

His 1714 publication [6], of about 1,000 pages, was intended to demonstrate the existence of God by teleological arguments. Never before had this been tried on such a scale, and none among Nieuwentijt's numerous imitators equaled his completeness. It is not clear, however, whether or to what degree he depended on William Derham, whose *Physico-Theology* [10] (see also [11]) appeared almost simultaneously. Nieuwentijt may have known of Derham's lectures of 1711–1712, which were the nucleus for the work.

It is an old idea that nature, by its purposiveness, betrays the existence of a creator; Nieuwentijt, however, was one of the first who, rather than relying on a few examples, reviewed the whole of natural sciences to show in detail how marvelously things fitted in the world. His work [6] looks like a manual of up-to-date science and as such it may have contributed to the propagation of knowledge. On the other hand, by the abundance of its argumentation, it is tiring reading and full of platitudes. Its fundamental shortcoming is its static world picture and its lack of any trace of the oncoming evolutionary ideas. Its background philosophy, however, is remarkably sound. Nieuwentijt opposed both chance and necessity as explanatory principles of nature. He preferred empiricist above rationalist arguments. Natural laws have, according to Nieuwentijt, factual rather than rational truth, and as such they must have been ordained by a lawgiver.

Nieuwentijt felt that rationalism led to Spinozism and other kinds of atheism. A more methodical struggle against rationalism was fought in his second major work [8]. This is, indeed, a methodology of science which surprises by a seemingly modern view.

In fact it is nothing but a philosophy of common sense, and this explains why it fell into oblivion amid more sophisticated philosophies. In this work [8] Nieuwentijt arrived at a clear distinction between what he called ideal and factual mathematics, and at the insight that both avail themselves of the same formal methods, that all ideal statements are conditional, and that the ultimate criterion for factual statements is corroboration by experience. Nieuwentijt distinguished himself from the British empiricists by his closeness to mathematics and exact sciences. Although his influence in philosophy was negligible, his position as a methodologist was unique up to modern times.

BIBLIOGRAPHY

1. *Disputatio medica inauguralis de obstructionibus,* 8 Feb. 1676, Ultraiecti.

2. *Considerationes circa analyseos ad quantitates infinitè parvas applicatae principia, et calculi differentialis usum in resolvendis problematibus geometricis* (Amsterdam, 1694).

3. *Analysis infinitorum, seu curvilineorum proprietates ex polygonorum natura deductae* (Amsterdam, 1695).

4. *Considerationes secundae circa calculi differentialis principia; et responsio ad virum nobilissimum C. G. Leibnitium* (Amsterdam, 1696).

5. "Nouvel usage des tables des sinus au moyen de s'en servir sans qu'il soit nécessaire de multiplier et de diviser," in *Journal littéraire,* **5** (1714), 166–174.

6. *Het regt gebruik der wereltbeschouwingen ter overtuiginge van ongodisten en ongelovigen, aangetoont door . . .* (Amsterdam, 1714).

6a. *The Religious Philosopher, or the Right Use of Contemplating the Works of the Creator*: (I) *In the Wonderful Structure of Animal Bodies,* (II) *In the Formation of the Elements,* (III) *In the Structure of the Heavens, Designed for the Conviction of Atheists,* trans. by J. Chamberlayne, 3 vols. (London, 1718).

6b. *L'existence de Dieu démontrée par les merveilles de la nature, en trois parties, où l'on traite de la structure des corps de l'homme, des élémens, des astres et de leurs divers effets,* trans. by P. Noguez (Paris, 1725).

6c. *Rechter Gebrauch der Weltbetrachtung zur Erkenntnis der Macht, Weisheit und Güte Gottes, auch Überzeugung der Atheisten und Ungläubigen,* trans. by J. A. v. Segner (Jena, 1747).

7. "Brief aen den Heer J. Bernard, zynde een antwoord op de Aenmerkingen van den Heer Bernard, omtrent de werelt-beschouwingen, in de Nouv. de la Repub. 1716, 252," in *Maandelijke Uittreksels, of Boekrael der Geleerde Werelt* (1716), 673–690.

8. *Gronden van zekerheid of de regte betoogwyze der wiskundigen so in het denkbeeldige als in het zakelijke: ter weerlegging van Spinosaas denkbeeldig samenstel;*

en ter aanleiding van eene sekere sakelyke wysbegeerte, aangetoont door . . . (Amsterdam, 1720).

9. G. G. L. [Leibniz], "Responsio ad nonnullas difficultates, a Dn. Bernardo Nieuwentijt circa methodum differentialem seu infinitesimalem motas," in *Acta eruditorum 1695,* pp. 310–316.

10. William Derham, *Physico-Theology, or a Demonstration of the Being and Attributes of God From His Works of Creation* (London, 1713).

11. William Derham, *Astro-Theology, or a Demonstration of the Being and Attributes of God From a Survey of the Heavens* (London, 1715).

12. *Nieuw Nederlandsch Biographisch Woordenboek,* **6** (1924), 1062–1063.

13. A. J. J. Van der Velde, "Bijdrage tot de bio-bibliographie van Bernard Nieuwentyt (1654–1718)," in *Bijdragen en Mededelingen Koninklijke Vlaamsche Academie van Taal- en Letterkunde 1926,* 709–718.

14. E. W. Beth, "Nieuwentyt's Significance for the Philosophy of Science," in *Synthese,* **9** (1955), 447–453.

15. H. Freudenthal, "Nieuwentijt und der teleologische Gottesbeweis," in *Synthese,* **9** (1955), 454–464.

16. J. Vercruysse, "La fortune de Bernard Nieuwentyd en France au 18e siècle et les notes marginales de Voltaire," in *Studies on Voltaire and the 18th Century,* **30** (1964), 223–246.

17. J. Vercruysse, "Frans onthaal voor een Nederlandse apologeet: Bernard Nieuwentyd—1654–1718," in *Tijdschrift van de Vrije Universiteit te Brussel,* **11** (1968–1969), 97–120.

HANS FREUDENTHAL

NIEUWLAND, JULIUS ARTHUR (*b.* Hansbeke, Belgium, 14 February 1878; *d.* Washington, D.C., 11 June 1936), *organic chemistry.*

Nieuwland was the son of poor Flemings who in 1880 immigrated to South Bend, Indiana, where they joined a settlement of Flemish speakers from the Ghent region. Nieuwland was educated in a German school. He graduated from Notre Dame University in 1899, and studied for the priesthood at the Congregation of the Holy Cross in South Bend and then at Holy Cross College of the Catholic University of America in Washington, D.C. He was ordained in 1903. Meanwhile, he studied botany and chemistry at the Catholic University, gaining a Ph.D. in 1904 with a thesis—which contained the germ of much of his later work—on the reactions of acetylene. His discovery of the reaction between acetylene and arsenic trichloride (which he did not pursue because of the noxious nature of the product) led to the development of the poison gas and vesicant lewisite (named after W. Lee Lewis) in World War I.

For several years Nieuwland almost abandoned chemistry and taught botany at Notre Dame, an interest which he maintained throughout his life. Botanical excursions were one of his favorite relaxations, and he published many papers on the subject—although none of them seem to be of importance. In 1909 Nieuwland founded the journal *American Midland Naturalist*; he edited it until near the end of his life.

In 1918 Nieuwland became professor of organic chemistry at Notre Dame and, with a series of junior collaborators, resumed his work on acetylene. He was able to polymerize acetylene under controlled conditions, using a cuprous chloride–ammonium chloride catalyst, to give a mixture of which the main constituent was divinyl-acetylene (hex-1,5-diene 3-yne). In 1925 a chance encounter at a scientific meeting led to the collaboration of Nieuwland with the firm of Du Pont, which was interested in this reaction. The Du Pont chemists modified the polymerization to produce good yields of vinyl-acetylene (but-1-ene 3-yne), which on treatment with hydrogen chloride formed 2-chlorobutadiene ("Chloroprene"). This in turn could be polymerized to the first really successful synthetic rubber, which Du Pont marketed in the early 1930's as "Duprene" or neoprene.

Nieuwland died suddenly of a heart attack while visiting his old university in Washington.

BIBLIOGRAPHY

I. ORIGINAL WORKS. A complete list of Nieuwland's papers is given in *American Midland Naturalist*, **17**, no. 4 (1936), vii–xv. There are ninety-seven biological articles, eighty-eight articles on chemistry, including, in addition to his acetylene studies, much pioneer work on the catalytic properties of boron trifluoride. His most important paper, "A New Synthetic Rubber: Chloroprene and Its Polymers," in *Journal of the American Chemical Society*, **53** (1931), 4198, was followed by a companion paper from the Du Pont team, *ibid.*, 4203. See also *The Chemistry of Acetylene* (New York, 1945), written with R. R. Vogt, which has a portrait as frontispiece.

II. SECONDARY LITERATURE. Nieuwland is noticed in *Dictionary of American Biography*, supp. 2 (1958), 488–489, and in *National Cyclopedia of American Biography* XXVI. The best account of his life and work is in a memorial ed. of the Notre Dame house journal, *Catalyzer* (February 1937), 39–44.

W. V. FARRAR

NIFO, AGOSTINO (*b*. Sessa Aurunca, Italy, *ca.* 1469–1470; *d.* Sessa Aurunca, 18 January 1538), *medicine, natural philosophy, psychology.*

The son of Giacomo Nifo and Francesco Gallione, Nifo received his early education at Naples before attending the University of Padua. After receiving his degree around 1490, he taught at Padua from about 1492 until 1499, when he returned to his native city. He was involved in controversies at Padua with his teacher, Nicoletto Vernia, as well as with his lifelong rival, Pietro Pomponazzi, and the Franciscan theologian Antonio Trombetta. In the south he became a member of the circle of the famed humanist Giovanni Pontano, and he himself wrote and published humanistic treatises. He had learned Greek by 1503. Nifo appears to have been professor of philosophy at Naples and at Salerno during the first decade of the sixteenth century and also to have practiced medicine. He served as physician to Gonsalvo Hernández de Córdoba in 1504–1505.

Subsequently, Leo X invited him to teach at the University of Rome, where he was ordinary professor of philosophy in 1514. In 1520 Leo made Nifo a count palatine, granted him the right to use the Medici name, and authorized him to grant degrees in his own name. Nifo had openly attacked Pomponazzi in his *De immortalitate animae* (1518), which was dedicated to Leo. He served as ordinary professor of philosophy at the University of Pisa from 1519 until 1522. He then departed for Salerno, where he appears to have taught from 1522 until 1535, except for the academic year 1531–1532, when he taught philosophy and medicine at Naples. Although the Florentines attempted to lure him back to Pisa in 1525 and Paul III asked him to return to Rome to teach natural philosophy in 1535, Nifo declined both invitations. He was elected mayor of Sessa and extended the formal welcome to Emperor Charles V during his visit there on 24 March 1536.

Nifo wrote commentaries on almost all the works of Aristotle, usually providing his own translation. In some cases he wrote a second, revised commentary. While he held to the doctrine of Averroës (Ibn Rushd) of the unity of the intellect in two early works, the commentary on Averroës' *Destructio destructionum* and the early commentary on the *De anima*, he rejected this as the true interpretation of Aristotle in his *De intellectu*, published in 1503. In later works he emphasized that the true interpretation of Aristotle is reached through reading the Greek text. Nifo also came to prefer the Greek commentators over Averroës. This shift is especially noticeable in his psychological and logical writings. He did not, however, give up his interest in establishing the true interpretation of Averroës himself.

In his early commentary on the *Physics*, book I, t.c. 4 (Venice, 1508; fols. 7v–8), Nifo held that through a *negotiatio* the intellect could grasp the cause of an effect and thus formulate a *propter quid* demonstration in natural philosophy, whereas in his posthumously published *Recognitiones* on the *Physics*, after having studied Aristotle and the Greek commentators more carefully, he rejected the notion of a *negotiatio* (see Venice ed., 1569, pp. 13–14) and proposed instead that the cause of an effect is learned through a merely hypothetical syllogism *(syllogismus coniecturalis)*. To the possible objection that natural science would then cease to be science, he replied that while it is not science *simpliciter*, like mathematics, it is still a science *propter quid*, but one that remains conjectural, insofar as in it the knowledge of the cause can never be as certain as the knowledge of the effect, since the latter is based on sense experience. Although Nifo allowed in his commentary on the *Posterior Analytics* (1526), book I, t.c. 21, that there could be *demonstratio simpliciter* in natural science, he gave no examples of such a demonstration. The other form of demonstration, that "from hypothesis," was now called *demonstratio coniecturalis* and appeared to dominate in science.

Nifo also attempted in his early commentary on the *Physics*, book VIII, t.c. 81 (fols. 236–236v), to reconcile Aristotle with the impetus theory by making the impetus the principal mover, and the medium and its properties only auxiliary causes or passive dispositions. Later, in his commentary on the *De caelo*, book III, t.c. 28 (Naples, 1517; fols. 22v–23), he added the interesting refinement that a *vis impressa* is communicated not only to the projectile but also to the air or medium. Nifo developed Averroës' doctrine of natural minima by further refining the notion of qualitative minima, which explain qualitative changes, and by using the theory of minima to explain physical structure and chemical reactions. His medical interests are clearly evident in his *De ratione medendi* and his unpublished commentary on Hippocrates' *Aphorisms*. They are also occasionally reflected in remarks found in his *De pulchro et amore*, in which he proposed a sexual theory of love, and in his commentaries on the *Parva naturalia* and the *De animalibus*.

BIBLIOGRAPHY

I. Original Works. There is no collected ed. of Nifo's works. His commentaries on Aristotle, some of which contain a commentary on Ibn Rushd, include *Super tres libros De anima* (Venice, 1503), repr. with rev. commentary (Venice, 1522, 1523, 1544, 1549, 1552, 1553, 1554, 1559); *Aristotelis De generatione et corruptione liber Augustino Nipho philosopho suessano interprete et expositore* (Venice, 1506), repr. with *Recognitiones* and *Quaestio de infinitate*

primi motoris (Venice, 1526, 1543, 1550, 1557, 1577); *Aristotelis Physicarum acroasum hoc est naturalium auscultationum liber interprete atque expositore Eutyco Augustino Nypho phylotheo suessano* (Venice, 1508, 1519), to which *Recognitiones* was later added (Venice, 1540, 1543, 1549, 1552, 1558, 1559, 1569); *In quattuor libros De caelo et mundo et Aristotelis et Averrois expositio* (Naples, 1517); *Parva naturalia Augustini Niphi Medices philosophi suessani* (Venice, 1523); *Suessanus super posteriora cum tabula, Eutychi Augustini Nyphi Medices philosophi suessani commentaria in libris posteriorum Aristotelis* (Venice, 1526, 1538, 1539, 1544, 1548, 1552, 1553, 1554, 1565; Paris, 1540); and *Expositiones in omnes Aristotelis libros De historia animalium, De partibus animalium et earum causis, ac De generatione animalium* (Venice, 1546), which was completed in 1534 and published posthumously.

Besides his comments on Ibn Rushd in these works, Nifo wrote commentaries on Ibn Rushd's *Destructio destructionum* (Venice, 1497); *De animae beatitudine* (Venice, 1508); *De substantia orbis* (Venice, 1508); and a short opusculum, *Averrois de mixtione defensio* (Venice, 1505). There is also a commentary on Ptolemy, *Ad Apotelesmata Ptolemaei eruditiones* (Naples, 1513); and one on the *Aphorisms* of Hippocrates, Biblioteca Lancisiana, Rome, Codex 158, fols. 55 ff. Other works of interest are his *De demonibus*, printed with his *De intellectu* (Venice, 1503); *De diebus criticis* (Venice, 1504); *De nostrarum calamitatum causis liber* (Venice, 1505); *De immortalitate animae* (Venice, 1518); *De falsa diluvii prognosticatione* (Naples, 1519); *De figuris stellarum helionoricis* (Naples, 1526); *De pulchro et amore* (Rome, 1531); and *De ratione medendi* (Naples, 1551), which was completed in 1528.

II. Secondary Literature. See Leopoldo Cassese, "Agostino Nifo a Salerno," in *Atti del Centro di studi di medicina medioevale*, **3** (1958), app. to *Rassegna storica salernitana*, **19** (1958), 3–17; Angelo Crescini, *Le origini del metodo analitico, Il Cinquecento* (Trieste, 1965), 141–144, 181, 187; E. J. Dijksterhuis, *The Mechanization of the World Picture*, C. Dikshoorn, trans. (Oxford, 1961), 236–237, 278; Giovanni di Napoli, *L'immortalità dell'anima nel Rinascimento* (Turin, 1963), 203–217, 309–314; Pierre Duhem, *Études sur Léonard de Vinci*, III (Paris, 1955), 115–120; Eugenio Garin, *La cultura filosofica del Rinascimento italiano* (Florence, 1961), 114–118, 295–303; and *Storia della filosofia italiana*, II (Turin, 1966), 523–527, 535–538, 572–573; Michele Giorgiantonio, "Un nostro filosofo dimenticato del'400 (Luca Prassicio e Agostino Nifo)," in *Sophia* (Naples), **16** (1948), 212–214, 303–312; Gustav Hellmann, *Beiträge zur Geschichte der Meteorologie*, I, Nr. 1 (Berlin, 1914), 40–44, 79–83; Edward P. Mahoney, "Agostino Nifo's *De sensu agente*," in *Archiv für Geschichte der Philosophie*, **53** (1971), 119–142; "Agostino Nifo's Early Views on Immortality," in *Journal of the History of Philosophy*, **8** (1970), 451–460; "A Note on Agostino Nifo," in *Philological Quarterly*, **50** (1971), 125–132; "Nicoletto Vernia and Agostino Nifo on Alexander of Aphrodisias: An Unnoticed Dispute," in *Rivista critica di storia della filosofia*, **23** (1968), 268–296; "Pier Nicola Castellani and Agostino Nifo on Averroës' Doctrine of the Agent Intellect," *ibid.*, **25** (1970), 387–409; and Anneliese Maier, *Zwei Grundprobleme der scholastischen Naturphilosophie*, 2nd ed. (Rome, 1951), 61, 295–297.

See also Bruno Nardi, *Saggi sull'aristotelismo padovano dal secolo XIV al XVI* (Florence, 1958); and *Sigieri di Brabante nel pensiero del Rinascimento italiano* (Rome, 1945), *passim*; Antonino Poppi, *Causalità e infinità nella scuola padovana dal 1480 al 1513* (Padua, 1966), 222–236; and *Saggi sul pensiero inedito di Pietro Pomponazzi* (Padua, 1970), 97–101, 121–144; John Herman Randall, *The School of Padua and the Rise of Modern Science* (Padua, 1961), 42–47, 57 ff., 74 ff.; Wilhelm Risse, *Die Logik der Neuzeit*, I (Stuttgart, 1964), 218–229; Lynn Thorndike, *A History of Magic and Experimental Science*, V (New York, 1941), 69–98, 162 ff., 182–188; Giuseppe Tommasino, *Tra umanisti e filosofi* (Maddaloni, 1921), pt. I, 123–147; Pasquale Tuozzi, "Agostino Nifo e le sue opere," in *Atti e memorie della R. Accademia di scienze, lettere ed arti* (Padua), n.s. **20** (1904), 63–86; Andreas G. M. van Melsen, *From Atomos to Atom: The History of the Concept "Atom"* (Pittsburgh, 1952), 64–76; and William A. Wallace, *Causality and Scientific Explanation*, I (Ann Arbor, 1972), 139–153.

EDWARD P. MAHONEY

NIGGLI, PAUL (*b.* Zofingen, Switzerland, 26 June 1888; *d.* Zurich, Switzerland, 13 January 1953), *crystallography, mineralogy, petrology, geology, chemistry.*

Niggli's father was a teacher and principal of the technical high school at Zofingen. Both his father and his Gymnasium teacher Fritz Mühlberg sparked his lifelong enthusiasm for the natural sciences and for geological-mineralogical problems in particular. As early as his high school years he participated in the mapping of his home canton; and at the age of nineteen he wrote his first scientific paper, "Die geologische Karte von Zofingen" (1913).

In the fall of 1907 Niggli enrolled in the Section for Natural Science Teachers at the Eidgenössische Technische Hochschule in Zurich. His early interest in both the descriptive and the analytic aspects of research prompted him to choose a petrologic topic for his M.S. thesis under Ulrich Grubenmann. In 1911 he received his teacher's diploma and, after a brief stay in the department of physical chemistry of the Technical University at Karlsruhe, received his Ph.D. in 1912 at the University of Zurich. His thesis, which became famous, was entitled *Die Chloritoidschiefer des nordöstlichen Gotthardmassivs.* It showed the traits of Niggli's style of research, the combination of a fundamentalist approach with a strong trend toward the integration of broad aspects.

Shortly after receiving his doctorate, Niggli qualified as a lecturer; and in 1913 he was at the Geophysical Laboratory of the Carnegie Institution in Washington, D.C., where he worked with Norman L. Bowen on phase diagrams of petrology, especially those with a volatile component. From the end of 1915 to 1918 he was a professor at Friedrich Rinne's Institute at Leipzig; he then taught for two years at Tübingen. In 1920 Niggli succeeded Grubenmann as professor of mineralogy and petrography at the University of Zurich and the Swiss Federal Institute of Technology. He held these positions until his death, receiving but declining offers from well-known foreign universities.

Niggli's influence is still felt in virtually all fields of applied and pure crystallography, mineralogy, and petrology. To those who did not know him Niggli appeared at times to be dry or even unfriendly; this was a result of his intense dedication to his work and his modest, simple, and direct way of dealing with people. To a majority of his students, co-workers, and friends he was by no means authoritative or despotic; he was warm and interested in their scientific problems and education and in their more personal affairs. His terms as rector of the Eidgenössische Technische Hochschule (1927–1931) and the University of Zurich (1940–1942) are proof of his talent for organization and his interest in public affairs. Nevertheless, the intensity of his scientific disputes, particularly over transformism, spilled over to affect personal relationships and the careers of two generations of Swiss geologists.

Niggli took a great and continuing interest in his teaching. During his thirty-two years at Zurich he constantly sought to improve all aspects of instruction in his field, in order to offer the students a well-rounded education. His courses were rather condensed and often hard to follow because of the wealth of material presented in a short time. The advanced student or co-worker could, however, gain much from taking the "same" course a second time, not only because of the density and breadth of material but also because Niggli hardly ever gave the same lecture in the same way or with the same content.

Although Niggli was mainly interested in theory, he devoted considerable time to the field and to the field training of his students. The excursions were always well prepared; and Niggli attempted to offer a balanced program covering igneous, metamorphic, sedimentary, and applied petrology.

Niggli's significant accomplishments range from theoretical considerations of crystal lattices through many facets of petrology and geochemistry to the very practical problems of avalanche prevention through snow petrology and mineralogy. His crystallographic accomplishments were summarized by P. P. Ewald:

Crystallographers will remember him as the author of *Geometrische Kristallographie des Diskontinuums* (1919) in which he transformed the theory of space groups from the mathematical skeleton left by Schoenflies (1891), E. S. Fedorov (1891), and Harold Hilton (1903), to a helpful friend and advisor of the modern crystallographer. This first of Niggli's books testifies well to his aim of achieving convergence of previously separate fields. Once this was accomplished, he stopped; he never made an all-out attempt at structure determination, the details of which he may have felt likely to divert him from his main course. He kept, however, a profound interest in extending morphological methods to account for the inner structure of crystals. His two papers "Atombau und Kristallstruktur" (1921) contain a detailed survey of atomic and ionic volumes in the solid state throughout the periodic system and discuss the importance of similarity of volumes for the crystallographic properties of salts. His book *Kristallographische und Strukturtheoretische Grundbegriffe* (1928) is an attempt to arrive at a more refined classification of the translation lattices of structures and to connect to it the external morphology of the crystals. His papers "Topologische Strukturanalyse" and "Stereochemie der Kristallverbindungen" (1928–1933) serve as preliminary study for his book *Grundlagen der Stereochemie* (1945), which, by its treatment of the internal morphology of crystals, is a counterpart to his textbook *Spezielle Mineralogie* (1924). Niggli's urge for unifying, condensing and classifying knowledge so as to make it applicable to ever wider fields also stands out in papers on "Charaktertafeln" (1950–1951) in which a method is developed for symbolizing each space group so as to make any further reference to tables unnecessary.

Even in view of the infiltration of detailed wave-mechanical bond theory into the realm of crystallography Niggli remained convinced of the lasting power of morphological methods. Morphology was the central theme of his interest and philosophy and his last large book *Probleme der Naturwissenschaften erläutert am Begriffe der Mineralart* (1949) is, in this sense, his testament ["Paul Niggli," p. 240].

During the twenty years Niggli edited the *Zeitschrift für Kristallographie*, the journal acquired an international reputation. During these two first decades of X-ray analysis he strove to maintain a reasonable unified system of crystallographic terminology. It is also characteristic of his analytical mind that as early as 1919 Niggli recognized, in his *Geometrische Kristallographie des Diskontinuums*, the difference between "real" and "ideal" crystals, at that time speaking of the "pathology of crystals." In 1934

Zeitschrift für Kristallographie published, at his urging, a double issue on ideal and real crystals. Niggli published over sixty papers on crystal structures and summarized this field in the two-volume *Lehrbuch der Mineralogie und Kristallchemie* (1941–1944); a third volume (dealing with crystal chemistry) was destroyed by fire in Berlin during the last months of World War II. These three volumes were actually the third edition of his *Lehrbuch der Mineralogie* (1920). Significantly, volume I contains the foundation of a statistical morphological science, the principles of which, largely original with Niggli, have only recently begun to find application. In his last years Niggli showed that even the wave-mechanical approach to crystal physics required the assistance of morphological concepts, and he presented papers and lectures on the vibration symmetries and degrees of freedom of vibrations of atomic complexes.

In petrography and petrology Niggli also tended to be the integrating, unifying spirit. Almost a century of work had to be pulled together. It should be stressed that throughout his career Niggli used and emphasized the importance of physical chemistry and that phase diagrams formed an essential part of his courses on petrology and mineral deposits. But he did not mistake bare experimental results for proof of natural processes; rather, he was aware that experiments are bound to be oversimplifications, which are designed by man and may miss some essential factors present in natural processes. When experimental results did not match observations of nature, he suggested the possibility of misdirection or of missing parameters in the experiment.

Soon after his stay at the Geophysical Laboratory of the Carnegie Institution, Niggli recognized that his métier was integrating available experimental results instead of duplicating or adding to them. He accomplished this task in long, close teamwork with Bowen at the Geophysical Laboratory, A. Smits at Amsterdam, and especially with colleagues at the Eidgenössische Technische Hochschule and the University of Zurich.

Niggli applied the theory of phase equilibria for the first time to the role of volatile fractions in magmas, as also to problems of metamorphic petrology. The experimental criteria acquired during his work at the Geophysical Laboratory were soon applied to an understanding of the pneumatolytic and auto-hydrothermal alterations of the Eibenstock granite near Dresden, work done in cooperation with F. Rinne. The result was his first book, *Die leichtflüchtigen Bestandteile im Magma* (1920), which received an award from the Fürstlich Jablonowitsche Gesellschaft in Leipzig and became a standard work.

The 1937 book *Das Magma und seine Produkte* (*mit besonderer Berücksichtigung der leichtflüchtigen Bestandteile*) can be considered the second edition of the 1920 work.

Niggli and his students undertook the gigantic task of petrographic classification and interpretation of the chemical analyses of the world's rocks. The original CIPW-norm procedures proved to be inadequate for the project, so he modified them, creating "molecular values," which soon were used throughout the world and were known as Niggli values. These two fundamentally new principles inherent in these "values" calculated from the weight percentages of metal oxides are (1) conversion to atomic percentages instead of weight percentages, which blur the crystal-chemical relationships, and (2) the immediate juxtaposition of the atomic abundance of the basic oxides (aluminum, iron, magnesium, manganese, calcium, sodium, and potassium) with that of silicon dioxide since most of these elements are contained in silicates. This allows the rapid calculation of the possible mineralogical composition of a rock.

Combined with variation diagrams and normative calculation schemes designed by Niggli and his co-workers (1922–1945), the new methods proved far superior to any others. Some of the basic ideas which he developed are (1) the principle of magmatic crystallization and the influence of the volatile fraction; (2) the principle of gravitative crystallization differentiation in magmas; (3) the principle of petrographic-geochemical provinces; and (4) the importance of a calculation and a comparison of the normative and the modal composition of rocks. His principal works on the mineralogical composition of rocks were "Das Magma und seine Produkte" (*Naturwissenschaften*, **9** [1921], also published as a book in 1937); *Gesteins- und Mineralprovinzen* (1923); "Die komplexe gravitative Kristallisationsdifferentiation" (*Schweizerische mineralogische und petrographische Mitteilungen*, **18** [1938]); "Die Magmentypen," written with A. H. Stutz (*ibid.*, **16** [1936]); and, with C. Burri, the two-volume *Die jungen Eruptivgesteine des mediterranen Orogens* (1945–1948). A late but important work was "Gesteins-chemismus und Magmenlehre."

Niggli's 1948 book *Gesteine und Minerallagerstätten* summarized his previous work. In it he developed petrologic science from the level of the crystal structure (including the role of trace elements in petrology) to the mineral, the rock specimen as the mineral aggregate, the outcrop, the regional, and the global levels, in a synthesis of geochemical and geometric problems. The set of rock fabric patterns in this book illustrates his knowledge of petrographic-petrologic processes and fabric possibilities.

Niggli also applied his molecular values and norm calculations to metamorphic and sedimentary rocks to show that extreme transformist or relatively migrationist interpretations were oversimplifications. Extreme migrationist views were challenged about 1970–1972, when new global comparisons showed the average composition of sediments to be basaltic rather than granitic. This revived a differentiation concept of anatexis, similar to that proposed by Niggli. The principles of metamorphic transformation were well known to him; but extensions of local processes to a regional scale were unacceptable to his critical mind.

Niggli insisted on the application of physico-chemical principles, as is exemplified in all of his work on metamorphic rocks starting in 1913–1914 (some papers with J. Johnston) and in the *Gesteinsmeta-morphose* (1924), prepared with Grubenmann.

Niggli's work is sometimes considered to be that of an "extreme orthodox magmatist." On examination it is obvious that he was aware of the importance of exchange reactions in metamorphism. Nevertheless, before invoking a deus ex machina for the majority of igneous rocks, he insisted that the physicochemical aspects of the development of both igneous and metamorphic processes had to be understood. He was aware that his theory of magmatic processes and provinces was not the final answer to all problems of field petrology and that additional work was needed. This is perhaps best expressed in a statement from his 1952 paper "Gesteinschemismus und Magmenlehre," in which he explicitly states that there may be various ways of interpreting regional variations.

Niggli's interest in sedimentary rocks led to original papers on clastic sediments, especially the morphological aspects of grains. His classification of shapes, developed with dal Vesco, is probably still the most widely used one. The general principles of Niggli's petrology-petrography, especially the close ties with his crystallographic interests, are summarized in volume I of *Gesteine und Minerallagerstätten* (1948), and he offered as complete a summary on sedimentary rocks in volume II. His papers on snow research, which had a profound impact in the field, also concern sedimentation, diagenesis, and metamorphism.

Niggli exerted a strong influence on applied petrology, especially through his work on rock weathering and other aspects of building stone petrology, most of it done with F. de Quervain, who headed the Geotechnische Prüfstelle and the Geotechnische Kommission, both created at Niggli's suggestion. Although never directly active in consulting work—he was too dedicated to fundamental science—Niggli nevertheless was in constant touch with applied fields of his science. He knew very well how often new "pure" aspects emerge from technological applications, and he also realized that most of his students had to prepare for work in applied fields. Also in this vein he promoted the publication of the geotechnical map of Switzerland and was active in the foundation of the Schweizerische Mineralogische und Petrographische Gesellschaft, of which he was president from 1928 to 1930.

Niggli wrote several papers and one booklet on ore deposits. Here his main point of departure was the accumulation of metals during the gravitative crystallization differentiation and the accumulation of volatiles in later magmatic stages. His booklet *Versuch einer natürlichen Klassifikation der im weiteren Sinne magmatischen Lagerstätten* (1925) is a classic synopsis of magmatic ore deposits. In this work Niggli devoted less space to experimental problems and to direct observations of ore deposits, being more concerned to classify synopses of the published results and observations. This approach proved to be generally acceptable in crystallography, mineralogy, and petrology but rather negative with regard to ore deposits, the descriptive terminology of which was filled with preconceived genetic concepts. He did, however, recognize many physicochemical relationships within magmatic and hydrothermal ore deposits for the first time, especially the role of the volatile fraction in the accumulation of metals in a magma. His booklet was internationally quoted and used, especially the English translation.

BIBLIOGRAPHY

I. ORIGINAL WORKS. A summary of Niggli's works can be found in R. L. Parker, "Memorial of Paul Niggli" (see below); a complete list of all publications is given by J. Marquard and I. Schroeter, in *Schweizerische mineralogische und petrographische Mitteilungen*, 33 (1953), 9–20.

Niggli's major works include *Die Chloritoidschiefer des nordöstlichen Gotthardmassivs* (diss., University of Zurich; Bern, 1912); *Geometrische Kristallographie des Diskontinuums* (Leipzig, 1918–1919); *Lehrbuch der Mineralogie* (Berlin, 1920), 2nd ed., 2 vols. (Berlin, 1924–1926), 3rd ed., entitled *Lehrbuch der Mineralogie und Kristallchemie*, 3 vols. (Berlin, 1941–1944), vol. III destroyed by fire; "Das Magma und seine Produkte," in *Naturwissenschaften*, 9 (1921), 463–471; *Gesteins- und Mineralprovinzen* (Berlin, 1923), written with P. I. Beger; *Die Gesteinsmetamorphose* (Berlin, 1924), written with U. Grubenmann; *Versuch einer natürlichen Klassifikation der im weiteren Sinne magmatischen Lagerstätten* (Halle, 1925), trans. by Thomas Murby as *Ore Deposits of Magmatic Origin. Their Genesis and Natural Classification* (London, 1929); *Tabellen zur allgemeinen und speziellen Mineralogie* (Berlin, 1927); *Kristallographische und strukturtheoretische Grundbegriffe,* which is *Handbuch der Experimentalphysik*, VII, pt. 1 (Leipzig, 1928); "Chemismus schweizerischer Gesteine," which is *Beiträge zur geologischen Karte der Schweiz*, Geotechnische Reihe, 8, no. 14 (1930), written with F. de Quervain and R. U. Winterhalter; *Geotechnische Karte der Schweiz 1:200 000*, 4 sheets (Bern, 1934–1938), prepared with F. de Quervain, M. Gschwind, and R. U. Winterhalter; *Internationale Tabellen zur Bestimmung von Kristallstrukturen* (Berlin, 1935), written with E. Brandenberger; and "Die Magmentypen," in *Schweizerische mineralogische und petrographische Mitteilungen*, 16 (1936), 335–399, written with A. H. Stutz.

Also see *Das Magma und seine Produkte* (*mit besonderer Berücksichtigung der leichtflüchtigen Bestandteile*), (Leipzig, 1937), which is the 2nd ed. of *Die leichtflüchtigen Bestandteile im Magma* (Leipzig, 1920); "Die komplexe gravitative Kristallisationsdifferentiation," in *Schweizerische mineralogische und petrographische Mitteilungen*, 18 (1938), 610–664; *La loi des phases en minéralogie et pétrographie* (Paris, 1938); *Die Mineralien der Schweizeralpen*, 2 vols. (Basel, 1940), written with J. Koenigsberger and R. L. Parker; *Grundlagen der Stereochemie* (Basel, 1945); *Schulung und Naturerkenntnis* (Erlenbach–Zurich, 1945); *Die jungen Eruptivgesteine des mediterranen Orogens*, 2 vols. (Zurich, 1945–1948), written with C. Burri; "Krystallogia von J. H. Hottinger (1698)," in *Veröffentlichungen der Schweizerischen Gesellschaft für Geschichte der Medizin und der Naturwissenschaften*, no. 14 (1946); *Gesteine und Minerallagerstätten*, 2 vols. (Basel, 1948–1952), written with E. Niggli; *Probleme der Naturwissenschaften erläutert am Begriff der Mineralart* (Basel, 1949); "Gesteinschemismus und Magmenlehre," in *Geologische Rundschau*, 39 (1951), 8–32; and *International Tables for X-ray Crystallography* (Birmingham, 1952), written with E. Brandenberger.

II. SECONDARY LITERATURE. See E. Brandenberger, "Paul Niggli (1888–1953). Seine Verdienste um die Lehre des festen Körpers," in *Zeitschrift für angewandte Mathematik und Physik*, 4 (1953), 415–418; P. P. Ewald, "Paul Niggli," in *Acta crystallographica*, 6 (Mar. 1953), 225–226; P. Karrer and E. Brandenberger, *Prof. Dr. Paul Niggli. Ansprachen zu seinem Gedenken* (Zurich, 1953); F. Laves, "Paul Niggli," in *Experientia*, 9 (1953), 197–202; F. Laves and A. Niggli, "In Memoriam: Paul Niggli's Crystallographic Oeuvre," in *Zeitschrift für Kristallographie und Mineralogie*, 120 (1964), 212–215; H. O'Daniel, K. H. Scheumann, and H. Schneiderhöhn, "Paul Niggli," in *Neues Jahrbuch für Mineralogie* (1953), 51–67; R. L. Parker, "Memorial of Paul Niggli," in *American Mineralogist*, 39 (1954), 280–283; F. de Quervain, "Prof. Dr. Paul Niggli," in *Schweizerische mineralogische und petrographische Mitteilungen*, 33 (1953), 1–20; K. H. Scheumann, "Paul Niggli und sein Werk," in *Geologie*, 2 (1953), 124–130; and A. Streckeisen, "Paul Niggli," in *Mitteilungen der Naturforschenden Gesellschaft in Bern*, n.s. 11 (1954), 109–113.

G. C. AMSTUTZ

NIKITIN, SERGEY NIKOLAEVICH (*b.* Moscow, Russia, 4 February 1851; *d.* St. Petersburg, Russia, 18 November 1909), *geology.*

Nikitin's father was a dissector in the department of anatomy at Moscow University. While still a Gymnasium student the boy was attracted to the natural sciences, especially botany and geology. In 1867 he entered the natural sciences section of the Faculty of Physics and Mathematics at Moscow University. After graduating in 1871, Nikitin taught botany and geography in secondary schools. He was one of the organizers of the Moscow Natural History Courses for Women, where he lectured in mineralogy and geology. At the same time he studied Paleozoic and Mesozoic deposits of the Russian platforms. In 1878 Nikitin was awarded the master's degree for work on the ammonites.

In 1882, when the Russian Geological Survey was founded, he was elected its senior geologist. Concerned with the stratigraphy of the Russian platform, he investigated the coal deposits in the Moscow Basin and the Permian deposits of the Ural foothills. He suggested the name "Tatar layer" for the Upper Permian horizons; divided Jurassic deposits, according to the ammonites, into seven paleontological zones; and established a phylogenetic series of Kelloveyskikh and Oxford ammonites. Nikitin was a Darwinian who introduced evolutionary theory into invertebrate paleontology. He compiled a stratigraphic scheme of the Russian Upper Cretaceous deposits, comparing them with corresponding deposits in Western Europe, and determined the northern limit of the distribution of the Upper Cretaceous remains. Nikitin assigned great importance to the study of Quaternary deposits, distinguishing ten sorts of regions in Russia according to geological types of glacial deposits. Regarding the origin of loess he advocated the eolian hypothesis.

Nikitin laid the foundation for systematic hydrogeological and hydrological research in Russia. Participating in the expeditions organized by the Geological Survey to study the southern arid regions through investigations of the sources of Russian rivers, he generalized the material obtained and published several works. These investigations were of great importance for the development of agriculture. He studied the conditions of occurrence of underground water in the Moscow region and showed the possibility of using artesian wells for supplying the capital. From 1907 through 1909 he was president of the Hydrological Committee. He presented his conclusions on the conditions for artesian water supply to cities and on the hydrogeological conditions for railroad regions.

From 1905 through 1907 Nikitin headed an expedition that studied the geological structure of the Mugodzhar Hills. During the last years of his life, at the request of the imperial mining department, he was concerned with ways to prevent the flooding of the salt mines in the Urals. Nikitin was well acquainted with the geological literature and published a bibliographical guide, *Russkaya geologicheskaya biblioteka* ("Russian Geological Library," 1886–1900), and surveys of Russian and general geology.

In 1883 the St. Petersburg Academy of Sciences awarded Nikitin the Helmersen Prize for his paleontological works, and in 1894 the Russian Geographical Society awarded him the Medal of Constantine. In 1902 he was elected a corresponding member of the St. Petersburg Academy of Sciences.

BIBLIOGRAPHY

I. ORIGINAL WORKS. Nikitin's most important writings are "Ammonity gruppy *Amaltheus funiferus*" ("Ammonites of the Group *Amaltheus funiferus*"), in *Bulletin de la Société impériale des naturalistes de Moscou,* n.s. **3** (1878), 81–160; "Darvinizm i vopros o vide v oblasti sovremennoy paleontologii" ("Darwinism and the Question of Form in Contemporary Paleontology"), in *Mysl* (St. Petersburg), no. 8 (1881), 144–170; no. 9 (1881), 229–245; "Yurskie obrazovania mezhdu Rybinskom, Mologoy i Myshkinym" ("Jurassic Formations Between Rybinsk, Mologa and Myshkin"), in *Materialy dlya geologii Rossii,* **10** (1881), 199–331; "Posletretichnye otlozhenia Germanii v ikh otnoshenii k sootvetstvennym obrazovaniam Rossii" ("Post-Tertiary Deposits of Germany In Their Relations to the Corresponding Formations of Russia"), in *Izvestiya Geologicheskago komiteta,* **5,** no. 3–4 (1886), 133–185; "Sledy melovogo perioda v Tsentralnoy Rossii" ("Traces of the Cretaceous Period in Central Russia"), in *Trudy Geologischeskago komiteta,* **5,** no. 2 (1888), 1–205; and "Ukazatel literatury po burovym na vodu skvazhinam v Rossii" ("A Guide to Literature on Wells Drilled for Water in Russia"), supp. to *Izvestiya Geologicheskago komiteta,* **29** (1911).

II. SECONDARY LITERATURE. See F. N. Chernyshev, "Sergey Nikolaevich Nikitin," in *Izvestiya Imperatorskoi akademii nauk,* 6th ser., **3,** no. 18 (1909), 1171–1173; F. N. Chernyshev, A. A. Borisyak, N. N. Tikhonovich, and M. M. Prigorovsky, "Pamyati Sergeya Nikolaevicha Nikitina" ("Recollections of . . . Nikitin"), in *Izvestiya Geologicheskago komiteta,* **28,** no. 10 (1909), 1–51; and N. N. Karlov, "S. N. Nikitin i znachenie ego rabot dlya razvitia otechestvennykh geologicheskikh nauk" ("S. N. Nikitin and the Importance of His Work for the Development of Native Geological Sciences"), in *Ocherki po istorii geologicheskikh znany,* no. 1 (1953), 157–180.

IRINA V. BATYUSHKOVA

NĪLAKAṆṬHA (*b*. Tṛ-k-kaṇṭiyūr [Kuṇḍapura], Kerala, *ca*. 14 June 1444; *d*. after 1501), *astronomy*.

Nīlakaṇṭha, a Nampūtiri Brahman, was born in the house *(illam)* called Keḷallūr (Keralasadgrāma), which is said to be identical with the present Eṭamana *illam* in Tṛ-k-kaṇṭiyūr, a village near Tirur, Kerala. His father was named Jātavedas, and the family belonged to the Gārgyagotra and followed the Āśvalāyanasūtra of the *Ṛgveda*; Nīlakaṇṭha was a Somasutvān (performer of the Soma sacrifice). He studied Vedānta and some astronomy under Ravi, but his principal instructor in *jyotiḥśāstra* was Dāmodara (*fl*. 1417), the son of the famous Parameśvara (*ca*. 1380–1460), whom he also met at the Dāmodara house in Ālattūr (Aśvatthagrāma), Kerala. His younger brother, Śaṅkara, studied astronomy under his tutelage and in turn professed that science. It is possible, but not certain, that Nīlakaṇṭha is identical with the father of the Rāma who wrote a *Laghurāmāyaṇa* in Malayālam.

Nīlakaṇṭha was a follower of Parameśvara's *dṛgganita* system (see essay in Supplement), although he gives various parameters in his several works (see D. Pingree, in *Journal of the Oriental Institute, Baroda* 21 [1971–1972], 146–148). These works include the following:

1. The *Golasāra*, in fifty-six verses, gives the parameters of his planetary system, a description of the celestial spheres, and a description of the principles of computation used in Indian mathematical astronomy. It was edited by K. V. Sarma (Hoshiarpur, 1970).

2. The *Siddhāntadarpaṇa*, in thirty-two verses, gives another set of parameters and a description of (impossible) planetary models. It also was edited by K. V. Sarma (Madras, 1955). Nīlakaṇṭha's commentary *(vyākhyā)* on the *Siddhāntadarpaṇa* has not been published.

3. The *Candracchāyāgaṇita* describes, in thirty-one verses, the computation of the moon's zenith distance. Neither it nor Nīlakaṇṭha's commentary *(vyākhyā)* has been published.

4. The *Tantrasaṅgraha* is an elaborate treatise on *dṛgganita* astronomy, composed in 1501. It consists of eight chapters:

a. On the mean motions of the planets.

b. On the true longitudes of the planets.

c. On the three questions relating to the diurnal rotation of the sun.

d. On lunar and solar eclipses.

e. Particulars of solar eclipses.

f. On the *pātas* of the sun and moon.

g. On the first visibilities of the moon and planets.

h. On the horns of the moon.

The *Tantrasaṅgraha* was edited with the commentary, *Laghuvṛtti*, of Śaṅkara Vāriyar (*fl*. 1556) by S. K. Pillai (Trivandrum, 1958).

5. The *Āryabhaṭīyabhāṣya* is an extensive and important commentary on the *Āryabhaṭīya* composed by Āryabhaṭa I in 499. Nīlakaṇṭha's patron for this work was the religious head of the Nampūtiri Brahmans, Netranārāyaṇa. In his commentary on Kālakriyā 12–15 he states that he observed a total eclipse of the sun on 6 March 1467 (Oppolzer no. 6358) and an annular eclipse at Anantakṣetra on 28 July 1501 (not in Oppolzer). The *Āryabhaṭīyabhāṣya* was published in three volumes by K. S. Sastri (volumes I and II) and S. K. Pillai (volume III), (Trivandrum 1930–1957).

6 and 7. In the *Āryabhaṭīyabhāṣya*, Nīlakaṇṭha refers to his *Grahanirṇaya* on eclipses and to his *Sundararājapraśnottara*, in which he answers questions posed by Sundararāja, the author of a commentary on the *Vākyakaraṇa*. Neither of these works is extant.

8. An untitled prose work on eclipses by Nīlakaṇṭha is included in a manuscript of the *Siddhāntadarpaṇavyākhyā*; it refers to the *Āryabhaṭīyabhāṣya*, and thus is his last known work.

BIBLIOGRAPHY

Nīlakaṇṭha's method of computing π is discussed by K. M. Marar and C. T. Rajagopal, "On the Hindu Quadrature of the Circle," in *Journal of the Bombay Branch of the Royal Asiatic Society*, n.s. **20** (1944), 65–82. A general survey of his life and works (now superseded by the introductions to Sarma's latest eds.) is given by K. V. Sarma, "Gārgya-Kerala Nīlakaṇṭha Somayājin: The Bhāṣyakāra of the Āryabhaṭīya (1443–1545)," in *Journal of Oriental Research* (Madras), **26** (1956–1957), 24–39; and by K. K. Raja, "Astronomy and Mathematics in Kerala," in *Brahmavidyā*, **27** (1963), 118–167, esp. 143–152.

DAVID PINGREE

NILSSON-EHLE, HERMAN (*b*. Skurup, Sweden, 12 February 1873; *d*. Lund, Sweden, 29 December 1949), *genetics, plant breeding*.

Nilsson-Ehle was the son of Nils Nilsson, a farmer, and his wife, Elin. After first studying in Malmö, he enrolled in 1891 at the University of Lund, where he received the candidate's degree in 1894, the licentiate degree in 1901, and the Ph.D. in 1909. He began his scientific research in 1894, at first concentrating on plant taxonomy and plant physiology. In 1900 he became an assistant at the Swedish Seed Association in Svalöf (near Lund), and thereafter devoted himself to the new science of genetics and its practical applications in plant breeding.

Nilsson-Ehle realized the fundamental importance of Mendel's principles of heredity, which had just been rediscovered, and he was especially impressed by Mendel's clarification of the mechanism of genetic recombination. Nilsson-Ehle was the first to demonstrate that economically important properties in cultivated plants are inherited according to Mendel's laws and may be recombined in a specific way. In a now-classic paper of 1906 he recommended artificial crosses as the best method of obtaining a recombination of various desirable properties. He cited a series of examples from his own experience and pointed out that part of the offspring of experimentally produced hybrids combined the valuable properties of the parents. At the same time he obtained, as expected, other offspring which represented a combination of the undesirable properties of the parents.

In three papers published in 1908–1911 Nilsson-Ehle demonstrated that quantitative characters (size, earliness, resistance to disease) are inherited in the same Mendelian way as the qualitative characters (differences in flower color, etc.) with which Mendel and the early Mendelists had been working. As a rule, however, the quantitative characters were found to be conditioned by a relatively high number of polymeric (or multiple) genes. After recombination these genes may give rise to numerous quantitative gradations of the characters involved in the crosses. This finding was a very important contribution to the development of basic genetics and a solid basis for its practical application to plant breeding.

In 1915 Nilsson-Ehle was appointed to the chair of physiological botany at the University of Lund. Two years later he moved to the chair of genetics. From 1925 until his retirement in 1939 Nilsson-Ehle was director of the Swedish Seed Association. During this period as an active administrator, he encouraged the development of new fields of research and in 1931 organized a new department for chromosome investigations and the production of new types of polyploid cultivated plants. He realized that induced mutations would be important in plant breeding, and advocated mutation research.

As professor emeritus, Nilsson-Ehle became actively interested in forestry and horticulture, and he was helpful in the founding and development of organizations which sought to improve the stock for forests and orchards through breeding.

Nilsson-Ehle was a member of many academies and received several honorary doctorates. His life was marked by a cyclical mental state: periods of ill health, extraordinary activity, deep depression, and great optimism. He was a fascinating combination of creative fantasy and sober realism; and he combined a farmer's intimate practical knowledge of soil and crops with the theoretical education and logical acumen of a university professor. Moreover, he was talented musically and often entertained himself and his guests at the piano. He was married and had a daughter and two sons.

BIBLIOGRAPHY

I. ORIGINAL WORKS. Among Nilsson-Ehle's numerous publications the following ones are of especial importance: "Einige Ergebnisse von Kreuzungen bei Hafer und Weizen," in *Botaniska notiser* (1908), 257–294; "Kreuzungsuntersuchungen an Hafer und Weizen," in *Acta Universitatis lundensis*, n.s. 2, **5**, no. 2 (1909), 1–122; **7**, no. 6 (1911), 1–84; and "Mendélisme et acclimatation," in *IV*[e] *Conférence international génétique Paris* (Paris, 1911), 1–22.

II. SECONDARY LITERATURE. See A. Müntzing, "Lebensbeschreibung von H. Nilsson-Ehle," in *Zeitschrift für Pflanzenzüchtung*, **29**, no. 1 (1950), 110–114. In *Sveriges utsädesförenings tidskrift* (1950), no.1, there are articles in Swedish about Nilsson-Ehle and his importance for the development of genetics and plant breeding. See also A. Müntzing, "Minnesteckning över Professor H. Nilsson-Ehle," in *Kungliga Fysiografiska Sällskapets förhandlingar*, **20** (1950), 1–7; Å. Gustafsson, "Herman Nilsson-Ehle, minnesteckning," in *Levnadsteckningar över Kungliga Svenska Vetenskapsakademiens ledamöter*, **175** (1971), 279–293; A. Müntzing, "Om aktualiteten av Herman Nilsson-Ehles teoretiska forskning," in *Sveriges Utsädesförenings tidskrift* (1973), no. 2–3, 159–168; and E. Åkerberg, "Om aktualiteten i Herman Nilsson-Ehles insatser i växtförädling och jordbruksforskning," *ibid.*, 169–178.

ARNE MÜNTZING

NISSL, FRANZ (*b.* Frankenthal, Germany, 9 September 1860; *d.* Munich, Germany, 11 August 1919), *psychiatry, neuropathology.*

Nissl was the son of Theodor Nissl and Maria Haas. He is known for the discovery of a granular basophilic substance, now called Nissl's bodies, that is found in the nerve cell body and the dendrites. In connection with this discovery he classified changes in the distribution and number of these granules following disease or the severing of the axon; coined the term *nervöses grau*, or gray nerve network, a misleading concept of a diffuse interconnection of all nerve processes; and, both alone and with Alois Alzheimer, made a detailed study of dementia paralytica, paying special attention to the behavior of microglia (rod cells).

Nissl's father, who taught Latin in a Catholic school, intended his son to become a priest but, against his parents' wishes, Nissl studied medicine at Munich University. His interest in the nervous system was firmly established by his first scientific effort. He entered a competition for a prize in neurology offered by the Medical Faculty at Munich. The judge was Bernard von Gudden, a scientist and psychiatrist. Gudden was so impressed by Nissl's work on the pathological changes of cortical neurons that he offered him an assistantship in 1884. Gudden drowned in the lake of Starnberg with his patient, King Ludwig of Bavaria, in 1886, and Nissl then became an assistant at the psychiatric hospital in Frankfurt. There he met the comparative neurologist Ludwig Edinger and the neuropathologist Carl Weigert. Nissl worked for seven years with Alzheimer, a psychiatrist and an outstanding neuropathologist. Together they edited the *Histologische und Histopathologische Arbeiten über die Grosshirnrinde* (1904–1921). In 1895 Nissl moved to the University of Heidelberg, becoming university lecturer in 1896, associate professor of psychiatry in 1901, and full professor and director of the department of psychiatry in 1904.

The burden of teaching and administration, combined with poor research facilities, forced Nissl to leave many scientific projects unfinished. He also suffered from a kidney disease. World War I proved to be an even greater burden for he was commissioned to administer a large military hospital as well. In 1918 Nissl moved to Munich to take a research position at the Deutsche Forschungsanstalt für Psychiatrie. He died a year later, before deriving any benefit from these new opportunities.

The present stage of development of neurohistological techniques, including electron microscopy, makes possible an appraisal of Nissl's real scientific achievements. The granular basophilic Nissl's substance is an important ultrastructure of nerve cells, composed of ribosomes and the membranes of the endoplasmic reticulum. The Nissl bodies, in reacting to injury and toxins, mirror the life cycle of a neuron very closely, and the importance Nissl attached to them was fully justified. The gray nerve network, however, proved to be untenable and demonstrates that the use of only one technique leads to faulty interpretation. Nissl and his contemporaries fought for recognition of their respective views. His only monograph, *Die Neuronenlehre und ihre Anhänger* (1903), is a sad and depressing account of a speculative mind incapable of listening to other scientists'

arguments. His attitude was rigid, and his stature in German neurology was one of the reasons that after the magnificent start given to German neuropathology by the studies of Nissl, W. Spielmeyer, Alzheimer, Weigert, and others, no real advances were made by fully experimental methods. Such studies would have early revealed that many observations of cellular changes or of the gray nerve network resulted from bad fixation and application of only one technique.

Nevertheless, Nissl's arguments against the neuron doctrine, which is based on areas of contact between neurons, survived in many quarters until recent times, when electron microscopy amply confirmed the early observations of his contemporary Ramón y Cajal, which Nissl had rejected in vitriolic terms. While Nissl's studies on dementia paralytica are still valid, his admirable attempt to discover a neuropathological cause for mental diseases failed. This is not to detract from Nissl, however, for the cause of these mainly functional diseases is neurobiochemical, not morphological.

BIBLIOGRAPHY

I. ORIGINAL WORKS. Nissl's writings include "Resultate und Erfahrungen bei der Untersuchung der pathologischen Veränderungen der Nervenzellen in der Grosshirnrinde" his diss. (unpublished) (1884); "Über die Veränderungen der Ganglienzellen am Facialiskern des Kaninchens nach Ausreissung des Nerven," in *Versammlungen des Südwestdeutschen Psychiatervereins in Karlsruhe*, **22** (1890); "Über experimentell erzeugte Veränderungen an den Vorderhornzellen des Rückenmarkes bei Kaninchen," in *Zeitschrift für Neurologie*, **48** (1892), 675–682; "Über die sogenannten Granula der Nervenzellen," in *Neurologisches Zentralblatt*, **13** (1894), 676–685, 781–789, 810–814; "Die Beziehungen der Nervenzellsubstanzen zu den tätigen, ruhenden und ermüdeten Zellzuständen," in *Allgemeine Zeitschrift für Psychiatrie*, **52** (1896), 1147–1154; "Mitteilungen zur pathologischen Anatomie der Dementia paralytica," in *Archiv für Psychiatrie*, **28** (1896), 987–992; "Über die Veränderungen der Nervenzellen nach experimentell erzeugter Vergiftung. Autoreferat," in *Neurologisches Zentralblatt*, **15** (1896); "Über einige Beziehungen zwischen Nervenzellerkrankungen und gliösen Erscheinungen bei verschiedenen Psychosen," in *Archiv für Psychiatrie*, **32** (1899), 656–676; *Die Neuronenlehre und ihre Anhänger* (Jena, 1903); "Zur Histopathologie der paralytischen Rindenerkrankung," in *Histologische und histopathologische Arbeiten über die Grosshirnrinde*, **1** (Jena, 1904); "Diskussionsbemerkung zu Alzheimer: Die syphilitischen Geistesstörrungen," in *Neurologisches Zentralblatt*, **32** (1909), 680; and "Zur Lehre der Lokalisation in der Grosshirnrinde des Kaninchens," in *Sitzungsberichte der*

Heidelberger Akademie der Wissenschaften, Math.-naturwiss. Kl. (1911).

II. SECONDARY LITERATURE. Biographies are A. Jakob, "Franz Nissl," in *Deutsche medizinische Wochenschrift,* XLV, (1919), 1087; E. Kraepelin, "Franz Nissl," in *Münchener medizinische Wochenschrift* (1919); and "Lebensschicksale deutscher Forscher," *ibid.* (1920); H. Marcus, "Franz Nissl," in *Minnesood i Svenska läkaresällskapet* (1919); P. Schröder, "Franz Nissl," in *Monatsschrift für Psychiatrie . . .,* **46** (1919), 294; H. Spatz, "Franz Nissl," in *Berliner klinische Wochenschrift* (1919), 1006; and "Nissl und die theoretische Hirnanatomie," in *Archiv für Psychiatrie und Nervenkrankheiten,* **87** (1929), 100–125; and W. Spielmeyer, "Franz Nissl," in *Kirchhoffs deutsche Irrenärzte,* **2** (1924), 288.

Specialized works are J. E. C. Bywater and P. Glees, *Der Einfluss der Fixationsart und des Intervalles zwischen Tod und Fixation auf die Struktur der Motoneurone des Affen* (Zurich, 1959), offprint from *Verhandlungen der Anatomischen Gesellschaft* (Jena), **55** (1959), 194–200; P. Glees, "Ludwig Edinger," in *Journal of Neurophysiology,* **15** (1952), 251–255; and "Neuere Ergebnisse auf dem Gebiet der Neurohistologie: Nissl-Substanz, corticale Synapsen, Neuroglia und intercellulärer Raum," in *Deutsche Zeitschrift für Nervenheilkunde,* **184** (1963), 607–631; P. Glees and J. E. C. Bywater, "The Effect of the Mode of Fixation and the Interval Between Death and Fixation on Monkeys' Motoneurones," in *Proceedings of the Physiology Society Journal of Physiology,* **149** (1959), 3–4P; and P. Glees and K. Meller, "The Finer Structure of Synapses and Neurones. A Review of Recent Electronmicroscopical Studies," in *Paraplegia,* **2,** no. 2 (1964), 77–95.

PAUL GLEES

NOBEL, ALFRED BERNHARD (*b.* Stockholm, Sweden, 21 October 1833; *d.* San Remo, Italy, 10 December 1896), *chemistry.*

Nobel's father, Immanuel Nobel the younger, was a builder, industrialist, and inventor; his great-great-great-grandfather, Olof Rudbeck, was one of the most important Swedish scientists of the seventeenth century. His mother was Andrietta Ahlsell.

Nobel attended St. Jakob's Higher Apologist School in Stockholm in 1841–1842. The family then moved to St. Petersburg, where he and his brothers were tutored privately from 1843 to 1850 by Russian and Swedish tutors. They were also encouraged to be inventive by their energetic father.

After a two-year study trip to Germany, France, Italy, and North America, Nobel had improved his knowledge in chemistry and was an excellent linguist, with a mastery of German, English, French, Swedish, and Russian. During the Crimean War (1853–1856) Nobel worked in St. Petersburg in his father's firm,

which produced large quantities of war matériel. After the war the new Russian government canceled all delivery agreements; Immanuel Nobel had to declare bankruptcy, and he returned to Sweden in 1859.

Immanuel had long experimented with powder-charged mines, and his attention had been drawn by Nikolai Zinin and Yuli Trapp to the explosive substance nitroglycerin. Both he and Alfred worked with it independently, using different methods. In 1862 Immanuel was the first to demonstrate a comparatively simple way of producing nitroglycerin on a factory scale, using Ascano Sobrero's method with some modifications.

In 1863 Alfred Nobel, back in Sweden, developed his first important invention, the Nobel patent detonator, constructed so that detonation of the liquid nitroglycerin explosive charge was effected by a smaller charge placed in a metal cap charged with detonating mercury (mercury fulminate). The "initial ignition principle," using a strong shock rather than heating, was thus introduced into the technique of blasting.

In 1865 the world's first true factory for producing nitroglycerin was put into operation by the Nobel company, Nitroglycerin Ltd., in an isolated area outside Stockholm. Young Alfred Nobel was not only managing director but also works engineer, correspondent, traveling salesman, advertising manager, and treasurer. This responsibility marked the beginning of his life as an inventor and industrialist and led to the establishment of many factories throughout the world and to the development of new production methods. Accidents in factories and in the handling of nitroglycerin made Nobel aware of the danger of fluid nitroglycerin. After a long period of experimentation, in 1867 he patented dynamite in Sweden, England, and the United States. It was an easily handled, solid, and ductile explosive that consisted of nitroglycerin absorbed by kieselguhr, a very porous diatomite. The invention aroused great interest among users of explosives. Nitroglycerin, the fundamental discovery of Sobrero (1847), had been transformed into a useful explosive. A worldwide industry was built up by Alfred Nobel himself.

Guhr dynamite, as it was known, had certain technical weaknesses. Continuing his research, Nobel in 1875 created blasting gelatin, a colloidal solution of nitrocellulose (guncotton) in nitroglycerin which in many respects proved to be an ideal explosive. Its force was somewhat greater than that of pure nitroglycerin, it was less sensitive to shock, and it was strongly resistant to moisture and water. It was called Nobel's Extra Dynamite, Express Dynamite, Blasting Gelatin,

Saxonite, and Gelignite. As early as 1875 it was put into production in most of Nobel's dynamite factories.

The problem of improving blasting powder had quite early occupied the Nobels. In 1879 Alfred Nobel was working on less smoky military explosive charges for artillery missiles, torpedoes, and ammunition. In 1887 he produced a nearly smokeless blasting powder —Ballistite, or Nobel's blasting powder—a mixture of nitroglycerin and nitrocellulose plus 10 percent camphor, which upon ignition burned with almost mathematical precision in concentric layers.

Nobel's last discovery in the realm of explosives was progressive smokeless powder (Swedish primary patent no. 7552, in 1896). A further product of Ballistite for special purposes, it was developed in his laboratory at San Remo during the last years of his life.

Nobel's interests as an inventor were by no means confined to explosives; his later work covered electrochemistry, optics, biology, and physiology. The list of his patents runs to no fewer than 355 in different countries. His pioneer work helped later inventors solve many problems in the manufacture of artificial rubber, leather, and silk, of semiprecious and precious stones from fused alumina, and of other products.

Through his skill as an industrialist and his fundamental patents on explosives, Nobel became a multimillionaire. By his last will and testament, dated Paris, 27 November 1895, he left his total fortune, over 33 million Swedish crowns (over two million pounds sterling), to a foundation that would award prizes "to those who, during the preceding year, shall have conferred the greatest benefit on mankind." The prize-awarding institutions are the Royal Swedish Academy of Science (physics, chemistry), the Royal Caroline Medical Institute (medicine, physiology), the Swedish Academy (literature), and a committee of the Norwegian Parliament (peace). The prize for economic sciences is a separate entity established by the Swedish Riksbank.

BIBLIOGRAPHY

Nobel's only writing is *On Modern Blasting Agents* (Glasgow, 1875).

The three main biographies are *Alfred Nobel och hans släkt* (Stockholm, 1926); E. Bergengren, *Alfred Nobel, the Man and His Work* (London–New York, 1962), with a supp. on the Nobel Institutions and the Nobel prizes by N. K. Ståhle; and H. Schück and R. Sohlman, *The Life of Alfred Nobel* (London, 1929).

Other works to be consulted are *The Book of High Explosives* (Birmingham, 1908); A. P. Gelder and H. Schlatter, *History of the Explosives Industry in America* (New York, 1927); *The History of Nobel's Explosives Co. Ltd. and Nobel Industries Ltd. 1871–1926*, which is vol. I of *Imperial Chemical Industries Ltd. and Its Founding Companies* (Birmingham, 1938); F. D. Miles, *A History of Research in the Nobel Division of ICI* (Birmingham, 1955); R. Moe, *Le Prix Nobel de la paix et l'Institut Nobel norvégien* (Oslo, 1932); H. de Mosenthal, "The Inventor of Dynamite," in *Nineteenth Century*, no. 260 (Oct. 1898), 567–581; H. Schück et al., *Nobel, the Man and His Prizes* (Amsterdam, 1962), a discussion of the role of the Nobel prizes in the development of the prize fields during the first sixty years by eminent representatives of the five prize sections, with a biographical sketch of Nobel by H. Schück and a sketch of Nobel and the Nobel Foundation by R. Sohlman; and B. von Suttner, *Memoirs* (Stuttgart–Leipzig, 1909).

TORSTEN ALTHIN

NOBERT, FRIEDRICH ADOLPH (*b.* Barth, Pomerania [now German Democratic Republic], 17 January 1806; *d.* Barth, 21 February 1881), *clockmaking, optical instruments.*

Nobert was the elder son of Johann Friedrich Nobert, a clockmaker. After a scanty education he took up his father's occupation. In 1827, at a Berlin trade exhibition, Nobert entered a pocket chronometer that was commended and brought him to the attention of the astronomer Johann Encke. To check the going of his timepieces Nobert constructed a telescope and, with Encke's encouragement, made a number of astronomical measurements. In the spring of 1833, realizing that his ambitions would never be achieved without further education, he applied for a scholarship to study at the Technical Institute at Charlottenburg in Berlin. During the academic year 1833–1834 Nobert learned dividing methods and made a circle-dividing engine. This was most probably the machine with which he later ruled diffraction gratings, test objects, and micrometers; marked out scales for optical instruments; and cut chronometer gears. In 1835 Nobert was appointed *Universitätsmechaniker* at the University of Greifswald. During the 1840's he began to develop the fine ruling techniques that brought him some measure of fame. Not long after his father's death in 1846 Nobert returned to the family house in Barth, where he remained, working alone, until his death.

The key to Nobert's production of scientific instruments is his use of the circle-dividing engine. The art of dividing straight lines, arcs, and circles into various numbers of equal parts is the most important and difficult aspect of the work of the mathematical instrument maker. At the end of a paper on improvements in dividing the circle (1845), Nobert described his newly discovered technique for using the dividing engine in ruling parallel lines on glass that was to be used in micrometers by microscopists. In 1846 he described an extension of the technique to the production of a resolution test plate for the microscope. At this time the optical microscope was undergoing rapid improvements in the design of the objective lens systems, and a standard physical test was needed for judging improvements in resolution. Nobert's first test plate consisted of ten bands, each comprising a number of lines a definite distance apart, the first and tenth bands being ruled to a spacing of 2.25 and 0.56 microns respectively. Whenever the finest band on one of the test plates had been resolved by improved objectives, Nobert made another plate containing even finer bands; in all he made a series of seven different test plates between 1845 and 1873. The last plate in the series contained bands ruled to a spacing of 0.11 micron, well below the resolution limit of the optical microscope. Plates of this type were sold in London in 1880 for £15 apiece.

Nobert used his circle-dividing engine to produce rulings on glass for other purposes. The most important of these were the diffraction gratings made for V. S. M. van der Willigen in Haarlem and Ångström in Uppsala. For his celebrated measurements of the wavelengths of the elements in the spectrum of the sun, Ångström used four gratings made by Nobert. Other devices made by ruling were a micrometer for the astronomical telescope and a "spectrum plate" designed to prove that the velocity of light is greater in air than in glass. Nobert also made compound microscopes and objective lenses.

BIBLIOGRAPHY

I. ORIGINAL WORKS. Nobert's writings include "Ueber Kreistheilung im Allgemeinen und über einige, bei einer Kreistheilmaschine angewendete, verfahren zur Erzielung einer grossen Vollkommenheit der Theilung derselben," in *Verhandlungen des Vereins zur Beförderung des Gewerbfleisses in Preussen* (1845), 202–212; "Ueber die Prüfung und Vollkommenheit unserer jetzigen Mikroskope," in *Annalen der Physik und Chemie,* **67** (1846), 173–185; "Ueber Glas-Skalen von Herren Universitäts-Mechanicus Nobert in Greifswalde," in *Ergänzungs-Heft zu den Astronomischen Nachrichten* (Altona, 1849), cols. 93–96; "Ueber eine Glasplatte mit Theilungen zur Bestimmung der Wellenlänge und relativen Geschwindigkeit des Lichts in der Luft und im Glase," in *Annalen der Physik und Chemie,* **85** (1852), 83–92; "Ein Ocularmikrometer mit leuchtenden farbigen Linien im dunkel Gesichtsfelde," *ibid.,* 93–97; "Die höchste Leistung des heutigen Mikroskops, und seine Prüfung durch künstliche und natürliche Objekte," in *Mittheilungen aus dem naturwissenschaftlichen Vereine von Neu-Vorpommern und Rügen in Greifswald,* **13** (1882), 92–105. For a list of papers on astronomical observations and other matters see the Royal Society's *Catalogue of Scientific Papers,* IV, 628.

Some letters from Nobert to Ernst Abbe, as well as two microscopes made by Nobert, are in the optical museum of VEB Carl Zeiss, Jena, German Democratic Republic. Nobert's circle-dividing engine is preserved in the Smithsonian Institution, Washington, D.C.

II. SECONDARY LITERATURE. See W. Rollman, "Friedrich Adolph Nobert," in *Mittheilungen aus dem naturwissenschaftlichen Vereine von Neu-Vorpommern und Rügen in Greifswald,* **15** (1884), 38–58; G. L'E. Turner, "The Microscope as a Technical Frontier in Science," in *Historical Aspects of Microscopy,* S. Bradbury and G. L'E. Turner, eds. (Cambridge, 1967), 175–199; "The Contributions to Science of Friedrich Adolph Nobert," in *Bulletin of the Institute of Physics and the Physical Society,* **18** (1967), 338–348; and "F. A. Nobert's Invention of Artificial Resolution Tests for the Optical Microscope," in *Actes du XIᵉ Congrès international d'histoire des sciences,* III (Warsaw, 1968), 435–440; and G. L'E. Turner and S. Bradbury, "An Electron Microscopical Examination of Nobert's Finest Test-Plate of Twenty Bands," in *Journal of the Royal Microscopical Society,* **85** (1966), 435–447.

G. L'E. TURNER

NOBILI, LEOPOLDO (*b.* Trassilico, Italy, 1784; *d.* Florence, Italy, 5 August 1835), *physics.*

Although Nobili received a university training, he did not immediately become a physicist. He passed the first years of his adult life as an artillery captain in Modena and Brescia. He resigned from the military to become a professor of physics at Florence, where he conducted most of his experimental and theoretical work.

Nobili was primarily interested in the electrical current. His education in the school of Ampère had taught him to view currents as phenomena that could be analyzed by supposing the existence of a central, action-at-a-distance force between their parts. Ampère had never given a precise definition of what the current was or how it was connected with the electrical fluid of Coulomb and Poisson. This imprecision troubled Nobili, and his earliest papers attempted to clarify the nature of electrical currents (1, 2).

Nobili was impressed by the existence of what seemed to be two distinct types of electrical current. Those currents that occur whenever there is a temperature gradient across a conductor he termed "thermoelectric currents." The currents that are generated in processes involving wet conductors, as in a Voltaic apparatus, he called "hydroelectric currents." Although both currents exerted forces on magnets and on other current-bearing conductors in the same manner, in accordance with Ampère's law, Ampère gave no explanation of how currents originating in two seemingly distinct fashions could produce the same range of effects. Nobili resolved this dichotomy by deciding that there was actually only one type of current: thermoelectric. He believed that the currents produced with wet conductors did not result from direct chemical action, as Volta's followers thought, but were created by the heat generated in the chemical action.

Once Nobili became convinced that all currents resulted from the release of heat, he thought that the conjunction of heat flux and electrical current was more than coincidental; he concluded that the current is a flow of heat or caloric. Nobili firmly believed in his identification of caloric flow as electrical current, and it was this belief that led him to oppose the contemporary theory of the Voltaic decomposition of water (3).

By the 1820's and 1830's Volta's successors thought that Voltaic decomposition of water depended on the generation of electromotive force produced by the contact of dissimilar metals. If a strip of zinc is joined end to end to a strip of copper, and the ends of the compound strip are placed in a vessel of water (see Figure 1), then the water is decomposed into oxygen and hydrogen, and the zinc is oxidized. Volta's group held that there were two successive processes involved here. First the contact of the zinc and the copper produced a potential difference between the zinc and copper terminals of the compound strip. Since the strip was not closed, the electrical potential could not be neutralized. A state of electrostatic tension thus resulted between the ends of the strip, in which the zinc became positively electrified and the copper became negatively electrified. Since water is composed of two elements, one of which bears negative electrical fluid (oxygen) and the other positive fluid (hydrogen),

the water particles will line up between the submerged ends of the copper-zinc strip.

This configuration produces a constant stress on the water particles and induces some of them to dissociate into electronegative and electropositive components. This dissociation initiates the second stage, in which the dissociated components are attracted to the electrified zinc and copper terminals where they are held, thereby neutralizing the potential for electrical contact which occurred when the zinc and copper were joined. It is this neutralization, a quasi-static relaxation of tension in the copper-zinc strip, that is the electrical current conceptualized by Volta's disciples.

The Voltaic explanation was rejected by Nobili and indeed by every follower of Ampère. They believed that the forces of currents were unrelated to the forces of statical electricity; the Ampèrean forces pertained uniquely to the electrical current that was thought of as an entity distinct from the statical fluid of Coulomb. Nobili believed, furthermore, that the current was a flow of caloric, not a relaxation of statical tension; and he felt called upon to explain Voltaic decomposition in Ampèrean terms. He began by criticizing the Voltaic schema.

He reasoned that the current in the strip, according to Volta, occurred as a result of the neutralizing process of oxidation; it should therefore follow, and not precede, the commencement of oxidation. But in fact the current preceded oxidation. Nobili reasoned that when the ends of the copper-zinc strip were submerged in water, a true electrical current or caloric flow was immediately engendered between the submerged ends. As Ampère had shown, the adjacent parts of any electrical current repel one another; that is why, for example, a closed conducting loop bearing a current tends to expand. Nobili suspected that this principle was operative in the Voltaic apparatus. He thought that the current, while attempting to increase in length because of the self-repulsion of its parts, exerted a powerful force at the zinc and copper terminals between which it occurs. It was this force that wrenched the water particles apart. Nobili thus concluded that the generation of current by Voltaic means was not due to the release of statical electricity by electrochemical decomposition; rather, the release was itself the result of the presence of an electrical current.

Nobili's explanation of the Voltaic apparatus was influential in eliminating the assumption of any direct connection between statical electricity and the electrical current. His conclusion was also essential to the ultimate acceptance of Maxwellian electrodynamics and, more significantly on the Continent, to the

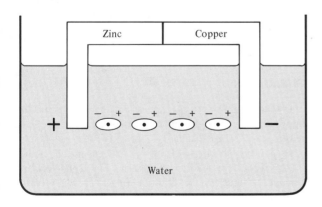

FIGURE 1

ability of Weber and his group to concentrate their attention on the electrical forces necessary to explain the phenomena discovered by Faraday.

BIBLIOGRAPHY

Works by Nobili referred to above are:

1. "Sur une nouvelle classe de phénomènes électrochimiques," in *Annales de chimie*, **34** (1827), 280–297.

2. "Sur la nature des courans électriques," in *Bibliothèque universelle*, **37** (1828), 118–144, 180–184.

3. "De la distribution et des effets des courans électriques dans les masses conductrices," *ibid.*, **49** (1835), 263–281, 416–436.

JED Z. BUCHWALD

NODDACK, WALTER (*b.* Berlin, Germany, 17 August 1893; *d.* Bamberg, West Germany, 7 December 1960), *chemistry*.

Noddack attended the secondary school in his native city and then entered the University of Berlin in 1912 to study chemistry, physics, and mathematics. World War I interrupted his studies and he therefore did not receive his doctorate until 1920. His dissertation, completed under the direction of W. Nernst, examined Einstein's law of photochemical equivalence. Noddack then worked for two years with Nernst at the Physical Chemistry Institute of the University of Berlin, and in 1922 he became director of the chemical laboratory of the Physikalisch-Technische Reichsanstalt under Nernst.

In 1927 Noddack became director at the newly founded Photochemistry Laboratory in Berlin, but he subsequently accepted an offer from the University of Freiburg to become chairman of the department of physical chemistry (1935). In 1941 he became director of both the physical chemistry department and the Research Institute for Photochemistry at the University of Strasbourg. Following World War II, Noddack offered his services to the Philosophisch-Theologische Hochschule in Bamberg, where instruction in chemistry was being introduced. From 1956 to 1960 he directed the newly established research institute for geochemistry in Bamberg.

Noddack's principal achievement was the discovery of element seventy-five of the periodic table, which he called rhenium (after the Rhine). He conducted this research in Berlin with his co-worker, Ida Tacke, whom he married in 1926. (Their joint research continued until his death.) They discovered rhenium by X-ray spectroscopy in columbite that had been systematically enriched (*Naturwissenschaften*, **13** [1925], 567). O. Berg also assisted in the discovery. Although they succeeded in obtaining two milligrams of rhenium from various ores, it was not until 1926, when they produced the first gram of rhenium, that they were able to examine the chemical properties of the new element.

Simultaneously with this discovery, the Noddacks claimed that they had discovered a second new element, element forty-three of the periodic table, which they named masurium. This element was discussed for years in the literature until E. Segrè and C. Perrier discovered that it could be produced only artificially; they named the element technetium.

In the field of geochemistry Noddack studied the abundance of individual elements in the crust of the earth and in the universe. This research was based on the evaluation of 1,600 mineral assays. Noddack believed that every element was present in every mineral, but could not be detected in its existing concentrations with the analytic methods then available. He thought that for each element there was a threshold concentration, beyond which the element could be recognized in all minerals; he named this *Allgegenwartskonzentration* ("omnipresent concentration") and calculated it for various elements. Noddack also took considerable interest in theoretical and practical questions concerning the rare earths and their separation.

Noddack's second major field of research was photochemistry. In 1920 he found, on a photographic plate, that under suitable conditions an absorbed quantum $h\nu$ of blue or ultraviolet radiation corresponds to a silver atom. He then investigated the photographic quantum sensitivity of X and α radiation. In his studies on photographic sensitizing Noddack gave particular attention to the physical properties of the sensitizing coloring substances, and his treatment of photochemical problems in the human eye led him to a new demonstration of the three visual pigments.

BIBLIOGRAPHY

I. ORIGINAL WORKS. Noddack and his wife published approximately 100 papers in various periodicals. Their major work is *Das Rhenium* (Leipzig, 1933).

II. SECONDARY LITERATURE. H. Meier and E. Ruda, "Zum Tode von Walter Noddack," in *Zeitschrift für Chemie*, **2** (1962), 33; and O. Bayer *et al.*, "Walter Noddack," in *Chemische Berichte*, **96** (1963), xxvii.

FERENC SZABADVÁRY

NOEL, ÉTIENNE (*b.* Bassigny, Haute-Marne, France, 29 September 1581; *d.* La Flèche, France, 16 October 1659), *physics.*

Noel entered the Society of Jesus in 1599. He taught in several colleges, although principally at La Flèche, where he was *répétiteur* of philosophy when the young Descartes was studying there prior to 1612; he later became rector of the college. He served as vice-provincial of the Society in 1645–1646. At the end of 1646, when he became rector of the Collège de Clermont in Paris, Noel sent to Descartes his first two published works: *Aphorismi physici* (1646) and *Sol flamma* (1646). The double perspective that characterized all of Noel's later work is already present: adherence to Aristotelian physics and receptiveness to new ideas.

Well disposed towards Descartes, Noel had several disputes with Pascal, a more radically modern physicist than Descartes. In 1646 Pascal took part in the first performance of Torricelli's experiment in France. It raised a problem: what remained above the mercury? The traditional philosophers, opposed to the existence of a vacuum, suggested that it was either air or vapors of mercury, while Descartes, also a partisan of the Aristotelian universe, proposed the idea of a subtle matter. When Pascal published *Expériences nouvelles touchant le vide* (1647), in which he disputed the Aristotelian concept of a full universe, Noel sent the young scientist a letter containing his objections. He asserted, in particular, that the upper portion of Torricelli's barometer was not empty, but was filled with a refined air that had entered through the pores in the glass—a notion similar to Descartes's. Pascal's reply was a lesson in method, which exposed the lack of rigor in Noel's principles and arguments. Noel made a few concessions, but refused to admit the existence of a vacuum. He attacked this idea in a work with the baroque title *Le plein du vide* (1648), which provoked new criticisms from Pascal in *Lettre à Le Pailleur.*

Noel remained unembittered toward Pascal and the following summer, in *Gravitas comparata*, honored Pascal for his role in developing an experiment to produce a vacuum within a vacuum. Noel left Paris a short time later and returned to La Flèche, where he published several further works of minor importance.

BIBLIOGRAPHY

A bibliography of Noel's writings is given in Sommer-vogel, ed., *Bibliothèque de la Compagnie de Jésus*, V (Brussels–Paris, 1894), cols. 1789–1790. The Noel–Descartes correspondence is found in Descartes's *Oeuvres de Descartes*, C. Adam and P. Tannery, eds., IV (Paris, 1901), 498, 567, 584–586; V (1903), 101, 117–118, 119–120, 549–552; XII (1910), 556; and Descartes's *Correspondance*, C. Adam and G. Milhaud, eds., I (Paris, 1936), 374–375; II (1939), 29–30; VII (1960), 171, 221, 238–240, 411; VIII (1963), 7.

For a discussion of Noel or his work see *Oeuvres de Descartes*, C. Adam and P. Tannery, eds., I (Paris, 1897), 382–384, 454–456; Dupont-Ferrier, *Du collège de Clermont au lycée Louis-le-Grand*, III (Paris, 1925), 7; Pascal, *Oeuvres*, Brunschvicg and Boutroux, eds., II (Paris, 1908), 77–125, 158, 174–214, 253–282, 291–294, and *Oeuvres complètes*, J. Mesnard, ed., II (Paris, 1971), 509–540, 556–576, 584–602, 633–639; C. de Waard, *L'expérience baro-métrique, ses antécédents et ses applications* (Thouars, 1936); and J. Lewis, "Pascal's Physical Science," unpub. diss. (Princeton, 1968).

JEAN MESNARD

NOETHER, AMALIE EMMY (*b.* Erlangen, Germany, 23 March 1882; *d.* Bryn Mawr, Pennsylvania, 14 April 1935), *mathematics.*

Emmy Noether, generally considered the greatest of all female mathematicians up to her time, was the eldest child of Max Noether, research mathematician and professor at the University of Erlangen, and Ida Amalia Kaufmann. Two of Emmy's three brothers were also scientists. Alfred, her junior by a year, earned a doctorate in chemistry at Erlangen. Fritz, two and a half years younger, became a distinguished physicist; and his son, Gottfried, became a mathematician.

At first Emmy Noether had planned to be a teacher of English and French. From 1900 to 1902 she studied mathematics and foreign languages at Erlangen, then in 1903 she started her specialization in mathematics at the University of Göttingen. At both universities she was a nonmatriculated auditor at lectures, since at the turn of the century girls could not be admitted as regular students. In 1904 she was permitted to matriculate at the University of Erlangen, which granted her the Ph.D., *summa cum laude,* in 1907. Her sponsor, the algebraist Gordan, strongly influenced her doctoral dissertation on algebraic invariants. Her divergence from Gordan's viewpoint and her progress in the direction of the "new" algebra first began when she was exposed to the ideas of Ernst Fischer, who came to Erlangen in 1911.

In 1915 Hilbert invited Emmy Noether to Göttingen. There she lectured at courses that were given under his name and applied her profound invariant-theoretic knowledge to the resolution of problems which he and Felix Klein were considering. In this connection she was able to provide an elegant pure mathematical formulation for several concepts of Einstein's general theory of relativity. Hilbert repeatedly tried to obtain her appointment as *Privatdozent*, but the strong prejudice against women prevented her "habilitation" until 1919. In 1922 she was named a *nichtbeamteter ausserordentlicher Professor* ("unofficial associate professor"), a purely honorary position. Subsequently, a modest salary was provided through a *Lehrauftrag* ("teaching appointment") in algebra. Thus she taught at Göttingen (1922–1933), interrupted only by visiting professorships at Moscow (1928–1929) and at Frankfurt (summer of 1930).

In April 1933 she and other Jewish professors at Göttingen were summarily dismissed. In 1934 Nazi political pressures caused her brother Fritz to resign from his position at Breslau and to take up duties at the research institute in Tomsk, Siberia. Through the efforts of Hermann Weyl, Emmy Noether was offered a visiting professorship at Bryn Mawr College; she departed for the United States in October 1933. Thereafter she lectured and did research at Bryn Mawr and at the Institute for Advanced Study, Princeton, but those activities were cut short by her sudden death from complications following surgery.

Emmy Noether's most important contributions to mathematics were in the area of abstract algebra, which is completely different from the early algebra of equation solving in that it studies not so much the results of algebraic operations (addition, multiplication, etc.) but rather their formal properties, such as associativity, commutativity, distributivity; and it investigates the generalized systems that arise if one or more of these properties is not assumed. Thus, in classical algebra it is postulated that the rational, the real, or the complex numbers should constitute a "field" with respect to addition and multiplication, operations assumed to be associative and commutative, the latter being distributive with respect to the former. One of the traditional postulates, namely the commutative law of multiplication, was relinquished in the earliest example of a generalized algebraic structure (William Rowan Hamilton's "quaternion algebra" of 1843) and also in many of the 1844 Grassmann algebras. The entities in such systems and in some of the research of Emmy Noether after 1927 are still termed numbers, albeit hypercomplex numbers. In further generalization the elements of an algebraic system are abstractions that are not necessarily capable of interpretation as numbers, and the binary operations are not literally addition and multiplication, but merely laws of composition that have properties akin to the traditional operations.

If Hamilton and Grassmann inspired Emmy Noether's later work, it was Dedekind who influenced the abstract axiomatic "theory of ideals" which

Noether developed from 1920 to 1926. The Dedekind ideals—which are not numbers but sets of numbers— were devised in order to reinstate the Euclidean theorem on unique decomposition into prime factors, a law which breaks down in algebraic number fields. Two of the generalized structures which Noether related to the ideals are the "group" and the "ring."

A group is more general than a field because it involves only a single operation (either an "addition" or a "multiplication") which need not be commutative. It is, then, a system $\{S, \bigcirc\}$ where S is a set of elements, $\bigcirc$ is a closed associative binary operation, and S contains a unit element or identity as well as a unique inverse for every element. A ring is a system $\{S, \oplus, \otimes\}$ which is a commutative group with respect to $\oplus$, an "addition," and which is closed under a "multiplication," that is, a second binary associative operation $\otimes$, which is distributive with respect to the first operation. Finally, a subset of a ring with a commutative multiplication $\otimes$ is called an "ideal" if it is a subgroup of the additive group of the ring—for this it is sufficient that the difference of any two elements of the subset belong to that set—and if it contains all products of subset elements by arbitrary elements of the ring. In a ring with a noncommutative multiplication, there are left ideals and right ideals.

Emmy Noether showed that the ascending chain condition is important for ideal theory. A ring satisfies that condition if every sequence of ideals C_1, C_2, C_3, · · ·, in the ring—such that each ideal is a proper part of its successor—has only a finite number of terms. Noether demonstrated that for a commutative ring with a unit element the requirement is equivalent to each of two other requirements: namely, that every ideal in the ring have a finite basis, that is, that the ideal consist of the set of all elements

$$x_1 a_1 + x_2 a_2 + \cdot \cdot \cdot + x_n a_n,$$

where the a_i are fixed elements of the ring and the x_i are any elements whatsoever in the ring, and that, given any nonempty set of ideals in the ring, there be at least one ideal which is "maximal" in that set.

Having formulated the concept of primary ideals—a generalization of Dedekind's prime ideals—Noether used the ascending chain condition in order to prove that an ideal in a commutative ring can be represented as the intersection of primary ideals. Then she studied the necessary and sufficient conditions for such an ideal to be the product of "prime power ideals." A somewhat different aspect of ideal theory was her use of polynomial ideals to rigorize, generalize, and give modern pure mathematical form to the concepts and methods of algebraic geometry as they had first been developed by her father and subsequently by the Italian school of geometers.

In another area Emmy Noether investigated the noncommutative rings in linear algebras like the Hamilton and Grassmann systems. An "algebra" is a ring in which the two binary operations are supplemented by a unary operation, an external or scalar multiplication, that is, a multiplication by the elements (scalars) of a specified field. From 1927 to 1929 Emmy Noether contributed notably to the theory of representations, the object of which is to provide realizations of noncommutative rings (or algebras) by means of matrices or linear transformations in such a way that all relations which involve the ring addition and/or multiplication are preserved; in other words, to study the homomorphisms of a given ring into a ring of matrices. From 1932 to 1934 she was able to probe profoundly into the structure of noncommutative algebras by means of her concept of the *verschränktes* ("cross") product. In a 1932 paper that was written jointly with Richard Brauer and Helmut Hasse, she proved that every "simple" algebra over an ordinary algebraic number field is cyclic; Weyl called this theorem "a high water mark in the history of algebra."

Emmy Noether wrote some forty-five research papers and was an inspiration to Max Deuring, Hans Fitting, W. Krull, Chiungtze Tsen, and Olga Taussky Todd, among others. The so-called Noether school included such algebraists as Hasse and W. Schmeidler, with whom she exchanged ideas and whom she converted to her own special point of view. She was particularly influential in the work of B. L. van der Waerden, who continued to promote her ideas after her death and to indicate the many concepts for which he was indebted to her.

BIBLIOGRAPHY

I. ORIGINAL WORKS. Among Emmy Noether's many papers are "Invarianten beliebiger Differentialausdrücke," in *Nachrichten von der Gesellschaft der Wissenschaften zu Göttingen* (1918), 37–44; "Moduln in nichtkommutativen Bereichen, insbesondere aus Differential- und Differenzenausdrücken," in *Mathematische Zeitschrift*, **8** (1920), 1–35, written with W. Schmeidler; "Idealtheorie in Ringbereichen," in *Mathematische Annalen*, **83** (1921), 24–66; "Abstrakter Aufbau der Idealtheorie in algebraischen Zahlund Funktionenkörpern," *ibid.*, **96** (1927), 26–61; "Über minimale Zerfällungskörper irreduzibler Darstellungen," in *Sitzungsberichte der Preussischen Akademie der Wissenschaften zu Berlin* (1927), 221–228, written with R. Brauer; "Hyperkomplexe Grössen und Darstellungstheorie," in *Mathematische Zeitschrift*, **30** (1929), 641–692;

"Beweis eines Hauptsatzes in der Theorie der Algebren," in *Journal für die reine und angewandte Mathematik*, **167** (1932), 399–404, written with R. Brauer and H. Hasse; and "Nichtkommutative Algebren," in *Mathematische Zeitschrift*, **37** (1933), 514–541.

II. SECONDARY LITERATURE. For further information about Noether and her work, see A. Dick, "Emmy Noether," in *Revue de mathématiques élémentaires*, supp. 13 (1970); C. H. Kimberling, "Emmy Noether," in *American Mathematical Monthly*, **79** (1972), 136–149; E. E. Kramer, *The Nature and Growth of Modern Mathematics* (New York, 1970), 656–672; B. L. van der Waerden, "Nachruf auf Emmy Noether," in *Mathematische Annalen*, **111** (1935), 469–476; and H. Weyl, "Emmy Noether," in *Scripta mathematica*, **3** (1935), 201–220.

EDNA E. KRAMER

NOETHER, MAX (*b.* Mannheim, Germany, 24 September 1844; *d.* Erlangen, Germany, 13 December 1921), *mathematics.*

Max Noether was the third of the five children of Hermann Noether and Amalia Würzburger. Noether's father was a wholesaler in the hardware business—a family tradition until 1937 when it was "Aryanized" by the Nazis. Max attended schools in Mannheim until an attack of polio at age fourteen made him unable to walk for two years and left him with a permanent handicap. Instruction at home enabled him to complete the Gymnasium curriculum; then, unassisted, he studied university-level mathematics.

After a brief period at the Mannheim observatory, he went to Heidelberg University in 1865. There he earned the doctorate in 1868 and served as *Privatdozent* (1870–1874) and as associate professor (*extraordinarius*) from 1874 to 1875. Then he became affiliated with Erlangen as associate professor until 1888, as full professor (*ordinarius*) from 1888 to 1919, and as professor emeritus thereafter. In 1880 he married Ida Amalia Kaufmann of Cologne. She died in 1915. Three of their four children became scientists, including Emmy Noether, the mathematician.

Noether was one of the guiding spirits of nineteenth-century algebraic geometry. That subject was motivated in part by problems that arose in Abel's and Riemann's treatment of algebraic functions and their integrals. The purely geometric origins are to be found in the work of Plücker, Cayley, and Clebsch, all of whom developed the theory of algebraic curves—their multiple points, bitangents, and inflections. Cremona also influenced Noether, who in turn inspired the great Italian geometers who followed him—Segre, Severi, Enriques, and Castelnuovo. In another direction Emmy Noether and her disciple B. L. van der Waerden made algebraic geometry rigorous and more general. Lefschetz, Weil, Zariski, and others later used topological and abstract algebraic concepts to provide further generalization.

In both the old and the new algebraic geometry, the central object of investigation is the algebraic variety, which, in n-dimensional space, is the set of all points $(x_1, x_2, \cdots, x_n)$ satisfying a finite set of polynomial equations,

$$f_i(x_1, x_2, \cdots, x_n) = 0, \quad i = 1, 2, \cdots, r$$

with $r \leqslant n$ and coefficients in the real (or complex) field or, in modern algebraic geometry, an arbitrary field. Thus, in the plane, the possible varieties are curves and finite sets of points; in space there are surfaces, curves, and finite point sets; for $n > 3$ there are hypersurfaces and their intersections.

Following Cremona, Noether studied the invariant properties of an algebraic variety subjected to birational transformations; that is, one-to-one rational transformations with rational inverses, those of lowest degree being the collineations or projective transformations. Next in order of degree come the quadratic transformations, for which Noether obtained a number of important theorems. For example, any irreducible plane algebraic curve with singularities can be transformed by a finite succession of standard quadratic transformations into a curve whose multiple points are all "ordinary" in the sense that the curve has multiple but distinct tangents at such points.

In 1873 Noether proved what came to be his most famous theorem: Given two algebraic curves

$$\Phi(x, y) = 0, \quad \Psi(x, y) = 0$$

which intersect in a finite number of isolated points, then the equation of an algebraic curve which passes through all those points of intersection can be expressed in the form $A\Phi + B\Psi = 0$ (where A and B are polynomials in x and y) if and only if certain conditions (today called "Noetherian conditions") are satisfied. If the intersections are nonsingular points of both curves, the desired form can readily be achieved. The essence of Noether's theorem, however, is that it provides necessary and sufficient conditions for the case where the curves have common multiple points with contact of any degree of complexity.

Although Noether asserted that his results could be extended to surfaces and hypersurfaces, it was not until 1903 that the Hungarian Julius König actually generalized the Noether theorem to n dimen-

sions by providing necessary and sufficient conditions for the

$$A_1 f_1 + A_2 f_2 + \cdots + A_n f_n = 0$$

form to be possible for the equation of the surface or hypersurface through the finite set of points of intersection of n surfaces ($n = 3$) or n hypersurfaces ($n > 3$),

$$f_1(x_1, x_2, \cdots, x_n) = 0, \cdots, f_n(x_1, x_2, \cdots, x_n) = 0.$$

Noether himself derived a theorem that gives conditions for the equation of a surface passing through the curve of intersection of the surfaces $\Phi(x, y, z) = 0$ and $\Psi(x, y, z) = 0$ to have the form $A\Phi + B\Psi = 0$. Generalization turned out to be complicated and difficult, but Emanuel Lasker, the chess champion, saw that the issue could be simplified by the use of the theory of polynomial ideals which he and Emmy Noether had developed. Thus he was able to derive Noetherian conditions for the

$$A_1 f_1 + \cdots + A_r f_r = 0$$

form to be possible for a hypersurface through the intersection of

$$f_1(x_1, x_2, \cdots, x_n) = 0, \cdots, f_r(x_1, x_2, \cdots, x_n) = 0$$

with $n > 3$, $r < n$, in which case the intersection will, in general, be a curve or a surface or a hypersurface.

The Noether, König, and Lasker theorems all start with a set of polynomial equations that defines a variety the nature of which varies in the different propositions. The objective in every case is the same, namely to see under what conditions a polynomial that vanishes at all points of the given variety can be expressed as a linear combination of the polynomials originally given. Since those polynomials play a basic role, it is especially significant that the representation of a variety as the intersection of other varieties, that is, by a set of polynomial equations, is not unique. Thus a circle in space might be described as the intersection of two spheres, or as the intersection of a cylinder and a plane, or as the intersection of a cone and a plane, and so forth. Hence, in general, the only impartial way to represent a given variety

$$f_i(x_1, x_2, \cdots, x_n) = 0, \quad i = 1, 2, \cdots, r$$

where $r \leqslant n$ and the f_i are polynomials with real or complex coefficients, is not by this one system of equations, but rather in terms of all polynomial equations which points on the variety satisfy. Now if $f(x_1, x_2, \cdots, x_n)$ and $g(x_1, x_2, \cdots, x_n)$ are any two polynomials that vanish at all points of the given variety, then the difference of the polynomials also vanishes at those points, as does the product of either polynomial by an arbitrary polynomial, $A(x_1, x_2, \cdots, x_n)$. By the definition of an ideal, these two facts are sufficient for the set of all polynomials that vanish at every point of the variety to be a polynomial ideal in the ring of polynomials with real (complex) coefficients, and it is that ideal which is considered to represent the variety. The linear combinations $A_1f_1 + A_2f_2 + \cdots + A_rf_r$ obviously vanish at all points of the given variety and hence belong to the representative polynomial ideal. These are the linear combinations that were the subject of the special criteria developed in the Noether, König, and Lasker theorems. Other important results related to the representative polynomial ideal are contained in a famous proposition of Hilbert, namely his basis theorem.

BIBLIOGRAPHY

I. ORIGINAL WORKS. Noether's most important papers include "Zur Theorie des eindeutigen Entsprechens algebraischer Gebilde von beliebig vielen Dimensionen," in *Mathematische Annalen*, 2 (1870), 293–316; "Über einen Satz aus der Theorie der algebraischen Functionen," *ibid.*, 6 (1873), 351–359; "Über die algebraische Functionen und ihre Anwendung in der Geometrie," *ibid.*, 7 (1874), 269–310, written with A. W. von Brill; and "Die Entwicklung der Theorie der algebraischen Functionen in älterer und neurer Zeit," in *Jahresbericht der Deutschen Mathematiker-vereinigung*, 3 (1894), 107–566, written with A. W. von Brill.

II. SECONDARY LITERATURE. On Noether and his work, see A. W. von Brill, "Max Noether," in *Jahresbericht der Deutschen Mathematiker-vereinigung*, 32 (1923), 211–233; A. Dick, "Emmy Noether," in *Revue de mathématiques élémentaires*, supp. 13 (1970), 4–8, 19, 53–56, 67–68; W. Fulton, *Algebraic Curves, an Introduction to Algebraic Geometry* (New York–Amsterdam, 1969), 119–129; J. König, *Einleitung in die Allgemeine Theorie der Algebraischen Grössen* (Leipzig, 1903), 385–398; E. Lasker, "Zur Theorie der Moduln und Ideale," in *Mathematische Annalen*, 60 (1905), 44–46, 51–54; F. S. Macaulay, "Max Noether," in *Proceedings of the London Mathematical Society*, 2nd ser., 21 (1920–1923), 37–42; and C. A. Scott, "A Proof of Noether's Fundamental Theorem," in *Mathematische Annalen*, 52 (1899), 593–597.

See also J. G. Semple and L. Roth, *Introduction to Algebraic Geometry* (Oxford, 1949), 94–99, 391; R. J. Walker, *Algebraic Curves* (Princeton, 1950), 120–124; H. Wieleitner, *Algebraische Kurven*, II (Berlin–Leipzig, 1919), 18–20, 45, 88; and the editors of *Mathematische Annalen*, "Max Noether," in *Mathematische Annalen*, 85 (1922), i–iii.

EDNA E. KRAMER

NOGUCHI, (SEISAKU) HIDEYO (*b.* Sanjogata, Okinashima-mura, Fukushima, Honshu, Japan, 24 November 1876; *d.* Accra, Gold Coast, 21 May 1928), *microbiology*.

Despite humble origins and a physical handicap, Noguchi attained extraordinary fame during his lifetime. He discovered *Treponema pallidum* in the brain of general paralytics, and he proved that either Oroya fever or verruga peruana might be produced by *Bartonella bacilliformis*. But his technique and conclusions were often faulty, and his work on *Leptospira icteroides* as the causal agent of yellow fever was gravely misleading.

Noguchi was the second child and only son of Sayosuke, a thriftless peasant, and his illiterate but industrious wife, Shika; he was given the name Seisaku. As an infant he was burned by an indoor brazier, and his left hand was seriously injured. After rapid elementary schooling, he attended secondary school at Inawashiro, a distance of three miles from his home, and graduated with honors at age seventeen. While working as dispenser to a local surgeon (who restored partial function to his crippled hand) and as janitor at a dental college, he studied medicine from borrowed books. Helped financially by friends, he briefly attended a proprietary medical school in Tokyo, receiving his practitioner's diploma in 1897.

Various temporary appointments then followed, including an assistantship at S. Kitasato's Institute for Infectious Diseases, where advancement was slow. Noguchi then replaced the name Seisaku by Hideyo ("to excel in the world"). When Simon Flexner visited the Institute in 1899, leading a medical commission from the Johns Hopkins University, Noguchi expressed a desire to study pathology and bacteriology in the United States. Flexner cautiously endorsed the wish. In December 1900, having borrowed passage money, Noguchi arrived unannounced and penniless at the University of Pennsylvania, where Flexner had become professor of pathology. With Weir Mitchell's modest support, Noguchi began snake venom investigations under Flexner's tutelage.

Noguchi amassed data for a dozen papers, and grants were forthcoming. In 1903 the Carnegie Institution appointed Noguchi research assistant and awarded him a one-year fellowship at the Statens Seruminstitut, Copenhagen. He was befriended by Thorvald Madsen, who stressed quantitative accuracy and physicochemical concepts in their immunologic studies, mainly of venoms and potent antivenins. Late in 1904 Noguchi began an assistantship at the Rockefeller Institute for Medical Research, inaugurated under Flexner's direction. He and Flexner were the first scientists in America to confirm F. Schaudinn's discovery of *Spirochaeta pallida* (1905). Following A. Wassermann's publication of his complement-fixation test (1906), Noguchi became preoccupied with problems of syphilis and produced twenty papers and a book on serodiagnostic methods. Between 1909 and 1913 he cultured *Sp. pallida* and various other spirochetes in artificial media, described specific cutaneous reactions in latent and tertiary syphilitics after intradermal injection of emulsified spirochetes ("luetin"), and detected *Sp. pallida* in the brain of paretics—the sole enduring accomplishment of this prolific period. He also reported on cultivable bodies as probable causal agents of poliomyelitis, rabies, and trachoma. Late in 1913 his lecture-demonstrations were received triumphantly in European medical centers. His promotion to membership in the Institute ensued from these researches.

In 1915 Noguchi visited his ailing mother in Japan and received the Order of the Rising Sun and an Imperial Prize. He also learned of *Spirochaeta icterohaemorrhagiae*, recently identified by R. Inada and Y. Ido as the causative agent of hemorrhagic jaundice (Weil's disease). Noguchi made extensive studies of this microorganism, whose generic name he revised to *Leptospira*. Soon after reaching Guayaquil in 1918 with the Rockefeller Foundation yellow fever commission to Ecuador, he isolated an organism resembling *Leptospira icterohaemorrhagiae* from several allegedly classic cases of yellow fever. Convalescent patients' serums showed positive "Pfeiffer reactions" (specific bacteriolysis) with this organism, which he named *Leptospira icteroides*; outbreaks of the disease in Yucatan, Peru, and Brazil yielded similar evidence. Guinea pigs injected with *L. icteroides* developed lesions like those of yellow fever. Numerous reports from Noguchi implicating *L. icteroides* as the causative agent of yellow fever appeared between 1919 and 1922, in the *Journal of Experimental Medicine*, then under Flexner's editorship. Yellow fever prophylactic vaccine and therapeutic antiserum were prepared from this organism and distributed experimentally by the Rockefeller Institute until 1926. Noguchi's inconclusive experimental data had been superimposed on the fallacious presumptions that leptospiral jaundice and yellow fever were distinguishable by regional physicians and were caused by kindred agents. Criticism was silenced initially by overenthusiasm, but in 1924, at a Jamaican conference on tropical medicine, the leptospiral theory was disputed on various grounds. By then, yellow fever had practically vanished from the Western Hemisphere, and further tests therefore awaited transfer of the campaign to West Africa.

Noguchi had meanwhile solved the long-standing enigma concerning the relationship between Oroya fever (Carrión's disease) and verruga peruana. Using his special *Leptospira* medium, he isolated *Bartonella bacilliformis* (previously uncultivated) from an Oroya fever patient's blood, and also from verruga nodules. This microorganism, administered intravenously to macaques, provoked an acute febrile anemia, whereas intradermal inoculations caused local verruga formation. Thus the etiologic unity of these diseases was established. He simultaneously resumed enquiries into trachoma among Arizona Indians and isolated *Bacterium granulosis*, which induced progressive granular conjunctivitis when injected into the conjunctiva of monkeys and chimpanzees. Notwithstanding his stimulating monograph on trachoma (1928), this bacillus gained little credence.

In 1927 the leptospiral theory was finally discredited by careful reports that *L. icteroides* and *L. icterohaemorrhagiae* were indistinguishable, and by investigations sponsored by the Rockefeller Foundation International Health Board in Nigeria since late 1925. After negative bacteriologic findings in sixty-seven typical cases, a filterable virus was implicated, undetected by guinea pigs, but producing characteristic fatal lesions in rhesus monkeys. Convalescent human serums protected such animals, but *L. icteroides* vaccine and antiserum did not. In September 1927, before these results were published, the senior author, Adrian Stokes, died from yellow fever. Noguchi sailed for Africa in October. He established himself at Accra, Gold Coast, in the Medical Research Institute; the director, W. A. Young, collaborated closely with Noguchi. By frenzied, often solitary work, day and night, Noguchi apparently confirmed the viral findings, but also isolated a banal bacillus that he considered significant. When about to depart for New York, after six unhappy months, he fell ill and in nine days died of yellow fever. One week later, the same fate befell Young. A marble memorial to their joint research was erected in the Institute's compound at Accra. Noguchi's tomb, surmounted by natural rock, is in Woodlawn Cemetery, New York.

Speculation centered upon the manner of Noguchi's fatal infection. To some, increasing despondency and ill health suggested that he courted infection, fulfilling a youthful motto, "Success or suicide"; others, particularly in Japan, viewed his death as martyrdom, worthy of veneration. Noguchi's quick perceptivity and remarkable energy often permitted him to correct or amplify the more original discoveries of others. Unfortunately, he applied bacteriologic techniques to many viral diseases. In the laboratory he was deft and ingenious, but disorderly and extravagant; at Accra, for example, he accumulated over 500 monkeys. These qualities were magnified by his propensity for working alone upon multiple projects. His small stature, fine head, and oriental manners could be very appealing, despite his frequent unpredictability and moodiness. An unannounced, childless marriage to Mary Dardis in 1912 moderated a tempestuous life.

Noguchi's spoken English was difficult to follow, and his writings needed editing; but he understood many languages. Various universities conferred honorary doctorates, and several countries granted honors and decorations. He was awarded the John Scott Medal (1920) and the Kober Medal (1925). Noguchi's tragedy lay not in lack of recognition, but arose from insatiable ambition, reckless industry, and shrewd intelligence so overlaid with disarming modesty and charm that mentors and benefactors in Japan and America minimized his faults and overestimated his capabilities. He owed an incalculable debt to the Rockefeller Institute for Medical Research, which fostered his activities for a quarter-century. In its library stands a striking bronze bust of Noguchi done in his last year of life.

BIBLIOGRAPHY

I. ORIGINAL WORKS. A list of Noguchi's publications, comprising 186 titles, was prepared and bound with 81 collected reprints by the Rockefeller Institute for Medical Research (now the Rockefeller University). In 1935 some 20 sets of this bibliography were distributed to selected libraries in the United States, Japan, and Europe. The list has minor inaccuracies and omissions. Most of his writings are in English, but a few appear only in German, French, or Spanish. Many articles had multiple publication, sometimes in two or more foreign-language journals.

Noguchi's books are *Snake Venoms* (Washington, 1909); *Serum Diagnosis of Syphilis and the Butyric Acid Test for Syphilis* (Philadelphia, 1910; 2nd ed., 1911; 3rd ed., 1912); and *Laboratory Diagnosis of Syphilis* (New York, 1923). He also contributed chapters to various texts, of which the more noteworthy are "Snake Venoms," in W. Osler and T. McCrae, eds., *System of Medicine*, I (London, 1907), 247–265; "Serodiagnostic de la syphilis," in A. Gilbert and M. Weinberg, eds., *Traité du sang* (Paris, 1921); "Yellow Fever," in R. L. F. Cecil, ed., *Textbook of Medicine* (Philadelphia, 1927); and "The Spirochetes," in E. O. Jordan and I. S. Falk, eds., *The Newer Knowledge of Bacteriology and Immunology* (Chicago, 1928), 452–497.

His early contributions to the knowledge of snake venoms and antivenins and of hemolysins, include "Snake Venom in Relation to Haemolysis, Bacteriolysis, and Toxicity," in *Journal of Experimental Medicine*, 6 (1902), 277–301, and "On the Plurality of Cytolysins in Snake Venom," in *Journal of Pathology and Bacteriology*, 10 (1905), 111–124, both written with S. Flexner; "The Photodynamic Action of Eosin and Erythrosin Upon Snake Venom," in *Journal of Experimental Medicine*, 8 (1906), 252–267; "The Influence of Temperature Upon the Rate of Reaction (Haemolysis, Agglutination, Precipitation)," *ibid.*, 337–364, written with T. Madsen and L. Walbum; and "Toxins and Antitoxins—Snake Venoms and Antivenins," *ibid.*, 9 (1907), 18–50, written with T. Madsen.

Among his reports on the serodiagnosis of syphilis are "The Relation of Protein, Lipoids and Salts to the Wassermann Reaction," *ibid.*, 11 (1909), 84–99; "A New and Simple Method for the Serum Diagnosis of Syphilis," *ibid.*, 392–401; "The Present Status of the Noguchi System of Serodiagnosis of Syphilis," in *Interstate Medical Journal*, 18 (1911), 11–25; "Biochemical Studies on So-called Syphilis Antigen," in *Journal of Experimental Medicine*, 13 (1911), 43–68, and "The Comparative Merits of Various Complements and Amboceptors in the Serum Diagnosis of Syphilis," *ibid.*, 78–91, both written with J. Bronfenbrenner; "A Cutaneous Reaction in Syphilis," *ibid.*, 14 (1911), 557–568; "Experimental Research in Syphilis, With Especial Reference to Spirochaeta pallida (Treponema pallidum)," in *Journal of the American Medical Association*, 58 (1912), 1163–1172, the Fenger-Senn Memorial Address; "A Homohemolytic System for the Serum Diagnosis of Syphilis," in *Journal of Experimental Medicine*, 28 (1918), 43–67.

His wide-ranging studies of spirochetes include "On the Occurrence of *Spirochaeta pallida*, Schaudinn, in Syphilis," in *Medical News*, 86 (1905), 1145, written with S. Flexner; "A Method for the Pure Cultivation of Pathogenic Treponema pallidum (Spirochaeta pallida)," in *Journal of Experimental Medicine*, 14 (1911), 99–108; "The Pure Cultivation of Spirochaeta duttoni, Spirochaeta kochi, Spirochaeta obermeieri, and Spirochaeta novyi," *ibid.*, 16 (1912), 199–210; "A Demonstration of Treponema pallidum in the Brain in Cases of General Paralysis," *ibid.*, 17 (1913), 232–238, written with J. W. Moore; "Spirochaetes," in *Journal of Laboratory and Clinical Medicine*, 2 (1917), 365–400, 472–499, the Harvey lecture; "Spirochaeta icterohaemorrhagiae in American Wild Rats and Its Relation to the Japanese and European Strains," in *Journal of Experimental Medicine*, 25 (1917), 755–763; "Morphological Characteristics and Nomenclature of Leptospira (Spirochaeta) icterohaemorrhagiae (Inada and Ido)," *ibid.*, 27 (1918), 575–592; and "The Survival of Leptospira (Spirochaeta) icterohaemorrhagiae in Nature; Observations Concerning Microchemical Reactions and Intermediary Hosts," *ibid.*, 609–625.

Noguchi's fallacious claims respecting yellow fever involve over 30 papers. The laboratory data, embodied mainly in a series of 18 reports in the *Journal of Experimental Medicine*, include "Etiology of Yellow Fever. I. Symptomatology and Pathological Findings of the Yellow Fever Prevalent in Guayaquil," in *Journal of Experimental Medicine*, 29 (1919), 547–564; "II. Transmission Experiments on Yellow Fever," *ibid.*, 565–584; "III. Symptomatology and Pathological Findings in Animals Experimentally Infected," *ibid.*, 585–596; "VI.

Cultivation, Morphology, Virulence, and Biological Properties of Leptospira icteroides," *ibid.*, **30** (1919), 13–29; "VII. Demonstration of Leptospira icteroides in the Blood, Tissues, and Urine of Yellow Fever Patients and of Animals Experimentally Infected With the Organism," *ibid.*, 87–93; "IX. Mosquitoes in Relation to Yellow Fever," *ibid.*, 401–410; and "X. Comparative Immunological Studies in Leptospira icteroides and Leptospira icterohaemorrhagiae," *ibid.*, **31** (1920), 135–158. Among four papers written with I. J. Kligler are "Immunological Studies With a Strain of Leptospira Isolated From a Case of Yellow Fever in Mérida, Yucatan," *ibid.*, **32** (1920), 627–637, and "Experimental Studies on Yellow Fever in Northern Peru," *ibid.*, **33** (1921), 239–252.

Other key publications are "Prophylactic Inoculation Against Yellow Fever," in *Journal of the American Medical Association*, **76** (1921), 96–99, written with W. Pareja; "Prophylaxis and Serum Therapy of Yellow Fever," *ibid.*, **77** (1921), 181–185; *Experimental Studies of Yellow Fever in Northern Brazil*, Monograph no. 20, Rockefeller Institute for Medical Research (New York, 1924), written with H. R. Muller and others; "The Pfeiffer Reaction in Yellow Fever," in *American Journal of Tropical Medicine*, **4** (1924), 131–138; and "Yellow Fever Research, 1918–1924: A Summary," in *Journal of Tropical Medicine and Hygiene*, **28** (1925), 185–193.

Noguchi's earliest searches for the causal agent of trachoma appeared as "The Relationship of the So-called Trachoma Bodies to Conjunctival Affections," in *Archives of Ophthalmology*, **40** (1911), 1–9, and culminated in his monograph "The Etiology of Trachoma," in *Journal of Experimental Medicine*, **48** (1928), supp. no. 2. Misleading reports on the causal agents of rabies and poliomyelitis are "Contribution to the Cultivation of the Parasite of Rabies," in *Journal of Experimental Medicine*, **18** (1913), 314–316; "Experiments on the Cultivation of the Microorganism Causing Epidemic Poliomyelitis," *ibid.*, 461–485, written with S. Flexner; and "Concerning Survival and Virulence of the Microorganism Cultivated From Poliomyelitis Tissues," *ibid.*, **21** (1915), 91–102, written with S. Flexner and H. L. Amoss.

Characteristic papers on miscellaneous researches include "Pure Cultivation in Vivo of Vaccine Virus Free From Bacteria," *ibid.*, 539–570; "Bacteriological and Clinical Studies of an Epidemic of Koch-Weeks Bacillus Conjunctivitis Associated With Cell Inclusion Conjunctivitis," *ibid.*, **22** (1915), 304–318, written with M. Cohen; "Immunity Studies of Rocky Mountain Spotted Fever. II. Prophylactic Inoculation in Animals," *ibid.*, **38** (1923), 605–626; "The Isolation and Maintenance of Leishmania on the Medium Employed for the Cultivation of Organisms of the Leptospira Group of Spirochetes," in *American Journal of Tropical Medicine*, **5** (1925), 63–69, written with A. Lindenberg; and "Comparative Studies of Herpetomonads and Leishmanias. I. Cultivation of Herpetomonads From Insects and Plants," in *Journal of Experimental Medicine*, **44** (1926), 307–325, written with E. B. Tilden.

Noguchi's last series of publications, on the causal agent of Oroya fever and verruga peruana, comprises 17 reports in the *Journal of Experimental Medicine*, of which three appeared posthumously in 1929. The more important are "Etiology of Oroya Fever. I. Cultivation of Bartonella bacilliformis," *ibid.*, **43** (1926), 851–864, written with T. S. Battistini; "III. The Behaviour of Bartonella moniliformis in Macacus rhesus," *ibid.*, **44** (1926), 697–713; "The Etiology of Verruga Peruana," *ibid.*, **45** (1927), 175–189; "VIII. Experiments on Cross-Immunity Between Oroya Fever and Verruga Peruana," *ibid.*, 781–786; and "XIV. The Insect Vectors of Carrión's Disease," *ibid.*, **49** (1929), 993–1008, written with R. C. Shannon *et al.*

Correspondence with Flexner and others is in the Simon Flexner Papers at the American Philosophical Society Library, Philadelphia. Relevant material is also among the Philip S. Hench Collection, Walter Reed Yellow Fever Archive, at the Alderman Library, University of Virginia.

II. SECONDARY LITERATURE. Memorial addresses delivered at the New York Academy of Medicine on 20 December 1928 are T. Smith, "Hideyo Noguchi, 1876–1928," in *Bulletin of the New York Academy of Medicine*, 2nd ser., **5** (1929), 877–884; and W. Welch, *ibid.*, 884–886. The chief obituary is S. Flexner, "Hideyo Noguchi. A Biographical Sketch," in *Science*, **69** (1929), 653–660, repr. with portrait in *Report of the Smithsonian Institution* (1929), pp. 595–608. G. Eckstein, *Noguchi* (New York–London, 1931), a vivid but awkward biography, lacks an index and authenticating details.

Other references to Noguchi's life and work are S. Benison, *Tom Rivers. Reflections on a Life in Medicine and Science* (Cambridge, Mass., 1967), pp. 93–98; A. R. Burr, *Weir Mitchell. His Life and Letters* (New York, 1929), pp. 293–296; P. F. Clark, "Hideyo Noguchi, 1876–1928," in *Bulletin of the History of Medicine*, **33** (1959), 1–20, with portrait; H. Hanson, *The Pied Piper of Peru* (Jacksonville, Fla., 1961), pp. 83–85; P. de Kruif, *The Sweeping Wind* (New York, 1962), pp. 17–18; K. Morishita, "Dr. Noguchi's Last Photo," in *Tokyo-iji-shinski*, no. 3143 (1939), 1920–1921, in Japanese; W. A. Sawyer, "A History of the Activities of the Rockefeller Foundation in the Investigation and Control of Yellow Fever," in *American Journal of Tropical Medicine*, **17** (1937), 35–50; M. G. Schultz, "A History of Bartonellosis (Carrión's disease)," in *American Journal of Tropical Medicine and Hygiene*, **17** (1968), 503–515; A. Takahashi, ed., *Hideyo Noguchi, November 9, 1876–May 21, 1928* (Tokyo, 1961), a booklet published by the Doctor Noguchi Memorial Association; and G. Williams, *The Plague Killers* (New York, 1969), pp. 215–249.

Crucial reports that finally discredited *L. icteroides* include A. Agramonte, "Some Observations Upon Yellow Fever Prophylaxis," in *Proceedings of the International Conference on Health Problems in Tropical America, Held at Kingston, Jamaica, July 22-August 1, 1924* (Boston, 1924), 201–227; W. Schüffner and A. Mochtar, "Gelbfieber und Weilsche Krankheit," in *Archiv für Schiffs- u. Tropen-Hygiene*, **31** (1927), 149–165; A. W. Sellards, "The Pfeiffer Reaction With Leptospira in Yellow Fever," in *American Journal of Tropical Medicine*, **7** (1927), 71–95; A. Stokes *et al.*, "Experimental Transmission of Yellow Fever to Laboratory Animals," *ibid.*, **8** (1928), 103–164; and M. Theiler and A. W. Sellards, "The Immunological Relationship of Yellow Fever as it Occurs in West Africa and in South America," in *Annals of Tropical Medicine and Parasitology*, **22** (1928), 449–460.

CLAUDE E. DOLMAN

NOLLET, JEAN-ANTOINE(*b*. Pimprez, near Noyon, France, 19 November 1700; *d*. Paris, France, 24 April 1770), *physics*.

Nollet's rise from the semiliterate peasantry to the top of the aristocratic Paris Academy of Sciences was a *chef d'oeuvre* of the Age of Reason. His village curé had recognized his intelligence and recommended him for the Church; his father, a stranger to learning, reluctantly consented; Jean-Antoine, having completed the humanities course in the provincial *collège* of Clermont, went to Paris to study theology. The capital opened his mind. The range of commodities, industries, and techniques particularly took his fancy; and soon he was devoting more time to the processes later pictured in the *Encyclopédie* of Diderot than to the system of St. Thomas. He supported himself by tutoring while inertia carried him to a master's degree in theology (1724) and the diaconate (*ca*. 1728); but there he suspended his clerical career, withdrew with the equivocal title "abbé," and cast about for a livelihood in the unpromising borderland between science and art.

Nollet had become acquainted with a few like-minded individuals who, with the financial backing of the Comte de Clermont, had constituted themselves a Société des Arts dedicated to bringing science to the artisan. In 1728 Nollet joined this group, which included Clairaut, La Condamine, and Grandjean de Fouchy—all of whom were to be his colleagues in the Academy—and Pierre Polinière, a public lecturer on natural philosophy, who was to leave him an example and an audience. The Société des Arts disbanded in the early 1730's, partly because of the disparity between its purpose and its purse and partly because of the opposition of the Academy of Sciences. The short-lived association, however, probably decided Nollet's future, for it was doubtless through contacts made there that he came to the attention of two leading academicians, C. F. Dufay and R. A. F. de Réaumur. From 1731 or 1732 to about 1735 Nollet assisted them in investigations of extraordinary range, touching the anatomy of insects, the fertilization of frogs, thermometry, pneumatics, phosphorescence, magnetism, and what was to become Nollet's special subject, electricity. From his masters the abbé learned

—besides an ocean of facts—the best contemporary laboratory technique and a useful, moderate Cartesian approach to physical theory. Moreover, through them, especially Dufay, who took him on a *Gelehrtenreise* to England and Holland, Nollet came to know a number of men of science, including the two most successful expositors of Newton, J. T. Desaguliers and W. J. 'sGravesande.

On returning from Holland in 1735, Nollet decided to take up the calling of Polinière, who had died the preceding year, and to follow the methods, if not the theories, of the expositors of Newton. But he found the requisite apparatus so expensive that he could finance it only by building and selling duplicates. "I wielded the file and scissors myself [he wrote of that time]; I trained and hired workmen; I aroused the curiosity of some gentlemen, who placed my products in their studies; I levied a kind of voluntary tribute; in a word (I will not hide it) I have often made two or three instruments of the same kind in order to keep one for myself."[1] By 1738 Nollet could handle an order from Voltaire for instruments costing over 10,000 livres, equivalent to about as many dollars today.[2]

Nollet's *cours de physique* was perhaps the most popular exhibition of its kind ever given.[3] With carefully orchestrated demonstrations performed on some 350 different instruments, the abbé entertained his enthusiastic auditors as, in the spirit of the Enlightenment, he undertook to dispel their "vulgar errors, extravagant fears and faith in the marvelous."[4] These were not mere shows, as one sees from their expanded syllabus, the famous *Leçons de physique*, which appeared in six volumes between 1743 and 1748 and was often reprinted. The presentations are lively, comprehensive, and up-to-date, with full directions for realizing the effects under study and excellent illustrations of apparatus. Nollet strove ceaselessly to perfect his technique; and his last work, *L'art des expériences* (1770), offers the "amateur of physics" the distillation of forty years of attention to the "choice, construction, and use of instruments." The establishment as well as the literate and leisured public rewarded the abbé. In 1739 he entered the Academy as "adjunct mechanician" and went to Turin to instruct the heir to the kingdom of Sardinia; in 1741 the Académie Royale de Bordeaux invited him to lecture before it, and three years later he enlightened the dauphin and the queen at Versailles. Eventually Nollet collected the newly created chair of physics at the Collège de Navarre (the first such post at the University of Paris), an annual lectureship at the technical schools of La Fère and Mézières (where Coulomb attended his course), the succession to

Réaumur as pensionary in the Academy's class of "mechanics," and appointment as preceptor to the royal family.

Nollet's repertoire always included electricity. Until 1745 the electrical demonstrations offered nothing beyond the results of Hauksbee, Stephen Gray, and Dufay, while the accompanying patter probably referred the phenomena to the vague vortical theories of Dufay and Fontenelle. In February of that year, however, word reached Nollet of the first fundamentally new experiments since those of his master: the antics of G. M. Bose and ignition of spirits by sparks. These colorful effects interested Nollet both as showman and as physicist; he threw himself into their study, from which he emerged, three months later, with the elements of the ill-fated theory of simultaneous effluence and affluence.

The theory is a compound of Cartesian common sense, bits and pieces of earlier hypotheses, the results of the Germans, and immediate experience. From the last—the sparks, pricklings, hissings, snappings, and smells surrounding a working electric—Nollet inferred, as had most electricians before him, that electricity consists in the action of a particular matter in motion. From the German experiments he deduced that, contrary to the opinion of Dufay, the matters of electricity and light are fundamentally the same and that, consequently, one can safely infer from the appearance of the brush discharge that the electrical matter leaves a charged body in divergent conical jets. Such jets, in their entirety, make up the body's "effluence." In answer to it, as suggested by the earlier theories of Cabeo, Hauksbee, and Privat de Molières, environing objects and even the air return an "affluence" to the body. According to Nollet the two currents, which differ only in direction, not in kind, nearly or exactly balance, so that a body can never be emptied of its electrical matter. Finally, in accordance with the principles of Descartes, Nollet insisted that all "attractions" and "repulsions" arise from the direct impact of the electrical matter in motion: "mechanical explanations are the only ones capable of advancing experimental physics."[5] Since the effluent flow is divergent and the affluent roughly homogeneous, one understands that local imbalances always exist; and, if one can accept certain ancillary hypotheses about the distribution of the imbalances, one may perceive why Bose, Musschenbroek, and many other physicists agreed that (in the words of Réaumur) "a more probable and natural explanation [of electrical phenomena] can scarcely be expected."[6] Nollet immediately became the chief of the European electricians. In the late 1740's he consolidated his position with several papers and two books—which,

among other things, tried to apply the theory to the Leyden jar—and with a trip across the Alps, undertaken at the request of his colleagues and at the expense of his government, to examine electrical cures advertised by Italian physicians. His expert, tactful, decisive debunking of these claims won him a kind word from Benjamin Franklin.

Shortly after his return from Italy, at the height of his reputation, Nollet found himself the quarry of Buffon, who was promoting the translation of a book by an unknown printer from Philadelphia. The abbé at first believed this American to be a fabrication of his enemies, and in this he was not far wrong; for Buffon, whose raging feud with Réaumur had reached a new stage of ferocity with the publication of the first volumes of the *Histoire naturelle* (1749), pushed Franklin in an effort to embarrass his enemy's favorite and most successful disciple. The plot worked far better than Buffon could have hoped. In the spring of 1752 his henchman, the naturalist Dalibard, issued the translation, prefaced by a "short history of electricity" that found space for third-rate contributors and none for Nollet; and while contemporaries puzzled over the slap, the plotters announced that Franklin's views about lightning had been proved by experiments they had set up in Marly-la-Ville, a small town outside Paris where Dalibard had earlier botched a geologizing errand for Buffon. No one remembered that in the fourth volume of his *Leçons* (1748) Nollet had stressed the analogy between electricity and lightning. Franklin's name was on everyone's lips. "The abbé Nollet," Buffon wrote in evident satisfaction, "is dying of chagrin from it all."[7] Worst of all, from Nollet's point of view, the apparent success of the lightning experiment lent support to the truly menacing aspects of Franklin's scheme, with which in fact it had nothing to do.

The first menace was the Philadelphia theory of the Leyden jar, which unfortunately for Nollet had been discovered just after the system of effluence and affluence. The new theory required the novel assumption that glass was impermeable to the electrical matter, a proposition in manifest disagreement with the patent fact that a feather in a sealed bottle can be drawn by an external electrified object. Franklin, concerned to elucidate the Leyden jar, accepted impenetrability and with it macroscopic action at a distance; Nollet, eager to retain the standard theory of electrical motions, insisted on transparency and mechanical action. The second threat was the doctrine that electricity (but not the electrical matter) came in two qualitatively different, opposite, and mutually destructive types. In Nollet's system only quantitative differences can obtain; it could never handle the disappearance of electricity in the discharge of the Leyden jar.

Nollet recognized these menaces and replied in an amusing set of *Lettres sur l'électricité* (1753), containing a wealth of counterexamples which drew their strength from Franklin's occasional obscurities, imprecisions, exaggerations, and inappropriate appeals to traditional effluvial models. Buffon's group was unable to respond and seized with relief the reply of Franklin's first European paladin, Giambatista Beccaria, which they issued in French before they left the field. Within the Academy, Franklin found a supporter in J. B. Le Roy, who had learned electricity from Nollet. But Le Roy was not a match for his mentor, whose tireless ingenuity, expressed in seven memoirs and two more volumes of *Lettres*, kept the Academy bamboozled until his death in 1770.

France had no electrician of stature again before Coulomb. One must not conclude, however, that Nollet's attack on Franklinists had no positive results. Under prodding from Paris the Philadelphia system was progressively refined into classical electrostatics. In particular, the need to come to terms with Nollet colored the reforms of Aepinus (1759); and Nollet himself, by spreading the dualistic theory of Robert Symmer in Italy, set in train developments that culminated in the invention of the electrophorus (1775), which in turn forced the excision of the last vestiges of the traditional theories (the "electrical atmospheres") from Franklin's system.

For the rest Nollet was by no means the ignorant and friendless recluse of Franklinist mythology. Among his important work outside electricity and pedagogy are his discovery and clear explanation of osmotic pressure (1748) and his account of the hatmaking trade (1765). Among his immediate disciples were M. J. Brisson and J. A. Sigaud de la Fond, and, among his correspondents, Bergman, Bose, Musschenbroek, William Watson, and Benjamin Wilson. His friends included Réaumur, the permanent secretary of the Academy, Grandjean de Fouchy, and the portrait painter Quentin de La Tour. He was one of the few people acceptable at both Cirey and Versailles. Despite his success he retained close ties with his family, whom he often helped financially. "No one [according to Grandjean] ever saw him lose his composure or his unfailing consideration; he only became excited when he talked about physics."[8]

NOTES

1. *Programme ou idée générale d'un cours de physique expérimentale* (Paris, 1738), xviii–xix.

2. Estimated from letters from Voltaire to B. Moussinot, June and July 1738, in Voltaire, *Correspondance*, 107 vols., T. Besteman, ed. (Geneva, 1953–1965), VII, *passim*.

3. It attracted some 500 auditors in 1760. Bengt Ferrner, *Resa i Europa 1758–1762*, S. G. Lindroth, ed. (Uppsala, 1956), xliii.

4. *Programme*, pp. xxxv–xxxvi.

5. Nollet to Bergman, 20 September 1766, in *Torbern Bergman's Foreign Correspondence*, G. Carlid and J. Nordström, eds., I (Stockholm, 1965), 285.

6. Réaumur to J. F. Séguier, 25 May 1747, in *Lettres inédites de Réaumur*, G. Musset, ed. (La Rochelle, 1886), 60.

7. Buffon to de Ruffey, 22 July 1752, in *Correspondance de Buffon de 1729 à 1788*, N. de Buffon, ed., 2nd ed., 2 vols., I (Paris, 1885), 84.

8. *Histoire de l'Académie . . . des sciences* (1770), 135. Bošković also testified to Nollet's wisdom and kindness; see Elizabeth Hill, in L. L. Whyte, ed., *Roger Joseph Boscovich* (London, 1961), 61.

BIBLIOGRAPHY

I. ORIGINAL WORKS. Nollet's chief works are *Programme ou idée générale d'un cours de physique expérimentale* (Paris, 1738); *Leçons de physique expérimentale*, 6 vols. (Paris, 1743–1748), often repr. and once trans. into Spanish (Madrid, 1757); "Conjectures sur les causes de l'électricité des corps," in *Mémoires de l'Académie des sciences* for 1745, 107–151; *Essai sur l'électricité des corps* (Paris, 1746; 4th ed., 1764); "Recherches sur les causes du bouillonnement des liquides," in *Mémoires de l'Académie des sciences* for 1748, 57–109; *Recherches sur les causes particulières des phénomènes électriques* (Paris, 1749; 2nd ed., 1754); *Lettres sur l'électricité*, 3 vols. (Paris, 1753–1767); "Nouvelles expériences d'électricité faites à l'occasion d'un ouvrage publié depuis peu en Angleterre, par M. Robert Symmer," in *Mémoires de l'Académie des sciences* for 1761, 244–258; *L'art de faire les chapeaux* (Paris, 1765); and *L'art des expériences ou avis aux amateurs de la physique*, 3 vols. (Paris, 1770; 3rd ed., 1784). Nollet published a great many papers in the volumes of the Paris Academy; the content of most of them appears in his books, the chief exception being the reports of his Italian trip published in the Academy's *Mémoires* for 1749 and 1750.

The best bibliography of both Nollet's works and secondary literature is in J. Torlais, *Un physicien au siècle des lumières, l'abbé Nollet 1700–1770* (Paris, 1954), 251–262. Less complete is the entry in *Nouvelle table des articles contenus dans les volumes de l'Académie royale des sciences de Paris depuis 1666 jusqu'en 1770* (Paris, 1775); Poggendorff is quite inadequate.

The most important MS remains are letters to Étienne-François Dutour (1711–1789), a corresponding member of the Academy and Nollet's staunchest supporter; the correspondence, which covers 25 years and includes drafts of Dutour's replies, is preserved at the Burndy Library, Norwalk, Connecticut. The MS of Nollet's Italian travel diary is in the Bibliothèque Municipale, Soissons.

II. SECONDARY LITERATURE. Information about Nollet's career may be collected from his books, from his dossier at the Académie des Sciences, from the *éloge* by Grandjean de Fouchy in *Histoire de l'Académie . . . des sciences* for 1770 (1771), 121–137, and from the published correspondence of Bergman, Buffon, Mme du Châtelet, Franklin, Montesquieu, Réaumur, and Voltaire. The best biography is the work of Torlais cited above; see also his "Une grande controverse scientifique au xviiie siècle, l'abbé Nollet et Benjamin Franklin," in *Revue d'histoire des sciences*, **9** (1953), 339–349; "Une rivalité célèbre, Réaumur et Buffon," in *Presse médicale*, **66**, no. 2 (1958), 1057–1058; and *Un esprit encyclopédique en dehors de "l'Encyclopédie."* *Réaumur d'après les documents inédits*, 2nd ed. (Paris, 1961). Important additional data is given by R. Hahn, *The Anatomy of a Scientific Institution, The Paris Academy of Sciences, 1666–1803* (Berkeley, 1971), esp. 108–110; V. Lecot, *L'abbé Nollet de Pimprez* (Noyon, 1856); and G. H. Quignon, *L'abbé Nollet, physicien. Son voyage en Piémont et en Italie* (Amiens, 1905). For details about Nollet's physics see the works of Torlais; J. A. Sigaud de la Fond, *Précis historique et expérimental des phénomènes électriques* (Paris, 1781); J. C. Poggendorff, "Über die Entdeckung der Diffusion tropfbarer Flüssigkeiten," in *Annalen der Physik*, **139** (1884), 350–351; and the unfriendly Franklinist histories, such as J. Priestley, *The History and Present State of Electricity*, 3rd ed., rev. (London, 1775); and I. B. Cohen, *Franklin and Newton* (Philadelphia, 1956).

JOHN L. HEILBRON

NORDENSKIÖLD, (NILS) ADOLF ERIK (*b.* Helsinki, Finland, 18 November 1832; *d.* Dalbyö, Sweden, 12 August 1901), *geography, geology, mineralogy, history of cartography.*

Nordenskiöld came from a distinguished family of soldiers, administrators, and scientists who, originally Swedish, had long been settled in southern Finland. He was educated at the University of Helsinki, where his outspoken liberalism brought him into conflict with the Russian administration of the country; he was therefore compelled to leave Finland shortly after his graduation. In 1858 he departed for Sweden, where a reputation based upon his first publications in mineralogy had preceded him. In the same year, at the age of twenty-six, he was appointed chief of the mineralogy division of Sweden's National Museum, a post that he held for the rest of his life.

Nordenskiöld's career as an Arctic explorer had begun even earlier. In 1857 he made his first Arctic voyage, accompanying Otto Torrell to Spitsbergen. He either participated in or led four more voyages to Spitsbergen in the course of the next fifteen years, as well as leading eight other expeditions between 1864 and 1886. His explorations culminated in the voyage of the *Vega*; this expedition, carried out under his command in 1878–1879, penetrated the seas north of Asia to reach the Pacific, thus achieving the long-sought northeastern passage to the Orient. (For this accomplishment, King Oscar of Sweden created Nordenskiöld a baron.)

Nordenskiöld was responsible for making scientific work an integral part of Arctic exploration. The expeditions that he conducted were distinguished by careful planning; scientific equipment was meticulously prepared and a well-qualified staff selected to aid in the collection of data and observations. Nordenskiöld himself contributed to the extensive series of papers that resulted from these voyages; that of the *Vega* was reported in five volumes that dealt with the zoological, botanical, geodetic, geomagnetic, geophysical, oceanographic, and anthropological aspects of the regions investigated. These volumes marked the beginning of serious polar studies.

Nordenskiöld also wrote on a wide range of subjects within his chief fields of interest, geology and mineralogy. Many of his descriptive publications remain valuable but, for the most part, his theoretical contributions are now of only historical interest. He did more important work in the history of science. He was interested in Swedish science of the eighteenth century from an early age, and the preparations for his Arctic voyages led him to study historic maps. He published a fundamental work on Scheele, then the two magnificent folio volumes, published simultaneously in Swedish and English, that laid the foundations of the history of cartography. These were *Facsimile-atlas to the Early History of Cartography* (1889) and *Periplus—An Essay on the Early History of Charts and Sailing Directions* (1897).

The *Facsimile-atlas* is a survey of map making, from the Alexandrine cartographer Ptolemy to the beginnings of scientific surveying and mapping in the seventeenth century. *Periplus*, its companion piece, offered a collection of historically important charts and documents, assembled for the first time. While there had been attempts to compose histories of cartography earlier in the nineteenth century—most notably the works of Santarém and Jomard—they had not met the criteria of careful, truly scientific inquiry; it was Nordenskiöld who first applied the critical approach of the historian to this field of study.

Nordenskiöld exerted considerable influence on two generations of Scandinavian natural scientists, offering them ample field experience and generous support. He helped to establish the study of the earth sciences and promoted polar research in northern Europe, and was persuasive in urging others to report their findings in these fields.

BIBLIOGRAPHY

The basic bibliography of Nordenskiöld's complete works was published in *Ymer*, **21**, no. 2 (1902), 277–302. The same issue contains extensive accounts of Nordenskiöld's life and statements on his work as a polar explorer, geologist, mineralogist, and historian of geography and cartography. His successor at the Swedish National Museum, H. Sjögren, published a detailed memorial of Nordenskiöld's scientific accomplishments in *Geologiska Föreningens i Stockholm Förhandlingar*, **34** (1912), 45–100. Popular biographies were published by Sven Hedin, *Adolf Erik Nordenskiöld—en levnadsbeskrivning* (Stockholm, 1926); and Henrik Ramsay, *Nordenskiöld Sjöfararen* (Stockholm, 1950). An extensive biography in English is George Kish, *Northeast Passage: Adolf Erik Nordenskiöld, His Life and Times* (Amsterdam, 1973).

GEORGE KISH

NORDENSKIÖLD, NILS ERIK

NORDENSKIÖLD, NILS ERIK (*b.* Frugård, Nyland [now Uusimaa], Finland, 23 November 1872; *d.* Stockholm, Sweden, 28 April 1933), *zoology, history of biology*.

Nordenskiöld belonged to a well-known Swedish-Finnish family which had for generations produced outstanding government officials, military men, and scientists. He grew up on the family estate in southern Finland and in the 1890's studied biology at the University of Helsinki. He did graduate work at Padua and Leipzig and in 1899 was appointed lecturer in zoology at Helsinki, where he taught invertebrate anatomy until 1915. His zoological studies, some of which were done at foreign universities and marine biological stations, were devoted almost entirely to the systematics, anatomy, histology, and spermatogenesis of the Acarina (especially hydrachnids and ticks).

Nordenskiöld's most important contribution were in the history of biology. His series of lectures at the University of Helsinki (1916–1917) constituted an extensive survey of the development of biology and was published in Swedish in three volumes as *Biologiens historia* (1921–1924). This work, which was soon translated into German and English, received great international acclaim. Grounded on basic studies in the botanical and zoological sources, it shows as a rule sound judgment in its final analyses, though Nordenskiöld shows, by treating Darwin's theory of natural selection as obsolete and discredited, that he was dependent on the general evaluation of the biologists of his generation. Throughout his work he always considers the relation of biological theories to philosophy and general cultural development.

In 1917 Nordenskiöld moved to Sweden and became a citizen. In 1926 he was appointed lecturer in the history of zoology at the University of Stockholm. He became a member of the Finnish Academy of Science and Letters in 1908.

BIBLIOGRAPHY

Nordenskiöld's *Biologiens historia* was published in German (Jena, 1926) and in English as *The History of Biology* (New York, 1928; new ed., 1935; London, 1929).

A secondary source is Tor Carpelan and L. O. T. Tudeer, *Helsingfors universitet. Lärare och tjänstemän från år 1828*, II (Helsinki, 1925), 669–670, and supp. II (Helsinki, 1940), 586–587.

STEN LINDROTH

NORMAN, ROBERT

NORMAN, ROBERT (*fl.* England, late sixteenth century), *navigation, magnetism*.

Little is known of Norman, an English instrument maker of the late sixteenth century, other than that he was for a considerable time a sailor and later had a house at Radcliffe, where he sold navigational instruments. At a time when sailing and the construction of good compasses were of primary importance, Norman established his reputation not only as a maker of superior instruments but also as one interested in their irregularities.

In making magnetic compasses, Norman noticed that the needle did not remain parallel to the earth's surface but that the north-seeking pole dipped toward the earth. He constructed his compasses with a wax counterbalance on the south-seeking pole to counteract this dip; when, by accident, he found that the attached wax did not serve as an equalizer if the needle was shortened, he became interested in the theory of the phenomenon.

In *The Newe Attractive* (1581), a treatise on the lodestone, Norman discussed the known properties of the magnet; suggested that the orientation of the compass was due to its turning toward, rather than its being attracted to, a certain point; and related his newly discovered deviation of the needle from the horizontal. He measured this deviation to be 71°50′ at London and was interested in finding its value at other points on the earth's surface.

This work appears to have been well known and was one of the few writings on magnetism favorably referred to by William Gilbert. In *De magnete* (1600), Gilbert credited Norman with the discovery of the dip of the magnetic needle and suggested that this property could be used to measure latitude on the earth's surface.

Norman also published *The Safegarde of Saylers* (1590), a book of sailing directions which he translated from Dutch.

BIBLIOGRAPHY

Norman's writings are *The Newe Attractive, Containying a Short Discourse of the Magnes or Lodestone, and Amongst Other His Vertues, of a Newe Discovered Secret and Subtill Properties, Concerning the Declinying of the Needle Touched Therewith Under the Plaine of the Horizon* (London, 1581); and *The Safegarde of Saylers* (London, 1590).

SUZANNE KELLY

NORTON, JOHN PITKIN (*b.* Albany, New York, 19 July 1822; *d.* Farmington, Connecticut, 5 September 1852), *agriculture, agricultural chemistry.*

Norton was encouraged by his father, John Treadwell Norton, a wealthy Connecticut farmer, to study "scientific" farming. To this end, Norton attended the lectures given by Benjamin Silliman, Sr., Denison Olmsted, and other faculty members at Yale College from 1840 to 1842, although he never formally matriculated. In addition he learned experimental chemistry and mineralogy in Benjamin Silliman, Jr.'s private laboratory. In 1842 and 1843 Norton enrolled for lectures at the Harvard Law School and attended many scientific lectures in Boston. From 1843 to 1844 he again enrolled in the younger Silliman's laboratory, where his progress was so rapid and his interest in "scientific" agriculture so great that the Sillimans arranged for him to spend two years, from 1844 to 1846, in Scotland with James F. W. Johnston, whose work in agricultural chemistry was well known in the United States. While in Scotland, Norton won a prize of £50 from the Highland Agricultural Society of Scotland for his essay on the chemical constitution of oats.

Together with the two Sillimans, Norton devised a plan for professorships in agricultural and practical chemistry at Yale College. After initial reluctance from the Yale Corporation, this plan was approved on 19 August 1846, although formal instruction in these sciences did not begin until 1 November 1847. These professorships, in what was known informally as the Yale School of Applied Chemistry, evolved into the Sheffield Scientific School. Norton, named professor of agricultural chemistry, was probably the first in the United States to hold such a special position. He was also awarded an honorary M.A. by Yale College at this time, his only academic degree. Following his election to this professorship, Norton spent the winter of 1846 and the spring of 1847 in Gerardus Johannes Mulder's chemistry laboratory in Utrecht, analyzing plant proteins.

In addition to his full schedule of lectures and laboratory instruction at Yale College, Norton campaigned vigorously throughout the Northeast for a new scientific approach to agriculture and agricultural education. In keeping with this emphasis he organized laboratory instruction at the School of Applied Chemistry around analytical chemistry, believing that accurate soil analysis was essential to improved farming.

Norton was a well-trained chemist capable of research of high quality, as his work in Scotland and Utrecht showed. He is chiefly remembered for his inspirational leadership of the scientific farming movement in the United States, for his part in founding a leading American scientific institution, and for an excellent textbook on scientific farming. Many of his students made substantial contributions to scientific teaching and research, although only a handful remained in scientific agriculture. Of these the best-known were Samuel W. Johnson and William H. Brewer, both of whom later joined the faculty of the Sheffield Scientific School. Johnson became director of the Connecticut Agricultural Experiment Station (the first of its kind in the nation), and Brewer taught agricultural sciences.

BIBLIOGRAPHY

I. ORIGINAL WORKS. Norton's most important publication was *Elements of Scientific Agriculture* (Albany, 1850). He wrote a series of articles for *Cultivator*, n.s. **1–9** (1844–1852), as well as for *American Agriculturist* (1844–1846). His major research papers were "On the Analysis of the Oat," in *American Journal of Science*, 2nd ser., **3** (1847), 222–236, 318–333; "Account of Some Researches on the Protein Bodies of Peas and Almonds, and a Body of a Somewhat Similar Nature Existing in Oats," *ibid.*, **5** (1848), 22–33; "On the Value of Soil Analysis, and the Points to Which Special Attention Should Be Directed," in *Proceedings of the American Association for the Advancement of Science* (1850), 199–206, written with William J. Craw.

The Yale Memorabilia Room in the Sterling Memorial Library at Yale University holds a sizable collection of Norton MS and printed material, including unpublished diaries and letters.

II. SECONDARY LITERATURE. There is an extensive, although short and fragmentary, literature on Norton. One of the most recent accounts is Louis I. Kuslan, "The Founding of the Yale School of Applied Chemistry," in

Journal of the History of Medicine and Allied Sciences, **24**, no. 4 (1969), 430–451. See also *Memorials of John Pitkin Norton* (Albany, 1853), a collection of contemporary periodical accounts of his work, which particularly stresses his deep religious faith; and Russell H. Chittenden, *History of the Sheffield Scientific School of Yale University*, I (New Haven, 1928), ch. 2. Margaret W. Rossiter, "Justus Liebig and the Americans" (Ph.D. dissertation, Yale University, 1971), will also be useful.

Louis I. Kuslan

NORWOOD, RICHARD (*b.* Stevenage, Hertfordshire, England, 1590; *d.* Bermuda, 1665), *mathematics, surveying, navigation.*

Norwood's family were gentlefolk who apparently had fallen upon hard times; he attended grammar school, but at the age of fifteen was apprenticed to a London fishmonger. The many seamen he met in London aroused his interest in learning navigation and seeing the world. Eventually he was able to switch his apprenticeship to a coaster plying between London and Newcastle. He tells in his *Journal* how, while forced to lay over for three weeks at Yarmouth, he went through Robert Record's treatise on arithmetic, *The Ground of Arts*. So involved was he in studying mathematics that he almost forgot to eat and caught "a spice of the scurvy." During the following years Norwood made several voyages to the Mediterranean and on his first trip was fortunate to find a fellow passenger with an extensive mathematical library, among which was Leonard Digges's *Pantometria*. On following trips Norwood himself took along mathematical books, including Euclid's *Elements* and Clavius' *Algebra*.

To retrieve a piece of ordnance that had fallen into the harbor at Lymington, Norwood devised a kind of diving bell, descended in it to the bottom, and was able to attach a rope to the lost piece. This exploit brought him to the attention of the Bermuda Adventurers, a company that planned to finance its colonization of Bermuda by exploiting the oyster beds that supposedly surrounded the islands. In 1616 Norwood joined them and sailed for Bermuda. It soon became evident that very few pearls were to be found, and Norwood was then offered the task of surveying the islands. He made several surveys between 1614 and 1617, and upon their completion he returned to London. In 1622 he married Rachel Boughton, and in the same year his map of Bermuda was published by Nathaniel Newbery. No copy of this map is now known to exist, but in 1624–1625 Samuel Purchas reprinted the Newbery version.

Upon his return to London, Norwood taught

mathematics and wrote a number of books on mathematics and navigation, which went through many editions. His *Trigonometrie, or, The Doctrine of Triangles* (1631), based on the logarithms of Napier and Briggs as well as on works by Wright and Gunter, was intended essentially as a navigational aid to seamen. In it Norwood explained the common logarithms, the trigonometrical functions, the spherical triangles, and their applications to the problems confronting the navigator. He posed practical problems of increasing complexity; his explanations were clear; and he enabled the navigator to determine his course with the aid of a plane or Mercator chart and the logarithmic and trigonometric formulas. He emphasized great circle navigation by giving the formulas involved and thus facilitated the calculations. In his *The Seaman's Practice* (1637), he set out a great circle course between the Lizard (the southernmost point in Great Britain) and Bermuda.

Norwood was the first to use consistently the trigonometric abbreviations s for sine, t for tangent, sc for sine complement, tc for tangent complement, and sec for secant.

The Seaman's Practice was especially concerned with the length of a degree and improvements in the log line. In 1635 Norwood measured the length of a degree along the meridian between London and York. His degree was 367,167 English feet, a surprisingly good measurement in view of the crude tools he used. Based on this volume, he reknotted the log line, putting a knot every fifty feet. Running this with a half-minute glass gave sixty sea miles to a degree.

Norwood was a convinced nonconformist, and because of Archbishop Laud's oppressive actions he decided to leave England. He returned to Bermuda in 1638 and established himself as a schoolmaster; planted olive trees and shipped olive oil to London; and made a new survey in 1663. He also corresponded with the newly founded Royal Society.

BIBLIOGRAPHY

I. Original Works. The British Museum has a copy of Norwood's chart of Bermuda, which, together with his *Description of the Sommer Islands*, was repr. in John Speed, *A Prospect of the Most Famous Parts of the World* (London, 1631). His other works include *Trigonometrie, or, The Doctrine of Triangles* (London, 1631); *The Seaman's Practice; Containing a Fundamental Problem in Navigation, Experimentally Verified* (London, 1637); *Fortification, or Architecture Military* (London, 1639); *Table of the Sun's True Place, Right Ascension, Declination, etc.* (London, 1657); and *A Triangular Canon Logarithmicall* (London, 1665[?]). *The Journal of Richard Norwood, Surveyor of Bermuda; With Introductions by Wesley F. Craven and*

Walter B. Hayward (New York, 1945). Norwood wrote this account of his early life when he was 49 years old, but it ends with the year 1620. It is concerned with his religious conversion. The intros. are excellent and the book also contains a biblio. of Norwood's writings (pp. lix–lxiv).

II. SECONDARY LITERATURE. Norwood's contributions to mathematics and navigation are extensively discussed in E. G. R. Taylor, *The Mathematical Practitioners of Tudor and Stuart England* (Cambridge, 1954); and David W. Waters, *The Art of Navigation in England in Elizabethan and Early Stuart Times* (London, 1958).

LETTIE S. MULTHAUF

NOSTRADAMUS, MICHAEL (latinized form of **NOSTREDAME, MICHEL DE**) (*b.* Saint-Rémy, France, 14 December 1503; *d.* Salon, France, 2 July 1566), *medicine, astrology.*

More than any other writer in modern times Nostradamus knew how to titillate the deep-seated craving, felt by potentate and plebeian alike, to foresee the future, near and remote.

After receiving his early education in the liberal arts at the University of Avignon, he proceeded to the University of Montpellier to study medicine. When a plague broke out in southern France, many of the local licensed physicians cravenly fled from the epidemic, whereas the student Nostradamus courageously enlisted in the struggle to combat it. After traveling about for four years in this intensive and dangerous effort, he returned to Montpellier when the pestilence abated and was officially matriculated on 23 October 1529. He was, however, labeled an apothecary, accused of having slandered doctors, and was struck from the list of students by Guillaume Rondelet, who was the procurator of students during that year.[1] Nevertheless, the jealous and hostile faculty was coerced into co-opting Nostradamus by strong pressure from a grateful populace and a student body eager to learn from his experience. Yet at this stage of his life he was not satisfied to settle down in the humdrum routine of a university professor of medicine, surrounded by unfriendly colleagues. In 1532 Nostradamus left Montpellier with no definite destination in mind.

During the course of his travels he was invited by a prominent intellectual, Julius Caesar Scaliger, to join his circle in Agen.[2] There Nostradamus married and became the father of two children. But when the Inquisition came to Agen, he deemed it prudent to leave. After the uproar subsided, he returned to Agen, only to have a recurrence of the plague wipe out the three members of his family.

Once more alone in the world, Nostradamus resumed the life of the wandering physician. When the plague again ravaged Aix-en-Provence, that stricken city persuaded him in 1546 to help fight the dread disease and, in gratitude for his labors, awarded him a pension for life. The following year he settled in Salon, a small town halfway between Avignon and Marseilles. On 11 November 1547 he married a wealthy widow, who bore him six children, of whom the eldest, César (perhaps named after Scaliger), became the first local historian of Provence. In the Salon cadastral survey of 1552 Nostradamus acknowledged acquisition of a house after his marriage.[3] The marked improvement in his financial situation freed him from the necessity of continuing his medical practice for the sake of the income. Nevertheless, on 23 September 1555 he was consulted at Salon by Felix Platter and some German fellow students from Montpellier, and on 20 October 1559 he gave medical advice to Bishop Laurent Strozzi at Béziers.[4] But for years he had devoted his major energies to an entirely different pursuit.

Wrapping himself in the mantle of the ancient Hebrew prophets, to whose religion his ancestors had adhered until his grandfather's compulsory conversion to Roman Catholicism,[5] and claiming divine inspiration for his astrological forecasts, Nostradamus dedicated the first edition of his *Prophecies* to his infant son César on 1 March 1555. This opening salvo also contained the first three centuries, or groups of 100 quatrains of rhyming iambic pentameters, plus century IV, quatrains 1–53. The numerous allusions to heavily veiled persons, places, and events were strewn about in no discernible arrangement, either chronological or geographical. These deliberately vague forebodings, promulgated in a France trembling on the verge of a religious civil war, were an instantaneous success. Nostradamus was promptly summoned to the capital in 1556 to cast the horoscopes of the royal children. Encouraged by such favorable responses and ignoring his harsh critics, Nostradamus published his first seven centuries in 1557 (I–VI and VII, 1–40), and on 27 June 1558 dedicated to King Henry II centuries VIII–X (issued posthumously in 1568).

In 1560 Pierre de Ronsard (1524–1585), prince of poets and poet of princes, aligned himself with Nostradamus:

> By the ambiguous words of his prophetic voice,
> Like an ancient oracle, he has for many years
> Predicted the greatest part of our destiny.
> I would not have believed him, had not Heaven,
> Which separates good from evil for humans,
> Been on his side.[6]

On 17 October 1564 the young King Charles IX sought out the seer at Salon. But the rationalistic philosopher Pierre Gassendi examined a horoscope cast by Nostradamus, his fellow Provençal, for the father of a personal friend and showed it to have been totally wrong in numerous details.[7]

By the same token, the adversaries and supporters of Nostradamus have continued until our time respectively to denounce him as a charlatan and to predict retrospectively such portentous crises as the French Revolution and World War II.

NOTES

1. Marcel Gouron, "Documents inédits sur l'Université de médecine de Montpellier (1495–1559)," in *Montpellier médical*, 3rd ser., **50** (1956), 374–375; and Gouron, ed., *Matricule de l'Université de médecine de Montpellier 1503–1599*, Travaux d'Humanisme et Renaissance, XXV (Geneva, 1957), 58.
2. Vernon Hall, "Life of Julius Caesar Scaliger," in *Transactions of the American Philosophical Society*, **40** (1950), 117.
3. Edgar Leroy, "Nostradamus, médecin de la Faculté de médecine de Montpellier," in *Histoire de la médecine*, **4** (Mar. 1954), 10, with a facs. of Nostradamus' oath of allegiance to the University of Montpellier on p. 7.
4. *Beloved Son Felix, the Journal of Felix Platter*, Seán Jennett, trans. (London, 1961), 107; Gouron, "Documents," pp. 375–377.
5. Paul Masson, *Dictionnaire biographique*, which is *Les Bouches-du-Rhône Encyclopédie départementale*, IV, pt. 2 (Paris–Marseilles, 1931), 357.
6. Pierre de Ronsard, *Oeuvres* (Paris, 1560), III, *Poèmes*, bk. V, "Élégie à Guillaume des Autels," ll. 184–188; repr. in Ronsard's *Oeuvres complètes*, Paul Laumonier, ed., 2nd ed., X (Paris, 1939), 359.
7. Pierre Gassendi, *Syntagma philosophicum*, pt. 2 (physics), sec. 2, bk. 6, ch. 5, in Gassendi's *Opera omnia* (Lyons, 1658; repr. Stuttgart–Bad Cannstatt, 1964), I, 745–746. The relevant passage was trans. into English in P. Gassendus, *The Vanity of Judiciary Astrology* (London, 1659), 139–141.

BIBLIOGRAPHY

See Edgar Leoni, *Nostradamus: Life and Literature* (New York, 1961), 77–89 for the original works and 89–101 for the secondary literature. See also H. Noll-Husum, "Nostradamus und die Astronomie," in *Vierteljahrsschrift der astronomischen Gesellschaft*, **71** (1936), 242–249; Nostradamus, *Interprétation des hiéroglyphes de Horapollo*, Pierre Rollet, ed. (Aix-en-Provence, 1968); Michel Chomarat, *Nostradamus entre Rhône et Saône* (Lyons, 1971); and Pierre Guérin, *Le véritable secret de Nostradamus* (Paris, 1971).

EDWARD ROSEN

NOVARA, DOMENICO MARIA (*b.* Ferrara, Italy, 1454; *d.* Bologna, Italy, 1504), *astronomy.*

As is indicated by Novara's surname (Novara or da Novara), that city in northwestern Italy had been

the home of his ancestors. One of them, however, had been invited to move eastward to Ferrara, where Domenico Maria was born.[1] Hence he was variously known as Maria (as Kepler always cited him), Novara (or da Novara), and Ferrariensis (of Ferrara).[2] In his own publications he usually called himself Domenico Maria da Novara of Ferrara.

In his publications Novara described himself as holding two academic degrees, Doctor of Arts and Doctor of Medicine. It is not yet known when and where he pursued these studies, but from 1483 to 1504 he taught at Bologna University.[3] As professor of astronomy,[4] he was, in addition to his teaching duties, required to publish a prognostication for every year. Such a slender and ephemeral forecast, of which only a relatively small number of copies was printed, has often perished without a trace. In Novara's case, however, his writings were available as late as 1619.[5] At present twelve of his twenty-one prognostications still survive.[6]

After the return of Columbus' crew from his first voyage to America, the outbreak of syphilis in southern Europe stimulated widespread discussion. According to a contemporary Bolognese writer, "The astrologers assert that the cause of this disease was the conjunction of Jupiter and Saturn on 9 November 1484, and they base this [date] on the very accurate observation of Professor Domenico Maria of Ferrara, this being the city where he was born but he has become a citizen of Bologna by virtue of his accomplishments and work."[7] The foregoing statement has been misunderstood to mean "The astrologers, particularly Dominicus Maria of Ferrara, attributed this new disease to the conjunction of 1484."[8] What the astrologers took from the professor of astronomy, however, was the date of the conjunction, not the etiology of syphilis.

In his prognostication published in 1489, Novara declared that the latitude of Cádiz and of places in Italy was found in his own time to exceed by 1°10′ the corresponding latitude reported in Ptolemy's *Geography*. Since this discrepancy occurred too often to be attributed to scribal error, Novara concluded that northern latitudes in general had been increasing imperceptibly since antiquity. This systematic displacement he ascribed to a gradual shift of the terrestrial north pole toward the zenith in a slow motion requiring 395,000 years to complete the circuit. Novara's thesis was quoted in Giovanni Antonio Magini's widely consulted planetary tables, from which it was repeated by William Gilbert, Willebrord Snel, Pierre Gassendi, and Giovanni Battista Riccioli.[9] While Novara's greatest pupil, Copernicus, did not accept his teacher's argument that the terrestrial pole had changed its direction, that mistaken view may nevertheless have encouraged him to doubt the traditionally asserted absolute immobility of the earth.[10]

Novara's tombstone was erected by one of the two heirs to whom he had bequeathed all his modest worldly goods in the absence of a wife, children, and servants.[11]

NOTES

1. Lorenzo Barotti, ed., *Memorie istoriche di letterati ferraresi*, II (Ferrara, 1793), 26–27.
2. Johannes Kepler, *Gesammelte Werke* (Munich, 1937–   ), II, 135:29–30; VII, 147:12; XIII, 114:63; XIV, 16:347, 26:191, 27:219, 55:515, 347:218, 352:389; XV, 308:94; XVII, 339:13, 353:82; Kepler, *Opera omnia*, Christian Frisch, ed., VIII (Frankfurt–Erlangen, 1871), 235:25. The "Ferrariensis" who is cited three times in Galileo's student papers is someone other than Domenico Maria Novara of Ferrara, whom Galileo discussed in a marginal note in his copy of William Gilbert's *Magnet*: Galileo Galilei, *Opere*, nat. ed. (Florence, 1890–1909; repr. 1968), I, 32:6, 76:33, 105:27; VIII, 625.
3. Umberto Dallari, ed., *I rotuli dei lettori legisti e artisti dello studio bolognese dal 1384 al 1799*, I (Bologna, 1888), 121–185.
4. Not astrology, as in Lynn Thorndike, *A History of Magic and Experimental Science*, V (New York, 1941), 234. The name of the regular course was changed from astrology to astronomy a decade before Novara was born: Dallari, I, 18, 21; with the single exception of 1463–1464, see I, 64.
5. Kepler, *Gesammelte Werke*, XVII, 339:9–12.
6. Gustav Hellmann, "Versuch einer Geschichte der Wettervorhersage im XVI. Jahrhundert," in *Abhandlungen der Preussischen Akademie der Wissenschaften*, Phys.-math. Kl. (1924), no. 1, 34.
7. Bartholomeus Cocles, *Chyromantie ac physionomie anastasis* (Bologna, 1504), bk. VI, ch. 248, sig. T2r.
8. Thorndike, *op. cit.*, V, 62–63.
9. More recently by Curtze, "Ueber . . . Schriften . . . Ferrara," 519–520; by Boncompagni, "Sopra alcuni scritti . . . Ferrara," 146–148; and by Antonio Favaro, ed., *Carteggio inedito di Ticone Brahe, Giovanni Keplero . . . con Giovanni Antonio Magini* (Bologna, 1886), 80–81.
10. Edward Rosen, *Three Copernican Treatises*, 3rd ed. (New York, 1971), 323.
11. The tombstone no longer survives, and a transcription of it inadvertently postponed Novara's death by 10 years in Roman numerals, an error corrected by Silvestro Gherardi, *Di alcuni materiali per la storia della facoltà matematica nell'antica Università di Bologna* (Bologna, 1846), 37–38, offprinted from R. Accademia delle scienze dell'Istituto di Bologna, *Nuovi annali delle scienze naturali*, 2nd ser., 5 (1846), 161–187, 244–268, 321–356, 401–436, and trans. into German by Maximilian Curtze, in *Archiv der Mathematik und Physik*, 52 (1871), 106–107. According to the transcription of the tombstone, Novara died on 1 Sept., whereas the university's payroll records report his death on 17 Aug. and on 20 Aug.—Carlo Malagola, *Della vita e delle opere di Antonio Urceo* (Bologna, 1878), 350–351. The inventory of Novara's bequeathed property was found and published by Lino Sighinolfi, "Domenico Maria Novara e Nicolò Copernico allo Studio di Bologna," in *Studi e memorie per la storia dell'Università di Bologna*, 5 (1920), 213–215, 235.

BIBLIOGRAPHY

Novara's writings are listed and discussed in the following (presented in chronological order): Maximilian Curtze, "Ueber einige bis jetzt unbekannte gedruckte Schriften des Domenico Maria Novara da Ferrara," in *Altpreussische Monatsschrift*, 7 (1870), 515–521; Baldassarre Boncompagni, "Sopra alcuni scritti stampati, finora non conosciuti, di Domenico Maria Novara da Ferrara," in *Bullettino di bibliografia e di storia delle scienze matematiche e fisiche*, 4 (1871), 140–149, 340–341; Domenico Berti, *Copernico e le vicende del sistema copernicano in Italia* (Rome, 1876), 34–42, 179–184; Gustav Hellmann, *Beiträge zur Geschichte der Meteorologie*, in Veröffentlichungen des K. Preussischen Meteorologischen Instituts no. 296 (Berlin, 1917), 217; and Pietro Riccardi, *Biblioteca matematica italiana*, enl. ed., II (Milan, 1952), 205–207.

See also Luigi Napoleone Cittadella, "Domenico Maria Novara," in *Buonarroti*, 11 (1876), 157–163; Ferdinando Jacoli, "Intorno alla determinazione di Domenico Maria Novara dell'obliquità dell'eclittica," in *Bullettino di bibliografia e di storia delle scienze matematiche e fisiche*, 10 (1877), 75–88; Paul J. Melchior, "Sur une observation faite par Copernic et Dominique Maria," in *Bulletin de l'Académie r. de Belgique. Classe des sciences*, 5th ser., 40 (1954), 416–417; and Edward Rosen, "Copernicus and His Relation to Italian Science," forthcoming under the auspices of the Accademia dei Lincei.

EDWARD ROSEN

NOVY, FREDERICK GEORGE (*b.* Chicago, Illinois, 9 December 1864; *d.* Ann Arbor, Michigan, 8 August 1957), *microbiology.*

Novy's father, a tailor, and his mother, a milliner, emigrated from Bohemia to the United States in 1864. A high school teacher stimulated Novy's interest in chemistry, and he received a bachelor's degree in that subject from the University of Michigan in 1886. Following graduation he remained at Michigan— where he was to spend his entire career—to work as an assistant to the organic chemist Albert Prescott and to pursue graduate studies. He received a master's degree in 1887 for his research on cocaine and its derivatives. In that year the direction of his interests began to shift from organic chemistry to physiological chemistry and bacteriology when he accepted an instructorship in the department of hygiene and physiological chemistry, headed by Victor Vaughan. He continued his graduate work, receiving the D.Sc. in 1890 and a medical degree in 1891. He was promoted to assistant professor in the latter year and to junior professor in 1893. In 1902 he became professor and chairman of the newly founded department of bacteriology, a post that he held until his retirement in 1935. He also served as dean of the medical school from 1933 to 1935.

Novy's strong commitment to truth and to meticulous scientific work was immortalized in the person of Max Gottlieb in Sinclair Lewis' novel *Arrowsmith*. Paul de Kruif, one of Novy's students, served as a technical advisor to Lewis on the book and helped to create the character of Gottlieb, the dedicated scientist, who represented a blend of Novy and Jacques Loeb. Honors received by Novy during his lifetime included membership in the National Academy of Sciences and the American Philosophical Society and honorary degrees from the University of Cincinnati and the University of Michigan. He was married in 1891 and was the father of five children.

Novy was one of the pioneers in bacteriology in the United States. He and Vaughan spent their vacation in 1888 at Robert Koch's Berlin laboratory, learning the techniques and concepts of the new science of bacteriology. In January 1889 they instituted at Michigan a course that may well have been the first systematic laboratory instruction in bacteriology offered at an American medical school, altho gh lectures and occasional experiments in the subject iad apparently entered the medical curriculum of some American universities by that time. The course, which consisted of three months of intensive laboratory work, was so successful that it was made a required part of the medical curriculum in 1890. Novy was also one of the charter members of the Society of American Bacteriologists, founded in 1899, and served as its president in 1904.

Novy's early work in microbiology dealt with the toxic products produced by bacteria. In 1888 he collaborated with Vaughan on a book on this subject which expressed the view that pathogenic bacteria cause disease by decomposing complex substances in the body to produce poisonous alkaloids. By the fourth edition of the work (1902), the authors had adopted a view more in accord with current thought: that the bacterial toxins involved in disease are usually complex proteins which are synthesized by the microorganisms. They still overemphasized, however, the importance of toxins in infectious diseases. For example, the symptoms of anthrax and pneumonia were assumed to be due to toxins produced by the bacteria involved, whereas the ability of these and many other bacteria to produce disease actually appears to be due to their invasiveness (their ability to invade tissues and spread and multiply).

Novy devoted a significant amount of attention to anaerobic bacteria and developed apparatus for the cultivation and study of these organisms, such as the Novy jar, an anaerobic culture method in which the air in the jar is removed by a vacuum pump and replaced by an inert gas such as nitrogen. In 1894 he discovered and isolated the organism now known as *Clostridium novyi*, a species of gas gangrene bacillus.

Novy is probably best known for his extensive studies on trypanosomes and spirochetes, and he was apparently the first to cultivate a pathogenic protozoan (the trypanosome) in an artifical culture medium. *Spirochaeta novyi*, the organism that causes the American variety of relapsing fever, was discovered in his laboratory in 1906.

Among Novy's other research contributions were his studies in microbial respiration (especially on the respiration of the tubercle bacillus) and his investigation of anaphylaxis.

BIBLIOGRAPHY

I. ORIGINAL WORKS. For bibliographies of Novy's publications, see Esmond Long, "Frederick George Novy," in *Biographical Memoirs. National Academy of Sciences*, **33** (1959), 342–350; and S. E. Gould, "Frederick George Novy, Microbiologist," in *American Journal of Clinical Pathology*, **29** (1958), 305–309. For his views on bacterial toxins, see *Ptomaines and Leucomaines, or the Putrefactive and Physiological Alkaloids* (Philadelphia, 1888), written with V. C. Vaughan. The 4th ed., which was considerably revised, is entitled *Cellular Toxins, or the Chemical Factors in the Causation of Disease* (Philadelphia–New York, 1902). For a review of his work on trypanosomes, see his "On Trypanosomes," *Harvey Lectures*, **1** (1905–1906), 33–72. On spirochetes, see "Relapsing Fever and Spirochetes," in *Transactions of the Association of American Physicians*, **21** (1906), 456–464, written with R. E. Knapp. His most important studies on anaphylaxis were reported in a series of papers in *Journal of Infectious Diseases*, **20** (1917). Two important papers entitled "Microbic Respiration" appeared *ibid.*, **36** (1925), 109–232.

II. SECONDARY LITERATURE. Two substantial biographical articles about Novy have been cited above: Long, pp. 326–350; and Gould, pp. 297–309. There is list of eight biographical sketches in Genevieve Miller, ed., *Bibliography of the History of Medicine of the United States and Canada, 1939–1960* (Baltimore, 1964), 80–81. See also Thomas Francis, Jr., "Frederick George Novy, 1864–1957," in *Transactions of the Association of American Physicians*, **71** (1958), 35–37; and the article on Novy in *National Cyclopedia of American Biography*, XVI (1918), 93. Paul de Kruif, *The Sweeping Wind: A Memoir* (New York, 1962), makes several references to Novy—including pp. 93–94, 96, 102–103, 109—and discusses the aspects of Novy's character that were portrayed in the person of Max Gottlieb in Sinclair Lewis' *Arrowsmith*.

JOHN PARASCANDOLA

NOYES, ARTHUR AMOS (*b*. Newburyport, Massachusetts, 13 September 1866; *d*. Pasadena, California, 3 June 1936), *chemistry*.

Noyes's father, Amos Noyes, was an able and scholarly lawyer. One of his forebears, Nicolas Noyes, had come from England in 1633 and had settled in the town (then called Newbury) in 1635. Noyes's mother, Anna Page Andrews Noyes, was interested in literature, especially poetry. After her husband's death in 1896 she became a close companion to her son, who never married.

As a boy Noyes carried out chemical experiments at home. When he graduated from high school he found that he could not attend the Massachusetts Institute of Technology because of lack of money. At home he studied all of the first-year subjects except drawing and was able to enter the sophomore class at M.I.T. the following year, when he was granted the Wheelright Scholarship, which had been established for Newburyport students. He received his bachelor's degree in 1886, with a thesis on the action of heat on ethylene. He continued his research in organic chemistry, and after receiving the M.S. in 1887 he was appointed assistant in analytical chemistry. During this period he became a close friend of one of his students, George Ellery Hale, who was later to play an important part in his life.

In the summer of 1888 Noyes, accompanied by two other M.I.T. graduates in chemistry, went to Europe for advanced study in organic chemistry under Adolf von Baeyer at Munich. On their arrival in Rotterdam they received word that there would be no space for them in Baeyer's laboratory, and Noyes elected Leipzig as the alternative. There Wilhelm Ostwald had just begun to present lectures in the new subject of physical chemistry, and Noyes became interested in this field. He carried out an investigation of deviations from van't Hoff's laws of perfect solutions, for which he received his doctorate in 1890. On his return to M.I.T. he was for a number of years engaged in teaching analytical chemistry, organic chemistry, and physical chemistry. Noyes wrote a book on each of these subjects: *A Detailed Course of Qualitative Chemical Analysis* (1895), following a preliminary edition, *Notes on Qualitative Analysis* (1892); *Laboratory Experiments on the Class Reactions and Identification of Organic Substances* (1898), written with S. P. Mulliken; and *The General Principles of Physical Science* (1902). His textbook on qualitative analysis, which has gone through many editions, was widely used and of great importance in introducing concepts of physical chemistry into that field. His first book on physical chemistry was later expanded, with the collaboration of Miles Sherrill, into a textbook,

at first entitled *The General Principles of Chemistry* and in later editions *A Course of Study in Chemical Principles*, which has been of much value in bringing precision into the teaching of this subject in the United States. A characteristic of *Chemical Principles* was the use of problems so phrased as to lead the student to derive the basic equations. These two books have been described as revolutionizing the teaching of analytical chemistry and physical chemistry in America.

One of Noyes's important contributions to chemistry, carried out with many collaborators, was his thorough study of the chemical properties of the rarer elements and the development of a complete system of chemical analysis including these elements. This work, which extended over a period of twenty-five years, was summarized in *A System of Qualitative Analysis for the Rare Elements* (1927), written with W. C. Bray.

Noyes was one of the first chemists to surmise that the large deviations from unity of the activity coefficients of ions might be ascribed to the interaction of the electric charges of the ions. He carried out extensive studies of the properties of solutions of electrolytes, over a wide range of temperatures and pressures. Around 1920 this work culminated in the testing of the theory of electrostatic interactions of ions that was proposed by S. R. Milner in 1911 and by P. Debye and E. Hückel in 1923.

In 1903 Noyes became director of the Research Laboratory of Physical Chemistry at M.I.T., which was set up under a provision that half of the support would be provided by Noyes himself. He was director of this laboratory for sixteen years. He also served as acting president of M.I.T. for two years, beginning in 1907.

In 1913, at the request of George Ellery Hale, Noyes became associated on a part-time basis with the California Institute of Technology (then called Throop College of Technology), and in 1919 he resigned his post at M.I.T. and moved to California. During the remaining years of his life he devoted himself to developing the California institution into a great center of education and research in science and engineering. He and Hale, who was a member of the board of trustees, succeeded in bringing the physicist Robert Andrews Millikan from Chicago to Pasadena to develop the physics program and to serve as chief administrative officer of the Institute.

Noyes was a very good chemist. He diligently carried on research throughout his life and made some significant discoveries. But he was a great teacher of chemistry, and it is as a teacher of chemistry that he will be long remembered. He believed that students of chemistry should be introduced to research as early as possible. He was always on the watch for "carefully selected seeds," and he was a good judge of young people. In Boston he had been fond of sailing, and he made trips on his yacht with young friends. In Pasadena this interest was largely replaced by camping. He had a large touring car, and he liked to drive with the top down. It was his custom in the 1920's to invite new graduate students in chemistry to go with him on a camping trip to the desert, or to stay for a day with him in his beach house. These trips gave him an opportunity to size them up. The time was spent partly in enjoying nature and partly in discussions of scientific interest. In the evening he would often recite poetry at length, with evident pleasure and enthusiasm. He was also fond of tennis.

Noyes's personality was reserved, but he was not at all withdrawn from the general activities of the California Institute of Technology nor of American scientists as a whole. He never sought publicity and was rarely mentioned publicly in connection with innovations or changes in policy that led to the progress of the California Institute of Technology, although he was often the one who was responsible for the policies. It seems likely that Noyes was primarily responsible for the emphasis on pure rather than applied science, the limitation of the number of undergraduate students to 160 (later 180) per annual class, and the emphasis on the humanities and on undergraduate, graduate, and postdoctorate research.

In 1895 Noyes founded a journal, *Review of American Chemical Research*, which in 1907 became *Chemical Abstracts*. He was president of the American Chemical Society in 1904—the youngest man ever to hold office. During World War I he served as chairman of the National Research Council, an organization set up through the efforts of Noyes, Hale, and Millikan to aid the National Academy of Sciences in advising the government on scientific questions. He was president of the American Association for the Advancement of Science in 1927; and he was awarded the Humphry Davy Medal by the Royal Society in 1927, the Willard Gibbs Medal by the Chicago Section of the American Chemical Society in 1915, and the Theodore William Richards Medal by the Northeastern Section of the American Chemical Society in 1932 (first recipient). He was a member of a number of scientific societies.

Despite his reserved personality, which was perhaps due to shyness, Noyes had a great influence on students. He inspired them by his own unselfish devotion to science, his high principles, and his idealism, which was sometimes expressed in poetic selections that he read in class. He believed in the importance of a broad basic education. He strove to discover the most talented among his students as early as possible, and to encourage them by the provision of special instruction and other opportunities for rapid growth, such as scholarships permitting summer travel in Europe. His estate was left to the California Institute of Technology for the support of research in chemistry.

The qualities of Noyes that impressed themselves most strongly on his associates were his gentlemanliness, integrity, and unselfishness. His effectiveness in his work is attested by the great number of able scientists who came under his influence and received part of their training from him.

BIBLIOGRAPHY

A bibliography of Noyes's writings is given in Linus Pauling, "Arthur Amos Noyes, a Biographical Memoir," in *Biographical Memoirs. National Academy of Sciences*, **31** (1958), 322–346.

Other biographical notices are Frederick G. Keyes, "Arthur Amos Noyes," in *Nucleus* (Boston) (Oct. 1936), 28–33; R. A. Millikan, "Arthur Amos Noyes," in *Science*, **83** (1936), 613; and Miles S. Sherrill, "American Contemporaries: Arthur Amos Noyes," in *Industrial and Engineering Chemistry*, **23** (Apr. 1931), 443; and "Arthur Amos Noyes (1866–1936)," in *Proceedings of the American Academy of Arts and Sciences*, **74** (1940), 150–155.

LINUS PAULING

NOYES, WILLIAM ALBERT (*b.* Independence, Iowa, 6 November 1857; *d.* Urbana, Illinois, 24 October 1941), *chemistry.*

The youngest son of Spencer W. and Mary Packard Noyes, William Albert grew up in a farm environment, which did not lend itself to the study of chemistry. Although he enrolled at Grinnell College in classical studies, he read chemistry on the side and earned both the A.B. and B.S. degrees in 1879. He continued at Grinnell, teaching and studying analytical chemistry until January 1881, when he entered Johns Hopkins to study with Ira Remsen. In June 1882 he received not only the Ph.D. from Johns Hopkins, for work on benzene oxidation with chromic acid, but also an A.M. from Grinnell.

Noyes spent a year at Minnesota as an instructor and then, in 1883, went to the University of Tennessee as professor of chemistry. He married Flora Collier in December 1884. The couple had three children— Ethel and Helen, who both died in early childhood, and William Albert, Jr. His first wife died, and in 1902 Noyes married Mattie Elwell; they had one son,

Charles Edward. In 1886 Noyes began a seventeen-year career at the Rose Polytechnic Institute, Terre Haute, Indiana, where most of his work on camphor derivatives, especially camphoric acid, was performed. In 1889 he spent several months in Munich at the laboratory of Adolf von Baeyer.

In 1903 Noyes left the Institute to become chief chemist at the National Bureau of Standards, where he was engaged in atomic weight determinations. Burning hydrogen over palladium in pure oxygen and weighing the resultant water, he obtained a value of 1.00787:16 for the critical hydrogen:oxygen weight ratio, which still stands as one of the most precise chemical determinations ever made.

In 1907 Noyes became director of the chemical laboratories at the University of Illinois. He held this post until his retirement in 1926. Noyes married his third wife, Katherine Macy, in 1915; they had two sons, Richard Macy and Henry Pierre.

Besides his determination of the hydrogen:oxygen ratio, Noyes studied the structure of camphor and its derivatives and conducted early applications of the valence theory.

Noyes published his own numerous works and edited the papers of his colleagues while serving as editor of the *Journal of the American Chemical Society* from 1902 to 1917. He was the first editor of the following publications: *Chemical Abstracts* (1907–1910), *Chemical Reviews* (1924–1926), and the *American Chemical Society Scientific Monographs* (1919–1941).

BIBLIOGRAPHY

I. ORIGINAL WORKS. The majority of Noyes's papers appeared in *Journal of the American Chemical Society* between 1900 and 1941; his earlier works appeared in *American Chemistry Journal*. A key paper illustrating his researches in camphor chemistry is "Confirmation of Bredt's Formula. Some Derivatives of Inactive Camphoric Acid," in *American Chemistry Journal*, **27** (1902), 425, written with A. Patterson. His more important books are *Elements of Qualitative Analysis* (first published privately in 1887; 5th ed., 1926); *A Textbook of Organic Chemistry* (New York, 1913); and *Modern Alchemy* (Springfield, 1932), written with W. A. Noyes, Jr.

II. SECONDARY WORKS. Two excellent biographical sketches are Austin M. Patterson, "William Albert Noyes," in *Science*, **94** (1941), 477–479; and B. S. Hopkins, "William Albert Noyes," in *Journal of the American Chemical Society*, **66** (1944), 1045–1056, which includes a bibliography.

GERALD R. VAN HECKE

NUMEROV, BORIS VASILIEVICH (*b.* Novgorod, Russia, 17 January 1891; *d.* 19 March 1943), *astronomy, gravimetry.*

Numerov graduated in 1909 from the Novgorod Gymnasium and entered the faculty of physics and mathematics at St. Petersburg University. On graduating in 1913 he remained in the department of astronomy to prepare for a scientific career. In 1913–1915 he was a supernumerary astronomer at Pulkovo Observatory, where he observed on the zenith telescope. From 1915 to 1925 Numerov was astronomer-observer at the University's astronomical observatory. In 1924 he was appointed professor of practical astronomy, higher geodesy, and the technology of computation at the university and professor of mathematics at the Mining Institute.

In 1919 Numerov organized the Computation Bureau, the aim of which was to compile an astronomical yearbook. The following year a subdivision was established, the State Computation Institute (in 1924 renamed the Leningrad Astronomical Institute and now the Institute of Theoretical Astronomy of the Soviet Academy of Sciences). From 1920 through 1936 Numerov directed the Institute. In 1926–1928 he was also director of the Leningrad geophysical observatory. From 1931 to 1935 he also headed the section of applied mathematics of the State Optical Institute. In 1929 he was elected corresponding member of the USSR Academy of Sciences. In 1934 he received a doctorate in physical and mathematical sciences.

In 1930–1934 Numerov headed the Astronomical Committee of the People's Commissariat of Education, created to plan and organize astronomical institutions in the Soviet Union and to coordinate their work. Its successor was the Astronomical Council of the USSR Academy of Sciences. In connection with the work of the committee, Numerov traveled to Holland, France, England, Germany, and the United States, visiting astronomical observatories and observing geophysical methods of prospecting for useful minerals. In 1920–1926 Numerov was president of the Russian Astronomical Society.

Numerov's scientific career was devoted to practical astronomy and astrometry, celestial mechanics, and gravimetry. He was notable in Soviet astronomy for having organized the construction and manufacture of gravimetric and astronomical instruments and equipment. For this purpose in 1928 he created a mechanical workshop at the Leningrad Astronomical Institute and, later, a construction bureau that produced a number of new and improved gravimeters and the first Soviet telescope—a reflector with a thirty-two-centimeter mirror installed at the first mountain astronomical observatory in the Soviet Union, at Abastumani (now the Abastumani Astrophysical Observatory of the Academy of Sciences of the Georgian S.S.R.). In 1931, under the presidency of Numerov, the Commission of Astronomical Instruments was created in the All-Union Cooperative of Optical-Mechanical Production. It laid the foundation for the industrial manufacture of large astronomical instruments.

Numerov's new program and method (1916) of analyzing zenith telescope observations was used in determining variations in latitude and was later adopted at Pulkovo. He developed a complete theory of the zenith telescope and introduced formulas for the influence of instrumental errors, proposed a new method of studying the forms of pivots of transit instruments, and developed a theory of universal and photographic transit instruments. At the beginning of the 1920's Numerov organized the compilation and publication of astronomical yearbooks, necessary for the observatories and numerous expeditions; later the Astronomical Institute also compiled and published *Morskoi ezhegodnik* ("Marine Yearbook") and *Aviatsionny ezhegodnik* ("Aviation Yearbook"). In the astronomy of ephemerides, Numerov developed useful tables and charts for computing geographical and Gauss-Kruger rectangular coordinates.

Numerov's new method of computing planetary perturbation was widely used in compiling the annual reference book founded by Numerov, *Efemeridi malykh planet* ("Ephemerides of Asteroids"), which acquired an international reputation. In 1923, for large-scale computation and improvements in the calculation of the orbits of asteroids, he proposed an original and effective method of integrating differential equations of celestial mechanics (the method of extrapolation). The application of this method allowed the computation in 1930 of a new and very precise ephemerides of the eighth satellite of Jupiter. After 1923 the satellite was not sighted again until it was rediscovered on 22 November 1930 by astronomers at the Lick Observatory in California.

Numerov gave a theoretical basis to the analysis of star catalogs by means of observational data on asteroids, and he proposed an original plan for international cooperation in determining the constants that characterize star catalogs. This plan was approved by the International Astronomical Union in 1935 and is now used in working the catalog of faint stars.

Numerov introduced into practice the pendulum gravimeter and the variograph for studying the upper layers of the earth's crust in geological prospecting. He participated in about ten gravimetrical expeditions to the Urals, the Donets Basin, the Kazakh S.S.R., and other areas, testing the new instruments developed under his direction: a light quarter-second pendulum apparatus, a half-second pendulum apparatus, a gravitation torsion balance with three levers, and many others. Numerov's plan for a general gravimetrical survey of the Soviet Union provided extremely valuable results.

BIBLIOGRAPHY

I. ORIGINAL WORKS. Numerov's earlier works include "Nouveau programme pour le zénith-télescope," in *Izvestiya Pulkovskoi observatorii*, **7**, no. 1 (1916), 1–20; "Teoria universalnogo instrumenta" ("Theory of the Universal Instrument"), in *Astronomichesky ezhegodnik na 1923* ("Astronomical Yearbook for 1923"; Petrograd, 1923), app. 3, 239–272; "Novy metod opredelenia orbit i vychislenia efemerid s uchetom vozmushcheny" ("A New Method for Determining Orbits and Computing Ephemerides That Takes Into Account Perturbations"), in *Trudy Glavnoi rossiiskoi astrofizicheskoi observatorii*, **2** (1923), 188–288; "A Method of Extrapolation of Perturbations," in *Monthly Notices of the Royal Astronomical Society*, **84** (1924), 592–601; "Chislennoe integrirovanie uravneny nevozmushchennogo dvizhenia v polyarnykh koordinatakh" ("Numerical Integration of Equations of Unperturbed Motion in Polar Coordinates"), in *Byulleten Astronomicheskogo Instituta*, no. 2 (1924), 7–107; "Résultats du calcul des éphémerides et des perturbations approchées des coordonnées rectangulaires de 99 planètes pour l'époque 1921–1925," in *Izvestiya Glavnoi astronomicheskoi observatorii v Pulkove*, **10**, no. 94 (1924), 58–155; "Teoreticheskie osnovania primenenia gravimetricheskogo metoda v geologii" ("Theoretical Bases of the Application of the Gravimetric Method in Geology"), in *Izvestiya Geologicheskogo komiteta*, **14**, no. 3 (1925), 331–347; "Calcul des éphémerides pour une excentricité arbitraire," in *Journal des Observateurs*, **7** (1926), 125–130; "Berechnung der gestörten Ephemeriden nach der Extrapolationsmethode," in *Byulleten Astronomicheskogo Instituta*, no. 12 (1926), 109–120; "Hilfstafeln zur Bahnbestimmung und gestörten Ephemeridenrechnung nach der Extrapolationsmethode," *ibid.*, no. 13 (1926), 121–152; *Programma sposoba Talkotta dlya opredelenia shiroty* ("Program of the Talcott Method for Determining Latitude"; Leningrad, 1927); and "Zavisimost mezhdu mestnymi anomaliami sily tyazhesti i proizvodnymi ot potentsiala" ("The Relation Between the Local Anomalies in the Force of Gravitation and the Derivatives of Potential"), in *Doklady Akademii nauk SSSR*, ser. A (1929), 101–105, and in *Zeitschrift für Geophysik*, **5**, no. 2 (1929), 58–62.

Later works include "Gravitatsionny variometr s tremya rychagami" ("A Gravitational Torsion Balance With Three Levers"), in *Byulleten Astronomicheskogo Instituta*, no. 30 (1931), 103–108; "K voprosu opredelenia sistematicheskikh oshibok skloneny fundamentalnykh zvezd" ("On the Problem of Determining Systematic Errors in the Declination of Fundamental Stars"), *ibid.*, no. 32 (1932), 139–147; "Konstruirovanie i izgotovlenie astronomicheskikh priborov" ("The Construction and Manufacture of Astronomical Instruments"), in *Astronomia* ("Nauka v SSSR za 15 let") "Astronomy" ("Science in the USSR During 15 Years"); (Moscow–Leningrad, 1932), 207–215; "Svetosilny fotografichesky meridianny krug" ("An Efficient Photographic Meridian Circle"), in *Astronomichesky zhurnal*, **12** (1935), 349–355, and in *Doklady Akademii nauk SSSR*, **3** (1935), 201–204; "Primenenie metoda ekstrapolirovania k tochnomy vychisleniyu vozmushchennogo dvizhenia malykh planet" ("The Use of the Method of Extrapolation for the Exact Computation of the Perturbational Motion of Asteroids"), in *Astronomichesky zhurnal*, **12** (1935), 455–475; "K voprosu o sovmestnom opredelenii popravok elementov planety i Zemli" ("On the Problem of the Simultaneous Correction of the Elements of a Planet and of the Earth"), in *Astronomichesky zhurnal*, **12** (1935), 584–593, and in *Astronomical Journal*, **45**, no. 12 (1936), 105–111; "K voprosu ob opredelenii geoida na osnovanii gravitatsionnykh nablyudeny" ("On the Problem of Determining the Geoid on the Basis of Gravitational Observations"), in *Astronomichesky zhurnal*, **12**, no. 1 (1935), 47–59; "K voprosu opredelenia sistematicheskikh oshibok zvezdnykh polozheny" ("On the Problem of Determining Systematic Errors in Stellar Positions"), in *Doklady Akademii nauk SSR*, **2** no. 7 (1935), 451–457, in *Astronomichesky zhurnal*, **12**, no. 4 (1935), 339–348, and in *Journal des Observateurs*, **18**, no. 4 (1935), 57–64, in French; "K voprosu o postroenii fundamental'nogo kataloga slabykh zvezd" ("On the Problem of Compiling a Fundamental Catalog of Faint Stars"), in *Doklady Akademii nauk SSSR*, **12** (1936), 261–263, and in *Astronomische Nachrichten*, **260** (1936), 305–322; "Ob opredelenii figury geoida na osnovanii nablyudeny sily tyazhesti" ("On Determining the Figure of the Geoid on the Basis of Observations of the Force of Gravity"), in *Doklady Akademii nauk SSSR*, **12** (1936), 265–268, written with D. N. Khramov; "On the Problem of the Stability of the Motion of Trojans," in *Byulleten Astronomicheskogo Instituta*, no. 41 (1936), 1–4; and "Absolute Perturbations of Polar Coordinates of Asteroids From Outer Planets," *ibid.*, no. 42 (1937), 37–57, in English and Russian.

II. SECONDARY LITERATURE. On Numerov and his work, see S. I. Seleshnikov, in *Astronomichesky kalendar* for 1966 (Moscow, 1965), 211–214; and N. S. Yakhontova, "Boris Vasilievich Numerov, 1891–1943," in *Byulleten Instituta teoreticheskoi astronomii*, **9**, no. 3 (1963), 213–215, with portrait.

P. G. KULIKOVSKY

NUÑEZ SALACIENSE, PEDRO (*b.* Alcácer do Sol, Portugal, 1502; *d.* Coimbra, Portugal, 11 August 1578), *mathematics, cosmography.*

Nuñez's parents are believed to have been Jewish, since he was registered as a "new Christian." He was married at Salamanca in 1523 to Giomar de Arias, daughter of a Spanish Christian, Pedro Fernández de Arias; they had six children. The earliest information on his education places him as an independent student at the University of Salamanca in 1521 and 1522. He moved to Lisbon in 1524 or 1525, at which time he received a bachelor's degree in medicine while simultaneously extending his knowledge of mathematics and studying astrology. This excellent preparation served as a basis for his appointment as royal cosmographer on 16 November 1529. In recognition of his abilities as a practical researcher, he was named on 4 December 1529 to the professorship of moral philosophy at the University of Lisbon, then to the chair of logic (15 January 1530); during 1531 and 1532 he also held the chair of metaphysics. At the same time Nuñez was pursuing his own studies, and on 16 February 1532 he graduated as licentiate in medicine from the University of Lisbon.

The professorship of mathematics at Lisbon was moved to Coimbra in 1537; and on 16 October 1544 Nuñez was named to the post, which he occupied until his retirement on 4 February 1562. On 22 December 1547 he was named chief royal cosmographer and fulfilled the duties of the office until his death.

Nuñez was called to court on 11 September 1572 by his former student Sebastian, grandson of John III. He remained in Lisbon for two years as adviser for the projected reform of weights and measures, which was promulgated in 1575. He was also appointed professor of mathematics for the instruction of pilots, navigators, and cartographers. After the reform of weights and measures he returned to Coimbra, where he remained until his death.

Considered the greatest of Portuguese mathematicians, Nuñez reveals in his discoveries, theories, and publications that he was a first-rate geographer, physicist, cosmologist, geometer, and algebraist. In addition to works in Portuguese (*Tratado da sphera*), he wrote and published several works in Latin so that his discoveries might be utilized by educated people of other nations. His writings are rigorously scientific and usually contain a profusion of drawings and figures so that they may be understood more easily.

Among Nuñez's students in Lisbon were the brothers of John III, Louis and Henry, the latter the future king and cardinal. While at Coimbra he taught Clavius, known as the sixteenth-century Euclid. Also among his outstanding students were Nicolas Coelho de Amaral, who succeeded Nuñez in his professorship; Manuel de Figueredo, who became chief royal cosmographer; and João de Castro, viceroy of India, and one of the greatest Portuguese navigators.

Nuñez made important contributions in the design of instruments. In astronomical observations the impossibility of precisely measuring small portions of an arc was an impediment, and to overcome this difficulty he conceived the idea of the nonius. In its original form this instrument, consisting of forty-four concentric auxiliary circles, was attached to an astrolabe for measuring fractions of a degree. Upon each circle and upon their quadrants were equal divisions, ranging from eighty-nine on the circle of greatest diameter to forty-six on the circle of least diameter. Each circle had one division less than the one outside it and one division more than the one inside, making it possible to take a reading from the circle that gave the most accurate approximation.

This instrument has not been modified during the four centuries since it was devised, but it has been refined. In 1593 Clavius reduced the auxiliary circles to one divided into sixty-one parts and divided the limb of the astrolabe into sixty; and in 1631 Pierre Vernier let the auxiliary arc move freely by attaching it to the alidade of the astrolabe. (The latter variation is called a vernier in some countries.) With the nonius exceedingly small measures may be read on any scale or system of division, either circular or rectilinear.

As a navigator Nuñez made a significant discovery based on observations reported to him in 1533 by Admiral Martim Afonso de Sousa. They relate to rhumb line sailing and to great circle sailing. The former is the course of the ship while sailing on a single bearing (always oblique to the meridian in the direction of one and the same point of the compass), subsequently (1624) called "loxodrome" by Willebrord Snell. The latter, which is the shortest distance between any two terrestrial points, has been called "orthodrome"; in it the bearing varies. Until that time pilots had considered them equivalent; but Nuñez demonstrated their dissimilarity, an important discovery that exerted great influence on the making of charts for navigation. For this purpose he conceived and drew curved rhumb lines (1534–1537), several years before Mercator made a loxodromic terrestrial globe with rhumb lines for eight sea routes in each quadrant, drawn from various points in different latitudes (1541).

Another of Nuñez's contributions to navigation was his technique for determining latitude by means of two readings of the sun's altitude and the azimuth, with solutions that were quite interesting and ingenious but of little practical use on shipboard; they relate more to the concerns of a scientist in the observatory than to the needs of a practical navigator and therefore have fallen into disuse.

In physics and seamanship Nuñez wrote a commentary on Aristotle's mechanical problem of propulsion by oars. It is a contribution to the geometry of motion —an attempt to determine, at each moment and in every circumstance, the deviation of the boat in relation to the oars.

Nuñez's cosmological theories relating to solar and lunar motions are important, as are his inquiries into the duration of day and night, the transformation of astronomical coordinates, and other problems concerning the motions of celestial bodies. He commented on the planetary theories of Georg Peurbach; worked on the problem of determining the duration of twilight; and solved the problem of afterglow or second twilight.

Nuñez also exhibited mathematical ability in geometry with his original solutions to the problems of spherical triangles. He demonstrated the errors made by Oronce Fine, professor at the Collège de France, in his attempt to solve three problems by means of ruler and compass: trisecting an angle, doubling a cube, and squaring a circle.

Finally, Nuñez was a poet; his highly regarded sonnets were collected and published by Joaquín Ignacio de Fraitas (Coimbra, 1826).

BIBLIOGRAPHY

I. ORIGINAL WORKS. *Tratado da sphera* (Lisbon, 1537) consists of three parts: (1) annotated translations by Nuñez from Sacrobosco's *Tractatus de sphaera*, writings on the theory of the sun and moon by Georg Peurbach, and the first book of Ptolemy's *Geography*; (2) two writings by Nuñez, a treatise on certain difficulties in navigation and a treatise in defense of his navigation chart and tables of the movements of the sun and its declination; (3) an epigram in Latin written to Nuñez by Jorge Coelho. The first part of this work was reprinted at Lisbon in 1911 and 1912, the second part in 1913, and a facs. ed. was published at Munich in 1915. There is an ed. of a French trans. prior to 1562, published in France. The Latin version, *Opera quae complectuntur, primum duos libros . . .* , was published at Basel in 1566 and in subsequent, much improved, eds. in 1573 and 1592. It is in this work that the theory of loxodromic curves is first set forth.

Other works are *De crepusculis liber unus* (Lisbon, 1542; 2nd ed., Coimbra, 1571), which treats the afterglow and the nonius; *Astronomici introductorii De Sphaera epitome* (n.p., n.d. [1543?]), with 12 folios thought to be an introduction to *Tratado da sphera*; *De erratis Promtii Orontii Finaei, regii mathematicarum Lutetice professoris* (Coimbra, 1546; 2nd ed. 1571); and *Libro de álgebra en arithmética y geometría* (Antwerp, 1567).

In *De crepusculis*, Nuñez mentions MS treatises, now believed lost, on the geometry of spherical triangles, on the astrolabe, on the geometrical representation of the sphere on a plane surface, on proportions in measurement, and on the method of delineating a globe for the use of navigators. Another MS mentioned is a work on the sea routes to Brazil. In catalog no. 508, item no. 15, of Maggs Bros. bookstore in London, there is a reference to "Codice de circa 1560 de Nunes (Pedro) y Vaz Fraguoso (Pedro)," containing the elements of navigation and routes to the East, which is believed to have been compiled by Vaz Fraguoso.

II. SECONDARY LITERATURE. See the following, listed chronologically: *Diccionario enciclopédico hispano-americano*, XIII (Barcelona, 1813), 1190–1198; Rodolfo Guimaräes, *Sur la vie et l'oeuvre de Pedro Nunes* (Coimbra, 1915); Luciano Pereira da Silva, *As obras de Pedro Nunes, sua cronologia bibliográfica* (Coimbra, 1925); and A. Fontoura da Costa, *Pedro Nunes (1502–1578)* (Lisbon, 1938); and *Quarto centenârio da publicaçao de Tratado de sphera de Pedro Nunes* (Lisbon, 1938).

J. M. LÓPEZ DE AZCONA

NUSSELT, ERNST KRAFT WILHELM (*b.* Nuremberg, Germany, 25 November 1882; *d.* Munich, Germany, 1 September 1957), *heat transfer, thermodynamics.*

Nusselt was the first significant contributor to the subject of analytical convective heat transfer. He completed his schooling at a time when the problems of heating and cooling in the increasingly high-performance power equipment of the early twentieth century finally demanded accurate analysis. For a century Fourier's mathematical theory of heat conduction in rigid media had provided the only analytical attack on the problem, but it was inadequate to predict the heat flux in a flowing fluid. In 1915 Nusselt cut the Gordian knot. Although analytical solutions to the appropriate fluid-flow equations were so intrinsically complicated that they had to await the more fundamental work of others, Nusselt used dimensional analysis to show, in a single stroke, the functional form that such solutions would have to take. He thus made it possible to generalize limited experimental data.

Nusselt was the son of Johannes Nusselt, a factory owner, and Pauline Fuchs Nusselt. He completed his early education in Nuremberg in 1900 and then enrolled at the Technische Hochschule in Munich to study mechanical engineering. After six semesters he transferred to the Technische Hochschule of Charlottenburg, in Berlin, where he completed his studies. He then returned to Munich and passed his mechanical engineering diploma examination there.

Nusselt began his studies toward a doctorate in mechanical engineering in Munich, and from 1906 through 1907 he served as an assistant to Oskar Knoblauch, who was also the teacher of another early heat transfer luminary, Ernst Schmidt. He completed the degree in August 1907, and from then until 1925 he moved about Germany from post to post. From September 1907 to June 1909 he was assistant to the well-known thermodynamicist Richard Mollier at the Technische Hochschule in Dresden. He then worked in the heat technology division of the Sulzer brothers' firm in Switzerland (1909–1911). He returned to the mechanical laboratory in Dresden in 1913 and held indefinite teaching appointments until 1917. From January 1918 through March 1919 he returned to industry and worked at the Badische Anilin- und Soda-Fabrik in Ludwigshafen. In April 1920 he was appointed professor at the Technische Hochschule in Karlsruhe. In 1925, Nusselt was named to the chair in theoretical mechanics at the Technische Hochschule in Munich. He retired from this post in 1952 and was succeeded by Schmidt.

Two of Nusselt's most important works were completed during his years in Dresden. His paper on the similitude of convective heat transfer, "The Basic Law of Heat Transfer" (1915), followed his earlier work on the thermal conductivity of insulating materials and some work with heat convection coefficients. The scope of his 1915 paper, however, was far broader; in this work he set up the dimensionless functional equations for both natural and forced convection. He thus reduced the large number of physical variables that appear in the boundary layer equations to the familiar dimensionless groups that today bear the names "Nusselt number," "Reynolds number," "Prandtl number," and "Grashof number." He also noted additional groups that are needed when physical properties vary or when the full equations of motion are used to define natural convection. It was thus possible for experimentalists to reduce limited data into these few parameters and to form simple empirical equations among them. Such correlations have, in most cases, preceded heat transfer theory down to the present day.

His other major contribution during this period was a paper entitled "The Film Condensation of Steam" (1916), in which he provided a clear-headed and simple description of the film condensation of any liquid by linearizing the temperature profile and ignoring inertia in the liquid. Subsequent efforts to refine this heat transfer prediction have failed to alter his numerical results, except for liquid metals and the most extreme heat fluxes.

Nusselt's later works branched into radiant heat transfer, combustion, and a variety of applications of heat transfer and thermodynamics to power equipment. In 1930 he provided an important description of the similarity between heat and mass transfer, and in 1934 and 1944 he published the first and second volumes, respectively, of a book on technical thermodynamics.

Nusselt was married on December 12, 1917, while teaching at Dresden, to Susanne Thurmer. The couple had two daughters and one son. Nusselt was an energetic man, strongly inner-directed, soft-spoken, and self-contained. He was an avid mountain climber throughout his life, and he appears to have equated the methodical assault of a mountain to the kind of assault a man should make on the problems that beset him. He brought this same kind of energy and concentration to his technical work. He was, however, circumspect and, perhaps, even cautious.

During the 1930's and 1940's German scientists made great advances in heat transfer. But Nusselt did not wield great influence within the peer group that controlled this field. It was probably not in his makeup to do so, and he is known to have suffered from a chronic internal ailment during these years. Although he was an exacting taskmaster with his students, he apparently lacked charisma and he was not a good lecturer.

In 1947 Nusselt's son, Dietrich, also a mountaineer, fell to his death on the east wall of the Riffelkopf in the Wetterstein Gebirge. Nusselt did little more in his remaining years, and upon his retirement he left the university completely and lived out his life in relative seclusion.

BIBLIOGRAPHY

I. Original Works. G. Lück and G. Kling (see below) both provide a bibliography of over 50 major works. Nusselt's most important writings include "Das Grundgesetz des Wärmeüberganges," in *Gesundheits Ingenieur*, **38** (1915), 872; "Die Oberflächenkondensation des Wasserdampfes," in *Zeitschrift des Vereines deutscher Ingenieure*, **60** (1916), 541, 569; "Wärmeübergang, Diffusion und Verdunstung," in *Zeitschrift für angewandte Mathematik und Physik*, **10** (1930), 105; and *Technische Thermodynamik*, 2 vols. (Berlin, 1934, 1944). Nusselt's autobiographical deposition for the American occupation force after World War II provides a wealth of personal detail.

II. Secondary Literature. Poggendorff, VIIa, 455, lists several biographical articles; the most extensive is G. Kling in *Chemie-Ingenieur-Technik*, **24** (1952), 597–608, which includes a bibliography of works by both Nusselt and his co-workers. G. Lück's article on Nusselt's retirement in *Gesundheits Ingenieur*, **74** (1953), 7–8, also provides a similar bibliography. *Allgemeine Warmetechnik*, **3** (1952), 161–163, includes a bibliography and a list of Nusselt's doctoral students and their theses.

JOHN H. LIENHARD

NUTTALL, THOMAS (*b*. Long Preston, near Settle, Yorkshire, England, 5 January 1786; *d*. Nut Grove Hall, near St. Helens, Lancashire, England, 10 September 1859), *botany, ornithology, natural history.*

Very little is known of the early life of Nuttall. A bachelor throughout his life, he was extremely reticent about his personal affairs. Through careful frugality while in America, he was able to make numerous field trips collecting botanical specimens.

His father, James Nuttall, married Mary Hardacre in January 1785. He died before Thomas was twelve years old, and his profession is unknown. The family was not prosperous, and at the age of fourteen Thomas was apprenticed to an uncle to learn the printing trade. At the conclusion of his apprenticeship, he sought other employment. In 1808 he sailed for Philadelphia, and shortly after his arrival in America, he became a friend of and plant collector for Benjamin Smith Barton.

With Barton's encouragement Nuttall began to take a serious interest in American flora, teaching himself the principles of botany. In 1809 he made two field trips, collecting botanical specimens for Barton. The next year Barton outlined and financed a more ambitious collecting program, which was designed to take Nuttall through hazardous Indian country into Canada. Unable to complete Barton's itinerary, Nuttall joined an expedition of John Jacob Astor's Pacific Fur Company. The English botanist John Bradbury was also a member of this party. Traveling up the Missouri River, the two Englishmen collected new species of plants from lands that were botanically unexplored. At the conclusion of the expedition, Nuttall sailed for England in the fall of 1811. The War of 1812 prevented his return to America until 1815.

Nuttall published the results of his first western trip in *The Genera of North American Plants, and a Catalogue of the Species, to the Year 1817* (1818). As the first comprehensive study of American flora, this work established his reputation as a botanist. Although he classified his plants by the Linnaean system, Nuttall nevertheless discussed the natural relationships of the different genera he described. He thus provided American naturalists with an introduction to the merits of A. L. de Jussieu's natural system of classification. *Genera* described many western species new to botany and helped to stimulate an interest in the study of the plant life of the western United States.

From 1818 to 1820 Nuttall journeyed west again, collecting plants on the Arkansas River in Indian territory. In May 1820 he presented a paper describing the geology and fossils of the Mississippi Valley to the Academy of Natural Sciences of Philadelphia. His memoir anticipated modern geological techniques of stratigraphical correlations by suggesting a similarity between the geological formations of America and Europe.

In late 1822 Nuttall received his first professional appointment when he was named curator of the botanic garden at Cambridge, Massachusetts, and lecturer in natural history at Harvard. He remained at Harvard for eleven years, occasionally absenting himself for collecting trips. For the use of his students, he published *An Introduction to Systematic and Physiological Botany* (1827). The second edition of this work introduced new materials on plant physiology; and in its descriptions of the cellular composition of plants, Nuttall partially anticipated Schleiden's cell theory. While in Cambridge, Nuttall developed an interest in ornithology and began to gather data for a guide to North American birds. Between 1832 and 1834 he published his only major ornithological study, *A Manual of the Ornithology of the United States and Canada*. This inexpensively priced study demonstrated Nuttall's intimate familiarity with the literature on the subject and his personal observations of birds in their natural habitats. One of the most original features of this work was his careful attempt to describe the songs of birds through syllabic patterns.

Claiming that he was "vegetating at Harvard," Nuttall desired to return to the virgin flora of the West. His discovery of many new species of plants on his Arkansas trip convinced him that the study of western plant life was still in its initial stages. Resigning his position at Harvard, he won a chance to go west once more when he joined Nathaniel Jarvis Wyeth's second expedition to Oregon in 1834. Nuttall invited the young ornithologist John Kirk Townsend to accompany this party. Arriving safely in Oregon, Nuttall was the first experienced botanist to have traveled across the continent collecting specimens. On the Pacific coast he gathered not only plants but also mollusks and crustaceans. After spending two winters in Hawaii collecting, he returned to Boston in September 1836.

Nuttall's remaining years in America (1836–1841) were spent primarily in Philadelphia, where he began to work up the recently acquired western specimens. He included some of his data on western plants in his contribution to Torrey and Gray's *Flora of North America*. His last major activity in America was the preparation of a three-volume appendix for a new edition of François André Michaux's *North American Sylva*. This appendix, which was also published

separately, contained extensive information on the sylva of the western United States.

In 1842 Nuttall returned to England; and except for a six-month visit to America in 1847–1848, he remained there until his death in 1859. His last years in England were not notable for any major botanical studies and were not scientifically productive. He did become interested in the rhododendrons of Assam, but published only a brief paper on the subject.

Nuttall's greatest scientific strength was his meticulous skill as a fieldworker and his detailed knowledge of plants in their native habitats. His taxonomy was at times marred by use of the Linnaean system of classification. Since he was forced by the circumstances of wilderness travel to collect specimens at random seasons, his data about the seasonal development of plants were often insufficient for correct taxonomic determination. Nuttall willingly shared specimens that he obtained on his expeditions with more skilled and specialized co-workers in other fields of natural history. He provided materials for further study to Audubon, Say, Pursh, Gambel, Torrey, Gray, and John Bachman. Nevertheless, Nuttall was the preeminent figure in the discovery of the flora of the American West.

BIBLIOGRAPHY

I. ORIGINAL WORKS. In addition to papers in *Transactions of the American Philosophical Society* and *Journal of the Academy of Natural Sciences of Philadelphia*, Nuttall published the following works: *The Genera of North American Plants, and a Catalogue of the Species, to the Year 1817* (Philadelphia, 1818); *An Introduction to Systematic and Physiological Botany* (Cambridge, Mass., 1827; 2nd ed., enl., 1830); *A Journal of Travels Into the Arkansa Territory, During the Year 1819* (Philadelphia, 1821), repr. as vol. XIII of Reuben G. Thwaites, ed., *Early Western Travels, 1748–1846* (Cleveland, 1905); *A Manual of the Ornithology of the United States and Canada: The Land Birds* (Cambridge, Mass., 1832) and *The Water Birds* (Boston, 1834); and *The North American Sylva*, 3 vols. (Philadelphia, 1842–1849).

II. SECONDARY WORKS. The only extensive study of Nuttall's life and career is Jeannette E. Graustein, *Thomas Nuttall Naturalist, Explorations in America 1808–1841* (Cambridge, Mass., 1967). Carefully documented, this volume contains numerous references to Nuttall's correspondence and scientific papers as well as contemporary biographical notices. Additional information is in Richard G. Beidleman, "Some Biographical Sidelights of Thomas Nuttall, 1786–1859," in *Proceedings of the American Philosophical Society*, **104**, no. 1 (Feb. 1960), 86–100; Jeannette E. Graustein, "Nuttall's Travels Into the Old Northwest, an Unpublished 1810 Diary," in *Chronica* botanica, **14**, nos. 1–2 (1950–1951), 1–85; and Francis W. Pennell, "Travels and Scientific Collections of Thomas Nuttall," in *Bartonia*, **18** (1936), 1–51.

PHILLIP DRENNON THOMAS

NYLANDER, FREDRIK (*b.* Uleåborg [now Oulu], Russia [now Finland], 9 September 1820; *d.* Contrexéville, Vosges, France, 2 October 1880), *botany, medicine.*

Nylander was the son of Anders Nylander, a merchant, and the former Margareta Magdalena Fahlander, and the great-grandson of Johan Nylander, bishop of Borgå (now Porvoo) and of Åbo (now Turku). In 1853 he married Ida Babette Hummel, of Frankfurt.

Nylander received his secondary education at the Gymnasium in Åbo and graduated in 1836. He matriculated at the University of Helsinki in the same year and was awarded a master's degree in 1840; he remained at the university to specialize in botany and medicine, taking his examination and receiving his candidate's degree in medicine in 1843. He was lecturer in botany at the University of Helsinki from 1843 to 1853. In 1844 he received his doctorate in botany. He spent 1843–1846 at the St. Petersburg Botanic Garden, during which time he became fluent in Russian. On his return he studied for several months at the University of Uppsala. In 1853 he received the M.D. from the University of Helsinki and was then appointed assistant to the municipal physician of Uleåborg. He became the municipal physician there in 1865, holding that post until his death.

Nylander was the first to study the flora of Finland critically. He made many botanical expeditions and published five important papers on the Finnish-Russian flora. He pioneered in the botanical exploration of the then almost unknown Kola Peninsula. With Johan Ångström (1813–1879) he explored eastern Finland, Russian Karelia to the White Sea, and Russian Lapland in the summer of 1843. The following summer they explored Russian and Norwegian Lapland.

Nylander later abandoned botanical pursuits to concentrate on medicine. In addition to administering health services in Uleåborg, he was active in the political life of Finland, being elected by the Socialist party to the House of Burghers of Uleåborg and representing the party in the Diet convened in 1872. Throughout his later life Nylander was a staunch promoter of the Finnish language for all official use and a strong partisan of Finnish autonomy. Reports of his appointment as professor by the city of Uleåborg in 1877 are unclear; the title must have been honorary, since the University of Oulu was not founded until 1959.

BIBLIOGRAPHY

Biographical notices include A. Oswald Kairamo, "Fredrik Nylander," in *Kansallinen Elamakerrasto*, IV (Porvoo, 1927), 251; Sextus Otto Lindberg, "Fredrik Nylander, 1820–1880," in *Meddelanden af Societas pro fauna et flora fennica*, **6** (1881), 260; N. J. S[cheutz], "Fredrik Nylander," in *Botaniska notiser* (1880), no. 6, 199; and Theodor Saelan, "Nylander, Fredrik," in *Acta Societatis pro fauna et flora fennica*, **43** (1928), 354–355, which contains a bibliography.

GEORGE H. M. LAWRENCE

NYLANDER, WILLIAM (*b.* Uleåborg [now Oulu], Russia [now Finland], 3 January 1822; *d.* Paris, France, 29 March 1899), *botany.*

Nylander, brother of Fredrik Nylander, was the son of Anders Nylander, a merchant, and Margareta Magdalena Fahlander. He never married. For much of his life he was a world authority on the identification of lichens.

Nylander graduated from the gymnasium in Åbo (now Turku) in 1839 and matriculated at the University of Helsinki the same year. He passed examinations as a candidate in philosophy in 1843 and continued his studies at the university, where he received the M.D. in 1847. He never established medical practice, and his interests thereafter were limited to natural history.

An ardent naturalist, Nylander traveled throughout Finland in 1847 and 1848, collecting plant and insect specimens. His early publications dealt with entomology, especially with the identification of Finnish ants and bees. In 1848 Nylander went to Paris, where he studied lichens at the Muséum d'Histoire Naturelle, under Charles Tuslane. During most of the following decade he published much about lichens, primarily their classification and identification, and his work was acclaimed in Europe and America.

In 1857 Nylander became the first professor of botany at the University of Helsinki. Unhappy with his treatment there, he resigned in 1863 and emigrated permanently to France, where he had neither academic affiliation nor gainful employment.

Through his abundant, if often trivial, publications Nylander became known as the one who had acquired the reputation of being able to identify lichens from any part of the world. Specimens that he identified became his personal property, and he subsequently amassed the world's richest and largest private lichen herbarium. In 1868 the French government awarded Nylander the Prix des Mazières for his contributions to lichenology. Somewhat earlier the Portuguese had conferred on him the Ordre du Christ. He was elected to honorary membership in learned societies and stood at the pinnacle of his career.

The decade following 1868 witnessed revolutionary discoveries about the origin and biology of lichens, concepts accepted internationally by leading scientists.

Nylander held to the earlier theory that the green cells in lichens were primitive prototypes of algae. In 1867 Schwendener proposed that the green cells were themselves true algae, parasitized and imprisoned by fungal hyphae, and that the two separate and unrelated organisms lived together by obligative symbiosis. This was proved by Rees in England (1871), by Bornet in France (1872), and by du Bary in Germany (1873). Summarily dismissing the new ideas, Nylander became one of a shrinking handful who held to the earlier but scientifically untenable view. His vitriolic abuse of fellow botanists in France and elsewhere closed the doors of many institutions, including the Muséum d'Histoire Naturelle, to Nylander. Most editors then denied him access to publication in their journals. He became a paranoid recluse who considered all who disagreed with him to be his enemies.

About 1879, in poor financial circumstances, Nylander made an agreement with the University of Helsinki whereby, in return for a lifetime annual pension of 1,200 francs, he would bequeath it his lichen herbarium, library, notebooks, and papers.

Nylander's full bibliography contains 314 papers. Less than a score of them, published before 1875, continue to be recognized as major contributions to botanical science; but so great was the impact he made during the first fifteen years of his professional life that he will always be counted as the dominant lichenologist of the mid-nineteenth century.

BIBLIOGRAPHY

I. ORIGINAL WORKS. Nylander's major scientific works include *Conspectus florae Helsingforsiensis* (Helsinki, 1852); "Essai d'une classification des lichens. I–II," in *Mémoires de la Société impériale académique des sciences naturelles de Cherbourg* (1854), 5–16 and (1855), 161–202; "Énumération générale des lichens, avec l'indication sommaire de leur distribution géographique," *ibid.* (1857), 85–146, 332–339; "Prodromus lichenographie Galliae et Algeriae," in *Actes de la Société linnéenne de Bordeaux*, **21** (1857), 249–467, also pub. as a separate vol. with the same title (Bordeaux, 1857); *Synopsis methodica lichenum omnium hucusque cognitorum, praemissa introductione lingua gallica*

tracta, 2 vols. (I, Paris, 1858–1860; II, Paris [1861], 1869), never completed; "Lichenes Scandinaviae," in *Notiser ur sällskapets pro fauna et flora fennica förhandlingar*, **5** (1861), 1–312; and "Lichenes Lapponiae orientalis," *ibid.*, n.s. **8** (1882), 101–192—a few preprints correctly dated 1866 are known.

II. SECONDARY LITERATURE. Biographical notices include Alphonse Boistel, "Le Professeur William Nylander," in *Revue générale de botanique*, **11** (1899), 218–237, with partial bibliography; Auguste Hue, "William Nylander," in *Bulletin. Société botanique de France*, **47** (1899), 152–165, with portrait; Thorgny Krok, "Nylander, William," in *Bibliotheca botanica suecana* (Uppsala–Stockholm, 1925), 559–560; and Theodor Saelan, "Nylander, William," in *Acta Societatis pro fauna et flora fennica*, **43** (1928), 355–379, with complete bibliography.

GEORGE H. M. LAWRENCE

OBRUCHEV, VLADIMIR AFANASIEVICH (*b.* Klepenino, Rzhev district, Tver [now Kalinin] guberniya, Russia, 10 October 1863; *d.* Moscow, U.S.S.R., 19 June 1956), *geology, geography*.

Obruchev was the son of Afanasy Aleksandrovich Obruchev, a personnel officer in the Russian army, and Paulina Hertner, the daughter of a German pastor. After attending elementary school in Brest, he graduated from the technical high school in Vilna (now Vilnyus), where he showed a special interest in geography and the natural sciences, especially chemistry. In 1881 he won admission to the St. Petersburg Mining Institute, from which he graduated in 1886. It was there that he first became strongly interested in geology.

Obruchev showed outstanding abilities during his first Transcaspian expedition (1886–1888). He was assigned the task of studying the Transcaspian depression, discovering the conditions of the mobility of quicksand in the regions where railroads were being constructed, seeking water-bearing levels in the sands, and making observations of the Tedzhen and Murgab rivers and the ancient Amu-Darya river bed. His study of the action of the wind as a geological agent inspired a lifelong interest in wind processes, particularly in the production of loess. Obruchev's interest in dynamic geology also dates from this expedition.

Contrary to the then prevalent opinion that the Transcaspian sands were of exclusively oceanic origin, Obruchev discovered convincing evidence that they were of triple origin—marine, continental, and fluviatile. Explaining the conditions under which immobile or slightly mobile sands become mobile, he suggested practical measures that were subsequently implemented for combating migrating sands.

166

In 1888 Obruchev accepted Mushketov's offer to go to eastern Siberia as staff geologist of the Irkutsk Administration of Mines, with supervision of a vast territory that comprised part of Irkutsk and Yenisey provinces, and the Yakutsk and Transbaikal regions. The geology of Siberia subsequently remained his main scientific topic. Obruchev's area of responsibility included the study of the geological structure of the area and distribution of useful minerals, especially gold, which was mined in the Olekma-Vitim district and the area surrounding Lake Baikal.

In 1889 Obruchev completed an expedition across the Pribaikal Mountains, studied mica deposits on the Slyudyanka River, looked for lapsis lazuli and lazurite in the Khamar-Daban range and graphite on the Baikal island of Olkhon, investigated mineral springs in the Nilova Desert, and prospected for brown coal on the banks of the Oka River (a tributary of the Angara). In 1890 and 1891 he inspected in detail the goldfields in the Olekma-Vitim basins. Goldfields in these districts—in contrast to those in other areas of Siberia—were generally covered with a coating of loose glacial deposits to a depth of 60–180 feet. These layers prevented the gold-bearing ones from rewashing and preserved their extraordinary richness.

Having confirmed Kropotkin's views on the preglacial history of the gold-bearing layers, Obruchev explained their origin not by the erosion of thick quartz lodes, as had been asserted, but from the gold disseminated in thin quartz veins and in pyrite, dispersed through certain layers of bedrock. The destruction of this rock *in situ*, at the bottom of valleys under river beds, endowed these deposits with their unique structure and enriched their gold, which was chemically extracted from pyrite. This explanation of the origin of gold deposits, given by Obruchev in 1900, has retained its importance.

Summarizing his predecessors' results in "Geologichesky ocherk Irkutskoy gubernii" ("A Geological Sketch of Irkutsky Province," 1890), Obruchev expressed his own views on the current question of the origin of the depression of Lake Baikal.

> When one stands on an elevation at the edge of the majestic depression of Baikal, it is impossible to agree with Chersky's opinion that this depression is the result of the combination of prolonged erosion and slow crustal folding. It is too deep and wide, and its slopes are too steep and precipitous. Such a depression could have been created only by faulting, and comparatively recently; otherwise its steep slopes would have been smoothed by erosion and the lake would have been filled with its products [*Moi puteshestvia po Sibiri* ("My Travels Through Siberia"; 1948), p. 35].

Obruchev's views on the origin of the Baikal depression were supported by Suess. Obruchev published more than thirty works during his four years with the mining administration. During the winters he worked in the eastern Siberian section of the Russian Geographical Society as director of affairs and curator of its museum.

From 1892 to 1895 Obruchev traveled through Mongolia and China as a member of the central Asian expedition. His work gained him a worldwide reputation as an explorer and geologist: the Russian Geographical Society awarded him the Przhevalsky Prize and the Great Medal of Constantine, and the Paris Academy of Sciences honored him for his contributions. His research was based on Suess's synthetic work on the geology of central Asia. From Kyakhta to Kuldja, Obruchev investigated the steppe, the Gobi Desert, and the quicksands of the Ordos Desert; traveled throughout the loess area of northern China; spent time in the Alashan range; investigated the Nan Shan and eastern Kunlun ranges; visited the shores of Lake Koko Nor; traveled through all the oases of Kansu Province; traced the course of the Edsin Gol River; crossed the mountainous southwestern region of the Gobi and central Mongolia; and thus extended Richthofen's research deep into central Asia to the north, northwest, and west. From along the eastern Tien Shan Mountains he came out into Kuldja. His two-volume diary of the expedition (1900–1901) has remained the only source material on certain areas of central Asia.

Through his work with the expedition Obruchev disproved Richthofen's ideas about the Tertiary Lake Khanka in central Asia, showing that the multicolored deposits of the Khanka suite are continental. In addition he noted that continental conditions had prevailed there since the Mesozoic Era. He introduced significant corrections and additions into Richthofen's theory of the formation and distribution of loess in China and central Asia. Contrary to Richthofen's views, Obruchev asserted that there is no loess in those depressions of central Asia that are part of the area of weathering and wind erosion. Obruchev considered that it was precisely from this area that the loess was carried by wind to the borders of central Asia, mainly into northern China, where it was deposited, preserving and smoothing the forms of the ancient topography.

Returning to Irkutsk in 1895 as head of a special mining party, Obruchev spent the next three years studying the geology of Selenga Dauria (western Transbaikalia) along the route of the main line of the Trans-Siberian railway, then under construction. The material gathered on this expedition formed the basis for the conclusions presented in his *Orografichesky i geologichesky ocherk Yugo-Zapadnogo Zabaykalya* ("Orographical and Geological Sketch of Southwestern Transbaikalia"), for which the Russian Academy of Sciences awarded him the Helmersen Prize.

Advancing new ideas about the tectonics of Siberia, Obruchev believed that the Transbaikal, composed of huge stretches of granites and crystalline slates, was part of the oldest dry land of Eurasia, "of the ancient shield of Asia." Around this skeletal nucleus, he believed, further growth of the continent had occurred in more recent periods, from the Paleozoic to the Quaternary. The concept of the "ancient shield of Asia," raised in the works of Ivan Chersky and Obruchev, was accepted by Suess. Obruchev continued to develop it throughout his life.

Obruchev also developed the idea of the origin of a series of large depressions of the Transbaikal, filled with Mesozoic and Cenozoic deposits. In his opinion these depressions are grabens, which appeared as a result of faulting of the rigid blocks of the ancient shield of Asia, and they are similar to the depressions of Lake Baikal. Obruchev subsequently developed a concept of the prime role of faulting in the formation of the surface and geological structure of Siberia.

From 1898 to 1901 Obruchev worked in St. Petersburg on the material from his expeditions to central Asia and Transbaikalia. In 1899 he studied geology in Germany, Switzerland, and Austria, where he became acquainted with Suess; and in 1900 he participated in the Eighth International Geological Congress in Paris. In 1901 he again returned to Siberia, to study the goldfields on the Bodaybo River.

In 1901, on Mushketov's recommendation, Obruchev was invited to the Tomsk Technological Institute to organize a department of mining and to teach general geology. An outstanding teacher (1902–1912), he also founded the Siberian school of geology.

Continuing his research expeditions, Obruchev traveled to the border regions of Dzungaria in the summers of 1905, 1906, and 1909, with the aim of clarifying the interplay of the Altay and Tien Shan mountain systems. In his opinion the distinguishing feature of the topography is that the mountain heights are remains of an ancient plateau, broken by faults. The flat peaks are horsts (tectonically raised blocks), while the hollows dividing them—the grabens—are zones of tectonic sinking along faults, filled in by large amounts of lake alluvium. Obruchev discovered traces in the mountains of two glaciations and established the presence of vertical zonation of vegetation.

After retiring from Tomsk in 1912, Obruchev settled in Moscow, continued writing up the results of his previous research, and conducted geological fieldwork by contract with private firms. In 1914 he traveled through the Russian Altay, and the following year his *Altayskie etyudy* ("Altay Studies") appeared; the second sketch, "O tektonike Russkogo Altaya" ("On the Tectonics of the Russian Altay"), is of special interest for his analysis of the views of previous investigators, notably Helmersen, P. A. Chikhachev, G. E. Shchurovsky, Karl Ritter, Cotta, Chersky, and Suess.

In Suess's presentation the Altay are folded mountains convex to the south, formed by tangential stresses of the earth's crust. Obruchev, however, after studying the adjoining Dzungaria, concluded that the recent topography was primarily the result of faulting:

> It was not ancient folds that caused this topography; they have long since been worn down and reduced to almost a plain. It was, rather, faults that turned the entire area into a combination of horsts and grabens. Such land cannot be called plicate; it was such in Paleozoic times but has long since lost its characteristic peculiarities; what now dominate here are more or less extensive stepped plateaux, broad plains, frequently arranged inconsistently with a stretch of Paleozoic sedimentary rock, of which the planed-off ends come to the surface [*Izbrannye trudy*, V (Moscow, 1963), 34].

On his trips through the Altay, Obruchev was especially concerned with relating the topography to the geological structure. He established the existence of a severe discrepancy in orographical maps in relation to the actual position of mountain ranges. In describing the topography, Obruchev counted three mountain chains: "The Russian Altay in its topography has little similarity to the system of narrow and long mountain chains of folded origin. It is, rather, an ancient plateau, a highland, broken down by faulting into more or less broad and long parts, frequently consisting of two or more ledges of different heights and divided by deep and wide fault valleys" (*ibid.*, p. 43).

Obruchev was close to the truth; according to the latest data the mountainous Altay is a complex block-fault structure, formed as a result of an arched uplift, faulting, and uneven vertical displacements of separate blocks of an ancient peneplained surface of folded Paleozoic formations. The uplift occurred at the end of the Tertiary and was especially forceful in the mid- and upper Tertiary periods. The raised parts of the peneplain form mountain systems; the lowered parts are the hollows between mountains. Obruchev's original treatment of the geomorphology of the Altay and Siberia was subsequently developed as an independent branch of science—neotectonics.

At the request of the Higher Council of National Economy, in 1918 Obruchev went to the Donets Basin to prospect for fire-clays and marls. Cut off from central Russia by the civil war, he was obliged to accept the post of professor of geology at the University of the Crimea, in Simferopol. In 1920 he returned to Moscow and the following year was appointed to the chair of applied geology at the Moscow Mining Academy. For the next eight years he taught advanced courses on ore deposits and field geology; his lecture material formed the basis for the texts *Polevaya geologia* ("Field Geology") and *Rudnye mestorozhdenia* ("Ore Deposits"). The former is the best-known Soviet handbook for beginning geologists and covers the entire work cycle of the geologist-prospector.

In 1928 Obruchev reported on Chinese loess to the All-Union Geological Congress in Tashkent, of which he was also president. Elected a member of the Soviet Academy of Sciences in 1929, he subsequently headed its Geological Institute and Committee for the Study of Permafrost. Working with materials gathered in China and Dzungaria, he began the compilation of the five-volume *Istoria geologicheskogo issledovania Sibiri* ("History of Geological Research in Siberia").

Obruchev retained a lifelong interest in the geography of Siberia, especially in the former glaciation of the northern region, already suggested by Kropotkin after the expedition of 1866. The formation of the topography of Siberia also occupied an important place in Obruchev's geographical works. Defending the necessity of the geomorphological regionalization of Siberia, Obruchev wrote in the first volume of *Geologia Sibiri* ("The Geology of Siberia," 1935) that earlier characterizations of its regions, given by Kropotkin, Chersky, and Suess, had become outdated and required "certain more or less essential changes and additions on the basis of new data." His delineation and characterizations of ten geomorphological regions have retained their importance.

Referring to the research of Hans Stille and W. H. Bucher, Obruchev wrote in "Molodost relefa Sibiri" ("The Youth Stage of the Topography of Siberia"), "At present, on the basis of the numerous investigations of the past decade, it is possible to assert with full justification that the topography not only of the ancient shield but of almost all Siberia is young and was formed by movements during the Tertiary and post-Tertiary periods that attained in places quite a substantial amplitude."

Obruchev also investigated the conditions of the origin and development of permafrost, its geographical distribution, and its influence on agriculture. For many years he was head of the Institute of Permafrost Management of the Soviet Academy of Sciences.

In 1937 Obruchev was head of the Soviet delegation to the Seventeenth International Geological Congress in Moscow, and in 1939 he became editor of the geological series of *Izvestiya Akademii nauk SSSR*. During World War II he was secretary of the Geological Sciences Section of the Soviet Academy of Sciences. A recipient of many Soviet medals and awards, he was corresponding member of the Royal Geographical Society and a member of the Russian Geographical Society, the American Geological and Geographical Societies, the Geological Society of China, the Hungarian Geographical Society, and the Deutsche Geophysikalische Gesellschaft.

A volcano in Transbaikalia, a glacier in the Mongolian Altay, a peak in the Russian Altay, and a steppe between the Murgab and Amu-Darya rivers are named for him. In 1941 a prize in his name was established for work on the geology of Siberia.

BIBLIOGRAPHY

I. ORIGINAL WORKS. Obruchev's published work comprises more than 2,000 pages of scientific works and more than 3,000 reviews for foreign journals, as well as teaching materials, classical works on geology and mining, and several science fiction novels. Only the basic works reflecting his scientific activity are given here.

His selected works were published as *Izbrannye trudy*, 6 vols. (Moscow, 1958–1964). His account of his journey completed at the request of the Imperial Russian Geographical Society was published as *Tsentralnaya Azia, Severny Kitay i Nan-Shian* ("Central Asia, Northern China, and the Nan Shan"), 2 vols. (St. Petersburg, 1901). Subsequent fundamental works include *Rudnye mestorozhdenia* ("Ore Deposits"), 2 vols. (Moscow–Leningrad, 1928–1929); *Istoria geologicheskogo issledovania Sibiri* ("History of Geological Research on Siberia"), 5 pts. (Moscow–Leningrad, 1931–1949); *Polevaya geologia* ("Field Geology"), 2 vols. (Moscow–Leningrad, 1927; 4th ed., 1932); and *Izbrannye raboty po geografii Azii* ("Selected Works on the Geography of Asia"), 3 vols. (Moscow, 1951). *V staroy Sibiri* ("In Old Siberia"; Irkutsk, 1958), contains articles, recollections, and letters from 1888 to 1955.

His science fiction works include *Plutonia. Neobychaynoe puteshestvie v nedra Zemli* ("Plutonia. An Extraordinary Journey to the Depths of the Earth"; Leningrad, 1924; repr., Moscow, 1958); *V debryakh Tsentralnoy Azii. Zapiski kladoiskatelya* ("In the Depths of Central Asia. Notes of a Treasure Hunter"), 3rd ed. (Moscow, 1955); and *Zemlya Sannikova ili poslednie onkilony* ("The Land of Sannikov or the Last Onkilons"; Moscow, 1958).

II. Secondary Literature. On Obruchev and his work, see *V. A. Obruchev* (Moscow–Leningrad, 1946), materials for a bibliography published by the Soviet Academy of Sciences; and the notice in *Bolshaya sovetskaya entsiklopedia* ("Great Soviet Encyclopedia"), 2nd ed., XXX, 390–392. See also A. N. Granina, "Deyatelnost V. A. Obrucheva v Vostochno-Sibirskom otdele Geograficheskogo obshchestva SSSR" ("The Career of V. A. Obruchev in the Eastern Siberian Section of the Geographical Society of the U.S.S.R."), in *Izvestiya Vsesoyuznogo geograficheskogo obshchestva*, **89**, no. 2 (1957), 123–130; L. G. Kamanin and B. A. Fedorovich, "V. A. Obruchev—issledovatel Sredney i Tsentralnoy Azii i Sibiri" ("V. A. Obruchev—Investigator of Middle and Central Asia and Siberia"), in *Voprosy geomorfologii i paleogeografii Azii* ("Questions of Asian Geomorphology and Paleogeography"; Moscow, 1955); E. M. Murzaev *et al.*, *Vladimir Afanasevich Obruchev. Zhizn i deyatelnost* (Moscow, 1959), on his life and work; and V. V. Obruchev and G. N. Finashina, *Vladimir Afanasevich Obruchev* (Moscow, 1965), with bibliography.

G. V. Naumov

OCAGNE, PHILBERT MAURICE D' (*b.* Paris, France, 25 March 1862; *d.* Le Havre, France, 23 September 1938), *mathematics, applied mathematics, history of mathematics.*

D'Ocagne was a student and then *répétiteur* at the École Polytechnique. He then became a civil engineer and a professor at the École des Ponts et Chaussées. In 1912 he was appointed professor of geometry at the École Polytechnique. He was elected to the Académie des Sciences on 30 January 1922.

Active both as researcher and teacher, d'Ocagne published a great many articles, mostly on geometry, in mathematical journals and in the *Comptes rendus . . . de l'Académie des sciences.* His name, however, remains linked especially with graphical calculation procedures and with the systematization he gave to that field under the name of nomography. Graphical calculation consists in the execution of graphs employing straight-line segments representing the numbers to be found. This discipline was reduced to an autonomous body of principles chiefly through the work of Junius Massau (1852–1909). Nomography, on the other hand, consists in the construction of graduated graphic tables, nomograms, or charts, representing formulas or equations to be solved, the solutions of which were provided by inspection of the tables.

The overwhelming majority of formulas and equations encountered in practice can be represented graphically by three systems of converging straight lines. By making a dual transformation on the nomograms d'Ocagne obtained nomograms on which the relationship among the variables consisted in the alignment of numbered points. Hence this type of nomogram is called an aligned-point nomogram.

In a pamphlet published in 1891 d'Ocagne presented the first outline of a rationally ordered discipline embracing all the individual procedures of nomographic calculation then known. Pursuing this subject, he succeeded in defining and classifying the most general modes of representation applicable to equations with an arbitrary number of variables. The results of all these investigations, along with a considerable number of applications, were set forth in *Traité de nomographie* (1899), which was followed by other more or less developed expositions. This material appeared in fifty-nine partial or entire translations in fourteen languages.

D'Ocagne retained a lifelong interest in the history of science and published many articles on the subject, some of which were collected.

BIBLIOGRAPHY

D'Ocagne published many articles in *Comptes rendus . . . de l'Académie des sciences, Revue de mathématiques spéciales, Nouvelles annales de mathématiques, Annales des ponts et chaussées, Bulletin de la Société mathématique de France, Enseignement mathématique, Mathésis,* and other journals. His books include *Nomographie, les calculs usuels effectués au moyen des abaques* (Paris, 1891); *Le calcul simplifié par les procédés mécaniques et graphiques* (Paris, 1893; 2nd ed., 1905; 3rd ed., 1928); *Traité de nomographie. Théorie des abaques, applications pratiques* (Paris, 1899; 2nd ed., 1921); *Calcul graphique et nomographie* (Paris, 1908; 2nd ed., 1914); *Souvenirs et causeries* (Paris, 1928); *Hommes et choses de science,* 3 vols. (Paris, 1930–1932); and *Histoire abrégée des sciences mathématiques,* René Dugas, ed. (Paris, 1955).

Jean Itard

OCHSENIUS, CARL (*b.* Kassel, Germany, 9 March 1830; *d.* Marburg, Germany, 9 December 1906), *geology, sedimentology.*

The son of an administrator at the court of Hessen-Kassel, Ochsenius attended the Gymnasium and the Polytechnische Schule in Kassel, where he studied mining engineering and geology. In 1851 he accompanied his professor, Rudolf Amandus Philippi, on an expedition to Chile, where he remained for twenty years. During his stay in South America, Ochsenius investigated coal, salt, guano, and sulfur deposits; served in various administrative and directorial positions; and traveled widely as a German consul.

In 1879 he married Rau von Holzhausen; they had four children. After 1871 Ochsenius settled in Marburg, where he was a private scientist and promoter of potash mining near Hannover. He began to publish reports on the observations made during his twenty years abroad. In 1884 the University of Marburg awarded him an honorary doctorate.

Ochsenius is best known for his book *Die Bildung der Steinsalzlager und ihrer Mutterlaugensalze* (1877). This work was outstanding for the great amount of direct observations reported, for the accuracy with which the depositional sequence of salt formation was presented, and for the vigor with which a relatively new idea on the origin of salt deposits was presented (Bischof had offered some preliminary ideas pointing in this direction in the second edition of his *Lehrbuch der chemischen und physikalischen Geologie* [1863–1871]). This new idea, the "bar theory" to explain thick deposits of salt, gypsum, and other evaporites, assumes lagoons separated by bars from the ocean proper. As water is lost by evaporation, evaporites precipitate in the lagoon and additional seawater is fed into the lagoon from the open ocean. With increasing evaporation, the salinity in the almost closed basin increases to the point where gypsum, rock salt, and other evaporites are deposited. The best examples, in Ochsenius' opinion, are the basins of Kara-Bogaz-Gol and Adzhi Darya on the eastern rim of the Caspian Sea. The Stassfurt sequence of the German Zechstein also appeared to confirm his theory. The physicochemical results of van't Hoff's work were welcomed by Ochsenius as confirmations of his observations in nature.

The bar theory was opposed by Johannes Walther, whose "desert theory" proposed a formation of salt basins as closed evaporation basins. Both theories were confirmed by observation of present-day processes, but the bar theory was preferred by more geoscientists. Ochsenius published his last revision of this theory in 1906, the year of his death.

Ochsenius contributed other models and theories to the earth sciences, but none was as successful as his bar theory. Of his theory on petroleum formation only the close association of petroleum and salt provinces has remained confirmed. Equally well confirmed was his theory on partial uplift zones of continents, which was based on numerous observations on the Pacific coast of South America. On the other hand, his theory on coal formation was based on a too restricted observation and, consequently, today applies only to local, special modes of origin. He had tried to apply his bar theory to coal genesis in an attempt to explain the facies change coal / sandstone or coal / claystone in soft-water basins.

Ochsenius' contributions to science are based on an enormous wealth of keenly remembered and recorded observations and on his independent, undogmatic approach. If a theory appeared to be confirmed by observations, he was not afraid to stand alone in its defense.

BIBLIOGRAPHY

I. Original Works. Ochsenius' writings include *Die Bildung der Steinsalzlager und ihrer Mutterlaugensalze unter besonderer Berücksichtigung der Flöze von Douglashall in der Egelnschen Mulde* (Halle, 1877); *Chile, Land und Leute* (Leipzig, 1884); "Bedeutung des orographischen Elementes 'Barre' in Hinsicht auf Bildungen und Veränderungen von Lagerstätten und Gesteinen," in *Zeitschrift für praktische Geologie*, **1** (1893), 189–201, 217–233; and "Theorien über die Entstehung der Salzlager," in *Deutschlands Kaliindustrie*, supp. to the newspaper *Industrie*, 2nd ed. (1906), 1–8.

II. Secondary Literature. See the unsigned "Dr. Carl Ochsenius, der Forscher und Mensch," in *Festschrift zum 100 jährigen Geburtstage* (Chemnitz, 1931), pp. 67–161; Kurt Ochsenius, "Zum 100. Geburtstag von Dr. Carl Christian Ochsenius," in *Zeitschrift Kali und verwandte Salze*, **24**, no. 5 (1930), 68–70; and W. Weissermel, "Zum 100. Geburtstag von Carl Ochsenius," in *Zeitschrift der Deutschen geologischen Gesellschaft*, **82**, no. 4 (1930), 229–236.

G. C. Amstutz

OCKENFUSS, LORENZ. See **Oken, Lorenz.**

OCKHAM, WILLIAM OF (*b.* Ockham, near London, England, *ca.* 1285; *d.* Munich, Germany, 1349), *philosophy, theology, political theory.*

Traditionally regarded as the initiator of the movement called nominalism, which dominated the universities of northern Europe in the fourteenth and fifteenth centuries and played a significant role in shaping the directions of modern thought, William of Ockham ranks, with Thomas Aquinas and Duns Scotus, as one of the three most influential Scholastic philosophers. Of his early life nothing is known; but it is supposed that he was born in the village of Ockham, Surrey, between 1280 and 1290 and that he became a Franciscan friar at an early age. He entered Oxford around 1310 as a student of theology and completed his formal requirements for the degree by lecturing on Peter Lombard's *Sentences* in the years 1318–1319, thereby becoming a *baccalaureus formatus*, or *inceptor*. During the next four years, while awaiting the teaching license which would have made him a *magister actu regens,* or doctor of theology, Ockham took part in quodlibetal disputations, revised his lectures on the first book of the *Sentences* for public circulation, and wrote some philosophical and theological treatises.

In this period his teachings, recognized for their power and originality, became a center of controversy and aroused opposition from partisans of Duns Scotus, whose doctrines Ockham criticized, as well as from most of the Dominican masters and some of the secular teachers. In 1323 one of the latter, John Lutterell, went to the papal court at Avignon to press charges of heretical teaching against Ockham, who was summoned to Avignon to answer these accusations early in 1324. Because his academic career was cut short by these events, so that he never received his license to teach, he came to be known as "the venerable inceptor"—that is, candidate who never received the doctoral degree he had earned.

At Avignon, Ockham stayed at the Franciscan convent while awaiting the outcome of the process against him; and during this period he probably wrote several of his theological and philosophical works. A commission of six theologians was appointed by Pope John XXII to examine the charges against his teaching; and although this commission drew up two lists of suspect doctrines, no action appears to have been taken on the charges. Meanwhile Ockham became actively involved in the dispute then raging between Michael of Cesena, general of the Franciscan order, and Pope John XXII over the question of evangelical poverty; and he gave his support to Cesena.

When, in May 1328, it became apparent that the pope was about to issue an official condemnation of their position, Cesena, Ockham, and two other Franciscan leaders fled by night from Avignon and sought the protection of the German emperor, Louis of Bavaria. Louis, whose claim to the imperial crown was contested by Pope John, welcomed the support of Ockham in his cause, as well as that of Marsilius of Padua. The pope, enraged by this defection, excommunicated Ockham and his companions, not for heretical doctrines but for disobedience to his authority. During the ensuing years Ockham remained at Munich and devoted his energies to writing a series of treatises and polemical works directed against John XXII, some of which contained carefully argued discussions of the powers and functions of the papal office, the church, and the imperial or civil authority. When Louis of Bavaria died in 1347, the contest with the Avignon papacy became a lost cause; and there is some evidence that Ockham sought to reconcile himself with the Franciscan faction that had remained loyal to the pope. It is thought that he died in 1349, a victim of the Black Plague, and that he was buried in the Franciscan church at Munich.

Ockham's writings, as preserved, fall into three main groups: philosophical, theological, and political. The philosophical works include commentaries and sets of questions on Aristotle's *Physics* and commentaries on Porphyry's *Predicables* and Aristotle's *Categoriae, De interpretatione,* and *De sophisticis elenchis.* Ockham wrote an independent work on logic, entitled *Summa logicae,* that gave full expression to his own philosophy of language and logical doctrines. An incomplete treatise, published under the title *Philosophia naturalis,* dealt with the concepts of motion, place, and time in an original and independent manner. Of his theological writings the most important is the set of questions on book I of the *Sentences,* edited by Ockham for publication and therefore known as his *ordinatio,* along with the questions on the other three books, which are in the form of *reportata* (stenographic versions of the lectures as actually delivered). The *Quodlibeta septem,* containing 172 questions on theological and philosophical topics divided among seven quodlibetal disputations, are of great value as an expression of Ockham's distinctive philosophical positions.

Of logical as well as theological interest are the treatise *De praedestinatione et de praescientia dei et de futuris contingentibus* and the work known as *De sacramento altaris,* which seems to consist of two distinct treatises and which is devoted chiefly to arguing that the doctrine of transubstantiation does not require the assumption that quantity is an entity distinct from substances or qualities. One other theological work, the authenticity of which has been questioned, is the *Centiloquium theologicum,* consisting of 100 conclusions directed mainly to showing that doctrines of natural theology cannot be proved by evident reason or experience.

The third group of Ockham's writings is made up of the polemical and political works written in his Munich period. Many of these are of interest only in connection with the historical events of the time; but some of them contain important discussions of moral, legal, and political concepts and issues developed in connection with the controversies over the powers of pope and emperor, of church and state. Such are the lengthy *Dialogus inter magistrum et discipulum de imperatorum et pontificum potestate,* the *Octo quaestiones super potestate et dignitate papali,* and the shorter but eloquent *Tractatus de imperatorum et pontificum potestate,* written in 1347. Modern critical editions of the political works are well under way; but editions of the philosophical and theological writings are very much needed, since the early printed editions are both rare and not fully reliable, while some important works (those on Aristotle's *Physics*) have never been printed at all.

Ockham was a thinker of profound originality, independence, and critical power. Although he had scarcely any acknowledged disciples, and did not found a school in the sense of having followers committed to defense of his teachings (as did Thomas Aquinas and Duns Scotus), the actual influence exerted by Ockham's thought, in his own time and into the seventeenth century, was of a significance and breadth that may well have surpassed that of Aquinas or Scotus. This influence is clearly discernible in the empiricist doctrines of Locke and Hume, in the controversies concerning faith and merit associated with the Reformation, and in the political theories that found expression in the Conciliar Movement and in seventeenth-century constitutional liberalism. Although some historians have portrayed Ockham as an innovator who revolted against the traditional values and standards of medieval Christendom, it is nearer the truth to say that he was very much a product of the medieval culture and educational system, who sought to resolve problems that were generated by that culture and that had reached critical dimensions in his own time.

The condemnations of strict Aristotelianism that took place in 1277 were symptomatic of a crisis in the Scholastic effort to harmonize Greek metaphysics with the Christian creed; while the conflict between Philip the Fair and Boniface VIII, followed by the controversy between Louis of Bavaria and John XXII, brought to the surface issues concerning the sources of political and ecclesiastical authority that were becoming acute with the decline of the feudal system. It was to save the values threatened by these conflicts, rather than to destroy them, that Ockham subjected the prevailing Scholastic positions to criticism, and sought more adequate and powerful principles of analysis. His chief contributions to philosophy, lying in the areas of philosophy of language, metaphysics, and theory of knowledge, were the direct result of his effort, as a theologian, to meet the twofold commitment to reason and experience, on the one hand, and to the articles of the faith, on the other.

This dual commitment to faith and reason finds expression in two maxims that are constantly invoked in Ockham's writings. The first is that God can bring about anything whose accomplishment does not involve a contradiction. Although this principle is accepted on the basis of the Christian creed, it is equivalent to the philosophical principle that whatever is not self-contradictory is possible, so that what is actually the case cannot be established on a priori

grounds but must be ascertained by experience. The second maxim, known as Ockham's Razor because of his frequent use of it, is the methodological principle of economy in explanation, frequently expressed in the formula "What can be accounted for by fewer assumptions is explained in vain by more." Ockham often expressed it, however, in this longer form: "Nothing is to be assumed as evident, unless it is known per se, or is evident by experience, or is proved by the authority of Scripture" (*Sentences* I, d. 30, qu. 1).

These maxims are equivalent in force and constitute the unifying principle of Ockham's doctrine, whether viewed in its theological or philosophical aspect. They determine a view of the universe as radically contingent in its being, a theory of knowledge that is thoroughly empiricist, and a rejection of all realist doctrines of common natures and necessary relations in things— all of which constitute what is called Ockham's nominalism. They also eliminate every form of determinism in Ockham's metaphysics and psychology, by associating the principle of divine omnipotence with that of divine liberty and freedom of choice and by making the liberty of the human will basic to moral and legal theory.

A first consequence of these principles is the elimination of various metaphysical "distinctions" that played a dominant role in late thirteenth-century Scholasticism and that derived in large measure from the interpretation of Aristotle made by the Islamic philosopher Ibn Sīnā. The real distinction between essence and existence, held to be a doctrine of St. Thomas Aquinas, supposed that in an existing thing its essence or nature, although not separable from its existence, is nevertheless really distinct from it. Ockham argued that if essence and existence are distinct realities, then it is not self-contradictory for one to exist without the other; but since it is self-contradictory to suppose that an essence exists without existence, it follows that there cannot be a real distinction between the two. By a similar argument it is shown that there cannot be a real distinction between individuals and their natures, as the theory of common natures existing in individuals supposes.

Ockham directed his main critique against the Scotist theory that the common nature differs from the individuating principle by a formal distinction that is less than a real distinction but more than a distinction of reason. To show that this involves a contradiction, Ockham argued as follows: Let the common nature be indicated by the letter a and the individuating difference by the letter b. Then, according to Duns Scotus, a is formally distinct from b. But Scotus must concede that a is not formally distinct from a. Yet,

Ockham argued, wherever contradictory predicates are verified of two things, those two things must be really distinct. Hence *b* and *a* cannot be really identical if they are formally distinct, as Scotus claimed; and by the same argument it can be shown that if they are really identical, they cannot be formally distinct.

The notion of a common nature in individuals, really or formally distinct from them, is therefore self-contradictory; and it remains that universality is a property of terms, or of concepts expressed by general nouns, and is simply their capacity to be used to signify or denote many individuals. In denying that there is any universality in things, Ockham does not deny that the basis for universal predication of general terms is objectively present in individual things; he only denies that the fact that Socrates and Plato, for example, are similar in that each is a man entails that there is some entity common to both and distinct from each. Ockham's nominalism is not to be construed as a doctrine that denies any foundation in things for the generality of terms, and his theory of human cognition rests squarely on the assumption that direct experience of existing things gives rise to concepts of universal character that directly signify things as they are or can be.

Since whatever exists is individual, Ockham holds that our knowledge of things is based on a direct and immediate awareness of what is present to our senses and intellect, which he calls intuitive cognition. He defines this type of awareness as one which enables us to form an evident judgment of contingent fact—that is, that the object apprehended exists, or that it is qualified in a certain way, or is next to another object, and so forth. Such cognition gives rise only to singular contingent propositions that are evident; hence it does not yield scientific knowledge in Aristotle's sense, in which premises and conclusions must be of universal character. Every intuitive cognition, however, can give rise to an abstractive cognition of the same object, which Ockham defines as the cognition of an object which does not suffice for an evident judgment concerning the existence of the object or concerning a contingent fact about the object. Thus, while I am observing Socrates and hearing him talk, I can judge evidently that Socrates exists and that he is talking; but if I depart from the spot and then form the proposition that Socrates exists, or that he is talking, my statement is not evident and may in fact be false.

But Ockham insists that there is no distinction between intuitive and abstractive cognition with respect to objects cognized, but only with respect to their capacity to yield evident judgments of existence and contingent fact. In the natural course of events, every abstractive cognition presupposes an intuitive

cognition of an object understood by it; but Ockham says that since the cognitions are distinct from each other and from their objects, it is logically possible for God to cause an intuitive cognition of an object which is not present or not presently existing. In such a case, Ockham says, the intuitive cognition will yield a judgment that the object is not present or that it does not exist; for it would be self-contradictory to hold that one can have an evident judgment that an object exists, if it does not exist.

The general propositions which serve as premises of scientific knowledge, in the strict sense, are established by inductive generalization from singular judgments evident by experience. But Ockham holds that such scientific statements, being formed from abstractive cognitions of their objects, cannot have absolute evidence, or necessary truth, as categorical propositions; they must be construed as necessary propositions concerning the possible, or as conditional statements. Except for premises of mathematics, which are known per se by the meanings of the terms, the principles of the natural sciences are held by Ockham to be evident by experience but not as necessary in the absolute sense, although they may be said to be necessary in the conditional sense of presupposing the common course of nature without divine interference.

Ockham's empirical theory of knowledge and his nominalist doctrine of the relation of discourse to reality are reinforced by a remarkably original and thoroughgoing use of the *logica moderna* of the arts faculties, with its theory of the supposition of terms, which takes the form of a fully developed philosophy of language. Ockham's *Summa logicae* gives the most complete expression to this semantically oriented logic.

Ockham's treatment of theology is consistent with his treatment of philosophy and natural science, in the sense that absolute evidence for theological propositions cannot be had in this life and only a positive theology based on acceptance of the testimony of Christ and the saints is possible. The order established by God and revealed in the laws of the church, which Ockham ascribes to God's *potentia ordinata*, is freely established by divine choice but is not necessary, since God, by his absolute power, could have ordained a different order. In moral and political philosophy Ockham applies these same criteria of divine freedom and omnipotence to refute the claims of pope and emperor alike to absolute power and dominion over members of the church or citizens of the state. The dignity of man is found in his freedom of choice; and Ockham reiterates that the law of God is a law of liberty, not to be degraded and corrupted into absolutism and coercive tyranny.

BIBLIOGRAPHY

I. ORIGINAL WORKS. Individual works include *Quodlibeta septem* (Paris, 1487; Strasbourg, 1491); *Summa logicae* (Paris–Bologna, 1498; Venice, 1508, 1522, 1591; Oxford, 1675), modern ed. of *Pars prima* and *Pars IIa et tertiae prima*, P. Boehner, ed., 2 vols. (St. Bonaventure, N.Y., 1951–1954); *De sacramento altaris et De corpore christi* (Strasbourg, 1491), with *Quodlibeta*, new ed. by T. B. Birch, with English trans., *The De sacramento altaris of William of Ockham* (Burlington, Iowa, 1930); *Summulae in libros Physicorum* (Bologna, 1494; Venice, 1506; Rome, 1637), also known as *Philosophia naturalis; Super quatuor libros Sententiarum . . . quaestiones* (Lyons, 1495), with *Centiloquium theologicum*, modern critical ed. of *Sentences I, Prologus* and *Dist.* I, Gedeon Gal, O.F.M., ed. (St. Bonaventure, N.Y., 1967); *Expositio aurea . . . super artem veterem* (Bologna, 1496), modern ed. of the *Proemium* and *Expositio super librium Porphyrii*, Ernest A. Moody, ed. (St. Bonaventure, N.Y., 1965); *Tractatus de praedestinatione et de praescientia Dei et de futuris contingentibus*, P. Boehner, ed. (St. Bonaventure, N.Y., 1945), English trans. by Marilyn McCord Adams and Norman Kretzmann (New York, 1969); *Dialogus inter magistrum et discipulum* (Lyons, 1495); and *The De imperatorum et pontificum potestate of William of Ockham*, C. K. Brampton, ed. (Oxford, 1927).

Collections are *Guillelmi de Ockham Opera politica*, vol. I, J. G. Sikes, ed. (Manchester, 1940), vol. III, H. S. Offler, ed. (Manchester, 1956), other vols. in preparation or in course of publication; and *Ockham: Philosophical Writings*, P. Boehner, ed. (Edinburgh, 1957), selections with English trans.

II. SECONDARY LITERATURE. On Ockham and his work see Nicola Abbagnano, *Guglielmo di Ockham* (Lanciano, 1931); Léon Baudry, *Le Tractatus de principiis theologiae attribué à G. d'Occam* (Paris, 1936); *Guillaume d'Occam*, I, *L'homme et les oeuvres* (Paris, 1950), with an excellent bibliography; and *Lexique philosophique de Guillaume d'Occam* (Paris, 1958); P. Boehner, *Collected Articles on Ockham* (St. Bonaventure, N.Y., 1956); Franz Federhofer, *Die Erkenntnislehre des Wilhelm von Ockham* (Munich, 1924); Martin Gottfried, *Wilhelm von Ockham* (Berlin, 1949); Robert Guelluy, *Philosophie et théologie chez Guillaume d'Ockham* (Louvain–Paris, 1947); Erich Hochstetter, *Studien zur Metaphysik und Erkenntnislehre Wilhelms von Ockham* (Berlin, 1927); Georges de Lagarde, *La naissance de l'esprit laïque au déclin du moyen âge*, IV–VI (Paris, 1942–1946); Ernest A. Moody, *The Logic of William of Ockham* (New York–London, 1935); Simon Moser, *Grundbegriffe der Naturphilosophie bei Wilhelm von Ockham* (Innsbruck, 1932); Richard Scholz, *Wilhelm von Ockham als politischer Denker und sein Breviloquium de principatu tyrannico* (Leipzig, 1944); Herman Shapiro, *Motion, Time and Place According to William Ockham* (St. Bonaventure, N.Y., 1957); Cesare Vasoli, *Guglielmo d'Occam* (Florence, 1953), which contains a good bibliography; Paul Vignaux, *Justification et prédestination au XIVᵉ siècle* (Paris, 1934); *Le nominalisme au XIVᵉ siècle* (Montreal, 1948); and "Nominalisme" and "Occam," in *Dictionnaire de théologie catholique*, 15 vols. (Paris, 1903–1950), XI, cols. 733–789, 864–904; Damascene Webering, *The Theory of Demonstration According to William Ockham* (St. Bonaventure, N.Y., 1953); and Sytse Zuidema, *De Philosophie van Occam in zijn Commentaar op de Sententien*, 2 vols. (Hilversum, 1936).

ERNEST A. MOODY

ODDI, RUGGERO (*b.* Perugia, Italy, 20 July 1864; *d.* Tunis, Tunisia, 22 March 1913), *medicine*.

The son of Filippo Oddi and Zelinda Pampaglini, Oddi spent four years at the University of Perugia, one at Bologna, and one at Florence, where he graduated in medicine and surgery on 2 July 1889. He remained as an assistant at the Physiology Institute in Florence (directed by L. Luciani) and made a study trip to the Experimental Pharmacological Institute at the University of Strasbourg (directed by Oswald Schmiedeberg), during which he isolated chondroitin sulfate from the amyloid substance. In January 1894 Oddi was appointed head of the Physiology Institute at the University of Genoa, from which he resigned on 1 April 1900 as the result of a complex series of events (reconstructed in 1965 by L. Belloni). This was followed by a short period as physician in the Belgian Congo, during which time his mental condition became more unbalanced, partly as a result of his using narcotics.

Oddi's main contribution is the discovery of the sphincter of the choledochus, made at Perugia as a fourth-year medical student (1886–1887). Intent on studying *in vivo* the action of bile on the digestion, he had the idea of obtaining an uninterrupted flow of bile into the duodenum by removing the reservoir. In a dog that had been cholecystectomized some time before, he was surprised to observe a marked dilatation of the bile ducts, which led him to suppose "that at the outlet of the choledochus into the duodenum there was a special device which allowed the flow of bile only at certain times, preventing it at others, so that the bile, no longer accumulating in the gallbladder, but compelled to create a space in the larger bile ducts, thus caused their enormous dilatation."

A subsequent series of refined morphological researches in various animal species allowed him to demonstrate, both at the outlet of the choledochus and at the outlet of Wirsüng's duct, that there is a special sphincteral device that is largely independent of the muscular layers of the intestine.

Oddi also measured the tone of the sphincter of the choledochus by perfecting an experimental device substantially identical with that used today for the intraoperative manometry of the biliary ducts.

BIBLIOGRAPHY

I. Original Works. Oddi's writings include "Di una speciale disposizione a sfintere allo sbocco del coledoco," in *Annali dell'Università Libera di Perugia*, **2** (1886–1887), vol. I, Facoltà medico-chirurgica, 249–264 and pl. IX; "Effetti dell'estirpazione della cistifellea," in *Bullettino delle scienze mediche*, 6th ser., **21** (1888), 194–202; "Sulla tonicità dello sfintere del coledoco," in *Archivio per le scienze mediche*, **12** (1888), 333–339; "Sul centro spinale dello sfintere del coledoco," in *Lo Sperimentale*, sec. biologica, **48** (1894), 180–191; "Sulla esistenza di speciali gangli nervosi in prossimità dello sfintere del coledoco," in *Monitore zoologico italiano*, **5** (1894), 216–219 and pl. IV; "Ueber das Vorkommen von Chondroïtinschwefelsäure in der Amyloidleber," in *Archiv für experimentelle Pathologie und Pharmakologie*, **33** (1894), 376–388; "Sulla fisiopatologia delle vie biliari," in *Conferenze cliniche italiane dirette dal Prof. Achille de Giovanni . . .*, 1st ser., I (Milan, n.d.), 77–124; *L'inibizione dal punto di vista fisio-patologico, psicologico e sociale* (Turin, 1898); and *Gli alimenti e la loro funzione nella economia dell'organismo individuale e sociale* (Turin, 1902). For other publications by Oddi, see the works by L. Belloni below.

II. Secondary Literature. See Luigi Belloni, "Sulla vita e sull'opera di Ruggero Oddi (1864–1913)," in *Rendiconti dell'Istituto lombardo di scienze e lettere*, Classe di scienze (B), **99** (1965), 35–50; and "Über Leben und Werk von Ruggero Oddi (1864–1913), dem Entdecker des Schliessmuskels des Hauptgallenganges," in *Medizinhistorisches Journal*, **1** (1966), 96–109.

Luigi Belloni

ODIERNA (or **Hodierna**), **GIOANBATISTA** (*b*. Ragusa, Sicily, 13 April 1597; *d*. Palma di Montechiaro, Sicily, 6 April 1660), *astronomy, meteorology, natural history.*

A self-taught scholar, Odierna was born into a modest artisan family, and, apart from a journey to Rome and Loreto, spent all his life in Sicily. He taught mathematics and astronomy at the school in Ragusa and later studied theology at Palermo. He observed the three comets of 1618–1619, which spurred the famous polemic resolved in 1623 by Galileo in his *Saggiatore*. Odierna's observations were published many years later, when he was at the peak of his career.

After having read Galileo's *Sidereus nuncius*, Odierna wrote an enthusiastic appraisal of it, in which he mentions that Galileo had presented him with a telescope of moderate focal distance. He served the barons of Montechiaro as chaplain and parish priest of their newly founded town of Palma di Montechiaro, in the province of Agrigento. They gave him an apartment on a high floor of their palace for his astronomical observations and later named him archpriest and court mathematician.

Odierna's observations were aimed principally at determining the period of revolution of the four satellites of Jupiter. Like Galileo he tried to predict their eclipses, which would have helped to solve the long-standing, important problem of determining longitudes at sea; lacking sufficient knowledge of celestial mechanics, neither he nor Galileo was successful. Odierna's pamphlet on the subject, *Medicaeorum ephemerides* (1656), was dedicated to Grand Duke Ferdinand II of Tuscany.

With Galileo's telescope, Odierna made careful observations of Saturn but did not comprehend the true shape of its ring. In 1656 he published a pamphlet on it, *Protei caelestis seu Saturni systema*, and sent it to Huygens, the discoverer of the nature of the ring. Huygens replied, encouraging him to continue his useful observations, and sent him a drawing of his pendulum clock to assist him in his research.

After studying the passage of light through prisms Odierna offered a vague explanation of the rainbow and of the spectrum. His *Thaumantia Junonis nuntia praeconium pulchritudinis* (1647), on the nature of the iris and its colors, was followed in 1652 by *Thaumantiae miraculum*.

Odierna's interest in meteorology resulted in some research on cyclones. In natural history his explanation of the structure and function of the retractile poison fangs of vipers anticipated the work of Redi. In his studies on the eyes of flies and of other insects, he used a microscope and a camera obscura.

Odierna's numerous works were almost all published at Palermo and are now in the Municipal Library of Palermo and in the University Library of Catania. Although they cannot be said to be of real scientific value, Odierna must be considered among the pioneers of the experimental method.

BIBLIOGRAPHY

Odierna's "L'occhio della mosca" was repub. with a commentary by C. Pighetti, in *Physis*, **3** (1961), 309–335, with a complete bibliography.

See also G. Abetti, "Don Giovanni Battista Odierna," in *Celebrazioni siciliane* (Urbino, 1939), 3–28, and "Onoranze a D. Gioanbatista Hodierna della città di Ragusa in Sicilia," in *Physis*, **3** (1961), 177–179.

Giorgio Abetti

ODINGTON. See **Walter of Odington.**

ODLING, WILLIAM (*b.* Southwark, London, England, 5 September 1829; *d.* Oxford, England, 17 February 1921), *chemistry*.

Before Odling became Waynflete professor of chemistry at Oxford, where he was a conscientious teacher but personally uninterested in the emergence of a research school, he had been (with A. W. Williamson, B. C. Brodie, and E. Frankland) one of England's leading theoretical chemists during the exciting renaissance of British chemistry between 1850 and 1870. The only son of George Odling, a London doctor with a long family tradition of medicine, he received his elementary schooling at Stockwell and then at the interesting Nesbit's Chemical Academy and Agricultural College, where he gave his first public lecture in 1844.[1] He entered Guy's Hospital at the age of sixteen and was one of the hospital's first students to take the London University M.D. in 1851.

Although Odling attended A. W. Hofmann's course at the Royal College of Chemistry for a semester in 1848, his principal chemistry teacher, who initially biased him toward toxicological studies, was Alfred Swaine Taylor (1806–1880), the Guy's lecturer in chemistry and medical jurisprudence. Odling held several teaching positions at Guy's while he was medical officer of health for Lambeth from 1856 to 1862. From 1863 to 1870 he taught chemistry at St. Bartholomew's Hospital and, on Faraday's death in 1867, became Fullerian professor of chemistry at the Royal Institution, whereupon he abandoned applied medicine except for remunerative work on water analysis. On succeeding Brodie at Oxford in 1872 he married Elizabeth Mary Smee; they had three sons. He retired in 1912 but remained active until the end of his life.

Through his long and influential association with the Chemical Society, which he joined in 1848, Odling became a close friend and overmodest "follower" of Brodie and of the older and authoritative Williamson. Through the latter he met Kekulé during his *Wanderjahre* in London (1854–1855). All these men had been stimulated by the revolutionary French chemistry of C. Gerhardt and A. Laurent. On the latter's death in 1853 Williamson recommended Odling to Biot (Laurent's editor) as the English translator of Laurent's posthumous masterpiece, *Méthode de chimie*. The few months Odling spent in Paris with Gerhardt in 1854 completed his chemical education. Thereafter he and Williamson became the formidable British spokesmen for the type theory and for two-volume formulas (such as H_2O for water instead of the prevailing confusion of HO or H_4O_2).

Laurent and Gerhardt also induced Odling's lifelong interest in the problems of classifying chemical compounds and of exploiting their analogies. Unlike Williamson, however, who believed in the existence of atoms, Odling was sufficiently influenced by Brodie's skepticism to remain uncommitted to atomism per se. Like Gerhardt he preferred to regard formulas as heuristic devices, and he had only harsh words for the pictorial "fancies" involved in the graphic formulas of A. Crum Brown and Frankland. The radical theories of J. J. Berzelius—and more recently those of H. Kolbe and Frankland—were irrational, Odling believed, because they involved the real existence of hypothetical components; on the other hand, Gerhardt's two-volume types (hydrogen, hydrogen chloride, water, and ammonia) merely used chemical analogies that were based upon facts which did not involve a commitment to unattainable absolute structures.

This positivism was prominent in Odling's first paper to the Chemical Society in 1853.[2] In it he extended Williamson's use of the multiple water type (which classified compounds as substitution products in a double water molecule) and showed how all salts, however complex, could be reduced to "the types of one or more atoms of water." ("Atom" was here being used in a conventionalist sense.) For example, the problematic phosphoric acids were construed as

$$\left.\begin{array}{l}2H'\\2H'\end{array}\right\}2O'' \qquad \left.\begin{array}{l}PO'''\\H'\end{array}\right\}2O'' \quad \text{metaphosphoric acid}$$

$$\left.\begin{array}{l}3H'\\3H'\end{array}\right\}3O'' \qquad \left.\begin{array}{l}PO'''\\3H'\end{array}\right\}3O'' \quad \text{orthophosphoric acid}$$

while alum was a quadruple water type:

$$\left.\begin{array}{l}2SO_2''\\KAl_2'''\end{array}\right\}4O''$$

Superscript single, double, and triple vertical lines were introduced by Odling to indicate the equivalence, or "replaceable value," of the element or group within the type formula compared with hydrogen. (He recognized, and allowed for, elements with variable equivalence, such as Fe' in ferrous and Fe''' in ferric salts; CO'' in carbonic and CO' in oxalic acid.) This useful notation was rapidly adopted by other chemists, and by the 1860's the vertical lines were recognized as denoting the valence of particular atoms. Odling also introduced "mixed types" for molecules like sodium thiosulfate:

$$\left.\begin{matrix} H \\ H \end{matrix}\right\}O'' \quad \left.\begin{matrix} Na' \\ \end{matrix}\right\}O'' \\ \left.\begin{matrix} H \\ H \end{matrix}\right\}O'' \quad \left.\begin{matrix} SO_2'' \\ Na' \end{matrix}\right\}S''$$

Such types were later used extensively by Kekulé.

In an important lecture on hydrocarbons in 1853, Odling extended these ideas and argued, against the radical school, that in hydrogen compounds, such as the hydrocarbons, there were as many potential radicals as there were parts of hydrogen.[3] Thus

1 HCl
2 $H \cdot OH$ H^2O
3 $H \cdot NH^2$ $H^2 \cdot NH$ H^3N
4 $H \cdot CH^3$ $H^2 \cdot CH^2$ $H^3 \cdot CH$ H^4C methane

The methane example was made famous by Kekulé in 1857 as the "marsh gas type." But unlike Kekulé, who possessed an offprint of the lecture, Odling failed to extend the type to several carbon compounds and thus failed to exploit the unifying possibilities implicit in his own notation, $C^{iv}H^4$: the quadrivalence of carbon. Here the historical problem of Odling's precise influence on Kekulé, as opposed to Gerhardt's, is particularly puzzling.

Odling was one of the secretaries at the international conference on a rational system of combining or atomic weights held at Karlsruhe in 1860. Between 1853 and 1863 he was an enthusiastic propagandist for Gerhardt's partial revision of atomic weights,[4] and although he did not immediately see the need for Cannizzaro's more sweeping revision, he accepted and publicized them beginning in 1864. Like Laurent, Odling had a passion for learned, and often impracticable, neologisms. In 1864 he introduced the terms "monad," "dyad," and "tetrad" for variable units of valence, and "artiads" and "perissads" for elements with even and odd valences, respectively. The terms were used widely in British textbooks and examinations until about 1900.

Odling's interest in classification inevitably led him to examine the natural relationships between chemical elements and to publish several prescient schemes between 1857 and 1865. He was perhaps unique among Mendeleev's many predecessors in placing more emphasis on the physical and chemical analogies between elements and their compounds, rather than indulging in numerical speculations about the atomic weights of the elements.

NOTES

1. His lecture on chemical affinity was reported extensively in *Maidstone and South Eastern Gazette* (June 1844), quoted in entirety by Freeman, pp. 190–202.
2. "On the Constitution of Acids and Salts as Substitution Products Formed on the Water Type," in *Quarterly Journal of the Chemical Society*, **7** (1855), 1–21 (read 7 Nov. 1853 and probably extensively revised).
3. "On the Constitution of the Hydrocarbons," in *Proceedings of the Royal Institution of Great Britain*, **2** (1854–1858), 63–66 (read 16 Mar. 1855).
4. "On the Atomic Weights of Oxygen and Water," in *Quarterly Journal of the Chemical Society*, **11** (1859), 107–129; and "On the Molecule of Water," in *Chemical News*, **8** (1863), 147–152.

BIBLIOGRAPHY

I. ORIGINAL WORKS. An unpublished bibliography of 150 publications is given by P. J. Freeman in "The Life and Times of William Odling (1829–1921), Waynflete Professor of Chemistry, 1872–1912" (B.Sc. thesis, Oxford, 1963). This supplements the published bibliography by John L. Thornton and Anna Wiles, "William Odling, 1829–1921," in *Annals of Science*, **12** (1956), 288–295. To their lists, the following significant items should be added: "On the Basis of Chemical Notation," in *Nature*, **1** (1869–1870), 600–602; "On the Unit Weight and Mode of Constitution of Compounds," in *Chemical News*, **45** (1882), 63–65; and "The Whole Duty of a Chemist," in *Nature*, **33** (1885–1886), 99, a reply to an editorial attack (*ibid.*, 73–77) on Odling's presidential address to the Royal Institute of Chemistry.

Odling's textbooks, which are easily confused, are *A Course of Practical Chemistry, Arranged for the Use of Medical Students* (London, 1854; 2nd ed., 2 pts., 1863–1865; 3rd ed., 1865 [*sic*], trans. into Russian by R. Savtschenkoff [St. Petersburg, 1867] and read by Mendeleev; 4th ed., 1869; 5th ed., 1876, trans. into French by A. Naquet [1876]); *A Manual of Chemistry, Descriptive and Theoretical*, pt. I (London, 1861)—pt. II never appeared, but see MSS—trans. into Russian (St. Petersburg, 1863), German (Erlangen, 1865), and French (Paris, 1868)—for the O_3 formula for ozone, see 1861 ed., pp. 93–94; *Tables of Chemical Formulae* (London, 1864), 8 leaves, the first list of Cannizzaro's atomic weights in English; *Lectures on Animal Chemistry, Delivered at the Royal College of Physicians* (London, 1866), also trans. into Russian (St. Petersburg, 1867); *A Course of Six Lectures on the Chemical Changes of Carbon*, W. Crookes, ed. (London, 1869), also trans. into French (Paris, 1870); *Outlines of Chemistry, or Brief Notes of Chemical Facts* (London, 1870 [published Nov. 1869]), for the influential bleaching powder formula, see p. 24; and *Science Primers for the People*, no. 2, *Chemistry* (London, 1883).

Odling translated Auguste Laurent, *Chemical Method, Notation, Classification and Nomenclature* (London, 1855; some copies released in 1854). Odling's final book was the extraordinary *The Technic of Versification: Notes and Illustrations* (Oxford–London, 1916), which Marsh (below) aptly described as "a kind of type theory of verse with a symbolic notation almost chemical."

The principal archival sources are, in London: Chemical Society (B Club and Roscoe papers, photographs), Imperial College Archives, Royal Institute of Chemistry (several MSS, including drafts of pt. II of the *Manual*), Royal Institution, Royal Society (referee reports); at Harpenden: Rothamsted Experimental Station (J. H. Gilbert papers); at Oxford: Museum of History of Science (Rev. F. J. J. Smith papers).

II. SECONDARY LITERATURE. There are two good obituaries: J. E. Marsh, in *Journal of the Chemical Society*, **119** (1921), 553–564, with portrait; and H. B. D.[ixon], in *Proceedings of the Royal Society*, **100A** (1922), i–vii, with portrait. An Oxford student's caricature of Odling lecturing is reproduced in R. T. Gunther, *Early Science at Oxford*, XI (Oxford, 1937), 293. To Freeman, and Thornton and Wiles (above), add K. R. Webb, "William Odling, Third President 1883–88," in *Journal of the Royal Institute of Chemistry*, **81** (1957), 728–733; and J. R. Brown and J. L. Thornton, "William Odling as Medical Officer of Health at Lambeth," in *Medical Officer*, **102** (1959), 77–78.

No detailed study of Odling's influence on theoretical chemistry in the 1850's and 1860's has yet been made. For some indications see, on the problem of Odling's influence on Kekulé, R. Anschütz, *August Kekulé*, I (Berlin, 1929), *passim*; on Odling's contribution to valence, C. A. Russell, *History of Valency* (Leicester, 1970), *passim*; on Odling's attitude toward Brodie and atomism, W. H. Brock, *The Atomic Debates* (Leicester, 1967), *passim*, which includes four letters; on Odling and the periodic law, J. W. van Spronsen, "William Odling wegbereider en ontdekker van het periodiek systeem der elementen 1864–1964," in *Chemisch Weekblad*, **60** (1964), 683–686; and his *Periodic System of Chemical Elements* (Amsterdam–London–New York, 1969), 87–90, 112–116, 349–350; and, for general orientation, J. R. Partington, *A History of Chemistry*, IV (London, 1964), *passim*. Finally, for a glimpse of Odling the administrator, see R. B. Pilcher, *The Institute of Chemistry of Great Britain and Ireland. History of the Institute, 1877–1914* (London, 1914), *passim*.

W. H. BROCK

OENOPIDES OF CHIOS (*b.* Chios; *fl.* fifth century B.C.), *astronomy, mathematics.*

The notice of Pythagoras in Proclus' summary of the history of geometry is followed by the sentence,[1] "After him Anaxagoras of Clazomenae touched many questions concerning geometry, as also did Oenopides of Chios, being a little younger than Anaxagoras, both of whom Plato mentioned in the *Erastae*[2] as having acquired a reputation for mathematics." This fixes the birthplace of Oenopides as the island of Chios and puts his active life in the second third of the fifth

century B.C.[3] Anaxagoras was born about 500 B.C. and died about 428 B.C. There is confirmation from Oenopides' researches into the "great year" (see below), which suggest that he could not have differed greatly in date from Meton, who proposed his own Great Year in 432. Like Anaxagoras, Oenopides almost certainly conducted his researches in Athens.

In the opening words of the *Erastae*, to which Proclus refers, Socrates is represented as going into the school of Dionysius the grammarian, Plato's own teacher,[4] and seeing two youths earnestly discussing some astronomical subject. He could not quite catch what they were saying, but they appeared to be disputing about Anaxagoras or Oenopides, and to be drawing circles and imitating some inclinations with their hands. In the light of other passages in Greek authors, this is a clear reference to the obliquity of the ecliptic in relation to the celestial equator. Eudemus in his history of astronomy, according to Dercyllides as transmitted by Theon of Smyrna, related that Oenopides was the first to discover the obliquity of the zodiac,[5] and there appears to have been a widespread Greek belief to that effect. Macrobius,[6] for example, drawing on Apollodorus, notes that Apollo was given the epithet $\Lambda o \xi i a s$ because the sun moves in an oblique circle from west to east, "as Oenopides says." Aëtius[7] says that Pythagoras was the first to discover the obliquity of the ecliptic, and that Oenopides claimed the discovery as his own, while Diodorus[8] says that it was from the Egyptian priests and astronomers that he learned the path of the sun to be oblique and opposite to the motion of the stars (that is, fixed stars). He is not recorded as having given any value to the obliquity, but it was probably he who settled on the value of 24°, which was accepted in Greece until refined by Eratosthenes.[9] Indeed, if Oenopides did not fix on this or some other figure, it is difficult to know in what his achievement consisted, for the Babylonians no less than the Pythagoreans and Egyptians must have realized from early days that the apparent path of the sun was inclined to the celestial equator.

In the same passage as that already mentioned, Theon of Smyrna[10] attributes to Oenopides the discovery of the period of the Great Year. This came to mean a period in which all the heavenly bodies returned to their original relative positions, but in early days only the motions of the sun and moon were taken into account and the Great Year was the least number of solar years which coincided with an exact number of lunations. Before Oenopides it was calculated that the sun and the moon returned to the same relative positions after a period of eight years, the *octaëteris*, in which three years of thirteen months or

384 days were distributed among five years of twelve months or 354 days, giving the solar year an average of $365\frac{1}{4}$ days and making the lunar month a shade over $29\frac{1}{2}$ days. Oenopides appears to have been the first to give a more exact rendering, possibly in an attempt to take account also of the planetary motions. Aelian records that he set up at Olympia a bronze inscription stating that the Great Year consisted of fifty-nine years, and Aëtius confirms the period,[11] while Censorinus[12] states that he made the year to be $365\frac{22}{59}$ days, which implies a Great Year of 21,557 days. Oenopides no doubt fixed upon a period of fifty-nine years, as P. Tannery[13] first showed, by taking the figures of $29\frac{1}{2}$ days for a lunar month and 365 days for a solar year, and deducing that in fifty-nine years there would on this basis be exactly 730 lunations. Observation would have established, Tannery argued, that in 730 lunar months there were 21,557 days, from which it follows that the year consists of $365\frac{22}{59}$ or 365.37288 days and the month of 29.53013 days. The cycle of nineteen years that Meton and Euctemon proposed in 432 B.C., on which the present ecclesiastical calendar is ultimately based, gives a year of $365\frac{5}{19}$ or 365.26315 days and a month of 29.53191 days. The modern value for the sidereal year is 365.25637 days and for the mean synodic month is 29.53059 days.

Oenopides' figure for the lunar month is, therefore, if Tannery is right, more exact than that of Meton (indeed, very exact, for the error does not exceed a third of a day in the whole fifty-nine years), but his figure for the year is considerably less exact, amounting to seven days for the whole period.

But could Oenopides have calculated at that date so exact a figure for the mean synodic month (which requires a long period of observation) when he had so inaccurate a figure for the solar year (to establish which as about $365\frac{1}{4}$ days would require only a few consecutive observations of the times of the solstices)? In a private communication G. J. Toomer is skeptical. He believes that Oenopides did not assign any specific number of days to the Great Year, and the year-length of $365\frac{22}{59}$ days attributed to him by Censorinus is a later reconstruction. Someone at this later date asked himself what is the length of the year according to Oenopides. He answered the question by taking the standard length of the mean synodic month of his own time, namely (expressed sexagesimally) 29; 31, 50, 8, 20 days. This is found in Geminus as well as the *Almagest* and was a fundamental Babylonian parameter adopted by Hipparchus. The hypothetical investigator multiplied this by the 730 months of Oenopides' period and obtained 21,557 days and a fraction of a day. Dividing 21,557 by the

59 years of the cycle, he declared that Oenopides' year consisted of $365\frac{22}{59}$ days—that is to say, the figure is a later deduction using a completely anachronistic value for the month. This is credible. The critical question is whether Oenopides could have had at his disposal records extending over more than his own adult life showing that in 730 lunations there were 21,557 days; if he did, it would be strange for him not to have known a more exact figure for the year.

Tannery[14] holds that Oenopides' Great Year was intended to cover the revolutions of the planets and of the sun and moon, but he is forced to conclude that Oenopides could not have taken them all into account. The ancient cosmographers gave the time for Saturn to traverse its orbit as thirty years, for Jupiter twelve years, and for Mars two years, which would allow two revolutions for Saturn in the Great Year, five for Jupiter, and thirty or thirty-one for Mars. If the latter figure is taken as the more correct, and the figure of 21,557 days in the Great Year is divided by these numbers, we get values for the revolutions of the three planets which do not differ by more than one percent from the correct values. Tannery considers that the degree of inaccuracy ought rather to be judged by the error in the mean position of the heavenly body at the end of the period; this would be only 2° in the case of Saturn and 9° for the sun, but 107° for Mars. If Oenopides had indicated in which sign of the zodiac the planet would be found at the end of the period, the error would have been obvious when the time came.

According to Achilles Tatius,[15] Oenopides was among those who believed that the path of the sun was formerly the Milky Way; the sun turned away in horror from the banquet of Thyestes and has ever since moved in the path defined by the zodiac.

Two propositions in geometry were discovered by Oenopides according to Eudemus as preserved by Proclus. Commenting on Euclid I.12 ("to a given infinite straight line from a given point which is not upon it to draw a perpendicular straight line") Proclus[16] says: "Oenopides was the first to investigate this problem, thinking it useful for astronomy. But, in the ancient manner, he calls the perpendicular 'a line drawn gnomon-wise,' because the gnomon is at right angles to the horizon." When he comes to Euclid I.23 ("on a given straight line and at a given point on it, to construct a rectilineal angle equal to a given rectilineal angle") Proclus[17] comments: "This problem is rather the discovery of Oenopides, as Eudemus relates." Heath[18] justly observes that the geometrical reputation of Oenopides can hardly have rested on such simple propositions as these, nor could he have been the first to draw a perpendicular in practice. Possibly he was the first to draw a perpendicular to a straight line by means of a ruler and compass (instead of a set-square), and it may have been he who introduced into Greek geometry the limitation of the use of instruments in all plane constructions—that is, in all problems equivalent to the solution of algebraic equations of the second degree—to the ruler and compasses. He also may have been the first to give a theoretical construction to Euclid I.23.

This question bears on an interesting problem to which Kurt von Fritz[19] has devoted much attention. According to Proclus,[20] "Zenodotus, who stood in the succession of Oenopides but was one of the pupils of Andron, distinguished the theorem from the problem by the fact that the theorem seeks what is the property predicated of its subject-matter, but the problem seeks to find what is the cause of what effect" (as translated by Heath,[21] but Glenn R. Morrow[22] translates τίνος ὄντος τί ἐστιν as "under what conditions something exists"). The meaning was probably no clearer to Proclus than it is to us, but it may be that Oenopides was one of those who helped to create the distinction between theorems and problems. Taken in conjunction with what was said in the previous paragraph, it would appear that he made a special study of the methodology of mathematics.

Oenopides had an original theory to account for the Nile floods. He held that the water beneath the earth is cold in the summer and warm in the winter, a phenomenon proved by the temperature of deep wells. In winter, when there are no rains in Egypt, the heat that is shut up in the earth carries off most of the moisture, but in summer the moisture is not so carried off and overflows the Nile. Diodorus Siculus, who recorded the theory, reasonably objected that other rivers of Libya, similar in position and direction to the Nile, are not so affected.[23]

It is related that Oenopides, seeing an uneducated youth who had amassed many books, observed, "Not in your coffer but in your breast."[24] Sextus Empiricus[25] says that Oenopides laid special emphasis on fire and air as first principles. Aëtius[26] says that Diogenes (of Apollonia), Cleanthes, and Oenopides made the soul of the world to be divine. Cleanthes left a hymn to Zeus in which the universe is considered a living being with God as its soul, and if Aëtius is correct then Oenopides must have anticipated these views by more than a century. Diogenes is known to have revived the doctrine of Anaximenes that the primary substance is air, and presumably Oenopides in part shared this view but gave equal primacy to fire as a first principle.

NOTES

1. Proclus: *Procli Diadochi in primum Euclidis, Elementorum librum commentarii*, G. Friedlein, ed. (Leipzig, 1873, repr. 1967), pp. 65.21–66.4.
2. Plato, *Erastae (Amatores)*, 132 A.B, in J. Burnet, ed., *Platonis opera*, II (Oxford, 1901, repr. 1946). The Platonic authorship of the *Erastae* has been denied, but this does not affect its evidence for Oenopides.
3. The "Vita Ptolemaei e schedis Savilianis descripta" found in a Naples MS (Erwin Rohde, *Kleine Schriften*, I [Tübingen–Leipzig, 1901], p. 123, n. 4) is therefore in error in saying that Oenopides lived "towards the end of the Peloponnesian war" but more accurate in adding "at the same time as Gorgias the orator and Zeno of Elea and, as some say, Herodotus, the historian, of Halicarnassus." Diogenes Laërtius IX.41 (H.S. Long, ed., II [Oxford, 1964], 450. 23–25) says that Democritus "would be a contemporary of Archelaus, the pupil of Anaxagoras, and of the circle of Oenopides"; and he adds that Democritus makes mention of Oenopides—presumably in a work that has not survived.
4. Diogenes Laërtius III. 4 (H. S. Long, ed., I [Oxford, 1964], 122.13).
5. Theon of Smyrna, *Expositio rerum mathematicarum ad legendum Platonem utilium*, E. Hiller, ed. (Leipzig, 1878), 198.14–16. H. Diels's conjecture λόξωσιν ("obliquity") for διάζωσιν ("girdle") is almost certainly correct.
6. Macrobius, *Saturnalia* I.17.31, F. Eyssenhardt, ed., 2nd ed. (Leipzig, 1893), 93.28–94.2.
7. Aëtius, II.12, 2, Ps.-Plutarch, *De placitis philosophorum*, B. N. Bernardakis, ed. (*Plutarchi Chaeronensis Moralia*, Teubner, V [Leipzig, 1893]), 284.8–9.
8. Diodorus Siculus, *Bibliotheca historica*, I.98.3, C. H. Oldfather, ed., I (London–New York, 1933), pp. 334.29, 337.4.
9. Proclus, *In primum Euclidis*, Friedlein, ed., p. 269.11–21, states that Euclid IV.16 (which shows how to construct a regular polygon of fifteen sides in a circle, each side therefore subtending an angle of 24° at the center) was inserted "in view of its use in astronomy." Erastosthenes found the distance between the tropical circles to be 11/83 of the whole meridian, giving a value for the obliquity of 23°51′20″ as Ptolemy records in *Syntaxis*, J. L. Heiberg, ed., I.12 (Leipzig, 1898), p. 68.3–6.
10. Theon of Smyrna, *op. cit.*, p. 198.15.
11. Aelian, *Varia historia*, X.7, C. G. Kuehn ed., II (Leipzig, 1780), 65–67; Aëtius, II.32.2, *op. cit.*, 316.1–7.
12. Censorinus, *De die natali* 19.2, F. Hultsch, ed. (Leipzig, 1867), 40.19–20.
13. Paul Tannery, *Mémoires scientifiques*, II (Toulouse–Paris, 1912), 359.
14. *Ibid.*, 358, 362–363.
15. Achilles Tatius, *Introductio in Aratum* 24, E. Maass ed., *Commentariorum in Aratum reliquiae* (Berlin, 1898), p. 55.18–21. Aristotle, *Meteorologica* I.8, 345A, 13–25, Fobes, ed. (Cambridge, Mass. 1919, repr. Hildesheim, 1967), notes that certain of the so-called Pythagoreans held the same view and pointedly asks why the zodiac circle was not scorched in the same way.
16. Proclus, *In primum Euclidis*, Friedlein, ed., 283.7–10.
17. *Ibid.*, 333.5–6.
18. Thomas Heath, *A History of Greek Mathematics*, I (Oxford, 1921), 175.
19. Kurt von Fritz, "Oinopides" in Pauly-Wissowa, **17** (Stuttgart, 1937), cols. 2267–2271.
20. Proclus, *In primum Euclidis*, Friedlein, ed., p. 80.15–20.
21. Thomas L. Heath, *The Thirteen Books of Euclid's Elements*, 2nd ed., I (Cambridge, 1926; New York, 1956), 126.
22. Glenn R. Morrow, *Proclus: A Commentary on the First Book of Euclid's Elements* (Princeton, 1970), p. 66.
23. Diodorus Siculus I. 41.1–3, *op. cit.*, vol. 1, pp. 144.23–147.17.

24. *Gnomologium Vaticanum* 743, L. Sternbach, ed. (Berlin, 1963), n. 420.
25. Sextus Empiricus, *Pyrrhoniae hypotyposes*, iii. 30.
26. Aëtius, I.7, 17, *op. cit.*, 284.8–9.

BIBLIOGRAPHY

No works by Oenopides have survived, nor are the titles of any known. The ancient references to him are collected in Diels-Kranz, *Die Fragmente der Vorsokratiker*, 6th ed. (Dublin–Zurich, 1969), 41(29), 393–395. The most useful modern studies are Paul Tannery, "La grande année d'Aristarque de Samos," in *Mémoires de la Société des sciences physiques et naturelles de Bordeaux*, 3rd ser., **4** (1888), 79–96, reprinted in *Mémoires scientifiques*, J. L. Heiberg and H. G. Zeuthen, ed., **2** (Paris–Toulouse, 1912), 345–366; Thomas Heath, *Aristarchus of Samos. The Ancient Copernicus* (Oxford, 1913), 130–133; Kurt von Fritz, "Oinopides," in Pauly-Wissowa-Kroll, *Real-Encyclopädie der classischen Altertumswissenschaft*, **17** (Stuttgart, 1937), cols. 2258–2272; D. R. Dicks, *Early Greek Astronomy to Aristotle* (London, 1970), 88–89, 157, 172; Jürgen Mau, "Oinopides," in *Der Kleine Pauly*, IV (Stuttgart, 1972), cols. 263–264.

IVOR BULMER-THOMAS

OERSTED, HANS CHRISTIAN (*b.* Rudkøbing, Langeland, Denmark, 14 August 1777; *d.* Copenhagen, Denmark, 9 March 1851), *physics.*

Oersted was the elder son of an apothecary, Søren Christian Oersted, and his wife, the former Karen Hermansen. The demands of his father's business and his mother's superintendence of a large family forced his parents to place Hans Christian and his younger brother, Anders Sandøe, with a German wigmaker and his wife while they were still young boys. It was there that Oersted learned German by translating a German Bible and speaking with the couple. The brothers' intellectual abilities were soon apparent, and neighbors did what they could to stimulate and educate them. In this way they picked up the rudiments of Latin, French, and mathematics. When Oersted was eleven, he began to serve as his father's assistant in the pharmacy, thereby gaining a practical knowledge of the fundamentals of chemistry.

This was not much formal education; but when the two brothers arrived in Copenhagen in 1794, they were able to pass the entrance examination for the university with honors. At this point they parted intellectual company; Anders went on to become a jurist and Hans Christian pursued a career in natural philosophy. The most important of Oersted's courses for his intellectual development was that offered on Kant and the critical philosophy. Oersted became a passionate Kantian and defender of Kant's philosophical views, which were to be of fundamental importance to his scientific development. They were even to be the agent that led him to his most important discovery, electromagnetism.

At the University of Copenhagen, Oersted studied astronomy, physics, mathematics, chemistry, and pharmacy. In 1797 he received his pharmaceutical degree with high honors. The following year he became a member of the editorial staff of a new periodical, *Philosophisk repertorium for faedrelandets nyeste litteratur*, which was devoted to the propagation and defense of Kantian philosophy. Although short-lived, the journal provided Oersted with an opportunity to mature his philosophical thinking. An unpublished article that he wrote for it served as the starting point for his doctoral dissertation. In 1799 he received his doctorate with a thesis entitled "Dissertatio de forma metaphysices elementaris naturae externae," which states Oersted's appreciation of the importance of Kantian philosophy for natural philosophy and, in addition, provides a clue to the two areas in which he was to apply his scientific training: electromagnetism and research on the compressibility of gases and liquids.

After a brief stint as the manager of a pharmacy, Oersted set out in the summer of 1801 on a journey that was to complete his scientific education. The scientific world was in ferment over the recently announced discovery of the voltaic pile (1800), and Oersted eagerly pursued information relating to galvanism and its relation to chemistry. A small voltaic battery of his own invention gained him entry to others' laboratories, and he gathered knowledge and ideas as he visited Berlin, Göttingen, and Weimar. Again the influences at work on him were twofold. At Göttingen he was given an introduction to Johann Ritter, who was then publishing on the chemical effects of current electricity. Ritter focused Oersted's attention on the forces of chemical affinity and their relationship to electricity. Ritter's highly unorthodox ideas on matter and force also stimulated Oersted to develop his own concepts. At Berlin he attended lectures on *Naturphilosophie* and met such *Naturphilosophen* as Henrik Steffens and Franz von Baader. He read Schelling and heard Friedrich Schlegel. As a result he developed his philosophical insights by comparing his own metaphysics with those of the *Naturphilosophen*. Since both Oersted and the *Naturphilosophen* drew their inspiration and basic ideas from Kant, it is no coincidence that Oersted's later philosophy closely resembled *Naturphilosophie.*

Oersted was saved from the extravagances of a Schelling by his basic respect for empirical fact. Nevertheless, during this trip it was his philosophical penchant that dominated, for although he was suspicious of Schelling's system-building, he swallowed as fact what were only wild guesses by Ritter and the Hungarian chemist J. J. Winterl. Indeed, it was as a defender of Winterl and Ritter that Oersted made his scientific debut in Paris.

The result was disastrous. Winterl's "system" rested on two archetypal substances—Andronia and Thelycke—the essences of acidity and basicity. From these Winterl developed a chemistry of conflicting opposites which, because of its philosophical beauty, completely seduced Oersted. The French chemists, however, were scornful; and Oersted was blasted in the *Annales de chimie et de physique.* It was a valuable lesson. Henceforth, Oersted tended increasingly to hold his philosophical enthusiasms in check at least until he had some evidence for their plausibility. The lesson was driven home by his championing of Ritter's work. To his dismay, he discovered that many of the experimental results his friend reported in the journals were, like Winterl's Andronia, mere figments of his imagination. The pain of having made a scientific fool of himself taught Oersted the critical attitude necessary for the successful pursuit of scientific knowledge.

Oersted returned to Denmark in 1804, preceded by his reputation as an uncritical enthusiast. He had hoped for a professorship in physics but was disappointed by the failure of the warden of the University of Copenhagen to nominate him. He turned, instead, to public lectures, which became so popular that he finally gained an extraordinary professorship in 1806. He then began his own scientific work in earnest. A series of sober publications, among which was an excellent paper on acoustical figures (1810), gradually erased his earlier reputation. He began a steady advance in the academic hierarchy and in reputation. In 1824 he founded the Society for the Promotion of Natural Science and in 1829 became the director of the Polytechnic Institute in Copenhagen, a position that he held until his death. Oersted was a superb teacher and, almost single-handed, raised the level of Danish science to that of the major countries of Europe. He was also an ardent popularizer of science, writing articles and reviews for popular journals. Some of these writings, collected and published as *The Soul in Nature,* reveal his deepest philosophical and scientific beliefs.

There is a unity in Oersted's scientific work that is rarely found in the results of someone whose researches ranged from the forces of chemical affinity, electromagnetism, and the compressibility of fluids and gases to the new phenomenon of diamagnetism. This unity was drawn from Oersted's philosophy, inspired by his reading of Kant. Most Kantian scholars today would insist that Oersted totally misread Kant and came to conclusions to which Kant would have objected. That charge is probably correct; but what is important is that Oersted, and a number of other philosophers and scientists of the time, misread Kant in the same way. Basically, what Oersted thought Kant was saying was that science was not merely the *dis*-covery of Nature; that is, the scientist did not just record empirical facts and sum them up in mathematical formulas. Rather, the human mind imposed patterns upon perceptions; and the patterns were scientific laws. That those patterns were not arbitrary was guaranteed by the existence of Reason. Human reason corresponded to the Divine Reason, for man was made in the image of God. And, inasmuch as God had created Nature, it too shared in the Divine Reason. Thus human reason, unaided, could construct the laws of nature by virtue of its congruence with the Divine Reason. "Was der Geist versprecht, leistet die Natur" is a misquotation from Schiller's *Columbus*—"Mit dem Genius steht die Natur in ewigem Bunde, Was der Eine Verspricht, leistet die andre gewiss"—that Oersted used more than once in *The Soul in Nature.* It represents the basic position of *Naturphilosophie.*

Oersted's reading of Kant led him to more than an attitude toward nature. It also gave him what he felt was a firm metaphysical foundation for his beliefs. In a now neglected treatise, *Metaphysische Anfangsgründe der Naturwissenschaft* (1786), Kant had abandoned some of his agnosticism expressed in the antinomies in the *Critique of Pure Reason.* More particularly, whereas in the *Critique* he had argued that it was impossible for reason to decide between an atomistic or a plenist concept of matter, in the *Metaphysische Anfangsgründe* he came down on the side of the antiatomists. He argued that we experience only force; that force manifests itself in matter as the force of attraction that defines the limits of a body and the force of repulsion that gives a body the property of impenetrability. These two forces Kant called *Grundkräfte* (basic forces). Other forces, such as electricity, magnetism, heat, and light, he hinted, were merely modifications of the *Grundkräfte* under different conditions.

Oersted read both the *Critique of Pure Reason* and the *Metaphysische Anfangsgründe* while still at the university. His doctoral dissertation is a defense of the *Metaphysische Anfangsgründe* and an attempt to have it accepted in Denmark as a basic philosophical treatise. As early as 1800 it is possible to discern the two elements that were fundamental in Oersted's later scientific work: the clear enunciation of the doctrine of forces and the disbelief in atoms. The first was to lead him, through the convertibility of forces, to the discovery of electromagnetism; the second seems to have been the stimulus behind his work on compressibility, for if solid, incompressible atoms existed, there ought to come a point when further compression of a gas or fluid was impossible.

In 1800, however, Oersted's ideas were only half-formed. He was far more *au courant* in philosophy than he was in science. This is why his journey to Germany and France was so crucial. It acquainted him with men who were at the frontiers of science and forced him to bring his philosophical speculations down to earth.

The reentry was a difficult one. The "new" chemistry of Lavoisier and the other French chemists left him unmoved because it turned its back on the very questions, such as elective affinity and the true nature of acids and bases, that fascinated Oersted. Winterl's system, on the other hand, was just what he was looking for. Instead of some thirty-odd elements, defined only empirically as the last products of a laboratory analysis, Winterl offered two fundamental and opposed substances. Andronia and Thelycke could be viewed as materializations of the *Grundkräfte* and chemistry could then, it was hoped, be seen as a Kantian science. Similarly, Ritter's work in electrochemistry appeared to Oersted as a development of Kantian thought and all of a piece with his own philosophy of forces. It was only when his philosophical theories and the empirical facts refused to fit together in repeatable experiments that Oersted's critical faculties were awakened. It is significant that, at this point, he did not reject his philosophical faith. Instead, he rejected the physical systems of Winterl and Ritter. His first real scientific achievement was to create his own system, based upon his own experiments. The results appeared in German in 1812 and in a French translation in 1813. The title of the latter, *Recherches sur l'identité des forces chimiques et électriques*, indicates its purpose. From the *Grundkräfte*, Oersted hoped to deduce a system of chemistry that would be in accordance with the results of experiment.

The *Recherches* is an undeservedly neglected work of theoretical chemistry. The standard histories of chemistry barely mention it, yet it tried to come to grips with some of the major problems of the day. Specifically, it sought to make some sense of the various chemical reactions involved in combustion and the neutralization of acids and bases. By 1813 Lavoisier's theory of acids and of combustion could be severely criticized. Humphry Davy's work on chlorine showed that oxygen was not the only supporter of combustion. The fact that hydrochloric acid contained no oxygen also proved that oxygen was not, as Lavoisier had claimed, the principle of acidity. Oersted now tried to show how one could create a new chemistry based on forces, not elements.

According to Oersted, the Kantian *Grundkräfte* of attraction and repulsion manifest themselves in chemistry as combustibles and combusters. These forces are in conflict and when allowed, in combustion, to come to grips with one another, so to speak, produce the light and heat that are so preeminently the effects of combustion. But these two forces do not annihilate one another chemically; instead, they produce a higher synthesis—the acids and bases. Acidity and basicity, in turn, are opposites which unite to form the neutral salts. The supposedly Hegelian triad of thesis-antithesis-synthesis is here clear and is a standard aspect of *Naturphilosophie*. Although this analysis of fundamental chemical processes did provide a conceptual unity where before there was chaos, it left little impression upon Oersted's colleagues. Nor did his final chapters, in which he examined the convertibility of other forces. By 1813 everyone admitted the chemical role of electricity, but Oersted's treatment of it seemed to be of little help. What is of interest, at least to historians of science, is his discussion of the possibility of the conversion of electricity into magnetism.

It is important to stress that electromagnetism was not an effect to be expected according to the orthodox, corpuscular theories of the day. Coulomb seemingly had proved in the 1780's that electricity and magnetism were two entirely different species of matter whose laws of action were mathematically similar but whose natures were fundamentally different. The conversion of one into the other was, literally, unthinkable. Hence, those who accepted Coulomb's findings simply did not look for a magnetic effect.

For Oersted the situation was quite different. The Kantian doctrine of *Grundkräfte* led directly to the idea of conversion of forces. All that was necessary was to discover the conditions under which such conversions took place. The particular conditions for the conversion of electricity into magnetism were deduced by Oersted from the nature of electricity. Electricity to him was a conflict of the positive and negative aspects of magnetism, which conflict spread out in wave fashion in space. When the electric conflict was confined in a rather narrow-gauge wire, the result was heat. When the conflict was restricted still further by decreasing the diameter of the wire, light was produced. So, Oersted suggested in his treatise on the identity of chemical and electrical forces, the magnetic force should be produced when the electrical conflict is still further confined in a very narrow-gauge wire. In 1813, therefore, he had already predicted the existence of the electromagnetic effect. He was wrong, of course, on the conditions; and this error, together with his increasing teaching duties in the years that followed, prevented him from bringing his prediction to reality. The actual discovery was made in the early spring of 1820 and may best be given in Oersted's own words.

Electromagnetism itself was discovered in the year 1820, by Professor Hans Christian Oersted, of the University of Copenhagen. Throughout his literary career, he adhered to the opinion, that the magnetical effects are produced by the same powers as the electrical. He was not so much led to this, by the reasons commonly alleged for this opinion, as by the philosophical principle, that all phenomena are produced by the same original power. . . . His researches upon this subject, were still fruitless, until the year 1820. In the winter of 1819-20, he delivered a course of lectures upon electricity, galvanism, and magnetism, before an audience that had been previously acquainted with the principles of natural philosophy. In composing the lecture, in which he was to treat of the analogy between electricity and magnetism, he conjectured, that if it were possible to produce any magnetical effect by electricity, this could not be in the direction of the current, since this had been so often tried in vain, but that it must be produced by a lateral action. This was strictly connected with his other ideas; for he did not consider the transmission of electricity through a conductor as an uniform stream, but as a succession of interruptions and reestablishments of equilibrium, in such a manner that the electrical powers in the current were not in quiet equilibrium, but in a state of continual conflict. . . . The plan of the first experiment was, to make the current of a little galvanic trough apparatus, commonly used in his lectures, pass through a very thin platina wire, which was placed over a compass covered with glass. The preparations for the experiments were made, but some accident having hindered him from trying it before the lecture, he intended to defer it to another opportunity; yet during the lecture, the probability of its success appeared stronger, so that he made the first experiment in the presence of the audience. The magnetical needle, though included in a box, was disturbed; but as the effect was very feeble, and must, before its law was discovered, seem very irregular, the experiment made no strong impression on the audience ["Thermo-electricity," in *Edinburgh Encyclopaedia* (1830), XVIII, 573–589; repr. in Oersted's *Scientific Papers*, II, 356].

Oersted could not be sure that the effect was the one he had anticipated, and therefore he deferred working on it for some three months. In July he resumed his researches and made certain that a current-carrying wire is surrounded by a circular magnetic field. The results appeared in a short paper, written in Latin, sent to the major scientific journals in Europe. The "Experimenta circa effectum conflictus electrici in acum magneticam," dated 21 July 1820, opened a new epoch in the history of physics. From it followed the creation of electrodynamics by Ampère and Faraday's *Experimental Researches in Electricity*.

Oersted's second major area of research involved the compressibility of gases and fluids. It may be, as his biographer Kirstine Meyer implies, that he became interested in this problem by noting inconsistencies in the experiments of previous investigators. There may also, however, be a matter of theoretical importance involved. In all his experiments on compressibility, especially the compressibility of fluids, Oersted was intent upon proving that the reduction in volume was proportional to the pressure. If this were so, then the law of compressibility would provide a smooth pv curve. The existence of incompressible atoms, occupying space, would force a discontinuity in this curve if and when the point could be reached when the atoms were packed tightly together. Oersted's system of forces permitted continual compression, and it seems plausible that his experiments on compressibility were intended to test the atomic hypothesis. The results were inconclusive, but his apparatus and critical acumen in detecting sources of error were of basic importance for later investigations of compressibility.

Oersted's last scientific researches were on the phenomena of diamagnetism. He tried to account for diamagnetic substances by assuming reverse polarity and reverse inductive effects in substances that were repelled from, rather than attracted to, a magnetic pole. This work, in the late 1840's, was made obsolete by Faraday's investigations, which showed that the concept of polarity could not be applied to diamagnetics.

In his last years Oersted returned to his first love, philosophy. In a series of articles, published together in *The Soul in Nature*, he considered the relation between beauty and science. He still saw the hand of God in both. Beauty in art and music was the Divine Reason manifested in the harmonies of sight and sound. "Spirit and nature are one, viewed under two different aspects. Thus we cease to wonder at their harmony." Oersted's last work, *The Soul in Nature*, was left unfinished when he died on 9 March 1851. It was intended to express, in final form, the faith that had guided his entire scientific career.

BIBLIOGRAPHY

I. Original Works. There is an autobiography, in Danish, in *Kofod's Konversationslexikon*, XXVIII (Copenhagen, 1828), but it deals only with Oersted's earlier years. The published primary sources are *H. C. Ørsted, Scientific Papers. Collected Edition With Two Essays on His Work* by Kirstine Meyer, 3 vols. (Copenhagen, 1920); and *Correspondance de H. C. Orsted avec divers savants*, H. C. Harding, ed., 2 vols. (Copenhagen, 1920). There is also a considerable amount of unpublished MS material at the Royal Academy of Sciences in Copenhagen. Oersted's views on philosophy, nature, and aesthetics are found in *The Soul in Nature* (London, 1852; repr. 1966).

II. Secondary Literature. The only biography to deal with Oersted's entire scientific life is that by Kirstine Meyer, which introduces the *Scientific Papers*.

There are a number of specialized studies on Oersted. Bern Dibner, *Oersted and the Discovery of Electromagnetism* (Norwalk, Conn., 1961), is a study of Oersted's most important work. Robert C. Stauffer's "Speculation and Experiment in the Background of Oersted's Discovery of Electromagnetism," in *Isis*, **48** (1957), 33 ff.; and "Persistent Errors Regarding Oersted's Discovery of Electromagnetism," *ibid.*, **44** (1953), 307 ff., first drew scholarly attention to the importance of *Naturphilosophie* for an understanding of Oersted's scientific career.

L. Pearce Williams

OHM, GEORG SIMON (*b.* Erlangen, Bavaria, 16 March 1789; *d.* Munich, Bavaria, 6 July 1854), *physics.*

Ohm was the oldest son of Johann Wolfgang Ohm, master locksmith, and Maria Elisabeth Beck, daughter of a master tailor. Of the Protestant couple's seven children, only two others survived childhood: Martin the mathematician and Elisabeth Barbara. The father, a self-sacrificing autodidact, gave his sons a solid education in mathematics, physics, chemistry, and the philosophies of Kant and Fichte; their considerable mathematical ability was recognized in 1804 by the Erlangen professor Karl Christian von Langsdorf, who enthusiastically likened them to the Bernoullis. Of considerably less importance than his father's tutoring was Ohm's attendance (1800–1805) at the Erlangen Gymnasium, where the predominantly classical instruction stressed recitation, translation, and interpretation of texts. On 3 May 1805 he matriculated at the University of Erlangen, where he studied for three semesters until his father's displeasure at his supposed overindulgence in dancing, billiards, and ice skating forced him to withdraw in virtual exile to rural Switzerland. In September 1806 Ohm began a two-and-a-half-year stint teaching mathematics at one

Pfarrer Zehender's *Erziehungsinstitut* in Gottstadt bei Nydau, Bern canton; in March 1809, he went to Neuchâtel for two years as a private tutor. Just before this move he had expressed to Langsdorf the desire to follow him to Heidelberg; but he was dissuaded with the advice that he would be better off studying Euler, Laplace, and Lacroix on his own.

By Easter of 1811 Ohm was back at the University of Erlangen, where on 25 October, after having passed the required examinations, he received the Ph.D. He subsequently taught mathematics for three semesters as a *Privatdozent*, his only university affiliation until near the end of his life. Lack of money and the poor prospects for advancement at Erlangen forced Ohm to seek other employment from the Bavarian government; but the best he could obtain was a post as a teacher of mathematics and physics at the low-prestige, poorly attended *Realschule* in Bamberg, where he worked with great dissatisfaction from January 1813 until the school's dissolution on 17 February 1816. From 11 March 1816 until his release from Bavarian employ on 9 November 1817, he was assigned, in the capacity of an auxiliary instructor, to teach a section of mathematics at the overcrowded Bamberg *Oberprimärschule*.

On 11 September 1817 Ohm had been offered the position of *Oberlehrer* of mathematics and physics at the recently reformed Jesuit Gymnasium at Cologne, and he began work there (evidently) sometime before the end of the year. The ideals of *wissenschaftliche Bildung* had infused the school with enthusiasm for learning and teaching; and this atmosphere—which appears later to have waned—coupled with the requirement that he teach physics and the existence of a well-equipped laboratory, stimulated Ohm to concern himself for the first time avidly with physics. He studied the French classics—at first Lagrange, Legendre, Laplace, Biot, and Poisson, later Fourier and Fresnel—and, especially after Oersted's discovery of electromagnetism in 1820, did experimental work in electricity and magnetism. It was not until early in 1825, however, that he undertook research with an eye toward eventual publication. On 10 August 1826 Ohm was granted a year's leave of absence, at half pay, to go to Berlin to continue this work. When his leave ended in September 1827, he had not yet attained his fervently sought goal of a university appointment.

Not wishing to return to Cologne, Ohm formally severed his connections there in March 1828 and accepted a temporary job to teach three recitation classes of mathematics a week at the Allgemeine Kriegsschule in Berlin. Sometime during 1832 he also took on a class at the Vereinigte Artillerie- und Ingenieurschule there. Continuing to find all higher academic doors closed to him in Prussia, Ohm hoped to have better luck in Bavaria; but although his ample qualifications were duly recognized, he could elicit no better offer (18 October 1833) than the professorship of physics at the Polytechnische Schule in Nuremberg, a job that brought him no improvement over his previous circumstances except the desirable title of professor.

Finally Ohm began to receive belated official recognition of the importance of his earlier work: he became a corresponding member of the Berlin (1839) and Turin (1841) academies, and on 30 November 1841 he received the Royal Society's Copley Medal. He became a full member of the Bavarian Academy in 1845 and was called to Munich on 23 November 1849 to be curator of the Academy's physical cabinet, with the obligation to lecture at the University of Munich as a full professor. He did not receive the chair of physics until 1 October 1852, less than two years before his death.

Ohm's first work was an elementary geometry text, *Grundlinien zu einer zweckmässigen Behandlung der Geometrie als höheren Bildungsmittels an vorbereiter len Lehranstalten* (Erlangen, 1817), which embodied his ideas on the role of mathematics in education. The student, he believed, should learn mathematics as if it were the free product of his own mind, not as a finished product imposed from without. Ideally, by fostering the conviction that the highest life is that devoted to pure knowledge, education should create a self-reliance and self-respect capable of withstanding all vicissitudes in one's external circumstances. One detects in these sentiments the reflection not only of his own early education but also of the years of isolation in Switzerland and of personal and intellectual deprivation at Bamberg. The resulting inwardness of Ohm's character and the highly intellectualized nature of his ideals of personal worth were an essential aspect of the man who would bring the abstractness of mathematics into the hitherto physical and chemical domain of galvanic electricity.

Ohm's decision in 1825 to undertake, and publish, the original research that was to immortalize his name was made only after he had become convinced that his life had run into a dead end, that he must extricate himself from what had become a stultifying situation at Cologne. Overburdened with students, finding little appreciation for his conscientious efforts, and realizing that he would never marry, he turned to science both to prove himself to the world and to have something solid on which to base his petition for a position in a more stimulating environment. (Similarly, the occasion for the publication of his geometry book had been the desire to leave Bamberg.)

Ohm's first scientific paper was "Vorläufige Anzeige des Gesetzes, nach welchem Metalle die Contakt-elektricität leiten" (May 1825).[1] In it he sought a functional relationship between the decrease in the electromagnetic force exerted by a current-carrying wire and the length of the wire. A brief discussion of his procedure is necessary to understand his results and their implications for his further work. From the zinc and copper poles of a voltaic pile he ran two wires, *A* and *B*, the free ends of which terminated in small mercury-filled cups, *M* and *N*; between *M* and another cup, *O*, he ran a third wire, *C*. Together *A*, *B*, and *C* formed what he called the "invariable conductor," to distinguish it from one of the seven wires of different lengths that, when placed in the circuit between *O* and *N*, constituted the "variable conductor." Among the latter was one "very thick" wire, four inches long, and six thinner ones, 0.3 line (.025″) in diameter, ranging in length from one foot to seventy-five feet. Finally, over wire *C* hung the magnetic needle of a Coulomb torsion balance, which served to measure the electromagnetic force exerted when one of the variable conductors completed the circuit.

Ohm referred all his force readings to the so-called normal force produced by the short, thick wire and chose as his variable the loss in force (*Kraftverlust*) brought about by one of the six longer and thinner test wires. This loss in force was equal to the difference between the normal force and the lesser force occasioned by one of the other wires, divided by the normal force. Tabulating these values against the lengths of the wires, he found that his data were well represented by the formula $v = 0.41 \log (1 + x)$, where v is the loss in force and x is the length of the wire in feet. (This seems to have been a purely empirical fit to his data.) Differentiating this equation—whereby he apparently forgot he was using common logarithms—to get $dv = m [dx/(1 + x)]$, Ohm then speculated that its general form might be $dv = m [dx/(a + x)]$; a would represent the equivalent length of the invariable conductor (which in the previous case by chance had been equal to 1). Hence the general equation, ignoring an additive constant, is $v = m \log (1 + x/a)$, which he found quite well confirmed by subsequent experiments and took as the sought-for law. Ohm believed that the coefficient m was a function of the normal force, the thickness of the wire, the value of a, and the "electric tension of the force." He seems actually to have believed that the loss in force would be total (that is, $v = 1$) for a sufficiently long conductor, as required by his formula. One of the striking features of this and Ohm's other early papers was their direct foundation on experiment. Indeed, several could be taken as models of inductive derivation of mathematical laws from empirical data. In his mature work of 1827, however, Ohm, under the influence of Fourier, adopted a highly abstract theoretical mode of presentation that obscured the theory's close relationship with experiment.

It is not obvious why Ohm chose to measure the loss in force and not the force itself. It should be noted, however, that he nowhere spoke of measuring the current; rather, he wanted to find out by what amount the electromagnetic force exerted by a given conductor was weakened when another, longer conductor was placed in the same circuit. From the beginning he sought a law that would elucidate the complex relationship between battery and conductor, and it is possible that he regarded the progressive attenuation of the battery's force by ever longer conductors as the central phenomenon to be explained. In this regard it is significant that three of Ohm's cryptic references to his formula's applicability were to the behavior of different forms of the pile; the other reference was to a series of experiments in which Poggendorff had shown that the magnifying effect of a multiplier eventually reached a limit as the number of turns— and thereby also the length of the conductor—was increased.[2]

In the same month that Ohm's first paper was published (May 1825) there appeared an extract in Férussac's *Bulletin des sciences mathématiques* of A.-C. Becquerel's and Barlow's work on the electric conductibility of metals.[3] Becquerel, like Davy before him, was primarily interested in comparing the "conducting powers" of different wires.[4] Their findings were similar: Becquerel said that to obtain the same conductibility with wires of the same metal, their lengths should be in the same ratio as their cross sections; Davy had said that the conducting powers of wires of the same metal varied directly with their mass (per unit length) and inversely with their length. Each also determined the relative conductibility of different metals, although their results differed markedly. Whereas neither Becquerel nor Davy actually measured anything like the current or the electromagnetic effect—both preferring an equilibrium or null-effect type of experiment —Barlow sought a direct relationship between current intensity, as measured by the deflection of a magnetic needle, and the length and diameter of the conductor. He found that this intensity varied roughly with the inverse square root of the length of the wire and that, for wires all of the same length, it increased with their diameters only up to a certain point, after which any further increase in the diameter of the wire had no effect on the intensity.

Additional experiments by both Barlow and Becquerel had corroborated that the electromagnetic effect did not vary sensibly at different points along the same wire, thereby proving that something having to do with the current remained constant throughout the circuit. Barlow had expected to find a steady diminution of effect either from the positive pole to the negative or from both poles toward the center, and thereby to be able to decide in favor of either the one-fluid or the two-fluid theory of electricity; hence the apparent inconclusiveness of this experiment puzzled him. Becquerel, however, used the same observation, in conjunction with his finding that conductibility decreased with length, in his explanation of the nature of the electric current. He conceived of it as a double stream, going in opposite directions, of positive and negative electricity, such that the intensity or quantity of each—Becquerel was not precise in his distinctions —decreased arithmetically from its pole of origin, resulting in a constant net current at all points. This conjecture, along with Becquerel's original observation that the electromagnetic effect did not vary over the length of the conductor, may have influenced Ohm's subsequent work. In it Ohm clarified with mathematical precision exactly what remained constant (the current) and what gradually decreased (the tension, or electroscopic force) along a conducting wire. At the least Ohm now took it upon himself to eliminate the discrepancies among these related findings. His suspicion, subsequently disproved, that conductibility varied with the strength of the current, made it all the more natural for him to incorporate the force into the relationship for conductibilities.

In February and April 1826, Ohm published two important papers that dealt separately with the two major aspects of his ultimately unified theory of galvanic electricity. The first, "Bestimmung des Gesetzes, nach welchem Metalle die Contaktelektricität leiten, nebst einem Entwurfe zu einer Theorie des Voltaischen Apparates und des Schweiggerschen Multiplicators," announced a comprehensive law for electric current that brought order into the hitherto confused collection of phenomena pertaining to the closed circuit, including the solution to the problem of conductibility as he and others had conceived of it.[5] The second paper, "Versuch einer Theorie der durch galvanische Kräfte hervorgebrachten elektroskopischen Erscheinungen," broke new ground in associating an electric tension with both open and closed galvanic circuits.[6]

Ohm's experimental procedure in the first of these papers was analogous to that which he had used earlier but was modified in several significant ways. First, at Poggendorff's suggestion he now used a thermoelectric pile in order to eliminate the fluctuations in current strength accompanying the voltaic pile, fluctuations that Ohm attributed to changes produced by the current in the distribution (*Vertheilung*) of the components of the liquid conductor. Second, he sought a direct relationship between the electromagnetic force of the current and the entire length of the connecting wire. Although there is some evidence that Ohm may have been in possession of his new, correct law before he undertook this later series of experiments, he presented it as if it were a straightforward induction from his data and later consistently referred to it as having been derived from his experiments.

Be that as it may, in the paper in question Ohm simply observed that the data from each of his several series of experiments were very closely represented by the formula $X = a/(b + x)$, where X is the strength of the electromagnetic effect—which he took as a measure of the electric current—of a conductor of length x on the magnetic needle of a Coulomb torsion balance, and where a and b are constants the exact nature of which he proposed to determine from additional series of carefully controlled experiments. The observation that b remained constant for all series of experiments, whereas a varied with temperature, led Ohm to conclude that a depended solely on the electromotive force (*erregende Kraft*) of the pile and b solely on the resistance (*Leitungswiderstand* or, more commonly, *Widerstandslänge*) of the remaining portion of the circuit, in particular that of the pile itself. He also observed that the electromotive force of the thermoelectric pile appeared to be exactly proportional to the temperature difference at its end points. This process of reasoning back and forth between the experimental data and their mathematical representation, through which he was able to discover the physical significance of the terms, is a characteristic of Ohm's methodology.

After reconfirming the validity of his law by further series of experiments, Ohm exhibited its explanatory powers on some of the chief unsolved problems which had occupied scientists working on the pile; and he showed how it also cast light on a number of other previously reported but poorly understood experimental findings. For example, he was able to explain the apparent differences in behavior between voltaic and thermoelectric pile by pointing out that although both the electromotive force a and the resistance b are normally much greater in the voltaic pile than in the thermoelectric pile, the current in a circuit composed solely of a thermoelectric element bent back upon itself—for which $x = 0$ in the expression $a/(b + x)$ —could exert just as great an electromagnetic effect as the voltaic pile. According to Ohm's formula, however, the introduction of another conductor into each circuit would result in a relatively much greater diminution in the electromagnetic effect of the thermoelectric circuit than of the hydroelectric circuit, which was known to be the case. It had previously seemed anomalous that of two piles capable of registering the same electromagnetic action, one, the thermoelectric, should be incapable of producing either chemical actions or the ignition of fine wires. Such differences had either been attributed to a qualitative difference between electricities stemming from different sources or had been explained by saying that the electricity produced by the thermoelectric pile was greater in quantity but lower in intensity relative to that of the hydroelectric, or voltaic, pile. In addition, Ohm developed a simple mathematical theory of the multiplier that enabled him to say under exactly what conditions it would either amplify or diminish the electromagnetic effect, why this amplification eventually reached a maximum, and why the multiplier usually seemed to weaken the electromagnetic effect of a thermoelectric circuit, whereas it markedly strengthened that of a hydroelectric circuit. The fruitful application of Ohm's simple law to existing problems was an explanatory tour de force.

Ohm's second major paper of 1826 announced the beginnings of a comprehensive theory of galvanic electricity based, he said, on the fact that the contact of heterogeneous bodies produced and maintained a constant electric tension (*Spannung*). He deferred the systematic exposition of this theory to a later work, however, and limited himself to stating without derivation the two equations that constituted its heart: $X = kw(a/l)$ and $u - c = \pm(x/l)a$, where X is the strength of the electric current in a conductor of length l, cross section w, and conductibility (*Leitungsvermögen*) k produced by a difference in electric tension a at its end points; where u is the electroscopic force at a variable point x of the conductor; and where c is a constant independent of x. By means of the first equation one can, with respect to overall conducting power (or resistance), reduce the actual length of a wire of whatever cross section and conductibility to the equivalent length of one wire chosen arbitrarily as a standard. Letting l now be this equivalent length—called the reduced length (*reducirte Länge*) of the conductor—Ohm wrote his first law in the simpler form $X = a/l$, the expression which has become known as Ohm's law.

After pointing out briefly how this law, which corresponded to the one he had developed in his previous paper, embraced his and others' findings on the conductibility of different wires, Ohm devoted the rest of the paper to developing the implications of the second, electroscopic law and to comparing these implications with previously known facts. In this work he showed that his formula successfully explained those experiments which measured the electroscopic force at different points (especially the poles) of open and closed, and grounded and ungrounded, circuits. Here again the explanatory power of his law was impressive.

The fully developed presentation of his theory of electricity appeared in Ohm's great work, *Die galvanische Kette, mathematisch bearbeitet* (Berlin, 1827). Hoping to make the book more accessible to the mathematically unsophisticated, he devoted the first third of it to an introduction in which he attempted an essentially geometric presentation of his theory. The introduction, which contained a discussion of the theory's success in explaining the property of conductibility, the phenomena of the pile, and the behavior of the electromagnetic multiplier, was virtually the only part of the book in which he referred explicitly to the theory's very close connections with experiment. But in neither the introduction nor in the body of the work, which contained the more rigorous development of the theory, did Ohm bring decisively home either the underlying unity of the whole or the connections between fundamental assumptions and major deductions. For example, although his theory was conceived as a strict deductive system based on three fundamental laws (*Grundgesetze*), he nowhere indicated precisely which of their several mathematical and verbal expressions he wished to be taken as the canonical form. The following exposition, although simplified by the omission of steps in the derivation and of the theory's more specialized developments, follows the letter of Ohm's work as it attempts to provide a clearer synopsis than is sometimes afforded by the book.

As a preliminary to the formulation of his fundamental laws, Ohm defined the electroscopic force operationally as that force the presence of which was detected by means of an electroscope, and the quantity of electricity of a body as the product of the magnitude of its electroscopic force times its volume. These definitions, in the context of the larger theory, gave the previously vague but universally used notions of intensity and quantity of electricity a precise interpretation.

Ohm's first *Grundgesetz* pertained to the communication of electricity from one body to another, and it involved the explicit assumption that the quantity of electricity communicated was proportional to the difference in the bodies' electroscopic force, an assumption the validity of which would be proved by the subsequent correspondence between theory and experiment. This hypothesis, coupled with the definition of conductibility as the quantity of electricity

transferred per unit time across a unit distance, led directly to the expression

$$(1) \qquad \frac{\kappa(u' - u)\, dt}{s}$$

for the quantity of electricity communicated in time dt between two bodies of electroscopic force u' and u, separated by a distance s, where κ is the conductibility relative to these bodies. This may be taken as the mathematical expression of his first fundamental law.

Ohm's second *Grundgesetz*—which he based on the results of experiments Coulomb had done on the loss into the surrounding air of the electricity of a charged body—declared that, for an infinitesimal slice of thickness dx of a current-carrying conductor of circumference c, this loss across the surface in the time interval dt was proportional to that time, to the electroscopic force of the slice, and to its surface area, or to

$$(2) \qquad bcu\,dx\,dt,$$

where b is a constant dependent only on the condition of the air. As Ohm himself observed, this law has little or no applicability to galvanic phenomena; it was included for the sake of completeness and to maintain the desired parallelism between the fundamental equations of electricity and heat.

Ohm's third *Grundgesetz* embodied the fundamental tenet of the contact theory of electricity by asserting that heterogeneous bodies in contact maintain a constant difference in electroscopic force (tension) across their common surface. Mathematically,

$$(3) \qquad (u) - (u') = a,$$

where the parentheses simply indicate that the quantities they enclose are to be evaluated at the common surface between the two conductors, and where a is the magnitude of the constant difference. This fact he considered to be the basis (*Grundlage*) of all galvanic phenomena.

Ohm derived several important results directly from the first fundamental law. Applying it to three infinitesimal slices M', M, and $M_{,}$ of a homogeneous prismatic current-carrying conductor, the quantities of electricity transferred from M' to M, and from $M_{,}$ to M, are

$$(4) \qquad \frac{\kappa(u' - u)\, dt}{dx} \quad \text{and} \quad \frac{\kappa(u_{,} - u)\, dt}{dx},$$

respectively, where u', u, and $u_{,}$ are the electroscopic force and $x + dx$, x, and $x - dx$ are the abscissas of M', M, and $M_{,}$. Hence the total increase in the quantity of electricity of slice M is $[\kappa(u' + u_{,} - 2u)dt]/dx$, which, by means of the Taylor series expansions for u' and u, , can be written as

$$(5) \qquad \kappa\omega \frac{d^2u}{dx^2} \, dxdt,$$

where the conductibility κ has now been referred to unit cross section, ω being the cross section of the conductor. Furthermore, observing that each of the expressions in (4) is individually equal to $\kappa\omega(du/dx)dt$, Ohm defined the electric current S as the quantity of electricity passing through a given cross section of the conductor in unit time, and wrote

$$(6) \qquad S = \kappa\omega \frac{du}{dx},$$

which related the current directly to the (change in) electroscopic force. He then used this equation as the basis of the important condition for the continuity of current between two conductors,

$$(7) \qquad \kappa\omega \left(\frac{du}{dx} \right) = \kappa'\omega' \left(\frac{du'}{dx} \right),$$

where the parentheses have the same meaning as in (3).

The total change in the quantity of electricity of an infinitesimal slice of conductor is found by adding expressions (2) and (5). But, from the definition of quantity of electricity, this change is just equal to $\omega(du/dt)dxdt$—which quantity must, however, be multiplied by a factor γ, analogous to the coefficient for heat capacity, if equal changes in electroscopic force are not always accompanied by equal changes in the quantity of electricity. From these considerations Ohm derived the important general equation

$$(8) \qquad \gamma \frac{du}{dt} = \kappa \frac{d^2u}{dx^2} - \frac{bc}{\omega} u.$$

Although Ohm solved this equation in its full generality, as well as for the steady-state case when $b \neq 0$ (that is, when the influence of the air may not be ignored), the only really useful solution was for the steady-state case when $b = 0$. Under these conditions the equation reduces to $0 = d^2u/dx^2$, the general solution of which is

$$(9) \qquad u = fx + c.$$

For the idealized case of a simple circuit composed of a conductor of length l, bent back upon itself so that the cross sections at $x = 0$ and $x = l$ are in contact, and of a single source of tension (*Erregungsstelle*) located at this common point, equation (3), taken in conjunction with (9), implies that

$$(u)_{x=l} - (u)_{x=0} = f \cdot l - f \cdot 0 = a.$$

Hence $f = a/l$; and for this simple circuit

$$(10) \qquad u = (a/l)x + c,$$

where the constant c is determined whenever the electroscopic force at any one point is known—as, for example, by the circuit's being grounded.

In a derivation too lengthy to recapitulate here, Ohm showed that equation (10) can be generalized to circuits composed of any number of different conductors and sources of electromotive force, for which

$$(11) \qquad u = (A/L)y - O + c,$$

where A is the sum of the tensions of all sources of electromotive force; L is the total reduced length of the entire circuit; y is the so-called reduced abscissa, equal to the reduced length of that portion of the circuit between the origin and the point in question; and O is the sum of the tensions of all sources lying between the origin and that point.

Now from equations (6) and (11) one has

$$S = \kappa\omega \frac{du}{dy} \cdot \frac{dy}{dx} = \kappa\omega \frac{A}{L} \cdot \frac{dy}{dx}.$$

As Ohm showed from the (here omitted) derivation of equation (11), dy/dx, which simply relates the change in reduced length to the change in real length of the conductor, is just equal to $1/\kappa\omega$. Hence

$$(12) \qquad S = A/L.$$

This equation—which is, again, Ohm's law as we know it—states that the current in a galvanic circuit is constant across all cross sections and is equal to the sum of all the tensions divided by the total reduced length of the circuit.

Equations (11) and (12) epitomize the theory as it pertains to the electroscopic and current manifestations of the galvanic circuit, respectively. Ohm's major conceptual originality lay in explicating the intrinsic relationship between tension and current, and in associating a varying electric tension, or electroscopic force, with each point of a current-carrying wire. The relationship between these two classes of phenomena had at best been obscure when, as was often the case, they were not regarded as mutually exclusive. This belief was, however, not without foundation, since in general one had been able to measure the electric tension of a pile only when no current flowed. Earlier experiments of Erman, Ritter, and C. C. F. Jäger, to which Ohm referred, had demonstrated not only the presence of an electroscopic force at the poles of a pile closed by means of a poor conductor (such as water) but also the progressive decrease in this force from the poles toward the center of the connecting conductor.[7] To the extent to which these experiments had not simply been forgotten, however, they were thought inapplicable to the case of metallic conduction because of the traditional classification of substances into perfect, imperfect, and nonconductors, each with its own peculiar characteristics. To Ohm, who had the mathematical physicist's tendency to regard properties less as an "either-or" of some quality than as a "more-or-less" of some quantity, such distinctions could have no intrinsic validity; and he did not hesitate to apply to metals findings originally restricted to imperfect conductors.

It was not a matter of casual importance that Ohm regarded the force arising at the contact surface of heterogeneous substances as the cardinal fact and starting point of his theory, for his acceptance of the contact theory of electricity was probably crucial to the genesis of his own theory. It was the contact theory that asserted the existence of an impulsive electromotive force, and it was this electromotive force (of the closed pile) which Ohm identified conceptually with the electroscopic force (of the open pile). Measurement of the electric tension of the open pile (while no current flowed and no chemical activity took place) by means of an electroscope was one of the foundation stones of the contact theory, as was the fact that this tension increased as the number of metallic couples was increased. Indeed, the very existence of such an additive electromotive force was an acute embarrassment to the defenders of the chemical theory of the pile, who consequently tended to play down the very phenomena from which Ohm borrowed one of his central concepts.

Ohm structured his theory in conscious imitation of Fourier's *Théorie analytique de la chaleur* (1822), a fact that may have induced him to deemphasize its experimental side in favor of an abstract deductive rigor, in striking contrast with the inductivist tone of his earliest papers. In particular his basic expressions for the conduction of electricity through a solid (1) and for the loss of electricity from the surface into the air (2), as well as his resulting general equation (8), are exactly analogous to Fourier's equations for the motion of heat. Although he did not spell out just how, Ohm wished the analogy between electricity and heat to be taken seriously, not as something merely coincidental but as revealing some underlying relationship. It is possible that Seebeck's thermoelectric pile had powerfully suggested the intimate relationship between the two phenomena that Ohm endeavored to exploit in his own theory.

Although Ohm's work was not immediately and universally appreciated even within Germany—largely because the majority of German physicists in 1827 represented a soon-to-be-superseded nonmathematical approach to physics—already by the early 1830's it was beginning to be used by all the younger physicists working in electricity: Gustav Theodor Fechner gave Ohm's theory a prominent place in his *Lehrbuch des Galvanismus und der Elektrochemie* (Leipzig, 1829) and subjected it to rigorous experimental testing (and confirmation) in his *Massbestimmungen über die galvanische Kette* (Leipzig, 1831); Heinrich Friedrich Emil Lenz used it in his first paper on electromagnetic induction, "Über die Gesetze nach welchen der Magnet auf eine Spirale einwirkt wenn er ihr plötzlich genähert oder von ihr entfernt wird und über die vortheilhafteste Construction der Spiralen zu magneto-electrischem Behufe," read on 7 November 1832;[8] Wilhelm Eduard Weber and Karl Friedrich Gauss used it from 1832–1833 in connection with their investigations on terrestrial magnetism and their construction of precision instruments; and Moritz Hermann Jacobi became familiar with it sometime after 1833 and used it in his first appreciable publication, *Mémoire sur l'application de l'Électro-Magnétisme au Mouvement des Machines* (Potsdam, 1835). On the other hand, the question of how fast Ohm's work became known and appreciated by the majority of scientists who were not particularly concerned with that branch of physics has still to be answered. One would like to know, for instance, how soon it entered the textbooks; suggesting its rather quick adoption was its inclusion in the *Supplementband* (Vienna, 1830–1831) to Andreas Baumgartner's *Naturlehre* (a popular text that went through eight editions between 1824 and 1845), although it remains to be seen whether this example was typical. English and French physicists seem not to have become aware of Ohm's work and its profound implications for electrical science until the late 1830's and early 1840's.[9]

It has been repeatedly asserted ever since the middle of the last century that Ohm's work had to await the recognition of foreign scientists around 1840 before it became well known in Germany. Insofar as his fame among the larger scientific and nonscientific community is concerned, there may be some truth to that assertion. However, by then his work had already been used by those working in electricity who should have appreciated it, at least among the scientists born after 1800. Nor does that traditional explanation gain plausibility from the observation that in the nineteenth century the notion had become a commonplace in Germany that Germans only esteemed what came from abroad, hence the uncritical commentator had a familiar and convenient dictum ready at hand to explain a complex situation.[10] The issue of the acceptance of Ohm's work by contemporary scientists has been further confounded with his lack of success in securing an academic appointment. In connection with the latter, to make matters worse, the fact that his chief adversaries in Berlin—Johannes Schulze, a powerful figure in the ministry of education, and Georg Friedrich Pohl, professor of physics at the Friedrich-Wilhelms-Gymnasium—were followers of Hegel and of *Naturphilosophie* has wrongly been taken as characteristic of the general situation in German physics. And even this confrontation was not simply a matter of ideologies: Martin Ohm, several years before, had incurred Schulze's dislike and had gained the reputation in Berlin of being a dangerous revolutionary because of his criticisms of the educational system; among his suggestions for reform had been the use of his brother's geometry text, which did not find favor in Berlin.

NOTES

1. In Schweigger's *Journal für Chemie und Physik*, **44** (1825), 110–118. Also in Poggendorff's *Annalen der Physik und Chemie*, **4** (1825), 79–88.
2. J. C. Poggendorff, "Physisch-chemische Untersuchungen zur nähern Kenntniss des Magnetismus der voltaischen Säule," in *Isis von Oken* (1821), **2** (**9** in the series), no. 8, cols. 687–710.
3. A.-C. Becquerel, "Du pouvoir conducteur de l'électricité dans les métaux, et de l'intensité de la force électro-dynamique en un point quelconque d'un fil métallique qui joint les deux extrémités d'une pile lui à l'Académie royale des sciences le 31 Janvier 1825," in *Annales de chimie et de physique*, **32** (Aug. 1826), 420–430; and Peter Barlow, "On the Laws of Electro-Magnetic Action, as Depending on the Length and Dimensions of the Conducting Wire, and on the Question, Whether Electrical Phenomena Are Due to the Transmission of a Single or of a Compound Fluid ?" in *Edinburgh Philosophical Journal*, **12**, no. 23 (Jan. 1825), 105–114. Extracts of these, which Ohm saw, appeared in *Bulletin des sciences mathématiques, astronomiques, physiques et chimiques*, **3**, no. 5 (May 1825), 293–296 and 296–298, respectively.
4. Humphry Davy, "Farther Researches on the Magnetic Phaenomena Produced by Electricity; With Some New Experiments on the Properties of Electrified Bodies in Their Relations to Conducting Powers and Temperature," in *Philosophical Transactions of the Royal Society*, **111** (1821), 425–439. Ohm knew the German trans. in Gilbert's *Annalen der Physik*, **71** (1822), 241–261.
5. In Schweigger's *Journal für Chemie und Physik*, **46** (1826), 137–166.
6. In Poggendorff's *Annalen der Physik und Chemie*, **6** (1826), 459–469; *ibid.*, **7** (1826), 45–54, 117–118.
7. Paul Erman, "Ueber die electroskopischen Phänomene der Voltaischen Säule," in Gilbert's *Annalen der Physik*, **8** (1801), 197–209; and "Ueber die electroskopischen Phänomene des Gasapparats an der Voltaischen Säule," *ibid.*, **10** (1802), 1–23; J. W. Ritter, "Versuche und Bemerkungen über den Galvanismus der Voltaischen Batterie. . . . Dritter Brief," *ibid.*, **8** (1801), 385–473; C. C. F. Jäger, "Ueber die electroskopischen Aeusserungen der Voltaischen Ketten und Säulen," *ibid.*, **13** (1803), 399–433. Even the

recent experiment of Ampère and Becquerel had left open the question of whether tension was associated with complete conduction by metals, since they too detected a tension only at the poles of a pile closed by means of a so-called incomplete conductor; see "Note sur une Expérience relative à la nature du courant électrique, faite par MM. Ampère et Becquerel," in *Annales de chimie et de physique*, **27** (Sept. 1824), 29–31.
8. *Mémoires de l'Académie impériale des sciences de St.-Pétersbourg*, 6th ser. Sciences mathématiques, physiques et naturelles, **2** (1833), 427–457; repr. in Poggendorff's *Annalen der Physik und Chemie*, **34** (1835), 385–418; and trans. in Taylor's *Scientific Memoirs*, **1** (1837), 608–630.
9. The first exposition of Ohm's work in French that I know of was Élie Wartmann, "Des travaux et des opinions des Allemands sur la pile voltaïque," in *Archives de l'électricité*, **1** (1841), 31–66, followed by Auguste de la Rive, "Observations sur l'article de M. Wartmann . . .," *ibid.*, 67–73.
10. See, for example, Schweigger's *Journal für Chemie und Physik*, **10** (1814), 355; **23** (1818), 372; **33** (1821), 20; and Poggendorff's *Annalen der Physik und Chemie*, **3** (1825), 191. Leibniz' comment on his countrymen, "nil nisi aliena mirantur," was often cited to support the generality of this supposed nationality trait.

BIBLIOGRAPHY

I. ORIGINAL WORKS. A nearly complete list of Ohm's scientific papers is found in the Royal Society *Catalogue of Scientific Papers*, IV, 665–666. Also useful is Poggendorff, II, cols. 316–318, which lists books as well as papers. All but two of Ohm's papers, plus the book *Die galvanische Kette*, are collected in *Gesammelte Abhandlungen von G. S. Ohm*, edited with an intro. by E. Lommel (Leipzig, 1892). Ohm also wrote a textbook, *Grundzüge der Physik als Compendium zu seinen Vorlesungen*, 2 vols. (Nuremberg, 1853–1854). Two of his earlier papers—"Vorläufige Anzeige . . ." and "Bestimmung des Gesetzes . . ."—are reprinted in *Das Grundgesetz des elektrischen Stromes. Drei Abhandlungen von Georg Simon Ohm (1825 und 1826) und Gustav Theodor Fechner (1829)*, C. Piel, ed. (Leipzig, 1938), which is Ostwald's Klassiker der exakten Wissenschaften, no. 244. An English trans. of the second of these papers has been published by Niels H. de Vaudrey Heathcote as "A Translation of the Paper in Which Ohm First Announced His Law of the Galvanic Circuit, Prefaced by Some Account of the Work of His Predecessors," in *Science Progress*, **26**, no. 101 (July 1931), 51–75.

The 1st ed. of *Die galvanische Kette* has been reprinted in facs. (Brussels, 1969). It was translated into English by William Francis as "The Galvanic Circuit Investigated Mathematically," in R. Taylor, J. Tyndall, and W. Francis, eds., *Scientific Memoirs, Selected From the Transactions of Foreign Academies and Learned Societies and From Foreign Journals*, II (London, 1841), 401–506, and later reprinted (New York, 1891), no. 102 in the Van Nostrand Science Series. There is a French ed., *Théorie mathématique des courants électriques*, translated with preface and notes by Jean-Mothée Gaugain (Paris, 1860); and an Italian one (not seen), "Teoria matematica del circuito galvanico," in *Cimento* (Pisa), **3** (1845), 311–348; **4** (1846), 85–96, 169–183, 246–266. A very useful and informative source is *Aus Georg Simon Ohms handschrift-lichem Nachlass. Briefe, Urkunden und Dokumente*, Ludwig Hartmann, ed. (Munich, 1927), which contains much of the MS material on Ohm in the Deutsches Museum in Munich.

II. SECONDARY LITERATURE. The fullest biography is Heinrich von Füchtbauer, *Georg Simon Ohm. Ein Forscher wächst aus seiner Väter Art* (Berlin, 1939), which contains extracts of letters not available elsewhere. Also very informative is the article by Carl Maximilian von Bauernfeind in *Allgemeine deutsche Biographie*, XXIV (Leipzig, 1887), 187–203. Useful for some aspects of his background and life is Ernst G. Deuerlein, *Georg Simon Ohm 1789–1854. Leben und Wirken des grossen Physikers* (Erlangen, 1939; 2nd ed., enl., 1954). Two contemporary eulogies are valuable: Friedrich von Thiersch, "Rede zur Feier des hohen Geburtsfestes Sr. Majestät des Königs Maximilian II. von Bayern," in *Gelehrte Anzeigen der k. bayerischen Akademie der Wissenschaften*, **40** (Jan.–June 1855), Bulletins der drei Classen, nos. 3–4 (Jan. 5–8), cols. 26–32, 33–35 (indicated here are only those portions of the speech dealing with Ohm; the bulk of the published article is a long footnote on Ohm by Philipp Ludwig Seidel, cols. 29–35); and Johann von Lamont, *Denkrede auf die Akademiker Dr. Thaddäus Siber und Dr. Georg Simon Ohm . . .* (Munich, 1855).

The best account of Ohm's electrical work is Morton L. Schagrin, "Resistance to Ohm's Law," in *American Journal of Physics*, **31**, no. 7 (July 1963), 536–547. Not to be trusted, especially in its translations, is Henry James Jacques Winter, "The Reception of Ohm's Electrical Researches by His Contemporaries," in *London, Edinburgh and Dublin Philosophical Magazine and Journal of Science*, 7th ser., **35**, no. 245 (June 1944), 371–386. Worth consulting are two articles by John L. McKnight: "The Intellectual Development of Georg Simon Ohm," in *Actes du XIe Congrès international d'histoire des sciences, Varsovie-Toruń-Kielce-Cracovie, 24–31 août 1965*, III (Wrocław-Warsaw-Cracow, 1968), 318–322; and "Laboratory Notebooks of G. S. Ohm: A Case Study in Experimental Method," in *American Journal of Physics*, **35**, no. 2 (Feb. 1967), 110–114, although his account is rather too Baconian. See also Eugen Lommel, *Georg Simon Ohm's wissenschaftliche Leistungen. . . .* (Munich, 1889), English trans. by William Hallock, "The Scientific Work of George Simon Ohm," in *Annual Report of the Board of Regents of the Smithsonian Institution, . . . 1891* (Washington, 1893), 247–256.

There were several contemporary reviews of *Die galvanische Kette*: Georg Friedrich Pohl, in *Jahrbücher für wissenschaftliche Kritik* (Berlin) (1828), **1**, nos. 11/12–13/14, Jan., cols. 85–96, 97–103; Ludwig Friedrich Kämtz, in *Allgemeine Literatur-Zeitung* (Halle–Leipzig) (1828), **1**, nos. 13–14, Jan., cols. 97–104, 105–109; and *Leipziger Literatur-Zeitung* (18 Dec. 1828), cols. 2562–2565, anonymous, although possibly written by Heinrich Wilhelm Brandes, professor of physics at Leipzig and an editor of the journal. Problematical is the review that appeared anonymously in the obscure journal edited by K. W. G.

Kastner, *Proteus. Zeitschrift für Geschichte der gesammten Naturlehre*, **1**, no. 2 (1828), 349–377. Ohm referred to it as if it had been written by Johann Wilhelm Andreas Pfaff, but Füchtbauer says Ohm himself was really the author, as evidenced by a letter from Martin Ohm to his brother; in fact it seems as if Kastner, Pfaff, and Ohm all had a hand in it. It is less a review than a complete recapitulation and reformulation of the full mathematical theory, an undertaking which probably only Ohm would have ventured. As such it should perhaps be numbered among Ohm's works.

KENNETH L. CANEVA

OKEN (or **Okenfuss**), **LORENZ** (*b.* Bohlsbach bei Offenburg, Baden, Germany, 1 August 1779; *d.* Zurich, Switzerland, 11 August 1851), *natural science, philosophy, scientific congresses.*

The son of poor farmers in the Black Forest, Oken studied at the universities of Freiburg, Würzburg, and Göttingen. In 1803, at the age of twenty-four, he published a system of *Naturphilosophie*, thereby marking his adherence to the school of thought founded by Schelling a few years earlier. Throughout his life he remained faithful to this way of thinking, which he outlined in 1805 in a small book of methodological importance, *Die Zeugung*. Oken was a prolific writer whose works record his growing erudition and developing conceptions about nature. After graduating from the University of Freiburg in 1804, he held various teaching posts at Göttingen, Jena, Munich, and Erlangen. The frequent changes in his place of employment were occasioned by the boldness of the ideas he taught; the violence of the scientific polemics in which he engaged; and his political activities in revolutionary youth movements, which the German principalities severely repressed. Finally, in 1832 he secured a post at the recently founded University of Zurich, where he was a respected teacher until his death in 1851.

Oken took an active interest in all branches of natural history and of human knowledge in general, including optics, mineralogy, and even military science. His contributions to anatomy deal with the osteology of the skull and the vertebrae, the organogenesis of the intestinal tract and the umbilical cord, and, more generally, with the subject of comparative anatomy. Oken's importance lies far more in the formulation of a number of fundamental concepts, which constitute the guiding threads of his many publications. A good example of his treatment of one of the major themes of his corpus is offered by *Die Zeugung*, in which he discusses the elementary units of living organisms, "the infusoria." In this work Oken contended that all flesh can be broken down into infusoria and that all higher animals consist of constituent animalcules. "For this reason," he wrote, "we shall call them primal animals (*Urthiere*)." From the semantic point of view, this word is crucial; it was long used to designate the protozoans, and the prefix *Ur-* is one of the key elements of *Naturphilosophie* and of Romantic thought in general.

According to Oken, these primal animals constitute the original material not only of the animals as we know them but also of the plants; they may thus be called the primal material of all organized beings. The primal animals are subordinated to a higher organism in which they facilitate a unique common function, or in which they carry out this function by realizing their own potentialities. When the entities are combined they form another entity, the organism, which is a fusion of primal beings, each element having lost its individuality in favor of a higher unity.

In his treatise on the philosophy of nature Oken postulated the existence of a primal slime. It results from a combination of various processes that—when they reach equilibrium—must produce a sphere; for the organism is the image of the planet and therefore possesses an analogous spherical form. This primal plasm supposedly formed along the boundary of the seas and the earth. Oken held that a primal mucous follicle emerged from this plasm or infusorian, and that the genesis of the organism is merely the accumulation of an infinite number of mucous particles. According to Oken, organisms are not preformed; no organism is created that is larger than an infusorial particle; no organism is created or has ever been created that is not microscopic. Everything that is larger has not been created but has developed. Man was not created; he developed. These aphoristic pronouncements were repeatedly reprinted in the re-editions of the book until 1843, that is, until five years after the appearance of Schwann's work on the cell theory. These formulations of Oken's prefigure some of the fundamental concepts of nineteenth-century natural science.

In the preceding paragraphs we have summarized one aspect of Oken's thought; his own prose has become virtually impenetrable to the modern reader unfamiliar with the enthusiastic outpourings of Romantic philosophy. Throughout his scientific career Oken devoted his greatest efforts to fostering the study of natural science, a subject then at the height of its development. In communicating his enthusiasm for it, he was, of course, also attempting to promote his own views. He was one of the first to stress the pedagogical value of natural history at all levels of instruction, and he wrote many books for both students and adults.

Sometimes these were modest works, but often they were major treatises in several volumes. Further, Oken founded his own journal, *Isis oder enzyklopädische Zeitung von Oken*, which for three decades (1817–1847) published popular scientific articles of a very high caliber. Oken himself wrote the majority of the articles; and in them he set forth his basic views, particularly the theory that the skull is composed of several vertebrae. In claiming priority for this theory he became involved in a long and bitter polemic with Goethe, a complicated affair that has been the subject of many historical studies. Several times Oken discussed the traditional explanations of the origin of the first man. Oken often gave free vent to his anger regarding the contemporary political situation and thus ran afoul of the official censors. By the variety of its contents, *Isis* offers a remarkable picture of the development of the field of natural history in the first half of the nineteenth century.

Oken made a lasting contribution to science through his role in the creation of scientific congresses organized outside the framework of the universities. Accordingly, at the Congrès Scientifique de France, held at Strasbourg in 1842, Oken, despite his absence, was acclaimed the father of scientific meetings. He was the founder of the Gesellschaft Deutscher Naturforscher und Aerzte, which first met at Leipzig in 1822, and the 107th meeting of which was held in Munich in October 1972. The proceedings of this organization constitute a precious record of the fruitful union of biology and medicine in the first half of the nineteenth century. After a very agitated university and scientific career, Oken found a peaceful life and a definite appointment at the University of Zurich, and as the rector of this newly created university he had to receive the young Georg Büchner, who now figures prominently in the history of European thought by virtue of the recent vogue for his literary works. By profession Büchner was a naturalist who had received his doctorate at Strasbourg under Georges Louis Duvernoy and was appointed to the University of Zurich after giving an inaugural lecture that recently has been republished. Unfortunately, Büchner died before he was able to take up his post.

Contemporary judgments of Oken's work and personality varied considerably, but there was general agreement on the importance of his contributions to comparative anatomy, which were well known even outside the German-speaking countries. One of his shorter books was published in French in 1821, when Oken was already known among scientists in Paris. For example, Cuvier's lecture notes contain analyses of Oken's writings. In 1830 Oken's name was mentioned again during the famous controversy at the

Académie des Sciences, in which Goethe played an important, although indirect role.

Oken has never been completely forgotten, but during his lifetime sharply differing assessments were made of his role. Claude Bernard, in his *Introduction à l'étude de la médecine expérimentale*, mentioned Oken along with Goethe, Carus, Geoffroy Saint-Hilaire, and Darwin. On the hundredth anniversary of his birth (1879) he was the subject of an article in *Die Gartenlaube*, one of the most widely read popular German newspapers. A. Ecker (1880) wrote a fervent biography of Oken, which was translated into English. Then, with the rise of experimental natural science—the founders of which were hostile to *Naturphilosophie*—Oken's reputation was temporarily eclipsed. Interest in him revived around 1930 when historians of science began to study the Romantic period more closely. Indeed, their opinions of Oken tend to reflect their overall assessment of Romantic biology and its repercussions in modern biology.

Oken has also been exploited for ideological purposes. In 1939 J. Schuster interpreted certain passages in Oken's writings as favorable to German nationalism. The philosopher Ernst Bloch has treated Oken from the opposite point of view in his treatise on the problem of materialism.

The centennial of Oken's death was commemorated by a colloquium held in 1951 in his native region, at Freiburg im Breisgau. The few quotations given above show that despite his obscurity and combativeness, Oken remains a subject of considerable interest for the historian of modern biology.

BIBLIOGRAPHY

I. ORIGINAL WORKS. A complete list of Oken's publications can be found in the works of Ecker (1880) and Pfannenstihl (1953), both of which are cited below. The following works are referred to in the text: *Die Zeugung* (Bamberg, 1805); *Beiträge zur vergleichenden Zoologie, Anatomie und Physiologie*, 2 pts. (Bamberg–Würzburg, 1806–1807), written with D. G. Kieser; *Über Bedeutung der Schädelknochen* (Bamberg, 1807); *Erste Ideen zur Theorie des Lichts, der Finsternis, der Farben und der Wärme* (Jena, 1808); *Grundzeichnung des natürlischen Systems der Erze* (Jena, 1809); *Über den Wert der Naturgeschichte besonders für die Bildung der Deutschen* (Jena, 1809); *Lehrbuch der Naturphilosophie* (Jena, 1809, 1831; 3rd ed., Zurich, 1843); "Über die Bedeutung der Schädelknochen . . .," in *Isis*, **1** (1817), 1204–1208; and "Oken, wie er zur Bedeutung der Schädelknochen gekommen," in *Isis*, **2** (1818), 511–512.

See also "Entstehung des ersten Menschen," in *Isis*, **5** (1819), 1117–1123; *Esquisse du système d'anatomie, de physiologie et d'histoire naturelle* (Paris, 1821), *Natur-*

geschichte für Schüler (Bamberg, 1821); "Vergleichung alter Sagen und Überlieferungen mit Okens Ansicht der Entstehung des Menschen aus dem Meere," in *Isis*, **9** (1821), 1113–1115; *Allgemeine Naturgeschichte für alle Stände*, 13 vols. (Stuttgart, 1839–1842); *Elements of Physiophilosophy*, trans. from the German by A. Tulk (London, 1847); and "Über die Schädelwirbel gegen Hegel und Göthe," in *Isis* (1847), 557–560.

II. SECONDARY LITERATURE. See E. Bloch, *Das Materialismusproblem. Seine Geschichte und Substanz* (Frankfurt am Main, 1972), esp. 258–260; H. Bräuning-Oktavio, *Oken und Goethe im Lichte neuer Quellen. Beiträge zur deutschen Klassik* (Weimar, 1959); G. Büchner, "Mémoire sur le système nerveux du barbeau," in *Mémoires de la Société du Muséum d'histoire naturelle de Strasbourg*, **2** (1835), 1–57; *Über Schädelnerven. Probevorlesung* (Zurich, 1836), in Büchner, *Sämmtliche Werke und Briefe*, W. R. Lehmann, ed. (Hamburg, 1971); *Congrès scientifique de France*, I (1843), esp. 87, 581; A. Ecker, *Lorenz Oken* (Stuttgart, 1880), trans. into English by A. Tulk (London, 1883); E. Gagliardi, H. Nabholz, J. Strohl, *Die Universität Zürich 1833–1933 und ihre Vorläufer* (Zurich, 1938), esp. 262–276; M. Klein, *Histoire des origines de la théorie cellulaire* (Paris, 1936), esp. 18–22; "Sur les résonances de la philosophie de la nature en biologie moderne et contemporaine," in *Revue philosophique*, **144** (1954), 514–543; and "Goethe et les naturalistes français," in *Goethe et l'esprit français* (Paris, 1958), 169–191, esp. 177; D. Kuhn, *Empirische und ideelle Wirklichkeit. Studien über Goethes Kritik des französischen Akademiestreites* (Graz, 1967); A. Lang, "Oken, Lorenz (eigentlich Okenfuss)," in *Allgemeine deutsche Biographie*, XXIV (Leipzig, 1881), 216–226; E. T. Nauk, "Lorenz Oken und die medizinische Fakultät Freiburg im Breisgau," in *Oken Heft* (1951), 21–74; E. Nordenskiöld, *Die Geschichte der Biologie* (Jena, 1926), esp. 290–294; and "Oken-Heft," in *Berichte der Naturforschenden Gesellschaft zu Freiburg im Breisgau*, **41**, no. 1 (1951).

See also J. L. Pagel, "Oken," in *Biographisches Lexikon der hervorragenden Aerzte aller Zeiten und Völker*, **4** (Vienna–Leipzig, 1886), 416; M. Pfannenstiehl, "Lorenz Oken," in *Oken-Heft* (1951), 7–20; "Schriften und Varia über Lorenz Oken von 1806 bis 1951," *ibid.*, 101–118; and "Lorenz Oken. Sein Leben und Werken," in *Freiburger Universitätsreden*, n.s. **14** (1953); J. Schuster, "Oken, Welt und Wesen, Werk und Wirkung," in *Archiv für Geschichte der Mathematik, der Naturwissenschaften und der Technik*, n.s. **3** (1929), 54–70; J. Schuster, ed., *Laurentius Oken gesammelte Schriften, Programme zur Naturphilosophie* (Berlin, 1939), esp. 320–328, Oken Geistesgeschichtliche Stellung; C. Sterne [Ernst Krause], "Ludwig Lorenz Oken. Zum hundertjährigen Geburtstag eines Vielgeschmäheten," in *Gartenlaube* (1879), 518–520; J. Strohl, *Oken und Büchner. Zwei Gestalten aus der Uebergangszeit von Naturphilosophie zu Naturwissenschaft* (Zurich, 1936); and G. von Wyss, *Die Hochschule Zürich in den Jahren 1833–1883* (Zurich, 1883).

MARC KLEIN

OLAUS MAGNUS (*b.* Linköping, Sweden, October 1490; *d.* Rome, Italy, 1557), *geography, ethnology.*

Olaus Magnus was born to a middle-class family. He attended school in Linköping and in 1510 traveled abroad to prepare himself for a career in the Swedish church. With the support of a canonry he studied for almost seven years on the Continent, among other places at the University of Rostock, where, probably in 1513, he received his baccalaureate. Upon his return to Sweden, Olaus became in 1518 a deputy to Arcimboldi, the papal seller of indulgences; and in that capacity he traversed the wilderness of Norrland and the high mountains of Norway. He may have reached Lofoten and southern Finnmark before returning to Sweden and proceeding south via Torneå (Tornio, Finland). Later he was a vicar in Stockholm and cathedral dean in Strängnäs. In 1523 the Church ordered him to Rome and he never returned to Sweden. The Lutheran Reformation erupted in Sweden, but Olaus, together with his brother, Archbishop Johannes Magnus, remained loyal to the Catholic faith. Both brothers spent several years as refugees in Danzig, but in 1537 they traveled to Italy, where with gusto they took part in the game of church politics. In 1544, after his brother's death, Olaus was appointed archbishop of Sweden and in that capacity attended some of the meetings of the Council of Trent. He continued to live in Rome for the rest of his life.

During his long exile Olaus Magnus published two scientific works which give him a pioneering position in the geographic research of Scandinavia. The first was the monumental map of the Scandinavian countries, *Carta marina* (Venice, 1539), of which only two copies are extant. It is executed in woodcut and also shows the Atlantic Ocean with its islands from Scotland to Iceland and Greenland. It was the first fairly reliable map of northern Europe and was based on older maps (Claudius Clavus, Ziegler) and Olaus' own notes—and perhaps also on some no longer extant nautical charts. Olaus used a few astronomical latitude determinations. His vivid illustrations of wild animals, skiing Lapps, and tumbling sea monsters gave the *Carta marina* life and movement.

In 1555, at Rome, Olaus published his great description of the Scandinavian peoples, *Historia de gentibus septentrionalibus*. Originally this work was planned as a detailed description of the more noteworthy features of the map. Olaus dealt with nature and the life of the people in Scandinavia—especially Sweden—the Lapps and Finns, the climate and physical geography, agriculture and mining, the wild animals, and the Swedish people in their daily occupations. The work has to be used with care, since large parts were simply copied from older European literature. But he also builds upon his own memories and experiences in this description of an entire country at the beginning of a new period. The basic thread is primitivistic, and Olaus Magnus, a warm patriot, praises the freezing winter cold and the harsh Scandinavian virtues.

Both of Olaus' works were of great influence. The *Carta marina* was indispensable for later cartographers; and the historical work, which was published in many editions, for generations informed the educated European about Scandinavia.

BIBLIOGRAPHY

I. ORIGINAL WORKS. The *Carta marina* has been published in many facs. eds., including Lychnosbibliotek, XI, 1 (Malmö, 1949). A copperplate engraving on a reduced scale was produced by Antonio Lafreri (Rome, 1572). The *Historia* is available in Latin, French, Italian, Dutch, and German; facs. ed. (Copenhagen, 1972). A modern Swedish trans., *Historia om de nordiska folken*, 4 vols. (Uppsala, 1909–1925), has a vol. of commentary by John Granlund (Uppsala, 1951).

II. SECONDARY LITERATURE. The most recent biography is Hjalmar Grape, *Olaus Magnus* (Stockholm, 1970), in Swedish. A basic work is Herman Richter, *Olaus Magnus Carta marina 1539*, Lychnosbibliotek, XI, 2 (Lund, 1967). See also Karl Ahlenius, *Olaus Magnus och hans framställning af Nordens geografi* (Uppsala, 1895).

STEN LINDROTH

OLBERS, HEINRICH WILHELM MATTHIAS (*b.* Arbergen, near Bremen, Germany, 11 October 1758; *d.* Bremen, 2 March 1840), *medicine, astronomy.*

Olbers was the eighth of the sixteen children of Johann Jürgen Olbers, a Protestant minister. He became interested in astronomy when he was about fourteen, but the Gymnasium in Bremen which he attended was a typical humanistic institution of that time where almost no mathematics or science was taught. In order to understand astronomy Olbers taught himself mathematics and tried to compute the solar eclipse of 1774. In 1777 he began the study of medicine in Göttingen under Blumenbach and Ernst Baldinger, and also attended lectures in physics and mathematics by G. C. Lichtenberg and, especially, A. G. Kästner, who was in charge of the small observatory at Göttingen. But mainly he studied astronomy on his own. His lifelong concern with comets dates from January 1779, when he used his observations of Bode's comet to calculate its orbit according to Euclid's method. In 1780 he independently discovered a comet that was simultaneously observed by Montaigne. Meanwhile, Olbers continued his medical studies, concentrating on a problem that involved the application of mathematics to physiology. His dissertation, *De oculi mutationibus* (Göttingen, 1780), explains how the eye adapts to a change in focus by changing the shape of the eyeball; only much later was it discovered that only the lens changes shape. Later, as a practicing physician he specialized in ophthalmology, a field hardly recognized at that time.

In 1781, after receiving his medical degree at Göttingen, Olbers went on a study trip to Vienna, where he visited hospitals during the day, enjoyed the aristocratic social life of the city in the evenings and spent the nights at the Vienna observatory. Throughout his life he profited from needing only four hours of sleep, so that after a long and busy day of practicing medicine he could "relax" by observing the sky. In Vienna he was thus able to follow the course of the recently discovered planet Uranus.

At the end of 1781 Olbers settled in Bremen and soon acquired an extensive medical practice. It was mainly through his efforts that inoculation was introduced in the city, and he was highly praised for his work during several cholera epidemics. When the "magnetic cures" of Mesmer started a great controversy, Olbers published an article admitting the reality of some of them but also expressed the opinion that future understanding of physiology would explain them without the assumption of a special power.

In 1785 Olbers married Dorothea Köhne, who died a year later at the birth of their daughter. In 1789 he married Anna Adelheid Lurssen, by whom he had one son. After the death of his daughter in 1818 and of his second wife in 1820, he retired from active medical practice to devote the rest of his life to astronomy.

Olbers installed an observatory on the second floor of his house, using its two large bay windows for his telescopes. At various times he possessed two achromatic Dollond refractors, a Schröter reflector, a heliometer and refractor from Fraunhofer's workshop, and three comet seekers, made by Hofmann, Weickhardt, and Fraunhofer. He had no transit instrument or fixed instrument of any kind. His library became one of the best private astronomical collections in Europe. For over fifty years he carefully gathered astronomical literature and assembled a collection in the field of cometography that was practically complete. After Olbers' death, F. G. W. Struve bought this library for the new Pulkovo observatory, near St. Petersburg. Struve's new catalog of the collection listed 4,361 items, consisting of 39 sky charts, 1,607 monographs, and 2,715 articles.

Busy with his new medical practice when he first moved to Bremen, Olbers had less time for astronomy; but in 1786 he met J. H. Schröter, whose private observatory in nearby Lilienthal was one of the best-equipped on the Continent, and they worked closely together for many years. In 1796 Olbers discovered a comet and calculated its parabolic orbit with a new method, simpler than that used by Laplace. In a letter to F. X. von Zach, director of the newly founded observatory on the Seeberg, near Gotha, Olbers asked whether his treatise on this method should be printed, and if so, how this could best be done. After reading the treatise and using it with excellent results to compute the orbit of the comet of 1779, which had presented great difficulties to many astronomers, von Zach decided to see it through the press himself. It appeared at Weimar in 1797 under the title *Über die leichteste und bequemste Methode, die Bahn eines Kometen aus einigen Beobachtungen zu berechnen*. This work immediately established Olbers among the foremost astronomers of his time, and his method was used throughout the nineteenth century.

Despite the work of Newton and Lambert, the computation of cometary orbits had until then been a very laborious process. Laplace had given formulas for the computation of a parabola through successive approximations, but the procedure was cumbersome and unsatisfactory. It had been assumed that when three observations of a comet had been obtained within a short period of time, the radius vector of the middle observation would divide the chord of the orbit of the comet from the first to the last observation in relation to the traversed time. The finding that this assumption could be applied with equal advantage to the three positions of the earth in its orbit was Olbers' contribution. This basic idea led to a rapidly converging process of calculation, and Olbers worked out simple and easily calculated formulas.

The space between the planets Mars and Jupiter, shown mathematically by Bode's law, had long intrigued astronomers. The first asteroid was discovered by G. Piazzi at the Palermo observatory on 1 January 1801. He noticed a starlike object that moved during the succeeding days. He communicated this news to other astronomers; and although it was soon realized that this must be a new planet, named Ceres by Piazzi, it disappeared before more observations could be made. At that time it was still impossible to compute an orbit from such a small arc without assuming the eccentricity. Then the twenty-three-year-old Gauss was able to determine the orbit by a new method; and it was Olbers who, on 1 January 1802, found the new planet very near where Gauss had calculated it would be. This episode was the beginning of their lifelong friendship; and when Gauss visited Olbers in 1803, each had his portrait painted to give to the other. The two portraits now hang in the Göttingen observatory. While following Ceres, Olbers discovered a second asteroid, Pallas, on 28 March 1802; a third, Juno, was discovered by Harding at Lilienthal in 1804. The orbits of these small planets suggested to Olbers that they had a common point of origin and might have originated from one large planet. Accordingly, for years he searched the sky where the orbits of Ceres, Pallas, and Juno approached each other; the result was the discovery of Vesta on 29 March 1807.

The search for comets remained Olbers' main interest, and his industry was rewarded with the discovery of four. Of particular interest is the comet that he discovered on 6 March 1815, which has an orbit of seventy-two years, similar to Halley's. Olbers also calculated the orbits of eighteen other comets. Noticing that comets consist of a starlike nucleus and a parabolic cloud of matter, he supposed that this matter was expelled by the nucleus and repelled by the sun. In "Über die Durchsichtigkeit des Weltraums," published in 1823 in *Berliner astronomisches Jahrbuch für das Jahr 1826,* Olbers discussed the paradox that now bears his name: If we accept an infinite, uniform universe, the whole sky would be covered by stars shining as brightly as our sun. Olbers explained the paradox of the dark night sky by assuming that space is not absolutely transparent and that some interspace matter absorbs a very minute percentage of starlight. This effect is sufficient to dim the light of the stars, so that they are seen as points against the dark sky. The idea was not absolutely new; Halley had written about it and a young Swiss astronomer, Jean Philippe Loys de Chéseaux, had published an essay in 1744 using a very similar argument.

Olbers was also interested in the influence of the moon on weather, the origin of meteorite showers, and the history of astronomy. He was a member of Museum, the scientific society in Bremen, and through the years gave over eighty lectures there (of which only one was on a medical subject).

Although Olbers usually declined official posts, he felt it his duty to participate in the government during the time that Bremen was part of the French empire (1811–1813). This commitment forced him to spend time in Paris, where he met some of the French astronomers.

Olbers was held in great esteem by his contemporaries. He conducted an extensive correspondence with Gauss, Bessel, Encke, Schröter, and other astronomers. He also encouraged many young astronomers with good advice and made great efforts to obtain positions for them at various observatories. One of them, Friedrich Wilhelm Bessel, a twenty-year-old apprentice in a merchant's office, had approached Olbers in 1804 with his calculation of the orbit of Halley's comet. Olbers was so impressed with his work that, after suggesting some additions, he recommended it for publication and sought to obtain the directorship of the new observatory at Königsberg for Bessel.

A very modest man, Olbers later claimed that his greatest contribution to astronomy had been to lead Bessel to become a professional astronomer. Bessel's eulogy, written in 1845, ended: "He was to me the most noble friend. With wise and fatherly counsel he guided my youth; 171 letters which I possess from him are written proof of my right to extend my devotion beyond the limits of science."

BIBLIOGRAPHY

I. ORIGINAL WORKS. Olbers' article on Mesmer's cures is "Erklärung über die in Bremen durch den sogenannten Magnetismus vorgenommenen Kuren," in *Deutsches Museum* (Oct. 1787), 296–312. A complete listing of his almost 200 articles is in vol. I of C. Schilling, *Wilhelm Olbers, sein Leben und seine Werke,* 2 vols. in 3 pts. and supp. (Berlin, 1894–1909); this work also contains the complete correspondence with Gauss. Olbers' correspondence with Bessel was published by A. Erman, *Briefwechsel zwischen W. Olbers und F. W. Bessel,* 2 vols. (Leipzig, 1852). The Staatsbibliothek in Bremen has a collection of Olbers' papers.

II. SECONDARY LITERATURE. *Von Bremer Astronomen und Sternfreunden,* W. Stein, ed. (Bremen, 1958), contains six papers on various aspects of Olbers' career and a partial listing of his works. Bessel's obituary is in *Astronomische Nachrichten,* **22** (1845), cols. 265–270. Another, unsigned obituary is in *Proceedings of the Royal Society,* **4** (1837–1843), 267–269. Struve reported on the purchase of Olbers' library for the observatory at Pulkovo in *Astronomische Nachrichten,* **19** (1842), 307–312. The paradox of the dark night sky is extensively treated by Stanley L. Jaki in *The Paradox of Olbers' Paradox* (New York, 1969). See also Otto Struve, "Some Thoughts on Olbers' Paradox," in *Sky and Telescope,* **25** (1963), 140–142; and Stanley L. Jaki, "New Light on Olbers' Dependence on Chéseaux," in *Journal for the History of Astronomy,* **1** (1970), 53–55. F. X. von Zach describes Olbers' observatory and instruments in "Auszug aus einem astronomischen Tagebuche, geführt auf einer Reise nach Celle, Bremen und Lilienthal in Sept. 1800," in *Monatliche Correspondenz . . .,* **3** (1801), 113–145. There is a biographical notice in *Allgemeine deutsche Biographie,* XXIV, 236–238.

LETTIE S. MULTHAUF

OLDENBURG, HENRY (*b.* Bremen, Germany, *ca.* 1618; *d.* London, England, 5 September 1677), *scientific administration.*

There were three eminent secretaries of seventeenth-century scientific societies: Lorenzo Magalotti, Henry Oldenburg, and J. B. du Hamel. Although both the Italian and the Frenchman left behind substantial memorials of the societies with which they were associated, Oldenburg alone made a profession of scientific administration. In his fifteen years of service to the Royal Society he founded a complete system of records (still extant), created an international correspondence among scientists, and furnished a monthly account of scientific developments.

The Oldenburg family, which had moved to Bremen from Münster in the sixteenth century, was long associated with education. Henry's father, for whom he was named, taught from about 1610 to 1630 at the Paedagogium in Bremen; his last years were spent in the new university that Gustavus II founded at Dorpat (now Tartu), Estonia, where he died in 1634. The year of his son's birth can only be deduced from the facts that the boy entered the Gymnasium Illustre of Bremen in May 1633 and, after proceeding to the degree of Master of Theology in November 1639, went on to the University of Utrecht in 1641. Moreover, it seems likely that at his second marriage in 1668 Oldenburg described himself as "about fifty" years old. There are no details concerning his early life, and the only known means for his support during his minority was a lease of some ecclesiastical property, acquired by his grandfather, which he retained all his life—although apparently not as a useful source of income. His studies at the Gymnasium were largely theological, with Hebrew, Latin, Greek, rhetoric, logic, and mathematics as other subjects. Again, there is no evidence to show how he acquired his mastery of modern languages.

After a brief appearance at the University of Utrecht, where he possibly became acquainted with the philosophy of Descartes, Oldenburg vanishes for twelve years. It is likely that he followed the plan indicated in his letter to G. J. Vossius (August 1641) of acting as a private tutor, for in these years he acquired, apparently, a wide knowledge of France, Italy, Switzerland, Germany, and possibly England, and command of their respective languages. It also seems likely that when Oldenburg reappears he had already been tutor to a number of young Englishmen (among them Edward Lawrence, Robert Honywood, and William Cavendish, later first duke of Devonshire); he had now (as Milton testified) a perfect knowledge of English. Hence it is likely that he had spent some period in England, a deduction

confirmed by the next certain event in his life—his selection by the city government of Bremen, at a moment when he had returned to his birthplace, to go on a diplomatic mission to Oliver Cromwell, with the object of protecting the maritime interests of Bremen. Oldenburg arrived in England at the end of July 1653 and presented a memorial to Cromwell in December without achieving much result (partly owing to the state of confusion in the English government) before the conclusion of peace between England and Holland brought an end to the seizures at sea. Oldenburg remained in England, where he made new acquaintances or revived old ones, until a second call arrived from Bremen in August 1654 asking him to enlist Cromwell's friendship in aid of Bremen's resistance against a Swedish onslaught. This time Oldenburg (despite his laments of lack of money) achieved a partial diplomatic success. When this business was done—and possibly even before—Oldenburg returned to his tutorial employment, although there is no positive evidence of it before March 1656, when he was negotiating with the two Boyle families of Cork and Ranelagh.

By this time Oldenburg was certainly acquainted with John Dury, Samuel Hartlib, John Milton, Thomas Hobbes, the learned and pious Lady Ranelagh and her more famous brother, Robert Boyle, and no doubt many others who moved in the circles of persons who as yet were more inclined toward religion than toward philosophy or science and, insofar as they hoped for material progress in this life, saw it as dependent on the mysteries of technical invention. Like Boyle, Oldenburg did not figure in the Gresham College group; but his tutorship of Boyle's nephew Richard Jones (later third viscount and first earl of Ranelagh, 1641–1712) took him to Oxford in 1656, and so to acquaintance with John Wilkins and, no doubt, others residing at the university and constituting its Philosophical Club. In the summer of 1657 Oldenburg took young Jones for a long stay in France, with excursions into Germany. In several cities, but in Paris above all, Oldenburg and his pupil participated in learned societies, while, under the simultaneous promptings of Boyle and Hartlib, his interests and his acquaintanceships moved steadily toward science and medicine, especially chemistry. On these travels Oldenburg began to learn his trade as a scientific intelligencer and to become the friend of scientists. His return to London with Jones slightly preceded that of Charles II, and he soon began to develop his correspondence with the Continent. On 29 November 1660, at the famous meeting of the Gresham College group, he was listed as a candidate member of a formal scientific society—later to be the

Royal Society of London for Improving Natural Knowledge—which he joined in January 1661.

Thereafter Oldenburg's whole life was devoted to the Royal Society. He made one more trip abroad, to Bremen in the summer of 1661; returning through Holland, he met both Huygens and Spinoza. He was twice married. On 22 October 1663 he married Dorothy West, a woman not much younger than he, who possessed an estate of £400 with which (in part) his house in Pall Mall, near Lady Ranelagh's, was bought; she died early in 1665. In August 1668 Oldenburg married (with her father's consent) Dora Katherina Dury, aged about sixteen, who had been his ward for some years. She brought him a small property in Kent, near Charlton, which was their summer home. There were two children of this marriage, Rupert, born *ca.* 1673, and a younger daughter, Sophia.

In the first royal charter granted to the Royal Society (15 July 1662), as in the second (1663), Oldenburg was named one of its two secretaries, although he had not hitherto played a great part in its affairs. Probably, like Hooke, he owed his position to Robert Boyle, his constant friend and occasional employer over many years. His obvious chief qualifications for this honorary office were his industry, knowledge of languages, and literary gifts. Few Englishmen at this time possessed close contacts with the learned men of the Continent or knew much about work being done abroad. Certainly neither John Wilkins nor any of his successors in the titular first secretaryship hesitated to leave all conduct of the Society's affairs in Oldenburg's hands.

Oldenburg thus defined (British Museum MS Add 4441, fol. 27) the secretary's business as it had matured by the spring of 1668:

> He attends constantly the Meetings both of ye Society and Councill; noteth the Observables, said and done there; digesteth ym in private; takes care to have ym entred in the Journal- and Register-books; reads over and corrects all entrys; sollicites the performances of taskes recommended and undertaken; writes all Letters abroad and answers the returns made to ym, entertaining a correspondence wth at least 30. persons; employes a great deal of time, and takes much pain in inquiring after and satisfying forrain demands about philosophicall matters, dispenseth farr and near store of directions and inquiries for the society's purpose, and sees them well recommended etc.
>
> Query. Whether such a person ought to be left unassisted?

Besides the journal book, which held notes of meetings, and the register book, in which copies of impor-

tant contributions were entered, Oldenburg kept the Council minutes and a letter book in which all the more important incoming and outgoing letters were extracted, as well as the files of original letters and papers, records of membership, and a cipher record of discoveries. Since he could pursue no regular career—although he did have earnings as an editor and translator, especially from the *Philosophical Transactions*—Oldenburg suffered increasing impoverishment until the Society allowed him an annual salary of £40 beginning in April 1668; also about this time he gained the assistance of an amanuensis.

Between them the Royal Society's two permanent officers, Robert Hooke and Henry Oldenburg, provided a great part of the matter discussed at the weekly meetings, Oldenburg drawing upon his correspondence and the books presented to the Society through himself. Apart from the week-by-week business of the Society this correspondence was Oldenburg's greatest burden, involving as it did receiving and answering an average of probably six or seven letters a week during the working period of the year, some of them long and difficult documents. The postal service to many points abroad was nonexistent or unreliable and, in any case, expensive; but in the last ten years of his service Oldenburg was able to exploit diplomatic channels by enlisting young men in embassies as his correspondents and agents. From 1666 he instructed correspondents to write by post to "Grubendol, London." This was a code address; it seems that Grubendol letters were delivered to the office of the secretary of state (and there paid for); in return Oldenburg reported any news of events abroad his letters might contain. Despite the Royal Society's privileges, maintaining correspondence with foreigners could be perilous, especially in time of war—as Oldenburg discovered when he was thrown in the Tower for some weeks during the summer of 1667. The probability is that in a letter to some foreigner that was seized he had expressed a patriotic but injudicious resentment that the English government had fallen down in its measures to protect England from the Dutch fleet. Thereafter he was very scrupulous in sticking to scientific matters.

From 6 March 1665 selected portions of the letters submitted to him were published in the *Philosophical Transactions: Giving Some Accompt of the Present Undertakings, Studies and Labours of the Ingenious in Many Considerable Parts of the World*, which were interrupted only twice in Oldenburg's lifetime: once by the plague, when a few issues were printed at Oxford although Oldenburg remained in London, attentive to the Society's concerns, and again when Oldenburg was imprisoned. The *Philosophical*

Transactions formed the first purely scientific journal containing both formal contributions and short notes about work in progress, as well as book reviews that were often long and of critical value. They became the principal vehicle of interchange between English and Continental science, supplementing Oldenburg's correspondence; and for some investigators—of whom Leeuwenhoek is the most obvious example—they were their sole vehicle of publication. Oldenburg also encouraged the Royal Society to undertake, and personally managed, the publication of separate works: those of Malpighi are best known in this category. He both translated and published Steno's *Prodromus*; and generally through his letters and the open pages of the *Transactions* he gave encouragement to all the younger English scientists of the decade 1667–1677, including Isaac Newton, as well as many on the Continent.

As a scientific journalist and administrator of the Royal Society, Oldenburg has been accused of over-enthusiasm. At a time when the line between a private letter and a paper for publication was dubious, he committed some errors of discretion; but he was never guilty of a breach of confidence. He tended to regard everything disclosed at an ordinary meeting of the Royal Society as public, unless a special request was made—and, indeed, the Society was opposed to secrecy about discoveries. He can hardly be censured for communicating accounts of meetings to absent fellows, whether native or, like Huygens and Hevelius, foreign. Oldenburg perhaps had an excessive faith in the power of the process of critique-and-rebuttal to elicit truth, but this was often for the sake of enhancing English prestige in a manner that his contemporaries expected of him. There is no evidence that Oldenburg (who was careful not to claim English nationality, which he sought only in the last months of his life) favored foreigners; the accusations on this score leveled against him by Hooke were without foundation, and he was fully vindicated by the Royal Society's Council. Hooke was Oldenburg's sole enemy, and then only after the Huygens' spring-balance watch patent application of 1675.

Oldenburg's conception of the Royal Society's function was consistently and simply Baconian; it was the task of the learned and the well-endowed of his age to compile an authentic natural history from which posterity would elucidate a sound natural philosophy. "Natural history" included, besides passive investigation of flora and fauna, minerals, the heavens, and even wonders and prodigies, active experimentation, such as blood transfusion and medical injection, with which he was much concerned. All this Oldenburg regarded as an international enterprise, in which the efforts of established societies should be strengthened by individual zeal in every nation. He regretted the poverty of the Royal Society, contrasting it with the lavish resources enjoyed by the Académie Royale des Sciences. Like Boyle and Newton he distrusted a priorist systems of nature and sometimes, like Bacon, spoke of the amelioration of human life as a major object of the scientific movement. But in practice he gave a warm welcome to any piece of solid work, whether in scientific description, pure mathematics, experimental physics, or astronomical calculation. As an editor he had a sound instinct, although (like his age) distorted by an excessive preoccupation with medical curiosities and teratology. If Oldenburg's approach to the advancement of science was not greatly ahead of that general in his time, it did not lag behind.

Oldenburg remained steadily active until the last months of his life. He died on 5 September 1677 after a brief illness and was buried at Bexley, Kent; his wife died on 17 September. Since Oldenburg was intestate, letters of administration were taken out to make provision for the children; Boyle probably had a hand in the arrangements. Rupert Oldenburg, then serving as a lieutenant, committed suicide in 1724; of the fate of Sophia no trace remains. Oldenburg's considerable library was bought by the earl of Anglesey, whose vast collection was in turn dispersed in 1686. Some of Oldenburg's books are now in the British Museum, and others appear on the antiquarian market.

BIBLIOGRAPHY

I. ORIGINAL WORKS. Besides the *Philosophical Transactions* and the literary activities already mentioned, Oldenburg translated several of Boyle's books into Latin and probably acted as a literary assistant to John Evelyn. He also published an English translation of François Bernier's *History of the Late Revolution of the Empire of the Great Mogul* and was possibly the translator of some other works published over the initials "H.O." For his correspondence see A. Rupert Hall and Marie Boas Hall, eds., *The Correspondence of Henry Oldenburg*, I–IX (Madison–Milwaukee–London, 1965–1973), a work that is still continuing.

II. SECONDARY LITERATURE. Friedrich Althaus in the Munich *Beilage zur Allgemeinen Zeitung*, no. 212 (2 August 1889), pp. 1–3, gave an account of Oldenburg's family and early life in Bremen. For the rest, see A. Rupert Hall and Marie Boas Hall, "Why Blame Oldenburg?" in *Isis*, **53** (1962), 482–491; "Some Hitherto Unknown Facts About the Private Career of Henry Oldenburg," in *Notes and Records of the Royal Society of London*, **18** (1963), 94–103; "Further Notes on Henry Oldenburg," *ibid.*, **23** (1968), 33–42; M.

B. Hall, "Henry Oldenburg and the Art of Scientific Communication," in *British Journal for the History of Science*, **2** (1964–1965), 277–290; and A. R. Hall, "Henry Oldenburg et les relations scientifiques au XVIIᵉ siècle," in *Revue d'histoire des sciences*, **23** (1970), 285–304. See also T. Sprat, *History of the Royal Society* (London, 1667); T. Birch, *History of the Royal Society* (London, 1756–1757; repr. 1968), and Robert Hooke, *Diary, 1672–80*, H. W. Robinson and W. Adams, eds. (London, 1935).

A. RUPERT HALL

OLDHAM, RICHARD DIXON (*b*. Dublin, Ireland, 31 July 1858; *d*. Llandrindod Wells, Wales, 15 July 1936), *geology, seismology*.

Oldham was the third son of Thomas Oldham, a distinguished geologist who was professor of geology at Trinity College, Dublin, and then a director of the geological surveys of Ireland and India. He was educated in England, first at Rugby and then at the Royal School of Mines. He followed in his father's footsteps by joining the staff of the Geological Survey of India in 1879. He devoted much energy to completing the unfinished work of his father, who died in 1878, notably an extensive investigation of a great earthquake in Cachar in 1869.

Oldham became superintendent of the Geological Survey of India and wrote some forty of its publications, chiefly on earthquakes in India, the hot springs of India, the geology of the Son Valley, and the structure of the Himalayas and the Ganges plain, taking account of geodetic observations. He developed a great interest in the then emerging science of seismology and is now noted more for his contributions to seismology than to geology. He left India in 1903, partly because of ill health, and returned to England, spending some time working with the seismologist John Milne on the Isle of Wight. Later, for health reasons, he lived in the Rhone Valley and then in Wales; but he remained an active contributor to science until about eight years before his death. He was awarded the Lyell Medal of the Geological Society of London in 1908 and was elected to the Royal Society in 1911.

Oldham became famous for his report on the great Assam earthquake of 12 June 1897, one of the most violent of modern times, which caused complete devastation over 9,000 square miles and was felt over 1.75 million square miles. It far surpassed in quality all reports on previous earthquakes, describing the remarkable Chedrang fault, with a thirty-five-foot uplift at one point; gave evidence of the occurrence of fractures without apparent rock displacement; showed that in some places accelerations of the ground

motion had exceeded the vertical acceleration of gravity; and reported the results of the first resurvey ever carried out after a large earthquake. From the point of view of seismology, the most far-reaching result was the first clear identification on seismograms of the onsets of the primary (P), secondary (S), and tertiary (surface) waves, previously predicted in long-standing mathematical theory. This identification showed that the earth could be treated as perfectly elastic to good approximation in studying seismic waves, a result of supreme importance to the further development of seismology.

Oldham also supplied the first clear evidence that the earth has a central core (1906). Others had suspected its existence but had not succeeded in obtaining direct evidence. In the course of analyzing some of Milne's records of large earthquakes, Oldham invariably found delays in the arrival of P waves at points on the earth diametrically opposite to earthquake sources; and he showed that the delays could be interpreted only in terms of the presence of a sizable core inside which the average P velocity is substantially less than in the surrounding shell.

Oldham was an original and independent thinker whose writings, whatever the subject, were always interesting and suggestive. One account describes him as "a little too independent sometimes for those in authority." There is a suggestion that he was impatient with the red tape of administrators less brilliant than himself. Above all, he is noted as a pioneer in the application of seismology to the study of the interior of the earth.

BIBLIOGRAPHY

I. ORIGINAL WORKS. Oldham's "Report on the Great Earthquake of 12th June 1897" was published in *Memoirs of the Geological Survey of India*, **29** (1899), i–xxx, 1–379, along with a supplementary report in **30** (1900), 1–102. "On the Propagation of Earthquake Motion to Great Distances," in *Philosophical Transactions of the Royal Society*, **194A** (1900), 135–174, includes his work on P, S, and surface waves. Most of his other papers were published in *Memoirs of the Geological Survey of India*.

II. SECONDARY LITERATURE. See the accounts of Oldham's life by C. Davison, in *Obituary Notices of Fellows of the Royal Society of London*, **2** (1936–1938), 111–113; and by P. L., in *Nature*, **138** (Aug. 1936), 316–317.

K. E. BULLEN

OLDHAM, THOMAS (*b*. Dublin, Ireland, 4 May 1816; *d*. Rugby, England, 17 July 1878), *geology*.

Oldham was educated privately in Dublin and

received his B.A. from Trinity College, Dublin, in 1836. Next, at Edinburgh he studied engineering; also geology and mineralogy under Robert Jameson, professor of natural history. On his return to Ireland in 1839, Oldham became chief geological assistant to J. E. Portlock, who was in charge of the Ordnance Survey in Ireland. Oldham supplied the mineral identifications for Portlock's *Report on the Geology of Londonderry* . . . (London, 1843). In 1844 he was appointed assistant professor of engineering at Trinity College, and a year later he became professor of geology there. In 1846 he also became local director of the Irish branch of the Geological Surveys of the United Kingdom, but continued to occupy the chair of geology.

During the next four years Oldham carried out much geological work. His noteworthy discovery in 1849 of hitherto unnoticed radiating fanlike impressions in the Cambrian rocks of Bray Head, County Wicklow, aroused intense interest; and the paleontologist Edward Forbes gave the name *Oldhamia* to the presumed fossil. The nature of this fossil has been disputed, but it is now thought to be a trace fossil—that is, a sedimentary structure caused by a living creature.

In November 1850 Oldham was appointed, on a five-year agreement, as geological surveyor to the East India Company. Although he succeeded another surveyor, D. H. Williams, he took no narrow view of his new post, immediately describing himself as the "Superintendent of the Geological Survey of India," and began to recruit other geologists to his staff. His office was renewed every five years until his retirement in 1876.

Oldham is justifiably regarded as the architect of the Geological Survey of India; under his guidance a remarkable amount of work was carried out, and large areas of India were surveyed geologically. Particular attention was given to a survey of the Indian coalfields, and in 1864 Oldham issued an elaborate report, *On the Coal Resources of India*. At the same time, under his supervision several serial publications were begun: *Annual Reports*, *Records*, *Memoirs*, and the important *Palaeontologia Indica*. Oldham initiated the scientific study of earthquakes in India and published a catalog of earthquakes. He also brought to the attention of European geologists much new information on the Cretaceous rocks. A vast collection of Indian rocks and fossils was accumulated, and shortly before Oldham's retirement it was transferred to the Indian Museum in Calcutta.

Oldham was elected a fellow of the Royal Society in 1848, and in 1875 the Society awarded him a Royal Medal.

BIBLIOGRAPHY

Oldham's scientific papers are listed in the Royal Society *Catalogue of Scientific Papers*, IV, 672; VIII, 528. His geological work in India was published by the Geological Survey of India.

There is no biography of Oldham, but details of his career are given by T. G. Bonney, in *Dictionary of National Biography*, XLII (1895), 111, which is based partly on an obituary notice in *Quarterly Journal of the Geological Society of London*, **35** (1879), "Proceedings," 16. The circumstances relating to his appointment in India are given by Sir Cyril S. Fox, "The Geological Survey of India, 1846 to 1947," in *Nature*, **160** (1947), 889. For a brief appraisal of his work there, see Sir Lewis Fermor, "Geological Survey of India, Centenary Celebrations," *ibid.*, **167** (1951), 10.

JOAN M. EYLES

OLIVER, GEORGE (*b.* Middleton-in-Teesdale, Durham, England, 13 April 1841; *d.* Farnham, Surrey, England, 27 December 1915), *physiology*.

Oliver was the second son of W. Oliver, a surgeon. He prepared at Gainford School, Yorkshire, for medical studies at University College, London, qualifying for membership in the Royal College of Surgeons in 1863 and receiving the M.B. in 1865. After brief periods of practice at Stockton-on-Tees and Redcar, he won the gold medal in obtaining his M.D. (London) in 1873. Oliver settled in Harrogate, where he practiced medicine from 1876 to 1908, then retired to Farnham. His first wife, Alice Hunt, died in 1898. Two years later he married Mary Ledyard, who survived him. Winter residence in London, afforded him by the seasonal nature of his Harrogate practice, allowed Oliver to be active in a number of medical and scientific societies. He was a member of the Physiological Society and of the Medical Society of London and a fellow of the Royal College of Physicians of London, the Royal Society of Medicine, and the Royal Microscopical Society.

Oliver was one of the many medical students influenced by William Sharpey, professor of anatomy and physiology at University College, to devote himself to the development of more scientific methods of diagnosis and therapy. With extensive clinical experience, knowledge of physiology and chemistry, and considerable technical ingenuity, he devised accurate and convenient techniques for, among other things, the analysis of blood and of urine, the measurement of circulatory phenomena, and the assessment of the therapeutic effects of medicinal waters. Notable examples of his contributions in this area are his introduction of urinary testing papers and of his hemacytometer, hemoglobinometer, arteriometer, and sphygmomanometer.

Oliver's interest in the circulation and his facility with instruments led to his most important scientific contribution, a collaboration with Edward A. Schäfer (later Sir Edward Sharpey-Schafer) in 1893–1895, in which the two elucidated the cardiovascular effects of the administration of extracts of the adrenal medulla and of the pituitary. Oliver administered glycerin extracts of a number of different organs to his son, noting their various effects with particular reference to the caliber of the peripheral arteries, as measured by his arteriometer. In Schäfer's words:

> Dr. George Oliver had been making a large number of clinical observations upon the effect of various organ extracts upon the circulation, but had been unable to arrive at any very definite conclusions regarding them. Amongst these was extract of suprarenal capsule, extract of thyroid gland, extract of brain and so on. He consulted me as to what steps might be taken to arrive at a clearer understanding in regard to their action, and I invited him to investigate their physiological action along with me upon animals in the laboratory. This we proceeded to do; and the result of the investigation was that the majority of the extracts from which he supposed that he had obtained definite results in man gave no indications of physiological activity; whereas on the other hand, the extract of suprarenal capsule gave such manifest indications of activity that it was quite clear that a very important principle was contained within this organ. The properties of this principle we then proceeded to work out . . . [from Schäfer's testimony before the second Royal Commission on Vivisection, in *British Parliamentary Papers*, **57** (1908), 430].

Addison in 1849 had associated a diseased state of the adrenal glands with the set of clinical symptoms characteristic of the disease that now bears his name. Brown-Séquard (1856) showed that excision of the entire adrenal glands of animals was inevitably fatal. Oliver and Schäfer's experiments demonstrated conclusively that intravenous injection of small quantities of aqueous extract of adrenal gland into various animals produced striking effects: a sharp increase in blood pressure owing to contraction of the arterioles, cardiac inhibition, shallower respiration, and prolongation of muscular contractions. They showed that the extract took effect through direct action on the peripheral arterioles; that the activity of the extract was preserved through digestion; that the active principle was produced by the medulla and not by the cortex of the gland; and that the active principle was absent in extracts of glands from patients with advanced Addison's disease. Oliver and Schäfer identified the active principle which they had dem-

onstrated in the adrenals with a substance described by Vulpian (1856) in his distinction between the cortex and the medulla of the gland. They contrasted their results with those of Paolo Pellacani (1874) and Pio Foà and Pellacani (1884), who had found that injection of adrenal extract into animals was generally fatal. In related work they demonstrated that extract of pituitary in relatively large quantities caused a somewhat smaller rise in blood pressure due to contraction of arterioles and augmentation of heart action.

Oliver and Schäfer's accomplishment was the first detailed study of the effect of the active principle of a ductless gland. By explicitly rejecting the autointoxication theory, which held that fatalities following excision of the adrenals were due to the accumulation in the blood of toxins that it was the normal function of the adrenals to destroy, they helped to shape the endocrine doctrine. They pointed out that the production of a specific active principle, diffused through the blood, appeared to be the essential function of certain ductless glands, notably the thyroid and the adrenals. Their work was the basis for subsequent research in which J. J. Abel (1899) isolated and named the active principle of the adrenal medulla, epinephrine, and Thomas Bell Aldrich (1901) and Jokichi Takamine (1901) prepared it in crystalline form.

BIBLIOGRAPHY

I. ORIGINAL WORKS. Oliver's many medical writings are listed in *Index medicus*. His endocrinological researches are dealt with in four papers written with E. A. Schäfer: "On the Physiological Action of Extract of the Suprarenal Capsules," in *Journal of Physiology*, **16** (1894), i–iv, and **17** (1894–1895), ix–xiv; "The Physiological Effects of Extracts of the Suprarenal Capsules," *ibid.*, **18** (1895), 230–276; and "On the Physiological Action of Extracts of Pituitary Body and Certain Other Glandular Organs," *ibid.*, 277–279. See also "The Croonian Lectures: A Contribution to the Study of the Blood and the Circulation. Lecture II," in *British Medical Journal* (1896), **1**, 1433–1437; and "The Action of Animal Extracts on the Peripheral Vessels," in *Journal of Physiology*, **21** (1897), xxii–xxiii; and the book *Pulse-Gauging. A Clinical Study of Radial Measurement and Pulse-Pressure* (London, 1895). A few letters by Oliver are in the Wellcome Institute of the History of Medicine, London, and in the library of the Royal College of Physicians, London. His MS notes of William Jenner's lectures in medicine for the session of 1862–1863 are in the library of University College Hospital, London.

II. SECONDARY LITERATURE. Sources for Oliver's life and work are the notices in *Lancet* (1916), **1**, 105; and

British Medical Journal (1916), **1**, 73; *Munk's Roll* (London, 1955), IV, 324; *Presidential Address to the Royal College of Physicians of London* (London, 1916), 27–29; and T. R. Elliot, "Sir William Jenner and Dr. George Oliver," in *University College Hospital Magazine*, **19** (1934), 159–163. On Oliver's work in the context of early endocrinology, see E. A. Schäfer, "Internal Secretions," in *Lancet* (1895), **2**, 321–324, an important theoretical discussion, and "On the Present Condition of Our Knowledge Regarding the Functions of the Suprarenal Capsules," in *British Medical Journal* (1908), **1**, 1277–1281, 1346–1351. See also the following books: L. F. Barker, ed., *Endocrinology and Metabolism* (London, 1922); A. Biedl, *The Internal Secretory Organs: Their Physiology and Pathology*, L. Forster, trans. (London, 1913); C. McC. Brooks, J. L. Gilbert, H. A. Levey, and D. R. Curtis, *Humors, Hormones and Neurosecretions* (New York, 1962); J. F. Fulton and L. G. Wilson, eds., *Selected Readings in the History of Physiology*, 2nd ed. (Springfield, Ill., 1966); E. Gley, *The Internal Secretions. Their Physiology and Application to Pathology*, M. Fishberg, trans. (New York, 1917), H. D. Rolleston, *The Endocrine Organs in Health and Disease, With an Historical Review* (London, 1936); E. A. Schäfer, ed., *Text-Book of Physiology*, I (Edinburgh–London, 1898); and *The Endocrine Organs. An Introduction to the Study of Internal Secretion* (London, 1916); and S. Vincent, *Internal Secretion and the Ductless Glands* (London, 1912).

RICHARD D. FRENCH

OLSZEWSKI, KAROL STANISŁAW (*b.* Broniszow, Poland, 29 January 1846; *d.* Cracow, Poland, 24 March 1915), *chemistry, physics.*

Olszewski was a pioneer in the field of low-temperature phenomena who became famous, along with Z. von Wroblewski, for achieving the liquefaction of air. His father, a Polish landowner, was killed during a peasants' uprising a few months after the birth of his son; and Olszewski was brought up by relatives. From 1866 to 1872 he studied natural science at Cracow and at Heidelberg, from which he received the doctorate in 1872. He then became assistant to Emil Czyrnianski, professor of chemistry at the Jagiellonian University in Cracow; in 1891 he was appointed professor of chemistry there, a post he held until his death. Olszewski was a member of the Cracow Academy of Sciences.

In 1883 Olszewski and Wroblewski liquefied air, oxygen, nitrogen, and carbon monoxide. Their successes owed much to Olszewski's previous work on the liquefaction of carbon dioxide. After Wroblewski's death Olszewski was the only expert in Poland on the liquefaction of gases. He determined the inversion temperatures of oxygen and nitrogen and, in 1902,

that of hydrogen. He also liquefied argon and fluorine. Olszewski and Wroblewski were able to liquefy hydrogen only in its dynamic state; it appeared as a cloud of fog in the midst of escaping hydrogen gas. Olszewski attempted to liquefy hydrogen in its static state, but the first to do so was James Dewar (1898), who used the new procedure of air liquefaction developed by Linde and Hampson: the cooling of gases by means of their internal efficiency, using the counterflow principle. Olszewski, however, improved Dewar's methods and adapted them to practical laboratory work.

Olszewski worked on the liquefaction of helium as early as 1895, but without success; the existing methods were not applicable because of the low critical temperature of helium, and Linde's process was unavailable to him because of its high cost. (The liquefaction of helium was achieved in 1908 by H. Kamerlingh Onnes.) Olszewski was a thorough researcher with great manual dexterity and experimental intuition. His devices for air and hydrogen liquefaction were very highly regarded and were manufactured under license by the Cracow mechanic L. Grodzicki.

BIBLIOGRAPHY

I. ORIGINAL WORKS. Olszewski's numerous scientific papers include "Ueber die Verflüssigung des Sauerstoffs, Stickstoffs und Kohlenoxyds," in *Annalen der Physik und Chemie*, n.s. **20** (1883), 243–257, written with Z. von Wroblewski; "Ueber die Dichte des flüssigen Methans, sowie des verflüssigten Sauerstoffs und Stickstoffs," *ibid.*, **31** (1887), 58–74; "Ueber das Absorptionsspectrum des flüssigen Sauerstoffs und der verflüssigten Luft," *ibid.*, **33** (1888), 570–575; "Bestimmung des Siedepunkts des Ozons und der Erstarrungstemperatur des Aethylens," *ibid.*, **37** (1889), 337–340; "Bestimmung der kritischen- und der Siedetemperatur des Wasserstoffs," *ibid.*, **56** (1895), 133–143; "Liquefaction of Gases," in *Philosophical Magazine*, 5th ser., **39** (1895), 188–213; "Ein Versuch, das Helium zu verflüssigen," in *Annalen der Physik und Chemie*, n.s. **59** (1896), 184–192; "Experimentelle Bestimmung der Inversionstemperatur der Kelvinschen Erscheinung," in *Annalen der Physik*, 4th ser., **7** (1902), 818–823; "Apparate zur Verflüssigung von Luft und Wasserstoff," *ibid.*, **10** (1903), 768–782; "Ein neuer Apparat zur Verflüssigung des Wasserstoffs," *ibid.*, **12** (1903), 196–201; "Ein Beitrag zur Bestimmung des kritischen Punktes des Wasserstoffs," *ibid.*, **17** (1905), 986–993; "Weitere Versuche, das Helium zu verflüssigen," *ibid.*, 994–998; and "On the Temperature of Inversion of the Joule-Kelvin Effect for Air and Nitrogen," in *Philosophical Magazine*, 6th ser., **13** (1907), 722–724.

Bibliographies of Olszewski's writings are in Academy of Sciences, Cracow, *Katalog der Akademischen Publika-*

tionen seit 1873 bis 1909 (Cracow, 1910); and Poggendorff, IV, 1095; and V, 922–923.

II. Secondary Literature. M. von Smoluchowski, "Karl Olszewski—ein Gelehrtenleben," in *Naturwissenschaften*, **5** (1917), 738–740, includes a biographical note on Wroblewski; see also H. Kamerlingh Onnes, "Karol Olszewski," in *Chemikerzeitung*, **39** (1915), 517–519.

A chronological list of publications in Polish—courtesy of Dr. I. Stroński, Cracow—includes the following: *Kronika Uniwersytetu Jagiellońskiego 1864–1887* (Cracow, 1887), 83–86, 184; E. Kurzyniec, "O pierszeństwie skroplenia wodoru w stanie dynamicznym" ("On the Priority of the Liquefaction of Hydrogen in the Dynamic State"), in *Prace Komisji historii medycyny*, **3** (1953), 303–315; K. Adwentowski, A. Pasternak, and Z. Wojtaszek, "Dewar czy Olszewski?" ("Dewar or Olszewski?"), in *Kwartalnik historii nauki . . .*, **1** (1956), 539–561, including letters from M. Pattison Muir and Sir William Ramsay to Olszewski; A. Pasternak, "Karol Olszewski (1846–1915) i Zygmunt Wroblewski (1845–1888)," in *Polscy badacze przyrody* ("Polish Investigators of Nature"; Warsaw, 1959), 174–203; K. Adwentowski, A. Pasternak, and Z. Wojtaszek, "Karol Olszewski jako uczony i nauczyciel" ("Karol Olszewski as Teacher and Scientist"), in *Studia i materiały z dziejów nauki polskiej*, ser. C, **3** (1959), 193–229, including a report on Olszewski's laboratory by his former co-worker K. Adwentowski; and Z. Wojtaszek, "O działalności naukowej Karola Olszewskiego poza dziedzina kriogeniki" ("On Olszewski's Scientific Work Outside Cryogenics"), *ibid.*, **9** (1964), 135–173 (which includes Olszewski's researches on the chemistry of water and a bibliography of these papers); and "Zarys historii katedr chemicznych Uniwersytetu Jagiellońskiego" ("Compendium of the History of the Chairs of Chemistry in the Jagiellonian University of Cracow"), in *Studia ad universitatis Iagellonicae Cracoviensis facultatis mathematicae, physicae, chemiae cathedrarum historiam pertinentia* (Cracow, 1964), 133–219.

Hans-Günther Körber

OLUFSEN, CHRISTIAN FRIIS ROTTBØLL (*b.* Copenhagen, Denmark, 15 April 1802; *d.* Copenhagen, 29 May 1855), *astronomy.*

Olufsen was the son of the Danish political economist Christian Olufsen. During his studies at the University of Copenhagen he was awarded a gold medal for a mathematical treatment of eclipses, and he later spent two years with Bessel at Königsberg. In 1829 he became senior astronomer at the University of Copenhagen observatory and, three years later, was promoted to professor of astronomy and director of the observatory. In 1840 he received his doctorate with a dissertation on the derivation of the lunar parallax.

In 1829 Schumacher had suggested that the Royal Danish Academy of Sciences and Letters produce new tables for the sun, some preparatory work having already been done by Bessel; but more than twenty years elapsed before the tables were completed through cooperation between Hansen, in Gotha, and Olufsen. The former derived the perturbations in the movement of the earth, and the latter made comparisons with a long series of observations and the final determination of the mean motion of the earth. Olufsen's investigation in 1831 of the systematic errors in the observations made with the Greenwich mural quadrant when Maskelyne was astronomer royal was a prerequisite for the use of the Greenwich observations of the sun for his work with Hansen.

For the series of star maps covering the declinations −15° to +15°, which were initiated and published by the Berlin Academy (*Akademische Sternkarten*), Olufsen took over the right ascension 1^h; and he gave a detailed report on the course of the total solar eclipse of 28 July 1851, as observed from Kalmar, Sweden.

Olufsen worked in several fields, but he was often hampered by illness. His main contribution was made in connection with the work of his contemporaries, particularly that of Bessel, toward reforming and improving the foundation of astronomy.

BIBLIOGRAPHY

Olufsen's memoirs include "Untersuchungen über den Greenwicher Mauerquadranten während Maskelynes Direction der dortigen Sternwarte," in *Astronomische Nachrichten*, **9** (1831), 85–106; "Untersuchungen über den Werth der Mondsparallaxe, die aus den in der Mitte des vorigen Jahrhunderts angestellten correspondirenden Beobachtungen abgeleitet werden kann," *ibid.*, **14** (1837), 209–226; "Ueber die Sonnenfinsterniss am 7ten Juli 1842," *ibid.*, **22** (1844), 217–230, 232–242; and "Beobachtung der totalen Sonnenfinsterniss am 28sten Juli 1851 in Calmar," *ibid.*, **33** (1851), 219–222. Among his books are *Disquisitio de parallaxi lunae* (Copenhagen, 1840), his dissertation; *Tentamen de longitudine speculae Havniensis. Praemittuntur considerationes de conaminibus, quae initio seculi octavi decimi ad astronomiam practicam reformandam instituit inclytissimus Roemerus* (Copenhagen, 1840); *Begyndelsesgrunde af astronomien med anvendelse paa den mathematiske Geographie* (Copenhagen, 1848); and *Tables du soleil, exécutées d'après les ordres de la Société royale des sciences de Copenhague* (Copenhagen, 1853; supp., 1857), written with P. A. Hansen.

There is an obituary by P. Pedersen in *Oversigt over det K. Danske Videnskabernes Selskabs Forhandlinger* (1856), 96–103.

Axel V. Nielsen

OLYMPIODORUS (*b.* Thebes, Egypt, *ca.* 360–385; *d.* after 425), *history, alchemy.*

The earliest known event in the life of Olympiodorus is a mission in 412 for Emperor Honorius to Donatus, leader of the Huns. About 415 he was in Athens; and about 423 he went to Egypt, where he visited Nubia, Thebes, Talmis, Syene (now Aswan), the oasis of Siwa, and the priests of Isis at Philae. He probably lived at times in Byzantium, Ravenna, and Rome; and he knew the latter city well. He was not a Christian. At Athens, Olympiodorus associated with the Sophists and was a friend of the grammarian Philtatius. He was personally acquainted with Valerius, the prefect of Thrace. He called himself a poet (ποιητής), a word that is sometimes interpreted as "alchemist."

Olympiodorus is known primarily for his Greek history, *Materials for History*, a continuation of the work of Eunapius (*d.* after 414). The original work, covering the period from 407–425, is preserved only in fragments in the *Bibliotheca* of Photius, the ninth-century patriarch of Constantinople. Olympiodorus' history is dedicated to the Emperor Theodosius II and describes in twenty-two books the history of the Western Empire from the seventh consulship of Honorius to the accession of Valentinian III. The work is an impartial and interesting commentary by an educated observer who had firsthand knowledge of the troubled decades of the early fifth century.

Certain authorities, such as Berthelot and Lippmann, credit Olympiodorus of Thebes with being the author of a Greek work on alchemy entitled variously "The Philosopher Olympiodorus to Pelasius, King of Armenia, on the Divine and Sacred Art" and "The Alexandrian Philosopher Olympiodorus on the Book of Deeds by Zosimus and on the Sayings of Hermes and the Philosophers." The work is quite extensive, with a wealth of disconnected quotations; some of those from Zosimus of Panopolis (late third century) are new. The author presents a very confused and poor explanation of alchemy and displays little practical understanding of his subject, although there is considerable alchemical imagery with Gnostic and Egyptian influence and language. He attempts to draw parallels between the views of the great alchemists and the views of such philosophers as Thales, Anaximander, Anaximenes, Parmenides, and Xenophanes on the origin of matter. He cites the many books of the ancients that were to be found in the Ptolemaic library at Alexandria, written in allegory, with the words having a mystical, double sense which only the initiate can understand. There is little mention of alchemical apparatus. Among his alchem-ical predecessors he mentions Agathodaemon, Chimes, Maria the Jewess, and Synesius.

Other authorities, especially Hammer-Jensen, consider the author of this alchemical work to have been a Neoplatonic philosopher of the sixth century known as Olympiodorus of Alexandria. The author of commentaries on the works of Plato and Aristotle, this Olympiodorus is much esteemed as an interpreter of Plato.

BIBLIOGRAPHY

I. ORIGINAL WORKS. The work on alchemy is found in Marcellin P. E. Berthelot, *Collection des anciens alchimistes grecs*, 3 vols. (Paris, 1887–1888; Osnabrück, 1967), II, 69–106, III, 75–115. The excerpts from Olympiodorus' historical work, as preserved by Photius, are published in Ludwig A. Dindorf, *Historici graeci minores*, I (Leipzig, 1870), 450–472.

II. SECONDARY LITERATURE. See M. P. E. Berthelot, *Les origines de l'alchimie* (Paris, 1885), 191–199 and *passim*; and *Introduction à l'étude de la chimie des anciens et du moyen âge* (Paris, 1889; Brussels, 1966), *passim*; Walter Haedicke, "Olympiodoros" no. 11, in Pauly-Wissowa, *Real-Encyclopädie der classischen Altertumswissenschaft*, 1st ser.; XVIII, pt. 1 (Stuttgart, 1939), cols. 201–207; Ingeborg Hammer-Jensen, *Die älteste Alchemie, Meddelelser fra den K. Danske Videnskabernes Selskab, Hist.-fil. Meddel.*, IV, no. 2 (Copenhagen, 1921); Arthur J. Hopkins, *Alchemy, Child of Greek Philosophy* (New York, 1967), 77; Edmund O. von Lippmann, *Entstehung und Ausbreitung der Alchemie*, I (Berlin, 1919), 96–102; Riess, "Alchemie," in Pauly-Wissowa, 1st ser., I (Stuttgart, 1894), col. 1349; and George Sarton, *Introduction to the History of Science*, I (Baltimore, 1927), 389.

KARL H. DANNENFELDT

OMALIUS D'HALLOY, JEAN BAPTISTE JULIEN D' (*b.* Liège, Belgium, 16 February 1783; *d.* Brussels, Belgium, 15 January 1875), *geology.*

D'Omalius d'Halloy, who played a major role in the transition from the stratigraphic systems of Werner or Guettard to those of de la Beche and Murchison, was the only son of Jean Bernard d'Omalius d'Halloy, the son of an old and wealthy family, and Sophie de Thier de Skeuvre. Following his parents' wishes, he was educated in the family tradition of law and public service. In 1801 he was sent to Paris, where they expected him to become acquainted with literature, art, and theater. But Paris was also the scientific center of Europe, and d'Omalius was attracted to the natural sciences. In 1803, over parental protests, he began serious scientific study, attending the lectures of Lacépède, the zoologist; Antoine de Fourcroy, the chemist; and Cuvier.

D'Omalius made his first geological tour in the Ardennes and Lorraine in 1804. In 1805–1806 he traveled throughout France, including the Belgian provinces, making the observations for his first important paper, "Essai sur la géologie du nord de la France" (1808), which established his scientific reputation. In this publication d'Omalius began, on the Continent, stratigraphic subdivision of the major Wernerian classes by superposition and paleontological criteria. This type of subdivision was later associated in England with the work of Bakewell and Smith and in America with that of Maclure and Eaton. The success of the essay led Coquebert de Montbret, head of the Bureau of Statistics of France, to engage d'Omalius to prepare a geological map of the Empire. This work was begun in 1809 and completed in 1813, but new administrative duties prevented d'Omalius from preparing the map for publication until 1823. In 1813 he presented to the Institut de France a "Mémoire sur l'étendue géographique du terrain des environs de Paris," extending and significantly modifying the work on the Paris basin begun by Cuvier and Brongniart.

Political events ended the first period of d'Omalius' scientific career. From 1813 to 1830, with his father's urging, d'Omalius served in a succession of public offices: mayor of Brabant (1813); superintendent of Dinant (1814), then secretary general of Liège; and governor of the province of Namur, Netherlands (1815–1830). A notable achievement of his administration was the *Code administratif de la province de Namur* (1827), on which he worked for several years. The establishment of Belgian independence in 1830 ended his governorship of Namur and allowed him to resume his scientific career. He never again entirely gave up science for public service, although in 1848 he was elected to the Belgian senate from Dinant, holding office until his death. From 1851 to 1870 he was vice-president of the senate.

In the first period of his scientific career, from 1804 to 1813, d'Omalius worked in stratigraphy and mineralogy. His *Essai* of 1808 opposed Wernerian geology by arguing that the inclination of strata is not due to deposition and that, in the same basin, inclined strata are older than horizontal strata. He also distinguished ten terrains among the strata of northern France. In his *Observations sur un essai de carte géologique de la France, des Pays-Bas et des contrées voisines* (1823), d'Omalius brought the local descriptions of the geology of France into a uniform and sophisticated stratigraphic column, one that, in conjunction with the parallel efforts of Alexandre

Brongniart, enjoyed wide acceptance and formed the basis for the development of Continental stratigraphy in the first half of the nineteenth century.

After returning to geology in 1830, d'Omalius was more speculative than in the earlier period. He also wrote about ethnology and defended the theory of evolution. His controversial views grew out of his conservative refusal to accept complete uniformitarianism in geology. In papers and in his textbook, *Éléments de géologie* (1831), d'Omalius argued that contemporary geological processes are not capable of having produced all formations. Reasoning from the traditional hypothesis that the earth was originally a hot mass cooling slowly, he insisted that the deepest structures—Werner's primitive terrain, which he renamed plutonic terrain—had been formed by heat agencies no longer intensely active. Even in later epochs, when upper strata were formed by deposition in water, heat remained a secondary cause. Thus d'Omalius believed that many deposits of sand in Belgium had been ejected from the hot interior. These views, which he strongly defended in the 1840's and 1850's, when uniformitarianism was being accepted by the scientific community, were d'Omalius' resolution of the contest between Werner's and Hutton's theories, which influenced his early career. In 1833, with the aim of completing an introduction to the science that he called "inorganic natural history," he published a 900-page tome on astronomy, meteorology, and mineralogy, the *Introduction à la géologie*.

While d'Omalius nominally eschewed hypotheses, he early adopted the catastrophic idea of craters of elevation, and he was one of the first to accept glacial concepts. D'Omalius was, in 1831, an early defender of the theory of organic evolution, rejecting Cuvier's theory of successive creations as a "purely gratuitous hypothesis" (*Éléments de géologie*, pp. 526–527). He believed that species are not absolutely fixed, but change in response to changes in environment. Domestication, in which man alters species by controlling nutrition, for instance, is strong analogical evidence for similar processes in nature. While he rejected the notion that man had developed from a polyp, he did believe that the human species had evolved to some extent, suggesting that if man had existed at the beginning of the coal age, then at that time he must have possessed lungs permitting him to live in an atmosphere with more carbon dioxide than his lungs now allow. Contemporary man's racial differentiation similarly resulted from changes in environment.

In his later years d'Omalius was reluctant to accept Charles Darwin's theory of the origin of species. He agreed that natural selection occurs and alters species to a small degree, but he did not think natural selection is powerful enough to explain the major developments in paleontological series. He continued to believe that only environmental changes were sufficient to make major alterations in species.

D'Omalius' evolutionary views were undoubtedly inspired by Lamarck and Geoffroy St.-Hilaire, whose famous debate with Cuvier over evolution had occurred in 1830, but they also derived from a fundamental belief in vital forces. D'Omalius thought that the hypothesis of physical-chemical forces was unable to explain living phenomena; rather, he believed that "each form of living being is determined by a special force" ("Quatrième note sur les forces naturelles," in *Bulletin de l'Académie royale des sciences . . . de Belgique*, **32** [1871], 48–49). He conceived of the vital forces as analogous to the director of an industrial plant who oversees the assembly of a product according to his design. The vital forces thus directed organic responses to environmental change, thereby making evolution possible. This concept of vital force was compatible with the concept of an immortal soul—a matter of importance to d'Omalius, who was a practicing Catholic.

D'Omalius d'Halloy was a member of the Royal Academy of Sciences, Letters, and Fine Arts of Belgium and a foreign member of the Academy of Sciences (Paris).

BIBLIOGRAPHY

I. ORIGINAL WORKS. Most of d'Omalius' articles were published in the *Journal des mines* and the *Annales des mines*, its successor; and the *Bulletin* and *Mémoires* of the Royal Academy of Sciences, Letters, and Fine Arts of Belgium. Scattered pieces of correspondence are listed in the *Catalogue générale des manuscrits des bibliothèques publiques en France*, **48**, **55**; and "Paris: Tome II," *passim*.

His most important works are "Essai sur la géologie du nord de la France," in *Journal des mines*, **24** (1808), 123–158, 271–318, 345–392, 439–466; "Observations sur un essai de carte géologique de la France, des Pays-Bas, et des contrées voisines," in *Annales des mines*, **7** (1823), 353–376; *Éléments de géologie* (Paris, 1831); *Introduction à la géologie ou première partie des éléments d'histoire naturelle inorganique, comprenant des notions d'astronomie, de météorologie et de minéralogie* (Paris, 1833); and *Coup d'oeil sur la géologie de la Belgique* (Brussels, 1842).

II. SECONDARY LITERATURE. The best biographical memoir is J. Guequier, "Omalius d'Halloy," in *Biographie nationale . . . de Belgique*, **16** (1901), 157–166, with partial bibliography. There is a detailed biography by Jules Gosselet, in *Bulletin de la Société géologique de France*, **6** (1878), 453–467, which succeeds the major study by

E. Dupont, "Notice sur la vie et les travaux de J. B. J. d'Omalius d'Halloy," in *Annuaire de l'Académie royale de Belgique*, **42** (1876), 181–296, with a complete bibliography.

RONALD C. TOBEY

OMAR KHAYYAM. See **al-Khayyāmī.**

OMORI, FUSAKICHI (*b.* Fukui, Japan, 30 October 1868; *d.* Tokyo, Japan, 8 November 1923), *seismology*.

Omori entered the College of Science of the Imperial University, Tokyo, in 1886. After graduating in physics in 1890, he turned his attention to the then rapidly emerging science of seismology. He became a lecturer at the university in 1893; and after some further study in Italy and Germany he became professor of seismology in 1897, a post which he held until his death. During this period he was secretary of the Japanese Committee for the Prevention of Earthquake Disasters, becoming noted as Japan's foremost seismologist of the time and one of the world's great early seismologists.

Omori's work was inspired by Seikei Sekiya and by John Milne, who in Tokyo had become one of the great pioneers of modern seismology. Under Milne's encouragement Omori made the first precise studies of earthquake aftershocks and published an important memoir on this subject in 1894. His studies, principally of a great earthquake in the Japanese provinces of Mino and Owari in 1891, led him to evolve a formula, still quoted, for the rate of falloff of aftershocks following major earthquakes.

Omori is probably most noted today for his work in designing seismological instruments. One of these, a horizontal-pendulum-type seismograph, was used in many countries and, with certain modifications, is still in use in some observatories. Omori was the first to experiment with the tiltmeter, an instrument designed to measure small tilting of geological blocks before, during, and after large earthquakes. An important innovation, this instrument led to the gathering of much information useful in predicting earthquakes.

Omori carried out pioneering work on earthquake zoning—the division of a region into areas of greater and less earthquake risk. He showed, incidentally, that destructive Japanese earthquakes were centered predominantly under the steeply sloping ocean floors on the Pacific side of Japan, a result of some importance to modern theories of earthquake occurrence.

Omori's contributions touched on practically all aspects of seismology, and his published papers are numerous. Further topics treated by him include the characteristics of earthquake motions as recorded on seismograms; detailed measurements of periods, displacements, and accelerations of the motions; the location of earthquake sources from seismograph records; the evolution of earthquake intensity scales based on acceleration measurements; experiments on the overturning of brick columns on shaking tables designed to simulate earthquakes; measurements of vibrations of buildings, bridges, chimneys, and towers during earthquakes; and the compilation of earthquake catalogs. He was also interested in the mechanism of volcanoes and used seismic methods in studying them. Omori also applied his ideas in investigations of large earthquakes in India, California, Sicily, and Formosa.

Omori's approach was that of the practical physicist. It has been stated that his achievements, important as they are, could have been greatly enhanced had he been more mathematically minded. But that judgment does not detract from his central importance in maintaining unbroken the distinguished reputation of Japanese seismological research since Milne's time.

On 1 September 1923 Omori, who had gone to Australia to attend a Pan-Pacific Science Congress, visited the Riverview Observatory in Sydney. While he was there, the seismographs started to trace out records of a large distant earthquake. This event proved to be a great earthquake in the province of Kanto, Japan, which caused the loss of 140,000 lives and left Tokyo in ruins. During Omori's return by sea to Japan, his health declined sharply. He died shortly after his return in the university hospital close by the wrecked buildings where he had carried out his lifework.

BIBLIOGRAPHY

Omori's many seismological papers, some of which were written in English, appeared mainly in Japanese journals, especially *Publications of the Imperial Earthquake Investigation Committee, Transactions of the Seismological Society of Japan*, and *Journal of the College of Science, Imperial University of Tokyo*. Other papers were published in *Bollettino della Società sismologica italiana*. Omori's major papers include "On the Aftershocks of Earthquakes," in *Journal of the College of Science, Imperial University of Tokyo*, **7** (1895), 111–200; and "Materials for the Earthquake History of Japan From the Earliest Times Down to 1866," which is *Publications of the Imperial Earthquake Investigation Committee*, **46**, nos. 1 and 2 (1904), written in Japanese with S. Sekiya.

On Omori's life and work, see Charles Davison, *The Founders of Seismology* (Cambridge, 1927), ch. 11; publication details (without titles) of about 100 of Omori's papers are given.

See also *Who's Who in Japan* (Tokyo, 1912), 691.

K. E. BULLEN

ONNES, HEIKE KAMERLINGH. See **Kamerlingh Onnes, Heike.**

OPPEL, ALBERT (*b.* Hohenheim, Württemberg, Germany, 19 December 1831; *d.* Munich, Germany, 22 December 1865), *paleontology, biostratigraphy*.

Oppel was the son of a professor at the agricultural college in Hohenheim, near Stuttgart. He spent most of his school years in Stuttgart, where he was introduced to geology and mineralogy by J. G. von Kurr. In 1851 he entered the University of Tübingen, where he became one of Quenstedt's most talented students. Oppel was a passionate and gifted collector, and even as a student he amassed a first-rate collection of fossils of the Württemberg Jurassic.

Oppel received his doctorate in 1853 with the dissertation "Über den Mittleren Lias in Schwaben." In the following years he visited the Jurassic exposures in Germany, France, England, and Switzerland and met the most important investigators of the Jurassic in these countries. He formed a particularly close friendship with d'Orbigny in Paris. In 1858 he became an assistant to Andreas Wagner at the Bavarian State Paleontological Collection in Munich. Oppel became an assistant professor there in 1860 and, following Wagner's death in 1861, was appointed full professor of paleontology and curator of the paleontological collections at the University of Munich—posts he held until his death. In 1861 he married Anna Herbort, a friend of his sister; they had two children. Their younger child died at the beginning of December 1865. Soon afterward Oppel fell ill and died of typhoid fever at the age of thirty-four.

With his dissertation Oppel laid the foundation for his scientific lifework, the investigation of the Jurassic system. His fundamental work was *Die Juraformation Englands, Frankreichs und des südwestlichen Deutschlands* (1856–1858). Previously the Jurassic deposits of these countries had been subdivided according to local, and frequently lithological, features. Oppel showed, however, that a subdivision may be based solely on paleontological content—that is, on certain faunal species or assemblages—even when the lithological character of the sediments involved is quite varied. By means of fossils he divided the Jurassic formation into thirty-three sections, which he called zones. Each zone was characterized by a number of typical animal species, mostly ammonites. Thus the Jurassic deposits of western Europe were correlated independently of their lithology.

Like d'Orbigny, whose methodology he followed, Oppel based his stratigraphic division on the acceptance of sharply delineated faunal assemblages or faunal species that suddenly appear and disappear. This approach presupposed Linnaeus' concept of the immutability of species and Cuvier's catastrophism. When Darwin's work on the origin of species appeared in 1859, Oppel experienced a great inner conflict. He accepted the theory of evolution only hesitatingly, in the last years of his life. Nevertheless, his concept of the zone is an indispensable resource of modern biostratigraphy, despite the altered theoretical foundations.

Following the appearance of his comparative studies on the Jurassic, Oppel began publishing *Paläontologische Mittheilungen* at Munich. The first five essays, which he himself wrote, dealt chiefly with the invertebrates of the Jurassic and demonstrated his taxonomic acuity. The nearness of the Alps directed his attention to problems of Alpine Jurassic stratigraphy, and in his last work (1865) he distinguished the Tithonian stage. In this designation he included the boundary layers between the Jurassic and Cretaceous in the Alpine and transalpine regions and characterized them through the ammonites they contained. Oppel devoted much time and energy to his collections, and he enriched them to an extraordinary degree; further expanded by his successor Zittel, they became world famous.

BIBLIOGRAPHY

I. ORIGINAL WORKS. Oppel's writings include *Die Juraformation Englands, Frankreichs und des südwestlichen Deutschlands* (Stuttgart, 1856–1858); *Paläontologische Mittheilungen . . .*, 5 pts. (Stuttgart, 1862–1865); and "Die tithonische Etage," in *Zeitschrift der Deutschen geologischen Gesellschaft*, **17** (1865), 535–558.

II. SECONDARY LITERATURE. See F. von Hochstetter, "Zur Erinnerung an Dr. Albert Oppel," in *Jahrbuch der Geologischen Reichsanstalt*, **16** (1866), 59–67; J. G. von Kurr, "Nekrolog des Professor Dr. Albert Oppel," in *Jahreshefte des Vereins für vaterländische Naturkunde in Württemberg*, **23** (1867), 26–30; K. Lambrecht and W. and A. Quenstedt, "Palaeontologi. Catalogus bio-bibliographicus," in *Fossilium catalogus*, **2**, pt. 72 (1938), 320; and the obituaries by K. F. P. von Martius, in *Sitzungsberichte der Bayerischen Akademie der Wissenschaften zu München*, **1** (1866), 380–386, with bibliography; W. W. Smyth, in *Quarterly Journal of the Geological Society of London*, **23** (1867), "Proceedings," 48–49; and H. Woodward, in *Geological Magazine*, **3** (1866), 95–96, with bibliography.

HEINZ TOBIEN

OPPENHEIM, SAMUEL (*b*. Braunsberg, Moravia [now Brušperk, Czechoslovakia], 19 November 1857; *d*. Vienna, Austria, 15 August 1928), *astronomy*.

After leaving the Gymnasium at Teschen, Austrian Silesia, Oppenheim began his studies of mathematics, physics, and astronomy in 1875 at the University of Vienna. His teachers included Boltzmann, Petzval, Stefan, and Weiss. In 1878 he had to undergo a year of military service. He obtained his teaching diploma in mathematics and physics in 1880 and was employed as a teacher at the Akademisches Gymnasium in Vienna. From 1883 onward, he also worked at the university observatory. After receiving the Ph.D., he became assistant astronomer and in 1889 lecturer in astronomy. From 1888 he worked for some time as associate astronomer at Kuffner's private observatory at Ottakring, then a suburb of Vienna. In order to have a safe economic basis, Oppenheim again accepted employment as a teacher in secondary schools: in Vienna (1891), Arnau, Bohemia (1896), and Karolinenthal, near Prague (1899). He also gave lectures in astronomy at Charles University, where in 1902 he became associate professor. In 1911 he was finally called to Vienna and appointed full professor at the university. Oppenheim became a member of the Astronomische Gesellschaft in 1889 and of the Austrian Academy of Sciences in 1920.

The major part of Oppenheim's work was devoted to theoretical astronomy. He studied the influence of rotation on the shape of heavenly bodies, and he published valuable contributions to the three-body and *n*-body problem and to the theory of gravitation. A considerable part of his work dealt with the motions of the stars and with stellar statistics. Oppenheim also performed many numerical calculations of the orbits of comets and minor planets, and he also promoted astrophysics by a great number of visual and photographic observations. After 1917 he was editor of the astronomy volumes of the *Encyklopädie der Mathematischen Wissenschaften*.

BIBLIOGRAPHY

I. ORIGINAL WORKS. Oppenheim's works include "Eine neue Integration der Differential-Gleichungen der Planetenbewegung," in *Sitzungsberichte der Akademie der Wissenschaften in Wien*, **87** (1883); "Rotation und Präcession eines flüssigen Sphäroids," *ibid.*, **92** (1885); "Eine

Gleichung, deren Wurzeln die mittleren Bewegungen im *n*-Körperproblem sind," in *Publikationen der von Kuffnerschen Sternwarte*, **1** (1889); "Bahnbestimmung des Kometen 1846," in *Sitzungsberichte der Akademie der Wissenschaften in Wien*, **99** (1890); "Bahnbestimmung des Planeten (290) Bruna," *ibid.*, **100** (1891); and "Ausmessung des Sternhaufens G.C. Nr. 1166," in *Publikationen der von Kuffnerschen Sternwarte*, **3** (1894).

See also "Bestimmung der Kräfte, durch welche die Bewegung dreier Körper in gegebenen Curven erzeugt werden," *ibid.*, **3** (1894); *Zur Lehre von den Bewegungen der Doppelsterne* (Vienna, 1894); *Fortpflanzungsgeschwindigkeit der Gravitation* (Vienna, 1895); "Specielle periodische Lösungen im Problem der drei Körper," in *Publikationen der von Kuffnerschen Sternwarte*, **4** (1896); *Kritik des Newton'schen Gravitationsgesetzes* (Prague, 1903); "Bestimmung der Periode einer periodischen Erscheinung nebst Anwendung auf die Theorie des Erdmagnetismus," in *Sitzungsberichte der Akademie der Wissenschaften in Wien*, **118** (1909); "Die Eigenbewegungen der Fixsterne," in *Denkschriften der Akademie der Wissenschaften*, **87** (1912); **92** (1916); **93** (1917); **97** (1921); "Zur Frage nach der Fortpflanzungsgeschwindigkeit der Gravitation," in *Annalen der Physik*, **53** (1917); "Theorie der Gleichgewichtsfiguren der Himmelskörper," in *Encyklopädie der Mathematischen Wissenschaften* (Leipzig, 1919); and *Das astronomische Weltbild im Wandel der Zeit* (Leipzig, 1920).

Other works include "Die scheinbare Verteilung der Sterne," in *Sitzungsberichte der Akademie der Wissenschaften in Wien*, **130** (1921); "Statistische Untersuchungen über die Bewegung der kleinen Planeten," in *Denkschriften der Akademie der Wissenschaften*, **97** (1921); *Weltuntergang in Sage und Wissenschaft* (1921), written with K. Ziegler; *Kometen* (Vienna, 1922); "Perioden der Sonnenflecken," in *Sitzungsberichte der Akademie der Wissenschaften in Wien*, **137** (1928). Besides, there are about thirty papers that are mainly concerned with the determination of orbits of planets, comets, stellar statistics and proper motions, and with theoretical mechanics; these papers are published in the *Astronomische Nachrichten*, **113** (1886), to **232** (1928), and in other periodicals.

II. SECONDARY LITERATURE. See W. E. Bernheimer, in *Beiträge zur Geophysik*, **20** (1928), in *Forschungen und Fortschritte*, **4** (1928), and in *Nature*, **122** (London, 1928), 657; K. Graff, in *Almanach. Österreichische Akademie der Wissenschaften*, **79** (1929), 183–186; J. Rheden, in *Astronomische Nachrichten*, **233** (1928), 295; C. Wirtz, in *Vierteljahrsschrift der Astronomischen Gesellschaft*, **64** (1929), 20–30, with a portrait of Oppenheim; and Poggendorff, vols. III, 988; IV, 1096–1097; V, 923; VI, 1913.

KONRADIN FERRARI D'OCCHIEPPO

OPPENHEIMER, J. ROBERT (*b.* New York, N.Y., 22 April 1904; *d.* Princeton, New Jersey, 18 February 1967), *theoretical physics.*

Robert Oppenheimer achieved great distinction in four very different ways: through his personal research, as a teacher, as director of Los Alamos, and as the elder statesman of postwar physics. These different activities belong to different periods, except that his role as teacher overlaps in time with several of these periods. We may therefore review these different contributions separately, while following a chronological order.

J. Robert Oppenheimer was the son of Julius Oppenheimer, who had immigrated as a young man from Germany.[1] The father was a successful businessman, and the family was well-to-do. His mother, the former Ella Freedman, was a painter of near professional standard, and both parents had taste for art and music.

As a boy Oppenheimer showed a wide curiosity and the ability to learn quickly. He went to the Ethical Culture School in New York, a school with high academic standards and liberal ideas. He went as a student to Harvard in 1922, and in spite of following a very broad curriculum, which included classical languages as well as chemistry and physics, he completed the four-year undergraduate course in three years and graduated *summa cum laude* in 1925.

With all the breadth of his interests, Oppenheimer was quite clear that his own subject was physics. During his undergraduate course he profited much from the contact with Percy Bridgman, an eminent physicist who himself had wide-ranging interests and whose publications dealt with topics far beyond the field of his own experiments; they included philosophical questions.

After graduating, Oppenheimer went to Europe; and during his four years of travel he established himself as a theoretical physicist.

Research in Quantum Mechanics. The year 1925 marked the beginning of an exhilarating period in theoretical physics. During that year Heisenberg's first paper on the new quantum mechanics appeared, and Dirac started to develop his own version of Heisenberg's theory in a paper which appeared in the same year. Schrödinger's first paper on his wave equation was published early in 1926. Up to that time the principles of the quantum theory had been grafted onto the classical equations of mechanics, with which they were not consistent. The resulting rules sometimes gave unique predictions which agreed with observation; sometimes the answers were ambiguous; and sometimes the rules could not be applied at all. The new ideas showed the way of obtaining a logically consistent and mathematically clear description, and it looked as if all the old paradoxes of atomic theory would resolve themselves.

This started a period of intense activity, during which all atomic phenomena had to be reexamined in the light of the new ideas. Oppenheimer's quickness in grasping new ideas helped him to play a part in this process. His first paper was submitted for publication in May 1926, less than four years from his entering Harvard and less than a year after Heisenberg's first paper on quantum mechanics.[2] It shows him in full command of the new methods, with which he showed that the frequencies and intensities of molecular band spectra could be obtained unambiguously from the new mechanics. A second paper, submitted in July, is concerned with the hydrogen atom;[3] by this time he was making use of the full apparatus of matrix mechanics developed by Born, Heisenberg, and E. P. Jordan, of the alternative techniques of Dirac, and of Schrödinger's wave mechanics. These two papers were written in Cambridge, and he acknowledged help from Ralph H. Fowler and Paul Dirac.

In the second paper Oppenheimer raises the question of the continuous spectrum and discusses the question of how to formulate the normalization of the wave functions for that case. This was the beginning of his interest in a range of problems which were to occupy him for some time.

In 1926 Max Born invited Oppenheimer to come to Göttingen, where he continued his work on transitions in the continuous spectrum, leading to his first calculations of the emission of X rays. He also developed, jointly with Born, the method for handling the electronic, vibrational, and rotational degrees of freedom of molecules, now one of the classical parts of quantum theory, referred to as the "Born-Oppenheimer method."[4] He obtained his Ph.D. degree in the spring of 1927.

Oppenheimer remained in Europe until 1929, spending some time with Paul Ehrenfest in Leiden and with Wolfgang Pauli in Zurich; the influence of both these men helped further to deepen his understanding of the subject. He continued with the work on radiative effects in the continuous spectrum, which he recognized as one of the important and difficult problems of the time, and found ways of improving the approximations used, which still serve as a pattern for work in this field. Among his minor papers, one deals with electron pickup by ions, a problem which requires the use of nonorthogonal wave functions.[5]

In 1929 Oppenheimer accepted academic positions both at the University of California, Berkeley, and at the California Institute of Technology; and between 1929 and 1942 he divided his time between these two institutions. The list of his papers during this period might almost serve as a guide to what was important

in physics at that time. He was now at the top of his form in research work, and he knew what was important, so that he did not waste his time on pedantic detail. In some of these papers Oppenheimer struggled with key problems which were not yet ripe for solution, such as the difficulties of the electromagnetic self-energy, or the paradox of the "wrong" statistics of the nitrogen nucleus (wrong because, before the discovery of the neutron, nuclei were believed to consist of protons and electrons).[6] But on others he was able to take important steps forward. He saw the importance of Dirac's idea to avoid the difficulty of negative energy states for electrons by assuming them all filled except for a few holes, which were then positively charged particles. He showed, however, that Dirac could not be right in identifying these as protons, since they would have to have the same mass as electrons.[7] Thus he practically predicted the positron three years before its discovery by Carl Anderson.

When cosmic-ray experiments showed serious contradiction with theory, Oppenheimer studied the possibility that this might indicate a breakdown of the accepted quantum theory of radiation.[8] When the discovery of the meson resolved the paradox, he took great interest in the properties of the new particle. He also developed, in a paper with J. F. Carlson, an elegant method for investigating electron-photon showers in cosmic rays.[9] In the 1930's the cyclotron and other accelerators opened up the atomic nucleus to serious study, and Oppenheimer participated in asking important questions and in answering some of them. His paper with G. Volkoff shows a very early interest in stars with massive neutron cores.[10]

During the California period Oppenheimer proved to be an outstanding teacher of theoretical physics. He attracted many pupils, both graduate students and more senior collaborators, many of whom, under his inspiration, became first-rate scholars. His important qualities as a teacher were those which characterized his research: his flair for the key question, his quick understanding, and his readiness to admit ignorance and to invite others to share his struggle for the answer. His influence on his pupils was enhanced by his perceptive interest in people and by his habit of informal and charming hospitality. After his marriage in 1940 his wife, the former Katherine Harrison, helped maintain this easy and warm hospitality.

Oppenheimer still maintained a great breadth of interests, adding even Sanskrit to the languages he could, and did, read. At first his interests were exclusively academic; and he showed little interest in political questions, or in the national and world events of the day. But in the mid-1930's he became acutely aware of the disturbing state of the world— unemployment at home, Hitler, Mussolini, and the Spanish Civil War in Europe. He became interested in politics and, like many liberal intellectuals of the day, became for a time involved with the ideas of left-wing groups.

The list of publications by Oppenheimer and his group shows a break in 1941, and this marks almost the end of his personal research (the exception being three papers published after the war) but by no means of his influence on the development of physics.

Atomic Energy: Los Alamos. The change was the result of Oppenheimer's involvement with atomic energy. After the discovery of fission he, like many others, had started thinking about the possibility of the practical release of nuclear energy. With his quick perception he was aware of the importance of fast neutrons for any possible bomb. In 1940 and 1941 the idea of releasing nuclear energy was beginning to be taken seriously. A number of groups in different universities were working on the feasibility of a nuclear reactor, and others on methods for separating uranium isotopes. The latter would ultimately lead to the production of the light isotope (U^{235}) in nearly pure form, and this is capable of sustaining a chain reaction with fast neutrons. The reactor work led to the production of plutonium, which can be used for the same purpose. While these efforts were well under way by the beginning of 1942, there was no coordinated work on the design of an atomic weapon, its critical size, methods of detonating it, and so on. Oppenheimer had attended some meetings at which such matters were discussed, and early in 1942 he was asked to take charge of the work on fast neutrons and on the problem of the atomic bomb.

On the theoretical side Oppenheimer assembled at Berkeley a conference of first-rate theoreticians, including Edward Teller, who on that occasion first suggested the possibility of a thermonuclear explosion. The work continued in a theoretical group led by Oppenheimer at Berkeley. The experimental determination of the relevant nuclear data was divided between a large number of small nuclear physics laboratories; this hampered progress, since it was difficult for these groups to maintain adequate contact, particularly in view of the secrecy with which the whole project had to be treated.

When, therefore, the United States government brought the atomic energy work under the auspices of the army and put Colonel (later General) Leslie Groves in charge of the project under the code name "Manhattan District," Oppenheimer suggested to Groves that the weapon development be concentrated in a single laboratory. This should include the theory and the nuclear physics work as well as the chemical, metallurgical, and ordnance aspects of the project. In this way the different groups could work together effectively.

Groves accepted the proposal, and on Oppenheimer's advice chose the site of a boys' boarding school at Los Alamos, New Mexico, a region Oppenheimer knew and loved—he had a ranch there. The remoteness of the site made access and transport problems difficult but seemed to have an advantage in reducing contacts with the outside—and therefore the risk of leakage of information.

Groves not only followed Oppenheimer's advice in the creation and location of the laboratory, but he selected Oppenheimer as its director. This was a bold decision, since Oppenheimer was a theoretician with no experience of administration or of organizing experimental work. Events proved Groves right, and the work of the laboratory was extremely effective. In the view of most of the wartime members of Los Alamos, its success owed much to Oppenheimer's leadership.

He attracted a strong team of first-rate scientists, who came because of their respect for Oppenheimer as a scientist and because of his evident sense of purpose. Inside the laboratory he was able to maintain completely free exchange of information between its scientific members; in other words, in exchange for the isolation of the laboratory and the restrictions on travel which its members had to accept, there was none of the "compartmentalization" favored in other atomic energy laboratories for the sake of security. Oppenheimer was able to delegate responsibility and to make people feel they were being trusted. At the same time his quick perception enabled him to remain in touch with all phases of the work. When there were major problems or major decisions to be taken, he guided the discussions of the people concerned in the same spirit of a joint search for the answer in which he had guided the discussions with his students. In the work he did not spare himself, and in response he obtained a sustained effort from all his staff.

It seems that the laboratory was set up just in time, because when the design of the plutonium bomb was ready, enough plutonium was available for the first bomb. The plutonium bomb required a greater design and development effort than the uranium bomb, since the more intense neutron background required a much more rapid assembly from subcritical conditions to the final, highly critical configuration. Failing this, a stray neutron is likely to set off the chain reaction when the assembly is only just critical, giving an explosion of very poor efficiency.

When the test of the first bomb at Alamogordo demonstrated the power of the new weapon, all spectators felt a terrified awe of the new power, mixed with pride and satisfaction at the success of their endeavors. Initially some were more conscious of the one emotion, some of the other. Oppenheimer, whose attitude to his own faults was as unmerciful as to those of others, if not more so, admitted later that he could not resist feeling satisfaction with the key part he had played in the work. Many accounts have quoted the verses from his Sanskrit studies of the *Bhagavad-Gita* which went through his mind at the time of the test, the first referring to the "radiance of a thousand suns" and the other saying, "I am become Death, the destroyer of worlds." Besides the awareness of the technical achievement, Oppenheimer clearly did not lose sight of the seriousness of the implications.

None of this was public knowledge until 6 August 1945, when the first uranium bomb was dropped on Hiroshima. The implications of the decision to use the bomb to destroy a city will continue to occupy historians for a long time. Oppenheimer played some part in this decision: he was one of a panel of four scientists (the others being A. H. Compton, E. Fermi, and E. O. Lawrence) who were asked in May 1945 to discuss the case for the military use of the bomb on Japan. They were told that it would be impossible to cancel or delay the planned invasion of Japan, which was sure to be very costly in lives, unless Japan surrendered beforehand. Their opinion, which Oppenheimer supported, was that a demonstration on an uninhabited island would not be effective, and that the only way in which the atom bomb could be used to end the war was by actual use on a "military" target in a populated area. Today, in retrospect, many people, including many scientists, deplore this advice and the use of the bomb. Oppenheimer commented in 1962: "I believe there was very little deliberation. ... The actual military plans at that time ... were clearly much more terrible in every way and for everyone concerned than the use of the bomb. Nevertheless, my own feeling is that if the bombs were to be used there could have been more effective warning and much less wanton killing. . . ."[11] He remained for the rest of his life acutely conscious of the responsibility he bore for his part in developing the weapon and in the decision to use it.

The Aftermath of the Bomb: Princeton. At the end of 1945 Oppenheimer returned to California. This did not mean, however, returning to an ivory tower. He was by now a national figure, and his advice much in demand; he was also very seriously concerned with the issues raised by the invention of atomic weapons.

He took part in the drafting of the "Acheson-Lilienthal Report," which proposed the international control of atomic energy. Most of the language of this report is undoubtedly Oppenheimer's and so, probably, are many of its ideas. The authors of this report wrote it in a generous spirit: international control of the new weapons would be used to ensure peace and to prevent any nation's threatening another with the formidable new weapons. It probably never had much chance of becoming a political reality. A proposal embodying the outline of the report, but hardly its spirit, was presented to the United Nations by Bernard Baruch as the "Baruch Plan," but nothing came of it.

In 1946 the Atomic Energy Commission was set up under the McMahon Act, which provided for civilian control of atomic energy. The first proposal, the May-Johnson Bill, which would have led to military control, was defeated very largely because of the opposition from scientists, although Oppenheimer was prepared to accept it. The commission appointed a General Advisory Committee, with Oppenheimer as chairman; and he served in that capacity until 1952. The committee did more than give technical advice; it had great influence on the policy of the commission. Oppenheimer's role as chairman was not to dominate opinion but to clarify the issues and to formulate people's thoughts. In addition to the General Advisory Committee, he served on numerous other committees concerned with policy questions relating to atomic weapons and defense.

In October 1947, Oppenheimer moved to Princeton, New Jersey, to become director of the Institute for Advanced Study. Until then the Institute had been a kind of retreat for great scientists and scholars who wanted to get on with their studies in peace. Under Oppenheimer's regime the population of the Institute grew in number, and it included many young scientists, mostly as short-term members for a year or two. They included many visitors from other countries. Oppenheimer was an active member of the physics department and usually presided at seminar meetings.

Under Oppenheimer's influence the physics group became one of the centers at which the current problems of modern physics were most clearly understood. Many colleagues came to discuss their ideas with Oppenheimer, and to do so meant exposing one's thoughts to penetrating scrutiny and sometimes to withering criticism. Oppenheimer now had less time for physics than in the prewar days, and he had to form his judgments more rapidly. He was fallible, and there were occasions when he violently and effectively attacked some unfortunate speaker whose ideas were perhaps not proved but were worth debating; there were other instances when he hailed as very promising ideas which later proved barren.

The early Princeton years were a time when there was again a buoyant optimism in physics. The theory of electrons and their electromagnetic field had been stagnant for many years because of the infinities predicted by quantum theory for the field energy of a point charge. The discovery of the "Lamb shift" in the hydrogen spectrum showed that there were some questions to which theoretical answers were needed, and the attempts to find the answers showed how one could bypass the troublesome infinities. S. Tomonaga, J. Schwinger, R. P. Feynman, and F. J. Dyson developed consistent formulations for the new form of the theory, and it was hoped that they could be extended to the proton and neutron and their interactions with the newly discovered meson field. It was a time of intense debate and discussion, and much of this took place at small ad hoc meetings of theoreticians, at which Oppenheimer was at his best in guiding discussion and in helping people to understand each other (and sometimes themselves). The phrase he used in an interview to describe the work at the Institute, "What we do not know we try to explain to each other," is very appropriate for these sessions. He had always had a remarkable gift for finding the right phrase, and he had now become an absolute master of the epigram.

While he did not resume personal research on any substantial scale (he was coauthor of three papers on physics after the war, one of them being a criticism of somebody else's theory), Oppenheimer's participation in meetings at the Institute and elsewhere was still a major factor in the development of ideas in physics.

As director of the Institute, Oppenheimer was responsible also for the policy in other fields, including pure mathematics and history. Here the breadth of his knowledge was a unique qualification. He did not, of course, take part in the work of the other groups as he did in physics, but he could understand what was being done and could comment in a manner respected by the experts.

Throughout the postwar period Oppenheimer wrote and lectured much. At first the subject was predominantly atomic energy and its implications, and the scheme for its international control. Later he became more concerned with the relations between the scientist and society and, from this, with the problem of conveying an adequate understanding of science to the layman. In his Reith lectures on the B.B.C., "Science and the Common Understanding," he attempted to set out what science is about.[12] The language of such lectures was probably not easily followed in detail by the nonscientist, but it had a poetic quality which to many listeners brought the subject closer.

The "Oppenheimer Case." In December 1953, Oppenheimer was informed that his security clearance—that is, his access to secret information—was being withdrawn, because of accusations that his loyalty was in doubt. He exerted his right to ask for hearings, and he was exposed to the grueling experience of over three weeks' quasi-judicial hearings, in which all his past was exposed to detailed scrutiny. The charges were in part his opposition in 1949 to a crash program for developing the hydrogen bomb, and in part his contacts or associations in the late 1930's and early 1940's with Communists and fellow travelers, contacts which had been known to the A.E.C. many years before and had then not been considered sufficiently derogatory to impede his clearance.

It is impossible to understand how these charges could be raised without remembering the atmosphere of hysterical fear of Communism of the Joseph McCarthy era and also without noting that Oppenheimer had made many enemies, who were delighted at this opportunity of curbing his influence. Some of these enemies were people he had bested in public debate, whom his devastating logic had not only shown to be wrong but also made to appear ridiculous. Others were people interested in military policy who feared his influence, which could act contrary to their interests.

The hearings before the three-man Personnel Security Board were originally intended to be confidential, but eventually the transcript was published.[13] It remains an interesting historical document. The board found that Oppenheimer was "a loyal citizen" but, by a two-to-one majority, that he was to blame for opposing the hydrogen-bomb program and later was lacking in enthusiasm for it.

The report of the board went to the Atomic Energy Commission. The commissioners did not uphold the board's (majority) decision censuring Oppenheimer for his views on the hydrogen bomb—this would have caused a powerful reaction in the scientific community—but confirmed the withdrawal of his clearance, in a majority verdict, mainly on grounds of "defects of character." This was opposed by one of the commissioners, the physicist Henry Smyth, who wrote a minority report in favor of Oppenheimer and criticizing the arguments of his colleagues.[14]

Oppenheimer continued as director of the Institute and with his writing and lecturing. On many occasions audiences at his lectures gave him ovations clearly intended to express their sympathy for him and their indignation at the treatment he had received.

In 1963, when the McCarthy era was an embarrassing memory, when many of the people who had conducted the Oppenheimer investigation and made decisions had been succeeded by others, and when tempers had cooled, it was decided to make a gesture of reconciliation. Oppenheimer was given the Enrico Fermi Award for 1963, a prize of high prestige awarded by the Atomic Energy Commission. The award is usually conferred by the president, and John F. Kennedy had the intention of doing so when he was assassinated. It was then conferred by Lyndon Johnson, and Oppenheimer acknowledged it with the words he had intended to say to President Kennedy: "I think it is just possible . . . that it has taken some charity and some courage for you to make this award today."

Oppenheimer knew for almost a year that he had throat cancer, and he could contemplate this fact and talk about it as lucidly as about a conclusion in physics.

NOTES

1. There has been controversy whether in "J. Robert" the "J" stood for "Julius." P. M. Stern (footnote at the beginning of ch. 2 of the book cited in the bibliography) quotes evidence that this was the case. We use the style Oppenheimer used, with the explanation that the letter J "stood for nothing."
2. Oppenheimer, "On the Quantum Theory of Vibration-Rotation Bands," in *Proceedings of the Cambridge Philosophical Society*, **23** (1926), 327–335.
3. Oppenheimer, "On the Quantum Theory of the Problem of the Two Bodies," *ibid.*, 422–431.
4. Max Born and Oppenheimer, "Zur Quantentheorie der Molekeln," in *Annalen der Physik*, 4th ser., **84** (1927), 457–484.
5. Oppenheimer, "On the Quantum Theory of the Capture of Electrons," in *Physical Review*, **31** (1928), 349–356.
6. Oppenheimer, "Note on the Theory of the Interaction of Field and Matter," *ibid.*, **35** (1930), 461–477; P. Ehrenfest and Oppenheimer, "Note on the Statistics of Nuclei," *ibid.*, **37** (1931), 333–338.
7. Oppenheimer, "On the Theory of Electrons and Protons," *ibid.*, **35** (1930), 562–563.
8. Oppenheimer, "Are the Formulas for the Absorption of High Energy Radiation Valid?" *ibid.*, **47** (1935), 44–52.
9. Oppenheimer and J. F. Carlson, "On Multiplicative Showers," *ibid.*, **51** (1937), 220–231.
10. Oppenheimer and G. Volkoff, "On Massive Neutron Cores," *ibid.*, **55** (1937), 374–381.
11. Oppenheimer, *The Flying Trapeze*, the Whidden lectures for 1962 (London, 1964), pp. 59–60.
12. Oppenheimer, *Science and the Common Understanding*, Reith lectures, British Broadcasting Corporation, Nov. 1953 (New York, 1953; London, 1954).
13. United States Atomic Energy Commission, *In the Matter of J. Robert Oppenheimer. Transcript of Hearings Before the Personnel Security Board* (Washington, D.C., 1954).
14. United States Atomic Energy Commission, *In the Matter of J. Robert Oppenheimer. Text of Principal Documents* (Washington, D.C., 1954).

BIBLIOGRAPHY

I. ORIGINAL WORKS. A full list of Oppenheimer's writings can be found in the article by H. A. Bethe, in *Biographical Memoirs of Fellows of the Royal Society*, **14** (1968), 391–416.

II. SECONDARY LITERATURE. There is as yet no book-length biography of Oppenheimer. Among his obituary notices the most important are the one by Bethe, cited above, and the record of speeches at a memorial meeting by R. Serber, V. F. Weisskopf, A. Pais, and G. T. Seaborg, in *Physics Today*, **20**, no. 10 (Oct. 1967), 34–53. The dual biography by Nuel Pharr Davis, *Lawrence and Oppenheimer* (New York, 1968), has been strongly criticized by many reviewers—for instance, F. Oppenheimer, in *Physics Today*, **22**, no. 2 (Feb. 1969), 77–80.

Numerous books are primarily concerned with the "Oppenheimer case" but bring in much biographical material. The most scholarly of these is P. M. Stern, *The Oppenheimer Case; Security on Trial* (New York, 1969). In addition there are C. P. Curtis, *The Oppenheimer Case. The Trial of a Security System* (New York, 1955); and J. Major, *The Oppenheimer Hearing* (London, 1971). H. Chevalier, *Oppenheimer, The Story of a Friendship* (New York, 1965), criticizes Oppenheimer for his conduct when questioned on security; it also contains many interesting facets of Oppenheimer's life at Berkeley.

There is also a considerable literature on the history of the Manhattan Project, including Oppenheimer's part in it. The official record is *A History of the United States Atomic Energy Commission*, I, R. G. Hewlett and D. E. Anderson, Jr., *The New World* (University Park, Pa., 1962), II, R. G. Hewlett and F. Duncan, *Atomic Shield* (University Park, Pa., 1969). Other examples are Leslie R. Groves, *Now It Can Be Told* (New York, 1962); Lewis L. Strauss, *Men and Decisions* (New York, 1962); D. E. Lilienthal, *Journals*, II, *The Atomic Energy Years 1945–1950* (New York, 1964), and III, *The Venturesome Years 1950–1955* (New York, 1966); and L. Giovanetti and F. Freed, *The Decision to Drop the Bomb* (New York, 1965).

RUDOLF PEIERLS

OPPOLZER, THEODOR RITTER VON (*b.* Prague, Bohemia [now Czechoslovakia], 26 October 1841; *d.* Vienna, Austria, 26 December 1886), *astronomy, geodesy.*

His father, Johann von Oppolzer, was a leader of the Vienna school of medicine and professor at the universities of Prague, Leipzig, and Vienna. Oppolzer's first teacher, Franz Jahne, discovered and encouraged his outstanding mathematical abilities. After attending the Piaristen-Gymnasium in Vienna from 1851 to 1859 he studied medicine—in accordance with his father's wishes—and received the M.D. in 1865. Having also studied astronomy, he built a private observatory in the Josephstadt, a recently incorporated suburb of

Vienna. His main instrument, a seven-inch refracting telescope, was then probably the largest in the Austrian empire. By 1866 he had published more than seventy papers on astronomy, comprising observations, computations of the orbits of comets and asteroids, and analytical investigations of related problems. In March 1866 he became lecturer on astronomy at the University of Vienna. In 1868 he participated in the Austrian expedition to Aden to observe a solar eclipse, and in 1874 he observed the transit of Venus at Iaşi, Rumania.

Elected to the Imperial Academy of Sciences of Vienna in 1869, Oppolzer subsequently became a member of nearly every European and American learned society. In 1870 he was appointed associate professor and, in 1875, full professor of astronomy and geodesy at the University of Vienna. In 1873 he became director of the Gradmessungs-Bureau, the Austrian geodetic survey, which was very active under his direction. At the eighth conference of the Internationale Erdmessung, held at Berlin in 1886, he was elected vice-president of the International Geodetic Association. He died a few months later, after having revised the major part of the proofs of his last work, "Canon der Finsternisse."

On 1 June 1865 Oppolzer married Coelestine Mautner von Markhof, daughter of a prominent Austrian industrialist; they had six children. Three planetoids are named for two of their three daughters, Hilda and Agatha, and for his wife. A son, Egon Ritter von Oppolzer, founded the astronomical observatory at Innsbruck.

The great majority of Oppolzer's more than 300 papers deal with the determination and improvement of the orbits of comets and asteroids—sometimes based on Oppolzer's own observations—and with the computation of ephemerides derived from the orbital elements. Dissatisfied with merely routine work, Oppolzer improved existing methods: as early as 1864, for example, he developed new formulas for calculating the differential correction of planetary or cometary orbital elements directly from the deviations from the computed positions. His two-volume *Lehrbuch zur Bahnbestimmung der Cometen und Planeten* (1870–1880) comprises all the materials then necessary for understanding and determining both preliminary and definitive orbits: the basic concepts, mathematical tools, practically arranged formulas, extensive auxiliary tables, and examples drawn from the author's own experience.

About 1868 Oppolzer began to study the computation of ancient and modern eclipses, intending to compile a catalog of the relevant data of all eclipses from the beginning of reliable history, whether observations of the eclipses were actually known by him.

These data were to be computed on the basis of modern knowledge of the exact laws of solar and lunar motion. After several years of discouraging setbacks, Oppolzer realized the impossibility of completing the work within a reasonable time by using Hansen's tables, then the best available. Instead of abandoning the work, however, he devised new methods and tables that, despite their greater accuracy, were much easier to use. They were published in 1881 as "Syzygien-Tafeln für den Mond nebst ausführlicher Anweisung. . . . "

Oppolzer then organized—partly at his own expense—the immense project that resulted in the "Canon der Finsternisse." The "Canon" contains, with minor exceptions, the relevant data of every lunar and solar eclipse, with charts of the central paths of the latter, from 1207 B.C. to A.D. 2163. Oppolzer also planned a fundamental improvement of the lunar theory. Left unfinished at his death, the expansions of the derivatives necessary for this purpose were completed under the supervision of his collaborator R. Schram, who followed Oppolzer's ideas.

In his work in geodesy Oppolzer revealed uncommon administrative ability. He introduced technical improvements in the registration of time signals and in the use of the reversible pendulum for gravimetry. Many differences of longitude between primary stations of the European triangulation frame were determined by Oppolzer or under his supervision. He represented Austria with distinction at international conferences and soon won esteem for his profound knowledge. Apart from his admirable scientific qualities, he showed great social responsibility. Beloved for his generous liberality, he devoted his last public speech to the association for the welfare of sick students founded by his father.

BIBLIOGRAPHY

I. Original Works. Oppolzer's works include "Entwickelung von Differentialformeln zur Verbesserung einer Planeten- oder Cometenbahn," in *Sitzungsberichte der Akademie der Wissenschaften in Wien*, **49** (1864), 271–288; "Definitive Bahnbestimmung des Planeten (58) Concordia," *ibid.*, **57** (1868), 343–383; *Lehrbuch zur Bahnbestimmung der Cometen und Planeten*, 2 vols. (Leipzig, 1870–1880; 2nd ed., rev. and enl., 1882), French trans. by Ernest Pasquier, *Traité de la détermination des orbites des comètes et des planètes* (Paris, 1886); "Über den Venusdurchgang des Jahres 1874," in *Sitzungsberichte der Akademie der Wissenschaften in Wien*, **61** (1870), 515–599; "Das Schaltbrett der österreichischen Gradmessung," *ibid.*, **69** (1874), 379–398; "Entwickelung der Differentialquotienten der wahren Anomalie und des

Radiusvectors nach der Excentricität in nahezu parabolischen Bahnen," in *Monatsberichte der Deutschen Akademie der Wissenschaften zu Berlin* (1878), 852–859; and "Syzygien-Tafeln für den Mond nebst ausführlicher Anweisung . . .," in *Publikationen der Astronomischen Gesellschaft*, **16** (1881).

See also "Ermittlung der Störungswerthe in den Coordinaten durch die Variation entsprechend gewählter Constanten," in *Denkschriften der Akademie der Wissenschaften*, **46** (1882), 45–75; "Tafeln für den Planeten (58) Concordia," *ibid.*, **47** (1883), 149–159; "Tafeln zur Berechnung der Mondesfinsternisse," *ibid.*, 243–275; "Bestimmung der Schwere mit Hilfe verschiedener Apparate," in *Zeitschrift für Instrumentenkunde*, **4** (1884), 303–316, 379–387; "Entwurf einer Mondtheorie," in *Denkschriften der Akademie der Wissenschaften*, **51** (1885), 69–105; "Canon der Finsternisse," *ibid.*, **52** (1887), 1–376, repr. as *Canon of Eclipses* (New York, 1962), with trans. of text and pref. by Owen Gingerich and Donald H. Menzel; and "Astronomische Refraction," *ibid.*, **53** (1887), 1–52.

II. SECONDARY LITERATURE. Robert Schram, "Nekrolog Theodor von Oppolzer," in *Vierteljahrsschrift der Astronomischen Gesellschaft*, **22** (1887), 177–208, contains a bibliography of Oppolzer's work. See also Eduard Suess, "Bericht," in *Almanach der Akademie der Wissenschaften in Wien*, **37** (1887), 183–189, with partial bibliography.

KONRADIN FERRARI D'OCCHIEPPO

ORBELI, LEON ABGAROVICH (*b*. Tsakhkadzor, Russia, 7 July 1882; *d*. Leningrad, U.S.S.R., 9 December 1958), physiology.

Orbeli was the son of Abgar Iosifovich Orbeli, a well-known jurist in Transcaucasia. After graduating from the Gymnasium in Tbilisi in 1899, he entered the Military Medical Academy in St. Petersburg. Orbeli's general biological views were formed under the influence of the lectures of the zoologist N. A. Kholodkovsky and the histologist M. A. Lavdovsky. While still a student in the second course, Orbeli studied physiology and began to work in the laboratory of I. P. Pavlov. Here he carried out his first experimental research—the activity of pepsin iron before and after the severing of the vagus nerves (1903). For the next thirty-five years, Orbeli's life and scientific career were closely connected with the work of Pavlov.

Following his graduation from the Military Medical Academy (1904), Orbeli became an intern at the Nikolai Hospital in Kronshtadt. His move to the Naval Hospital in St. Petersburg gave him the opportunity to continue his experimental research in Pavlov's laboratory. Orbeli joined Pavlov at the very height of Pavlov's research on conditioned reflexes.

In Orbeli's dissertation, "Conditioned Reflexes of the Eye in Dogs" (1908), he showed that a change in the intensity of light could serve as a conditioned stimulus for a dog, even though the dog could not distinguish the color. On Pavlov's recommendation Orbeli worked for two years with Hering in Germany, Langley and Barcroft in England, and at the Marine Biological Station in Naples.

Orbeli's scientific career was spent in the leading Russian physiological centers. He worked in the physiology departments of the Institute of Experimental Medicine; the Military Medical Academy, where he was chairman of the department from 1925 to 1950; and the First Leningrad Medical Institute, where he was also chairman from 1920 to 1930. He also worked in the physiology laboratory of P. F. Lesgaft at the Petrograd Scientific Institute in the Biological Station in Koltushakh.

After Pavlov's death Orbeli was the most prominent physiological scientist in the U.S.S.R.; he directed the I. P. Pavlov Institute of Physiology of the Academy of Sciences of the U.S.S.R., the Institute of Evolutionary Physiology and Pathology of Higher Nervous Activity in Koltushakh (now Pavlovo). In 1956 Orbeli organized the I. M. Sechenov Institute of Evolutionary Physiology (now the Institute of Evolutionary Physiology and Biochemistry of the Academy of Sciences of the U.S.S.R.).

Orbeli wrote more than 200 works on experimental and theoretical science. They embrace a varied range of problems in physiology and theoretical medicine. Most of them are grouped around the following subjects: physiology of the higher nervous activity and the sense organs; of the regularities of cerebrospinal coordination; physiology of the autonomic nervous system, the rapid development of which led to his creation of the theory of the adaptive-trophic role of the sympathetic nervous system in the organism (the classic Orbeli–Ginedinsky phenomenon, 1923); the theory of the physiological role of the cerebellum as regulator of the autonomic functions; the physiology of kidney activity and the problem of pain; and environmental physiology particularly in deep-sea diving.

Orbeli was responsible for the development of evolutionary physiology, and he proposed special experimental methods of research to study regularities of the evolution of functions. Orbeli's basic ideas in this direction are generalized in his program report "Basic Problems and Methods of Evolutionary Physiology" (1956). He created a large physiological school, and among his students were Y. M. Kreps, A. G. Ginetsinsky, A. V. Lebedinsky, and A. V. Tonkikh.

Orbeli's scientific and organizational talent was highly appreciated by the Soviet government. In 1935 he was elected an active member of the Academy of Sciences of the U.S.S.R., and in 1945 he was awarded the title of Hero of Socialist Labor. He was also a member of various foreign academies.

BIBLIOGRAPHY

Orbeli's major writings are included in *Izbrannye trudy* ("Selected Works") 5 vols. (Moscow–Leningrad, 1961–1968).

On Orbeli and his work see L. G. Leybson, in L. A. Orbeli, *Izbrannye trudy*, **1**, 13–55; and *Leon Abgarovich Orbeli* (Leningrad, 1973).

N. A. GRIGORIAN

ORBIGNY, ALCIDE CHARLES VICTOR DES-SALINES D' (*b.* Couëron, Loire-Atlantique, France, 6 September 1802; *d.* Pierrefitte-sur-Seine, near Saint-Denis, France, 30 June 1857), *paleontology.*

D'Orbigny's father, Charles-Marie Dessalines d'Orbigny, came from Santo Domingo. After serving as a naval doctor, he practiced medicine in Couëron and finally in La Rochelle on the Atlantic coast. He was an enthusiastic scientist and often took his sons Alcide and Charles collecting with him on excursions along the coast near their house. As a result of this experience, Alcide decided on a career in science. In 1819 he began systematic zoological research, studying in Paris under Cordier. In June 1826 he left for South America on a commission for the Muséum d'Histoire Naturelle and did not return until March 1834. During these eight years he traveled through the entire continent, making extensive scientific studies under difficult and often dangerous conditions. At the time, much of the continent had been explored only slightly or not at all.

Following his return d'Orbigny spent the rest of his life in Paris, except for brief periods of travel. He was but little concerned about his career. Because of his novel, unorthodox ideas and hypotheses, he had many opponents among his French colleagues. The zoologists dismissed his taxonomic and systematic works and his views concerning the geographic distribution of animals. Many geologists opposed his stratigraphic conceptions. His division of the history of the earth into stages evoked repeated criticism. Lastly, zoologists and geologists were united against him; they thought that paleontology was not an independent science, but that it was only the zoology and botany of fossil organisms.

Consequently, d'Orbigny's initial attempt to obtain a professorship in Paris was unsuccessful. In 1853 a government decree finally created—especially for him—a chair of paleontology at the Muséum d'Histoire Naturelle; the position still exists. Although with his appointment to this chair he had attained his life's goal, he had become embittered by the years of hostility and criticism from his colleagues. He thought that he could forget this adversity by working harder, but the increased activity helped undermine his health and he died of a heart ailment, which had caused him much pain during the last year of his life. D'Orbigny was survived by a wife and children. He was a member of many scientific societies and academies in France and abroad, and on two occasions he won the Wollaston Medal of the Geological Society of London.

D'Orbigny's first scientific publications were brief studies of recent and Jurassic gastropods and of the masticatory apparatus of the nautiluses. From 1819 much of his early research was devoted to recent and fossil Foraminifera. After seven years of work in this field, he published *Tableau méthodique de la classe des Céphalopodes* (1826). In Lamarck's classification the protozoans were still grouped under the cephalopods. D'Orbigny accepted this view, but he separated—under the name Foraminifera—the microscopic forms from the other cephalopods. His classification encompassed five classes, fifty-three genera, and 600 species. He based his classification on the number and arrangement of the chambers of the shell. Among the forms that he described were living taxa from South America, the Canary Islands, Cuba, and the Antilles. He also described Cretaceous fossils from the Paris Basin and Tertiary fossils from the Vienna Basin. Although d'Orbigny knew the entire group of Foraminifera better than anyone else at the time, he did not grasp their true systematic position. This was first perceived in 1835 by Dujardin, who discovered their protozoan nature and grouped them with the infusorians.

Between 1834 and 1847 d'Orbigny published in ten volumes the results of his eight-year expedition to South America. The material—which extended to zoology, geography, geology, paleontology, ethnography, and anthropology—constituted the most detailed description of a continent ever made. While supervising this publication, d'Orbigny was occupied with many other projects. With Férussac he published *Histoire naturelle des Céphalopodes vivants et fossiles* (1839–1848). He wrote the sections on mollusks, echinoderms, sponges, and Foraminifera for Webb and Berthelot's *Histoire naturelle des îles Canaries* (1839–1840) and the sections on ornithology, Foraminifera, and mollusks for Ramon de la Sagra's *Histoire naturelle de Cuba et des Antilles* (1839–1843). His *Histoire naturelle des Crinoides* appeared in 1840 and his *Galérie ornithologique des Oiseaux d'Europe* in 1836–1838 in fifty-two installments. He was the author of a series of paleontological monographs, including one on the Foraminifera of the Upper Cretaceous of the Paris Basin (1840) and one on the Mesozoic and Tertiary fossils of European Russia and of the Ural Mountains (in the great work by Murchison, P. E. de Verneuil, and Keyserling [1845]). Finally, he published a series of brief studies on Cretaceous and Jurassic ammonites, belemnites, and gastropods, on Cretaceous Rudista, and on Tertiary Sepioideans.

From 1840 until his death, d'Orbigny was involved with the publication of his principal work, *Paléontologie française*. In this work he set forth the paleontology and the stratigraphic distribution of all the known forms of mollusks, echinoderms, brachiopods, and bryozoans found in the French Jurassic and Cretaceous deposits. For many years this critical catalog was of great assistance to French geologists and stratigraphers. A notable feature of the work was its treatment of the bryozoans. With the exception of isolated works by other authors, there had previously existed no comprehensive survey of the bryozoans. D'Orbigny provided a critical synthesis of all the living and fossil forms of this phylum, which embraced 1,929 species, of which 879 species came from the Cretaceous period alone. D'Orbigny did not complete *Paléontologie française*. After his death the work was continued, with the aid of the French Geological Society, by Cotteau, Deslongchamps, Piette, De Loriol, and Fromentel, but it was never finished.

D'Orbigny published a still more comprehensive paleontological work, *Prodrome de paléontologie stratigraphique universelle* (1850–1852). It consisted of critical lists of all the fossil mollusks and of other invertebrate groups, which were arranged according to their stratigraphic distribution. D'Orbigny made consistent use of this novel approach and divided the sediments and their fossil contents into twenty-seven stages (*étages*). The stages were named for localities or regions and all were spelled with the same *-ian* ending (*-ien* in French)—Silurian, Callovian, Aptian, Cenomanian, and so forth. Furthermore, the stages were designated by characteristic fossils, and the 18,000 species under consideration were divided into twenty-seven stages.

In this manner d'Orbigny obtained twenty-seven successive extinct faunas. He examined the faunas and ascertained that most species in any given stage no longer appeared in the next younger one; rather, they were replaced by new species. He therefore arrived at a conception of successive destructions and creations of animals in the course of the earth's history. This conception corresponded to the views that Cuvier had set forth in his theory of catastrophism. The theory of evolution has put an end to all such ideas about new creations, including d'Orbigny's. Nevertheless, the term and important elements of the concept of the stage are still valid, and many of the names that d'Orbigny created for the stages remain in use. D'Orbigny may thus be considered one of the founders of modern biostratigraphy. In *Prodrome* d'Orbigny presented a great number of new species, but did not illustrate them. Between 1906 and 1937 the *Annales de paléontologie* (volumes 1–26) published the expanded diagnoses and illustrations of a large portion of the types first described by d'Orbigny.

With the publication of *Prodrome* and of another basic work with similar aims, *Cours élémentaire de paléontologie et de géologie stratigraphiques* (1849–1852), d'Orbigny established the close and enduring connection between invertebrate paleontology and stratigraphic geology that has proved so fruitful for both disciplines.

BIBLIOGRAPHY

I. ORIGINAL WORKS. D'Orbigny's works include *Tableau méthodique de la classe des Céphalopodes* (Paris, 1826); *Voyage dans l'Amérique méridionale*, 10 vols. (Paris, 1834–1847); *Galérie ornithologique des Oiseaux d'Europe* (Paris, 1836–1838); *Histoire naturelle générale et particulière des Céphalopodes acétabulifères vivants et fossiles* (Paris, 1839–1848), with Férussac; *Histoire naturelle générale et particulière des Crinoides vivants et fossiles, comprenant la description zoologique et géologique de ces animaux* (Paris, 1840); *Paléontologie française. Description zoologique et géologique de tous les animaux mollusques et rayonnés fossiles de France*, 8 vols. (Paris, 1840–1856); *Cours élémentaire de paléontologie et de géologie stratigraphiques*, 3 vols. (Paris, 1849–1852); *Prodrome de paléontologie stratigraphique universelle des animaux mollusques et rayonnés*, 3 vols. (Paris, 1850–1852).

II. SECONDARY LITERATURE. On d'Orbigny and his work, see *Notice analytique sur les travaux de Géologie, de Paléontologie et de Zoologie de M. Alcide d'Orbigny, 1823–1856* (Paris, 1856); P. Fischer, "Notice sur la vie et sur les travaux d'Alcide d'Orbigny," in *Bulletin de la Société géologique de France*, ser. 3, **6** (Paris, 1878), 434–453, which contains a bibliography; A. Gaudry, "Alcide d'Orbigny, ses voyages et ses travaux," in *Revue des Deux Mondes* (Paris, 15 February 1859); C. L. V. Monty, "D'Orbigny's Concepts of Stage and Zone," in *Journal of Paleontology*, **42**, no. 3 (1968), 689–701; J. E. Portlock, "Obituary," in *Quarterly Journal of the Geological Society of London*, **14** (London, 1858), lxxiii–lxxix; and K. A. Zittel, *Geschichte der Geologie und Paläontologie bis Ende des 19. Jahrhunderts* (Munich–Leipzig, 1899), pp. 297, 441, 669–670, 692, 696, 705–706, 777, 796, 800, 811.

HEINZ TOBIEN

ORESME, NICOLE (*b.* France, *ca.* 1320; *d.* Lisieux, France, 1382), *mathematics, natural philosophy.*

Oresme was of Norman origin and perhaps born near Caen. Little is known of his early life and family. In a document originally drawn in 1348, "Henry Oresme" is named along with Nicole in a list of masters of arts of the Norman nation at Paris. Presumably this is a brother of Nicole, for a contemporary manuscript[1] mentions a nephew of Nicole named Henricus *iunior.* A "Guillaume Oresme" also appears in the records of the College of Navarre at Paris as the holder of a scholarship in grammar in 1352 and in theology in 1353; he is later mentioned as a bachelor of theology and canon of Bayeux in 1376.

Nothing is known of Nicole Oresme's early academic career. Apparently he took his arts training at the University of Paris in the 1340's and studied with the celebrated master Jean Buridan, whose influence on Oresme's writing is evident. This is plausible in that Oresme's name appears on a list of scholarship holders in theology at the College of Navarre at Paris in 1348. Moreover, in the same year he is listed among certain masters of the Norman nation, as was noted above. After teaching arts and pursuing his theological training, he took his theological mastership in 1355 or 1356; he became grand master of the College of Navarre in 1356.

His friendship with the dauphin of France (the future King Charles V) seems to have begun about this time. In 1359 he signed a document as "secretary of the king," whereas King John II had been in England since 1356 with the dauphin acting as regent. In 1360 Oresme was sent to Rouen to negotiate a loan for the dauphin.

Oresme was appointed archdeacon of Bayeux in 1361. He attempted to hold this new position together with his grand-mastership, but his petition to do so was denied and he decided to remain in Navarre. Presumably he left Navarre after being appointed canon at Rouen on 23 November 1362. A few months later (10 February 1363) he was appointed canon at Sainte-Chapelle, Paris, obtaining a semiprebend. A year later (18 March 1364) he was appointed dean of the cathedral of Rouen. He held this dignity until his appointment as bishop of Lisieux in 1377, but he does not appear to have taken up residency at Lisieux until 1380. From the occasional mention of him in university documents it is presumed that from 1364 to 1380 Oresme divided his time between Paris and Rouen, probably residing regularly in Rouen until 1369 and in Paris thereafter. From about 1369 he was busy translating certain Aristotelian Latin texts into French and writing commentaries on them. This was done at the behest of King Charles V, and his appointment as

bishop was in part a reward for this service. Little is known of his last years at Lisieux.

Scientific Thought. The writings of Oresme show him at once as a subtle Schoolman disputing the fashionable problems of the day, a vigorous opponent of astrology, a dynamic preacher and theologian, an adviser of princes, a scientific popularizer, and a skillful translator of Latin into French.

One of the novelties of thought associated with Oresme is his use of the metaphor of the heavens as a mechanical clock. It has been suggested that this metaphor—which appears to mechanize the heavenly regions in a modern manner—arises from Oresme's acceptance of the medieval impetus theory, a theory that explained the continuance of projectile motion on the basis of impressed force or impetus. Buridan, Oresme's apparent master, had suggested the possibility that God could have impressed impetuses in the heavenly bodies, and that these, acting without resistance or contrary inclination, could continue their motion indefinitely, thus dispensing with the Aristotelian intelligences as the continuing movers. A reading of several different works of Oresme, ranging from the 1340's to 1377, all of which discuss celestial movers, however, shows that Oresme never abandoned the concept of the intelligences as movers, while he specifically rejected impetuses as heavenly movers in his *Questiones de celo.*[2] In these discussions he stressed the essential differences between the mechanics governing terrestrial motion and that involved in celestial motions. In two passages of his last work, *Livre du ciel et du monde d'Aristote,*[3] he suggests (1) the possibility that God implanted in the heavens at the time of their creation special forces and resistances by which the heavens move continually like a mechanical clock, but without violence, the forces and resistances differing from those on earth; and (2) that "it is not impossible that the heavens are moved by a power or corporeal quality in it, without violence and without work, because the resistance in the heavens does not incline them to any other movement nor to rest but only [effects] that they are not moved more quickly." The latter statement sounds inertial, yet it stresses the difference between celestial resistance and resistance on the earth, even while introducing analogues to natural force and resistance. In other treatments of celestial motions Oresme stated that "voluntary" forces rather than "natural" forces are involved, but that the "voluntary" forces differ from "natural" ones in not being quantifiable in terms of the numerical proportionality theorems applicable to natural forces and resistances.[4] In addition to his retention of intelligences as movers, a further factor prevents the identification of any of Oresme's treatments of celestial movers with the proposal of Buridan. For Buridan, *impetus* was a thing of permanent nature (*res natura permanens*) which was corruptible by resistance and contrary inclination. But Oresme seems to hold in his *Questiones de celo*[5] that impetus is not permanent, but is self-expending by the very fact that it produces motion. If this is truly what Oresme meant, it would be obviously of no advantage to use such impetuses in the explanation of celestial motions, for unless such impetuses were of infinite power (and he would reject this hypothesis for all such powers) they would have to be renewed continually by God. One might just as well keep the intelligences as movers. An even more crucial argument against the idea that Oresme used the impetus theory to explain heavenly motion is that he seems to have associated impetus with accelerated motion, and yet insisted on the uniform motion of the heavens. Returning to the clock metaphor, it should be noted that in the two places in which the metaphor is employed, Oresme did not apply it to the whole universe but only to celestial motions.

One of these passages in which the clock metaphor is cited leads into one of Oresme's most intriguing ideas—the probable irrationality of the movements of the celestial motions. The idea itself was not original with Oresme, but the mathematical argument by which he attempted to develop it was certainly novel. This argument occurs in his treatise *Proportiones proportionum* ("*The Ratios of Ratios*"). His point of departure in this tract is Thomas Bradwardine's fundamental exponential relationship, suggested in 1328 to represent the relationships between forces, resistances, and velocities in motions:

$$\frac{F_2}{R_2} = \left(\frac{F_1}{R_1}\right)^{\frac{V_2}{V_1}}.$$

Oresme went on to give an extraordinary elaboration of the whole problem of relating ratios exponentially. It is essentially a treatment of fractional exponents conceived as "ratios of ratios."

In this treatment Oresme made a new and apparently original distinction between irrational ratios of which the fractional exponents are rational, for example, $\left(\frac{2}{1}\right)^{\frac{1}{2}}$, and those of which the exponents are themselves irrational, apparently of the form $\left(\frac{2}{1}\right)^{\sqrt{1/2}}$. In making this distinction Oresme introduced new significations for the terms *pars*, *partes*, *commensurabilis*, and *incommensurabilis*. Thus *pars* was used to stand for the exponential part that one ratio is of another. For example, starting with the ratio $\left(\frac{2}{1}\right)^{\frac{1}{2}}$, Oresme would say, in terms of his exponential calculus, that this irrational ratio is "one half part" of the ratio $\frac{2}{1}$—

meaning, of course, that if one took the original ratio twice and composed a ratio therefrom, $\frac{2}{1}$ would result. Or one would say that the ratio $\frac{2}{1}$ can be divided into two "parts" exponentially, each part being $\left(\frac{2}{1}\right)^{\frac{1}{2}}$, or more succinctly in modern representation:

$$\frac{2}{1} = \left[\left(\frac{2}{1}\right)^{\frac{1}{2}}\right]^2.$$

Furthermore, Oresme would say that such a ratio as $\left(\frac{3}{1}\right)^{\frac{2}{3}}$ is "two third parts" of $\frac{3}{1}$, meaning that if we exponentially divided $\frac{3}{1}$ into

$$\left(\frac{3}{1}\right)^{\frac{1}{3}} \cdot \left(\frac{3}{1}\right)^{\frac{1}{3}} \cdot \left(\frac{3}{1}\right)^{\frac{1}{3}},$$

then $\left(\frac{3}{1}\right)^{\frac{2}{3}}$ is two of the three "parts" by which we compose the ratio $\frac{3}{1}$, again representable in modern symbols as

$$\left(\frac{3}{1}\right) = \left(\frac{3}{1}\right)^{\frac{2}{3}} \cdot \left(\frac{3}{1}\right)^{\frac{1}{3}}.$$

This new signification of *pars* and *partes* also led to a new exponential treatment of commensurability. After this detailed mathematical treatment, Oresme claimed (without any real proof) that as we take a larger and larger number of the possible whole number ratios greater than one and attempt to relate them exponentially two at a time, the number of irrational ratios of ratios (that is, of irrational fractional exponents relating the pairs of whole number ratios) rises in relation to the number of rational ratios of ratios. From such an unproved mathematical conclusion, Oresme then jumps to his central theme, the implications of which reappear in a number of his works: it is probable that the ratio of any two unknown ratios, each of which represents a celestial motion, time, or distance, will be an irrational ratio. This then renders astrology—the predictions of which, he seems to believe, are based on the precise determinations of successively repeating conjunctions, oppositions, and other aspects—fallacious at the very beginning of its operations. A kind of basic numerical indeterminateness exists, which even the best astronomical data cannot overcome. It should also be noted that Oresme composed an independent tract, the *Algorism of Ratios*, in which he elucidated in an original way the rules for manipulating ratios.

Oresme's consideration of a very old cosmological problem, the possible existence of a plurality of worlds, was also novel. Like the great majority of his contemporaries, he ultimately rejected such a plurality in favor of a single Aristotelian cosmos, but before doing so he stressed in a cogent paragraph the possibility that God by His omnipotence could so create such a plurality.[6]

All heavy things of this world tend to be conjoined in one mass [*masse*] such that the center of gravity [*centre de pesanteur*] of this mass is in the center of this world, and the whole constitutes a single body in number. And consequently they all have one [natural] place according to number. And if a part of the [element] earth of another world was in this world, it would tend towards the center of this world and be conjoined to its mass. . . . But it does not accordingly follow that the parts of the [element] earth or heavy things of the other world (if it exists) tend to the center of this world, for in their world they would make a mass which would be a single body according to number, and which would have a single place according to number, and which would be ordered according to high and low [in respect to its own center] just as is the mass of heavy things in this world. . . . I conclude then that God can and would be able by His omnipotence [*par toute sa puissance*] to make another world other than this one, or several of them whether similar or dissimilar, and Aristotle offers no sufficient proof to the contrary. But as it was said before, in fact [*de fait*] there never was, nor will there be, any but a single corporeal world. . . .

This passage is also of interest in that it reveals Oresme's willingness to consider the possible treatment of all parts of the universe by ideas of center of gravity developed in connection with terrestrial physics.

The passage also illustrates the technique of expression used by Oresme and his Parisian contemporaries, which permitted them to suggest the most unorthodox and radical philosophical ideas while disclaiming any commitment to them.

The picture of Oresme's view of celestial physics and its relationship to terrestrial phenomena would not be complete without further mention of his well-developed opposition to astrology. In his *Questio contra divinatores* with *Quodlibeta annexa* we are told again and again that the diverse and apparently marvelous phenomena of this lower world arise from natural and immediate causes rather than from celestial, incorporeal influences. Ignorance, he claims, causes men to attribute these phenomena to the heavens, to God, or to demons, and recourse to such explanations is the "destruction of philosophy." He excepted, of course, the obvious influences of the light of the sun on living things or of the motions of celestial bodies on the tides and like phenomena in which the connections appear evident to observers. In the same work he presented a lucid discussion of the existence of demons. "Moreover, if the Faith did not pose their existence," he wrote, "I would say that from no natural effect can they be proved to exist, for all things [supposedly arising from them] can be saved naturally."[7]

In examining his views on terrestrial physics, we should note first that Oresme, along with many fourteenth-century Schoolmen, accepted the conclusion that the earth could move in a small motion of translation.[8] Such a motion would be brought about by the fact that the center of gravity of the earth is constantly being altered by climatic and geologic changes. He held that the center of gravity of the earth strives always for the center of the world; whence arises the translatory motion of the earth. The whole discussion is of interest mainly because of its application of the doctrine of center of gravity to large bodies. Still another question of the motion of the earth fascinated Oresme, that is, its possible rotation, which he discussed in some detail in at least three different works. His treatment in the *Du ciel*[9] is well known, but many of its essential arguments for the possibility of the diurnal rotation of the earth already appear in his *Questiones de celo*[10] and his *Questiones de spera*.[11] These include, for example, the argument on the complete relativity of the detection of motion, the argument that the phenomena of astronomy as given in astronomical tables would be just as well saved by the diurnal rotation of the earth as by the rotation of the heavens, and so on. At the conclusion of the argument, Oresme says in the *Questiones de spera* (as he did in the later work): "The truth is, that the earth is not so moved but rather the heavens." He goes on to add, "However I say that the conclusion [concerning the rotation of the heavens] cannot be demonstrated but only argued by persuasion." This gives a rather probabilistic tone to his acceptance of the common opinion, a tone we often find in Oresme's treatment of physical theory. The more one examines the works of Oresme, the more certain one becomes that a strongly skeptical temper was coupled with his rationalism and naturalism (of course restrained by rather orthodox religious views) and that Oresme was influenced deeply by the probabilistic and skeptical currents that swept through various phases of philosophy in the fourteenth century. He twice tells us in the *Quodlibeta* that, except for the true knowledge of faith, "I indeed know nothing except that I know that I know nothing."[12]

In discussing the motion of individual objects on the surface of the earth, Oresme seems to suggest (against the prevailing opinion) that the speed of the fall of bodies is directly proportional to the time of fall, rather than to the distance of fall, implying as he does that the acceleration of falling bodies is of the type in which equal increments of velocity are acquired in equal periods of time.[13] He did not, however, apply the Merton rule of the measure of uniform acceleration of velocity by its mean speed, discovered at Oxford in the

1330's, to the problem of free fall, as did Galileo almost three hundred years later. Oresme knew the Merton theorem, to be sure, and in fact gave the first geometric proof of it in another work, but as applied to uniform acceleration in the abstract rather than directly to the natural acceleration of falling bodies. In his treatment of falling bodies, despite his different interpretation of *impetus*, he did follow Buridan in explaining the acceleration of falling bodies by continually accumulating impetus. Furthermore, he presented (as Plutarch had done in a more primitive form) an *imaginatio*—the device of a hypothetical, but often impossible, case to illustrate a theory—of a body that falls through a channel in the earth until it reaches the center. Its impetus then carries it beyond the center until the acquired impetus is destroyed, whence it falls once more to the center, thus oscillating about the center.[14]

The mention of Oresme's geometrical proof of the Merton mean speed theorem brings us to a work of unusual scope and inventiveness, the *Tractatus de configurationibus qualitatum et motuum*, composed in the 1350's while Oresme was at the College of Navarre. This work applies two-dimensional figures to hypothetical uniform and nonuniform distributions of the intensity of qualities in a subject and to equally hypothetical uniform and nonuniform velocities in time.

There are two keys to our proper understanding of the *De configurationibus*. To begin with, Oresme used the term *configuratio* in two distinguishable but related meanings, that is, a primitive meaning and a derived meaning. In its initial, primitive meaning it refers to the fictional and imaginative use of geometrical figures to represent or graph intensities in qualities and velocities in motions. Thus the base line of such figures is the subject when discussing linear qualities or the time when discussing velocities, and the perpendiculars raised on the base line represent the intensities of the quality from point to point in the subject, or they represent the velocity from instant to instant in the motion (Figs. 1–4). The whole figure, consisting of all the perpendiculars, represents the whole distribution of intensities in the quality, that is, the quantity of the quality, or in case of motion the so-called total velocity, dimensionally equivalent to the total space traversed in the given time. A quality of uniform intensity (Fig. 1) is thus represented by a rectangle, which is its *configuratio*; a quality of uniformly nonuniform intensity starting from zero intensity is represented as to its configuration by a right triangle (Fig. 3), that is, a figure where the slope is constant ($GK/EH = CK/GH$). Similarly, motions of uniform velocity and uniform acceleration are represented,

respectively, by a rectangle and a right triangle. There is a considerable discussion of other possible configurations.

Differences in configuration—taken in its primitive meaning—reflect for Oresme in a useful and suitable fashion internal differences in the subject. Thus we can say by shorthand that the external configuration represents some kind of internal arrangement of intensities, which we can call its essential internal *configuratio*. So we arrive at the second usage of the term configuration, in which the purely spatial or geometrical meaning is abandoned, since one of the variables involved (namely intensity) is not essentially spatial, although, as Oresme tells us, variations in intensity can be represented by variations in the length of straight lines. He suggests at great length how differences in internal configuration may explain many physical and even psychological phenomena, which are not simply explicable on the basis of the primary elements that make up a body. Thus two bodies might have the same amounts of primary elements in them and even in the same intensity, but the configuration of their intensities may well differ, and so produce different effects in natural actions.

The second key to the understanding of the configuration doctrine of Oresme is what we may call the suitability doctrine. It pertains to the nature of configurations in their primitive meaning of external figures and, briefly, holds that any figure or configuration is suitable or fitting for description of a quality, when its altitudes (ordinates, we would say in modern parlance) on any two points of its base or subject line are in the same ratio as the intensities of the quality at those points in the subject. The phrase used by Oresme to describe the key relationship of intensities and altitudes occurs at the beginning of Chapter 7 of the first part, where he tells us that:

> Any linear quality can be designated by every plane figure which is imagined as standing perpendicularly on the linear extension of the quality and which is proportional in altitude to the quality in intensity. Moreover a figure erected on a line informed with a quality is said to be "proportional in altitude to the quality in intensity" when any two lines perpendicularly erected on the quality line as a base and rising to the summit of the surface or figure have the same ratio to each other as do the intensities at the points on which they stand.

Thus, if you have a uniform linear quality, it can be suitably represented by *every* rectangle erected on the given base line designating the extension of the subject (for example, either *ADCB* or *AFEB*, or any other rectangle on *AB* in Fig. 1), because any rectangle on that base line will be "proportional in altitude to the

$$\frac{MK}{IG} = \frac{LK}{HG} = 1$$

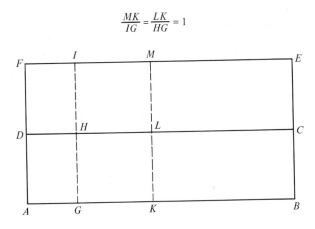

FIGURE 1

intense as the first one, we would have a rectangle whose altitude is everywhere twice as high as that of the rectangle specifying the first uniform quality.

The essential nature of this suitability doctrine was not present in the *Questiones super geometriam Euclidis*, and in fact it is specifically stated there that some specific quality must be represented by a specific

$$\frac{GK}{EH} = \frac{CK}{GH}$$

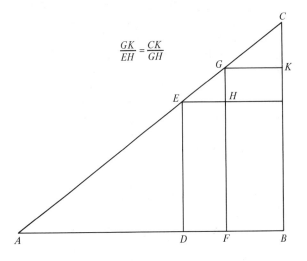

FIGURE 3

quality in intensity," the ratio of any two intensities always being equal to one (that is, $MK/IG = LK/HG = 1$). Similarly, a uniformly difform quality will be represented by *every* right triangle on the given base line, since two altitudes on any one right triangle will have the same ratio to each other as the corresponding two altitudes over the same points of the base line of any other triangle (that is, in Fig. 2, $DB/FE = CB/GE$).

figure rather than a specific kind of figure; that is, a quality represented by a semicircle (Fig. 4) is representable only by that single semicircle on the given base line. But in the *De configurationibus* (pt. 1, ch. 14) Oresme decided in accordance with his fully developed suitability doctrine that such a quality that is representable by a semicircle can be represented by any curved

$$\frac{DB}{FE} = \frac{CB}{GE}$$

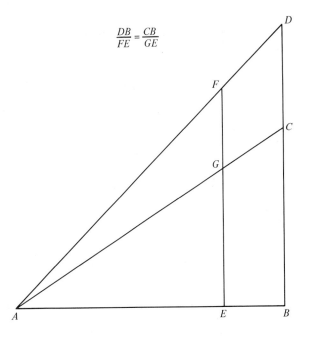

FIGURE 2

$$\frac{CD}{EF} = \frac{HD}{FG} = \frac{JD}{IF}$$

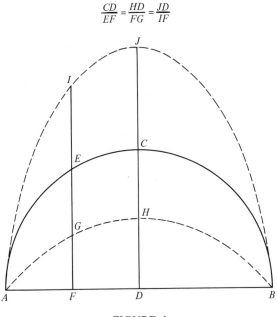

FIGURE 4

The only proviso is, of course, that when we compare figures—say, one uniform quality with another—we must retain some specific figure (say rectangle) as the point of departure for the comparison. Thus, in representing some uniform quality that is twice as

figure on the same base whose altitudes (ordinates) would have any greater or lesser constant ratio with the corresponding altitudes (ordinates) of the semicircle (for example, in Fig. 4, $CD/EF = HD/FG = JD/IF$). He was puzzled as to what these higher or lower figures would be. For the figures of higher altitudes, he definitely rejected their identification with segments of circles, and he said he would not treat the figures of lower altitudes. Unfortunately, Oresme had little or no knowledge of conic sections. In fact the conditions he specified for these curves comprise one of the basic ways of defining ellipses: if the ordinates of a circle $x^2 + y^2 = a^2$ are all shrunk (or stretched) in the same ratio b/a, the resulting curve is an ellipse whose equation is $x^2/a^2 + y^2/b^2 = 1$. Oresme, without realizing it, has given conditions that show that the circle is merely one form of a class of curves that are elliptical. It is quite evident that Oresme arrived at the conclusion of this chapter by systematically applying the basic and sole criterion of suitability of representation, which he has already applied to uniform and uniformly difform qualities; namely, "that the figure be proportional in altitude to the quality in intensity," which is to say that any two altitudes on the base line have the same ratios as the intensities at the corresponding points in the subject. He had not adequately framed this doctrine in the *Questiones super geometriam Euclidis,* and in fact he denied it there, at least in the case of a quality represented by a semicircle or of a uniform or uniformly difform quality formed from such a difform quality. In this denial he confused the question of sufficiently representing a quality and that of comparing one quality to another.

While the idea of internal configuration outlined in the first two parts of the book had little effect on later writers and is scarcely ever referred to, the third part of the treatise—wherein Oresme compared motions by the external figures representing them, and particularly where he showed (Fig. 5) the equality of a right triangle representing uniform acceleration with a rectangle

representing a uniform motion at the velocity of the middle instant of acceleration—was of profound historical importance. The use of this equation of figures can be traced successively to the time of its use by Galileo in the third day of his famous *Discorsi* (Theorem I). And indeed the other two forms of the acceleration law in Galileo's work (Theorem II and its first corollary) are anticipated to a remarkable extent in Oresme's *Questiones super geometriam Euclidis.*[15]

The third part of the *De configurationibus* is also noteworthy for Oresme's geometric illustrations of certain converging series, as for example his proof in chap. 8 of the series

$$1 + \frac{1}{2} \cdot 2 + \frac{1}{4} \cdot 3 \cdots + \frac{1}{2^{n-1}} \cdot n \cdots = 4.$$

He had showed similar interest in such a series in his *Questions on the Physics* and particularly in his *Questiones super geometriam Euclidis.* In the latter work he clearly distinguished some convergent from divergent series. He stated that when the infinite series is of the nature that to a given magnitude there are added "proportional parts to infinity" and the ratio a/b determining the proportional parts is less than one, the series has a finite sum. But when $a > b$, "the total would be infinite," that is, the series would be divergent. In the same work he gave the procedure for finding the following summation:

$$1 + \frac{1}{3} + \frac{1}{9} + \frac{1}{27} + \cdots + \frac{1}{3^n} + \frac{1}{3^{n+1}} + \cdots = \frac{3}{2}.$$

In doing so, he seems to imply a general procedure for the summation of all series of the form:

$$1 + \frac{1}{m} + \frac{1}{m^2} + \frac{1}{m^3} + \cdots + \frac{1}{m^n} + \frac{1}{m^{n+1}} + \cdots.$$

His general rule seems to be that the series is equal to y/x when, $(1/m^i - 1/m^{i+1})$ being the difference of any two successive terms,

$$m^i \left(\frac{1}{m^i} - \frac{1}{m^{i+1}} \right) = \frac{x}{y}.$$

As we survey Oresme's impressive accomplishments, it is clear that his natural philosophy lay within the broad limits of an Aristotelian framework, yet again and again he suggests subtle emendations or even radical speculations.

NOTES

1. MS Paris, BN lat. 7380, 83v: cf. MS Avranches, Bibl. Munic. 223, 348v.
2. Bk. II, quest. 2.

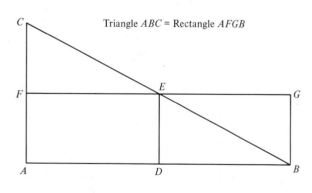

Triangle ABC = Rectangle $AFGB$

FIGURE 5

3. Menut text, 70d–71a; 73d.
4. *Questiones de spera*, quest. 9; *Questiones de celo*, bk. II, quest. 2.
5. Bk. II, quest. 13.
6. *Du ciel*, 38b, 39b–c.
7. MS Paris, BN lat. 15126, 127v.
8. *Questiones de spera*, quest. 3.
9. 138b–144c; see also Clagett, *Science of Mechanics*, 600–608.
10. Bk. II, quest. 13.
11. Quest. 6[8]; see also Clagett, *Science of Mechanics*, 608, n. 23.
12. BN lat. 15126, 98v, 118v.
13. *Questiones de celo*, bk. II, quest. 7.
14. *Questiones de celo*, ibid.; *Du ciel*, 30a–b; Clagett, *Science of Mechanics*, 570.
15. Clagett, *Nicole Oresme and the Medieval Geometry of Qualities*, etc., ch. 2, pt. A.

BIBLIOGRAPHY

I. ORIGINAL WORKS. Oresme's scholarly writings reflect a wide range of interests and considerable originality. He was the author of more than thirty different writings, the majority of which are unpublished and remain in manuscript. They can be conveniently grouped into five categories:

1. Collections of, or individual, *questiones*. These include questions on various works of Aristotle: *Meteorologica* (perhaps in two versions, with MS St. Gall 839, 1–175v being the most complete MS of the vest version); *De sensu et sensato* (MS Erfurt, Amplon. Q. 299, 128–157v); *De anima* (MSS Bruges 514, 71–111v; Munich, Staatsbibl. Clm 761, 1–40v; a different version with an *expositio* in Bruges 477, 238v–264r, may also be by Oresme); *De generatione et corruptione* (MS Florence, Bibl. Naz. Centr., Conv. Soppr. H. ix. 1628, 1–77v; a different version in MS Vatican lat. 3097, 103–146; and Vat. lat. 2185, 40v–61v, may be by him); *Physica* (MS Seville, Bibl. Colomb. 7–6–30, 2–79v); and *De celo* (MSS Erfurt, Amplon. Q. 299, 1–50; Q. 325, 57–90). These also include questions on the *Elementa* of Euclid (edit. of H. L. L. Busard [Leiden, 1961]; additional MS Seville, Bibl. Colomb. 7–7–13, 102v–112) and on the *Sphere* of Sacrobosco (MSS Florence, Bibl. Riccard. 117, 125r–135r; Vat. lat. 2185, 71–77v; Venice Bibl. Naz. Marc. Lat. VIII, 74, 1–8; Seville, Bibl. Colomb. 7–7–13; a different version is attributed to him in Erfurt, Amplon. Q. 299, 113–126). There are other individual questions that are perhaps by him: *Utrum omnes impressiones* (MS Vat. lat. 4082, 82v–85v; edit. of R. Mathieu, 1959), *Utrum aliqua res videatur* (MS Erfurt, Amplon. Q. 231, 146–150), *Utrum dyameter alicuius quadrati sit commensurabilis coste eiusdem* (MS Bern A. 50, 172–176; H. Suter, ed., 1887; see *Isis*, **50** [1959], 130–133), and *Questiones de perfectione specierum* (MS Vat. lat. 986, 125–133v). This whole group of writings seems to date from the late 1340's and early 1350's, that is, from the period when Oresme was teaching arts.

2. A group of mathematico-physical works. This includes a tract beginning *Ad pauca respicientes* (E. Grant, ed., 1966), which is sometimes assigned the title *De*

motibus sperarum (MS Brit. Mus. 2542, 59r); a *De proportionibus proportionum* (E. Grant, ed. [Madison, Wisc., 1966]); *De commensurabilitate sive incommensurabilitate motuum celi* (E. Grant, ed. [Madison, Wisc., 1971]); *Algorimus proportionum* (M. Curtze, ed. [Thorn, 1868], and a partial ed. by E. Grant, thesis [Wisconsin, 1957]); and *De configurationibus qualitatum et motuum* (M. Clagett, ed. [Madison, Wisc., 1968]). These works also probably date from the period of teaching arts, although some may date as late as 1360.

3. A small group of works vehemently opposing astrology and the magical arts. Here we find a *Tractatus contra iudiciarios astronomos* (H. Pruckner, ed., 1933; G. W. Coopland, ed., 1952); a somewhat similar but longer exposition in French, *Le livre de divinacions* (G. W. Coopland, ed., 1952); and a complex collection commonly known as *Questio contra divinatores* with *Quodlibeta annexa* (MS Paris, BN lat. 15126, 1–158; Florence, Bibl. Laurent. Ashb. 210, 3–70v; the *Quodlibeta* has been edited by B. Hansen in a Princeton University diss. of 1973). The first two works almost certainly date before 1364; the last is dated 1370 in the manuscripts but in all likelihood is earlier.

4. A collection of theological and nonscientific works. This includes an economic tract *De mutationibus monetarum* (many early editions; cf. C. Johnson, ed. [London, 1956]; this work was soon translated into French, cf. E. Bridrey's study), a *Commentary on the Sentences of Peter Lombard* (now lost but referred to by Oresme); a short theological tract *De communicatione ydiomatum* (E. Borchert, ed., 1940); *Ars sermonicinandi*, i.e., on the preaching art (MSS Paris, BN lat. 7371, 279–282; Munich, Clm 18225); a short legal tract, *Expositio cuiusdam legis* (Paris, BN lat. 14580, 220–222v); a *Determinatio facta in resumpta in domo Navarre* (MS Paris, BN lat. 16535, 111–114v); a tract predicting bad times for the Church, *De malis venturis super Ecclesiam* (Paris, BN 14533, 77–83v); a popular and oft-published *Sermo coram Urbano V* (delivered in 1363; Flaccus Illyricus, ed. [Basel, 1556; Lyons, 1597]), a *Decisio an in omni casu* (possibly identical with a *determinatio* in MS Brussels, Bibl. Royale 18977–81, 51v–54v); a *Contra mendicacionem* (MSS Munich, Clm 14265; Kiel, Univ. Bibl. 127; Vienna, Nat.-bibl. 11799); and finally some 115 short sermons for Sundays and Feast Days, *Sacre conciones* (Paris, BN lat. 16893, 1–128v). The dating of this group is no doubt varied, but presumably all of them except the *Commentary on the Sentences* postdate his assumption of the grand-mastership at Navarre.

5. A group of French texts and translations. This embraces a popular tract on cosmology, *Traité de l'espere* (L. M. McCarthy, ed., thesis [Toronto, 1943]), which dates from about 1365; a translation and commentary, *Le livre de ethiques d'Aristote* (A. D. Menut, ed., [New York, 1940]), completed in 1372; a similar translation and commentary of the *Politics—Le livre de politique d'Aristote* (Vérard, ed. [Paris, 1489; cf. Menut's ed., in *Transactions of the American Philosophical Society*, n.s. **60**, pt. 6 (1970)]), completed by 1374; the *Livre de ycono-*

mique d'Aristote (Vérard, ed. [Paris, 1489]; A. D. Menut, ed. [Philadelphia, 1957]), completed about the same time; and finally, *Livre du ciel et du monde d'Aristote* (A. D. Menut and A. J. Denomy, eds. [Toronto, 1943], new ed., Madison, Wisc., 1968), completed in 1377. To these perhaps can be added a translation of *Le Quadripartit de Ptholomee* (J. F. Gossner, ed., thesis [Syracuse, 1951]), although it is attributed to G. Oresme.

6. Modern editions. These comprise "De configurationibus qualitatum et motuum," in M. Clagett, ed., *Nicole Oresme and the Medieval Geometry of Qualities* (Madison, Wisc., 1968); E. Grant, ed., "*De proportionibus proportionum*" and "*Ad pauca respicientes*" (Madison, Wisc., 1966); *Nicole Oresme and the Kinematics of Circular Motion* (Madison, Wisc., 1971); A. D. Menut, ed., *Le livre de ethiques d'Aristote* (New York, 1940); A. D. Menut and M. J. Denomy, eds., *Le livre de ciel et du monde d'Aristote*, in *Mediaeval Studies*, 3–5 (1941–1943), rev. with English trans. by Menut (Madison, Wisc., 1968).

II. Secondary Literature. Only a brief bibliography is given here because the extensive literature on Oresme appears in full in the editions of Grant, Clagett, and Menut listed above. These editions include full bibliographical references to the other editions mentioned in the list of Oresme's works.

Works on Oresme include E. Borchert, "Die Lehre von der Bewegung bei Nicolaus Oresme," in *Beiträge zur Geschichte der Philosophie und Theologie des Mittelalters*, **31**, no. 3 (1934); M. Clagett, *The Science of Mechanics in the Middle Ages* (Madison, Wisc., 1959, 1961); M. Curtze, *Die mathematischen Schriften des Nicole Oresme (ca. 1320–1382)* (Berlin, 1870); P. Duhem, *Études sur Léonard de Vinci*, 3 vols., (Paris, 1906–1913); *Le système du monde*, VI–X (Paris, 1954–1959). See also the following works by A. Maier, *An der Grenze von Scholastik und Naturwissenschaft*, 2nd ed. (Rome, 1952); *Die Vorläufer Galileis im 14. Jahrhundert* (Rome, 1949); *Zwei Grundproblem der scholastischen Naturphilosophie*, 2nd ed. (Rome, 1952); and O. Pederson, *Nicole Oresme, og hans Naturfilosofiske System. En undersøgelse af hans skrift* "Le livre du ciel et du monde" (Copenhagen, 1956).

Marshall Clagett

ORIBASIUS (*fl.* Pergamum, fourth century), *medicine.*

The life of Oribasius (or Oreibasius, the correct form of his name is not certain) is described by Eunapius in his *Lives of the Philosophers and Sophists.* This article follows Schröder's presentation, which is based on Eunapius and other sources. Since Oribasius is mentioned among the Sophists, he was an "iatrosophist"—a concept which appeared before the fourth century and which referred to a physician of a particular rhetorical and philosophical orientation.[1] He came from a prominent family in Pergamum, where he was born at the beginning of the fourth century. He may have studied medicine there, but

most of his medical education was obtained at Alexandria. In the late Hellenistic period the study of medicine at Alexandria had become "scholastic," as Galen termed it—it was divorced from practice and was purely theoretical.[2] Oribasius, however, dissociated himself from physicians who were overly concerned with rhetoric and philosophy.

In Pergamum, Oribasius belonged to the circles representing the intellectual elite of the age; there he met the future Emperor Julian the Apostate, who later made Oribasius his physician in ordinary and head of his library. The relationship between the two plainly was very close, and Oribasius' political influence was correspondingly great. He also was a political official, quaestor of Constantinople. In addition he was closely associated with the emperor's cultural program, including the restoration of pagan religion. Oribasius' notes (a *hypomnema*) on the emperor's life have not survived, but they served as an essential source for Eunapius' biography of Julian and, evidently, as a source for some parts of the historical writings of Ammianus Marcellinus. Banished after Julian's death along with other of his supporters, Oribasius was later rehabilitated. He was married and had four children; a son named Eustathius was also a physician.

The initial stimulus for Oribasius' work as a medical writer was a suggestion by Julian that he prepare abstracts *(epitomai)* of Galen's works; this composition has not survived. His most extensive surviving work (although it was not transmitted intact) is *Iatrikai synagogai* (or *Collectiones medicae*), which contains excerpts from the writings of the more important Greek physicians. These extracts are primarily, but not entirely, verbatim.[3] From this large work he produced *Synopsis for Eustathius* (also extant), a kind of abridged edition or vade mecum for his son. There still exists *For Eunapius*, a collection of easily procured medicines compiled for the layman. The known lost works, in addition to the historical account already mentioned, are *To the Perplexed Physicians*, *On Diseases*, *Anatomy of the Intestines*, and, outside the field of medicine, *On Royal Rule* (only the title of the last work is known). Two spurious writings are extant: *Introductions to Anatomy* and a commentary on the *Aphorisms* of Hippocrates. (The authorship of the commentary, which is preserved only in Latin translation, should be carefully examined since it contains material of interest for the history of medicine.)[4]

Oribasius' encyclopedic medical writings became the model for such authors as Aëtius of Amida. They also found a large audience in the Latin West, as the early (fifth century [?]) Latin translations of them testify. The Arabs also drew freely on Oribasius' works. For the historian of medicine Oribasius is especially important for his role in preserving earlier, more important medical authors, whom we know about, in part, only through his excerpts.

NOTES

1. See F. Kudlien, "The Third Century A. D.—a Blank Spot in the History of Medicine?" in L. G. Stevenson and R. P. Multhauf, eds., *Medicine, Science and Culture. Historical Essays in Honor of Owsei Temkin* (Baltimore, 1968), 32 ff.
2. See F. Kudlien, "Medical Education in Classical Antiquity," in C. D. O'Malley, ed., *The History of Medical Education* (Berkeley–Los Angeles–London, 1970), 23 ff.
3. For the rather complicated situation, see F. Kudlien, *Die handschriftliche Überlieferung des Galenkommentars zu Hippokrates De articulis* (Berlin, 1960), 49–54.
4. See F. Kudlien, "Pelops no. 5," in Pauly-Wissowa, *Real-Encyclopädie*, supp. X (Stuttgart, 1965), col. 531.

BIBLIOGRAPHY

The best ed. of the genuine writings is by H. Raeder in *Corpus medicorum Graecorum*, VI, 1–3 (Amsterdam, 1964). The best survey is H. O. Schröder, "Oreibasios," in Pauly-Wissowa, supp. VII (Stuttgart, 1940), cols. 797–812.

Fridolf Kudlien

ORLOV, ALEKSANDR YAKOVLEVICH (*b.* Smolensk, Russia, 6 April 1880; *d.* Kiev, Ukrainian S.S.R., 28 January 1954), *astronomy, gravimetry, seismology.*

The thirteenth child of a priest, Orlov graduated from the Voronezh Gymnasium in 1898. He revealed a strong interest in astronomy while attending the mathematical section of St. Petersburg University; a student work published in 1901 on the total solar eclipse of 1907 was awarded a prize by the Society of Natural Scientists. He graduated with distinction in 1902 and the following year broadened his scientific background by studying at the Sorbonne, at Lund with C. L. W. Charlier, and at Göttingen with Johann Wiechert.

In 1905–1906 Orlov was an assistant at the Yurev [now Tartu, Estonian S.S.R.] observatory and inspected seismic stations in Transcaucasia for the Permanent Central Seismic Commission. At Pulkovo observatory (1906–1908) he made observations on the zenith telescope, analyzed observations, and improved existing methods. In 1908 he returned to Yurev to undertake seismic research, and in 1909 he became director of the seismic station of the Yurev observatory. He was an active member of the Perma-

nent Central Seismic Commission in his capacity as representative of Yurev University, where he also lectured on seismology, the theory of seismic instruments, celestial mechanics, and geodesy. His remarkable series of observations on tidal lunar-solar deformations of the earth, made with the horizontal pendulum, provided the subject of his thesis, defended in 1910 at St. Petersburg, for the degree of master of astronomy and geodesy. In July 1911 Orlov was one of three Russian delegates to the congress of the International Seismological Association at Manchester, at which he was elected a member of the international commission for the study of tidal deformations of the earth. In 1911, on the recommendation of the congress, he organized a new seismic station in Tomsk. He also visited Yerkes Observatory in connection with his research on the motion of matter in comet tails. In the summer of 1912 Orlov took part in a major gravimetric expedition to western Siberia. In December of that year he was named extraordinary professor at Novorossysk University in Odessa and director of the Odessa observatory. After defending his doctoral dissertation in 1915, he became professor at Odessa University, where he taught spherical and theoretical astronomy, celestial mechanics, and advanced geodesy.

After the 1917 Revolution, Orlov took part in renewing the destroyed triangulation network of southern Russia from the Dnieper to the Dniester rivers. In 1921 he organized computation and publication of the marine astronomical yearbook, issued until 1924. Remaining as director of the Odessa observatory, Orlov was elected rector of Kiev University in 1919 and from 1920 to 1923 was academician of the Academy of Sciences of the Ukrainian S.S.R. In 1924 a gravimetric observatory was organized, at Orlov's suggestion, in Poltava, to produce gravimetric maps of the Ukrainian S.S.R. and to study tidal deformations of the earth and polar perturbation. In 1926 this observatory began to conduct regular work at Poltava and gravimetric expeditions.

In 1927 Orlov was elected corresponding member of the Soviet Academy of Sciences; from 1934 to 1938 he was professor of astronomy at the P. K. Sternberg Astronomical Institute in Moscow. From 1938 to 1951 he again headed the Poltava observatory, and in 1939 he became a member of the Academy of Sciences of the Ukrainian S.S.R. While he was director of the Main Astronomical Observatory of the Academy (1944–1950) construction was begun on a new observatory in Goloseevo, near Kiev. At Orlov's initiative the first All-Union Conference on Latitude was held in October 1939; on its recommendation a commission on latitude of the Astronomical Council of the Soviet Academy of Sciences was created. Orlov was its president until 1952.

Orlov's scientific work touched on several areas: (1) the motion of the poles and variations in latitude; (2) tidal deformations of the earth; (3) seismology; and (4) geodesy and geophysics. He also studied comets, the rotation of the sun, precalculations of the circumstances of solar eclipses, problems of theoretical astronomy, analysis of curves of brightness of variable stars, and curves of radial velocities of spectroscopic binary stars.

Work in the first area inaugurated a yearlong series (1906–1907) of extremely precise observations on the zenith telescope at Pulkovo. Orlov developed a graphic method of selecting pairs of stars for determining latitudes (so-called latitude pairs) from observations by the method of equal zenith angles. A group of works beginning in 1916 was completed with the publication of articles that were of great importance in interpreting the results of observations of the International Latitude Service. Orlov introduced a new definition of the concept of "mean latitude" and, using harmonic analysis of the variations of altitude of the pole, confirmed Chandler's indication that there were variations of latitude besides the fourteen-month period in the motion of the pole.

At the Poltava gravimetric observatory, extensive observations were begun on the two bright stars α Persei and α Ursae Majoris, which could be observed there at their highest culmination near the zenith both day and night. Orlov proposed the establishment of stations at the same latitude, at Blagoveshchensk (on the Amur River) and Winnipeg, Canada.

Orlov discovered slow nonpolar variations of latitude and devised a method for excluding them and for correcting the polar coordinates for the period during which the International Latitude Service existed (1892–1952). His numerous investigations led him to develop a new method of determining the polar coordinates from latitude observations at an isolated station.

Orlov's last works deal with an explanation of semimonthly changes in latitude and with the determination of the coefficients of the principal member of nutation based on declination. A posthumously published work on secular polar motion contains a vast amount of observational and theoretical material (259,000 observations of latitude) on the basis of which Orlov discovered the annual secular motion of the pole to be 0.004″ per year, proceeding along a meridian of 69° W.

Orlov's son Boris (1906–1963) was an astrometrist at Pulkovo. His son Aleksandr (b. 1915) became a specialist in celestial mechanics at the P. K. Sternberg Astronomical Institute.

BIBLIOGRAPHY

I. ORIGINAL WORKS. Orlov's selected works were published as *Izbrannye trudy*, 3 vols. (Kiev, 1961). His writings include "O polnom zatmenii Solntsa 1907 goda" ("On the Total Solar Eclipse of 1907"), in *Izvestiya Russkago astronomicheskago obshchestva*, no. 9 (1901), 48–52, with map; and *ibid.*, no. 10 (1903), 131–145; "Nablyudenia potoka Perseid v 1901 g., sdelannye v Pulkove" ("Observations at Pulkovo on the Stream of Perseids in 1901"), in *Izvestiya Imperatorskoi akademii nauk*, **36** (1902), 45–52; "Sur la théorie des appareils seismiques," in *Bulletin astronomique*, **23** (1906), 286–291; "Über die Untersuchungen der Schwankungen der Erdrinde," in *Protokoly Obshchestva estestvoispytatelei pri Imperatorskom yurevskom universitete*, **15** no. 3 (1906), 147–162; "Über die von Fürst Golitzin angestellte Versuch mit einem nahezu aperiodischen Seismographen," *ibid.*, 167–173; "Beobachtungen am grossen Zenitellescop von 7 Februar 1907 bis zum 28 Februar 1908," in *Publications de l'Observatoire central (Nicolas) à Poulkova*, 2nd ser., **18** (1908), 1–66; "Ob opredeleny postoyannykh k i n uravnenia $d^2\theta/dt^2 + 2k(d\theta/dt) + n^2\theta = 0$" ("On the Determination of the Constants k and n of the Equation . . ."), in *Protokoly Obshchestva estestvoispytatelei pri Imperatorskom yurevskom universitete*, **10**, pt. 4 (1908), 243–258; "Graphische Methode zur Auswahl der Sternpaare für die Breitenbestimmung nach der Methode gleicher Zenitdistanzen," in *Trudy Astronomicheskoi observatorii, Yurevskogo universitet*, **21**, pt. 2 (1909), 3–12; "Beobachtungen über die Deformation des Erdkörpers unter dem Attractionseinfluss des Mondes an Zöllner'schen Horizontalpendeln," in *Astronomische Nachrichten*, **186** (1910), 81–88; and "Novy sposob opredelenia velichiny ottalkivatelnoy sily Solntsa" ("A New Method for Determining the Value of the Repulsive Force of the Sun"), in *Izvestiya Imperatorskoi akademii nauk*, no. 7 (1910), 517–522.

Subsequent works include "Pervy ryad nablyudeny s gorizontalnymi mayatnikami v Yurieve nad deformatsiami Zemli pod vlianiem lunnogo prityazhenia" ("First Series of Observations with the Horizontal Pendulums at Yurev on the Deformation of the Earth Under the Influence of Lunar Gravity"), in *Trudy Astronomicheskoi observatorii, Yurevskogo universiteta*, **23**, pt. 1 (1911), his master's diss.; "Sur la théorie des queues des comètes," in *Astronomische Nachrichten*, **196** (1914), 231; "Über der ursprüngliche Bredichinsche Teorie des Kometenschweife," *ibid.*, **198** (1914), 161; "Opredelenie sily tyazhesti v Zapadnoy Sibiri" ("Determination of the Force of Gravity in Western Siberia"), in *Trudy Astronomicheskoi observatorii* (Odessa), no. 1 (1914), 1–22; "Rezultaty yurievskikh, tomskikh i potsdamskikh nablyudeny nad lunno-solnechnymi deformatsiami Zemli" ("Results of Yurev, Tomsk, and Potsdam Observations on the Lunar-Solar Deformation of the Earth"), *ibid.*, no. 2 (1915), 1–281, his doctoral diss.; "Rezultaty 18-letnego ryada nablyudeny solnechnykh pyaten, proizvedennogo v Odesse s konstantinovskim geliografom" ("Results of an Eighteen-Year Series of Observations of Sunspots Made at Odessa With a Constantinov Heliograph"), in *Izvestiya Imperatorskoi akademii nauk*, no. 2 (1915), 135–144; *Teoreticheskaya astronomia s prilozheniem tablits* ("Theoretical Astronomy With Appended Tables"; Odessa, 1920); Odessa Astronomical Observatory, "Harmonic Analysis of the Latitude Observations. I. Kazan, Carlo-forte, Greenwich" (Odessa, 1925), 7–30; "Harmonic Tables for Spectroscopic Binaries," *ibid.*, (1930), 1–8; "Über die Dreiachsigkeit des Trägheitsellipsoids der Erde aus Breitenbeobachtungen," in *Verhandlungen der siebenten Tagung der Baltischen geodetischen Kommission* (1934), 319–339; and "Opredelenie lunnykh geomagnitnykh variatsy pri pomoshchi schetnykh mashin" ("Determination of the Lunar Geomagnetic Variations With the Aid of Computing Machines"), in *Izvestiya akademii nauk SSSR*, Ser. Geograf. i geofiz., no. 2 (1937), 195-206.

His later works include "O deformatsiakh Zemli po nablyudeniam v Tomske i Poltave s gorizontalnymi mayatnikami" ("On the Deformations of the Earth According to Observations in Tomsk and Poltava With Horizontal Pendulums"), *ibid.*, no. 1 (1939), 3–29; *Kurs teoreticheskoy astronomii* ("Course of Theoretical Astronomy"; Moscow, 1940), written with his son, Boris; "Dvizhenie zemnogo polyusa po nablyudeniam shiroty v Pulkovo . . ." ("The Motion of the Earth's Pole According to Latitude Observations at Pulkovo . . ."), in *Byulleten Gosudarstvennogo astronomicheskogo instituta imeni Shternberga*, no. 7 (1941), 5–26; "Dvizhenie mgnovennogo polyusa Zemli otnositelno srednego polyusa za 46 let" ("Motion of the Instantaneous Pole of the Earth Relative to the Mean Pole for Forty-Six Years"), *ibid.*, no. 8 (1941), 5–34; "On Variations of Greenwich Mean Latitude," in *Doklady Akademii nauk SSSR*, **42**, no. 9 (1944), 377–381; "On the 'Ellipcity' of the Earth's Equator," *ibid.*, **43**, no. 8 (1944), 327–328; "The Mean Annual Motion of the Earth's Principal Axes of Inertia," *ibid.*, **51**, no. 7 (1946), 509; "O vekovom dvizheny polyusov" ("On the Secular Motion of the Poles"), in *O zadachakh i programme nablyudeny Mezhdunarodnoy Sluzhby Shiroty* ("On the Problems and Program of Observations of the International Latitude Service"; Moscow, 1954), 13–18; "Analiz pulkovskikh nablyudeny na zenit-teleskope s 1915 po 1928" ("An Analysis of Observations at Pulkovo on the Zenith Telescope from 1915 to 1928"), in Orlov's *Izbrannye trudy*, I (Kiev, 1961), 234–261; and "Sluzhba shiroty" ("Latitude Service"), *ibid.*, 270–334.

II. SECONDARY LITERATURE. On Orlov and his work, see Z. N. Aksentieva, "Ocherk zhizni i tvorchestva Orlova" ("Sketch of the Life and Work of Orlov"), in *Izbrannye trudy*, I, 3–37; and the obituary by Z. N. Aksentieva and V. P. Fedorov, in *Trudy Glavnoi astronomicheskoi observatorii v Pulkove*, no. 146 (1954).

P. G. KULIKOVSKY

ORLOV, SERGEY VLADIMIROVICH (*b.* Moscow, Russia, 18 August 1880; *d.* Moscow, 12 January 1958), *astronomy, astrophysics.*

Orlov was the son of a physician. After graduating in 1899 from the First Gymnasium in Moscow, he entered the department of physics and mathematics of Moscow University. Orlov began his scientific career while a student, when, influenced by the lectures of Vitold Ceraski, he became a non-staff assistant at the university's observatory and conducted observations with the transit instrument. Following graduation in 1904, Orlov continued to work in the observatory, began teaching, and served as an artillery officer in the Russo-Japanese War. Since his former position in the Moscow University observatory was filled, from 1906 to 1918 Orlov taught at the First Gymnasium and began his study of comets. During 1914–1917 Orlov again was on active duty in the army. After recovering from a compound fracture of the leg, he was demobilized in 1917 and returned to the Gymnasium, where he taught mathematics and physics and served as vice-director until 1920, meanwhile continuing his study of comets. In 1917 he had received the right to teach at the university level; and in 1920–1922 he was professor at Perm University, where he headed the department of astronomy and physics.

In 1922 Orlov was again in Moscow, where he became a staff member of the State Astrophysical Institute (acting director 1923–1931) and at the Moscow University Astronomical-Geodesical Scientific Research Institute, which in 1931, after merging with the university observatory, became the P. K. Sternberg Astronomical Institute. From 1931 to 1935 Orlov was its vice-director and from 1943 to 1952 its director.

In 1926 Orlov became professor at Moscow University, where he gave courses in astrophysics and comet astronomy; in 1935 he received the doctorate in physical and mathematical sciences, and from 1938 he was head of the department of comet astronomy. From 1935 to 1957 he was president of the Commission on Comets and Meteors of the Astronomical Council of the USSR Academy of Sciences. In 1943 Orlov was elected associate member of the Academy of Sciences and was awarded the state prize for scientific work on comet and meteor astronomy. His other honors include two Orders of Lenin, two Orders of the Red Banner of Labor, and several state medals. As early as 1908 he had photographed and studied photographs of Morehouse's comet, and in 1910 he had observed Halley's comet. This work formed the beginning of investigations that led to more than seventy publications on the astronomy of comets.

Bredikhin's research in the mechanical theory of

comet forms was further developed by Orlov and his school. Orlov faced the problem of creating a theory that would embrace the mechanistic properties of motion, as well as the physical peculiarities of comets and their changes through time. At first he examined the mechanistic theory and put its formulas into a form more convenient for calculation. He then gave an improved method of determining the values of the repulsive accelerations of the action of the sun on the particles of comets' tails; these accelerations were multiples of 22.3 ($I + M = 22.3 \cdot n$, where n can take a value from 1 to 9). This method was based on careful study of the displacements of separate details of the tails as functions of time.

Enlarging on Bredikhin's ideas, Orlov examined and returned to the improvement of Bredikhin's classification of comet forms and developed his own classification of the forms of comet heads. In connection with the latter he produced a theory of the head of the comet, based on the proposition that both the sun and the nucleus of the comet are centers of repulsion forces acting on molecules that separate from the body of the comet's nucleus as the comet approaches the sun and are pushed out into the tail. According to this theory, cross sections of the nuclei of comets were on the order of several kilometers, a dimension confirmed by further research. This theory was also proved by analysis of the structure of the envelopes of comet heads and calculation of the masses of comets.

While continuing Bredikhin's work Orlov also pioneered in the astrophysical study of comets, taking into consideration the mechanism of luminescence of a comet and variations in its spectra. He repeatedly returned to the study of cometary luminescence and the laws of its variation in a periodic comet from one appearance to another. Orlov provided a method of determining the parameters of comet's light and for many comets established the integral, or absolute stellar, magnitudes (the stellar magnitudes reduced to so-called standard conditions, for which one takes the distance of the comet from the sun and its distance from the earth as equal to astronomical unity). As Orlov showed, the law of variation of the parameters of a comet's light is of great importance for the study of the origin and evolution of comets. In particular he noted that a comet's brightness, related to standard conditions, depends on the phase of solar activity.

Orlov's cosmogonic hypothesis suggested that the formation of comets was a result of accidental collisions between two asteroids, which led to explosions that destroyed these small bodies. Fragments acquired varied orbits, usually elliptical. As they approached the sun and came under the influence of its radiation, the fragments released gases that later formed the envelopes of the heads and of the tails of comets. Research on spectra enabled Orlov to determine the gas composition of the straight-tail (type I) comets; less was known at that time about tails of types II and III—he took them to be dust. He was the first to identify lines of nickel in the spectrum of comets.

Orlov's construction of new astronomical instruments was related to the organization of observations of comets and meteors in the Soviet Union. He created a special camera for photographing comets and developed a method for their photogrammetry. Orlov was widely known as an excellent lecturer.

BIBLIOGRAPHY

I. ORIGINAL WORKS. Orlov's writings include "Issledovanie ochertany golovy komety" ("Research on Outlining the Head of a Comet"), in *Zhurnal fizikomatematicheskogo obshchestva pri Permskom Gosudarstvennom Universitete* (1919), no. 2, 139–144; "Opredelenie ottalkivatelnykh sil Solntsa v khvoste komety po dvum nablyudeniam polozhenia oblachnogo obrazovania" ("Determination of the Repulsive Force of the Sun on the Tail of a Comet From Two Observations of the Position of the Cloud Formation"), in *Trudy Glavnoi rossiiskoi astrofizicheskoi observatorii*, **1** (1922), 231–236; "O svyazi mezhdu yarkostyu komet i deyatelnostyu na poverkhnosti Solntsa" ("On the Relations Between the Brightness of Comets and Activity on the Solar Surface"), *ibid.*, **2** (1923), 150; "Opredelenie ottalkivatelnykh sil Solntsa–kometa Galleya (1910 II)" ("Determination of the Repulsive Forces of the Sun and Halley's Comet . . ."), in *Russkii astronomicheskii zhurnal*, **2**, no. 3 (1925), 4–21; "The Series of Carbon Monoxide in the Spectrum of Comets. 1908 III (Morehouse)," in *Astronomische Nachrichten*, **225** (1925), 397–400; "The Spectrum of the Comet 1882 II," in *Russkii astronomicheskii zhurnal*, **4**, no. 1 (1927), 1–9; and "Oblachnye obrazovania v khvoste komety 1908 III (Morehouse), 14–17 oktyabrya" ("Cloud Formations in the Tail of Comet 1908 III [Morehouse] 14–17 October"), in *Astronomicheskii zhurnal*, **5**, no. 4 (1928), 193–202.

Other works are "Mekhanicheskaya teoria kometnykh form" ("Mechanical Theory of Comet Forms"), in *Trudy Gosudarstvennogo astrofizicheskogo instituta*, **3**, no. 4 (1928), 3–79, also in *Astronomischeskii zhurnal*, **6**, no. 2 (1929), 180–186; "Priroda ottalkivatelnykh sil Solntsa v khvostakh komet" ("The Nature of the Repulsive Forces of the Sun in the Tails of Comets"), in *Astronomicheskii zhurnal*, **8**, no. 3–4 (1931), 199–205; "Stereoskopichesky metod fotografirovania komet" ("The Stereoscopic Method of Photographing Comets"), *ibid.*, **9**, nos. 1–2 (1932), 71–81; "O dvizhenii oblachnykh obrazovany v khvoste komety 1908 III (Morehouse)" ("On the Motion of the Cloud Formation in the Tail of Comet 1908 III . . ."), *ibid.*, nos. 3–4 (1932), 163–165; *Komety*

("Comets"; Moscow, 1935); "Spectroskopia komet" ("Spectroscopy of Comets"), in *Uspekhi astronomicheskikh nauk*, **4** (1935), 46–60; "Stroenie golovy komety" ("Structure of the Head of a Comet"), in *Astronomicheskii zhurnal*, **12**, no. 1 (1935), 1–20; and "Origin of Sporadic Meteors," in *Observatory*, **59**, no. 743 (1936), 132–135.

Subsequent writings are "Vidimye radianty kosmicheskikh meteornykh potokov" ("Visible Radiants of Meteor Showers in Space"), in *Astronomicheskii zhurnal*, **13** (1936), 388–396; "Mnogoyarusnye obolochki komet s khvostami I tipa" ("Many-Layered Envelopes of Comets With Tails of Type I"), *ibid.*, **14**, no. 2 (1937), 130–134; "Evolyutsia i proiskhozhdenie komet" ("Evolution and Origin of Comets"), *ibid.*, **16**, no. 1 (1939), 3–27; "Istoria Gosudarstvennogo astrofizicheskogo instituta, 1922–1931" ("History of the State Astrophysical Institute . . ."), in *Uchenye zapiski Moskovskogo gosudarstvennogo universiteta*, **58** (1940), 121–136; "Proiskhozhdenie komet" ("Origin of Comets"), in *Uspekhi astronomicheskikh nauk* (1941), no. 2, 101–121; "Isklyuchitelnye komety. Bolshaya sentyabrskaya kometa 1882 II" ("Exceptional Comets. The Great September Comet 1882 II"), in *Astronomicheskii zhurnal*, **21**, no. 5 (1944), 201–202; *Priroda komet* ("The Nature of Comets"; Moscow, 1944); *Golova komety i novaya klassifikatsia kometnykh form* ("Head of a Comet and a New Classification of Comet Forms"; Moscow, 1945); "Moshchnost i svetosila astrografa i spektrografa" ("Power and Optical Efficiency of the Astrograph and Spectrograph"), in *Astronomicheskii zhurnal*, **22**, no. 1 (1945), 1–10; and "Sinkhrony v khvostakh komet" ("Synchrones in the Tails of Comets"), *ibid.*, no. 4, 202–214.

Later works are "Rol' F. A. Bredikhina v razvitii mirovoy nauki" ("The Role of F. A. Bredikhin in the Development of World Science"), in *Uchenye zapiski Moskovskogo gosudarstvennogo universiteta*, no. 91 (1947), 157–185; *Fedor Aleksandrovich Bredikhin, 1871–1904* (Moscow, 1948); "Komety," in *Astronomia v SSSR za tridtsat let. 1917–1947* ("Astronomy in the USSR for Thirty Years . . ."; Moscow, 1948), 83–88; "Asteroidy i meteority" ("Asteroids and Meteorites"), in *Meteoritika*, no. 5 (1949), 3–13; "Meteornye potoki i komety" ("Meteor Showers and Comets"), in *Astronomicheskii zhurnal*, **17**, no. 1 (1940), 4–7; "Reflektory Maksutova i Shmidta" ("The Reflectors of Maksutov and Schmidt"), *ibid.*, **30**, no. 5 (1953), 546–551; "Rol bazisa pri opredelenii meteornykh orbit" ("The Role of the Basis for the Determination of Meteor Orbits"), in *Byulleten Komissii po kometam i meteoram Astrosoveta, Akademiya nauk SSSR*, no. 1 (1954), 24–28; and "Komety," in *Astronomia v SSSR za 40 let* ("Astronomy in the USSR for Forty Years"; Moscow, 1960), written with S. M. Poloskov.

II. Secondary Literature. See "Chestvovanie chlena-korrespondenta AN SSSR S. V. Orlova" ("Celebration in Honor of Associate Member . . . Academy of Sciences . . ."), in *Vestnik Akademii nauk SSSR* (1951), no. 6, 81–83; "Orlov, Sergey Vladimirovich," in *Bolshaya Sovetskaya entsiklopedia* ("Great Soviet Encyclopaedia"), 2nd ed., XXXI (1955), 203; S. M. Poloskov, "Vydayu-

shchiysya issledovatel' komet S. V. Orlov (k 70-letiyu so dnya rozhdenia)" ("Outstanding Investigator of Comets S. V. Orlov [on His Seventieth Birthday]"), in *Priroda* (1951), no. 11, 73–75; "S. V. Orlov (k 75-letiyu so dnya rozhdenia" (". . . on His Seventy-Fifth Birthday"), in *Byulleten Stalinabadskoi astronomicheskoi observatorii*, no. 14 (1955), 3–4; "S. V. Orlov (Nekrolog)," in *Astronomicheskii tsirkulyar Akademii nauk SSSR*, no. 190 (1958), 1–3; "Semidesyatiletie S. V. Orlova" ("Seventieth Birthday of S. V. Orlov"), in *Byulleten Vsesoyuznogo astronomo-geodezicheskogo obshchestva*, no. 10 (1951), 3–4; "Sergey Vladimirovich Orlov (k 40-letiyu nauchnoy i pedagogicheskoy deyatelnosti)" (". . . on the Fortieth Anniversary of His Scientific and Teaching Career"), in *Tsirkulyar Stalinabadskoi astronomicheskoi observatorii*, nos. 64–65 (1948), 1; and "Sergey Vladimirovich Orlov (nekrolog)," in *Astronomicheskii zhurnal*, **35**, no. 3 (1958), 321–322.

P. G. Kulikovsky

ORNSTEIN, LEONARD SALOMON (*b.* Nijmegen, Netherlands, 12 November 1880; *d.* Utrecht, Netherlands, 20 May 1941), *physics*.

At the suggestion of his teacher, H. A. Lorentz, Ornstein took as the subject of his doctoral dissertation (Leiden, 1908) the application of J. W. Gibbs's general methods in statistical mechanics (published in 1902) to various concrete problems in molecular theory, including the determination of the equation of state of a nonideal gas.

Ornstein generalized Boltzmann's definition of the probability of a macroscopically definable state of a gas to include systems in which the interactions between the molecules could no longer be neglected. He then calculated the probability that in a homogeneous system a given spatial distribution of density arises. This was the basis of virtually all of Ornstein's subsequent work as a theoretical physicist, and, in particular, of his collaboration with F. Zernike from 1914 to 1917 on the formation of molecular "swarms." They were thus able to add an important correction to the Einstein-Smoluchowski theory of opalescence of a fluid at its critical point, taking account of the correlation of the density fluctuations in volume elements separated by distances comparable to the wavelength of the scattered light.

From 1909 to 1914 Ornstein was lecturer in theoretical physics at the University of Groningen, where Hermanus Haga tried unsuccessfully to interest him in the experimental work of the physical institute. Late in 1914 Ornstein succeeded Peter Debye in the chair of theoretical physics at the University of Utrecht, where W. H. Julius provided him with a room in the physical institute. There, largely through contact with and in collaboration with W. J. H. Moll,

Ornstein began to experiment on liquid crystals, regarded as an exemplification of his theory of molecular swarms.

It was, however, only in 1920, when Ornstein became acting director of the Utrecht physical institute (substituting for the ailing Julius, whom he succeeded officially in 1925), that his work shifted decisively into experiment, and his organizational talent began to unfold. In the preceding dozen years he had published about fifty papers; in the following twenty years over two hundred bore his own name, and almost five hundred additional papers were published from his institute, which underwent three substantial enlargements in this period. Among these publications were eighty-eight doctoral dissertations: about one per year in the early 1920's, increasing to three per year in the middle and late 1920's, and reaching a peak of about seven per year in the mid-1930's.

With great acumen Ornstein had recognized around 1918 that the advance of atomic physics would require that the exceedingly precise measurements of spectral frequencies be supplemented by quantitative measurements of intensities. Immediately upon taking charge of the Utrecht institute, which had had under Julius a strong tradition in spectroscopy and radiation intensity measurements in the service of solar physics, Ornstein charted an ambitious, coordinated program for the systematic investigation of techniques of intensity measurement—especially the blackening of photographic plates—for exploitation in the service of atomic physics. Although generally regarded by his colleagues as a difficult man, within his own institute "the boss" maintained a harmonious collaboration of all the staff upon his tightly integrated program.

In the period 1923–1925 the publications from his institute on the simple integral relations ("sum rules") between the intensities of spectral lines originating in transitions out of or into a complex spectral term, and also between the intensities of the several components into which a spectral line is split in a magnetic field, were an important stimulus and a unique source of data for theoretical atomic physicists seeking a quantum mechanics via Bohr's correspondence principle.

In the late 1920's and early 1930's, after the development of a quantum mechanics whose arbitrary character he found unsatisfying, Ornstein and his co-workers turned these same techniques—for which his institute was then world-famous—to a wide variety of problems: the investigation of electric arcs; the Raman effect; liquid crystals; the determination of isotopic ratios; and purely technical problems of illumination engineering and lightbulb lifetimes.

From the mid-1920's Ornstein, although an active Zionist, cultivated the closest relations with Dutch industry, providing space in his institute and his personal supervision for technical chores in electrical and heat, as well as light, engineering. "The physicist *in* society" was one of his watchwords.

In the late 1930's, on the proposal and with the support of the Rockefeller Foundation, Ornstein moved into biophysics, concentrating upon bacterial luminescence and photosynthesis. Far more would have emerged from this effort had not the war and German occupation intervened. In November 1940 Ornstein was forbidden entrance to his laboratory; he died six months later.

BIBLIOGRAPHY

A complete annotated bibliography of Ornstein's publications and the publications from his institute to 1933 is given in *L. S. Ornstein; A Survey of His Work From 1908 to 1933 Dedicated to Him by His Fellow-Workers and Pupils* (Utrecht, 1933), reviewed by W. Gerlach in *Naturwissenschaften*, **22** (1934), 111–112. The text of this publication, pp. 1–86, gives semipopular expositions of the various areas of research conducted by Ornstein and his institute. The publications of the "Biophysical Group Utrecht–Delft," 1936–1940, are listed in A. F. Kamp *et al.*, eds., *Albert Jan Kluyver. His Life and Work* (Amsterdam, 1959), 548–553. Ornstein's style in the direction of research is also described, pp. 30–31.

Valuable obituary notices are H. A. Kramers, "Levensbericht van L. S. Ornstein," in *Jaarboek van het Koninkl. Akademie van Wetenschappen, Amsterdam* (1940–1941), 225–231; F. Zernike, "Ornsteins Levenswerk," in *Nederlandsch Tijdschrift voor Naturkunde*, **8** (1941), 253–265; and R. C. Mason, "Leonard Salomon Ornstein," in *Science*, **102** (1945), 638–639.

Additional biographical data is given in *Wie is dat*, 4th ed. (The Hague, 1938), pp. 314–315. Some twenty-five letters by Ornstein are listed in T. S. Kuhn *et al.*, *Sources for History of Quantum Physics* (Philadelphia, 1967), 71–72.

PAUL FORMAN

ORTA, GARCIA D' (or **da Orta**) (*b*. Castelo de Vide, Portugal, *ca.* 1500; *d*. Goa, India, *ca.* 1568), *botany, pharmacology, tropical medicine, anthropology*.

Among the Portuguese voyagers and travelers in Asia, d'Orta was the first to use his position to add to the knowledge in Europe of South Asian flora. He thus showed how inadequate were the inherited Greek and Arabic sources on Indian botany and pharmacology. His work provided Western scholars with their introduction to tropical medicine and with their basic data on almost all of the major cultivated plants of the region.

His parents, Fernão and Leonor d'Orta, were Spanish Jews who had taken refuge at Castelo de Vide, in the Portuguese province of Alentejo, when the Jews were banished from Spain in 1492. Faced again in 1497 with the choice between conversion or exile, they became Christians. Probably their eldest son, Garcia, was born soon afterward. The family presumably retained cultural links with Spain, for d'Orta was sent to study at the universities of Salamanca and Alcalá de Henares; he returned to his native town in 1523. He was officially examined and became qualified to practice medicine in 1526 and then moved to Lisbon. In 1530, after two unsuccessful applications, he was appointed to lecture on natural philosophy at the University of Lisbon, where he was elected to the council in 1533. Despite this apparently rapid progress in his career, he sailed for Goa on 12 March 1534, as personal physician to M. A. de Sousa, who had been appointed "captain general by sea" of the Portuguese in India and was later viceroy. From 1534 to 1538 d'Orta traveled extensively along the western coast of India and Ceylon attending de Sousa on his campaigns. He thus met and treated some of the leading Indian princes of the Deccan, notably Burhān Nizām Shah, the sultan of Ahmadnagar, who became a personal friend.

In 1538 he settled permanently in Goa and acquired a country estate on the island of Bombay. Once established, he began to bring over his sisters, who had already been in the hands of the Inquisition and might have hoped that they would be safe from persecution in Goa. In 1541 he married a distant relative, Brianda de Solis; the couple had several daughters. After de Sousa's return to Portugal, d'Orta remained in India as vice-regal physician. It has been assumed that, as the most eminent doctor of Goa, he must also have been a physician at the Goa hospital and prison, but documentary evidence is lacking.

D'Orta participated in a public disputation on philosophy and medicine in 1559, when he was described as very old—indeed "already decrepit"—and learned. Feeling that old age was pressing in on him, he decided to make known the information he had collected during his thirty years in India. This work, *Coloquios dos simples e drogas he cousas medicinais da India* (Goa, 1563), is in the form of dialogues between d'Orta and a colleague, newly arrived in Goa and anxious to know about the materia medica of India. Most of the simples discussed were of vegetable origin, but amber, ivory, and pearls were also among his topics. For each specimen he provided the names in the local languages as well as the names in Greek and Arabic. He then described the size and form of the plant, its leaves, flowers, and fruit; what parts were used; the methods of cultivation and preparation; and the exact location where each plant was grown.

Although d'Orta's central concern was medicinal substances, he often digressed to include other edible plants unknown in Europe. Besides verbal inquiry he used his mercantile contacts to procure specimens, which he attempted to cultivate in his garden (for example, *Eugenia malaccensis* from Malaya). Although he made his own mistakes, he corrected many more, and first reported on several important local food plants, notably mangoes, mangosteens, durians, and jakfruit, which shares a chapter with three other fruit trees then new to European botany. He also described accurately other plants formerly known only as processed commodities or from garbled texts. He was a pioneer in the study of Indian diseases then new to European medicine. His description of the symptoms of Asian cholera became a standard reference, and he carefully observed the effects of chronic dysentery, cobra bite, and datura poisoning.

Besides the plants, drugs, and diseases he found in India, d'Orta was greatly interested in Indian sociology. He described the caste system, the Parsee religion, and the social role of such practices as betel chewing and the consumption of *bangue* (cannabis). D'Orta established friendships with both Muslims and Hindus, and he learned much from their medicine. Although he wrote of Portuguese achievements patriotically, he could appreciate other cultures. He had after all assumed Portuguese Catholicism as a mask. He also reported his discussions with Arab and Jewish merchants from the Middle East. One of these last, who styled himself Isaac of Cairo, and so a Turkish subject, was really a kinsman from Castelo de Vide.

D'Orta was one of the first European scholars to express admiration for the civilization of China, and believed that Western medicine would benefit from closer contact. He realized, too, that the medieval Arabic authors on materia medica knew more about India than the Greeks, and he did not hesitate to challenge the authority of classical texts. This cultural relativism and skepticism toward Western tradition may be attributed in part to his origins.

These origins at last caught up with him. Investigations by the Holy Office brought up his name, and perhaps only his influential position protected his family, for soon after his death his sister Catarina was arrested. From her interrogations it appears that d'Orta had secretly encouraged his family to honor the sabbath and the fasts of Judaism as faithfully as possible. Catarina was martyred in 1569, and the rest of the family was deported to Portugal. In 1580 d'Orta's remains were exhumed and burned.

D'Orta's book may then have been suppressed in Goa. The Flemish botanist L'Écluse came across a copy in Lisbon. He extracted the essential information on the characteristics and properties of the economic and medicinal plants of India, and published an epitome in Latin as *Aromatum et simplicium . . . historia* (Antwerp, 1567); Italian and French translations were also published. Much of d'Orta's material later reappeared in a Spanish work. Although his entertaining dialogue and thoughtful comments were lost, his contributions to botany and tropical medicine were thus saved and absorbed into the mainstream of European natural history.

BIBLIOGRAPHY

D'Orta's work was entitled *Coloquios dos simples e drogas he cousas medicinais da India e assi dalgũas frutas achadas . . .* (Goa, 1563). Two nineteenth-century eds. appeared; the standard one was edited and annotated by de Ficalho, 2 vols. (Lisbon, 1891–1895). This ed. was translated into English by C. Markham as *Colloquies on the Simples and Drugs of India* (London, 1913), but without the introductory material or notes.

L'Écluse's epitome appeared as *Aromatum et simplicium aliquot medicamentorum apud Indos nascentium historia ante biennium quidem Lusitanica lingua . . . conscripta, D. Garcia ab Horto auctore* (Antwerp, 1567); five eds. appeared between 1567 and 1605, and a facs. ed. was produced in 1963. An Italian trans. was made in 1576 (later eds. appeared between 1582 and 1616), and a French trans. was made in 1602 (2nd ed., 1619). C. Acosta, *Tractado de las drogas y medicinas de las Indias orientales* (Burgos, 1588), corrects, epitomizes, and illustrates d'Orta's work.

There are two biographies of d'Orta: A. de Silva Carvalho, "Garcia d'Orta," in *Revista da Universidade de Coimbra*, **12** (1934), 61–246; and Count de Ficalho, *Garcia da Orta e o seu tempo* (Lisbon, 1886).

The journal *Garcia de Orta. Revista da Junta das missões geográficas e de investigacões do ultramar* has been named in his honor; **10**, no. 4 (1963), is a special issue commemorating the publication of *Coloquios dos simples*; an extensive bibliography by J. Walter is included.

A. G. KELLER

ORTEGA, JUAN DE (*b.* Palencia, Spain, *ca.* 1480; *d. ca.* 1568), *mathematics.*

Ortega was a member of the Order of Preachers and was assigned to the province of Aragon. He taught arithmetic and geometry in Spain and Italy. Ortega followed the classical tradition and drew

inspiration, like his Spanish contemporaries, from the arithmetic of Boethius. His work reveals the influence of the more important mathematicians of the thirteenth and fourteenth centuries, but he was apparently unfamiliar with fifteenth-century works.

Ortega wrote *Cursus quattuor mathematicarum artium liberalium* (Paris, 1516) and *Tractado subtilisimo d'aritmética y de geometria* (Barcelona, 1512). The first part of the latter was devoted to commercial arithmetic and contains many examples, practical rules, and conversion tables for the various currencies then in use in the different regions of Spain. The second part gives instruction in practical rules of geometry "whereby anybody can measure any figure."

This work is of historical interest mainly for the numerical values that he obtained in extracting square roots, which appear in some of the geometric applications in the second part of the book. Almost identical editions were published in Seville in 1534, 1537, and 1542 (each published by Ortega himself), in which he modified the roots extracted in the first edition. He replaced them with values satisfying the Pell equation ($x^2 - Ay^2 = 1$); these values thereby gave the best approximation of square roots. Mathematicians have wondered how Ortega managed to evolve a method enabling him to find such closely approximate values, when a general solution of the Pell equation was presumably not achieved before Fermat (1601–1665).

Ortega's *Aritmética* became famous throughout Europe; the work was published in Lyons (1515), Rome (1515), Messina (1522), and Cambray (1612). It was also published in Seville (1552), probably posthumously, as it contained inadmissible changes. (This publication was later corrected.) The Lyons edition was the first book on commercial arithmetic to be published in French.

BIBLIOGRAPHY

Works that discuss Ortega and his work are Cantor, *Vorlesungen über die Geschichte der Mathematik*, II (Leipzig, 1908), 388; J. E. Hofmann, *Geschichte der Mathematik*, trans. into Spanish as *Historia de la Matemática* (Mexico City, 1960), I, 109–110; J. Rey Pastor, "Los matemáticos españoles del siglo XVI," in *Biblioteca scientia*, no. 2 (1926), 67, and "Las aproximaciones de Fr. Juan de Ortega," in *Revista matemática hispano-americana*, **7** (1925), 158.

MARÍA ASUNCIÓN CATALÁ

ORTELIUS (or **Oertel**), **ABRAHAM** (*b.* Antwerp, Brabant [now Belgium], 14 April 1527; *d.* Antwerp, 4 July 1598), *cartography, geography.*

With the exception of his friend Mercator, Ortelius was the principal cartographer of the sixteenth century. He was born to a Catholic family whose origins were in Augsburg. At the age of twenty he was admitted as an illuminator of maps into the guild of St. Luke in his native town. Soon he was able to earn his living by buying, coloring, and selling maps produced by map makers in various countries. Ortelius traveled widely in his profession; he went regularly to the Frankfurt Fair and visited Italy several times before 1558. In the period 1559–1560 he traveled through Lorraine and Poitou in the company of Mercator, who encouraged him to become a cartographer and to draw his own maps. The first product of this new activity was an eight-sheet map of the world published in 1564. In 1565 he published a map of Egypt (two sheets), in 1567 a map of Asia (two sheets), and in 1570 a map of Spain (six sheets).

The growing demand for maps of distant countries, caused by the rapidly expanding colonization and the development of commerce, had already led to the production of large collections of maps of various size and provenance, for instance, Lafreri's atlas published *ca.* 1553. At the suggestion of the Dutch merchant and map collector Hooftman, and of his friend Radermacher, Ortelius undertook the publication of a comprehensive atlas of the world. It appeared in May 1570 in the form of a single volume, in folio, entitled *Theatrum orbis terrarum,* published by Egidius Coppens Diesth and printed by Plantin in Antwerp. It contained fifty-three sheets with a total of seventy copperplate maps, most of them engraved by Frans Hoogenberg, and thirty-five leaves of text.

The atlas clearly reveals that Ortelius was more an editor of maps than an original mathematical cartographer. Unlike Mercator, Ortelius never devised new projections; but acquiring the rights to utilize maps produced by others, he reduced them to a uniform size and brought their geographic contents up-to-date. To cite a single example, his map of Denmark, entitled *Daniae regni typus,* had as its immediate prototype a map with the same title made in 1552 by Marcus Jordan. But Ortelius' map also included a number of features taken from *Caerte van oostland,* drawn *ca.* 1543 by Cornelis Anthoniszoon, and from Niccolò Zeno's map published in 1558. The map was, however, an unmistakable improvement on previous maps of the region, which was also true for most of his other maps, although Ortelius still relied to a certain extent on material later shown to be legendary, for example the travels of Prester John.

A unique feature of the *Theatrum* was the *Catalogus cartographorum,* in which Ortelius listed eighty-seven map makers as authorities for his own work. Much of what we know of the minor cartographers of the fifteenth and sixteenth centuries derives from this catalog.

The *Theatrum* was an immediate success and the Plantin press published a long series of editions and epitomes in Latin between 1570 and 1624. In 1625 the copyright was acquired by Willem Blaeu, who in 1631 published an appendix to the work and then edited his own *Theatrum orbis terrarum sive atlas novus* in 1634. Ortelius' atlas was translated into Dutch (*Toonneel des aertbodems*) in 1571 and 1598; German (*Schawplatz des erdbodems*) in 1572, 1573, 1580, 1602; French (*Théâtre de l'univers*) in 1572, 1574, 1578, 1581, 1587, 1598; Spanish (*Theatro de la tierra universal*) in 1588, 1600, 1602, 1612; Italian (*Theatro del mondo*) in 1608 and 1612; and English (*Abraham Ortelius his Epitome of the Theatre of the Worlde*) in 1603 and 1606. Ortelius continually revised the new editions, adding new maps and reediting the old. In 1573 he published the first *Additamentum theatri orbis terrarum.* This work was later incorporated into the *Theatrum* (1601), which contained no fewer than 161 maps and 183 authorities.

The *Theatrum* won for Ortelius the title of geographer to King Philip II of Spain. (Arias Montanus vouched for Ortelius' orthodox faith, which had been under suspicion.) It also secured for him a substantial income, enabling him to continue his travels to collect new material. In 1577 he visited England and Ireland, making the personal acquaintance of John Dee, Camden, Hakluyt, and other British geographers. The report of a similar journey in 1575 appeared as *Itinerarium per nonnullas Galliae Belgicae partes,* written in collaboration with J. Vivianus and published in 1584 by Plantin.

During the later part of his life, Ortelius spent much time on classical studies. His large collection of ancient coins and other antiquities was described in the *Deorum dearumque capita ex vetustis numismatibus . . . effigiata et edita ex museo A. Ortelii,* published by P. Gallaeus in 1573 (later editions are 1582, 1602, 1680, 1683, 1699). An edition of *C. J. Caesaris omnia quae extant* (1593) appeared in Leiden and *Aurei saeculi imago, sive Germanorum veterum vita, mores, ritus et religio iconibus delineata* (1596) was published in Antwerp. Of particular interest are Ortelius' works on ancient geography, which began with his *Synonymia geographica,* published by Plantin in 1578 and later revised as *Thesaurus geographicus* (1587, 1596). In 1584 he published *Nomenclator Ptolemaicus,* which dealt with place names in Ptolemy's geography, and *Parergon,* a collection of maps illustrating ancient history, printed by Plantin. The *Nomenclator* and the

Parergon were incorporated into several of the later editions of the *Theatrum*; thus the 1601 edition contained forty maps from the *Parergon*. Ortelius also collaborated with Marcus Welser on his edition of the *Tabula Peutingeriana* (Venice, 1591), a fourth-century Roman military itinerary map.

BIBLIOGRAPHY

On Ortelius and his work, see L. Bagrow, *Abrahami Ortelii Catalogus Cartographorum* (Gotha, 1928–1930); J. Denucé, *Oud–Nederlandsche Kaartmakers in betrekking met Plantin*, II (Antwerp 1913), 1–252, which contains a good biography of Ortelius and a complete bibliography of the *Theatrum*; J. H. Hessels, "Abrahami Ortelii . . . et virorum eruditorum . . . epistolae," in J. H. Hessels, *Ecclesiae Londino-Batavae archivum*, I (Cambridge, 1887); and H. E. Wauermans, *Histoire de l'école cartographique belge et anversoise du XVIe siècle* (Brussels, 1895).

OLAF PEDERSEN

ORTON, JAMES (*b.* Seneca Falls, N.Y., 21 April 1830; *d.* Lake Titicaca, Bolivia, 24 September 1877), *natural history, exploration.*

James Orton was the fifth of eight sons of Azariah Giles Orton, preacher, poet, and classicist, and Minerva Squire Orton. He wrote *The Miner's Guide and Metallurgist's Directory* at nineteen, in the year of the gold rush; but he did not go to California. After graduating from Williams College (B.A. 1855), he attended Andover Theological Seminary and held three pastorates, but Mark Hopkins, president of Williams College, turned Orton permanently toward natural history.

Under the auspices of Williams College and with a loan of instruments from the Smithsonian Institution, Orton directed an Andean expedition in 1867 to determine whether deposits in the upper Amazon Valley were of marine or, as Louis Agassiz insisted, glacial origin. He crossed the Ecuadorian Andes and by canoe descended the Rio Napo, "a steaming vapor-bath." He found marine shells at Pebas, Peru. Only the botanist William Jameson and the zoologist Gaetano Osculati had preceded Orton's party across the Guamani Pass on a scientific expedition. Orton's *Andes and Amazon* (1870) was dedicated to Charles Darwin.

In 1869 Orton, a staunch supporter of coeducation, introduced natural history instruction at Vassar College and recounted his experiences in *Liberal Education of Women* (1873). His *Comparative Zoology, Structural and Systematic* (1876), expounding Agassiz's functional approach, was an influential text. In 1873

Orton directed a second Andean expedition, from Pará to Yurimaguas, across the Andes and down to Lima, collecting for specialists in a wide number of fields and telling of these experiences in the third edition of *Andes and Amazon* (1876).

In 1876 Orton set out on a third expedition, traveling to the trans-Andean rain forests. Although the expedition seemed well planned, the hired porters and much of the escort provided by the Bolivian government mutinied, leaving the small party to make its way through most difficult terrain to Lake Titicaca. Orton had never enjoyed good health, and he succumbed from exhaustion while crossing the lake. He was buried on Estaves Island. Unfortunately, Orton's collections and notes from this last expedition were lost during shipment to New York.

BIBLIOGRAPHY

I. ORIGINAL WORKS. Orton's principal writings include *The Miner's Guide and Metallurgist's Directory* (New York–Cincinnati, 1849); *The Proverbialist and the Poet: Proverbs Illustrated by Parallel or Relative Passages From the Poets, to which are Added Latin, French, Spanish, and Italian Proverbs* (Philadelphia, 1852); *Andes and the Amazon; or, Across the Continent of South America* (New York, 1870); repr., 1871; 2nd ed., 1876); *Underground Treasures, How and Where to Find Them* (Hartford, 1872); and *Comparative Zoology, Structural and Systematic, for Use in Schools and Colleges* (New York, 1876).

Orton published a number of short papers in the *American Journal of Science, Geological Magazine, Proceedings of the American Association for the Advancement of Science, Annals and Magazine of Natural History*, and *American Naturalist*, including one in the latter, "The Great Auk, *Alca impennis*," concerning the former model for Audubon's drawing in the Vassar College collection, **3** (1869), 539–542.

He edited *The Liberal Education of Women, the Demand and the Method, Current Thoughts in America and England* (New York, 1873), to which he contributed 8 chapters. Four of Orton's letters, written between 1867 and 1868, are preserved in the S. F. Baird correspondence, Smithsonian Institution. Manuscript "Notes for New Edition" of *Andes and Amazon*, including 251 queries and references, is preserved in De Golyer Library, University of Oklahoma.

II. SECONDARY LITERATURE. There is no published bibliography of his writings. The essential sketch is Susan R. Orton, "A Sketch of James Orton," in *Vassar Quarterly*, **1** (1916), 1–8, in which the date of death accepted here appears. See also E. D. Cope, "An Examination of the Reptilia and Batrachia Obtained by the Orton Expedition to Ecuador and the Upper Amazon, With Notes on Other Species," in *Proceedings of the Academy of Natural Sciences of Philadelphia*, **20** (1868), 96–140, and "On Some Batrachia and Nematognathi Brought From the Upper Amazon by Professor Orton," *ibid.*, **26** (1874), 120–137;

Philip Reese Uhler, "Notices of the Hemiptera Obtained by the Expedition of Prof. James Orton in Ecuador and Brazil," in *Proceedings of the Boston Society of Natural History*, **12** (1869), 321–327; George Dale Smith, "List of Coleoptera Collected by Professor James Orton in Ecuador and Brazil," *ibid.*, 327–330; and Samuel Hubbard Scudder, "Notes on Orthoptera Collected by Professor James Orton on Either Side of the Andes of Equatorial South America," *ibid.* 330–345. Ruth D. Turner, "James H. Orton. His Contributions to the Field of Fossil and Recent Mollusks," in *Revista del Museo argentino de ciencias naturales "Bernardino Rivadavia." Ciencias zoológicas*, **8** (1962), 89–99—the title included an erroneous middle initial.

Henry Morris Myers and Philip Van Ness Myers, *Life and Nature Under the Tropics* (New York, 1871), 194–323, relates to a contingent of the first Andean expedition.

JOSEPH EWAN

OSBORN, HENRY FAIRFIELD (*b.* Fairfield, Connecticut, 8 August 1857; *d.* Garrison, New York, 6 November 1935), *vertebrate paleontology.*

Osborn was the eldest son of William Henry Osborn, president of the Illinois Central Railroad, and of Virginia Reed Sturges. He spent his early life in the vicinity of New York City. He attended the College of New Jersey (now Princeton University), where he was much influenced by President James McCosh and Arnold Guyot, director of the museum. At Princeton he began a lifelong friendship with William Berryman Scott. In their junior year, Scott and Osborn became intensely interested in the fossil remains of extinct reptiles and mammals. The young men accordingly organized their first paleontological expedition. They spent the summer of 1877 in Colorado and Wyoming, still a wild land inhabited by less than friendly Indians and by some of the "old mountain men." In 1878 there was a second expedition, and it was at this time that they met and became disciples of Edward Drinker Cope of Philadelphia, the rival of O. C. Marsh of Yale.

After completing their undergraduate work at Princeton, Scott and Osborn went abroad for postgraduate studies. Osborn studied under T. H. Huxley and Francis Maitland Balfour in London. He also met Charles Darwin, an encounter that he never forgot.

He returned to join the faculty at Princeton, and in 1881 married Lucretia Perry; they had five children. In 1891 he was called to Columbia University to found a department of biology and to the American Museum of Natural History to found a department of mammalian paleontology (soon to become the department of vertebrate paleontology). He spent the remainder of his life in New York City, where he was actively associated with Columbia until 1910 and with the American Museum of Natural History until his death.

In addition to his career as first head of the biology department at Columbia, Osborn was first dean of the graduate faculty, and for many years was Da Costa professor of zoology, in which capacity he trained numerous students, many of whom became distinguished zoologists and paleontologists. At the same time he served as head of the department of vertebrate paleontology, where he was instrumental in building a collection of worldwide importance. For twenty-five years he was also president of the American Museum of Natural History and was largely responsible for making it probably the largest natural history museum in the world.

In spite of his involvement with these several concurrent careers, Osborn was primarily a research scientist. He continually studied fossil vertebrates, and with the aid of assistants and colleagues, who did much of the detailed work for him, he published some 600 papers, books, and monographs.

Although Osborn was concerned with the details of vertebrate evolution—particularly that of reptiles and mammals—he was especially interested in the larger problems of life. He was a theorist and proposed various explanations for many aspects of evolution. His important contributions to the knowledge of evolution within many groups of mammals and reptiles were, nonetheless, based upon the fossil evidence. He had a grand concept of the adaptive radiation of life; yet in spite of his penetrating mind, he never seemed to appreciate fully the significance of genetic studies to the modern concept of evolution. Osborn was also a master of synthesis, a capacity illustrated by his enormous monographs on the titanotheres and the proboscideans.

BIBLIOGRAPHY

A full bibliography of Osborn's works (exclusive of newspaper articles, abstracts, and some popular articles) will be found in William K. Gregory, "Biographical Memoir of Henry Fairfield Osborn 1857–1935," in *Biographical Memoirs. National Academy of Sciences*, **19** (1938), 53–119. See also George Gaylord Simpson, "Henry Fairfield Osborn," in *Dictionary of American Biography*, **11**, supp. 1 (New York, 1944), 584–587, which includes a bibliography.

EDWIN H. COLBERT

OSBORNE, THOMAS BURR (*b.* New Haven, Connecticut, 5 August 1859; *d.* New Haven, 29 January 1929), *protein chemistry.*

Osborne was the son of Arthur Dimon Osborne, who was educated in law but subsequently became the president of a local bank, and Frances Louisa Blake. His ancestors on both sides can be traced in the history of New Haven to its earliest years. Osborne prepared for college at the Hopkins Grammar School in New Haven and was graduated from Yale in 1881. A year was spent in the study of medicine before he entered the Yale graduate school, where he studied chemistry under W. G. Mixter. His doctoral dissertation (1885) described the analytical determination of niobium in columbite, the mineral occurring in Connecticut in which niobium (originally named columbium) had been discovered in 1801 by Charles Hatchett. In 1886, at the invitation of Samuel W. Johnson, director of the recently established Connecticut Agricultural Experiment Station and professor of agricultural chemistry of the Sheffield Scientific School of Yale University, Osborne joined the staff of the experiment station as an analytical chemist. In the same year he married Elizabeth Anna Johnson, his director's daughter.

While a graduate student at Yale, and during a year spent as an instructor, Osborne published several papers on analytical methods. At the experiment station he developed what became known as the Osborne beaker method for the mechanical analysis of soils, a method still in use. In 1889, owing to the passage of the Hatch Act of 1887, which provided for additional funds, Osborne began the investigations of the proteins of plant seeds, which became his lifelong work. The initial suggestion to begin this research was made by Johnson, impressed with the related work of Heinrich Ritthausen in Germany.

Johnson's former teacher at Yale, J. P. Norton, had studied the proteins of the oat kernel some forty years earlier, and Johnson, noting that Ritthausen had not investigated this seed, suggested that further work was desirable, Although in later years Osborne stated that no seed that he subsequently worked with ever presented such difficulties as had the oat kernel, he succeeded in preparing what appeared to be a homogeneous alcohol-soluble protein and also a globulin that was obtained in crystalline form.

This success led to broadly planned research into the proteins of seeds used as human or animal food. Within two years he had obtained crystalline globulins from six different seeds, and from 1889 to 1901 he examined no less than thirty-two species, including a number of legumes, many common nuts, and the most important cereal grains. His skillful use of saline solvents, his control of acidity by the intelligent use of indicators long before the theory of pH had been developed, the use of temperature gradients, dialysis,

and in certain instances of alcohol to precipitate the components of the extracts, demonstrate an instinctive appreciation of the physicochemical properties of proteins, which was many years in advance of theoretical explanation of these matters. He established what since have become the classical methods for the isolation of proteins from plant seeds.

Osborne was somewhat restricted during this early period of research by two considerations. Liebig's dictum of fifty years earlier that there are only four kinds of protein in nature (albumin, casein, fibrin, and gelatin), although shown to be greatly oversimplified by Ritthausen and others, was still a dominating principle. Second, the motive for many studies was to show that identical proteins could be prepared from analogous tissues of different species. Inasmuch as the main criterion for the differentiation of preparations from different species was the comparison of the content of carbon, hydrogen, nitrogen, and sulfur, it can be understood how Osborne was at first frequently misled into believing that he had obtained the same protein from two or more different seeds. Thus, in 1894, he stated that, since their ultimate composition was essentially the same, the globulins of hempseed, castor bean, squashseed, flax, wheat, maize, and cottonseed are identical. For this widely distributed protein he suggested the name edestin (from the Greek for "edible"), a name later applied only to the globulin of hempseed.

Several other instances of apparent identity between pairs or small groups of proteins of different origin were later encountered. Nevertheless, as his experience broadened, Osborne became increasingly suspicious of the validity of such conclusions, and, at the turn of the century, he began to subject his extensive collection of proteins to detailed chemical study. He examined such properties as the solubility in saline solution, coagulation temperature, specific rotation, heat of combustion, color tests, the behavior of sulfur when the protein was heated with alkali in the presence of lead, and the quantitative behavior of protein toward acid and alkali, in which he sought for differences between proteins that seemed to be identical. Osborne's closest attention, however, was given to the determination of the different forms of nitrogen in the products of complete acid hydrolysis of the proteins and to the determination of the basic amino acids by the recently published method of Kossel and Kutcher. Glutamic and aspartic acids were also determined by direct isolation.

Osborne soon came to the conclusion that the detection of differences was the fundamental problem. To obtain more complete characterizations, for about

242

five years beginning in 1906, he devoted the full resources of his laboratory to the determination of the amino acid composition of many of the most important proteins. He used the ester distillation method of Emil Fischer for these determinations. The outcome of this labor was that, with only a few exceptions, proteins that closely resembled each other in ultimate composition could be distinguished from each other in terms of amino acid composition. When the highly sensitive biologic test dependent upon the anaphylaxis reaction became available, Osborne, from 1911 to 1916, collaborated with H. Gideon Wells of the University of Chicago in a comprehensive study of the seed proteins. Their studies showed that only two or three doubtful instances remained where proteins of different origin could not be distinguished from each other. This demonstration of the specificity of vegetable proteins with respect to source remains one of Osborne's fundamental contributions to protein chemistry.

In 1909, Osborne invited Lafayette B. Mendel of Yale to join him in a study of the nutritive properties of the seed proteins. It was widely believed that, with a few exceptions, all proteins are alike in nutritive effect. The striking differences in the amino acid composition that had been found for a number of common food proteins raised the question of the validity of this view. The collaboration with Mendel continued from 1909 until Osborne's retirement in 1928. They developed a technique for feeding rats that enabled them to measure the food intake, and within a few years obtained convincing proof that the amino acids tryptophan and lysine are essential in the diet. Although it was clear that the rat can synthesize some of the amino acids, this capacity is strictly limited. The study of the effect of lysine was especially rewarding since it showed that the growth of a young rat could be quantitatively controlled by the supply of lysine, either as such or combined in the protein of the diet. Animals stunted by low levels of lysine intake could be induced to grow at any age by increasing the supply.

The outstanding accomplishment of the first few years of this collaboration was the discovery of what became known as vitamin A. This discovery resulted from the comparison of the growth of rats on diets consisting of purified components, of which one contained dried whole milk and the other only the lactose and inorganic salts of milk. The substitution of butter for some of the lard in the second diet prevented the loss of weight and eventual death of the animals. This observation was made early in 1913. The conclusion was obvious that butter contains a trace amount of some fat-soluble organic substance that is essential in nutrition. Unfortunately the submission to a journal of a similar observation by E. V. McCollum of the University of Wisconsin preceded by three weeks the receipt of the Osborne and Mendel paper, and McCollum is accordingly regarded as the discoverer of the first vitamin to be recognized. Although Osborne and Mendel devoted considerable study to the natural distribution of the fat-soluble vitamin (notably finding that cod liver oil is a rich source) and the later-discovered water-soluble vitamin, their main interest during the extremely active period from 1911 to 1924 was in the phenomena of growth and in the nutritive properties of various proteins. They studied the effects of high-protein diets, low-carbohydrate and low-fat diets, and variations in the supply of inorganic salts. They obtained rational explanations for many empirical practices in animal feeding that had been found advantageous, and they cleared up the relation between nutritional ophthalmia and vitamin A. They also contributed to the demonstration of the nutritional origin of rickets; the common use of cod liver oil and orange juice in the diets of children stems largely from their work.

Although the main interest of the laboratory continued to be in nutrition, Osborne, with the aid of his assistants, also devoted much effort to the many purely chemical problems that arose. In 1919 he and Alfred J. Wakeman prepared the first vitamin-rich concentrate from an extract of brewer's yeast. The concentrate was used for many years in the laboratory and was marketed successfully by a former assistant, Isaac F. Harris, who had become a manufacturing chemist. The observation that the alfalfa plant is rich in vitamins led to attempts to prepare the proteins from green leaves. Only moderate success attended these efforts, but A. C. Chibnall of the Imperial College, London, who joined Osborne's group in 1923 and 1924, was later successful. In his last years Osborne also stimulated the investigations by his assistants of the simpler nitrogenous substances present in plants, a field of study that had been neglected since the early work of Ernst Schulze in Switzerland in the last decades of the nineteenth century.

Unlike his collaborator Mendel, Osborne did not have a large group of loyal and devoted students to keep his memory alive. To those who worked with him he was a rare stimulus, a formidable opponent in argument, and an ever genial but just critic. His major, in fact almost his only, interest was in the work of the laboratory. He served for many years as a director of the local bank of which his father had been president, but this and the group of close friends at his club, together with his interest in the birds of Connecticut, upon which he was an authority, provided the major relief from his daily work at the laboratory bench.

BIBLIOGRAPHY

I. ORIGINAL WORKS. Osborne's bibliography published in Vickery's memoir (see below) lists titles of 252 papers that appeared in various chemical journals between 1884 and 1929. A nearly complete bound collection of his work is in the Osborne Library at the Connecticut Agricultural Experiment Station in New Haven. The papers on the preparation of proteins appeared in the *American Chemical Journal* or *Journal of the American Chemical Society* until 1904; nearly all were reprinted in the annual *Report of the Connecticut Agricultural Experiment Station.*

Most of the papers from 1891 to 1897 were translated into German by V. Griessmayer in *Die Proteide der Getreidarten, Hülsenfrüchte und Ölsamen sowie einiger Steinfruchte* (Heidelberg, 1897). Griessmayer continued to translate and publich most of Osborne's papers, which appeared up to 1908, in *Zeitschrift für das landwirtschaftliche Versuchswesen in Österreich* or, after 1904, in *Zeitschrift für analytische Chemie.*

From 1904 to 1910 Osborne's papers were published in the *American Journal of Physiology*; subsequent papers appeared in the *Journal of Biological Chemistry*, to which Osborne and Mendel contributed most of their collaborative papers on nutrition between 1912 and 1927. Including annual reports to the Carnegie Institution of Washington there were 111 of these. The six papers on the anaphylaxis reactions of the seed proteins, written with H. Gideon Wells, appeared in the *Journal of Infectious Diseases* between 1911 and 1916.

Osborne's works also include *The Proteins of the Wheat Kernel*, Carnegie Institution of Washington Publication no. 84 (Washington, D.C., 1907); *The Vegetable Proteins*, in R. H. Plimmer and F. G. Hopkins, eds., Monographs on Biochemistry (London, 1909; 2nd ed., rev., 1924); and *Feeding Experiments with Isolated Food-Substances, Parts I and II*, Carnegie Institution of Washington Publication no. 156 (Washington, D.C., 1911), written with Mendel.

II. SECONDARY LITERATURE. Hubert Bradford Vickery has written three articles on Osborne: "Thomas Burr Osborne, 1859–1929," in *Biographical Memoirs. National Academy of Sciences*, **14** (1931), 261–304; "Thomas B. Osborne, a Memorial," in *Bulletin. Connecticut Agricultural Experiment Station*, **312** (1930); and "Thomas Burr Osborne," in *Journal of Nutrition*, **59** (1956), 1–26.

The bulletin published by the experiment station contains several obituary notices, the records of the presentation of an honorary degree by Yale University in 1910, the presentation of the John Scott Medal by the board of directors of City Trusts of Philadelphia in 1922, and the presentation of the Thomas Burr Osborne Medal by the American Association of Cereal Chemists in 1928. It also contains reprints of Osborne's addresses on protein

chemistry to several organizations, a previously unpublished paper on bird migration, and a complete bibliography of his papers.

HUBERT BRADFORD VICKERY

OSGOOD, WILLIAM FOGG (*b.* Boston, Massachusetts, 10 March 1864; *d.* Belmont, Massachusetts, 22 July 1943), *mathematics.*

Osgood was the son of William Osgood and Mary Rogers Gannett. After preparing for college at the Boston Latin School, he entered Harvard College in 1882 and was graduated second in his class in 1886. He remained at Harvard for a year of graduate work in mathematics and was awarded the A.M. in 1887. Osgood spent much of his first two years at Harvard studying the classics but was largely influenced by the mathematical physicist Benjamin Osgood Peirce, one of his favorite teachers, and by Frank Nelson Cole. Cole had attended Felix Klein's lectures on function theory and lectured on the subject, following Klein's ideas, at Harvard during 1885–1887. Osgood went to the great German center of mathematics at Göttingen in 1887, largely because of Klein's presence there.

In 1887 there was great mathematical activity in Europe, brought about especially by the introduction of rigor into current research. Under the influence of Klein, Osgood embraced this tendency, which remained a commitment throughout his life. Osgood went to Erlangen in 1889 to continue his graduate work. His dissertation, a study of Abelian integrals of the first, second, and third kinds, was based on previous work by Klein and Max Noether. The topic was part of the theory of functions, to which Osgood devoted much of his later life. After receiving his Ph.D. at Erlangen in 1890, Osgood married Anna Terese Ruprecht of Göttingen and returned to the United States. He then joined the Harvard department of mathematics, where he remained for forty-three years. He brought with him the spirit of research, then new in the United States, as well as that of rigor. A year later Maxime Bôcher returned to Harvard, and the two were influential in fostering the new attitude there.

Osgood's main research papers concerned convergence of sequences of continuous functions, solutions of differential equations, Riemann's theorem on the mapping of a simply connected region, the calculus of variations, and space-filling curves. These topics are classical, and Osgood's results are important and deep. Klein invited Osgood to write an article for the *Encyklopädie* on the theory of functions; the writing of it (1901) gave Osgood an unparalleled knowledge of the field and its history. His *Lehrbuch der Funktionentheorie* (1907) subsequently became the standard treatise. Osgood was one of the world's outstanding mathematics teachers through that work and through others on analytic geometry, calculus, and advanced calculus. Over the years he instilled ideals and habits of careful and accurate thought in hundreds of elementary as well as advanced students. After his retirement from Harvard in 1933, he lectured for two years at the National University of Peking.

Osgood's favorite recreations were travel by car, smoking cigars, and occasional games of tennis and golf. He was kindly although somewhat reserved, but warm to those who knew him. He and his first wife had two sons and a daughter. He married Celeste Phelps Morse in 1932.

BIBLIOGRAPHY

Personal recollections; Harvard Class of 1886 *Reports* for 1886, 1889, 1894, 1898, 1901, 1906, 1911, 1926, 1936; and clippings in Harvard University Archives. See also *Dictionary of American Biography*, supp. 3, 574–575.

J. L. WALSH

OSIANDER, ANDREAS (*b.* Gunzenhausen, Bavaria, Germany, 19 December 1498; *d.* Königsberg, Germany [now Kaliningrad, U.S.S.R.], 17 October 1552), *theology, astronomical and mathematical publishing.*

On 9 July 1515 Osiander was admitted to the University of Ingolstadt as a "cleric of the Eichstätt" diocese.[1] Without obtaining a degree he moved to Nuremberg, where he taught Hebrew and was ordained a priest in 1520. He enthusiastically embraced the new Lutheran movement and soon became one of its most militant spokesmen. When Nuremberg accepted the pro-Catholic Augsburg Interim, Osiander left and joined the Protestant Duke Albert of Prussia. On 27 January 1549 he arrived in Königsberg, where the recently founded university appointed him professor of theology.[2] His doctrinal views were bitterly opposed by the more orthodox followers of Martin Luther in the "Osiander Controversy," which continued after Osiander's death.

In 1538 Rheticus obtained a leave of absence from Wittenberg University in order to visit German astronomers. In Nuremberg he met Osiander, whose hobby was the mathematical sciences. Hence, when Rheticus' *Narratio prima*, the first printed discussion of the Copernican astronomy, was published in 1540, a copy was sent to Osiander, who was shocked by the claim of the new system to be true; he regarded divine revelation as the sole source of truth. In similar letters to Rheticus and Copernicus on 20 April 1541, when Rheticus was waiting in Frombork (Frauenburg) for Copernicus to put the final touches on the manuscript of *De revolutionibus orbium coelestium*, Osiander urged the inclusion in the introduction of the statement that even if the Copernican system provided a basis for correct astronomical computations, it might still be false. Copernicus firmly rejected Osiander's recommendation.

Nevertheless, subsequent events enabled Osiander to impose his fictionalist philosophy of science on *De revolutionibus*, while its author lay helpless and dying in far-off Frombork. Copernicus had entrusted the printing of *De revolutionibus* to Rheticus, who supervised the early stages of the process in the shop of Johannes Petreius (Hans Peter) in Nuremberg. When Rheticus had to go to the University of Leipzig, which had just appointed him professor of mathematics, he was replaced as editor of *De revolutionibus* by Osiander, who surreptitiously slipped into the authentic front matter an unsigned preface composed by himself and expounding his anti-Copernican fictionalism.[3]

When copies of *De revolutionibus* reached Rheticus in Leipzig, he became enraged and sent to the City Council of Nuremberg a sharp protest that was written by Tiedemann Giese, the closest friend of Copernicus, who had died in the meantime. Petreius replied that he had received the false preface in a form undifferentiated from the rest of the material. Whereas Osiander never publicly acknowledged his authorship of the interpolated preface, he did so privately,[4] and thus finally in 1609 Kepler's *Astronomia nova* was able to identify Osiander as the culprit.

Osiander was more sympathetic to the mathematician Cardano. Both of them were astrologers, and they exchanged letters about horoscopes for some five years before Cardano on 9 January 1545 dedicated *Artis magnae sive de regulis algebraicis liber unus*— which initiated the theory of algebraic equations—to Osiander, who edited the work for Petreius.[5]

NOTES

1. Götz F. v. Pölnitz, ed., *Die Matrikel der Ludwig-Maximilians-Universität Ingolstadt-Landshut-München*, I (Munich, 1937), 381.
2. His son Lucas was admitted to the university in the summer semester of 1549 (Georg Erler, ed., *Die Matrikel der Universität Königsberg in Preussen*, I [Leipzig, 1908–1910], 10).
3. Osiander's preface was translated into English by Edward Rosen, *Three Copernican Treatises*, 3rd ed. (New York, 1971), pp. 24–25.

4. Ernst Zinner, *Entstehung und Ausbreitung der copperni-canischen Lehre* (Erlangen, 1943), p. 453.

5. Cardano's dedication was translated into English by T. Richard Witmer, *The Great Art or the Rules of Algebra by Girolamo Cardano* (Cambridge, Mass., 1968), p. 2.

BIBLIOGRAPHY

I. ORIGINAL WORKS. Osiander's works are chronologically enumerated (1522–1552) in Gottfried Seebass, *Das reformatorische Werk des Andreas Osiander* (Nuremberg, 1967), pp. 6–58, with nine portraits of Osiander as frontispiece and supplement.

II. SECONDARY LITERATURE. On Osiander and his work, see Wilhelm Möller, *Andreas Osiander* (Elberfeld, 1870; repr. Nieuwkoop, 1965), and his article, "Osiander," in *Allgemeine deutsche Biographie*, XXIV (1887; 1970), 473–483; and G. Seebass, *op. cit.*, pp. xi–xviii.

EDWARD ROSEN

OSMOND, FLORIS (*b*. Paris, France, 10 March 1849; *d*. St. Leu, Seine-et-Oise, France, 18 June 1912), *metallography.*

Osmond studied metallurgy under Samson Jordan at the École Centrale des Arts et Manufactures. After a short period with the Fives-Lille machine shop he joined Denain et Anzin, where he worked with Bessemer and open-hearth installations. From 1880 to 1884 Osmond was chief of the chemical laboratory of Schneider, Creusot, where he began his microscopic study of iron and steel in collaboration with a colleague in the physical testing laboratories. After 1884 Osmond, who was of a retiring disposition, left active business and returned to Paris, where he continued his research, corresponding with professional friends and publishing some eighty papers before his death.

Osmond's earliest interests concerned the effects of tempering and hardening cast steel and, particularly, the phenomena that occur during the heating and cooling of steel. The Le Chatelier pyrometer became available in 1886; and with the help of it Osmond took up the studies suggested by Tschernoff in 1868, by W. F. Barrett in 1873, and by Le Chatelier and others. Osmond proceeded to determine the so-called critical points at which the abnormal retardation or acceleration in the temperature drop occurs during the cooling of an iron sample—effects which indicate a liberation or an absorption of heat. From these investigations he concluded that allotropic β iron is the principal cause of the new properties communicated to steel by hardening. Osmond's experiments with tungsten steel showed that variations in the hardness of steel could be obtained by altering the initial temperature of heating and the rate of cooling; he did not publish this finding, which, in a sense, anticipated the Taylor-White process (1898).

By 1890 Osmond recognized three modifications of iron: α, β, and γ. His research led to the allotropic theory, the subject of much argument in the 1890's. It was opposed by the "carbonists," including John Oliver Arnold, who maintained that all the phenomena observed in the hardening of steel are explained by changes in the condition of the carbon and are in no way due to allotropic modifications of the iron. The Iron and Steel Institute (London) recognized the merits of both arguments by awarding the Bessemer Medal to Arnold in 1905 and to Osmond in 1906.

Osmond made substantial contributions to microscopical investigations of the structure of iron and steel. Although his interest may have been derived from the work of Hermann Vogelsang of Delft, he started with H. C. Sorby's methods, which he developed, especially in the preparation of samples. In the final polishing Osmond developed a method of "polish attack," in which the sample was rubbed on a sheet of parchment covered with calcium sulfate moistened with an infusion of licorice by which some of the constituents of the steel were colored.

Osmond's observations led him to identify and name sorbite, austenite, and troostite, commemorating Sorby, Sir W. C. Roberts-Austen, and Troost, an early associate of Osmond's who presented the latter's early papers to the Académie des Sciences in 1886–1887. Osmond rechristened H. M. Howe's hardenite "martensite" in honor of Adolf Martens, another pioneer in metallography. His own name was commemorated in osmondite, a term now obsolete in the nomenclature.

Osmond was awarded prizes by the Société d'Encouragement pour l'Industrie Nationale in 1888 and 1895, and the Lavoisier Medal in 1897.

BIBLIOGRAPHY

I. ORIGINAL WORKS. Among Osmond's more than 80 papers are: "Théorie cellulaire des propriétés de l'acier," in *Annales des mines* (Mémoires), 8th ser., **8** (1885), 5–84, written with Jean Werth; "Sur les phénomènes qui se produisent pendant le chauffage et le refroidissement de l'acier fondu," in *Comptes rendus . . . de l'Académie des sciences*, **103** (1886), 743–746, 1135–1137; "Rôle chimique du manganèse," *ibid.*, **104** (1887), 985–987; "Sur les residues que l'on extrait des aciers," *ibid.*, 1800–1812, written with J. Werth; "Die Metallographie als Untersuchungsmethode," in *Stahl und Eisen*, **17** (1897), 904–913; "Metallography as a Testing Method," in *Metallographist*, **1** (1898), 5–27; "What is the Inferior Limit of the Critical Point A_2?" *ibid.*, **2** (1899), 169–186; "On the Crystallography of Iron," *ibid.*, **3** (1900), 181–219; 275–290; *The Microscopic Analysis of Metal*, J. E. Stead, ed. (London, 1904); "Les expériences du Prof. Heyn sur la trempe et le revenu des aciers," in *Revue de métallurgie* (Mémoires), **3** (1906), 621–632; and "Crystallization of Iron," in *Journal of the Iron and Steel Institute*, **71**, no. 3 (1906), 444–492, written with G. Cartaud.

II. SECONDARY LITERATURE. See John O. Arnold and A. McWilliams, "The Diffusion of Elements in Iron," in *Engineering*, **68** (1899), 249; Henry M. Howe, *The Metallurgy of Steel* (New York, 1890), 163 ff.; and "The Heat Treatment of Steel: Note on Osmond's Theory," in *Transactions of the American Institute of Mining Engineers*, **23** (1893), 520; and the unsigned obituary in *Engineering*, **94** (1912), 56–58.

P. W. BISHOP

OSTROGRADSKY, MIKHAIL VASILIEVICH (*b.* Pashennaya [now in Poltava oblast], Russia, 24 September 1801; *d.* Poltava [now Ukrainian S.S.R.], 1 January 1862), *mathematics, mechanics.*

Ostrogradsky was born on the estate of his father, Vasily Ivanovich Ostrogradsky, a landowner of modest means; his mother was Irina Andreevna Sakhno-Ustimovich. After he had spent several years at the Poltava Gymnasium, the question of his future arose. Ostrogradsky hoped to become a soldier; but the life of an officer was expensive, the salary alone would not support him, and the family had little money to spare. It was decided to prepare him for the civil service and to give him a university education, without which his career would be limited. In 1816 Ostrogradsky enrolled in the physics and mathematics department of Kharkov University, where he received a good mathematical education under A. F. Pavlovsky and T. F. Osipovsky. He was especially influenced by the latter, an outstanding teacher and author of the three-volume *Kurs matematiki* (1801–1823), which was well known in its time, and also of philosophical papers in which he criticized Kant's apriorism from the materialistic point of view. In 1820 Ostrogradsky passed the examinations for the candidate's degree, and the university council voted to award it to him. But the minister of religious affairs and national education refused to confirm the council's decision and proposed that Ostrogradsky take the examinations again if he wished to receive his degree. Ostrogradsky rejected this proposal, and therefore did not obtain a university diploma.

The true reason for the arbitrary reversal of the council's decision was the government's struggle with the nonconformist and revolutionary attitudes prevalent among the Russian intelligentsia. The national educational system was headed by conservative bureaucrats who encouraged a combination of piety and mysticism at the universities. In the autumn of 1820 Osipovsky was suspended after having been rector of Kharkov University for a number of years. The animosity felt toward him was extended to Ostrogradsky, his best and favorite pupil, who, according to his own account later, was at that time a complete materialist and atheist. The ground for the refusal to grant him a diploma was that, under the influence of Osipovsky, he and the other students of mathematics did not attend lectures on philosophy and theology.

Ostrogradsky continued his mathematical studies in Paris, where Laplace and Fourier, Legendre and Poisson, Binet and Cauchy worked, and where outstanding courses were offered at the École Polytechnique and other educational institutions. Ostrogradsky's rapid progress gained him the friendship and respect of the senior French mathematicians and of his contemporaries, including Sturm. The Paris period of his life (1822–1827) was for Ostrogradsky not only "years of traveling and apprenticeship" but also a period of intense creative work. Between 1824 and 1827 he presented to the Paris Academy several papers containing important new discoveries in mathematical physics and integral calculus. Most of these discoveries were incorporated in his later papers; a memoir on hydrodynamics was published by the Paris Academy in 1832, and individual results in residue theory appeared, with his approval, in the works of Cauchy.

In the spring of 1828 Ostrogradsky arrived in St. Petersburg. There, over a period of several months, he presented three papers to the Academy of Sciences. In the first, on potential theory, he gave a new, more exact derivation of Poisson's equation for the case of a point lying within or on the surface of an attracting mass. The second was on heat theory, and the third on the theory of double integrals. All three appeared in *Mémoires de l'Académie impériale des sciences de St.-Péterbourg*, 6th ser., **1** (1831). On 29 December 1828 Ostrogradsky was elected a junior academician in the section of applied mathematics. In 1830 he was elected an associate and in 1832 a full academician. His work at the Academy of Sciences restored to it the brilliance in mathematics that it had won in the eighteenth century but had lost in the first quarter of the nineteenth.

Ostrogradsky's activity at the Academy was manifold. He contributed some eighty-odd reports in mathematics and mechanics, delivered public lectures, wrote detailed reviews of papers submitted to the Academy, and participated in the work of commissions on the introduction of the Gregorian calendar and the decimal system of measurement. At the behest of the government he also investigated exterior ballistics problems. Ostrogradsky also devoted a great deal of time to teaching and did much to improve mathematical instruction in Russia. From 1828 he lectured at the Naval Corps (later the Naval Academy); from 1830, at the Institute of Means of Communication; and from 1832, at the General Pedagogical Institute. Later he also lectured at the General Engineering College and at the General Artillery College.

From 1847 Ostrogradsky accomplished a great deal as chief inspector for the teaching of the mathematical sciences in military schools. His textbooks on elementary and higher mathematics include a very interesting course on algebra and an exposition of the theory of numbers. Ostrogradsky's educational views were ahead of their time in many respects, particularly his program for the education of children between the ages of seven and twelve, which is expounded in *Considérations sur l'enseignement* (St. Petersburg–Paris, 1860), written with I. A. Blum.

It was mainly Ostrogradsky who established the conditions for the rise of the St. Petersburg mathematical school organized by Chebyshev, and who was the founder of the Russian school of theoretical mechanics. His direct disciples included I. A. Vyshnegradsky, the creator of the theory of automatic regulation, and N. P. Petrov, the author of the hydrodynamic theory of lubricants. Ostrogradsky's services were greatly appreciated by his contemporaries. He was elected a member of the American Academy of Arts and Sciences in 1834, the Turin Academy of Sciences in 1841, and the Rome Academy of Sciences in 1853; in 1856 he was elected a corresponding member of the Paris Academy of Sciences.

Ostrogradsky's scientific work closely bordered upon the developments originating in the École Polytechnique in applied mathematics and in directly related areas of analysis. In mathematical physics he sought a grandiose synthesis that would embrace hydromechanics, the theory of elasticity, the theory of heat, and the theory of electricity by means of a unique homogeneous method. The realization of this plan was beyond the capacity of one man and beyond the resources of the nineteenth century; it remains uncompleted to date.

Ostrogradsky contributed significantly to the development of the method of separating variables that was so successfully applied by Fourier in his work on the conduction of heat (1822). In "Note sur la théorie de la chaleur," presented in 1828 and published in 1831 (see his *Polnoe sobranie trudov*, I,

62–69), Ostrogradsky was the first to formulate a general schema of the method of solving boundary-value problems, which Fourier and Poisson had applied to the solution of individual problems.

For linear partial differential equations with constant coefficients Ostrogradsky established the orthogonality of the corresponding system of proper functions (eigenfunctions). Auxiliary means of calculation in this determination were Ostrogradsky's theorem for the reduction of certain volume integrals to surface integrals and the general formula for arbitrary conjugate linear differential operators with constant coefficients for a three-dimensional space, generally called Green's theorem. In terms of modern vector analysis Ostrogradsky's theorem states that the volume integral of the divergence of a vector field A taken over any volume v is equal to the surface integral of A taken over the closed surface s surrounding the volume v:

$$\iiint (\nabla \cdot A)\, dv = \iint A d\bar{s}.$$

(Ostrogradsky himself expressed this proposition in terms of ordinary integral calculus.) This theorem is also called Gauss's theorem, Green's theorem, or Riemann's theorem.

Ostrogradsky next applied his general results to the theory of heat, deriving formulas for the coefficients a_k in the expansion of an arbitrary function $f(x, y, z)$

into a series $\sum_{k=0}^{\infty} a_k u_k$ of eigenfunctions $u_k(x, y, z, \theta_k)$

of the corresponding boundary-value problem—a generalized Fourier series. He noted the difficulty connected with investigating the convergence of this type of series expansion and only touched on the problem of the existence of eigenvalues of θ_k; satisfactory solutions to these questions were not found until the turn of the twentieth century, by Poincaré and V. A. Steklov, among others.

A large part of these discoveries was contained in two memoirs presented by Ostrogradsky to the Paris Academy of Sciences in 1826–1827. In the second of these he solved the problem of the conduction of heat in a right prism with an isosceles right triangle as a base; Fourier and Poisson had previously examined the cases of a sphere, a cylinder, and a right rectangular parallelepiped. Lamé mentioned this solution, which was not published during Ostrogradsky's lifetime, in an 1833 paper. General results in the theory of heat analogous to Ostrogradsky's (but without his integral theorem) were also obtained by Lamé and Duhamel, who presented their papers to the Paris Academy of Sciences in 1829 (published in 1833).

At first Ostrogradsky investigated heat conduction in a solid body surrounded by a medium having a constant temperature. In "Deuxième note sur la théorie de la chaleur," presented in 1829 and published in 1831 (see *Polnoe sobranie trudov*, I, 70–72), he reduced this problem to the case when the temperature of the surrounding medium is a given function of the coordinates of space and time. Finally, in "Sur l'équation relative à la propagation de la chaleur dans l'intérieur des liquides," presented in 1836 and published in 1838 (*ibid.*, pp. 75–79), he derived the corresponding differential equation for an uncompressed moving liquid free of internal friction, thereby confirming Fourier's results by more thorough analysis.

At the same time Ostrogradsky studied the theory of elasticity; in this field his work meshed with Poisson's parallel investigations. Starting from the work of Poisson, who was the first to establish precisely the necessary condition of the extremum of a double integral with variable limits (1833), Ostrogradsky obtained important results in the calculus of variations. In "Mémoire sur le calcul des variations des intégrales multiples," presented in 1834 and published in 1838 (*ibid.*, III, 45–64), he derived equations containing the necessary conditions of the extremum of an integral of any multiplicity. To accomplish this he had to develop substantially the theory of multiple integrals. He generalized the integral theorem which he had found earlier, that is, reduced an n-tuple integral from an expression of the divergent type taken over any hypervolume to an $(n-1)$-tuple integral taken over the corresponding boundary hypersurface; derived a formula for the substitution of new variables in an n-tuple integral (independently of Jacobi, who published it in 1834); and described in detail the general method for computing an n-tuple integral by means of n consecutive integrations with respect to each variable.

In "Sur la transformation des variables dans les intégrales multiples," presented in 1836 and published in 1838 (*ibid.*, pp. 109–114), Ostrogradsky was the first to derive in a very modern manner (with a geometrical interpretation) the rule of the substitution of new variables in a double integral; he later extended this method to triple integrals. His work in the calculus of variations was directly related to his work in mechanics.

Ostrogradsky made two important discoveries in the theory of ordinary differential equations. In "Note sur la méthode des approximations successives," presented in 1835 and published in 1838 (*ibid.*, pp. 71–75), he proposed a method of solving nonlinear equations by expanding the unknown quantity into a power series in α, where α is a small parameter, in order to avoid "secular terms" containing the independent variable outside the sign of trigonometric functions. This important idea received further development in the investigations of H. Gylden (1881), Anders Lindstedt (1883), Poincaré, and Lyapunov. In "Note sur les équations différentielles linéaires," presented in 1838 and published in 1839 (*ibid.*, pp. 124–126), Ostrogradsky derived, simultaneously with Liouville, a well-known expression for Wronski's determinant, one of the basic formulas in the theory of differential linear equations.

Ostrogradsky also wrote several papers on the theory of algebraic functions and their integrals (*ibid.*, pp. 13–44, 175–179). The foundation of this theory was laid in 1826 by Abel, whom Ostrogradsky may have met in Paris. From Ostrogradsky's general results there follows the transcendency of a logarithmic function and of the arc tangent. His investigations were parallel to Liouville's work in the same area; they were continued in Russia by Chebyshev and his pupils. In "De l'intégration des fractions rationnelles," presented in 1844 and published in 1845 (*ibid.*, pp. 180–214), Ostrogradsky proposed a method for finding the algebraic part of an integral of a rational function without preliminary expansion of the integrand into the sum of partial fractions. This algebraic (and rational) part is calculated with the aid of rational operations and differentiations. Hermite rediscovered this method in 1872 and included it in his textbook on analysis (1873). It is sometimes called Hermite's method.

In "Mémoire sur les quadratures définies," written in 1839 and published in 1841 (*ibid.*, pp. 127–153), which grew out of his work in ballistics, Ostrogradsky gave a new derivation of the Euler-Maclaurin summation formula with a remainder term in the form in which it is now often presented (Jacobi published an equivalent result in 1834) and applied the general formulas to the approximation calculus of definite integrals. Several articles are devoted to probability theory—for example, one on the sample control of production, presented in 1846 and published in 1848 (*ibid.*, pp. 215–237), and to algebra. In general, however, as a mathematician Ostrogradsky was always an analyst.

Ostrogradsky's memoirs in mechanics can be divided into three areas: the principle of virtual displacements; dynamic differential equations; and the solution of specific problems.

Ostrogradsky's most important investigations in mechanics deal with generalizations of its basic principles and methods. He made a substantial contribution to the development of variational principles. The fundamental "Mémoire sur les équations différentielles relatives au problème des isopérimètres," presented in 1848 and published in 1850 (*ibid.*, II, 139–233), belongs in equal measure to mechanics and the calculus of variations. Because of his mathematical approach Ostrogradsky's investigations significantly deepened the understanding of variational principles.

In the paper just cited Ostrogradsky examined the variational problem in which the integrand depends on an arbitrary number of unknown functions of one independent variable and their derivatives of an arbitrary order and proved that the problem can be reduced to the integration of canonical Hamiltonian equations, which can be viewed as the form into which any equations arising in a variational problem can be transformed. This transformation requires no operation other than differentiation and algebraic operations. The credit for this interpretation of the dynamics problem belongs to Ostrogradsky. He also eased the restrictions on constraints, which had always been considered stationary, and thus significantly generalized the problem. Therefore the variational principle formulated by Hamilton in 1834–1835 might more accurately be called the Hamilton-Ostrogradsky principle. Jacobi also worked in the same direction, but his results were published later (1866).

At the same time Ostrogradsky prepared the important paper "Sur les intégrales des équations générales de la dynamique," also presented in 1848 and published in 1850 (*ibid.*, III, 129–138). In it he showed that even in the more general case, when the constraints and the force function depend on time (this case was not considered by Hamilton and Jacobi), the equations of motion can be transformed into Hamiltonian form. Generally, the development of the classical theory of the integration of canonical equations was carried out by Hamilton, Jacobi, and Ostrogradsky.

Ostrogradsky's results related to the development of the principle of virtual displacements are stated in "Considérations générales sur les moments des forces," presented in 1834 and published in 1838 (*ibid.*, II, 13–28). This paper significantly broadened the sphere of application of the principle of virtual displacements, extending it to the relieving constraints.

In "Mémoire sur les déplacements instantanés des systèmes assujettis à des conditions variables," presented and published in 1838 (*ibid.*, pp. 32–59), and "Sur le principe des vitesses virtuelles et sur la force d'inertie," presented in 1841 and published in 1842 (*ibid.*, pp. 104–109), Ostrogradsky gave a rigorous proof of the formula expressing the principle of virtual displacements for the case of nonstationary constraints.

"Mémoire sur la théorie générale de la percussion," presented in 1854 and published in 1857 (*ibid.*, pp. 234–266), presents Ostrogradsky's investigations of the impact of systems, in which he assumed that the constraints arising at the moment of impact are preserved after the impact. The principle of virtual displacements is extended here to the phenomenon of inelastic impact, and `the basic formula of the analytical theory of impact is derived.

Ostrogradsky also wrote papers containing solutions to particular problems of mechanics that had arisen in the technology of his time. A series of his papers on ballistics deserves special mention: "Note sur le mouvement des projectiles sphériques dans un milieu résistant" and "Mémoire sur le mouvement des projectiles sphériques dans l'air," both presented in 1840 and published in 1841; and "Tables pour faciliter le calcul de la trajectoire que décrit un mobile dans un milieu résistant," presented in 1839 and published in 1841 (*ibid.*, pp. 70–94). In the first two papers Ostrogradsky investigated the motion of the center of gravity and the rotation of a spherical projectile the geometrical center of which does not coincide with the center of gravity; both topics were important for artillery at that time. The third paper contains tables, computed by Ostrogradsky, of the function $\Phi(\theta) = 2 \int d\theta/\sin^3 \theta$, used in ballistics. These papers stimulated the creation of the Russian school of ballistics in the second half of the nineteenth century.

BIBLIOGRAPHY

I. Original Works. Most of Ostrogradsky's papers appeared in French in publications of the St. Petersburg Academy of Sciences. The most complete bibliography of his works and of writings concerning him is by M. G. Novlyanskaya in Ostrogradsky's *Izbrannye trudy* ("Selected Works"), V. I. Smirnov, ed. (Moscow, 1958), 540–581. Other collections of Ostrogradsky's writings are *Polnoe sobranie sochineny* ("Complete Collected Works"), I, pt. 2, *Lektsii po analiticheskoy mekhanike, 1834* ("Lectures on Analytic Mechanics"), and II, *Lektsii algebraicheskogo i transtsendentnogo analiza, 1837* ("Lectures on Algebraic and Transcendental Analysis"; Moscow–Leningrad, 1940–1946), never completed; and *Polnoe sobranie trudov* ("Complete Collected Works"), I. Z. Shtokalo, ed., 3 vols. (Kiev, 1959–1961), which contains commentaries and articles by I. Z. Shtokalo, I. B. Pogrebyssky, E. Y. Remez, Y. D. Sokolov, S. M. Targ, and others but does not include the 1834 and 1837 works above or the two articles that follow; and "Dokazatelstvo odnoy teoremy integralnogo ischislenia" ("Proof of One Theorem in the Integral Calculus") and "Memuar o rasprostranenii tepla vnutri tverdykh tel" ("Memoir on the Conduction of Heat Within Solid Bodies"), in *Istoriko-matematicheskie issledovaniya*, **16** (1965), 49–96, Russian translations of two previously unpublished articles presented to the Paris Academy in 1826–1827, with an introduction by A. P. Youschkevitch.

II. SECONDARY LITERATURE. See Y. L. Geronimus, *Ocherki o rabotakh korifeev russkoy mekhaniki* ("Essays on the Work of the Leading Figures in Russian Mechanics"; Moscow, 1952), 13–57; B. V. Gnedenko and I. B. Pogrebyssky, *Mikhail Vasilievich Ostrogradsky (1801–1862). Zhizn i rabota. Nauchnoe i pedagogicheskoe nasledie* (". . . Life and Work. Scientific and Pedagogical Heritage"; Moscow, 1963), the most complete work on his life and accomplishments; A. T. Grigorian, *Mikhail Vasilievich Ostrogradsky (1801–1862)* (Moscow, 1961); and *Ocherki istorii mekhaniki v Rossii* ("Essays on the History of Mechanics in Russia"; Moscow, 1961), see index; *Istoria otechestvennoy matematiki* ("History of Russian Mathematics"), I. Z. Shtokalo, ed.-in-chief, II (Kiev, 1967), see index; A. I. Kropotov and I. A. Maron, *M. V. Ostrogradsky i ego pedagogicheskoe nasledie* ("Ostrogradsky and His Pedagogical Heritage"; Moscow, 1961); *Mikhail Vasilievich Ostrogradsky. 1862–1962. Pedagogicheskoe nasledie. Dokumenty o zhizni i deyatelnosti* (". . . Pedagogical Heritage. Documents on His Life and Activity"), I. B. Pogrebyssky and A. P. Youschkevitch, eds. (Moscow, 1961), a supp. to *Polnoe sobranie trudov* containing a Russian trans. of Ostrogradsky and Blum's *Considérations sur l'enseignement* (St. Petersburg–Paris, 1860), and Ostrogradsky's "Zapiski integralnogo ischislenia" ("Lectures on Integral Calculus"); E. Y. Remez, "O matematicheskikh rukopisyakh akademika M. V. Ostrogradskogo" ("On the Mathematical Manuscripts of Academician M. V. Ostrogradsky"), in *Istoriko-matematicheskie issledovaniya*, **4** (1951), 9–98; S. P. Timoshenko, *History of Strength of Materials* (New York–Toronto–London, 1953); I. Todhunter, *A History of the Progress of the Calculus of Variations During the Nineteenth Century* (Cambridge, 1861); P. I. Tripolsky, ed., *Mikhail Vasilievich Ostrogradsky. Prazdnovanie stoletia dnya ego rozhdenia* (". . . Celebration of the Centenary of His Birth"; Poltava, 1902), which contains short sketches on his life and scientific and educational activities—of special interest are an article by Lyapunov on his work in mechanics (pp. 115–118) and one by Steklov on Ostrogradsky's paper in mathematical physics (pp. 118–131); A. Youschkevitch, *Michel Ostrogradski et le progrès de la science au XIXe siècle* (Paris, 1967); and *Istoria matematiki v Rossii do 1917 goda* ("History of Mathematics in Russia to 1917"; Moscow, 1968), see index; and N. E. Zhukovsky, "Uchenye trudy M. V. Ostrogradskogo po mekhanike" ("Ostrogradsky's Scientific Works in Mechanics"), in Zhukovsky's *Polnoe sobranie sochineny* ("Complete Collected Works"), VII (Moscow–Leningrad, 1950), 229–246.

A. P. YOUSCHKEVITCH

OSTWALD, CARL WILHELM WOLFGANG (*b.* Riga, Latvia, Russia, 27 May 1883; *d.* Dresden, Germany, 22 November 1943), *colloid chemistry, zoology.*

Ostwald, the second child of Wilhelm Ostwald, was a founder of colloid chemistry. He attended the Realgymnasium in Leipzig and at the age of fifteen composed a scientific work on the cases of the larvae of the caddis fly. After completing his secondary education he studied zoology at Leipzig under Carl Chun. From 1904 to 1906 he was a research assistant to Jacques Loeb at Berkeley, California. There he became friendly with the physiologist and physician M. H. Fischer, with whom he worked on the theory of fertilization. He qualified as a lecturer in biology at Leipzig in 1907 and he became professor of colloid chemistry in 1915. In 1907 he became editor of *Zeitschrift für Chemie und Industrie der Kolloide* and, beginning in 1909, he also edited *Kolloidchemische Beihefte.* Through these journals and through the Kolloid Gesellschaft, founded in 1922 at his suggestion —he was its first president and held that post for two decades—Ostwald organized and encouraged research in colloid chemistry. In 1923 he was appointed director of the colloid chemistry division of the physical-chemical institute at the University of Leipzig. He became a full professor there in 1935 and had a large circle of students.

In his zoological studies Ostwald explained the suspension of plankton and described the process of fertilization as a colloidal phenomenon. He established that there are no sharp differences between mechanical decompositions and colloidal and molecular solutions. He also defined colloids as disperse systems that are generally polyphasic and that possess particles 1–100 millimicrons in size. Ostwald worked on colloid chemistry problems involving, for example, bread and rubber. In addition he discovered the rule of color dispersion in the optics of colloidal systems and explained the irregular flow behavior of colloids, their textural viscosity, and their textural turbulence. He also worked on the law governing precipitation in saturated colloidal solutions, electrolytic coagulation, and other colloidal properties. In addition he developed a method of foam analysis. Through the lectures he gave outside Leipzig, especially in the United States, and through his books Ostwald made an essential contribution to obtaining international recognition of colloid chemistry as an independent field.

BIBLIOGRAPHY

I. ORIGINAL WORKS. Ostwald's approximately 200 scientific papers include "Kolloidwissenschaft, Elektrotechnik und heterogene Katalyse," in *Kolloidchemische Beihefte*, **32** (1930), 1–48; "Über mesomorphe und kolloide Systeme," in *Zeitschrift für Kristallographie . . .*, **79** (1932), 222–254; "Über Osmose und Solvation disperser Systeme," in *Zeitschrift für physikalische Chemie*, **159A** (1932), 375–392; "Elektrolytkoagulation und Elektrolytaktivitätskoeffizient," in *Kolloidzeitschrift*, **73** (1935)–**87** (1939), and **94** (1941), 169–184 (12 papers on this topic); "Metastrukturen der Materie," in *Kolloidchemische Beihefte*, **42** (1935), 109–124; "Über die andere geschichtliche Wurzel der Kolloidwissenschaft," in *Kolloid-Zeitschrift*, **84** (1938), 258–265; and "Physikalisch-chemische Metastasen," *ibid.*, **100** (1942), 2–57.

Ostwald's books include *Neue theoretische Betrachtungsweise in der Planktologie* (Stuttgart, 1903); *Grundriss der Kolloidchemie* (Leipzig, 1909); *Die Welt der vernachlässigten Dimensionen* (Leipzig, 1914), which consists of lectures given in the United States; *Praktikum der Kolloidchemie* (Leipzig, 1920); and *Licht und Farbe in Kolloiden* (Dresden–Leipzig, 1924).

Bibliographies of his works are in: Poggendorff, IV, 1103; V, 930–931; VI, 1929–1931; and VIIa, pt. 3, 484–486; and in the biographical article by Lottermoser.

II. Secondary Literature. See the following, listed chronologically: A. Lottermoser, "Wolfgang Ostwald 60 Jahre alt," in *Kolloidzeitschrift*, **103**, no. 2 (1943), 89–94 (bibliography 91–94); G. F. Hüttig, "Wolfgang Ostwald," in *Forschungen und Fortschritte*, **20** (1944), 118–119; and other obituaries by R. E. Oesper, in *Journal of Chemical Education*, **22** (1945), 263; by H. Ebring, in *Kolloidzeitschrift*, **115** (1949), 3–5; by E. A. Hauser, in *Journal of Chemical Education*, **32** (1955), 2–9; and by M. H. Fischer, "Wolfgang Ostwalds Weg zur Kolloidchemie," in *Kolloidzeitschrift*, **145** (1956), 1–2.

Hans-Günther Körber

OSTWALD, FRIEDRICH WILHELM (*b.* Riga, Latvia, Russia [now Latvian S.S.R.], 2 September 1853; *d.* Leipzig, Germany, 4 April 1932), *chemistry, color science.*

For a detailed study of his life and work, see Supplement.

OTT, ISAAC (*b.* Northampton County, Pennsylvania, 30 November 1847; *d.* Easton, Pennsylvania, 1 January 1916), *physiology.*

Discoverer of the heat-regulating center of the brain in 1887, Ott received the B.A. and M.A. from Lafayette College and the M.D. from the University of Pennsylvania in 1869, with a dissertation on typhoid fever. Following an internship at St. Mary's Hospital in Philadelphia, he did postgraduate study at Leipzig, Würzburg, and Berlin. In 1873 he was appointed demonstrator of experimental physiology

at the University of Pennsylvania, where he organized a physiological laboratory and lectured on physiology until 1878. He became a fellow in biology at the Johns Hopkins University in 1879, lecturing the same year in physiology at the Medico-Chirurgical College of Philadelphia while it was still a society. In 1894 Ott was appointed professor of physiology at the Medico-Chirurgical College, filling the chair until the college merged with the University of Pennsylvania about the time of his death in 1916. He served as dean in 1895, and each year he selected five of his most promising students as members of the American Physiological Society.

In 1876 Ott settled at Easton. He wrote more than fifty scientific papers, the last one in 1910 on internal secretions, and wrote the book *The Action of Medicines* (1878).

Ott performed experiments demonstrating that there are areas in the brain which exert considerable control over the body temperature and pinpointed the center for temperature regulation in the region of the corpora striata. From his pioneering work in neurophysiological technique have come a multitude of studies. He also devoted considerable study to the physiological action of drugs and discovered the path and decussation of the sudorific, sphincter-inhibitory and thermo-inhibitory fibers in the spinal cord and the innervation of the sphincters.

He served as president of the American Neurological Association. In his opening address as president of the Section on Physiology at the first Pan American Medical Congress in 1895, he reviewed work in physiology in the United States and noted that research required special commitment because it was exhausting financially as well as physically.

BIBLIOGRAPHY

I. Original Works. Ott's article "The Relation of the Nervous System to the Temperature of the Body," in *Journal of Nervous and Mental Diseases*, **11** (1884), 141–152, is item 1416 in F. H. Garrison and L. T. Morton, *A Medical Bibliography* (London, 1943), with a statement that Ott wrote important papers on the nervous regulation of body temperature. His papers on the heat center in the brain and on the thermo-inhibitory apparatus were published in the same journal, **14** (1887), 150–162, 428–438; and **15** (1888), 85–104. His book *Fever: Its Thermotaxis and Metabolism* (New York, 1914) is listed in Garrison and Morton as item 2115. He also published *Cocaine, Veratria and Gelsemium: Toxicological Studies* (Philadelphia, 1874). His works on lobelia, thebaine, lycotomia, poisonous mushrooms, ethyl bromide, Jamaica dogwood, loco weed, lily of the valley, rattlesnake venom, copperhead snake venom, absinthism epilepsy, antipyretics, heroin, and adrenalin are listed in the *Surgeon General's Catalogue.* Further writings are *Textbook of Physiology* (Philadelphia, 1904; 2nd ed., 1907; 3rd ed., 1909; 4th ed., 1913); *The Parathyroid Glandules From a Physiological and Pathological Standpoint* (Philadelphia, 1910); *Internal Secretions From a Physiological and Therapeutical Standpoint* (Easton, Pa., 1910); and *Contributions From the Physiological Laboratory of the Medico-Chirurgical College of Philadelphia* (Philladelphia, 1914), written with John C. Scott.

II. SECONDARY LITERATURE. A good biographical sketch is presented in Howard A. Kelly and Walter L. Burrage, eds., *American Medical Biographies* (Baltimore, 1920), 869. There is a contemporary sketch by W. B. Atkinson in *Physicians and Surgeons of the United States* (Philadelphia, 1878), 172–173. Ott's photograph is reproduced on plate 69, facing p. 343, accompanied by a sketch and an excerpt from his classic article "The Heat-Center in the Brain" (in *Journal of Nervous and Mental Diseases*, **14** [1887], 150–162), in John F. Fulton and Leonard G. Wilson, *Selected Readings in the History of Physiology*, 2nd ed. (Springfield, Ill., 1966), 337. There is a biographical sketch in *Appleton's Cyclopaedia of American Biography*, IV (1888), 608; and obituaries in *Journal of the American Medical Association*, **26** (1916), 206; by Joseph McFarland, in *Journal of Nervous and Mental Diseases*, **43** (1916), 201; in *Medical Record*, **89** (1916), 72; and in *New York Medical Journal*, **103** (1916), 80. G. Clark, Magoun, and Ranson refer to his work in "Hypothalamic Regulation of Body Temperature," in *Journal of Neurophysiology*, **2** (1939), 61–80.

SAMUEL X. RADBILL

OUDEMANS, CORNEILLE ANTOINE JEAN ABRAM (*b.* Amsterdam, Netherlands, 7 December 1825; *d.* Arnhem, Netherlands, 29 August 1906), *medicine, botany, mycology.*

Oudemans was the son of Anthonie Cornelis Oudemans, an educator, and Jacoba Adriana Hammecker. Two of their other children became prominent scientists: Jean Abraham Crétien Oudemans, an astronomer, and Antoine Corneille Oudemans, a chemist. Oudemans received his elementary education in Weltevreden, Java, where his father was the principal of a grammar school; at the age of fourteen he was sent back to the Netherlands to study Latin and Greek in preparation for admission to a university. Two years later he became a medical student at Leiden, where he was granted the M.D. on 5 November 1847. A subsequent study trip to Paris and Vienna was cut short by the March Revolution of 1848. Soon after his return on 9 August 1848, Oudemans was appointed lecturer in botany, materia medica, and natural history at the clinical school of Rotterdam, where he also set up a practice. While in Rotterdam he was very active in the field of public health and also published the results of his pharmacological investigations.

In 1859 Oudemans was offered the chair of medicine and botany at the Athenaeum of Amsterdam, vacant after Miquel moved to the University of Utrecht. He gave his inaugural lecture on 21 November. When the Athenaeum obtained university status in 1877, Oudemans became its first *rector magnificus.* In the same year his teaching duty was reduced to systematic botany and pharmacognosy; Hugo de Vries was appointed lecturer in plant physiology and anatomy. After his retirement in 1896 he settled in Arnhem.

While at Amsterdam, Oudemans became increasingly interested in the fungi of the Netherlands, a subject on which he became the national expert. His *Révision des champignons* (1892–1897) and *Catalogue raisonné* (1904) are still standard works on Dutch mycology, as is his posthumously published *Enumeratio systematica fungorum*, on which he worked for twenty-five years. In this book he described all the known European parasitic fungi. The work was published under the supervision of J. W. Moll, professor of botany at Groningen, to which university Oudemans left his collection of parasitic fungi.

BIBLIOGRAPHY

I. ORIGINAL WORKS. The library of the University of Amsterdam has the following MS notes by Oudemans: "Hebra's Klinik über Hautkrankheiten. Angefangen 20 März (1848). Aufgeschrieben in einem Privatkurs von C. A. J. A. Oudemans."

His earliest published works are *De fluxu menstruo* (Leiden, 1847), his dissertation; *Algemeen verslag der subcommissie voor den Aziatischen braakloop, geheerscht hebbende te Rotterdam* (Rotterdam, 1849); *Systematisch overzicht der geneeskundige gewassen* (Rotterdam, 1851); *Aanteekeningen op het systematisch- en pharmacognostisch botanische gedeelte der Pharmacopoea Neerlandica*, 2 vols. (Rotterdam, 1854–1856); *Bijdrage tot de kennis van de morphologische en anatomische structuur van de vrucht en het zaad des kamferbooms (Dryobalanops camphora, Colebr.) van Sumatra* (Rotterdam, 1855); *Brief van de openbare gezondheidscommissie te Rotterdam omtrent het planten van boomen aldaar* (n.p., 1855); *Flora van Nederland*, 3 vols. and atlas (Haarlem, 1859–1862; 2nd ed., Amsterdam, 1872–1874); and *Over de plantkunde, beschouwd in hare trapsgewijze ontwikkeling van de vroegste tijden* (Amsterdam, 1859), his inaugural lecture.

Writings from the 1860's and 1870's are *Brief over de hervorming en uitbreiding van het natuur- en geneeskundig onderwijs aan het Athenaeum Illustre te Amsterdam . . .* (Amsterdam, 1860), written with C. E. V. Schneevoogt; *Ueber den Sitz der Oberhaut bei den Luftwurzeln der Orchideen* (Amsterdam, 1861); *Annotationes criticae in Cupuliferas nonnulas Javanicas* (Amsterdam, 1865); *Hand-*

leiding tot de pharmacognosie van het planten- en dierenrijk (Haarlem, 1865); *Leerboek der plantenkunde*, 2 vols. (Utrecht–Amsterdam, 1866–1870); *Eerste beginselen der plantenkunde* (Amsterdam–Rotterdam–Utrecht, 1868); and *Rede ter herdenking van den sterfdag van Carolus Linnaeus, eene eeuw na diens verscheiden* (Amsterdam, 1878).

His latest works were *Leerboek der plantenkunde, ten gebruike bij het hooger onderwijs,* I, *Vormleer en rang-schikking der planten* (Zaltbommel, 1883; 2nd ed., Nijmegen, 1896)—vols. II and III written by Hugo de Vries; *Revisio pyrenomycetum in regno Batavorum, hujusque detectorum* (Amsterdam, 1884); *Révision des champignons, tout supérieurs qu'inférieurs, trouvés jusqu'à ce jour dans les Pays-Bas,* 2 vols. (Amsterdam, 1892–1897); *Beteekenis der geslachtsnamen van de phanerogamen en de vaat kryptogamen* (Bussum, 1899); *Catalogue raisonné des champignons des Pays-Bas* (Amsterdam, 1904); and *Enumeratio systematica fungorum,* J. W. Moll, R. de Boer, and L. Vuyck, eds., 5 vols. (The Hague, 1919–1924).

In addition Oudemans wrote a large number of papers, a list of more than eighty is in the obituary by J. W. Moll and in the Royal Society *Catalogue of Scientific Papers,* IV, 715; VIII, 543–544; X, 970–971; XII, 552; and XVII, 657.

II. SECONDARY LITERATURE. See P. J. Lotsy, "Corneille Antoine Jean Abram Oudemans," in *Nieuw nederlandsch biografisch woordenboek,* I (Leiden, 1911), 1396–1397; J. W. Moll, "C. A. J. A. Oudemans," in *Jaarboek van de K. Akademie van wetenschappen . . . Amsterdam,* **62** (1909), 57–105, with bibliography; W. F. R. Suringar, "C. A. J. A. Oudemans," in *Eigen Haard,* **21** (1895), 773–775; and J. S. Theissen, "Corneille Antoine Jean Abram Oudemans," in *Gedenkboek van het Athenaeum en de Universiteit van Amsterdam, 1632–1932,* I (Amsterdam, 1932), 649–650.

PETER W. VAN DER PAS

OUGHTRED, WILLIAM (*b.* Eton, Buckinghamshire, England, 5 March 1575; *d.* Albury, near Guildford, Surrey, England, 30 June 1660), *mathematics.*

Oughtred's father was a scrivener who taught writing at Eton and instructed his young son in arithmetic. Oughtred was educated as a king's scholar at Eton, from which he proceeded to King's College, Cambridge, at the age of fifteen. He became a fellow of his college in 1595, graduated B.A. in 1596, and was awarded the M.A. in 1600.

Ordained a priest in 1603, Oughtred at once began his ecclesiastical duties, being presented with the living of Shalford, Surrey. Five years later he became rector of Albury and retained this post until his death. Despite his parochial duties he continued to devote considerable time to mathematics, and in 1628 he was called upon to instruct Lord William Howard, the young son of the earl of Arundel. In carrying out this task he prepared a treatise on arithmetic and algebra.

This slight volume, of barely 100 pages, contained almost all that was then known of these two branches of mathematics; it was published in 1631 as *Clavis mathematicae.*

Oughtred's best-remembered work, the *Clavis* exerted considerable influence in England and on the Continent and immediately established him as a capable mathematician. Both Boyle and Newton held a very high opinion of the work. In a letter to Nathaniel Hawes, treasurer of Christ's Hospital, dated 25 May 1694 and entitled "A New Scheme of Learning for the Mathematical Boys at Christ's Hospital," Newton referred to Oughtred as "a man whose judgment (if any man's) may be relyed on." In Lord King's *Life of Locke* we read: "The best Algebra yet extant is Oughtred's" (I, 227). John Aubrey, in *Brief Lives,* maintained that Oughtred was more famous abroad for his learning than at home and that several great men came to England for the purpose of meeting him (II, 471).

John Wallis dedicated his *Arithmetica infinitorum* (1655) to Oughtred. A pupil of Oughtred, Wallis never wearied of sounding his praises. In his *Algebra* (1695) he wrote: "The *Clavis* doth in as little room deliver as much of the fundamental and useful parts of geometry (as well as of arithmetic and algebra) as any book I know," and in its preface he classed Oughtred with the English mathematician Thomas Harriot.

The *Clavis* is not easy reading. The style is very obscure, and rules are so involved as to make them difficult to follow. Oughtred carried symbolism to excess, using signs to denote quantities, their powers, and the fundamental operations in arithmetic and algebra. Chief among these were X for multiplication, ⊐ for "greater than"; ⊏ for "less than"; and ∼ for "difference between." Ratio was denoted by a dot; proportion, by ::. Thus the proportion $A : B = \alpha : \beta$ was written $A \cdot B :: \alpha \cdot \beta$. Continued proportion was written $\div$. Of the maze of symbols employed by Oughtred, only those for multiplication and proportion are still used. Yet, surprisingly, there is a complete absence of indices or exponents from his work. Even in later editions of the *Clavis,* Oughtred used *Aq, Ac, Aqq, Aqc, Acc, Aqqc, Aqcc, Accc, Aqqcc,* to denote successive powers of *A* up to the tenth. In his *Géométrie* (1637) Descartes had introduced the notation x^n but restricted its use to cases in which *n* was a positive whole number. Newton extended this notation to include fractional and negative indices. These first appeared in a letter to Oldenburg for transmission to Leibniz—the famous *Epistola prior* of June 1676—in which Newton illustrated the newly discovered binomial theorem.

In *La disme*, a short tract published in 1585, Simon Stevin had outlined the principles of decimal fractions. Although a warm admirer of Stevin's work, Oughtred avoided his clumsy notation and substituted his own, which, although an improvement, was far from satisfactory. He did not use the dot to separate the decimal from the whole number, undoubtedly because he already used it to denote ratio; instead, he wrote a decimal such as 0.56 as 0|56.

Oughtred is generally regarded as the inventor of the circular and rectilinear slide rules. Although the former is described in his *Circles of Proportion and the Horizontal Instrument* (1632), a description of the instrument had been published two years earlier by one of his pupils, Richard Delamain, in *Grammelogie, or the Mathematical Ring*. A bitter quarrel ensued between the two, each claiming priority in the invention. There seems to be no very good reason why each should not be credited as an independent inventor. Oughtred's claim to priority in the invention of the rectilinear slide rule, however, is beyond dispute, since it is known that he had designed the instrument as early as 1621.

In 1657 Oughtred published *Trigonometria*, a work of thirty-six pages dealing with both plane and spherical triangles. Oughtred made free use of the abbreviations *s* for sine, *t* for tangent, *se* for secant, *sco* for sine of the complement (or cosine), *tco* for cotangent, and *seco* for cosecant. The work also contains tables of sines, tangents, and secants to seven decimal places as well as tables of logarithms, also to seven places.

It is said that Oughtred, a staunch royalist, died in a transport of joy on hearing the news of the restoration of Charles II.

BIBLIOGRAPHY

I. ORIGINAL WORKS. Oughtred's chief writing is *Arithmeticae in numeris et speciebus institutio . . . quasi clavis mathematicae est* (London, 1631); 2nd ed., *Clavis mathematicae* (London, 1648). English translations were made by Robert Wood (1647) and Edmond Halley (1694). Subsequent Latin eds. appeared at Oxford in 1652, 1667, and 1693.

His other works are *The Circles of Proportion and the Horizontal Instrument*, W. Forster, trans. (London, 1632), a treatise on navigation; *The Description and Use of the Double Horizontal Dial* (London, 1636); *A Most Easy Way for the Delineation of Plain Sundials, Only by Geometry* (1647); *The Solution of All Spherical Triangles* (Oxford, 1651); *Description and Use of the General Horological Ring and the Double Horizontal Dial* (London, 1653); *Trigonometria* (London, 1657), trans. by R. Stokes as *Trigonometrie* (London, 1657); and *Canones sinuum, tangentium, secantium et logarithmorum* (London, 1657).

A collection of Oughtred's papers, mainly on mathematical subjects, was published posthumously under the direction of Charles Scarborough as *Opuscula mathematica hactenus inedita* (Oxford, 1677).

II. SECONDARY LITERATURE. On Oughtred or his work, see John Aubrey, *Brief Lives*, Andrew Clark, ed. (Oxford, 1898), II, 106, 113–114, 471. W. W. R. Ball, *A History of the Study of Mathematics at Cambridge* (Cambridge, 1889); Florian Cajori, *William Oughtred, a Great Seventeenth-Century Teacher of Mathematics* (Chicago–London, 1916); Moritz Cantor, *Vorlesungen über Geschichte der Mathematik*, 2nd ed., II (Leipzig, 1913), 720–721; Charles Hutton, *Philosophical and Mathematical Dictionary*, new ed. (London, 1815), II, 141–142; and S. J. Rigaud, ed., *Correspondence of Scientific Men of the Seventeenth Century*, I (Oxford, 1841), 11, 16, 66.

J. F. SCOTT

OUTHIER, RÉGINALD (*b.* La Marre-Jousserans, near Poligny, France, 16 August 1694; *d.* Bayeux, France, 12 April 1774), *astronomy, cartography.*

Outhier, for many years canon of the cathedral of Bayeux, was one of the many provincial amateur scientists who supplied the academicians in Paris with somewhat raw observations, which they, in turn, used in order to support their more general theories and treatises. His scientific observations covered astronomy, meteorology, and cartography, both terrestrial and celestial.

Outhier's scientific communications began in 1727, when he presented a celestial globe of his own invention to the Académie Royale des Sciences. In addition to the positions of the stars, this globe, moved by clockwork, indicated the apparent path of the sun along the ecliptic and various motions of the moon. On 1 December 1731 Outhier was named correspondent of Jacques Cassini and, twenty-five years later, correspondent of Cassini de Thury.

In preparation for an exact map of France, Cassini in 1733 drew a line perpendicular to the meridian of Paris westward from Paris to the sea. Outhier, then secretary to Paul d'Albert de Luynes, the very scientific bishop of Bayeux, joined the surveying party from Caen to St.-Malo. After the triangulation was accomplished, the party went to Bayeux to make some celestial observations. Cassini was impressed by the large sundial with lines at five-minute intervals that Outhier had traced on the cathedral library. Around this time Outhier drew a map of the diocese of Bayeux, published in 1736, and others of the bishopric of Meaux and of the archbishopric of Sens.

In 1736–1737 the Academy sponsored an expedition to Lapland to measure the length of a degree of latitude near the North Pole, in order to determine the actual figure of the earth; and Outhier, "dont la capacité dans l'ouvrage que nous allions faire, etoit connuë..." (Maupertuis, *La figure de la terre*, p. xv), was invited to participate. He assisted in the astronomical observations, drew eighteen maps of the lands through which they passed, and studied the religious and social customs of the Lapps. His detailed journal of the voyage was published in 1744.

In 1752 Outhier drew, and presented to the Academy, a map of the Pleiades that was by far the most accurate map of the region. It included ninety-nine stars of the third through the tenth magnitudes, thirty-five of which had been measured by Le Monnier; coordinates were given for every ten minutes of celestial latitude and longitude and every twenty minutes of right ascension and declination. Other reports to the Academy concerned the weather at Bayeux, the transit of Venus of 1761, six lunar eclipses, and two solar eclipses.

BIBLIOGRAPHY

I. ORIGINAL WORKS. Outhier's account of his journey to Lapland is *Journal d'un voyage au nord, en 1736 et 1737* (Paris, 1744; repub. Amsterdam, 1746), English trans. in John Pinkerton, ed., *A General Collection of the Best and Most Interesting Voyages and Travels*, I (London, 1808), 259–336. Two of his maps appeared as *Carte topographique du diocèse de Bayeux, divisé en ses quatre archidiaconés et ses dix sept doyenés . . . par l'Abbé Outhier*, 2 sheets (1736); and *Cartes de l'évêché de Meaux et de l'archévêché de Sens*.

His earlier articles include "Globe mouvant inventé par M. l'Abbé Outhier, prestre," in *Machines et inventions approuvées par l'Académie royale des sciences*, V (Paris, 1735), 15–17; "Le mesme globe perfectionné et presenté en MDCCXXXI," *ibid.*, pp. 19–20; "Addition au globe mouvant, par M. l'Abbé Outhier," *ibid.*, 21–22; "Observations de l'éclipse de Jupiter & de ses satellites par la lune, faites à Sommervieux près de Bayeux par M. l'Évêque de Bayeux le 17 juin 1744, par M. Cassini," in *Mémoires de l'Académie royale des sciences . . .* (1744), 415–416; "Extrait des observations de l'éclipse de lune, faites à Bayeux le 2 novembre 1743 au matin, & communiquées à l'Académie, par M. le Monnier fils," *ibid.* (1745), 511; "Observation de l'éclipse du soleil, du 25 juillet 1748, faite à Bayeux par M. l'Abbé Outhier," in *Mémoires . . . présentés par divers sçavans*, 2 (1755), 307–308; "Observation de l'éclipse de lune, du 8 août 1748, faite à Bayeux, dans l'évêché par M. l'Abbé Outhier," *ibid.*, pp. 309–310; "Observation de l'éclipse de lune, du 23 décembre 1749, faite à Bayeux, par M. l'Abbé Outhier, correspondent de l'Académie," *ibid.*, pp. 311–312; "Observation de l'éclipse du soleil du 8 janvier 1750, faite à Bayeux, par M. l'Abbé Outhier," *ibid.*, pp. 313–314; "Sur une nouvelle quadrature par approximation, par M. l'Abbé Outhier . . .," *ibid.*, p. 333; and "Cartes des Pléyades . . .," *ibid.*, pp. 607–608 and pl. XXV; "Observations météorologiques faites à Bayeux en 1756," *ibid.*, 4 (1763), 612–613; "Autre observation du passage de Vénus, faite à Bayeux le 6 juin 1761, avec une lunette de 34 pouces garnis d'un micromètre dont chaque tour de vis est divisé en 42 parties," *ibid.*, 6 (1764), 133–134; "Observation de l'éclipse de lune, faite à Bayeux le 18 mai 1761," *ibid.*, p. 134; and "Observation de l'éclipse de lune du 8 mai 1762, au matin, faite à Bayeux," *ibid.*, p. 176.

II. SECONDARY LITERATURE. See C. F. Cassini de Thury, "De la carte de la France et de la perpendiculaire à la méridienne de Paris," in *Mémoires de l'Académie royale des sciences . . .* (1733), 389–405; H. F., "Outhier (Réginald ou Regnauld)," in *Nouvelle biographie générale*, XXXVIII (Paris, 1864), cols. 982–983; and P. Maupertuis, *La figure de la terre* (Paris, 1738).

DEBORAH JEAN WARNER

OVERTON, CHARLES ERNEST (*b.* Stretton, Cheshire, England, 25 February 1865; *d.* Lund, Sweden, 27 January 1933), *cell physiology, pharmacology.*

Overton was the son of the Reverend Samuel Charlesworth Overton and Harriet Jane Fox, daughter of the Reverend W. Darwin Fox, a second cousin of Charles Darwin. He was educated at Newport Grammar School until 1882, when his mother, for health reasons, moved with her children to Switzerland. He studied biology, especially botany, at the University of Zurich, where in 1889 he obtained the Ph.D. and, in 1890, was appointed *Dozent* in biology. From Zurich, Overton moved to the University of Würzburg in 1901 as assistant to Max von Frey in the physiology department. In 1907 he accepted the chair of pharmacology at the University of Lund, where he remained until his retirement in 1930. In 1912 he married Dr. Louise Petrén. Overton published his most important papers between about 1893 and 1902. His productivity subsequently decreased considerably, owing to impaired health.

As early as 1890–1893, before finding his final field of research, Overton had done pioneering work in plant cytology, in which he showed that the haploid chromosome number is characteristic not only of the sex cells themselves but also of the whole gametophyte.

At about this time Overton became interested in the fundamental problem of how living cells, isolated from their surroundings so that the solutes in the sap are prevented from diffusing out, are nevertheless able to take up nutrients from without and to throw off the waste products of their metabolism. In the 1890's living cells were commonly thought to be virtually impermeable to the great majority of solutes but readily permeable to water. Overton, however, observed that there is a whole series of intermediate cases between substances totally unable to penetrate living protoplasts and those that do so as rapidly as water. Moreover, he found that all the widely different kinds of plant and animal cells are surprisingly similar in their permeability properties. In 1899 Overton pointed out a striking parallel between the permeating powers of different substances and their relative fat solubility—that is, their partition coefficient in a system composed of fat and water. The smaller this coefficient, the more difficult the passage of the substance through the protoplast. This was at first sight a very surprising result, but Overton explained it by assuming that the invisible plasma membranes, already theoretically postulated by Pfeffer, are "impregnated" with fatlike substances, such as cholesterol or phosphatides.

This hypothesis, now universally known as Overton's lipoid (or lipide) theory of plasma permeability, was first published in a preliminary form, his intention being to present the detailed basis for it in a later extensive publication. The larger work containing definite proof of the theory was, however, never finished. Thus, it is understandable that, although the theory aroused a great deal of interest, it also met with doubt and even violent opposition, especially since Overton never replied to the attacks on his views. Apart from minor modifications, however, later experiments have confirmed his results.

In 1896 Overton pointed out that both plant and animal cells can transport solutes against the concentration gradient. Such an active transport carried out at the expense of energy set free by metabolic processes is a phenomenon quite different from the simple diffusion of substances through the protoplasts. Active transport, as Overton anticipated, has proved to be of fundamental importance to living cells.

In carrying out permeability experiments with muscle cells, Overton found that their irritability is reversibly lost when the sodium ions that are normally present between them diffuse out from the muscles. To explain this and other related observations, he tentatively proposed the hypothesis that for an extremely short interval the surface of the contracting muscle fibers becomes permeable to sodium and potassium ions. This fundamental idea in the theory of propagation of impulses in nerves and muscles was worked out almost fifty years later by A. L. Hodgkin and A. F. Huxley, for which they were awarded the Nobel Prize in physiology or medicine in 1963.

In studying the permeability properties of plant and animal cells, Overton observed that those substances which, owing to their great lipide solubility, penetrate the protoplasts most rapidly also have the ability to produce narcosis. It was only natural that he assumed their narcotizing effect to be in some way dependent on their lipide solubility. Almost simultaneously with Overton but independently of him, the pharmacologist Hans Horst Meyer reached much the same conclusion. Although the Meyer-Overton theory does not offer a complete explanation of the mechanism of narcosis, it remains an important starting point for newer, more elaborate theories of this phenomenon.

A gentle and placid man, Overton had a striking intuitive ability to recognize the great, fundamental problems and to envision a means of solving them without recourse to complicated apparatus. He never founded a school in the proper sense of the word, and his publications, almost all of which were written in German, do not seem to have been widely read in the original, especially in English-speaking countries. Nevertheless, his influence on the development of cell physiology and pharmacology has been strong and long-lasting. He was one of those scientists whose stature is more obvious after their death than it was during their lifetime.

BIBLIOGRAPHY

Overton's most important publications are "On the Reduction of the Chromosomes in the Nuclei of Plants," in *Annals of Botany*, **7** (1893), 139–143; "Über die allgemeinen osmotischen Eigenschaften der Zelle, ihre vermutlichen Ursachen und ihre Bedeutung für die Physiologie," in *Vierteljahrsschrift der Naturforschenden Gesellschaft in Zürich*, **44** (1899), 88–135; *Studien über die Narkose* (Jena, 1901); "Beiträge zur allgemeinen Muskel- und Nervenphysiologie," in *Pflügers Archiv für die gesamte Physiologie*, **92** (1902), 346–386; and "Über den Mechanismus der Resorption und Sekretion," in W. Nagel, ed., *Handbuch der Physiologie des Menschen*, II (Brunswick, 1907), 744–898.

For a more complete biography and bibliography, see P. R. Collander, "Ernest Overton (1865–1933), a Pioneer to Remember," in *Leopoldina*, 3rd ser., **8–9** (1962–1963), 242–254.

RUNAR COLLANDER

OWEN, DAVID DALE (*b.* New Lanark, Scotland, 24 June 1807; *d.* New Harmony, Indiana, 13 November 1860), *geology.*

Owen was the son of Robert Owen, the utopian philanthropist and progressive mill owner, and Anne

Caroline Dale Owen. He was educated at home in the classics, mechanics, and architectural drawing and, from the age of seventeen to twenty, at P. E. von Fellenberg's "progressive school" in Hofwyl, near Bern, Switzerland, at which he studied the classics, music, drawing, chemistry, and natural history. He then spent a year in Glasgow, studying principally chemistry under Andrew Ure at the Andersonian Institution. In 1825 Owen's father and William Maclure purchased the village, factories, and lands of New Harmony, Indiana, from George Rapp. In 1828, Owen came to New Harmony, which remained his home for the rest of his life. With Henry D. Rogers he went back to London in 1831 to study chemistry, then returned to New Harmony in 1833 and studied medicine in Cincinnati at various times from 1835 to 1837. He graduated in 1837 and used the title of doctor, although he never practiced. His medical training was to gain more scientific background, especially for his developing interest in geology. He spent part of the summer of 1836 as assistant to Gerard Troost, state geologist of Tennessee. He married Caroline Neef in 1837.

When the Indiana Geological Survey was established in 1837, Owen was immediately appointed state geologist. Always mindful of the practical application of science, he made a regional survey to determine the major rock divisions, the limits of the coal-bearing rocks, the iron ore deposits, and building stones. He was the first American to use the term "Carboniferous" in the present restricted sense. He also recognized the Cincinnati arch just east of Indiana, a structural axis that controlled the westward-dipping strata in Indiana. Owen's Indiana reports led to his appointment in 1839 to explore the United States mineral lands of the Dubuque lead district in southwestern Wisconsin, southeastern Iowa, and northeastern Illinois, then the most important lead mining area in the country. Aided by John Locke and a corps of 139 assistants, in two months Owen covered 11,000 square miles; he presented maps in February 1840 and a report in June 1840. The expertly colored geologic maps and sections, the sketches of topographic features, and the lithographic fossil plates, all by Owen, added greatly to the value of the report.

In 1847 Owen was appointed to survey the mineral lands of the Chippewa land district, an area extending from northeastern Iowa and southern Wisconsin to Lake Superior. In his report, presented the following year, he correctly analyzed the stratigraphy and structure and paid particular attention to the economic geology. Owen and his assistants continued explorations into Minnesota and to Lake Winnipeg,

then into Iowa and the South Dakota Badlands. The resulting report (1852) included an atlas of maps and plates engraved from sketches by Owen and his brother Richard. The most sumptuous American geological publication to that time, it is still of great significance.

Appointed state geologist of Kentucky in 1854, Owen not only made detailed geologic, chemical, economic, and soil studies but also constructed base maps. His medical training enabled him to relate certain diseases to soil and mineral types. After Owen's part in the Kentucky fieldwork was completed and the third volume of the surveys was published, he accepted an appointment as state geologist of Arkansas in 1857, at very little salary, for the opportunity to examine unknown territory.

In 1859 the Indiana Geological Survey was reactivated and Owen was appointed state geologist, with the understanding that his brother Richard, who had long been associated with him, could begin the fieldwork while Owen was completing the second volume of the Arkansas survey. But Owen, who had long suffered from recurrent malaria, was seized by other ailments, including acute rheumatism, and soon became practically immobilized. Nevertheless, he dictated the last of the Arkansas report to two secretaries and completed the work three days before he died.

A superb field geologist, Owen attracted and retained capable assistants who could lead his field parties and could also contribute important parts of the final reports. In a day when verbosity was not uncommon he wrote in a lucid, well-outlined, compact manner and completed the writing and editing of his reports in remarkably short time. He was a talented artist; and his works contain hundreds of maps, sections, and diagrams and scores of lithographic plates, some of which are today sought by collectors.

Owen's reports contain meticulous and accurate descriptions, reasonable analysis of origin, and wide correlation with American and foreign strata; they also introduced to America some of the terminology for Paleozoic systems used today. A skilled chemist with a knowledge of mechanics and a naturalist-physician, he produced geological writings ranging through paleontology, stratigraphy, mineralogy, and structure. Above all he related economic resources to geology in a way that endeared him to "practical" men and to legislators.

Many of his assistants and associates, some of whom received their first geological experience under him, become important geologists: Robert Peter, F. B. Meek, Richard Owen, John Evans, J. G. Norwood, E. T. Cox, C. C. Parry, Benjamin F. Shumard,

G. C. Swallow, Peter Lesley, Charles Whittlesey, and John Locke.

BIBLIOGRAPHY

I. ORIGINAL WORKS. Owen's major works are *Report of a Geological Reconnaissance of the State of Indiana*, 2 pts. (Indianapolis, 1838–1839); *Report of a Geological Exploration of Part of Iowa, Wisconsin and Illinois* . . . (Washington, 1840–1844); *Report of a Geological Reconnaissance of the Chippewa Land District* (Washington, 1848); *Report of a Geological Survey of Wisconsin, Iowa and Minnesota*, 2 vols. (Philadelphia, 1852); *Report of the Geological Survey in Kentucky*, 4 vols. (Frankfort, Ky., 1856–1861); and *Report of a Geological Reconnaissance . . . of Arkansas*, 2 vols. (Little Rock, 1858; Philadelphia, 1860). One of the most significant of his many short papers is that read before the Geological Society of London in 1842, "On the Geology of the Western States of North America," in *Quarterly Journal of the Geological Society of London*, **2** (1846), 433–447, with an important map and correlation of English and American Paleozoic rocks. Scores of papers, reviews, and short reports that form additional records of his travels and activities are listed most completely in Hendrickson (see below). Each larger report contains a long introduction describing the establishment, organization, associates, progress of the fieldwork, and publications. The body of most reports gives an account of day-to-day activities, which together form a detailed "scientific biography."

In addition to the bibliography of Owen's publications in Hendrickson see those in J. M. Nickles, "Geologic Literature of North America 1785–1918," in *Bulletin of the United States Geological Survey*, no. 746 (1923), 804–805; and detailed and annotated lists in Max Meisel, *A Bibliography of American Natural History* (New York, 1929), see III, 633, for the many entries. J. B. Marcou, "Writings of D. D. Owen," in *Bulletin. United States National Museum*, **30** (1885), 247–251, presents a partially annotated list of fossil genera and species described by Owen and his associates.

II. SECONDARY LITERATURE. An excellent biography, with a portrait of Owen and extensive bibliographies of his publications, of source materials, and of related documents and publications is W. B. Hendrickson, *David Dale Owen, Pioneer Geologist of the Middle West* (Indianapolis, 1943). Caroline Dale Snedecker, a granddaughter of Owen's, relates much personal history and includes interesting illustrations in *The Town of the Fearless* (Garden City, N.Y., 1931). Various obituary notices, most of them listed by Hendrickson, were published at intervals after Owen's death. William E. Wilson, in *The Angel and the Serpent, the Story of New Harmony* (Bloomington, Ind., 1964), gives the best account, with many illustrations and portraits, of Robert Owen and William Maclure at New Harmony and the later activities of Robert Owen's family. The "Obituary Notice," in *Fourth Report of the Geological Survey in Kentucky* (1861), 323–330, is of especial interest and value because it formed the basis for later obituaries. The unsigned obituary by one of the editors (Benjamin Silliman, Jr.?) in *American Journal of Science*, **31** (1861), 153–155, has information on scientific associates and an evaluation of Owen's work by an editor who published many of his papers. A very brief sketch by N. H. Winchell(?) with a portrait is in *American Geologist*, **4** (1889), 65–72. The account by W. J. Youmans in *Pioneers of Science in America* (New York, 1896), 500–508, is mostly derived from Winchell. That in H. A. Kelly and W. L. Burrage, *Dictionary of American Medical Biography* (New York–London, 1928), 927–928, adds little new material.

G. P. Merrill, in *First One Hundred Years of American Geology* (New Haven, 1924; New York, 1962), 194–200, 217–218, 271–275, 321–323, 365–367, has summarized Owen's geological contributions and reproduced important illustrations. Merrill's article in *Dictionary of American Biography*, XIV, 116–117, is a summary of these longer notes. See also *National Cyclopaedia of American Biography*, VIII, 113. A number of biographies of geologists contain considerable information on Owen's association with them (see list in Hendrickson, pp. 160–164). Charles Keyes, "The Transplantation of English Terranal Classification to America by David Dale Owen," in *Pan-American Geologist*, **34** (1923), 81–94, is a fulsome description with some minor inaccuracies of Owen's introduction of English names for Paleozoic periods in which the strata of the Mississippi Valley were deposited. Sir Charles Lyell, *A Second Visit to the United States of North America*, II (London, 1849), 269–274, recounts his visit to Owen at New Harmony and his excursions to see Wabash Valley geology.

GEORGE W. WHITE

OWEN, GEORGE (*b.* Henllys, Pembrokeshire, Wales, 1552; *d.* 1613), *geology.*

A member of an old and distinguished South Wales family, Owen became vice admiral of the maritime counties of Pembroke and Cardigan and was twice sheriff of Cardigan. He was eminent as a local historian and topographer. In geology he is important not so much for the few paragraphs he wrote that happen to come within that subject as for the historical context in which he wrote them.

The description of the geology of Britain can hardly be said to have been begun at any definite time. There are, first of all, the casual remarks of the medieval writers and those made by John Leland in his *Itinerary* (*ca.* 1538) and by William Camden in his *Britannia* (1586). But in 1603 Owen included in his manuscript "Description of Pembrokeshire" an account of the occurrence of the (Carboniferous) limestone and coal measures of South Wales. He did so for the practical guidance of those wishing to exploit these materials; but in detailing the course of the limestone, he established the geological fact of bands of outcrop traceable across country. His account is thus the first attempt to "map" a British geological formation, if only verbally. He prepared a topographical map to accompany his description of Pembrokeshire; had he delineated his information on it, he would have provided a true geological map some two centuries before any other was made. Owen described the course of the limestone as being in two separate "veins"; these are really both the same limestone, outcropping on the north and south sides of the syncline of the South Wales coalfield. Owen clearly had no idea of geological structure, and his remarks cannot be said to form part of a continuous evolution of geological knowledge. It was not until the second half of the seventeenth century that the scientific spirit really came alive and produced a band of naturalists who, among their wide-ranging scholarly researches, collected, described, and discussed truly geological matters.

BIBLIOGRAPHY

Owen's most important work is *The Description of Pembrokeshire*, written in 1603; the authoritative ed. is that by his descendant, Henry Owen, published as no. 1 in Cymmrodorion Record Series (London, 1892), with geological commentary in the footnotes.

Owen's biography is given by Henry Owen in the intro. to his ed. of . . . *Pembrokeshire* (see above). See also D. Lleufer Thomas, in *Dictionary of National Biography*, XLII (1895), 408–410. Detailed commentaries on Owen's geological observations are made by A. Ramsay, in *Passages in the History of Geology*, pt. 2 (London, 1849), 8–11; by F. J. North, "From Giraldus Cambrensis to the Geological Map," in *Transactions of the Cardiff Naturalists' Society*, **64** (1931), 20–97, see 24–29; and by J. Challinor, "The Early Progress of British Geology—I," in *Annals of Science*, **9** (1953), 124–153, see 127–129.

JOHN CHALLINOR

OWEN, RICHARD (*b*. Lancaster, England, 20 July 1804; *d*. Richmond Park, London, England, 18 December 1892), *comparative anatomy, vertebrate paleontology, geology*.

Owen, the younger son of Richard and Catherine Parrin Owen, lost his father in 1809. When six years old Owen was enrolled at the Lancaster Grammar School, where he was a younger schoolmate of William Whewell. In 1820 Owen was apprenticed to the first of three Lancaster surgeons under whom he studied. As an apprentice he had access to postmortems and dissections at the local jail, which sparked an early interest in anatomy and started him collecting anatomical specimens. Before completing his apprenticeship, Owen matriculated in October 1824 at the University of Edinburgh, where he attended the anatomical lectures of Alexander Munro Tertius. More importantly, Owen was able to attend the extramural lectures on anatomy given by John Barclay, from whom Owen gained considerable knowledge of comparative anatomy. In April 1825 Barclay recommended that Owen go to London to study at St. Bartholomew's Hospital with John Abernethy, to whom Barclay addressed a letter of introduction on Owen's behalf. Abernethy immediately appointed Owen to be his prosector. After qualifying as a member of the Royal College of Surgeons in August 1826, Owen set up practice in Lincoln's Inn Fields.

Abernethy had recognized Owen's dissecting ability and knowledge of comparative anatomy. As president of the Royal College of Surgeons, Abernethy had Owen appointed assistant to the conservator, William Clift, of the Hunterian Collection. Owen soon became engaged to Clift's only daughter, Caroline, whom he married in 1835. His primary task was to assist Clift in the preparation of the long-needed catalogue of John Hunter's wide-ranging collection, which had been purchased for the College of Surgeons and served as the nucleus of the College's Museum. Since most of Hunter's notes concerning the specimens had been lost, Owen was obliged to perform many fresh dissections in order to identify the specimens. His general assistance to Clift included serving as Georges Cuvier's guide around the Museum in 1830. This encounter led to an invitation to visit Cuvier in Paris, which Owen did the following year. He later considered the experiences of that trip and the contact with Cuvier a major influence on his work. In 1836 Owen was appointed Hunterian professor at the Royal College of Surgeons, an appointment that necessitated his presenting annually a course of twenty-four lectures based on some aspect of the Hunterian Collection.

Owen succeeded Clift as conservator of the Museum and continued in that position until his appointment in 1856 as superintendent of the natural history departments of the British Museum. At that time these departments were still housed with all the other departments of the British Museum in Bloomsbury. In 1859 Owen sent a forceful report to the trustees of the Museum detailing his views and plans for a new building in a separate location to house the natural history departments. There was much talk and little action, until Gladstone took an interest in Owen's scheme and introduced a bill into Parliament. Finally, in 1871 work was begun on the new Natural History Museum in South Kensington, with the galleries laid out after the design Owen had submitted in 1859. Owen continued as superintendent of the Natural History Museum until after it was fully installed in the new building. He retired in 1884 and was then made K.C.B. After leaving the Royal College of Surgeons, Owen was free to accept the Fullerian lectureship in physiology at the Royal Institution. He also lectured at the Royal School of Mines and on many natural history topics in London and throughout Great Britain. During his career he received most major awards in his fields, including both the Royal and Copley medals from the Royal Society; he was also a member of many British and foreign scientific societies. He served on several royal commissions that dealt with aspects of public health and was president of the British Association for the Advancement of Science in 1858. After a lengthy decline in his health, he died on 18 December 1892 at Sheen Lodge (in Richmond Park), the use of which Queen Victoria had granted him in 1851.

Unfortunately Owen is principally remembered as T. H. Huxley's antagonist at the 1860 meeting of the British Association and in the ensuing debates over Darwin's *On the Origin of Species*. This view neglects his authorship of massive quantities of detailed monographs and papers, which made known many new organisms (both recent and fossil), helped to delineate several natural groups, and laid the bases for much later work by many investigators. The attention of the scientific community was first focused on Owen in 1832 when he published *Memoir on the Pearly Nautilus* (*Nautilus Pompilius, Linn.*), which was based on a single specimen of this delicate organism that had previously been known only by its shell. In this superb piece of descriptive anatomy he also modified Cuvier's Cephalopoda and proposed two orders that were considered valid until 1894. He reviewed the Cephalopoda in an 1836 article for Robert Todd's *Cyclopaedia of Anatomy and Physiology*. Among many other works on invertebrates, Owen in 1835 described the parasite that causes trichinosis.

In 1828 Owen began dissecting the animals that died in the gardens of the Zoological Society of London and soon after helped to organize the evening scientific meetings, the publication of which became *Proceedings of the Zoological Society of London*. Of all the exotic forms to which he thus had access probably none interested him more than the monotremes and marsupials. Before his work the means of generation and of feeding the young of these groups was very much a matter for discussion. Through specimens from the Zoological Society and the many specimens

collected for him in Australia and New Zealand, Owen was able to establish in a series of papers both the mammalian nature and the egg-laying mode of reproduction of the monotremes. Similarly he was able to present details of the reproductive processes of the marsupials. This work was brought together in the articles "Monotremes" and "Marsupials" in Todd's *Cyclopaedia*. Later in his career he was sent and described numerous fossil monotremes and marsupials, which further supported his argument that these forms compose two distinct groups within the Mammalia and that they have long been geographically isolated.

Primates in general and anthropoid apes in particular were of early and lasting interest to Owen, especially in their relation to man. He published many accounts of the anatomy of various primates from the aye-aye to the gorilla. In 1839 Owen began a series, "Contributions to the Natural History of the Anthropoid Apes," which at first was concerned with the osteology of the orangutans but was broadened to include the other apes as specimens became available, often being sent to him by African explorers. Owen separated man from the anthropoid apes into a separate subclass of Mammalia, the Archencephala, primarily on the basis of several supposed differences in the gross structure of their brains.

Lyell introduced Owen to Charles Darwin in October 1836, and thus began a long friendship. The following year Darwin turned over to Owen, for description, his South American fossils. Up to this time Owen had not published on any fossils but did have a broad knowledge of the anatomy, especially the osteology, of recent vertebrates. He described Darwin's *Toxodon platensis* (1837), and his description of Darwin's fossils from South America was published as the first volume of *The Zoology of the Voyage of H.M.S. "Beagle"* (1840). The teeth of some of these fossils intrigued Owen and led him into a major study of the structure of teeth. He addressed a report to the British Association in 1838, which served as the nucleus of his *Odontography* (1840–1845) and his article, "Odontology," in *Encyclopaedia Britannica* (1858). This work on teeth contained a great deal of new information and presented a uniform nomenclature for the teeth and their parts that was of considerable service to zoologists.

In addition to the monographs on comparative anatomy, Owen published several general works. Certain of his Hunterian lectures were published as separate volumes: on invertebrates (1843), on fishes (1844, 1846), and again on invertebrates (1855). Between 1866 and 1868 Owen published the massive *On the Anatomy and Physiology of Vertebrates*, the conclusion of which contains some of Owen's views on Darwin's hypothesis. All of these works were based on the prodigious amount of dissection and observation performed by Owen in preparing the five-volume *Descriptive and Illustrative Catalogue of the Physiological Series of Comparative Anatomy* of the Royal College of Surgeons' Museum, of his nearly twenty courses of Hunterian lectures, and of his many research papers.

Owen's paleontological work began in 1837 with Darwin's South American fossils and especially with the monograph on the *Toxodon* that Owen recognized as an intermediate type, with anatomical characteristics normally identified with rodents, cetaceans, and pachyderms. His works on Darwin's other fossils, for example, *Glyptodon* and *Macrauchenia*, are no less important. Owen also published on the marsupial characteristics of a group of fossils from the Stonesfield Slate (1838) and the first part of his major *Report on British Fossil Reptiles* to the British Association (1839, part two in 1841). This two-part *Report* was the framework on which Owen developed his exhaustive four-volume *History of British Fossil Reptiles* (1849–1884), which is a separate publication of his collected papers issued principally by the Palaeontographical Society. Also in 1839 Owen received a fragment of a femur from New Zealand, which he identified as belonging to a previously unknown giant terrestrial bird. This first paper on the New Zealand moa developed into a major series of publications on *Dinornis* and similar flightless birds. In addition, Owen paid particular attention to fossils from South Africa and Australia, the latter in relation to his interest in marsupials, and published many new species and new descriptions of these faunae. Also of interest are his 1842 studies of English Triassic labyrinthodonts and his description of the Jurassic bird *Archaeopteryx* (1863).

Despite the quantity of Owen's paleontological work before 1856, his career can be divided in two segments, not only by his place of employment but also by the different emphases of his work. While at the College of Surgeons the principal thrust of his work, and also of that museum's collections, was comparative-anatomical. After his transfer to the British Museum with its rich collections of fossils, which Owen further enriched, he naturally changed the emphasis of his work to paleontology. One aspect of this changed emphasis was his series of lecture courses on paleontology at the Royal School of Mines beginning in 1857; they were well-received and later compiled in his popular text *Paleontology* (1860).

Owen made a number of contributions to taxonomy, often modifying and clarifying one or another taxon in the course of his anatomical investigations and in describing many previously unknown species and genera. An excellent example of this work is his recognition of the marsupials as a natural, geographically defined group. Owen did undertake a classification of the Mammalia, in his Rede lecture in May 1859, in which he gave primal import to certain characteristics of the cerebral hemispheres. By these criteria he divided the Mammalia into four subclasses of equivalent value: Lyencephala, Lissencephala, Gyrencephala, and Archencephala (in order of increasing complexity). A strength of Owen's classification was the close association of the monotremes and marsupials in his Lyencephala. The other three subclasses graded imperceptibly into one another. The Archencephala, moreover, contained just one species, man. Owen believed that his cerebral criteria—anterior and posterior extension of the cerebral hemispheres, the posterior horn of the lateral ventricle, and the presence of a *hippocampus minor*—separated man further from the anthropoid apes than the latter were separated from the most primitive primates.

Owen's views on the transmutation of species are not entirely clear, partly on account of his writing style. In 1848 he claimed to have no idea of what the secondary causes may have been by which the Creator introduced new species, and he refrained from publishing on the subject. He did think that there were six possible ways in which the Creator might have acted but would not enumerate them. This, of course, was soon after the publication of Chambers' *Vestiges of the Natural History of Creation*, a much talked-about book. He had no objection to the notion that the Creator may have worked through secondary causes and recognized that, among animals, there had been an ascent and progression. Owen responded very vigorously to Darwin's *Origin* in a long, anonymous attack in the *Edinburgh Review* for April 1860. He was totally unable to accept the possibility that selective action of external circumstances might cause new species to arise. He observed that no effects of any of the hypothetical transmuting influences had been recorded. His objections were not to evolution's having occurred but rather that Darwin's mechanism, natural selection, had not been demonstrated as adequate. He thought "an innate tendency to deviate from parental type" the most probable way that secondary causes have produced one species from another (*On the Anatomy and Physiology of the Vertebrates*, III, p. 807).

The basic ideas Owen put forth in his Rede lecture had been presented previously in London meetings, one of which T. H. Huxley attended. Huxley doubted the validity of Owen's subclass Archencephala, investigated the matter to his own satisfaction, and incorporated his opposing findings in his teaching without publishing them. When Owen repeated his views in a discussion following another's paper at the 1860 meeting of the British Association in Oxford, Huxley was prepared to contradict Owen directly and publicly, stating that he would give evidence to support his contradiction in a more appropriate place. This Huxley did, with the assistance of others, particularly Flower, in a series of publications from 1860 to 1863. Owen simply failed to see certain anatomical structures and relationships. He appears to have operated on the assumption that man possessed unique mental capabilities and that any such unique capabilities must be based in some unique anatomical structure or structures; therefore man could be distinguished from the anthropoid apes by just such structures. Not to be ignored is the fact that many considered Owen to be the preeminent anatomist of his time, and he had held two prominent positions in the British scientific community. In contrast, Darwin had not held any similar position and had already retired, seemingly, to the country; and Huxley, who had backed the argumentative Owen into a corner, was a relative youngster in 1860. These personal factors must have played a role in this whole controversy.

Owen's comparative-anatomical and paleontological work is in the best Cuvierian tradition and perhaps comparable only to that of Cuvier. At the same time Owen was guided by a strong affinity to that school of thought which strongly repelled Cuvier—German *Naturphilosophie*. This affinity is amply evidenced by Owen's further development of the idea of a vertebral archetype, promulgated by Goethe and Carus, in his *On the Archetype and Homologies of the Vertebrate Skeleton* (1848), his *Anatomy of Fishes* (1846), and *On the Nature of Limbs* (1849). It was hardly coincidence that Owen was instrumental in having Oken's *Lehrbuch der Naturphilosophie* translated and published in London in 1847. Also, Owen wrote the article "Oken" for the eighth edition of *Encyclopaedia Britannica*. Through his elaboration of his theory of archetypes, Owen provided a major assist to the much-needed standardization of anatomical nomenclature and greatly clarified the distinction between the anatomical concepts of homology and analogy. In addition, from the *Naturphilosophen* Owen acquired the notion of a specific character resulting from the interaction of two opposing forces working within the developing embryo. His view that one species might develop from another by "an innate tendency to deviate from the parental type" meshes well with the developmental forces he saw operative in each individual.

BIBLIOGRAPHY

I. ORIGINAL WORKS. There are more than 600 titles in Owen's bibliography. The greatest bulk of his papers are in the collections of the British Museum (Natural History). These include correspondence, notebooks, drafts of papers, and interleaved copies of most of his own books. The Royal College of Surgeons has a smaller but important collection of Owen's papers, mostly dating from the period when he was there. In addition, letters by Owen are to be found in the papers of his many correspondents.

The most important of Owen's separate publications include *Memoir on the Pearly Nautilus* (*Nautilus Pompilius, Linn.*) (London, 1832); *Descriptive and Illustrative Catalogue of the Physiological Series of Comparative Anatomy*, 5 vols. (London, 1833–1840); *Fossil Mammalia*, pt. 1 of *The Zoology of the Voyage of H.M.S "Beagle"* (London, 1840); *Odontography; or a Treatise on the Comparative Anatomy of the Teeth; Their Physiological Relations, Mode of Development, and Microscopic Structure in the Vertebrate Animals* (London, 1840–1845); *On the Archetype and Homologies of the Vertebrate Skeleton* (London, 1848); *On the Classification and Geographical Distribution of the Mammalia, Being the Lecture on Sir Robert Rede's Foundation* (London, 1859); *Palaeontology, or a Systematic Summary of Extinct Animals and Their Geological Relations* (Edinburgh, 1860; 2nd ed., Edinburgh, 1861); and *On the Anatomy and Physiology of Vertebrates*, 3 vols. (London, 1866–1868). *The Life of Sir Richard Owen*, cited below, contains an exhaustive chronological bibliography of about 650 items; this is the most complete listing of his works.

II. SECONDARY LITERATURE. The most important source for Owen's life and work is by his grandson, Rev. Richard Owen, *The Life of Sir Richard Owen* (London, 1894), which contains an essay by Thomas Henry Huxley, "Owen's Position in the History of Anatomical Science," II, 273–332, and the bibliography cited above, II, 333–382. See also William Henry Flower, "Richard Owen," in *Dictionary of National Biography*, XIV (1894–1895), 1329–1338.

WESLEY C. WILLIAMS

OZANAM, JACQUES (*b.* Bouligneux, Bresse, France, 1640; *d.* Paris, France, 3 April 1717 [?]), *mathematics*.

Ozanam came from a Jewish family that had converted to Catholicism. As the younger of two sons he was educated for the clergy, but chemistry and mechanics interested him more than theology. He was said to be generous, witty, and gallant; and probably he was too tolerant to have made a good churchman of his day. Except for a tutor who may have helped him slightly, Ozanam taught himself mathematics.

Four years after Ozanam had begun studying for the church, his father died; he then devoted himself to mastering mathematics, with considerable success.

He taught mathematics at Lyons without charge until the state of his finances led him to charge a fee. A lucky circumstance took him to Paris, where his teaching brought him a substantial income. Being young and handsome, his gallantry as well as his penchant for gambling drained his resources; Ozanam sought a way out by marrying a modest, virtuous young woman without means. Although his financial problems remained unsolved, the marriage was happy and fruitful; there were twelve children, most of whom died young. After his marriage Ozanam's conduct was exemplary; always of a mild and cheerful disposition, he became sincerely pious and shunned disputes about theology. He was wont to say that it was the business of the Sorbonne doctors to discuss, of the pope to decide, and of a mathematician to go straight to heaven in a perpendicular line.

Following the death of his wife in 1701, misfortune quickly befell Ozanam. In the same year the War of the Spanish Succession broke out; and many of his students, being foreign, had to leave Paris. From then on, the income from his professional activities became small and uncertain. The last years of his life were melancholy, relieved only by the dubious satisfaction of being admitted as an *élève* of the Academy of Sciences. Ozanam never regained his customary health and spirits, and died of apoplexy, probably on 3 April 1717, although there is some reason to believe that it may have been between 1 April and 6 April 1718.

By almost any criterion Ozanam cannot be regarded as a first-rate mathematician, even of his own time. But he had a flair for writing and during his career wrote a number of books, some of which were very popular, passing through many editions. According to Montucla:

> He promoted mathematics by his treatise on lines of the second order; and had he pursued the same branch of research, he would have acquired a more solid reputation than by the publication of his *Course, Récréations*, or *Dictionnaire mathématique*; but having to look to the support of himself and family, he wisely consulted the taste of his purchasers rather than his own [*Histoire des mathématiques*, II, 168].

In short, his contributions consisted of popular treatises and reference works on "useful and practical mathematics," and an extremely popular work on mathematical recreations; the latter had by far the more lasting impact. Ozanam's *Récréations* may be regarded as the forerunner of modern books on mathematical recreations. He drew heavily on the works of Bachet de Méziriac, Mydorge, Leurechon, and Daniel Schwenter; his own contributions were

somewhat less significant, for he was not a particularly creative mathematician. The work was later augmented and revised by Montucla and, still later, was translated into English by Hutton (1803).

Ozanam is not to be confounded with a contemporary geometer, Sébastien Leclerc (1637–1714), who upon occasion used the pseudonym Ozonam.

BIBLIOGRAPHY

I. ORIGINAL WORKS. Ozanam's writings include *Méthode pour tracer les cadrans* (Paris, 1673, 1685, 1730); *La géométrie pratique du sr Boulenger* (Paris, 1684, 1689, 1691, 1736, 1764); *Tables de sinus, tangentes et sécantes; et des logarithmes* . . . (Paris, 1685, 1697, 1720, 1741); *Traité de la construction des équations pour la solution des problèmes indéterminez* (Paris, 1687); *Traité des lieux géométriques, expliquez par une méthode courte et facile* (Paris, 1687); *Traité des lignes du premier genre, expliquées par une méthode nouvelle et facile* (Paris, 1687); *Usage du compas de proportion* . . . *augmenté d'un traité de la division des champs* (Paris, 1688, 1691, 1700, 1736, 1748, 1794); *Usage de l'instrument universel.* . . . (Paris, 1688, 1700, 1748); and *Méthode de lever les plans et les cartes de terre et de mer, avec toutes sortes d'instrumens, et sans instrumens.* . . . (Paris, 1693, 1700, 1750, 1781).

His major works are *Dictionnaire mathématique, ou, idée générale des mathématiques.* . . . (Amsterdam–Paris, 1691), translated and abridged by Joseph Raphson (London, 1702); *Cours de mathématique, qui comprend toutes les parties les plus utiles et les plus necessaires à un homme de guerre, & à tous ceux qui se veulent perfectionner dans les mathématiques*, 5 vols. (Paris, 1693), also 3 vols. in 1 (Amsterdam, 1697), translated as *Cursus mathematicus: Or a Compleat Course of the Mathematicks*. . ., 5 vols. (London, 1712); and *Récréations mathématiques et physiques* . . ., 4 vols. (Paris, 1694, 1696, 1698, 1720, 1725, 1735, 1778, 1790; Amsterdam, 1698), translated as *Recreations Mathematical and Physical* . . . (London, 1708); as *Recreations in Mathematics and Natural Philosophy* . . . *First Composed by M. Ozanam* . . . *Lately Recomposed by M. Montucla, and Now Translated into English* . . . *by Charles Hutton* (London, 1803, 1814), rev. by Edward Riddle (London, 1840, 1844); and as *Recreations for Gentlemen and Ladies, or, Ingenious Amusements* . . . (Dublin, 1756).

Among his other works are *Traité des fortifications* . . . (Paris, 1694), translated by J. T. Desaguliers as *Treatise of Fortification* . . . (Oxford, 1711, 1727); *Nouveaux élémens d'algèbre* . . ., 2 vols. (Amsterdam, 1702); *Géographie et cosmographie* (Paris, 1711); *La perspective, théorique et pratique* (Paris, 1711, 1720); *La méchanique* . . . *tirée du cours de mathématique de M. Ozanam* (Paris, 1720); *La gnomonique* . . . *tirée du cours de mathématique de M. Ozanam* (Paris, 1746); and *Traité de l'arpentage et du toisé, nouvelle édition, mise dans un nouvel ordre par M. Audierne*

(Paris, 1779). Ozanam also published several articles in the *Journal des sçavans*, including a proof of the theorem that neither the sum nor the difference of two fourth powers can be a fourth power.

His translations or editions of works by others include a revised and enlarged ed. of Adriaan Vlacq, *La trigonométrie rectiligne et sphérique . . . avec tables* (Paris, 1720, 1741, 1765); and *Les élémens d'Euclide du R. P. Dechalles . . . et de M. Ozanam . . . démontrés d'une manière . . . par M. Audierne* (Paris, 1753).

II. Secondary Literature. See Heinrich Zeitlinger, ed., *Bibliotheca chemico-mathematica* (London, 1921), I, 171, and II, 643; Moritz Cantor, *Vorlesungen über die Geschichte der Mathematik*, 2nd ed. (Leipzig, 1913), II, 770, and III, 102–103, 270, 364; Fontenelle, "Éloge . . .," in *Oeuvres diverses*, III (The Hague, 1729), 260–265; Charles Hutton, *Philosophical and Mathematical Dictionary*, II (London, 1815), 144; J. E. Montucla, *Histoire des mathématiques*, II (Paris, 1799), 168; *The Penny Cyclopaedia of the Society for the Diffusion of Useful Knowledge*, XVII (London, 1840), 111–112; and Edward Riddle, *Dr. Hutton's Philosophical Recreations* (London, 1840), v–vii.

William L. Schaaf

OZERSKY, ALEKSANDR DMITRIEVICH (*b.* Chernigov guberniya, Russia, 21 September 1813; *d.* St. Petersburg, Russia, 1 October 1880), *mining engineering, geology.*

Ozersky's father, Dmitry Nikitich Ozersky, owned a small estate and had the rank of state councillor. His mother, Varvara Aleksandrovna, came from a noble family. In 1831 Ozersky, who had graduated from the Mining Cadet Corps (now the Leningrad G. V. Plekhanov Mining Institute), returned to the corps as a tutor in chemistry; lectured on mining statistics and mineralogy from 1833 to 1857; and from 1848 to 1851 was school inspector. From 1857 to 1864 he was the head of the Altay mines and for several years during the period was civilian governor of Tomsk. Upon his return to St. Petersburg, Ozersky worked in the Mining Department and until almost the end of his life was a member of the Committee on Mining Science. In 1857 he was given the rank of major general and in 1866 that of lieutenant general. He married Sofia Semenovna Gurieva and had two daughters, Olga and Sofia.

Ozersky's extremely varied scientific interests at the beginning of his career included the chemical analysis of minerals, rocks, and alloys. Through these precise investigations he established the composition of a number of Russian minerals, pointing out a number of cases in which new names had been suggested for already known minerals. Commissioned by the Free Economic Society to systematize its collection of natural stones according to use, he distinguished thirteen groups of minerals, rocks, and ores.

In his study of ore deposits—following the ideas on the origin of ores then current—Ozersky accepted the sublimation theory: that metalliferous veins are formed by a cooling of "metallic sublimates" that penetrate into cavities, fissures, and pores of a rock. During his expeditions he tried not only to study outcrops of ore and mineral deposits but also to deduce their genesis, in order to assist further prospecting.

While working in Transbaikalia, Ozersky established that ore deposits do not depend upon the enclosing strata but are directly associated with intrusive igneous rocks. He determined a pattern according to which all the deposits of Transbaikalia could be grouped into several isolated stretches. This regularity was later confirmed and was of great practical importance in prospecting.

Ozersky was also interested in the origin of sulfur, saltpeter, and other nonmetallic minerals. He believed in the organic origin of oil, assuming that it could have an animal beginning, particularly a molluscan one. In his regional investigations he attached great importance to the problems of stratigraphic subdivision and solved a number of complicated problems of geological age.

While working in the Baltic provinces in 1843, Ozersky was the first to compile a detailed sequence of Silurian strata of this area, which is now regarded as a classic example of the Lower Paleozoic of northern Europe. Minor subdivisions that he distinguished according to paleontological and lithological data are still valid. Although at that time there were no adequate tables for making paleontological determinations, the stratigraphic scheme worked out by Ozersky proved so accurate that all subsequent studies have confirmed it without introducing any vital changes. Later, in Transbaikalia, he discovered Jurassic deposits which had long been overlooked, an omission resulting in the compilation of erroneous tectonic and paleogeographical schemes of that vast territory. Not until the middle of the twentieth century were his conclusions fully confirmed.

In tectonics Ozersky was a plutonist, believing that all uplifts are determined by injections of a liquid magma into a sedimentary shell. At the same time he admitted the possibility of alternating ascending and descending movements, suggesting that they be called oscillation movements, a term later accepted in geological literature. He indicated that vertical crustal movements could be divided into "local," involving only small portions of the crust, and "general," resulting in an uplift or subsidence of an entire continent.

Ozersky's Russian translation (1845) of Murchison's *The Geology of Russia in Europe and the Ural Mountains* included much new data obtained through extensive research conducted during the four years following the appearance of Murchison's book. He also supplied many footnotes with references to the studies of Russian geologists that had served as a basis for Murchison's work.

Ozersky was a materialist, stating that only experience and its practical application can be depended upon to determine the laws of nature. In public lectures he discussed the interrelations between material objects and natural phenomena, the cycle of matter (the circulation of substances through chemical change), the process of development as reflected in everything that surrounds us, and the fact that light can be emitted only by existing bodies.

Advocating the development of industry and the national economy, Ozersky urged the expansion of railway transport in Russia, the construction of canals to connect the major rivers, the use of hard coal instead of charcoal by industrial enterprises, and the introduction of modern methods in metallurgy.

BIBLIOGRAPHY

Ozersky's major writings are "Geognostichesky ocherk severo-zapadnoy Estlyandii" ("Geognostic Outline of Northwestern Estonia"), in *Gornyi zhurnal*, **2** (1844), 157–208, 285–338; "Vstupitelnye lektsii i kurs prikladnoy mineralogii" ("Introductory Lectures of the Course in Applied Mineralogy"), in *Zhurnal Ministerstva narodnogo prosveschenia*, **46**, sec. 2 (1845), 1–38, 87–111, 161–224; and "Ocherk geologii, mineralnykh bogatstv i gornogo promysla Zabaykalya" ("Outline of the Geology, Mineral Reserves, and Mining Industry of Transbaikalia"), in *Izdaniya SPb. Mineral. Obshchestva*, **8** (1867), 89c.

On Ozersky and his work, see V. V. Tikhomirov and T. A. Sofiano, "Zabyty russky geolog A. D. Ozersky" ("A. D. Ozersky, Forgotten Russian Geologist"), in *Byulleten Moskovskogo obshchestva ispytatelei prirody*, geological ser., **29**, no. 1 (1954).

V. V. TIKHOMIROV

PACCHIONI, ANTONIO (*b.* Reggio nell' Emilia, Italy, 13 June 1665; *d.* Rome, Italy, 5 November 1726), *medicine.*

Pacchioni studied in his native town and obtained his degree in medicine on 25 April 1688. In 1689 he moved to Rome, where at first he attended the Santo Spirito hospital. He was assistant physician at the Ospedale della Consolazione from 26 May 1690 to 3 June 1693, and then remained for six years in Tivoli as the town doctor. In 1699 Pacchioni returned to Rome and established a successful medical practice; he later became head doctor at the Hospital of San Giovanni in Laterano and then at the Ospedale della Consolazione. Interested in anatomy, he was guided by Malpighi (who lived in Rome from 1691 to 1694) and collaborated with Lancisi.

Among Pacchioni's dissertations, from 1701 on, dealing with the structure and functions of the dura mater, the *Dissertatio epistolaris de glandulis conglobatis durae meningis humanae* (1705) is particularly well known and contains his description of the arachnoidal, or so-called Pacchioni, granulations. Pacchioni attributed to these bodies the faculty of secreting lymph for lubricating the sliding movement between the meninges and the brain. He believed that the contraction of the dura mater, then considered to be muscular in nature, served to compress the glands, which, according to Malpighi's doctrine, constituted the cerebral cortex. Pacchioni also collaborated with Lancisi on the explanatory text to Eustachi's *Tabulae anatomicae* (1714).

BIBLIOGRAPHY

I. ORIGINAL WORKS. Pacchioni's principal work is *Dissertatio epistolaris de glandulis conglobatis durae meningis humanae, indeque ortis lymphaticis ad piam meningem productis* (Rome, 1705), reprinted in his *Opera* (Rome, 1741).

II. SECONDARY LITERATURE. On Pacchioni and his work, see Enrico Benassi, "Carteggi inediti fra il Lancisi, il Pacchioni ed il Morgagni," in *Rivista di storia delle scienze mediche e naturali*, **23** (1932), 145–169; Maria Bertolani del Rio, "Antonio Pacchioni 1665–1726," in Luigi Barchi, ed., *Medici e naturalisti Reggiani* (Reggio nell' Emilia, 1935), 659–667; Pietro Capparoni, "Lo stato di servizio di Antonio Pacchioni all'Ospedale della Consolazione in Roma ed un suo medaglione onorario," in *Rivista di storia critica delle scienze mediche e naturali*, **2** (1914), 241–245; Jacopo Chiappelli, "Notizie intorno alla vita di Antonio Pacchioni da Reggio," in *Raccolta d'opuscoli scientifici e filologici*, **3** (1730), 79–102; and Girolamo Tiraboschi, *Biblioteca Modenese*, III (Modena, 1783), 415–419.

LUIGI BELLONI

PACINI, FILIPPO (*b.* Pistoia, Italy, 25 May 1812; *d.* Florence, Italy, 9 July 1883), *anatomy, histology.*

Pacini was the son of Francesco Pacini, a cobbler, and Umiltà Dolfi. He was educated, with public assistance, at the Pistoia episcopal seminary and later at the classical academy. In 1830 he entered the medical school attached to the Ospedale del Ceppo; he completed his studies at the University of Pisa, where he graduated in surgery in 1839 and in medicine in 1840. In the latter year Pacini was also appointed assistant at the Institute of Comparative Anatomy in Pisa; he assumed a similar post at the Institute of Human Anatomy in 1843, and became a substitute teacher there the following year.

In 1847 Pacini began to teach descriptive anatomy at the Lyceum in Florence; he subsequently (1849) became director of the anatomical museum and professor of topographical anatomy at the medical school there, and from 1859 also teacher of microscopical anatomy. (Throughout Pacini's career at the Florence medical school, the professor of descriptive anatomy was Luigi Paganucci.) As a teacher Pacini, convinced of the fundamental importance of the biological sciences to medical education, initiated a number of new programs; he was, however, occasionally frustrated and embittered by the antagonism of Bufalini, director of the department of internal medicine.

Pacini was primarily interested in microscopical research; as early as 1833 he had access to a primitive instrument, and in 1843 was given a good one by the Pistoian philanthropist Niccolò Puccini. The following year Pacini designed his own microscope, which he constructed the next year with the help of Amici; this was the best to which he ever had access. In 1868 he constructed another compound (which he called "inverted") instrument for photographic and chemical use; this, together with the 1845 microscope, is preserved in the Museo di Storia della Scienza in Florence.

Pacini saw the corpuscles that are now named for him early in his career; indeed, he discovered them in a hand that he was dissecting as a student in the Pistoia hospital in 1831, when he was nineteen. He first saw the corpuscles around the digital branches of the median nerve, and suggested that they were "nervous ganglia of touch"; but he soon found them also in the abdominal cavity. Although he studied these corpuscles microscopically from 1833 on, Pacini published his research only in 1840, when his *Nuovi organi scoperti nel corpo umano* appeared. The name "Pacini's corpuscles" was proposed in 1844 by Koelliker, who had confirmed their existence; in 1862, however, the Viennese anatomist Carl Langer claimed priority for Abraham Vater—although Vater's work, published in 1741, had been forgotten and was certainly unknown to Pacini. At all events, Pacini was the first to describe the distribution of the corpuscles in the body, their microscopic structure, and their

nerve connections; he also interpreted the function of the corpuscles as being concerned with the sensation of touch and deep pressure.

Pacini made another important observation in 1854, when, in the midst of an epidemic in Florence, he discovered the cholera vibrio. He microscopically examined the blood and feces of those afflicted with the disease and the intestines of those dead from it. He published his findings in a report, *Osservazioni microscopiche e deduzioni patologiche sul cholera asiatico*, in which he stated that cholera is a contagious disease, characterized by destruction of the intestinal epithelium, followed by extreme loss of water from the blood (for which condition he later recommended, in 1879, the therapeutic intravenous injection of saline solution). Pacini went on to declare that the intestinal injuries common to the disease were caused by living microorganisms (which he called "vibrions"); he further provided drawings of the vibrions that he had observed microscopically in abundance in the intestines of cholera victims.

Despite the significance of his researches, Pacini was overlooked when, following the epidemic of 1866, the Italian government distributed medals for meritorious work against cholera. In 1884 Koch rediscovered the cholera vibrio, which he isolated in pure culture, and named it "Komma Bacillus"; by applying his rigorous postulates, he was further able to prove that the bacillus was the sole cause of the disease. Koch presented his findings to the Cholera Commission of the Imperial Health Office in Berlin; the commission also recognized Pacini's priority in discovering the microorganism.

In addition to conducting his own histological research, Pacini enthusiastically advocated the teaching of microscopic anatomy. He himself gave a course in practical microscopy as early as 1843, while he was still at Pisa; in 1847 he published a plea for the teaching of histology, and in 1861 he presented a collection of selected microscopical preparations to the first Italian Exposition, held at Florence. He published further notes on histological technique as late as 1880. His specific contributions include a description of the *membrana limitans interna* of the human retina (1845) and reports on the electric organ of the Nile *Silurus* (1846 and 1852) and on the structure of bone (1851). He also published work in practical anatomy, including a study of the muscular mechanics of respiration in man (1847); he later (1870) developed a method of artificial respiration based upon a rhythmic movement of the shoulders of the unconscious subject.

Pacini was a pious and charitable man. He never married, and his work was generally unrecognized; he died in a poorhouse, and was buried in the cemetery of the Misericordia in Florence. In 1835 his remains were transferred, with the remains of two other anatomists, Atto Tigri and Filippo Civinini (Castaldi), to the church of Santa Maria delle Grazie in Pistoia.

BIBLIOGRAPHY

I. ORIGINAL WORKS. For a complete bibliography of Pacini's fifty-five works, see Castaldi, below. Works of particular interest are "Sopra un particulare genere di piccoli corpi globulari scoperti nel corpo umano da Filippo Pacini," in *Archivio delle scienze medico-fisiche*, **8** (1835), and in *Nuovo giornale dei letterati*, parte scientifica, **32** (1836), 109–114; *Nuovi organi scoperti nel corpo umano* (Pistoia, 1840); "Nuove ricerche microscopiche sulla tessitura intima della retina," in *Nuovi annali delle scienze naturali* (July-Aug. 1845), and separately repr. (Bologna, 1845); "Sopra l'organo elettrico del Siluro del Nilo," *ibid.* (July 1846); "Sulla questione della meccanica dei muscoli intercostali," in *Gazzetta toscana delle scienze medico-fisiche*, **5** (1847), 153–156; "Cosa è ed a che è buona l'anatomia microscopica del corpo umano," *ibid.*, 193–199; "Nuovo ricerche microscopiche sulla tessitura intima delle ossa," in *Gazzetta medica italiana federativa* (Nov. 1851); "Osservazioni microscopiche e deduzioni patologiche sul colera asiatico," *ibid.* (Dec. 1854), and repr. in *Sperimentale*, **78** (1924), 277–282; "Della natura del colera asiatico," in *Cronaca medica di Firenze* (10 Aug. and 10 Nov. 1866); and "Il mio metodo di respirazione artificiale per la cura dell'asfissia," in *Imparziale*, **10** (1870), 481–486.

See also "Dei fenomeni e delle funzioni di trasudamento nell'organismo animale," in *Sperimentale*, **28** (1874), 436–438, 537–563, 681–722; "Del processo morboso del colera asiatico del suo stadio di morte apparente e della legge matematica da cui è regolato," *ibid.*, **33** (1879), 355–369, 466–499, 573–597; "Di alcuni metodi di preparazione e di conservazione degli elementi microscopici dei tessuti animali o vegetali," in *Giornale internazionale delle scienze mediche*, **2** (1880), 337–350; and *Nuove osservazioni microscopiche sul colera* (Milan, 1885).

II. SECONDARY LITERATURE. On Pacini and his work see A. Bianchi, *Relazione e catalogo dei manoscritti di Filippo Pacini esistenti nella R. Biblioteca Nazionale di Firenze* (Florence, 1889); L. Castaldi, "Filippo Pacini nel quarantesimo anniversario della sua morte," in *Rivista di storia delle scienze mediche e naturali*, **14** (1923), 182–212, with complete bibliography; "Filippo Pacini," in *Sperimentale*, **78** (1924), 275–283; "Un manoscritto inedito di Filippo Pacini sull'ordinamento degli studi anatomici," in *Rivista di storia delle scienze mediche e naturali*, **16** (1925), 13–17; "Discorso per la translazione delle salme di Filippo Civinini, Filippo Pacini ed Atto Tigri nella Chiesa di S. Maria delle Grazie presso l'Ospedale del Ceppo. Letto in 29 Settembre 1935 nel Palazzo Comunale di Pistoia," *ibid.*, **26** (1935), 289–310; G. Chiarugi, "Corpuscoli lamellosi del Pacini," in *Istituzioni di anatomia dell'uomo*, IV (Milan, 1921), 789–793; A. Filippi, "Filippo Pacini," in *Sperimentale*, **37** (1883), 109–111; P. Franceschini, "Filippo Pacini e il colera," in *Physis*, **13** (1971), 324–332; J. Herrick, *Introduction to Neurology* (Philadelphia, 1928), 89; A. Koelliker, *Ueber die Pacinischen Körperchen des Menschen und der Säugethiere* (Zurich, 1844); C. Langer, "Zur Anatomie und Physiologie der Haut," in *Sitzungsberichte der Akademie der Wissenschaften in Wien*, Math.-naturwiss. Klasse, **44** (1861), 19–46, and **45** (1862), 133–188; and G. Sanarelli, *Il Colera* (Milan, 1931), 73, 74, 80.

PIETRO FRANCESCHINI

PACINOTTI, ANTONIO (*b.* Pisa, Italy, 17 June 1841; *d.* Pisa, 24 March 1912), *electrophysics*.

Although Pacinotti spent most of his life as a physics professor at various Italian universities, his subsequent reputation was due largely to his invention of a new form of armature used in electric motors and generators. The armature design that Pacinotti first described in a paper published in *Nuovo cimento* (June 1864) became a key element in the evolution from the magnetoelectric generator to the commercial self-excited dynamo during the next decade. The Pacinotti armature consisted of an iron ring with projecting teeth interspersed with coils which formed a closed series circuit with connections to a commutator. In his paper Pacinotti pointed out that his new machine could be used as either a direct-current motor or a generator. A similar ring armature design was developed, apparently independently, by Gramme by 1869.

Pacinotti developed his armature while a student at the University of Pisa. His electrical investigations were encouraged by his father, Luigi Pacinotti, a professor of mathematics and physics at Pisa who had himself engaged in electrical studies during the 1840's. A laboratory notebook kept by the younger Pacinotti indicates that he began his armature experiments in 1858. His work was interrupted by a year of service in the corps of engineers during the war for Italian independence. Pacinotti later claimed to have conceived the idea for radial teeth on his armature after having seen radially stacked muskets during the war. He returned to Pisa and resumed his electrical experiments, which culminated in a small test machine described in his 1864 paper. Following his graduation from Pisa in 1861, Pacinotti taught at Florence until 1864. He then taught physics at the Royal Institute of Technology in Bologna and at the University of Cagliari before returning to the University of Pisa, where he spent the rest of his life.

In the wake of the widespread publicity given the Gramme dynamo, Pacinotti called attention to his own earlier work in a note published in *Comptes rendus . . . de l'Académie des sciences* in 1871. Belated recognition followed at the Vienna Exposition of 1873 and the Paris Electrical Exhibition of 1881. From the perspective of the historian of science, Pacinotti provides an interesting example of a scientist whose contribution was not especially impressive but whose reputation ultimately derived from his recognition by the new electrical engineering profession for his invention of a device that was largely ignored until its apparently independent invention some years later.

BIBLIOGRAPHY

I. Original Works. See the Royal Society *Catalogue of Scientific Papers*, IV, 733; VIII, 549; X, 977–978; XII, 553; XVII, 668; for lists of Pacinotti's published papers, most of which appeared in *Nuovo cimento*. His paper on armature design, "Correnti elettriche generate dall'azione del calorico e della luce," in *Nuovo cimento*, **19** (1864), 378–384, was reprinted as *Descrizione di una macchina elettro-magnetica* (Bergamo, 1912), with accompanying French, English, German, and Latin translations.

II. Secondary Literature. There is a three-part biographical essay based on original sources by Franklin L. Pope, in *Electrical Engineer* (New York), **14** (1892), 259–262, 283–284, 339–341. See S. P. Thompson, *Dynamo-Electric Machinery*, 3rd ed. (London, 1888), for a detailed discussion of Pacinotti's design and the results of Thompson's experiments comparing the electrical performance of the Pacinotti and Gramme armatures. An obituary appeared in *Electrical World*, **59** (1912), 732–733. See also G. Polvani, *Antonio Pacinotti: la vita, l'opera* (Milan, 1932).

James E. Brittain

PACIOLI, LUCA (*b.* Sansepolcro, Italy, *ca.* 1445; *d.* Sansepolcro, 1517), *mathematics, bookkeeping.*

Luca Pacioli (Lucas de Burgo), son of Bartolomeo Pacioli, belonged to a modest family of Sansepolcro, a small commercial town in the Tiber valley about forty miles north of Perugia. All we know of his early life is that he was brought up by the Befolci family of Sansepolcro. It has been suggested that he may have received part of his early education in the atelier of his older compatriot Piero della Francesca (1410–1492). As a young man he entered the service of Antonio Rompiansi, a Venetian merchant who lived in the fashionable Giudecca district. Pacioli lived in Rompiansi's house and helped to educate his three sons. While doing so he studied mathematics under Domenico Bragadino, who held classes in Venice, probably at the school that the republic had established

near the Church of San Giovanni di Rialto for those who did not want to go to Padua. The experience Pacioli gained in Rompiansi's business and the knowledge he gathered at Bragadino's school prompted him to write his works on arithmetic, the first of which he dedicated to the Rompiansi brothers in 1470. Their father was dead by then and Pacioli's employment probably had ended. He then stayed for several months in Rome as the guest of the architect Leone Battista Alberti.

Sometime between 1470 and 1477 Pacioli was ordained as a friar in the Franciscan order in fulfillment of a vow. After completing his theological studies he began a life of peregrination, teaching mathematics in various cities of Italy. From 1477 to 1480 he gave lessons in arithmetic at the University of Perugia and wrote a treatise on arithmetic for the benefit of his students (1478). In 1481 he was in Zara (now Zadar, Yugoslavia), then under Venetian rule, where he wrote another work on arithmetic. After teaching mathematics successively at the universities of Perugia, Naples, and Rome in 1487–1489, Pacioli returned to Sansepolcro. In 1494 his major work, *Summa de arithmetica, geometria, proportioni et proportionalita*, was ready for the publisher and he went to Venice to supervise the printing. He dedicated the book to the young duke of Urbino, Guidobaldo da Montefeltro (1472–1508), who, it is believed, was his pupil. The dedicatory letter suggests that Pacioli had been closely associated with the court of Urbino. This is confirmed by the altarpiece painted by Piero della Francesca for the Church of San Bernardino in Urbino (now in Milan), in which the figure of St. Peter the Martyr is portrayed by Pacioli. The painting shows Duke Federigo (Guidobaldo's father) praying before the Virgin and Child surrounded by angels and saints. A painting by Jacopo de' Barbari in the Naples Museum shows Pacioli demonstrating a lesson in geometry to Guidobaldo.

In 1497 Pacioli was invited to the court of Ludovico Sforza, duke of Milan, to teach mathematics. Here he met Leonardo da Vinci, who was already in Sforza's employment. That Leonardo consulted Pacioli on matters relating to mathematics is evident from entries in Leonardo's notebooks. The first part of Pacioli's *Divina proportione* was composed at Milan during 1496–1497, and it was Leonardo who drew the figures of the solid bodies for it. Their stay in Milan ended in 1499 with the entry of the French army and the consequent capture of Sforza. Journeying through Mantua and Venice, they arrived in Florence, where they shared quarters. Leonardo's stay in Florence, which lasted until the middle of 1506, was interrupted by a short period in the service of Cesare Borgia.

In 1500 Pacioli was appointed to teach Euclid's *Elements* at the University of Pisa, which had been transferred to Florence because of the revolt of Pisa in 1494. The appointment was renewed annually until 1506. In 1504 he made a set of geometrical figures for the Signoria of Florence, for which he was paid 52.9 lire. He was elected superior of his order for the province of Romagna and shortly afterward (1505) was accepted as a member of the monastery of Santa Croce in Florence. During his stay in Florence, Pacioli also held an appointment at the University of Bologna as *lector ad mathematicam* (1501–1502). At this time the University of Bologna had several *lectores ad arithmeticam*, one of whom was Scipione dal Ferro, who was to become famous for solving the cubic equation. It has been suggested that Pacioli's presence in Bologna may have encouraged Scipione to seek a solution of the cubic equation, but there is no evidence to support this apart from Pacioli's statement in the *Summa* that the cubic equation could not be solved algebraically.

Since his arrival in Florence, Pacioli had been preparing a Latin edition of Euclid's *Elements* and an Italian translation. He had also written a book on chess and had prepared a collection of recreational problems. On 11 August 1508 Pacioli was in Venice, where he read to a large gathering in the Church of San Bartolomeo in the Rialto an introduction to book V of Euclid's *Elements*. A few months later, on a supplication made by him to the doge of Venice, he was granted the privilege that no one but he could publish his works within the republic for fifteen years. The works listed were the fifteen books of Euclid, *Divina proportione*, "De viribus quantitatis," "De ludo scachorum," and *Summa de arithmetica*. The Latin edition of Euclid and the *Divina proportione* were published in 1509. Pacioli was called once more to lecture in Perugia in 1510 and in Rome in 1514.

On several occasions Pacioli came into conflict with the brethren of his order in Sansepolcro. In 1491, on a complaint made to the general of the order, he was prohibited from teaching the young men of the town; but this did not prevent his being called to preach the Lenten sermons there in 1493. It is likely that certain minor privileges granted to him by the Pope had aroused enmity or jealousy. Although a petition had been sent to the general of the order in 1509, he was shortly afterward elected commissioner of his convent in Sansepolcro. A few years later Pacioli renounced these privileges and in 1517, shortly before his death, his fellow townsmen petitioned that he be appointed minister of the order for the province of Assisi.

The commercial activity of Italy in the late Middle Ages had led to the composition of a large number of treatises on practical arithmetic to meet the needs of merchant apprentices. Evidence of this is found in the extant works of the *maestri d'abbaco* of central and northern Italy. Some of them even contained chapters devoted to the rules of algebra and their application, no doubt influenced by the *Liber abbaci* of Leonardo Fibonacci. The first printed commercial arithmetic was an anonymous work that appeared at Treviso in 1478. By the end of the sixteenth century about 200 such works had been published in Italy. Pacioli wrote three such treatises: one at Venice (1470), one at Perugia (1478), and one at Zara (1481). None of them was published and only the second has been preserved.

Pacioli's *Summa de arithmetica . . .* (1494) was more comprehensive. Unlike the practical arithmetics, it was not addressed to a particular section of the community. An encyclopedic work (600 pages of close print, in folio) written in Italian, it contains a general treatise on theoretical and practical arithmetic; the elements of algebra; a table of moneys, weights, and measures used in the various Italian states; a treatise on double-entry bookkeeping; and a summary of Euclid's geometry. He admitted to having borrowed freely from Euclid, Boethius, Sacrobosco, Leonardo Fibonacci, Prosdocimo de' Beldamandi, and others.

Although it lacked originality, the *Summa* was widely circulated and studied by the mathematicians of the sixteenth century. Cardano, while devoting a chapter of his *Practica arithmetice* (1539) to correcting the errors in the *Summa*, acknowledged his debt to Pacioli. Tartaglia's *General trattato de' numeri et misure* (1556–1560) was styled on Pacioli's *Summa*. In the introduction to his *Algebra*, Bombelli says that Pacioli was the first mathematician after Leonardo Fibonacci to have thrown light on the science of algebra—"primo fu che luce diede a quella scientia."[1] This statement, however, does not mean that algebra had been neglected in Italy for 300 years. Another edition of Pacioli's *Summa* was published in 1523.

Pacioli's treatise on bookkeeping, "De computis et scripturis," contained in the *Summa*, was the first printed work setting out the "method of Venice," that is, double-entry bookkeeping. Brown has said, "The history of bookkeeping during the next century consists of little else than registering the progress of the *De computis* through the various countries of Europe."[2]

The *Divina proportione*, written in Italian and published in 1509, was dedicated to Piero Soderini, perpetual gonfalonier of Florence. It comprised three books: "Compendio de divina proportione," "Tractato de l'architectura," and "Libellus in tres partiales tractatus divisus quinque corporum regularium." The first book, completed at Milan in 1497, is dedicated to Ludovico Sforza. Its subject is the golden section or divine proportion, as Pacioli called it, the ratio obtained by dividing a line in extreme and mean ratio. It contains a summary of Euclid's propositions (including those in Campanus' version) relating to the golden section, a study of the properties of regular polyhedrons, and a description of semi-regular polyhedrons obtained by truncation or stellation of regular polyhedrons. Book 2 is a treatise on architecture, based on Vitruvius, dedicated to Pacioli's pupils at Sansepolcro. To this he added a treatise on the right proportions of roman lettering. The third book is an Italian translation, dedicated to Soderini, of Piero della Francesca's *De corporibus regularibus*.

Also in 1509 Pacioli published his Latin translation of Euclid's *Elements*. The first printed edition of Euclid (a Latin translation made in the thirteenth century by Campanus of Novara from an Arabic text) had appeared at Venice in 1482. It was severely criticized by Bartolomeo Zamberti in 1505 when he was publishing a Latin translation from the Greek. Pacioli's edition is based on Campanus but contains his own emendations and annotations. It was published in order to vindicate Campanus, apparently at the expense of Ratdolt, the publisher of Campanus' translation.

Among the works that Pacioli had intended to publish is "De viribus quantitatis," a copy of which, in the hand of an amanuensis, is in the University Library of Bologna.[3] The name of the person to whom the work was dedicated has been left blank. It is an extensive work (309 folios) divided into three parts: the first is a collection of eighty-one mathematical recreational problems, a collection larger than those published a century later by Bachet de Méziriac and others; the second is a collection of geometrical problems and games; the third is a collection of proverbs and verses. No originality attaches to this work, for the problems are found scattered among earlier arithmetics and, in fact, a collection is attributed to Alcuin of York. Pacioli himself called the work a compendium. Some of the problems are found in the notebooks of Leonardo da Vinci, and the work itself contains frequent allusions to him.

Pacioli's Italian translation of Euclid's *Elements* and his work on chess, "De ludo scachorum," dedicated to the marquis of Mantua, Francesco Gonzaga, and his wife, Isabella d'Este, were not published and there is no trace of the manuscripts.

Vasari, in writing the biography of Piero della Francesca, accused Pacioli of having plagiarized the work of his compatriot on perspective, arithmetic, and geometry.[4] The accusations relate to three works by Piero—*De prospectiva pingendi*, "Libellus de quinque corporibus regularibus," and *Trattato d'abaco*, all of which have been published only since the turn of the twentieth century.[5] In 1908 Pittarelli came to the defense of Pacioli, pointing out that any accusation of plagiarism in regard to *De prospectiva* was unjust, since Pacioli had acknowledged Piero's work in both the *Summa* and the *Divina proportione*.[6] As for the *Libellus*, it has been established by Mancini that Pacioli's work is a translation of it that lacks the clarity of the original.[7] In the case of the *Trattato*, although Piero can claim no originality for it, it has been possible to find in it at least 105 problems of the *Summa*.[8]

The writings of Pacioli have provided historians of the Renaissance with important source material for the study of Leonardo da Vinci. The numerous editions and translations of the *De computis et scripturis* are evidence of the worldwide esteem in which Pacioli is held by the accounting profession. Pacioli made no original contribution to mathematics; but his *Summa*, written in the vulgar tongue, provided his countrymen, especially those not schooled in Latin, with an encyclopedia of the existing knowledge of the subject and enabled them to contribute to the advancement of algebra in the sixteenth century.

NOTES

1. Rafael Bombelli, *Algebra* (Bologna, 1572), d 2v.
2. Brown, *History of Accounting*, p. 119.
3. An ed. of the MS by Paul Lawrence Rose of New York University is in press.
4. Vasari, *Vite*, pp. 360, 361, 365.
5. Codex Palat. Parma, published by C. Winterberg (1899); Codex Vat. Urb. lat. 632, published in 1915 by Mancini; Codex Ash. 280, published in 1971 by Arrighi.
6. Pittarelli, "Luca Pacioli . . .."
7. Mancini, "L'opera 'De corporibus regularibus' . . .."
8. Jayawardene, "The *Trattato d'abaco* of Piero della Francesca."

BIBLIOGRAPHY

I. ORIGINAL WORKS. Pacioli's writings include *Summa de arithmetica, geometria, proportioni et proportionalita* (Venice, 1494; 2nd ed. Toscolano, 1523)—there are several eds. of the treatise on bookkeeping, "De computis et scripturis," contained in the *Summa*, fols. 197v–210v, in the original Italian and in trans.; *Divina proportione* (Venice, 1509)—there are two extant MSS containing the "Compendio de divina proportione," one in the University of Geneva Library (Codex 250) and the other in the Biblioteca Ambrosiana, Milan (Codex 170, parte superiore), the second published as no. XXXI of Fontes Ambrosiani, Giuseppina Masotti Biggiogero, ed. (Verona, 1956); *Euclid megarensis opera . . . a Campano . . . tralata. Lucas Paciolus emendavit* (Venice, 1509); "De viribus quantitatis" (University of Bologna Library, Codex 250), described by Amedeo Agostini in "Il 'De viribus quantitatis' di Luca Pacioli," in *Periodico di matematiche*, **4** (1924), 165–192, and by Carlo Pedretti in "Nuovi documenti riguardanti Leonardo da Vinci," in *Sapere* (15 Apr. 1952), 65–70.

An unpublished arithmetic, written in Perugia (1478), is in the Vatican Library (Codex Vat. lat. 3129).

II. SECONDARY LITERATURE. Studies of Pacioli's life and work are listed by G. Masotti Biggiogero in "Luca Pacioli e la sua 'Divina proportione,'" in *Rendiconti dell'Istituto lombardo di scienze e lettere*, ser. A, **94** (1960), 3–30.

The earliest biographical sketch, written by Bernardino Baldi in 1589, was not published. Baldassare Boncompagni made a critical study of it with the help of archival documents: "Intorno alle vite inedite di tre matematici . . . scritte da Bernadino Baldi," in *Bullettino di bibliografia e di storia delle scienze mathematiche e fisiche*, **12** (1879), 352–438, 863–872. Other archival documents were published by D. Ivano Ricci in *Luca Pacioli, l'uomo e lo scienziato* (Sansepolcro, 1940). R. E. Taylor, *No Royal Road: Luca Pacioli and His Times* (Chapel Hill, N. C., 1942), is a lively narrative but unreliable as a biography. Pacioli's work is discussed by L. Olschki in *Geschichte der neusprachlichen wissenschaftlichen Literatur*, I (Leipzig, 1919), 151–239.

Stanley Morison, *Fra Luca Pacioli of Borgo San Sepolcro* (New York, 1933), contains a study of that part of the *Divina proportione* dealing with roman lettering. The history of bookkeeping is discussed by Richard Brown in *A History of Accounting and Accountants* (London, 1905), 108–131.

On the accusations of plagiarism see Giorgio Vasari's life of Piero della Francesca in *Le vite de' più eccellenti architetti, pittori e scultori italiani* (Florence, 1550); G. Pittarelli, "Luca Pacioli usurpò per se stesso qualche libro di Piero de' Franceschi," in *Atti, IV Congresso internazionale dei matematici, Roma, 6–11 aprile 1908*, III (Rome, 1909), 436–440; G. Mancini, "L'opera 'De corporibus regularibus' di Pietro Franceschi detto Della Francesca usurpata da fra Luca Pacioli," in *Memorie della R. Accademia dei Lincei*, classe di scienze morali, storiche e filologiche, ser. 5, **14** (1915), 446–477, 488–580; and Gino Arrighi's ed. of Piero della Francesca's *Trattato d'abaco* (Pisa, 1971), 24–34. See also S. A. Jayawardene, "The *Trattato d'abaco*" of Piero della Francesca," in *Studies in the Italian Renaissance: A Collection in Honour of P. O. Kristeller* (in press).

S. A. JAYAWARDENE

PACKARD, ALPHEUS SPRING, JR. (*b.* Brunswick, Maine, 19 February 1839; *d.* Providence, Rhode Island, 14 February 1905), *entomology*.

Packard was the son of Alpheus Spring Packard, professor of Greek and Latin at Bowdoin College, and Frances Elizabeth Appleton, daughter of Jesse Appleton, president of Bowdoin College. His boyhood was spent exploring nature, a pastime in which he was encouraged by his father. By the time he had matriculated at Bowdoin College in 1857, he had investigated most branches of natural history. Packard's attention early turned to entomology, and in 1860 he was invited to join an expedition to Labrador arranged by Chadbourne. After graduating in 1861, he spent the summer with the Maine Geological Survey and published his first two scientific papers, on entomology and on geology.

In the fall of 1861 Packard went to the Lawrence Scientific School to study under Louis Agassiz. Packard was simultaneously able to pursue his medical studies, mainly human anatomy and medicine, and was awarded the M.D. from the Maine Medical School at Bowdoin College in 1864. By now he appeared so well established in his assistantship that there was every reason to believe he would be appointed one of the permanent curators of the new Museum of Comparative Zoology. Unfortunately a smoldering quarrel between the junior assistants and Agassiz relating to their duties and obligations to the museum led to a complete break; Packard, A. Hyatt, E. S. Morse, F. W. Putnam, S. H. Scudder, and A. E. Verrill left Cambridge in 1864.

Immediately after leaving Cambridge, Packard was again invited to join an expedition to Labrador, this time under the direction of the artist William Bradford. On his return to the United States in the fall of 1864 he received a commission as assistant surgeon of the First Maine Veteran Volunteers. But Packard's Civil War experience was of short duration, and it appears that his medical practice was equally brief, being confined entirely to these few months.

For the next thirteen years Packard led an almost itinerant existence of writing and editing, interrupted by short appointments at various institutions and agencies. He was acting librarian and custodian of the Boston Society of Natural History for a year. In 1867 he was one of the curators of the Peabody Academy of Science, Salem, Massachusetts, and, in 1877–1878, was its director. He lectured on economic entomology at the Maine State College of Agriculture and the Mechanical Arts (now the University of Maine) for a year and for several years at the Massachusetts Agricultural College (now University of Massachusetts). At Bowdoin he lectured first on entomology and later on comparative anatomy. During the winter of 1869–1870 he studied marine life at Key West and in the Dry Tortugas, then did similar work in Charleston, South Carolina. In March 1867 he joined forces with E. S. Morse, A. Hyatt, and F. W. Putnam to found *American Naturalist*, a popular scientific monthly.

Packard was its editor for some twenty years, writing many of the articles.

In 1874 he was associated with the Kentucky Geological Survey, investigating the fauna of Mammoth Cave; and in 1875–1876 he was with the U.S. Geological Survey of the Territories under F. V. Hayden. In 1877 Packard was appointed secretary to the U.S. Entomological Commission, headed by C. V. Riley.

With his appointment as professor of zoology and geology at Brown University, a greater maturity may be seen in his approach to scientific matters. During these years he produced some of his finest writings: *The Cave Fauna of North America* (1888), *The Labrador Coast* (1891), *Textbook of Entomology* (1898), and the three-volume *Monograph of the Bombycine Moths of North America* (1895–1914). He was a foreign member of the Royal Entomological and the Linnean societies of London and held an honorary Ph.D. and an honorary LL.D. from Bowdoin.

By inclination Packard was a naturalist in the early nineteenth-century tradition at a time when specialization was the fashion. His bibliography therefore appears miscellaneous, and the quality of his writings is varied. In his own time he was considered a general zoologist and geologist; today he is generally thought of as an entomologist, since with only a few exceptions it is his entomological publications that have real currency. As an educator, he provided a sound basis for training a new generation of professional entomologists with his *Half Hours With Insects* (1873), *Our Common Insects* (1873), *Guide to the Study of Insects* (1869), and *Textbook of Entomology* (1898).

Much of Packard's reputation now rests in the more enduring area of taxonomy, and it has been estimated that he described as new over 50 genera and about 580 species of invertebrates. About a quarter of these have now been placed in synonymy. Packard's descriptive methods varied greatly in quality. He turned out a good many inadequately analyzed and artificial descriptions, which accounts in part for a high share of the synonyms noted; yet, as with the bombycid moths, he delineated careful morphological features. Packard's reputation therefore rests with *A Monograph of the Geometrid Moths or Phalaenidae of the United States* (1876) and *Monograph of the Bombycine Moths of North America* (1895–1914). The latter work in particular more nearly corresponds to present-day expectations.

Attention is frequently drawn to Packard's place in the history of economic entomology; but although he contributed an impressive list of shorter papers, bulletins, and books on injurious insects, he was not an original or really professional applied entomologist.

The value of these writings consisted in providing sound life history studies and calling attention to potential problems. *Insects Injurious to Forest and Shade Trees* (1881, 1890), a compilation with some original observations, represents his best and most useful effort in this field.

Packard's work in marine invertebrate zoology was divided among taxonomic treatments, both living and fossil, one of his best being *A Monograph of the Phyllopod Crustacea of North America* (1883); and embryological and anatomical investigations, typical of which is *On the Embryology of Limulus polyphemus* (1871).

Packard's embryological studies, perhaps reflecting the influence of Agassiz, were fairly inclusive of the invertebrates, embracing both primitive and higher insects, crustaceans, and some anomalous forms, such as *Peripatus*. The first serious American student of insect embryology, he was one of the first in the United States to introduce the concept of comparative embryology. His preliminary *Life Histories of Animals, Including Man; or, Outlines of Comparative Embryology* (1876) was followed by *Outlines of Comparative Embryology* (1878); these pioneer efforts, however, were soon superseded by Francis Maitland Balfour's masterly two-volume *A Treatise on Comparative Embryology* (1880–1881).

An undoubted pioneer masterpiece of Packard's was *The Cave Fauna of North America, With Remarks on the Anatomy of the Brain and Origin of the Blind Species* (1888), combining the disciplines of taxonomy, anatomy, and evolution.

Packard's sustained interest in evolution was more Lamarckian than Darwinian, more teleologic than mechanistic. His interest in Lamarck's zoological philosophy led to *Lamarck, the Founder of Evolution; His Life and Work, With Translations of His Writings on Organic Evolution* (1901) and, with E. D. Cope and A. Hyatt, virtually to found the neo-Lamarckian movement, which influenced the writings of later nineteenth- and early twentieth-century American taxonomists.

BIBLIOGRAPHY

A rather complete listing of the many biographical notices of Packard is in Mathilde M. Carpenter, "Bibliography of Biographies of Entomologists," in *American Midland Naturalist*, **33** (1945), 76–77. Of those listed, J. S. Kingsley, "Sketch of Alpheus Spring Packard," in *Popular Science Monthly*, **33** (1888), 260–267, and A. D. Mead, "Alpheus Spring Packard," *ibid.*, **67** (1905), 43–48, are especially good; but the definitive reference is T. D. A. Cockerell, "Biographical Memoir of Alpheus Spring

Packard," in *Biographical Memoirs. National Academy of Sciences*, **9** (1920), 181–236, which quotes extensively from Packard's unpublished diaries and includes a complete bibliography of his writings. No modern evaluation of Packard's impact on American science has been published.

CALVERT E. NORLAND

PADOA, ALESSANDRO (*b*. Venice, Italy, 14 October 1868; *d*. Genoa, Italy, 25 November 1937), *mathematical logic, mathematics*.

Padoa attended a secondary school in Venice, the engineering school in Padua, and the University in Turin, from which he received a degree in mathematics in 1895. He taught in secondary schools at Pinerolo, Rome, and Cagliari, and (from 1909) at the Technical Institute in Genoa.

Padoa was the first to devise a method for proving that a primitive term of a theory cannot be defined within the system by the remaining primitive terms. This method was presented in his lectures at Rome early in 1900 and was made public at the International Congress of Philosophy held at Paris later that year. He defined a system of undefined symbols as irreducible with respect to the system of unproved propositions when no symbolic definition of any undefined symbol can be deduced from the system of unproved propositions. He also said:

> To prove that the system of undefined symbols is irreducible with respect to the system of unproved propositions, it is necessary and sufficient to find, for each undefined symbol, an interpretation of the system of undefined symbols that verifies the system of unproved propositions and that continues to do so if we suitably change the meaning of only the symbol considered ["Essai . . .," p. 322].

Although it took the development of model theory to bring out the importance of this method in the theory of definition, Padoa was already convinced of its significance. (A proof of Padoa's method was given by Alfred Tarski in 1926 and, independently, by J. C. C. McKinsey in 1935.)

In lectures at the universities of Brussels, Pavia, Bern, Padua, Cagliari, and Geneva, Padoa was an effective popularizer of the mathematical logic developed by Giuseppe Peano's "school," of which Padoa was a prominent member. He was also active in the organization of secondary school teachers of mathematics and participated in many congresses of philosophy and mathematics. In 1934 he was awarded the ministerial prize in mathematics by the Accademia dei Lincei.

BIBLIOGRAPHY

I. ORIGINAL WORKS. A list of 34 of Padoa's publications in logic and related areas of mathematics (about half of all his scientific publications) is in Antonio Giannattasio, "Due inediti di Alessandro Padoa," in *Physis* (Florence), **10** (1968), 309–336. To this may be added three papers presented to the Congrès International de Philosophie Scientifique at Paris in 1935 and published in *Actualités scientifiques et industrielles* (1936): "Classes et pseudo-classes," no. 390, 26–28; "Les extensions successives de l'ensemble des nombres au point de vue déductif," no. 394, 52–59; and "Ce que la logique doit à Peano," no. 395, 31–37.

Padoa's method was stated in "Essai d'une théorie algébrique des nombres entiers, précédé d'une introduction logique à une théorie déductive quelconque," in *Bibliothèque du Congrès international de philosophie, Paris, 1900*, III (Paris, 1901), 309–365. An English trans. (with references to Padoa's method) is in Jean van Heijenoort, ed., *From Frege to Gödel: A Source Book in Mathematical Logic 1879–1931* (Cambridge, Mass., 1967), 118–123. Padoa's major work is "La logique déductive dans sa dernière phase de développement," in *Revue de métaphysique et de morale*, **19** (1911), 828–832; **20** (1912), 48–67, 207–231, also published separately, with a preface by G. Peano (Paris, 1912).

II. SECONDARY LITERATURE. There is no biography of Padoa. Some information on his life and work may be found in the obituaries in *Bollettino dell'Unione matematica italiana*, **16** (1937), 248; and *Revue de métaphysique et de morale*, **45** (1938), Apr. supp., 32; and in F. G. Tricomi, "Matematici italiani del primo secolo dello stato unitario," in *Memorie della Accademia delle scienze di Torino*, 4th ser., no. 1 (1962), 81.

HUBERT C. KENNEDY

PAGANO, GIUSEPPE (*b*. Palermo, Sicily, 21 September 1872; *d*. Palermo, 9 August 1959), *physiology*.

For a detailed study of his life and work, see Supplement.

PAINLEVÉ, PAUL (*b*. Paris, France, 5 December 1863; *d*. Paris, France, 29 October 1933), *mathematics*.

Painlevé's father, Léon Painlevé, and grandfather, Jean-Baptiste Painlevé, were lithographers. Through his grandmother, Euphrosine Marchand, he was a descendant of Napoleon I's valet. As gifted in literature as in the sciences, Painlevé received excellent marks in secondary school.

After hesitating between a career as a politician, engineer, and researcher Painlevé chose the last, which had been offered him by the École Normale Supérieure. Admitted in 1883, he received his *agrégation* in mathematics in 1886. He worked for a time at Göttingen, where Schwarz and Klein were teaching, and at the same time completed his doctoral dissertation (1887). Painlevé became professor at Lille in 1887. In 1892 he moved to Paris, where he taught at the Faculty of Sciences and the École Polytechnique, the Collège de France (1896), and the École Normale Supérieure (1897).

Painlevé received the Grand Prix des Sciences Mathématiques (1890), the Prix Bordin (1894), and the Prix Poncelet (1896); and was elected a member of the geometry section of the Académie des Sciences in 1900. In 1901 he married Marguerite Petit de Villeneuve, niece of the painter Georges Clairin; she died at the birth of their son Jean (1902), who became one of the creators of scientific cinematography.

Painlevé was interested in the infant field of aviation, and as the passenger with Wilbur Wright and Henri Farman he even shared for a time the record for duration of biplane flights (1908). He was a professor at the École Supérieure d'Aéronautique (1909) and president of several commissions on aerial navigation.

In 1910 Painlevé turned to politics. Elected a deputy from the fifth arrondissement of Paris, the "Quartier Latin," he headed naval and aeronautical commissions established to prepare for the country's defense. In 1914 he created the Service des Inventions pour les Besoins de la Défense Nationale, which became a ministry in 1915. Minister of war in 1917, Painlevé played an important role in the conduct of military operations: he supported the efforts of the Army of the Near East in the hope of detaching Austria-Hungary from the German alliance. He conducted the negotiations with Woodrow Wilson over the sending of American combat troops to France. He also had Foch appointed as head of the allied chiefs of staff.

In 1920 Painlevé was commissioned by the Chinese government to reorganize the country's railroads. From 1925 to 1933 he was several times minister of war and of aviation, president of the Council of Ministers, and an active participant in the League of Nations and in its International Institute of Intellectual Cooperation.

As a mathematician Painlevé always considered questions in their greatest generality. After his first works concerning rational transformations of algebraic curves and surfaces, in which he introduced biuniform transformations, he was remarkably successful in the study of singular points of algebraic differential equations. His goal was to obtain general propositions on the nature of the integral considered as a function of the variable and of the constants, par-

ticularly through distinguishing the "perfect integrals," definable throughout their domain of existence by a unique development.

In old problems in which the difficulties seemed insurmountable, Painlevé defined new transcendentals for singular points of differential equations of a higher order than the first. In particular he determined every equation of the second order and first degree whose critical points are fixed. This work was presented in notes published in the *Comptes rendus . . . de l'Académie des sciences* beginning in 1887.

The results of these studies are applicable to the equations of analytical mechanics which admit rational or algebraic first integrals with respect to the velocities. Proving, in the words of Hadamard's *éloge*, that "continuing [the work of] Henri Poincaré was not beyond human capacity," Painlevé extended the known results concerning the n-body problem. He also corrected certain accepted results in problems of friction and of the conditions of certain equilibriums when the force function does not pass through a maximum.

BIBLIOGRAPHY

I. Original Works. Painlevé's mathematical writings are "Sur les lignes singulières des fonctions analytiques," in *Annales de la Faculté des sciences de Toulouse* (1888), his doctoral dissertation; "Sur la transformation des fonctions harmoniques et les systèmes triples de surfaces orthogonales," in *Travaux et mémoires de la Faculté des sciences de Lille*, **1** (Aug. 1889), 1–29; "Sur les équations différentielles du premier ordre," in *Annales scientifiques de l'École normale supérieure*, 3rd ser., **8** (Jan.–Mar. 1891), 9–58, 103–140; (Aug.–Sept. 1891), 201–226, 267–284; **9** (Jan. 1892), 9–30; (Apr.–June 1892), 101–144, 283–308; "Mémoire sur la transformation des équations de la dynamique," in *Journal de mathématiques pures et appliquées*, 4th ser., **10** (Jan. 1894), 5–92; "Sur les mouvements et les trajectoires réels des systèmes," in *Bulletin de la Société mathématique de France*, **22** (Oct. 1894), 136–184; *Leçons sur l'intégration des équations de la dynamique et applications* (Paris, 1894); *Leçons sur le frottement* (Paris, 1895); *Leçons sur l'intégration des équations différentielles de la mécanique et applications* (Paris, 1895); *Leçons sur la théorie analytique des équations différentielles professées à Stockholm . . .* (Paris, 1897); "Sur les équations différentielles dont l'intégrale générale est uniforme," in *Bulletin de la Société mathématique de France*, **28** (June 1900), 201–261; "Sur les équations différentielles du second ordre et d'ordre supérieur dont l'intégrale générale est uniforme," in *Acta mathematica*, **25** (Sept. 1900), 1–80; and contributions to Émile Borel, *Sur les fonctions de variables réelles et les développements en série de polynomes* (Paris, 1905), 101–147; and Pierre Boutroux, *Leçons sur les fonctions* *définies par les équations différentielles du premier ordre* (Paris, 1908), 141–187.

Painlevé's other works include *L'aviation* (Paris, 1910; 2nd ed., 1911), written with Émile Borel; *Cours de mécanique de l'École polytechnique*, 2 vols. (Paris, 1920–1921); *Les axiomes de la mécanique. Examen critique et note sur la propagation de la lumière* (Paris, 1922); *Cours de mécanique* (Paris, 1929), written with Charles Platrier; *Leçons sur la résistance des fluides non visqueux*, 2 vols. (Paris, 1930–1931); and *Paroles et écrits* (Paris, 1936).

II. SECONDARY LITERATURE. See the collection made by the Société des Amis de Paul Painlevé, *Paroles et écrits de Paul Painlevé* (Paris, 1936), with prefaces by Paul Langevin and Jean Perrin; and Jean Painlevé, *Textes inédites et analyse des travaux scientifiques jusqu'en 1900*.

LUCIENNE FÉLIX

PAINTER, THEOPHILUS SHICKEL (*b.* Salem, Virginia, 22 August 1889; *d.* Fort Stockton, Texas, 5 October 1969), *genetics, cytogenetics.*

Painter is best known for introducing the use of the giant salivary gland chromosomes of the fruit fly, *Drosophila melanogaster*, into cytogenetic studies. With this material he was able to demonstrate in 1933 what until then had been only an assumption: that Mendelian genes could be identified with specific bands on physical structures in cell nuclei, the chromosomes.

The son of the Reverend Franklin Verzelius Newton Painter, professor of modern languages at Roanoke College, Virginia, and Laura Shickel, Painter received his early education from tutors. He entered Roanoke College in 1904 and received the B.A. in 1908. He then went to Yale on a fellowship in chemistry but soon found biology more to his liking. Working under A. Petrunkevitch and R. G. Harrison, Painter received the Ph.D. in 1913, then studied during 1913–1914 under Theodor Boveri at Würzburg. Returning to Yale in 1914, he served as an instructor in zoology for two years (1914–1916); during the summers of 1914 and 1915 he was an instructor in the invertebrate zoology course at the Marine Biological Laboratory, Woods Hole, Massachusetts. In 1916 he accepted a post as adjunct professor of zoology at the University of Texas, beginning a long and distinguished association with that university. In 1922 he was promoted to full professor and in 1939 became a distinguished professor in the graduate school. From 1944 to 1946 he was acting president of the university and from 1946 until 1952 was president. In 1952 he resigned the presidency to return to full-time teaching and research. In 1966 he retired but continued his research and participated regularly in graduate seminars until his death.

By the 1930's most geneticists were convinced that Mendel's genes had an actual physical existence and were arranged in a linear fashion on the cell's chromosomes (the common analogy was to beads on a string). At that time, however, there was no direct proof for the validity of this assumption. The linkage maps for *Drosophila* prepared by T. H. Morgan, C. B. Bridges, and their associates between 1915 and 1925 were only formalisms derived almost wholly from analysis of crossover frequencies (that is, breeding data). Positions on these maps represented only relative distances between the various genes. Several studies in the 1920's had suggested that a point-by-point correspondence between linkage maps and the structure of actual chromosomes could be determined. In 1929, however, Painter and H. J. Muller (both at the University of Texas) cast doubt on this idea by showing that while the linear sequence determined by crude cytological methods and that determined by crossover data were the same, the spatial correspondences were not. That is, there appeared to be long areas of the chromosome in which no crossovers occurred, and other, shorter areas where a great deal seemed to occur.

The major problem in determining the correspondence between linkage maps and the structure of chromosomes was the small size of the latter. Workers before the 1930's studied various types of somatic (body) or oögonial cells, in which the chromosomes were so small that detailed observation, particularly of the banding pattern, was impossible. Larger chromosomes had been observed in the 1880's and, in the 1920's, in salivary gland cells of young larval dipterans; but these cells had been found difficult to work with and observational studies with them had not been carried very far. In 1930 Painter found that if older larvae (almost ready to pupate) were used, large and easily observable chromosomes could be obtained. They offered ideal material for the study of small chromosomal segments and thus for the detection of modifications of chromosomal structure that could be correlated with variations in linkage maps. Painter's paper of December 1933 established a method that made possible the detailed analysis of *Drosophila* and other insect chromosomes and provided the long-awaited confirmation of the chromosomal theory of heredity—the idea that genes are located on chromosomes.

In addition to studying the chromosomes of insects, Painter pioneered in the structural analysis of human chromosomes. He provided new techniques for studying human karyotypes (the full complement of chromosomes from a species, observed in squash or other preparations) and suggested ways of relating chromosomal aberrations to disease. He was also interested in the relationship of heterochromatin and chromosome puffing to ribonucleic acid (RNA) synthesis, and at the time of his death he was studying the nucleic acids found in the royal jelly of the honeybee. It was Painter's original count of the human karyotype, in 1929 and subsequently, that established the erroneous total of 48 chromosomes (rather than 46), believed for many years to be the actual number for the human species.

Painter was a member of the U.S. National Academy of Sciences (1938), the American Philosophical Society (1939), the American Society of Zoologists, the American Genetics Society, the American Society of Naturalists, and Sigma Xi. He received the David Girard Elliott Medal of the National Academy of Sciences (1934) in recognition of his work on the giant chromosomes of *Drosophila*, and the M. D. Anderson Award from the University of Texas, for his "scientific creativity and teaching" (1969). He was also awarded an honorary D.Sc. from Yale (1936) and an LL.D. from Roanoke College (1942).

In 1917 Painter married Anna Mary Thomas, whom he had met at Woods Hole in 1914. They had four children.

BIBLIOGRAPHY

Painter's last scientific publication, "The Origin of the Nucleic Acid Bases Found in the Royal Jelly of the Honeybee," appeared posthumously in *Proceedings of the National Academy of Sciences . . .*, **64** (Sept. 1969), 64–66. His major work on the giant salivary gland chromosomes in *Drosophila* is "A New Method for the Study of Chromosome Rearrangements and Plotting of Chromosome Maps," in *Science*, **78** (1933), 585–586.

No full-scale biographical study exists, although presumably one will be issued by the National Academy of Sciences. My major sources of information have been a brochure from the University of Texas, announcing the M. D. Anderson Award in 1969 (supplied by Mrs. T. S. Painter), and several personal communications from Mrs. Painter in 1970.

GARLAND E. ALLEN

PALEY, WILLIAM (*b.* Peterborough, England, July 1743; *d.* Lincoln, England, 25 May 1805), *natural theology*.

Paley was the eldest son of Elizabeth and William Paley. His father, a graduate of Christ's College, Cambridge, was a vicar and minor canon of the Church of England, and was headmaster of the grammar school at Giggleswick. Paley was educated at his father's school and entered Christ's College, Cambridge, in 1759, receiving his B.A. in January 1763. Following graduation he began to teach at an academy in Greenwich, but in June 1766 he was elected a fellow of his college and returned to Cambridge. Paley's last formal connection with Christ's College was as tutor from 1771 to 1774.

While an undergraduate, Paley had shown promise in mathematics; he continued his interest in that field during his tenure as tutor, when he corrected the proofs of *Miscellanea analytica*, written by Edward Waring, the Lucasian professor of mathematics. Paley had been ordained a deacon in the Church of England by 1766. While a fellow of Christ's College, he gave a lecture entitled "Metaphysics, Morals and the Greek Testament" and discussed the *Being and Attributes of God*, written by the Reverend Samuel Clarke. It was at this time that Paley became friendly with a number of other fellows interested in natural theology, including John Jebb, and joined the Hyson Club. His interest in metaphysics and in Clarke's work led him to support an attempt of other latitudinarians to relax the stringency of the church's organization and government.

In 1775 Paley was presented with the rectorship of Musgrave in Cumberland, the first of several ecclesiastical posts he was to hold. By 1782 he had become archdeacon of Carlisle, and his financial position was assured. Written at this time, his *Principles of Moral and Political Philosophy* was taken from his lectures and enjoyed wide success. His rise in the church continued, and by the end of 1785 he had become chancellor of the diocese of Carlisle. Having become interested in the abolition of the slave trade, he lectured against slavery and became, on the local level, very much of a public figure.

Paley's abandonment of a purely academic career may be seen in his refusal of the mastership of Jesus College, Cambridge, in 1792, for financial reasons. Instead of returning to the university he continued to accumulate increasingly lucrative ecclesiastical holdings and continued to publish. By 1794 his writings advanced his religious career, for his *Evidences of Christianity* was warmly regarded by the church and he was rewarded with new benefices. In 1795 Paley received the Doctorate of Divinity at Cambridge and the rectorship of Bishop-Wearmouth, a post worth £1,200 a year. He remained in residence at Carlisle and was appointed a justice of the peace for the region.

Although much in demand as a public speaker, illness in 1800 forced Paley to give up this aspect of his career. In 1802 he published his most significant book, *Natural Theology; or, Evidences of the Existence and*

Attributes of the Deity. He died three years later. Paley was married twice: to Jane Hewitt, who died in 1791; and in 1795 to a Miss Dobinson. His son by his first marriage, Edmund, wrote a life of his father.

Paley's fame is as a writer of textbooks. His works were used at Cambridge for nearly half a century after his death. His own religious views inclined toward liberalism; and while he never embraced the Unitarian point of view, as did so many of his friends, he was not hostile toward Arianism or Unitarianism. *Natural Theology* is perhaps most significant for Paley's efforts to reconcile liberal orthodox Christianity with divine providence. As an undergraduate at Cambridge, Charles Darwin read much of Paley's writings:

> In order to pass the B.A. examination, it was also necessary to get up Paley's *Evidences of Christianity*, and his *Moral Philosophy*. This was done in a thorough manner, and I am convinced that I could have written out the whole of the *Evidences* with perfect correctness, but not of course in the clear language of Paley. The logic of this book, and, as I may add, of his *Natural Theology*, gave me as much delight as did Euclid. The careful study of these works, without attempting to learn any part by rote, was the only part of the academical course which, as I then felt and as I still believe, was of the least use to me in the education of my mind. I did not at that time trouble myself about Paley's premises; and taking these on trust, I was charmed and convinced by the long line of argumentation [*Charles Darwin's Autobiography*, Sir Francis Darwin, ed. (New York, 1961), 34–35].

Paley's underlying belief, expressed in *Natural Theology*, was that the world is essentially a happy place. Nature was God and God was good; the proof of the goodness of God and Nature could be found in day-to-day experiences. In the most often quoted passage of his work Paley says:

> It is a happy world after all. The air, the earth, the water teem with delighted existence. In a spring noon, or a summer evening, on whichever side I turn my eyes, myriads of happy beings crowd upon my view. "The insect youth are on the wing." Swarms of new-born flies are trying their pinions in the air. Their sportive motions, their wanton mazes, their gratuitous activity, testify their joy and the exultation which they feel in their lately discovered faculties. . . . The whole winged insect tribe, it is probable, are equally intent upon their proper employments, and under every variety of constitution, gratified, and perhaps equally gratified, by the offices which the author of their nature has assigned to them [*Natural Theology*, p. 236].

The work itself, written as a treatise against atheism and teleological in the extreme, shows how the workings of the body, the functions of animals and plants, and the arrangement of the human frame all manifest the workings of the deity. Throughout, Paley dwells on how things could not possibly have been organized otherwise, using mechanical examples to illustrate further the existence of the deity. In his description of the muscles Paley attempted to work out a dynamic approach to muscular action (*Natural Theology*, in *The Works of William Paley*, I [Cambridge, 1830], 71; hereafter cited by page only). In his treatment of organisms Paley cited symmetry as a further proof of divinity (pp. 101, 166). He also used the relations of "parts one to another" to show the works of God; his classic example was that the sexes are "manifestly made for each other" (p. 143).

Believing that the various parts of animals complement each other, Paley used the term "compensation" —which, to him, was "a relation, when the defects of one part, or one organ, are supplied by the structure of another part, or of another organ" (p. 146). He showed how the "short, unbending neck of the elephant" is compensated by the trunk, and in the course of this description he took issue with the ideas of Erasmus Darwin.

> If it be suggested, that this proboscis may have been produced in a long course of generations, by the constant endeavour of the elephant to thrust out his nose, (which is the general hypothesis by which it has been lately attempted to account for the forms of animated nature), I would ask, how was the animal to subsist in the mean time; during the process; *until* this prolongation of snout were completed? What was to become of the individual, whilst the species was perfecting? (pp. 146–147)

Paley took no stand on evolution as such, believing that "our business is simply to point out the relation which an organ bears to the peculiar figure of the animal to which it belongs" (p. 147).

After dealing with animate matter Paley turned to a very brief treatment of the elements—a highly simplistic and almost Aristotelian one. His comments on astronomy sum up his idea of the purpose of scientific studies: "My opinion of [it] has always been, that it is *not* the best medium through which to prove the agency of an intelligent Creator; but that, this being proved, it shows, beyond all other sciences, the magnificence of his operation" (p. 197). Paley's conclusion seems quite clear. "In the observable portion of nature organisms are formed one beneath another . . . the Deity can mould and fashion the parts in material nature so as to fulfill any purpose whatsoever which he is pleased to appoint" (p. 280).

Paley's significance in the history of science is twofold. His writings on natural theology clearly reveal the changed framework of the late eighteenth and early nineteenth centuries as opposed to the late seventeenth century. The purely physical universe no longer could suffice to furnish proof for God's existence, but emphasis had turned to biological evidence to show the beneficence of the deity's workings. On the whole Paley's universe was a benevolent one. This very benevolence, coupled with Paley's popularity as a textbook writer, helped to create the atmosphere so hostile to Charles Darwin in the 1850's and 1860's.

BIBLIOGRAPHY

I. ORIGINAL WORKS. Paley's published works were collected as *The Works of William Paley, D.D.*, 6 vols. (Cambridge, 1825; 2nd ed., 1830). Included are *Natural Theology; or, Evidences of the Existence and Attributes of the Deity, Sermons on Various Subjects, Horae Paulinae, Clergyman's Companion, The Young Christian Instructed, Principles of Moral and Political Philosophy*, and *A View of the Evidences of Christianity*. In 1820 *Natural Theology* was reprinted for the twentieth time, and in 1835–1839, it formed the core of a work entitled *Natural Theology* by Lord Brougham. Paley's work was translated into Spanish, French, and Italian.

II. SECONDARY LITERATURE. G. W. Meadley, "Memoirs of William Paley, D.D.," in *The Works of William Paley, D.D.*, I, is the best treatment of the subject. Meadley had been a close friend of Paley, and his work is more detailed than Edmund Paley's "Life of William Paley," which is prefixed to the 1825 ed. of Paley's *Works*. Other treatments are derivative and depend on Meadley's. Passing reference is paid to Paley by L. E. Elliott-Binns, *Religion in the Victorian Era* (London–Redhill, 1946). Although the most extensive treatment is to be found in Leslie Stephen's article in *Dictionary of National Biography*, XV (1967–1968), 101–107, Stephen is concerned less with Paley's influence on science than in the recounting of personal anecdotes.

JOEL M. RODNEY

PALISA, JOHANN (*b.* Troppau, Austrian Silesia [now Czechoslovakia], 6 December 1848; *d.* Vienna, Austria, 2 May 1925), *astronomy*.

While a student Palisa became known for his skill in mathematics, and in 1866 he entered the University of Vienna to study that subject. He soon became attracted to astronomy, the science to which he devoted the rest of his life. In 1870 he was appointed assistant astronomer at the Vienna observatory. There he performed routine observations: at night, of positions of stars with the meridian circle; during the day, of the spots on the sun's disk. In 1871 Palisa became associate astronomer at the Geneva observatory. A few months later, at the age of twenty-three, he was appointed director of the Austro-Hungarian naval observatory at Pola, with the rank of commander. His main task was precise timekeeping by astronomical observations. For this purpose a new meridian circle was acquired; and he himself invented the "Chronodeik," a small instrument for determining time by measuring equal heights of stars east and west of the meridian. Palisa was also eager to promote scientific research. Inspired by Oppolzer, he began to observe asteroids systematically with the small telescope at Pola; he obtained a great many positions of asteroids already known and discovered twenty-eight new ones between 1874 and 1880.

Meanwhile, a splendid new observatory had been built at Vienna with a refracting telescope of twenty-seven inches aperture, the largest at that time. Palisa agreed to join the observatory as associate astronomer after being assured that the large telescope would always be at his disposal. From 1883 to the fall of 1924 he used it to the utmost, an undertaking made arduous because there was no automation: the telescope, the observer's stage, and the dome were moved manually. The discovery of 120 asteroids is Palisa's best-known work but not his greatest. About 1893 Max Wolf at Heidelberg had begun to discover asteroids photographically; and the two scientists, after a short period of rivalry, achieved an effective collaboration: Wolf continued to discover asteroids near their opposition, while Palisa followed them to great distances with his powerful telescope so that their orbits could be determined with greater precision. Valuable products of Palisa's work are two catalogs containing the positions of 4,696 stars and, from his collaboration with Wolf, the 210 sheets of the Palisa-Wolf photographic charts of the sky. Undoubtedly, Palisa was the most effective observer of the Austrian astronomers.

BIBLIOGRAPHY

I. ORIGINAL WORKS. Palisa's writings are listed in Poggendorff, III, 1000–1001; IV, 1113; V, 937–938; and VI, 1941. They include "Das Meridian-Instrument zu Pola," in *Repertorium für Experimental-Physik physikalische Technik . . .*, **13** (1877); "Beobachtungen während der Sonnenfinsternis 6. Mai 1883," in *Sitzungsberichte der Akademie der Wissenschaften in Wien*, Abt. IIa, **88** (1884), 1018–1031; "Katalog von 1238 Sternen," in *Denkschriften der Akademie der Wissenschaften* (Vienna), **67** (1899), written with F. Bidschof; "Sternlexikon von −1° bis +19° Deklination," in *Annalen der Universitätssternwarte in Wien*, 4th ser., **17** (1902); "Über einen Plan zur Herstellung von Ekliptikal-Sternkarten," in *Vierteljahrsschrift der Astronomischen Gesellschaft* (Leipzig), **39** (1904) and **41** (1906); and "Katalog von 3458 Sternen," in *Annalen der Universitätssternwarte in Wien*, 4th ser., **19** (1908). There are also many observational notes in *Astronomische Nachrichten*, **76–222** (1870–1924).

II. SECONDARY LITERATURE. See J. Hepperger, "J. Palisa," in *Astronomische Nachrichten*, **225** (1925); S. Oppenheim, "J. Palisa," in *Vierteljahrsschrift der Astronomischen Gesellschaft* (Leipzig), **60** (1925); and J. Rheden, *Johann Palisa* (Vienna, 1925), a pamphlet.

KONRADIN FERRARI D'OCCHIEPPO

PALISSY, BERNARD (*b.* La Capelle Biron, France, *ca.* 1510; *d.* Paris, France, *ca.* 1590), *natural history, hydrology.*

Palissy was first trained in the manufacture and decoration of stained glass windows. As his profession became less in demand, however, he took up land surveying in order to support his wife and children (of whom there were at least six). Some time around 1539 he became interested in enameled pottery and, after sixteen years of tireless experimentation (during which, by his own account, he burned his furniture and floorboards to fuel his kiln), perfected a technique for making a "rustic" enameled earthenware that brought him fame and a modest fortune. Some of his works are preserved in the Louvre and the Cluny Museum in Paris and in the Victoria and Albert Museum and the British Museum in London. These extant pieces are molded and decorated with modeling or applied ornaments, often in patterns derived from contemporary engravings; Palissy probably never used the potter's wheel, and no identifying mark of his is known. The governor of Saintes, where Palissy settled, was the constable Anne, Duc de Montmorency, who had a keen interest in the fine arts and became Palissy's patron.

Palissy converted to Protestantism in about 1546. He was one of the first Huguenots in Saintes, and was much persecuted for his religion. He was imprisoned in Bordeaux around 1559 and, had it not been for Anne de Montmorency, who took his case directly to the queen mother, Catherine de Médicis, he would almost certainly have been executed. The queen mother appointed him *inventeur des rustiques figulines du roy*, and commissioned him to decorate the new Tuileries palace. Palissy thus became established in Paris, where in 1575 he began to give public lectures on natural history. Despite his lack of formal education Palissy's lectures, according to Désiré Leroux, attracted the most learned men in the capital.

Palissy wrote two major books, *Recepte véritable*, published in 1563, and *Discours admirables*, published in 1580. (A small pamphlet describing the building of a grotto for Anne de Montmorency was also published in 1563.) The form of the two works is similar; each is a dialogue, between "Demande" and "Réponse" in the *Recepte*, and between "Théorique" and "Pratique" in the *Discours*. "Réponse" and "Pratique" give voice to Palissy's own ideas and concepts.

In *Recepte véritable*, Palissy discussed a wide variety of topics, including agriculture (for which he proposed better methods for farming and for the use of fertilizers), geology (in which he touched upon the origin of salts, springs, precious stones, and rock formations), mines, and forestry. He also suggested plans for an ideal garden, to be decorated with his earthenware and with biblical quotations, and discussed the founding and persecution of the Protestant church at Saintes. As part of this ecclesiastic history he included plans for a spiral fortress, which he claimed would be invincible and which would presumably offer a refuge for Protestants in time of war.

The second book, *Discours admirables*, probably incorporates Palissy's Paris lectures. It, like the earlier work, deals with an impressive array of subjects: agriculture, alchemy, botany, ceramics, embalming, engineering, geology, hydrology, medicine, metallurgy, meteorology, mineralogy, paleontology, philosophy, physics, toxicology, and zoology. The book is divided into several chapters, the first and longest of which is concerned with water. The others take up metals and their nature and generation; drugs; ice; different types of salts and their nature, effects, and methods of generation; characteristics of common and precious stones; clay and marl; and the potter's art.

Palissy's views on hydrology and paleontology, as expressed in the *Discours*, are of particular interest. He was one of the few men of his century to have a correct notion of the origins of rivers and streams, and he stated it forcefully, denying categorically that rivers can have any source other than rainfall. An early advocate of the infiltration theory, he refuted, with great skill and logic, the old theories that streams came from seawater or from air that had condensed into water. He also wrote on the principles of artesian wells, the recharging of wells from nearby rivers, and forestation for the prevention of soil erosion, and presented plans for constructing "fountains" for domestic water supply.

Palissy discussed fossils extensively. Like Xenophanes of Colophon, he believed them to be remnants of animals and plants. He firmly rejected the idea that they were detritus of the biblical flood, suggesting that inland fossils are found on site as the result of the congelation of a lake. He recognized the relation between these fossils and living species and, in some cases, extinct ones. He was one of the first to hold a reasonably correct view of the process of petrification.

(Duhem in *Études sur Léonard de Vinci* has pointed out that all these ideas may well be derived from Cardano's *De subtilitate*, with which Palissy was familiar, and hence from the thought of Leonardo da Vinci.)

Palissy held other advanced views. From experimentation he concluded that all minerals with geometric crystal forms must have crystallized in water; his classification of salts was nearly correct; and he suggested the concept of superposition for the development of sedimentary rocks. In his writings on medicine he demonstrated that potable gold was neither potable nor beneficial, and he showed that mithridate, a remedy composed of some 300 ingredients, was useless and probably harmful. He presented observations in support of his scientific ideas, and scathingly denounced established authorities if their findings did not agree with his own data. While there is some question concerning his originality—La Rocque discussed his dependence on thirty-one other writers on earth sciences whose works were available in the sixteenth century, and Thorndike charged him with plagiarizing Jacques Besson's *L'art et la science de trouver les eaux* of 1567—there is little doubt that Palissy was probably one of the first men in France to teach natural sciences from facts, specimens, and demonstrations rather than hypotheses.

Although he was well known as a potter, Palissy's scientific work was not widely recognized in his lifetime. In 1588, soon after religious warfare once more broke out in France, Palissy was again imprisoned. He was taken to the Conciergerie, then transferred to the Bastille, where he died.

BIBLIOGRAPHY

I. ORIGINAL WORKS. Palissy's works are *Recepte véritable par laquelle tous les hommes de la France pourront apprendre à multiplier et augmenter leurs trésors* (La Rochelle, 1563); *Architecture et ordonnance de la grotte rustique de Monseigneur le duc de Montmorency* (La Rochelle, 1563; repr. Paris, 1919); and *Discours admirables de la nature des eaux et fontaines* (Paris, 1581), translated by Aurèle La Rocque as *The Admirable Discourses of Bernard Palissy* (Urbana, Ill., 1957).

Collected eds. of Palissy's works include those of B. Faujas de Saint-Fond and N. Gobet (Paris, 1777), which contains incorrectly attributed works and a dedication to Benjamin Franklin; and of Anatole France (Paris, 1880).

II. SECONDARY LITERATURE. See C. L. Brightwell, *Palissy the Potter; or the Huguenot, Artist and Martyr* (New York, 1835); H. Morley, *Palissy the Potter*, 2 vols. (London, 1852); E. Dupuy, *Bernard Palissy, l'homme, l'artiste, le savant, l'écrivain* (Paris, 1894); Désiré Leroux, *La vie de Bernard Palissy* (Paris, 1927); Lynn Thorndike,

A History of Magic and Experimental Science, V (New York, 1941), 441, 465, 596–599; H. R. Thompson, "The Geographical and Geological Observations of Bernard Palissy, the Potter," in *Annals of Science*, **10**, no. 2 (1954), 149–165; and A. K. Biswas, *History of Hydrology* (Amsterdam, 1970), 149–155.

MARGARET R. BISWAS
ASIT K. BISWAS

PALLADIN, VLADIMIR IVANOVICH (*b*. Moscow, Russia, 23 July 1859; *d*. Petrograd [now Leningrad], U.S.S.R., 3 February 1922), *biochemistry, plant physiology*.

Palladin attended the Gymnasium in Moscow and in 1883 graduated from Moscow University, where he remained to prepare for a career in teaching. In 1886 he defended his thesis, "The Meaning of Oxygen for Plants," for the M.A. From 1886 Palladin was an instructor and then a professor of botany at the Institute of Agriculture and Forestry in Novaya Aleksandriya, and from 1889 he was professor of botany at Kharkov University. In 1889 he defended his doctoral dissertation, "The Influence of Oxygen on the Decomposition of Proteins in Plants," at the University of Warsaw, to which he had transferred in 1897. In 1901 Palladin was appointed to the chair of physiology at St. Petersburg University and in the Higher Courses for Women. Here Palladin began his lengthy teaching career; among his students were the physiologists S. P. Kostychev, N. A. Maksimov, and D. A. Sabinin.

Palladin was the author of two well-known textbooks: *Fisiologia rasteny* ("Plant Physiology"), which for more than thirty years was used as the basic text in all Russian higher educational institutions, and *Anatomia rasteny* ("Plant Anatomy"). In 1906 he was elected corresponding member of the Academy of Sciences, and in 1914 academician. He then retired from St. Petersburg University and conducted his scientific work at the academy.

In his two graduate dissertations Palladin showed the existence of a close bond between the two most important biological processes: respiration and protein metabolism. He established that the carbons formed in plants are the products of the incomplete oxidation of proteins, for which the assimilation of oxygen from the air is necessary. Palladin also studied the process of evaporation of water, the content of proteins and mineral substances, the process of respiration in green and etiolated plants, and the conditions under which etiolated plants become green. He then advanced to questions concerning the chemical physiology of plants, first studying the transformation of nitrogenous substances and the energy processes in plants, and then studying the process of plant respiration, to which he devoted the last year of his life.

Three basic stages can be noted in Palladin's research into plant respiration. First, basing his work on the discovery of oxidizing enzymes at the end of the nineteenth century, he concluded that the oxidation-reduction processes, which represent a chain of strictly coordinated enzyme reactions, are the basis of respiration in plants. His monograph "The Respiration of Plants as the Sum of the Enzyme Processes" (1907) attracted much attention from Russian and foreign scientists; it clarified the details of the anaerobic and aerobic phases of respiration from the point of view of the activity of specific enzymes, which transform in succession the intermediate products of respiration. A careful study of the action of the oxidase and peroxidase enzymes showed that their oxidizing energy is limited and cannot have an oxidizing effect on the respiratory substrate: carbohydrates or the products of carbohydrate decomposition. In the second stage of his research into respiration, Palladin sought to discover the intermediaries between oxidases and carbohydrates. They proved to be aromatic compounds of the polyphenol type, which he called respiratory chromogens. Palladin formulated the role of the respiratory chromogens in "The Significance of Respiratory Pigments in the Oxidizing Processes of Plants and Animals" (1912). A new point in this work was the discovery of intermediate agents: carriers of oxygen. But the process of respiration was still understood in accordance with Lavoisier's hypothesis of a process analogous to burning. In the last stage of his research Palladin showed that this hypothesis was false. He stated that respiratory chromogens do not activate the oxygen in the air; instead, they activate the hydrogen in carbohydrates with the aid of the enzyme reductase (dehydrogenase). Respiratory chromogens were thus carriers of hydrogen, not oxygen. He discovered that simultaneously with the decomposition of water, the oxygen of which goes into the oxidation of the respiratory substrate and forms carbonic acid, the hydrogen is temporarily bonded to the respiratory pigment. This work predated that of H. O. Wieland, with whom the phenomenon is usually associated.

This first phase of respiration is the most basic. It occurs under anaerobic conditions and the carbonic acid separated out is formed by the respiratory substrate (carbohydrates), not by atmospheric oxygen. The oxygen in the air takes part only in the second phase of respiration, interacting with hydrogen in the pigments and restoring their activity. Palladin's

theory of respiration brought him an international reputation. His concept of the active role of hydrogen was new, as well as his theory of the active participation of water in the oxidation-reduction process of respiration.

BIBLIOGRAPHY

I. ORIGINAL WORKS. Palladin's major works include "Bedeutung des Sauerstoffes für die Pflanzen," in *Byulletin' Moskovskogo obschestva ispytalelii prirody*, **62**, no. 3 (1882), 44–126; *Vliyanie kisloroda na raspadenie belkovykh veshchestv v rasteniyakh* ("The Influence of Oxygen on the Decomposition of Albuminous Substances in Plants"; Warsaw, 1889), which also appeared in *Bericht der Deutschen botanischen Gesellschaft*, **5** (1887), 326–328, **6** (1888), 205–212, and **7** (1889), 126–130; *Fiziologia rasteny* ("Plant Physiology," Kharkov, 1891; 9th ed., Petrograd, 1922); "Fiziologicheskie issledovania nad etiolirovannymi listyami" ("Physiological Research on Etiolated Leaves"), in *Trudy Obschestva ispytatelei prirody pri Imperatorskom khar'kovskom universite*, **26** (1892), 67–68; this article also appeared in *Bericht der Deutschen botanischen Gesellschaft*, **9** (1891), 194–198, 229–232; "Recherche sur la respiration des feuilles vertes et étiolées," in *Revue générale de botanique*, **5** (1893), 449–473; and *Anatomia rasteny* ("Plant Anatomy," Kharkov, 1895; 7th ed. Petrograd, 1924).

Palladin's subsequent writings include "Dykhanie rasteny kak summa fermentativnykh protsessov" ("Breathing of Plants as the Sum of Enzyme Processes"), in *Zapiski Imperatorskoi akademii nauk*. Fiziko–Matematicheskomu, ser. 8, **20** (1907), no. 5, 5–64; "Die Atmungspigmente der Pflanzen," in *Hoppe–Seyler's Zeitschrift fur physiologische Chemie*, **55** (1908), 207–222; "Znachenie vody v protsesse spirtovogo brozhenia i dykhania" ("The Importance of Water in the Process of Alcohol Fermentation and Breathing"), in F. N. Krasheninnikov, ed., *Sbornik statey; posvyashchenny K. A. Timiryazevy ego uchenikami v oznamenovanie Semidesyatogo dlya ego rozhdenia* ("A collection of articles offered to K. A. Timiryazev by his pupils in honor of his seventieth birthday," Moscow, 1916), pp. 1–34; and *Izbrannye trudy* ("Selected Works"; Moscow, 1960), which gives a bibliography of Palladin's works.

II. SECONDARY LITERATURE. For works on Palladin and his work, see S. Kostychev, "V. I. Palladin. Nekrolog" ("V. I. Palladin, Obituary"), in *Zhurnal Russkago botanicheskogo obshchestva*, **7** (1922), 173–186; C. N., "W. Palladin," in *Biochemische Zeitschrift*, **130** (1922) 321–322; S. D. Lvov, "V. I. Palladin kak osnovopolozhnik sovremennogo uchemia o dykhanii" ("V. I. Palladin as the Founder of the Contemporary Theory of Breathing"), in *Vestnik Leningradskogo gosudarstvennogo universiteta*, nos. 4, 5, 50–71; "Zesedanie i doklady, posvyashchennye pamyati V. I. Palladin" ("Session and Reports Devoted to the Memory of V. I. Palladin"), in *Biokhimiya*, **17**, no. 2 (1952), 246–254; B. A. Rubin, "Idei V. I. Palladina i sovremennoe sostoyanie uchenia o dykhanii rasteny" ("V. I. Palladin's Ideas and the Present State of the Theory of Breathing"), in *Vestnik sel'skokhozyaistvennoi nauki*, no. 9 (1960), 39–49; and E. M. Senchenkova, "Vydayush-chysya russky biokhimik i fiziolog rasteny. K stoletiyu so dyna rozhdenia V. I. Palladin" ("Distinguished Russian Biochemist and Plant Physiologist. For the Hundredth Anniversary of the Birthday of V. I. Palladin"), in *Voprosy istorii estestvoznaniya i tekhniki*, no. 9 (1960), 134–138.

E. M. Senchenkova

PALLAS, PYOTR SIMON (*b.* Berlin, Germany, 3 October 1741; *d.* Berlin, 20 September 1811), *natural science, geography*.

Pallas was the son of a professor at the Berlin Medical-Surgical Academy. He received his early education at home and from 1754 to 1759 studied at the Medical-Surgical Academy and the universities of Halle, Göttingen, and Leiden. In his dissertation for the doctorate in medicine, which he defended at Leiden, Pallas refuted the Linnaean classification of worms. From 1761 to 1766 he studied collections of marine animals in England and Holland. In *Elenchus zoophytorum* (1766) he gave a detailed classification of corals and sponges, which had just been transferred by zoologists from the plant kingdom to the animal. In 1763 Pallas was elected a member of the Royal Society of London and the Academia Caesarea Leopoldina.

In 1767 Pallas was invited to work at the St. Petersburg Academy of Sciences. He was elected ordinary academician and had the rank of acting state councillor. For more than forty years Pallas was associated exclusively with the development of Russian science. During his first years there he studied nature and the peoples of the Russian empire, participating in the "Academic expeditions" of 1768–1774. His research as leader of the first Orenburg detachment of the expeditions covered both European Russia and Asia. Pallas and his companions journeyed from St. Petersburg to Moscow; crossed the Volga at Simbirsk (now Ulyanovsk); and explored the Zhiguli Mountains and the southern Urals, the steppes of western Siberia and the Altay, Lake Baikal, and the mountains of Transbaikalia. The easternmost regions visited were the basins of the Shilka and Argun rivers. On his way back to St. Petersburg, Pallas studied the Caspian depression and the lower reaches of the Volga. His results were published in *Reise durch verschiedenen Provinzen des russischen Reichs . . .* (1771–1776), which later appeared in Russian (1773–1778) and in French, English, and Italian. Pallas' writings and the other materials of the "Academic expeditions" enriched natural history by providing massive amounts of empirical data which made it possible to generalize on the geographical distribution of plants and animals and to gain knowledge about the orography, climate, population, and economy of varied and little-studied regions of Russia.

In St. Petersburg Pallas published a series of works, including a monograph on rodents (1778) and on the genus *Astragalus* (1780); assembled a collection of botanical, zoological, and mineralogical specimens; and was the Admiralty historiographer and teacher of the future Emperor Alexander I and his brother Constantine. At the Academy of Sciences he proposed bold projects for new expeditions to northern and eastern Siberia. Pallas' discussions of the formation of mountains (1777) and the variability of animals (1780) are of great importance. Pallas offered a paleogeographic interpretation of fossil animal remains found in the frozen strata of Siberia, although he was influenced by ideas that explained these phenomena in terms of the sudden catastrophic incursion of oceanic water from the south. In an illustrated collection, *Flora Rossia*, he described 283 species of ancient trees and began work on the description of the fauna of Russia.

In 1793–1794 Pallas studied the southern provinces of Russia—the steppes near the Caspian Sea, the northern Caucasus, and the Crimea. The natural beauty of the Crimea and its healthy climate made him decide to live there permanently. Catherine II granted him two estates on the shore and a house in Simferpol, as well as a subsidy to establish a school of horticulture and enology. In 1795 Pallas moved to the Crimea, where he studied nature and developed gardens and vineyards in the Sudak and Koz valleys. He published *Fizicheskoe i topograficheskoe opisanie Tavricheskoy gubernii* ("Physical and Topographical Description of Taurida Province"; 1795) and wrote articles on the agricultural technology of the warm areas of the Crimea. His main efforts were devoted to preparing materials of the 1793–1794 trip for publication and to compiling a complete description of the fauna of Russia. In 1799–1801 he published an account of the trip that included an important description of the Crimea.

The writing of a zoological geography of the Russian empire, which was the main goal of Pallas' life, took much work and money; and its preparation for publication went slowly. Because of his declining health, and his wish to hasten the appearance of his work in print, Pallas moved in 1810 to Berlin, where he died a year later. The St. Petersburg Academy of Sciences, without waiting to prepare the drawings for publication, began in 1811 to publish Pallas'

Zoographia Rosso-Asiatica . . ., the last volume of which appeared in 1831.

A versatile scientist, Pallas was in many ways reminiscent of the scientific encyclopedists of antiquity. Among his contemporaries he was a peer of Linnaeus and Buffon; in zoology, he was a predecessor of Cuvier. As a geographer he may be considered a predecessor of Humboldt. Pallas sought to advance from merely describing nature to finding the causal interrelationships and hidden regularities of natural phenomena. Using the comparative method, he laid the bases of a new natural history that excluded the metaphysical approach. Pallas' achievements in zoology and botany were especially important. He was one of the first to use anatomical characteristics in classifying animals. His research in comparative anatomy provided the foundations for animal taxonomy. He described hundreds of species of animals and plants; expressed interesting ideas on their relationships to the environment; and noted the boundaries and areas of their distribution, which led to the development of the science of biogeography.

Pallas' views on the evolution of animals and plants reflected the contradictions in the science of his age and underwent changes during his lifetime. In the 1760's and 1770's he assumed the unity of origin and historical development of the organic world. In 1766 he proposed the first known scheme to express the sequential development of animal organisms in terms of a family tree. Later he spoke as a metaphysician and catastrophist, recognizing the constancy and nonvariability of species. In 1780 Pallas showed that all known species arose at one general time. He denied Buffon's idea that food, climate, and way of life influence the variation of species and Linnaeus' idea that species vary through the process of hybridization.

Pallas' contribution to geology and geography was great. From his descriptions of the natural features of Russia later generations of scientists drew much that was new and useful. He formulated the first general hypothesis of the formation of mountains. In his opinion, granite constituted the skeleton of the earth and its nucleus. Emerging after some time in the form of marine islands, the granite appeared framed with slate, the product of the disintegration of the granite. Limestones containing organic remains and constituting a Secondary formation are even younger. The friable rocks of adjacent foothills were separated out into a Tertiary formation. The raising of the mountains and the receding of the seas occurred, in Pallas' opinion, as a result of volcanic processes. These processes caused the inclined position of layers, especially of the steep position of the most ancient rocks. Pallas' ideas on the structure and origin of mountains played an important role in the further development of theoretical geology, as Cuvier pointed out.

The progressive significance of Pallas' views consisted in the recognition of a prolonged geological history of the earth and of the important role of both volcanic (inner) and external forces and their mutual influence in the development of the earth. In many ways, however, he shared the opinions of seventeenth- and eighteenth-century diluvialists. His work was influential in the development of evolutionary ideas of nature, as was acknowledged by Charles Darwin in England and K. F. Rulye in Russia.

Pallas left a deep impression in paleogeography, medicine, ethnography, the history of geography, and philology. His impressive capacity for work resulted in 170 published writings, including dozens of major reports on research. He was an active member of many Russian and foreign scientific societies, institutes, and academies. Plants and animals—including the plant genus *Pallasia* (the name given by Linnaeus) and the Crimean pine *Pinus Pallasiana*—were named in honor of Pallas. Stony meteorites are called pallasite; and a volcano (Pallasa) in the Kuril Islands and a reef in New Guinea bear his name.

BIBLIOGRAPHY

I. ORIGINAL WORKS. Pallas' writings include *Reise durch verschiedenen Provinzen des russischen Reichs in den Jahren 1768–1773,* 2 vols. (St. Petersburg, 1771–1776); "Observations sur la formation des montagnes et sur les changements arrivés au globe," in *Acta academiae scientiarum imperialis Petropolitanae,* pt. 1 (1777), 21–64; *Novae species quadrupedum et glirium ordine* (Erlangen, 1778); *Species astragalorum* (Leipzig, 1780); "Mémoire sur la variation des animaux," in *Acta academiae scientiarum imperialis Petropolitanae,* 4, pt. 2 (1780), 69–102; "O Rossyskikh otkrytiakh na moryakh mezhdu Aziey i Amerikoy" ("On Russian Discoveries in the Seas Between Asia and America"), in *Mesyatseslov istorichesky i geografichesky* (1781), 1–150, also in German (1782) and Danish (1784); *Flora Rossia . . .* (St. Petersburg, 1784–1788), also in Russian (St. Petersburg, 1786); *Bemerkungen auf einer Reise in die südlichen Statthalterschaften des russischen Reichs in den Jahren 1793 und 1794,* 2 vols. (Leipzig, 1799–1801); and *Zoographia Rosso-Asiatica,* 3 vols. (St. Petersburg, 1811–1831).

II. SECONDARY LITERATURE. See V. V. Belousov, "Pallas–puteshestvennik i geolog" ("Pallas—Traveler and Geologist"), in *Priroda* (1941), no. 3, 111–116; G. P. Dementev, "Pyotr Simon Pallas (1741–1811)," in *Lyudi russkoy nauki. Biologia . . .* ("People of Russian Science. Biology . . ."; Moscow, 1963), 34–44, with bibliography;

Y. K. Efremov, "Pyotr Simon Pallas," in N. N. Baransky *et al.*, eds., *Otechestvennye fiziki-geografy i puteshestvenniki* ("Native Physical Geographers and Travelers"; Moscow, 1950), 132–144, with bibliography; V. Marakuev, *Pyotr Simon Pallas, ego zhizn, uchenye trudy i puteshestvia* (". . . His Life, Scientific Works and Travels"; Moscow, 1877); "Peter Simon Pallas (1741–1811)," in *Lomonosov. Schlözer. Pallas. Deutsch-Russische Wissenschaftsbeziehungen im 18. Jahrhundert* (Berlin, 1962), 245–317; and B. E. Raykov, "Russkie biologi-evolyutsionisty do Darvina" ("Russian Evolutionist Biologists Before Darwin"), in *Materialy k istorii evolyutsionnoy idei v Rossii* ("Material for a History of the Idea of Evolution in Russia"), I (Moscow–Leningrad, 1952), 42–105, with bibliography.

VASILIY A. ESAKOV

PALMER, EDWARD (*b.* Hockwold cum Wilton, England, 12 January 1831; *d.* Washington, D.C., 10 April 1911), *natural history*.

Although Palmer had little formal education and was never robust in health, he was a gifted collector and made significant contributions to knowledge, especially from about 1860 to 1880. During this period he worked primarily in the western United States and mostly in areas that were still sparsely or not at all occupied by Europeans. His notes and observations on the manners and customs of the western Indians, and his collections of ancient and modern Indian artifacts, are among the more important sources of information on the ethnology and archaeology of the tribes in question. Throughout his career he collected botanical and zoological specimens, and also those of anthropological interest, but in his later years he increasingly restricted himself to the collection of herbarium specimens. His most-quoted paper is "Food Products of the North American Indians," in *Report of the Commissioner of Agriculture* (1871), and most of his other publications are short papers on the same or similar subjects.

Palmer came to the United States in 1849. He was introduced to natural history and to the practice of medicine by serving as hospital steward on a naval expedition to Paraguay (1853–1855). He attended lectures at the Homeopathic College in Cleveland, Ohio (1856–1857) and, having thus qualified himself, earned much of his living for the next eleven years as a physician and surgeon, while at the same time collecting and distributing biological specimens. He served as a contract surgeon at various army posts in Colorado, Kansas, and Arizona (1862–1867), and as medical officer at an Indian agency in what is now Oklahoma (1868). After this he gave up medicine and devoted himself exclusively to collecting in Mexico and the western United States, with intermittent support from government agencies, including the U.S. Department of Agriculture, the Smithsonian Institution, the Bureau of Ethnology, and the Commission of Fish and Fisheries. Private sponsors included the Peabody Museum of Archaeology and Ethnology, and other biologically oriented museums at Harvard University. In 1878 and thereafter, Palmer helped to finance several of his own trips to Mexico by selling subscriptions to his sets of exsiccatae. Asa Gray and his successors at Harvard supplied determinations for the botanical specimens that Palmer collected and helped with the sale and distribution of his duplicates. In Palmer's later years J. N. Rose of the U.S. National Herbarium performed a similar service for him.

Palmer's zoological collections include representatives of most of the major groups of macroscopic animals, but he seems never to have collected extensively in any group except the insects. His botanical collections, which were widely distributed, included an estimated 100,000 specimens, or about 20,000 different gatherings, of which about 2,000 represented species new to science. Although lacking the unique quality of his early ethnological collections, his botanical specimens were often from areas seldom or never before visited by experienced collectors. They provided a basis for modern taxonomic and phytogeographic studies, especially of northern and western Mexico, and they stimulated further exploration in western North America.

BIBLIOGRAPHY

I. ORIGINAL WORKS. Palmer's own publications comprise 25 papers published in scientific periodicals and government documents. Most of the papers deal with the uses of plants or other articles by the American Indians; all are cited in full in McVaugh's book on Palmer, which is mentioned below.

About 650 of Palmer's letters (1852–1911), and many handwritten documents pertaining to specimens, are extant. The richest sources of these are the governmental archives in Washington, D.C. (especially those of the Smithsonian Institution), and the archives and libraries of Harvard University.

Palmer's personal collection of MSS and correspondence was sold at auction in 1914, and much of it has never been located since that time (McVaugh, *Edward Palmer*, p. 407, item 45 of "References and Sources").

II. SECONDARY LITERATURE. Rogers McVaugh, *Edward Palmer: Plant Explorer of the American West* (Norman,

1956), includes a detailed listing of Palmer's itineraries by date and locality, copies of some early field notes, and a bibliography of all known books and papers written by or about Palmer to 1955.

ROGERS McVAUGH

PAMBOUR, FRANÇOIS MARIE GUYONNEAU DE (*b.* Noyon, France, 1795), *civil engineering.*

Pambour attended the École Polytechnique (1813–1815) and upon being commissioned he entered the artillery; later he transferred to the general staff. His principal contributions were to the theory and practice of steam engines and steam locomotives.

His *Théorie de la machine à vapeur*, which went through several editions and translations, had a fundamental and mathematical approach such as might be expected from an applied physicist rather than a practical engineer. As late as 1876 the work was authoritatively referred to as "the most celebrated treatise of Pambour . . . published in 1844, then far superior to other works and still in many respects one of the best standards on the subject." In his definitive treatise, R. H. Thurston frequently refers to Pambour's work in the highest terms. He points out that much of his original work has been "demonstrated anew by a certain number of modern writers who appear to ignore the works of Pambour." Pambour was certainly not a mere empiricist. His original researches were reported in papers communicated to *Comptes rendus* of the Academy of Sciences. In addition to his treatise on the steam engine, he wrote an equally successful practical work, *Traité théorique et pratique des machines locomotives* (1835).

Despite his productivity, Pambour was never a member of the Academy. A candidate for the Section de Mécanique in 1837, 1840, and 1843, he failed to be elected, quite possibly as a result of scientific differences with Poncelet, probably intensified by a personal clash. There was a polemic between Pambour and Poncelet's protégé, Morin, over the choice of "frictional coefficients" for calculating the work of piston expansion. This quantity relates the volumes occupied by a unit weight of steam in the boiler and in the cylinder. According to the older theory of Poncelet and Morin, the ratio can be determined by a constant coefficient. In the new theory of Pambour, the coefficient is no longer constant, but varies with operating conditions. Pambour's denunciation of Poncelet's earlier work elicited a sharp response. Not only was Pambour rejected by the Academy, but his name does not appear in the *École Polytechnique, Livre du centenaire*, which includes biographical sketches of considerably less distinguished graduates.

Pambour's most important contribution to the theory of the steam engine dealt with the calculation of the work obtained under a given set of operating conditions. An earlier treatment, developed by Poncelet and Morin, calculated the work of expansion by the use of Boyle's law but did not take into account the drop in steam temperature upon expansion. Pambour, on the other hand, assumed that the steam remained saturated throughout the engine; and since the temperature of saturated steam varies with pressure, Boyle's law is quite incapable of representing this situation. Instead, Pambour used an empirical formula that involved two experimentally determined constants. He assumed various pressure situations between the boiler and the cylinder and with his formula calculated the work for these cases. The assumption that steam is saturated during expansion is not used today; the Rankine cycle—where the expansion is from superheated steam into wet steam—has replaced Pambour's scheme. The improvement over the assumption of Boyle's law was tremendous and represented a major advance. Clapeyron used Pambour's formula in his 1851 lectures but recognized the shortcomings inherent in the assumption that the steam was "dry."

BIBLIOGRAPHY

There are forty-four papers by Pambour on the steam engine in *Comptes rendus hebdomadaires des séances de l'Académie des sciences,* **4–21** (1837–1845). His subsequent interest in hydraulic turbines was reflected in seventeen papers in vols. **32–75** (1851–1872). *Théorie de la machine à vapeur* appeared in French (Brussels, 1837; Paris, 1839, 1844; Liège, 1847, 1848), in English (London, 1839; Philadelphia, 1840), and in German, with intro. by A. L. Crelle (Berlin, 1849).

On Pambour's candidature at the Academy, see *Comptes rendus,* **4** (1837), 556; **10** (1840), 504; and **17** (1843), 1310.

On Pambour and his work, see James Renwick, *Treatise on the Steam Engine,* 3rd ed. (New York, 1848), pref.; R. S. McCullock, *Treatise on the Mechanical Theory of Heat and Its Application to the Steam Engine* (New York, 1876), pp. 24, 261–263; R. H. Thurston, *Traité de la machine à vapeur,* I (Paris, 1893), 260; and F. Zernikow, *Die Theorie der Dampfmaschinen* (Brunswick, 1857), 14–23.

MILTON KERKER

PANDER, CHRISTIAN HEINRICH (*b.* Riga, Latvia, Russia, 24 July 1794; *d.* St. Petersburg, Russia, 22 September 1865), *embryology, anatomy, paleontology.*

Pander was the son of a wealthy banker of German descent. After studying in the local schools of Riga, he entered the University of Dorpat in 1812. Dorpat had been refounded in 1798, and its faculty was German trained. Prior to this the Baltic gentry had traditionally sent their sons to German universities. Although his father had wanted him to study medicine, Pander was more interested in natural history, but he attempted to combine the two. At Dorpat he came under the influence of the anatomist Karl Friedrich Burdach, who had also taught Karl Ernst von Baer; Baer later continued Pander's embryologic researches.

In 1814 Pander left Dorpat for Berlin and from there he went on to Göttingen. In March 1816, at a congress of Baltic students resident in Germany, he renewed his acquaintance with Baer, who persuaded him to come to the University of Würzburg and study under Ignaz Döllinger. In his autobiography Baer states that he, Pander, and Döllinger had discussed Döllinger's hope that someone would study anew the development of the chick embryo. Pander took on the task and he received his M.D. at Würzburg in 1817. His dissertation, "Historia metamorphoseos quam ovum incubatum prioribus quinque diebus subit," was amplified and then published in German (1817) with illustrations by the elder E. J. d'Alton.

Pander discovered the trilaminar structure of the chick blastoderm, a term he also coined. He stated that he used the term blastoderm, from the Greek *blastos*, germ, and *derma*, skin, because the embryo chose it as "its seat and its domicile, contributing much to its configuration out of its own substance, therefore in the future we shall call it blastoderm." He described the three layers as the serous or outer, the vascular or middle, and the mucous or inner. In the twelfth hour of embryonic development he reported that the blastoderm consisted of two entirely separate layers: an inner layer, thick and opaque; and an outer layer, thin, smooth, and transparent. Between these two a third layer developed, in which blood vessels formed and from which "events of the greatest importance subsequently occur." When Baer received a copy of Pander's work in 1818 at Königsberg University, where he was serving as prosector to his old Dorpat professor, Burdach, he began his own investigations, which ultimately revolutionized embryology. Baer's first treatise on the subject includes an introduction styled as a personal letter to Pander, explaining his differences with his old friend.

Pander, for reasons that are not entirely clear, never pursued his early research, although he regarded his studies as incomplete and had expressed himself only briefly on the subsequent transformations that took place in the embryo. After receiving his degree, Pander traveled in a leisurely manner through Germany, France, Spain, Holland, and England with d'Alton as a companion, visiting anatomical museums and making various paleontological, geological, and biological observations. In 1821 he began publishing a series of papers on comparative osteology with illustrations by d'Alton. In these osteological studies Pander developed an evolutionary theory of the development of animal forms which had strong Lamarckian overtones. Goethe endorsed his transformist ideas, and Darwin was aware of them through secondary sources.

In 1821 he became a member of the Academy of Sciences in St. Petersburg. He also traveled extensively in Russia at this time and wrote an account of the natural history of Bukhara in central Asia. In 1826 he became a fellow of the Zoological Academy in St. Petersburg, but in 1827 he returned to his estate near Riga, where he remained until 1842. In that year he returned to St. Petersburg as a member of the Mining Institute. He then resumed his travels in Russia, observing geological characteristics and gathering paleontological materials. Although he wrote almost exclusively on geological matters, he began a zoological collection and surveyed geological formations around St. Petersburg. Little of his later work has received the attention paid his earlier studies.

BIBLIOGRAPHY

I. ORIGINAL WORKS. Pander's major works include *Dissertatio inauguralis, sistens historiam metamorphoseos, quam ovum incubatum prioribus quinque diebus subit* (Würzburg, 1817); *Beiträge zur Entwickelungsgeschichte des Hühnchens im Eie* (Würzburg, 1817); *Der vergleichende Osteologie*, 12 vols. (Bonn, 1821–1831), written with E. d'Alton, and divided into individually titled sections *Riesenfaulthier* (1821), *Pachydermata* (1821), *Raubthiere* (1822), *Widerkauer* (1823), *Nagathiere*, 2 vols. (1823–1824), *Vierhänder* (1825), *Zahnlose Thiere* (1826), *Robben und Lamantine* (1826), *Cetaceen* (1827), *Beutelthiere* (1828), and *Chiropteren und Insectivoren* (1831); "Naturgeschichte der Bukharei," in George Meyendorf, *Reise von Orenburg nach Buchara* (Jena, 1826), trans. into French as *Voyage d'Orenbourg à Boukhara* (Paris, 1826); *Beiträge zur Geognosie des russischen Reichs* (St. Petersburg, 1830); *Monographie der Fossilen Fische des silurischen Systems der Russisch–Baltischen Gouvernements* (St. Petersburg, 1856); *Über die Placodermen des devonischen Systems* (St. Petersburg, 1857); *Über die Ctenodipterinen des devonischen Systems* (St. Petersburg, 1858); and *Über die Saurodipterinen, Dendrodonten, Glyptolepiden und Cheirolepiden des devonischen Systems* (St. Petersburg, 1860).

II. SECONDARY LITERATURE. For Pander's contributions to embryology, see Erik Nordenskiöld, *The History of Biology* (New York, 1928), 368–369; Jane M. Oppenheimer,

"The Non-Specificity of the Germ Layers," in her *Essays in the History of Embryology and Biology* (Cambridge, Mass., 1967), 256–294, which is reprinted with additional bibliography from the *Quarterly Review of Biology*, **15** (1940), 1–27; and Alexander Vucinich, *Science in Russian Culture: A History to 1860* (Stanford, 1963), 206, 362.

Personal information about Pander is found in the autobiography of Karl Ernst von Baer, *Nachrichten über Leben und Schriften des Geheimrathes Dr. Karl Ernst von Baer, mitgetheilt von ihm selbst. Veröffentlicht bei Gelegenheit seines fünfzigjährigen Doctor-Jubiläums, am 29. August 1864, von der Ritterschaft Ehstlands* (St. Petersburg, 1866); and in L. Steidt's article in *Allgemeine Deutsche Biographie*, XXV (Leipzig, 1875–1901), 117–119.

VERN L. BULLOUGH

PANETH, FRIEDRICH ADOLF (*b*. Vienna, Austria, 31 August 1887; *d*. Mainz, Germany, 17 September 1958), *radiochemistry, inorganic chemistry.*

Paneth was the second of three sons of Joseph and Sophie Schwab Paneth. His father was a noted physiologist who discovered the histological cells that bear his name. Paneth completed his education at the universities of Munich and Glasgow, and received the Ph.D. from the University of Vienna in 1910. On 6 December 1913 he married Else Hartmann; the couple had two children, Eva and Heinrich Rudolph. From 1912 to 1918 Paneth served as an assistant to Stefan Meyer at the Vienna Institute for Radium Research. In 1913 he spent a short period of time with Soddy at the University of Glasgow and visited Rutherford's laboratory in Manchester. After a brief stay at the Prague Institute of Technology, he spent three years at Hamburg University (1919–1922). He then worked at the University of Berlin, where he remained until 1929. He was invited to give the George Fisher Baker lectures at Cornell University in 1926–1927.

From 1929 to 1933 Paneth was professor and director of the chemistry laboratories at the University of Königsberg. The growth of the Nazi movement, however, was a factor in his decision to leave Königsberg for good and become a guest lecturer at the Imperial College of Science and Technology in London, a position he held for five years. After serving one year as a reader in atomic chemistry at the University of London, he accepted a professorship at the University of Durham, where he remained from 1939 until 1953, when he returned to Germany as director of the Max Planck Institute for Chemistry in Mainz. From 1943 to 1945 he was head of the chemistry division of the Joint British-Canadian Atomic Energy Team in Montreal. He served as president of the Joint Commission on Radioactivity, an organization of the International Council of Scientific Unions, from 1949 to 1955. He received the Lavoisier, Stas, Liebig, and Auer von Welsbach medals from the chemical societies of France, Belgium, Germany, and Austria.

Paneth's intellectual career manifested a progressively broader and deeper range of interests as his professional competence increased. He contributed significantly to the development of radioactive tracer techniques, synthesized and characterized new metal hydrides, and experimentally verified the existence of free radicals in the thermal decomposition of organic compounds. In subsequent research he developed methods for determining the age of rocks and meteorites by measuring the helium content, and he applied exceedingly sensitive methods of helium measurement to determine the composition of the stratosphere as a function of altitude up to 45 miles. In addition to his purely experimental research, Paneth was keenly interested in the philosophical, cosmological, and historical aspects of science. His involvement in radiochemical studies soon led him to an active interest in the history of alchemy, while his experimental work in the quantitative determination of trace amounts of helium led him to a consuming, lifelong interest in the study of meteorites, from a historical as well as a cosmological perspective. After his death a trust fund was established to administer his meteorite collection housed at the Max Planck Institute, to further and encourage research concerned with meteorites, and to augment the collection.

One of Paneth's first papers in chemistry dealt with the acid-catalyzed rearrangement of quinidine and cinchonidine. But his work at the Radium Institute soon involved him in the study of radioactive substances. All of Paneth's future experimental work progressed from this early research. Several unsuccessful efforts to separate radium D and thorium B from lead and its compounds gradually led him to the realization that radium D and thorium B must be "isotopes" of lead. These studies, carried out in collaboration with the Hungarian radiochemist Georg von Hevesy, developed into the exploration of radium D and thorium B as indicators to determine the solubility of the slightly soluble compounds lead sulfide and lead chromate. A similar attempt to separate the radioactive products of thorium led to the preparation and isolation of BiH_3 and the realization that thorium C and radium E were isotopes of bismuth. The yield of bismuth hydride was exceedingly low when produced by the ordinary methods of preparation; it was thus undetected for a long time, and only the use of radioactive isotopes permitted the detection of the minute quantities formed.

While studying the metal hydrides of bismuth, lead, tin, and polonium, Paneth employed the mirror deposition technique to decompose the hydride and concentrate the corresponding metal. Although it seems a short step from a study of the unstable metal hydrides to a study of the metal alkyls, this step took twelve years. During his stay at the University of Berlin, Paneth's classic paper, written with Wilhelm Hofeditz, appeared (1929), announcing the preparation and identification of the free methyl radical from lead tetramethyl.

During the intervening period of time (1917–1929), Paneth worked at developing the sensitive methods for determining trace amounts of helium for which he is justly famous. Using spectroscopic techniques first, and mass spectrometry later, he successfully applied these techniques (1) to the determination of the content of natural gas from various sources, (2) to the quantitative determination of the rate of diffusion through glass, (3) to several unsuccessful efforts to measure helium produced by the attempted transmutation of various lighter elements into helium, and (4) to the quantitative determination of helium in rocks, artificial glasses, and meteorites. From 1929 to the end of his life the study of meteorites increasingly dominated his interests. He refined ever more accurately his techniques for determining the age of rocks by their helium content and the helium : radium ratio. These determinations were further refined when it was discovered that part of the ^{4}He is converted into ^{3}He by cosmic-ray bombardment in space. From these measurements he estimated the ages of iron meteorites to be in the range of 10^8 to 10^9 years and speculated that they were formed within the solar system. Until 1935 Paneth also continued his studies of organic free radicals. (This field was later developed by the physical and organic chemists.) About this same time he succeeded in his efforts to induce artificial transmutation by obtaining measurable amounts of helium from the neutron bombardment of boron.

In 1935 Paneth began to investigate trace components of the stratosphere. In an interesting series of papers he determined the He, O_3, and NO_2 content of the atmosphere and investigated the extent of gravitational separation of the components of the atmosphere. His basic finding was that there is no appreciable gravitational separation below 40 miles, while there appears to be a measurable change in relative concentration above 40 miles. Paneth subsequently returned to the field of free radical chemistry and explored the use of radioactive isotopes for the study of free radicals with the mirror removal technique. His last paper, on meteorites, appeared posthumously in *Geochimica et cosmochimica acta* as an introduction to a series of studies on the Breitscheid meteorite, which fell in Breitscheid, Dillkreis, West Germany, on 11 August 1956.

BIBLIOGRAPHY

I. ORIGINAL WORKS. Paneth's principal works are *Lehrbuch der Radioaktivität* (Leipzig, 1923; 2nd ed., 1931), written with G. von Hevesy, trans. into English, by R. W. Lawson, as *A Manual of Radioactivity* (London, 1926; 2nd ed., 1938), and also into Russian and Hungarian (1924–1925); *Radioelements as Indicators and Other Selected Topics in Inorganic Chemistry* (New York, 1928), the George Fisher Baker lectures; "Über die Darstellung von freiem Methyl," in *Berichte der Deutschen chemischen Gesellschaft*, **62B** (1929), 1335–1347, written with W. Hofeditz; *The Origin of Meteorites* (Oxford, 1940), the Halley lectures, and "Der Meteorit von Breitscheid," in *Geochimica et cosmochimica acta*, **17** (1959), 315–320. *Chemistry and Beyond*, H. Dingle and G. R. Martin, eds. (New York, 1964), is a selection of Paneth's writings, and also contains an extensive bibliography.

II. SECONDARY LITERATURE. See Otto Hahn, "Friedrich A. Paneth," in *Zeitschrift für Electrochemie*, **61** (1957), 1121, written on the occasion of Paneth's seventieth birthday; and K. Peters' bibliographical appendix to one of Paneth's last articles, "Hat Chladni das Pallas-Eisen in Petersburg gesehen," in *Österreichische Chemikerzeitung*, **59** (1958), 289–291.

ERNEST G. SPITTLER, S.J.

PANNEKOEK, ANTONIE (*b.* Vaassen, Netherlands, 2 January 1873; *d.* Wageningen, Netherlands, 28 April 1960), *astronomy*.

Pannekoek was the son of Johannes Pannekoek and Wilhelmina Dorothea Beins. In 1903 he married Johanna Maria Nassau Noordewier, a teacher of Dutch literature. His family belonged to the rural middle class, and through his wife Pannekoek entered literary and musical circles. He was a member of the Royal Netherlands Academy of Sciences and an honorary member of the American Astronomical Society. In addition he received an honorary doctorate from Harvard University and the gold medal of the Royal Astronomical Society.

An amateur astronomer since his youth, Pannekoek studied astronomy at Leiden University. He began his career in 1895 as a geodesist and became observer at the Leiden observatory in 1898, but he grew disenchanted with the old-fashioned meridian work, which he considered of little scientific use. A teacher of Marxist theory at the Socialist party school in

Berlin from 1905, and later in Bremen, he came to oppose the increasing opportunism in the German Socialist party.

At the outbreak of World War I, Pannekoek returned to the Netherlands, where he became a high-school teacher. Since leaving the observatory he had followed the progress of astronomy and had written several scientific papers. He now finished his book *De wonderbouw der wereld* (1920), an excellent and original historical introduction to astronomy. Long interested in Babylonian astronomy, he published several papers on this subject, while continuing his lectures on Marxism at Leiden. His nomination as a vice-director of the Leiden observatory was rejected by the minister of education; but the city of Amsterdam, not dependent on the state, appointed him lecturer in mathematics and astronomy at its municipal university, where he founded a modest but very active astronomical institute. Named professor in 1925, he was dismissed by the German occupation government in 1941.

A chance readings of Saha's paper on ionization in stellar atmospheres prompted Pannekoek to begin work in astrophysics, and he became the founder of modern astrophysics in the Netherlands. His investigations of the structure of the galaxy that includes our solar system extended over sixty years. He made careful and detailed drawings, with isophotes, of the northern and southern Milky Way, and later repeated this work on extrafocal photographs. Improving on the work of Kapteyn, he studied our galaxy as a function of galactic longitude as well as latitude. Dissatisfied by smoothed mean values, he gave full attention to star clouds and dark nebulae. He discovered the typical groups of early stars that were later called associations.

In the area of ionization theory and the composition of stellar atmospheres, Pannekoek was the first to modify Henry Norris Russell's work and to assume a huge preponderance of hydrogen, a view subsequently confirmed. He was also the first to apply "detailed analysis" to stellar atmospheres, taking into account the change in physical properties of the successive layers. With M. Minnaert he published the first quantitative analysis of the flash spectrum during a solar eclipse. Calling attention to the surprisingly low value of gravitation that may be deduced from the spectra of giant stars, he interpreted the brightness maxima of Cepheids as the ejection of gaseous shells.

Pannekoek's work on the history of astronomy, culminating in his *History of Astronomy*, emphasized the broad lines of the evolution of the disciplines and the relation between astronomy and society. Of still wider scope is his *Anthropogenesis*, in which he traced the origin of man and the development into *Homo sapiens.*

BIBLIOGRAPHY

I. ORIGINAL WORKS. Among Pannekoek's earlier writings are *Untersuchungen über den Lichtwechsel Algols* (Leiden, 1902), his diss.; *De astrologie en hare betekenis voor de ontwikkeling der sterrekunde* (Leiden, 1916), his inaugural lecture; "Die nördliche Milchstrasse," in *Annalen van de Sterrewacht te Leiden*, **11**, no. 3 (1920); *De wonderbouw der wereld* (Amsterdam, 1920); *Researches on the Structure of the Universe*, Publications of the Astronomical Institute of the University of Amsterdam, nos. 1–2 (1924–1929); "The Ionization Formula for Atmospheres Not in Thermodynamic Equilibrium," in *Bulletin of the Astronomical Institutes of the Netherlands*, **3** (1926), 207–209; *Results of Observations of the Solar Eclipse of June 29, 1927. Photometry of the Flash Spectrum* (Amsterdam, 1928), written with M. Minnaert; "Die südliche Milchstrasse," in *Annalen van der Boss\-Sterrewacht* (Lembang), **2**, no. 1 (1929), 1–73; "Die Ionisation in den Atmosphären der Himmelskörper," in *Handbuch der Astrophysik*, III, pt. 1 (Berlin, 1930), 256–350; "The Theoretical Contours of Absorption Lines," in *Monthly Notices of the Royal Astronomical Society*, **91** (1930), 139–169, 519–531; *Photographische Photometrie der nördlichen Milchstrasse*, Publications of the Astronomical Institute of the University of Amsterdam, no. 3 (1933); "Theoretical Colour Temperatures," in *Monthly Notices of the Royal Astronomical Society*, **95** (1935), 529–535; *The Theoretical Intensities of Absorption Lines in Stellar Spectra*, Publications of the Astronomical Institute of the University of Amsterdam, no. 4 (1935); and "Ionization and Excitation in the Upper Layers of an Atmosphere," in *Monthly Notices of the Royal Astromical Society*, **96** (1936), 785–793.

Later works include "The Hydrogen Lines Near the Balmer Limit," *ibid.*, **98** (1938), 694–709; *A Photometric Study of Some Stellar Spectra*, Publications of the Astronomical Institute of the University of Amsterdam, no. 6, 2 pts. (1939–1946), written with G. B. van Albada; *Investigations on Dark Nebulae*, *ibid.*, no. 7 (1942); "Anthropogenese, een studie over het ontstaan van den mens," in *Verhandelingen der K. akademie van wetenschappen*, **42**, no. 1 (1945), translated as *Anthropogenesis, a Study of the Origin of Man* (Amsterdam, 1953); "The Line Spectra of Delta Cephei," in *Physica*, **12** (1946), 761–767; "Planetary Theories," in *Popular Astronomy*, **55** (1947), 422–438, and **56** (1948), 2–13 (Copernicus), 63–75 (Kepler), 177–192 (Newton), 300–312 (Laplace); *Photographic Photometry of the Southern Milky Way*, Publications of the Astronomical Institute of the University of Amsterdam, no. 9 (1949), written with D. Koelbloed; "Line Intensities in Spectra of Advanced Types," in *Publications of the Dominion Astrophysical Observatory*,

Victoria, B.C., **8** (1950), 141–223; *De groei van ons wereld-beeld* (Amsterdam–Antwerp, 1951), translated as *A History of Astronomy* (London, 1961); and "The Origin of Astronomy," in *Monthly Notices of the Royal Astronomical Society*, **111** (1951), 347–356.

II. SECONDARY LITERATURE. Two short biographies are G. B. van Albada, "Ter nagedachtenis van Prof. Panne-koek," in *Hemel en dampkring*, **58** (1960), 105; and B. J. Bok, "Two Famous Dutch Astronomers," in *Sky and Telescope*, **20** (1960), 74–76.

M. MINNAERT

PAPANICOLAOU, GEORGE NICHOLAS (*b.* Kími, Greece, 13 May 1883; *d.* Miami, Florida, 19 February 1962), *anatomy.*

The son of a physician, George Nicholas undertook the study of medicine and received the M.D. from the University of Athens in 1904. After postgraduate work in biology at the universities of Jena, Freiburg, and Munich, from which he received his doctorate in 1910, he returned to Greece and married Mary Mavroyeni, the daughter of a high-ranking military officer.

Papanicolaou decided to forgo the practice of medicine in favor of an academic career, in which his wife served as his lifelong associate. En route to Paris, Papanicolaou stopped for a visit at the Oceanographic Institute of Monaco and accepted an unexpected offer to join its staff. He worked for one year as a physiologist and then returned to Greece upon the death of his mother. After serving for two years as second lieutenant in the medical corps of the Greek army during the Balkan War, he immigrated to the United States.

In 1913 Papanicolaou was appointed assistant in the pathology department of New York Hospital, and in 1914 he became assistant in anatomy at Cornell Medical College. Until 1961 he conducted ali of his scientific research, devoted almost exclusively to the physiology of reproduction and exfoliative cytology, at these two affiliated institutions, each of which named a laboratory in his honor. He was designated professor emeritus of clinical anatomy at Cornell in 1951. In November 1961 Papanicolaou moved to Florida and became director of the Miami Cancer Institute, but died three months later of acute myocardial infarction. The institute was renamed the Papanicolaou Cancer Research Institute in November 1962. An indefatigable worker, Papanicolaou is said never to have taken a vacation.

Papanicolaou is best known for his development of the technique, eponymically termed the Papanicolaou smear, or "Pap test," for the cytologic diagnosis of cancer, especially cancer of the uterus—second only to the breast as the site of origin of fatal cancers in American women.

The history of cancer cytology dates from 1867, when Beale observed tumor cells in the smears of sputum from a patient with carcinoma of the pharynx. He suggested the microscopic examination of desquamated cells for the detection of cancer of other organs, including the uterus and urinary tract.[1] Friedlaender noted, in his subsequent microscopic examination of fluid exuding from ulcerating cancers of the uterus, distinctive cellular elements that helped establish the diagnosis.[2] In 1908 Königer called attention to the striking differences in the size and shape of cancer cells obtained from serous cavities, the abundance of vacuoles and fatty droplets in the cytoplasm, the enlargement of the nucleus, and the presence of multiple nucleoli within it.[3]

Papanicolaou was invited by Charles R. Stockard, chairman of the Cornell Medical School department of anatomy, to join him in his work in experimental genetics. In 1917 he began a study of the vaginal discharge of the guinea pig, with the hope of finding an indicator of the time of ovulation; he would thus be able to obtain ova at specific stages of development. He sought traces of blood, as seen during estrus in certain other species, such as the cow and bitch, and in the menstrual discharge of primates and women. In the course of his daily examination of the guinea pig vaginal fluid, obtained through a small nasal speculum, Papanicolaou saw no blood. He noted instead a diversity in the forms of the epithelial cells in a sequence of cytologic patterns recurring in a fifteen- to sixteen-day cycle, which he was able to correlate with the cyclic morphologic changes in the uterus and ovary. Papanicolaou thus established the technique that became the standard for studying the sexual (estrous) cycle in other laboratory animals, especially the mouse and rat, and for measuring the effect of the sex hormones.

In 1923 Papanicolaou extended his studies to human beings in an effort to learn whether comparable vaginal changes occur in woman in association with the menstrual cycle. His first observation of distinctive cells in the vaginal fluid of a woman with cervical cancer gave Papanicolaou what he later described as "one of the most thrilling experiences of my scientific career" and soon led to a redirection of his work.

His early reports on cancer detection, however, which appeared from 1928, failed to arouse the interest of clinicians. Cytologic examination of the vaginal fluid seemed an unnecessary addition to the proven procedures for uterine cancer diagnosis—cervical biopsy and endometrial curettage. In 1939, while collaborating with the gynecologist Herbert Traut, Papanicolaou began to concentrate his studies on human beings. Their research culminated in the publication of *Diagnosis of Uterine Cancer by the Vaginal Smear*. This monograph encompassed a variety of physiologic and pathologic states, including the menstrual cycle, puerperium, abortion, ectopic pregnancy, prepuberty, menopause, amenorrhea, endometrial hyperplasia, vaginal and cervical infections, and 179 cases of uterine cancer (127 cervical and 52 corporeal). The work was instrumental in gaining clinical acceptance of the smear as a means of cancer diagnosis, for superficial lesions could thus be detected in their incipient, preinvasive phase, before the appearance of any symptoms.

The Papanicolaou smear soon achieved wide application as a routine screening technique. The death rate from cancer of the uterus among women aged thirty-five to forty-four who were insured under industrial policies by the Metropolitan Life Insurance Company was almost halved in the decade from 1951 to 1961, decreasing from 16.0 to 8.2 per 100,000; while the corresponding reduction in the death rate from cancer of all sites was from 74.0 to 66.0.

Although the *Atlas of Exfoliative Cytology* lists the criteria for malignancy in the shed cells, Papanicolaou used to state that he could not explain how he recognized a smear as positive for malignancy any more than he could explain how to recognize an acquaintance by describing his facial expression. Yet he taught thousands of students how to detect cancer cells under the microscope, and they carried his teachings to all parts of the world. Papanicolaou's technique was rapidly extended to the diagnosis of cancer of other organs from which scrapings, washings, or exudates could be obtained. The principal value of the Papanicolaou smear lies in cancer screening, but it is also applied to the prediction of cancer radiosensitivity, the evaluation of the effectiveness of radiotherapy, and the detection of recurrence after treatment.

It has been suggested that Papanicolaou's work ranks with the discoveries of Roentgen and Marie Curie in reducing the burden of cancer. Cancer of the uterine cervix is nearly 100 percent curable when recognized in its incipiency.

NOTES

1. Beale, L. S., *The Microscope in Its Application to Practical Medicine*, 3rd ed. (Philadelphia, 1867), p. 197.
2. Friedlaender, C., *The Use of the Microscope in Clinical and Pathological Examinations*, 2nd ed., trans. by H. C. Coe (New York, 1885), pp. 168–169.
3. Königer, H., *Die Zytologische Untersuchungsmethode, ihre Entwicklung und ihre Klinische Verwerthung an den Ergüssen Seröser Höhlen* (Jena, 1908), pp. 99–100.

BIBLIOGRAPHY

I. ORIGINAL WORKS. Papanicolaou's works include "The Existence of a Typical Oestrous Cycle in the Guinea Pig—With a Study of its Histological and Physiological Changes," in *American Journal of Anatomy*, **22** (1917), 225–283, written with C. Stockard; "New Cancer Diagnosis," in *Proceedings. Third Race Betterment Conference, January 2–6, 1928* (1928), 528–534; *Diagnosis of Uterine Cancer by the Vaginal Smear* (New York, 1943), written with H. Traut; and *Atlas of Exfoliative Cytology* (Cambridge, Mass., 1954).

II. SECONDARY LITERATURE. On Papanicolaou and his work, see "Dedication of the Papanicolaou Cancer Research Institute," in *Journal of the American Medical Association*, **182** (1962), 556–559; H. Speert, *Obstetric and Gynecologic Milestones* (New York, 1958), 286; and D. E. Carmichael, *The Pap Smear: Life of George N. Papanicolaou* (Springfield, Ill., 1973).

HAROLD SPEERT

PAPIN, DENIS (*b*. Blois, France, 22 August 1647; *d*. London [?], England, *ca*. 1712), *technology*.

Papin was the son of Denys Papin and Magdaleine Pineau. He studied medicine at the University of Angers, from which he received the M.D. in 1669. He was apparently early intent upon a scientific career, since shortly after graduation he went to Paris, where he began working as an assistant to Christiaan Huygens. Papin was a skillful mechanic; he constructed an air pump, with which he performed a number of experiments under Huygens' direction. These were eventually published (1674), and included some attempts at preserving food in a vacuum that testify to Papin's utilitarian bent of mind.

In 1675 Papin went to London. He took with him letters of introduction to Henry Oldenburg, but it was with Robert Boyle that he soon established himself. In *A Continuation of New Experiments*, published by Boyle in 1680, Papin described both the investigations that he had made with Boyle (chiefly on the air pump) and those that he had conducted himself. In Boyle's scientific household Papin also invented his "steam digester," a pressure cooker for which he invented a safety valve that was to be technologically important in the development of steam power. He demonstrated the digester to the Royal Society, under the auspices of Robert Hooke, in May 1679. In the latter part of the same year, he was employed by Hooke to write letters for the society, at two shillings each. He was not elected a fellow until late in 1680.

Papin was again in Paris with Huygens at some time in 1680; in 1681 he went to Venice, where he was director of experiments at Ambrose Sarotti's academy.

He remained there for three years; among his duties was the performance of diverse experiments for the entertainment of the members, who periodically gathered in Sarotti's library. Papin returned to London in 1684 to serve as temporary curator of experiments to the Royal Society, at a salary of £30 a year. He sought the more lucrative post of secretary, but Halley was elected in his stead. His own work at this time consisted primarily of experiments in hydraulics and pneumatics, a number of which were published in the Royal Society's *Philosophical Transactions*.

In 1687 Papin went to Germany and joined a number of his fellow Huguenots at the University of Marburg, where he had been appointed professor of mathematics. He married and acquired a large family, which further strained his always inadequate finances. At this time Papin's interests in air pumps and steam pressure merged to provide an innovative solution to the widespread need for raising water. He considered first a piston ballistic pump using gunpowder, the idea for which he had earlier discussed with Huygens but claimed as his own (in a letter of 6 March 1704 to Leibniz). Papin met the problem of a 20 percent residue of elastic air remaining in the chamber after combustion by substituting steam for the gunpowder. In 1690 he published an account of a single cylinder engine in which water was both boiled and condensed in a tube beneath a piston. Atmospheric pressure forced the piston down again. While not immediately practical in actual operation, the piston arrangement had the advantage, Papin noted, of requiring steam at pressure low enough to be accommodated by vessels artisans of the time could make. Thomas Newcomen independently achieved great success following this line.

Papin remained in correspondence with Huygens during these years, and at one point, having tired of his heavy teaching load and low salary, appealed to him for help in finding a new position. Huygens could offer him nothing in The Hague, however, and Papin in 1695 was given a place in the court of the landgrave of Hesse, in Kassel. Here he devised a number of pumps and other practical inventions that intermittently interested his patron. He was made counsellor to the landgrave, and received recognition for his work in raising water from the Fulde. In 1705 Leibniz sent him a diagram of T. Savery's high-pressure steam pump. Papin designed a modification to this engine, published in *Ars nova ad aquam ignis adminiculo efficacissime elevandam* (1707), which though workable was not to prove as fruitful as the original piston model.

In 1707 Papin returned to England, but his old friends were gone, and he received no permanent appointment from the Royal Society. He drifted into obscurity and died, probably about 1712, but certainly at some date before 1714.

BIBLIOGRAPHY

Papin's more important writings include: *Nouvelles expériences du vuide* (Paris, 1674), repr. in Huygens, *Oeuvres complètes*, XIX (The Hague, 1937), 231; *A New Digester or Engine for Softening Bones* . . . (London, 1681;) *A Continuation of the New Digester of Bones, Together With Some Improvements and New Uses of the Air Pump* (London, 1687); "Nova methodus ad vires motrices validissimas levi pretio comparandas," in *Acta Eruditorum* (1690), and *Ars nova ad aquam ignis adminiculo efficacissime elevandam* (Kassel, 1707). A complete bibliography of Papin's writings, together with a biography, was published by Ernst Gerland, *Leibnizens und Huygens' Briefwechsel mit Papin, nebst der Biographie Papins* (Berlin, 1881), which contains the letter from Papin to Leibniz of 6 March 1704. See also Bannister, *Denis Papin, sa vie et ses écrits* (Blois, 1847); and Louis de Saussaye, *La vie et les ouvrages de Denis Papin* (Lyons, 1869).

For evaluation of Papin's place in the development of steam power see R. Thurston, *A History of the Growth of the Steam Engine* (New York, 1878); R. L. Galloway, *The Steam Engine and Its Inventors* (London, 1881); and H. W. Dickenson, *A Short History of Steam Power* (Cambridge, 1938).

Patricia P. MacLachlan

PAPPUS OF ALEXANDRIA (*b.* Alexandria, *fl.* A.D. 300–350), *mathematics, astronomy, geography.*

In the silver age of Greek mathematics Pappus stands out as an accomplished and versatile geometer. His treatise known as the *Synagoge* or *Collection* is a chief, and sometimes the only, source for our knowledge of his predecessors' achievements.

The *Collection* is in eight books, perhaps originally in twelve, of which the first and part of the second are missing. That Pappus was an Alexandrian is affirmed by the titles of his surviving books and also by an entry in the *Suda Lexicon*.[1] The dedication of the seventh and eighth books to his son Hermodorus[2] provides the sole detail known of his family life. Only one of Pappus' other works has survived in Greek, and that in fragmentary form—his commentary on Ptolemy's *Syntaxis* (the *Almagest*). A commentary on book X of Euclid's *Elements*, which exists in Arabic, is generally thought to be a translation of the commentary that Pappus is known to have written, but some doubts may be allowed. A geographical work, *Description of the World*, has survived in an early Armenian translation.

The dates of Pappus are approximately fixed by his

reference in the commentary on Ptolemy to an eclipse of the sun that took place on the seventeenth day of the Egyptian month Tybi in the year 1068 of the era of Nabonasar. This is 18 October 320 in the Christian era, and Pappus writes as though it were an eclipse that he had recently seen.[3] The *Suda Lexicon*, which is followed by Eudocia, would make Pappus a contemporary of Theon of Alexandria and place both in the reign of Theodosius I (A.D. 379–395), but the compiler was clearly not well informed. The entry runs: "Pappus, of Alexandria, philosopher, lived about the time of the Emperor Theodosius the Elder, when Theon the Philosopher, who wrote on the *Canon* of Ptolemy, also flourished. His books are: *Description of the World, Commentary on the Four Books of Ptolemy's Great Syntaxis, Rivers of Libya, Interpretation of Dreams.*" The omission of Pappus' chief work and the apparent confusion of the *Syntaxis* with the *Tetrabiblos* of Ptolemy[4] does not inspire confidence. The argument that two scholars could not have written in the same city, on the same subject, at the same time, without referring to each other may not be convincing, for that is precisely what scholars are liable, deliberately or inadvertently, to do. But detailed examination shows that when Theon wrote his commentary on the *Syntaxis* he must have had Pappus' commentary before him.[5] A scholium to a Leiden manuscript of chronological tables, written by Theon, would place Pappus at the turn of the third century, for opposite the name Diocletian (A.D. 284–305) it notes: "In his time Pappus wrote."[6] This statement cannot be reconciled with the eclipse of A.D. 320, but it is more than likely that Pappus' early life was spent under Diocletian, for he would certainly have been older than fifteen when he wrote his commentary on the *Syntaxis.*

The several books of the *Collection* may well have been written as separate treatises at different dates and later brought together, as the name suggests. It is certain that the *Collection*, as it has come down to us, is posterior to the *Commentary on the Syntaxis*, for in book VIII Pappus notes that the rectangle contained by the perimeter of a circle and its radius is double the area of the circle, "as Archimedes showed, and as is proved by us in the commentary on the first book of the *Mathematics* [*sc.*, the *Syntaxis mathematica* of Ptolemy] by a theorem of our own."[7] A. Rome concludes that the *Collection* was put together about A.D. 340, but K. Ziegler states that a long interval is not necessary, and that the *Collection* may have been compiled soon after A.D. 320.[8] It has come down to us from a single twelfth-century manuscript, Codex Vaticanus Graecus 218, from which all the other manuscripts are derived.[9]

T. L. Heath judiciously observes that the *Collection*, while covering practically the whole field of Greek geometry, is a handbook rather than an encyclopedia; and that it was intended to be read with the original works, where extant, rather than take their place. But where the history of a particular topic is given, Pappus reproduces the various solutions, probably because of the difficulty of studying them in many different sources. Even when a text is readily available, he often gives alternative proofs and makes improvements or extensions.[10] The portion of book II that survives, beginning with proposition 14, expounds Apollonius' system of large numbers expressed as powers of 10,000. It is probable that book I was also arithmetical.

Book III is in four parts. The first part deals with the problem of finding two mean proportionals between two given straight lines, the second develops the theory of means, the third sets out some "paradoxes" of an otherwise unknown Erycinus, and the fourth treats of the inscription of the five regular solids in a sphere, but in a manner quite different from that of Euclid in his *Elements*, XIII.13–17.

Book IV is in five sections. The first section is a series of unrelated propositions, of which the opening one is a generalization of Pythagoras' theorem even wider than that found in Euclid VI.31. In the triangle

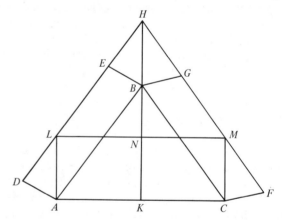

FIGURE 1

ABC let any parallelograms *ABED, BCFG* be drawn on *AB, AC* and let *DE, FG* meet in *H*. Join *HB* and produce it to meet *AC* in *K*. The sum of the parallelograms *ABED, BCFG* can then be shown to be equal to the parallelogram contained by *AC, HB* in an angle equal to the sum of the angles *BAC, DHB*. (It is, in fact, equal to the sum of *ALNK, CMNK*; that is, to the figure *ALMC*, which is easily shown to be a parallelogram having the angle *LAC* equal to the sum of the angles *BAC, DHB*.)

The second section deals with circles inscribed in the figure known as the ἄρβηλος or "shoemaker's knife." It is formed when the diameter *AC* of a semi-

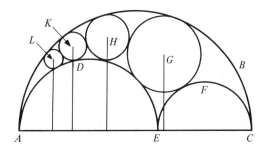

FIGURE 2

circle *ABC* is divided in any way at *E* and semicircles *ADE*, *EFC* are erected. The space between these two semicircles and the semicircle *ABC* is the ἄρβηλος. In a series of elegant theorems Pappus shows that if a circle with center *G* is drawn so as to touch all three semicircles, and then a circle with center *H* to touch this circle and the semicircles *ABC*, *ADE*, and so on *ad infinitum*, then the perpendicular from *G* to *AC* is equal to the diameter of the circle with center *G*, the perpendicular from *H* to *AC* is double the diameter of the circle with center *H*, the perpendicular from *K* to *AC* is triple the diameter of the circle with center *K*, and so on indefinitely. Pappus records this as "an ancient proposition" and proceeds to give variants. This section covers as particular cases propositions in the *Book of Lemmas* that Arabian tradition attributes to Archimedes.

In the third section Pappus turns to the squaring of the circle. He professes to give the solutions of Archimedes (by means of a spiral) and of Nicomedes (by means of the conchoid), and the solution by means of the quadratrix, but his proof is different from that of Archimedes. To the traditional method of generating the quadratrix (see the articles on Dinostratus and Hippias of Elis), Pappus adds two further methods "by means of surface loci," that is, curves drawn on surfaces. As a digression he examines the properties of a spiral described on a sphere.

The fourth section is devoted to another famous problem in Greek mathematics, the trisection of an angle. Pappus' first solution is by means of a νεῦσις or verging—the construction of a line that has to pass through a certain point—which involves the use of a hyperbola. He next proceeds to solve the problem directly, by means of a hyperbola, in two ways; on one occasion he uses the diameter-and-ordinate property (as in Apollonius), and on another he uses the focus-directrix property. This property is proved in book VII. Pappus then reproduces the solutions by means of the quadratrix and the spiral of Archimedes; he also gives the solution of a νεῦσις, which he believes

Archimedes to have unnecessarily assumed in *On Spirals*, proposition 8.

In the preface to book V, which deals with isoperimetry, Pappus praises the sagacity of bees who make the cells of the honeycomb hexagonal because of all the figures which can be fitted together the hexagon contains the greatest area. The literary quality of this preface has been warmly praised. Within the limits of his subject, Pappus looks back to the great Attic writers from a world in which Greek had degenerated into Hellenistic. In the first part of the book Pappus appears to be reproducing Zenodorus fairly closely; in the second part he compares the volumes of solids that have equal surfaces. He gives an account of thirteen semiregular solids, discovered and discussed by Archimedes (but not in any surviving works of that mathematician) that are contained by polygons all equilateral and equiangular but not all similar. He then shows, following Zenodorus, that the sphere is greater in volume than any of the regular solids that have surfaces equal to that of the sphere. He also proves, independently, that, of the regular solids with equal surfaces, that solid is greater which has the more faces.

Book VI is astronomical and deals with the books in the so-called *Little Astronomy*—the smaller treatises regarded as an introduction to Ptolemy's *Syntaxis*. In magistral manner he reviews the works of Theodosius, Autolycus, Aristarchus, and Euclid, and he corrects common misrepresentations. In the section on Euclid's *Optics*, Pappus examines the apparent form of a circle when seen from a point outside the plane in which it lies.

Book VII is the most fascinating in the whole *Collection*, not merely by its intrinsic interest and by what it preserves of earlier writers, but by its influence on modern mathematics. It gives an account of the following books in the so-called *Treasury of Analysis* (those marked by an asterisk are lost works): Euclid's *Data* and *Porisms*,* Apollonius' *Cutting Off of a Ratio*, *Cutting Off of an Area*,* *Determinate Section*,* *Tangencies*,* *Inclinations*,* *Plane Loci*,* and *Conics*. In his account of Apollonius' *Conics*, Pappus makes a reference to the "locus with respect to three or four lines" (a conic section); this statement is quoted in the article on Euclid (IV, 427 *ad fin.*). He also adds a remarkable comment of his own. If, he says, there are more than four straight lines given in position, and from a point straight lines are drawn to meet them at given angles, the point will lie on a curve that cannot yet be identified. If there are five lines, and the parallelepiped formed by the product of three of the lines drawn from the point at fixed angles bears a constant ratio to the parallelepiped formed by the product of the

295

other two lines drawn from the point and a given length, the point will be on a certain curve given in position. If there are six lines, and the solid figure contained by three of the lines bears a constant ratio to the solid figure formed by the other three, then the point will again lie on a curve given in position. If there are more than six lines it is not possible to conceive of solids formed by the product of more than three lines, but Pappus surmounts the difficulty by means of compounded ratios. If from any point straight lines are drawn so as to meet at a given angle any number of straight lines given in position, and the ratio of one of those lines to another is compounded with the ratio of a third to a fourth, and so on (or the ratio of the last to a given length if the number of lines is odd) and the compounded ratio is a constant, then the locus of the point will be one of the higher curves. Pappus had, of course, no symbolism at his disposition, nor did he even use a figure, but his meaning can be made clearer by saying that if $p_1, p_2, \ldots, p_n$ are the lengths of the lines drawn at fixed angles to the lines given in position, and if (a having a given length and k being a constant)

$$\frac{p_1}{p_2} \cdot \frac{p_3}{p_4} \ldots \frac{p_{n-1}}{p_n} = k \text{ when } n \text{ is even, or}$$

$$\frac{p_1}{p_2} \cdot \frac{p_3}{p_4} \ldots \frac{p_n}{a} = k \text{ when } n \text{ is odd,}$$

then the locus of the point is a certain curve.

In 1631 Jacob Golius drew the attention of Descartes to this passage in Pappus, and in 1637 "Pappus' problem," as Descartes called it, formed a major part of his *Géométrie*.[11] Descartes begins his work by showing how the problems of conceiving the product of more than three straight lines as geometrical entities, which so troubled Pappus, can be avoided by the use of his new algebraic symbols. He shows how the locus with respect to three or four lines may be represented as an equation of degree not higher than the second, that is, a conic section which may degenerate into a circle or straight line. Where there are five, six, seven, or eight lines, the required points lie on the next highest curve of degree after the conic sections, that is, a cubic; if there are nine, ten, eleven, or twelve lines on a curve, one degree still higher, that is, a quartic, and so on to infinity. Pappus' problem thus inspired the new method of analytical geometry that has proved such a powerful tool in subsequent centuries. (See the article on Descartes, IV, 57.)

In his *Principia* (1687) Newton also found inspiration in Pappus; he proved in a purely geometrical manner that the locus with respect to four lines is a conic section, which may degenerate into a circle. It is impossible to avoid seeing in Newton's conclusion to lemma XIX, cor. ii, a criticism of Descartes: "Atque ita Problematis veterum de quatuor lineis ab *Euclide* incaepti et ab *Apollonio* continuati non calculus, sed compositio Geometrica, qualem Veteres quaerebant, in hoc Corollario exhibetur."[12] But in this instance it was Descartes, and not Newton, who had the forward vision. Pappus observes that the study of these curves had not attracted men comparable to the geometers of previous ages. But there were still great discoveries to be made, and in order that he might not appear to have left the subject untouched, Pappus would himself make a contribution. It turns out to be nothing less than an anticipation of what is commonly called "Guldin's theorem."[13] Only the enunciations, however, were given, which state

> Figures generated by complete revolutions of a plane figure about an axis are in a ratio compounded (*a*) of the ratio [of the areas] of the figures, and (*b*) of the ratio of the straight lines similarly drawn to [*sc.* drawn to meet at the same angles] the axes of rotation from the respective centers of gravity. Figures generated by incomplete revolutions are in a ratio compounded (*a*) of the ratio [of the areas] of the figures and (*b*) of the ratio of the arcs described by the respective centers of gravity; it is clear that the ratio of the arcs is itself compounded (1) of the ratio of the straight lines similarly drawn [from the respective centers of gravity to the axis of rotation] and (2) of the ratio of the angles contained about the axes of rotation by the extremities of these straight lines.

Pappus concludes this section by noting that these propositions, which are virtually one, cover many theorems of all kinds about curves, surfaces, and solids, "in particular, those proved in the twelfth book of these elements." This implies that the *Collection* originally ran to at least twelve books.

Pappus proceeds to give a series of lemmas to each of the books he has described, except Euclid's *Data*, presumably with a view to helping students to understand them. (He was half a millennium from Apollonius and elucidation was probably necessary.) It is mainly from these lemmas that we can form any knowledge of the contents of the missing works, and they have enabled mathematicians to attempt reconstructions of Euclid's *Porisms* and Apollonius' *Cutting Off of an Area, Plane Loci, Determinate Section, Tangencies,* and *Inclinations.* It is from Pappus' lemmas that we can form some idea of the eighth book of Apollonius' *Conics.*

The lemmas to the *Cutting Off of a Ratio* and the *Cutting Off of an Area* are elementary, but those to the *Determinate Section* show that this work amounted to a theory of involution. The most interesting lemmas concern the values of the ratio $AP \cdot PD : BP \cdot PC$, where (A, D), (B, C) are point-pairs on a straight line and P is another point on the straight line. Pappus investigates the "singular and least" values of the ratio and shows what it is for three different positions of P.

The lemmas to the *Inclinations* do not call for comment. The lemmas to the second book of the *Tangencies* are all concerned with the problem of drawing a circle so as to touch three given circles, a problem that Viète and Newton did not consider it beneath their dignity to solve.[14] The most interesting of Pappus' lemmas states: Given a circle and three points in a straight line external to it, inscribe in the circle a triangle, the sides of which shall pass through the three points.

The lemmas to the *Plane Loci* are chiefly propositions in algebraic geometry, one of which is equivalent to the theorem discovered by R. Simson, but generally known as Stewart's theorem:[15] If A, B, C, D are any four points on a straight line, then

$$AD^2 \cdot BC + BD^2 \cdot CA + CD^2 \cdot AB + BC \cdot CA \cdot AB = 0.$$

The remarkable proposition that Pappus gives in his description of Euclid's *Porisms* about any system of straight lines cutting each other two by two has already been set out in modern notation in the article on Euclid (IV, 426–427). The thirty-eight lemmas that he himself provides to facilitate an understanding of the *Porisms* strike an equally modern note. Lemma 3, proposition 129 shows that Pappus had a clear understanding of what Chasles called the anharmonic ratio and is now generally called the cross-ratio of four points. It proves the equality of the cross-ratios that

are made by any two transversals on a pencil of four lines issuing from the same point. The transversals are, in fact, drawn from the same point on one of the straight lines—in Figure 3 they are *HBCD* and *HEFG*, cutting the lines *AH*, *AL*, *AF*, and *AG*—but it is a simple matter to extend the proof, and Pappus proves that

$$\frac{HE \cdot GF}{HG \cdot FE} = \frac{HB \cdot DC}{HD \cdot BC},$$

that is to say, the cross-ratio is thus invariant under projection.

Lemma 4, proposition 130 shows, even more convincingly than the lemmas to the *Determinate Section*, that Pappus had an equally clear grasp of involution. In Figure 4, *GHKL* is a quadrilateral and *ABCDEF* is

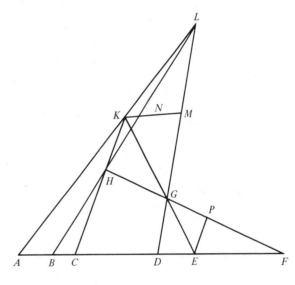

FIGURE 4

any transversal cutting pairs of opposite sides and the diagonals in (A,F), (C,D), (B,E). Pappus shows that

$$\frac{AF \cdot BC}{AB \cdot CF} = \frac{AF \cdot DE}{AD \cdot EF}.$$

(Strictly, what Pappus does is to show that if, in the figure, which he does not set out in detail, this relationship holds, then F, G, H lie on a straight line, but this is equivalent to what has been said above.) This equation is one of the ways of expressing the relationship between three pairs of conjugate points in involution. That Pappus gives these propositions as lemmas to Euclid's *Porisms* implies that they must have been assumed by Euclid. The geometers living just before Euclid must therefore have had an understanding of cross-ratios and involution, although these properties were not named for 2,250 years.

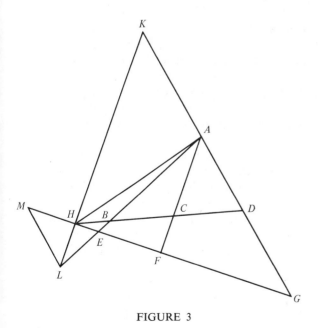

FIGURE 3

Lemma 13, proposition 139 has won its way into text books of modern geometry as "The Theorem of Pappus."[16] It establishes that if, from a point C, two transversals CE, CD cut the straight lines AN, AF, AD (see Figure 5) so that A, E, B and C, F, D are two sets of collinear points, then the points G, M, K are collinear. GMK is called the "Pappus line" of the two sets of collinear points.

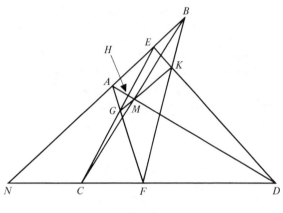

FIGURE 5

In the second of the two lemmas that Pappus gives to the *Surface Loci*, he enunciates and proves the focus-directrix property of a conic, which, as we have seen, he had already once employed. There is only one other place in any surviving Greek text in which this property is used—the fragment of Anthemius' *On Remarkable Mechanical Devices*. G. L. Toomer, however, has recently discovered this property in an Arabic translation of Diocles' treatise *On Burning Mirrors* in Mashhad (Shrine Library, MS 392/5593) and Dublin (Chester Beatty Library, Arabic MS 5255). But Pappus' passage remains the only place in ancient writing in which the property is proved.

Book VIII is devoted mainly to mechanics, but it incidentally gives some propositions of geometrical interest. In a historical preface Pappus justifies the claim that mechanics is a truly mathematical subject as opposed to one of merely utilitarian value. He begins by defining "center of gravity"—the only place in Greek mathematics where it is so defined—gives the theory of the inclined plane; shows how to construct a conic through five given points; solves the problem of constructing six equal hexagons around the circumference of a circle so as to touch each other and a seventh equal hexagon at the center; discourses on toothed wheels; and in a final section (which may be wholly interpolated) gives extracts from Heron's description of the five mechanical powers: the wheel and axle, the lever, the pulley, the wedge, and the screw.

Commentary on the Almagest. A commentary by Pappus on book V (with lacunae) and book VI of Ptolemy's *Syntaxis* exists in the Florentine manuscript designated L (ninth century) and in various other manuscripts. But this commentary is only part of a larger original. In the *Collection* Pappus refers to his commentary (*scholion*) on the first book of the *Almagest*, and in the surviving sixth book he makes the same reference, repeating a proof of his own for Archimedes' theorem about the area of a circle which, he says, he had given in the first book. In the compilation of uncertain authorship known as the *Introduction to the Almagest* there is a reference to a method of division "according to the geometer Pappus," which would seem to hark back to the third book.[17] In the fifth book of the commentary Pappus refers to a theorem in connection with parallax proved in his fourth book.[18] Although there is no direct reference to the second book, there is sufficient evidence that he commented on the first six books, and he may have written on all thirteen. The date of the commentary, as we have seen, must be soon after 320.

At the outset of his fifth book Pappus gives a summary of Ptolemy's fourth book, and at the beginning of his sixth book he summarizes Ptolemy's fifth book, which suggests that his commentary was a course of lectures. This theory is borne out by the painstaking and methodical way in which he explains, apparently for an audience of beginners, the details of Ptolemy's theory.

Ptolemy's fourth book introduces his lunar theory, and he explains the "first or simple anomaly" (irregularity of the movements of the moon) by postulating that the moon moves uniformly round the circumference of a circle (the epicycle), the center of which is carried uniformly round a circle concentric with the ecliptic. Pappus, following Ptolemy closely, explains in his fifth book that this needs correction for a second anomaly, which disappears at the new and full moons but is again noticeable when the moon is at the quadratures—provided that it is not then near its apogee or perigee, an irregularity later called evection. He also explains in detail Ptolemy's hypothesis that the circle on which the epicycle moves (the deferent) is eccentric with the ecliptic, and that the center of the eccentric circle itself moves uniformly round the center of the earth. To account for certain irregularities not explained by these anomalies, Ptolemy postulates a further correction which he calls prosneusis (that is, inclination or verging). In this context prosneusis means that the diameter of the epicycle which determines apogee and perigee is not directed to the center of the ecliptic but to a point on the line joining the center of the eccentric and the center of the ecliptic produced,

298

and as far distant from the latter as the latter is from the former. After a gap in the manuscript, Pappus begins his comment again in the middle of this subject and proceeds to deal with a further complication. He states that the true position of the moon may not be where it is seen in the heavens on account of parallax, which may be neglected for the sun but not for the moon. He gives details for the construction of a "parallactic instrument" (an alidade) used for finding the zenithal distances of heavenly bodies when crossing the meridian. He had previously given details of "an astrolabe" (really an armillary sphere) described by Ptolemy.[19] He also follows Ptolemy closely in his deduction of the sizes and distances of the sun and moon, the diameter of the shadow of the earth in eclipses, and the size of the earth.

In the sixth book, again following Ptolemy closely, Pappus explains the conditions under which conjunctions and oppositions of the sun and moon occur. This explanation leads to a study of the conditions for eclipses of the sun and moon and to rules for predicting when eclipses will occur. The book closes with a study of the points of first and last contact during eclipses.

Pappus, like Theon after him, not only follows Ptolemy's division into chapters but enumerates theorems as Ptolemy does not. It is clear that Theon had Pappus' commentary before him when he wrote over a century later, and in some cases Theon lifted passages directly from Pappus.

Commentary on Euclid's Elements. Eutocius[20] refers to a commentary by Pappus on the *Elements* of Euclid and it probably extended to all thirteen books. In Proclus' commentary on book I there are three references to Pappus,[21] and it is reasonable to believe that they relate to Pappus' own commentary on the *Elements* as they do not relate to anything in the *Collection.* Pappus is said to have pointed out that while all right angles are equal to one another, it is not true that an angle equal to a right angle is always

a right angle—it may be an angle formed by arcs of circles and thus cannot be called a right angle. He is also alleged to have added a superfluous axiom: If unequals are added to equals, the excess of one sum over the other is equal to the excess of one of the added quantities over the other. He also added a complementary axiom about equals added to unequals, as well as certain axioms that can be deduced from the definitions. He gave a neat alternative proof of Euclid I.5 (the angles at the base of an isosceles triangle are equal) by comparing the triangle *ABC* with the triangle *ACB*, that is, the same triangle with the sides taken in reverse order (Figure 6).

Eutocius states that Pappus, in his commentary on the *Elements*, explains how to inscribe in a circle a polygon similar to a polygon inscribed in another circle. This would doubtless be in his commentary on book XII, and Pappus probably solved the problem in the same manner as a scholiast to XII.1, that is, by making the angles at the center of the second circle equal to the angles at the center of the first.[22]

If Pappus wrote on books I and XII it is likely that he also commented on the intermediate books, and the fact that he commented on book X is attested by a scholiast to Euclid's *Data*[23] and by the *Fihrist*, in which it is stated that the commentary was in two parts.[24]

A two-part commentary on the tenth book of Euclid's *Elements* does actually exist in Arabic,[25] and it is usually identified with that of Pappus. It was discovered in a Paris manuscript by F. Woepcke in 1850, but the manuscript lacks diacritical marks and Woepcke himself read the consonantal skeleton of the author's name as Bls, which he interpreted as meaning Valens, probably Vettius Valens, an astronomer of the age of Ptolemy.[26] Heiberg showed this interpretation to be impossible, and was the first scholar to identify the commentary with that which Pappus was known to have written.[27] H. Suter pointed out that the Arabic for Bls could easily be confused with Bbs, and as there is no P in Arabic, Pappus would be the author indicated.[28] This was accepted by T. L. Heath,[29] and indeed generally, but when Suter's translation of Woepcke's text was published in 1922[30] he raised the question whether the prolixity and Neoplatonic character of the treatise did not indicate Proclus as the author. In the latest study of the subject (1930) William Thomson denied the charges of prolixity and mysticism and accepted the authorship of Pappus.[31] It must be admitted that the commentary is in a wholly different style from the severely mathematical nature of the *Collection*, or even of the more elementary commentary on the *Almagest*, and the question of authorship cannot be regarded as entirely free from doubt.

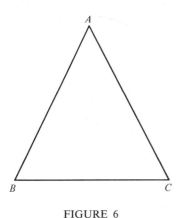

FIGURE 6

The superscription to the first part of the commentary and the subscription to the second part state that the Arabic translation is the work of Abū 'Uthman al-Dimishqī (*fl. ca.* 908–932), who also translated the tenth book of Euclid's *Elements*. The postscript to the second part adds that the copy of the commentary was written in 969 by Aḥmad ibn Muḥammad ibn 'Abd al-Jalīl, that is, the Persian geometer generally known as al-Sizjī (*ca.* 951–1029).

Some two dozen passages in the commentary have parallels in the scholia to Euclid's book X, sometimes remarkably close parallels. The simplest explanation is that the scholiast made his marginal notes with Pappus' commentary in front of him.

Euclid's book X is a work of immense subtlety, but there is little in the commentary that calls for comment. The opening section has an interest for the historian of mathematics as it distinguishes the parts played by the Pythagoreans, Theaetetus, Euclid, and Apollonius in the study of irrationals. It also credits Theaetetus with a classification of irrationals according to the different means.[32] He is said to have assigned the medial line to geometry ($\sqrt{xy}$ is the geometric mean between x, y), the binomial to arithmetic ($\frac{1}{2}[x + y]$ is the arithmetic mean between x, y), and the apotome to harmony (the harmonic mean $[2xy]/[x + y]$ between x, y is $[(2xy)/(x^2 - y^2)] \cdot [x - y]$, which is the product of a binomial and an apotome.)

Other Mathematical Works. Marinus, in the final sentence of his commentary on Euclid's *Data*,[33] reveals that Pappus also commented on the *Data*. Pappus apparently showed that Euclid's teaching followed the method of analysis rather than synthesis. Pappus also mentions a commentary that he wrote on the *Analemma* of Diodorus, in which he used the conchoid of Nicomedes to trisect an angle.[34]

The *Fihrist* includes among Pappus' works "A commentary on the book of Ptolemy on the *Planisphaerium*, translated by Thābit into Arabic." The entry leaves it uncertain whether Thābit ibn Qurra (*d.* 901) translated Ptolemy's work or Pappus' commentary, but Hājjī Khalīfa states that Ptolemy was the author of a treatise on the *Planisphaerium* translated by Thābit. He also adds that Ptolemy's work was commented on by "Battus al Roûmi [that is, late Greek], an Alexandrian geometer." "Battus" is clearly "Babbus," that is Pappus.[35]

Geography. The *Description of the World* mentioned in the *Suda Lexicon* has not survived in Greek, but the *Geography* bearing the name of the Armenian Moses of Khoren (although some scholars see in it the work of Anania Shirakatsi) appears to be a translation, or so closely based on Pappus' work as to be

virtually a translation. The *Geography*, if correctly ascribed to Moses, was written about the beginning of the fifth century. The archetype has not survived, and the manuscripts contain both a long and a short recension. The character of Pappus' work may be deduced from two passages of Moses, or the pseudo-Moses, which may be thus rendered:[36] "We shall begin therefore after the *Geography* of Pappus of Alexandria, who has followed the circle or the special map of Claudius Ptolemy" and "Having spoken of geography in general, we shall now begin to explain each of the countries according to Pappus of Alexandria." From these and other passages J. Fischer[37] deduced that Pappus' work was based on the world map and on the special maps of Ptolemy rather than on the text itself, and as Pappus flourished only a century and a half after Ptolemy it is a fair inference that the world map and the special maps date back to Ptolemy himself. Pappus appears to have written with Ptolemy's maps as his basis, but about the world as he knew it in the fourth century.

Nothing is known of the second geographical work, *Rivers of Libya*, mentioned in the *Suda Lexicon*, or of the *Interpretation of Dreams*. The interpretation of dreams is akin to astrology, and there would be nothing surprising in a work on the subject by an ancient mathematician.

Music. It is possible that the commentary on Ptolemy's *Harmonica*, which was first edited by Wallis as the work of Porphyry, is, from the fifth chapter of the first book on, the work of Pappus. Several manuscripts contain the first four chapters only, and Lucas Holstein found in the Vatican a manuscript containing a definite statement that Porphyry's commentary was confined to the first four chapters of the first book and that Pappus was responsible for the remainder. Montfaucon also noted the same manuscript under the title "Pappi De Musica." Wallis did not accept the attribution because the title of the whole work and the titles of the chapters imply that it is wholly the work of Porphyry and because he could detect no stylistic difference between the parts. But the titles prove nothing, as Porphyry no doubt did comment, or intended to comment, on the whole work, and only missing parts would have been taken from another commentary, and arguments based on differences of style, especially in a technical work, are notoriously difficult. Hultsch and Jan were satisfied that Pappus was the author, but Düring was emphatically of the opinion that the whole is the work of Porphyry, and Ver Eecke agreed.[38] It must be left an open question.

Hydrostatics. An Arabic manuscript discovered in Iran by N. Khanikoff and published in 1860 under the title *Book of the Balance of Wisdom, an Arabic Work on the Water Balance, Written by al-Khazini in the Twelfth Century*[39] attributes to Pappus an instrument for measuring liquids and describes it in detail. The instrument is said by Khanikoff to be nearly identical with the volumeter of Gay-Lussac. If the attribution is correct—and there seems no reason to doubt it—the instrument may have been described in the missing part of the eighth book of the *Collection* or it may have had a place in a separate work on hydrostatics, of which no other trace has survived.

An Alchemical Oath. An oath attributed to "Pappus, philosopher" in a collection of alchemical writings may be genuine—if not *vero*, it is at least *ben trovato*—and if so it may tell us something of Pappus' syncretistic religious views in an age when paganism was retreating before Christianity. It is an oath that could have been taken equally by a pagan or a Christian, and it would fit in with the dates of Pappus. It could be gnostic, it has a Pythagorean element in it, there may be a veiled reference to the Trinity, and there is a Byzantine ring to its closing words. It reads: "I therefore swear to thee, whoever thou art, the great oath, I declare God to be one in form but not in number, the maker of heaven and earth, as well as the tetrad of the elements and things formed from them, who has furthermore harmonized our rational and intellectual souls with our bodies, who is borne upon the chariots of the cherubim and hymned by angelic throngs."[40]

A Vatican manuscript containing Ptolemy's *Handy Tables* has on one folio a short text about the entry of the sun into the signs of the zodiac, which F. Boll has shown must refer to the second half of the third century and which E. Honigmann attributes to Pappus. But this is no more than an unsubstantiated guess, which Boll himself refrained from making.[41]

A Florentine manuscript catalogued by Bandini notices Ἡμεροδρόμιον Πάππου τῶν διεπόντων καὶ πολευόντων, that is, daily tables of governing and presiding stars compiled by Pappus.[42]

NOTES

1. *Suda Lexicon*, Adler, ed., Vol. I, Pars IV (Leipzig, 1935), P 265, p. 26.
2. Pappus, *Collectio*, III.1, F. Hultsch, ed., I, 30.4; VII.1, Hultsch, ed., II, 634.1. Nothing more is known of Hermodorus or of Pandrosion and Megethion, to whom the third and fifth books are dedicated; or of his philosopher-friend Hierius, who pressed him to give a solution to the problem of finding two mean proportionals (Hultsch, ed., III, 3–8). A phrase in Proclus, *In primum Euclidis*, Friedlein, ed. (Leipzig, 1873; repr. Hildesheim, 1967), p. 429.13, οἱ . . . περὶ Πάππον, implies that he had a school.
3. A. Rome, *Commentaires de Pappus et de Théon d'Alexandrie sur l'Almageste*, I (Rome, 1931), 180.8–181.23, *Studi e Testi*, no. 54 (1931). The eclipse is no. 402 in F. K. Ginzel, *Spezialler Kanon der Sonnen und Mond Finsternisse* (Berlin, 1899), p. 87, and no. 3642 in T. von Oppolzer, *Canon der Finsternissen* (Vienna, 1887), repr. translated by Owen Gingerich (New York, 1962), p. 146. Rome, who first perceived the bearing of this eclipse on the date of Pappus, argues that if the total, or nearly total, eclipse of A.D. 346 had taken place, Pappus would certainly have chosen it for his example, and that the better eclipse of A.D. 291 was already too distant to be used (A. Rome, *op. cit.*, pp. x–xiii).
4. So A. Rome, *op. cit.*, I, xvii, note 1, suggests. This is more convincing than the conjecture of F. Hultsch, *op. cit.*, III, viii, note 3, that Δ is a copyist's error for ΙΓ.
5. A. Rome, *op. cit.*, II, lxxxiii, *Studi e Testi*, no. 72 (1936).
6. Leiden MS, no. 78, of Theon's ed. of the *Handy Tables*, fol. 55. This was first noted by J. van der Hagen, *Observationes in Theonis Fastos Graecos priores* (Amsterdam, 1735), p. 320, and his view was followed by H. Usener, "Vergessenes III," in *Rheinisches Museum*, n.s. **28** (1873), 403–404, and F. Hultsch, *op. cit.*, III, vi–vii, but none of these scholars realized the significance of Pappus' reference to the eclipse of A.D. 320.
7. Pappus, *Collectio* VIII.46, *op. cit.*, III, 1106.13–15. Rome, *op. cit.*, I, 254, note 1, gives reasons for thinking that the third theorem of book V of the *Collectio* is a fragment, all that now survives, of book I of the *Commentary on the Syntaxis*, and that it is an interpolation by an ed.
8. A. Rome, see previous note; K. Ziegler, in Pauly-Wissowa, XVIII (Waldsee, 1949), col. 1094.
9. F. Hultsch, *op. cit.*, I, p. vii–xiv.
10. Thomas Heath, *A History of Greek Mathematics*, II (Oxford, 1921), 357–358. A full and excellent conspectus of the *Collection* is given by Heath, *loc. cit.*, pp. 361–439; Gino Loria, *Le scienze esatte nell'antica Grecia*, 2nd ed. (Milan, 1914), pp. 658–700; and Paul Ver Eecke, *Pappus d'Alexandrie: La Collection mathématique*, I (Paris–Bruges, 1933), xiii–cxiv.
11. René Descartes, *Des matières de la géométrie* (Leiden, 1637), book I, 304–314, book II, 323–350; David Eugene Smith and Marcia C. Latham, *The Geometry of René Descartes With a Facsimile of the First Edition* (New York, 1925; repr. 1954), book I, 17–37, book II, 59–111.
12. Isaac Newton, *Philosophiae naturalis principia mathematica* (London, 1687; repr. London, 1953), "De motu corporum," lib. 1, sect. 5, lemma XIX, pp. 74–75.
13. Pappus, VII.42, *op. cit.*, II, 682.7–15. The whole passage in which this occurs is attributed by Hultsch to an interpolator, but without reasons given, and by Ver Eecke (*op. cit.*, I, xcvi) for unconvincing stylistic reasons and lack of connection with the context. But Heath pertinently observes (*A History of Greek Mathematics*, II, 403) that no Greek after Pappus would have been capable of framing an advanced proposition. Ver Eecke (*op. cit.*, I, xcv, cxxiii) observes that Paul Guldin (1577–1643) could not have been inspired by the passage in Pappus as Commandino did not include it in his first ed. (Pesaro, 1588) and he could not have seen the second ed. (Bologna, 1660), augmented with this passage by Manolessius. But this conclusion is an error; the passage is in the first no less than the second ed. See also the article on Guldin.
14. F. Vieta, *Apollonius Gallus* (Paris, 1600), problem x, pp. 7–8; Isaac Newton, *Arithmetica universalis* (Cambridge, 1707), problem xli *ad finem*, pp. 181–182, 2nd ed. (London, 1722), problem xlvii *ad finem*, p. 195; *Principia* (London, 1687; repr. London, 1953), lemma XVI, pp. 67–68.
15. Robert Simson, *Apollonii Pergaei locorum planorum libri II restituti* (Glasgow, 1749), pp. 156–221; Matthew Stewart, *Some General Theorems of Considerable Use in the Higher Parts of Mathematics* (Edinburgh, 1746), pp. 1–2. See also

Moritz Cantor, *Vorlesungen über Geschichte der Mathematik*, III (Leipzig, 1898), 523–528.

16. For example, E. A. Maxwell, *Geometry For Advanced Pupils* (Oxford, 1949), p. 97. The term "Pappus' Theorem" is thus used by Renaissance and modern geometers in two different ways.

17. C. Henry, *Opusculum de multiplicatione et divisione sexagesimalibus, Diophanto vel Pappo tribuendum* (Halle, 1879), p. viii; A. Rome, *op. cit.*, **1**, xvi.

18. A. Rome, *op. cit.*, I, 76.19–77.1.

19. For a reconstruction of the astrolabe and parallactic instrument as described by Pappus, with illustrations, see A. Rome, *Annales de la Société scientifique de Bruxelles*, **47** (1927), 77–102, 129–140, and *op. cit.*, **1**, 3–5, 69–77.

20. Eutocius, *Commentarii in libros Archimedis De Sphaera et cylindro*, p. 1.13, *ad init.*, *Archimedis opera omnia*, J. L. Heiberg, ed., 2nd ed., III (Leipzig, 1915), corr. repr. Evangelos S. Stamatis (Stuttgart, 1972), p. 28.19–22: ὅπως μὲν οὖν ἔστιν εἰς τὸν δοθέντα κύκλον πολύγωνον ἐγγράψαι ὅμοιον τῷ ἐν ἑτέρῳ ἐγγεγραμμένῳ, δῆλον, εἴρηται δὲ καὶ Πάππῳ εἰς τὸ ὑπόμνημα τῶν Στοιχείων.

21. Proclus, *In primum Euclidis*, Friedlein, ed., pp. 189.12–191.4, 197.6–198.15, 249.20–250.19.

22. *Euclidis opera omnia*, J. L. Heiberg and H. Menge, eds., V (Leipzig, 1888), scholium 2, 616.6–617.21.

23. *Ibid.*, VI (Leipzig, 1896), scholium 4 *ad definitiones*, 262.4–6: δύναται δὲ καὶ ῥητὸν καὶ ἄλογον δεδομένον εἶναι, ὡς λέγει Πάππος ἐν ἀρχῇ τοῦ εἰς τὸ ι΄ Εὐκλείδου.

24. H. Suter, "Das Mathematiker Verzeichniss im Fihrist des Ibn abî Jaʿkûb an-Nadîm," in *Zeitschrift für Mathematik und Physik*, **37** (1892), suppl. (or *Abhandlungen zur Geschichte der Mathematik*, **6**), p. 22. The whole entry runs, in English: "Pappus the Greek. His writings are: A Commentary on the book of Ptolemy concerning the representation of the sphere in a plane, translated by Thābit into Arabic. A commentary on the tenth book of Euclid, in two parts."

25. Bibliothèque Nationale (Paris), MS no. 2457 (Supplément arabe de la Bibliothèque impériale no. 952.2). The manuscript contains about fifty treatises, of which nos. 5 and 6 constitute the two books of the commentary.

26. Woepcke described the manuscript and translated four passages into French in his "Essai d'une restitution de travaux perdus d'Apollonius sur les quantités irrationelles," in *Mémoires présentés par divers savants à l'Académie des sciences*, **14** (1856), 658–720. He developed his theory about the authorship in *The Commentary on the Tenth Book of Euclid's Elements by Bls*, which he published anonymously and without date or place of publication. Woepcke read the name of the author in the title of the first book of the commentary as B.los (the dot representing a vowel) and in other manuscripts as B.lis, B.n.s, or B.l.s.

27. J. L. Heiberg, *Litterärgeschichtliche Studien über Euklid* (Leipzig, 1882), pp. 169–170. Heiberg points out that one of the manuscripts cited by Woepcke states that "B.n.s le Roumi" (that is, late Greek) was later than Claudius Ptolemy, while the *Fihrist* says that "B.l.s le Roumi" wrote a commentary on Ptolemy's *Planisphaerium*. As Vettius Valens lived under Hadrian, he was therefore older than Ptolemy—an elder contemporary. Moreover the *Fihrist* gives separate entries to B.l.s and Valens. See also Suter, *op. cit.*, p. 54, note 92.

28. H. Suter, "Das Mathematiker Verzeichniss im Fihrist," pp. 22, 54, note 92.

29. T. L. Heath, *The Thirteen Books of Euclid's Elements*, 2nd ed., III (Cambridge, 1905, 1925; repr. New York, 1956), 3; Heath, *A History of Greek Mathematics*, I (Oxford, 1921), 154–155, 209, II, 356.

30. H. Suter, "Der Kommentar des Pappus zum X Buche des Eukleides," in *Abhandlungen zur Geschichte der Naturwissenschaften und der Medizin*, **4** (1922), 9–78; see p. 78 for the question of authorship.

31. William Thomson and Gustav Junge, *The Commentary of Pappus on Book X of Euclid's Elements* (Cambridge, Mass., 1930; repr. New York, 1968), pp. 38–42.

32. There is nothing in the opening section about the rational and irrational being "given," as Pappus is stated by a scholiast (see note 23) to have maintained at the beginning of his commentary. This may be evidence against the ascription of the Arabic treatise to Pappus.

33. *Euclidis opera omnia*, J. L. Heiberg and H. Menge, eds., VI, 256.22–25.

34. F. Hultsch, *op. cit.*, I, 246.1–3. Ἀνάλημμα, as in Ptolemy's work with that title, means the projection of the circles of a celestial sphere on the plane. Neither the work of Diodorus nor the commentary of Pappus has survived. In Ptolemy's work certain segments of a semicircle are required to be divided into six equal parts, and it is easy to see how Pappus would need to trisect an arc or angle.

35. H. Suter, "Das Mathematiker Verzeichniss im Fihrist," p. 22 (see note 24, supra); Hājjī Khalīfa, *Lexikon bibliographicum et encyclopaedicum*, G. Fluegel, ed., V (London, 1850), 61–62, no. 9970, *s.v.* Kitab testih el koret. The *Planisphaerium* is a system of stereographic projection by which points on the heavenly sphere are represented on the plane of the equator by projection from a pole.

36. Translated from the French of P. Arsène Soukry, *Géographie de Moïse de Corène d'après Ptolémée* (Venice, 1881), p. 7.

37. J. Fischer, "Pappus und die Ptolemaeus Karten," in *Zeitschrift der Gesellschaft für Erdkunde zu Berlin*, **54** (1919), 336–358.

38. John Wallis, *Claudii Ptolemaei Harmonicorum libri III* (Oxford, 1682), reprinted in *Opera mathematica*, III (Oxford, 1699); the commentary is on pp. 183–355 of the latter work, and the authorship is discussed on p. 187. It has been edited in modern times by Ingemar Düring as *Porphyrios' Kommentar zur Harmonienlehre des Ptolemaios* (Göteborg, 1932). His discussion of the authorship is on pp. xxxvii–xxxix. Lucas Holstenius, *Dissertatio De vita et scriptis Porphyrii* (Rome, 1630), c. vi, p. 55: Neque tamen in universum ἁρμονικῶν opus scripsit Porphyrius, sed in quatuor duntaxat prima capita: cetera dein Pappus pertexuit. Ita enim in alio manuscripto Vaticano titulus indicat: Πορφυρίου ἐξήγησις εἰς δ΄ πρῶτα κεφάλαια τοῦ πρώτου τῶν ἁρμονικῶν Πτολεμαίου. Sequitur deinde, Πάππου ὑπόμνημα εἰς τὰ ἀπὸ τοῦ ε΄ κεφαλαίου καὶ ἐφεξῆς. Bernard de Montfaucon, *Bibliotheca bibliothecarum manuscriptorum nova*, I (Paris, 1739), 11B. Paul Ver Eecke, *Pappus d'Alexandrie: La Collection mathématique*, I (Paris-Bruges, 1933), cxv–cxvi. F. Hultsch, *op. cit.*, III, xii. C. Jan, *Musici scriptores graeci* (Leipzig, 1895; repr. Hildesheim, 1962), p. 116 and note 1.

39. See *Journal of the American Oriental Society*, **6** (1860), 40–53; and the article on al-Khāzinī, IV, 338–341; and bibliography, 349–351.

40. C. G. Grumer, *Isidis, Christiani et Pappi philosophi Iusjurandum chemicum nunc primum graece et latine editum* (Jena, 1807); M. Berthelot and C. E. Ruelle, *Collection des anciens alchimistes grecs* (Paris, 1888), pp. 27–28, traduction, pp. 29–30; Paul Tannery, "Sur la réligion des derniers mathématiciens de l'antiquité," in *Annales de philosophie chrétienne*, **34** (1896), 26–36, repr. in *Mémoires scientifiques*, **2** (1912), 527–539, esp. pp. 533–535. Tannery seems inclined to think that the oath may be correctly attributed to Pappus the mathematician, and he speculates that he may have been a gnostic.

41. Vaticanus Graecus 1291, fol. 9r. F. Boll, "Eine illustrierte Prakthandschrift der astronomischen Tafeln des Ptolemaios," in *Sitzungsberichte der Königliche Bayerische Akademie der Wissenschaften, philosophisch-philologischen und historischen Classe*, **29** (1899), 110–138. E. Honigmann, *Die sieben Klimata und die πόλεις ἐπίσημοι* (Heidelberg, 1929), p. 73.

42. Codex Laurentianus XXXIV plut. XXVIII; A. M. Bandini, *Catalogus Bibliothecae Laurentianae*, II (Florence, 1767), 61.

BIBLIOGRAPHY

I. ORIGINAL WORKS. *Collection.* The only complete ed. of the extant Greek text is F. Hultsch, *Pappi Alexandrini Collectionis quae supersunt e libris manu scriptis edidit latina interpretatione et commentariis instruxit Fridericus Hultsch,* 3 vols. (Berlin, 1876–1878). The Greek text is accompanied by a Latin translation on the opposite page and there are invaluable introductions, notes, and appendixes. Apart from a tendency to invoke interpolators too readily, the work is a model of scholarship.

The only translation of the whole extant text into a modern language is that of Paul Ver Eecke, *Pappus d'Alexandrie, La Collection Mathématique: oeuvre traduite pour la première fois du grec en français avec une introduction et des notes,* 2 vols. (Paris–Bruges, 1933). A German translation of books III and VIII is given by C. J. Gerhardt, *Die Sammlung des Pappus von Alexandrien, griechisch und deutsch herausgegeben,* 2 vols. (Halle, 1871).

The *Collection* first became known to the learned world when Commandino included Latin translations of various extracts in his editions of Apollonius (Bologna, 1566) and Aristarchus (Pesaro, 1572). After Commandino's death, his complete Latin trans. of the extant Greek text, except the unknown fragment of book II, appeared as *Pappi Alexandrini Mathematicae Collectiones a Federico Commandino Urbinate in Latinum conversae et commentariis illustratae* (Pesaro, 1588). Reprints appeared in 1589 (Venice) and 1602 (Pesaro) and a second ed. was published by C. Manolessius in 1660 (Bologna); despite the editor's claims it was inferior to the first ed.

Extracts from the Greek text were published in works by Marc Meiboom (1655), John Wallis (1688; first publication of the missing fragment of book II, which he found in a MS in the Savilian Library at Oxford), David Gregory (1703), Edmond Halley (1706, 1710), Robert Simson (1749), Joseph Torelli (1769), Samuel Horsley (1770), J. G. Camerer (1795), G. G. Bredow (1812), Hermann J. Eisenmann (1824), C. J. Gerhardt (1871).

Commentary on Ptolemy's Syntaxes (*Almagest*). The only complete ed. of the extant Greek text (part of book V and book VI) is Adolphe Rome, *Commentaires de Pappus et de Théon d'Alexandrie sur l'Almageste, texte établi et annoté par A. Rome*; vol. I is *Pappus d'Alexandrie: Commentaire sur les livres 5 et 6 de l'Almageste* (Rome, 1931), *Studi e Testi* no. 54. The work lacks only the indexes that would have been published at the end of the commentaries if Rome's design had not been interrupted by the destruction of his papers in the war.

The extant Greek text of book V was printed, with numerous errors and together with Theon's commentary, at the end of the *editio princeps* of the *Almagest.* This ed. was published by Grynaeus and Camerarius (Basel, 1538), but contained no mention of Pappus on the title page. F. Hultsch began, but was not able to complete, an ed. of the commentary by Pappus and Theon; see his "Hipparchos über die Grosse und Entfernung der Sonne," in *Berichte über die Verhandlungen der königlich sachsischen Gesellschaft der Wissenschaften,* Philologisch-Historische Klasse, **52** (1900), 169–200. This work is vitiated by a fundamental error—what he thought was a working over of Pappus' text by Theon was really the same text—an error he would undoubtedly have recognized had he been able to continue his research.

Commentary on Euclid's Elements. The text of Abū 'Uthman al-Dimishqī's Arabic translation of a Greek commentary on the tenth book of Euclid's *Elements,* generally believed to be part of Pappus' commentary on the *Elements,* is published—with an English trans. and notes—in William Thomson and Gustav Junge, *The Commentary of Pappus on Book X of Euclid's Elements,* VIII in Harvard Semitic Series (Cambridge, Mass., 1930; repr. New York, 1968), 189–260. This supersedes the first printed version of the text by F. Woepcke, *The Commentary on the Tenth Book of Euclid's Elements by Bls* (Paris, 1855)—published without indication of author, place, or date. The Arabic text and trans. are the work of William Thomson. There is a German translation in H. Suter, "Der Kommentar des Pappus zum X Buche des Eukleides," in *Abhandlungen zur Geschichte der Naturwissenschaften und der Medizin* (1922), 9–78.

Commentary on Ptolemy's Harmonics. John Wallis, *Claudii Ptolemaei Harmonicorum libri III* (*graece et latine*), *Joh. Wallis recensuit, edidit, versione et notis illustravit, et auctarium adjecit* (Oxford, 1682). The commentary, which follows Ptolemy's text, is the work of Porphyry for the first four chapters, but possibly of Pappus from the fifth chapter on. The work was reprinted in *Johannis Wallis S.T.D. Operum mathematicorum Vol. III* (Oxford, 1699), 183–355, as *Porphyrii in Harmonica Ptolemaei commentarius nunc primum ex Codd. MSS.* (*Graece et latine*) *editus,* with (?) Pappus' share of the commentary on pp. 266–355. There is a modern text, with copious notes, by Ingemar Düring, "Porphyrios' Kommentar zur Harmonielehre des Ptolemaios," in *Göteborgs högskolas årskrift,* **38** (1932), i–xliv, 1–217; see also Bengt Alexanderson, *Textual Remarks on Ptolemy's Harmonica and Porphyry's Commentary,* which is *Studia Graeca et latina Gothoburgensia,* XXVII (Göteborg, 1969).

Geography. What is believed to be essentially an early Armenian trans. of Pappus' *Geography* is given, with a French rendering, in P. Arsène Soukry, *Géographie de Moïse de Corène d'après Ptolémée, texte arménien, traduit en français* (Venice, 1881).

II. SECONDARY LITERATURE. *General.* Konrat Ziegler, "Pappos 2," in Pauly-Wissowa, XVIII (1949), cols. 1084–1106; the prefatory matter, notes, and appendices to the works by Hultsch, Rome, and Ver Eecke are cited above; Moritz Cantor, *Vorlesungen über Geschichte der Mathematik,* I, 3rd ed. (Leipzig, 1907), 441–455; Gino Loria, *Le scienze esatte nell'antica Grecia,* 2nd ed. (Milan, 1914), pp. 656–703; T. L. Heath, *A History of Greek Mathematics,* II (Oxford, 1921), 355–439.

Collection. The works cited in the previous paragraph are helpful. See also Robert Simson, *Apollonii Pergaei locorum planorum libri II restituti* (Glasgow, 1749); "De porismatibus tractatus," in *Opera quaedam reliqua*

(Glasgow, 1776), pp. 315–594; Michel Chasles, *Les trois livres de porismes d'Euclide, rétablis pour la première fois, d'après la notice et les lemmes de Pappus* (Paris, 1860); Paul Tannery, "L'arithmetique des Grecs dans Pappus" in *Mémoires de la Société des sciences physiques et naturelles de Bordeaux*, **3** (1880), 351–371, repr. in *Mémoires scientifiques*, **1** (1912), pp. 80–105; "Note sur le problème de Pappus" ("Pappus' problem" in the sense used by Descartes), in *Oeuvres de Descartes*, C. Adam and P. Tannery, eds., VI (Paris, 1902), 721–725, repr. in *Mémoires scientifiques*, **3** (1915), 42–50; J. S. MacKay, "Pappus on the Progressions" (a translation of Pappus on means), in *Proceedings of the Edinburgh Mathematical Society*, **6** (1888), 48–58; J. H. Weaver, "Pappus," in *Bulletin of the American Mathematical Society*, **23** (1916–1917), 127–135; "On Foci of Conics," *ibid.*, 361–365; N. Khanikoff, "Analysis and Extracts of *Book of the Balance of Wisdom*, an Arabic Work on the Water Balance, Written" by al-Khāzinī in the Twelfth Century, in *Journal of the American Oriental Society*, **6** (1860), lecture 1, ch. 7, 40–53; and al-Khāzinī, *Kitāb mīzān al-ḥikma* (Hyderabad, Deccan, A.H. 1359 [A.D. 1940–1941]). For further references see Bibliography to article on al-Khāzinī, in *Dictionary of Scientific Biography*, IV, 349–351. An article by Malcolm Brown, "Pappus, Plato and the Harmonic Mean," is promised for *Phronesis*.

Commentary on Ptolemy's Syntaxis (Almagest). F. Hultsch, *Hipparchos über die Grosse und Entfernung der Sonne* as above; S. Gunther, "Über eine merkwürdige Beziehung zwischen Pappus und Kepler," in *Bibliotheca mathematica*, n.s. **2** (1888), 81–87; A. Rome, "L'astrolabe et le météoroscope, d'après le commentaire de Pappus sur le 5e livre de l'Almageste," in *Annales de la Société scientifique de Bruxelles*, **47** (1927), 77–102; "L'instrument parallactique d'après le commentaire de Pappus sur le 5e livre de l'Almageste," *ibid.*, 129–140; *Pappus d'Alexandrie: Commentaire sur les livres 5 et 6 de l'Almageste* as above.

Commentary on Euclid's Elements. F. Woepcke, "Essai d'une restitution de travaux perdus d'Apollonius sur les quantités irrationelles," in *Mémoires présentés par divers savants à l'Académie des sciences de l'Institut de France*, **14** (1856), 658–720; J. L. Heiberg, *Litterärgeschichtliche Studien uber Euklid* (Leipzig, 1882), pp. 169–170; H. Suter, *Der Kommentar des Pappus zum X Buche des Eukleides* as above; William Thomson and Gustav Junge, *The Commentary of Pappus on Book X of Euclid's Elements* as above.

Geography. J. Fischer, "Pappus und die Ptolemaeus Karten," in *Zeitschrift der Gesellschaft für Erdkunde zu Berlin*, **54** (1919), 336–358; *Claudii Ptolemaei Geographiae Codex Urbinas Graecus 82, Tomus prodromus* (Leiden–Leipzig, 1922), 419–436; E. Honigmann, *Die sieben Klimata und die πόλεις ἐπίσημοι* (Heidelberg, 1929), c.x., "Pappus und Theon," pp. 72–81.

Commentary on Ptolemy's Harmonics. Ingemar Düring, "Die Harmonielehre des Klaudios Ptolemaios," in *Göteborgs högskolas årskrift*, **36** (1930); and "Ptolemaios und Porphyrios uber die Musik," *ibid.*, **40** (1934), 1–293.

                    IVOR BULMER-THOMAS

PARACELSUS, THEOPHRASTUS PHILIPPUS AUREOLUS BOMBASTUS VON HOHENHEIM (*b.* Einsiedeln, Switzerland, *ca.* 1493 [or 1 May 1494(?)]; *d.* Salzburg, Austria, 24 September 1541), *chemistry, medicine, natural philosophy, cosmology, theology, occultism, iatrochemistry.*

"Paracelsus," a nickname dating from about 1529, may denote "surpassing Celsus"; it might also represent a latinization of "Hohenheim," or even refer to his authorship of "para [doxical]" works that overturned tradition. Paracelsus was the son of William of Hohenheim, a member of the Bombast (Banbast) family of Swabia, who practiced medicine from 1502 to 1534 at Villach, in Carinthia; his mother was a bondswoman of the Benedictine abbey at Einsiedeln. Paracelsus received his early education–particularly in mining, mineralogy, botany, and natural philosophy—from his father. He was later taught by several bishops and apparently by Johannes Trithemius, abbot of Sponheim and a famous exponent of the occult, who was also in contact with Heinrich Cornelius Agrippa von Nettesheim. Paracelsus did practical work in the Fugger mines of Hutenberg, near Villach, and in those of Siegfried Fueger at Swaz.

In addition, Paracelsus probably studied at various Italian universities, perhaps including that of Ferrara. It is not certain that he received the doctorate; the only documentation would seem to be a personal deposition made before a magistrate in Basel. (This deposition was accepted in lieu of an oath by a witness in a lawsuit between two Strasbourg burghers, one of whom had been a patient of Paracelsus.) At any rate, in his laudatory preface to Paracelsus' *Grosse Wundartzney* (Augsburg, 1536), Wolfgang Thalhauser, municipal physician at Augsburg, called Paracelsus "Doctor of both medicines." It is possible that Paracelsus took a lower medical degree at Ferrara, as may be borne out by his subsequent service as a military surgeon, first in the service of Venice, then elsewhere on his early travels (including those to Scandinavia and probably to the Middle East and Rhodes). His work as a surgeon reflected his nonconformity; in reply to the traditional separation of medicine and surgery he coined the phrase "*In judicando*, a physician; *in curando*, a surgeon."

In 1525 Paracelsus was in Salzburg for a short time. While there he barely escaped prosecution for exhibiting sympathy for the Peasants' War. Following some abortive attempts to establish himself in southern Germany and Switzerland, he set up a successful practice in Strasbourg. Called in consultation to Basel, he saved the life of the influential humanist and publisher Johannes Froben. His conservative and cautious treatment of Froben and the medical advice that he gave Erasmus, who was then staying in Froben's house, won Paracelsus the post of municipal physician and professor of medicine at Basel in March 1527. His appointment was sponsored by Strasbourg and Basel church reformers, especially Johannes Oecolampadius, and was not approved by the academic authorities. The latter refused to admit into their company a man who not only failed to submit qualifying documents and declined to take the required oath, but also issued instead an iconoclastic document, the *Intimatio*. In this work (which, published as a broadside, is extant in the 1575 edition of Michael Toxites' *Libri paragraphorum*, although it has been lost in its original form since 1616) he professed disagreement with Galenic medicine and promised to introduce a new syllabus based upon his own firsthand experience as a naturalist and in treating patients at the sickbed.

Paracelsus next offended by burning a copy of Ibn Sīnā's *Canon* at a student rag on St. John's Day (24 June). He further lectured in German, contrary to academic tradition, and admitted barber-surgeons to his courses. When his patron Froben died suddenly in October 1527 his opponents (who had at first been subdued by the enthusiastic response to Paracelsus' teaching by the students, humanists, and reformed churchmen) gained ground. Chief among his enemies were the professors and, especially, the apothecaries, who objected to his control of the pharmacies and his criticism of the profits that they made. The public mood began to turn against Paracelsus, and his protests to the town council—directed against the apothecaries and against the lampoons that the fickle students had begun to publish about him—were ignored.

The final clash came in February 1528, when Paracelsus publicly denounced a magistrate who had found against him in a lawsuit against a church dignitary who had had Paracelsus called in for consultation in an acute abdominal emergency. He promised Paracelsus the enormous fee that he demanded; having been cured by a few of Paracelsus' laudanum pellets, the patient refused to pay, and Paracelsus charged him with default. Having made his denunciation, Paracelsus had to leave Basel, thereby relinquishing all his property. He never obtained academic preferment or contractual employment again.

Paracelsus then paid a short visit to Lorenz Fries at Colmar, stopped for a brief time at Esslingen, and reached Nuremberg in 1529. He remained there until 1530, leaving after a series of altercations, the chief of which developed from his stated disapproval of the treatment of syphilis by guaiacum and poisonous doses of mercury; his criticism of Lutheran orthodoxy; and his assertion of his right to publish (which was followed by censure elicited from the Leipzig Medical Faculty). He proceeded to Beratzhausen, where he worked on his *Buch paragranum*, then spent an uneasy period of about two years at St. Gall. There Paracelsus completed his main medical work, the *Opus paramirum*, which he dedicated to Joachim de Watt (Vadianus), humanist, Zwinglian churchman, geographer, and acting mayor. He left despite influential friends and patients and good chemical laboratory facilities.

Paracelsus then embarked upon a career as a wandering lay preacher, appearing in "beggar's garb" in Appenzell, Innsbruck, and Sterzing (Vipiteno). He failed to secure the medical work he hoped for, but was able to study miners' diseases—silicosis and tuberculosis—at Solbad Hall in Tirol and at Schwaz, to which he returned in 1533. He found better luck in Merano, St. Moritz, and especially Pfäfers-Bad Ragaz, where in August 1535 he was consulted by the abbot John Jacob Russinger. The following summer he was at Augsburg and Ulm, where he supervised the printing of his *Grosse Wundartzney*. He traveled on through Bavaria (where he wrote a work on tartar) to Kromau in Bohemia, where he wrote a work on the occult-metaphysical *philosophia sagax* and continued chemical laboratory work. In Austria he had audiences with Ferdinand of Bohemia, brother of Charles V, and thus temporarily recovered much of his former status.

In 1538 Paracelsus reached Villach, where his father had died four years before. Here he completed his *Carinthian Trilogy*, which included a *Chronica* of the land, his last work *On Tartar*, the *Labyrinth of Doctors Perplexed*, and the *Seven Defenses* against his critics. Although the Carinthian authorities promised him publication, the work was not published until 1564, and then in Cologne. Paracelsus again practiced medicine, although his own health was failing. Finally he was called to Salzburg by Ernest of Wittelsbach, suffragan of that city; he died there and was buried in the almshouse of St. Sebastian. The site of his grave was a place of pilgrimage for the sick for a long time after.

Paracelsus' difficult personality may have been formed from his resentment of his father's illegitimate birth and of his mother's status as a bondswoman. He was an angry man, and his career followed a pattern of initial triumphs followed by losing battles, in the course of which he alienated even his best friends and patrons. His wholesale condemnation of traditional science and medicine found its parallel in his rough behavior and in his unwillingness to make concessions to custom and authority. He sought to learn new cures and remedies from the common people, and spent many hours drinking with them in low taverns; his expertise concerning wines and vintages is apparent in some of his medical writings. Paracelsus was prepared to treat the poor without any reward, but required high fees of the rich and reviled them if they defaulted (or, indeed, if he even thought that they had). Some of his cures were probably not as brilliant as they appeared to be at the time, and some may even have done more harm than good in the long run (although some of these criticisms must be discounted as the canards of more traditional practitioners).

A portrait of Paracelsus painted when he was in his thirties shows him as beardless, but with a shock of hair, and almost pathologically obese. Pictures made a decade later show him still beardless and mostly bald, while in his older years he is depicted as looking haggard and ill. The best eyewitness account of him is that of his amanuensis, Johannes Oporinus, which was published in Daniel Sennert's *De chymicorum cum Aristotelicis et Galenicis consensu et dissensu* (Wittenberg, 1619). Oporinus was one of the few men for whom Paracelsus himself had words of praise, and he in his turn admired Paracelsus and his medicine although, as a young and timid scholar, he was often overawed by his master's unconventional behavior.

Oporinus found fault with Paracelsus' rejection of organized religion and classical scholarship, as well as his addiction to drink, noting that he had been averse to wine until his twenty-fifth year, but later challenged peasants to drinking contests from which he emerged victorious. He gave a vivid report of Paracelsus' astonishing resilience, describing him dictating, late at night after a drinking bout, with perfect coherence and sense, in a manner that could not have been bettered by a sober person. Then Paracelsus threw himself on his bed with his long sword (which he said he had got from a hangman) still girded about him, and, suddenly leaping up, brandished his sword like a madman, frightening his famulus to distraction. Oporinus further states that Paracelsus was busy all day in his laboratory, but lived luxuriously, never short of money. As part of his treatment for ulcers, in which he ignored the usual restrictive diets, he dined lavishly with his patients, curing them "with a full stomach." He was not interested in women, and never had sexual intercourse. He liked to buy expensive new clothes, and tried to give his old ones away, although no one would accept them because they were extremely dirty. (This last comment is at variance with Paracelsus' announced scorn of academic robes

and with the modest and practical dress shown in his portraits.)

In his own writings Paracelsus dealt in paradoxes, interlarded with undisguised obscenities and endless outbursts against traditional doctrines and their professors. His works might have at times appeared to be the ravings of a megalomaniac, enjoining the whole learned world to follow him in new paths, away from deceitfully wrong and "excrementitious" humoral lore. Nonetheless, he created a new style and a refreshing and witty language, perfectly suited to the ideas that he wished to convey. These ideas—those of a naturalist physician, spiritualist and symbolist thinker, and passionately religious and charitable fighter against perceived evil—are reflected in the contradictory interpretations that posterity has placed upon Paracelsus' work. He was, for example, extolled in the early years of the nineteenth century, the era of Romanticism and *Naturphilosophie*, and reviled before and after, at the beginning of the age of scientific medicine.

Only a few of Paracelsus' works were published during his lifetime. Among these were some astrological-mantic forecasts, including the *Practica gemacht auf Europen in dem nechstkunfftigen Dreyssigtsen Jahr* (Nuremberg, 1529); *Usslegung des Commeten erschynen zu mitlem Augsten anno 1531*; *Practica teutsch auf das MDXXXV Jar*; and *Prognostication auff XXIIII jar*. More important were his critical appraisal of the treatment of syphilis by guaiacum (*Vom Holtz Guaiaco gründlicher heylung* [Nuremberg, 1529]); a related treatise on the "impostures" committed therein (*Von der Französischen kranckheit Drey Bücher* [Nuremberg, 1530]); a booklet on Bad Pfäfers (*Von dem Bad Pfeffers Tugenden, Krefften und würckung, ursprung und herkommen, Regiment und ordinantz* [1535]); and the *Grosse Wundartzney* (Augsburg, 1536). Most of his writings came to light in the decades following his death, and their publication reached a peak in about 1570 with the *Archidoxis*, a handbook of Paracelsian chemistry that went through many editions after the first one, issued at Cracow in 1569. The *Archidoxis* was edited or translated by Adam Schröter, Adam of Bodenstein, Michael Toxites, Gerard Dorn, Balthassar Flöter, and G. Forberger, among others. The work of all of them predated that of John Huser, who edited the first definitive collected editions—ten quarto volumes in 1589–1591, folio editions of 1603 and 1605, and the surgical folio of 1605.

Among Paracelsus' practical achievements was his management of wounds and chronic ulcers. These conditions were overtreated at the time, and Paracelsus' success lay in his conservative, noninterventionist approach, which was based upon his belief in natural healing power and *mumia*, an active principle in tissues. He thus continued the tradition of Theodoric Borgognoni of Lucca and his pupil Henry of Mondeville, both of whom had advocated, in the thirteenth and fourteenth centuries, aseptic and pus-preventing treatment, the method of *rara vulnerum medicatio* (as Cesare Magati named it in 1616). Chemical therapy had been used chiefly externally by the ancients, but Paracelsus recognized the superiority of chemicals taken internally over the traditional, mostly herbal, internal medicines. He imposed strict controls upon their use, however, holding that chemicals must be given only in moderate doses (in contrast to the toxic doses of mercury then used in treating syphilis) and only in detoxified form, achieved by washing the chemical substance with water and alcohol, "to cleanse the sharpness," or by oxidation and the induction of solubility (as, for example, in heating white crystalline arsenic with saltpeter or in converting harmful iron sulfides into therapeutic sulfates).

Paracelsus also knew of the diuretic action of mercury in the treatment of dropsy and of the narcotic and sedative properties of ether-like preparations that he obtained from the interaction of alcohol and sulfur. He demonstrated the latter in an early pharmacological experiment on chickens, which he probably selected because of the well-known narcotic effect of *Hyoscyamus* (henbane) upon them (an effect also called *mort aux poules*).

Paracelsus' description of miners' diseases was the first to identify silicosis and tuberculosis as occupational hazards. He was also the first to recognize the congenital form of syphilis, and to distinguish it from postpartum infection. He studied visceral—notably osseous and nervous—syphilis in its protean manifestations, and differentiated it from hydrargyrosis, the morbid syndrome caused by toxic doses of mercury. (He did, however, regard syphilis as a specific modification of other diseases, rather than as a separate entity.) Paracelsus also gave the first purely medical account of dancing mania and chorea, proposing a natural explanation in place of previous supernatural theories (including possession by demons). He described the symptoms of hysteria, hysterical blindness among them.

Paracelsus further drew the connection between cretinism and goiter, which he identifed as being endemic and related to the mineral content of drinking water. He recognized the significance of acid in mineral waters as a powerful aid to gastric digestion and wrote of the "hungry acid" *(acetosum esurinum)* in the stomach of some animals as permitting them to digest metals and stone. He was not, however, aware of the essential role of acid in the gastric digestion of all animals—this was recognized and studied thoroughly only by J. B. van Helmont, between 1624 and 1644, and to a lesser degree by slightly earlier workers, including Quercetanus (Duchesne), Petrus Castelli, Fabius Violet (perhaps Quercetanus writing under another name), and Johannes Walaeus. Paracelsus also observed the precipitation of albumin in urine after acid (rennet or vinegar) is added.

Chief among Paracelsus' contributions to medical theory was his new concept of disease. He demolished the ancients' notion of disease as an upset of humoral balance—either an excess *(hyperballonta)* or an insufficiency *(elleiponta)*—or as a displacement or putrefaction of humors. Each of these conditions depended on the constitution (or *physis*, or "temperament") of the individual, as determined by the humoral variations appropriate to him. There were therefore as many diseases as there were individuals, and no disease was considered to be a classifiably separate entity, having a specific agent and specific anatomical effects.

Paracelsus completely reversed this concept, emphasizing the external cause of a disease, its selection of a particular locus, and its consequent seat. He sought and found the causes of diseases chiefly in the mineral world (notably in salts) and in the atmosphere, carrier of star-born "poisons." He considered each of these agents to be a real *ens*, a substance in its own right (as opposed to humors, or temperaments, which he regarded as fictitious). He thus interpreted disease itself as an *ens*, determined by a specific agent foreign to the body, which takes possession of one of its parts, imposing its own rules on form and function and thereby threatening life. This is the parasitistic or ontological concept of disease—and essentially the modern one. It was substantially elaborated by Helmont. The significance of specific disease-semina, its connection with imagination, ideas, and passions, and the bodily manifestation of spiritual impulses, as inculcated by Helmont, is clearly anticipated in Paracelsus' concept of disease.

Paracelsus' new idea of disease led him to new modes of therapy. He directed his treatment specifically against the agent of the disease, rather than resorting to the general anti-humoral measures (such as sweating, purging, bloodletting, and inducing vomiting) that had been paramount in ancient therapy for "removal of excess and addition of the deficient." For Galenic remedies derived from plants, he substituted specifics, often applied on homeopathic principles. Here his notion of "signatures" came into play, in the selection of herbs that in color and shape resembled the affected organ (as, for example, a yellow plant for the liver or an orchid for the testicle). Paracelsus' search for such specific medicines led him to attempt to isolate the efficient kernel (the *quinta essentia*) of each substance. His method was thus one of separating drugs into their component parts, rather than compounding them as the ancients had done.

In nontherapeutic chemistry Paracelsus described new products arising from the combination of metals and devised a method of concentrating alcohol by freezing out its watery component. He also developed a new way to prepare aquafortis, and demonstrated its transformation into an oil at the bottom of its container when laminated metals were dissolved in it. In the *Archidoxis* he grouped chemicals according to their susceptibility to similar chemical processes, although it has been stated (and disputed) that many of the chemicals that Paracelsus believed to be discrete entities were in fact identical distillates containing nitric or hydrochloric acid. It is difficult to reproduce some of the processes that Paracelsus described, partly because he made deliberate omissions in the interest of secrecy (as, for example, in his instructions for producing the *arbor Dianae*). It is no easier to decide what in his work is truly original or to what degree it had been anticipated by the Lullists and by Johannes de Rupiscissa, the latter in the middle of the fourteenth century, particularly in regard to the preparation of potable metals and *quintae essentiae*. Certainly Paracelsus was the first to devise such advanced laboratory techniques as the use of detoxification and freezing to concentrate alcohol and invented new preparations (including those of the ether group and probably tartar emetic); he was, moreover, the first to attempt to construct a chemical system.

Much has been written about Paracelsus as an alchemist, but he was not really interested in the classical alchemists' problems of transmutation, the philosopher's stone, or making gold. Rather, "alchemy" meant to him the invention of new and nontoxic metals for medicinal uses. He was, however, perceptibly influenced by the medieval alchemists and by a number of contemporary herbalist-distillers, including Hieronymus Brunschwig and Ulstadt.

While the observations and achievements in medicine and chemistry cited may be regarded as stepping-stones to modern science, it must be realized that they are selected from the larger body of Paracelsus' writings, which, in their totality, evoke a world of *magia naturalis* far removed from the modern spirit of independent inquiry. Paracelsus, on the whole, was as much (and on occasion more) of a seer, a cosmosophic, and religious metaphysician, than he was a naturalist and scientist. Having turned against book learning, formal logic, and complacent human reasoning, he espoused the study of nature, which must be "read" by traveling from land to land—"one land, one page." He thus sought to find the invisible, spiritual forces that make visible bodies act.

Paracelsus believed that these invisible forces achieved their purpose through what he called "knowledge," which is not of the observer but rather of the object observed—the vital principle that, for instance, insures that the seed of a pear tree will grow into a pear tree, and not any other kind. Knowledge therefore lies in the object itself and in its specific function; man can acquire this knowledge only through union with the object—a meeting of the spirit of the inquirer with the spirit of the object. It is through this traffic between astral bodies (which Paracelsus defined as ethereal spirits of finest corporality, the star-born carriers of vital principles—"souls"—on their descent from heaven) that knowledge is obtained in the "light of nature," which represents the total of the specific forces embedded in the visible objects of nature and their grasping by the mind through union with them. These forces, the divine arcana, are specific for every object and are visualized as volatile. Knowledge through union with the object is possible to man because all the substances and objects of the ambient world are somewhere and somehow represented in him. This follows from the close parallelism between the macrocosm and the microcosm, and from the attraction of like to like (universal sympathy).

Paracelsus believed that the microcosmic state of man permitted the study of the universe, so that science and knowledge are possible. He called the study of nature and medicine "astronomy," and urged that every physician be an astronomer—that is, that he study the *astra*. Paracelsus' term *astra* denoted not so much the stars themselves and their influences on sublunary objects (so important in traditional astrology) as the essential virtues and functions of individual objects and their correspondences within all realms of nature, including the stars.

Indeed, each object (or part thereof) has its vital principles—*astra*—as well as its celestial star, with which it shares specific properties. In the realm of plants, for example, a certain herb represents a certain star, as well as a corresponding mineral, organ, part of the body, disease, and remedy. A certain fungus is thus a "fruit" of the earth; its equivalent is vitriol, a "fruit" of water. Arsenic, a mineral product of water, emerges as a terrestrial "fruit" in the form of another kind of fungus. *Vitriolum terrae album* is equivalent to *Pfifferling*, a kind of chanterelle; *vitriolum aquae* is the copper of water.

In explaining the curative power of mercury in the treatment of dropsy, Paracelsus was able to reach a rational and protoscientific statement: the cure results from the expulsion of a "wrong" salt, which tends to be dissolved, and its replacement by one that remains in a fixed, coagulated state. Less scientifically, however, this "life of salts" depends upon celestial impression—a "shot" fired by the star that represents the salt in its own world "divides" the bodily salt in the same way that the sun melts snow. The salt is then restored by the celestial virtue of the mercury. Paracelsus thus retained much of traditional astrology, although giving it an original, astrosophical interpretation and denying the exclusive power of the stars over human life. Moreover he corrected or dismissed a great deal of traditional lore, and argued generally against complacent human reasoning, which is "from the stars" and comes to us with the astral body.

Paracelsus' notion of knowledge led him to advocate humanitarian treatment of imbeciles; he posited that since man's reason and pseudo-knowledge are "astral," the simpleton, unaffected by the astral snares of human reasoning, retains a closeness to God (which incidentally allows him to make accurate predictions). Paracelsus also—although not unequivocally—recommended humane treatment of the insane, stating that their illness is perfectly natural and not simply the work of demons. There are nevertheless certain retrogressive features in his system of therapy—among them many popular superstitions, including *Dreck-Apotheke*—that are the result of his adherence to *magia naturalis* and the idea of cosmic sympathy.

Paracelsus' general natural philosophy is spiritualist. The important forces in nature are the invisible "spirits," such as the *quintae essentiae*; these are the life substances of objects, and the *magus* may know how to extract them, particularly from herbs and chemicals. These "spirits" may not be defined in terms of the merely passive ("female") elements, qualities, humors, and elemental mixtures of the ancients; they are rather the specific, active ("male") *arcana* and primordial seeds (*semina*) that emanate immediately from God and direct and inform nature. Each contains its own *archeus* (*vulcanus*), which determines individual form and function. The *archeus* is also called the "internal alchemist," the digestive principle that, acting in the "kitchen" ("stomach") of each organ (most notably the stomach proper), separates nourishment from poison, the pure from the impure, and the assimilable from waste. Its failure leads to the deposit of sediment which constitutes pathological change, the Paracelsian tartar. The *semina* are preformed in an invisible, nonmaterial

foreworld, the *iliastrum*, or prime matter. This has nothing to do with matter in any modern sense of the word; rather, it denotes in themselves uncreated (divine) impulses (*logoi*), the word *Fiat* that directs an original watery "matter" to form individual objects. This matter is an archetype of water, an invisible "creative" water, that is linked to ordinary water (that is, the material, elementary world) through bearing the *semina* of all things; it is thus the mother element, the mother of all things. Earth, air, and fire also occur in the Paracelsian world, not in the ancient sense of elements, but rather in the Platonic and cabalistic (*sepher jezirah*) sense of mothers. They are the wombs that give birth to groups of objects, each specific to its source; thus, minerals and metals are the "fruit" of water, and plants and animals—including man—the "fruit" of earth. Each of these "fruits" has its corresponding "fruit" in each of the mothers.

The three Paracelsian principles—salt, sulfur, and mercury—likewise do not replace the elements of the ancients, nor are they matter of any kind. They are rather principles within matter that condition the state in which matter can occur. There is thus in every object a principle (salt) responsible for its solid state; a principle (sulfur) responsible for its inflammable, or "fatty," state; and a third (mercury) responsible for its smoky (vaporous) or fluid state. Of these, salt is of particular importance in medicine. It represents a state of fixation, or coagulation and sedimentation, and appears in the pathological form of tartar. According to Paracelsus, tartar is any pathological change that can be interpreted as a deposit, and the remains of unassimilable nourishment for an organ; it is coagulated under the influence of acid and affects the function and vitality of all parts of the body. It is the opposite pole of spirit, that is, of what is active and alive; it reflects the pain of hell and its curse in the life of man. It is for this reason named "tartar" (*cagastrum*), which also denotes the useless and troublesome deposits in wine vats, comparable with the stones and calcifications in the channels of the kidneys, bladder, bronchi, and other organs. (In the lungs such deposits cause pulmonary tuberculosis through bronchial obstruction.) A universal solvent, the *liquor alcahest*, was therefore a medical necessity, and one was eagerly sought.

Of the Paracelsian principles, mercury denotes the highest spiritual state and sulfur an intermediate one. Sulfur is also called the soul, and forms the link between the spirit (*geist*, mercury) and the solid body (salt). It is, of course, the spirit that animates an object—that is, that accounts for its specific form and function—and each object is an "essential thing" by virtue of its specific spirit. This spirit gives life, a "spiritual, invisible, incomprehensible thing"; it makes dead things "male" (*männisch*)—that is, alive and responsive.

Paracelsus was conversant with traditional academic medicine and naturalism, both ancient and medieval. Although he was in principle intent on destroying tradition, he retained much of it in his system, often merely changing its emphasis or giving it an original interpretation. For example, during the Middle Ages salt as a third principle was known as *faex*, or earth; its admixture to sulfur and mercury had been thought to influence proportionally the nature of a metal (by Michael Scott) or any other substance (as in the *Seven Hermetic Tracts*, Archelaus [Arisleus], and the Lullists). Paracelsus assigned salt a much wider, cosmological significance. For him it stood for not only a bodily admixture to a mineral or metal, determining its baseness or nobility, but also for anything that makes an object appear in solid form— that is, for whatever causes a deposit in a nonsolid medium or promotes any change in normal appearance, as in an organ. In the latter sense, salt can denote any morbid anatomical change, although it can at the same time mean any particular salt, such as sodium chloride, that has caused this change. (Paracelsus referred to an ulcer as a salt mine.)

Obviously, then, Paracelsus was not the first to introduce the concept of the third principle, but he did give it an original meaning. Similarly, he adapted the *quintae essentiae* and their use in medicine, merely studying them further and applying them in greater depth and compass. He did the same thing with the notion of the astral body, a Neoplatonic concept that he probably owed to Marsilio Ficino, whose influence is apparent in the work of Paracelsus as a whole, as well as in detail in his medical system (most notably in his ideas on the cause of the plague, the role of the stars, and *magia naturalis*). Natural magic meant, to Paracelsus, capturing and demonstrating heavenly "gifts," rather than cultivating demons, in which parallels between his work and that of Agrippa von Nettesheim may be seen.

Among the medieval antecedents of Paracelsus' work, Konrad of Megenberg's *Buch der Natur*, of about 1350, would seem to be of importance. Paracelsus used the phrase "Buch der Natur," recommending that it be studied instead of the printed books of scholars and professors. Doctrines that are common to Paracelsus and Konrad of Megenberg may also be traced to earlier sources. Some of these ideas bear the clear stamp of gnosticism, as transmitted in the esoteric and largely suppressed medieval literature of the "prohibited arts" of alchemy and astrology. These ideas include the notion of the lower, "incompetent," astral "administrators" responsible for the creation of our evil world, which emerges in Paracelsian (and notably pseudo-Paracelsian) treatises as the concept of the *vulcani* or *archei*. Likewise gnostic are the ideas of prime and uncreated matter (and its offspring, the original "water"); of the elemental and material world as the workshop of the devil; of missiles from the stars as the agents of insanity, aggression, delinquency, and physical malformations; of the splitting off of a female principle in the person of God; and of the ultimate return of the immortal soul to God and of the fine pneumatic shell of the soul (the astral body) to the stars.

Paracelsus made a distinct contribution to medicine and chemistry in his nomenclature for substances that were already known—for example, he substituted the word "alcohol" (which had previously meant any subtle substance in dispersion) for "spirit of wine" and "synovia" for the "gluten" contained within the joints. His more general reforms were, however, less scientific in character (although protoscientific elements are prominent in them). Chief among his posthumous accomplishments was the introduction of Paracelsian chemicals into the *London Pharmacopoeia* of 1618—some of these, calomel among them, owed their inclusion to Croll, Turquet de Mayerne, and other Paracelsists. But however his specific contributions may be qualified, there is no doubt of Paracelsus' essential influence, both direct and indirect, on medical reform, as there can be no doubt of his truly naturalist empiricism and skepticism toward the prevalent Galenic tradition.

Medicine was the chief focus of Paracelsus' labors, the center of his anthropocentric world. Nonetheless, it was by no means his only concern—he was further committed to the reformation of religion and society. At heart neither Protestant nor Catholic, he opposed any closed church (*Mauerkirche*) and fought against what he considered to be the fraudulent rationalism of dogma and formal juridical logic, the man-made snares "sold" by jurists as divine laws and institutions. He belonged to the group of spiritualist and individualist reformers that included Sebastian Franck, whom he had probably met at Nuremberg and Augsburg. Like Franck, he wrote in paradoxical language to advocate a return to the ideals of early Christianity, including poverty, the redistribution of wealth, and regeneration ("glorification of the flesh") through the sacraments.

In sum, Paracelsus was a great doctor and an able chemist. That he achieved little in his lifetime (apart from his success in his practice and in the laboratory) may be attributed in part to his uncompromisingly destructive attitude toward tradition. His views encompassed both astrological superstitions and quite conspicuously modern descriptions of diseases, together with shrewd appraisals of their nature and causes. He remained ignorant of a number of important surgical methods that were practiced widely by his contemporaries and, although repudiating astrological beliefs in many instances, he nonetheless incorporated them into his own work and added multifarious mantic lore of his own. It must be noted, however, that his credulousness was part of his unprejudiced and empirical attitude; he was intent on testing all reported observations, no matter how unlikely they might be. All these factors are reflected in the varied quality of his writings—a mixture of pansophic and religious parables with naturalism and medicine, the mixture of the "medieval" and the modern. Nevertheless, Paracelsus is basically consistent in theory and practice. A number of apparent contradictions in his work disappear when they are considered in their proper context, while others are clearly the result of developmental changes in his life and general outlook.

What is in the end most remarkable in Paracelsus' work is that he achieved real advances in chemistry and medicine through the revival and original development of lore that had been kept alive only at a very low level (or had, indeed, been suppressed as heresy). This lore—alchemy, astrology, and the "prohibited arts"—can be traced to Hellenistic and Oriental Neoplatonism, gnosticism, and syncretism; in Paracelsus' hands it became, if not scientific, at least protoscientific. It is difficult to overrate the effect of Paracelsus' achievement on the development of medicine and chemistry. Some thirty years after his death a powerful Paracelsian movement began to agitate naturalists and physicians all over Europe. It was set in motion by the need to find new and immediately effective medicines, and even orthodox traditionalists joined in in some form (usually in attempting to devise a conciliatory and eclectic synopsis of Paracelsian and Galenic practice). Despite opposition and vilification, the influence of Paracelsus and the Paracelsians is apparent in the work of Van Helmont, Boyle, Willis, Sylvius, Stahl, Boerhaave, and others, well into the eighteenth century.

BIBLIOGRAPHY

I. ORIGINAL WORKS. The few works appearing during Paracelsus' lifetime and the first definitive collected ed. by Huser are mentioned in the text. The complete modern and critical ed. of the medical, chemical, metaphysical, and mantic works is Karl Sudhoff, *Theophrast von Hohenheim, genannt Paracelsus, sämtliche Werke*, 14 vols. (Munich, 1922–1933). An index to the Sudhoff ed. is found in Martin Müller, *Registerband zu Sudhoffs Paracelsus Gesamtausgabe* (Einsiedeln, 1960), which is *Nova acta paracelsica*, supp. (1960). Of the planned *Die theologischen und religionswissenschaftlichen Schriften*, only 1 vol. appeared, Wilhelm Matthiessen, ed. (Munich, 1923). It was resumed in a new critical ed. by Kurt Goldammer, of which 6 vols. (of 14 planned) have appeared (Wiesbaden, 1955–1973). An annotated version of the Huser quarto ed. in modern German appeared under the name of B. Aschner, 4 vols. (Jena, 1926–1932); a "study ed." of the most important works in slightly modernized form by Will-E. Peuckert, 5 vols. (Basel–Stuttgart, 1965–1968); and an annotated digest by J. Strebel, 8 vols. (St. Gall, 1944–1949).

Modern individual eds. and translations of single or several treatises include A. E. Waite, *The Hermetic and Alchemical Writings of Paracelsus the Great*, 2 vols. (London, 1894; repr. 1966); Franz Strunz, *Das Buch Paragranum* (Jena, 1903), and *Volumen und Opus Paramirum* (Jena, 1904); H. E. Sigerist *et al.*, *Four Treatises of Theophrastus of Hohenheim* (Baltimore, 1941); K. Leidecker, *Volumen Medicinae Paramirum*, which is *Bulletin of the History of Medicine*, supp. 3 (Baltimore, 1949); Kurt Goldammer, *Paracelsus. Sozial-ethische und sozialpolitische Schriften* (Tübingen, 1952); K. Goldammer *et al.*, *Die Kärntner Schriften* (Klagenfurt, 1955); Robert Blaser, *Theophrastus von Hohenheim, Liber de nymphis, sylphis, pygmaeis et salamandris et caeteris spiritibus* (Bern, 1960), Altdeutsche Übungstexte, XVI; K. Goldammer and K.-H. Weimann, *Paracelsus vom Licht der Natur und des Geistes* (Leipzig, 1960), *Labyrinthus medicorum* and selected theological treatises with bibliography and critical notes; Paul F. Cranefield and W. Federn, "Paracelsus on Goitre and Cretinism," in *Bulletin of the History of Medicine*, **37** (1963), 463–471; K. Goldammer, *Das Buch der Erkanntnus des Theophrast von Hohenheim. Aus der Handschrift mit einer Einleitung* (Berlin, 1964), Texte des Späten Mittelalters, XVIII; Paul F. Cranefield and W. Federn, "The Begetting of Fools. An Annotated Translation of Paracelsus, De generatione Stultorum," in *Bulletin of the History of Medicine*, **41** (1967), 56–74, 161–174.

II. SECONDARY LITERATURE. Biographical material may be found in W. Artelt *et al.*, *Theophrastus Paracelsus*, F. Jaeger, ed. (Salzburg, 1941); K. Bittel, *Paracelsus-Museum, Stuttgart. Paracelsus-Dokumentation. Referatenblätter* (Stuttgart, 1943); R. H. Blaser, "Neue Erkenntnisse zur Basler Zeit des Paracelsus," in *Nova acta Paracelsica*, supp. VI (Einsiedeln, 1953); R. J. Hartmann, *Theophrast von Hohenheim* (Stuttgart–Berlin, 1905); E. Schubert and K. Sudhoff, *Paracelsus-Forschungen*, 2 vols. (Frankfurt, 1887–1889); K. Sudhoff, *Paracelsus. Ein deutsches Lebensbild aus den Tagen der Renaissance* (Leipzig, 1936); and E. Wickersheimer, "Paracelse à Strasbourg," in *Centaurus*, **1** (1951), 356–365, which includes documentation concerning Paracelsus' doctorate (?) at Ferrara.

Literary criticism and a bibliography of the corpus of Paracelsian writings, together with a discussion of questions of authenticity, can be found in K. Sudhoff, *Versuch einer Kritik der Echtheit der Paracelsischen Schriften*, 2 vols. (Berlin, 1894–1899). Vol. I, *Bibliographia Paracelsica* (repr. Graz, 1958), contains a complete annotated catalog and collation of all works issued under Paracelsus' name from 1527 to 1893. Vol. II, *Paracelsus Handschriften*, contains an analysis and collation of letters, documents, and MSS of treatises on medical, chemical, alchemical, theological, and magical subjects (there are no autographs other than of letters, recipes, and documents).

A bibliography of secondary Paracelsian literature is given in K. Sudhoff, *Nachweise zur Paracelsus Literatur* (Munich, 1932), continued in K.-H. Weimann, *Paracelsus-Bibliographie 1932–1960. Mit einem Verzeichnis neuentdecker Paracelsus Handschriften* (Wiesbaden, 1963), which is vol. II of K. Goldammer, ed., *Kosmosophie*.

Paracelsus' doctrines and sources are discussed in K. Goldammer, "Der Beitrag des Paracelsus zur neuen wissenschaftlichen Methodologie und zur Erkenntnislehre," in *Medizin-historisches Journal*, **1** (1966), 75–95; "Die Paracelsische Kosmologie und Materietheorie in ihrer wissenschaftsgeschichtlichen Stellung und Eigenart," *ibid.*, **6** (1971), 5–35; and "Bemerkungen zur Struktur des Kosmos und der Materie bei Paracelsus," in *Medizingeschichte in unserer Zeit. Festgabe für Edith Heischkel und Walter Artelt* (Stuttgart, 1971), 121–144; C. G. Jung, *Paracelsica* (Zurich-Leipzig, 1942); Walter Pagel, "Religious Motives in the Medical Biology of the XVIIth Century," in *Bulletin of the Institute of History of Medicine, Johns Hopkins University*, **3** (1935), 97–312; *Paracelsus—Introduction to Philosophical Medicine in the Era of the Renaissance* (Basel–New York, 1958), trans. into French by M. Deutsch (Paris, 1963); "Paracelsus and Techellus the Jew," in *Bulletin of the History of Medicine*, **34** (1960), 274–277; "Paracelsus and the Neoplatonic and Gnostic Tradition," in *Ambix*, **8** (1960), 125–166; "The Prime Matter of Paracelsus," *ibid.*, **9** (1961), 117–135; *Das medizinische Weltbild des Paracelsus. Seine Zusammenhänge mit Neuplatonismus und Gnosis* (Wiesbaden, 1962), which is vol. I of K. Goldammer, ed., *Kosmosophie*; "The Wild Spirit (Gas) of J. B. Van Helmont (1579–1644) and Paracelsus," in *Ambix*, **10** (1962), 1–13; "Paracelsus' aether-ähnliche Substanzen und ihre pharmakologische Auswertung an Hühnern. Sprachgebrauch (*henbane*) und Konrad von Megenbergs *Buch der Natur* als mögliche Quellen," in *Gesnerus*, **21** (1964), 113–125: "Paracelsus: Traditionalism and Mediaeval Sources," in *Medicine, Science and Culture. Historical Essays in Honor of Owsei Temkin* (Baltimore, 1968), 51–75; and "Van Helmont's Concept of Disease—to be or not to be? The Influence of Paracelsus," in *Bulletin of the History of Medicine*, **46** (1972), 419–454.

See also Walter Pagel and P. Rattansi, "Vesalius and Paracelsus," in *Medical History*, **8** (1964), 309–328; Walter Pagel and Marianne Winder, "Gnostisches bei Paracelsus und Konrad von Megenberg," in *Fachliteratur des Mittelalters. Festschrift für Gerhard Eis* (Stuttgart,

1968), 359–371; "The Eightness of Adam and Related 'Gnostic' Ideas in the Paracelsian Corpus," in *Ambix*, **16** (1969), 119–139; and "The Higher Elements and Prime Matter in Renaissance Naturalism and the Paracelsian Corpus," in *Ambix*, **21** (in press); and O. Temkin, "The Elusiveness of Paracelsus," in *Bulletin of the History of Medicine*, **26** (1952), 201–217.

The moral philosophy, sociology, and theology of Paracelsus are discussed in K. Goldammer, "Paracelsische Eschatologie," in *Nova acta Paracelsica*, **5** (1948), 45–85, and **6** (1954), 3–37; *Paracelsus. Natur und Offenbarung* (Hannover, 1953); *Paracelsus-Studien* (Klagenfurt, 1954); "Das theologische Werk des Paracelsus," in *Nova acta Paracelsica*, **7** (1954), 78–102; and "Friedensidee und Toleranzgedanke bei Paracelsus und den Spiritualisten," in *Archiv für Reformationsgeschichte*, **47** (1956), 20–46, 180–212. See also Bodo Sartorius von Waltershausen, *Paracelsus am Eingang der Deutschen Bildungsgeschichte* (Leipzig, 1936).

Critical assessments of Paracelsus' achievements in medicine and chemistry are in E. Darmstaedter, *Arznei und Alchemie. Paracelsus-Studien* (Leipzig, 1931), which is vol. XX in the series Studien zur Geschichte der Medizin; and "Paracelsus 'De natura rerum,' " in *Janus*, **37** (1933), 1–18, 48–62, 109–115, 323–324; F. Dobler, "Chemische Arzneibereitung bei Paracelsus am Beispiel seiner Antimonpräparate," in *Pharmaceutica acta helvetiae*, **37** (1957), 181–193, 226–252; W. Ganzenmüller, "Paracelsus und die Alchemie des Mittelalters," in *Angewandte Chemie*, **54** (1941), 417–431, repr. in his *Beiträge zur Geschichte der Technologie und Alchemie* (Weinheim, 1956), 300–314; R. Hooykaas, "Die Elementenlehre des Paracelsus," in *Janus*, **39** (1935), 175–187; "Die Elementenlehre der Iatrochemiker," *ibid.*, **41** (1937), 1–28; and "Chemical Trichotomy Before Paracelsus?," in *Archives internationales d'histoire des sciences*, **28** (1949), 1063–1074; R. Multhauf, "Medical Chemistry and the Paracelsians," in *Bulletin of the History of Medicine*, **28** (1954), 101–126; "J. B. Van Helmont's Reformation of the Galenic Doctrine of Digestion," *ibid.*, **29** (1955), 154–163; and "The Significance of Distillation in Renaissance Medical Chemistry," *ibid.*, **30** (1956), 329–346; W. Pagel, "J. B. Van Helmont's Reformation of the Galenic Doctrine of Digestion—and Paracelsus," *ibid.*, **29** (1955), 563–568; and "Van Helmont's Ideas on Gastric Digestion and the Gastric Acid," *ibid.*, **30** (1956), 524–536; J. R. Partington, *A History of Chemistry*, II (London, 1961), 115–151; J. K. Proksch, *Paracelsus über die venerischen Krankheiten und die Hydrargyrose* (Vienna, 1882); *Paracelsus als medizinischer Schriftsteller* (Vienna–Leipzig, 1911); and *Zur Paracelsus-Forschung* (Vienna–Leipzig, 1912); W. Schneider, "Der Wandel des Arzneischatzes im 17. Jahrhundert und Paracelsus," in *Sudhoffs Archiv für Geschichte der Medizin und der Naturwissenschaften*, **45** (1961), 201–215; "Grundlagen für Paracelsus' Arzneitherapie," *ibid.*, **49** (1965), 28–36; and "Paracelsus und die Entwickelung der pharmazeutischen Chemie," in *Archiv der Pharmazie*, **299** (1967), 737–746; and *Geschichte der pharmazeutische Chemie* (Weinheim, 1973); T. P. Sherlock, "The Chemical

Work of Paracelsus," in *Ambix*, **3** (1948), 33–63; G. Urdang, "How Chemicals Entered the Official Pharmacopoeias," in *Archives internationales d'histoire des sciences*, **7** (1954), 303–314; and P. Walden, "Paracelsus als Chemiker," in *Angewandte Chemie*, **54** (1941), 421–427.

Paracelsus' influence is discussed in Allen G. Debus, "The Paracelsian Compromise in Elizabethan England," in *Ambix*, **8** (1960), 71–97; "Solution Analyses Prior to Robert Boyle," in *Chymia*, **8** (1962), 41–61; "The Paracelsian Aerial Niter," in *Isis*, **55** (1964), 43–61; "An Elizabethan History of Medical Chemistry," in *Annals of Science*, **18** (1962), 1–29; and *The English Paracelsians* (London, 1965); P. M. Rattansi, "Paracelsus and the Puritan Revolution," in *Ambix*, **11** (1963), 24–32; Dietlinde Goltz, *Studien zur Geschichte der Mineralnamen in Pharmazie, Chemie und Medizin von den Anfängen bis auf Paracelsus* (Wiesbaden, 1972), which was supp. in *Sudhoffs Archiv für Geschichte der Medizin und der Naturwissenschaften*, **14**; and Audrey B. Davis, *Circulation Physiology and Medical Chemistry in England 1650–1680* (Lawrence, Kansas, 1973).

WALTER PAGEL

PARAMEŚVARA (*b.* Ālattūr, Kerala, India, *ca.* 1380; *d. ca.* 1460), *astronomy.*

Parameśvara was born into a learned Nampūtiri Brāhmaṇa family of Kerala, which belonged to the Bhṛgugotra and followed the Āśvalāyanasūtra of the *Ṛgveda*. His father remains obscure, but his grandfather studied under the astrologer Govindabhaṭṭa of Ālattūr (1236–1314). The family resided in an *illam* ("house") called Vaṭaśśeri (Vaṭaśreṇi) in the village of Ālattūr (Aśvatthagrāma) on the north bank of the river Nilā at its mouth in Kerala. Parameśvara states that this place lies eighteen *yojanas* west of the meridian of Ujjain, and that the sine of its latitude is 647 (with R = 3,438); its latitude, then, is 10°51′N.

Parameśvara names Rudra as his teacher. Nīlakaṇṭha (*b.* 1444), the pupil of his son Dāmodara, states that Parameśvara studied under Nārāyaṇa and Mādhava; the latter was a well-known astronomer of Saṅgamagrāma in Kerala who lived between *ca.* 1340 and *ca.* 1425. Parameśvara's dates are fixed not only by the epochs of his several astronomical works, but also by his eclipse observations which extended from 1393 to 1432 (see D. Pingree, in *Journal of the American Oriental Society*, **87** [1967], 337–339). His latest recorded observation was made in 1445, although he states in a verse cited by Nīlakaṇṭha that he made observations for fifty-five years—that is, until 1448 if the observations commenced in 1393. Since Nīlakaṇṭha, who was born in 1444, knew him personally, Parameśvara could not have died much before 1460.

Parameśvara's greatest achievements were the revisions of the accepted parameters of planetary motions, the *parahita* that were based on the *Āryabhaṭīya* of Āryabhaṭa I (*b.* 476), and the accepted procedure of eclipse-computations on the basis of his observations. He called this new system the *dṛggaṇita* (see essay in Supplement). He was also active in the composition of commentaries on the standard astronomical texts that were in use in Kerala.

BIBLIOGRAPHY

I. ORIGINAL WORKS. Parameśvara's works include the following. *Pārameśvara* (*ca.* 1408), B. D. Āpaṭe, ed. (Poona, 1946), is a commentary on the *Laghubhāskarīya* of Bhāskara I (*fl.* 629); *Grahaṇamaṇḍana*, K. V. Sarma, ed. (Hoshiarpur, 1965), is a treatise on eclipses, of which an earlier version contained 87 verses, and a later 100; the epoch is 15 July 1411. The *Dṛggaṇita* (1431), K. V. Sarma, ed. (Hoshiarpur, 1963), gives his new parameters, which modify those of the *parahita* system. The work contains new parameters of mean motions of the planets, of their mean longitudes at the beginning of the Kaliyuga, and of their two equations, and a table of their equations at intervals of 6° of argument. It also mentions the *Grahaṇamaṇḍana*. Nīlakaṇṭha in the *Āryabhaṭīyabhāṣya* written after 1501 understood the fifty-five years of Parameśvara's observations to antedate the *Dṛggaṇita*, but this would make him nearly a century old in Nīlakaṇṭha's own youth.

The *Siddhāntadīpikā*, published by T. S. Kuppanna Sastri (Madras, 1957), is a commentary on the *Bhāṣya*, written by Govindasvāmin (*fl. ca.* 800–850) on the *Mahābhāskarīya* of Bhāskara I (*fl.* 629). In this work Parameśvara cites the series of eclipse observations (including one at Nāvākṣetra in 1422 and two at Gokarṇa in 1425 and 1430), which extended from 1393 to 1432. The *Grahaṇanyāyadīpikā*, K. V. Sarma, ed. (Hoshiarpur, 1966), discusses eclipse theory in eighty-five verses and cites both the *Grahaṇamaṇḍana* and the *Siddhāntadīpikā*. The first *Goladīpikā* (1443) contains four chapters that deal respectively with the armillary sphere, the motions of the planets, geography, and gnomon-problems. It was edited with Parameśvara's own commentary, *Vivṛti*, by K. V. Sarma (Madras, 1957). *Grahaṇāṣṭaka*, a short treatise in ten verses, gives the fundamental information required for the computation of eclipses. It was edited by K. V. Sarma, in *Journal of Oriental Research, Madras*, **28** (1958–1959), 47–60.

Other works include *Vākyakaraṇa*, an unpublished treatise on the *vākya* system of astronomy (see essay in Supplement); *Bhaṭadīpikā*, H. Kern, ed. (Leiden, 1874), a commentary on the *Āryabhaṭīya* of Āryabhaṭa I (*b.* 476); *Vivaraṇa*, an unpublished commentary on the *Lilāvatī* of Bhāskara II (*b.* 1115); and *Karmadīpikā*, B. Āpaṭe, ed. (Poona, 1945), is a commentary on the *Mahābhāskarīya* of Bhāskara I (*fl.* 629), in which Parameśvara mentions his *Siddhāntadīpikā*, his *Vākyadīpikā* (= *Vākyakaraṇa*), his *(Grahaṇa)nyāyadīpikā*, his *Goladīpikā*, and his *Bhaṭadīpikā*, and also two lost works: a *Muhūrtāṣṭakadīpikā* on astrology and a *Bhādīpikā*.

Vivaraṇa is a commentary on the *Sūryasiddhānta*, K. S. Shukla, ed. (Lucknow, 1957), in which the amount of precession is reckoned for 1432. This *Vivaraṇa* refers to his *Pārameśvara* on the *Laghubhāskarīya*, his *Siddhāntadīpikā*, his *Līlāvatīvivaraṇa*, and his *Karmadīpikā*. The *Pārameśvara*, B. D. Āpaṭe, ed. (Poona, 1952), is a commentary on the *Laghumānasa* of Muñjāla (*fl.* 932). A second *Goladīpikā*, T. G. Śāstrī, ed. (Trivandrum, 1916), consists of 302 verses and discusses a number of problems that relate to the celestial spheres. In this work Parameśvara refers to his *Siddhāntadīpikā*, to his first *Goladīpikā*, and to his *Karmadīpikā*. A *Jātakapaddhati*, K. S. Menon, ed. (Trivandrum, n.d.), is on horoscopes; and an unpublished commentary, *Vṛtti*, is on the *Vyatīpātāṣṭaka*, which is a work on the *pātas* of the sun and moon. A number of astrological works by Parameśvara exist in MSS in South India: *Ācārasaṅgraha*, a commentary on the *Muhūrtaratna* of Govindabhaṭṭa (1236–1314), the teacher of Parameśvara's grandfather; a commentary on the *Jātakapaddhati* of Śrīpati (*fl.* 1040); and a commentary on the *Ṣaṭpañcāśikā* of Pṛthuyaśas (*fl. ca.* 575).

II. SECONDARY LITERATURE. The best source of information on Parameśvara is in the introductions to K. V. Sarma's works. Unfortunately, there is as yet no study of how Parameśvara's observations affected his astronomy. A brief summary of what was then known about him is given by K. K. Raja, "Astronomy and Mathematics in Kerala," in *Brahmavidyā*, **27** (1963), 118–167, esp. 136–143.

DAVID PINGREE

PARDIES, IGNACE GASTON (*b.* Pau, France, 5 September 1636; *d.* Paris, France, 21 April 1673), *physics.*

A Jesuit, Pardies deserves a place in the history of physics for having intervened in the debate on the ideas of Newton and of Huygens at certain decisive moments. His work, which is not extensive, is characteristic of a transitional period. The establishment of the Jesuits at Pau in 1622 determined the course of Pardies's life. His Christian name resulted from the friendship of his father, a royal counselor at the Parlement of Navarre, for the Jesuits. It was at their *collège* that he began his studies.

After finishing his secondary education in 1652, Pardies decided to become a Jesuit. He entered into the novitiate and with it the remarkably well-organized network of Jesuit studies and schools. From 1654 to 1656 at Toulouse he completed the philosophical phase of the curriculum by the study of logic and physics. From 1656 to 1660 he taught humanities at Bordeaux. After studying theology, he was ordained a priest in 1663 and was admitted to the order in 1665.

During the remaining eight years of his life Pardies taught at La Rochelle (1666–1668), Bordeaux (1668–1670), and the Collège Clermont in Paris (1670–1673). During this period he demonstrated his ability to conduct scientific research while teaching—without neglecting his clerical duties. Indeed, it was while carrying out his ministry at the hospital of Bicêtre during the Easter season of 1673 that he contracted a fatal illness.

"As good a cleric as a scientist," according to the Jesuit chronicler of the *Mémoires de Trévoux* (1726), Pardies nevertheless presented a problem to his order. From the time of his appointment at La Rochelle his superiors distrusted him because he was known "to pursue strange opinions avidly," and until his death he was continually obliged to compromise his true views on philosophy and science, to the point that they cannot be established with certainty.

Pardies's first work, *Horologium thaumanticum duplex*, dates from 1652; it is not known whether it was ever published. He discussed the subject more completely in a treatise published in 1673. His correspondence from 1661 to 1665 with Kircher offers insight into the source of his initial inspiration, the influence of Maignan on him, and reveals the originality of his "two marvelous clocks." This originality consists not only in the optical device—a cone receiving light on its base that transformed a stopped-down pencil of solar rays into a luminous plane perpendicular to this pencil; its originality is also reflected in the use of this plane, which rotated with the sun, with either a sundial or a translucent terrestrial sphere. Thus, while still a student of theology, Pardies was engaged in scientific research at a level indicating that he possessed a good education, keen intelligence, and skill in handling instruments.

These qualities also marked his subsequent work. In *Discours du mouvement local* (1670), *Élémens de géométrie* (1671), and *La statique ou la science des forces mouvantes* (1673) he presented his material tersely and suggestively. Despite certain inadequacies (notably his account of the laws of impact) it is easy to see in these writings a striving for the most economical axioms obtained through rational reflection on empirical data. During this period Pardies also wrote works of a more philosophical character; the titles indicate criticism of Descartes and reveal how greatly their author felt the need to clarify his position. *Discours du mouvement local* was published with additional remarks designed to counter the charge of Cartesianism. Yet although Pardies clearly derived a great deal from the Cartesian heritage and felt obliged to defend himself in this regard, Cartesianism was not his principal source. Through the *Philosophical Transactions of the Royal Society* he was directly informed of the advances of English science (quadrature of the hyperbola and the competition concerning the laws of impact); and in Paris he closely followed the work of the newly established Académie Royale des Sciences and especially that of Huygens, with whom he was in personal contact.

Pardies's last book offers more of his ideas, which were based on knowledge derived from widely different sources. It contains, for example, his critical study of Descartes's letter to Beeckman on the speed of light (1634) and his demonstration of the tautochronism of the cycloidal pendulum.

Hindered by the philosophical climate, Pardies made his most important scientific contribution not in his writings, but in his correspondence. It is there that we find the objections that Pardies expressed to Newton concerning his theory of colors and the *experimentum crucis*—objections that enabled Newton to clarify certain difficult points. Pardies's unpublished manuscripts contain a theory of waves and vibrations that—judging from the fragments presented by Pierre Ango in 1682—might well have played an important role in the development of physics.

Although Pardies did not have the time to devote the full measure of his abilities to science, he was undoubtedly one of those vigorous intellects that science always needs, along with great discoverers, especially in an age of transition. That he was just such an intellect is evident from his pedagogical writings and his contacts with the pioneers of physics. Leibniz' impression of him confirms this view. A member of the great line of Jesuit scientists that persisted throughout the seventeenth century, he was, to a greater degree than his predecessors, embroiled in philosophical disputes. Beneath the Aristotelian language that he sometimes sought to preserve, new meanings emerge. His notions, as bold as they were naïve, purported to demonstrate the spirituality of the soul by virtue of its capacity to understand the infinite through the "clear and distinct ideas" of certain geometric arguments.

BIBLIOGRAPHY

I. Original Works. Pardies's writings include *Horologium thaumanticum duplex* (? Paris, 1662); *Dissertatio de motu et natura cometarum* (Bordeaux, 1665); *Theses mathematicae ex mechanica* (Bordeaux, 1669); *Discours du mouvement local* (Paris, 1670), also translated into English (London, 1670); *Élémens de géométrie* (Paris, 1671), also translated into Dutch (Amsterdam, 1690),

Latin (Jena, 1693), and English (London, 1746); *Discours de la connaissance des bestes* (Paris, 1672); *Lettre d'un philosophe à un cartésien de ses amis* (Paris, 1672); *La créance des miracles* (Paris, 1673); *Deux machines propres à faire les quadrans avec une très grande facilité* (Paris, 1673); *La statique ou la science des forces mouvantes* (Paris, 1673); *Atlas céleste* (Paris, 1674); and *Oeuvres de mathématiques* (Paris, 1691, 1694, 1701, 1721).

Letters written by Pardies or concerning him can be found in *The Correspondence of Isaac Newton*, H. W. Turnbull, ed., I (Cambridge, 1959); *Oeuvres complètes de Huygens*, VI-VIII, *passim*; and *The Correspondence of Henry Oldenburg*, R. Hall and M. Hall, eds., VIII. Portions of his correspondence remain unpublished.

II. Secondary Literature. See Pierre Ango, S.J., *L'optique* (Paris, 1682), which draws on Pardies's MSS; and the unsigned article in *Mémoires pour servir à l'histoire des sciences et des beaux-arts (Mémoires de Trévoux)* (Apr. 1726), 667–668. August Ziggelaar, S.J., *Le physicien Ignace Gaston Pardies S.J. (1636–1673)*, vol. XXVI of Bibliotheca Universitatis Havniensis (Odense, 1971), contains the most complete documentation available and eliminates the need to present a more detailed listing of secondary works here.

Pierre Costabel

PARÉ, AMBROISE (*b.* Laval, Mayenne, France, 1510 [?]; *d.* Paris, France, 22 December 1590), *surgery*.

Paré was the son of an artisan. He served an apprenticeship to a barber-surgeon in the provinces (probably at Angers or Vitré), then went to Paris, where he became house surgical student at the Hôtel-Dieu, a post that provided him a valuable opportunity to study anatomy by dissection. About 1536 Paré became a master barber-surgeon and entered military service under Maréchal Montejan; he accompanied the army on an expedition to Italy, where he spent two years. He returned to Paris in 1539, but intermittently participated in military campaigns throughout most of the next three decades. In 1552 Henry II appointed him one of his *chirurgiens ordinaires*; he became *premier chirurgien* to Charles IX in 1562 and served Henry III in the same capacity. He had a flourishing practice at court and in Paris, and, as a military surgeon, treated the wounded of both sides during the Wars of Religion. (Although often reported to have been a Huguenot, Paré remained a Roman Catholic throughout his life.)

Military practice afforded Paré experience in treating a wide variety of injuries. In particular, he revolutionized the treatment of gunshot wounds, which had been considered to be poisonous and were routinely cauterized with boiling oil. At the siege of Turin in 1536, Paré (according to his own account, published almost half a century later) ran out of hot oil and instead used a "digestive" dressing composed of egg yolk, oil of roses, and turpentine. The following day, he noted that the soldiers who had had their wounds dressed in this improvised manner were recovering better than the soldiers treated by the conventional method; they were free from pain, and their wounds were neither inflamed nor swollen. Paré then experimented with a number of different dressings (including some containing *aqua vitae*, which would, together with turpentine, have acted as a topical antiseptic) and concluded that gunshot wounds were not in themselves poisonous, and did not require cautery. He reported his discovery in his first treatise, *La méthode de traicter les playes faites par les arquebuses et aultres bastons à feu*, published in 1545. This treatise, written in the vernacular because Paré knew no Latin, brought him immediate fame.

Paré also rejected cautery as a method of achieving hemostasis and advocated the ligature of blood vessels to control hemorrhage during amputations. He devised a new instrument for this purpose—the "crow's beak," a sort of hemostat that he used to grasp the vessels to be ligated. His obstetrical surgery was also innovative, and he revived the ancient technique of podalic version for difficult deliveries. (Paré's method was widely used after his own time; one of his chief disciples, Jacques Guillemeau, was primarily an obstetrician, and his influence on French surgeon-obstetricians extended throughout the seventeenth century.)

Paré's motto was "Je le pensai, Dieu le guarist"— "I dressed him, God cured him." Many of the details of his surgery are no longer of scientific interest; despite his innovations, he labored under the humoral theories and superstitions common to sixteenth-century surgery and was ignorant of such considerations as circulation of the blood and asepsis. Nonetheless, he saved many patients who would be the despair of a modern surgeon, and he came to represent the ideal practitioner, both for his technical competence and for his humanitarian concern for his patients. He had a vague, reasoned anticipation of some form of transmissible infection and a crude appreciation of public health measures. Although his knowledge of pathology was at best rudimentary, he advocated and practiced (and left records of) autopsy investigations of fatal illnesses.

Paré's theories and writings were often in opposition to those of the university authorities. He desired to spread anatomical knowledge among his fellow barber-surgeons, and to this end performed dissections and wrote a number of works on anatomy. His use of the vernacular and the advent of the printing press assured his books a wide distribution, although they were published, for the most part, only after legal conflicts with the members of the Paris Faculty of Medicine, who wished to suppress them. Paré was, however, widely supported by his noble clientele, and their support brought him the acceptance of professional associations that were usually closed to all except university graduates. (He was, for example, invited to join the guild of academic surgeons of Paris in the Collège de Saint-Côme, despite his lack of Latin —although the physicians and the Faculty of Medicine always scorned and snubbed him.)

Paré's personal life was marked by honesty, piety, and concern for the poor and defenseless. His last recorded act is his having, at the age of eighty, stopped a religious procession in the streets of Paris so that he might plead with its leader, the archbishop of Lyons, to come to terms with Henry of Navarre (later Henry IV), who was then besieging the city. Paré hoped thus to alleviate the lot of the starving Parisians, and whether or not his unprecedented act exerted any influence, the siege was lifted about a week later. He died shortly thereafter, in the house in which he had lived throughout most of his active career. He was buried in his parish church, St. André-des-Artes (destroyed in 1807, in which year Paré's bones were transferred to the catacombs). Twice married, Paré had nine children, of whom three daughters survived him.

Paré left a powerfully reactivated surgical tradition at his death. His many publications, which were translated into both Latin and modern languages, circulated throughout Europe, and had considerable influence during his life and well into the following century. But in France itself surgeons were again under the Hippocratic yoke within two generations, and the art of surgery had reverted to about the level at which Paré had found it. Part of the blame for this must be attached to the reactionary character of French academic medicine, and to the prevailing moral and religious antipathy toward the scientific principle; the reforms for which Paré struggled did not come until both the Collège de Saint-Côme and the once powerful and privileged Faculty of Medicine were abolished in the French Revolution.

BIBLIOGRAPHY

Paré's writings have been collated and described in Janet Doe, *A Bibliography of the Works of Ambroise Paré; Premier Chirurgien et Conseiller du Roy* (Chicago, 1937). His best books are in J. F. Malgaigne, *Oeuvres complètes d'Ambroise Paré*, 3 vols. (Paris, 1840). Accounts of Paré's life and works are available in W. B. Hamby, *The Case Reports and Autopsy Records of Ambroise Paré* (Springfield, Ill., 1960), *Surgery and Ambroise Paré* (Norman, Okla., 1965), trans. and ed. from J. F. Malgaigne, *op cit.*, and

Ambroise Paré; Surgeon of the Renaissance (St. Louis, Mo., 1967). See also G. Keynes, ed., *The Apologie and Treatise of Ambroise Paré, Containing the Voyages Made Into Divers Places With Many of His Writings Upon Surgery* (Chicago, 1952); F. P. Packard, ed., *The Life and Times of Ambroise Paré* (New York, 1926); and Stephen Paget, *Ambroise Paré and His Times: 1510–1590* (New York–London, 1897).

WALLACE B. HAMBY

PARENAGO, PAVEL PETROVICH (*b.* Ekaterinodar [now Krasnodar], Russia, 20 March 1906; *d.* Moscow, U.S.S.R., 5 January 1960), *astronomy.*

The son of a physician and surgeon, Parenago lived in Moscow from 1912. He graduated from secondary school in 1922 and, seven years later, from the Faculty of Physics and Mathematics of the University of Moscow. The sudden appearance in 1920 of a nova in the constellation Cygnus aroused Parenago's interest in the observation of variable stars. He began his scientific work as a serious observer of variable stars in 1921–1922. In 1925, while still a student, Parenago became a computer at the Astrophysical Institute, which in 1931 became part of the P. K. Sternberg State Astronomical Institute at the University of Moscow. From 1932 he was a senior scientific worker.

Parenago began teaching in 1930 as an assistant at the Steel Institute, then was a lecturer in the department of mathematics. In 1935 he was awarded a doctorate in the physical and mathematical sciences without defense of a dissertation. In 1937 Parenago became a lecturer, and in 1938 professor, in the department of astronomy of the Faculty of Mechanics and Mathematics of the University of Moscow. From 1940 he was the head of the department of stellar astronomy there, which had been organized on his initiative. He gave courses in stellar astronomy, spherical astronomy, probability theory and mathematical analysis of observations, and stellar dynamics. In 1953 he was elected corresponding member of the Soviet Academy of Sciences.

Parenago began his scientific career with photometric research on variable stars—first visually and then from photographs in the collection at the Moscow and Simeiz observatories—in all he investigated more than 600 variable stars. He subsequently advanced to a study of all the properties of variable stars, including their motion. He discovered statistical relationships of their physical and kinematic properties to their spatial distribution in the galaxy, and investigated the use of variable stars in studying its structure. For these purposes he collected and systematized a vast amount of observational material that reflected all aspects of contemporary knowledge of the stars. The foremost Soviet stellar astronomer, Parenago was head of a school of specialists in this field and the organizer of the first department of stellar astronomy in the Soviet Union. His textbook *Kurs zvezdnoy astronomii* ("Course of Stellar Astronomy") went through three editions (1938, 1946, and 1954) and was translated into some foreign languages.

The numerous investigations of variable stars by Parenago and other Moscow astronomers and their rich collection of references to the existing world literature led the Executive Committee of the International Astronomical Union in 1946 to commission them to name new variable stars and to compile and edit *Obshchy katalog peremennykh zvezd* ("General Catalog of Variable Stars"). Parenago was coauthor of its two editions (1948, 1958) and of the first edition of *Katalog zvezd, zapodozrennykh v peremennosti* ("Catalog of Stars Suspected of Being Variable"; 1951). His long monograph, *Fizicheskie peremennye zvezdy* ("Physical Variable Stars"; 1937), written with B. V. Kukarkin, and the popular *Peremennye zvezdy i sposoby ikh nablyudenia* ("Variable Stars and Methods of Observing Them"; 1938; 2nd ed., 1947), were of great value for amateur astronomers.

His interest in the structure of the galaxy and in using variable stars as indicators of distance led Parenago to develop a method of taking into account the absorption of light in galactic space by particles of interstellar dust. This method substantially increased the precision of determining galactic distances. A number of important works by Parenago concern the Hertzsprung-Russell diagram (the "spectrum-luminosity" relation and the "mass-radius-luminosity" relation). Parenago also studied the kinematics of stars and stellar dynamics. With Kukarkin he developed and published a new, evolutionary meaning for the concept of subsystems of various objects in the galaxy; and, using extensive statistical material, he obtained fundamental quantitative properties of these subsystems and provided a method for evaluating the total number of objects in each of them. Parenago studied the law of rotation of the galaxy and its spiral structure. He also developed a new theory of galactic potential (that is, of the law of the variation of the force of galactic attraction with the distance from the center of the galaxy) and the theory of the galactic orbit of stars and the sun. He published 225 works, more than 40 reviews, and more than 150 popular articles and books.

From 1947 to 1951 Parenago was president of the Moscow Astronomical and Geodetic Society. In 1949 he received the Bredikhin Prize of the Soviet

Academy of Sciences for a series of works on the structure of the galaxy. He was awarded the Order of Lenin in 1951. Parenago was a member of the International Astronomical Union and participated in a number of international and Soviet congresses and scientific conferences.

BIBLIOGRAPHY

I. ORIGINAL WORKS. Parenago's writings of the 1930's are "O periode i krivoy izmenenia yarkosti SW Cygni" ("On the Period and Curve of the Variation of Brightness of SW Cygni"), in *Astronomicheskii zhurnal*, **8**, nos. 3–4 (1931), 229–239; "Shkaly zvezdnykh velichin" ("Stellar Magnitude Scales"), in *Uspekhi astronomicheskikh nauk*, **2** (1933), 104–122; "The Catalogue of Parallaxes of Variable Stars," in *Astronomicheskii zhurnal*, **11**, no. 1 (1934), 29–39; "Untersuchungen über veränderliche Sterne mit unbekannten Lichtwechsel," in *Peremennye zvezdy*, **4**, no. 9 (1934), 301–317; "O vrashchenii Urana vokrug osi" ("On the Rotation of Uranus Around Its Axis"), in *Astronomicheskii zhurnal*, **11**, no. 5 (1934), 487–496; "The Shapes of Light Curves of Long-Period Cepheids," in *Zeitschrift für Astrophysik*, **11**, no. 5 (1936), 337–355, written with B. V. Kukarkin; "The Mass-Luminosity Relation," in *Astronomicheskii zhurnal*, **14**, no. 1 (1937), 33–48; "Standard Light Curves of Cepheids," *ibid.*, **14**, no. 3 (1937), 181–193, written with B. V. Kukarkin; "Issledovania izmeneny bleska 208 peremennykh zvezd (1920–1937)" ("Research on the Variations in Brightness of 208 Variable Stars"), in *Trudy Gosudarstvennogo astronomicheskogo instituta im. P. K. Shternberga*, **12**, iss. 1 (1938), 1–132; "Obobshchennaya zavisimost massa-svetimost" ("Generalized Mass-Luminosity Relation"), in *Astronomicheskii zhurnal*, **16**, no. 6 (1939), 7–14; and "Opredelenie galakticheskoy orbity Solntsa" ("Determination of the Galactic Orbit of the Sun"), *ibid.*, **16**, no. 4 (1939), 18–24.

During the 1940's Parenago published "Issledovania, osnovannye na svodnom kataloge zvezdnykh parallaksov" ("Researchs Based on the Summary Catalog of Stellar Parallaxes"), in *Trudy Gosudarstvennogo astronomicheskogo instituta im. P. K. Shternberga*, **13**, pt. 1 (1940), 59–117; "O temnykh tumannostyakh i pogloshchenii sveta v galaktike" ("On Dark Nebulae and the Absorption of Light in the Galaxy"), in *Astronomicheskii zhurnal*, **17**, no. 4 (1940), 1–22, also in *Byulleten Gosudarstvennogo astronomicheskogo instituta im. P. K. Shternberga*, **4** (1940), 3–24; "O diagramme Rassela-Khertsshprunga" ("On the Hertzsprung-Russell Diagram"), in *Astronomicheskii zhurnal*, **21**, no. 5 (1944), 223–229; "O mezhzvezdnom pogloshchenii sveta" ("On the Interstellar Absorption of Light"), *ibid.*, **22**, no. 3 (1945), 129–150; "Some Works on the Structure of the Galaxy," in *Popular Astronomy*, **53** (1945), 441–446; "Fizicheskie kharakteristiki subkarlikov" ("Physical Characteristics of Subdwarfs"), in *Astronomicheskii zhurnal*, **23**, no. 1 (1946), 31–39; "O dvizheniakh sharovykh skopleny" ("On the Motions of Globular Clusters"), *ibid.*, **24**, no. 3 (1947), 167–176; "Prostranst-

vennoe dvizhenie peremennykh zvezd tipa RR Lyrae" ("Spatial Motion of Variable Stars of Type RR Lyrae"), in *Peremennye zvezdy*, **6**, no. 2 (1948), 79–88; "Shkaly i katalogi zvezdnykh velichin" ("Scales and Catalogs of Stellar Magnitudes"), in *Uspekhi astronomicheskikh nauk*, 2nd ser., **4** (1948), 257–287; and "Stroenie Galaktiki" ("Structure of the Galaxy"), *ibid.*, pp. 69–171, translated into German in *Abhandlungen aus der Sowjetischen Astronomie*, **3** (1953), 7–113.

Writings by Parenago that appeared in the 1950's were "O gravitatsionnom potentsiale Galaktiki" ("On the Gravitational Potential of the Galaxy"), 2 pts., in *Astronomicheskii zhurnal*, **27**, no. 6 (1950), 329–340, and **29**, no. 3 (1952), 245–287; "Issledovanie prostranstvennykh skorosteyzvezd" ("Research on the Spatial Velocities of Stars"), *ibid.*, no. 3 (1950), 150–168, also in *Trudy Gosudarstvennogo astronomicheskogo instituta im. P. K. Shternberga*, 20 (1951), 26–80; "Issledovanie zavisimosti massa-radius-svetimost. I. Opredelenie empiricheskoy zavisimosti massa-radius-svetimost. II. Teoreticheskaya interpretatsia empiricheskikh zavisimostey" ("Research on the Mass-Radius-Luminosity Relation. I. Determination of the Empirical Mass-Radius-Luminosity Relation. II. Theoretical Interpretation of Empirical Relations"), in *Trudy Gosudarstvennogo astronomicheskogo instituta im. P. K. Shternberga*, **20** (1951), 81–146, written with A. G. Masevich, also in *Astronomicheskii zhurnal*, **27**, no. 3 (1950), 137–149, and no. 4 (1950), 202–210; "O gravitatsionnom potentsiale galaktiki. II" ("On the Gravitational Potential of the Galaxy. II"), in *Astronomicheskii zhurnal*, **29**, no. 3 (1952), 245–287; "Issledovanie zvezd v oblasti tumannosti Oriona" ("Research on Stars in Areas of Nebulae of Orion"), *ibid.*, **30**, no. 3 (1953), 249–264, also in *Byulleten* "*Peremennye zvezdy*," **9**, no. 2 (1953), 89–93, and in *Trudy Gosudarstvennogo astronomicheskogo instituta im. P. K. Shternberga*, **25** (1954), 1–547; "O spiralnoy strukture galaktiki po radionablyudeniam na volne 21 sm." ("On the Spiral Structure of the Galaxy According to Radio Observations on a Wave of 21 cm."), in *Astronomicheskii zhurnal*, **32**, no. 3 (1955), 226–238; "Plan kompleksnogo izuchenia izbrannykh oblastey Mlechnogo Puti" ("Plan of Complex Study of Selected Areas of the Milky Way"), *ibid.*, **33**, no. 5 (1956), 749–755; "O kinematike razlichnykh posledovatelnostey na diagramme spektr-svetimost" ("On the Kinematics of Various Sequences in the Spectrum-Luminosity Diagram"), *ibid.*, **35**, no. 3 (1958), 488–490; and "The Hertzsprung-Russell Diagram From Photoelectric Observations of Nearby Stars," in J. Greenstein, ed., *The Hertzsprung-Russell Diagram* (Paris, 1959), 11–18.

In 1960 he published "Zvezdnaya astronomia" ("Stellar Astronomy"), in *Astronomia v SSSR za 40 let. 1917–1957* ("Astronomy in the U.S.S.R. for Forty Years . . ."; Moscow, 1960), 227–259.

II. SECONDARY LITERATURE. See *Astronomichesky tsirkulyar SSSR*, no. 169 (1956), 23, on Parenago's 50th birthday; *Astronomicheskii zhurnal*, **37**, no. 1 (1960), 191–192, an obituary; *Bolshaya sovetskaya entsiklopedia*, 2nd ed., XXXII (1955), 88; B. V. Kukarkin, "P. P. Parenago

(1906–1960)," in *Peremennye zvezdy*, **13**, no. 1 (1960), 3–5; D. Y. Martynov, *Astronomichesky kalendar na 1951 g.* ("Astronomical Calendar for 1951"; Gorky, 1951), 144–145, on Parenago's receiving the Bredikhin Prize; "P. P. Parenago (1906–1960)," in *Istoriko-astronomicheskie issledovania*, no. 7 (1961), 335–394, which consists of articles by B. A. Vorontsov-Velyaminov, B. V. Kukarkin, A. S. Sharov, and F. A. Tsitsin, and a bibliography of Parenago's works; A. S. Sharov, "On the Photometric Catalogue of Stars in the Region of Orion Nebulae, Compiled by P. P. Parenago," in *Astronomical Journal*, **66**, no. 2 (1961), 103; and W. Zonn, "Pavel Petrowich Parenago," in *Postępy astronomii*, **8**, no. 3 (1960), 175.

P. G. KULIKOVSKY

PARENT, ANTOINE (*b.* Paris, France, 16 September 1666; *d.* Paris, 26 September 1716), *physics.*

Parent was the son of an *avocat au conseil.* His mother's uncle, Antoine Mallet, took charge of the boy's education when he was only three; Fontenelle, our source for this information, does not indicate why the father relinquished the duty. An elderly and pious man, Mallet pointed Parent toward a career in law. After dutifully completing that study, Parent turned to mathematics, for which he had independently acquired a taste. He attended the lectures of La Hire and Sauveur—the latter considered Parent a rare genius. For a short time Parent accompanied the Marquis d'Alègre on military campaigns, studying fortifications, and then devoted his time exclusively to science.

In 1699, when Gilles Filleau des Billettes was elected to the Académie des Sciences, he selected Parent as his *élève.* Parent carried the title until his death. His failure to advance was due to a lack of clarity in his writing, his antipathy to Cartesian science, and his aggressive, tactless, critical, and uncompromising candor in dealing with colleagues. Fontenelle declared that he had "goodness without showing it," scarcely a generous remark in an official eulogy. Parent, who never married, lived alone according to an austere, disciplined regimen. In 1716 he contracted a fatal case of smallpox.

Parent's interests were very wide-ranging, although such broad scope was not uncommon in his day, before the branches of science were carefully delineated. He wrote on astronomy, cartography, chemistry, biology, sensationalist psychology and epistemology, music, practical and abstract mathematics, and various mechanical phenomena, particularly those of the strength of materials and the effects of friction on motion. He often reviewed and commented on the works of others. He read many papers to the Académie des Sciences but few were published in the *Mémoires.* His most frequent avenues of publication were the *Journal des sçavans* and the *Journal de Trévoux.* In 1705 he launched his own periodical, *Essais et recherches de mathématiques et de physique,* which, although short-lived, provided a means of publishing much of his completed work. A three-volume sequel (1713) remains his best-known and most comprehensive collection.

Some of Parent's work was clearly original. The work of 1705 contained a memoir, originally delivered to the Academy in 1700, on the description of a sphere according to the techniques of analytic geometry. Although the treatment is awkward by modern standards, the clear understanding and use of space coordinates was not to become routine for many years. Typically, Parent aimed at extending the power both of geometry and of the new calculus, although he was far from moving to pure analysis. Some of his earliest work was in cartography, and it seems to have left a permanent mark on his style. The ever-present correspondence between the descriptive device of geometry and the physical space being described is characteristic of mapping. Parent never went the full journey to pure abstraction. He anticipated neither Lagrange nor Laplace but, rather, Coulomb, the engineer who absorbed enough science to emerge as an early physicist. The similarity is confirmed in Parent's work. The early study of fortifications has been mentioned. The publication of 1713 included an article describing the conditions of stress on a loaded beam, in which Parent first recognized the existence of a shear stress. Certain aspects of the analysis were not extended or even repeated for over half a century, by Coulomb.

Parent's contributions to science are not best characterized by the listing of "firsts." The power of the new mathematics was such that originality became common during the first decade of the eighteenth century. Parent learned from La Hire and Sauveur, and he shared the stage during his prime with Varignon, Hermann, Jakob I Bernoulli, and others. If Parent had a particular characteristic, it is perhaps his sense of the practical. The utilitarian aspect, seldom absent in his work, is noticeable from his first paper, on the calculation of frictional forces in machines, to his last, on the theoretical and practical applications of arithmetic.

Also remarkable is the degree to which Parent's criticisms of scientific work extended into the thematic or paradigmatic foundations of science. This tendency is evident even in the first volume of the 1713 publication; the bulk of that book is devoted to an attack on Descartes's *Principia philosophiae,* a work then nearly seventy years old. Parent went through it almost paragraph by paragraph; the reviewer in the *Journal des sçavans* needed three pages just to list the points upon which Parent and Descartes disagreed. For Parent, who was an atomist, motion could not produce hardness in objects, nor could it account for their specific shapes; Cartesian laws of motion were entirely incorrect. Parent refuted Cartesian notions on the formation of the elements, on comets, weight, the nature and effects of fire, on air, winds, hail and lightning, subterranean heat, light and color, the nature of ideas and the principles of music. Parent did not, as might be suspected, consider Newton to be Descartes's principal adversary, although Parent and Newton clearly shared many concepts. But Parent seems to have viewed Descartes as having been refuted by the whole range of seventeenth-century mechanical philosophers who embraced atomism. His attack on Descartes seems also to have been aimed at the Academy, most of whose members still clung to Cartesian science.

Parent emerges from the skimpy historical record as rather stiff, pious, solitary, independent, hardworking, and intelligent. He foreshadowed the Enlightenment in his unflagging critical spirit, his attempt to develop the scientific view of nature, and his conviction that the seemingly esoteric nature of mathematics had a very real utility.

BIBLIOGRAPHY

I. ORIGINAL WORKS. Parent's most comprehensive single work is *Essais et recherches de mathématiques et de physique*, 3 vols. (Paris, 1713). Also valuable are his reviews and articles in the *Journal des sçavans* and the *Journal de Trévoux*.

II. SECONDARY LITERATURE. Parent has not received much attention. His official *éloge* by Fontenelle appeared in the *Histoire de l'Académie Royale des Sciences* for 1716 and constitutes the bulk of the available biographical information. An account of his mathematics can be found in C. B. Boyer, *History of Analytic Geometry* (New York, 1956), 156 ff. The most extensive treatment of his physics is in Clifford Truesdell, *The Rational Mechanics of Flexible or Elastic Bodies*, in Euler's *Opera omnia*, 2nd ser., XI, pt. 2 (Lausanne, 1950), 109–114. Truesdell repeats some of his comments in "The Creation and Unfolding of the Concept of Stress," in *Essays in the History of Mechanics* (New York, 1968), 184–238. See also C. S. Gillmor, *Coulomb and the Evolution of Physics and Engineering in Eighteenth-Century France* (Princeton, 1971), *passim*; Isaac Todhunter, *A History of the Theory of Elasticity and of the Strength of Materials* (New York, 1960), *passim*; and Hunter Rouse and Simon Ince, *History of Hydraulics* (New York, 1963), *passim*.

J. MORTON BRIGGS JR.

PARKHURST, JOHN ADELBERT (*b*. Dixon, Illinois, 24 September 1861; *d*. Williams Bay, Wisconsin, 1 March 1925), *astronomy*.

Parkhurst was the son of Sanford and Clarissa J. Hubbard Parkhurst. Upon the death of his mother, when he was five years old, he was adopted by his aunt and uncle Dr. and Mrs. Abner Hagar, who lived in Marengo, Illinois. After completing public schools there, he attended Wheaton College, Wheaton, Illinois, from 1878 to 1881. Parkhurst graduated from the Rose Polytechnic Institute, Terre Haute, Indiana, in 1886 with a B.S. in mechanical engineering and remained there as instructor of mathematics for the next two years; in 1897 he received an M.S. from Rose.

He married Anna Greenleaf of Terre Haute, Indiana, in 1888; their only child died in infancy. From childhood, when he walked with crutches, until his death, caused by a cerebral hemorrhage, Parkhurst suffered from poor health but nevertheless worked diligently. He was a member of the American Astronomical Society, the British Astronomical Association, and the Astronomische Gesellschaft of Hamburg, and was a fellow of the Royal Astronomical Society. For many years he was active in the Congregational Church in Williams Bay and was elected the first town supervisor.

Parkhurst's interest in astronomy was stimulated by reading the works of Thomas Dick. Although his time in Marengo was devoted principally to business, he spent his leisure hours making astronomical observations, mainly of variable stars. Within a decade that part-time research had led to approximately fifty published papers.

After occasionally serving as a nonresident computer for the Washburn Observatory, Parkhurst made an important change in his professional life with the opening in 1897 of the University of Chicago's Yerkes Observatory. In 1898 he became a volunteer research assistant at Yerkes; and in 1900, with his appointment as assistant, he began working full time on astronomy. He remained at Yerkes until his death twenty-five years later, having progressed to the rank of associate professor in 1919.

Parkhurst's first published paper at Yerkes, "The Spectra of Stars of Secchi's Fourth Type," was written with George E. Hale and Ferdinand Ellerman. His specialty was stellar photometry, both visual and photographic. In 1906 the Carnegie Institution of Washington published his longest work, *Researches in Stellar Photometry During the Years 1894 to 1906, Made Chiefly at the Yerkes Observatory*. Perhaps his most important paper, however, was "Yerkes Actinometry," published in 1912. It contained his determinations of the visual and photographic magnitudes, color indexes, and spectral classes of all stars brighter than apparent magnitude 7.5 between $+73°$ north declination and the celestial north pole.

As Yerkes' representative, Parkhurst began collaborating in 1900 with Harvard, Lick, and McCormick observatories in a comparison, published in 1923, of the brightnesses of faint stars with those of known bright stars. He also helped prepare no. XII of the appendix to J. G. Hagen's *Atlas stellarum variabilium*; and he determined the photographic magnitudes and color indexes of 1,500 stars in twenty-four Kapteyn Fields, the report of which was published posthumously.

Parkhurst participated in three solar eclipse expeditions to measure coronal brightness but encountered clear skies for only one—that of 24 January 1925, shortly before his death.

BIBLIOGRAPHY

I. ORIGINAL WORKS. Parkhurst published nearly 100 papers, principally in *Astronomical Journal*, *Astrophysical Journal*, *Astronomische Nachrichten*, and *Popular Astronomy*. His key works include *Researches in Stellar Photometry During the Years 1894 to 1906, Made Chiefly at the Yerkes Observatory* (Washington, D.C., 1906); "Yerkes Actinometry," in *Astrophysical Journal*, **36** (1912), 169–227; "Photometric Magnitudes of Faint Standard Stars Measured Visually at Harvard, Yerkes, Lick and McCormick Observatories," in *Memoirs of the American Academy of Arts and Sciences*, **14**, no. 4 (1923), 209–307, written with S. A. Mitchell *et al.*; "Methods Used in Stellar Photographic Photometry at the Yerkes Observatory Between 1914 to 1924," in *Astrophysical Journal*, **62** (1925), 179–190, written with Alice Hall Farnsworth; and "Zone $+45°$ of Kapteyn's Selected Area: Photographic Photometry for 1,550 Stars," in *Publications of the Yerkes Observatory*, **4**, pt. 6 (1927), 230–289.

II. SECONDARY LITERATURE. Biographical information is given in J. McKeen and Dean R. Brimell, eds., *American Men of Science* (Lancaster, Pa., 1921), 526; and in Poggendorff, VI, 1951. A biographical sketch by Raymond S. Dugan is in *Dictionary of American Biography*, XIV, 246–247. Obituaries are by R. G. Aitken and E. B. Frost, in *Publications of the Astronomical Society of the Pacific*, **37** (1925), 85–88; by Storrs B. Barrett, in *Popular Astronomy*, **33** (1925), 280–284, which includes a portrait from 1923; and by E. B. Frost in *Astrophysical Journal*, **61** (1925), 454. There are unsigned obituaries in *Astronomische Nachrichten*, **224** (1925), 147–148; in *Observatory*, **48** (1925), 120; and in *Monthly Notices of the Royal Astronomical Society*, **86** (1926), 185–186.

RICHARD BERENDZEN

PARKINSON, JAMES (*b.* Hoxton Square, London, England, 11 April 1755; *d.* London, 21 December 1824), *medicine, paleontology.*

Parkinson's father, John Parkinson, was a surgeon. Where James studied is not known, but in 1784 his name appeared on a list of surgeons approved by the Corporation of London, and in 1785 he attended a series of lectures by John Hunter. On 21 May 1783 he married Mary Dale of Hoxton Square; they had six children.

Parkinson's early career was overshadowed by his involvement in a variety of social and revolutionary causes. This involvement was mainly through pamphlets that he wrote anonymously or under the pseudonym "Old Hubert." He advocated reform and representation of the people in the House of Commons, the institution of annual parliaments, and universal suffrage. Parkinson joined the London Corresponding Society for Reform of Parliamentary Representation in 1792, and it was between then and 1795 that he was most often heard from with regard to social and political change.

In 1780 Parkinson published, anonymously, *Observations on Dr. Hugh Smith's Philosophy of Physic*, a critical appraisal of Smith's theories. With the exception of a brief account of the effects of lightning (1789), Parkinson published nothing more in the sciences until his political and social activities lessened near the end of the century. His medical practice continued to flourish, however, and during this period he became interested in geology and paleontology. In 1799 his *Chemical Pocket-Book*, a guide for the student and layman, was published; it reflected his interests in medicine, geology, and fossils. In 1799 a work called *Medical Admonitions* was also published. It was the first in a series of popular medical works by Parkinson aimed toward the improvement of the general health and well-being of the population. It is likely that these works represented a continuation of the same zeal for the welfare of the people that was expressed by his political activism. His humanitarianism appeared again in 1811, when he crusaded for better safeguards in regulating madhouses and for legal protection for the mental patients, their keepers, doctors, and families.

Parkinson was the author of several medical treatises of particular interest to the profession. These included a work on gout (1805) and a report on a perforated and gangrenous appendix with peritonitis (1812). The latter is probably the earliest description of that condition in the English medical literature. Parkinson's most important medical work was *An Essay on the Shaking Palsy* (1817). In this short essay Parkinson established the disease as a clinical entity. Sorting through a variety of palsied conditions, which he had observed throughout his career, Parkinson gave the classic, albeit in modern terms limited, clinical description of the illness: "Involuntary tremulous motion, with lessened muscular power, in parts not in action and even when supported; with a propensity to bend the trunk forwards, and to pass from a walking to a running pace: the senses and intellect being uninjured." Symptoms that had been assumed to be characteristic of distinct illnesses, such as tremulous agitans and the violent propensity to run, were shown to be part of a single ailment. A study of several cases and a sorting-out of the symptoms comprises most of the work. Parkinson made no decision concerning the cause but suggested that it arose from "a disordered state of that part of the medulla which is contained in the cervical vertebrae." The illness described by Parkinson, now called Parkinson's disease, is understood today as one form of several clinical events.

Sometime in the late eighteenth century, Parkinson began to collect and study fossils. This was a pleasant avocation for him, and he enjoyed making short trips with his children and his friends to collect or observe fossil plants and animals. In the second edition of the *Chemical Pocket-Book* (1801) he made a public appeal for information on fossils. As he attempted to learn more about their identification and interpretation he discovered that there was little help available in English works. He decided, therefore, to write an introduction to the study of fossils. The first of three volumes of *Organic Remains of a Former World* was published in 1804, the second in 1808, and the third in 1811. Parkinson wanted these volumes to be useful to the beginning student as well as to the advanced collector. Volume I discusses the plant kingdom. The work is somewhat more theoretical than either of the other volumes and is also the least interesting; indeed, this volume met with only moderate success and was soundly criticized for its dullness and poor grammar. At a time when new discoveries in geology were causing concern among the theologians, the volume was also criticized for its failure to offer any mode of reconciliation between geology and theology. Much of the book is devoted to the question of whether coal, peat, and other bituminous products are vegetable in origin; a small portion is devoted to fossil woods, ferns, and other plants. There was considerable disagreement on this issue among Parkinson's contemporaries. Parkinson believed these products originated from plants, and he developed a theory of "bituminous fermentation" to explain the transformation. This fermentation, one of several kinds he recognized as normal to the vegetable kingdom, operated in the absence of external air and under conditions that prohibited the escape of volatile principles in the vegetation. A fluid was thus created. A modification of this fluid occurred through the "oxygenizement" of carbon by the mixture of earthy and metallic salts.

Volume II (on the fossil zoophytes) and volume III (on the fossil starfish, echini, shells, insects, amphibia, and mammals) are more descriptive and met with a better reception. In volume III Parkinson introduced the discoveries of Lamarck, Cuvier, and William Smith. From Smith he adopted the use of fossils as stratigraphic markers, from Lamarck information on shells, and from Cuvier knowledge of the amphibia and land mammals. These volumes, though descriptive, give insight into Parkinson's basic position with regard to geological theory.

He was opposed to the Huttonian theory of the earth. Although probably not a strict adherent of Werner's neptunism, he favored it. In studying the relation of fossils to their strata he was convinced that the creation of life had taken a long time and had proceeded in an orderly fashion, in keeping with scriptural history. After the creation of primary rocks, vegetables were created, then animals of the water and air, followed by land animals and man. He emphasized the Biblical Flood in some cases, but creation and extinction were continuing processes guided by the hand of God. To reconcile his concept of geological time with theology, he adopted from some of his contemporaries the notion that each day of creation represented a long period of time. Parkinson was adamantly opposed to any theory of gradual, natural evolution. The "creative power," he argued, worked continually through new creations.

The volumes of *Organic Remains* were well illustrated with many plates (some in color) done by Parkinson. The plates were later republished in Gideon Mantell's *Pictorial Atlas of Fossil Remains* (1850). The work was a major contribution to the development of British paleontology, particularly as a thorough and usable compilation of information on British fossils.

On 13 November 1807 Parkinson met with several of his friends, including Sir Humphry Davy and George Greenough, at the Freemason's Tavern. Together they formed the Geological Society of London. Parkinson was a contributor to the first volume (1811) of the society's *Transactions* with a detailed study of the London basin entitled "Observations on Some of the Strata in the Neighbourhood of London, and on the Fossil Remains Contained in Them."

In 1822 Parkinson published *Outlines of Oryctology,*

which he considered a supplement to Conybeare and Phillips' *Outlines of the Geology of England and Wales, With an Introductory Compendium of the General Principles of That Science, and Comparative Views of the Structure of Foreign Countries*. It is similar to *Organic Remains*, with some additions and changes based on newer developments in geology. He adopted catastrophism and viewed the creation of life in a sequence and manner like that outlined by Cuvier.

BIBLIOGRAPHY

I. ORIGINAL WORKS. Parkinson's major works are *The Chemical Pocket-Book, or Memoranda Chemica: Arranged in a Compendium of Chemistry: With Tables of Attractions, etc. Calculated as Well for the Occasional Reference of the Professional Student, As to Supply Others With a General Knowledge of Chemistry* (London, 1799); *Organic Remains of a Former World. An Examination of the Mineralized Remains of the Vegetables and Animals of the Antediluvian World Generally Termed Extraneous Fossils* (London, 1804–1811); *An Essay on the Shaking Palsy* (London, 1817); and *Outlines of Oryctology: An Introduction to the Study of Fossil Organic Remains; Especially of Those Found in the British Strata: Intended to Aid the Student in His Inquiries Respecting the Nature of Fossils and Their Connection With the Formation of the Earth* (London, 1822).

Parkinson's notes on J. Hunter's lectures were transcribed by his son and published as *Hunterian Reminiscences* (London, 1833).

II. SECONDARY LITERATURE. On Parkinson and his work see W. R. Bett, "James Parkinson: Practitioner, Pamphleteer, Politician and Pioneer in Neurology," in *Medical Press*, **234** (1955), 148; G. S. Boulger, "James Parkinson," in *Dictionary of National Biography*; J. Challinor, "Beginnings of Scientific Paleontology in Britain," in *Annals of Science*, **6** (1948), 46–53; M. Critchley, ed., *James Parkinson (1755–1824). A Bicentenary Volume of Papers Dealing with Parkinson's Disease Incorporating the Original Essay on the Shaking Palsy* (London, 1955), contains a biography of Parkinson by W. H. McMenemey and a bibliography; J. M. Eyles, "James Parkinson (1755–1824)," in *Nature*, **176** (1955), 580–581; and L. G. Rowntree, "James Parkinson," in *Johns Hopkins Hospital Bulletin*, **23** (1912), 33–45.

PATSY A. GERSTNER

PARKINSON, SYDNEY (*b.* Edinburgh, Scotland, *ca.* 1745; *d.* at sea 26 January 1771), *natural history drawing*.

Parkinson, a young Scottish artist, who died on the return voyage of the *Endeavour* after the disastrous stay for refitting at Batavia (where nearly everyone on board contracted malaria or dysentery, or both), was an extremely gifted and versatile draftsman and colorist. His beautiful and accurate drawings of the plants and animals collected on Cook's first voyage round the world, his studies of exotic landscapes, their peoples, and artifacts, together with Joseph Banks's collections and manuscripts, combined to make that voyage one of the most memorable in the annals of scientific discovery.

Sydney was the younger son of Joel Parkinson, an Edinburgh brewer, and his wife Elizabeth. Nothing is known of his education—not a very formal one since he signed himself both Sydney and Sidney—but in his teens he was apprenticed to a woolen draper. He seems to have had innate aptitude for drawing. His brother Stanfield wrote in the introduction to his posthumously published journal that from an early age "taking a delight in drawing flowers, fruits and other objects of natural history, he soon became as proficient in that stile of painting as to attract the notice of the most celebrated botanists and connoisseurs in that study."

When Parkinson was about twenty his widowed mother moved to London, where he exhibited with the Free Society in 1765 and 1766. About that time he was engaged by another Scot, James Lee of the Vineyard Nursery, Hammersmith, to teach his favorite daughter Ann, some thirteen years old. It was to her that Parkinson bequeathed his "utensils" and some botanical paintings that remained in the possession of the Lee family until 1970. The paints and brushes have been lost, but the paintings are now in the National Library, Canberra; Ann's paintings, considered by the great Danish entomologist J. C. Fabricius to be the best British natural history drawings of the day, are in the library of the Royal Botanic Gardens, Kew. Early in 1767 Lee introduced Parkinson to Banks, and from then on the young artist worked extensively on Banks's collections, first on the plates (now in the British Museum [Natural History]) of the Loten collection from Ceylon, then on the invertebrates, fishes, and birds collected by Banks in Newfoundland and Labrador in 1766. These drawings of Newfoundland animals, and of other exotic species in Banks's possession, many of which are still unidentified, are in the Print Room, British Museum.

Banks himself wrote in warm terms of Parkinson's industry in the *Endeavour*. During the voyage the young Scot made nearly 1,000 drawings of plants, about 300 of animals ranging from pellucid coelenterates to tropical birds, all of which may be seen in the British Museum (Natural History); he also executed some 200 topographical and ethnographical drawings, now in the Manuscript Room, British Museum.

Parkinson also recorded Polynesian and other vocabularies whenever the opportunity arose. In some cases his lists exceeded those compiled by Banks, who, more cautiously, recorded only ninety-one Tahitian words against Parkinson's 300. Parkinson also listed eighty-one Tahitian plants, with their economic uses as fish poisons, dyes, medicines, textiles, and as building material. These lists appeared in the illegally published edition (1773) of his completed journal (which was lost) when his brother Stanfield attempted to forestall the official account of the voyage. A second edition appeared in 1784. A self-portrait is owned by the British Museum (Natural History); an engraving of him at an earlier age forms the frontispiece to his journal.

BIBLIOGRAPHY

The illegally published edition of Parkinson's journal is Sydney Parkinson, *A Journal of a Voyage to the South Seas in His Majesty's Ship the Endeavour* (London, 1773, 2nd ed. 1784). See also F. C. Sawyer, "Some Natural History Drawings Made During Captain Cook's First Voyage Round the World," in *Journal of the Society for the Bibliography of Natural History*, **2** (1950), 190–193; J. C. Beaglehole, ed., *The Journals of Captain James Cook*, I, *The Voyage of the Endeavour* (Cambridge, 1955); and *The Endeavour Journal of Joseph Banks*, 2 vols. (Sydney, 1962); A. M. Lysaght, *Joseph Banks in Newfoundland and Labrador, 1766* (London, 1971); A. M. Lysaght and D. L. Serventy, "Some Erroneous Distribution Records in Parkinson's Journal of a Voyage to the South Seas," in *Emu*, **56** (1956), 129–130; W. F. Miller, "Sydney Parkinson and His Drawings," in *Journal of the Friends' Historical Society*, **8** (1911), 123–127; and E. J. Willson, *James Lee and the Vineyard Nursery* (London, 1961).

AVERIL M. LYSAGHT

PARMENIDES OF ELEA (*b. ca.* 515 B.C.; *d.* after 450 B.C.), *natural philosophy.*

Parmenides' dates are inferred from the dramatic situation in Plato's *Parmenides*. A different chronology, which ultimately comes from Apollodorus of Athens, should be rejected, since it is based on an attempt to place Parmenides' birth in 540/539 B.C., the year of the founding of Elea. (See Tarán, *Parmenides*, pp. 3–4.)

Parmenides puts forward his philosophy in a poem in dactylic hexameter. Mainly through Simplicius, who cites it in his commentaries on Aristotle's *Physics* and *De Caelo*, we have approximately 155 lines of it; there are, in addition, six lines extant only in a Latin translation. More information on his thought may be obtained from Plato, Aristotle, Theophrastus, and the doxographers who depend mainly upon Aristotle and Theophrastus. Because one must depend on others for his thought and because of Aristotle's notorious tendency to interpret his predecessors in the light of his own philosophy, the indirect information about Parmenides cannot be taken at face value but must be analyzed critically. (This has been done by Cherniss in *Aristotle's Criticism of Presocratic Philosophy*. On the secondary sources of information about Parmenides, see Tarán, *Parmenides*, pp. 269–295.)

To emphasize the objectivity of his method, Parmenides presents his doctrine through a nameless goddess whom the poet reaches after traveling in a chariot driven by the daughters of the sun. According to this goddess, only two ways of inquiry can be conceived of: that which asserts the existence of being and that which accepts as necessary the existence of not-being. Since it is impossible to think that which in no way is, only the first way can be pursued. Nevertheless the goddess, with a probable allusion to Heraclitus' doctrine of the unity of contraries, attacks as the extreme of folly those mortals who believe that being and not-being are the same and yet not the same. She asks her hearer to judge her argument by reason and not to let himself be led astray by the senses; for there is but one way, that which maintains that only being exists. By assuming that there is no *tertium quid* between being and absolute not-being, the goddess construes a tight and cogent reasoning which shows that that which is (being) must be ungenerated, imperishable, homogeneous, changeless, immovable, complete, and unique. These characteristics are meant to emphasize from the negative side the unique and unalterable existence of being, for it is implied that if being did not have any one of these characteristics, one would have to admit the existence of something different from being; and such an admission, given the original assumption, would be tantamount to accepting the existence of not-being. Consequently mortals' opinions about the phenomenal world are meaningless, since they refer to something that has no existence whatever. Yet the pupil must learn the opinions of men, for it is essential to know the source of error so that no one should outstrip him. The minimal error is the belief, conscious or not, that difference is real. Since the minimal difference implies the existence of two things, the goddess shows how from it a whole world of difference and change can be derived. In accordance with this purpose of offering a cosmogony or cosmology as a model of reference, the goddess describes doctrines that were more or less current at Parmenides' time.

Parmenides' basic mistake is his misapplication of the law of the excluded middle to the disjunction being:: not-being. Otherwise his reasoning is flawless, and none of the philosophers who came immediately after him was able to refute him. The refutation was reserved for Plato, especially in his *Sophist*; but Plato recognized the importance of Parmenides' attempt to apply the exigencies of logical proofs to thought and its object.

BIBLIOGRAPHY

The fragments of Parmenides' poem and most of the evidence from secondary sources are in H. Diels and W. Kranz, eds., *Die Fragmente der Vorsokratiker*, 6th ed., I (Berlin, 1952), 217–246.

Modern works dealing with Parmenides are P. Albertelli, *Gli Eleati. Testimonianze e frammenti* (Bari, 1939); K. Bormann, *Parmenides. Untersuchungen zu den Fragmenten* (Hamburg, 1971); J. Burnet, *Early Greek Philosophy*, 4th ed. (London, 1930), 169–196; G. Calogero, *Studi sull'eleatismo* (Rome, 1932); H. Cherniss, *Aristotle's Criticism of Presocratic Philosophy* (Baltimore, 1935); H. Diels, *Parmenides Lehrgedicht. Griechisch und Deutsch* (Berlin, 1897); H. Fränkel, *Wege und Formen frühgriechischen Denkens*, 2nd ed. (Munich, 1960); W. K. C. Guthrie, *A History of Greek Philosophy*, II (Cambridge, 1965), 1–80; U. Hölscher, *Parmenides. Vom Wesen des Seiendes* (Frankfurt, 1969); J. B. McDiarmid, "Theophrastus on the Presocratic Causes," in *Harvard Studies in Classical Philology*, **61** (1953), 85–156; J. Mansfeld, *Die Offenbarung des Parmenides und die menschliche Welt* (Assen, 1964); A. Mourelatos, *The Route of Parmenides* (New Haven, 1970); G. Reale, in his ed. of E. Zeller's *Die Philosophie der Griechen, La filosofia dei greci nel suo sviluppo storico*, I, pt. 3 (Florence, 1967); K. Reinhardt, *Parmenides und die Geschichte griechischen Philosophie* (Bonn, 1916); L. Tarán, *Parmenides. A Text With Translation, Commentary, and Critical Essays* (Princeton, 1965); M. Untersteiner, *Parmenide. Testimonianze e frammenti* (Florence, 1958); W. J. Verdenius, *Parmenides. Some Comments on His Poem* (Groningen, 1942); and E. Zeller, *Die Philosophie der Griechen*, 7th ed., W. Nestle, ed., I (Leipzig, 1923), 679–741.

LEONARDO TARÁN

PARMENTIER, ANTOINE-AUGUSTIN (*b.* Montdidier, France, 12 or 17 August 1737; *d.* Paris, France, 17 December 1813), *chemistry, nutrition, agriculture, public health, pharmacy.*

Born into a bourgeois family of modest means, Parmentier was apprenticed at an early age to an apothecary in Montdidier. In 1755 he left for Paris to continue his apprenticeship; but two years later, during the Seven Years War, he joined the French army in Germany as *apothicaire sous-aide.* Wounded in action and captured five times by the Prussians, he nevertheless returned safely to Paris in 1763. To support himself he worked in an apothecary shop and in his spare time attended lectures given by Nollet, Bernard de Jussieu, and G.-F. Rouelle. In 1766 he competed successfully for the post of *apothicaire gagnant-maîtrise* at the Hôtel Royal des Invalides and in 1772 was commissioned *apothicaire-major* of that institution. Two years later the Sisters of Charity, the nursing order in charge of the pharmacy service at the Invalides since 1676, caused Parmentier's commission to be revoked. Despite this temporary setback, he carved out a brilliant career in military pharmacy, eventually achieving the rank of inspector general in the army health service.

Parmentier's earliest investigation, dating from about 1771, concerned the chemical and nutritive constituents of the potato. This research was soon broadened to include a large number of indigenous plants which he recommended as food in times of scarcity and famine, ascribing their nutritive value to their starch content. These early efforts resulted in a published memoir (1773), which was awarded a prize by the Besançon Academy of Sciences, Belles-Lettres, and Arts and later formed the basis of a greatly expanded work, *Recherches sur les végétaux nourissants qui, dans les temps de disette, peuvent remplacer les alimens ordinaires* (1781).

Of all these plants it was the potato that most interested Parmentier, and it is unfortunate that his long and successful campaign to popularize the cultivation and use of the potato in France as a cheap and abundant source of food has tended to obscure his other accomplishments in food chemistry and nutrition. Typical and worthy of note are his chemical analyses of wheat and flour (1776), chestnuts (1780), milk (1790 and 1799, in collaboration with Nicolas Deyeux), and chocolate (1786 and 1803). Parmentier devoted considerable time to formulating cheap and nutritious soups for the poor and to the technology of bread-making. In 1780 he was instrumental in founding, with his colleague Cadet de Vaux, the first government-sponsored school of baking in France. During France's economic warfare with England (1806–1812), Parmentier achieved some success in fostering the production of grape syrup as a substitute for cane sugar, which had become scarce and expensive.

A member of the prestigious Royal Society of Agriculture in Paris and an *agronome* of repute,

Parmentier conducted far-ranging investigations that included preservation of grain and flour; improvements in milling; cultivation of corn; and preservation of vinegar, wine, and meat, as well as methods for detecting their adulteration. He contributed articles to the twelve-volume *Cours complet d'agriculture*, launched by Abbé François Rozier in 1781; he collaborated in the writing of the twenty-four-volume *Nouveau dictionnaire d'histoire naturelle* (Paris, 1803–1804); and he was the author of *Économie rurale et domestique*, of which six volumes of the projected eight appeared (Paris, 1788–1793).

Parmentier also evinced a strong interest in public health, reflected in his publications on the quality of water from the Seine (1775 and 1787), chemical studies with Deyeux of pathological changes in the blood (1791 and 1794), and his collaboration with Laborie and Cadet de Vaux on cesspools (1778) and with them and Hecquet on exhumations (1783). Parmentier was active in the movement to provide free smallpox vaccinations to the poor, and in 1802 he was appointed to the newly created Council of Health for the Department of the Seine.

Frankly utilitarian in his scientific orientation, Parmentier in his life and work personified the best sentiments and aspirations of the Enlightenment. In addition to his close association with Cadet de Vaux and Deyeux, Parmentier numbered among his collaborators Bertrand Pelletier, Chaptal, Huzard, Rozier, Thouin and d'Ussieux. A member of many learned societies, he was admitted to the Academy of Sciences in 1795 and in 1801 was one of the founding members of the Société d'Encouragement pour l'Industrie Nationale.

BIBLIOGRAPHY

I. ORIGINAL WORKS. An annotated bibliography of Parmentier's publications is given in A. Balland, *La chimie alimentaire dans l'oeuvre de Parmentier* (Paris, 1902), 377–426. See also J.-M. Quérard, *La France littéraire ou dictionnaire bibliographique* (Paris, 1827–1839), VI, 603–606; and the Royal Society *Catalogue of Scientific Papers (1800–1863)*, IV, 762–763.

II. SECONDARY LITERATURE. For older material covering the period 1781–1897, see Balland (above), pp. 427–434, which lists 34 references. Recent sources include Arthur Birembaut, "L'école gratuite de boulangerie," in René Taton, ed., *Enseignement et diffusion des sciences en France au XVIIIᵉ siècle* (Paris, 1964), 493–509; A. J. Bourde, *Agronomie et agronomes en France au XVIIIᵉ siècle*, 3 vols. (Paris, 1967), II, 637–643, 913; and III, 1291, 1331, 1533–1534; Maurice Bouvet, "Hommage à Parmentier," in *Revue d'histoire de la pharmacie*, **12**, no. 151 (Dec. 1956),

478–480; Maurice Javillier, "Antoine Parmentier," in *Figures pharmaceutiques françaises* (Paris, 1953), 29–34; and R. Massy, "À l'apothicairerie de l'Hôtel royal des Invalides: Le conflit de 1772 entre l'administration de l'hôtel et les Filles de la Charité," in *Revue d'histoire de la pharmacie*, **11**, no. 142 (Sept. 1954), 315–324.

ALEX BERMAN

PARNAS, JAKUB KAROL (*b.* Tarnopol, Poland [now Ukrainian S.S.R.], 16 January 1884; *d.* Moscow, U.S.S.R., 29 January 1949), *biochemistry.*

Parnas studied chemistry in the universities of Berlin, Strasbourg, Zurich, and Munich, where he received the Ph.D. in 1907. He was associate professor of chemistry at Strasbourg in 1913 and professor of physiological chemistry at Warsaw (1916–1919) and Lvov (1920–1941). From 1943 he was head of the Biological and Medical Chemistry Institute of the Soviet Academy of Medical Sciences in Moscow, where he also established a Laboratory of Physiological Chemistry as part of the Soviet Academy of Sciences.

In the course of his career Parnas educated a large number of biochemists and exerted an important influence on the development of biochemistry, both in Poland and throughout the world. As a researcher, his chief fields of investigation were the biochemistry of muscles, especially the interdependence of the metabolism of carbohydrates and that of phosphorus; ammonia production in its relationship to the function of muscles; and the connection between nitrogen metabolism and the metabolism of adenosine monophosphate, including its deamination and dephosphorylation. He discovered the phosphorolysis of glycogen, and, by establishing reaction sequences linking the metabolism of carbohydrates with that of phosphorus, initiated the method of studying life processes now characteristic of molecular biology. In 1937, in collaboration with the Niels Bohr Institute in Copenhagen, Parnas became one of the first to apply P^{32} to biochemical investigations, particularly to that of the metabolism of muscles in vitro. He thus attained a detailed picture of the functional metabolism of muscles; the enzymatic pathway that he thereby established is sometimes known as the Embden, Meyerhof, and Parnas (EMP) scheme.

Parnas was a member of the Polish Academy of Sciences, the Soviet Academy of Sciences, the Soviet Academy of Medical Sciences, the Academy of Medicine in Paris, and the Leopoldina. He received honorary degrees from the universities of Athens and Paris.

BIBLIOGRAPHY

I. ORIGINAL WORKS. Parnas published about 120 scientific papers and a number of reviews, of which a list of twenty may be found in Dorothy M. Needham, *Machina carnis* (Cambridge, 1971), p. 706. A complete list of Parnas' works has been compiled by Irena Mochnacka in *Acta Biochimica Polonica*, **3** (1956), 3–39. His textbooks include *Chemja Fizjologiczna* ("Physiological Chemistry"; Warsaw–Lvov, 1922).

II. SECONDARY LITERATURE. An article on Parnas and his work appears in *Wielka Encyklopedia Powszechna* (Warsaw, 1966), and J. Heller and W. Mozołowski have described his teaching activity in *Postępy Biochemii*, **4** (1958), 5–16, where there is also a bibliography.

T. W. KORZYBSKI

PARSEVAL DES CHÊNES, MARC-ANTOINE

(*b.* Rosières-aux-Salines, France, 27 April 1755; *d.* Paris, France, 16 August 1836), *mathematics.*

Little is known of Parseval's life or work. He was a member of a distinguished French family and described himself as a squire; his marriage in 1795 to Ursule Guerillot soon ended in divorce. An ardent royalist, he was imprisoned in 1792 and later fled the country when Napoleon ordered his arrest for publishing poetry against the regime. He was nominated for election to the Paris Academy of Sciences in 1796, 1799, 1802, 1813, and 1828; but the closest he came to being elected was to place third to Lacroix in 1799.

Parseval's only publications seem to have been five memoirs presented to the Academy of Sciences. The second of these (dated 5 April 1799) contains the famous Parseval theorem, given here in his own notation:

If there are two series

$$A + Bf + Cf^2 + Ff^3 + \cdots = T$$

$$a + b\tfrac{1}{f} + c\tfrac{1}{f^2} + f\tfrac{1}{f^3} + \cdots = T'$$

as well as the respective sums T, T', then we obtain the sum of the series

$$Aa + Bb + Cc + Ff + \cdots = V$$

by multiplying T by T' and, in the new function $T \times T'$, substituting

$$\cos u + \sqrt{-1} \sin u$$

for the variable f, which will yield the function V'. Then for f substitute

$$\cos u - \sqrt{-1} \sin u$$

which will yield the new function V''. We then obtain

$$V = \frac{1}{u} \int \frac{V' + V''}{2} \, du,$$

u being made equal to $180°$ after integrating.

According to Parseval, the theorem was suggested by a method of summing special cases of series of products, presented by Euler in his *Institutiones calculi differentialis* of 1755. He believed the theorem to be self-evident, suggesting that the reader multiply the two series and recall that $(\cos u + i \sin u)^m = \cos mu + i \sin mu$, and gave a simple example that would "confirm its validity." He noted that it could be used only if the imaginaries in V' and V'' cancel one another, and he hoped to overcome this inconvenience. This hope was realized in a note appended to his next memoir (dated 5 July 1801), in which he gave a simplified version of the theorem. In modern notation the theorem states:

If, in the series $M = A + Bs + Cs^2 + \cdots$ and $m = a + bs + cs^2 + \cdots$, s is replaced by $\cos u + i \sin u$, and the real and imaginary parts are separated so that

$$M = P + Qi$$

and

$$m = p + qi,$$

then

$$\frac{2}{\pi} \int_0^\pi Pp \, du = 2Aa + Bb + Cc + \cdots.$$

(There is an error in Parseval's statement: the 2 in the right-hand side of the last equation is missing.)

In his memoirs, which were not published until 1806, Parseval applied his theorem to the solution of certain differential equations suggested by Lagrange and d'Alembert. The theorem first appeared in print in 1800, in Lacroix's *Traité des différences et des séries* (p. 377). By 1810 Delambre, in his *Rapport historique sur les progrès des sciences mathématiques depuis 1789, et sur leur état actuel,* could report that Prony had given, and published, lectures at the École Polytechnique taking Parseval's procedure into account and that Poisson had used a method dependent on an equation of this type. Since then dozens of equations have been called Parseval equations, although some only remotely resemble the original. Although Parseval's method involves trigonometric series, he never tried to find a general expression for the series coefficients; and hence he did not contribute directly to the theory of Fourier series. It should be noted that although Parseval viewed his theorem as a formula for summing infinite series, it was taken up at the end of the century as defining properties in more abstract treatments of analysis.

BIBLIOGRAPHY

I. ORIGINAL WORKS. Parseval's five memoirs appeared in *Mémoires présentés à l'Institut des Sciences, Lettres et Arts, par divers savans, et lus dans ses assemblées. Sciences mathématiques et physiques.* (*Savans étrangers.*), **1** (1806): "Mémoire sur la résolution des équations aux différences partielles linéaires du second ordre" (5 May 1798), 478–492; "Mémoire sur les séries et sur l'intégration complète d'une équation aux différences partielles linéaires du second ordre, à coefficiens constans" (5 Apr. 1799), 638–648; "Intégration générale et complète des équations de la propagation du son, l'air étant considéré avec ses trois dimensions" (5 July 1801), 379–398; "Intégration générale et complète de deux équations importantes dans la mécanique des fluides" (16 Aug. 1803), 524–545; and "Méthode générale pour sommer, par le moyen des intégrales définies, la suite donnée par le théorème de M. Lagrange, au moyen de laquelle il trouve une valeur qui satisfait à une équation algébrique ou transcendente" (7 May 1804), 567–586.

II. SECONDARY LITERATURE. A brief biography is in *Généalogies et souvenirs de famille; les Parseval et leurs alliances pendant trois siècles, 1594–1900,* I (Bergerac, 1901), 281–282. The memoirs are described in Niels Nielsen, *Géomètres français sous la Révolution* (Copenhagen, 1929), 192–194. The relation of Parseval's theorem to the work of Fourier is discussed in Ivor Grattan-Guinness, *Joseph Fourier, 1768–1830* (Cambridge, Mass., 1972), 238–241, written with J. R. Ravetz.

HUBERT C. KENNEDY

PARSONS, WILLIAM, Third Earl of Rosse (*b.* York, England, 17 June 1800; *d.* Monkstown, Ireland, 31 October 1867), *astronomy.*

William Parsons was the eldest son of Lawrence Parsons, second Earl of Rosse, and a descendant of the Sir William Parsons who had gone to Ireland in the sixteenth century. Prior to the death of his father, in 1841, he held the title Lord Oxmantown, under which style some of his scientific papers were published. (His own eldest son, Lawrence Parsons, held the same succession of titles; since he too was an astronomer, this has given rise to some confusion.) He received his early education privately at Birr Castle, the family seat, then in 1818 went for a year to Trinity College, Dublin. He next attended Magdalen College, Oxford, matriculating in 1821 and graduating with first-class honors in mathematics in 1822.

Since the nobility of his time took their responsibilities seriously, it was expected that Oxmantown would follow his father's example and take his place in the Irish government. As an undergraduate, in 1821, he was returned as a member of parliament for King's County, a seat that he held until 1834. He proved to be an able political economist and an effective committee member. He was appointed to further civil duties; in 1831 he was named lord lieutenant of County Offaly, in which Birr is situated, and in 1834 he became colonel of the local militia. In 1845, as the earl of Rosse, he was elected Irish representative peer and sat in the English House of Lords.

Rosse's scientific achievements were all the more remarkable in light of his activities as an administrator and public servant. His chief contributions were to astronomical instrumentation, particularly the design and construction of large telescopes. He early realized the need for instruments of greater aperture and light-grasp than were provided by William Herschel's forty-eight-inch aperture telescope of 1789; his own experiments, which he began in about 1826, were first concentrated on instruments incorporating the new optically excellent small-aperture Fraunhofer refractors. Having without success investigated the possibility of devising large fluid lenses, he was soon convinced that large apertures could be achieved only with reflectors. He therefore took up the search for an appropriate material for casting large mirrors and, after a number of experiments, decided to use an alloy of four parts of copper and one of tin. This alloy was both harder and more brittle than steel, it crystallized easily, and thus casting it was difficult. Rosse first tried making sectional mirrors composed of annular rings surrounding a central disk, all soldered to a brass disk having the same coefficient of expansion, but these proved ineffective in instruments of greater than eighteen-inch aperture.

Rosse was thus forced to develop a technique for casting solid disks. Having designed a mold ventilated so as to permit the mirror to cool evenly all over in an annealing oven, he finally achieved his aim. He completed a sectional thirty-six-inch speculum in 1839 and a superior solid mirror of the same size in 1840. In 1842 he cast the first seventy-two-inch disk, which, mounted in the meridian between two brick walls nearly sixty feet high, became known as the "Leviathan of Parsonstown" (Parsonstown being an old name of Birr). The telescope was completed in 1845; it had a focal length of fifty-four feet (with a nominal maximum magnification of 6,000), and a tube about seven feet in diameter. The mirror itself weighed nearly four tons, and its flexure under gravitation was controlled by twenty-seven felt-covered cast iron platforms. While it was not completely maneuverable, it was mounted so that a considerable portion of the sky was visible, and Rosse and his collaborator, the Reverend Thomas Romney Robinson, were able to utilize it to carry out, especially between 1848 and 1878, a number of important observations of nebulae.

With the new telescope Rosse and his co-workers were able to see hitherto unsuspected detail in many hundreds of nebulae, and to resolve many of these nebulae into stars. They abolished some of the existing distinctions (annular/planetary, for example) and added some new classes. Rosse himself was the first to detect the spiral nature of some nebulae, of which he published a number of fine drawings that clearly demonstrated the value of a large reflector of high optical quality.

In addition to overcoming the problems inherent in casting large solid mirrors (he eventually cast one of eighty-four inches), Rosse devised improvements in grinding and polishing techniques. Although he had initially believed that only hand finishing would be delicate enough to give a good conformation, he found this to be incorrect. He then designed an apparatus in which the mirror was rotated horizontally in a water bath (for constant temperature) beneath a grinding and polishing tool that could be moved in either a straight line or an ellipse of any eccentricity. The mirror could be tested *in situ* by observing its image in a watch dial fixed some fifty feet above it. The machine was driven by steam and was widely copied. Rosse also designed and executed a simple but effective clockwork drive for a large (eighteen-inch) equatorial. His interests extended further to the building of iron-armored ships, on which some correspondence is printed in his *Scientific Papers*. He took some of the earliest lunar photographs.

Rosse was married in 1836 to a Yorkshirewoman, Mary Field; they had four sons. He was president of the British Association at its 1843 meeting in Cork and president of the Royal Society from 1848 to 1854. In 1852 he served as chancellor of Trinity College, Dublin, and he was a member of the board of visitors of both Greenwich Observatory and Maynooth College. Following the potato famine of 1846 Rosse devoted the major part of the rents from his Irish properties to alleviating the poverty of the local inhabitants; he was held in great affection by his tenants, some 4,000 of whom attended his funeral.

BIBLIOGRAPHY

I. ORIGINAL WORKS. All of Rosse's scientific papers have been reprinted in Charles Parsons, ed., *The Scientific Papers of William Parsons, Third Earl of Rosse 1800–1867* (London, 1926). Rosse's chief publications are "An Account of Experiments on the Reflecting Telescope," in *Philosophical Transactions of the Royal Society*, **130** (1840), 503–528; "Observations of Some of the Nebulae," *ibid.*, **134** (1844), 321–323; "Observations of the Nebulae," *ibid.*, **140** (1850), 499–514; and "On the Construction of Specula of Six-Feet Aperture; and a Selection From the Observations of Nebulae Made With Them," *ibid.*, **151** (1861), 681–745.

II. SECONDARY LITERATURE. Two useful notes on Rosse are in *Proceedings of the Royal Society*, **16** (1868), xxxvi–xlii; and *Monthly Notices of the Royal Astronomical Society*, **29** (1869), 123–130. See also J. D. North, *The Measure of the Universe* (Oxford, 1965), esp. ch. 1.

J. D. NORTH
COLIN A. RONAN

PARTINGTON, JAMES RIDDICK (*b.* Bolton, Lancashire, England, 20 June 1886; *d.* Weaverham, Cheshire, England, 9 October 1965), *chemistry, dissemination of knowledge.*

Partington studied chemistry at the University of Manchester and, after a short period of research in organic chemistry under Arthur Lapworth, received an 1851 Exhibition scholarship. He worked under Nernst in Berlin on the specific heats of gases, continuing his research after his appointment as lecturer in chemistry at Manchester in 1913. During World War I he carried out investigations with E. K. Rideal for the Ministry of Munitions on the purification of water and the oxidation of nitrogen; he was subsequently knighted for this work. From 1919 to 1951 he was professor of chemistry at Queen Mary College, London University, where he continued his research on the specific heats of gases.

Remembered primarily as a historian of chemistry, Partington was gifted with an encyclopedic mind and a great facility for writing. His chief work, *A History of Chemistry*, is an outstanding accomplishment that surpasses any work on the subject since Hermann Kopp's *Geschichte der Chemie* (1843–1847). Its four volumes deal with the history of chemistry from antiquity to the present. Although at his best in describing the personalities and contributions of the great pioneers, Partington also included accounts of their less important contemporaries. His method consisted of summarizing the successive accomplishments of contributors to chemistry, rather than of organizing the history of the subject around a given sequence of themes or topics. The vein is biographical, not narrative, but the comprehensive accounts permit the reader to constitute his own narrative.

In 1965 Partington was awarded the Sarton Medal of the American History of Science Society during the Eleventh International Congress of the History of Science held in Warsaw and Cracow.

BIBLIOGRAPHY

I. ORIGINAL WORKS. Partington published several historical papers in *Annals of Science*. They include "Joan Baptista van Helmont," **1** (1936), 359–384; "Historical Studies on the Phlogiston Theory," **2** (1937), 361–404; **3** (1938), 1–58, 337–371; **4** (1939), 113–149, written with D. McKie; "The Origins of the Atomic Theory," **4** (1939), 245–282; and "Jeremias Benjamin Richter and the Law of Reciprocal Proportions," **7** (1951), 173–198; **9** (1953), 289–314.

His early books are *Higher Mathematics for Chemical Students* (London, 1911; 4th ed., 1931); *A Textbook of Thermodynamics* (London, 1913); *The Alkali Industry* (London, 1918); *A Textbook of Inorganic Chemistry for University Students* (London, 1921; 6th ed., 1950); *The Nitrogen Industry* (London, 1922), written with L. H. Parker; *Chemical Thermodynamics* (London, 1924; 4th ed., rev. and enl., 1950); *The Specific Heats of Gases* (London, 1924), written with W. G. Shilling; *Calculations in Physical Chemistry* (London–Glasgow, 1928), written with S. K. Tweedy; *The Composition of Water* (London, 1928); and *Everyday Chemistry* (London, 1929; 3rd ed., 1952).

Subsequent works are *A School Course of Chemistry* (London, 1930); *Origins and Development of Applied Chemistry* (London, 1935); *A Short History of Chemistry* (London, 1937; 3rd ed., rev. and enl., 1965); *A College Course of Inorganic Chemistry* (London, 1939); and *Intermediate Chemical Calculations* (London, 1939), written with K. Stratton.

His later writings are *General and Inorganic Chemistry for University Students* (London, 1946; 4th ed., 1966); *An Advanced Treatise on Physical Chemistry*, 4 vols. (London, 1949–1953); *A History of Greek Fire and Gunpowder* (Cambridge, 1960); *The Life and Work of William Higgins, Chemist* (*1763–1825*) (New York, 1960), written with T. S. Wheeler; and *A History of Chemistry*, 4 vols. (London–New York, 1961–1970).

II. SECONDARY LITERATURE. See the obituary notice in *The Times* (11 Oct. 1965), p. 12.

HAROLD HARTLEY

PASCAL, BLAISE (*b.* Clermont-Ferrand, Puy-de-Dôme, France, 19 June 1623; *d.* Paris, France, 19 August 1662), *mathematics, mechanical computation, physics, epistemology.*

Varied, original, and important, although often the subject of controversy, Pascal's scientific work was intimately linked with other aspects of his writings, with his personal life, and with the development of several areas of science. Consequently a proper understanding of his contribution requires a biographical framework offering as precise a chronology as possible.

Pascal's mother, Antoinette Begon, died when he was three; and the boy was brought up by his father, Étienne, who took complete charge of his education.

In 1631 the elder Pascal left Clermont and moved to Paris with his son and two daughters, Gilberte (1620–1687), who married Florin Périer in 1641, and Jacqueline (1625–1661), who entered the convent of Port-Royal in 1652.

The young Pascal began his scientific studies about 1635 with the reading of Euclid's *Elements*. His exceptional abilities, immediately and strikingly apparent, aroused general admiration. His sister Gilberte Périer left an account, more doting than objective, of her brother's life and, in particular, of his first contacts with mathematics. According to her, Pascal accompanied his father to the meetings of the "Académie Parisienne" soon after its founding by Mersenne in 1635 and played an important role in it from the first. This assertion, however, is not documented; and it appears more likely that it was at the beginning of 1639 that Pascal, not yet sixteen, began to participate in the activities of Mersenne's academy. In that year Girard Desargues had just published his *Brouillon project d'une atteinte aux événemens des rencontres du cone avec un plan*; but his originality, his highly personal style and vocabulary, and his refusal to use Cartesian algebraic symbols baffled most contemporary mathematicians. As the only one to appreciate the richness of this work, which laid the foundations of projective geometry and of a unified theory of conic sections, Pascal became Desargues's principal disciple in geometry.

Projective Geometry. Grasping the significance of Desargues's new conception of conics, Pascal adopted the basic ideas of the *Brouillon project:* the introduction of elements at infinity; the definition of a conic as any plane section of a cone with a circular base; the study of conics as perspectives of circles; and the involution determined on any straight line by a conic and the opposite sides of an inscribed quadrilateral. As early as June 1639 Pascal made his first great discovery, that of a property equivalent to the theorem now known as Pascal's "mystic hexagram"; according to it, the three points of intersection of the pairs of opposite sides of a hexagon inscribed in a conic are collinear.[1] He also soon saw the possibility of basing a comprehensive projective study of conics on this property. (The property amounts to an elegant formulation, in geometric language, of the condition under which six points of one plane belong to a single conic.) Next he wrote *Essay pour les coniques* (February 1640), a pamphlet, of which only a few copies were published [1].[2] A plan for further research, illustrated with statements of several typical propositions that he had already discovered, the *Essay* constituted the outline of a great treatise on conics that he had just conceived and begun to prepare.

Pascal seems to have made considerable progress by December 1640, having deduced from his theorem most of the propositions contained in the *Conics* of Apollonius.[3] Subsequently, however, he worked only intermittently on completing the treatise. Although Desargues and Mersenne alluded to the work in November 1642 and 1644, respectively, it was apparently not until March 1648 that Pascal obtained a purely geometric definitive general solution to the celebrated problem of Pappus, which had furnished Descartes with the principal example for illustrating the power of his new analytic geometry (1637).[4] Pascal's success marked an important step in the elaboration of his treatise on conics, for it demonstrated that in this domain projective geometry might prove as effective as the Cartesian analytic methods. Pascal therefore reserved the sixth, and final, section of his treatise, "Des lieux solides" (geometric loci composed of conics), for this problem.

In 1654 Pascal indicated that he had nearly completed the treatise [12], conceived "on the basis of a single proposition"—a work for which he had "had the idea before reaching the age of sixteen" and which he then "constructed and put in order." He also mentioned some special geometric problems to which his projective method could usefully be applied: circles or spheres defined by three or four conditions; conics determined by five elements (points or tangents); geometric loci composed of straight lines, circles, or conics; and a general method of perspective.

Pascal made no further mention of this treatise, which was never published. It seems that only Leibniz saw it in manuscript, and the most precise details known about the work were provided by him. In a letter [23] of 30 August 1676 to E. Périer, one of Pascal's heirs, Leibniz stated that the work merited publication and mentioned a number of points concerning its contents, which he divided into six parts: (1) the projective generation of conics; (2) the definition and properties of the "mystic hexagram"—Pascal's theorem and its applications; (3) the projective theory of poles and polars and of centers and diameters; (4) various properties related to the classic definitions of conics on the basis of their axes and foci; (5) *contacts coniques*, the construction of conics defined by five elements (points or tangents); and (6) solid loci (the problem of Pappus). Besides reading notes on a number of passages of Pascal's treatise [15], Leibniz's papers preserve the text of the first part, "Generatio conisectionum" [14].

The content and inspiration of this introductory chapter are readily apparent from the full title: "The Generation of Conics, Tangents, and Secants; or the Projection of the Circumference, Tangents, and Secants of the Circle for Every Position of the Eye and of the Plane of the Figure." The text presents in an exceptionally elegant form the basic ideas of projective geometry already set forth, in a much less explicit fashion, in Desargues's *Brouillon project*.[5] Although these few elements of Pascal's treatise preserved by Leibniz do not provide a complete picture of its contents, they are sufficient to show the richness and clarity of Pascal's conceptions once he had become fully aware of the power of projective methods. It is reasonable to assume that publication of this work would have hastened the development of projective geometry, impeded until then by the obscurity of Desargues's writings and by their limited availability. Despite the efforts of Philippe de la Hire,[6] the ultimate disappearance of the treatise on conics and the temporary eclipse of both *Essay pour les coniques* (which was not republished until 1779) and Desargues's *Brouillon project* (rediscovered in 1864) hindered the progress of projective geometry. It was not truly developed until the nineteenth century, in the work of Poncelet and his successors. Poncelet, in fact, was one of the first to draw attention to the importance of Pascal's contribution in this area.

Pascal was soon obliged to suspend the contact with the "Académie Parisienne" that had encouraged the precocious flowering of his mathematical abilities. In 1640 he and his sisters joined their father, who since the beginning of that year had been living in Rouen as a royal tax official. From the end of 1640 until 1647 Pascal made only brief and occasional visits to Paris, and no information has survived concerning his scientific activity at the beginning of this long provincial interlude. Moreover, in 1641 he began to suffer from problems of health that several times forced him to give up all activity. From 1642 he pursued his geometric research in a more or less regular fashion; but he began to take an interest in a new problem, to the solution of which he made a major contribution.

Mechanical Computation. Anxious to assist his father, whose duties entailed a great deal of accounting, Pascal sought to mechanize the two elementary operations of arithmetic, addition and subtraction. Toward the end of 1642 he began a project of designing a machine that would reduce these operations to the simple movements of gears. Having solved the theoretical problem of mechanizing computation, it remained for him to produce such a machine that would be convenient, rapid, dependable, and easy to operate. The actual construction, however, required relatively complicated wheel arrangements and proved to be extremely difficult with the rudimentary and inaccurate techniques available. In this venture Pascal displayed remarkable practical sense, great concern for efficiency, and undeniable stubbornness. Supervising a team of workers, he constructed the first model within a few months but, judging it unsatisfactory, he decided to modify and improve it. The considerable problems he encountered soon discouraged him and caused him to interrupt his project. At the beginning of 1644 encouragement from several people, including the chancellor of France, Pierre Séguier, induced Pascal to resume the development of his "arithmetic machine." After having constructed, in his words, "more than fifty models, all different," he finally produced the definitive model in 1645. He himself organized the manufacture and sale of the machine.

This activity is the context of Pascal's second publication, an eighteen-page pamphlet [2] consisting of a "Lettre dédicatoire" to Séguier and a report on the calculating machine—its purpose, operating principles, capabilities, and the circumstances of its construction ("Avis nécessaire à ceux qui auront curiosité de voir ladite machine et de s'en servir"). The text concludes with the announcement that the machine can be seen in operation and purchased at the residence of Roberval. Pascal's first work of this scope, the pamphlet is both a valuable source of information on the guiding ideas of his project and an important document on his personality and style.

It is difficult to estimate the success achieved by Pascal's computing machine, the first of its kind to be offered for sale—an earlier one designed by W. Schickard (1623) seems to have reached only the prototype stage. Although its mechanism was quite complicated, Pascal's machine functioned in a relatively simple fashion—at least for the two operations to which it was actually applied.[7] Its high price, however, limited its sale and rendered it more a curiosity than a useful device. It is not known how many machines were built and sold; seven still exist in public and private collections.[8] For a few years Pascal was actively involved in their manufacture and distribution, for which he had obtained a monopoly by royal decree (22 May 1649) [22]. In 1652 he demonstrated his machine during a lecture before fashionable audience and presented one to Queen Christina of Sweden. For some time, however, he had been directing his attention to problems of a very different kind.

Raised in a Christian milieu, Pascal had been a practicing Catholic throughout his youth but had never given any special consideration to problems of faith. Early in 1646, however, he became converted to the austere and demanding doctrine of Saint-Cyran (Jean Duvergier de Hauranne), whose views were close to those of the Jansenists. This event profoundly marked the rest of Pascal's life. The intransigence of his new convictions was underscored at Rouen between February and April 1647, when Pascal and two friends denounced certain bold theological positions defended by Jacques Forton de Saint-Ange. This change in attitude did not, however, prevent Pascal from embarking on a new phase of scientific activity.

Fluid Statics and the Problem of the Vacuum. To understand and evaluate Pascal's work in the statics of gases and liquids, it is necessary to trace the origins of the subject and to establish a precise chronology. In his *Discorsi* (1638) Galileo had noted that a suction pump cannot raise water to more than a certain height, approximately ten meters. This observation, which seemed to contradict the Aristotelian theory that nature abhors a vacuum, was experimentally verified about 1641 by R. Maggiotti and G. Berti. V. Viviani and E. Torricelli modified the experiment by substituting mercury for water, thereby reducing the height of the column to about seventy-six centimeters. Torricelli announced the successful execution of this experiment in two letters to M. Ricci of 11 and 28 June 1644. Describing the experiment in detail, he gave a correct interpretation of it based on the weight of the external column of air and the reality of the existence of the vacuum.[9] Mersenne, informed of the work of the Italian scientists, attempted unsuccessfully to repeat the experiment, which for some time fell into neglect.

In October 1646 Mersenne's friend P. Petit, who was passing through Rouen, repeated the experiment with the assistance of Étienne and Blaise Pascal. At the end of November 1646 Petit described the event in a letter to Pierre Chanut. Meanwhile, Pascal, seeking to arrive at firm conclusions, had repeated the experiment in various forms, asserting that the results contradicted the doctrine of the *horror vacui*. Profiting from the existence at Rouen of an excellent glassworks, Pascal conducted a series of further experiments in January and February 1647. He repeated Torricelli's experiment with water and wine, using tubes of different shapes, some as long as twelve meters, affixed to the masts of ships. These experiments became known in Paris in the spring of 1647. Gassendi wrote the first commentary on them, and Mersenne and Roberval undertook their own experiments. The first printed account of the entire group of Pascal's experiments was *Discours sur le vide* by P. Guiffart, of Rouen, written in April 1647 and published in August of that year. Just as it was published, word reached Paris that a barometric experiment had been conducted at Warsaw in July 1647 by V. Magni, who implicitly claimed priority. Roberval responded on 22 September with a Latin *Narratio* (published at Warsaw in December), in which he established the priority of Torricelli's and Pascal's experiments and revealed new details concerning the latter.

Pascal soon intervened directly in the debate. During the summer of 1647 his health had deteriorated; and he left Rouen with his sister Jacqueline to move to Paris, where their father joined them a year later. Henceforth, Pascal maintained contacts both with the Jansenists of Port-Royal and with the secular intellectuals of Paris, who were greatly interested in the interpretation of the experiments with the vacuum. He had two discussions on this topic with Descartes (23 and 24 September), who may have suggested that he compare barometric observations made at different altitudes.[10] This idea was also proposed by Mersenne in his *Reflexiones physico-mathematicae* (beginning of October 1647).[11] At this time Pascal wrote a report of his experiments at Rouen, a thirty-two-page pamphlet published in October 1647 as *Expériences nouvelles touchant le vide* [3]. In this "abridgment" of a larger work that he planned to write, Pascal admitted that his initial inspiration derived from the Italian barometric experiment and stated that his primary goal was to combat the idea of the impossibility of the vacuum. From his experiments he had deduced the existence of an apparent vacuum, but he asserted that the existence of an absolute vacuum was still an unconfirmed hypothesis. Consequently his pamphlet makes no reference to the explanation of the barometric experiment by means of the weight of the air, proposed by Torricelli in 1644.[12] According to his sister Jacqueline, however, Pascal had been a firm proponent of this view from 1647.[13] In any case his concern was to convince his readers; he therefore proceeded cautiously, affirming only what had been irrefutably demonstrated by experiment.

Despite his moderate position, Pascal's rejection of the theory of the impossibility of the vacuum involved him in vigorous debate. With the publication of the *Expériences nouvelles*, a friend of Descartes's, the Jesuit Estienne Noël, declared in a letter to Pascal that the upper portion of Torricelli's tube was filled with a purified air that had entered through the pores of the glass.[14] In a dazzling reply (29 October 1647) [4] Pascal clearly set forth the rules of his scientific method and vigorously upheld his position. Several days later Noël reaffirmed the essence of his views but expressed a desire to end the dispute.[15] It was indirectly resumed, however, after Noël published a new and violent critique of the *Expériences nouvelles*.[16] In a letter to his friend F. Le Pailleur [5], Pascal refuted Noël's second letter and criticized his recent publication. In April 1648 Étienne Pascal entered the debate against Noël.[17] The dispute soon ended, however,

when Noël published a much more moderate Latin version of his short treatise.[18]

During this controversy scientists in Paris had become interested in the problem of the vacuum, devoting many experiments to it and proposing a number of hypotheses to explain it. Having participated in discussions on the topic, Pascal conceived one of the variants of the famous experiment of the vacuum within the vacuum, designed to verify the hypothesis of the column of air.[19] He seems, however, to have expected a still better confirmation of the hypothesis from a program of simultaneous barometric observations at different altitudes (at Clermont-Ferrand and at the summit of Puy de Dôme), the execution of which he entrusted to his brother-in-law, Périer. One of these observations, now known as the "Puy de Dôme experiment," was carried out on 19 September 1648. Pascal immediately published a detailed, twenty-page account of it, *Récit de la grande expérience de l'équilibre des liqueurs . . .* [6], consisting principally of Périer's letter and report. In a short introduction he presented the experiment as the direct consequence of his *Expériences nouvelles*, and the text of a letter of 15 November 1647 to Périer, in which he explained the goal of the experiment and the principle on which it was based. He concluded by pointing out his analogous experiment at the Tour St. Jacques in Paris and by announcing his conversion to the principles of the existence of the absolute vacuum and of the weight of air.

The *Récit*, which marks an important phase of Pascal's research on the vacuum, gave rise to two heated controversies.[20] The first arose at the end of the seventeenth century, when several authors denied Pascal's priority with regard to the basic principle of the Puy de Dôme experiment. This question, however, is of only secondary importance. While it appears that the principle was formulated simultaneously—on the basis of different presuppositions—by Pascal, Descartes, and Mersenne, only Pascal tested it and integrated it into an exceptionally cogent chain of reasoning.

The second controversy was launched in 1906–1907 by F. Mathieu, who challenged both Pascal's scientific originality and his honesty. He accused Pascal of having fabricated the letter to Périer of 15 November 1647 after completion of the event, in order to take credit for the experiments of the vacuum within the vacuum and of Puy de Dôme. Although the heated debate that ensued did not produce any unanimously accepted conclusions, it did stimulate research that brought to light many unpublished documents. In an assessment of the question J. Mesnard, after examining the arguments and clarifying many points, suggests that Pascal probably did send the contested letter to Périer on 15 November 1647 but may have altered the text slightly for publication. This compromise judgment is probably close to the truth.

At the beginning of 1649 Périer, following Pascal's instructions, began an uninterrupted series of barometric observations designed to ascertain the possible relationship between the height of a column of mercury at a given location and the state of the atmosphere. The *expérience continuelle*, which was a forerunner of the use of the barometer as an instrument in weather forecasting, lasted until March 1651 and was supplemented by parallel observations made at Paris and Stockholm.[21] Pascal continued working on a major treatise on the vacuum; but only a few fragments, dating from 1651, have survived: a draft of a preface [7] on the relationships between reason and authority and between science and religion, and two short passages published by Périer in 1663.[22] In June 1651 a Jesuit accused Pascal of claiming credit for Torricelli's experiment. In two letters [9, 10], of which only the first was printed, Pascal recounted—with several serious errors—the history of that experiment and laid claim to the idea of the Puy de Dôme experiment.

Pascal soon put aside his great treatise on the vacuum in order to write a shorter but more synthetic work. Divided into two closely related parts, this work is devoted to the laws of hydrostatics and to the demonstration and description of the various effects of the weight of air. It was completed about the beginning of 1654 and marked the end of Pascal's active research in physics. It was published posthumously by Périer, along with several appendices, in 1663 as *Traités de l'équilibre des liqueurs et de la pesanteur de la masse de l'air* . . . [13]. The fruit of several years of observations, experiments, and reflection, it is a remarkable synthesis of new knowledge and theories elaborated since the work of Stevin and Galileo. The highly persuasive *Traités*, assembling and coordinating earlier results and recent discoveries, are characterized above all by their rigorous experimental method and by the categorical rejection of Scholasticism. In hydrostatics Pascal continued the investigations of Stevin, Galileo, Torricelli, and Mersenne. He clearly set forth the basic principles of the science, although he did not fully succeed in demonstrating them satisfactorily. In particular he provided a lucid account of the fundamental concept of pressure.

The untoward delay in the publication of the *Traités* obviously reduced its timeliness; for in the meanwhile the study of the weight of air and the existence of the vacuum had been profoundly affected by the work of Otto von Guericke and Robert Boyle.[23] In this area, in fact, the *Traités* essentially systematized, refined, and

developed experiments, concepts, and theories that, for the most part, had already been discussed in the *Expériences nouvelles* and the *Récit*. Pascal's influence, therefore, must be measured as much by the effect of these preliminary publications and the contemporary writings of Mersenne and Pecquet, which reflect his thinking, as by the posthumous *Traités*.[24] This influence was certainly considerable, for it partially conditioned all subsequent research on the subject; but it cannot easily be separated from that of, for instance, Roberval and Auzout, who participated in the rapid progress of research on the vacuum at Paris in 1647 and 1648. Nevertheless, for their synthetic treatment of the subject, clarity, and rigor, the *Traités* are indisputably a classic of seventeenth-century science.

Although from October 1646 Pascal had been deeply interested in problems of the vacuum, he was often impeded in his research by poor health and by religious concerns. The death of his father in September 1651 and the entry of his sister Jacqueline into the convent of Port-Royal in January 1652 marked a turning point in his life. In better health and less preoccupied with religious problems, he pursued his scientific work while leading a more worldly existence. Beginning in the summer of 1653 he frequently visited the duke of Roannez. Through the duke he met the Chevalier de Méré, who introduced him to the problems of games of chance. At the beginning of 1654, in an address [12] to the Académie Parisienne de Mathématique, which was directed by F. Le Pailleur, Pascal listed the works on geometry, arithmetic, and physics that he had already completed or begun writing and mentioned, in particular, his recent research on the division of stakes.[25]

Calculus of Probabilities. The Arithmetical Triangle. The year 1654 was exceptionally fruitful for Pascal. He not only did the last refining of his treatises on geometry and physics but also conducted his principal studies on arithmetic, combinatorial analysis, and the calculus of probability. This work can be seen in his correspondence with Fermat [16] and his *Traité du triangle arithmétique* [17].

Pascal's correspondence with Fermat between July and October 1654 marks the beginning of the calculus of probability. Their discussion focused on two main problems. The first concerned the probability that a player will obtain a certain face of the die in a given number of throws. The second, more complex, consisted in determining, for any game involving several players, the portion of the stakes to be returned to each player if the game is interrupted. Fermat succeeded in solving these problems by using only combinatorial analysis. Pascal, on the other hand, seems gradually to have discovered the advantages of the systematic application of reasoning by recursion. This recourse to mathematical induction, however, is not clearly evident until the final section of the *Traité du triangle arithmétique*, of which Fermat received a copy before 29 August 1654.

The *Traité* was printed in 1654 but was not distributed until 1665. Composed partly in French and partly in Latin, it has a complex structure; but the discovery of a preliminary Latin version of the first part makes it easier to trace its genesis.[26] Although the principle of the arithmetical triangle was already known,[27] Pascal was the first to make a comprehensive study of it. He derived from it the greatest number of applications, the most important and original of which are related to combinatorial analysis and especially to the study of the problems of stakes. Yet it is impossible to appreciate Pascal's contribution if it is considered solely from the perspective of combinatorial analysis and the calculus of probability. Several modern authors have shown that Pascal's letters to Fermat and the *Traité du triangle arithmétique* can be fully understood only when they are seen as preliminary steps toward a theory of decision.[28]

As E. Coumet has pointed out, Pascal's concern, beyond the purely mathematical aspect of the problems, was to link decisions and uncertain events. His aim was not to define the mathematical status of the concept of probability—a term that he did not employ —but to solve the problem of dividing stakes. This innovative effort must therefore be viewed in the context of the discussions conducted by jurists, theologians, and moralists in the sixteenth and seventeenth centuries on the implications of chance in the most varied circumstances of individual and community life. Unrecognized until recently, this aspect of Pascal's creative work is revealed in its full significance in the light of recent ideas on game theory and decision theory.

On the other hand, Pascal's research on combinatorial analysis now appears much less original. Considered in the context of the vigorous current of ideas on the subject in the sixteenth and seventeenth centuries, it is noteworthy less for the originality of its results than for the clarity, generality, and rigor with which they are presented.[29] Pascal's contribution to the calculus of probability is much more direct and indisputable: indeed, with Fermat he laid the earliest foundations of this discipline.[30] The *Traité du triangle arithmétique* contains only scattered remarks on the subject; in addition, only a part of the correspondence with Fermat [16] has been preserved, and its late publication (1679 and 1779) certainly reduced its direct influence. Fortunately, through Huygens the original contribution of Pascal and Fermat in this area became quickly known. During a stay in Paris in 1655 Huygens was informed in detail of their work, and he recast their ideas in the light of his own conceptions in his *Tractatus de ratiociniis in aleae ludo*. With its publication in 1657 the essential elements of the new science were revealed.[31] Nevertheless, the calculus of probability did not experience further development until the beginning of the eighteenth century, with Jakob I Bernoulli, P. R. de Montmort, and A. de Moivre.

Unsatisfied by his worldly life and intense scientific activity, Pascal was again drawn to religious concerns. Following a second conversion, during the famous "nuit de feu" of 23 November 1654, he abandoned his scientific work in order to devote himself to meditation and religious activity and to assist the Jansenists in their battle against many enemies, particularly the Jesuits. Working anonymously, between 13 January 1656 and 24 March 1657 Pascal composed the eighteen *Lettres provinciales* with the assistance of his friends from Port-Royal, Antoine Arnauld and Pierre Nicole. A masterpiece of polemic, this eminent contribution to the debate then agitating Christian doctrine was first published as a collection in 1657 under the pseudonym Louis de Montalte. Although Pascal produced other polemical writings, he worked primarily on preparing a defense of Christianity directed to nonbelievers. This unfinished project was the source of several posthumously published writings, the most important being the *Pensées*, published in 1670. The object of numerous commentaries and penetrating critical studies, this basic work fully displays Pascal's outstanding philosophical and literary talents.

Although concerned above all with meditation and religious activities during this period, Pascal was not totally estranged from scientific life thanks to his friends, particularly Carcavi. Around 1657, at the request of Arnauld, Pascal prepared a work entitled *Éléments de géométrie*, of which there remain only a few passages concerning methodology: the brief "Introduction à la géométrie," preserved among Leibniz's papers [18]; and two fragments, "De l'esprit géométrique" and "De l'art de persuader" [19]. Finally, in 1658 Pascal undertook a brilliant, if short-lived, series of scientific studies.

The Calculus of Indivisibles and the Study of Infinitesimal Problems. During 1658 and the first months of 1659 Pascal devoted most of his time to perfecting the "theory of indivisibles," a forerunner of the methods of integral calculus. This new theory enabled him to study problems involving infinitesimals: calculations of areas and volumes, determinations of centers of gravity, and rectifications of curves.

From the end of the sixteenth century many authors, including Stevin (1586), L. Valerio (1604), and Kepler

(1609 and 1615), had tried to solve these fundamental problems by using simpler and more intuitive methods than that of Archimedes, which was considered a model of virtually unattainable rigor.[32] The publication in 1635 of Cavalieri's *Geometria* marked the debut of the method of indivisibles;[33] its principles, presentation, and applications were discussed and elaborated in the later writings of Cavalieri (1647 and 1653) and in those of Galileo (1638), Torricelli (1644), Guldin (1635–1641), Gregory of Saint-Vincent (1647), and A. Tacquet (1651). (The research of Fermat and Roberval on this topic remained unpublished.)[34] The method, which assumed various forms, constituted the initial phase of development of the basic procedures of integral calculus, with the exception of the algorithm.

Pascal first referred to the method of indivisibles in a work on arithmetic of 1654, "Potestatum numericarum summa."[35] He observed that the results concerning the summation of numerical powers made possible the solution of certain quadrature problems. As an example he stated a known result concerning the integral of x^n for whole n, $\int_0^a x^n \, dx = a^{n+1}/(n+1)$, in modern notation.[36] This arithmetical interpretation of the theory of indivisibles permitted Pascal to give a sufficiently precise idea of the order of infinitude[37] and to establish the natural relationship between "la mesure d'une grandeur continue" and "la sommation des puissances numériques." In the fragment "De l'esprit géométrique" [19], composed in 1657, he returned to the notion of the indivisible in order to specify its relationship to the notions of the infinitely small and of the infinitely large and to refute the most widespread errors concerning it.

At the beginning of 1658 Pascal believed that he had perfected the calculus of indivisibles by refining his method and broadening its field of application. Persuaded that in this manner he had discovered the solution to several infinitesimal problems relating to the cycloid or *roulette*, he decided to challenge other mathematicians to solve these problems.[38] Although rather complicated, the history of this contest is worth a brief recounting because of its important repercussions during a crucial phase in the birth of infinitesimal calculus. In an unsigned circular distributed in June 1658, Pascal stated the conditions of the contest and set its closing date at 1 October [20a]. In further unsigned circulars and pamphlets [20], issued between July 1658 and January 1659, he modified or specified certain of the conditions and announced the results. He also responded to the criticism of some participants and sought to demonstrate the importance and the originality of his own solutions.

Most of the leading mathematicians of the time followed the contest with interest, either as participants (A. de Lalouvère and J. Wallis) or as spectators working on one or several of the questions proposed by Pascal or on related problems—as did R. F. de Sluse, M. Ricci, Huygens, and Wren.[39] Their solutions having been judged incomplete and marred by errors, Lalouvère and Wallis were eliminated. Their heated reactions to this decision were partially justified by the bias it displayed and the commentaries that accompanied it.[40] This bias, which also appears in certain passages of Pascal's *Histoire de la roulette* [20b, 20d], was the source of intense polemics concerning, in particular, the importance of Torricelli's original contribution.[41] At the end of the contest Pascal published his own solutions to some of the original problems and to certain problems that had been added in the meantime. In December 1658 and January 1659 he brought out, under the pseudonym A. Dettonville, four letters setting forth the principles of his method and its applications to various problems concerning the cycloid, as well as to such questions as the quadrature of surfaces, cubature of volumes, determination of centers of gravity, and rectification of curved lines. In February 1659 these four pamphlets were collected in *Lettres de A. Dettonville contenant quelques-unes de ses inventions de géométrie . . .* [21].

This publication of some 120 pages has a very complex structure. The first of the *Lettres* consists of five sections with independent paginations, and the three others appear in inverse order of their composition.[42] Thus only by returning to the original order is it possible to understand the logical sequence of the whole, follow the development of Pascal's method, and appreciate the influence on it of the new information he received and of his progress in mastering infinitesimal problems.[43]

When he began the contest, Pascal knew of the methods and the chief results of Stevin, Cavalieri, Torricelli, Gregory of Saint-Vincent, and Tacquet; but he was not familiar with the bulk of the unpublished research of Roberval and Fermat. Apart from this information, and in addition to the arithmetical procedures that he applied, starting in 1654, to the solution of problems of the calculus of indivisibles, Pascal possessed a new method inspired by Archimedes. It was elaborated on a geometric foundation, its point of departure being the principle of the balance and the concepts of static moment and center of gravity. Pascal learned of the importance of the results obtained by Fermat and Roberval—notably in the study of the cycloid—at the time he issued his first circular. This information led him to modify the subject of the contest and to develop his own method further. Similarly, in August 1658, when he was informed of the result of the rectification of the cycloid, Pascal extended rectification to other arcs of curves and then undertook to determine the center of gravity of these arcs, as well as the area and center of gravity of the surfaces of revolution generated by their revolution about an axis. Consequently the *Lettres* present a method that is in continual development, appearing increasingly complex as it becomes more precise and more firmly based. The most notable characteristics of this work, which remained unfinished, are the importance accorded to the determination of centers of gravity, the crucial role of triangular sums and statical considerations, its stylistic rigor and elegance, and the use of a clear and precise geometric language that partially compensates for the absence of algebraic symbolism.[44] Among outstanding contributions of the work are the discovery of the equality of curvature of the generalized cycloid and the ellipse; the deepening of the concept of the indivisible; a first step toward the concept of the definite integral and the determination of its fundamental properties; and the indirect recourse to certain methods of calculation, such as integration by parts.

Assimilated and exploited by Pascal's successors, these innovations contributed to the elaboration of infinitesimal methods. His most productive contribution, however, appears to have been his implicit use of the characteristic triangle.[45] Indeed, Leibniz stated that Pascal's writings on the characteristic triangle were an especially fruitful stimulus for him.[46] This testimony from one of the creators of infinitesimal calculus indicates that Pascal's work marked an important stage in the transition from the calculus of indivisibles to integral calculus. Pascal was unable, however, to transcend the overly specific nature of his conceptions. Neither could he utilize to full effect the power and generality of the underlying methods nor develop the results he obtained. This partial failure can be attributed to two causes. First, his systematic refusal to adopt Cartesian algebraic symbolism prevented him from realizing the necessity of the formalization that permitted Leibniz to create the integral calculus. Second, his preoccupation with mystic concerns led him to interrupt his research only a short time after he had begun it.

Early in 1659 Pascal again fell gravely ill and abandoned almost all his intellectual undertakings in order to devote himself to prayer and to charitable works.[47] In 1661 his desire for solitude increased after the death of his sister Jacqueline and a dispute with his friends from Port-Royal. Paradoxically, it was at this time that Pascal participated in a project to establish a public transportation system in Paris, in the form of carriages charging five *sols* per ride—a

scheme that went into effect in 1662.[48] Some writers have asserted that Pascal's doctrinal intransigence had diminished in this period to such a point that at the moment of his death he renounced his Jansenist convictions, but most of the evidence does not support this interpretation.

Pascal was a complex person whose pride constantly contended with a profound desire to submit to a rigorous, Augustinian insistence on self-denial. An exceptionally gifted polemicist, moralist, and writer, he was also a scientist anxious to help solve the major problems of his day. He did not, it is true, produce a body of work distinguished by profound creativity, on the model of such contemporaries as Descartes, Fermat, and Torricelli. Still, he was able to elucidate and systematize several rapidly developing fields of science (projective geometry, the calculus of probability, infinitesimal calculus, fluid statics, and scientific methodology) and to make major original contributions to them. In light of this manifold achievement Pascal, a leading opponent of Descartes, was undoubtedly one of the outstanding scientists of the mid-seventeenth century.

NOTES

1. The first known formulation of this theorem was as lemma 1 of *Essay pour les coniques*. It clearly differs from the modern statement by not referring explicitly to the inscribed hexagon and by apparently being limited to the case of the circle (even though the corresponding figure illustrates the case of the ellipse). According to remarks made by Leibniz, it seems that this theorem, in its hexagonal formulation and under the name "hexagramme mystique," held a central place in Pascal's treatise on conics, now lost. The fact that the *Essay pour les coniques* contains only statements without demonstrations makes it impossible to ascertain the precise role Pascal assigned to this theorem in 1640.

2. The numbers in square brackets refer to the corresponding works listed in sec. 1 of the bibliography. For a more detailed study of the *Essay*, see R. Taton, in *Revue d'histoire des sciences*, **8** (1955), 1–18, and in *L'oeuvre scientifique de Pascal* (Paris, 1964), 21–29; and J. Mesnard, ed., *Blaise Pascal. Oeuvres complètes*, II (1971), 220–225 (cited below as Mesnard).

3. See Mersenne's letter to Theodore Haak of 18 Nov. 1640, in Mesnard, II, 239.

4. On the references by Desargues and Mersenne, see *ibid.*, 279–280, 299. On the problem of Pappus, see Mersenne's letter to Constantijn Huygens of 17 Mar. 1648 in C. Huygens, *Oeuvres complètes de Christiaan Huygens, publiées par la Société Hollandaise des Sciences*, II (1888), 33, and in Mesnard, II, 577–578. On Descartes, see Taton, in *L'oeuvre scientifique de Pascal*, 45–50; and M. S. Mahoney, "Descartes: Mathematics and Physics," in *DSB*, IV, 56.

5. See Taton, in *L'oeuvre scientifique . . .*, 55–59 (for "Generatio conisectionum") and 53–72 (for the treatise as a whole). See also his "Desargues," in *DSB*, IV, 46–51.

6. See Taton, "La Hire, Philippe de," in *DSB*, VII, 576–578.

7. See D. Diderot, "Arithmétique (Machine)," in *Encyclopédie*, I (1751), 680–684.

8. See J. Payen, in *L'oeuvre scientifique de Pascal*, 229–247.

9. See, in particular, C. De Waard, *L'expérience barométrique, ses antécédents et ses explications* (Thouars, 1936), 110–123; M. Gliozzi, "Origine e sviluppi dell'esperienza torricelliana," in *Opere de Evangelista Torricelli*, G. Loria and G. Vassura, eds., IV (Faenza, 1944), 231–294; and W. E. K. Middleton, *The History of the Barometer* (London, 1964).

10. Jacqueline Pascal gave some details of these meetings in a letter to her sister Gilberte of 25 Sept. 1647. (See Mesnard, II, 478–482.) In a letter to Mersenne of 13 Dec. 1647 and in two letters to Carcavi of 11 June and 17 Aug. 1649 (see *ibid.*, 548–550, 655–658, 716–719) Descartes stated that he had suggested this idea, which was the origin of the celebrated Puy de Dôme experiment of 19 Sept. 1648, to Pascal.

11. See *ibid.*, 483–489.

12. Torricelli held that the space above the column of mercury was empty. Considering the horizontal plane determined by the exterior level of the mercury, he asserted that the weight of the column of mercury equaled the weight of a column of air of the same base, which implied simultaneously the existence of the vacuum, the weight of the air, and the finiteness of the terrestrial atmosphere. In 1651 Pascal admitted that he was aware of Torricelli's explanation as early as 1647 (see *ibid.*, 812), but he insisted that at that time the explanation was only a conjecture; it had yet to be verified by experiment, and for this reason he undertook the experiment of Puy de Dôme.

13. Letter to Gilberte Pascal of 25 Sept. 1647 (see *ibid.*, 482).

14. See *ibid.*, 513–518.

15. See *ibid.*, 528–540.

16. It was a brief work with the picturesque title *Le plein du vide* (Paris, 1648); see Mesnard, II, 556–558. This work was reprinted by Bossut in *Oeuvres de Blaise Pascal*, C. Bossut, ed., IV (The Hague, 1779), 108–146.

17. See Mesnard, II, 584–602.

18. E. Noël, *Plenum experimentis novis confirmatum* (Paris, 1648); see Mesnard, II, 585.

19. This experiment is mentioned without details in Pascal's *Récit . . .* (see Mesnard, II, 678). The reality of the experiment is confirmed by the quite precise description of it that Noël gave in his *Gravitas comparata* (Paris, 1648); on this point see Mesnard, II, 635–636, which presents the Latin text, a French translation, and an explanatory diagram derived from an earlier study by P. Thirion. The principle of this experiment consists of conducting Torricelli's experiment in an environment where the pressure can be varied from atmospheric pressure to zero. Other variants were devised at almost the same time by Roberval (Mesnard, II, 637–639) and by Auzout (*ibid.*, 767–771). A fourth variant, easier to carry out in practice, is described in Pascal's *Traités de l'équilibre des liqueurs et de la pesanteur de la masse de l'air . . .* (*ibid.*, 1086–1088).

20. See *ibid.*, 653–676.

21. F. Périer published an account of them in 1663 as an appendix to Pascal's *Traités de l'équilibre . . .* (pp. 195–209); see Mesnard, II, 738–745. The fact that the first observations made at Stockholm were carried out by Descartes appears to indicate that he had become reconciled with Pascal.

22. The preface was not published until 1779, when it appeared under the title "De l'autorité en matière de philosophie" (Bossut, II, 1–12). The passages published by Périer appear at the end of Pascal's *Traités de l'équilibre . . .* (pp. 141–163).

23. See F. Krafft, "Guericke," in *DSB*, V, 574–576; and C. Webster, "The Discovery of Boyle's Law and the Concept of the Elasticity of Air in the Seventeenth Century," in *Archive for History of the Exact Sciences*, **2** (1965), 441–502, esp. 447–458.

24. M. Mersenne, *Reflectiones physico-mathematicae* (Paris, 1647); and J. Pecquet, *Experimenta nova anatomica* (Paris, 1651). To these works should be added publications by Noël, already cited, as well as those of Roberval and of V. Magni (see Webster, *op. cit.*), and, above all, the corre-spondence of scientists from Italy, France, England, Poland, and other European countries.

25. The word used in French to designate this problem, *parti*, is the past participle (considered as the noun form) of the verb *partir*, understood in the sense of "to share." The problem consists in finding, for a game interrupted before the end, the way of dividing the stakes among the players in proportion to their chances of winning at the time of interruption.

26. See Mesnard, II, which provides an introduction to the texts (pp. 1166–1175) and the texts themselves, both of the first printing, in Latin with French translation (pp. 1176–1286), and of the second, with translation of the Latin passages (pp. 1288–1332).

27. This figure, in more or less elaborated forms that were equivalent to lists of coefficients of the binomial theorem, appeared as early as the Middle Ages in the works of Naṣīr al-Dīn al Ṭūsī (1265) and Chu Shih-chieh (1303). The arithmetical triangle reappeared in the sixteenth and seventeenth centuries in the writings of Apian (1527), Stifel, Scheubel, Tartaglia, Bombelli, Peletier, Trenchant, and Oughtred. But Pascal was the first to devote to it a systematic study linked to many questions of arithmetic and combinatorial analysis.

28. See, for example, E. Coumet, "La théorie du hasard est-elle née par hasard?" in *Annales. Économies, sociétés, civilisations* (1970), 574–598, as well as the studies of G.-T. Guilbaud (1952) and the other works on operational research, cybernetics, game theory, and other fields cited in Coumet's article (p. 575, notes 1 and 2).

29. See E. Coumet, "Mersenne, Frénicle et l'élaboration de l'analyse combinatoire dans la première moitié du XVIIe siècle" (a typescript thesis, Paris, 1968), and "Mersenne: Dénombrements, répertoires, numérotations de permutations," in *Mathématiques et sciences humaines*, **10** (1972), 5–37.

30. See I. Todhunter, *A History of the Mathematical Theory of Probability From the Time of Pascal to That of Laplace* (Cambridge–London, 1865; repr. New York, 1949), 7–21.

31. See F. Van Schooten, *Exercitationum mathematicarum libri quinque* (Leiden, 1657), 519–534, and H. J. M. Bos's article on Huygens in *DSB*, VI, 600.

32. On Archimedes see the article by M. Clagett in *DSB*, I, 213–231, esp. 215–222, for his infinitesimal methods and 229 for the diffusion of his writings in the sixteenth and seventeenth centuries. It should be noted that at this period mathematicians were aware only of his rigorous method of presentation, which Gregory of Saint-Vincent termed the "method of exhaustion." Archimedes' much more intuitive method of discovery did not become known until the rediscovery of his *Method* in 1906. On the infinitesimal work of Stevin, Valerio, and Kepler, see C. B. Boyer, *The Concept of Calculus* (New York, 1949), 98–111.

33. B. Cavalieri, *Geometria indivisibilibus continuorum nova quadam ratione promota* (Bologna, 1635). On this subject see Boyer, *op. cit.*, pp. 111–123; A. Koyré, in *Études d'histoire de la pensée scientifique* (Paris, 1966), 297–324; and the article on Cavalieri by E. Carruccio in *DSB*, III, 149–153.

34. See Boyer, *op. cit.*, pp. 123–147, 154–165.

35. Reprinted in Mesnard, II, 1259–1272; see esp. 1270–1272. This work is the next to last—but also one of the earliest written—of the brief treatises making up the *Traité du triangle arithmétique* [17].

36. "The sum of all the lines of any degree whatever is to the larger line and to the higher degree as unity is to the exponent of the higher degree" (Mesnard, II, 1271). On Pascal's infinitesimal work see H. Bosmans, in *Archivio di storia della scienza*, **4** (1923), 369–379; Boyer, *op. cit.*, pp. 147–153; F. Russo, in *L'oeuvre scientifique de Pascal* (Paris, 1964), 136–153; and P. Costabel, *ibid.*, 169–206.

37. "In the case of a continuous magnitude (*grandeur continue*), magnitudes of any type (*genre*), when added in any number desired to a magnitude of higher type, do not increase it at all. Thus, points add nothing to lines, [nor] lines to surfaces, [nor] surfaces to solids, or, to use the language of numbers in a treatise devoted to numbers, roots do not count with respect to squares, [nor] squares with respect to cubes . . . . Therefore, lower degrees should be neglected as possessing no value" (Mesnard, II, 1271–1272).

38. The cycloid is the curve generated by a point M of the circumference of a circle (C) that rolls without sliding on a straight line D. AB, the base of the cycloid, is equal to $2\pi r$ (where r is the radius of the circle C). Derived curves are obtained by the displacement of a point M' situated on the interior (curtate cycloid) or M'' on the exterior (prolate cycloid) of the moving circle. Defined by Roberval in 1637,

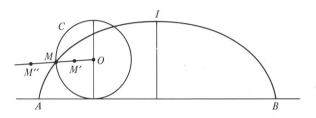

FIGURE 1

these curves had served since that year—under the name of *roulettes*, trochoids, or cycloids—as key examples for the solution of various problems pertaining to the infinitesimal calculus. These problems included the construction of tangents to plane curves by the use of the method of indivisibles, the determination of plane areas, the calculation of volumes, and the determination of centers of gravity. The cycloid thus played an important role in the patient efforts that resulted in the transition from the method of indivisibles to the infinitesimal calculus. Between 1637 and 1647 Roberval, then Fermat and Descartes, and finally Torricelli were particularly interested in the solution of infinitesimal problems associated with the cycloid; and bitter priority disputes broke out between Roberval and Descartes and then between Roberval and Torricelli. But in June 1658, when Pascal distributed his first circular, it appears that he had only a very imperfect knowledge of prior work on this subject.

The practice of setting up a contest was very common at the time. A similar contest, initiated by Fermat in January 1657 on questions of number theory, continued to set Fermat against some of the participants, notably Wallis. See O. Becker and J. E. Hofmann, *Geschichte der Mathematik* (Bonn, 1951), 192–194.

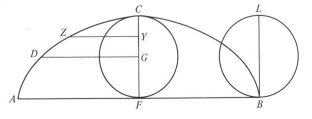

FIGURE 2

The contest problem was the following: Given an arch of the cycloid of base AB and of axis CF, one considers the semicurvilinear surface CZY defined by the curve, the axis, and a semichord ZY parallel to the base. The problem is to find (1) the area of CZY and its center of gravity; (2) the volumes of the solids V_1 and V_2 generated by the revolution of CZY about CY and about ZY, as well as their centers of gravity; and (3) the centers of gravity of the semisolids obtained by cutting V_1 and V_2 by midplanes.

39. In his *Histoire de la roulette* [20b], Pascal mentions the results sent to him by these four authors and notes, in particular, the rectification of the arch of the cycloid communicated to him by Wren. He points out that he has extended this operation to an arbitrary arc AZ originating at the summit of the cycloid and that he has determined the center of gravity of this arc AZ, as well as the areas and centers of gravity of the surfaces of revolution generated by the rotation of AZ about the base or about the axis of the cycloid. Carcavi, the president of the jury, also mentioned the results sent by Fermat, particularly those on the areas of the surfaces of revolution.

40. See A. Lalouvère, *Veterum geometria promota in septem de cycloide libris* (Toulouse, 1660); and J. Wallis, *Tractatus duo, prior de cycloide, posterior de cissoide* (Oxford, 1659). On the latter publication see K. Hara, "Pascal et Wallis au sujet de la cycloïde," in two parts: the first in *Annals of the Japanese Association for the Philosophy of Science*, **3**, no. 4 (1969), 36–57, and the second in *Gallia* (Osaka), nos. 10–11 (1971), 231–249.

41. See, in particular, C. Dati, *Lettera della vera storia della cicloide* (Florence, 1663).

42. This question is raised by K. Hara, in "Quelques additions à l'examen des textes mathématiques de Pascal," in *Gallia* (Osaka), no. 7 (1962); by P. Costabel, in *L'oeuvre scientifique de Pascal*, 169–198; and by J. Mesnard, in Mesnard, I, 31-33.

43. The original order is reproduced in vol. III of Mesnard's ed. of Pascal's works (in preparation).

44. See Bosmans, *op. cit.*; Boyer, *op. cit.*, pp. 147–153; Russo, *op. cit.*, pp. 136–153; and Costabel, *ibid.*, pp. 169–206.

45. See Russo, *op. cit.*, pp. 149–151. It should be noted that the expression "characteristic triangle" was introduced not by Pascal but by Leibniz. See also Boyer, *op. cit.*, pp. 152-153; Boyer points out that this figure had previously been used by Torricelli and Roberval and even by Snell (1624). In modern notation, the characteristic triangle at a point $M(x_0, y_0)$ of a plane curve (C) of equation $y = f(x)$ is a right triangle, the first two sides of which, parallel to the axes Ox and Oy, are of length dx and dy; its diagonal, of length ds, is parallel to the tangent to the curve (C) at M.

46. See a letter from Leibniz to Jakob I Bernoulli of Apr. 1703, in Leibniz, *Mathematische Schriften*, C. I. Gerhardt, ed., III (Halle, 1856), 72–73. This letter is reproduced by J. Itard in *Histoire générale des sciences*, 2nd ed., II (Paris, 1969), 245–246. For other statements by Leibniz concerning his knowledge of Pascal's writtings, see P. Costabel, in *L'oeuvre scientifique de Pascal*, 201–205.

47. Pascal wrote again to Fermat (10 Aug. 1660), met Huygens (5 and 13 Dec. 1660), and conversed with the duke of Roannez on the force of rarefied air and on flying. These are the few indications that we have regarding Pascal's scientific activity during the last three years of his life.

48. See M. Duclou, *Les carrosses à cinq sols* (Paris, 1950).

BIBLIOGRAPHY

I. ORIGINAL WORKS. There have been many complete eds. of Pascal's works. The most important from the point of view of scholarship are the following:

a. *Oeuvres de Blaise Pascal*, C. Bossut, ed., 5 vols. (The Hague, 1779), abbrev. as Bossut.

b. *Oeuvres de Blaise Pascal publiées selon l'ordre chronologique*, L. Brunschvicg, P. Boutroux, and F. Gazier, eds., 14 vols. (Paris, 1904–1914), part of the Collection des Grands Écrivains de la France, abbrev. as G.E.

c. *Oeuvres complètes de Blaise Pascal*, J. Chevalier, ed. (Paris, 1954), in Bibliothèque de la Pléiade," abbrev. as PL.

d. *Blaise Pascal. Oeuvres complètes*, J. Mesnard, ed., 2 vols. to date (Paris, 1964–1971); abbrev. as Mesnard. This last ed., which surpasses all previous ones, so far comprises only vol. I (*Introduction générale* and *Documents généraux*) and vol. II (*Oeuvres diverses, 1623–1654*). It has been used in preparing this article.

Each reference to a passage in one of these eds. will consist of the abbreviation, the volume number, year of publication of the volume, and page number. The list below includes most of Pascal's surviving scientific writings, cited in the order in which they were written. For each writing there is the title, its presumed date of composition, and its various eds.: the first (indicated as "orig." if published during Pascal's lifetime and as "1st ed." if posthumous) and the chief subsequent eds. (in separate vols. and in the sets of complete works cited above, as well as any other ed. containing important original material).

1. *Essay pour les coniques* (1639–1640). Orig. (Paris, Feb. 1640); Bossut, IV (1779), 1–7; G.E., I (1908), 243–260 and XI (1914), 347; PL (1954), 57–63, 1380–1382; R. Taton, "L' 'Essay pour les coniques' de Pascal," in *Revue d'histoire des sciences*, **8** (1955), 1–18; Mesnard, II (1971), 220–235.

2. *Lettre dédicatoire à Monseigneur le Chancelier sur le sujet de la machine nouvellement inventée par le sieur B. P. pour faire toutes sortes d'opérations d'arithmétique par un mouvement réglé sans plume ni jetons avec un avis nécessaire à ceux qui auront curiosité de voir ladite machine et de s'en servir* (1645). Orig. (Paris, 1645); Bossut, IV (1779), 7–24; G.E., I (1908), 291–314; PL (1954), 347–358; Mesnard, II (1971), 329–341.

3. *Expériences nouvelles touchant le vide. . . . Avec un discours sur le même sujet . . . dédié à Monsieur Pascal, conseiller du roi . . . par le sieur B. P. son fils. Le tout réduit en abrégé et donné par avance d'un plus grand traité sur le même sujet* (Sept.–early Oct. 1647). Orig. (Paris, Oct. 1647); Bossut, IV (1779), 51–68; G.E., II (1908), 53–76; PL (1954), 359–370; Mesnard, II (1971), 493–508.

4. Pascal's correspondence with Noël (late Oct.–early Nov. 1647). 1st ed., Bossut, IV (1779), 69–108; G.E., II (1908), pp. 77–125; PL (1954), 370–377, 1438–1452; Mesnard, II (1971), 509–540.

5. Pascal's letter to Le Pailleur (Feb. 1648). 1st ed., Bossut, IV (1779), 147–177; G.E., II (1908), 177–211; PL (1954), 377–391; Mesnard, II (1971), 555–576.

6. *Récit de la grande expérience de l'équilibre des liqueurs projetée par le sieur B. P. pour l'accomplissement du traité qu'il a promis dans son abrégé touchant le vide et faite par le sieur F. P. en une des plus hautes montagnes d'Auvergne* (autumn 1648). Orig. (Paris, 1648), repr. in facs. with intro. by G. Hellmann (Berlin, 1893) and in *Traités de l'équilibre des liqueurs et de la pesanteur de la masse de l'air . . .* (Paris, 1663; repr. 1664, 1698); Bossut, IV (1779), 345–369; G.E., II (1908), 147–162, 349–358, and 363–373; PL (1954), 392–401; Mesnard, II (1971), 653–690.

7. Preface to the treatise on the vacuum ("De l'autorité en matière de philosophie") (1651). 1st ed., Bossut, II (1779), 1–12; G.E., II (1908), 127–145, and XI (1914), 348–349; PL (1954), 529–535; Mesnard, II (1971), 772–785.

8. Fragments of "Traité du vide" (1651). 1st ed., in *Traités de l'équilibre des liqueurs et de la pesanteur de la masse de l'air . . .* (Paris, 1663), 141–163; Bossut, IV (1779), 326–344; G.E., II (1908), 513–529; PL (1954), 462–471; Mesnard, II (1971), 786–798.

9. *Lettre de M. Pascal le fils adressante à M. le Premier Président de la Cour des aides de Clermont-Ferrand . . .* (July 1651). Orig. (Clermont-Ferrand, 1651); Bossut, IV (1779), 198–214; G.E., II (1908), 475–495; PL (1954), 402–409; Mesnard, II (1971), 799–813.

10. Continuation of the correspondence with M. de Ribeyre (July–Aug. 1651). 1st ed., Bossut, IV (1779), 214–221; G.E., II (1908), 496–502; PL (1954), 409–411; Mesnard, II (1971), 814–818.

11. Letter from Pascal to Queen Christina of Sweden (June 1652). 1st ed., in F. Granet and P. N. Desmolets, eds., *Recueil de pièces d'histoire et de littérature*, III (Paris, 1738), 117–123; Bossut, IV (1779), 25–29; G.E., III (1908), 23–34; PL (1954), 502–504; Mesnard, II (1971), 920–926.

12. "Celeberrimae matheseos academiae Parisiensis" (Paris, 1654). 1st. ed., Bossut, IV (1779), 408–411; G.E., III (1908), 293–308; PL (1954), 71–74, 1400–1404 (French trans.); Mesnard, II (1971), 1121–1135 (with French trans.).

13. *Traités de l'équilibre des liqueurs et de la pesanteur de la masse de l'air. Contenant l'explication des causes de divers effets de la nature qui n'avaient point été bien connus jusques ici, et particulièrement de ceux que l'on avait attribués à l'horreur du vide* (completed at the latest in 1654). 1st ed. (Paris, Nov. 1663). The text of the *Traités* corresponds to pp. 1–140; pp. 141–163 reproduce the only two fragments known of the great treatise on the vacuum prepared by Pascal in 1651; the rest of the volume contains (pp. 164–194) a repr. of the *Récit de la grande expérience . . .* and (pp. 195–232) texts by F. Périer and others. Important subsequent eds. are those of 1664 and 1698; Bossut, IV (1779), 222–325; G.E., III (1908), 143–292, and IX (1914), 352; PL (1954), 412–471; Mesnard, I (1964), 679–689 (preface by F. Périer), and II (1971), 739–745 (account of Périer's observations), 787–798 (two "Fragments d'un traité du vide"), and 1036–1101 (the actual *Traités*).

14. "Generatio conisectionum" (completed about 1654). 1st ed. in *Sitzungsberichte der K. Preussischen Akademie der Wissenschaften zu Berlin*, **1** (1892), 197–202 (edited by C. I. Gerhardt); G.E., II (1908), 234–243; PL (1954), 66–70, 1382–1387 (French trans.); Mesnard, II (1971), 1108–1119.

15. Leibniz's notes on Pascal's treatise on conics (the notes date from 1676, but the treatise was finished about

1654). 1st ed. (partial) in *Sitzungsberichte der K. Preussischen Akademie der Wissenschaften zu Berlin*, **1** (1892), 195–197, edited by C. I. Gerhardt; G.E., II (1908), 227–233; P. Costabel, in *L'oeuvre scientifique de Pascal* (1964), 85–101 (with French trans.); Mesnard, II (1971), 1120–1131 (with French trans.).

16. Correspondence with Fermat (July–Oct. 1654). 1st ed., P. Fermat, *Varia opera mathematica . . .* (Toulouse, 1679), 179–188 (for the three letters by Pascal; for the other four see Bossut); Bossut, IV (1779), 412–445; Fermat, *Oeuvres*, P. Tannery and C. Henry, eds., II (1894), 288–314, and III (1896), 310–311; PL (1954), 74–90; Mesnard, II (1971), 1132–1158.

17. *Traité du triangle arithmétique, avec quelques petits traités sur la même matière* (1654). 1st ed. (Paris, 1665). Without the first four pages (title page, foreword, and table of contents) and the plate, this work was printed during Pascal's lifetime (1654) but was not distributed. It consists of four parts: the "Traité du triangle arithmétique" itself; two papers devoted to various applications of the triangle; and a fourth paper on numerical orders, powers, combinations, and multiple numbers that is formed of seven sections, the first in French and the rest in Latin. J. Mesnard has identified a preliminary Latin version of the part of this treatise that was published in French.

Subsequent eds.: Bossut, V (1779), 1–134; *Oeuvres complètes de Pascal*, C. Lahure, ed., II (1858), 415–494 (with French trans. of the Latin passages); G.E., III (1908), 311–339, 341–367, 433–598, and XI (1914), 353, 364–390; PL (1954), 91–171, 1404–1432 (translations); Mesnard, II (1971), 1166–1332—repr. with French trans. of the entire preliminary Latin ed., *Triangulus arithmeticus*, followed by the new sections of the *Traité*.

18. "Introduction à la géométrie" (written about the end of 1657). 1st ed. in *Sitzungsberichte der K. Preussischen Akademie der Wissenschaften zu Berlin*, **1** (1892), 202–204 (C. I. Gerhardt, ed.); G. E., IX (1914), 291–294; PL (1954), 602–604, 1476; J. Itard, in *L'oeuvre scientifique de Pascal* (1964), 102–119.

19. "De l'esprit géométrique" and "De l'art de persuader" (written about 1657–1658). 1st ed. (partial), P. N. Desmolets in *Continuation des mémoires de littérature et d'histoire*, V, pt. 2 (Paris, 1728), 271–296; Bossut, II (1779), 12–38, 39–57; G.E., IX (1914), 240–290; PL (1954), 574–602.

20. Various items pertaining to the cycloid competition (June 1658–Jan. 1659).

a. Three circulars addressed to the contestants: the first in Latin (June 1658); the second in Latin (July 1658); the third in French and Latin (dated 7 Oct. in the French text and 9 Oct. in the Latin version).

b. *Histoire de la roulette . . .* (10 Oct. 1658), also in Latin, *Historia trochoidis* (same date).

c. *Récit de l'examen et du jugement des écrits envoyés pour les prix proposés publiquement sur le sujet de la roulette . . .* (25 Nov. 1658).

d. *Suite de l'histoire de la roulette . . .* (12 Dec. 1658, with an addition on 20 Jan. 1659); the Latin version exists only in MS.

A more detailed description of this group of writings is provided by L. Scheler, in *L'oeuvre scientifique de Pascal* (1964), 30–31, and in Mesnard, I (1964), 163–167. Subsequent eds. are Bossut, V (1779), 135–213; G.E., VII (1914), 337–347, and VIII (1914), 15–19, 155–223, 231–246, 289–319; PL (1954), 180–223, 1433–1435 (French trans. of the circulars of June and July 1658).

21. *Lettres de A. Dettonville contenant quelques-unes de ses inventions de géométrie . . .* (Paris, Feb. 1659). This vol. contains a title page (written after the rest of contents), four sheets of plates, and four letters published between Dec. 1658 and Jan. 1659 (in the order 1, 4, 3, 2).

Letter no. 1: *Lettre de A. Dettonville à Monsieur de Carcavy, en lui envoyant: Une méthode générale pour trouver les centres de gravité de toutes sortes de grandeurs. Un traité des trilignes et de leurs onglets. Un traité des sinus du quart de cercle. Un traité des solides circulaires. Et enfin un traité général de la roulette, contenant la solution de tous les problèmes touchant la roulette qu'il avait proposés publiquement au mois de juin 1658*. Orig. (Paris, 1658).

Letter no. 2: *Lettre de A. Dettonville à Monsieur A. D. D. S. en lui envoyant: La démonstration à la manière des anciens de l'égalité des lignes spirale et parabolique*. Orig. (Paris, 1658).

Letter no. 3: *Lettre de A. Dettonville à Monsieur de Sluze, chanoine de la cathédrale de Liège, en lui envoyant: La dimension et le centre de gravité de l'escalier. La dimension et le centre de gravité des triangles cylindriques. La dimension d'un solide formé par le moyen d'une spirale autour d'un cône*. Orig. (Paris, 1658).

Letter no. 4: *Lettre de A. Dettonville à Monsieur Huggyens [sic] de Zulichem, en lui envoyant: La dimension des lignes de toutes sortes de roulettes, lesquelles il montre être égales à des lignes elliptiques*. Orig. (Paris, 1659).

Later eds.: Bossut, V (1779), 229–452; G.E., VIII (1914), 247–288, 325–384, and IX (1914), 1–149, 187–204; PL (1954), 224–340, 1436–1437; a facs. of the original ed. has recently appeared (London, 1966).

Two other important documents relating to Pascal's scientific work are the following:

22. The license for his calculating machine (22 May 1649). 1st ed. in *Recueil de diverses pièces pour servir à l'histoire de Port-Royal* (Utrecht, 1740), 244–248; Bossut, IV (1779), 30–33; G.E., II (1908), 399–404; Mesnard, II (1971), 711–715.

23. Letter from Leibniz to Étienne Périer of 30 Aug. 1676 concerning Pascal's treatise on conics. 1st ed., Bossut, V (1779), 459–462; G.E., II (1908), 193–194; PL (1954), 63–65; J. Mesnard and R. Taton, in *L'oeuvre scientifique de Pascal* (1964), 73–84.

II. SECONDARY LITERATURE. A very complete bibliography of studies on Pascal's scientific work published before 1925 can be found in A. Maire, *Bibliographie générale des oeuvres de Pascal*, 2nd ed., I, *Pascal savant* (Paris, 1925). Most of the more recent works on the subject (except for those dealing with the cycloid) are cited in the bibliographies in Mesnard, II (1971)—geometry, 227–228, 1108; combinatorial theory and the calculus of probability, 1135, 1175; the calculating machine, 327–

328; physics, 349, 459, 513, 675–676, 777, 804, 1040; miscellaneous, 1031.

Two general studies in particular should be mentioned: P. Humbert, *L'oeuvre scientifique de Pascal* (Paris, 1947), a survey written for a broad audience; and *L'oeuvre scientifique de Pascal* (Paris, 1964), a joint effort that restates the main aspects of Pascal's career and scientific work (with the exception of the theory of combinations and the calculus of probability). Other recent studies worth consulting are A. Koyré, "Pascal savant," in *Blaise Pascal, l'homme et l'oeuvre* (Paris, 1956), pp. 259–285; K. Hara, "Examen des textes mathématiques dans les oeuvres complètes de Pascal d'après les Grands Écrivains de la France," in *Gallia* (Osaka), no. 6 (1961); "Quelques additions à l'examen des textes mathématiques de Pascal," *ibid.*, no. 7 (1962); and "Pascal et Wallis au sujet de la cycloïde, I," in *Annals of the Japan Association for Philosophy of Science*, 3, no. 4 (1969), 166–187; "Pascal et Wallis . . . , II," in *Gallia*, nos. 10–11 (1971), 231–249; and "Pascal et Wallis . . . , III," in *Japanese Studies in the History of Science*, no. 10 (1971), 95–112; N. Bourbaki, *Éléments d'histoire des mathématiques*, 2nd ed. (Paris, 1969), see index; M. E. Baron, *The Origins of the Infinitesimal Calculus* (London, 1969), esp. 196–205; and E. Coumet, "La théorie du hasard est-elle née par hasard?" in *Annales. Économies, sociétés, civilisations*, 5 (May–June 1970), 574–598.

RENÉ TATON

PASCAL, ÉTIENNE (*b.* Clermont-Ferrand, France, 2 May 1588; *d.* Paris, France, 24 September 1651), *mathematics.*

The son of Martin Pascal, treasurer of France, and Marguerite Pascal de Mons, Pascal married Antoinette Begon in 1616. They had three children: Gilberte (1620–1687), who in 1641 married Florin Périer; Blaise (1623–1662), the philosopher and scientist; and Jacqueline (1625–1661), who in 1652 entered the convent of Port-Royal.

Elected counselor for Bas-Auvergne in 1610, Pascal became president of the Cour des Aides in 1625. His wife died in 1626, and in 1631 he left Clermont to settle in Paris with his children. He devoted himself to his son's education while gaining a reputation as a talented mathematician and musician. In 1634 Pascal was one of five commissioners named to examine J. B. Morin's "invention" for the determination of longitudes. As early as 1635 he frequented "Mersenne's academy" and was in contact with Roberval, Desargues, and Mydorge.

In November 1635 Mersenne dedicated to Pascal the "Traité des orgues" of his *Harmonie universelle* (1636). Roberval communicated to Pascal his first discoveries concerning the cycloid and intervened on his side in the debate concerning the nature of gravity

(interpreting it in terms of attraction—letter to Fermat of 16 August 1636; Fermat's response of 23 August). At the beginning of 1637 Fermat wrote his "Solution d'un problème proposé par M. Pascal." At about the same time Pascal introduced a special curve, the conchoid of a circle with respect to one of its points, to be applied to the problem of trisecting an angle. Roberval called it the "limaçon de M. Pascal" and determined its tangent by his kinematic method. In February 1638 Roberval joined Pascal in defending Fermat's *De maximis et minimis*, which had been attacked by Descartes.

Having been obliged to return to Auvergne from March 1638 to April 1639, Pascal then moved to Rouen, where he was appointed intendant of the province, a post he held until 1648. He had given his son Blaise a solid foundation in mathematics, and he now fostered the development of his work, mainly through his contacts with many scientists. In October 1646 Pascal participated with his son and P. Petit in the first repetition in France of Torricelli's experiment. In April 1648 he joined in the debate between Blaise and the Père E. Noël concerning the problem of the vacuum. He returned to Paris in August 1648, was in Auvergne from May 1649 to November 1650, then spent his last months in Paris.

BIBLIOGRAPHY

I. ORIGINAL WORKS. The rare documents concerning Pascal's scientific work are reproduced in the major eds. of his son's complete works: *Oeuvres de Blaise Pascal publiées selon l'ordre chronologique*, L. Brunschvicg, P. Boutroux, and F. Gazier, eds., 14 vols. (Paris, 1908–1914), in the collection Grands Écrivains de la France (hereafter cited as G.E.); and *Blaise Pascal. Oeuvres complètes*, J. Mesnard, ed., I, II (Paris, 1964, 1971) (hereafter cited as Mesnard).

They include "Jugement porté par les commissaires Étienne Pascal, Mydorge, Beaugrand, Boulanger, Hérigone sur l'invention du sieur J. B. Morin," in G.E., I, 194–195, and Mesnard, II, 82–99; "Lettre d'Étienne Pascal et Roberval à Fermat, samedi 16 août 1636," in *Oeuvres de Fermat*, P. Tannery, C. Henry, and C. de Waard, eds., 5 vols. (Paris, 1891–1922), II, 35–50 (hereafter cited as Fermat), also in G.E., I, 177–193, and Mesnard, II, 123–140; "Lettre de Fermat à Étienne Pascal et Roberval, 23 août 1636," in Fermat, II, 50–56, and in Mesnard, II, 140–146; "Solutio problematis a Domino de Pascal propositi" (Jan. or Feb. 1637), in Fermat, I, 70–74, also in G.E., I, 196–201, and Mesnard, II, 148–156, also translated into French as "Solution d'un problème proposé par M. de Pascal," in Fermat, III, 67–71, and Mesnard, II, 149–156; "Réponse de Descartes à un écrit des amis de M. de Fermat" (1 Mar. 1638), in *Oeuvres de Descartes*, C. Adam and P. Tannery, eds., II (Paris, 1898), 1–15, also in *Descartes. Correspondance*, C. Adam and G. Milhaud, eds., II (Paris, 1939), 143–153, *Correspondance du P. Marin Mersenne*, C. De Waard and B. Rochot, eds., VII (Paris, 1962), 64–73, and Mesnard, II, 164–174; and "Lettre de M. Pascal le Père au R. P. Noël" (Apr. 1648), in G.E., II, 255–282, and in Mesnard, II, 584–602.

II. Secondary Literature. Documents, notices, and details concerning the life and work of Pascal can be found in G.E., I, 5–28, 170–176, and II, 533–562; Mesnard, I, 459–464, 510–515, 571–576, 721–722, 727–729, 754–771, 1077–1079, 1091–1100, and II, 119–123, 157–163, 174–188, 217, 253–254, 841–863; the ed. of Descartes's *Oeuvres* cited above, index, V, 607; the ed. of Descartes's correspondence cited above, II, 379–381 and index; and Mersenne's correspondence cited above, vols. IV–VII, see index.

The catalog of a commemorative exhibition held at the Bibliothèque Nationale in 1962, *Blaise Pascal, 1623–1662* (Paris, 1962), furnishes references to many documents concerning Étienne Pascal: nos. 1, 9, 10, 14, 17, 18, 22–27, 31, 32, 34, 36, 38, 41, 60, 67, 69–72, 76, 77, 168. Other references are in A. Maire, *Pascal savant* (Paris, 1925), 270–275 and index.

Additional details are in M. Cantor, *Vorlesungen über Geschichte der Mathematik*, 2nd ed., II (Leipzig, 1900), 675, 679, 681, 875, 881, 882; J. Mesnard, *Pascal et les Roannez*, 2 vols. (Paris, 1965), see index; and P. Tannery, *Mémoires scientifiques*, X (Paris, 1930), 372, 382–383, and XIII (Paris, 1934), 337–338.

René Taton

PASCH, MORITZ (*b.* Breslau, Germany [now Wrocław, Poland], 8 November 1843; *d.* Bad Homburg, Germany, 20 September 1930), *mathematics*.

Pasch studied chemistry at Breslau but changed to mathematics at the suggestion of Heinrich Schröter, to whom, along with Kambly, his teacher at the Elisabeth Gymnasium, he dedicated his dissertation (1865). Later, at Berlin, he was influenced by Weierstrass and Kronecker. He maintained his mathematical activity with scarcely a break for sixty-five years, for the first seventeen years in algebraic geometry and later in foundations, the work on which his fame rests. His first two papers were written in collaboration with his lifelong friend J. Rosanes. Except for rapid promotion, Pasch's career at the University of Giessen was not unusual: in 1870, *Dozent*; in 1873, extraordinary professor; in 1875, after an offer of an extraordinary professorship from the University of Breslau, ordinary professor. In 1888 he obtained the chair left vacant by the death of Heinrich Baltzer. He was also active in administration, becoming dean in 1883 and rector in 1893–1894. In order to dedicate himself more fully to his scientific work, he retired in 1911. In celebration of his eightieth birthday Pasch received honorary Ph.D.'s from the universities of Frankfurt and Freiburg. He was a member of the Deutsche Mathematiker-Vereinigung. His name is perpetuated in Pasch's axiom, which states that in a plane, if a line meets one side of a triangle, then it meets another. His outward life was simple, although saddened by the early death of his wife and one of two daughters. He died while on a vacation trip away from Giessen.

The axiomatic method as it is understood today was initiated by Pasch in his *Vorlesungen über neuere Geometrie* (Leipzig, 1882; 2nd ed., Berlin, 1926). It consists in isolating from a given study certain notions that are left undefined and are expressly declared to be such (the *Kernbegriffe*, in Pasch's terminology of 1916), and certain theorems that are accepted without proof (the *Kernsätze*, or axioms). From this initial fund of notions and theorems, the other notions are to be defined and the theorems proved using only logical arguments, without appeal to experience or intuition. The resulting theory takes the form of purely logical relations between undefined concepts.

To be sure, there are preliminary explanations, and a definite philosophy is disclosed for choosing the axioms. According to Pasch the initial notions and theorems should be founded on observations. Thus the notion of point is allowed but not that of line, since no one has ever observed a complete (straight) line; rather, the notion of segment is taken as primitive. Similarly, a planar surface, but not a plane, is primitive.

Pasch's analysis relating to the order of points on a line and in the plane is both striking and pertinent to its understanding. Every student can draw diagrams and see that if a point *B* is between point *A* and point *C*, then *C* is not between *A* and *B*, or that every line divides a plane into two parts. But no one before Pasch had laid a basis for dealing logically with such observations. These matters may have been considered too obvious; but the result of such neglect is the need to refer constantly to intuition, so that the logical status of what is being done cannot become clear. According to Pasch, the appeal to intuition formally ceases once the *Kernbegriffe* and *Kernsätze* are stated.

The higher geometry of Pasch's day was projective geometry that used real numbers as coordinates. Pasch therefore considered that the foundation would be laid once the coordinates had been introduced. In doing this he presented notions of congruence, which were nonprojective. This is somewhat disappointing in view of Staudt's 1847 program for founding projective geometry solely on projective terms (though we may emphasize that the congruence axioms are original with Pasch). But F. Klein had uncovered some nonrigorous thinking in Staudt's proof that a one-to-one mapping between two lines that sends harmonic quadruples into harmonic quadruples is uniquely determined by the images of three points—without, however, obtaining notable success in clarifying this matter. Pasch proved this fundamental theorem on the basis of the Archimedean character of the ordering on the line, and not on its completeness, as Klein proposed to do. The congruence notions were introduced, at least in part, in order to state Archimedes' axiom. Once the fundamental theorem was proved, the introduction of coordinates could be easily accomplished—as M. Dehn remarks in a historical appendix to Pasch's *Vorlesungen*—on the basis of the Eudoxian theory of book V of Euclid's *Elements*. But for Pasch this procedure was complicated by his empiricist point of view.

It would be easy to overlook the significance of Pasch's foundational achievements for several reasons. First, it is now a commonplace to present theories in an axiomatic way, so that even logic itself is presented axiomatically. Thus Pasch's innovation achieves the status of being a trifle.

There are also widespread misconceptions as to what is in book I of the *Elements*: it is thought that Euclid had an axiomatic way of presenting geometry. This view is further confounded by a lack of clarity as to what the axiomatic method is and what geometry is. Anyone who looks at book I of the *Elements* with modern hindsight sees that something is wrong, but it would take delicate historical considerations to place the source of the faults in a correct light.

The Greeks of Euclid's time had the axiomatic method; Aristotle's description of it can be considered a close approximation to the modern one. Or, better yet, one may consider Eudoxus' theory of magnitude as presented in book V of the *Elements*. Except for style (which, however, may indicate a difference in point of view), the procedure presented there coincides with Pasch's. It is known, however, that the *Elements* is a compilation of uneven quality, so that even with the definitions, postulates, and common notions of book I, it is unwarranted to assume that book I is written from the same point of view as book V.

In some versions of book I, as it has come down to us, there are five "common notions" and five postulates. T. L. Heath considers it probable that common notions 4 and 5 were interpolations; and P. Tannery maintains that they were none of them authentic. The first three postulates are the "postulates of construction," the fourth states that all right angles are equal, and the fifth is the parallel postulate. It has been argued that the first three postulates were meant to help meet the injunction to limit the means of construction to "straightedge and compass"; that there was no intention to say anything about space. One could eliminate these three, as well as the fourth, without changing the rigor of the book or the points of view disclosed relative to geometry or to the axiomatic method. The fifth does not appear until proposition 29, so that the first twenty-eight propositions (minus the constructions) are, from a modern axiomatic point of view, based on nothing.

Although deduction is a prominent feature of book I of the *Elements*, the contents of the book and the history of the parallel postulate show that geometry was conceived as the study of a definite object, "external space." With the invention of non-Euclidean geometry around 1800, it began to dawn on mathematicians that their concern is with deduction, and not with a supposed external reality. Applications, if any, may be left to the physicist. With G. Fano's miniature projective plane of just seven points and seven lines (1892), the revolution may be considered to have been completed. Hilbert, through his work in geometry and logic, consolidated it.

Pasch initiated the axiomatic method, although the foundational developments of his time were against this point of view. Thus Cantor's striking discoveries were based, from an axiomatic point of view, on nothing. Dedekind rightly contrasted his own treatment of magnitude with Eudoxus'; Dedekind's was constructive, whereas Eudoxus' was axiomatic. As late as 1903 Frege poked fun at the axiomatic method as presented in Hilbert's *Grundlagen der Geometrie*.

The Italian geometers, particularly Peano, continued Pasch's work. In 1889 Peano published both his exposition of geometry, following Pasch, and his treatment of number. It is tempting to see in the former work a source of the latter, although there were many other sources.

Pasch played a crucial role in the innovation of the axiomatic method. This method, with contributions from logic and algebra, is a central feature of twentieth-century mathematics.

BIBLIOGRAPHY

See M. Dehn and F. Engel, "Moritz Pasch," in *Jahresbericht der deutschen Mathematiker-Vereinigung*, **44** (1934), 120–142; P. Tannery, "Sur l'authenticité des axiomes d'Euclide," in *Bulletin des sciences mathématiques et astronomiques* (1884), 162; G. Fano, "Sui postulati fondamentali della geometria proiettiva," in *Giornale di matematiche* (Naples), **30** (1892), 106–132; G. Frege, "Über die Grundlagen der Geometrie," in *Jahresbericht der deutschen Mathematiker-Vereinigung*, **12** (1903), 319–324,

368–375; H. Freudenthal, "Zur Geschichte der Grundlagen der Geometrie," in *Nieuwe Archieff voor Wiskunde*, **5** (1957), 105–142; T. L. Heath, *The Thirteen Books of Euclid's Elements*, 2nd ed. (New York, 1956), esp. 221, 225, 232; and A. Seidenberg, "The Ritual Origin of Geometry," in *Archive for History of Exact Sciences*, **1** (1962), 488–527, esp. 497 f.

A. SEIDENBERG

PASCHEN, LOUIS CARL HEINRICH FRIEDRICH

(*b.* Schwerin, Mecklenburg, 22 January 1865; *d.* Potsdam, Germany, 25 February 1947), *experimental physics.*

Friedrich Paschen, "probably the greatest experimental spectroscopist of his time,"[1] was born into a Lutheran family of scientifically inclined Mecklenburg officers, military and civil. His paternal grandfather, H. C. Friedrich Paschen (1804–1873), was director of the Mecklenburg geodetic survey and a noted astronomer. An uncle, Carl Paschen (1835–1911), who rose to the rank of admiral in the German navy, was a well-known hydrographer. Paschen himself, although he incorporated some official virtues—and although, as was customary, he became a lieutenant in the reserve[2]—seems nonetheless to have reacted against the authoritarian structure and the social-political attitudes of the German officer class. Apparently against the wishes of his family, and with no independent income, he resolved to accept the greater hazards of an academic career.[3]

After completing his secondary education in Schwerin, Paschen began his university studies in 1884 at Strasbourg, where he joined the group of disciples around August Kundt, the most charismatic and influential professor of physics in Germany, who happened also to be a native of Mecklenburg and a graduate of the Schwerin Gymnasium. After two years at Strasbourg, Paschen studied for one year at the University of Berlin, then returned to Kundt for supervision of his doctoral research. The degree was conferred in September 1888.[4] Paschen's doctoral research, suggested to him by Kundt, established "Paschen's law": that the sparking voltage depends only on the product of the gas pressure and the distance between the electrodes—one of the first and most important of the numerous scaling laws in this field.[5]

From October 1888 to April 1891 Paschen was Wilhelm Hittorf's last assistant in the Physical Institute of the Catholic Academy (subsequently University) of Münster. In keeping with Hittorf's interests and the experimental facilities available in consequence of them, Paschen turned to electrolytic solutions and, giving the first proofs of his enormous energy, in two

years published seven papers on a variety of investigations of electrolytic potentials. It was here, under Hittorf, that Paschen learned the value and technique of precision measurements.[6] Hittorf retired in 1890, and at Easter 1891 Paschen became Heinrich Kayser's teaching assistant in the Physical Institute of the Technische Hochschule at Hannover. Paschen continued in this position until Easter 1901, serving under Conrad Dieterici after Kayser's departure for Bonn in the summer of 1894.

Paschen had qualified as lecturer (*habilitiert*) at Hannover in the spring of 1893; and in March 1895, in return for his refusal of a position at Aachen, the Prussian government created a permanent lectureship (*etatsmässige Dozentur*) in physics and photography for him. This post gave him a relatively comfortable income of 3,300 marks in 1895, rising to about 4,000 marks by 1900, and a relatively large amount of time for research.[7]

In the late 1880's Kirchhoff's function (the universal but unknown temperature and wavelength dependence of the ratio of the radiant energy emitted to that absorbed by a body in thermal equilibrium—and, consequently also the dependence of the radiant energy in a cavity in thermal equilibrium upon those two variables) was beginning to draw considerable attention from theorists following the bolometric measurements of S. P. Langley and others.[8] Kayser proposed that Paschen improve upon Langley's results by more extensive measurements using the reflection gratings with which the institute had been well supplied for Kayser's and Carl Runge's spectroscopic work.[9] Gratings should have the advantage that their spectra, in contrast to prismatic spectra, may be chosen "normal" —that is, to disperse equal wavelength intervals into equal angular intervals—so that intensity measurements will give the desired distribution function directly. Only after a year spent building up the necessary apparatus—including the most sensitive galvanometer (of the Thomson astatic type) constructed until then or for years afterward—did Paschen discover late in 1892 that the gratings were unusable because the metal on which they were ruled showed irregular selective reflection in the infrared.[10]

But now committed to this problem, which would absorb most of his research efforts for the ten years at Hannover, Paschen turned to prismatic spectra and made a very accurate investigation of the dispersion of fluorite (correcting Rubens' measurements). At the same time he determined the infrared absorption by carbon dioxide and water vapor (obtaining results which, although presumed to have been superseded by Rubens, remained in 1913 the strongest evidence in favor of Bjerrum's quantum theory of molecular absorption).[11] Paschen also investigated the much mooted question of whether heat alone could bring gases to radiate, demonstrating—in contrast with the results of Ernst Pringsheim—the existence of infrared spectral lines produced by merely heating the gas.[12]

Paschen spent much of 1894 observing the deflections of his galvanometer attached to a very delicate bolometer irradiated by heated platinum with various surfacings; but the reduction of this data to emissive power as a function of λ and T, especially the transformation from a prismatic to a normal spectrum, required extremely laborious computations. By the summer of 1894 he had strong indications that $\lambda_{max} \cdot T =$ constant, "or: the frequency of the main thermal vibrations of the molecular parts of an absolutely black body is proportional to the absolute temperature."[13] Paschen, who was never willing to take time from the laboratory to find out in the library what others had done—nor, indeed, ever to write a review article in any of the fields in which he was to become an authority—seems to have been ignorant of W. Wien's publication of thermodynamic deductions of this "displacement law" in 1893 and 1894. Paschen was, however, aware of the general growth of interest and activity in this field, especially in Berlin (Rubens, Pringsheim, Lummer, among others); and in the summer of 1895, in view of the growing competition, Paschen published his result.[14]

But to complete the computation of all those "hundreds of curves" from which Paschen hoped to induce Kirchhoff's function would take many months more. It was thus not difficult for Carl Runge, excited by Ramsay's announcement in March 1895 of the discovery of terrestrial helium and bereft of his experimental collaborator Kayser, to persuade Paschen that the investigation of the spectrum of this new element was far more important to science. Compared with Paschen's infrared researches, this optical spectroscopy proved to be child's play: in one day he obtained for Runge the yellow D_3 helium line, and within three months they brought out two papers giving an astonishingly accurate inventory of the helium lines and an astonishingly successful arrangement of them into series.[15] Overnight Paschen acquired an international reputation; and accompanying Runge to England in September 1895, he was received most warmly by British physicists.[16]

In the fall of 1895 Paschen returned to the calculation of his emissive power curves, and by the spring of 1896 he had found $I = c_1 \lambda^{-\alpha} \exp(-c_2/\lambda T)$ with $\alpha = 5.5$, for an iron oxide surface.[17] The formulas for this curve previously proposed by V. A. Michelson and H. F. Weber, involving exponential dependence, had undoubtedly been suggestive; and Runge's suggestion that the energy curves be plotted on logarithmic scales was also helpful.[18] Wien, who had evidently gotten into contact with Paschen after the publication of Paschen's preliminary announcement in the summer of 1895, was informed in advance of these results by mail. Wien replied that he had deduced exactly this formula, but with $\alpha = 5$, some time earlier. The derivation of what soon became known as Wien's law—doubtless previously withheld from publication because of its highly arbitrary character—now appeared immediately following Paschen's paper in the *Annalen*.[19]

In the laboratory, however, Paschen had been obliged by lack of space and funds to forgo entirely his bolometric work from mid-1895 to the end of 1896. Instead he joined Runge in the quest for spectral series, first unsuccessfully with argon and then successfully with the homologous oxygen-sulfur-scandium-tellurium spectra. Through this collaboration Paschen became fully familiar with the field that would occupy him almost exclusively after the question of the blackbody radiation formula had finally been settled. That question certainly could not be settled by measurements on heated surfaces; to assert definitely that his (and Wien's) formula was the true one, and to fix the value of α, would require a far more perfect realization of ideal blackbody radiation.

For this purpose Paschen's competitors, enjoying the ample resources of the Physikalisch-Technische Reichsanstalt at Berlin, had in 1895 begun the construction of elaborate "thermostatic cavities." In the spring of 1897 Paschen, with a grant from the Berlin Academy and with the enlargement of the Hannover Institute, began to construct similar equipment.[20] By the summer of 1898 he had found $\alpha = 5$ and was "convinced that there can remain no further doubt about the correctness of the formula itself and of its constants."[21] Indeed, he was so convinced that when deviations between theory and experiment began to appear in 1899, he took them as indicative of undetected sources of experimental error. Only in 1900, when the Berlin experimentalists, working with much larger values of $\lambda \cdot T$, found clear deviations from Wien's law, did Paschen (like Planck) reconceptualize his experiments as a search for the limits of the law's range of validity. Paschen construed the results positively as strong evidence for the validity of Planck's formula in the intermediate region between Wien's and Rayleigh's formulas.[22] In striking and curious contrast with the Berlin experimentalists, who were literally enraged at him, throughout his work on the blackbody radiation problem Paschen the pure experimentalist showed himself to be more than ready to enlist experiment in the service of theory.

This continued to be characteristic of the man, of his work, and of some of his greatest successes.

In these years, despite the volume and importance of his work, it looked very much as if Paschen, like Runge, would remain stuck at Hannover.[23] Finally, however, in February 1901, the University of Tübingen, after failing successively to secure Paul Drude, Philipp Lenard, and Hermann Ebert, appointed Paschen professor of physics at a salary of 3,500 marks, including living quarters in a moderately good institute.[24] Here Paschen brought a wife in September 1901; here he raised a daughter and married her to his student Hermann Schüler in August 1920;[25] here he kept a guest room "always standing ready for the physicists." In 1908 the institute was substantially enlarged to accommodate the growing number of advanced students, including eventually many from abroad.[26] In July 1915 Paschen refused a call to Göttingen as Eduard Riecke's successor, gaining in return 9,000 marks for his institute and 3,000 marks per annum salary supplement.[27] In 1919 he was offered and actually accepted the Bonn chair of physics in succession to Kayser; but he reneged early in 1920 because he was persuaded that, in consequence of the political and above all the economic situation, the move would be to the disadvantage of his research work.[28]

Paschen had lost a year or two of research in transferring to Tübingen, and the diverse resources of the institute there led to forays into other fields—radioactivity and the nature of X rays, canal rays, and the mechanism of light emission. By 1908, however, he was focusing again on the problem of spectral series. Important forces bringing his attention back to this problem, to which he would devote himself exclusively for the remainder of his career, were, on the one hand, the presence of Walther Ritz at Tübingen during the winter of 1907–1908 and, on the other hand, Arno Bergmann's discovery in 1907 of new series in the infrared spectra of the alkalies (the so-called f series). Paschen now had, as he had not had at Hannover in the early 1890's, the facilities necessary for a systematic bolometric search for infrared spectral lines, above all the large-capacity high-voltage storage batteries to maintain intense, steady discharges. Returning to helium, in the spectrum of which he had previously detected bolometrically (June 1895) a few lines predicted by Runge's series formulas, he found in the spring of 1908 additional lines that did not fit in that series system. Paschen was looking everywhere for the impurity responsible for these lines when a letter arrived from Ritz announcing his newly invented combination principle and suggesting that helium

lines might exist at precisely those wavelengths Paschen had observed. Following this striking confirmation, Ritz suggested that Paschen look for hydrogen lines at frequencies $\nu = N\ (1/3^2 - 1/m^2)$, $m = 4, 5 \ldots$, and this "Paschen series" was soon found.[29]

In 1899 Thomas Preston had presented evidence that the magnetic splitting of spectral lines (Zeeman effect) was characteristic for the series to which they belonged, and in 1900 Runge and Paschen had begun a very careful investigation of Preston's rule. Paschen had been able to do no more than make the requisite photographs before he left for Tübingen; the striking quantitative results published under their names were entirely Runge's.[30] By 1905, however, Paschen had begun to equip himself to continue this work; and from 1907 he and his students employed the Zeeman effect extensively and with great success as an aid in identifying series lines.

At the same time, however, they found a large number of apparent exceptions to Preston's rule. In the simplest case those were very narrow doublet or triplet line groups showing the "normal" splitting pattern characteristic of a single line rather than the anticipated superposition of the "anomalous" splittings of the individual components of the group. Paschen, investigating this general circumstance with his student Ernst Back, and basing himself upon Ritz's conception of a spectral line as the combination of two independently subsisting terms, showed in 1912 that in sufficiently strong magnetic fields—i.e., fields strong enough for the magnetic splitting to be large compared with the separation of the components of the line group—all the splitting patterns transform themselves into the "normal" pattern. This "Paschen-Back effect" was immediately seized upon as potentially one of the most revealing clues to atomic structure and the mechanism of emission of spectral lines.[31]

In the last days of July 1914 the intense activity at Paschen's institute ceased abruptly. The German students and the institute staff rushed to the colors, and the foreign students fled over the Swiss border; Paschen himself seems not to have had the least inclination to participate in the war effort.[32] In the summer of 1915, with the aid of a single technician recalled from the army, Paschen again took up the most interesting problem upon which he had been working in 1914—"Bohr's helium lines," the lines previously construed as the sharp series of hydrogen but now ascribed by Bohr to ionized helium. In the course of this work, which was initially intended to check Bohr's prediction of a small difference between the Rydberg constant, N, for hydrogen and helium, and which was hampered by the diffuseness of the lines, Paschen discovered that a particular layer in the negative glow inside the common cylindrical-cathode Geissler tube gave especially sharp and complete spectra. Following up this observation, he developed the Paschen hollow-cathode discharge tube, in which, under the right conditions, the glow discharge retreats entirely into the largely field-free interior of a boxlike cathode. This device, which was the basis of much of the subsequent work on series and multiplet structure by Paschen and his students, showed the fine structure of Bohr's helium lines with extraordinary clarity and completeness.[33] (Here the series structure is the fine structure.)

Paschen had already reached this point when, late in 1915, Arnold Sommerfeld wrote inquiring about data with which to compare the relativistic fine structure demanded by the extension of Bohr's theory that he was then developing. Paschen was impressed; enlisting himself in the service of Sommerfeld's theory, he spent all his free time in the following six months confirming its predictions in detail, so far as possible.[34] The single paper presenting his results was immediately recognized as the tremendous advance in knowledge that it was.[35]

After the war spectroscopic activity in Paschen's institute increased rapidly. During the six years before his departure in 1924 Tübingen was unquestionably the most important center of atomic spectroscopy in Germany, at a time when this technique was far and away the most important for the advance of theoretical atomic physics. Paschen's own outstanding successes were the ordering of the neon spectrum—almost 1,000 lines—into spectral series; the evocation of the missing combinations between complex spectral terms by magnetic fields of appropriate strength (violation of the selection rule for the total angular momentum quantum number); and the first analysis of the spectra of an atom in its doubly ionized, as well as its neutral, and singly ionized states.[36] At the same time his associates Ernst Back and Alfred Landé were in the forefront of research on multiplet structure and Zeeman effects.

Paschen accepted, as few other experimentalists did, the priorities and guidance of atomic theorists. In 1922, persuaded that Landé could do more for him in this respect than any other available young theorist, he fought recklessly, tenaciously, and ultimately successfully against the social-political prejudices of his university for Landé's appointment to Tübingen's professorship (Extraordinariat) for theoretical physics.[37]

In July 1924 Paschen was looking forward to spending that fall and winter in the United States as the second German physicist (after Sommerfeld) to receive the compliment of a visiting professorship there since the war. But September found Paschen in Berlin, not Ann Arbor, for the month before he had been offered the presidency of the Physikalisch-Technische Reichsanstalt, as successor to Nernst.[38] Paschen accepted the post at the urging of the leading Berlin physicists, and with the intent of restoring basic research as a principal function of the institution. In this endeavor he had only very limited success because of budgetary constraints, bureaucratic resistance, and Paschen's own limitations as administrator and politician. After scarcely a year in the post he was letting it be known that he would be glad to return to a university chair.[39] He stayed on, however, gradually building up a spectroscopic laboratory to continue his previous line of research. As honorary professor, Paschen lectured at the University of Berlin on some topic in spectroscopy or physical optics for two hours a week every term. As a member of the Berlin Academy (from July 1925) and of various committees and commissions, and as president of the Deutsche Physikalische Gesellschaft (1925–1927), he played a prominent role—although not a key role—in the lively scientific life in Berlin during the later Weimar period.

Paschen's post, the highest to which a German experimental physicist could aspire, was *ipso facto* coveted by Johannes Stark; immediately following the Nazi seizure of power, Stark had himself appointed to it, effective 1 May 1933.[40] Forced out of his office and into retirement, Paschen was still able to continue working for a few years in his laboratory—at the cost of considerable difficulty and personal humiliation. Finally he withdrew to his home in Charlottenburg, where he confined himself to the evaluation of his spectrographs.[41] In November 1943 his home and all his possessions went up in flames in a bombing raid. Paschen then moved to Potsdam, where, weakened by postwar deprivations, he died of pneumonia early in 1947.

NOTES

1. S. Tolansky, "Friedrich Paschen," p. 1040.
2. Paschen, "Vita," Dissertation (1888); University of Tübingen Archives, 128/Paschen.
3. W. Gerlach, "Friedrich Paschen," p. 277; Paschen to Kayser, 14 Nov. 1895.
4. Paschen, "Vita," Dissertation (1888); "Antrittsrede," in *Sitzungsberichte der Deutschen Akademie der Wissenschaften zu Berlin* (1925), cii.
5. Paschen, "Ueber die zum Funkenübergang . . . erforderliche Potentialdifferenz," in *Annalen der Physik und Chemie*, n.s. **37** (1889), 69–96; J. J. Thomson, *The Conduction of Electricity Through Gases*, 2nd ed. (Cambridge, 1906), pp. 451 ff.

6. Paschen, "Antrittsrede" (1925), p. cii.
7. Paschen to Kayser, 17 Mar. 1895; 18 July 1895; 14 Nov. 1895; 8 Feb. 1898.
8. H. Kangro, *Vorgeschichte des Planckschen Strahlungsgesetzes.*
9. H. Kayser, "Erinnerungen aus meinem Leben" (1936), pp. 162–163 (typescript). Copy in the Library of the American Philosophical Society, Philadelphia.
10. Kangro, *Vorgeschichte* . . ., pp. 60–73.
11. Kangro, "Ultrarotstrahlung . . .," p. 181.
12. Kangro, *Vorgeschichte* . . ., *loc. cit.*
13. Paschen, "Über Gesetzmässigkeiten in den Spectren fester Körper und über eine neue Bestimmung der Sonnentemperatur," in *Nachrichten der Gesellschaft der Wissenschaften zu Göttingen*, Math.-phys. Kl. (1895), 294–304, dated June 1895.
14. *Ibid.*
15. "Spielerei": Paschen to Kayser, 18 July 1895; Iris Runge, *Carl Runge*, pp. 73–74. Henry Crew, visiting Hannover at this time, found Paschen reminded him of J. E. Keeler, while his "laboratory like Nernst's in apparent dissorder—using concave grating without any mounting, simply sets up the three parts in a room." Crew, "Diary," 12 July 1895 (American Institute of Physics, New York).
16. Paschen to Kayser, 14 Nov. 1895; Runge, *op. cit.*, p. 76.
17. Paschen, "Ueber Gesetzmässigkeiten in den Spectren fester Körper. Erste Mittheilung," in *Annalen der Physik*, 3rd ser., **58** (1896), 455–492, dated May 1896; Kangro, *Vorgeschichte* . . ., pp. 74–89.
18. Paschen, "Ueber Gesetzmässigkeiten in den Spectren fester Körper. Zweite Mittheilung," in *Annalen der Physik*, **60** (1897), 663–723, dated Jan. 1897, see 723.
19. Paschen to Kayser, 4 June [1896]; H. Kangro, "Das Paschen-Wiensche Strahlungsgesetz."
20. Paschen to Kayser, 2 Aug. 1896; *Sitzungsberichte der Preussischen Akademie der Wissenschaften* (8 Apr. 1897), 453; and (18 May 1899), 438.
21. Paschen to Kayser, 17 July 1898. This result was not published, however, until almost one year later in *Sitzungsberichte der Akademie der Wissenschaften zu Berlin* (1899), 405–420.
22. Kangro, *Vorgeschichte* . . ., pp. 165–179, 223.
23. Paschen to Kayser, 19 Jan. 1901.
24. University of Tübingen Archives, 128/Paschen; Paschen to L. Graetz, 22 July 1901 (Deutsches Museum, Munich); Paschen to Kayser, 18 Feb. 1901. "Der Neubau des physikalischen Instituts für die kgl. württemb. Landes-Universität Tübingen," in *Deutsche Bauzeitung*, **24** (1890), 213, 217.
25. Paschen to Sommerfeld, 25.8.20 (SHQP mf33); Paschen to E. Wiedemann, 10.6.13 (Darmst.).
26. Württemberg, Landtag, Kammer der Abgeordneten, "Begründung einer Exigenz von 125000 Mk. zur Erweiterung des physikalischen Instituts der Universität Tübingen," in *Verhandlungen*, 37. Landtag (1907), Beilagenband 1, Heft 15, pp. 16–18. H. M. Randall, who spent the year 1910-1911 at Tübingen, recalled that "Paschen offered to show me how each element of his entire infrared setup was constructed . . . . By 1914 a complete infrared installation of the Paschen type had been set up at Michigan. . . ." "Infrared Spectroscopy at the University of Michigan," in *Journal of the Optical Society of America*, **44** (1954), 97–103. Many examples could be given of the imitation of Paschen's installations by former students.
27. University of Tübingen Archives, 128/Paschen and 117/904; Akten der Naturwissenschaftlichen Fakultät, Tübingen.
28. Paschen to Kayser, 18 June 1919; Paschen to Sommerfeld, 25 Jan. 1919 [*sic*; actually 1920], Mar. 1920 (SHQP mf 33).
29. Paschen, "Zur Kenntnis ultraroter Linienspektra. I. (Normalwellenlängen bis 27000 Å.-E)," in *Annalen der Physik*, **27** (1908), 537–570, received 12 Aug. 1908; W. Ritz, *Gesammelte Werke*, Pierre Weiss, ed. (Paris, 1911), 521–525.
30. Runge, *op. cit.*, p. 108.

31. Paschen and E. Back, "Normale und anomale Zeemaneffekte," in *Annalen der Physik*, **39** (1912), 897–932; Paul Forman, "Back," in DSB, I, 370–371; J. B. Spencer, *Zeeman Effect*, 1896–1913.
32. Paschen to Kayser, 4 Feb. 1916.
33. H. Schüler, "Erinnerungen eines Spektrokopikers . . .," in H. Leussink *et al.*, *Studium Berolinense* (Berlin, 1960), 816–826.
34. Paschen to Sommerfeld, 32 letters Nov. 1915–Aug. 1916 (Archive for History of Quantum Physics).
35. Paschen, "Bohr's Heliumlinien," in *Annalen der Physik*, 4th ser., **50** (1916), 901–940, received 1 July 1916.
36. Paschen, "Das Spektrum des Neon," *ibid.*, **60** (1919), 405–453, and "Nachtrag," *ibid.*, **63** (1920), 201–220; Paschen and E. Back, "Liniengruppen magnetisch vervollständigt," in *Physica* (Eindhoven), **1** (1921), 261–273; and Paschen, "Die Funkenspektren des Aluminiums," in *Annalen der Physik*, 4th ser., **71** (1923), 142–161, 537–571.
37. Paul Forman, *Environment and Practice of Atomic Physics in Weimar Germany*, Ph.D. diss., Univ. of California, Berkeley, 1967 (Ann Arbor, Mich., 1968), 455–489.
38. University of Tübingen Archives, 128/Paschen; Paschen to Bohr, 11 Jan. 1924; 10 July 1924 (Archive for History of Quantum Physics).
39. Paschen to Sommerfeld, 14 Dec. 1924 (Archive for History of Quantum Physics); W. Wien to Ministerialrat [?], 14 Jan. 1926 (University of Munich Archives, Personalakten W. Wien, EII-698); H. Schüler, in *Physikalische Blätter*, **3** (1947), 232–233.
40. Armin Hermann, "Albert Einstein und Johannes Stark," in *Sudhoffs Archiv* . . ., **50** (1966), 267–286, see 283, which includes material on Paschen's role in the consideration of Stark for membership in the Berlin Academy, Dec. 1933–Jan. 1934.
41. Gerlach, *op. cit.*, p. 279.

BIBLIOGRAPHY

I. ORIGINAL WORKS. The only lists of Paschen's publications are the Royal Society *Catalogue of Scientific Papers*, XVII, 721–722; and Poggendorff, IV, 1121, 1286–1287 (under Runge); V, 618 (under Kayser), 946, 1078–1079 (under Runge), 1349 (under Weinland); VI, 1956–1957, 2291 (under R. A. Sawyer); VIIa, pt.3, 506–507. The following additional items have come to my attention: *Ueber die zum Funkenübergang in Luft, Wasserstoff und Kohlensäure bei verschiedenen Drucken erforderliche Potentialdifferenz*, his doctoral dissertation at Strasbourg (Leipzig, 1889), differs significantly from the version in *Annalen der Physik*, **37** (1889), 69–96; "Terrestrial Helium," in *Nature*, **52** (6 June 1895), 128, *Chemical News* . . ., **71** (14 June 1895), 286; and *Chemiker-zeitung*, **19** (1895), 977, written with C. Runge; "Ueber das Strahlungsgesetz des schwarzen Körpers," in *Annalen der Physik*, 4th ser., **4** (1901), 277–298; "Eine neue Bestimmung der Dispersion des Flusspates im Ultrarot," *ibid.*, 299–303; "Bestimmung des selectiven Reflexionsvermögens einer Planspiegel," *ibid.*, 304–306; "Erweiterung des Seriengesetzes der Linienspectra auf Grund genauer Wellenlängenmessungen im Ultraroth," in *Comptes rendus du Congrès international de radiologie et électricité, Brussels 1910*, I (Brussels, 1911), 588–600, also in *Jahrbuch der Radioaktivität und Elektronik*, **8** (1911), 174–186;

and "Antrittsrede," in *Sitzungsberichte der Preussischen Akademie der Wissenschaften* (2 July 1925), cii–civ.

Paschen's MSS apparently were destroyed with his home in 1943. Some 4 letters from Paschen to N. Bohr, 4 to W. Gerlach, 1 to S. A. Goudsmit, 1 to L. Graetz, 37 to H. Kayser, 1 to J. Königsberger, 14 to A. Landé, 1 to A. G. Shenstone, 87 to A. Sommerfeld, 2 to J. R. Swinne, and 1 to E. Wiedemann are listed or cited in T. S. Kuhn *et al.*, *Sources for History of Quantum Physics* (Philadelphia, 1967), 72–73. The items in the Darmstädter Collection, F1c(4) 1893, cited there, particularly the important collection of 37 letters to Kayser—17 Mar. 1895; 24 Mar. (1895); 18 July (1895), 1 Aug. 1895; 14 Nov. 1895; 25 Nov. 1895; 30 Dec. 1895; 7 Jan. 1896; 13 Feb. 1896; 6 May 1896; 4 June (1896); 19 July 1896; 25 July 1896; 2 Aug. 1896; 8 Feb. 1898; 23 Feb. 1898; 17 July 1898; 30 Dec. 1900; 3 Jan. 1901; 9 Jan. 1900 [*sic*; actually 1901]; 19 Jan. 1901; 22 Jan. 1901; 18 Feb. 1901; 7 June 1902; 5 July 1903; 7 July 1903; 9 June 1905; 16 June 1905; 19 Nov. 1910; 3 Oct. 1912; 14 Oct. 1913; 4 Feb. 1916; 2 July 1915 [*sic*; actually 1916]; 29 Mar. 1919; 18 June 1919; 11 Oct. 1921; 14 Sept. 1923—are now in the Staatsbibliothek Preussischer Kulturbesitz, Berlin-Dahlem. The Nachlass Stark in the same depository includes 25 letters from Paschen to Stark: 11 Jan. 1905; 28 Jan. 1905; 6 Mar. 1905; 19 May (1906); 10 July 1906; 2 Aug. 1906; 29 Sept. 1906; 3 Oct. 1906; 10 Nov. (1906); 12 Feb. 1907; 3 June 1907; 19 June (1907); 21 June (1907); 27 June (1907); 29 Feb. 1908; 18 Mar. 1911; 15 Oct. 1911; 18 Oct. 1911; 20 Oct. 1911; 22 Oct. 1911; 3 Oct. 1918; 20 Oct. 1918; 4 May 1927; 6 July 1927; 2 Oct. 1927. There are 3 additional letters to N. Bohr—11 Jan. 1924; 30 Mar. 1924; 10 July 1924—in the Bohr Collection, Niels Bohr Institutet, Copenhagen; and at least 3 letters to W. F. Meggers —6 Sept. 1921; 14 June 1924; 15 Oct. 1925—in the Meggers Papers, American Institute of Physics, New York.

II. SECONDARY LITERATURE. Paschen's work on Kirchhoff's emission function is discussed in detail by Hans Kangro, *Vorgeschichte des Planckschen Strahlungsgesetzes. Messungen und Theorien der spektralen Energieverteilung . . .* (Wiesbaden, 1970), summarized in Kangro's "Das Paschen-Wiensche Strahlungsgesetz und seine Abänderung durch Max Planck," in *Physikalische Blätter*, **25** (1969), 216–220, and touched upon in his "Ultrarotstrahlung bis zur Grenze elektrisch erzeugter Wellen: Das Lebenswerk von Heinrich Rubens," in *Annals of Science*, **26** (1970), 235–259, and **27** (1971), 165–200. Paschen's collaboration and personal relations with Carl Runge are described in Iris Runge, *Carl Runge und sein wissenschaftliches Werk* (Göttingen, 1949). The magneto-optical work of Paschen and his school up through the discovery of the Paschen-Back effect is discussed in detail in James Brooks Spencer, *An Historical Investigation of the Zeeman Effect, 1896–1913*, Ph.D. diss., U. of Wisconsin, 1964 (Ann Arbor, 1964). Some of their later work is discussed by P. Forman, "Alfred Landé and the Anomalous Zeeman Effect, 1919–1921," in *Historical Studies in the Physical Sciences*, **2** (1970), 153–262.

There are no biographical studies of Paschen apart from the very few and spare obituary notices. The best of these is Walther Gerlach, "Friedrich Paschen," in *Jahrbuch der Bayerischen Akademie der Wissenschaften* (1944–1948), 277–280—Paschen had been elected a corresponding member in 1922. Others are S. Tolansky, "Friedrich Paschen," in *Proceedings of the Physical Society of London*, **59** (1947), 1040–1041; H. Schüler, "Friedrich Paschen," in *Physikalische Blätter*, **3** (1947), 232–233; R. Seeliger, "Nachruf auf Friedrich Paschen," in *Jahrbuch der Deutschen Akademie der Wissenschaften zu Berlin* (1946–1949), 199–201; W. Heisenberg et al., "Friedrich Paschen," in *Annalen der Physik*, 6th ser., **1** (1947), 137–138. A notice by Carl Runge in honor of Paschen's sixtieth birthday, "Friedrich Paschen," in *Naturwissenschaften*, **13** (1925), 133–134, gives reminiscences of the origin of their collaboration in 1895; Niels Bohr, "Friedrich Paschen zum siebzigsten Geburtstag," *ibid.*, **23** (1935), 73, testifies to Paschen's "happy intuition, by which he always has pursued experimentally those problems the investigation of which proved to be of decisive significance for the extension of general theoretical conceptions."

PAUL FORMAN

PASTEUR, LOUIS (*b.* Dole, Jura, France, 27 December 1822; *d.* Chateau Villeneuve-l'Étang, near Paris, France, 28 September 1895), *crystallography, chemistry, microbiology, immunology.*

Outline of Pasteur's Career

1829–1831	Student at École Primaire, Arbois
1831–1839	Student at Collège d'Arbois
1839–1842	Student at Collège Royal de Besançon
1842–1843	Student at Barbet's School and Lycée St.-Louis, Paris
1843–1846	Student at École Normale Supérieure (Paris)
1846–1848	Préparateur in chemistry, École Normale
1849–1854	Professor of chemistry, Faculty of Sciences, Strasbourg suppléant, 1849–1852 titulaire, 1852–1854
1854–1857	Professor of chemistry and dean of the Faculty of Sciences, Lille
1857–1867	Administrator and director of scientific studies, École Normale
1867–1874	Professor of chemistry, Sorbonne
1867–1888	Director of the laboratory of physiological chemistry, École Normale
1888–1895	Director of the Institut Pasteur (Paris)

In addition:

Sept.–Dec. 1848 Professor of physics, Lycée de Dijon
1863–1868 Professor of geology, physics, and chemistry in their application to the fine arts, École des Beaux-Arts (Paris)

List of Pasteur's Major Prizes and Honors

1853 Chevalier of the Imperial Order of the Legion of Honor
1853 Prize on racemic acid, Société de Pharmacie de Paris
1856 Rumford Medal, Royal Society (for work in crystallography)
1859 Montyon Prize for Experimental Physiology, Académie des Sciences
1861 Zecker Prize, Académie des Sciences (chemistry section)
1862 Alhumbert Prize, Académie des Sciences
1862 Elected member of the Académie des Sciences (mineralogy section)
1866 Gold Medal, Comité Central Agricole de Sologne (for work on diseases of wine)
1867 Grand Prize Medal of the Exposition Universelle (Paris), for method of preserving wine by heating
1868 Honorary M.D., University of Bonn (returned during Franco-Prussian War, 1870–1871)
1868 Promoted to commander of the Legion of Honor
1869 Elected fellow of the Royal Society
1871 Prize for silkworm remedies, Austrian government
1873 Commander of the Imperial Order of the Rose, Brazil
1873 Elected member of the Académie de Médecine
1874 Copley Medal, Royal Society (for work on fermentation and silkworm diseases)
1874 Voted national recompense of 12,000 francs
1878 Promoted to grand officer of the Legion of Honor
1881 Awarded Grand Cross of the Legion of Honor
1882 Grand Cordon of the Order of Isabella the Catholic
1882 National recompense augmented to 25,000 francs
1882 Elected to Académie Française
1886 Jean Reynaud Prize, Académie des Sciences
1887 Elected perpetual secretary, Académie des Sciences (resigned because of illness in January 1888)
1892 Jubilee celebration at the Sorbonne

Chronological Outline of Pasteur's Major Research Interests

1847–1857 Crystallography: optical activity and crystalline asymmetry
1857–1865 Fermentation and spontaneous generation; studies on vinegar and wine
1865–1870 Silkworm diseases: *pébrine* and *flacherie*
1871–1876 Studies on beer; further debates over fermentation and spontaneous generation
1877–1895 Etiology and prophylaxis of infectious diseases: anthrax, fowl cholera, swine erysipelas, rabies

Pasteur and His Place in History. If Pasteur was a genius, it was not through ethereal subtlety of mind. Although often bold and imaginative, his work was characterized mainly by clearheadedness, extraordinary experimental skill, and tenacity—almost obstinacy—of purpose. His contributions to basic science were extensive and very significant, but less revolutionary than his reputation suggests. The most profound and original contributions are also the least famous. Beginning about 1847 Pasteur carried out an impressive series of investigations into the relation between optical activity, crystalline structure, and chemical composition in organic compounds, particularly tartaric and paratartaric acids. This work focused attention on the relationship between optical activity and life and provided much inspiration and several of the most important techniques for an entirely new approach to the study of chemical structure and composition. In essence, Pasteur opened the way to a consideration of the disposition of atoms in space, and his early memoirs constitute founding documents of stereochemistry.

From crystallography and structural chemistry Pasteur moved to the controversial and interrelated topics of fermentation and spontaneous generation. If he did more than anyone to promote the biological theory of fermentation and to discredit the theory of spontaneous generation, his effect was due less to profound conceptual originality than to experimental ingenuity and polemical virtuosity. He did broach and contribute fundamentally to important questions in microbial physiology—including the relationship between microorganisms and their environment—but he was readily distracted from such basic issues by more practical concerns—the manufacture of wine, vinegar, and beer, the diseases of silkworms, and the etiology and prophylaxis of diseases in general.

To an extent, Pasteur's interest in practical problems evolved naturally from his basic research, especially that on fermentation, for the biological theory of fermentation contained obvious implications for industry. By insisting that each fermentative process could be traced to a specific living microorganism, Pasteur not only drew attention to the purity of the causative organism and the amount of oxygen employed, but also suggested that the primary industrial product could be preserved by appropriate sterilizing procedures, called "pasteurization" almost from the outset. Furthermore, the old and widely accepted analogy between fermentation and disease made any theory of the former immediately relevant to the latter. Pasteur's biological theory of fermentation virtually implied a biological or "germ" theory of disease. This implication was more rapidly developed by others, particularly Joseph Lister; but Pasteur also perceived it from the first and devoted his last twenty years almost exclusively to the germ theory of disease.

No one insisted more strongly than Pasteur himself on the degree to which his pragmatic concerns grew out of his prior basic research. He saw the progression from crystallography through fermentation to disease as not only natural but virtually inevitable; he had been "enchained," he wrote, by the "almost inflexible logic of my studies."[1] This view, however enduring and widely accepted, has not gone entirely unchallenged. René Dubos has emphasized how Pasteur's work could have taken many other directions with equal fidelity to the internal logic of his research.[2] To some extent Pasteur chose, or at least allowed himself to pursue, the practical consequences of his work at the expense of his potential contributions to basic science. Without disputing the immense value and fertility of the basic research he did accomplish, it is fascinating to speculate on what might have been. Late in life, Pasteur indulged in similar speculation and expressed regret that he had abandoned his youthful researches before fully resolving the relationship between asymmetry and life. Had he contributed as much as he had once hoped toward this problem, he would surely have fulfilled his ambition of becoming the Newton or Galileo of biology.

By taking another direction, however, Pasteur revealed the enormous medical and economic potential of experimental biology. He himself developed only one treatment directly applicable to a human disease—his treatment for rabies—but his widely publicized and highly successful efforts on behalf of the germ theory were immediately credited with saving much money and many lives. It is for this reason above all that he was recognized and honored during his lifetime and that his name remains a household word.

As his letters make clear, Pasteur chose his path under the impulse of complex and mixed motives.

Apart from the internal logic of his research, these motives included ambition for fame and imperial favor, his wish to serve his country and humanity, and his concern for financial security (more for the sake of his work and his family than for himself). In the highly competitive academic life of mid-nineteenth-century France, he was unabashedly ambitious and opportunistic. Not yet thirty, he consoled his rather neglected wife by telling her that he would "lead her to posterity."[3] Pasteur's correspondence is filled with references to academic politics and with appeals for support from his influential friends—notably Biot and Dumas at the outset of his career, and later a number of important ministers and government officials, including Emperor Louis Napoleon and Empress Eugénie.

Pasteur sometimes complained bitterly of the neglect of science by the French state; but once his concern with practical problems became manifest, he had remarkable success in getting what he sought—a new laboratory, additional personnel, a larger research budget, a national pension for himself, even railroad passes for himself and his assistants. His support, although not spectacular in comparison with that provided to some scientists in German universities, was unusually generous by French governmental standards.

To supplement it, Pasteur competed actively for awards from private societies and foreign governments. Here too he enjoyed considerable success. For his work on racemic acid, for example, he received a prize of 1,500 francs from the Société de Pharmacie de Paris in 1853; and for his efforts to aid the silkworm industry, he was awarded 5,000 florins by the Austrian government. By far the most spectacular award for which Pasteur competed—in this case unsuccessfully—was a prize of 625,000 francs offered in 1887 by the government of New South Wales for practical measures to reduce the rabbit population. As unpublished correspondence makes clear,[4] Pasteur sought this fortune partly for the sake of his family and partly to support the projected Institut Pasteur, toward the creation of which a widely publicized and highly successful drive had been launched. The fame of Pasteur's treatment for rabies attracted donors throughout the world, and the value of their contributions surpassed 2 million francs by November 1888,[5] when the Institut Pasteur was officially inaugurated. The French National Assembly had already voted him two national recompenses—one in 1874 with an annual value of 12,000 francs and another in 1883 that increased his life annuity to 25,000 francs and made it transferable upon his death to his wife and then to his children.[6]

Pasteur secured yet other revenues from patents or

licenses for products and processes that resulted from his research. In 1861 he patented his method of making vinegar; and he later received patents or licenses for his methods of preserving wine and manufacturing beer, for a bacterial filter (the Chamberland-Pasteur filter), and for his vaccines against fowl cholera, anthrax, and swine erysipelas. No adequate account exists of the fate of these patents and licenses, but some were allowed to enter the public domain or were otherwise unexploited, while those for the filter and vaccines apparently yielded large revenues, most of which seem to have gone to the state or to the Institut Pasteur. Apparently at the urging of his wife and family, Pasteur accepted some unknown amount of the income from his patents. His will reveals only that he left his wife "all that the law allows."[7] Apparently Pasteur amassed no large personal fortune. Although exaggerated, his insistence that he worked solely for the love of science and country and the standard portrayal of him as a "savant désintéressé" carry more conviction than attempts to depict him as a scientific prostitute. Compared, for example, with Liebig, he was a model of commercial restraint.[8]

Pasteur displayed no comparable restraint in controversy. Combative and enormously self-assured, he could be devastating to the point of cruelty. He so offended one opponent, an eighty-year-old surgeon, that the latter challenged him to a duel.[9] Although often counseled to spend his energy more productively, Pasteur was constitutionally incapable of suffering criticism in silence. A few debates, notably on spontaneous generation, did stimulate valuable work and produce important clarifications, but most were barren. Sharing with many contemporary scientists a zealous concern for his intellectual property, he spent considerable time and effort to establish the priority of concepts and discoveries, particularly his process for preserving wines. Pasteur generally gave credit to others only grudgingly and mistrusted those who claimed to have reached similar views independently. He also shared a rather simpleminded and absolutist notion of scientific truth, rarely conceding the possibility of its being multifaceted and relative. By appeal to public demonstrations—notably in the sensational vaccination experiments at Pouilly-le-Fort—and by frequent recourse to "judiciary" commissions of the Académie des Sciences, Pasteur almost invariably won public and quasi-official sanction for his views.[10]

Although in some ways unfortunate, Pasteur's polemical inclinations and talents were a major factor in his success. Intuitively at least, he perceived that the essential measure of a scientist's achievement is the degree to which he can persuade the scientific community of his views. By this measure, Pasteur was enormously successful, thanks in part to his tendency toward self-advertisement. The most obvious factor contributing to his success was his tremendous capacity for work; equally important was his ability to concentrate intensely on one problem for remarkably long periods. Other factors, especially obvious in his early work in crystallography, were his powerful visual imagination and highly developed aesthetic sense. Perhaps the most surprising factor invoked to explain Pasteur's success was his myopia, which reportedly enhanced his close vision so that, in an object under the microscope or between his hands, he saw things hidden to those around him.[11]

His father's constant concern for his health suggests that Pasteur had never been robust, and excessive physical and mental exertion further undermined his constitution. On 19 October 1868, in the midst of silkworm studies, Pasteur suffered a cerebral hemorrhage that completely paralyzed his left side. Treated with leeches and later by electricity and mineral waters, he improved somewhat but retained a lifelong hemiplegia that impaired his speech and prevented his performing most experiments. He continued to design and direct experiments with his usual care and ingenuity, but their execution was often left to collaborators. For nearly twenty years Pasteur's health remained fairly stable; but in the autumn of 1886 he began to experience cardiac deficiency and in October 1887 he suffered another stroke that further impaired his speech and mobility. His strength fading steadily, Pasteur was visibly feeble when he moved into the Institut Pasteur in 1889. In 1892 he expressed a brief enthusiasm for Charles Brown-Séquard's controversial testicular injections, but in 1894 he suffered what was probably a third stroke.[12] At his death he was almost completely paralyzed.

Virtually obsessed with science and its applications, Pasteur devoted little thought to political, philosophical, or religious matters. His beliefs in these areas were basically visceral or instinctive. His close association with the Second Empire reflects his political instincts. Despite a youthful flirtation with republicanism during the Revolution of 1848, Pasteur was essentially conservative, not to say reactionary. He considered strong leadership, firm law enforcement, and the maintenance of domestic order more important than civil liberty or even democracy, which he distrusted lest it lead to national mediocrity or vulgar tyranny. Yearning for the past glory of France, which he traced to Napoleon, he believed that Louis Napoleon might somehow restore it.[13]

From the coup d'état of 2 December 1851, by which Louis Napoleon dissolved the Constituent Assembly, Pasteur declared himself a "partisan" of the new leader.[14] Partly through Dumas, whom Napoleon III named a senator, Pasteur developed personal relations with the imperial household, to which he sent copies of his works on fermentation and spontaneous generation. Especially after 1863, when Dumas presented him to Louis Napoleon, Pasteur openly sought to attract imperial interest to his research. He dedicated his book on wines (1866) to the emperor and his book on silkworm diseases (1870) to the empress, who had encouraged him during the difficult early stages of this work.[15]

Louis Napoleon's deposition in 1870 nullified an imperial decree of 27 July 1870 by which Pasteur would have been awarded a national pension and made a senator. In 1868 the emperor had promoted Pasteur to commander of the Legion of Honor, and in 1865 had invited him to Compiègne, the most elegant imperial residence. During a week there Pasteur, in giddy letters to his wife, betrayed his awe of, and fascination with, imperial power, pomp, and wealth.[16] No mere political opportunist, however, he continued to acknowledge his association with and indebtedness to the empress after the abdication—in the face of advice that it could be politically imprudent to do so.[17]

However firm Pasteur's loyalty to the Second Empire, his general patriotism was even stronger. In 1871, despite tempting offers from Milan and Pisa, Pasteur remained in France, partly because of his wife's unwillingness to expatriate but especially because he felt it would be an act of desertion to leave his country in the wake of its crushing defeat by Prussia.[18] That defeat and the excesses of the Prussian army so aroused Pasteur that he vowed to inscribe all of his remaining works with the words, "Hatred toward Prussia. Revenge! Revenge!"[19] Also in 1871 he returned in protest an honorary M.D. awarded in 1868 by the University of Bonn. In an exchange of letters with the dean of the faculty of medicine there, which he published as a brochure, Pasteur cried out in rage at the "barbarity" being visited upon his country by Prussia and its king. In another brochure of 1871, "Some Reflections on Science in France," Pasteur emphasized the disparity between the state support of science in France and in Germany, and traced the defeat of France in the war to its excessive tolerance toward the "Prussian canker [*chancre*]" and to its neglect of science during the preceding half-century.

During the war and, later, the Commune, Pasteur withdrew to the provinces and launched his studies on beer—his explicit object being to bring France into competition with the superior German breweries. In 1873, when he patented the process that resulted from these studies, Pasteur stipulated that beer made by his method should bear in France the name "Bières de la revanche nationale" and abroad the name "Bières françaises."[20] Chauvinism undoubtedly played some part in his refusal to grant permission to translate his *Études sur la bière* into German and in his bitter and protracted controversy with Robert Koch in the 1880's. Even on the eve of his death, Pasteur's memories of the war remained so strong that he declined the Prussian Ordre Pour le Mérite.[21]

In 1875 Pasteur was asked by friends in Arbois to run for the Senate. Saying that he had no right to a political opinion because he had never studied politics, he nonetheless consented to run as a conservative. Presenting himself as the candidate of science and patriotism, he rehearsed his published explanations for the fall of France in the Franco-Prussian War and made his central political pledge "never [to] enter into any combinations the goal of which is to upset the established order of things."[22] Although Pasteur's strong commitment to scientific professionalism probably struck some as elitist, the main issues against him were his conservatism, his links with the Second Empire, and his suspected Bonapartist loyalties. In response, Pasteur reported that the emperor had died owing him 4,000 francs and disclaimed any link with organized Bonapartist groups. He was soundly defeated, receiving only 62 votes, nearly 400 less than each of the two successful candidates (both republicans). Although asked at least twice during the 1880's to run again for the Senate, Pasteur declined while his strength for scientific work remained. By then he referred to politics as ephemeral and sterile compared with science, a view that can only have been reinforced by his hostile reception on a visit to Arbois in 1888.[23] In 1892, no longer strong enough for research, Pasteur began soliciting support for a place in the Senate but eventually withdrew.[24]

At the center of Pasteur's public views on religion and philosophy lay his insistence on an absolute separation between matters of science and matters of faith or sentiment.[25] Although he was reared and died a Catholic, religious ritual and sectarian doctrine held little attraction for him. He cared as little for formal philosophy. By 1865 he had read only a few "absurd passages" in Comte, and he described his own philosophy as one "entirely of the heart."[26] Throughout his life he disdained materialists, atheists, freethinkers, and positivists. In 1882, in his inaugural address to the Académie Française, Pasteur found wanting the positivistic philosophy of Émile Littré, whom he was replacing. For Pasteur, the failures of positivism included its lack of real intellectual novelty, its confusion of the true experimental method with the "restricted method" of observation, and above all its disregard for "the most important of positive notions, that of the Infinite," one form of which is the idea of God. Pasteur never doubted the existence of the spiritual realm or of the immortal soul. In that sense, and in his opposition to philosophical materialism, he was a spiritualist. Indeed, in his inaugural address he spoke of the service his research had rendered to the "spiritualistic doctrine, much neglected elsewhere, but certain at least to find a glorious refuge in your ranks."[27]

Pasteur's chief contribution to the "spiritualist doctrine" was his campaign against spontaneous generation, the religiophilosophical consequences of which he emphasized in an address at the Sorbonne in 1864 while fervently denying that these broader issues had influenced his actual research. To the extent that any question was truly scientific, he argued, neither spiritualism nor any other philosophical school had a place in it. The "experimental method" alone could arbitrate scientific disputes. And while limited hypotheses played an essential role in the experimental method, speculation on the ultimate origin and end of things was beyond the realm of science. Despite this public posture, Pasteur sometimes speculated on the origin of life and attempted to create it experimentally, as he finally confessed in 1883.[28] And while the results of his work on fermentation, spontaneous generation, and disease may point toward a vitalistic rather than a mechanistic position, it would be misleading if not erroneous to label Pasteur a vitalist.

Pasteur was frank, stubborn, prodigiously self-confident, intensely serious—almost somber—and rather aloof toward those outside his select circle. Obsessed with his work, he brooked no interference with it. Sincerely kind to children, he could be insensitive and exploitative to others. His passion for tidiness and cleanliness approached the eccentric, and fear of infection allegedly made him wary of shaking hands or of eating without first wiping the dinnerware and scrutinizing his food.[29] Pasteur tended to be highly secretive about the general direction of his current work, even with his most trusted assistants; and his insistence on absolute control of his laboratory reportedly extended even to the recording of experimental notes and the labeling of animal cages.[30]

An innovative administrator and fastidious organizer, Pasteur showed a legendary devotion to detail. As director of scientific studies at the École Normale Supérieure, he proposed procedural and structural reforms, notably with regard to the *agrégés-préparateurs* (laboratory assistants who were graduates of the school); founded a journal, *Annales scientifiques de l'École normale supérieure*; and raised the standards and reputation of the scientific section so that it began to challenge the École Polytechnique. On the other hand, Pasteur's handling of student discipline betrayed an inflexible and rather authoritarian spirit. His relations with students were described as "hardly frequent" but "often disagreeable."[31] He dealt summarily, unsympathetically, and sometimes arbitrarily with student complaints about food and rules; and by 1863 he was openly appalled by what he considered student insubordination. In 1867 Pasteur was removed from his post as administrator and director of scientific studies precisely because of his rigid and unpopular stand against a student protest involving free speech and anti-imperial sentiment.[32]

Pasteur was considered an excellent teacher, and his lectures were beautifully organized if not spellbinding. During the last two decades of his life, however, he taught only by precept and example in the laboratory and only those few who could simultaneously contribute to his own work and meet his exacting standards. He therefore trained very few students directly, but several of them—notably Émile Duclaux and Émile Roux—transmitted the spirit of his work to others who established and staffed the more than 100 medical institutes and scientific centers that now bear Pasteur's name. Despite his tendency to be as demanding of others as he was of himself, Pasteur inspired tremendous loyalty. If any assistant-collaborator felt that his contributions were being unduly appropriated to Pasteur's name, none ever expressed that feeling publicly.

In realizing most if not all of his ambitions, Pasteur became a national hero and "benefactor of humanity" to many while arousing the envy and hostility of others. A portion of the medical profession and of what he denigrated as the "so-called scientific press" vilified him as an intolerant representative of "official" science, an egomaniacal and greedy opportunist, and a would-be suppressor of dissident views. Some fervently denied that his work had brought the immense industrial, agricultural, or medical benefits claimed for it. In addition to debates over the safety and efficacy of Pasteur's treatment for rabies, there were questions about the degree of success of his other vaccines, preservative processes, and remedies.[33] These questions deserve more detailed examination; but the contemporary attacks on Pasteur were generally so exaggerated and badly argued that they fail now, as then, to persuade others of the residue of truth they contain.

Early Life and Education. Until the late seventeenth century the Pasteurs were simple laborers or tenant farmers in the Franche-Comté, on the eastern border of France. Then, for two generations, Pasteur's ancestors were millers at Lemuy, in service to the count of Udressier. About the middle of the eighteenth century his great-grandfather migrated to Salins-les-Bains, where he became a tanner and, by payment to and "special grace" of the count of Udressier, achieved independence for himself and his posterity. Pasteur's grandfather, Jean-Henri Pasteur (1769–1796), moved to Besançon, where he too worked as a tanner. His only son, Jean-Joseph Pasteur, was Louis Pasteur's father.

Born in 1791, Jean-Joseph Pasteur was drafted into the French army in 1811. As a member of the celebrated Third Regiment of Napoleon's army, he served with distinction in the Peninsular War during 1812–1813. By 1814, when he was discharged, he had attained the rank of sergeant major and had been awarded the cross of the Legion of Honor. Upon his return to civilian life, Jean-Joseph also became a tanner, initally at Besançon. In 1816 he married Jeanne-Étiennette Roqui, daughter of a gardener from a family of the Franche-Comté. They moved to Dole, where the first four of their five children were born. Louis, their third child, was preceded by a son who died in infancy and by a daughter born in 1818; two daughters were born later. About 1826 the family moved to Marnoz, the native village of the Roqui family, and in 1827 to the neighboring town of Arbois, on the Cuisance River, where a tannery had become available for lease. It was in Arbois, a town of about 8,000 inhabitants, that Louis grew up and to which he returned periodically.

From his parents Louis absorbed the traditional *petit bourgeois* values: familial loyalty, moral earnestness, respect for hard work, and concern for financial security. Jean-Joseph, who had received little education, wished only that his son should join the faculty of a local *lycée*. Louis, who at one time apparently shared this goal, gradually directed his vision toward the scientific elite in Paris. Jean-Joseph's modest ambitions for his son seem entirely in keeping with Louis's early performance at school. In 1831, after two years in the associated École Primaire, Louis entered the Collège d'Arbois as a day pupil; he was for several years considered only a slightly better-than-average student. Until quite near the end of his secondary schooling, nothing in his record presaged his later success and fame. Only his genuine, if immature, artistic talent seemed to promise anything exceptional. Several early portraits of friends, teachers, and acquaintances have been preserved; two sensitive character sketches of his parents reveal a talent quite beyond the ordinary.

If Louis ever seriously considered an artistic career, he was dissuaded by his pragmatic father and by Bousson de Mairet, a family friend and headmaster of the Collège d'Arbois until 1837. Under Bousson and his successor, Romanet, Louis's scholarly enthusiasm was at last aroused; and he swept the school prizes

during the academic year 1837–1838. Bousson and Romanet also awakened his ambition to prepare for the École Normale Supérieure. Apparently with this end in view, it was arranged that he enter the preparatory school in Paris headed by M. Barbet, himself a Franc-Comtois. Louis arrived in Paris in October 1838; less than a month later, overwhelmed by homesickness, he returned to Arbois. His superb performance that year at the Collège d'Arbois inspired him to prepare again for the École Normale.

Because the Collège d'Arbois had no class in philosophy leading to the baccalaureate in letters, Louis was compelled to continue his studies elsewhere. On 29 August 1840 he received his bachelor's degree in letters from the Collège Royal de Besançon. He received a mark of "good" in all subjects except elementary science, in which he received a "very good." Consumed with the ambition of entering the science section of the École Normale, he had first to obtain a bachelor's degree in science. His family's financial burdens were eased by his appointment as "preparation master" or tutor at the Collège Royal de Besançon. After two years in the class of special mathematics there, Pasteur received his baccalaureate in science on 13 August 1842, although in physics he was considered merely "passable," and in chemistry "mediocre." Two weeks later he was declared admissible to the École Normale, but he was dissatisfied with his rank of fifteenth among twenty-two candidates and declined admittance. Having also considered a career as a civil engineer, Pasteur took the entrance examination for the École Polytechnique in September but failed.[34] He decided to spend another year preparing for the École Normale. In letters to his parents he emphasized the importance of study in Paris; and in October 1842 he returned to Barbet's boarding school.

Like all students at the Barbet school, Pasteur attended the classes of the Lycée St.-Louis; but he also went to hear Jean-Baptiste Dumas, professor of chemistry at the Sorbonne, whose fervent admirer he quickly became. At the end of the academic year 1842–1843, he took first prize in physics at the Lycée St.-Louis, sixth "accessit" in physics in the annual general competition, and was admitted fourth on the list of candidates to the science section of the École Normale, which he entered in the autumn of 1843.

Until November 1848 Pasteur studied and worked at the École Normale. Before he could join even a secondary school faculty, he had to pass the license examination and to compete in the annual *agrégation*. In the license examination, which he took in 1845, Louis placed seventh. In September 1846 he placed third in the annual *agrégation* in the physical sciences. His appointment in October as *préparateur* in chemistry to Antoine Jérome Balard at the École Normale enabled Pasteur to continue toward his doctorate, which he received in August 1847 with dissertations in both physics and chemistry. While awaiting an appropriate post, he continued to work as *préparateur* at the École Normale and launched those studies on optical activity which were to make his early reputation.

Optical Activity, Asymmetry, Crystal Structure. By the time he completed his dissertation in physics, Pasteur's interest in optical activity had emerged. Already attracted to crystallography by the lectures of Gabriel Delafosse, professor of mineralogy at the École Normale, he found his interest intensified by his association with Auguste Laurent, who worked in the same laboratory from late in 1846 until April 1847. In his dissertation Pasteur also expressed indebtedness to Biot, whose own polarimeter Pasteur had used and whose pioneering papers on the optical activity of organic liquids had served as a guide. Essentially a preliminary methodological study, Pasteur's dissertation focused in part on the relation between isomorphism and optical activity. The results, based on two pairs of isomorphic substances, supported Laurent's view that substances of the same crystalline form possess the same optical activity in solution. One of these isomorphic pairs belonged to the tartrates, and Pasteur's other references to the tartrates and paratartrates suggest that he had already begun a systematic study of them.

Ordinary tartaric acid had been known since the eighteenth century. Prepared from salts of the tartar deposited as a by-product in wine vats, it had become especially important in medicine and in dyeing. Racemic or paratartaric acid had come to the attention of chemists only in the 1820's, when Gay-Lussac established that it possessed the same chemical composition as ordinary tartaric acid. Because of their importance for the emerging concept of isomerism, the two acids had thereafter attracted considerable notice. The studies of Biot and Eilhard Mitscherlich had established that aqueous solutions of tartaric acid and its derivatives rotated the plane of polarized light to the right, while aqueous solutions of racemic acid and its derivatives exerted no effect on it. Indeed, in a brief note of 1844 Mitscherlich had claimed that in one case—the sodium-ammonium double salts—the tartrates and paratartrates were identical in every respect, including crystalline form and atomic arrangement, except for this difference in optical activity.

Pasteur later emphasized the seminal role of Mitscherlich's note in his work. He had been deeply disturbed, he said, by the notion that "two substances could be as similar as claimed by Mitscherlich without

being completely identical."[35] Pasteur's approach to the problem reflects his tutelage under Delafosse, who had made a special study of hemihedrism and naturally emphasized it in his lectures.[36] Through him Pasteur learned of the earlier work of Haüy, Biot, and John Herschel on crystallized quartz. Haüy had shown that some quartz crystals are hemihedral to the left, while others are hemihedral to the right. Biot had shown that some quartz crystals rotate the plane of polarized light to the left, while others of the same thickness rotate it an equal amount to the right. Herschel in 1820 had established a causal connection between the asymmetrical crystalline forms and the direction of optical activity. Because quartz displays optical activity only in the crystallized state and loses it when dissolved, it had been recognized that only the quartz crystal as a whole, and not its constituent molecules, is asymmetrical. But Biot had also found a number of natural organic substances—oil of turpentine, camphor, sugar, tartaric acid—that were optically active in aqueous solutions or in the fluid state. As he emphasized, optical activity in such cases—unlike that of quartz—must depend on an asymmetry in the form of the constituent molecules.

Obviously prepared in part by the ideas of Delafosse and Laurent, Pasteur became convinced that the molecular asymmetry of optically active liquids ought to find expression in an asymmetry or hemihedrism in their crystalline form. In May 1848—having published several related papers on isomorphism and dimorphism in various compounds—Pasteur announced the discovery of small hemihedral facets on the crystals of all nineteen tartrate compounds he had studied. In all of them the hemihedral facets inclined in the same direction, and the direction of optical activity was the same. In the optically inactive paratartrates Pasteur expected to find perfectly symmetrical crystals. This expectation was confirmed with the notable exception of the sodium-ammonium paratartrate on which Mitscherlich's claims specifically rested. At first disappointed when he found hemihedrism in these crystals, Pasteur soon noticed that certain crystals inclined to the right, others to the left. Pasteur meticulously separated them by hand, dissolved them, and found that solutions of the right-handed crystals rotated the plane of polarized light in one direction while solutions of the left-handed crystals rotated it in the opposite direction to approximately the same degree. When equal weights of the two kinds of crystals were dissolved separately and then combined, the result was an optically inactive sodium-ammonium paratartrate.

Similar results were obtained with the acids from which the sodium-ammonium salts had been derived.

Right-handed salts gave a right-handed acid identical to ordinary tartaric acid. Left-handed salts gave a hitherto unknown acid identical to tartaric acid except for the left-handed direction of both its hemihedrism and optical activity. Combinations of equal weights of the left-handed and right-handed acids yielded an acid identical to racemic or paratartaric acid. Pasteur now concluded that the optical inactivity of paratartaric acid (and hence of its derivatives) resulted from its being a combination of two optically active acids that were mirror images of each other, the separate optical activities of which, in opposite directions, compensated for or canceled each other.

These results, quickly confirmed by Biot and further developed by Pasteur in a series of papers between 1848 and 1850, bear striking testimony to the fertility of an admittedly a priori conception. Indeed, so powerful was Pasteur's conviction that tartrates and other optically active substances must possess hemihedral facets that he was able not only to see subtle distinctions that had eluded earlier observers, but in a sense even to produce them by appropriate adjustments in the conditions of crystallization. The hemihedral forms of sodium-ammonium paratartrate appear only under quite special and delicate conditions, especially with regard to temperature, a circumstance that leads some to assign luck a rather large role in Pasteur's first great discovery.[37] His decision to begin with the tartrates and paratartrates seems at least as fortunate, for in no other optically active compounds is the relationship between molecular asymmetry and crystalline structure so clear or straightforward; and Pasteur soon had to contend with several "exceptions" to his "law of hemihedral correlation."

Meanwhile, his credentials having been established, Pasteur was appointed professeur suppléant in chemistry at the Faculty of Sciences in Strasbourg on 29 December 1848. On 29 May 1849 he married Marie Laurent, daughter of the rector of the Strasbourg Academy. Devoted to her husband and his career, tolerating his intense absorption in his work,[38] and often serving as his stenographer or secretary, she bore him three daughters who died before reaching maturity, a son, Jean-Baptiste (b. 1851), who became a diplomat, and a fourth daughter, Marie-Louise (b. 1858), who in 1879 married René Vallery-Radot, later Pasteur's biographer.

At Strasbourg, Pasteur continued and greatly extended his work on optical activity and molecular asymmetry despite expanding teaching duties. During 1850 and 1851 he turned to asparagine and its derivatives (aspartic acid, malic acid, the aspartates and malates), which were among the very few optically active compounds from which crystals could be ob-

tained in sizes and amounts adequate for his investigations. Most of these compounds, too, display hemihedral facets as well as optical activity and, at least in this respect, fulfilled Pasteur's expectations. Indeed, malic acid shares so many analogies with tartaric acid (with which it occurs naturally in the grape) that Pasteur was led to postulate a common atomic grouping for the two and to predict the existence of a hitherto unknown left-handed malic acid and of an optically inactive malic acid analogous to, and appearing naturally with, racemic acid. Several of the asparates and malates do not, however, conform to Pasteur's conclusions concerning the tartrates and paratartrates. Certain compounds, for example, rotate the plane of polarized light in a direction contrary to the direction of their hemihedrism. A few display hemihedrism in the absence of optical activity, while others display optical activity in the absence of hemihedral crystals. Even in cases where the relationship between optical activity and crystalline form does seem to conform to Pasteur's "law," the evidence is more ambiguous. Similar difficulties emerged when other groups of optically active compounds were investigated.

Some of these difficulties escaped Pasteur, who naturally sought confirmation and not refutation of his earlier conclusions. His response to those "exceptions" which he did recognize was sometimes brilliant, sometimes evasive, but always ingenious. For cases of hemihedrism in the absence of optical activity, he had a ready explanation derived from the case of quartz. Like quartz, he argued, such substances must possess not true molecular asymmetry but merely a fortuitous asymmetry in the form of their crystal as a whole. More generally and more importantly, he suggested that minor aspects of the conditions of crystallization could mask the existence of a clear and consistent correlation between molecular and crystalline asymmetry; and he even managed in several cases to adjust the crystallizing medium and conditions so as to produce the "hidden" hemihedral facets he sought.[39] This bold achievement perhaps accounts for the confidence with which Pasteur announced as late as 1856 that the only legitimate exception to his law was one which he himself had discovered: amyl alcohol, which shared with a few other compounds the property of being optically active in the absence of crystalline asymmetry but which also displayed in its mode of crystallization unique features that convinced Pasteur that any "hidden" asymmetry could never be revealed.[40]

By 1860, as the number of apparent "exceptions" multiplied, Pasteur had subtly shifted the emphasis of his position so that optical activity became the primary index of molecular asymmetry, while crystalline form was relegated to a secondary although still important position.[41] Never, it seems, did he fully and openly abandon his basic conviction that optical activity (and hence molecular asymmetry) must somehow find expression in crystalline form.[42]

In speculating on the kind of atomic arrangements that could produce molecular asymmetry, Pasteur suggested tentatively in 1860 that the atoms of a right-handed compound, for example, might be "arranged in the form of a right-handed spiral, or . . . situated at the corners of an irregular tetrahedron."[43] But he never developed these suggestions, and it was left to others—notably Le Bel and van't Hoff in 1874—to link his work with Kekulé's theory of the tetrahedral carbon atom. From this linkage emerged the concept of the asymmetrical carbon atom, which underlies all subsequent developments in stereochemistry. Besides adding precision and clarity to Pasteur's earlier investigations of molecular asymmetry, these developments in stereochemistry raised further doubts about the validity of some of his principles.

Asymmetry and Life. In the meantime, Pasteur's preconceptions had opened a fertile new territory to him. But scarcely had he entered it when he committed himself firmly and permanently to another guiding idea—that optical activity was somehow intimately associated with life and could not be produced artificially by ordinary chemical procedures. The precise origin and basis of this idea are the subject of some controversy;[44] but Pasteur's commitment to it seems undeniable by 1852, and he may have held it implicitly from the outset of his career. Even then, evidence existed (especially from the work of Biot and Laurent) that optical activity was generally present in organic products and uniformly absent from inorganic substances. In any case, Pasteur's conviction of an association between life and optical activity ultimately became far more fundamental to him than his belief in a correlation between molecular and crystalline asymmetry.

Quite probably because of this conviction, Pasteur reacted dramatically to the work of Victor Dessaignes, who announced in 1850 that he had prepared aspartic acid by heating optically inactive starting materials (maleic and fumaric acids). Since the only known aspartic acid was optically active, Dessaignes's discovery seemed to constitute the artificial creation of optical activity. Upon hearing of this work, Pasteur went immediately to Dessaignes's laboratory in Vendôme to obtain samples of the new acid. As he expected, it proved to be a hitherto unknown inactive aspartic acid, as did the malic acid prepared from it. The possibility remained, however, that these newly

discovered inactive acids were "racemic"—that is, that they owed their optical inactivity to a compensation between left-handed and right-handed forms. Initially, in a memoir of 1852, Pasteur rejected this possibility on the ground that such "racemic" acids could be synthesized only from "racemic" starting materials, while the available evidence suggested that neither the maleic nor the fumaric acid with which Dessaignes had begun could possess such a constitution.

Having rejected this explanation for the inactivity of Dessaignes's aspartic and malic acids, Pasteur boldly suggested that they belonged to an entirely new class of compounds—those the original asymmetry of which had been "untwisted" so that they had become inactive by total absence of any asymmetry, "inactive by nature" rather than "inactive by compensation." The existence of such compounds (subsequently designated by the prefix "meso") was quickly confirmed by Pasteur's preparation of "mesotartaric" acid, a compound he predicted on the basis of his belief that all forms of malic acid should have counterpart forms of tartaric acid. His hypothetical "mesomalic" acid has never been found, however, and it now seems certain that Dessaignes's synthetic malic acid was in fact "racemic" or "inactive by compensation." That Pasteur did not recognize it as such has led to the assumption that he operated under the sway of preconceived ideas—an assumption that gains immense force from Pasteur's remark of 1860 that if Dessaignes's malic acid were inactive by compensation between left-handed and right-handed forms, he would have performed the remarkable feat of producing not just one but two optically active substances from inactive starting materials.[45]

A similar interpretation can be given to Pasteur's trip of October 1852 through the tartaric acid factories of Germany and Austria. His explicit aim was to find the origin of and new sources for paratartaric or racemic acid, which had become scarce and which resisted attempts to produce it in the laboratory. For these reasons the Société de Pharmacie in 1851 had established a prize of 1,500 francs for the resolution of two questions: Does racemic acid preexist in certain tartrates? How can racemic acid be produced from tartaric acid?

Another chemist might have sought the answers solely within the laboratory, but Pasteur's conviction that asymmetry could not be produced chemically suggested another approach. Since racemic acid is a combination of right- and left-handed tartaric acids, its production from ordinary (right-handed) tartaric acid implied the transformation of a portion of right-handed tartaric acid into its left-handed form. By 1852 Pasteur had become convinced that such a transformation was chemically impossible and that racemic acid might best be sought by tracing it to its natural origin. He therefore visited the tartaric acid factories where racemic acid had once appeared or was now believed to appear, in order to compare the sources and natures of the tartars they used as well as their modes of manufacture. A survey of factories in Saxony, Vienna, and Prague revealed a correlation between the appearance of racemic acid and the use of crude tartars, especially from Italy. Pasteur concluded that racemic acid preexisted naturally to varying degrees in crude tartars and resulted not from some accidentally discovered industrial procedure. Since most manufacturers used semirefined rather than crude tartars, Pasteur asked one of them to switch back to crude Italian tartars, with the expected result that racemic acid soon reappeared in the factory. In addition he persuaded two manufacturers to seek racemic acid by treating the mother liquids left from the initial purification of their semirefined tartars, and this effort too had rapid success.

During this journey Pasteur met a German industrial chemist who claimed to have achieved what Pasteur then considered impossible—the chemical transformation of tartaric into racemic acid. Although he soon confirmed his belief that this particular claim was inaccurate, Pasteur unexpectedly achieved the transformation in May 1853 by heating cinchonine tartrate at 170° C. for five to six hours. This procedure also yielded a small amount of inactive "mesotartaric" acid, the existence of which Pasteur had predicted the year before and in search of which he had apparently undertaken the experiment. In the memoir (1 August 1853) in which he announced these two discoveries, Pasteur disclosed a new method for separating racemic acid into its left- and right-handed components. His original method, involving the manual separation of the crystals, was laborious and extremely limited in applicability. The central feature of the new method was the chemical combination of racemic acid with optically active bases. Under appropriate conditions they affected the solubility of the resulting paratartrates in such a way as to favor the crystallization of only one of the two forms that together compose the paratartrate. Although introduced by Pasteur only for the case of racemic acid, this new method clearly had wider applicability and was soon used to separate the left- and right-handed components in other "racemic" substances (substances inactive by compensation).

In November 1852, immediately after his foreign tour, Pasteur was promoted to *professeur titulaire* at Strasbourg. For his work on racemic acid and crystallography he received the prize of 1,500 francs from the Société de Pharmacie (1853), membership in the

Legion of Honor, and the Rumford Medal of the Royal Society (1856). In December 1857, having moved to Lille and become deeply involved in the study of fermentation, Pasteur announced in preliminary fashion the discovery of a third method for separating racemic acid. If the first method is considered as manual and the second as chemical, then the new method was biological or physiological. In essence, it depended on the capacity of certain microorganisms to "discriminate" between left- and right-handed forms and selectively to metabolize one or the other.

The particular example that Pasteur described grew out of his study of the fermentation of ammonium paratartrate. Following this fermentation with a polarimeter, he found that the fermenting fluid displayed increasing optical activity to the left. Eventually, the fluid yielded only left-handed ammonium tartrate. The right-handed form originally present in the paratartrate had been selectively attacked during the fermentation, while the left-handed form had been left alone. Pasteur linked this discriminatory action with the nutritional needs of a living microorganism presumed to be responsible for the fermentation. Initially vague about its nature, he showed in 1860 that a specific mold, *Penicillium glaucum*, selectively metabolized the right-handed form in a solution of ammonium paratartrate containing a little phosphate. Later qualified, modified, and generalized by others, Pasteur's new method became applicable to the separation of left- and right-handed forms in a number of compounds. Another method of wide applicability was discovered in 1868 by Désiré Gernez, one of Pasteur's assistants. He showed that a single crystal of either the left- or the right-handed form, when sown into a supersaturated solution of a paratartrate, induced the selective crystallization of the form sown.

Pasteur retained a lifelong conviction that asymmetry and life are intimately associated. To do so, however, he had to refine, qualify, and even deny some aspects of his original position, especially in the face of accumulating evidence that racemic acid and other racemic substances could be produced from optically inactive compounds by ordinary chemical procedures. Ultimately Pasteur merely insisted that the artificial production of racemic substances should in no way be compared with the production of a single active substance unaccompanied by its inverse form. Ascribing the latter faculty to nature alone, he perceived in it the last barrier between organic and inorganic phenomena.[46]

If this mode of thought seems to stamp Pasteur as a vitalist, a slightly different perspective can make him seem a mechanist, for he spoke not of "vital forces" but of "asymmetrical forces." While emphasizing that these asymmetrical forces were not deployed in ordinary chemical procedures, he nonetheless connected them with, and sought them among, physical forces at work in the cosmos. In particular, he suggested that the earth is asymmetrical, in the sense that when it turns on its axis, its mirror image rotates in a different direction. And if an ether moving with the rotating earth presides over electrical and magnetic phenomena, the latter must be considered asymmetrical in the same sense. Solar light, too, presents an asymmetrical aspect, for it strikes the earth (and its organisms) at an angle which would be inverted in a mirror. Somehow, Pasteur believed, these or other asymmetrical forces must generate asymmetry (and thus life) in matter.

Not content merely to harbor such boldly speculative ideas, Pasteur sought experimental evidence for them. As early as 1853, while still at Strasbourg, he tried to bring asymmetrical forces to bear upon crystallization by means of powerful magnets built to his specifications. At Lille he tried to modify the normal character of optically active substances by using a large clockwork mechanism to rotate a plant continuously in alternate directions and by using a reflector-and-heliostat arrangement to reverse the natural movement of solar rays directed on a plant from its moment of germination. Biot and others discouraged such experiments as a waste of physical and mental resources, and Pasteur admitted that he must have been a "little mad" to undertake them.[47] Nonetheless, despite his lack of success, Pasteur never abandoned hope that life might someday be created, or at least profoundly modified, in the laboratory under the influence of such asymmetric forces. It seems a remarkable paradox that he could retain this hope while attacking all attempts by others to achieve spontaneous generation.[48] In any case, all subsequent research has supported Pasteur's convictions that optical activity and life are somehow intimately associated and that the production of a single active substance unaccompanied by its mirror image is indeed nature's prerogative except under highly exceptional and basically "asymmetrical" conditions.

Fermentation: The Background. In December 1854 Pasteur was named professor of chemistry and dean of the newly established Faculty of Sciences at Lille. Located at the center of the most flourishing industrial region in France, it was designed in part to bring science to the service of local industry. While resisting any emphasis on applied subjects at the expense of basic science, Pasteur strongly supported this goal and sought to link industry and the Faculty of Sciences in his own courses and activities. For instance, he taught the principles and techniques of bleaching, of extracting and refining sugar, and especially of fermentation and the manufacture of beetroot alcohol, an important local industry. During 1856, he went regularly to the beetroot alcohol factory of M. Bigo, seeking the cause of and remedies for recent disappointments in the quality of that product. For this reason especially, Pasteur's interest in fermentation has often been traced to the brewing industry in Lille.

Pasteur, however, traced his interest to 1849, when Biot informed him that amyl alcohol displayed optical activity.[49] For a brief period during that year, he apparently tried to study the compound, but the problem of securing pure amyl alcohol in adequate quantities led him to abandon the topic. His transfer to Lille may well have reactivated his intention to continue these studies, for amyl alcohol was readily available as a by-product of several industrial fermentations. By August 1855 Pasteur had published a paper showing that the crude amyl alcohol found in industrial fermentations was composed of two isomeric forms, one optically active and the other optically inactive. A careful study of the two forms and their derivatives convinced him by June 1856 that he had found the first legitimate exception to his "law of hemihedral correlation." His determination to investigate this exception thoroughly probably helped to direct his attention to fermentations.

Once attracted to the study of fermentation, Pasteur naturally pondered the source of asymmetry in its optically active products, notably amyl alcohol. The prevailing view traced the optical activity of amyl alcohol to the sugar (also optically active) that served as the starting material in fermentations. Pasteur, however, believed that the molecular structure of amyl alcohol differed too greatly from that of sugar for its optical activity to originate there. His tendency to associate asymmetry and optical activity with life may then have brought him to the view that fermentation depends on the activity of living microorganisms. In taking this view Pasteur defied the dominant chemical theory of fermentation, but his basic position was by no means novel or obscure. Since 1837 several observers—notably Charles Cagniard de Latour and Theodor Schwann—had insisted that alcoholic fermentation depended on the vital activity of brewer's yeast. This view had been ridiculed and eventually overwhelmed by Liebig and Berzelius, who insisted that the process was chemical rather than vital or biological. Their position drew impressive support from indisputably chemical processes considered analogous to fermentation—most notably the action of the soluble digestive "ferments" (enzymes) diastase and pepsin. But the alternative biological theory had also been founded and developed on the basis of persuasive evidence that must have given Pasteur enormous comfort when he launched his campaign against the chemical theory.

Lactic Fermentation. The opening salvo in that campaign was a short memoir on lactic fermentation, presented in August 1857 to the Society of the Sciences, Agriculture, and the Arts in Lille. Émile Duclaux has suggested that two factors induced Pasteur to focus first on the relatively unimportant lactic fermentation (most familiar as the process producing sour milk) rather than alcoholic fermentation: (1) a large quantity of amyl alcohol is produced during lactic fermentation and (2) alcoholic fermentation had already been thoroughly investigated without seriously threatening the dominant chemical theory. In a sense, unless and until living organisms were implicated in other fermentations, advocates of the chemical theory could continue to doubt the essential role of living yeast in alcoholic fermentation.[50]

Pasteur's memoir expressed the basic approach and point of view which informed all of his subsequent work on fermentation. After a historical introduction he began by claiming that "just as an alcoholic ferment exists—namely, brewer's yeast—which is found wherever sugar breaks down into alcohol and carbonic acid—so too there is a special ferment—a lactic yeast—always present when sugar becomes lactic acid." In an ordinary lactic fermentation, this "lactic yeast" appeared as a gray deposit the central role of which could be demonstrated by isolating and purifying it. To do this Pasteur took the soluble extract from brewer's yeast, added to it some sugar and some chalk, and then sprinkled in a trace of the gray deposit from an ordinary lactic fermentation. In this way he invariably produced a lively and indisputably lactic fermentation, with the gray deposit increasing in amount as the fermentation progressed. Viewed macroscopically, this deposit resembled ordinary pressed or drained brewer's yeast. Under the microscope it seemed to be composed of "little globules or very short segmented filaments, isolated or in clusters, which form irregular flakes resembling those of certain amorphous precipitates." An extremely small amount of the deposit sufficed to decompose a large amount of sugar. Although smaller and harder to see than brewer's yeast, the lactic ferment seemed to Pasteur so analogous to it that he supposed the two "yeasts" might belong to closely related species or families.

Throughout the memoir Pasteur more nearly assumed than proved that lactic yeast "is a living organism, . . . that its chemical action on sugar corresponds to its development and organization," and that the nitrogenous substances in the fermenting medium served merely as its food. Nonetheless, his

convictions were firm and his conception of fermentation was already remarkably complete. Nothing demonstrates this more forcefully than his discussion of the conditions essential for good fermentations, which include not only a pure and homogeneous ferment but also an appropriate nutrient medium, well adapted to the "individual nature" of the ferment. "In this respect," he wrote, "it is important to realize that the circumstances of neutrality, alkalinity, acidity, or chemical composition of the liquids play a great part in the predominant growth of . . . a ferment, for the life of each does not adapt itself to the same degree to different states of the environment." Acidity, for example, favors the development of the alcoholic over the lactic fermentation, while in neutral or slightly alkaline media the situation is reversed. Furthermore, the purity of the fermentation is greatly enhanced by protecting it from air and by the method of sowing pure ferments, for both prevent the invasion of "foreign vegetation or infusoria." An unsown fermentable medium, like an unseeded plot of land, "soon becomes crowded with various plants and insects that are mutually harmful." Pasteur even referred to the capacity of "the essential oil of onion" to inhibit the development of both brewer's yeast and infusoria without affecting the growth of the lactic ferment—a remark to which some have traced the concept of antibiotics.

With two striking exceptions this memoir contains the central theoretical and methodological features of all of Pasteur's work on fermentation—the biological conception of fermentation as the result of the activity of living microorganisms; the view that the substances in the fermenting medium serve as food for the causative microorganism and must therefore be appropriate to its nutritional requirements; the notion of specificity, according to which each fermentation can be traced to a specific microorganism; the recognition that particular chemical features of the medium can promote or impede the development of any one microorganism in it; the notion of competition among different microorganisms for the aliments contained in the media; the assumption that air might be the source of the microorganisms that appear in fermentations; and the technique of directly and actively sowing the microorganism presumed responsible for a given fermentation in order to isolate and purify it. The two missing features, which soon completed Pasteur's basic conception, were the technique of cultivating microorganisms (and thereby producing fermentations) in a medium free of organic nitrogen and his notion of fermentation as "life without air."

In October 1857, two months after presenting his memoir on the lactic fermentation, Pasteur left Lille for the École Normale in Paris, where he had been named director of scientific studies and administrator, his duties including "the surveillance of the economic and hygienic management, the care of general discipline, intercourse with the families of the pupils and the literary or scientific establishments frequented by them."[51] Because these positions included neither laboratory nor allowance for research expenses, Pasteur was obliged to make frequent appeals to governmental agencies for financial support. Although he considered such appeals "antipathetic to the character of a scientist worthy of the name,"[52] he made them with sufficient success to secure his own research laboratory, which consisted at first of two rooms in an attic of the École Normale. By December 1859 he had gained possession of a small pavilion, which was expanded considerably in 1862. In these surroundings Pasteur pursued his study of fermentation and quickly extended his basic conclusions on lactic fermentation to various others, notably the tartaric, butyric, and acetic as well as alcoholic.

Alcoholic Fermentation. In December 1857 Pasteur published the first in a series of abstracts, notes, and letters on alcoholic fermentation that culminated in a long and classic memoir of 1860. Divided into two major sections, dealing respectively with the fate of sugar and of yeast in alcoholic fermentation, it inflicted on the chemical theory what Duclaux called "a series of blows straight from the shoulder, delivered with agility and assurance."[53] Pasteur established that alcoholic fermentation invariably produces not only carbonic acid and ethyl alcohol—as was well known—but also appreciable quantities of glycerin and succinic acid as well as trace amounts of cellulose, "fatty matters," and "indeterminate products." On the basis of these results, Pasteur emphasized the complexity of alcoholic fermentation and attacked the tendency of chemists since Lavoisier to depict it as the simple conversion of sugar into carbonic acid and alcohol. If the alleged simplicity of the process had formerly been seen as evidence of its chemical nature, he argued, then its actual complexity ought now to be seen as evidence of its dependence on the activity of a living organism. In truth, the complexity of alcoholic fermentation was such as to prevent the writing of a complete equation for it, a fact which was only to be expected, since chemistry was "too little advanced to hope to put into a rigorous equation a chemical act correlative with a vital phenomenon."

However impressive this line of attack against the chemical theory, an even more decisive mode of argument derived from Pasteur's ability to produce yeast and alcoholic fermentation in a medium free of organic nitrogen. To a pure solution of cane sugar he added only an ammonium salt and the minerals obtained by

incineration of yeast, then sprinkled in a trace of pure brewer's yeast. Although the experiment was difficult and not always successful, this method could produce an alcoholic fermentation accompanied by growth and reproduction in the yeast and the evolution of all the usual products. If any one constituent of this medium were eliminated, no alcoholic fermentation took place. Obviously, argued Pasteur, the yeast must grow and develop in this mineral medium by assimilating its nitrogen from the ammonium salt, its mineral constituents from the yeast ash, and its carbon from the sugar. In fact, it is precisely the capacity of yeast to assimilate combined carbon from sugar that explains why it can decompose sugar into carbonic acid and alcohol. Above all, there is in this medium none of the "unstable organic matter" required by Liebig's theory.

When this memoir on alcoholic fermentation appeared, Pasteur had already begun to exploit more widely his new method of cultivating microorganisms in a medium free of organic nitrogen. Described initially in a note of December 1858, this method had been applied to the lactic ferment by February 1859. Indeed, "Pasteur's fluid"—a solution of sugar, yeast ash, and ammonium salt—proved far more conducive to the growth of the lactic ferment than to that of brewer's yeast. Sometimes, the lactic fermentation appeared "spontaneously" in this medium, even when only brewer's yeast had been sown. From similar events in ordinary crude alcoholic fermentations, some chemists had concluded that lactic acid was a normal by-product of alcoholic fermentation. Pasteur showed, however, that the appearance of lactic acid in such cases could be associated with an accidental contamination of the fermenting medium by the lactic ferment. To ensure the uncontaminated growth of the lactic ferment itself, it was necessary only to add calcium carbonate to the solution of sugar, yeast ash, and ammonium salt.

Fermentation and Putrefaction as "Life Without Air." In November 1860 Pasteur described the successful cultivation of *Penicillium* "or any mucedinous fungus" in a medium of pure water, cane sugar, phosphates, and an acid ammonium salt. By February 1861 he had isolated a specific butyric ferment and had produced butyric fermentation in a similar medium. In two respects this new butyric ferment greatly suprised him: (1) unlike brewer's yeast and the lactic ferment, it was motile and thus, presumably, a member of the animal kingdom; and (2) while examining microscopically the liquid from a butyric fermentation, he noticed that the rodlike "infusoria" lost their motility and vitality at the margins of the slide glass but remained active in the center. Assuming that this phenomenon depended on the presence of atmospheric air at the margins of the slide glass, Pasteur passed a current of ordinary air through a butyric fermentation. Within an hour or two the butyric fermentation had ceased and all the motile rods had been killed. Carbonic acid gas, on the other hand, exerted no appreciable effect on their life and reproduction. Pasteur concluded that the butyric ferment is an infusorium and that this infusorium lives without free oxygen gas. This was, he believed, the first known example of an animal ferment and of an animal capable of living without free oxygen.

From the beginning naturalists challenged Pasteur's belief that the butyric ferment was an animal, because for many of them motility had ceased to be an automatic index of animality. More specifically, the genus *Vibrionia*, to which Pasteur assigned his new ferment, had been identified as vegetable in 1854 by Ferdinand Cohn, who had linked it with the algae and bacteria. Not surprisingly, then, an English translator immediately suggested that Pasteur's supposed butyric "infusorium" probably belonged instead among the algae.[54] Cohn later placed it among the bacteria (*Bacillus subtilis*).[55] Although Pasteur quickly qualified his assertion of the animality of the new ferment, he demonstrated little concern about the taxonomic issue and little serious interest in the literature of the naturalists, whom he seemed sometimes to despise. This attitude and Pasteur's inadequacies as a naturalist led to some confusion about and hostility toward his work—and by no means solely in the case of the butyric ferment. In later years Pasteur became somewhat more sensitive to taxonomic issues and emphasized the importance of physiological characters as a taxonomic criterion; but his generally casual attitude toward microbial morphology and nomenclature helped to exacerbate some of the debates over spontaneous generation, the transformation of microbial species, and the germ theory of disease. Cohn and Koch, among others, chastised Pasteur severely for his lack of rigor in these areas.

Pasteur's discovery of the butyric ferment and of its death in air gave a new direction to his studies on fermentation. He quickly investigated the effect of free oxygen on other ferments and moved gradually toward a new definition of fermentation as "life without air." In June 1861 he reported that the activity of brewer's yeast depended fundamentally on the degree of free oxygen available to it. Like ordinary fungi or infusoria, it grew and reproduced with great vigor in the presence of air. As a ferment, however, it was virtually powerless under such circumstances; only in the absence of free oxygen did it display a significant capacity to ferment sugar. For Pasteur the explanation was obvious: when deprived of free oxygen, the yeast

of necessity attacked the sugar in order to extract its combined oxygen.

In March 1863 Pasteur announced that calcium tartrate fermented in a medium free of organic nitrogen by the action of a motile infusorium analogous to the butyric ferment. Like the butyric, the new ferment lived only in the absence of air and belonged to the genus *Vibrionia*, although its external form differed greatly from the butyric ferment. In a medium exposed to the air, the new ferment developed only when protected by organisms that consumed free oxygen at the surface of the medium, while the ferment lived and developed at lower, oxygen-free levels. Fermentation, Pasteur now suggested, is merely "nutrition without the consumption of free oxygen gas." In this conception, he believed, lay the key to "the secret and mysterious character of all true fermentations and, possibly, that of many normal and abnormal actions in the organization of living things."

Among these "normal and abnormal actions" was putrefaction, generally defined as the decomposition of vegetable or animal matter with the evolution of fetid gases. In April and June 1863, on the basis of rather sketchy evidence, Pasteur extended to the phenomena of putrefaction the central conclusions of his work on fermentation. Like fermentation, he insisted, putrefaction can be traced to the vital activity of living ferments. Indeed, except for the action of microorganisms, the constituents of dead plants and animals could be considered "relatively indestructible." To express the matter in more poetic terms, "life takes part in the work of death in all its phases," for the decomposition associated with death depends on the development and multiplication of microorganisms. Moreover, death is as essential to the cycle of life as life is to the phenomena of death. For it is only as a consequence of death and putrefaction that carbon, nitrogen, and oxygen become available as nutrients to support the life of other organisms. Thus, in an eternal cycle, life stems from death and death from life.

Within this cosmic perspective, reformulated and reemphasized on other occasions, Pasteur developed a more prosaic analysis of the nature and action of the microorganisms involved in the decomposition of dead substances. These organisms are of two kinds: (1) the oxidative microorganisms—the mycodermas and their relatives—which in the course of their vital activity transfer atmospheric oxygen to the dead organic substances and thereby enormously increase the rate of combustion and (2) the putrefactive ferments per se, which (like the butyric ferment) belong to the genus *Vibrionia* and live only in the absence of air. Putrefaction and fermentation are, therefore, analogous processes, for both involve the decomposition of substances by organisms living in the absence of air. In fact, putrefaction is merely the fermentation of substances containing a relatively high proportion of sulfur, and the release of this sulfur in gaseous form produces the fetid odors commonly associated with putrefaction.

In other words, Pasteur emphasized, a putrescible liquid exposed to atmospheric air experiences two distinct sorts of chemical decomposition correlative with the life and development of two distinct sorts of microscopic organisms. On the one hand the anaerobes—putrefactive ferments living below the surface, in the absence of air—determine "acts of fermentation": they transform nitrogenous materials into simpler but still complex substances. On the other hand the aerobes—oxidative microorganisms living at the surface, in the presence of air—can assimilate these intermediate products and transform them into the "simplest binary combinations"—water, ammonia, and carbonic acid. Bulloch has traced the concept of anaerobism or "life without air" to Leeuwenhoek and Spallanzani.[56] Their work in this regard having been completely forgotten, however, Pasteur has always been recognized as the architect of the idea.

Studies on Acetic Fermentation and Vinegar. By the time he published his papers on putrefaction, Pasteur was deeply involved in the study of acetic fermentation and the manufacture of vinegar. Beginning in July 1861, he produced a series of papers on acetic fermentation that linked theory with industrial practice and culminated in a long memoir (1864) and in *Études sur le vinaigre* (1868). When he began this work, acetic fermentation was widely viewed as a chemical, catalytic process, comparable with the well-known oxidation of alcohol to aldehyde and acetic acid in the presence of finely divided platinum. This conception seemed in accord with the German method of manufacturing vinegar, in which the fermenting medium consisted of a dilute alcohol solution, a trace of acetic acid, and some "unstable organic matter" such as sharp wine or acid beer. When this liquid trickled through a hollow column of wine casks containing loosely piled beechwood shavings, the alcohol was oxidized to acetic acid with the release of heat and the production of an upward current of air that constantly renewed the supply of oxygen. As Liebig interpreted this method, the "unstable organic matter" initiated fermentation and the beechwood shavings facilitated the oxidation process while remaining unaltered (that is, they acted as a catalyst). In all of this there was no hint of biological action.

Pasteur approached acetic fermentation fully confident that he would find in this case, too, that a microorganism was essential to the process. The rela-

tive ease with which he succeeded can be partly ascribed to the character of the French method of vinegar production, for which the leading center was Orléans. This method differed markedly from the German method. In Orléans vinegar was produced by the slow oxidation of wine in covered casks stacked on end, about one-third empty and exposed to the air by an opening or "window" above the surface of the fermenting liquid. On the surface of the liquid, which consisted of a mixture of finished vinegar and new wine, there appeared a delicate pellicle—long known as "mother of vinegar"—the presence of which was recognized as essential to the process. From his earlier studies of fermentation, Pasteur knew that such pellicles could be formed by microorganisms. Moreover, several observers had already suggested that the "mother of vinegar" consisted of living organisms. In 1822 Persoon had named it "mycoderma" precisely to suggest that it was a fungal skin. And in 1837 Friedrich Kützing had drawn a connection between the life of this skin and the production of vinegar—as indeed, in the same year he had also connected the life of yeast with the production of alcohol in ordinary alcoholic fermentation. As in alcoholic fermentation, Pasteur could draw inspiration from a tradition that viewed fermentation as a vital process. His task was to present this case so persuasively as to override the dominant authority and arguments of Liebig.

To do so, Pasteur resorted again to media free of organic nitrogen. By July 1862 he had succeeded in cultivating *Mycoderma aceti* in a medium of dilute alcohol, ammonia, and mineral salts. When sown into such a medium, *Mycoderma aceti* consistently produced acetic acid; and Pasteur again emphasized the ability of a microorganism to yield fermentation in the absence of the "unstable organic matter" required by Liebig's theory. Moreover, he was able to detect a thin film of *Mycoderma aceti* on the beechwood shavings so important in the German method of vinegar production. Protected from or deprived of the *Mycoderma*, the shavings lost their capacity to produce acetic acid. Their only role, Pasteur insisted, was to provide a site for the growth and development of *Mycoderma aceti*. In his view *Mycoderma aceti* acted by transmitting the "combustive action" of atmospheric oxygen to alcohol and thus oxidizing it to acetic acid. If no alcohol remained in the fermenting medium, the *Mycoderma* could attack the acetic acid it had produced and complete the oxidation to water and carbonic acid. *Mycoderma aceti* also ceased to produce acetic acid if submerged; only at the surface of the fermenting medium, in the presence of abundant oxygen, did it support acetic fermentation. Although this latter fact posed an obvious difficulty for his concept of fermentation as "life without air," Pasteur made no attempt to resolve the issue at the time.

From the beginning of his research on acetic fermentation, Pasteur recognized its industrial significance; and in July 1861 he took out a patent "for the manufacture of vinegar or acetic acid by means of molds, in particular *Mycoderma vini* and *Mycoderma aceti*."[57] To a considerable extent Pasteur's interpretation of acetic fermentation merely provided a rationale for industrial practices that had already been introduced empirically, although it did allow somewhat greater confidence in and control over them. Perhaps the most important advantage that Pasteur ascribed to his method of manufacturing vinegar was that it permitted the process to be directed at will. No longer was it necessary to await the "spontaneous" appearance of the mycodermic pellicle, which sometimes took several weeks. Manufacturers could now produce acetification quickly and reliably by direct sowing of *Mycoderma aceti*. Moreover, he claimed, his method produced acetic acid three to five times as rapidly as the Orléans method and greatly reduced the losses by evaporation experienced in the German method. By 1868, when he published his *Études sur le vinaigre*, Pasteur could appeal by analogy to his recent studies on the "diseases" of wine in order to discuss the diseases of vinegar, all of which (like the diseases of wine) could be prevented by heating finished vinegar to about 55° C.

Studies on Wine. In December 1863 Pasteur published the first of the papers that culminated in his *Études sur le vin* (1866; 2nd ed. 1873). In that first paper, dealing with the role of atmospheric oxygen in vinification, he sought to establish that the aging of wine resulted from the slow penetration of atmospheric oxygen through the porous wood casks into which new wine was decanted. By virtue of this slow oxidation, he claimed, new wine grows less harsh and acid to the taste as it becomes clearer and lighter from the precipitation of dark coloring matters. In his second paper (January 1864) Pasteur examined the "alterations" or "diseases" of wine, especially wine from the Jura, his native department. Reviewing the familiar diseases of "turned," "acid," "ropy," or "oily" wine, he associated each with a microscopic organism. He summarized the results of his first two papers by noting that "wine, which is produced by a cellular vegetation acting as a ferment [namely, yeast], is altered only by the influence of other vegetations of the same order; and once removed from the effects of their parasitism, it is made or matured principally by the action of atmospheric oxygen penetrating slowly through the staves of the casks."

Since the diseases of wine are due to the develop-

ment of foreign organisms, which are present before the wine becomes sensibly "sick" and the germs of which are bottled with the wine, the crucial task was to find a way of killing these germs without damaging the taste or other qualities of wine. On 1 May 1865 Pasteur told the Académie des Sciences that his attempts to cure diseased wines with chemical antiseptics had been less than satisfying, but that he had found a perfectly reliable and practical procedure for preserving healthy wine: by heating it in closed vessels for an hour or two at a temperature between 60° and 100° C. As a result of small-scale preliminary trials, Pasteur progressively lowered the temperature to between 50° and 60° C. Within this range, he claimed, wine could be perfectly protected from disease at minimum risk to its taste, bouquet, and color.

As soon as Pasteur publicly disclosed this method, which he patented in April 1865, alternative claims began to appear. In a series of letters and notes published between 1865 and 1872, nearly all of which were reproduced or incorporated into the two editions of his *Études sur le vin*, Pasteur repeatedly defended his priority rights, even as he became increasingly informed of the long history of "empirical" attempts to preserve wine. Eventually he admitted that he had been anticipated by Nicolas Appert, who had specifically proposed the application to wine of his method of preserving foodstuffs by heating them in closed vessels. Nonetheless, he insisted that he had rescued from oblivion and established on the basis of rigorous scientific experiments what had been only a poorly tested and entirely empirical technique.

In support of the practicability of his method, Pasteur cited a series of commissions the members of which generally preferred the taste of heated to untreated wine. One commission was appointed by the French navy in 1868 to test the feasibility of applying Pasteur's process to wines destined for the fleet and the French colonies. The results were impressive enough for the navy to adopt Pasteur's process. Further evidence of the value of his method was reflected in the grand prize awarded Pasteur by the jury of the Exposition Universelle (1867); the use abroad of the word "pasteurization" to denote the heating of wine; and the prizes from agricultural societies and from the Société d'Encouragement pour l'Industrie Nationale for the best apparatus for heating wine (fifteen examples of which Pasteur described and illustrated in the second edition of *Études sur le vin*).

Spontaneous Generation: The Background. Almost from the beginning of his work on fermentation and despite attempts by Biot and Dumas to dissuade him, Pasteur became embroiled in the controversial issue of spontaneous generation. Although advanced in several more or less sophisticated versions, the doctrine of spontaneous generation rests at bottom on the notion that living organisms can arise independently of any immediate living parent, whether from inorganic substances (abiogenesis) or from organic debris (heterogenesis). In his classic paper of 1861, "Mémoire sur les corpuscules organisés qui existent dans l'atmosphère . . .," Pasteur included a fairly substantial historical introduction, which seems greatly to have influenced subsequent histories of the debate. To account for the modern rise of the doctrine—following its apparent destruction in the seventeenth century by Francesco Redi's experiments on the generation of insects—Pasteur emphasized the influence of the microscope. By revealing a teeming world of hitherto unseen living organisms of dubious or unknown parentage, the microscope gave the doctrine a new lease on life and led to a celebrated eighteenth-century dispute between Spallanzani and Needham. Spallanzani seemed largely to carry the day by showing that infusions boiled for forty-five minutes in closed vessels (to destroy any organisms they might already contain) thereafter remained free of alteration and microbial life. But his technique was open to the objection that the air in his sealed flasks might have been altered in such a way as to render spontaneous generation impossible. In the early nineteenth century this objection took special force from Gay-Lussac's study of the role of oxygen in fermentation and putrefaction. Having found that oxygen was absent from substances preserved by Appert's canning process, and that grapes crushed under mercury in a bell jar fermented only upon the introduction of air, Gay-Lussac concluded that oxygen was essential to the onset of fermentation and putrefaction (and hence to the appearance of any microorganisms associated with these alterations).

As Pasteur emphasized, Gay-Lussac's experiments made it imperative to remove any doubts about the possible alteration of air in Spallanzani's flasks. Toward this end Theodor Schwann made "a great step forward" in 1837 by showing that boiled meat infusions could be preserved from alteration in flasks in which the air was continually renewed, provided only that the added air had been heated or "calcined" before entering the flasks. Schwann's experiment extended that of Franz Schulze, who in 1836 had achieved similar results by drawing the added air through potassium hydroxide and sulfuric acid; it also helped to set the stage for the work of Heinrich Schröder and Theodor von Dusch, who in the 1850's exposed alterable substances to ordinary air filtered through cotton. By these means Schwann, Schulze, Schröder, and Dusch prevented putrefaction, fermentation, and microbial life in many alterable substances—including meat infusions, beer, must, starch paste, and the constituents of milk taken separately—and tended to suppose that they had done so by eliminating airborne germs. But in the case of other substances—notably milk, egg yolk, and dry meat—their experiments often failed and helped to sustain the view that something like spontaneous generation could occur.

Moreover, Pasteur insisted, even those experiments which seemed to contradict spontaneous generation did so only in the sense of showing that an unknown something in atmospheric air was essential to life in organic infusions. This unknown principle seemed often to be eliminated by heat, cotton, or certain chemical reagents; but insofar as Schwann and others tended to suppose that atmospheric germs had thus been killed or eliminated, they "had no more proofs for their opinion," wrote Pasteur, "than those who believed that [the unknown principle] might be a gas, fluid, noxious effluvia, etc., and who consequently were inclined to believe in spontaneous generation." There, according to Pasteur, the issue lay when Félix Pouchet launched his attempt to establish the doctrine of spontaneous generation on the basis of irrefutable experiments. Pouchet, a respected naturalist from Rouen and a corresponding member of the Académie des Sciences, published in 1859 his long and controversial *Hétérogénie ou traité de la génération spontanée*, which created a sensation in France and probably stimulated the Académie des Sciences to institute the Alhumbert Prize in 1860 for the best "attempt, by well conducted experiments, to throw new light on the question of so-called spontaneous generation."

Pasteur won this competition with his "Mémoire sur les corpuscules. . . ." By ignoring a wide range of other factors that helped to discredit spontaneous generation (notably studies of cell division and the debate, by then resolved, over the origin of parasitic worms),[58] it magnified the importance of his own contributions; and nearly all subsequent accounts have followed suit. At the end of his historical introduction, Pasteur traced his interest in spontaneous generation to his work on fermentation, and particularly to his recognition that the ferments were living organisms:

> Then, I said to myself, one of two things must be true. The true ferments being living organisms, if they are produced by the contact of albuminous materials with oxygen alone, considered merely as oxygen, then they are spontaneously generated. But if these living ferments are not of spontaneous origin, then it is not just the oxygen as such that intervenes in their production—the gas acts as a stimulant to a germ carried

with it or already existing in the nitrogenous or fermentable materials. At this point, to which my study of fermentation brought me, I was thus obliged to form an opinion on the question of spontaneous generation. I thought I might find here a powerful support for my ideas on those fermentations which are properly called fermentations.

As this passage suggests, it is perhaps artificial to separate Pasteur's study of spontaneous generation from his work on fermentation, especially since some of his adversaries contended that microorganisms could appear as a result of fermentation rather than as its cause. The question of the origin of the ferments was therefore crucial, and Pasteur's concern with it is apparent from his earliest paper on fermentation.

In 1858, in his initial paper on fermentation, Pasteur wrote that the lactic ferment "originates spontaneously, with as much facility as brewer's yeast, whenever conditions are favorable," but immediately emphasized in a footnote that he used the word "spontaneously" merely to "describe the fact, leaving entirely aside any judgment on the question of spontaneous generation." In February 1859 he addressed the issue

somewhat more directly, asserting that in his experiments the lactic ferment always came "uniquely by way of the atmospheric air." If he boiled his medium and then removed it from all contact with air or exposed it only to previously calcined air, no microbial life or fermentation of any kind appeared. "On this point," he wrote, "the question of spontaneous generation has made an advance."

Spontaneous Generation, 1860–1861. Beginning in February 1860, Pasteur presented to the Académie des Sciences a series of notes focusing specifically on spontaneous generation. In the first and most important of these papers, he began by examining the solid particles of the air, which he collected by aspirating atmospheric air through a tube plugged with guncotton. When this guncotton was dissolved in a sedimentation tube containing an alcohol-ether mixture, the solid particles trapped by it settled at the bottom. Although this method killed any germs or microorganisms in the trapped particles, microscopic examination always revealed a variable number of corpuscles, the form and structure of which closely resembled those of living organisms. But were these

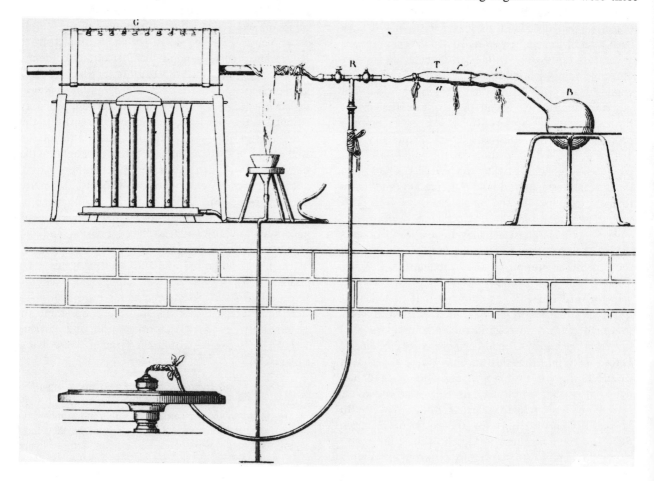

FIGURE 1

"organized corpuscles" in fact the "fecund germs" of the microorganisms which appeared in alterable media exposed to the air? In search of an answer, Pasteur employed three distinct methods. With the first, involving the use of a pneumatic trough filled with mercury, he obtained somewhat dubious or inconsistent results and abandoned it in favor of a second method, which he characterized as "unassailable and decisive." In a flask of about 300 cubic centimeters, he placed 100 to 150 cubic centimeters of sugared yeast water, which he boiled for a few minutes. After the flask had cooled, he filled it with calcined air (by means of a neck connected to a red-hot platinum tube) and then sealed it in a flame. The liquid in such a flask, deposited in a stove at 28–32° C., could remain there indefinitely without alteration.

Having thus far only repeated the experiments of Schwann and others, Pasteur now introduced an important modification. After a month to six weeks he removed the flask from the stove and connected it to an elaborate apparatus so arranged that a small wad of guncotton previously charged with atmospheric dust could be made to slide into the hitherto sterile liquid in the flask (see Figure 1). In twenty-four to thirty-six hours, the liquid swarmed with familiar microorganisms. Thus, Pasteur concluded, the dust of the air, sown in an otherwise sterile medium, produces organisms of the same sort and in the same period of time as would appear if the liquid were freely exposed to ordinary air. Finally, to counter the objection that these microorganisms arose not from germs in the atmospheric dust but "spontaneously" from the organic matter in the guncotton, Pasteur replaced the guncotton with dust-charged asbestos, a mineral substance, and obtained the same results. With dust-free or precalcined asbestos, on the other hand, no growths appeared in the flask.

To confirm and extend these conclusions on the role of atmospheric dust, Pasteur employed a third method, perhaps the most influential by virtue of its elegant simplicity: the famous "swan-necked" flask. After preparing a series of flasks in the same manner as in the second method, he drew their necks out into very narrow extensions, curved in various ways and exposed to the air by an opening one to two millimeters in diameter (see Figure 2). Without sealing these flasks, he boiled the liquid in most of them for several minutes, leaving three or four unboiled to serve as controls. If all the flasks were then placed in calm air, the unboiled liquids became covered with various molds in twenty-four to forty-eight hours, while the boiled flasks remained unaltered indefinitely despite their exposure. Moreover, if one of the curved necks were detached from a hitherto sterile flask and placed up-

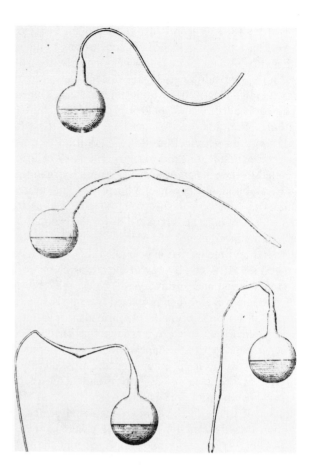

FIGURE 2

right in it, vegetative growths appeared in a day or two. Pasteur concluded that the "sinuosities and inclinations" of his swan-necked flasks protected the liquids from growths by capturing the dusts that entered with the air. In fact, Pasteur insisted, nothing in the air—whether gases, fluids, electricity, magnetism, ozone, or some unknown or occult agent—constitutes a condition of microbial life except the germs carried by atmospheric dusts.

According to Duclaux, the swan-necked flask method was suggested to Pasteur by Balard; and Pasteur admitted that Chevreul had already done "similar experiments" in his chemistry lectures.[59] But if in this case, as in his experiments with calcined air, Pasteur borrowed importantly from the techniques of his predecessors, he also developed and exploited them with greater effect and influence. By the force of its conclusions and the variety and ingenuity of its experimental techniques, his paper of 6 February 1860 propelled Pasteur to preeminence among the opponents of spontaneous generation. All of his subsequent work in this field can be seen as an extension, elaboration, and defense of the principles and methods set forth here.

By May 1860, as promised at the end of his February paper, Pasteur had extended his conclusions to media other than albuminous sugar water—namely, to urine and milk, two substances highly susceptible to alteration in air. Deprived of atmospheric dust, Pasteur claimed, boiled urine could be stored indefinitely without alteration, even at the temperature most favorable to its putrefaction. But the addition of dust-charged asbestos to a previously sterile flask of urine resulted in the appearance of various microorganisms and an abundant deposit of phosphates and urates. One of the microorganisms could be identified as the "true ferment of urine," responsible for the production of ammoniacal urine. Its germ, like those of the infusoria and molds that appeared with it, could have entered the flask only by way of the atmospheric dust.

Unlike urine and sugared yeast water, milk boiled for two minutes and then protected from atmospheric dusts did not remain unaltered. Instead, it invariably coagulated within three to ten days, this coagulation being associated with the appearance and development of vibrios. By no means, however, did this alteration imply that spontaneous generation had taken place. For if the duration of boiling were increased, the number of flasks in which milk coagulated decreased proportionately. And if the temperature were increased to 110° or 112° C., no vibrios appeared and the milk did not coagulate. Obviously, Pasteur concluded, a temperature of 100° C. does not entirely destroy the fecundity of the vibrio germs, while a temperature of 110° to 112° C. does.

In September and November 1860, Pasteur described another famous set of experiments in which he exposed alterable liquids to the natural atmosphere of different locations and altitudes, hoping thereby to discredit the belief that any quantity of ordinary air, however minute, is sufficient for the production of organized growths in any kind of infusion. In his view this belief enjoyed currency chiefly because of Gay-Lussac's analysis of Appert's preserves and his experiment with grapes crushed under mercury, for these studies led him to associate fermentation or putrefaction with the presence of oxygen, even in minute quantities. On this basis the partisans of spontaneous generation had elaborated a seemingly impressive argument against the notion of airborne germs. For if the most minute quantity of air can produce the microorganisms appropriate to any infusion, and if these organisms are supposed to derive from pre-existent germs, then the air must be so loaded with a multitude of different germs as to be foggy at least, if not as dense as iron.

Pasteur's approach to this problem was deceptively simple. After boiling sugared yeast water in sealed flasks, he broke the necks to admit the surrounding air, immediately resealed the flasks in a flame, and stored them in a stove at a temperature favorable to the development of microorganisms. Under these conditions the liquids in the flasks sometimes remained entirely unaffected, a "simple and unobjectionable proof" that a limited quantity of ordinary air does not invariably produce infusorial growths. On the other hand, the result accorded well with the notion of the variable dissemination of germs in the air. The latter notion received further support from the fact that it was easy to alter the proportion of flasks in which microbial life appeared merely by exposing them to the air in various locations or altitudes. In the vaults of the Paris observatory, for example, the proportion of exposed flasks that later showed infusorial growths was much lower than in Pasteur's laboratory at the École Normale. This proportion also decreased with increased altitude. Thus, of twenty flasks opened at the foot of the Jura plateau, eight later showed vegetative growths; of twenty exposed on one of the Jura mountains, 850 meters above sea level, five produced growths; and of twenty opened on a glacier at Montanvert, 2,000 meters above sea level, only one flask underwent subsequent alteration. For Pasteur such results authorized the conclusion that germs are variably disseminated in the air, their relative abundance depending on locality, altitude, and other environmental circumstances.

In January 1861, in his fifth paper on spontaneous generation, Pasteur described the influence of temperature on the fecundity of fungal spores. Spallanzani had found that fungal spores could survive boiling in water at 100° C. and—without assigning a precise upper limit—had claimed that they could even resist the heat of a furnace when dry. Pasteur denied that this upper limit was as high as Spallanzani had supposed and criticized his experimental technique for its failure to ensure that any observed fungi derived solely from the spores he had sown and not from additional spores in the air or on the experimental apparatus. His own method, which seemed to Pasteur "beyond reproach," was a modification of the technique he had used to sow dust-charged asbestos into sterile media in an atmosphere of calcined air. In this case the asbestos was charged with fungal spores and then heated to a determined temperature before sowing. Pasteur found that in a vacuum or in dry air, such spores could remain fecund even after being heated at 120–125° C. for as long as an hour. On the other hand, their fecundity was completely destroyed by heating them at a temperature of 127–130° C. for twenty to thirty minutes. These results also offered a means of proving that fungal spores exist in the atmospheric dust, for the sowing of such dust at 120–125° C. produced fungi, while none appeared when the dust was sown at 125–130° C.

The Memoir of 1861. In May 1861, at a meeting of the Société Chimique de Paris, Pasteur presented the major results of his work on spontaneous generation in a lecture later expanded into his prize-winning memoir. Although this memoir is essentially a restatement of his earlier papers on the topic, it is richer in detail and contains some new material, including the historical introduction. The appearance under the microscope of atmospheric dust and of the organisms found in infusions received considerable attention, as did the role of contaminated mercury as a source of error in the experiments of Pouchet and others. Pasteur had barely hinted at the latter possibility in his initial paper of 6 February 1860 and had made it explicit in a note of September 1860. Pouchet's experimental case for spontaneous generation rested chiefly on his ability to produce microbial life by adding germ-free air to boiled hay infusions under mercury. Pasteur admitted that Pouchet's precautions seemed to eliminate every source of possible contamination by living germs with one exception—the mercury. But this exception was crucial, Pasteur argued, since ordinary laboratory mercury often contains germs. As proof he cited the following comparative experiments. If a globule of ordinary mercury is dropped into an alterable liquid in an atmosphere of calcined (and hence germ-free) air, microbial life appears within two days. But if the mercury is previously calcined, not a single living organism will appear. Indeed, so thoroughly did Pasteur mistrust experiments with the mercury trough that he insisted that this mode of experimentation be banished from the field.

The most important new material in the 1861 memoir concerned the effect of the alkalinity of a medium on the heat-resistance of germs in it. Pasteur identified the alkalinity of milk as the chief reason why boiling at 100° C. failed to protect it from subsequent alteration. As evidence he noted that sugared yeast water—ordinarily protected by boiling at 100° C.—must be heated at 105–110° if its alkalinity is increased by the addition of chalk.

This memoir seriously damaged the doctrine of spontaneous generation, but the blow was far from fatal and many unresolved issues remained. In Pasteur's mind the most obvious weakness of his work on spontaneous generation was its exclusive reliance on experiments involving heated substances—"organic matters which are not only dead but which have also been carried to the temperature of boiling." To all such experiments, partisans of spontaneous generation could object that so high a temperature profoundly modified organic substances and perhaps destroyed a "vegetative force" or some other condition essential for spontaneous generation. For this reason Pasteur long sought to extend his conclusions to "natural organic substances, not previously heated"—in short, to "natural substances such as life elaborates them." In April 1863 he announced that he had found a way to take fresh blood and urine directly from healthy, living organisms and to preserve both substances from putrefaction without preliminary boiling. He immediately asserted that these results "carry a final blow to the doctrine of spontaneous generation," and he attached enormous importance to them in all subsequent debates over spontaneous generation and the germ theory of disease.

The Pasteur-Pouchet Debate. Pasteur's work on spontaneous generation created as great a sensation in France as had Pouchet's *Hétérogénie* (1859), and in neither case was the sensation confined to scientific circles. The wide public interest in the debate stemmed from its presumed religiophilosophical and even political implications, for the issue of spontaneous generation formed part of the general debate raging in France between materialism and spiritualism. Pouchet's results were invoked in support of materialism, evolutionism, and radical politics, while Pasteur's opposing results were used to support spiritualism, the Biblical account of creation, and conservative politics. In April 1864, in a lecture at the Sorbonne, Pasteur emphasized that the doctrine of spontaneous generation (like materialism in general) threatened the very concept of God the Creator. And although he insisted that he had approached the issue without preconceived ideas, and would willingly have announced in favor of spontaneous generation had "experiment imposed the view on me," there is reason to believe that he wanted a priori to deny the existence of spontaneous generation at least as fervently as Pouchet wanted to affirm it.[60] For Pasteur's position in the debate was in keeping both with his conservative religious and political convictions and with certain aspects of his concept of fermentation—notably the idea of specificity, which implied the transmission of hereditary characters and led to a belief in an ordinary kind of generation among microorganisms. That Pasteur was influenced by such a priori convictions seems clear from his tendency automatically to suspect error in any experiment—including his own—which might be used in support of spontaneous generation and from the eagerness with which he accused Pouchet and other heterogeneticists of technical errors without having repeated their experiments carefully.

Perhaps partly for this reason, as well as the public notoriety of Pasteur's experiments on the glacier at

Montanvert, Pouchet decided to expose his usual hay infusions to the atmosphere at high altitude, following Pasteur's procedure and without using mercury. In November 1863 Pouchet and two collaborators, Nicolas Joly and Charles Musset, announced that the results of their experiments, conducted in the Spanish Pyrenees, contradicted Pasteur's results at Montanvert. For when they exposed their flasks to the air, all subsequently showed microbial growths, as one would expect if the organic material in infusions required only oxygen to organize itself spontaneously into living organisms. In his contemptuous reply to this announcement, Pasteur criticized Pouchet and his collaborators for using a short file instead of long pincers to break the necks of their flasks and for limiting their flasks to so small a number as eight.

In January 1864 the Académie des Sciences named a commission to adjudicate the dispute. When the commission proposed that the participants in the debate repeat their principal experiments before it in March, Pouchet and his collaborators asked that the meeting be delayed until the summer, on the ground that warm weather was conducive to the success of their experiments. In June the commission met with Pasteur and his adversaries, but the latter objected to the program as arranged by the commission and withdrew without repeating their experiments. The commission then observed a series of Pasteur's experiments and verified their exactitude in a report that scarcely veiled its contempt for the opposite side.[61]

As Duclaux has emphasized, this episode might have had a different outcome had Pouchet and his collaborators maintained their nerve in the face of Pasteur's self-assurance and the contempt of the commission.[62] For although no one seemed to realize it immediately, there was a crucial difference between the experiments of Pasteur and those of Pouchet—namely, that Pasteur used yeast water as his alterable medium, while Pouchet used hay infusions. And while boiling easily kills the microorganisms common to yeast water, decoctions of desiccated hay often contain heat-resistant bacilli endospores which can survive high heat and subsequently develop in the presence of oxygen. For this reason Pouchet's flasks could have given microbial life and could have been used in support of spontaneous generation. Only after 1876, especially as a result of the work of Ferdinand Cohn and John Tyndall, did the heat-resistant hay bacillus endospore become fully recognized. Ironically, Pasteur had briefly considered a possibility of this sort in his Sorbonne lecture of 1864, but his attention seems to have been diverted by his zeal to ascribe technical errors to Pouchet. Thus, even though Pasteur had examined the heat resistance of fungal spores and had recognized

the role of heat resistance in other cases, the full complexity and importance of the issue became clear to him only during his debate with Henry Charlton Bastian in the 1870's and in the wake of work by Cohn, Tyndall, Koch, and others.

The Silkworm Problem: The Background. On 8 December 1862, three weeks before he won the Alhumbert Prize, Pasteur had been elected to membership in the mineralogy section of the Académie des Sciences, succeeding in his third formal campaign for the honor. His often active participation in the weekly meetings of the Academy regularly took him away from his laboratory and administrative tasks. So did his lectures at the École des Beaux-Arts, where from November 1863 to October 1867 he was the first professor of geology, physics, and chemistry in their application to the fine arts, and where he introduced laboratory procedures oriented toward the problems of art and its materials. Pasteur also found time to write historical articles on Lavoisier in 1865 and on his friend Claude Bernard in 1866. But the most exhausting demand on his time from 1865 through 1870 was the silkworm problem, which took him away from Paris for several months each year.

By 1865 French sericulturists had become almost frantic about a blight which had afflicted their silkworms for the past fifteen to twenty years—a disease so disastrous as to reduce silk production over this period by a factor of six. In Alais [now Alès] alone, the center of French sericulture, the revenue loss was estimated at 120 million francs for the fifteen years before 1865.[63] The gravity of the situation aroused the concern of the ministry of agriculture and of Dumas, Pasteur's mentor and patron, who was from Alais. In May 1865 Dumas asked Pasteur to study the silkworm blight. Confessing utter ignorance of the problem and noting that he had never even touched a silkworm,[64] Pasteur nonetheless acceded to Dumas's request and immersed himself in the relevant literature, notably Quatrefages's 1859 work.

According to most authorities, the blight resulted from a disease called *pébrine* (pepper) by Quatrefages because of the small black spots frequently seen on sick worms. Its symptoms included stunted or interrupted growth, sluggishness, loss of appetite, and premature death. A general association had also been established between *pébrine* and the existence of microscopic "corpuscles" within the internal organs of diseased worms. Although considerable controversy surrounded the precise role and nature of these corpuscles, several authorities considered them to be the cause of the disease. Those who did tended to suppose that the corpuscles were living parasites, a position that drew support from Agostino Bassi's pioneering studies in the 1830's of another major silkworm disease, muscardine, which he had traced to a fungal parasite. Unfortunately the microscopic corpuscles of *pébrine* could sometimes be found in apparently healthy broods, while their absence failed to guarantee either healthy worms or good silk cocoons. Nonetheless, in 1859, after detecting the corpuscles even in silkworm eggs, where they increased in size and number as hatching time approached, Marco Osimo had tried to establish a preventive measure based on the rejection of corpuscular eggs and pupae. Preliminary trials of Osimo's method were unimpressive, however, and the problem remained obscure.

Pasteur's Early Silkworm Studies. Pasteur's initial firsthand experience with *pébrine* disposed him to doubt both its contagiousness and the causative role of the corpuscles. On his first trip to Alais, in June 1865, he observed two neighboring cultures or broods the opposite fates of which seemed to refute the supposed connection between *pébrine* and the internal corpuscles. The first brood, a successful one, had already spun its cocoons and had therefore entered the pupa stage of its life cycle; the second brood, which had proceeded sluggishly and poorly, as if diseased, had not yet made its passage from silkworms to pupae. Surprisingly, the pupae and moths of the successful brood contained corpuscles in abundance, while the worms of the poor brood contained almost none. Similar cases appeared in other silkworm nurseries around Alais. Some of the surprise abated as the second brood continued to pass through its life cycle. The previously rare corpuscles became increasingly frequent in the pupae, and eventually every moth contained them in profusion. Nonetheless, from the rarity of corpuscles in the sick worms of the second brood, Pasteur concluded that *pébrine* must be a constitutional, hereditary disease, existing prior to and independently of the corpuscles. These corpuscles he supposed to be products of the disease, perhaps resulting from tissue disintegration. Since both broods displayed corpuscles, both must have been diseased; but presumably the first brood had been attacked only late in its life cycle (and thus without serious damage to its silk crop), while the second brood had suffered more severely since an earlier stage.

This conception led Pasteur to essentially the same preventive remedy proposed by Osimo—the selection of eggs from noncorpuscular moths and the rejection of those from corpuscular moths. That Pasteur could advocate this method of egg selection while denying the causative role of the corpuscles becomes less paradoxical if the corpuscles are regarded as an index of the severity of the disease. In Pasteur's view corpuscular moths were obviously in an advanced state of

the disease; and although noncorpuscular moths might also be sick, they must be less seriously so and thus less likely to produce diseased offspring. This method of egg selection, which Pasteur announced only two weeks after his arrival in Alais, remained at the core of his remedial proposals even as his conception of the silkworm plague underwent a dramatic change.

For various reasons this change took place with almost agonizing slowness. In the first place, the tentative conclusions drawn from one year's silk culture could be overturned by the results of the next year, and no way existed to circumvent fully this prolonged natural delay. Moreover, if the material selected happened to fail for reasons unconnected with the prevailing blight, it became useless as a guide to the disease. Personal tragedies and burdens further frustrated Pasteur's efforts. During his first brief trip to Alais, toward the end of the silkworm season of 1865, his father died. The studies of the following year were briefly interrupted by the death of his two-year-old daughter. Immediately after the 1867 season he became the focus of the student protest which ended with his dismissal from the administration of the École Normale. His activities during the 1869 and 1870 seasons were restricted by his debilitating stroke of October 1868.

But the most fundamental obstacle lay in the inherent complexity of the task. Only gradually did it become clear that the silkworm plague involved at least two independent diseases, which differed in ways precisely calculated to confuse students of the problem. Under the weight of these burdens, Pasteur leaned heavily on the moral support of Dumas and Empress Eugénie, and—beginning in 1866—on the companionship and assistance of his loyal collaborators Désiré Gernez, Maillot, Jules Raulin, and Émile Duclaux. For about five months of every year through 1870, one or more of these collaborators joined Pasteur and his wife at Pont-Gisquet, near Alais, where in an abandoned orangery they arranged a makeshift laboratory and carried out the experiments which the master had designed.

From the outset Pasteur's basic experimental strategy was to compare carefully the results of cultures from relatively corpuscular moths with those from relatively noncorpuscular moths. These painstaking studies established the following general conclusions: (1) the more corpuscular the parent moths, the less successful the resulting crop of silk cocoons; (2) while the offspring from partially corpuscular moths sometimes gave a good first crop, they never gave a good second crop; (3) in any brood, however corpuscular the eggs from which it derived, some noncorpuscular moths could always be found. If this third result offered hope that healthy moths (and hence cultures) could always appear even in the midst of disease, the first two tended to emphasize the connection between the corpuscles and the disease and to reinforce the value of selecting eggs from noncorpuscular moths. Indeed, this method had so won Pasteur's confidence by the end of the 1866 season that he began to rely on it to make bold public prophecies. In a letter to the mayor of St.-Hippolyte-du-Fort he predicted the fate during the 1867 season of fourteen batches of eggs he had examined there the year before. In twelve of the fourteen cases, the results conformed closely to his predictions.[65]

None of these results, however, really demonstrated either that the corpuscles caused the disease or that they were living parasites. Nor did a clear answer emerge from preliminary feeding experiments conducted by Pasteur in 1866. For when he fed healthy worms mulberry leaves smeared with corpuscles—to see if *pébrine* could be transmitted in this way—many of the young worms died without becoming corpuscular. On the other hand, similar feeding experiments by Gernez seemed strongly to support the parasitic theory of *pébrine*. Besides establishing a general association between a corpuscular diet and *pébrine*, Gernez showed more precisely that the time at which the corpuscles of *pébrine* appeared in a brood depended directly on the time at which the corpuscular diet had been introduced. Pasteur, however, was not yet convinced. When he reported the results of Gernez's experiments in November 1866, he focused chiefly on those which showed that broods from noncorpuscular moths gave good silk crops—in other words, he reemphasized the practical value of his method of egg selection.

By now, it seems, Pasteur's collaborators were thoroughly convinced both that the corpuscles caused *pébrine* and that they were living parasites. His reluctance to accept this view greatly surprised them, and Duclaux went so far as to accuse him of obstinacy.[66] Pasteur's hesitation is indeed remarkable, not only in view of Gernez's persuasive results but more emphatically in view of his abiding faith in the pathological implications of the germ theory of fermentation—a faith which ought presumably to have disposed him toward a parasitic etiology for *pébrine*. Nonetheless, his initial observations in 1865, and the evidence which he knew best from his own research, conflicted in some respects with the parasitic theory. As late as January 1867, he listed four major objections to a parasitic etiology for *pébrine*: "(1) the disease is certainly constitutional in a number of circumstances and precedes the appearance of corpuscles; (2) the feeding of corpuscular matter often kills young worms without corpus-

cles appearing in their bodies; (3) I have been unable thus far to discover a mode of reproduction for the corpuscles; (4) their mode of appearance resembles a transformation of tissues."[67]

These objections depended in part on Pasteur's inadequate knowledge of protozoan reproduction and on his then defective technique for detecting corpuscles. But they derived in larger measure from a general confusion between *pébrine* and another disease, *morts-flats* or *flacherie*, the complex etiology of which was even more obscure than that of *pébrine*. Perhaps partly because he shared this confusion with so many other authorities on *pébrine*, Pasteur resisted a rapid and careless extension of the germ theory to the diseases of silkworms. Indeed, no other work by Pasteur displays greater sensitivity to the complex relationships between heredity, environment, and parasitism; and Duclaux—while accusing Pasteur of obstinacy—wrote that he did not know a "more beautiful example of scientific investigation" than Pasteur's study of the silkworm problem.[68]

The Silkworm Season of 1867. The silkworm season of 1867 marked a watershed in Pasteur's investigations. Before it ended, he had become a convert to the parasitic theory of *pébrine* and had come to recognize that *morts-flats* or *flacherie*—which most authorities linked with *pébrine*—was an independent disease, with its own character and etiology.[69] His conversion to the parasitic theory of *pébrine* depended chiefly on mounting evidence of its contagiousness. To establish this characteristic, it was necessary to discredit the notion that the disease arose in consequence of a mysterious epidemic environment. Pasteur rejected this notion on the ground that broods derived from noncorpuscular moths—of which he had secured a large supply—usually remained sound and noncorpuscular even in the midst of the allegedly epidemic environment. This result helped to clear the way for further feeding experiments, for it undermined the objection that worms which became sick on a corpuscular diet might owe their disease to an epidemic environment having no connection with their diet.

Against this background Pasteur and his collaborators repeated on a large scale Gernez's feeding experiments and supplemented them with experiments in which healthy silkworms were directly inoculated with corpuscles through surface punctures. In both ways, although especially by corpuscular diets, otherwise healthy worms contracted *pébrine* and became highly corpuscular. Having reached this point, Pasteur seems to have encountered little difficulty in discovering a mode of reproduction for the corpuscles, a mode strikingly different from the budding and binary fission of the microorganisms which he knew best but a mode familiar to protozoologists.

While these studies seemed, therefore, to establish the contagiousness of *pébrine*, with parasitic corpuscles as its cause, they raised another question: If *pébrine* is contagious, of what use is the method of egg selection as a remedy? In the first place, Pasteur replied, the corpuscles of diseased parent moths can be transmitted directly to their eggs; and in this sense *pébrine* is simultaneously hereditary and contagious. Moreover, if the offspring of noncorpuscular moths later contract *pébrine*—whether by eating corpuscular leaves or by inoculation—the incubation period is long enough to ensure that all, or virtually all, the worms will spin cocoons and yield a silk crop. And since the corpuscles lose their fecundity and pathogenicity from one silkworm season to the next, the only effective source of contagion in each season must be the corpuscles contained in the eggs produced by corpuscular moths. If, therefore, all the eggs from corpuscular moths are rejected, *pébrine* ought to disappear quickly. In this way Pasteur developed a new and more impressive rationale for his method of egg selection, but his hope that it could lead to total elimination of *pébrine* was doomed by the fact that the corpuscle enjoys hosts other than the silkworm.[70]

Studies on Flacherie, 1867–1870. Pasteur's recognition of *flacherie* as an independent disease can be traced at least in part to his confidence in the method of egg selection, undoubtedly reinforced by his new conviction of the parasitic nature of *pébrine*. For what especially alerted him to the independence of *flacherie* was the failure during the 1867 season of entire broods descended from noncorpuscular moths. Most of these unsuccessful broods, which appeared in Pasteur's cultures as well as those of several breeders to whom he had sent the eggs, displayed neither the corpuscles nor the black external spots of *pébrine*. Instead, nearly all the worms died with the familiar symptoms of *flacherie* —symptoms different enough from those of *pébrine* to have received a separate name, although most authorities (including Pasteur) had hitherto supposed that these symptoms merely represented a special stage or effect of *pébrine*. That *flacherie* represented a well-defined and independent hereditary disease now seemed clear not only from the absence of corpuscles in these diseased broods but also from the way it attacked all of the offspring of certain batches of eggs, even though these eggs had been cultivated in widely different environments.

Although these events aroused great practical concern, they also helped to clarify much of the apparently contradictory evidence. In the case of Pasteur's work, it now seemed clear that his initial observations at

Alais, as well as his preliminary experiments with corpuscular diets, had miscarried through the intervention of *flacherie*. Because, in both cases, he had observed death and disease in the absence of corpuscles, he had supposed that *pébrine* must be a constitutional disease. The events of the 1867 season strongly suggested that such a constitutional disease did exist, but that this disease was *flacherie* rather than *pébrine*. Compared with *pébrine*, *flacherie* had contributed rather little to the ruinous silkworm blight; but its character and etiology demanded great attention because it threatened the method of egg selection on which Pasteur had based his hopes for the rejuvenation of French sericulture. At first Pasteur merely advised the rejection of eggs from broods which displayed high mortality, languor, or any other symptom of *flacherie*. Then, during the silkworm seasons of 1868 to 1870, he sought to unravel the etiology of *flacherie* from that of *pébrine* and to find a prophylactic method for it as reliable as the method of egg selection he had devised for *pébrine*.

At the outset of these studies on *flacherie*, two striking phenomena arrested Pasteur's attention: (1) the strongly hereditary aspect of the disease, as revealed by the almost constant and devastating appearance of *flacherie* in descendants of broods which had shown some symptoms of the disease before spinning their cocoons and laying their eggs; and (2) the abundant presence of microorganisms in the intestinal canals of worms attacked by *flacherie*. Notable among these microorganisms, which were virtually absent from healthy worms, were vibrions (bacilli) and a "petit ferment en chapelets de grains" (a micrococcus), which resembled an organism he had already associated with certain fermentations.[71] As with the corpuscles of *pébrine*, Pasteur at first supposed that these microorganisms were a consequence of the disease rather than its cause, their chief significance being diagnostic rather than etiological.[72] More specifically, he conceived of *flacherie* as a sort of hereditary susceptibility to indigestion, in consequence of which ingested mulberry leaves underwent fermentation in the intestinal canal. On this view the microorganisms associated with intestinal fermentation, and especially the small ferment in chains, served as a physical index of a late stage in the disease. But even while thus denying the intestinal microorganisms a direct causative role in *flacherie*, Pasteur put them at the center of his efforts to develop a prophylactic measure against it. In brief, he counseled the rejection of pupae the stomachs of which contained the small ferment in chains, since they were certain to transmit the hereditary predisposition to their offspring.

During the silkworm season of 1869, Pasteur considerably modified his conception of *flacherie*.[73] As in the case of *pébrine*, feeding experiments seem to have been chiefly responsible for this shift. On a diet of leaves smeared with excrement from worms with *flacherie*, previously healthy worms fell sick with the disease. Thus *flacherie*, like *pébrine*, was contagious as well as hereditary. Unlike *pébrine*, however, *flacherie* owed its hereditary character not to the direct transmission of a microorganism from the parent moths to the eggs but to a constitutional weakness of which there was no immediately visible sign. While thus retaining part of his original conception of the disease, Pasteur now perceived—however dimly—that this hereditary weakness involved a susceptibility not so much to indigestion per se as to the germs of the microorganisms later seen in the intestinal canal. Henceforth he identified the intestinal microorganisms as the proximate cause of *flacherie*. Unlike the corpuscles of *pébrine*, these microorganisms are common and universally distributed. They must therefore become pathogenic in silkworms only under special circumstances. Hereditary susceptibility to them in certain silkworms clearly forms one of these special circumstances; but other such conditions must exist, for *flacherie* sometimes appears "accidentally" or "spontaneously" in a brood without any hereditary predisposition. In such cases, Pasteur suggested, unusual conditions of temperature, humidity, or ventilation in the nursery must either promote the multiplication of the causative microorganisms on the leaves or lower the resistance of the silkworms to the ingested germs. To prevent or reduce all forms of *flacherie*, therefore, it was necessary not only to reject infected pupae but also to monitor and to control as far as possible the environmental conditions in the nursery.

According to René Dubos, the etiology of *flacherie* is even more complex than Pasteur realized.[74] Among other things, the susceptibility of silkworms to the bacteria of *flacherie* seems to depend on the intervention of a filterable virus. However that may be, Pasteur had attained a remarkably keen insight into the essential features of *pébrine* and *flacherie*. He recognized the subtlety and importance of the questions this work raised about the interaction of parasite, host, and environment in the production of disease; and he later advised young physicians to study his *Études sur la maladie des vers à soie* (1870) as an introduction to such issues.[75] But Pasteur had grown increasingly tired of this work, especially as he became confident that he had provided the basis for a practical solution to the problems of French sericulture. In fact, between 1868 and 1870 study of the etiology of *flacherie* occupied him less fully and directly than his efforts to establish and proselytize his practical measures against the silk-worm blight. Toward this end he engaged in an enormous correspondence with sericulturists and their trade journals, distributed vast quantities of eggs for industrial trials, and became a practical sericulturist. These efforts brought Pasteur recognition and testimonials from commissions and sericulturists, many of whom adopted his methods. If his success was less than total, it was certainly considerable and he did not lose confidence even during a serious depression in French sericulture from 1879 to 1881, which he ascribed not to a failure of his methods but to bad weather and to the comparatively low prices of Oriental silk.[76]

Debates Over Fermentation, 1871–1876. From 1865 to 1870, while Pasteur was preoccupied with the silkworm problem, his theory of fermentation enjoyed increasing favor, especially abroad. What criticism did appear during that period failed to distract him from his central task. In 1871, however, the *Annales de chimie et de physique* published a French translation of a wide-ranging critique by Liebig, who had broken a long silence on the issue in two lectures (1868, 1869). In a reply of almost arrogant brevity, Pasteur discussed only two aspects of Liebig's critique, both of which involved direct challenges to experimental claims made a decade before by Pasteur: (1) that pure yeast and a simple alcoholic fermentation could be produced in a medium free of organic nitrogen and (2) that acetic fermentation required the intervention of *Mycoderma aceti*. Pasteur responded by challenging Liebig to submit the dispute to a commission of the Académie des Sciences. Before this commission, Pasteur boldly predicted, he would prepare, in a medium free of organic nitrogen, as much beer yeast as Liebig might reasonably demand and would demonstrate the existence of *Mycoderma aceti* on the surface of the beechwood shavings in the German method of acetification.[77]

Although Liebig died in 1873 without accepting Pasteur's challenge, some aspects of his critique were adopted in France by Edmond Frémy and Auguste Trécul, among others. These critics earned the scorn and ridicule of Pasteur who went so far as to impugn the patriotism of those who dared to defend a "German theory" against a "French theory" after the Franco-Prussian War.[78] Despite their often personal and repetitive character, the ensuing debates nonetheless contributed to Pasteur's understanding and articulation of the issues surrounding fermentation and spontaneous generation. Insofar as the debates concerned fermentation as such, their chief value was to induce Pasteur to clarify his views on the role of oxygen in the process, to extend to all living cells his theory of fermentation without air, and to begin at last to face directly and explicitly the ambiguities of his definition of fermentation.

In his papers of the 1860's Pasteur had implied that oxygen played no role in fermentation, unless to impede it. In fact, some brewers supposed that he advocated the total elimination of air during brewing, a natural enough conclusion from his theory of fermentation as "life without air."[79] Only gradually, under prodding from such critics as Frémy, Oscar Brefeld, and Moritz Traube, did Pasteur begin to emphasize that oxygen played an essential, if strictly limited, role. In the face of Frémy's repeated insistence that some contact with oxygen was essential to the fermentation of grape juice, Pasteur finally acknowledged in 1872 that this view contained a kernel of truth, in that yeast—the true agent of fermentation— did require some oxygen in order to germinate.[80] In 1875 he responded in essentially similar fashion to the objections of Brefeld and Traube, whose careful experiments suggested that yeast deprived of free oxygen either could not live at all or else provoked at most a very feeble and incomplete fermentation. In his reply Pasteur suggested that they had been misled by using contaminated yeast or yeast too old and "exhausted" to germinate in an oxygen-free environment.[81] In his *Études sur la bière* (1876), in which he also described a new and perfected method of preparing pure yeast, Pasteur emphasized that yeast occasionally required small quantities of oxygen in order to retain its "youth" and its capacity to germinate in oxygen-free environments. Having now achieved a new appreciation for the importance of oxygen in brewing, and especially the advantages of aerated wort, he insisted only that air should be carefully limited and freed of foreign germs rather than entirely eliminated.

In the meantime Pasteur had extended his theory of fermentation to all living cells, a development that Dumas believed might well mark "an epoch in the history of general physiology."[82] Beginning in October 1872, Pasteur set forth and elaborated the view that because fermentation is a manifestation of life in the absence of free oxygen, and because every living cell can survive at least temporarily under such conditions, "all living things are ferments in certain conditions of their life." As evidence he cited experiments showing that *Mycoderma vini* and *Penicillium glaucum*, ordinarily aerobic organisms that consume free oxygen, can live for a time in the absence of free oxygen—when forcibly submerged in a sugared liquid medium, for example. Under these anaerobic conditions the *Mycoderma* and *Penicillium* become ferments: they decompose the sugar in order to extract its combined oxygen and carbon, producing alcohol in the process. Similarly, intact grapes, prunes, plums, and other fruits give off a small quantity of alcohol in an oxygen-free environment. In the latter case the cells of the fruit decompose the sugar in the fruit to obtain carbon and oxygen and thus the heat (or energy) required for physiological processes. As early as 1861 Pasteur had described in preliminary fashion a converse phenomenon—the capacity of yeast, ordinarily an anaerobic organism, to become adapted to a more or less aerobic existence, in which case its power as a ferment decreased or disappeared. In August 1875 he specified the conditions under which yeast could become a fully aerobic plant, living exactly like common molds. The essential task was to germinate the yeast on a liquid of large surface area in the presence of abundant oxygen. Under these circumstances it consumed free oxygen and did not produce fermentation.[83]

In several respects these ideas confused Pasteur's contemporaries, and some of his opponents thought he had unwittingly exposed fundamental flaws in his germ theory of fermentation. By revealing the protean character of yeast and other lower organisms—which might live as either anaerobes or aerobes, as ferments or not—he seemed to undermine his insistence on the specificity and peculiarity of fermentative microorganisms. More directly, his suggestion that fruit cells could produce alcohol—without the intervention of living microorganisms—struck some as an outright contradiction of his earlier views and the entire germ theory of fermentation. While responding to such confusion and criticism, Pasteur finally emphasized and clarified several points hitherto largely implicit or otherwise submerged in his work. Above all, he revealed how much his theory of fermentation depended on his carefully circumscribed definition of the process.

The Circularity of Pasteur's Theory of Fermentation. From the beginning of his work on fermentation, Pasteur had restricted the germ theory to "fermentations proprement dites" ("fermentations properly so-called"). When, in February 1872, Frémy demanded to know what he meant by this expression, "so vague and so elastic," Pasteur said that he applied it to "the fermentations that I have studied and which include all the best characterized fermentations, those which are as old as the world, those which give bread, wine, beer, sour milk, ammoniacal urine, etc., etc., those in which the ferments are, according to my researches, living beings which arise and multiply during the act of fermentation."[84] On the other hand, processes such as the so-called diastatic fermentation, by which starch was converted into sugar, did not merit inclusion among the fermentations "properly so-called," because they involve a soluble chemical ferment (an enzyme) rather than a living microorganism. In other words, Pasteur excluded from his definition of fermentation those processes of decomposition which he admitted to be chemical rather than biological.

But Pasteur also excluded from the list of true fermentations certain processes of decomposition that he had identified as biological. He probably acted intentionally, for example, when he omitted acetic fermentation from the list of the "best characterized fermentations" that he had studied. This obvious omission can be explained by supposing that Pasteur defined as "fermentations properly so-called" only those processes associated with anaerobic microorganisms. Because acetification depended on *Mycoderma aceti*, an aerobic organism, it must be excluded from the true fermentations, even though it met two other fundamental criteria—it was microbial and it involved the decomposition of a weight of substance vastly greater than the weight of the responsible microorganism. Along somewhat similar lines, Pasteur differentiated between alcoholic fermentation "properly so-called" and the nonmicrobial production of alcohol by fruit cells in the absence of free oxygen. By itself, he insisted, the production of alcohol is no index of true alcoholic fermentation, for the latter process also yields glycerin, succinic acid, and other substances. This process is called "alcoholic fermentation" only by abbreviation; to be precise, one ought to designate it by its complete equation, the complexity of which reflects its dependence on living yeast.

As these examples make clear, Pasteur's theory of fermentation reduced to a virtual tautology, for any process which failed to conform to that theory in every respect automatically failed to qualify as a fermentation "properly so-called." In similar fashion Liebig might have maintained an unassailable chemical theory of fermentation had he been willing to exclude from his definition of fermentation those processes which Pasteur associated with microorganisms. In so doing, however, Liebig would have excluded many of the decomposition processes traditionally regarded as fermentations—most notably ordinary alcoholic fermentation, which had always been considered the archetypal fermentative process. The fertility and power of Pasteur's theory derived precisely from its applicability to these familiar processes, and he seemed remarkably unconcerned that it did not also apply to those processes associated with such soluble chemical ferments as diastase, emulsin, or pepsin. By admitting, or at least implying, that his theory also failed to apply to certain biological processes—including acetification and the nonmicrobial production of alcohol by fruit cells—Pasteur invited confusion and threatened his own attempt to generalize the theory to all living cells. That his study of fermentation nonetheless produced

valuable insights, both theoretical and practical, illustrates forcefully that not all circles are vicious.

The Issue of a Soluble Alcoholic Ferment. In retrospect, the most intriguing feature of the debate between Pasteur and Liebig is the extent to which they seemed ultimately to approach a mutually acceptable conception of fermentation. By 1869, at least, Liebig was prepared to admit the possibility that alcoholic fermentation depended in part on the life of yeast. Adopting a hypothesis by no means original with him, he suggested that living yeast cells might secrete a soluble chemical ferment, analogous to diastase or pepsin, which then induced the decomposition of sugar into alcohol and carbonic acid. This hypothesis drew particular support from the knowledge that yeast did produce at least one other soluble ferment, *ferment glycosique* or invertase, responsible for inverting cane sugar.

Pasteur made no immediate objection to Liebig's suggestion; indeed, in his memoir of 1860 on alcoholic fermentation, he had mentioned the possibility that yeast might act by secreting a soluble ferment. In 1875, two years after Liebig's death, Pasteur suggested that the processes resulting from soluble ferments might someday be reunited with the true fermentations "in some way as yet unknown."[85] In July 1876 he conceded that the ammoniacal fermentation of urine, which he had ascribed since 1860 to a living microorganism, could be traced more immediately to a soluble chemical ferment produced by the living ferment. When his opponents tried to exploit this concession, Pasteur emphasized that for twenty years he had devoted himself chiefly to demonstrating that the agents of fermentations were microorganisms. The precise mechanism by which these agents acted was a problem of a different order and required further investigation.[86]

From this perspective the debate between Pasteur and Liebig seems to have ended as an essentially semantic dispute, a disagreement born of their approach to the phenomena of fermentation at different levels, with Liebig seeking its proximate cause and Pasteur content to establish more remote correlations. As Duclaux emphasized, however, the two positions implied strikingly different experimental strategies.[87] Because this difference was reinforced by long-standing disagreements over experimental results, and by personal and national antagonisms, Pasteur and Liebig found it difficult to make concessions; and their potential rapprochement remained largely submerged in mutual hostility.

Even in the absence of these difficulties, Pasteur and Liebig might never have achieved a fully compatible conception of fermentation, for some of the issues which divided them reemerged in Pasteur's debate with Marcelin Berthelot, the leading French advocate of the modified chemical theory of fermentation. Like Liebig, Berthelot had initially opposed Pasteur's attempt to implicate living organisms in fermentation and had then moved to the view that living yeast might act by secreting a soluble alcoholic ferment. His views on fermentation derived particular authority from his having isolated from yeast the soluble ferment responsible for the inversion of cane sugar. In 1878 Berthelot arranged for the posthumous publication of manuscript notes in which Claude Bernard criticized Pasteur's theory of fermentation and claimed to have isolated a soluble ferment capable of producing alcoholic fermentation independently of living yeast.

The publication of this manuscript placed Pasteur in an awkward position, for Bernard had long contributed his immense authority and support to Pasteur's cause. To some extent Pasteur adopted the strategy of impugning Berthelot's motives rather than the work of the revered Bernard, who had neither authorized the publication of his manuscript notes nor described their contents to Pasteur. Nonetheless, in a full-length critique of Bernard's manuscript (1879), Pasteur attacked in devastating fashion the experiments by which Bernard believed he had destroyed Pasteur's theory of fermentation as life without air. By carefully repeating these experiments and comparing them with his own, Pasteur went a long way toward justifying his claim that Bernard's results were mistaken, dubious, or badly interpreted. In this task Pasteur benefited from the patently crude and preliminary character of Bernard's experiments (at least as they were represented in the manuscript notes) and from their author's inability to reply or defend himself. While expressing reluctance about taking advantage of these circumstances, Pasteur justified his action on methodological grounds. In his view Bernard's manuscript offered a dramatic example of the danger of "systems" and "preconceived ideas," a danger which Bernard himself had done so much to expose in his *Introduction à l'étude de la médecine expérimentale* (1865).

Saying that Bernard had somehow forgotten his own wise precepts, Pasteur suggested that he had been led astray by an a priori conviction of a fundamental opposition between organic syntheses, which he supposed to be peculiarly vital phenomena, and organic decompositions (including fermentation, combustion, and putrefaction), which he supposed to be physico-chemical rather than vital processes. Because his theory of fermentation linked life and organic syntheses with a process of organic decomposition, Pasteur continued, it conflicted with Bernard's general conception of life and thereby earned his rejection. From this perspective it was easy to understand why Bernard not only embraced the view that the immediate cause of fermentation was a soluble alcoholic ferment but also claimed that this soluble ferment existed—independently of yeast cells—in the juice of grapes at a certain stage of their maturity. By this claim Bernard sought to deny living yeast any role in fermentation, while even Liebig and Berthelot were willing to concede that it might be essential for the production of the hypothetical soluble ferment. Unfortunately for Bernard, said Pasteur, his claim was refuted by the fact that grapes of any degree of maturity never fermented when carefully protected from yeast germs.

If this version of Bernard's "preconceived ideas" was less than fair or accurate—as Duclaux suggests[88] —Pasteur may have been driven to it by the extravagance of Bernard's views. But it is clear from his critique of Bernard, and from the associated debate with Berthelot, that he was also suspicious of the more moderate attempts by Berthelot and Liebig to incorporate his "physiological" theory of fermentation into the modified chemical theory that yeast acted by secreting a soluble alcoholic ferment. Even as he insisted that he would be neither surprised nor disturbed by the discovery of such a chemical ferment—indeed, he reportedly sought it himself by grinding and plasmolyzing yeast cells[89]—Pasteur asserted that the role of soluble ferments would one day be eclipsed by that of life without air.[90]

Until his death Pasteur could retain this hope as he surveyed a long tradition of unsuccessful attempts to isolate a soluble alcoholic ferment. In 1897, however, while engaged in apparently unrelated immunological research, Eduard Buchner achieved this goal and thereby cast Pasteur's physiological theory of fermentation into the shade. Even then, however, the phenomenon known as the "Pasteur effect"—the inhibition of fermentation in the presence of free oxygen —remained as real as it was inexplicable. More recently, the physiological and chemical theories of fermentation have come to be seen as complementary rather than opposed. If, at some level, Pasteur perceived this possibility, he never explained precisely how the notion of a soluble alcoholic ferment could be reconciled with the doctrine of fermentation as life without air. He sought instead to defend the conclusions he had already reached and challenged the wisdom or necessity of invoking the concept of a soluble alcoholic ferment. For him this concept remained a gratuitous and unproved assumption. For his opponents, and particularly for Berthelot, the concept of life without air was an equally gratuitous, unproved and unnecessary hypothesis. In short, if our present conception of fermentation suggests that the debate was largely semantic and capable of easy resolution, its participants were unable to see it that way.

Studies on Beer. During the late 1860's the "pasteurization" of wine and vinegar became increasingly common. The process found a new application in Austria and Germany, where the practice of heating bottled beer to 55° C. became widespread following the publication of Pasteur's *Études sur le vin* (1866). Beginning in May 1871, largely under the stimulus of the Franco-Prussian War, Pasteur launched a study of beer in hopes of serving "a branch of industry in which Germany is superior to us."[91] This effort, begun in Émile Duclaux's laboratory at Clermont-Ferrand, led to a series of patents and to Pasteur's *Études sur la bière* (1876). Meanwhile, Pasteur had become embroiled in a series of debates over fermentation and spontaneous generation; and the book on beer consists for the most part of a sometimes oddly organized and largely tedious rehearsal of those debates.

Only two chapters in the book were directed specifically toward the practical problems of brewing. In the first chapter Pasteur sought to demonstrate that the alterations or "diseases" of beer depend on the appearance and development of foreign microorganisms, "not at this time a new idea," according to Duclaux.[92] In the last chapter Pasteur described his process for manufacturing beer, which emphasized the use of pure yeast and carefully limited quantities of pure air. As in his books on vinegar and wine, he gave considerable space to descriptions and drawings of the industrial apparatus his new method would require. Perhaps because of its wide adoption in the German brewing industry, the method of preserving beer by heat received only passing and skeptical attention.[93]

Among the advantages that Pasteur claimed for his new method of manufacturing beer, the most important were the elimination or reduction of costly cooling techniques (introduced empirically, but now explicable as a means of impeding the development of pathogenic organisms) and the protection of finished beer from disease. Nonetheless, Pasteur admitted that his process had "not yet been practically adopted,"[94] a result he ascribed chiefly to the costly retooling it would require. If attempts were ever made to exploit Pasteur's patents on beer, their fate has yet to be described. Nonetheless, his more general contributions to the study of brewing attracted the attention and admiration of some industrial brewers, notably J. C. Jacobsen, founder of the Carlsberg brewery in Denmark. In the late 1870's Jacobsen gave 1.5 million francs for the creation of a magnificent laboratory at his Carlsberg brewery. For this laboratory, which soon became a leading center of biochemical research, he commissioned a bust of Pasteur, who responded by dedicating his 1879 critique of Bernard to Jacobsen.[95]

Pasteur and Spontaneous Generation, 1871–1879. If Pasteur believed that his triumph over Pouchet would silence the partisans of spontaneous generation, he was soon disappointed. The issue remained a subject of lively debate, especially in England and Germany, where Pasteur's critics had rather less to fear from the judiciary proceedings of the Académie des Sciences and from the presumed association of spontaneous generation with Darwinian evolution and radical politics. When Pasteur rejoined the controversy in 1871 his chief French opponents were those who simultaneously challenged his theory of fermentation—notably Frémy and Trécul. In France the debate on spontaneous generation now focused on the origin of the alcoholic yeasts, although attention was also paid to the origin of the microorganisms found in putrefying eggs and in human abscesses. Bound up with these specific concerns were the broader issues of the transmutation of microbial species, the nature and distribution of germs, and the distinction between aerobic and anaerobic life. From July 1876 to July 1877 Pasteur also engaged in a celebrated controversy with the English naturalist H. Charlton Bastian, who claimed he could produce microorganisms spontaneously in neutral or alkaline urine. From this debate—the most productive in the series—Pasteur emerged with a firmer grasp of the relative distribution of germs in air, in water, and on solid objects, and—most important—with a greater appreciation for the heat resistance of certain microorganisms.

Pasteur rejoined the spontaneous generation controversy by attacking Frémy's claim that the yeasts of vinification arose internally and spontaneously from grape juice upon contact with the air. In 1872, in an attempt to make his point decisively, Pasteur showed that a drop of unheated natural grape juice, aspirated from the interior of a ripe grape, would neither ferment nor give yeasts in germ-free air.[96] He took great delight in this delicate experiment, which he often linked with his earlier demonstrations that natural urine and blood could be preserved in germ-free air even without preliminary heating. Although Frémy and Trécul managed to find objections against even this experiment, Pasteur disposed of them quite readily and continued to cite the experiment as definitive proof against the internal, spontaneous origin of yeasts.

When Frémy then sought support for spontaneous generation in Pasteur's demonstration that fruits could remain intact (hence closed to external germs) and yet produce alcohol in an oxygen-free environment, Pasteur was obliged to emphasize that no microorganisms participated in this process; it was a case of the fruit cells themselves acting as "ferments" under anaerobic conditions.

But Pasteur went much further. Denying that the yeasts of wine originated spontaneously within the grape, he sought to establish their precise external origin and to clarify their more general properties. By 1876, when he reported his results in his *Études sur la bière*, he felt confident that he had established the following generalizations about the alcoholic yeasts: (1) a great many yeasts exist, differing in form, physiological properties, and in the taste and other qualities that they impart to the fermenting liquid; (2) the yeasts of wine derive from germs that are particularly abundant on the wood of the grape cluster, somewhat less abundant on the surfaces of the grapes themselves, and rare in ordinary atmospheric air; (3) these germs gradually decrease in number and fecundity during the winter and are entirely absent from the surfaces of immature grapes; (4) these germs increase in number and fecundity as the grapes mature and as the time of the vintage approaches (so that when the ripe grapes are crushed, no yeasts need be sown, as they must in brewing beer); (5) these germs require oxygen to retain their vitality and thus their capacity to produce fermentation; and (6) the species of yeast are distinct, are not transformed one into the other, and do not represent special developmental forms of another plant.[97]

In 1878, after Bernard's manuscript on fermentation had revived the notion of an internal origin for the yeast of wine, Pasteur confirmed under natural conditions the central conclusions of his *Études sur la bière*. In July of that year, immediately after reading Bernard's manuscript, he ordered the construction of several glass hothouses, with which he intended to cover some of the still immature vines in his own vineyard near Arbois. This plan had been executed by early August, before any yeast germs had appeared on the grape clusters. As a further precaution he wrapped some of the clusters within the hothouses in sterile cotton. By 10 October all the grapes had ripened and the time of vintage had arrived. As Pasteur expected, the exposed grapes easily and rapidly fermented when crushed, while those protected from yeast germs by the hothouses did not, except in one case. The grapes wrapped in cotton within the hothouses never fermented when proper precautions were taken. On the other hand, if these grapes were subsequently exposed in the open air, they soon fermented when crushed with the yeast germs that they had in the meantime received.[98]

By these experiments Pasteur went a long way toward a definitive demonstration of the external origin of the yeast of wine. He had by then reached an equally firm position on the issue of the transmutation of microbial species. From the 1840's to the early 1870's, an increasing number of botanists claimed that they had observed the transformation of one microbial species into another; and their claims had been enlisted in support of Darwinian evolution and spontaneous generation. In 1861 Pasteur specifically challenged several presumed cases of microbial transmutation—notably of *Penicillium glaucum* into beer yeast—and his more general opposition to the doctrine seems implicit in his work on fermentation and spontaneous generation, with its dependence on the specificity and hereditary continuity of microorganisms. Nonetheless, Pasteur gave little explicit attention to the issue before the 1870's; and his general position had been obscured by his claim of 1862 that he had observed the transformation of *Mycoderma vini* into the alcoholic yeast of wine under anaerobic conditions, more specifically when submerged in a fermentable liquid. For the next decade, as he continued to hold this view, Pasteur used it in support of his theory of fermentation as life without air. Then quite suddenly, in October and November 1872, he reconsidered and abandoned his earlier claim.[99]

By Pasteur's own account, this change of view had its origin in two sorts of observational evidence. First, even when he had sown only *Mycoderma vini* into the fermentable liquid, he sometimes found cells of *Mucor mucedo* or *racemosus* as well as yeast cells among the submerged mycodermic pellicle. Assuming that this *Mucor* could have entered the medium only from the surrounding air, he began to wonder if the air could not also be the source of the yeast cells he had hitherto supposed to be the transformed cells of *Mycoderma vini*. Second, yeast cells sometimes failed to appear in the submerged pellicle, even when the experimental conditions seemed identical. Why should the presumed transmutation fail to take place in these cases? To resolve his doubts Pasteur modified his swan-necked flasks in such a way as to permit the comparative study of the same microorganism under anaerobic conditions (when submerged) and under aerobic conditions (on the surface of a shallow liquid) without exposing the liquid medium to the ambient air or to any other external source of germs. Under these conditions Pasteur never again observed the supposed transformation of *Mycoderma vini* into yeast, and he never again wavered in his opposition to the notion of direct microbial transmutation.

If Pasteur felt any embarrassment about rejecting his original belief in the transmutability of *Mycoderma vini*, he probably found more than adequate consolation in the circumstance that his theory of fermentation not only remained intact but also acquired a new extension and generality. For his rejection of the transmutability of *Mycoderma vini* coincided with, and perhaps depended upon, the extension to all living cells of his theory of fermentation as life without air. In the light of this generalized version of his theory, he could and did ascribe fermentative power directly to the cells of *Mycoderma vini* under anaerobic conditions, without needing to suppose that they acquired this power by virtue of a transformation into yeast cells. Pasteur may also have been encouraged to take this position and to reassert his general opposition to microbial transformism by the influential work of the German botanists Anton de Bary and Ferdinand Cohn. Certainly he was not alone in his opposition to immediate microbial transformism; and de Bary, Cohn, and others contributed more than he to the general rejection of the doctrine.

With regard, finally, to Pasteur's general position on the transmutation of species, it should be emphasized that he did not directly and explicitly repudiate Darwinian evolutionary theory per se. Although clearly skeptical of the theory and suspicious of its popularity—which he ascribed to its failure to require "rigorous experimentation" or "profound observations"[100]—Pasteur insisted only that no one had demonstrated the immediate transformation of one microbial species into another.

The Pasteur-Bastian Debate. By July 1876, when Pasteur locked horns with H. Charlton Bastian, that influential English advocate of spontaneous generation had already established his reputation through his long and controversial *The Beginnings of Life* (1872) and had engaged the attention and opposition of the English physicist John Tyndall. Although Bastian's advocacy of spontaneous generation depended on a wide range of experimental evidence and theoretical considerations, his dispute with Pasteur focused very narrowly on one issue: whether microorganisms can originate spontaneously in neutral or alkaline urine. Pasteur seems publicly to have ignored Bastian's work until the latter sent a note to the Académie des Sciences in which he claimed that microorganisms appeared under carefully specified conditions in urine that had been boiled and subsequently protected from atmospheric germs. According to Bastian, the requisite physicochemical conditions were the intervention of potash and oxygen and a storage temperature of 50° C. On the assumption that the boiling killed any organism in the urine, Bastian claimed to have produced spontaneous generation.

Within a week Pasteur had repeated Bastian's experiment and had confirmed in most cases his central result—boiled urine rendered alkaline by aqueous potash did indeed yield microbial life in germ-free air.

With Bastian's interpretation of this result, however, Pasteur profoundly disagreed. In his view Bastian's result merely proved "that certain inferior germs resist 100° C. in neutral or slightly alkaline media, no doubt because their envelopes are not penetrated by water under these conditions as they are . . . [in] slightly acid media."[101] He referred Bastian to his memoir of 1861 on organized corpuscles in the atmosphere, in which he had discussed the heat resistance of microorganisms in alkaline media, and challenged him to repeat his experiments using potash—whether solid or in aqueous solution—that had been previously heated to 110° C. Under these conditions, Pasteur asserted, the urine would remain sterile and Bastian's "spontaneous generation" would cease to exist.

The terms of Pasteur's challenge imply his belief that Bastian had unwittingly introduced the germs of microbial life into his urine flasks by using germ-charged potash or germ-charged water. Over the next several months Bastian refused to abandon his claim, insisting on the absurdity of the notion that germs could resist so caustic a substance as potash, demanding a direct demonstration of the heat resistance of germs, and complaining that his experimental procedures (including the exact neutralization of urine by potash) had not been faithfully reproduced by Pasteur and his collaborators, Jules Joubert and Charles Chamberland. As they fended off these objections, Pasteur and his collaborators sought to establish more precisely the external origin and degree of heat resistance of the germs supposedly introduced into Bastian's flasks. Pasteur and Joubert launched a study of the distribution of germs in water, reinforcing and extending the earlier results of the English physiologist Burdon-Sanderson concerning the enormous quantity of bacteria in ordinary streams and the presence of germs even in distilled water unless it was stored in vessels rendered germ-free by flaming. They also noted the absence of germs in water from deep sources, where surface germs could not penetrate, and insisted on the extreme minuteness of the bacterial germs, which passed through all ordinary filters and required the invention of a new method for their collection (presumably a prototype of Chamberland's porcelain bacterial filter).

In July 1877, having accepted Pasteur's challenge to submit their dispute to a commission of the Académie des Sciences, Bastian went to Paris to repeat his experiment in the presence of this commission. Like Pouchet, however, Bastian eventually withdrew after a long and confusing dispute with the commissioners.[102] Once again facing a commission on spontaneous generation without an opponent, Pasteur reaffirmed his claim that neutral urine could be kept sterile if all proper precautions were observed. By this time he had clearly identified three possible sources of germ contamination in Bastian's experiments: to the potash solution originally suspected he added the experimental apparatus (even when carefully washed, since all water contains germs) and the urine, which can from the outset harbor germs capable of surviving boiling at 100° C. He did not yet fully appreciate the latter possibility, however, believing that the acidity of the normal urine with which Bastian began would prevent the appearance of these heat-resistant germs, and he chose instead to indict contaminated apparatus as the source of germs in Bastian's experiments.

Only after Cohn, Koch, Tyndall, and others had established the existence of highly resistant bacterial endospores; only when it became clear that certain microorganisms could survive a temperature of 100° C. even in acid media; and only as microbial life continued to appear in certain liquids (notably urine and infusions of hay or cheese) despite every precaution to eliminate germs from the experimental apparatus—only then did Pasteur begin fully to perceive the possibility that the liquids used by Pouchet, Bastian, and other advocates of spontaneous generation may sometimes have harbored microbial life from the beginning rather than having subsequently acquired it through careless experimental technique.

By this time other aspects of Pasteur's doctrine had come under open and serious challenge, especially in England.[103] Some challenged his evidence that the "organized corpuscles" in the air were living organisms, for that evidence was largely indirect and failed to establish a direct link between any particular living microorganism and its presumed antecedent germ or corpuscle. Others asked how germs living in the air— and thus presumably aerobic—could be responsible for processes that Pasteur ascribed to the activity of anaerobic microorganisms. To meet this argument, Pasteur suggested that germs possessed only latent life while in the air and therefore should not be called aerobic organisms in the ordinary sense of the word.[104]

More or less convergent with these challenges were two doubts shared even by those who fully accepted his claim that the air contained living ferments. One doubt concerned whether the atmosphere in fact carried as much microbial life as Pasteur supposed. Pasteur himself had contributed toward this question by showing that microbial life was variably disseminated in the atmosphere and was certainly not so widespread as to exist in every sample of air. He had also drawn attention to the relatively high concentration of microbial life on grape clusters and in water as compared with the atmosphere, but some of his contemporaries advocated an even greater shift of emphasis to liquids and solid surfaces. The second doubt concerned the precise meaning of Pasteur's often casual use of "germ." In 1877 Burdon-Sanderson argued that Pasteur's "organized corpuscles" were in fact finished, adult microorganisms and not their "germs" or precursors.[105]

In both cases subsequent research has tended to confirm the doubts. Indeed, Pasteur himself emphasized in 1878 that surgeons had far more to fear from germs on their instruments or hands than from germs in the air,[106] and he seems not to have disputed the growing belief that many of his "germs" were adult microorganisms. Insofar as the word "germs" is used for adult microorganisms today, it is merely a perpetuation of Pasteur's vague designation.

On the other hand, none of these doubts and criticisms really undermined Pasteur's central positions on spontaneous generation and on "the infinite role of infinitely small" organisms. If some of his earlier views now required modification, and if Cohn, Tyndall, and others ultimately contributed as much as he to the still dominant sentiment against the doctrine of spontaneous generation, he had nonetheless laid the groundwork. Nor did Pasteur fail to derive practical benefit from the new attention to bacterial spores and to liquids and solids as the main vehicles of germ contamination. In Pasteur's laboratory, and almost certainly under his watchful eye, Chamberland pursued some of the issues arising from the dispute with Bastian. In his doctoral dissertation (1879) Chamberland established the basic rules of modern bacteriological technique by showing that temperatures of at least 115° C. were required to ensure the destruction of heat-resistant microorganisms in liquids, while temperatures of at least 180° C. were required to achieve the same result on dry surfaces. Especially in the wake of Chamberland's work, the autoclave and the flaming of glassware became standard in microbiological equipment and technique.[107]

Pasteur and Medicine: The Background. Almost from the beginning of his work on fermentation and spontaneous generation, Pasteur made frequent reference to its potential medical implications. Sharing the common belief that fermentation and disease were analogous processes, he naturally supposed that the germ theory could apply to disease as well as to fermentation—as Theodor Schwann, among others, had supposed before him. In fact, in the late 1850's, when Pasteur began his study of fermentation and spontaneous generation, the status of the germ theory of disease paralleled almost precisely the status of the germ theory of fermentation. In both cases the germ theory held less favor than alternative theories, but serious claims had been made for it on the basis of solid and highly suggestive evidence. Advocates of the germ theory of fermentation appealed chiefly to evidence that yeast was a living organism; the germ theory of disease drew its most impressive support from accumulating evidence of the important role played by living parasites in a number of plant and animal diseases, including such human maladies as trichinosis, scabies, and the fungal skin diseases, notably scalp favus.

At the same time, however, critics of the germ theory could cite apparently contradictory evidence, could insist that any microorganisms associated with disease or fermentation were merely epiphenomenal products of these processes rather than their cause, and could argue that the alleged examples of microbial processes were atypical or unimportant. With regard to disease, even those who accepted the pathogenic role of microscopic parasites in certain diseases often doubted or denied their role in the major killer diseases of man or other vertebrates. The notion that tiny living agents could kill vastly larger organisms struck many as absurd. Moreover, the complexity of disease, and the peculiarity of its expression in each patient, impressed most physicians with the seeming irregularity, spontaneity, and mystery of the process. From this perspective the germ theory of disease seemed too simplistic, inflexible, and remote, particularly because it emphasized the role of agents possessing a life and origin independent of the organisms in which disease became manifest.

In contrast with the emphasis of the germ theory on the "exteriority" of disease, the dominant concepts of the process stressed the internal state and quality of the affected organism. When external agents found a place in these schemes—and the existence of epidemics virtually required their inclusion—they were generally denied a life of their own and accorded a distinctly secondary role. In traditional medical doctrine these external agents—whether meteorological conditions, "cosmic-telluric" forces, subtle fluids, noxious effluvia, chemical poisons, or inanimate particles—acted chiefly as contributors to, or as transmitters of, pathological states the proximate genesis of which was internal and spontaneous. In Pasteur's view the future of medicine depended on a literally life-and-death struggle against this traditional doctrine of the interiority and spontaneity of disease, a doctrine which found capsule expression in the slogan "Disease is in us, of us, by us."[108]

Through his efforts on behalf of the germ theory of fermentation and against spontaneous generation, Pasteur became a highly influential, if largely indirect, participant in this struggle during the two decades after 1857. His studies on the silkworm diseases may seem to represent his most direct and important contribution to the germ theory of disease, but persuasive evidence of microbial participation in certain insect diseases had long existed without transforming medical theory. Vastly more influential in this regard were two medical contributions immediately inspired and encouraged by Pasteur's work on fermentation. The more familiar and dramatic of these was antiseptic surgery, introduced in the 1860's by Lister, who openly saluted Pasteur for having provided in the germ theory of fermentation "the sole principle" upon which the antiseptic system had been built.[109] Although only gradually and rather reluctantly accepted, especially in England, Lister's method eventually created a revolution in surgery and enormously advanced the cause of the germ theory of disease.

Almost simultaneously the French pathologist Casimir Joseph Davaine sought to establish a microbial etiology for anthrax or splenic fever, taking as his point of departure a paper by Pasteur on the fermentation of butyric acid. Struck by the similarity between Pasteur's butyric ferment and some rods he had observed more than a decade before in anthrax blood, Davaine in 1863 launched his attempt to demonstrate experimentally that anthrax was caused by these rodlike organisms or "bacteridia." Ultimately the path from Pasteur's work on fermentation to Davaine's on anthrax carried traffic both ways, for anthrax became the subject of Pasteur's first excursion into medical research per se. Twice in 1865 Pasteur took part in discussions on anthrax at the Académie des Sciences. On both occasions he gave qualified support to Davaine's basic position, but the tone of his remarks betrayed his belief that anthrax remained obscure in many respects.[110] As early as 1867 he specifically identified anthrax as the disease he hoped soon to study.[111] Not until 1877, however, did he publish the first of his papers on anthrax.

For a man of his bold readiness to tackle the major problems of the day, and for a man whose research had so long approached the medical domain, Pasteur seems to have hesitated a surprisingly long time before entering the struggle against traditional medical doctrine. His hesitation is all the more surprising because it persisted despite his expressed desire to undertake specifically medical research and his possessing ample opportunity, adequate resources, and—in a sense—an imperial mandate to do so. This mandate—along with the resources and facilities to carry it out—followed a remarkable appeal that Pasteur addressed simultaneously (on 5 September 1867) to Louis Napoleon and to the minister of public instruction.

Having just removed Pasteur from his administrative posts at the École Normale, the Ministry of Public

Instruction had offered him a professorship in chemistry at the Sorbonne and a position as *maître de conférences* in organic chemistry at the École Normale, with the right to retain his old apartment and laboratory there. Pasteur submitted a counter proposal. He agreed fully with his appointment at the Sorbonne but objected to the proposed position at the École Normale on several grounds, including his concern that two teaching posts might impede his research. Instead, he proposed the construction at the École Normale of a new, spacious, and well-endowed laboratory of physiological chemistry in which he would not teach but would continue his research. He supported his proposal by referring to "the necessity of maintaining the scientific superiority of France against the efforts of rival nations" and by projecting studies of immense practical importance on infectious diseases in general and on anthrax in particular.[112]

The emperor immediately expressed his support for Pasteur's project in a letter to the minister of public instruction. Construction began in August 1868, the cost of 60,000 francs being shared equally by the Ministry of Public Instruction and the Ministry of the House of the Emperor. The new laboratory, thirty meters long, was to be linked by a gallery with the pavilion Pasteur had occupied since 1859. Largely because of the Franco-Prussian War, however, the laboratory remained incomplete as late as 1871. In September of that year, following the departure from Paris of the Prussian troops and the Communards, Pasteur returned from the provinces and immediately asked to be relieved of his remaining teaching duties at the Sorbonne because of his health. Claiming thirty years in university service (including his days as a tutor at Besançon), he requested a retirement pension as well as a separate national recompense in recognition of his contributions.[113]

By 1874, when Pasteur achieved the last of these goals, he had still taken no direct steps toward the study of anthrax projected in 1867. For the first four of the intervening years, his attention had been diverted by his desire to complete the silkworm studies under way since 1865 and by the Franco-Prussian War. By late 1871, however, he had solved the silkworm problem to his satisfaction and had at his disposal the new and presumably disease-oriented laboratory, as well as an annual research allowance of 6,000 francs. And yet, instead of turning to the direct study of disease, he continued through 1876 to devote his energies and the resources of his laboratory to his studies on beer and to the persistent controversies over spontaneous generation and his germ theory of fermentation.

To a degree Pasteur considered the solution of these problems—and especially the destruction of the doctrine of spontaneous generation—a prerequisite to the direct and effective study of disease.[114] But in the paper of 1877 that marked his full-fledged entry into the medical arena, Pasteur offered another explanation for his prior absence. Although long "tormented" with the desire of tackling the great medical problems of the day, he wrote, he had hesitated until now for two reasons: (1) he had needed a "courageous and devoted collaborator," a requirement at last fulfilled in Jules Joubert, and (2) being "a stranger to medical and veterinary knowledge," he had needed to overcome his fear of his own "insufficiency."[115]

That Pasteur required assistance to undertake the experimental study of disease can scarcely be denied—not only because of his partial paralysis but also because of his attitude toward vivisection. As one who found vivisection personally repugnant,[116] and yet considered it essential to his task, he needed collaborators who were willing and able to undertake the animal experiments he designed. But in view of the seeming ease with which he attracted such assistants—not only Joubert but also Duclaux, Chamberland, Émile Roux, Louis Thuillier, and Adrien Loir—one wonders whether he could not have found them long before 1877 had he really tried. On the other hand, Pasteur's fear of "insufficiency"—while scarcely in keeping with his usual self-assurance and his bold excursions into other fields in which he could claim no professional competence—does find some echo in his general ambivalence toward holders of the M.D. degree. Although he tended to disdain doctors for their traditionalism, their pretensions to scientific knowledge, and their preference for ritual and oratory over experiment, he envied their social status, their clinical experience, and their immediate, dramatic utility.

Much of this ambivalence emerged during meetings of the Académie de Médecine, where Pasteur became a frequent and controversial participant after his election to membership in 1873. Besides repeatedly defending his views on fermentation, putrefaction, and spontaneous generation, he occasionally ventured into discussions of more strictly medical topics even before 1877—most notably on urinary disorders and the use of cotton wool dressings in surgery. As might be expected, he linked ammoniacal urine with the "true ferment of urine" which he had discovered in the early 1860's and which had since been studied in great detail by van Tieghem. For the treatment of such disorders he proposed the injection into the bladder of antiseptics, particularly dilute boric acid, the destructive action of which on the ammoniacal ferment he examined and the therapeutic efficacy of which he later affirmed on the basis of clinical reports from those willing to adopt his suggestion.[117]

With regard to cotton wool dressings, Pasteur argued that their efficacy depended not on the exclusion of air, as many surgeons supposed, but on the capacity of cotton wool to trap germs without impeding the circulation over the wound of presumably beneficial pure oxygen.[118] If physicians and surgeons were annoyed by these unsolicited incursions into their domain by a "mere chemist," they were incensed by his implicit charge that they often produced disease by carrying pathogenic microorganisms into their patients on contaminated hands or instruments. As early as 1874, in a passage reflecting his deep commitment to the germ theory of disease before he had entered strictly medical research, Pasteur wrote: "If I had the honor of being a surgeon, I would never introduce any instrument into the human body without having passed it through boiling water, or better yet through a flame, immediately before the operation."[119]

The Etiology of Anthrax: The Background. By the time Pasteur finally did undertake his study of anthrax, its etiology had been largely resolved. Perhaps because it was a well-defined, economically important, and often fatal epidemic disease of large animals—particularly of cattle and sheep, although it could occur in humans in the form of "the malignant pustule"—anthrax had long been the subject of intense study and controversy. Davaine's work of the 1860's therefore aroused great interest, and anthrax quickly became a major focus for the debate between advocates and opponents of the germ theory of disease. Advocates of the germ theory emphasized Davaine's claim that bacteridia always appeared in the blood of animals afflicted with anthrax but never in that of animals free of its symptoms, while opponents of the theory denied this invariable association and insisted that Davaine had failed in any case to prove the causative role of the bacteridia. If he had shown that anthrax blood could transmit the disease from one animal to another, he had not fully demonstrated that the bacteridia were the agents of this transmission. Moreover, Davaine's conception of anthrax scarcely helped to explain its behavior under natural conditions—its appearance or frequency in any given season or why it should selectively attack certain herds or fields while sparing others. His attempt to implicate flies as vectors of the infection failed to account persuasively for these and other features of the disease.[120]

Into this breach stepped Robert Koch, whose classic study of 1876 unraveled the complete life cycle of Davaine's bacteridia (Koch's *Bacillus anthracis*) and established the existence of an endospore phase. These anthrax spores, which preserved the virulence of the rods, could form in the blood and tissues of an animal after death and, once formed, resisted subsequent putrefaction or drying. Koch immediately recognized that these resistant spores held the key to understanding the natural behavior of anthrax, for they could retain their pathogenicity from one season to the next and could produce a recurrence of the disease in specific localities under appropriate conditions of temperature and moisture. Suggesting that natural infection probably took place through the food, he proposed preventive measures against the disease. In addition he developed new techniques for cultivating the anthrax bacillus and showed that successive cultures remained virulent despite repeated dilution.

Pasteur on the Etiology of Anthrax and Septicemia. Despite these achievements, which attracted widespread attention and acclaim, Pasteur believed that some doubts remained. As evidence he cited Paul Bert's claim of January 1877 that anthrax blood could produce death even after its bacteridia had been killed by compressed oxygen. Since death occurred "without any trace" of bacteridia, Bert concluded that the latter were "neither the cause nor the necessary effect of anthrax." Instead, he ascribed the disease to a "virus," by which he meant a soluble chemical poison or some other inanimate agent. In his first memoir on anthrax (April 1877), Pasteur challenged Bert's hypothesis by extending Koch's successive dilution experiments. By greatly increasing the number of cultures (Koch had stopped at eight) and by using a much larger volume of cultural liquid each time, Pasteur diluted an initial drop of anthrax blood to the point of virtual disappearance. Nonetheless, each successive culture retained the original virulence. In his view this result persuasively established the dependence of anthrax on a living microorganism, for no other agent could have retained its power through so drastic a dilution. Only an agent which reproduced itself in each successive culture—almost certainly a living organism—could be responsible for the continued virulence of the original drop of blood. If Bert's hypothetical chemical poison did exist, it must be capable of self-reproduction or must be continuously secreted by the multiplying bacteridia. But these possibilities, remote in any case, became even more so in view of the fact that the filtered liquid from each culture (which ought to contain any soluble poison) produced no effect when injected.

In his next paper on anthrax (July 1877), Pasteur offered his own interpretation of Bert's experiment and applied it as well to the most damaging earlier evidence against Davaine's work. In essence, he argued that the architects of this earlier evidence (notably Leplat and Jaillard), and probably Bert as well, had confused anthrax with a form of septicemia. As early as 1865, when Davaine made a similar charge, Pasteur had lent credence to it by reporting that Leplat and Jaillard's supposed anthrax blood contained putrefactive microorganisms foreign to anthrax.[121] What he sought now to do was to develop this argument and, more generally, to clarify the relationship between anthrax and septicemia. In this relationship, he insisted, the crucial factor is the time that elapses between death and the extraction of blood. At first, for perhaps eighteen hours, the blood of an animal dead of anthrax contains only the anthrax bacteridia. Eventually, however, this blood undergoes putrefaction and the bacteridia progressively disappear. Despite this disappearance, or despite the destruction of the bacteridia by compressed oxygen, the blood can remain virulent and can produce death in another animal.

One reason for the continued virulence, Pasteur suggested, was that such blood might continue to harbor anthrax bacteridia in the endospore phase, for the anthrax spores not only resist putrefaction (as Koch had already shown) but also survive the action of compressed oxygen. In such cases, however, the spores ought to germinate when injected into another animal, reproducing ordinary anthrax with its familiar rods. But Leplat and Jaillard, and apparently Bert as well, had insisted that their injections of anthrax blood had produced death in the absence of any microorganisms whatever. In these cases, Pasteur argued, the blood must have ceased to carry anthrax and must have become putrid or septic instead. More important, he claimed that the most familiar effects of putrid or septic injections also depended upon a microorganism —the hitherto unknown *vibrion septique*—and not an inanimate septic "virus," as was commonly believed. To explain how the *vibrion septique* had previously escaped detection, Pasteur focused on the preoccupation of earlier observers with the blood, a concern that had distracted them from a systematic search for pathogenic microorganisms in other parts of the body. In an animal dying of septicemia, microorganisms could be found in abundance in the muscles and in the abdominal serosities near the intestinal canal, but not in the blood until just before death. And when these organisms finally did enter the bloodstream, they became peculiarly long and translucent and easily escaped detection.

But even if such organisms did exist, and even if they had invaded the blood used by Bert, how could they have retained their virulence after being subjected to compressed oxygen? Pasteur's answer hinged on the assertion that the new septic vibrio, like the anthrax bacteridium, had a resistant spore phase. In fact, he

insisted, this spore phase appears within hours of the application of compressed oxygen to septic blood. Preserved in this immobile, resistant phase from further attack by oxygen, the septic vibrio can return to its motile, filamentary phase upon injection into another animal, producing death with the usual symptoms of septicemia. With only slightly less confidence, Pasteur suggested that the septic vibrio was one of the putrefactive vibrios found in the intestinal canal. If so, septicemia might properly be called "putrefaction on the living." And since various putrefactive vibrios exist, one could expect a corresponding range of septic infections from the inoculation of putrid materials.

On the way to this new interpretation of Bert's experiment—which Bert soon adopted—Pasteur offered some novel views on the physiological properties and modus operandi of the anthrax bacteridia. Having observed that filtered anthrax serum produced agglutination of the blood, he suggested that this familiar symptom of the disease might be due to a soluble ferment produced by the bacteridia. But this suggestion produced no shift in his basic conviction that the bacteridia themselves, and not any soluble ferment, were responsible for death from anthrax. In search of a mechanism by which the bacteridia might kill, he began by insisting on their aerobic character. Once in the blood, he supposed, these aerobic bacteridia would compete with the red blood cells, "those aerobic beings *par excellence*," for oxygen. If the bacteridia won this struggle for existence, the animal would die of asphyxia, as suggested by the black color of the blood and viscera. In support of this notion, Pasteur reported that other aerobic microorganisms could impede the development of the bacteridia in cultural liquids or in animal bodies. Most remarkably, even animals highly susceptible to anthrax could survive an injection of bacteridia so long as the latter were accompanied by competing aerobic microorganisms. By his suggestion that these facts "authorize the greatest hopes from the therapeutic point of view," Pasteur has won credit as a prophet of bacteriotherapy, in the development of which he played no direct or substantial role.[122]

In March 1878 Pasteur described a remarkable new experiment on which he placed great importance for both the etiology and the treatment of anthrax. He showed that it was possible to transmit anthrax to hens, which are ordinarily refractory, merely by lowering their body temperature a few degrees. Aware that anthrax bacteridia could not develop in otherwise appropriate media at a temperature above 44° C., Pasteur had wondered whether the natural immunity of hens to anthrax might be due to the naturally elevated temperature of their blood. By plunging the legs of a chicken in an ice bath, he lowered its blood temperature several degrees; previously injected bacteridia were then able to develop and to induce death from anthrax. Conversely, he was able to prevent anthrax in rabbits, which are ordinarily susceptible, by raising their blood temperature several degrees. On this basis he hoped that it might prove possible to cure humans of "malignant pustule" by placing them in a bath warm enough to maintain a blood temperature of 41–42° C. By July 1878 he had cured a chilled hen of advanced anthrax by warming it. When some members of the Académie de Médecine raised objections against these experiments, Pasteur effectively demolished them in dramatic confrontations before the full Academy and by demanding a judiciary commission, which verified the exactitude of his results. In the light of subsequent research, Pasteur's interpretation of these results, as well as the therapeutic hopes he based on them, seem somewhat naïve, for such drastic changes in body temperature produce effects far more general and profound than those bearing directly on the anthrax bacteridia. Nonetheless, his results offered striking experimental evidence that receptivity to disease depends on factors beyond the mere presence of pathogenic agents.

Pasteur on the Etiology of Natural Anthrax. Beginning with reports to the minister of agriculture in September and October 1879, and more fully in a memoir of July 1880, Pasteur extended and refined Koch's views on the etiology of natural anthrax. Through feeding experiments on large domestic animals (which Koch had not used), he confirmed Koch's suggestion that the natural mode of transmission was the food. More specifically, he showed that sheep could contract anthrax by ingesting bacteridia spores, especially when the spores were mixed with a prickly diet of thistle leaves or short barbs of oats and barley. The resulting lesions strongly suggested that the disease began in the mouth and back of the throat. To explain how sheep and cattle came upon anthrax spores under natural conditions, Pasteur recalled that these spores withstood putrefaction and could therefore persist for months or even years in soil where diseased animals had been buried. Indeed, these spores could be found on the soil above such graves—where grazing animals might ingest them—while no spores could be found on the soil just a few meters away. In this way the existence of "infected fields" could be readily understood; they were fields in which animals dead of anthrax had been buried.

Pasteur's most original contribution to the problem concerned the mechanism by which the immotile anthrax spores were brought from animal graves to the surface of the earth. The agent of this transfer, he insisted, was the common earthworm. After several

days in soil containing anthrax spores, earthworms carried the spores in their intestinal canals. When they rose to the surface, they ejected these spores along with their earth castings. Once on the surface, the anthrax spores could attach to the plants on which sheep and cattle grazed or—as Pasteur recognized in January 1881[123]—could be inhaled. These conclusions authorized a fairly obvious and simple prophylactic measure: animals dead of anthrax must never be buried in fields intended for grazing or the growing of fodder, at least not unless the soil in such fields was inimical to earthworms. If this measure were followed, Pasteur rather extravagantly predicted, anthrax could be a thing of the past, for the disease is never spontaneous and can be found only where its germs have been disseminated "by the innocent complicity of earthworms."[124]

The Extension of the Germ Theory to Other Diseases. Although Pasteur's work on the etiology of anthrax and septicemia was largely a confirmation and extension of Koch's work, it helped to raise anthrax to its special status as the first major killer disease of large animals widely admitted to be parasitic. Besides lending credence and interest to a series of earlier but inconclusive attempts to implicate microorganisms in major vertebrate and human diseases, this achievement ushered in what came to be known as the golden age of bacteriology. In less than two decades the microbial theory of disease was extended to tuberculosis, cholera, diphtheria, typhoid, gonorrhea, pneumonia, tetanus, and plague. Surprisingly, Pasteur and the French school contributed only minimally. The vast majority of these pathogenic microorganisms were isolated and studied by Koch and the German school, thanks in part to Koch's mastery of microscopic morphology, classification, and technique and more particularly to his method of pure solid cultures, which Pasteur praised without adopting. For the most part Pasteur and the French school focused instead on the problems of immunity from and prophylaxis against microbial diseases—in a word, on vaccination.[125] But only after 1880 did these differences become dramatically clear, to be quickly reinforced by national and personal rivalries. Between 1878 and 1880 Pasteur and Koch seemed to be aiming toward similar goals: the elucidation of septicemia in its various forms and the extension of the germ theory to diseases other than anthrax.

Pasteur's contributions toward these goals include a lecture of April 1878, "La théorie des germes et ses applications à la médecine et la chirurgie," and a memoir of May 1880 on the extension of the germ theory to the etiology of certain common diseases. In the 1878 lecture, delivered before the Académie de Médecine, Pasteur described the results of studies undertaken on the *vibrion septique* since discovering it the year before. To a large degree these results merely gave more explicit, elaborate, and confident form to his original conception of the septic vibrio. After several unsuccessful attempts to cultivate this organism by ordinary means, Pasteur and his collaborators had decided that it might be an obligate anaerobe, incapable of living in the presence of the oxygen dissolved in ordinary cultural liquids. They therefore switched to cultures in a vacuum or an atmosphere of carbon dioxide, with immediate success. But this obligate anaerobism applied only to the motile, filamentary phase of the vibrio. In its spore phase, the *vibrion septique* could obviously live in oxygen; it even survived the compressed oxygen used by Bert. The persistent virulence of the septic blood in Bert's experiment—as well as the existence of natural septicemia in any form—depended absolutely on this spore phase. Only in the form of resistant spores could the otherwise anaerobic vibrio exist in ordinary air, ready to germinate and to produce septicemia if the spores penetrated a portion of an animal where oxygen was absent or nearly so. Until they reached such a site, the spores could not germinate and thus remained harmless.

In other words, Pasteur argued, the *vibrion septique* may be harmless or pathogenic according to environmental circumstances, just as its form, reproductive capacity, and virulence vary in different artificial media. Similarly, one of the most common bacteria resembles the anthrax bacillus in its physiological properties—including obligate aerobism—and yet is harmless because it cannot live at the temperature of the animal body. Yet another vibrio—the hitherto unrecognized "microbe of pus"—resembles yeast in its capacity to live either aerobically or anaerobically. And, like any solid body, the microbe of pus produces an abscess, or pocket of pus, upon injection into a guinea pig or rabbit. But the inordinate size of the resulting abscess clearly depends on the vital activity of the new microbe; if killed by heat before injection, it produces a much smaller abscess. Although far less dangerous than the anthrax bacillus or the *vibrion septique*, the microbe of pus can sometimes produce metastatic abscesses, purulent infection, and death. It can also modify the action and virulence of those more dangerous microorganisms when associated with them. More generally, the nature and relative proportions of specific microbes determine a richly varied set of pathological states.

If this is the central thrust of Pasteur's lecture of April 1878, the message must be extracted from a diffuse and atypically obscure presentation of his views. In the same lecture, and more or less haphazardly, he also described methods for separating aerobic from anaerobic microorganisms; mentioned the difficulties his results posed for microbial classification; offered hygienic advice to surgeons; insisted that the acquired knowledge of anthrax and septicemia upset the doctrine of spontaneity; and argued that the *vibrion septique* was the true cause (rather than a product) of septicemia, probably without the intervention of any soluble ferment. In his advice to surgeons, probably the most famous section of the lecture, Pasteur repeated and embellished the counsel he had given in 1874, before his entry into the medical arena. After the same opening phrase ("if I had the honor to be a surgeon"), he stated: "Impressed as I am with the dangers to which the patient is exposed by the germs of microbes scattered over the surface of all objects, particularly in hospitals, not only would I use none but perfectly clean instruments, but after having cleansed my hands with the greatest care and subjected them to a rapid flaming . . . I would use only lint, bandages and sponges previously exposed to air of a temperature of 130 to 150° C.; I would never use any water which had not been subjected to a temperature of 110 to 120° C." Finally, Pasteur quoted with pride from a lecture given at the Académie des Sciences several weeks earlier by the distinguished surgeon Sédillot, who introduced the word "microbe" for microorganism and enthusiastically supported the germ theory and the new "Listerian" surgery arising from it.

In his paper of May 1880, "De l'extension de la théorie des germes à l'étiologie de quelques maladies communes," Pasteur implicated microbes in furuncles (boils), osteomyelitis, and puerperal fever. He reached his views on boils and on osteomyelitis after studying a single case of each. The case of boils belonged to one of Pasteur's own assistants (Duclaux). When pricked and submitted to culture, these boils gave a unique aerobic microbe of the form later called staphylococcus, to which Pasteur ascribed the local inflammation and consequent pus. When he found this same microbe in pus taken from the infected bone of a girl, he boldly asserted that "osteomyelitis is the boil of the bone marrow." In his discussion of puerperal fever, Pasteur described seven cases of the disease, in each of which he found highly presumptive evidence of microbial participation, and then developed the views he had already expressed during a debate on puerperal fever at the Académie de Médecine in March 1879.[126] During that debate Pasteur asserted that a microbe shaped like strings of beads caused most childbed infections, and he joined Semmelweiss in charging doctors themselves with the transmission of these infections. With his usual tone of disdain toward those who tried to classify microbes, he noted that some German authors had given the Latin name "micrococcus" to organisms having the form of the new puerperal microbe—including the ferment of ammoniacal urine and the microbe of *flacherie*—and had even tried to implicate micrococci in puerperal fever. Despite these attempts the etiology of puerperal fever remained obscure, chiefly because various microbes associated with pus could intervene to modify the symptoms and course of the disease. Finally, for the treatment of puerperal infections, Pasteur advocated the use of sterile water and bandages and, more specifically, the application to the infected genital tissues of a 4 percent solution of boric acid, which combined the advantages of known destructiveness toward at least one micrococcus (the ferment of ammoniacal urine) and inoffensiveness to mucous membranes.

Immunity, Virus Diseases, and the Discovery of Vaccines: The Background. From the outset of his work on anthrax, and even as his study of septic infections converged with the German effort to identify new pathogenic microbes, Pasteur pondered what were then known in France as the "virus diseases." Typified by smallpox and presumed to be nonmicrobial, their most striking feature was that they did not recur (or recurred in milder form) in the same individual. The strength of the "virus" (or poison) considered responsible for each of these diseases was usually assumed to be fixed and uniform for any given species but variable from one species to another. In particular the cowpox virus, which maintained a constant virulence through hundreds of transfers from man to man, clearly declined in strength when passed from cow to man. Thus "humanized," the cowpox virus became the "vaccine" introduced by Edward Jenner at the end of the eighteenth century to protect man from attacks of smallpox. Most authorities adopted Jenner's belief that his vaccine was simply a milder form of the smallpox virus, modified by passage through the cow. But others claimed that smallpox and cowpox were independent diseases, due to distinct viruses of inherently different strengths. Variations in severity between different smallpox epidemics and in the course of the same epidemic, as well as variations in the duration of the immunity produced, further obscured the nature of the virus diseases. Moreover, attempts to find "vaccines" or modified viruses against other diseases had produced nothing, and Jenner's vaccine remained unique.

As early as June 1877, Pasteur announced that he had begun to study the virus diseases, and particularly the cowpox virus that served as Jenner's vaccine.[127] He clearly hoped to isolate a cowpox microbe (a vain hope), and he may already have perceived some con-

nection between the virus diseases and the microbial diseases of anthrax and septicemia. In fact, Davaine and others had already shown that anthrax and septicemia shared one property of the virus diseases: their virulence could be modified by passage through living animals. But the meaning of this isolated fact was obscure, and the possibility that such variations might be related to variations in the microbes of anthrax and septicemia ran afoul of the doctrine of microbial specificity. Although Pasteur had done much to establish this doctrine, and continued to deny the transmutability of microbial species, his position was somewhat more flexible than that of Cohn or Koch. During his study of septicemia, he noticed that different cultures of the *vibrion septique* varied in virulence when injected into animals. At first, in keeping with the doctrine of microbial specificity, he supposed that these variable virulences depended on different species or varieties of septic vibrio. In April 1878, however, in his lecture on the germ theory to the Académie de Médecine, he suggested that these variations should be ascribed to the effects of different cultural media on the properties of a single *vibrion septique*. To suspect a connection between microbial diseases and the virus diseases, and to recognize that the virulence of the *vibrion septique* could be artificially modified, were to take some preliminary steps toward the concept of attenuated viruses and the technique of vaccination. But these early, almost instinctive steps gained real force and direction only through Pasteur's study of fowl cholera.

Fowl Cholera and the Discovery of Vaccines. In December 1878 Toussaint, a professor at the Alfort Veterinary School, sent Pasteur some blood from a cock dead of fowl cholera.[128] The symptoms of this disease, which has no relation to human cholera, include weakness, loss of coordination, droopy wings, erect feathers, and somnolence usually ending in death. Its progress through an infected poultry yard can be extremely swift, with most of the hens dead or dying in a few days. Like a few others before him, Toussaint linked the disease with a microbe, which he found in the blood of all hens having the disease. Beginning with the blood sent him by Toussaint, Pasteur immediately sought to isolate the microbe in a state of perfect purity[129] and to demonstrate by the method of successive cultures that it was the true and sole cause of fowl cholera. He soon found that this nonmotile microbe—in the form of a figure eight but so tiny as to resemble isolated dots—developed much more readily in neutral chicken broth than in the neutral urine used by Toussaint. By March 1879 he had found that a culture almost uniformly fatal for chickens was relatively benign for guinea pigs, and he drew an analogy between the guinea pig and yeast extract, both being cultural media ill-suited to the development of the fowl cholera microbe.[130]

In February 1880 Pasteur announced that although the fowl cholera microbe retained its virulence through successive cultures in chicken broth, he had found a way of decreasing its virulence "by certain changes in the mode of culture." In this milder form the microbe usually produced disease, but not death, in chickens. More important, the chickens that recovered from this less virulent form of the microbe became relatively immune to the highly virulent form. Unlike ordinary chickens they did not die from an injection of the microbe in its usual form. In other words, Pasteur concluded, "The disease is its own preventive. It has the character of the virus diseases, which do not recur." What gave this result special importance and novelty was the demonstrably microbial nature of fowl cholera. Preventive inoculations were not new, but they had never been used against a disease known to be caused by a microorganism that might be cultivated outside of living organisms. Never had it been known that the property of nonrecurrence, associated with the so-called virus diseases, could belong to a microbial disease. Fowl cholera thus formed the first clear link between microbial diseases and diseases "in the virus of which life has never been recognized." Although many difficulties remained before the attenuated microbe of fowl cholera could be properly compared with Jenner's vaccine—in particular, its constancy through a series of inoculations had yet to be assured —it offered hope that every "virus" might be artificially cultivated and that "vaccines" might be obtained against the infectious diseases "which afflict humanity, and which are the greatest scourge of agriculture in the rearing of domestic animals."

In announcing these dramatic results, Pasteur declined to reveal the method by which he had obtained the attenuated form of the fowl cholera microbe, saying that he wished to assure independence in his studies. Despite complaints from members of the Académie de Médecine in particular, he persisted in this course for nine months, during which period he reported the results of his subsequent studies. In April 1880 he admitted that inoculation with the attenuated form of the fowl cholera microbe produced very different results in different hens, but he insisted that the procedure always conferred some benefit. Even when two or more inoculations were required for complete protection against the disease, each acted in some measure to impede its course. He emphasized that "vaccinated" chickens, as well as species naturally resistant to the disease, must represent cultural media somehow ill-suited for the development of the microbe and suggested that this immunity probably resulted from the absence of some substance essential to the life of the microbe.

This suggestion drew support from the fact that cultivations of whatever sort (whether ordinary plants, parasites, or microbes) modify a given medium (or "soil") in such a way as to make subsequent cultivations of the same species difficult or impossible. Thus, after four days as a medium for the fowl cholera microbe, chicken broth will not support a new inoculation of the microbe. After the second day it will do so, but less readily than at first, which suggests that some essential substance is progressively withdrawn from the medium by the microbe. The same effects might be explained by supposing that the developing microbe produces some substance which is toxic to itself, but Pasteur rejected this hypothesis on the ground that cultural extracts developed easily in new chicken broth, although such extracts should contain any self-toxic substance secreted by the fowl cholera microbe.

In May 1880 Pasteur suggested that the fowl cholera microbe produces a soluble narcotic responsible for the characteristic somnolence of the disease. This suggestion echoes his earlier proposal that the anthrax bacillus produces a soluble substance responsible for the agglutination of the blood in that disease. Now, as then, he made the soluble substance responsible for only one symptom of the disease and ascribed death chiefly to asphyxia, citing the violet-tinged combs of diseased chickens and the aerobic character of the fowl cholera microbe, which implied a struggle for oxygen with the red blood cells. As evidence that somnolence and death had independent causes, he reported that vaccination prevented death but not extract-induced somnolence.

From late May to early October 1880, Pasteur participated in heated debates at the Académie de Médecine over the significance of his work on fowl cholera vis-à-vis smallpox and Jenner's vaccine. Having shown that the fowl cholera "vaccine" was only modified fowl cholera "virus" (or microbe), he felt confident that Jenner's vaccine was only modified smallpox virus. His opponents denied the relevance to this question of experiments on fowl cholera, claimed that physicians had long held the view that Pasteur was now needlessly repeating, and criticized him for keeping secret his method of attenuating the fowl cholera microbe. In return Pasteur insisted on the importance of experimental evidence, accused his opponents of failing to grasp the real issue in dispute (the relation between smallpox and vaccine, not between cowpox and vaccine), defended his "reserve" on the method of attenuation, and ridiculed one opponent's surgical procedures so viciously that the latter had to be restrained from physically assaulting Pasteur, whom he soon challenged to a duel.[131]

Finally, in October 1880, Pasteur described his method of attenuating the fowl cholera microbe. The first step was to procure the microbe in its most virulent form by taking it from a chicken dead of the chronic form of the disease. In successive cultures made at brief intervals, this virulence remained constant; but attenuation set in when the intervals reached two or three months. In general, the longer the intervals, the weaker the virulence became, although the results defied mathematical regularity. Throughout these changes in virulence, the microbe remained essentially constant in form. Furthermore, a virus (or microbe) of any given virulence retained this degree of virulence so long as successive cultures were made at brief intervals. To explain attenuation, Pasteur invoked the effect on the microbe of prolonged exposure to atmospheric oxygen. As proof he reported that no attenuation occurred in closed tubes, however long the intervals between cultures might be. He suggested that oxygen might have a similar effect on other viruses or microbes and might even be responsible for the natural limits characteristic of great epidemics. Neither here nor anywhere else did Pasteur specify why oxygen should weaken microbes, especially those aerobic microbes (including the anthrax bacillus and the fowl cholera microbe) which ordinarily depended on it for life.

At one point in this memoir, Pasteur alluded again to his prior silence on the method of attenuation. The "true reason" for that silence, he said, ought now to be clear: "Time was an element in my researches." What he did not reveal even now was the remarkable manner in which the crucial role of time had become known. In this case, as in his discovery of hemihedrism in the paratartrates, Pasteur seemed to enjoy extremely good luck.[132] During his early experiments on fowl cholera, he followed his usual practice of making fresh cultures of the microbe every day or so. From late July to October 1879, however, the cultures were allowed to lie idle while he vacationed at Arbois.[133] During this period nearly all the cultures had become sterile and resisted attempts to restore their fecundity by inoculation into chickens. The seemingly useless cultures were about to be discarded when Pasteur proposed that the chickens in which they had produced no apparent effect be subjected to a fresh inoculation from a fecund, virulent culture. The chickens survived. "With [this] one blow," wrote Duclaux, "fowl cholera passed to the list of virus diseases and vaccination was discovered!"[134] Against those who might call this discovery mere luck, Duclaux insisted that some "secret instinct," some "spirit of divination" had led

his master to it, while Pasteur himself might have repeated his famous phrase, "Chance favors only the prepared mind."[135]

In any case Pasteur immediately recognized that he had found a technique capable of extension to other diseases, and he moved toward this goal even as he kept secret his method of attenuating the fowl cholera microbe. Anthrax, the disease he knew best, served naturally as his first choice in the effort to find other vaccines. One may wonder how far this effort would have gone or how successful it would have been without a major expansion in Pasteur's facilities and resources. In May 1880, shortly after the discovery of the fowl cholera vaccine, the city of Paris gave him access to some unoccupied land near his laboratory. On this site, which belonged to the old Collège Rollin, he made extensive provisions for the care and shelter of the many animals used in his experiments. Simultaneously the annual budget for his laboratory—fixed at 6,000 francs since 1871—was supplemented by an annual credit of 50,000 francs from the Ministry of Agriculture.[136] As he surveyed his new domain and as his team of assistants grew larger, Pasteur found new scope for those qualities that led Duclaux to compare him to "a chief of industry who watches everything, lets no detail escape him, wishes to know everything, to have a hand in everything, and who, at the same time, puts himself in personal relation with all his clientele. . . ."[137]

Pasteur and the Discovery of Anthrax Vaccine. In his attempt to place anthrax among the virus or non-recurring diseases and to find a vaccine against it, Pasteur faced competitors, notably Auguste Chauveau and Toussaint. As early as September 1879, Chauveau undertook to explain the relative immunity of Algerian sheep from anthrax and to reinforce that immunity by preventive inoculations. In July 1880 Toussaint announced that he had obtained an effective vaccine against anthrax. In opposition to Pasteur, Chauveau and Toussaint shared an essentially chemical theory of immunity, ascribing it to a soluble substance released by and noxious to the developing anthrax bacilli. Toussaint's proposed vaccine reflects this view: he used filtered and defibrinated anthrax blood heated for ten minutes at 55° C. He supposed that these procedures freed a soluble vaccine from its microbial companions and claimed that sheep injected with this serum survived inoculations of virulent anthrax.

Toussaint's announcement clearly shook Pasteur, whose biological theory of immunity and vaccination it directly threatened. Immediately upon hearing of the announcement, while on vacation at Arbois, he wrote Chamberland and Roux, his collaborators throughout his studies on anthrax, and asked them to join him for experiments designed to examine Toussaint's claims.[138] They soon found that Toussaint's proposed vaccine did indeed provide protection in most cases; but they rejected his interpretation of how this process took place, criticized his experimental technique on several grounds, and disputed the general safety and practicability of his method of vaccination. While doing so, they defended Pasteur's alternative conception of immunity and developed a different anthrax vaccine, based on fundamentally the same principles and techniques employed in the discovery of the fowl cholera vaccine.

As a matter of fact, Pasteur briefly considered the possibility that the fowl cholera vaccine might also serve as an anthrax vaccine. In August 1880, on the basis of preliminary experiments, he claimed that chickens inoculated with the fowl cholera vaccine became simultaneously immune from anthrax. Unlike ordinary chickens, they did not contract anthrax when injected with its bacilli and subsequently chilled. Pasteur noted that this result, if established, would constitute the creation of immunity from anthrax by means of an entirely different parasitic disease. If applicable to other virulent diseases, it gave hope of immense therapeutic consequences, even in human diseases. Since he made no further mention of this result, one can only surmise that subsequent experiments failed to corroborate these preliminary claims.

Slightly earlier, in July 1880, Pasteur had made brief and passing reference to another possible mode of vaccination against anthrax: the gradual and moderate feeding of anthrax spores. He claimed that this idea had first occurred to him in the late summer of 1878, during his experiments on the etiology of natural anthrax. Having noticed that some sheep fell sick but did not die from the ingestion of anthrax spores, he injected eight of them with virulent anthrax blood. Of these eight sheep all but one survived the virulent injection, leading Pasteur to conclude that their recovery from diet-induced anthrax had rendered them immune to subsequent attacks of the disease. In reporting these results, nearly two years after they had been achieved, Pasteur recalled that Toussaint, who had just announced the discovery of a new anthrax vaccine, had witnessed these experiments, initially with skepticism but ultimately with conviction as to their accuracy.[139]

Pasteur drew additional attention to these experiments and extended their basic result to cows in a letter of 27 September 1880. The occasion was a report to the minister of agriculture on a proposed empirical treatment for anthrax in cows. He reported that no valid judgment could be made of the proposed treatment because cows inoculated with anthrax sometimes

succumbed despite the treatment, while others recovered in the absence of any treatment whatever. A far more interesting and significant conclusion—based on experiments conducted in August 1879 and in mid-September 1880—was that recovery from an initial attack of anthrax preserved cows subsequently injected with virulent anthrax blood. Thus in cows, as in sheep, anthrax does not recur and inoculations that do not kill act as preventives. In the same report Pasteur defended his biological theory of immunity against Chaveau's chemical theory. Contrary to Chaveau, he insisted that no toxic substance need be invoked to explain the relative immunity from anthrax of Algerian sheep and the reinforcement of the immunity by preventive inoculations. Instead, Algerian sheep ought to be compared with chickens, which are naturally and inherently resistant to anthrax without the intervention of any substance toxic to the anthrax bacillus. The proof of this contention lay in the fact that the mere act of chilling (which could hardly destroy any such substance) permitted the development of the anthrax bacillus in otherwise refractory hens. Moreover, the reinforcement of immunity by preventive inoculations could be likened to the progressive sterility of successive cultures of the bacilli in a given medium.

Pasteur undoubtedly realized that defending his theory of immunity or claiming priority for the discovery of nonrecurrence in anthrax was quite different from producing a safe and effective vaccine. After his "accidental" discovery of the fowl cholera vaccine, the path to such a vaccine must have seemed fairly direct: by increasing the interval between successive cultures of the anthrax microbe, and thus prolonging its exposure to atmospheric oxygen, he could hope to attenuate it. However, the extension of this method to the anthrax microbe was neither so obvious as he might have feared nor so rapid and straightforward as he might have hoped. Not until February 1881 did Pasteur announce the production of the new anthrax vaccine; the resistant spore phase of the anthrax bacillus (a phase which the fowl cholera microbe does not possess) had formed the chief obstacle because it undergoes no alteration upon exposure to atmospheric oxygen.

To attenuate the anthrax microbe, therefore, it was necessary to prevent the production of spores without simultaneously killing the microbe. This feat could be accomplished only by a quite delicate application of heat during cultivation. More specifically, in a medium of neutral chicken broth, the bacillus could live and grow without forming spores at a temperature between 42° and 44° C., while a temperature of 45° C. killed it. Once obtained, however, this asporogenous culture underwent rapid attenuation. After only eight days at 42–44° C., the culture proved harmless to guinea pigs, rabbits, and sheep, three species otherwise highly susceptible to anthrax. Most important, the microbe could be cultivated and conserved in this harmless state, as well as in each degree of attenuation achieved during the previous eight days; and each of these attenuated strains acted as a preventive or vaccine for the less attenuated strain that immediately preceded it. Pasteur claimed that he had already had great success in protecting sheep from anthrax with these vaccines and announced that the method would be given a large-scale trial when the sheep-penning season arrived in the Beauce district.

Also in the memoir of February 1881, Pasteur described the results of experiments in which animals of various ages and species had been injected with variously attenuated strains of the anthrax microbe. Despite the almost random character these experiments must sometimes have presented, they led to a general conclusion of great theoretical and practical importance—that attenuated viruses (or microbes) could return to their original virulence after successive cultures in appropriate animals. Thus a one-day-old guinea pig might succumb to an anthrax microbe that had been attenuated to the point of harmlessness for an adult of the species. If passed from this one-day-old guinea pig to progressively older ones, the microbe gained steadily in virulence until it reached its original capacity to kill adult guinea pigs and even sheep. Unless subjected anew to the attenuation procedure, the microbe would retain this original virulence. In similar fashion a fowl cholera microbe attenuated to harmlessness for chickens might remain virulent for canaries or other small birds and might regain its original virulence by passage through them. This progressive return to original virulence not only offered a means of preparing vaccines of all intermediate degrees but also suggested a possible explanation for new eruptions of old epidemic diseases and for the occasional appearance of entirely new epidemic diseases. By progressive passage through other species, a microorganism might regain a virulence once lost through natural attenuation or might become virulent to a species for which it had hitherto been harmless. With remarkable prescience Pasteur thus broached the question of the evolutionary relationship between parasites and their hosts. In essence, he had perceived that different animal species, including man, can serve as reservoirs of infection for each other; and he recognized that there was virtually no hope of a complete and final victory over epidemic diseases by preventive measures of any sort, including his own.

When this memoir appeared, Toussaint had not yet published the results obtained with his proposed anthrax vaccine. But Pasteur already felt confident that Toussaint's "uncertain" method would compare poorly with his own, which rested on the existence of vaccines producible at will and without resort to anthrax blood. A month later, in March 1881, he subjected Toussaint's proposed vaccine to a probing critique. In the first place, he argued, whatever success Toussaint had achieved resulted not from the death of the anthrax microbe (and consequent isolation of a presumed soluble vaccine) but from its unintentional attenuation by heat. Unfortunately, this protective modification of the anthrax bacillus was only one of three possible effects of Toussaint's unreliable method of heating anthrax blood to 55° C. In certain cases the microbe might survive without modification and thus retain its original virulence upon injection. In still other cases it might indeed be killed, as Toussaint supposed; but its injection would then fail to protect the animal from a subsequent attack of anthrax. Nor did the use of filtration improve the reliability of the method. Filtered anthrax blood might retain all its original virulence; more commonly, it would fail to act at all and would thus confer no protection against a subsequent attack. In short, no anthrax vaccine could be produced by successive filtrations or dilutions of an original quantity of anthrax blood. Moreover, even when Toussaint's method did attenuate the anthrax microbe, and even if it could be made to do so consistently and reliably, it still presented serious practical difficulties. Unlike Pasteur's fowl cholera vaccine or his new anthrax vaccine, Toussaint's heat-modified anthrax microbe could not be reproduced in culture so as to preserve its modified virulence. His method therefore required a large and continually renewed supply of anthrax blood.

In a separate paper of the same day (21 March 1881), Pasteur reported that he and his collaborators had produced an anthrax vaccine so attenuated that it failed to kill even newborn guinea pigs. This vaccine, the product of forty-three days of attenuation at 42–43° C., could therefore regain its original virulence only through some new species even more susceptible to anthrax. Nonetheless, the new vaccine displayed no appreciable morphological differences from the most virulent form of the bacillus and grew with equal facility in artificial media. Most important, this fully attenuated microbe (and all others of intermediate virulence) shared with the original, fully virulent culture the capacity to form spores that preserved the virulence of the anthrax rods. This meant that anthrax vaccines of whatever degree of virulence could be fixed in that state by passage into the spore phase and could then be stored or transported over long distances without fear of alteration.

The Experiments at Pouilly-le-Fort. Unlike fowl cholera—which was quite rare and local in its effects—anthrax posed a severe economic threat to French agriculture and animal husbandry. According to Pasteur, estimates of the annual loss from anthrax ranged from 20 to 30 million francs.[140] His announcement of an effective anthrax vaccine therefore excited great interest, and the Agricultural Society of Melun quickly proposed a public field test of the new method. Much of the initiative came from H. Rossignol, a veterinarian who had earlier satirized the growing deification of the germ theory and of Pasteur as its "pontiff" and "prophet," and who now produced a list of about 100 subscribers willing to underwrite the costs of a field trial of Pasteur's anthrax vaccine.[141] At the end of April 1881, Pasteur and the Agricultural Society of Melun agreed upon a course of experiments, to be arranged and supervised by Rossignol, who gave the program wide publicity by sending copies throughout the world. The program captured international attention as much by its uncompromising and boldly prophetic character as by its inherent importance—so much so that the *Times* of London sent its Paris correspondent to Rossignol's farm at Pouilly-le-Fort to provide a serial eyewitness account.[142]

As initially agreed upon, the program called for the injection of virulent anthrax culture into fifty sheep of any age, variety, or sex, of which half were to be unvaccinated while the other half were to be previously vaccinated by separate inoculations of two unequally attenuated anthrax cultures. Pasteur predicted that all twenty-five unvaccinated sheep would die from the virulent injection, while all twenty-five vaccinated sheep would recover completely and would be indistinguishable from ten additional sheep kept apart as an index of normalcy. At the request of the Agricultural Society of Melun, Pasteur later agreed to substitute two goats for two of the fifty sheep and to extend the trial to ten cows, of which six were to be vaccinated and four unvaccinated. Although somewhat less confident of the results on cows, he predicted that the six vaccinated cows would remain healthy when injected with the virulent culture, while the four unvaccinated cows would die or at least become very ill.

The experiments began on 5 May 1881 with the injection of an attenuated anthrax culture into twenty-four sheep, one goat, and six cows. On 17 May each of these animals was inoculated with a second attenuated culture, somewhat more virulent than the first. On 31 May Pasteur and his assistants—Chamberland, Roux, and Thuillier—injected a fully virulent anthrax culture into each of these thirty-one vaccinated animals and into twenty-nine unvaccinated animals

—twenty-four sheep, one goat, and four cows. They inoculated the vaccinated and unvaccinated animals alternately, "to render the experiments more comparative," and set 2 June as the date on which the crowd should reassemble to observe the results. In the meantime some of the vaccinated animals became feverish and Pasteur's faith wavered briefly; indeed, it has been asserted that he temporarily feared the possibility of public ridicule and, in an overwrought state, accused Roux of carelessness and thought of sending him to face the crowd alone.[143] But a telegram from Rossignol informed him on the morning of 2 June that he would find a "stunning success" when he arrived at Pouilly-le-Fort that afternoon.[144] When he and his collaborators made their triumphant arrival at two o'clock, all of the vaccinated sheep were alive and apparently healthy; all but three of the unvaccinated sheep were dead, and they were failing rapidly. Two dropped before the spectators' eyes, and the third died at the end of the day. The six vaccinated cows were also perfectly healthy, while the four unvaccinated ones were swollen and feverish. Upon seeing Pasteur, the crowd burst into applause and congratulations. It was perhaps the single most dramatic moment in a singularly dramatic scientific career.

In his published account of these experiments (13 June 1881), Pasteur reported that one of the vaccinated sheep (a ewe) had died on 3 June. But an autopsy revealed that this ewe had been pregnant and that her fetus had been dead for two weeks. Rossignol and a fellow veterinarian, who had jointly conducted the autopsy, therefore linked the ewe's death with that of her fetus, a diagnosis that aroused acrimonious but inconclusive debate. In the same report Pasteur insisted that the vaccine should be prepared and controlled in his laboratory, at least for the time being, lest a poor application of the method compromise its future. Finally, he emphasized that the new vaccine, as an artificial product of the laboratory, marked a great advance over Jenner's smallpox vaccine, which was a unique and mysterious natural product. On 22 June 1881 he developed this distinction further, noting that the smallpox microbe remained unknown, if indeed it existed at all, and that the preservative powers of the Jennerian vaccine gradually deteriorated, presumably because it could not be conserved in the form of spores.[145]

Anthrax Vaccination After 1881. Controversy and Triumph. In the wake of the dramatic success at Pouilly-le-Fort, Pasteur and his laboratory received a flood of requests for supplies of the new anthrax vaccine. On Christmas Day 1881, in a private note to the president of the Council of Ministers, Pasteur proposed the creation of a state factory for the manufacture of anthrax vaccine, of which he should be the director, assisted by Chamberland and Roux. By its support for this project, the French state would gain prestige and gratitude as the disease disappeared. In return Pasteur asked only that he and his family "be freed of material preoccupations."[146] Ultimately the government rejected Pasteur's proposal, and his laboratory remained the center for the manufacture of anthrax vaccine; one annex of the laboratory, under Chamberland's supervision, was given over entirely to the production of this and other vaccines discovered subsequently.

As efforts were made to meet the growing demand for the anthrax vaccine, Pasteur noted that his achievement raised at least one important new problem—the duration of immunity conferred by the vaccine. By June 1881 his experiments suggested that protection against injections of highly virulent anthrax culture lasted at least six months, leading him to suppose that it would last at least a year under normal conditions of field exposure. If it proved necessary, annual revaccination should pose no serious obstacles, for the procedure took little time and the vaccine cost very little to produce.[147] In late January 1882 Pasteur injected a new virulent anthrax culture into the sheep vaccinated nearly eight months before at Pouilly-le-Fort. All survived. In his view this result solved the question from a practical point of view, since the normal anthrax season ran only from April to October. Animals vaccinated in April of each year would therefore acquire complete protection from the disease.[148] By March 1883 Pasteur realized that the duration of immunity followed no general law, varying from animal to animal, so that annual revaccination was indeed indicated.[149]

In the meantime Pasteur basked in the fame and general success of his method of anthrax vaccination, while seeking to explain and to minimize those failures or "accidents" which occurred as the procedure became increasingly common and increasingly distant from his direct control. In August 1881 he went in triumph to London, where he addressed the International Congress of Medicine on vaccination. While summarizing his earlier achievements, Pasteur reported that a commission of doctors and veterinarians had asked him to repeat the Pouilly-le-Fort experiments using infected anthrax blood in place of a virulent anthrax culture as a test of the preservative powers of the attenuated vaccine. These experiments, conducted at Chartres, produced equally decisive and favorable results. Pasteur characterized vaccination as a great advance in "microbiology" (the word he preferred to the more restrictive and "Germanic" word "bacteriology")[150] and emphasized that his extension of the word "vaccination" to include preventive inoculations of any sort of attenuated culture was meant as homage to Jenner. What he had seen and heard during the Congress (including Koch's technique of solid culture) struck him as evidence not merely of the advance but of the triumph of the germ theory of disease.[151]

Late in January 1882, during his return to Pouilly-le-Fort for experiments on the duration of immunity, Pasteur received three medals commemorating his original experiments there. At the festive meeting of the Agricultural Society of Melun, where this honor was bestowed, Pasteur reported that more than 32,000 sheep had already been vaccinated, with a mortality rate about one-tenth that of unvaccinated sheep under ordinary conditions of field exposure. In fact, about 400 sheep had been saved, and the number would have been even greater had the vaccinations been made earlier in the season. As for those deaths which did occur immediately after vaccination, only a portion should be ascribed to accidents in the procedure itself; the others should be charged to the disease having already invaded the animal before its vaccination.

By June 1882, however, as reports of accidents increased, Pasteur admitted that the vaccines supplied by his laboratory from November 1881 to March 1882 had been less than adequate, despite their being direct cultural descendants of earlier, completely successful vaccines. Experience revealed that the vaccines gradually deteriorated (like Jenner's vaccine), leading to two sorts of unfortunate accidents: (1) the first of the two preventive inoculations might be made with a culture too weak compared with the second, so that the latter produced death upon injection; and (2) both vaccines might be too weak to act as a preventive against the natural disease. When these problems and their causes became clear, effective new vaccines were developed and sent free of charge to all who requested them. Pasteur also insisted again that accidents could occur through no fault of the vaccine itself: not even Jenner's vaccine could prevent smallpox once the disease had become established. Moreover, because of interspecific or interracial differences in susceptibility to anthrax, a vaccine perfectly appropriate for, say, one race of sheep might be entirely unsuited to another. Therefore vaccination should be extended to a new race of sheep or cattle only after preliminary tests had determined the appropriate degree of attenuation. In any case, occasional accidents should not be allowed to obscure the demonstrable overall value of anthrax vaccination. To encourage its general adoption, Pasteur proposed that farmers be reimbursed for any losses suffered from accidents in the procedure, with the revenues for this guarantee to be raised by a surcharge of ten centimes on each vaccination.[152]

At the end of this paper of June 1882, in response to a remark from the floor, Pasteur charged the veterinary school at Turin with a careless experimental error that had undermined confidence in his method of vaccination. A commission from that school had found that his vaccines failed to prevent death from the injection of virulent anthrax blood. In a bold assertion from afar, Pasteur ascribed their failure to the inadvertent use of anthrax blood contaminated with septicemia. For nearly a year thereafter, Pasteur and the Turin school exchanged charges and invective in open forum. When the Turin school denied his assertion and accused him of arrogance for his diagnosis-at-a-distance, Pasteur offered to come to Turin to demonstrate that anthrax blood becomes partially septic within a day. The Turin school replied that no such simple and restricted demonstration could decide the real issues in dispute and compared Pasteur to a "duelist who challenges all those who dare to contradict him . . . but who has the habit of choosing the weapons and of obliging his adversaries to fight with their hands tied."[153]

In his rejoinder Pasteur continued to limit the debate to the narrow confines within which he had placed it. He reported that Roux had confirmed the point he wished to demonstrate before the Turin school by showing that the blood of an anthrax victim dead for twenty-six hours contained both the anthrax bacillus and the *vibrion septique*, which could be separated by appropriate methods of culture (the bacillus grew in air; the vibrio *in vacuo*). Besides implying that his adversaries feared a direct confrontation with him, Pasteur impugned their motives by citing a passage in which they had distorted his views by quoting him out of context. According to Pasteur Vallery-Radot, the Turin school never admitted defeat; nevertheless anthrax vaccination soon became as widespread in Italy as elsewhere.[154]

More disturbing criticism of Pasteur's work came from Germany, where Koch and his school contributed their impressive authority to the cause. Scarcely concealed beneath the scientific and methodological issues dividing Pasteur and Koch were powerful personal and national antagonisms. As one whose basic training lay in chemistry, and whose attitude toward naturalists and physicians sometimes approached the contemptuous, Pasteur belonged to a tradition different from that of Koch, whose training had been in medicine and whose career owed so much to the botanist Ferdinand Cohn. A more immediate source of their later confrontation lay in Pasteur's tendency to minimize the originality and decisiveness of Koch's work on anthrax. Indeed, from 1877 he repeatedly claimed priority for the discovery of

resistant bacilli endospores, citing passages in which he had described the formation of resistant "corpuscules brillants" or "corpuscules-germs" in *flacherie*.[155] To him Koch's discovery of a resistant endospore phase for the anthrax microbe amounted to merely a confirmation and extension of this earlier discovery. With a convenient disregard for the difference between his rather brief, ambiguous description of "corpuscules brillants" in *flacherie* and Koch's precise and full-fledged account of the anthrax endospore, Pasteur implied that the special character and full significance of these "corpuscles" had always been clear to him and ought therefore to have been clear to Koch and other naturalists. In fact, as Koch well knew, the existence and significance of bacilli endospores had received little attention before 1875, when Ferdinand Cohn recognized their crucial place in the life cycle of *Bacillus subtilis*.

However deep and long-standing these tensions between Koch and Pasteur may have been, they remained largely suppressed until 1881, when the German Sanitary Office published the first volume of its journal, *Mittheilungen aus dem Kaiserlichen Gesundheitsamt*. In this volume Koch and his students attacked Pasteur's work on disease on several grounds, of which perhaps the most damning was their charge that his liquid media (as opposed to Koch's solid media) failed to guarantee pure cultures. In fact, the German school alleged that Pasteur's supposedly "attenuated" anthrax cultures or "vaccines" were merely contaminated cultures.[156] They also accused him of confusing several other diseases with septicemia and of unacknowledged dependence on Koch and others for the most accurate and valuable portions of his work. Koch disputed Pasteur's claims that earthworms play a central role in the spread of anthrax and that domestic animals ordinarily contract it through lesions of the mouth and throat caused by prickly diets.

Pasteur was apparently unaware of these charges when he met Koch at the International Congress of Medicine in August 1881 and described the latter's solid media as a "great progress." But in September 1882, during the International Congress of Hygiene and Demography at Geneva, he mounted a vigorous defense of his work against Koch and his pupils, blaming their "inexperience" for the "multitude of errors" they had committed.[157] Fortified by the results of his experiments at Pouilly-le-Fort, he virtually demanded a response from Koch, who sat among the audience. Considering the Congress an inappropriate forum for such a discussion, Koch had little to say, although he promised to respond in print to Pasteur's address. Three months later he kept his promise with

Ueber die Milzbrandimpfung. Eine Entgegnung auf den von Pasteur in Genf gehaltenen Vortrag. In an abrupt shift of position, Koch hailed the discovery of attenuation as a major achievement but gave Toussaint, rather than Pasteur, priority for it and justified his own earlier skepticism on the ground that his French rival had failed at first to provide a complete and explicit account of his method of attenuating the anthrax microbe. Moreover, he continued to condemn Pasteur's method of vaccination from a practical point of view, citing the experiments of the veterinary school of Turin as well as other "accidents" and unresolved issues, including the duration of immunity. He continued also to cast aspersions on the purity of Pasteur's cultures, on his secrecy, and on his more general knowledge of medicine and pathological bacteriology. Pasteur's rejoinder took the form of a long open letter to Koch dated Christmas Day 1882. Combining heavy sarcasm with considerable persuasion, he refuted Koch's critique point by point until it seemed to contain nothing of value but belated and grudging concessions to Pasteur's point of view.

For several years thereafter, the Pasteur-Koch dispute remained mostly in the shadows, as Pasteur's method of anthrax vaccination spread throughout Europe with striking success. In April 1883 Pasteur could insist that the new anthrax vaccines—introduced in November 1882—were so safe that not a single animal had fallen victim to a vaccination accident in the meantime, while their efficacy was so great that he could not have been consoled had attenuation been other than a "French discovery."[158] One month later Pasteur reported that a field trial of his anthrax vaccine had been conducted in two regions of Germany, under the auspices of the Prussian minister of agriculture, and that the results of the first year, released that month in Berlin, had been so favorable that the farmers of those regions had decided to adopt the procedure.[159] In August 1887, after Pasteur announced that the "Berlin school" had been converted, Koch denied that he had modified his views on the practical value of vaccination and insisted that no guarantee existed as to the accuracy of Pasteur's glowing statistics on the procedure. To Pasteur this position represented blind obstinacy in the face of the testimony of veterinarians, whose reports he promised to submit to the forthcoming International Congress of Hygiene and Demography in Vienna. These reports, like all that followed, can only have embarrassed Koch. By 1894 Chamberland could report that 3,400,000 sheep and 438,000 cattle had been vaccinated against anthrax, with respective mortality rates of 1 and 0.3 percent. Comparing these rates with earlier mortalities among unvaccinated animals, he estimated a saving through vaccination of five million francs for sheep and two million francs for cattle.[160]

The Attenuation of the "Saliva Microbe" and of a Microbe Found in "Horse Typhoid." Pasteur's work on anthrax vaccines reinforced his belief in the general applicability of the method of attenuation discovered for the fowl cholera microbe. As early as June 1881, soon after the Pouilly-le-Fort experiment, he reported the extension of this method to a third microbe: the "saliva microbe" (later recognized to be a pneumococcus), first obtained in December 1880 from the saliva of a child dead of rabies. This saliva produced rapid death upon injection into dogs or rabbits, the blood of which became infested by the new microbe, similar morphologically (a figure eight) but not physiologically to the fowl cholera microbe. The origin of the saliva raised the possibility that the new microbe might play some role in rabies; and Pasteur spent several weeks investigating this possibility, while carefully refraining from publishing any definite conclusions. In March 1881, having found the new microbe in the saliva of young victims of other diseases and in healthy adults, he denied any connection between it and rabies. Indeed, by the time he announced his success in attenuating this new microbe, again by prolonged exposure to atmospheric oxygen, he suggested that it might be entirely harmless to man, however lethal its effects when injected into rabbits or dogs.[161]

In September 1882, at the International Congress of Hygiene and Demography in Geneva, Pasteur gave a much fuller account of his work on the saliva microbe and disclosed the discovery of a fourth example of attenuation by atmospheric oxygen—that of a microbe obtained from the nasal discharges of a horse dead of "horse typhoid." Rabbits injected with these discharges died in less than twenty-four hours of a "veritable typhoid fever," accompanied by the appearance in their blood of a new microbe—once again in the form of a figure eight. Like the microbes of fowl cholera, anthrax, and saliva, the aerobic character of which it shared, this new microbe underwent no change in virulence in closed tubes but became progressively less virulent (or more attenuated) upon exposure to the air. As in his early work on the saliva microbe, Pasteur carefully avoided any conclusion as to the possible role of this microbe in horse typhoid.[162] Despite such caution, his adversaries accused him of trying to forge an etiological link between these microbes and the diseases of the subjects from which they had been taken, especially in the case of the microbe taken from the rabid youth. Koch referred sarcastically to Pasteur's fondness for microbes in the form of a figure eight and suggested that the animals allegedly killed by these suspicious new microbes had merely died of different forms of septicemia.

In part Koch's objections reflect a more general difference of emphasis between him and Pasteur (or between the German and French schools). For while they basically agreed on the specificity of microbes, they differed as to the range of variability within a given species and as to the relative importance of morphological and physiological properties in microbial identification. Probably because of his mastery of technique in the naturalist tradition, Koch gave pride of place to morphology—to careful, detailed descriptions and pictorial representations of microbial form. The relative reliability and constancy of form in microbes grown in his solid cultural media tended naturally to reinforce this morphological bias. Pasteur's lack of training in and relative disdain for the naturalist tradition led him to focus instead on the physiological properties of microbes, and this functional bias drew additional force from observations of morphological variability in the richly varied liquid media he used.

This is not to say that Koch ignored physiological considerations, or that Pasteur ignored morphology,[163] but merely to assert a difference in emphasis. From this perspective some of their specific disagreements can be more readily understood. Pasteur's claim that microbes of similar form (notably the figure eight) had radically different functions—produced different diseases—was bound to arouse Koch's skepticism. But Pasteur's physiological bias made him rather more sensitive to the variable behavior of a given microbe in different environments, and he early recognized that different animal species constitute different cultural media or "terrains" for the microbe to which they are host. Not surprisingly, therefore, he tended to identify microbes by virtue of their biological action when injected into a given animal species. Thus, his claim that the "saliva microbe" differed from the *vibrion septique* rested above all on the fact that guinea pigs, which were strikingly susceptible to septicemia, proved entirely refractory to injections of the new microbe.[164]

For somewhat similar reasons Pasteur was more disposed than Koch to suppose that a given microbe could undergo intrinsic changes in its properties by successive passages through the same or different animal species. That the virulence of a microbe could be increased by successive passages within a species was known before Pasteur began his work. Indeed, Koch had drawn attention to this fact in the cases of anthrax and traumatic infectious diseases. But Koch emphasized the effect of such passages on microbial purity—he described the technique in 1878 as "the best and surest method of pure cultivation"—and did

not suppose that the intrinsic properties of the microbe had thereby been changed.[165] This helps to explain his assumption that Pasteur's "attenuated" anthrax cultures must have been impure. Pasteur, by contrast, insisted that his attenuated anthrax cultures were pure and that they resulted from real changes in the properties of the microbe itself. These changes could be reversed and the microbe returned to its original virulence by passage through animals of different ages and species.[166]

In his address at the Geneva Congress in 1882, Pasteur extended these conclusions to the saliva microbe and to the microbe found in "horse typhoid." He reported that the virulence of the former microbe in guinea pigs could be increased by successive passages through that species, while the virulence of the latter in rabbits could be increased by successive passages through that species. More important, he now recognized that successive passages through one species could reduce the virulence of a microbe toward another species. Thus the saliva microbe became increasingly less virulent to rabbits by successive passages through guinea pigs, and the microbe found in "horse typhoid" became progressively less virulent to guinea pigs by successive passages through rabbits. In effect this amounted to the discovery of a new method of attenuation, which Pasteur was soon to exploit against swine erysipelas and rabies.

Discovery of the Vaccine Against Swine Erysipelas. Although Pasteur's attention had been drawn to swine erysipelas (*rouget du porc* or hog cholera) as early as 1877 by Achille Maucuer, a veterinarian in the township of Bollène, in Vaucluse,[167] he was then too preoccupied with other work to give it any serious attention. In the summer of 1881, he sent Chamberland to Bollène to study the disease, but nothing seems to have come of that effort. Six months later Louis Thuillier, another of Pasteur's assistants, went to Peux, in Vienne, where in March 1882 he isolated a new microbe (now called *Erysipelothrix insidiosa*), which he implicated in swine erysipelas. Almost immediately Thuillier returned with cultures of this new microbe to Pasteur's laboratory at the École Normale, where they began searching for a means of attenuating it. Early in April, however, Thuillier was sent to Germany to supervise a field trial of anthrax vaccination on the model of the Pouilly-le-Fort experiments, as he had done in Hungary the year before. For the next two months Thuillier remained in Germany as Pasteur pursued the search for a vaccine against swine erysipelas. By mid-October, Pasteur apparently had made considerable progress; and in November he, Thuillier, and Adrien Loir went to Bollène to conduct preliminary small-scale trials. From there, on

3 December 1882, Pasteur sent J.-B. Dumas a letter, to be read at the Académie des Sciences, in which he outlined the basic results to date of their hitherto unpublished studies. He included Thuillier's new microbe among those having the form of a figure eight and reported that it killed rabbits and sheep as well as hogs but had no effect on chickens. More important, he announced that they had proved the nonrecurrence of swine erysipelas and had prepared an attenuated form of the microbe, inoculation with which made hogs refractory to the disease. While noting that additional confirmatory experiments needed to be done, Pasteur expressed confidence that the new vaccine would be ready by the next spring to save hogs from this seasonal blight, which in 1882 had claimed an estimated 20,000 animals in the departments of the Rhône Valley alone and in 1879 an estimated 900,000 hogs in the United States.[168]

In November 1883, after further successful testing of the new vaccine, Pasteur gave the Académie des Sciences a more extended account of the studies on swine erysipelas. He began with a warm tribute to Thuillier, who had died of cholera in September, at the age of twenty-seven. His death affected Pasteur deeply, in part because it came while Thuillier served on the ill-fated French Cholera Commission, sent to Egypt at Pasteur's urging and under his guidance to study the very disease that killed him. As if to intensify the tragedy, the German Cholera Commission, in Egypt at the same time under Koch's leadership, made considerable progress and eventually isolated a comma-shaped bacillus to which Koch definitely and triumphantly ascribed cholera in the early months of 1884.

Pasteur took consolation in the heroic quality of Thuillier's death and in the outcome of their joint study of swine erysipelas, the results of which he presented in both their names. He reported that the immunity conferred by the new vaccine lasted at least a year, but that its general diffusion faced practical difficulties owing to wide variations in the susceptibility of different breeds of hogs to the disease. Studies were already under way, however, to prepare vaccines of a strength appropriate to each breed; and while absolutely definitive results could not yet be claimed, he decided to disclose the method by which the microbe had been attenuated. By way of introduction, Pasteur recalled his earlier discovery that the saliva microbe became attenuated for rabbits by successive passages through guinea pigs.

Having learned that pigeons and rabbits, as well as hogs, suffered severely from infectious disease in the department of Vaucluse, Pasteur and his team wondered whether these species might share with hogs a susceptibility to the microbe of swine erysipelas—and

if so, what effects its successive passage through them might have. They quickly established that pigeons and rabbits did indeed die from injections of the microbe; and while successive passages through pigeons increased the virulence of the microbe for hogs, successive passages through rabbits had the opposite effect. In fact, several passages through rabbits so attenuated its virulence in hogs that it became harmless to them. At this point inoculation of the cultures protected hogs from the effects of somewhat less attenuated cultures. By injecting hogs with a series of progressively more virulent cultures, they could be rendered immune to the natural disease. According to Bulloch, this method of vaccination was used on more than 100,000 hogs in France between 1886 and 1892, and on more than 1 million hogs in Hungary from 1889 to 1894.[169]

In revealing this new method of attenuation, Pasteur emphasized the variability of viruses or microbes in different media and made an arresting comparison between their variability and that of higher organisms. In fact, he suggested, microbes are no more variable than higher organisms; they seem to be only because they reproduce so rapidly, with an immense number of generations succeeding each other in short order. By contrast, higher organisms require thousands or millions of years to achieve the same number of generations. Thus, even though higher organisms, no less than microbes, display "plasticity" under the influence of the environmental conditions in which successive generations live, they seem static to us. As usual, Pasteur said nothing about the possible implications of such ideas for the transmutability of species or for Darwinian evolutionary theory.

The Search for a Rabies Vaccine, 1881–1884. Because rabies is so rare in man (in France its victims probably never reached more than 100 in any year) and can be quite readily controlled by muzzling and quarantine of dogs, many observers of Pasteur's career have been somewhat puzzled by his interest in it. Some have traced his concern to a traumatic childhood experience. In October 1831 a rabid wolf bit several Arboisiens and terrorized the entire region. The standard treatment, then as since antiquity, was to cauterize the wounds immediately with a red-hot iron; and the youthful Pasteur reportedly saw a man submit to this excruciating procedure at a blacksmith's shop near his home. Despite all efforts some of the wolf's victims died, including at least one whose name and circumstances Pasteur recalled more than half a century later.[170]

As a result of this episode, Pasteur may long have shared the popular horror of the disease. Indeed, in several ways rabies was precisely suited to inspire terror and a sense of mystery. Its rarity made it seem that its victims had been perversely singled out, especially since they were often children. Its usual victim and agent was man's favorite pet. Its long incubation period, ordinarily a month, at least, produced suspense and dread in any victim of an animal bite, especially because medical care was utterly powerless and death absolutely certain once the symptoms became manifest. Above all, the symptoms were believed to embody the ultimate in agony and degradation, stripping the victims of their sanity and reducing them to quivering, convulsive, animal-like shadows of their former selves. Although this conception of rabies depended more on observations of "mad" dogs than on clinical evidence, it so gripped the public imagination that the short, dry cough of human victims was compared to the bark of a dog. Few realized that the disease had a quite peaceful "paralytic" form as well as a "mad" form, or that the supposed fear of water—which gave the disease its other popular name, hydrophobia—stemmed from difficulty in swallowing and not from a fear of water per se.

Thus, however unimportant rabies may have been in terms of vital statistics, Pasteur must have realized that its conqueror would be hailed as a popular savior. And indeed he was. For if the anthrax experiments at Pouilly-le-Fort had created public confidence in the germ theory of disease, his treatment for rabies set off an international chorus of cheers the tangible echo of which was the Institut Pasteur. Not even Pasteur could have hoped for such a result from the outset, however; and before he had fully achieved it, he offered an additional explanation for his interest in rabies—an explanation at once more prosaic and plausible than the others. Speaking at Copenhagen in 1884, he emphasized that the extension of vaccination to human diseases presented special difficulties, notably because "experimentation, [if] allowable on animals, is criminal on man." For this reason vaccination could be extended to man only on the basis of a deep knowledge of animal diseases, "in particular those which affect animals in common with man." As the oldest, most familiar, and most striking example of such a disease, rabies was a natural choice to satisfy Pasteur's "desire to penetrate further" into the problem.[171]

Initially, from December 1880 through March 1881, Pasteur's work on rabies was bound up with that on the "saliva microbe." Once convinced that this microbe had no connection with rabies, and finding himself unable to implicate any other microbe in the disease, Pasteur approached rabies rather differently from the way in which he had so successfully attacked fowl cholera and anthrax. The central feature of his

work on these diseases, as he often insisted, was the cultivation and attenuation of the implicated microbe in sterile cultural media, outside the animal economy. With a flexibility born partly of necessity, Pasteur now made the living organism the sole cultural medium for the rabies virus. In this, as in so much of his work, a thread of continuity runs through the seemingly dramatic shift in approach. For he had long conceived of living organisms as cultural media, and he already knew that the microbes of fowl cholera and anthrax could vary in virulence in different living media. Moreover, believing that Jenner's still mysterious vaccine was merely attenuated smallpox virus, he had additional reason to hope that any "virus," including that of rabies, might be altered in virulence by passage through appropriate animals, even though it resisted attempts to cultivate it *in vitro*. In fact, as we now know, Pasteur could have accomplished what he did toward the conquest of rabies only by this rather indirect approach. Of the "virus" diseases that he studied, only rabies is a virus disease in the modern sense; its agent is a filterable virus, invisible under the ordinary microscope, the *in vitro* cultivation of which has not yet been achieved. In this connection it is interesting that Robert Koch studied rabid brains during the 1880's.[172] That this work was apparently fruitless may well have been partly due to Koch's tendency to emphasize the visible and tangible aspects of disease agents over their physiological behavior in different media.

Although fortunate and essential, Pasteur's decision to proceed in the absence of an *in vitro* rabies culture did not lead far by itself. The lengthy incubation period, as well as the uncertainty of the standard modes of transmission, made new techniques imperative. Neither the injection of rabid saliva nor the bite of a rabid animal produced rabies consistently, and neither method reduced the incubation period. Similar objections applied to the subcutaneous inoculation of rabid nerve tissue, a method that seemed well chosen in view of the patently neuropathic symptoms of the disease. In May 1881, in his first memoir on rabies per se, Pasteur described a new experimental method for transmitting the disease with certainty and with a greatly reduced incubation period. The new method, perhaps suggested by Roux,[173] involved the extraction of cerebral matter from a rabid dog under sterile procedures and its subsequent inoculation directly onto the surface of the brain of a healthy animal, under the dura mater, after trephining. Under these conditions the inoculated animal invariably contracted rabies after an incubation period of about two weeks.

In December 1882, Pasteur reported that rabies could also be transmitted (usually in paralytic form) by the intravenous injection of its virus, the character of which remained obscure. Whether transmitted by this intravenous method or by the intracranial method announced earlier, the incubation period had now been reduced to six to ten days, although Pasteur declined to reveal how, "leaving aside for the moment all details." Among the other results of his 200 experiments, perhaps the most important was the discovery that a few dogs were "accidentally" or inherently resistant to injections of the virulent virus. After recovering from the effects of one such injection, these dogs became immune to subsequent injections. This result established that rabies shared the distinguishing feature of the other "virus" diseases—it did not recur in an animal that had survived an attack. That rabies shared this feature had been far from certain, since death so consistently claimed its victims. Only with the removal of this doubt did it become entirely reasonable to hope that the search for a vaccine might eventually succeed.

Fortified by this assurance and armed with their new techniques of transmission, Pasteur and his collaborators pressed toward a rabies vaccine. In February 1884 Pasteur announced that they had reached their goal. By now they had returned to the method of intracranial inoculation, described as easy to learn and almost always successful. Although the virus continued to resist all attempts at artificial cultivation, Pasteur held to his assumption that a microbe of rabies did exist. At the very least, he maintained, a rabid brain could easily be distinguished from a normal one, for the medulla of the former contained numerous fine granules, resembling simple dots and suggesting a microbe of extreme tenuity. Whether or not further research established that these granules were "actually the germ of rabies," Pasteur and his team had made what seemed to him a vastly more important discovery: the rabies virus (like the microbes of fowl cholera, anthrax, saliva, and "rabbit typhoid") could be prepared in varying degrees of virulence by successive passages through different animal species. In any given species a series of passages led eventually to a fixed degree of virulence, measured by the number of days of incubation for a given quantity of inoculated virus. This maximum or "fixed" virulence varied in different animals and had already been reached naturally in the dog by virtue of countless transfers by bites through past ages.

As this suggestion implies, Pasteur emphatically rejected the notion that rabies could arise "spontaneously" in the absence of the virus. But the really important consequence of the varying states of virulence was that they allowed "a method of rendering dogs refractory to rabies in numbers as large as desired." Like his earlier methods of vaccination, this method involved the serial inoculation of progressively virulent cultures, beginning with one attenuated to the point of harmlessness. By this method Pasteur and his team had already produced twenty-three dogs capable of sustaining the most virulent rabies virus. Indirectly the problem of prophylaxis in man had also thus been essentially solved, for he ordinarily contracted rabies only from dogs. Moreover, the lengthy incubation period of the disease offered hope that a victim might be rendered refractory before the symptoms became manifest.

In his fourth memoir on rabies (May 1884), Pasteur elaborated very briefly on the methods by which the rabies virus had been prepared in varying degrees of virulence. To weaken or attenuate the virus, it was passed from dog to monkey and then successively from monkey to monkey. After just a few such passages it had become so attenuated that its hypodermic injection into dogs never resulted in rabies; indeed, even intracranial inoculation usually produced no effect. On the other hand, the virulence of ordinary canine rabies could be increased by successive passages through guinea pigs or rabbits; in the latter it achieved its maximum fixed virulence only after a considerable number of passages. By these means, Pasteur noted, one can prepare and keep on hand a series of viruses of various strengths, the most attenuated of which are nonlethal from the outset but protect the inoculated animal from the effects of somewhat more virulent viruses, which in their turn act as a vaccine against still more virulent strains, until eventually the animal is always rendered refractory to even the most virulent and ordinarily fatal virus. If all dogs were vaccinated in this way, rabies could eventually be eliminated; but until that "distant period" it seemed important to search for a means of preventing the disease during the long incubation that followed the bite of a rabid animal. Indeed, Pasteur believed that the method was already at hand to render bitten patients refractory before the disease became manifest. "But," he emphasized, "proofs must be collected from different animal species, and almost *ad infinitum*, before human therapeutics can make bold to try this mode of prophylaxis on man himself."

Toward this end Pasteur requested the convening of a commission, to be appointed by the minister of public instruction, to which he could submit his present results and future experiments. Two sorts of experiments seemed to him best calculated to carry conviction. First, twenty of his vaccinated dogs should be placed with twenty unvaccinated dogs, and all forty should then be subjected to the bites of rabid dogs. Second, the same experiment should be made, except that the forty dogs should sustain the intracranial inoculation of ordinary canine rabies instead of the bites of rabid dogs. "If the facts announced by me are real," Pasteur predicted, "not one of my twenty [vaccinated] dogs will contract rabies, while the twenty control animals will." The proposed commission was duly appointed that very month and issued its initial report early in August 1884. After two months of experiments conducted under its scrutiny, none of Pasteur's twenty-three vaccinated dogs had contracted rabies—whether from the bites of rabid dogs or from inoculation of the rabies virus. By contrast, two-thirds of the unvaccinated control dogs had already become rabid.

Later in August, in a major address to the International Congress of Medicine at Copenhagen, Pasteur proudly repeated these results and finally described in considerable detail the method of intracranial inoculation and his process of preparing the rabies virus in varying degrees of virulence. He reported that the search for an organism which would act as an attenuating medium for the virus had been long and frustrating. Through a great number of experiments, the animals selected as candidates for this role proved to increase rather than to attenuate the virulence of the virus. Not until December 1883 did they happen upon the proper "attenuating" organism—the monkey. Toward the end of this address, Pasteur again raised the issue of the rabies microbe: "You must be feeling, gentlemen, that there is a great blank in my communication; I do not speak of the microorganism of rabies. We have not got it. . . . Long still will the art of preventing diseases have to grapple with virulent diseases, the microorganic germs of which escape our investigation."

Despite the encouraging initial results, it gradually became clear that Pasteur's proposed method was not infallible—no more than fifteen or sixteen dogs in twenty could be rendered refractory to rabies with absolute certainty. Furthermore, the results of the method could be ascertained only after three or four months, a circumstance that would have severely limited its scope in human practice, particularly in emergency cases. For these reasons Pasteur undertook to discover a new method of prophylaxis that would be both more rapid and more certain.

In doing so, Pasteur could look forward to yet another major government-financed expansion of his facilities. Very early in its deliberations, the rabies commission recommended the establishment of a large kennel yard for the housing and observation of Pasteur's experimental dogs. The site initially chosen, in the Bois de Meudon, was quickly abandoned in the face of vigorous protests from inhabitants of the neighborhood. Similar local protests erupted upon the selection of a second site—in the park of Villeneuve l'Étang, near St.-Cloud, a state domain that had once belonged to Louis Napoleon. Although these protests helped to delay an appropriation of 100,000 francs promised to Pasteur, they ultimately proved ineffectual. By May 1885 the old stables of the château of St.-Cloud had been converted into a large paved kennel with accommodations for sixty dogs. A laboratory was also established, and living quarters nearby were renovated for Pasteur's private use.[174]

Rabies Vaccination, 1884–1886: Its Extension to Man. Awaiting the completion of this new complex—which eventually became a branch of the Institut Pasteur and was the site of his death—Pasteur pursued his quest for a perfected method of preventing rabies. In December 1884 he reluctantly declined to treat a bitten child by the means at his disposal, noting that he had not yet established that his method would work on dogs after they had been bitten and confessing that even if he proved successful at that, his hand would "tremble" before applying the treatment to humans, "for what is possible on the dog may not be so on man."[175] By March 1885, however, he had begun to test his method on dogs already bitten;[176] and on 6 July 1885 he decided to treat nine-year-old Joseph Meister, from Alsace, "not without feelings of utmost anxiety," even though he had been assured by two sympathetic physicians that the boy was otherwise "doomed to inevitable death" and even though his new method of prophylaxis had never failed in dogs.

In a memoir of 26 October 1885, Pasteur described this new method and the circumstances under which he had made his fateful decision. In essence the new method involved *in vitro* attenuation rather than the earlier method of passage through monkeys. It depended first on the preparation of a virus both pure and perfectly consistent in its virulence. This had been accomplished by using a virus passed successively through rabbits over a period of three years. Now in its ninetieth passage, this virus invariably produced an incubation period of seven days and had done so for nearly forty consecutive passages. In two earlier memoirs Pasteur had reported that a given rabies virus retained its virulence for weeks in the encephalon and spinal cord of the infected animal, as long as these tissues were preserved from putrefaction by storage at 0–12° C. He now revealed that this technique could be modified in such a way as to attenuate the virus. Adopting a technique introduced by Roux, who prevented putrefaction of rabbit spinal cords by suspending them in a dry atmosphere instead of by cooling,[177] Pasteur excised strips of spinal cord from rabbits dead of the seven-day "fixed" virus and suspended them in flasks in which the atmosphere was kept dry by addition of caustic potash. He found that the virulence of the virus in these strips gradually diminished and eventually disappeared. The time required for this process depended somewhat on the thickness of the strips but more importantly on atmospheric temperature. Up to a point, the higher the temperature, the more quickly attenuation was achieved. Ordinarily the virus became attenuated to the point of harmlessness in about two weeks.

Using a spinal strip that had been drying for some two weeks, the first step in the actual treatment was to mash a portion of it in a sterile broth and then to inject the resulting paste into the animal to be protected. On successive days the injections came from progressively fresher marrows and eventually from a highly virulent strip that had been drying for only a day or two. By this method, Pasteur reported, he had rendered fifty dogs of all ages and types refractory to rabies when young Meister appeared unexpectedly at his laboratory, accompanied by his mother and the owner of the dog responsible for the attack. Two days before, on 4 July, Meister had been bitten in fourteen places on his hands, lower legs, and thighs. These wounds, some so deep that he could scarcely walk, had been cauterized with carbolic acid by a local physician twelve hours after the attack. The dog had been killed by its owner, whom Pasteur sent home after having been assured that his skin had not been broken by the dog's fangs. That the dog was indeed rabid seemed certain from its behavior and from the presence in its stomach of hay, straw, and wood chips.

Pasteur immediately consulted Alfred Vulpian, a member of the rabies commission, and Jacques Joseph Grancher, who worked in his laboratory. Both considered young Meister doomed; and after Pasteur told them of his new results, both urged him to use the new method on the boy. The treatment, begun that evening, lasted ten days, during which Meister received thirteen abdominal injections derived from progressively more virulent rabbit marrows. By the end of the treatment, Meister was being inoculated with the most virulent rabies virus known—that of a mad dog augmented by a long series of passages through rabbits. Nonetheless, he had remained healthy during the nearly four months since he had been bitten, and his recovery therefore seemed assured. According to Dubos, Meister eventually became a concierge at the Institut Pasteur and lived until 1940, when he chose to commit suicide rather than open Pasteur's burial crypt to the advancing German army.[178]

At the end of his memoir of 26 October 1885, Pasteur announced that a week earlier he had begun to treat a second boy, a fifteen-year-old shepherd named Jean-Baptiste Jupille, who had been viciously bitten while killing a rabid dog that threatened the lives of six younger comrades. He had not arrived at Pasteur's laboratory for treatment until six days after having been bitten (as compared with two days for Meister), prompting Pasteur to emphasize that the length of time that could safely be allowed to pass between bites and treatment presented the "most anxious question for now." During the brief and uniformly laudatory discussion that followed Pasteur's memoir, Vulpian proposed the founding of a special service for the treatment of rabies by Pasteur's method (a proposal ultimately realized in the Institut Pasteur) and the president of the Académie des Sciences predicted that the date of this meeting would "remain forever memorable in the history of medicine and forever glorious for French science." When, a day later, Pasteur read the same memoir to the Académie de Médecine, its president expressed the nearly identical sentiment that the date of the meeting would "remain one of the most memorable, if not the most memorable, in the history of the conquests of science and in the annals of the Academy."[179]

In March 1886, Pasteur reported that young Jupille remained well (like Meister, Jupille ultimately joined the staff of the Institut Pasteur, where he served until his death in 1923)[180] and that 350 patients had now submitted to his rabies treatment. One had died despite the treatment, but Pasteur defended his method by emphasizing that ten-year-old Louise Pelletier had not arrived for treatment until thirty-seven days after being attacked and by showing that the fatal virus had an incubation period characteristic of dog-bite virus and not of the virus used in his prophylactic treatment.

In the same memoir Pasteur referred briefly to the problem of reliable statistics, which remained at the center of all subsequent debate over his antirabies treatment. Admitting his surprise at the large number of people who came for treatment, he suggested that the frequency of rabid bites had previously been underestimated out of reluctance to inform victims that they might have contracted a fatal disease. Moreover, he emphasized that he had drawn up a very rigorous catalog of the cases, insisting where possible that the victims bring certificates from veterinarians or doctors testifying to the rabid state of the attacking animal. Although it nonetheless proved necessary to treat cases in which dogs were merely suspected of being rabid, Pasteur declined to treat anyone whose clothes had not been visibly penetrated. Supporting himself particularly on statistics giving an average of one death per six bitten victims in the department of the Seine from 1878 to 1883, he insisted that his treatment was "henceforth an established fact" and deserved a special new institution. In the discussion which followed, he argued that one such center in Paris would suffice for all of Europe if those who would eventually apply the treatment abroad came there for training.[181]

Pasteur and the Rabies Treatment After 1886. Even a few of Pasteur's disciples and collaborators opposed his quickness in applying the prophylactic treatment to human cases. Indeed, Roux broke with Pasteur over the issue, refusing to sign the first report on the treatment and leaving the laboratory for several months.[182] Naturally opposition was far more severe outside Pasteur's circle, particularly among traditional medical men, antivivisectionists, and antivaccinationists, and especially as others died after receiving the treatment. Some critics charged that these deaths occurred not despite the treatment but because of it, in effect accusing Pasteur of involuntary manslaughter, and the father of one dead child actually filed suit against him.[183] Despite overwhelming statistical evidence as to the safety of his treatment, Pasteur was not allowed to forget the occasional failures. By May 1886 he complained, with some justice, that his efforts had made him the target of a "hostile press" and of "malevolent persons" in the Académie de Médecine.[184] Nonetheless, it was impossible to claim absolute safety and efficacy for his treatment, and Pasteur's attempts to perfect it may have done more to exacerbate doubts than to dispel them.

In Meister's case the last injection in the series had been prepared from spinal marrow only one day old, but soon afterward Pasteur decided that five-day-old marrow should suffice for the final injection. For certain severe cases, however, he developed a more intensive version of the treatment, which he had begun to apply by September 1886. In these cases he returned to one-day-old marrow for the last injection in each series and increased the number of injections per day so that the patient went through three series of injections in the same period of time (ten days) that Meister and other victims had gone through one series.[185] Besides raising some doubt as to Pasteur's full confidence in his treatment, these modifications failed to eliminate occasional failures. In a very few cases death occurred under circumstances suggesting that the "intensive" method of treatment may have been responsible,[186] and Pasteur abandoned it within a year of its introduction. To some this indecision served as evidence that Pasteur's method was empirical rather than truly "scientific," as did the rather casual leap from animal experiments to human therapeutics and Pasteur's practice of keeping certain details of the method secret.[187] Against the latter criticism, however, Pasteur could appeal to the need for quality control

and could produce a list of those to whom every detail of the method had been taught in his laboratory.[188]

Ultimately the chief and most persuasive criticism of Pasteur's treatment concerned the statistical evidence of its efficacy. In its report of 1887, probably the most judicious contemporary evaluation of the treatment, the English Rabies Commission emphasized the unreliability of statistics on rabies. Besides the frequent difficulty or impossibility of establishing that the attacking animal had in fact been rabid, immense uncertainty surrounded the exact influence of the character and location of bites, of interracial and interspecific differences among attacking animals, and of cauterization and other treatments applied before Pasteur's vaccine. The uncertainty of these and other factors helped to explain why previous estimates of the mortality from the bites of rabid dogs varied from 5 percent to 60 percent. Despite its testimony as to the exactitude of Pasteur's experiments and its conviction that his treatment had saved a considerable number of lives, the English Rabies Commission recommended the less dramatic course of enacting and enforcing more stringent police regulations on dogs. By this approach, already operating with striking success in Australia and Germany, rabies was virtually eliminated from England by the turn of the century.[189]

Meanwhile, however, English citizens were among those making the pilgrimage to Paris in hope of being saved from rabies. If Pasteur's treatment evoked strong opposition from certain quarters, it won lavish praise and gratitude from nearly all who submitted to it; and centers for the treatment quickly spread to other nations. Despite the cavils of unbitten and unthreatened adversaries, a steady stream of fearful victims came to Paris and offered Pasteur living testimony of the value of his achievement. By November 1886, about a year after the first treatment, nearly 2,500 persons had been treated in Paris alone. Of 1,726 French citizens treated, only 12 had died; and Pasteur refused to acknowledge failure in two of these cases, including that of Louise Pelletier. On Pasteur's reckoning, therefore, the mortality rate after his treatment amounted to about 0.6 percent, as compared with the most optimistic estimate of 5 percent in the absence of his treatment.[190]

At Pasteur's death about 20,000 persons had undergone his rabies treatment at centers throughout the world, with a mortality rate of less than 0.5 percent.[191] By 1905 this number had reached 100,000; and by 1935, 51,057 persons had been treated at the Institut Pasteur alone, with only 151 deaths—a mortality rate of 0.29 percent.[192] Despite these statistics, controversy continued to surround the safety and efficacy of Pasteur's original method of vaccination and of the modified versions introduced subsequently. By the mid-twentieth century it had become clear that the repeated injection of rabid nerve tissue could sometimes produce paralysis and that such accidents could be strikingly reduced by resort to dead vaccines in place of Pasteur's living, attenuated vaccine. Even better results were achieved with live vaccine cultivated in duck eggs rather than in nerve tissue.[193] By 1973, another rabies vaccine had been developed in the hope that a single preventive injection could replace the long and painful series of abdominal injections.[194] If most epidemiologists now doubt that Pasteur's treatment has saved as many lives as once believed, if others deny its value under present social circumstances, and if all agree that muzzling and quarantine of dogs is a preferable approach to the rabies problem, Pasteur's achievement nonetheless had an impact and importance not fully represented in statistical terms—not only for those whose lives or peace of mind were saved by it but also for the promise and foundation it gave to the immensely successful campaign to extend immunization to other human diseases.[195]

Pasteur on Pest Control. On the basis of a few isolated passages in his work, Pasteur has been called a prophet not only of bacteriotherapy but also of chemotherapy. If this was indeed true, he was scarcely a toiler in the vineyard of either discipline. A rather similar judgment attaches to his few scattered remarks on biological methods of pest control, the prophetic character of which may seem more compelling in an era when ecology is in the ascendant and insecticides under suspicion. One pest of particular concern to him was phylloxera, a plant louse which, by its ruinous effects on vineyards in France and elsewhere, interfered for several years with the adoption and spread of his process for preserving wine. "In a time of famine," wrote Duclaux, "no one need consider how to keep grapes, and the heating of wines was little practiced except for those which must be shipped under bad conditions as to keeping, for example, in the commissariat of the Navy."[196] In 1882 Pasteur suggested that if phylloxera were subject to some contagious disease, and if the causative microbe of this disease could be isolated and cultivated, then the pest might be controlled by introducing the microbe into infested vineyards. This suggestion seems not to have been seriously pursued, however, and the phylloxera plague eventually declined as mysteriously as it had arisen.[197]

Under the stimulus of a 625,000-franc prize offered by the afflicted countries, Pasteur sought far more seriously a practical means of reducing the destructively large rabbit population in Australia and New Zealand. Having observed the remarkable susceptibility of rabbits to fowl cholera, he proposed that their food be contaminated with the fowl cholera bacillus, in hopes of establishing an epizootic outbreak of the disease among them. In 1888, following a highly successful preliminary trial of this method on an estate in Rheims, he sent a team of his collaborators to Sydney, Australia, where they were to organize and launch the antirabbit campaign. Ultimately, however, the Australian government refused to authorize a full-scale field trial and Pasteur failed to win the prize. He ascribed this outcome chiefly to the irrational fear aroused in Australia by the word "cholera," even though fowl cholera has nothing in common with human cholera.[198] Subsequent attempts along similar lines, however, have made it clear that Pasteur underestimated the difficulty of establishing a progressive epizootic disease in any animal population. Whether they arise naturally or are produced artificially, epizootics and epidemics are limited in their spread by factors which Pasteur did not fully appreciate and which remain to some degree obscure.[199]

Pasteur and Chemical Theories of Immunity. Like some of Pasteur's contemporary critics, Dubos has characterized his work on vaccination as largely "empirical" rather than "scientific."[200] Compared with the time and energy he invested in the search for effective vaccines, his efforts to establish a theoretical basis for attenuation and immunity were rather casual and undeveloped. Nonetheless, his concepts of immunity are interesting, the more so because they underwent a dramatic shift as a result of his work on rabies. Throughout his work on fowl cholera, anthrax, and swine erysipelas, Pasteur linked immunity with the biological, and particularly the nutritional, requirements of the pathogenic organism. In the case of animals inherently immune to a given disease, he supposed either that their natural body temperature was inimical to the development of the appropriate microbe or that they lacked some substance(s) essential to its life and nutrition. In animals rendered immune by recovery from a prior attack or by preventive inoculations, he supposed that each invasion by a given microbe (even in the attenuated state) removed a portion or all of some essential nutritional element(s), thereby rendering subsequent cultivation difficult or impossible. In January 1880, during a discussion of his work on fowl cholera, Pasteur illustrated his conception by applying it to cases of long-lasting immunity. Such cases could be explained by supposing that elements as rare as cesium or rubidium were essential to the life of the appropriate microbe and present only in trace amounts in the tissues of the invaded animal. Under these circumstances the initial invasion of the microbe could exhaust the supply of the essential element(s), rendering the animal refractory for as long as it took it to recoup a sufficient supply of the rare substance(s).[201]

At some point during his work on rabies, however, Pasteur began to doubt the validity of this biological "exhaustion" theory in the case of immunity against rabies. By his own account, he converted to a chemical "toxin" theory for rabies early in 1884;[202] but he gave no public indication of his conversion until his memoir of 26 October 1885, where it paled into insignificance beside the drama of young Meister and Jupille. Moreover, his conversion remained tentative and undeveloped, with the details of the supporting experiments reserved for a later paper. For the time being, Pasteur merely asserted that the vaccinal properties of the desiccated rabbit marrows seemed to result not from a decrease in the intrinsic virulence of the rabies virus but from a progressive quantitative decrease in the amount of living virus contained in the marrows. Then, citing other evidence that microbes could produce substances toxic to themselves (evidence that he had minimized while holding the "exhaustion" theory), Pasteur suggested that the virus might be composed of two distinct substances, "the one living and capable of multiplying in the nervous system, the other not living but nonetheless capable in suitable proportion of arresting the development of the former."

In January 1887, in the first issue of the *Annales de l'Institut Pasteur*, Pasteur gave a somewhat fuller account of the considerations that had led him to adopt a chemical theory of immunity for rabies. He noted that in rabbits the same rabies virus could give either a prolonged incubation period or a minimum incubation of seven days, depending on the manner and hence the quantity in which it was injected. This finding upset Pasteur's earlier assumption that length of incubation depended only on the intrinsic virulence and not on the quantity of the virus. Even more remarkable, large quantities of a given vaccine generally seemed to produce immunity more readily than smaller quantities. If immunity depended only on the action of an attenuated, living virus capable of self-reproduction, then small quantities ought to work just as effectively as large ones. Finally, Pasteur cited cases in which vaccination rendered animals immediately refractory to rabies without their showing any prior symptoms of an attenuated form of the disease.

To Pasteur these results seemed explicable only on the assumption that the rabies virus (or microbe) produced a nonliving vaccinal substance inimical to its own development. If this were so, the result of any given injection would depend on the relative proportions of living virus and vaccinal substance at the time of injection. The situation was complicated by the fact that the quantity of vaccinal substance depended to some extent on the amount of living virus from which it was derived. If the quantity of living virus introduced was small, as in intracranial inoculation or animal bites, the quantity of associated vaccinal substance would also necessarily be small. In such cases the supply of vaccinal substance might be inadequate to prevent the multiplication of the virus, and rabies would appear. Such a conception helped to explain why intracranial inoculation invariably produced rabies and why the bites of rabid dogs never conferred immunity. On the other hand, if the preexisting supply of vaccinal substance were large compared with the amount of living virus, then it might prevent the virus from developing at all—as seemed to be the case in animals rendered refractory to rabies without showing any prior symptoms of an attenuated form of the disease. The progressive loss of virulence in desiccated rabbit marrows could be explained by supposing that the drying process destroyed the living virus more rapidly than it destroyed the nonliving vaccinal substance. On such grounds, Pasteur reported, he had sought marrows in which the rabies virus had been entirely destroyed but in which some vaccinal substance remained. Although his search for such nonliving vaccines had been inconclusive, he continued to hope they would be found, for their discovery would constitute "both a first-rate scientific fact and a priceless improvement on the present method of prophylaxis against rabies." On 20 August 1888 he announced encouraging results with injections of rabid spinal cord heated at 35° C. for forty-eight hours to kill the virus—results that led him to predict that a chemical vaccine against rabies would soon be found and utilized.[203]

In January 1888, Pasteur had thrown his unqualified support behind Roux and Chamberland's claim that they had found a soluble chemical vaccine against septicemia in guinea pigs. At the same time he described his own preliminary attempts to find a chemical vaccine against anthrax.[204] Toward this end he used anthrax blood heated at 45° C. for several days to kill the anthrax microbe—a technique strikingly similar to that which Toussaint had proposed in 1880 and which Pasteur had criticized severely. Thus, at the very end of his scientific life Pasteur proved willing to modify profoundly the biological point of view that underlay his most celebrated achievements in the study of fermentation, putrefaction, and disease. He did, however, retain the notion that a living virus or microbe was essential to the production of the chemical vaccine. And he did reveal a continuing sympathy for biological theories of immunity—most notably by drawing early and favorable attention to Élie Metchnikoff's "phagocytic" theory.[205] But in doing so he no longer insisted on the inviolability of his earlier views.

In short, the aging Pasteur demonstrated a remarkable flexibility of mind. Moreover, as Dubos suggests, he seemed to end by groping "towards the new continent where the chemical controls of disease and immunity were hidden."[206] How far he might have progressed toward chemical theories of disease and chemotherapy must remain unknown. But without fully acknowledging it—perhaps without fully recognizing it—he seemed to draw ever closer not only to his erstwhile opponents Liebig, Bernard, and Chaveau but also to his own roots in chemistry. Given just a little more time, he might have closed the circle.

Honored Life, Honored Death. The final decade of Pasteur's life brought him additional honors, of which the most tangible were the second national recompense of 1883 and the establishment of the Institut Pasteur in 1888. Perhaps the most cherished were his election on 8 December 1881 to the Académie Française; an official celebration in July 1883 at Dole, where a commemmorative plaque was placed on the house in which he was born; and, above all, the moving jubilee celebration in the grand amphitheater of the Sorbonne on 27 December 1892, depicted in the painting by Rixens.

So frail that he had to be led in on the arm of Sadi Carnot, president of the Third Republic, Pasteur found the huge amphitheater filled to overflowing with students from the French *lycées* and universities, with his former pupils and assistants, with delegations from all the major French scientific schools and societies, and with government officials, foreign ambassadors and dignitaries. Of the many speakers who honored his life and work, the surgeon Sir Joseph Lister was perhaps the most notable and certainly the best qualified to testify to the direct influence of Pasteur's work. Unable to deliver his own brief speech of appreciation, Pasteur delegated this task to his son. In it he counseled the young students to "live in the serene peace of laboratories and libraries" and spoke to the foreign delegates of his "invincible belief that Science and Peace will triumph over Ignorance and War, that nations will unite, not to destroy, but to build, and that the future will belong to those who have done most for suffering humanity."[207]

It was Pasteur's last public appearance but far from his last honor. By the time of his death, his name had been given to the *collège* in Arbois, to a village in Algeria, to a district in Canada, and to streets and schools throughout France and the world, not to mention the proliferating Pasteur institutes. On 5 October 1895 France honored Pasteur's passing with a state funeral at Nôtre Dame, complete with full military honors. Temporarily placed in one of the chapels at Nôtre Dame, his body was moved in January 1896 to the resplendent funeral crypt in the Institut Pasteur where it now reposes, and where his wife was interred in 1910.

NOTES

Pasteur's scientific papers have been cited only when it seemed that the information provided in the text (notably the dates of memoirs), combined with that given by the editor of Pasteur's *Oeuvres*, might be insufficient to guide the reader to the pertinent paper. Similarly, for more strictly biographical material, references have been provided only when that material seemed sufficiently unfamiliar or controversial to warrant documentation.

1. Pasteur, *Oeuvres*, I, 376.
2. See Dubos, *Louis Pasteur*, 359–362, 377–384.
3. Pasteur, *Correspondance*, I, 228.
4. Pasteur to John Tyndall, 8 Mar. 1888, Archives of the Royal Institution of Great Britain.
5. *Ibid.* Cf. Pasteur, *Correspondance*, IV, 229–231 and *passim*.
6. René Vallery-Radot, *The Life of Pasteur*, 245–246, 374–376. Cf. Pasteur, *Correspondance*, II, 552, 565–570, 573–574, 580–583; III, 350, 364, 373, 384.
7. See Pasteur, *Correspondance*, IV, 365.
8. See Cuny, *Louis Pasteur*, 15–18; and Dubos, *op. cit.*, 80–82. For a contemporary attempt to portray Pasteur as greedy and unscrupulous, see Auguste J. Lutaud, *M. Pasteur et la rage* (Paris, 1887), esp. 405–431. Fanatically opposed to Pasteur, Lutaud often distorted and misused the documents he adduced in support of his claims, but those documents are suggestive enough to deserve a more dispassionate reexamination.
9. See Pasteur, *Oeuvres*, VI, 489; and Pasteur, *Correspondance*, III, 173 ff.
10. For a valuable but rather uncritical survey of Pasteur's involvement in controversy, see Pasteur Vallery-Radot, *Pasteur inconnu*, 62–131.
11. See Roux, "The Medical Work of Pasteur," 384; Dubos, *op. cit.*, 80, 370; and Louis Chauvois, D. Wrotnowska, and E. Perrin, "L'optique de Pasteur," in *Revue d'optique théorique et instrumentale*, **45** (1966), 197–213, esp. 197 f.
12. See Pasteur, *Correspondance*, IV, 336, 341, and 357, n. 1.
13. On Pasteur's general political views, see Cuny, *Louis Pasteur*, 19–25; Pasteur Vallery-Radot, *op. cit.*, 175–220; and Pasteur, *Correspondance*, II, 355, 459, 461–463, 489, 523, 534, 593, 600, 611–630; III, 424, 436–437; IV, 268–270, 300.
14. See Pasteur, *Correspondance*, I, 228, 230.
15. On the origins of Pasteur's study of wine, see especially *ibid.*, II, 128–129; and Pasteur, *Oeuvres*, III, 481–482. On his relationship with the imperial house, see *Correspondance*, II, 62, 215–235, 245–246, 268, 286–287, 297, 345–346, 355, 385, 387–388, 407–408, 451, 459, 461–463, 471, 484–485, 489, 586, 627; and his correspondence with Col. Fave, Louis Napoleon's aide-de-camp, *ibid.*, 98–100, 110–111, 120–121, 125–126, 146–148, 160–161, 236–238. In an unpub. letter of 7 Aug. 1863, announced for sale in *The Month at Goodspeed's* (May 1965), 249–250, Pasteur refers to a request from the emperor that he "take care of the aged and their illnesses," an invitation that Pasteur thought "might be useful to me with public officials whose help I might have to ask."
16. Pasteur, *Correspondance*, II, 216–236.
17. See Pasteur Vallery-Radot, ed., *Pages illustres de Pasteur*, 8.

18. See Pasteur, *Correspondance*, II, 502–503, 511–514, 517–519.
19. *Ibid.*, 491–492.
20. See "Fabrication de la bière," in *Journal de pharmacie*, **17** (1873), 330–331.
21. See Pasteur, *Correspondance*, III, 115–116, 313–314, 335–346, 430–431; IV, 209, 213–214, 358–359. More generally on Pasteur as "patriote," see Pasteur Vallery-Radot, *Pasteur inconnu*, 175–192.
22. Pasteur, *Correspondance*, II, 612. More generally on his campaign and the election, see *ibid.*, 611–630; and Pasteur Vallery-Radot, *Pasteur inconnu*, 203–215.
23. See Ledoux, *Pasteur et la Franche-Comté*, 55 ff.
24. Pasteur, *Correspondance*, IV, 340.
25. On Pasteur's general philosophical and religious positions, see René Vallery-Radot, *op. cit.*, 242–245, 342–343; Dubos, *op. cit.*, 385–400; Pasteur Vallery-Radot, *Pasteur inconnu*, 221–238; André George, *Pasteur* (Paris, 1958); and Pasteur, *Oeuvres*, II, 328–346; VI, 55–58; VII, 326–339.
26. Pasteur, *Correspondance*, II, 213–214.
27. Pasteur, *Oeuvres*, VII, 326–339, quote on 326. For an English trans. of Pasteur's inaugural address, see Eli Moschcowitz, "Louis Pasteur's Credo of Science: His Address When He Was Inducted Into the French Academy," in *Bulletin of the History of Medicine*, **22** (1948), 451–466.
28. Pasteur, *Oeuvres*, I, 376. For the Sorbonne address of 1864, see *ibid.*, II, 328–346. More generally, see Farley and Geison, "Science, Politics and Spontaneous Generation . . ." (in press).
29. See Adrien Loir, "L'ombre de Pasteur," **15** (1938), 508; and Dubos, *op. cit.*, 79–80, 370.
30. See Duclaux, *Pasteur*, 147–148, 174–175; Roux, "The Medical Work of Pasteur," 384–387; Loir, *op. cit.*, esp. **14** (1937), 144–146, 659–664; and Dubos, *op. cit.*, 370.
31. See Victor Glachant, "Pasteur disciplinaire: Un incident à l'École normale supérieure (novembre 1864)," in *Revue universitaire*, **47** (1938), 97–104, quote on p. 97.
32. *Ibid.*; Pasteur Vallery-Radot, *Pasteur inconnu*, 36–58; and Pasteur, *Correspondance*, II, 136–142, 332–339.
33. See esp. Lutaud, *op. cit.*, 418–430.
34. Ledoux, *op. cit.*, 33.
35. Pasteur, *Oeuvres*, I, 370. Cf. *ibid.*, 323 ff.
36. Pasteur later wrote that he might never have discovered hemihedrism in the tartrates had not Delafosse given such "particular development and special attention" to hemihedrism in his lectures. See Pasteur, *Correspondance*, IV, 386. For other expressions of his indebtedness to Delafosse, see Pasteur, *Oeuvres*, I, 66, 322, 398.
37. See J. R. Partington, *A History of Chemistry*, IV (London, 1964), 751–752; and Aaron Ihde, *The Development of Modern Chemistry* (New York, 1964), 322–323.
38. In 1884 Marie wrote to her daughter: "Your father, always very busy, says little to me, sleeps little, gets up at dawn—in a word continues the life that I began with him 35 years ago today." Pasteur, *Correspondance*, III, 418.
39. See esp. Pasteur, *Oeuvres*, I, 203–241.
40. *Ibid.*, 275–279, 284–288.
41. *Ibid.*, 314–344, esp. 331.
42. *Ibid.*, 369–380, 391–394.
43. *Ibid.*, 327.
44. See esp. Dorian Huber, "Louis Pasteur and Molecular Dissymmetry: 1844–1857" (M.A. thesis, Johns Hopkins University, 1969).
45. Pasteur, *Oeuvres*, I, 334–336. But see Huber, *op. cit.*, esp. 40–58.
46. See Pasteur, *Oeuvres*, I, 345–350, 360–365, 369–386.
47. Pasteur, *Correspondance*, I, 325–326. Cf. Pasteur Vallery-Radot, ed., *Pages illustres de Pasteur*, 10–13.
48. This paradox is explored in Farley and Geison, *op. cit.*
49. See Pasteur, *Oeuvres*, I, 275; II, 3.

50. Duclaux, *Pasteur*, 69. Cf. Pasteur, *Oeuvres*, II, 85–86.
51. René Vallery-Radot, *op. cit.*, 84. Cf. Pasteur, *Oeuvres*, VII, 186.
52. Pasteur, *Correspondance*, II, 183.
53. Duclaux, *op. cit.*, 73.
54. See *Annals and Magazine of Natural History*, 3rd ser., **7** (1861), 343, n.
55. See Ferdinand Cohn, "Untersuchungen über Bacterien, II.," in *Beiträge zur Biologie der Pflanzen*, **1**, no. 3 (1875), 141–207, esp. 194–196. More generally on Cohn, see Gerald L. Geison, "Ferdinand Cohn," in *Dictionary of Scientific Biography*, III, 336–341.
56. Bulloch, *History of Bacteriology*, 232.
57. Pasteur, *Oeuvres*, II, 13–14, n. 3.
58. See John Farley, "The Spontaneous Generation Controversy (1700–1860): The Origin of Parasitic Worms," in *Journal of the History of Biology*, **5** (1972), 95–125.
59. See Duclaux, *op. cit.*, 107; and Pasteur, *Oeuvres*, II, 190.
60. See Farley and Geison, *op. cit.*
61. This report is reproduced in Pasteur, *Oeuvres*, II, 637–647.
62. Duclaux, *op. cit.*, 109–111.
63. Pasteur, *Correspondance*, II, 198.
64. *Ibid.*, 193–195.
65. Pasteur, *Oeuvres*, IV, 86.
66. Duclaux, *op. cit.*, 163.
67. Pasteur, *Oeuvres*, IV, 465.
68. See Duclaux, *op. cit.*, 147, 158–159, 165, quote on 147.
69. See esp. Pasteur, *Oeuvres*, IV, 100–134, 188–199, 500–510.
70. See Duclaux, *op. cit.*, 172–173.
71. See Pasteur, *Oeuvres*, IV, 544.
72. *Ibid.*, 564–571.
73. *Ibid.*, 590–595.
74. Dubos, *op. cit.*, 225.
75. Pasteur, *Oeuvres*, VI, 132.
76. *Ibid.*, IV, 729–732.
77. *Ibid.*, II, 361–366.
78. *Ibid.*, 379, 396.
79. *Ibid.*, V, 265, n. 1.
80. *Ibid.*, II, 403–404, n. 3. Cf. *ibid.*, 427–429.
81. *Ibid.*, 430–435, 443–444.
82. *Ibid.*, 387.
83. *Ibid.*, 440–442.
84. *Ibid.*, 374–380, quote on 375.
85. *Ibid.*, VI, 33.
86. *Ibid.*, 80–86.
87. Duclaux, *op. cit.*, 128–131.
88. *Ibid.*, 206–209.
89. See Robert Kohler, "The Background to Eduard Buchner's Discovery of Cell-Free Fermentation," in *Journal of the History of Biology*, **4** (1971), 35–61, on 39–40.
90. See Pasteur, *Oeuvres*, II, 353, 538, 588–593, esp. 535.
91. *Ibid.*, V, 5.
92. Duclaux, *op. cit.*, 188.
93. See Pasteur, *Oeuvres*, V, 18, n. 1.
94. *Ibid.*, 307.
95. See Pasteur, *Correspondance*, III, 71, 78–79. For facsimiles of these letters, as well as several others exchanged between Pasteur and Jacobsen (including two previously unpublished letters from Pasteur), see *Lettres échangées entre J. C. Jacobsen et Louis Pasteur au cours des années 1878–1882* (Copenhagen, 1964).
96. Pasteur, *Oeuvres*, II, 385–386.
97. *Ibid.*, 453–455; V, *passim*; and Duclaux, *op. cit.*, 214–218.
98. Pasteur, *Oeuvres*, II, 541–545, 559–567.
99. *Ibid.*, 150–158, 373, 383, 389 n. 1, 407–408, esp. n. 2; V, 98–101; and Duclaux, *op. cit.*, 192–197.
100. Pasteur, *Oeuvres*, V, 101; II, 411. Pasteur only once used Darwin's name in print—while pointing out that the belief in microbial transformism was losing ground by 1876, "in spite of the growing favor of Darwin's system." *Ibid.*, V, 79.
101. *Ibid.*, II, 461.

102. For Bastian's account of his disagreements with the commission, see Henry Charlton Bastian, "The Commission of the French Academy and the Pasteur-Bastian Experiments," in *Nature*, **16** (1877), 277–279. Cf. Pasteur, *Oeuvres*, II, 459–473. More generally, see Glenn Vandervliet, *Microbiology and the Spontaneous Generation Debate During the 1870's* (Lawrence, Kan., 1971), 55–64; and J. K. Crellin, "The Problem of Heat Resistance of Microorganisms in the British Spontaneous Generation Controversy of 1860–1880," in *Medical History*, **10** (1966), 50–59.

103. See Vandervliet, *op. cit., passim*; and J. K. Crellin, "Airborne Particles and the Germ Theory; 1860–1880," in *Annals of Science*, **22** (1966), 49–60.

104. Pasteur, *Oeuvres*, II, 478–481.

105. J. Burdon-Sanderson, "Bacteria," in *Nature*, **17** (1877), 84–87, on 84. More generally, see Crellin, "Airborne Particles and the Germ Theory."

106. Pasteur, *Oeuvres*, VI, 124.

107. See Duclaux, *op. cit.*, 119; Bulloch, *op. cit.*, 109; and Vandervliet, *op. cit.*, 63–64.

108. See Pasteur, *Oeuvres*, VI, 167, 188, 590.

109. See Pasteur, *Correspondance*, II, 577.

110. Pasteur, *Oeuvres*, VI, 161–163. Cf. *ibid.*, 469.

111. Pasteur, *Correspondance*, II, 350.

112. *Ibid.*, 346–351.

113. *Ibid.*, 551–552, 564–568, 570, 573–574, 580–584.

114. See Pasteur, *Oeuvres*, VI, 166–167, 188; II, 465.

115. *Ibid.*, VI, 167.

116. See Roux, *op. cit.*, 382. Pasteur's work made him the target of antivivisectionists and antivaccinationists, especially in England. For his contemptuous attitude toward these movements, see Pasteur, *Correspondance*, IV, 86, 109, 143, 193, 232, 294, 296–297.

117. Pasteur, *Oeuvres*, VI, 71–84, 138, 140, 157, 543.

118. *Ibid.*, 89–103.

119. *Ibid.*, 71.

120. See Duclaux, *op. cit.*, 237–241; and Bulloch, *op. cit.*, 179–182, 207, and *passim*.

121. Pasteur, *Oeuvres*, VI, 162–163.

122. *Ibid.*, 178; and Dubos, *op. cit.*, 309–310, 380–381.

123. Pasteur, *Oeuvres*, VI, 271–272.

124. *Ibid.*, 262.

125. Bulloch, *op. cit.*, 213, 236–238; and Dubos, *op. cit.*, 261.

126. Pasteur, *Oeuvres*, VI, 131–138.

127. *Ibid.*, 470.

128. *Ibid.*, 495.

129. Claiming, in private, that Toussaint had failed to do so. See *ibid.*, VII, 49.

130. *Ibid.*, VI, 132–133.

131. *Ibid.*, 471–489, esp. 489. Cf. Pasteur, *Correspondance*, III, 173–179.

132. But, more generally on luck and chance in Pasteur's work, see Dubos, *op. cit.*, 100–101, 219, 327, 340, 342.

133. Throughout that period all of Pasteur's correspondence is addressed from Arbois. See Pasteur, *Correspondance*, III, 98–115.

134. Duclaux, *op. cit.*, 281.

135. First used by Pasteur in his 1854 inaugural address at Lille. See Pasteur, *Oeuvres*, VII, 131. He used it again in 1871, in an address at Lyons, and in 1881, while discussing his famous Pouilly-le-Fort experiments on anthrax vaccination. See *ibid.*, 215; VI, 348.

136. See Pasteur, *Correspondance*, III, 121, 138–140.

137. Duclaux, *op. cit.*, 178.

138. Pasteur, *Correspondance*, III, 158–171, esp. 159, 166.

139. Pasteur, *Oeuvres*, VI, 256, n. 2.

140. *Ibid.*, 371.

141. See René Vallery-Radot, *op. cit.*, 313–315.

142. See Pasteur, *Oeuvres*, VI, 350, 710–711.

143. Charles Nicolle, *Biologie de l'invention* (Paris, 1932), 62–65. Cf. Roux, *op. cit.*, 379.

144. See Pasteur, *Correspondance*, III, 196–199.

145. Pasteur, *Oeuvres*, VI, 358, 360, 363–364.

146. See Pasteur, *Correspondance*, III, 271, n. 2.

147. Pasteur, *Oeuvres*, VI, 365–366.

148. *Ibid.*, 385.

149. *Ibid*, 441.

150. See Pasteur, *Correspondance*, IV, 262–263, 286–287.

151. Pasteur, *Oeuvres*, VI, 370.

152. *Ibid.*, 386–390.

153. *Ibid.*, 454.

154. *Ibid.*, 458, n. 1.

155. See *ibid.*, 115, 165, 174, 424.

156. Perhaps with some justice, since Chamberland reportedly added cultures of *Bacillus subtilis* to some tubes of anthrax vaccine. See Dubos, *op. cit.*, 341.

157. Pasteur, *Oeuvres*, VI, 403–411.

158. See *ibid.*, 446–447, 450.

159. *Ibid.*, 459.

160. See Duclaux, *op. cit.*, 293.

161. Pasteur, *Oeuvres*, VI, 367–368. Cf. 398–399, 570–571.

162. *Ibid.*, esp. 402, n. 1.

163. Cf. Dubos, *op. cit.*, 193–194.

164. Pasteur, *Oeuvres*, VI, 555.

165. See Bulloch, *op. cit.*, 226.

166. Pasteur, *Oeuvres*, VI, 332–338, esp. 335–336.

167. *Ibid.*, 525. More generally on Pasteur's work on this disease, see Frank and Wrotnowska, *Correspondence of Pasteur and Thuillier*, 53 ff.

168. See Pasteur, *Oeuvres*, VI, 523, 525.

169. See Bulloch, *op. cit.*, 246–247.

170. Ledoux, *op. cit.*, 16–17. Cf. Dubos, *op. cit.*, 332.

171. See Pasteur, *Oeuvres*, VI, 591.

172. See Frank and Wrotnowska, *op. cit.*, 137.

173. See Dubos, *op. cit.*, 264.

174. See Pasteur, *Correspondance*, III, 421, 425–428, 441–445; and René Vallery-Radot, *op. cit.*, 398, 406, 410–411.

175. Pasteur, *Correspondance*, III, 445–446.

176. *Ibid.*, IV, 14–15.

177. See Dubos, *op. cit.*, 333–334.

178. *Ibid.*, 336. But if Dubos's romantic version of Meister's suicide is true, it seems remarkable that his death should have been reported so briefly and casually in *Isis*, **37** (1947), 183, where no mention is made of suicide or the German army and where the year of his death is given as 1941 rather than 1940.

179. See Pasteur, *Oeuvres*, VI, 611–612.

180. See *Journal of the American Medical Association*, **81** (1923), 1445.

181. Pasteur, *Oeuvres*, VI, 621.

182. See Dubos, *op. cit.*, 335, 347.

183. *Ibid.*, 347.

184. Pasteur, *Oeuvres*, VI, 626–627.

185. *Ibid.*, 633–634.

186. See *ibid.*, 875–876.

187. See *ibid.*, 836–844, esp. 841–842.

188. Pasteur, *Correspondance*, IV, 75–76.

189. See Pasteur, *Oeuvres*, VI, 870–877; and *Black's Medical Dictionary*, 27th ed. (London, 1967), 743. Rabies did return to England, however, between 1918 and 1940, presumably because of violations of the quarantine regulations. See H. J. Parish, *A History of Immunization* (Edinburgh, 1965), 56–57.

190. Pasteur, *Oeuvres*, VI, 628–629.

191. Duclaux, *op. cit.*, 299.

192. Bulloch, *op. cit.*, 251.

193. See Parish, *op. cit.*, 58.

194. See *New York Times*, 21 Aug. 1973, 1:4.

195. Cf. Dubos, *op. cit.*, 350–353.

196. Duclaux, *op. cit.*, 143.

197. See Dubos, *op. cit.*, 310; and Pasteur, *Oeuvres*, VII, 32–35.

198. See Pasteur, *Oeuvres*, VII, 86–93; and Pasteur, *Correspondance*, IV, 227–229, 231–232, 237, 240–251, 257–260, 270–273, 278–280, 287–290.

199. Dubos, *op. cit.*, 312. Cf. Bulloch, *op. cit.*, 243.
200. Dubos, *op. cit.*, 379–382 and *passim*.
201. Pasteur, *Oeuvres*, VI, 290–291.
202. *Ibid.*, 463.
203. *Ibid.*, 550.
204. *Ibid.*, 464–466.
205. *Ibid.*, 645, n. 2.
206. Dubos, *op. cit.*, 357–358.
207. See René Vallery-Radot, *op. cit.*, 447–451. For the text of Pasteur's address, see Pasteur, *Oeuvres*, VII, 426–428.

BIBLIOGRAPHY

I. ORIGINAL WORKS. Virtually every word that Pasteur published during his lifetime, including all of his books, monographs, and scientific papers, has been reproduced in the monumental and magnificent *Oeuvres de Pasteur*, Pasteur Vallery-Radot, ed., 7 vols. (Paris, 1922–1939). This work also contains a number of letters, notes, and MSS that were not published during Pasteur's lifetime and a number of documents by others relating to his work, including several reports by commissions of the Académie des Sciences. Each volume has a brief introduction by Pasteur Vallery-Radot, who adds helpful editorial notes and comments throughout. The volumes are organized topically as follows: I, molecular asymmetry; II, fermentations and spontaneous generation; III, studies on vinegar and wine; IV, studies on the silkworm disease; V, studies on beer; VI, infectious diseases, virus vaccines, and rabies prophylaxis; VII, scientific and literary miscellania. Vol. VII also contains a complete index of names cited in all of the volumes, a complete chronological bibliography of Pasteur's publications, and a masterful "analytic and synthetic" subject index. In every way *Oeuvres de Pasteur* is a triumph of careful and diligent scholarship.

Of the approximately 500 communications published by Pasteur during his lifetime, many cover essentially similar ground, and the great majority are very brief notes or letters. Many others are nonscientific in content. The Royal Society *Catalogue of Scientific Papers* lists nearly 200 papers. Only the most important and influential of his published works can be listed here. His published books and monographs were *Études sur le vin. Ses maladies, causes qui les provoquent. Procédés nouveaux pour le conserver et pour le vieillir* (Paris, 1866; 2nd ed., rev. and enl., 1873); *Études sur le vinaigre, sa fabrication, ses maladies, moyens de les prévenir; nouvelles observations sur la conservation des vins par la chaleur* (Paris, 1868); *Études sur la maladie des vers à soie. Moyen pratique assuré de la combattre et d'en prévenir le retour*, 2 vols. (Paris, 1870); *Études sur la bière. Ses maladies, causes qui les provoquent, procédé pour la rendre inaltérable, avec une théorie nouvelle de la fermentation* (Paris, 1876), trans. by Frank Faulkner and D. Constable Robb as *Studies on Fermentation; the Diseases of Beer, Their Causes and the Means of Preventing Them* (London, 1879); and *Examen critique d'un écrit posthume de Claude Bernard sur la fermentation* (Paris, 1879).

Of Pasteur's papers on molecular asymmetry, the most important are "Recherches sur le dimorphism," in *Annales de chimie et de physique*, 3rd ser., 23 (1848), 267–294; "Recherches sur les relations qui peuvent exister entre la forme cristalline, la composition chimique et le sens de la polarisation rotatoire," *ibid.*, 24 (1848), 442–459; "Recherches sur les propriétés spécifiques des deux acides qui composent l'acide racémique," *ibid.*, 28 (1850), 56–99; "Nouvelles recherches sur les relations qui peuvent exister entre la forme cristalline, la composition chimique et le phénomène de la polarisation rotatoire," *ibid.*, 31 (1851), 67–102; "Mémoire sur les acides aspartique et malique," *ibid.*, 34 (1852), 30–64; "Nouvelles recherches sur les relations qui peuvent exister entre la forme cristalline, la composition chimique et le phénomène rotatoire moléculaire," *ibid.*, 38 (1853), 437–483; "Notice sur l'origine de l'acide racémique," in *Comptes rendus hebdomadaires . . . de l'Académie des sciences* (hereafter *Comptes rendus*), 36 (1853), 19–26; "Transformation des acides tartriques en acide racémique. Découverte de l'acide tartrique inactif. Nouvelle méthode de séparation de l'acide racémique en acides tartriques droit et gauche," *ibid.*, 37 (1853), 162–166; "Sur le dimorphisme dans les substances actives. Tetartoédrie," *ibid.*, 39 (1854), 20–26; "Mémoire sur l'alcool amylique," *ibid.*, 41 (1855), 296–300; "Isomorphisme entre les corps isomères, les uns actifs les autres inactifs sur la lumière polarisée," *ibid.*, 42 (1856), 1259–1264; "Études sur les modes d'accroissement des cristaux et sur les causes des variations de leurs formes secondaires," in *Annales de chimie et de physique*, 3rd ser., 49 (1857), 5–31; and "Note relative au *Penicillium glaucum* et à la dissymétrie moléculaire des produits organiques naturels," in *Comptes rendus*, 51 (1860), 298–299.

Pasteur's views on molecular asymmetry and optical activity, as they stood at the end of his active research on the problem, are admirably summarized in "Recherches sur la dissymétrie moléculaire des produits organiques naturels," in *Leçons de chimie professées en 1860* (Paris, 1861), 1–48, trans. by George Mann Richardson as "On the Asymmetry of Naturally Occurring Organic Compounds," in *The Foundations of Stereochemistry; Memoirs by Pasteur, van't Hoff, Lebel, and Wislicenus* (New York, 1901), 3–33. Pasteur's continuing interest in the relationship between asymmetry and life can be seen in ["Observations sur les forces dissymétriques"], in *Comptes rendus*, 78 (1874), 1515–1518; "Sur une distinction entre les produits organiques naturels et les produits organiques artificiels," *ibid.*, 81 (1875), 128–130; "La dissymétrie moléculaire," in *Revue scientifique*, 3rd ser., 7 (1884), 2–6; and "Réponses aux remarques de MM. Wyrouboff et Jungfleisch sur 'La dissymétrie moléculaire,'" in *Bulletin de la Société chimique de Paris*, n.s. (1884), 215–220. For a projected volume that would gather his earlier works on molecular asymmetry Pasteur wrote a preface, an introduction, and a historical note (1878); published posthumously by Pasteur Vallery-Radot in *Oeuvres de Pasteur*, I, 389–412, they serve as an excellent introduction to Pasteur's mature views on the subject.

Excluding his works on vinegar, wine, and beer (discussed below), the longest and most important of Pasteur's papers on fermentation and spontaneous generations is

"Mémoire sur les corpuscules organisés qui existent dans l'atmosphère, examen de la doctrine des générations spontanées," in *Annales des sciences naturelles*. Zoologie, 4th ser., **16** (1861), 5–98, significant portions of which are trans. into English in Conant, *Harvard Case Histories* (see below), 494–504, 509–516. Also important are "Mémoire sur la fermentation appelée lactique," in *Mémoires de la Société des sciences, de l'agriculture et des arts de Lille*, 2nd ser., **5** (1858), 13–26, trans. into English with the omission of three paragraphs in Conant, *Harvard Case Histories*, 453–460; "Mémoire sur la fermentation alcoolique," in *Annales de chimie et de physique*, 3rd ser., **58** (1860), 323–426; "De l'origine des ferments. Nouvelles expériences relatives aux générations dites spontanées," in *Comptes rendus*, **50** (1860), 849–854, of which an English trans. appeared in *Quarterly Journal of Microscopical Science*, **8** (1860), 255–259; "Recherches sur le mode de nutrition des Mucédinées," in *Comptes rendus*, **51** (1860), 709–712, of which an English trans. appeared in *Quarterly Journal of Microscopical Science*, 2nd ser., **1** (1861), 213–215; "Animalcules infusoires vivant sans gaz oxygène libre et déterminant des fermentations," in *Comptes rendus*, **52** (1861), 344–347, of which an English version appeared in *Annals and Magazine of Natural History*, 3rd ser., **7** (1861), 343–344; "Expériences et vues nouvelles sur la nature des fermentations," in *Comptes rendus*, **52** (1861), 1260–1264; and "Quelques faits nouveaux au sujet des levûres alcooliques," in *Bulletin de la Société chimique de Paris* (1862), 66–74.

Also see "Nouvel exemple de fermentation déterminée par des animalcules infusoires pouvant vivre sans gaz oxygène libre, et en dehors de tout contact avec l'air de l'atmosphère," in *Comptes rendus*, **56** (1863), 416–421, of which an English version appeared in *Annals and Magazine of Natural History*, 3rd ser., **11** (1863), 313–317; "Examen du rôle attribué au gaz oxygène atmosphérique dans la destruction des matières animales et végétales après la mort," in *Comptes rendus*, **56** (1863), 734–740, of which an English version appeared in *Chemical News and Journal of Physical (Industrial) Science*, **7** (1863), 280–282; "Recherches sur la putréfaction," in *Comptes rendus*, **56** (1863), 1189–1194; "Des générations spontanées," in *Revue des cours scientifiques*, **1** (1864), 257–265; "Note sur un mémoire de M. Liebig, relatif aux fermentations," in *Comptes rendus*, **73** (1871), 1419–1424; "Sur la nature et l'origine des ferments," *ibid.*, **74** (1872), 209–212; "Réponse à M. Frémy," *ibid.*, 403–404; "Faits nouveaux pour servir à la connaissance de la théorie des fermentations proprement dites," *ibid.*, **75** (1872), 784–790, of which an English trans. appeared in *Quarterly Journal of Microscopical Science*, 2nd ser., **13** (1873), 351–356; "Nouvelles expériences pour démontrer que le germe de la levure qui fait le vin provient de l'extérieur des grains de raisin," in *Comptes rendus*, **75** (1872), 781–782; "Note sur la fermentation des fruits et sur la diffusion des germes des levures alcooliques," *ibid.*, **83** (1876), 173–176; "Note sur l'altération de l'urine, à propos d'une communication du Dr Bastian, de Londres," *ibid.*, 176–180; "Sur l'altération de l'urine. Réponse à M. le

Dr Bastian," *ibid.*, 377–378; "Note sur l'altération de l'urine, à propos des communications récentes du Dr Bastian," *ibid.*, **84** (1877), 64–66, written with Jules François Joubert; "Réponse à M. le Dr Bastian," *ibid.*, 206; "Sur les germes des bactéries en suspension dans l'atmosphère et dans les eaux," *ibid.*, 206–209, written with J. F. Joubert; and "Note au sujet de l'expérience du Dr Bastian, relative à l'urine neutralisée par la potasse," *ibid.*, **85** (1887), 178–180. Also of interest are Pasteur's papers on Claude Bernard's posthumous MS on fermentation. These papers, published in 1878 and 1879, are reproduced in the appendix to Pasteur's *Examen critique d'un écrit posthume de Claude Bernard* (cited above).

The most important of Pasteur's papers on vinegar and acetic acid fermentation are "Sur la fermentation acétique," in *Bulletin de la Société chimique de Paris* (1861), 94–96, repr. in *Oeuvres*, III, 3–5; "Études sur les mycoderms. Rôle de ces plantes dans la fermentation acétique," in *Comptes rendus*, **54** (1862), 265–270; "Suite à une précédente communication sur les mycodermes. Nouveau procédé industriel de fabrication du vinaigre," *ibid.*, **55** (1862), 28–32; and "Mémoire sur la fermentation acétique," in *Annales scientifiques de l'École normale supérieure*, **1** (1864), 115–158. The last paper was combined with an otherwise unpublished lecture (delivered at Orléans in Nov. 1867) to yield Pasteur's *Études sur le vinaigre* (cited above).

Of Pasteur's papers on wine, the most important are "Études sur les vins. Première partie: De l'influence de l'oxygène de l'air dans la vinification," in *Comptes rendus*, **57** (1863), 936–942; "Études sur les vins. Deuxième partie: Des altérations spontanées ou maladies des vins, particulièrement dans le Jura," *ibid.*, **58** (1864), 142–150; "Procédé pratique de conservation et d'amélioration des vins," *ibid.*, **60** (1865), 899–901; "Note sur les dépôts qui se forment dans les vins," *ibid.*, 1109–1113; and "Nouvelles observations au sujet de la conservation des vins," *ibid.*, **61** (1865), 274–278. These papers form the basis of and were developed into Pasteur's *Études sur le vin* (cited above). Apart from his book *Études sur la bière* (cited above), Pasteur's only important publication on beer is "Études sur la bière; nouveau procédé de fabrication pour la rendre inaltérable," in *Comptes rendus*, **77** (1873), 1140–1148.

In *Études sur la maladie des vers à soie* (cited above), Pasteur reproduced all his important papers on silkworm disease written between 1865 and 1869 with the exception of "Sur la nature des corpuscules des vers à soie. Lettre à M. Dumas," in *Comptes rendus*, **64** (1867), 835–836. Of the post-1869 papers on silkworm disease, the most important are "Rapport adressé à l'Académie sur les résultats des éducations pratiques de vers à soie, effectuées au moyen de graines préparées par les procédés de sélection," *ibid.*, **71** (1870), 182–185; "Note sur l'application de la méthode de M. Pasteur pour vaincre la pébrine," in *Annales scientifiques de l'École normale supérieure*, 2nd ser., **1** (1872), 1–9, written with Jules Raulin; and "Note sur la flacherie," *ibid.*, 11–21, written with Jules Raulin.

The most important of Pasteur's communications on anthrax and septicemia are "Charbon et septicémie," in *Comptes rendus*, **85** (1877), 101–115, written with J. F. Joubert; ["Discussion sur l'étiologie du charbon"], in *Bulletin de l'Académie de médecine*, 2nd ser., **6** (1877), 921–926; ["Discussion sur l'étiologie du charbon. Poules rendues charbonneuses"], *ibid.*, **7** (1878), 253–255, 259–261; "Sur le charbon des poules," in *Comptes rendus*, **87** (1878), 47–48, written with J. F. Joubert and C. E. Chamberland; "Recherches sur l'étiologie et la prophylaxie de la maladie charbonneuse dans le département d'Eure-et-Loir," in *Recueil de médecine vétérinaire*, **56** (1879), 193–198; "Sur l'étiologie de l'affection charbonneuse," in *Bulletin de l'Académie de médecine*, 2nd ser., **8** (1879), 1063–1065, written with C. E. Chamberland and Émile Roux; "Étiologie du charbon [discussion]," *ibid.*, 1152–1157, 1159, 1183–1186, 1222–1234; and "Sur l'étiologie du charbon," in *Comptes rendus*, **91** (1880), 86–94, written with C. E. Chamberland and E. Roux, of which an English version appeared in *Chemical News*, **42** (1880), 225–227.

See also "Sur l'étiologie des affections charbonneuses," in *Comptes rendus*, **91** (1880), 455–457; "Sur la non-récidive de l'affection charbonneuse," *ibid.*, 531–538, written with C. E. Chamberland; "Sur la longue durée de la vie des germes charbonneux et sur leur conservation dans les terres cultivées," *ibid.*, **92** (1881), 209–211, written with C. E. Chamberland and E. Roux; "Note sur la constatation des germes du charbon dans les terres de la surface des fosses où on a enfoui des animaux charbonneux," in *Bulletin de l'Académie de médecine*, 2nd ser., **10** (1881), 308–311, written with C. E. Chamberland and E. Roux; "De la possibilité de rendre les moutons réfractaires au charbon par la méthode des inoculations préventives," in *Comptes rendus*, **92** (1881), 662–665, written with C. E. Chamberland and E. Roux; "Le vaccin du charbon," *ibid.*, 666–668, written with C. E. Chamberland and E. Roux; "Compte rendu sommaire des expériences faites à Pouilly-le-Fort, près Melun, sur la vaccination charbonneuse," *ibid.*, 1378–1383, written with C. E. Chamberland and E. Roux; "La vaccination charbonneuse. Réponse à un mémoire de M. Koch," in *Revue scientifique*, 3rd ser., **5** (1883), 74–84; "Sur la vaccination charbonneuse," in *Comptes rendus*, **96** (1883), 967–982; "Les doctrines dites microbiennes et la vaccination charbonneuse," in *Bulletin de l'Académie de médecine*, 2nd ser., **12** (1883), 509–514; and "La commission de l'École vétérinaire de Turin," in *Comptes rendus*, **96** (1883), 1457–1462.

Of Pasteur's communications on fowl cholera, the most important are "Sur les maladies virulentes, et en particulier sur la maladie appelée vulgairement choléra des poules," in *Comptes rendus*, **90** (1880), 239–248, of which an English version appeared in *Chemical News*, **42** (1880), 4–7; "Sur le choléra des poules; études des conditions de la non-récidive de la maladie et de quelques autres de ses caractères," in *Comptes rendus*, **90** (1880), 952–958, of which an English version appeared in *Chemical News*, **42** (1880), 321–322, and **43** (1881), 5–6; "Expériences tendant à démontrer que les poules vaccinées pour le choléra sont réfractaires au charbon," in *Comptes rendus*, **91** (1880), 315; and "De l'atténuation de virus du choléra des poules," *ibid.*, 673–680, of which an English version appeared in *Chemical News*, **43** (1881), 179–180.

Of Pasteur's papers on rabies, the most important are "Sur une maladie nouvelle provoquée par la saliva d'un enfant mort de la rage," in *Comptes rendus*, **92** (1881), 159–165, written with C. E. Chamberland and E. Roux; "Sur la rage," *ibid.*, 1259–1260, written with C. E. Chamberland, E. Roux, and Louis Ferdinand Thuillier; "Nouveaux faits pour servir à la connaissance de la rage," *ibid.*, **95** (1882), 1187–1192, written with C. E. Chamberland, E. Roux, and L. F. Thuillier; "Nouvelle communication sur la rage," *ibid.*, **98** (1884), 457–463, written with C. E. Chamberland and E. Roux; "Sur la rage," *ibid.*, 1229–1231, written with C. E. Chamberland and E. Roux; "Méthode pour prévenir la rage après morsure," *ibid.*, **101** (1885), 765–773, 774; "Nouvelle communication sur la rage," *ibid.*, **103** (1886), 777–784; and "Lettre sur la rage," in *Annales de l'Institut Pasteur*, **1** (1887), 1–18. These communications, as well as two others, are trans. into English in Jean R. Suzor, *Hydrophobia: An Account of M. Pasteur's System* (London, 1887). The two additional communications trans. by Suzor are "Résultats de l'application de la méthode pour prévenir la rage après morsure," in *Comptes rendus*, **102** (1886), 459–466, 468–469; and "Note complémentaire sur les résultats de l'application de la méthode de prophylaxie de la rage après morsure," *ibid.*, 835–838.

Pasteur also published a number of papers on infectious diseases other than anthrax, fowl cholera, and rabies. The most important of these are ["Discussion sur la peste en Orient"] in *Bulletin de l'Académie de médecine*, 2nd ser., **8** (1879), 176–182; "Septicémie puerpérale," *ibid.*, 256–260, 271–274, 488–493; "Commission dite de la peste," *ibid.*, **9** (1880), 386–390; "De l'extension de la théorie des germes à l'étiologie de quelques maladies communes [I. Sur les furoncles. II. Sur l'ostéomyélite. III. Sur la fièvre puerpérale]," in *Comptes rendus*, **90** (1880), 1033–1044; "Note sur la péripneumonie contagieuse des bêtes à cornes," in *Recueil de médecine vétérinaire*, **59** (1882), 1215–1223; and "La vaccination du rouget des porcs à l'aide du virus mortel atténué de cette maladie," in *Comptes rendus*, **97** (1883), 1163–1169, written with L. F. Thuillier.

For Pasteur's more general views on infectious diseases and vaccines, see especially "Observations verbales, à l'occasion du rapport de M. Gosselin," in *Comptes rendus*, **80** (1875), 87–95; "Discussion sur la fermentation," in *Bulletin de l'Académie de médecine*, 2nd ser., **4** (1875), 247–257, 265–282, 283, 284–290; "La théorie des germes et ses applications à la médecine et la chirurgie," in *Comptes rendus*, **86** (1878), 1037–1043, written with J. F. Joubert and C. E. Chamberland; "De l'atténuation des virus et de leur retour à la virulence," *ibid.*, **92** (1881), 429–435, written with C. E. Chamberland and E. Roux; "Vaccination in Relation to Chicken-Cholera and Splenic Fever," in *Transactions of the International Medical Congress, Seventh Session Held in London . . . 1881*,

4 vols. (London, 1881), I, 85–90; "Des virus-vaccins," in *Revue scientifique*, 3rd ser., **3** (1881), 225–228; "De l'atténuation des virus," *ibid.*, **4** (1882), 353–361, written with C. E. Chamberland, É. Roux, and L. F. Thuillier; and "Microbes pathogènes et vaccins," in *Semaine médicale*, **4** (1884), 318–320.

Of Pasteur's nonscientific writings the most interesting are "Lavoisier," in *Moniteur universel*, 4 Sept. 1865, 1198; "Claude Bernard. Idée de l'importance de ses travaux, de son enseignement et de sa méthode," *ibid.*, 7 Nov. 1866, 1284–1285; *Quelques réflexions sur la science en France* (Paris, 1871); *Discours de réception à l'Académie française* (Paris, 1882); *Réponse au discours de M. J. Bertrand à l'Académie française* (Paris, 1885); "Discours prononcé à l'inauguration de l'Institut Pasteur, le 14 novembre 1888," in *Inauguration de l'Institut Pasteur* (Paris, 1888), 26–30; and "Discours prononcé par Pasteur, le 27 décembre 1892, à l'occasion de son jubilé," in *Jubilé de M. Pasteur* (Paris, 1893), 24–26.

A major portion of Pasteur's vast correspondence was assembled and published in his *Correspondance*, Pasteur Vallery-Radot, ed., 4 vols. (Paris, 1940–1951). Arranged chronologically over the period 1840–1895, these letters provide a detailed account of Pasteur's activities and vividly illuminate every aspect of his life and career. Pasteur's own letters dominate the collection, but many letters to him and many by members of his family are included. For published versions of nearly 100 additional letters to or by Pasteur, as well as several other previously unpublished documents, see *Pages illustres de Pasteur*, Pasteur Vallery-Radot, ed. (Paris, 1968), 7–55; and *Correspondence of Pasteur and Thuillier Concerning Anthrax and Swine Fever Vaccination*, translated and edited by Robert M. Frank and Denise Wrotnowska with a preface by Pasteur Vallery-Radot (University, Ala., 1968).

Pasteur's grandson, Pasteur Vallery-Radot, spent his life seeking and collecting his grandfather's letters, MSS, and papers. In 1964 he gave most of his collection to the Bibliothèque Nationale. It comprises 7 file boxes of correspondence to Pasteur, 6 file boxes of correspondence by Pasteur, 1 file box of letters about him, and 22 packets containing MSS of his works and his laboratory and course notebooks. Although this material became generally accessible upon Pasteur Vallery-Radot's death in 1971, the collection is not yet classified for use. Apart from this material, some letters or MSS by and relating to Pasteur are deposited in the Reynolds collection at the University of Alabama at Birmingham, the Institut Pasteur in Paris, the Maucuer family in Paris, the Carlsberg Foundation in Copenhagen, the Bayerische Staatsbibliothek in Munich, the Laboratoire Arago in Banyuls-sur-Mer, the Royal Institution and the Wellcome Institute of the History of Medecine in London, the National Library of Medicine in Bethesda, Maryland, and the Burndy Library in Norwalk, Connecticut. Still other letters or MSS may be deposited in other libraries or may be privately owned. A number of official and administrative documents by and about Pasteur are deposited in French national and provincial archives. Many such documents have been extracted or otherwise put to use in the articles by Denise Wrotnowska (see below).

In addition to its small collection of Pasteur's personal letters and the resplendent funeral chapel where Pasteur and his wife are interred, the Institut Pasteur houses the Musée Pasteur, which includes the following: Pasteur's personal apartment, preserved as it was when he lived there; Pasteur's personal library, including annotated volumes of his communications to the Académie des Sciences; about 1,000 pieces of Pasteur's laboratory instruments and equipment, including chemical products with his labels, microscopes, wood models of crystals, flasks, and bottles; Pasteur's medals, diplomas, and other personal souvenirs; several of the portraits and pastel drawings he did as a youth (including the superb portraits of his parents); an iconography of about 5,000 photographs, drawings, and portraits of Pasteur, his disciples, and the Institut Pasteur; as yet uncataloged MS material on Émile Roux, Alexandre Yersin, Élie Metchnikoff, and Albert Calmette; documents concerning the Institut Pasteur; and a historical library. Pasteur museums also exist in Arbois, Dole, and Strasbourg.

II. SECONDARY LITERATURE. Perhaps no life in science has been so minutely described as Pasteur's, and rarely does a biographer or historian have access to such a wealth of carefully preserved primary material. Nonetheless, no fully adequate scientific biography of Pasteur exists, and the vast majority of the literature on him is derivative and essentially useless. Only the most important and valuable of the literature can be listed here.

There are three basic sources for Pasteur's life and work. The standard biography is René Vallery-Radot, *La vie de Pasteur* (Paris, 1900), trans. by Mrs. R. L. Devonshire as *The Life of Pasteur*, 2 vols. (London, 1901; 2nd, abr. ed., 1906)—references in the notes are to the 2nd ed. This biography, from which most of the literature on Pasteur derives, is distinguished for its extensive use of Pasteur's correspondence (including some still not published) and for its extraordinary detail. But it is without scholarly apparatus, occasionally obscure about dates, weak on historical background, and too exclusively concerned with Pasteur as an isolated genius at the expense of the more general context of his scientific work. Moreover, Vallery-Radot, who was Pasteur's son-in-law, is often openly hostile toward Pasteur's opponents and is so devoid of critical judgment as to approach hagiography. A second basic source is Émile Duclaux, *Pasteur: Histoire d'un esprit* (Paris, 1896), trans. by Erwin F. Smith and Florence Hedges as *Pasteur: The History of a Mind* (Philadelphia, 1920). Although also virtually devoid of scholarly apparatus, this book provides a lucidly brilliant and critical analysis of Pasteur's work by one of his most celebrated students. Illuminated by a prescient historiography, it remains one of the most impressive and perceptive books ever written on the development of a scientist's thought. The third basic source is René Dubos, *Louis Pasteur: Free Lance of Science* (Boston, 1950), trans. by Elisabeth Dussauze as *Louis Pasteur: Franc-tireur de la science*, with a preface by Robert Debré (Paris, 1955). Although sometimes almost embarrassingly dependent on Duclaux, and although virtually undocumented, this book is more sensitive to the larger context of Pasteur's work and surpasses Duclaux's by interweaving the evolution of Pasteur's scientific thought with his other activities and attitudes. Distinguished by a lucid and graceful style, it offers insights and perspectives unavailable to Duclaux so soon after Pasteur's death.

Of the remaining full-scale general accounts of Pasteur's life and work, the most valuable are Hilaire Cuny, *Louis Pasteur: L'homme et ses théories* (Paris, 1963), trans. by Patrick Evans as *Louis Pasteur: The Man and His Theories* (London, 1965); Jacques Nicolle, *Un maître de l'enquête scientifique, Louis Pasteur* (Paris, 1953), trans. as *Louis Pasteur; a Master of Scientific Enquiry* (London, 1961); and *Pasteur: sa vie, sa méthode, ses découvertes* (Paris, 1969); and Percy F. and Grace C. Frankland, *Pasteur* (New York, 1898). René Dubos, *Pasteur and Modern Science* (New York, 1960), is essentially an elegantly spare reworking of Dubos's earlier full-bodied biography. René Vallery-Radot foreshadowed his later full-scale biography in the anonymously published *Pasteur, histoire d'un savant par un ignorant* (Paris, 1883), which appeared twelve years before Pasteur's death. Also of interest is René Vallery-Radot, *Madame Pasteur* (Paris, 1941), a brief panegyric written in 1913 and eventually released for publication by Pasteur Vallery-Radot, whose own *Louis Pasteur: A Great Life in Brief*, trans. by Alfred Joseph (New York, 1958), is perhaps the best short biography of Pasteur.

Of the remaining books on Pasteur, several deserve mention on rather more specialized grounds. François Dagognet, *Méthodes et doctrines dans l'oeuvre de Pasteur* (Paris, 1967), is a highly suggestive, sometimes brilliant, but essentially ahistorical account of Pasteur's scientific work. Particularly valuable for their insights into the modern consequences of Pasteur's program are Henri Simonnet, *L'oeuvre de Louis Pasteur* (Paris, 1947); and Albert Delaunay, *Pasteur et la microbiologie* (Paris, 1967). For a valuable account of some of the less familiar and essentially nonscientific aspects of Pasteur's career, see Pasteur Vallery-Radot, *Pasteur inconnu* (Paris, 1954). Pasteur's early life receives detailed scrutiny in E. Ledoux, *Pasteur et la Franche-Comté; Dole, Arbois, Besançon* (Besançon, 1941), an appealing attempt to elucidate the influences on him of the land, climate, and demography of his native region. Louis Blaringhem, *Pasteur et le transformisme* (Paris, 1923), approaches Pasteur's work from an interesting perspective and seeks to trace to his work on fermentation, the genetic technique of "pure lines" and other intervening developments in biology. *The Pasteur Fermentation Centennial, 1857–1957* (New York, 1958) contains the contributions of Pasteur Vallery-Radot and René Dubos to a symposium held on the centennial of the publication of Pasteur's first memoir on lactic fermentation. Among the books that reprint extracts or selections from Pasteur's works are *Les plus belles pages de Pasteur*, Pasteur Vallery-Radot, ed. (Paris, 1943); *Pasteur: Pages choisies*, Ernest Kahane, ed. (Paris, 1957);

Louis Pasteur: Choix de textes, bibliographie, portraits, fac-similés, Hilaire Cuny, ed. (Paris, 1963); *Louis Pasteur; recueil de travaux*, Pasteur Vallery-Radot, ed. (Paris, 1966); and *Louis Pasteur: Extraits de ses oeuvres*, R. Dujarric de la Rivière, ed. (Paris, 1967).

Of the multitude of articles on Pasteur, the most generally valuable are those written by his students. Particularly informative with regard to Pasteur's personality and interaction with his assistants are Émile Roux, "L'oeuvre médicale de Pasteur," in *Agenda du chimiste* (Paris, 1896), trans. by Erwin F. Smith as "The Medical Work of Pasteur," in *Scientific Monthly*, **21** (1925), 365–389, to which version the notes refer; and Adrien Loir, "L'ombre de Pasteur," in *Mouvement sanitaire*, **14** (1937), 43–47, 84–93, 135–146, 188–192, 269–282, 328–348, 387–399, 438–445, 487–497, 572–573, 619–621, 659–664; **15** (1938), 179–181, 370–376, 503–508.

See also Émile Duclaux, "Le laboratoire de M. Pasteur à l'École normale," in *Revue scientifique*, 4th ser., **15** (Apr. 1895), 449–454; and "Le laboratoire de M. Pasteur," in *Le centenaire de l'École normale, 1795–1895* (Paris, 1895), 458 ff., and repro. in the centenary volume sponsored by the Institut Pasteur, *Pasteur, 1822–1922* (Paris, 1922), 39–54. Also repro. in the centenary volume are Roux's paper of 1896, "L'oeuvre médicale de Pasteur" (55–87); his "L'oeuvre agricole de Pasteur" (89–101), originally delivered to the Société Nationale d'Agriculture on 22 Mar. 1911; and his "Madame Pasteur" (102–104), a speech originally delivered on 28 Sept. 1910, when she was interred in the Pasteur crypt at the Institut Pasteur. See also Élie Metchnikoff, "Recollections of Pasteur," in *Ciba-Symposium*, **13** (1965), 108–111; and *The Founders of Modern Medicine: Pasteur, Koch, Lister* (New York, 1939). For a lengthy list of obituary notices on Pasteur, see the Royal Society *Catalogue of Scientific Papers*, XVII, 726–727.

Of the more narrowly focused literature on Pasteur (including that cited in full in the notes above), several works deserve special mention. Pasteur's religious position is explored in great detail in George (n. 25). On his handling of student discipline, see Glachant (n. 31). Various aspects of his career are explored in the articles of Denise Wrotnowska, among which the most significant are "Pasteur, professeur à Strasbourg (1849–1854)," in *92rd Congrés national des sociétés savantes*, I (Strasbourg–Colmar, 1967), 135–144; "Candidatures de Pasteur à l'Académie des sciences," in *Histoire de la médecine*, spec. no. (1958), 1–23; "Pasteur et Lacaze–Duthiers, professeur d'histoire naturelle à la Faculté des sciences de Lille," in *Histoire des sciences médicales* (1967), no. 1, 1–13; "Pasteur, précurseur des laboratoires auprès des musées," in *Bulletin du Laboratoire du Musée du Louvre* (1959), no. 4, 46–61; and "Recherches de Pasteur sur le rouget du porc," in *90th Congrès nationale des sociétés savantes*, III (Nice, 1965), 147–159. Pasteur's work on crystallography and molecular asymmetry is explored in admirable detail in Huber (n. 44). Seymour Mauskopf, *Crystals and Compounds* (forthcoming), examines the French crystallographic tradition from which Pasteur emerged, and offers a novel interpretation of his discovery of optical isomerism, emphasizing Laurent's influence and the issue of isomorphism. See also J. D. Bernal, "Molecular Asymmetry," in *Science and Industry in the Nineteenth Century* (London, 1953), 181–219; and Nils Roll-Hansen, "Louis Pasteur—a Case Against Reductionist Historiography," in *British Journal for the Philosophy of Science*, **23** (1972), 347–361.

For an English trans. of nearly all of Pasteur's first memoir on fermentation, together with a brief account of its genesis and impact, see James Bryant Conant, "Pasteur's Study of Fermentation," in *Harvard Case Histories in Experimental Science*, II (Cambridge, Mass., 1957), 437–485. On the relationship of Pasteur's work on fermentation to Buchner's discovery of zymase, see Kohler (n. 89). For an English trans. of significant portions of Pasteur's prize-winning memoir of 1861 on organized particles in the atmosphere, together with a more general discussion of the controversy over spontaneous generation, see Conant, "Pasteur's and Tyndall's Study of Spontaneous Generation," *op cit.*, 487–539. For an attempt to show that Pasteur's work on and public posture toward spontaneous generation were motivated in part by political factors, see John Farley and Gerald L. Geison, "Science, Politics and Spontaneous Generation in Nineteenth-Century France: The Pasteur-Pouchet Debate," in *Bulletin of the History of Medicine*, **48** (1974). The same debate is treated at length by Pouchet's disciple Georges Pennetier, *Un débat scientifique: Pouchet et Pasteur, 1858–1868* (Rouen, 1907). More generally on spontaneous generation, see Crellin (n. 102 and n. 103) and Vandervliet (n. 102). On the larger historical context of Pasteur's biological work, see William Bulloch, *The History of Bacteriology* (London, 1938); and William D. Foster, *A History of Medical Bacteriology and Immunology* (London, 1970).

GERALD L. GEISON

PASTOR, JULIO REY. See **Rey Pastor, Julio.**

PATRIZI, FRANCESCO (also **Patrizzi** or **Patricio**; Latin form, **Franciscus Patricius**) (*b.* Cherso, Istria, Italy, 25 April 1529; *d.* Rome, Italy, 7 February 1597), *mathematics, natural philosophy.*

Patrizi studied at Ingolstadt, at the University of Padua (1547–1554), and at Venice. While in the service of various noblemen in Rome and Venice he made several trips to the East, where he perfected his knowledge of Greek, and to Spain. He lived for a time at Modena and at Ferrara, before being appointed to a personal chair of Platonic philosophy at the University of Ferrara by Duke Alfonso II d'Este in 1578. He remained there until 1592, when Pope Clement VIII summoned him to a similar professorship in Rome, a post he held until his death.

Patrizi had interests in many different intellectual fields; he published works on poetry, history, rhetoric, literary criticism, metaphysics, ethics, natural philosophy, and mathematics, besides translating a number of Greek works into Latin. His thought is a characteristic blend of Platonism (in the widest sense in which the word is used when referring to the Renaissance) and natural philosophy, with a very strong anti-Aristotelian bent. The latter critical tendency is developed in his *Discussiones peripateticae* (Venice, 1571; much enlarged edition, Basel, 1581).

Patrizi's importance in the history of science rests primarily on his highly original views concerning the nature of space, which have striking similarities to those later developed by Henry More and Isaac Newton. His position was first set out in *De rerum natura libri II priores, alter de spacio physico, alter de spacio mathematico* (Ferrara, 1587) and was later revised and incorporated into his *Nova de universis philosophia* (Ferrara, 1591; reprinted Venice, 1593), which is his major systematic work. Rejecting the Aristotelian doctrines of *horror vacui* and of determinate "place," Patrizi argued that the physical existence of a void is possible and that space is a necessary precondition of all that exists in it. Space, for Patrizi, was "merely the simple capacity (*aptitudo*) for receiving bodies, and nothing else." It was no longer a category, as it was for Aristotle, but an indeterminate receptacle of infinite extent. His distinction between "mathematical" and "physical" space points the way toward later philosophical and scientific theories.

The primacy of space (*spazio*) in Patrizi's system is also seen in his *Della nuova geometria* (Ferrara, 1587), the essence of which was later incorporated into the *Nova de universis philosophia*. In it Patrizi attempted to found a system of geometry in which space was a fundamental, undefined concept that entered into the basic definitions (point, line, angle) of the system.

The full impact of Patrizi's works on later thought has yet to be evaluated.

BIBLIOGRAPHY

Lega Nazionale di Trieste, *Onoranze a Francesco Patrizi da Cherso: Catalogo della mostra bibliografica* (Trieste, 1957), presents the most complete listing of primary and secondary works to 1957. Other general works are B. Brickman, *An Introduction to Francesco Patrizi's Nova de universis philosophia* (New York, 1941); P. O. Kristeller, *Eight Philosophers of the Italian Renaissance* (Stanford, 1964), ch. 7; and G. Saitta, *Il pensiero italiano nell'umanesimo e nel Rinascimento*, 2nd ed. (Florence, 1961), II, ch. 9.

Works on Patrizi's concept of space are B. Brickman, "Francesco Patrizi on Physical Space," in *Journal of the History of Ideas*, **4** (1943), 224–245; E. Cassirer, *Das Erkenntnisproblem*, 3rd ed. (Berlin, 1922), I, 260–267; W. Gent, *Die Philosophie des Raumes und der Zeit*, 2nd ed. (Hildesheim, 1962), 81–83; and M. Jammer, *Concepts of Space* (Cambridge, Mass., 1954), 84–85.

CHARLES B. SCHMITT

PAUL OF AEGINA (*b.* Aegina; *fl.* Alexandria, A.D. 640), *medicine.*

The details of Paul of Aegina's life are meager. He was born on the island of Aegina in the Saronic Gulf and studied and practiced medicine at Alexandria, where he remained after the Arabic invasion of 640.

Paul's most important and only extant work is his seven-book medical encyclopedia, *Epitome medicae libri septem.* According to Islamic sources, he also wrote two other works, a volume on gynecology and one on toxicology. Muslim physicians considered him one of the most eminent of Greek medical authorities, and he is frequently quoted in their works. In the preface to his work, Paul indicated that he prepared his review of earlier Greek medical practices in order that physicians, regardless of where they found themselves, could have a brief synopsis of pertinent medical procedures. He did not claim to be original; and, indeed, he noted that he had added only a few practices of his own. His study was based primarily on Oribasius' seventy-volume medical encyclopedia. Through Oribasius, Paul acquired and transmitted many of the Galenic medical concepts. Although he used other sources, unlike Oribasius, he did not cite them.

Paul divided the *Epitome* into the following sections:

Book I. Hygiene and regimen
Book II. Fevers
Book III. Bodily afflictions arranged topically
Book IV. Cutaneous complaints and intestinal worms
Book V. Toxicology
Book VI. Surgery
Book VII. Properties of medicines.

In the first book Paul examined in some detail the general principles of hygiene. Beginning with an analysis of the problems of pregnant women, he proceeded to a review of the problems of hygiene in the successive ages of man. He was interested in the establishment of the proper regimen for every stage of human development. In the Galenic tradition he subscribed to the earlier Greek humoral pathology of the four elements with their respective qualities. Paul contended that through various forms of dietary, medical, and physical manipulations, a proper balance could be achieved in the body and man would thus enjoy good health. He provided instructions for the care of the eyes and teeth, the retention of hearing, and the problems of impotence. His attitudes toward the role of the temperaments is clearly based on Oribasius' interpretation of Galen's thoughts on this subject. Paul maintained that man is in his best temperament when he exists in a middle position between all extremes—leanness and obesity, softness and hardness, hot and cold, and wet and dry. Individuals have particular attributes as their bodies vary from the mean. Bodies with hot and dry temperaments differ substantially from those with cold and moist temperaments. Depending upon their constituency of humors, internal organs also have different temperaments. The numerous permutations of possible temperaments and humors both explain the diverse medical conditions of men and necessitate the numerous varieties of medicines and treatments. Since food is vital to sound health and to the balance of the humors, he presented a sustained discussion of dietary therapeutics with a description of numerous foods and their medicinal virtues.

In Book II, Paul analyzed the nature and manifestations of fevers as characteristics of particular diseases. He utilized the duration and degree of fever as one of the prognoses for the course of a disease. High fevers indicate an acute illness; low fevers, a chronic sickness. The pulse is another important prognostic tool, and he classified sixty-two varieties of pulse. He defined pulse

> . . . as a movement of the heart and arteries, taking place by a diastole and systole. Its object is two fold; for, by the diastole, which is, as it were, an unfolding and expansion of the artery, the cold air enters, ventilating and resuscitating the animal vigour, and hence the formation of vital spirits; and by the systole, which is, as it were, a falling down and contraction of the circumference of the artery towards the centre, the evacuation of the fuliginous superfluities is effected [Adams, *Seven Books*, I, 202].

Paul also utilized alvine discharges, urine, and sputa as indications of the body's conditions.

In Book III Paul surveyed ailments that affect the body. Beginning with afflictions of the hair, he proceeded through diseases of the head (eye, ear, nose, and throat) to mental problems and then to internal ailments (heart, stomach, kidney, liver, and uterus). He concluded with comments on corns, calluses, and nails. Paul's topical approach enabled him to critique the general medical complications of the body's organs and their respective treatments. He recommended bleeding for cephalalgia, hemicrania, phrenitis, erysipelas of the brain, and lethargy, and he encouraged diverse medicines and select bleedings for the control of epilepsy, melancholy, apoplexy, and nervous diseases. His review of the kidneys, liver, and spleen embodies the best traditions of classical medical thought. Kidney stones are formed by thick earthy humors that are heated by the body. Baths and compound medicines are methods of expelling these stones. Diseases and afflictions of the uterus, and complicated labors are examined thoroughly in the final passages of this book. Paul maintained that when the fetus is in a preternatural position it should be restored to its natural position.

> . . . sometimes drawing it down, sometimes pressing it back, sometimes rectifying the whole. If a hand or foot protrude we must seize upon the limb and drag it down, for thereby it will be more wedged in, or may be dislocated or fractured; but fixing the fingers about the shoulders or the hip joint of the foetus, the part that had protruded is to be restored to its proper position. If there be a wrong position of the whole foetus, attended with impaction, we must first push it upwards from the mouth of the womb, then lay hold of it, and direct it properly to the mouth of the uterus [Adams, *Seven Books*, I, 648].

Paul did not describe podalic version, and Islamic surgeons followed his example and consequently failed to include this in their medical procedures. His comments on complicated labors were closely studied by Muslim medical thinkers.

Cutaneous afflictions and their treatments are outlined in Book IV. Some diseases, such as elephantiasis, leprosy, and cancer, could not be healed because it was impossible to find medicines that were stronger than the ailments; but it was possible in certain cases to control the progress of the disease. Paul's description of cancer is abridged from Galen. According to Galenic theory, cancers are formed by the overheating of black bile. Because of the thickness of the humor that precipitated cancer, it was incurable. His description of the three types of intestinal worms (round, broad, and ascarids) is rather curious. The round worms were generated in the small intestinal membrane from bilious humors; the broad worm was converted from the intestinal membrane into a living animal; and ascarids, formed by bad diet, arose in the region near the rectum. All of these worms were to be treated with bitter astringents.

Toxicology was of interest to classical medical authorities; and in Book V Paul summarized the principal comments of ancient authors upon this theme. Information is provided for the treatment of bites or stings of vipers, mad dogs, spiders, scorpions, and crocodiles. This section terminates with a series of antidotes for

henbane, fleawort, hemlock, wolfsbane, smilax, gypsum, arsenic, and lead.

Paul's most important and original contributions are in Book VI, on surgery. He divided this book into a section that examines manual operations on the flesh and into passages that review treatment of fractures and dislocations. The work contains one of the most detailed descriptions of ophthalmic surgery in antiquity and describes procedures for the removal of cataracts, and operations for trichiasis, ectropion, cysts, symblepharon, and staphyloma. Surgical techniques for tracheotomies, tonsilectomies, nasal polyps, abdominal paracentesis, catheterization, hemorrhoidectomies, and lithotomies are outlined. Since bleeding was an important aspect of his medical procedures, he spared few details in his descriptions of venesection, cupping, cauterization, and ligation for bleeding vessels. In Book III Paul had sought to alleviate the problems of difficult labor with drugs and repositioning of the fetus, but in Book VI he offered surgical techniques for cases in which the fetus must be removed to save the mother's life. This book concludes with a survey of useful methods for the treatment of fractures and dislocations.

His concluding book is a summary of simple and compound medicines used in the practice of the healing art. The majority of this information was derived from the Dioscoridian tradition, for Paul utilized ninety minerals, 600 plants, and 168 animals from Dioscorides' *De materia medica*.

Paul's *Epitome* provided Islamic physicians with their most substantial account of Greek surgical procedures. Al-Zahrāwī and al-Razi used it extensively in their works, and Fabrici based much of his surgery on the techniques detailed in Paul's sixth book. The *Epitome* also transmitted the whole range of classical Greek medical thought to the Islamic world.

BIBLIOGRAPHY

I. Original Works. The *editio princeps* of Paul's medical encyclopedia was the Aldine ed. (Venice, 1528). Francis Adams' very satisfactory English trans. of Paul's work, prepared for the Sydenham Society, *The Seven Books of Paulus Aegineta*, 3 vols. (London, 1844–1847), contains an excellent commentary on Paul's relationship with Greek and Arabic medical traditions. René Briau prepared a Greek ed. and French trans. of Paul's Book VI, on surgery, *La chirurgie de Paul d'Égine* (Paris, 1855); there is a German trans. by J. Berendes (Leiden, 1914).

II. Secondary Literature. For discussions of Paul's contributions and thought see E. Gurlt, "Paulus von Aegina," in *Geschichte der Chirurgie*, I (Berlin, 1898), 558–590; Signorelli Remo, "Ostetricia e ginecologia nel bizantino Paolo d'Egina e nell' arabo Albucasi," in *Minerva medica*, **58** (24 Nov. 1967), 4118–4131; and Konrad Straubel, "Zahn- und Mundleiden und deren Behandlung bei Paulos von Aigina" (diss., University of Leipzig, 1922).

PHILLIP DRENNON THOMAS

PAUL OF ALEXANDRIA (*fl.* Alexandria, *ca.* A.D. 378), *astrology.*

Paul composed an elementary textbook, *Εἰσαγωγικά*, which was designed to instruct students in the fundamental concepts of astrology. The second edition of this brief text, addressed to his son Cronamon, is extant; in chapter twenty Paul gives as an example for the determination of the weekday the computation for "today, 20 Mecheir 94 Diocletian," or 14 February A.D. 378. No further biographical details are known.

Paul names as his sources Ptolemy, Apollinarius, Apollonius of Laodicea, the *Panaretus* (of Hermes Trismegistus), the wise men of the Egyptians, and Hermes Trismegistus himself. In addition, relations of his text to a number of other astrological texts—for example, those of Firmicus Maternus and Rhetorius —can be discerned. Astronomically, Paul was not incompetent but never became profound. He discussed the planets' heliacal risings and settings (ch. 14) and their stationary points (ch. 15; he referred the reader desiring accurate computations to Ptolemy's *Handy Tables*); and he treated the moon's phases (ch. 16), the sun's longitude for any day (ch. 28), and the establishment of the ascendant (ch. 29) and the midheaven (ch. 30).

Paul's work became reasonably popular. It was used as the basis for a course of lectures delivered at Alexandria between May and July A.D. 564—probably by Olympiodorus (*Heliodori, ut dicitur, in Paulum Alexandrinum Commentarium*, E. Boer, ed. [Leipzig, 1962]; compare L. G. Westerink, "Ein astrologisches Kolleg aus dem Jahre 564," in *Byzantinische Zeitschrift*, **64** [1971], 6–21). Chapters 1 and 2 (p. 1, line 1– p. 10, line 8 in E. Boer's edition) were translated into Armenian by Ananias of Shirak in the seventh century (A. G. Abrahamyan, ed., item 21 of Ananias' collected works [Yerevan, 1944], pp. 327–330. I owe this reference to Prof. R. C. Thompson of Harvard). A summary of Paul's work was included in an important Byzantine treatise on astrological authorities (pp. xxi– xxiv in E. Boer's ed.), and the text was illuminated by numerous scholia (pp. 102–134, in E. Boer's ed.), at least some of which are of the twelfth century (O. Neugebauer, in E. Boer's ed., pp. 136–137).

Modern interest in Paul has largely centered on two problems. Al-Bīrūnī alleged that the Indian astronomer Pauliśa (or Puliśa) was a Greek, Paulus of Alexandria. Although al-Bīrūnī later corrected his error, many more recent scholars have continued to repeat it. The reasons for the rejection of the identification will be found in O. Neugebauer and D. Pingree, *The Pañcasiddhāntikā of Varāhamihira*, I (Copenhagen, 1970), 12–13 (Pauliśa's peculiar Greco-Babylonian astronomy is summarized by Varāhamihira in his *Pañcasiddhāntikā*, I, 11–13; III; VI–VII; and XVII, 65–80 [?]); and in D. Pingree, "The Later Pauliśasiddhānta," in *Centaurus*, **14** (1969), 172–241, where it is shown that al-Bīrūnī's *Pauliśasiddhānta* was written at Sthāneśvara in the eighth century, and that it follows the *ārdharātrikapakṣa* that was founded by Āryabhaṭa.

Several scholars have contended that there is a relation of direct dependence between a geographical list in Acts of the Apostles and the astrological geography in Paul's *Εἰσαγωγικά*. This relation has been disproved by B. M. Metzger, "Ancient Astrological Geography and Acts 2: 9–11," in W. W. Gasque and R. P. Martin, eds., *Apostolic History and the Gospel* (Exeter, 1970), 123–133.

BIBLIOGRAPHY

The standard ed. of Paul is E. Boer, *Pauli Alexandrini Elementa apotelesmatica* (Leipzig, 1958). The articles on Paul by W. Gundel, in Pauly-Wissowa, *Real-Encyclopädie der classischen Altertumswissenschaft*, XVIII, pt. 2, cols. 2376–2386; and W. Gundel and H. G. Gundel, *Astrologumena* (Wiesbaden, 1966), 236–239, are no longer of much value.

DAVID PINGREE

PAUL OF VENICE (*b.* Udine, Italy, *ca.* 1370; *d.* Padua, Italy, 15 June 1429), *natural philosophy, logic.*

Christened Paolo Nicoletti da Udine, Paul of Venice was a highly respected scholar and leader of the Hermits of St. Augustine. He was the son of Nicoletto di Venezia, a noble citizen of Udine, and his wife Elena. Paul received his early religious and literary training at the monastery of St. Stephen in Venice. In 1390 the Augustinian order sent him for university training to Oxford, where he studied both natural philosophy and terminist logic and seems to have been influenced by Ockhamism. But his sympathies seem to have lain primarily with the Averroists, although he also adopted some doctrines of earlier Augustinians, especially Gregory of Rimini. After a fairly brief period at Oxford, Paul apparently studied at Paris, where he likely knew and studied with Pierre d'Ailly, a leading nominalist of the period.

Paul returned to Italy about 1395, and although little is known of his activities for nearly twelve years after that date, he must have been occupied with preaching and lecturing, since by 1408 he had already acquired a considerable reputation. In that year he was listed among the masters at Padua.

In 1413 Paul served briefly as Venetian ambassador to the king of Poland, and during the next two years, he lectured at Siena, Bologna, and Paris. In 1415 he was summoned before the Venetian Council of Ten, apparently on a charge of having interrupted his lectures at Padua in order to lecture elsewhere. He was ordered not to travel outside Venice for a year. In 1416 he was allowed to leave Venice, on the condition that he not attend the Council of Constance. He returned to Padua, where he remained for three years.

Most of Paul's work was written between 1409 and 1417; and because of his growing reputation as a philosopher, in 1417 the friars of his convent received the rare honor of being entitled to wear the black beret reserved for patricians of Venice. In 1420, when he was elected prior provincial of Siena and of the province of Marche Tarvisine, he was at the height of his fame. He was accorded such honorific titles as *monarcha sapientiae, summus Italiae philosophus,* and *Aristotelis genius.*

Paul's fame apparently carried with it a certain immunity. In 1420 a dispute with a Friar Francesco Porcerio led to a trial for heresy, and in that same year he was again summoned before the Council of Ten and exiled to Ravenna, where he was ordered to remain for at least five years. Neither of these difficulties seems to have affected his fortunes, and he simply ignored the sentence of the Council. In 1421 he was reelected prior provincial of Marche Tarvisine, and in 1422 he became regent at the Siena convent. He was deputed to lecture at Bologna in May 1424, moved to Perugia in November of that year, and was granted a faculty to visit Rome in 1426. In 1427 he was a professor at Siena, and he was rector of the university during 1428.

On 16 June 1428 Paul's petition to return to Padua was granted, and a year later he died and was buried there. The cause of his death is unknown.

Although Paul was widely known as a prominent rationalist with Averroist tendencies, and although his work in natural philosophy was widely read, his real importance seems to have been primarily in the field of logic. In the late fifteenth and early sixteenth centuries, his *Logica* was inscribed in the list of required texts at Venice, Padua, and Ferrara. And his logic remained widely read in many parts of Italy even until near the end of the seventeenth century, when it was still used as a text in Jesuit schools.

Paul's four logical works—although not markedly original—probably constitute the most thorough and encyclopedic exposition of the so-called terminist logic written during the Middle Ages. He seems to have read and thoroughly digested the most important logical work since Peter of Spain, and one finds in his writings, presented with admirable clarity, order, and understanding, almost all of the important concepts, problems, and proposed solutions of problems of terminist logic. His *Logica* is undoubtedly his most important and enduring contribution.

Paul's work in natural philosophy, on the other hand, is much less impressive in every respect. Although he wrote extensively in both natural philosophy and geology, his work seems to be wholly derivative and eclectic in a not very discriminating fashion.

In approaching Paul's work in natural philosophy, Duhem has shown the importance of distinguishing clearly between the early *Expositio super octo phisicorum* (completed in 1409) and the later *Summa naturalium.*[1] In the *Expositio,* Paul revealed himself as an orthodox Averroist not only on the question of the unicity of the agent intellect, but also on other issues. By the time of the *Summa,* he had moved away from Averroës in important respects and located his primary influences among the Parisian natural philosophers of the fourteenth century. In part this move away from Averroës can probably be explained as an attempt to maintain the orthodox position on God's omnipotence, as expressed in the 1277 condemnations. Thus when in the *Summa* Paul admitted the logical possibility of an actually infinite magnitude, while still maintaining that such a magnitude cannot occur in nature, he did not merely abandon Averroës to follow Albert of Saxony, but he also affirmed an accepted condition of divine omnipotence.[2] The same must be said of his admission that God could move the entire universe.[3] Paul often attempted unsuccessfully to reconcile Averroistic positions with the assertion of divine omnipotence. For example, he continued to uphold a notion of absolute place (he calls it *locus situalis*) as a relation of objects to the center of the universe, even though he had abandoned the notion of an immobile earth as the center of the universe and treated the center merely as a geometrical point.[4]

At least some of the shifts in Paul's positions should be treated as genuine changes of opinion, not merely as efforts at orthodoxy. A good example is the change in his view of projectile motion. In the *Expositio,* he takes the view that a projectile, after losing contact with the projecting instrument, is carried by successive waves of air.[5] For this view Paul found support not only in Aristotle and Averroës, but also in Walter Burley, a realist and terminist who remained with the standard Peripatetic position. In the *Summa* he had come to accept an account of projectile motion in terms of an "impetus" imparted to the object by the projecting instrument, a theory most importantly linked with Jean Buridan.[6] While this does seem to reflect a genuine change of opinion, it is also a good example of Paul's eclecticism. Although he supported his new view in language reminiscent of Buridan and Albert of Saxony, and although he followed Buridan's and Albert's arguments in extending the theory of the impetus to account for the acceleration of freely falling bodies,[7] the version of the theory that he accepted is not Buridan's, but a version usually associated with the Scotist Francis of Marchia. For Buridan the impetus transferred to the projectile would keep it moving indefinitely, were it not for the resistance of the air. Paul followed Francis of Marchia in explaining the tendency of the projectile to lose velocity by the view that the impetus, since it is not natural to the projectile but is impressed on it by violence, is gradually lost as the motion continues.

As Duhem has shown, Paul's work is repeatedly marred by elementary confusions. Thus in the *Expositio*, after attributing projectile motion to the push of air, he wrote about how much further a projectile would move in a void.[8] In attempting to defend Aristotle against Ockham's view that motion is identical with the thing moved, he used an argument based on the possibility of God's removing all form from prime matter and then moving the prime matter, hardly a defense that Aristotle would have appreciated.[9]

When we turn to Paul's geological theories in his *De compositione mundi*, the judgment must be much the same. Despite the fact that the intervening century and a half had witnessed both important theoretical advances and a number of significant discoveries and empirical observations, Paul's work is heavily dependent on the *Composizione del mondo* of Ristoro (written in 1282). Duhem goes so far as to characterize Paul as no more than a plagiarist of Ristoro.[10] Thus he not merely copied Ristoro's accounts of the origin of mountains and rivers and of the Mediterranean Sea, failing to take account of the discoveries of Marco Polo and others, but he even failed to include some of Ristoro's most interesting observations, such as the presence of fossils high on mountains.[11] Furthermore, Paul continued to rely heavily on astrological arguments and failed to take account of the strong antiastrological arguments of earlier philosophers such as Oresme.

Although Paul undoubtedly aided in the dissemi-

nation of Parisian natural philosophy in Italy, he should probably not be accounted an important figure in medieval science.

NOTES

1. Duhem, *Études sur Léonard de Vinci*, vol. III, p. 104.
2. Paul of Venice, *Summa naturalium Aristotelis*, pt. II (*De caelo et mundo*), 7.
3. *Ibid.*, pt. VI (*Metaphysica*), sec. 37.
4. Paul of Venice, *Expositio super octo phisicorum libros Aristotelis*, book IV, tract I, ch. 3, pt. 2, note 6.
5. *Ibid.*, pt. 1.
6. Paul of Venice, *Summa naturalium Aristotelis*, pt. II (*De caelo et mundo*), sec. 22.
7. *Ibid.*, pt. I (*Physica*), sec. 32.
8. Paul of Venice, *Expositio super octo phisicorum libros Aristotelis*, book VII, tract II, ch. 2, pt. 1.
9. *Ibid.*, book III, tract I, ch. 3, *dubium secundum*.
10. P. Duhem, *Le système du monde*, vol. IV, pp. 199–210, esp. pp. 209–210, where Duhem compares a number of passages from Paul and Ristoro. See also L. Thorndike, *Science and Thought in the Fifteenth Century*, pp. 195–232.
11. Paul of Venice, *De compositione mundi*, esp. chs. 18–27.

BIBLIOGRAPHY

I. ORIGINAL WORKS. Paul's most important works on logic and natural philosophy are *Logica* (Bologna, [?], 1472; Venice, 1475, 1478, 1480, 1485, 1488, 1492, 1493, 1498, 1565; Milan, 1474, 1478, 1484); *Expositio super libros de generatione et corruptione* (Perugia, 1475 [?]; Venice, 1498); *Summa naturalium Aristotelis* (Venice, 1476, 1503; Milan, 1476; Paris, 1514, 1521); *Expositio in libros Posteriorum Aristotelis* (Venice, 1477, 1481, 1486, 1491, 1494, 1518); *Quadratura* (Pavia, 1483; Venice, 1493; Paris, 1513); *Sophismata* (Pavia, 1483; Venice, 1493; Paris, 1514); *Universalia predicamenta sexque principia* (Venice, 1494); *De compositione mundi* (Venice, 1498); *Expositio super octo phisicorum libros Aristotelis* (Venice, 1499); *Logica magna* (Venice, 1499); and *In libros de anima* (Venice, 1504).

II. SECONDARY LITERATURE. On Paul and his work, see I. M. Bochenski, *History of Formal Logic* (Notre Dame, Indiana, 1961); P. Duhem, *Études sur Léonard de Vinci*, II (Paris, 1955), 319–327, and index in vol. III; P. Duhem, *Le système du monde*, IV (Paris, 1954), 199–210, and vol. X (Paris, 1959), 377–439; A. B. Emden, *A Biographical Register of Oxford University*, III (Oxford, 1959), 1944–1945, which contains a bibliography; A. Maier, *An der Grenze von Scholastik und Naturwissenschaft* (Essen, 1943), p. 207; A. Maier, *Zwei Grundprobleme der scholastischen Naturphilosophie*, 2nd ed. (Rome, 1951), 273–274; F. Momigliano, *Paolo Veneto e le correnti de pensiero religioso e filosofico nel tempo suo* (Udine, 1907); D. A. Perini, *Bibliographia Augustiniana*, III (Florence, 1929–1938), 29–46, contains biographical note and bibliography; and L. Thorndike, *Science and Thought in the Fifteenth Century* (New York, 1929), 195–232.

T. K. SCOTT, Jr.

PAULI, SIMON. See **Paulli, Simon.**

PAULI, WOLFGANG (*b.* Vienna, Austria, 25 April 1900; *d.* Zurich, Switzerland, 14 December 1958), *physics.*

Wolfgang Pauli's father, a distinguished and original scholar, was professor of colloid chemistry at the University of Vienna and was also named Wolfgang. Thus his son, in his early work, called himself Wolfgang Pauli, Jr. The child was baptized a Catholic, his godfather being Ernst Mach, the physicist and critical philosopher. Pauli went to school in Vienna. Toward the end of his high school studies he became acquainted with Einstein's general theory of relativity, which at that time was completely new. He read it secretly during dull classroom hours. He was truly proficient in higher mathematics, for he had previously studied Jordan's *Cours d'analyse* in the same manner. Einstein's papers had made a deep impression on him. It was, he said, as if scales had fallen from his eyes; one day, so it appeared to him, he suddenly understood the general theory of relativity.

After finishing high school Pauli decided to study theoretical physics. He went to Arnold Sommerfeld in Munich, who was then the most imposing teacher of theoretical physics, in Germany or elsewhere. Many outstanding theoreticians were his pupils, including Heisenberg and Bethe. Here Pauli further perfected his analytical skills, which he later again and again masterfully put to use. Felix Klein was then publishing the *Encyklopädie der mathematischen Wissenschaften*, a monumental compilation that was to examine the current state of science from all sides. Leading scholars—mathematicians and physicists—were contributors. Klein had requested Sommerfeld to write an article on relativity theory for the *Encyklopädie*. Sommerfeld ventured to entrust the task to Pauli, who although scarcely twenty years old had published several papers on the subject. (Sommerfeld revealed admirable courage and insight in letting a student in his fourth semester write this important article.)

Pauli soon completed a monograph of about 250 pages, which critically presented the mathematical foundations of the theory as well as its physical significance. He took thorough account of the already very considerable literature on the subject but at the same time clearly put forth his own interpretation. Despite the necessary brevity of discussion, the monograph is a superior introduction to the special and general theories of relativity; it is in addition a first-rate historical document of science, since, together with H. Weyl's *Raum, Zeit, Materie* ("Space, Time, and Matter"), it is the first comprehensive presentation of the mathematical and physical ideas of Einstein, who himself never wrote a large work about his theory.

Sommerfeld was elated by this performance and wrote to Einstein that Pauli's article was "simply masterful"—and so it has remained to the present day. Pauli showed here for the first time his art of presenting science, which marks everything he wrote.

In Sommerfeld's institute Pauli also became acquainted with the quantum theory of the atom. He wrote in his Nobel lecture:

> While, in school in Vienna, I had already obtained some knowledge of classical physics and the then new Einstein relativity theory, it was at the University of Munich that I was introduced by Sommerfeld to the structure of the atom, somewhat strange from the point of view of classical physics. I was not spared the shock which every physicist, accustomed to the classical way of thinking, experienced when he came to know of Bohr's "basic postulate of quantum theory" for the first time.

It is a modest expression when Pauli speaks of "some knowledge of classical physics and the . . . Einstein relativity theory." This must be taken into account to understand what it means for a "physicist, accustomed to the classical way of thinking," to experience a shock from Bohr's postulate. There were, to be sure, few students scarcely twenty years of age who had penetrated the classical way of thinking as deeply as Pauli had. At this age the shock must have been great.

In 1922 Pauli obtained the doctorate with the thesis "Über das Modell der Wasserstoffmolekülions." Soon thereafter he began to work on the anomalous Zeeman effect. As he reports in his Nobel lecture, these studies finally culminated in the discovery of the exclusion principle, announced in "Ueber den Zusammenhang des Abschlusses der Elektronengruppen im Atom mit der Komplexstruktur der Spektren" (*Zeitschrift für Physik*, **31** [1925], 765). The markedly complicated title shows that here Pauli had solved an intricate problem. Landé, Sommerfeld, and Bohr among others believed, particularly in the case of the alkali metals, that the atomic core around which the valence electron moved possessed an angular momentum and that this was the cause of the magnetic anomaly. Why the atomic core should possess a half-integral angular momentum and a magnetic moment was, to be sure, unclear. Even more incomprehensible was the situation regarding the alkaline earths which possess both a singlet and a triplet system; these two systems should also be explained from the properties of the core. Indeed, the core should always possess the same electron configuration; but in the two cases it would interact differently with the valence electrons. No one could say how this would happen; and Bohr spoke of a *Zwang*, or constraint, which had no mechanical analogue. Now because the core, the closed noble gas configuration, should possess such peculiar properties, it was further believed that the core could not be characterized by the quantum numbers of the individual electrons: the "permanence of the quantum numbers" would have to be given up.

Pauli now proposed that the magnetic anomaly be understood as a result of the properties of the valence electron: in it appears, as he wrote, "a classically nondescribable two-valuedness in the quantum-theoretic properties of the electron." The atomic core, on the other hand, possesses no angular momentum and no magnetic moment. This assumption meant that the "permanence of the quantum numbers," Bohr's *Aufbauprinzip*, could be retained: each electron, even in a closed shell, could in principle be described by quantum numbers. In addition to the already known n, l, and m, one now needed a fourth, which is denoted today by the spin quantum number s. After such a strong foundation was laid, Pauli went on to study the structure of the core, which had previously been considered by E. C. Stoner (*Philosophical Magazine*, **48** [1924], 709). Pauli was able to explain Stoner's rule by means of his famous exclusion principle:

> There can never be two or more equivalent electrons in an atom, for which in a strong field the values of all the quantum numbers n, k_1, k_2 and m are the same. If an electron is present, for which these quantum numbers (in an external field) have definite values, then this state is "occupied."

In this formulation the atom is first considered in a strong external field (Paschen-Back effect), since only then can the quantum numbers for single electrons be defined. However, on thermodynamic grounds (the invariance of the statistical weights during an adiabatic transformation of the system) the number of possible states in strong and weak fields must, as Pauli observed, be the same. Thus the number of possible configurations of the various unclosed electron shells could now be ascertained.

The discovery of the exclusion principle builds the crowning conclusion to the old quantum theory based on the correspondence principle, which Pauli described in *Handbuch der Physik*, XXIII (1926). When the article was published, new developments had already occurred; in rapid succession the fundamental work of Heisenberg, Dirac, and Schrödinger appeared, leading to a proper, mathematically consistent quantum mechanics.

Following Dirac's precedent, Jordan, Heisenberg,

and Pauli developed the relativistic quantum electrodynamics. This theory occupied physicists for a good twenty years before it became clear that, in spite of all the doubts and disappointments, one of the most precise physical theories had been discovered. Disappointment and doubt had arisen primarily from the following circumstances: It was known for a long time that in the quantum theory of light and the electron, the Sommerfeld fine structure constant $e^2/hc = \alpha$ plays an exceptional role: α is a dimensionless quantity and has the value $1/137$. In it three areas of theoretical physics are symbolically united: electromagnetism, which is represented by e; relativity, represented by c; and quantum theory, represented by h. It was therefore believed that if a relativistic quantum electrodynamics was successfully developed, it would at the same time yield a theory of α. Thereby, so it was further hoped, a natural solution would be found for the problem of the infinite self-energy of the electron, an insurmountable problem in the classical electron theory. These hopes have not been fulfilled.

In order to accommodate the new developments, Pauli wrote an article on wave mechanics for the second edition of the *Handbuch der Physik* (XXIV, pt.1 [1933]), "Die allgemeinen Prinzipien der Wellenmechanik." A student at the time, the author well remembers meeting Hermann Weyl on the street and his saying, "What Pauli has written on wave mechanics is again completely outstanding!" This judgment of a connoisseur is still valid today: the same article, twenty-five years later, was used unchanged in the new handbook (1958). Pauli's presentation was thoroughly modern and well thought out, considering that such articles frequently become outdated after only a few years.

While the work on the Pauli principle and the first *Handbuch* article—"the Old Testament"—was done in Hamburg, the second article—"the New Testament"— was written in Zurich. After finishing his thesis under Sommerfeld's guidance, Pauli had gone to Göttingen as an assistant to Max Born. Here he met Niels Bohr, who invited him to Copenhagen. From there he soon went to Hamburg, where he held an assistantship under Wilhelm Lenz and gave his inaugural lecture as *Privatdozent*. In 1928 the Swiss Board of Education appointed him Debye's successor as professor at the Eidgenössische Technische Hochschule, where he remained until his death in 1958. At the same time Schrödinger had left the University of Zurich, where wave mechanics was developed, and he was succeeded by Gregor Wentzel. Both professors were very young and brought a rich and active scientific life to Zurich. For many years Pauli and Wentzel organized a seminar together, in which the more important new work from practically all areas of theoretical physics was critically discussed.

By today's standards facilities at both schools were at that time rather limited. At the Technical University, Pauli was the only lecturer for theoretical physics, and students specializing in this field were practically nonexistent. But Pauli—in contrast to Wentzel at the university—did have an assistantship at his disposal. This was a research position, and he always filled it with someone who had already attained the doctorate. These assistants became his true pupils: R. Kronig, Rudolf Peierls, H. B. G. Casimir, and V. F. Weisskopf were his assistants during his first ten years in Zurich, and all were scholars who later became well-known in the field.

Pauli was never what one would call a good lecturer. He mumbled to himself, and his writing on the blackboard was small and disorganized. Above all, though, he had the tendency during the lecture to think over the subject at hand—which, as Wilhelm Ostwald remarked in *Great Men*, hinders teaching. And so his lectures were difficult to understand—but nevertheless his students were fascinated and greatly stimulated. On the whole he radiated a very strong personal force. One was immediately impressed by his sharp and critical judgment. In discussions he was in no way willing, and perhaps completely unable, to accept unclear formulations. He seemed hard to convince, or he reacted in a sharply negative manner. Thereby he forced his partner in discussion to self-criticism and to a more logical organization of his thoughts. If, however, one succeeded in convincing Pauli of an idea, then at the same time one's own thoughts were brought to a greater clarity. In this sense he was a truly Socratic teacher who helped in the birth of the ideas of others.

The great influence that Pauli exerted on students and colleagues cannot be ascribed to his imposing critical understanding alone. Nor did the respect that one had for him originate solely from his often caustic way of jumping at his discussion partner, which put many into disarray. Such attacks, although occasionally malicious, were not intended to be mean and had a humorous, ironic side. It was the daemon of the man that one sensed. Theoretical physics surely appears quite rational, but it rises from irrational depths. And so it rests on a daemonic background that can lead to serious conflicts. Pauli had experienced and endured this deep within himself. He had, as few others, earnestly endeavored to master this conflict rationally. Since mathematics and theoretical physics are creations of the human soul, and since they come out of the structure of the soul, he took up the ideas of C. G. Jung in order to better understand

the meaning of scientific activity. The results of these efforts are numerous essays and lectures, and particularly his study "Der Einfluss archetypischer Vorstellungen auf die Bildung naturwissenschaftlicher Theorien bei Kepler." It appeared—and Pauli attached importance to this—in the book *Naturerklärung und Psyche* (1952), which he published with C. G. Jung.

It appears that Pauli's colleagues did not always understand how earnestly he wrestled with the philosophical foundations of science and how strongly he experienced their irrational origin. But in some obscure manner they felt it and realized it in outward experiences. These experiences took form in the strange phenomena known as the "Pauli effect": Pauli's mere presence in a laboratory would cause all sorts of misfortunes. So believed critical scholars, such as Otto Stern, who was friendly with Pauli, and so Pauli himself believed. The great impression that his personality made on all who came in contact with him can be correctly assessed only when this mysterious side of his complex being is taken into account.

One of Pauli's most significant accomplishments in physics while in Zurich is the neutrino hypothesis. With it he correctly explained the continuous β spectrum, at that time very puzzling. In a lecture before the Naturforschende Gesellschaft in Zurich in 1957 he presented the history of this discovery. Niels Bohr was of the opinion that in the case of β decay the conservation of energy should be only statistically valid. If this were conceded, then the conservation of angular momentum and the statistical laws for particles of spin 1/2 would be violated. In the early days of the development of atomic theory, Bohr was ready to sacrifice the *Aufbauprinzip* and the permanence of the quantum numbers and to introduce a mechanically unexplainable *Zwang*; and he was now also prepared to give up the classical conservation laws. He was always "ready and willing" to discover the unexpected in the realm of atomic dimensions. Pauli, on the other hand, resolved only with great difficulty to let fall natural laws that had previously been confirmed everywhere. Just as he held on to the permanence of the quantum numbers in his theory of the closing of shells in atoms, which led him to the exclusion principle, so it appeared to him right to retain the conservation laws. Thus he proposed in a letter of 4 December 1930 to Lise Meitner and associates "the continuous β-spectrum would be understandable under the assumption that during β-decay a neutron is emitted along with the electron. . . ."

Since the letter was written before Chadwick had discovered the neutron in the nucleus, the discussion here involved another particle, which Fermi then christened "neutrino." At the Solvay Congress in 1933, Pauli again extensively justified his proposal, which was published in the Congress report. Shortly thereafter, in 1934, Fermi worked out his theory of β decay, which, in spite of unsolved basic difficulties, has been confirmed amazingly well.

During the war Pauli was active at the Institute for Advanced Study in Princeton; but later, after careful consideration, he returned to Zurich. He lived happily with his wife in Zollikon, near great forests that invited meditative strolls. Consistent as he was, he now earned Swiss citizenship.

This article has intentionally avoided giving even an approximately complete review of Pauli's scientific work, for there is practically no area of theoretical physics in which he did not decisively take part. The aim has been to make clear, in connection with his most important contributions, the manner in which he worked. A last example to be mentioned is his important work on discrete symmetries in field theory. He dedicated it to Niels Bohr on his seventieth birthday under the title "Exclusion Principle, Lorentz Group and Reflection of Space-time and Charge." Starting from investigations by Schwinger and Lüders, Pauli showed that every Lorentz invariant Lagrangian field theory is invariant under the operation CTP, whereas C, T, and P separately do not have to be symmetries of the theory. This study had greatly occupied him, as he occasionally told me, and I guessed that he had hidden thoughts about the matter which he did not express. So I asked him if in this work there was not in fact another problem between the lines and if he might not say something about it. But he denied my conjecture: he was interested in these symmetries in their own right.

Not much later it was discovered that in weak interactions—for example, in β decay—the parity (P) is not conserved (Lee and Yang, 1956). Pauli was greatly stirred by this discovery. It seemed to him at first extraordinarily repugnant that in nature right and left should not enjoy equal status. But then he realized that the symbolic, to some extent natural-philosophic, concept which he saw in this symmetry did indeed remain: for as he had made clear one year earlier, CTP must be a valid symmetry if only the natural laws are Lorentz invariant. Thus, guided by his own genius, he had meaningfully prepared for the coming developments.

Just as Pauli received a shock when, as a student, he first became acquainted with the strange laws of quantum theory, so did he receive a shock from the nonconservation of parity. For it was always his hope

that physics would indicate the mysterious harmony of God and Nature. This hope was not illusory. Precisely in his most important work he had shown how apparently paradoxical phenomena could be explained through a harmonious extension of the previously confirmed theory. And so theoretical physics since Kepler, Galileo, and Newton appeared to him as a great house the foundations of which, despite many changes, would never be shaken. It was because he felt this way, and because he considered himself a representative of a great tradition, that he reacted so sharply against obscure arguments and superficial speculation. He expressed himself thus concerning his position to a colleague: "In my youth I believed myself to be a revolutionary; now I see that I was a classicist."

In December 1958, Pauli became violently and seriously ill, and on December 14 he died. At the funeral Viktor Weisskopf said he was "the conscience of theoretical physics." This is truly the shortest statement that can render the impression which this rare man made on all who knew him.

BIBLIOGRAPHY

A complete list of Pauli's books, articles, and studies is in *Theoretical Physics in the Twentieth Century, a Memorial Volume to Wolfgang Pauli* (New York, 1960). Collections include his scientific papers (New York, 1964) and *Aufsätze und Vorträge über Physik und Erkenntnistheorie* (Brunswick, 1961).

M. FIERZ

PAULIŚA (*fl.* India, fourth or fifth century), *astronomy.*

Pauliśa was the author of a textbook *(siddhānta)* on astronomy in Sanskrit. The work was largely based on the Greek adaptations of Mesopotamian astronomy that began to be introduced into India in the third century by Sphujidhvaja (*fl.* 269/270) and perhaps earlier, in the second century, by Yavaneśvara (*fl.* 149/150) (see essay in Supplement). Since the *Pauliśasiddhānta* was revised by Lāṭadeva (*fl.* 505), the original must have been written between *ca.* 300 and *ca.* 450; it is probably, then, to be associated with the patronage of the Guptas, of whom the one most noted for his interest in literary efforts is Candragupta II (*fl. ca.* 375–415). Of Pauliśa himself we can say nothing save that his name may be a transliteration of the Greek Παῦλος. His identification with Paul of Alexandria (*fl.* 378)—at which al-Bīrūnī first hinted—is certainly false, as it is based on a misreading of the place-name Tanaysar (Sthāneśvara or Sthāṇvīśvara) in a later *Pauliśasiddhānta* that was written in the eighth century and that followed the *ārdharātrikapakṣa* of Āryabhaṭa I (*b.* 476) (see D. Pingree, "The Later Pauliśasiddhānta," in *Centaurus*, **14** [1969], 172–241).

The original *Pauliśasiddhānta*, as revised by Lāṭadeva, is known to us only through the *Pañcasiddhāntikā* of Varāhamihira (*fl. ca.* 550). From that work we learn of Pauliśa's method of computing the days lapsed since epoch (I, 11–13); his solar and lunar equations, the former computed from a Greek model, the latter going back to Babylonian techniques (III; 1–3, 5–8); his method of computing oblique ascensions, longitudinal differences, and the daily motion of the sun (III, 10–17); his rules relating to the Indian time-units called *karaṇas* (sixtieths of a synodic month), *tithis* (thirtieths of a synodic month), and *ṛtus* (seasons of two synodic months), and to the *pātas* of the sun and moon, the *ṣaḍaśītimukhas* (ecliptic arcs of 86° beginning from Libra 0°), and the *saṅkrāntis* (entries of the sun into the several zodiacal signs) (III, 18–27); his computation of lunar latitude (III, 28–29); his theory of lunar and solar eclipses (VI, VII); and his planetary theory, based on a Greek adaptation of Babylonian astronomy in which the synodic arcs of the planets and their elongations from the sun at the occurrence of the "Greek-letter" phenomena are utilized (XVII, 64–80).

BIBLIOGRAPHY

All of the material relevant to Pauliśa will be found in O. Neugebauer and D. Pingree, *The Pañcasiddhāntikā of Varāhamihira*, 2 vols. (Copenhagen, 1970–1971).

DAVID PINGREE

PAULLI, SIMON (*b.* Rostock, Mecklenburg, 6 December 1603; *d.* Copenhagen, Denmark, 23 April 1680), *botany, anatomy.*

Paulli was the son of Heinrich Paulli, a professor at Rostock and physician in ordinary at the Danish court. He studied anatomy at Rostock and Leiden, and later at Paris under Jean Riolan. After a trip to England he received his medical degree at Wittenberg in 1630. He practiced medicine at Rostock and Lübeck from 1634 to 1639 and from 1639 to 1648 was professor of medicine at Rostock. In 1648 Paulli was appointed professor of anatomy, surgery, and botany at Copenhagen. Simultaneously he became physician in ordinary to the Danish king, who granted him the revenue from the bishopric of Aarhus. In 1655 he gave a series of botany lectures in Rostock.

Paulli made notable contributions to the technical literature of anatomy and botany. His botanical writings were discussed in detail by Albrecht von Haller, who praised him not only for compiling existing botanical knowledge but also for comparing it with information derived from his own experiments. More a practitioner than a theoretician, he recommended the use of simple medications. His biography is included in the posthumous Frankfurt edition of *Quadripartitum botanicum* (1708).

BIBLIOGRAPHY

I. ORIGINAL WORKS. Paulli's major work is *Quadripartitum botanicum de simplicium medicamentorum facultatibus . . .* (Rostock, 1640; Strasbourg, 1667–1668; Frankfurt, 1708), in which he arranges plants according to the seasons, in the form of a floral almanac; within each season the plants are listed in alphabetical order. Along with the uses and effects of the vegetal medicines, he provides bibliographical information. The star-thistle (*Centaurea calcitrapa*) is discussed here, apparently for the first time. An appendix reprints his Rostock inaugural lecture, "De officio medicorum, pharmacopoeorum et chirurgorum," which contains the first mention of the use of a cow's bladder in giving enemas. The lecture was printed separately at Rostock in 1639.

He also wrote *Flora Danica, det er Dansk Urtebog . . .* (Copenhagen, 1648), which is also arranged according to the seasons and, besides descriptions of plants, includes information on their synonyms and medicinal properties and 393 illustrations, some original and some taken from Matthias de Lobel and Joannes Moretus (Moerentorf); *Viridaria varia regia et academica publica . . .* (Copenhagen, 1653), which consists of catalogs of the botanical gardens of Copenhagen, Paris, Warsaw, Oxford, Leiden, and Groningen, as well as catalogs of exotic and native plants, listed according to location; *Parekbasis seu digressio de . . . causa febrium . . . Appendix, seu Historica relatio de . . . anatomico et chirurgico casu ad . . . Johannem Riolanum . . . anno 1652* (Frankfurt, 1660), which includes a description of scurvy and venereal diseases; *Miscella antiquae lectionis cujus quatuor monumenta in praefatione enumerata in publicam lucem reduxit . . .* (Strasbourg, 1664), a historical work; *Commentarius de abusu tabaci Americanorum veteri et herbae Theé Asiaticorum in Europa novo . . .* (Strasbourg, 1665, 1681), also in English trans. by Robert James (London, 1746)—according to Haller, Paulli also published *Libellum de usu et abusu tabaci et herbae Theae* in 1635; *Orbis terraqueus in tabulis geographicis et hydrographicis descriptus . . .* (Strasbourg, 1670), a geographical work; and *Historia litteraria, sive Dispositio librorum omnium facultatum . . .* (Strasbourg, 1671), an encyclopedic work.

Miscella . . ., Orbis . . ., and *Historia . . .* are cited in *Catalogue général de la Bibliothèque Nationale*, CXXXI (Paris, 1935), cols. 671 ff. Haller mentions a letter entitled

"De gramine ossifrago epistola ad Th. Bartholinum a Beughemio citatur."

II. Secondary Literature. See A. Blanck, *Die mecklenburgischen Aerzte* (Schwerin, 1874), 30; A. von Haller. *Bibliotheca botanica*, 2 vols. (Zurich, 1771–1772; repr, Hildesheim–New York, 1969): I, 459, and II, 333; C. Krause, in *Allgemeine deutsche Biographie*, XXV (1885), 274; J. Krey, *Andenken an die Rostocker Gelehrten*, VI, 8 f.; Linnaeus, *Bibliotheca botanica* (Amsterdam, 1736), 36, 48, 69–71, 74, 78–79, 86, 92, 97, 143; and H. Schelenz, *Geschichte der Pharmazie* (Berlin, 1904), 495, 526.

Karin Figala

PAULY, AUGUST (*b*. Munich, Germany, 13 March 1850; *d*. Munich, 9 February 1914), *zoology, entomology.*

Pauly's development as a scientist was completely self-motivated and self-directed. His father, Cölestin Pauly, was from the south of France; formerly a farrier, and during Pauly's childhood a wine merchant and innkeeper in Munich, he was known for his hot temper. Pauly's mother, Johanna Riehle, a Bavarian, was more sympathetic toward the boy, but neither parent understood his deep longing for a good education. They intended him for a career in commerce, but in the depths of the wine cellar Pauly secretly read to provide himself with the equivalent of the Gymnasium studies. Adolf Bayersdorfer, a lifelong friend of Pauly and later his brother-in-law, helped him in his plans to pass the examinations and enter the University of Munich. Whenever Pauly could, he attended lectures on a wide variety of subjects at the Konservator an der alten Pinakothek in Munich. He had a sensitive disposition, was interested in art, and thought that he might become a painter.

Pauly took his examinations in 1873, and in 1877 received the doctorate with a dissertation in zoology. His interest in biology had most probably been influenced by his teacher Carl Theodor Ernst von Siebold, professor of zoology and comparative anatomy, whose assistant Pauly became. From 1877 to 1885 Pauly edited an ornithological journal; since he had access to a large store of histological material, he became an expert in avian pathological anatomy. He gave lectures on both the theoretical and practical aspects ·of forest entomology, and donated considerable material for the study of insects, insect damage, and forest zoology to the Royal Institute for Experimental Forestry in Munich. He was appointed extraordinary professor of applied zoology at the University of Munich in 1896. His wife, Mathilde von Portheim, gave him invaluable help in his work as he suffered progressive difficulties with his vision, including retinal detachment.

Since Pauly set zoology within the framework of a philosophically oriented outlook that combined the love of nature with a broad interest in art, pressing biological questions were frequently discussed within the group of his friends, which included artists and a poet, and later, the scientists Boveri and Spemann when he came to Munich.

Pauly's zoological lectures reflected his dissatisfaction with Darwin's explanation of the evolutionary process, for Pauly thought it highly unlikely that chance variations could accumulate and coincide to account for the adaptation and correlation of organs. Pauly looked back, rather, to Lamarck, discounting the emphasis that had been placed on Lamarck's view of the role of use and disuse and stressing the psychic factor that entered into the Lamarckian doctrine. *Darwinismus und Lamarckismus. Entwurf einer psychophysischen Teleologie*, published in 1905, represented the sum of Pauly's thirty years' work in evolutionary theory.

Pauly conceived of evolution as being the result of an "inner teleology," a capacity for change in response to a consciously apprehended need within the organism itself. Adaptation was a "discovery," a change in response to this necessity. Adaptive changes were inherited but could be maintained only through use; the changed organ or organism tended to revert to its former condition should the function cease. Pauly referred the underlying psychic circumstances not only to the brain but also to each organ and cell; variation on this psychological basis was, of course, present in the plant as well as the animal world. He maintained this neo-Lamarckian evolutionary theory—a vitalistic viewpoint that he believed was derived from Lamarck, but which was actually uniquely his own—throughout his lifetime.

BIBLIOGRAPHY

I. Original Works. Pauly's most important work was *Darwinismus und Lamarckismus. Entwurf einer psychophysischen Teleologie* (Munich, 1905). He also published his aphorisms in that year and wrote short essays on his evolutionary beliefs.

II. Secondary Literature. On Pauly and his work, see Fritz Baltzer, *Theodor Boveri, Life and Work of a Great Biologist 1862–1915*, trans. by Dorothea Rudnick (Berkeley–Los Angeles, 1967), 8–10, 36, 48, 130–131, 141–142, which describes their friendship and correspondence and the milieu in Munich. See also Friedrich Wilhelm Spemann, ed., *Hans Spemann, Forschung und Leben* (Stuttgart, 1948), 145–150, 157–164, which presents a valuable reminiscence of Pauly and his background, as well as an excellent account of his theory.

Obituaries are M. Merk-Buchberg, "Zum Gedächtnis August Paulys," in *Zoologischer Beobachter* (1914), 87–88; K. Escherich, "August Pauly," in *Zeitschrift für angewandte Entomologie*, **1** (1914), 370–373; Max Friedemann, "Psychobiologie. Zum andenken an August Pauly," in *Berliner klinische Wochenschrift*, **51**, pt. 2 (1914), 1441–1443; Adolf Leiber, "August Pauly," in *Süddeutsche Monatshefte*, **11**, no. 2 (1914), 161–166; and the *Deutsche biographisches Jahrbuch, 1914–1916* (Berlin, 1925), 303, which has further biographical references.

GLORIA ROBINSON

PAVLOV, ALEKSEI PETROVICH (*b.* Moscow, Russia, 13 November 1854; *d.* Bad Tölz, Germany, 9 September 1929), *geology*.

Pavlov was the son of a retired military man. He entered the Moscow Gymnasium in 1866, then in 1874 enrolled in the Faculty of Physics and Mathematics of Moscow University. He was talented in both art and music, and his eventual choice of a scientific career may have been influenced by his Gymnasium teachers. At the university he attended the lectures of the distinguished geologists G. E. Shchurovsky and M. A. Tolstopiatov; his diploma topic, ammonites, was suggested to him by Shchurovsky, and his work won him a gold medal. After graduating from the university, Pavlov taught natural history in the secondary schools of Tver (now Kalinin) from 1878 to 1880; in the latter year he went to Moscow, at Shchurovsky's invitation, to become curator of the geological and mineralogical collections of the university. At the same time he began to study for the master's degree, make practical studies in mineralogy, and teach in the Higher Courses for Women.

In 1883 Pavlov, at the request of the St. Petersburg Mineralogical Society, conducted field research in the lower and middle Volga regions; this research formed the basis for his master's thesis, *Nizhnevolzhskaya yura* ("The Jurassic Period of the Lower Volga"), which he defended the following year. He then traveled abroad, first to Paris and the Auvergne, then to Vienna, where he attended the lectures of Suess. While in Paris he met M. V. Illich-Shishatskaya, a young widow who was auditing lectures on geology and paleontology there; they were married in 1886. Pavlov returned to the Volga to do further fieldwork in the summer of 1885; he studied Cretaceous deposits and made important observations regarding the stratigraphy of Cretaceous and Tertiary deposits. In January 1886 he became a professor at the Uni-

versity of Moscow—a post that he held for the rest of his life—and in May of that year he defended a doctoral dissertation on the *Aspidcervas acanthicum* of eastern Russia.

Pavlov's teaching was inseparable from his scientific work. His course on introductory geology, which he gave for about ten years, was extremely popular, as were the field excursions that he conducted for his students. His courses at the Moscow Archaeological Institute and at the Moscow Mining Academy, together with those at the university, brought him a large number of pupils who formed the nucleus of the Moscow school of geologists that he trained. His concern for the reform of secondary education in Russia culminated in a book, published in 1905, in which he stressed the need for the teaching of science at that level. He later devoted a number of articles to the subject.

Pavlov's purely scientific works comprised a wide range of topics, including stratigraphy, paleontology, tectonics, Quaternary geology, and practical geology. A single, early work is devoted to vertebrate paleontology; in an article on *Archaeopteryx*, published in 1884, Pavlov suggested that this genus, having achieved its greatest development in the Jurassic period, was an evolutionary side branch, destined to extinction through poor adaptation to life.

Pavlov spent a number of years studying the Mesozoic deposits of the Russian platform and the Boreal phases of the Mesozoic era throughout northern Europe. In his master's thesis of 1884 he had traced the upper and lower boundaries of the Volga Jurassic deposits and had studied their fauna; he later established a discontinuity in the Jurassic deposits of the same region, associated with the perturbations of the Jurassic sea, and showed the sharp line of contact between the Jurassic and Cretaceous deposits. In 1888, while attending the Fourth International Geological Congress in London, Pavlov studied the local Jurassic and Cretaceous profiles and examined the collections of Jurassic and Cretaceous fossils in English museums. He utilized this new material in a comparative stratigraphic analysis of the Jurassic deposits of England and of the central part of European Russia. His conclusion was that these deposits were possibly equivalent.

Pavlov drew upon later studies of the profile of the province of Boulogne to make still wider generalizations and to compare deposits. He thus discovered a great similarity between the Jurassic fauna of the Volga region and that of Europe. By 1896 he had synthesized an enormous amount of material into a comprehensive classification of the Upper Jurassic and Lower Cretaceous deposits of Europe and Russia and had completed a paleogeographic survey of these areas. He presented the results of these researches to the International Geological Congresses of 1897 and 1900.

In his study of the stratigraphy of the Lower Cretaceous Pavlov used materials drawn from his investigations of the Russian plains and the Pechersky caves, as well as from collections gathered in northern Siberia. He established that there had been two Boreal periods in the Lower Cretaceous and that the sea which had flooded the lower Volga in the Albian stage had been connected to the sea of Western Europe. He proposed a series of paleogeographic maps and described the character of these Lower Cretaceous deposits; he also studied the Upper Cretaceous deposits of the same area and incorporated them into a stratigraphical scheme in which he noted a number of new paleontologically distinct horizons. He further established the distribution of the Lower Tertiary deposits of the Volga and differentiated them paleographically.

Pavlov's work in Quaternary geology also began early; indeed, his interest in the period dated from Shchurovsky's lectures at the university. He studied Quaternary deposits in his first expedition to the Volga, then, during the Third International Geological Congress, held in 1885 in Berlin, investigated the glacial deposits of Germany. After several years of investigating and comparing Quaternary deposits Pavlov was able to reach a number of conclusions concerning the genetic types of continental deposits, the number of glaciations that had caused them, and the genesis of modern topography. He summed these up in a paper of 1888, "Geneticheskie tipy materikovykh obrazovany lednikovoy i poslelednikovoy epokhi" ("Genetic Types of Continental Formations of Glacial and Post-glacial Epochs").

Pavlov defined two types of glacial deposits. The first, the talus, consists of deposits formed by the weathering and decomposition of bedrock which have formed a slope at the bottom of a steeper declivity. This process played a large part in the formation of the topography of the Russian platform; Pavlov included a number of different types of rocks within this concept and assigned an aqueous origin to all of them, even loess. He studied the loess of the Volga region, Turkestan, and Western Europe to conclude that the process by which the Turkestan loess had been formed was very similar to that of the talus, save only that the loess had been formed by torrential mountain deluges running through the valleys, rather than by rain. He proposed to call these deposits, his second type, "proluvium." Pavlov emphasized the importance of talus and proluvial deposits in his

studies of ancient continental deposits, suggesting that these were the result of weathering processes that had taken place in the earliest geological age. He integrated his findings into his investigations of contemporary topology.

In his researches on the history of the glacial epoch Pavlov compared Russian and Western European Neogene and Quaternary deposits to establish threefold glaciation. He noted two waves of cold in the Pliocene period, then a first glaciation, covering a large part of Europe, in the Quaternary. Following the moderate and moist Chellian and Achellian periods, a second glacier covered the whole of northern Europe, developing a glacial cover almost equal to the first. The characteristic morainic landscape of Europe was formed during the second interglacial epoch; the third glaciation was less widespread than the first two, and was followed by the present warm and moist period.

Pavlov conducted paleontological research as an adjunct to his stratigraphic work. He was particularly interested in Mesozoic ammonites and belemnites and devoted several works to the description of belemnites from the Spiton deposits, in which he showed the similarity of their forms and established their genetic series and natural classification. In a report given to the Eighth International Geological Congress, which met in Paris in 1900, Pavlov proposed a new genetic classification for fossil organisms, arguing that the morphological classification then accepted did not correspond to evolutionary theory. His own system was based upon phylogenetic properties; he suggested that the terms "genetic series," "genetic line," "phyletic branch," "generation," and "species and variety" be used for more detailed subdivision. A convinced Darwinian, he drew upon the example of the ammonites to analyze questions of phylogeny and ontogeny and thereby discovered the phenomenon of phylogenetic acceleration. He applied his own classification to aucella, comparing examples from Russian and Western European deposits, and described Pliocene paludinas as part of his analysis of Quaternary material. Pavlov's wife assisted him actively in his paleontological work, and several of his students continued it.

In the course of his work on the Volga Pavlov also became concerned with tectonic phenomena. In 1887 he suggested the existence of faulting in the northern border region of Zhigulaya. An adherent of the contractionist theory, he ascribed this faulting to that cause and noted further that petroleum deposits might be associated with this dislocation, a prediction that was later confirmed. Pavlov also discovered a fault on the right bank of the Volga and thus accounted for the general tectonic features of that area. He further found a new element in the structure of the Russian platform, the great gentle down-warpings that he called "synclines." He interpreted these as local uplifts and depressions in the crust of the earth.

In the field of theoretical tectonics Pavlov, as early as the end of the 1890's, began to study the topography of the moon and its genesis. He later made a comparative analysis of the topography of the moon and that of the earth, which he reported in a paper of 1908, "Lik zemli i lik luni" ("The Face of the Earth and the Face of the Moon"), and in another read to the Astronomical Society in 1922. He emphasized the importance of the study of lunar topography for understanding terrestrial processes. In his investigation of the basic morphology of the earth he concluded that its structure was determined at an early stage of its development, but that its fundamental topography was then obscured by its massive sedimentary cover and the action of the hydrosphere. The moon, which because of its weak gravity lacks an air and water cover, may therefore, in its continents and in its depressions, serve as a model of the first stages of the development of the surface of the earth. Beginning with the contraction hypothesis, Pavlov suggested that lunar forms were shaped by the solidification of the molten moon and by volcanic action; such forms, he added, had previously existed on earth, but had been transformed by the forces of contraction and the processes of weathering.

Although he was primarily interested in theoretical geology, Pavlov did not neglect its practical aspects. He investigated the landslides on the shores of the Volga and concluded that they resulted from the geological structure of the slope of the river bank, its steepness, the activity of underground water, and the leaching-out effect of the river itself. He divided landslides into two types—gravity slides, embracing the lower part of the slope; and pushing slides, whose movement begins at the top of the slope and embraces it almost in its entirety. (Some landslides may partake of both types.) Pavlov also treated the distribution of forces acting in massive landslides and suggested preventive measures. He provided a classification of rocks for engineering purposes and was frequently consulted about the construction of railroads and bridges, work related to the then new field of engineering geology.

Pavlov was, in addition, often consulted on hydrogeological problems, including the irrigation of arid areas and the reasons for the hardness of water. He was interested in soil and emphasized the relationship between soil and bedrock and topology and stressed the geological processes of soil formation. He thus contributed a good deal of the basic geological research upon which the Russian discipline of soil science was founded.

In addition to his geological works, Pavlov published a number of books in the history of science and a number of popular scientific works. He gave well-attended popular lectures, too, and several of his books went through multiple editions. He was an active member of several scientific societies and received many honors. He died at the spa of Bad Tölz, where he went with his wife to recover from a serious illness. He was active until the last days of his life, investigating the mineral springs of the resort.

BIBLIOGRAPHY

I. ORIGINAL WORKS. Pavlov's writings include *Nizhne-volzhskaya yura. Klassifikatsia otlozheny i spiski isko-paemykh* ("The Jurassic of the Lower Volga. Classification of Deposits and Notes on Fossils"; Moscow, 1884); "Notes sur l'histoire géologique des oiseaux," in *Bulletin de la Société impériale des naturalistes de Moscou*, **60** (1884), 100–123; "Samarskaya luka i Zheguli" ("The Samara Bend and the Zhiguli Hills"), in *Trudy Geologicheskago komiteta*, **2**, no. 5 (1885–1887), 1–63; "Geneticheskie tipy materikovykh obrazovany lednikovoy i poslelednikovoy epokhi" ("Genetic Types of Continental Formations of Glacial and Post-glacial Epochs"), in *Izvestiya Geologicheskago komiteta*, **7** (1889), 243–261; "Études sur les couches jurassiques et crétacées de la Russie," in *Bulletin de la Société impériale des naturalistes de Moscou*, n.s. **3** (1890), 61–127, 176–179; "Argiles de Speeton et leurs équivalents," *ibid.*, **5** (1892), 214–276, 455–570; "On the Classification of the Strata Between the Kimheridgian and Aptian," in *Quarterly Journal of the Geological Society of London*, **52** (1896), 542–554; *Polveka v istorii nauki ob iskopaemykh · organizmakh* ("Half a Century in the History of Science of Fossil Organisms"; Moscow, 1897); "O reliefe ravnin i ego izmeneniakh pod vlianiem raboty podzemnykh i poverkhnostnykh vod" ("On the Topography of Plains and Its Changes Under the Influence of Underground and Surface Waters"), in *Zemlevedenie*, **5** (1898), 91–147; and *Vulkany na Zemle i vulkanicheskie yavlenia vo vselennoy* ("Volcanoes on the Earth and Volcanic Phenomena in the Universe"; St. Petersburg, 1899).

Works published in the twentieth century include *Kratky ocherk istorii geologii* ("A Brief Sketch of the History of Geology"; Moscow, 1901); "Ob izmeneniakh v geografii Rossii v yurskoe i melovoe vremya" ("On the Changes in the Geography of Russia in the Jurassic and Cretaceous Eras"), in *Nauchnoe slovo*, **1** (1903), 143–145; *Opolzni Simbirskogo i Saratovskogo povolzhya* ("Landslips in the Simbirsk and Saratov Volga Region"; Moscow,

1903); *Geologichesky ocherk okrestnostey Moskvy* ("Geological Sketch of the Surroundings of Moscow"), 5th ed. (Moscow, 1907). "Enchaînement des aucelles et aucellines du crétacé russe," in *Nouveaux mémoires de la Société des naturalistes de Moscou*, **17**, no. 1 (1907), 1–92; *Geologia nastoyashchego vremeni* ("Geology of the Present Time"; Moscow, 1914); "Yurskie i nizhnemelovye Cephalopods severnoy Sibiri" ("The Jurassic and Lower Crétaceous Cephalopoda of Northern Siberia"), in *Zapiski Imperatorskoi akademii nauk*, 8th ser., **21**, no. 4 (1914), 1–68; *Ocherki istorii geologicheskikh znany* ("Sketches of the History of Geological Knowledge"; Moscow, 1921); and *Neogenovye i posletretichnye otlozhenia Yuzhnoy i Vostochnoy Evropy* ("Neocene and Post-Tertiary Deposits of Southern and Eastern Europe"; Moscow, 1925).

II. SECONDARY LITERATURE. On Pavlov and his work, see A. P. Mazarovich, *Aleksey Petrovich Pavlov* (Moscow, 1948); N. S. Shatsky, "O sineklizakh A. P. Pavlova" ("On Pavlov's Syneclises"), in *Byulleten Moskovskogo obshchestva ispytatelei prirody*, Otdel. geolog., **18**, nos. 3–4 (1940), 39–45; and V. A. Varsanofieva, *Aleksey Petrovich Pavlov i ego rol v razvitii geologii* ("Aleksey Petrovich Pavlov and His Role in the Development of Geology"; Moscow, 1947).

IRINA V. BATYUSHKOVA

PAVLOV, IVAN PETROVICH (*b.* Ryazan, Russia, 27 September 1849; *d.* Leningrad, U.S.S.R., 27 February 1936), *physiology, psychology.*

Pavlov was the son of a priest, Pyotr Dmitrievich Pavlov, and his wife, Varvara Ivanova. He was sent at the age of eleven to the religious school in Ryazan and, after graduating, entered the seminary of that town, where he studied the current literature on natural science, including I. M. Sechenov's *Refleksy golovnogo mozga* ("Reflexes of the Brain") and the popular works of D. I. Pisarev. He did not complete his studies there, but in 1870 entered the natural sciences section of the Faculty of Physics and Mathematics at St. Petersburg University. While he was a third-year student the lectures and experimental work of E. F. Cyon decisively stimulated his interest in physiology and he carried out experimental research on the influence of the nerves on the circulation of the blood. Pavlov was awarded a gold medal for a student work on the nerves that govern the pancreas (1875), written with M. I. Afanasiev.

To broaden his knowledge of physiology Pavlov entered the third-year course at the Military Medical Academy after graduating from the university in 1875. His studies were directed primarily toward theoretical medicine. In the physiology laboratory of the veterinary section of the academy, directed by K. N.

Ustimovich, Pavlov conducted the research on the physiology of the circulation of the blood that brought him into contact with S. P. Botkin. He subsequently organized and headed the physiology laboratory of Botkin's clinic (1878–1890) and conducted investigations on the physiology of circulation and of digestion. On 19 December 1879 he received the degree of doctor of medicine; in 1881 he married Serafima Vasilievna Karchevskaya.

In Botkin's laboratory Pavlov was exposed to an atmosphere of "nervism," which "extended the influence of the nervous system to the greatest possible amount of an organism's activity."[1] During this period Pavlov wrote his doctoral dissertation, on the efferent nerves of the heart, which he defended on 21 May 1883. In 1884–1886 he worked in the laboratories of Karl Ludwig in Leipzig and Rudolf Heidenhain in Breslau and, at the latter, carried out his only research in the physiology of invertebrates, published in 1885.

In 1883 Pavlov became *Privatdozent* in physiology at the Military Medical Academy and in 1890 was appointed professor in the department of pharmacology. At the same time he became director of the physiology section of the Institute of Experimental Medicine and conducted research on the physiology of digestion that was summarized in a work published in 1897. In 1895, after the retirement of I. R. Tarkhanov, Pavlov moved to the department of physiology, which he headed until 1925. For the rest of his life his activity was concentrated at three institutes: the Institute of Physiology of the Soviet Academy of Sciences that now bears his name, the Institute of Experimental Medicine, and the biological station at Koltushy (now Pavlovo), near Leningrad.

Pavlov's scientific work received worldwide recognition. In 1904 he was awarded the Nobel Prize in physiology or medicine for his research on digestion. In 1907 he was elected an academician of the Russian Academy of Sciences. In August 1935 he presided over the Fifteenth International Physiological Congress, held at Leningrad and Moscow.

Pavlov enriched physiology and the natural sciences with a new method and a new methodology. The latter derived from his general biological thought, which was directed toward the study of the whole organism under the conditions of its normal activity. For Pavlov the living organism was a complex system, the study of which—like that of any system—demanded the use of both the analytic and synthetic methods of scientific research. He considered the main problem of experimental research in physiology to be the study of reciprocal influence and reciprocal action within the organism, and the relation of the organism to its environment. In his first study of circulation he emphasized that such work was possible only by a method that allowed the systematic investigation of "those mutual relationships in which the separate constituent parts of the complex hemodynamic machine are found during its life activity."[2] Research must be conducted under normal conditions on unprepared animal specimens.

Toward the end of the nineteenth century the essential problem of physiology was becoming the replacement of the traditional, vivisectional method with a long-term, environmental one. Such replacement was called for by the logic of the development of physiology; a vast amount of data had been accumulated by means of the vivisectional method, but it was becoming increasingly apparent that the entire organism must be studied in its natural conditions. On the limitations of vivisection Pavlov said:

> Strict experiment . . . can serve the aims of physiological analysis—that is, the general clarification of the functions of a given part of an organism and its conditions—more successfully. But when, how, and to what degree the activity of the separate parts is connected . . . constitutes the content of physiological synthesis, and . . . is frequently difficult or simply impossible to deduce from the data of strict experiment, for the setting of the experiment (narcosis, curarization, operations) is inevitably linked to a certain amount of destruction of the normal processes of the organism.[3]

Pavlov conceived the method of long-term experiment, which he introduced into the laboratory, not only as a technique of experimental research but also as a way of thinking. The continuous method inaugurated a new era in the physiology of digestion and led to new work and concepts, especially in experimental surgery and in the physiology of the brain. In his first lecture on the physiology of digestion, Pavlov said, "Science moves in spurts, depending on progress made in its methods. With each step forward in methods we rise, so to speak, to a higher step, from which a wider horizon opens to us, with subjects previously unseen."[4] He therefore developed a synthetic physiology designed to "determine precisely the actual course of particular physiological phenomena in a whole and normal organism."[5]

The object of Pavlov's research was both the organism as a system and any of its separate organs that fulfilled a definite function. He was not concerned with the basic principles and foundations of life, believing them to be the proper subjects of not physiological but rather physicochemical research. Characterizing his approach, he wrote:

> I would prefer to remain a pure physiologist, that is, an investigator who studies the functions of separate organs, the conditions of their activity, and the synthesis of their function in the total mechanism of a part or in the whole of the organism; and I am little interested in the ultimate, deep basis for the function of an organ or of its tissues, for which primarily chemical or physical analysis is required.[6]

His devotion to the synthetic approach did not, however, hinder Pavlov from analytical study of the organism, "going into the depths of cellular and molecular physiology."[7] Emphasizing the problems and goals of physiological analysis, he pointed out its role in elucidating the functional mechanisms of the organs. He distinguished four levels, or degrees, of experimental physiological research—organismic, organic, cellular, and molecular—all of which must, in the final analysis, reflect the properties of a living substance. Pavlov was well aware of the necessity of a definite, regular relationship between the holistic and analytical (or organicist and reductionist) approaches of scientific research. As a founder of organicism he clearly foresaw the advent of the cellular and molecular physiology that would greatly alter the course of organic physiology.

Pavlov stated his notion of the levels of physiological research in a speech dedicated to the memory of Heidenhain (1897), in which he said that "organic physiology . . . began its study with the middle of life; its principle, the basis of life, is in the cell."[8] He considered Heidenhain "a cellular physiologist, a representative of that physiology which must replace . . . contemporary organic physiology and which must be considered the forerunner of the last step in the science of life—the physiology of the living molecule."[9]

The greatest part of Pavlov's research is devoted to three major areas: the physiology of the circulation of the blood (1874–1888), the physiology of digestion (1879–1897), and the physiology of the brain and of higher nervous activity (1902–1936). His earliest research in the physiology of circulation was devoted to the mechanisms that regulate blood pressure. He described the role of the nerve mechanism in the adaptive activity of the blood vessels, specifying the role of the vagus nerve as a regulator of blood pressure. In his doctoral dissertation he showed that cardiac function is governed by four nerves which respectively inhibit, accelerate, weaken, and intensify it. (Prior to his work and that of Gaskell it was believed that the influence of the nerves on the heart was limited to changing its rhythm.) Pavlov's research in this area culminated with the publication in 1888 of his work on the intensifying nerve, in which he proposed that its influence be understood as trophic. In the 1920's he returned to trophic innervation, the idea upon which

L. A. Orbeli had based his theory of the adaptive-trophic role of the sympathetic nervous system.

Pavlov's research on the physiology of digestion (1897, 1906, 1911) required him to devise new techniques and thereby marked a turning point in his work. His method for studying the action of the digestive organs involved surgical intervention on the entire digestive tract, performed under conditions of strict asepsis and antisepsis, which allowed him to observe the normal activity of a particular digestive gland in a healthy animal. (A mastery of surgery was, for Pavlov, as necessary to the physiologist as a knowledge of physical and chemical methods of research.) His surgical procedures included the formation of various types of fistulas from the salivary glands, the stomach, and the pancreas to the body surface, known as esophagotomy; "imaginary feeding," carried out with E. O. Shumova-Simanovskaya (1889); the operation on the small ventricle of the stomach, formed with P. P. Khizhin (1894); and the severing of two branches of the vagus nerve and the application of the fistula of Eck (1892). He was thus enabled to investigate, more or less directly, the mechanisms governing the salivary glands, stomach, pancreas, kidneys, and intestines.

Pavlov's experiments proceeded from contemporary ideas about the neural and humoral regulation of the digestive process and of its consequences in various parts of the digestive tract. He showed that there is a close connection between the properties of salivary secretion and the kind of food consumed (the Pavlovian curves of salivary secretion). He elucidated the role of enzymes in digestion and, with N. P. Shepovalnikov, discovered enterokinase—which he called "the enzyme of enzymes"—in the intestinal secretion (1894). His theoretical conclusions were of broad biological significance. His theory of specific irritability was of particular importance—in showing that the concept of general irritability is scientifically untenable, he demonstrated specific irritability in various parts of the digestive tract. The Pavlovian theory of digestion was of great value in the clinical pathology of the stomach and intestines.

Following his work on the physiology of digestion, Pavlov turned to the physiology of behavior. By the beginning of the twentieth century many physiologists, zoologists, and psychologists had already undertaken experiments to study the function of the brain, but had assembled only fragmentary data. Pavlov drew upon Darwin's theory of evolution—which stressed psychological as well as physiological continuity—and Sechenov's reflexology to create his own theory of behavior. Pavlov thus described the genesis of his behaviorism: "The time is ripe for the transition to experimental analysis of the subject from the objective, external side, as in all the other natural sciences. This transition has made possible the recently born [study of] comparative physiology, which itself arose as one of the results of the influence of evolutionary theory."[10]

Pavlov investigated the activity of the cortex and the cerebral hemispheres, basing his work on fundamental facts, concepts, and terminology of the physiology of the nervous system. He chose to approach these areas through studying the salivary glands, which had attracted his attention because of their modest role in the organism and because their activity could be subjected to strict quantitative measurement. He had, moreover, already encountered the phenomenon of "psychic" salivation in the course of his investigations on the physiology of digestion, and wished to study it further. Subjective psychology held that saliva flowed because the dog wished to receive a choice bit of meat, but Pavlov, "an experimenter from head to foot," rejected this method as fallacious and chose to pursue the investigation objectively.

Pavlov could not help but see the "psychic" stimulation of the salivary glands as a phenomenon analogous to the normal digestive reflex. Both digestion and salivation were reflexive; only the external agents that evoked the reflexes were different. The digestive reflex was triggered by the essential mechanical and chemical properties of the food; the salivary by nonphysiological "signals," including the form and odor of the food. Using the concept of the reflex as an elementary response of the organism to external stimulus, Pavlov termed the normal digestive reaction an unconditioned reflex, and the activity of the salivary glands, stimulated by various environmental agents, a conditioned reflex.

Pavlov described the formation of the conditioned reflex, showing it to be based, like the unconditioned reflex, on the innate activity of the organism. He demonstrated that any environmental factor can enter into a temporary relation with the natural activity of the organism through combination with the unconditioned reflex. He noted that the chief characteristics of conditioned reflexes are that they are developed throughout the life of an organism (and are therefore extraordinarily subject to change, depending on the environment) and that they are provoked by stimuli that act as signals. Taken together, these qualities ensure the organism a completely individual adaptive activity. Pavlov saw in the conditioned reflex a mechanism through which the ameliorative potentialities of the organism are increased.

Pavlov made his first public statement on the conditioned reflex in 1903, in a paper presented to the

Fourteenth International Medical Congress in Madrid. He expanded upon the subject three years later, when he wrote that

> . . . with the general biological point of view before us we find in this conditioned reflex an improved adaptive mechanism or, in other words, a more precise mechanism for counterbalance with the environment. The organism reacts with natural phenomena that are vital to it in the most sensible and most precautionary way, since all other, even the smallest phenomena . . ., although accompanying the first only temporarily, present themselves as signals of the first—signals of the stimulus. The subtlety of the procedure makes itself known in the formation of the conditioned reflex as well as in its suppression, when it ceases to be a correct signal. Here, we must think, lies one of the main mechanisms of progress in the more finely differentiated nervous system. . . . The concept of the conditioned stimulus must be seen as the fruit of the previous work of biologists. . . .[11]

Pavlov found in the conditioned reflex a mechanism of individual adaptation which, he held, exists throughout the entire animal world. "A temporary nervous connection is a universal physiological phenomenon in the animal world and exists in us ourselves."[12]

Pavlov localized conditioned-reflex activity in the cerebral hemispheres of the brain, demonstrating that the center for such activity is to be found in the cortex, among the cortical agents of innate reflexes. Pavlov considered the possibility that subcortical formations may be responsible for the placement of the conditioned-reflex centers, but did not offer any direct evidence for this. He showed that with the formation of conditioned reflexes in the functional state of nerve centers displacements occur in the form of increases in irritability. The cells of the higher sections of the central nervous system, and their branches, he suggested, must therefore undergo definite subtle structural and physicochemical changes. "The locking-in, the formation of new connections," he wrote, "we relate to the function of the separating membrane, if it exists, or simply to the fine branching between neurons [that is], between the separate nerve cells."[13] Pavlov's hypothesis has been verified by more recent neurophysiological data, which have demonstrated the plastic character of the changes in the synactial apparatus as a result of excitation. Through work on the conditioned reflex investigators were able to establish that the activity of the cerebral hemispheres is based on the processes of excitation and inhibition. Further experiments, designed to elucidate the dynamics and mutual relationships of these processes, revealed a definite regularity in their development.

An important concomitant of Pavlov's experimental work was his creation of experimental neuroses, which arose when contradictory stimuli were offered the subject. Such neuroses may serve as a rough model for functional disease of the human nervous system; Pavlov and his co-workers attributed them to the disturbance of balance between the cortical processes of excitation and inhibition. In 1924, in Pavlov's laboratory, I. P. Razenkov, investigating the induced conflict of basic nerve processes in the activity of the cerebral cortex, observed the same phase states as N. E. Vvedensky had observed in the nerve fiber. It was shown that disturbance of cortical activity passed through four stages: inhibiting, characterized by the absence of all reflexes; paradoxical, in which strong stimuli produce little or no effect, while weak stimuli induce greater effects; equalizing, in which all conditioned stimuli, regardless of their intensity, produce the same effect; and intermediate to the norm, in which stimuli of average intensity produce the greatest effect, and strong or weak conditioned stimuli induce little or no effect. Pavlov applied Razenkov's work, which provided the first description of phase states of the central nervous system, toward understanding the nature of human psychic illness. From 1918 on, he regularly visited a psychiatric clinic in Udelnaya, near Leningrad, to study the patients.

A. G. Ivanov-Smolensky, V. V. Rikman, I. S. Rosenthal, I. O. Narbutovich, and L. N. Fedorov were active in the creation of the theory of experimental neuroses. Pavlov's student M. K. Petrova was able to induce deliberately various specific neuroses in animals and subsequently to suppress them. This study of experimental neuroses was closely related to the development of the theory of types of behavior. In 1909 Pavlov reported his pioneering work on behavior to the Society of Russian Physicians in St. Petersburg; his paper *Dalneyshie shagi obektivnogo analiza slozhnonervnykh yavleny* ("Further Steps in the Objective Analysis of Complex Nerve Phenomena") discussed carefully conditioned, "weak-nerved" dogs, in which it was difficult to induce inhibition.

In addition to studying animal behavior and the accumulating experimental material, Pavlov and his co-workers made the first attempt to provide a scientific basis for the ancient Hippocratic classification of temperaments. They established the existence of four basic types of behavior, which they classified according to the strength, mobility, and constancy of the basic nerve processes.

During the 1920's two ramifications emerged from Pavlov's basic theory: the study of comparative physiology of behavior and the theory of human behavior. Pavlov had experimented on dogs, mice, and monkeys; his students expanded the range of animal subjects, E. M. Kreps working with Ascidia, Y. P. Frolov with fish, N. A. Popov and B. I. Bayandurov with doves, P. M. Nikiforovsky and E. A. Asratyan with amphibians and reptiles, and G. A. Vasiliev and A. N. Promptov with birds. In the next decade Pavlov himself took up the idea of the genetic study of behavior; a biological station was established for this purpose at Koltushy, near Leningrad. On the basis of comparative physiological data an attempt was made, chiefly by Pavlov's student L. A. Orbeli, to create an evolutionary physiology of behavior.

Pavlov attributed decisive importance to the signals that characterize conditioned-reflex activity. He assumed the existence of two signal systems, of which one, the primary system, is found in both animals and man, whereas the secondary system is peculiar to man, and it is this system that makes possible the distinctively human activities of abstract thought and speech. In recent years he has come to be regarded as a mechanist who saw complex behavior as the sum of individual conditioned reflexes. This is a profound error, since in Pavlov's view the brain, through its capacity for subtle analysis and complex synthesis, integrates a vast range of conditioned reflexes into coherent behavior corresponding to the specific circumstances and needs of the organism. If in the early stages of his work Pavlov and his students were chiefly concerned with the study of elementary conditioned reflexes, they later turned to purposeful study of the more complex forms.

A distinguished scientific administrator, Pavlov created a large research school that, at various times, employed about 300 physiologists and physicians. He also organized a number of major research centers, including the physiological section of the Institute of Experimental Medicine, the Institute of Physiology of the Soviet Academy of Sciences, and the biological station at Koltushy. With Pavlov's active cooperation the Russian Physiological Society (now the I. P. Pavlov All-Union Physiological Society) was organized in 1917. He was an active member of the Society of Russian Physicians in St. Petersburg, and his services were highly valued by the Soviet government.

NOTES

1. I. P. Pavlov, *Polnoe sobranie sochineny* ("Complete Collected Works"), 2nd ed. (Moscow–Leningrad, 1951–1952), I, 197.
2. *Ibid.*, I, 82.
3. *Ibid.*, VI, 321.
4. *Ibid.*, II, bk. 2, p. 22.
5. *Ibid.*, 36.
6. A. F. Samoylov, *Izbrannye trudy* ("Selected Works"), V. V. Parin, ed. (Moscow, 1967), 301.

7. I. P. Pavlov, *Polnoe sobranie sochineny*, I, 574.
8. *Ibid.*, VI, 104.
9. *Ibid.*, 107.
10. *Ibid.*, IV, 19.
11. *Ibid.*, II, bk. 1, p. 71.
12. *Ibid.*, II, bk. 2, p. 182.
13. *Ibid.*, II, bk. 2, p. 61.

BIBLIOGRAPHY

I. ORIGINAL WORKS. Pavlov's writings were collected as *Polnoe sobranie trudov* ("Complete Collected Works"), 5 vols. (Moscow, 1940–1949). The 2nd ed. is *Polnoe sobranie sochineny* ("Complete Collected Works"), 6 vols. (Moscow–Leningrad, 1951–1952). There is a German trans. of this ed., *Sämtliche Werke*, L. Pickenhain, ed., 6 vols. (Berlin, 1953–1956).

Works referred to in the text are *O nervakh, zavedyvayushchikh rabotoy v podzheludochnoy zheleze* ("On the Nerves That Govern the Pancreas"; 1875), written with M. I. Afanasiev; *O tsentrobezhnykh nervakh serdtsa* ("On the Efferent Nerves of the Heart"; St. Petersburg, 1883), his doctoral diss.; "Kak bezzubka raskryvaet svoi stvorki" ("How the Anodonta Opens Its Valves"; *Polnoe sobranie sochineny*, 1, 466–493, also in *Pflügers Archiv*, **37** [1885], 6–31); *Lektsii o rabote glavnykh pishchevaritelnykh zhelez* ("Lectures on the Function of the Main Food-Digesting Glands"; 1897); *Eksperimentalnaya psikhologia i psikhopatologia na zhivotnykh* ("Experimental Psychology and Psychopathology in Animals"; 1903), his first public statement on the conditioned reflex; "Vneshnyaya rabota pishchevaritelnykh zhelez i ee mekhanizm" ("The External Function of the Digestive Glands and Its Mechanism"), in W. Nagel, ed., *Handbuch der Physiologie des Menschen*, II (Brunswick, 1907), 666–743; "Operativnaya metodika izuchenia pishchevaritelnykh zhelez" ("An Operative Method of Studying the Digestive Glands"; in Tigerstedt's *Handbuch der physiologischen Methodik*, Band II, IH, Leipzig, 1911); *Dvadtsatiletny opyt obektivnogo izuchenia vysshey nervnoy deyatelnosti (povedenia) zhivotnykh* ("Twenty Years of Experiments in the Objective Study of Higher Nervous Activity [Behavior] of Animals"; Moscow–Petrograd, 1923); and *Lektsii o rabote bolshikh polushary golovnogo mozga* ("Lectures on the Function of the Cerebral Hemispheres"; Moscow, 1927), Pavlov edited vols. **1–6** (1924–1936) of *Trudy fiziologicheskikh laboratorii imeni I. P. Pavlova*.

Important English translations of his works include *The Work of the Digestive Glands*, W. H. Thompson, trans. (London, 1902; 2nd ed., 1910); *Conditioned Reflexes*, G. V. Anrep, trans. and ed. (London, 1927; repr. New York, 1960), a trans. of the 1923 work cited above; and *Lectures on Conditioned Reflexes*, W. H. Gantt, trans. (New York, 1928).

II. SECONDARY LITERATURE. On Pavlov's life and work, see P. K. Anokhin, *Ivan Petrovich Pavlov Zhizn, deyatelnost i nauchnaya shkola* (". . . Life, Work, and Scient¹ School"; Moscow–Leningrad, 1949); E. A. Asraty; *Ivan Petrovitch Pavlov, Work* (Moscow, 1974), in Eng¹

B. P. Babkin, *Pavlov* (Chicago, 1949); Y. P. Frolov, *Pavlov and His School* (London, 1937), written by a student of Pavlov; and E. M. Kreps, ed., *I. P. Pavlov v vospominaniakh sovremennikov* ("Pavlov Recalled by His Contemporaries"; Leningrad, 1967).

N. A. Grigorian

PAYEN, ANSELME (*b.* Paris, France, 17 January 1795; *d.* Paris, 13 May 1871), *industrial chemistry, agricultural chemistry.*

Payen's father owned a factory in Grenelle, a suburb of Paris, in which sal ammoniac was made from animal waste. He would not permit his son to go to school and himself took charge of his education; the boy grew up well-informed on scientific matters but rather unsociable—traits that persisted throughout his life. The turmoil of the Hundred Days and its aftermath prevented him from entering the École Polytechnique, but he studied chemistry privately with Vauquelin and Chevreul. Payen's first industrial venture was the manufacture of borax, which until then had been imported; but more significant was his advocacy of animal charcoal (the carbonaceous residues from the Grenelle works) as superior to wood charcoal for decolorizing purposes in the recently established beet-sugar industry. He also had his own beet-sugar factory at Vaugirard.

In 1829 Payen began to teach industrial chemistry at the École Centrale des Arts et Manufactures; ten years later he was also appointed to a similar chair at the Conservatoire des Arts et Métiers, although he did not abandon his industrial interests. He wrote a large number of papers, mostly on technological matters of local and temporary concern. The subjects included manures (some in collaboration with Boussingault), sugar refining, rubber and gutta-percha, water supply, potato blight, and phylloxera. Such extensive publication would imply some dilution of quality, and Berzelius had an uncharitably low opinion of Payen's rank as a chemist. "I know the man so well," Berzelius wrote to Wöhler on 12 January 1847, "that I never rely on him where accuracy is concerned. But when it is a matter of writing pamphlet-fodder for the general public, then he is in his element."

Payen is remembered mainly for his work on carbohydrates, some of it done with Persoz. In 1833 they found that starch was hydrolyzed to sugar by a substance contained in malt, which they called diastase, now known to be a mixture of extracellular enzymes. (Their priority in this matter was disputed.) Payen later showed that starch has the same chemical composition, regardless of the species of plant from

which it is prepared. In 1838 he distinguished two components in woody tissue, an isomer of starch for which he coined the name cellulose and the "true woody material," later called lignin; the two could be separated chemically.

Payen spent all his life in the same poor quarter of Paris, much respected by his working-class neighbors. Of his five children only one daughter survived him. He is reputed never to have missed a lecture until, toward the end of 1869, he collapsed in front of his class. Even so, during the siege of Paris he devoted himself to attempts to make various unusual materials edible. He died during the Commune and was given an unpublicized and perfunctory funeral to the distant rattle of musketry.

BIBLIOGRAPHY

I. Original Works. Approximately 200 papers written by Payen or in collaboration with others are listed in the Royal Society *Catalogue of Scientific Papers*, IV, 783–789; VIII, 574–575; XII, 563. His books, in their various eds., are listed in Bibliothèque Nationale *Catalogue général des livres imprimés*, CXXXI, cols. 947–964. His most important work is the printed version of his lecture course, *Manuel du cours de chimie organique appliquée aux arts industriels et agricoles*, J. J. Garnier, ed., 2 vols. (Paris, 1842–1843).

II. Secondary Literature. The most informative obituary of Payen is by J.-A. Barral, in *Mémoires publiés par la Société centrale d'agriculture de France* (1873), 67–87. A. Girard, in *Annales du Conservatoire des arts et métiers*, **9** (1870) [*sic*], 317–331, admits to knowing little of Payen's private life and confines himself to an account of his work. An anonymous notice in *Revue scientifique*, **8** (1871), 94–96, is no more than a list of some of Payen's papers with brief comments.

W. V. Farrar

PAYKULL, GUSTAF (*b.* Stockholm, Sweden, 21 August 1757; *d.* Vallox-Säby, Sweden, 28 January 1826), *entomology.*

Paykull, a civil servant, became a chamberlain at the royal court in 1796 and was appointed master of the royal household in 1815. He received the title of baron in 1818.

In his twenties Paykull had a moderately successful career in literature, and wrote dramas and satirical plays. Later, however, his interest turned to natural history, where the possibilities for original contributions were decidedly greater than in belles lettres. On his estate, Vallox-Säby, in Uppland, he amassed the largest private zoological collection ever assembled in Scandinavia. The bird collections reportedly occupied

1,362 drawers, and the insects were said to have included 8,600 species. The shellfish and fish collections were comparatively large, but the collection of animals preserved in alcohol was unimportant. The extensive group of the larger tropical mammals—lion, leopard, camel, zebra—was noteworthy. Some of Paykull's animals were purchased during foreign travels—in Holland and France, for instance. The entire collection was donated in 1819 to the state and became the nucleus of the present National Museum of Natural History in Stockholm.

Although Paykull was a general zoologist, his primary interest was in entomology, and it was there that he made his lasting contributions. His first publications (1785–1786) concerned moths; he later concentrated on beetles, although he did produce a short article on the Lepidoptera in 1793. His main work, *Fauna Suecica*, appeared in three parts (1798–1800). Intending from the outset to publish a complete Swedish fauna, Paykull began with the insects—"to finish them off." His aim was not realized, however, and the three published volumes concern only beetles. An extraordinarily fine and well executed work, it clearly shows Paykull's taxonomical competence. In his *Fauna*, as well as in smaller monographs on various beetle families, Paykull accurately described a great number of new species.

BIBLIOGRAPHY

Paykull's major writings are *Monographia staphylinorum Sueciae* (Uppsala, 1789); *Monographia caraborum Sueciae* (Uppsala, 1790); *Monographia curculionum Sueciae* (Uppsala, 1792); *Fauna Suecica. Insecta*, 3 vols. (Uppsala, 1789–1800); and *Monographia histeroidum* (Uppsala, 1811). For the best account of Paykull and his work see Sten Lindroth, *Kungl. Svenska Vetenskapsakademiens Historia 1759–1818*, II (Stockholm, 1967), *passim*.

Bengt-Olof Landin

PEACOCK, GEORGE (*b.* Denton, near Darlington, Durham, England, 9 April 1791; *d.* Ely, England, 8 November 1858), *mathematics*.

Peacock is known for his role in the reform of the teaching of mathematics at Cambridge and his writings on algebra. His father, Thomas, was perpetual curate at Denton; and Peacock was educated at home. He entered Trinity College, Cambridge, in 1809 and received the B.A. in 1813, as second wrangler; the M.A. in 1816; and the D.D. in 1839. In 1815 he was named lecturer at Trinity and was a tutor from 1823 to 1839. He was a moderator of the tripos examination in 1817,

1819, and 1821. Peacock was a member of the Analytical Society, founded by Charles Babbage for the purpose of revitalizing mathematical studies at Cambridge. Toward this end Peacock, Babbage, and John Herschel published a translation of an elementary calculus text by Lacroix (1816). In 1820 Peacock published a collection of examples in differential and integral calculus. These works, and his influence as moderator, tutor, and lecturer, were major factors in replacing the fluxional notation and the geometric methods, which had been entrenched at Cambridge since the time of Newton, with the more fruitful analysis and Leibnizian notation.

In 1837 Peacock became Lowndean professor of geometry and astronomy at Cambridge, but in 1839 he was appointed dean of Ely. (He had been ordained in 1822.) Although he moved to Ely and no longer lectured, he remained active in the affairs of Cambridge. In 1841 he published a book on the statutes of the university in which he urged reform, and he served on two government commissions dealing with the question. Peacock was a member of the Cambridge Philosophical Society, the Royal Astronomical Society, the Geological Society of London, and the British Association for the Advancement of Science. He was elected a fellow of the Royal Society in 1818. He married Frances Elizabeth Selwyn in 1847. They had no children.

Peacock's mathematical work, although not extensive, is significant in the evolution of a concept of abstract algebra. In the textbook *A Treatise on Algebra* (1830), revised in 1842–1845, he attempted to put the theory of negative and complex numbers on a firm logical basis by dividing the field of algebra into arithmetical algebra and symbolic algebra. In the former the symbols represented positive integers; in the latter the domain of the symbols was extended by his principle of the permanence of equivalent forms. This principle asserts that rules in arithmetical algebra, which hold only when the values of the variables are restricted, remain valid when the restriction is removed. Although it was a step toward abstraction, Peacock's view was limited because he insisted that if the variables were properly chosen, any formula in symbolic algebra would yield a true formula in arithmetical algebra. Thus a noncommutative algebra would not be possible.

Peacock's other works include a survey on the state of analysis in 1833, prepared for the British Association for the Advancement of Science. It is an invaluable source for a contemporary view of the important problems at that time. Peacock also wrote a biography of Thomas Young and was one of the editors of his miscellaneous works.

BIBLIOGRAPHY

Early works are Sylvestre Lacroix, *An Elementary Treatise on the Differential and Integral Calculus*, translated by Charles Babbage, George Peacock, and John Herschel, with notes by Peacock and Herschel (Cambridge, 1816); and *A Collection of Examples of the Differential and Integral Calculus* (Cambridge, 1820). *A Treatise on Algebra* (Cambridge, 1830) is rare, but there is a rev. ed., 2 vols. (Cambridge, 1842–1845; repr. New York, 1940). Other writings include "Report on the Recent Progress and Present State of Certain Branches of Analysis," in *Report of the British Association for the Advancement of Science* (1834), 185–352; "Arithmetic," in *Encyclopaedia Metropolitana*, I (London, 1845); 369–523; *The Life of Thomas Young* (London, 1855); *Miscellaneous Works of the Late Thomas Young*, vols. I and II, George Peacock, ed., vol. III, John Leitch, ed. (London, 1855); and *Observations on the Statutes of the University of Cambridge* (London, 1841).

A complete bibliography of his writings can be found in Daniel Clock, "A New British Concept of Algebra: 1825–1850" (Ph.D. diss., U. of Wisconsin, 1964), 10–12; this work also contains an extensive discussion of Peacock's life and work.

ELAINE KOPPELMAN

PEALE, CHARLES WILLSON (*b.* Queen Anne's County, Maryland, 15 April 1741; *d.* Philadelphia, Pennsylvania, 22 February 1827), *museum direction.*

Eldest of the five children of Charles Peale, sometime clerk in the General Post Office, London, and Margaret Triggs of Annapolis, Maryland, Peale grew up in Chestertown, Maryland, where his father was master of the Kent County school. Deciding at the age of twenty-one that painting might be more profitable than saddlemaking, for which he had been trained, he sought instruction from John Hesselius in Maryland and from John Singleton Copley in Boston. Thereafter he displayed such skill in his portraits of the Maryland gentry that in 1767 several joined to send him abroad for two years to study under Benjamin West, who was later historical painter to George III. In 1776 Peale settled in Philadelphia, the largest and wealthiest city in the British colonies. During the American Revolution, Peale, a zealous patriot, served as a militia officer in the campaign of Trenton and Princeton (1776–1777) and the defense of Philadelphia (1777–1778); and as a "furious Whig" he took a prominent role in political controversies of the period. All the while he continued to paint—Washington, his officers, and the men of the Revolution are known today largely through Peale's eyes. With the return of peace Peale's artistic genius, mechanical skill, and patriotic vision of America's future found expression in designs for grand public displays, such as Philadelphia's Federal Procession of 4 July 1788.

A commission to make drawings of bones of a prehistoric creature from the banks of the Ohio River gave Peale the idea of establishing a natural history museum. It was the first in the United States, for the American Museum of Pierre Eugène du Simitière, although containing natural history specimens, was primarily historical. The museum was opened in Peale's house at Third and Lombard Streets in 1786; thereafter, without entirely giving up painting, he devoted his principal energies to it. It was moved in 1794 to the hall of the American Philosophical Society, the members of which took a warm interest in it and seemed thereby to endorse Peale's plans; and in 1802 it was established in the State House (now Independence Hall). Most of the exhibits were gifts or deposits (for instance, specimens collected by Lewis and Clark presented by President Jefferson); others were secured by Peale himself, most notably the nearly complete skeletons of two mastodons from Orange County, New York, in 1801. When the museum collections were sold in 1848 and 1854 the catalog listed 1,824 birds, 250 quadrupeds, 650 fish, 135 reptiles, lizards, and tortoises, 269 portraits, and thirty-three cases of shells.

Peale's museum demonstrated sound principles of scientific exposition. The exhibits were arranged according to the modified Linnaean system, from fossils and insects to "animal man" (although Peale never obtained a preserved specimen of *Homo sapiens*); and as far as possible they were shown in their natural forms, attitudes, and backgrounds, which Peale as painter readily provided. Special attention was directed to likenesses between species and to distinctive features: a rattlesnake, for example, was mounted with its jaws open and a glass was placed so that all might see the fangs and venom sacs.

Peale viewed the museum as an element in a reformed system of education suitable for virtuous republicans and hoped it would become a national institution, comparable with the Muséum National d'Histoire Naturelle in Paris, with salaried staff, lecturers, laboratories, and publications. This would have required greater support than the federal government was prepared to give. Peale had, therefore, to depend on admission charges, which, although they produced a satisfactory income, constantly exposed the museum—especially when Peale was not personally in charge—to pressures to compromise with scientific integrity. The museum was a school for such young naturalists as Alexander Wilson, John D. Godman, and Richard Harlan, who drew upon its collections for their earliest researches. Similar institutions were established in Baltimore and New York by Peale's sons Rubens and Rembrandt; and catchpenny imitations which did not scorn, as Peale's museum did, to "catch the eye of the gaping multitude," sprang up throughout the country after 1820. These cabinets of jumbled curiosities gave a few young men an introduction to science, but they misled many as to science's nature and scope and fell far short of Peale's goal.

Peale was married three times: to Rachel Brewer of Annapolis in 1762; to Elizabeth DePeyster of New York, in 1791; and to Hannah Moore of Philadelphia, in 1805. From the two first marriages he had seventeen children, of whom eleven reached maturity. Most of them displayed artistic talents of high order; two—Titian Ramsay and Franklin—achieved distinction in natural history and mechanics, respectively.

BIBLIOGRAPHY

Peale's paintings have been listed and illustrated in Charles Coleman Sellers, "Portraits and Miniatures by Charles Willson Peale" and "Charles Willson Peale With Patron and Populace," in *Transactions of the American Philosophical Society*, n.s. **42**, pt. 1 (1952), and **59**, pt. 3 (1969), respectively. The contents of the museum are described in Peale and A. F. M. J. Palisot de Beauvois, *A Scientific and Descriptive Catalogue of Peale's Museum* (Philadelphia, 1796); and Peale's *Guide to the Philadelphia Museum* (Philadelphia, 1804). Peale explained his purpose and philosophy in *Introduction to a Course of Lectures on Natural History* (Philadelphia, 1800). His letter books, diary, autobiography, and other MSS are in the American Philosophical Society library.

Charles Coleman Sellers, *Charles Willson Peale* (New York, 1969), based upon a lifetime of study, is a sufficient introduction to, as well as the last authority on, the subject and its sources.

WHITFIELD J. BELL, JR.

PEALE, REMBRANDT (*b.* near Richboro, Pennsylvania, 22 February 1778; *d.* Philadelphia, Pennsylvania, 3 October 1860), *biology.*

For a detailed study of his life and work, see Supplement.

PEALE, TITIAN RAMSAY (*b.* Philadelphia, Pennsylvania, 2 November 1799; *d.* Philadelphia, 13 March 1885), *natural history.*

Titian Peale, youngest son of Charles Willson Peale and his second wife, Elizabeth DePeyster Peale, knew Philadelphia's scientific men from childhood.

His formal education ended at age thirteen. At sixteen he was sketching for volume I of Thomas Say's *American Entomology* (1824). At eighteen he was elected to the Academy of Natural Sciences of Philadelphia.

Peale's first natural history collecting expedition was in 1817–1818, to Florida and the Sea Islands of Georgia, with Thomas Say, George Ord, and William Maclure. In 1819–1820 he was assistant naturalist with Stephen Long's expedition to the Rocky Mountains, making 122 sketches and drawings.

During 1822–1838 Peale was employed chiefly at the Philadelphia Museum. In the winter of 1824–1825 he collected in Florida for Charles Lucien Bonaparte and then drew all but one of the plates for volume I of Bonaparte's *American Ornithology* (1825); many of the specimens from which the plates were drawn were of Peale's collecting. He visited Maine in 1829 and returned from a trip to Colombia (1830–1832) with 500 bird skins, as well as drawings and butterflies, for exhibition in the Philadelphia Museum. In 1833 he issued a prospectus for what he hoped would be his most important publication, *Lepidoptera Americana*, temporarily abandoned because it was too expensive. He was elected to the American Philosophical Society in 1833.

Peale's great opportunity came when he was appointed as a naturalist on the United States South Seas Surveying and Exploring Expedition (the Wilkes expedition) of 1838–1842. On the homeward journey, in June 1841, one of the expedition's ships, the *Peacock*, was wrecked. A large proportion, and the best, of Peale's bird and animal specimens, all of his butterflies, and an extensive collection of native artifacts were lost—the results of three years' collecting. Still other specimens, which had been shipped back, were improperly handled. Bureaucratic restrictions, lack of library facilities, and Peale's sometimes difficult temper and financial problems combined with quarrels over the quality of engravings to present difficulties during the preparation of his book, which Charles Wilkes titled *Mammalia and Ornithology* (1848).

Wilkes objected to Peale's preface, in which he said that although the government specified that only new species should be described, he felt it would have been more appropriate also to record times and places of observations of known species. There was some criticism of Peale's nomenclature, and therefore Wilkes suppressed the volume shortly after its publication. In 1852 John Cassin, a brilliant taxonomist, was appointed to rewrite it. In his *Mammalogy and Ornithology* (1858) the classifications and names of the species are often different, but Peale's field observations are quoted extensively. According to Harley Harris Bartlett, "Cassin went too far afield to find species to which Peale's might be reduced, and the more modern conception of geographic species might justify the reinstatement of [a number of] Peale's species" ("Reports of the Wilkes Expedition," p. 689). It was a crushing professional defeat for Peale.

Peale was later an examiner in the U.S. Patent Office (1849–1872), did amateur photography, wrote occasional articles, painted, and worked on his manuscript on butterflies. He was a passionate and careful field observer and collector, rather than a "closet naturalist" or skilled taxonomist, at a time when questions of synonymy and nomenclature were deemed of increasing importance. Consequently, he often observed and collected species that others subsequently recorded and described.

BIBLIOGRAPHY

I. ORIGINAL WORKS. Peale's papers are in the collections of the American Philosophical Society, the Historical Society of Pennsylvania, the Academy of Natural Sciences of Philadelphia, and the Library of Congress. The American Museum of Natural History, New York, possesses Peale's unpublished MS "The Butterflies of North America, Diurnal Lepidoptera, Whence They Come; Where They Go; and What They Do," with 3 vols. of accompanying drawings and paintings. His *Lepidoptera Americana. Prospectus* (Philadelphia, 1833) and *Mammalia and Ornithology*, vol. VIII of the Scientific Reports of the U.S. Exploring Expedition of 1838–1842 (Philadelphia, 1848), are extremely rare.

II. SECONDARY LITERATURE. Jessie Poesch, *Titian Ramsay Peale, 1799–1885, and His Journals of the Wilkes Expedition*, which is *Memoirs of the American Philosophical Society*, **52** (1961), an extensive bibliography. See also Harley Harris Bartlett, "The Reports of the Wilkes Expedition, and the Work of the Specialists in Science," in *Proceedings of the American Philosophical Society*, **82** (1940), 601–705; Mary E. Cooley, "The Exploring Expedition in the Pacific," *ibid.*, 707–719; Clifford Merrill Drury, *Diary of Titian Ramsay Peale* (Los Angeles, 1957); Daniel C. Haskell, *The United States Exploring Expedition, 1838–1842, and Its Publications 1844–1874—a Bibliography* (New York, 1942); and Asa Orrin Weese, ed., "The Journal of Titian Ramsay Peale, Pioneer Naturalist," in *Missouri Historical Review*, **41** (1947), 147–163, 266–284.

JESSIE POESCH

PÉAN DE SAINT-GILLES, LÉON (*b.* Paris, France, 4 January 1832; *d.* Cannes, France, 22 March 1862), *analytical chemistry.*

Péan de Saint-Gilles was born into an old and very rich family. His father, like his ancestors, was a notary.

Sickly and weak, he never attended public schools but received private tutoring instead. At the age of seventeen he earned his bachelor of letters degree. Departing from family tradition, he chose scientific research as a profession, but his poor health made it impossible for him to pursue university studies on a regular basis. Thus, he gained his knowledge of chemistry by himself and acquired practical laboratory experience under the guidance of Pelouze, a student and successor of Gay-Lussac at the École Polytechnique. Independently wealthy, Péan de Saint-Gilles later had a laboratory built for himself in which he carried out chemical investigations. He had already achieved some success when death ended his very promising career. The symptoms of consumption appeared at the beginning of 1861; he moved to Cannes to aid his cure, but it was no longer of any help. He was survived by his widow and two children.

Péan de Saint-Gilles's most important work was in the field of titrimetry. In 1846 Frédéric Marguerite introduced the standard solution of potassium permanganate (then called chameleon solution) into the volumetric analysis employed in the determination of iron. Pelouze applied this method to other determinations. Péan de Saint-Gilles extended the use of potassium permanganate as a titrimetric solution for the quantitative determination of nitrite and iodide, as well as of oxalic acid and other organic substances. All of these procedures are still used. He also worked on the identification of the oxidation products of organic substances. His investigations in the area of inorganic chemistry are of no particular importance.

In physical chemistry he examined, in collaboration with Berthelot, the esterification of alcohols with acids. They found that the reaction was never complete but reached a state of equilibrium. This state was independent of the quality of the alcohol and acid. Finding that "the amount of ester formed in each moment is proportional to the product of the reacting substances," they attempted to give a mathematical formulation of the phenomenon. It was the crucial reaction to which Guldberg and Waage referred in their enunciation of the law of mass action in 1864.

BIBLIOGRAPHY

There is a bibliography of Péan de Saint-Gilles's works in Poggendorff, III, 1010–1011. See also the Royal Society *Catalogue of Scientific Papers*, which lists eighteen works, seven of them written with Berthelot. The latter include the important "Recherches sur les affinités," in *Annales de chimie et de physique*, 3rd ser., **65** (1862), 385–422; **66** (1862), 5–110; **68** (1863), 225–359.

On his life and work, see M. Berthelot, "Nécrologie," in *Bulletin. Société chimique de France*, **A5** (1863), 226–227; J. R. Partington, *A History of Chemistry*, IV (London, 1964), 584–585; and F. Szabadváry, *History of Analytical Chemistry* (Oxford, 1966), 251.

F. Szabadváry

PEANO, GIUSEPPE (*b.* Spinetta, near Cuneo, Italy, 27 August 1858; *d.* Turin, Italy, 20 April 1932), *mathematics, logic.*

Giuseppe Peano was the second of the five children of Bartolomeo Peano and Rosa Cavallo. His brother Michele was seven years older. There were two younger brothers, Francesco and Bartolomeo, and a sister, Rosa. Peano's first home was the farm Tetto Galant, near the village of Spinetta, three miles from Cuneo, the capital of Cuneo province, in Piedmont. When Peano entered school, both he and his brother walked the distance to Cuneo each day. The family later moved to Cuneo so that the children would not have so far to walk. The older brother became a successful surveyor and remained in Cuneo. In 1974 Tetto Galant was still in the possession of the Peano family.

Peano's maternal uncle, Michele Cavallo, a priest and lawyer, lived in Turin. On this uncle's invitation Peano moved to Turin when he was twelve or thirteen. There he received private lessons (some from his uncle) and studied on his own, so that in 1873 he was able to pass the lower secondary examination of the Cavour School. He then attended the school as a regular pupil and in 1876 completed the upper secondary program. His performance won him a room-and-board scholarship at the Collegio delle Provincie, which was established to assist students from the provinces to attend the University of Turin.

Peano's professors of mathematics at the University of Turin included Enrico D'Ovidio, Angelo Genocchi, Francesco Siacci, Giuseppe Basso, Francesco Faà di Bruno, and Giuseppe Erba. On 16 July 1880 he completed his final examination "with high honors." For the academic year 1880–1881 he was assistant to D'Ovidio. From the fall of 1881 he was assistant and later substitute for Genocchi until the latter's death in 1889. On 21 July 1887 Peano married Carola Crosio, whose father, Luigi Crosio (1835–1915), was a genre painter.

On 1 December 1890, after regular competition, Peano was named extraordinary professor of infinitesimal calculus at the University of Turin. He was promoted to ordinary professor in 1895. In 1886 he had been named professor at the military academy, which was close to the university. In 1901 he gave up his position at the military academy but retained his professorship at the university until his death in 1932, having transferred in 1931 to the chair of complementary mathematics. He was elected to a number of scientific societies, among them the Academy of Sciences of Turin, in which he played a very active role. He was also a knight of the Order of the Crown of Italy and of the Order of Saint Maurizio and Saint Lazzaro. Although he was not active politically, his views tended toward socialism; and he once invited a group of striking textile workers to a party at his home. During World War I he advocated a closer federation of the allied countries, to better prosecute the war and, after the peace, to form the nucleus of a world federation. Peano was a nonpracticing Roman Catholic.

Peano's father died in 1888; his mother, in 1910. Although he was rather frail as a child, Peano's health was generally good. His most serious illness was an attack of smallpox in August 1889. After having taught his regular class the previous afternoon, Peano died of a heart attack the morning of 20 April 1932. At his request the funeral was very simple, and he was buried in the Turin General Cemetery. Peano was survived by his wife (who died in Turin on 9 April 1940), his sister, and a brother. He had no children. In 1963 his remains were transferred to the family tomb in Spinetta.

Peano is perhaps most widely known as a pioneer of symbolic logic and a promoter of the axiomatic method, but he considered his work in analysis most important. In 1915 he printed a list of his publications, adding: "My works refer especially to infinitesimal calculus, and they have not been entirely useless, seeing that, in the judgment of competent persons, they contributed to the constitution of this science as we have it today." This "judgment of competent persons" refers in part to the *Encyklopädie der mathematischen Wissenschaften*, in which Alfred Pringsheim lists two of Peano's books among nineteen important calculus texts since the time of Euler and Cauchy. The first of these books was Peano's first major publication and is something of an oddity in the history of mathematics, since the title page gives the author as Angelo Genocchi, not Peano: *Angelo Genocchi, Calcolo differenziale e principii di calcolo integrale, publicato con aggiunte dal D.*^r *Giuseppe Peano.* The origin of the book is that Bocca Brothers wished to publish a calculus text based on Genocchi's lectures. Genocchi did not wish to write such a text but gave Peano permission to do so. After its publication Genocchi, thinking Peano lacked regard for him, publicly disclaimed all credit for the book, for which Peano then assumed full responsibility.

Of the many notable things in this book, the *Encyklopädie der mathematischen Wissenschaften* cites theorems and remarks on limits of indeterminate expressions, pointing out errors in the better texts then in use; a generalization of the mean-value theorem for derivatives; a theorem on uniform continuity of functions of several variables; theorems on the existence and differentiability of implicit functions; an example of a function the partial derivatives of which do not commute; conditions for expressing a function of several variables with Taylor's formula; a counterexample to the current theory of minima; and rules for integrating rational functions when roots of the denominator are not known. The other text of Peano cited in the *Encyklopädie* was the two-volume *Lezioni di analisi infinitesimale* of 1893. This work contains fewer new results but is notable for its rigor and clarity of exposition.

Peano began publication in 1881 with articles on the theory of connectivity and of algebraic forms. They were along the lines of work done by D'Ovidio and Faà di Bruno. Peano's work in analysis began in 1883 with an article on the integrability of functions. The article of 1890 contains original notions of integrals and areas. Peano was the first to show that the first-order differential equation $y' = f(x, y)$ is solvable on the sole assumption that f is continuous. His first proof dates from 1886, but its rigor leaves something to be desired. In 1890 this result was generalized to systems of differential equations using a different method of proof. This work is also notable for containing the first explicit statement of the axiom of choice. Peano rejected the axiom of choice as being outside the ordinary logic used in mathematical proofs. In the *Calcolo geometrico* of 1884 Peano had already given many counterexamples to commonly accepted notions in mathematics, but his most famous example was the space-filling curve that was published in 1890. This curve is given by continuous parametric functions and goes through every point in a square as the parameter ranges over some interval. Some of Peano's work in analysis was quite original, and he has not always been given credit for his priority; but much of his publication was designed to clarify and to make rigorous the current definitions and theories. In this regard we may mention his clarification of the notion of area of a surface (1882, independently discovered by H. A. Schwarz); his work with Wronskians, Jacobians, and other special determinants, and with Taylor's formula; and his generalizations of quadrature formulas.

Peano's work in logic and in the foundations of mathematics may be considered together, although he never subscribed to Bertrand Russell's reduction of mathematics to logic. Peano's first publication in logic

was a twenty-page preliminary section on the operations of deductive logic in *Calcolo geometrico secondo l'Ausdehnungslehre di H. Grassmann* (1888). This section, which has almost no connection with the rest of the text, is a synthesis of, and improvement on, some of the work of Boole, Schröder, Peirce, and McColl. The following year, with the publication of *Arithmetices principia, nova methodo exposita*, Peano not only improved his logical symbolism but also used his new method to achieve important new results in mathematics; this short booklet contains Peano's first statement of his famous postulates for the natural numbers, perhaps the best known of all his creations. His research was done independently of the work of Dedekind, who the previous year had published an analysis of the natural numbers, which was essentially that of Peano but without the clarity of Peano. (This was the only work Peano wrote in Latin.) *Arithmetices principia* made important innovations in logical notation, such as $\in$ for set membership and a new notation for universal quantification. Indeed, much of Peano's notation found its way, either directly or in a somewhat modified form, into mid-twentieth-century logic.

In the 1890's he continued his development of logic, and he presented an exposition of his system to the First International Congress of Mathematicians (Zurich, 1897). At the Paris Philosophical Congress of 1900, Peano and his collaborators—Burali-Forti, Padoa, and Pieri—dominated the discussion. Bertrand Russell later wrote, "The Congress was a turning point in my intellectual life, because I there met Peano."

In 1891 Peano founded the journal *Rivista di matematica*, which continued publication until 1906. In the journal were published the results of his research and that of his followers, in logic and the foundations of mathematics. In 1892 he announced in the *Rivista* the *Formulario* project, which was to take much of his mathematical and editorial energies for the next sixteen years. He hoped that the result of this project would be the publication of a collection of all known theorems in the various branches of mathematics. The notations of his mathematical logic were to be used, and proofs of the theorems were to be given. There were five editions of the *Formulario*. The first appeared in 1895; the last was completed in 1908, and contained some 4,200 theorems. But Peano was less interested in logic as a science per se than in logic as used in mathematics. (For this reason he called his system "mathematical logic.") Thus the last two editions of the *Formulario* introduce sections on logic only as it is needed in the proofs of mathematical theorems. The editions through 1901 do contain separate, well-organized sections on logic.

The postulates for the natural numbers received minor modifications after 1889 and assumed their definitive form in 1898. Peano was aware that the postulates do not characterize the natural numbers and, therefore, do not furnish a definition of "number." Nor did he use his mathematical logic for the reduction of mathematical concepts to logical concepts. Indeed, he denied the validity of such a reduction. In a letter to Felix Klein (19 September 1894) he wrote: "The purpose of mathematical logic is to analyze the ideas and reasoning that especially figure in the mathematical sciences." Peano was neither a logicist nor a formalist. He believed rather that mathematical ideas are ultimately derived from our experience of the material world.

In addition to his research in logic and arithmetic, Peano also applied the axiomatic method to other fields, notably geometry, for which he gave several axiom systems. His first axiomatic treatment of elementary geometry appeared in 1889 and was extended in 1894. His work was based on that of Pasch but reduced the number of undefined terms from four to three: point and segment, for the geometry of position (1889), and motion, also necessary for metric geometry (1894). (This number was reduced to two by Pieri in 1899.)

The treatise *Applicazioni geometriche del calcolo infinitesimale* (1887) was based on a course Peano began teaching at the University of Turin in 1885 and contains the beginnings of his "geometrical calculus" (here still influenced by Bellavitis' method of equipollences), new forms of remainders in quadrature formulas, new definitions of length of an arc of a curve and of area of a surface, the notion of a figure tangent to a curve, a determination of the error term in Simpson's formula, and the notion of the limit of a variable figure. There is also a discussion of the measure of a point set, of additive functions of sets, and of integration applied to sets. Peano here generalized the notion of measure that he had introduced in 1883. Peano's popularization of the vectorial methods of H. Grassmann—beginning with the publication in 1888 of the *Calcolo geometrico secondo l'Ausdehnungslehre di H. Grassmann*—was of more importance in geometry. Grassmann's own publications have been criticized for their abstruseness. Nothing could be clearer than Peano's presentation, and he gave great impetus to the Italian school of vector analysis.

Peano's interest in numerical calculation led him to give formulas for the error terms in many commonly used quadrature formulas and to develop a theory of "gradual operations," which gave a new method for the resolution of numerical equations. From

1901 until 1906 he also contributed to actuarial mathematics, when as a member of a state commission he was asked to review a pension fund.

Peano also wrote articles on rational mechanics (1895–1896). Several of these articles dealt with the motion of the earth's axis and had their origin in the famous "falling cat" experiment of the Paris Academy of Sciences in the session of 29 October 1894. This experiment raised the question: "Can the earth change its own orientation in space, using only internal actions as animals do?" Peano took the occasion to apply his geometrical calculus in order to show that, for example, the Gulf Stream alone was able to alter the orientation of the earth's axis. This topic was the occasion of a brief polemic with Volterra over both priority and substance.

By 1900 Peano was already interested in an international auxiliary language, especially for science. On 26 December 1908 he was elected president of the Akademi Internasional de Lingu Universal, a continuation of the Kadem Volapüka, which had been organized in 1887 by the Reverend Johann Martin Schleyer in order to promote Volapük, the artificial language first published by Schleyer in 1879. Under Peano's guidance the Academy was transformed into a free discussion association, symbolized by the change of its name to Academia pro Interlingua in 1910. (The term "interlingua" was understood to represent the emerging language of the future.) Peano remained president of the Academia until his death. During these years Peano's role as interlinguist eclipsed his role as professor of mathematics.

Peano's mathematical logic and his ideography for mathematics were his response to Leibniz' dream of a "universal characteristic," whereas Interlingua was to be the modern substitute for medieval Latin, that is, an international language for scholars, especially scientists. Peano's proposal for an "interlingua" was *latino sine flexione* ("Latin without grammar"), which he published in 1903. He believed that there already existed an international scientific vocabulary, principally of Latin origin; and he tried to select the form of each word which would be most readily recognized by those whose native language was either English or a Romance language. He thought that the best grammar was no grammar, and he demonstrated how easily grammatical structure may be eliminated. His research led him to two areas: one was the algebra of grammar, and the other was philology. The latter preoccupation resulted most notably in *Vocabulario commune ad latino-italiano-français-english-deutsch* (1915), a greatly expanded version of an earlier publication (1909). This second edition contains some 14,000 entries and gives for each the form to be adopted in Interlingua, the classic Latin form, and its version in Italian, French, English, and German (and sometimes in other languages), with indications of synonyms, derivatives, and other items of information.

In his early years Peano was an inspiring teacher; but with the publication of the various editions of the *Formulario*, he adopted it as his text, and his lectures suffered from an excess of formalism. Because of objections to this method of teaching, he resigned from the military academy in 1901 and a few years later stopped lecturing at the Polytechnic. His interest in pedagogy was strong, and his influence was positive. He was active in the Mathesis Society of school teachers of mathematics (founded in 1895); and in 1914 he organized a series of conferences for secondary teachers of mathematics in Turin, which continued through 1919. Peano constantly sought to promote clarity, rigor, and simplicity in the teaching of mathematics. "Mathematical rigor," he wrote, "is very simple. It consists in affirming true statements and in not affirming what we know is not true. It does not consist in affirming every truth possible."

As historian of mathematics Peano contributed many precise indications of origins of mathematical terms and identified the first appearance of certain symbols and theorems. In his teaching of mathematics he recommended the study of original sources, and he always tried to see in his own work a continuation of the ideas of Leibniz, Newton, and others.

The influence of Peano on his contemporaries was great, most notably in the instance of Bertrand Russell. There was also a school of Peano: the collaborators on the *Formulario* project and others who were proud to call themselves his disciples. Pieri, for example, had great success with the axiomatic method, Burali-Forti applied Peano's mathematical logic, and Burali-Forti and Marcolongo developed Peano's geometrical calculus into a form of vector analysis. A largely different group was attracted to Peano after his shift of interest to the promotion of an international auxiliary language. This group was even more devoted; and those such as Ugo Cassina, who shared both the mathematical and philological interests of Peano, felt the closest of all.

It has been said that the apostle in Peano impeded the work of the mathematician. This is no doubt true, especially of his later years; but there can be no question of his very real influence on the development of mathematics. He contributed in great measure to the popularity of the axiomatic method, and his discovery of the space-filling curve must be considered remarkable. While many of his notions, such as area and integral, were "in the air," his originality is undeniable. He was not an imposing person, and his

gruff voice with its high degree of lallation could hardly have been attractive; but his gentle personality commanded respect, and his keen intellect inspired disciples. Much of Peano's mathematics is now of historical interest; but his summons to clarity and rigor in mathematics and its teaching continues to be relevant, and few have expressed this call more forcefully.

BIBLIOGRAPHY

I. ORIGINAL WORKS. See Ugo Cassina, ed., *Opere scelte*, 3 vols. (Rome, 1957–1959), which contains half of Peano's articles and a bibliography (in vol. I) that lists approximately 80 percent of Peano's publications. A more complete list is in Hubert C. Kennedy, ed., *Selected Works of Giuseppe Peano* (Toronto, 1972). The fifth ed. of the *Formulario mathematico* has been reprinted in facsimile (Rome, 1960).

II. SECONDARY LITERATURE. The most complete biography is Hubert C. Kennedy, *Giuseppe Peano* (Basel, 1974). Ten articles on the work of Peano are in Ugo Cassina, *Critica dei principî della matematica e questioni di logica* (Rome, 1961) and *Dalla geometria egiziana alla matematica moderna* (Rome, 1961). Also see Alessandro Terracini, ed., *In memoria di Giuseppe Peano* (Cuneo, 1955), which contains articles by eight authors. A list of these and other items is in *Selected Works of Giuseppe Peano*.

HUBERT C. KENNEDY

PEARL, RAYMOND (*b.* Farmington, New Hampshire, 3 June 1879; *d.* Hershey, Pennsylvania, 17 November 1940), *biology, genetics.*

Pearl was the only child of Frank Pearl and Ida May McDuffee. He attended public schools in Farmington and nearby Rochester. In 1899 he earned the A.B. in biology at Dartmouth College and the Ph.D. in 1902 at the University of Michigan, where his dissertation was on the behavior of a flatworm (*Planaria*). He also studied at the University of Leipzig in 1905 and at University College, London, from 1905–1906. In London he studied under Karl Pearson, whose influence led Pearl to apply statistics to population studies.

Pearl was an instructor in zoology at the University of Pennsylvania (1906–1907) until he became chairman of the department of biology at the Maine Agricultural Experiment Station (1907–1918), where he studied the heredity and reproduction of poultry and cattle. As chief of the statistical division of the U.S. Food Administration from 1917 to 1919, he studied the relationship of food to population. Pearl's long association with the Johns Hopkins University began in 1918, when he became professor of biometry and vital statistics in the School of Hygiene and Public Health (1918–1925). He was also professor of biology in the School of Medicine (1923–1940), research professor and director of the Institute of Biological Research (1925–1930), and statistician at the Johns Hopkins Hospital (1919–1935).

A prodigious researcher and a voluminous and articulate writer, Pearl achieved renown as a pioneer in world population changes, birth and death rates, and longevity. He founded and edited the *Quarterly Review of Biology* from 1926 and *Human Biology* from 1929. On 29 June 1903 he married Maud Mary DeWitt, who assisted his researches and writing; the couple had two daughters, Ruth DeWitt and Penelope Mackey.

Pearl attracted public attention in 1920 with a mathematical equation for determining population to the year 2100. His predictions deviated only 3.7 percent in the 1940 census. His other research findings, often controversial, led him to believe that the length of life varied inversely with the tempo or pace of living, that heredity predominated over environment in the length of life and in shaping one's destiny, that moderate drinkers lived longer than total abstainers, and that intellectuals had a better chance to live longer than did manual workers. In one study he analyzed the reproductive histories; the use of contraception; and the social, economic, educational, health, and religious histories of 30,949 mothers.

Pearl received many honorary degrees for his work relating biology to the social sciences. He was president of the International Union for Scientific Investigation of Population Problems (1928–1931), the American Association of Physical Anthropologists (1934–1936), and the American Statistical Association (1939). He was also made a knight (1920) and an officer (1929) of the Crown of Italy.

Obituary accounts called Pearl "a statistician of the human race" and a "biologist-philosopher." H. L. Mencken praised his lucid writing style, his wide knowledge and interests, his scientific creativity, and his delight in playing the French horn.

BIBLIOGRAPHY

I. ORIGINAL WORKS. Pearl's writings include *Modes of Research in Genetics* (New York, 1915); *Diseases of Poultry* (New York, 1915), written with F. M. Surface and M. R. Curtis; *The Nation's Food* (Philadelphia, 1920); *The Biology of Death* (Philadelphia, 1922); *Introduction to Medical Biometry and Statistics* (Philadelphia–London, 1923; 3rd ed. 1940); *Studies in Human Biology* (Baltimore, 1924); *The Biology of Population Growth* (New York, 1925);

Alcohol and Longevity (New York, 1926); *To Begin With* (New York, 1927; rev. 1930); *The Rate of Living* (New York, 1928); *Constitution and Health* (London–New York, 1933); *The Ancestry of the Long-Lived* (Baltimore, 1934), written with Ruth DeWitt Pearl; and *The Natural History of Population* (London–New York, 1939). He was editorial associate of *Biometrika, Journal of Agricultural Research, Genetics, Metron, Biologia generalis,* and *Acta biotheoretica.*

II. SECONDARY LITERATURE. Biographical and obituary accounts are A. W. Freeman, "Raymond Pearl, 1879–1940," in *American Journal of Public Health*, **31**, no. 1 (1941), 81–82; H. S. Jennings, "Raymond Pearl, 1879–1940," in *Biographical Memoirs. National Academy of Sciences*, **22** (1943), 295–347, which includes a full bibliography; H. L. Mencken, in Baltimore *Sun* (24 Nov. 1940); J. R. Miner and J. Berkson, "Raymond Pearl, 1879–1940," in *Scientific Monthly*, **52** (1941), 192–194; "News and Notes," in *American Journal of Sociology*, **16**, no. 4 (1941), 604; and *Dictionary of American Biography*, XXII, supp. 2 (1958), 521–522.

FRANKLIN PARKER

PEARSON, GEORGE (*b.* Rotherham, England, 1751 [baptized 4 September]; *d.* London, England, 9 November 1828), *chemistry.*

Pearson was one of the first chemists in Britain to accept the "antiphlogistic" theories of Lavoisier. He is best known for his role in introducing into Britain the nomenclature devised by Lavoisier and other leading French chemists. He studied medicine at Edinburgh University from 1770 to 1774 and received the M.D. in 1773; he also studied chemistry under Joseph Black. After a brief period at St. Thomas's Hospital in London, Pearson spent about two years in Europe before establishing a medical practice in Doncaster, where he stayed for about six years. He eventually moved to London. In 1787 he became chief physician at St. George's Hospital. He was admitted to the Royal Society in 1791 and for many years served on the council.

On hearing of Edward Jenner's successful inoculation against smallpox, in which he used matter from the pustule of a cowpox patient, Pearson became interested in the subject. He published a number of articles and pamphlets and eventually set up an institution to provide vaccinations. His program, however, delayed, rather than hastened, the general adoption of Jenner's method, for Pearson used a defective vaccine which frequently produced severe eruptions resembling smallpox. Ill feeling thus developed between Jenner and Pearson. Pearson tended to belittle Jenner's achievements and opposed his successful claim for remuneration from the government.

A glimpse of Pearson as a lecturer was afforded by the American chemist Benjamin Silliman, who, when planning his visit to Europe in 1805–1806, had been given a letter of introduction to Pearson as "the greatest chemist in England." Silliman said Pearson lectured on chemistry, materia medica, and therapeutics for two and a quarter hours without a break. "There was no interval for breathing or for a gentle transition to a new subject. This mental repletion was not favorable to intellectual digestion" (see G. P. Fisher, *Life of Benjamin Silliman*, I [New York, 1866], 144–145).

Nevertheless Pearson seems to have been a competent chemist. He investigated the composition of "James's powder," a popular febrifuge which made a fortune for Robert James. He found that it was a mixture of bone ash and antimony oxide. In 1792 he extended the work of Smithson Tennant, who had shown that carbon was obtained when powdered marble was heated with phosphorus (S. Tennant, "On the Decomposition of Fixed Air," in *Philosophical Transactions of the Royal Society*, **81** [1791], 182–184). Pearson showed that sodium carbonate could be similarly decomposed, and he discovered calcium phosphide by heating phosphorus with quicklime. He noted the reaction of calcium phosphide with water and the spontaneous combustion of "phosphoric air" (phosphine). With J. Stodart he investigated the composition of Indian ("wootz") steel (see R. A. Hadfield, *Faraday and His Metallurgical Researches* [London, 1931], pp. 36–37 and *passim*) and made a useful contribution to the history of metallurgy by analyzing some ancient weapons and utensils.

In 1789 the Dutch chemists A. Paete van Troostwijk and J. R. Deiman succeeded in decomposing water by frictional electricity, although they were unable to show conclusively that hydrogen and oxygen were formed. They were assisted by J. Cuthbertson, who constructed the apparatus; Cuthbertson also collaborated with Pearson in a series of experiments over two years in the 1790's in which a more convincing demonstration of the formation of the two constituents of water was effected. The amount of the gases actually obtained was, however, very small; and an entirely successful decomposition of water by electricity was not possible until the invention of the voltaic cell.

Pearson also published a number of papers of mainly medical content. He investigated a number of body tissues and fluids and showed, for example, that the blackening of lung tissue is caused by the absorption of carbon from the atmosphere. These researches were continued in an unpublished Bakerian lecture.

A feature of Lavoisier's *Méthode de nomenclature chimique* (Paris, 1787) had been a large folding sheet which presented, in columns, the names of all known substances, classified according to the tenets of the new chemistry. Lavoisier gave the proposed new names in adjoining columns. It was this sheet that Pearson translated. He included both English and Latin equivalents, an explanatory text, and many additions. Pearson also adopted the term "nitrogen," which was first coined by Chaptal as *nitrogène*. Pearson considered the original French *azote* unsuitable because it was based on a purely negative characteristic.

BIBLIOGRAPHY

I. ORIGINAL WORKS. Pearson's M.D. dissertation was *Disputatio physica inauguralis de putridine animalibus post mortem quam superveniente* (Edinburgh, 1773). His investigation of the waters from the springs in Buxton, conducted while he was living in Doncaster, is embodied in *Observations and Experiments for Investigating the Chymical History of the Tepid Springs of Buxton . . .*, 2 vols. (London, 1784) and in a short pamphlet, *Directions for Impregnating the Buxton Water, With Its Own and Other Gases, and for Composing Artificial Buxton Water* (London, 1785). His most important pamphlets on vaccination are *An Inquiry Concerning the History of the Cow Pox Principally With a View to Supersede and Extinguish the Small Pox* (London, 1789), repr. in E. M. Crookshank, ed., *History and Pathology of Vaccination*, II (London, 1889), 34–91; and *An Examination of the Report of the Committee of the House of Commons on the Claims of Remuneration for the Vaccine Pock Inoculation: Containing a Statement of the Principal Historical Facts of the Vaccina* (London, 1802). He also published *Heads and Notes of a Course of Chemical Lectures* (London, 1806).

Pearson's work on nomenclature is *A Translation of the Table of Chemical Nomenclature, Proposed by De Guyton, Formerly de Morveau, Lavoisier, Berthollet & de Fourcroy; With Additions & Alterations, Prefixed by an Explanation of the Terms, and Some Observations on the New System of Chemistry* (London, 1794). A 2nd, enl. ed. was published in 1799, in which Pearson added tables of chemical affinity; the new symbols of J. H. Hassenfratz and P. A. Adet, which had appeared in the *Méthode de nomenclature chymique*; and symbols used by T. Bergman and C. J. Geoffroy. He also included objections that had been made to the new nomenclature by various chemists.

An incomplete list of Pearson's papers is in the Royal Society *Catalogue of Scientific Papers*, IV, 795. His writings include "Experiments & Observations to Investigate the Composition of James's Powder," in *Philosophical Transactions of the Royal Society*, **81** (1791), 317–367; "Experiments Made With the View of Decompounding Fixed Air, or Carbonic Acid," *ibid.*, **82** (1792), 289–308;

"Experiments to Investigate the Nature of a Kind of Steel, Manufactured at Bombay and There Called Wootz; With Remarks on the Properties and Composition of the Different States of Iron," *ibid.*, **85** (1795), 322–346; "Observations on Some Ancient Metallic Arms & Utensils; With Experiments to Determine Their Composition," *ibid.*, **86** (1796), 395–451; "Experiments & Observations Made With a View of Ascertaining the Nature of the Gas Produced by Passing Electric Discharges Through Water," *ibid.*, **87** (1797), 142–158, full paper in Nicholson's *Journal of Natural Philosophy, Chemistry & the Arts*, **1** (1797), 241–248, 299–305, 349–355; "On the Colouring Matter of the Black Bronchial Glands, and of the Black Spots of the Lungs," in *Philosophical Transactions of the Royal Society*, **103** (1813), 159–170; and "Researches to Discover the Faculties of Pulmonary Absorption, With Respect to Charcoal," MS in the Royal Society Archives, A.P. 13 (1827–1829), no. 21, read 20 Dec. 1827.

II. SECONDARY LITERATURE. No informative biography of Pearson exists. A short account, with a list of his publications, is in *Gentleman's Magazine*, **99**, pt. 1 (1829), 129–131. A partial account of Pearson's involvement with Jenner is given by D. Fisk, *Dr. Jenner of Berkeley* (London, 1959), 148 and *passim*.

E. L. SCOTT

PEARSON, KARL (*b.* London, England, 27 March 1857; *d.* Coldharbour, Surrey, England, 27 April 1936), *applied mathematics, biometry, statistics.*

Pearson, founder of the twentieth-century science of statistics, was the younger son and the second of three children of William Pearson, a barrister of the Inner Temple, and his wife, Fanny Smith. Educated at home until the age of nine, he was sent to University College School, London, for seven years. He withdrew in 1873 for reasons of health and spent the next year with a private tutor. He obtained a scholarship at King's College, Cambridge, in 1875, placing second on the list. At Cambridge, Pearson studied mathematics under E. J. Routh, G. G. Stokes, J. C. Maxwell, Arthur Cayley, and William Burnside. He received the B.A. with mathematical honors in 1879 and was third wrangler in the mathematical tripos that year.

Pearson went to Germany after receiving his degree. At Heidelberg he studied physics under G. H. Quincke and metaphysics under Kuno Fischer. At Berlin he attended the lectures of Emil du Bois-Reymond on Darwinism. With his father's profession no doubt in mind, Pearson went up to London, took rooms in the Inner Temple in November 1880, read in Chambers in Lincoln's Inn, and was called to the bar in 1881. He received an LL.B. from Cambridge University in 1881 and an M.A. in 1882, but he never practiced.

Pearson was appointed Goldsmid professor of applied mathematics and mechanics at University College, London, in 1884 and was lecturer in geometry at Gresham College, London, from 1891 to 1894. In 1911 he relinquished the Goldsmid chair to become the first Galton professor of eugenics, a chair that had been offered first to Pearson in keeping with Galton's expressed wish. He retired in 1933 but continued to work in a room at University College until a few months before his death.

Elected a fellow of the Royal Society in 1896, Pearson was awarded its Darwin Medal in 1898. He was awarded many honors by British and foreign anthropological and medical organizations, but never joined and was not honored during his lifetime by the Royal Statistical Society.

In 1890 Pearson married Maria Sharpe, who died in 1928. They had one son, Egon, and two daughters, Sigrid and Helga. In 1929 he married a co-worker in his department, Margaret Victoria Child.

At Cambridge, Pearson's coach under the tripos system was Routh, probably the greatest mathematical coach in the history of the university, who aroused in Pearson a special interest in applied mathematics, mechanics, and the theory of elasticity. Pearson took the Smith's Prize examination, which called for the very best in mathematics. He failed to become a prizeman; but his response to a question set by Isaac Todhunter was found, on Todhunter's death in 1884, to have been incorporated in the manuscript of his unfinished *History of the Theory of Elasticity*, with the comment "This proof is better than De St. Venant's."[1] As a result, in the same year Pearson was appointed by the syndics of the Cambridge University Press to finish and edit the work.

Pearson did not confine himself to mathematics at Cambridge. He read Dante, Goethe, and Rousseau in the original, sat among the divinity students listening to the discourse of the university's regius professor of divinity, and discussed the moral sciences tripos with a fellow student. Before leaving Cambridge he wrote reviews of two books on Spinoza for the *Cambridge Review*, and a paper on Maimonides and Spinoza for *Mind*.

Although intensely interested in the basis, doctrine, and history of religion, Pearson rebelled at attending the regular divinity lectures, compulsory since the founding of King's in 1441, and after a hard fight saw compulsory divinity lectures abolished. He next sought and, with the assistance of his father, obtained release from compulsory attendance at chapel; after which, to the astonishment and pique of the authorities, he continued to attend as the spirit moved him.

Pearson's life in Germany, as at Cambridge, involved much more than university lectures and related study. He became interested in German folklore, in medieval and renaissance German literature, in the history of the Reformation, and in the development of ideas on the position of women. He also came into contact with the ideas of Karl Marx and Ferdinand Lassalle, the two leaders of German socialism. His writings and lectures on his return to England indicate that he had become both a convinced evolutionist and a fervent socialist, and that he had begun to merge these two doctrines into his own rather special variety of social Darwinism. His given name was originally Carl; at about this time he began spelling it with a "K." A King's College fellowship, conferred in 1880 and continued until 1886, gave Pearson financial independence and complete freedom from duties of any sort, and during these years he was frequently in Germany, where he found a quiet spot in the Black Forest to which he often returned.

In 1880 Pearson worked for some weeks in the engineering shops at Cambridge and drew up the schedule in Middle and Ancient High German for the medieval languages tripos. In the same year he published his first book, a literary work entitled *The New Werther*, "by Loki," written in the form of letters from a young man wandering in Germany to his fiancée.

During 1880–1881 Pearson found diversion from his legal studies in lecturing on Martin Luther at Hampstead, and on socialism, Marx, and Lassalle at workingmen's clubs in Soho. In 1882–1884 he gave a number of courses of lectures around London on German social life and thought from the earliest times up to the sixteenth century, and on Luther's influence on the material and intellectual welfare of Germany. In addition he published in the *Academy*, *Athenaeum*, and elsewhere a substantial number of letters, articles, and reviews relating to Luther. Many of these were later republished, together with other lectures delivered between 1885–1887, in his *The Ethic of Freethought* (1888).

During 1880–1884 Pearson's mathematical talent was not entirely dormant. He gave University of London extension lectures on "Heat" and served as a temporary substitute for absent professors of mathematics at King's College and University College, London. At the latter Pearson met Alexander B. W. Kennedy, professor of engineering and mechanical technology, who was instrumental in securing Pearson's appointment to the Goldsmid professorship.

During his first six years in the Goldsmid chair, Pearson demonstrated his great capacity for hard work and extraordinary productivity. His professorial duties included lecturing on statics, dynamics, and mechanics, with demonstrations and proofs based on geometrical and graphical methods, and conducting practical instruction in geometrical drawing and projection. Soon after assuming the professorship, he began preparing for publication the incomplete manuscript of *The Common Sense of the Exact Sciences* left by his penultimate predecessor, William Kingdon Clifford; and it was issued in 1885. The preface, the entire chapter "Position," and considerable portions of the chapters "Quantity" and "Motion" were written by Pearson. A far more difficult and laborious task was the completion and editing of Todhunter's unfinished *History of the Theory of Elasticity*. He wrote about half the final text of the first volume (1886) and was responsible for almost the whole of the second volume, encompassing several hundred memoirs (1893). His editing of these volumes, along with his own papers on related topics published during the same decade, established Pearson's reputation as an applied mathematician.

Somehow Pearson also found the time and energy to plan and deliver the later lectures of *The Ethic of Freethought* series; to complete *Die Fronica* (1887), a historical study that traced the development of the Veronica legend and the history of the Veronica-portraits of Christ, written in German and dedicated to Henry Bradshaw, the Cambridge University Librarian; and to collect the material on the evolution of western Christianity that later formed much of the substance of *The Chances of Death* (1897). In these historical studies Pearson was greatly influenced and guided by Bradshaw, from whom he learned the importance of patience and thoroughness in research. In 1885 Pearson became an active founding member of a small club of men and women dedicated to the discussion of the relationship between the sexes. He gave the opening address on "The Woman's Question," and addressed a later meeting on "Socialism and Sex." Among the members of the group was Maria Sharpe, whom he married in 1890.

In the 1890's the sole duty of the lecturer in geometry at Gresham College seems to have been to give three courses per year of four lectures to an extramural audience on topics of his own choosing. Pearson's aim in applying for the lectureship was apparently to gain an opportunity to present some of his ideas to a fairly general audience. In his first two courses, delivered in March and April 1891 under the general title "The Scope and Concepts of Modern Science," he explored the philosophical foundations of science. These lectures, developed and enlarged, became the first edition of *The Grammar of Science* (1892), a remarkable book that influenced the scientific thought of an entire generation.

Pearson outlined his concept of the nature, scope, function, and method of science in a series of articles in the first chapter of his book. "The material of science," he said, "is coextensive with the whole physical universe, not only . . . as it now exists, but with its past history and the past history of all life therein," while "The function of science" is "the classification of facts, the recognition of their sequence and their relative significance," and "The unity of all science consists alone in its method, not its material . . . It is not the facts themselves which form science, but the method in which they are dealt with." In a summary of the chapter he wrote that the method of science consists of "(a) careful and accurate classification of facts and observation of their correlation and sequence; (b) the discovery of scientific laws by aid of the creative imagination; (c) self-criticism and the final touchstone of equal validity for all normally constituted minds." He emphasized repeatedly that science can only describe the "how" of phenomena and can never explain the "why," and stressed the necessity of eliminating from science all elements over which theology and metaphysics may claim jurisdiction. The *Grammar of Science* also anticipated in many ways the revolutionary changes in scientific thought brought about by Einstein's special theory of relativity. Pearson insisted on the relativity of all motion, completely restated the Newtonian laws of motion in keeping with this primary principle, and developed a system of mechanics logically from them. Recognizing mass to be simply the ratio of the number of units in two accelerations as "expressed briefly by the statement that mutual accelerations are *inversely* as masses" (ch. 8, sec. 9), he ridiculed the current textbook definition of mass as "quantity of matter." Although recognized as a classic in the philosophy of science, the *Grammar of Science* is little read today by scientists and students of science mainly because its literary style has dated it.

Pearson was thus well on the way to a respectable career as a teacher of applied mathematics and philosopher of science when two events occurred that markedly changed the direction of his professional activity and shaped his future career. The first was the publication of Galton's *Natural Inheritance* in 1889; the second, the appointment of W. F. R. Weldon to the Jodrell professorship of zoology at University College, London, in 1890.

Natural Inheritance summed up Galton's work on correlation and regression, concepts and techniques that he had discovered and developed as tools for measuring the influence of heredity;[2] presented all that he contributed to their theory; and clearly reflected his recognition of their applicability and value in studies of all living forms. In the year of its appearance, Pearson read a paper on *Natural Inheritance* before

the aforementioned small discussion club, stressing the light that it threw on the laws of heredity, rather than the mathematics of correlation and regression. Pearson became quite charmed by the concept and implications of Galton's "correlation," which he saw to be a "category broader than causation . . . of which causation was only the limit, and [which] brought psychology, anthropology, medicine and sociology in large parts into the field of mathematical treatment," which opened up the "possibility . . . of reaching knowledge—as valid as physical knowledge was then thought to be—in the field of living forms and above all in the field of human conduct."[3] Almost immediately his life took a new course: he began to lay the foundations of the new science of statistics that he was to develop almost single-handed during the next decade and a half. But it is doubtful whether much of this would have come to pass had it not been for Weldon, who posed the questions that impelled Pearson to make his most significant contributions to statistical theory and methodology.[4]

Weldon, a Cambridge zoologist, had been deeply impressed by Darwin's theory of natural selection and in the 1880's had sought to devise means for deriving concrete support for it from studies of animal and plant populations. Galton's *Natural Inheritance* convinced him that the most promising route was through statistical studies of variation and correlation in those populations. Taking up his appointment at University College early in 1891, Weldon began to apply, extend, and improve Galton's methods of measuring variation and correlation, in pursuit of concrete evidence to support Darwin's "working hypothesis." These undertakings soon brought him face to face with problems outside the realm of the classical theory of errors: How describe asymmetrical, double-humped, and other non-Gaussian frequency distributions? How derive "best"—or at least "good"—values for the parameters of such distributions? What are the "probable errors" of such estimates? What is the effect of selection on one or more of a number of correlated variables? Finding the solution of these problems to be beyond his mathematical capacity, Weldon turned to Pearson for help.

Pearson, in turn, seeing an opportunity to contribute, through his special skills, to the improvement of the understanding of life, characteristically directed his attention to this new area with astonishing energy. The sudden change in his view of statistics, and the early stages of his rapid development of a new science of statistics are evident in the syllabuses of his lectures at Gresham College in 1891–1894 and in G. Udny Yule's summaries of Pearson's two lecture courses on the theory of statistics at University College during the sessions of 1894–1895 and 1895–1896,[5] undoubtedly the first of their kind ever given. Pearson was an enthusiast for graphic presentation; and his Gresham lectures on "Geometry of Statistics" (November 1891–May 1892) were devoted almost entirely to a comprehensive formal treatment of graphical representation of statistical data from the biological, physical, and social sciences, with only brief mention of numerical descriptive statistics. In "Laws of Chance" (November 1892–February 1893) he discussed probability theory and the concept of "correlation," illustrating both by coin-tossing and card-drawing experiments and by observations of natural phenomena. The term "standard deviation" was introduced in the lecture of 31 January 1893, as a convenient substitute for the cumbersome "root mean square error" and the older expressions "error of mean square" and "mean error"; and in the lecture of 1 February, he discussed whether an observed discrepancy between a theoretical standard deviation and an experimentally determined value for it is "sufficiently great to create suspicion." In "The Geometry of Chance" (November 1893–May 1894) he devoted a lecture to "Normal Curves,"[6] one to "Skew Curves," and one to "Compound Curves."

In 1892 Pearson lectured on variation, and in 1893 on correlation, to research students at University College, the material being published as the first four of his *Philosophical Transactions* memoirs on evolution. At this time he worked out his general theory of normal correlation for three, four, and finally n variables. Syllabuses or summaries of these lectures at University College are not available, but much of the substance of the four memoirs is visible in Yule's summaries. Those of the lectures of November 1895 through March 1896 reveal Pearson's early groping toward a general theory of skew correlation and non-linear regression that was not published until 1905. His summary of Pearson's lecture of 14 May 1896 shows that considerable progress had already been made on both the experimental and theoretical material on errors of judgment, measurement errors, and the variation over time of the "personal equations" of individual observers that constituted Pearson's 1902 memoir on these matters.

These lectures mark the beginning of a new epoch in statistical theory and practice. Pearson communicated some thirty-five papers on statistical matters to the Royal Society during 1893–1901. By 1906 he had published over seventy additional papers embodying further statistical theory and applications. In retrospect, it is clear that Pearson's contributions during this period firmly established statistics as a discipline in its own right. Yet, at the time, "the main purpose of all this work" was not development of statistical theory and techniques for their own sake but, rather, "development and application of statistical methods for the study of problems of heredity and evolution."[7]

In order to place the whole of Pearson's work in proper perspective, it will be helpful to examine his contributions to distinct areas of theory and practice. Consider, for example, his "method of moments" and his system of wonderfully diverse frequency curves. Pearson's aim in developing the method of moments was to provide a general method for determining the values of the parameters of a frequency distribution of some particular form selected to describe a given set of observational or experimental data. This is clear from his basic exposition of the subject in the first (1894) of his series of memoirs entitled "Contributions to the Mathematical Theory of Evolution."[8]

The foundations of the system of Pearson curves were laid in the second memoir of this series, "Skew Variation in Homogeneous Material" (1895). Types I–IV were defined and applied in this memoir; Types V and VI, in a "Supplement . . ." (1901); and Types VII–XII in a "Second Supplement . . ." (1916). The system includes symmetrical and asymmetrical curves of both limited and unlimited range (in either or both directions); most are unimodal, but some are U-, J-, or reverse J-shaped. Pearson's purpose in developing them was to provide a collection of frequency curves of diverse forms to be fitted to data as "*graduation curves*, mathematical constructs to describe more or less accurately what we have observed."[9] Their use was facilitated by the central role played by the method of moments: (1) the appropriate curve type is determined by the values of two dimensionless ratios of centroidal moments,

$$\beta_1 = \frac{\mu_3^2}{\mu_2^3} \quad \text{and} \quad \beta_2 = \frac{\mu_4}{\mu_2^2},$$

defined in the basic memoir (1894); and (2) values of the parameters of the selected types of probability (or frequency) curve are determined by the conditions $\mu_0 = 1$ (or $\mu_0 = N$, the total number of observations), $\mu_1 = 0$, and the observed or otherwise indicated values of $\mu_2 (= \sigma^2)$, β_1, and β_2. The acceptance and use of curves of Pearson's system for this purpose may also have been aided by the fact that all were derived from a single differential equation, to which Pearson had been led by considering the slopes of segments of frequency polygons determined by the ordinates of symmetric and asymmetric binomial and hypergeometric probability distributions. That derivation may well have provided some support to Pearson curves as probability or frequency curves, rather than as purely arbitrary graduation curves. Be that as it may, the fitting of Pearson curves to observational data was extensively practiced by biologists and social scientists in the decades that followed. The results did much to dispel the almost religious acceptance of the normal distribution as the mathematical model of variation of biological, physical, and social phenomena.

Meanwhile, Pearson's system of frequency curves acquired a new and unanticipated importance in statistical theory and practice with the discovery that the sampling distributions of many statistical test functions appropriate to analyses of small samples from normal, binomial, and Poisson distributions—such as χ^2, s^2, t, s_1^2/s_2^2, and r (when $\rho = 0$)—are represented by particular families of Pearson curves, either directly or through simple transformation. This application of Pearson curves, and their use to approximate percentage points of statistical test functions whose sampling distributions are either untabulated or analytically or numerically intractable, but whose moments are readily evaluated, have now transcended their use as graduation curves; they have also done much to ensure the value of Pearson's comprehensive system of frequency curves in statistical theory and practice. The use of Pearson curves for either purpose would, however, have been gravely handicapped had not Pearson and his co-workers prepared detailed and extensive tables of their ordinates, integrals, and other characteristics, which were published principally in *Biometrika* beginning in 1901, and reprinted, with additions, in his *Tables for Statisticians and Biometricians* (1914; Part II, 1931).

As statistical concepts and techniques of correlation and regression originated with Galton, who devised rudimentary arithmetical and graphical procedures (utilizing certain medians and quartiles of the data in hand) to derive sample values for his "regression" coefficient, or "index of co-relation," r. Galton was also the first, though he had assistance from J. D. Hamilton Dickson, to express the bivariate normal distribution in the "Galtonian form" of the frequency distribution of two correlated variables.[10] Weldon and F. Y. Edgeworth devised alternative means of computation, which, however, were somewhat arbitrary and did not fully utilize all the data. It was Pearson who established, by what would now be termed the method of maximum likelihood, that the "best value of the correlation coefficient" (ρ) of a bivariate normal distribution is given by the sample product-moment coefficient of correlation,

$$r = \frac{\Sigma xy}{N s_x s_y} = \frac{\Sigma xy}{\sqrt{\Sigma(x^2) \cdot \Sigma(y^2)}},$$

where x and y denote the deviations of the measured values of the x and y characteristics of an individual sample object from their respective arithmetic means (m_x and m_y) in the sample, Σ denotes summation over all N individuals in the sample, and s_x and s_y are the sample standard deviations of the measured values of x and y, respectively.[11] The expression "coefficient of correlation" apparently was originated by Edgeworth in 1892,[12] but the value of r defined by the above equation is quite properly known as "Pearson's coefficient of correlation." Its derivation may be found in section 4b. of "Regression, Heredity, and Panmixia" (1896), his first fundamental paper on correlation theory and its application to problems of heredity.

In the same memoir Pearson also showed how the "best value" of r could be evaluated conveniently from the sample standard deviations s_x, s_y and either s_{x-y} or s_{x+y}, thereby avoiding computation of the sample product moment ($\Sigma xy/N$); gave a mistaken expression for the standard deviation of the sampling error[13] of r as a measure of ρ in large samples—which he corrected in "Probable Errors of Frequency Constants . . ." (1898); introduced the term "coefficient of variation" for the ratio of a standard deviation to the corresponding mean expressed as a percentage; expressed explicitly, in his discussion of the trivariate case, what are now called coefficients of "multiple" correlation and "partial" regression in terms of the three "zero-order" coefficients of correlation (r_{12}, r_{13}, r_{23}); gave the partial regression equation for predicting the (population) mean value of trait X_1, say, corresponding to given values of traits X_2 and X_3, the coefficients of X_2 and X_3 being expressed explicitly in terms of r_{12}, r_{13}, r_{23} and the three sample standard deviations (s_1, s_2, s_3); gave the formula for the large-sample standard error of the value of X_1 predicted by this equation; restated Edgeworth's formula (1892) for the trivariate normal distribution in improved determinantal notation; and carried through explicitly the extension to the general case of a p-variate normal correlation surface, expressed in a form that brought the computations within the power of those lacking advanced mathematical training.

In this first fundamental memoir on correlation, Pearson carried the development of the theory of multivariate normal correlation as a practical tool almost to completion. When the joint distribution of a number of traits X_1, X_2, . . ., X_p, ($p \geqq 2$) over the individuals of a population is multivariate normal, then the population coefficients of correlation, ρ_{ij}, ($i, j = 1, 2, \ldots, p$; $i \neq j$), completely characterize the degrees of association among these traits in the population—traits X_i and X_j are independent if and only if $\rho_{ij} = 0$ and completely interdependent if and only if ρ_{ij} equals ± 1—and the regression in the population of each one of the traits on any combination of the others is linear. It is clear from footnotes to section 5 of this memoir that Pearson was fully aware that linearity of regressions and this comprehensive feature of population (product-moment) coefficients of correlation do not carry over to multivariate skew frequency distributions, and he recognized "the need of [a] theory of skew correlation" which he proposed to treat "in a memoir on skew correlation."[14] The promised memoir, *On the General Theory of Skew Correlation and Non-Linear Regression*, appeared in 1905.

Pearson there dealt with the properties of the correlation ratio, $\eta(=\eta_{yx})$, a sample measure of correlation that he had introduced in a paper of 1903 to replace the sample correlation coefficient, r, when the observed regression curve of y on x (obtained by plotting the means of the y values, $\bar{y}_{x_i}$, corresponding to the respective x values, $x_1, x_2, \ldots$, as a function of x) exhibits a curvilinear relationship and showed that η is the square root of the fraction of the variability of the N y values about their mean, $\bar{y}$, that is ascribable to the variability of the y means $\bar{y}_{x_i}$ about $\bar{y}$; that $1 - \eta^2$ is the fraction of the total variability of the y values about their mean $\bar{y}$ contributed by the variability of the y values within the respective x arrays about their respective mean values, $\bar{y}_{x_i}$, within these arrays; and that $\eta^2 - r^2$ is the fraction ascribable to the deviations of the points $(\bar{y}_{x_i}, x_i)$ from the straight line of closest fit to these points, indicating the importance of the difference between η and r as an indicator of the departure of regression from linearity.[15] He also gave an expression for the standard deviation of the sampling error of η in large samples that has subsequently been shown to be somewhat inaccurate; classified the different forms of regression curves and the different patterns of within-array variability that may arise when the joint distribution of two traits cannot be represented by the bivariate normal distribution, terming the system "homoscedastic" or "heteroscedastic" according to whether the within-array variability is or is not the same for all arrays, respectively; gave explicit formulas for the coefficients of parabolic, cubic, and quartic regression curves, in terms of $\eta^2 - r^2$ and other moments and product moments of the sample values of x and y; and listed the conditions in terms of $\eta^2 - r^2$ and the other sample moments and product moments that must be satisfied for linear, parabolic, cubic, and quartic regression equations to be adequate representations of the observed regression of y on x.

In a footnote to the section "Cubical Regression," Pearson noted that he had pointed out previously[16]

that when a polynomial of any degree, p $(p \leq n)$, is fit to all of n distinct observational points by the method of moments, the curve determined by "the method of moments becomes identical with that of least squares"; but, he continued, "the retention of the method of moments . . . enables us, without abrupt change of method, to introduce the need for η, and to grasp at once the application of the proper SHEPPARD'S corrections [to the sample moments and product moments of x and y when the measurements of either or both are coarsely grouped]."

Pearson clearly favored his method of moments; but the method of least squares has prevailed. However, use of the method of least squares to fit polynomial regression curves in a bivariate correlation situation involves an extension beyond the original formulation and development of the method of least squares by Legendre, Gauss, Laplace, and their followers in the nineteenth century. In this classical development of the method of least squares, one of the variables—x, for example—was a quantity that could be measured with negligible error, and the other, y, a quantity of interest functionally related to x, the observed values of which for particular values of x, Y_x, were, however, subject to nonnegligible measurement errors. The problem was to determine "best" values for the parameters of the functional relation between y and x despite the measurement errors in the observed values of Y_x. The method of least squares as developed by Gauss gave a demonstrably optimal solution when the functional dependence of y upon x was expressible with negligible error in a form in which the unknown parameters entered linearly—for instance, as a polynomial in x. In the Galton-Pearson correlation situation, in contrast, the traits X and Y may both be measurable with negligible error with respect to any single individual but some population of individuals have a joint frequency or probability distribution. The regression of y on x is not an expression of a mathematical functional dependence of the trait Y on the trait X but, rather, an expression of the mean of values of Y corresponding to values of $X = x$ as a function of x—for example, as a polynomial in x. In the classical least-squares situation, the aim was to obtain the best possible approximation to the correct functional relation between the variables despite variations introduced by unwanted errors of measurement. In the Galton-Pearson correlation situation, on the other hand, the aim of regression analysis is to describe two important characteristics of the joint variation of the traits concerned. Pearson's development of the theory of skew correlation and nonlinear regression was, therefore, not merely an elaboration on the work of Gauss but a major step in a new direction.

Pearson did not pursue the theory of multiple and partial correlation beyond the point to which he had carried it in his basic memoir on correlation (1896). The general theory of multiple and partial correlation and regression was developed by his mathematical assistant, G. Udny Yule, in two papers published in 1897. Yule was the first to give mathematical expressions for what are now called partial correlation coefficients, which he termed "net correlation coefficients." What Pearson had called coefficients of double regression, Yule renamed net regressions; they are now called partial regression coefficients. The expressions "multiple correlation" and "partial correlation" stem from the paper written with Alice Lee and read to the Royal Society in June 1897.[17]

In order to see whether the correlations found in studies of the heredity of continuously varying physical characteristics held also for the less tractable psychological and mental traits, Pearson made a number of efforts to extend correlation methods to bivariate data coarsely classified into two or more ordered categories with respect to each trait. Thus, in "On the Correlation of Characters Not Quantitatively Measurable" (1900), he introduced the "tetrachoric" coefficient of correlation, r_t, derived on the supposition that the traits concerned were distributed continuously in accordance with a bivariate normal distribution in the population of individuals sampled, though not measured on continuous scales for the individuals in the sample but merely classified into the cells of a fourfold table in terms of more or less arbitrary but precise dichotomous divisions of the two trait scales. The derived value of r_t was the value of the correlation coefficient (ρ) of the bivariate normal distribution with frequencies in four quadrants corresponding to a division of the x, y plane by lines parallel to the coordinate axes that agreed exactly with the four cell frequencies of the fourfold table. Hence the value of r_t calculated from the data of a particular fourfold table was considered to be theoretically the best measure of the intensity of the correlation between the traits concerned. Pearson gave a formula for the standard deviation of the sampling error of r_t in large samples. He corrected two misprints in this formula and gave a simplified approximate formula in a paper of 1913.[18]

To cope with the intermediate case, in which one characteristic of the sample individuals is measured on a continuous scale and the other is merely classified dichotomously, Pearson, in a *Biometrika* paper of 1909, introduced (but did not name) the "biserial" coefficient of correlation, say r_b.

The idea involved in the development of the "tetrachoric" correlation coefficient, r_t, for data classified in a fourfold table was extended by Pearson in 1910 to cover cases in which "one variable is given by alternative and the other by multiple categories." The sample measure of correlation introduced but not named in this paper became known as "biserial η" because of its analogy with the biserial correlation coefficient, r_b, and the fact that it is defined by a special adaptation of the formula for the correlation ratio, η, based on comparatively nonrestrictive assumptions with respect to the joint distribution of the two traits concerned in the population sampled. The numerical evaluation of "biserial η," however, involves the further assumption that the joint variation of the traits is bivariate normal in the population; and its value for a particular sample, say r_η, is taken to be an estimate of the correlation coefficient, ρ, of the assumed bivariate normal distribution of the traits in the population sampled. The sampling variation of r_η as a measure of ρ was unknown until Pearson published an expression for its standard error in large samples from a bivariate normal population in 1917.[19] It is not known how large the sample size N must be for this asymptotic expression to yield a satisfactory approximation.

Meanwhile, Charles Spearman had introduced (1904) his coefficient of rank-order correlation, say r', which, although first defined in terms of the rank differences of the individuals in the sample with respect to the two traits concerned, is equivalent to the product-moment correlation coefficient between the paired ranks themselves. Three years later Pearson, in "On Further Methods of Determining Correlation," gave the now familiar formula, $\hat{\rho} = 2 \sin(\pi r'/6)$, for obtaining an estimate, $\hat{\rho}$, of the coefficient of correlation (ρ) of a bivariate normal population from an observed value of the coefficient of rank-order correlation (r') derived from the rankings of the individuals in a sample therefrom with respect to the two traits concerned; he also presented a formula for the standard error of $\hat{\rho}$ in large samples.

The "tetrachoric" and "biserial" coefficients of correlation and "biserial η" played important parts in the biometric, eugenic, and medical investigations of Pearson and the biometric school during the first two decades of the twentieth century. Pearson was fully aware of the crucial dependence of their interpretation upon the validity of the assumed bivariate normality and was circumspect in their application; his discussions of numerical results are full of caution. (A sample product-moment coefficient of correlation, r, always provides a usable determination of the product-moment coefficient of correlation, ρ, in the population sampled, bivariate normal or otherwise. On the other hand, when the joint distribution of the two traits concerned is continuous but not bivariate normal in the population sampled, exactly what interpretations are to be accorded to observed values of r_t, r_b, and r_n is not at all clear; and if assumed continuity with respect to both variables is not valid, their interpretation is even less clear—they may be virtually meaningless.) The crucial dependence of the interpretation of these measures on the uncheckable assumption of bivariate normality of the joint distribution of the traits concerned in the population sampled, together with their uncritical application and incautious interpretation by some scholars, brought severe criticism; and doubt was cast on the meaning and value of "coefficients of correlation" thus obtained. In particular, Pearson and one of his assistants, David Heron, ultimately became embroiled in a long and bitter argument on the matter with Yule, whose paper embodying a theory and a measure of association of attributes free of any assumption of an underlying continuous distribution Pearson had communicated to the Royal Society in 1899. Despite this skepticism, r_t, r_b, and r_n have survived and are used today as standard statistical tools, mainly by psychologists, in situations where the traits concerned can be logically assumed to have a joint continuous distribution in the population sampled and the at least approximate normality of this distribution is not seriously questioned.

Pearson did not attempt to investigate sampling distributions of r or η in small samples from bivariate normal or other population distributions because he saw no need to do so. He and his co-workers in the 1890's and early 1900's saw their mission to be the advancement of knowledge and understanding of "variation, inheritance, and selection in Animals and Plants" through studies "based upon the examination of *statistically large numbers* of specimens," and the development of statistical theory, tables of mathematical functions, and graphical methods needed in the pursuit of such studies.[20] They were not concerned with the analysis of data from small-scale laboratory experiments or with comparisons of yield from small numbers of plots of land in agricultural field trials. It was the need to interpret values of r obtained from small-scale industrial experiments in the brewing industry that led "Student" (W. S. Gosset) to discover in 1908 that r is symmetrically distributed about 0 in accordance with a Pearson Type II curve in random samples of any size from a bivariate normal distribution when $\rho = 0$; and, when $\rho \neq 0$, its distribution is skew, with the longer tail toward 0, and cannot be represented by any of Pearson's curves.[21]

In another paper published earlier in 1908 ("The Probable Error of a Mean"), "Student" had dis-covered that the sampling distribution of s^2 (the square of a sample standard deviation), in random samples from a normal distribution, can be represented by a Pearson Type III curve. Although these discoveries stemmed from knowledge and experience that "Student" had gained at Pearson's biometric laboratory in London and were published in the journal that Pearson edited, they seem to have awakened no interest in Pearson or his co-workers in developing statistical theory and techniques appropriate to the analysis of results from small-scale experiments. This indifference may have stemmed from preoccupation with other matters, from recognition that establishment of the small trends or differences for which they were looking required large samples, or from a desire "to discourage the biologist or the medical man from believing that he had been supplied with an easy method of drawing conclusions from scanty data."[22]

In September 1914 Pearson received the manuscript of the paper in which R. A. Fisher derived the general sampling distribution of r in random samples of any size $n \geq 2$ from a bivariate normal population with any degree of correlation, $-1 \leq \rho \leq +1$, and pointed out the extreme skewness of the distribution for large positive or negative values of ρ even for large sample sizes.[23] Pearson responded with enthusiasm, congratulated Fisher "very heartily on getting out the actual distribution form of r," and stated that "if the analysis is correct which seems highly probable, [he] should be delighted to publish the paper in *Biometrika*."[24] A week later he wrote to Fisher: "I have now read your paper fully and think it marks a distinct advance . . . I shall be very glad to publish it . . . [it] shall appear in the next issue [May 1915] . . . I wish you had had the leisure to extend the last pages a little . . . I should like to see some attempt to determine at what value of n and for what values of ρ we may suppose the distribution of r practically normal."[25]

In the "last pages" of the paper, Fisher introduced two transformations of r, $r/\sqrt{1 - r^2}$ and $\tanh^{-1} r$, his aim being to find a function of r whose sampling distribution would have greater stability of form as ρ varied from -1 to $+1$, would be more nearly symmetric, or would have an approximately constant standard deviation, for all values of ρ. The first of these two transformations he considered in detail. Denoting the transformed variable by t, and the corresponding transformation of ρ by τ, he showed that the mean value of t was proportional to τ, the constant of proportionality increasing toward unity with increasing sample size. He also gave exact formulas for $\sigma^2(t)$, $\beta_1(t)$, and $\beta_2(t)$, and tables of their numerical values for selected values of τ^2 from .01 to 100 (that is, ρ from .0995 to .995) and sample sizes n from 8 to 53. Although the distribution of t was, by design, much less asymmetric and of more stable form than the distribution of r—this became unmistakably clear when the corresponding values of $\beta_1(r)$ and $\beta_2(r)$ became known in the "Cooperative Study" (see below)—the transformation was not an unqualified success: its distribution was not close to normal except in the vicinity of $\rho = 0$, and $\sigma^2(t)$ was not approximately constant but nearly proportional to $1/(1 - \rho^2)$. In the final paragraph Fisher dismissed the second transformation for the time being with the comment (with respect to the aims mentioned above): "It is not a little attractive, but so far as I have examined it, it does not tend to simplify the analysis . . . ." (He later found it very much to his liking.)

Reasoning about a function of sample values, such as r, in terms of a transform of it, instead of in terms of the function itself, seems to have been foreign to Pearson's way of thinking. He wrote to Fisher:

> I have rather difficulties over this r and t business—not that I have anything to say about it from the theoretical standpoint—but there appear to me difficulties from the everyday applications with which we as statisticians are most familiar. Let me indicate what I mean.
>
> A man finds a correlation coefficient r from a small sample n of a population; often the material is urgent and an answer on the significance has to be given at once. What he wants to know, say, is whether the true value of $r(\rho)$ is likely to exceed or fall short of his observed value by, say, .10. It may be for instance the correlation between height of firing a gun and the rate of consumption of a time fuse, or between a particular form of treatment of a wound and time of recovery. . . . For example, suppose that $\rho = .30$, and I want to find what is the chance that in 40 observations the resulting r will lie between .20 and .40. Now what we need practically are the β_1 and β_2 for $\rho = .30$ and $n = 40$, and if they are not sufficiently Gaussian for us to use the probability integral, we need the frequency curve of r for $\rho = .30$ and $n = 40$ to help us out. . . . Had I the graph of t I could deduce the graph of r, and mechanically integrate to determine the answer to my problem, but you have not got the ordinates of the t-curve and the practical problem remains it seems to me unsolved. It still seems to me essential (i) to determine β_1 and β_2 accurately for r . . . and (ii) determine a table of frequencies or areas (integral curve) of the r distribution curve for values of ρ and n which do not provide approximately Gaussian results. Of course you may be able to dispose of my practical difficulties, which do not touch your beautiful theory.[26]

Pearson then proposed a specific program of tabulation of the ordinates of the frequency curves for r for selected values of ρ and n to be executed by his trained calculators "unless you really want to do them yourself." The letter in which Fisher is said to have "welcomed the suggestion" that the computations of these ordinates be carried out at the Galton laboratory "seems to have been lost through the disturbance of papers during the 1939–45 war."[27] On the other hand, Fisher seems to have agreed (in this missing, or some other, letter) to undertake the evaluation of the integral of the distribution of r for a selection of values of ρ and n. In a May 1916 letter to Pearson he comments, "I have been very slow about my paper on the probability integral."

When not engaged in war work, Pearson and several members of his staff took on the onerous task of developing reliable formulas for the moments of the distribution of r and calculating tables of its ordinates for ρ from 0.0 to 0.9 and selected values of n. In May 1916, Pearson wrote to Fisher: ". . . the *whole* of the correlation business has come out quite excellently . . . . By [$n =$] 25 my curves [curves of the Pearson system] give the frequency very satisfactorily, but even when $n = 400$, for high values of ρ the normal curve is really not good enough . . . ."[28] It is quite clear from this correspondence between Pearson and Fisher during 1914–1916 that the relationship was entirely friendly, and the implication in some accounts of Fisher's life and work[29] that this venture was carried out without his knowledge is far from correct.

The results of this joint effort of Pearson and his staff were published as ". . . A Cooperative Study" in the May 1917 issue of *Biometrika*. Included were tables of ordinates of the distribution of r for $\rho = 0.0(0.1)0.9$ and $n = 3(1)25, 50, 100, 400$; values of $\beta_1(r)$ and $\beta_2(r)$ for the same ρ when $n = 3, 4, 25, 50, 100, 400$; and of the normal approximation to the ordinates for $n = 100$, $\rho = 0.9$, and $n = 400$, $\rho = 0.7(0.1)0.9$. There were also photographs of seven cardboard models showing, for example, the changes in the distribution of r from U-shaped through J-shaped to skew "cocked hat" forms with increasing sample size for $n = 2(1)25$ for $\rho = 0.6, 0.8$, and illustrating the rate of deviation from normality and increasing skewness with increase of ρ from 0.0 to 0.9 in samples of 25 and 50. This publication represented a truly monumental undertaking. Unfortunately, it had little long-range impact on practical correlation analysis, and it contained material in the section "On the Determination of the 'Most Likely' Value of the Correlation in Sampled Population" that contributed to the widening of the rift that was beginning to develop between Pearson and Fisher.

In his 1915 paper Fisher derived (pp. 520–521), from his general expression for the sampling distribution of r in samples of size n from a bivariate normal population, a two-term approximation,

$$\hat{\rho} = r \Big/ \left(1 + \frac{1 - r^2}{2n}\right),$$

to the "relation between an observed correlation of the sample and the *most probable value* of the correlation of the whole population" [emphasis added]. He referred to his 1912 paper "On an Absolute Criterion for Fitting Frequency Curves" for justification of this procedure.[30] Inasmuch as Pearson had shown in his 1896 memoir that an observed sample from a bivariate normal population is "the most probable" when $\rho = r$ ($\mu_x = m_x$, $\sigma_x = s_x$, $\mu_y = m_y$, and $\sigma_y = s_y$), Fisher's proposed adjustment must have been puzzling to him. The result Fisher obtained is the same as what would be obtained, via the sampling distribution of r, by the method of inverse probability, using Bayes's theorem and an assumed uniform a priori distribution of ρ from -1 to $+1$. This, and Fisher's use of the expression "most probable value," evidently led Pearson, who presumably drafted the text of the "Cooperative Study,"[31] to state mistakenly (pp. 352, 353) that Fisher had assumed such a uniform a priori distribution in deriving his result. Pearson may have been misled also by a "Draft of a Note"[32] that he had received from Fisher in mid-1916, commenting on a paper by Kirstine Smith that had appeared in the May 1916 issue of *Biometrika*, in which Fisher had written: "There is nothing at all 'arbitrary' in the use of the method of moments for the normal curve; as I have shown elsewhere it flows directly from the absolute criterion ($\Sigma \log f$ a maximum) derived from the Principle of Inverse Probability."

Not realizing that Fisher had not only not assumed a uniform a priori distribution of ρ but had also considered his procedure (which he later termed the method of "maximum likelihood") to be completely distinct from "inverse probability" via Bayes's theorem with an assumed a priori distribution, Pearson proceeded to devote over a page of the "Study" to pointing out the absurdity of such an "equal distribution of ignorance" assumption when estimating ρ from an observed r. Several additional pages contain a detailed consideration of alternative forms for the a priori distribution of ρ, showing that with large samples the assumed distribution had little effect on the end result but in small samples could dominate the sample evidence, from which he concluded that "in problems like the present indiscriminate use of Bayes' Theorem is to be deprecated" (p. 359). All of this amounted to flogging a dead horse, so to speak, because Fisher was as fully opposed as Pearson to using Bayes's theorem in such problems. Unfortunate-

ly, Fisher probably was totally unaware of this offending section before proofs became available in 1917. Papers such as the "Study" were not readily typed in those days, so that there would have been only a single manuscript of the text and tables prior to typesetting. Had Fisher, who was then teaching mathematics and physics in English public schools, been in closer touch with Pearson, these misunderstandings might have been resolved before publication of the offending passages.

In August 1920 Fisher sent Pearson a copy of his manuscript "On the 'Probable Error' of a Coefficient of Correlation Deduced From a Small Sample," in which he reexamined in detail the $\tanh^{-1} r$ transformation and, denoting the transformed variable by z and the corresponding transformation of ρ by ζ, showed that z can be taken to be approximately normally distributed about a mean of $\zeta + \dfrac{\rho}{2(n-1)}$ with a standard deviation equal to $1/\sqrt{n-3}$, the normal approximation being extraordinarily good even in very small samples—of the order of $n = 10$. This transformation thus made it possible to answer questions of the types that Pearson had raised without recourse to tables of the integral of the distribution of r, and obviated the immediate need for the preparation of such tables. (It was not until 1931 that Pearson suggested to Florence N. David the computation of tables of the integral. Values of the integral obtained by quadrature of the ordinates given in the "Cooperative Study" were completed in 1934. Additional ordinates and values of the integral were calculated to facilitate interpolation. These improved tables, together with four charts for obtaining confidence limits for ρ given r, were published in 1938.[33])

In his discussion of applications, Fisher took pains to point out that the formula he had given in his 1915 paper for what he then "termed the 'most likely value,' which [he] now, for greater precision, term[ed] the 'optimum' value of ρ, for a given observed r" involved in its derivation "no assumption whatsoever as to the probable distribution of ρ," being merely that value of ρ for which the observed r occurs with greatest frequency." He also noted that one is led to exactly the same expression for the optimum value of ρ in terms of an observed r if one seeks the optimum through the z distribution rather than the r distribution and he commented that the derivation of this optimum cannot, therefore, be inferred to depend upon an assumed uniform prior distribution of ζ and upon an assumed uniform prior distribution of ρ, since these two assumptions are mutually inconsistent. Then, "though . . . reluctant to criticize the distinguished statisticians who put their names to the Cooperative

Study," Fisher went on to criticize with a tone of ridicule some of the illustrative examples of the application of Bayes's theorem considered on pp. 357–358 of the "Study," without noting the authors' conclusions from these, and other examples considered, that such "use of Bayes' Theorem is to be deprecated" (p. 359) and when applied to "values observed in a small sample may lead to results very wide from the truth" (p. 360). Fisher concluded his paper with a "Note on the Confusion Between Bayes' Rule and My Method of the Evaluation of the Optimum."

Pearson returned the manuscript to Fisher with the following comment:

> . . . I fear if I could give full attention to your paper, which I cannot at the present time, I should be unlikely to publish it in its present form, or without a reply to your criticisms which would involve also a criticism of your work of 1912—I would prefer you publish elsewhere. Under present printing and financial conditions, I am regretfully compelled to exclude all that I think erroneous on my own judgment, because I cannot afford controversy.[34]

Fisher therefore submitted his paper to *Metron*, a new journal, which published the work in its first volume.[35]

The cross criticism, at cross purposes, conducted by Pearson and Fisher over the use of Bayes's theorem in estimating ρ from r was multiply unfortunate: it was unnecessary and ill-timed; it might have been avoided; and it fostered ill will and fueled the innately contentious temperament of both parties at an early stage of their argument over the relative merits of the method of moments and method of maximum likelihood. This argument was started by Fisher's "Draft of a Note," which Pearson took to be a criticism not only of the minimum chi-square technique that Kirstine Smith had propounded but also of his method of moments, and refused to publish it in both original (1916) and revised (1918) forms on the grounds of its being controversial and liable to provoke a quarrel among contributors.[36] The argument, which grew into a raging controversy, was fed by later developments on various fronts and continued to the end of Pearson's life—and beyond.[37]

In 1922 Fisher found the sampling distribution of η^2 in random samples of any size from a bivariate normal population in which the correlation is zero ($\rho = 0$), and later (1928) derived the distribution of η^2 in samples of any size when the x values are fixed and the y values are normally distributed with a common standard deviation σ about array means $\mu_{y|x}$ which may be different for different values of x, thereby giving rise to a nonzero value of the "population" correlation ratio. In particular, it was found that for any value of the population correlation ratio different from zero, the sampling distribution of η tends in sufficiently large samples to be approximately normal about the population value with standard error given by Pearson's formula; but when the correlation ratio in the population is exactly zero—that is, when sampling from uncorrelated material—the sampling distribution of η does not tend to normality with increasing sample size for any finite number of arrays. This led to formulation of new procedures, since become standard, for testing the statistical significance of an observed value of η and of $\eta^2 - r^2$ as a test for departure from linearity.

In 1926 Pearson showed that the distribution of sample regression coefficients, that is, of the slopes of the sample regression of y on x and of x on y, respectively, is his Type VII distribution symmetrical about the corresponding population regression coefficient. It tends to normality much more rapidly than the distribution of r with increasing sample size, so that the use of Pearson's expressions for the standard error of regression coefficients is therefore valid for lower values of n than in the case of r. It is, however, not of much use in small samples, since it depends upon the unknown values of the population standard deviations and correlation, σ_y, σ_x, and ρ_{xy}. Four years earlier, however, in response to repeated queries from "Student" in correspondence, Fisher had succeeded in showing that in random samples of any size from a general bivariate normal population, the sampling distribution of the ratio $(b - \beta)/s_{b-\beta}$, where β is the population regression coefficient corresponding to the sample coefficient b, and $s_{b-\beta}$ is a particular sample estimate of the standard error of their difference, does not depend upon any of the population parameters other than β and is given by a special form of Pearson's Type VII curve now known as "Student's" t-distribution for $n - 2$ degrees of freedom. Consequently, it is this latter distribution, free of "nuisance parameters," that is customarily employed today in making inferences about a population regression coefficient from an observed value of the corresponding sample coefficient.

Although the final steps of correlation and regression analyses today differ from those originally advanced by Pearson and his co-workers, there can be no question that today's procedures were built upon those earlier ones; and correlation and regression analysis is still very much indebted to those highly original and very difficult steps into the unknown taken by Pearson at the turn of the century.

Derivation of formulas for standard errors in large samples of functions of sample values used to estimate parameters of the population sampled did not, of course, originate with Pearson. It dates from Gauss's derivation (1816) of the standard errors in large samples of the respective functions of successive sample absolute moments that might be used as estimators of the population standard deviation. Another early contribution was Gauss's derivation (1823) of a formula comparable with that derived by Pearson in 1903 for the standard error in large samples of the sample standard deviation as estimator of the standard deviation of an arbitrary population having finite centroidal moments of fourth order or higher. Subsequent writers treated these matters somewhat more fully and made a number of minor extensions, but the first general approach to the problem of standard errors and intercorrelations in large samples of sample functions used to estimate values of population parameters is that given in "On the Probable Errors of Frequency Constants . . . ," written by Pearson and his young French mathematical demonstrator, L. N. G. Filon, and read to the Royal Society in November 1897. In section II there is the first derivation of the now familiar expressions for the asymptotic variances and covariances of sample estimators of a group of population parameters in terms of mathematical expectations of second derivatives of the logarithm of what is now called the "likelihood function," but without recognition of their applicability only to maximum likelihood estimators, a limitation first pointed out by Edgeworth (1908).[38] Today these formulas are usually associated with Fisher's paper "On the Mathematical Foundations of Theoretical Statistics" (1922)—and perhaps rightly so, because, although the expressions derived by Pearson and Filon, and by Fisher, are of identical mathematical form, what they meant to Pearson and Filon in 1897 and continued to mean to Pearson may have been quite different from what they meant to Fisher.[39] (This may have been a major obstacle to their conciliation.)

Specific formulas derived by Pearson and Filon included expressions for the standard error of a coefficient of correlation r; the correlation between the sample means m_x and m_y of two correlated traits; the correlation between the sample standard deviations, s_x and s_y; the correlation between a sample coefficient of correlation r and a sample standard deviation s_x or s_y; the standard errors of regression coefficients, and of partial regression coefficients, for the two- and three-variable cases, respectively; and the correlations between pairs of sample correlation coefficients (r_{12}, r_{13}), (r_{12}, r_{34})—all in the case of large samples from a correlated normal distribution. In the process it was noted that in the case of large samples from a correlated normal distribution, the errors of sample

means are uncorrelated with the errors of sample standard deviations and sample correlation coefficients; and that through failure to recognize the existence of correlation between the errors of sample standard deviations and a sample correlation coefficient, the formula given previously for the large sample standard error of the sample correlation coefficient r was in error, because it was appropriate to the case in which the population standard deviations, σ_x and σ_y, are known exactly. Large sample formulas were found also for the standard errors and correlations between the errors of sample estimates of the parameters of Pearson Type I, III, and IV distributions, making this the first comprehensive study of such matters in the case of skew distributions.

Pearson returned to this subject in a series of three editorials in *Biometrika*, "On the Probable Errors of Frequency Constants," prepared in response to a need expressed by queries from readers. The first (1903) deals with the standard errors of, and correlations between, (i) cell frequencies in a histogram and (ii) sample centroidal moments, in terms of the centroidal moments of a univariate distribution of general form. Some of the results given are exact and some are limiting values for large samples. In some instances a "probable error" ($= 0.6745 \times$ standard error) is given, but the practice is deprecated: "The adoption of the 'probable error' . . . as a measure of . . . exactness must not, however, be taken as equivalent to asserting the validity of the normal law of errors or deviations, but merely as a purely conventional reduction of the standard deviation. It would be equally valid provided it were customary to omit this reduction or indeed to multiply the standard deviation by any other conventional factor" (p. 273).

The extension to samples from a general bivariate distribution was made in "Part II" (1913), reproduced from Pearson's lecture notes. Formulas were given for the correlation of errors in sample means; the correlation of errors in sample standard deviations; the standard error of the correlation coefficient r (in terms of the population coefficient of correlation ρ and the β_2's of the two marginal distributions); the correlation between the random sampling deviations of a sample mean and a sample standard deviation for the same variate; correlation between the random sampling deviations of sample mean of one variate and the standard deviation of a correlated variate; the correlation between a mean and a sample coefficient of correlation; the correlation between the sampling deviations of a sample standard deviation and sample coefficient of correlation; and the standard errors of coefficients of linear regression lines and of the means of arrays. In this paper it is also shown that in the case of all symmetric distributions, there is no correlation between the sample mean and sample standard deviation. "Part III" (1920) deals with the standard errors of, and the correlations between, the sampling variations of the sample median, quartiles, deciles, and other quantiles in random samples from a general univariate distribution. The relative efficiency of estimating the standard deviation of a normal population from the difference between two symmetrical quantiles of a large sample therefrom is discussed, and the "optimum" is found to be the difference between the seventh and ninety-third percentiles.

The results given in these three editorials are derived by a procedure considerably more elementary than that employed in the Pearson-Filon paper. Some of the results given are exact; others are limiting values for large samples; and many have become more or less standard in statistical circles.

The July 1900 issue of *Philosophical Magazine* contained Pearson's paper in which he introduced the criterion

$$\chi^2 = \Sigma \frac{(f_i - F_i)^2}{F_i}$$

as a measure of the agreement between observation and hypothesis overall to be used as a basis for determining the probability with which the differences $f_i - F_i$, $(i = 1, 2, \ldots, k)$, collectively might be due solely to the unavoidable fluctuations of random sampling, where f_i denotes the observed frequency (the observed number of observations falling) in the ith of k mutually exclusive categories, and F_i is the corresponding theoretical frequency (the number expected in the ith category in accordance with some particular true or hypothetical frequency distribution), with $\Sigma f_i = \Sigma F_i = N$, the total number of independent observations involved. To this end he derived the sampling distribution of χ^2 in large samples as a function of k, finding it to be a specialized form of the Pearson Type III distribution now known as the "χ^2 distribution for $k - 1$ degrees of freedom," the $k - 1$ being explained by the remark (in our notation) "only $k - 1$ of the k errors are variables; the kth is determined when the first $k - 1$ are known"; he also gave a small table of the integral of the distribution for χ^2 from 1 to 70 and k from 3 to 20. Of Pearson's many contributions to statistical theory and practice, this χ^2 text for goodness of fit is certainly one of his greatest; and in its original and extended forms it has remained one of the most useful of all statistical tests.

Four years later, in *On the Theory of Contingency and Its Relation to Association and Normal Correlation*, Pearson extended the application of his χ^2 criterion to the analysis of the cell frequencies in a "contingency table" of r rows and c columns resulting from the partitioning of a sample of N observations into r distinct classes in terms of some particular characteristic, and into c distinct classes with respect to another characteristic; showed how the χ^2 criterion could be used to test the independence of the two classifications; termed $\phi^2 = \chi^2/N$ the "mean square contingency" and

$$C = \sqrt{\frac{\chi^2}{N + \chi^2}}$$

the coefficient of mean square contingency; showed that, if a large sample from a bivariate normal distribution with correlation coefficient ρ is partitioned into the cells of a contingency table, then C^2 will tend to approximate ρ^2 as the number of categories in the table increases, the correct sign of ρ then being determined from the order of the two classifications and the pattern of the cell frequencies within the $r \times c$ table; and that, when $r = c = 2$, ϕ^2 is equal to the square of the product-moment coefficient of correlation computed from the observed frequencies in the fourfold table with purely arbitrary values (for instance, 0, 1) assigned to the two row categories and to the two column categories.

Pearson made much of the fact that the value of χ^2 and of C is unaffected by reordering either or both of the marginal categories, so that χ^2 provides a means of testing the independence of the two characteristics (such as eye color and occupation) in terms of which the marginal classes are defined without, and independently of, any additional assumptions as to the nature of the association, if any. In view of the above-mentioned relation of C to ρ under the indicated circumstances, C would seem to be a generally useful measure of the degree or intensity of the association when a large value of χ^2 leads to rejection of the hypothesis of independence; and Pearson proposed its use for this purpose. It is, however, not a very satisfactory measure of association—for example, the values of C obtained from an $r \times c$ classification and an $r' \times c'$ classification of the same data will usually be different. Also, some fundamental objections have been raised to the use of C, or any other function of χ^2, as a measure of association. Nonetheless, C played an important role in its day in the analysis of data classified into $r \times c$ tables when the categories for both characteristics can be arranged in meaningful orders— if the categories for either characteristic cannot be put into a meaningful order, then there can be no satisfactory measure of the intensity of *the* association; and a large value of χ^2 may simply be an indication of some fault in the sampling procedure.

In a 1911 *Biometrika* paper, Pearson showed how

his χ^2 criterion could be extended to provide a test of the hypothesis that "two independent distributions of frequency [arrayed in a $2 \times c$ table] are really samples from the same population." The theoretical proportions in the respective cells implied by the presumed common population being unknown, they are estimated from the corresponding proportions of the two samples combined. Illustrative examples show that to find P, the probability of a larger value of χ^2, the "Tables for Testing Goodness of Fit" are to be entered with $n' = c$, signifying that there are $c - 1$ "independent variables" ("degrees of freedom") involved, which agrees with present practice. In a *Biometrika* paper, "On the General Theory of Multiple Contingency . . ." (1916), Pearson gave a new derivation of the χ^2 distribution, as the limiting distribution of the class frequencies of a multinomial distribution as the sample size $N \to \infty$; pointed out (pp. 153–155) that if q linear restraints are imposed on the n' cell frequencies in addition to the usual $\Sigma f_i = N$, then to find P one must enter the tables with $n' - q$; and extended the χ^2 technique to testing whether the frequencies arrayed in two $(2 \times c)$ contingency tables can be considered random samples from the same bivariate population. In this application of "partial χ^2," Pearson considers the c column totals of each table to be fixed, thereby imposing $2c$ linear restraints on the $4c$ cell frequencies involved. The theoretical proportion, p_{1j}, in the presumed common population, corresponding to the cell in the top row and jth column of either table being unknown, it is taken as equal to the corresponding proportion in this cell of the two tables combined, $(j = 1, 2, \ldots, c)$, thereby imposing c additional linear restraints (p_{2j} is, of course, simply $1 - p_{1j}$, $[j = 1, 2, \ldots, c]$). Hence there remain only $4c - 2c - c = c$ "independent variables"; and Pearson notes that the χ^2 tables are to be entered with $n' = c + 1$. These two papers clearly contain the basic elements of a large part of present-day χ^2 technique.

In section 5 of his 1900 paper on χ^2, Pearson pointed out that one must distinguish between a value of χ^2 calculated from theoretical frequencies F_i derived from a theoretical probability distribution completely specified a priori and values of χ_s^2, say, calculated from theoretical frequencies $\tilde{F}_i$ derived from a theoretical probability distribution of specified form but with the values of one or more of its parameters left unspecified so that "best values" for these had to be determined from the data in hand. It was clear that χ_s^2 could never exceed the "true" χ^2. From a brief, cursory analysis Pearson concluded that the difference $\chi^2 - \chi_s^2$ was likely to be negligible. Evidently he did not realize that the difference might depend on the number of constants the values of which were determined from

the sample and that, if k constants were fit, χ_s^2 might be zero.

Ultimately Fisher showed in a series of three papers (1922, 1923, 1924) that when the unknown parameters of the population sampled are efficiently estimated from the data in such a manner as to impose c additional linear restraints on t cell frequencies, then, when the total number of observations N is large, χ_s^2 will be distributed in accordance with a χ^2 distribution for $(t - 1 - c)$ degrees of freedom. Pearson had recognized this in the cases of the particular problems discussed in his 1911 and 1916 papers considered above; but he never accepted Fisher's modification of the value of n' with which the "Tables of Goodness of Fit" were to be entered in the original 1900 problem of testing the agreement of an observed and a theoretical frequency distribution when some parameters of the latter were estimated from the observed data, or in the 1904 problem of testing the independence of the two classifications of an $r \times c$ contingency table.

During Pearson's highly innovative decade and a half, 1891–1906, in addition to laying the foundations of the major contributions to statistical theory and practice reviewed above, he also initiated a number of other topics that later blossomed into important areas of statistics and other disciplines. Brief mention was made above of "On the Mathematical Theory of Errors of Judgment . . ." (1902). This investigation was founded on two series of experiments in which three observers each individually (a) estimated the midpoints of segments of straight lines; and (b), estimated the position on a scale of a bright line moving slowly downward at the moment when a bell sounded. The study revealed that the errors of different observers estimating or measuring the same series of quantities are in general correlated; that the frequency distributions of such errors of estimation or measurement certainly are not always normal; and that the variation over a period of time of the "personal equation" (the pattern of the systematic error or bias of an individual observer) is not explainable solely by the fluctuations of random sampling. The investigation stemmed from Pearson's observation that when three observers individually estimate or measure a series of physical quantities, the actual magnitudes of which may or may not be known or determinable, then, on the assumption of independence of the judgments of the respective observers, it is possible to determine the standard deviations of the distributions of measurement errors of each of the three observers from the observed standard deviations of the differences between their respective measurements of the same quantities. The investigation reported in this memoir is thus the forerunner of the work carried out by Frank E. Grubbs during the 1940's on methods for determining the individual precisions of two, three, four, or more measuring instruments in the presence of product variability.

A second example is provided by Pearson's "Note on Francis Galton's Problem" (August 1902), in which he derived the general expression for the mean value of the difference between the rth and the $(r + 1)$th individuals ranked in order of size in random samples of size n from any continuous distribution. This is one of the earliest general results in the sampling theory of order statistics, a very active subfield of statistics since the 1930's. Pearson later gave general expressions for the variances of, and correlations between, such intervals in random samples from any continuous distribution in a joint paper with his second wife, "On the Mean . . . and Variance of a Ranked Individual, and . . . of the Intervals Between Ranked Individuals, Part I . . ." (1931).

A third example is the theory of "random walk," a term Pearson coined in a brief letter, "The Problem of the Random Walk," published in the 17 July 1905 issue of *Nature*, in which he asked for information on the probability distribution of the walker's distance from the origin after n steps. Lord Rayleigh replied in the issue of 3 August, pointing out that the problem is formally the same as that of "the composition of n isoperiodic vibrations of unit amplitude and of phases distributed at random" (p. 318), which he had considered as early as 1880, and indicated the asymptotic solution as $n \to \infty$. The general solution for finite n was published by J. C. Kluyver in Dutch later the same year and, among other applications, provides the basis for a test of whether a set of orientation or directional data is "random" or tends to exhibit a "preferred direction." With John Blakeman, Pearson published *A Mathematical Theory of Random Migration* (1906), in which various theoretical forms of distribution were derived that would result from random migration from a point of origin under certain ideal conditions and solutions to a number of subsidiary problems were given, results that have found various other applications. Today "random walks" of various kinds, with and without reflecting or absorbing barriers, play important roles not only in the theory of Brownian motion but also in the treatment of random phenomena in astronomy, biology, physics, and communications engineering; in statistics, they are used in the theory of sequential estimation and of sequential tests of statistical hypotheses.

Pearson's involvement in heredity and evolution dates from his first fundamental paper on correlation and regression (1896), in which, to illustrate the value of these new mathematical tools in attacking problems of heredity and evolution, he included evaluations of partial regressions of offspring on each parent for sets of data from Galton's *Record of Family Faculties* (London, 1884) and considerably extended Galton's collateral studies of heredity by considering types of selection, assortative mating, and "panmixia" (suspension of selection and subsequent free interbreeding). Galton's formulation, in *Natural Inheritance* (1889), of his law of ancestral heredity was somewhat ambiguous and imprecise because of his failure to take into account the additional mathematical complexity involved in the joint consideration of more than two mutually correlated characteristics. Pearson supposed him to mean (p. 303) that the coefficients of correlation between offspring and parent, grandparent, and great-grandparent, . . . were to be taken as $r, r^2, r^3, \ldots$. This led him to the paradoxical conclusion that "a knowledge of the ancestry beyond the parents in no way alters our judgment as to the size of organ or degree of characteristic probable in the offspring, nor its variability" (p. 306), a conclusion that he said in a footnote "seems especially noteworthy" inasmuch as it is quite contrary to what "it would seem natural to suppose."

In "On the Reconstruction of the Stature of Prehistoric Races" (1898), Pearson used multiple regression techniques to predict ("reconstruct") average measurements of extinct races from the sizes of existing bones and known correlations among bone lengths in an extant race, as a means of testing the accuracy of predictions in evolutionary problems in the light of certain evolutionary theories.

Meanwhile, Galton had formulated (1897) his "law" more precisely. After some correspondence Pearson, in "On the Law of Ancestral Heredity" (1898), subtitled "A New Year's Greeting to Francis Galton, January 1, 1898," expressed what he christened "Galton's Law of Ancestral Heredity" in the form of a multiple regression equation of offspring on midparental ancestry

$$x_0 = \frac{1}{2} \frac{\sigma_0}{\sigma_1} x_1 + \frac{1}{4} \frac{\sigma_0}{\sigma_2} x_2 + \frac{1}{8} \frac{\sigma_0}{\sigma_3} x_3 + \cdots,$$

where x_0 is the predicted deviation of an individual offspring from the mean of the offspring generation, x_1 is the deviation of the offspring's "midparent" from the mean of the parental generation, x_2 the deviation of the offspring's "midgrandparent" from the mean of the grandparental generation, and so on, and $\sigma_0 \sigma_1 \ldots$ are the standard deviations of the distributions of individuals in the respective generations. In order that this formulation of Galton's law be unambiguous, it was necessary to have a precise definition of "sth midparent." The definition that Pearson adopted "with reservations" was "[If] a father is a first parent, a grandfather a second parent, a great-grandfather a third parent, and so on, [then] the mid sth parent or the sth mid-parent is derived from [is the mean of] all 2^s individual sth parents" (footnote, p. 387).

From this formulation Pearson deduced theoretical values for regression and correlation coefficients between various kin, tested Galton's stature data against these expectations, and suggested generalizing Galton's law by substituting $\gamma\beta$, $\gamma\beta^2$, $\gamma\beta^3$, ... for Galton's geometric series coefficients 1/2, 1/4, 1/8, ... to allow "greater scope for variety of inheritance in different species" (p. 403). In the concluding section Pearson claims: "If either [Galton's Law], or its suggested modification be substantially correct, they embrace the whole theory of heredity. They bring into one simple statement an immense range of facts, thus fulfilling the fundamental purpose of a great law of nature" (p. 411). After noting some difficulties that would have to be met and stating, "We must wait at present for further determinations of hereditary influence, before the actual degree of approximation between law and nature can be appreciated," he concluded with the sweeping statement: "At present I would merely state my opinion that, with all due reservations it seems to me that . . . it is highly probable that [the law of ancestral heredity] is the simple descriptive statement which brings into a single focus all the complex lines of hereditary influence. If Darwinian evolution be natural selection combined with *heredity*, then the single statement which embraces the whole field of heredity must prove almost as epoch-making to the biologist as the law of gravitation to the astronomer" (p. 412).

These claims were obviously too sweeping. Neither the less nor the more general form of the law was founded on any clear conception of the mechanism of heredity. Also, most unfortunately, some of the wording employed—for instance, "I shall now proceed to determine . . . the correlation between an individual and any sth parent from a knowledge of the regression between the individual and his mid-sth parent" (p. 391) —tended to give the erroneous idea that the law expressed a relation between a particular individual and his sth parents, and thus to mislead biologists of the period, who had not become fully conscious that regression equations merely expressed relationships that held on the average between the generic types of "individuals" involved, and not between particular individuals of those types.

During the summer vacations of 1899 and 1900 Pearson, with the aid of many willing friends and colleagues, collected material to test a novel theory of "homotyposis, which if correct would imply that the correlation between offspring of the same parents should on the average be equal to the correlation between undifferentiated like organs of an individual." The volume of data collected and reduced was far greater than Pearson had previously attempted. The result was a joint memoir by Pearson and several members of his staff, "On the Principle of Homotyposis and Its Relation to Heredity . . . Part I. Homotyposis in the Vegetable Kingdom," which was "received" by the Royal Society on 6 October 1900. William Bateson, biologist and pioneer in genetics, who had just become a convert to Mendel's theory, was one of those chosen to referee the memoir, which was "read"—presumably only the five-page abstract[40] and certainly in highly abridged form—at the meeting of 15 November 1900. In the discussion that followed the presentation, Bateson sharply criticized the paper, its thesis being, in his view, mistaken; and other fellows present added criticism of both its length and its content.

The next day (16 November 1900) Weldon wrote to Pearson: "The contention 'that numbers mean nothing and do not exist in Nature' is a very serious thing, which will have to be fought. Most other people have got beyond it, but most biologists have not. Do you think it would be too hopelessly expensive to start a journal of some kind?. . ."[41] Pearson was enthusiastically in favor of the idea—on 13 December 1900 he wrote to Galton that Bateson's adverse criticism "did not apply to this memoir only but to all my work, . . . if the R. S. people send my papers to Bateson, one cannot hope to get them printed. It is a practical notice to quit. This notice applies not only to *my* work, but to most work on similar statistical lines."[42] On 29 November Weldon wrote to him: "Get a better title for this would-be journal than I can think of!"[43] Pearson replied with the suggestion that "the science in future should be called Biometry and its official organ be *Biometrika*."[44]

A circular was sent out during December 1900 to solicit financial support and resulted in a fund sufficient to support the journal for a number of years. Weldon, Pearson, and C. B. Davenport were to be the editors; and Galton agreed to be "consulting editor." The first issue appeared in October 1901, and the editorial "The Scope of *Biometrika*" stated:

> *Biometrika* will include (a) memoirs on variation, inheritance, and selection in Animals and Plants, based upon the examination of statistically large numbers of specimens (this will of course include statistical investigations in anthropometry); (b) those developments of statistical theory which are applicable to biological problems; (c) numerical tables and graphical solutions tending to reduce the labour of

statistical arithmetic; (d) abstracts of memoirs, dealing with these subjects, which are published elsewhere; and (e) notes on current biometric work and unsolved problems.

In the years that followed, *Biometrika* became a major medium for the publication of mathematical tables and other aids to statistical analysis and detailed tables of biological data.

The memoir on homotyposis was not published in the *Philosophical Transactions* until 12 November 1901, and only after a direct appeal by Pearson to the president of the Royal Society on grounds of general principle rather than individual unfairness. Meanwhile, Bateson had prepared detailed adverse criticisms. Under pressure from Bateson, the secretary of the Royal Society put aside protocol and permitted the printing of Bateson's comments and their issuance to the fellows at the meeting of 14 February 1901— before the full memoir by Pearson and his colleagues was in their hands, and even before its authors had been notified whether it had been accepted for publication. Then, with the approval of the Zoological Committee, Bateson's full critique was published in the *Proceedings of the Royal Society* before the memoir criticized had appeared.[45] One can thus appreciate the basis for the acerbity of Pearson's rejoinder, which he chose to publish in *Biometrika*[46] because he had been "officially informed that [he had] a right to a rejoinder, but only to such a one as will not confer on [his] opponent a right to a further reply!" (footnote, p. 321).

This fracas over the homotyposis memoir was but one manifestation of the division that had developed in the 1890's between the biometric "school" of Galton, Weldon, and Pearson and certain biologists— notably Bateson—over the nature of evolution. The biometricians held that evolution of new species was the result of gradual accumulation of the effects of small continuous variations. In 1894 Bateson published a book in which he noted that deviations from normal parental characteristics frequently take the form of discontinuous "jumps" of definite measurable magnitude, and held that discontinuous variation of this kind—evidenced by what we today call sports or mutations—is necessary for the evolution of new species.[47] He was deeply hurt when Weldon took issue with this thesis in an otherwise very favorable review published in *Nature* (10 May 1894).

When Gregor Mendel's long-overlooked paper of 1866 was resurrected in 1900 by three Continental botanists, the particulate nature of Mendel's theory of "dominance" and "segregation" was clearly in keeping with Bateson's views; and he became a totally committed Mendelist, taking it upon himself to convert all English biologists into disciples of Mendel.

Meanwhile, Weldon and Pearson had become deeply committed adherents to Galton's law of ancestral heredity, to which Bateson was antipathetic. There followed a heated controversy between the "ancestrians," led by Pearson and Weldon, and the "Mendelians," led by Bateson. Pearson and Weldon were not, as some supposed, unreceptive to Mendelian ideas but were concerned with the too ready acceptance of Mendelism as a complete gospel without regard to certain incompatibilities they had found between Mendel's laws of "dominance" and "segregation" and other work. Weldon, the naturalist, regarded Mendelism as an unimportant but inconvenient exception to the ancestral law. Pearson, the applied mathematician and philosopher of science, saw that Mendelism was not incompatible with the ancestral law but in some circumstances could lead directly to it; and he sought to bring all heredity into a single system embodying both Mendelian and ancestrian principles, with the latter dominant. To Bateson, Mendel's laws were the truth and all else was heresy. The controversy raged on with much mutual incomprehension, and with great bitterness on both sides, until Weldon's death in April 1906 removed the most committed ancestrian and Bateson's main target.[48] Without the help of Weldon's biologically trained mind, Pearson had no inclination, nor the necessary training, to keep in close touch with the growing complexity of the Mendelian hypothesis, which was coming to depend increasingly on purely biological discoveries for its development; he therefore turned his attention to unfinished business in other areas and to eugenics.

During the succeeding decades Mendelian theory became firmly established—but only after much testing on diverse material, clarification of ideas, explanation of "exceptions," and tying in with cytological discoveries. Mendel's laws have been shown to apply to many kinds of characters in almost all organisms, but this has not entirely eliminated "biometrical" methods. Quite the contrary: multiple regression techniques are still needed to cope with the inheritance of quantitative characters that presumably depend upon so many genes that Mendelian theory cannot be brought to bear in practice. For example, coat color of dairy cows depends upon only a few genes and its Mendelian inheritance is readily verified; but the quantitative trait of milk production capacity is so complex genetically that multiple regression methods are used to predict the average milk-production character of offspring of particular matings, given the relevant ancestral information.

In fact, geneticists today ascribe the reconciliation of the "ancestral" and "Mendelian" positions, and definitive synthesis of the two theories, to Fisher's first genetical paper, "The Correlations to be Expected Between Relatives on the Supposition of Mendelian Inheritance" (1918), in which, in response to new data, he improved upon the kinds of models that Pearson, Weldon, and Yule had been considering 10–20 years before, and showed clearly that the correlations observed between human relatives not only could be interpreted on the supposition of Mendelian inheritance, but also that Mendelian inheritance must lead to precisely the kind of correlations observed.

Weldon's death was not only a tremendous blow to Pearson but also removed a close colleague of high caliber, without whom it was not possible to continue work in biometry along some of the lines that they had developed during the preceding fifteen years. Yet Pearson's productivity hardly faltered. During his remaining thirty years his articles, editorials, memoirs, and books on or related to biometry and statistics numbered over 300; he also produced one in astronomy and four in mechanics and about seventy published letters, reviews, and prefatory and other notes in scientific publications, the last of which was a letter (1935) on the aims of the founders of *Biometrika* and the conditions under which the journal had been published.

Following Weldon's death, Pearson gave increasing attention to eugenics. In 1904 Galton had provided funds for the establishment of a eugenics record office, to be concerned with collecting data for the scientific study of eugenics. Galton kept the office under his control until late in 1906, when, at the age of eighty-four, he turned it over to Pearson. With a change of name to eugenics laboratory, it became a companion to Pearson's biometric laboratory. It was transferred in 1907 to University College and with a small staff carried out studies of the relative importance of heredity and environment in alcoholism, tuberculosis, insanity, and infant mortality.[49] The findings were published as Studies in National Deterioration, nos. 1–11 (1906–1924) and in Eugenics Laboratory Memoirs, nos. 1–29 (1907–1935). Thirteen issues of the latter were devoted to "The Treasury of Human Inheritance" (1909–1933), a vast collection of pedigrees forming the basic material for the discussion of the inheritance of abnormalities, disorders, and other traits.

Pearson's major effort during the period 1906–1914, however, was devoted to developing a postgraduate center in order "to make statistics a branch of applied mathematics with a technique and nomenclature of its own, to train statisticians as men of science . . . and in general to convert statistics in this country from being the playing field of *dilettanti* and controversialists into

a serious branch of science, which no man could attempt to use effectively without adequate training, any more than he could attempt to use the differential calculus, being ignorant of mathematics."[50] At the beginning of this period Pearson was not only head of the department of applied mathematics, but also in charge of the drawing office for engineering students, giving evening classes in astronomy, directing the biometric and eugenics laboratories, and editing their various publications, and *Biometrika*, a tremendous task for one man. In the summer of 1911, however, he was able to cut back somewhat on these diverse activities by relinquishing the Goldsmid chair of applied mathematics to become the first Galton professor of eugenics and head of a new department of applied statistics in which were incorporated the biometric and eugenics laboratories. But he also assumed a new task about the same time: soon after Galton's death in 1911, his relatives had asked Pearson to write his biography. The first volume of *The Life, Letters and Labors of Francis Galton* was published in 1914, the second volume in 1925, and the third volume (in two parts) in 1930. It is an incomparable source of information on Galton, on Pearson himself, and on the early years of biometry. Although the volume of Pearson's output of purely statistical work was somewhat reduced during these years by the task of writing this biography, it was still immense by ordinary standards.

Pearson was the principal editor of *Biometrika* from its founding to his death (vols. 1–28, 1901–1936), and for many years he was the sole editor. Under his guidance it became the world's leading medium of publication of papers on, and mathematical tables relating to, statistical theory and practice. Soon after World War I, during which Pearson's group was deeply involved in war work, he initiated the series Tracts for Computers, nos. 1–20 (1919–1935), many of which became indispensable to computers of the period. In 1925 he founded *Annals of Eugenics* and served as editor of the first five volumes (1925–1933). Some of the tables in *Tables for Statisticians and Biometricians* (pt. I, 1914; pt. II, 1931) appear to be timeless in value; others are no longer used. *The Tables of the Incomplete Beta-Function* (1934), a compilation prepared under his direction over a period of several decades, remains a monument to him and his co-workers.

In July 1932 Pearson advised the college and university that he would resign from the Galton professorship the following summer. The college decided to divide the department of applied statistics into two independent units, a department of eugenics with which the Galton professorship would be associated, and a new department of statistics. In October 1933 Pearson was established in a room placed at his disposal by the zoology department; his son, Egon, was head of the new department of statistics; and R. A. Fisher was named the second Galton professor of eugenics. Pearson continued to edit *Biometrika* and had almost seen the final proofs of the first half of volume 28 through the press when he died on 27 April 1936.

NOTES

1. Quoted by E. S. Pearson in *Karl Pearson: An Appreciation* . . ., p. 4 (*Biometrika*, **28**, 196).
2. Galton discovered the statistical phenomenon of regression around 1875 in the course of experiments with sweet-pea seeds to determine the law of inheritance of size. Using 100 parental seeds of each of 7 different selected sizes, he constructed a two-way plot of the diameters of parental and offspring seeds from each parental class. Galton then noticed that the median diameters of the offspring seeds for the respective parental classes fell nearly on a straight line. Furthermore, the median diameters of offspring from the larger-size parental classes were less than those of the parents; and for the smaller-size parental classes, they were greater than those of the parents, indicating a tendency of the "mean" offspring size to "revert" toward what might be described as the average ancestral type. Not realizing that this phenomenon is a characteristic of any two-way plot, he first termed it "reversion" and, later, "regression."

   Examining these same data further, Galton noticed that the variation of offspring size within the respective parental arrays (as measured by their respective semi-interquartile ranges) was approximately constant and less than the similarly measured variation of the overall offspring population. From this empirical evidence he then inferred the correct relation, variability of offspring family $= \sqrt{1 - r^2} \times$ variability of overall offspring population, which he announced in symbolic form in an 1877 lecture, calling r the "reversion" coefficient.

   A few years later Galton made a two-way plot of the statures of some human parents of unselected statures and their adult children, noting that the respective marginal distributions were approximately Gaussian or "normal," as Adolphe Quetelet had noticed earlier from examination of each of these variables separately, and that the frequency distributions along lines in the plot parallel to either of the variate axes were "apparently" Gaussian distributions of equal variation, which was less than, and in a constant ratio $\sqrt{1 - r^2}$ to, that of the corresponding marginal distributions. To obtain a numerical value for r, Galton expressed the deviations of the individual values of both variates from their respective medians in terms of their respective semi-interquartile ranges as a unit, so that r became the slope of his regression line.

   In 1888 Galton made one more great and far-reaching discovery. Applying the techniques that he had evolved for the measurement of the influence of heredity to the problem of measuring the degree of association between the sizes of two different organs of the same individual, he reached the conception of an "index of co-relation" as a measure of the degree of relationship between two such characteristics and recognized r, his measure of "reversion" or "regression," to be such a coefficient of co-relation or correlation, suitable for application to all living forms.

   Galton, however, failed to recognize and appreciate the additional mathematical complexity necessarily involved in the joint consideration of more than two mutually correlated characteristics, with the result that his efforts to formulate and implement what became known as his law of ancestral heredity were somewhat confused and imprecise. It remained for Pearson to provide the necessary generalization and precision of formulation in the form of a multiple regression formula.

   For fuller details, see Pearson's "Notes on the History of Correlation" (1920).
3. *Speeches . . . at a Dinner . . . in [His] Honour*, pp. 22–23; also quoted by E. S. Pearson, *op. cit.*, p. 19 (*Biometrika*, **28**, 211).
4. An examination of *Letters From W. S. Gosset to R. A. Fisher 1915–1936*, 4 vols. (Dublin, 1962), issued for private circulation only, reveals that Gosset (pen name "Student"), played a similar role with respect to R. A. Fisher. When and how they first came into contact is revealed by the two letters of Sept. 1912 from Gosset to Pearson that are reproduced in E. S. Pearson's "Some Early Correspondence . . ." (1968).
5. E. S. Pearson, *op. cit.*, apps. II and III.
6. Pearson was not the first to use this terminology: "Galton used it, as did also Lexis, and the writer has not found any reference which seems to be its first use" (Helen M. Walker, *Studies . . .*, p. 185). But Pearson's consistent and exclusive use of this term in his epoch-making publications led to its adoption throughout the statistical community.
7. E. S. Pearson, *op. cit.*, p. 26 (*Biometrika*, **28**, 218).
8. The title "Contributions to the Mathematical Theory of Evolution" or "Mathematical Contributions . . ." was used as the general title of 17 memoirs, numbered II through XIX, published in the *Philosophical Transactions* or as Drapers' Company Research Memoirs, and of 8 unnumbered papers published in the *Proceedings of the Royal Society*. "Mathematical" became and remained the initial word from III (1896) on. No. XVII was announced before 1912 as a forthcoming Drapers' . . . Memoir but has not been published to date.
9. From Pearson, "Statistical Tests," in *Nature*, **136** (1935), 296–297, see 296.
10. Pearson, "Notes on the History of Correlation," p. 37 (Pearson and Kendall, p. 197).
11. Pearson did not use different symbols for population parameters (such as μ, σ, ρ) and sample measures of them (m, s, r) as has been done in this article, following the example set by "Student" in his first paper on small-sample theory, "The Probable Error of a Mean" (1908). Use of identical symbols for population parameters and sample measures of them makes Pearson's, and other papers of this period, difficult to follow and, in some instances, led to error.
12. Pearson, "Notes on the History of Correlation," p. 42 (Pearson and Kendall, p. 202).
13. In the rest of the article, the term "standard error" will be used instead of "standard deviation of the sampling error." Pearson consistently gave formulas for, and spoke of the corresponding "probable error" (or "p.e.") defined by,

    probable error = 0.674489 . . . × standard error,

    the numerical factor being the factor appropriate to the normal distribution, and reserved the term "standard deviation" (and the symbol σ) for description of the variation of individuals in a population or sample.
14. Footnote, p. 274 (*Early . . . Papers*, p. 134).
15. There are always two sample η's, η_{yx} and η_{xy}, corresponding to the regression of y on x and the regression of x on y, respectively, in the sample. When these regressions are both exactly linear, $\eta_{yx} = \eta_{xy} = r$; otherwise η_{yx} and η_{xy} are different.

    In this memoir Pearson defines and discusses the correlation ratio, η_{yx}, and its relation to r entirely in terms of a sample of N paired observations, (x_i, y_i), $(i = 1, 2,..., N)$. The implications of various equalities and inequalities between the correlation ratio of a trait X with respect to a trait Y in some general (nonnormal) bivariate population and ρ, the product-moment coefficient of correlation of X and Y in this population, are discussed, for example, in W. H. Kruskal, "Ordinal Measures of Association," in *Journal of American Statistical Association*, **53** (1958), 814–861.

16. In Pearson, "On the Systematic Fitting of Curves to Observations and Measurements," in *Biometrika*, **1**, no. 3 (Apr. 1902), 264–303, see p. 271.

17. Pearson and Alice Lee, "On the Distribution of Frequency (Variation and Correlation) of the Barometric Height at Diverse Stations," in *Philosophical Transactions of the Royal Society*, **190A** (1898), 423–469, see 456 and footnote to 462, respectively.

18. Pearson, "On the Probable Error of a Coefficient of Correlation as Found From a Fourfold Table," in *Biometrika*, **9**, nos. 1–2 (Mar. 1913), 22–27.

19. Pearson, "On the Probable Error of Biserial η," *ibid.*, **11**, no. 4 (May 1917), 292–302.

20. *Ibid.*, **1**, no. 1 (Oct. 1901), 2. Emphasis added.

21. Student, "Probable Error of a Correlation Coefficient," *ibid.*, **6**, nos. 2–3 (Sept. 1908), 302–310. In a 1915 letter to R. A. Fisher (repro. in E. S. Pearson, "Some Early Correspondence . . .," p. 447, and in Pearson and Kendall, p. 470), Gosset tells "how these things came to be of importance [to him]" and, in particular, says that the work of "the Experimental Brewery which concerns such things as the connection between analysis of malt or hops, and the behaviour of the beer, and which takes a day to each unit of the experiment, thus limiting the numbers, demanded an answer to such questions as 'If with a small number of cases I get a value r, what is the probability that there is really a positive correlation of greater than (say) 25?' "

22. E. S. Pearson, "Some Reflexions . . .," pp. 351–352 (Pearson and Kendall, pp. 349–350).

23. R. A. Fisher, "Frequency Distribution of the Values of the Correlation Coefficient in Samples From an Indefinitely Large Population," in *Biometrika*, **10**, no. 4 (May 1915), 507–521.

24. Letter from Pearson to Fisher dated 26 Sept. 1914, repro. in E. S. Pearson, "Some Early Correspondence . . .," p. 448 (Pearson and Kendall, p. 408).

25. Letter from Pearson to Fisher dated 3 Oct. 1914, partly repro. *ibid.*, p. 449 (Pearson and Kendall, p. 409).

26. Letter from Pearson to Fisher dated 30 Jan., 1915, partly repro. *ibid.*, pp. 449–450 (Pearson and Kendall, pp. 409–410).

27. *Ibid.*, p. 450 (Pearson and Kendall, p. 410).

28. Letter from Pearson to Fisher dated 13 May 1916, repro. *ibid.*, p. 451 (Pearson and Kendall, p. 411).

29. J. O. Irwin, in *Journal of the Royal Statistical Society*, **126**, pt. 1 (Mar. 1963), 161; F. Yates and K. Mather, in *Biographical Memoirs of Fellows of the Royal Society*, **9** (Nov. 1963), 98–99; P. C. Mahalanobis, in *Biometrics*, **20**, no. 2 (June 1964), 214.

30. R. A. Fisher, "On an Absolute Criterion for Fitting Frequency Curves," in *Messenger of Mathematics*, **41** (1912), 155–160.

This paper marks Fisher's break away from inverse probability reasoning via Bayes's theorem but, although evident in retrospect, the "break" was not clear-cut: not having yet coined the term "likelihood," he spoke (p. 157) of "the probability of any particular set of θ's" (that is, of the parameters involved) being "proportional to the chance of a given set of observations occurring"— which appears to be equivalent to the proposition in the theory of inverse probability that, assuming a uniform a priori probability distribution of the parameters, the ratio of the a posteriori probability that $\theta = \theta_o + \xi$ to the a posteriori probability that $\theta = \theta_o$ is equal to the ratio of the probability of the observed set of observations when $\theta = \theta_o + \xi$ to their probability when $\theta = \theta_o$. He also described (p. 158) graphical representation of "the inverse probability system." On the other hand, he did stress (p. 160) that only the relative (not the absolute) values of these "probabilities" were meaningful and that it would be "illegitimate" to integrate them over a region in the parameter space.

Fisher introduced the term "likelihood" in his paper "On the Mathematical Foundations of Theoretical Statistics," in *Philosophical Transactions of the Royal Society*, **222A** (19 Apr. 1922), 309–368, in which he made clear for the first time the distinction between the mathematical properties of "likelihoods" and "probabilities," and stated:

> I must plead guilty in my original statement of the Method of Maximum Likelihood to having based my argument upon the principle of inverse probability; in the same paper, it is true, I emphasized the fact that such inverse probabilities were relative only . . . . Upon consideration . . . I perceive that the word probability is wrongly used in such a connection: probability is a ratio of frequencies, and about the frequencies of such [parameter] values we can know nothing whatever (p. 326).

31. E. S. Pearson, "Some Early Correspondence . . .," p. 452 (Pearson and Kendall, p. 412).

32. Repro. *ibid.*, pp. 454–455 (Pearson and Kendall, pp. 414–415).

33. F. N. David, *Tables of the Ordinates and Probability Integral of the Distribution of the Correlation Coefficient in Small Samples* (London, 1938).

34. Letter from Pearson to Fisher dated 21 Aug. 1920, repro. in E. S. Pearson, "Some Early Correspondence . . .," p. 453 (Pearson and Kendall, p. 413).

35. R. A. Fisher, "On the 'Probable Error' of a Coefficient of Correlation Deduced From a Small Sample," in *Metron*, **1**, no. 4 (1921), 1–32.

36. Letters from Pearson to Fisher dated 26 June 1916 and 21 Oct. 1918, repro. in E. S. Pearson, "Some Early Correspondence . . .," pp. 455, 456, respectively (Pearson and Kendall, pp. 415, 416).

37. Pearson, "Method of Moments and Method of Maximum Likelihood," in *Biometrika*, **28**, nos. 1–2 (June 1936), 34–59; R. A. Fisher, "Professor Karl Pearson and the Method of Moments," in *Annals of Eugenics*, **7**, pt. 4 (June 1937), 303–318.

38. F. Y. Edgeworth, "On the Probable Error of Frequency Constants," in *Journal of the Royal Statistical Society*, **71** (1908), 381–397, 499–512, 652–678.

39. The identical mathematical form of expressions derived by the method of maximum likelihood and by the method of inverse probability, if a uniform prior distribution is adopted, has been a source of continuing confusion. Thus, the "standard errors" given by Gauss in his 1816 paper were undeniably derived via the method of inverse probability and, strictly speaking, are the standard deviations of the a posteriori probability distributions of parameters concerned, given the observed values of the particular functions of sample values considered. On the other hand, by virtue of the above-mentioned equivalence of form, Gauss's 1816 formulas can be recognized as giving the "standard errors," that is, the standard deviations of the sampling distributions, of the functions of sample values involved for fixed values of the corresponding population parameters. Consequently, speaking loosely, one is inclined today to attribute to Gauss the original ("first") derivation of these "standard error" formulas, even though he may have had (in 1816) no conception of the "sampling distribution," for fixed values of a population parameter, of a sample function used to estimate the value of this parameter. In contrast, the result given in his 1821 paper almost certainly refers to the sampling

distribution of *s*, and not to the a posteriori distribution of *σ*.

Edgeworth's discussion is quite explicitly in terms of inverse probability. Pearson-Filon asymptotic formulas are derived afresh in this context and are said to be applicable only to "solutions" obtained by "the genuine inverse method," the "fluctuation of the *quaesitum*" so determined "being less than that of any other determination" (pp. 506–507).

The correct interpretation of the formulas derived by Pearson and Filon is somewhat obscured by their use of identical symbols for population parameters and the sample functions used to estimate them, and by the fact that their choice of words is such that their various summary statements can be interpreted either way. On the other hand, their derivation starts (p. 231) with consideration of a ratio of probabilities, introduced without explanation but for which the explanation may be the "proposition in the theory of Inverse Probability" mentioned in note 30 above; and Pearson says, in his letter of June 1916 to Fisher (see note 32), "In the first place you have to demonstrate the logic of the Gaussian rule . . . I frankly confess I approved the Gaussian method in 1897 (see *Phil. Trans.* Vol. 191, A, p. 232), but I think it logically at fault now." These facts suggest that Pearson and Filon may have regarded the "probable errors" and "correlations" they derived as describing properties of the joint a posteriori probability distribution of the population parameters, given the observed values of the sample functions used to estimate them.

40. *Proceedings of the Royal Society*, **68** (1900), 1–5.
41. Quoted by Pearson in his memoir on Weldon, in *Biometrika*, **5**, no. 1 (Oct. 1906), 35 (Pearson and Kendall, p. 302).
42. Letter from Pearson to Galton, quoted in Pearson's *Life . . . of Francis Galton*, IIIA, 241.
43. Quoted by Pearson in his memoir on Weldon, in *Biometrika*, **5**, no. 1 (Oct. 1906), 35 (Pearson and Kendall, p. 302).
44. *Ibid.*
45. W. Bateson, "Heredity, Differentiation, and Other Conceptions of Biology: A Consideration of Professor Karl Pearson's Paper 'On the Principle of Homotyposis,'" in *Proceedings of the Royal Society*, **69**, no. 453, 193–205.
46. Pearson, "On the Fundamental Conceptions of Biology," in *Biometrika*, **1**, no. 3 (Apr. 1902), 320–344.
47. W. Bateson, *Materials for the Study of Variation, Treated With Especial Regard to Discontinuity in the Origin of Species* (London, 1894).
48. For fuller details, see either of the articles by P. Froggatt and N. C. Nevin in the bibliography; the first is the more complete.
49. These studies were not without a price for Pearson: he became deeply involved almost at once in a hot controversy over tuberculosis and a fierce dispute on the question of alcoholism. See E. S. Pearson, *Karl Pearson . . .*, pp. 59–66 (*Biometrika*, **29**, 170–177).
50. From a printed statement entitled *History of the Biometric and Galton Laboratories*, drawn up by Pearson in 1920; quoted in E. S. Pearson, *Karl Pearson . . .*, p. 53 (*Biometrika*, **29**, 164).

BIBLIOGRAPHY

I. ORIGINAL WORKS. A bibliography of Pearson's research memoirs and his articles and letters in scientific journals that are on applied mathematics, including astronomy, but not statistics, biometry, anthropology, eugenics, or mathematical tables, follows the obituary by L. N. G.

Filon (see below). A bibliography of his major contributions to the latter five areas is at the end of P. C. Mahalanobis, "A Note on the Statistical and Biometric Writings of Karl Pearson" (see below). The individual mathematical tables and collections of such tables to which Pearson made significant contributions in their computation or compilation, or through preparation of explanatory introductory material, are listed and described in Raymond Clare Archibald, *Mathematical Table Makers* (New York, 1948), 65–67.

Preparation of a complete bibliography of Pearson's publications was begun, with his assistance, three years before his death. The aim was to include all of the publications on which his name appeared as sole or part author and all of his publications that were issued anonymously. The result, *A Bibliography of the Statistical and Other Writings of Karl Pearson* (Cambridge, 1939), compiled by G. M. Morant with the assistance of B. L. Welch, lists 648 numbered entries arranged chronologically under five principal headings, with short summaries of the contents of the more important, followed by a sixth section in which a chronological list, "probably incomplete," is given of the syllabuses of courses of lectures and single lectures delivered by Pearson that were printed contemporaneously as brochures or single sheets. The five major categories and the number of entries in each are the following:

I. Theory of statistics and its application to biological, social, and other problems (406);

II. Pure and applied mathematics and physical science (37);

III. Literary and historical (67);

IV. University matters (27);

V. Letters, reviews, prefatory and other notes in scientific publications (111).

Three omissions have been detected: "The Flying to Pieces of a Whirling Ring," in *Nature*, **43**, no. 1117 (26 Mar. 1891), 488; "Note on Professor J. Arthur Harris' Papers on the Limitation in the Applicability of the Contingency Coefficient," in *Journal of the American Statistical Association*, **25**, no. 171 (Sept. 1930), 320–323; and "Postscript," *ibid.*, 327.

The following annotated list of Pearson's most important publications will suffice to reveal the great diversity of his contributions and their impact on the biological, physical, and social sciences. The papers marked with a single asterisk (*) have been repr. in *Karl Pearson's Early Statistical Papers* (Cambridge, 1948) and those with a double asterisk (**), in E. S. Pearson and M. G. Kendall, eds., *Studies in the History of Probability and Statistics* (London–Darien, Conn., 1970), referred to as Pearson and Kendall.

"On the Motion of Spherical and Ellipsoidal Bodies in Fluid Media" (2 pts.), in *Quarterly Journal of Pure and Applied Mathematics*, **20** (1883), 60–80, 184–211; and "On a Certain Atomic Hypothesis" (2 pts.), in *Transactions of the Cambridge Philosophical Society*, **14**, pt. 2 (1887), 71–120, and *Proceedings of the London Mathematical Society*, **20** (1888), 38–63, respectively. These early papers on the motions of a rigid or pulsating atom in an infinite incompressible fluid did much to increase Pearson's stature in applied mathematics at the time.

William Kingdon Clifford, *The Common Sense of the Exact Sciences* (London, 1885; reiss. 1888), which Pearson edited and completed.

Isaac Todhunter, *A History of the Theory of Elasticity and of the Strength of Materials From Galilei to the Present Time*, 2 vols. (Cambridge, 1886–1893; reiss. New York, 1960), edited and completed by Pearson.

The Ethic of Freethought (London, 1888; 2nd ed., 1901), a collection of essays, lectures, and public addresses on free thought, historical research, and socialism.

"On the Flexure of Heavy Beams Subjected to a Continuous Load. Part I," in *Quarterly Journal of Pure and Applied Mathematics*, **24** (1889), 63–110, in which for the first time a now-much-cited exact solution was given for the bending of a beam of circular cross section under its own weight, and extended to elliptic cross sections in ". . . Part II," *ibid.*, **31** (1899), 66–109, written with L. N. G. Filon.

The Grammar of Science (London, 1892; 3rd ed., 1911; reiss. Gloucester, Mass., 1969; 4th ed., E. S. Pearson, ed., London, 1937), a critical survey of the concepts of modern science and his most influential book.

* "Contributions to the Mathematical Theory of Evolution," in *Philosophical Transactions of the Royal Society*, **185A** (1894), 71–110, deals with the dissection of symmetrical and asymmetrical frequency curves into normal (Gaussian) components and marks Pearson's introduction of the method of moments as a means of fitting a theoretical curve to experimental data and of the term "standard deviation" and σ as the symbol for it.

* "Contributions to the Mathematical Theory of Evolution. II. Skew Variation in Homogeneous Material," *ibid.*, **186A** (1895), 343–414, in which the term "mode" is introduced, the foundations of the Pearson system of frequency curves is laid, and Types I–IV are defined and their application exemplified.

* "Mathematical Contributions to the Theory of Evolution. III. Regression, Heredity, and Panmixia," *ibid.*, **187A** (1896), 253–318, Pearson's first fundamental paper on correlation, with special reference to problems of heredity, in which correlation and regression are defined in far greater generality than previously and the theory of multivariate normal correlation is developed as a practical tool to a stage that left little to be added.

The Chances of Death and Other Studies in Evolution, 2 vols. (London, 1897), essays on social and statistical topics, including the earliest adequate study ("Variation in Man and Woman") of anthropological "populations" using scientific measures of variability.

* "Mathematical . . . IV. On the Probable Errors of Frequency Constants and on the Influence of Random Selection on Variation and Correlation," in *Philosophical Transactions of the Royal Society*, **191A** (1898), 229–311, written with L. N. G. Filon, in which were derived the now-familiar expressions for the asymptotic variances and covariances of sample estimators of a group of population parameters in terms of derivatives of the likelihood function (without recognition of their applicability only to maximum likelihood estimators), and a number of particular results deduced therefrom.

* "Mathematical . . . V. On the Reconstruction of the Stature of Prehistoric Races," *ibid.*, **192A** (1898), 169–244, in which multiple regression techniques were used to reconstruct predicted average measurements of extinct races from the sizes of existing bones, given the correlations among bone lengths in an extant race, not merely as a technical exercise but as a means of testing the accuracy of predictions in evolutionary problems in the light of certain evolutionary theories.

"Mathematical . . . On the Law of Ancestral Heredity," in *Proceedings of the Royal Society*, **62** (1898), 386–412, a statistical formulation of Galton's law in the form of a multiple regression of offspring on "mid-parental" ancestry, with deductions therefrom of theoretical values for various regression and correlation coefficients between kin, and comparisons of such theoretical values with values derived from observational material.

"Mathematical . . . VII. On the Correlation of Characters not Quantitatively Measurable," in *Philosophical Transactions of the Royal Society*, **195A** (1901), 1–47, in which the "tetrachoric" coefficient of correlation r_t was introduced for estimating the coefficient of correlation, ρ, of a bivariate normal distribution from a sample scored dichotomously in both variables.

* "On the Criterion That a Given System of Deviations From the Probable in the Case of a Correlated System of Variables Is Such That It Can Be Reasonably Supposed to Have Arisen From Random Sampling," in *London, Edinburgh and Dublin Philosophical Magazine and Journal of Science*, 5th ser., **50** (1900), 157–175, in which the "χ^2 test of goodness of fit" was introduced, one of Pearson's greatest single contributions to statistical methodology.

"Mathematical . . . IX. On the Principle of Homotyposis and Its Relation to Heredity, to the Variability of the Individual, and to That of Race. Part I. Homotyposis in the Vegetable Kingdom," in *Philosophical Transactions of the Royal Society*, **197A** (1901), 285–379, written with Alice Lee *et al.*, a theoretical discussion of the relation of fraternal correlation to the correlation of "undifferentiated like organs of the individual" (called "homotyposis"), followed by numerous applications; the paper led to a complete schism between the biometric and Mendelian schools and the founding of *Biometrika*.

* "Mathematical . . . X. Supplement to a Memoir on Skew Variation," *ibid.*, 443–459; Pearson curves Type V and VI are developed and their application exemplified.

* "On the Mathematical Theory of Errors of Judgment With Special Reference to the Personal Equation," *ibid.*, **198A** (1902), 235–299, a memoir still of great interest and importance founded on two series of experiments, each with three observers, from which it was learned, among other things, that the "personal equation" (bias pattern of an individual observer) is subject to fluctuations far exceeding random sampling and that the

errors of different observers looking at the same phenomena are in general correlated.

"Note on Francis Galton's Problem," in *Biometrika*, **1**, no. 4 (Aug. 1902), 390–399, in which Pearson found the general expression for the mean value of the difference between the rth and the $(r + 1)$th ranked individuals in random samples from a continuous distribution, one of the earliest results in the sampling theory of order statistics —similar general expressions for the variances of and correlations between such intervals are given in his joint paper of 1931.

"On the Probable Errors of Frequency Constants," in *Biometrika*, **2**, no. 3 (June 1903), 273–281, an editorial that deals with standard errors of, and correlations between, cell frequencies and sample centroidal moments, in terms of the centroidal moments of a univariate distribution of general form. The extension to samples from a general bivariate distribution was made in pt. II, in *Biometrika*, **9**, nos. 1–2 (Mar. 1913), 1–19; and to functions of sample quantiles in pt. III, *ibid.*, **13**, no. 1 (Oct. 1920), 113–132.

* Mathematical . . . XIII. On the Theory of Contingency and Its Relation to Association and Normal Correlation, Drapers' Company Research Memoirs, Biometric Series, no. 1 (London, 1904), directed toward measuring the association of two variables when the observational data take the form of frequencies in the cells of an $r \times c$ "contingency table" of qualitative categories not necessarily meaningfully orderable, an adaptation of his χ^2 goodness-of-fit criterion, termed "square contingency," being introduced to provide a test of overall departure from the hypothesis of independence and the basis of a measure of association, the "coefficient of contingency" $c = \sqrt{\chi^2/(\chi^2 + n)}$, which was shown to tend under certain special conditions to the coefficient of correlation of an underlying bivariate normal distribution.

On Some Disregarded Points in the Stability of Masonry Dams, Drapers' Company Research Memoirs, Technical Series, no. 1 (London, 1904), written with L. W. Atcherley, in which it was shown that the assumptions underlying a widely accepted procedure for calculating the stresses in masonry dams are not satisfied at the bottom of the dam, the stresses there being in excess of those so calculated, with consequent risk of rupture near the base—still cited today, this paper and its companion *Experimental Study . . .* (1907) caused great concern at the time, for instance, with reference to the British-built Aswan Dam.

* Mathematical . . . XIV. On the General Theory of Skew Correlation and Non-Linear Regression, Drapers' Company Research Memoirs, Biometric Series, no. 2 (London, 1905), dealt with the general conception of skew variation and correlation and the properties of the "correlation ratio" η (introduced in 1903) and showed for the first time the fundamental importance of the expressions $(1 - \eta^2) \sigma_y^2$ and $(\eta^2 - r^2) \sigma_y^2$ and of the difference between η and r as measures of departure from linearity, as well as those conditions that must be satisfied for linear, parabolic, cubic, and other regression equations to be adequate.

"The Problem of the Random Walk," in *Nature*, **72** (17 July 1905), 294, a brief letter containing the first explicit formulation of a "random walk," a term Pearson coined, and asking for information on the probability distribution of the walker's distance from the origin after n steps—Lord Rayleigh indicated the asymptotic solution as $n \to \infty$ in the issue of 3 Aug., p. 318; and the general solution for finite n was published by J. C. Kluyver in Dutch later the same year.

Mathematical . . . XV. A Mathematical Theory of Random Migration, Drapers' Company Research Memoirs, Biometric Series, no. 3 (London, 1906), written with John Blakeman. Various theoretical forms of distribution were derived that would result from random migration from an origin under certain ideal conditions, and solutions to a number of subsidiary problems were given—results that, while not outstandingly successful in studies of migration, have found various other applications.

** "Walter Frank Raphael Weldon, 1860–1906," in *Biometrika*, **5**, nos. 1–2 (Oct. 1906), 1–52 (repr. as paper no. 21 in Pearson and Kendall), a tribute to the man who posed the questions that impelled Pearson to some of his most important contributions, with additional details on the early years (1890–1905) of the biometric school and the founding of *Biometrika*.

Mathematical . . . XVI. On Further Methods of Determining Correlation, Drapers' Company Research Memoirs, Biometric Series, no. 4 (London 1907), dealt with calculation of the coefficient of correlation, r, from the individual differences $(x - y)$ in a sample and with estimation of the coefficient of correlation, ρ, of a bivariate normal population from the ranks of the individuals in a sample of that population with respect to each of the two traits concerned.

An Experimental Study of the Stresses in Masonry Dams, Drapers' Company Research Memoirs, Technical Series, no. 5 (London, 1907), written with A. F. C. Pollard, C. W. Wheen, and L. F. Richardson, which lent experimental support to the 1904 theoretical findings.

A First Study of the Statistics of Pulmonary Tuberculosis, Drapers' Company Research Memoirs, Studies in National Deterioration, no. 2 (London, 1907), and *A Second Study . . .: Marital Infection, . . .* Technical Series, no. 3 (London, 1908), written with E. G. Pope, the first two of seven publications by Pearson and his co-workers during 1907–1913 on the then-important and controversial subjects of the inheritance and transmission of pulmonary tuberculosis.

"On a New Method of Determining Correlation Between a Measured Character A, and a Character B, of which Only the Percentage of Cases Wherein B Exceeds (or Falls Short of) a Given Intensity Is Recorded for Each Grade of A," in *Biometrika*, **6**, nos. 1 and 2 (July–Oct. 1909), 96–105, in which the formula for the biserial coefficient of correlation, "biserial r," is derived but not named, and its application exemplified.

"On a New Method of Determining Correlation When One Variable Is Given by Alternative and the Other by Multiple Categories," *ibid.*, **7**, no. 3 (Apr. 1910), 248–257, in which the formula for "biserial η" is derived but not named, and its application exemplified.

A First Study of the Influence of Parental Alcoholism on the Physique and Ability of the Offspring, Eugenics Laboratory Memoirs, no. 10 (London, 1910), written with Ethel M. Elderton, gave correlations between drinking habits of the parents and the intelligence and various physical characteristics of the offspring, and examined the effect of parental alcoholism on the infant death rate.

A Second Study . . . Being a Reply to Certain Medical Critics of the First Memoir and an Examination of the Rebutting Evidence Cited by Them, Eugenics Laboratory Memoirs, no. 13 (London, 1910), written with E. M. Elderton.

A Preliminary Study of Extreme Alcoholism in Adults, Eugenics Laboratory Memoirs, no. 14 (London, 1910), written with Amy Barrington and David Heron. The relations of alcoholism to number of convictions, education, religion, prostitution, mental and physical conditions, and death rates were examined, with comparisons between the extreme alcoholic and the general population.

"On the Probability That Two Independent Distributions of Frequency Are Really Samples From the Same Population," in *Biometrika*, **8**, nos. 1–2 (July 1911), 250–254, in which his χ^2 goodness-of-fit criterion is extended to provide a test of the hypothesis that two independent samples arrayed in a $2 \times c$ table are random samples from the same population.

Social Problems: Their Treatment, Past, Present and Future..., Questions of the Day and of the Fray, no. 5 (London, 1912), contains a perceptive, eloquent plea for replacement of literary exposition and folklore by measurement, and presents some results of statistical analyses that illustrate the complexity of social problems.

The Life, Letters and Labours of Francis Galton, 3 vols. in 4 pts. (Cambridge, 1914–1930).

Tables for Statisticians and Biometricians (London, 1914; 2nd ed., issued as "Part I," 1924; 3rd ed., 1930), consists of 55 tables, some new, the majority repr. from *Biometrika*, a few from elsewhere, to which Pearson as editor contributed an intro. on their use.

"On the General Theory of Multiple Contingency With Special Reference to Partial Contingency," in *Biometrika*, **11**, no. 3 (May 1916), 145–158, extends the χ^2 method to the comparison of two ($r \times 2$) tables and contains the basic elements of a large part of present-day χ^2 technique.

"Mathematical Contributions . . . XIX. Second Supplement to a Memoir on Skew Variation," in *Philosophical Transactions of the Royal Society*, **216A** (1916), 429–457, in which Pearson curves Types VII–XI are defined and their applications illustrated.

"On the Distribution of the Correlation Coefficient in Small Samples. Appendix II to the Papers of 'Student' and R. A. Fisher. A Cooperative Study," in *Biometrika*, **11**, no. 4 (May 1917), 328–413, written with H. E. Soper, A. W. Young, B. M. Cave, and A. Lee, an exhaustive study of the moments and shape of the distribution of r in samples of size n from a normal population with correlation coefficient ρ as a function of n and ρ, and of its approach to normality as $n \to \infty$, with special attention to determination, via inverse probability, of the "most likely value" of ρ from an observed value of r—the paper that initiated the rift between Pearson and Fisher.

"De Saint-Venant Solution for the Flexure of Cantilevers of Cross-Sections in the Form of Complete and Curtate Circular Sectors, and the Influence of the Manner of Fixing the Built-in End of the Cantilever on Its Deflection," in *Proceedings of the Royal Society*, **96A** (1919), 211–232, written with Mary Seegar, a basic paper giving the solution regularly cited for cantilevers of such cross sections—Pearson's last paper in mechanics.

** "Notes on the History of Correlation. Being a Paper Read to the Society of Biometricians and Mathematical Statisticians, June 14, 1920," in *Biometrika*, **13**, no. 1 (Oct. 1920), 25–45 (paper no. 14 in Pearson and Kendall), deals with Gauss's and Bravais's treatment of the bivariate normal distribution, Galton's discovery of correlation and regression, and Pearson's involvement in the matter.

Tables of the Incomplete Γ-Function Computed by the Staff of the Department of Applied Statistics, University of London, University College (London, 1922; reiss. 1934), tables prepared under the direction of Pearson, who, as editor, contributed an intro. on their use.

Francis Galton, 1822–1922. A Centenary Appreciation, Questions of the Day and of the Fray, no. 11 (London, 1922).

Charles Darwin, 1809–1922. An Appreciation. . . ., Questions of the Day and of the Fray, no. 12 (London, 1923).

"Historical Note on the Origin of the Normal Curve of Errors," in *Biometrika*, **16**, no. 3 (Dec. 1924), 402–404, announces the discovery of two copies of a long-overlooked pamphlet of De Moivre (1733) which gives to De Moivre priority in utilizing the integral of essentially the normal curve to approximate sums of successive terms of a binomial series, in formulating and using the theorem known as "Stirling's formula," and in enunciating "Bernoulli's theorem" that imprecision of a sample fraction as an estimate of the corresponding population proportion depends on the inverse square root of sample size.

"On the Skull and Portraits of George Buchanan," *ibid.*, **18**, nos. 3–4 (Nov. 1926), 233–256, in which it is shown that the portraits fall into two groups corresponding to distinctly different types of face, and only the type exemplified by the portraits in the possession of the Royal Society conforms to the skull.

"On the Skull and Portraits of Henry Stewart, Lord Darnley, and Their Bearing on the Tragedy of Mary, Queen of Scots," *ibid.*, **20B**, no. 1 (July 1928), 1–104, in which the circumstances of Lord Darnley's death and the history of his remains are discussed, anthropometric characteristics of his skull and femur are described and shown to compare reasonably well with the portraits, and the pitting of the skull is inferred to be of syphilitic origin.

"Laplace, Being Extracts From Lectures Delivered by Karl Pearson," *ibid.*, **21**, nos. 1–4 (Dec. 1929), 202–216, an account of Laplace's ancestry, education, and later life that affords necessary corrections to a number of earlier biographies.

Tables for Statisticians and Biometricians, Part II (London, 1931), tables nearly all repr. from *Biometrika*, with pref. and intro. on use of the tables by Pearson, as editor.

"On the Mean Character and Variance of a Ranked Individual, and on the Mean and Variance of the Intervals Between Ranked Individuals. Part I. Symmetrical Distributions (Normal and Rectangular)," in *Biometrika*, **23**, nos. 3–4 (Dec. 1931), 364–397, and ". . . Part II. Case of Certain Skew Curves," *ibid.*, **24**, nos. 1–2 (May 1932), 203–279, both written with Margaret V. Pearson, in which certain general formulas relating to means, standard deviations, and correlations of ranked individuals in samples of size *n* from a continuous distribution are developed and applied (in pt. I) to samples from the rectangular and normal distributions, and (in pt. II) to special skew curves (Pearson Types VIII, IX, X, and XI) that admit exact solutions.

Tables of the Incomplete Beta-Function (London, 1934), tables prepared under the direction of and edited by Pearson, with an intro. by Pearson on the methods of computation employed and on the uses of the tables.

"The Wilkinson Head of Oliver Cromwell and Its Relationship to Busts, Masks and Painted Portraits," in *Biometrika*, **26**, nos. 3–4 (Dec. 1934), 269–378, written with G. M. Morant, an extensive analysis involving 107 plates from which it is concluded "that it is a 'moral certainty' drawn from circumstantial evidence that the Wilkinson Head is the genuine head of Oliver Cromwell."

"Old Tripos Days at Cambridge, as Seen From Another Viewpoint," in *Mathematical Gazette*, **20** (1936), 27–36.

Pearson edited two scientific journals, to which he also contributed substantially: *Biometrika*, of which he was one of the three founders, always the principal editor (vols. **1–28**, 1901–1936), and for many years the sole editor; and *Annals of Eugenics*, of which he was the founder and the editor of the first 5 vols. (1925–1933). He also edited three series of Drapers' Company Research Memoirs: Biometric Series, nos. 1–4, 6–12 (London, 1904–1922) (no. 5 was never issued), of which he was sole author of 4 and senior author of the remainder; Studies in National Deterioration, nos. 1–11 (London, 1906–1924), 2 by Pearson alone and as joint author of 3 more; and Technical Series, nos. 1–7 (London, 1904–1918), 1 by Pearson alone, the others with coauthors. To these must be added the Eugenics Laboratory Memoirs, nos. 1–29 (London, 1907–1935), of which Pearson was a coauthor of 4. To many others, including the 13 issues (1909–1933) comprising "The Treasury of Human Inheritance," vols. I and II, he contributed prefatory material; the Eugenics Laboratory Lecture Series, nos. 1–14 (London, 1909–1914), 12 by Pearson alone and 1 joint contribution; Questions of the Day and of the Fray, nos. 1–12 (London, 1910–1923), 9 by Pearson alone and 1 joint contribution; and Tracts for Computers, nos. 1–20 (London, 1919–1935), 2 by Pearson himself, plus a foreword, intro., or prefatory note to 5 others.

Pearson has given a brief account of the persons and early experiences that most strongly influenced his development as a scholar and scientist in his contribution to the volume of *Speeches . . .* (1934) cited below; fuller accounts of his Cambridge undergraduate days, his teachers, his reading, and his departures from the norm of a budding mathematician are in "Old Tripos Days" above. His "Notes on the History of Correlation" (1920) contains a brief account of how he became involved in the development of correlation theory; and he gives many details on the great formative period (1890–1906) in the development of biometry and statistics in his memoir on Weldon (1906) and in vol. IIIA of his *Life . . . of Francis Galton*.

A very large number of letters from all stages of Pearson's life, beginning with his childhood, and many of his MSS, lectures, lecture notes and syllabuses, notebooks, biometric specimens, and data collections have been preserved. A large part of his scientific library was merged, after his death, with the joint library of the departments of eugenics and statistics at University College, London; a smaller portion, with the library of the department of applied mathematics.

Some of Pearson's letters to Galton were published by Pearson, with Galton's replies, in vol. III of his *Life . . . of Francis Galton*. A few letters of special interest from and to Pearson were published, in whole or in part, by his son, E. S. Pearson, in his "Some Incidents in the Early History of Biometry and Statistics" and in "Some Early Correspondence Between W. S. Gosset, R. A. Fisher, and Karl Pearson," cited below; and a selection of others, from and to Pearson, together with syllabuses of some of Pearson's lectures and lecture courses, are in E. S. Pearson, *Karl Pearson: An Appreciation . . .*, cited below.

For the most part Pearson's archival materials are not yet generally available for study or examination. Work in progress for many years on sorting, arranging, annotating, cross-referencing, and indexing these materials, and on typing many of his handwritten items, is nearing completion, however. A first typed copy of the handwritten texts of Pearson's lectures on the history of statistics was completed in 1972; and many dates, quotations, and references have to be checked and some ambiguities resolved before the whole is ready for public view. Hence we may expect the great majority to be available to qualified scholars before very long in the Karl Pearson Archives at University College, London.

II. SECONDARY LITERATURE. The best biography of Pearson is still *Karl Pearson: An Appreciation of Some Aspects of His Life and Work* (Cambridge, 1938), by his son, Egon Sharpe Pearson, who stresses in his preface that "this book is in no sense a Life of Karl Pearson." It is a reissue in book form of two articles, bearing the same title, published in *Biometrika*, **28** (1936), 193–257, and **29** (1937), 161–248, with two additional apps. (II and III in the book), making six in all. Included in the text are numerous instructive excerpts from Pearson's publications, helpful selections from his correspondence, and an outline of his lectures on the history of statistics in the seventeenth and eighteenth centuries. App. I gives the syllabuses of the 7 public lectures Pearson gave at Gresham

College, London, in 1891, "The Scope and Concepts of Modern Science," from which *The Grammar of Science* (1892) developed; app. II, the syllabuses of 30 lectures on "The Geometry of Statistics," "The Laws of Chance," and 'The Geometry of Chance" that Pearson delivered to general audiences at Gresham College, 1891–1894; app. III, by G. Udny Yule, repr. from *Biometrika*, **30** (1938), 198–203, summarizes the subjects dealt with by Pearson in his lecture courses on "The Theory of Statistics" at University College, London, during the 1894–1895 and 1895–1896 sessions; app. VI provides analogous summaries of his 2 lecture courses on "The Theory of Statistics" for first- and second-year students of statistics at University College during the 1921–1922 session, derived from E. S. Pearson's lecture notes; and apps. IV and V give, respectively, the text of Pearson's report of Nov. 1904 to the Worshipful Company of Drapers on "the great value that the Drapers' Grant [had] been to [his] Department" and an extract from his report to them of Feb. 1918, "War Work of the Biometric Laboratory."

The following publications by E. S. Pearson are useful supps. to this work: "Some Incidents in the Early History of Biometry and Statistics, 1890–94," in *Biometrika*, **52**, pts. 1–2 (June 1965), 3–18 (paper 22 in Pearson and Kendall); "Some Reflexions on Continuity in the Development of Mathematical Statistics, 1885–1920," *ibid.*, **54**, pts. 3–4 (Dec. 1967), 341–355 (paper 23 in Pearson and Kendall); "Some Early Correspondence Between W. S. Gosset, R. A. Fisher, and Karl Pearson, With Notes and Comments," *ibid.*, **55**, no. 3 (Nov. 1968), 445–457 (paper 25 in Pearson and Kendall); *Some Historical Reflections Traced Through the Development of the Use of Frequency Curves*, Southern Methodist University Dept. of Statistics THEMIS Contract Technical Report no. 38 (Dallas, 1969); and "The Department of Statistics, 1971. A Year of Anniversaries . . ." (mimeo., University College, London, 1972).

Of the biographies of Karl Pearson in standard reference works, the most instructive are those by M. Greenwood, in the *Dictionary of National Biography*, *1931–1940* (London, 1949), 681–684; and Helen M. Walker, in *International Encyclopedia of the Social Sciences*, XI (New York, 1968), 496–503.

Apart from the above writings of E. S. Pearson, the most complete coverage of Karl Pearson's career from the viewpoint of his contributions to statistics and biometry is provided by the obituaries by G. Udny Yule, in *Obituary Notices of Fellows of the Royal Society of London*, **2**, no. 5 (Dec. 1936), 73–104; and P. C. Mahalanobis, in *Sankhyā*, **2**, pt. 4 (1936), 363–378, and its sequel, "A Note on the Statistical and Biometric Writings of Karl Pearson," *ibid.*, 411–422.

Additional perspective on Pearson's contributions to biometry and statistics, together with personal recollections of Pearson as a man, scientist, teacher, and friend, and other revealing information are in Burton H. Camp, "Karl Pearson and Mathematical Statistics," in *Journal of the American Statistical Association*, **28**, no. 184 (Dec. 1933), 395–401; in the obituaries by Raymond Pearl, *ibid.*,

31, no. 196 (Dec. 1936), 653–664; and G. M. Morant, in *Man*, **36**, no. 118 (June 1936), 89–92; and in Samuel A. Stouffer, "Karl Pearson—An Appreciation on the 100th Anniversary of His Birth," in *Journal of the American Statistical Association*, **53**, no. 281 (Mar. 1958), 23–27. S. S. Wilks, "Karl Pearson: Founder of the Science of Statistics," in *Scientific Monthly*, **53**, no. 2 (Sept. 1941), 249–253; and Helen M. Walker, "The Contributions of Karl Pearson," in *Journal of the American Statistical Association*, **53**, no. 281 (Mar. 1958), 11-22, are also informative and useful as somewhat more distant appraisals. L. N. G. Filon, "Karl Pearson as an Applied Mathematician," in *Obituary Notices of Fellows of the Royal Society of London*, **2**, no. 5 (Dec. 1936), 104–110, seems to provide the only review and estimate of Pearson's contributions to applied mathematics, physics, and astronomy. Pearson's impact on sociology is discussed by S. A. Stouffer in his centenary "Appreciation" cited above; and Pearson's "rather special variety of Social-Darwinism" is treated in some detail by Bernard Semmel in "Karl Pearson: Socialist and Darwinist," in *British Journal of Sociology*, **9**, no. 2 (June 1958), 111–125. M. F. Ashley Montagu, in "Karl Pearson and the Historical Method in Ethnology," in *Isis*, **34**, pt. 3 (Winter 1943), 211–214, suggests that the development of ethnology might have taken a different course had Pearson's suggestions been put into practice.

The great clash at the turn of the century between the "Mendelians," led by Bateson, and the "ancestrians," led by Pearson and Weldon, is described with commendable detachment, and its after-effects assessed, by P. Froggatt and N. C. Nevin in "The 'Law of Ancestral Heredity' and the Mendelian-Ancestrian Controversy in England, 1889–1906," in *Journal of Medical Genetics*, **8**, no. 1 (Mar. 1971), 1–36; and "Galton's 'Law of Ancestral Heredity': Its Influence on the Early Development of Human Genetics," in *History of Science*, **10** (1971), 1–27.

Notable personal tributes to Pearson as a teacher, author, and friend, by three of his most distinguished pupils, L. N. G. Filon, M. Greenwood, and G. Udny Yule, and a noted historian of statistics, Harald Westergaard, have been preserved in *Speeches Delivered at a Dinner Held in University College, London, in Honour of Professor Karl Pearson, 23 April 1934* (London, 1934), together with Pearson's reply in the form of a five-page autobiographical sketch. The centenary lecture by J. B. S. Haldane, "Karl Pearson, 1857–1957," published initially in *Biometrika*, **44**, pts. 3–4 (Dec. 1957), 303–313, is also in *Karl Pearson, 1857–1957. The Centenary Celebration at University College, London, 13 May 1957* (London, 1958), along with the introductory remarks of David Heron, Bradford Hill's toast, and E. S. Pearson's reply.

Other publications cited in the text are Allan Ferguson, "Trends in Modern Physics," in British Association for the Advancement of Science, *Report of the Annual Meeting, 1936*, 27–42; Francis Galton, *Natural Inheritance* (London–New York, 1889; reissued, New York, 1972); R. A. Fisher, "The Correlation Between Relatives on the Supposition of Mendelian Inheritance," in *Transactions of the Royal*

Society of Edinburgh, **52** (1918), 399–433; H. L. Seal, "The Historical Development of the Gauss Linear Model," in *Biometrika*, **54**, pts. 1–2 (June 1967), 1–24 (paper no. 15 in Pearson and Kendall); and Helen M. Walker, *Studies in the History of Statistical Method* (Baltimore, 1931).

CHURCHILL EISENHART

[Contribution of the National Bureau of Standards, not subject to copyright.]

PEASE, FRANCIS GLADHELM (*b.* Cambridge, Massachusetts, 14 January 1881; *d.* Pasadena, California, 7 February 1938), *astronomy.*

Pease, the son of Daniel and Katherine Bangs Pease, received his education in Illinois. He attended high school in Highland Park and the Armour Institute of Technology in Chicago (now part of the Illinois Institute of Technology), where in 1901 he received a B.S. in mechanical engineering. Armour also awarded Pease an honorary M.S. in 1924 and D.Sc. in 1927, and he received another honorary D.Sc. in 1934 from Oglethorpe University. In 1922 he was elected a fellow of the Royal Astronomical Society.

Upon completing his formal education in 1901, Pease became a staff member of the Yerkes Observatory. There, with G. W. Ritchey, he studied problems in optics and instrument design, and carried out astronomical observations with the twenty-four-inch reflector. In 1904 he went to the Mount Wilson Observatory, where he remained as an instrument designer and astronomer until his death. He did, however, spend 1918 as chief draftsman in the engineering section of the National Research Council.

Pease's combination of instrumental expertise and observational experience made him immensely important to the Mount Wilson Observatory, which was embarking upon the construction of several large instruments. He helped design the 60-inch and 100-inch reflectors, the 60-foot and 150-foot towers, and the 20-foot and 50-foot interferometers, as well as much of the auxiliary equipment used with these instruments. He was also responsible for solving many of the design problems of the 200-inch telescope. Aware of the immense potential of larger telescopes, Pease spent considerable time developing plans and publishing papers on the need for, and uses and difficulties of, such instruments.

Pease was associated with A. A. Michelson in the first determination of stellar diameters, using interferometers for the difficult measurement of fringes as a function of mirror separation. He also assisted Michelson and F. Pearson in redetermining the velocity of light and in repeating the Michelson-Morley experiment.

Pease's direct spectrographic study with W. S. Adams and M. L. Humason of nebulae and star clusters enabled him to continue the measurements of rotations and radial velocities of spirals, which V. M. Slipher had begun in 1914 at the Lowell Observatory. In astronomical research Pease made important, although not pioneering, contributions; in instrument design, however, he was a leading figure of the twentieth century.

BIBLIOGRAPHY

I. ORIGINAL WORKS. Pease's key papers include "Radial Velocities of Six Nebulae," in *Publications of the Astronomical Society of the Pacific*, **27** (1915), 239–240; "The Rotation and Radial Velocity of the Spiral Nebula N.G.C. 4594," in *Proceedings of the National Academy of Sciences*, **2** (1916), 517–521; "Photographs of Nebulae With the 60-Inch Reflector 1911–1916," in *Astrophysical Journal*, **46** (1917), 24–55; "Interferometer Observations of Star Diameters," in *Publications of the Astronomical Society of the Pacific*, **34** (1921), 183; "On the Design of Very Large Telescopes," *ibid.*, **38** (1926), 195–207; "The Ball-Bearing Support System for the 100-Inch Mirror," *ibid.*, **44** (1932), 257, 308–312; and "Measurement of the Velocity of Light in a Partial Vacuum," in *Astrophysical Journal*, **82** (1935), 26–61, written with A. A. Michelson and F. Pearson. Some of his correspondence with G. E. Hale about designs for the 200-inch telescope are in the George Ellery Hale papers (1882–1937), Mount Wilson Observatory library, Pasadena, California.

II. SECONDARY LITERATURE. Obituaries of Pease are W. S. Adams, in *Publications of the Astronomical Society of the Pacific*, **50** (1938), 119–121, with portrait; G. Stromberg, in *Popular Astronomy*, **46** (1938), 357–359; and by an anonymous writer in *Monthly Notices of the Royal Astronomical Society*, **99** (1938), 312. For additional, related material see A. Pannekoek, *A History of Astronomy* (London, 1961). Pease's contributions to Mount Wilson are discussed in Helen Wright's biography of Hale, *Explorer of the Universe* (New York, 1966).

RICHARD BERENDZEN
RICHARD HART

PECHAM, JOHN (*b.* Sussex, England, *ca.* 1230–1235; *d.* Mortlake, Surrey, England, 8 December 1292), *optics, cosmology, mathematics.*

Pecham was probably born in the vicinity of Lewes in Sussex, possibly in or near the village of Patcham, and received his elementary education at the priory of Lewes.[1] He later matriculated in the arts faculties at Paris and Oxford, probably in that order. He became a Franciscan in the late 1240's or in the 1250's and was sent to Paris to undertake theological studies between 1257 and 1259.[2] In 1269 he received the doctorate in theology and for the next two years served as regent master in theology. Pecham returned to Oxford in 1271 or 1272 as eleventh lecturer in theology to the Franciscan school, a position he held until his appointment as provincial minister of the order in 1275. Two years later he was called to Italy as master in theology to the papal curia, and in 1279 he was elected archbishop of Canterbury. During his thirteen years as archbishop, Pecham maintained a zealous program of reform. He conscientiously endeavored to improve the administration of his province and persistently fought the practices of plurality and nonresidence; he called two reform councils and opposed, at every opportunity, the spread of "dangerous" philosophical novelties.

Of Pecham's intellectual development we know very little, although the major forces shaping his outlook probably came from within his own order. In the thirteenth century the Franciscan Order was a stronghold of Augustinianism and, consequently, of opposition to the new Aristotelian and Averroist ideas penetrating Europe. It is thus no surprise that Pecham became one of the leaders in the resistance against heterodox Aristotelian, and even more moderate Thomist, innovations.[3] But the Franciscan Order could provide more than antagonism toward philosophical and theological novelties. Among the English Franciscans a tradition of mathematical science had been initiated by Robert Grosseteste (who lectured to the Franciscans at Oxford and probably bequeathed his library to them at his death) and advanced by Roger Bacon. There can be little doubt that this tradition influenced Pecham: there is ample evidence that he and Bacon were personally acquainted and, indeed, resided together in the Franciscan friary at Paris during the period when Bacon was writing his principal scientific works. Nevertheless, this should not be taken to mean that Pecham was Bacon's student or protégé (there is no evidence for either) or that the influences on Pecham were limited to the Franciscan Order; Pecham's optical works, for example, reveal the influence not only of Augustine, Grosseteste, and Bacon but also of Aristotle, Euclid, al-Kindī, Ibn al-Haytham, Moses Maimonides, and perhaps Ptolemy and Witelo; and the primary influence in this instance was not Augustine or Grosseteste or Bacon, but Ibn al-Haytham.

Works. Pecham's indisputably genuine works on natural philosophy and mathematical science are *Tractatus de numeris* (or *Arithmetica mystica*); *Tractatus de perspectiva*; *Perspectiva communis*, extant in both an original and a revised version; and *Tractatus de sphera*. In addition to these, a treatise entitled *Theorica planetarum* is attributed to Pecham in several manuscripts and has commonly been regarded as genuine, although the question of its authenticity has in fact never been explored with care. Material of considerable scientific import is also contained in Pecham's treatises on the soul, *Tractatus de anima* and *Questiones de anima*, and his *Questiones de beatitudine corporis et anime*.[4] Two other scientific treatises have also been attributed to Pecham, *Perspectiva particularis* and *Tractatus de animalibus*, but there is no evidence supporting either attribution.

Of Pecham's scientific works only those on optics have been subjected to serious scrutiny; nevertheless, it is possible to make a few remarks about several of the others. The *Tractatus de sphera* was apparently a rival to, rather than a commentary on, Sacrobosco's *De sphaera*.[5] In this work Pecham presents an elementary discussion of the sphericity (or circularity) of the principal bodies of the world (for instance, the heavens, raindrops, and solar radiation passing through noncircular apertures); the rotation of the heavens; the equality and inequality of days; the climatic zones of the terrestrial sphere; the origin of eclipses; and other topics of a cosmologic nature.

The *Tractatus de numeris* begins with the classification of number into abstract and concrete; concrete number is further subdivided into corporeal and spiritual number, spiritual number is divided into five additional categories, and so on. After further discussion of the elementary properties of numbers (odd and even, equality and inequality) and the perceptibility of number by the external senses, Pecham turns to the mystical properties of numbers: he employs number to elucidate the mysteries of the Trinity and concludes with an analysis of the mystical meanings of the numbers 1 to 30, 36, 40, 50, 100, 200, 300, and 1,000.

The earliest of Pecham's optical works was the *Tractatus de perspectiva*, probably written for the Franciscan schools during Pecham's years as a teacher at Paris or Oxford (1269–1275) or possibly during his provincial ministership (1275–1277). It is a rambling piece of continuous prose, not divided into propositions like the later *Perspectiva communis*, that treats the full range of elementary optical matters. Like the *Tractatus de numeris*, and unlike the *Perspectiva communis*, it is filled with quotations from the Bible and patristic sources, especially Augustine, that give it a theological and devotional flavor. With a few exceptions the *Tractatus de perspectiva* and *Perspectiva communis* are identical in theoretical content, although each includes certain topics that the other omits.

The work on which Pecham's fame has chiefly rested

is the *Perspectiva communis*, probably written between 1277 and 1279 during Pecham's professorship at the papal curia.[6] In the first book Pecham discussed the propagation of light and color, the anatomy and physiology of the eye, the act of visual perception, physical requirements for vision, the psychology of vision, and the errors of direct vision. In book II he discussed vision by reflected rays and presented a careful and sophisticated analysis of image formation by reflection. Book III was devoted to the phenomena of refraction, the rainbow, and the Milky Way.

The central feature of Pecham's optical system and the dominant theme of book I of the *Perspectiva communis* is the theory of direct vision. Here, as elsewhere, Pecham endeavored to reconcile all the available authorities—Aristotle, Euclid, Augustine, al-Kindī, Ibn al-Haytham, Ibn Rushd, Grosseteste, and Bacon. Following Ibn al-Haytham, Pecham argued that the emission of visual rays from the observer's eye is neither necessary nor sufficient as an explanation of sight; the primary agent of sight is therefore the ray coming to the eye from a point on the visible object. But in an attempt to follow Aristotle, al-Kindī, and Grosseteste as well, Pecham argued that visual rays do nevertheless exist and perform the important, but not always necessary, function of moderating the luminous rays from the visible object and making them "commensurate with the visual power." Thus Pecham, like Bacon, resolved the age-old debate between the emission and intromission theories of vision in favor of a twofold radiation, although, to be sure, priority was given to rays issuing from the visible object.

The rays issuing from points on the visible object fall perpendicularly onto the cornea and penetrate without refraction to the sensitive ocular organ, the glacial humor (or crystalline lens); nonperpendicular rays are weakened by refraction and therefore can be largely ignored. Since only one perpendicular ray issues from each point of the visible object and the collection of such perpendicular rays maintains a fixed order between the object and the eye, a one-to-one correspondence is established between points on the object and points on the glacial humor, and unconfused perception of the visual field is thus achieved. Vision is not "completed," however, in the glacial humor. There is a further propagation of the rays (or species) through the vitreous humor and optic nerve to the common nerve, where species from the two eyes combine, and eventually to the anterior part of the brain and the "place of interior judgment."

Pecham's optical system included significantly more than a theory of direct vision. He briefly discussed the doctrine of species; treated at length the propagation of rays; and developed a theory to explain how solar radiation, when passing through noncircular apertures, gives rise to circular images. He expressed the full law of reflection and applied it to image formation by plane, spherical, cylindrical, and conical mirrors; in this analysis he revealed an implicit understanding of the nature of the focal point of a concave mirror. Although he did not possess a mathematical law of refraction, he successfully applied the general qualitative principles of refraction to the images that result from refraction at plane and circular interfaces between transparent media of various densities. In his discussion of the rainbow Pecham again attempted to reconcile different theories. He argued that all three kinds of rays (rectilinear, reflected, and refracted) concur in the generation of the rainbow.

Significance and Influence. Pecham saw himself primarily not as a creative scientific thinker but as an expositor of scientific matters in elementary terms. He remarked at the beginning of the *Tractatus de sphera*:

> In the present opusculum, I intend to explain the number, figure, and motion of the principal bodies of the world (as well as related matters) insofar as is sufficient for an understanding of the words of Holy Scripture. And certain of these matters I have found treated in other works, but because of their difficulty, brevity, and in some cases falsity, they are useless for the elementary students that I intend to serve.[7]

In the *Tractatus de perspectiva* he remarked that he had undertaken to discuss light and number "for the sake of my simpler brothers," and in the preface to the *Perspectiva communis* he indicated that his goal was to "compress into concise summaries the teachings of perspective, which [in existing treatises] are presented with great obscurity."[8] Pecham's significance in the history of science is principally the result of his success in achieving this goal. He is most notable not as one who formulated new theories and interpretations, although on many occasions he did, but as one who skillfully presented scientific knowledge to his contemporaries and posterity by writing elementary textbooks.

Pecham's success was greatest in the case of the *Perspectiva communis*. This text is still extant in more than sixty manuscripts and went through twelve printed editions, including a translation into Italian, between 1482 and 1665. It was used and cited by many medieval and Renaissance natural philosophers, including Dominicus de Clavasio, Henry of Langenstein, Blasius of Parma, Lorenzo Ghiberti, Leonardo da Vinci, Albert Brudzewski, Francesco Maurolico, Giambattista della Porta, Girolamo Fabrici, Johannes Kepler, Willebrord Snellius, and G. B. Riccioli. It was lectured upon, in the late Middle Ages, at the universities of Vienna, Prague, Paris, Leipzig, Cracow, Würzburg, Alcalá, and Salamanca.[9] The *Perspectiva communis* was the most widely used of all optical texts from the early fourteenth until the close of the sixteenth century, and it remains today the best index of what was known to the scientific community in general on the subject.

NOTES

1. The evidence for both claims is a letter written by Pecham in 1285, in which he refers to his "nourishment from childhood" in the vicinity of the priory of Lewes and the comforts and honors he has received from its teachers; see *Registrum epistolarum*, III, 902. Several historians have argued that Pecham was born in Kent rather than Sussex.
2. In assigning the latter dates, I am following Douie, *Archbishop Pecham*, p. 8.
3. On Pecham's position vis-à-vis Averroism and Thomism, see Fernand van Steenberghen, *The Philosophical Movement in the Thirteenth Century* (Edinburgh, 1955), 94–104. Van Steenberghen calls Pecham "the true founder of neo-Augustinianism" (p. 103).
4. Scientific content is especially evident in the *Questiones de beatitudine corporis et anime*, in *Johannis Pechami Quaestiones tractantes de anima*, Hieronymus Spettman, ed., which is *Beiträge zur Geschichte der Philosophie des Mittelalters*, XIX, pts. 5–6 (Münster, 1918), although Pecham's psychology is apparent in all of them. On Pecham's psychology see Sharp, *Franciscan Philosophy*, 185–203; and *Die Psychologie des Johannes Pecham*, Spettman, ed., which is *Beiträge zur Geschichte der Philosophie des Mittelalters*, XX, pt. 6 (Münster, 1919).
5. According to Thorndike, *Sphere of Sacrobosco*, 24–25.
6. The dating of the *Perspectiva communis* is discussed in Lindberg, *Pecham and the Science of Optics*, 14–18; and in Lindberg, "Lines of Influence in Thirteenth-Century Optics: Bacon, Witelo, and Pecham," in *Speculum*, **46** (1971), 77–83.
7. Latin text in Thorndike, *op. cit.*, 445.
8. See Lindberg's eds. of these two treatises for the texts.
9. For a fuller account of the influence of the *Perspectiva communis*, see Lindberg, *Pecham and the Science of Optics*, 29–32.

BIBLIOGRAPHY

I. ORIGINAL WORKS. The *Perspectiva communis* (in both the original and the revised versions) is available in a recent ed. and English trans. by David C. Lindberg, *John Pecham and the Science of Optics* (Madison, Wis., 1970). The known extant MSS and eleven early printed eds. are listed in the intro. to this ed. Pecham's other optical work, the *Tractatus de perspectiva*, is also available in a modern critical version, David C. Lindberg, ed., in Franciscan Institute Publications, Text Ser. no. 16 (St. Bonaventure, N.Y., 1972).

No other complete scientific work of Pecham has been printed. The first five chapters of the *Tractatus de numeris* have been edited from four MSS and published as an appendix to *Tractatus de anima Ioannis Pecham*, Gaudentius Melani, ed. (Florence, 1948), 138–144. Lynn

Thorndike has published the opening paragraphs and incipits of later paragraphs of the *Tractatus de sphera* in *The Sphere of Sacrobosco and Its Commentators* (Chicago, 1949), 445–450; and Pierre Duhem has published the section on pinhole images from this same work in *Le système du monde*, III (Paris, 1915), 524–529. The *Theorica planetarum* is extant only in MS.

For a full listing of Pecham's works, including extant MSS and eds., see Victorinus Doucet, "Notulae bibliographicae de quibusdam operibus Fr. Ioannis Pecham O.F.M.," in *Antonianum*, **8** (1933), 207–228, 425–459; Palémon Glorieux, *Répertoire des maîtres en théologie de Paris au XIIIᵉ siècle*, II (Paris, 1933), 87–98; and *Fratris Johannis Pecham quondam archiepiscopi Cantuariensis Tractatus tres de paupertate*, C. L. Kingsford *et al.*, eds. (Aberdeen, 1910), 1–12.

II. SECONDARY LITERATURE. The best biography of Pecham is Decima L. Douie, *Archbishop Pecham* (Oxford, 1952). Other valuable sources on Pecham's life and thought are David Knowles, "Some Aspects of the Career of Archbishop Pecham," in *English Historical Review*, **57** (1942), 1–18, 178–201; Hieronymus Spettman, "Quellenkritisches zur Biographie des Johannes Pecham," in *Franziskanische Studien*, **2** (1915), 170–207, 266–285; and D. E. Sharp, *Franciscan Philosophy at Oxford in the Thirteenth Century* (Oxford, 1930), 175–207. For a short biographical sketch and additional bibliography, see Lindberg, *Pecham and the Science of Optics*, 3–11.

Pecham's optical work has been most fully analyzed in the following works by David C. Lindberg: *John Pecham and the Science of Optics*; "The *Perspectiva communis* of John Pecham: Its Influence, Sources, and Content," in *Archives internationales d'histoire des sciences*, **18** (1965), 37–53; "Alhazen's Theory of Vision and Its Reception in the West," in *Isis*, **58** (1967), 321–341; and "The Theory of Pinhole Images From Antiquity to the Thirteenth Century," in *Archive for History of Exact Sciences*, **5**, no. 2 (1968), 154–176. Brief descriptions of Pecham's other scientific works are found in *Registrum epistolarum fratris Johannis Peckham archiepiscopi Cantuariensis*, Charles T. Martin, ed., III (London, 1885), lvi–cxlv; and Lynn Thorndike, "A John Peckham Manuscript," in *Archivum Franciscanum Historicum*, **45** (1952), 451–461.

DAVID C. LINDBERG

PECQUET, JEAN (*b.* Dieppe, France, 9 May 1622; *d.* Paris, France, February 1674), *anatomy*.

Pecquet spent his youth in Normandy, first in Dieppe and then in Rouen, where he met Blaise Pascal. In 1642 he went to Paris, where he was a member of the various scientific circles that preceded the Académie des Sciences. He joined the entourage of the Fouquet brothers, François, bishop of Agde, and Nicolas, superintendent of finance. Pecquet enrolled at the Paris Faculty of Medicine around 1646, at the age of twenty-four. Finding the atmosphere unfavorable, he matriculated at Montpellier on 15 July 1651, received his *licence* on 16 February 1652, and defended his doctoral thesis on 23 March 1652.

Pecquet subsequently returned to Paris, where he had both worldly and scientific careers. He was physician to Nicolas Fouquet, as well as to the Marquise de Sévigné, her daughter, and her grandchildren; his name occasionally appears in the marquise's correspondence. Pecquet was also friendly with the Paris scientists Jacques Mentel, Louis Gayant, Adrien Auzout, and Claude Perrault. He probably knew Steno from the time of the latter's visit to Paris (1664–1665), as well as other, less notable foreign physicians, including Martin Bogdan, municipal physician of Bern.

The quantity of Pecquet's scientific production was slight. He participated in experiments on the transfusion of blood performed in 1666–1667 at the Académie des Sciences, as did his friends Gayant (provost of the *communauté* of the surgeons of Paris and consulting surgeon to the royal army) and Perrault.

The *Mémoires de l'Académie royale des sciences* for 1666 to 1669 (**10** [Paris, 1730], 476–477) mentions a note on liver parasites. Pecquet also debated the question of the agent of vision with Mariotte (1669), who contended that it was in the choroid coat; Pecquet believed the retina to be the sensory membrane. Pecquet's only important accomplishment was the discovery of the chyle reservoir, which he called the *receptaculum chyli*—not *cisterna*, a word introduced into anatomical nomenclature by his friend Thomas Bartholin.

To understand the genesis of Pecquet's investigations, it is important to remember that when he began them, the great discovery dividing and preoccupying physicians was that of the circulation of the blood. Harvey had announced it in 1628 and returned to it in 1649 in his two letters to Jean Riolan, dean of the Paris Faculty of Medicine. Rejected by the Paris medical officials, Harvey's discovery was taught at the Jardin du Roi and furnished dissenting physicians with subjects for study, such as the circulation of various body fluids, as well as methodology, *anatomia animata*, the ancestor of experimental physiology. The experimenting physician, who actively examined nature instead of passively contemplating it, was Pecquet's ideal.

Pecquet was probably introduced to the study of the lymphatic system by Mentel, a Harveian physician who had received his doctorate at Paris in 1632. Tradition relates that Mentel had observed human lymphatics around 1629, but it was Aselli's discovery in 1622 of the chyliferous vessels in the dog that drew the attention of researchers to the "white vessels."

Aselli's discovery had also propagated the erroneous idea—accepted by Vesling (1647), among others—that the chyliferous vessels terminate in the liver after traversing the pancreas. Aselli's "pancreas" included both the true pancreas and the groups of ganglia situated behind it in the mesentery.

Harvey believed that the resorption of the chyle occurred in the mesenteric veins and that the liver was the site of hematopoiesis. His chief opponent, Jean Riolan, did not admit the existence of the "white vessels," even though Falloppio had probably seen the lymphatics of the liver and Eustachi had observed the thoracic duct of the horse (*vena alba thoracis*) in 1564. In 1642, the year of Pecquet's arrival in Paris, Johann Georg Wirsung discovered the duct that bears his name; but yet instead of identifying it with the excretory canal of the pancreas, he considered it to be a chyliferous vessel, emerging from the intestine and ending in the pancreas.

It was in these circumstances that Pecquet, while still a student, defied the reigning conceptions and engaged not in the "mute and frozen science" of cadaver anatomy, but in *anatomia animata* on dogs, cattle, pigs, and sheep. Using a dog that was digesting, he showed the following:

1. If the heart has been resected, pressure on the mesenteric root causes the chyle to spurt into the superior vena cava.

2. The chyle is directed toward the subclavian veins by two paravertebral canals that swell when their distal extremities are ligatured.

3. The origin of the ascending chyliferous ducts is situated in a prevertebral and subdiaphragmatic ampulla—"this sought-after sanctuary of the chyle, this reservoir sought with so much difficulty."

4. The posterior part of Aselli's pancreas is composed of lymphatic ganglia.

5. No mesenteric chyliferous vessel goes to the liver (a fact confirmed by Glisson in 1654), and the inferior vena cava, incised above the liver, reveals no trace of chyle.

The human thoracic duct was rediscovered by Thomas Bartholin, Rudbeck, and Gayant. With Perrault and Gayant, Pecquet eventually observed the communications of the human thoracic duct with the lumbar veins.

Pecquet's discovery was received with great interest and provoked sharp debate, particularly with Riolan. It was warmly welcomed by Bartholin, who distinguished the chylous vessels from the other "white vessels," which he called lymphatics, and by Rudbeck. The latter saw the lymphatic system as a new type of vascular system, long unrecognized because it can be made visible only by special preparations. Nevertheless, the significance of the lymphatic system was still far from clear. Lympho-neural anastomoses were described in terms of Cartesian "nerve tubes." Wharton and Glisson thought that the glands received juices excreted by the nerves and that they eliminated them through the lymphatics. Pecquet, like Bartholin, believed that communications existed between the *cisterna chyli* and the urinary tracts; these structures would short-circuit the renal tubule system and would explain the rapid filling of the bladder after a copious intake of liquid.

Pecquet's friend Perrault likened the thoracic duct to a glandular canal, thus showing that he still believed in humors carried by special vessels and not in "juices" (*sucs*) synthesized by the glands from the blood.

BIBLIOGRAPHY

I. ORIGINAL WORKS. Pecquet's major writing, the *Experimenta nova anatomica*, went through several eds. and versions: *Experimenta nova anatomica, quibus incognitum hactenus chyli receptaculum et ab eo per thoracem in ramos usque subclavios vasa lactea deteguntur . . .* (Paris, 1651); a copy of this ed., with illustrations (Harderwijk, n.d.); *Experimenta nova academica . . . Huic secundae editioni . . . accessit de thoracicis lacteis dissertatio in qua Jo. Riolani responsio ad eadem experimenta nova anatomica reputatur . . .* (Paris, 1654); *Experimenta nova anatomica, quibus incognitum hactenus chyli receptaculum et ab eo, per thoracem, in ramos usque subclavios vasa lactea deteguntur. . . . Accedunt clarissimorum virorum epistolae tres ad auctorem,* in J. A. Munierus, *De venis tam lacteis thoracicis quam lymphaticis . . .* (Genoa, 1654); and *Experimenta nova anatomica . . . chyli motu* (Amsterdam, 1661). An English trans. is *New Anatomical Experiments by Which the Hitherto Unknown Receptacle of the Chyle and the Transmission From Thence to the Subclavial Veines by the Now Discovered Lacteal Chanels of the Thorax Is Plainly Made Apear in Brutes . . . Being an Anatomical Historie Publickly Propos'd by Thomas Bartoline to Michael Lysere, Answering* (London, 1653).

Other works are *Brevis destructio, seu litura responsionis Riolani ad ejusdem Pecqueti esperimenta per Hyginum Thalassium* (Paris, 1655; Amsterdam, 1661), which is also found in Siboldus Hemsterhuys, ed., *Messis aurea, seu collectanea anatomica . . .* (Leiden, 1654; Heidelberg, 1659); in Daniel Le Clerc and J. J. Manget, eds., *Bibliotheca anatomica* (Geneva, 1685); and in Thomas Bartholin, *Anatomia . . . tertium ad sanguinis circulationem reformata . . .* (Laon, 1651); "Lettre de M. Pecquet à M. de Carcavi touchant une nouvelle découverte de la communication du canal thoracique avec la veine émulgente," in *Journal des sçavans* (4 Apr. 1667), 53–56; and *Réponse . . . à la lettre de Mr. l'Abbé Mariotte sur une nouvelle découverte touchant la vueüe* (Paris, 1668); and "Lettres écrites par MM. Mariotte, Pecquet et Perrault sur le sujet d'une nouvelle découverte touchant la vueüe faite par M. Mariotte," in *Recueil de plusieurs traitez de mathématiques de l'Académie royale des sciences* (Paris, 1676).

II. SECONDARY LITERATURE. There are unsigned articles on Pecquet in A. L. Bayle and A. J. Thillaye, eds., *Biographie médicale*, II (Paris, 1855), 13–14; J.-E. Dezeimeris, *Dictionnaire historique de la médecine ancienne et moderne*, III (Paris, 1836), 689; *Dictionnaire historique de la médecine ancienne et moderne*, III (Mons, 1778), 507–508; Michaud, ed., *Biographie universelle ancienne et moderne*, XXXIII, 247–249; and *Nouvelle biographie médicale depuis les temps les plus reculés jusqu'au nos jours*, XXXIX (Paris, 1863), 443–444.

Bartholin's fundamental investigations on the human lymphatics in 1652–1653 were completed by those made independently in 1651 by Olof Rudbeck. On this subject see V. Maar, "Thomas Bartholinus," in *Janus*, **21** (1916), 273–301, and Axell Garböe, "Thomas Bartholin," I–II. On Rudbeck see *Annals of Medical History* (1928).

See also Berchon, "Victor Hugo et la découverte de Pecquet," in *Chronique médicale*, **21** (1914), 429–431; A. Chéreau, "Pecquet," in *Dictionnaire encyclopédique des sciences médicales*, XXII (Paris, 1886), 202; J. Delmas, "Pecquet," in *Médecins célèbres* (Paris, 1947), 94; R. Desgenettes, "Pecquet," in *Dictionnaire des sciences médicales—biographie médicale*, VI (Paris, 1824), 384–385; P. Gilis, "Pecquet," in *Bulletin de la Société des sciences médicales et biologiques de Montpellier*, **3** (1921–1922), 32–60, with portrait; in *Montpellier médical*, **43** (1921), 627–628; and *Normandie médicale*, **32** (1922), 141–156, 177–191; E. Hintzsche, "Anatomia animata," in *Revue Ciba*, **64** (1948), 2398–2399; Georges Laux, "Tricentenaire de la thèse de J. Pecquet. L'oeuvre anatomique," in *Séance publique de la section montpellieraine de la Société française d'histoire de la médecine, 31 Mars 1952* (Montpellier, 1952); and "Jean Pecquet. Son oeuvre anatomique," in *Monspelliensis Hippocrates*, no. 37 (1967), 7–12, with portraits and facsimiles; Jean Lucq, "Jean Pecquet. 1622–1674," a thesis at the University of Paris (1925, no. 22); Pagel, "Pecquet," in *Biographisches Lexicon der hervorragenden Ärzte*, IV (Berlin–Vienna, 1932), 543; the anonymous "Jean Pecquet," in *Progrès médical*, ill. supp. no. 5 (1926), 39–40; P. Rabier, "Le centenaire de Pecquet," in *Paris médical*, **46** (supp.) (1922), 171–173; and C. Webster, "The Discovery of Boyle's Law and the Concept of the Elasticity of Air in the Seventeenth Century," in *Archive for History of Exact Sciences*, **2** (1965), 441–502, with illustrations and references.

PIERRE HUARD
MARIE-JOSÉ IMBAULT-HUART

PEIRCE, BENJAMIN (*b.* Salem, Massachusetts, 4 April 1809; *d.* Cambridge, Massachusetts, 6 October 1880), *mathematics, astronomy.*

In an address before the American Mathematical Society during the semicentennial celebration of its founding in 1888 as the New York Mathematical Society, G. D. Birkhoff spoke of Benjamin Peirce as having been "by far the most influential scientific personage in America" and "a kind of father of pure mathematics in our country."

Peirce's background and training were completely American. The family was established in America by John Peirce (Pers), a weaver from Norwich, England, who settled in Watertown, Massachusetts, in 1637. His father, Benjamin Peirce, graduated from Harvard College in 1801, and served for several years as representative from Salem in the Massachusetts legislature; he was Harvard librarian from 1826 until 1831, prepared a printed catalog of the Harvard library (1830–1831), and left a manuscript history of the university from its founding to the period of the American Revolution (published 1833). Peirce's mother, Lydia Ropes Nichols of Salem, was a first cousin of her husband. On 23 July 1833 Peirce married Sarah Hunt Mills, daughter of Harriette Blake and Elijah Hunt Mills of Northampton, Massachusetts. They had a daughter, Helen, and four sons: James Mills Peirce, professor of mathematics and an administrator at Harvard for fifty years; Charles Sanders Peirce, geodesist, mathematician, logician, and philosopher; Benjamin Mills Peirce, a mining engineer who wrote the U.S. government report on mineral resources and conditions in Iceland and Greenland; and Herbert Henry Davis Peirce, a diplomat who served on the staff of the legation in St. Petersburg and who later arranged for the negotiations between Russia and Japan that led to the Treaty of Portsmouth on 5 September 1905.

Peirce attended the Salem Private Grammar School, where Henry Ingersoll Bowditch was a classmate. This relationship influenced the entire course of Peirce's life, since Ingersoll Bowditch's father, Nathaniel Bowditch, discovered Peirce's unusual talent for mathematics. During Peirce's undergraduate career at Harvard College (1825–1829), the elder Bowditch enlisted Peirce's aid in reading the proof-sheets of his translation of Laplace's *Traité de mécanique céleste*. Peirce gave evidence of his own mathematical powers in his revision and correction of Bowditch's translation and commentary on the first four volumes (1829–1839), and also with his proof (in 1832) that there is no odd perfect number that has fewer than four prime factors.

Peirce taught at Bancroft's Round Hill School at Northampton, Massachusetts, from 1829 until 1831, when he was appointed tutor in mathematics at Harvard College; he received his M.A. from that institution in 1833. At Harvard he became University professor of mathematics and natural philosophy (1833–1842), then Perkins professor of astronomy and mathematics (1842–1880). During the early days of his teaching at Harvard, Peirce published a popular series of textbooks on elementary branches of mathematics.

Peirce's continued interest in the theory of astronomy was apparent in his study of comets. Around 1840 he made observations in the old Harvard College observatory; his 1843 Boston lectures on the great comet of that year stimulated the support that led to the installation of the new telescope at the Harvard Observatory in June 1847. Since 1842 Peirce had also supervised the preparation of the mathematics section of the ten-volume *American Almanac and Repository of Useful Knowledge*, and in 1847 he published therein a list of known orbits of comets. In 1849 Charles Henry Davis, a brother-in-law of Peirce's wife, was appointed superintendent of the newly created *American Ephemeris and Nautical Almanac*, and Peirce was appointed consulting astronomer (1849–1867).

Peirce was not only helpful to Davis in planning the general form of the *Ephemeris*, but he also began a revision of the theory of planets. He had become deeply interested in the work of Le Verrier and John Couch Adams that had permitted Galle's discovery of the planet Neptune on 23 September 1846. In cooperation with Sears Walker, Peirce determined the orbit of Neptune and its perturbation of Uranus. Simon Newcomb wrote in his *Popular Astronomy* (1878) that the investigation of the motion of the new planet was left in the hands of Walker and Peirce for several years, and that Peirce was "the first one to compute the perturbations of Neptune produced by the action of the other planets." Peirce was led to believe that Galle's "happily" discovered Neptune and Le Verrier's calculated theoretical planet were not the same body and that the latter did not exist—an opinion that led to considerable controversy.

In conjunction with his work on the solar system, Peirce became interested in the mathematical theory of the rings of Saturn. In 1850 George Phillips Bond, assistant in the Harvard College observatory, discovered Saturn's dusky ring and on 15 April 1851 announced to a meeting of the American Academy of Arts and Sciences his belief that the rings were fluid, multiple, and variable in number. Peirce published several mathematical papers on the constitution of the rings in which he reached the same conclusion concerning their fluidity. His review of the problem at that time led to a most unfortunate priority dispute.

Peirce also enjoyed a distinguished career in the U.S. Coast Survey. In 1852 he accepted a commission —at the request of Alexander Dallas Bache, who was then superintendent—to work on the determination of longitude for the Survey. This project involved Peirce in a thorough investigation of the question of errors of observation; his article "Criterion for the Rejection of Doubtful Observations" appeared in B. A. Gould's *Astronomical Journal* in July 1852. The criterion was designed to determine the most probable hypothesis whereby a set of observations might be divided into normal and abnormal, when "the greater part is to be regarded as normal and subject to the ordinary law of error adopted in the method of least squares, while a smaller unknown portion is abnormal and subject to some obscure source of error." Some authorities regarded "Peirce's criterion"—which gave good discrimination and acceptable practical results—as one of his most important contributions, although it has since been demonstrated to be invalid.

After Bache's death Peirce became superintendent of the Coast Survey (1867–1874), while maintaining his association with Harvard. He arranged to carry forward Bache's plans for a geodetic system that would extend from the Atlantic to the Gulf. This project laid the foundation for a general map of the country independent of detached local surveys. Peirce's principal contribution to the development of the Survey is thought to have been the initiation of a geodetic connection between the surveys of the Atlantic and Pacific coasts. He superintended the measurement of the arc of the thirty-ninth parallel in order to join the Atlantic and Pacific systems of triangulation.

Peirce also took personal charge of the U.S. expedition that went to Sicily to observe the solar eclipse of 22 December 1870, and, as a member of the transit of Venus commission, sent out two Survey parties— one to Nagasaki and the other to Chatham Island —in 1874. Peirce also played a role in the acquisition of Alaska by the United States in 1867, since in that year he sent out a reconnaissance party, whose reports were important aids to proponents of the purchase of that region. In 1869 he sent parties to observe the eclipse of the sun in Alaska and in the central United States.

Peirce's eminence made him influential in the founding of scientific institutions in the United States. In 1847 the American Academy of Arts and Sciences appointed him to a committee of five in order to draw up a program for the organization of the Smithsonian Institution. From 1855 to 1858 he served with Bache and Joseph Henry on a council to organize the Dudley observatory at Albany, New York, under the direction of B. A. Gould. In 1863 he became one of the fifty incorporators of the National Academy of Sciences.

Despite his many administrative obligations, Peirce continued to do mathematics in the 1860's. He read before the National Academy of Sciences a number of papers on algebra, which had resulted from his interest in Hamilton's calculus of quaternions and finally led to Peirce's study of possible systems of multiple algebras. In 1870 his *Linear Associative Algebra* appeared as a memoir for the National Academy and was lithographed in one hundred copies for private circulation. The opening sentence states that "Mathematics is the science which draws necessary conclusions." George Bancroft received the fifty-second copy of the work, and in an accompanying letter (preserved in the manuscript division of the New York Public Library) Peirce explained that

> This work undertakes the investigation of all possible single, double, triple, quadruple, and quintuple Algebras which are subject to certain simple and almost indispensable conditions. The conditions are those well-known to algebraists by the terms of *distributive* and *associative* which are defined on p. 21. It also contains the investigation of all sextuple algebras of a certain class, i.e., of those which contain what is called in this treatise an *idempotent* element.

D. E. Smith and J. Ginsburg, in their *History of Mathematics Before 1900*, speak of Peirce's memoir as "one of the few noteworthy achievements in the field of mathematics in America before the last quarter of the century." It was published posthumously in 1881 under the editorship of his son Charles Sanders Peirce (*American Journal of Mathematics*, **4**, no. 2, 97–229).

In *A System of Analytic Mechanics* (1855) Peirce again set forth the principles and methods of the science as a branch of mathematical theory, a subject he developed from the idea of the "potential." The book has been described as the most important mathematical treatise that had been produced in the United States up to that time. Peirce's treatment of mechanics has also been said, by Victor Lenzen, to be "on the highest level of any work in the field in English until the appearance of Whittaker's *Analytical Dynamics*" in 1904. Peirce was widely honored by both American and foreign scholarly and scientific societies.

BIBLIOGRAPHY

I. ORIGINAL WORKS. Peirce's works include *An Elementary Treatise on Sound* (Boston, 1836); *An Elementary Treatise on Algebra* (Boston, 1837), to which are added exponential equations and logarithms; *An Elementary Treatise on Plane and Solid Geometry* (Boston, 1837); *An Elementary Treatise on Plane and Spherical Trigonometry, . . . Particularly Adapted to Explaining the Construction of Bowditch's Navigator and the Nautical Almanac* (Boston, 1840); *An Elementary Treatise on Curves, Functions, and Forces,* 2 vols. (Boston, 1841, 1846); and *Tables of the Moon* (Washington, D.C., 1853) for the *American Ephemeris and Nautical Almanac. Tables of the*

Moon was used in taking the *Ephemeris* up to the volume for 1883 and was constructed from Plana's theory, with Airy's and Longstreth's corrections, Hansen's two inequalities of long period arising from the action of Venus, and Hansen's values of the secular variations of the mean motion and of the motion of the perigee.

Later works are *A System of Analytic Mechanics* (Boston, 1855); *Linear Associative Algebra* (1870), edited by C. S. Peirce, which appeared in *American Journal of Mathematics*, **4** (1881), 97–229, and in a separate vol. (New York, 1882); and James Mills Peirce, ed., *Ideality in the Physical Sciences*, Lowell Institute Lectures of 1879 (Boston, 1881).

Peirce's unpublished letters are in the National Archives, Washington, D. C., and in the Benjamin Peirce and Charles S. Peirce collections of Harvard University.

II. Secondary Literature. On Peirce and his work, see reminiscences by Charles W. Eliot, A. Lawrence Lowell, W. E. Byerly, Arnold B. Chace, and a biographical sketch by R. C. Archibald, in *American Mathematical Monthly*, **32** (1925), repr. as a monograph, with four new portraits and addenda (Oberlin, 1925), which contains in sec. 6 a listing with occasional commentary of Peirce's writings and massive references to writings about him. See also Bessie Zaban Jones and Lyle Gifford Boyd, *The Harvard College Observatory* (Cambridge, Mass., 1971), esp. the chap. entitled "The Two Bonds," which gives a detailed description of the unhappy relationship that developed between Peirce and George and William Bond.

See further R. C. Archibald, in *Dictionary of American Biography* (New York, 1934); A. Hunter Dupree, "The Founding of the National Academy of Sciences—A Reinterpretation," in *Proceedings of the American Philosophical Society*, **101**, no. 5 (1957), 434–441; M. King, ed., *Benjamin Peirce . . . A Memorial Collection* (Cambridge, Mass., 1881); Victor Lenzen, *Benjamin Peirce and the United States Coast Survey* (San Francisco, 1968); Simon Newcomb, *Popular Astronomy* (New York, 1878), esp. pp. 350 (on the rings of Saturn), 363 (on the perturbation of Neptune), and 403 (on comets); H. A. Newton, "Benjamin Peirce," in *Proceedings of the American Academy of Arts and Sciences*, 16, n.s., **8**, pt. 2 (1881), 443–454, repr. in *American Journal of Science*, 3rd ser., **22**, no. 129 (1881), 167–178; James Mills Peirce, in *Lamb's Biographical Dictionary of the United States*, VI (Boston, 1903), 198; and Poggendorff, II (1863), 387–388; and III (1858–1883), 1012–1013. See also F. C. Pierce, *Peirce Genealogy* (Worcester, Mass., 1880).

Carolyn Eisele

PEIRCE, BENJAMIN OSGOOD, II (*b.* Beverly, Massachusetts, 11 February 1854; *d.* Cambridge, Massachusetts, 14 January 1914), *mathematics, physics.*

Peirce's father, who bore the same names, was by 1849 a merchant in the South African trade, having previously been professor of chemistry and natural philosophy at Mercer University, Macon, Georgia. His mother was Mehetable Osgood Seccomb of Salem, Massachusetts. Peirce and his father were close companions, and in 1864 they traveled together to the Cape of Good Hope. They shared a love of music; Peirce's father played the flute and Peirce himself frequently sang in Oratorio and Choral Society performances. Later in his professional career at Harvard, Peirce served as a member of the committee on honors and higher degrees in music.

In 1872, after a two-year apprenticeship as a carpenter (during which he read extensively and perfected the Latin his father had taught him) Peirce was admitted to Harvard College. He became the first research student of John Trowbridge and published, during his junior year, a paper that revealed a "remarkable knowledge of Becquerel, Rowland, Maxwell, and Thomson; a remarkable use of electromagnetic equipment; a remarkable application of mathematics." Under Trowbridge's influence he investigated magnetization; he later developed an interest in problems in heat conduction, and wrote a number of papers on those subjects.

Peirce was graduated in 1876 with highest honors in physics. During the next year, he served as laboratory assistant to Trowbridge and then studied under Wiedemann in Leipzig, where he took the Ph.D. (1879). In 1880 he worked in Helmholtz' laboratory in Berlin, where he met Karl Pearson, who became his lifelong friend. He also met Isabella Turnbull Landreth, a student in the conservatory of music, and they were married in her native Scotland in 1882. They had two daughters.

Peirce's research efforts in Germany were in a sense unrewarding. Edwin Hall wrote of the "unhappy turn of fate" that led Peirce to devote "a year or more of intense labor on gas batteries at a time when physical chemistry was floundering through a bog of experimentation . . . misdirected by the false proposition that the electromotive force of a battery should be calculable from the heat yielded by the chemical operations occurring in it." Peirce exercised the greatest care in testing some 400 batteries, of six different types, and found no data to support this principle, which had been advocated by Wiedemann and by William Thomson. Although he regretfully recorded his findings, he did not openly challenge such authorities, and Wiedemann and Thomson's theorem was only later disproved by J. Willard Gibbs and Helmholtz.

In 1880 Peirce returned to the United States and taught for one year at the Boston Latin School. He began his teaching career at Harvard University as an instructor in 1881, and in 1888, following Lovering's retirement, was appointed Hollis professor of mathematics and natural philosophy. He soon established himself as an able administrator.

In 1883 Peirce was one of the first scientists to study retinal sensitivity by means of the spectrum instead of revolving discs. But his 1889 work, "Perception of Horizontal and of Vertical Lines," was essentially psychological. The full extent of his mathematical talent was first revealed in 1891, in a paper entitled "On Some Theorems Which Connect Together Certain Line and Surface Integrals." His *Short Table of Integrals*, which eventually became an indispensable reference tool for scientists and mathematicians, was first published as a pamphlet in 1889.

Peirce was a member of various American and foreign societies. In 1913 he served as president of the American Physical Society, which he had helped to organize, and as vice-president of the American Mathematical Society. He also served as an editor of the *Physical Review*. He was a cousin, at several removes, of Charles Sanders Peirce.

BIBLIOGRAPHY

I. ORIGINAL WORKS. Poggendorff, III, col. 1013; IV, cols. 1128–1129; and V, cols. 952–953, gives a detailed bibliography. Peirce's major works are *Elements of the Theory of the Newtonian Potential Function* (Boston, 1888); *A Short Table of Integrals*, issued as a pamphlet in 1889, but subsequently published in Byerly, ed., *Elements of the Integral Calculus* (Boston, 1889) and enlarged in many later eds.; and *Mathematical and Physical Papers, 1903–1913* (Cambridge, Mass., 1926), which contains 56 papers. Peirce's papers and correspondence are preserved in the archives of Harvard College.

II. SECONDARY LITERATURE. For works on Peirce and his work, see *American Men of Science*, 2nd ed. (Lancaster, Pa., 1910), p. 364; R. Archibald, in *Dictionary of American Biography*, XIV, 397–398; *Boston Transcript* (14 Jan. 1914); Edwin Hall, *et al.*, "Harvard University Minute on the Life and Services of Professor Benjamin Osgood Peirce," in the university archives, repr. from *Harvard University Gazette* (21 Feb. 1914); Edwin Hall, "Biographical Memoir of Benjamin Osgood Peirce," in *Biographical Memoirs. National Academy of Sciences*, **8** (1919), 437–466, which also contains a complete bibliography of his mathematical and physical papers; *Lamb's Biographical Dictionary of the United States*, VI (Boston, 1903), 198; J. Trowbridge, "Benjamin Osgood Peirce," in *Harvard Grads's Magazine* (Mar. 1914); A. G. Webster, "Benjamin Osgood Peirce," in *Science* (1914), repr. in *Nation* (23 Apr. 1914); and *Who's Who in America, 1912–1913*.

CAROLYN EISELE

PEIRCE, CHARLES SANDERS (*b.* Cambridge, Massachusetts, 10 September 1839; *d.* Milford, Pennsylvania, 19 April 1914), *logic, geodesy, mathematics, philosophy, history of science.*

Peirce frequently asserted that he was reared in a laboratory. His father, Benjamin Peirce, was professor of mathematics and natural philosophy at Harvard University at the time of Charles's birth; he personally supervised his son's early education and inculcated in him an analytic and scientific mode of thought. Peirce attended private schools in Cambridge and Boston; he was then sent to the Cambridge High School, and, for a term, to E. S. Dixwell's School, to prepare for Harvard. While at college (1855–1859), Peirce studied Schiller's *Aesthetische Briefe* and Kant's *Kritik der reinen Vernunft*, both of which left an indelible mark on his thought. He took the M.A. at Harvard (1862) and the Sc.B. in chemistry, *summa cum laude*, in the first class to graduate from the Lawrence Scientific School (1863). Despite his father's persistent efforts to encourage him to make a career of science, Peirce preferred the study of methodology and logic.

Upon graduation from Harvard, Peirce felt that he needed more experience in methods of scientific investigation, and he became a temporary aide in the U.S. Coast Survey (1859). For six months during the early 1860's he also studied, under Louis Agassiz, the techniques of classification, a discipline that served him well in his logic research. Like Comte, Peirce later set up a hierarchy of the sciences in which the methods of one science might be adapted to the investigation of those under it on the ladder. Mathematics occupied the top rung, since its independence of the actualities in nature and its concern with the framing of hypotheses and the study of their consequences made its methodology a model for handling the problems of the real world and also supplied model transforms into which such problems might be cast and by means of which they might be resolved.

Peirce was appointed a regular aide in the U.S. Coast Survey on 1 July 1861 and was thereby exempted from military service. On 1 July 1867 he was appointed assistant in the Survey, a title he carried until his resignation on 31 December 1891. In the early days his assignments were diverse. He observed in the field the solar eclipse of 1869 in the United States and selected the site in Sicily from which an American expedition—headed by his father and including both himself and his wife—observed the solar eclipse of 22 December 1870. He was temporarily in charge of the Coast Survey Office in 1872, and on 30 November of that year his father appointed him to "take charge of the Pendulum Experiments of the Coast Survey." Moreover he was to "investigate the law of deviations of the plumb line and of the azimuth from the spheroidal theory of the earth's figure." He was further directed to continue under Winlock the astronomical work that he had begun in 1869, while an assistant at the Harvard College Observatory; his observations, completed in 1875, were published in 1878 in the still important *Photometric Researches*. He was an assistant computer for the nautical almanac in 1873, and a special assistant in gravity research from 1884 to 1891. During the 1880's, however, Peirce found it increasingly difficult, under the changing administration of the Survey, to conform to the instructions issued him; in 1891 he tendered a forced resignation and left government service. (In 1962 a Coast and Geodetic Survey vessel was named for him, in somewhat belated recognition of his many contributions.)

Peirce's astronomical work, which he began in 1867, was characterized as "pioneer" by Solon I. Bailey, director of the Harvard Observatory in 1920. Peirce attempted to reform existing scales of magnitudes with the aid of instrumental photometry, and he investigated the form of the galactic cluster in which the sun is situated, the determination of which was "the chief end of the observations of the magnitude of the stars."

From April 1875 to August 1876 Peirce was in Europe to learn the use of the new convertible pendulum, "to compare it with those of the European measure of a degree and the Swiss Survey," and to compare his "invariable pendulums in the manner which has been usual by swinging them in London and Paris." In England he met Lockyer, Clifford, Stokes, and Airy; and in Berlin, Johann Jacob Baeyer, the director of the Prussian Geodetic Institute, where Peirce compared the two standards of the German instrument and the American one. He was invited to attend the meetings of the European Geodetic Association held in Paris during the summer of 1875, and there made a name as a research geodesist. His discovery of an error in European measurement, which was due to the flexure of the pendulum stand, led to the important twenty-three-page report that Plantamour read for him at Geneva on 27 October 1877. The first Peirce pendulum was invented in June 1878 and superseded the Repsold model used in the Coast and Geodetic Survey. Although the United States did not become a member of the International Geodetic Association until 1889, Peirce's geodetic work was widely recognized. His paper on the value of gravity, read to the French Academy on 14 June 1880, was enthusiastically received, and he was invited to attend a conference on the pendulum of the Bureau des Longitudes.

In 1879 Peirce succeeded in determining the length

of the meter from a wavelength of light. Benjamin Peirce described this feat, an adumbration of the work of Michelson, as "the only sure determination of the meter, by which it could be recovered if it were to be lost to science." By 1882 Peirce was engaged in a mathematical study of the relation between the variation of gravity and the figure of the earth. He claimed that "divergencies from a spherical form can at once be detected in the earth's figure by this means," and that "this result puts a new face on the relation of pendulum work to geodesy."

Peirce's mathematical inventiveness was fostered by his researches for the Coast Survey. His theory of conformal map projections grew out of his studies of gravity and resulted in his quincuncial map projection of 1876, which has been revived by the Coast Survey in chart no. 3092 to depict international air routes. This invention represented the first application of elliptic functions and Jacobian elliptic integrals to conformal mapping for geographical purposes. Peirce was further concerned with topological mapping and with the "Geographical Problem of the Four Colors" set forth by A. B. Kempe. The existential graphs that he invented as a means of diagrammatic logical analysis (and which he considered his *chef d'oeuvre*) grew out of his experiments with topological graphic elements. These reflect the influence on his thought of Tait's historic work on knots and the linkage problems of Kempe, as well as his own belief in the efficacy of diagrammatic thinking.

Peirce's interest in the linkage problem is first documented in the report of a meeting of the Scientific Association at the Johns Hopkins University, where Peirce was, from 1879 to 1884, a lecturer in logic and was closely associated with members of the mathematics department directed by J. J. Sylvester. (It was Sylvester who arranged for the posthumous republication, with addenda and notes by Charles Peirce, of Benjamin Peirce's *Linear Associative Algebra*.) Peirce had persuaded his father to write that work, and his father's mathematics influenced his own. J. B. Shaw has pointed out that two other lines of linear associative had been followed besides the direct one of Benjamin Peirce, one by use of the continuous group first announced by Poincaré and the other by use of the matrix theory first noted by Charles Peirce. Peirce was the first to recognize the quadrate linear associative algebras identical with matrices in which the units are letter pairs. He did not, however, regard this combination as a product, as did J. W. Gibbs in his "Elements of Vector Analysis" of 1884. Gibbs's double-dot product, according to Percey F. Smith, "is exactly that of C. S. Peirce's vids, and accordingly the algebra of dyadics based upon the double-dot law of multiplication is precisely the matricular algebra" of Peirce. In his *History of Mathematics*, Florian Cajori wrote that "C. S. Peirce showed that of all linear associative algebras there are only three in which division is unambiguous. These are ordinary single algebra, ordinary double algebra, and quaternions, from which the imaginary scalar is excluded. He showed that his father's algebras are operational and matricular." Peirce's work on nonions was to lead to a priority dispute with Sylvester.

By the time Peirce left the Johns Hopkins University, he had taken up the problem of continuity, a pressing one since his logical analysis and philosophical interpretation required that he deal with the infinite. In his 1881 paper "Logic of Number," Peirce claimed to have "distinguished between finite and infinite collections in substantially the same way that Dedekind did six years later." He admired the logical ingenuity of Fermat's method of "infinite descent" and used it consistently, in combination with an application of De Morgan's syllogism of transposed quantity that does not apply to the multitude of positive integers. Peirce deduced the validity of the "Fermatian method" of reasoning about integers from the idea of correspondence; he also respected Bolzano's work on this subject. He was strongly impressed by Georg Cantor's contributions, especially by Cantor's handling of the infinite in the second volume of the *Acta Mathematica*. Peirce explained that Cantor's "class of *Mächtigkeit* aleph-null is distinguished from other infinite classes in that the *Fermatian inference* is applicable to the former and not to the latter; and that generally, *to any smaller class some mode of reasoning is applicable which is not applicable to a greater one.*" In his development of the concept of the orders of infinity and their aleph representations, Peirce used a binary representation (which he called "secundal notation") of numbers. He eventually developed a complete algorithm for handling fundamental operations on numbers so expressed. His ingenuity as an innovator of symbolic notation is apparent throughout this work.

Peirce's analysis of Cantor's *Menge* and *Mächtigkeit* led him to the concept of a supermultitudinous collection beyond all the alephs—a collection in which the elements are no longer discrete but have become "welded" together to represent a true continuum. In his theory of logical criticism, "the temporal succession of ideas is continuous and not by discrete steps," and the flow of time is similarly continuous in the same sense as the nondiscrete superpostnumeral multitudes. Things that exist form an enumerable collection, while those *in futuro* form a denumerable collection (of multitude aleph-null). The possible different courses of the future have a first abnumeral multitude (two raised to the exponent aleph-null) and the possibilities of such possibilities will be of the second abnumeral multitude (two raised to the exponent "two raised to the exponent aleph-null"). This procedure may be continued to the infinitieth exponential, which is thoroughly potential and retains no relic of the arbitrary existential—the state of true continuity. Peirce's research on continuity led him to make an exhaustive study of topology, especially as it had been developed by Listing.

Peirce's philosophy of mathematics postulated that the study of the substance of hypotheses only reveals other consequences not explicitly stated in the original. Mathematical procedure therefore resolves itself into four parts: (1) the creation of a model that embodies the condition of the premise; (2) the mental modification of the diagram to obtain auxiliary information; (3) mental experimentation on the diagram to bring out a new relation between parts not mentioned in its construction; and (4) repetition of the experiment "to infer inductively, with a degree of probability practically amounting to certainty, that every diagram constructed according to the same precept would present the same relation of parts which has been observed in the diagram experimented upon." The concern of the mathematician is to reach the conclusion, and his interest in the process is merely as a means to reach similar conclusions, whereas the logician desires merely to understand the process by which a result may be obtained. Peirce asserted that mathematics is a study of what is or is not logically possible and that the mathematician need not be concerned with what actually exists. Philosophy, on the other hand, discovers what it can from ordinary everyday experience.

Peirce characterized his work in the following words: "My philosophy may be described as the attempt of a physicist to make such conjecture as to the constitution of the universe as the methods of science may permit. . . . The best that can be done is to supply a hypothesis, not devoid of all likelihood, in the general line of growth of scientific ideas, and capable of being verified or refuted by future observers." Having postulated that every additional improvement of knowledge comes from an exercise of the powers of perception, Peirce held that the observation in a necessary inference is directed to a sort of diagram or image of the facts given in the premises. As in mathematics, it is possible to observe relations between parts of the diagram that were not noticed in its construction. Part of the business of logic is to construct such diagrams. In short, logical truth has the same source as mathematical truth, which is derived from the observation of diagrams. Mathematics uses the language of imagery to trace out results and the language of abstraction to make generalizations. It was Peirce's claim to have opened up the subject of abstraction, where Boole and De Morgan had concentrated on studies of deductive logic.

In 1870 Peirce greatly enlarged Boolean algebra by the introduction of a new kind of abstraction, the dyadic relation called "inclusion"—"the connecting link between the general idea of logical dependence and the idea of sequence of a quantity." The idea of quantity is important in that it is a linear arrangement whereby other linear arrangements (for example, cause and effect and reason and consequent) may be compared. The logic of relatives developed by Peirce treats of "systems" in which objects are brought together by any kind of relations, while ordinary logic deals with "classes" of objects brought together by the relation of similarity. General classes are composed of possibilities that the nominalist calls an abstraction. The influence of Peirce's work in dyadic relations may be seen in Schröder's *Vorlesungen über die Algebra der Logik*, and E. V. Huntington included Peirce's proof of a fundamental theorem in his "Sets of Independent Postulates for the Algebra of Logic" and in *The Continuum* referred to a statement that Peirce had published in the *Monist*. Peirce's contribution to the foundations of lattice theory is widely recognized.

In describing multitudes of systems within successive systems, Peirce reached a multitude so vast that the individuals lose their identity. The zero collection represents germinal possibility; the continuum is concrete-developed possibility; and "The whole universe of true and real possibilities forms a continuum upon which this universe of Actual Existence is a discontinuous mark like a point marked on a line."

The question of nominalism and realism became for Peirce the question of the reality of continua. Nature syllogizes, making inductions and abductions—as, for example, in evolution, which becomes "one vast succession of generalizations by which matter is becoming subjected to ever higher and higher laws." Laws of nature in the present form are products of an evolutionary process and logically require an explanation in such terms. In the light of the logic of relatives, Peirce maintained, the general is seen to be the continuous and coincides with that opinion the medieval Schoolmen called realism. Peirce's Scotistic stance—in opposition to Berkeley's nominalism—caused him to attack the nominalistic positions of Mach, Pearson, and Poincaré. Peirce accused the positivists of confusing psychology with logic in mistaking sense impressions, which are psychological inferences, for logical data. Joseph Jastrow tells of

being introduced by Peirce "to the possibility of an experimental study of a psychological problem," and they published a joint paper, "On Small Differences in Sensation," in the *Memoirs of the National Academy of Sciences* (1884).

William James was responsible for Peirce's worldwide reputation as the father of the philosophical doctrine that he originally called pragmatism, and later pragmaticism. Peirce's famous pragmatic maxim was enunciated in "How to Make Our Ideas Clear," which he wrote (in French) on shipboard before reaching Plymouth on the way to the Stuttgart meetings of the European Geodetic Association in 1877. The paper contains his statement of a laboratory procedure valid in the search for "truth"—"Consider what effects, that might conceivably have practical bearings, we conceive the object of our conception to have. Then, our conception of these effects is the whole of our conception of the object." In a letter to his former student Christine Ladd-Franklin, Peirce emphasized that "the meaning of a *concept* . . . lies in the manner in which it could *conceivably* modify purposive action, and *in this alone*." Moreover "pragmatism is one of the results of my study of the formal laws of signs, a study guided by mathematics and by the familiar facts of everyday experience and by no other science whatever." John Dewey pointed out that reality, in Peirce's system, "means the object of those beliefs which have, after prolonged and cooperative inquiry, become stable, and 'truth,' the quality of these beliefs, is a logical consequence of this position." The maxim underlies Peirce's epistemology, wherein the first procedure is a guess or hypothesis (abductive inference) from which are set up subsidiary conclusions (deductive inference) that can be tested against experimental evidence (inductive inference).

The results of the inductive process are ratios and admit of a probability error, abnormal occurrences corresponding to a ratio of zero. This is valid for infinite classes, but for none larger than the denumeral. Consequently, induction must always admit the possibility of exception to the law, and absolute certainty is unobtainable. Every boundary of a figure that represents a possible experience ought therefore to be blurred, and herein lies the evidence for Peirce's claim to priority in the enunciation of a triadic logic.

Morris Cohen has characterized Peirce's thought as germinal in its initiation of new ideas and in its illumination of his own "groping for a systematic view of reason and nature." Peirce held that chance, law, and continuity are basic to the explanation of the universe. Chance accounts for the origin of fruitful ideas, and if these meet allied ideas in a mind prepared for them, a welding process takes place—a process called the law of association. Peirce considered this to be the one law of intellectual development.

In his educational philosophy Peirce said that the study of mathematics could develop the mind's powers of imagination, abstraction, and generalization. Generalization, "the spilling out of continuous systems of ideas," is the great aim of life. In the early 1890's he was convinced that modern geometry was a rich source of "forms of conception," and for that reason every educated man should have an acquaintance with projective geometry (to aid the power of generalization), topology (to fire the imagination), and the theory of numbers (to develop the power of exact reasoning). He kept these objectives in view in the mathematics textbooks that he wrote after his retirement from the Coast Survey; these works further reflect the influence of Arthur Cayley, A. F. Möbius, and C. F. Klein. Peirce's adoption of Cayley's mathematical "absolute" and his application of it to his metaphysical thought is especially revealing. "The Absolute in metaphysics fulfills the same function as the absolute in geometry. According as we suppose the infinitely distant beginning and end of the universe are *distinct*, *identical*, or *nonexistent*, we have three kinds of philosophy, hyperbolic, parabolic, or elliptic." Again "the first question to be asked about a continuous quantity is whether the two points of its absolute coincide." If not, are they in the real line of the scale? "The answers will have great bearing on philosophical and especially cosmogonical problems." For a time Peirce leaned to a Lobachevskian interpretation of the character of space.

Peirce once wrote to Paul Carus, editor of the *Monist*, "Few philosophers, if any, have gone to their work as well equipped as I, in the study of other systems and in the various branches of science." In 1876, for example, Peirce's thought on the "economy of research" was published in a Coast and Geodetic Survey report. It became a major consideration in his philosophy, for the art of discovery became for him a general problem in economics. It underlay his application of the pragmatic maxim and became an important objective in his approach to problems in political economy, in which his admiration of Ricardo was reflected in his referring to "the peculiar reasoning of political economy" as "Ricardian inference." Peirce's application of the calculus approach of Cournot predated that of Jevons and brought him recognition (according to W. J. Baumol and S. W. Goldfeld) as a "precursor in mathematical economics."

Peirce also sought systems of logical methodology in the history of logic and of the sciences. He became known for his meticulous research in the scientific and logical writings of the ancients and the medieval Schoolmen, although he failed to complete the book on the history of science that he had contracted to write in 1898. For Peirce the history of science was an instance of how the law of growth applied to the human mind. He used his revised version of the Paris manuscript of Ptolemy's catalogue of stars in his astronomical studies, and he included it for modern usage in *Photometric Researches*. He drew upon Galileo—indeed, his abductive inference is identical twin to Galileo's *il lume naturale*—and found evidence of a "gigantic power of right reasoning" in Kepler's work on Mars.

Peirce spent the latter part of his life in comparative isolation with his second wife, Juliette Froissy, in the house they had built near Milford, Pennsylvania, in 1888. (His second marriage, in 1883, followed his divorce from Harriet Melusina Fay, whom he had married in 1862.) He wrote articles and book reviews for newspapers and journals, including the *Monist*, *Open Court*, and the *Nation*. As an editorial contributor to the new *Century Dictionary*, Peirce was responsible for the terms in logic, metaphysics, mathematics, mechanics, astronomy, and weights and measures; he also contributed to the *Dictionary of Philosophy and Psychology*. He translated foreign scientific papers for the Smithsonian publications, served privately as scientific consultant, and prepared numerous papers for the National Academy of Sciences, to which he was elected in 1877 and of which he was a member of the Standing Committee on Weights and Measures. (Earlier, in 1867, he had been elected to the American Academy of Arts and Sciences.) Peirce also lectured occasionally, notably at Harvard (where he spoke on the logic of science in 1865, on British logicians in 1869–1870, and on pragmatism in 1903) and at the Lowell Institute. None of his diverse activities was sufficient to relieve the abject poverty of his last years, however, and his very existence was made possible only by a fund created by a group of friends and admirers and administered by his lifelong friend William James.

BIBLIOGRAPHY

I. Original Works. Bibliographies and works by Peirce include Carolyn Eisele, ed., *The New Elements of Mathematics by Charles S. Peirce*, 4 vols. (The Hague, 1974); Charles Hartshorne and Paul Weiss, eds., *The Collected Papers of Charles Sanders Peirce*, I–VI (Cambridge, Mass., 1931–1935); Arthur W. Burks, ed., VII–VIII (Cambridge, Mass., 1958), with a bibliography in vol. VIII—supp. 1 to this bibliography is by Max Fisch, in Philip Wiener, and Harold Young, eds., *Studies in the Philosophy of Charles Sanders Peirce*, 2nd ser. (1964), 477–485, and supp. 2 is in *Transactions of the Charles S. Peirce Society*, **2**, no. 1 (1966), 51–53. Also see Max Fisch, "A Draft of a Bibliography of Writings About C. S. Peirce," in *Studies*, 2nd ser., 486–514; supp. 1 is in *Transactions of the Charles S. Peirce Society*, **2**, no. 1 (1966), 54–59. Papers in the Houghton Library at Harvard University are listed in Richard S. Robin, *Annotated Catalogue of the Papers of Charles S. Peirce* (Amherst, 1967). In addition, see Richard S. Robin, "The Peirce Papers: A Supplementary Catalogue," in *Transactions of the Charles S. Peirce Society*, **7**, no. 1 (1971), 37–58. Unpublished MSS are in the National Archives, the Library of Congress, the Smithsonian Archives, and in the Houghton Library, Harvard University.

For Peirce's work during 1879–1884, see *Johns Hopkins University Circulars*, esp. "On a Class of Multiple Algebras," **2** (1882), 3–4; "On the Relative Forms of Quaternions," **13** (1882), 179; and "A Communication From Mr. Peirce [On nonions]," **22** (1883), 86–88.

In the period 1870–1885, Peirce published fourteen technical papers as appendices to *Reports of the Superintendent of the United States Coast and Geodetic Survey.* See "Notes on the Theory of Economy of Research," **14** (1876), 197–201, repr. in W. E. Cushen, "C. S. Peirce on Benefit-Cost Analysis of Scientific Activity," in *Operations Research* (July–Aug., 1967), 641–648; and "A Quincuncial Projection of the Sphere," in **15** (1877), published also in *American Journal of Mathematics*, **2** (1879), and in Thomas Craig, *A Treatise on Projections* (1882). See also "Photometric Researches," in *Annals of Harvard College Observatory* (Leipzig, 1878); and preface: "A Theory of Probable Inference"; note A: "Extension of the Aristotelian Syllogistic"; and note B: "The Logic of Relatives," in Peirce, ed., *Studies in Logic. By Members of the Johns Hopkins University* (Boston, 1883).

See also *Charles S. Peirce Über die Klarheit unserer Gedanken* (*How to Make Our Ideas Clear*), ed., trans., and with commentary by Klaus Oehler (Frankfurt am Main, 1968); Edward C. Moore, ed., *Charles S. Peirce: The Essential Writings* (New York, 1972); *Charles S. Peirce Lectures on Pragmatism* (*Vorlesungen über Pragmatismus*), ed., trans., and annotated by Elisabeth Walther (Hamburg, 1973); and Morris R. Cohen, ed., *Chance, Love, and Logic* (New York, 1923).

II. SECONDARY LITERATURE. The *Transactions of the Charles S. Peirce Society* contain a large number of papers on Peirce and his work. There are also interesting biographical notices in a number of standard sources—see esp. those by N. Bosco, in *Enciclopedia filosofica* (Florence, 1967); Murray G. Murphey, in *The Encyclopedia of Philosophy* (New York–London, 1967); Paul Weiss, in *Dictionary of American Biography* (New York, 1934); and Philip P. Wiener, in *International Encyclopedia of the Social Sciences* (New York, 1968). See also *Lamb's Biographical Dictionary of the United States* (Boston, 1903); and *American Men of Science* (1906), which contains Peirce's own list of his fields of research.

Recent books, not necessarily listed in the bibliographies cited above, include John F. Boler, *Charles S. Peirce and Scholastic Realism* (Seattle, 1963); Hanna Buczynska-Garewicz, *Peirce* (Warsaw, 1965); Douglas Greenlee, *Peirce's Concept of Sign* (The Hague–Paris, 1973); Edward C. Moore and Richard S. Robin, eds., *Studies in the Philosophy of Charles Sanders Peirce*, 2nd ser. (Amherst, 1964); Murray G. Murphey, *The Development of Peirce's Philosophy* (Cambridge, Mass., 1961); Francis E. Reilly, *Charles S. Peirce's Theory of Scientific Method* (New York, 1970); Don D. Roberts, *The Existential Graphs of Charles S. Peirce* (The Hague–Paris, 1973); Elisabeth Walther, *Die Festigung der Überzeugung und andere Schriften* (Baden-Baden, 1965); Hjamer Wennerberg, *The Pragmatism of C. S. Peirce* (Uppsala, 1962); and Philip P. Wiener and Frederic H. Young, eds., *Studies in the Philosophy of Charles Sanders Peirce* (Cambridge, Mass., 1952).

Especially pertinent to this article are J. C. Abbott, *Trends in Lattice Theory* (New York, 1970); Oscar S. Adams, "Elliptic Functions Applied to Conformal World Maps," in *Department of Commerce Special Publication No. 112* (1925); and "The Rhombic Conformal Projection," in *Bulletin géodésique*, **5** (1925), 1–26; Solon I. Bailey, *History and Work of the Harvard College Observatory* (1931); W. J. Baumol and S. M. Goldfeld, *Precursors in Mathematical Economics* (London, 1968); Max Bense and Elisabeth Walther, *Wörterbuch der Semiotik* (Cologne, 1973); Garrett Birkhoff, *Lattice Theory* (New York, 1948); Rudolf Carnap, *Logical Foundations of Probability* (London, 1950); Clarence I. Lewis, *A Survey of Symbolic Logic* (Berkeley, 1918); James Byrnie Shaw, *Synopsis of Linear Associative Algebra* (Washington, D.C., 1907); Percey F. Smith, "Josiah Willard Gibbs," in *Bulletin of the American Mathematical Society* (Oct. 1903), 34–39; and Albert A. Stanley, "Quincuncial Projection," in *Surveying and Mapping* (Jan.–Mar., 1946).

See also the section "Charles Sanders Peirce," in *Journal of Philosophy, Psychology and Scientific Methods*, **13**, no. 26 (1916), 701–737, which includes Morris R. Cohen, "Charles S. Peirce and a Tentative Bibliography of His Published Writings"; John Dewey, "The Pragmatism of Peirce"; Joseph Jastrow, "Charles Peirce as a Teacher"; Christine Ladd-Franklin, "Charles S. Peirce at the Johns Hopkins"; and Josiah Royce and Fergus Kernan, "Peirce as a Philosopher."

More recently published essays by Carolyn Eisele, not necessarily listed above, include "The *Liber abaci* Through the Eyes of Charles S. Peirce," in *Scripta mathematica*, **17** (1951), 236–259; "Charles S. Peirce and the History of Science," in *Yearbook. American Philosophical Society* (Philadelphia, 1955), 353–358; "Charles S. Peirce, American Historian of Science," in *Actes du VIIIᵉ Congrès international d'histoire des sciences* (Florence, 1956), 1196–1200; "The Charles S. Peirce-Simon Newcomb Correspondence," in *Proceedings of the American Philosophical Society*, **101**, no. 5 (1957), 410–433; "The Scientist-Philosopher C. S. Peirce at the Smithsonian," in *Journal of the History of Ideas*, **18**, no. 4 (1957), 537–547; "Some Remarks on the Logic of Science of the Seventeenth

Century as Interpreted by Charles S. Peirce," in *Actes du 2ᵉᵐᵉ Symposium d'histoire des sciences* (Pisa–Vinci, 1958), 55–64; "Charles S. Peirce, Nineteeth-Century Man of Science," in *Scripta mathematica*, **24** (1959), 305–324; "Poincaré's Positivism in the Light of C. S. Peirce's Realism," in *Actes du IXᵉ Congrès international d'histoire des sciences* (Barcelona–Madrid, 1959), 461–465; "The Quincuncial Map-Projection of Charles S. Peirce," in *Proceedings of the 10th International Congress of History of Science* (Ithaca, 1962), 687; and "Charles S. Peirce and the Problem of Map-Projection," in *Proceedings of the American Philosophical Society*, **107**, no. 4 (1963), 299–307.

Other articles by Carolyn Eisele are "Fermatian Inference and De Morgan's Syllogism of Transposed Quantity in Peirce's Logic of Science," in *Physis. Rivista di storia della scienza*, **5**, fasc. 2 (1963), 120–128; "The Influence of Galileo on the Thought of Charles S. Peirce," in *Atti del Simposio su Galileo Galilei nella storia e nella filosofia della scienza* (Florence–Pisa, 1964), 321–328; "Peirce's Philosophy of Education in His Unpublished Mathematics Textbooks," in Edward C. Moore and Richard S. Robin, eds., *Studies in the Philosophy of Charles Sanders Peirce*, 2nd ser., (1964), 51–75; "The Mathematics of Charles S. Peirce," in *Actes du XIᵉ Congrès international d'histoire des sciences* (Warsaw, 1965), 229–234; "C. S. Peirce and the Scientific Philosophy of Ernst Mach," in *Actes du XIIᵉ Congrès international d'histoire des sciences* (Paris, 1968), 33–40; and "Charles S. Peirce and the Mathematics of Economics," in *Actes du XIIIᵉ Congrès international d'histoire des sciences* (Moscow, 1974).

Essays by Max H. Fisch include "Peirce at the Johns Hopkins University," in Philip P. Wiener and Frederic H. Young, eds., *Studies in the Philosophy of Charles Sanders Peirce* (Cambridge, 1952), 277–312, written with Jackson I. Cope; "Alexander Bain and the Genealogy of Pragmatism," in *Journal of the History of Ideas*, **15** (1954), 413–444; "A Chronicle of Pragmaticism, 1865–1897," in *Monist*, **48** (1964), 441–466; "Was There a Metaphysical Club in Cambridge?," in Edward C. Moore and Richard S. Robin, eds., *Studies in the Philosphy of Charles Sanders Peirce*, 2nd ser. (Amherst, 1964), 3–32; "Peirce's Triadic Logic," in *Transactions of the Charles S. Peirce Society*, **2** (1966), 71–86, written with Atwell Turquette; "Peirce's Progress from Nominalism Toward Realism," in *Monist*, **51** (1967), 159–178; and "Peirce's Ariste: The Greek Influence in His Later Philosophy," in *Transactions of the Charles S. Peirce Society*, **7** (1971), 187–210.

Essays by Victor F. Lenzen include "Charles S. Peirce and *Die Europaische Gradmessung*," in *Proceedings of the XIIth International Congress of the History of Science* (Ithaca, 1962), 781–783; "Charles S. Peirce as Astronomer," in Edward C. Moore and Richard S. Robin, eds., *Studies in the Philosophy of Charles Sanders Peirce*, 2nd ser. (Amherst, 1964), 33–50; "The Contributions of Charles S. Peirce to Metrology," in *Proceedings of the American Philosophical Society*, **109**, no. 1 (1965), 29–46; "Development of Gravity Pendulums in the 19th Century,"

in United States Museum Bulletin 240: *Contributions From the Museum of History and Technology, Smithsonian Institution*, paper 44 (Washington, 1965), 301–348, written with Robert P. Multhauf; "Reminiscences of a Mission to Milford, Pennsylvania," in *Transactions of the Charles S. Peirce Society*, **1** (1965), 3–11; "The Role of Science in the Philosophy of C. S. Peirce," in *Akten des XIV Internationalen Kongresses für Philosophie* (Vienna, 1968), 371–376; "An Unpublished Scientific Monograph by C. S. Peirce," in *Transactions of the Charles S. Peirce Society*, **5** (1969), 5–24; "Charles S. Peirce as Mathematical Geodesist," *ibid.*, **8** (1972), 90–105; and "The Contributions of C. S. Peirce to Linear Algebra," in Dale Riepe, ed., *Phenomenology and Natural Existence* (*Essays in Honor of Martin Farber*) (New York, 1973), 239–254.

Carolyn Eisele

PEIRESC, NICOLAS CLAUDE FABRI DE (*b.* Belgentier, Var, France, 1 December 1580; *d.* Aix-en-Provence, France, 24 June 1637), *astronomy, scientific patronage.*

Peiresc was the son of Raynaud de Fabri, sieur de Callas and *conseiller* in the Parlement of Provence, and Marguerite de Bompar de Magnan. Originally from Pisa, the Fabri family had lived in Provence for many years, acquiring property and social standing. The name that Nicolas Claude assumed formally in 1624 was derived from an estate in his mother's dowry, the hamlet of Peiresc high in the Alpes de Provence. Peiresc's education began at Aix and Avignon and continued at the Jesuit *collège* at Tournon, where he made his first contact with astronomy.

In 1599 Peiresc went to Padua, where he met the erudite and generous jurist, numismatist, and antiquarian Giovanni Vincenzo Pinelli (1535–1602). Here also he met Galileo, then a professor at the university. From some time in 1600 he traveled in Italy, Switzerland, and France, visiting galleries and libraries and meeting learned men; he finally settled down to serious legal studies at Montpellier under the rigorous and inspiring teaching of Julius Pacius (1550–1635), a learned Protestant who had taught in Hungary, at Heidelberg, and Sedan before going to Aix, Padua, and Valence. The influences of Pinelli and Pacius stimulated in Peiresc a curiosity about antiquity, the arts and sciences, and the diversity of the natural world; and he viewed these two men as living examples of the Renaissance *virtuoso*: the man of taste and intelligence who communicates his knowledge as he offers his books and instruments for the use and satisfaction of his contemporaries.

Having received his degree in law, Peiresc returned to Aix and was admitted *conseiller* (1604) in the

Parlement of Provence, taking over the seat of an uncle. In 1605 he went to Paris as secretary to Guillaume du Vair, president of the Parlement of Provence, and in 1606 accompanied ambassador Le Fèvre de La Boderie to England, where he met L'Obel, the botanist of James I, and numerous learned amateurs of the arts and sciences, among them William Camden and Henry Savile. He returned to France by way of the Netherlands, meeting other antiquarians and scientists, including L'Écluse, to whom he later sent seeds of thirty-six plants native to Provence and names of others provided by friends in Aix and its environs.

Peiresc's attitude toward natural phenomena was exhibited in early July 1608, when mysterious splashes of red appeared suddenly on walls and trees in and around the city. Popularly attributed to a "rain of blood," the phenomenon was regarded with superstitious awe. Peiresc considered the circumstances and concluded that the red substance had been excreted by the chrysalides of the butterfly *Vanessa*, which was numerous in the summer of that year.

In 1610 Peiresc read Galileo's *Sidereus nuncius* and learned of the latter's discoveries made with the newly developed telescope. Peiresc's patron, du Vair, had already acquired one of the new instruments and with it the astronomer Joseph Gaultier[1] and Peiresc were the first in France (24 and 25 November) to see the four satellites of Jupiter. Galileo named these satellites Sidera Medicea. In 1611 Peiresc observed Venus and Mercury in the morning sky after sunrise, distinguished the crescent phases of Venus, and was the first to see the nebula in the sword of Orion, announced and described by Huygens in 1658.

During these years Peiresc's main interest was the recording of the times of planetary events. The journal he kept from 24 November 1610 to 21 June 1612 preserves a record largely of observations of the relative positions of the satellites of Jupiter, gradually establishing their period of revolution. In this work he had several assistants, the most helpful being Jean Lombard, who made an expedition to Marseilles, Malta, Cyprus, and Tripoli (Lebanon), in each place recording the positions of the satellites in local time. These observations, later collated with time recorded in Aix, permitted Peiresc to calculate terrestrial longitudinal differences. This interest was maintained by Peiresc and in 1635 led finally to a more systematic and successful operation: determining the length of the Mediterranean with a good deal of accuracy.

In 1616 Peiresc again went to Paris in the service of du Vair and remained there for about seven years. Now a mature scholar, he met intellectual circles in Paris on even terms and was soon introduced to the "Cabinet" of the Dupuy brothers, learned librarians and students of law and history, in whose quarters meetings were held weekly for erudite discussions and exchange of news. Through the Dupuys, Peiresc met many men with whom he maintained contact for the rest of his life. Most important for the scientific movement was Mersenne, the Minorite father in whose cell a more specialized group met at frequent intervals. Mersenne and Peiresc regularly exchanged letters, discussing news of books, experiments, observations, and the theories and opinions that were opening fresh perspectives on knowledge of the natural world.[2]

While in Paris, Peiresc made a simple telescopic observation of the comet of 1618; but without mathematical instruments he could take no angular measurements, and he left no record of what motion he had perceived. In the same year Louis XIII granted him the abbacy of a monastic house at Guîtres, north of Bordeaux, the income from which permitted the employment of a priest for ecclesiastical duties and also funds which he could use for the purchase of books. His position as *abbé* was regularized when he took the tonsure in 1624. After du Vair died in 1621, Peiresc remained in Paris until the summer of 1623, when he returned to Provence and remained there for the rest of his life. During his stay in Paris he sponsored or assisted in the publication of important books, representative of his own erudite and scientific interests. These works included the *Epistolae mathematicae de divinatione* (1623) of George of Ragusa (1579–1622), the *Histoire des grands chemins de l'empire romain* (1622) by Nicolas Bergier (1567–1623), as well as the much read satiric novel *Argenis* by John Barclay (1582–1621). One notes that these books were all by men recently deceased or in failing health.

Peiresc's long and fruitful association with Gassendi began about 1624. Gassendi had been teaching in Aix since 1616. When Peiresc returned from Paris he joined Gaultier in urging Gassendi to continue his philosophic writings against Aristotelianism and, later, to develop his discussion of the atomistic philosophy of Epicurus. Peiresc could lay no claim to profound philosophic insight, but he was as discontented as Gassendi with the stagnation of traditional physics. He was fully aware of the changes in intellectual perspective demanded by the accumulation of new facts in every field of human interest. His outlook was that of a collector, rather than of a systematist, who was content to accumulate artifacts and data, books and manuscripts of many kinds, plants and animals, and by correspondence to make his collections and knowledge available to innumerable friends, including many whom he would never meet.

With his associates Peiresc again timed celestial

happenings on the occasion of the lunar eclipse of 20 June 1628, observed in Aix with Gassendi and Gaultier, and in Paris by Mersenne and the mathematician Mydorge. These observations permitted the calculation of the Paris–Aix longitudinal differential with much greater accuracy than had been possible before. Parhelia were observed in 1629 and a solar eclipse in 1631; but the transit of Mercury anticipated by Kepler for 7 November 1631 was missed by Gaultier and Peiresc, who had to admit sheepishly that they had taken too long over Mass and that when they had climbed to the observatory toward noon the sky was clear, the sun spotless, and the transit was over.[3] Thus Gassendi, in Paris, made the only serious observation of the first predicted transit of a planet across the disk of the sun.

From this time on, celestial phenomena were studied with vigor. In 1633 sunspots attracted attention; Gassendi's suggestion that they were actually spots on the solar surface and not small satellites close to the sun was verified. In 1634 an observatory was constructed on the roof of the Hôtel Callas and observations of Jupiter, Mercury, and Saturn were made, mostly by Gassendi. Expeditions to Tycho Brahe's observatory at Uraniborg and to Alexandria had established the meridians of those places, and now a network was ready for larger operations. The lunar eclipse of 28 August 1635 was more widely observed than any other to date, largely as a result of the many priests, merchants, and secretaries of embassies (trained under Gaultier, Peiresc, and Gassendi) who were able to use instruments supplied by Peiresc and to establish more or less effective stations in Rome, Naples, Aleppo, Cairo, and Tunis. Reports from these scattered points, taken with observations made at Aix, Digne, and Paris, permitted reasonably accurate longitudinal distances covering most of the Mediterranean, from Marseilles to the Levant, particularly Aleppo. Results concerning the dimensions of the sea were checked by consultation with experienced pilots in the port of Marseilles.

The work done by Gaultier, Gassendi, and Peiresc in determining the true length of the Mediterranean depended on the development of the telescope, more accurate timekeeping, and the presence of observers capable of using modern instruments at appropriate points on or near the coasts of the sea. The mapping of reference points on the lunar surface and the use of positions of the satellites of Jupiter permitted closer approximations to the true length of about 41°30′ of longitude as opposed to 60° given in the Ptolemaic maps and to the generally exaggerated dimensions of the portolans.

Innumerable personal interests also filled the life of Peiresc. His duties as a member of the sovereign court of Provence, as a *sénateur*, and as a priest of the church, were considerably less exacting than the needs of his gardens and collections, his correspondence, and the call of science as he understood it. He could be deemed a dilettante were it not that the activities he shared or sponsored achieved a degree of success. Besides his work in astronomy, for which Gaultier and Gassendi must be given much credit, Peiresc collected and studied fossils and crystals, as well as ancient coins and medals. He was well aware of the importance of the latter for establishing historical sequences. After Gassendi had sent him Aselli's *De lactibus* (1627), Peiresc sponsored in his house the dissection by local surgeons of a cadaver, finding the chyliferous vessels in the human body as Aselli had found them in other mammals. Similarly, as the fame of Harvey's *De motu cordis et sanguinis* spread abroad, Peiresc planned to trace in the heart the channels in the septum—which Harvey had not found but which a local surgeon, one Payen, claimed to have exhibited to Peiresc and Gassendi.

Peiresc was told of Harvey's *De motu cordis* in early August 1629, a full year after its publication at Frankfurt. In a letter to the Dupuys, 11 August 1629,[4] he told of his interest in the book and on 15 September he thanked them for obtaining a copy for him, asking that it be sent by the post. In the meantime Gassendi had written that he had seen the book before leaving for Germany. He expressed his views in a letter to Mersenne,[5] saying that the circulation through the arteries and veins seems "fort vraysemblable et establye"; but that he finds that Harvey imagines that the blood cannot pass from the right ventricle to the left by way of the septum "là où il me souvient que le Sieur Payen nous a fait voir autrefois qu'il y a non seulement des pores mais des canaux très ouverts."[6]

On 17 January 1630[7] Peiresc wrote to the Dupuys that the book had come; but that he had not been able to read it—"mais à ce peu que j'en ay veu, je le trouve bien agréable." He regretted the death of Payen, the local surgeon whom Gassendi claimed had made a curious observation that Harvey could have used.

Gassendi returned to this subject in the *Vita*,[8] saying that he had informed Peiresc of this excellent new book by outlining its argument and adding that Peiresc had wished to obtain a copy in order to investigate the valves in the veins and to observe other things, including the wanderings (*maeandros*) of the channels of the heart, which Harvey denied but of which "I assured him" (*quos Harvaeus est inficiatus, et de quibus ipse feceram securum*).

There is no reference in either context to anatomical research resulting from the reading of Harvey's book and nothing corresponding to the work on the lymphatics later done under Peiresc's guidance. There is merely a reference, somewhat vague, to a dead local surgeon, who claimed to have discovered certain passages in the septum that Harvey did not find.

In these investigations Peiresc was a sponsor and in some cases the originator of such trials; but the actual work, even on occasion the astronomical observations he recorded, was performed by his staff and associates. There is little reason to believe he himself was sufficiently skilled to perform operations of any delicacy. He was a patron and amateur of science, the arts, and erudition, better equipped to write letters to his friends than to record concisely and effectively the investigations carried out in the Hôtel Callas.

Peiresc's interest in lenses and concave mirrors led him in 1634 to speculate about vision and to study the structure and function of different parts of the eye. With a local surgeon, Cayre, and with his own assistant, Lombard, and occasionally with Gaultier and Gassendi, Peiresc dissected the eyes of a small shark, dolphin, tuna, ox, sheep, owl, and an eagle and eaglet. The results of these investigations are recorded in MS 1877 at the Bibliothèque Inguimbertine, Carpentras. He also recorded personal observations of the behavior of his own eyes, for example, the persistence of afterimages. None of these records were published, although there are references to them in the correspondence; nor did this work lead to theoretical or practical results.[9]

From about 1634 Gassendi lived more or less continuously as a guest of Peiresc while working on the philosophy of Epicurus. Peiresc's health, never robust, declined in early 1637; and it is related that he died on 24 June in the arms of Gassendi, the pattern of whose life was now seriously disturbed. The philosopher sought other refuges and finally turned to the congenial task of writing his widely read and influential book *Viri illustris Nicolai Ciaudii Fabricii de Peiresc . . . vita*, published in Paris by Cramoisy in 1641 after critical reading by François Luillier, Jean Chapelain, and perhaps others. Through this book Peiresc and his work came to be known to many who had neither visited his collections and library at Belgentier and Aix nor exchanged letters with him. Translated under the title *The Mirrour of True Nobility and Gentility* (1657) by William Rand and dedicated to the English virtuoso John Evelyn, this record of a patron of the sciences takes its place in the literature associated with the rise of organized natural philosophy in England.

An understanding of Peiresc's intellectual position must be derived from his activity taken as a whole rather than from any personal statement. A practical man, he found little reason to think one kind of knowledge superior to another. He believed that an intelligent person can link experience in one discipline with what is learned in another and that cooperation and free communication are the basis on which sound knowledge—that is, science—and therefore human wisdom can advance. It is a mistake to look at merely one aspect of Peiresc's career or to consider it from a special point of view, for in his life of service to learning in all its forms, he exemplified much of what Francis Bacon proposed in his utopian Salomon's House. Peiresc's protest to Cardinal Francesco Barberini (31 January 1635) on behalf of Galileo is typical of his foresight: he saw that in the long run an adverse judgment would profit no man, neither the cause of religion nor the cause of truth, and that Galileo would be a martyr, as Socrates had been, to forces of darkness and ignorance if he were right and to a gospel professing mercy if he were wrong. Like many men of science, Peiresc may be described as a skeptic, which indicates merely that he reserved judgment, awaiting truth as time reveals it. He was a product of the Renaissance in his comprehensiveness, his delight in beauty, and his spontaneous vitality. He took pleasure in old books and coins and in collecting plants and animals.

Peiresc's gardens at Belgentier were in their day the third largest in France, surpassed only by those of the king at Paris and at Montpellier. Peiresc is known to have had jasmine from India, guaiac from South America, Persian lilacs, Egyptian papyrus, varieties of myrtle, ginger, lentiscus, and *polianthes tuberosa*, as well as foreign grapes. He was fond of cats, introducing the Angora cat to Europe, and briefly possessed an elephant and a type of antelope described as an "alzaron."

Gabriel Naudé, Mazarin's bibliophile librarian, who described Peiresc's house at Aix as a "marché très fréquenté," where one could see "des marchandises très précieuses des deux Indes, Éthiopie, Grèce, Allemagne, Italie, Espagne, Angleterre . . . aucun navire n'entrait dans les ports de France sans apporter pour Peiresc des statues, des manuscrits samaritains, coptes, arabes, hébreux, chinois, grecs, les restes de l'antiquité la plus reculée."[10]

Although Peiresc's ideas were often vague and their theoretical basis imprecise, the spontaneity of his reactions to events and observations led to questions that Gassendi, in particular, deemed worthy of consideration; and he often developed these ideas in directions that Peiresc could neither foresee nor exploit. Bloch has suggested that this combination of two very different intellects was fruitful not only in Gassendi's thought but also in the evolution of science in France. It is probable that the organization of the amateurs of science in the house of Habert de Montmor, where Gassendi spent his last years, was a by-product of the extended periods during which Gassendi participated in or witnessed the intense and sometimes ill-coordinated investigations carried out at the Hôtel Callas and at Belgentier.

Perhaps not skillful himself, Peiresc did not withdraw, as some do, into bookish speculation, but rather drew on the talents of the skilled. His work for science was a natural extension of his taste for the arts and erudition. Two very different men summed up his work: soon after Peiresc's death, J.-L. Guez de Balzac wrote: "Dans une fortune médiocre, il avait les pensées d'un grand seigneur"; and at the end of the century, Pierre Bayle stated: "Jamais homme ne rendit plus de services à la République des Lettres que celui-ci."[11]

NOTES

1. P. Humbert, "Joseph Gaultier de la Villette, astronome provençal," in *Revue d'histoire des sciences et de leurs applications*, **1** (1948), 314–342.
2. *Correspondance du P. Marin Mersenne*, Mme Paul Tannery et al., eds., I–VII (Paris, 1932–   ), contains much well-documented information on Peiresc and his interests.
3. P. Humbert, "A propos du passage de Mercure, 1631," in *Revue d'histoire des sciences et de leurs applications*, **3** (1950), 27 ff., discusses part of the text of a letter of Peiresc to Gassendi, 22 Dec. 1631, that Tamizey de Larroque left unpublished, doubtless as "trop scientifique."
4. *Correspondance du Mersenne*, III, 156.
5. *Correspondance du Mersenne*, II, 132 ff.
6. *Ibid.*, IV, 208.
7. *Correspondance du Mersenne*, III, 216–217.
8. Also in Gassendi, *Opera*, V, 300–301.
9. Cf. P. Humbert, "Les études de Peiresc sur la vision," in *Archives internationales d'histoire des sciences*, **4** (1951), 654–659.
10. Quoted without source by Isaac Uri, *François Guyet* (Paris, 1886), 41.
11. Pierre Bayle, *Dictionnaire historique et critique*, III (Paris, 1720), 2217. Jean-Louis Gues de Balzac to F. Luillier, 15 Aug. 1640, in *Oeuvres*, L. Moreau, ed., I (Paris, 1854), 474–478.

BIBLIOGRAPHY

No published works by Peiresc are known to exist. His correspondence has been collected in *Lettres de Peiresc*, P. Tamizey de Larroque, ed., Documents Inédits sur l'Histoire de France, 7 vols. (Paris, 1888–1898); 10 vols. were originally planned. This ed. is difficult to use because the editor omitted many passages and sometimes whole letters as "trop scientifique"; also, the classification by correspondents does not facilitate the establishment of historical or biographical sequence. Letters received by Peiresc, the originals of which are scattered, appeared in twenty-one separately annotated publications; they have been reprinted in *Les correspondants de Peiresc*, Tamizey

de Larroque, ed., 2 vols. (Geneva, 1972). See also *Correspondants de Peiresc dans les anciens Pays-bas*, R. Lebègue, ed. (Brussels, 1943).

Francis W. Gravit, *The Peiresc Papers*, University of Michigan Contributions in Modern Philology no. 14 (Ann Arbor, Mich., 1950), lists 193 separate items, ten MSS now lost, and sixty-two secondary MSS, mostly copies. It lists neither the 200 or more ancient and medieval MSS nor the 5,000 books in Peiresc's library. Of the MSS that Gravit lists, nos. 18, 47, 65, 76, 113, 129, 132, 133, 145, and 146 seem to be the most valuable for the historians of various sciences. Gravit also wrote a substantial unpublished diss., "Peiresc, Patron of Scholars" (Ph.D. diss., Univ. of Michigan, 1939).

The basis of any biographical study must be Gassendi's *Viri illustris Nicolai Claudii Fabricii de Peiresc . . . vita* (Paris, 1641; The Hague, 1651, 1655), trans. by W. Rand as *The Mirrour of True Nobility and Gentility* (London, 1657); this work also appeared in Gassendi's *Opera omnia*, V (Lyons, 1658) and was abridged unfaithfully by J. B. Requier, *Vie de N. Peiresc, conseiller au Parlement de Provence* (Paris, 1770). Pierre Borel, a physician from Castres, added factual material to Gassendi's *Vita* in *Auctorium ad vitam Peirescii* (The Hague, 1655).

Pierre Humbert, *Un amateur: Peiresc* (Paris, 1933), and G. Cahen-Salvador, *Un grand humaniste: Peiresc* (Paris, 1951), are the most extensive studies in recent times. Each has a bibliography with reference to original documents and to studies of detail.

Olivier René Bloch, in *La philosophie de Gassendi: nominalisme, matérialisme et métaphysique*, International Archives of the History of Ideas no. 38 (The Hague, 1971), remarks on both the importance of Peiresc's cosmopolitan outlook and his emphasis on observation and experiment in the development of Gassendi's thinking. Seymour L. Chapin, "Astronomical Activities of Nicolas Claude Fabri de Peiresc," in *Isis*, **48** (1957), 13–29, is a good survey of its field, based on the printed material but apparently without fresh contact with the MSS.

Other articles are listed by Alexandre Cioranescu in his *Bibliographie de la littérature française du 17ᵉ siècle*, III (Paris, 1965–1966), nos. 53.790–53.925.

Harcourt Brown

PEKELHARING, CORNELIS ADRIANUS (*b.* Zaandam, Holland, 19 July 1848; *d.* Utrecht, Holland, 18 September 1922), *physiological chemistry, medicine.*

Pekelharing was the son of Cornelis Pekelharing, a physician, and Johanna van Ree. In 1866 he became a medical student at the University of Leiden, where his interests ranged from chemistry and physics to social and religious questions. The well-known physiologist Adriaan Heynsius appointed him his assistant from 1871 to 1876. A skilled and conscientious laboratory worker, Pekelharing became a licensed physician in 1872 and established a practice in Leiden. In 1873 he married Willemina Geertruida Campert; they had five children. Pekelharing received the M.D. degree in 1877 after a masterful defense of his dissertation on the determination of urea in blood and tissues.

Pekelharing returned to the laboratory after his appointment in 1878 as instructor in physiology and anatomy at the School of Veterinary Medicine in Utrecht. Here he developed his research interests, many of which he maintained throughout his life, without neglecting his teaching duties. Protein digestion in the stomach and its end product, which he called pepton, were thoroughly investigated by Pekelharing's group. His many publications on this subject became the center of much controversy in Europe. He also elucidated the protein nature of pepsin. The study of anthrax led him to take up bacteriology. He went to Leipzig to study with Julius Cohnheim, a student of Virchow's, and in 1886 visited Koch in Berlin. In 1881 Pekelharing was appointed professor of pathology and anatomy at the University of Utrecht. In 1888 his assignment was changed to physiological chemistry and histology. With the neurologist C. Winkler and a growing number of medical students, Pekelharing studied the role of leukocytes in inflammation and phagocytosis. He continued his work with anthrax bacilli and spores. Subsequent investigations included the role of calcium in blood clotting, arteriosclerosis, urine pigments, enzyme precursors, intestinal iron absorption, and hemoglobin and glycogen in oysters.

Pekelharing became interested in nutrition in 1886, when he and Winkler were sent to the Dutch East Indies to investigate the cause of beriberi. They were joined by Christiaan Eijkman. In accordance with the general conviction of his time, Pekelharing looked for and found a microorganism that he believed to be the causative agent of beriberi. Nevertheless, he was not totally satisfied with his findings; and upon returning to Holland in 1887, he convinced the Dutch government of the need for a medical research laboratory in Batavia (now Djakarta). Eijkman was appointed its first director. With the assistance of Grijns, Eijkman subsequently discovered the involvement of a dietary factor in the development of beriberi.

In 1905 Pekelharing reported his efforts to maintain mice on a diet of purified nutrients. Only with the addition of milk as a dietary supplement was he able to do so; and he thus concluded that milk contains an unknown substance which, even though in very small amounts, is essential to the diet. Without these substances, he stated, the organism lacks the ability to metabolize the major nutrients. Since his reports were published in Dutch, Pekelharing's findings received little attention. Nevertheless, he became increasingly

interested in the new science of nutrition, and in 1908 his monograph on proteins as food appeared. During World War I he focused his attention on problems of mass feeding. He was also instrumental in the founding of the Netherlands Institute of Nutrition.

In 1918 Pekelharing reached retirement age, but he continued studying and writing. An outstanding teacher and scholar, he was active in the Royal Netherlands Association for the Advancement of Medicine, served twice as its president, and was editor of its journal for many years. Pekelharing was in the forefront of the fight against alcoholism. His many honors included membership in the Royal Netherlands Academy of Sciences.

BIBLIOGRAPHY

I. ORIGINAL WORKS. A complete bibliography of Pekelharing's papers and monographs is given in J. M. Baart de la Faille et al., "Leven en werken van Cornelis Adrianus Pekelharing 1848–1922," in N.V. A. Oosthoek's Uitgeversmij (Utrecht, 1948), 211–217. His major works include "Sur le dosage de l'urée," in Archives néerlandaises des sciences exactes et naturelles, 10 (1875), 56; "Recherches sur la nature et la cause du beriberi et sur les moyens de la combattre," in Baillière et Fils (1888), written with Winkler; "Ueber eine neue bereitungsweise des Pepsins," in Zeitschrift physiologische Chemie, 22 (1896), 233; "On the Proteins of the Glandular Thymus," in Proceedings of the Section of Sciences. K. Nederlandse akademie van wetenschappen, 3 (1901), 383; and "Der Eiweiszverbrauch im Tierkorper," in Zentralblatt für die gesamte Physiologie und Pathologie des Stoffwechsels, n.s. 4 (1909), 289.

II. SECONDARY LITERATURE. On Pekelharing and his work, see A. M. Erdman, "Cornelis Adrianus Pekelharing," in Journal of Nutrition, 83 (1964), 1–9; and H. Zwaardemaker, "L'oeuvre de C. A. Pekelharing jusqu'à son septuagenaire," in Archives néerlandaises de physiologie de l'homme et des animaux, 2 (1918), 451–464.

ANNE MARIE ERDMAN

PELETIER, JACQUES (b. Le Mans, France, 25 July 1517; d. Paris, France, July 1582), *mathematics, medicine.*

Peletier was the ninth of fifteen children born to Pierre Peletier, a barrister in Le Mans, and Jeanne le Royer. His family, educated in theology, philosophy, and law, wanted him to pursue these diciplines. He therefore studied philosophy at the Collège de Navarre (Paris) and read law for five years in Le Mans. But when he became secretary in the late 1530's to René du Bellay, bishop of Le Mans, he decided that his interests were not in philosophy or law.

In 1541 Peletier published *L'art poëtique d'Horace, traduit en vers François*, the preface of which pleaded for a national language, thus anticipating the ideas of the later Pléiade. He also studied Greek, mathematics, and later medicine, always as an autodidact. In 1543 he became rector of the Collège de Bayeux in Paris, a post that soon bored him. He therefore left Paris in 1547 and lived as a vagabond. Among the cities he visited were Bordeaux, Poitiers, Lyons, Paris, and Basel. Working alternately as a teacher in mathematics and as a surgeon, he devoted his life to poetry and science. Peletier shared with the Pléiade, a group of seven poets whose leader was Pierre de Ronsard, a desire to create a French literature. He also stated that French was the perfect instrument for the sciences and planned to publish mathematical books in the vernacular. Temporarily, however, he published only in Latin (1557–ca. 1572) because no one would accept his somewhat peculiar French orthography. Peletier's poetry had scientific aspects, especially the second part of *L'amour des amours* (1555), in which he published descriptive-lyric verses on nature, natural phenomena, and astronomy which revealed the influence of Lucretius. He also published two minor works on medicine.

In 1545 Peletier published a short comment on Gemma Frisius' *Arithmeticae practicae methodus facilis*. In 1549 the *Arithmétique* appeared. In this work Peletier tried to satisfy both the theoretical requirements and the practical needs of the businessman. This topic had been previously discussed in Latin by C. Tunstall and Gemma Frisius, but Peletier was the first to combine both in a textbook in the vernacular. Peletier wrote *L'algèbre* (1554) in French in his own orthographic style. In this work he adopted several original and ingenious ideas from Stifel's *Arithmetica integra* (1544) and showed himself to have been strongly influenced by Cardano. Peletier's work presented the achievements already reached in Germany and Italy, and he was the first mathematician to see relations between coefficients and roots of equations.

In the *In Euclidis elementa demonstrationum* (1557) Peletier rejected the method of superposition as nongeometric. His arguments for this opinion, however, were used for the contrary view by Petrus Ramus. A long note on the angle of contact—in Peletier's view not a finite quantity and not an angle at all—was the starting point for various disputes, especially with C. Clavius. This work was vehemently criticized by J. Buteo.

Translations into French or Latin and several reprints, especially of the French editions, indicate that Peletier's works were quite successful. His other mathematical publications were devoted to such topics as the measurement of the circle, contact of straight lines and curves with curves, and duplication of the cube. The basic ideas in these publications often originated in Peletier's discussions with Buteo and Clavius.

BIBLIOGRAPHY

I. ORIGINAL WORKS. An incomplete bibliography of Peletier's works is given by C. Jugé in *Jacques Peletier du Mans* (Paris, 1907). His major works include *Arithmeticae practicae methodus facilis per Gemmam Frisium, huc accesserunt Peletarii annotationes* (Paris, 1545), subsequent eds. between 1549 and 1557; *L'arithmétique departie en quatre livres* (Poitiers, 1549), with later eds. between 1552 and 1969; *L'algebre departie en deus livres* (Lyons, 1545; 3rd ed., 1620), with a Latin trans. as *De occulta parte numerorum* (Paris, 1560); *In Euclidis elementa geometrica demonstrationum libri sex* (Lyons, 1557; 2nd ed., Geneva, 1610), with a French trans. (Geneva, 1611); *Commentarii tres, primus de dimensione circuli, secundus de contactu linearum, tertius de constitutione horoscopi* (Basel, 1563), an ed. of the second part also appeared (Paris, 1581); *Disquisitiones geometricae* (Lyons, 1567); and *In C. Clavium de contactu linearum apologia* (Paris, 1579).

Peletier's letter *ad Razallium* against Buteo was published at the end of *De occulta parte numerorum*. The *In Euclidis elementa* of 1573, mentioned by Jugé, is not Peletier's work but one of the many eds. "cum praefatione St. Gracilis."

II. SECONDARY LITERATURE. C. Jugé (see above) provides a biography of Peletier. For works on his poetry, see A. Boulanger, *L'art poétique de Jacques Peletier* (Paris, 1830); *Dictionnaire des lettres françaises, le seizième siècle* (Paris, 1951), 561–563; F. Letessier, "Un humaniste Manceau: Jacques Peletier (1517–1582)," in *Lettres d'humanité. Bulletin de l'Association Guillaume Budé*, supp. 9 (1950), 206–263; H. Staub, *Le curieux désir. Scève et Peletier du Mans poètes de la connaissance* (Geneva, 1967); and D. B. Wilson, "The Discovery of Nature in the Work of Jacques Peletier du Mans," in *Bibliothèque d'humanisme et renaissance*, **16** (1954), 298–311. His contacts with the Pléiade are discussed by H. Chamard in *Histoire de la Pléiade* (Paris, 1961–1963), *passim*, and in L. C. Porter's intro. to the repr. of Peletier's *Dialogue de l'ortografe e prononciation françoese* (Poitiers, 1550; repr. Geneva, 1966).

Peletier's mathematics is discussed by H. Bosmans, "L'algèbre de J. Peletier du Mans," in *Revue des questions scientifiques*, **61** (1907), 117–173, which uses the 1556 ed. of *L'algèbre*; N. Z. Davis, "Sixteenth-century French Arithmetics on the Business Life," in *Journal of the History of Ideas*, **21** (1960), 18–48; V. Thebault, "A French Mathematician of the Sixteenth Century: Jacques Peletier (1517–1582)," in *Mathematics Magazine*, **21** (1948), 147–150; M. Thureau, "J. Peletier, mathématicien manceau au XVIᵉ siècle," in *La province du Maine*, 2nd ser., **15** (1935), 149–160, 187–199; and J. J. Verdonk, *Petrus Ramus en de wiskunde* (Assen, 1966), 264–268; on his contacts with P. Nunez, see L. de Matos, *Les Portugais en France au XVIᵉ siècle* (Coimbra, 1952), 123–125.

J. J. VERDONK

PELL, JOHN (*b*. Southwick, Sussex, England, 1 March 1611; *d*. London, England, 12 December 1685), *mathematics*.

Pell was the son of John Pell, vicar of Southwick, and Mary Holland, who both died when he was a child. In 1624 he left Steyning School in Sussex for Trinity College, Cambridge. He received the B.A. in 1629 and the M.A. in 1630. By the latter year he was assistant master at Collyer's School in Horsham, and then at Samuel Hartlib's short-lived Chichester academy. On 3 July 1632 he married Ithamaria Reginalds, the second daughter of Henry Reginalds of London. In 1638 the Comenian group, of which Hartlib was a leading member, arranged his move to London; and he soon won a reputation for his knowledge of mathematics and languages. The success of the group was thwarted by political developments; not wanting to take a church living, Pell had to emigrate to secure a mathematical post. In December 1643 he became professor of mathematics at Amsterdam and, in 1646, at the newly opened academy in Breda. From 1654 to 1658 he was a Commonwealth agent in Zurich. After the Restoration, Pell became rector of Fobbing in Essex, vicar of Laindon, and then chaplain to Gilbert Sheldon, bishop of London.[1] For a time he lived with a former pupil at Brereton Hall. He died in London in poverty.[2]

Opinions about Pell's significance as a mathematician have always varied, and a full assessment will be impossible until his writings have been collected and analyzed. Houzeau and Lancaster, and others, have suggested that his "Description and Use of the Quadrant" (1628) and other works were printed. His first publication was undoubtedly *Idea of Mathematics*, which appeared anonymously after circulating in manuscript in an early version before 1630. The work was published in Latin and in English in 1638 and republished as part of John Dury's *The Reformed Librarie-Keeper* in 1650. The *Idea* won Pell "a great deal of repute both at home and abroad" and led to his post at Amsterdam. His arguments are clearly very close to those of Bacon, Comenius, and their followers but also have a large personal element. The tract stressed the importance of mathematics and proposed "the writing of a *Consilarius Mathematicus*, the establishment of a public library of all mathematical

books, and the publication of three new treatises." A copy was sent by Pell's patron, Theodore Haak, to P. Mersenne, who circulated the work; Descartes replied approvingly.[3]

At Amsterdam, Pell's fame was enhanced by his *Controversiae de vera circuli mensura* (1647), which attacked C. S. Longomontanus and earned the approbation of Roberval, Hobbes, Cavendish, Cavalieri, Descartes, and others.[4] In 1647 Pell read his *oratio inauguralis* at Breda and was praised by an eyewitness[5] for the excellence of his delivery and his explanation of "the use and dignity" of mathematics.

Most mathematicians know of Pell through his equation[6] $x^2 = 1 + Ay^2$. Some suggest that Euler mistakenly attributed to Pell some work of William Lord Brouncker, but the equivalent equation $x = 12yy - zz$ occurs in Thomas Brancker's 1668 translation, *An Introduction to Algebra*,[7] of J. H. Rahn's *Teutsche Algebra oder algebraische Rechenkunst* (Zurich, 1659). Pell edited the latter part of the translation. Aubrey, however, stated that "Rhonius was Dr. Pell's scholar at Zurich and came to him every Friday night after he had writt his post-lettres" and claimed that the *Algebra* was essentially Pell's work.[8] If this statement is accepted, Pell should also be credited with innovations in symbolism (particularly $\div$) and with setting out equations in three columns (two for identification and one for explanation), otherwise credited to Rahn. Without further evidence, it is best to assume that there was joint responsibility for these innovations and that Pell's contemporary reputation as a mathematician, and particularly as an algebraist, was not unearned.

NOTES

1. His academic reputation is indicated by his D.D. at Lambeth and election as a fellow of the Royal Society in 1663.
2. Some of his books and manuscripts were acquired by Richard Busby, master of Westminster School, which still has some of his books. The MSS came to the British Museum via Thomas Birch; other manuscripts were left at Brereton.
3. Wallis, "An Early Mathematical Manifesto," *passim*.
4. Dijksterhuis, "John Pell," p. 293.
5. Edward Norgate, quoted by D. Langedijk in " ' De illustre schole ende Collegium Auriacum ' te Brede," p. 131.
6. Whitford, *The Pell Equation*, p. 2. Cajori does not accept or even refer to Whitford's argument.
7. *Loc. cit.*, p. 143, no. 34. The relation between the 1659 and 1668 eds. is discussed in more detail in a forthcoming article by C. J. Scriba.
8. Aubrey's biography was partly checked by Pell himself and later supplemented by Haak.

BIBLIOGRAPHY

I. ORIGINAL WORKS. For a 1967 repr. of the 1638 *Idea* and the 1682 and 1809 versions, see Wallis. Two other anonymous works not cited in the text are *Easter Not*

Mistimed (London, 1664) and *Tabula numerorum quadratorum* (London, 1672). See notes for a reference to his many MSS, often mistakenly said to have been published.

II. SECONDARY LITERATURE. Writings on Pell and his work are J. Aubrey's biography of Pell, Bodleian MS 6 f.53, printed in *Brief Lives*, A. Clark, ed., II (Oxford, 1898), 121–131, and in O. L. Dick's 1949–1950 ed.; P. Bayle, in *A General Dictionary, Historical and Critical*, J. P. Bernard et al., eds., VIII (London, 1739), 250–253; T. Birch, *The History of the Royal Society of London*, IV (London, 1757), 444–447; F. Cajori, "Rahn's Algebraic Symbols," in *American Mathematical Monthly*, **31** (1924), 65–71; E. J. Dijksterhuis, "John Pell in zijn strijd over de rectificatie van den cirkel," in *Euclides*, **8** (1932), 286–296; J. C. Houzeau and A. Lancaster, *Bibliographie générale de l'astronomie* (Brussels, 1882–1887, repr., 1964); and D. Langedijk, " 'De illustre schole ende Collegium Auriacum' te Brede," in G. C. A. Juten, ed., *Taxandria: Tijdschrift voor Noordbrabentsche geschiedenis en volkskunde xlii*, III (Bergen op Zoom, 1932), 128–132.

For additional information see C. de Waard's biography of Pell in *Nieuw Nederlandsch biografisch woordenboek*, III (1914), cols. 961–965; and "Wiskundige bijdragen tot de pansophie van Comenius," in *Euclides*, **25** (1950), 278–287; P. J. Wallis, "An Early Mathematical Manifesto —John Pell's *Idea of Mathematics*," in *Durham Research Review*, no. 18 (1967), 139–148; E. E. Whitford, *The Pell Equation* (New York, 1912); and A. Wood, in *Fasti Oxonienses*, P. Bliss, ed., I (London, 1815), cols. 461–464, and in 1967 fasc., repr. (New York–London).

P. J. WALLIS

PELLETIER, BERTRAND (*b.* Bayonne, France, 31 July 1761; *d.* Paris, France, 21 July 1797), *chemistry*.

The son of Bertrand Pelletier, a pharmacist, and Marie Sabatier, Pelletier was apprenticed to his father until 1778. He then continued his pharmaceutical training with Bernard Coubet in Paris, where he was befriended by Jean d'Arcet, an acquaintance of his father, and Pierre Bayen. In 1782 he became d'Arcet's assistant and lecture demonstrator at the Collège de France, and soon published his first paper, an account of the preparation and properties of arsenic acid.

In 1783, on d'Arcet's recommendation, H. M. Rouelle's widow appointed Pelletier manager of her pharmacy in the rue Jacob. He qualified as a master pharmacist in 1784, the year of his marriage to Marguerite Sédillot, and then bought the Rouelle pharmacy (which is still called the Pharmacie Pelletier). Chemical investigations took up much of his time, however, so he assigned the management of the business to his elder brother, Charles, also a master pharmacist. From 1783 Pelletier was registered as a student in the Paris Faculty of Medicine, but he did not graduate there. In 1790, however, he made two journeys to Rheims, where he passed the examinations for his doctorate in medicine. The Paris Académie des Sciences elected Pelletier in 1792, the year before its suppression.

Pelletier was a skillful chemist, concerned more with experiment than with theory. He spent much time checking and extending the researches of others, and therefore made few original contributions to chemistry. His experimental skill was demonstrated in 1784, when, at the suggestion of the crystallographer J.-B. Romé de l'Isle, he prepared crystals of several very soluble or deliquescent salts by using the techniques of slow evaporation and seeding. In 1785 he confirmed C. W. Scheele's discovery that the gas now called chlorine is obtained from the reaction between marine (hydrochloric) acid and manganese calx (dioxide) and, independently of C. L. Berthollet, he came to the incorrect conclusion that it was a compound of marine acid and oxygen. But unlike Berthollet, Pelletier did not yet accept Lavoisier's antiphlogistic theory; and he followed Scheele in describing chlorine as "dephlogisticated marine acid," even though he thought it was a compound containing marine acid. By the action of chlorine on alcohol Pelletier obtained a product that he regarded as marine ether (ethyl chloride), but it must have been chloral. He considered other "ethers"—sulfuric (diethyl ether), nitrous (ethyl nitrite or nitrate), and acetous (ethyl acetate)—to be formed by the action on alcohol of the dephlogisticated air present in the various acids. He continued to call oxygen "dephlogisticated air" until late in 1787; and it seems that, like his mentor d'Arcet, he accepted Lavoisier's theory and the new nomenclature only after some hesitation.

Pelletier's important series of researches on phosphorus (1785–1792) included the preparation, for the first time, of the phosphides of most metals. The slow oxidation of phosphorus over water yielded an acid that he thought was phosphorous acid, but he did not complete his examination of its salts; in 1816 P. L. Dulong showed that it was in fact hypophosphoric acid.

Copper was scarce during the French Revolution, and in 1790 Pelletier devised a process for recovering it from bell metal (an alloy of copper and tin) by oxidation with manganese dioxide, which attacked the tin before the copper; a good yield was obtained, but A. F. Fourcroy's method of atmospheric oxidation proved to be cheaper and was generally preferred. Pelletier became a member of the Bureau de Consultation des Arts et Métiers and of the Commission Temporaire des Arts, and for both organizations he helped to prepare reports on crafts and industries of national importance. The best-known of these was the report recommending Nicolas Leblanc and Michel-Jean Dizé's process for soda manufacture; but Pelletier also contributed to reports on M. E. Janety's malleable platinum, Armand Seguin's method for tanning leather, the production of soap, and the repulping of waste paper.

Pelletier was appointed assistant professor when the École Polytechnique opened in 1794, and he helped Guyton de Morveau with the course on mineral chemistry. In 1795 he was elected to the Institut de France. He served on a commission of the Institut that investigated methods of refining and analyzing saltpeter for gunpowder manufacture, but he was already suffering from pulmonary tuberculosis and died before the work was finished.

BIBLIOGRAPHY

I. ORIGINAL WORKS. Most of Pelletier's publications are listed in Poggendorff, II, col. 392. His reports on the extraction of copper from bell metal, the manufacture of soda, and the production of soap were written jointly with Jean d'Arcet and others; details are given in the bibliography of d'Arcet, in *Dictionary of Scientific Biography*, III, 561. These and other reports were reprinted with all of Pelletier's scientific articles in a collected ed. published by his brother Charles Pelletier and his brother-in-law Jean Sédillot, as *Mémoires et observations de chimie de Bertrand Pelletier*, 2 vols. (Paris, 1798). As a member of various commissions of the Institut de France, Pelletier contributed to several reports that were eventually published in *Procès-verbaux des séances de l'Académie [des Sciences]*, I (Hendaye, 1910), 32–33 (examination of some minerals with d'Arcet and Claude Lelievre); 71–75 (analysis of an alloy, with P. Bayen *et al.*); 228–230 (report on a memoir by N. Deyeux, with L. B. Guyton de Morveau *et al.*); 244–256 (report on the refining and analysis of salpeter, with Guyton *et al.*).

In 1792 Pelletier joined the editorial board of *Annales de chimie*; he was also a coeditor of the short-lived *Journal d'histoire naturelle*, which was founded by Lamarck and others in 1792 and ceased publication in the same year.

II. SECONDARY LITERATURE. See Sédillot, "Éloge de B. Pelletier," in *Mémoires et observations de chimie de Bertrand Pelletier*, I (Paris, 1798), vii–xxvii. Further information, including the correct date of birth, is given by P. Dorveaux, "Bertrand Pelletier," in *Revue d'histoire de la pharmacie*, **6** (1937), 5–24.

W. A. SMEATON

PELLETIER, PIERRE-JOSEPH (*b*. Paris, France, 22 March 1788; *d*. Paris, 19 July 1842), *chemistry, pharmacy*.

Following the example of his distinguished father, Bertrand Pelletier, Pierre-Joseph chose pharmacy and chemistry as his lifework. In 1810 he qualified as a pharmacist after achieving a brilliant scholastic record at the École de Pharmacie. Pelletier earned his *docteur ès sciences* in 1812 and in 1815 was named assistant professor of natural history of drugs at the École de Pharmacie, but he lectured mainly on mineralogy, which he had studied under R. J. Haüy. In 1825 he was promoted to full professor of natural history, succeeding Pierre Robiquet, and in 1832 he became assistant director of the school. In addition to his academic responsibilities and research commitments, Pelletier directed a pharmacy on the rue Jacob and a chemical plant at Clichy.

Pelletier's early scientific efforts, mostly concerned with the analysis of gum resins and coloring matter in plants, culminated in 1817 with a brilliant work on the isolation of emetine in an impure form. His collaborator in this work was François Magendie. The investigation of gum resins, begun by Pelletier in a report on opopanax in 1811, was followed in the next two years by publications on sagapenum, asafetida, bdellium, myrrh, galbanum, and caranna gum. From 1813 to 1817 his articles dealt mainly with natural products (sarcocolla, toad venom, amber, olive gum) and coloring matter contained in red sandalwood, alkanet, and curcuma (written with H. A. Vogel). In 1817 the discovery by Pelletier and Magendie of the "matière vomitive" in ipecac root, named emetine by Pelletier and verified by animal experiments, was announced in a paper read before the Académie des Sciences.

The period from 1817 to 1821 was remarkably productive for Pelletier, who had acquired a new collaborator in Joseph-Bienaimé Caventou, a gifted young pharmacy intern attached to the Saint-Antoine Hospital in Paris. Their keen interest in the chemistry of natural products led them in 1817 to study the action of nitric acid on the nacreous material of human biliary calculi and the green pigment in leaves, which they were the first to name chlorophyll. In 1818 they obtained crotonic acid from croton oil and analyzed carmine in cochineal. Two years later, in 1820, Pelletier and Caventou isolated ambrein from ambergris. But it was their discovery of a number of plant alkaloids that brought them international fame: strychnine (1818); brucine (1819); veratrine (1819), independently of Karl Meissner; cinchonine, first obtained by B. A. Gomes in 1810 but again isolated and more extensively studied by Pelletier and Caventou (1820); quinine, the most important of their discoveries from a therapeutic standpoint (1820); and caffeine (1821), independently of Robiquet and Runge. Pelletier also published his own research on gold compounds (1820), piperine (1821), and various species of cinchona (1821).

For the remaining two decades of his life Pelletier continued his alkaloid and phytochemical investigations. He also studied the decomposition products of pine, resin, amber, and bitumen. In 1823 the results of a combustion analysis of nine alkaloids, undertaken by Pelletier and J.-B. Dumas, provided conclusive evidence for the presence of nitrogen in alkaloids, a fact that Pelletier and Caventou had earlier failed to ascertain. In 1832 Pelletier reported his discovery of narceine, a new opium alkaloid, to the Academy. The following year he published, with J. P. Couerbe, a study of picrotoxin. But relations between the two collaborators became embittered when Couerbe, Pelletier's former student and *chef des travaux* at his chemical plant, refused to support Pelletier's claim to priority in the isolation of thebaine, another opium alkaloid. Instead, Couerbe gave credit for the discovery of thebaine (called paramorphine by Pelletier) to Thiboumery, who had been employed by Pelletier as *directeur des travaux*. A happier association was subsequently established with Philippe Walter. This collaboration led to the publication of a number of interesting papers on oily hydrocarbons obtained from the destructive distillation of amber and bitumen. One of these studies (1837–1838), of an oily by-product of pine resin used in the manufacture of illuminating gas, resulted in their discovery of a substance that they designated as "rétinnaphte," now known as toluene (C_7H_8).

After 1821, Pelletier and Caventou still conducted a few investigations jointly, including further researches on strychnine and procedures for its extraction from nux vomica (1822); chemical examination of upas (1824); the manufacture of quinine sulfate (1827); and the isolation of cahinca acid, the bitter crystalline substance in cahinca root (1830), with André François. Although ably carried out, this work was overshadowed by their earlier discoveries. Among Pelletier's other collaborative efforts were an analysis, undertaken with Corriol, of a species of cinchona (*Cinchona cordifolia*), which enabled them to isolate aricine in 1829; a chemical examination of curare, with Petroz, in 1829; and a posthumous memoir on guaiacum with H. Sainte-Claire Deville, published in 1844.

Pelletier was named a member of the Paris Académie de Médecine in 1820 and was elected to the Académie des Sciences in 1840. In 1827 Pelletier and Caventou were awarded the Montyon Prize of 10,000 francs by the latter academy in recognition of their discovery of quinine. Both men were also honored in 1900 by an impressive statue erected on the boulevard St.-Michel, which was destroyed during the German occupation of Paris in World War II but was replaced by another monument dedicated in 1951. A firm defender of the established political order, Pelletier was, in Caventou's words, "partisan aussi sincère qu'éclairé de nos institutions monarchiques et constitutionnelles."

BIBLIOGRAPHY

I. Original Works. Pelletier's earlier papers include "Analyse de l'opopanax," in *Annales de chimie*, **79** (1811), 90–99; "Analyse du galbanum," in *Bulletin de pharmacie*, **4** (1812), 97–102; "Réflexions sur le tannin et sur quelques combinaisons nouvelles de l'acide gallique avec des substances végétales," in *Annales de chimie*, **87** (1813), 103–108, 218–219; "Examen chimique de quelques substances colorantes de nature résineuse," in *Bulletin de pharmacie*, **6** (1814), 432–453; "Mémoire sur la gomme d'olivier," in *Journal de Pharmacie*, **2** (1816), 337–343; and "Recherches chimiques et physiologiques sur l'ipécacuanha," in *Annales de chimie et de physique*, **4** (1817), 172–185, written with F. Magendie. For the most important joint publications of Pelletier and Caventou, see Alex Berman, "Caventou," in *Dictionary of Scientific Biography*, III, 160.

Other representative works by Pelletier, written alone or in collaboration, are "Faits pour servir à l'histoire de l'or," in *Annales de chimie et de physique*, **15** (1820), 5–26, 113–127; "Examen chimique du poivre (*Piper nigrum*)," *ibid.*, **16** (1821), 337–351; "Recherches sur la composition élémentaire et sur quelques propriétés caractéristiques des bases salifiables organiques," *ibid.*, **24** (1823), 163–191, written with J.-B. Dumas; "Note sur la caféine," in *Journal de pharmacie*, **12** (1826), 229–233; "Notice sur une nouvelle base salifiable organique venant du Pérou," *ibid.*, **15** (1829), 565–568, written with Corriol; "Examen chimique du curare," in *Annales de chimie et de physique*, **40** (1829), 213–219, written with Petroz; "Nouvelles recherches sur l'opium," *ibid.*, **50** (1832), 240–280; "Nouvelle analyse de la coque du Levant," *ibid.*, **54** (1833), 178–208, written with Couerbe; and "Examen chimique des produits provenant du traitement de la résine pour l'éclairage au gaz," *ibid.*, **67** (1838), 269–303, written with P. Walter. For a more complete listing of Pelletier's articles, see Royal Society *Catalogue of Scientific Papers*, IV, 806–810.

II. Secondary Literature. See A. Bussy, "Discours prononcé à la distribution des prix de l'École de pharmacie pour l'année 1842, suivi d'une notice sur feu Pelletier, par A. Bussy, secrétaire de l'École," in *Journal de pharmacie et de chimie*, 3rd ser., **3** (1843), 48–58; J. B. Caventou, "Discours prononcé sur la tombe de M. Pelletier," in *Bulletin de l'Académie royale de médecine*, **7** (1841–1842), 1011–1016; *Centenaire de l'École supérieure de pharmacie, 1803–1903* (Paris, 1904), 264–265, 283–284, 295–297, 354; Marcel Delépine, "Joseph Pelletier and Joseph Caventou," in *Journal of Chemical Education*, **28** (1951), 454–461; J.-B. Dumas, *Discours prononcé aux funérailles de M. Pelletier, le 22 juillet 1842* (Paris, 1842);

M. M. Janot, "Joseph Pelletier, 1788–1842," in *Figures pharmaceutiques françaises* (Paris, 1953), 59–64; J. R. Partington, *A History of Chemistry*, IV (London–New York, 1964), 244–245, 558, and *passim*; and Horst Real and Wolfgang Schneider, "Wer entdeckte Chinin und Cinchonin?" in *Beiträge zur Geschichte der Pharmazie*, **22**, no. 3 (1970), 17–19.

ALEX BERMAN

PELOUZE, THÉOPHILE-JULES (*b.* Valognes, Manche, France, 26 February 1807; *d.* Paris, France, 31 May 1867), *chemistry*.

Pelouze was the son of Edmond Pelouze, whose interests in industrial technology and invention were reflected in many publications. Pelouze decided originally on a career in pharmacy and after serving apprenticeships in pharmacies in La Fère and Paris, he was appointed to a hospital pharmacy internship at the Salpêtrière in Paris. An accidental meeting with Joseph-Louis Gay-Lussac, whose student and laboratory assistant he later became, changed the course of his life. Pelouze, undaunted by financial hardship, so impressed Gay-Lussac by his zeal and talents that Gay-Lussac became a lifelong patron and friend of the young chemist. In 1830 Pelouze secured a post teaching chemistry in Lille and shortly thereafter competed successfully for the position of assayer at the Paris mint. Further recognition and success came rapidly: he was elected to the Académie des Sciences (1837); he taught and was professor of chemistry at the École Polytechnique (1831–1846) and at the Collège de France (1831–1850); he was president of the Commission of the Mint (1848); he was a member of the Paris Municipal Council (1849); and he succeeded Gay-Lussac as consulting chemist at the Saint-Gobain glassworks (1850).

Beginning in 1830, Pelouze quickly established himself as an outstanding analytical and experimental chemist. His early investigations included studies of salicin (1830), with Jules Gay-Lussac; sugar beet (1831); fermentation (1831), with Frédéric Kuhlmann; conversion of hydrocyanic acid into formic acid; and decomposition of ammonium formate into hydrocyanic acid and water (1831). Later he investigated pyrogallic acid (1833); ethyl phosphoric acid (1833); discovered ethyl cyanide (1834); and found the correct formula for potassium dinitrosulfite (1835). In 1836 Pelouze and Liebig, with whom Pelouze had worked in Giessen, published a long memoir dealing with a number of organic substances, including their discovery of oenanthic ester and the corresponding acid. Noteworthy, too, were Pelouze's discovery of nitrocellulose (1838); oxidation of borneol to obtain camphor

(1840); synthesis of butyrin (1843), with Amédée Gélis; production of glycerophosphoric acid (1845); work on curare (1850), with Claude Bernard; and investigation of American petroleum (1862–1864), with Auguste Cahours. Interested mainly in empirical facts, Pelouze was, unfortunately, indifferent to the seminal chemical theories of his time.

In Paris, Pelouze founded the most important private laboratory school of chemistry in France. He trained many students and made his laboratory facilities available for the personal research of Bernard and other French and foreign chemists.

BIBLIOGRAPHY

I. ORIGINAL WORKS. Pelouze published at least 90 papers, alone or with other eminent chemists, most of which appeared in *Annales de chimie et de physique* or in *Comptes rendus . . . de l'Académie des sciences* and were repr. in other periodicals. For listings of these articles, see A. Goris *et al.*, *Centenaire de l'internat en pharmacie des hôpitaux et hospices civils de Paris* (Paris, 1920), 531–532; Poggendorff, II, 394–396, and III, 1015; and Royal Society *Catalogue of Scientific Papers*, IV, 810–814, and VIII, 583.

A major work by Pelouze was his *Traité de chimie générale*, 3 vols. and atlas (Paris, 1848–1850), written with E. Frémy. In later eds. the work was expanded, and it also appeared in a number of abridged versions.

II. SECONDARY LITERATURE. On Pelouze and his work, see J.-B. Dumas, *Discours et éloges académiques*, I (Paris, 1885), 127–198; C. von Martius, "Nekrolog auf Th. Julius Pelouze," in *Neues Repertorium für Pharmacie*, **17** (1868), 506–510; J. R. Partington, *A History of Chemistry*, IV (London–New York, 1964), 395 and *passim*; Warren De la Rue, "Proceedings of the Chemical Society," in *Journal of the Chemical Society*, **21** (1868), xxv–xxix; and Marc Tiffeneau, in A. Goris *et al.*, *Centenaire de l'internat en pharmacie des hôpitaux et hospices civils de Paris* (Paris, 1920), 615.

For Pelouze's relations with Bernard, see Joseph Schiller, *Claude Bernard et les problèmes scientifiques de son temps* (Paris, 1967), 63–64.

ALEX BERMAN

PELTIER, JEAN CHARLES ATHANASE (*b.* Ham, France, 22 February 1785; *d.* Paris, France, 27 October 1845), *physics*.

Peltier was born to a poor family; his father earned a living as a shoemaker. A quick intelligence and perseverance were displayed at an early age, as were mechanical skills. His formal education, however, was limited to the local schools. At the age of fifteen he was apprenticed to a German clockmaker named Brown in Saint-Quentin. He was refused permission to study and was generally ill-treated; after two years, in 1802, his father removed him from this position and apprenticed him in Paris to another clockmaker, named Métra, who had worked for A.-L. Bréquet. After an attempt to enter the army, which was prevented by his mother's disapproval, Peltier attracted the attention of Bréquet and entered his employ in 1804. In 1806 Peltier established his own shop and married a Mlle Dufant. The death of his wife's mother in 1815 brought him a modest inheritance, which was sufficient for their needs, and he retired.

Even while working at his trade, Peltier read broadly; when he retired, he devoted his attention to a wide range of studies and began to compose a Latin grammar. He then became interested in the phrenology of Franz Gall and was inspired, at age thirty-six, to study anatomy in order to obtain a more complete knowledge of the structure of the brain. He attended a number of vivisection demonstrations by Magendie, in which electricity was used to stimulate nerves. These demonstrations led Peltier to the study of electricity, which he pursued for the last twenty years of his life.

Peltier's first scientific paper was delivered to the Académie des Sciences in 1830. In it he showed that chemical effects can be obtained from a dry pile if the surface area of the plates is sufficiently large. This work also showed that Peltier had some understanding of the difference between current and voltage, with which electricians were to struggle for another ten years.

Stimulated by the work of Nobili, Peltier constructed a sensitive galvanometer to measure the conductivities of antimony and bismuth for small currents. Peltier's use of small samples of these nonductile materials was fortunate because the anomalous behavior of these materials led him to construct a thermoelectric thermoscope and to measure the temperature distribution along a series of thermocouple circuits. He discovered that a cooling effect can take place at one junction and excessive heating at the other. He then confirmed this discovery by using an air thermometer in place of the thermoscope.

Peltier did not pursue the effect he had discovered, and its importance was not fully recognized until after the thermodynamic work of William Thomson twenty years later. He did, however, write a paper on thermoelectric piles, and he spent some time studying the relations between static and dynamic electricity.

Peltier's remaining scientific endeavors fell into two major categories: microscopy and meteorology. His work in microscopy was an outgrowth of his anatomical and physiological interests; most of his observations were on various animalcules. In meteorology he made numerous measurements of electrical charges in the atmosphere and developed a theory that accounted for various cloud and storm formations on the basis of charge distribution. In 1842 he conducted a field trip to obtain such measurements. A cold resulting from this trip left him in a weakened condition, from which he never recovered.

BIBLIOGRAPHY

I. ORIGINAL WORKS. A bibliography of more than 60 papers by Peltier is contained in the Royal Society *Catalogue of Scientific Papers*, IV, 814–817. His discovery of the "Peltier effect" appears in "Nouvelles expériences sur la caloricité des courants électriques," in *Annales de chimie*, **56** (1834), 371–386.

II. SECONDARY LITERATURE. A memoir by Peltier's son, F. A. Peltier, *Notice sur la vie et les travaux scientifiques de J. C. A. Peltier* (Paris, 1847), was translated by M. L. Wood in *Report of the Board of Regents of the Smithsonian Institution* (1867), 158–202.

BERNARD S. FINN

PEMBERTON, HENRY (*b.* London, England, 1694; *d.* London, 9 March 1771), *physics, mathematics, physiology, medicine.*

Little is known of Pemberton's family or youth beyond the significant fact that he was introduced to mathematics at grammar school. He read, independently, Halley's editions of Apollonius and then traveled to Leiden to study medicine with Boerhaave. In Leiden he was further introduced to the work of Newton, the decisive event of his intellectual life. Pemberton interrupted his stay in Leiden to study anatomy in Paris and then returned to London about 1715 to attend Saint Thomas's Hospital. Although he took his degree at Leiden in 1719, he never practiced medicine extensively because of his delicate health. He did, however, serve for several years as professor of physics at Gresham College.

Pemberton's thesis, on the mechanism by which the eye accommodates to objects at different distances (1719), was his most important independent work. Treating the crystalline lens as a muscle, he argued that it accommodates to vision at varying distances by changes in shape. Students of physiological optics in the eighteenth century knew the work, and Pemberton ranks as one of the precursors of Thomas Young.

Pemberton's work on the mechanism of accommodation was nearly his last independent work, for he was determined to join the circle of Newton's epigones. He attempted, unsuccessfully, to approach the master through John Keill. But Richard Mead, Newton's friend and physician, showed Newton a paper in which Pemberton refuted Leibniz' measure of the force of moving bodies—an obsequious essay larded with references to "the great Sir Isaac Newton." Although the measure of the force of moving bodies was not an issue germane to Newtonian mechanics, Newton was apparently pleased with the attack on Leibniz. He made Pemberton's acquaintance; and Pemberton sought to cement the relation by contributing another obsequious essay on muscular motion, which converted itself into a panegyric on Newtonian method, to Mead's edition of Cowper's *Myotomia reformata*, completed in 1723 and published in 1724. When work on the third edition of Newton's *Principia* began late in 1723, Pemberton was the editor.

Pemberton devoted the major portion of his attention to the edition during the following two and a half years. He was a conscientious editor who carefully attended to the details of style and consistency, but nothing more substantive in the edition bears his stamp. The third edition of the *Principia* (1726) is the primary vehicle by which Pemberton's name has survived. The meagerness of his contribution, in comparison with the promise of his thesis at Leiden, suggests how deadening the role of sycophant can be.

Pemberton had labored assiduously to earn Newton's favor; apparently he intended to make his position near Newton the foundation of a career. Already he was at work on a popularization of Newtonianism for those without mathematics—*A View of Sir Isaac Newton's Philosophy*, which finally appeared in 1728 with prefatory assurances that Newton had read and approved it. He had also announced an English translation of the *Principia* and a commentary on it. In 1728 he received the Gresham position. Other aspiring young men had also courted Newton, however, and they chose to dispute the inheritance. John Machin, secretary of the Royal Society, sponsored and aided Andrew Motte's rival translation, which beat Pemberton's work to the press. Discouraged, he abandoned the commentary and virtually ended his career as a scientist.

Pemberton was thirty-five years old when Motte's translation appeared in 1729. Although he lived more than forty years more, he did almost nothing further to fulfill his earlier promise. During the 1730's, he was drawn into the fringes of the *Analyst* controversy on the foundations of the calculus. In 1739 the College of Physicians engaged him to reedit and translate their pharmacopoeia—*The Dispensatory of the Royal College of Physicians* (1746). He spent the following seven years on the project, attempting, he said, to purge it of the trifles that disgraced it. From the point of view of medical science, the job was undertaken too soon, and it had to be repeated again before the end of the century. At Gresham College he delivered courses of lectures on chemistry and physiology, which his friend James Wilson later published; both were minor works. Toward the end of his life he returned to his early love of mathematics and published four papers in the *Philosophical Transactions of the Royal Society*.

Pemberton was a man of deep friendships and broad learning. His first publication was a mathematical letter addressed to James Wilson, to whom, fifty years later, he left his papers. In his *View of Newton's Philosophy* he published a poem on Newton by a young friend, Richard Glover, whose continuing poetic efforts evoked pamphlets written by Pemberton praising Glover's work with a show of literary erudition. Glover's political connections led Pemberton to write an essay on political philosophy, which remained unpublished. He also wrote on weights and measures. He was known as a lover of music who never missed a performance of a Handel oratorio.

BIBLIOGRAPHY

I. Original Works. Pemberton's major works include *Dissertatio physica-medica inauguralis de facultate oculi qua ad diversas rerum conspectarum distantias se accommodat* (Leiden, 1719); *Epistola ad amicum de Cotesii inventis, curvarum ratione, quae cum circulo & hyperbola comparationem admittunt* (London, 1722); "Introduction. Concerning the Muscles and Their Action," in William Cowper, *Myotomia reformata*, Richard Mead, ed. (London, 1724); "A Letter to Dr. Mead . . . Concerning an Experiment, Whereby It Has Been Attempted to Shew the Falsity of the Common Opinion, in Relation to the Force of Bodies in Motion," in *Philosophical Transactions of the Royal Society*, **32** (1722), 57; *A View of Sir Isaac Newton's Philosophy* (London, 1728); *Observations on Poetry, Especially the Epic* (London, 1738); *The Dispensatory of the Royal College of Physicians* (London, 1746); *Some Few Reflections on the Tragedy of Boadicia* (London, 1753); *A Course of Chemistry* (London, 1771); and *A Course of Physiology* (London, 1773).

II. Secondary Literature. See I. Bernard Cohen, "Pemberton's Translation of Newton's *Principia*, With Notes on Motte's Translation," *Isis*, **54** (1963), 319–351; and *Introduction to Newton's 'Principia'* (Cambridge, Mass., 1971), 265–286; and the biographical sketch published by James Wilson as the preface to Pemberton's *Course of Chemistry*.

Richard S. Westfall

PENCK, ALBRECHT (*b.* Reuditz [near Leipzig], Germany, 25 September 1858; *d.* Prague, Czechoslovakia, 7 March 1945), *geomorphology, geology, paleoclimatology, hydrology, cartography.*

Born near the outer limit of the maximum southward advance of the Quaternary Scandinavian ice sheet, Penck took a lifelong interest in glacial deposits. In 1875 he entered the University of Leipzig to study natural sciences. In the same year Otto Torell delivered a forceful lecture at Berlin, which persuaded his audience that the boulder clay of the north European plain had been carried by a continental ice sheet and not by floating ice; he thus vindicated and perpetuated the ideas of A. Bernhardi (1832). Shortly thereafter Penck found and wrote about a northern "basalt" erratic embedded in the diluvium near Leipzig.

At Leipzig, Penck studied chemistry under Adolf Kolbe, geology under Hermann Credner, mineralogy and petrography under Ferdinand Zirkel, and botany under August Schenk. In 1877 Credner chose him to assist in a geological survey of Saxony, and Penck mapped on a scale of 1:25,000 the Grimma-Colditz area southeast of Leipzig. In 1879, from the detailed analysis of a sequence of glacial sedimentation (*Geschiebeformation*) that showed alternations of unbedded glacial clay (*Geschiebelehm*) and laminated sands and clays, he postulated at least three main ice advances, or glacial phases, interspersed with two interglacial periods during which rivers had laid down normal, bedded deposits. In the following year Penck worked under the geologist Karl von Zittel at Munich, which was near the outer (northern) limit of the maximum advance of the Alpine ice sheets. Penck's subsequent investigations into Quaternary geology were especially concerned with Alpine glaciation.

In 1882 Penck summarized his local fieldwork in *Die Vergletscherung der deutschen Alpen . . .*, which soon became a standard reference and was his *Habilitationsschrift* as *Privatdozent* in geography at the University of Munich (1883). Two years later Penck was elected to the chair of physical geography at the University of Vienna, where he stayed for nearly twenty years, developing a well-equipped geographical institute and achieving an international reputation. When not lecturing or writing, he collaborated with Eduard Brückner on extensive field studies in the Alpine valleys undertaken with a view to perfecting a chronology of ice sheet advances and retreats. The immediate result was a series of articles on the influence of glaciers on valley development and valley forms; the ultimate result was the classic three-volume *Die Alpen im Eiszeitalter* (1901–1909). During this period Penck traveled to England several times from 1883, to the Pyrenees (1884), and to Norway (1892) to study glacial features and other landform types for a general work on morphology. Reflected in numerous articles dealing with erosion and denudation, these travels culminated in his two-volume *Morphologie der Erdoberfläche* (1894). Penck subsequently made several journeys through Western Europe and visited Canada and the United States in 1898, the Balkans and Australia in 1900, and the United States and Mexico in 1904. Before leaving Vienna, he had taught many German and foreign scholars.

In 1906 Penck succeeded Ferdinand von Richthofen in the chair of geography at the Geographisches Institut of the University of Berlin. His inaugural lecture dealt with the fundamental importance of fieldwork in geographical studies ("Beobachtung als Grundlage der Geographie"), and as director of the institute for the next twenty years he set a fine example. In the winter of 1908–1909 he and his family visited the United States. Penck taught at Columbia University and lectured at Yale and other universities; he also met G. K. Gilbert in California. They returned to Germany via Hawaii, Japan, North China, and Siberia. (In the same scholar exchange program, W. M. Davis lectured on landforms at Berlin.) The last part of *Die Alpen im Eiszeitalter* appeared in 1909; up to this time, and for a few more years, his work at Berlin was virtually an extension of his studies at Vienna.

The outbreak of World War I was a turning point in Penck's thought rather than in his life. Apart from Quaternary problems, which had always interested him, his thinking became more geographical and less geomorphological. Directing more effort to sociopolitical themes, he showed an increasing interest in ethnographic, cultural, and nationalistic topics. In 1917–1918 he served as rector of the University of Berlin; and his inaugural discourse, "Über politische Grenzen," was a study of frontiers, especially European. The best frontiers, he thought, coincided with the living space (*Lebensraum*) indispensable to the life and security of a state. Germany had in part acquired *Lebensraum* but unfortunately had failed to retain the entire mineral basin of Lorraine. Now in 1917 Penck hoped that Germany would keep all the territories currently occupied so far as they were indispensable, and that it would further acquire colonies to furnish essential raw materials.

These and similar views led his friend Davis to write in a review of the *Festband* (1918) that was presented to Penck by his former students on his sixtieth birthday: "He used to be liked as much as admired but during the war some of his statements have lessened the esteem formerly felt for him" (*Geographical Review*, **10** [1920], 249). Penck played a considerable role in the revival of German nationalism after World War I; he was, for example, one of the chief advocates of the foundation of the Berlin *Volkshochschule*. The *Lebensraum* concepts (*Reichboden*, *Sprachboden*, *Volksboden*, and *Kulturboden*) and the ethnographic, cultural, and social surveys fostered and undertaken by Penck and others later proved disastrous but were then highly popular in Germany and had honored antecedents in the work of Friedrich Ratzel. Penck thus enjoyed great national esteem and achieved membership in the Berlin Mittwochsgesellschaft. His success was dimmed when his brilliant son Walther died of cancer in September 1923 at the age of 35. He supervised the publication (1924–1928) of his son's literary remains, including four articles, and *Die morphologische Analyse* (Stuttgart, 1924), an important contribution to the study of landforms.

About this time Penck's interests in oceanography yielded their best results. As director of the Institut für Meereskunde he was responsible for extending the oceanographic museum at the University of Berlin and was involved in the arrangements for the Meteor Expedition (1925), which, under A. Merz, made several sounding traverses in the South Atlantic. In 1926 Penck retired from the chair of geography at Berlin and was succeeded by his former student Norbert Krebs. Penck continued to live in Berlin, however, where he worked on geographical and editorial problems in connection with the geographical institute of the university.

By 1927 much of the wartime breach of friendship with Davis had been healed—largely owing to the death of Walther Penck, whom Davis greatly admired—and Penck spent some time lecturing in the United States, with the University of Arizona as his base. In 1928 he presided with distinction over both the centennial celebrations of the Berlin Gesellschaft für Erdkunde and the meetings of the Oceanic Conference. Most of his biographers consider 1928 "the peak of his career," but from a scientific viewpoint there can be no doubt that he reached his peak in 1909 or 1910. His Austrian work was full of scientific innovations and included his concepts for an international map on the scale of 1:1,000,000; his Berlin work was full of the less scientific branches of geography. In fact, for the last thirty years of his life he was more a regional geographer and demographer than an earth scientist.

The majority of Penck's approximately sixty-five articles and books written after his retirement concern Quaternary chronology, cartography, and population problems. There remained withal more than a hint of *Lebensraum*—evident in his description in 1934 of Krebs's important atlas *Deutsches Lebensraum in Mitteleuropa*, which Penck had initiated, and in his associated interest in political boundaries—as well as a tinge of regional geography (*Länderkunde*). Penck also wrote several competent biographies, including

those of Brückner (1928), Gilbert (1929), J. Partsch (1928), Richthofen (1930, 1933), and F. von Wieser (1929). His last projects involved the study, with a group of students, of the relationship between the potential productivity and possible number of inhabitants per unit area of land mass. During World War II, his house was damaged by bombs and he moved to Prague.

The assessment of Penck's contributions to the earth sciences is complicated by the change in his views. He did not hesitate to accept new theories or to recant his ideas. This development can be illustrated clearly from three facets of his work. First, in his concepts of regional geography he was an early follower of Richthofen. Thus, his "Das deutsche Reich" (1887) superimposed spatial distribution of various phenomena upon a detailed physical base, with the use of new physiographic terms such as *Alpenvorland* (foreland). But after 1914 his regional concepts changed rapidly to unit areas of landscape in which the visible repercussions of the natural and sociocultural environment allowed the establishment of core and fringe areas. Man's activities and his acquired traits and inherited characteristics entered more strongly into the spatial relationships. The concept of *Lebensraum* loomed large with what might be considered a regrettable chauvinistic veneer, and with strong hints at possible expansion and regrets at the noncoincidence of political, social, economic, and cultural distributions. Second, Penck changed his views considerably on the descriptive analysis of landforms. At first his elaborate empirical descriptions lacked any notable sequential development among the individual forms; but under the influence of Davis, Penck recognized the value of a "cyclic" or sequential progress. After 1918, he rejected Davis' theory; and, with his son Walther, he placed the rate and nature of uplift as dominant factors in the analysis of certain landforms. Third, Penck quite early agreed with Suess on the leading principle that secular variations in the relative altitude of land and sea were due to worldwide fluctuations of sea level (eustasism) rather than to crustal movements. By 1900 Penck had modified his views and had accepted independent crustal movement (regional or local) as a concomitant factor in elevating or depressing coastlines.

Assessing his contributions is complicated also by the wide range of geographical topics that he discussed. He published more than 400 books and articles, and many of the latter were issued separately in book form. Yet his chief scientific writings concerned four branches of the natural sciences: Quaternary geology and chronology, geomorphology, hydrology, and cartography.

In his Quaternary studies Penck's early work on the superficial deposits of the north German lowlands and of the Alpine valleys and piedmont plains increased the number of distinct epochs of glaciation to three or four. Prior to Penck's work only two such epochs were commonly accepted in continental Europe. Although James Geikie had enumerated five ice advances and James Croll (on climatic theories) had enumerated seven, Penck's suggestions were the first to be based on firm geological evidence. Following the publication of *Die Alpen im Eiszeitalter* the sedimentation evidence for at least four main ice advances in the Alps was indisputable. They were named Günz, Mindel, Riss, and Würm; the first three being right-bank alpine tributaries of the Danube and the last a tributary of the Isar River near Munich. Penck used for reference the capital letters in a wide-spaced alphabetical sequence, which could, if necessary, incorporate future discoveries of ice advances in a mnemonic order.

For nearly half a century this scheme provided a nomenclature and a time scale for European Pleistocene studies. *Die Alpen im Eiszeitalter* was a milestone in the history of the investigation of the Quaternary; its results, according to Davis (*Geographical Journal*, **34** [1909], 651), formed "an indispensable guide for all future students of the subject, a standard from which all future progress must be measured." The findings revealed the great length of the Riss-Würm interglacial and its mildness as compared with the present, as shown by the plant-bearing Hötting breccia near Innsbruck. The Würm (or last ice advance) had, on moraine evidence, experienced at least three significant pauses or stages of retreat. During glaciation, the permanent snowline of the Alps had advanced 1,200 meters; and its lowering was caused, Penck and Brückner believed, by a moderate decrease in the mean annual temperature and consequent increase in proportion of snowfall to total precipitation, rather than by an increase in the total precipitation.

Penck lived to see significant modifications to his scheme: an older (Donau) advance was subsequently added, which allowed the scheme to conform to postulated variations in insolation; various local terminologies not based on the Alps were adopted for Scandinavian and British ice advances; and the main glaciations, particularly the Würm, were more rigorously divided into stadials and interstadials. Among Penck's other contributions to glacial geology was the term "tillite," which he coined in 1906 for the ancient Dwyka moraines of the Permo-Carboniferous glaciation in South Africa.

Penck's main contributions in geomorphology were to the general classification of landforms, to knowledge of individual landform types, and to the significance of climatic change in landform-analysis. *Morphologie der Erdoberfläche*, the first unified text of geomorphology, followed in the tradition of Suess's *Das Antlitz der Erde* (1883–1909) and Richthofen's *Führer für Forschungsreisende* (1886); Penck acknowledged his debt to each and also to James Dana. Penck's work is divided into three parts. The first part deals mathematically—with the aid of numerous formulas and equations—with general surface morphology and with Penck's concepts of morphography and morphometry. The second part describes in detail the various forms (landforms) that are recognizable upon the surface of the earth and the various processes, endogenous and exogenous, at work in their genesis. Most of the principles stated here were basically familiar to students; but much of the information was new and the presentation was unified, ingenious, and scientific. The third includes the study of oceans, coastlines, and islands. The arrangement of particular sections in this work is similar to Richthofen's *Forschungsreisende*; but the battalion of facts, evidence, and computations more closely resembles Dana's geological manuals. As Charles Lapworth wrote admiringly (*Geographical Journal*, **5** [1895], 580), "the work is an encyclopaedia of facts and conclusions, admirably classified and digested; and affords, at the same time, a complete index to the literature of the subject."

Penck's attempt to construct a unified system of landform analysis and classification was of outstanding importance. He emphasized form or shape in relation to genetic processes rather than to functional processes, and he stated that fundamental types could be formed by many different processes. This Penckian system created or nurtured the German, as distinct from the American (Davisian), system of landform analysis. Penck's system was expounded more widely in 1895 at the Sixth International Geographical Congress in London, with a masterly summary in English. It stated that changes on the earth's surface result from erosion (true erosion and denudation), accumulation, and dislocation, which cause the formation of new surfaces and the destruction and alteration of existing surfaces. Thus the character of the surface relief depends partly on the geological structure, including disturbances and dislocations. These structures may be of a stratified nature (practically horizontal, undulating or warped, intensely folded, fractured) or of an igneous nature (extrusive, or volcanic, and intrusive). Erosion, denudation, and accumulation affect these geological structural types and result in the creation of six fundamental forms: the plain, the escarpment, the valley, the mount, the cup-shaped hollow, and the cavern. The forms are differentiated by their slopes; and the "form-elements" combine to build up fundamental forms, which usually occur in association or groups to compose a special landscape. The form-elements, fundamental forms, and landscape are the three minor morphological elements of the earth's surface. The three higher categories are the extended area of equal elevation (a combination of landscapes); the system (a grouping of such areas); and, finally, the continental block and abyssal deep. Between the surface forms and agencies of change there is one relation: the major forms are due exclusively to dislocations, while the minor forms arise in a variety of ways.

The same fundamental form can arise from either erosion, accumulation, or dislocation; and, with one exception, each process can result in the six fundamental forms. Penck suggested the term "homoplastic" for forms with the same shape and "homogenetic" for those with the same origin. But a uniform and clear terminology, he believed, as well as a knowledge of the genesis of landforms superior to that existing already, had to be acquired. Penck stated that each fundamental form (except the plain) includes three groups of homogenetic features and that each group falls into various subdivisions according to the special kind of erosion, accumulation, or dislocation that has operated. By naming the homogenetic members of each fundamental form according to its genesis, two nouns could express both the plastic and genetic relations (plain of accumulation). The definition could be made more explicit by the addition of adjectives (plain of marine accumulation).

One of the most significant features of geomorphology is the contrast between this system of landform description and the cyclic concept of Davis, who postulated a sequence of development in each landscape and based landform analysis mainly on structure, process, and stage. Penck recognized and used some of Davis' sequential ideas; but as Walther Penck, from 1912 on, became intrigued with the intense folding of the Andes, he and his father increasingly emphasized the importance of rate of uplift on valley-side slopes.

In 1919 Penck published an important article on the summit levels of the Alps, "Die Gipfelflur der Alpen." In direct opposition to Davis' theory that peaks were eroded uplifted peneplains, Penck developed the concept that mechanical disintegration rapidly increases with altitude and that in each region there exists a maximum altitude above which the highest relief will not rise. He explained the existence of mature, or flattened, surfaces at great heights in the interior ranges of the Alps by assuming that the massif was uplifted slowly at first and then more rapidly. The spacing of the valley dissection and the nature of the valley-side slopes reflect this accelerating uplift. At an intermediate stage the sharp ridge crests (where the steep valley-side slopes intersect) will maintain a constant, absolute altitude and constant relief because the rate of upheaval and rate of deepening of the master valleys are balanced. Davis replied at length in "The Cycle of Erosion and the Summit Level of the Alps" (*Journal of Geology*, **31** [1923], 1–41). Although he admitted Penck's exceptional stature as a geographer and the "unquestionably large" value of the "Gipfelflur" essay, he considered it necessary to "correct" Penck's so-called corrections of the Davisian system. Influenced by his German colleagues, Penck moved increasingly away from Davis' cyclic concept, and by 1928 he had already abandoned the idea of the sequential development of landforms in favor of a scheme based on the ratio between rates of erosion and uplift ("Die Geographie unter den erdkundlichen Wissenschaften," in *Naturwissenschaften*, **16** [1928], 33–41).

Penck's chief contributions to knowledge of individual landform types were to glacial forms, especially in *Die Alpen im Eiszeitalter*. In this work, as well as in earlier articles, he stressed the importance of the overdeepening of glacial valleys and the significance of glacial through valleys. He was also instrumental in pointing out the general association between till sheets, terminal moraines, and outwash gravels and sands that develop on bordering lowlands outside the overdeepened piedmont basin at the end of an Alpine valley. This association was later equally applied to the peripheries of continental ice sheets.

To climatic geomorphology, as distinct from paleoclimatology, Penck made two significant contributions. He recognized an areal classification of surface morphology based on correlations with humid, subhumid, semiarid, arid, and nival (glacial) climatic areas. He was one of the first to insist that "we see on the earth's surface not only the features of the present climate but also those of a past climate" (*American Journal of Science*, **19** [1905], 169).

Penck's detailed accounts of the Danube River, "Die Donau" (1891), and the Oder River (1899) were among the earliest scientific analyses of the water budget and the flow regime of Central European rivers. His interest in cartography was responsible for initiating many distribution maps and at least one influential atlas on sociocultural themes, Krebs's *Deutsches Lebensraum in Mitteleuropa*. He advocated the production of Prussian maps on the scale of 1:100,000 for general purposes and of a standard series of global maps on a scale of 1:1,000,000. Penck introduced this idea in 1891 at the Fifth International Geographical Congress in Berne. The matter was raised at each successive congress; and at the eighth congress in Washington (1904), Penck again addressed the delegates and presented maps compiled on that scale by the French, Germans, and British. International conferences were subsequently held in 1908 and 1913 to resolve outstanding problems with regard to standard specifications, spelling, and production of the 1:1,000,000 world sheets (IMW). Of the estimated 840 sheets needed to cover the land areas of the world, only 97 had been published by 1931. Within a few years of Penck's death the greater part of the land areas had been covered by standard IMW maps.

BIBLIOGRAPHY

I. ORIGINAL WORKS. Penck's published works comprise about 410 books and articles. The selective list given here includes works referred to in the text and others that exemplify his contributions to major themes. His works on Pleistocene geology include *Die Vergletscherung der deutschen Alpen* . . . (Leipzig, 1882); *Die Alpen im Eiszeitalter*, 3. vols (Leipzig, 1901–1909), written with E. Brückner; and "Europa im Eiszeitalter," in *Geographische Zeitschrift*, **43** (1937), pt. 1. On geomorphology, see *Morphologie der Erdoberfläche*, 2 vols. (Stuttgart, 1894); "Die Geomorphologie als genetische Wissenschaft," in *Report of the Sixth International Geographical Congress, London, 1895* (1896), 735–757; and "Die Gipfelflur der Alpen," in *Sitzungsberichte der Preussischen Akademie der Wissenschaften*, **17** (1919), 256–263. On hydrography, see "Die Donau," in *Schriften des Vereins zur Verbreitung naturwissenschaftlicher Kenntnisse in Wien*, **31** (1891), 1–101; and "Der Oderstrom," in *Geographische Zeitschrift*, **5** (1899), 19–47, 84–94; and on cartography, "The Construction of a Map of the World on a Scale of 1:1,000,000," in *Geographical Journal*, **1** (1893), 253–261.

Other works on geography include "Das deutsche Reich," in A. Kirchhoff's *Länderkunde von Europa*, I (Leipzig, 1887), 115–596; "Die österreichische Alpengrenze," in *Zeitschrift der Gesellschaft für Erdkunde zu Berlin* (1915), 329–368, 417–448; *Über politische Grenzen* (Berlin, 1917); "Die Stärke der Verbreitung des Menschen," in *Mitteilungen der Geographischen Gesellschaft in Wien* (1942), 241–269; *Beobachtung als Grundlage der Geographie* (Berlin, 1906); and "Geography Among the Earth Sciences," in *Proceedings of the American Philosophical Society*, **66** (1927), 621–644.

II. SECONDARY LITERATURE. The chief biographies and bibliographies of Penck are: *1877–1903: Druckschriften von Albrecht Penck* (Vienna, 1903), a list of 162 items compiled by A. E. Forster; Erich Wunderlich, "Albrecht Penck: Zu seinem 70 Geburtstag am 25. September 1928," in *Geographischer Anzeiger*, **29** (1928), 297–306; *1877–1928. Druckschriften von Albrecht Penck* . . . (Berlin, 1928), with bibliography of 350 items to early 1928; Norbert Krebs, "Nachruf auf Albrecht Penck," in *Jahrbuch der Deutschen Akademie der Wissenschaften zu Berlin* (1946–1949), 202–212; Johann Sölch, "Albrecht Penck," in *Mitteilungen der Geographischen Gesellschaft in Wien*, **89** (1946), Heft 7–12, 88–122; Walter Behrmann, "Albrecht Penck 25.9.1858–7.3.45," in *A. Petermanns Mitteilungen aus J. Perthes Geographischer Anstalt*, **92** (1948), 190–193; H. Spreitzer, "Albrecht Penck," in *Quartär* (1951), 109–139; Edgar Lehmann, "Albrecht Penck," in *Deutsche Akademie der Wissenschaften zu Berlin*, **64** (1959); Herbert Louis, "Albrecht Penck und sein Einfluss auf Geographie und Eiszeitforschung," in *Die Erde*, **89** (1958), Heft 3–4, 161–182 (extends bibliography of *1928 Druckschriften* to a total of 406 items); and G. Englemann, "Bibliographie Albrecht Penck," in *Wissenschaftliche Veröff d. Deutschen Institut für Länderkunde* (1960), 331–447. For controversy over landform analysis between the Pencks and W. M. Davis see R. J. Chorley *et al.*, *The History of the Study of Landforms*, II (London, 1973). For Penck's contributions to cartography see Walter Behrmann, "Die Bedeutung Albrecht Pencks für die Kartographie," in *Blätter d. Dt. Kartogr. Ges.*, no. 2 (1938), 22 pp.

ROBERT P. BECKINSALE

PENCK, WALTHER (*b*. Vienna, Austria, 30 August 1888; *d*. Stuttgart, Germany, 29 September 1923), *geology.*

Penck's interest in natural science developed under the tutelage of both his father, the geologist and geomorphologist Albrecht Penck, and his teacher, Paul Pfurtscheller. When the elder Penck moved to the University of Berlin, Penck began his undergraduate studies there, but these studies were soon interrupted when, in 1908–1909, he accompanied his father to the United States, where the latter was an exchange professor at Columbia University. During this year, he traveled widely with his father and met many geologists, including G. K. Gilbert. After returning to Berlin via Hawaii, Japan, China, and Siberia, Penck enrolled at the University of Heidelberg, from which he graduated; he subsequently continued his studies in Vienna.

In 1912 he was appointed geologist to the Dirección General de Minas in Buenos Aires, where he was responsible for geological surveying and topographic mapping in northwest Argentina. Aided by his mountaineering ability, he mapped some 4,500 square miles of territory in less than two years and made a reconnaissance across the Andes. It was during these years (1912–1914) that Penck formalized his ideas regarding the pattern of tectonic movements. His studies of the Upper Cretaceous and Tertiary sediments flanking the Calchaqui mountains, Sierra de Famatina, and Sierra de Fiambalá in the Puna de Atacama led him, like his father, to posit temporal patterns of uplift much more varied than the pattern of rapid uplift followed by long quiescence, which was accepted by W. M. Davis.

Penck believed that most tectonic movements began and ended slowly, and that the common pattern of such movements involved a slow initial uplift, an accelerated uplift, a deceleration in uplift, and, finally, quiescence. There can be no doubt that much of Penck's geomorphic work was an attempt to provide physiographic support for the general pattern of uplift that he had previously inferred from stratigraphical evidence. The importance that Penck placed on identifying the movements of the source area from the record of sedimentation is clearly stated in the first chapter of his *Die morphologische Analyse*. Few geologists would now attempt more than to suggest the occurrence of some generalized uplift on the sole evidence of the sedimentary record, and even fewer would infer the pattern of uplift in any great detail. In 1917 Joseph Barrell showed that much of the character of the sedimentary record is determined by the subsidence of the basin of sedimentation, as distinct from the behavior of the adjacent source area. Although these behaviors are often so closely linked that it is difficult to distinguish between them, the work of Barrell began to cast doubt on the simple association between the nature of sedimentation and the pattern of uplift of the source area.

The major results of Penck's work in Argentina were not published until the end of World War I. The war broke out while he was in Germany on leave and, although his South American work qualified him for a geological post at the University of Leipzig, he served for a while in the German army in Alsace. At the end of 1915 he was appointed professor of mineralogy and geology at the University of Constantinople. For the next two and a half years he made tectonic observations in Anatolia (where he visited the Bithynian Olympus) and did varied geological work in the region of the Sea of Marmara (where he studied the coal strata of the Dardanelles). He also served as a professor at the Agricultural College of Halkaly. Malaria forced him to return to Germany in the summer of 1918; shortly thereafter he published the two substantial works that summarized his studies in Turkey.

Penck was unable to return to Turkey after the end of the war, and he became an unsalaried professor at the University of Leipzig, where he also held a lectureship in topographical and geological surveying. Refusing, despite straitened financial circumstances, more lucrative posts that would have inhibited his researches, Penck studied the terrain of the German highlands, and in particular that of the Black Forest. In 1921 he recovered some of his Turkish assets. Shortly afterward he died of cancer, survived by his wife and two small sons.

During the last years of his career, Penck developed his most influential ideas on the interpretation of landforms through analysis of the relationships between endogenetic (diastrophic) and exogenetic (erosional) processes. Of the three major publications that embodied his views, only the least important, "Wesen und Grundlagen der morphologischen Analyse" (1920), was published before his death. "Die Piedmontflächen des südlichen Schwarzwaldes" (1925) was based upon two lectures that he gave at Leipzig in December 1921; his book *Die morphologische Analyse* (1924) was only part of a contemplated larger work and was assembled and edited by his father. This last, posthumously published work was not only fragmentary but also hurriedly written, full of obscure terminology, and often unclear. Apart from J. E. Kesseli's mimeographed translation (1940) of an abstract of chapter 6, which discussed the development of slopes, *Die morphologische Analyse* was not translated into English until 1953. Simons, one of Penck's later translators, wrote "I have hardly ever met more difficult and obscure language. Quite often it was difficult to tell whether he said yes or no."

It is unfortunate that, for a period of more than twenty years, the only English interpretation of Penck's geomorphic ideas was that available in a highly critical article published in 1932 by his major opponent, W. M. Davis. Davis concentrated on Penck's Black Forest paper of 1925 and, besides seizing on the obvious difficulties of interpreting topographic discontinuities as the product of continuous crustal uplift, grossly misrepresented Penck's ideas, particularly in attributing to him the postulate of the parallel retreat of one major slope element which leaves beneath itself a surface of less declivity (compare fig. 4 of Davis' 1932 article with fig. 4 of Penck's 1925 publication). By World War II the Davis-Penck controversy, as it was carried out in the English-speaking world, had foundered in a doctrinaire and depressingly semantic morass.

Penck believed that landforms could be interpreted through the ratios that might be expected to occur between exogenetic processes (which he believed to be of uniform type but developed at different rates in different climates) and a wide spectrum of endogenetic processes. He also thought that diastrophic movements were of two major types, which could occur independently or together. He named the first type *Grossfalt* ("great" or "broad" fold) and stated that it was produced by lateral compression with flanking synclines; this fold became narrower with time and was superficially faulted and thrusted in later stages. Penck interpreted "basin and range" structures as belonging to this type. He treated these in detail in *Die morphologische Analyse*, in which he tried unsuccessfully to show that the facies of the sediments derived from these folds do not indicate intermittent uplift. He viewed the whole summit area of such a range as a deformed primary peneplain that was formed during slow initial uplift and correlated with unconformities in the basin.

The second type of movement defined by Penck was regional arching. He stated that this movement was produced by differential uplift, thus generating domes (*Gewölbes*) that progressively expanded their area with time but were not necessarily associated with flanking down-warps. Penck slighted the physiographic results of this type of movement in *Die morphologische Analyse*, but described them in detail in his 1925 paper on the Black Forest. All popular expositions of Penck's geomorphic views were based to some extent on his description of the landforms that might be developed on such a dome, the surface of which forms a series of stepped erosional benches (*Piedmonttreppen*) of differing age.

Where the two types of crustal movements occur together, as in the Alps, Penck thought that a more complex deformation was produced in which the regional doming, often outlasting the *Grossfalt*, was responsible for the general relief. He believed that regional up-doming began with a major phase of waxing development (*aufsteigende Entwicklung*) in which the accelerating uplift rates were generally in excess of stream degradation and the resulting landforms were dominated by the crustal instability. This development was followed by a general decline in the rate of uplift, during which a short period of uniform development (*gleichförmige Entwicklung*), in which the rate of erosion by streams overtook those of uplift, was succeeded by a dominantly waning phase (*absteigende Entwicklung*), during which the rate of uplift decreased, becoming stable as the landscape became progressively dominated by the erosional processes of valley widening. In this model the initially slow uplift would result in the formation and subsequent elevation of a primary peneplain (*Primärrumpf*), with convex valley-side slopes. As the uplift accelerated, the peneplain would be surrounded by a series of *Piedmonttreppen*, each of which had

originated as a piedmont flat (*Piedmontfläche*) on the slowly rising dome margin. Penck believed convex breaks of slope (*Knickpunkte*) to form on the radially draining river courses during accelerating uplift, leaving "one convex nick after the other . . ., below each one there begins a narrow, steep course reach with convex valley slopes, above each there is a broader reach with concave slope profiles" ("Die Piedmontflächen des südlichen Schwarzwaldes," p. 90). The concave stream-reaches between the convex nicks are formed in association with the *Piedmonttreppen*; each tends to act as an independent local baselevel for the subsequent valley widening on either side of the stream course. Penck made no clear distinction between continuous acceleration of uplift and continuous but intermittently accelerated uplift; the mechanisms that he evoked for the production of *Piedmonttreppen* and *Knickpunkte* also lacked clarity. Davis made much of these points and the modern geomorphologist can only find it difficult to understand how topographic discontinuities can develop during the waxing phase of Penck's model.

Penck's imaginative work was nonetheless of particular value in repairing the omission of diastrophic causes in much of the classic geomorphic literature.

BIBLIOGRAPHY

I. Original Works. The more important of Penck's 34 works include *Die tektonischen Grundzüge Westkleinasiens* (Stuttgart, 1918); "Grundzüge der Geologie des Bosporus," in *Veröffentlichungen des Instituts für Meereskunde an dem Universität*, n.s. **4** (1919), 1–71; "Der Südrand der Puna de Atacama (Nordwestargentinien). Ein Beitrag zur Kenntnis des andinen Gebirgstypus und zu der Frage der Gebirgsbildung," in *Abhandlungen der Sächsischen Akademie der Wissenschaften*, Math.-Phys. Kl., **37**, no. 1 (1920), 1–420; "Wesen und Grundlagen der morphologischen Analyse," in *Bericht Sächsischen Akademie der Wissenschaften*, Math.-nat. Kl., **72** (1920), 65–102; "Über die Form Andiner Krustenbewegungen und ihre Beziehung zur Sedimentation," in *Geologische Rundschau*, **14** (1923), 301–315; "Die morphologische Analyse. Ein Kapitel der physikalischen Geologie," in *Geographische Abhandlungen*, 2nd ser., **2** (1924), 1–283; this work was subsequently published separately (Stuttgart, 1924), and trans. by H. Czech and K. C. Boswell as *Morphological Analysis of Landforms* (London, 1953). This ed. contains a short biography of Penck (pp. vii–viii) and a list of his publications (pp. 352–353). See also "Die Piedmontflächen des südlichen Schwarzwaldes," in *Zeitschrift der Gesellschaft für Erdkunde zu Berlin* (1925), 83–108, with mimeographed trans. by M. Simons, "The Piedmont-flats of the Southern Black Forest" (1961).

II. Secondary Literature. On Penck and his works, are O. Ampferer, "Walther Penck," in *Verhandlungen der*

Geologischen Bundesanstalt, **4** (1924), 81–82; H. G. Backlund, "Walther Penck," in *Geologiska Foreningins I Stockholm Förhandlinger*, **45**(5) (1923), 445–447; J. Barrell, "Rhythms and the Measurement of Geologic Time," in *Bulletin of the Geological Society of America*, **28** (1917), 745–904; H. Baulig, "Sur les gradins de piedmont," in *Journal of Geomorphology*, **2** (1939), 281–304, a somewhat misguided criticism of Penck's concept of slope development; I. Bowman, "The Analysis of Landforms: W. Penck on the Topographic Cycle," in *Geographical Review*, **16** (1926), 122–132, a critical article on *Die morphologische Analyse* written with the approval and help of Davis; R. J. Chorley, "The Diastrophic Background to Twentieth-Century Geomorphological Thought," in *Bulletin of the Geological Society of America*, **74** (1963), 953–970; R. J. Chorley *et al.*, *The History of the Study of Landforms*, (Methuen–London, 1973), *passim*, which presents the important personal correspondence between Penck and Davis; W. M. Davis, "Piedmont Benchlands and the Primärrumpfe," in *Bulletin of the Geological Society of America*, **43** (1932), 399–440, a detailed attack on Penck's 1925 publication; G. K. Gilbert, "The Convexity of Hilltops," in *Journal of Geology*, **17** (1909), 344–350; and J. E. Kesseli, *The Development of Slopes* (Berkeley, Calif., 1940), mimeographed; F. Kossmat, "Walther Penck," in *Centralblatt für Mineralogie, Geologie und Paläontologie*, **25** (1924), 123–127.

Additional works include H. Lautensach, "Albrecht und Walther Penck," in *Zeitschrift für Geomorphologie*, n.s. **2** (1958), 245–250; A. G. Ogilvie, "Argentine Physiographical Studies: A Review," in *Geographical Review*, **13** (1923), 112–121, a review of "Der Südrand der Puna de Atacama" and other works; A. Penck, "Biography of Walther Penck," Foreword to *Die morphologische Analyse* (1924), VII–XVIII; A. Penck, "Letter Regarding 'Die morphologische Analyse,'" in *Geographical Review*, **16** (1926), 350–352, a reply to Bowman (1926); C. O. Sauer, "Landforms in the Peninsular Range of California as Developed About Warner's Hot Springs and Mesa Grande," in *University of California Publications in Geography*, **3**, no. 4 (1929), 199–290, an attempt to apply Penck's geomorphic notions in North America; M. Simons, "The Morphological Analysis of Landforms: A New Review of the Work of Walther Penck," in *Transactions of the Institute of British Geographers*, no. 31 (1962), 1–14, a penetrating review of many of Penck's ideas; and indispensable in the preparation of this biographical note; H. Spreitzer, "Die Piedmonttreppen in der regionalen Geomorphologie," in *Erdkunde*, **5**, no. 4 (1951), 294–304; Symposium, "Walther Penck's Contribution to Geomorphology," in *Annals of the Association of American Geographers*, **30** (1940), 219–284; Y.-F. Tuan, "The Misleading Antithesis of Penckian and Davisian Concepts of Slope Retreat in Waning Development," in *Proceedings of the Indiana Academy of Science*, **67** (1958), 212–214; and O. D. von Engeln, *Geomorphology* (New York, 1942), 256–268, an exposition based on Davis (1932).

RICHARD CHORLEY

PENNANT, THOMAS (*b.* Downing, near Holywell, Flintshire, Wales, 14 June 1726; *d.* Downing, 16 December 1798), *natural history.*

Pennant was the eldest son of David and Arabella Mytton Pennant. His first schooling was under the Reverend W. Lewis. In 1744 he matriculated at Queen's College, Oxford, but left without an undergraduate degree, probably because he had been active in troubles between undergraduates and faculty. In 1759 he married Elizabeth Falconer; they had a daughter, Arabella, and a son, David. His wife died in 1764, and Pennant married Anne Mostyn in 1777; two children, Thomas and Sarah, were born to them. He subsequently inherited his father's property, where he discovered a rich lead mine.

Calling himself a "moderate Tory," Pennant was active in politics and served as sheriff of Flintshire. He enjoyed excellent health throughout the first seventy years of his life, ascribing it to traveling on horseback and avoiding supper, which he called "the meal of excess." He kept a strict schedule, retiring at ten and rising at seven, and concentrated seriously when he worked. The recognition that he received included election to the Royal Society of Uppsala (1757), the Royal Society of London (1767), and various foreign societies.

Pennant's passion for natural history began in 1738, when he received a copy of Willughby's *Ornithology* as a gift. In 1746, while still at Queen's College, Pennant toured Cornwall and met the geologist William Borlase, who encouraged his interest in minerals and fossils. Pennant's first publication, a description of an earthquake at Downing in 1750, appeared in the *Philosophical Transactions of the Royal Society.* He later admitted a "rage" to become an author.

At Pennant's suggestion Gilbert White began writing the letters that became *The Natural History and Antiquities of Selborne*, and forty-four of the 110 letters in it are addressed to Pennant. Pennant also had a talent for observation and organization. He was able to combine his own observations with information from Thomas Hutchins, Ashton Blackburn, Alexander Garden, Benjamin Smith Barton, and Peter Simon Pallas, and thus to produce his classic work, *Arctic Zoology.* He also corresponded with the leading naturalists of his day.

Although Pennant had little ability for theorizing, he did contribute to organizing, popularizing, and promoting the study of natural history. His writings tended to emphasize the goodness and usefulness of nature, which he considered a reflection of a sanctified creation. In classification he supported the views of his countryman John Ray and, later, those of Linnaeus. Pennant was a representative of the best of the gentleman-naturalists who flourished in the late eighteenth century and who sought to comprehend all of nature.

BIBLIOGRAPHY

I. ORIGINAL WORKS. Pennant's published writings include *The British Zoology. Class 1. Quadrupeds. 2. Birds* (London, 1766), enl. to 4 vols., *British Zoology* (London, 1768–1770), a standard text; *Indian Zoology* (London [?], 1769); *Synopsis of Quadrupeds* (Chester, 1771), with subsequent eds. entitled *History of Quadrupeds.* The first of his travel books, *A Tour in Scotland, 1769* (Chester, 1771), was reissued in 1772, 1774, 1775, and 1790; Pennant also published many travel accounts and guidebooks for the British Isles, but one, the *Tour on the Continent, 1765* (London, 1948), remained unpublished until it was edited by G. P. de Beer for the Ray Society.

Subsequent writings are *Genera of Birds* (Edinburgh, 1773); *Arctic Zoology,* 2 vols. (London, 1784–1785) and its *Supplement* (London, 1787); *Catalogue of My Works* (London, 1786); *The Literary Life of the Late Thomas Pennant, Esq. by Himself* (London, 1793); *The History of the Parishes of Whiteford and Holywell* (London, 1796); *Outlines of the Globe,* 4 vols. (London, 1798–1800). The last title, planned to reach 14 vols., describes imaginary travels to many parts of the world. Several of Pennant's books were translated for foreign-language eds. and many appeared in several English eds.

II. SECONDARY LITERATURE. There is no sufficient biography of Pennant in existence; and his autobiographical work, cited above, is incomplete at best. Information about him can be found in Georges Cuvier, "Thomas Pennant," in *Biographie universelle,* XXXIII (Paris, 1823), 315–318; R. W. T. Gunther, *Early Science in Oxford,* XI (Oxford, 1937), 131–132, for information about Pennant's college experiences; Sir William Jardine, *The Natural History of Humming-Birds,* II (Edinburgh, 1833), 1–39; W. L. McAtee, "The North American Birds of Thomas Pennant," in *Journal of the Society for the Bibliography of Natural History,* **4,** pt. 2 (January 1963), 100–124; W. L. McAtee, "Thomas Pennant," in *Nature Magazine,* **45** (Feb. 1952), 98, 108; John Nichols, *Literary Anecdotes of the Eighteenth Century,* VIII (London, 1815), *passim;* Peter Simon Pallas, *A Naturalist in Russia; Letters From Peter Simon Pallas to Thomas Pennant,* Carol Urness, ed. (Minneapolis, 1967), with a biography of Pennant on 169–175; and Warwick Wroth, "Thomas Pennant," in *Dictionary of National Biography,* XLIX (1895), 320–323.

CAROL URNESS

PENNY, FREDERICK (*b.* London, England, 10 April 1816; *d.* Glasgow, Scotland, 22 November 1869), *analytical chemistry, toxicology.*

The third son of Charles Penny, a wholesale stationer, Penny was educated at schools in Sherborne (Dorset) and Tooting, London. He was then apprenticed (1833–1838) to the pharmacist and analytical chemist Henry Hennell at the Apothecaries' Hall. Penny described himself as a "pupil" at the lectures of W. T. Brande and M. Faraday at the Royal Institution in 1836 and 1837.

In 1839, on Thomas Graham's strong recommendation, Penny suceeded W. Gregory in the unremunerative chair of chemistry at Anderson's College, Glasgow. Classes were small and brought him few fees, and this, together with the rents he assumed for his laboratory and classrooms, forced Penny to exploit his brilliant analytical talents in legal and commercial consultancy. Consequently, he published little and failed to fulfill the scientific promise he had shown in London. He did, however, visit Liebig at the University of Giessen in 1843 and was awarded the Ph.D. there on the basis of his published work.

Penny was dwarfed by a crooked spine caused when a governess threw him to the ground as a child. He married a Miss Perry in 1851 and had one daughter.

Penny's most important paper appeared in 1839. While trying to assay potassium nitrate in crude saltpeter, he found that the actual, as opposed to the theoretical, quantities of potassium chloride that are produced by the reaction of the nitrate with hydrochloric acid are different. Suspecting that the received chemical equivalents were at fault, he undertook a polished reappraisal of the equivalent weights (oxygen = 8) of the key elements: chlorine, nitrogen, potassium, sodium, and silver.

This elegant work, which was highly praised by, and later influential on, J. S. Stas, was of twofold significance. First, for the practical techniques involved: the use of special apparatus and a counterpoised balance, the use of carefully prepared reagents, and the exploitation of the nitrate-chloride and chlorate-chloride conversions. Second, because the results, Penny thought, confirmed those that E. Turner had published in 1833, and implied that "the favourite hypothesis [Prout's], of all equivalents being simple multiples of hydrogen, is no longer tenable" (*Philosophical Transactions,* **129** [1839], 32). Stas came to the same conclusion in 1860. Although Penny was clearly Turner's successor and Stas's predecessor in matters of atomic weight determinations—Penny even suspected the correctness of the published combining weight of carbon—he was unable to pursue these investigations in Scotland.

All of Penny's Scottish publications related to practical problems. In 1850 he introduced a volumetric determination of iron by the reduction of potassium chromate, or bichromate, using potassium ferricyanide as an external indicator. He also extended this method to the estimation of tin and iodine. Although A. W. von Hofmann once professed never to have heard of him, Penny had a high commercial reputation in Scotland; and, undoubtedly, he restored Anderson College's dormant reputation for medical, and especially technical, chemistry. He also became widely known throughout British medical and legal circles for the brilliance and composure of his Crown evidence in murder trials, notably those of the celebrated Madeleine Smith in 1857 (arsenic, nonproven) and Dr. Edward Pritchard in 1865 (aconite, guilty). The last months of Penny's life were made bitter by James Young's tactless endowment of an additional chair of technical chemistry at Anderson College, which appeared to Penny to threaten his livelihood and reputation.

BIBLIOGRAPHY

I. ORIGINAL WORKS. Penny's two most important papers are "On the Application of the Conversion of Chlorates and Nitrates Into Chlorides, and of Chlorides Into Nitrates, to the Determination of Equivalent Numbers," in *Philosophical Transactions of the Royal Society*, **129** (1839), 13–33; and "On a New Method for the Determination of Iron in Clay-band and Black-band Ironstone," in *Chemical Gazette*, **8** (1850), 330–337. Twelve other papers are listed in the Royal Society *Catalogue of Scientific Papers*, IV, 819–820; and VIII, 587. Penny's pamphlets are *Testimonials in Favour of Frederick Penny, Ph.D. . . . Candidate for the Professorship of Chemistry in the University of Edinburgh* [Glasgow, 1843]; *The Public Wells of Glasgow, With Analytical Reports by R. D. Thomson, M.D. and Dr. Penny* (Glasgow, 1848); and *Chemical Report on the Examination of the Water of Loch Katrine* (Glasgow, 1854).

Subsequent works include *Glasgow Water Supply Question* (Glasgow, 1855); *Report on the Experimental Operations at Loch Katrine* (London, 1855); and the important *Dr. Penny's Remonstrance and Appeal Against the Nomination and Appointment of an Additional Professor of Chemistry in Anderson's University* (Glasgow, 1869); and four analyses of Glasgow waters, 1854–1855, in J. Burnet, ed., *History of the Water Supply to Glasgow* (Glasgow, 1869).

Adams (below) mentioned "a large quantity of unfinished manuscripts," but these have not been located. There are several letters at the University of Glasgow and the Andersonian Library at the University of Strathclyde, Glasgow.

II. SECONDARY LITERATURE. Replies to Penny's 1869 pamphlet are *Retraction and Apology to Dr. Penny with Reference to Evidence Given by Mr. Mayer Before a Select Committee of the House of Commons* (Glasgow, 1869); and J. Adams, *Reasons of Protest* (Glasgow, 1869). These pamphlets may be found in Glasgow and Edinburgh libraries.

Obituary notices are James Adams, in *Glasgow Medical Journal*, **2** (1870), 258–270, concerned mainly with defending Penny's reputation; James Bryce, "President's Address," in *Proceedings of the Glasgow Philosophical Society*, **7** (1871), 364–371; A. H. Sexton, *The First Technical College* (London, 1894), 50; and A. W. Williamson, in *Journal of the Chemical Society*, **23** (1870), 301–306.

See also A. J. Berry, "Frederick Penny. A Forgotten Worker on Equivalent Weights," in *Chemistry and Industry*, **51** (1932), 453–454; J. Butt, "James Young, Scottish Industrialist and Philanthropist" (Ph.D. thesis, University of Glasgow, 1964), *passim*; and H. Irvine, "The Centenary of Penny's [Volumetric] Process," in *Science Progress*, **39** (1951), 63–66.

W. H. BROCK

PENSA, ANTONIO (*b*. Milan, Italy, 15 September 1874; *d*. Pavia, Italy, 17 August 1970), *anatomy, histology, embryology*.

For a detailed study of his life and work, see Supplement.

PERCY, JOHN (*b*. Nottingham, England, 23 March 1817; *d*. London, England, 19 June 1889), *metallurgy*.

Percy was the third son of Henry Percy, a solicitor. Persuaded, against his inclination, to prepare for a medical career, he studied in Paris (where he met Gay-Lussac, Thenard, and Jussieu) and Edinburgh, where he graduated M.D. in 1838. His thesis, on the presence of alcohol in the brain, won a gold medal. Although he obtained a hospital post in Birmingham in 1839, he never established a practice; instead, his early interest in chemistry was reawakened by the local metal industries. In the same year he married Grace Piercy, who died in 1880.

In 1846 Percy studied the nature of slags; he later turned to the extraction of silver from its ores by a process dependent upon the solubility of silver chloride in sodium thiosulfate (a phenomenon discovered by Herschel in 1819). He was elected to the Royal Society in 1847. In 1851 he was appointed lecturer at the Metropolitan School of Science (later the Royal School of Mines), which was then under the direction of Sir Henry de la Beche. He subsequently became professor and thus exerted a profound influence on the progress of British metallurgy; many of his pupils achieved great distinction. His teaching was both methodical and innovative; and Percy transformed metallurgy from a repertoire of practices into a scientific discipline. The inventions of his pupils (for example, the Thomas-Gilchrist process for making iron from phosphorus-rich ores) were, however, more important than Percy's own.

Using the results of a large number of chemical analyses, Percy made a survey of the national resources of iron ore. This survey was incorporated into his large, unfinished work on metallurgy; perhaps the first writer since the Renaissance to attempt to achieve the comprehensiveness of Agricola and Ercker. Percy held many official lectureships, including one at the Royal Military Academy in Woolwich, and was called upon for technical advice on many military defense questions. He disapproved of the removal by the government of the Royal School of Mines to South Kensington, and he resigned in 1879.

Percy made two personal collections during his life: one of watercolors and engravings, which was dispersed by sale after his death, and one of metallurgical specimens of historical interest, which has fortunately survived intact and is now in the Science Museum at South Kensington. He was a lifelong student of political and social questions, often forcefully expressing himself in public, both in speech and writing, although he was not always sensitive to the appropriateness of the occasion.

BIBLIOGRAPHY

Percy's major work was *A Treatise on Metallurgy*, 4 vols. (1864–1880). See also J. F. Blake, *Catalogue of the Collection of Metallurgical Specimens Formed by the Late John Percy, Esq.*, . . . (London, 1892).

Obituary notices are found in *Athenaeum*, **1** (1889), 795; *Journal of the Iron and Steel Institute*, **1** (1889), 210; and *Proceedings of the Geological Society*, **46** (1890), 45.

FRANK GREENAWAY

PEREIRA (or **Pererius**), **BENEDICTUS** (*b*. Ruzafa [near Valencia], Spain, 1535; *d*. Rome, Italy, 6 March 1610), *physics, mechanics, astrology.*

Little is known of Pereira's early life before his admission to the Society of Jesus in 1552. After joining the order, Pereira was sent to Sicily, and then Rome, to complete his education. In Rome he taught various disciplines and arts and became known also as an exponent of scripture, on which he left several commentaries.

Pereira's most important work was his treatise on natural philosophy, *De communibus omnium rerum naturalium*, known also as *Physicorum . . . libri*. First published in Rome in 1562, this Aristotelian commen-

tary went through many subsequent European editions and was used as a philosophy textbook in the flourishing Jesuit schools. It was widely read and is cited in several of the writings of the young Galileo. The section on dynamics (book XIV) is staunchly Aristotelian. Although various theories of violent motion were described, most were rejected, particularly the Parisian impetus theory. In his dislike for Parisian dynamics, Pereira belonged to a strong Italian tradition, upheld also by Girolamo Cardano, Gasparo Contarini, Andrea Cesalpino, and Girolamo Borro.

The *De communibus* was quoted frequently in the Renaissance debate on the nature of mathematics. Like some of his fellow Aristotelians, Pereira was reluctant to allow Aristotle's admission that abstract mathematical demonstrations were of the greatest certainty. Pereira took the extreme position that neither mathematics nor any other science could satisfy Aristotle's very strict criteria for certainty.

Pereira's *Adversus fallaces et superstitiosas artes* (1591) was an outright attack on the occult arts, including alchemy and natural magic, the interpretation of dreams, and astrology. Like the *De communibus*, this treatise enjoyed a wide circulation, although for more notorious reasons. In denouncing magic, Pereira began with the paradoxical premise that natural magic did exist, and was indeed the noblest part of physics, mathematics, and medicine. Because of this exalted status, however, natural magic was accessible to only a very few learned and good men. The evil, therefore, lay in the pretensions of the ignorant and wicked to such knowledge. Such pretensions resulted in abuses, deception, and poverty. Despite its intrinsic goodness, Pereira advocated that the pursuit of natural magic, and of alchemy in particular, be banned.

In the section on the interpretation of dreams, Pereira reverted to Aristotle and concluded that dreams ought neither to be heeded nor disregarded to excess. Those who accepted fixed rules in this matter should be denounced as followers of superstition.

The final section, which dealt with astrology, seems to have been inspired by Sixtus V's bull of 1586 condemning judicial astrology. Pereira used arguments from Giovanni Pico della Mirandola and other sources to show that the heavens do not manifest portents and that the rules of astrology are absurd— any fulfillment of predictions was ascribed to the work of demons. Pereira passed over in silence the acceptance of astrology by Thomas Aquinas and Albertus Magnus. In his earlier *De communibus*, however, the Jesuit had cited Aquinas' favorable opinion without adding a condemnation thereof.

512

PEREIRA

BIBLIOGRAPHY

I. ORIGINAL WORKS. Pereira's major works are *De communibus omnium rerum naturalium principiis et affectionibus, libri quindecim* (Rome, 1562), later reprinted in Rome (1576, 1585), Venice (1586, 1592, 1609), Paris (1579, 1585, 1589), Lyons (1585, 1588, 1603), Cologne (1595, 1598, 1601, 1603, 1609), and Ingolstadt (1590); and *Adversus fallaces et superstitiosas artes, id est de magia, de observatione somniorum et de divinatione astrologica, libri tres* (Ingolstadt, 1591), later reprinted in Venice (1591, 1592), Lyons (1592, 1602, 1603), Paris (1616), and Cologne (1598, 1612). An English trans. by Percy Enderbie, *The Astrologer Anatomised*, was issued in London in 1661 and again in 1674. Some of Pereira's commentaries on Aristotle's *Physics* are in the Nationalbibliothek, Vienna, MSS 10476, 10478, 10491, and 10509.

II. SECONDARY LITERATURE. Pereira's religious and scientific works are listed in A. De Backer, ed., *Bibliothèque de la Compagnie de Jésus*, VI (Brussels–Paris), 499–507. Pereira is discussed briefly in Pierre Duhem, *Études sur Léonard de Vinci*, III (Paris, 1913), 203–204; Lynn Thorndike, *A History of Magic and Experimental Science*, VI (New York, 1914), 409–413; and Neal W. Gilbert, *Renaissance Concepts of Method* (New York, 1960), 91. For Galileo's citations of Pereira, see Galileo Galilei, *Opere*, A. Favaro, ed., I (Florence, 1890), 24, 35, 145, 318, 411.

PAUL LAWRENCE ROSE

PEREIRA, DUARTE PACHECO

PEREIRA, DUARTE PACHECO (*b.* Santarém [?], Portugal, *ca.* 1460; *d.* Lisbon [?], Portugal, 1533), *navigation.*

Although not all of the voyages that the Portuguese sailor and pilot Pereira made during his lifetime are known, it is known that he did not participate in several expeditions cited by historians. For example, he would have been very young in 1471 to have taken part in the assault on the north African fortress of Arzila. It is generally believed that he was entrusted by King Manuel I with an expedition in 1498 to America, where he supposedly sailed along the coast of Brazil for the first time. But this assertion, based solely on one obscure passage in his book, is doubtful. It is certain, however, that in 1488 Pereira was on Prince's Island, southwest of the Cameroons, when Bartholomeu Dias was returning to Europe. Pereira was very ill, and the discoverer of the Cape of Good Hope brought him back to Lisbon. In 1503 Pereira was in India with Alfonso de Albuquerque. Remaining there to defend the weak king of Cochin against the powerful King Samorim of Calicut, he succeeded in driving back the fierce and repeated attacks of the Samorim against a small band of Portuguese. Pereira thus became a national hero, and the fame of his exploit at Cochin spread throughout Europe.

In 1505 Pereira returned to Lisbon, where he began his *Esmeraldo de situ orbis*, a title that is still unexplained. He never completed the work and it is known only from incomplete copies. He subsequently undertook missions for the king along the coasts of Portugal and North Africa. In 1519 he became governor of the fortress and commercial entrepôt of São Jorge da Mina, in the Gulf of Guinea. Three years later he was arrested and imprisoned in Lisbon, presumably as the result of irregularities he had committed while in office. After regaining the king's confidence, he was freed and was awarded a lifelong pension.

Pereira's *Esmeraldo* may be considered a routier, or collection of sailing directions, with an introduction on contemporary seamanship. It has several novel aspects, and departs from the style of the medieval routiers, for example that by Pierre Garcie. The introduction contains such interesting elements as the "rules of the sun" for determining latitudes and information concerning tides.

BIBLIOGRAPHY

I. ORIGINAL WORKS. The three published editions of the *Esmeraldo de situ orbis* are all based on surviving eighteenth-century copies. The 1st ed. (Lisbon, 1892) was published by Azevedo Bastos, in celebration of the fourth centenary of the discovery of America. The 2nd (Lisbon, 1905) contains philological comments by the editor A. E. Silva Dias. The 3rd ed. (Lisbon, 1954) was sponsored by the Academia Portuguesa de Historia and includes notes by Damião Peres.

II. SECONDARY LITERATURE. J. Barradas de Carvalho has published a series of studies on Pereira and the significance of his book in University of São Paulo, Brazil, *Revista de história* (1966–1970).

LUÍS DE ALBUQUERQUE

PÉRÈS, JOSEPH JEAN CAMILLE

PÉRÈS, JOSEPH JEAN CAMILLE (*b.* Clermont-Ferrand, France, 31 October 1890; *d.* Paris, France, 12 February 1962), *mathematics, mechanics.*

The son and son-in-law of distinguished philosophers, Pérès entered the École Normale Supérieure in 1908, became *agrégé* in mathematics in 1911, and was immediately awarded a scholarship to enable him to earn a doctorate. Introduced by Émile Borel to Vito Volterra, he left for Italy to prepare his dissertation under the latter's supervision. He defended the dissertation *Sur les fonctions permutables de Volterra* in 1915, while teaching *mathématiques spéciales* at the *lycée* of Montpellier. After brief stays at the faculties of Toulouse and Strasbourg, he was from 1921 to 1932 professor of rational and applied mechanics at Marseilles, where in 1930 he founded an institute of fluid mechanics. Called to the Sorbonne in 1932, he devoted his scientific efforts primarily to developing the field of fluid mechanics. But his personal qualities led to his being burdened with ever more numerous and demanding duties. He taught at all the *grandes écoles* and from 1954 to 1961 was dean of the Paris Faculty of Sciences during a difficult time of expansion and profound transformation. Moreover, he fulfilled extensive responsibilities in several major national and international research organizations, notably the Centre National de la Recherche Scientifique and the International Committee of Scientific Unions.

Pérès won prizes from the Académie des Sciences in 1932, 1938, and 1940 and was elected a member in 1942. He was a foreign member of the Accademia Nazionale dei Lincei, Accademia delle Scienze, and the National Academy of Sciences, as well as an active member of the Académie Internationale d'Histoire des Sciences from 1948. Pérès's positions and honors testify to his exceptionally fruitful life, devoted to the combination of teaching and research.

Volterra's initial influence on Pérès and their warm thirty-year friendship account to a large degree for the course of Pérès's research, which was at first oriented toward pure analysis and then toward mechanics. The events of his career simply accentuated a development the outlines of which were determined at the outset.

Pérès's results on integral equations extended those of Volterra, notably regarding composition products of permutable functions with a given function and, later, the composition of functions of arbitrary order. These findings are now considered classical, as is his theory of symbolic calculus, which is more general than Heaviside's. Work of this type in analysis harmonized with the needs of fluid mechanics. In the latter domain, which experienced great progress in France through Pérès's efforts, his work was linked in large part to that of other researchers. Aiming at various applications, especially in aeronautics, Pérès conducted studies on the dynamics of viscous fluids, on the theory of vortices, and on movements with slip streams while refining the method of electrical analogies. In constructing his "wing calculator," as well as analogous devices—for measuring the pressure of lapping waves on jetties, for example—Pérès remained in close contact with those testing the equipment. To his scientific colleagues he remained a circumspect theorist, animator, and promoter.

At the beginning of his career Pérès obtained two results, now bearing his name, that are not connected with the fields mentioned above. One concerned Levi-

Civita parallelism (1919); the other, impact with friction (1924). In the second area he achieved one of the last great successes of rational mechanics. The gift for theoretical speculation manifested in these investigations remained the mainspring of his work and of his influence, and the fruitfulness of both is explained by his openness to new ideas.

BIBLIOGRAPHY

I. Original Works. Pérès's books include *Sur les fonctions permutables de Vito Volterra* (Paris, 1915), his diss.; *Leçons sur la composition et les fonctions permutables* (Paris, 1924); *Les sciences exactes* (Paris, 1930); *Cours dé mécanique des fluides* (Paris, 1936); *Tables numériques pour le calcul de la répartition des charges aérodynamiques suivant l'envergure d'une aile* (Paris, 1936), written with L. Malavard and L. Romani; *Théorie générale des fonctionnelles* (Paris, 1936); *Notice sur les titres et travaux scientifiques* (Paris, 1942), submitted with his candidacy to the Academy; and *Mécanique générale* (Paris, 1953).

Among his memoirs published in the *Comptes rendus* of the Academy are "Actions d'un fluide visqueux sur un obstacle," **188** (1929), 310–312, 440–441; "Sur le mouvement limite d'Oseen," **192** (1931), 210–212; "Sur les analogies électriques en hydrodynamique," **194** (1932), 1314–1316, written with L. Malavard; "Sur le calcul analogique des effets de torsion," **211** (1940), 131–133, written with L. Malavard; "Sur le calcul expérimental," *ibid.*, 275–277; and "Calcul symbolique d'Heaviside et calcul de composition de V. Volterra," **217** (1943), 517–520.

His other noteworthy works include the editing of *Leçons sur les fonctions de lignes de Vito Volterra* (Paris, 1913); "Le parallélisme de M. Levi-Civita et la courbure Riemannienne," in *Rendiconti. R. Accademia dei Lincei* (June 1919); "Choc avec frottement," in *Nouvelles annales de mathématiques*, **2** (1924); Pérès edited this journal, with R. Brocard and H. Villat, from 1923 to 1927; "Une application nouvelle des mathématiques à la biologie, la théorie des associations biologiques," in *Revue générale des sciences* (1927); and "Les divers aspects de la mécanique. Quelques notions concernant son enseignement," in *Mécanique*, no. 322 (Feb. 1944), 27-29.

II. Secondary Literature. On Pérès and his work, see the notices by P. Costabel, in *Archives internationales d'histoire des sciences*, **15** (1962), 137–140; H. Villat, in *Comptes rendus . . . de l'Académie des sciences*, **254** (1962); and M. Zamansky, in *Revue de l'enseignement supérieur*, no. 2 (1962), 95–97.

Pierre Costabel

PÉREZ DE VARGAS, BERNARDO (*b.* Madrid, Spain, *ca.* 1500–1533), *astronomy, biology, metallurgy.*

Few biographical data are known of Pérez de Vargas. His parents were of distinguished lineage, hence the title of "magnífico" which he appended to his name. From Madrid he moved to the province of Málaga; in one of his works he described himself as a resident of Coín, a town of that province. In 1563 he published the *Repertorio perpetuo o fábrica del universo*. There are extant copies of only the second part of this work. Most Spanish bibliographers, including Colmeiro and Navarrete, believed that both parts were published, although Tamayo gave the opinion that the first part remained as a manuscript in folio, the whereabouts of which is unknown. Palau has suggested that the *Sumario de cosas notables*, published in 1560, is the aforementioned first part.

The *Repertorio perpetuo*, as the second half of its title indicates, dealt with the "structure of the universe." It discussed such subjects as the nature of matter, the age of the globe and man, time and its measurements, astrology, the proper times for purges and bloodletting, and lunar and solar eclipses.

Pérez de Vargas' most important work, however, was on metallurgy, *De re metalica*; this work was published in 1568, although it bore a royal license dated 1564. It was composed of nine books, varying from five to twenty-five chapters, and contained thirteen illustrations. Starting from a philosophical discussion of the form and matter of metals, Pérez de Vargas admitted the possibility of alchemy. The work then proceeded to discuss mining and the smelting and development of metals and minerals. Diego de Meneses, who had owned and worked mines in the New World for thirty years, wrote the preface; he recommended the circulation of *De re metalica* among the miners of Peru and in other parts of America, where he had observed that much gold and silver was lost owing to a lack of adequate knowledge concerning refining processes.

Despite Pérez de Vargas' assertion in the introduction that the contents of his book were culled from the works of many famous authors and that the greater part of it had been subjected to experimentation, *De re metalica* was largely copied, with some paragraphs lifted in full, from Vannuccio Biringuccio's more meritorious *Pirotechnia* (1540). Pérez de Vargas referred to some mines in Spain not mentioned by Biringuccio, but his dependence on the latter would explain the absence of any mention of the development of quicksilver and the process of amalgamation, already known in Spain and America; amalgamation was probably of Spanish origin.

Nonetheless, *De re metalica* was useful because it was the first extensive book on metallurgy in Spanish. Although there was a Spanish translation of Glanville's *De proprietatibus rerum* in the fifteenth century, the scope of this work was not as extensive as *De re metalica*.

BIBLIOGRAPHY

I. ORIGINAL WORKS. Pérez de Vargas' works are *Sumario de cosas notables* (Toledo, 1560); *Repertorio perpetuo o fábrica del universo* (Toledo, 1563); *De re metalica* (Madrid, 1568); and "De los edificios y máquinas que pertenecen al arte de laborar los metales," a work (probably a MS) referred to by Pérez de Vargas in one of his works but of which there are no extant copies. There is a two-volume French trans. of *De re metalica* entitled *Traité singulier de métallique* (Paris, 1743).

II. SECONDARY LITERATURE. On Pérez de Vargas and his work, see Eugenio Maffei and Ramón Rua Figueroa, *Apuntes para una biblioteca española*, 2 vols. (Madrid, 1873); and Felipe Picatoste y Rodríquez, *Apuntes para una biblioteca científica española del siglo XVI* (Madrid, 1891).

VICENTE R. PILAPIL

PERKIN, WILLIAM HENRY (*b.* London, England, 12 March 1838; *d.* Sudbury, England, 14 July 1907), *synthetic organic chemistry, physical organic chemistry.*

Perkin was the son of George Fowler Perkin, a builder and contractor. He became interested in chemistry at an early age and in 1851 was sent to the City of London school, where—although science was not part of the curriculum—he was able to attend the weekly lectures on chemistry given by one of the classmasters during the dinner hour. Perkin's father was opposed to his making a career in chemistry, but he was encouraged by his master, Thomas Hall, through whose intercession he was enrolled in the Royal College of Science when he was fifteen. Perkin attended the lectures of the German chemist A. W. von Hofmann and, by the end of his second year at the college, was appointed Hofmann's assistant.

Perkin established his own laboratory at home at about the same time; one of his first pieces of private research was concerned with a coloring material. With Arthur H. Church he began to investigate the reduction products of dinitrobenzene and dinitronaphthalene. From the latter, Perkin and Church obtained a colored substance that they named "nitrosonaphthalene," which proved to be one of the first of the azo-dyes derived from naphthalene to be manufactured. They subsequently patented their process. Perkin's major discovery, that of mauve, the first synthetic dyestuff, occurred shortly thereafter, during the Easter vacation of 1856, when Perkin was only eighteen.

Hofmann had previously remarked to Perkin on the desirability of synthesizing quinine. Taking up the problem Perkin (basing his experiments on the idea, now understood to be unsound, that the structure of a chemical compound could be determined from the molecular formula alone) first treated toluidine with bichromate of potash, then repeated the process with an aniline salt. From the latter he obtained not quinine but a dirty, dark-colored precipitate. Some special instinct caused him to examine this precipitate further, and he discovered it to have coloring properties. From it he succeeded in isolating mauve, or aniline purple, the first dyestuff to be produced commercially from coal-tar. Almost immediately he sent a sample to a firm of dyers in Perth, with the request that they try it for coloring silk. In reply he received a letter that said, "If your discovery does not make the goods too expensive, it is decidedly one of the most valuable that has come out for a long time."

Perkin thus decided to patent his method for manufacturing the new dyestuff. His father agreed to provide financial support, although Hofmann had tried to discourage the venture, and a factory building was begun at Greenford Green in June 1857. Among the initial problems that the manufacturers faced was the refining of suitable raw materials; the eighteen-year-old Perkin had to work out a method of converting nitrobenzene to aniline and to devise not only a new technique but also a new apparatus. Nonetheless Perkin's "Tyrian purple" was being used in London dyehouses within six months, and shortly thereafter other firms in England and France were engaged in its production. Many other procedures for making mauve were soon patented. These represented only slight modifications of Perkin's original process, but fortunately for Perkin none of these newer methods yielded mauve as cheaply as his "bichromate method."

Perkin's discovery gave impetus to a new coal-tar dyestuffs industry. Perkin was able to keep his factory working at a profit in spite of the discovery of a number of other new coloring materials by a number of other chemists; in 1864 he himself introduced a new method for the alkylation of magenta, which allowed him to compete with the manufacturers of other violet dyes.

In 1868 the German chemists Graebe and Liebermann announced that they had synthesized alizarin, the natural coloring matter of madder; their process, however, was too expensive to be of more than scientific interest. Within a year Perkin worked out two new methods to manufacture alizarin more cheaply; both used coal-tar products, one being based upon dichloroanthracene and the other upon the sulfonic acid of anthraquinone. Synthetic alizarin soon replaced rose madder as the prime red dye, both in England and on the Continent. By the end of 1869 Perkin's company had made a ton of alizarin, and by 1871 they were manufacturing 220 tons a year.

Perkin had always hoped to devote himself completely to pure science, and by 1873 he found that his factory and patents could guarantee him the means for a modest retirement. The following year, when he was thirty-six, he sold his factory and turned full time to the research in pure chemistry that he had conducted concurrently with his industrial work. He had already made significant contributions to organic chemistry, even while burdened by commerce; in 1858, a year after his factory had opened, he had discovered that aminoacetic acid could be obtained by heating bromoacetic acid with ammonia. By 1860, in collaboration with B. F. Duppa, he had established the relationships between tartaric, fumaric, and maleic acids and had accomplished the synthesis of cinnamic acid from dibromo succinic acid. About 1867 he began to investigate the action of acetic anhydride on aromatic aldehydes, which led him to the method of synthesizing unsaturated acids by what is now known as "Perkin's synthesis"—a method that he applied, within a year of its discovery, to synthesizing coumarin. This line of investigation culminated, after Perkin's retirement from the dyestuffs industry, in his discovery that cinnamic acid could be synthesized from benzaldehyde—a discovery that made possible the first synthesis of indigo by Baeyer and Caro.

Upon retiring from business Perkin had a new house built at Sudbury and converted the old, adjacent one into a laboratory, where he continued to work almost until the time of his death. In 1881 he became interested in the magnetic rotatory polarization of certain organic compounds and so developed his investigations that the examination of this property became an important tool in considering questions of molecular structure. Perkin devoted the last twenty-five years of his life to this physical aspect of organic chemistry; he was commended for his work by Professor Bruehl, himself one of the pioneers of the application of optical methods to the determination of chemical constitutions, who wrote to him in 1906, "Before you began work there was little, almost nothing, known of this subject, certainly nothing of practical use to the chemist. You created a new branch of science. . . ."

Perkin's personal life was essentially uneventful. His devotion to his work and his family was so complete that, aside from participating in the activities of several scientific societies, he took no part in outside affairs. He was married twice: in 1859 to Jemima Harriet Lissett, who died in 1862, then in 1866 to a Polish girl, Alexandrine Caroline Mollwo, who survived him. He had two sons, both of whom became distinguished professors of chemistry, from his first marriage and one son, Frederick, and four daughters from his second. Perkin was of a retiring disposition and chose to avoid publicity; although colleagues in pure chemistry accorded him considerable recognition for his work, he was less honored by his co-workers in the field of commercial dyestuffs manufacture. In 1906, however, jubilee celebrations were held in England and the United States in commemoration of Perkin's discovery of mauve; distinguished scientists and industrialists from all over the world attended them, and Perkin was knighted upon this occasion. He died of pneumonia, perhaps weakened by the strain attendant upon celebrity, shortly thereafter.

BIBLIOGRAPHY

A complete list of Perkin's work is in Sidney M. Edelstein, "Sir William Henry Perkin," in *American Dyestuff Reporter*, **45** (1956), 598–608.

For further information on Perkin's life and work, see B. Harrow, *Eminent Chemists of Our Times* (New York, 1927); R. Meldola, *Jubilee of the Discovery of Mauve and of the Foundation of Coal-Tar Industry by Sir W. H. Perkin* (London, 1906), and "Obituary Notice," in *Journal of the Chemical Society*, **93** (1908), 2214; and M. Reiman, "On Aniline and Its Derivatives," a treatise on the manufacture of aniline colors, to which is added an appendix, "The Report on the Colouring Matters Derived from Coal Tar," shown at the French Exhibition (1867) by A. W. von Hofmann, Mme G. DeLair, and C. Girard; William Crookes revised and edited the whole work (London, 1868).

SIDNEY EDELSTEIN

PERKIN, WILLIAM HENRY, JR. (*b.* Sudbury, Middlesex, England, 17 June 1860; *d.* Oxford, England, 17 September 1929), *organic chemistry.*

Perkin was the eldest son of William Henry Perkin, the pioneer of synthetic dyestuffs, and Jemima Lissett. He studied at the Royal College of Chemistry in London and then under Johannes Wislicenus at the University of Würzburg (1880–1882) and Adolf von Baeyer at the University of Munich (1882–1886). Soon after his return to Britain, he married, became professor of chemistry at the new Heriot-Watt College in Edinburgh (1887), and was elected a fellow of the Royal Society (1890). In 1892 he succeeded Carl Schorlemmer in the chair of organic chemistry at Owens College in Manchester. Following the retirement (1912) of William Odling from the Waynflete professorship, Perkin moved to Oxford to take charge of an almost moribund department, which, in spite of the stringencies of World War I, he quickly transformed.

Perkin was much influenced by the personality of von Baeyer, and he himself became almost the archetype of a German professor of the best sort—cultured (he was an accomplished musician), fond of walking holidays in the Alps, and devoted to a rather narrow field of research, which nevertheless infused and inspired all his teaching. He was a skilled practical worker, and at no time in his life did he desert the laboratory for long. As a young man in Germany he made the first derivatives of cyclopropane and cyclobutane, thus disproving a widely held opinion that only carbon rings of five or six members could exist. Von Baeyer subsequently used these results in his "strain theory." Perkin's later work was concerned almost entirely with the elucidation of the structures of natural products by degradation and synthesis. He and his students worked on camphor and the terpenes; the natural dyestuffs brazilin and haematoxylin; and a long series of alkaloids, including berberine, harmine, cryptopine, strychnine, and brucine.

When Perkin died complaints were voiced that he had never been given a Nobel Prize. His fundamental shortcoming, however, was that he always applied himself to problems with well-defined solutions, which he studied by established techniques. His solutions, when found, were incorporated into chemistry with little remark. Consequently, he is now less well remembered than his colleagues and brothers-in-law, Lapworth and Kipping, who preferred to break fresh ground.

BIBLIOGRAPHY

I. ORIGINAL WORKS. Perkin wrote more than 200 papers, mostly with collaborators, and a number of textbooks, of which the most successful was *Organic Chemistry* (London, 1894; 2nd ed., 1929), written with F. S. Kipping.

II. SECONDARY LITERATURE. The Chemical Society has published an elaborate obituary entitled *The Life and Work of Professor William Henry Perkin* (London, 1932), which contains a personal memoir by A. J. Greenaway and accounts of Perkin's scientific work by his former colleagues J. F. Thorpe and R. Robinson.

Obituary notices are in *Chemistry and Industry*, **48** (1929), 1008–1012; *Nature*, **124** (1929), 623–627; and *Proceedings of the Royal Society*, **130A** (1930), i–xii. Perkin is also noticed in J. R. H. Weaver, ed., *Dictionary of National Biography 1922–1930* (London, 1967), 665–667.

W. V. FARRAR

PÉRON, FRANÇOIS (*b.* Cérilly, France, 22 August 1775; *d.* Cérilly, 14 December 1810), *zoology, natural history.*

Péron came from a family of modest means. He was intellectually gifted, and began the study of theology in 1791. But he abandoned theology in 1792 and enrolled at the Allier Battalion. He fought in Alsace, where he was wounded and taken prisoner. He was exchanged in 1794 and returned home blind in one eye. Pierre-Lazare Petit-Jean, a notary in Cérilly, financed his studies for three years at the École de Médecine in Paris. Péron was very interested in anthropology, and after learning that an expedition to New Holland (Australia) needed a physician-naturalist, he volunteered for the position. With the support of the naturalists Jussieu and Lacépède, he was accepted as a zoologist.

On 19 October 1800 Péron sailed from Le Havre on the *Géographe*, bound for Tasmania and Australia, with stops scheduled at the Canary Islands, Mauritius, and Timor. Aboard the ship he made the acquaintance of Lesueur and the botanist Leschenault de la Tour. Dysentery broke out on board; and only a few of the scientists and crew survived. The voyage ended at Port Jackson (Sydney Harbor) on 20 June 1802. Péron, Leschenault, and Lesueur had been left behind on King Island (in Bass Strait), when the ship sailed to evade a storm. After twelve days on the island they were rescued by an English vessel.

On 25 March 1804 the expedition arrived at Lorient, France. They returned with about a hundred live animals that had never been seen in Europe, which were intended for the Empress Josephine's château, Malmaison; they also brought plants for her garden and a sizable herbarium. Péron became a frequent guest at Malmaison and was even the empress' reader, which was only a pretext so that he could be supported in his research. Péron, however, was weakened by the voyage, and he did not live long enough to complete his proposed works.

BIBLIOGRAPHY

I. ORIGINAL WORKS. Péron's works are *Histoire générale et particulière de tous les animaux qui composent la famille des médulles* (n.p., n.d.), written with Lesueur; *Mémoire sur le nouveau genre Pyrosoma* (n.p., n.d.); *Mémoire sur quelques faits zoologiques applicables à la théorie du globe, lu à la classe des sciences physiques et mathématiques de l'Institut national* (n.p., n.d.); *Notice sur l'habitation des animaux marins* (n.p., n.d.), written with Lesueur; *Observations sur la dysenterie des pays chauds et sur l'usage du bétel* (n.p., n.d.); *Précis d'un mémoire lu à l'Institut national sur la température des eaux de la mer, soit à sa surface, soit à diverses profondeurs* (Paris, 1816); and *Sur les Méduses du genre Eguorée* (Paris, 1816), 55–99, written with Lesueur.

See also *Voyage de découvertes aux terres australes exécuté sur les corvettes le Géographe, le naturaliste et la goëlette la Casuarina pendant les années 1800–1804 . . .*,

I (Paris, 1807), with a preface by Cuvier; *Voyage de découvertes aux terres australes. Historique*, II (Paris, 1816), continued by M. Louis Freycinet; *Voyage de découvertes aux terres australes . . . navigation et géographie . . . par L. Freycinet* (Paris, 1815); *Voyage de découvertes aux terres australes, fait par ordre du gouvernement sur les corvettes le Géographe, le Naturaliste et la goëlette Casuarina pendant les années 1800–1804 rédigé par F. Péron, et continué par Louis de Freycinet*, 2nd ed., 4 vols. (Paris, 1824).

II. SECONDARY LITERATURE. On Péron and his work, see in L. G. Michaud, ed., *Biographie universelle*, new ed. (Paris, 1854–1865), XXXII; and Maurice Girard, *François Péron, naturaliste, voyageur aux terres australes. Sa vie, appréciation de ses travaux, analyse raisonnée de ses recherches sur les animaux vertébrés et invertébrés d'après ses collections déposées au Muséum d'Histoire naturelle . . .* (Paris, 1857), with a portrait.

P. JOVET
J. MALLET

PÉROT, JEAN-BAPTISTE GASPARD GUSTAV ALFRED (*b.* Metz, France, 3 November 1863; *d.* Paris, France, 28 November 1925), *physics.*

Alfred Pérot's most important work involved experiments in optical interferometry and in electricity. He was ingenious with apparatus and as a teacher emphasized the importance of direct contact with experimentation.

Pérot studied at the *lycée* in Nancy and then at the École Polytechnique in Paris. In 1884 he returned to Nancy and worked under Blondlot in the physics laboratory of the university. In 1888 he received the *docteur ès sciences* for measurements of the specific volumes of saturated vapors and for determining the mechanical equivalent of heat. (Pérot's result for the mechanical equivalent was good, but the work was interesting mainly because of the indirect method that Pérot used, which was based on the equation of Clapeyron.)

In 1888 Pérot joined the University of Marseilles as *maître de conférences* and became much involved in problems of electricity. He studied dielectric properties and electromagnetic waves, and he was considered an expert on topics relating to the emerging electrical industry. (In 1894 a special chair in industrial electricity was created for him at the university.) He collaborated with Charles Fabry (1894–1901) in developing and using a new method of optical interferometry. In 1901 Pérot accepted an invitation to be the first director of the *laboratoire d'essais* of the Conservatoire des Arts et Métiers—a difficult administrative task. In 1908 he returned to research at the Meudon Observatory and to teaching at the

École Polytechnique. He studied properties of the solar atmosphere (using the interference and spectroscopic techniques he had developed earlier) and problems associated with the triode and with telegraphy.

The initial inspiration for developing interferometry came when Fabry was asked to help in measuring the distance between metallic surfaces about a micron apart. Fabry thought of using interference between rays of light that have undergone different numbers of reflections between the two surfaces, and he and Pérot began to study the very fine fringes produced by reflections between silvered films. This work led to development of the "Fabry-Pérot interferometer." According to Fabry, he and Pérot complemented each other nicely in this work; Fabry was more theoretically inclined, while Pérot imagined the actual mechanical arrangements that would make the technique succeed. Fabry, Pérot, and Macé de Lepinay (who already in 1885 used optical methods to determine thicknesses) used the new interferometer to determine the mass of a cubic centimeter of water. Later, Fabry and Pérot used the silver-film interferometer as a spectroscopic analyzer and measured the wavelengths of the black lines in the solar spectrum—thus making it possible to correct small errors in Rowland's wavelengths.

Pérot's analysis of solar spectra involved some interesting problems, since small shifts in wavelength are produced by a variety of causes (for example, pressure effects, convection currents, and the Doppler effect due to rotation). Pérot was inspired to separate some of these effects, and from 1920 to 1921 he tried to verify the gravitational red shift of Einstein's general theory of relativity. (Experimental verification of the gravitational red shift is difficult, and conclusive measurements were not made until 1960; see R. H. Dicke, *The Theoretical Significance of Experimental Relativity* [New York, 1964], 25–27.)

BIBLIOGRAPHY

I. ORIGINAL WORKS. Perot's articles include "Sur la mesure du volume spécifique des vapeurs saturées et la détermination de l'équivalent mécanique de la chaleur," in *Journal de physique*, 7 (1888), 129–148; "Les applications industrielles d'électricité," in *Annales de la Faculté des Sciences de Marseille*, 4 (1895); "Mesure de petites épaisseurs en valeur absolue," in *Comptes rendus hebdomadaires des séances de l'Académie des sciences*, 123 (1896), 802–805, written with C. Fabry; "Sur une nouvelle méthode de spectroscopie interférentielle," *ibid.*, 126 (1898), 34–36, written with C. Fabry; and "Mesure de la pression de l'atmosphère solaire dans la couche du magnésium et vérification du principe de relativité," *ibid.*, 172 (1921), 578–581. More of Perot's papers are listed in Royal Society *Catalogue of Scientific Papers*, 4th ser., XVII, 798, and in Poggendorff, IV, 1140–1141; V, 958–959; and VI, 1984.

II. SECONDARY LITERATURE. On Pérot's life and work, see the essays by Charles Fabry in *Bulletin de la Société astronomique de France*, 40 (1926), 40–43 (with the announcement on pp. 2–3), and in *Astrophysical Journal*, 64 (1926), 209–214, which includes information about the initial development of the silver-film interferometer. See also the bibliography to the article on C. Fabry.

SIGALIA DOSTROVSKY

PERRAULT, CLAUDE (*b*. Paris, France, 25 September 1613; *d*. Paris, 11 October 1688), *zoology, medicine, plant and animal physiology, architecture, mechanical engineering.*

Perrault was the son of Pierre Perrault, originally from Tours and an advocate at the Parlement de Paris, and Paquette Leclerc. It was a talented, versatile, and close-knit family; his brothers were the fairy-tale writer Charles Perrault and the hydrologist Pierre Perrault. As boys, the brothers collaborated in such things as writing mock-heroic verse, and in adult life each aided the career of the other. Perrault was educated at the Collège de Beauvais and then trained as a physician; he presented his thesis at the University of Paris in 1639. He then practiced quietly for the next twenty years, acquiring a reputation, but publishing nothing until he was invited to become a founding member of the Académie des Sciences in 1666. He may have owed this invitation, in part, to the influence of his brother Charles, who was then assistant to the chief minister, Colbert, patron of the Academy.

In June 1667 the Academy was invited to dissect a thresher shark and a lion which had died at the royal menagerie. The reports on these dissections were the first of a long series of anatomical descriptions, which ultimately included those of twenty-five species of mammals, seventeen birds, five reptiles, one amphibian, and one fish. These were eventually assembled in 1676 as memoirs toward a natural history of animals and first appeared anonymously. The anatomists worked as a team and every description had to be accepted by all. Nevertheless, Perrault's name has always been attached to the descriptions, and, in the early years at least, he was undoubtedly the leader of the group.

In general the reports followed a traditional pattern: the anatomists first compared the species with the accounts given by the ancient naturalists, then investigated any legends attached to the species, primarily to dispel them. The authors then proceeded to examine the external appearance of the head, the main internal organs, and the skeleton. Although problems of respiration in birds, fish, and aquatic mammals were of interest to them, the Parisian anatomists (like most naturalists of their day) considered the mechanisms of unusual anatomical features to be particularly worthy of investigation. Perrault discussed the structure of bird feathers and their adaptation to flight, and in his examination of ostrich feathers suggested why they were unsuited for this purpose. In the group's initial dissections Perrault stressed the mechanical functions of the spiral intestine of the shark and the mechanism that retracts the claws of the lion.

In the rationalist atmosphere of the day, it was the debunking of old and popular myths that most attracted public attention. The group tested whether salamanders lived in fire, whether pelicans fed their young with their own blood by stabbing their breasts, and whether chameleons could live on air and change their color to match that of their surroundings; in each case they found the old belief false. Perrault and his group did not, however, spend as much time on these points as has been supposed, and Perrault appears to have been prouder of his positive observations, as, for example, his careful description of the protrusion of the tongue of the chameleon (which he falsely attributed to vascular pressure) and the independent swiveling motion of its eyes. Although some of the discoveries on which the Parisians most prided themselves—including the nictitating membrane that Perrault first observed in a cassowary, the external lobation of the kidneys in the bear, and the castoreal glands of the beaver—had been observed earlier, no such detailed and exact descriptions and illustrations had been published before.

The Parisian dissections were made over several years as specimens became available, usually by the death of some animal at the menagerie. During this time, Perrault was certainly thinking about wider problems of comparative anatomy and physiology and botany. He claimed to have conceived independently and expounded to the Academy two theories, which, although subsequently shown to be erroneous, were in his lifetime, and for many years thereafter, highly influential. These theories concerned the circulation of sap in plants and the embryonic growth from preformed germs, which Perrault thought to be present in all parts of the body. He stated that his botanical theory was first proposed to the Academy in January 1667; it was not, however, a strictly circulatory theory. Perrault thought that there were two fluids at work, one conveying nourishment absorbed from the air through the branches and bark of the trunk to the roots, and a second transporting nourishment absorbed from the earth up to the branches through internal channels. His arguments,

which were supported by a number of experiments, had to be reevaluated by later workers, including Hales, who in the eighteenth century refuted this general hypothesis. Perrault's preformation theory, first stated in 1668, was somewhat overshadowed by the similar but more detailed expositions of his contemporaries.

Not until 1680 did Perrault begin to publish an all-embracing natural philosophy which comprehended these theories, together with his other researches in anatomy, various aspects of animal and plant physiology, and acoustics. The influence of Descartes, although scarcely acknowledged, is patent in this work. Accepting the concept of an atmosphere composed of coarser and subtler parts of the air and of a still finer "ethereal body," Perrault claimed that this assumption allowed him to explain the phenomena of elasticity and hardness. These two key ideas then enabled him to account for almost anything else, from metallurgical phenomena to the sounds of different musical instruments. He also thought that peristaltic motion explained the action of arteries and the contraction of muscles.

Perrault's longest essay was devoted to sound (or noise, as he preferred to call it), which he attempted to explain as an agitation of the air. This agitation, however, affects only the ear, which is not touched by wind or other motions of the air. Perrault rejected the concept of sound waves for he thought that sound should be understood as an agitation that occurs in a restricted space and is produced by the impact of particles in a narrow rectilinear beam. He also discussed the comparative anatomy of the organs of hearing in the various animals he had dissected, and discovered that the lower larynx is the organ of sound in birds. In order to establish the difference between sight and hearing, he made similarly detailed comparisons of different organs of vision.

Perrault's basic ideas had probably been developed well before their publication, but he lacked the leisure to write them up. In fact, at the height of his researches in natural history, he was even more active as an architect than as an anatomist. In 1667 he was invited to join the committee that eventually produced a plan for the completion of the Louvre. Much of his time over the next few years must have been devoted to this task (and to the intrigue that went with it), for the colonnade of the Louvre largely follows his plans. In the same year he produced designs for the observatory, which both he and Colbert hoped would be a center for all the activities of the Academy. When it was objected that Perrault's plans were not well suited for astronomical observations, they were modified, but the observatory, when completed, was still mainly his work. He also designed a triumphal arch, built a house for Colbert in Sceaux in 1673, and worked on two Paris churches from 1674 to 1678. The journal of his journey to Bordeaux in the autumn of 1669 contained mainly architectural notes.

In connection with his work on the Louvre, Perrault became interested in the problem of friction in machines. Several of the machines he designed to overcome this problem were used at the Louvre and then, in 1691, at the Invalides. These designs appeared with other inventions, among them a pendulum-controlled water clock and a pulley system to rotate the mirror of a reflecting telescope, in a posthumous collection published by his brother Charles. Perrault also included among his essays one on ancient music, to show its inferiority to that of his own day; but he was also enough of a classicist to translate Vitruvius.

After Colbert's death, the position of the Perrault family declined. Claude Perrault's house was among those torn down to make room for the Place des Victoires and he seems to have spent his last years writing his essays, possibly at his brother's house. But he was a keen academician until his death. He died of an infection received at the dissection of a camel. Although the extraordinary breadth of his interests and his ability to make significant discoveries in so many fields may have prevented him from achieving complete mastery in any one of them, Perrault was nevertheless an original and highly influential figure. Few of his predecessors described so many species in such detail, or with such clarity and precision.

BIBLIOGRAPHY

I. ORIGINAL WORKS. Many of Perrault's reports are included in *Mémoires pour servir à l'histoire naturelle des animaux* (Paris, 1671); for the complex publication history of this work, and of the individual *Descriptions anatomiques* that preceded it, see E. J. Cole, *A History of Comparative Anatomy* (London, 1944), 396–401. Subsequent works are *Essais de physique, ou recueil de plusieurs traites touchant les choses naturelles*, 4 vols. (Paris, 1680, 1688), republished with some minor works as *Oeuvres diverses de physique et de méchanique*, 2 vols. (Leiden, 1721); and *Recueil de plusieurs machines de nouvelle invention* (Paris, 1700).

II. SECONDARY LITERATURE. On Perrault and his work, see Charles Perrault, *Mémoires de ma Vie* (published with Claude Perrault), *Voyage à Bordeaux*, P. Bonnefon, ed. (Paris, 1909) and *Les hommes illustres qui ont paru en France, pendant ce siècle*, I (Paris, 1696), 67–68; J. Colombe, "Portraits d'ancêtres: III. Claude Perrault," in *Hippocrate*, **16**, nos. 4–5 (1949), 1–47; Marquis de Condorcet, *Éloges des academiciens de l'Académie Royal des Sciences* (Paris, 1773), 83–103; and A. Hallays, *Les Perrault* (Paris, 1926).

Perrault's anatomical descriptions are analyzed by E. J.

Cole (see above), 393–458; his architectural work is discussed in L. Hautecoeur, *Histoire de l'architecture classique en France*, III (Paris, 1948), 441–461; and the "Essais de physique" are discussed in J. Leibowitz, *Claude Perrault, physiologiste* (Paris, 1930).

The Perrault papers at the Academy are listed in a descriptive catalog (not seen by the author) prepared by Alan Gabbey. A copy is deposited in the Archives.

A. G. KELLER

PERRAULT, PIERRE (*b.* France, 1611; *d.* France, 1680), *natural history*.

Very little is known about the personal life of Pierre Perrault, who was rather overshadowed, at least during his lifetime, by his three younger brothers: Claude (1613–1688), a physician, scientist, and the architect of the Louvre; Nicholas (1624–1662), a noted theologian; and Charles (1628–1703), a critic and the author of the Mother Goose fairy tales. Pierre, following in his father's footsteps, became a lawyer and joined the government service as an administrator. He bought the post of receiver-general of finances for Paris. His timing was rather unfortunate, for Louis XIV soon remitted the *tailles* due for the previous ten years. Perrault, along with other tax collectors, encountered financial difficulties. He borrowed on the current year's (1664) revenue and was caught in the act by Colbert, who was then at the height of his power. He was dismissed and was forced to sell his post at a loss. The affair left him almost penniless.

It is not known how Perrault earned his living after the dismissal. He did, however, make a very poor translation of Alessandro Tassoni's *Secchia rapita*.

The book *De l'origine des fontaines* was published anonymously at Paris in 1674. Its authorship, the subject of considerable controversy in the past, has been variously attributed to André Félibien, Denis Papin, and finally to its true author, Perrault. In it he reviewed the various earlier hypotheses on the origin of springs and proposed an experimental investigation to prove that rainfall alone is sufficient to sustain the flow of springs and rivers throughout the year. Perrault considered the Seine River, from its source to Aynay-le-Duc (now Aignay-le-Duc), determining the total drainage area for that portion and making observations of annual rainfall. Using the average annual rainfall, he estimated the total volume of water that precipitated over the drainage area. The losses due to "feeding the trees, herbs, vapours, [and] extraordinary swellings of the river when it rains" were deducted from this figure to obtain sustained runoff. The total annual flow of the Seine was estimated by comparing its flow with that of the Gobelins River near Versailles, which had been measured previously. Having determined the total annual rainfall over the entire drainage area of the River Seine and the total annual flow of the river itself, Perrault experimentally demonstrated that only one-sixth of the annual rainfall was necessary to sustain the river flow. Thus for the first time it was scientifically proven that rainfall is more than adequate to supply river flow.

But Perrault did not believe in the general infiltration of rainwater and, thereby, recharge of groundwater. He went to great lengths to find evidence of general infiltration, and from his observations he concluded that it was only an occasional and local phenomenon. In the beginning of the second part of his book, Perrault differentiated his own view from that held by Vitruvius, Gassendi, Pallisy, and Jean François, which he referred to as "general opinion." He objected to their concept of infiltration of rainwater into the earth and said that he did not believe that there is enough precipitation for the earth to be soaked with it as much as necessary, and still leave over a sufficient quantity to cause rivers and springs.

Perrault's experimental work on the rainfall and runoff of the upper Seine is a milestone in the history of hydrology. Admittedly his experimental techniques were somewhat crude and his figures could have been more refined, but his reasoning was flawless, his method irrefutable, and his was the first experimentation to prove categorically that rivers originated from rainfall. Edmé Mariotte later used more sophisticated measuring techniques to confirm Perrault's findings. The second half of the concept of the hydrologic cycle, that enough water evaporated from oceans and rivers to come down as rainfall, was experimentally proved by Edmond Halley. Biswas, in his *History of Hydrology*, has suggested that Perrault, Mariotte, and Halley should be considered as cofounders of experimental hydrology. But when it is considered that Mariotte and Halley were familiar with Perrault's work and may have been considerably influenced by it, Perrault's contributions to hydrology become all the more important.

An international symposium on the history of hydrology was held at Paris in 1974, by the International Hydrological Decade, to mark the tricentenary of the publication of Perrault's trailblazing book.

BIBLIOGRAPHY

I. ORIGINAL WORKS. Perrault's major writing is *De l'origine des fontaines* (Paris, 1674), trans. by A. LaRocque as *On the Origin of Springs* (New York, 1967). He also translated Alessandro Tassoni's *La secchia rapita* as *Le seau enlevé* (Paris, 1678),

II. SECONDARY LITERATURE. See Asit K. Biswas, "Beginning of Quantitative Hydrology," in *Journal of the Hydraulics Division, American Society of Civil Engineers*, **94** (1968), 1299–1316; and *History of Hydrology* (Amsterdam, 1970), 208–213; and S. Delorme, "Pierre Perrault," in *Archives internationales d'histoire des sciences*, **27**, no. 3 (1948), 388–394.

MARGARET R. BISWAS
ASIT K. BISWAS

PERRIER, EDMOND (*b.* Tulle, France, 9 May 1844; *d.* Paris, France, 31 July 1921), *zoology*.

The son of a school principal in Tulle, Perrier began his education in that city and completed his secondary education in Paris. In 1864 he scored well on the competitive entrance examinations to both the École Polytechnique and the science section of the École Normale Supérieure. At the suggestion of Pasteur, who was then its director, he chose the latter institution, where his teacher was Lacaze-Duthiers. He earned his *licence ès sciences* in mathematics and physics in 1866, passed the *agrégation* in physics in 1867, and began teaching physics at the *lycée* of Agen. A few months later he returned to Paris where Lacaze-Duthiers had secured his appointment as *aide-naturaliste* at the Muséum d'Histoire Naturelle. He obtained his *licence* in natural science in 1868 and the following year defended a dissertation, "Recherches sur les pédicellaires et les ambulacres des astéries et des oursins," that earned him the *doctorat ès sciences naturelles*. Named *maître de conférences* in zoology at the École Normale Supérieure in 1872, he held similar posts several years later at the *écoles normales* of Sèvres and of St.-Cloud. He was appointed professor-administrator at the Muséum d'Histoire Naturelle in 1876 and in 1900 became its director, a post he held for some twenty years. Elected to the Paris Academy of Sciences in 1892, he became its president in 1913.

The major portion of Perrier's zoological work is devoted to the anatomy, physiology, and taxonomy of the invertebrates. He participated in oceanographic expeditions in the Atlantic in 1881 and in the Mediterranean in 1883 and established the Museum's marine biology laboratory at St.-Vaast-la-Hougue in 1887. He also was called upon to classify the material obtained on several expeditions, notably the starfish collected under the direction of Alexander Agassiz.

Perrier, who declared his acceptance of the theory of evolution in 1879, was always particularly interested in the study of the oligochaetes and echinoderms, since these groups represented the two major types of animal organization: segments arranged in linear series and segments radiating from a center. His richest theoretical work, and perhaps his most original, is *Les colonies animales* (1881). In it he attempted to comprehend the evolutionary formation of groups of organisms, starting with the simplest creatures, which are favored in their evolution by their ability to reproduce by division or budding. Certain of these creatures remained independent, while others, Perrier explained, agglomerated into colonies in which, at first, they preserved their individuality. In a subsequent stage this individuality was eliminated by the effect of a division of physiological labor implying a reciprocal dependence as well as a differentiation of forms. The later transformations of the linear or irregular (simple or coalescent) colonies were, on this view, the origin of the major taxonomic divisions.

At the time of its publication (1902) another work by Perrier, "Tachygenèse" (written in collaboration with Charles Gravier), was also of theoretical interest. In it Perrier sought to explain apparent difficulties of the biogenetic law by the phenomenon of embryogenic acceleration (according to which in a series of organisms, "the higher the given organism is in the series, the more rapid, in general, is the development and the more advanced is the stage of development at which hatching occurs") and by the occurrence of embryonic adaptations that modify the subsequent development of the organism.

Perrier quickly became one of the principal defenders in France of the theory of evolution; but he was never a Darwinian in the strict sense and was one of those mainly responsible for the revival of Lamarckism in France. Perrier was interested in the history of his discipline and wrote, in addition to his book on Lamarck, a long preface to Quatrefage's *Émules de Darwin* and *La philosophie zoologique avant Darwin*, which is still a useful source for the study of biology in the nineteenth century.

The author of a substantial *Traité de zoologie* and several textbooks, Perrier was also director of the *Annales des sciences naturelles* (zoologie) from 1900 until his death. His many popular articles appeared mainly in *Revue scientifique* and in the newspaper *Le temps*.

BIBLIOGRAPHY

I. Original Works. A chronological list of Perrier's scientific publications is at the end of the article by R. Anthony cited below. Perrier himself prepared, on the occasion of his candidacies for the Paris Academy of Sciences, *Notice sur les travaux scientifiques de H. O. Edmond Perrier* (Paris, 1875, 1886, 1892), containing an analysis of his works.

The following works are of particular importance: "Recherches sur les pédicellaires et les ambulacres des astéries et des oursins," in *Annales des sciences naturelles*, 5th ser., **12** (1869), 197–304; **13** (1870), 5–81; "Recherches pour servir à l'histoire des lombriciens terrestres," in *Archives (nouvelles) du Muséum d'histoire naturelle*, **8** (1872), 5–198; "Révision de la collection des stellérides du Muséum d'histoire naturelle," in *Archives de zoologie expérimentale et générale*, **4** (1875), 265–450 and **5** (1876), 1–104, 209–304; "Le transformisme et les sciences physiques," in *Revue scientifique*, **16** (1879), 890–895; "Rôle de l'association dans le regne animal," *ibid.*, **17** (1879), 553–559; *Les colonies animales et la formation des organismes* (Paris, 1881); "Sur l'appareil circulatoire des étoiles de mer," in *Comptes rendus . . . de l'Académie des sciences*, **94** (1882), 658–661, written with J. Poirier; *La philosophie zoologique avant Darwin* (Paris, 1884); *Le transformisme* (Paris, 1888); *Traité de zoologie*, 10 fascs. (Paris, 1893–1932); "La tachygenèse ou accélération embryogénique, son importance dans les modifications des phénomènes embryogéniques, son rôle dans la transformation des organismes," in *Annales des sciences naturelles* (zoologie), **16** (1902), 133–374, written with C. Gravier; *La terre avant l'histoire* (Paris, 1921); and *Lamarck* (Paris, 1925).

The bulk of Perrier's MSS are in the archives of the Paris Academy of Sciences and the Muséum National d'Histoire Naturelle.

II. Secondary Literature. To date no full study of Perrier's work exists, but the following articles are useful: R. Anthony, "Edmond Perrier, 1844–1921," in *Archives du Muséum national d'histoire naturelle*, 6th ser., **1** (1926), 1–14; C. Gravier, "En souvenir de M. Edmond Perrier," in *Bulletin du Muséum national d'histoire naturelle* (1921), no. 7; and M. Phisalix, "Edmond Perrier (1844–1921)," in *Bulletin de l'Association des élèves de Sèvres* (Jan. 1922).

C. Limoges

PERRIER, GEORGES (*b.* Montpellier, France, 28 October 1872; *d.* Paris, France, 16 February 1946), *geodesy.*

Perrier was the son of François Perrier, who revived French geodesy and created the Service Géographique de l'Armée. He graduated from the École Polytechnique in 1894 as an artillery officer, intending to continue his family's military and scientific tradition. For more than half a century his activities centered on geodesy, interrupted only by World War I and certain peacetime requirements of his military career. In addition to achieving important results as an officer in the Service Géographique de l'Armée, he concentrated on two major geodesic tasks.

First, as a young officer he played a major role in preparing and executing a scientific mission sent to Peru and Ecuador to measure an arc of meridian at low latitudes. Lasting from 1901 to 1906, the project was particularly difficult because of the topography and climate of the Andean cordilleras. Perrier was the only geodesist who participated from beginning to end. After returning to France he was placed in charge of evaluating and processing all the measurements and of publishing the results. This overwhelming assignment, delayed by the two world wars, dragged on and eventually diminished in significance as a result of the development of new geodetic techniques.

Second, from 1919 until his death Perrier served as secretary-general of the organization created in 1919 and several years later named the Association Internationale de Géodésie. This body replaced the Association Géodésique Internationale, which had had its headquarters in Potsdam and the activities of which had been interrupted in 1914 by the war. Transforming the general secretariat of the new association into a center of activity in geodesy, Perrier suggested and encouraged many undertakings, improved existing publications, and created new ones.

BIBLIOGRAPHY

I. Original Works. Perrier produced the following portions of Ministère de l'Instruction Publique, *Mission du Service géographique de l'armée pour la mesure d'un arc de méridien équatorial en Amérique du Sud, sous le contrôle scientifique de l'Académie des sciences (1899–1906)*, pt. B, *Géodésie et astronomie*: II, fasc. 1, *Notices sur les stations (Atlas)* (Paris, 1913); II, fasc. 1, *Notices sur les stations (Atlas). Appendice: Origine, notation et sens des noms géographiques de l'Atlas; vocabulaires espagnol–français et quichua–français*" (Paris, 1918); III, fasc. 1, *Angles azimutaux* (Paris, 1910); III, fasc. 2, *Compensation des angles, calcul des triangles* (Paris, 1912); III, fasc. 7, *Latitudes astronomiques observées aux théodolites à microscopes*, pt. 1, "Considérations générales," preceded by "Introduction historique" (also found in III, fasc. 8, *Latitudes astronomiques observées aux astrolabes à prisme*) (Paris, 1925); and III, fasc. 7, pts. 2 and 3, *Tableaux numériques des observations et conclusions* (Paris, 1911).

His other writings include *Pascal* (Paris, 1901), written with A. Hatzfeld—"Travaux scientifiques," pp. 113–191, is by Perrier; "La figure de la terre, les grandes opérations géodésiques, l'ancienne et la nouvelle mesure de l'arc méridien de Quito," in *Revue annuelle de géographie*, 2nd ser., **2** (108), 201–508; "Les académiciens au Pérou, 1735–1744," in *L'astronomie* (Mar.–Apr. 1911); *La géodésie militaire française, historique et travaux actuels et La géodésie moderne à l'étranger* (Paris, 1912), two lectures given at the Service Géographique de l'Armée; *Union géodésique et géophysique internationale: Première assemblée générale, Rome 1922, section de géodésie* (Toulouse, 1922); *Bibliographie des oeuvres de géographie mathématique publiées en France de 1910 à 1920 inclus, précédé d'un projet de classification et d'indexation pour la géo-*

graphie mathématique, which is **19**, sec. 2, no. 1 of *Bibliographie scientifique française* (1922); *Comptes rendus de la première assemblée générale de la section de géodésie de l'union géodésique et géophysique internationale, réunie à Rome en mai 1922, rédigés et publiés avec 18 annexes par . . . G. Perrier* (Toulouse, 1923)—the appendixes contain, under Perrier's signature, the secretary-general's administrative and financial report (79–83), a note on the junction of the French and Italian triangulations, (95–100), and a note on the electric recording of the oscillations of a pendulum (148–157), written with Gustave Ferrié.

Additional works are "Où en est la géodésie? Les problèmes et les travaux actuels," in *L'astronomie*, **37** (1923), 433–457, 505–526; "La deuxième assemblée générale de l'Union géodésique et géophysique internationale, section de géodésie, Madrid, . . . 1924," in *Bulletin géodésique*, no. 4 (Oct.–Dec. 1924), 241–278; "Les raisons géodésiques de l'isostasie terrestre," in *Annuaire du Bureau des longitudes* for 1926, Sect. B., *Tables de l'ellipsoïde de référence internationale* (Paris, 1928), written with E. Hasse; "La coopération internationale en géodésie et en géophysique. Troisième assemblée générale de l'Union internationale de géodésie et géophysique, Prague, 1927," in *Annuaire du Bureau des longitudes* for 1928, sect. C; "Triangulation de détail des régions andines centrale et septentrionale," in *Géographie*, **49**, nos. 5–6 (May–June 1928), 365–385, and **50**, nos. 1–2 (July–Aug. 1928), 26–49; "L'Académie des sciences, le Bureau des Longitudes et les grandes missions scientifiques," in *Annuaire du Bureau des longitudes* for 1933, Sect. C; and *Petite histoire de la géodésie* (Paris, 1939).

II. Secondary Literature. See Élie Cartan, "Information nécrologique . . .," in *Comptes rendus . . . de l'Académie des sciences*, **222**, no. 8 (18 Feb. 1946), 421–423; M. Delhau, "Le général Georges-François Perrier," in *Bulletin des séances. Institut r. colonial belge*, **18**, no. 1 (1947), 127–164, with photograph; "Le général Georges Perrier," in *Bulletin géodésique*, n.s. no. 1 (July 1946), 7–21, with photograph; "Notice nécrologique sur le général Perrier," in *Comptes rendus du Comité National français de géodésie et géophysique* for 1946, 17-21; and *Notice sommaire sur les titres et travaux scientifiques de M. Georges Perrier* (Toulouse, 1926).

A. Gougenheim

PERRIN, JEAN BAPTISTE (*b*. Lille, France, 30 September 1870; *d*. New York, New York, 17 April 1942), *physical chemistry.*

Along with his two sisters, Perrin was raised in modest circumstances by his mother, after his father, an army officer, died of wounds received in the Franco-Prussian War. Perrin obtained his secondary education in Lyons and at the Lycée Janson-de-Sailly in Paris, where his special preparation in mathematics enabled him, in 1891, after serving one year in the army, to gain entrance into the École Normale Supérieure. In 1895 he became *agrégé-préparateur* at the École Normale, and two years later he completed his doctorate.

Perrin's years as a student at the École Normale were exceptionally formative, owing primarily to the influence of his teacher, Marcel Brillouin, who was an outspoken advocate of Boltzmann's "statistical mechanics" and an outspoken adversary of Ostwald's and Mach's "energetics." It is possible to detect Perrin's atomistic biases even in his first paper of 1895, in which he reported experiments demonstrating that cathode rays are negatively charged by collecting them in a "Faraday cup" (an open-ended metal cylinder with appropriate electrical connections). In 1896 Perrin won the Joule Prize of the Royal Society for his experiments on cathode rays and for certain preliminary studies on Röntgen's recently discovered X rays; this formed the basis for his doctoral thesis the following year.

Soon after receiving his degree, Perrin married Henriette Duportal; they had a daughter, Aline, and a son, Francis. Perrin was placed in charge of developing a course in physical chemistry at the Sorbonne, for which he wrote his pro-Boltzmannian *Traité de chimie physique. Les principes* (1903). For Perrin, these years, in general, were years of transition. The focus of his research shifted from cathode rays and X rays, and a general concern with the atomic hypothesis (in 1901 he suggested, for example, that the atom was like a miniature solar system), to experiments on ion transport and the whole problem of how an electrolyte transfers its charge to the walls of a container (*électrisation par contact*). It was out of these studies, in turn, and the stimulation provided by Siedentopf and Zsigmondy's 1903 invention of the "slit ultramicroscope," that Perrin's interest arose in the behavior of colloidal particles and, in particular, in their Brownian motion. By 1906 this problem had already begun to attract his attention, and in 1908 he inaugurated his classic series of experiments on the subject.

It struck Perrin that colloidally suspended particles undergoing Brownian motion (as a result of collisions with the molecules of the surrounding fluid) should distribute themselves vertically in a definite way at equilibrium. Only after finding experimentally that their number decreases exponentially with increasing height, and only after proving that this variation (and hence a definite value of Avogadro's number) follows from kinetic theory, did Perrin learn, through Langevin, of Einstein's and Smoluchowki's 1905–1906 theoretical papers on Brownian motion, and sub-sequently understand that his work was also consistent with theirs.

In auxiliary experiments Perrin proved that Stokes's law (and hence his calculation of the particle's mass) was valid for particles as small as 0.1 micron. In 1909 Perrin's student Chaudesaigues also demonstrated the accuracy of Einstein's prediction that the mean displacement of a given particle undergoing Brownian motion is proportional to the square root of the time of observation, a result that undercut earlier criticisms of Einstein's work by Svedberg and others. During the same year Perrin continued to refine and extend his experiments (for example, he verified Einstein's formula for rotational Brownian motion). Perrin's work brought him a great deal of formal recognition over the years: in 1909 he was awarded the Prix Gaston Planté; he was appointed to a chair of physical chemistry especially created for him at the Sorbonne, which he held for three decades (1910–1940); in 1911 and 1921 he received invitations to the extremely influential Solvay conferences; in 1923 he was elected to the Académie des Sciences (he became its president in 1938); he was awarded eight honorary degrees, several prizes, and membership in seven foreign academies of science; and in 1926 he received the Nobel Prize for physics. His most fundamental conclusion—that he had finally uncovered irrefutable proof for the real existence of atoms—contrary to the assertions and expectations of Ostwald, Mach, and others—was soon universally accepted and popularized in his book *Les atomes* (1913), which went through many editions and translations.

As an army officer in World War I, Perrin worked on acoustic detection devices for submarines and other military equipment. Between 1918 and 1921 he studied the phenomenon of fluorescence and the interaction between light and matter. He simultaneously demonstrated his insight into current problems of nuclear physics by offering essentially correct, albeit qualitative, speculations on the origin of solar energy and on the nature of nuclear reactions. In subsequent years, as a convinced socialist, Perrin became increasingly involved with the institutional development of science in France. In the late 1930's, for example, he was primarily responsible both for establishing the Centre National de la Recherche Scientifique and for founding the Palais de la Découverte in Paris. In 1940 his well-known and outspoken antifascism made it necessary for him to emigrate from France. He came to the United States, where he helped establish the French University of New York (École Libre des Hautes Études). His son, Francis, was then teaching at Columbia University, where Perrin himself had been an exchange professor in 1913.

Perrin died in New York, but after the war his remains were returned to his homeland and buried in the Panthéon.

BIBLIOGRAPHY

I. ORIGINAL WORKS. Perrin's initial papers were "Nouvelles propriétés des rayons cathodiques," in *Comptes rendus hebdomadaires des séances de l'Académie des sciences*, **121** (1895), 1130; and his doctoral thesis, "Rayons cathodiques et rayons de Roentgen," in *Annales de chimie et de physique*, **11** (1897), 496–554. Two of his early Sorbonne papers were "Mécanisme de l'électrisation de contact et solutions colloïdales," in *Journal de chimie physique*, **2** (1904), 601–651; **3** (1905), 50–110. He offered a summary of his 1908–1909 work (especially as it had appeared earlier in *Comptes rendus*) in "Mouvement Brownien et réalité moléculaire," in *Annales de chimie et de physique*, **18** (1909) 1–114, trans. into English by Frederick Soddy (London, 1910) and into German by J. Donau (Dresden–Leipzig, 1910). Two of Perrin's postwar papers were "La fluorescence," in *Annales de physique*, **10** (1918), 133–159, and "Matière et lumière," *ibid.*, **11** (1919), 1–108. These papers, and later papers directed at institutional concerns, have been collected under the title *Oeuvres scientifiques de Jean Perrin* (Paris, 1950). Perrin's two most important books are *Traité de chimie physique*, I, *Les principes* (Paris, 1903), and *Les atomes* (Paris, 1913).

II. SECONDARY LITERATURE. The most comprehensive study of Perrin's life and work is Mary Jo Nye, *Molecular Reality* (London, 1972), the bibliography of which lists all important secondary sources on Perrin, certain primary source documents not contained in *Oeuvres scientifiques de Jean Perrin*, and other primary source documents, for example, Einstein's papers on Brownian motion.

ROGER H. STUEWER

PERRINE, CHARLES DILLON (*b.* Steubenville, Ohio, 28 June 1867; *d.* Villa General Mitre, Argentina, 21 July 1951), *astronomy.*

After a brief career in business, Perrine, who was skilled in photography, became professionally interested in astronomy. In 1893 he joined the staff at the Lick Observatory as secretary; shortly thereafter he became widely known for his discovery, observation, and calculation of orbits of comets and for his determination of solar parallax from observations of the asteroid Eros. He was made acting astronomer and then astronomer. In 1901 he discovered motion in the nebulosity surrounding a nova in Perseus.

Between 1900 and 1909 Perrine accompanied four eclipse expeditions, directing one in 1901 from the Lick to Sumatra. He also spoke on eclipses at the International Congress of Sciences (1904) in St. Louis. Perrine's most widely acclaimed scientific achievement occurred in 1904 and 1905, when he discovered the sixth and seventh satellites of Jupiter. His subsequent research, initiated by J. A. Keeler, with the Crossley reflector at the Lick was less publicized but of considerable importance; it led to most of the work contained in volume VIII of the *Publications of the Lick Observatory*, which dealt with nebulae and star clusters.

In 1909 Perrine was appointed director of the Argentine National Observatory at Cordoba. He was also responsible for the establishment of the astrophysical station in Bosque Alegre. The sixty-inch reflector at this station made it one of the principal observatories in the Southern Hemisphere. Sixteen volumes of the *Resultades del Observatorio Nacional Argentino* were published during Perrine's directorship.

Despite his scientific achievements, Perrine and his office became a target for nationalist politicians and he was attacked verbally by deputies in the Argentine Congress. In 1931 he was barely missed by a sniper's bullet and in 1933 the Argentine Congress passed legislation removing authority from the director of the observatory. He retired, under duress, in 1936.

Perrine received several honors, including the Lalande Prize from the Paris Academy of Sciences in 1897; the Gold Medal from the Sociedad Astronómica de Mexico; and an honorary Sc.D. from the University of Santa Clara, California, in 1905. He was president of the Astronomical Society of the Pacific in 1907, and a member of various American and foreign societies.

BIBLIOGRAPHY

I. ORIGINAL WORKS. Perrine published nearly 200 papers. His most important articles include "Comet c 1895," in *Publications of the Astronomical Society of the Pacific*, **7** (1895), 342–343; "Preliminary Report of Observations of the Total Solar Eclipse of 1901, May 17, 18," in *Astrophysical Journal*, **14** (1901), 349–359; "Motion of the Faint Nebula Surrounding Nova Persei," *ibid.*, 359–362; "Some Total Eclipse Problems," *ibid.*, **20** (1904), 331–337; "Experimental Determination of the Solar Parallax From Negatives of Eros Made With the Crossley Reflector," in *Publications of the Astronomical Society of the Pacific*, **16** (1904), 267; "Discovery of a Sixth Satellite to Jupiter," *ibid.*, **17** (1905), 22–23; and "The Seventh Satellite of Jupiter," *ibid.*, 62–63.

II. SECONDARY LITERATURE. Biographical information on Perrine is given in Jacques Cattel, ed., *American Men of Science*, 8th ed. (Lancaster, Pa., 1949), 1928; and a brief obituary by Jorge Bobone appears in *Publications of the Astronomical Society of the Pacific*, **63** (1951), 259.

A collection of some of Perrine's correspondence with Hale is contained in "The George Ellery Hale Papers, 1882–1937," microfilm ed., California Institute of Technology (Pasadena, Calif., 1968).

RICHARD BERENDZEN
DANIEL SEELEY

PERRONCITO, EDOARDO (*b.* Viale d'Asti, Italy, 1 March 1847; *d.* Pavia, Italy, 4 November 1936), *parasitology, bacteriology.*

Perroncito studied at the University of Turin and obtained a degree in veterinary medicine in 1867. In 1873 he won a public competition for the chair of veterinary pathological anatomy at Turin. His professorship was confirmed the following year; he was only twenty-seven.

His scientific interests were mainly in parasitology and bacteriology, subjects which although relevant to human pathology had been very little studied. The emphasis of his research shifted from the purely veterinary to the human level. Some of his most important research was on *Echinococcus* and *cysticercosis*, and on other parasitic infections in animals that are easily transmitted to man by infected food. Perroncito extended his interest to the prophylaxis and cure of these infections. He emphasized the importance of hygiene and advocated stricter and more complete supervision of meat and a more practical approach to the construction of slaughterhouses. He also campaigned for the adoption of refrigeration to preserve food products. He carried out important studies on bovine tuberculosis, which he demonstrated to be identical to human tuberculosis; and he was active in making known in Italy Pasteur's studies on the prophylaxis and cure of rabies.

There was almost no aspect of the infective pathology and parasitology of animals (and of some plants) that Perroncito did not carefully investigate, in many cases discovering their etiopathogenetic cause and searching for means of prevention and cures. These activities led to the creation in 1875 of the first chair of parasitology in Italy. The post, established at Turin, was created for Perroncito.

Perroncito's name is especially connected with his efforts to identify the cause of the anemia that was killing large numbers of the miners (mainly Italian) working on the St. Gotthard Tunnel. His research was also relevant to the health of thousands of other workers in mines, furnaces, and similar environments. Perroncito discovered that the fatal illness was caused by the presence in the human body of the worm *Anchylostoma duodenale*, already described by Dubini in 1843. He studied its complete biological cycle and

means of diffusion, and found that it could be eliminated by means of a medicine based on the oil of the male fern.

The priority of Perroncito's discovery was disputed by others involved in the same research. The polemic lasted many years, and Perroncito became very embittered, but his priority was finally acknowledged. In 1932, on the fiftieth anniversary of the completion of the St. Gotthard Tunnel, the Institut de France awarded him the Prix Montyon. His other awards included honorary degrees from academies, universities, and societies throughout the world.

BIBLIOGRAPHY

I. ORIGINAL WORKS. Perroncito's writings include *Sugli echinococchi negli animali domestici* (Turin, 1871); and "La tubercolosi in rapporto alla economia sociale e rurale," in *Annali della Accademia d'agricoltura di Torino*, **18** (1875). See also *L'anemia dei minatori in Ungheria* (Turin, 1886); *Sulla trasmissione della rabbia dalla madre al feto attraverso la placenta e per mezzo del latte* (Turin, 1887); *Studi sull'immunità pel carbonchio* (Turin, 1889); *Sulla utilizzazione delle carni degli animali da macello affetti da tubercolosi* (Turin, 1892); and *La maladie des mineurs . . . une question résolue* (Turin, 1912).

II. SECONDARY LITERATURE. See P. Ghisleni, "Edoardo Perroncito," in *Giornale dell'Accademia di medicina di Torino*, **100** (1937), 39–47, with bibliography; and V. Marzocchi, "Edoardo Perroncito," in *Patologia comparata della tubercolosi*, **3** (1937), 96–98.

CARLO CASTELLANI

PERRONET, JEAN-RODOLPHE (*b.* Suresnes, France, 8 October 1708; *d.* Paris, France, 27 February 1794), *civil engineering.*

Perronet was the son of a Swiss officer in French service. His maternal uncle, the mathematician Crousaz, encouraged his early interest in mathematics; but Perronet gave up his plan to join the Corps du Génie Militaire (he had passed the entrance examinations) when his father died. Instead, at the age of seventeen he entered the office of Debeausire, architect to the city of Paris. He soon carried out assignments of considerable responsibility. About 1735 he was named *sous-ingénieur* of the Corps des Ponts et Chaussées in the administrative district (*généralité*) of Alençon, where he designed roads. Perronet returned to Paris in 1747, when the Corps des Ponts et Chaussées appointed him head of its newly founded Bureau Central des Dessignateurs; this institution was designed mainly as a training center for young engineers and was later renamed École des Ponts et

Chaussées. Henceforth Perronet's engineering activities concentrated on the design and construction of bridges. In 1750 he became inspector general, and in 1763 head, of the Corps des Ponts et Chaussées, with the title *premier ingénieur du roi*. In addition, from 1757 until 1786 he served as inspector general of France's saltworks. He received many honors, and his fame remained undiminished to his death.

Apart from his institutional role as the highest-ranking civil engineer of the French state during the last decades of the *ancien régime*, Perronet's significance is twofold. As director of the École des Ponts et Chaussées, from the school's inception to his death in 1794, he was the founder of one of the world's first engineering schools. He graduated some 350 students, outstanding among whom were Antoine de Chézy, Emiland Marie Gauthey, and Riche de Prony.

In the design of bridges Perronet developed the classical stone arch bridge to its ultimate perfection. He increased the span of the individual arches, reduced the width of the piers, and shaped the arches in curves composed of several circle segments, which combined aesthetic elegance with ease of construction. This design not only minimized the interference of the bridge with the flow pattern of the river below, at normal level and in floods, but it also reduced the weight of the bridge, and hence its load upon the foundations. The best-known bridges—among the thirteen that Perronet designed—are the Pont de Neuilly (completed 1774), the Pont Sainte-Maxence (1785), and the Pont de la Concorde (1791), still standing.

BIBLIOGRAPHY

I. Original Works. Perronet's chief work is *Description des projets et de la construction des ponts de Neuilly, de Mantes, d'Orléans . . .* 3 vols. (Paris, 1782–1789). He also wrote a number of memoirs, most of which are on civil engineering; a bibliography compiled by W. Hoffmann (see below) lists eleven items.

II. Secondary Literature. Biographical works by Perronet's contemporaries are Pierre-Charles Lesage, *Notice pour servir à l'éloge de Perronet* (Paris, 1805); and G.-C.-F.-M. Riche de Prony, *Notice historique sur Jean-Rodolphe Perronet* (Paris, 1829). Useful articles on Perronet are in *Nouvelle biographie générale*, XXXIX (Paris, 1865), cols. 650–652; and, especially, W. Hoffmann, in J. S. Ersch and J. G. Gruber, eds., *Allgemeine Encyklopädie der Wissenschaften und Künste*, pt. 17, sec. 3 (Leipzig, 1842), 272–280. Perronet's bridges are discussed in James Kip Finch, "The Master of the Stone Arch," in *Consulting Engineer* (London), **18**, no. 4 (Apr. 1962), 128–132.

OTTO MAYR

PERROTIN, HENRI JOSEPH ANASTASE (*b.* St. Loup, Tarn-et-Garonne, France, 19 December 1845; *d.* Nice, France, 29 February 1904), *astronomy.*

Perrotin was the son of an employee in the telegraph service. His scholastic ability earned him scholarships to the *lycée* at Pau and the Faculté des Sciences of Toulouse. His professor at the latter, F. Tisserand, invited him to work at the Toulouse observatory (of which he was the director); and in 1873 he appointed Perrotin astronomer there.

During his career Perrotin made many observations, both astrometric (double stars, planets, satellites, asteroids, comets) and astrophysical (sunspots, study of planetary surfaces). He discovered five asteroids between 1874 and 1878 and a sixth in 1885. Perrotin turned to celestial mechanics and in his doctoral dissertation established the first precise theory of the asteroid Vesta (1879). In this work he expanded the perturbing function as far as the eighth order relative to the eccentricities and inclinations, an achievement that has been applied by astronomers to verify recent theories of Vesta.

In 1879 Perrotin was engaged by Raphaël Bischoffsheim, the banker who built the Nice observatory, to become director of the observatory and to install its equipment. Perrotin devoted himself to this task until his death. The most important of the instruments that he put into service was the seventy-six-centimeter refractor (1886), which was then the world's largest. An inspiring leader of men, he obtained a great deal from his collaborators; and the first years of the observatory were marked by important projects, notably in spectroscopy and work on the asteroids. Perrotin's measurements of the speed of light were based on the slotted-wheel method applied to beams sent between the observatory and surrounding hilltops. The value of 299,880 kilometers per second, obtained in 1902 utilizing a combined trajectory of ninety-two kilometers, was considered the best estimate for more than thirty years.

In 1892 Perrotin was elected a corresponding member of the Académie des Sciences, which had twice awarded him a prize; and in 1894 he became a corresponding member of the Bureau des Longitudes.

BIBLIOGRAPHY

I. Original Works. Perrotin's most important papers are "Théorie de Vesta," in *Annales de l'Observatoire astronomique magnétique et météorologique de Toulouse*, **1** (1880), B1–B90, also in *Annales de l'Observatoire de Nice*, **3** (1890), B1–B118, and **4** (1895), A3–A71; "Détermination des différences de longitudes entre Nice, l'Île Rousse et Ajaccio," in *Annales de l'Observatoire de Nice*,

8 (1904), 3–242, written with P. Hatt and L. Driencourt; and "Détermination de la vitesse de la lumière . . . ," *ibid.*, **11** (1908), A3–A98, written with A. Prim.

Perrotin published his astronomical observations, principally in about fifty "Notes," in *Comptes rendus . . . de l'Académie des sciences* from 1875 to 1903 and, between 1875 and 1889, in *Astronomische Nachrichten*. Also worthy of mention is "Parallaxe solaire déduite des observations d'Éros," in *Bulletin astronomique*, **20** (1903), 161–165.

Perrotin founded the *Annales de l'Observatoire de Nice*, directed the publication of the first 10 vols., and wrote several of them; among the latter are **2** (1887), devoted to various of his astronomical works, and "Description de l'Observatoire de Nice," in **1** (1899), 1–152. With a view toward the establishment of the Nice observatory, Perrotin made an extensive inquiry, the main findings of which he published in *Visite à divers observatoires d'Europe* (Paris, 1881), which describes the equipment of 31 European observatories.

II. Secondary Literature. The most important obituary notices are the unsigned "Todes-Anzeige," in *Astronomische Nachrichten*, **165** (1904), 254–255; and "M. Henri Perrotin," in *Nature*, **69** (1904), 468; and E. Stephan, "J. A. Perrotin," in *Annales de l'Observatoire de Nice*, **8** (1904), i–iv.

Jacques R. Lévy

PERSEUS (*fl.* third century B.C. [?]), *mathematics.*

Perseus is known only from two passages in Proclus. In one passage his name is associated with the investigation of "spiric" curves as that of Apollonius of Perga is with conics, Nicomedes with the conchoids, and Hippias of Elis with the quadratrices.[1] In the second passage, derived from Geminus, Proclus says that Perseus wrote an epigram upon his discovery, "Three lines upon five sections finding, Perseus made offering to the gods therefor."[2]

In another place Proclus says that a spiric surface is thought of as generated by the revolution of a circle standing upright and turning about a fixed point that is not its center; wherefore it comes about that there are three kinds of spiric surface according as the fixed point is on, inside, or outside the circumference.[3] The spiric surface is therefore what is known today as a "tore"; in antiquity Hero of Alexandria gave it the name "spire" or "ring."[4]

These passages throw no light on the provenance of Perseus and leave wide room for conjecture about his dates. He must have lived before Geminus, as Proclus relies on that author; and it is probable that the conic sections were well advanced before the spiric curves were tackled. Perseus therefore probably lived between Euclid and Geminus, say between 300 and 70 B.C., with a preference for the earlier date.

What Perseus actually discovered is also uncertain. In rather more precise language than that of Proclus, a spiric surface may be defined as the surface generated by a circle that revolves about a straight line (the axis of revolution) always remaining in a plane with it. There are three kinds of spiric surfaces, according as the axis of revolution is outside the circle, tangential to it, or inside it (which are called by Proclus the "open," "continuous," and "interlaced"; and by Hero the "open," "continuous," and "self-crossing").

A spiric section on the analogy of a conic section would be a section of a spiric surface by a plane, which it is natural to assume is parallel to the axis in the first place. Proclus says that the sections are three in number corresponding to the three types of surface, but this is difficult to understand or to reconcile with the epigram. G. V. Schiaparelli showed how three different spiric curves could be obtained by a section of an open tore according as the plane of section was more or less distant from the axis of revolution,[5] and Paul Tannery entered upon a closer mathematical analysis that led him to give a novel interpretation to the epigram.[6] If r is the radius of the generating circle, a the distance of its center from the axis, and d the distance of the cutting plane from the axis, in the case of the open tore (for which $a > r$), the following five cases may be distinguished:

$$a + r > d > a \tag{1}$$
$$d = a \tag{2}$$
$$a > d > a - r \tag{3}$$
$$d = a - r \tag{4}$$
$$a - r > d > 0 \tag{5}$$

Of these the curve produced by (4) is Proclus' first spiric curve, the "hippopede" or "horse-fetter," which is like a figure eight and had already been used by Eudoxus in his representation of planetary motion; (1) is Proclus' second, broad in the middle; (3) is his third, narrow in the middle; (2) is a transition from (1) to (3); and (5) produces two symmetrical closed curves. If the tore is "continuous" ("closed" in modern terminology), $a = r$, the forms (1), (2), and (3) remain as for the "open" tore, but (4) and (5) disappear and there is no new curve. If the tore is "interlaced" ("reentrant"), $a < r$, and the forms (4) and (5) do not exist; but there are three new curves corresponding to (1), (2), and (3), each with an oval inside it.

Tannery deduced that what the epigram means is that Perseus found three spiric curves in addition to the five sections. In this deduction he has been followed by most subsequent writers, Loria even finding support in Dante.[7] Although the interpretation is not impossible, it puts a strain upon the Greek. It is simpler to suppose that Tannery has correctly identified the five sections, but that Perseus ignored (2) and (5) as not really giving new curves. Thus he found "three curves in five sections." If we suppose that he took one of his curves from the five sections of the "open" tore, one from the five sections of the "continuous," and one from the five sections of the "interlaced," we could reconcile Proclus' statement also, but it is simpler to suppose that Proclus, writing centuries later, made an error.

NOTES

1. Proclus, *In primum Euclidis*, G. Freidlein, ed. (Leipzig, 1873; repr. Hildesheim, 1967), p. 356.6–12.
2. *Ibid.*, pp. 111.23–112.2.
3. *Ibid.*, p. 119.9–13.
4. Heron, Definitiones 97, in J. L. Heiberg, ed., *Heronis Alexandrini opera quae supersunt omnia*, IV (Leipzig, 1912), pp. 60.24–62.9.
5. G. V. Schiaparelli, *Le sfere omocentriche di Eudosso, di Calippo e di Aristotele* (Milan, 1875), pp. 32–34.
6. Paul Tannery, *Mémoires scientifiques*, II (Toulouse–Paris, 1912), pp. 26–28.
7. Gino Loria, *Le scienze esatte nell'antica Grecia*, 2nd ed. (Milan, 1914), p. 417, n. 2.

BIBLIOGRAPHY

On Perseus or his works, see T. L. Heath, *The Thirteen Books of Euclid's Elements*, 2nd ed. (Cambridge, 1926; repr. New York, 1956), I, 162–164; *A History of Greek Mathematics*, II (Oxford, 1921), 203–206; G. V. Schiaparelli, *Le sfere omocentriche di Eudosso, di Calippo e di Aristotele* (Milan, 1875), 32–34; and Paul Tannery, "Pour l'histoire des lignes et de surfaces courbes dans l'antiquité," in *Bulletin des sciences mathématiques et astronomiques*, 2nd ser., **8** (Paris, 1884), 19–30; repr. in *Mémoires scientifiques*, **2** (Toulouse–Paris, 1912), 18–32.

IVOR BULMER-THOMAS

PERSONNE, JACQUES (*b.* Saulieu, Côte-d'Or, France, 17 October 1816; *d.* Paris, France, 11 December 1880), *chemistry, pharmacy.*

Orphaned at an early age when his father, a lime-burner, died in an accident, Personne experienced much hardship. After completing an apprenticeship in pharmacy, he enrolled in the Paris School of Pharmacy and later competed successfully for an internship in a hospital pharmacy. From 1849 until his death in 1880, Personne served as chief pharmacist in three Paris municipal hospitals: Midi (1849–1857), Pitié (1857–1878), and Charité (1878–1880). From 1843 onward his connection with the Paris School of Pharmacy was continuous, first as *préparateur*, then as *chef des travaux*, and finally in 1877 as instructor of a newly established course in analytical chemistry. In 1875 Personne was admitted to the Academy of

Medicine. In 1877 at the age of sixty-one, he earned his *docteur ès sciences physiques* and in 1878 became a member of the Council on Public Hygiene and Health of the department of the Seine.

Despite his heavy hospital and teaching responsibilities, Personne carried on almost four decades of unremitting research. Among his most important investigations was a long chemical and botanical study of lupulin (1854). Later he produced the first experimental evidence that red phosphorus was safer than and superior to regular phosphorus in the production of hydrobromic and hydriodic acids and their esters (1861). In his researches on chloral hydrate (1869–1870)—which earned him the Barbier Prize of the Academy of Sciences—Personne not only developed standards of identity and purity for chloral hydrate, but also discovered and investigated chloral alcoholate. Personne believed that he had experimentally confirmed Liebreich's view that chloral hydrate owed its hypnotic effect to the release of chloroform in the blood. The Barbier Prize commission shared this belief, although it is now known that the *in vivo* effect is due to the release of trichloroethanol.

Personne's work includes such diverse subjects as acids and oxides of manganese, with Michel Lhermite (1851); fermentation of acetic acid (1853); oxidation of oil of turpentine (1856); chemical analysis of cannabis (1857); compounds formed by the interaction of iodine and tin (1862); and the determination of quinine in urine (1878).

BIBLIOGRAPHY

I. ORIGINAL WORKS. For listings of Personne's publications, see A. Goris *et al.*, *Centenaire de l'internat en pharmacie des hôpitaux et hospices civils de Paris* (Paris, 1920), 536–538; Poggendorff, III, 1024; and Royal Society *Catalogue of Scientific Papers*, IV, 837; VIII, 596; X, 1035.

II. SECONDARY LITERATURE. On Personne and his work, see E. C. Jungfleisch, "Discours prononcé aux obsèques de M. J. Personne," in *Journal de pharmacie et de chimie*, 5th ser., **3** (1881), 109–112; C. Méhu, "Discours prononcé aux obsèques de M. Personne," in *Bulletin de l'Académie de médecine*, **9** (1880), 1320–1322; and A. Villiers, "J. Personne," in *Centenaire de l'École supérieure de pharmacie de l'Université de Paris, 1803–1903* (Paris, 1904), 226–232.

ALEX BERMAN

PERSOON, CHRISTIAAN HENDRIK (*b.* Cape of Good Hope, South Africa, 31 December 1761; *d.* Paris, France, 15 November 1836), *botany, mycology.*

Persoon was the son of Christiaan Daniel Persoon (originally Persohn), a native of the island of Usedom [now Uznam] on the coast of Prussian Pomerania but a Dutch citizen at the time his son was born, and Elisabeth Wilhelmina Groenewald. The belief that his mother was a Hottentot has long since been disproved. Sent to Europe in 1775 for further education, he was orphaned the following year by the death of his father. Since Persoon and his two sisters were minors, their guardianship fell to the orphan masters at the Cape; and Persoon was to receive a sizable sum of money if he continued his studies. This legacy was sufficient to provide him with a modest annual income. Various adverse circumstances, however, caused him to live under the most dire conditions during a considerable period of his later life in Paris. He never accepted a paid position.

After attending the Gymnasium at Lingen, on the River Ems, he studied theology at Halle (1783–1786), medicine for a brief period at Leiden (1786), and medicine and the natural sciences at Göttingen (1787–1802). He never completed his university studies but in 1799 was awarded an honorary Ph.D. by the Kaiserlich-Leopoldinisch-Karolinische Deutsche Akademie der Naturforscher, then at Erlangen. He was elected a foreign or corresponding member of a number of learned societies. Almost nothing is known about his life while he was in Germany. At Halle he met F. W. von Leysser and other botanists, who may have been responsible for his turning to that field. At Göttingen he met J. A. Murray, professor of botany, and, undoubtedly, many young botanists who later became famous. Certainly half a dozen of them contributed important works on mycology. He must have known G. F. Hoffmann, then famous, who introduced Goethe to cryptogamy.

In 1802, for an unknown reason, Persoon moved to Paris, where he resided until his death. During much of this time his financial distress was great. In 1828 the Dutch government granted him a pension in exchange for his botanical collections, which are now in the Rijksherbarium at Leiden. As a gesture of gratitude, at his death Persoon donated his newly accumulated herbarium and library to the Dutch government. He maintained a wide correspondence with many botanists; a very extensive set of letters is now in the possession of the University Library at Leiden. He sought, unsuccessfully, to return to the Cape by invoking the assistance of James E. Smith of London.

Persoon is known in particular for his mycological publications, which culminated in the *Synopsis fungorum* (1801), rightly considered the basis of modern mycology. His classification was later elaborated by E. M. Fries and P. A. Saccardo. A modification of his system of the "macromycetes," now known as the Friesian tradition, still plays an important role in mycology, although it is gradually being replaced by a radically different system based mainly on microscopic characters. The influence of the *Synopsis* during the decades following its publication was enormous, for it made possible an unprecedented growth of the number of described genera and species of fungi. Persoon began a greatly revised version of the *Synopsis* under the title *Mycologia europaea*. Three volumes were published (1822–1826), but it remained incomplete.

At almost the same time the Swedish botanist and mycologist E. M. Fries began a rival work, the *Systema mycologicum* (1821–1832), which soon replaced Persoon's *Synopsis*. The latter has been accepted as the starting point for the nomenclature of the Gasteromycetes, Uredinales, and Ustilaginales; Fries's *Systema* became the starting point for the nomenclature of "fungi caeteri." In France one of Persoon's correspondents, J. B. Mougeot, kept the Persoonian tradition alive; and in the Vosges and the French Jura a flourishing group of mycologists included Lucien Quélet, Émile Boudier, and N. T. Patouillard. The mycological department of the Rijksherbarium publishes *Persoonia. A Mycological Journal*, and he also has been commemorated by a number of generic names, including *Persoonia* J. E. Smith (Proteaceae).

Persoon's importance as a phanerogamist is firmly based on his *Synopsis plantarum* (1805–1807), which sought to describe briefly all the phanerogams then known. Earlier he had reedited Murray's fifteenth edition of Linnaeus' *Systema vegetabilium* (1797). He advised F. W. Junghuhn to go to the Dutch East Indies and was instrumental in obtaining a post for him in the service of the Dutch government.

BIBLIOGRAPHY

I. ORIGINAL WORKS. Persoon's *Tentamen dispositionis methodicae fungorum* (Leipzig, 1797) contains the first draft of his classification of fungi. Knowledge of this group was further developed in *Synopsis fungorum* (Göttingen, 1801), which became the basis of mycological taxonomy, and *Mycologia europaea*, 3 vols. (Erlangen, 1822–1826). His fame as a phanerogamist is due mainly to his *Synopsis plantarum* (Paris–Tübingen, 1805–1807).

II. SECONDARY LITERATURE. An article by A. L. A. Fée on Persoon in *Giornale botanico italiano* (1846), translated into French by M. Rousseau as "Notice sur Persoon," in *Bulletin de la Société royale de botanie belgique*, **30** (1891), 50–60, and into Dutch by C. E. Destrée as "Aanteekeningen betreffende C. H. Persoon," in *Nederlandsch kruidkundig Archief*, **2**, no. 6 (1894), 366–377, is not altogether reliable. Other important publications (listed chronologically) are G. Schmid, "Eine unbekannte mykologische Arbeit Persoons (1793) zugleich ein Beitrag zur Lebensgeschichte des Verfassers," in *Zeitschrift für Pilzkunde*, **12** (1933), 54–60; J. Ramsbottom, "C. H. Persoon and James E. Smith," in *Proceedings of the Linnean Society of London*, **146** (1934), 10–21; and J. L. M. Franken, "Uit die lewe van 'n beroemde Afrikaner, Christiaan Hendrik Persoon," in *Annale van die Universiteit van Stellenbosch*, **15B** (1937), 1–102.

M. A. DONK

PERSOZ, JEAN-FRANÇOIS (*b.* Cortaillod, Neuchâtel, Switzerland, 9 June 1805; *d.* Paris, France, 12 or 18 September 1868), *chemistry.*

After working in several pharmacies in Neuchâtel, Persoz went to Paris, to study under Thenard at the Collège de France. From 1826 to 1832 he was *préparateur* to Thenard, and in 1833 he earned his *docteur ès sciences physiques* from the Paris Faculty of Sciences. Persoz then spent seventeen years (1833–1850) in Strasbourg in scientific, teaching, and administrative posts, including those of professor of chemistry at the Faculty of Sciences, assayer of the Mint, and professor of chemistry and director of the Strasbourg School of Pharmacy. Returning to Paris in 1850, Persoz became a *suppléant* to Dumas, who was then teaching chemistry at the Sorbonne; he was appointed *maître de conférences* at the École Normale Supérieure at about the same time. In 1852 he received a professorship at the Conservatoire des Arts et Métiers, where he lectured on dyeing and printing of textiles.

Among Persoz's most important accomplishments were two collaborative works published in 1833. The first, written with Anselme Payen, reported the isolation of diastase from malt extract; this research revealed that diastase converts starch into sugar. The second work, written with Biot, showed that the partial hydrolysis of starch with mineral acids yields a substance (dextrin) that proved dextrorotatory on the plane of polarized light, and that a similar effect on polarized light can be obtained by boiling cane sugar with dilute acid.

Persoz later published an influential book on the chemistry of molecular combinations, *Introduction à l'étude de la chimie moléculaire* (Paris–Strasbourg, 1839), and a four-volume treatise with an atlas on textile printing, *Traité théorique et pratique de l'impression des tissus* (Paris, 1846). Noteworthy, too, among his numerous investigations were his discovery of the production of methane by heating an acetate with caustic alkali (1839), his method for combining sulfur dioxide and phosphorus pentachloride to obtain thionyl chloride (1849), and his work on tungsten compounds (1863). Many of his publications dealt with analytical chemistry and chemical technology.

BIBLIOGRAPHY

I. ORIGINAL WORKS. In addition to his treatises on the chemistry of molecular combinations (1839) and on the printing of textiles (1846) already mentioned, Persoz published many papers in the *Comptes rendus hebdomadaires des séances de l'Académie des Sciences,* and in leading chemical journals of the day. For listings of Persoz's articles written alone or with others, see Poggendorff, II, 408–410, and III, 1024; and the Royal Society *Catalogue of Scientific Papers*, IV, 838–840, and VIII, 596.

II. SECONDARY LITERATURE. See Fritz Ferchl, *Chemisch-Pharmazeutisches Bio- und Bibliographikon* (Mittenwald, 1937), 404–405; Gabriel Humbert, *Contribution à l'histoire de la pharmacie strasbourgeoise* (Mulhouse, 1938), 217–219; E. V. McCollum, *A History of Nutrition* (Boston, 1957), 13–14, and *passim*; J. R. Partington, *A History of Chemistry*, IV (London–New York, 1964), 429 and *passim*; and René Sartory, "Jean-François Persoz," in *Figures pharmaceutiques françaises* (Paris, 1953), 95–100.

ALEX BERMAN

PETER ABANO. See **Abano, Pietro d'.**

PETER ABELARD. See **Abailard, Pierre.**

PETER OF AILLY. See **Ailly, Pierre d'.**

PETER BONUS. See **Petrus Bonus.**

PETER OF DACIA. See **Peter Philomena of Dacia.**

PETER PEREGRINUS, also known as **Pierre de Maricourt** (*fl. ca.* 1269), *magnetism.*

Other than that he was the author of the first extant treatise on the properties and applications of magnets, virtually nothing is known of Peregrinus. Two sources provide what little data we have: (1) his famous letter, or treatise, on the magnet, *Epistola Petri Peregrini de Maricourt ad Sygerum de Foucaucourt, Militem, De Magnete,*[1] "Letter on the Magnet of Peter Peregrinus of Maricourt to Sygerus of Foucaucourt, Soldier"; and (2) the *Opus tertium* of Roger Bacon.

Only one date in Peregrinus' life is fixed with certainty. At the conclusion of the *Epistola*, he added

"Completed in camp, at the siege of Lucera, in the year of our Lord 1269, eighth day of August."[2] From this account it would appear that Peregrinus was a member of the army of Charles of Anjou, King of Sicily, who was at that time personally directing an assault on Lucera (a city in Apulia, approximately twelve miles west of Foggia). Given Peregrinus' apparent interest in mechanical devices and instruments, Schlund has suggested[3] that he may have served in some technical capacity, perhaps as an engineer.

Peter may have received the appellation "Peregrinus" in connection with one or more of the assaults on Lucera. Under the control of the Hohenstaufens, Lucera had been besieged three times between 1255 and 1269, when it fell for the last time. The Papacy had declared these assaults against the Hohenstaufens and their Saracen allies official crusades. Since during the twelfth and thirteenth centuries the honorific title "Peregrinus" could be awarded not only to those who went on pilgrimages to the Holy Land but also to those who fought in recognized crusades in the Holy Land and elsewhere, Peregrinus may have thus earned it by participating in the siege.[4]

The manuscripts of the *Epistola* indicate that Peregrinus' full name was Petrus de Maharncuria, or Pierre de Maricourt, signifying that he probably came from the town of Méharicourt in Picardy.[5] Although there is evidence that Peregrinus was of noble birth,[6] the suggestion that he was a theologian is unconvincing[7] and the assertion that he was a Franciscan is baseless.[8] The lack of biographical data has even prompted an unsuccessful attempt to determine whether Peregrinus might be identical with one of his better-known thirteenth-century namesakes.[9]

Apart from the little that is revealed by the *Epistola* itself, further knowledge of Peregrinus as a scientist and investigator of natural phenomena depends heavily on the authenticity of certain statements in Roger Bacon's *Opus tertium* and whether in that same treatise Peregrinus is intended in references to a "Master Peter" (*Magister Petrus*). In chapter 11 of that treatise, which was written during 1267,[10] Bacon spoke of the need for good mathematicians and declared that "there are only two perfect mathematicians, Master John of London and Master (*Magister*) Peter de Maharn-curia, a Picard."[11] Since two of the five manuscripts of the *Opus tertium*, including the oldest, carry this statement in the margin, it is possible that it was added by a scribe and subsequently incorporated into the text of other manuscripts.[12] It is therefore difficult to give full credence to the claim that Bacon had Peregrinus[13] in mind when, a few lines below, he stated "Nor can any one

obtain their services [that is, of good mathematicians] unless he be the Pope or another great prince, especially the services of that one who is the best of all of them, of whom I have written in the *Opus minus* and shall write again in [the proper] place."[14] Since Peregrinus is neither mentioned nor alluded to in the single fragmentary manuscript of the *Opus minus*[15] known thus far, the claim is further eroded. Whoever may have been the author of the statement citing Peregrinus as one of two "perfect mathematicians," it is of interest that he referred to him as "Magister," probably signifying that Peregrinus had earned a Master of Arts degree, perhaps at the University of Paris.

In chapter 13 of the *Opus tertium*, Bacon praised a "Magister Petrus" as the only Latin writer to realize that experience rather than argument is the basis of certainty in science.[16] Later in the same chapter, following a discussion of burning mirrors, Bacon declared that "I know of only one person who deserves praise in the works of this science." At this point a marginal notation in one[17] of the five manuscripts used in Brewer's edition of the *Opus tertium* reads: "Notandum de magistro Petro de Maharne Curia" ("It should be noted that this is about Peter of Maharne Curia"). If the glossator is correct, it becomes highly probable that Peregrinus was intended by the earlier reference to "Magister Petrus," and Bacon's laudatory description in the lines that follow must also refer to Peregrinus. Bacon's description would reveal an idealistic and indefatigable scientist:

... for he does not trouble about discourses or quarrels over words, but follows the works of wisdom and keeps quietly to them. And so, though others strive blinkingly to see, as a bat in the twilight, the light of the sun, he himself contemplates it in its full splendour, on account of which he is a master of experiments (*dominus experimentorum*) and thus by experience he knows natural, medical, and alchemical things, as well as all things in the heavens and beneath them: indeed he is shamed if any layman, or grandam, or soldier, or country bumpkin knows anything that he himself does not know. Wherefore he has inquired into all operations of metal-founding, and the working of gold and silver and other metals, and of all minerals; and he knows all things pertaining to the army and to arms and the chase: and he has examined all that relates to agriculture, the measurement of land, and the works of farmers; and he has also reflected upon the experiments, devices, and incantations of witches and magicians, and likewise the illusions and tricks of all jugglers; so that nothing is hidden from him which he ought to know, and he knows how to reprobate[18] all things false and magical. And so without him it is impossible that philosophy could be completed, or be treated usefully or with certainty. But just as he cannot be valued with respect to price, so he does not estimate his own worth. For should he wish to stand well with kings and princes, he would find those who would honour him and enrich him. Or, if he were to show in Paris by his works of wisdom all that he knows, the whole world would follow him: yet because either way he would be hindered from the bulk of his experiments in which he most delights, so he neglects all honour and enrichment, the more since he might, whenever he wished it, attain to riches by his wisdom.[19]

On the basis of this extraordinary encomium, written in 1267, Bacon is conjectured to have met, or to have come to know about, Peregrinus during the 1260's. Had Bacon known him earlier, it is likely that he would have mentioned him in the *Communia mathematica*, written in the late 1250's, in which Robert Grosseteste, Adam March, and John Bandoun are singled out as praiseworthy mathematicians.[20]

Despite the paucity of information about Peregrinus, it seems evident that he was greatly, and perhaps primarily, interested in the construction of instruments and devices. According to him the *Epistola* was to form "part of a treatise in which we shall show how to construct physical instruments."[21] Similar interests are reflected in part 2, chapter 2, of the *Epistola*, where Peregrinus declares his intention to explain "how iron is held suspended in air by virtue of the stone"[22] in "the book on the action of mirrors" (*in libro de operibus speculorum*), which he was writing or planned to write.

Although no such work has yet been found, Bacon, immediately following the lengthy passage quoted above and still, presumably, speaking of Peregrinus, stated that "He [that is, Peregrinus] has already labored three years on one burning mirror [set for?] a fixed distance and, by the grace of God will soon complete it. Although we have books on the construction of such mirrors, the Latins are ignorant as to how to build them, nor has any among them attempted it."[23] Twice again in the *Opus tertium* (chaps. 33 and 36)[24] lavish praise is heaped upon the constructor of this same burning mirror (or so it seems) and the mirror itself, now mentioned as actually completed. A treatise on the construction of an astrolabe (*Nova compositio astrolabii particularis*),[25] in which the year 1261 is mentioned, has been attributed to Peregrinus and bears further witness to his overriding interest in the fabrication of instruments.

Scientific Thought. Since the *Epistola* is the sole authentic work attributed to Peregrinus that has been edited and made generally known, it alone must serve for the present as the basis of any evaluation of his scientific achievement. The *Epistola* is a brief treatise in two parts; the first, in ten chapters, describes the properties and effects of the lodestone, while the second, in three chapters, is devoted to the construction of three instruments utilizing the special properties and powers of the magnet.

The scope of the work and the essential prerequisites for conducting an investigation into magnetism are outlined in the first two chapters. Since the *Epistola* was to constitute part of a larger treatise on the construction of instruments, Peregrinus explicitly confined his attention to the manifest properties of the magnet, leaving aside all consideration of its occult powers. An investigator into the properties of magnetism must not only be knowledgeable about nature and the celestial motions, but also be clever in the use of his hands.

Turning to the magnet or lodestone itself, in part 1, chapter 3, Peregrinus named four characteristics: color (it should resemble polished iron exposed to the tarnishing effect of the air); homogeneity (although a magnet is rarely completely homogeneous, the more homogeneous it is, the more efficiently it performs); weight (a function of homogeneity and density—a heavier magnet is a better magnet); and virtue, or power to attract iron. Although the north-south orientation properties of a magnetized needle had been described and utilized in magnetic compasses since the eleventh century (and probably earlier) in China and since the twelfth century in the Latin West,[26] and although it had been known from antiquity that magnets could attract and repel iron, Peregrinus left the first extant account of magnetic polarity and methods for determining the poles of a magnet (pt. 1, chap. 4; Peregrinus may also have been the first to apply the term *polus* to a magnetic pole).[27] Just as the celestial sphere has a north and south pole, so also does every magnet.

A celestial analogy aided Peregrinus in his description of the first of two methods for locating the poles of a magnet. Since the meridian circles of the celestial sphere converge and meet at the poles, the lines drawn on a spherical magnet (called a *terrella* by William Gilbert[28] but perhaps first shaped and used by Peregrinus) will similarly meet at the poles when the investigator adheres to the following procedure:

> Let a needle or elongated piece of iron, slender like a needle, be placed on the stone, and a line be drawn along the length of iron dividing the stone in the middle. Then let the needle or iron be placed in another position on the stone and mark the stone with a line in a similar manner according to that position.[29]

All the lines drawn in this fashion will converge in the two opposite points or poles. The poles may also be found by noting at what point on the spherical magnet a needle or piece of iron clings with the greatest force. To render this method more precise, Peregrinus recommended that a small, oblong needle, or piece of iron, of approximately two finger nails in length, be located on or near the poles until it lies perpendicular to the stone (that is, stands upright). The marks representing these points should lie diametrically opposite. The two methods described here were also employed by Gilbert.[30]

In distinguishing north and south poles (pt. 1, chap. 5), Peregrinus presented a qualitative description of the fundamental law of magnetic polarity. If a lodestone is laid in a plate or cup, which in turn is placed in a vessel filled with water so that "the stone may be like a sailor in a ship"[31]—that is, free to turn in any direction without colliding into the sides of the vessel—then the north pole of the lodestone (*polus septentrionalis lapidis*) will face toward the north celestial pole and the south pole of the stone will face to the south celestial pole. Peregrinus observed that whenever the lodestone is forcibly turned away from its north-south orientation, it will always return to that orientation upon removal of the constraint.

The effect that a hand-held magnet will have upon a floating magnet serves as a paradigm for the general effect that one magnet has upon another (pt. 1, chap. 6). If the north pole of a hand-held magnet is brought in close proximity to the south pole of a floating magnet, the latter will seek to adhere to the former, an effect that will be repeated when the south pole of the hand-held magnet is brought near the north pole of the floating magnet. After formalizing this behavior in a general rule, Peregrinus observed that when the like poles of these magnets are brought close together, "the stone which you hold in your hand will appear to flee the floating stone."[32] To explain attraction and repulsion between the poles of magnets, Peregrinus resorted (pt. 1, chap. 9) to the agent-patient relationship so popular in medieval natural philosophy. He observed that if a magnet is broken in two each part will function as a magnet with north and south poles. If the opposite poles of the parts are then brought together, they will seek to unite and rejoin into a single magnet, since "an active agent strives not only to join its patient to itself but to unite with it, so that out of the agent and the patient there may be made one."[33] Indeed, if the two parts were cemented at the point of contact, the opposite poles would become unified and the resulting magnet would have a north and south pole

and be identical in every way with the original magnet. The union of agent and patient, which involves an attraction and union of opposite poles, is accounted for by a "likeness" or "similitude" (*similitudo*) between them. Peregrinus does not explain how a "similitude" between opposite poles is to be understood. John of St. Amand (*fl.* 1261–1298), a medical commentator and seeming contemporary of Peregrinus, likewise sought to account for the attraction between magnets and between a magnet and iron by saying that "it [that is, the magnet] does it by multiplying its like (*similitudo*) and, without any evaporation, exciting the active power which exists incomplete in iron, which is born to be completed by the form of the magnet, nay is moved towards it."[34]

On the question of which of two mutually attracting magnets is the agent and which the patient Peregrinus provided no answer. Presumably, if one magnet were assumed to be stationary (say, held in the hand) and the other free to move, it would be plausible to expect Peregrinus to designate the former as agent and the latter as patient. Otherwise the choice seems wholly arbitrary. Should the two north (or south) poles be brought into proximity, the two magnets could not be reunited into a single magnet, since the "identity or similitude of the parts would not be conserved."[35] The single magnet formed from the joining of like poles would possess two north (or south) poles and would differ in species from the original magnet, which possessed two opposite poles.

The ability of a magnet to orient itself with the celestial poles in a north-south direction is transmissible to iron upon contact (pt. 1, chap. 7). Let a magnetized iron needle (whether by "iron needle" Peregrinus meant iron or steel is left unspecified; if iron, the needle would have required repeated remagnetization)[36] be placed upon a piece of wood or straw that floats upon water. The end of the needle that had been touched by the region around the north pole of the magnet will turn toward the southern part of the heavens; and, conversely, the end touched by the area around the south pole of the magnet will orient itself toward the north celestial pole (but not the pole star). Since the magnetized needle takes on the polar properties of a magnet, it will behave like a magnet. Consequently, the south pole of the needle will be attracted to the north pole of the magnet and repelled by its south pole; and the north pole of the needle will be attracted to the south pole of the magnet and repelled by its north pole (pt. 1, chap. 8). The polarity of a magnetized needle is reversible, however, when, as Peregrinus (and also John of St. Amand)[37] observed, similar poles of a magnet and magnetized needle are brought into contact. When the north pole of a magnet is made to touch the north pole of a needle, it converts the latter to a south pole. "And the cause of this," Peregrinus explained, "is the impression of the last agent, confounding and changing the virtue of the first."[38] Given the agent-patient relationship discussed in part 1, chapter 9, Peregrinus would undoubtedly have accounted for this as the striving of an agent, the magnet, to unite with its patient, the iron needle. To achieve this objective the agent transforms the patient by altering its north pole to a south pole.

It was almost inevitable that Peregrinus should have inquired about the source of magnetic force (pt. 1, chap. 10). First, he disposed of the popular view that mines of magnetic stone in northern regions were the cause of the north-south orientation of a magnet. To support his position, Peregrinus stated that (1) magnetic stone is found in many parts of the world; (2) the polar regions are uninhabitable and thus could not be the source of magnetic stone; and (3) a magnet, or magnetized iron, orients to the south as well as the north. In rightly rejecting this notion, however, Peregrinus overlooked the fruitful concept, developed later by Gilbert, that the earth itself is a large spherical magnet. Instead, Peregrinus looked to the heavens in the belief that the poles of a magnet receive their virtue from the celestial poles.[39]

Although knowledge of magnetic declination (apparently already known in China in the eleventh century[40]) might have dissuaded Peregrinus from his opinion, there was reasonable evidence in its favor. Peregrinus was convinced that the poles of a magnet orient themselves in the meridian and that all meridians converge at the celestial poles; he was also aware that Polaris, the pole star, does not rest at the celestial north pole, but revolves around it—a fact virtually unknown to astronomers or seamen, which Columbus discovered for himself.[41] From this knowledge Peregrinus concluded that the poles of a magnet, or magnetized needle, always point directly to the celestial poles rather than to the pole star, as commonly believed.[42] From this conclusion it was an easy and perhaps irresistible inference that the poles of a magnet received their power to attract and repel directly from the celestial poles. Indeed, Peregrinus thought that every part of a spherical magnet received its power from the corresponding part of the celestial sphere.[43]

As a test for this claim, he suggested the construction of a spherical magnet with fixed pivots at its poles, which would leave the magnet free to rotate. The sphere should be positioned on the meridian circle "so that it moves in the manner of armillaries in such

a way that the elevation and depression of its poles may correspond with the elevation and depression of the poles of the heavens in the region where you may be."[44] If these instructions are followed faithfully, the spherical magnet, receiving magnetic virtue from every part of the celestial sphere, should commence to turn on its axis round the pivots,[45] thus simulating the daily celestial motion and functioning as a perfect clock. Although Peregrinus did not claim to have constructed such a perpetual motion machine, there is the hint that an abortive attempt was made, for he stated that failure of the sphere to perform as described could only be attributed to lack of skill in the contriver rather than deficiency in the theory, which he judged wholly sound. Peregrinus thus insulated his theory from the practical consequences that he himself deduced from it. Both Gilbert, who referred to this passage and mentioned Peregrinus by name,[46] and Galileo[47] rejected such claims.

Magnetic power as a source of perpetual motion is taken up again at the conclusion of the *Epistola* (pt. 2, chap. 3), where Peregrinus described construction of a continually moving toothed wheel powered by an oval magnet. The latter is so positioned that each tooth of the wheel will, in turn, be attracted to the north pole of the magnet. Under the influence of the attraction, the tooth acquires sufficient momentum to move beyond the north pole and into the vicinity of the south pole, by which it is repelled toward the north pole. As each tooth is alternately attracted and repelled, the wheel maintains a perpetual motion. Thus Peregrinus joined Villard de Honnecourt (*fl.* 1225–1250)[48] in proposing perpetual motion wheels in defiance of medieval Scholastic theory, which generally denied the possibility of inexhaustible forces in nature.

If Peregrinus' attempt to apply magnetic force to perpetual motion was misconceived, his use of it in the improvement of the compass was surely not. He described two compasses, one wet and one dry. The first (pt. 2, chap. 1), a floating compass, represents a considerable improvement over those that had been in use: an oval magnet is encapsulated in a wooden case and floated on water in a large rounded vessel. The rim of the vessel is divided into four quadrants according to the cardinal points of the compass. Each quadrant is then subdivided into ninety equal parts. A rule with sighting pins, positioned perpendicularly at each end, is placed on the encapsulated magnet. This rule extends to diametrically opposed points on the graduated rim. With this instrument, perhaps the first mariner's compass with divisions, not only could the direction of a ship be determined, but also the azimuth of the sun, moon, and stars.

Although the Chinese used magnetic compasses with geomantic divisions centuries before Peregrinus,[49] it is not clear whether they used them in the mariner's compass, of which clear mention is made in the eleventh century.[50]

The second compass (pt. 2, chap. 2), dry and pivoted, was deemed by Peregrinus an improvement over the floating compass. A vessel in the shape of a jar (which may be made from any solid material, preferably transparent) is constructed with a transparent lid of glass or crystal on which are marked the cardinal points. After subdividing each quadrant into ninety parts or degrees, a movable rule with perpendicular sights is fastened to the top of the lid. An axis of brass or silver is positioned at the center of the vessel between the bottom side of the lid and the bottom of the vessel. In the center of the axis, and at right angles to it, two needles—one of iron, the other of brass or silver—are inserted perpendicular to each other. Upon magnetizing the iron needle, the vessel, with its lid, is turned until the north-south points of the lid are aligned with the magnetized needle (as an obvious consequence, the silver or brass needle becomes aligned in an east-west direction). Azimuthal readings of the sun and stars may now be taken by rotating the movable rule on the lid. Peregrinus appears to have been the first to describe such a compass.[51]

The *Epistola* ranks as one of the most impressive scientific treatises of the Middle Ages. Although much of what Peregrinus included may have been known and expressed earlier in a vague and incomplete manner, the *Epistola* was the first extant treatise devoted exclusively to magnetism. Not only did Peregrinus bring together virtually all the relevant, contemporary knowledge on magnetism, but he obviously added to it and, of the greatest importance, organized the whole into a science of magnetism. He formulated rules for the determination of magnetic polarity, which then enabled him to enunciate rules for attraction and repulsion, all of which would today form the basis of an introductory lesson on magnetism.[52] As the two magnetic compasses and perpetual motion devices for clock and wheel testify, Peregrinus was also seriously concerned with the practical application of magnetic force. The subsequent influence of his treatise was considerable. The existence of at least thirty-one manuscript versions of it bears witness to its popularity during the Middle Ages. Of greater significance, however, was its eventual impact on Gilbert, who, in his famous *De Magnete* (1600), not only mentioned Peregrinus by name, but also drew upon the *Epistola* to build upon and add to the solid empirical rules on magnetic

polarity and induction formulated by Peregrinus more than three centuries earlier.

NOTES

1. I have cited the title as given in Bertelli's ed., *Bullettino*, I (Rome, 1868), 70. For variant titles, see Bertelli, *ibid.*, 4–7; E. Schlund, "Petrus Peregrinus von Maricourt," in *Archivum Franciscanum historicum*, **5** (1912), 22–39; and S. P. Thompson, "Petrus Peregrinus de Maricourt," in *Proceedings of the British Academy*, **2** (1905–1906), 400–407.
2. From the trans. of the *Epistola* by H. D. Harradon, "Some Early Contributions to the History of Geomagnetism—I," 17. Although the date 1269 is recorded in only three of thirty-one known MSS, it appears in what may be the oldest of them (see Schlund, "Petrus Peregrinus," in *Archivum Franciscanum historicum*, **4** [1911], 450, and **5** [1912], 23.
3. Schlund, *ibid.*, **4**, 455.
4. For details on the crusades against Lucera and the significance of the term "peregrinus," see Schlund, *ibid.*, 450–455.
5. Schlund, *ibid.*, 449.
6. Schlund, *ibid.*, 451; based on a Picard family "de Maricourt" listed in the *Dictionnaire de la Noblesse* by De la Chenaux-Desbois.
7. F. Picavet, *Essais sur l'histoire générale et comparée des théologies et des philosophies médiévales*, 240–242, 252.
8. Stewart Easton, *Roger Bacon and His Search for a Universal Science* (Oxford, 1952), 120–121.
9. Schlund, "Petrus Peregrinus," in *Archivum Franciscanum Historicum*, **4**, 441–448.
10. Bacon himself mentions the year. See 277, 278 of J. S. Brewer's ed. of Bacon's *Opus tertium*, *Opus minus*, and *Compendium studii philosophie* in *Fr. Rogeri Bacon Opera quaedam hactenus inedita* (*Rerum Britannicarum Medii Aevi*, no. 15; London, 1859).
11. Brewer, *ibid.*, lxxv and 35.
12. See Schlund, "Petrus Peregrinus," in *Archivum Franciscanum historicum*, **4**, 445–446.
13. Brewer, *Fr. Rogeri Bacon*, xxxvii.
14. My translation from Brewer, *ibid.*, 35. Brewer gives two variant translations on xxxvii and lxxv.
15. Brewer, *ibid.*, xxxvii. Even if an alleged marginal gloss in the *Opus minus* mentioning a "Master Peter" is a correct reading—and this is dubious—there is no good reason to assume that Peregrinus is the "Peter" intended (see Schlund, "Petrus Peregrinus," in *Archivum Franciscanum historicum*, **4**, 446–447).
16. Brewer, *ibid.*, 43.
17. Oxford, Bodleian, "e Musaeo" 155–3705; for the Latin text, see Brewer, *ibid.*, 46.
18. At this point, "Petrus de Maharne Curia" appears as a marginal gloss in British Museum, Cotton MSS, Tiberius C. V.
19. Thompson's trans., 380. I have slightly altered the trans., which was made from Brewer's ed., 46–47.
20. Easton, *Roger Bacon*, 88.
21. Pt. 1, chap. 1, as translated by H. D. Harradon in *Terrestrial Magnetism and Atmospheric Electricity*, **48** (1943), 6.
22. Harradon, *ibid.*, 16. On the claims to suspend iron in air by magnets, see Dorothy Wyckoff's trans., *Albertus Magnus Book of Minerals* (Oxford, 1967), 148 and Bertelli, *Bullettino* I, 87, n. 6.
23. My trans. from Brewer's ed., 47; see also Thompson's trans., 379.
24. Brewer, *Fr. Rogeri Bacon*, 112–116; see also F. Picavet, *Essais sur l'histoire générale et comparée des théologies et des philosophies médiévales*, 247, 252.
25. See Bertelli, *Bullettino*, I, 5, and bibliography, below.
26. Joseph Needham, *Science and Civilisation in China*, IV (Cambridge, 1962), pt. 1, 246, 249–250, and 274.

27. Schlund, "Petrus Peregrinus," in *Archivum Franciscanum historicum*, **4**, 636, n. 5.

28. *De Magnete*, bk. I, chap. 3, in Thompson's English trans., *On the Magnet*, 2nd ed. (New York, 1958; 1st ed., London, 1900), 13.

29. Harradon trans., 7.

30. *De Magnete*, bk. I, chap. 3 (13–14 of Thompson's trans.). Gilbert employed a third method using a versorium, that is, "a piece of iron touched with a loadstone, and placed upon a needle or point firmly fixed on a foot so as to turn freely about" (*ibid.*).

31. Harradon trans., 8.

32. *Ibid.*

33. *Ibid.*, 10. On medieval explanations of the causes of magnetic attraction, see W. James King, "The Natural Philosophy of William Gilbert and His Predecessors," in *Contributions From the Museum of History and Technology*, Smithsonian Institution Bulletin 218 (Washington, D.C., 1959), 125–129, and Harry A. Wolfson, *Crescas' Critique of Aristotle* (Cambridge, Mass., 1929), 90–92.

34. See Lynn Thorndike, "John of St. Amand on the Magnet," in *Isis*, **36** (1945), 156.

35. Harradon trans., 10.

36. E. Gerland, *Geschichte der Physik* (Berlin, 1913), 213.

37. Thorndike, "John of St. Amand," 157.

38. Harradon trans., 9.

39. Gilbert (*De Magnete*, bk. III, ch. 1; Thompson's trans., 116), citing Peregrinus by name, emphatically rejects this explanation.

40. Needham, *Science and Civilisation in China*, IV, pt. 1, 250.

41. See Samuel Eliot Morison, *Admiral of the Ocean Sea, A Life of Christopher Columbus*, 2 vols. (Boston, 1942), I, 271.

42. See Duane H. D. Roller, *The "De Magnete" of William Gilbert*, 36, 39.

43. A similar view was expressed by John of St. Amand; see Thorndike, "John of St. Amand," 156–157.

44. Harradon trans., 11–12.

45. Without mention of either Peregrinus or the title of the treatise, Nicole Oresme, in bk. II, question 3, of his *Questiones super De Celo*, makes a probable reference to this device. See Claudia Kren, "The 'Questiones super De Celo' of Nicole Oresme" (Ph.D. diss., University of Wisconsin, 1965), 474–476.

46. *De Magnete*, bk. VI, chap. 4; Thompson trans., 223.

47. Stillman Drake, trans., *Dialogue Concerning the Two Chief World Systems* (Berkeley, Calif., 1962), 413–414.

48. See Theodore Bowie, *The Sketchbook of Villard de Honnecourt* (Bloomington, Ind., 1959), 134.

49. Needham, *Science and Civilisation in China*, IV, pt. 1, 262–263, 296–297.

50. *Ibid.*, 279–280.

51. Thompson, 388. The Chinese did not learn of the dry, pivoted compass until the sixteenth century (Needham, *Science and Civilisation in China*, IV, pt. 1, 290).

52. E. J. Dijksterhuis, *The Mechanization of the World Picture* (Oxford, 1961), 153.

BIBLIOGRAPHY

I. ORIGINAL WORKS. The most complete list of MSS of the *Epistola* has been compiled by Erhard Schlund, O.F.M., in "Petrus Peregrinus von Maricourt, sein Leben und seine Schriften (ein Beitrag zur Roger Baco-Forschung)," in *Archivum Franciscanum historicum*, **5** (1912), 22–35. Of the 31 extant MSS described, 29 are Latin (for easier identification, the opening and closing lines [that is, *incipits* and *explicits*] are often supplied) and two, located in Vienna, represent two versions of a single Italian trans. made during the Middle Ages or Renaissance. In addition, five Latin MSS that may once have existed but the fate of which are unknown are also briefly cited and discussed. Another list of MSS, not quite as complete as Schlund's but including an English trans., possibly of the late sixteenth or early seventeenth century (in Gonville and Caius College, Cambridge), has been published by Silvanus P. Thompson, F.R.S., "Petrus Peregrinus de Maricourt and his Epistola De Magnete," in *Proceedings of the British Academy*, **2** (1905–1906), 400–404. Included are MSS owned by Thompson as well as five MSS the previous existence of which is plausibly conjectured. In the same article (404–408), Thompson presents the most comprehensive list yet produced of the printed eds. and trans. of the *Epistola* (11 partial and complete versions in all are cited; a useful but less extensive list appears in Schlund's article on 36–40; the original basis of both lists was probably furnished by Baldassare Boncompagni, "Intorno alle edizioni della *Epistola De Magnete* di Pietro Peregrino de Maricourt," in *Bullettino di bibliografia e di storia delle scienze matematiche e fisiche*, IV [Rome, 1871], 332–339).

The first published ed. was that of Achilles P. Gasser, *Petri Peregrini Maricurtensis De Magnete seu Rota perpetui motus libellus . . .* (Augsburg, 1558), which was followed by a few inadequate and truncated versions. Not until 1868 did the first critical text appear. Working from 7 MSS and the 1558 ed., Timoteo Bertelli, a Barnabite monk, published a new ed. of the *Epistola* in "Sulla Epistola di Pietro Peregrino di Maricourt e sopra alcuni trovati e teorie magnetiche del secolo XIII," in *Bullettino di bibliografia e di storia delle scienze matematiche e fisiche*, I (Rome, 1868), 70–89. A few years later, in an article entitled "Intorno a due Codici Vaticani della *Epistola De Magnete* di Pietro Peregrino di Maricourt," in *Bullettino*, IV (Rome, 1871), 303–331, Bertelli listed additional variant readings (see especially, 315–319) to his ed. of 1868. Using Bertelli's ed. and incorporating some of the later variants, G. Hellmann published another ed. of the *Epistola* in his *Rara Magnetica, Neudrucke von Schriften und Karten über Meteorologie und Erdmagnetismus* (Berlin, 1898), no. 10. By collating at least nine additional MSS (seven from Oxford) with printed eds., especially Bertelli's, Silvanus P. Thompson, on 390–398 of his article cited above, subsequently published a large number of additional variants, cuing them to the page and line numbers of Bertelli's 1868 ed. Thus despite five printed Latin eds., as well as a plagiarized version by Joannes Taisnier (1562), a facs. repr. by Bernard Quaritch (1900), and an ed. and English trans. promised by Charles Sanders Peirce (*Prospectus of an Edition of 300 Numbered Copies [150 for America] of the Earliest Work of Experimental Science: The Epistle of Pierre Pelerin de Maricourt to Sygur de Foucaucourt, Soldier, On the Lodestone* [New York, 1892], 16 pp., of which pp. 12–13 contain a sample Latin text based on Bibliothèque Nationale, fonds Latin, 7378A; see Thompson's article, pp. 406–407 for a full description and *Collected Papers of Charles Sanders Peirce*, VII, VIII, Arthur W. Burks, ed. [Cambridge, Mass., 1966], 280–282, for a lengthy quotation from 1–6 of the *Prospectus*, which Burks dates *ca.* 1893), there is as yet no single definitive Latin ed. based on all or most of the MSS.

Leaving aside two early printed English trans. of 1579 (?) and 1800 (see Thompson, 405), there now exist three major modern English trans. (1) Silvanus P. Thompson, *Epistle of Petrus Peregrinus of Maricourt, to Sygerus of Foucaucourt, Soldier, Concerning the Magnet* (London, 1902), based upon the eds. of Gasser, Bertelli, and Hellmann; (2) *The Letter of Petrus Peregrinus On the Magnet, A.D. 1269,* translated by Brother Arnold [Joseph Charles Mertens] with introductory notice by Brother Potamian [M. F. O'Reilly] (New York, 1904), made from the Gasser ed.; and (3) H. D. Harradon, "Some Early Contributions to the History of Geomagnetism-I," in *Terrestrial Magnetism and Atmospheric Electricity* (now the *Journal of Geophysical Research*), **48** (1943), 3–17. The title of the trans., which actually appears on 6–17, is "The Letter of Peter Peregrinus de Maricourt to Sygerus de Foucaucourt, Soldier, Concerning the Magnet." Although Harradon makes no mention of the ed., or eds., on which his trans. was based, one may conjecture that Hellmann's ed. was used. Included in Harradon's trans. is a prologue consisting solely of chapter titles, which Thompson (384 of the article cited above) believes is a scribal interpolation compiled from the original chapter headings that precede each chapter.

An as yet unexamined and unpublished work on the construction of an astrolabe is assigned to Peregrinus in the title of a treatise in Latin MS codex Vatican Palatine 1392, which reads: *Petri Peregrini Nova Compositio Astrolabii Particularis (Peter Peregrinus' New Composition [or Construction] of a Special Astrolabe)*; no folio numbers are given by T. Bertelli, who mentions the MS in his article "Sopra Pietro Peregrino di Maricourt e la sua *Epistola De Magnete,*" in *Bullettino,* I (Rome, 1868), 5. Since reference is made to certain astronomical tables completed by Campanus of Novara in 1261, the treatise was probably written after that date (see Bertelli, *ibid.,* 5, n. 1). A second MS is reported (without folio numbers) in the Library of Genoa ("à la Bibl. de Gênes") by J. G. Houzeau and A. Lancaster, *General Bibliography of Astronomy to the Year 1880,* I, pts. 1 and 2, new ed. with intro. and author index by D. W. Dewhirst (London, 1954), 640, col. 1, nr. 3197. Some suspicion is cast on this reference, however, since the authors cite the very page in Bertelli's memoir where the Vatican MS is listed. Whether they intended to identify a second manuscript "à la Bibl. de Gênes" (that is, Genoa) or merely to report the existence of the Vatican MS, which erroneously became a Genoa MS, is unclear. A possible third MS appears among Schlund's list of MSS of Peregrinus' *Epistola* ("Petrus Peregrinus von Maricourt," **5** [1912], 32, nr. 27). On fol. 20r–22v and 25v–36r of Latin codex Österreichische Nationalbibliothek, Vienna, 5184 (sixteenth century), treatises, or parts of treatises, appear,

titled, respectively, *Tractatus De Compositione Instrumenti Horarum Diei et Noctis (Treatise On the Construction of an Instrument for [Determining] the Hours of the Day and the Night)* and *Tractatus De Compositione Astrolabii (Treatise On the Construction of an Astrolabe).* Schlund conjectured that both were parts of the *Epistola,* but the second might well be all or part of the *Nova Compositio Astrolabii Particularis.*

II. SECONDARY LITERATURE. The most extensive study of Peregrinus and his *Epistola* consists of two memoirs by Timoteo Bertelli in *Bullettino di bibliografia e di storia delle scienze matematiche e fisiche,* I (Rome, 1868). The first ("memoria prima"), "Sopra Pietro Peregrino di Maricourt e la sua *Epistola De Magnete,*" 1–32, is concerned with Peregrinus' life, MSS of the *Epistola,* and contemporary and later authors, down to 1868, who spoke of Peregrinus, used his work, or edited and translated his treatise. The second memoir ("memoria seconda"), "Sulla *Epistola* di Pietro Peregrino di Maricourt e sopra alcuni trovati e teorie magnetiche del secolo XIII," is in three parts. Part 1, 65–89, includes the Latin ed. of the *Epistola* and a description of the MSS used; part 2, 90–99, 101–139, considers other medieval authors who discussed magnetism; and part 3, 319–420, further analyzes the content of the *Epistola* and traces its subsequent influence. A careful reexamination and evaluation of the life and works of Peregrinus, as well as a summary of the contents of the *Epistola* and an attempt to place Peregrinus in the context of scholastic thought, was published by Erhard Schlund, O.F.M., in "Petrus Peregrinus von Maricourt, sein Leben und seine Schriften (ein Beitrag zur Roger Baco-Forschung)," *in Archivum Franciscanum historicum,* **4** (1911), 436–455, 633–643. Peregrinus' life and *Epistola* are sketchily summarized by Silvanus P. Thompson, "Petrus Peregrinus de Maricourt and his *Epistola De Magnete,*" in *Proceedings of the British Academy,* **2** (1905–1906), 377–390 (the lists of variants, MSS, eds., and trans. of the *Epistola* mentioned above, follow on 390–408; see also Thompson's *Peregrinus and his Epistola* [London, 1907]). For an examination of Roger Bacon's alleged remarks about Peregrinus, and an attempt to demonstrate that Peregrinus was a theologian, see François Picavet, *Essais sur l'histoire générale et comparée des théologies et des philosophies médiévales* (Paris, 1913), chap. 11 ("Le maître des expériences, Pierre de Maricourt, l'exégète et le théologien vantés par Roger Bacon"), 233–254.

Among numerous summaries of the *Epistola,* see Park Benjamin, *The Intellectual Rise of Electricity* (London, 1895); Jean Daujat, *Origines et formation de la théorie des phénomènes électriques et magnétiques,* I (Paris, 1945); Paul Fleury Mottelay, *Bibliographical History of Electricity and Magnetism* (London, 1922), 45–54 (bibliography on 54); Duane H. D. Roller, *The "De Magnete" of William Gilbert* (Amsterdam, 1959), 39–42 (see bibliography, 186–190); and George Sarton, *Introduction to the History of Science,* II, pt. 2 (Baltimore, 1927–1948), 1030–1032, with bibliography. On the specific problem of declination, see Heinrich Winter, "Petrus Peregrinus von Maricourt und die magnetische Missweisung," in *Forschungen und Fortschritte*, **11** (1935), 304–306.

EDWARD GRANT

PETER PHILOMENA OF DACIA, also known as **Petrus Dacus, Petrus Danus, Peter Nightingale** (*fl.* 1290–1300), *mathematics, astronomy.*

Originally a canon of the cathedral in Roskilde, Denmark, Peter Nightingale first appears as the recipient of a letter from Hermann of Minden (provincial of the German Dominicans, 1286–1290) thanking him for the gift of some astronomical instruments and proposing to him that he leave Italy for Germany.[1] In 1291–1292 he is listed as a member of the University of Bologna,[2] where he taught mathematics and astronomy to pupils who included the astrologer Magister Romanus.[3] During 1292 Peter went to Paris, where in that and the following year he produced many writings. After that the sources are silent about him until 4 July 1303, when a letter from Pope Boniface VIII shows that he had returned to Denmark, in his former position as a canon of Roskilde.[4] The years of Peter's birth and death are unknown; and since he is not mentioned in the necrology of his cathedral, it is probable that he died abroad. Although he was a canon regular, he has often been considered a Dominican[5] and confused with the Swedish Dominican author of the same name. This mistake was corrected by H. Schück in 1895 but nevertheless persists in more recent literature.[6] His identification with another Petrus de Dacia, who in 1327 was rector of the University of Paris, has also been shown to be incorrect.[7]

A recent survey has revealed that there are more than 200 extant manuscripts of Peter's numerous works.[8] These can be divided into two groups, the first of which comprises the following writings:

Commentarius in Algorismum vulgarum (10 MSS). This commentary to Sacrobosco's well-known text-book of arithmetic was completed on 31 July 1291 at Bologna and is the only work of Peter Nightingale that has been edited and printed.[9] It contains some original contributions, notably a new and better method of extracting cube roots.[10]

Tabula multiplicationis (2 MSS). A multiplication table in the sexagesimal system and, accordingly, destined for use by astronomers.

Declaratio super Compotum (2 MSS). A commentary on the twelfth-century *Compotus metricus manualis* of Gerlandus of Besançon. It has not yet been examined.

Kalendarium with *canones* (56 MSS). This calendar for the period 1292–1369 was computed in Paris as a substitute for the much-used calendar of Robert Grosseteste, which had run out.[11] The appended *canones* give rules for adjusting the calendar for a new seventy-six-year period. Such adjustments were made in 1369 and around 1442. This calendar was intended to give more precise times of the phases of the moon than Grosseteste's work, with which it has often been confused.[12]

Tractatus eclipsorii (2 MSS). This newly found treatise describes the construction and use of a volvelle or equatorium for determining eclipses. It was written in Paris but contains a reference to Roskilde and is presumably the first evidence of Peter's interest in devising astronomical computers. It is followed by:

Tabulae coniunctionum solis et lune, that is, a table of mean conjunctions of the sun and moon;

Tabula temporis diurni, a table giving the length of the day as a function of the declination of the sun, calculated for the middle of the seventh climate (approximately the latitude of Paris);

Tabula diversitatis aspectuum lune ad solem, a table of the lunar parallax in longitude and latitude, for the same latitude as the preceding table, and meant to be used in connection with the *Tractatus eclipsorii*; and

Tabula equacionis dierum, a table of the equation of time as a function of the longitude of the sun.

Tabula lune with *canones* (68 MSS). This was Peter's most popular work. It exists in two versions: a numerical table and a diagram by which the approximate positions of the moon can be rapidly found from its age and the months of the year.

Tabula planetarum with *canones* (8 MSS). A diagram showing the governing planet for each day of the week and each hour of the day.

All the above works are well-authenticated writings by Peter Nightingale, but it is more difficult to ascertain the authorship of the treatises of the second group:

Tractatus de semissis (10 MSS). A long treatise on the construction and use of an equatorium for calculating planetary longitudes, written in Paris in 1293. No specimen of this instrument has survived, but a modern reconstruction based on the text was published in 1967.[13]

Tractatus novi quadrantis (18 MSS). This work was written in 1293 in Paris and describes the "new quadrant" invented some years earlier by Jacob ben Māḥir ibn Tibbon (Profatius Judaeus).[14] Peter's text seems to be a translation from the Hebrew original, provided with a careful introduction explaining the construction of this curious device, in which the astrolabe is transformed into a quadrant. It is not yet clear whether the other later Latin version dating from 1299 and attributed to Armengoud of Montpellier has anything to do with Peter's treatise.

540

Tractatus eclipsis solis et lune (1 MS). This is a brief treatise on how the problem of computing eclipses can be solved by geometrical construction.

In many manuscripts the three writings of the second group are attributed to a Petrus de Sancto Audomaro, or Peter of St.-Omer. But two manuscripts of the *Tractatus de semissis* are stated to be by Petrus Danus of St. Audomaro, while another simply calls the author Petrus Danus. Internal evidence and a comparison of astronomical parameters prove the three texts to be works by the same author, who accordingly must have been a very competent astronomer working in Paris at exactly the same time as Peter Nightingale. The latter is a definitely historical person, while it has been impossible to find any other records of the former in contemporary sources. Therefore, there are good reasons to agree with the hypothesis, proposed by E. Zinner in 1932, that the two authors are identical. In that case all the works mentioned above must be attributed to Peter Nightingale, whose possible connection with St.-Omer remains to be explained.

Apart from his works in pure mathematics, Peter Nightingale made two important contributions to medieval science. One was his work on astronomical computing instruments, for which he occupies a very important position in the history of astronomical computing machines. He was not the first Latin writer in this field, which in the later Middle Ages increasingly attracted the attention of astronomers. About 1260 the Paris astronomer Campanus of Novara had constructed a set of six equatoria for calculating longitudes.[15] Peter, however, was the first to invent a computer that solved this problem for all the planets with a single instrument. This device reduced the number of graduated circles and facilitated the construction of the instrument, the main principle of which was later adopted by John of Lignères and Chaucer.[16] The *Tractatus de semissis* also contains Peter's efforts to correct traditional astronomical parameters by new observations.

Peter's second achievement was in the field of astronomical tables, in which his calendar remained in constant use for 150 years. This calendar had the peculiar feature that for each day of the year it listed both the declination of the sun and the length of the day. The same features are found in a contemporary calendar by the Paris astronomer Guillaume de St.-Cloud, who seems to have collaborated with Peter Nightingale during the latter's sojourn in Paris.[17] The prehistory of this calendar was put into perspective by A. Otto, who in 1933 drew attention to a passage in the partly extant *Liber daticus* of Roskilde cathedral. It appears that in 1274 an unnamed astronomer belonging to the chapter made a series of observations,

unique for his time, of the altitude of the sun at noon, from which he calculated the length of the day by a *kardagas sinuum* (a trigonometrical diagram replacing a sine table).[18] Both the altitude and the length of the day were tabulated in the now lost calendar of the cathedral. In this respect the Roskilde calendar may be considered the prototype of the calendar calculated by Peter Nightingale in Paris. This is not to say that he was identical with the unknown Roskilde astronomer of 1274; but there is no doubt that it was he who brought the principle from Denmark to France, thus creating a hitherto unknown link between Scandinavian astronomy and European science in general.

NOTES

1. Published in Paul Lehmann, "Skandinaviens Anteil an der lateinischen Literatur und Wissenschaft des Mittelalters," in *Sitzungsberichte der Bayerischen Akademie der Wissenschaften zu München*, Phil.-hist. Abt. (1936), 53–54.
2. Ellen Jørgensen, "Om nogle middelalderlige forfattere der naevnes som hjemmehørende i Dacia," in *Historisk tidsskrift*, 8th ser., **3** (1910–1912), 253–260.
3. Lynn Thorndike, *History of Magic and Experimental Science*, III (New York, 1934), 647–649.
4. A. Krarup, in *Bullarium danicum*, no. 947 (1932), 834–835.
5. J. Quétif and J. Echard, *Scriptores ordinis Praedicatorum*, II (Paris, 1721).
6. H. Schück, *Illustrerad Svensk literaturhistoria*, I (Stockholm, 1895), 343; G. Sarton, *Introduction to the History of Science*, II (Baltimore, 1931), 996–997.
7. C. E. Bulaeus, *Historia Universitatis Parisiensis*, II (Paris, 1668), 210, 982; cf. H. Denifle and A. Chatelain, *Chartularium Universitatis Parisiensis*, II (Paris, 1891), nos. 863, 955.
8. This survey, by Olaf Pedersen, is not yet completed. It supersedes previous inventories by G. Eneström, "Anteckningar om matematikern Petrus de Dacia och hans skrifter," in *Öfversigt af K. Vetenskapsakademiens förhandlingar* (1885), 15–27, 65–70, and (1886), 57–60; and E. Zinner, *Verzeichnis der astronomischen Handschriften des deutschen Kulturgebietes* (Munich, 1925), nos. 2055–2082.
9. Maximilian Curtze, *Petri Philomeni de Dacia in Algorismum vulgarem Johannis de Sacrobosco commentarius una cum algorismo ipso* (Copenhagen, 1897).
10. G. Eneström, "Über die Geschichte der Kubikwurzelausziehung im Mittelalter," in *Bibliotheca mathematica*, 3rd ser., **14** (1914), 83–84; cf. M. Cantor, *Geschichte der Mathematik*, 2nd ed., II (Leipzig, 1899–1900), 90.
11. E. Zinner, "Petrus de Dacia, en middelalderlig dansk astronom," in *Nordisk astronomisk tidsskrift*, **13** (1932), 136–146; German trans. in *Archeion*, **18** (1936), 318–329.
12. First by J. Langebek, in *Scriptores rerum Danicarum*, IV (Copenhagen, 1786), 260 f., where Grosseteste's calendar was edited and attributed to Petrus de Dacia.
13. O. Pedersen, "The Life and Work of Peter Nightingale," in *Vistas in Astronomy*, **9** (1967), 3–10; cf. O. Pedersen, "Peder Nattergal og hans astronomiske regneinstrument," in *Nordisk astronomisk tidsskrift*, **44** (1963), 37–50.
14. This text has been edited in an unpublished thesis by Lydik Garm, "Profatius Judaeus' traktat om kvadranten" (Aarhus, Institute for the History of Science, 1966).
15. F. J. Benjamin and G. J. Toomer, *Campanus of Novara and Medieval Planetary Theory* (Madison, Wis., 1971).

16. D. J. de Solla Price, *The Equatorie of the Planetis* (Cambridge, 1955), 17 f. (Chaucer) and 188 f. (John of Lignères).
17. P. Duhem, *Le système du monde*, new ed., IV (Paris, 1954), 14 f.; cf. Zinner, *loc. cit.*
18. A. Otto, *Liber daticus Roskildensis* (Copenhagen, 1933), 32–33. The importance of the Roskilde astronomer was first pointed out by A. A. Bjørnbo, "Die mathematischen S. Marco-Handschriften in Florenz," in *Bibliotheca mathematica*, 3rd ser., **12** (1912), 116.

OLAF PEDERSEN

PETERS, CARL F. W. (*b.* Pulkovo, Russia, 16 April 1844; *d.* Königsberg, Germany [now Kaliningrad, R.S.F.S.R.], 2 November 1894), *astronomy, geodesy.*

Peters was the son of the astronomer Christian A. F. Peters. Between 1862 and 1866 he studied at Kiel, Berlin, and Munich, became adjunct at the Hamburg observatory in 1867, and received the Ph.D. at Göttingen in 1868. He then became his father's assistant at the Altona observatory. In this post he determined the length of the seconds pendulum for Altona, Berlin, and Königsberg and in 1870–1871 executed new observations with Bessel's pendulum apparatus between Königsberg and Güldenstein, a castle in Holstein. He was appointed observer in 1872 and the following year moved to Kiel, where the Altona observatory had been relocated. Here he became academic lecturer in astronomy in 1876.

After his father's death Peters edited three volumes of the *Astronomische Nachrichten* (**97–99**) during 1880–1881, and in 1882 he became assistant professor of astronomy. A year later he assumed the directorship of the chronometer *Observatorium* of the imperial navy, a post conferred on him because of his extremely careful investigations of the rate of chronometers. He determined that they were influenced not only by temperature but also by humidity and magnetism. He also reduced the existing observations of the double star 61 Cygni, deriving an accurate orbit, and edited the German version of A. N. Sawitsch's *Practical Astronomy.*

In 1888 Peters was appointed professor at the University of Königsberg and director of its observatory, where he began observations with the meridian circle. His early death, after a long illness, prevented any major achievements in this new field. Peters was a man of great kindness and cordiality, and of an unusually humane temperament.

BIBLIOGRAPHY

Peters' writings include *Astronomische Tafeln und Formeln* (Hamburg, 1871); *Entfernung der Erde von der Sonne* (Berlin, 1873); *Beobachtungen mit dem Besselschen Pendelapparat in Königsberg und Güldenstein* (Hamburg, 1874); "Einige Bemerkungen über die Vorbestimmung des Chronometerstandes," in *Annalen der Hydrographie . . .*, **5** (1877), 207–214; *Die Fixsterne*, in the series Wissenschaft der Gegenwart (Leipzig, 1883); "Magnetische Einflüsse auf den Gang der Chronometer," in *Annalen der Hydrographie . . .*, **12** (1884), 316–318; "Bestimmung der Bahn des Doppelsterns 61 Cygni," in *Astronomische Nachrichten*, **113** (1885), 321–340; and "Einfluss der Feuchtigkeit der Luft auf den Gang der Chronometer," in *Annalen der Hydrographie*, **15** (1887), 505–512.

An obituary is J. Franz, in *Vierteljahrsschrift der Astronomischen Gesellschaft* (Leipzig), **30** (1895), 12–16.

H.-CHRIST. FREIESLEBEN

PETERS, CHRISTIAN AUGUST FRIEDRICH (*b.* Hamburg, Germany, 7 September 1806; *d.* Kiel, Germany, 8 May 1880), *astronomy.*

Peters' father, a merchant, saw to it that his son, who did not regularly attend secondary school, obtained a good knowledge of mathematics and astronomy. He was so successful that H. C. Schumacher, the editor of the *Astronomische Nachrichten*, learned of Peters and induced him to study geodesy and astronomy. Peters subsequently entered the University of Königsberg, where he received the Ph.D. under Bessel. From 1834 to 1838, as assistant at the Hamburg observatory Peters observed mainly with the transit instrument. In 1839 he was appointed assistant at the new Pulkovo observatory, where he worked for nearly ten years, finally becoming assistant director under F. G. W. Struve. He observed the polestar, the newly discovered planet Neptune, and parallaxes of fixed stars.

In 1849 Peters returned to Königsberg to become professor of astronomy. This post was not connected with the directorship of the observatory, but he did have access to Bessel's famous heliometer. In 1854 Peters moved to Altona as director of the observatory and editor of the *Astronomische Nachrichten*, of which he edited fifty-eight volumes (**40–97**), from 1855 till the end of his life. In 1872 the Altona observatory was moved to Kiel and reconstructed on a larger scale, a plan that Peters had favored since 1864. In 1874 Peters became ordinary professor at the University of Kiel.

Both a student and a successor of Bessel, Peters sought to ascertain the base of spherical astronomy. His investigations concerning nutation, the proper motion of Sirius, and the parallaxes of fixed stars are his main achievements.

BIBLIOGRAPHY

Peters' writings include *Numerus constans nutationis . . . in specula Dorpatensi annis 1832–1838 observatis deductus* (St. Petersburg, 1842); "Resultate aus den Beobachtungen des Polarsterns an der Pulkowaer Sternwarte," in *Mémoires de l'Académie impériale des sciences de St.-Pétersbourg*, 6th ser., **3** (1844); "Über die eigene Bewegung des Sirius," in *Astronomische Nachrichten*, **32** (1851), 1–58; "Recherches sur la parallaxe des étoiles fixes," in *Mémoires de l'Académie impériale des sciences de St.-Pétersbourg*, 6th ser., **5** (1853); "Über die Länge des einfachen Sekundenpendels auf dem Schlosse Güldenstein," in *Astronomische Nachrichten*, **40** (1855), 1–152; *Bestimmung des Längenunterschiedes Altona–Schwerin* (Altona, 1861); "Ein Repsoldsches Äquatorial zu Altona," in *Astronomische Nachrichten*, **58** (1862), 271–352; *Das Land Swante–Wustrow oder das Fischland* (Wustrow, 1866); and *Bestimmung des Längenunterschiedes Göttingen–Altona* (Kiel, 1880).

An obituary is A. Winnecke, in *Vierteljahrsschrift der Astronomischen Gesellschaft* (Leipzig), **16** (1881), 5–8.

H. C. FREIESLEBEN

PETERS, CHRISTIAN HEINRICH FRIEDRICH

(*b.* Coldenbüttel, Schleswig, Denmark [now Schleswig-Holstein, Germany], 19 September 1813; *d.* Clinton, New York, 18 July 1890), *astronomy.*

After attending the Gymnasium in Flensburg Peters studied mathematics and astronomy with Encke at the University of Berlin, where he took the Ph.D. in 1836, then with Gauss at Göttingen. From 1838 to 1843 he worked on a private survey of Mount Etna, then was appointed director of the government trigonometric survey of Sicily. He held this post until 1848, when he was deported for actively supporting the Sicilian revolutionaries. In 1849, following the fall of Palermo, Peters went to Constantinople, where he remained for the next five years. Although political circumstances—including the Crimean War—precluded expeditions, Peters was able to learn Arabic and Turkish.

In 1854, carrying introductions from Humboldt, Peters immigrated to the United States. He was employed by the Coast Survey, and detailed first to the Cloverden Observatory in Cambridge, and then to the Dudley Observatory in Albany. In 1858 he was appointed professor of astronomy and director of the Litchfield Observatory at Hamilton College, Clinton, N.Y. Although funds were short (he often went for months without salary) he remained at Hamilton for the rest of his life.

Peters' primary scientific interest was observational positional astronomy. While in Naples, long before Carrington took up such work, Peters charted the latitudinal and longitudinal proper motions and internal developments of sunspots. At Hamilton College Peters attempted to chart, without photog-

raphy, all the stars down to (and even below) the fourteenth magnitude situated within 30° on either side of the ecliptic. He coincidentally discovered forty-eight asteroids and computed their orbits. In 1869 he organized a party to observe a total eclipse of the sun, and in 1874 he led one of eight U.S. government expeditions to observe the transit of Venus; he was a member of the International Astrophotographic Congress held in Paris in 1887. Drawing on both his linguistic ability and his astronomical knowledge Peters collated the star catalogs in various Continental manuscript copies of Ptolemy's *Almagest*. E. B. Knobel collated the British manuscripts and issued a revised Ptolemaic catalog in 1915, after Peters' death.

Peters was a member or a fellow of a number of scientific societies, including the American Academy of Arts and Sciences (1856), the National Academy of Sciences (1876), the American Philosophical Society (1878), and the Royal Astronomical Society (1879); he received the French Legion of Honor in 1887.

BIBLIOGRAPHY

I. ORIGINAL WORKS. Peters' writings include *De principio minimae actionis dissertatio* (Berlin, 1836); *Report . . . on the Longitude of Elmira* (Albany, 1864); *Report . . . on the Longitude and Latitude of Ogdensburg* (Albany, 1865); and *Celestial Charts Made at the Litchfield Observatory* (Clinton, N.Y., 1882), with 20 charts each covering 5° dec. and 20^m r.a., another 20 were finished, but unpublished, at his death. Posthumous works are E. B. Frost, ed., *Heliographic Positions of Sun-spots Observed at Hamilton College from 1860–1870*, Carnegie Institution of Washington Publication no. 43 (Washington, D.C. 1907); and *Ptolemy's Catalogue of Stars*, Carnegie Institution of Washington Publication no. 86 (Washington, D.C., 1915), written with E. B. Knobel. The Royal Society of London *Catalogue of Scientific Papers* lists 144 papers by Peters; his correspondence with his friend G. P. Bond is at the Harvard College Observatory.

II. SECONDARY LITERATURE. On Peters and his work, see *Christian Henry Frederick Peters, September 19, 1813, July 18, 1890* (Hamilton, N. Y., 1890), a memorial vol. printed for private circulation. Obituary notices are in *Sidereal Messenger*, **9** (1890), 439–442; *Monthly Notices of the Royal Astronomical Society*, **51** (1890–1891), 199–202; and *Astronomische Nachrichten* (Aug. 1890). For Peters' controversy with C. A. Borst over ownership of a research MS, see "Dr. Peters' Star Catalogue," in *Sidereal Messenger*, **8** (1889), 138–139, 455–458; *Utica Morning Herald* (9 Nov. 1889); and Simon Newcomb, *Reminiscences of an Astronomer* (Boston, 1903), 372–381.

DEBORAH JEAN WARNER

PETERSEN, JULIUS (*b.* Sorø, Denmark, 16 June 1839; *d.* Copenhagen, Denmark, 5 August 1910), *mathematics.*

Petersen's interest in mathematics was awakened at school, where his main occupation was solving problems and attempting the trisection of the angle. At the age of seventeen he entered the College of Technology in Copenhagen; but after some years of study he transferred to the University of Copenhagen, from which he graduated in 1866 and received the doctorate in 1871. His dissertation treated equations solvable by square roots with applications to the solution of problems by ruler and compass. During his university years and after graduation Petersen taught in secondary schools. In 1871 he was appointed docent at the College of Technology and, in 1887, professor at the University of Copenhagen, a post he held until the year before his death.

Through his terse, well-written textbooks Petersen has exerted a very strong influence on mathematical education in Denmark. Several of his books were translated into other languages. Worthy of particular mention is his *Methods and Theories for the Solution of Problems of Geometrical Constructions* (Danish, 1866; English, 1879; German, 1879; French, 1880; Italian, 1881; Russian, 1892). His other writings cover a wide range of subjects in algebra, number theory, analysis, geometry, and mechanics. Perhaps his most important contribution is his theory of regular graphs, inspired by a problem in the theory of invariants and published in *Acta mathematica* in 1891.

BIBLIOGRAPHY

Petersen's works are listed in Niels Nielsen, *Matematiken i Danmark 1801–1908* (Copenhagen–Christiania, 1910).

There are obituaries by H. G. Zeuthen, in *Oversigt over det K. Danske Videnskabernes Selskabs Forhandlinger 1910* (1910–1911), I, 73–75; C. Juel and V. Trier, in *Nyt Tidsskrift for Matematik, A,* **21** (1910), 73–77, in Danish; and C. Juel, "En dansk Matematiker," in *Matematisk Tidsskrift, A* (1923), 85–95.

BØRGE JESSEN

PETERSON, KARL MIKHAILOVICH (*b.* Riga, Russia [now Latvian S.S.R.], 25 May 1828; *d.* Moscow, Russia, 19 April 1881), *mathematics.*

Peterson was the son of a Latvian worker, a former serf named Mikhail Peterson, and his wife, Maria Mangelson. In 1847 he graduated from the Riga Gymnasium and enrolled at the University of Dorpat. The lectures of his scientific tutor Ferdinand Minding

provided an occasion for Peterson's writing his thesis "Über die Biegung der Flächen" (1853), for which he received the degree of bachelor of mathematics.

Later Peterson moved to Moscow where he worked first as a private teacher then, from 1865 until his death, as a mathematics teacher at the German Peter and Paul School. Becoming intimately acquainted with scientists close to N. D. Brashman and A. Y. Davidov, Peterson took an active part in the organization of the Moscow Mathematical Society and in its work. He published almost all of his writings in *Matematicheskii sbornik,* issued by the society.

In 1879 the Novorossiiskii University of Odessa awarded Peterson an honorary doctorate in pure mathematics for his studies on the theory of characteristics of partial differential equations, in which, by means of a uniform general method, he deduced nearly all the devices known at that time for finding general solutions of different classes of equations. These studies were to a certain extent close to the works of Davidov (1866) and N. Y. Sonin (1874). However, Peterson's principal discoveries are connected with differential geometry.

In the first part of his thesis Peterson established certain new properties of curves on surfaces and in the second part he continued Gauss's and Minding's works on the bending of surfaces. Here he for the first time obtained equations equivalent to three fundamental equations of Mainardi (1856) and Codazzi (1867–1869), which involve six coefficients of the first and the second quadratic differential forms of a surface. Peterson also proved—in different expression—the theorem usually bearing the name of Bonnet (1867): the geometrical form of the surface is wholly determined if the coefficients of both quadratic forms are given. Minding found the thesis excellent, but these results were not published during Peterson's lifetime and found no development in his articles which were printed after 1866. Brief information on Peterson's thesis was first given by P. Stäckel in 1901; a complete Russian translation of the manuscript, written in German and preserved in the archives of the University of Tartu, was published in 1952.

In his works Peterson elaborated new methods in the differential geometry of surfaces. Thus, he introduced the notion of bending on a principal basis, namely, bending under which a certain conjugate congruence of curves on the surface remains conjugate; such congruence is called the principal basis of a surface. Peterson established numerous general properties of conjugate congruences and studied in depth the bending on a principal basis of surfaces of second order, surfaces of revolution, minimal and translation surfaces. All these surfaces and some others constitute a class of surfaces, quite interesting in its properties, named after Peterson.

Although Peterson did not teach at the university, his ideas initiated the studies of B. K. Mlodzeevsky and, later, of his disciples Egorov, S. P. Finikov, and S. S. Bushgens. Peterson's discoveries also found a somewhat belated reputation and extension in other countries, for example, in the works of Darboux and Bianchi. Outside the Soviet Union, however, his remarkable studies on the theory of surfaces are still mentioned but rarely in the literature on the history of mathematics.

BIBLIOGRAPHY

I. ORIGINAL WORKS. Peterson's writings include "Ob otnosheniakh i srodstvakh mezhdu krivymi poverkhnostyami" ("On Relationships and Kinships Between Surfaces"), in *Matematicheskii sbornik*, **1** (1866), 391–438; "O krivykh na poverkhnostiakh" ("On Curves on Surfaces"), *ibid.*, **2** (1867), 17–44; *Über Kurven und Flächen* (Moscow–Leipzig, 1868); "Ob integrirovanii uravnenii s chastnymi proizvodnymi" ("On the Integration of Partial Differential Equations"), in *Matematicheskii sbornik*, **8** (1877), 291–361; **9** (1878), 137–192; **10** (1882), 169–223; and *Ob integrirovanii uravnenii s chastnymi proizvodnymi po dvum nezavisimym peremennym* ("On the Integration of Partial Differential Equations With Two Independent Variables"; Moscow, 1878). For a French trans. of Peterson's works, see *Annales de la Faculté des sciences de l'Université de Toulouse*, 2nd ser., **7** (1905), 5–263. See also "Ob izgibanii poverkhnostei" ("On the Bending of Surfaces"), his dissertation, in *Istoriko-matematicheskie issledovaniya*, **5** (1952), 87–112, with commentary by S. D. Rossinsky, pp. 113–133.

II. SECONDARY LITERATURE. On Peterson and his work, see (listed chronologically) P. Stäckel, "Karl Peterson," in *Bibliotheca mathematica*, 3rd ser., **2** (1901), 122–132; B. K. Mlodzeevsky, "Karl Mikhailovich Peterson i ego geometricheskie raboty" ("Karl Mikhailovich Peterson and His Geometrical Works"), in *Matematicheskii sbornik*, **24** (1903), 1–21; D. F. Egorov, "Raboty K. M. Petersona po teorii uravnenii s chastnymi proizvodnymi" ("Peterson's Works on Partial Differential Equations"), *ibid.*, 22–29— the last two appear in French trans. in *Annales de la Faculté des sciences de l'Université de Toulouse*, 2nd ser., **5** (1903), 459–479; D. J. Struik, "Outline of a History of Differential Geometry," in *Isis*, **19** (1933), 92–120; **20** (1933), 161–191; S. D. Rossinsky, "Karl Mikhailovich Peterson," in *Uspekhi matematicheskikh nauk*, **4**, no. 5 (1949), 3–13; I. Y. Depman, "Karl Mikhailovich Peterson i ego kandidatskaya dissertatsia" ("Peterson and His Candidature Dissertation"), in *Istoriko-matematicheskie issledovaniya*, **5** (1952), 134–164; I. Z. Shtokalo, ed., *Istoria otechestvennoy matematiki* ("History of Native Mathematics"), II (Kiev, 1967); and A. P. Youschkevitch, *Istoria matematiki v Rossii do 1917 goda* ("A History of Mathematics in Russia to 1917"; Moscow, 1968).

A. P. YOUSCHKEVITCH
A. T. GRIGORIAN

PETIT, ALEXIS THÉRÈSE (*b.* Vesoul, France, 2 October 1791; *d.* Paris, France, 21 June 1820), *physics.*

Petit was an outstanding pupil at the École Centrale in Besançon and later at a private school in Paris that was staffed by teachers from the École Polytechnique. He had fulfilled the entrance requirements for the École Polytechnique by the time he was ten-and-a-half and he enrolled there in 1807, when he was sixteen, the minimum age for entry. He was first in his entering class; when he graduated, in 1809, he was placed *hors de ligne*, and the next student in the year was designated "first." Petit was immediately taken onto the staff as a teaching assistant.

In 1810 Petit also became professor of physics at the Lycée Bonaparte in Paris. As a teacher he was both popular and successful, and when he succeeded to J.-H. Hassenfratz's chair of physics at the École Polytechnique in 1815, after a year as assistant professor, he extended and improved the courses in his subject. His last years, however, were clouded by grief and illness; shortly after the death of his young wife, in 1817, he contracted tuberculosis, the disease from which he died. He was a member of the Société Philomatique from February 1818 but was never elected to the Académie des Sciences.

Petit's most important work was done in collaboration with his close friend Pierre Dulong. (This collaborative work is discussed in detail in the article on Dulong.) Their association began in 1815, probably in response to the prize competition on thermometry and the laws of cooling that was then set by the first class of the Institute. By 1818, when the prize was awarded to them, Petit and Dulong had conducted a classic experimental investigation, which established the gas thermometer as the only reliable standard and put the approximate nature of Newton's law of cooling beyond all doubt. It was after a further year of intense activity, devoted mainly to the measurement of the specific heats of solids, that Petit and Dulong discovered their law of atomic heats. Since the discovery was made, suddenly and quite by chance, only one week before it was announced to the Académie des Sciences on 12 April 1819, it is not surprising that the evidence for their categorical statement, "the atoms of all elementary substances have exactly the same capacity for heat," was inadequate. In fact the exactness of the law was in doubt from the start and was never to be established.

Petit's comments on theoretical issues were characterized by his receptiveness to new ideas. He received a thoroughly conventional education in physics at the École Polytechnique, where the customary emphasis was placed on such doctrines as the corpuscular theory of light, the caloric theory of heat, and the other theories of imponderable fluids. Thus, not surprisingly, when he himself began to teach there, his teaching was completely orthodox, as may be seen in some manuscript notes of the lectures that he gave in the winter of 1814–1815. But in December 1815, as a result of some experiments on the refraction of light in gases—which he had performed with his brother-in-law Dominique Arago—Petit openly rejected the corpuscular theory and became one of the earliest supporters of the wave theory, which had just been revived in France by Fresnel.

The Petit-Dulong paper of April 1819 on atomic heats was likewise marked by a skepticism toward established doctrine. In it Petit and Dulong rejected the caloric theory and, almost certainly under the influence of Dulong's close friend Berzelius, substituted for it the electrical explanation of heats of chemical reaction. The 1819 paper also contained a statement of support for the chemical atomic theory, which, owing largely to the opposition of Berthollet and his followers, had made little headway in France.

Although he is best known for his experimental work, Petit had an equal, if not greater, talent for mathematics. Evidence of this is found in his brilliant doctoral thesis of 1811 on the theory of capillary action (treated in the manner of Laplace) and in a paper on the theory of machines written in 1818.

BIBLIOGRAPHY

I. Original Works. In the absence of an ed. of Petit's collected works, his papers have to be consulted in the journals in which they originally appeared. The *Annales de chimie et de physique* between 1816 and 1819 is the most useful source. A partial bibliography is given in Poggendorff, II, 415–416.

II. Secondary Literature. The standard biographical sketch of Petit is the obituary notice by J.-B. Biot, published in *Annales de chimie et de physique*, **16** (1821), 327–335, and *Journal de physique, de chimie, d'histoire naturelle et des arts*, **92** (1821), 241–248. On his work with Dulong, see R. Fox, "The Background to the Discovery of Dulong and Petit's Law," in *British Journal for the History of Science*, **4** (1968–1969), 1–22; J. Jamin, "Études sur la chaleur statique. Dulong et Petit," in *Revue des deux mondes*, 2nd ser., **11** (1855), 375–412; and J. W. van Spronsen, "The History and Prehistory of the Law of Dulong and Petit as Applied to the Determination of Atomic Weights," in *Chymia*, **12** (1967), 157–169. See also R. Fox, *The Caloric Theory of Gases From Lavoisier to Regnault* (Oxford, 1971), especially pp. 227–248. Petit's paper on the history of machines is discussed in C. C. Gillispie, *Lazare Carnot savant* (Princeton, 1971), 107–111.

Robert Fox

PETIT, PIERRE (*b*. Montluçon, France, 8 December 1594 or 31 December 1598; *d*. Lagny-sur-Marne, France, 20 August 1677), *physics, astronomy*.

The son of a minor provincial official, Petit spent his early adult life as *contrôleur de l'election* in Montluçon. In 1633 he traveled to Paris and was appointed Commissaire Provincial de l'Artillerie by Richelieu; he became Intendant Général des Fortifications in 1649. Petit's governmental career was complemented by an active role in French science for more than four decades.

A member of the group of savants meeting at Marin Mersenne's lodgings in the Place Royale, he exemplified those investigators who, in contrast to the increasingly doctrinaire Cartesians, emphasized the importance of accurate experimental observation in validating scientific theories. Petit criticized the lack of adequate astronomical facilities, which he thought had prevented the French from keeping abreast of observations made elsewhere in Europe, and urged the establishment of a royal observatory. His private collection of telescopes and instruments was among the best in Paris and included a number of his own inventions, most notably a perfected filar micrometer later used by Cassini I.

Petit worked with or knew many of the major scientists of the period. In 1646 he collaborated with Blaise Pascal in Rouen and repeated Torricelli's experiment on the barometric vacuum. A regular correspondent of Henry Oldenburg, Petit was keenly interested in the scientific studies pursued in England and played a central role in facilitating the exchange of ideas and inventions between the two national communities. His *Dissertation sur la nature des comètes* (1665) was praised in England and on the Continent for the accuracy and completeness of its observations and discussion; his studies on magnetic declination were equally well-praised. A leading member of the Montmor Academy, Petit was a forceful advocate for the creation of an official science organization. He was, however, ignored by Colbert in the initial selection of members of the Académie Royale des Sciences in

1666; this surprising disappointment was partially compensated for by Petit's election as one of the first foreign fellows of the Royal Society of London in April 1667.

BIBLIOGRAPHY

I. ORIGINAL WORKS. Petit's major scientific writings include *L'usage ou le moyen de pratiquer par une règle toutes les opérations du compas de proportion* (Paris, 1634); *Dissertation sur la nature des comètes . . . avec un discours sur les prognostiques des éclipses et autres matières curieuses* (Paris, 1665); *Dissertations académiques sur la nature du froid et du chaud . . . avec un discours sur la construction et l'usage d'un cylindre arithmétique, inventé par le même autheur* (Paris, 1671).

Petit gave a model account of his and Pascal's experiments in a letter to Pierre Chanut (French Ambassador to Sweden) dated 26 November 1646; the letter is reprinted in Blaise Pascal, *Oeuvres complètes*, Léon Brunschvicg and Pierre Boutroux, eds., I (Paris, 1908), 325–345. The account was published the following year as *Observation touchant le vuide faite pour la première fois en France* (Paris, 1647).

Petit gave an account of his filar micrometer, in which he acknowledged the simultaneous development of the same instrument by Auzout and Picard, in "Extrait d'une Lettre de M. Petit Intendant des Fortifications . . . touchant une nouvelle machine pour mesurer exactement les diamètres des astres. Du 12 Mars 1667," in *Journal des sçavans*, no. 9 (16 May 1667). Petit's complete bibliography is somewhat confused, since a number of his works have been attributed by some to another Pierre Petit of Paris, a prolific medical writer and historian as well as a contemporary of Petit. Thus Petit's *Observationes aliquot eclipsium solis et lunae, cum dissertationibus de latitudine Lutetiae, et declinatione magnetis, necnon de novo systemate mundi quod anonymus dudum proposuit*, published with Jean-Baptiste Du Hamel's *Astronomia physica* (Paris, 1660), is incorrectly attributed by the British Museum *Catalogue* to Petit the medical writer and historian. Finally, Petit's letters to Oldenburg are reprinted in A. Rupert Hall and Marie Boas Hall, eds., *The Correspondence of Henry Oldenburg* (Madison, 1965–   ), *passim*.

II. SECONDARY LITERATURE. Details concerning Petit's life and work are given in *Biographie universelle*, J. Michaud, ed., XXXII (repr. Graz, 1968), 588–589; and Jean Pierre Nicéron, *Mémoires pour servir à l'histoire des hommes illustres dans la république des lettres*, **42** (Paris, 1741), 191–195. On Petit's collaboration with Pascal, see Pierre Humbert, *L'oeuvre scientifique de Blaise Pascal* (Paris, 1947), pp. 73 ff. A useful account of Petit's activities in Parisian scientific circles is given by Harcourt Brown, *Scientific Organizations in Seventeenth Century France* (Baltimore, 1934), *passim*.

MARTIN FICHMAN

PETOSIRIS, PSEUDO- (*fl.* Egypt, second and first centuries B.C.), *astrology*.

During antiquity several texts relating to divination and astrology circulated under the names Petosiris and Nechepso. Nechepso is the name of a king whom Manetho included in the twenty-sixth Egyptian dynasty (*ca.* 600 B.C.); and the most famous Petosiris was the high priest of Thoth (*ca.* 300 B.C. [?]),[1] although many others bore this name signifying "Gift of Osiris." Whether the author of the works circulating under their names had these two individuals, or some others, in mind we cannot know.

The fragments of these works, which were collected by E. Riess,[2] fall into four main groups: (1) those using astral omens as developed by the Egyptians in the Achemenid and Ptolemaic periods from Mesopotamian prototypes to give general indications; (2) those derived from a revelation-text in which Nechepso the king, guided by Petosiris, sees a vision that grants him a knowledge of horoscopic truth; (3) a treatise on astrological botany for medical purposes and another on decanic medicine; and (4) treatises on numerology.

(1) The fragments of texts employing astral omina are largely from authors of late antiquity: Hephaestio of Thebes (*fl. ca.* 415), Proclus (410–485), and John Lydus (*fl. ca.* 560). As preserved to us, the fragments represent radical reworkings of the original texts. It is those fragments, and especially fragment 6 (Riess), which C. Bezold and F. Boll[3] saw to be related to Mesopotamian texts and that allowed Kroll[4] to date the original to the second century B.C. The fragments belonging to this text[5] use as omens eclipses, the heliacal rising of Sirius, and comets. Fragment 6[6] uses as omens the color of the eclipsed body; the simultaneous occurrence of winds blowing from the several directions and of shooting stars, halos, lightning, and rain; and the presence of the eclipsed body in each of the signs of the zodiac (a substitution for Egyptian months). Fragment 6 also divides the day or night into four periods, each of which has three seasonal hours. Most of these elements are found in the demotic papyrus published by R. A. Parker,[7] and many of them in the relevant tablets of the Sin and Shamash sections of the Babylonian astral omen series *Enūma Anu Enlil*.

Fragment 8[8] summarizes a similar treatment of eclipse omens from Campestrius, "who follows the Petosirian traditions." Fragment 7,[9] also on eclipse omens, seems to be from another but still ancient source in which the scheme of geographical references was rather strictly limited to Egypt and its neighbors in contrast to fragment 6, where the eclipses affect the whole Eurasian continent.

Fragment 12[10] gives annual predictions based on the situation at the heliacal rising of Sirius, including the positions of the planets and the color of the star and direction of the winds; it is to be compared to the demotic papyrus published by G. R. Hughes[11] and also with "Eudoxius"[12] and Pseudo-Zoroaster.[13] In the middle of Hephaestio, I, 23, is a description of the manner in which the effective force of the planets is transmitted through the spheres to the sublunar sphere. This passage presupposes both Aristotelian physical theories and a planetary system based on epicycles, eccentrics, or both. If the passage is a genuine quotation from a text written in the second century B.C., it is of the greatest interest as providing the earliest evidence known to us of a theory of astral influence. The fragment contains other elements of interest to a historian of horoscopy—for example, a categorization of the planets as malefic or benefic and the use of aspects. But these elements may have been added by Hephaestio or some unknown predecessor, or the whole chapter may have nothing to do with the work published under the names of Nechepso and Petosiris.

Very doubtful indeed is the attribution to that work of fragments 9,[14] 10,[15] and 11.[16] The ominous bodies are the comets, of which there was originally one type associated with each of the planets. Such comets of the planets are found also in early Sanskrit astral omen texts (for example, in the *Gargasaṃhitā*), but we have as yet no cuneiform tablets that would give us a common source. In any case, there is little reason to assign these specific fragments to Nechepso and Petosiris.

Perhaps also forming a part of the astral omen texts are two other sets of fragments dealing with problems that interested the earliest men who attempted to convert general omens into ones significant for individuals and who used Babylonian techniques. These two problems are the date of a native's conception[17] and the computation of the length of his life based on the rising times between the ascendent and the nonagesimal.[18]

(2) The horoscopic text includes all of the passages from Valens' *Anthologies* (I give the references to the edition by W. Kroll [Berlin, 1908]) and some from Firmicus Maternus. In it Nechepso saw a vision,[19] which included a perception of the motions of the planets that is redolent of pre-Ptolemaic astronomy. He described what he had learned from this revelation in at least thirteen books of very obscure iambic senarii. As we know the ideas there expressed only through the dim intellect of Vettius Valens, we are not surprised to find the "mysteries" largely either self-

contradictory or too fragmentary to be comprehended fully. Some passages in Valens[20] indicate that he knew of a separate work of Petosiris (entitled *Definitions*) in addition to that of Nechepso, to whom he usually refers as "the king," although in another place[21] he speaks of "the king and Petosiris" together. Several passages[22] contain quotations from "the king's" thirteenth book.

Among the principal astrological doctrines discussed by Nechepso and Petosiris in the poetic work (or works) available to Valens are the computation of the length of life of the native;[23] the calculation of the Lot of Fortune, which is also used in computing the length of life;[24] the determination of good and bad times during the native's life, based on various methods of continuous horoscopy (the planetary periods, the lord of the year, and the revolution of the years of nativities);[25] dangerous or climacteric times;[26] and various aspects of the native's life: travel,[27] injury,[28] children,[29] and death.[30] It is probable that Firmicus Maternus drew upon this same collection for his references to Petosiris' and Nechepso's geniture of the universe,[31] his statement that Petosiris only lightly touched upon the doctrine of the decans,[32] and his denial that Petosiris and Nechepso dealt with the *Sphaera barbarica*. Add also the discussion of initiatives in Julian.[33]

(3) Nechepso is known as an authority on materia medica (plants and stones) under astral influence.[34]

(4) The numerological treatises are of two sorts, both explained in a letter of Petosiris to King Nechepso, which is extant in numerous recensions. The simpler form utilizes only the numerical equivalent of the Greek letters in the querist's name; the second form utilizes the day of the lunar month and the "Circle of Petosiris."[35] Another numerological text, which is based on the zodiacal signs, occurs in a letter addressed to Nechepso.[36]

The significance of Pseudo-Petosiris' works (esp. 1 and 2) is their illumination of—although in a very fragmentary form—two important processes of Ptolemaic science: the development of the astral omens that the Egyptians of the Achemenid period had derived from Mesopotamia, and the invention of a new science of astrology based on Greek astronomy and physics in conjunction with Hellenistic mysticism and Egypto-Babylonian divination from astral omens. The effect of their teachings on their successors was profound, although the primitiveness of their methods meant that only their heirs of a mystic (Valens) or antiquarian (Hephaestio and Lydus) bent cite them in detail. That influence is acknowledged not only in the fragments mentioned above, but also at various places in the important *Epitome Parisina*.[37]

NOTES

1. G. Lefebvre, *Le tombeau de Petosiris*, 3 vols. (Cairo, 1923–1924).
2. E. Riess, "Nechepsonis et Petosiridis fragmenta magica," in *Philologus*, Supplementband **6** (1892), 327–394, to which many more fragments could be added.
3. C. Bezold and F. Boll, *Reflexe astrologischer Keilinschriften bei griechischen Schriftstellern* (Heidelberg, 1911).
4. W. Kroll, "Aus der Geschichte der Astrologie," in *Neue Jahrbücher für das Klassische Altertum, Geschichte und Deutsche Literatur*, **7** (1901), 559–577, esp. 573–577.
5. Frs. 6–12 in Riess, some of which are very dubious.
6. Hephaestio, I, 21, who attributes the material to the ancient Egyptians; another version, using Roman months rather than zodiacal signs, was published by F. Boll in *Catalogus Codicum Astrologorum Graecorum*, VII (Brussels, 1908), 129–151.
7. R. A. Parker, *A Vienna Demotic Papyrus on Eclipse and lunar-omina* (Providence, 1959).
8. Lydus, *De ostentis*, 9.
9. Hephaestio, I, 22.
10. Hephaestio, I, 23, who attributes it to the ancient, wise Egyptians; cf. *Catalogus Codicum Astrologorum Graecorum*, V, pt. 1 (Brussels, 1904), 204.
11. G. R. Hughes, "A Demotic Astrological Text," in *Journal of Near Eastern Studies*, **10** (1951), 256–264.
12. *Catalogus Codicum Astrologorum Graecorum*, VII, 181–187.
13. *Geoponica*, I, 8 and I, 10 = fr. 0,40; and fr. 0,41 in J. Bidez and F. Cumont, *Les mages hellénisés*, II (Paris, 1938), 178–183.
14. John Lydus, *De ostentis*, 11–15, from Campestrius.
15. Hephaestio, I, 24.
16. Servius, *In Aeneidem*, X, 272, who follows Avienus, but also mentions Campestris (sic!) and Petosiris.
17. Fr. 14; cf. Achinapolus in Vitruvius, *De architectura*, IX, 6,2; Pseudo-Zoroaster, fr. 0,14 Bidez-Cumont, II, 161–162; and A. Sachs, in *Journal of Cuneiform Studies*, **6** (1952), 58–60.
18. Frs. 16 and 17 and also fr. 5 (Valens, III, 16) and Valens, III, 3, and VIII, 6; cf. Berosus, frs. 32 and 33 in P. Schnabel, *Berossos und die babylonisch-hellenistische Literatur* (Leipzig–Berlin, 1923), 264. Also see Hephaestio, II, 18, 72 (this quotation does not include the important fragment of the *Salmeschoeniaca*, II, 18, 74–75), and Pliny's report of their computation of the distances of the planetary spheres (fr. 2). This last may belong to 2.
19. Fr. 1; Valens, VI, preface.
20. Valens, II, 3; VIII, 5; IX, 1.
21. *Ibid.*, VII, 5; cf. III, 10.
22. *Ibid.*, II, 3; III, 14; IX, preface; IX, 1.
23. *Ibid.*, III, 10 = fr. 18, which gives a computation based on a point computed similarly to a Lot and entirely different from the method employed in the passages we have assigned to 1.
24. This is given in Nechepso's thirteenth book and in Petosiris, *Definitions*; Valens, II, 3; III, 14 = fr. 19; IX, 1.
25. Valens, V, 6 = fr. 20 and VII, 5 = fr. 21; cf. III, 14; VI, 1.
26. *Ibid.*, III, 11 = fr. 23.
27. *Ibid.*, II, 28.
28. *Ibid.*, II, 36; cf. fr. 27 from Firmicus.
29. *Ibid.*, II, 39.
30. *Ibid.*, II, 41 = fr. 24.
31. Fr. 25, where they are correctly stated to be drawing on an Hermetic source; cf. test. 6.
32. Fr. 13, but cf. fr. 28.
33. *Catalogus Codicum Astrologorum Graecorum*, I (Brussels, 1898), 138. (I doubt the authenticity of the brief statement about quartile and trine aspect published in *Catalogus Codicum Astrologorum Graecorum*, VI [Brussels, 1903], 62.)
34. Frs. 28–32 and 35–36; the latter two, drawn from the work of Thessalus, should now be consulted in the edition of H.-V. Friedrich (Meisenheim am Glan, 1968); cf. also *Catalogus Codicum Astrologorum Graecorum*, I, 126.
35. Frs. 37–42; see also *Catalogus Codicum Astrologorum Graecorum*, I, 128; IV (Brussels, 1903), 120–121; XI, pt. 2 (Brussels, 1934), 152–154, 163–164; Pseudo-Bede in *Patrologia Latina*, XC, cols. 963–966; and cf. Psellus in a letter published in *Catalogus Codicum Astrologorum Graecorum*, VIII, pt. 1 (Brussels, 1929), 131.
36. *Catalogus Codicum Astrologorum Graecorum*, VII, 161–162.
37. *Ibid.*, VIII, pt. 3 (Brussels, 1912), 91–119.

BIBLIOGRAPHY

Aside from Riess's collection of fragments, the main study of Pseudo-Petosiris is C. Darmstadt, *De Nechepsonis-Petosiridis Isagoge quaestiones selectae* (Leipzig, 1916); unfortunately, he attributes to Nechepso-Petosiris far more than the evidence of the fragments warrants. Rather unsatisfactory articles are W. Kroll in Pauly-Wissowa's *Real-Encyclopädie der classischen Altertumswissenschaft*, **16** (1935), cols. 2160–2167; **19** (1938), col. 1165; and W. Gundel and H. G. Gundel, *Astrologumena* (Wiesbaden, 1966), 27–36.

DAVID PINGREE

PETRIE, (WILLIAM MATTHEW) FLINDERS

(*b*. Charlton, Kent, England, 3 June 1853; *d*. Jerusalem, Palestine [now Israel], 28 July 1942), *Egyptology, archaeology.*

Petrie's delicate health in childhood prevented him from going to school or university. He was taught at home by his parents and developed a keen interest in antiquities and surveying. His father encouraged him in particular to make surveys of British earthworks, which led to his interest in measurement. Petrie's first book, *Inductive Metrology*, appeared in 1877, and three years later he published a field survey of Stonehenge.

In 1881 Petrie's father, inspired by the notions of Charles Piazzi Smyth regarding the pyramids, planned a trip to Egypt with his son; but in the end, Petrie went alone to survey them. He published *The Pyramids and Temples of Gizeh* in 1883, following two years' work. In this book Petrie disproved and abandoned Smyth's esoteric theories. In 1883 the Egypt Exploration Fund (later Society) was founded, and Petrie was appointed its first field director. He wrote to Miss Edwards, the secretary of the Fund, in 1883 that "The prospect of excavating in Egypt is a most fascinating one to me, and I hope that the results may justify my undertaking such a work." This hope was brilliantly realized. Petrie began work at Tanis in 1884 but subsequently quarreled with the Fund. In 1894 he founded the Egyptian Research Account, renamed the British

School of Archaeology in Egypt in 1905. He directed this until 1926, when, disillusioned by Egypt, he transferred his attention to Palestine, where he worked until his death.

For a period of ten years (1896–1906) Petrie worked again with the Egypt Exploration Society, but on the whole his career was devoted to his own excavations and publications. His annual excavations were followed by immediate publications and exhibitions in London. In 1892 he was elected professor of Egyptology at the University of London, a post he held until 1933. He was also elected a fellow of both the Royal Society (1902) and the British Academy (1912) and was knighted in 1923. Petrie was essentially an individualist and a free-lance worker; he was an authoritarian who brooked no opposition or criticism, and, as Woolley has stated, he had "a dogmatic assurance of his own rightness." During his lifetime, he trained at least two generations of Egyptologists and Near Eastern archaeologists.

Among the most remarkable of his finds were the early royal tombs at Abydos; the Tell el-Amarna correspondence and the numerous relics at Tell el-Amarna itself; the discoveries of Mycenaean and Pre-Mycenaean pottery at Ghurob and Kahun; and the discovery of the predynastic cultures of Egypt, particularly those of Nakada and Ballas (which he examined in 1894–1895) and at Diospolis Parva (in 1898–1899). Nakada was revealed to be a prehistoric cemetery of more than 2,000 graves, and gave its name to the Nakada period. The British Museum declined Petrie's offer of the type series from this cemetery on the ground that they had been advised it was "unhistoric rather than prehistoric." In his memoir *Diospolis Parva* (1901) Petrie systematically arranged the predynastic Egyptian material for the first time and invented the technique of sequence-dating, which he further described in *Methods and Aims in Archaeology* (1904).

When he began working in Egypt, Petrie was sharply critical of the methods of his predecessors, and wrote that "The true line lies as much in the careful noting and comparison of small details as in more wholesale and off-hand clearance." The advances and developments in techniques and methods that he made, as well as his actual discoveries of dynastic and predynastic Egypt, made the last quarter of the nineteenth century the "heroic age" of Egyptian archaeology. In 1892 he published *Ten Years' Diggings*, a remarkable record of his work. He continued to dig in Egypt for more than thirty years, and then in Palestine for nearly twenty years. His fieldwork, diggings, and devotion to archaeology were later chronicled in his autobiography, *Seventy Years in Archaeology* (London, 1931).

In 1889, at Ghurob, Petrie found Mycenaean pottery among the remains of the late eighteenth dynasty; he also found what he called Proto-Greek or Aegean pottery. The following year, at Kahun, he found painted Aegean or Proto-Greek pottery mixed with that of the twelfth dynasty. No pottery of this Proto-Greek style had hitherto been found in the Aegean. Petrie, however, was not content to regard the Kahun pottery as "foreign" ware; he unhesitatingly cataloged it as Aegean, a splendid example of skilled guesswork. In 1891 he visited Mycenae to verify the dating of the Ghurob and Kahun sites. He recognized examples of Egyptian influence and actual imports of Egyptian objects in Mycenae that dated to the eighteenth dynasty period. He thus established two synchronisms; one between the Aegean or Proto-Greek ware and the twelfth dynasty of Egypt, and the second between Mycenae and the eighteenth dynasty. On this basis Petrie declared that an Aegean civilization had begun about 2500 B.C. and that the dates of the late Mycenaean civilization were between 1500 and 1000 B.C. He dated the Mycenaean "treasuries" between 1400 and 1200 B.C., the Vaphio cups at about 1200 B.C., and the shaft graves at 1150 B.C. It was a remarkably fine use of cross-dating and one of the first demonstrations of this method of extending historical chronology to primitive regions.

Ernest Gardner, then director of the British School of Archaeology in Athens, writing of Petrie's chronological structuring, said that he had "done more in a week than the Germans had done in ten years to clear up the matter from an Egyptian basis." Petrie himself, writing in 1931 and looking back to his early conclusions on Greek chronology, said "there seems little to alter in the outline reached then, though forty years have since passed."

Petrie's contributions to the development of archaeology included not only his substantive work in Egypt and Palestine, but also his revolutionary techniques, which are of paramount importance to the study of antiquity.

BIBLIOGRAPHY

Petrie's major works are cited above. See also Margaret Murray, *My First Hundred Years* (London, 1966); Sidney Smith, obituary notice in *Proceedings of the British Academy*, **28** (1942), 307–324; Leonard Woolley, in *Dictionary of National Biography 1941–1950* (London, 1959), 666–667; J. D. Wortham, *The Genesis of British Egyptology 1549–1906* (Norman, Okla., 1971), 115–126; and W. R. Dawson and E. P. Uphill, *Who Was Who in Egyptology* (London, 1972), 228–230, which has a complete bibliography.

GLYN DANIEL

PETROV, NIKOLAY PAVLOVICH (*b.* Trubchevsk, Orlovskaya Oblast, Russia, 25 May 1836; *d.* Tuapse, U.S.S.R., 15 January 1920), *mechanics, engineering.*

The son of a military man, Petrov graduated in 1855 from the Konstantinovsky Military School in St. Petersburg with the rank of ensign. He immediately enrolled at the Nikolaevskaya Engineering Academy, where Vyshnegradsky guided his studies in applied mechanics and where he was especially influenced by Ostrogradsky. After graduating in 1858, Petrov was retained in the mathematics department, where he worked under Ostrogradsky. On the latter's death in 1862 Petrov began teaching a course in higher mathematics at the Academy, while from 1866 he taught applied mechanics at the St. Petersburg Technological Institute. In 1867 he became adjunct professor of applied mechanics at the Engineering Academy.

In 1876 Petrov was invited to the U.S. Centennial Exposition in Philadelphia. On the basis of material collected during this trip, he published a work in 1882 on mechanical equipment for ports and cargo deposits at railroad stations; it presented methods for designing elevators and mechanical devices used in shipping grain and coal. In a paper published in 1878 on continuous braking systems of trains, he was the first to arrive at equations for the motion of wheels in the presence and absence of braking, as well as an equation for the motion of the center of gravity of a train during braking. As a result of his theoretical investigations, he found the true maximum speed for any possible braking system.

In 1888 Petrov was elected an honorary member of the Moscow Polytechnical Society and was awarded the gold medal of the Russian Technical Society for inventing an instrument to determine the internal and external friction of liquids. As chairman of the Department of State Railroads (1888–1892) he was active in the construction of the Trans-Siberian Railroad. In 1892 he became chairman of the Engineering Council of the Ministry of Means of Communication, and from 1893 to 1900 he was deputy minister of means of communication. In 1894 Petrov became an honorary member of the St. Petersburg Academy of Sciences, and from 1896 to 1905 he was chairman of the Russian Technical Society.

Petrov's most important work was in the hydrodynamic theory of lubricants. His investigations were characterized by a masterful command of mathematical method and an outstanding gift for experiment. Petrov's results were based on broad experimental study of friction in liquids and machines, the most important part of which he conducted himself.

Petrov accurately formulated the physical laws that might provide a basis for calculating elementary frictional forces. He demonstrated that the frictional force developed within a viscous liquid is proportional to the velocity of relative motion and to the surface area of contact; it does not depend on pressure, and the coefficient of proportionality depends only on the properties of the liquid.

Petrov also confirmed experimentally Newton's formula for the force of resistance of a viscous liquid with one layer in motion relative to another layer, $F = \mu(\partial v/\partial n)$, where v is the velocity, n is the distance between layers along the normal to the surface of constant velocity, and μ is the coefficient of viscosity. He determined that the coefficient of internal friction of a liquid varies significantly with temperature; this relationship was experimentally determined for each type of lubricant.

Petrov also studied the effects of friction in a bearing in which the lubricating layer fills the intermediate space between two cylinders having a common geometrical axis. During rotation friction causes the shaft to carry along with it the nearest adjacent layer of lubricant. This layer moves more slowly than the shaft, since it is impeded by friction with the next adjacent layer of lubricant. The last layer, carried along by friction with the penultimate layer, is in turn retarded by friction with the surface of the bearing. Using this model, Petrov examined bearings that operate with only a thin layer of lubricant. Computing the equations of motion of a viscous liquid, integrating them, and using Newton's formula presented above, he discovered the law of friction that bears his name.

Petrov's works have formed the basis of the hydrodynamic theory of friction in the presence of lubrication and provided an impetus for the further development of theoretical and experimental research in this field.

BIBLIOGRAPHY

I. ORIGINAL WORKS. The basic ideas of Petrov's hydrodynamic theory of lubricants were stated in the following works: "Trenie v mashinakh i vlianie na nego smazyvayushchey zhidkosti" ("Friction in Machines and the Influence of a Lubricating Liquid on It"), in *Inzhenernyi zhurnal*, **24** (1883), no. 1, 71–140; no. 2, 227–279; no. 3, 377–436; no. 4, 535–641; also published separately (St. Petersburg, 1883); *Opisanie i rezultaty opytov nad treniem zhidkostey i mashin* ("Description and Results of Experiments on the Friction of Liquids and Machines"; St. Petersburg, 1886); "Prakticheskie rezultaty opytov" ("Practical Results of Experiments"), in *Inzhenernyi zhurnal*, **26** (1887), also published separately; and "Frottement dans les machines," in *Zapiski Imperatorskoi akademii nauk*, **10**, no. 4 (1900), 1–84.

His other writings include "Ochertanie zubtsov kruglykh tsilindricheskikh koles dugami kruga" ("Configuration of the Teeth of Round Cylindrical Wheels With Circular Arcs"), in *Inzhenernyi zhurnal*, **12** (1871), which was his first scientific paper; *O nepreryvnykh tormoznykh sistemakh* ("On Continuous Braking Systems"; St. Petersburg, 1878); *Peregruzka i khranenie khlebnogo zerna. Peregruzka kamennogo ugla* ("Shipment and Storage of Cereal Grain. Shipment of Coal"; St. Petersburg, 1882); *Résultats les plus marquants de l'étude théorique et expérimentale sur le frottement médiat* (St. Petersburg, 1889); "Sur le frotte-ment des liquides," in *Izvestiya Imperatorskoi akademii nauk*, **5** (1896), 365–373; *Un moyen de déterminer les déformations du rail soutenu par des supports mobiles sous la pression d'une roue usée en mouvement* (St. Peters-burg, 1910); and *Davlenie koles na relysy. Prochnost relsov i ustoychivost puti* ("The Pressure of Wheels on Rails. Strength of the Rails and Stability of the Track"; St. Petersburg, 1915), a summary of his investigations.

Petrov's writings are also included in L. S. Leybenzon, ed., *Gidrodinamicheskaya teoria smazki* ("The Hydro-dynamic Theory of Lubricants"; Moscow–Leningrad, 1934); and L. S. Leybenzon, ed., *Gidrodinamicheskaya teoria smazki*, in the series Klassiki Nauki (Moscow, 1948), an anthology.

II. SECONDARY LITERATURE. On Petrov and his work, see A. S. Akhmatov, "Nikolay Pavlovich Petrov," in *Lyudi russkoy nauki. Tekhnika* ("Men of Russian Science"; Moscow, 1965), 240–246; A. T. Grigorian, *Ocherki istorii mekhaniki v Rossii* ("Sketches of the History of Mechanics in Russia"; Moscow, 1961), 213–217; V. M. Kostomarov and A. G. Burgvits, *Osnovopolozhnik teorii gidrodinamicheskogo trenia v mashinakh N. P. Petrov* ("Petrov, the Founder of the Theory of Hydrodynamic Friction in Machines"; Moscow, 1952); and L. S. Leybenzon and N. I. Glagolev, "Vydayushchysya ucheny i inzhener N. P. Petrov" ("The Outstanding Scientist and Engineer N. P. Petrov"), in *Izvestiya Akademii nauk SSSR*, no. 7 (1946), 929–933.

A. T. GRIGORIAN

PETROV, VASILY VLADIMIROVICH (*b.* Oboyan, Russia, 19 July 1761; *d.* St. Petersburg, Russia, 15 August 1834), *physics, chemistry.*

The son of a parish priest, Petrov graduated from the Kharkov Collegium in 1785 and studied at the Teacher's Gymnasium in St. Petersburg. He taught physics, mathematics, Latin, and Russian at the mining school of Barnaul (Altay) from 1788 to 1791, then taught in St. Petersburg at the Izmaylov Cadets School (1791–1797) and the Main Medical School. In 1795 Petrov became extraordinary professor and, in 1800, professor at the Medical-Surgical Academy. There he created a first-class *cabinet de physique* and

at the beginning of the nineteenth century did basic research in physical chemistry, electrostatics, and galvanism.

From 1802 Petrov was corresponding member, from 1809 extraordinary, and from 1815 ordinary academi-cian of the St. Petersburg Academy of Sciences. He was elected honorary member of the Erlangen Physics-Medical Society (1810) and the University of Vilna (1829).

Petrov was an active follower of Lavoisier not only in the promotion and application of the oxygen theory of combustion but also in the treatment of heat and light as chemical elements, in which he included electrical and galvanic fluids.

In *Sobranie fiziko-khimicheskikh novykh opytov i nablyudeny* ("Collection of New Physical-Chemical Experiments and Observations," 1801) and in a series of articles later published in *Umozritelnye issledovania Sankt-Peterburgskoy Akademii nauk* ("Speculative Research of the St. Petersburg Academy of Sciences") Petrov described his experiments on the possibility of burning organic and inorganic substances in a vacuum and in some gases that do not sustain combustion (carbon dioxide gas, hydrogen chloride, sulfur dioxide). He showed that even in the absence of air, substances containing oxygen can burn, whereas the transformation of metals into oxides is impossible.

Petrov also examined various forms of phosphores-cence from the viewpoint of the oxygen theory of combustion, showing that luminescence of rotten wood can occur only in the presence of oxygen and is a "very slow form of combustion." By demonstrating that the phosphorescence of minerals does not depend on the presence of oxygen, he distinguished photo-luminescence from chemiluminescence. He also investigated the relationship of the luminescence of crystal phosphorus to temperature.

In *Izvestie o galvani-voltovskikh opytakh* ("News of Galvanic-Voltaic Experiments," 1803) Petrov described experiments carried out in the spring of 1802 with a battery of 2,100 copper-zinc elements. He also related its structure and principles of operation—the use of sealing wax and wax to insulate the wires, chemical indicators for observing the oxidation-reduction processes at the electrodes, and methods for eliminating oxidization on the surfaces of the metal disks. Of great importance is the description of the stable arc discharge and the indication of its possible use in artificial lighting, melting metals, obtaining pure metallic oxides, and reduction of metals from oxides mixed with powdered carbon and oils.

Petrov observed discharges at low pressures and in particular described a decomposing discharge. He investigated the relation of electrolysis of various substances to temperature and electromotive force, and used parallel tubes filled with electrolytes to observe the relation of the current to the cross section of the conductor.

In *Novye elektricheskie opyty* ("New Electrical Experiments," 1804) Petrov described experiments that showed the possibility of electrifying metals by friction. He used the bell glass of an air pump for studying electrostatic discharges in a vacuum, in hydrogen, in nitrogen, and in carbon dioxide.

After his election to the Academy of Sciences, Petrov made meteorological observations in St. Petersburg and also processed the observations sent from other Russian cities. He conducted a special study of the relation of velocity of vaporization of ice and snow to atmospheric pressure, temperature, and wind force.

The decline in Petrov's scientific activity during the last twenty-five years of his life was caused by financial difficulties related to the War of 1812 and later by his impaired health.

BIBLIOGRAPHY

Petrov's "Izvestie o galvani-voltovskikh opytakh" ("News of Galvanic-Voltaic Experiments") was reprinted in *Sbornik k stoletiyu so dnya smerti pervogo russkogo electrotekhnika akademika Vasilia Vladimirovicha Petrova* ("Collection on the Centenary of the Death of the First Russian Electrotechnician Academician Vasily Vladimirovich Petrov"; Moscow–Leningrad, 1936), i–viii, 1–194.

There is an annotated bibliography of his works in S. I. Vavilov, ed., *K istorii fiziki i khimii v Rossii v nachale XIX v.* ("Toward a History of Physics and Chemistry in Russia at the Beginning of the Nineteenth Century"; Moscow–Leningrad, 1940), 193–210. Other works on Petrov's life and work are A. A. Eliseev, *Vasily Vladimirovich Petrov* (Moscow–Leningrad, 1949), with bibliography on pp. 172–179; O. A. Lezhneva, "Die Entwicklung der Physik in Russland in der ersten Hälfte des 19. Jahrhunderts," in *Beiträge zur Geschichte der Naturwissenschaften* (1960), 203-225; and Y. A. Shneyberg, "O bataree V. V. Petrova i ego opytakh s elektricheskoy dugoy i razryadom v vakuume" ("On Petrov's Battery and His Experiments With the Electric Arc and Discharge in a Vacuum"), in *Elektrichestvo*, no. 11 (1953), 72–75.

OLGA A. LEZHNEVA

PETROVSKY, IVAN GEORGIEVICH (*b.* Sevsk, Orlov guberniya, Russia, 18 January 1901; *d.* Moscow, U.S.S.R., 15 January 1973), *mathematics*.

Petrovsky's major works dealt with the theory of partial differential equations, the topology of algebraic curves and surfaces, and the theory of probability. After graduating from the technical high school in Sevsk in 1917, he worked in various Soviet institutions until 1922, when he entered Moscow University. He graduated from the division of physics and mathematics in 1927 and remained at Moscow until 1930 as a graduate student of D. F. Egorov.

From 1929 to 1933 Petrovsky was assistant professor and *dozent* at Moscow. In 1933 he became professor and, in 1935, doctor of physical-mathematical sciences. From 1951 he was head of the department of differential equations. During World War II he was dean of the faculty of mechanics and mathematics; and from 1951 until his death he was rector of the university.

Petrovsky combined his work at the university with activity at various scientific and teaching institutions. From 1943 he worked at the V. A. Steklov Institute of Mathematics at the Academy of Sciences, of which he was vice-director from 1947 to 1949. For many years he was editor-in-chief of *Matematicheskii sbornik*.

In 1943 Petrovsky was elected corresponding member and, in 1946, full member of the Soviet Academy of Sciences. From 1949 to 1951 he was academician-secretary of the division of physical and mathematical sciences of the Academy, and from 1953 until his death he was a member of the Presidium of the Academy. He was twice awarded the State Prize of the U.S.S.R., and he received the title of Hero of Socialist Labor. He was also a member of the Soviet Committee for the Defense of Peace and vice-president of the Institute of Soviet-American Relations.

Petrovsky's first research dealt with the investigation of the Dirichlet problem for Laplace's equation (1928) and the theory of functions of a real variable (1929). In the early 1930's he began research on the topology of algebraic curves and surfaces in which he achieved fundamental results and methods. In 1933 he proved Hilbert's hypothesis that a curve of the sixth order cannot consist of eleven ovals lying outside each other. The method that Petrovsky devised for this proof was useful in solving the more general problem of embedding components of algebraic curves of any order in a projective plane. In 1949 he generalized certain of his results to include algebraic surfaces in *n*-dimensional space.

The results of Petrovsky's work (1934) on the solvability of the first boundary-value problem for the heat equation were widely applied in the theory of probability, especially in research connected with the Khinchin-Kolmogorov law of the iterated logarithm. Petrovsky's article on the theory of random processes (1934) considerably influenced an investigation of limit laws for the sum of a large number of random variables with the aid of the transition to random processes with continuous time. The work also contains the so-called method of upper and lower sums, which became the basic analytical method of research in the field.

A second work, also published in 1934, examined the behavior, near the origin of the coordinates of integral curves, of a system of equations of the form

$$\frac{dx_i}{dt} = \sum_{k=1}^{n} \alpha_{ik} x_k + \phi_i(x_1, \cdots, x_n).$$

This work was, essentially, the first full investigation of a neighborhood of a singular point in the three-dimensional case.

In 1937–1938 Petrovsky distinguished and studied classes of systems of partial differential equations, which he first identified as either elliptical, hyperbolic, or parabolic. In 1937 he published his proof that the Cauchy problem for nonlinear systems of differential equations, which Petrovsky called hyperbolic, is well-posed. In a work on the Cauchy problem for systems of linear partial differential equations in nonanalytic functions (1938) Petrovsky studied systems for which the Cauchy problem is uniformly well-posed relative to the variation of the surface, for which the original data are given. Petrovsky introduced the concept of parabolic systems and investigated the problem of the analyticity of the solution of such systems in space dimensions. In 1937 he introduced his famous notion of elliptical systems and showed that when the functions are analytic, all sufficiently smooth solutions will be analytic, thereby giving a more complete solution of Hilbert's nineteenth problem. Petrovsky's results were the starting point for numerous investigations, including those of J. Leray and L. Gårding; and they determined the basic direction of the development of the theory of systems of partial differential equations.

In widely known works on the qualitative theory theory of hyperbolic equations (1943–1945) Petrovsky introduced the concept of lacunae and obtained necessary and sufficient conditions for the existence of stable lacunae for uniform hyperbolic equations with constant coefficients. He also solved completely the question of lacunae for linear hyperbolic systems with variable coefficients in the case of two independent variables.

In 1945 Petrovsky investigated the extent of the discontinuitites of the derivatives of the displacements on the surface of a nonuniform elastic body that is free from the influence of external forces; and in 1954 he examined the character of lines and surfaces of a discontinuity of solutions of the wave equations.

Many well-known specialists in the theory of differential equations were students of Petrovsky, whose seminar (the Petrovsky seminar) is a leading center for the study of the theory of partial differential equations. His course texts are widely known.

BIBLIOGRAPHY

I. ORIGINAL WORKS. Petrovsky's major works are "Über das Irrfahrtproblem," in *Mathematische Annalen*, **109** (1934), 425–444; "Ueber das Verhalten der Integralcurven eines Systemes gewöhnlicher Differentialgleichungen in der Nahe lines singularen Punktes," in *Matematicheskii sbornik*, **41** (1934), 107–156; "Zur ersten Randwertaufgabe der Warmeleitungsgleichung," in *Compositia Mathematica*, **1** (1935), 383–419; "Über das Cauchysche Problem für Systeme von partiellen Differentialgleichungen," in *Matematicheskii sbornik*, **2** (1937), 815–870; "On the Topology of Real Plane Algebraic Curves," in *Annals of Mathematics*, **39**, no. 1 (1938), 197–209; "Sur l'analyticité des solutions des systèmes d'équations différentielles," in *Matematicheskii sbornik*, **5** (1939), 3–70; *Lektsii po teorii obyknovennykh differentsialnykh uravneny* ("Lectures on the Theory of Ordinary Differential Equations"; Moscow–Leningrad, 1939); "On the Diffusion of Waves and the Lacunas for Hyperbolic Equations," in *Matematicheskii sbornik*, **17** (1945), 289–370; *Lektsii po teorii integralnykh uravneny* (Moscow–Leningrad, 1948), trans. into German as *Vorlesungen über die Theorie der Integralgleichungen* (Würzburg, 1953); and *Lektsii ob uravneniakh s chastnymi proizvodnymi* (Moscow–Leningrad, 1948), trans. into English as *Lectures in Partial Differential Equations* (New York, 1954) and into German as *Vorlesungen über partielle Differentialgleichungen* (Leipzig, 1955).

II. SECONDARY LITERATURE. On Petrovsky's work and its influence, see P. S. Aleksandrov *et al.*, "Ivan Georgievich Petrovsky," in *Uspekhi matematicheskikh nauk*, **26**, no. 2 (1971), 3–22, with bibliography, pp. 22–24; *Matematika v SSSR za 40 let* ("Mathematics in the U.S.S.R. for the Last Forty Years"), II (Moscow, 1959), 538–540, for a bibliography of 51 of his publications; and *Matematika v SSSR za 50 let* ("Mathematics in the U. S. S. R. for the Last Fifty Years"), II, pt. 2 (Moscow, 1970), 1035.

S. DEMIDOV

PETRUS. See also **Peter.**

PETRUS BONUS, also known as **Bonus Lombardus** or **Buono Lombardo of Ferrara** (*fl. ca.* 1323–1330), *alchemy.*

Petrus Bonus is known through the extant texts of his alchemical work. He has been differentiated from the Petrus Bonus who was doctor of laws in the

University of Ferrara (1396–1402), and from Petrus Bonus Advogarius, who taught astronomy or astrology at Ferrara and issued astrological predictions in the late fifteenth century.[1] According to the explicit of his alchemical treatise, Master Petrus Bonus, *phisicus*, or doctor of medicine, in 1323 discussed a *quaestio* on alchemy in Traù, Dalmatia.[2] In 1330 (1338, 1339, or 1350), in Pola, Istria, he composed *The Precious New Pearl*,[3] a scholastic exposition of the arguments, with the names of appropriate authorities, for and against the validity of the alchemical art.

Since Petrus Bonus approached the problem of the transmutation of baser metals into gold "with a pen in his hand rather than an alembic and with volumes of the past literature . . . rather than metals and chemicals,"[4] his accomplishment is significant not for any new chemical or scientific data but, rather, for the light that it sheds on current practices, theories, and authorities in alchemy and in other areas of natural science in the early fourteenth century. Like his earlier contemporaries, principally Arnald of Villanova, Ramón Lull, Roger Bacon, and Albertus Magnus, to whom alchemical writings of about the same era are attributed, but whose names Petrus Bonus did not mention,[5] he believed that the adept alchemist could produce gold from baser metals with the use of the philosophers' stone. He held that the substance and material cause of this stone was quicksilver alone. In this view Petrus Bonus agreed with Arnald of Villanova, whom he probably knew through the *Lilium*, one of the few works of Latin origin that he cited.[6]

He rejected the alternative view advanced in other writings that both quicksilver and sulfur are essential, that is, that quicksilver is the matter of, and sulfur is the active agent that shapes and forms, the philosophers' stone. According to Petrus Bonus, gold, the most perfect of metals, had been purified of sulfur. It had thus attained in nature the stage of perfection toward which the baser metals are striving.[7] Therefore, according to Petrus Bonus, the process of transmutation consists principally in the separation from the baser metals of the sulfur that blackens, corrodes, or discolors the metals. This separation goes on at a slow and protracted rate in nature in the bowels of the earth. The skilled operator, who is conversant with the literature of the past and adept in the use of the philosophers' stone, can greatly hasten the separation in the laboratory. The form of gold can be introduced in the twinkling of an eye, but only with divine assistance.[8]

Although Petrus Bonus insisted upon an acquaintance with the past literature of alchemy—chiefly that by authors of Greek and Arabic or Muslim origin,

such as Aristotle, Hermes, Morienus, Jabir ibn Hayyān (Geber), Rasis (probably al-Razī), Ibn Rushd, Ibn Sīnā, and the authors of the *Turba philosophorum* —and also upon the need to check speculative thought by practice and experience,[9] he nonetheless was finally obliged to admit that he had not been able to penetrate into the secret of the philosophers' stone and that he had to fall back upon the mystery of the divine will. Petrus Bonus accepted "alchemy as possessing a divine as well as natural character." He held that "sufficient natural reasons for the philosophers' stone" could not be given. One must believe it, as one believes in "the miracles of Christianity." To him the "art is a divine secret transcending both natural reason and experience."[10] Thus *The Precious New Pearl* lucidly portrays the direction that alchemists were to take in the fourteenth century and thereafter.

Petrus Bonus shed light not only on the practices and perplexities of the alchemists but also on other items of natural interest in his time. In seeking analogies to the transmutation of metals, he set forth the belief of his contemporaries in the spontaneous generation of frogs, ants, and flies from dust and clouds or refuse. He insisted that animals so generated were the same as those generated in the usual fashion.[11] Hence he affirmed that gold, which would stand up to various tests, was the same as gold generated in nature. Further, he revealed that it was current knowledge that in Aristotle's time there was opposition to the geocentric theory, with alternative views that, like the planets, the earth moved in circular motion.[12] He mentioned the discovery of gold in the silver mines of Serbia and noted the existence of silver mines in Germany and alum mines in the vicinity of Constantinople.[13]

Although *The Precious New Pearl* is an extremely detailed and even repetitive work, it is nonetheless very informative. This accounts for the interest in it of its first editor, Janus Lacinius, and the frequency of its publication in the sixteenth century and thereafter.

NOTES

1. Lynn Thorndike, *History of Magic and Experimental Science*, III (New York, 1934), 147 ff.
2. *Ibid.* Thorndike has quoted the citation by Tiraboschi, *Storia della letteratura italiana*, V (Milan, 1823), 332, of the Este MS, "Quaestio . . . per Magistrum Bonum Ferrariensem physicum sub MCCCXXIII anno . . . tunc temporis salariatum in civitate Traguriae de provincia Dalmatiae." Other MSS that contain similar citations regarding the *quaestio* in 1323 are British Museum, Harley 672, XV cent., f. 169; Orléans 289 (243), XV cent., f. 187v; and 290 (244), XVI cent., f. 96.
3. MS, British Museum, Harley 672, XV cent., f. 169, "Explicit Preciosa novella margarita edita a magistro Bono

Lombardo de Ferraria phisico introducens ad artem alkimie. Composita 1330 in civitate Polle in provincia Istrie"; similarly MSS Orléans 289 (243), XV cent., f. 187v, with the date 1338; Orléans 290 (244), XVI cent., ff. 1–96, with the date 1350. Of the printed eds., that of Zetzner, *Theatrum chemicum*, V (1660), has the dates 1330 and 1339, as does Manget, *Bibliotheca chemica curiosa*, II (1702), 1–80.

4. Thorndike, III, 153.
5. Thorndike, III, 150–151; for these authors see Thorndike, II and III, *passim*.
6. *Ibid.*, III, 160; Zetzner, V, 546, 559, 568. For Arnald of Villanova and the *Lilium*, see Thorndike, III, 62 ff.
7. Thorndike, III, 160; Zetzner, V, 508–509, 546, 567–568, 580.
8. Thorndike, III, 158 ff.; Zetzner, V, 550 ff., 632 ff.
9. Zetzner, V, 568 ff.; for the numerous authors cited, see esp. Julius Ruska, "L'alchimie à l'époque de Dante," in *Annales Guébhard-Séverine*, 10 (1934), 410–417, esp. pp. 415–416, where some twenty authors are counted without those of the *Turba philosophorum*.
10. Thorndike, III, 159; Zetzner, V, 580–588, caps. 6–8.
11. Thorndike, III, 162; Zetzner, V, 647.
12. Thorndike, III, 161–162; Zetzner, V, 642.
13. Thorndike, III, pp. 160–161; Zetzner V, 681, 682, also 549.

BIBLIOGRAPHY

I. Original Works. For MS texts, see British Museum, Harley 672, XV cent., ff. 1–169 (D. W. Singer, *Catalogue of Latin and Vernacular Alchemical Manuscripts in Great Britain and Ireland Dating From Before the XVI Century*, I [Brussels, 1928], no. 276); Orléans 289 (243), XV cent., ff. 1–187v (J. Corbett, *Catalogue des manuscrits alchimiques latins: II. Manuscrits des bibliothèques publiques des départements français* [Brussels, 1951], no. 39); Orléans 290 (244), XVI cent., ff. 1–96 (Corbett, II, no. 40); Bodleian, Ashmole 1426, XV cent., ff. 103–115, a gloss or commentary on *The Precious New Pearl* (Singer, I, no. 277); and Paris, Bibliothèque nationale 14006, XV cent., ff. 44r–45v, an abridgment of *The Precious New Pearl* (J. Corbett, *Catalogue des manuscrits alchimiques latins: I. Manuscrits des bibliothèques publiques de Paris* [Brussels, 1939], no 53).

The earliest printed ed. of *The Precious New Pearl* is Janus Lacinius, ed., *Pretiosa margarita, novella de thesauro, ac pretiosissimo philosophorum lapide (Petro Bono Ferrariensi autore) Artis huius diuinae typus, et methodus: collectanea ex Arnaldo, Rhaymundo, Rhasi, Alberto, et Michaele Scoto . . . apud Aldi filios* (Venice, 1546). For the above and other eds. of the 16th and early 17th centuries see Thorndike, III, 148, 151, notes 8 and 18.

The full text is in Eberhard Zetzner, *Theatrum chemicum praecipuos . . . tractatus de chemiae et lapidis philosophici antiquitate, veritate jure, praestantia et operationibus continens*, V (Strasbourg, 1660), 507–713; and J. J. Manget, *Bibliotheca chemica curiosa*, I (Geneva, 1702), 1–80. The Janus Lacinius ed. was translated into German by W. G. Stollen, *Pretiosa margarita, oder Neu-erfundene köstliche Perle, von dem unvergleichlichen Schatz und höchst-kostbahren Stein der Weisen . . .* (Leipzig, 1714); and into English, in abridged form, by A. E. Waite, *The New Pearl of Great Price: A Treatise Concerning the Treasure and Most Precious Stone of the Philosophers, or the Method and Procedure of This Divine Art* (London, 1894; 1963), 49–184.

II. Secondary Literature. The major critical study on Petrus Bonus is Lynn Thorndike, *History of Magic and Experimental Science*, III (New York, 1934), Ch. 9. Briefer accounts are Robert Multhauf, *The Origins of Chemistry* (London, 1966), 191–193; Julius Ruska, "L'alchimie à l'époque du Dante," in *Annales Guébhard-Séverine*, 10 (1934), 410–417 (translated by A. Kuenzi); George Sarton, *Introduction to the History of Science*, III (Baltimore, 1927–1948), 750–752; and J. M. Stillman, "Petrus Bonus and Supposed Chemical Forgeries," in *Scientific Monthly*, 16 (1923), 318–325.

Pearl Kibre

PETTENKOFER, MAX JOSEF VON (*b.* Lichtenheim [near Neuburg], Germany, 3 December 1818; *d.* Munich, Germany, 10 February 1901), *chemistry, hygiene, epidemiology.*

Although Pettenkofer initiated advances in many fields of science, his fame rests chiefly upon his pioneering accomplishments in experimental hygiene, a discipline that he founded. He was the fifth of eight children born in a solitary, converted custom house at a former border crossing in the Danubian marshes of lower Bavaria. His father, Johann Baptist Pettenkofer, the youngest of four sons of a customs official who had bought the property, farmed the peat bog unsuccessfully. His industrious mother, nineteen years her husband's junior, was of Oberpfalz peasant stock.

A parental uncle, the pharmacist Franz Xaver Pettenkofer, became court apothecary to Ludwig I of Bavaria in 1823. As his marriage was childless, he helped to educate four of his brother's children, who went successively to Munich to live in their uncle's official apartment in the royal residence; Max arrived there in 1827. While yearning for country-life simplicities, he gradually overcame the limits of his early parish schooling, made prizewinning progress through Latin school, and in 1837 matriculated with distinction from the humanistic Gymnasium. He entered the University of Munich with a predilection for philology, but accepted his uncle's insistence upon two years in philosophy and natural sciences, followed by apprenticeship at the court pharmacy.

On completing the prescribed university courses, of which he liked chemistry most, Pettenkofer worked diligently under tutelage, and in one year (instead of the usual three) was appointed assistant. But his uncle's strict regimen and demanding nature provoked the high-spirited youth to abandon his post and leave Munich. Under the pseudonym "Tenkof" he took minor parts at the Augsburg theater. Neither uncomplimentary reviews nor emissaries from the scandalized apothecary altered his course. He became captivated, however, by Helene Pettenkofer, the daughter of another uncle, Josef Pettenkofer, and she consented to marry Max, but only if he resumed steady ways. Accordingly, he left the stage, became engaged to Helene, and received a prodigal's welcome from his Munich uncle, who nevertheless forbade employment in the court pharmacy, advising that the medical profession was more suited to a former actor. In 1841 he returned to university studies and by June 1843 passed with honors the state qualifying examinations for both pharmacist and physician. His doctoral dissertation, "Ueber Mikania Guaco," concerned a plant of the *Eupatorium* genus, native to Mexico and Colombia, the sap of which reputedly had medicinal value in snakebite, rabies, and cholera. Self-experiments with a resin chemically extracted from the leaves induced vomiting, quickened pulse, and profuse sweating.

Pettenkofer was unwilling to practice either profession. While still an undergraduate, he had improved the sensitivity and specificity of Marsh's test for arsenic. The mineralogist Johann von Fuchs now advised additional training in medical chemistry, foreseeing a chair in this specialty at Munich. With Fuchs's influence Pettenkofer secured a bursary to attend the 1843–1844 winter semester at the University of Würzburg under Josef Scherer, a former pupil of Justus Liebig. There he detected large amounts of hippuric acid in the urine of a child whose diet consisted of only bread and apples—a notable example of the influence of diet upon urinary composition—and he developed the specific color reaction for bile, a test which still carries his name. He spent the summer at the University of Giessen, where he became an enthusiastic disciple of Liebig, who was then at the zenith of his powers and fame. Assigned to investigate the chemistry of meat, Pettenkofer discovered in human urine a new amino acid, creatinine. But a lack of funds ended this brief interlude.

At Munich, ministerial indifference dashed his hopes for a university post in medical chemistry. Unemployed and frustrated, he turned to poetry, composing romantic poems (published over forty years later as *Chemische Sonette*) glorifying chemistry and its adepts, from Roger Bacon to Liebig. In 1845 an appointment as assistant at the Royal Mint sufficiently relieved his financial problems, and he at last married his cousin Helene. Pettenkofer's duties, official and unofficial, agreeably challenged his resourcefulness. The costs of Bavarian currency conversion were minimized by retrieving precious metals from the old silver thalers.

The gold that was separated during refining contained about 3 percent silver, which was tenaciously combined with small amounts of platinum. Pettenkofer devised methods for separating the gold from the silver and recovering the platinum. By a slow-cooling process in the mint ovens he reproduced the beautiful bloodred ("haematinon") coloration of a specimen of *porporino antico* that was possessed by the king. He also reported the presence of sulfocyanic acid in human saliva. In 1846 he was elected an extraordinary member of the Bavarian Academy of Sciences. In the following year a ministerial change reopened the possibility of a chair in medical chemistry at Munich. On Fuchs's urging, Pettenkofer became a candidate. With royal support he was appointed extraordinary professor in November 1847, four months before the king abdicated.

His lecture course was developed slowly, with frequent changes of title and content. Pathologic chemistry was lightly treated, but the applications of chemistry to nutrition, public health, sanitation, and forensic medicine were emphasized. Another component, at first unformulated, concerned the role of chemistry in analyzing man's personal environment. His first lectures were given in 1853 under the title "Dietetic Physical-Chemistry," and not until 1865, when a new chair of hygiene was created for him, did he adopt the title "Lectures on Hygiene." In 1850 a remarkable address to the Royal Academy of Sciences, "Ueber die regelmässigen Abstände der Aequivalentzahlen der sogenannten einfachen Radicale," revealed his reflective mastery of chemical principles and presaged the periodic law of the elements. As verification of the alleged mathematical relationships between certain elements depended upon accurate atomic weights, Pettenkofer sought vainly for modest support from the Academy to make the necessary precise determinations. His classic report, published in an obscure journal, was reprinted in Liebig's *Annalen der Chemie und Pharmacie* (1858) when his priority was threatened. In 1899 the German Chemical Society awarded him a gold medal commemorating the fiftieth anniversary of the historic address.

Pettenkofer's versatility in solving practical problems was nearly inexhaustible. He observed that German hydraulic lime equaled English Portland cement in hardening properties if the calcining temperature of the marl was reduced. He also discovered means of improving the illuminating power of wood gas, so that (until coal became cheaper) several German cities and the Munich railway station were lit thereby. Such accomplishments enhanced his widespread reputation for chemical inventiveness and excited new demands. In the early 1860's his reports on the restoration of oil paintings in the Alte Pinakothek emphasized proper heating of art galleries to prevent the dampness that opacified many varnishes. He was also recognized professionally, being nominated to the Chief Medical Commission (Obermedizinalausschuss) in 1849. Upon his uncle's death (1850), he was appointed part-time director of the royal pharmacy, with tenure of the apartment in which his earlier years were spent. His brother Michael became court apothecary and managed the operating details. Under their administration, a model profitable institution developed.

When the new king, Maximilian II, informed the Commission about the uncomfortable dryness and possible health hazards from heating the palace by circulated hot air instead of stoves, Pettenkofer was delegated to resolve the problem. His investigations disclosed that the dryness was caused by the faster-flowing heated air desiccating the room walls. This project showed how such vague expressions as "salubrity of dwellings" could be substantiated by explicit data; the project was also his starting point for the new science of experimental hygiene. He realized that determinations of the physical properties of building materials were extensible to adjacent soil, personal clothing, and more remote factors concerning human health.

In 1852 he helped the king arrange Liebig's appointment to the chair of chemistry at Munich. Although Pettenkofer had become ordinary professor of medical chemistry (1852), he never competed with the acknowledged master, but instead attained preeminence in the discipline created by himself. At Pettenkofer's request, Liebig permitted his name to be attached to the well-known meat extract that Pettenkofer was manufacturing in the court pharmacy.

In 1857 better accommodation became available in the new Physiological Institute. Pettenkofer investigated air exchange in Munich hospitals and studied artificial ventilation in Paris. A simplified method of carbon dioxide assay was used to evolve standards of atmospheric purity for occupied rooms. He cast new light on air permeation through room walls, on the composition of ground air and its penetrability into dwellings, and on the vitiation of air by heating and lighting arrangements. Other studies were facilitated after 1860 by an airtight metallic respiratory apparatus, invented by himself and paid for by the king, who patronized science as his father had patronized the arts. This unique structure comfortably housed a human subject or large experimental animal for a given period while the gaseous exchange and all bodily gains or losses were measured exactly. Thus, in collaboration with Carl Voit, one of his earliest pupils, Pettenkofer established many basic nutritional facts,

such as the dietetic requirements of normal people at rest and in various activities, the vital necessity of adequate protein intake, the protein-sparing properties of carbohydrate and fat during starvation, and the need of the diabetic for extra protein and fat to replace unused carbohydrate.

These advances captured international attention. Voit became director of the Physiological Institute in 1863. Two years later Pettenkofer was made ordinary professor of hygiene and elected university rector at Munich. During an audience with the young King Ludwig II, Pettenkofer promoted hygiene so effectively that chairs were created at the universities of Würzburg and Erlangen, and the subject was made compulsory in state medical examinations. The new discipline had taken firm root. In 1865, with Voit and two associates, Pettenkofer founded and for eighteen years coedited the *Zeitschrift für Biologie*, which published many of his reports. The collaborative studies on nutrition continued until Pettenkofer moved into his own institute. His special interest in the hygienic importance of air focused on its relationship to clothing and particularly to soil. He discussed the functions of clothing in the first volume of the *Zeitschrift*; and in 1872, in three popular lectures, he dealt with air in relation to clothing, dwelling, and soil.

Pettenkofer believed that soil had fundamental sanitary relationships, for "an impurity cleaves longest and most tenaciously to the soil, which suffers no change of place, like air and water." An unexpected, late offshoot of his work on soil aeration was his observation of coal gas poisoning that resulted from seepage into houses from leaking mains (1883). But his initial and continuing preoccupation with this field stemmed mainly from his determination to explain recurrent cholera epidemics—the grimmest challenge to hygiene in the nineteenth century. His introduction to cholera came as member of a commission which had been appointed to review scientifically the 1854 epidemic in Bavaria. He studied ten outbreaks, including one in detail in Munich, where he and his young daughter contracted the disease and their cook died of it. As cholera was neither purely contagious (like smallpox) nor purely miasmatic (like malaria), Pettenkofer concluded that it must be contagio-miasmatic. He developed the hypothesis that choleraic excreta embodied a contagious factor, the dissemination of which required soils of a particular constitution: moist, porous, and polluted. Penetrating into such terrain, "the cholera germ-bearing excrements . . . modify the existing process of decay and decomposition," so that, besides normal putrefactive gases, "a specific Cholera-Miasma is developed, which is then spread along with other exhalations into the houses" (1855). From his survey in Munich, Pettenkofer concluded that drinking water did not cause the outbreak. This conclusion was then generalized and extended to typhoid fever (highly endemic in Munich). Thus his epidemiologic doctrines were directly opposed to those of John Snow and William Budd in England.

He then investigated the varied regional incidence of cholera in Bavaria. Its recurrence along certain river valleys was attributed to their being areas of natural soil drainage rather than routes of travel; and he deduced that the decisive factor in the genesis of cholera was the moisture content of the local soil, which was indicated by the groundwater level. When the water table fell, as in summer, a larger soil volume became available for production of the toxic miasma. This hypothesis of the mode of spread of cholera became an *idée fixe*. Pettenkofer expounded his theory in the *Zeitschrift*, and his pupils L. Buhl and L. Seidel each discussed the etiology of typhoid fever from similar theoretic standpoints (1865). Their position was a compromise between the "contagionist" view of Snow and Budd, which stated that the intestines of cholera and typhoid patients were the primary origin of the specific infective agents, and the extreme "localist" concept developed by James Cuningham and his Anglo-Indian associates in Calcutta, which held that cholera epidemics arose because of properties of the localities concerned. Pettenkofer's localist leanings did not diminish when disinfection of excreta failed to halt the 1866 cholera outbreak in Germany. Moreover, his travels in 1868 revealed that the comparative immunity of Lyons to cholera, and the recent outbreaks in Malta and Gibraltar, were all explicable by the nature of their underlying terrain. In an extensive review, *Boden und Grundwasser in ihren Beziehungen zu Cholera und Typhus* (1869), he proposed that an "x" factor, dependent on human intercourse, and a "y" factor, derived from soil, were essential for production of the true cholera poison "z." The reaction of the germ and the soil factor might occur in the soil itself, in the air of the dwelling, or perhaps in the human body. But the germ alone could no more cause cholera than swallowing yeast cells could produce intoxication.

Despite denials that cholera or typhoid might be waterborne, Pettenkofer's hygienic philosophy required that the water supply of a city be free from impurities. Between 1867 and 1883 he initiated measures whereby the drinking water supply of Munich became one of the finest in Europe. Further, since cholera tends to rage where the greatest filth prevails, he set about providing the city with good drainage and sewerage. After the 1854 epidemic, the possibility was mooted of replacing the unsanitary night cartage of excreta with float-canalization (*Schwemmkanalisation*), in which all excreta and household wastes were piped into the natural streamlets and canals traversing Munich and thence into the swift-flowing Isar River. This system, initiated in 1858, publicly backed in 1870, and completed in 1892, testified to Pettenkofer's vision, persistence, scientific authority, and persuasiveness.

Munich acknowledged these contributions to its cleanliness by granting him honorary citizenship in 1872. Two public lectures entitled "Ueber den Werth der Gesundheit für eine Stadt" (1873) illustrated well his talent for inspiring further hygienic advances. These advances were evident from the essay "Munich a Healthy Town" (1889) and from the changes reported to the Epidemiological Society of London by C. Childs, an English physician, in "The History of Typhoid Fever in Munich" (1898).

A flattering call to Vienna in 1872 was rejected by Pettenkofer after the German government assured him that a hygienic institute would be erected for him in Munich. In 1876 he also refused Chancellor Bismarck's invitation to direct the new *Reichsgesundheitsamt* (Imperial Health Office). He was honorary president of the second International Congress of Hygiene and Demography, held in 1878 in Paris. In 1879 the first Institute of Hygiene, with abundant space and equipment for teaching and research, was formally opened under his direction. In 1882, with H. W. von Ziemssen's collaboration, a *Handbuch der Hygiene* appeared; and with two colleagues Pettenkofer founded and coedited the *Archiv für Hygiene* in 1883. In that year he was granted hereditary nobility.

When Koch first announced his isolation of the "comma bacillus" from cholera cases (1883), Pettenkofer readily accepted this vibrio as his "x" factor, but refused to modify his views on the paramountcy of the telluric "y" factor. Perhaps his attitude hardened because he was not invited to the first cholera conference in Berlin in 1884, organized by the *Reichsgesundheitsamt*, of which he was an extraordinary member. Koch visited him at Munich later that year, but there was no rapprochement. Pettenkofer attended the second cholera conference (1885), but Koch seemed unaware or disdainful of Pettenkofer's thirty years of careful epidemiologic enquiries and of his major role at earlier cholera conferences in Weimar (1867), Berlin (1873), and Vienna (1874). Koch and Carl Flügge now founded and coedited the *Zeitschrift für Hygiene* (1885), while Pettenkofer reiterated and amplified his doctrines in the *Archiv für Hygiene* (1886–1887). The two schools of inflexible proponents and loyal adherents polarized their irreconcilable opinions in Munich and Berlin. In 1892, after a classic waterborne epidemic killed some 8,000 inhabitants of Hamburg within three months, the seventy-four-year-old Pettenkofer defiantly drank a cubic centimeter of culture of a recently isolated *Vibrio cholerae*. His colleague R. Emmerich subsequently repeated the experiment. They both developed diarrhea and excreted cholera vibrios for several days. These were proclaimed negative reactions and hence dramatic vindications of Pettenkofer's doctrine. The possibilities of immunity from previous exposure to the infecting agent or of the low virulence of the culture, however, were not raised. But the tide of evidence and opinion was not reversed, and little more was heard (outside the Munich school) of the localist or groundwater theory.

Honors and weariness now descended upon the intrepid hygienist. His seventieth birthday celebrations were highlighted by an address signed by over 100 artists, and by Munich and Leipzig jointly creating a Pettenkofer Foundation to provide prizes for scientific accomplishments in hygiene. For his doctoral jubilee in 1893, a *Festschrift* and a volume of *Archiv für Hygiene* were dedicated to him; Munich presented its golden Citizen's Medal, and many other German and foreign honors were bestowed. In 1896 he was granted the title *Excellenz*, and in 1897 received the Harben Medal from the Royal Institute of Public Health in England as "founder of scientific hygiene." Honorary doctorates from universities and honorary memberships in foreign medical and hygienic associations were very numerous. After his last public address, "Ueber Selbstreinigung der Flüsse" (1891), he continued academic lectures for another three years. In 1894 he became professor emeritus, and within the next two years resigned his editorial duties and retired from the court pharmacy. In 1899 he terminated a decade as president of the Bavarian Academy of Sciences.

Pettenkofer retained his apartment in the royal residence for winter living, but was happiest in his country home at Seeshaupt on the Starnberger See. He restored the land, pruned trees, planted crops, and rowed on the lake. His only contacts with hygiene were through occasional visitors and lively correspondence with former students and disciples, many of whom now held chairs of hygiene in Germany and other countries. Although physically strong and mentally alert, he complained of tiredness, loss of memory, and inability to concentrate. His wife had died in 1890; and of their five children, two sons and a daughter had predeceased her. Pettenkofer especially mourned the loss of his gifted, eldest son, a medical student, who died from tuberculosis in 1869. Despite the care and concern of surviving relatives, he feared for his reason and

threatened suicide. At the end of January 1901, a septic throat caused him much pain and insomnia, aggravating his depression. He bought a revolver and one night, two hours after being put to bed by his daughter-in-law, shot himself in the head. He was buried in South Munich cemetery. A seated statue of him was erected in a main public square. In 1944 the Hygienic Institute was destroyed by bombs and fire, but a new "Max v. Pettenkofer-Institut für Hygiene und Medizinische Microbiologie" was completed on a neighboring site in 1961. An adjacent street still bears his name.

Pettenkofer was genial and sociable, carrying his culture and honors lightly, yet fully aware of the significance of his theories. A leonine countenance, self-confident bearing, and skillful delivery helped to make his lectures unforgettable. He worked prodigiously, rising early and often retiring late. His character was truly benevolent and without pettiness or spite, but he could be scornful of pretentiousness, angry at injustice, and stubborn in self-defense. His disciples revered him as father-figure and wise visionary. As an investigator the mark of genius appeared in the discrimination with which he selected mundane phenomena for delicate observation, and in the unexcelled virtuosity applied to determining their cause. Although his approach was highly original, he firmly believed in the practical aims of scientific hygiene, and all his research and exhortative efforts were directed toward promoting public health in its broadest sense. Indirectly, by elevating hygiene to an accepted discipline in medical training, he stimulated a new social outlook that encouraged the physician to provide not only relief from disease, but counsel and leadership toward healthful living.

A lenient view may be taken of Pettenkofer's misconceptions about cholera and enteric infections. Although often biased in the selection of data supporting his contentions, the sincerity of his motives was never impugned. His deeply rooted obsession with soil as the major source of certain ailments yielded great benefits to Munich when its soil was cleansed by the introduction of effective drainage and sewerage. If the hubristic tendencies of some early bacteriologists were tempered by his skepticism, he was not antagonistic to their emerging science. At the 1885 conference, he magnanimously termed Koch's discovery "a great enrichment of our pathologic knowledge about cholera." His attitude might have been less obdurate against a more tactful adversary. After all, Koch was his junior by a quarter-century, and likewise even Pasteur by four years.

In honor of his eighty-first birthday, the citizens of Munich presented another gold medal "as a sign of unlimited veneration, gratitude and love," to "the High Priest of Hygiene . . . the remover of pernicious diseases from the home soil."

BIBLIOGRAPHY

I. Original Works. Pettenkofer wrote over 20 monographs and more than 200 separate articles in German scientific and medical journals between 1842 and 1898. There is no collected ed. E. E. Hume's biographical review (see below) contains a bibliography of 227 items, which is fairly complete but sprinkled with inaccuracies.

Among Pettenkofer's reports on hygienic topics available in English are "Ueber den Respirations- und Perspirations-Apparat im physiologischen Institute zu München," in Erdmann's *Journal für praktische Chemie*, **82** (1861), 40–50, trans. by A. Ten Brook as "Description of Apparatus for Testing the Results of Perspiration and Respiration in the Physiological Institute of Munich," in *Smithsonian Report for 1864*, (Washington, 1865), pp. 235–239; *Beziehungen der Luft zu Kleidung, Wohnung, und Boden* (Brunswick, 1872), 3 lectures, trans. by A. Hess as *The Relations of the Air to the Clothes We Wear, the House We Live in, and the Soil We Dwell On* (London, 1873); *Ueber den Werth der Gesundheit für eine Stadt* (Brunswick, 1873), 2 lectures, trans. by H. E. Sigerist as *The Value of Health to a City* (Baltimore, 1941), repr. from *Bulletin of the History of Medicine*, **10** (1941), 473–503, 593–613; and "Der Boden und sein Zusammenhang mit der Gesundheit des Menschen," in *Deutsche Rundschau*, **29** (1881), 217–234, trans. as "Sanitary Relations of the Soil," in *Popular Science Monthly*, **20** (1882), 332–340; and "Munich a Healthy Town" (Munich, 1889), written with H. W. von Ziemssen.

English trans. or expressions of Pettenkofer's views on cholera include "Observations on Dr. Buchanan's Lecture on Professor von Pettenkofer's Theory of the Propagation of Cholera and Enteric Fever," in *Medical Times and Gazette*, 1 (1870), 629–632, 661–663, 687–689; "On the Recent Outbreak of Cholera in Munich," *ibid.*, 1 (1874), 582–583; "On the Probability of an Invasion of Cholera in Europe," in *Sanitary Record*, n.s. 5 (1883–1884), 47–51; "Professor Max von Pettenkofer on Cholera Afloat," in *Lancet* (1884), **2**, 338–340; and "Cholera," *ibid.*, 769–771, 816–819, 861–864, 904–905, 992–994, 1042–1043, 1086–1088. His contributions to the 1874 Conférence Sanitaire Internationale in Vienna were freely translated by T. W. Hime as *Cholera: How to Prevent and Resist It* (London, 1875; 2nd ed. 1883).

Other such works are *Künftige Prophylaxis gegen Cholera* . . . (Munich, 1875), with trans. and abstr. by A. Rabaglioti as "Artificial Prophylaxis Against Cholera," in *Sanitary Record*, 3 (1875), 35–39; "Die Choleraepidemie in der königlichen bayerischen Gefangenenanstalt Laufen an der Salzach," in *Bericht der Choleracommission für das Deutsche Reich* (Berlin, 1875), with trans. as *Outbreak of Cholera Among Convicts* . . . (London, 1876); "Die Cholera in Syrien und die Choleraprophylaxe in Europa," in *Zeitschrift für Biologie*, **12** (1876), 102–128, with trans.

as "The Cholera in Syria and the Prophylaxis of Cholera in Europe," in *Practitioner*, **16** (1876), 401–415; "Neun ätiologische und prophylaktische Sätze aus den amtlichen Berichten über die Choleraepidemien in Ostindien und Nordamerica," in *Deutsche Vierteljahrsschrift für öffentliche Gesundheitspflege*, **9** (1877), 177–224, with trans. as "Nine Propositions Bearing on the Aetiology and Prophylaxis of Cholera, Deduced From the Official Reports of the Cholera-Epidemic in East India and North America," in *Practitioner*, **18** (1877), 135–160, 204–240; *Zum gegenwärtigen Stand der Cholerafrage* (Munich, 1887), repr. from *Archiv für Hygiene*, **4** (1886), 249–354, 397–546; **5** (1886), 353–445; **6** (1887), 1–84, 129–233, 303–358, 373–441; **7** (1887), 1–81, with abstr. and trans. by H. Koplik as "The Present Aspect of the Cholera Question," in *Lancet* (1886), **2**, 29–30, 89–91; and "Ueber Cholera mit Berücksichtigung der jüngsten Cholera-Epidemie in Hamburg," in *Münchener medizinische Wochenschrift*, **39** (1892), 807–817, discussion 826–828, with trans. and abstr. as "On Cholera, With Reference to the Recent Epidemic at Hamburg," in *Lancet* (1892), **2**, 1182–1185.

His monographs or booklets include *Ueber Chemie in ihrem Verhältnisse zur Physiologie und Pathologie* (Munich, 1848); *Ueber Oelfarbe und Conservirung der Gemälde durch das Regenerations-Verfahren* (Brunswick, 1850); *Untersuchungen und Beobachtungen über die Verbreitungsart der Cholera* . . . (Munich, 1855); *Ueber den Luftwechsel in Wohngebäuden* (Munich, 1858); *Ueber die Verlegung der Gottesäcker in Basel* (Basel, 1864); *Die Cholera vom Jahre 1866 in Weimar* (Weimar, 1867); *Ueber die Ursachen und Gegenwirkung von Cholera-Epidemien in Erfurt* (Erfurt, 1867); *Boden und Grundwasser in ihren Beziehungen zu Cholera und Typhus* (Munich, 1869), repr. from *Zeitschrift für Biologie*, **5** (1869), 171–310; *Das Canal- oder Siel-System in München* (Munich, 1869); *Verbreitungsart der Cholera in Indien* (Brunswick, 1871); *Ueber die Aetiologie des Typhus* (Munich, 1872); *Ueber den gegenwärtigen Stand der Cholera-Frage* . . . (Munich, 1873); *Vorträge über Canalisation und Abfuhr* (Munich, 1876); *Populäre Vorträge* (Munich, 1877); *Die Cholera* (Breslau, 1884); *Chemische Sonette* (Munich, 1886); *Der epidemiologische Theil des Berichtes über die Thätigkeit der zur Erforschung der Cholera im Jahre 1883 nach Aegypten und Indien entsandten deutschen Commission* (Munich, 1888); and *Die Verunreinigung der Isar durch das Schwemmsystem von München* (Munich, 1890). He was coeditor, with H. W. von Ziemssen, of *Handbuch der Hygiene* . . . (Leipzig, 1882). He also wrote several obituaries and commemorative tributes, of which the most important is *Dr. Justus Freiherrn von Liebig zum Gedächtnis* (Munich, 1874).

Characteristic chemical reports are "Sichere und einfache Methode das Arsenik mittelst des Marsh'schen Apparates entwickelt von allen andern ähnlichen Erscheinungen augenfällig zu unterscheiden," in *Repertorium für die Pharmacie*, **76** (1842), 289–307; "Nachtrag zur Arsenikprobe . . ." *ibid.*, **83** (1844), 328–336; "Ueber Mikania Guaco," *ibid.*, **86** (1844), 289–323, his inaugural diss. at the University of Munich (1844); "Ueber das Vorkommen einer grossen Menge Hippursäure im Menschenharne," in Liebig's *Annalen der Chemie und Pharmacie*, **52** (1844), 86–90; "Notiz über eine neue Reaction auf Galle und Zucker," *ibid.*, 90–96; "Vorläufige Notiz über einen neuen stickstoffhaltigen Körper im Harne," *ibid.*, 97–100; "Ueber die Affinirung des Goldes und über die grosse Verbreitung des Platins," in *Münchner gelehrte Anzeiger*, **24** (1847), 589–598; "Bemerkungen zu Hopfgartner's Analyse eines englischen und eines deutschen hydraulischen Kalkes," in Dingler's *Polytechnisches Journal*, **113** (1849), 357–371; "Ueber die regelmässigen Abstände der Aequivalentzahlen der sogenannten einfachen Radicale," in *Münchner gelehrte Anzeiger*, **30** (1850), 261–272; and "Ueber das Haematinon der Alten und über Aventuringlas," in Erdmann's *Journal für praktische Chemie*, **72** (1857), 50–53.

Works on various aspects of hygiene include "Ueber den Unterschied zwischen Luftheizung und Ofenheizung in ihrer Einwirkung auf die Zusammensetzung der Luft der beheizten Räume," in Dingler's *Polytechnisches Journal*, **119** (1851), 40–51, 282–290; "Ueber die wichtigsten Grundsätze der Bereitung und Benützung des Holzleuchtgases," in Erdmann's *Journal für praktische Chemie*, **71** (1857), 385–393; "Berichte über Ventilations-Apparate," in *Abhandlungen der naturwissenschaftlich-technischen Commission bei der königlichen baierischen Akademie*, **2** (1858), 19–68; "Besprechung allgemeiner auf die Ventilation bezüglicher Fragen," *ibid.*, 69–126; "Ueber die Bestimmung der freien Kohlensäure im Trinkwasser," in Erdmann's *Journal für praktische Chemie*, **82** (1861), 32–40; "Ueber eine Methode die Kohlensäure in der atmosphärischen Luft zu bestimmen," *ibid.*, **85** (1862), 165–184; "Ueber die Respiration," in Liebig's *Annalen der Chemie und Pharmacie*, supp. 2 (1862–1863), 1–52; "Untersuchungen über die Respiration," *ibid.*, 52–70, written with C. Voit; "Ueber die Wahl der Begräbnissplätze," in *Zeitschrift für Biologie*, **1** (1865), 45–68; "Ueber die Funktion der Kleider," *ibid.*, 180–194; "Ueber Kohlensäureausscheidung und Sauerstoffaufnahme während des Wachens und Schlafens beim gesunden und kranken Menschen," in Liebig's *Annalen der Chemie und Pharmacie*, **141** (1867), 295–322; "Ueber Kohlensäuregehalt der Luft im Boden (Grundluft) von München in verschiedenen Tiefen und zu verschiedenen Zeiten," in *Zeitschrift für Biologie*, **7** (1871), 395–417; **9** (1873), 250–257; "Beleuchtung des königliche Residenztheaters in München mit Gas und mit elektrischem Licht," in *Archiv für Hygiene*, **1** (1883), 384–388; "Ueber Vergiftung mit Leuchtgas," in *Festschrift des aerztlichen Vereins München . . .* (Munich, 1883), pp. 68–74; "Der hygienische Unterricht an Universitäten und technischen Hochschulen," opening speech, *VI. International Congress for Hygiene and Demography* (Vienna, 1887); and "Ueber Selbstreinigung der Flüsse," in *Deutsche medizinische Wochenschrift*, **17** (1891), 1277–1281.

Researches on the physiology of nutrition include "Ueber die Produkte der Respiration des Hundes bei Fleischnahrung, und über die Gleichung der Einnahmen und Ausgaben des Körpers dabei," in Liebig's *Annalen der Chemie und Pharmacie*, supp. 2 (1862–1863), 361–377; "Untersuchungen über den Stoffverbrauch des normalen Menschen," in *Zeitschrift für Biologie*, **2** (1866), 459–573; "Ueber den Stoffverbrauch bei der Zuckerharnruhr," *ibid.*, **3** (1867), 380–444; "Respirationsversuche am Hunde bei Hunger und ausschliesslicher Fettzufuhr," *ibid.*, **5** (1869), 369–392; and "Ueber die Zersetzungsvorgänge im Thierkörper bei Fütterung mit Fleisch und Fett," *ibid.*, **9** (1873), 1–40, all written with C. Voit. Pettenkofer's report on Liebig's meat extract is "Ueber Nahrungsmittel im Allgemeinen und über den Werth des Fleischextracts als Bestandtheil der menschlichen Nahrung insbesonders," in Liebig's *Annalen der Chemie und Pharmacie*, **167** (1873), 271–292.

His views on the etiology and epidemiology of cholera and typhoid fever are illustrated further in "Die Bewegung des Grundwassers in München von März 1856 bis März 1862," in *Sitzungsberichte der königlichen baierischen Akademie der Wissenschaften zu München*, **1** (1862), 272–290; "Ueber die Verbreitungsart der Cholera," in *Zeitschrift für Biologie*, **1** (1865), 322–374; "Ueber den gegenwärtigen Stand des Grundwassers in München," *ibid.*, 375–377; "Die sächsische Choleraepidemie des Jahres 1865," *ibid.*, **2** (1866), 78–144; "Ueber die Schwankungen der Typhussterblichkeit in München von 1850 bis 1867," *ibid.*, **4** (1868), 1–39; "Die Immunität von Lyon gegen Cholera und das Vorkommen der Cholera auf Seeschiffen," *ibid.*, 400–490; "Prof. Dr. Hallier über den Einfluss des Trinkwassers auf dem Darmtyphus in München," *ibid.*, 512–530; "Die Choleraepidemie des Jahres 1865 in Gibraltar," *ibid.*, **6** (1870), 95–119; "Die Choleraepidemien auf Malta und Gozo," *ibid.*, 143–203; "Typhus und Cholera und Grundwasser in Zürich," *ibid.*, **7** (1871), 86–103; and "Ueber Cholera auf Schiffen und den Zweck der Quarantänen," *ibid.*, **8** (1872), 1–70.

Additional writings include "Auszug aus den Untersuchungen von Dr. Douglas Cunningham in Ostindien, über die Verbreitungsart der Cholera," *ibid.*, 251–293; "Ist das Trinkwasser Quelle von Typhusepidemien ?", *ibid.*, **10** (1874), 439–526; "Aetiologie des Abdominal-Typhus," in *Archiv für öffentliche Gesundheitspflege*, **9** (1884), 92–100; "Ueber Desinfection der ostindischen Post als Schutzmittel gegen Einschleppung der Cholera in Europa," in *Archiv für Hygiene*, **2** (1884), 35–45; "Die Cholera in Indien," *ibid.*, 129–146; "Rudolf Virchow's Choleratheorie," in *Berliner klinische Wochenschrift*, **21** (1884), 488–490, with Virchow's reply, *ibid.*, 490–491; "Die Trinkwassertheorie und die Cholera-Immunität des Forts William in Calcutta," *ibid.*, **3** (1885), 147–182; "M. Kirchner, Ueber Cholera mit Berücksichtigung der jüngsten Choleraepidemie in Hamburg," in *Centralblatt für Bakteriologie*, **12** (1892), 898–904, in response to Kirchner's review, *ibid.*, 828–836; "Ueber die Cholera von 1892 in Hamburg und über Schutzmaassregeln," in *Archiv für Hygiene*, **18** (1893), 94–132; and "Choleraexplosionen und Trinkwasser," in *Münchener medizinische Wochenschrift*, **48** (1894), 221–224, 248–251.

A small collection of his reprints and decorations, and a fine oil portrait, survive in the rebuilt Max v.

Pettenkofer Institute of the University of Munich. The bulk of his papers are in the Bavarian State Museum in Munich.

II. Secondary Literature. Obituaries in German include "Nachruf Max von Pettenkofer gewidmet," in *Archiv für Hygiene*, **39** (1901), 313–320; R. Emmerich, "Erinnerungen an Max v. Pettenkofer," in *Deutsche Revue*, **27** (1902), 81–92; F. Erismann, "Max von Pettenkofer," in *Deutsche medizinische Wochenschrift*, **27** (1901), 209–211, 253–255, 285–287, 299–302, 323–327; M. Gruber, "Max v. Pettenkofer (1818–1901)," in *Wiener klinische Wochenschrift*, **14** (1901), 213–218; G. W. A. Kahlbaum, "Worte des Gedenkens an Max von Pettenkofer," in *Abhandlungen der Naturforschenden Gesellschaft zu Basel*, **13** (1901), 326–337; K. B. Lehmann, "Max von Pettenkofer," in *Münchener medizinische Wochenschrift*, **48** (1901), 464–473; L. Pfeiffer, "Zum Gedächtniss für Max von Pettenkofer," in *Hygienische Rundschau*, **11** (1901), 717–732; M. Rubner, "Zum Andenken an Max v. Pettenkofer," in *Berliner klinische Wochenschrift*, **38** (1901), 268–270, 301–303, 321–326; C. Voit, "Max von Pettenkofer, dem Physiologen, zum Gedächtnis," in *Zeitschrift für Biologie*, **41**, n.s. **23** (1901), I–VIII, and *Max von Pettenkofer zum Gedächtniss* (Munich, 1902).

Obituaries in English are by C. Childs, "Geheimrath Max von Pettenkofer, of Munich," in *Transactions of the Epidemiological Society*, n.s. **20** (1901), 118–125; J. S. Haldane, "The Work of Max von Pettenkofer," in *Journal of Hygiene*, **1** (1901), 289–294; and an editorial, with brief additional tributes by Sir John Simon and W. H. C. Corfield, "Max Josef von Pettenkofer, M.D.," in *British Medical Journal* (1901), **1**, 489–490.

Other German references bearing on Pettenkofer's life and work are L. Buhl, "Ein Beitrag zur Aetiologie des Typhus," in *Zeitschrift für Biologie*, **1** (1865), 1–25; E. Ebstein, "Max Pettenkofer als junger Professor—und Komödiant," in *Deutsche medizinische Wochenschrift*, **54** (1928), 1813–1814; R. Emmerich, "Max von Pettenkofer," in *Centralblatt für allgemeine Gesundheitspflege*, **12** (1893), 207–217, and *Max Pettenkofer's Bodenlehre der Cholera Indica* (Munich, 1910); H. Eyer, *100 Jahre Lehrstuhl für Hygiene an der Ludwig-Maximilians-Universität München* (Munich, 1965); K. M. Finkelnburg *et al.*, *Festschrift des niederrheinischen Vereins für öffentliche Gesundheitspflege und des Centralblattes für allgemeine Gesundheitspflege, zur Feier des 50jährigen Doctor-Jubiläums Max von Pettenkofers am 30. Juni 1893* (Bonn, 1893); I. Fischer, "Max von Pettenkofer," in *Biographisches Lexikon der hervorragender Aerzte*, 2nd ed., IV (1932), 576–577; *Jubelband dem Herrn Geh. Rath Prof. Dr. M. von Pettenkofer zu seinem 50jährigen Doctor-Jubiläum gewidmet von seinen Schülern* (*Archiv für Hygiene*, *17*) (1893); K. Kisskalt, "Max von Pettenkofer," in *Grosse Naturforscher*, H. W. Frickhinger, ed. (Stuttgart, 1948); A. Kohut, "Ein Gedenkblatt zu seinem 80. Geburtstage (3. Dezember)," in *Pharmazeutische Zeitung*, **43** (1898), 853–855; K. B. Lehmann, "Max v. Pettenkofer und seine Verdienste um die wissenschaftliche und praktische

Hygiene," in *Deutsche Vierteljahrsschrift für öffentliche Gesundheitspflege*, **25** (1893), 361–385; O. Neustatter, "Max Pettenkofer," in M. Neuberger, ed., *Meister der Heilkunde*, VII (Vienna, 1925), 1–89; "Die Pettenkofer-Feier," in *Münchener medizinische Wochenschrift*, **40** (1893), 536–538; E. Roth, "Von Pettenkofer als populärer Schriftsteller," in *Deutsche Vierteljahrsschrift für öffentliche Gesundheitspflege*, **25** (1893), 386–396; L. Seidel, "Ueber den numerischen Zusammenhang, welcher zwischen der Häufigkeit der Typhus-Erkrankungen und dem Stande des Grundwassers während der letzten 9 Jahre in München hervorgetreten ist," in *Zeitschrift für Biologie*, **1** (1865), 221–236; H. E. Sigerist, "Max Pettenkofer (1818–1901)," *Grosse Aerzte* (Munich, 1932), pp. 288–292; G. Sticker, *Abhandlungen aus der Seuchengeschichte und Seuchenlehre, II. Die Cholera* (Giessen, 1912), pp. 133–150, 290–299; J. Soyka, "Zur Aetiologie des Abdominal Typhus," in *Archiv für Hygiene*, **6** (1887), 257–302; and R. Virchow, "Erwiderung an Herrn von Pettenkofer," in *Berliner klinische Wochenschrift*, **21** (1884), 490–491.

Among English references to Pettenkofer's theories are G. Buchanan, "On Prof. Pettenkofer's Theory of the Propagation of Cholera," in *Medical Times and Gazette*, **1** (1870), 283–285; W. Budd, *Typhoid Fever . . .* (London, 1874; repr. New York, 1931); C. Childs, "The History of Typhoid Fever in Munich," in *Lancet* (1898), **1**, 348–354; J. M. Cuningham, "Recent Experience of Cholera in India," *ibid.* (1874), **1**, 477–479; H. M. Dietz, "The Activity of the Hygienist F. Erismann in Moscow (Unpublished Documents and Letters to M. v. Pettenkofer)," in *Clio Medica*, **4** (1969), 203–210; E. E. Hume, *Max von Pettenkofer . . .* (New York, 1927); E. McClellan, "A Reply to an Address of Prof. Max von Pettenkofer of Munich . . .," in *Practitioner*, **19** (1877), 61–78, 148–160; C.-E. A. Winslow, "Pettenkofer—The Last Stand," in *The Conquest of Epidemic Disease* (Princeton, N.J., 1944), chap. 15, pp. 311–336.

Claude E. Dolman

PETTERSSON, HANS (*b.* Kälhuvudet, Marstrand, Bohuslän, Sweden, 26 August 1888; *d.* Göteborg, Sweden, 25 January 1966), *oceanography.*

Like his father, Pettersson became one of the most outstanding oceanographers of his period. After matriculating in Stockholm, and graduating at Uppsala, he studied physics under K. Ångstrom and Sir William Ramsay, working on problems in optics and radioactivity. In 1913 he was appointed to the staff of the Svenska Hydrografiska-Biologiska Kommissionen and was soon publishing papers on the tides and currents of the Kattegat. Like his father, he was particularly interested in the differences of flow in stratified water, waves on internal boundary surfaces, and improved methods of measuring water density and flow. Throughout this work he had a compelling interest in changes of sea level brought about by meteorological factors and in the effect of oceans on climate. He shared his father's urge to demonstrate that the dominant meteorological features of northwest Europe are determined to a large extent by the heat capacity and water transport of the North Atlantic Ocean. Following up his earlier interests he studied radioactivity in seawater and sediments, and penetration of light into the sea.

In general, he seemed to prefer spectacular explanations for oceanic phenomena, invoking, for example, a cosmic origin rather than more disciplined terrestrial sources to explain the relatively high content of nickel in some deep sea sediments. He offered a catastrophe of a volcanic nature as an alternative to turbidity-current deposition to explain the presence of organic material on the equational West Atlantic plain. He held tenaciously to the idea that abyssal plains were formed by vast outpourings of lava. One of his aims was to promote discussion and rouse interest; in this he was very successful. He was a lecturer at the Göteborgs Högskola from 1914 to 1930, when he was appointed professor. He was largely responsible for persuading wealthy businessmen to finance the Oceanografiska Institutet in 1939, and was director of this laboratory until 1956.

He is best known for his round-the-world, Swedish deep-sea expedition in the *Albatross* (1946–1948), for which he again obtained the funds and loan of the vessel from a Swedish shipping combine and private donors. One of the main achievements was the sampling of deep-sea sediments with a new piston core-sampler, developed largely by his colleague Dr. Kullenberg. Radioactivity and optical studies were also prominent. Within four years of the return of the expedition he was honored by many universities and learned societies, mainly in Europe, but also in the United States. In 1956 he was elected a foreign member of the Royal Society of London.

After his retirement in 1956 he held a research professorship in geophysics at the University of Hawaii, and he continued as a prominent figure in oceanography and geophysics until his death in 1966.

BIBLIOGRAPHY

See M. Sears, ed., *Progress in Oceanography*, III (Elmsford, 1965), which commemorated his seventy-fifth birthday; it contains an appreciation of his work by one of his former pupils and a full bibliography of some 180 publications covering fifty-one years. His semipopular book *Westward Ho With the Albatross* (London–New York, 1953) covered much of his life story and many of his interests.

G. E. R. DEACON

PETTY, WILLIAM (*b.* Romsey, Hampshire, England, 26 May 1623; *d.* London, England, 16 December 1687), *economics, demography, geography.*

Petty was a prominent virtuoso in an age of the many-sided genius. His investigations ranged over anatomy, geodesy, the design of ships, and the application of mathematics to diverse natural and practical problems. His major contributions were in what he called "political arithmetic."

Petty was the eldest surviving child of Anthony Petty, a cloth worker and tailor who owned his home and probably some farm land, and Francesca Denby Petty. Petty's chief childhood amusement, he later told John Aubrey, was "looking on the artificiers, e.g. smyths, the watchmakers, carpenters, joiners, etc." After the usual education in Latin, Greek, and arithmetic, he became at the age of thirteen a cabin boy on a merchant ship. He studied navigation while aboard, although he was extremely nearsighted. After ten months at sea he broke his leg and was set ashore near Caen.

There he impressed the Jesuit fathers, who admitted him to their college. More than a year later he returned to England and joined the navy. He left after the outbreak of civil war but always retained an interest in ships and the sea. In 1643 Petty went to the Netherlands to study medicine at Utrecht, Leiden, and Amsterdam. In 1645 he continued on to Paris and studied anatomy with Hobbes, who introduced him to Mersenne and his famous circle of friends who studied natural philosophy.

In 1646 Petty returned to England and met Samuel Hartlib, who persuaded Petty to write a tract on education. The result was *The Advice of W.P. to Mr. S. Hartlib for the Advancement of Some Particular Parts of Learning* (1648). Following Bacon's teachings, Petty recommended the establishment of a college "for the advancement of all mechanical arts and manufactures," where practical and theoretical questions would be investigated and a history of the trades would be written.

Petty resumed his anatomical studies, and by 1649 he was at Oxford University. In that year John Wilkins and other members of the university formed a club for experimental philosophy, which was an antecedent to the Royal Society of London. Because of the proximity to an apothecary shop, the club soon began holding its meetings in Petty's lodgings.

Petty's professional standing advanced rapidly. In 1650 he received the doctorate of physic from Oxford and became a candidate at the College of Physicians in London. He was soon appointed professor of anatomy and vice-principal of Brasenose College, Oxford, and then professor of music at Gresham College, London.

Nevertheless, in 1652 he left these positions to become physician-general to Cromwell's army in Ireland. Having found it necessary to pay the troops with land from the defeated Irish, Cromwell needed a land survey. The appointed supervisor proved inept, and in 1655 Petty volunteered to carry out the survey in thirteen months. His proposal was accepted, and he engaged about a thousand men for the task. In March 1656 he completed the survey as scheduled. Although it was an outstanding achievement, there were errors of underestimation of 10–15 percent, which correspondingly lessened his pay.

As a supplement to the Down Survey, Petty also undertook the complete mapping of Ireland. A general map and thirty-five county and barony maps were printed around 1685—the most detailed maps ever published for a whole country. Even more detailed maps, some of which have since been destroyed by fire, remained unpublished. Another supplement to the survey was the first census of Ireland, which was taken around 1659; the extent of Petty's involvement is unknown. The only copy of the results was among his papers, but he never made use of it in his writings.

Petty's payment for his survey enabled him to buy cheaply forfeited and mortgaged lands, thus acquiring considerable property, which he continued to augment throughout his life. Having acquired his wealth from other men's misfortunes, Petty endured hostility and litigation for the rest of his life.

Petty abandoned the practice of medicine and henceforth devoted a major part of his time and thought to managing his property. He was a friend of the Cromwells, both before and after the collapse of the Commonwealth; but he acquiesced in the Restoration, and Charles II knighted him in April 1661. Petty aspired to, but never received, an important government office. One of his hopes was to be placed in charge of an Irish statistical office. He was also on good terms with James II, although disappointed in his support of the Irish Catholics.

In 1667 Petty married Elizabeth Fenton, the attractive and intelligent widow of Sir Maurice Fenton. She was the daughter of Sir Hardress Waller, a prominent Irish Protestant, who had provided strong support for Petty's land survey. Two sons and a daughter survived infancy. Although Petty twice declined a baronetcy, James II made his widow Baroness Shelburn in 1688.

By the end of 1659 Petty had returned from Ireland to London, where he rejoined those members of the philosophical club who had moved back from Oxford and had begun meeting at Gresham College. When this group became incorporated in 1662 as the Royal Society of London for Promoting Natural Knowledge,

Petty was a charter member of its council. In 1673 he was elected vice-president. His interest in the society never waned, but his participation was curtailed by his becoming nearly blind in 1670 and by his being in Ireland managing his property from 1666 to 1673 and from 1676 to 1685.

A thoroughgoing Baconian, Petty emphasized the collection of information and the practical application of scientific principles and knowledge. Before the Royal Society he read instructions concerning the technical processes that were involved in his father's trade of cloth making and dyeing. In a discourse read before the Royal Society in 1674, he expressed interests in the physical sciences that were typical for his time. He urged the study of number, weight, and measurement; and he provided diverse examples, such as the relationship between scale and strength of structures; calculations of the distance that sounds, odors, and light travel in relation to the magnitude of the source; and formulas to predict prices and human longevity. He also suggested an explanation of elasticity and an atomic theory of matter. Atoms, he believed, were tiny magnets of different sexes.

Petty's most notable application of physical principles to a practical problem was in the development of his famous twin-hulled ships. He designed and constructed three of them between 1662 and 1664 and another in 1684. The first two ships were rather successful, but the third sank in a storm with all hands aboard, and the fourth was a complete failure. His concept was nevertheless valid and is exemplified in the modern catamaran.

Petty's writings that are relevant to the social sciences were his most enduring achievement. The time spent managing his Irish holdings undoubtedly reduced the time he could devote to scholarly writings. On the other hand, those holdings provided the incentive and orientation for his writings on political arithmetic.

Inseparable from Petty's writings is John Graunt's important *Natural and Political Observations Mentioned in a Following Index, and Made Upon the Bills of Mortality*; published in January 1662, it started the sciences of demography and statistics. In spite of Graunt's name on the title page and his election to the Royal Society on the book's merits, seven contemporaries referred to Petty as the author. The available evidence does not establish their claim. Petty and Graunt became friends around 1650 and remained so until Graunt's death. It seems likely that Petty discussed with Graunt the analysis of the London bills of mortality, and perhaps the book would not have been published without Petty's assistance and encouragement.

Petty's writings most comparable to Graunt's *Observations* are ten brief essays (1682–1687) on the population of London, Dublin, Paris, Rome, and other cities. These essays reveal Petty's appreciation of the importance of population as an economic factor. Because there were few vital statistics for any city except London, he relied upon unreliable indexes of population, such as the number of chimneys.

Although his figures were sometimes questionable, Petty was the first economic theorist to make a significant attempt to base economic policy upon statistical data. His *Treatise of Taxes and Contributions* (1662), possibly his best work, discussed the economy of Ireland and England and attempted to formulate sound policies for promoting wealth and collecting taxes. He was the first to emphasize the value of labor as a part of national wealth, and he also provided an important analysis of rents. His short *Verbum sapienti*, written about 1665 and published in 1691, is more quantitative and contains the first estimate of national incomes and the first discussion of the velocity of money. *The Political Anatomy of Ireland*, written about 1672 and published in 1691, also on economic policy, is supported by economic geography. In *Political Arithmetick*, written 1671–1676 and published in 1690, Petty extended his discussion to a comparison of the wealth and economic policies of England and France. He argued the benefits of a division of labor and the gains from foreign trade in *Another Essay in Political Arithmetick Concerning the Growth of the City of London* (1683). He also left other manuscripts on the Irish and English economy that have been published only recently. These include *A Treatise of Ireland* (1899), written in 1687 as advice for James II—advice neither requested nor heeded.

Petty's writings were, nevertheless, influential in England, and as an economic theorist he was not surpassed before 1750.

BIBLIOGRAPHY

I. ORIGINAL WORKS. Excellent detailed bibliographies of Petty's published writings and some of his MSS are Charles Henry Hull, ed., *The Economic Writings of Sir William Petty Together With the Observations Upon the Bills of Mortality, More Probably by Captain John Graunt*, II (Cambridge, 1899; facs. ed., New York, 1963), 633–657; Yann Morvran Goblet [pseudonym for Louis Tréguiz], *La transformation de la géographie politique de l'Irelande au XVIIᵉ Siècle dans les cartes et essais anthropogéographiques de Sir William Petty*, I (Paris, 1930), ix–xxiii; and Geoffrey Keynes, *A Bibliography of Sir William Petty F.R.S. and Observations on the Bills of Mortality by John Graunt F.R.S.* (Oxford, 1971).

Twelve of Petty's treatises have been reprinted with introductions and notes by Henry Hull, in the work mentioned above. Other works by Petty include *The History of the Survey of Ireland, Commonly Called the Down Survey, A.D. 1655–6*, Thomas Aiskew Larcom, ed. (Dublin, 1851; facs. ed., New York, 1967); and *A Census of Ireland, circa 1659*, Séamus Pender, ed. (Dublin, 1939). Lord Lansdowne has edited three important collections of Petty's MSS: *The Petty Papers: Some Unpublished Writings of Sir William Petty*, 2 vols. (London–Boston, 1927; facs. ed., New York, 1966); *The Petty-Southwell Correspondence, 1676–1687* (London, 1928; facs. ed., New York, 1967); and *The Double Bottom or Twin-Hulled Ship* (Oxford, 1931).

II. SECONDARY LITERATURE. The most important contemporary accounts of Petty are by John Aubrey and John Evelyn; both are reprinted in Keynes' *Bibliography*, 85–95. The most detailed, and still essential, biography is Edmund George Petty Fitzmaurice, *The Life of Sir William Petty, 1623–1687* (London, 1895). Fitzmaurice also wrote the account in the *The Dictionary of National Biography*, new ed., XV, 999–1005. A very interesting recent biography is Eric Strauss, *Sir William Petty, Portrait of a Genius* (London–Glencoe, Ill., 1954), which includes a discussion of Petty's work.

Bacon and Hartlib's influence on Petty is admirably discussed in Walter E. Houghton, "The History of Trades: Its Relation to Seventeenth-Century Thought, as Seen in Bacon, Petty, Evelyn, and Boyle," in *Journal of the History of Ideas*, **2** (1941), 33–60, repr. in Philip P. Wiener and Aaron Noland, eds., *Roots of Scientific Thought, a Cultural Perspective* (New York, 1957), 354–381.

On Petty's participation in the beginnings of the Royal Society, see Thomas Sprat, *The History of the Royal-Society of London, for the Improving of Natural Knowledge* (London, 1667), facs. ed., Jackson I. Cope and Harold Whitmore Jones, eds. (St. Louis–London, 1958); Thomas Birch, *A History of the Royal Society for Improving of Natural Knowledge, From Its First Rise*, I (London, 1756); Irvine Masson and A. J. Youngson, "Sir William Petty, F.R.S. (1623–1687)," in *The Royal Society, Its Origins and Founders*, Harold Hartley, ed., (London, 1960), 79–90; and Margery Purver, *The Royal Society: Concept and Creation* (Cambridge, Mass., 1967).

On Petty's *Discourse Made Before the Royal Society the 26. of November 1674. Concerning the Use of Duplicate Proportion in Sundry Important Particulars: Together With a New Hypothesis of Springing or Elastique Motions*, see A. Wolf, F. Dannemann, A. Armitage, and Douglas McKie, *A History of Science, Technology and Philosophy in the 16th & 17th Centuries*, 2nd ed. (New York, 1950), 484; and Robert H. Kargon, "William Petty's Mechanical Philosophy," in *Isis*, **56** (1965), 63–66. On Petty's ships, see Lord Lansdowne's intro. to *The Double Bottom or Twin-Hulled Ship of Sir William Petty* (Oxford, 1931). On the relationship between science and technology in Petty's work, Hessen's famous arguments concerning Newton would for the most part apply; see Boris M. Hessen, "The Social and Economic Roots of Newton's 'Principia,' " in *Science at the Cross Roads: Papers Presented to the International Congress of the History of Science and Technology in London From June 20th to July 3rd, 1931 by the Delegates of the U.S.S.R.* (London, 1931), 151–176; repr. as separate book (New York, 1971).

On the Down Survey, see Petty's account mentioned above. On the census, see Pender's intro. to the volume cited above. The Down Survey, the maps, Petty's contributions to geography, and his career in Ireland are thoroughly and admirably discussed by Goblet (Tréguiz), *La transformation de la géographie politique*. See also his *A Topographical Index of the Parishes and Townlands of Ireland in Sir William Petty's MSS. Barony Maps (c. 1655–9)* (*Bibliothèque Nationale de Paris, fonds anglais, nos. 1 & 2*) and *Hiberniae Delineatio (c. 1672)* (Dublin, 1932); and Séan O'Domhnaill, "The Maps of the Down Survey," in *Irish Historical Studies*, **3** (1943), 381–392.

The authorship of Graunt's *Natural and Political Observations* has been discussed in C. H. Hull, ed., *Economic Writings of Sir William Petty*, I (Cambridge, 1899), xxxix–liv; Wilson Lloyd Bevan, "Sir William Petty. A Study in English Economic Literature," in *Publications of the American Economic Association*, **9**, no. 4, (1894), 42–46; Lansdowne, *The Petty Papers*, II, 273–284, and *The Petty-Southwell Correspondence*, xxiii–xxxii; Major Greenwood, *Medical Statistics From Graunt to Farr* (Cambridge, 1948), 36–39; D. V. Glass, "John Graunt and His *Natural and Political Observations*," in *Notes and Records. Royal Society of London*, **19** (1964), 78–89, 97–100; P. D. Groenewegen, "Authorship of the *Natural and Political Observations Upon the Bills of Mortality*," in *Journal of the History of Ideas*, **28** (1967), 601–602; and Geoffrey Keynes, *A Bibliography of Sir William Petty* (Oxford, 1971), 75–77.

On Petty's contributions to demography and statistics, see Harald Westergaard, *Contributions to the History of Statistics* (London, 1932; facs. ed., New York, 1968), 28–31; Greenwood, *Medical Statistics*, 2–27; Wolf *et al.*, *A History of Science*, 429–430, 598–602; Tréguiz, *La transformation*, II, bk. 7; James Bonar, *Theories of Population From Raleigh to Arthur Young* (London, 1931; facs. ed., London, 1966), ch. 3; and Charles F. Mullett, "Sir William Petty on the Plague," in *Isis*, **28** (1938), 18–25. Petty's interest in the application of his demographic ideas in the United States is discussed in James H. Cassedy, *Demography in Early America: Beginnings of the Statistical Mind* (Cambridge, Mass., 1969).

On Petty's contributions to economics, see Bevan, *Sir William Petty*; Phyllis Deane, "William Petty," in *International Encyclopedia of the Social Sciences*, XII (1968), 66–68; William Letwin, *The Origins of Scientific Economics* (London, 1963; Garden City, N.Y., 1964), ch. 5; Walter Müller, *Sir William Petty als politischer Arithmetiker: Eine soziologisch-statistische Studie* (Gelnhausen, 1932); Maurice Pasquier, *Sir William Petty, ses idées économiques* (Paris, 1903); and Eric Roll, *A History of Economic Thought*, 2nd ed. (New York, 1942), 99–114.

FRANK N. EGERTON III

PEUERBACH, GEORG VON (*b.* Purbach, Austria, 30 May 1423; *d.* Vienna, Austria, 8 April 1461), *mathematics, astronomy.*

For a detailed study of his life and work, see Supplement.

PEYER, JOHANN CONRAD (*b.* Schaffhausen, Switzerland, 26 December 1653; *d.* Schaffhausen, 29 February 1712), *physiology.*

Born into a patrician family, Peyer studied medicine in Basel before becoming the pupil of J. Guichard du Verney at Paris and later of Vieussens at Montpellier. He then returned to his native Schaffhausen, where, with his teacher Johann Jakob Wepfer and the latter's son-in-law Johann Conrad Brunner, he formed the "Schaffhausen trio," whose important contributions to the new methodology of medical research were the explanation of symptoms by connecting them with the lesion in the body, considered as the site of the disease, and experimentation on animals (*anatomia animata*) to study either the functioning of organs or the effects of medicines on the organism. It is not always easy to determine the individual contribution of each member of the trio from their joint publications. In any case, after a ten-year collaboration through the 1680's Peyer quarreled with Wepfer and Brunner and spent the rest of his life as professor of logic, rhetoric, and medicine at the local Gymnasium.

A member of the Academia Naturae Curiosorum Sacra, Peyser sent, under the pseudonym Pythagoras, many letters to the Academy on the antiperistaltic movements of the intestine, cinchona bark, and other subjects.

The iatrochemical theories that were then dominant oriented scientists toward the study of digestion and the digestive system. Hence in 1682 Peyer described the lymphatic nodules and masses located in the walls of the ileum that now bear his name (or sometimes that of Johannes Nicolaus Pechlin, who in 1672 had described them in a work on the anatomy of the intestinal tract). According to William Cole, Severino had observed follicular regions in the small intestine of animals (1645).

Adenography was then in fashion, but the interpretation of the dissections is somewhat confusing today. In fact, the chemical phenomena of digestion were unknown, and the theory of the lymphatic system had barely been outlined. It was thought that the excretory canal of the pancreas was a chyliferous vessel originating in the intestine and entering the gland (Moritz Hofmann, 1641; J. G. Wirsung, 1642). C. Brunner attributed to the duodenal glands that now bear his name a role in the secretion of lymph. Thus it is not astonishing that Peyer supposed the ileal follicles to be glands excreting the digestive juices (E. Hintzsche, pp. 2401–2402).

Peyer also studied the stomach of ruminants in his *Merycologia* (1685) and the anatomy of normal and abnormal fetuses. He probably assisted Brunner in his pancreatectomies on dogs. These operations resulted in diabetes, but neither Peyer nor Brunner could interpret the results obtained (1683).

Wepfer—apparently the leader of the Schaffhausen trio—had written the first book on experimental toxicology, *Cicutae aquaticae historia et noxae commentario illustrata* (1679). He had been led to believe that the activity of the heart could be explained only by its special structure. Peyer attempted to confirm this hypothesis, either with his Schaffhausen friends or with his Basel friend J. J. Harder. In 1681 he succeeded in making the hearts of dead animals—and even of human beings who had been hanged—beat by blowing air into the veins or by utilizing other stimuli. Despite his imperfect technique he achieved an artificial cardiac activity lasting several hours. These experiments were repeated in 1702 by Baglivi, who studied the tonicity of the *fibra motrix*. He concluded from them that the heart died not from lack of nervous fluid but from lack of blood (M. D. Grmek, pp. 313–314). Later, supporters of the doctrine of irritability investigated this problem, taken up by Bidder. In his letters to Harder, Peyer revealed some of his other discoveries: hermaphroditism of the pulmonate snail and an epizootic disease that may have been foot-and-mouth disease.

BIBLIOGRAPHY

I. ORIGINAL WORKS. Peyer's writings include *Exercitatio anatomico-medica de glandulis intestinorum, earumque usu et affectionibus. Cui subjungitur anatome ventriculi gallinacei* (Schaffhausen, 1677); *Methodus historiarum anatomico-medicarum exemplo ascitis vitalium organorum vitio ex pericardii coalitu cum corde nato illustrata* (Paris, 1678); *Meditatio de valetudine humana* (Basel, 1681); *Parerga anatomica et medica septem* (Geneva, 1681); *Paeonis et Pythagorae exercitationes anatomicae et medicae familiares bis quinquaginta, Hecatombe non Hecatae sed illustri Academiae naturae curiosorum sacra* (Basel, 1682), 100 letters dated 12 Jan. 1677 to 8 July 1681 written by Peyer (Pythagoras) and Harder (Paeonis); *Merycologia sive de ruminantibus et ruminatione commentarius* (Basel, 1685); and *Observatio circa urachum in foetu humano pervium*, edited by his son Johan Jacob Peyer (Lyons, 1721).

II. SECONDARY LITERATURE. See E.-H. Ackerknecht, *Grands médecins suises de 1500 à 1900, Conférence du Palais de la Découverte D109* (Paris, 1966); Conrad Brunner and Wilhelm von Muralt, *Aus den Briefen hervorragender schweizer Ärzte des 17. Jahrhunderts* (Basel, 1919), 153–226; M. D. Grmek, "La notion de fibre vivante chez les médecins de l'école iatrophysique," in *Clio medica*, **5**, no. 4 (Dec. 1970), 297–318; E. Hintzsche, "Anatomia animata," in *Revue Ciba*, **69** (1948), 2395–2412; *Recherches, découvertes et inventions des médecins suisses* (Basel, n.d.); Robert Lang, *Das Collegium humanitatis in Schaffhausen*, I, *1648–1727* (Leipzig, 1893), 29 ff.; F. V. Mandach, "Über das klassische Werk des schweizer Arztes Joh. Conr. Peyer *De glandulis intestinorum*," in *Korrespondenzblatt für schweizer Ärzte*, **33** (1903), 445–450, 479–482; Bernhard Peyer and Heinrich Peyer, "Bildnis und Siegel des Arztes Johann Conrad Peyer," supp. to *Veröffentlichungen der Schweizerischen Gesellschaft für Geschichte der Medizin . . .*, **13** (1943); Bernhard Peyer-Amsler, "Johann Conrad Peyer 1653–1712," in Reinhard Frauenfelder, ed., *Geschichte der Familie Peyer mit den Wecken 1410–1932* (Zurich, 1932), 299–346—notes 1–7, pp. 333 f., list all the works dealing with Peyer's life and there is a facs. of a letter by Peyer on p. 323 (note [I] on Peyer also appeared as a supp. to *Veröffentlichungen der Schweizerischen Gesellschaft für Geschichte der Medizin . . .*, **8** [1932]); and H. E. Sigerist, "Die Verdienste zweier Schaffhauser Ärzte (J. C. Peyer und J. C. Brunner) um die Erforschung der Darmdrüsen," in *Verhandlungen der Schweizerischen naturforschenden Gesellschaft*, **102** (1921), 153 f.

PIERRE HUARD
M. J. IMBAULT-HUART

PEYSSONNEL, JEAN ANDRÉ (*b.* Marseilles, France, 19 June 1694; *d.* Guadeloupe, 24 December 1759), *botany*, *zoology*.

The eighth child of a physician at the Hôtel-Dieu in Marseilles, Peyssonnel visited the Antilles at the age of fifteen and Egypt three years later. He studied medicine at the University of Aix, where he defended his dissertation before a jury headed by his father. He began practice in Marseilles, which in 1720 suffered a severe epidemic of plague; Peyssonnel's efforts on behalf of the stricken were rewarded by a royal pension. During his year in Marseilles, Peyssonnel and some friends founded an academy devoted to belles lettres, in the tradition of the "ancient academy" of Marseilles.

A naturalist by inclination, Peyssonnel was interested in marine natural history. He observed the Mediterranean currents and studied corals, confirming the "flowering" established twenty years earlier by Count Luigi Marsigli. How the plant produced the red, stony portion, however, remained a mystery. Peyssonnel informed the president of the Paris Academy of Sciences, the Abbé Jean-Paul Bignon, of the progress of his research. In 1723 he became a correspondent of the Academy, to which he presented a paper on coral, which he then considered to be a flowering plant, in 1724.

Sent by the king to the Barbary Coast "to make discoveries in natural history" and to visit the country around Tunis and Algiers, Peyssonnel described his travels in letters to Bignon and in 1726 sent him two papers concerning his research on madrepores and other corals. The essence of his findings was that corals are not plants, but animals. The so-called flowers retract on contact with air but reappear when returned to the sea. It is animal matter, soft and milky, that covers the organism's stony parts. Further, Peyssonnel believed that "insects," such as the sea anemone, are also corals and madrepores. Peyssonnel's text, transmitted by Bignon to Réaumur, was read at the Academy on 8 and 28 June and 3 July 1726. Réaumur did not mention the author's name for fear of subjecting a distinguished person to ridicule: the idea seemed unacceptable to everyone, as Peyssonnel was informed in letters from Réaumur and Bernard de Jussieu.

Undaunted, Peyssonnel continued his studies in marine natural history. His pamphlet of 1726, *Mémoire sur les courans de la Méditerranée*, was the first work written on the Mediterranean currents. Distributed by the municipal magistrates of Marseilles to ship captains, it stimulated them to communicate their own observations.

In 1727 Peyssonnel departed for Guadeloupe as "royal botanist to the American islands." He spent the rest of his life on that island, where he married and fathered a son and several daughters. Although prevented by administrative difficulties from mining the sulfur of the Grande Soufrière volcano, which he had explored, he continued his research on marine life. His results provided him with complete confirmation of his earlier assertions, a fact he communicated in a letter of 1733 to Antoine de Jussieu.

In 1740 Trembley discovered the green hydra that Réaumur called a polyp, and the analogy with "marine plants" was cited. Guettard and Bernard de Jussieu confirmed, on the French coast, Peyssonnel's findings, and Réaumur admitted his own error (*Mémoires pour servir à l'histoire naturelle des insectes*, VI [1742], preface). Although he regained the esteem of the Academy, of which he was a corresponding member, it was to the Royal Society of London that Peyssonnel sent a manuscript on coral (in French). A résumé of it was presented to the Society on 7 May 1752 by W. Watson, who stressed the high quality of Peyssonnel's work (*Philosophical Transactions* [1753]). In 1756 Peyssonnel published in London a translation of Watson's article, supplemented by various writings on a plan to establish an annual prize at the Academy of Marseilles, to be awarded for the best paper on a subject in marine natural history. From 1756 until his death in 1759 he presented ten further articles recording his scientific observations to the Royal Society.

BIBLIOGRAPHY

I. ORIGINAL WORKS. Peyssonnel's published works are *La contagion de la peste expliquée et les moyens de s'en préserver, par le S***, Docteur en Médecine* (Marseilles, 1723); *Mémoire sur les courans de la Méditerranée* (Marseilles, 1726), 600 copies of which were printed at the expense of the Chambre de Commerce de Marseille, repr. in the 1756 collection under the title "Essai de physique ou conjectures fondées sur quelques observations qui peuvent conduire à la connaissance et à l'explication des courans de la mer Méditerranée"; and *Traduction d'un article des Transactions Philosophiques sur le Corail—Projet proposé à l'Académie de Marseille pour l'établissement d'un Prix . . . et réponse de l'Académie. Diverses observations sur les courants de la Mer, faites en différents endroits* (London, 1756).

II. SECONDARY LITERATURE. For the coral controversy see *Registre de l'Académie des sciences*, 8 and 28 June and 3 July 1725; "Sur le corail," in *Histoire de l'Académie royale des sciences* (1727), 37–39; Réaumur, "Observations sur la formation du corail et des autres productions appelées plantes pierreuses," in *Mémoires de l'Académie royale des sciences* (1727), 269–281; Bernard de Jussieu, "Examen de quelques productions marines qui ont été mises au nombre des plantes et qui sont l'ouvrage d'une sorte d'insectes de mer," *ibid.*, (1742), 290–302; Pierre Flourens, "Analyse d'un ouvrage manuscrit intitulé Traité du Corail," in *Annales Sc. nat. Zoologie*, ser. 2, **9** (1838), 334; and "Analyse d'un ouvrage manuscrit intitulé Traité du Corail par le Sieur de Peyssonnel," in *Journal des Savants* (Feb. 1838). See also Dureau de la Malle, Peyssonnel, and Desfontaines, *Voyages dans les Régences de Tunis et d'Alger* (Paris, 1838); Alfred Lacroix, "Notice historique sur les cinq de Jussieu," in *Mémoires de l'Institut*, **63** (1941), 24; and Noël Duval, "La solution d'une énigme: les Voyageurs Peyssonnel et Giménez à Sbeitla en 1724," in *Bull. Soc. Nat. Antiquaires France* (2 June 1965).

LUCIEN PLANTEFOL

PEZARD, ALBERT (*b.* Neuflize, Ardennes, France, 1 April 1875; *d.* Paris, France, 21 November 1927), *endocrinology.*

Pezard's parents were farmers. He studied at the École Normale of Charleville, then won admission to the École Normale of St.-Cloud. Several professors from the Sorbonne who taught at St.-Cloud soon found that Pezard was one of their most brilliant science students. He obtained his master's degree from the University of Paris with honors in both physical and natural sciences. His success in the difficult natural sciences *agrégation* made him eligible for a professorship at two renowned Paris *collèges*: Colbert and Jean-Baptiste Say. Pezard's teaching duties were time-consuming but he fulfilled them even after 1908, when he was active in biological investigation. His attraction to research was noticed by one of his former teachers at St.-Cloud, Charles Gravier, professor of zoology at the Muséum d'Histoire Naturelle, who recommended him to Émile Gley, professor at the Collège de France, one of the founders of endocrinology. Gley directed the Station Physiologique du Collège de France, and Pezard was soon elected associate director of the station, where he worked for nearly twenty years.

Pezard first studied the physiological conditioning of the secondary sexual characters in birds. Full publication of his findings, in his doctoral dissertation, was delayed until 1918 by his military duties during World War I. The dissertation contained many new results, such as Pezard's establishment of the true secondary sexual characters in the male bird. He showed their strict dependence upon an endocrine secretion, since castration brings their immediate regression, and gave a simple but accurate algebraic expression of this regression. He also deduced from the effects of total castration of females that the ovary normally secretes a substance that inhibits the evolution of plumage and the development of spurs. This was the first demonstration of the inhibitory action of an endocrine secretion.

In his pioneer study on the experimental production of sexual inversion and hermaphroditism Pezard defined the conditions required for the grafting of testicular transplants. With this procedure he showed that a minimal, or threshold, amount of the specific hormone must be secreted in order to elicit a given secondary sexual character, each character responding to a different threshold amount. Thus Pezard introduced into endocrinology the "all or none" law, classical in other domains of physiology. In 1922 Pezard began the most active years of his scientific life, working closely with Fernand Caridroit and with Knud Sand of Copenhagen. Their papers, treating the experimental production of gynandromorphic hens, demonstrate the important implication of their results for genetics. They concluded that hormones obviously do not interfere with the fundamental genotypic complex, but act as regulators of dominances. These hormones operate, as in sex, not as creating agents but as exteriorization factors.

They neither destroy existing genes nor provide new genes.

Pezard's work, which exerted a marked influence on biologists, found artistic expression in his creation of strange and beautiful birds, such as hens decorated with a cock's brilliant feathers and proud comb. His reputation was great outside France, which failed to give him a university professorship and left him to teach high school pupils.

BIBLIOGRAPHY

Pezard's writings include "Sur la détermination des caractères sexuels secondaires chez les Gallinacés," in *Comptes rendus hebdomadaires des séances de l'Académie des sciences*, **153** (1911), 1027–1030; "Sur la détermination des caractères sexuels secondaires chez les Gallinacés, greffe de testicule et castration post-pubérale," *ibid.*, **154** (1912), 1183–1186; "Développement expérimental des ergots et croissance de la crête chez les femelles des Gallinacés," *ibid.*, **158** (1914), 513–516; "Transformation expérimentale des caractères sexuels secondaires chez les femelles des Gallinacés," *ibid.*, **160** (1915), 260–263; "Loi numérique de la régression des organes érectiles consécutive à la castration post-pubérale," *ibid.*, **164** (1917), 234–236; *Le conditionnement physiologique des caractères sexuels secondaires chez les Oiseaux* (Paris 1918); Édition du Bulletin Biologique de la France et de la Belgique; "Facteur modificateur de la croissance normale et loi de compensation," in *Comptes rendus hebdomadaires des séances de l'Académie des sciences*, **169** (1919), 997-1000; "Castration alimentaire chez les Coqs soumis au régime carné exclusif," *ibid.*, 1177–1180; "Secondary Sexual Characteristics and Endocrinology," in *Endocrinology*, **4** (1920), 527–541; "Castration intrapubéale et généralisation de la loi parabolique de régression," in *Comptes rendus hebdomadaires des séances de l'Académie des sciences*, **171** (1920), 1081–1084; "Loi du 'tout ou rien' ou de constance fonctionelle du testicule considéré comme glande à sécrétion interne," *ibid.*, **172** (1921), 89–91; and "Temps de latence dans les expériences de transplantation testiculaire et loi du 'tout ou rien,' " *ibid.*, 176–179.

See also "Numerical Law of Regression of Certain Secondary Sex Characters," in *Journal of General Physiology*, **9** (1921), 271–283; "Notion de 'seuil différentiel' et explication humorale du gynandromorphisme des Oiseaux bipartits," in *Comptes rendus hebdomadaires des séances de l'Académie des sciences*, **174** (1922), 1573–1576; "Notion de seuil différentiel et masculinisation progressive de certaines femelles d'oiseaux," *ibid.*, **175** (1922), 236–239; "Interpénétration testiculaire chez les Coqs castrés incomplètement," *ibid.*, 284–287, written with Caridroit; "L'hérédité sex-linked chez les Gallinacés. Interprétation fondée sur l'existence de la forme neutre et sur les propriétés de l'hormone ovarienne," *ibid.*, 910–913, written with Caridroit; "L'action de l'hormone testiculaire sur la valence relative des facteurs allélomorphes chez les Ovins," *ibid.*, 1099–1102, written with Caridroit; "La loi du 'tout ou rien.' Exposé général," in *Journal de physiologie et de pathologie générale,* **20** (1922), 200–211; "La loi du 'tout ou rien' et le gynandromorphisme endocrinien," *ibid.*, 495–508; "Caractères sexuels secondaires et tissus interstitiels," in *Comptes rendus de la Société de biologie,* **88** (1923), 245–248; "Critique de la théorie de Bouin et Ancel," *ibid.*, 333–336; "Féminisation d'un Coq adulte Leghorn doré," *ibid.*, **89** (1923), 947–950, written with Sand and Caridroit; "Le gynandromorphisme biparti experimental," *ibid.*, 1103–1106, written with Sand and Caridroit; and "Gynandromorphisme biparti fragmentaire d'origine mâle," *ibid.*, 1271–1274, written with Sand and Caridroit.

See also "Productions expérimentales du gynandromorphisme biparti chez les Oiseaux," in *Comptes rendus hebdomadaires des séances de l'Académie des sciences,* **176** (1923), 615–618, written with Sand and Caridroit; "Les modalités du gynandromorphisme chez les Oiseaux," *ibid.*, **177** (1923), 76–79, written with Caridroit; "Modifications raciales par greffe ovarienne chez les Coqs," in *Comptes rendus de la Société de biologie,* **40** (1924), 737–740, written with Sand and Caridroit; "Poecilandrie d'origine endocrinienne chez les Gallinacés," *ibid.*, **90** (1924), 676–679, written with Sand and Caridroit; "Potentialités homologues et potentialités hétérologues chez la Poule domestique," *ibid.*, 737–740; "Les effets de la castration sur le plumage du Coq domestique," *ibid.*, 677–680; "Remarques au sujet de l'hérédité sex-linked chez les Gallinacés," *ibid.*, 935–938, written with Caridroit; "Survie d'un transplant testiculaire actif en présence d'un ovaire producteur d'oeufs mûrs chez la Poule domestique," *ibid.*, 1459–1462; "Évolution et fonction d'un transplant ovarien chez un Coq adulte Leghorn doré," *ibid.*, **91** (1924), 1075–1078, written with Sand and Caridroit; and "Le gynandromorphisme en mosaïque et les dysharmonies endocriniennes chez les Gallinacés," *ibid.*, 1116–1119, written with Sand and Caridroit.

See also "Modifications hormono-sexuelles chez les Gallinacés adultes et théorie de la forme spécifique," in *Comptes rendus hebdomadaires des séances de l'Académie des sciences,* **178** (1924), 2011–2014, written with Sand and Caridroit; "Le gynandromorphisme biparti expérimental. Récurrences raciales dictées par la mue automnale et caractéres transitoires de certaines modifications pigmentaires," *ibid.*, **177** (1924), 1087–1090, written with Sand and Caridroit; "Le gynandromorphisme expérimental et le non antagonisme des glandes sexuelles chez les Gallinacés adultes," in *Comptes rendus de la Société de biologie,* **92** (1925), 427–430, written with Sand and Caridroit; "L'évolution des potentialités chez la Poulette," *ibid.*, 495–498; "Une notion nouvelle, l'existence d'un seuil différentiel racial dans certains complexes hybrides des Gallinacés," *ibid.*, 566–569, written with Sand and Caridroit; "Le gynandromorphisme périodique chez un Coq adulte," *ibid.*, 1034–1037, written with Sand and Caridroit; "Inversion sexuelle du plumage observées chez nos sujets lors de la récente mue et notion du seuil hormonique," *ibid.*, **93** (1925), 1094–1096, written with Sand and

Caridroit; "Quelques faits nouveaux concernant les greffes d'ovaires effectuées sur le Coq domestique," *ibid.*, **94** (1926), 520–523, written with Sand and Caridroit; "Analyse de quelques déviations sexuelles secondaires chez les Gallinacés," *ibid.*, 741–744, written with Caridroit; "La bipartition longitudinale de la plume, faits nouveaux concernant le gynandromorphisme élémentaire," *ibid.*, 1074–1078, written with Caridroit; and "La présence de l'hormone testiculaire dans le sang d'un Coq normal. Démonstration directe fondée sur la greffe autoplastique de crétillons," *ibid.*, **95** (1926), 296–300, written with Caridroit.

See further "Les hormones sexuelles et le gynandromorphisme chez les Gallinacés," in *Archives de biologie,* 36 (1927), 541–647, written with Sand and Caridroit; "Les caractères sexuels secondaires," in Masson's *Traité de physiologie,* XI (Paris, 1927), 125–163; "Inversion sexuelle autonome d'une Cane de Rouen," in *Comptes rendus de la Société de biologie,* **96** (1927), 1295–1298, written with Caridroit; "Classement des seuils différentiels du plumage chez la poule domestique," *ibid.*, 1372–1375, written with Caridroit; "Remarques concernant la loi du 'tout ou rien,' " *ibid.*, **97** (1927), 442–446; "État actuel de la théorie de l'interstitielle," *ibid.*, 620–623; "Die Bestimmung des Geschlechtsfunktion bei den Hühnern," in *Ergebnisse der Physiologie,* **27** (1928), 552–656; and *La détermination de la fonction sexuelle chez les Gallinacés,* Masson, ed. (Paris, 1930).

A. M. MONNIER

PEZENAS, ESPRIT (*b.* Avignon, France, 28 November 1692; *d.* Avignon, 4 February 1776), *hydrography, astronomy, physics.*

Pezenas was the son of E. F. Pezenas, a notary and court clerk, and Gabrielle de Rivières. He entered the Jesuit *collège* in Avignon at the age of ten and began his novitiate when he was seventeen. At the conclusion of his studies he taught first in Lyons and then at the Jesuit *collège* in Aix. (He taught physics in 1724 and 1726, logic in 1725, and metaphysics in 1727.) In 1728 Pezenas was named professor of hydrography at the École Royale d'Hydrographie at Marseilles and was assigned to teach the galley officers. He held this post until the school was closed in 1749, when he was appointed director of the Observatoire Ste.-Croix in Marseilles, a Jesuit establishment that later became Observatoire Royal de la Marine. In 1763, upon the suppression of the Jesuit order in France, Pezenas retired to Avignon, where he died at the age of eighty-three. He was a member of the Académie de Marine and, from 1750, a corresponding member of the Académie Royale des Sciences.

Besides giving highly regarded lectures, Pezenas published a number of works on hydrography and sailing. Notable among these were treatises on piloting, a work on gauging barrels and ships, and several treatises on nautical astronomy in which he set forth innovations in the use of such instruments as the vernier, the azimuth compass, and the octant. His interest in the determination of longitudes led to several papers on the subject. Assigned to design and estimate the cost of a canal that would supply Marseilles with water taken from the Durance River, he conducted the necessary surveying. Meanwhile, he had also become increasingly interested in astronomical observation. He paid for some of his observatory equipment, and in 1749 the king granted him the services of two associate astronomers, both Jesuits. He presented his first observations (including the determination of the latitude of Marseilles) in 1731 in the *Mémoires de Trévoux* (*Mémoires pour servir à l'histoire des sciences et des beaux arts*); the others appeared in *Histoire de l'Académie Royale des Sciences*.

Pezenas played a major role in the diffusion in France of important works by English scientists, especially in mathematics and optics. Although his scientific achievements were modest, Pezenas was an effective popularizer who made skillful use of the scientific and technical knowledge of his time.

BIBLIOGRAPHY

I. ORIGINAL WORKS. Pezenas's main publications are listed in J. M. Quérard, *La France littéraire*, VII (Paris, 1835), 111–112; Poggendorff, II, 422–423; and A. de Backer, A. de Backer, and C. S. Sommervogel, *Bibliothèque de la Compagnie de Jésus*, VI (Brussels–Paris, 1845), 647.

His principal works are *Éléments de pilotage* (Marseilles, 1733); *Pratique du pilotage* (Marseilles, 1741); *La théorie et la pratique du jaugeage* (Avignon, 1749); *Astronomie des marins . . .* (Avignon, 1766); and *Histoire critique de la découverte des longitudes* (Avignon, 1775). In addition, several of his papers are in *Mémoires de mathématique et de physique rédigés à l'Observatoire de Marseille*, 2 vols. (Avignon, 1755–1756).

Among the most important translations by Pezenas are C. Maclaurin's *Traité des fluxions*, 2 vols. (Paris, 1749), and *Éléments d'algèbre* (Paris, 1750); J. T. Desaguliers's *Cours de physique expérimentale*, 2 vols. (Paris, 1751); H. Baker's *Traité du microscope* (Paris, 1754); John Ward's *Guide des jeunes mathématiciens* (Paris, 1756); J. Harrison's *Principes de la montre . . .* (Avignon, 1767); R. Smith's *Cours complet d'optique*, 2 vols. (Avignon, 1767); and W. Gardiner's *Tables de logarithmes* (Avignon, 1770).

II. SECONDARY LITERATURE. Besides the bibliographical articles cited above, the principal studies of Pezenas's life and work are the following, listed chronologically:

J. Lalande, "Éloge du P. Pezenas," in *Journal des sçavans* (1779), 569–571, repr. in *Bibliographie astronomique* (Paris, 1803), see index; J. B. Delambre, in *Biographie universelle*, XXXIII (1823), 563–565, also in new ed., XXXII (1861), 663–664; P. Levot, in F. Hoefer, ed., *Nouvelle biographie générale*, XXXIX (1862), cols. 791–792; and B. Aoust, in *Mémoires de l'Académie des sciences, lettres et beaux-arts de Marseille*, ser. 2, **20** (1870), 1–16. Details concerning various aspects of Pezenas's work can be found in R. Taton, ed., *Enseignement et diffusion des sciences en France au XVIIIe siècle* (Paris, 1964), see index.

JULIETTE TATON

PFAFF, JOHANN FRIEDRICH (*b.* Stuttgart, Germany, 22 December 1765; *d.* Halle, Germany, 21 April 1825), *mathematics.*

Pfaff came from a distinguished family of Württemberg civil servants. His father, Burkhard Pfaff, was chief financial councillor and his mother was the only daughter of a member of the consistory and of the exchequer; Johann Friedrich was the second of their seven sons.

The sixth son, Christoph Heinrich (1773–1852), did work of considerable merit in chemistry, medicine, and pharmacy. He also investigated "animal electricity" with Volta, Humboldt, and others. Pfaff's youngest brother, Johann Wilhelm Andreas (1774–1835), distinguished himself in several areas of science, especially in mathematics, and became professor of mathematics at the universities of Würzburg and Erlangen; but the rapid changes in his scientific interests prevented him from attaining the importance of Johann Friedrich.

As the son of a family serving the government of Württemberg, Pfaff went to the Hohe Karlsschule in Stuttgart at the age of nine. The school, which was well-administered but subject to a harsh military discipline, served chiefly to train Württemberg's government officials and superior officers. Pfaff completed his legal studies there in the fall of 1785.

On the basis of mathematical knowledge that he acquired by himself, Pfaff soon progressed to reading Euler's *Introductio in analysin infinitorum*. In the fall of 1785, at the urging of Karl Eugen, the duke of Württemberg, he began a journey to increase his scientific knowledge. He remained at the University of Göttingen for about two years, studying mathematics with A. G. Kaestner and physics with G. C. Lichtenberg. In the summer of 1787 he traveled to Berlin, in order to improve his skill in practical astronomy with J. E. Bode. While in Berlin, on the recommendation of Lichtenberg, Pfaff was admitted to the circle of followers of the Enlightenment around Friedrich Nicolai. In the spring of 1788 he traveled to Vienna by way of Halle, Jena, Helmstedt, Gotha, Dresden, and Prague.

Through the recommendation of Lichtenberg, Pfaff was appointed full professor of mathematics at the University of Helmstedt as a replacement for Klügel, who had been called to Halle. Pfaff assumed the rather poorly paid post with the approval of the duke of Württemberg.

At first Pfaff directed all his attention to teaching, with evident success: the number of mathematics students grew considerably. Gauss, after completing his studies at Göttingen (1795–1798), attended Pfaff's lectures and, in 1798, lived in Pfaff's house. Pfaff recommended Gauss's doctoral dissertation and, when necessary, greatly assisted him; Gauss always retained a friendly memory of Pfaff both as a teacher and as a man.

While in Helmstedt, Pfaff aided students whose talents he recognized. For example, he was a supporter of Humboldt following his visit to Helmstedt and he recommended him to professors at Göttingen. During this period he also formed an enduring friendship with the historian G. G. Bredow. Their plan to edit all the fragments of Pappus of Alexandria progressed no further than a partial edition (Book 4 of the *Collectio*) done by Bredow alone.

In 1803 Pfaff married Caroline Brand, a maternal cousin. Their first son died young; the second, Carl, who edited a portion of his father's correspondence, became an historian, but his career was abbreviated by illness.

A serious threat to Pfaff's academic career emerged at the end of the eighteenth century, when plans were discussed for closing the University of Helmstedt. This economy measure was postponed—in no small degree as a result of Pfaff's interesting essay "Über die Vorteile, welche eine Universität einem Lande gewährt" (Häberlins *Staatsarchiv* [1796], no. 2)—but in 1810 the university was in the end closed. The faculty members were transferred to Göttingen, Halle, and Breslau. Pfaff went to Halle at his own request, again as professor of mathematics. After Klügel's death in 1812 he also took over the direction of the observatory there.

Pfaff's early work was strongly marked by Euler's influence. In his *Versuch einer neuen Summationsmethode . . .* (1788) he uncritically employed divergent series in his treatment of Fourier expansions. In editing Euler's posthumous writings (1792) and in the inaugural essay traditionally presented by new professors at Helmstedt—"programma inaugurale, in quo peculiaris differentialis investigandi ratio

ex theoria functionum deducitur" (1788)—as well as in 1795, Pfaff investigated series of the form

$$\sum_{k=1}^{n} \arctan \frac{f(k+1) - f(k)}{1 + f(k) \cdot f(k+1)}.$$

A friend of K. F. Hindenburg, the leader of the German combinatorial school, Pfaff prepared a series of articles between 1794 and 1800 for *Archiv der reinen und angewandten Mathematik* and *Sammlung combinatorisch-analytischer Abhandlungen*, which were edited by Hindenburg. The articles consistently reflect the long-winded way of thinking and expression of Hindenburg's school, with the single exception of "Analysis einer wichtigen Aufgabe des Herrn La Grange" (1794), which sought to free the Taylor expansion (with the remainder in Lagrange's form) from the tradition that embedded it in the theory of combinations and instead to present it as a primary component of analysis.

In 1797 Pfaff published at Helmstedt the first and only volume of an introductory treatise on analysis written in the spirit of Euler: *Disquisitiones analyticae maxime ad calculum integralem et doctrinam serierum pertinentes*. In 1810 he participated in the solution of a problem originating with Gauss that concerned the ellipse of greatest area that can be inscribed in a given quadrilateral. This led him to investigate conic pencils of rays.

Pfaff presented his most important mathematical achievement, the theory of Pfaffian forms, in "Methodus generalis, aequationes differentiarum partialium, necnon aequationes differentiales vulgares, utrasque primi ordinis, inter quotcunque variabiles, complete integrandi," which he submitted to the Berlin Academy on 11 May 1815. Although it was printed in the *Abhandlungen* of the Berlin Academy (1814–1815) and received an exceedingly favorable review by Gauss, the work did not become widely known. Its importance was not appreciated until 1827, when it appeared with a paper by Jacobi, "Über Pfaff's Methode, eine gewöhnliche lineare Differentialgleichung zwischen 2 *n* Variabeln durch ein System von *n* Gleichungen zu integrieren" (*Journal für die reine und angewandte Mathematik*, **2**, 347 ff.).

Pfaff's "Methodus" constituted the starting point of a basic theory of integration of partial differential equations which, through the work of Jacobi, Lie, and others, has developed into the modern Cartan calculus of extreme differential forms. (On this subject see, for example, C. Carathéodory, *Variationsrechnung und partielle Differentialgleichungen 1. Ordnung*, I [Leipzig, 1956].)

The core of the method that Pfaff made available

can be described as follows: In the title of the "Methodus" the expression "aequationes differentialis vulgares" appears; by this Pfaff meant equations of the form

$$\sum_{i=1}^{n} \varphi_i(x_1, x_2, \cdots, x_n)\, dx_i = 0,$$

the left side of which, in modern terminology, is a differential form in n variables (Pfaffian form). The equation itself is called a Pfaffian equation. Now, by means of a first-order partial differential equation in $n + 1$ variables,

$$F(x_1, x_2, \cdots, x_n; z; p_1, p_2, \cdots, p_n) = 0,$$

where the partial derivatives

$$p_i = \frac{\partial z}{\partial x_i}, \qquad i = 1, 2, \cdots, n,$$

one can easily transform the equation

$$dz - \sum_{i=1}^{n} p_i\, dx_i = 0$$

into a Pfaffian equation in $2n$ variables by eliminating dz.

The significance of the reduction of a partial differential equation to a Pfaffian equation had previously been recognized by Euler and Lagrange. The reduction could not be exploited, however, for lack of an integration theory of the Pfaffian forms which would be valid for all n; it was this deficiency that Pfaff's "Methodus" in large measure remedied. Gauss justifiably emphasized this aspect of Pfaff's work in his review in *Göttingische gelehrte Anzeigen* (1815).

Pfaff's theory is based on a transformation theorem that in current terminology, and going a little beyond Pfaff, can be stated in the following manner: A Pfaffian form $\sum_{i=1}^{n} \varphi_i\, dx_i$ with an even number of variables can be transformed, by means of a factor $\rho(x_1, x_2, \cdots, x_n)$ into a Pfaffian form of $n - 1$ variables. Moreover, for the case $n = 2$, ρ is simply the Euler multiplier or integrating factor of the differential equation $\varphi_1\, dx_1 + \varphi_2\, dx_2 = 0$. For $\sum_{i=1}^{4} \varphi_i\, dx_i$, therefore, there is a multiplier ρ, so that $\rho \sum_{i=1}^{4} \varphi_i\, dx_i$ can be written in the form $\sum_{i=1}^{3} \psi_i(y_1, y_2, y_3)\, dy_i$ and the y_i's are independent functions of $x_1, \cdots, x_4$.

For a Pfaffian form with an odd number of variables there is in general no corresponding multiplier that will enable one to reduce the number of variables.

In the 1827 article cited above, Jacobi later provided a suitable method of reduction: a Pfaffian form $\sum_{i=1}^{n} \varphi_i\, dx_i$ with an odd number of variables can,

through subtraction of a differential dw, which is always reducible by means of the transformation $x_i = f_i(y_1, y_2, \cdots, y_{n-1}, t)$ $i = 1, 2, \cdots, n$, be brought to the form $\sum_{i=1}^{n-1} \psi_i\, dy_i$, where the ψ_i's are functions of the y_i's.

Through alternately employing transformation (following Pfaff) and reduction (following Jacobi) one can finally bring every Pfaffian form with arbitrary number of variables into a canonical form: for $n = 2p$, into the form $z_1\, dz_2 + z_3\, dz_4 + \cdots + dz_{2p-1}\, dz_{2p}$; and for $n = 2p + 1$, into the form $dz_1 + dz_2, dz_3 + \cdots + z_{2p}\, dz_{2p+1}$.

Lie later gave the relationship between partial differential equations and Pfaffian forms a geometrical interpretation that possessed a greater intuitive clarity than the analytic approach.

BIBLIOGRAPHY

I. ORIGINAL WORKS. There is a list of Pfaff's writings in Poggendorff, II, cols. 424–425. They include *Versuch einer neuen Summationsmethode nebst anderen damit zusammenhängenden analytischen Bemerkungen* (Berlin, 1788); "Analysis einer wichtigen Aufgabe des Herrn La Grange," in Hindenburg's *Archiv der reinen und angewandten Mathematik*, **1** (1794), 81–84; *Disquisitiones analyticae maxime ad calculum integralem et doctrinam serierum pertinentes* (Helmstedt, 1797); "Methodus generalis, aequationes differentiarum partialium, necnon aequationes differentiales vulgares, utrasque primi ordinis, inter quotcunque variabiles complete integrandi," in *Abhandlungen der Preussischen Akademie der Wissenschaften* (1814–1815), 76–135, also translated into German by G. Kowalewski as no. 129 in Ostwald's Klassiker der exakten Wissenschaften (Leipzig, 1902); and *Sammlung von Briefen, gewechselt zwischen Johann Friedrich Pfaff . . .*, Carl Pfaff, ed. (Leipzig, 1853).

II. SECONDARY LITERATURE. See G. Kowalewski, *Grosse Mathematiker*, 2nd ed. (Munich–Berlin, 1939), 228–247; and Carl Pfaff's biographical introduction to his ed. of his father's correspondence, pp. 1–35. Also see articles on Pfaff in *Neuer Nekrolog der Deutschen*, **3** (1825), 1415–1418; and *Allgemeine deutsche Biographie*, XXV (Leipzig, 1887), 592–593.

H. WUSSING

PFEFFER, WILHELM FRIEDRICH PHILIPP (*b.* Grebenstein, near Kassel, Germany, 9 March 1845; *d.* Leipzig, Germany, 31 January 1920), *botany, chemistry.*

Pfeffer was the son of Wilhelm Pfeffer, an apothecary in Grebenstein, and Luise Theobald, whose family was associated with the clergy. His parents intended him to be the third successor to the apothecary shop founded by his great-grandfather; and even after several years of university study, Pfeffer still planned to enter the profession of pharmacy. He thus made sure of a strong background in chemistry, physics, and botany; and his theoretical studies and practical experience in preparation for such a career provided him with broad insight as well as techniques that would become useful in fine experimentation when he turned his full attention to botany, and especially to plant physiology.

Pfeffer attended the grammar school in Grebenstein until he was twelve; then, after additional private instruction, he entered the electoral Gymnasium near Kassel. Three years later (before he had completed his Gymnasium studies) his father took him into the apothecary shop as an apprentice. Pfeffer's father, a man of broad scientific interests, had a large herbarium and had gathered extensive collections in many fields of the natural sciences; he also wrote a textbook (unpublished) of pharmaceutical chemistry and corresponded with scientists in Germany and abroad. He supplemented his son's education in many ways and imbued him with his own enthusiasm for the study of nature. Pfeffer received an early introduction to botany by accompanying his father on expeditions into the nearby countryside, and at the age of six he was pressing flowers and collecting specimens for several of his own collections. The scope of Pfeffer's explorations was considerably widened when, at the age of twelve, he began to make excursions into the high Alps with his uncle Gottfried Theobald, whose geological and botanical trips kindled the boy's enthusiasms still further while also developing his ability as an alpinist. Pfeffer was fearless about searching in difficult locations for certain mosses and rare plant specimens, and he became one of the earliest to climb the Matterhorn. After his marriage (1884) to Henrika Volk, Pfeffer avoided such hazardous mountaineering, although visiting the Alps was always his favorite vacation.

As an apprentice to his father, Pfeffer prepared plants and herbs and ground ingredients, made up chemicals and medicinal preparations, and was responsible for various analyses. He also maintained the shop, which dispensed not only pharmaceutical chemicals but sold homemade candies, cleansers, and even shoe polish prepared on the premises. Pfeffer also used his father's microscope to examine the fine structure of various specimens of seeds, fibers, and starches; these observations were aided by texts on the microscope and Mohl's *Grundzüge der Anatomie und Physiologie der vegetabilischen Zelle*. After passing the examination for apothecary's assistant, Pfeffer entered the University of Göttingen to study chemistry and prepare for a career in pharmacy.

At Göttingen Pfeffer attended lectures in physics, chemistry, zoology, and botany; and although his atypical preparation for the university had left deficiencies which he now strove to remedy, he was able to begin a dissertation in chemistry shortly after his matriculation at the age of eighteen. He submitted this work, "Über einige Derivative des Glyzerins und dessen Überführung in Allylen," and received his doctorate in chemistry and botany in early February 1865, having spent barely four semesters at Göttingen.

In the summer of the same year, at Marburg, Pfeffer continued his pharmaceutical studies and for the time being set aside consideration of an academic career, thus following his father's wishes. He then received an assistantship at an apothecary in Augsburg and later in Chur, where his uncle still taught in the canton school. Pfeffer resumed alpine climbing with his uncle and gathered the mosses that were to be the subjects of his first papers. He subsequently returned to the University of Marburg and in December 1868 passed the examination qualifying him for the profession of apothecary, as his father had desired. He now became more certain, however, of his preference for an academic career in botany. Encouraged by the botanist Albert Wigand, Pfeffer studied the development of blossoms; he then went to Berlin at the end of the next summer and obtained a much sought-after place in Pringsheim's private laboratory, where the activity centered about investigations of the developmental history of plants. Here Pfeffer began his studies of the germination of *Selaginella*. He continued this work in Würzburg under the plant physiologist Julius von Sachs, who encouraged Pfeffer to direct his researches to problems in plant physiology. He thus studied the effects of light from different parts of the spectrum on the decomposition of carbon dioxide in plants and analyzed some of the effects of external stimuli on the growth of plants. The chemistry and physics he had previously studied were already of great help to Pfeffer, and he presented some of this work for his habilitation at Marburg.

In March 1871 Pfeffer was appointed *Privatdozent* at Marburg, where he investigated protein metabolism in plants and was especially concerned with the formation and diffusion of asparagine. He also began extensive researches, which continued for many years, on irritability in plants, studying movements due to irritability in the "sensitive plant," *Mimosa pudica*, and the staminal filaments of Cynareae. From his observations of irritability phenomena and his understanding of their broad implications, Pfeffer undertook his classic investigations of osmosis, with a view toward explaining the basic causes and mechanisms of such manifestations. Pfeffer's osmotic investigations were first pursued at Marburg and continued at the University of Bonn, where, in 1873, he was appointed professor extraordinarius of pharmacy and botany. He made the first direct measurements of osmotic pressures in plants during these studies, which, published in his *Osmotische Untersuchungen* (Leipzig, 1877), were to provide van't Hoff with the values for his calculations. Thus Pfeffer's work was invaluable in the development of the theory of solution, itself a landmark in the history of physical chemistry.

At Bonn, where he was also custodian of the Botanical Institute, Pfeffer's various researches included studies of the periodic movements of leaf organs. In 1877 he was appointed professor of botany at the University of Basel, but in the fall of 1878 he accepted a post at the University of Tübingen. During nine years here he investigated plant irritability and respiration. His work on chemotaxis, which evolved from this period, demonstrated the attraction of certain specific substances for the small, free-swimming organisms he was studying: it appeared that malic acid drew the spermatozoids of ferns and *Selaginella* to the archegonium, for they were as readily drawn to the capillary tubes that Pfeffer filled with this solution. In certain mosses Pfeffer discovered that a similar stimulus was exerted by a cane-sugar solution. He also found chemotropism in bacteria, flagellates, and various other organisms.

In 1881 Pfeffer published his comprehensive *Pflanzenphysiologie*, which became a well-known reference work. It presented not only a mine of information but also a view of Pfeffer's aims and philosophy as a physiologist studying fundamental life processes with plants as his subjects. His investigations probed all implications indefatigably, and he was ingenious at devising apparatus for acute measurements and various laboratory experiments. Pfeffer followed specific phenomena in his search for ultimate causes. He was convinced that changes in energy underlay the processes of plant life and that the phenomena of life, if penetrated, would be understood "as the natural consequences of given conditions." Yet he stressed the complexities of life processes: Within each cell there were constellations of still finer organizations and chains of interrelated reactions, and within each organ there were intricately correlated relationships. In the living organism there was a cooperation and self-regulation of the whole; following disturbances, this relationship tended to restore the previous equilibrium or establish a new state.

Students from both Germany and abroad entered Pfeffer's botanical laboratory at Tübingen, where they found a unique combination in his personal stimulus to their researches and the chance to work with a teacher whose knowledge of physics and chemistry had enabled him to pursue a range of microchemical and other investigations and to construct fine physical apparatus. Pfeffer was unmatched in setting up instrumentation to measure plant movements, growth, and osmotic pressure, and to pursue other observations toward the solution of complex problems of plant physiology. He had a penchant for theorizing, for exploring the range of possibilities, and for exactitude. His students' papers and some of his own appeared in the *Untersuchungen aus dem botanischen Institut zu Tübingen* (1881–1888).

In 1887 Pfeffer was appointed professor of botany at the University of Leipzig, where he was again active in teaching and busy with administrative duties and was the director of the Botanical Institute. His responsibilities were increased in 1895 when, following the death of Pringsheim, he became coeditor with Eduard Strasburger of the *Jahrbuch für wissenschaftliche Botanik*.

Pfeffer was a member of many leading societies, both German and foreign; and his honors included degrees from the universities of Halle, Königsberg, and Oslo. Cambridge University awarded him the honorary degree of doctor in science in 1898.

At Leipzig the number of Pfeffer's students grew; they came from many countries, and many were from the United States. A stimulating teacher in the laboratory, he enlivened his lectures with demonstrations and made use of projection apparatus. In a container for the purpose, students placed slips with questions he would answer at botanical club meetings, and surreptitiously Pfeffer added challenging questions of his own. In 1915, at the age of seventy, a *Festschrift* of the *Jahrbücher für wissenschaftliche Botanik* was dedicated to Pfeffer. (Among his former students named in the volume were Carl Correns and Wilhelm Johannsen.) In the same year a special issue of *Naturwissenschaften* commemorated Pfeffer's scientific contributions.

The last years of Pfeffer's life, however, were deeply troubled and unhappy: He had become increasingly affected by feelings of depression; the brutality of the war haunted him, and he was apprehensive about the political and social changes he witnessed in Germany. His only son was killed less than two months before the armistice.

Many of Pfeffer's investigations were basic to plant physiology. His work on the role of asparagine was controversial but nevertheless an advance in the study of plant metabolism when viewed in a historical context. He made extensive contributions to the study of irritability in plants by investigating the movements of leaves, the opening and closing of flowers, the influences of variations of light and of temperature, the effects of tactile stimuli, and the physiology of transmission in irritability phenomena. He also studied the sleep movements of plants. Using aniline dyes, he pioneered the method of vital staining and followed the assimilation and accumulation of various dyes within the living cell.

Abbé Nollet, in 1748, was the first to explain osmosis; and as Pfeffer noted, the phenomenon had been "rediscovered" several times. Moritz Traube had even constructed semipermeable precipitation membranes and used them in studying the cell. While searching for the causes of the extremely high osmotic pressures he had observed in plants, Pfeffer improved upon Traube's membranes. After trying methods that similarly produced membranes that were too fragile and burst under pressure, Pfeffer devised his *Pfeffer-Zelle*, his "pepper pots." Using unglazed, porous porcelain cells, he precipitated membranes of copper ferrocyanide within them. Tightly supported by the walls of the cells, they withstood increased pressures; and Pfeffer was able to make direct measurements of solutions of various substances at different concentrations and temperatures. Pfeffer considered this cell a model of the plant cell, or in his terms, the protoplast with its surrounding membranes. His results showed proportionate relationships between the concentrations of the solutions in the cells and the osmotic pressures, and temperature likewise proved to be directly related to osmotic pressure.

Pfeffer, then teaching at Bonn, communicated his findings to the physicist Clausius. The values for the pressures within a plant cell seemed inordinately high at first, and Pfeffer later recalled that Clausius had thought that such pressures must be impossible and was convinced of their accuracy only when Pfeffer gave him further proof. Pfeffer acknowledged, however, that Clausius did not examine the question closely at the time.

Some time afterward, at Amsterdam, the botanist De Vries, who was engaged in researches related to osmotic pressures, met van't Hoff and told him of the values for osmotic pressures that Pfeffer had published in *Osmotische Untersuchungen*. Van't Hoff then referred to Pfeffer's work and drew his broad analogies between osmotic pressures and gas pressures. Thus Pfeffer's determinations were the values on which van't Hoff based his theoretical considerations. In his classic paper of 1887 van't Hoff outlined the experimental method that Pfeffer had used in obtaining his determinations. Pfeffer's technical achievement in making these measurements proved difficult to match.

Pfeffer suggested in *Osmotische Untersuchungen* that the "threads" he had permitted to fall might be a point of departure for the physicist; later he described his work to Clausius. He recalled, "I repeatedly expressed in conversations with him that some kind of connection had to exist between the osmotic effect on one hand and the size and number of the molecules on the other hand" (E. Cohen, p. 119). Pfeffer was therefore aware of the wider implications of his measurements, even though he never pursued them. Nevertheless, the development of the theory of solution has assured Pfeffer an enduring role in the history of physical chemistry. He was one of the triumvirate who guided German botany in their time; the obituary in *Nature* stated, "With his death the three outstanding figures of the older German botany —Sachs, Strasburger, and Pfeffer—have all passed away."

BIBLIOGRAPHY

I. ORIGINAL WORKS. Pfeffer's major works are *Osmotische Untersuchungen. Studien zur Zellmechanik* (Leipzig, 1877) and *Pflanzenphysiologie. Ein Handbuch des Stoffwechsels und Kraftwechsels in der Pflanze*, 2 vols. (Leipzig, 1881; 2nd ed., 1897–1904), with English trans. by A. J. Ewart as *The Physiology of Plants, a Treatise Upon the Metabolism and Sources of Energy in Plants*, 3 vols. (Oxford, 1900–1906). A complete bibliography of 100 titles is given in Fitting (see below).

II. SECONDARY LITERATURE. A biographical sketch by G. Haberlandt, "Wilhelm Pfeffer," in *Naturwissenschaften*, **3** (1915), 115–118, is followed by contributions describing various aspects of Pfeffer's work: Ernst Cohen, "Wilhelm Pfeffer und die physikalische Chemie," pp. 118–120; Friedrich Czapek, "Die Bedeutung von W. Pfeffer's physicalischen Forschungen für die Pflanzenphysiologie," pp. 120–124; L. Jost, "Die Bedeutung Wilhelm Pfeffers für die pflanzenphysiologische Technik und Methodik," pp. 129–131; H. Kniep, "Wilhelm Pfeffer's Bedeutung für die Reizphysiologie," pp. 124–129.

See also Albert Charles Chibnall, *Protein Metabolism in Plants* (New Haven–London–Oxford, 1939), 6–34, 37, 42–45, 65, 122, 170–171, for Pfeffer's views on asparagine and his work on protein metabolism in plants; and Harry Clay Jones, trans. and ed., *The Modern Theory of Solution. Memoirs by Pfeffer, van't Hoff, Arrhenius and Raoult* (New York–London, 1899), pp. v–vi, 3–10, 14–16, with excerpts from *Osmotische Untersuchungen* and from van't Hoff's paper and its reference to Pfeffer's determinations of osmotic pressure.

For biographical material see Frank M. Andrews, "Wilhelm Pfeffer," in *Plant Physiology*, **4** (1929), 285–288; Hans Fitting, "Wilhelm Friedrich Philipp Pfeffer," in

Deutsches Biographisches Jahrbuch, II (1917–1920), 578–582, 750–751, and "Wilhelm Pfeffer," in *Berichte der Deutschen botanischen Gesellschaft*, **38** (1920), 30–63, with a comprehensive bibliography; Wilhelm Ostwald, "Wilhelm Pfeffer," in *Chemiker-Zeitung*, **44** (1920), 145; G. J. P. [George James Peirce], "Wilhelm Pfeffer," in *Science*, n.s. **51** (1920), 291–292; Hans and Ernst G. Pringsheim, "Wilhelm Pfeffer," in *Berichte der Deutschen chemischen Gesellschaft*, **53A** (1920), 36–39; V. H. R., "Prof. Wilhelm Pfeffer, For. Mem. R. S.," in *Nature*, **105** (1920), 302; and Wilhelm Ruhland, "Wilhelm Pfeffer," in *Berichte über die Verhandlungen der Sächsischen Akademie der Wissenschaften zu Leipzig. Math.–Phys. Kl.*, **75** (1923), 107–124.

GLORIA ROBINSON

PFEIFFER, PAUL (*b.* Elberfeld [now Wuppertal], Germany, 21 April 1875; *d.* Bonn, Germany, 4 March 1951), *chemistry*.

Pfeiffer was the son of Hermann Pfeiffer, a head clerk and later factory owner, and Emilie Willmund. He studied for two semesters at the University of Bonn under Kekulé and Anschütz before entering the University of Zurich (1894), where he became Werner's best-known student, protégé, and eventually "chief of staff." After receiving his doctorate in 1898, for a paper "Molekülverbindungen der Halogenide des 4-wertigen Zinns und der Zinnalkyle" (published with Werner as coauthor in *Zeitschrift für anorganische Chemie*, **17** [1898], 82–110), he studied for one semester each with Ostwald at Leipzig and Hantzsch at Würzburg. On 14 August 1901 he married his cousin Julie Hüttenhoff. In the same year, with the acceptance of his *Habilitationsschrift*, "Beitrag zur Chemie der Molekülverbindungen," he became *Privatdozent* at the University of Zurich and in 1908 associate professor of theoretical chemistry. In 1916, as a result of personal and political conflicts with Werner, Pfeiffer left Zurich for the University of Rostock, even though Werner was ill at the time and Pfeiffer was certain to be appointed his successor. In 1919 Pfeiffer moved to the Technische Hochschule in Karlsruhe. Three years later he was appointed to the directorship of the chemical institute at the University of Bonn, Kekulé's old chair, where he remained until his retirement in 1947.

Pfeiffer's work encompassed both inorganic and organic chemistry as well as the borderland between these disciplines. As the intellectual—but not academic—successor to Werner, Pfeiffer's main interest was in coordination compounds, particularly those of chromium. He investigated their constitution, configuration, isomerism, acid-base and hydrolysis reactions, and their relationships to double salts and salt hydrates. He was the first to apply Werner's coordina-

tion theory to crystals. He also studied both inorganic and organic tin compounds, inner complexes, metal organic compounds, and the chemistry of dyes. He was a pioneer in the field of halochromism—the formation of colored substances from colorless organic bases by the addition of acids or solvents. His contributions to pure organic chemistry include studies of cyclic compounds, quinhydrones, stilbene compounds, unsaturated acids, and the relationship of ethylene compounds to ethane and acetylene compounds.

BIBLIOGRAPHY

I. ORIGINAL WORKS. Most of Pfeiffer's work appeared in *Berichte der Deutschen chemischen Gesellschaft* and *Zeitschrift für anorganische Chemie*. His monograph *Organische Molekülverbindungen* (Stuttgart, 1922) was reprinted in 1927. A detailed bibliography is given in Poggendorff, VIIA, 552–553.

II. SECONDARY LITERATURE. A summary of Pfeiffer's early career appeared in the *Festschrift* compiled for the opening of the chemical institute of the University of Zurich, *75 Jahre chemischer Forschung an der Universität Zürich* (Zurich, 1909). There are biographical data and evaluations of Pfeiffer's work by his former student R. Wizinger, "P. Pfeiffers Beitrag zur Entwicklung der Komplexchemie," in *Angewandte Chemie*, **62A** (1950), 201–205; and "In memoriam Paul Pfeiffer, 1875–1951," in *Helvetica chimica acta*, **36** (1953), 2032–2037. An unpublished autobiography, "Mein Lebenslauf" (1947), is available at the University of Bonn. *Angewandte Chemie*, **62**, nos. 9–10 (20 May 1950), was devoted to Pfeiffer on the occasion of his seventy-fifth birthday. Biographical data can be found in G. B. Kauffman, "Crystals as Molecular Compounds: Paul Pfeiffer's Application of Coordination Theory to Crystallography," in *Journal of Chemical Education*, **47** (1970), 277–278; and R. E. Oesper, "Paul Pfeiffer," *ibid.*, **28** (1951), 62.

GEORGE B. KAUFFMAN

PFLÜGER, EDUARD FRIEDRICH WILHELM (*b.* Hanau, Germany, 7 June 1829; *d.* Bonn, Germany, 16 March 1910), *physiology*.

Pflüger's father, Johann Georg Pflüger, began his career as a businessman and commercial traveler. Later he became a passionate politician and leader of the democrats in Hanau. He was a combative person and repeatedly came into conflict with government agencies; in the period 1841–1843 he became involved in a high treason trial. In 1848 he was active in the Frankfurt *Vorparlament*. He studied law in his later years in order to participate in political proceedings, and in 1848 he founded his own political journal. His wife, Charlotte Wilhemine Richter, died in 1855. On his mother's side Pflüger was descended from

Huguenots who had immigrated to Hanau from Dauphiné. He married Christine Marc of Wiesbaden in 1869. They had three daughters, Anna (who later married the chemist Richard Anschütz), Rosa, and Hildegard.

Pflüger, who usually signed his name simply as Eduard, spent his childhood years in Hanau and attended the Gymnasium there. His youth was overshadowed by political developments. He himself became a passionate democrat in 1848, but only for a short time. When he was arrested with rebellious Heidelberg students in 1849, he abandoned politics and law, which he had been studying at Heidelberg since the summer term of 1849, and took up the study of medicine.

In the summer term of 1850 he became a medical student at Marburg, and in 1851 he continued his medical studies at Berlin. There he became an admirer and student of Johannes Müller. In his student years he worked in Müller's laboratory and attended lectures given by Müller and Emil du Bois-Reymond. He once witnessed a demonstration of the inhibitory effect of the vagus nerve on the heart in Müller's laboratory. As a result he had the auspicious idea of seeking similar inhibitory effects on the intestine, which he discovered in the rabbit. On the basis of this work he received the M.D. under Müller in 1855. He had already earned such a degree in 1851 at Giessen for a dissertation, dedicated to Müller, dealing with the reflex and psychical capacities of the spinal cord of the frog. This work, written at the age of twenty-two, was his first scientific publication. Pflüger studied this question several more times during his career.

Pflüger's work on electrotonus qualified him as a university lecturer under du Bois-Reymond at the end of 1858, a few months after Müller's death. Pflüger worked mostly at his own residence, although under the supervision of du Bois-Reymond, the founder of scientific electrophysiology. Electrotonus was, methodologically considered, a very difficult topic. Pflüger was able to determine the basic laws of the changes in sensitivity that take place in a section of nerve subjected to a direct current from a cathode and from an anode which, due to polarization, spreads "extrapolarly." The laws proved to be dependent on the polarity, the direction, and the strength of the direct current. This finding, known as Pflüger's law of convulsion, was required learning for German medical students until the middle of the twentieth century. The principles of the diagnostic and therapeutic applications of the galvanic current in medicine are based on it.

Pflüger's early researches revealed his exceptional sagacity, perseverance, and experimental exactness.

Because of these qualities, on 28 February 1859 he received the chair of physiology at Bonn, succeeding Helmholtz. Helmholtz had recommended him to the Bonn faculty by stating: "Concerning physiology, among the younger pure physiologists, Pflüger in Berlin appears to me to be the most talented and promising." In his new post Pflüger was given little in the way of space or resources. Helmholtz had been responsible for both anatomy and physiology, but Pflüger assumed only the physiological duties. He used not the old anatomy building in the Hofgarten (built in 1824) but the so-called pavilion, a cramped building in the northeast corner of the university, where a provisional laboratory was constructed for his use. The initially modest facilities caused Pflüger to direct his research to histology. He studied the embryonal development of the ovary (1861–1863), the nerve endings in the salivary glands (1866), and the gas exchange in the blood and in the cells. He developed a gas evacuation pump for this purpose (1864–1865).

In his investigation of the ovary Pflüger also studied the question of whether the egg cells originate by division or "in free cell formation." Pflüger saw the embryonal ovary built up from hollow tubular sacs ("Pflüger's Tubes"), in the lumina of which were closely packed, bright vesicles, which were probably cells. In these sacs he thought the oogonia could be seen. The endings of secretory nerves, he believed, entered directly into the secreting cells.

Pflüger's studies of gas exchange continued for many years and encountered extraordinarily great methodological difficulties. Before Pflüger's time it had not been established whether the oxidizing processes of combustion take place in the lungs (Lavoisier), in the blood (Müller), or in the cells. He succeeded in demonstrating that cellular activity, or the energy requirement of the cell, determines the magnitude of the oxygen consumption; the blood respires comparatively little. In related work, Pflüger's experiment on the "table-salt frog" became famous: a frog in which the blood was replaced by a physiological salt solution displayed no significant decrease in gas exchange.

The concept of the "respiratory quotient" was also developed by Pflüger. He conducted new studies to show that the exchange of gas in the lungs and tissues results exclusively from the fall in partial pressure of the gases and cannot be considered as a secretion. That the respiratory action is stimulated by a surplus of carbon dioxide and a lack of oxygen was another definitive result of his work. His famous work, "Über die physiologische Verbrennung in den lebenden Organismen" (Pflüger's *Archiv*, **10** [1875], 251–367), consolidated these studies, which subsequently led Pflüger to consider the maintenance of heat in cold surroundings. He found that cold was the most effective stimulus to the increase of metabolism. He localized this process—through the application of curare—in the musculature. This finding also remains valid today.

Another field that claimed Pflüger's attention for several decades was the metabolism of the nutritive substances: protein, fat, and carbohydrates. He published more than sixty works on glycogen alone. Yet, despite the greatest exactitude in his methods, his determination of nitrogen-bearing substances had no unqualified success. He erroneously saw in protein alone the "source of the muscular force," and he considered protein in general to be the only nourishing substance. In fact he held this position on the basis of the hypothesis that the secret of living matter lies in the molecular structure of protein. His finding that the quantity of protein ingested determines the amount of decomposition of protein in the body has proved to be correct. Thus he did not believe that glycogen could be produced from protein. He became involved in violent controversies over this matter, especially with Voit's school in Munich. Shortly before his death, however, he became convinced of his error.

Pflüger expended much time, energy, and inventiveness on the improvement of methods for determining glycogen. In opposition to most other researchers, he believed that pancreatic diabetes was a nervous disturbance. In a letter (1896) to his daughter we find the complaint: "Glycogen is still the dream of my days and nights. It's dreadful." The result of this work was a large book, *Das Glykogen und seine Beziehungen zur Zuckerkrankheit* (2nd ed., 1905). Containing over five hundred pages, it included a review of the entire world literature on this subject, which was cited and utilized precisely. He died of a liver carcinoma at the age of eighty-one.

Pflüger was only of medium stature but powerfully built. His large beard gave him a commanding appearance. He was humane, rather reticent, and even shy when away from the Institute or his family. Pflüger attended few congresses, but he exercised a strong influence on his contemporaries while serving as editor of *Archiv für die gesamte Physiologie des Menschen und der Tiere* from 1868. Few people went as far as Pflüger in public criticism of the works of others, and he was the most feared critic of his time. His remarks could even degenerate into personal abuse; and when it was a matter of eliminating errors, he knew no restraint. He demanded the highest standards of diligence, exactness, and conclusiveness in his demonstrations. His polemic against Voit, Hermann Munk, Hermann Senator, and others was caustic beyond all measure. "Criticism is the most important motive of progress;

for that reason I practice it," he stated (see *Archiv*, **222** [1929], 561). His lectures were original and stimulating but not easy.

Pflüger loved his family above all else. He enjoyed taking long hikes with his wife and daughters and was exceedingly concerned about his health and theirs. He was a Christian but did not belong to any church. He claimed countless animals for experimental purposes —156 dogs in the last year of his life alone. Despite his respect for living creatures, the desire for knowledge took precedence. His wife once sighingly remarked that she had constantly had a very powerful rival, science.

As much as Pflüger regarded hypotheses as necessary in his discipline and sometimes even made far-reaching use of them, to that same degree he confined his work to physiology and to the questions of the life sciences that are open to experiment. He viewed the living being as a great unity, full of both teleology and mechanism (*Archiv*, **15** [1877], 57–103). He reserved his greatest reverence for Aristotle, whose likeness hung over his desk. Pflüger attributed the unifying forces in the organism to the nervous system. He thus rejected the neuron theory and upheld the "continuity theory" of the nervous system. In his article "Über den elementaren Bau des Nervensystems" (*Archiv*, **112** [1906], 1–40) he expressed himself emphatically on this point. His articles constantly demonstrated an intensive study of the literature and even a careful study of original sources already historical in his day. For this reason his works are frequently useful reading even today.

Pflüger was strongly opinionated. He called himself a student of Müller, but not of du Bois-Reymond, with whom he had worked the longest. His relations with du Bois-Reymond were not always untroubled. He seems to have had no friend among the physiologists, and his only lifelong friend was the botanist N. Pringsheim.

Pflüger was a member of the Leopoldina and honorary member of many foreign academies; he also was awarded the order *Pour le mérite*. He was rector at Bonn in the academic year 1889–1890.

BIBLIOGRAPHY

I. ORIGINAL WORKS. Pflüger's works include "De functionibus medullae oblongatae et spinalis psychicis" (M.D. diss., Giessen University, 1851); *Die sensorischen Funktionen des Rückenmarks der Wirbeltiere nebst einer neuen Lehre von den Leitungsgesetzen der Reflexionen* (Berlin, 1853); "De nervorum planchnicorum functione" (M.D. diss., University of Berlin, 1855); *Das Hemmungsnervensystem für die peristaltische Bewegung der Gedärme* (Berlin, 1857); *Untersuchungen über die Physiologie des Elektrotonus* (Berlin, 1859); *Über die Eierstöcke der*

Säugethiere und des Menschen (Leipzig, 1863); *Über die Kohlensäure des Blutes* (Bonn, 1864); "Beschreibung meiner Gaspumpe," in *Untersuchung aus dem physiologischen Laboratorium zu Bonn* (Berlin, 1865), 183–188; *Die Endigungen der Absonderungsnerven in den Speicheldrüsen* (Bonn, 1866); "Uber die physiologische Verbrennung in den lebenden Organismen," in *Pflügers Archiv für die gesamte Physiologie des Menschen und der Tiere*, **10** (1875), 251–367.

Subsequent writings are "Die teleologische Mechanik der lebendigen Natur," in *Pflügers Archiv für die gesamte Physiologie des Menschen und der Tiere*, **15** (1877), 57–103; "Die Physiologie und ihre Zukunft," *ibid.*, 361–365; "Wesen und Aufgaben der Physiologie," *ibid.*, **18** (1878), 427–442, the inaugural address at the Physiological Institute at Bonn–Poppelsdorf, 9 Nov. 1878; *Die allgemeinen Lebenserscheinungen* (Bonn, 1889), his rectorial address at Bonn; *Über die Kunst der Verlängerung des menschlichen Lebens* (Bonn, 1890), oration delivered on the Kaiser's birthday; "Das Glykogen und seine Beziehungen zur Zuckerkrankheit," in *Pflügers Archiv für die gesamte Physiologie des Menschen und der Tiere*, **96** (1903), 1–398, published as a book (Bonn, 1905); and "Über den elementaren Bau des Nervensystems," in *Pflügers Archiv für die gesamte Physiologie des Menschen und der Tiere*, **112** (1906), 1–40.

II. SECONDARY LITERATURE. On Pflüger and his work see A. Bethe, "E. Pflüger als Begründer dieses Archives," in *Pflügers Archiv für die gesamte Physiologie des Menschen und der Tiere*, **222** (1929), 569–572; W. Bleibtreu, "Pflügers Persönlichkeit," *ibid.*, 562–568; W. Haberling and S. Pagel, "Pflüger, Ed. Fr. W.," in *Biographisches Lexikon der hervorragenden Ärzte aller Zeiten und Völker*, 2nd ed., IV (Berlin–Vienna, 1932), 586–587; E. Heischkel, "Eduard Pflüger (1829–1910). Physiologe," in *Lebensbilder aus Kurhessen und Waldeck, 1830–1930*, I. Schnack, ed., IV (Marburg, 1950), 253–263—*Lebensbilder* . . . is no. 20 of Veröffentlichungen der Historischen Kommission für Hessen und Waldeck; M. Nussbaum, *E. F. W. Pflüger als Naturforscher* (Bonn, 1909), with partial bibliography; R. Rosemann, "Pflügers Lebenswerk," in *Pflügers Archiv für die gesamte Physiologie des Menschen und der Tiere*, **222** (1929), 548–562; K. E. Rothschuh, *Entwicklungsgeschichte physiologischer Entdeckungen in Tabellenform* (Munich, 1952), nos. 137, 140, 221, 225, 227, 250, 603, 640, 677, 680, 721, 724, 748, 782, 941, 1000, 1032, 1317, 1392; and *Geschichte der Physiologie* (Berlin, 1953), esp. 133–134; F. Runkel, "Eduard Pflügers Vorfahren und Jugendzeit," in *Pflügers Archiv für die gesamte Physiologie des Menschen und der Tiere*, **222** (1929), 572–574; and K. Schmiz, *Die medizinische Fakultät der Universität Bonn 1818–1918. Ein Beitrag zur Geschichte der Medizin* (Bonn, 1920), 26–29. Obituaries are by René du Bois-Reymond in *Berliner klinische Wochenschrift*, **47** (1910), 658–659; H. Boruttau in *Deutsche medizinische Wochenschrift*, **36** (1910), 851–852; E. von Cyon in *Pflügers Archiv für die gesamte Physiologie des Menschen und der Tiere*, **132** (1910), 1–19; J. A. F. Dastre in *Comptes rendus des séances de la Société de biologie*, **68** (1910), 648–650; H. Leo in *Münchener medizinische Wochenschrift*, **57** (1910), 1128–

1129; F. Schenck in *Naturwissenschaftliche Rundschau* (1910), 340; and A. D. Waller in *Nature*, **83** (1910), 314.

Additional material may be found in the archives of the Institut für Geschichte der Medizin, Münster University.

K. E. ROTHSCHUH

PHILINUS OF COS (*b.* Cos; *fl. ca.* 250 B.C.), *medicine.*

Philinus of Cos, about whom very little information has survived (we do not know the titles of any of his works), played an interesting role in the beginnings of the empirical school of medicine and of medical skepticism. As a native of Cos he certainly was acquainted with the Hippocratic tradition. He studied under Herophilus, who originally belonged to the Coan school of physicians. It is not known whether, like his teacher, Philinus spent time in Alexandria; but it is known that by severing relations with Herophilus he helped to establish the empirical school of medicine (see Deichgräber, fr. 6), which had close contacts with the philosophy of skepticism. Since Herophilus also had such contacts (see F. Kudlien, "Herophilos und der Beginn der medizinischen Skepsis," in *Gesnerus*, **21** [1964], 1–13), the difference between him and his pupil may not be evident. The answer is that Philinus was fundamentally more rigorous than his teacher. He evidently transformed the latter's etiological skepticism (the chief causes of disease are not known) into an etiological nihilism (it is impossible to know the causes). This transformation had consequences in medical diagnostics: in opposition to his teacher, Philinus denied the utility of reading the pulse and "thereby shut medicine's diagnostic eyes" (see Deichgräber, fr. 77). On this point at least, Philinus' medical ideas emerge more clearly than H. Diller supposed; but beyond it nothing can be affirmed concerning Philinus.

BIBLIOGRAPHY

Accounts of Philinus are collected in K. Deichgräber, *Die griechische Empirikerschule*, 2nd enlarged ed. (Berlin–Zurich, 1965). See also H. Diller's short article "Philinos no. 9," in Pauly-Wissowa, *Real-Encyclopädie der classischen Altertumswissenschaft*, XIX (Stuttgart, 1938), cols. 2193–2194.

FRIDOLF KUDLIEN

PHILIP, ALEXANDER PHILIPS WILSON (*b.* Shieldhall, Scotland, 15 October 1770; *d.* Boulogne, France, 1851 [?]), *medicine, physiology.*

Wilson Philip was christened Alexander Philips Wilson but in 1811 changed his name to Alexander

Philips Wilson Philip; his writings that were published before 1807 bear the name of A. P. Wilson. He received his early education in Edinburgh and studied medicine at the medical school of the University of Edinburgh, under William Gregory, Alexander Monro (Secundus), Joseph Black, and William Cullen; he graduated M.D. on 25 June 1792 with a thesis entitled "De dyspepsia." After studying in London he returned to Edinburgh and was admitted a fellow of the Royal College of Physicians of Edinburgh on 3 February 1795. Probably for reasons of health, he left Edinburgh. In 1798 he was appointed physician to the Winchester County Hospital and in 1802 to the Worcester General Infirmary. On account of friction with local colleagues, he resigned the latter position in 1818 and in 1820 went to London, where he soon became a leading physician.

Wilson Philip was made a licentiate of the Royal College of Physicians of London on 22 December 1820, a fellow on 25 June 1834, and on 11 May 1826 a fellow of the Royal Society. Meantime he built up a large and lucrative practice, especially among the aristocracy who lived near his large home in fashionable Cavendish Square. About 1842 Wilson Philip retired from practice, and in 1843 or 1844 he moved to Boulogne (it has been said, to avoid imprisonment for insolvency). Nothing is known of this episode, and although it appears that he died in France, the precise date of his death is unknown. It has been suggested that W. M. Thackeray based his character Dr. Brand Firmin in *The Adventures of Philip* on Wilson Philip, but McMenemy thinks this unlikely.

Wilson Philip was a man of great energy and diligence, and of kindliness toward his patients. Although he himself possessed a critical outlook, he was temperamentally unable to tolerate criticism from others. He was frequently involved in vitriolic polemics and violent arguments concerning his clinical and experimental work, but he eschewed medical politics. Although he wrote profusely, his literary style is a difficult and tedious one, and he is often guilty of repetition and self-glorification.

In addition to a busy medical practice, Wilson Philip also undertook physiological research. Concerning clinical medicine he wrote on urinary gravel (1792), fevers (1799–1804, 1807), indigestion (1821, 1824), and on many other topics. From these works he may be seen to have been a critical and accurate observer of disease, but his reasoning—like that of many of his contemporaries— was based on unproven hypotheses and relics of the eighteenth-century systems of disease. No doubt his therapeutic advice, although mostly ill-founded, was often valuable, but his many publications contributed little or nothing to the advancement of internal medicine as a science. Nevertheless, his works were popular in their time, and the four-volume *Treatise on Febrile Diseases* (1799–1804), for example, went through four editions and was translated into German and French.

Wilson Philip investigated experimentally the action of opium (1795), mercury (1805), galvanism (1817), and Malvern waters (1805). His book on indigestion (1821), a condition that interested him throughout his career, was well received by the profession—unlike some of his later publications, including that *On the More Obscure Diseases of the Brain* (1835), which was described, according to McMenemy, as "the mixture as before" or a "shameless piece of self-exaltation."

In his early years Wilson Philip carried out important physiological research on the nervous system and capillary circulation, and today his reputation rests upon the results of this research. He was one of a small group of British physicians who contributed to the field of physiology, which at that time was dominated by the French and Germans. His book *An Experimental Enquiry Into the Laws of the Vital Functions* (1817) is especially important. In some famous experiments he showed that digestion ceased with section of the vagus nerve, that gastric secretion could be decreased by damage or removal of parts of the brain or spinal cord, and that movement of the gut could be independent of brain control. These results stimulated a great deal of further research.

Wilson Philip's studies of nervous influences on the cardiovascular system were equally significant. Albrecht von Haller maintained that the capillaries were unable to contract, but Wilson Philip, along with others, refuted this conclusion and went on to demonstrate that cardiac acceleration and inhibition were produced by stimulation of the nervous system. He also showed that although the heart and blood vessels act independently of the brain and spinal cord, as Haller and the French physiologist J. J. C. Legallois had maintained, they could be affected by drugs that acted on the nervous system, since they are centrally regulated (but not primarily centrally controlled). In his experiments Wilson Philip used the microscope to detect changes in the caliber of blood vessels. This marked a very early use of this instrument in physiological research in England.

BIBLIOGRAPHY

I. ORIGINAL WORKS. Pettigrew (1840, see below) lists many of Wilson Philip's publications, including several in journals; and McMenemy (1958, see below) cites the important works.

Works published before 1807 appear under the style "A. P. Wilson"; these include *An Inquiry Into the Remote Causes of Urinary Gravel* (Edinburgh, 1792; German trans., 1795); *An Experimental Essay on the Manner in Which Opium Acts on the Living Animal Body* (Edinburgh, 1795); *Treatise on Febrile Diseases*, 4 vols. (Winchester, 1799–1804; 4th ed., London, 1820; German trans., Leipzig, 1804–1812; and French trans., Paris, 1819); *Observations on the Use and Abuse of Mercury* (Winchester, 1805); *An Analysis of the Malvern Waters* (Worcester, 1805); and *An Essay on the Nature of Fever* (Worcester, 1807).

Works published after 1807 under the name "A. P. W. Philip," or "A. P. Wilson Philip" are *Treatise on Indigestion and Its Consequences* (London, 1821; 7th ed., 1833; German trans., Leipzig, 1823; and Dutch trans., Amsterdam, 1823; 8th ed. as *A Treatise on Protracted Indigestion and Its Consequences, etc.* [London, 1842]). His most important work is *An Experimental Inquiry Into the Laws of the Vital Functions* (London, 1817; 2nd ed., 1818; 3rd. ed., 1826; and 4th ed., 1839; with German trans., Stuttgart, 1828; and Italian trans., in F. Tantini, *Opusculi scientifici*, II [Pisa, 1822]).

II. SECONDARY LITERATURE. The best account of Wilson Philip and his work is by W. H. McMenemy, "Alexander Philips Wilson Philip (1770–1847), Physiologist and Physician," in *Journal of the History of Medicine and Allied Sciences*, **13** (1958), 289–328, with portrait. McMenemy adds useful details to the official accounts, such as those of W. Munk, *The Roll of the Royal College of Physicians of London*, III (London, 1878), 227–238; and J. F. Payne in *Dictionary of National Biography*, XV (London, 1921–1922), 1041–1042.

A contemporary assessment is that of T. J. Pettigrew, in *Medical Portrait Gallery. Biographical Memoirs of the Most Celebrated Physicians, Surgeons, etc.*, III (London, 1840), 16, with portrait. Wilson Philip's work in the context of the history of neurophysiology is discussed by M. Neuburger, in *Die historische Entwicklung der experimentellen Gehirn- und Rückenmarksphysiologie vor Flourens* (Stuttgart, 1897), 225, 251–255, 266–267, 270–271, 275. *Extracts From an Experimental Inquiry* (1817) is included in J. F. Fulton and L. G. Wilson, *Selected Readings in the History of Physiology*, 2nd ed. (Springfield, Ill., 1966), 82–85.

EDWIN CLARKE

PHILLIPS, JOHN (*b*. Marden, Wiltshire, England, 25 December 1800; *d*. Oxford, England, 24 April 1874), *geology, paleontology.*

Phillips was the son of John Phillips of Blaen-y-Ddol, Carmarthenshire, an excise officer, and Elizabeth Smith, daughter of John Smith of Churchill, Oxfordshire. He remained unmarried, and for many years at York and at Oxford his sister Anne was hostess at his home. Phillips was left an orphan at an early age and entered into the care of his maternal uncle, the geologist and land surveyor William Smith. He was educated at a school near Bath and also spent a year in the house, and under the instruction, of the Reverend Benjamin Richardson of Farleigh Hungerford; but his formal education ended before he reached the age of fifteen when he returned to live with his uncle in London. There, after Smith's fossil collection had been bought for the nation in 1815, Phillips assisted in preparing a catalog and in arranging the specimens "according to Linnaeus" before their delivery to the British Museum.

One of the few to practice the art of lithography in England, Phillips was entrusted by Smith with copying and lithographing some of his reports. For the next nine years he acted as assistant and amanuensis to his uncle and was almost constantly his companion. During this time, apart from professional assignments, they were occupied in compiling the series of county geological maps; and both together and independently they made many geological traverses throughout the north of England.

In 1824, following an invitation to Smith to lecture to the Yorkshire Philosophical Society, Phillips was engaged to arrange the fossil collections at the museum at York. Shortly afterward he was appointed curator of the society's museum, a post he held until the end of 1840.

In 1831 Phillips played a leading part in organizing at York the general meeting of British scientists that became the British Association for the Advancement of Science. This society was founded "to give a stronger impulse and a more systematic direction to scientific inquiry; to promote the intercourse of those who cultivate science in different parts of the British empire with one another, and with foreign philosophers; to obtain a more general attention to the objects of science, and a removal of any disadvantages of a public kind which impede its progress." He was the executive officer of the Association until 1859, and throughout this period he arranged the venue for its annual assembly, maintained close links with leading British scientists, edited the society's annual reports and, through his efficiency and cordial relationships, was a major contributor to its success.

Phillips was an accomplished lecturer and teacher and gave courses in geology and zoology in many towns in the north of England under the auspices of the local scientific and philosophical societies. In 1831 he began similar courses in London, and in 1834 he was elected a fellow of the Royal Society and was appointed professor of geology at King's College, London. He gave up this post in 1840, when he joined the Geological Survey under De la Beche. In 1844 he became professor of geology at Trinity College, Dublin. But an expected appointment to a senior position in the Irish branch of the Geological Survey, which could have been held concurrently, did not materialize; and the following year he relinquished his Dublin post. Phillips' assignment with the Geological Survey involved work in Southwest England and detailed geological mapping around the Malvern Hills and in South Wales.

In 1853 Phillips was appointed deputy reader in geology at Oxford, and on the death of William Buckland in 1856 he became reader and subsequently professor. He also played an important part in the building of the new University Museum and was its keeper until his death, which resulted from a fall.

By his early training and by inclination, Phillips was a practical field geologist, skilled in the making of geological maps; and his most important contributions to stratigraphical geology were descriptive. In his volumes on the geology of Yorkshire he recorded the stratigraphy and structure of the "Mountain" (Carboniferous) Limestone and the Jurassic and Cretaceous strata of a large area of the north of England. He also introduced the term "Yoredale series" (1836, p. 37) for sediments showing a rhythmic succession of shales, sandstones, and limestones, together constituting a special facies of the uppermost zone of the Carboniferous Limestone series in this region. He traced the changes in these sediments when followed laterally and interpreted them as due to changes in the depositional environment.

In his work on the fossils of Southwest England (1841, p. 160), Phillips introduced the term "Mesozoic" to identify the geological era between the Paleozoic and Cenozoic and to include the "New Red" (Triassic), "Oolitic" (Jurassic), and Cretaceous periods. This book was an essential supplement to De la Beche's first official Geological Survey memoir, *Report on the Geology of Cornwall, Devon, and West Somerset* (1839). A similar style of presentation was continued in Phillips' volume on the geology of the Paleozoic rocks in the vicinity of the Malvern Hills. By these descriptions and illustrations of the stratigraphy and characteristic fossils of particular formations, Phillips contributed notably to the background of knowledge by which progress in stratigraphical classification and correlation was made possible.

In 1852 John Phillips brought mature geological experience to his own personal observation of the physical features of the surface of the moon, using at first the great telescope belonging to the Earl of Rosse. These investigations arose out of the appointment of an ad hoc committee by the British Association charged with the task of procuring a new series of drawings or surveys of selected parts of the lunar disk. The drawings were to be made under a set of standard conditions of representation and on a uniform scale. Since few observers were willing to undertake the investigations, it was left to Phillips to pursue, virtually alone, the queries enumerated in the committee's prospectus. By 1853 he was recording his observations photographically on collodion plates and employing his great artistic skill in accurate and detailed drawings. After he reached Oxford there was an interval of several years before he resumed the study using an up-to-date telescope provided by the Royal Society.

Phillips emphasized the need for continuous observation of selected areas and the recording of each one at different times of the lunar day so that the configurations could be accurately determined. In a summary of his findings published in 1868, Phillips drew vivid analogies between many of the features seen on the surface of the moon and those known to him intimately by observation and measurement of the earth.

BIBLIOGRAPHY

I. ORIGINAL WORKS. A complete bibliography of Phillips' works is given in *Proceedings of the Yorkshire Geological Society* (see below). Works on regional geology include *Illustrations of the Geology of Yorkshire*, I (York, 1829; 2nd ed., London, 1835; 3rd ed., London, 1875), II (London, 1836); *Figures and Descriptions of the Palaeozoic Fossils of Cornwall, Devon, and West Somerset* (London, 1841); and *Geology of Oxford and the Valley of the Thames* (Oxford, 1871). His lunar work, submitted in papers to the Royal Society, is summarized in "Notices of Some Parts of the Surface of the Moon," in *Philosophical Transactions of the Royal Society*, **158** (1868), 333–346. Other astronomical researches are mentioned in "The Planet Mars," in *Quarterly Journal of Science*, **2** (1865), 369–381. Phillips included certain autobiographical material in his *Memoirs of William Smith* (London, 1844).

II. SECONDARY LITERATURE. A number of obituary notices are mentioned in T. Sheppard, "John Phillips," in *Proceedings of the Yorkshire Geological Society*, **22** (1933), 153–187, which contains a full bibliography of Phillips' scientific papers and a number of portraits with the location of the originals.

J. M. EDMONDS

PHILLIPS, THEODORE EVELYN REECE (*b.* Kibworth, Leicestershire, England, 28 March 1868; *d.* Headley, Surrey, England, 13 May 1942), *astronomy*.

Phillips was the son of the Reverend Abel Phillips, formerly of Barbados and a missionary in West Africa. He was educated at Yeovil Grammar School and in 1891 graduated B.A. from St. Edmund Hall, Oxford. He was ordained in the same year and became curate at the Church of the Holy Trinity in Taunton.

In 1896, while curate at Hendford (near Yeovil), Phillips began systematic observation of the planets, especially Jupiter and Mars, with a nine-and-a-quarter-inch altazimuth reflector. He continued his observations, with that instrument when he moved to Croydon and later, in Ashstead, when he acquired a twelve-and-a-quarter-inch equatorial reflector. From 1911 he used an eight-inch reflector loaned by the Royal Astronomical Society; and when he became rector of Headley in 1916, he used an eighteen-inch reflector loaned by the British Astronomical Association, of which he had been president from 1914 to 1916.

Phillips directed the Jupiter section of the association from 1900 to 1933 and the Saturn section from 1935 to 1940. From 1896 to 1941 he submitted more than 400 drawings to the Mars section.

Phillips' work on Jupiter followed that of A. S. Williams, W. F. Denning, and others who had observed the drift of the surface markings of Jupiter in different latitudes. These markings were charted by timing their passage over the central meridian; the *Memoirs of the British Astronomical Association* (1897–1898) included Phillips' tables with deduced rotation periods for different latitudes. The *Memoirs* for 1932–1933, which were not published until 1939, recorded his last observations.

The observations published under Phillips' care are perhaps the only satisfactory continuous records of the movements of Jupiter during his career. They include a complete history of the appearance and movement of the red spot and the south tropical disturbance, which was first recorded in 1901, and of disturbances heralding the return (1920, 1928) of the south equatorial belt and other south tropical characteristics. Phillips is known to have recorded more than 30,000 spot transits.

Following a suggestion by H. H. Turner, Phillips conducted a harmonic analysis of the light curves of about eighty stars; and he made protracted observations of double stars. He was the president of the International Astronomical Union, Commission Sixteen; and he represented the Church of England at Geneva in 1922, when it was proposed that Easter become a fixed, rather than a movable, feast. He was president of the Royal Astronomical Society from 1927 to 1929. Among his many observational records is a continuous set of rainfall records for Headley that covers twenty-five years; and he made a harmonic analysis of annual temperature curves for several places in Great Britain. Phillips was also

an amateur botanist and a university extension lecturer for many years. He received an honorary D.Sc. from Oxford University shortly before his death. In 1906 he married Mellicent Kynaston of Croydon. Their only son, the Reverend John E. T. Phillips, became an amateur astronomer.

BIBLIOGRAPHY

Phillips was a regular contributor to the publications of the British Astronomical Association. He also contributed the articles "Jupiter," "Mercury," "Neptune" (in part), "Saturn," and "Venus," in the 14th ed. of the *Encyclopaedia Britannica* (1929); revised R. S. Ball's *A Popular Guide to the Heavens*, 4th ed. (London, 1925); and collaborated with W. H. Steavenson in editing *Hutchinson's Splendour of the Heavens*, 2 vols. (London, 1923–1926).

For brief accounts of his life, see the obituary by B. M. Peek in *Journal of the British Astronomical Association*, **52** (1942), 203–208. See also M. Davidson, "Honour for Rev. T. E. R. Phillips," in *Observatory*, **64** (1942), 228–231, written shortly before Phillips' death.

J. D. NORTH

PHILLIPS, WILLIAM (*b.* London, England, 10 May 1775; *d.* London, 2 April 1828), *geology.*

Phillips inherited his father's printing and book-selling business in London but also devoted much leisure time to studying the natural sciences. In this pursuit he found a kindred spirit in his younger brother Richard (1778–1851), who became a successful chemist. They were the grandsons of Catherine Phillips, a noted Quaker, and were lifelong members of the Society of Friends.

Phillips became increasingly interested in geology and in November 1807 was one of the founding members of the London Geological Society. He subsequently spent most of his spare time on mineralogy and stratigraphy. In 1825 a crystal of the zeolite family—a hydrated aluminosilicate of calcium, potassium, and sodium—was named "phillipsite" after him. He was elected a fellow of the Royal Society in 1827.

Phillips contributed about twenty-seven papers, mainly on mineralogy and with an emphasis on Cornish minerals, to the *Transactions of the Geological Society of London* and to other scientific journals. He also published three influential, standard textbooks, two on mineralogy and one mainly on stratigraphy, that enjoyed a great popularity. His *Outlines of Mineralogy and Geology* (1815) had gone into a fourth edition by 1826, and his *Elementary Introduction to the Knowledge of Mineralogy* (1816) was largely rewritten and reissued as a fifth edition in 1852. This work is illustrated with numerous woodcuts of crystals. In 1829 the *Quarterly Journal of the Geological Society of London* stated that:

> It was after the invention of Dr. Wollaston's reflective Goniometer, that his [Phillips'] assiduity and success in the use of that beautiful instrument enabled him to produce his most valuable Crystallographic Memoirs; and the third edition of his elaborate work on Mineralogy [*An Elementary Introduction to Mineralogy* (London, 1823)] contains perhaps the most remarkable results ever yet produced in Crystallography, from the application of goniometric measurement, without the aid of mathematics.

Phillips' most influential work was that on the stratigraphy of England and Wales, which began as a digest of English geology, entitled *A Selection of Facts From the Best Authorities, Arranged so as to Form an Outline of the Geology of England and Wales* (London, 1818). Phillips then collaborated with Conybeare on an enlarged version, *Outlines of the Geology of England and Wales, With an Introductory Compendium of the General Principles of That Science, and Comparative Views of the Structure of Foreign Countries* (London, 1822). In this work Phillips incorporated some of his own fieldwork—for example, his study of the chalk cliffs on each side of the Strait of Dover. The book was received with great enthusiasm. Almost a century later (*Encyclopaedia Britannica*, 11th ed. [1910], XXI, 408) it was still extolled "as a model of careful original observation, of judicious compilation, of succinct description, and of luminous arrangement, it has been of the utmost service in the development of geology in Britain."

The stratigraphical information presented in this work actually depended largely on the ideas of William Smith. The book publicized Smith's concepts and methods and showed convincingly that a secure foundation for the comparative study of sedimentary strata can be obtained only with the assistance of fossil content. The clear geological cross sections and the fairly complete exposition of the contemporary geological knowledge of sedimentary rocks were based, with certain modifications, on Smith's terminology and stratigraphy.

Not surprisingly, this classic work had a marked influence on British geology, and the stratigraphical framework popularized in it became of worldwide significance. The book was highly praised in the United States, in an anonymous review, probably by Benjamin Silliman; it was, however, weak in its geomorphological concepts. Although the authors derided Wernerian theories, they still believed that valleys were formed by violent currents when the landmass was upheaved from beneath the sea.

BIBLIOGRAPHY

There is an obituary notice in *Quarterly Journal of the Geological Society of London, Proceedings*, **1** (1829), 113. Conybeare and Phillips' textbook is reviewed anonymously in *American Journal of Science*, **7** (1824), 203–240.

See also R. J. Chorley *et al.*, *The History of the Study of Landforms*, I (London, 1964), *passim*; and H. B. Woodward, *A History of the Geological Society of London* (London, 1907), *passim*.

ROBERT P. BECKINSALE

PHILO OF BYZANTIUM (*fl. ca.* 250 B.C.), *physics, mechanics, pneumatics.*

Little is known of Philo's life. Vitruvius includes him, with Archytas, Archimedes, Ctesibius, and others, in a list of inventors,[1] while Hero of Alexandria mentions a work on an automatic theater by him[2] and Eutocius cites his work on the duplication of the cube.[3] These constitute the only references to Philo in antiquity; what else is known of him must be inferred from the hints that exist in his few extant writings. These writings are fragments of a large book on mechanics, Philo's only known work. From references to the bronze-spring catapult, recently invented by Ctesibius (*fl. ca.* 270 B.C.), it is possible to reach an approximate date for Philo's career. It is also clear that Philo was able to travel to Rhodes and Alexandria to study catapults, which suggests that he may have been wealthy, or have had a wealthy patron—perhaps his friend Ariston, to whom each of the surviving books of the larger mechanics is dedicated. (Nothing else is known of Ariston, who would seem to have been a man of position and of some mathematical sophistication; in Arabic versions of the text his name is rendered as Māristūn.)[4] Diels states that, like Hero, Philo was a mere artisan, but he gives no reason for his opinion.[5]

Philo chose to write his textbook on mechanics, the Σύνταξις τῆς μηχανικῆς in κοινή, the vernacular common to the whole Greek-speaking world of the time. Certainly, this straightforward language is better suited to his practical purpose than is the intricate Attic literary prose; the book is full of technical detail that would be of service to an architect, contractor, or—in its sections on war machines and fortifications—a general, and perhaps Ariston was one of these.

It is possible, through studying the extant parts of Philo's *Mechanics*, to surmise what the contents of the whole work must have been, especially since Philo often refers both to what he has already written and to what he intends to write. One can thus reconstruct a list of nine books:

1. Introduction
2. On the lever (Μοχλικά)
3. On the building of seaports (Λιμενοποιικά)
4. On catapults (Βελοποιικά)
5. On pneumatics (Πνευματικά)
6. On automatic theaters (Αὐτοματοποιικά)
7. On the building of fortresses (Παρασκευαστικά)
8. On besieging and defending towns (Πολιορκετικά)
9. On stratagems.

Of these, book 4 (the *Belopoeica*), book 5 (the *Pneumatica*), book 7 (the *Paraskeuastica*), and book 8 (the *Poliorcetica*) are extant. The Greek text exists for the *Belopoeica* and for parts of the *Paraskeuastica* and *Poliorcetica*; the *Pneumatica* was for many years known only through a Latin translation of an Arabic version of the first sixteen chapters. At the beginning of the twentieth century, however, B. Carra de Vaux found three manuscripts (one in the Bodleian Library and two in the library of Hagia Sophia), which provided a fuller Arabic text; previous to this discovery, the *Paraskeuastica* and the *Poliorcetica* were together known as book 5.

Book 4, the *Belopoeica*, is concerned with the construction of catapults. It is from this book that Philo's travels may be discovered. In chapter 3, he remarks that his application of mathematics to the building of these weapons resulted from the interest that the Alexandrian kings took in the technical arts. (This would indicate that he was in Alexandria in the time of the first Ptolemies.) In chapter 4, Philo states that he has talked with catapult experts in Alexandria, and has examined catapults in Rhodes; in chapter 39, he goes on to mention that although he himself has not seen Ctesibius' bronze-spring catapult, it has been described to him in detail by people who have. He displays a great interest in experimentation, and would seem to have invented improved catapults himself.

Philo gives the rules for constructing a catapult from a module derived from either the length of the arrow or the weight of the projectile, a method first worked out by Alexandrian technicians. Philo then approaches the problem of designing a catapult that will deliver a missile of twice the weight of one launched by another catapult; in establishing the module by which such a catapult should be constructed it is necessary for him to double the cube, since the respective weights of the two missiles are in the proportion of the cubes of their diameters.

This cannot be done by a Euclidean construction with ruler and compass alone; the method is to find two mean proportionals. Philo writes (chapter 7):[6] "the reduplication of the cube, which I have explained in the first Book, but which I do not hesitate to write

here also" (see fig. 1). "Let there then be given a certain straight line *A*, for which, for example, we have to find the double cube; I then place the double of it, *B*, at right angles to it, and from the other end of *B* I draw at right angles a straight line of indefinite length

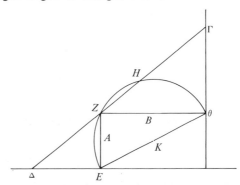

FIGURE 1.   Philo's figure for the reduplication of the cube.

towards *Γ* ⟨and from the end of *A* I draw at right angles a straight line of indefinite length towards *Δ*⟩; from the angle marked *Θ* ⟨to the angle marked *E*⟩ I draw the straight line *K*, and divide it in two, and let the middle be the point *K*, and with *K* as centre and *KΘ* as radius I draw a half circle, which goes also through the point *Z*. Then I take an accurately fashioned ruler and place it so that it cuts both the straight lines, taking care that one point of it touches the angle (*Z*), and turning it till I get the part of the ruler from the intersection marked *Γ* to the part that falls on the intersection with the circle, marked *H*[*Z*], to be equal to the length from the intersection marked to that which falls on the angle marked *Z*. And then *ΔE* is the double of the cube of *EZ*, *ΘΓ* of that of *EΔ*, *ΘZ* of that of *ΘΓ*. The diameter of the circle which has to take the spring is found in this way." Additions to the text have been made by August Brinkmann, and they are so obvious that it is possible that Philo himself omitted them, since this is just an example, an application, of the proof given in the first book.

Eutocius (in his commentary on Archimedes, *De sphaera*, Book 2)[7] reviews the different ways of reduplicating the cube. He gives Hero's solution, and then proceeds to Philo's solution, using Philo's figure, but with lettering of his own and adding two lines from Hero's figure, from *K* to *Δ* and *Z* (see fig. 2).

The construction is the same as that of Philo, but Eutocius uses fewer words. When he has placed the ruler, he writes: "Let it then be assumed that the ruler has the position taken by *ΔBEZ*, where, as stated, *ΔB* is equal to *EZ*. I say then that *ΑΔ* and *ΓZ* are middle proportions to *AB* and *BΓ*."

He then gives the proof in this way: "Let us assume

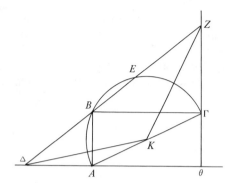

FIGURE 2. Eudocius' figure for Philo's reduplication of the cube.

that $\varDelta A$ and $Z\varGamma$ are prolonged and intersect at $\varTheta$; then it is evident that, since BA and $Z\varTheta$ are parallel, the angle at $\varTheta$ is right, and that the half-circle $AE\varGamma$ if it is filled out also will go through $\varTheta$. Now since $\varDelta B$ and EZ are equal, $E\varDelta$ by EB will be equal BZ by BE. But $E\varDelta$ by $\varDelta B$ is equal to $\varTheta A$ by $\varDelta A$, for both are equal to the square of the tangent from $\varDelta$; but BZ by ZE is equal to $\varTheta Z$ by $Z\varGamma$, for both are equal to the tangent from Z; so also $\varTheta \varDelta$ by $\varDelta A$ is equal to $\varTheta Z$ by $Z\varGamma$, and from this it follows that $\varDelta \varTheta$ to $\varTheta Z$ is equal to $B\varGamma$ to $\varDelta A$. But as $\varDelta \varTheta$ is to $\varTheta Z$, so is both $B\varGamma$ to $\varGamma Z$ and $\varDelta A$ to AB, for in the triangle $\varDelta \varTheta Z$ $B\varGamma$ is parallel to $\varDelta \varTheta$, and BA is parallel to $\varTheta Z$. So then, as $B\varGamma$ to $\varGamma Z$, so $\varGamma Z$ to $\varDelta A$ and $\varDelta A$ to AB, which was to be proved."

This construction is almost the same as that of Hero; for the $B\varTheta$ parallelogram is the same as that assumed in Hero's construction, and also the prolonged sides $\varTheta A$ and $\varTheta \varGamma$, and the ruler turned on the point B. The only difference is that (in Hero) we move the ruler around B until the lines from the middle of $A\varGamma$, that is K, become equal as they are intersected by it where they reach $\varTheta \varDelta$ and $\varTheta Z$, as $K\varDelta$ and KZ, but here (in Philo) till $\varDelta B$ becomes equal to EZ. In either construction the same thing follows, but the latter is more easy in practice, for to see that $\varDelta B$ and EZ are equal can be done if the ruler is divided into equal parts, and very much more easily than by trying to find out by compasses from K if the lines from K to $\varDelta$ and to Z are equal."

It appears that the first part of this construction was written by Eutocius in his own way, but the middle part, the proof, may be copied from Philo. The comparison of the two constructions is Eutocius' own contribution; when the figures are drawn, Hero's method is easier.

Philo goes on to describe a method of improving catapults by using wedges to tighten the sinews. He also gives a design for a catapult with bronze springs,

which he notes was inspired by the report that Ctesibius had made such a machine. Philo did not know the details of Ctesibius' catapult at the time he devised his own, and when he was able to compare the two weapons, he discovered that they were different in several respects. Philo also describes Ctesibius' air-driven catapult, and discusses an automatic catapult invented at Rhodes by a man named Dionysius, about whom nothing else is known. None of these devices is mentioned by either Vitruvius (in 25 B.C.) or Hero (in A.D. 62).

Philo's fifth book, the *Pneumatica*, begins with a series of introductory chapters that incorporate a number of experiments almost certainly taken from Ctesibius, the founder of the science of pneumatics. These chapters were copied by Hero. The rest of the book (like Hero's work on the same subject) consists of descriptions of pneumatic toys—trick jars, inexhaustible bowls, and other apparatus for parlor magic. Many of these were probably reconstructed or reinvented by Hero and others.[8] All the chapters are illustrated in the extant Arabic manuscripts, but the illustrations have never been published.

The surviving parts of books 7 and 8, the *Paraskeuastica* and the *Poliorcetica*, indicate that each book consisted of a large number of short chapters. These were considered together as one book in earlier editions of Philo's work, and may be divided into four sections, of which the first two constitute the *Paraskeuastica*, and the second two the *Poliorcetica*. The first section contains eighty-seven chapters devoted to the techniques of fortifying a town—constructing walls and towers, digging moats, setting up palisades, and placing catapults. The second section consists of fifty-seven chapters on provisioning a town against siege. In it Philo describes the proper construction of storerooms and lists the foodstuffs to be kept therein, together with methods for keeping stored foods fresh. He gives several recipes for "iron rations" for the besieged and recommends poisoning supplies that might otherwise be used by the enemy. He promises a full treatment of poisons in a later book (presumably one of the lost chapters of book 8). Philo also gives a list of materials and tools to be prepared or procured against a siege, and describes an optical telegraph, which employs a clepsydra as a mechanism, for maintaining communications with allies beyond the perimeter of the besieging forces.

Section 3 has seventy-five chapters, primarily devoted to a number of different ways of defending town walls against attack and to the means of defense against an attack from the sea. Philo also discusses the importance of the availability of good doctors during time of war, and states that invalids should be made certain of pensions, that the dead should be buried with honor, and that widows should be provided for.

The last section is 111 chapters long; it deals with how to lay siege to a town, how to use catapults, testudos, and other engines of war, and how to capture a town through starvation or bribery. There are also discussions of secret messages (Philo promises a more detailed discussion of this subject in a following book that is now lost) and cryptography, and an account of how to attack a town from the sea. Philo again mentions his longer treatment of poisons, this time as something that he has already written; it may therefore be assumed that this chapter is one of the missing fragments of book 8. He further states that he has drawn figures to illustrate all of the kinds of fortifications that he has discussed,[9] but these drawings have also been lost.

Since only fragments remain, it is impossible to judge Philo's textbook on mechanics as a whole. His thoroughness and attention to detail, however, are evident in every part that remains. He is not identical with a later Philo of Byzantium who wrote a work on the seven wonders of the world.

NOTES

1. Vitruvius, *De architectura*, VII, intro., 14.
2. Hero, *Automata*, XX, 1 and 3. *Heronis Alexandrini Opera quae supersunt omnia*, W. Schmidt, ed., I (Leipzig 1899).
3. Eutocius, in *Archimedis Opera omnia cum commentariis Eutocii*, J. L. Heiberg, ed., 2nd ed., III (Leipzig, 1915), 60.
4. B. Carra de Vaux, "L'invention de l'hydraulis," in *Revue des études grecques*, **21** (1908), 332–340.
5. Hermann Diels, "Über das physikalische System des Straton," in *Sitzungsberichte der Preussischen Akademie der Wissenschaften zu Berlin*, **9** (1893), 110, n. 3.
6. Philo, *Belopoeica*, VII.
7. Eutocius, *loc. cit.*, n. 3 above.
8. A. G. Drachmann, *Ktesibios, Philon and Heron*, no. 4 in the series Acta Historica Scientiarum Naturalium et Medicinalium (Copenhagen, 1948).
9. Philo, *Mechanics*, 7–8, sec. 1, ch. 87.

BIBLIOGRAPHY

Editions of Philo's works are the Greek texts of the *Belopoeica* and the *Poliorcetica*, both edited by R. Schoene: *Philonis mechanicae syntaxis libri IV et V* (Berlin 1892), where IV is the *Belopoeica* and V the *Poliorcetica*, later known as bks. 7–8.

A later ed. of the *Belopoeica* is Philon's *Belopoiika* (*4. Buch der Mechanik*), H. Diels and E. Schramm, eds., in *Abhandlungen der Preussischen Akademie der Wissenschaften* for 1918, Phil.-hist. Kl., no. 16 (Berlin 1919), with the Greek text, the MS figures, a German trans., and reconstructions of all the engines.

The book on *Pneumatics*, found in Arabic trans. only, is edited by Carra de Vaux: "Le livre des appareils pneu-

matiques et des machines hydrauliques, par Philon de Byzance," in *Notices et extraits des manuscrits de la Bibliothèque nationale*, **38** (1903), 27–235; here is the Arabic text and a French trans., with intro. (all figures redrawn by the editor).

A Latin trans. of an Arabic trans. of the first 16 chs. of the *Pneumatics* is edited by V. Rose, in *Anecdota graeca et graecolatina*, no. 2 (Berlin, 1870), 297–314, with 13 figs. copied from the MS. This ed. is reprinted in *Heronis Alexandrini Opera quae supersunt omnia*, I (Leipzig, 1899), 458–489, with a German trans. The original of the Latin trans. is slightly different from the Arabic text edited by Carra de Vaux.

The *Paraskeuastica* and *Poliorcetica* have been published as *Exzerpte aus Philons Mechanik B. VII und VIII* (*vulgo fünftes Buch*), Greek text with German trans. by H. Diels and E. Schramm, in *Abhandlungen der preussischen Akademie der Wissenschaften* for 1919 (1920), Phil.-hist. Kl., no. 12; with drawings by the editors.

A. G. DRACHMANN

PHILOLAUS OF CROTONA (*fl.* second half of fifth century B.C.), *philosophy, astronomy, medicine.*

Like the majority of the Pythagoreans living in the Pythagorean centers of Crotona and Metapontum in the middle of the fifth century B.C., Philolaus fled after the oubreak of the democratic rebellion, during which the meeting houses of the Pythagoreans were burned down. He first went to Thebes but later settled in Tarentum, the only remaining center of Pythagorean political activity in southern Italy. There he is said to have taught the mathematician Archytas.

The ancient tradition concerning Philolaus' writings and doctrine is so confused and contradictory that some modern scholars have questioned his very existence, others have doubted that he ever wrote or published anything, and still others have tried to show that none of the fragments that have come down to us under his name are genuine. His existence, however, is established beyond reasonable doubt by Plato's *Phaedo* (61e). That there existed at least one genuine work by Philolaus is attested by Menon, a disciple of Aristotle and a conscientious historian who lived only about a century later than Philolaus. In his history of medicine, of which fragments have been discovered on a papyrus, Menon quotes Philolaus' medical doctrines.

According to a widespread ancient tradition, Philolaus had written, but did not publish, a comprehensive work of Pythagorean doctrine, the manuscript of which was purchased by Plato at a very high price and then copied. Thus Philolaus' work became public. After the book had become famous, other books, not written by Philolaus, appear to have been published under his name. Since all later Pythagoreanism was strongly influenced by Plato, however, it is possible to distinguish between pre-Platonic and post-Platonic Pythagorean literature. This criterion has been used especially by W. Burkert to separate the probably genuine from the spurious fragments of Philolaus' work.

According to this criterion, the fragments that can be considered genuine contain many rather abstruse speculations concerning the relation between certain numbers and certain gods of traditional Greek mythology, as well as speculations about numbers, places, and things that were more noble or more deserving of honor ($\tau\iota\mu\iota\acute{\omega}\tau\epsilon\rho\alpha$) than others. The fragments also reveal a tendency to combine elements of various doctrines that were prevalent in the second half of the fifth century–although not all were specifically Pythagorean—into a rather muddled whole. This approach is hardly "scientific," and there would be little reason to mention Philolaus in the context of the history of science if it were not for the considerable influence of his astronomical system on the development of astronomy, even after Copernicus.

Knowledge of this system has evolved from three sources: Aetius, a late Greek doxographer; Achilles, a late Greek commentator on the *Phaenomena* of Aratus; and Aristotle. The first two, who differ from each other in some details, especially in the explanation of the light emanating from the sun, mention Philolaus by name. Aristotle, describing essentially the same system, does not mention Philolaus by name but attributes the system to "some of the Pythagoreans, who had lived in Southern Italy," Aristotle's reason for this attribution may be that, according to one tradition, Philolaus' system was further elaborated by a certain Hicetas.

Contrary to all earlier astronomical systems, and also to those accepted later by Plato and Aristotle, Philolaus' system states that the center of the universe is occupied not by the earth but by a central fire (not identical with the sun), around which the earth moves at considerable speed. That we do not see this central fire is explained by assuming that the earth always turns the same side to the central fire and that we live on the opposite side. A "counter-earth," or "anti-earth" ($\dot{\alpha}\nu\tau\acute{\iota}\chi\theta\omega\nu$), exists on the opposite side of the central fire but at a lesser distance. The "anti-earth" revolves around the central fire simultaneously with the earth, which is why we can never see it. Above the earth and at greater distances from the central fire the other heavenly bodies likewise revolve around the central fire—first the moon, then the sun, and then the five planets: Mercury, Venus, Mars, Jupiter, and Saturn. Outermost is the fiery sphere of the fixed stars. The sun, according to Aetius, is a hollow mirror collecting the light from below (the central fire?) and reflecting and focusing it toward the earth. According to Achilles, it is, on the contrary, similar to a convex lens, collecting the light that emanates from the outer sphere and likewise focusing it in the direction of the earth.

To understand the position of this strange system within the history of astronomy, the following factors must be considered. From the early fifth century B.C. Greek philosophers and cosmologists were puzzled by the apparently irregular motions of certain heavenly bodies. Anaxagoras, among others, tried to explain the movements of the stars by adopting mechanics as an analogue. Others, like the early Pythagoreans, considered the heavenly bodies as living beings endowed with a higher reason than humans. These philosophers thought it unworthy of such divine beings to move as irregularly as animals on this earth, rather than moving with a constant speed in the most beautiful and simple curve—the circle. Plato, who adopted this view, asked the famous mathematician Eudoxus of Cnidus if the apparent movements of the celestial bodies could be explained on the assumption that they were combinations of circular movements with constant velocity. Eudoxus actually succeeded in constructing a model of twenty-seven spheres, one within the other and each moving at a constant speed. The inner spheres were directed by the outer spheres, which moved in different directions. Thus the apparent movements of the celestial bodies could be approximately accounted for. Because of careful observations, some disagreements remained; and the number of spheres within this system was later increased to thirty-three by the astronomer Callippus. Aristotle increased the number to fifty-five. Within all these systems, however, the earth remained in the center of the universe.

By removing the earth from the center of the universe, Philolaus anticipated modern astronomical theory and, specifically, the Copernican system. It is natural to assume that Philolaus must have tried also to account mathematically for the apparent movements of the heavenly bodies, although it is difficult to see what the assumption of an invisible central fire and an invisible anti-earth could have contributed to such an explanation. Since the system, as reported by both Aetius and Achilles, did not present such an account, B. L. Van der Waerden suspected that Philolaus' system must have differed from that described by ancient tradition. He states that either the most important step in the development of early Greek astronomy was taken by a muddleheaded speculator (*ein Wirrkopf*) or Philolaus' system must have been different.

Van der Waerden, in a most ingenious fashion, constructed a Philolaic system, accounting to some extent for the observed phenomena. Aristotle, however, when speaking of the Pythagoreans of southern Italy, who invented the system otherwise ascribed to Philolaus, states expressly that they did not try to account for the observed phenomena by means of mathematical constructions. Instead, they removed the earth from the center of the universe for the sole reason that only fire, the most noble (τιμιώτατον) of the elements, was worthy of occupying that place. The anti-earth was added to bring the number of the celestial bodies to ten, because ten was the "perfect number." If this is so—and Aristotle, after all, is not a contemptible witness—one will have to admit that it was an unscientific, muddleheaded speculator and not a great mathematician or astronomer who first suggested, against all appearances, that the earth is not in the center of the universe and that it is not at rest but moving along at great speed. This speculator also encouraged the inventors of the heliocentric system, Aristarchus of Samos and Copernicus, who greatly admired Philolaus.

BIBLIOGRAPHY

On Philolaus and his influence, see A. Boeckh, *Philolaus des Pythagoreers Lehren nebst den Bruchstücken seines Werkes* (Berlin, 1819); W. Burkert, *Weisheit und Wissenschaft* (Nuremburg, 1962); H. Diels, "Über die Excerpte von Menons Iatrica in dem Londoner Papyrus 137," in *Hermes*, **28** (1893), 428–434; K. von Fritz, *Grundprobleme der Geschichte der antiken Wissenschaften* (Berlin, 1970), 157–166; R. Mondolfo, "Sui frammenti di Filolao. Contribuzione a una revisione del processo di falsità," in *Rivista di filologia e di istruzione classica*, n.s. **15** (1937), 225–245; B. L. Van der Waerden, "Die Astronomie der Pythagoreer," in *Verhandlingen der K. nederlandsche akademie van wetenschappen. Afdeeling natuurkunde*, **20**, pt. 1 (1951), 1–136; and K. von Fritz, "Philolaos" in *Pauly-Wissowa*, supp. XIII (1974), 453–483.

Kurt von Fritz

PHILOPONUS. See **John Philoponus.**

PIANESE, GIUSEPPE (*b*. Civitanova del Sannio, Campobasso, Italy, 19 March 1864; *d*. Naples, Italy, 22 March 1933), *pathological anatomy.*

For a detailed study of his life and work, see Supplement.

PIAZZI, GIUSEPPE (*b*. Ponte in Valtellina, Italy [now Switzerland], 16 July 1746; *d*. Naples, Italy, 22 July 1826), *astronomy.*

As a young man, Piazzi entered the Theatine Order in Milan. He completed his studies there and in Rome, obtaining the doctorate in philosophy and mathematics. From 1769 until 1779 Piazzi taught mathematics in a number of Italian cities; in 1780 he was summoned by the prince of Caramanico, Bourbon viceroy of Sicily, to fill the chair of higher mathematics at the Academy of Palermo. The viceroy encouraged Piazzi in his wish to establish an astronomical observatory in Palermo, and toward the end of the 1780's Piazzi went to England in order to obtain the best possible equipment.

In England, Piazzi met Maskelyne, William Herschel, and Ramsden. He investigated Herschel's large telescopes (indeed, he fell and broke his arm while examining one of them that was mounted outdoors) and, with Maskelyne, observed at Greenwich the solar eclipse of 3 June 1788. His first astronomical work, a study of the difference in longitude between various observatories, based on that of Greenwich, was published in the *Philosophical Transactions of the Royal Society* in 1789. The most important result of Piazzi's English visit, however, was the great five-foot vertical circle, a masterpiece of eighteenth-century technology, that he commissioned from Ramsden. It was installed in the new observatory in the Santa Ninfa tower of the royal palace of Palermo in 1789, and is still preserved there. The Palermo observatory opened in 1790, and Piazzi was appointed its director, a post that he retained for most of the rest of his life.

Having returned to Palermo, Piazzi took up the problem of the precise determination of the astronomical coordinates (direct ascension and declination) of the principal stars. Palermo was then the southernmost European observatory, and the favorable climatic conditions that it provided allowed him to study more stars than had been previously cataloged, with a greater degree of accuracy. In the course of his regular observations Piazzi, on the night of 1 January 1801, was searching a region in Taurus in which he hoped to see a star of the seventh magnitude, listed in Lacaille's catalog, which he had previously observed. Before that star appeared, however, he noticed the passage of a somewhat fainter body that Lacaille had not listed. Piazzi continued to observe the new body on the following evenings and ascertained from its movement that it must be a planet or comet. He watched it regularly until 11 February 1801, during which period its retrograde motion ceased and it began to advance, until it had moved near enough to the sun that it could not be seen at its passage to the meridian. It was therefore necessary to calculate its orbit around the sun to find it again.

Piazzi communicated his discovery almost immediately to his friend Barnaba Oriani, director of the Brera observatory in Milan, and to J. E. Bode, director of the Berlin observatory. As early as 24 January 1801, he wrote to Oriani:

> I have announced this star as a comet, but since it is not accompanied by any nebulosity and, further, since its movement is so slow and rather uniform, it has occurred to me several times that it might be something better than a comet. But I have been careful not to advance this supposition to the public. I will try to calculate its elements when I have made more observations.

In his reply to Piazzi, Oriani wrote, "I congratulate you on your splendid discovery of this new star. I do not think that others have noticed it, and because of its smallness, it is unlikely that many astronomers will see it."

Piazzi had made observations for a period of forty-one days, over a geocentric arc of only 3°. He published his observations in 1801 as *Risultati delle osservazioni della nuova stella scoperta il 1° gennaio 1801 nell'Osservatorio di Palermo.* Other astronomers were eager to rediscover the new body; if it were a planet, it should be possible, on the example of Uranus, to compute from Piazzi's observations a circular orbit, even if the arc of the presumably elliptical orbit were to prove short. In December 1801 Gauss calculated both such an orbit and an ephemeris for the new body. He communicated his calculations to F. X. von Zach, director of the Gotha observatory, who employed them to rediscover the body in almost exactly the position that Gauss had predicted. It was thus apparent that it was a planet, and in a publication of 1802, *Della scoperta del nuovo pianeta "Cerere Ferdinandea,"* Piazzi named it for Ceres, the patron goddess of Sicily.

Piazzi's discovery involved him in a genteel polemic with William Herschel; on 22 May 1802, following Olbers' discovery of Pallas, Herschel wrote to Piazzi from Slough to argue that these new planets could not be so called in the same sense as the planets within the solar system. He proposed that Ceres and Pallas be called "asteroids," since they are intermingled with, and similar to, the small fixed stars. He further suggested that these bodies were not worthy of the name of planets, since they did not occupy the space betweeen Mars and Jupiter "with sufficient dignity." Herschel went on to advocate three forms of celestial bodies—planets, asteroids, and comets—a hierarchy

that led Piazzi to gloss his letter with the remark, "Soon we shall also be seeing counts, dukes, and marquesses in the sky!" (In the course of time, the question has been resolved in Piazzi's favor, since "asteroid" has fallen into disuse, while the term "small planet" has become standard.)

In 1792 Piazzi returned to making precise determinations of the coordinates of the fixed stars. After ten years of intense and fatiguing work, he published at Palermo, in 1803, a catalog of the medial positions of 6,748 stars, under the title *Praecipuarum stellarum inerrantium positiones mediae ineunte saeculo decimonono ex observationibus habitis in specula panoromitana ab anno 1792 ad annum 1802.* This catalog was more accurate than any of its predecessors; Zach pronounced it epochal, and the Institut de France awarded it the Lalande prize for the best astronomical work published in 1803.

Piazzi then carried his work a step further. Doubting the value of the precession of the equinoxes used at that time, Piazzi undertook to determine the right ascension of a number of basic stars, relating them directly to the sun, in order to improve on earlier observations (including those made at Greenwich). Since he was at that time in poor health, he enlisted the aid of Niccolò Cacciatore as his collaborator. Piazzi's *Praecipuarum stellarum inerrantium positiones mediae ineunte saeculo decimonono ex observationibus habitis in specula panormitana ab anno 1792 ad annum 1813,* published in Palermo in 1813, cataloged the mean position of 7,646 stars. It was widely esteemed among astronomers, and the Institut de France again awarded Piazzi a prize.

In 1817 Piazzi published (again at Palermo) the two-volume *Lezioni elementari di astronomia,* which he sent to Oriani, who discussed it appreciatively. Oriani was also anxious to persuade Piazzi to republish his earlier observations, but he did not do so. (In 1845 L. von Littrow incorporated Piazzi's observations of 1792 to 1795 in *Annalen der K. K. Sternwarte in Wien.*)

In March 1817 Piazzi was summoned to Naples by King Ferdinand I, who wished him to supervise the completion of the observatory already under construction on the hill at Capodimonte. The building had been begun under the direction of Joachim Murat, but had remained unfinished because of the unstable political circumstances of the kingdom. Although he received an enthusiastic reception, Piazzi was reluctant to leave Palermo. He stated his feelings in a letter to Oriani: "I shall never yield to the invitations and kindnesses that are showered upon me so that I might remain in Naples. I would thus stain the last years of my life with the vilest ingratitude." The king nevertheless appointed him director general of the observatories of both Sicily and Naples, with the freedom to remain in whichever of the two kingdoms that he preferred, and Piazzi subsequently divided his time between the two. He took considerable pains in the building and equipping of the Naples observatory, and, on Oriani's recommendation, secured Carlo Brioschi, of the Istituto Geografico Militare Lombardo, as its director. Brioschi gratified Piazzi by publishing, in 1824, *Comentari astronomici della specola reale di Napoli.*

Piazzi returned to settle in Naples in 1824, his health weakened. The king had commissioned him to reform the system of weights and measures, and he had been elected president of the Neapolitan Academy of Sciences. He died in Naples of an acute disease.

BIBLIOGRAPHY

I. ORIGINAL WORKS. Piazzi's major works are cited in the text. In addition, a partial bibliography—beginning with his first publication, "Results of Calculations of the Observations Made at Various Places of the Eclipse of the Sun . . . on June 3, 1788," in *Philosophical Transactions of the Royal Society*, **79** (1789), 55–61—is in the Royal Society *Catalogue of Scientific Papers*, IV, 897.

II. SECONDARY LITERATURE. See G. Abetti, *Storia dell'astronomia* (Florence, 1963); F. Angelitti, "Per il centenario della morte dell'astronomo Giuseppe Piazzi," in *Memorie della Società astronomica italiana*, **3** (1925), 369–395; A. Bemporad, "Giuseppe Piazzi—commemorazione tenuta nella R. Università di Napoli," *ibid.*, 396–413; and Rudolf Wolf, *Biographien zur Kulturgeschichte der Schweiz*, IV (Zurich, 1862), 275–292.

GIORGIO ABETTI

PICARD, CHARLES ÉMILE (*b.* Paris, France, 24 July 1856; *d.* Paris, 11 December 1941), *mathematics.*

Picard's father, the director of a silk factory, was of Burgundian origin; his mother was the daughter of a doctor from northern France. At the death of her husband, during the siege of Paris in 1870, she was obliged to seek employment in order to care for her two sons. Picard was a brilliant student at the Lycée Henri IV and was especially interested in literature, Greek, Latin, and history. An avid reader with a remarkable memory, he acquired a rare erudition. For many years he retained a liking for physical exercise—gymnastics and mountain climbing—and an interest in carefully planned travel. He chose his vocation after reading a book on algebra at the end of his secondary studies. In 1874, after only one year of preparation, he was accepted as first candidate by the École Normale Supérieure and as second candidate by the École Polytechnique. After a famous interview with Pasteur, he chose the former, where he would be permitted to devote himself entirely to research. He placed first in the competition for the *agrégation* in 1877 but had already made several important discoveries and had received the degree of *docteur ès sciences.*

From 1877 to 1878 Picard was retained as an assistant at the École Normale Supérieure. Appointed professor at the University of Toulouse in 1879, he returned to Paris in 1881 as lecturer in physical and experimental mechanics at the Sorbonne and as lecturer in mechanics and astronomy at the École Normale Supérieure. Although he accepted these teaching posts outside his preferred field, Picard continued his work in analysis, and the first of the two famous theorems that bear his name dates from 1879, when he was twenty-three. In 1885 he was unanimously elected to the chair of differential and integral calculus at the Sorbonne, where he served as his own *suppléant* before reaching the prescribed age of thirty for the post. In 1897, at his own request, he exchanged this chair for that of analysis and higher algebra, where he was able to train students for research.

Nominated in 1881 by the section of geometry for election to the Académie des Sciences, he was elected in 1889. In 1886 he received the Prix Poncelet and in 1888 the Grand Prix des Sciences Mathématiques for a memoir that was greatly admired by Poincaré. Picard's mathematical activity during the period 1878–1888 resulted in more than 100 articles and notes. A member of the Académie Française (1924), he received the Grande Croix de la Légion d'Honneur in 1932 and the Mittag-Leffler Gold Medal from the Swedish Academy of Sciences in 1937. He received an honorary doctorate from five foreign universities and was a member of thirty-seven academies and learned societies.

Picard was chairman of numerous commissions, including the Bureau des Longitudes; and his administrative abilities and his sincere and resolute character earned him great prestige. As permanent secretary of the Académie des Sciences from 1917 to his death in 1941, he wrote an annual notice on either a scientist or a subject of current interest. He also wrote many prefaces to mathematical books and participated in the publication of works of C. Hermite and G.-H. Halphen.

An outstanding teacher, Picard was devoted to the young, and from 1894 to 1937 he trained more than 10,000 engineers at the École Centrale des Arts et Manufactures. He was responsible, with extraordinary success, for choosing pupil-teachers at the École Normale Supérieure de Jeunes Filles de Sèvres (1900–1927). He was director of the Société des Amis des Sciences, founded by Pasteur to look after needy scholars and their families.

In 1881 Picard married the daughter of his mentor and friend Charles Hermite. His life of uninterrupted professional success was clouded by the death of his daughter and two sons in World War I. His grandsons were wounded and captured in World War II, and the invasion and occupation of France darkened the last two years of his life. He died in the Palais de l'Institut, where he lived as *secrétaire perpétuel* of the Academy.

Picard's works were mostly in mathematical analysis and algebraic geometry. As early as 1878 he had studied the integrals of differential equations by making successive substitutions with equations having suitable partial derivatives. The following year he discovered the first of the two well-known theorems that bear his name. The first states: Let $f(z)$ be an entire function. If there exist two values of A for which the equation $f(z) = A$ does not have a finite root, then $f(z)$ is a constant. From this theorem it follows that if $f(z)$ is an entire function that is not a constant, there cannot be more than one value of A for which $f(z) = A$ has no solution.

Picard's second theorem, which extended a result stated by Weierstrass, states: Let $f(z)$ be a function, analytic everywhere except at a, where it has an essential isolated singularity; the equation $f(z) = A$ has in general an infinity of roots in any neighborhood of a. Although the equation can fail for certain exceptional values of the constant A, there cannot be more than two such values (1880). This result led to a classification of regular analytic functions; and it was the origin of important work carried out especially by Émile Borel and Otto von Blumenthal. The latter established generalizations that he called Picard's little theorem and Picard's big theorem. Picard's theorems revealed the fruitfulness of the idea of introducing, in the terms of a problem, a restriction bearing on the case of an exception that can be shown to be unique.

From 1883 to 1888 Picard extended Poincaré's investigations on automorphic functions to functions of two complex variables, which he called hypergeometric and hyperfuchsian (1883, 1885). These functions led Picard to the study of algebraic surfaces (1901). Setting himself the task of studying the analogies between the theory of linear differential equations and the theory of algebraic equations, Picard took up Galois's theory and obtained for a

linear differential equation a group of transformations now called the Picard group.

Picard's method for demonstrating the existence of the integrals of differential equations by successive approximations at first appears very simple. The introduction of n functions $u_1, u_2, \ldots, u_n$ reestablishes the system

$$\frac{du_i}{dx} = f_i(x, u_1, u_2, \cdots, u_n), \qquad i = 1, 2, \cdots, n,$$

with the initial conditions $x = x_0$ gives $u_i = a_i$. There is then resolved by n quadratures the system

$$\frac{dv_i}{dx} = f_i(x, a_1, a_2, \cdots, a_n),$$

the v_i satisfying the initial conditions and the same being true of

$$\frac{dw_i}{dx} = f_i(x, v_1, v_2, \cdots, v_n)$$

and so forth. It remains to prove—and this is the essential point—that under certain conditions (identified by Cauchy) the functions that are successively introduced tend toward limits that are precisely the desired integrals in the neighborhood of x_0. Picard himself extended his method to numerous cases, particularly to the equations of complex variables and also to integral equations. He, as well as his successors, thus demonstrated the preeminence of his method. Integral equations became of considerable importance in mathematical physics, with much of the genuine progress due to Fredholm. By completing the earlier works, Picard made more precise the necessary conditions for the existence of the various types of equations.

These works, as well as many dispersed results found in notes, were assembled in Picard's three-volume *Traité d'analyse*, which immediately became a classic and was revised with each subsequent edition. The work was accessible to many students through its range of subjects, clear exposition, and lucid style. Picard examined several specific cases before discussing his general theory.

In theoretical physics Picard applied analysis to theories of elasticity, heat, and electricity. He was particularly successful in achieving an elegant solution to the problem of the propagation of electrical impulses along cables (*équation des télégraphiques*). This research was to have been collected in a fourth volume of his treatise on analysis; but it appeared instead in four fascicles of *Cahiers scientifiques*.

After 1900 Picard published several historical and philosophical reflections, in particular *La science moderne et son état actuel* (1905), and speeches and reports. When he was more than eighty years old he presented considerations on the questions of homogeneity and similarity encountered by physicists and engineers.

Throughout his life Picard supported the innovations of other mathematicians, including the early work of Lebesgue. With Poincaré he was the most distinguished French mathematician of his generation.

BIBLIOGRAPHY

I. ORIGINAL WORKS. Picard's writings include "Sur la forme des équations différentielles du second ordre dans le voisinage de certains points critiques," in *Comptes rendus hebdomadaires des séances de l'Académie des sciences*, **87** (1878), 430–432, 743–746; "Mémoire sur les fonctions entières," in *Annales scientifiques de l'École normale supérieure*, 2nd ser., **9** (1880), 145-166; "Sur la réduction du nombre des périodes des intégrales abéliennes," in *Bulletin de la Société mathématique de France*, **11** (1883), 25–53; "Sur les fonctions hyperfuchsiennes provenant des séries hypergéométriques de deux variables," in *Annales scientifiques de l'École normale supérieure*, 3rd ser., **2** (1885), 357–384; "Mémoire sur la théorie des fonctions algébriques de deux variables indépendantes," in *Journal de mathématiques pures et appliquées*, 4th ser., **5** (1889), 135–319; *Traité d'analyse*, 3 vols. (Paris, 1891–1896); and *Théorie des fonctions algébriques de deux variables indépendantes*, 2 vols. (Paris, 1897–1906), written with Georges Simart.

Subsequent writings include "Sur la résolution de certaines équations à deux variables," in *Bulletin de la Société mathématique de France*, **25** (1901); *Sur le développement de l'analyse et ses rapports avec diverses sciences* (Paris, 1905); *La science moderne et son état actuel* (Paris, 1905); *L'histoire des sciences et les prétentions de la science allemande* (Paris, 1916); *Les sciences mathématiques en France depuis un demi-siècle* (Paris, 1917); *Discours et mélanges* (Paris, 1922); and *Mélange de mathématiques et de physique* (Paris, 1924).

His later writings are "Leçons sur quelques types simples d'équations aux dérivées partielles avec des applications à la physique mathématique," in *Cahiers scientifiques*, fasc. 1 (1925); "Leçons sur quelques équations fonctionnelles avec des applications à divers problèmes d'analyse et de physique mathématique," *ibid.*, fasc. 3 (1928); "Leçons sur quelques problèmes aux limites de la théorie des équations différentielles," *ibid.*, fasc. 5 (1930); "Leçons sur quelques équations fonctionnelles," *ibid.*, fasc. 6 (1930); *Un coup d'oeil sur l'histoire des sciences et des théories physiques* (Paris, 1930); "Quelques applications analytiques de la théorie des courbes et des surfaces algébriques," in *Cahiers scientifiques*, fasc. 9 (1931); and *Discours et notices* (Paris, 1936).

II. SECONDARY LITERATURE. An early biography of Picard is Ernest Lebon, *Émile Picard, biographie, bibliographie* (Paris, 1910), which has details of 256 of his works. See also René Garnier, ed., *Centenaire de la naissance d'Émile Picard* (Paris, 1957), which has reports of speeches by colleagues and pupils.

His mathematical discoveries are discussed in Émile Borel, *Leçons sur les fonctions méromorphes* (Paris, 1903), ch. 3; and Otto Blumenthal, *Principes de la théorie des fonctions entières d'ordre infini* (Paris, 1910), ch. 7.

LUCIENNE FÉLIX

PICARD, JEAN (*b.* La Flèche, France, 21 July 1620; *d.* Paris, France, 12 October 1682), *astronomy, geodesy.*

Nothing is known about Picard's youth. According to Esprit Pezenas in *Histoire critique de la découverte des longitudes* (1775), it was the astronomer Jacques de Valois who led Picard, a gardener for the duke of Créqui, to make astronomical observations and who may also have encouraged him to enter a seminary, where he seems to have taken religious orders; at the end of his life Picard may have been prior of the abbey of Rillé in Anjou. On 21 August 1645 Picard assisted Gassendi in the observation of a solar eclipse and remained with him for some time. Some biographers have thought that he substituted for and then replaced Gassendi in his chair at the Collège Royal, but archival evidence contradicts this hypothesis.

In 1666 Picard was named a founding member of the Académie Royale des Sciences and, even before its opening, participated in several astronomical observations. In collaboration with Adrien Auzout he perfected the movable-wire micrometer and utilized it to measure the diameters of the sun, the moon, and the planets. During the summer of 1667 he applied the astronomical telescope to the instruments used in making angular measurements—quadrants and sectors—and was aware that this innovation greatly expanded the possibilities of astronomical observation. The making of meridian observations by the method of corresponding heights, which he suggested in 1669, was not put into practice until after his death. Yet when the Academy decided to remeasure an arc of meridian in order to obtain a more accurate figure for the earth's radius, Picard was placed in charge of the operation. He employed the method of skeleton triangulation devised by Snell (1617) but greatly improved the associated observational techniques.

Picard decided to measure the distance between two localities at approximately the same meridian (Sourdon, near Amiens, and Malvoisine, near Corbeil-Essonnes), to determine the difference in their latitudes, and to deduce from these results the length of a degree of meridian. The project lasted from 1668 to 1670. The base he selected was very long (5,663 toises, as against the 168 used by Snell) and his triangulation consisted of thirteen triangles. But it was primarily through the use of instruments fitted with telescopes, quadrants, and sectors for angular measurements that Picard attained a precision thirty to forty times greater than that achieved previously. Consequently, from his measurements of the various distances and angles, Picard was able to obtain the notable result of 57,060 toises for the terrestrial degree (Lacaille in 1740 and Delambre in 1798 obtained 57,074 toises). This increased precision made possible a great advance in the determination of geographical coordinates and in cartography, and enabled Newton in 1684 to arrive at a striking confirmation of the accuracy of his principle of gravitation.

The results of this undertaking were scarcely published in Picard's *Mesure de la terre* (1671) when he undertook another project that he had proposed to the Academy in 1669: to determine the coordinates of Uraniborg in order to compare the important observations that Tycho Brahe had made there between 1576 and 1597 with those that would be obtained at the Paris observatory. During eight months in Denmark in 1671–1672, Picard, aided by one assistant and a young Danish astronomer, Ole Römer—whose outstanding ability he recognized—obtained the data he was seeking. He convinced Römer to return with him to France, and Römer carried out brilliant work at Paris until 1681. In compiling his registers of observations, Picard noted an annual displacement of the polestar that was later explained by the combination of aberration and nutation; he also brought back a copy of Tycho Brahe's registers of observations.

In 1673 Picard moved into the Paris observatory and collaborated with Cassini, Römer, and, later, Philippe de La Hire on the institution's regular program of observations. He also joined many missions away from the observatory. The first of these enabled him to provide more precise data on the coordinates of various French cities (1672–1674); others, conducted from 1679 to 1681 with La Hire, had the purpose of establishing the bases of the principal triangulation of a new map of France. The results of these geodesic observations were published in 1693 by La Hire. The new map that was drawn from them contained corrections of as much as 150 kilometers' longitude and 50 kilometers' latitude.

Meanwhile, in 1674–1675 Picard played a very active role in the important surveying operations undertaken by the Academy to supply the châteaux of Marly and Versailles with water. He perfected the existing methods, notably through use of the telescope

level, and prepared a treatise on this subject published by La Hire in 1684. Picard was not able to put the finishing touches on the book; for, although he was still making observations on 12 September 1682, he died on 12 October after a brief illness.

Picard was also responsible for many improvements in procedures and instruments and for observations that stimulated later discoveries. For example, he pointed out the phenomenon of "barometric glow," the first observation of electric discharge in a rarefied gas. He was also one of the first to observe the fixed stars in full daylight. In addition, he noted the influence of temperature on atmospheric refraction and participated in the measurement of the parallax of Mars in 1672. But it was not he who published the first volumes of *Connaissance des temps* (1679–1684), as has been supposed, but Joachim Dalencé, who obtained the royal privilege for them.

Considered in his time a scientist of the first rank, Picard has been eclipsed by several of his contemporaries—for instance, Cassini. Nevertheless, he helped to advance several branches of science, exhibiting a remarkable flair for observation and a very refined sense of the practical.

BIBLIOGRAPHY

I. ORIGINAL WORKS. Picard wrote two books, *Mesure de la terre* and *Traité du nivellement* (see below), and a number of memoirs that appeared in the following compendia published by the Académie Royale des Sciences: *Recueil de plusieurs traitez de mathématiques* . . . (Paris, 1676), abbrev. as *Recueil de plusieurs traitez*; *Divers ouvrages de mathématiques et de physique* . . . (Paris, 1693), 337–422: "Divers ouvrages de M. Picard," P. de La Hire, ed., abbrev. as *Divers ouvrages*; *Recueil d'observations faites en plusieurs voyages par ordre de sa majesté pour perfectionner l'astronomie et la géographie avec divers traitez astronomiques* . . . (Paris, 1693), no. 3, P. de La Hire, ed., abbrev. as *Recueil*; *Mémoires de l'Académie royale des sciences depuis 1666 jusqu'en 1699*, VI (Paris, 1730), 479–707, "Divers ouvrages de M. Picard," abbrev. as *Mémoires*, VI; *Mémoires de l'Académie royale des sciences depuis 1666 jusqu'en 1699*, VII (Paris, 1729), 191–411, abbrev. as *Mémoires*, VII.

Mesure de la terre (Paris, 1671) was repr. without change in 1676 (*Recueil de plusieurs traitez*, no. 2) and repub. in 1729 (*Mémoires*, VII, 133–190) and in 1740 (in *Degré du méridien entre Paris et Amiens déterminé par la mesure de M. Picard et par les observations de MM. de Maupertuis, Clairaut, Camus, Le Monnier* [Paris, 1740], 1–106).

Traité du nivellement . . . avec une relation de quelques nivellements faits par ordre du roy et un abrégé de la Mesure de la terre . . . (Paris, 1684), which included additional material supplied by the editor, P. de La Hire, was repub.

in 1728, 1730 (*Mémoires*, VI, 631–707), and 1780; it was also trans. into Italian (Florence, 1723) and German (Berlin, 1749).

Picard's other writings were published in 1693 by P. de La Hire—generally on the basis of MSS, some of which were incomplete—and were brought out again in 1729–1730 in vols. VI and VII of the *Mémoires*. These include "De la pratique des grands cadrans par le calcul" (*Divers ouvrages*, 341–365; *Mémoires*, VI, 481–531); "De mensuris" (*Divers ouvrages*, 366–368; *Mémoires*, VI, 532–537); "De mensura liquidorum & aridorum" (*Divers ouvrages*, 370–374; *Mémoires*, VI, 540–549); "Fragmens de dioptrique" (*Divers ouvrages*, 375–412; *Mémoires*, VI, 550–627); "Voyage d'Uranibourg ou observations astronomiques faites au Danemarck" (*Recueil*, 1–29; *Mémoires*, VII, 191–230)—the title page of the first version of this work seems to indicate that it was published separately in 1680; "Observations astronomiques faites en divers endroits du royaume en 1672, 1673, 1674" (*Recueil*, 33–46; *Mémoires*, VII, 327–347); "Observations faites à Brest et Nantes en 1679," written with La Hire (*Recueil*, 47–56; *Mémoires*, VII, 377–390); "Observations faites à Bayonne, Bordeaux et Royan en 1680," written with La Hire (*Recueil*, 57–64; *Mémoires*, VII, 391–398); and "Observations faites aux côtes septentrionales de France en 1681," written with La Hire (*Recueil*, 65–76; *Mémoires*, VII, 399–411).

Picard's astronomical observations made at Paris between 1666 and 11 Sept. 1682 were incorporated by P. C. Le Monnier in his *Histoire céleste* . . . (Paris, 1741), 1–263. The registers that Picard kept of his observations are at the Paris observatory.

II. SECONDARY LITERATURE. The few biographical accounts devoted to Picard are quite short: J. A. N. de Concordet, in *Éloges des académiciens* . . . (Paris, 1773), 36–48; J. B. Delambre, in Michaud, ed., *Biographie universelle*, XXXIV (Paris, 1823), 253–256, and new ed., XXXIII (Paris, 1861), 173–175; F. Arago, in *Notices biographiques*, III (Paris, 1854), 313–314; E.-M., in F. Hoefer, ed., *Nouvelle biographie générale* (Paris, 1862), cols. 48–49; E. Doublet, in *Revue générale des sciences* . . ., **31** (1920), 561–564; F. Boquet, in *Histoire de l'astronomie* (Paris, 1925), 362–364; and E. Armitage, in *Endeavour*, **13**, no. 49 (1954), 17–21.

More precise details are in Fontenelle, *Histoire de l'Académie royale des sciences*, I (Paris, 1733), see secs. on "Astronomie" for 1666–1682, and II (1733), 202, 222–223, 353–354; and C. Wolf, *Histoire de l'observatoire de Paris de sa fondation à 1793* (Paris, 1902), see index. Picard's astronomical and geodesic work is analyzed in some detail by J.-S. Bailly in *Histoire de l'astronomie moderne*, II (Paris, 1779), 335–354; J. de Lalande, in *Bibliographie astronomique* (Paris, 1803), see index; J. B. Delambre, in *Histoire de l'astronomie moderne*, II (Paris, 1821), 597–632; and G. Bigourdan, *L'astronomie* . . . (Paris, 1916), see index. His work in cartography was treated by L. Gallois, "L'Académie des sciences et les origines de la carte de Cassini," in *Annales de géographie*, **18** (1909), 193–204.

Picard's collaboration with Auzout is discussed in a dissertation by R. M. McKeon, "Établissement de l'astronomie de précision et oeuvre d'Adrien Auzout" (Paris, 1965), see index.

JULIETTE TATON
RENÉ TATON

PICCARD, AUGUSTE (*b.* Basel, Switzerland, 28 January 1884; *d.* Lausanne, Switzerland, 24 March 1962), *physics*.

With his twin brother, Jean Félix (*d.* Minneapolis, Minnesota, 23 January 1960), Auguste Piccard achieved fame and distinction as a scientist and explorer of the stratosphere and the ocean depths. Sons of Jules and Hélène Haltenhoff Piccard, the Piccards were members of a prominent Vaudois family: their grandfather was *commissaire général* of the canton; their father, head of the chemistry department at the University of Basel; and their uncle Paul, designer of the first Niagara Falls–type turbines, which he manufactured and sold through the Piccard-Pictet Company he founded in Geneva. They attended the local *Oberrealschule* before entering the Federal Institute of Technology in Zurich, from which Auguste received a degree in mechanical, and Jean Félix in chemical engineering. Obtaining doctorates in their respective fields, the brothers served for many years as university professors, Auguste in Zurich and Brussels, Jean Félix at Munich (where he was assistant to Adolph von Baeyer), Lausanne, Chicago, the Massachusetts Institute of Technology, and the University of Minnesota. Jean also held positions with the Hercules Powder Company in Delaware and the Bartol Foundation of the Franklin Institute, Philadelphia; he became a U.S. citizen in 1931.

Auguste Piccard attracted world attention when, on 27 May 1931, ascending with Paul Kipfer from Augsburg, Germany, in a free balloon, he achieved a new altitude record of 51,775 feet. The sixteen-hour flight, which ended on an Austrian glacier, marked the first use of a pressurized cabin for manned flight. It was followed on 18 August 1932 by an ascent with Max Cosyns from Zurich that attained 53,153 feet and ended near Lake Garda, Italy. After his last flight, in 1937, Piccard devoted himself for the better part of ten years to studies aimed at realizing his youthful dream to "plunge into the sea deeper than any man before." Although interrupted by World War II, his research resulted in 1948 in an unsuccessful first trial of his bathyscaphe (from the Greek for "deep" and "boat")—a self-propelled, untethered metal sphere designed on balloon principles and intended to withstand pressures of 12,000 pounds per square inch at a depth of 12,000 feet, off the Cape Verde Islands. Despite this failure, a second model in 1953 carried Auguste and his son Jacques to a depth of 10,168 feet, thereby trebling the 1934 record of William Beebe off the Bermuda coast. Piccard and his son built a third bathyscaphe, *Trieste*, which he sold to the U.S. navy in the late 1950's. He lived to see his son, with Lt. Don Walsh, USN, set a new world record of 35,800 feet with the *Trieste* in the Marianas Trench of the Pacific Ocean on 23 January 1960. Father and son were working on a new ship called a mesoscaphe at the time of Piccard's death.

BIBLIOGRAPHY

I. ORIGINAL WORKS. Piccard wrote three books: *Au-dessus des nuages* (Paris, 1933); *Entre ciel et terre* (Lausanne, 1946); and *Au fond des mers en bathyscaphe* (Paris, 1954), the last published simultaneously in German as *Über den Wolken. Unter den Wellen* (Wiesbaden, 1954). A bibliography of his articles is in Latil and Rivoire (below).

II. SECONDARY LITERATURE. Works on Piccard are Adelaide Field, *Auguste Piccard, Captain of Space, Admiral of the Abyss* (Boston, 1969); Alan Honour, *Ten Miles High, Two Miles Deep; The Adventures of the Piccards* (New York, 1957); Pierre de Latil and Jean Rivoire, *Le professeur Auguste Piccard* (Paris, 1962); Alida Malkus, *Exploring the Sky and Sea; Auguste and Jacques Piccard* (Chicago, 1961); and Kurt R. Stehling and William Beller, "The First Space-Gondola Flight," in *Skyhooks* (New York, 1962), ch. 13.

MARVIN W. MCFARLAND

PICCOLOMINI, ARCANGELO (*b.* Ferrara, Italy, 1525; *d.* Rome, Italy, 19 October 1586), *medicine, anatomy, physiology.*

Little is known of Piccolomini's early life. He received his doctorate in philosophy and medicine at Ferrara, probably in the late 1540's, and then taught philosophy at the University of Bordeaux. In 1556 he published at Paris a commentary on Galen's *De humoribus*; Piccolomini dedicated the work to Bishop Michele della Torre, the papal nuncio in France. Under the latter's patronage Piccolomini went to Rome, where he was eventually named physician to Pope Pius IV (1559–1565). He retained this office until his death, serving in turn Pius V, Gregory XIII, and Sixtus V. In 1575 he was given the chair in medical practice at the Sapienza, where he also lectured on anatomy. In 1582 he became general *protomedicus* for the Papal States.

In 1586 Piccolomini published his course of anatomical lectures, *Anatomicae praelectiones*, to forestall an unauthorized edition based on the notes of students. The course itself was supplemented by a series of anatomical demonstrations by the prosector Leonardo Biondini. This collaboration may explain why Piccolomini did not stress thorough morphological description in his own lectures, except in individual instances where he regarded the descriptions of earlier anatomists as faulty. Among his more noteworthy descriptions were those of the abdominal muscles, the termination of the acoustic nerve, the anastomoses of the fetal heart, and the differences between the male and female pelvis. He was the first anatomist after Salomon Alberti (1585) to describe the venous valves as a general phenomenon, although, like Alberti, Piccolomini probably learned of the valves from Fabrici, who had publicly announced the discovery at Padua in 1578 or 1579. Piccolomini also related a number of interesting pathological observations.

On the whole, though, the element of descriptive anatomy in the *Praelectiones* is quite subordinate to the discussion of highly abstract questions of psychology and physiology. In its theoretical orientation the work resembles the "Physiologia" (1542) of Jean Fernel, who was the one contemporary to receive Piccolomini's unqualified praise. Piccolomini followed Fernel in maintaining a strongly Neoplatonist view of the soul and in stressing the importance of "supra-elementary" powers as causes of vital phenomena. He thought, for example, that the formative powers of animal semen were due to three supraelementary powers derived from the parents, namely celestial heat, celestial spirit, and the vegetative soul—"instructed and taught by the most wise God." But these powers could only begin the process of generation, the completion of which required the direct infusion of a substantial form from the heavenly bodies for each new creature.

While Piccolomini derived much of his physiology from Aristotle and Galen, his system had many characteristic features of its own. Most notably, where Aristotle regarded the heart as the one ruling organ of the body and Galen emphasized the independence of the brain and heart in controlling different aspects of bodily function, Piccolomini upheld the supreme hegemony of the brain, on which even the heart depends for its vivifying and pulsatile faculties. Piccolomini also rejected other Aristotelian and Galenic doctrines in favor of supposedly Hippocratic views: that the semen of both parents is drawn from all parts of their bodies, that the heart is a muscle, and that the peculiar "celestial heat" of animals is not innate but is inhaled together with the air. On numerous other points of detail Piccolomini introduced his own functional doctrines, most of them derived speculatively; and his opposition to several of Colombo's physiological theories included a lengthy refutation of the pulmonary circulation. *Anatomicae praelectiones* was not reprinted after 1586, but it left a number of discernible marks on the detailed texture of physiological thought during the late sixteenth and early seventeenth centuries.

BIBLIOGRAPHY

I. ORIGINAL WORKS. Piccolomini's works include *In librum Galeni de humoribus commentarii*, with Greek and Latin texts (Paris, 1556; Venice, 1556); and *Anatomicae praelectiones . . . explicantes mirificam corporis humani fabricam: et quae animae vires, quibus corporis partibus, tanquam instrumentis, ad suas obeundas actiones, utantur; sicuti tota anima, toto corpore* (Rome, 1586).

II. SECONDARY LITERATURE. On Piccolomini and his work, see Francesco Pierro, *Arcangelo Piccolomini Ferrarese (1525–1586) e la sua importanza nell'anatomia postvesaliana*, which is Quaderni di storia della scienza e della medicina, no. 6 (Ferrara, 1965).

JEROME J. BYLEBYL

PICKERING, EDWARD CHARLES (*b.* Boston, Massachusetts, 19 July 1846; *d.* Cambridge, Massachusetts, 3 February 1919), *astronomy*.

Pickering, the elder son of Edward Pickering and Charlotte Hammond, was a descendant of one of New England's oldest and most distinguished families. John Pickering, his first American ancestor, had emigrated from Yorkshire and settled in Salem, Massachusetts, in 1636, bringing with him the family coat of arms (a lion rampant) and the motto *Nil desperandum*. Timothy Pickering, Edward's great-grandfather, served in the cabinets of Washington and John Adams.

Pickering attended the Boston Latin School for five years, where he "studied little and learnt less." Forced to memorize long passages of such works as Xenophon's *Anabasis*, he acquired a great distaste for the classics. He did, however, find time to read mathematical works on his own, for example Charles Davies' *Legendre's Elements of Geometry and Trigonometry*. Finding these more to his liking, he proposed entering the Lawrence Scientific School at Harvard, only to be informed by his schoolmaster that the "only requisite would be to know enough to come in when it rained."

Pickering entered the chemical department of the Lawrence Scientific School in 1862, largely on the advice of Charles William Eliot, then assistant professor of mathematics and chemistry. There he "found studies hard but delightful and enjoyed work exceedingly." That spring, although offered a position as assistant instructor, he declined it on Eliot's advice and entered the engineering department. He graduated *summa cum laude* in 1865, on his nineteenth birthday.

After two years as assistant instructor of mathematics at the Lawrence Scientific School, Pickering was appointed assistant professor of physics at the recently founded Massachusetts Institute of Technology. During his ten years there, he revolutionized the teaching of physics. With the encouragement of the Institute's founder and president, William Barton Rogers, Pickering established the first physical laboratory in America specifically designed for student instruction. He devised a series of experiments on the construction and use of apparatus, the properties of gases, and the mechanics of solids, and wrote instructions to enable students to perform them. The students were further encouraged to design experiments and to publish their original research. Pickering later compiled the instructions he had written and published them as *Elements of Physical Manipulation* (2 vols. [Boston, 1873–1876]), thereby producing the first American laboratory manual of physics.

On 10 October 1876 Eliot, who had been elected president of Harvard in 1869, appointed Pickering director of the Harvard College Observatory. Pickering and his wife, Elizabeth Wadsworth Sparks (the daughter of Jared Sparks, a noted historian and former Harvard president), moved to Observatory Hill on 1 February 1877; and he began work that very day. Eliot's selection of a physicist for the post rather than an observational astronomer evoked considerable criticism. Eliot, however, had sound reasons for his appointment: he had known Pickering both as an undergraduate at the Lawrence Scientific School and as a colleague on the MIT faculty, and was thoroughly familiar with his unusual scientific and administrative abilities. It is also possible that Eliot was aware that the direction of astronomical research was undergoing a crucial change.

Pickering realized at once that the greatest opportunities in astronomical research lay in the new field of astrophysics rather than in the astronomy of position and motion, which had occupied the chief place in the programs of other observatories. This is not to say that he ignored the "old" astronomy—several members of the observatory staff spent twenty years in preparing each of the two Harvard zones of the Astronomische Gesellschaft's star catalog—but the astrophysical work accomplished under his directorship was of incomparably greater volume and importance. He was a pioneer in three main fields of astronomical research: visual photometry, stellar spectroscopy, and stellar photography.

Before Pickering began his photometric measurements, he made two important decisions: (1) he adopted the magnitude scale suggested by Norman Pogson in 1854, whereby a change of one magnitude represents a change of a factor of 2.512 in brightness, and (2) he chose α Ursae Minoris (Polaris), then thought to be of constant brightness, as the standard and arbitrarily assigned a magnitude of 2.1 to it. Working in close cooperation with George B. Clark, Pickering designed and had the firm of Alvan Clark and Sons construct several new models of photometers. This culminated in the development of a revolutionary new instrument, the meridian photometer, in which the image of a star crossing the meridian is brought alongside the image of Polaris by suitable arrangement of mirrors and prisms. Thus each star could be measured at its point of highest visibility.

This photometric work continued for nearly a quarter of a century. Pickering never tired of the routine work involved and made more than 1.5 million photometric readings. The brightness of every visible star was measured and remeasured to obtain the greatest possible accuracy. The photometric studies culminated in 1908 with the publication of the *Revised Harvard Photometry*. Printed as volumes 50 and 54 of the *Annals of Harvard College Observatory*, it lists the magnitudes of more than 45,000 stars brighter than the seventh magnitude and remained the standard reference until photographic methods largely supplanted visual ones.

Pickering's researches in stellar spectroscopy were made possible largely through the establishment of the Henry Draper Fund in 1886. Under the terms of this fund, Mrs. Draper supplied money to the Harvard observatory for Pickering and his assistants to photograph, measure, and classify the spectra of the stars and to publish the resulting catalog in the *Annals* as a memorial volume to Henry Draper. The program consisted of three main parts: a general survey of stellar spectra for all stars north of $-25°$ and brighter than the sixth magnitude; a study of the spectra of the fainter stars; and a detailed investigation of the spectra of the brighter stars.

The spectra were produced by placing a large prism in front of the telescope's objective. While this did not give the definition attainable by use of a spectroscopic slit, it allowed a large number of spectra to be photographed on a single plate and gave sufficient definition. The principal investigators on the project were Williamina P. Fleming, Annie Jump Cannon, Antonia C. Maury (Henry Draper's niece), and a large corps of women computers. This led some contempo-

rary wags to refer to the Harvard team as "Pickering and his harem."

These investigations culminated with the publication of *The Henry Draper Catalogue*, printed between 1918 and 1924 as volumes 91–99 of the *Annals*. In this work nearly a quarter of a million stellar spectra were measured by Annie Jump Cannon and placed into one of twelve main spectral classes (P, O, B, A, F, G, K, M, R, N, Continuous, and Peculiar). This system was unanimously adopted by the International Union for Cooperation in Solar Research but was later modified slightly by the Committee of the International Astronomical Union on Spectral Classification.

The third principal field of Pickering's research was stellar photography. As early as 1883 he decided to chart all of the visible stars by means of photography. At the International Astrophotographic Congress organized in 1887 at Paris, however, it was decided that several observatories would participate in preparing a photographic atlas of the sky. Although the Congress intended to publish a map of the heavens in about five years, progress was so slow that Pickering maintained his desire to issue his own photographic map of the sky (the *Carte du ciel* is still incomplete). The acquisition of a substantial fund enabled him to carry out his plan. In 1889 Catherine Wolf Bruce, responding to a circular Pickering had issued, donated money for a large photographic telescope. The Bruce telescope, employing a photographic doublet with a twenty-four-inch aperture ground by the Clarks, was completed in 1893; and after a year and a half of testing, it was shipped to Harvard's Boyden Station at Arequipa, Peru, where it was used routinely in photographing the heavens.

In 1903 Pickering issued a *Photographic Map of the Entire Sky*, the first such map ever published, but it was not made with the Bruce telescope. Instead, two 2.5-inch doublets were used in Cambridge and Arequipa to map stars down to the twelfth magnitude on fifty-five plates. In addition, Pickering's habit of routinely photographing as large a portion of the visible sky as possible on every clear night resulted in the Harvard Photographic Library, which provides a photographic history, on some 300,000 glass plates, of all stars down to the eleventh magnitude. This record, duplicated nowhere else, is heavily relied on today by astronomers everywhere.

Other investigations that Pickering undertook were in photographic photometry. This was one of the chief interests of his later years, and an increasing part of the work of the observatory was devoted to establishing a standard system of stellar photographic magnitudes. The study of variable stars was also a marked feature of the observatory's work during his administration, and Pickering was instrumental in the founding of the American Association of Variable Star Observers. In 1889 he discovered that the brighter component of ζ Ursae Majoris (Mizar) was a spectroscopic binary—that is, a double star that is not resolvable in the telescope but which can be detected by the periodic doubling of its spectral lines. In 1886 he observed three systems of lines in ζ Puppis, the third of which formed a series closely resembling the Balmer lines of hydrogen. Pickering thought they represented hydrogen under some unknown conditions of temperature or pressure; Niels Bohr later showed that the "Pickering series" was actually due to ionized helium.

Pickering thoroughly believed in the advantages of broad associations in astronomy. One of his most cherished hopes was to organize a centralized institution to distribute funds to astronomers of all nations. He published several pamphlets on this subject, but his plan met with little success. For a short period Pickering was enabled to administer $500 gifts to American and European astronomers through another donation from Catherine Wolf Bruce. The establishment of the Carnegie Institution in 1902 raised his hopes, but he became bitterly disappointed when its executive committee made it clear that it preferred to support established observatories and other enterprises of its own creation rather than individual scientists. In 1906 he approached the Rockefeller Foundation for funds to implement another of his plans, the establishment of an international southern telescope at some favorable site, preferably in South Africa. Again he met with no success.

During his lifetime Pickering received numerous awards and honors. Six American and two European universities bestowed honorary doctorates upon him, and he was made a knight of the Prussian Ordre Pour le Mérite. Besides being a member of the American scientific societies, he was either a member or a foreign associate of the royal or national societies of England, France, Germany, Italy, Ireland, Sweden, Mexico, and Russia. He was awarded the Henry Draper Gold Medal of the National Academy of Sciences, the Rumford Gold Medal of the American Academy of Arts and Sciences, the Bruce Gold Medal of the Astronomical Society of the Pacific, and the gold medal of the Royal Astronomical Society on two occasions.

In all of his astronomical investigations, Pickering was not a speculator or theorizer but was content to be, in his words, "a collector of astronomical facts." Whereas theoretical reasoning not based on well-established data had little attraction for him, the posthumous value of the work of William Herschel and Friedrich Argelander appealed strongly to him.

Recognizing that the best service he could render to astronomy was the accumulation of facts, he instituted great research projects, often of a considerably routine nature, so that a sufficient basis in fact could be established for the solution of stellar problems by future astronomers.

When Pickering died in 1919, the Harvard College Observatory had been in operation for eighty years and he had been its director for forty-two of those years. Such a vast network of correspondence was established with observatories and astronomers throughout the world that the Harvard observatory under Pickering became the major distributing house of astronomical news. There was virtually no astronomer active during this period who did not benefit in one way or another from Pickering's interest and assistance. Indeed, many of his fellow astronomers thought of him as "the dean of American science."

BIBLIOGRAPHY

I. ORIGINAL WORKS. An extensive bibliography of Pickering's works numbering 266 items, prepared by Jenka Mohr, is appended to Solon I. Bailey, "Biographical Memoir of Edward Charles Pickering," in *Biographical Memoirs. National Academy of Sciences*, **15** (1934), 169–178. This bibliography does not include Pickering's contributions to the *Bulletin. Astronomical Observatory, Harvard College* or to *Circular. Astronomical Observatory of Harvard College* (many of which are unsigned), nor does it include the annual *Report. Astronomical Observatory of Harvard College*.

Pickering's papers are in the Harvard University Archives. The collection, which totals sixty-eight linear feet, includes personal and official correspondence, an autobiography, an autobiographical and personal notebook, and other notebooks and scrapbooks.

II. SECONDARY LITERATURE. There is as yet no full-length biography of Pickering. Bailey's "Memoir" cited above is taken, with only slight alterations, from his *History and Work of the Harvard Observatory* (New York, 1931). A recent and extremely valuable account of Pickering's observatory directorship can be found in Bessie Zaban Jones and Lyle Gifford Boyd, *The Harvard College Observatory. The First Four Directorships, 1839–1919* (Cambridge, Mass., 1971), more than half of which is devoted to the Pickering years.

HOWARD PLOTKIN

PICKERING, WILLIAM HENRY (*b.* Boston, Massachusetts, 15 February 1858; *d.* Mandeville, Jamaica, 16 January 1938), *astronomy*.

The younger brother of the astronomer E. C. Pickering, William graduated from Massachusetts

Institute of Technology in 1879. He taught there for a time and was appointed an assistant professor at Harvard observatory in 1887. In 1891 he set up Harvard's Boyden Station at Arequipa, Peru. Around 1900 he led expeditions to Jamaica, and from 1911 he was in charge of a permanent Harvard observing station there. Upon his retirement in 1924 the station became Pickering's private observatory.

Pickering was a pioneer in dry-plate celestial photography, and the Harvard photographic sky survey was undertaken at his suggestion. He took some of the earliest photographs of Mars (1888), and the lunar photographs he obtained in Jamaica (1900) were long the finest and most complete.

In 1899 Pickering discovered Phoebe on photographic plates taken, at his request, for possible new satellites of Saturn and demonstrated that it has a retrograde orbit. Saturn was the first planet known to possess both direct and retrograde satellites.

Pickering also made extensive visual observations of the planets and their satellites, discovering the "oases" on Mars (1892), recording apparent changes on the lunar surface (which he attributed to hoarfrost and vegetation), and claiming short rotation periods (now known to be incorrect) for Jupiter's Galilean satellites.

From 1907 Pickering paid considerable attention to predicting the location of trans-Neptunian planets; and after Pluto was discovered, faint images of it were located on plates taken for him in 1919. Percival Lowell, for whom Pickering had helped set up the observatory near Flagstaff, Arizona, in 1894, is generally accorded greater credit for the discovery, although Pickering's prediction was quite independent and more accurate in many respects; in any case, it has since become clear that the discovery was completely accidental.

BIBLIOGRAPHY

I. ORIGINAL WORKS. Among Pickering's principal writings are "Investigations in Astronomical Photography," in *Annals of Harvard College Observatory*, **32** (1895), 1–115; *The Moon* (New York, 1903); "The Ninth Satellite of Saturn," in *Annals of Harvard College Observatory*, **53** (1905), 45–73; "Researches of the Boyden Department," *ibid.*, **61** (1908), 1–103; "A Search for a Planet Beyond Neptune," *ibid.*, **61** (1909), 113–373; "Reports on Mars," nos. 1–44, in *Popular Astronomy*, **22–38** (1914–1930); *Mars* (Boston, 1921); and "Early Observations of the Elliptical Disks of Jupiter's Satellites," in *Annals of Harvard College Observatory*, **82** (1924), 61–74.

II. SECONDARY LITERATURE. Obituary notices are L. Campbell, in *Publications of the Astronomical Society of the Pacific*, **50** (1938), 122–125; and E. P. Martz, in *Popular Astronomy*, **46** (1938), 299–310. For a comparison of the conclusions by Lowell and Pickering concerning Mars, see W. W. Campbell, "The Problems of Mars," in *Publications of the Astronomical Society of the Pacific*, **30** (1918), 133–146. On Pickering's study of the Galilean satellites, see J. Ashbrook, "W. H. Pickering and the Satellites of Jupiter," in *Sky and Telescope*, **26** (1963), 335–336. On the nonpredictability of Pluto, see E. W. Brown, "On a Criterion for the Prediction of an Unknown Planet," in *Monthly Notices of the Royal Astronomical Society*, **92** (1931), 80–101.

BRIAN G. MARSDEN

PICTET, MARC-AUGUSTE (*b.* Geneva, Switzerland, 23 July 1752; *d.* Geneva, 19 April 1825), *physics*.

The son of Charles Pictet and Marie Dunant, Pictet came of an old and respected Genevan family. After a private education he studied at the Law Faculty of the Academy of Geneva, and it was only after qualifying as a lawyer in 1774 that he turned his attention seriously to science. His first mentor was the astronomer J. A. Mallet-Favre; but the greatest influence on him during his early years was that of H. B. de Saussure, who fostered Pictet's interest in physics and meteorology and secured his appointment to the chair of philosophy at the Academy of Geneva when age and ill health brought about his own retirement in 1786.

Pictet was always prominent in public life. At the time of Geneva's annexation to France in 1798, he did much to protect the interests of his city and of the Protestant religion, to which he ardently subscribed. Respected by Napoleon and a frequent visitor to Paris during the Consulate and Empire, he served as a member of the Tribunate from 1802 until 1807 and as one of the inspectors of the Imperial University from 1808 to 1815. He was prominent also in the scientific circles of Paris at this time; Berthollet and other French scientists were among his closest friends, and he often attended meetings of the first class of the Institut de France, of which he became a nonresident associate in 1802 and a corresponding member in 1803. His many other honors included membership of the Legion of Honor (1804) and fellowships of the Royal Society of London and of Edinburgh (1791 and 1796, respectively). By his marriage to Susanne Françoise Turrettini, which lasted from 1776 until her death in 1811, he had three daughters.

Pictet's most important research, a series of experiments on heat and hygrometry, was described in his *Essai sur le feu* (1790). Although widely read and even translated into English and German, the *Essai* broke little new ground. For example, his demonstration that radiant heat is reflected in the same way as light merely confirmed a conclusion already reached by J. H. Lambert and Saussure. And his experiments on the refraction and velocity of radiant heat, which could have led to results of great interest, were inconclusive. Although Pictet's research was mainly experimental, he was not uninterested in theory. In the *Essai*, for instance, he discussed the competing views of the nature of heat at some length and, although unwilling to commit himself, declared a slight preference for the material theory in the form given by Lavoisier. Hence he chose to consider heat as the fluid caloric rather than as a vibration either of the ordinary particles of matter or, in the manner of several Swiss scientists, of an all-pervading subtle fluid. He also opposed many of his compatriots by taking up Lavoisier's cause in the struggle for the acceptance of the new chemistry, and it was in a series of lectures given by Pictet in 1790 that Lavoisier's views were first given public support in Geneva.

Although Pictet earned a considerable reputation by his own research, which embraced geology, geodesy, astronomy, and meteorology as well as physics, he was even better known for his help and encouragement to others. In Geneva he worked tirelessly as a leading member of the Société des Arts, the Société de Physique et d'Histoire Naturelle, and the Société Helvétique des Sciences Naturelles; and he made what were probably his most important contributions to science as the joint founder and editor of two scientific journals: the *Journal de Genève* (published by the Société des Arts from 1787 to 1791) and the *Bibliothèque britannique*, founded in 1796 in collaboration with his younger brother Charles and his friend F. G. Maurice. Established originally to inform Continental readers of British publications and research, the *Bibliothèque britannique* was especially important in the maintenance of communications between Britain and the Continent during the Napoleonic Wars. After 1815, when it abandoned its special concern for British science and was renamed the *Bibliothèque universelle*, the journal continued to serve as a source of information that took little account of national barriers, although after 1815 Swiss contributions increased in number and importance.

Through his extensive international correspondence and his travels—mainly in France, Italy, and Britain—Pictet won many friends. As a zealous worker for science, a discerning editor, a patriot who showed no trace of chauvinism, and a man of great gentleness and modesty, he fully deserved the high esteem in which he was held throughout his life.

BIBLIOGRAPHY

I. ORIGINAL WORKS. Pictet's *Essai sur le feu* (Geneva, 1790), trans. into English by W. B[elcombe] as *An Essay on Fire* (London, 1791), was intended to be the first volume in a work entitled *Essais de physique*, but no further volumes were published. His *Voyage de trois mois en Angleterre, en Écosse, et en Irlande pendant l'été de l'an IX* (*1801 v. st.*) (Geneva, an XI [1802]), first published as a series of letters to the *Bibliothèque britannique*, is a very interesting account. His numerous contributions to the *Bibliothèque britannique* and the *Bibliothèque universelle* are listed in the Royal Society *Catalogue of Scientific Papers*, IV, 902–903; but his articles in the *Mémoires de la Société des arts* and the *Mémoires de la Société de physique et d'histoire naturelle de Genève* are not mentioned. Some of Pictet's correspondence with other Genevans is preserved at the Bibliothèque Publique et Universitaire of Geneva, but most of his personal papers and other correspondence are in the possession of the Rilliet family of Geneva. A copy, by Edmond Pictet, of a diary kept by Pictet while in Paris between 1802 and 1804 has been published as "Journal d'un genevois à Paris," in *Mémoires et documents publiés par la Société d'histoire et d'archéologie de Genève*, 2nd ser., **5** (1893–1901), 98–133.

II. SECONDARY LITERATURE. The standard biographical source is the obituary by J. P. Vaucher, in *Bibliothèque universelle*, sec. "Sciences et Arts," **29** (1825), 65–88. Other useful sketches are R. Wolf, *Biographien zur Kulturgeschichte der Schweiz*, III (Zurich, 1860), 373–394; Michaud, ed., *Biographie universelle*, new ed., XXXIII, 208–210; and A. de Montet, *Dictionnaire biographique des genevois et des vaudois*, II (Lausanne, 1878), 296–298. A MS of Edmond Pictet's "Dates des principales fonctions publiques, distinctions scientifiques et autres, dans la vie de Marc-Auguste Pictet," in the Bibliothèque Publique et Universitaire in Geneva, contains useful information not available elsewhere. Accounts of Pictet's connections with Mme de Staël and of his part in the introduction of Lavoisier's ideas in Geneva appear in P. Kohler, *Madame de Staël et la Suisse* (Lausanne–Paris, 1916), 408–412; and J. Deshusses, "Le physicien Marc-Auguste Pictet et l'adoption de la doctrine chimique de Lavoisier par les savants genevois," in *Bulletin de l'Institut national genevois*, **61** (1961), 100–112. His work for the *Bibliothèque britannique* is discussed in D. M. Bickerton, "A Scientific and Literary Periodical, the *Bibliothèque britannique* (1796–1815): Its Foundation and Early Development," in *Revue de littérature comparée*, **4** (1972), 527–547.

ROBERT FOX

PICTET, RAOUL-PIERRE (*b.* Geneva, Switzerland, 4 April 1846; *d.* Paris, France, 27 July 1929), *low-temperature physics.*

Pictet, son of Auguste Pictet-de Bock, a military

officer in various foreign services, was descended from a prominent Geneva family. After studying physics and chemistry in Geneva and Paris (1868–1870), he returned to his native city and devoted himself to experimentation in the physics of low temperatures, with an eye to the fast-growing and lucrative refrigeration industry. A compression refrigeration system that he developed, with sulfur dioxide as its cooling medium, functioned at a much lower pressure than competing systems; contact with water, however, often turned the refrigerant into corrosive sulfurous acid. This system, protected by a number of patents, was marketed with some success.

It was Pictet's researches that led to a scientific achievement which at once made him internationally famous. In December 1877, when Louis Paul Cailletet was about to report his liquefaction of oxygen to the Paris Academy of Sciences, Pictet cabled from Geneva that he had achieved the same feat. Cailletet and Pictet had worked independently and by different methods. While Cailletet's method had been to compress, cool, and expand the gas to be liquefied, Pictet had employed the "cascade" process, in which the refrigeration cycles of three different cooling media with successively lower critical temperatures were arranged in series, so that the gas liquefied first would act as a coolant in the liquefaction of the next. Pictet used sulfur dioxide in the first cycle, carbon dioxide in the second, and oxygen in the last. Although Cailletet could establish a priority of a few weeks, Pictet has been allowed to share the credit for the first liquefaction of an atmospheric gas. His claim also to have liquefied hydrogen was later shown to be based on error (Carl Linde, *Aus meinem Leben und von meiner Arbeit* [Munich, n.d. (1916?)], 68–72; Kurt Mendelssohn, *The Quest for Absolute Zero* [London, 1966], 41–42).

In 1879 Pictet was given a chair of "industrial physics" at the University of Geneva, which he held for seven years. In 1886 he left academic life to establish an industrial research laboratory in Berlin and to market his inventions. The chief feature of his refrigeration system now became a patented refrigerant, *liquide Pictet* (sulfur dioxide plus carbon dioxide), which involved him in controversy because he had claimed it to be exempt from the second law of thermodynamics. Although, as before, his machines were prone to disintegrate because of corrosion unless carefully shielded from moisture, he enjoyed some commercial success. During the later part of his life, spent in Paris, he continued to publish scientific papers; but his death in 1929 went virtually unnoticed.

BIBLIOGRAPHY

I. ORIGINAL WORKS. Pictet described his 1877 experiment in *Mémoire sur la liquéfaction de l'oxygène et la liquéfaction et solidification de l'hydrogène* (Geneva, 1878). An extensive bibliography of his works is in Poggendorff, III, 1040; IV, 1163; V, 975; and VI, 2014.

II. SECONDARY LITERATURE. Biographical data are found in an obituary by C.-E. Guye, in *Comptes rendus des séances de la Société de physique et d'histoire naturelle de Genève*, **47** (1930), 18–20; and in *Dictionnaire historique et biographique de la Suisse*, V (Neuchâtel, 1930). For discussions of his work see, besides Linde and Mendelssohn (cited in text), Ferdinand Rosenberger, *Geschichte der Physik*, III (Brunswick, 1887–1890), 416, 652–653; and W. R. Woolrich, *The Men Who Created Cold* (New York, 1967), 171–173.

OTTO MAYR

PIERCE, GEORGE WASHINGTON (*b.* Webberville, Texas, 11 January 1872; *d.* Franklin, New Hampshire, 25 August 1956), *applied physics.*

Pierce was the second of three sons of G. W. Pierce, a farmer and cattleman; his mother was Mary Gill Pierce. Academic talent manifested itself early: despite the limitations of rural schools in central Texas, he entered the University of Texas at eighteen with sufficiently advanced standing to graduate in three years. His first publication was written with his professor, Alexander Macfarlane, during his senior year. Pierce then taught in secondary schools and held various odd jobs for four years. In 1898 he won a fellowship to Harvard, where he remained for the rest of his scientific career. In 1900 he received the Ph.D. with a thesis on measurements of short radio waves.

After a postdoctoral year spent partly in Ludwig Boltzmann's laboratory at Leipzig, Pierce was appointed assistant in physics at Harvard and progressed steadily to a full professorship (1917); in 1921 he succeeded E. H. Hall as Rumford professor of physics. During these years he worked out much of the scientific underpinnings of electrical communications. He wrote basic papers of great lucidity on the resonant circuits and crystal detectors used in early radiotelegraphy, extended the use of semiconductor crystals to electroacoustics, and showed how mercury-vapor discharge tubes could be used for current control and sound recording.

In 1912 Pierce collaborated with A. E. Kennelly on measurements of the electric characteristics of telephone receivers, in the course of which work they discovered the concept of motional impedance. Pierce's work on submarine detection during World War I led to his offering the first postgraduate course anywhere on underwater sound signaling, to which the U.S. Navy for many years sent an annual contingent of student officers. Together with undergraduate courses on the applications of electromagnetic phenomena, it led to Harvard's pioneering position in radio communications, a position Pierce consolidated by writing the two classic American textbooks on the subject and by becoming the first director of Harvard's famed Cruft Laboratory in 1914. At Cruft he was associated for thirty years with E. L. Chaffee, who succeeded him as director.

Pierce is best remembered for bridging the gap between phenomenological knowledge and technological application of two similar physical effects: piezoelectricity and magnetostriction. The first effect led to the development of the quartz-crystal "Pierce oscillator" used in circuits that control the frequency of radio transmitters, standards, and meters; the second, to generators of underwater sound used in sonar and ultrasonic devices. The elucidation of physical phenomena by a professor, the elaboration of the principles of their applications by his doctoral students, and the technological utilization of the principles (and sometimes patents) outside the university—the sequence that came to characterize postgraduate education at the best American schools of engineering and applied science—thus had its inception at the Cruft Laboratory under Pierce and Chaffee. Less typically, Pierce became wealthy through his patents, some of which he exploited vigorously, and usually successfully, in the face of interference suits by large corporations. He was an exceedingly warm and droll individual, much revered by his students.

Pierce's work in ultrasound led to his later interests in sound generation by bats and insects, which persisted past his retirement in 1940 and led to a book published when he was seventy-six. His many honors included election to the National Academy of Sciences in 1920, the Medal of Honor of the Institute of Radio Engineers (of which he was president in 1918 and 1919) in 1929, and the Franklin Medal in 1943.

BIBLIOGRAPHY

I. ORIGINAL WORKS. Pierce's books are *Principles of Wireless Telegraphy* (New York, 1910); *Electric Oscillators and Electric Waves* (New York, 1919); and *The Song of Insects* (Cambridge, Mass., 1948). He also wrote or was coauthor of some 30 scientific papers and received 53 patents.

II. SECONDARY LITERATURE. The article by Frederick A. Saunders and Frederick V. Hunt in *Biographical Memoirs. National Academy of Sciences*, **33** (1959), 351–380, includes a complete list of Pierce's publications and American patents. A memoir by David Rines, his patent attorney, in the form of a letter addressed to F. V. Hunt, is in the archives of Harvard University, together with some of Pierce's notebooks and correspondence. For a glimpse of Pierce's influence on his contemporaries, see B. F. Miessner, *On the Early History of Radio Guidance* (San Francisco, 1964), 14, 24–25.

CHARLES SÜSSKIND

PIERI, MARIO (*b.* Lucca, Italy, 22 June 1860; *d.* Sant' Andrea di Còmpito (Lucca), Italy, 1 March 1913), *projective geometry, foundations of geometry.*

Pieri's father, Pellegrino Pieri, was a lawyer; his mother was Erminia Luporini. He began his university studies in 1880 at Bologna, where Salvatore Pincherle was among the first to recognize his talent; but he obtained a scholarship to the Scuola Normale Superiore of Pisa in November 1881 and completed his university studies there, receiving his degree on 27 June 1884. After teaching briefly at the technical secondary school in Pisa he became professor of projective geometry at the military academy in Turin and also, in 1888, assistant in projective geometry at the University of Turin, holding both posts until 1900. He became *libero docente* at the university in 1891 and for several years taught an elective course in projective geometry there.

On 30 January 1900, following a competition, he was named extraordinary professor of projective and descriptive geometry at the University of Catania. In 1908 he transferred to Parma, where in the winter of 1911 he began to complain of fatigue. His fatal illness, cancer, was diagnosed a few months later.

For ten years following his first publication in 1884, Pieri worked primarily in projective geometry. From 1895 he studied the foundations of mathematics, especially the axiomatic treatment of geometry. Pieri had made a thorough study of Christian von Staudt's geometry of position, but he was also influenced by his colleagues at the military academy and the university, Giuseppe Peano and Cesare Burali-Forti. He learned symbolic logic from the latter, and Peano's axiom systems for arithmetic and ordinary geometry furnished models for Pieri's axiomatic study of projective geometry.

In 1895 Pieri constructed ordinary projective geometry on three undefined terms: point, line, and segment. The same undefined terms were used in 1896 in an axiom system for the projective geometry of hyperspaces, and in 1897 he showed that all of the geometry of position can be based on only two undefined terms: projective point and the join of two projective points. In the memoir "I principii della geometria di posizione composti in un sistema logico-deduttivo" (1898) Pieri combined the results reached thus far into a more organic whole. Here the same two undefined terms were used to construct projective geometry as a logical-deductive system based on nineteen sequentially independent axioms—each independent of the preceding ones—which are introduced one by one as they are needed in the development, thus allowing the reader to determine on which axioms a given theorem depends. Of this paper Bertrand Russell wrote: "This is, in my opinion, the best work on the present subject" (*Principles of Mathematics*, 2nd ed. [New York, 1964], 382), a judgment that Peano echoed in his report in 1903 to the judging committee for the Lobachevsky Award of the Société Physico-Mathématique de Kasan. (Pieri received honorable mention, the prize going to David Hilbert.)

In their axiom systems for ordinary geometry, Pasch had used four undefined terms, and Peano three. With Pieri's memoir of 1899, "Della geometria elementare come sistema ipotetico-deduttivo," the number was reduced to two—point and motion—the latter understood as the transformation of one point into another. Pieri continued to apply the axiomatic method to the study of geometry, and in several subsequent publications he investigated the possibility of using different sets of undefined terms to construct various geometries. In "Nuovi principii di geometria proiettiva complessa" (1905) he gave the first axiom system for complex projective geometry that is not constructed on real projective geometry.

Two brief notes published in 1906–1907 on the foundations of arithmetic are notable. In "Sur la compatibilité des axiomes de l'arithmétique" he gave an interpretation of the notion of whole number in the context of the logic of classes; and in "Sopra gli assiomi aritmetici" he selected as primitive notions "number" and "successor of a number," and characterized them with a system of axioms that from a logical point of view simplified Peano's theory. In 1911 Pieri may have been on the point of beginning a new phase of his scientific activity. He was then attracted by the vectorial calculus of Burali-Forti and Roberto Marcolongo, but he left only three notes on this subject.

Pieri became one of the strongest admirers of symbolic logic; and although most of his works are published in more ordinary mathematical language, the statements of colleagues and his own statements show that Pieri considered the use of Peano's symbolism of the greatest help not only in obtaining rigor but also in deriving new results.

Pieri was among the first to promote the idea of geometry as a hypothetical-deductive system. His address at the First International Congress of Philosophy in 1900 had the highly significant title "Sur la géométrie envisagée comme un système purement logique." Bertrand Russell wrote in 1903: "The true founder of non-quantitative Geometry is von Staudt. . . . But there remained one further step, before projective Geometry could be considered complete, and this step was taken by Pieri. . . . Thus at last the long process by which projective Geometry has purified itself from every metrical taint is completed" (*Principles of Mathematics*, 2nd ed. [New York, 1964], 421).

BIBLIOGRAPHY

I. ORIGINAL WORKS. A chronological list of Pieri's publications appears in Beppo Levi, "Mario Pieri," in *Bullettino di bibliografia e storia delle scienze matematiche*, 15 (1913), 65–74, with additions and corrections in 16 (1914), 32. The list includes 57 articles, a textbook of projective geometry for students at the military academy, a translation of Christian von Staudt's *Geometrie der Lage*, and four book reviews.

II. SECONDARY LITERATURE. Besides the obituary by Beppo Levi (cited above), see Guido Castelnuovo, "Mario Pieri," in *Bollettino della mathesis*, 5 (1913), 40–41; and [Giuseppe Peano], "Mario Pieri," in *Academia pro Interlingua, Discussiones*, 4 (1913), 31–35. On the centennial of Pieri's birth Fulvia Skof published "Sull'opera scientifica di Mario Pieri," in *Bollettino dell'Unione matematica italiana*, 3rd ser., 15 (1960), 63–68.

HUBERT C. KENNEDY

PIERO DELLA FRANCESCA. See **Francesca, Piero della.**

PIERRE. See **Peter.**

PIETTE, LOUIS-ÉDOUARD-STANISLAS (*b.* Aubigny, Ardennes, France, 1827; *d.* 1906), *archaeology, paleontology.*

Although trained as a lawyer and active as a magistrate, Piette is best known for his archaeological and paleontological research. He made major contributions to Paleolithic archaeology by his own discoveries, his championship of Paleolithic art, his special study of Paleolithic portable art, and his ideas on the classification of the Paleolithic. He discovered Gourdan, Lortet, Mas-d'Azil and Brassempouy, all sites of Paleolithic art, and excavated prehistoric barrows at Avezac-Prat, Bartres, Osun, and La Halliade, near Lourdes.

Mortillet's proposal for classifying the Paleolithic into Chellean, Mousterian, Solutrean, and Magdalenian was later modified by inserting the Acheulean between Chellean and Mousterian. Piette's scheme divided the Paleolithic into the Amygdalithic, Niphetic, and Glyptic periods. The Amygdalithic was characterized by hand axes and comprised the Chellean and Acheulean. The Niphetic was Mortillet's Mousterian. The Glyptic, or "âge des beaux-arts," was characterized by the presence of art. Piette divided it into three stages: Papalian or Eburnian, characterized by sculpture in relief and in the round; Gordanian, characterized by engravings and of a time when animals now extinct still existed; and Lorthetian, also characterized by engravings, particularly on reindeer bone, but not associated with any extinct fauna. Piette's scheme was never adopted; but his collection of portable art, now at the Musée des Antiquités Nationales at St.-Germain-en-Laye, is one of the most important collections of Paleolithic art in existence. His *L'art pendant l'âge du renne* (1907) was beautifully illustrated with 100 plates by J. Pilloy. Accepting the authenticity of the Altamira paintings, which had been disputed since their discovery in 1875, Piette claimed that they were Magdalenian in date and described them as authentic in his *Équides de la période quaternaire d'après les gravures de ce temps* (1887).

In 1887 Piette began digging at Mas-d'Azil (Ariège) in the foothills of the Pyrenees, about forty miles southwest of Toulouse. Here the Arise River tunnels through the rock for over a quarter of a mile; and in this great tunnel, on both banks of the river, Piette excavated two rock shelters and found, above a rich Magdalenian deposit, a thick layer containing flat harpoons of staghorn, and pebbles painted with red ochre mixed with bones of red deer and wild boar. To this post-Magdalenian industry Piette gave the name Azilian. He believed that the pebbles represented an early form of alphabetic writing.

In 1879 still another amateur archaeologist and lawyer, Edmond Vielle, found an industry of post-Paleolithic character in the Aisne, which he labeled the Tardenoisian. The Azilian and the Tardenoisian were the first industries of what is now called the Mesolithic. Piette established this period between Paleolithic and Neolithic but called it the Metabatic

Age, or Age of Transition. Like his other names, it was not widely adopted.

BIBLIOGRAPHY

See *Collection Piette: art mobilier préhistorique* (Paris, 1964), with a preface by Henri Breuil, introduction by André Varagnac, and a catalogue by Marthe Chollot. See also G. E. Daniel, *A Hundred Years of Archaeology* (London, 1950), 122–126, 131–132, 232.

GLYN DANIEL

PIGOTT, EDWARD (*b.* 1753; *d.* Bath, England, 1825), and **PIGOTT, NATHANIEL** (*b.* Whitton, Middlesex, England; *d.* 1804), *astronomy.*

Although there is little personal data extant concerning the lives of the gentleman astronomers Nathaniel and Edward Pigott, their careers cast interesting light on the early development of stellar astronomy in Great Britain. Nathaniel Pigott was the son of Ralph Pigott and his wife Alathea, the daughter of William, eighth Viscount Fairfax of Gilling Castle; while in France he married, at an unknown date, Anna Mathurina de Beriol. Edward Pigott was their second son, the first to survive infancy.

Nathaniel Pigott was a surveyor and landed proprietor as well as an amateur astronomer. He spent much of his life on the Continent, settling for a while at Caen. The *Philosophical Transactions of the Royal Society* for 1767 contains an account of his observations there of the solar eclipse of 16 August 1765. He also recorded observations of the transit of Venus of 1769, made a series of meteorological and longitudinal measurements in the Low Countries (between 1770 and 1778), and observed the transit of Mercury of 1786 from Louvain. He was elected to the Royal Society on 16 June 1772; to the Brussels Academy in 1773; and became a corresponding member of the Paris Academy of Sciences in 1776.

In September 1771 the Pigott family left Caen to return to England. They lived at Frampton, Glamorganshire for ten years, then moved to Bootham, Yorkshire, where they improvised an astronomical observatory in the garden. Edward Pigott had already become actively engaged in astronomy—he assisted his father in the observation of 1769—and at Bootham he made his first discovery, that of a nebula in the constellation Coma Berenices. In 1783 he discovered a new comet, an accomplishment later mistakenly ascribed to his father, along with a number of others. (A possible source of this confusion lies in the fact that although Edward Pigott kept a diary—much of it in French—comprising his work from 1770 until 1782, he failed to put his name to it.)

In about 1783 Edward Pigott struck up a friendship with John Goodricke, who had himself the year before, when he was eighteen, discovered the periodic variability of Algol. Not to be outdone, Edward Pigott noted that the star η Aquilae is periodically variable; he made his discovery on 10 September 1784, apparently the same night upon which Goodricke discovered the variability of yet another star, β Lyrae. Within the week, Goodricke had also determined the variability of δ Cephei. The happy partnership ended prematurely in April 1786, when Goodricke died, at the age of twenty-one, probably from pneumonia that he contracted while making observations. Edward Pigott's reaction to this loss must be left to conjecture, since his diary ceased before that time.

Having, in the latter part of 1786, accompanied his father to Louvain to observe the transit of Mercury, Edward Pigott may have extended his stay there. At any rate, he did not report any new variables until 1795, when, having returned to England and settled at Bath, he announced the variability of R Scuti and R Coronae Borealis. (These stars were later recognized as prototypes of certain classes of irregular variables; the cause of their variability is as yet uncertain.) He subsequently discovered another variable star in Scutum and two more comets, as well as determining the proper motions of several stars. Edward Pigott was a frequent contributor to the *Philosophical Transactions of the Royal Society*; his works published therein include a number of important papers on the method of observing stars with a transit instrument.

Following the treaty of Amiens in 1802, Edward Pigott took the first opportunity to return to the Continent. When hostilities again broke out in 1803, he was arrested and detained at Fontainebleau. He wrote to William Herschel of his melancholy at "being separated from my journals, books and instruments"; after a time, however, doubtless through the good offices of his friends in the French scientific community, he was supplied with materials to continue his work. A treatise that he wrote while in detention was published by the Royal Society in 1803; Sir Joseph Banks, president of that body, exerted his influence on Edward Pigott's behalf, and secured his release in 1806. In the meantime, Nathaniel Pigott had died while traveling abroad.

In 1807 Edward Pigott observed the great comet of that year; at this time he was at his home in Belvedere, Bath. The last record of him is a letter from John Herschel, dated from Slough, 8 May 1821. Herschel asked Edward Pigott if he and his father might propose him for membership in the recently formed London (later Royal) Astronomical Society; although Pigott could not have been unaware of the honor that such a proposal implied, and although he endorsed Herschel's letter as an invitation, there is no indication that he replied to it. He was sixty-eight, and his interest in the subject may have faded—or perhaps he was disappointed that his lifework had not received recognition from another quarter. At any rate, he was not elected, and nothing else is known of him until his death.

BIBLIOGRAPHY

The following works by Pigott appeared in *Philosophical Transactions of the Royal Society*: "Account of a Nebula in Comâ Berenices," **71** (1781), 82; "On the Discovery of a Comet in 1783," **74** (1784), 20; "Observations on the Comet in 1783," *ibid.*, 460; "Observations of a New Variable Star," **75** (1785), 127; "On Those Stars Which the Astronomer of the Last Century Suspected to be Changeable," **76** (1786), 189; "On the Transit of Mercury Over the Sun, Made at Louvain, in the Netherlands," *ibid.*, 389; "The Latitude and Longitude of York Determined From Astronomical Observations, With the Method of Determining the Longitudes of Places by Observation of the Moon's Transit Over the Meridian," *ibid.*, 409; "An Account of Some Luminous Arches," **80** (1790), 47; "Determination of the Longitudes and Latitudes of Some Remarkable Places Near the Severn," *ibid.*, 385; "On the Periodical Changes of Brightness of Two Fixed Stars," **87** (1797), 133; "On the Changes in the Variable Star in Sobieski's Shield, From Five Years Observations; With Conjectures Respecting Unenlightened Heavenly Bodies," **95** (1805), 131; "Observations on the Eclipse of the Sun, Aug. 11, 1765, at Caen in Normandy," **57** (1767), 402; "Observations on the Transit of Venus, Jan. 3, 1769, at Caen," **60** (1770), 257; "Meteorological Observations at Caen for 1765–69," **61** (1771), 274; "Astronomical Observations in the Austrian Netherlands," **66** (1776), 182; "Discovery of Double Stars in 1779, at Frampton House in Glamorganshire," **71** (1781), 84; "Astronomical Observations," *ibid.*, 347; "An Observation of the Meteor of August 18th, 1783, Made on Hewitt Common, Near York," **74** (1784), 457; "Observations of the Transit of Mercury Over the Sun's Disc, Made at Louvain, May 3, 1786," **76** (1786), 384.

ZDENĚK KOPAL

PILATRE DE ROZIER, JEAN FRANÇOIS (*b.* Metz, France, 30 March 1754; *d.* Wimille, near Boulogne, France, 15 June 1785), *education, aeronautics.*

The son of Mathurin Pilastre du Rosier, an innkeeper, and Madeleine Willemart, Pilatre was baptized

François but later added the name Jean and modified his surname. (He never used the form Pilâtre.) After studying pharmacy for three years in Metz, he attended scientific courses at Paris. About 1776 he taught a physics course in Paris and for a short time was professor of chemistry at the Société d'Émulation of Rheims. He returned to Paris about 1780 as keeper of the physics and natural history cabinets of the Comte de Provence, brother of Louis XVI.

Under his patronage Pilatre founded the Musée, a private institution for higher education that opened in 1781 on the rue St.-Avoye but soon moved to the rue de Valois. By 1785 it had 700 members, including the academicians Condorcet, Fourcroy, and Vicq d'Azyr, who gave their encouragement, as well as many ladies and gentlemen of society who attended lectures on a wide range of literary and scientific subjects.

Pilatre was skilled at arranging lecture demonstrations, but the research that he attempted was of little merit. Hoping to be elected to the Académie des Sciences, he submitted chemical and physical memoirs on several occasions between 1781 and 1784 but was never proposed as a candidate. His one useful invention was a respirator that enabled a man working in the noxious atmosphere of a deep well or cesspit to breathe fresh air, which was supplied by a flexible hose from the surface; and he courageously demonstrated it himself. It was praised in 1783 by the Société Royale de Médecine but was not generally adopted.

After being present at Versailles on 19 September 1783, when Étienne Montgolfier's hot-air balloon safely carried a sheep, a cock, and a duck for two miles, Pilatre took part in the trials of a new balloon constructed by Montgolfier in the garden of J. B. Réveillon, a paper manufacturer. This balloon had room for two in a gallery around the base, from which a brazier suspended under the opening of the balloon could be fed with fuel. On 15 October Pilatre rose to eighty feet in the tethered balloon, and he soon learned how to vary the altitude by controlling the fire. Accompanied by Marquis François Laurent d'Arlandes, an infantry major, he made the first human flight on 21 November 1783, taking off from the Château de la Muette, west of Paris, and traveling nearly six miles across the city in about twenty-five minutes, at about 3,000 feet. Pilatre made several other hot-air ascents, notably with Joseph Montgolfier and five others at Lyons on 19 January 1784 and with the chemist J. L. Proust on 23 June 1784, when they reached a height of about 11,000 feet above Versailles.

Pilatre hoped to make the first aerial crossing of the English Channel, and early in January 1785 he was at Boulogne. The wind was unfavorable, however, and the honor went to J. P. Blanchard and John Jeffries, who flew from Dover to Calais on 7 January. Pilatre accompanied Blanchard to Paris and made him a member of the Musée. He subsequently visited England, where he was present at ascents by Blanchard on 21 May and 3 June.

Pilatre attempted his own crossing from Boulogne on 15 June 1785, accompanied by Pierre Ange Romain, one of the constructors of the balloon, which was of a new type designed by Pilatre. Hot air being denser than hydrogen, a hot-air balloon had to be larger than a hydrogen balloon in order to carry the same weight; but its altitude could easily be changed by varying the size of the fire. Hoping to combine the advantages of both types, Pilatre attached a hydrogen balloon to the top of a small cylindrical hot-air balloon, apparently assuming that escaping hydrogen would rise and that there would be no danger of ignition from the fire below. Tragically, he was proved wrong. The balloon caught fire at about 1,700 feet and crashed near Boulogne, killing both occupants.

After Pilatre's death the Musée was reorganized and called the Lycée. Under its later names, Lycée Républicain (1792) and Athénée de Paris (1802), it played an important part in the scientific and cultural life of Paris until the 1840's.

BIBLIOGRAPHY

I. ORIGINAL WORKS. Seven memoirs by Pilatre were published in *Observations sur la physique* . . . between 1780 and 1782, and are listed in the collective index, *ibid.*, **29** (1786), 468–469. Some other writings were published posthumously by A. Tournon de la Chapelle (see below). Pilatre's only separate publication was *Première expérience de la montgolfière construite par ordre du roi* . . . (Paris, 1784), a description of his flight on 23 June 1784.

II. SECONDARY LITERATURE. There are a number of incorrect dates in Pilatre's first biography, Tournon de la Chapelle, *La vie et les mémoires de Pilatre de Rozier* (Paris, 1786). Later accounts of value are Léon Babinet, *Notice sur Pilatre de Rozier* (Metz, 1865); P. Dorveaux, "Pilatre de Rozier," in *Bulletin de la Société d'histoire de la pharmacie* (1920), 209–220, 249–258; and "Pilatre de Rozier et l'Académie des sciences," in *Cahiers lorrains*, **8**, (1929), 162–166, 182–185; and W. A. Smeaton, "Jean François Pilatre de Rozier, the First Aeronaut," in *Annals of Science*, **11** (1955), 349–355. Some dates in Smeaton's account are corrected in a review by A. Birembaut in *Archives internationales d'histoire des sciences*, **11** (1958), 100–101; further details of Pilatre's flights correcting statements by earlier authors are given by W. A. Smeaton, "The First and Last Balloon Ascents of Pilatre de Rozier," *ibid.*, 263–269.

For accounts of the early history of the Musée (after 1785, the Lycée), see C. Cabanes, "Histoire du premier musée autorisé par le gouvernement," in *Nature* (Paris), **65** (1937), pt. 2, 577–583; and W. A. Smeaton, "The Early Years of the Lycée and the Lycée des Arts: A Chapter in the Lives of A. L. Lavoisier and A. F. de Fourcroy. I. The Lycée of the Rue de Valois," in *Annals of Science*, **11** (1955), 257–267.

Pilatre's early balloon flights are described by B. Faujas de Saint-Fond, *Description des expériences de la machine aérostatique de MM. de Montgolfier . . .* (Paris, 1783; 2nd ed., 1784) and *Première suite de la description . . .* (Paris, 1784), of which no further volumes were published.

W. A. SMEATON

PINCHERLE, SALVATORE (*b.* Trieste, Austria [now Italy], 11 March 1853; *d.* Bologna, Italy, 10 July 1936), *mathematics*.

Born of a Jewish business family, Pincherle completed his preuniversity studies in Marseilles, where his family had migrated. The unusually sophisticated teaching of science there seems to have been a decisive factor in diverting his interest from the humanities to mathematics; and by 1869, when he entered the University of Pisa, the decision to study mathematics had already matured. His teachers at Pisa included Betti and Dini; Pincherle was greatly affected by both of them. After graduating in 1874, Pincherle became a teacher at a *liceo* in Pavia. A scholarship for study abroad enabled him to spend the academic year 1877–1878 in Berlin, where he met Weierstrass, who influenced all his subsequent work. In 1880 Pincherle became professor of infinitesimal analysis at the University of Palermo. He remained there only a few months, having been appointed to a chair at the University of Bologna. He retired in 1928.

Pincherle greatly improved the level of mathematics at the University of Bologna, which had badly deteriorated during the final years of papal domination. The university later acknowledged his contribution by naming the mathematics institute for him during his lifetime. In Bologna, Pincherle also founded (1922) the Italian Mathematical Union, of which he was the first president. At the Third International Congress of Mathematicians, held at Bologna in 1928, of which he was president, Pincherle restored the truly international character of international mathematical congresses by reopening participation to German and other mathematicians who had been excluded since World War I.

Pincherle's contributions to mathematics were mainly in the field of functional analysis, of which he was one of the principal founders, together

with Volterra. Remaining faithful to the ideas of Weierstrass, he did not take the topological approach that later proved to be the most successful, but tried to start from a series of powers of the D derivation symbol. Although his efforts did not prove very fruitful, he was able to study in depth the Laplace transformation, iteration problems, and series of generalized factors. He was the author of several textbooks, notably for secondary schools, at which he had had direct practical experience.

Pincherle was a member of the Accademia Nazionale dei Lincei and the Bayerische Akademie der Wissenschaften, which, despite the rise of Nazism, sent him a warm message on his eightieth birthday in 1934. In 1954 the city of Trieste held a solemn celebration of the centenary of his birth.

BIBLIOGRAPHY

There is an accurate bibliography of Pincherle's writings from 1874 to 1936, with 245 references, by Ettore Bortolotti, in *Bollettino dell'Unione matematica italiana*, **16** (1937), 37–60. On his life and work, see the notices by Ugo Amaldi, in *Annali di matematica pura ed applicata*, 4th ser., **17** (1938), 1–21; and Leonida Tonelli, in *Annali della Scuola normale superiore*, 2nd ser., **6** (1937), 1–10; and F. G. Tricomi, *Salvatore Pincherle nel centenario della nascità*, Pubblicazioni della Facoltà di scienze e d'ingegneria, Università di Trieste, ser. A, **60** (1954).

F. G. TRICOMI

PINCUS, GREGORY GOODWIN (*b.* Woodbine, New Jersey, 9 April 1903; *d.* Boston, Massachusetts, 22 August 1967), *endocrinology*.

Pincus is best known for his work with his associates in the development of the birth control pill. He received the B.S. degree at Cornell University in 1924 and a master's and doctorate in science at Harvard in 1927, working under the geneticist W. E. Castle and the animal physiologist W. J. Crozier. Pincus lived in Europe from 1929 to 1930, studying at Cambridge with F. H. A. Marshall and John Hammond, both pioneers in reproductive biology, and then at the Kaiser Wilhelm Institute with the geneticist R. B. Goldschmidt. In 1930 he returned to Harvard and was appointed an assistant professor in 1931. His pioneer work, *The Eggs of Mammals*, was published in 1936. He was at the University of Cambridge in 1937 and became a visiting professor in 1938 at Clark University, Worcester, Massachusetts.

Pincus conducted research on stress for the U.S. navy and air force during World War II. In

1944 he and Hudson Hoagland established the Worcester Foundation for Experimental Biology, which soon became internationally known as a center for the study of steroid hormones and mammalian reproduction. Pincus became a professor at Tufts Medical School in 1945 and at Boston University in 1951. In 1944 he organized the annual Laurentian Hormone Conference and edited the first twenty-three volumes of its proceedings, *Recent Progress in Hormone Research* (1946–1967).

Encouraged by Margaret Sanger, in 1951 Pincus and M. C. Chang started their studies on the effects of various newly synthesized hormones on reproduction in laboratory animals and found that several progestational compounds administered orally could prevent pregnancy, mainly by inhibition of ovulation. In collaboration with J. Rock and C. R. Garcia, Pincus immediately extended these studies to humans and perfected the oral contraceptive pill.

With his associates Pincus published about 350 papers on tropism in rats, genetics of mice, fatherless rabbits, fertilization and transplantation of eggs, diabetes, cancer, schizophrenia, adrenal hormones, and aging. He made significant contributions to knowledge of the effects, metabolism, and biosynthesis of steroid hormones. Pincus was coeditor of *The Hormones*, volumes 1–5, and his book, *Control of Fertility*, was published in 1965. He died of myeloid metaplasia, probably due to his early work with organic solvents.

Pincus was prominent in the study of mammalian reproductive physiology and endocrinology for more than thirty-five years. Some of his contributions in the early 1930's concerned processes involved in mammalian fertilization and development. With increasing knowledge of steroid hormones in the early 1940's, his attention became increasingly focused on the roles of these substances in general physiology and especially in reproduction. In the early 1950's, when powerful, orally active, synthetic hormonelike compounds were produced, Pincus and his associates seized the opportunity to develop an oral contraceptive. Their success was such that they produced that pharmaceutical rarity, a chemical agent that is virtually 100 percent effective. More important, the work of Pincus and his colleagues has transformed family planning in all the parts of the world in which it is systematically employed.

BIBLIOGRAPHY

Pincus' works include *The Eggs of Mammals* (New York, 1936); "The Comparative Behavior of Mammalian Eggs in vivo and in vitro," in *Proceedings of the American Philosophical Society*, **83** (1940), 631–646, written with H. Shapiro; "Studies of the Biological Activity of Certain 19-Nor Steroids in Female Animals," in *Endocrinology*, **59** (1956), 695–707, written with M. C. Chang *et al.*; *The Control of Fertility* (New York, 1965); and "Control of Conception by Hormonal Steroids," in *Science*, **153** (1966), 493–550.

M. C. CHANG

PINEL, PHILIPPE (*b*. Jonquières, near Castres, France, 20 April 1745; *d*. Paris, France, 25 October 1826), *medicine*.

Pinel was the son of a master surgeon who practiced in St.-Paul-Cap-de-Joux, a village between Castres and Toulouse. His mother, Élisabeth Dupuy, came from a family that had since the seventeenth century produced a number of physicians, apothecaries, and surgeons. Despite this medical heritage, Pinel's early education, first at the Collège de Lavaur and then at the Collège de l'Esquille in Toulouse, was an essentially literary one; he was greatly influenced by the Encyclopedists, particularly Rousseau. Having decided upon a career in religion, he enrolled in the Faculty of Theology at Toulouse in July 1767; in April 1770, however, he left it for the Faculty of Medicine, from which he received the M.D. on 21 December 1773. Simultaneously with his medical training, Pinel studied mathematics, an interest that is apparent in his medical writings.

In 1774 Pinel went to Montpellier, where for four years he frequented the medical school and hospitals. He there began to formulate and to practice the principles that he later recommended to his students: "Take written notes at the sickbed and record the entire course of a severe illness." He supported himself by giving mathematics lessons, conducting a private anatomy course, and writing theses for rich students. He also met Chaptal, who later acknowledged Pinel's influence upon his intellectual development. In 1777 Pinel presented two iatromechanical papers, on the application of mathematics to human anatomy, to the Société Royale des Sciences de Montpellier; he was named a corresponding member in July of that year.

In 1778 Pinel went to Paris. He carried with him letters of recommendation to the geometer Jacques Cousin, who advised him to give up medicine and devote himself to the exact sciences. He visited libraries and hospitals (particularly P. J. Desault's service at the Hôtel-Dieu) and frequented the *salon* of Mme Helvétius, into which he had been introduced by Cabanis, and where he met Franklin. Mme Helvétius's house in Auteuil was a gathering place for the school later called *idéologues*, and Pinel became acquainted with the sensationalist doctrines of Locke and

Condillac, which strongly influenced his work. As a graduate of Toulouse, however, he was unable to practice medicine in the capital.

In 1784 Pinel became editor of the *Gazette de santé*, in which he published a number of articles chiefly concerned with hygiene and mental disorders, a subject in which he had interested himself following the illness of a friend in 1783. In 1785 he translated William Cullen's *First Lines of the Practice of Physic* and three volumes of the *Philosophical Transactions of the Royal Society* into French. He also wrote articles on medicine for the daily *Journal de Paris* and, in 1788, published a new edition of Baglivi's *Opera omnia*.

Pinel took no active political role during the Revolution, but devoted himself to attempting to aid those who had been proscribed, among them Condorcet. On 25 August 1793 he was appointed, at the instance of his friends Cabanis and Jacques Thouret, *médecin des infirmeries* of the Hospice de Bicêtre, where he was able to begin implementing his ideas on the humane treatment of the insane. (He had previously been a frequent visitor at the Belhomme nursing home for the mentally ill, but had been unable to convince the director—who was primarily concerned with making a profit—to accept his therapeutic notions.) At the Bicêtre Pinel had the chains removed from his patients, an event commemorated in both paintings and popular prints. On 13 May 1795 he became chief physician of the Hospice de la Salpêtrière, a post that he retained for the rest of his life. Here he was in charge of 5,000 pensioners, aged women, and chronically ill patients; there was a 600-bed ward for the mentally ill, a 250-bed infirmary for acutely ill patients, and, at first, a small infirmary for sick orphans. Pinel was eventually assisted in his work by A. J. Landré-Beauvais, J. E. D. Esquirol, and C. J. A. Schwilgué.

On 4 December 1794 the Convention Nationale (three years after the dissolution of the medical guilds and faculties by the Legislative Assembly) established three *écoles de santé*, and Pinel, upon the recommendation of Fourcroy and Thouret, was named adjunct professor of medical physics at the school in Paris. In 1795 he became professor of medical pathology, a chair that he held for twenty years; he was briefly dismissed from this position in 1822, with ten other professors suspected of political liberalism, but reinstated as an honorary professor shortly thereafter. Pinel was elected to the Académie des Sciences in 1804 and was a member of the Academy of Medicine from its founding in 1820. In addition to working in hospitals and teaching, Pinel often served as a consulting physician, although he did not have the rich and influential patients that Corvisart or Portal did.

The difficult beginning and slow progress of his career neither discouraged nor embittered Pinel, and his eventual success did not diminish his modesty. Although he is properly considered one of the founders of psychiatry, Pinel's contemporaries regarded him as a master of internal medicine, a reputation based upon the authoritative classification of diseases that he set out in his *Nosographie philosophique*, published in 1798.

Pinel's nosological work should be viewed in the context of the great eighteenth-century concern with classification, of which the works of Linnaeus are exemplary. Specifically medical classifications had been offered by William Cullen and David McBride, in 1769 and 1787, respectively, while Erasmus Darwin's *Zoonomia* appeared in 1794–1796. Pinel was aware of the difficulties that his predecessors had faced, but he approached his task cheerfully, secure in his belief that a disease was "an indivisible whole from its commencement to its conclusion, a regular ensemble of characteristic symptoms." Since these symptoms could be observed and analyzed, a classification of disease was possible.

Pinel thus divided diseases into five classes—fevers, phlegmasias, hemorrhages, neuroses, and diseases caused by organic lesions. Nearly one third of the *Nosographie* is devoted to the first class, fevers, which Pinel subdivided into angiotenic, meningogastric, adenomeningic, adynamic, ataxic, and adenoneural forms, corresponding respectively to the inflammatory, bilious, mucous, putrid, malignant, and pestilential fevers of the ancient authors. Pinel subsequently added the order of hectic fevers, which had been described in 1803 by his then disciple Broussais; these six classes were further subdivided into eight genera and a number of species.

Pinel classified phlegmasias by the structure of the affected membranes (or tissues). He thus arrived at five orders: cutaneous phlegmasias, including eruptive fevers and dermatological diseases; mucous phlegmasias, classified by location, and including opthalmia, quinsy, gastritis, and enteritis; serous phlegmasias, including phrenitis, pleurisy, and peritonitis; parenchymatous and cellular phlegmasias; and phlegmasias of the muscle, fibrous, and synovial tissues. In his *Traité des membranes* of 1800, Bichat acknowledged the influence of Pinel's book on his own work.

Among Pinel's third class of diseases, hemorrhage, only those of the mucous membranes (epistaxis, hemoptysis, hematemesis, hemorrhoid, and metrorrhagia) seemed to him to have been studied sufficiently. Among the fourth class, neuroses, Pinel included not only psychiatric illnesses, but also diseases of the sense organs, spasmodic visceral disorders, and dysfunctions of the genital organs. His fifth class, which in the first edition of the *Nosographie* he called "diseases of which the seat is in the lymphatic system," comprised more generally systemic diseases, scurvy, syphilis, and cancer among them, as well as heart disease, dropsy, and kidney stone.

Pinel composed the *Nosographie* as a textbook. It went through several editions, among which important variations may be found. In the first, for example, Pinel refused to acknowledge the distinguishing features of scarlet fever and puerperal fever, although he later classified scarlet fever among the eruptive fevers, and remarked on the occurrence in the same epidemic of both simple quinsy and true scarlet fever. He continued to deny the existence of puerperal fever as an entity (as he continued to deny the existence of fevers concomitant to any stage of reproduction), and it was only in the last edition of his book that he recognized it as a special form of peritonitis. Although the *Nosographie* was a notable success among Pinel's students and disciplines, it also provoked a number of criticisms. Broussais, in particular, attacked Pinel's ideas on idiopathic fevers. Pinel chose to ignore his critics, however, and even forbade his followers to respond to them.

Pinel's other medical writings, from his first communications to the Montpellier Société Royale des Sciences, give evidence of his mathematical training. He drew up precise "tables synoptiques" to determine the frequency of occurrence of certain illnesses, together with their modes of development and their prognoses. He conducted rigorous experiments to measure the effectiveness of various medicines, and devised a numerical method of evaluation. His own therapy was conservative; he contented himself with a pharmacopoeia of only fifty-five vegetable substances and thirty-nine "chemical products," which he used sparingly. He recorded his "extreme distaste" for polypharmacy, objected to the use of bloodletting and purges, and proscribed the use of quinine (even for malaria) and opiates (even for severe pain). Nonetheless, Pinel easily accepted new discoveries, including Corvisart's technique of sounding by percussion and the use of the stethoscope for mediate auscultation, introduced by Laennec. Pinel created an inoculation clinic in his service at the Salpêtrière in 1799 and the first vaccination in Paris was given there in April 1800.

Pinel's psychiatric work effectively transformed the prison for the insane into a hospital. He did not merely initiate better treatment for the mentally ill, however, but rather concerned himself with establishing psychiatry as a discrete branch of medicine. He published a number of articles on the subject, beginning in 1784, then synthesized his findings in "Recherches et observations sur le traitement moral des aliénés" (1799) and *Traité medico-philosophique de l'aliénation mentale* (1801), to which his 1807 communication to the Institut de France is an important supplement.

Pinel's classification of mental diseases retained the old divisions of such illnesses as manic, melancholic, demented, and idiotic. He presented these classes (with a disclaimer—it was necessary to retain them "for the time being," since medicine was not advanced enough for subtler distinctions) as late as 1812. He nevertheless made finer distinctions, isolating mania from delirium, and pointing out that in this state the intellectual functions might be intact, and, in his description of idiocy, citing stupor, the first stage of some types of mental disease. Pinel recognized the relationship between periodic mania and melancholy and hypochondria and stressed the danger of suicide by the melancholic patient. He also mentioned the possibility of altruistic homicide.

In establishing the cause of mental illness, Pinel was wary of "metaphysical discussions or certain ideological ramblings," and he categorically rejected the notion of demonic possession or sorcery. Faithful to the doctrines of Locke and Condillac, he considered emotional disorders to be the primary factor in precipitating intellectual dysfunctions; he also took into account heredity, morbid predisposition, and what he called individual sensitivity.

Pinel's psychiatric therapeutics, his "traitement moral," represented the first attempt at individual psychotherapy. His treatment was marked by gentleness, understanding, and goodwill. He was opposed to violent methods—although he did not hesitate to employ the straitjacket or force-feeding when necessary. He recommended close medical attendance during convalescence, and he emphasized the need of hygiene, physical exercise, and a program of purposeful work for the patient. A number of Pinel's therapeutic procedures, including ergotherapy and the placement of the patient in a family group, anticipate modern psychiatric care.

Pinel was also concerned with the proper training of infirmary personnel and with the proper administration of an institution for the mentally ill. A generation of specialists in mental diseases, led by Esquirol, was educated at the Salpêtrière and disseminated Pinel's ideas throughout Europe.

Pinel was married in 1792 to Jeanne Vincent; of their three sons, one, Scipion, became a specialist in mental illness. Having been widowed in 1811, Pinel was married again, in 1815, to Marie-Madeleine Jacquelin-Lavallée.

BIBLIOGRAPHY

I. ORIGINAL WORKS. Pinel's writings include *Nosographie philosophique ou méthode de l'analyse appliquée à la médecine* (Paris, *an* VII [1798]; 6th ed., 1818); "Recherches et observations sur le traitement moral des aliénés," in *Mémoires de la Société médicale d émulation de Paris*, **2** (*an* VII [1799]), 215–255; *Traité médico-philosophique sur l'aliénation mentale ou la manie* (Paris, *an* IX [1801]; 2nd ed., 1809); *La médecine rendue plus précise et plus exacte par l'application de l'analyse* (Paris, 1802; 3rd ed., 1815); and "Resultats d'observations et construction de tables pour servir à déterminer le degré de probabilité de la guérison des aliénés," in *Mémoires de la classe des sciences mathématiques et physiques de l'Institut* (1807), 169–205.

II. SECONDARY LITERATURE. See E. H. Ackerknecht, *Medicine at the Paris Hospital 1794–1848* (Baltimore, 1967); H. Baruk, *La psychiatrie française de Pinel à nos jours* (Paris, 1967); F. J. V. Broussais, *Examen de la doctrine médicale généralement adoptée et des systèmes modernes de nosologie* (Paris, 1816); P. Chabbert, "Philippe Pinel à Paris (jusqu'à sa nomination à Bicêtre)," in *Comptes rendus du XIXᵉ Congrès international de l'histoire de la médecine* (Basel, 1966), 589–595; M. Foucault, *Histoire de la folie à l'âge classique* (Paris, 1961); and *Naissance de la clinique, une archéologie du regard médical* (Paris, 1963); W. H. Lechler, *Philippe Pinel. Seine Familie, seine Jugend- und Studienjahre* (Munich, 1960); W. Riese, "Philippe Pinel (1745–1826), His Views on Human Nature and Disease, His Medical Thought," in *Journal of Nervous and Mental Disease*, **114**, no. 4 (Oct. 1951), 313–323; and "An Outline of a History of Ideas in Psychotherapy," in *Bulletin of the History of Medicine*, **25**, no. 5 (Sept.–Oct. 1951), 442–456; R. Sémelaigne, *Aliénistes et philanthropes, les Pinel et les Tuke* (Paris, 1912); and J. Vinchon, "Philippe Pinel," in *Commentaires sur dix grands livres de la médecine française* (Paris, 1968), 89–106.

PIERRE CHABBERT

PINGRÉ, ALEXANDRE-GUI (*b*. Paris, France, 4 September 1711; *d*. Paris, 1 May 1796), *astronomy*.

There seem to be few details of Pingré's early life, but he is said to have been a somewhat precocious child with a great desire for knowledge. He was educated by the Congregation of Ste. Geneviève and in 1727, at the age of sixteen, entered the religious order of Ste. Geneviève de Senlis. There is no doubt of his intellectual abilities, for in 1735, when he was only twenty-four, he became professor of theology at the University of Ste. Geneviève. Like many French Roman Catholic clerics, Pingré followed the rather independent Augustinian opinions of the seventeenth-century bishop Cornelis Jansen; and when action was taken against the Jansenists in 1745, he was deprived of his chair and sent by his order to teach Latin in the schools outside Paris. Accused more than once of corrupting the minds of his young pupils, he was obliged to move from one place to another until the eminent surgeon Claude Le Cat decided to help him. In 1744 Le Cat had founded an academy of sciences in Rouen; and since the academy was still without an astronomer in 1749, he invited Pingré, who had recently moved to that city, to accept the post. It proved to be the turning point of Pingré's career. He was later recalled to Paris, where he settled permanently as an astronomer but with literature, history, music and, in later life, botany as his hobbies. Contemporaries spoke of him with affection, and he seems to have been a pious, kindly, and tolerant man.

Pingré was thirty-eight when he began a serious study of astronomy, but within a year he was able to calculate the lunar eclipse of 23 December 1749 well enough for his results to be submitted to the Académie des Sciences in Paris. Certain writers of Pingré's obituary notices state that through these calculations he found an error in the figures for the eclipse that Lacaille had prepared; but the evidence for this, and for the strong friendship between the two men, appears doubtful. Pingré's rapidly growing abilities as an astronomer need not be questioned, however; and after he had made observations at Rouen of the transit of Mercury across the sun's disk in 1753, his reputation was sufficient for the Academy to elect him a *correspondant*. Soon afterward his order recalled him to Paris and established a small observatory for him on the roof of the Abbey of Ste. Geneviève.

Also during 1753 Pingré began working with P. C. Le Monnier, who greatly encouraged him and whom he helped prepare *État du ciel à l'usage de la marine*, a nautical almanac giving hour angles of the moon for the purpose of determining longitude at sea by use of a method devised by Le Monnier. Although the work, which was complementary to the *Connaissance des temps*, did not find favor with mariners and appeared only for the years 1754 to 1757, it greatly enhanced Pingré's reputation as a computer.

Indeed, Pingré was becoming well-known as an astronomer; and in 1755 he was appointed a member of the commission established that year to examine the measurement of an arc of the meridian made some eighty years before by Jean Picard. In 1756 the Academy honored Pingré by electing him *associé libre*, the highest rank of membership open to a cleric. Also about this time he was invited by the provost of the Paris guilds to design a sundial for the corn market that would display the entry of the sun into the various zodiacal signs; this involved Pingré in an observing program as well as much computation, and he did not complete the commission until 1764.

Under the leadership of Delisle, the French took a great interest in the transit of Venus that was to occur in 1761, and Pingré became involved in the international arrangements made to ensure that observations of a phenomenon allowing the sun's distance to be precisely determined were carried out from points as widely scattered as possible. Commissioned by the Academy to observe from Rodriguez Island in the Indian Ocean, he left France early in January 1761.

This was during the Seven Years War; and since British naval supremacy might possibly present difficulties, Pingré armed himself with instructions from the British authorities commanding that he be unmolested and allowed to proceed without delay. His outward journey was uneventful until his ship met a damaged French vessel, the commander of which ordered Pingré's ship to stay with him. Only Pingré's dogged persistence got him and his assistant Denis Thuillier transshipped at last, and they arrived at Rodriguez with little time left to establish their observatory. At the transit on 6 June there was rain and cloud for a great part of the time, so that only a few observations could be made. The island was later sacked three times by the British; and the expedition's ship was attacked and boarded on the way home, despite Pingré's British instructions. Therefore, when the vessel reached Lisbon, Pingré decided to journey overland, reaching Paris late in May 1762. The expedition was not unsuccessful, for besides a few useful transit observations, Pingré had many longitude determinations, some of which led to a replotting of the charted position of the Cape Verde Islands. His analysis of the observations led him to the rather large value of 10.6 seconds of arc for the solar parallax, a figure that he later modified.

On his return Pingré also became engaged in preparing a second edition of Lacaille's *L'art de vérifier les dates*, originally designed to give sufficient details of eclipses during the previous 1,800 years to serve as a guide for dating historical events. He checked all the calculations and added additional eclipses up to A.D. 1900. The new edition appeared in 1770; but Pingré continued to work on the subject, computing eclipses back to 1000 B.C. and publishing the results in the *Mémoires de mathématique et physique* . . . of the French Academy. He also took a leading part in the preparations for observing the 1769 transit of Venus, and in 1766 and 1767 he presented two reports to the Academy about suitable observing stations. Undaunted by his previous experiences—the war was now over—Pingré set forth on voyages in 1767, 1768, and 1771. They were primarily intended to check chronometers by Ferdinand Berthoud and

Leroy, but on the 1768 voyage he visited Haiti, where he observed the 1769 transit. Later he recomputed the solar parallax from the complete observations; and in 1772 he announced a value of 8.8 seconds of arc, a figure extremely close to the present figure of 8.794.

Having become astronomer-geographer to the navy, Pingré in 1769 was appointed chancellor of his old university. In 1772 he became librarian at Ste. Geneviève; and although in his sixties, he continued with his computing and began to take an increasing interest in old observations. Pingré put his immense classical knowledge to use in translating and editing the *Astronomica* of Marcus Manilius and the earlier *Phaenomena* of Aratus of Soli, and especially in preparing his most important work, the two-volume *Cométographie ou traité historique et théorique des comètes* (1783–1784). This monumental work was divided into four parts, the first of which was a history of astronomy from Babylonian and Egyptian times, with particular reference to ideas about comets. The second part was a catalog of all comets observed since antiquity, with the orbital elements of 166 for which paths had been computed, 50 of them by Pingré himself. The third section discussed cometary returns, theories about the nature of comets, and the physical effects likely to ensue from their close approach to the earth. The fourth part concerned cometary orbits and methods for computing them. The high reputation of the *Cométographie* was deserved, and as recently as 1950 it was officially recommended as a source book of cometary information.

Pingré's other great book, the purely historical *Annales célestes du dix-septième siècle*, took him thirty years to complete and contained carefully checked and edited astronomical observations from the seventeenth century, both published and unpublished. In 1791 Le Monnier and Lalande persuaded the Academy to vote a large sum for its publication; but the printer was slow, and Pingré's death in 1796, coupled with devaluation the preceding year, led the printer to abandon the project and to sell the printed sheets as wastepaper. Worse still, the manuscript was lost. Almost a century later, however, a Parisian bibliophile found in a country town what turned out to be Le Monnier's set of sheets; and the remainder of the manuscript was discovered in the archives of the Paris observatory. In 1898, at the instigation of C. G. Bigourdain, the Academy again decided to publish; and the volume appeared in 1901. There is still a voluminous collection of Pingré's unedited manuscripts at the library of Ste. Geneviève. They do not seem to be astronomical, however, but to cover his other interests, ranging from translations of Spanish voyages, history and historical criticism, and literary sketches to liturgical hymns, musical satires, and a vast amount of French and Latin poetry. It is as an astronomer, however, that Pingré is remembered.

BIBLIOGRAPHY

I. Original Works. Pingré's main works were *Cométographie ou traité historique et théorique des comètes*, 2 vols. (Paris, 1783–1784); and *Annales célestes du dix-septième siècle*, C. G. Bigourdain, ed. (Paris, 1901).

II. Secondary Literature. The most complete biographical note is G. Riche de Prony, "Notice sur la vie et les ouvrages d'Alexandre-Gui Pingré," in *Mémoires de l'Institut national des sciences et arts. Sciences mathématiques et physiques* (*an* VI [1798]), **1**, xxvi–xlvi. There is also a reasonably full note on his astronomy, with some strictures on his accuracy as an observer, by J. B. Delambre in his *Histoire de l'astronomie du dix-huitième siècle* (Paris, 1827), 664–687. For a résumé of his work in connection with the transits of Venus, see H. Woolf, *The Transits of Venus* (Princeton, 1959), esp. 98–115.

Colin A. Ronan

PIRES, TOMÉ (*b*. Portugal, *ca*. 1470; *d*. China, *ca*. 1540), *pharmacology*.

Little is known of Pires' life before his arrival in India in 1511. He was the son of a royal apothecary and was himself "apothecary of prince D. Alfonso," perhaps the son of João II, king of Portugal. Pires was undoubtedly attracted to India by the prospect of the good career that apothecaries (frequently mentioned in documents of the period) could expect to make for themselves there. In 1511, aboard a fleet commanded by Garcia de Noronha, he went to Cochin, where it may be assumed that he practiced his trade. From a letter signed by Pires it can be deduced that he held the post of "factor of drugs."

In 1513 Pires was chosen to go to Malacca to help the factor put an end to troubles arising from trade duties. He was subsequently named registrar and checker of the Portuguese entrepôt, as well as factor. Profiting from his stay in Malacca, he took a position as clerk in a fleet bound for Java, where he visited the northern coast.

When Pires returned to Cochin in 1515, the first Portugese voyage to China was being organized; and Lopo Soares d'Albergaria, successor to Afonso de Albuquerque in the Portuguese government of India, chose him to be ambassador to China. Pires departed in February 1516 aboard a fleet of five ships under the command of Fernão Peres de Andrade. A year later he arrived in Canton but had to remain there for three years before reaching the court of the emperor in Peking. He was rejected by the Chinese nobles, and thus returned, disillusioned, to Canton, where he and three or four companions were imprisoned. Pires was later freed but was never able to leave China.

Pires' only known work is *Suma oriental*, a masterpiece on the geography, ethnography, and commerce of the Orient at the beginning of the sixteenth century.

BIBLIOGRAPHY

I. Original Works. Two different MS copies of *Suma oriental* have been preserved: at the National Library in Lisbon (MS 299), and at the library of the Chamber of Deputies in Paris. The first ed., an Italian trans. by G. B. Ramusio from a MS similar to the one in Paris, was *Sommario di tutti le regni, citta, & populi orientali, con li traffichi & mercantie, che iui si trovano, comenciando dal mar Rosso fino alli populi della China* (Venice, 1550). In 1944 Armando Cortesão prepared the Paris MS for publication, providing an English trans. and detailed intro., *The Suma Oriental of Tomé Pires, an Account of the East, From the Red Sea to Japan . . . Written in Malacca and India in 1512–1515 . . .*, 2 vols. (London, 1944).

II. Secondary Literature. On Pires and his work, see A. Cortesão, *A primeira embaixada portuguesa à China* (Lisbon, 1945); and "A propósito do ilustre boticário Tomé Pires," in *Revista portuguesa de farmácia*, **13** (1963). See also T'ien-Tse Chang, "Malacca and the Failure of the First Portuguese Embassy to Peking," in *Journal of Southeast Asian History*, **3** (1962).

Luís de Albuquerque

PIRÎ RAIS (or **Re'is**), **MUḤYÎ AL-DÎN** (*b*. Gelibolu [Gallipoli], Turkey, 1470; *d*. Egypt, 1554), *geography, cartography*.

Pirî Rais was the son of Hajjî Muḥammad Rais and the nephew of Kemal Rais, a famous Turkish admiral. From 1487 to 1493 he served in the Turkish navy and fought in several battles under the supervision of his uncle. After the death of his uncle in 1511, he left the navy and began work on his first map. Subsequently he entered the service of the Algerian corsair Khair al-Din Barbarossa (*ca*. 1483–1546).

In 1516–1517 Pirî was given command of several vessels that were involved in the Ottoman campaign against Egypt. He conquered Alexandria, a feat that enabled him to meet Sultan Selim I (1512–1520), to whom he presented the map of 1513, completed at Gelibolu.

After Egypt was joined to the Ottoman Empire, Pirî returned to Gelibolu and began to write his *Kitab-i Bahriye*. Because of the conflict in Egypt, he was appointed as a guide to Ibrahim Pasha of Parga

(1493–1536). On the way to Egypt, a storm forced the fleet to take refuge at Rhodes for a month. Pirī's frequent references to his records attracted the attention of Ibrahim Pasha, who encouraged him to complete his book so that it could be presented to the sultan. In 1526 Pirī was appointed admiral of the South Seas. His last official post was admiral of the Red and Arabian seas.

In 1929 a fragment of a map was discovered in the Topkapi Palace Museum (Figure 1). It depicts the Iberian Peninsula, the western bulge of North Africa, the Atlantic Ocean, and the coast and the islands of America. It is drawn with great care on gazelle hide and includes colored pictures and marginal notes about the countries, peoples, animals, and plants. The signature reveals that this is the map drawn by Pirī Rais in 1513 and presented to Sultan Selim I in 1517.

The map is a portolano chart, a design that was thought to be simple and to have no mathematical basis. There are no markings for latitude and longitude; instead there are lines radiating from centers. The assumption regarding mathematics is erroneous. The existence of a mathematical basis for Pirī's map was initially suggested by the five projection centers in the Atlantic Ocean. It was then easy to convert the portolano to modern coordinates of latitude and longitude. One can see two compass roses, one in the north and one in the south. Each is divided into thirty-two parts, and the division lines extend beyond the rose frames.

In one of the marginal notes, Pirī states that he used some twenty maps in constructing his own. Eight of these were of the world, drawn in the days of Alexander the Great; four were by Portuguese explorers and recorded the discoveries made before 1508 on the South American coast by Vespucci, Vicente Yáñez Pinzón (commander of the *Niña* in 1492–1493), and Juan Díaz de Solis (*d.* 1516); one by an Indian; and one, the most important, that had belonged to Columbus. The latter may have come into Pirī's possession during the fight against the Spanish in the western basin of the Mediterranean in 1501.

Pirī's map has all the important information that was on Columbus' map. For instance, Trinidad is spelled "Kalerot," which probably was derived from a point on the island that was named Galera by Columbus. Puerto Rico is called San Juan Bautista. The drawing of islands on the South American coast opposite Trinidad shows the influence of Columbus, who believed the newly discovered continent to be a group of islands. Haiti was called Hispaniola by Columbus and the Island of Spain by Pirī. The Antilles and Cuba are shown on the map as a continent, as

they were believed to be by Columbus. Hence Pirī called Central America "the coast of Antillia."

The fifth marginal note about America and its discovery states:

> These coasts are named the shores of Antillia. They were discovered in the year 896 of Hijra. It is reported that a Genoese infidel named Colombo discovered these places. A book fell into the hands of Colombo; and he found in it that at the end of the western side [of the world], there were coasts and islands and all sorts of metals and precious stones. Having studied this book thoroughly, Colombo explained these matters to the great of Genoa and said, "Give me two ships. Let me go and find these places." They said, "Can an end or a limit be found to the Western Sea? Its vapor is full of darkness." Colombo saw that no help was forthcoming from the Genoese. He went to the king of Spain and told his story in detail. The answer was that of the Genoese. Colombo petitioned for a long time, until finally the king of Spain gave him two ships, saw that they were well equipped, and said, "Colombo, if it happens as you say, we will make you an Admiral."

Cities and citadels are indicated on the map by red lines. Mountains are drawn in outline and rivers are marked with thick lines; rocky regions are indicated in black; shoals and shallow waters by reddish dots; and rocky areas in the sea by crosses. One of the remarkable aspects of Pirī's map is that the features on the Atlantic coast of Africa bear Turkish names: Babadağ (Father Mountain); Akburun (White Cape), now Cape Blanco; Yeşil Burun (Green Cape), now Cape Verde; this map is an original work based on various maps and the personal experience of Pirī Rais and his friends.

In 1528, in Gelibolu, Pirī Rais drew a second map (Figure 2). The upper left corner shows the northern part of the Atlantic Ocean and newly discovered regions of North and Central America. Greenland is in the north and the Azores in the south. The Azores include San Mikal, Santa Maria, Buriko, and San Jorjo. Two large pieces of land are depicted. The one in the north is called Baccolao; the other, Terra Nova. Pirī says that both were discovered by Portuguese. Terra Nova had not yet been fully explored, and only the known parts are shown on the map. He calls Florida "San Juan Bautisto," the name given to Puerto Rico on the 1513 map. Cuba, Haiti, the Bahamas, and the Antilles are drawn accurately.

In the notes near Labrador, Pirī says, "This is Baccolao; the Portuguese infidels discovered it. The coasts of Terra Nova were discovered by the Portuguese explorer Carlos Real in 1500, and his brother Miguel Real discovered Labrador a year later."

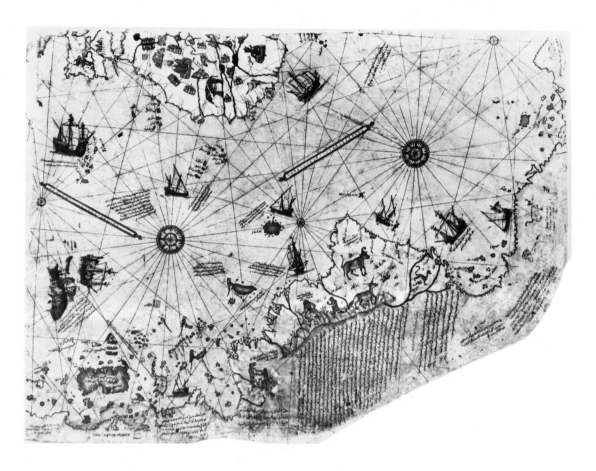

FIGURE 1.   The map of 1513 (above).

FIGURE 2.   The map of 1528 (below).

FIGURE 3.   Map from *Kitab-ī Bahriye*, 1521 (right).

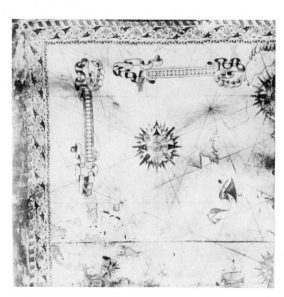

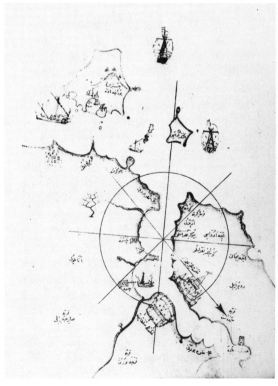

Pirī cites an explorer who planned to travel overland to reach the ocean. It is quite possible that he meant Balboa, who crossed the Isthmus of Panama and reached the Pacific Ocean in 1513.

By comparison of these two maps one can easily deduce that Pirī Rais continued to follow the new discoveries with great care. He showed only the parts of the world that had been discovered and left the unexplored areas blank. When Vespucci declared that South America was a new continent, that land drew the attention of the geographers. Consequently various maps of the new continent were drawn, and Pirī was the most important of the cartographers involved.

To make available all his own observations and all previous information that he could not fit onto the maps, Pirī collected them as *Kitab-ī Bahriye* ("On Navigation"; 1521). It is basically a naval guidebook with essential data on the most important coastal routes and large maps and detailed charts in different colors (Figure 3). The main portion of the book is devoted to the Mediterranean coast and islands.

The book is composed of twenty-one chapters. Pirī first gives historical and geographical information and then discusses the necessary practical navigational data. The accuracy of many of his statements is indisputable. In chapters 1 and 2 Pirī explains his aim in writing the book and describes his life at sea with Kemal Rais. In chapters 3–5 he gives information about storms, winds, and the compass. Chapters 6 and 7 concern maps and emblematic signs on maps. In chapter 8 Pirī discusses the continents and the seas. Chapter 9 is devoted to the geographic discoveries of the Portuguese. In chapter 21 Pirī mentions the Atlantic Ocean and tells the reader of a new continent, Antiliā, the mountains of which contain rich gold ores and in the seas, pearls. He says that it was discovered by sailors and gives information about the inhabitants, frightful creatures having flat faces and eyes a full span apart. The chapter on the Western Sea contains all that was known about the discovery of America at the time.

BIBLIOGRAPHY

Pirī Rais's only published work is *Kitab-ī Bahriye*, Şerafettin Yaltkaya, ed. (Istanbul, 1935).

Secondary literature includes A. Adnan Adivar, *Osmanli Türklerinde Ilim* (Istanbul, 1943); Afet I'nan, "Bir Türk amirali; XVI. astin büyük ceografi: Pirī Reis (Un amiral-géographe turc du XVI siècle—Pirī Reis, auteur de la plus ancienne carte de l'Amérique)," in *Belleten*, **1**, no. 2 (1937), 333–348; *America's Oldest Map Made by a Turkish Admiral: Pirī Reis*, trans. by Leman Yolaç (Ankara, 1950); and *Pirī Reis' in Amerika Haritasi, 1513–1528*

(Ankara, 1950); Yusuf Akçura, "Map Drawn by Pirī Reis," in *Illustrated London News* (23 July 1923); and "Pirī Reis haritasi hakkinda izahname (Die Karte des Pirī Reis. Pirī Reis Map. Carte de Pirī Reis)," in *Türk tarih karamu* (Istanbul, 1935); H. Alpagut, F. Kurtoğlu, "Mukaddime I-LV: Pirī Reis: *Kitab-ī Bahriye*," ibid., no. 2 (Istanbul, 1935); W. Y. Callien, "The Evolution of the Map of the Earth (Dünya haritasinin evrimi)," in *Ankara üniversitesi Dil ve tarih-coğrafya fakültesi dergisi*, **8**, no. 1 (1949), 149–153; H. Deismann, *Forschungen und Funde im Serai* (Berlin–Leipzig, 1933), 111–122; and Charles H. Hapgood, "Ancient Knowledge of America and Antarctica," in *Actes du dixième Congrès international d'histoire des sciences* (Ithaca, N.Y., 1962), 479–485.

See also P. Kahle, *Pirī Reis, Bahriye, Das türkisches Segelhandbuch für das mittelländische Meer von Jahre 1521*, 2 vols. (Berlin–Leipzig, 1926); "Importe colombiane in una carta turco del 1513," in *Cultura*, **1**, fasc. 10 (Milan–Rome, 1931), 1–13; *Die verscholtene Columbus Karte non 1498 in einer Türkisen Weltkarte von 1513* (Berlin–Leipzig, 1933), with trans. as "The Lost Columbus Map of 1498 Discovered in a Turkish Map of the World of 1513," in *Aligarh Muslim University Journal* (1935); Hans von Mžik, "Pirī Reis und seine *Bahriye*," in *Beiträge zur historischen Geographie* (Leipzig–Vienna, 1929), 60–76; Ibrahim Hakki Konyali, *Topkapi Sarayinda deri üzerine yapilmiş eski haritalar* (Istanbul, 1936); K. Kretchmer, *Die Entwicklung der Kartographie von America* (Gotha, 1891); Eugen Oberhummer, "Eine Turkische Karte zur Entdeckung Americas," in *Anzeiger der Akademie der Wissenschaften, Wien* (1931), 18–27; and "Eine Karte des Colombus in Türkische Überlieferung," in *Mitteilungen der Geographischen Gesellschaft, Wien*, **78** (1934), 115.

Additional works are Sadi Selen, "Pirī Reis' in Şimali Amerika Haritasi . . .," in *Belleten*, **1**, no. 2 (1937), 515–523; and Huseyin Yurdaydin, "*Kitab-ī Bahriye*'nin telif meselesi," in *Ankara üniversitesi Dil ve tarih-coğrafya fakultesi dergisi*, **10**, pts. 1–2 (1952), 143–146.

SEVIM TEKELI

PIROGOV, NIKOLAY IVANOVICH (*b.* Moscow, Russia, 25 November 1810; *d.* Vishnya, Ukraine, Russia, 5 December 1881), *surgery, anatomy.*

The son of a major in the commissary service, Pirogov grew up in fairly cultured surroundings, learned to read early, and was fluent in foreign languages as a child. In 1824 the family was left without means and the father died suddenly. Pirogov might have become a civil servant; but Efrem Mukhin, the family physician, who was professor of surgery and anatomy at Moscow University, arranged for him to be admitted to the Medical Faculty at Moscow, even though Pirogov was then only fourteen and the entrance age was sixteen.

Pirogov chose surgery as his specialty; but during his four years at the university he was present at only two operations and did not perform any himself. Nevertheless, he received a good general theoretical preparation. After graduating in 1828, he was sent, with Mukhin's advice and help, for a teaching career at Dorpat (now Tartu) University, where the professorial institute was being established. He studied surgery and anatomy under the direction of J. F. Moier and in 1832 defended his doctoral dissertation, on the ligation of the ventral aorta. In this important work, which was soon published in a German translation, Pirogov tried not only to improve the technical procedure of the operation but also to explain how the body reacts to it.

From 1833 to 1835 Pirogov visited the leading German clinics and observed the existing state of surgery. He became convinced that without special study of anatomy and physiology, surgery—even with the most advanced technique—could never rise to the level of a science but would remain an art. Upon his return to Russia, Pirogov found that the chair of surgery at Moscow University that he himself had hoped to win was occupied; thus, in 1836, he accepted the post of professor of surgery at Dorpat. Although he was only twenty-six, his reputation was already substantial. A work published the following year laid the foundation of surgical anatomy.

From 1841 to 1856 Pirogov headed the department of surgery and the surgical clinic, founded on his initiative, at the 1,000-bed hospital of the St. Petersburg Medical-Surgical Academy. He also taught pathological anatomy and founded a museum of anatomical pathology at the Academy. Working in an unheated, poorly lit basement that was the anatomical theater of the Academy, Pirogov lectured and performed countless operations and 12,000 dissections in anatomical pathology. During this time he spent about three years in military service, organizing and providing medical aid to the wounded. In 1847 he developed a theory of the action and use of anesthetic and, before using it on a patient, tested it on himself. He was the first to introduce anesthetic through the rectum, and in his clinic choloroform was first used in Russia. He also originated the intravenous administration of anesthetic ether. Pirogov was the first to use ether under battle conditions (1847); and in 1854–1855, during the siege of Sevastopol, he introduced the mass use of anesthetic in surgical operations at the front.

Pirogov's work on topographical anatomy (1851–1859) laid a firm foundation for that field as a special area of science having great practical significance for surgery. The work was followed by his discovery of new methods of anatomical research: the study of the forms and relative positions of the organs by dissecting frozen cadavers and removing organs from them. Both these extremely simple methods opened previously unknown possibilities for precisely determining the forms and positions of organs and tissues. Pirogov's work comprised four volumes of drawings of organs and tissues in their natural relative positions and an explanatory text. It immediately received widespread recognition and enhanced his reputation as a distinguished surgeon and anatomist.

During the Crimean War, Pirogov organized medical aid and developed the basic principles of field surgery. The first to use plaster casts, he conceived the technique in 1851 while observing the work of a sculptor. His experiences in field surgery, published in German in 1864, became a standard reference.

In 1856 Pirogov returned to St. Petersburg. Irritated by conditions at the Medical-Surgical Academy, he retired permanently from teaching and hospital work. In the same year Pirogov published a paper on the problems of pedagogy, which produced a great impression. He condemned the restrictions on education for the poor and for non-Russians and supported education for women. He also came out against early specialization and advocated the development of secondary schools. After the death of Nicholas I, Pirogov was appointed director of school affairs for the south of Russia. He came into conflict with the governor-general of Odessa and in 1858 was transferred to the same post in Kiev. He was forced to retire three years later and settled on his estate in the southern Ukraine. In 1862 he was named director of a group of young Russian scientists sent abroad to prepare for professorships. After Garibaldi had been severely wounded in the leg in August 1862 during the battle of Aspromonte, Pirogov attended him and recommended a successful method of cure. After his return to Russia in 1866, Pirogov lived almost exclusively on his estate, which he left for prolonged periods only twice: in 1870, when he traveled to the battlefields of the Franco-Prussian War as a representative of the Russian Red Cross; and in 1877, when he served as a surgeon in the Russo-Turkish War for the independence of Bulgaria.

Pirogov's other achievements include a procedure for amputation of the shin that retained the calcaneal bone; improved methods of tying the major blood vessels for hemostasis; a classic description of shock; the use—before the introduction of antisepsis—of spirit of camphor, aqueous solution of chlorine, or tincture of iodine to combat the festering of wounds;

and the demonstration of the importance of diet in treating the wounded.

Pirogov is considered a founder of contemporary surgery and topographical anatomy, and I. P. Pavlov credited him with placing surgery on a scientific basis.

BIBLIOGRAPHY

I. Original Works. Recent collections of Pirogov's writings are *Izbrannye pedagogicheskie sochinenia* ("Selected Pedagogical Works"; Moscow, 1953); and *Sobranie sochiney* ("Collected Works"), 8 vols. (Moscow, 1957–1962). Important works published during his lifetime are *Anatomia topographica. . .* , 4 vols. (Petropoli, 1851–1859); and *Grundzüge der allgemeinen Kriegschirurgie* (Leipzig, 1864), translated into Russian as *Nachala obshchey voennopolevoy khirurgii* ("The Principles of General Military Field Surgery"), 2 vols. (Dresden, 1865–1866).

II. Secondary Literature. See N. N. Burdenko, "N. I. Pirogov—osnovopolozhnik voenno-polevoy khirurgii" ("N. I. Pirogov—Founder of Military Field Surgery"), in Pirogov's *Nachala obshchey voenno-polevoy khirurgii* ("Beginnings of General Military Field Surgery"), I (Moscow, 1941), 9–42; A. M. Geselevich and Y. I. Smirnov, *N. I. Pirogov* (Moscow, 1960); A. N. Maksimenkov, *N. I. Pirogov* (Leningrad, 1961); and I. G. Rufanov, *N. I. Pirogov—veliky russky khirurg i ucheny* ("N. I. Pirogov—Great Russian Surgeon and Scientist"; Moscow, 1956).

S. R. Mikulinsky

PISANO. See **Fibonacci, Leonardo.**

PISO, WILLEM (*b.* Leiden, Netherlands, *ca.* 1611; *d.* Amsterdam, Netherlands, November 1678), *medicine, pharmacy.*

Piso was the son of Hermannus Piso van Cleef; his mother's name is unknown.[1] He matriculated as a medical student at the University of Leiden in 1623, at the age of twelve. He received the M.D. degree at Caen[2] on 4 July 1633 and subsequently established a practice in Amsterdam.

Piso's fame rests on his work as physician of the Dutch settlement in Brazil (1636–1644), which had Johan Maurits van Nassau as governor.[3] In Brazil he gathered the data for the books that made him famous.

Although he already had the M.D. degree, Piso matriculated at Leiden again on 3 March 1645, after his return from Brazil. He must soon have rejoined his former chief, however, for a letter dated September 1645 was sent by Piso to his friend Caspar van Baerle, professor of philosophy at the Amsterdam Athenaeum, from the "camp of Maurits van Nassau."[4]

Later Piso settled in Amsterdam, and on 1 September 1648 he married Constantia Spranger. He became a leading physician there, serving as *decanus* of the Collegium Medicum from 1656 to 1660 and again in 1670. His name is mentioned as a consultant in the works of Nicolaas Tulp (1593–1674) and Job van Meekren (*ca.* 1611–1666). Piso is remembered for his work in tropical medicine and pharmacy, but his chief contribution was perhaps his scientific approach to his work. Although he could not free himself from the Hippocratic and Galenic doctrines he had studied at the university and he explained some of his observations in terms of them, he studied the medical lore of the Brazilian natives very closely and felt free to adopt and recommend their methods if they proved effective. He was the first to point out that the health of Europeans in the tropics is best preserved by adopting the way in which the natives live. During the first decades of their settling in the tropics, the Dutch lived as they had at home. They built their houses of brick; they wore heavy, dark clothing; they swaddled their infants. These practices caused diseases, and the infant mortality was enormous. Piso told his compatriots to beware of the cool nights in Brazil; not to drink too much, especially sour beverages; and to take plenty of exercise. Piso recorded what he had learned in Brazil in *Historia naturalis Brasiliae*, a folio volume in twelve books, the first four of which are by Piso while the others, which deal chiefly with natural history, were written by Georg Markgraaff.[5] The latter was probably an assistant to Piso, for Piso says in several places that he ordered Markgraaff to make drawings or perform other chores in his free time. In a later edition, *De Indiae utriusque re naturali et medica*, the subject matter of the book is more consolidated, and in the process the distinction between contributions of the two authors has become somewhat vague; consequently, Piso has sometimes been accused of plagiarism. That Piso did not intend to take credit for what was not his is clear from the first edition, where the line is sharply drawn. His editing of *De Indiae* was done because the work was intended to be a handbook of tropical medicine, pharmacology, and natural history. His share covered the Americas, while the East Indies were represented by adding the complete works of Jacobus Bontius.[6]

Piso was the first to distinguish yaws, which he called *bubas*, from venereal disease, and he recommended the treatment used by the natives. He stated that defective nutrition was the cause of hemeralopia (day blindness),[7] fully discussed tropical intestinal disorders, and distinguished their various forms. Dysentery ("fluxus cum febre et sanguine") in particular was well researched; Piso recommended the native root ipecacuanha to cure this disease, advice followed for centuries. His description of the chigoe (*Pulex penetrans* or *Tunga penetrans*), the troubles it causes, and the treatment of these troubles has never been surpassed. Here again, he needed the experience of the Brazilian natives. On numerous expeditions into the back country, Piso searched for medicinal herbs. He was the first to bring ipecacuanha to the attention of the Western medical world and also discussed such American specifics as *Radix Chinae* (*Smilax Pseudo China* L.), sarsaparilla (*Smilax sarsaparilla* L.), *Radix mechoacan* (*Convolvulus brasiliensis* L.), sassafras (*Laurus sassafras* L.), and guaiacum (*Guaiacum officinale*). For these and other contributions, Piso deserves to be remembered as one of the pioneers of tropical medicine.

NOTES

1. Van Andel (intro. to "Capita") and Baumann state that the family name originally was Pies, while von Römer states that it was Lepois. A French physician, Charles Lepois (1563–1633), was also known as Carolus Piso. But Willem Piso's father was called "van Cleef," which indicates a German origin.
2. Not at Leiden in 1630, as von Römer states.
3. Johan Maurits van Nassau (1604–1679) served in the army of the States-General from 1621.
4. According to van Andel (intro. to "Capita"). This can be only Johan Maurits van Nassau; there was no Maurits van Nassau alive at the time. Upon his return from Brazil, Johan Maurits van Nassau became lieutenant general of the cavalry of the States-General (24 Oct. 1644) and commander of the fortress at Wesel (3 Nov. 1644). Hence this letter was probably sent from Wesel. See F. J. G. ten Raa and F. de Bas, *Het Staatsche leger*, IV (Breda, 1918), 332, 342.
5. Markgraaff appears to be unknown in the biographical literature. There is no M.D. after his name on the title page of the *Historia naturalis Brasiliae*. After his name is "de Liebstadt, Misnici Germani." Both Liebstadt and Misnia (Meissen) are in Saxony. According to van Andel ("Willem Piso"), Markgraaff died in 1643 on the coast of Guinea. Several variants of the name are found: Marcgraaf, Marggraaf, Marcgrav.
6. Jacobus Bontius (1592–1631), physician to the Dutch East India Company in Batavia, 1627–1631. See *Opuscula selecta Neerlandicorum de arte medica*, X (Amsterdam, 1931).
7. Not night blindness, as van Andel ("Willem Piso") states. See Vos, where a translation of Piso's remarks on tropical eye diseases is given.

BIBLIOGRAPHY

I. ORIGINAL WORKS. *Historia naturalis Brasiliae . . . in qua non tantum plantae et animalia, sed indigenarum morbi, ingenia mores describuntur et iconibus supra quingentas illustrantur* (Amsterdam, 1648) consists of 12 books, of which the first 4, by Piso, have the general title *De medicina Brasiliensi*. They are *De aere, aquis et locis*; *De morbis endemiis*; *De venenatis et antidotis*; *De facultatibus*

simplicium. The remaining 8 were written by Markgraaff under the general title *Historiae rerum naturalium Brasiliae*. These are *De plantis* (3 bks.); *De piscibus*; *De avibus*; *De quadripedibus et serpentibus*; *De insectis*; *De ipsa regione et illius incolis cum appendice de Tapuyis et de Chilensibus*.

In *De Indiae utriusque re naturali et medica* (Amsterdam, 1658), the subject matter is organized somewhat differently. Piso contributed *Historiae naturalis et medicinae Indiae Occidentalis libri quinque* and *Mantissa aromatica, sive de aromatum cardinalibus quator et plantis aliquot Indices in medicinum receptis, relatio nova*; Markgraaff wrote *Tractatus topographicus et meteorologicus Brasiliae, cum observatione eclipsis solaris, quibus additi sunt illius et aliorum commentarii de Brasiliensium et Chilensum indole et lingua*. This book also contains the complete works of Jacobus Bontius with some editing by Piso.

Oost- en West Indische warande. Vervattende aldaar de leef- en geneeskonst. Met een verhaal van de specerijen, boom- en aard gewassen, dieren, etc. Door Jac. Bontius, Gul. Piso en Geo. Markgraef (Amsterdam, 1694; 2nd ed. 1734) is not a trans. but an abstract of the writings of Bontius, Piso, and Markgraaff compiled for the use of naval and tropical surgeons.

A later Latin ed. of part of his main work is *Historia medica Brasiliae, novam editionem curavit et praefatus est Josephus Eques de Vering . . .* (Vienna, 1817); a Portuguese version is *Historia natural do Brasil illustrada. Edição comemorativa do primeiro cinquentenário du Museu Paulista* (São Paulo, 1948).

In the library of the University of Leiden are three letters by Piso: to J. van Wullen, a Lutheran minister in Amsterdam; to Nicolaas Heinsius (1620–1681), a Latin poet and diplomat; and to Caspar van Baerle (1584–1648), the author of *Rerum per octennium in Brasilia et alibi nuper gestarum* (Amsterdam, 1647, 1660, 1698; German trans., Kleve, 1659).

II. SECONDARY LITERATURE. See M. A. van Andel, "Willem Piso, een baanbreker der tropische geneeskunde," in *Bijdragen tot de geschiedenis der geneeskunde*, **4** (1924), 239–254; "Bontius en Piso over de dysenterie in de beide Indiën," *ibid.*, **11** (1931), 285–292; and intro. to "Capita nonnula de ventris fluxibus, de dysteneria, de lue indica, di ipecacuanha," in *Opuscula selecta Neerlandicorum de arte medica*, XIV (Amsterdam, 1937), xii–xxxviii, the best study on Piso, in Dutch and English, with extracts from Piso's works in the original language (Latin or Dutch) with English trans.; E. D. Baumann, *Uit drie eeuwen Nederlandsche geneeskunde* (Amsterdam, 1951), 102–104; L. S. A. M. von Römer, "Dr. Willem Piso," in *Nieuw Nederlandsch biografisch woordenboek*, IX (1933), 805–806, unreliable; J. van den Vondel, "Behoude reis aen Willem Pizo, Graef Maurits van Nassaus doctor, staende op sijn vertreck naer Brezijl," in H. Diferee, ed., *De volledige werken van Joost van den Vondel*, II (Utrecht, 1929), 375–376; and J. A. Vos, "De geneeskunde, in het bijzonder de oogheelkunde, bij Willem Piso," in *Bijdragen tot de geschiedenis der geneeskunde*, **39** (1959), 7–11.

PETER W. VAN DER PAS